AF601475

# ASTRONOMY AND ASTROPHYSICS ABSTRACTS

A Publication of the Astronomisches Rechen-Institut Heidelberg
Member of the International Council for Scientific and Technical Information
Astronomy and Astrophysics Abstracts is Prepared Under the Auspices of the International Astronomical Union

## Volume 38
## Literature 1984, Part 2

Edited by
S. Böhme W. Fricke H. Hefele I. Heinrich W. Hofmann
D. Krahn V. R. Matas L. D. Schmadel G. Zech

Springer-Verlag Berlin Heidelberg GmbH 1985

Astronomisches Rechen-Institut, Mönchhofstraße 12–14,
D-6900 Heidelberg 1, F.R. Germany
Telex: 461 336 ARIHD D
Director: Prof. Dr. Walter Fricke

Astronomy and Astrophysics Abstracts
Department head: Dr. Lutz D. Schmadel
Editors-in-Chief: Inge Heinrich, Dr. Lutz D. Schmadel

ISBN 978-3-662-12348-5 ISBN 978-3-662-12346-1 (eBook)
DOI 10.1007/978-3-662-12346-1


Originally published by Springer-Verlag Berlin Heidelberg New York Tokyo in 1985
Softcover reprint of the hardcover 1st edition 1985
Library of Congress Catalog Card Number 72-104650.

Media conversion: Daten- und Lichtsatz-Service, Würzburg.

2153/3130-543210

# Preface

*Astronomy and Astrophysics Abstracts* aims to present a comprehensive documentation of the literature concerning all aspects of astronomy, astrophysics, and their border fields. It is devoted to the recording, summarizing, and indexing of the relevant publications throughout the world. *Astronomy and Astrophysics Abstracts* is prepared by a special department of the *Astronomisches Rechen-Institut* under the auspices of the *International Astronomical Union.*

Volume 38 records literature published in 1984 and received before February 15, 1985. Some older documents which we received late and which are not surveyed in earlier volumes are included too. We acknowledge with thanks contributions of our colleagues all over the world. We also express our gratitude to all organizations, observatories, and publishers which provide us with complimentary copies of their publications. It is a pleasure to thank Ms. Ute Esser for her valuable contributions to this volume.

Starting with Volume 33, all the recording, correction, and data processing work was done by means of computers. The recording was done by our technical staff members Ms. Helga Ballmann, Ms. Mona El-Choura, Ms. Monika Kohl, Ms. Sylvia Matyssek. Ms. Karin Burkhardt, Ms. Susanne Schlötelburg, and Mr. Stefan Wagner supported our task by careful proofreading. It is a pleasure to thank them all for their encouragement.

Heidelberg, March 1985 **The Editors**

# Contents

# Introduction

## Astronomical Bibliographies

*Astronomy and Astrophysics Abstracts* started documentation and abstracting work in 1969 as the direct successor of the Astronomischer Jahresbericht. For information on astronomical literature before this date consultation of one of the following bibliographies is suggested:

(1) J. J. de Lalande, Bibliographie Astronomique, Paris 1803 (this work covers the time from 480 B. C. to the year 1803, VIII + 966 pages).

(2) J. C. Houzeau, A. Lancaster, Bibliographie générale de l'astronomie, Volume I (in two parts), Bruxelles 1887, 1889, Volume II, Bruxelles 1882. The complete title of Volume II is "Bibliographie générale de l'astronomie ou catalogue méthodique des ouvrages, des mémoires et des observations astronomiques, publiés depuis l'origine de l'imprimerie jusqu'en 1880". A new edition of these volumes was prepared by D. W. Dewhirst in 1964.

(3) Bibliography of Astronomy, 1881–1898. The literature of this period was recorded on standard slips by the Observatoire Royal de Belgique. From the material (some 52,000 items) a microfilm version was produced by University Microfilms Limited, Tylers Green, High Wycombe, Buckinghamshire, England, in 1970.

(4) Astronomischer Jahresbericht, 1899 gegründet von Walter Wislicenus, herausgegeben vom Astronomischen Rechen-Institut in Heidelberg (formerly in Berlin), Verlag W. de Gruyter, Berlin. For the period from 1899 to 1968 sixty-eight volumes were published, each of which, in general, covers the literature of one year.

## Concept of *Astronomy and Astrophysics Abstracts*

This abstracting service aims to present a comprehensive documentation of the literature in all fields of astronomy and astrophysics and their border fields. It appears in semi-annual volumes. Two of these volumes cover the literature of one calendar year. Every effort will be made to ensure that the average time interval between the receiving date of the original documents and publication of the abstracts will not exceed eight months. This time interval is near to that achieved by monthly abstracting journals, compared to which our system of accumulation of information over six months offers the advantage of greater convenience for the user.

The main characteristics of the concept of *Astronomy and Astrophysics Abstracts* may be summarized as follows:

(1) The subdivision of astronomy and its border fields into subject categories is facilitated by the fact that the astronomical objects appear to be particularly well suited for the formation of categories. It may be assumed that such subdivisions can be maintained for a long period. Experience shows, however, that progress in research might imply minor changes in the classification scheme.

(2) Each paper has been classified into one of 106 numbered subject categories and given a serial number within the category. In this way each item is numbered by six figures: the first three indicate the number of the category, the following three the serial number within the category. Reference to an abstract in Volume 1 is indicated by "01" before the number of the category; for example: 01.074.028, denotes Volume 1, category 074, abstract 028. A paper might be classified into more than one category. In this case, its abstract is placed only in one category, whereas in the other categories only cross references are given. These are listed at the end of each category.

(3) Authors' abstracts are used whenever possible. Popular articles are not abstracted.

(4) If possible, titles of papers and abstracts are given in English. A special reference is made to titles which we have not taken in the original language.

The whole material was recorded by means of modified ITT 3030 microcomputers. All text recording programs and other data processing software were developed by Multicom GmbH, Gröbenzell, F. R. Germany and by our staff members as well. The index computations were carried out on the IBM 3081 D computer of the University of Heidelberg.

## Classification Systems

The two most common and widely used classification systems in astronomy and astrophysics are given by Class 9 of the International Classification System for Physics, published by the International Council of Scientific Unions Abstracting Board (Second edition 1978. ICSU-AB, 17 Rue Mirabeau, 75017 Paris, France, ISSN 0305-9618), and the *Astronomy and Astrophysics Abstracts* classification. In order to facilitate literature searches, we introduce a concordance relation between these two very different systems. This solution is only a unilateral one. Starting from the fourth hierarchical level of the ICSU-AB system, the appropriate *Astronomy and Astrophysics Abstracts* chapter numbers are listed. This cannot imply an identical content of the respective chapters in both systems. In many cases there is only a rather partial concordance, and therefore the *Astronomy and Astrophysics Abstracts* numbers are enclosed in parentheses. Considering our objectives, only the astronomical part of Class 9 of the ICSU-AB scheme is covered.

## Transliteration Scheme for the Russian Alphabet

The transliteration of the Russian alphabet in use in *Astronomy and Astrophysics Abstracts* is presented here.

| | | | | | |
|---|---|---|---|---|---|
| А | а | a | П | п | p |
| Б | б | b | Р | р | r |
| В | в | v | С | с | s |
| Г | г | g | Т | т | t |
| Д | д | d | У | у | u |
| Е | е | e | Ф | ф | f |
| Ё | ё | e | Х | х | kh |
| Ж | ж | zh | Ц | ц | ts |
| З | з | z | Ч | ч | ch |
| И | и | i | Ш | ш | sh |
| Й | й | j | Щ | щ | shch |
| К | к | k | Ы | ы | y |
| Л | л | l | Ь | ь | ' |
| М | м | m | Э | э | eh |
| Н | н | n | Ю | ю | yu |
| О | о | o | Я | я | ya |

This transliteration was recommended by the Abstracting Board of the International Council of Scientific Unions in 1969. It corresponds essentially to the transliteration proposed by the Academy of Sciences, Moscow. In this case the letters can be read and printed by usual data processing machines. If the names of Russian authors in the literature are transliterated in a different scheme, we present the names as they are given in the references cited and in addition in brackets according to our transliteration table.

## Sources of Information

The majority of sources of information for this volume is given in category **001 Periodicals** and in category **008 Observatories, Institutes.** It may be noted that the titles of the periodicals are given in the original languages, and that Russian titles have been transliterated applying the transliteration scheme given above. Category 008 records publication series of observatories and astronomical institutes. Titles of the periodicals have been given following the recommendations of the "International List of Periodical Title Word Abbreviations" and its additions (see also **Abbreviations**). In most cases they permit recognition of the full title without recourse to the key in category 001. If other secondary sources have been consulted, we cite these papers and give reference to the respective services.

The total number of papers (some do not give names of authors) recorded in this volume amounts to 9,361.

## Author and Subject Indexes

The subject category and the serial number have been used as a reference both in the author index and the subject index. These references are more precise than page references and offer considerable advantages in indexing by means of computers.

The author index of this volume contains 10,863 names.

We consider the subject index as an approximation to an optimal index covering all fields of astronomy and astrophysics. Starting with Volume 18, the subject index was enlarged to a certain extent in order to provide a thesaurus of astronomical and astrophysical terms. At present, the *Astronomy and Astrophysics Abstracts Vocabulary* 07.84, containing some 2,200 key words, is in use. This is done not only for the users' convenience, but also with the intention to propose the use of special key words to authors and publishers.

While each volume is scheduled to contain an author index and a subject index, the magnetic tapes containing the index information will be used to produce separate index volumes (authors and subjects) at intervals of five years.

The sorting program for the author and subject indexes is based on the IBM SORT/MERGE Program. This program sorts blank before hyphen and before letters. Apostrophes are ignored by a special routine.

The users are requested to inform us on spelling errors within the author index in order to assist us in eliminating mistakes in future cumulative indexes.

---

*Astronomy and Astrophysics Abstracts* is prepared at the *Astronomisches Rechen-Institut,* Heidelberg under the auspices of the *International Astronomical Union* on a non-profit basis. The editors urge publishers of literature related to astronomy and astrophysics to provide our service in due time with complimentary copies of their material.

Publications should be mailed to:

Astronomy and Astrophysics Abstracts
Astronomisches Rechen-Institut
Moenchhofstrasse 12–14
D 6900 Heidelberg 1
F. R. Germany
Telex: 461 336 ARIHD D

# Concordance Relation

between the ICSU-AB International Classification System for Physics and the *Astronomy and Astrophysics Abstracts* Classification Scheme

| ICSU-AB International Classification System for Physics | Astronomy and Astrophysics Abstracts Classification Scheme |
|---|---|
| ***0 General*** | |
| **01.10 Announcements, news, and organizational activities** | |
| 01.10.C | 006 (010) |
| 01.10.F | 011 |
| 01.10.H | 013 (010) |
| **01.30 Physics literature and publications** | |
| 01.30.B | 012 (014) |
| 01.30.C | 012 |
| 01.30.E | 003 |
| 01.30.K | 002 (003) |
| 01.30.M | 003 |
| 01.30.P | 003 |
| 01.30.R | 014 |
| 01.30.T | 002 |
| **01.40 Education** | 014 |
| **01.50 Educational aids** | 014 |
| **01.60 Biographical, historical, and personal notes** | 004 (005, 006, 007) |
| **01.65 History of science** | 004 |
| **01.90 Other topics of general interest** | 015 |
| ***9 Geophysics, Astronomy, and Astrophysics*** | |
| **91.10 Geodesy and gravity** | |
| 91.10.B | 045 |
| 91.10.N | 044 |
| 91.10.Q | 081 |

| ICSU-AB International Classification System for Physics | Astronomy and Astrophysics Abstracts Classification Scheme |
|---|---|
| **91.25 Geomagnetism and paleomagnetism; geoelectricity** | 084 |
| **91.35 Earth's interior structure and properties** | 081 |
| **91.90 Other topics in solid Earth physics** | 081 |
| **92.60 Meteorology** | 082 |
| **92.65 Atmospheric optics** | 082 |
| **94.10 Physics of the neutral atmosphere** | |
| 94.10.B | 082 |
| 94.10.D | 082 |
| 94.10.F | 082 |
| 94.10.G | 082 (063) |
| 94.10.H | 082 |
| 94.10.L | 082 |
| 94.10.N | 106 |
| 94.10.Q | 082 |
| 94.10.S | 084 |
| **94.20 Physics of the ionosphere** | |
| 94.20.B | 083 |
| 94.20.D | 083 |
| 94.20.M | 083 (084) |
| 94.20.P | 083 (084) |
| 94.20.W | 083 (062) |
| 94.20.Y | 083 (084) |
| **94.30 Physics of the magnetosphere** | |
| 94.30.C | 084 |
| 94.30.D | 084 |
| 94.30.E | 084 |
| 94.30.F | 084 (062) |
| 94.30.G | 084 (062) |
| 94.30.H | 084 |
| 94.30.L | 084 |
| 94.30.M | 084 |
| 94.30.S | 084 |
| 94.30.V | 084 (074) |
| 94.30.W | 084 (078, 144) |

| ICSU-AB<br>International Classification System for Physics | Astronomy and Astrophysics Abstracts<br>Classification Scheme |
|---|---|
| **94.40 Cosmic rays** | |
| 94.40.C | 144 |
| 94.40.E | 144 (078, 106) |
| 94.40.H | 078 |
| 94.40.K | 144 (085) |
| 94.40.L | 144 (078) |
| 94.40.V | 105 (144) |
| **94.60 Interplanetary space** | |
| 94.60.D | 074 |
| 94.60.F | 074 |
| 94.60.G | 074 (062) |
| 94.60.K | 106 |
| 94.60.M | 106 |
| 94.60.Q | 074 (091, 094) |
| 94.60.R | 106 (062) |
| **94.80 Aerospace facilities and techniques, space research** | |
| 94.80.P | 053 (051) |
| 94.80.R | 053 (051) |
| 94.80.W | 035 |
| **95.10 Fundamental astronomy** | |
| 95.10.C | 042 (043, 052) |
| 95.10.E | 042 (052) |
| 95.10.G | 041 (079, 095, 096) |
| 95.10.J | 041 |
| **95.30 Fundamental aspects of astrophysics** | |
| 95.30.C | 061 (022) |
| 95.30.E | 022 |
| 95.30.G | 022 (061) |
| 95.30.J | 063 |
| 95.30.L | 062 |
| 95.30.Q | 062 |
| 95.30.S | 066 (161) |
| **95.45 Observatories** | 008 (009) |

| ICSU-AB International Classification System for Physics | Astronomy and Astrophysics Abstracts Classification Scheme |
|---|---|
| **95.55 Astronomical instruments** | |
| 95.55.B | 032 |
| 95.55.C | 032 (031) |
| 95.55.E | 032 |
| 95.55.J | 033 |
| 95.55.L | 035 |
| **95.65 Auxiliary and recording instruments** | 034 |
| **95.70 Other instrumentation and techniques (including clocks, frequency standards, etc.)** | 034 (036) |
| **95.75 Techniques of observation and reduction** | |
| 95.75.D | 036 |
| 95.75.F | 036 |
| 95.75.H | 036 |
| 95.75.K | 036 |
| 95.75.M | 036 |
| 95.75.P | 036 (021) |
| **95.80 Catalogues, atlases, etc.** | 002 (046) |
| **96.10 General, solar nebula, and cosmogony** | 107 (091) |
| **96.20 Moon** | |
| 96.20.B | 094 |
| 96.20.D | 094 |
| 96.20.J | 094 |
| **96.30 Planets and satellites (excluding the moon)** | |
| 96.30.D | 092 |
| 96.30.E | 093 |
| 96.30.G | 097 |
| 96.30.H | 098 |
| 96.30.K | 099 |
| 96.30.M | 100 |
| 96.30.T | 101 |

| ICSU-AB International Classification System for Physics | Astronomy and Astrophysics Abstracts Classification Scheme |
|---|---|
| **96.50 Other objects in the planetary system** | |
| 96.50.D | 106 |
| 96.50.G | 102 (103) |
| 96.50.K | 104 |
| 96.50.M | 105 |
| **96.60 Solar physics** | |
| 96.60.C | 080 (075) |
| 96.60.F | 071 (080) |
| 96.60.K | 080 |
| 96.60.M | 071 |
| 96.60.N | 073 (074) |
| 96.60.P | 074 |
| 96.60.Q | 072 |
| 96.60.R | 073 (076, 077) |
| 96.60.S | 073 |
| 96.60.V | 078 (074) |
| **97.10 Stellar characteristics and properties** | |
| 97.10.B | 131 |
| 97.10.C | 065 (061) |
| 97.10.E | 064 (063) |
| 97.10.F | 112 (064) |
| 97.10.H | 112 (064) |
| 97.10.K | 116 (065) |
| 97.10.L | 116 (065) |
| 97.10.N | 115 |
| 97.10.Q | 115 |
| 97.10.R | 115 (113, 114) |
| 97.10.T | 114 |
| 97.10.V | 111 |
| 97.10.W | 111 |
| **97.20 Normal stars (by class): general or individual** | |
| 97.20.D | 121 |
| 97.20.R | 126 |
| **97.30 Variable and peculiar stars (including novae)** | |
| 97.30.E | 114 (112) |
| 97.30.F | 116 |

| ICSU-AB<br>International Classification System for Physics | Astronomy and Astrophysics Abstracts<br>Classification Scheme |
|---|---|
| 97.30.G | 122 (123) |
| 97.30.J | 122 (123) |
| 97.30.K | 122 (123) |
| 97.30.N | 122 (123) |
| 97.30.Q | 124 (117, 122, 123) |
| 97.30.S | 122 (123) |
| **97.60 Late stages of stellar evolution (including black holes)** | |
| 97.60.B | 125 |
| 97.60.G | 126 (067) |
| 97.60.J | 067 |
| 97.60.L | 067 |
| 97.60.S | 067 (065) |
| **97.80 Binary and multiple stars (including extrasolar planetary systems)** | |
| 97.80.D | 118 (002) |
| 97.80.F | 120 |
| 97.80.H | 119 |
| 97.80.J | 142 (117) |
| 97.80.K | 118 |
| 97.80.M | 118 |
| **98.10 Stellar dynamics** | 151 |
| **98.20 Stellar clusters and associations** | |
| 98.20.C | 152 |
| 98.20.E | 153 |
| 98.20.H | 154 |
| **98.40 Interstellar matter and nebulae** | |
| 98.40.B | 131 |
| 98.40.C | 131 |
| 98.40.F | 131 |
| 98.40.H | 132 |
| 98.40.J | 133 |
| 98.40.K | 131 |
| 98.40.M | 134 |
| 98.40.N | 125 |

| ICSU-AB International Classification System for Physics | Astronomy and Astrophysics Abstracts Classification Scheme |
|---|---|
| **98.50 The Galaxy; extragalactic objects and systems** | |
| 98.50.C | 157 |
| 98.50.E | 157 (161) |
| 98.50.H | 157 (158, 161) |
| 98.50.K | 160 |
| 98.50.L | 155 |
| 98.50.M | 160 (156) |
| 98.50.R | 158 |
| 98.50.T | 160 |
| **98.70 Other objects and background radiations of unknown origin or distances** | |
| 98.70.D | 141 |
| 98.70.J | 159 |
| 98.70.L | 133 |
| 98.70.Q | 142 (143) |
| 98.70.S | 144 |
| 98.70.V | 161 (142, 143) |
| **98.80 Cosmology** | |
| 98.80.B | 161 |
| 98.80.D | 161 (066) |
| 98.80.F | 061 (161) |

# Abbreviations

Abbreviations used in *Astronomy and Astrophysics Abstracts* are primarily based on the 'International List of Periodical Title Word Abbreviations', prepared for the UNISIST/ICSU-AB Working Group on Bibliographic Descriptions (1970).

| | |
|---|---|
| **A.A.B.** | Associazione Astrofili Bolognesi |
| **Aarg.** | Aargang |
| **AAS** | American Astronomical Society |
| **AAVSO** | American Association of Variable Star Observers |
| **Abh.** | Abhandlung – |
| **Abstr.** | Abstract – |
| **Abt.** | Abteilung |
| **Acad.** | Academi –, Academy |
| **Accad.** | Accademi – |
| **Act.** | Active, Activit – |
| **Adm.** | Administr – |
| **Adv.** | Advanc – |
| **Aehron.** | Aehronomi – |
| **Aeron.** | Aeronom – |
| **Aeronaut.** | Aeronauti – |
| **Aerosp.** | Aerospace |
| **Afr.** | Africa – |
| **AG** | Astronomische Gesellschaft |
| **AIAA** | American Institute of Aeronautics and Astronautics |
| **AJB** | Astronomischer Jahresbericht |
| **Akad.** | Akadem – |
| **Ala.** | Alabama |
| **Alm.** | Almanac – |
| **Am.** | America –, Amerika –, Amerique – |
| **Amat.** | Amateur – |
| **Amst.** | Amsterdam |
| **An.** | Anais, Anale –, Anali –, Anals |
| **Anal.** | Analis –, Analit –, Analys –, Analyt – |
| **Angew.** | Angewandt – |
| **Ann.** | Annaes, Annal – |
| **Annu.** | Annu – |
| **Anst.** | Anstalt |
| **Anu.** | Anual –, Anuar – |
| **Anz.** | Anzeiger |
| **Appl.** | Applied |
| **Arb.** | Arbeit |
| **Arch.** | Archiv – |
| **Årg.** | Årgang |
| **Argent** | Argentin – |
| **Ariz.** | Arizona |
| **Ark.** | Arkiv – |
| **Arkh.** | Arkhiv – |
| **Artif.** | Artifici – |
| **ASA** | Astronomical Society of Australia |
| **Asoc.** | Asocia – |
| **ASP** | Astronomical Society of the Pacific |
| **ASSA** | Astronomical Society of Southern Africa |
| **Assem.** | Assembl – |
| **Assoc.** | Associ – |
| **Assoz.** | Assozi – |
| **Astrofis.** | Astrofisic – |
| **Astrofiz.** | Astrofizi – |
| **Astrometr.** | Astrometr – |
| **Astron.** | Astronom – |
| **Astronaut.** | Astronauti –, Astronauty – |
| **Astrophys.** | Astrophys – |
| **ASV** | Astronomical Society of Victoria |
| **ASWA** | Astronomical Society of Western Australia |
| **At.** | Atom – |
| **Atmos.** | Atmosf –, Atmosph – |
| **Aust.** | Australi – |
| **BAA** | British Astronomical Association |
| **Barc.** | Barcelona |
| **Bayer.** | Bayerisch – |
| **Beitr.** | Beitrag, Beiträge |
| **Belg.** | Belge –, Belgi – |
| **Beob.** | Beobacht – |
| **Beogr.** | Beograd – |
| **Ber.** | Bericht – |
| **Bibl.** | Bibliot – |
| **Bibliogr.** | Bibliograf –, Bibliograph – |
| **BIH** | Bureau International de l'Heure |
| **Bimest.** | Bimestr – |
| **Bl.** | Blatt, Blätter |
| **Bol.** | Boletin |
| **Boll.** | Bolletino |
| **Br.** | British |
| **Bras.** | Brasil – |
| **Brun.** | Brunens – |
| **Bruss.** | Brussel, s |
| **Brux.** | Bruxelles |
| **Bul.** | Buleten –, Buletin –, Bulten |
| **Bulg.** | Bulgar – |
| **Bull.** | Bulletin –, Bullettino |
| **Bur.** | Bureau – |
| **Byul.** | Byuleten –, Byuletin – |
| **Byull.** | Byulleten – |
| **C. R.** | Comptes Rendus |
| **Cah.** | Cahier – |
| **Calif.** | California |
| **Camb.** | Cambridge |
| **Can.** | Canadi –, Canada |
| **Carol.** | Carolina – |
| **Cas.** | Casopis |
| **Cat.** | Catalog – |
| **Celest.** | Celestial |
| **Cent.** | Center, Central, Centrale, Centrally, Centre |
| **Cercet.** | Cercetari |
| **Cesk.** | Ceskoslov – |
| **Chem.** | Chemi – |
| **Chim.** | Chimi – |
| **Chin.** | Chinese |
| **Chron.** | Chronic –, Chronik, Chronique |
| **Chronom.** | Chronometr – |
| **Cie.** | Compagnie |
| **Cienc.** | Ciencia – |
| **Cient.** | Cientific – |
| **Circ.** | Circolar –, Circolo, Circolaire –, Circular –, Circulo |
| **Cirk.** | Cirkulaer – |
| **Cl.** | Clasa, Classe – |
| **Co.** | Companies, Company |
| **Coll.** | College |
| **Collect.** | Collect – |
| **Colloq.** | Colloqui – |
| **Colo.** | Colorado |

| | |
|---|---|
| **Comet.** | Cometary |
| **Commentat.** | Commentat – |
| **Commun.** | Communica – |
| **Comput.** | Computation, Computer –, Computing |
| **Comun.** | Comunica – |
| **Conf.** | Conferen – |
| **Congr.** | Congres – |
| **Conn.** | Connecticut |
| **Contract.** | Contract – |
| **Contrib.** | Contribu – |
| **Copenh.** | Copenhagen |
| **Cosm.** | Cosmic – |
| **Cosmochim.** | Cosmochimi – |
| **COSPAR** | Committee on Space Research |
| **Crystallogr.** | Crystallograph – |
| **CSIRO** | Commonwealth Scientific and Industrial Research Organization |
| **Cult.** | Cultur –, Cultuur |
| **Curr.** | Current |
| **Czech.** | Czechoslovak – |
| **D. C.** | District of Columbia |
| **DDR** | Deutsche Demokratische Republik |
| **Del.** | Delaware |
| **Dep.** | Departament, Département, Department |
| **Dev.** | Development –, Développement – |
| **Dig.** | Digest |
| **Dir.** | Director – |
| **Diss.** | Disserta – |
| **Div.** | Divis – |
| **Doc.** | Document – |
| **Dok.** | Dokument – |
| **Dokl.** | Doklad – |
| **Dom.** | Dominion |
| **Dtsch.** | Deutsch |
| **Ed.** | Edit – |
| **Edinb.** | Edinburgh |
| **Ehksp.** | Ehksperiment – |
| **Eidg.** | Eidgenössisch – |
| **Eksp.** | Eksperiment – |
| **Electron.** | Electroni – |
| **Eng.** | Engineer – |
| **Environ.** | Environment – |
| **Equip.** | Equipement, Equipment |
| **Ergeb.** | Ergebnis – |
| **ESA** | European Space Agency |
| **ESO** | European Southern Observatory |
| **Espec.** | Especial – |
| **ESRO** | European Space Research Organization |
| **Eur.** | Europ – |
| **Eval.** | Evaluation – |
| **Exp.** | Experiment – |
| **Extraterr.** | Extraterrestr – |
| **F. R. Germany** | Federal Republic of Germany |
| **Fac.** | Facolt –, Faculd –, Facult – |
| **Fak.** | Fakult |
| **Fasc.** | Fascicul – |
| **Fenn.** | Fenni – |
| **Finn.** | Finni – |
| **Fis.** | Fisic –, Fisik – |
| **Fiz.** | Fizic –, Fizik –, Fizyk – |
| **Fla.** | Florida |
| **Fluid.** | Fluidi – |
| **Fond.** | Fondation –, Fondazione |
| **Fortschr.** | Fortschritt – |
| **Fotogr.** | Fotograf – |
| **Found.** | Foundation – |
| **Fr.** | Français – |
| **Freq.** | Frequen – |
| **Fundam.** | Fundamenta – |
| **Fys.** | Fysik –, Fysisch, Fysisk – |
| **Fyz.** | Fyzik – |

| | |
|---|---|
| **G.** | Giornale |
| **Ga.** | Georgia |
| **Gaz.** | Gazeta, Gazette |
| **Gazz.** | Gazzetta |
| **Gen.** | General |
| **Geochem.** | Geochem – |
| **Geochim.** | Geochim – |
| **Geod.** | Geodaes –, Geodaet –, Geodes –, Geodet –, Geodez – |
| **Geofis.** | Geofis – |
| **Geofiz.** | Geofiz – |
| **Geofys.** | Geofys – |
| **Geogr.** | Geograf –, Geograph – |
| **Geokhim.** | Geokhim – |
| **Geol.** | Geolog –, Geolosk – |
| **Geomagn.** | Geomagneti – |
| **Geophys.** | Geophys – |
| **Ges.** | Gesellschaft |
| **Gesch.** | Geschichte |
| **Gl.** | Glavno – |
| **Glas.** | Glasnik |
| **Gos.** | Gosudarst – |
| **Gov.** | Government – |
| **Grenzgeb.** | Grenzgebiet – |
| **GSFC** | Goddard Space Flight Center |
| **H.M.** | Her Majesty's, His Majesty's |
| **Hamb.** | Hamburg |
| **Handb.** | Handbook, Handbuch |
| **Heidelb.** | Heidelberg – |
| **Helv.** | Helveti – |
| **Her.** | Herald – |
| **Hist.** | History |
| **Hochsch.** | Hochschule |
| **Hoegsk.** | Hoegskol – |
| **HR-diagram** | Hertzsprung-Russell diagram |
| **Hung.** | Hungar – |
| **Hydrogr.** | Hydrograf –, Hydrograph – |
| **IAF** | International Astronautical Federation |
| **IAU** | International Astronomical Union |
| **IBM** | International Business Machines Corporation |
| **ICSTI** | International Council for Scientific and Technical Information |
| **ICSU** | International Council of Scientific Unions |
| **ICSU-AB** | International Council of Scientific Unions – Abstracting Board |
| **IEEE** | Institute of Electrical and Electronics Engineers |
| **Ill.** | Illinois |
| **Inc.** | Incorporated |
| **Ind.** | Industr – |
| **Inf.** | Informat –, Informaz –, Informe – |
| **Ing.** | Ingenieur |
| **INIS** | International Nuclear Information System |
| **INSPEC** | International Information Services for the Physics and Engineering Communities |
| **Inst.** | Institut –, Instytut – |
| **Instrum.** | Instrument – |
| **Int.** | International, Internazional – |
| **Intell.** | Intelligenc – |
| **Inter.** | Intérieur –, Interior |
| **Interplanet.** | Interplanetary |
| **Intez.** | Intezet – |
| **Invest.** | Investiga – |
| **Ionos.** | Ionosfer –, Ionospher – |
| **Ir.** | Irish |
| **Iskusstv.** | Iskusstvenn – |
| **Isr.** | Israel – |
| **Issled.** | Issledovan – |
| **Ist.** | Istitut |
| **Ital.** | Itali – |
| **Izd.** | Izdatel – |
| **Izv.** | Izvesti – |

**J.** Joernaal –, Jornal –, Journal –
**Jaarb.** Jaarboek –
**Jahr.** Jahrbuch, Jahrbücher
**Jahresber.** Jahresbericht –
**Jahresschr.** Jahresschrift
**Jahrg.** Jahrgang
**JPL** Jet Propulsion Laboratory
**Jpn.** Japan –

**K.** Königlich –, Koninkljik –, Kunglig –
**Kans.** Kansas
**Kartogr.** Kartograf –
**Kernforsch.** Kernforschung
**Kernphys.** Kernphysik –
**Khem.** Khemyi –
**Khim.** Khimi –
**Kim.** Kimija –, Kimya
**Kl.** Klass –
**Kolloq.** Kolloquium –
**Komet.** Kometnyj
**Komm.** Kommission –
**Konf.** Konfer –
**Kongr.** Kongress
**Kosm.** Kosmich –
**Kosmog.** Kosmogon –
**Kozp.** Kozponti
**KPNO** Kitt Peak National Observatory
**Ky.** Kentucky

**La.** Louisiana
**Lab.** Laborato –
**LEST** Large European Solar Telescope
**Lett.** Letter –, Lettra, Lettre
**Libr.** Librair –, Librar –

**Madr.** Madrid
**Mag.** Magasin, Magazin –
**Magn.** Magneti –, Magnitn –
**Mar.** Marin –
**Mass.** Massachusetts
**Mat.** Matemaat –, Matemat –
**Mater.** Material –
**Math.** Mathemat –
**Md.** Maryland
**Meas.** Measur –
**Mec.** Mecani –
**Mech.** Mechani –
**Medd.** Meddelande –, Meddelelse
**Meded.** Mededeeling, Mededeling –
**Mekh.** Mekhani –
**Mem.** Memento –, Memoir –, Memori –, Memory –, Memuary
**Memo.** Memorand –
**Mens.** Mensile, Mensual –, Mensuel –
**Messtech.** Messtechni –
**Meteorol.** Meteorolog –
**Mex.** Mexic –
**Mich.** Michigan
**Micromec.** Micromecaniq –
**Miner.** Mineral, Minerale –, Minerali –
**Mineral.** Mineralog –
**Minn.** Minnesota
**Miss.** Mississippi
**MIT** Massachusetts Institute of Technology
**Mitt.** Mitteilung –
**Mo.** Missouri
**Mod.** Modern –
**Mol.** Molecul –, Molekul –
**Mon.** Monat, Monatlich –, Month –
**Monogr.** Monograph –
**Mont.** Montana
**MPI** Max-Planck-Institut
**Mt.** Mount
**Muench.** Muenchen
**Mus.** Museum

**N. C.** North Carolina
**N. D.** North Dakota
**N. H.** New Hampshire
**N. J.** New Jersey
**N. M.** New Mexico
**N. Y.** New York
**N. Z.** New Zealand
**Nablyud.** Nablyudeni –
**Nac.** Nacion –
**Nachr.** Nachricht –
**NASA** National Aeronautics and Space Administration
**Nat.** Natur –
**Natl.** National –
**Naturforsch.** Naturforsch –
**Naturwiss.** Naturwissenschaft –
**Natuurkd.** Natuurkunde
**Nauchn.** Nauchny –
**Nauk.** Nauka, Naukite, Naukov –, Naukow –
**Naut.** Nautic –
**Nav.** Naval –
**Navig.** Navigat –
**Naz.** Nazion –
**Nebr.** Nebraska
**Ned.** Nederland –
**Nev.** Nevada
**Newsl.** Newsletter –
**Not.** Notationes, Notic –, Notise, Notizi –
**Nouv.** Nouveau –, Nouvell –
**Nov.** Novoe
**Nucl.** Nucléaire –, Nuclear –, Nucl –
**Nukl.** Nukle –
**Numer.** Numeri –

**O-va** Obshchestva
**O-vo** Obshchestvo
**Obs.** Observ –
**Österr.** Österreich –
**Off.** Offic –
**Okla.** Oklahoma
**Opt.** Optic –, Optik –, Optique
**Oreg.** Oregon
**Oss.** Osserva –

**Pa.** Pennsylvania
**Pac.** Pacific
**Paleontol.** Paleontolog –
**Pap.** Paper –, Papier
**Part.** Particle
**Pekin.** Pekinens –
**Perem.** Peremenn –
**Period.** Periodi –
**Petrol.** Petrolog –
**Philos.** Philosoph –
**Photogr.** Photograf –, Photograph –
**Photogramm.** Photogrammetr –
**Photom.** Photometr –
**Phys.** Physic –, Physik –, Physique –, Physisch
**Pict.** Picture –
**Planet.** Planetary
**Pol.** Polish, Polon –
**Pr.** Prac –
**Prelim.** Prelimin –
**Prepr.** Preprint
**Prib.** Pribor –
**Prikl.** Prikladn –
**Prilozh.** Prilozhen –
**Prir.** Prirodn –
**Prirodoved.** Prirodoved –

| | |
|---|---|
| **Probl.** | Problem – |
| **Proc.** | Proceedings |
| **Prod.** | Prodott –, Produc –, Produkt, |
| **Prog.** | Progres – |
| **Propag.** | Propagation |
| **Prospect.** | Prospecting |
| **Prov.** | Provinc –, Provints –, Provinz – |
| **Pubbl.** | Pubblicazion – |
| **Publ.** | Publicac –, Publicas –, Publicat –, Publikas –, Publikat – |
| **Q.** | Quarterly |
| **Quant.** | Quantit – |
| **R.** | Royal |
| **R. I.** | Rhode Island |
| **Radiat.** | Radiati – |
| **Radioact.** | Radioactiv –, Radioaktiv – |
| **Radioisot.** | Radioisotop – |
| **Rap.** | Raport – |
| **Rapp.** | Rapport – |
| **RAS** | Royal Astronomical Society |
| **Rec.** | Record – |
| **Rech.** | Recherche – |
| **Ref.** | Referat –, Reference –, Referieren |
| **Relat.** | Related, Relation – |
| **Relativ.** | Relativit – |
| **Rend.** | Rendicont – |
| **Rep.** | Report – |
| **Repr.** | Reprint – |
| **Repub.** | Republi – |
| **Res.** | Research – |
| **Result.** | Resultad –, Resultat – |
| **Rev.** | Review –, Revisio, Revista, Revue – |
| **Rezul't.** | Rezul'tat – |
| **Ric.** | Ricerca, Ricerche |
| **Riv.** | Rivist – |
| **Roum.** | Roumain – |
| **Rundsch.** | Rundschau |
| **S. Afr.** | South Africa |
| **S. C.** | South Carolina |
| **S. D.** | South Dakota |
| **SAF** | Société Astronomique de France |
| **SAI** | Società Astronomica Italiana |
| **Samml.** | Sammlung – |
| **SAO** | Smithsonian Astrophysical Observatory |
| **SAS** | Société Astronomique de Suisse |
| **Satell.** | Satellite |
| **Sb.** | Sbornik – |
| **Scand.** | Scandinavi – |
| **Sch.** | Schul – |
| **Schr.** | Schrift – |
| **Schriftenr.** | Schriftenreihe |
| **Schweiz.** | Schweizer – |
| **Sci.** | Scienc –, Scient –, Scienz – |
| **Scr.** | Scripta, Scritt – |
| **Secc.** | Seccion – |
| **Sect.** | Secti – |
| **Sekc.** | Sekci –, Sekcj – |
| **Sekt.** | Sektion –, Sektor – |
| **Sekts.** | Sektsi – |
| **Sel.** | Seleccion –, Select –, Selek –, Selezione |
| **Selsk.** | Selskab –, Selskap – |
| **Semin.** | Séminair –, Seminar – |
| **Sep.** | Separat – |
| **Ser.** | Seria –, Serie –, Seriya |
| **Serv.** | Servic –, Serviz – |
| **Sess.** | Sessi – |
| **Signal.** | Signalétique – |
| **Simp.** | Simpoz – |
| **Sin.** | Sinica |
| **Sitzungsber.** | Sitzungsbericht – |
| **Skr.** | Skrift – |
| **Smithson.** | Smithsonian |
| **Soc.** | Sociedad –, Societ – |
| **Sol.** | Solar |
| **Soln.** | Solnechn – |
| **Sonderdr.** | Sonderdruck – |
| **Soobshch.** | Soobshchen – |
| **South.** | Southern |
| **Spacecr.** | Spacecraft |
| **Spat.** | Spatial – |
| **Spec.** | Special – |
| **Spectrosc.** | Spectroscop – |
| **Spectrosk.** | Spectroskop – |
| **Spets.** | Spetsial – |
| **Spez.** | Spezial –, Speziell – |
| **SSR** | Sovetskaya Sotsialisticheskaya Respublika |
| **SSSR** | Soyuz Sovetskikh Sotsialisticheskikh Respublik |
| **St.** | Saint –, Sankt –, Sant – |
| **– St.** | – Straße, Street |
| **Stand.** | Standard –, Standart – |
| **Sternw.** | Sternwarte – |
| **Stiint.** | Stiintific – |
| **Stn.** | Station, Stazione |
| **Stud.** | Studia, Studie –, Studii |
| **Supl.** | Suplement –, Supliment – |
| **Suppl.** | Supplement – |
| **Surv.** | Survey – |
| **Syd.** | Sydney |
| **Symp.** | Sympos –, Sympoz – |
| **Syst.** | System – |
| **Sz.** | Szemle |
| **Teach.** | Teacher –, Teaching |
| **Tec.** | Tecni – |
| **Tech.** | Techni – |
| **Technol.** | Technolog – |
| **Tecnol.** | Tecnolog – |
| **Teh.** | Tehnic –, Tehnika, Tehnisk – |
| **Tehnol.** | Tehnolog –, Tehnolosk – |
| **Tek.** | Tekni – |
| **Tekh.** | Tekhni – |
| **Tekhnol.** | Tekhnolog – |
| **Teknol.** | Teknolog – |
| **Telesc.** | Telescop – |
| **Telev.** | Television – |
| **Tenn.** | Tennessee |
| **Teor.** | Teoret –, Teori – |
| **Terr.** | Terrestr – |
| **Test.** | Testing |
| **Tex.** | Texas |
| **TH** | Technische Hochschule |
| **Theor.** | Theoret –, Theori – |
| **Tidschr.** | Tidschrift – |
| **Tidskr.** | Tidskrift – |
| **Tidsskr.** | Tidsskrift – |
| **Top.** | Topic – |
| **Torun.** | Torunensis |
| **Tr.** | Trudy |
| **Trans.** | Transactions, Transazione |
| **Tsentr.** | Tsentral – |
| **Tsirk.** | Tsirkulyar – |
| **TU** | Technical University |
| **Uch.** | Uchen – |
| **Uchebn.** | Uchebn – |
| **UK** | United Kingdom |
| **Umsch.** | Umschau |
| **UN** | United Nations |
| **Univ.** | Universidad –, Universit –, Univerzitet – |
| **Ups.** | Upsaliens – |

**US** United States
**USA** United States of America
**USSR** Union of Soviet Socialist Republics
**Utr.** Utrecht

**Va.** Virginia
**Var.** Various
**Ver.** Verein –, Verenig –
**Veränderl.** Veränderlich –
**Verh.** Verhandl –
**Vermess.** Vermessung –
**Vermessungswes.** Vermessungswesen
**Veröff.** Veröffentlich –
**Vesn.** Vesnik
**Vestn.** Vestnik
**Vetensk.** Vetenskap –
**Vgl.** Vergleich –
**Vidensk.** Videnskab –, Videnskap
**Vierteljahresschr.** Vierteljahresschrift –
**Vierteljahrsschr.** Vierteljahrsschrift
**VLB** Very Long Baseline
**Volcanol.** Volcanolog –
**Vopr.** Vopros –
**Vortr.** Vorträge
**Vses.** Vsesoyuzn –
**Vt.** Vermont
**Vyp.** Vypusk –
**Vyssh.** Vyssh –
**Vyzk.** Vyzkum –

**W. Va.** West Virginia
**Wash.** Washington
**West.** Western
**Wet.** Wetenschap –, Wetenskap –
**Wis.** Wisconsin
**Wiss.** Wissenschaft –
**Wyo.** Wyoming

**Yad.** Yadern –

**Z.** Zeitschrift –
**ZA** Zero Age
**ZAED** Zentralstelle für Atomkernenergie-Dokumentation
**Zap.** Zapisk –, Zapyisk –
**Zaved.** Zaveden –
**Zent.** Zentral
**Zentralbl.** Zentralblatt
**Zesz.** Zeszyt
**Zh.** Zhurnal –
**Zirk.** Zirkular

# Periodicals, Proceedings, Books, Activities

## 001 Periodicals

**A.A.O. Newsl.**
A.A.O. Newsletter. Anglo–Australian Observatory, PO Box 296, Epping, N.S.W. 2121, Australia.

**AAS Photo–Bull.**
American Astronomical Society Photo–Bulletin. Working Group on Photographic Materials, 211 Space Sciences Building. University of Florida, Gainesville, Fla. 32611, USA. ISSN 0065–7433.

**Abastumanskaya Astrofiz. Obs., Byull.**
Abastumanskaya Astrofizicheskaya Observatoriya, Gora Kanobili. Byulleten'. Akademiya Nauk Gruzinskoj SSR. Izdatel'stvo Metsniereba, Tbilisi, USSR. ISSN 0375–6644.

**Abh. Akad. Wiss. DDR**
Abhandlungen der Akademie der Wissenschaften der DDR. Abteilung Mathematik, Naturwissenschaften, Technik. Akademie–Verlag, Berlin, German Democratic Republic.

**Abh. Hamb. Sternw.**
Abhandlungen aus der Hamburger Sternwarte. Hamburger Sternwarte, Universität Hamburg, Gojenbergsweg 2, D–2050 Hamburg 80, F.R. Germany. ISSN 0374–1583.

**Abstr. Submitted Pap.**
Abstracts of Submitted Papers. Institute of Astronomy, The Observatories, Madingley Road, Cambridge CB3 0HA, England.

**Acad. R. Belg., Bull. Cl. Sci.**
Académie Royale de Belgique, Bulletin de la Classe des Sciences (Koninklijke Academie van België, Mededelingen van de Klasse der Wetenschappen). 5e Série, Palais des Académies, Bruxelles, Belgium.

**Acad. R. Belg., Mém. Cl. Sci.**
Académie Royale de Belgique, Mémoires de la Classe des Sciences. Collection in 8°, 2e Série, Palais des Académies, Bruxelles, Belgium.

**Acad. Sci. Est. SSR, Div. Phys. Math. Tech. Sci., Prepr.**
Academy of Sciences of the Estonian SSR, Division of Physical, Mathematical and Technical Sciences, Preprints. Institute of Astrophysics and Atmospheric Physics, Academy of Sciences of the Estonian SSR, 202444 Tartu, Tôravere, Estonian SSR, USSR.

**Acta Acust.**
Acta Acustica. Science Press, Beijing, People's Republic of China. Subscription address: Guozi Shudian, PO Box 399, Beijing, People's Republic of China.

**Acta Astron.**
Acta Astronomica. An International Quarterly Journal. Polska Akademia Nauk, Komitet Astronomii, Państwowe Wydawnictwo Naukowe, Warszawa–Kraków, Poland. Subscription address: Ars Polona, 00–068 Warszawa, Krakowskie Przedmieście 7, Poland. ISSN 0001–5237.

**Acta Astron. Sin.**
Acta Astronomica Sinica. Purple Mountain Observatory, Academia Sinica. Nanjing, People's Republic of China. English translation in Chin. Astron. Astrophys. ISSN 0001–5245.

**Acta Astronaut.**
Acta Astronautica. Journal of the International Academy of Astronautics. Pergamon Press, Oxford – New York – Toronto – Paris – Frankfurt – Sydney. ISSN 0094–5765.

**Acta Astrophys. Sin.**
Acta Astrophysica Sinica. Beijing Astronomical Observatory, Academia Sinica, Beijing, People's Republic of China. Subscription address: Guozi Shudian, PO Box 399, Beijing, People's Republic of China. English translation in Chin. Astron. Astrophys.

**Acta Cosmologica**
Acta Cosmologica. Zeszyty Naukowe Universytetu Jagiellońskiego. Państwowe Wydawnictwo Naukowe, Warszawa–Kraków, Poland. ISSN 0137–2386.

**Acta Fac. Rerum Nat. Univ. Comenianae, Astron. Geophys.**
Acta Facultatis Rerum Naturalium Universitatis Comenianae, Astronomia et Geophysica. Published for the Komenský University by Slovenské pedagogické nakladatel'stvo, 89112 Bratislava, Czechoslovakia.

**Acta Geod. Geophys.**
Acta Geodaetica et Geophysica. Institute of Geodesy and Geophysics, Academia Sinica, Beijing, People's Republic of China. Subscription address: Guozi Shudian, PO Box 399, Beijing, People's Republic of China.

**Acta Geod. Geophys. Montan.**
Acta Geodaetica, Geophysica et Montanistica. Akademiai Kiado, H–1054 Budapest, Alkotmany utca 21, Hungary. ISSN 0374–1842.

**Acta Geophys. Pol.**
Acta Geophysica Polonica. Subscription address: Ars Polona, 00–068 Warszawa, Krakowskie Przedmieście 7, Poland. ISSN 0001–5725.

**Acta Geophys. Sin.**
Acta Geophysica Sinica. Department of Geophysical Research, Academia Sinica, Beijing, People's Republic of China. Subscription address: Guozi Shudian, PO Box 399, Beijing, People's Republic of China. ISSN 0001–5733.

**Acta Math. Sin.**
Acta Mathematica Sinica. Academia Sinica, Beijing, People's Republic of China. Subscription address: Guozi Shudian, PO Box 399, Beijing, People's Republic of China.

**Acta Mech. Sin.**
Acta Mechanica Sinica. Academia Sinica, Beijing, People's Republic of China. Subscription address: Guozi Shudian, PO Box 399, Beijing, People's Republic of China. ISSN 0254–3060.

**Acta Meteorol. Sin.**
Acta Meteorologica Sinica. Subscription address: Guozi Shudian, PO Box 399, Beijing, People's Republic of China. ISSN 0577–6619.

**Acta Phys. Acad. Sci. Hung.**
Acta Physica Academiae Scientiarum Hungaricae. Subscription address: Kultura, Hungarian Trading Co., H–1389 Budapest 62, PO Box 149, Hungary. ISSN 0001–6705.

**Acta Phys. Austriaca**
Acta Physica Austriaca. Springer–Verlag, Wien, Austria. ISSN 0001–6713.

**Acta Phys. Pol., Ser. A**
Acta Physica Polonica, Ser. A. (General Physics, Solid State Physics, Atomic and Molecular Spectroscopy, Applied Physics). Polska Akademia Nauk, Warszawa, Poland. Subscription address: Ars Polona, 00–068 Warszawa, Krakowskie Przedmieście 7, Poland. ISSN 0587–4246.

**Acta Phys. Pol., Ser. B**
Acta Physica Polonica, Ser. B (Elementary Particle Physics, Nuclear Physics, Theory of Relativity, Field Theory). Polska Akademia Nauk, Warszawa, Poland. Subscription address: Ars Polona, 00–068 Warszawa, Krakowskie Przedmieście 7, Poland. ISSN 0587–4254.

**Acta Phys. Sin.**
Acta Physica Sinica. Institute of Physics, Academia Sinica, Beijing, People's Republic of China. ISSN 0372–736X.

**Acta Phys. Slovaca**
Acta Physica Slovaca. VEDA Publishing House of the Slovak Academy of Sciences, 89530 Bratislava, Klemensova 19, Czechoslovakia. ISSN 0323–0465.

**Acta Polytech. Scand., Appl. Phys. Ser.**
Acta Polytechnica Scandinavia, Applied Physics Series. Finnish Academy of Technical Sciences, Kansakoulukatu 10A, SF–00100 Helsinki, Finland. ISSN 0355–2721.

**Acta Sci. Nat. Univ. Pekin.**
Acta Scientiarum Naturalium Universitatis Pekinensis. Beijing, People's Republic of China. ISSN 0479–8023.

**Acta Sci. Nat. Univ. Sunyatseni**
Acta Scientiarum Naturalium Universitatis Sunyatseni (Zhongshandaxue Xuebao). Canton Post Office, Canton, People's Republic of China.

**Acta Tech. Acad. Sci. Hung.**
Acta Technica Academiae Scientiarum Hungaricae. Akademiai Kiado, Budapest 1363, PO Box 24, Hungary. ISSN 0001–7035.

**Acta Tech. ČSAV**
Acta Technica Československá akademie věd. Academia, Publishing House of the Czechoslovak Academy of Sciences, Vodičkova 40, 112 29 Praha 1, Czechoslovakia. Subscription address: John Benjamins N.V., Periodical Trade, Warmoesstraat 54, Amsterdam, The Netherlands. ISSN 0001–7043.

**Acta Univ. Carol. Math. Phys.**
Acta Universitatis Carolinae. Mathematica et Physica. Fakulta matematicko–fyzikální, Karlova universita, Praha, Czechoslovakia. ISSN 0001–7140.

**A.D.I.O.N. Bull.**
A.D.I.O.N. Bulletin. Association pour le Développement International de l'Observatoire de Nice. Observatoire de Nice, B.P. 252, F–06007 Nice Cedex, France.

**Adv. Astronaut. Sci.**
Advances in the Astronautical Sciences. American Astronautical Society, Publicatons Office, PO Box 28130, San Diego, Calif. 92128, USA. ISSN 0065–3438.

**Adv. Phys.**
Advances in Physics. Taylor & Francis Ltd., London, England. ISSN 0001–8732.

**Adv. Space Res.**
Advances in Space Research. The official journal of the Committee on Space Research (COSPAR). Pergamon Press, Oxford – New York – Toronto – Sydney – Paris – Frankfurt. ISSN 0273–1177.

**Aéronaut. Astronaut.**
L'Aéronautique et l'Astronautique. Editions Air et Cosmos, 6 Rue Anatole de la Forge, F–75017 Paris, France. ISSN 0001–9275.

**AIAA J.**
AIAA Journal. American Institute of Aeronautics and Astronautics, 1290 Avenue of the Americas, New York, N.Y. 10019, USA. ISSN 0001–1452.

**AIP Conf. Proc.**
AIP Conference Proceedings. American Institute of Physics, 335 East 45th Street, New York, N.Y. 10017, USA. ISSN 0094–243X.

**Algoritm. Nebesnoj Mekh.**
Algoritmy Nebesnoj Mekhaniki (Materialy Matematicheskogo Obespecheniya EhVM). Institut Teoreticheskoj Astronomii Akademii Nauk SSSR. 191187 Leningrad, D–187, nab. Kutuzova, dom–10. USSR.

**Alta Freq.**
Alta Frequenza. Ufficio Centrale AEI–CEI, Viale Monza 259, I–20126 Milano, Italy. ISSN 0002–6557.

**Am. Assoc. Variable Star Obs. Bull.**
The American Association of Variable Star Observers Bulletin. The American Association of Variable Star Observers, 187 Concord Avenue, Cambridge, Mass. 02138, USA.

**Am. Assoc. Variable Star Obs. Rep.**
The American Association of Variable Star Observers Report. The American Association of Variable Star Observers, 187 Concord Avenue, Cambridge, Mass. 02138, USA.

**Am. J. Phys.**
American Journal of Physics. Published for the American Association of Physics Teachers by the American Institute of Physics, 335 East 45th Street, New York, N.Y. 10017, USA. ISSN 0002–9505.

**Am. Mineral.**
American Mineralogist. Mineralogical Society of America, 2000 Florida Avenue, N.W., Washington, D.C. 20009, USA. ISSN 0003–004X.

**Am. Sci.**
American Scientist. Society of Sigma XI, 345 Whitney Avenue, New Haven, Conn. 06510, USA. ISSN 0003–0996.

**An. Acad. Bras. Cienc.**
Anais da Academia Brasileira de Ciencias. Caixa Postal 229, ZC–00 Rio de Janeiro gb, Brazil. ISSN 0001–3765.

**An. Fis., Ser. B**
Anales de Fisica, Serie B (Aplicaciones, Metodes e Instrumentos). Real Sociedad Espanola de Fisica, Facultades de Ciencias, Ciudad Universitaria, Madrid 3, Spain. ISSN 0211–6251.

**Anal. Chem.**
Analytical Chemistry. American Chemical Society, 1155 16th Street N.W., Washington, D.C. 20036, USA. ISSN 0003–2700.

**Anglo–Aust. Obs., Prepr.**
Anglo–Australian Observatory, Preprints (AAO PP). Anglo–Australian Observatory, PO Box 296, Epping, N.S.W. 2121, Australia.

**Anglo–Aust. Telesc., Annu. Rep.**
Anglo–Australian Telescope, Annual Report. Anglo–Australian Observatory, PO Box 296, Epping, N.S.W. 2121, Australia.

**Ann. Acad. Sci. Fenn., Ser. A VI**
Annales Academiae Scientiarum Fennicae, Series A VI (Physica). Academia Scientiarum Fennica, Snellmaninkato 9–11, 00 170 Helsinki 17, Finland. ISSN 0066–2003.

**Ann. Fond. Louis de Broglie**
Annales de la Fondation Louis de Broglie. Fondation Louis de Broglie, 1 rue Montgolfier, F–75003 Paris, France.

**Ann. Geophys.**
Annales Geophysicae. European Geophysical Society. Gauthier–Villars. Subscription address: C.D.R. Centrale des Revues, 11 rue Gossin, F–92543 Montrouge Cedex, France.

**Ann. Inst. Henri Poincaré, Sect. A**
Annales de l'Institut Henri Poincaré, Section A (Physique Théoretique). 11 Rue Pierre–Curie, Paris 5, France. ISSN 0020–2339.

**Ann. Isr. Phys. Soc.**
Annals of the Israel Physical Society. c/o Department of Physics, Bar–Ilan University, Ramat–Gan, Israel. Adam Hilger Ltd., Bristol, England. ISSN 0309–8710.

**Ann. N.Y. Acad. Sci.**
Annals of the New York Academy of Sciences. The New York Academy of Sciences, 2 East 63rd Street, New York, N.Y., USA. ISSN 0077–8923.

**Ann. Nucl. Energy**
Annals of Nuclear Energy. Pergamon Press, Oxford – New York – Toronto – Paris – Frankfurt – Sydney. ISSN 0306–4549.

**Ann. Obs. Astron. Alger**
Annales de l'Observatoire Astronomique d'Alger. Observatoire Astronomique de l'Université d'Alger, Algiers, Algeria.

**Ann. Phys. (Leipzig)**
Annalen der Physik (Leipzig). Johann Ambrosius Barth, Salomonstr. 18B, Leipzig 701, German Democratic Republic. ISSN 0003–3804.

**Ann. Phys. (N.Y.)**
Annals of Physics (New York). Academic Press Inc., New York – London. ISSN 0003–4916.

**Ann. Phys. (Paris)**
Annales de Physique (Paris). Masson et Cie S.A., 120 Boulevard Saint–Germain, F–75280 Paris Cedex 06, France. ISSN 0003–4169.

**Ann. Sci.**
Annals of Science. Taylor & Francis Ltd., London, England. ISSN 0003–3790.

**Ann. Sci. Univ. Besançon, Phys.**
Annales Scientifiques de l'Université de Besançon, Physique. Institut des Sciences Naturelles, Place Leclerc, F–25000 Besançon, France. ISSN 0365–6543.

**Ann. Shanghai Obs., Acad. Sin.**
Annals of Shanghai Observatory, Academia Sinica. Shanghai Scientific and Technical Publishers, Shanghai, Rei Jing Er Street 450, People's Republic of China.

**Ann. Soc. Sci. Brux., Sér. I**
Annales de la Société Scientifique de Bruxelles, Série I (Sciences Mathématiques, Astronomiques et Physiques). Rue de Bruxelles 61, B–5000 Namur, Belgium. ISSN 0037–959X.

**Ann. Tokyo Astron. Obs., Second Ser.**
Annals of the Tokyo Astronomical Observatory, Second Series. Tokyo Astronomical Observatory, University of Tokyo, Mitaka, Tokyo 181, Japan. ISSN 0082–4704.

**Ann. Univ.–Sternw. Wien**
Annalen der Universitäts–Sternwarte Wien. Institut für Astronomie der Universität Wien, Türkenschanzstr. 17, A–1180 Wien, Austria. Published by Ferd. Dümmlers Verlag, Bonn, F.R. Germany. ISSN 0342–4030.

**Annu. Rep. Astron. Inst. Greece**
Annual Reports of the Astronomical Institutes of Greece. Published by the Greek National Committee for Astronomy, Athens, Greece.

**Annu. Rep. (B.I.H.)**
Annual Report (B.I.H.). Bureau International de l'Heure, 61, avenue de l'Observatoire, F–75014 Paris, France.

**Annu. Rep. Dir., Mt. Wilson Las Campanas Obs.**
Annual Report of the Director, The Mt. Wilson and Las Campanas Observatories. Mt. Wilson and Las Campanas Observatories of the Carnegie Institution of Washington, 813 Santa Barbara Street, Pasadena, Calif. 91101–1292, USA.

**Annu. Rep. Geophys. Obs.**
Annual Report of Geophysical Observations. The International Latitude Observatory of Mizusawa, Mizusawa–Shi, Iwate–Ken, Japan. ISSN 0579–5958.

**Annu. Rep. Int. Polar Motion Serv.**
Annual Report of the International Polar Motion Service. Central Bureau of the International Polar Motion Service, International Latitude Observatory of Mizusawa, Mizusawa–Shi, Iwate–Ken, Japan. ISSN 0074–7432.

**Annu. Rep. Meteorol. Obs. Int. Latitude Obs. Mizusawa**
Annual Report of the Meteorological Observations made at the International Latitude Observatory of Mizusawa. The International Latitude Observatory of Mizusawa, Mizusawa–Shi, Iwate–Ken, Japan. ISSN 0303–8378.

**Annu. Rev. Astron. Astrophys.**
Annual Review of Astronomy and Astrophysics. Annual Reviews Inc., 4139 El Camino Way, Palo Alto, Calif. 94306, USA. ISSN 0066–4146.

**Annu. Rev. Earth Planet. Sci.**
Annual Review of Earth and Planetary Sciences. Annual Reviews Inc., 4139 El Camino Way, Palo Alto, Calif. 94306, USA. ISSN 0084–6597.

**Annu. Rev. Fluid Mech.**
Annual Review of Fluid Mechanics. Annual Reviews Inc., 4139 El Camino Way, Palo Alto, Calif. 94306, USA. ISSN 0066–4189.

**Annu. Rev. Nucl. Part. Sci.**
Annual Review of Nuclear and Particle Science. Annual Reviews Inc., 4139 El Camino Way, Palo Alto, Calif. 94306, USA. ISSN 0163–8998.

**Annu. Univ. Sofia**
Annuaire de l'Université de Sofia. Faculté de Physique. Bibliothèque de l'Université, Sofia, Bulgaria. ISSN 0584–0279.

**Antenna**
L'Antenna. Via Monte Generoso 6/a, I–20155 Milano, Italy. ISSN 0003–5386.

**Anz. Österr. Akad. Wiss., Math.–Naturwiss. Kl.**
Anzeiger der Österreichischen Akademie der Wissenschaften, Mathematisch–Naturwissenschaftliche Klasse. Springer–Verlag, Wien, Austria. ISSN 0065–535X.

**Appl. Opt.**
Applied Optics. A monthly publication of the Optical Society of America. American Institute of Physics, 335 East 45th Street, New York, N.Y. 10017, USA. ISSN 0003–6935.

**Appl. Phys., B**
Applied Physics, B (Photophysics and Laser Chemistry). Springer–Verlag, Berlin – Heidelberg – New York – Tokyo. ISSN 0721–7269.

**Appl. Phys. Lett.**
Applied Physics Letters. American Institute of Physics, 335 East 45th Street, New York, N.Y. 10017, USA. ISSN 0003–6951.

**Appl. Spectrosc.**
Applied Spectroscopy. Society for Applied Spectroscopy, 428 East Preston Street, Baltimore, Md. 21202, USA. ISSN 0003–7028.

**Appl. Spectrosc. Rev.**
Applied Spectroscopy Reviews. Marcel Dekker Inc., 270 Madison Avenue, New York, N.Y. 10016, USA. ISSN 0570–4928.

**Arch. Hist. Exact Sci.**
Archive for History of Exact Sciences. Springer–Verlag, Berlin – Heidelberg – New York – Tokyo. ISSN 0003–9519.

**Arch. Int. Hist. Sci.**
Archives Internationales d'Histoire des Sciences. Istituto dell'Enciclopedia Italiana, fondata da Giovanni Treccani, Roma, Italy. ISSN 0003–9810.

**Arch. Sci.**
Archives des Sciences. Société de Physique et d'Histoire Naturelle de Genève. Subscription address: Librairie Payot, 6 Rue Grenus, CH–1211 Geneva 11, Switzerland. ISSN 0003–9705.

**Archaeoastronomy (U.K.)**
Archaeoastronomy. Supplement to Journal for the History of Astronomy. Science History Publications Ltd., Halfpenny Furze, Mill Lane, Chalfont St Giles, Bucks. England, HP8 4NR. ISSN 0142–7253.

**Archaeoastronomy (U.S.A.)**
Archaeoastronomy. The Bulletin of the Center for Archaeoastronomy, Space Sciences Building, University of Maryland, College Park, Md. 20742, USA. ISSN 0190–9940.

**Archaeometry**
Archaeometry. Research Laboratory for Archaeology and the History of Art, Oxford University, 6 Keble Road, Oxford OX1 3QJ, England. ISSN 0003–813X.

**Archenhold–Sternw. Berlin–Treptow, Sonderdr.**
Archenhold–Sternwarte Berlin–Treptow, Sonderdruck. Archenhold–Sternwarte, DDR–1193 Berlin, Alt Treptow 1, German Democratic Republic.

**Archenhold–Sternw. Berlin–Treptow, Vortr. Schr.**
Archenhold–Sternwarte Berlin–Treptow, Vorträge und Schriften. Archenhold–Sternwarte, DDR–1193 Berlin, Alt Treptow 1, German Democratic Republic. ISSN 0570–6262.

**Ark. Fys. Semin. Trondheim**
Arkiv for det Fysiske Seminar i Trondheim. c/o Institutt for Teoretisk Fysikk, Universitetet i Trondheim, NTH, N–7034 Trondheim, Norway. ISSN 0365–2459.

**Ark. Mat.**
Arkiv för Matematik. Institut Mittag–Leffler, Auravägen 17, S–182 62 Djursholm, Sweden. ISSN 0004–2080.

**Armagh Obs., Repr.**
Armagh Observatory, Reprints. Armagh Observatory, Armagh BT61 9DG, Northern Ireland.

**Artif. Satell.**
Artificial Satellites. Polish Academy of Sciences, Space Research Committee. Państwowe Wydawnictwo Naukowe, Warszawa–Łódź, Poland. Subscription address: Ars Polona, 00–068 Warszawa, Krakowskie Przedmieście 7, Poland. ISSN 0571–205X.

**Astrofiz. Issled. Izv. Spets. Astrofiz. Obs.**
Astrofizicheskie Issledovaniya. Izvestiya Spetsial'noj Astrofizicheskoj Observatorii. Akademiya Nauk SSSR. Izdatel'stvo Nauka, Leningradskoe Otdelenie, Leningrad. 199164 Leningrad, V–164, Mendeleevskaya l., 1, USSR. English translation in Bull. Spec. Astrophys. Obs. – North Caucasus. ISSN 0324–1459.

**Astrofizika**
Astrofizika. Izdatel'stvo Akademii Nauk Armyanskoj SSR, Erevan, USSR. English translation in Astrophysics. ISSN 0571–7132.

**Astrofys. Inst., Vrije Univ. Bruss., Overdruk**
Astrofysisch Instituut, Vrije Universiteit Brussel, Overdruk. Astrofysisch Instituut, Vrije Universiteit Brussel, Pleinlaan 2, B–1050 Brussel, Belgium.

**Astrometr. Astrofiz.**
Astrometriya i Astrofizika. Respublikanskij Mezhvedomstvennyj Sbornik. Akademiya Nauk Ukrainskoj SSR, Glavnaya Astronomicheskaya Observatoriya. Naukova Dumka, 252601 Kiev, Repina 3, USSR. ISSN 0582–8198.

**Astron. Astrophys.**
Astronomy and Astrophysics. A European Journal. Springer–Verlag, Berlin – Heidelberg – New York – Tokyo. ISSN 0004–6361.

**Astron. Astrophys., Suppl. Ser.**
Astronomy and Astrophysics, Supplement Series. A European Journal. Les Editions de Physique, Z.I. de Courtaboeuf, B.P. 112, F–91944 Les Ulis Cedex, France. ISSN 0365–0138.

**Astron. Bull., Carter Obs.**
Astronomical Bulletin, Carter Observatory. Carter Observatory, PO Box 2909, Wellington 1, New Zealand. ISSN 0373–7268.

**Astron. Circ.**
Astronomical Circular. Edited by the Chinese Astronomical Society. Compiled by the Editors of Acta Astronomica Sinica, Purple Mountain Observatory, Academia Sinica, Nanjing, People's Republic of China.

**Astron. Contrib. Univ. Manchester, Ser. II: Jodrell Bank Repr.**
Astronomical Contributions from the University of Manchester, Series II: Jodrell Bank Reprints. Nuffield Radio Astronomy Laboratories, Jodrell Bank, Macclesfield, Cheshire SK11 9DL, England.

**Astron. Contrib. Univ. Manchester, Ser. III**
Astronomical Contributions from the University of Manchester, Series III. Department of Astronomy, University of Manchester, Oxford Road, Manchester M13 9PL, England.

**Astron. Data Cent. Bull.**
AstronomicaL Data Center Bulletin. National Space Science Data Center/World Data Center A for Rockets and Satellites. National Aeronautics and Space Administration, Goddard Space Flight Center, Greenbelt, Md. 20771, USA.

**Astron. Express**
Astronomy Express. Cambridge University Press, London – New York – New Rochelle – Melbourne – Sydney. ISSN 0265–5365.

**Astron. Her.**
Astronomical Herald. Astronomical Society of Japan, Tokyo Astronomical Observatory, Oosawa Mitaka, Tokyo, Japan. ISSN 0374–2466.

**Astron. Inst. "Anton Pannekoek", Univ. Amst., Repr.**
Astronomical Institute "Anton Pannekoek", University of Amsterdam, Reprint.

**Astron. J.**
The Astronomical Journal. Published for the American Astronomical Society by the American Institute of Physics, 335 East 45th Street, New York, N.Y. 10017, USA. ISSN 0004–6256.

**Astron. Mitt. Wien**
Astronomische Mitteilungen Wien. Institut für Astronomie der Universität Wien, Türkenschanzstraße 17, A–1180 Wien, Austria.

**Astron. Nachr.**
Astronomische Nachrichten. Akademie–Verlag, DDR–108 Berlin, Leipziger Straße 3–4, German Democratic Republic. ISSN 0004–6337.

**Astron. Obs. Trieste, Publ.**
Astronomical Observatory Trieste, Publications. Osservatorio Astronomico di Trieste, Via G.B. Tiepolo 11, I–34131 Trieste, Italy.

**Astron. Pap.**
Astronomical Papers prepared for the use of the American Ephemeris and Nautical Almanac. Published by the Nautical Almanac Office, U.S. Naval Observatory by direction of the Secretary of the Navy and under the authority of Congress. U.S. Government Printing Office, Washington, D.C., USA.

**Astron. Q.**
The Astronomical Quarterly. Pachart Publishing House, 1130 San Lucas Circle, Tucson, Ariz. 85704, USA. ISSN 0364–9229.

**Astron. Raumfahrt**
Astronomie und Raumfahrt. Kulturbund der DDR, Zentrale Kommission Astronomie und Raumfahrt. Redaktionssitz: 9630 Crimmitschau, Pionier– und Jugendsternwarte "Johannes Kepler", Strasse der Jugend 8. Available from Zeitungsvertriebsamt, Abt. Export, 1004 Berlin, Strasse der Pariser Kommune 3–4, German Democratic Republic. ISSN 0587–565X.

**Astron. Rechen-Inst. Heidelb., Mitt., Ser. A**
Astronomisches Rechen–Institut Heidelberg, Mitteilungen, Serie A. Astronomisches Rechen–Institut, Mönchhofstraße 12–14, D–6900 Heidelberg, F.R. Germany.

**Astron. Rechen-Inst. Heidelb., Mitt., Ser. B**
Astronomisches Rechen–Institut Heidelberg, Mitteilungen, Serie B. Astronomisches Rechen–Institut, Mönchhofstraße 12–14, D–6900 Heidelberg, F.R. Germany.

**Astron. Rep.**
The Astronomical Reports. Polskie Towarzystwo Miłośników Astronomii, Polish Amateur Astronomical Society, ul. Ludwika Solskiego 30/8, PL–31–027 Kraków, Poland.

**Astron. Sch.**
Astronomie in der Schule. Verlag Volk und Wissen, DDR–1086 Berlin, Krausenstraße 50, Postfach 1213, German Democratic Republic. ISSN 0004–6310.

**Astron. Tidsskr.**
Astronomisk Tidsskrift. Astronomisk Selskab, Kobenhavn; Norsk Astronomisk Selskap, Oslo; Svenska Astronomiska Sällskapet, Stockholm. Subscription address: Svenska Astronomiska Sällskapet, Stockholms Observatorium, S–13300 Saltsjöbaden, Sweden. ISSN 0004–6345.

**Astron. Tsirk.**
Astronomicheskij Tsirkulyar. Izdavaemyj Byuro Astronomicheskikh Soobshchenij Akademii Nauk SSSR. ISSN 0365–7248.

**Astron. Vestn.**
Astronomicheskij Vestnik. Izdatel'stvo Nauka, 918, Moskva, K–9, USSR. English translation in Sol. Syst. Res. ISSN 0320–930X.

**Astron. Zh.**
Astronomicheskij Zhurnal. Akademiya Nauk SSSR. Izdatel'stvo Nauka, Moskva, 103717, GSP, Moskva, K–62, Podsosenskij per., 21, USSR. English translation in Soviet Astron. ISSN 0004–6299.

**Astronomia**
Astronomia. Periodico trimestrale dell'Unione Astrofili Italiani. Subscription address: L. Baldinelli, C.P. 1630, I–40100 Bologna A.D., Italy. ISSN 0392–2308.

**Astronomie**
L'Astronomie et Bulletin de la Société Astronomique de France. Société Astronomique de France, 3, Rue Beethoven, F–75016 Paris, France. ISSN 0004–6302.

**Astronomy**
Astronomy. Astro Media Corp., 625 E. St. Paul Avenue, PO Box 92788 Milwaukee, Wis. 53202, USA. ISSN 0091–6358.

**Astrophys. J.**
The Astrophysical Journal. Published by The University of Chicago Press for the American Astronomical Society. The University of Chicago Press, 5801 S. Ellis Avenue, Chicago, Ill. 60637, USA. ISSN 0004–637X.

**Astrophys. J., Lett. Ed.**
The Astrophysical Journal, Letters to the Editor. Published by The University of Chicago Press for the American Astronomical Society. The University of Chicago Press, 5801 S. Ellis Avenue, Chicago, Ill. 60637, USA. ISSN 0571–7248.

**Astrophys. J., Suppl. Ser.**
The Astrophysical Journal, Supplement Series. Published by The University of Chicago Press for the American Astronomical Society. The University of Chicago Press, 5801 S. Ellis Avenue, Chicago, Ill. 60637, USA. ISSN 0067–0049.

**Astrophys. Lett.**
Astrophysical Letters. Gordon and Breach Science Publishers Inc., New York – London – Paris. ISSN 0004–6388.

**Astrophys. Prepr. Ser.**
Astrophysics Preprint Series. Astronomy and Astrophysics Group, Faculty of Physics, Pontificia Universidad Católica de Chile, Casilla 6014, Santiago de Chile, Chile.

**Astrophys. Relativ., Prepr. Ser.**
Astrophysics and Relativity, Preprint Series. Department of Applied Mathematics and Astronomy, University College, PO Box 78, Cardiff CF1 1XL, England.

**Astrophys. Space Phys. Rev.**
Astrophysics and Space Physics Reviews. Soviet Scientific Reviews, Section E. Harwood Academic Publishers GmbH, PO Box 786, Cooper Station, New York, N.Y. 10276, USA. ISSN 0143–0432.

**Astrophys. Space Sci.**
Astrophysics and Space Science. An International Journal of Cosmic Physics. D. Reidel Publishing Company, Dordrecht – Boston. ISSN 0004–640X.

**Astrophysics**
Astrophysics. A cover–to–cover translation of Astrofizika of the Academy of Sciences of the Armenian SSR. Consultants Bureau, 227 West 17th Street, New York, N.Y. 10011, USA. ISSN 0004–6396.

**Atmos. Environ.**
Atmospheric Environment. Pergamon Press, Oxford – New York – Toronto – Paris – Frankfurt – Sydney. ISSN 0004–6981.

**Atomkernenerg. Kerntech.**
Atomkernenergie und Kerntechnik. Verlag Karl–Thiemig AG, Postfach 900740, D–8000 München 90, F.R. Germany. ISSN 0171–5747.

**Atti Accad. Naz. Lincei, Ser. Ottava, Rend.**
Atti della Accademia Nazionale dei Lincei. Serie Ottava, Rendiconti. Classe di Scienze fisiche, matematiche e naturali. Accademia Nazionale dei Lincei, Roma, Italy. ISSN 0001–4435.

**Atti Accad. Sci. Torino I**
Atti della Accademia delle Scienze di Torino I. Classe di Scienze fisiche, matematiche e naturali. Via Accademia delle Scienze 6, Via Maria Vittoria 3, Torino (208), Italy. ISSN 0001–4419.

**Atti Fond. Giorgio Ronchi**
Atti della Fondazione Giorgio Ronchi. Largo Enrico Fermi 1, I–50125 Arcetri–Firenze, Italy. ISSN 0015–606X.

**Aust. Comput. J.**
Australian Computer Journal. Australian Trade Publications, 28 Chippen Street, Chippendale, N.S.W. 2008, Australia. ISSN 0004–8917.

**Aust. J. Geod. Photogramm. Surv.**
Australian Journal of Geodesy, Photogrammetry, and Surveying. School of Surveying, University of New South Wales, PO Box 1, Kensington, NSW 2033, Australia. ISSN 0313–9220.

**Aust. J. Phys.**
Australian Journal of Physics. Commonwealth Scientific and Industrial Research Organization (CSIRO), 314 Albert Street, East Melbourne, Victoria 3002, Australia. ISSN 0004–9506.

**Aust. Phys.**
Australian Physicist. Australian Institute of Physics, Science Centre, 35–43 Clarence Street, Sydney, NSW 2000, Australia. ISSN 0004–9972.

**Austrian Pap. Asteroids**
Austrian Papers on Asteroids. Institut für Astronomie, Universitätsplatz 5, A–8010 Graz, Austria.

**Automatica**
Automatica. Pergamon Press, Oxford – New York – Frankfurt – Paris – Sydney – Toronto – Tokyo. ISSN 0005–1098.

**BAV Mitt.**
BAV Mitteilungen. Berliner Arbeitsgemeinschaft für Veränderliche Sterne e.V., Sternwarte, Munsterdamm 90, D–1000 Berlin 41, F.R. Germany.

**BAV Rundbrief**
BAV Rundbrief. Mitteilungsblatt der Berliner Arbeitsgemeinschaft für Veränderliche Sterne. BAV Berliner Arbeitsgemeinschaft für Veränderliche Sterne e.V., Sternwarte, Munsterdamm 90, D–1000 Berlin 41, F.R. Germany. ISSN 0405–5497.

**BBSAG Bull.**
Bedeckungsveränderlichen Beobachter der Schweizerischen Astronomischen Gesellschaft Bulletin. Available from K. Locher, Rebrain 39, 8624 Grüt, Switzerland.

**B.I.H. Circ.**
Bureau International de l'Heure (B.I.H.) Circulars A, D, E. 61, avenue de l'Observatoire, F–75014 Paris, France.

**Biul. Obs. Astron. Uniw. M. Kopernika Toruniu**
Biuletyn Obserwatorium Astronomicznego Uniwersytetu M. Kopernika w Toruniu. Institute of Astronomy, Nicolaus Copernicus University, Chopina 12/18, PL–87–100 Toruń, Poland.

**Bol. Acad. Cienc. Fis. Mat. Nat.**
Boletin de la Academia de Ciencias Fisicas Matematicas y Naturales. Academia de Ciencias Fisicas Matematicas y Naturales, Apartado de Correo 1421, Caracas, Venezuela. ISSN 0366–1652.

**Bol. Asoc. Argent. Astron.**
Boletin de la Asociación Argentina de Astronomía, La Plata, Argentina. ISSN 0571–3285.

**Bol. Astron.**
Boletin Astronômico. Observatório do Capricórnio, Prefeitura Municipal de Campinas–SP, Brazil.

**Bol. Astron. Obs. Madr.**
Boletín Astronómico del Observatorio de Madrid. Instituto Geografico Nacional, General Ibáñez de Ibero, 3. Madrid 3, Spain. ISSN 0373–7101.

**Bol. Astron. R Muscae**
Boletin Astronomico y R Muscae. Publicación Oficial de la Liga Ibero–Americana de Astronomía, LIADA. Apartado 700, Mérida 5101–A, Venezuela. Casilla de Correos Nro 51 Suc. 48B, 1448 Buenos Aires, Argentina.

**Bol. Inst. Tonantzintla**
Boletin del Instituto de Tonantzintla. Instituto Nacional de Astrofisica, Optica y Electronica, Apartados Postales Nos. 216 y 51, Puebla, Pue., Mexico. ISSN 0303–7584.

**Bol. Obs. Ebro**
Boletín del Observatorio del Ebro. Consejo Superior de Investigaciones Científicas, Observatorio del Ebro, Roquetas (Tarragona), Spain. ISSN 0211–5166.

**Boll. Geofis. Teor. Appl.**
Bollettino di Geofisica Teorica ed Applicada. Osservatorio Geofisico Sperimentale, I–34123 Trieste, Italy. ISSN 0006–6729.

**Boundary–Layer Meteorol.**
Boundary–Layer Meteorology. D. Reidel Publishing Company, Dordrecht – Boston – London. ISSN 0006–8314.

**Boyden Obs., Publ.**
Boyden Observatory, Publication. Boyden Observatory, Astronomy Department, University of the Orange Free State, PO Box 339, Bloemfontein 9300, South Africa.

**Boyden Obs., Repr.**
Boyden Observatory, Reprint. Boyden Observatory, Astronomy Department, University of the Orange Free State, PO Box 339, Bloemfontein 9300, South Africa.

**Br. Astron. Assoc. Circ.**
British Astronomical Association Circular. British Astronomical Association, Burlington House, Piccadilly, London W1V 0NL, England. ISSN 0264–4185.

**Br. J. Philos. Sci.**
British Journal for the Philosophy of Science. Cambridge University Press, Cambridge – London – New York – New Rochelle – Melbourne – Sydney. ISSN 0007–0882.

**Br. J. Photogr.**
British Journal of Photography. Henry Greenwood & Co. Ltd., 28 Great James Street, London WC1N 3HL, England. ISSN 0007–1196.

**Bulg. J. Phys.**
Bulgarian Journal of Physics. Bulgarian Academy of Sciences, Faculty of Physics, 5 Anton Ivanov Blvd., 1126 Sofia, Bulgaria. ISSN 0323–9217.

**Bull. Am. Astron. Soc.**
Bulletin of the American Astronomical Society. Published by the American Institute of Physics, 335 East 45th Street, New York, N.Y. 10017, USA. ISSN 0002–7537.

**Bull. Assoc. Fr. Obs. Etoiles Variables**
Bulletin de l'Association Française des Observateurs d'Etoiles Variables. Revue trimestrielle. A.F.O.E.V. Observatoire de Lyon, F–69230 Saint Genis Laval, France. ISSN 0153–9949.

**Bull. Astron.**
Bulletin Astronomique. Observatoire Royal de Belgique (Astronomisch Bulletin. Koninklijke Sterrenwacht van België). Observatoire Royal de Belgique, 3, avenue Circulaire, Uccle, B–1180 Bruxelles, Belgium.

**Bull. Astron. Inst. Czech.**
Bulletin of the Astronomical Institutes of Czechoslovakia. Academia, Publishing House of the Czechoslovak Academy of Sciences, Vodičkova 40, 11229 Praha 1, Czechoslovakia. ISSN 0004–6248.

**Bull. Astron. Soc. India**
Bulletin of the Astronomical Society of India. Astronomical Society of India, Osmania University, Hyderabad 500 007, India. ISSN 0304–9523.

**Bull. Aust. Math. Soc.**
Bulletin of the Australian Mathematical Society. University of Queensland Press, St. Lucia, Queensland 4067, Australia. ISSN 0004–9727.

**Bull. Cl. Sci., Acad. R. Belg.**
Bulletin de la Classe des Sciences, Académie Royale de Belgique. Académie Royale des Sciences des Lettres et de Beaux–Arts de Belgique, Brussels, Belgium. ISSN 0001–4141.

**Bull. Crimean Astrophys. Obs.**
Bulletin of the Crimean Astrophysical Observatory. A cover–to–cover translation of Izv. Krymskoj Astrofiz. Obs. Allerton Press, Inc. 150 5th Avenue, New York, N.Y. 10011, USA.

**Bull. Etoiles Tardives Spectre Particulier**
Bulletin sur les Etoiles Tardives à Spectre Particulier (Newsletter of Chemically Peculiar Late–type Stars). Observatoire de Strasbourg, 11, rue de l'Université, F–67000 Strasbourg, France.

**Bull. Géod.**
Bulletin Géodésique. The Journal of the International Association of Geodesy. Bureau Central de l'Association Internationale de Géodésie, 39 rue Gay–Lussac, F–75005 Paris, France. ISSN 0007–4632.

**Bull. Geogr. Surv. Inst.**
Bulletin of the Geographical Survey Institute. Geographical Survey Institute, Ministry of Construction, Kitasato–1, Yatabe–Machi, Tsukuba–Gun, Ibaraki–ken, Japan. ISSN 0373–7160.

**Bull. Geophys.**
Bulletin of Geophysics. National Central University, Chung–Li, Taiwan, Republic of China. ISSN 0253–4800.

**Bull. Inf. Cent. Données Stellaires**
Bulletin d'Information du Centre de Données Stellaires. Observatoire de Strasbourg, 11, rue de l'Université, F–67000 Strasbourg, France. ISSN 0242–6536.

**Bull. Inf. Etoiles Be**
Bulletin d'Information sur les Etoiles Be (Be Star Newsletter). Observatoire de Strasbourg, 11, rue de l'Université, F–67000 Strasbourg, France.

**Bull. Obs. Astron. Belgr.**
Bulletin de l'Observatoire Astronomique de Belgrade. Observatoire Astronomique de Belgrade, Beograd, Volgina 7, Yugoslavia. ISSN 0373–3734.

**Bull. Res. Inst. Sci. Meas., Tôhoku Univ.**
Bulletin of the Research Institute for Scientific Measurements, Tôhoku University, Sendai, Japan. ISSN 0040–8689.

**Bull. Signal., Sect. 120**
Bulletin Signalétique, Section 120: Géophysique Externe, Astronomie et Astrophysique. Centre National de la Recherche Scientifique, Centre de Documentation Scientifique et Technique, 26, rue Boyer, F–75971 Paris Cedex 20, France. ISSN 0240–894X.

**Bull. Soc. R. Sci. Liège**
Bulletin de la Société Royale des Sciences de Liège. L'Université, 15 Avenue des Tilleurs, Liège, Belgium. ISSN 0037–9565.

**Bull. Spec. Astrophys. Obs. – North Caucasus**
Bulletin of the Special Astrophysical Observatory – North Caucasus. A cover–to–cover translation of Astrofiz. Issled. Izv. Spets. Astrofiz. Obs. Allerton Press, Inc., 150 5th Avenue, New York, N.Y. 10011, USA.

**Bull., Time Serv. Mizusawa Obs.**
Bulletins, Time Service of the Mizusawa Observatory. The International Latitude Observatory of Mizusawa, Mizusawa–Shi, Iwate–Ken, Japan. ISSN 0580–6585.

**Bull. Tokyo Gakugei Univ., Ser. IV**
Bulletin of Tokyo Gakugei University, Series IV (Mathematics and Natural Sciences) 4–1–1 Nukui–kita–machi, Koganei, Tokyo, Japan. ISSN 0371–6813.

**Byull. Inst. Astrofiz.**
Byulleten' Instituta Astrofiziki. Akademiya Nauk Tadzhikskoj SSR. Izdatel'stvo Donish, Dushanbe, USSR. ISSN 0568–6865.

**Byull. Inst. Teor. Astron.**
Byulleten' Instituta Teoreticheskoj Astronomii. Akademiya Nauk SSSR. Leningradskoe Otdelenie, Izdatel'stvo Nauka, Leningrad V–164, Mendeleevskaya l., 1, USSR. ISSN 0002–3302.

**C. R. Acad. Sci., Sér. Gén., Vie Sci.**
Comptes Rendus de l'Académie des Sciences, Série Générale, La Vie des Sciences. Académie des Sciences, Paris. Subscription address: Gauthier–Villars, C.D.R., Centrale des Revues, 11, rue Gossin, F–92543 Montrouge Cedex, France.

**C. R. Acad. Sci., Sér. II**
Comptes Rendus de l'Académie des Sciences, Série II: Mécanique, Physique, Chimie, Sciences de l'Univers, Sciences de la Terre. Académie des Sciences, Paris. Subscription address: Gauthier–Villars, C.D.R., Centrale des Revues, 11, rue Gossin, F–92543 Montrouge Cedex, France. ISSN 0249–6305.

**Can. J. Earth Sci.**
Canadian Journal of Earth Sciences. National Research Council of Canada, Ottawa KIA OR6, Canada. ISSN 0008–4077.

**Can. J. Phys.**
Canadian Journal of Physics. National Research Council of Canada, Ottawa KIA OR6, Canada. ISSN 0008–4204.

**Carter Obs., Repr. Ser.**
Carter Observatory, Reprint Series. Carter Observatory, PO Box 2909, Wellington 1, New Zealand.

**Cartes Synoptiques**
Cartes Synoptiques de la Chromosphère Solaire et Catalogues des Filaments et des Centres d'Activité. Observatoire de Paris, Section d'Astrophysique, F–92190 Meudon, France.

**Celest. Mech.**
Celestial Mechanics. An International Journal of Space Dynamics. D. Reidel Publishing Company, Dordrecht – Boston. ISSN 0008–8714.

**Cent. Astron. Sci. Spat., Obs. Sol.**
Centre de l'Astronomie et des Sciences Spatiales, Observations Solaires. Centre de l'Astronomie et des Sciences Spatiales, Academiei Republicii Socialiste România, Bucuresti, Rumania.

**Cent. Astrophys., Prepr. Ser.**
Center for Astrophysics, Preprint Series. Center for Astrophysics, 60 Garden St., Cambridge, Mass. 02138, USA.

**Centaurus**
Centaurus. International Magazine of the History of Mathematics, Science, and Technology. Munksgaard Ltd., Copenhagen, Denmark. ISSN 0008–8994.

**Česk. Čas. Fyz., Sekce A**
Československý časopis pro fyziku, Sekce A. Academia Publishing House of the Czechoslovak Adademy of Sciences, Vodičkova 40, 11229 Praha 1, Czechoslovakia. ISSN 0009–0700.

**Chem. Phys.**
Chemical Physics. North–Holland Publishing Company, PO Box 211, 1000 AE Amsterdam, The Netherlands. ISSN 0301–0104.

**Chem. Phys. Lett.**
Chemical Physics Letters. North–Holland Publishing Company, PO Box 211, 1000 AE Amsterdam, The Netherlands. ISSN 0009–2614.

**Chin. Astron. Astrophys.**
Chinese Astronomy and Astrophysics. A Selected Translation of Acta Astron. Sin. and Acta Astrophys. Sin. Pergamon Press, Oxford – New York – Toronto – Paris – Frankfurt – Sydney. ISSN 0275–1062.

**Chin. J. Phys.**
Chinese Journal of Physics. Physical Society of the Republic of China, Physics Department, National Taiwan University, Taipei, Taiwan, Republic of China. ISSN 0577–9073.

**Chin. Phys.**
Chinese Physics. Selected translations from current issues of major Chinese physics and astronomy journals. American Institute of Physics, 335 East 45th Street, New York, N.Y. 10017, USA. ISSN 0273–429X.

**Ciel**
Le Ciel. Bulletin de la Société Astronomique de Liège. Société Astronomique de Liège, B–4410 Vottem, Belgium.

**Ciel Espace**
Ciel et Espace. Association Française d'Astronomie, 115 rue de Charenton, F–75012 Paris, France.

**Ciel Terre**
Ciel et Terre. Bulletin de la Société Royale Belge d'Astronomie, de Météorologie et de Physique du Globe. Société Royale Belge d'Astronomie, de Météorologie et de Physique du Globe, 3, avenue Circulaire, B–1180 Bruxelles, Belgium. ISSN 0009–6709.

**Circ. Czech. Obs., Time Latitude**
Circular of the Czechoslovak Observatories, Time and Latitude. Czechoslovak Academy of Sciences, Astronomical Institute, Prague, Czechoslovakia.

**Circ. Inf.**
Circulaire d'Information. Union Astronomique Internationale. Commission des Etoiles Doubles. Observatoire de Paris, F–92190 Meudon, France.

**Circ. Stn. Astron. Int. Latitudine, Carloforte–Cagliari**
Circolari della Stazione Astronomica Internazionale di Latitudine, Carloforte–Cagliari. Series A, B, C. Stazione Astronomica Internazionale di Latitudine, Carloforte–Cagliari, Italy.

**Circ. Time Latitude Serv.**
Circular Time and Latitude Service. Polish Academy of Sciences, Astronomical Latitude Observatory, Borowiec, Poland.

**Classical Quantum Gravity**
Classical and Quantum Gravity. Institute of Physics, 47 Belgrave Square, London SW1X 8QX, England.

**Clim. Change**
Climatic Change. D. Reidel Publishing Company, Dordrecht – Boston – London. ISSN 0165–0009.

**CODATA Bull.**
CODATA Bulletin. Committee on Data for Science and Technology of the International Council of Scientific Unions. Pergamon Press, Oxford – New York – Toronto – Sydney – Paris – Frankfurt. ISSN 0366–757X.

**Coelum**
Coelum. Periodico bimestrale per la Divulgazione dell'Astronomia. Osservatorio Astronomico Universitario di Bologna, Cassella Postale 596, I–40100 Bologna, Italy. ISSN 0366–7588.

**Comments Astrophys.**
Comments on Astrophysics. A Journal of Critical Discussion of the Current Literature. Comments on Modern Physics: Part C. Gordon and Breach Science Publishers, New York – London. ISSN 0146–2970.

**Comments At. Mol. Phys.**
Comments on Atomic and Molecular Physics. Gordon and Breach Science Publishers, New York – London. ISSN 0010–2687.

**Comments Nucl. Part. Phys.**
Comments on Nuclear and Particle Physics. Gordon and Breach Science Publishers, New York – London. ISSN 0010–2709.

**Comments Plasma Phys. Controlled Fusion**
Comments on Plasma Physics and Controlled Fusion. Gordon and Breach Science Publishers, New York – London. ISSN 0374–2806.

**Commun. Astron. Dep., Univ. Ankara**
Communications from the Astronomical Department, University of Ankara. Fen Fakültesi, Ankara, Turkey.

**Commun. Konkoly Obs.**
Communications from the Konkoly Observatory of the Hungarian Academy of Sciences, Budapest, Hungary. ISSN 0324–2234.

**Commun. Math. Phys.**
Communications in Mathematical Physics. Springer–Verlag, Berlin – Heidelberg – New York – Tokyo. ISSN 0010–3616.

**Commun. Theor. Phys.**
Communications in Theoretical Physics. Huazhong University of Science and Technology Press, Wuhan, People's Republic of China.

**Commun. Univ. Obs., St. Andrews**
Communications from the University Observatory, St. Andrews. University Observatory, Buchanan Gardens, St. Andrews, Fife KY16 9LZ, Scotland.

**Comput. Graphics**
Computer Graphics. Association for Computing Machinery, 1133 Avenue of the Americas, New York, N.Y. 10036, USA. ISSN 0097–8930.

**Comput. Methods Appl. Mech. Eng.**
Computer Methods in Applied Mechanics and Engineering. North–Holland Publishing Company, PO Box 211, 1000 AE Amsterdam, The Netherlands. ISSN 0045–7825.

**Comput. Phys. Commun.**
Computer Physics Communications. North–Holland Publishing Company, Amsterdam, The Netherlands. ISSN 0010–4655.

**Comput. Syst.**
Computer Systems. Techpress Publishing Company Ltd., Walton House, 93 High Street, Bromley BR1 1JW, England. ISSN 0264–4193.

**Computer**
Computer. The Institute of Electrical and Electronic Engineers Computer Society, 5855 Naples Plaza, Suite 301, Long Beach, Calif. 90803, USA. ISSN 0018–9162.

**Comun. Obs. Astron. Univ. Coimbra**
Comunicações do Observatório Astronomico da Universidade de Coimbra, Portugal.

**Contemp. Phys.**
Contemporary Physics. Taylor and Francis Ltd., 10–14 Macklin Street, London, WC2B 5NF, England. ISSN 0010–7514.

**Contrib. Astron. Obs. Skalnaté Pleso**
Contributions of the Astronomical Observatory Skalnaté Pleso. VEDA, vydavatel'stvo Slovenskej akadémie vied, Bratislava, Czechoslovakia.

**Contrib. Atmos. Phys.**
Contributions to Atmospheric Physics. (Beiträge zur Physik der Atmosphäre). Friedrich Vieweg & Sohn Verlagsgesellschaft mbH, Postfach 5829, D–6200 Wiesbaden, F.R. Germany. ISSN 0005–8173.

**Contrib. Bosscha Obs.**
Contributions from the Bosscha Observatory. Bosscha Observatory, Bandung Institute of Technology, Department of Science, Lembang, Java, Indonesia.

**Contrib. Dep. Astron., Univ. Kyoto**
Contributions from the Department of Astronomy, University of Kyoto. Department of Astronomy, Faculty of Science, University of Kyoto, Sakyo–ku, Kyoto 608, Japan. ISSN 0388–0230.

**Contrib. Dep. Astron., Univ. Tokyo**
Contributions from the Department of Astronomy, University of Tokyo. Department of Astronomy, University of Tokyo, Bunko–ku, Tokyo 113, Japan. ISSN 0563–8038.

**Contrib. Dep. Geod. Astron., Univ. Thessaloniki**
Contributions from the Department of Geodetic Astronomy, University of Thessaloniki. Department of Geodetic Astronomy, University of Thessaloniki, Thessaloniki, Greece.

**Contrib. Dom. Astrophys. Obs.**
Contributions from the Dominion Astrophysical Observatory. Dominion Astrophysical Observatory, National Research Council of Canada, 5071 West Saanich Road, Victoria, B.C. V8X 4M6, Canada.

**Contrib. Dunsink Obs.**
Contributions from the Dunsink Observatory. Dunsink Observatory, Castleknock, County Dublin, Republic of Ireland.

**Contrib. Geophys. Inst. Slovak Acad. Sci.**
Contributions of the Geophysical Institute of the Slovak Academy of Sciences. Slovak Academy of Sciences, Dubravská 9, 84228 Bratislava, Czechoslovakia. ISSN 0586–4607.

**Contrib. Inst. Argent. Radioastron.**
Contribuciones del Instituto Argentino de Radioastronomia. Instituto Argentino de Radioastronomia, Casilla de Correo No. 5, RA–1894 Villa Elisa, Prov. de Buenos Aires, Argentina.

**Contrib. Inst. Astron. Res.**
Contributions of the Institute for Astronomical Research. Institute for Astronomical Research, Inc., PO Box 15854, Baton Rouge, La. 70895, USA.

**Contrib. Kwasan Hida Obs., Univ. Kyoto**
Contributions from the Kwasan and Hida Observatories, University of Kyoto. Kwasan Observatory, University of Kyoto, Yamashina, Kyoto, 607, Japan. Hida Observatory, University of Kyoto, Kamitakara, Gifu–ken, 506–13, Japan. ISSN 0388–2349.

**Contrib. La. State Univ. Obs.**
Contributions of the Louisiana State University Observatory. Louisiana State University Observatory, Baton Rouge, La. 70803, USA.

**Contrib. Lick Obs.**
Contributions from the Lick Observatory. Lick Observatory, Board of Studies in Astronomy and Astrophysics, University of California at Santa Cruz, Santa Cruz, Calif. 95064, USA.

**Contrib. Nicholas Copernicus Obs. Planetarium Brno**
Contributions of the Nicholas Copernicus Observatory and Planetarium in Brno. Nicholas Copernicus Observatory and Planetarium Brno, Czechoslovakia.

**Contrib. Nizamiah Japal–Rangapur Obs.**
Contributions from Nizamiah and Japal–Rangapur Observatories. Centre of Advanced Study in Astronomy, Osmania University, Hyderabad–500 007, India.

**Contrib. Oss. Astron. Torino**
Contributi dell'Osservatorio Astronomico di Torino (Pino Torinese). Osservatorio Astronomico di Torino, I–10025 Pino Torinese (Torino), Italy.

**Contrib. Van Vleck Obs.**
Contributions from the Van Vleck Observatory. Van Vleck Observatory, Middletown, Conn., USA.

**Copenh. Univ. Obs., Repr.**
Copenhagen University Observatory, Reprints. University Observatory, Øster Voldgade 3, DK–1350 Copenhagen K, Denmark.

**Cosmic. Res.**
Cosmic Research. A cover–to–cover translation of Kosm. Issled. Consultants Bureau, 227 West 17th Street, New York, N.Y. 10011, USA. ISSN 0010–9525.

**Cracow Obs. Repr.**
Cracow Observatory Reprints. Cracow Observatory, Jagellonian University, Cracow, Poland.

**CSELT Rapp. Tec.**
CSELT Rapporti Tecnici. Centro Studi e Laboratori Telecomunicazioni, Via Guglielmo Reiss Romoli 274, Torino, Italy. ISSN 0390–1815.

**Curr. Sci.**
Current Science. Current Science Association, Raman Research Institute, Bangalore 560006, India. ISSN 0011–3891.

**Czech. J. Phys., Sect. B**
Czechoslovak Journal of Physics, Section B. Academia Publishing House of the Czechoslovak Academy of Sciences, Vodičkova 40, 11229 Praha 1, Czechoslovakia. ISSN 0011–4626.

**Data Rep. Hydrogr. Obs., Ser. Astron. Geod.**
Data Report of Hydrographic Observations, Series of Astronomy and Geodesy. Hydrographic Department of Japan, Tsukiji–5, Chuo–ku, Tokyo 104, Japan.

**Debrecen Heliophys. Obs. Hung. Acad. Sci., Repr.**
Debrecen Heliophysical Observatory of the Hungarian Academy of Sciences, Reprint. Heliophysical Observatory of the Hungarian Academy of Sciences, 4010 Debrecen, Hungary.

**Dějiny Věd Tech.**
Dějiny Věd a Techniky. Academia Praha, Praha, Czechoslovakia. ISSN 0300–4414.

**Dep. Astron. McDonald Obs. Univ. Tex., Repr.**
Department of Astronomy and McDonald Observatory of the University of Texas, Reprints. Astronomy Department, R.L.M. 15.220, University of Texas, Austin, Tex. 78712, USA.

**Dep. Astrophys., Univ. Oxford, Publ.**
Department of Astrophysics, University of Oxford, Publication. Department of Astrophysics, University of Oxford, South Parks Road, Oxford OX1 3RQ, England.

**Dokl. Akad. Nauk BSSR**
Doklady Akademii Nauk BSSR. Akademiya Nauk, Minsk, Belorussian SSR. ISSN 0002–354X.

**Dokl. Akad. Nauk SSSR. Ser. Mat. Fiz.**
Doklady Akademii Nauk SSSR. Seriya Matematika, Fizika. Izdatel'stvo Nauka, Moskva. Editorial address: 117874 GSP–7 Moskva, B–485, Profsoyuznaya Ul., 90 Kom. 533, USSR. ISSN 0002–3264.

**Dokl. Bolg. Akad. Nauk**
Doklady Bolgarskoj Akademii Nauk. Sofiya, Bulgaria. ISSN 0366–8681.

**Dtsch. Geod. Komm. Bayer. Akad. Wiss., Reihe A**
Deutsche Geodätische Kommission bei der Bayerischen Akademie der Wissenschaften, Reihe A: Theoretische Geodäsie. Deutsche Geodätische Kommission, Marstallplatz 8, D–8000 München 22, F.R. Germany. ISSN 0065–5309.

**Dtsch. Geod. Komm. Bayer. Akad. Wiss., Reihe B**
Deutsche Geodätische Kommission bei der Bayerischen Akademie der Wissenschaften, Reihe B: Angewandte Geodäsie. Deutsche Geodätische Kommission, Marstallplatz 8, D–8000 München 22, F.R. Germany. ISSN 0065–5317.

**Dtsch. Geod. Komm. Bayer. Akad. Wiss., Reihe C**
Deutsche Geodätische Kommission bei der Bayerischen Akademie der Wissenschaften, Reihe C: Dissertationen. Deutsche Geodätische Kommission, Marstallplatz 8, D–8000 München 22, F.R. Germany. ISSN 0065–5325.

**Dtsch. Geod. Komm. Bayer. Akad. Wiss., Reihe D**
Deutsche Geodätische Kommission bei der Bayerischen Akademie der Wissenschaften, Reihe D: Tafelwerke. Deutsche Geodätische Kommission, Marstallplatz 8, D–8000 München 22, F.R. Germany.

**Dtsch. Geod. Komm. Bayer. Akad. Wiss., Reihe E**
Deutsche Geodätische Kommission bei der Bayerischen Akademie der Wissenschaften, Reihe E: Geschichte und Entwicklung der Geodäsie. Deutsche Geodätische Kommission, Marstallplatz 8, D–8000 München 22, F.R. Germany. ISSN 0065–5341.

**Dtsch. Hydrogr. Inst. Hamb., Zeit–Breitendienst**
Deutsches Hydrographisches Institut Hamburg, Zeit– und Breitendienst. Deutsches Hydrographisches Institut, D–2000 Hamburg, F.R. Germany.

**Dudley Obs. Rep.**
Dudley Observatory Report. The Dudley Observatory, 69 Union Avenue, Schenectady, N.Y. 12308, USA.

**Dunsink Obs. Repr.**
Dunsink Observatory Reprints. Dunsink Observatory, Castleknock, County Dublin, Republic of Ireland.

**Earth, Moon, Planets**
Earth, Moon, and Planets. An International Journal of Comparative Planetology. D. Reidel Publishing Company, Dordrecht – Boston. ISSN 0167–9295. Formerly entitled The Moon and the Planets.

**Earth Orientation Bull.**
Earth Orientation Bulletin. U.S. Naval Observatory, Time Service Publications, Series 7. U.S. Naval Observatory, Time Service Division (62C), Washington, D.C. 20390, USA.

**Earth–Oriented Appl. Space Technol.**
Earth–Oriented Applications of Space Technology. Pergamon Press, Oxford – New York – Frankfurt. ISSN 0277–4488.

**Earth Planet. Sci. Lett.**
Earth and Planetary Science Letters. Elsevier Scientific Publishing Company, PO Box 211, 1000 AE Amsterdam, The Netherlands. ISSN 0012–821X.

**Earth–Sci. Rev.**
Earth–Science Reviews. Elsevier Scientific Publishing Company, PO Box 330, 1000 AH Amsterdam, The Netherlands. ISSN 0012–8252.

**Electron. Lett.**
Electronics Letters. Institution of Electrical Engineers, Savoy Place, London, WC2R 0BL, England. ISSN 0013–5194.

**Electronics**
Electronics. McGraw–Hill Publishing Company, 1221 Avenue of the Americas, New York, N.Y. 10020, USA. ISSN 0013–5070.

**Elektronik**
Elektronik. Franzis–Verlag GmbH, D–8000 München 37, Postfach 3701 20, Karlstraße 37, F.R. Germany. ISSN 0013–5658.

**Elettron. Telecomun.**
Elettronica e Telecomunicazioni. Via Arsenale 41, I–10121 Torino, Italy. ISSN 0013–6123.

**Endeavour New Ser.**
Endeavour New Series. Pergamon Press, Oxford – New York – Toronto – Paris – Frankfurt – Sydney. ISSN 0013–7162.

**EOS Trans. Am. Geophys. Union**
EOS Transactions of the American Geophysical Union. American Geophysical Union, 2000 Florida Avenue N.W., Washington, D.C. 20009, USA. ISSN 0096–3941.

**ESA Brochure**
ESA Brochure (ESA BR). European Space Agency, Scientific and Technical Publications Branch, ESTEC, Postbus 299, 2200 AG Noordwijk, The Netherlands. ISSN 0250–1589.

**ESA Bull.**
ESA Bulletin. European Space Agency, Scientific and Technical Publications Branch, ESTEC, Postbus 299, 2200 AG Noordwijk, The Netherlands. ISSN 0376–4265.

**ESA IUE Newsl.**
ESA IUE Newsletters. The ESA IUE Observatory. Apartado 540 65, Madrid, Spain. Subscription address: European Space Agency, 8–10 rue Mario–Nikis, F–75738 Paris Cedex 15, France.

**ESA J.**
ESA Journal. European Space Agency, Scientific and Technical Publications Branch, ESTEC, Postbus 299, 2200 AG Noordwijk, The Netherlands. ISSN 0379–2285.

**ESA SCI**
European Space Agency SCI. European Space Agency, Scientific and Technical Publications Branch, ESTEC, Postbus 299, 2200 AG Noordwijk, The Netherlands.

**ESA Spec. Publ.**
ESA Special Publication (ESA SP). European Space Agency, Scientific and Technical Publications Branch, ESTEC, Postbus 299, 2200 AG Noordwijk, The Netherlands. ISSN 0379–6566.

**ESO Annu. Rep.**
ESO Annual Report. European Southern Observatory, Karl–Schwarzschild–Straße 2, D–8046 Garching bei München, F.R. Germany. ISSN 0531–4496.

**ESO Sci. Prepr.**
ESO Scientific Preprint. European Southern Observatory, Karl–Schwarzschild–Straße 2, D–8046 Garching bei München, F.R. Germany.

**ESO Sci. Rep.**
ESO Scientific Report. European Southern Observatory, Karl–Schwarzschild–Straße 2, D–8046 Garching bei München, F.R. Germany.

**ESO Tech. Rep.**
ESO Technical Report. European Southern Observatory, Karl–Schwarzschild–Straße 2, D–8046 Garching bei München, F.R. Germany.

**Eur. J. Phys.**
European Journal of Physics. Institute of Physics, 47 Belgrave Square, London SW1X 8QX, England. ISSN 0143–0807.

**Europhys. News**
Europhysics News. European Physical Society, PO Box 69, CH–1213 Petit–Lancy 2, Switzerland. ISSN 0531–7479.

**EXOSAT Express**
EXOSAT Express. ESA, EXOSAT Observatory, ESOC, Robert–Bosch–Str. 5, D–6100 Darmstadt, F.R. Germany.

**Feingerätetechnik**
Feingerätetechnik. VEB Verlag Technik, Oranienburger Straße 13/14, DDR–1020 Berlin, German Democratic Republic. ISSN 0014–9683.

**Fis. Tecnol.**
Fisica e Tecnologia. Societa Italiana di Fisica, Via Loderingo Degli Andalo 2, I–40124 Bologna, Italy. ISSN 0391–9757.

**Fiz. Sz.**
Fizikai Szemle. Kiadja a Lapkiado Vallalat, Budapest VII, Lenin korut 9–11, Hungary. ISSN 0015–3257.

**Fizika**
Fizika. Mladost Export–Import, 41000 Zagreb, Ilica 30, Yugoslavia. ISSN 0015–3206.

**Folia Fac. Sci. Nat. Univ. Purkynianae Brun., Phys.**
Folia Facultatis Scientiarum Naturalium Universitatis Purkynianae Brunensis, Physica. University J. E. Purkyne, 61137 Brno–Kotlarska 2, Czechoslovakia. ISSN 0323–0287.

**Fortschr. Phys.**
Fortschritte der Physik. Akademie–Verlag, DDR–108 Berlin, Leipziger Straße 3–4, German Democratic Republic. ISSN 0015–8208.

**Found. Phys.**
Foundations of Physics. Plenum Publishing Corporation, 227 West 17th Street, New York, N.Y. 10011, USA. ISSN 0015–9018.

**Fra Fys. Verden**
Fra Fysikkens Verden. Fysisk Institutt, Universitetet i Trondheim, Norges Laererhogskole, N–7000 Trondheim, Norway. ISSN 0015–9247.

**Fujitsu Sci. Tech. J.**
Fujitsu Scientific and Technical Journal. Fujitsu Ltd., 1015 Kamikodanaka, Nakahara–ku, Kawasaki 211, Kanagawa, Japan. ISSN 0016–2523.

**Fundam. Cosmic Phys.**
Fundamentals of Cosmic Physics. Gordon and Breach Science Publishers, New York - London - Paris. ISSN 0094-5846.

**Funkschau**
Funkschau. Franzis-Verlag GmbH, Postfach 370120, Karlstr. 37, D-8000 München 2, F.R. Germany. ISSN 0016-2841.

**Fys. Tidsskr.**
Fysisk Tidsskrift. Subscription address: Jul. Gjellerups Boghandel, Solvgade 87, DK-1307 Kobenhavn, Denmark. ISSN 0016-3392.

**G. A.A.B.**
Giornale dell'A.A.B. Associazione Astrofili Bolognesi, Casella Postale 1630, I-40100 Bologna A.D., Italy. ISSN 0392-3932.

**G. Astron.**
Giornale Astronomia. Società Astronomica Italiana, Largo E. Fermi, 5, I-50125 Firenze, Italy. ISSN 0390-1106.

**Gemini**
Gemini. Newsletter of the Royal Greenwich Observatory. Royal Greenwich Observatory, Herstmonceux Castle, Hailsham, East Sussex BN27 1RP, England.

**Gen. Relativ. Gravitation**
General Relativity and Gravitation. Published under the auspices of the International Committee on General Relativity and Gravitation GRG. Plenum Publishing Corporation, 233 Spring Street, New York, N.Y. 10013, USA. ISSN 0001-7701.

**Geochim. Cosmochim. Acta**
Geochimica et Cosmochimica Acta. Journal of the Geochemical Society and the Meteoritical Society. Pergamon Press, New York - Oxford - Toronto - Paris - Frankfurt - Sydney. ISSN 0016-7037.

**Geod.-Geophys. Arb. Schweiz**
Geodätisch-Geophysikalische Arbeiten in der Schweiz. Schweizerische Geodätische Kommission, Institut für Geodäsie und Photogrammetrie, ETH-Hönggerberg, CH-8093 Zürich, Switzerland.

**Geod. Geophys. Veröff., Reihe II**
Geodätische und Geophysikalische Veröffentlichungen, Reihe II (Solar-terrestrische Beziehungen und Physik der Atmosphäre). Nationalkomitee für Geodäsie und Geophysik bei der Akademie der Wissenschaften der Deutschen Demokratischen Republik, DDR-1500 Potsdam, Telegrafenberg, German Democratic Republic.

**Geod. Geophys. Veröff., Reihe III**
Geodätische und Geophysikalische Veröffentlichungen, Reihe III (Physik der festen Erde). Nationalkomitee für Geodäsie und Geophysik bei der Akademie der Wissenschaften der Deutschen Demokratischen Republik, DDR-1500 Potsdam, Telegrafenberg, German Democratic Republic. ISSN 0435-6187.

**Geod. Kartogr.**
Geodezja i Kartografia. Komitet Geodezji Polskiej Akademii Nauk. Publisher: Państwowe Wydawnictwo Naukowe, Warszwa, Poland. ISSN 0016-7134.

**Geomagn. Aehron.**
Geomagnetizm i Aehronomiya. Akademiya Nauk SSSR. Izdatel'stvo Nauka, Moskva. English translation in Geomagn. Aeron. ISSN 0016-7940.

**Geomagn. Aeron.**
Geomagnetism and Aeronomy. A cover-to-cover translation of Geomagn. Aehron. American Geophysical Union, 2000 Florida Avenue, N.W., Washington, D. C. 20009, USA. ISSN 0016-7932.

**Geophys. Astrophys. Fluid Dyn.**
Geophysical and Astrophysical Fluid Dynamics. Gordon and Breach Science Publishers, New York - London - Paris. ISSN 0309-1929.

**Geophys. J. R. Astron. Soc.**
Geophysical Journal of the Royal Astronomical Society. Published for the Royal Astronomical Society by Blackwell Scientific Publications, Oxford - London - Edinburgh - Boston - Melbourne. ISSN 0016-8009.

**Geophys. Norv.**
Geophysica Norvegica. Universitetsforlaget, PO Box 2959, Toyen, Oslo 6, Norway. ISSN 0332-5903.

**Geophys. Prospect.**
Geophysical Prospecting. European Association of Exploration Geophysicists, PO Box 162, 2501 AN The Hague, The Netherlands. ISSN 0016-8025.

**Geophys. Res. Lett.**
Geophysical Research Letters. American Geophysical Union, 2000 Florida Avenue, N.W., Washington, D.C. 20009, USA. ISSN 0094-8276.

**Geophys. Surv.**
Geophysical Surveys. D. Reidel Publishing Company, Dordrecht - Boston - London. ISSN 0046-5763.

**Geophysics**
Geophysics. Society of Exploration Geophysicists, PO Box 3098, Tulsa, Okla. 74101, USA. ISSN 0016-8033.

**GEOS Circ.**
GEOS Circular. Series: EB (eclipsing binaries), RR (RR Lyrae type variables), SA (small-amplitude variables), SR (red variables). GEOS (Groupe d'Etude et d'Observation Stellaire), 12 rue Bezout, F-75014 Paris, France.

**Gerlands Beitr. Geophys.**
Gerlands Beiträge zur Geophysik. Akademische Verlagsgesellschaft Geest & Portig KG, DDR-7010 Leipzig, Sternwartenstraße 8, German Democratic Republic. ISSN 0016-8696.

**Glas. Mat., Ser. III**
Glasnik Matematički, Serija III. Društvo matematičara i fizičara SRH, Marulićev trg 19, YU-41001 Zagreb, P.P. 187, Yugoslavia. ISSN 0017-095X.

**Greenwich Time Rep.**
Greenwich Time Report. Royal Greenwich Observatory, Time and Latitude Service, Herstmonceux Castle, Hailsham, East Sussex BN27 1RP, England. ISSN 0264-4177.

**Hadronic J.**
Hadronic Journal. Hadronic Press Inc., Nonantum, Mass. 02195, USA. ISSN 0162-5519.

**Heavens**
The Heavens. The Oriental Astronomical Association, Ōtsu-shi, Shiga-ken, Japan. In Japanese.

**Helv. Phys. Acta**
Helvetica Physica Acta. Schweizerische Physikalische Gesellschaft. Birkhäuser Verlag, Elisabethenstraße 19, CH-4000 Basel 10, Switzerland. ISSN 0018-0238.

**HHI Sol. Data**
HHI Solar Data. Solar Radio Emission. Heinrich-Hertz-Institut, Solare Beobachtungsergebnisse. Akademie der Wissenschaften der DDR, Zentralinstitut für Solar-Terrestrische Physik (Heinrich-Hertz-Institut), DDR-1199 Berlin-Adlershof, German Democratic Republic. ISSN 0323-309X.

**HHI STP Rep.**
HHI Solar-Terrestrische Physik Reports. Heinrich-Hertz-Institut. Akademie der Wissenschaften der DDR, Zentralinstitut für Solar-Terrestrische Physik (Heinrich-Hertz-Institut), DDR-1199 Berlin-Adlershof, German Democratic Republic.

**HHI Suppl. Ser. Sol. Data**
HHI Supplement Series of Solar Data. Heinrich-Hertz-Institut. Akademie der Wissenschaften der DDR, Zentralinstitut für Solar-Terrestrische Physik (Heinrich-Hertz-Institut), DDR-1199 Berlin-Adlershof, German Democratic Republic. ISSN 0323-3197.

**H.M. Naut. Alm. Off., Libr. Repr.**
H.M. Nautical Almanac Office, Library Reprint. H.M. Nautical Almanac Office, Royal Greenwich Observatory, Herstmonceux Castle, Hailsham, East Sussex BN27 1RP, England.

**Hvar Obs. Bull.**
Hvar Observatory Bulletin. Faculty of Geodesy, University of Zagreb, Kačićeva 26, YU-41000 Zagreb, Yugoslavia. ISSN 0351-2651.

**I.A.P.P.P. Commun.**
International Amateur-Professional Photoelectric Photometry Communication. Fairborn Observatory, 1247 Folk Road, Fairborn, Ohio 45324, USA.

**IAU Circ.**
International Astronomical Union, Circular. Central Bureau for Astronomical Telegrams, Smithsonian Astrophysical Observatory, 60 Garden Street, Cambridge, Mass. 02138, USA. ISSN 0081-0304.

**IAU Inf. Bull.**
International Astronomical Union Information Bulletin. IAU-UAI Secrétariat, 61, Avenue de l'Observatoire, F-75014 Paris, France. Published by D. Reidel Publishing Company, Dordrecht - Boston - London.

**Icarus**
Icarus. International Journal of Solar System Studies. Academic Press Inc., New York - London. ISSN 0019-1035.

**IEE Proc. F**
IEE Proceedings F (Communications, Radar and Signal Processing). Institution of Electrical Engineers, PO Box 8, Southgate House, Stevenage, Herts. SG1 1HQ, England. ISSN 0143-7070.

**IEEE J. Solid-State Circuits**
IEEE Journal of Solid-State Circuits. The Institute of Electrical and Electronics Engineers, 345 East 47th Street, New York, N.Y. 10017, USA. ISSN 0018-9200.

**IEEE Spectrum**
IEEE Spectrum. The Institute of Electrical and Electronics Engineers, 345 East 47th Street, New York, N.Y. 10017, USA. ISSN 0018-9235.

**IEEE Trans. Aerosp. Electron. Syst.**
IEEE Transactions on Aerospace and Electronic Systems. The Institute of Electrical and Electronics Engineers, 345 East 47th Street, New York, N.Y. 10017, USA. ISSN 0018-9251.

**IEEE Trans. Antennas Propag.**
IEEE Transactions on Antennas and Propagation. The Institute of Electrical and Electronics Engineers, 345 East 47th Street, New York, N.Y. 10017, USA. ISSN 0018-926X.

**IEEE Trans. Commun.**
IEEE Transactions on Communications. The Institute of Electrical and Electronics Engineers, 345 East 47th Street, New York, N.Y. 10017, USA. ISSN 0090-6778.

**IEEE Trans. Ind. Electron.**
IEEE Transactions on Industrial Electronics. The Institute of Electrical and Electronics Engineers, 345 East 47th Street, New York, N.Y. 10017, USA. ISSN 0278-0046.

**IEEE Trans. Instrum. Meas.**
IEEE Transactions on Instrumentation and Measurement. The Institute of Electrical and Electronics Engineers, 345 East 47th Street, New York, N.Y. 10017, USA. ISSN 0018-9456.

**IEEE Trans. Microwave Theory Tech.**
IEEE Transactions on Microwave Theory and Techniques. The Institute of Electrical and Electronics Engineers, 345 East 47th Street, New York, N.Y. 10017, USA. ISSN 0018-9480.

**IEEE Trans. Nucl. Sci.**
IEEE Transactions on Nuclear Science. The Institute of Electrical and Electronics Engineers, 345 East 47th Street, New York, N.Y. 10017, USA. ISSN 0018-9499.

**IEEE Trans. Plasma Sci.**
IEEE Transactions on Plasma Science. The Institute of Electrical and Electronics Engineers, 345 East 47th Street, New York, N.Y. 10017, USA. ISSN 0093-3813.

**IMA J. Appl. Math.**
IMA Journal of Applied Mathematics. Academic Press Inc., London - New York. ISSN 0272-4960.

**Indian East. Eng.**
Indian and Eastern Engineer. "Piramal Mansion", 235 Dr. D. Naoroji Road, Bombay 400 001, India. ISSN 0019-4352.

**Indian J. Cryog.**
Indian Journal of Cryogenics. Indian Cryogenics Council, Jadavpur University, Calcutta 700 032, India. ISSN 0379-0479.

**Indian J. Hist. Sci.**
Indian Journal of History of Science. Indian National Science Academy, Bahadur Shah Zafar Marg, New Delhi 110 002, India.

**Indian J. Phys., Part B**
Indian Journal of Physics, Part B. Indian Association for the Cultivation of Science, 2 & 3 Raja Subodh Chandra Mallik Road, Calcutta 700 032, India. ISSN 0374-3330.

**Indian J. Pure Appl. Math.**
Indian Journal of Pure and Applied Mathematics. Indian National Science Academy, Bahadur Shah Zafar Marg, New Delhi 110 002, India. ISSN 0019-5588.

**Indian J. Pure Appl. Phys.**
Indian Journal of Pure and Applied Physics. Council of Scientific & Industrial Research. Publications & Information Directorate, Hillside Road, New Delhi 110 012, India. ISSN 0019-5596.

**Indian J. Radio Space Phys.**
Indian Journal of Radio and Space Physics. Council of Scientific & Industrial Research. Publications & Information Directorate, Hillside Road, New Delhi 110 012, India. ISSN 0367-8393.

**Indian J. Theor. Phys.**
Indian Journal of Theoretical Physics. Institute of Theoretical Physics, Bognan Kutir, 4–1 Mohan Bagan Lane, Calcutta 700 004, India. ISSN 0019–5693.

**Inf. Bull. Variable Stars**
Information Bulletin on Variable Stars. Commission 27 of the IAU. Konkoly Observatory, Budapest, Hungary. ISSN 0374–0676.

**Infrared Phys.**
Infrared Physics. An International Research Journal. Pergamon Press, Oxford – New York – Toronto – Sydney – Paris – Frankfurt. ISSN 0020–0891.

**Inst. Astron. Astrophys. Tech. Univ. Berlin, Mitt.**
Institut für Astronomie und Astrophysik der Technischen Universität Berlin, Mitteilungen. Institut für Astronomie und Astrophysik der Technischen Universität Berlin, Hardenbergstraße 36, D–1000 Berlin 12, F.R. Germany.

**Inst. Astron. Fis. Espacio, Tirada Aparte**
Instituto de Astronomía y Fisica del Espacio, Tirada Aparte. Instituto de Astronomía y Fisica del Espacio, Casilla de Correo 67, Sucursal 28, 1428 Buenos Aires, Argentina.

**Inst. Astron., Univ. Camb., Annu. Rep.**
Institute of Astronomy, University of Cambridge, Annual Report. The Observatories, Madingley Road, Cambridge CB3 0HA, England.

**Inst. Astrophys. Paris, Pré–Publ.**
Institut d'Astrophysique de Paris, Pré–Publication. Institut d'Astrophysique, 98 bis, Boulevard Arago, F–75014 Paris, France.

**Inst. Astrophys., Univ. Liège, Collect. 4°**
Institut d'Astrophysique, Université de Liège, Collection in 4°. Institut d'Astrophysique, Université de Liège, Avenue de Cointe, 5, B–4200 Cointe-Ougrée, Belgium.

**Inst. Astrophys., Univ. Liège, Collect. 8°**
Institut d'Astrophysique, Université de Liège, Collection in 8°. Institut d'Astrophysique, Université de Liège, Avenue de Cointe, 5, B–4200 Cointe–Ougrée, Belgium.

**Inst. Obs. Mar., Bol. Astron.**
Instituto y Observatorio de Marina, Boletin Astronomico. Instituto y Observatorio de Marina, San Fernando (Cadiz), Spain.

**Inst. Obs. Mar., Bol. Ser. B**
Instituto y Observatorio de Marina, Boletin Serie B. (Pequeños Planetas). Instituto y Observatorio de Marina, San Fernando (Cadiz), Spain. ISSN 0558–3993.

**Inst. Obs. Mar., Bol. Ser. C**
Instituto y Observatorio de Marina, Boletin Serie C. (Rotacion de la Tierra). Instituto y Observatorio de Marina, San Fernando (Cadiz), Spain. ISSN 0210–6485.

**Inst. Obs. Mar., Bol. Ser. D**
Instituto y Observatorio de Marina, Boletin Serie D. (Ocultaciones). Instituto y Observatorio de Marina, San Fernando (Cadiz), Spain.

**Inst. Obs. Mar., Mem. Act.**
Instituto y Observatorio de Marina, Memoria de las Actividades. Instituto y Observatorio de Marina, San Fernando (Cadiz), Spain.

**Inst. Teor. Astrofys., Blindern–Oslo, Småtrykk**
Institutt for Teoretisk Astrofysikk, Blindern–Oslo, Småtrykk. Institute of Theoretical Astrophysics, University of Oslo, PO Box 1029, Blindern, N–0315 Oslo 3, Norway.

**Inst. Theor. Astrophys., Blindern–Oslo, Rep.**
Institute of Theoretical Astrophysics, Blindern–Oslo, Reports. Institute of Theoretical Astrophysics, University of Oslo, PO Box 1029, Blindern, N–0315 Oslo 3, Norway. ISSN 0078–6780.

**Inst. Theor. Astrophys., Blindern–Oslo, Repr.**
Institute of Theoretical Astrophysics, Blindern–Oslo, Reprints. Institute of Theoretical Astrophysics, University of Oslo, PO Box 1029, Blindern, N–0315 Oslo 3, Norway.

**Inst. Theor. Astrophys., Univ. Oslo, Publ. Ser.**
Institute of Theoretical Astrophysics, University of Oslo, Publication Series. Institute of Theoretical Astrophysics, University of Oslo, PO Box 1029, Blindern, N–0315 Oslo 3, Norway. ISSN 0800–6652.

**Inst. Theor. Phys. Sternw. Univ. Kiel, Repr.**
Institut für Theoretische Physik und Sternwarte der Universität Kiel, Reprints. Institut für Theoretische Physik und Sternwarte der Universität Kiel, D–2300 Kiel, F.R. Germany.

**Int. Comet Q.**
The International Comet Quarterly. Department of Physics and Astronomy, Appalachian State University, Boone, N.C. 28608, USA. ISSN 0736–6922.

**Int. J. Gen. Syst.**
International Journal of General Systems. Gordon and Breach Science Publishers Inc., New York – London – Paris. ISSN 0308–1079.

**Int. J. Heat Mass Transfer**
International Journal of Heat and Mass Transfer. Pergamon Press, Oxford – New York – Toronto – Sydney – Paris – Frankfurt. ISSN 0017–9310.

**Int. J. Infrared Millimeter Waves**
International Journal of Infrared and Millimeter Waves. Plenum Publishing Corporation, 227 West 17th Street, New York, N.Y. 10011, USA. ISSN 0195–9271.

**Int. J. Mass Spectrom. Ion Processes**
International Journal of Mass Spectrometry and Ion Processes. Elsevier Scientific Publishing Company, PO Box 330, 1000 AH Amsterdam, The Netherlands. ISSN 0020–7381.

**Int. J. Non–Linear Mech.**
International Journal of Non–Linear Mechanics. Pergamon Press, Oxford – New York – Toronto – Paris – Frankfurt – Sydney. ISSN 0020–7462.

**Int. J. Theor. Phys.**
International Journal of Theoretical Physics. Plenum Publishing Corporation, 233 Spring Street, New York, N.Y. 10013, USA. ISSN 0020–7748.

**Interdisciplinary Sci. Rev.**
Interdisciplinary Science Reviews. Heyden & Son Ltd., Spectrum House, Hillview Gardens, London NW4 2JQ, England. ISSN 0308–0188.

**Ir. Astron. J.**
The Irish Astronomical Journal. A Half–Yearly Publication under the Auspices of the Observatories of Armagh and Dunsink. Armagh Observatory, Armagh BT61 9DG, Northern Ireland. ISSN 0021–1052.

**IRIS Bull. A**
IRIS Bulletin A (Earth Orientation). Subcommission International Radio Interferometric Surveying (IRIS) Steering Committee. Available from National Geodetic Survey, National Oceanic and Atmospheric Administration, N/CG114, Rockville, Md. 20852, USA.

**ISIS**
ISIS. An International Review devoted to the History of Science and its Cultural Influences. Department of History and Sociology of Science, University of Pennsylvania, Philadelphia, Pa. 19104, USA. ISSN 0021-1753.

**Issled. Solntsa Krasnykh Zvezd**
Issledovaniya Solntsa i Krasnykh Zvezd. (Investigations of the Sun and Red Stars). Akademiya Nauk Latvijskoj SSR, Radioastrofizicheskaya Observatoriya. Izdatel'stvo Zinatne, Riga, USSR. ISSN 0135-1303.

**Izv. Akad. Nauk Arm. SSR, Ser. Fiz.**
Izvestiya Akademii Nauk Armyanskoj SSR, Seriya Fizika. Akademiya Nauk Armyanskoj SSR, Erevan. 375019 Erevan, ul. Barekamutyan, 24, USSR. ISSN 0002-3035.

**Izv. Astron. Ehngel'gardt. Obs.**
Izvestiya Astronomicheskoj Ehngel'gardtovskoj Observatorii. Izdatel'stvo Kazanskogo Universiteta, Kazan, ul. Lenina, d. 4/5, USSR.

**Izv. Glav. Astron. Obs. Pulkovo**
Izvestiya Glavnoj Astronomicheskoj Observatorii v Pulkove. Akademiya Nauk SSSR. Leningradskoe Otdelenie, Izdatel'stvo Nauka, Leningrad. 199164 Leningrad, V-164, Mendeleevskaya l., 1, USSR. ISSN 0367-7966.

**Izv. Krymskoj Astrofiz. Obs.**
Izvestiya Ordena Trudovogo Krasnogo Znameni Krymskoj Astrofizicheskoj Observatorii. Akademiya Nauk SSSR. Izdatel'stvo 'Nauka', Moskva, USSR. English translation in Bull. Crimean Astrophys. Obs. ISSN 0367-8466.

**Izv. Vyssh. Uchebn. Zaved., Radiofiz.**
Izvestiya Vysshikh Uchebnikh Zavedenij, Radiofizika. Gor'kovskij Universitet. Gor'kij, ul. Lyadova 25, USSR. English translation in Radiophys. Quantum Electron. ISSN 0021-3462.

**J. Am. Assoc. Variable Star Obs.**
The Journal of the American Association of Variable Star Observers. The American Association of Variable Star Observers, 187 Concord Avenue, Cambridge, Mass. 02138, USA.

**J. Am. Chem. Soc.**
Journal of the American Chemical Society. American Chemical Society, 1155 16th Street N.W., Washington, D.C. 20036, USA. ISSN 0002-7863.

**J. Appl. Meteorol.**
Journal of Applied Meteorology. American Meteorological Society, 45 Beacon Street, Boston, Mass. 02108, USA. ISSN 0021-8952.

**J. Appl. Phys.**
Journal of Applied Physics. American Institute of Physics, 335 East 45th Street, New York, N.Y. 10017, USA. ISSN 0021-8979.

**J. Astron. Soc. Egypt**
Journal of the Astronomical Society of Egypt. Astronomical Society of Egypt, Astronomy Department, Faculty of Sciences, Cairo University, Cairo, Egypt.

**J. Astron. Soc. West. Aust.**
Journal of the Astronomical Society of Western Australia. Astronomical Society of Western Australia, PO Box S1460, Perth, W.A. 6000, Australia.

**J. Astronaut. Sci.**
Journal of the Astronautical Sciences. American Astronautical Society, 6060 Duke Street, Alexandria, Va. 22304, USA. ISSN 0021-9142.

**J. Astrophys. Astron.**
Journal of Astrophysics and Astronomy. Indian Academy of Sciences, PO Box 8005, Bangalore 560 080, India. ISSN 0250-6335.

**J. Atmos. Sci.**
Journal of the Atmospheric Sciences. American Meteorological Society, 45 Beacon Street, Boston, Mass. 02108, USA. ISSN 0022-4928.

**J. Atmos. Terr. Phys.**
Journal of Atmospheric and Terrestrial Physics. Pergamon Press, Oxford - New York - Frankfurt. ISSN 0021-9169.

**J. Aust. Math. Soc., Ser. B**
Journal of the Australian Mathematical Society, Series B. (Applied Mathematics). Department of Mathematics, University of Queensland, St. Lucia, QLD 4067, Australia. ISSN 0334-2700.

**J. Br. Astron. Assoc.**
Journal of the British Astronomical Association. The British Astronomical Association, Burlington House, Piccadilly, London, W1V 0NL, England. ISSN 0007-0297.

**J. Br. Interplanet. Soc.**
Journal of the British Interplanetary Society. The British Interplanetary Society, 27-29 South Lambeth Road, London, SW8 1SZ, England. ISSN 0007-084X.

**J. Chem. Educ.**
Journal of Chemical Education. American Chemical Society, 119 West 24th Street, New York, N.Y. 10011, USA. ISSN 0021-9584.

**J. Chem. Phys.**
Journal of Chemical Physics. American Institute of Physics, 335 East 45th Street, New York, N.Y. 10017, USA. ISSN 0021-9606.

**J. Comput. Phys.**
Journal of Computational Physics. Academic Press Inc., 111 5th Avenue, New York, N.Y. 10003, USA. ISSN 0021-9991.

**J. Differ. Equations**
Journal of Differential Equations. Academic Press Inc., New York - London. ISSN 0022-0396.

**J. Electron Spectrosc. Relat. Phenom.**
Journal of Electron Spectroscopy and Related Phenomena. Elsevier Scientific Publishing Company, PO Box 330, 1000 AH Amsterdam, The Netherlands. ISSN 0368-2048.

**J. Electrostat.**
Journal of Electrostatics. Elsevier Scientific Publishing Company, PO Box 330, 1000 AH Amsterdam, The Netherlands. ISSN 0304-3886.

**J. Environ. Sci.**
Journal of Environmental Sciences. Institute of Environmental Sciences, 940 East Northwest Highway, Mt. Prospect, Ill. 60056, USA. ISSN 0022-0906.

**J. Fluid Mech.**
Journal of Fluid Mechanics. Cambridge University Press, Bentley House, 200 Euston Road, London, NW1 2DB, England. ISSN 0022-1120.

**J. Geodyn.**
Journal of Geodynamics. Geophysical Press, Brouwersgracht 236, 1013 HE Amsterdam, The Netherlands. ISSN 0264-3707.

**J. Geol.**
Journal of Geology. The University of Chicago Press, 5801 S. Ellis Avenue, Chicago, Ill. 60637, USA. ISSN 0022–1376.

**J. Geomagn. Geoelectr.**
Journal of Geomagnetism and Geoelectricity. University of Tokyo Press, c/o Business Centre for Academic Societies, 4–16 Yayoi 2–chome, Bunkyo–ku, Tokyo 113, Japan. ISSN 0022–1392.

**J. Geophys.**
Journal of Geophysics. Springer–Verlag, Berlin – Heidelberg – New York – Tokyo. ISSN 0340–062X.

**J. Geophys. Res.**
Journal of Geophysical Research. Section A: Space Physics, Section B: Solid Earth and Planets, Section C: Oceans and Atmospheres. American Geophysical Union, 2000 Florida Avenue, N.W., Washington, D.C. 20009, USA. ISSN 0148–0227.

**J. Guid. Control Dyn.**
Journal of Guidance, Control, and Dynamics. American Institute of Aeronautics and Astronautics, 1290 Avenue of the Americas, New York, N.Y. 10104, USA. ISSN 0731–5090.

**J. Hist. Astron.**
Journal for the History of Astronomy. Published by Science History Publications Ltd., Halfpenny Furze, Mill Lane, Chalfont St Giles, Bucks. HP8 4NR, England. ISSN 0021–8286.

**J. Inst. Electron. Commun. Eng. Jpn.**
Journal of the Institute of Electronics and Communications Engineers of Japan. Institute of Electronics and Communications Engineers of Japan, Denshi Tsushin Gakkai, Kikai–Shinko–Kaikan, 5–8, Shibakoen 3 Chome Minato–ku, Tokyo 105, Japan. ISSN 0373–6121.

**J. Instr. Electron. Telecommun. Eng.**
Journal of the Institution of Electronics and Telecommunication Engineers. Institution of Electronics and Telecommunication Engineers, 2 Institutional Area, Lodi Road, New Delhi 110 003, India. ISSN 0377–2063.

**J. Korean Astron. Soc.**
The Journal of the Korean Astronomical Society. The Korean Astronomical Society, Seoul, Korea.

**J. Magn. Magn. Mater.**
Journal of Magnetism and Magnetic Materials. North–Holland Publishing Company, PO Box 211, 1000 AE Amsterdam, The Netherlands. ISSN 0304–8853.

**J. Math. Phys.**
Journal of Mathematical Physics. American Institute of Physics, 335 East 45th Street, New York, N.Y. 10017, USA. ISSN 0022–2488.

**J. Math. Phys. Sci.**
Journal of Mathematical and Physical Sciences. Indian Institute of Tech., Madras, India. ISSN 0047–2557.

**J. Mec. Theor. Appl.**
Journal de Mécanique Théorique et Appliquée. Centrale des Revues, 11 rue Gossin, F–92543 Montrouge Cedex, France. ISSN 0750–7240.

**J. Mech. Eng. Lab.**
Journal of Mechanical Engineering Laboratory. Namiki Sakura–mura, Niihari–gun, Ibaraki, Japan. ISSN 0388–4252.

**J. Meteor Res.**
The JournaL of Meteor Research. Space News Publishing Co., PO Box 66521, Baton Rouge, La. 70896, USA. ISSN 0277–6057.

**J. Microcomput. Appl.**
Journal of Microcomputer Applications. Academic Press Inc. Ltd., 24 – 28 Oval Road, London NW1 7DX, England. ISSN 0143–3792.

**J. Mol. Spectrosc.**
Journal of Molecular Spectroscopy. Academic Press Inc., 111 5th Avenue, New York, N.Y. 10003, USA. ISSN 0022–2852.

**J. Nanjing Univ.**
Journal of Nanjing University. (Natural Sciences Edition). Nanjing University, Nanjing, People's Republic of China.

**J. Navig.**
The Journal of Navigation. The Royal Institute of Navigation. Cambridge University Press, Cambridge – London – New York – New Rochelle – Melbourne – Sydney. ISSN 0020–3009.

**J. Opt. (Calcutta)**
Journal of Optics. Optical Society of India, Department of Applied Physics, University of Calcutta, 92 Acharya Prafulla Chandra Road, Calcutta 700 009, India. ISSN 0304–5811.

**J. Opt. (Paris)**
Journal of Optics. Masson Editeur, 120 Boulevard Saint–Germain, F–75280 Paris Cedex 06, France. ISSN 0150–536X.

**J. Opt. Soc. Am. A**
Journal of the Optical Society of America A. Optics and Image Science. Published for the Optical Society of America by the American Institute of Physics, 335 East 45th Street, New York, N.Y. 10017, USA. ISSN 0740–3232.

**J. Opt. Soc. Am. B**
Journal of the Optical Society of America B. Optical Physics. Published for the Optical Society of America by the American Institute of Physics, 335 East 45th Street, New York, N.Y. 10017, USA. ISSN 0740–3224.

**J. Phys.**
Journal de Physique. Les Editions de Physique, Z. I. de Courtabœuf, B.P. 112, F–91944 Les Ulis Cedex, France. ISSN 0302–0738.

**J. Phys. A**
Journal of Physics A. (Mathematical and General Physics). Institute of Physics, 47 Belgrave Square, London SW1X 8QX, England. ISSN 0305–4470.

**J. Phys. B**
Journal of Physics B. (Atomic and Molecular Physics). Institute of Physics, 47 Belgrave Square, London SW1X 8QX, England. ISSN 0022–3700.

**J. Phys. Chem.**
Journal of Physical Chemistry. American Chemical Society, 1155 16th Street N.W., Washington, D.C. 20036, USA. ISSN 0022–3654.

**J. Phys. Chem. Ref. Data**
Journal of Physical and Chemical Reference Data. American Chemical Society, 1155 16th Street, N. W., Washington, D. C. 20036, USA. ISSN 0047–2689.

**J. Phys. Colloq.**
Journal de Physique Colloque. Les Editions de Physique, Z.I. de Courtabœuf, B.P. 112, F–91944 Les Ulis Cedex, France. ISSN 0449–1947.

**J. Phys. D**
Journal of Physics D. (Applied Physics). Institute of Physics, 47 Belgrave Square, London SW1X 8QX, England. ISSN 0022–3727.

**J. Phys. E**
Journal of Physics E. (Scientific Instruments). Institute of Physics, 47 Belgrave Square, London SW1X 8QX, England. ISSN 0022–3735.

**J. Phys. Earth**
Journal of Physics of the Earth. University of Tokyo Press. Subscription address: Japan Publications Trading Company Ltd., C.P.O. 722, Tokyo, Japan. ISSN 0022–3743.

**J. Phys. F**
Journal of Physics F. (Metal Physics). Institute of Physics, 47 Belgrave Square, London SW1X 8QX, England. ISSN 0305–4608.

**J. Phys. G**
Journal of Physics G. (Nuclear Physics). Institute of Physics, 47 Belgrave Square, London SW1X 8QX, England. ISSN 0305–4616.

**J. Phys. Lett.**
Journal de Physique Lettres. Les Editions de Physique, Z.I. de Courtabœuf, B.P. 112, F–91944 Les Ulis Cedex, France. ISSN 0302–072X.

**J. Phys. Soc. Jpn.**
Journal of the Physical Society of Japan. Room 211, Kikai Shinko Building, 3–5–8 Shiba Koen, Minato–ku, Tokyo 105, Japan. ISSN 0031–9015.

**J. Plasma Phys.**
Journal of Plasma Physics. Cambridge University Press, Cambridge – London – New York – New Rochelle – Melbourne – Sydney. ISSN 0022–3778.

**J. Proc. R. Soc. N.S.W.**
Journal and Proceedings of the Royal Society of New South Wales. Science Centre, 35 Clarence Street, Sydney, N.S.W. 2000, Australia. ISSN 0035–9173.

**J. Quant. Spectrosc. Radiat. Transfer**
Journal of Quantitative Spectroscopy and Radiative Transfer. Pergamon Press, Oxford – New York – Toronto – Paris – Frankfurt – Sydney. ISSN 0022–4073.

**J. R. Astron. Soc. Can.**
The Journal of the Royal Astronomical Society of Canada. The Royal Astronomical Society of Canada, 124 Merton Street, Toronto, Ont. M4S 2Z2, Canada. ISSN 0035–872X.

**J. Radio Res. Lab.**
Journal of the Radio Research Laboratories. Radio Research Laboratories, Ministry of Posts & Telecommunications, Nukui–Kitamachi, Konganei–shi, Tokyo 184, Japan. ISSN 0033–8001.

**J. Res. Natl. Bur. Stand.**
Journal of Research of the National Bureau of Standards. Subscription address: US Government Printing Office, Washington, D.C. 20402, USA. ISSN 0091–0635.

**J. Sci. Ind. Res.**
Journal of Scientific and Industrial Research. Council of Scientific & Industrial Research, Publications & Information Directorate, Hillside Road, New Delhi 110 012, India. ISSN 0022–4456.

**J. Sci. Res. Banaras Hindu Univ.**
Journal of Scientific Research of the Banaras Hindu University. Department of Physics, Faculty of Science, PO Banaras Hindu University, Varnasi 221 005, India. ISSN 0447–9483.

**J. Soc. Instrum. Control Eng.**
Journal of the Society of Instrument and Control Engineers. Society of Instrument and Control Engineers, Hongo, 1–35–28–303, Bunkyo–ku, Tokyo 113, Japan. ISSN 0453–4662.

**J. Spacecr. Rockets**
Journal of Spacecraft and Rockets. American Institute of Aeronautics and Astronautics, 1290 Avenue of the Americas, New York, N.Y. 10104, USA. ISSN 0022–4650.

**J. Spectrosc. Soc. Jpn.**
Journal of the Spectroscopical Society of Japan. Spectroscopical Society of Japan, Room 301, Clean Building, 1–13, Kanda–Awaji–cho, Chiyoda–ku, Tokyo 101, Japan. ISSN 0038–7002.

**J. Stat. Phys.**
Journal of Statistical Physics. Plenum Publishing Corporation, 227 West 17th Street, New York, N.Y. 10011, USA. ISSN 0022–4715.

**J. Toyo Univ., Gen. Educ., Nat. Sci.**
Journal of the Toyo University, General Education, Natural Science. Toyo University, 28, Hakusan 5–chôme, Bunkyo–ku, Tokyo, Japan.

**Jenaer Rundsch.**
Jenaer Rundschau. Jenoptik Jena GmbH. VEB Verlag Technik, Oranienburger Str. 13/14, DDR–1020 Berlin, German Democratic Republic. ISSN 0368–203X.

**JETP Lett.**
JETP Letters. A cover–to–cover translation of Pis'ma v Zhurnal Ehksperimental'noj i Teoreticheskoj Fiziki. American Institute of Physics, 335 East 45th Street, New York, N.Y. 10017, USA. ISSN 0021–3640.

**Johns Hopkins APL Tech. Dig.**
Johns Hopkins APL Technical Digest. The Johns Hopkins University Applied Physics Laboratory. Johns Hopkins Road, Laurel, Md. 20707, USA. ISSN 0270–5214.

**Jpn. J. Appl. Phys., Part 1**
Japanese Journal of Applied Physics, Part 1 (Regular Papers and Short Notes). Publication Office, Daini Toyokaiji Building, 24–8, Shinbashi 4–chome, Minato–ku, Tokyo 105, Japan. ISSN 0021–2922.

**Jpn. J. Appl. Phys., Part 2**
Japanese Journal of Applied Physics, Part 2 (Letters). Publication Office, Daini Toyokaiji Building, 24–8, Shinbashi 4–chome, Minato–ku, Tokyo 105, Japan. ISSN 0021–4922.

**Kandilli Obs., Heliophys. Serv. Publ., Second Ser.**
Kandilli Observatory, Heliophysics Service Publications, Second Series. Kandilli Observatory, Istanbul, Turkey.

**Kapteyn Astron. Inst., Annu. Rep.**
Kapteyn Astronomical Institute, Annual Report. Department of Astronomy Rijksuniversiteit Groningen, Groningen, The Netherlands.

**KDD Tech. J.**
KDD Technical Journal. International Communications Research Institute, Tokyo, Japan. ISSN 0452–3431.

**Kexue Tongbao (Beijing)**
Kexue Tongbao. (In Chinese). Academia Sinica, Beijing. Subscription address: Science Press, No. 137, Chaayangmennei Street, Beijing, People's Republic of China. ISSN 0250–7862.

**Kexue Tongbao (London)**
Kexue Tongbao. (English translation of Kexue Tongbao). Scientific and Technical Books Service Ltd., PO Box 197, London WC2N 4DE, England.

**Kodaikanal Obs. Bull., Ser. A**
Kodaikanal Observatory Bulletins, Series A. Indian Institute of Astrophysics, Bangalore 560 034, India.

**Kodaikanal Obs. Repr.**
Kodaikanal Observatory Reprints. Indian Institute of Astrophysics, Bangalore 560 034, India.

**Komet. Tsirk.**
Kometnyj Tsirkulyar. Gruppa po Issledovaniyu Komet Sektsii 'Solnechnaya Sistema' Astronomicheskogo Soveta AN SSSR. Kievskij Universitet im. T.G. Shevchenko. Glavnaya Astronomicheskaya Observatoriya AN USSR.

**Komety Meteory**
Komety i Meteory. Akademiya Nauk Tadzhikskoj SSR. Astronomicheskij Sovet Akademii Nauk SSSR. Izdatel'stvo 'Donish', Dushanbe, 29, ul. Ajni, 121, korp. 2, USSR. ISSN 0568–6199.

**Kosm. Issled.**
Kosmicheskie Issledovaniya. Akademiya Nauk SSSR. Izdatel'stvo 'Nauka', Moskva. Editorial address: 103717 GSP, Moskva, K–62, Podsosenskij per., 21. USSR. English translation in Cosmic Res., Consultants Bureau, New York, N.Y., USA. ISSN 0023–4206.

**Kosm. Luchi**
Kosmicheskie Luchi. Rezul'taty Issledovanij po Mezhdunarodnym Geofizicheskim Proektam. Mezhduvedomstvennyj Geofizicheskij Komitet pri Prezidiume Akademii Nauk SSSR. Izdatel'stvo Sovetskoe radio, Moskva, Glavpochtamt, a/ya No. 693. USSR.

**Kozmos**
Kozmos. Popular Astronomical Journal. Slovenské ústredie amatérskej astronomie, Hurbanovo.

**Latitude Circ.**
Latitude Circular. Astronomic–Geodetical Observatory at Józefosław. Warsaw Technical University, Warsaw, Poland.

**LEST Found., Annu. Rep.**
Large European Solar Telescope Foundation, Annual Report. Institute of Theoretical Astrophysics, University of Oslo, PO Box 1029, Blindern, N–0315 Oslo 3, Norway.

**LEST Found., Tech. Rep.**
Large European Solar Telescope Foundation, Technical Report. Institute of Theoretical Astrophysics, University of Oslo, PO Box 1029, Blindern, N–0315 Oslo 3, Norway.

**Lett. Math. Phys.**
Letters in Mathematical Physics. D. Reidel Publishing Company, Dordrecht – Boston. ISSN 0377–9017.

**Lett. Nuovo Cimento**
Lettere al Nuovo Cimento. Società Italiana di Fisica. Editrice Compositori, viale XII Giugno, 1, I–40124 Bologna, Italy. ISSN 0024–1318.

**Lick Obs. Bull.**
Lick Observatory Bulletin. Lick Observatory, Board of Studies in Astronomy and Astrophysics, University of California at Santa Cruz, Santa Cruz, Calif. 95064, USA.

**Lohrmann–Obs., Tech. Univ. Dresden, Zirk.**
Lohrmann–Observatorium, Technische Universität Dresden, Zirkular. Lohrmann–Observatorium, Technische Universität Dresden, DDR–8027 Dresden, Mommsenstraße 13, German Democratic Republic.

**Lowell Obs. Bull.**
Lowell Observatory Bulletin. Lowell Observatory, Flagstaff, Ariz., USA.

**Madà**
Madà (Science). The Weizmann Science Press of Israel, Jerusalem, Israel. ISSN 0368–833X.

**Magy. Geofiz.**
Magyar Geofizika. Lapkiado Vallalat, H–1073 Budapest, Lenin korut 9–11, Hungary. ISSN 0025–0120.

**Marées Terr., Bull. Inf.**
Marées Terrestres, Bulletin d'Informations. Association Internationale de Géodesie, Commission Permanente des Marees Terrestres. Observatoire Royal de Belgique, 3, avenue Circulaire, B–1180 Bruxelles, Belgium.

**Marées Terr., Bull. Obs.**
Marées Terrestres, Bulletin d'Observations. Observatoire Royal de Belgique, 3, avenue Circulaire, B–1180 Bruxelles, Belgium.

**Math. Modelling**
Mathematical Modelling. Pergamon Press, New York – Oxford – Toronto – Paris – Frankfurt – Sydney. ISSN 0270–0255.

**Math. Proc. Camb. Philos. Soc.**
Mathematical Proceedings of the Cambridge Philosophical Society. Cambridge University Press, Cambridge – London – New York – New Rochelle – Melbourne – Sydney. ISSN 0305–0041.

**Max–Planck–Inst. Radioastron., Bonn, Sonderdr. Ser. A**
Max–Planck–Institut für Radioastronomie, Bonn, Sonderdrucke Serie A. Max–Planck–Institut für Radioastronomie, Auf dem Hügel 69, D–5300 Bonn, F.R. Germany.

**Meccanica**
Meccanica. Pitagora Editrice, via Zamboni 57, I–40126 Bologna, Italy. ISSN 0025–6455.

**Meded. K. Acad. Wet., Lett. Schone Kunsten Belg., Kl. Wet.**
Mededelingen van de Koninklijke Academie voor Wetenschappen, Letteren en Schone Kunsten van Belgie, Klasse der Wetenschappen. Koninklijke Academie voor Wetenschappen, Letteren en Schone Kunsten van Belgie. Paleis der Academien, Hertogsstraat 1, B–1000 Bruxelles, Belgium.

**Mem. Astron. Soc. India**
Memoirs of the Astronomical Society of India. Astronomical Society of India, Department of Astronomy, Osmania University, Hyderabad 500 007, India.

**Mem. Fac. Eng., Kyoto Univ.**
Memoirs of the Faculty of Engineering, Kyoto University. Faculty of Engineering, Kyoto University, Kyoto, Japan. ISSN 0023–6063.

**Mem. Fac. Eng., Osaka City Univ.**
Memoirs of the Faculty of Engineering, Osaka City University. Faculty of Engineering, Osaka City University, 459 Sugimoto–cho, Sumi Yoshi–kum, Osaka, Japan. ISSN 0078–6659.

**Mem. Fac. Sci., Kyoto Univ., Ser. Phys., Astrophys., Geophys., Chem.**
Memoirs of the Faculty of Science, Kyoto University, Series of Physics, Astrophysics, Geophysics, and Chemistry. Faculty of Science, University of Kyoto, Kyoto 608, Japan. ISSN 0368–9689.

**Mem. Jpn. Astron. Study Assoc.**
Memoirs of the Japan Astronomical Study Association. National Science Museum, Ueno Park, Taito–ku, Tokyo, Japan.

**Mem. Soc. Astron. Ital.**
Memorie della Società Astronomica Italiana. Società Astronomica Italiana, Largo Fermi, 5, I-50125 Firenze, Italy. ISSN 0037-8720.

**Mercury**
Mercury. The Journal of the Astronomical Society of the Pacific. The Astronomical Society of the Pacific, 1290 24th Avenue, San Francisco, Calif. 94122, USA. ISSN 0047-6773.

**Messenger**
The Messenger - El Mensajero. European Southern Observatory, Karl-Schwarzschild-Straße 2, D-8046 Garching bei München, F.R. Germany. ISSN 0722-6691.

**Meteoritics**
Meteoritics. The Journal of the Meteoritical Society. Center for Meteorite Studies, Arizona State University, Tempe, Ariz. 85287, USA. ISSN 0026-1114.

**Meteoritika**
Meteoritika. Akademiya Nauk SSSR. Komitet po Meteoritam. Izdatel'stvo 'Nauka', Moskva. 117864 GSP-7, Moskva V-485, Profsoyuznaya ul., d. 90, USSR. ISSN 0369-2507.

**Meteorol. Mag.**
Meteorological Magazine. General Meteorological Office, London Road, Bracknell, Berks. RG12 2SZ, England. ISSN 0026-1149.

**Metrologia**
Metrologia. Springer-Verlag, Berlin - Heidelberg - New York. ISSN 0026-1394.

**Micro Syst.**
Micro Systèmes. Société Parisienne d'Edition, 2 à 12 rue de Bellevue, F-75940 Paris Cedex 19, France. ISSN 0183-5084.

**Mikrocomput. Z.**
Mikrocomputer Zeitschrift. Franzis-Verlag GmbH, Karlstr. 37, D-8000 München 2, F.R. Germany. ISSN 0720-4442.

**Mikrowellen Mag.**
Mikrowellen Magazin. Sprechsaal-Verlag, PO Box 401, D-8630 Coburg, F.R. Germany. ISSN 0722-8244.

**Minor Planet Bull.**
The Minor Planet Bulletin. Bulletin of the Minor Planets Section of the Association of Lunar and Planetary Observers. Editorial Office: R. P. Binzel, Department of Astronomy, University of Texas, Austin, Tex. 78712, USA.

**Minor Planet Circ.**
The Minor Planet Circulars/Minor Planets and Comets. Minor Planet Center, Smithsonian Astrophysical Observatory, 60, Garden Street, Cambridge, Mass. 02138, USA. ISSN 0736-6884.

**Mitsubishi Denki Giho**
Mitsubishi Denki Giho. Mitsubishi Electric Corporation, Mitsubishi Denki Building, Marunouchi, Tokyo 100, Japan. ISSN 0369-2302.

**Mitt. Archenhold-Sternw. Berlin-Treptow**
Mitteilungen der Archenhold-Sternwarte Berlin-Treptow. Archenhold-Sternwarte, DDR-1193 Berlin, Alt Treptow 1, German Democratic Republic.

**Mitt. Astron. Ges.**
Mitteilungen der Astronomischen Gesellschaft, Hamburg. Subscription address: Astronomisches Institut der Universität Bochum, Postfach 102148, D-4630 Bochum, F.R. Germany. ISSN 0172-5483.

**Mitt. Astrophys. Obs. Potsdam**
Mitteilungen des Astrophysikalischen Wissenschaften der DDR, Zentralinstitut für Astrophysik. Astrophysikalisches Observatorium Potsdam, DDR-1500 Potsdam, Telegrafenberg, German Democratic Republic.

**Mitt. Inst. Astron. Phys. Geod. Tech. Hochsch. Münch.**
Mitteilungen aus dem Institut für Astronomische und Physikalische Geodäsie der Technischen Hochschule München. Institut für Astronomische und Physikalische Geodäsie der Technischen Hochschule München, D-8000 München, F.R. Germany.

**Mitt. Inst. Theor. Geod. Univ. Bonn**
Mitteilungen aus dem Institut für Theoretische Geodäsie der Universität Bonn. Institut für Theoretische Geodäsie der Universität Bonn, Nußallee 17, D-5300 Bonn 1, F.R. Germany. ISSN 0343-7361.

**Mitt. Karl-Schwarzschild-Obs. Tautenburg**
Mitteilungen des Karl-Schwarzschild-Observatoriums Tautenburg. Akademie der Wissenschaften der DDR, Zentralinstitut für Astrophysik. Karl-Schwarzschild-Observatorium Tautenburg, DDR-6901 Tautenburg, German Democratic Republic.

**Mitt. Lohrmann-Obs. Tech. Univ. Dresden**
Mitteilungen des Lohrmann-Observatoriums der Technischen Universität Dresden. Lohrmann-Observatorium, Technische Universität Dresden, DDR-8027 Dresden, Mommsenstraße 13, German Democratic Republic.

**Mitt. Satell.-Beobachtungsstn. Zimmerwald**
Mitteilungen der Satelliten-Beobachtungsstation Zimmerwald. Druckerei der Universität Bern, Switzerland.

**Mitt. Sonnenobs. Kanzelhöhe**
Mitteilungen des Sonnenobservatoriums Kanzelhöhe. Sonnenobservatorium Kanzelhöhe, A-9520 Sattendorf, Austria.

**Mitt. Sternw. Babelsberg, Neue Folge**
Mitteilungen der Sternwarte Babelsberg, Neue Folge. Akademie der Wissenschaften der DDR, Zentralinstitut für Astrophysik. Sternwarte Babelsberg, DDR-1502 Potsdam-Babelsberg, Rosa-Luxemburg-Straße 17a, German Democratic Republic.

**Mitt. Sternw. Münch.**
Mitteilungen der Sternwarte München. Institut für Astronomie und Astrophysik der Universität München, Universitäts-Sternwarte, Scheinerstr. 1, D-8000 München 80, F.R. Germany.

**Mitt. Sternw. Sonneberg**
Mitteilungen der Sternwarte zu Sonneberg. Akademie der Wissenschaften der DDR, Zentralinstitut für Astrophysik. Sternwarte Sonneberg, DDR-6400 Sonneberg, German Democratic Republic.

**Mitt. Univ.-Sternw. Innsb.**
Mitteilungen der Universitäts-Sternwarte Innsbruck. Institut für Astronomie der Universität Innsbruck, Universitätsstraße 4, A-6020 Innsbruck, Austria.

**Mitt. Univ.-Sternw. Jena**
Mitteilungen der Universitäts-Sternwarte zu Jena. Universitätssternwarte Jena, Schillergäßchen 2, DDR-6900 Jena, German Democratic Republic.

**Mitt. Universitätssternw. Graz**
Mitteilungen der Universitätssternwarte Graz. Institut für Astronomie, Universitätsplatz 5, A-8010 Graz, Austria.

**Mitt. Veränderliche Sterne**
Mitteilungen über Veränderliche Sterne. Akademie der Wissenschaften der DDR, Zentralinstitut für Astrophysik, Sternwarte Sonneberg, DDR–6400 Sonneberg, German Democratic Republic.

**Mitt. Zentralinst. Phys. Erde**
Mitteilungen des Zentralinstituts für Physik der Erde. Akademie der Wissenschaften der DDR, Zentralinstitut für Physik der Erde, DDR–1500 Potsdam, Telegrafenberg A17, German Democratic Republic.

**Mod. Geol.**
Modern Geology. Gordon and Breach Science Publishers, New York – London. ISSN 0026–7775.

**Mon. Not. R. Astron. Soc.**
Monthly Notices of the Royal Astronomical Society. Published for the Royal Astronomical Society by Blackwell Scientific Publications, Oxford – London – Edinburgh – Boston – Melbourne. ISSN 0035–8711.

**Mon. Notes Astron. Soc. S. Afr.**
Monthly Notes of the Astronomical Society of Southern Africa. South African Astronomical Observatory, PO Box 9, Observatory, 7935 Cape, South Africa. ISSN 0024–8266.

**Mon. Notes Int. Polar Motion Serv.**
Monthly Notes of the International Polar Motion Service. Central Bureau of the International Polar Motion Service, International Latitude Observatory of Mizusawa, Mizusawa-shi, Iwate-ken, Japan. ISSN 0020–8337.

**Mons Astrophys. Pap.**
Mons Astrophysical Papers. (Communications du Département d'Astrophysique de la Faculté des Sciences de Mons. Departement d'Astrophysique, Université de Mons, B–7000 Mons, Belgium.

**MPA Rep.**
Max–Planck–Institut für Physik und Astrophysik, Institut für Astrophysik, Reports. Max–Planck–Institut für Physik und Astrophysik, Institut für Astrophysik, Karl–Schwarzschild–Straße 1, D–8046 Garching bei München, F.R. Germany.

**MPE Contrib.**
Max–Planck–Institut für Physik und Astrophysik, Institut für Extraterrestrische Physik, Contributions. Max–Planck–Institut für Extraterrestrische Physik, Giessenbachstraße, D–8046 Garching bei München, F.R. Germany.

**MPE Prepr.**
Max–Planck–Institut für Physik und Astrophysik, Institut für Extraterrestrische Physik, Preprints. Max–Planck–Institut für Extraterrestrische Physik, Giessenbachstraße, D–8046 Garching bei München, F.R. Germany. ISSN 0340–8922.

**MPE Rep.**
Max–Planck–Institut für Physik und Astrophysik, Institut für Extraterrestrische Physik, Reports. Max–Planck–Institut für Extraterrestrische Physik, Giessenbachstraße, D–8046 Garching bei München, F.R. Germany. ISSN 0173–699X.

**MPE Repr.**
Max–Planck–Institut für Physik und Astrophysik, Institut für Extraterrestrische Physik, Reprints. Max–Planck–Institut für Extraterrestrische Physik, Giessenbachstraße, D–8046 Garching bei München, F.R. Germany.

**MPG Spiegel**
MPG Spiegel. Max–Planck–Gesellschaft zur Förderung der Wissenschaften, Residenzstraße 1a, D–8000 München 2, F.R. Germany. ISSN 0341–7727.

**MSN Microwave Syst. News**
MSN Microwave Systems News. EW Communications Inc., 1170 East Meadow Drive, Palo Alto, Calif. 94303, USA. ISSN 0164–3371.

**Mt. Stromlo Siding Spring Obs., Repr.**
Mount Stromlo and Siding Spring Observatories, Reprints. Mount Stromlo and Siding Spring Observatories, Research School of Physical Sciences, The Australian National University, Private Bag, Woden PO, ACT 2606, Australia.

**Nablyud. Iskusstv. Nebesn. Tel**
Nablyudeniya Iskusstvennykh Nebesnykh Tel. Published by Astronomicheskij Sovet Akademii Nauk SSSR, Moskva. Moskva Zh–17, ul. Pyatnitskaya, 48, Astronomicheskij Sovet AN SSSR.

**Nachr. Akad. Wiss. Göttingen. II**
Nachrichten der Akademie der Wissenschaften in Göttingen. II. Mathematisch–Physikalische Klasse. Vandenhoeck und Ruprecht, Göttingen, F.R. Germany. ISSN 0065–5295.

**Nachr. Karten–Vermessungswes., Reihe I**
Nachrichten aus dem Karten– und Vermessungswesen, Reihe I: Orginalbeiträge. Verlag des Instituts für Angewandte Geodäsie, Frankfurt a.M., F.R. Germany. ISSN 0469–4236.

**Nachr. Olbers–Ges. Bremen**
Nachrichten der Olbers–Gesellschaft Bremen. Dr. Walter Stein, Werderstr. 73, Bremen, F.R. Germany.

**Nakano wa Kangaeru noda**
Nakano wa Kangaeru noda. S. Nakano, PO Box No. 32, Sumoto Post Office, Hyogo–Ken, 656–91, Japan.

**Nanjing Univ. Obs., Publ.**
Nanjing University Observatory, Publications. Department of Astronomy and Astrophysics, Nanjing University, Nanjing, People's Republic of China.

**NASA Conf. Publ.**
NASA Conference Publication. National Aeronautics and Space Administration, Scientific and Technical Information Branch, Washington, D.C. 20546, USA. For sale by the National Technical Information Service, Springfield, Va. 22161, USA.

**NASA Contract. Rep.**
NASA Contractor Report. National Aeronautics and Space Administration, Scientific and Technical Information Branch, Washington, D.C. 20546, USA. For sale by the National Technical Information Service, Springfield, Va. 22161, USA. ISSN 0565–7059.

**NASA IUE Newsl.**
NASA IUE Newsletter. National Aeronautics and Space Administration, Goddard Space Flight Center, Greenbelt, Maryland 20771, USA. ISSN 0738–2677.

**NASA Ref. Publ.**
NASA Reference Publication. National Aeronautics and Space Administration, Scientific and Technical Information Branch, Washington, D.C. 20546, USA. For sale by the National Technical Information Service, Springfield, Va. 22161, USA.

**NASA Spec. Publ.**
NASA Special Publication. National Aeronautics and Space Administration, Scientific and Technical Information Branch, Washington, D.C. 20546, USA. For sale by the National Technical Information Service, Springfield, Va. 22161, USA. ISSN 0091–0805.

**NASA Tech. Brief**
NASA Technical Brief. NASA Technology Utilization Program, Technology Transfer Division, PO Box 8757, Baltimore/Washington International Airport, Md. 21240, USA. ISSN 0096–7494.

**NASA Tech. Memo.**
NASA Technical Memorandum. National Aeronautics and Space Administration, Scientific and Technical Information Branch, Washington, D.C. 20546, USA. For sale by the National Technical Information Service, Springfield, Va. 22161, USA. ISSN 0499–9320.

**NASA Tech. Note**
NASA Technical Note. National Aeronautics and Space Administration, Scientific and Technical Information Branch, Washington, D.C. 20546, USA. For sale by the National Technical Information Service, Springfield, Va. 22161, USA. ISSN 0499–9339.

**NASA Tech. Pap.**
NASA Technical Paper. National Aeronautics and Space Administration, Scientific and Technical Information Branch, Washington, D.C. 20546, USA. For sale by the National Technical Information Service, Springfield, Va. 22161, USA.

**Natl. Acad. Sci. Lett.**
National Academy Science Letters. National Academy of Sciences, 5–Lajpat Rai Road, Allahabad 211 002, India. ISSN 0250–541X.

**Natl. Astron. Ionos. Cent., Astron. Prepr.**
National Astronomy and Ionosphere Center, Astronomy Preprints. National Astronomy and Ionosphere Center, Space Sciences Building, Cornell University, Ithaca, N.Y. 14853, USA.

**Natl. Astron. Ionos. Cent., Astron. Publ.**
National Astronomy and Ionosphere Center, Astronomy Publications. National Astronomy and Ionosphere Center, Space Sciences Building, Cornell University, Ithaca, N.Y. 14853, USA.

**Natl. Astron. Ionos. Cent., Newsl.**
National Astronomy and Ionosphere Center, Newsletter. National Astronomy and Ionosphere Center, PO Box 995, Arecibo, P.R. 00613, Puerto Rico.

**Natl. Electron. Rev.**
National Electronics Reviews. National Electronics Council, Abell House, John Islip Street, London SW1P 4LN, England. ISSN 0305–2257.

**Natl. Geogr.**
National Geographic. National Geographic Society, 17th and M Sts. N.W., Washington, D.C. 20036, USA. ISSN 0027–9358.

**Natl. Radio Astron. Obs., Repr., Ser. A**
National Radio Astronomy Observatory, Reprints, Series A. National Radio Astronomy Observatory, PO Box 2, Green Bank, W.Va. 24944, USA.

**Natl. Radio Astron. Obs., Repr., Ser. B**
National Radio Astronomy Observatory, Reprints, Series B. National Radio Astronomy Observatory, PO Box 2, Green Bank, W.Va. 24944, USA.

**Nature**
Nature. Macmillan Journals Ltd., London, England. Subscription address: Nature, Circulation Dept., Brunel Road, Basingstoke, Hants RG21 2XS, England. ISSN 0028–0836.

**Naturwissenschaften**
Die Naturwissenschaften. Springer–Verlag, Berlin – Heidelberg – New York – Tokyo. ISSN 0028–1042.

**Nauchn. Inf.**
Nauchnye Informatsii. Astronomicheskij Sovet Akademii Nauk SSSR, Moskva, USSR. ISSN 0130–9773.

**Naučna Misao**
Naučna Misao (Scientific Idea). Society for Promotion and Propagation of Science, 41000 Zagreb, Pavleka Miškine 37, Yugoslavia. ISSN 0467–0468.

**Navigation (Paris)**
Navigation. Institut Française de Navigation, 3, avenue Octave–Greard, F–75340 Paris Cedex 07, France. ISSN 0028–1530.

**Navigation (Wash.)**
Navigation. Journal of the Institute of Navigation. Institute of Navigation, Suite 832. 815 15th Street, N.W., Washington, D.C. 20005, USA. ISSN 0028–1522.

**Ned. Tijdschr. Natuurkd. A**
Nederlands Tijdschrift voor Natuurkunde A. Nederlandse Natuurkundige Vereniging, Princetonplein 5, NL–3508 TA Utrecht, The Netherlands. ISSN 0378–6374.

**New Phys., Korean Phys. Soc.**
New Physics, Korean Physical Society. Korea Institute for Industrial Economics & Technology, PO Box 205, Chungryang, Seoul, Republic of Korea. ISSN 0374–4914.

**New Sci.**
New Scientist. New Science Publications, Commonwealth House, 1–19 New Oxford Street, London WC1A 1NG, England. ISSN 0028–6664.

**News Lett. Astron. Soc. N.Y.**
News Letter of the Astronomical Society of New York. Astronomial Society of New York, 1125 Oxford Place, Schenectady, N.Y. 12308, USA.

**Nizamiah Rangapur Obs. Dep. Astron. Osmania Univ., Repr.**
Nizamiah and Rangapur Observatories and Department of Astronomy of Osmania University, Reprint. Centre of Advanced Study in Astronomy, Osmania University, Hyderabad 500 007, India.

**Nova Acta Leopoldina**
Nova Acta Leopoldina. Abhandlungen der Deutschen Akademie der Naturforscher Leopoldina. Deutsche Akademie der Naturforscher Leopoldina, DDR–4010 Halle (Saale), German Democratic Republic. ISSN 0369–5034.

**Nova Acta Regiae Soc. Sci. Ups., Ser. V: A**
Nova Acta Regiae Societatis Scientiarum Upsaliensis, Ser. V: A (Astronomy and Mathematical Sciences). Royal Society of Sciences of Uppsala. Subscripton address: Almqvist & Wiksell International, Stockholm, Sweden. ISSN 0346–6253.

**NSSDC/WDC–A–R&S, Publ.**
NSSDC/WDC–A–R&S, Publication. National Space Science Data Center/World Data Center A for Rockets and Satellites, National Aeronautics and Space Administration, Goddard Space Flight Center, Greenbelt, Md. 20771, USA.

**Nucl. Instrum. Methods Phys. Res., Sect. A**
Nuclear Instruments and Methods in Physics Research, Section A. Accelerators, Spectrometers, Detectors and Associated Equipment. North–Holland Publishing Company, PO Box 211, 1000 AE Amsterdam, The Netherlands. ISSN 0167–5087.

**Nucl. Instrum. Methods Phys. Res., Sect. B**
Nuclear Instruments and Methods in Physics Research, Section B. Beam Interactions with Materials and Atoms. North–Holland Publishing Company, PO Box 211, 1000 AE Amsterdam, The Netherlands. ISSN 0168–583X.

**Nucl. Phys. A**
Nuclear Physics A. North–Holland Publishing Company, PO Box 211, 1000 AE Amsterdam, The Netherlands. ISSN 0029–5582.

**Nucl. Phys. B, Part. Phys.**
Nuclear Physics B, Particle Physics. North–Holland Publishing Company, PO Box 211, 1000 AE Amsterdam, The Netherlands. ISSN 0550–3213.

**Nucl. Sci. Appl., Sect. A**
Nuclear Science Applications, Section A. Harwood Academic Publishers GmbH, Poststraße 22, CH–7000 Chur, Switzerland. ISSN 0191–1686.

**Nucl. Tracks Radiat. Meas.**
Nuclear Tracks and Radiation Measurements. Pergamon Press, Oxford – New York – Toronto – Paris – Frankfurt – Sydney. ISSN 0191–278X.

**Numer. Math.**
Numerische Mathematik. Springer–Verlag, Berlin – Heidelberg – New York – Tokyo. ISSN 0029–599X.

**Nuovo Cimento A**
Il Nuovo Cimento A. Società Italiana di Fisica, Editrice Compositori, viale XII Giugno, 1, I–40124 Bologna, Italy. ISSN 0369–4097.

**Nuovo Cimento B**
Il Nuovo Cimento B. Società Italiana di Fisica, Editrice Compositori, viale XII Giugno, 1, I–40124 Bologna, Italy. ISSN 0369–4100.

**Nuovo Cimento C**
Il Nuovo Cimento C. Società Italiana di Fisica, Editrice Compositori, viale XII Giugno, 1, I–40124 Bologna, Italy.

**N.Z. J. Sci.**
New Zealand Journal of Science. New Zealand Department of Scientific and Industrial Research. Subscription address: Science Information Division, DSIR, PO Box 9741, Wellington, New Zealand. ISSN 0028–8365.

**Obs. Astron. Antares, Contrib. Cient.**
Observatório Astronômico Antares, Contribuição Cientifica. Universidade Estadual de Feira de Santana, Feira de Santana, Brazil.

**Obs. Astron. Córdoba, Tirada Aparte**
Observatorio Astronomico Córdoba, Tirada Aparte. Observatorio Astronomico, Laprida 854, 5000 Córdoba, Argentina.

**Obs. Astron. La Plata, Sep. Astron.**
Observatorio Astronómico La Plata, Separata Astronomica. Observatorio Astronómico, Universidad Nacional de La Plata, La Plata, Argentina.

**Obs. Astron. Univ. Nac. La Plata, Ser. Astron.**
Observatorio Astronómico de la Universidad Nacional de La Plata, Serie Astronómica. Observatorio Astronómico, Universidad Nacional de La Plata, La Plata, Argentina. ISSN 0325–3163.

**Obs. Astron. Univ. Nac. La Plata, Ser. Espec.**
Observatorio Astronómico de la Universidad Nacional de La Plata, Serie Especial. Observatorio Astronómico, Universidad Nacional de La Plata, La Plata, Argentina. ISSN 0325–3015.

**Obs. Astrophys. Lab., Univ. Helsinki, Rep.**
Observatory and Astrophysics Laboratory, University of Helsinki, Report. Observatory and Astrophysics Laboratory, University of Helsinki, Tähtitorninmäki, SF–00130 Helsinki 13, Finland. ISSN 0355–9289.

**Obs. Lyon Prepr.**
Observatoire de Lyon Preprints. Observatoire de Lyon, F–69230 St. Genis Laval, France.

**Obs. Lyon Repr.**
Observatoire de Lyon Reprint. Observatoire de Lyon, F–69230 St. Genis Laval, France.

**Obs. R. Belg., Commun., Sér. A**
Observatoire Royal de Belgique, Communications, Série A. (Koninklijke Sterrenwacht van België, Mededelingen, Reeks A). Observatoire Royale de Belgique, 3, avenue Circulaire, Uccle, B–1180 Bruxelles, Belgium.

**Obs. R. Belg., Commun., Sér. B**
Observatoire Royal de Belgique, Communications, Série B. (Koninklijke Sterrenwacht van België, Mededelingen, Reeks B). Observatoire Royal de Belgique, 3, avenue Circulaire, Uccle, B–1180 Bruxelles, Belgium.

**Obs. Satell.**
Observations of Satellites. Finnish Meteorological Institute. Subscription address: Government Printing Centre, Marketing Department, PO Box 516, SF–00101 Helsinki 10, Finland. ISSN 0355–2004.

**Obs. Variable Stars, Rep.**
Observations of Variable Stars, Report. Nederlandse Vereniging voor Weer–en Sterrenkunde. Kapteyn Astronomical Institute, Postbus 800, 9700 AV Groningen, The Netherlands.

**Observatory**
The Observatory. A Review of Astronomy. Royal Greenwich Observatory, Herstmonceux Castle, Hailsham Sussex, BN27 1RP England, ISSN 0029–7704.

**Occas. Rep. R. Obs., Edinb.**
Occasional Reports of the Royal Observatory, Edinburgh. Royal Observatory, Blackford Hill, Edinburgh EH9 3HJ, Scotland. ISSN 0309–099X.

**Occultation Newsl.**
Occultation Newsletter. International Occultation Timing Association (IOTA), PO Box 596, Tinley Park, Ill. 60477, USA.

**Österr. Z. Vermessungswes. Photogramm.**
Österreichische Zeitschrift für Vermessungswesen und Photogrammetrie. Österreichischer Verein für Vermessungswesen und Photogrammetrie, Friedrich–Schmidt–Platz 3, A–1082 Wien, Austria.

**Onsala Space Obs., Prepr.**
Onsala Space Observatory, Preprint. Chalmers University of Technology, Onsala Space Observatory, S–43900 Onsala, Sweden.

**Opt. Acta**
Optica Acta. Taylor and Francis Ltd., 4 John Street, London WC1N 2ET, England. ISSN 0030–3909.

**Opt. Commun.**
Optics Communications. North–Holland Publishing Company, PO Box 211, 1000 AE Amsterdam, The Netherlands. ISSN 0030–4018.

**Opt. Eng.**
Optical Engineering. Society of Photo–Optical Instrumentation Engineers (SPIE), PO Box 10, 405 Fieldston Road, Bellingham, Wash. 98225, USA. ISSN 0091–3286.

**Opt. Lett.**
Optics Letters. Published for the Optical Society of America by the American Institute of Physics, 335 East 45th Street, New York, N.Y. 10017, USA. ISSN 0146–9592.

**Opt. Pura Apl.**
Optica Pura y Aplicada. Sociedad Española de Optica, Serrano 121, Madrid 6, Spain. ISSN 0030–3917.

**Origins Life**
Origins of Life. An International Journal Devoted to the Scientific Study of the Origin of Life. D. Reidel Publishing Company, Dordrecht – Boston – London. ISSN 0302–1688.

**Orion**
Orion. Zeitschrift der Schweizerischen Astronomischen Gesellschaft (SAG). Zentralsekretariat, Hirtenhofstraße 9, CH–6005 Luzern, Switzerland. ISSN 0030–557X.

**Orione**
Orione. Rivista Trimestrale di Astronomia. Rivista Orione, via Roma 6, I–10025 Pino Torinese, Italy.

**Oss. Astrofis. Catania, Pubbl.**
Osservatorio Astrofisico di Catania, Pubblicazione. Osservatorio Astrofisico, città universitaria, I–95125 Catania, Italy.

**Oss. Astron. Milano–Merate, Contrib.**
Osservatorio Astronomico Milano–Merate, Contributo. Osservatorio Astronomico, Via Emilio Bianchi 46, I–22055 Merate, Como, Italy.

**Oss. Mem. Oss. Astrofis. Arcetri**
Osservazioni e Memorie dell'Osservatorio Astrofisico di Arcetri. Università Degli Studi di Firenze, Firenze, Italy.

**Oyo Buturi**
Oyo Buturi. Japan Society of Applied Physics, Room No. 209–2, Kikai–Shinko Building, 3 Shiba–Koen Minato–ku, Tokyo, Japan. ISSN 0369–8009.

**Patrika**
Patrika. Newsletter of the Indian Academy of Sciences. Indian Academy of Sciences, Bangalore 560 080, India.

**Perem. Zvezdy**
Peremennye Zvezdy (Variable Stars). Sbornik statej, izdavaemyj Astronomicheskim Sovetom Akademii Nauk SSSR. Editorial address: Sternberg State Astronomical Institute of the Moscow University, Universitetskij prospekt, 13, 117234 Moscow, USSR. ISSN 0373–7683.

**Perem. Zvezdy, Prilozh.**
Peremennye Zvezdy, Prilozhenie (Variable Stars, Supplement). Sbornik statej, izdavaemyj Astronomicheskim Sovetom Akademii Nauk SSSR. Astronomicheskij Sovet Akademii Nauk SSSR, Moskva, USSR.

**Philos. Trans. R. Soc. London, Ser. A**
Philosophical Transactions of the Royal Society of London, Series A: Mathematical and Physical Sciences. The Royal Society, 6 Carlton House Terrace, London SW1Y 5AG, England. ISSN 0080–4614.

**Photogr. Sci. Eng.**
Photographic Science and Engineering. Society of Photographic Scientists and Engineers, Suite 204, 1330 Massachusetts Avenue N.W., Washington, D.C. 20005, USA. ISSN 0031–8760.

**Photogramm. Eng. Remote Sensing**
Photogrammetric Engineering and Remote Sensing. American Society of Photogrammetry, 105 North Virginia Avenue, Falls Church, Va. 22046, USA. ISSN 0099–1112.

**Photonics Spectra**
Photonics Spectra. The Magazine of Optical/Electro–Optical/ Laser Technology. Published by the Optical Publishing Co., Inc., PO Box 1146, Pittsfield, Mass. 01201, USA. ISSN 0191–0647.

**Photorin**
Photorin, Mitteilungen der Lichtenberg–Gesellschaft e.V.

**Phys. Abstr.**
Physics Abstracts. Science Abstracts Series A. An INSPEC Publication published by the Institution of Electrical Engineers in association with the Institute of Electrical and Electronics Engineers Inc. Subscription address: INSPEC Marketing Department, IEE, Station House, Nightingale Road, Hitchin, Herts. SG5 1RJ, England. ISSN 0036–8091.

**Phys. Bl.**
Physikalische Blätter. Physik–Verlag GmbH, Postfach 1260/1280, D–6940 Weinheim, F.R. Germany. ISSN 0031–9279.

**Phys. Briefs**
Physics Briefs. Physikalische Berichte. Edited by Deutsche Physikalische Gesellschaft und Fachinformationszentrum Energie, Physik, Mathematik in cooperation with American Institute of Physics. Physik–Verlag GmbH, Postfach 1260/1280, D–6940 Weinheim, F.R. Germany. ISSN 0170–7434.

**Phys. Bull.**
Physics Bulletin. Institute of Physics, 47 Belgrave Square, London SW1X 8QX, England. ISSN 0031–9112.

**Phys. Chem. Miner.**
Physics and Chemistry of Minerals. Springer–Verlag, Berlin – Heidelberg – New York – Tokyo. ISSN 0342–1791.

**Phys. Earth Planet. Inter.**
Physics of the Earth and Planetary Interiors. Elsevier Scientific Publishing Company, PO Box 211, 1000 AE Amsterdam, The Netherlands. ISSN 0031–9201.

**Phys. Educ.**
Physics Education. Institute of Physics. 47 Belgrave Square, London SW1X 8QX, England. ISSN 0031–9120.

**Phys. Energ. Fortis Phys. Nucl.**
Physica Energiae Fortis et Physica Nuclearis. Science Press, Beijing. Subscription address: Guozi Shudian, PO Box 399, Beijing, People's Republic of China. ISSN 0254–3052.

**Phys. Fluids**
Physics of Fluids. American Institute of Physics, 335 East 45th Street, New York, N.Y. 10017, USA. ISSN 0031–9171.

**Phys. Lett. A**
Physics Letters A. (General, Atomic, and Solid State Physics). North–Holland Publishing Company, PO Box 211, 1000 AE Amsterdam, The Netherlands. ISSN 0031–9163.

**Phys. Lett. B**
Physics Letters B. (Nuclear, Elementary Particle, and High–Energy Physics). North–Holland Publishing Company, PO Box 211, 1000 AE Amsterdam, The Netherlands. ISSN 0031–9163.

**Phys. Rep.**
Physics Reports. North–Holland Publishing Company, PO Box 211, 1000 AE Amsterdam, The Netherlands. ISSN 0370–1573.

**Phys. Rev. A**
Physical Review A. (General Physics). Published for the American Physical Society by the American Institute of Physics, 335 East 45th Street, New York, N.Y. 10017, USA. ISSN 0556–2791.

**Phys. Rev. B**
Physical Review B. (Condensed Matter). Published for the American Physical Society by the American Institute of Physics, 335 East 45th Street, New York, N.Y. 10017, USA. ISSN 0163–1829.

**Phys. Rev. C**
Physical Review C. (Nuclear Physics). Published for the American Phyiscal Society by the American Institute of Physics, 335 East 45th Street, New York, N.Y. 10017, USA. ISSN 0556–2813.

**Phys. Rev. D**
Physical Review D. (Particles and Fields). Published for the American Physical Society by the American Institute of Physics, 335 East 45th Street, New York, N.Y. 10017, USA. ISSN 0556–2821.

**Phys. Rev. Lett.**
Physical Review Letters. Published for the American Physical Society by the American Institute of Physics, 335 East 45th Street, New York, N.Y. 10017, USA. ISSN 0031–9007.

**Phys. Scr.**
Physica Scripta. An International Journal for Experimental and Theoretical Physics. The Royal Swedish Academy of Sciences, Publications Department, Box 50005, S–104 05 Stockholm, Sweden. ISSN 0031–8949.

**Phys. Teach.**
Physics Teacher. American Institute of Physics, 335 East 45th Street, New York, N.Y. 10017, USA. ISSN 0031–921X.

**Phys. Technol.**
Physics in Technology. Institute of Physics, 47 Belgrave Square, London SW1X 8QX, England. ISSN 0305–4624.

**Phys. Today**
Physics Today. American Institute of Physics, 335 East 45th Street, New York, N.Y. 10017, USA. ISSN 0031–9228.

**Phys. Unserer Zeit**
Physik in unserer Zeit. Verlag Chemie GmbH, Postfach 1260/1280, D–6940 Weinheim, F.R. Germany. ISSN 0031–9252.

**Physica A**
Physica A. North–Holland Publishing Company, PO Box 211, 1000 AE Amsterdam, The Netherlands. ISSN 0378–4371.

**Physica B, C**
Physica B and C (Low Temperature and Solid State Physics; Atomic, Molecular and Plasma Physics; Optics). North–Holland Publishing Company, PO Box 211, 1000 AE Amsterdam, The Netherlands. ISSN 0378–4363.

**Physica D**
Physica D. North–Holland Publishing Company, PO Box 211, 1000 AE Amsterdam, The Netherlands. ISSN 0167–2789.

**Pis'ma Astron. Zh.**
Pis'ma v Astronomicheskij Zhurnal. Akademiya Nauk SSSR. Izdatel'stvo 'Nauka', Moskva. 103717, GSP, Moskva, K–62, Podsosenskij per., 21. USSR. English translation in Soviet Astron. Lett. ISSN 0320–0108.

**Planet. Space Sci.**
Planetary and Space Science. Pergamon Press, Oxford – New York – Paris – Frankfurt. ISSN 0032–0633.

**Plasma Phys. Controlled Fusion**
Plasma Physics and Controlled Fusion. Pergamon Press, Oxford – New York – Toronto – Paris – Frankfurt – Sydney. ISSN 0032–1028.

**Pokroky Mat., Fyz. Astron.**
Pokroky matematiky, fyziky a astronomie. Academia, Praha, Czechoslovakia. ISSN 0032–2423.

**Port. Phys.**
Portugaliae Physica. Sociedade Portuguesa de Fisica, Avenida de Republica, 37–40, P–1000 Lisboa, Portugal. ISSN 0048–4903.

**Postepy Astron.**
Postepy Astronomii. Czasopismo Poświecone Upowszechnianiu Wiedzy Astronomicznej. Polskie Towarzystwo Astronomiczne, Warszawa. Printed in Poland by Pánstwowe Wydawnictwo Naukowe, Warszawa–Lódź, Poland. ISSN 0032–5414.

**Postepy Fiz.**
Postepy Fiziyki. Polskie Towarzystwo Fizykczne, 00–681 Warszawa, ul. Hoza 69, Poland. ISSN 0032–5430.

**Pramāna**
Pramāna. Indian Academy of Sciences, Bangalore 560 006, India. ISSN 0304–4289.

**Prepr. Astron. Inst. Univ. Basel**
Preprint of the Astronomical Institute of the University of Basel. Astronomisches Institut der Universität Basel, Venusstraße 7, CH–4102 Binningen, Switzerland.

**Prepr. Steward Obs.**
Preprints of the Steward Observatory. Steward Observatory, The University of Arizona, Tucson, Ariz. 85721, USA.

**Priroda**
Priroda. Izdatel'stvo 'Nauka', Moskva. Editorial address: 117049, Moskva, GSP–1, Maronovskij per., 26, USSR. ISSN 0032–874X.

**Probl. Kosm. Fiz.**
Problemy Kosmicheskoj Fiziki. Respublikanskij Mezhvedomstvennyj Nauchnyj Sbornik. Izdatel'stvo pri Kievskom Gosudarstvennom Universitete Izdatel'skogo Obedineniya 'Vishcha Shkola'. Kiev, USSR. ISSN 0555–2796.

**Proc. Astron. Soc. Aust.**
Proceedings of the Astronomical Society of Australia. Astronomical Society of Australia, Sydney. Subscription address: Division of Radiophysics, PO Box 76, Epping, N.S.W. 2121, Australia. ISSN 0066–9997.

**Proc. Cosmic–Ray Res. Lab. Nagoya Univ.**
Proceedings of the Cosmic–Ray Research Laboratory of Nagoya University. Cosmic Ray Research Laboratory and Department of Physics, Nagoya University, Chikusa–ku, Nagoya, Japan. ISSN 0250–5444.

**Proc. IEEE**
Proceedings of the IEEE. The Institute of Electrical and Electronics Engineers, 345 East 47th Street, New York, N.Y. 10017, USA. ISSN 0018–9219.

**Proc. Indian Acad. Sci., Earth Planet. Sci.**
Proceedings of the Indian Academy of Sciences, Earth and Planetary Sciences. Indian Academy of Sciences, Bangalore 560 080, India. ISSN 0370–0089.

**Proc. Indian Acad. Sci., Eng. Sci.**
Proceedings of the Indian Academy of Sciences, Engineering Sciences. Indian Academy of Sciences, Bangalore 560 080, India. ISSN 0253–4096.

**Proc. Indian Natl. Sci. Acad., Part A**
Proceedings of the Indian National Science Academy, Part A. (Physical Sciences). The Indian National Science Academy, Bahadur Shah Zafar Marg, New Delhi 110 002, India. ISSN 0370–0046.

**Proc. Int. Latitude Obs. Mizusawa**
Proceedings of the International Latitude Observatory of Mizusawa. The International Latitude Observatory of Mizusawa, Mizusawa-Shi, Iwate-Ken, Japan. ISSN 0536-3403.

**Proc. Jpn. Acad., Ser. A**
Proceedings of the Japan Academy, Series A. (Mathematical Sciences). Japan Academy, Ueno Park, Tokyo 110, Japan. ISSN 0386-2194.

**Proc. Jpn. Acad., Ser. B**
Proceedings of the Japan Academy, Series B. (Physical and Biological Sciences). Japan Academy, Ueno Park, Tokyo 110, Japan. ISSN 0386-2208.

**Proc. K. Ned. Akad. Wet., Ser. B**
Proceedings of the Koninklijke Nederlandse Akademie van Wetenschappen, Series B. (Physical Sciences). North-Holland Publishing Company, PO Box 211, 1000 AE Amsterdam, The Netherlands. ISSN 0023-3366.

**Proc. Natl. Acad. Sci. India, Sect. A**
Proceedings of the National Academy of Sciences of India, Section A (Physical Sciences). National Academy of Sciences of India, 5 Lajpatrai Road, Allahabad-2, India. ISSN 0369-8203.

**Proc. Natl. Acad. Sci. U.S.A.**
Proceedings of the National Academy of Sciences of the United States of America. National Academy of Sciences, 2101 Constitution Avenue, Washington, D.C. 20418, USA. ISSN 0027-8424.

**Proc. R. Soc. Edinb., Sect. A**
Proceedings of the Royal Society of Edinburgh, Section A (Mathematical and Physical Sciences). Royal Society of Edinburgh, 22 George Street, Edinburgh EH2 2PQ, Scotland. ISSN 0308-2105.

**Proc. R. Soc. London, Ser. A**
Proceedings of the Royal Society of London, Series A. (Mathematical and Physical Sciences). The Royal Society, 6 Carlton House Terrace, London, SW1Y 5AG, England. ISSN 0080-4630.

**Proc. SPIE Int. Soc. Opt. Eng.**
Proceedings of the SPIE - The International Society for Optical Engineering. The International Society for Optical Engineering, PO Box 10, Bellingham, Wash. 98227, USA. ISSN 0277-786X.

**Prog. Part. Nucl. Phys.**
Progress in Particle and Nuclear Physics. Pergamon Press, New York - Toronto - Paris - Frankfurt - Sydney. ISSN 0146-6410.

**Prog. Theor. Phys.**
Progress of Theoretical Physics. Research Institute for Fundamental Physics and the Physical Society of Japan. Subscription address: Yukawa Hall, Kyoto University, 606 Kyoto, Japan. ISSN 0033-068X.

**Prog. Theor. Phys., Suppl.**
Progress of Theoretical Physics, Supplement. Research Institute for Fundamental Physics and the Physical Society of Japan. Subscription address: Yukawa Hall, Kyoto University, 606 Kyoto, Japan. ISSN 0375-9687.

**PTB Mitt.**
PTB Mitteilungen. Forschen + Prüfen. Amts- und Mitteilungsblatt der Physikalisch-Technischen Bundesanstalt Braunschweig - Berlin. Friedr. Vieweg & Sohn Verlagsgesellschaft mbH, Faulbrunnenstraße 13, D-6200 Wiesbaden, F.R. Germany. ISSN 0030-834X.

**Pubbl. Stn. Astron. Int. Latitudine, Carloforte-Cagliari, Nuov. Ser.**
Pubblicazioni della Stazione Astronomica Internazionale di Latitudine, Carloforte-Cagliari, Nuova Serie. Stazione Astronomica Internazionale di Latitudine, Carloforte-Cagliari, Italy.

**Pubbl. Varie Fuori Ser. Oss. Astron. Torino (Pino Torinese)**
Pubblicazioni Varie Fuori Serie dell'Osservatorio Astronomico di Torino (Pino Torinese). Osservatorio Astronomico di Torino, I-10025 Pino Torinese (Torino), Italy.

**Publ. Astron. Dep. Eötvös Univ.**
Publications of the Astronomy Department of the Eötvös University. Department of Astronomy, Eötvös Lorand University, Budapest, Hungary.

**Publ. Astron. Inst. Czech. Acad. Sci.**
Publications of the Astronomical Institute of the Czechoslovak Academy of Sciences. Astronomical Institute, Czechoslovak Academy of Sciences, Budečská 6, 120 23 Praha 2, Czechoslovakia.

**Publ. Astron. Obs. Sarajevo**
Publications of the Astronomical Observatory of Sarajevo. Astronomical Observatory of Sarajevo, Sarajevo, M. Tita 44, Yugoslavia. ISSN 0351-4587.

**Publ. Astron. Opservatorije Beogr.**
Publikacija Astronomske Opservatorije u Beogradu (Publications de l'Observatoire Astronomique de Beograd). Astronomska Opservatorija u Beogradu, YU-11050 Beograd, Volgina 7, Yugoslavia. ISSN 0373-3742.

**Publ. Astron. Soc. Jpn.**
Publications of the Astronomical Society of Japan. Astronomical Society of Japan, Tokyo Astronomical Observatory. Mitaka-shi, Tokyo 181, Japan. ISSN 0004-6264.

**Publ. Astron. Soc. Pac.**
Publications of the Astronomical Society of the Pacific. The Astronomical Society of the Pacific, 1290 24th Avenue, San Francisco, Calif. 94122, USA. ISSN 0004-6280.

**Publ. Astrophys. Obs. Potsdam**
Publikationen des Astrophysikalischen Observatoriums zu Potsdam. Akademie der Wissenschaften der DDR, Zentralinstitut für Astrophysik. Astrophysikalisches Observatorium Potsdam, DDR-1500 Potsdam, Telegrafenberg, German Democratic Republic.

**Publ. Beijing Astron. Obs.**
Publications of the Beijing Astronomical Observatory. Beijing Astronomical Observatory, Academia Sinica, Beijing, People's Republic of China.

**Publ. Bosscha Obs.**
Publications of the Bosscha Observatory. Bandung Institute of Technology, Department of Science, Lembang, Indonesia.

**Publ. Debrecen Heliophys. Obs.**
Publications of Debrecen Heliophysical Observatory. Hungarian Academy of Sciences, H-4010 Debrecen, Hungary. ISSN 0209-7567.

**Publ. Dep. Astron., Univ. Beogr.**
Publications of the Department of Astronomy, University of Beograd. Department of Astronomy, University of Beograd, Beograd, Yugoslavia. ISSN 0350-3283.

**Publ. Dep. Astron., Univ. Cape Town**
Publications of the Department of Astronomy of the University of Cape Town. Astronomy Department, University of Cape Town, Rondebosch, 7700 Cape, South Africa.

**Publ. Dep. Astron., Univ. Chile**
Publicaciones Departamento de Astronomía, Universidad de Chile. Observatorio Astronomico Nacional, Cerro Calan, Santiago de Chile, Chile.

**Publ. Dep. Geod. Astron., Univ. Thessaloniki**
Publications of the Department of Geodetic Astronomy, University of Thessaloniki. Department of Geodetic Astronomy, University of Thessaloniki, Thessaloniki, Greece.

**Publ. Dom. Astrophys. Obs.**
Publications of the Dominion Astrophysical Observatory. Dominion Astrophysical Observatory, National Research Center of Canada, 5071 West Saanich Road, Victoria, B.C. V8X 4M6, Canada. ISSN 0078-6950.

**Publ. Ege Univ. Obs.**
Publications of Ege University Observatory. Ege University Observatory, P.K. 21, Bornova-Izmir, Turkey.

**Publ. Espec. Obs. Astron. Quito**
Publicacion Especial del Observatorio Astronomico de Quito. Observatorio Astronomico de Quito, Escuela Politécnica Nacional, Quito, Ecuador.

**Publ. Finn. Geod. Inst.**
Publications of the Finnish Geodetic Institute. Finnish Geodetic Institute, Helsinki, Finland. ISSN 0085-6932.

**Publ. Inst. Geophys., Pol. Acad. Sci., Ser. F (Planet Geod.)**
Publications of the Institute of Geophysics, Polish Academy of Sciences, Series F (Planetary Geodesy). Państwowe Wydawnictwo Naukowe, Warszawa-Łódź, Poland. ISSN 0318-0222.

**Publ. Int. Latitude Obs. Mizusawa**
Publications of the International Latitude Observatory of Mizusawa. The International Latitude Observatory of Mizusawa, Mizusawa-shi, Iwate-ken, Japan. ISSN 0386-0779.

**Publ. Istanbul Univ. Obs.**
Publications of the Istanbul University Observatory. Istanbul University Observatory, Istanbul, Turkey.

**Publ. Lick Obs.**
Publications of the Lick Observatory. Lick Observatory, Board of Studies in Astronomy and Astrophysics, University of California at Santa Cruz, Santa Cruz, Calif. 95064, USA.

**Publ. Obs. Astron. Nac., Univ. Colombia**
Publicaciones del Observatorio Astronomico Nacional. Observatorio Astronomico Nacional, Universidad Nacional de Colombia, Apartado Aéreo 2584, Bogotá, Colombia.

**Publ. Obs. Astron. "Prof. Manuel de Barros", Fac. Cienc. Porto**
Publicações do Observatório Astronómico "Prof. Manuel de Barros", da Faculdade de Ciências do Porto. Observatório Astronómico "Prof. Manuel de Barros", Universidade do Porto, Monte da Virgem - Vila Nova de Gaia, Portugal.

**Publ. Obs. Astron. Univ. Cluj**
Publicatiile Observatorului Astronomic al Universitatii din Cluj. Astronomical Observatory, University Babes-Bolyai, Cluj-Napoca, Romania.

**Publ. Obs. Bordeaux, Nouv. Sér.**
Publications de l'Observatoire de Bordeaux, Nouvelle Série. Observatoire de l'Université de Bordeaux I, F-33270 Floirac, France.

**Publ. Obs. Ebro**
Publicaciones del Observatorio del Ebro (Miscelanea). Consejo Superior de Investigaciones Cientificas, Observatorio del Ebro, Roquetas (Tarragona), Spain.

**Publ. Obs. Genève, Sér. A**
Publications de l'Observatoire de Genève, Série A. Observatoire de Genève, CH-1290 Sauverny, Switzerland.

**Publ. Obs. Genève, Sér. B**
Publications de l'Observatoire de Genève, Série B. Observatoire de Genève, CH-1290 Sauverny, Switzerland.

**Publ. Purple Mt. Obs.**
Publications of the Purple Mountain Observatory. Purple Mountain Observatory, Academia Sinica, Nanking, People's Republic of China.

**Publ. R. Obs., Edinb.**
Publications of the Royal Observatory, Edinburgh. Science Research Council. Royal Observatory, Blackford Hill, Edinburgh EH9 3HJ, Scotland. ISSN 0305-2001.

**Publ. Semin. Astron. Geod. Univ. Complutense Madr.**
Publicaciones del Seminario de Astronomia y Geodesia de la Universidad Complutense de Madrid. Seminario de Astronomia y Geodesia, Facultad de Ciencias Matemáticas, Universidad Complutense, Madrid, Spain.

**Publ. Shaanxi Astron. Obs.**
Publications of the Shaanxi Astronomical Observatory. Shaanxi Astronomical Observatory, Academia Sinica, Lintong, Xian, People's Republic of China.

**Publ. Spéc. Cent. Données Stellaires**
Publication Spéciale du Centre de Données Stellaires. Observatoire de Strasbourg, 11, rue de l'Université, F-67000 Strasbourg, France.

**Publ. U.S. Nav. Obs., Second Ser.**
Publications of the United States Naval Observatory, Second Series. U.S. Government Printing Office, Washington, D.C. 20402, USA.

**Publ. Variable Star Sect., R. Astron. Soc. N.Z.**
Publications of Variable Star Section, Royal Astronomical Society of New Zealand. Astronomical Research Ltd., PO Box 3093, Greerton, Tauranaga, New Zealand.

**Publ. Warner Swasey Obs.**
Publications of the Warner and Swasey Observatory. Case Western Reserve University, 1975 Taylor Road, East Cleveland, Ohio 44112, USA. ISSN 0160-2500.

**Q. Appl. Math.**
Quarterly of Applied Mathematics. American Mathematical Society, PO Box 1571, Providence, R.I. 02901, USA. ISSN 0033-569X.

**Q. Bull. Sol. Act.**
Quarterly Bulletin on Solar Activity. International Astronomical Union. Published by the Tokyo Astronomical Observatory, University of Tokyo, Mitaka, Tokyo 181, Japan.

**Q. J. R. Astron. Soc.**
The Quarterly Journal of the Royal Astronomical Society. Published for the Royal Astronomical Society by Blackwell Scientific Publications, Oxford - London - Edinburgh - Boston - Melbourne. ISSN 0035-8738.

**Q. J. R. Meteorol. Soc.**
Quarterly Journal of the Royal Meteorological Society. Cromwell House, High Street, Bracknell, Berks., England. ISSN 0035-9009.

**R. Greenwich Obs., Annu. Rep.**
Royal Greenwich Observatory, Annual Report. Royal Greenwich Observatory, Herstmonceux Castle, Hailsham, East Sussex BN27 1RP, England. ISSN 0308-3322.

**R. Greenwich Obs., Bull.**
Royal Greenwich Observatory, Bulletins. Royal Greenwich Observatory, Herstmonceux Castle, Hailsham, East Sussex BN27 1RP, England. ISSN 0308–5074.

**R. Obs. Ann.**
Royal Observatory Annals. Royal Greenwich Observatory, Herstmonceux Castle, Hailsham, East Sussex BN27 1RP, England. ISSN 0080–4371.

**R. Obs. Edinb., Res. Facilities**
Royal Observatory Edinburgh, Research and Facilities. Royal Observatory, Blackford Hill, Edinburgh EH9 3HJ, Scotland.

**Radiat. Eff.**
Radiation Effects. Gordon and Breach Science Publishers Inc., New York – London – Paris. ISSN 0033–7579.

**Radio Sci.**
Radio Science. American Geophysical Union, 2000 Florida Avenue, N.W., Washington, D.C. 20009, USA. ISSN 0048–6604.

**Radiochem. Radioanal. Lett.**
Radiochemical and Radioanalytical Letters. Elsevier Sequoia S.A., PO Box 851, CH–1001 Lausanne, Switzerland. ISSN 0079–9483.

**Rapp. Act. Obs. Paris**
Rapport d'Activité de l'Observatoire de Paris. Observatoire de Paris, F–92190 Meudon, France.

**Rech. Aerosp.**
Recherche Aerospatiale. Office National d'Etudes et de Recherches Aerospatiales, 29 Avenue de la Division Leclerc, F–92320 Chatillon, France. ISSN 0034–1223.

**Recherche**
Recherche. Société d'Editions Scientifiques, 57, rue de Seine, F–75006 Paris, France. ISSN 0029–5671.

**Ref. Zh., 51. Astron.**
Referativnyj Zhurnal, 51. Astronomiya. Gosudarstvennyj Komitet SSSR po Nauke i Tekhnike. Vsesoyuznyj Institut Nauchnoj i Tekhnicheskoj Informatsii. Akademiya Nauk SSSR. Moskva. 125219, Moskva, A–219, Baltijskaya ul., 14, USSR. ISSN 0486–2236.

**Ref. Zh., 52. Geod. Aehrosemka**
Referativnyj Zhurnal, 52. Geodeziya i Aehrosemka. Gosudarstvennyj Komitet SSSR po Nauke i Tekhnike. Vsesoyuznyj Institut Nauchnoj i Tekhnicheskoj Informatsii. Akademiya Nauk SSSR. Moskva. 125219, Moskva, A–219, Baltijskaya ul., 14, USSR. ISSN 0375–9717.

**Ref. Zh., 62. Issled. Kosm. Prostranstva**
Referativnyj Zhurnal, 62. Issledovanie Kosmicheskogo Prostranstva. Gosudarstvennyj Komitet SSSR po Nauke i Tekhnike. Vsesoyuznyj Institut Nauchnoj i Tekhnicheskoj Informatsii. Akademiya Nauk SSSR. Moskva. 125219, Moskva, A–219, Baltijskaya ul., 14, USSR. ISSN 0034–2408.

**Rep. Finn. Geod. Inst.**
Reports of the Finnish Geodetic Institute. Finnish Geodetic Institute, Helsinki, Finland. ISSN 0355–1962.

**Rep. Inst. Phys. Chem. Res.**
Reports of the Institute of Physical and Chemical Research. Institute of Physical and Chemical Research, Rikagaku Kenkyushu, Wako–shi, Saitama 351, Japan. ISSN 0020–3084.

**Rep. Obs. Lund**
Reports from the Observatory of Lund. Institutionen för Astronomi, Lunds Universitet, Box 1107, 22104 Lund, Sweden. ISSN 0349–4217.

**Rep. Prog. Phys.**
Reports on Progress in Physics. Published by the Institute of Physics, 47 Belgrave Square, London SW1X 8QX, England. ISSN 0034–4885.

**Rep. Ser., Dep. Phys. Sci., Univ. Turku**
Report Series, Department of Physical Sciences, University of Turku. Department of Physical Sciences, University of Turku, SF–20500 Turku 50, Finland. ISSN 0356–9896.

**Res. Bull. Meisei Univ., Phys. Sci. Eng.**
Research Bulletin of Meisei University, Physical Sciences and Engineering. Meisei University, Hino–Shi, Tokyo, Japan. ISSN 0388–130X.

**Res. Cent. Astron. Appl. Math., Acad. Athens, Contrib. Ser. I**
Research Center for Astronomy and Applied Mathematics, Academy of Athens, Contributions Series I (Astronomy). Research Center for Astronomy and Applied Mathematics, Academy of Athens, Athens, Greece.

**Res. Lab. Electron. Onsala Space Obs., Res. Rep.**
Research Laboratory of Electronics and Onsala Space Observatory, Research Report. Chalmers University of Technology, Research Laboratory of Electronics, Fack, S–40220 Göteborg 5, Sweden.

**Rev. Astron.**
Revista Astronomica. Organo de la Asociación Argentina Amigos de la Astronomia. Asociación Argentina Amigos de la Astronomia, C. C. 369–C. Central, 1000–Buenos Aires, Argentina. ISSN 0044–9253.

**Rev. Bras. Fis.**
Revista Brasileira de Fisica. Sociedade Brasileira de Fisica, Caixa Postal 20553, Sao Paulo SP, Brazil. ISSN 0374–4922.

**Rev. Fac. Sci. Univ. Istanbul, Ser. C**
Review of Faculty of Science University of Instanbul, Serie C. (Üniversitesi fen Fakültesi Mecmuasi, Seri C). Fen Fakültesi Basimevi, Istanbul, Turkey. ISSN 0444–7298.

**Rev. Geophys. Space Phys.**
Reviews of Geophysics and Space Physics. American Geophysical Union, 2000 Florida Avenue, N.W., Washington, D.C. 20009, USA. ISSN 0034–6853.

**Rev. Hist. Sci.**
Revue d'Histoire des Sciences. Presses Universitaires de France, Paris, France. ISSN 0151–4105.

**Rev. Mex. Astron. Astrofis.**
Revista Mexicana de Astronomia y Astrofisica. Instituto de Astronomia, Universidad Nacional Autónoma de México, Apartado Postal 70–264, Mexico 20, D.F., Mexico. ISSN 0185–1101.

**Rev. Mex. Fis.**
Revista Mexicana de Fisica. Sociedad Mexicana de Fisica, Apartado Postal No. 20–364, Mexico 20, D.F., Mexico. ISSN 0035–001X.

**Rev. Mod. Phys.**
Reviews of Modern Physics. Published for the American Physical Society by the American Institute of Physics, 335 East 45th Street, New York, N.Y. 10017, USA. ISSN 0034–6861.

**Rev. Phys. Appl.**
Revue de Physique Appliquée. Editions de Physique, Z.I. de Courtabœuf, B.P. 112, F–91944 Les Ulis Cedex, France. ISSN 0035–1687.

**Rev. Quest. Sci.**
Revue des Questions Scientifiques. Société Scientifique de Bruxelles, 61, rue de Bruxelles, B-5000 Namur, Belgium. ISSN 0035-2160.

**Rev. Radio Res. Lab.**
Review of the Radio Research Laboratories. Ministry of Posts & Telecommunications, Nukui-Kitamachi, Konganei-shi, Tokyo 184, Japan. ISSN 0033-801X.

**Rev. Roum. Phys.**
Revue Roumaine de Physique. Academie Republicii Populare Romine, B.P. 134-135, Bucuresti, Rumania. ISSN 0035-4090.

**Rev. Roum. Sci. Tech., Sér. Mec. Appl.**
Revue Roumaine des Sciences Techniques, Série de Mécanique Appliquée. Academia R.S.R., Str. Constantin Mille 15, Bucuresti, Rumania. ISSN 0035-4074.

**Rev. Sci. Instrum.**
Review of Scientific Instruments. American Institute of Physics, 335 East 45th Street, New York, N.Y. 10017, USA. ISSN 0034-6748.

**Rezul't. Nablyud. Iskusstv. Sputnikov Zemli**
Rezul'taty Nablyudenij Iskusstvennykh Sputnikov Zemli. Published by Astronomicheskij Sovet Akademii Nauk SSSR, Ryazanskij Gosudarstvennyj Pedagogicheskij Institut, Ryazan'. 390000 GSP, Ryazan, ul. Svobody, 46, USSR. ISSN 0131-8586.

**Ric. Astron.**
Ricerche Astronomiche. Specola Vaticana, I-00120 Città del Vaticano.

**Rijksuniv. Gent, Sterrenkundig Obs., Meded.**
Rijksuniversiteit Gent, Sterrenkundig Observatorium, Mededeling. Astronomical Observatory, State University of Ghent, Krijgslaan 271, B-9000 Ghent, Belgium.

**Říše hvězd**
Říše hvězd. Czech popular astronomical journal. Panorama, Praha, Czechoslovakia.

**Riv. Nuovo Cimento**
La Rivista del Nuovo Cimento della Società Italiana di Fisica. Editrice Compositori, viale XII Giugno 1, I-40124 Bologna, Italy. ISSN 0035-5917.

**SAAO Newsl.**
SAAO Newsletter. South African Astronomical Observatory, PO Box 9, Observatory 7935, Cape, South Africa.

**Sac Peak Update**
Sac Peak Update. Sacramento Peak Observatory, Sunspot, N.M. 88349, USA.

**S.Afr. Astron. Obs., Annu. Rep.**
South African Astronomical Observatory, Annual Report. South African Astronomical Observatory, PO Box 9, Observatory, 7935 Cape, South Africa. ISSN 0250-0671.

**S.Afr. Astron. Obs., Circ.**
South African Astronomical Observatory, Circulars. South African Astronomical Observatory, PO Box 9, Observatory, 7935 Cape, South Africa.

**S.Afr. J. Phys.**
South African Journal of Physics. Bureau for Scientific Publications, PO Box 1758, Pretoria 0001, South Africa. ISSN 0379-4377.

**Schweiz. Tech. Z.**
Schweizerische Technische Zeitschrift. Schweizerischer Technischer Verband, Weinbergstraße 41, CH-8023 Zürich, Switzerland. ISSN 0040-151X.

**Sci. Am.**
Scientific American. Scientific American, Inc., 415 Madison Avenue, New York, N.Y. 10017, USA. ISSN 0036-8733.

**Sci. Atmos. Sin.**
Scientia Atmospherica Sinica. Science Press, Peking. Subscription address: Guozi Shudian, PO Box 399, Beijing, People's Republic of China.

**Sci. Dimension**
Science Dimension. National Research Council of Canada, Ottawa K1A OR6, Canada. ISSN 0036-830X.

**Sci. Pap. Inst. Phys. Chem. Res.**
Scientific Papers of the Institute of Physical and Chemical Research. Rikagaku Kenkyusho, Wako-shi, Saitama 351, Japan. ISSN 0020-3092.

**Sci. Prog.**
Science Progress. Blackwell Scientific Publications, Oxford - London - Edinburgh - Boston - Melbourne. ISSN 0036-8504.

**Sci. Rep. Tôhoku Univ., Eighth Ser.**
The Science Reports of the Tôhoku University, Eighth Series (Physics and Astronomy). Faculty of Science, Tôhoku University, Sendai 980, Japan. ISSN 0388-5607.

**Sci. Sin., Ser. A**
Scientia Sinica, Series A (Mathematical, Physical, Astronomical and Technical Sciences). Academia Sinica, Beijing. Science Press, No. 137, Chaoyangmennei Street, Beijing, People's Republic of China. Subscription address: Scientific and Technical Books Service Ltd., PO Box 197, London WC2N 4DE, England. ISSN 0253-5831.

**Science**
Science. American Association for the Advancement of Science, 1515 Massachusetts Avenue, N.W., Washington, D.C. 20005, USA. ISSN 0036-8075.

**Scr. Fac. Sci. Nat. Univ. Purkynianae Brun., Phys.**
Scripta Facultatis Scientiarum Naturalium Universitatis Purkynianae Brunensis, Physica. University J.E. Purkyne, 61137 Brno-Kotlarská 2, Czechoslovakia. ISSN 0231-6129.

**Semin. Astron. Geod., Univ. Madr., Publ.**
Seminario de Astronomia y Geodesia, Universidad Complutense, Facultad de Ciencias Matemáticas, Madrid, Publicación. Universidad Complutense, Madrid, Spain.

**Sendai Astron. Rap.**
Sendai Astronomiaj Raportoj. Astronomical Institute, Tôhoku University, Sendai 980, Japan.

**Shaanxi Astron. Obs., Repr.**
Shaanxi Astronomical Observatory, Reprints. Shaanxi Astronomical Observatory, Academia Sinica, Lintong, Xian, People's Republic of China.

**Shanghai Obs., Acad. Sin., Prepr.**
Shanghai Observatory, Academia Sinica, Preprint. Shanghai Observatory, Academia Sinica, 80 Nandan Road, Shanghai, People's Republic of China.

**Sitzungsber. Akad. Wiss. DDR, Math. Naturwiss. Tech.**
Sitzungsberichte der Akademie der Wissenschaften der DDR, Mathematik - Naturwissenschaften - Technik. Akademie-Verlag, DDR-1086 Berlin, Leipziger Straße 3-4, German Democratic Republic. ISSN 0138-3956.

**Sitzungsber. Heidelb. Akad. Wiss.**
Sitzungsberichte der Heidelberger Akademie der Wissenschaften. Mathematisch–Naturwissenschaftliche Klasse. Springer-Verlag, Berlin – Heidelberg – New York – Tokyo. ISSN 0371–0165.

**Sitzungsber., Österr. Akad. Wiss., Math.–Naturwiss. Kl. Abt. II**
Sitzungsberichte, Österreichische Akademie der Wissenschaften, Mathematisch–Naturwissenschaftliche Klasse, Abteilung II (Mathematik, Astronomie, Physik, Meteorologie und Technik). Springer–Verlag, Wien, Austria. ISSN 0029–8816.

**Sky Telesc.**
Sky and Telescope. Sky Publishing Corporation, 49 Bay State Rd., Cambridge, Mass. 02238–1290, USA. ISSN 0037–6604.

**Smithson. Astrophys. Obs., Spec. Rep.**
Smithsonian Astrophysical Observatory, Special Report. Available from the Publications Division, Distribution Section, Smithsonian Astrophysical Observatory, Cambridge, Mass. 02138, USA.

**Sol. Energy**
Solar Energy. Pergamon Press, Oxford – New York – Toronto – Paris – Frankfurt – Sydney. ISSN 0038–092X.

**Sol. Maps Act.**
Solar Maps and Activity. Manila Observatory, Solar Division, Manila, Philippines.

**Sol. Phys.**
Solar Physics. A Journal for Solar Research and the Study of Solar Terrestrial Physics. D. Reidel Publishing Company, Dordrecht – Boston. ISSN 0038–0938.

**Sol. Syst. Res.**
Solar System Research. Translation of Astron. Vestn. Consultants Bureau, 227 West 17th Street, New York, N.Y. 10011, USA. ISSN 0038–0946.

**Sol. Terr. Environ. Res. Jpn.**
Solar Terrestrial Environmental Research in Japan. Institute of Space and Aeronautical Science, University of Tokyo, 4–6–1, Komaba, Meguro–ku, Tokyo 153, Japan. ISSN 0386–5444.

**Solid State Phys.**
Solid State Physics. Agne Gijutsu Center, Kitamura Building, 5–1–25, Minamiayama, Minatoku, Tokyo, Japan. ISSN 0454–4544.

**Soln. Dannye, Byull.**
Solnechnye Dannye, Byulleten' (Solar Data). Akademiya Nauk SSSR. Astronomicheskij Sovet i Glavnaya Astronomicheskaya Observatoriya. Izdatel'stvo 'Nauka', Leningradskoe Otdelenie, Leningrad. 199164 Leningrad, V–164, Mendeleevskaya l. 1, USSR. ISSN 0552–5829.

**Sonne**
Sonne. Mitteilungsblatt der Amateursonnenbeobachter. Wilhelm–Foerster–Sternwarte, Munsterdamm 90, D–1000 Berlin 41, F.R. Germany. ISSN 0721–0094.

**Soobshch. Akad. Nauk Gruz. SSR**
Soobshcheniya Akademii Nauk Gruzinskoj SSR. Akademiya Nauk Gruzinskoj SSR. ISSN 0002–3167.

**Soobshch. Byurakan. Obs.**
Soobshcheniya Byurakanskoj Observatorii. Akademiya Nauk Armyanskoj SSR, Erevan. 375019 Erevan, Barekamutyan, 24, USSR. ISSN 0370–8691.

**Soobshch. Gos. Astron. Inst. Shternberg**
Soobshcheniya Gosudarstvennogo Astronomicheskogo Instituta im. P.K. Shternberga. Moskovskij Gosudarstvennyj Universitet im. M.V. Lomonosova. Izdatel'stvo Moskovskogo Universiteta, Moskva. Moskva, K–9, ul. Gertsena, 5/7, USSR. ISSN 0038–1489.

**Soobshch. Spets. Astrofiz. Obs.**
Soobshcheniya Spetsial'noj Astrofizicheskoj Observatorii. Akademiya Nauk SSSR. Izdanie Spetsial'noj Astrofizicheskoj Observatorii, USSR.

**South. Stars**
Southern Stars. Journal of the Royal Astronomical Society of New Zealand (Inc.). Royal Astronomical Society of New Zealand, PO Box 3181, Wellington, New Zealand. ISSN 0049–1640.

**Sov. Astron.**
Soviet Astronomy. A translation of Astron. Zh. American Institute of Physics, 335 East 45th Street, New York, N.Y. 10017, USA. ISSN 0038–5301.

**Sov. Astron. Lett.**
Soviet Astronomy Letters. A translation of Pis'ma Astron. Zh. American Institute of Physics, 335 East 45th Street, New York, N.Y. 10017, USA. ISSN 0360–0327.

**Space Educ.**
Space Education. A Publication of The British Interplanetary Society. The British Interplanetary Society, 27/29 South Lambeth Road, London SW8 1SZ, England. ISSN 0621–1813.

**Space Sci. Rev.**
Space Science Reviews. D. Reidel Publishing Company, Dordrecht – Boston. ISSN 0038–6308.

**Space Telesc. Sci. Inst., Newsl.**
Space Telescope Science Institute, Newsletter. Space Telescope Science Institute, 3700 San Martin Drive, Homewood Campus, Baltimore, Md. 21218, USA.

**Space Telesc. Sci. Inst., Prepr. Ser.**
Space Telescope Science Institute, Preprint Series. Space Telescope Science Institute, 3700 San Martin Drive, Homewood Campus, Baltimore, Md. 21218, USA.

**Spaceflight**
Spaceflight. A Publication of The British Interplanetary Society. The British Interplanetary Society, 27/29 South Lambeth Road, London SW8 1SZ, England. ISSN 0038–6340.

**Spectrosc. Lett.**
Spectroscopy Letters. Marcel Dekker Inc., 270 Madison Avenue, New York, N.Y. 10016, USA. ISSN 0038–7010.

**Speculations Sci. Technol.**
Speculations in Science and Technology. Elsevier Sequoia, S.A., PO Box 851, CH–1001 Lausanne 1, Switzerland. ISSN 0155–7785.

**Sterne**
Die Sterne. Zeitschrift für alle Gebiete der Himmelskunde. Johann Ambrosius Barth, DDR–7010 Leipzig, Salomonstraße 18b, German Democratic Republic. ISSN 0039–1255.

**Sterne Weltraum**
Sterne und Weltraum. Astronomische Monatsschrift. Verlag Sterne und Weltraum Dr. Vehrenberg GmbH, Portiastraße 10, D–8000 München 90, F.R. Germany, ISSN 0039–1263.

**Sternenbote**
Der Sternenbote. Österreichische Astronomische Monatsschrift. Astronomisches Büro, Hasenwartgasse 32, A–1238 Wien, Austria. ISSN 0039–1271.

**Stockholms Obs., Rep.**
Stockholms Observatorium, Report. Stockholms Observatorium, S-13300 Saltsjöbaden, Sweden.

**Strolling Astron.**
The Strolling Astronomer. The Journal of The Association of Lunar and Planetary Observers. The Strolling Astronomer, Box 3 AZ, University Park, N.M. 88003, USA. ISSN 0039-2502.

**Stud. Cercet. Fiz.**
Studii si Cercetari de Fizica. Academia Republicii Populare Romine, PO Box 134-5, Calca Victoriei 126, Bucuresti, Rumania, ISSN 0039-3940.

**Stud. Galileiani**
Studi Galileiani. Vatican Observatory Publications, Special Series. Specola Vaticana, I-00120 Città del Vaticano.

**Stud. Geophys. Geod.**
Studia Geophysica et Geodaetica. Geophysical Institute of the Czechoslovak Academy of Sciences. Published by Academia, Prague, Czechoslovakia.

**Stud. Hist. Philos. Sci.**
Studies in History and Philosophy of Science. Pergamon Press, Oxford - New York - Toronto - Paris - Frankfurt - Sydney. ISSN 0039-3681.

**Stud. Soc. Sci. Torun., Sect. F**
Studia Societatis Scientiarum Torunensis, Sectio F (Astronomia). Państwowe Wydawnictwo Naukowe, Warszawa - Poznań - Toruń, Poland. ISSN 0082-5573.

**Surv. High Energy Phys.**
Surveys in High Energy Physics. Harwood Academic Publishers GmbH, Poststraße 22, CH-7000 Chur, Switzerland. ISSN 0142-2413.

**Swarthmore Astron. Repr.**
Swarthmore Astronomical Reprints. Sproul Observatory, Swarthmore College, Swarthmore, Pa. 19081, USA.

**Syd. Obs. Pap.**
Sydney Observatory Papers. School of Physics, University of Sydney, Sydney, N.S.W. 2006, Australia.

**Syst. Int.**
Systems International. IPC Electrical Electronic Press Ltd., Quadrant House, The Quadrant, Sutton, Surrey SM2 5AS, England. ISSN 0309-1171.

**Tartu Astrofüüs. Obs. Publ.**
W. Struve nimelise Tartu Astrofüüsika Observatooriumi, Publikatsioonid. Eesti NSV Teaduste Akadeemia, Tartu.

**Tartu Astrofüüs. Obs. Teated**
Tartu Astrofüüsika Observatoorium Teated. Eesti NSV Teaduste Akadeemia W. Struve nim. Tartu Astrofüüsika Observatoorium. Valgus, Tallinn.

**Tech. Mess. - TM**
Technisches Messen - TM. Archiv für Technisches Messen. R. Oldenbourg Verlag GmbH, Rosenheimer Str. 145, D-8000 München 80, F.R. Germany. ISSN 0171-8096.

**Tech. Mitt. Krupp Forschungsber.**
Technische Mitteilungen Krupp Forschungsberichte. Fachbücherei Krupp, Postfach 917, D-4300 Essen 1, F.R. Germany. ISSN 0494-9382.

**Tectonophysics**
Tectonophysics. International Journal of Geotectonics and the Geology and Physics of the Interior of the Earth. Elsevier Publishing Company, PO Box 211, 1000 AE Amsterdam, The Netherlands. ISSN 0040-1951.

**Telecommun. J.**
Telecommunication Journal. (English Edition). International Telecommunications Union. Place des Nations, CH-1221 Genève 20, Switzerland. ISSN 0497-137X.

**Tellus, Ser. B**
Tellus, Series B. (Chemical and Physical Meteorology). Swedish Geophysical Society, Arrhenius Laboratoriet, S-10691 Stockholm, Sweden. ISSN 0280-6509.

**THEOCHEM**
THEOCHEM. Elsevier Scientific Publishing Company, PO Box 330, 1000 AH Amsterdam, The Netherlands. ISSN 0166-1280.

**Theor. Pap.**
Theoretic Papers. The Blindern Theoretic Research Team, P.B. 1029, Blindern, Oslo 3, Norway.

**Time Freq. Serv., Bull.**
Time and Frequency Services, Bulletin. Shaanxi Astronomical Observatory, Academia Sinica, Lintong, Xian, People's Republic of China.

**Time Serv. Annu. Rep.**
Time Service Annual Report. Xu-Jia-Hui Section, Shanghai Observatory, Academia Sinica. Subscription address: Shanghai Scientific and Technical Publishers, Shanghai, Rei Jing Er Street 450, People's Republic of China.

**Time Serv. Bull.**
Time Service Bulletin. Osservatorio Astronomico di Torino, I-10025 Pino Torinese (Torino), Italy.

**Tôhoku Geophys. J.**
Tôhoku Geophysical Journal. Science Reports of the Tôhoku University, Fifth Series. Faculty of Science, Tôhoku University, Sendai 980, Japan. ISSN 0040-8794.

**Tokyo Astron. Bull., Second Ser.**
Tokyo Astronomical Bulletin, Second Series. Tokyo Astronomical Observatory, University of Tokyo, Mitaka, Tokyo 181, Japan. ISSN 0082-4690.

**Tokyo Astron. Obs., Kiso Inf. Bull.**
Tokyo Astronomical Observatory, Kiso Information Bulletin. Tokyo Astronomical Observatory, University of Tokyo, Mitaka, Tokyo 181, Japan. ISSN 0286-1380.

**Tokyo Astron. Obs. Rep.**
Tokyo Astronomical Observatory Report. Tokyo Astronomical Observatory, University of Tokyo, Mitaka, Tokyo 181, Japan. ISSN 0374-4639.

**Tokyo Astron. Obs. Repr.**
Tokyo Astronomical Observatory Reprints. Tokyo Astronomical Observatory, University of Tokyo, Mitaka, Tokyo 181, Japan. ISSN 0082-4712.

**Tokyo Astron. Obs., Time Latitude Bull.**
Tokyo Astronomical Observatory, Time and Latitude Bulletins. Tokyo Astronomical Observatory, University of Tokyo, Mitaka, Tokyo 181, Japan. ISSN 0388-3701.

**Tr. Astrofiz. Inst. Alma-Ata**
Trudy Astrofizicheskogo Instituta. Akademiya Nauk Kazakhskoj SSR. Izdatel'stvo Nauka Kazakhskoj SSR, Alma-Ata. 480021, Alma-Ata, ul. Shevchenko, 28, USSR.

**Tr. Astron. Obs., Leningrad**
Trudy Astronomicheskoj Observatorii. Uchenye Zapiski Gosudarstvennogo Universiteta im. A.A. Zhdanova, Seriya matematicheskikh nauk = Izdatel'stvo Leningradskogo Gosudarstvennogo Universiteta A.A. Zhdanova, Leningrad. 199164, Leningrad V–164, Universitetskij nab., 7/9, USSR. ISSN 0136–8109; ISSN 0136–8141.

**Tr. Glav. Astron. Obs. Pulkovo, Ser. 2**
Trudy Glavnoj Astronomicheskoj Observatorii v Pulkove, Seriya 2. Akademiya Nauk SSSR. Leningradskoe Otdelenie, Nauka, Leningrad. 199164, Leningrad V–164, Mendeleevskaya l., 1, USSR.

**Tr. Gos. Astron. Inst. Shternberg**
Trudy Gosudarstvennogo Astronomicheskogo Instituta im. P.K. Shternberga. Izdatel'stvo Moskovskogo Universiteta, Moskva, USSR. ISSN 0371–6791.

**Tr. Inst. Teor. Astron., Leningrad**
Trudy Instituta Teoreticheskoj Astronomii, Akademiya Nauk SSSR. Izdatel'stvo Nauka, Leningrad. 199164, Leningrad V–164, Mendeleevskaya l., 1, USSR. ISSN 0568–6016.

**Tr. Kazan. Gorod. Astron. Obs.**
Trudy Kazanskoj Gorodskoj Astronomicheskoj Observatorii. Izdatel'stvo Kazanskogo Universiteta, Kazan. Kazan, ul. Lenina, 2, USSR. ISSN 0371–8247.

**Tr. Tashkent. Astron. Obs.**
Trudy Ordena Trudovogo Krasnogo Znameni Astronomicheskogo Instituta Akademii Nauk Uzbekskoj SSR. Izdatel'stvo FAN Uzbekskoj SSR, Tashkent. Tashkent, ul. Gololya, 70, USSR.

**Trans. Am. Nucl. Soc.**
Transactions of the American Nuclear Society. American Nuclear Society, 555 North Kensington Avenue, La Grange Park, Ill. 60525, USA. ISSN 0003–018X.

**Trans. Astron. Obs. Yale Univ.**
Transactions of the Astronomical Observatory of Yale University. Yale University Observatory, 260 Whitney Avenue, New Haven, Conn. 06511, USA.

**Trans. Bose Res. Inst.**
Transactions of the Bose Research Institute. Bose Research Institute, 93/1 Acharya Prafulla Chandra Road, Calcutta 9, India. ISSN 0006–7903.

**Trans. IAU**
Transactions of the International Astronomical Union. Published on behalf of the IAU by D. Reidel Publishing Company, Dordrecht - Boston - London.

**Trans. Inst. Electron. Commun. Eng. Jpn., Part C**
Transactions of the Institute of Electronic and Communication Engineers of Japan, Part C. Denshi Tsushin Gakkai, Kikai-Shinko-Kaikan, 5–8 Shibakoen 3 Chome, Minato-ku, Tokyo 105, Japan. ISSN 0373–6113.

**Transp. Theory Stat. Phys.**
Transport Theory and Statistical Physics. Marcel Dekker Inc., 270 Madison Avenue, New York, N.Y. 10016, USA. ISSN 0041–1450.

**Tsirk. Astron. Inst. Tashkent**
Tsirkulyar Astronomicheskogo Instituta Tashkent. Akademiya Nauk Uzbekskoj SSR. Izdatel'stvo FAN Uzbekskoj SSR, Tashkent. Tashkent, ul. Gololya, 70, USSR. ISSN 0373–7675.

**Tsirk., Astron. Obs. L'vov**
Tsirkulyar, Astronomicheskaya Observatoriya L'vov. L'vovskij Ordena Lenina Gosudarstvennyj Universitet im. Ivana Franko. Izdatel'stvo L'vovskogo Universiteta, L'vov. 290005, L'vov–5, ul. Lomonosova, 8, USSR. ISSN 0374–0722.

**Tsirk. Shemakh. Astrofiz. Obs.**
Tsirkulyar Shemakhinskoj Astrofizicheskoj Observatorii. Akademiya Nauk Azerbajdzhanskoj SSR. Izdatel'stvo Ehlm, Baku. 370143 Baku–143, prospekt Narimanova, 31. ISSN 0135–0420.

**Turku Univ. Obs., Informeto**
Turku University Observatory, Informeto. Turku University Observatory, Tuorla, SF–2150 Piikkiö, Finland.

**Turku Univ. Obs., Informo**
Turku University Observatory, Informo. Turku University Observatory, Tuorla, SF–21500 Piikkiö, Finland.

**UKIRT Rep.**
UKIRT Report. United Kingdom Infrared Telescope, 900 Leilani Street, Hilo, Hawaii 96720, USA. ISSN 0260–9983.

**Umschau**
Die Umschau. Das Wissenschaftsmagazin. Postfach 110262, Stuttgarter Straße 18–24, D–6000 Frankfurt am Main 11, F.R. Germany. ISSN 0722–8562.

**Univ. Barc., Dep. Fis. Tierra Cosmos, Publ.**
Universidad de Barcelona, Departamento de Fisica de la Tierra y del Cosmos, Publicacion. Departamento de Fisica de la Tierra y del Cosmos, Universidad de Barcelona, Spain.

**Univ. Hannover, Astron. Stn., Veröff.**
Universität Hannover, Astronomische Station, Veröffentlichungen. Universität Hannover, Astronomische Station, Hannover, F.R. Germany.

**Univ. Obs., St. Andrews, Repr.**
University Observatory, St. Andrews, Reprint. University Observatory, Buchanan Gardens, St. Andrews, Fife KY16 9LZ, Scotland.

**Univ. Tex., Monogr. Astron.**
The University of Texas, Monographs in Astronomy. Department of Astronomy, University of Texas, Austin, Tex. 78712, USA.

**Univ. Tex., Publ. Astron.**
University of Texas, Publications in Astronomy. Department of Astronomy, University of Texas, Austin, Tex. 78712, USA. ISSN 0276–1106.

**Upps. Astron. Obs. Ann.**
Uppsala Astronomiska Observatoriums Annaler. Astronomiska Observatoriet, Box 515, S–75120 Uppsala, Sweden.

**Upps. Astron. Obs. Rep.**
Uppsala Astronomical Observatory, Report. Astronomiska Observatoriet, Box 515, S–75120 Uppsala, Sweden.

**Urania (Barc.)**
Urania. Revista de Astronomía y Ciencias Afines. Organo de la Sociedad Astronómica de España y America y de la Union Nacional de Astronómia y Ciencias Afines. Sociedad Astronomica de España y America, Avenida Generalisimo, 337, Barcelona, Spain.

**Urania (Cracow)**
Urania. Miesiecznik Polskiego Towarzystwa Miłośników Astronomii. Polskie Towarzystwo Miłośników Astronomii, Zarzad Gl., 31–027 Kraków, Solskiego 30/8, Poland. ISSN 0042–0794.

**U.S. Nav. Obs., Circ.**
U.S. Naval Observatory, Circular. U.S. Naval Observatory, Washington, D.C. 20390, USA.

**U.S. Nav. Obs., Time Serv. Publ., Ser. 01**
U.S. Naval Observatory, Time Service Publications, Series 1 (Worldwide Primary Time and Frequency VLF and HF Transmissions). U.S. Naval Observatory, Time Service Division (62C), Washington, D.C. 20390, USA.

**U.S. Nav. Obs., Time Serv. Publ., Ser. 04**
U.S. Naval Observatory, Time Service Publications, Series 4 (Daily Time Differences and Relative Phase Values). U.S. Naval Observatory, Time Service Division (62C), Washington, D.C. 20390, USA.

**U.S. Nav. Obs., Time Serv. Publ., Ser. 06**
U.S. Naval Observatory, Time Service Publications, Series 6 (A.1–UT1 Data). U.S. Naval Observatory, Time Service Division (62C), Washington, D.C. 20390, USA.

**U.S. Nav. Obs., Time Serv. Publ., Ser. 10**
U.S. Naval Observatory, Time Service Publications, Series 10 (Astronomical Programs). U.S. Naval Observatory, Time Service Division (62C), Washington, D.C. 20390, USA.

**U.S. Nav. Obs., Time Serv. Publ., Ser. 11**
U.S. Naval Observatory, Time Service Publications, Series 11 (Time Service Report). U.S. Naval Observatory, Time Service Division (62C), Washington, D.C. 20390, USA.

**U.S. Nav. Obs., Time Serv. Publ., Ser. 14**
U.S. Naval Observatory, Time Service Publications, Series 14 (Time Service Announcement). U.S. Naval Observatory, Time Service Division (62C), Washington, D.C. 20390, USA.

**Utr. Sterrekundige Overdrukken**
Utrechtse Sterrekundige Overdrukken. Sonnenborgh Observatory, Zonnenburg 2, NL–3512 Utrecht, The Netherlands.

**Uttar Pradesh State Obs., Repr.**
Uttar Pradesh State Observatory, Reprint. Uttar Pradesh State Observatory, Manora Peak, Naini Tal 263 139, India.

**Vasiona**
Vasiona. Revue d'Astronomie. Bulletin de la Société Astronomique 'R. Bošković'. Vasiona, Narodna opservatorija, Kalemegdan, Gornji Grad, Beograd, Yugoslavia. ISSN 0506–4295.

**Vatican Obs. Publ.**
Vatican Observatory Publications. Specola Vaticana, I–00120 Città del Vaticano.

**Veröff. Archenhold–Sternw. Berlin–Treptow**
Veröffentlichungen der Archenhold–Sternwarte Berlin–Treptow. Archenhold–Sternwarte, DDR–1193 Berlin, Alt–Treptow 1, German Democratic Republic.

**Veröff. Astron. Inst. Bonn**
Veröffentlichungen der Astronomischen Institute Bonn. Ferd. Dümmler Verlag, Kaiserstraße 31–37, D–5300 Bonn 1, F.R. Germany. ISSN 0340–9821.

**Veröff. Astron. Rechen–Inst. Heidelb.**
Veröffentlichungen des Astronomischen Rechen–Instituts Heidelberg. Verlag G. Braun, Karl–Friedrich–Straße 14–18, D–7500 Karlsruhe 1, F.R. Germany.

**Veröff. Bayer. Komm. Int. Erdmessung Bayer. Akad. Wiss., Astron.–Geod. Arb.**
Veröffentlichungen der Bayerischen Kommission für die Internationale Erdmessung der Bayerischen Akademie der Wissenschaften, Astronomisch–Geodätische Arbeiten. Bayerische Akademie der Wissenschaften, Marstallplatz 8, D–8000 München 22, F.R. Germany. ISSN 0340–7691.

**Veröff. Forschungsinst. Dtsch. Mus. Gesch. Naturwiss. Tech., Reihe A**
Veröffentlichungen des Forschungsinstituts des Deutschen Museums für die Geschichte der Naturwissenschaften und der Technik, Reihe A. Forschungsinstitut des Deutschen Museums für die Geschichte der Naturwissenschaften und der Technik, D–8000 München, F.R. Germany. ISSN 0418–9949.

**Veröff. Remeis–Sternw. Bamberg**
Veröffentlichungen der Remeis–Sternwarte Bamberg. Astronomisches Institut der Universität Erlangen–Nürnberg. Remeis–Sternwarte, Sternwartstraße 7, D–8600 Bamberg, F.R. Germany.

**Veröff. Sternw. Pulsnitz**
Veröffentlichungen der Sternwarte Pulsnitz. Sternwarte Pulsnitz, DDR–8514 Pulsnitz, German Democratic Republic.

**Veröff. Sternw. Sonneberg**
Veröffentlichungen der Sternwarte in Sonneberg. Akademie der Wissenschaften der DDR, Zentralinstitut für Astrophysik, Sternwarte Sonneberg, DDR–6400 Sonneberg, German Democratic Republic. ISSN 0373–6199.

**Veröff. Wilhelm–Foerster–Sternw.**
Veröffentlichungen der Wilhelm–Foerster–Sternwarte. Wilhelm–Foerster–Sternwarte, Munsterdamm 90, D–1000 Berlin 41, F.R. Germany.

**Veröff. Zentralinst. Phys. Erde**
Veröffentlichungen des Zentralinstituts für Physik der Erde. Akademie der Wissenschaften der DDR, Zentralinstitut für Physik der Erde, DDR–1500 Potsdam, Telegrafenberg A17, German Democratic Republic.

**Vesmír**
Vesmír. Přírodovědecký časopis Čs. akademie věd. Publisher: Academia, Praha, Czechoslovakia.

**Vestn. Khar'kov. Univ.**
Vestnik Khar'kovskogo Universiteta. Izdatel'stvo pri Khar'kovskom Gosudarstvennom Universitete Izdatel'skogo Obedineniya Vishcha Shkola, Khar'kov. 310003, Khar'kov–3, ul. Universitetskaya, 16, USSR.

**Vestn. Kiev. Univ. Astron.**
Vestnik Kievskogo Universiteta. Astronomiya. Izdatel'stvo pri Kievskom Gosudarstvennom Universitete Izdatel'skogo Obedineniya Vishcha Shkola, Kiev. 252001, Kiev–1, Kreshchatik, 4, USSR. ISSN 0203–7319; ISSN 0321–3927.

**Vestn. Leningr. Univ., Mat. Mekh. Astron.**
Vestnik Leningradskogo Universiteta Matematika, Mekhanika, Astronomiya. Leningradskij Universitet, Universitetskaya Nab. 7/9 Leningrad B–164, USSR. ISSN 0024–0850.

**Vestn. Mosk. Univ., Ser. 3. Fiz. Astron.**
Vestnik Moskovskogo Universiteta, Seriya 3. Fizika, Astronomiya. Izdatel'stvo Moskovskogo Universiteta, 103009, Moskva, Ul.Gertsena 5/7, USSR. English translation in Mosc. Univ. Phys. Bull. ISSN 0579–9392.

**Villanova Univ. Obs. Contrib.**
Villanova University Observatory Contributions. Department of Astronomy, Villanova University, Villanova, Pa. 19085, USA.

**Vilniaus Astron. Obs. Biul.**
Vilniaus Astronomijos Observatorijos Biuletenis. Vilniaus Valstybinis V. Kapsuko Universitetas, LTSR Mokslu Akademijos Fizikos Institutas, Vilnius. 232031, Vilnius 31, Čiurlionio 29, Lithuania, USSR. ISSN 0136–3697.

**Vistas Astron.**
Vistas in Astronomy. An International Review Journal. Pergamon Press, Oxford – New York – Frankfurt. ISSN 0083–6656.

**Warner Swasey Obs., Repr.**
Warner and Swasey Observatory, Reprints. Warner and Swasey Observatory, Case Western Reserve University, 1975 Taylor Road, East Cleveland, Ohio 44112, USA.

**Warsaw Univ. Obs. Pol. Acad. Sci., N. Copernicus Astron. Cent., Repr.**
Warsaw University Observatory and Polish Academy of Sciences and N. Copernicus Astronomical Center, Reprint. Warsaw University Observatory and N. Copernicus Astronomical Center, Polish Academy of Sciences, Bartycka 18, PL–00–716 Warsaw, Poland.

**Weather**
Weather. James Glaisher House, Grenville Place, Bracknell, Berks, RG12 1BX, England. ISSN 0043–1656.

**Wiss. Z. Friedrich–Schiller–Univ. Jena, Math.–Naturwiss. Reihe**
Wissenschaftliche Zeitschrift der Friedrich–Schiller–Universität Jena, Mathematisch–Naturwissenschaftliche Reihe. Friedrich–Schiller–Universität Jena. Subscription address: Zeitungsvertriebsamt, Abt. Export, DDR–1004 Berlin, Straße der Pariser Kommune 3/4, German Democratic Republic. ISSN 0043–6836.

**Wiss. Z. Humboldt–Univ. Berlin, Math.–Naturwiss. Reihe**
Wissenschaftliche Zeitschrift der Humboldt–Universität zu Berlin, Mathematisch–Naturwissenschaftliche Reihe. Humboldt–Universität Berlin. Subscription address: Zeitschriftenvertriebsamt, Abt. Export, DDR–1004 Berlin, Straße der Pariser Kommune 3/4, German Democratic Republic. ISSN 0522–9863.

**Wrocław Astron. Obs., Repr.**
Wrocław Astronomical Observatory, Reprint. Wrocław University Observatory, Wrocław, Poland.

**Wuli**
Wuli. Science Press, Beijing. Subscription address: Guozi Shudian, PO Box 399, Beijing, People's Republic of China. ISSN 0379–4148.

**Yad. Fiz.**
Yadernaya Fizika. Izdatel'stvo Nauka, 103717 GSP Moskva K–62. Podsosenskij Per. 24, SSSR. English translation in Sov. J. Nucl. Phys. ISSN 0044–0027.

**Yamamoto Circ.**
Yamamoto Circular. Yamamoto Observatory, 289, Kamitanakami–kiryutyo, OTU, Sigaken, 520–21 Japan.

**Z. Angew. Math. Mech.**
Zeitschrift für angewandte Mathematik und Mechanik. Akademie–Verlag GmbH, DDR–108 Berlin, Leipziger Str. 3–4, German Democratic Republic. ISSN 0044–2267.

**Z. Angew. Math. Phys.**
Zeitschrift für Angewandte Mathematik und Physik. Verlag Birkhäuser, Postfach 4000, Basel 24, Switzerland. ISSN 0044–2275.

**Z. Naturforsch., A**
Zeitschrift für Naturforschung, Teil A (Physik, Physikalische Chemie, Kosmophysik). Verlag der Zeitschrift für Naturforschung, PO Box 2645, D–7400 Tübingen, F.R. Germany. ISSN 0340–4811.

**Z. Phys., A**
Zeitschrift für Physik, A. (Atoms und Nuclei). Springer–Verlag, Berlin – Heidelberg – New York – Tokyo. ISSN 0340–2193.

**Z. Phys., B**
Zeitschrift für Physik, B (Condensed Matter). Springer–Verlag, Berlin – Heidelberg – New York – Tokyo. ISSN 0722–3277.

**Z. Phys., C**
Zeitschrift für Physik, C (Particles and Fields). Springer–Verlag, Berlin – Heidelberg – New York – Tokyo. ISSN 0170–9739.

**Z. Vermessungswes.**
Zeitschrift für Vermessungswesen. Verlag Konrad Wittner KG, Postfach 147, D–7000 Stuttgart 1, F.R. Germany. ISSN 0513–9155.

**Zeiss Inf.**
Zeiss Information. Carl Zeiss, Oberkochen. F.R. Germany. ISSN 0044–2054.

**Zeit–Breitenbestimmungen**
Zeit– und Breitenbestimmungen, Empfangszeiten von Zeitsignalen, Präzisionszeitvergleiche. Akademie der Wissenschaften der DDR, Zentralinstitut für Physik der Erde, DDR–1500 Potsdam, Telegrafenberg A 17, German Democratic Republic.

**Zemlya Vselennaya**
Zemlya i Vselennaya. Astronomiya, Geofizika. Issledovaniya Kosmicheskogo Prostranstva. Nauchno–Populyarnyj Zhurnal Akademii Nauk SSSR. Izdatel'stvo Nauka, Moskva, USSR. ISSN 0044–3948.

**Zenit**
Zenit. Populair–wetenschappelijk maandblad over sterrenkunde, weerkunde, ruimtevaart, ruimte–onderzoek en aanverwante wetenschappen en technieken. Stichting De Koepel, Nachtegaalstraat 82 bis, Utrecht, The Netherlands. ISSN 0165–0211.

**Zentralbl. Math. Grenzgeb. – Math. Abstr.**
Zentralblatt für Mathematik und ihre Grenzgebiete – Mathematics Abstracts. Heidelberger Akademie der Wissenschaften und Fachinformationszentrum Energie Physik Mathematik GmbH, Karlsruhe. Springer–Verlag, Berlin – Heidelberg – New York – Tokyo. ISSN 0044–4235.

**Zh. Ehksp. Teor. Fiz.**
Zhurnal Ehksperimental'noj i Teoreticheskoj Fiziki. Akademiya Nauk SSSR, Leninskij Prosp. 14, Moskva, USSR. English translation in Sov. Phys.–JETP. ISSN 0044–4510.

**Zvaigžnota Debess**
Zvaigžnota Debess. Latvijas PSR Zinātnu Akadēmijas Radioastrofizikas Observatorijas Populārzinatnisks Rakstu Krāgums. Izdevnieciba Zinātne, Riga. 226004 Riga, Vienibas Gatve 11, USSR. ISSN 0135–129X.

# Journals Abstracted Completely

A selected number of journals listed in category 001 (periodicals) are central to the subject scope of *Astronomy and Astrophysics Abstracts*. Depending on their relevance, almost all papers of the journals listed below are abstracted in our service.

AAS Photo-Bull.
Acta Astron.
Acta Astron. Sin.
Acta Astrophys. Sin.
Acta Cosmologica
Algoritm. Nebesnoj Mekh.
Am. Assoc. Variable Star Obs. Bull.
Annu. Rev. Astron. Astrophys.
Archaeoastronomy (U.K.)
Archaeoastronomy (U.S.A.)
Astrofizika
Astrometr. Astrofiz.
Astron. Astrophys.
Astron. Astrophys., Suppl. Ser.
Astron. Circ.
Astron. Data Cent. Bull.
Astron. J.
Astron. Nachr.
Astron. Q.
Astron. Tidsskr.
Astron. Tsirk.
Astron. Vestn.
Astron. Zh.
Astronomia
Astronomie
Astrophys. J.
Astrophys. J., Lett. Ed.
Astrophys. J., Suppl. Ser.
Astrophys. Lett.
Astrophys. Space Sci.

BAV Rundbrief
BBSAG Bull.
Bol. Asoc. Argent. Astron.
Bol. Astron.
Bol. Astron. R Muscae
Br. Astron. Assoc. Circ.
Bull. Am. Astron. Soc.
Bull. Assoc. Fr. Obs. Etoiles Variables
Bull. Astron. Inst. Czech.
Bull. Astron. Soc. India
Bull. Inf. Cent. Données Stellaires

Celest. Mech.
Chin. Astron. Astrophys.
Ciel
Ciel Terre
Circ. Inf.
Coelum
Comments Astrophys.

Earth, Moon, Planets

Fundam. Cosmic Phys.

G. Astron.
Gen. Relativ. Gravitation
GEOS Circ.

I.A.P.P.P. Commun.
IAU Circ.
Icarus
Inf. Bull. Variable Stars
Int. Comet Q.
Ir. Astron. J.
Issled. Solntsa Krasnykh Zvezd

J. Am. Assoc. Variable Star Obs.
J. Astrophys. Astron.
J. Br. Astron. Assoc.
J. Hist. Astron.
J. Meteor Res.
J. R. Astron. Soc. Can.

Komet. Tsirk.
Komety Meteory

Mem. Astron. Soc. India
Mem. Soc. Astron. Ital.
Mercury
Messenger
Meteoritics
Meteoritika
Minor Planet Bull.
Minor Planet Circ.
Mitt. Astron. Ges.
Mon. Not. R. Astron. Soc.
Mon. Notes Astron. Soc. S.Afr.

Nablyud. Iskusstv. Nebesn. Tel
Nauchn. Inf.
News Lett. Astron. Soc. N.Y.

Observatory
Occultation Newsl.
Orion
Orione

Perem. Zvezdy
Perem. Zvezdy, Prilozh.
Pis'ma Astron. Zh.
Postępy Astron.
Probl. Kosm. Fiz.
Proc. Astron. Soc. Aust.
Publ. Astron. Soc. Jpn.
Publ. Astron. Soc. Pac.
Publ. Variable Star Sect., R. Astron. Soc. N.Z.

Q. J. R. Astron. Soc.

Rev. Astron.
Rev. Mex. Astron. Astrofis.

Sky Telesc.
Sol. Phys.
Soln. Dannye Byull.
South. Stars
Space Sci. Rev.
Sterne
Sterne Weltraum
Strolling Astron.

Trans. IAU

Urania (Barcelona)
Urania (Cracow)

Vasiona
Vestn. Khar'kov. Univ.
Vestn. Kiev. Univ. Astron.
Vistas Astron.

Yamamoto Circ.

Zenit
Zvaigžnota Debess

# Publications of Observatories and Astronomical Institutes

Reports, communications, publications, numbered series of reprints and preprints of observatories and astronomical institutes (listed in category 001) which are scanned completely in our service are listed below.

A.A.O. Newsl.
Abastumanskaya Astrofiz. Obs., Byull.
Abh. Hamb. Sternw.
Abstr. Submitted Pap.
Acad. Sci. Est. SSR, Div. Phys. Math. Tech. Sci., Prepr.
A.D.I.O.N. Bull.
Anglo-Aust. Obs., Prepr.
Anglo-Aust. Telesc., Annu. Rep.
Ann. Obs. Astron. Alger
Ann. Shanghai Obs., Acad. Sin.
Ann. Tokyo Astron. Obs., Second Ser.
Ann. Univ.-Sternw. Wien
Annu. Rep. Astron. Inst. Greece
Annu. Rep. (B.I.H.)
Annu. Rep. Dir., Mt. Wilson Las Campanas Obs.
Annu. Rep. Geophys. Obs.
Annu. Rep. Int. Polar Motion Serv.
Annu. Rep. Meteorol. Obs. Int. Latitude Obs. Mizusawa
Archenhold-Sternw. Berlin-Treptow, Sonderdr.
Archenhold-Sternw. Berlin-Treptow, Vortr. Schr.
Armagh Obs., Repr.
Astrofiz. Issled. Izv. Spets. Astrofiz. Obs.
Astrofys. Inst., Vrije Univ. Bruss., Overdruk
Astron. Bull., Carter Obs.
Astron. Contrib. Univ. Manchester, Ser. II: Jodrell Bank Repr.
Astron. Contrib. Univ. Manchester, Ser. III
Astron. Inst., "Anton Pannekoeck", Univ. Amst., Repr.
Astron. Mitt. Wien
Astron. Obs. Trieste, Publ.
Astron. Pap.
Astron. Rechen-Inst. Heidelb., Mitt., Ser. A
Astron. Rechen-Inst. Heidelb., Mitt., Ser. B
Astron. Zeit-Breitenbestimmungen
Astrophys. Prepr. Ser.
Astrophys. Relativ., Prepr. Ser.
Austrian Pap. Asteroids

BAV Mitt.
B.I.H. Circ.
Biul. Obs. Astron. Uniw. M. Kopernika Toruniu
Bol. Astron. Obs. Madr.
Bol. Inst. Tonantzintla
Bol. Obs. Ebro
Boyden Obs., Publ.
Boyden Obs., Repr.
Bull. Astron.
Bull. Etoiles Tardives Spectre Particulier
Bull. Inf. Cent. Données Stellaires
Bull. Inf. Etoiles Be
Bull. Obs. Astron. Belgr.
Bull., Time Serv. Mizusawa Obs.
Byull. Inst. Astrofiz.
Byull. Inst. Teor. Astron.

Carter Obs., Repr. Ser.
Cartes Synoptiques
Cent. Astron. Sci. Spat., Obs. Sol.
Cent. Astrophys., Prepr. Ser.
Circ. Czech. Obs., Time Latitude
Circ. Stn. Astron. Int. Latitudine, Carloforte-Cagliari
Circ. Time Latitude Serv.
Commun. Astron. Dep., Univ. Ankara
Commun. Konkoly Obs.
Commun. Univ. Obs., St. Andrews
Comun. Obs. Astron. Univ. Coimbra
Contrib. Astron. Obs. Skalnaté Pleso
Contrib. Bosscha Obs.
Contrib. Dep. Astron., Univ. Kyoto
Contrib. Dep. Astron., Univ. Tokyo
Contrib. Dep. Geod. Astron., Univ. Thessaloniki
Contrib. Dom. Astrophys. Obs.
Contrib. Dunsink Obs.
Contrib. Inst. Argent. Radioastron.
Contrib. Inst. Astron. Res.
Contrib. Kwasan Hida Obs., Univ. Kyoto
Contrib. La. State Univ. Obs.
Contrib. Lick Obs.
Contrib. Nicholas Copernicus Obs. Planetarium Brno
Contrib. Nizamiah Japal-Rangapur Obs.
Contrib. Oss. Astron. Torino
Contrib. Van Vleck Obs.
Copenh. Univ. Obs., Repr.
Cracow Obs. Repr.

Data Rep. Hydrogr. Obs., Ser. Astron. Geod.
Debrecen Heliophys. Obs., Hung. Acad. Sci., Repr.
Dep. Astron. McDonald Obs. Univ. Tex., Repr.
Dep. Astrophys., Univ. Oxford, Publ.
Dtsch. Hydrogr. Inst. Hamb., Zeit-Breitendienst
Dudley Obs. Rep.
Dunsink Obs. Repr.

Earth Orientation Bull.
ESO Annu. Rep.
ESO Sci. Prepr.
ESO Sci. Rep.
ESO Tech. Rep.

Gemini
Greenwich Time Rep.

HHI Sol. Data
HHI STP Rep.
HHI Suppl. Ser. Sol. Data
H.M. Naut. Alm. Off., Libr. Repr.
Hvar Obs. Bull.

Inst. Astron. Astrophys. Tech. Univ. Berlin, Mitt.
Inst. Astron. Fis. Espacio, Tirada Aparte
Inst. Astron., Univ. Camb., Annu. Rep.
Inst. Astrophys. Paris, Pré-Publ.
Inst. Astrophys. Univ. Liège, Collect. 4°
Inst. Astrophys. Univ. Liège, Collect. 8°
Inst. Obs. Mar., Bol. Astron.
Inst. Obs. Mar., Bol. Ser. B
Inst. Obs. Mar., Bol. Ser. C
Inst. Obs. Mar., Bol. Ser. D
Inst. Obs. Mar., Mem. Act.
Inst. Teor. Astrofys., Blindern-Oslo, Småtrykk
Inst. Theor. Astrophys., Blindern-Oslo, Rep.
Inst. Theor. Astrophys., Blindern-Oslo, Repr.
Inst. Theor. Astrophys., Univ. Oslo, Publ. Ser.
Inst. Theor. Phys. Sternw. Univ. Kiel, Repr.
Izv. Astron. Ehngel'gardt. Obs.
Izv. Glav. Astron. Obs. Pulkovo
Izv. Krymskoj Astrofiz. Obs.

Kandilli Obs., Heliophys. Serv. Publ., Second Ser.
Kapteyn Astron. Inst., Annu. Rep.
Kodaikanal Obs. Bull., Ser. A
Kodaikanal Obs. Repr.

Latitude Circ.
Lick Obs. Bull.
Lohrmann-Obs., Tech. Univ. Dresden, Zirk.
Lowell Obs. Bull.

Marées Terr., Bull. Inf.
Marées Terr., Bull. Obs.
Max-Planck-Inst. Radioastron., Bonn, Sonderdr. Ser. A
Messenger
Mitt. Archenhold-Sternw. Berlin-Treptow
Mitt. Astrophys. Obs. Potsdam
Mitt. Karl-Schwarzschild Obs. Tautenburg
Mitt. Lohrmann-Obs. Tech. Univ. Dresden
Mitt. Satell.-Beobachtungsstn. Zimmerwald
Mitt. Sonnenobs. Kanzelhöhe
Mitt. Sternw. Babelsberg, Neue Folge
Mitt. Sternw. Münch.
Mitt. Sternw. Sonneberg
Mitt. Universitätssternw. Graz
Mitt. Univ.-Sternw. Innsb.
Mitt. Univ.-Sternw. Jena
Mitt. Veränderliche Sterne
Mitt. Zentralinst. Phys. Erde
Mon. Notes Int. Polar Motion Serv.
Mons Astrophys. Pap.
MPE Contrib.
MPE Prepr.
MPE Rep.
MPE Repr.
Mt. Stromlo Siding Spring Obs., Repr.

Nanjing Univ. Obs., Publ.
NASA IUE Newsl.
Natl. Astron. Ionos. Cent., Astron. Prepr.
Natl. Astron. Ionos. Cent., Astron. Publ.
Natl. Astron. Ionos. Cent., Newsl.
Natl. Radio Astron. Obs., Repr., Ser. A
Natl. Radio Astron. Obs., Repr., Ser. B
Nizamiah Rangapur Obs. Dep. Astron. Osmania Univ., Repr.

Obs. Astron. Antares, Contrib. Cient.
Obs. Astron. Córdoba, Tirada Aparte
Obs. Astron. La Plata, Sep. Astron.
Obs. Astron. Univ. Nac. La Plata, Ser. Astron.
Obs. Astron. Univ. Nac. La Plata, Ser. Espec.
Obs. Astrophys. Lab., Univ. Helsinki, Rep.
Obs. Lyon Prepr.
Obs. Lyon Repr.
Obs. R. Belg., Commun., Sér. A
Obs. R. Belg., Commun., Sér. B
Obs. Variable Stars, Rep.
Occas. Rep. R. Obs., Edinb.
Onsala Space Obs., Prepr.
Oss. Astrofis. Catania, Publ.
Oss. Astron. Milano-Merate, Contrib.

Prepr. Astron. Inst. Univ. Basel
Prepr. Steward Obs.
Proc. Int. Latitude Obs. Mizusawa
Pubbl. Stn. Astron. Int. Latitudine Carloforte-Cagliari, Nuov. Ser.
Pubbl. Varie Fuori Ser. Oss. Astron. Torino (Pino Torinese)
Publ. Astron. Dep. Eötvös Univ.
Publ. Astron. Inst. Czech. Acad. Sci.
Publ. Astron. Obs. Sarajevo
Publ. Astron. Opservatorije Beogr.
Publ. Astrophys. Obs. Potsdam
Publ. Beijng Astron. Obs.
Publ. Bosscha Obs.
Publ. Debrecen Heliophys. Obs.
Publ. Dep. Astron. Univ. Beogr.
Publ. Dep. Astron. Univ. Cape Town
Publ. Dep. Astron., Univ. Chile
Publ. Dep. Geod. Astron., Univ. Thessaloniki
Publ. Dom. Astrophys. Obs.
Publ. Ege Univ. Obs.
Publ. Inst. Geophys., Pol. Acad. Sci., Ser. F (Planet. Geod.)
Publ. Int. Latitude Obs. Mizusawa
Publ. Istanbul Univ. Obs.
Publ. Lick Obs.
Publ. Obs. Astron. Nac., Univ. Colombia
Publ. Obs. Astron. "Prof. Manuel de Barros", Fac. Cienc. Porto
Publ. Obs. Astron. Univ. Cluj
Publ. Obs. Bordeaux, Nouv. Ser.
Publ. Obs. Ebro
Publ. Obs. Genève, Sér. A
Publ. Obs. Genève, Sér. B
Publ. Purple Mt. Obs.
Publ. R. Obs., Edinb.
Publ. Semin. Astron. Geod. Univ. Complutense Madr.
Publ. Shaanxi Astron. Obs.
Publ. Spéc. Cent. Données Stellaires
Publ. U.S. Nav. Obs., Second Ser.
Publ. Warner Swasey Obs.

Q. Bull. Sol. Act.

R. Greenwich Obs., Annu. Rep.
R. Greenwich Obs., Bull.
R. Obs. Ann.
R. Obs. Edinb., Res. Facilities
Rapp. Act. Obs. Paris
Rep. Obs. Lund
Rep. Ser., Dep. Phys. Sci., Univ. Turku
Res. Cent. Astron. Appl. Math., Acad. Athens, Contrib. Ser. I
Res. Lab. Electron. Onsala Space Obs., Res. Rep.
Ric. Astron.
Rijksuniv. Gent, Sterrenkundig Obs., Meded.

SAAO Newsl.
Sac Peak Update
S.Afr. Astron. Obs., Annu. Rep.
S.Afr. Astron. Obs., Circ.
Sci. Rep. Tôhoku Univ., Eighth Ser.
Semin. Astron. Geod., Univ. Madr., Publ.
Sendai Astron. Rap.
Shaanxi Astron. Obs., Repr.
Shanghai Obs., Acad. Sin., Prepr.
Smithson. Astrophys. Obs., Spec. Rep.
Sol. Maps Act.
Soobshch. Byurakan. Obs.
Soobshch. Gos. Astron. Inst. Shternberg
Soobshch. Spets. Astrofiz. Obs.
Space Telesc. Sci. Inst., Newsl.
Space Telesc. Sci. Inst., Prepr. Ser.
Stockholms Obs., Rep.
Stud. Galileiani
Stud. Soc. Sci. Torun., Sect. F
Swarthmore Astron. Repr.
Syd. Obs. Pap.

Tartu Astrofüüs. Obs. Publ.
Tartu Astrofüüs. Obs. Teated
Theor. Pap.
Time Freq. Serv., Bull.
Time Serv. Annu. Rep.
Time Serv. Bull.
Tokyo Astron. Bull., Second Ser.
Tokyo Astron. Obs., Kiso Inf. Bull.
Tokyo Astron. Obs. Rep.
Tokyo Astron. Obs. Repr.
Tokyo Astron. Obs., Time Latitude Bull.
Tr. Astrofiz. Inst. Alma-Ata
Tr. Astron. Obs., Leningrad
Tr. Glav. Astron. Obs. Pulkovo, Ser. 2
Tr. Gos. Astron. Inst. Shternberg
Tr. Inst. Teor. Astron., Leningrad
Tr. Kazan. Gorod. Astron. Obs.
Tr. Tashkent. Astron. Obs.
Trans. Astron. Obs. Yale Univ.

Tsirk. Astron. Inst. Tashkent
Tsirk. Astron. Obs. L'vov
Tsirk. Shemakh. Astrofiz. Obs.
Turku Univ. Obs., Informeto
Turku Univ. Obs., Informo

UKIRT Rep.
Univ. Barc., Dep. Fis. Tierra Cosmos, Publ.
Univ. Hannover, Astron. Stn., Veröff.
Univ. Obs., St. Andrews, Repr.
Univ. Tex., Monogr. Astron.
Univ. Tex., Publ. Astron.
Upps. Astron. Obs. Ann.
Upps. Astron. Obs., Rep.
U.S. Nav. Obs., Circ.
U.S. Nav. Obs., Time Serv. Publ., Ser. 1
U.S. Nav. Obs., Time Serv. Publ., Ser. 4
U.S. Nav. Obs., Time Serv. Publ., Ser. 6
U.S. Nav. Obs., Time Serv. Publ., Ser. 10
U.S. Nav. Obs., Time Serv. Publ., Ser. 11
U.S. Nav. Obs., Time Serv. Publ., Ser. 14
Utr. Sterrekundige Overdrukken
Uttar Pradesh State Obs., Repr.

Vatican Obs. Publ.
Veröff. Archenhold-Sternw. Berlin-Treptow
Veröff. Astron. Inst. Bonn
Veröff. Astron. Rechen-Inst. Heidelb.
Veröff. Remeis-Sternw. Bamberg
Veröff. Sternw. Pulsnitz
Veröff. Sternw. Sonneberg
Veröff. Wilhelm-Foerster-Sternw.
Veröff. Zentralinst. Phys. Erde
Villanova Univ. Obs. Contrib.
Vilniaus Astron. Obs. Biul.

Warner Swasey Obs., Repr.
Warsaw Univ. Obs. Pol. Acad. Sci.,
N. Copernicus Astron. Cent., Repr.
Wrocław Astron. Obs., Repr.

# 002 Bibliographical Publications, Documentation, Catalogues, Atlases

002.001 **The Case low-dispersion northern sky survey. II.**
N. Sanduleak, P. Pesch.
Astrophys. J., Suppl. Ser., Vol. 55, No. 3, p. 517 – 532 (1984).
Positions, estimated magnitudes, and finding charts (when needed) are provided for 125 blue and/or emission-line galaxies, probable H II regions in nine galaxies, 120 blue and/or emission-line stellar (unresolved) objects including QSO candidates, and 38 known and probable blue stars. Almost all of the objects are in the region $10^h50^m < \alpha < 12^h50^m$ and $+29.^\circ0 < \delta < +33.^\circ0$ (1950). The objects, whose blue magnitudes range from 11 to 18, were found on low-dispersion objective-prism plates taken with the Burrell Schmidt-type telescope at Kitt Peak.

002.002 **Subject and Author Indexes of Books and Papers prepared at the Main Astronomical Observatory of the Ukrainian SSR Academy of Sciences and published by "Naukova Dumka".**
I. S. Dovgalevskaya, O. V. Klimenko.
Astrometr. Astrofiz., Vyp. 52, p. 53 – 91 (1984).

002.003 **Catalogue of cataclysmic binaries, low-mass X-ray binaries and related objects. Third edition.**
H. Ritter.
Astron. Astrophys., Suppl. Ser., Vol. 57, No. 3, p. 385 – 418 (1984).
The catalogue lists coordinates, magnitudes, orbital parameters, stellar parameters of the components and other characterizing properties of 97 cataclysmic binaries, 20 low-mass X-ray binaries and 20 related objects together with a comprehensive selection of the relevant literature. In addition the catalogue contains a list of references to published finding charts for 131 of the 137 objects. A cross-reference list of object designations concludes the catalogue.

002.004 **Chemical evolution and the origin of life. Bibliography Supplement 1982.**
L. G. Pleasant, C. Ponnamperuma.
Origins Life, Vol. 15, No. 1, p. 55 – 59 (1984).

002.005 **Metal-deficient giants in the galactic field. Catalogue and some physical parameters.**
A. Bartkevičius.
Vilniaus Astron. Obs. Biul., Nr. 66, 89 pp. (1984). In Russian.
A catalogue of 408 known metal-deficient giants ([Fe/H] $\leqslant$ –0.8) in the galactic field compiled from various spectroscopic and photometric investigations is presented. $E_{B-V}$, $(B-V)_0$, $M_V$, and photometric spectral type, including metallicity, luminosity, and spectral class, are derived.

002.006 **The second Byurakan spectral sky survey. I. The results of the area centred at $\alpha = 11^h30^m$, $\delta = 59°00'$.**
B. E. Markarian, D. A. Stepanyan.
Astrofizika, Tom 19, Vyp. 4, p. 639 – 654 (1983). In Russian. English translation in Astrophysics, Vol. 19, No. 4.
The first list of objects of the second Byurakan survey covering a field of $4° \times 4°$ degrees and centred at $\alpha = 11^h30^m$, $\delta = 59°00'$ is presented. It contains data on 107, peculiar by their physical nature, galaxies and objects and 11 blue stars. 11 candidates for QSO, 7 Seyfert galaxies, 11 BSO, 29 galaxies with significant UV continuum and 56 emission galaxies without significant UV continuum were found.

002.007 **The second Byurakan spectral sky survey. II. The results of the area centred at $\alpha = 09^h50^m$, $\delta = +55°00'$.**
B. E. Markarian, D. A. Stepanyan.
Astrofizika, Tom 20, Vyp. 1, p. 21 – 34 (1984). In Russian. English translation in Astrophysics, Vol. 20, No. 1.
The second list of objects of the second Byurakan survey covering a field of $4° \times 4°$ degrees and centred at $\alpha = 09^h50^m$, $\delta = +55°00'$ is presented. It contains data on 110 peculiar, by their physical nature, galaxies and objects and 24 blue stars. In the given area were found: 20 candidates for QSO, 4 Seyfert galaxies, 27 BSO, 24 galaxies with significant UV continuum and 39 emission galaxies without significant UV continuum.

002.008 **Citations to federally-funded and unfunded research.**
H. A. Abt.
Publ. Astron. Soc. Pac., Vol. 96, No. 581, p. 563 – 565 (1984).
The 1409 papers published in 1974 in the three general American astronomical research journals produced 21,106 citations (references) during the years 1975 – 1979. Papers coming from most kinds of organizations (universities, NSF and NASA centers, foundations, and miscellaneous governmental and military organizations) have statistically the same average citation rates, whereas papers from commercial companies rank higher and foreign papers (in 1974) rank lower. Sloan Fellows und university authors receiving most forms of auxiliary outside funding (NASA, NSF) produce higher-cited papers than papers without such funding.

002.009 **Bibliography.**
Z. Kopal, M. Moutsoulas, F. B. Waranius (Editors).
Earth, Moon, Planets, Vol. 31, No. 1, p. 79 – 121 (1984).
The bibliography records articles concerned with the moon, the planets and other objects (asteroids, comets, meteorites, cosmic dust, other particles, etc.). The articles entered into the data base at the Lunar and Planetary Institute Library, January 1983 through April 1983.

002.010 **A catalogue of symbiotic stars.**
D. A. Allen.
Proc. Astron. Soc. Aust., Vol. 5, No. 3, p. 369 – 421 (1984).
More than 100 symbiotic stars are now known. A catalogue of them, complete to 1984 January 1, is presented. Finding charts are given for all examples, and optical spectra for the majority. A table summarises the observational material at X-ray, ultraviolet, infrared and radio wavelengths.

002.011 **Les catalogues astrométriques du C.D.S.**
C. Jaschek.
Astrométrie, géodésie, mécanique céleste, p. 219 – 221 (1984). – See Abstr. 012.021.

002.012 **Astronomical catalogues: definition, elements, and afterlife.**
C. Jaschek.
Q. J. R. Astron. Soc., Vol. 25, No. 3, p. 259 – 266 (1984).
Based on a look at the different meanings of the term catalogue (or catalog), a definition is proposed. In an analysis of the main elements, a number of requirements that catalogues should satisfy are pointed out. A section is devoted to problems connected with computer-readable versions of printed catalogues.

002.013 **International directory of amateur astronomical societies. – IDAAS.**
A. Heck, J. Manfroid (Editors).
André Heck, Observatoire Astronomique, 11, rue de l'Université, F-67000 Strasbourg, France. Jean Manfroid, Institut d'Astrophysique, avenue de Cointe 5, B-4200 Cointe-Ougrée, Belgium. 3+278 pp. Price BF 600.00, US$ 12.00 (1984).
Review in Sky Telesc., Vol. 68, No. 1, p. 37 (1984).

002.014 **An IUE high-resolution atlas of O-type spectra.**
N. R. Walborn, J. N. Heckathorn, R. J. Panek.
ESA Spec. Publ., ESA SP-218, p. 255 (1984). Abstract. – See Abstr. 012.024.

002.015 **IUE low-dispersion spectra reference atlas.**
B. J. M. Hassall, A. Heck, A. Talavera, D. Egret, C. Jaschek, M. Jaschek.
ESA Spec. Publ., ESA SP-218, p. 263 - 266 (1984). - See Abstr. 012.024.

The preparation of this publication has confirmed that MK classifications cannot simply be transferred to the ultraviolet range. A set of transparencies illustrates the reference sequences constructed from the ultraviolet data. A magnetic-tape copy of all the spectra pertaining to this volume is available from the Strasbourg Stellar Data Center.

002.016 **Catalogue of stellar magnitudes, colour indices, spectral classes of stars in the direction to the cluster NGC 6633.**
M. S. Kazanasmas, N. A. Mis'kin, L. F. Tomak.
Odes. univ. Odessa, 23 pp. (1983). In Russian. Abstr. in Ref. Zh., 51. Astron., 8.51.375 (1984).

002.017 **Interstellar travel and communication bibliography: 1984 update.**
Z. Paprotny, J. Lehmann, J. Prytz.
J. Br. Interplanet. Soc., Vol. 37, No. 11, p. 502 - 512 (1984).

002.018 **Charts for southern variables. Series No. 17.**
F. M. Bateson, M. Morel.
Astronomical Research Ltd., PO Box 3093, Greerton, Tauranga, New Zealand. 4 pp. + charts 751 - 800 (1984).

002.019 **Radial velocities and positions from objective prism plates. The catalogue.**
J. Stock.
Centro de Investigaciones de Astronomía, A. P. 264, Mérida 5101-A, Venezuela, 170 pp. (1984). See Abstr. 111.009.

002.020 **Phobos: close encounter imaging from the Viking Orbiters.**
T. C. Duxbury, J. D. Callahan, A. C. Ocampo.
NASA Ref. Publ., NASA RP-1109, 7 + 51 pp. (1984).

002.021 **UK National Report for the Quadrennium 1979 - 1983 to the International Association of Geomagnetism and Aeronomy (*Bibliography*).**
B. R. Leaton, K. M. Creer.
Geophys. J. R. Astron. Soc., Vol. 79, No. 2, p. 651 - 690 (1984).

002.022 **A comment on the authors' ages for the most-cited papers.**
A. R. Rao, M. N. Vahia.
Publ. Astron. Soc. Pac., Vol. 96, No. 582, p. 661 - 662 (1984).

In continuation of the work of Abt (1983) a study of Indian Astronomers' publication patterns is made. The pattern is found to be similar to that obtained by Abt. An explanation for the curve of variation of most cited work with age, based on individual and coauthored works (in later ages with junior colleagues), is suggested. The suggestion is further investigated by a study of the Indian Astronomers' principal and coauthor publications over their active working age.

002.023 **A library of stellar spectra.**
G. H. Jacoby, D. A. Hunter, C. A. Christian.
Astrophys. J., Suppl. Ser., Vol. 56, No. 2, p. 257 - 281 (1984).

Spectra for 161 stars having spectral classes O - M and luminosity classes V, III, and I have been incorporated into a library available on magnetic tape. The spectra extend from 3510 to 7427 Å at a resolution of $\sim$4.5 Å. The typical photometric uncertainty of each resolution element in the spectra is on the order of 1%, while broad-band variations are smaller than 3%.

002.024 **Voyager 1 and 2 atlas of six Saturnian satellites.**
R. M. Batson, E. M. Lee, K. F. Mullins, B. A. Skiff, P. M. Bridges, J. L. Inge, H. Masursky, M. E. Strobell.
NASA Spec. Publ., NASA SP-474, 8 + 175 pp. (1984).

The atlas contains preliminary maps of Mimas, Enceladus, Tethys, Dione, Rhea, and Iapetus and includes the pictures and other data gathered primarily by the Voyager 1 and 2 spacecraft that were used to compile the maps.

002.025 **Catalogue of right ascensions of near-polar stars.**
V. N. Ershov.
Glav. Astron. Obs. Akad. Nauk SSSR. Leningrad, 93 pp. (1984). In Russian. Abstr. in Ref. Zh., 51. Astron., 10.51.125 (1984).

002.026 **Fotografischer Mondatlas.**
W. Schwinge.
Johann Ambrosius Barth, Leipzig, GDR. 208 pp. Price DM 46.00 (1983).

002.027 **Citations to single and multiauthored papers.**
H. A. Abt.
Publ. Astron. Soc. Pac., Vol. 96, No. 583, p. 746 - 749 (1984).

It is shown that on the average, the number of citations to papers increases linearly with the number of authors. However, if team research takes substantially more time to perform than individual research, the advantage is lost. The larger citation rate for many-authored papers is not due to the higher citation rate for longer papers because they tend to be shorter than few-authored papers; it is also not due to certain more-active fields having more multiauthored papers. Longer papers yield substantially larger numbers of citations than shorter papers, but a single long paper still yields fewer citations than the total for several shorter ones of the same combined lengths.

002.028 **Astronomie im 20. Jahrhundert. Eine quantitative Analyse des "Source Book of Astronomy and Astrophysics".**
D. B. Herrmann.
Sterne, 60. Band, Heft 5, p. 284 - 290 (1984).

002.029 **A new edition of the Yale Parallax Catalogue.**
W. van Altena.
Publ. Astron. Soc. Pac., Vol. 96, No. 584, p. 797 (1984). Abstract. - See Abstr. 010.281.

002.030 **Viking orbiter mosaic status report.**
S. D. Winterhalter.
Bull. Am. Astron. Soc., Vol. 16, No. 3, Part I, p. 680 (1984). Abstract. - See Abstr. 010.061.

002.031 **Planetary imaging data archive-laser videodisc storage and retrieval.**
L. J. Pieri, M. D. Martin, R. E. Sailors.
Bull. Am. Astron. Soc., Vol. 16, No. 3, Part I, p. 705 (1984). Abstract. - See Abstr. 010.061.

002.032 **The ST Guide Star Catalog.**
H. Jenkner, B. M. Lasker, J. L. Russell.
Bull. Am. Astron. Soc., Vol. 16, No. 3, Part I, p. 723 (1984). Abstract. - See Abstr. 010.064.

002.033 **A catalogue of visual extinction in Taurus and Perseus.**
J. Cernicharo, R. Bachiller.
Astron. Astrophys., Suppl. Ser., Vol. 58, No. 2, p. 327 - 350 (1984).

A catalogue containing high angular resolution data on visual extinction in Taurus and Perseus has been compiled from star counts on the prints of the Palomar Observatory Sky Survey (POSS). This catalogue may serve as a finding list for infrared and radio molecular searches of protostellar objects and high density regions. It could be used as a statistical source of data for fragmentation and physical properties studies of dark clouds. The star counts procedure is discussed.

002.034 **Catalogue of dark nebulae and globules for galactic longitudes 240 to 360 degrees.**
J. V. Feitzinger, J. A. Stüwe.
Astron. Astrophys., Suppl. Ser., Vol. 58, No. 2, p. 365 - 386 (1984).

A catalogue of dark nebulae and globules has been compiled from a study of the ESO (*B*) and SRC-*J* Sky Atlas for galactic longitudes $240° < l^{II} < 360°$. This catalogue closes the great southern gap open since the work of Lynds (1962) for the northern hemisphere. The catalogue (with cross-references) contains positions, sizes, opacities and the van den Bergh (1972) classification on the filamentary morphology of 489 dark clouds and 331 globules.

002.035 **Radial velocities of 232 stars belonging to two fields of 4° × 4°. Third list.**
C. Fehrenbach, R. Burnage, A. Peton.
Astron. Astrophys., Suppl. Ser., Vol. 58, No. 2, p. 435 – 440 (1984). In French.
The authors give a radial velocity catalogue of 232 stars including exact position, spectral type and approximate magnitude. These stars belong to two fields of 4° × 4° field No. 9: $l$ = 58°.42, $b$ = −30°.95, field No. 10: $l$ = 169°.38, $b$ = 30°.97.

002.036 **A search for faint blue stars in high galactic latitudes. III. Twelve PSS fields at declinations −6° and −12° near the South Galactic Pole.**
J. Berger, A.-M. Fringant.
Astron. Astrophys., Suppl. Ser., Vol. 58, No. 3, p. 565 – 583 (1984).
This third paper completes the survey of faint blue objects near the South Galactic Pole: a catalogue of 2484 starlike or compact objects is presented with the 1950 positions, estimated magnitudes, colour classes and extensive identifications with stars of previous surveys and known QSOs.

002.037 **Spectroscopic binaries. 14th Complementary Catalogue.**
A. Pédoussaut, N. Ginestet, J. M. Carquillat.
Astron. Astrophys., Suppl. Ser., Vol. 58, No. 3, p. 601 – 655 (1984). In French.
This catalogue is a continuation of the "Catalogues Complémentaires des Binaires Spectroscopiques de l'Observatoire de Toulouse". It contains orbital elements for 380 spectroscopic binaries.

002.038 **A revised supernova catalogue.**
R. Barbon, E. Cappellaro, F. Ciatti, M. Turatto, C. T. Kowal.
Astron. Astrophys., Suppl. Ser., Vol. 58, No. 3, p. 735 – 750 (1984).
Essential data for 568 supernovae, discovered since 1885 up to 1983, and their parent galaxies are presented. This catalogue updates and revises previous listings, and some of its information is summarized in tabular and graphical form and briefly commented. An appendix gives the main references to observations of supernovae outside the optical range.

002.039 **Documentation for the machine-readable version of *The Revised AFGL Infrared Sky Survey Catalogue* (Price and Murdock 1983).**
W. H. Warren Jr.
NSSDC/WDC-A-R&S, Publ., No. 84-05, 5+17 pp. (1984).
A detailed description of the machine-readable catalog as it is currently being distributed from the Astronomical Data Center is given. The catalog contains a main data file of 2970 sources and a supplemental file of 3176 sources measured at wavelengths of 4.2, 11, 20 and 27 μm.

002.040 **Documentation for the machine-readable version of the *AGK3 Star Catalogue of Positions and Proper Motions North of −2°.5 Declination* (Dieckvoss and Collaborators 1975).**
W. H. Warren Jr.
NSSDC/WDC-A-R&S, Publ., No. 84-06, 5+16 pp. (1984).
A detailed description of the machine-readable catalog as it is currently being distributed from the Astronomical Data Center is given. The version described is that originally received from the Astronomisches Rechen-Institut, Heidelberg, with a few minor corrections found when the file was examined at the Astronomical Data Center, NASA Goddard Space Flight Center.

002.041 **Documentation for the machine-readable version of the *ANS Ultraviolet Photometry Catalogue of Point Sources* (Wesselius et al. 1982).**
W. H. Warren Jr.
NSSDC/WDC-A-R&S, Publ., No. 84-10, 5+12 pp. (1984).
The machine-readable version of the catalog is described in detail, with a byte-by-byte format description and characteristics of the data file given. The catalog is a compilation of ultraviolet photometry in five bands, within the wavelength range 155 nm to 330 nm, for 3573 mostly stellar objects. Additional cross reference data (object identifications, *UBV* photometry and MK spectral types) are included in the catalog.

002.042 **Documentation for the machine-readable version of *A Library of Stellar Spectra* (Jacoby, Hunter and Christian 1984).**
W. H. Warren Jr.
NSSDC/WDC-A-R&S, Publ., No. 84-11, 5+8 pp. (1984).
A detailed description of the machine-readable library as it is currently being distributed from the Astronomical Data Center is given. The library contains digital spectra for 161 stars of spectral classes O through M and luminosity classes I, III, and V in the wavelength range 3510 Å to 7427 Å. The documentation includes a format description, a table of the indigenous characteristics of the magnetic tape file, and a sample listing of logical records exactly as they are recorded on the tape.

002.043 **Compiled catalogue of positions and proper motions of 4949 geodetic stars from +90° to −90°. Part I.**
E. V. Khrutskaya.
Glav. Astron. Obs. Akad. Nauk SSSR. Leningrad, 152 pp. (1984). In Russian. Abstr. in Ref. Zh., 51. Astron., 11.51.90 (1984).

002.044 **Compiled catalogue of positions and proper motions of 4949 geodetic stars from +90° to −90°. Part II.**
E. V. Khrutskaya.
Glav. Astron. Obs. Akad. Nauk SSSR. Leningrad, 124 pp. (1984). In Russian. Abstr. in Ref. Zh., 51. Astron., 11.51.91 (1984).

002.045 **International register of stellar spectroscopists. Edition 3.**
W. K. Bonsack, A. Slettebak (Editors).
W. K. Bonsack, Institute for Astronomy, 2680 Woodlawn Drive, Honolulu, Hawaii 96822, USA. 34 pp. (1984).

002.046 **A bibliography of planetary geology and geophysics principal investigators and their associates, 1983 – 1984.**
N. E. Witbeck (Editor).
NASA Tech. Memo, NASA TM-87361, 3+53 pp. (1984).
The bibliography serves as a companion piece to NASA TM-86246 "Reports of Planetary Programs – 1983", see Abstr. 37.003.018.

002.047 **Faint blue objects at high galactic latitude. IV. Palomar Schmidt fields centered on Selected Areas 55 and 94.**
K.-l. Huang, P. D. Usher.
Astrophys. J., Suppl. Ser., Vol. 56, No. 3, p. 393 – 401 (1984). With plates 3 – 16.
Ultraviolet excess objects have been selected from Palomar Schmidt fields centered on Selected Areas 55 and 94. The authors list color classes, subclasses, and approximate positions and magnitudes of the selected objects.

002.048 **The HEAO A-1 X-ray source catalog.**
K. S. Wood, J. F. Meekins, D. J. Yentis, H. W. Smathers, D. P. McNutt, R. D. Bleach, E. T. Byram, T. A. Chubb, H. Friedman.
Astrophys. J., Suppl. Ser., Vol. 56, No. 4, p. 507 – 649 (1984). With plate 20.
A catalog of sources observed by the NRL Large Area Sky Survey Experiment on HEAO 1 (HEAO A-1) is presented. Positions and intensities are given for 842 X-ray sources in the energy range 0.25 – 25 keV, with a limiting flux of 250 nJy at 5 keV for

a Crab–like spectrum, or $\sim$0.25 UFU. About one–half of the sources remain unidentified.

002.049 **Compiling a photographic atlas–catalogue of reference stars from observations with the wide–angle astrograph.**
D. N. Ponomarev, Yu. A. Shokin.
Soobshch. Spets. Astrofiz. Obs., Vyp. 40, p. 27 – 31 (1983). In Russian. – See Abstr. 012.076.

Methods of making an atlas–catalogue of reference stars in the declination range $-8° < \delta < -3°$ with a limiting magnitude of $15^m$ are described.

002.050 **Composantes de systèmes doubles ou multiples de l'Index (1976.5) ayant un même No. BD mais figurant séparément dans l'AGK2/3.**
O. Nys.
Bull. Inf. Cent. Données Stellaires, No. 27, p. 63 – 90 (1984).

Components of double and multiple stars of the Index catalogue (1976.5) having the same BD number but existing separately in the AGK2/3 catalogue.

002.051 **A catalogue of classical (evolved) Algol–type binary candidate stars.**
E. Budding.
Bull. Inf. Cent. Données Stellaires, No. 27, p. 91 – 129 (1984).

A catalogue of some 414 classical Algol–type binary candidates is presented, together with a brief background review and remarks on what can be gathered from the listed parameter values. Apart from the entries on numerous well–known and studied Algols, the catalogue draws attention to a large number of probably similar but generally less well–known stars, as well as others which may have been cited as possible Algols, though for which the overall evidence appears weaker. Some problems surrounding the classification and interpretation of candidates which do not conform to the 'classical' picture too well are discussed.

002.052 **An up–to–date catalogue of the Vilnius UPXYZVTS photometry.**
P. North.
Bull. Inf. Cent. Données Stellaires, No. 27, p. 133 – 134 (1984).

This catalogue on magnetic tape contains the data in the UPXYZVTS system of Vilnius Observatory published up to mid–1984. There are 2848 published data for 2596 stars.

002.053 **A catalogue of equivalent widths of the interstellar 2200 Å band.**
J. Dorschner, C. Friedemann, J. Gürtler, R. Schielicke.
Bull. Inf. Cent. Données Stellaires, No. 27, p. 137 – 139 (1984).

Nous présentons ici un catalogue contenant la largeur équivalente de la bande interstellaire 2200 Å pour 709 étoiles O et B rougies différemment. Le matériel d'observation a été fourni, pour l'essentiel, par le satellite TD–1.

002.054 **Preliminary catalogue of photographic UBV photometry for stars in open clusters.**
J.–C. Mermilliod.
Bull. Inf. Cent. Données Stellaires, No. 27, p. 141 – 143 (1984).

This catalogue collects published UBV photographic photometry for stars in the field of open clusters. The first version contains 42,575 entries for more than 34,000 stars in 127 open clusters, representing about 75 percent of the published data. The main effort has been devoted to the work on cross–identification, in order to bring the various identifications existing for each cluster into the uniform numbering system already adopted for other catalogues.

002.055 **IUE meeting bibliography for peculiar stars.**
A. Heck.
Bull. Inf. Cent. Données Stellaires, No. 27, p. 145 – 157 (1984).

002.056 **Errata to the Bright Star Catalogue.**
W. H. Warren Jr.
Bull. Inf. Cent. Données Stellaires, No. 27, p. 159 (1984).

002.057 **Additions and corrections to the Bright Star Catalogue.**
D. Hoffleit.
Bull. Inf. Cent. Données Stellaires, No. 27, p. 161 – 195 (1984).

002.058 **News from SIMBAD.**
F. Ochsenbein.
Bull. Inf. Cent. Données Stellaires, No. 27, p. 203 – 207 (1984).

Recent improvements of the astronomical contents of SIMBAD since the last issue of this Bulletin are summarized in this paper.

002.059 **Catalogs currently available on microfiche.**
Bull. Inf. Cent. Données Stellaires, No. 27, p. 209 – 215 (1984).

002.060 **Catalogue of observations of total occultations of stars by the Moon for the years 1972 to 1980 and of grazing occultations for the years 1963 to 1980.**
G. M. Appleby, L. V. Morrison, M. T. White.
R. Greenwich Obs., Bull., No. 192, 12 pp. (1984). With 11 microfiches.

002.061 **Astronomy and Astrophysics Abstracts. Vol. 37. Literature 1984, Part 1.**
S. Böhme, W. Fricke, H. Hefele, I. Heinrich, W. Hofmann, D. Krahn, V. R. Matas, L. D. Schmadel, G. Zech (Editors).
Published for Astronomisches Rechen–Institut by Springer–Verlag, Berlin – Heidelberg – New York – Tokyo. 10+937 pp. Price DM 189.00, US$ 68.50 [Subscription Price DM 151.20, US$ 54.80]. ISBN 3–540–13937–0, ISBN 0–387–13937–0 (USA).

002.062 **Finding charts for field horizontal–branch A–type stars.**
A. G. D. Philip.
Contrib. Van Vleck Obs., No. 2, p. 1 – 17 (1984).

002.063 **Identification charts for F2 – G5 stars in a North Galactic Pole region.**
A. R. Upgren.
Contrib. Van Vleck Obs., No. 2, p. 19 – 26 (1984).

002.064 **Identification charts for selected stars of color class M in the NLTT catalogue.**
E. W. Weis.
Contrib. Van Vleck Obs., No. 2, p. 27 – 31 (1984).

002.065 **A search for southern M–type stars.**
P. K. Lu, D. J. MacConnell.
Contrib. Van Vleck Obs., No. 2, p. 33 – 49 (1984).

002.066 **A list of marginal barium stars.**
P. C. Keenan.
Bull. Etoiles Tardives Spectre Particulier, No. 1, p. 5 (1984).

002.067 **Bibliography U.R.S.S.**
A. Alksnis, H. R. Johnson.
Bull. Etoiles Tardives Spectre Particulier, No. 1, p. 9 – 11 (1984).

Bibliography of papers on carbon stars published 1983/84.

002.068 **Bibliography.**
D. Ballereau, A. Feinstein, A. M. Hubert–Delplace, H. Hubert, J. Jugaku, P. Koubsky, E. Mendoza, A. Slettebak.
Bull. Inf. Etoiles Be, No. 10, p. 69 – 72 (1984).

Bibliography of papers on Be stars published 1983/84.

002.069 **The 105 cm Schmidt plates (Nos. 3951 – 4250).**
Tokyo Astron. Obs., Kiso Inf. Bull., Vol. 2, No. 1, p. 1 – 7 (1984).

Fundamental data of the plates taken with the Kiso 105 cm Schmidt telescope are given.

002.070 **List of dissertations made at the Sternberg Astronomical Institute in 1983.**
L. N. Bondarenko.
Astron. Tsirk., No. 1310, p. 7 – 8 (1984). In Russian.

002.071 **Catalog of cluster C–M diagrams.**
R. E. White, A. G. D. Philip.
Prepr. Steward Obs., No. 533, 21 pp. (1984).
The cited references are believed to be a complete compilation through the latest published volume of the Astronomy and Astrophysics Abstracts, i.e. Vol. 33, which contains the citations for literature through the first–half of 1983.

002.072 **Triennial Report concerning the galactic globular star clusters.**
R. E. White.
Prepr. Steward Obs., No. 562, 25 pp. (1984).
The interval covered by the literature search reported here has been limited to 1 Januray 1981 to 30 June 1984; these dates include Vol. 29 – 34 of the "Astronomy and Astrophysics Abstracts". In this edition of the "Triennial Report" is found a noticeably greater abundance of information concerning clusters in galaxies other than our own Milky Way. Obviously, the impact of the high quantum–efficiency solid state detectors coupled to large aperture telescopes is being felt.

002.073 **Fourth Catalog of Orbits of Visual Binary Stars.**
C. E. Worley, W. D. Heintz.
Publ. U.S. Nav. Obs., Second Ser., Vol. 24, Part 7, 53 pp. (1983).
This catalog continues the series of catalogs of visual binary orbits previously published by W. S. Finsen (1934, 1938), C. E. Worley (1963), and jointly by Finsen and Worley (1970). The combination of the independent resources and opinions of two authors results in a more complete, accurate, and consistent catalog.

002.074 **Gesamtverzeichnis der deutschsprachigen astronomischen Hochschulschriften. Erster Band, Dissertationen bis 1983.**
M. Gossler.
Bibliographische Informationen 24, Universitätsbibliothek Graz, Graz, Austria. 17 + 192 pp. (1984).

002.075 **Catalogue of Metal–Deficient F – M stars. Part 1. Stars Classified Spectroscopically. Supplement 1.**
A. Bartkevičius.
Vilniaus Astron. Obs. Biul., Nr. 68, p. 3 – 32 (1984).
The first supplement (MDSPS1) to the Catalogue of Metal–Deficient F – M Stars Classified Spectroscopically (MDSP) (1980) includes 289 new metal–deficient stars selected from the literature sources published in 1977 – 1982. Criteria for the selection of stars are the same as in the MDSP Catalogue. The list and data of the new metal–deficient F – M stars are given in a table.

002.076 **The Second Byurakan Spectral Sky Survey. III. Results of the area centered at $\alpha = 08^h00^m$, $\delta = +59°00'$.**
B. E. Markarian, D. A. Stepanyan.
Astrofizika, Tom 20, Vyp. 3, p. 513 – 524 (1984). In Russian. English translation in Astrophysics, Vol. 20, No. 3.
The third list of objects of the Second Byurakan Survey in the area centred at $\alpha = 08^h00^m$, $\delta = +59°00'$ is presented. This list contains data on 94 objects and galaxies and 12 blue stars peculiar in their physical nature. In the given area six candidates for QSOs, 13 BSOs, 21 galaxies with significant UV continuum and 54 emission galaxies without significant UV continuum were found.

002.077 **Postwar growth in the length of astronomical and other scientific papers.**
V. Trimble.
Publ. Astron. Soc. Pac., Vol. 96, No. 586, p. 1007 – 1016 (1984).

002.078 **Cambridge Fotoatlas der Planeten.**
G. Briggs, F. Taylor.
Franckh'sche Verlagshandlung, Stuttgart, F.R. Germany. 255 pp. Price DM 84.00 (1984). For the English original see 32.002.144.
Review in Astron. Tidsskr., Årg. 17, Nr. 3, p. 141 (1984).

002.079 **Deep sky catalog and star atlas.**
R. G. Coutchie.
R. G. Coutchie, 22018 Ybarra Rd., Woodland Hills, Calif. 91364, USA. 17 pp. Price $ 7.50.
Review in Sky Telesc., Vol. 68, No. 2, p. 136 (1984).

002.080 **A bibliography of James Ferguson.**
J. R. Milburn.
Published by the author, 58 Bierton Road, Aylesbury, Bucks, England. 29 pp. + 2 microfiches. Price £ 4.75 (1983).
Review in J. Hist. Astron., Vol. 15, Part 3, No. 44, p. 218 – 219; 1984 (*M. Hoskin*).

002.081 **CRC handbook of high resolution infrared laboratory spectra of atmospheric interest.**
D. G. Murcray, A. Goldman (Editors).
CRC Press, Inc., Boca Raton, Florida, USA. 282 pp. Price $ 49.95 (1981). ISBN 0–8493–2950–7.
Review in Astrophys. Lett., Vol. 24, No. 2, p. 117; 1984 (*W. A. Traub*).

002.082 **Check a possible supernova (*Forty photographs of galaxies*).**
J. Salmi.
Juhani Salmi, Vesijar venkatu 36 C 40, SF–15140 Lahti 14, Finland. Obtainable from The Webb Society, 9 Fairacres Close, Herne Bay, Kent, CT6 6LL, England. Price £ 3.65, $ 4.80 (1984).
Reviews in J. Br. Astron. Assoc., Vol. 95, No. 1, p. 41 – 42; 1984 (*K. G. Jones*); Sky Telesc., Vol. 68, No. 3, p. 230 (1984); Strolling Astron., Vol. 30, Nos. 11 – 12, p. 254; 1984 (*W. S. Houston*).

002.083 **Statni Hvezdarna Inventar Archivniho Fondu. (Inventory of the Archive Group "State Astronomical Observatory").**
A. Slechtova.
Ustredni Archiv CSAV (Central Archive of Czechoslovak Academy of Sciences), Prague, Czechoslovakia. 22 + 202 pp. (1981).
Review in J. Hist. Astron., Vol. 15, Part 3, No. 44, p. 223 (1984).

002.084 **Catalog of 3539 zodiacal stars for the equinox 1950.0.**
Nautical Almanac Office. Arcturus Press, 95 Walton Rd., Sidcup, Kent DA14 4LL, England. 125 pp. Price $ 13.00 (1984). ISBN 0–907322–36–0.
Review in Sky Telesc., Vol. 68, No. 3, p. 230 (1984).

**Catalogue des étoiles cibles utilisées pour l'étude spectroscopique du milieu interstellaire diffus.**
See Abstr. 013.071.

**The Astrographic Catalogue and the Carte du Ciel.**
See Abstr. 013.072.

**Compiling technique of special reference catalogues of star positions and its application to comet Halley watch.**
See Abstr. 036.080.

**Comprehensive star catalogues.**
See Abstr. 036.151.

**Compilation du Catalogue d'Entrée. Programmes astrométriques sur Hipparcos.**
See Abstr. 041.015.

**Catalogue of declinations of 254 stars in the neighbourhood of 72 extragalactic radio sources.**
See Abstr. 041.020.

**Catalogue of right ascensions of 109 stars in the system of the compiled catalogue of the USSR Time Service (CTS).**
See Abstr. 041.029.

**Catalogue of right ascensions of 376 stars from the FK4 with declinations from −20° to +90° obtained from observations with the Tashkent meridian circle in 1976 – 1978, and systematic and random differences TMK–FK4 for these stars.**
See Abstr. 041.030.

**Systematic errors of Belgrade NPZT programme star catalogue.**
See Abstr. 041.032.

**Zum astrometrischen Nutzen von PZT–Katalogen.**
See Abstr. 041.045.

**Ein Katalog ausgewählter kompakter Radioquellen für die Konstruktion eines extragalaktischen radio/optischen Referenzsystems. I. Einführung und optische Daten. II. Radiodaten.**
See Abstr. 043.008.

**Der CPC2 Katalog, ein photographisches Referenzsternsystem hoher Genauigkeit am Südhimmel.**
See Abstr. 043.009.

**Ein photographischer PZT–Referenzsternkatalog für den Deklinationsgürtel +52°5.**
See Abstr. 043.010.

**Cartes synoptiques de la chromosphère solaire et catalogues des filaments et des centres d'activité.**
See Abstr. 073.089.

**A solar spectral line list between 10 and 200 Å modified for application to high spectral resolution X–ray astronomy.**
See Abstr. 076.011.

**Inventory of the Vatican Meteorite Collection.**
See Abstr. 105.081.

**Determination of relative proper motions and the membership of the open cluster NGC 1817.**
See Abstr. 111.015.

**Absolute proper motions of 774 stars with respect to 10 galaxies in 4 areas of the southern sky.**
See Abstr. 111.016.

**Verification of the calibration of Harold L. Johnson's 13–color photometry.**
See Abstr. 113.031.

**A search for ultraviolet–excess objects. II.**
See Abstr. 113.055.

**IUE low–dispersion spectra flux catalogue. Part 1. Normal stars (magnetic tape).**
See Abstr. 114.002.

**Investigation of the linearity of the Alma–Ata spectrophotometric star catalogue.**
See Abstr. 114.115.

***UBVRI* catalogue of dwarf novae.**
See Abstr. 117.020.

**Ein Atlas und Katalog nördlicher Zwergnovae.**
See Abstr. 117.245.

**Les aventures de Mira Ceti.**
See Abstr. 122.032.

**A spectral catalogue of southern hemisphere Mira variable stars.**
See Abstr. 122.073.

**Ein Katalog und Atlas galaktischer Novae.**
See Abstr. 124.006.

**Statistical studies of supernova remnants.**
See Abstr. 125.011.

**A new look at radio supernova remnants.**
See Abstr. 125.017.

**Dark cloud and globule distribution for galactic longitudes 230 to 360 degrees.**
See Abstr. 131.269.

**IUE UV spectra of extragalactic H II regions. I. The catalogue and the atlas.**
See Abstr. 132.006.

**A catalogue of expansion velocities in planetary nebulae.**
See Abstr. 134.043.

**A radio continuum survey of the galactic plane at 11 cm wavelength. I. The area 357°4 $\leqslant l \leqslant$ 76°, −1°5 $\leqslant b \leqslant$ 1°5.**
See Abstr. 155.057.

**The identification of HDE objects with Large Magellanic Cloud clusters and nebulae.**
See Abstr. 156.014.

**The Horologium region. I. The data.**
See Abstr. 160.003.

**A list of nearby groups of galaxies.**
See Abstr. 160.012.

**Morphological classification (revised RS) of Abell clusters in $D = 5$ and an analysis of observed correlations.**
See Abstr. 160.039.

# 003 Books

003.001 **Methods in radiative transfer.**
W. Kalkofen (Editor).
Cambridge University Press, Cambridge – London – New York – New Rochelle – Melbourne – Sydney. 8+450 pp. Price £ 25.00, US$ 49.50 (1984). ISBN 0-521-25620-8.
The individual contributions are included in their corresponding subject categories – see abstracts 063.001 – 063.015.

003.002 **Advances in photoelectric photometry, Vol. 2.**
R. C. Wolpert, R. M. Genet (Editors).
Fairborn Observatory, 1247 Folk Road, Fairborn, Ohio 45324, USA. 4+185 pp. (1984). ISBN 0-911351-05-1.
The individual contributions are included in their corresponding subject categories – see abstracts 009.006 – 009.008, 013.004 – 013.007, 021.003, 032.004, 082.017, 119.016.

003.003 **Reversals of the Earth's magnetic field.**
J. A. Jacobs.
Adam Hilger Ltd., Bristol, England; Accord, Mass., USA. 9+230 pp. Price £ 19.95, US$ 35.00 (1984). ISBN 0-85274-442-0.
Review in Nature, Vol. 310, No. 5976, p. 436; 1984 (*D. H. Tarling*).
Contents: The Earth's magnetic field. The magnetisation of rocks. The morphology of geomagnetic reversals. Geomagnetic excursions. Models for reversals. Magnetostratigraphy. The Earth's magnetic field and climate.

003.004 **Astrophysical techniques.**
C. R. Kitchin.
Adam Hilger Ltd., Bristol, England. 12+438 pp. Price £ 40.00 cloth, £ 15.00 paper (1984). ISBN 0-85274-461-7 cloth, ISBN 0-85274-484-6 paper.
Reviews in Astron. Express, Vol. 1, Issue 2, p. 81 (1984); Spaceflight, Vol. 26, No. 12, p. 476 (1984).
Contents: 1. Detectors: optical and infrared detection, radio and microwave detection, X-ray and gamma-ray detection, cosmic ray detectors, neutrino detectors, gravitational radiation. 2. Imaging: photography, electronic imaging, scanning, interferometry, speckle interferometry, occultations, radar. 3. Photometry: photometry, photometers. 4. Spectroscopy: spectroscopy, spectroscopes. 5. Other techniques: astrometry, polarimetry, solar studies, magnetometry. Appendices I – IX.

003.005 **Quantum theory of gravity. Essays in honor of the 60th birthday of Bryce S. DeWitt.**
S. M. Christensen (Editor).
Adam Hilger Ltd., Bristol, England. 16+483 pp. Price £ 30.00 (1984). ISBN 0-85274-755-1.
Reviews in Nature, Vol. 310, No. 5980, p. 804; 1984 (*P. Townsend*); Observatory, Vol. 104, No. 1062, p. 238; 1984 (G. Gibbons).
The individual contributions are included in their corresponding subject categories – see abstracts 061.027 – 061.035, 066.122 – 066.140, 067.035 – 067.037, 161.070, 161.071.

003.006 **Essays in theoretical physics. In honour of Dirk ter Haar.**
W. E. Parry (Editor).
Pergamon Press, Oxford – New York – Toronto – Sydney – Paris – Frankfurt. 8+337 pp. Price US$ 45.00 (1984).
The individual contributions within the subject scope of Astronomy and Astrophysics Abstracts are included in their corresponding categories – see abstracts 067.051, 067.052, 080.012, 144.013, 151.018.

003.007 **Spectrophotometric investigations of quiescent solar prominences.**
N. N. Morozhenko; edited by K. V. Alikaeva.
Akademiya Nauk Ukrainskoj SSR, Glavnaya Astronomicheskaya Observatoriya. Naukova Dumka, Kiev. 164 pp. Price 1 Rbl. 40 Kop. (1984). In Russian.
Review in Ref. Zh., 51. Astron., 10.51.69 (1984).

003.008 **Physics of gravitating systems. I. Equilibrium and stability.**
A. M. Fridman, V. L. Polyachenko, translated by A. B. Aries, I. N. Poliakoff.
Springer Verlag, New York – Berlin – Heidelberg – Tokyo. 17+468 pp. Price $ 97.30, DM 248.00 (1984). ISBN 0-387-11045-3 (USA), ISBN 3-540-11045-3 (F.R. Germany).
Review in Nature, Vol. 313, No. 6002, p. 507; 1985 (*D. Lynden-Bell*).
Contents: Introduction. Theory: Equilibrium and stability of a nonrotating flat gravitating layer. Equilibrium and stability of a collisionless cylinder. Equilibrium and stability of collisionless spherically symmetrical systems. Equilibrium and stability of collisionless ellipsoidal systems. Equilibrium and stability of flat gravitating systems.

003.009 **Physics of gravitating systems. II. Nonlinear collective processes: nonlinear waves, solitons, collisionless shocks, turbulence. Astrophysical applications.**
A. M. Fridman, V. L. Polyachenko, translated by A. B. Aries, I. N. Poliakoff.
Springer-Verlag, New York – Berlin – Heidelberg – Tokyo. 11+358 pp. Price $ 85.00, DM 218.00 (1984). ISBN 0-387-13103-5 (USA), ISBN 3-540-13103-5 (F.R. Germany).
Review in Nature, Vol. 313, No. 6002, p. 507; 1985 (*D. Lynden-Bell*).
Contents: Theory: Non-Jeans instabilities of gravitating systems. Problems of nonlinear theory. Astrophysical applications: General remarks. Spherical systems. Ellipsoidal systems. Disk-like systems. Spiral structure. Other applications. Appendix.

003.010 **Mathematical astronomy in Copernicus's De revolutionibus. Parts 1+2.**
N. M. Swerdlow, O. Neugebauer.
Studies in the History of Mathematical and Physical Sciences, Vol. 10. Springer-Verlag, New York – Berlin – Heidelberg – Tokyo. 16+537 pp (Part 1), 5+172 pp. (Part 2). Price $ 93.30, DM 238.00 (Parts 1+2). ISBN 0-387-90939-7 (USA), ISBN 3-540-90939-7 (F. R. Germany).
Contents of Part 1: General introduction. Trigonometry and spherical astronomy. The motions of the earth. Lunar theory and related subjects. Planetary theory of longitude. Planetary theory of latitude.
Contents of Part 2: Notation and symbols. Parameters. Tables. Figures. Bibliographical abbreviations. Subject index.

003.011 **Colours of the stars.**
D. Malin, P. Murdin.
Cambridge University Press, Cambridge – London – New York – New Rochelle – Melbourne – Sydney. 9+198 pp. Price £ 13.95, $ 27.95 (1984). ISBN 0-521-25714-X.
Reviews in Astron. Express, Vol. 1, Issue 3, p. 112 (1984); Nature, Vol. 313, No. 6002, p. 508; 1985 (*D. W. Hughes*); Spaceflight, Vol. 26, No. 11, p. 430 (1984).
Contents: Colours of the stars. Photography and colour. Dust clouds of Sagittarius. Changing colours of the stars. Nebulae from stars. Stars from nebulae. Starlight and dust. Colours of galaxies. Red-shift, blue-shift. Appendix I: Image manipulation. Appendix II: Additive colour photography.

003.012 **The early years of radio astronomy. Reflections fifty years after Jansky's discovery.**
W. T. Sullivan III (Editor).
Cambridge University Press, Cambridge – London – New York – New Rochelle – Melbourne – Sydney. 10 + 421 pp. Price £ 25.00 (1984). ISBN 0–521–25485–X. – See also 37.003.155.
The individual contributions are included in their corresponding subject categories – see abstracts 004.013 – 004.032, 005.009.

003.013 **Secrets of the sun.**
R. G. Giovanelli.
Cambridge University Press, Cambridge – London – New York – New Rochelle – Melbourne – Sydney. 11 + 116 pp. Price £ 11.95, US$ 19.95 (1984). ISBN 0–521–25521–X.
Reviews in Astron. Express, Vol. 1, Issue 2, p. 81 (1984); Astron. Tidsskr., Årg. 17, Nr. 4, p. 192 (1984); Nature, Vol. 313, No. 6002, p. 508; 1985 (*D. W. Hughes*); Spaceflight, Vol. 26, No. 11, p. 432 (1984).
Contents: The sun as we see it. Inside the sun. The sun's curious rotation. Gases in motion. Those notorious sunspots. Magnetic fields outside sunspots and what they do. The sunspot cycle – what and why. Coming out from the sun – first the chromosphere. Coming out from the sun – the corona and beyond. Beauty and action – prominences. Action and beauty – flares. Sunspots and the weather.

003.014 **Calculation of deformations of mechanical knots of a telescope.**
P. V. Shcheglov (Editor).
Izdatel'stvo Moskovskogo universiteta, Moskva. 134 pp. Price 35 Kop. (1983). In Russian.
The individual contributions are included in their corresponding subject categories – see abstracts 032.012 – 032.017.

003.015 **Physics of planetary interiors.**
G. H. A. Cole.
Adam Hilger Ltd, Bristol, England. 14 + 208 pp. Price £ 22.00, $ 39.00 cloth, £ 9.95, $ 16.00 paper (1984). ISBN 0–85274–444–7 cloth, ISBN 0–85274–445–5 paper.
Contents: Some observational data. The planetary body. Consequences of internal pressure: hydrostatic equilibrium. Observed shape and internal mass distribution. Thermal effects and thermodynamics. Magnetic fields of internal origin. Planet of large mass. Terrestrial-type planet. Planetary bodies composed mainly of ice.

003.016 **Die Physik der Zeitrichtung.**
H. D. Zeh.
Lecture notes in physics, Vol. 200. Springer–Verlag, Berlin–Heidelberg–New York–Tokyo. 5 + 86 pp. Price DM 18.00, $ 6.60 (1984). ISBN 3–540–13336–4 (F. R. Germany). ISBN 0–387–13336–4 (USA).

003.017 **Die farbigen Dämmerungserscheinungen.**
K. Bullrich.
Birkhäuser Verlag, Basel – Boston – Stuttgart. 104 pp. Price DM 26.80 (1982). ISBN 3–7643–1355–2.
Contents: Einführung in die Grundlagen zur Erklärung der Himmelsfarben. Der vertikale Aufbau der Atmosphäre. Die spektrale Lichtstreuung und Lichtabsorption in der Atmosphäre. Die Molekülatmosphäre und das atmosphärische Aerosol. Das Erscheinungsbild des Ablaufs der Dämmerung. Die farbarme Dämmerung und die Leuchtdichteverteilung am Himmel nach Sonnenuntergang. Die leuchtenden Nachtwolken, das Nordlicht, das Nachthimmelslicht, das Zodiakallicht und die Perlmutterwolken.

003.018 **Atmospheric electrodynamics.**
H. Volland.
Physics and Chemistry in Space, Vol. 11. Springer–Verlag, Berlin – Heidelberg – New York – Tokyo. 9 + 205 pp. Price DM 98.00, $ 35.70 (1984). ISBN 3–540–13510–3 (F.R. Germany), ISBN 0–387–13510–3 (USA).
Contents: Introduction. Plasma component of the air. Thunderstorms and related phenomena: global electric circuit, thunderstorms, thunderstorm electrification, lightning, Sferics. Tidal wind interaction with ionospheric plasma: electric *Sq* and *L* currents, ionospheric dynamo. Solar wind interaction with magnetosphere: coupling between solar wind and geomagnetic field, global magnetospheric electric fields and currents. Theory of wave propagation within the magnetosphere. Waves in the magnetosphere. Appendix.

003.019 **Orbits for amateurs with a microcomputer.**
D. Tattersfield.
Stanley Thornes Ltd., Old Station Drive, Leckhampton, Cheltenham GL53 0DN, England; Halsted Press, 605 Third Avenue, New York, New York 10158, USA. 13 + 171 pp. Price £ 15.95, $ 29.95 (1984). ISBN 0–85950–146–9 (England); ISBN 0–470–27502–2 (USA).
Reviews in Mon. Notes Astron. Soc. S.Afr., Vol. 43, Nos. 11 – 12, p. 96 – 97; 1984 (*G. Roberts*); Observatory, Vol. 104, No. 1063, p. 282 – 283; 1984 (*P. Duffett-Smith*); Space Educ., Vol. 1, No. 8, p. 383 (1984); Zenit, 11. Jaarg., No. 9, p. 390; 1984 (*D. van den Oudenalder*).
Contents: Construction of an ephemeris for an orbiting comet. Determination of the elements of a comet orbit. Perturbations of a comet orbit. The differential correction of a comet orbit.

003.020 **Kosmologie heute. Eine Einführung in ihre philosophischen und naturwissenschaftlichen Problemkreise.**
J. Meurers.
Wissenschaftliche Buchgesellschaft Darmstadt, F.R.Germany. 12 + 339 pp. Price DM 81.00 (1984). ISBN 3–534–07020–8.

003.021 **Explicando astronáutica. O homem e a conquista do espaço.**
R. de Freitas Mourão.
Editora Tecnoprint Ltda., Rua da Proclamação, 109–Rio de Janeiro, RJ, Brazil. 148 pp. (1984).

003.022 **O universo inflacionário.**
R. R. de Freitas Mourão.
Livraria Francisco Alves Editora S/A, Rua Sete de Setembro, 177–Centro 20050 – Rio de Janeiro, RJ, Brazil. 218 pp.(1983).

003.023 **Veränderliche Sterne.**
C. Hoffmeister, G. Richter, W. Wenzel.
Zweite, völlig überarbeitete Auflage. Springer–Verlag, Berlin – Heidelberg – New York – Tokyo. 335 pp. Price DM 66.00 (1984). ISBN 3–540–13396–8 (F. R. Germany), ISBN 0–387–13396–8 (USA).
Contents: Allgemeine Hinweise. Pulsierende Veränderliche. Eruptive Veränderliche. Bedeckungssterne. Ergänzungen zur Typologie. Entdeckung Veränderlicher Sterne. Bedeutung der Veränderlichen Sterne für die Erforschung des Baus der Galaxis und der Sternentwicklung. Beobachtungsmethoden und Organisation. Literatur.

003.024 **Annual Review of Astronomy and Astrophysics, Vol. 22, 1984.**
G. Burbidge, D. Layzer, J. G. Phillips (Editors).
Annu. Rev. Astron. Astrophys., Vol. 22. 10 + 635 pp. Price US$ 44.00 (USA), US$ 47.00 (elsewhere) (1984). ISBN 0–8243–0922–7.
The individual contributions are included in their corresponding subject categories – see abstracts 005.017, 033.022, 036.052, 067.079, 074.034, 076.012, 080.076, 080.077, 107.023, 117.124, 117.125, 125.040, 144.031, 155.039, 157.111, 157.112, 158.113, 160.038, 161.226.

003.025 **Understanding relativity. Origin and impact of a scientific revolution.**
S. Goldberg.
Birkhäuser, Boston – Basel – Stuttgart. 17 + 494 pp. Price DM 72.00 (1984). ISBN 0–8176–3150–X.

003.026 **Gamma ray astronomy.**
R. Hillier.
Oxford Studies in Physics. Clarendon Press, Oxford, England. 8+202 pp. Price DM 79.80 (1984). ISBN 0-19-851451-4.
Contents: Introduction. The generation of gamma rays. The problem of background radiation in gamma-ray astronomy. Low energy gamma-ray telescopes. Compton telescopes. Spark chamber telescopes for high energy gamma rays. The detection of extensive air showers from very high energy gamma rays. Solar gamma rays. Gamma rays from the Galactic disc. Discrete gamma-ray sources in the Galaxy. Extragalactic gamma-ray sources and the diffuse background radiation.

003.027 **Grundlagen der Ephemeridenrechnung. Eine Zusammenstellung der wichtigsten Formeln und Daten zur Berechnung der Bahnen von Körpern im Sonnensystem.**
O. Montenbruck.
Sterne und Weltraum, Taschenbuch 10. Verlag Sterne und Weltraum Dr. Vehrenberg GmbH, Portiastraße 10, D-8000 München 90, F.R. Germany. 166 pp. Price DM 28.00 (1984). ISBN 3-87973-913-0.
Contents: Koordinatensysteme. Zeitrechnung. Das Zweikörperproblem. Das Mehrkörperproblem. Die Mondbahn. Physische Ephemeriden.

003.028 **Leben und Werk des Königsberger Astronomen Friedrich Wilhelm Bessel.**
F. Schmeidler.
Prussia-Schriftenreihe, Werk 5. ILMA-Verlag, Kelkheim/T, F.R. Germany. 88 pp. Price DM 29.80 (1984). ISBN 3-9800-506-1-0.

003.029 **Schau mal in die Sterne. Himmelsbeobachtungen mit bloßem Auge und dem Feldstecher.**
H. Oberndorfer.
Kosmos. Gesellschaft der Naturfreunde. Franckh'sche Verlagshandlung, Stuttgart, F.R. Germany. 184 pp. Price DM 19.80 (1984). ISBN 3-440-05404-7.

003.030 **Astrophysik. Eine Einführung.**
H. Sautter.
2., neubearbeitete Auflage. Uni-Taschenbücher 1188. Gustav Fischer Verlag, Stuttgart, F.R. Germany. 8+243 pp. Price DM 26.80 (1984). ISBN 3-437-20278-2.

003.031 **Cold plasma waves.**
H. G. Booker.
Developments in Electromagnetic Theory and Applications, Vol. 2. Martinus Nijhoff Publishers, Dordrecht - Boston - Lancaster. 15+349 pp. Price Dfl. 165.00, US$ 59.50, £ 41.95 (1984). ISBN 90-247-2977-7.
Contents: Elementary properties of a plasma. Maxwell's equations. Isotropic plasma. Alternating current in a magnetoplasma. General properties of phase propagation in a magnetoplasma. General properties of group propagation in a magnetoplasma. Propagation of phase along the imposed magnetic field. Energy flow and group velocity for longitudinal propagation. Propagation of phase transverse to the imposed magnetic field. Elliptic polarization of the X wave for transverse propagation. Energy behaviour of the X wave for transverse propagation. Propagation at any angle to the imposed magnetic field. The radio approximation. The hydromagnetic approximation. The quasi-longitudinal and quasi-transverse approximations. Directional behaviour of group velocity in a magnetoplasma. The field of an antenna in a magnetoplasma. Directional behaviour of the power radiated by a dipole in a magnetoplasma.

003.032 **Relativity and engineering.**
J. Van Bladel.
Springer Series in Electrophysics, Vol. 15. Springer-Verlag, Berlin - Heidelberg - New York - Tokyo. 12+402 pp. Price DM 80.00 (1984). ISBN 3-540-12561-2 (F.R. Germany), ISBN 0-387-12561-2 (USA).
Contents: Kinematics in inertial axes. Dynamics in inertial axes. Vacuum electrodynamics in inertial axes. Fields in media in uniform translation. Boundary-value problems for media in uniform translation. Electromagnetic forces and energy. Accelerated systems of reference. Gravitation. Maxwell's equations in a gravitational field. Electromagnetism of accelerated bodies. Field problems in a gravitational field.

003.033 **Interstellar chemistry.**
W. W. Duley, D. A. Williams.
Academic Press, London - Orlando - San Diego - San Francisco - New York - Toronto - Montreal - Sydney - Tokyo - São Paulo. 8+251 pp. Price $ 45.00 (1984). ISBN 0-12-223360-3.
Contents: Introduction. Introduction: chemistry. Gas phase reactions. Photophysics of interstellar molecules. Grain surface chemistry. The molecular hydrogen problem. Modelling the chemistry in interstellar clouds. Interstellar chemistry in special situations. The present state and future prospects of interstellar chemistry.

003.034 **General relativity and relativistic astrophysics.**
N. Straumann.
Springer-Verlag, Berlin - Heidelberg - New York - Tokyo. Extended and revised edition. 13+459 pp. Price DM 112.00, $ 40.80 (1984). ISBN 3-540-13010-1 (F.R. Germany), ISBN 0-387-13010-1 (USA). German edition see 31.003.124.
Contents: Part I. Differential geometry: Differentiable manifolds. Tangent vectors, vector and tensor fields. The Lie derivative. Differential forms. Affine connections. Part II. General theory of relativity: The principle of equivalence. Einstein's field equations. The Schwarzschild solution and classical tests of general relativity. Weak gravitational fields. The post-Newtonian approximation. Part III. Relativistic astrophysics: Neutron stars. Rotating black holes. Binary X-ray sources. Accretion onto black holes and neutron stars.

003.035 **Handbuch der Physik. Band XLIX/7, Geophysik III, Teil VII.**
K. Rawer (Editor).
Springer-Verlag, Berlin - Heidelberg - New York - Tokyo. 9+544 pp. Price DM 448.00 [Subscription Price DM 358.40] (1984). ISBN 3-540-11425-4 (F.R. Germany), ISBN 0-387-11425-4 (USA).
The individual contributions are included in their corresponding subject categories - see abstracts 062.146, 076.018, 082.078.

003.036 **Planeten, Monde, Ringsysteme: Kamerasonden erforschen unser Sonnensystem.**
W. Engelhardt.
Birkhäuser Verlag, Basel - Boston - Stuttgart. 331 pp. Price DM 88.00 (1984). ISBN 3-7643-1618-7.
Contents: Prolog. Das Planetensystem wird entdeckt. Techniken der Planetenforschung. Merkur. Mariner 10. Venus. Pioneer-Venus. Mars. Mariner 4. Mariner 6 und 7. Mariner 9. Das Viking-Programm. Jupiter. Projekt Galileo. Saturn - der Ringplanet. Pioneer Jupiter/Saturn. Uranus. Das Voyager-Programm. Neptun. Pluto. Space Telescope. Kometen. Giotto.

003.037 **Meteoritic and meteor investigations.**
Yu. A. Dolgov (Editor).
Nauka, Novosibirsk. 240 pp. (1983).
Review in Ref. Zh., 51. Astron., 12.51.59 (1984).
See Abstr. 105.106.

003.038 **Solar activity investigations and the Prognoz space system. 1975 - 1978.**
R. Z. Sagdeev (Editor).
Nauka, Moskva. 237 pp (1984). In Russian.
Review in Ref. Zh., 62. Issled. Kosm. Prostranstva, 12.62.104 (1984).
The individual contributions are included in their corresponding subject categories - see abstracts 074.059, 074.060, 078.019, 085.036, 106.054 - 106.056, 143.042, 144.076.

003.039 **Geodynamics Branch Research Report – 1983.**
W. D. Kahn, S. C. Cohen (Editors).
NASA Tech. Memo., NASA TM-86123, 5+114 pp. (1984).
The individual contributions within the subject scope of Astronomy and Astrophysics Abstracts are included in their corresponding categories – see abstracts 044.069, 044.070, 052.042, 052.043, 081.055, 081.056.

003.040 **Das Sonnensystem.**
J. A. Wood, übersetzt und ergänzt von K. Hiller.
Geowissen kompakt, Band 7. Ferdinand Enke Verlag, PO Box 1304, D-7000 Stuttgart 1, F.R. Germany. 4+247 pp. Price DM 24.80 (1984). ISBN 3-432-94441-1. English edition see 25.003.170.

003.041 **Das intelligente Universum. Eine neue Sicht von Entstehung und Evolution.**
F. Hoyle.
Umschau Verlag, Frankfurt, F.R. Germany. 256 pp. Price DM 39.80 (1984). ISBN 3-524-69052-1. English edition see 37.003.099.

003.042 **Meyers Handbuch Weltall.**
K. Schaifers, G. Traving.
6., völlig neu bearbeitete Auflage. Bibliographisches Institut, Mannheim – Wien – Zürich. 656 pp. Price DM 68.00 (1984). ISBN 3-411-02155-1.

003.043 **Planetary interiors.**
W. B. Hubbard.
Van Nostrand Reinhold Company, New York – Wokingham – Melbourne – Agincourt. 9+334 pp. Price US$ 42.50, DM 195.50 (1984). ISBN 0-442-23704-9.
Review in Sky Telesc., Vol. 68, No. 6, p. 529 (1984).

003.044 **Urknall und Evolution des Kosmos – Fortschritte in der Kosmologie.**
W. Priester.
Rheinisch-Westfälische Akademie der Wissenschaften, Vorträge N 333. Westdeutscher Verlag, Opladen, F.R. Germany. 85 pp. Price DM 25.00 (1984). ISBN 3-531-08333-3.

003.045 **Magnetic fields in astrophysics.**
Ya. B. Zeldovich (*Ya. B. Zel'dovich*), A. A. Ruzmaikin (*A. A. Ruzmajkin*), D. D. Sokoloff (*D. D. Sokolov*).
The Fluid Mechanics of Astrophysics and Geophysics, Vol. 3. Gordon and Breach Science Publishers, New York – London – Paris – Montreux – Tokyo. 16+365 pp. Price US$ 80.00, DM 321.00 (1983). ISBN 0-677-06380-6 (USA), ISBN 0-677-05530-7 (UK).
Reviews in Astron. Express, Vol. 1, Issue 3, p. 111 (1984); Geophys. Astrophys. Fluid Dyn., Vol. 29, No. 4, p. 366 – 368; 1984 (*U. Frisch, D. Galloway*); Nature, Vol. 313, No. 6003, p. 606; 1985 (*L. Mestel*).
Contents: Introduction. Evidence of cosmic magnetism. Origin of magnetic fields. Dynamos. Conditions for magnetic field generation. The topology of flows in brief. Magnetic fields in stationary flows. Kinematic turbulent dynamos. The turbulent dynamo in a disk. Some topics in non-linear turbulent dynamo theory. Stellar cycles. The Galaxy and its magnetic field: observations. The galactic dynamo. The role of magnetic fields in star formation. Magnetic fields in cosmology. Accretion on black holes. Strong magnetic fields.

003.046 **Planets and their atmospheres. Origin and evolution.**
J. S. Lewis, R. G. Prinn.
International Geophysics Series, Vol. 33. Academic Press Inc., Orlando – San Diego – San Francisco – New York – London – Toronto – Montreal – Sydney – Tokyo – São Paulo. 10+470 pp. Price US$ 59.00 cloth, US$ 29.50 paper (1984). ISBN 0-12-446580-3 cloth, ISBN 0-12-446582-X paper.
Reviews in Icarus, Vol. 60, No. 3, p. 720 – 721; 1984 (*D. J. Stevenson*); Phys. Abstr., Vol. 87, No. 1238, Entry 73185 (1984).

003.047 **Physics of thermal gaseous nebulae (physical processes in gaseous nebulae).**
L. H. Aller.
Astrophysics and Space Science Library, Vol. 112. D. Reidel Publishing Company, Dordrecht – Boston – Lancaster. 10+350 pp. Price Dfl. 125.00, US$ 49.50, £ 31.75 (1984). ISBN 90-277-1814-8.
Contents: Types of gaseous nebulae. Spectra of gaseous nebulae. Radiative excitation, ionization, recombination, and fluorescence. Line and continuous spectra of hydrogen and helium. Collisionally excited lines and plasma diagnostics. Radiative equilibrium of a gaseous nebula. Nebular models. Dust and related phenomena in gaseous nebulae. Shock waves. Diffuse nebulae, wind-blown shells and planetaries. Chemical compositions of gaseous nebulae. Nebular compositions, element building, and stellar evolution.

003.048 **James E. Keeler. Pioneer American astrophysicist and the early development of American astrophysics.**
D. E. Osterbrock.
Cambridge University Press, Cambridge – London – New York – New Rochelle – Melbourne – Sydney. 12+411 pp. Price £ 35.00, US$ 39.50, DM 150.40 (1984). ISBN 0-521-26582-7.

003.049 **Spacelab. Research in earth orbit.**
D. Shapland, M. Rycroft.
Cambridge University Press, Cambridge – London – New York – New Rochelle – Melbourne – Sydney. 192 pp. Price £ 11.95, DM 52.00 (1984). ISBN 0-521-26077-9.
Contents: Space, our latest frontier. Spacelab – made in Europe. Into orbit aboard the Space Shuttle. Science and technology from Spacelab. Preparing for a Spacelab mission. The first Spacelab flight. Preliminary results of the first mission. Orbiting laboratories of the future.

003.050 **A workbook for astronomy.**
J. Waxman.
Cambridge University Press, Cambridge – London – New York – New Rochelle – Melbourne – Sydney. 12+356 pp. Price £ 15.00, US$ 19.95 (1984). ISBN 0-521-25312-8.

003.051 **Astronomical objects for southern telescopes. With an addendum for northern observatories. A handbook for amateur observers.**
E. J. Hartung.
Cambridge University Press, Cambridge – London – New York – New Rochelle – Melbourne – Sydney. 10+238 pp. Price £ 8.95, US $ 17.50 (1984). ISBN 0-521-31887-4. First published 1968.
Reviews in Astron. Express, Vol. 1, Issue 2, p. 86 (1984); J. Br. Astron. Assoc., Vol. 95, No. 1, p. 45 (1984).

003.052 **Saturn.**
T. Gehrels, M. S. Matthews (Editors).
The University of Arizona Press, Tucson, Arizona, USA. 11+968 pp. Price US$ 49.95 (1984). ISBN 0-8165-0829-1.
Review in Nature, Vol. 313, No. 6004, p. 716; 1985 (*D. J. Southwood*).
The individual contributions are included in their corresponding subject categories – see abstracts 100.129 – 100.149.

003.053 **Itogi Nauki i Tekhniki. Seriya Astronomiya. Tom 25. Physics of the sun.**
I. S. Shcherbina-Samojlova (Editor).
Vsesoyuznyj Institut Nauchnoj i Tekhnicheskoj Informatsii, Moskva. 194 pp. Price 1 Rbl.80 Kop. (1984). ISSN 0202-0742. In Russian.
The individual contributions are included in their corresponding subject categories – see abstracts 080.121, 073.114.

003.054 **Realm of the Universe.**
G. O. Abell.
Third edition. Saunders College Publishing, New York. 539 pp. Price $ 29.95 (1984). ISBN 0-03-058504-X.
Review in Sky Telesc., Vol. 68, No. 2, p. 136 (1984).

003.055 **Heroic efforts at meteor crater, Arizona: selected correspondence between Daniel Moreau Barringer and Elihu Thomson.**
H. J. Abrahams.
Fairleigh Dickinson University Press, Rutherford, N.J., USA. 322 pp. Price $ 35.00 (1983).
Review in J. Hist. Astron., Vol. 15, Part 3, No. 44, p. 219 – 220; 1984 (*W. G. Hoyt*).

003.056 **Meteorite – Boten aus dem Weltall.**
G. Aumann.
Veröffentlichungen aus dem Natur-Museum Coburg, Heft 22. Natur-Museum Coburg, Park 6, D-8630 Coburg, F.R. Germany. 84 pp. Price DM 6.00.
Review in Sterne Weltraum, 23. Jahrg., Nr. 7, p. 403; 1984 (*G. D. Roth*).

003.057 **Der Feldstecher, Wirkung und Leistung – dargestellt am Sternhimmel.**
H. Baderschneider.
Verlag Bode & Co. KG, Pforzheim, F.R. Germany. Available from "Neuen Optikerjournal", Postfach 1166, D-7530 Pforzheim, F.R. Germany. 45 pp. Price DM 16.00 (1983).
Review in Orion, 42. Jahrg., Nr. 205, p. 234; 1984 (*M. Zeller*).

003.058 **Quantum, space and time – the quest continues.**
A. O. Barut, A. van der Merwe, J.-P. Vigier (Editors).
Cambridge University Press, Cambridge – London – New York – New Rochelle – Melbourne – Sydney. 7 + 672 pp. Price £ 25.00 (1984). ISBN 0-521-31911-0.
From J. Br. Astron. Assoc., Vol. 94, No. 6, p. 287 (1984).

003.059 **Sternzeichen und Sternbilder. (Die Tafeln der ersten Ausgabe der "Uranometria" aus dem Jahr 1603).**
J. Bayer, mit einem Nachwort von K. Helmer.
Reihe "Die bibliophilen Taschenbücher". Harenberg Kommunikation Dortmund, F.R. Germany. 216 pp. Price DM 14.80 (1981).
Review in Sterne Weltraum, 23. Jahrg. Nr. 11, p. 602; 1984 (*G. D. Roth*).

003.060 **Man discovers the galaxies.**
R. Berendzen, R. Hart, D. Seeley.
Columbia University Press, 562 West 11th Street, New York, NY 10025. 228 pp. Price $ 30.00 cloth, $ 10.95 paper (1984).
From Strolling Astron., Vol. 30, Nos. 11 – 12, p. 255 (1984).

003.061 **Yardsticks of the Universe.**
O. Bishop.
Peter Bedrick Books, New York; Harper and Row, New York. 6 + 125 pp. Price $ 10.95 cloth, $ 5.95 paper (1984).
From Science, Vol. 225, No. 4659, p. 345 (1984).

003.062 **Modern numerical methods in problems of celestial mechanics.**
T. V. Bordovitsyna.
Nauka, Moskva. 136 pp. (1984). In Russian.
Review in Ref. Zh., 51. Astron., 8.51.26 (1984).

003.063 **Stephen Hawking's Universe.**
J. Boslough.
William Morrow, New York. 158 pp. Price $ 12.95 (1984).
Review in Nature, Vol. 312, No. 5991, p. 212 – 213; 1984 (*J. D. Barrow*).

003.064 **Fotografia astronomica.**
P. Bourge, J. Dragesco, Y. Dargery.
Cesco Capanna Editore. 240 pp. Price L. 18,900.
Review in Coelum, Vol. 52, N. 6, p. 386; 1984 (*R. Merighi*).

003.065 **Le ciel de vos vacances.**
S. Brunier, A. Cirou, P. Kohler, J. Lacroux.
Hachette, 79 Blvd. Saint-Germain, 75288 Paris 06, France. 205 pp. Price FF 78.00, $ 9.75 (1984). ISBN 2-01-010176-6.
From Sky Telesc., Vol. 68, No. 5, p. 426 (1984).

003.066 **Telescopes for the 1980s.**
G. Burbidge, A. Hewitt (Editors).
Mir, Moskva. 321 pp. (1984). In Russian.
Review in Ref. Zh., 51. Astron., 8.51.33 (1984). English original edition see 30.003.041.

003.067 **Prehistoric astronomy and ritual.**
A. Burl.
Shire Publications, Cromwell House, Church Street, Princes Risborough, Aylesbury, Bucks, England. 56 pp. Price £ 1.95 (1983).
Review in Archaeoastronomy (U.K.), No. 7, p. S152 (1984).

003.068 **The surface of Mars.**
M. H. Carr.
Yale University Press, New Haven – London. 11 + 232 pp. Price £ 19.95 paper (1984). ISBN 0-300-03242-0. For the hardcover edition see 31.003.026.
Reviews in J. Br. Astron. Assoc., Vol. 95, No. 1, p. 41; 1984 (*P. Moore*); Spaceflight, Vol. 26, No. 11, p. 430 – 431 (1984).

003.069 **Magnetic monopoles.**
R. A. Carrigan Jr., W. P. Tower (Editors).
Plenum Publishing Corporation, New York – London. 10 + 337 pp. Price $ 47.50 (1983). ISBN 0-306-41399-X.
From J. Br. Astron. Assoc., Vol. 95, No. 1, p. 45 (1984).

003.070 **Exploring the night sky with binoculars.**
D. Chandler.
David Chandler Co., PO Box 309, La Verne, Calif. 91750, USA. Price $ 5.95.
Review in Sci. Am., Vol. 251, No. 6, p. 25 (1984).

003.071 **Amateur astronomy: pocket skyguide.**
M. R. Chartrand III.
Newnes Books, 84-88 The Centre, Feltham, Middlesex TW13 4BH, England. 272 pp. Price £ 3.95 (1984). ISBN 0-600-35708-2.
Review in Observatory, Vol. 104, No. 1063, p. 279; 1984 (*R. W. Argyle*).

003.072 **Collected works of Meghnad Saha.**
S. Chatterjee (Editor).
Sangam Books, London. 8 + 591 pp. Price £ 11.25 (1984).
From J. Br. Astron. Assoc., Vol. 95, No. 1, p. 45 (1984).

003.073 **Ascent to orbit: a scientific autobiography.**
A. C. Clarke.
John Wiley & Sons, Chichester – New York – Brisbane – Toronto – Singapore. 10 + 226 pp. Price £ 14.95, $ 19.95 (1984). ISBN 0-471-87910-X.
Reviews in Astron. Tidsskr., Årg. 17, Nr. 3, p. 141 (1984); J. Br. Astron. Assoc., Vol. 95, No. 1, p. 44; 1984 (*C. A. Ronan*); Sky Telesc., Vol. 68, No. 3, p. 230 (1984).

003.074 **Rare halos, mirages, anomalous rainbows and related electro-magnetic phenomena: a catalog of geophysical anomalies.**
Compiled by W. R. Corliss.
Sourcebook Project, Glen Arm, Md., USA. 238 pp. (1984).
Review in Icarus, Vol. 60, No. 3, p. 719 – 720; 1984 (*A. Meinel, M. Meinel*).

003.075 **Die ersten Astronomen. Eine Einführung in die Ursprünge der Astronomie.**
J. Cornell, übersetzt von T. Westermayr.
Birkhäuser Verlag, Basel – Boston -Stuttgart. 259 pp. Price sfr. 32.00, DM 36.00 (1983). ISBN 3-7643-1379-X.
Reviews in Sterne, 60. Band, Heft 6, p. 376; 1984 (*J. Hamel*); Sterne Weltraum, 23. Jahrg., Nr. 7, p. 403; 1984 (*C. Leinert*).

003.076 **Time stands still: new light on megalithic science.**
K. Critchlow.
St. Martin's Press, New York, USA. 181 pp. Price $ 16.95 (1982).
Review in Archaeoastronomy (U.S.A.), Vol. 6, Nos. 1 – 4, p. 150 – 153; 1983 (*E. W. MacKie*).

003.077 **An introduction to experimental astronomy.**
R. B. Culver.
2nd edition. W. H. Freeman &Co., San Francisco – Oxford. 196 pp. Price £ 9.50 (1984).
Reviews in Observatory, Vol. 104, No. 1062, p. 237 – 238; 1984 (*D. McNally*); Space Educ., Vol. 1, No. 8, p. 382 (1984); Space Sci. Rev., Vol. 39, Nos. 1/2, p. 217; 1984 (*M. M. Dworetsky*).

003.078 **Astri collassati.**
A. Curir.
Il Castello Collane Tecniche, Via Ravizza 16, 20149 Milan, Italy. 79 pp. Price L. 9,500, $ 5.75 (1983).
Reviews in Orione, Vol. 4, N. 3, p. 136; 1984 (*S. Baroni*); Sky Telesc., Vol. 68, No. 3, p. 230 (1984).

003.079 **Beyond vision.**
J. Darius.
Oxford University Press, Oxford, England. 224 pp. Price £ 15.00 (1984). ISBN 0–19–853245–8.
Review in J. Br. Astron. Assoc., Vol. 95, No. 1, p. 44 – 45; 1984 (*C. Ronan*).

003.080 **The brightest stars.**
C. de Jager.
Mir, Moskva. 493 pp. (1984). In Russian. English edition see 28.003.060.
Review in Ref. Zh., 51. Astron., 8.51.31 (1984).

003.081 **Plurality of worlds: the origins of the extraterrestrial life debate from Democritus to Kant.**
S. J. Dick.
Cambridge University Press, Cambridge – London – New York – New Rochelle – Melbourne – Sydney. 10 + 246 pp. Price £ 7.50 paper (1984). For the hardcover edition see 31.003.035.
Reviews in Astron. Express, Vol. 1, Issue 2, p. 87 (1984); Astron. Tidsskr., Årg. 17, Nr. 4, p. 191 – 192 (1984).

003.082 **Dynamic astronomy.**
R. T. Dixon.
Fourth edition. Prentice–Hall Inc., Englewood Cliffs, N.J. 07632, USA. 555 pp. Price $ 25.95 (1984). ISBN 0–13–221333–8.
Reviews in Sky Telesc., Vol. 68, No. 5, p. 426 (1984); Strolling Astron., Vol. 30, Nos. 9 – 10, p. 207; 1984 (*J. R. Smith*).

003.083 **Planeten – Geschwister der Erde?**
J. Dorschner.
Akzent–Reihe, Band 27. 2. Aufl. Urania Verlag, Leipzig – Jena – Berlin. 128 pp. Price M 4.50 (1984).
Review in Astron. Raumfahrt, 22. Jahrg., Heft 6, p. 141 – 142; 1984 (*A. Dill*).

003.084 **Astronomia pratica.**
P. Duffet–Smith.
Editore Sansoni. 166 pp. Price L. 20,000.
Review in Orione, Vol. 4, N. 2, p. 82; 1984 (*M. Di Martino*).

003.085 **Gli asteroidi.**
P. Farinella, P. Paolicchi, V. Zappalà.
Il Castello Collane Tecniche, Via Ravizza 16, 20149 Milan, Italy. 127 pp. Price L. 11,500, $ 7.00 (1983).
Reviews in Orione, Vol. 4, N. 2, p. 83 – 84; 1984 (*S. Baroni*); Sky Telesc., Vol. 68, No. 3, p. 230 (1984).

003.086 **Guida alla osservazione del cielo.**
G. Favero.
Arnoldo Mondadori Editore. 146 pp. Price L. 12,000 (1984).
Review in Coelum, Vol. 52, N. 6, p. 385 – 386; 1984 (*S. Baroni*).

003.087 **Séparation des radiations par les filtres optiques.**
M. Françon.
Masson, Paris. 100 pp. (1984).
Review in Sol. Phys., Vol. 94, No. 2, p. 427; 1984 (*B. Valniček*).

003.088 **The amazing universe.**
H. Friedman.
National Geographic Society, Washington, D.C., 197 pp. (1984).
Review in Space Educ., Vol. 1, No. 8, p. 383 – 384 (1984).

003.089 **The light–hearted astronomer.**
K. Fulton.
AstroMedia Corporation, PO Box 92788, Milwaukee, Wis. 53202, USA. 115 pp. Price $ 6.95 (1984).
Reviews in Spaceflight, Vol 26, No. 12, p. 476 (1984); Strolling Astron., Vol. 30, Nos. 11 – 12, p. 253; 1984 (*R. Wend*).

003.090 **Radiation of gravitational waves by means of electrodynamical systems.**
D. V. Gal'tsov, Yu. V. Grats, V. I. Petukhov.
Moskovskij Gosudarstvennyj Universitet, Moskva. 128 pp. (1984). In Russian.
Review in Ref. Zh., 51. Astron., 8.51.27 (1984).

003.091 **Einführung in die Astronomie.**
R. H. Giese.
Bibliographisches Institut, Mannheim – Wien – Zürich. 9 + 440 pp. Price DM 44.00 (1984). See also 32.003.042.
From Phys. Bl., 40. Jahrg., Heft 11, p. 360 (1984).

003.092 **Einstein, the first hundred years.**
M. Goldsmith, A. Mackay, J. Woudhuysen (Editors).
Pergamon Press, Oxford – New York – Frankfurt – Paris – Sydney – Toronto – Tokyo. 14 + 200 pp. Price $ 19.95 (1980).
Review in J. Hist. Astron., Vol. 15, Part 3, No. 44, p. 222; 1984 (*O. Gingerich*).

003.093 **Artificial particle beams in space plasma studies.**
B. Grandal (Editor).
Plenum Publishing Corporation, New York, USA. 693 pp. Price $ 85.00 (1982). ISBN 0–306–40985–2.
Review in Astrophys. Lett., Vol. 24, No. 2, p. 127 – 128; 1984 (*S.–I. Akasofu*).

003.094 **Frozen star.**
G. Greenstein.
Freundlich Books, distributed by Scribner Book Companies, New York (available from Sky Publishing Corporation). 274 pp. Price $ 16.95 (1983). ISBN 0–88191–011–2.
Review in Sky Telesc., Vol. 68, No. 5, p. 421 – 422; 1984 (*T. Page*).

003.095 **Stellar chromospheres or the 2800 Å Mg II doublet in astrophysics.**
G. A. Gurzadyan.
Nauka, Moskva. 423 pp. (1984). In Russian.
Review in Ref. Zh., 51. Astron., 9.51.57 (1984).

003.096 **Early man and the cosmos.**
E. Hadingham.
Walker, New York. 280 pp. Price $ 22.50 (1984).
From Phys. Today, Vol. 37, No. 7, p. 73 (1984).

003.097 **Cosmic discovery. The search, scope, and heritage of astronomy.**
M. Harwit.
The MIT Press, Cambridge, Mass. 12 + 334 pp. Price $ 9.95 (1984). Reprint of the 1981 edition, see 30.003.071.
From Science, Vol. 225, No. 4666, p. 1018 (1984).

003.098 **New techniques in astronomy. Instruments and methods of investigation.**
Collection of scientific papers. Astron. sovet Akad. Nauk SSSR. Nauka, Leningrad. 140 pp. (1984). In Russian.
Review in Ref. Zh., 51. Astron., 11.51.39 (1984).

003.099 **Observing the universe. A "New Scientist" guide.**
N. Henbest (Editor).
Basil Blackwell, 108 Cowley Road, Oxford OX4 1JF, England. 16+288 pp. Price £ 5.50 (1984). ISBN 0-85520-726-4.
Reviews in Astron. Express, Vol. 1, Issue 3, p. 111 (1984); Astron. Tidsskr., Årg. 17, Nr. 3, p. 142 (1984); Space Educ., Vol. 1, No. 8, p. 382 (1984); Zenit, 11. Jaarg., No. 10, p. 442; 1984 (*G. Schilling*).

003.100 **Stonehenge. An ancient masonic temple.**
R. A. Herner.
Macoy Publishing and Masonic Supply Co., Richmond, Va., USA. 16+144 pp. Price $ 15.95 (1984).
From Science, Vol. 225, No. 4667, p. 1181 (1984).

003.101 **The history of astronomy from Herschel to Hertzsprung.**
D. B. Herrmann.
Cambridge University Press, Cambridge - London - New York - New Rochelle - Melbourne - Sydney. 220 pp. Price £ 12.50 (1984). ISBN 0-521-25733-6.
Reviews in Astron. Tidsskr., Årg. 17, Nr. 3, p. 142 (1984); Spaceflight, Vol. 26, Nos. 9-10, p. 384 (1984).

003.102 **Relativity and its roots.**
B. Hoffmann.
Scientific American Books. W. H. Freeman and Co., New York. 161 pp. Price $ 18.95 cloth, $ 9.95 paper (1983). ISBN 0-7167-1478-7 cloth, ISBN 0-7167-1510-4 paper.
Review in Sky Telesc., Vol. 68, No. 1, p. 34-35; 1984 (*E. Sheldon*).

003.103 **Materie - Bewegung - kosmische Entwicklung.**
W. Hollitscher, H. Horstmann.
Akademie-Verlag, Berlin, GDR. 233 pp. Price M 15.00 (1983).
Review in Astron. Raumfahrt, 22. Jahrg., Heft 6, p. 143-144; 1984 (*H. Meusinger*).

003.104 **Evolution of the expanding earth.**
M. Horai.
Translated from the Japanese edition. Nedra, Moskva, 109 pp. (1984). In Russian.
Review in Ref. Zh., 51. Astron., 12.51.61 (1984).

003.105 **Remote sounding of atmospheres.**
J. T. Houghton, F. W. Taylor, C. D. Rodgers.
Cambridge University Press, Cambridge - London - New York - New Rochelle - Melbourne - Sydney. 343 pp. Price £ 35.00 (1984).
Review in Spaceflight, Vol. 26, No. 11, p. 431 (1984).

003.106 **From grains to bacteria.**
F. Hoyle, C. Wickramasinghe.
University College Cardiff Press, Cardiff, England. 6+210 pp. Price £ 7.95 paper (1984).
Review in Astron. Express, Vol. 1, Issue 2, p. 88 (1984).

003.107 **Das Astrolabium und die Uhr.**
J. Hügin.
Verlag Wilhelm Kempter, Ulm, F.R. Germany. 168 pp. Price DM 48.00 (1978).
Review in J. Hist. Astron., Vol. 15, Part 3, No. 44, p. 223 (1984).

003.108 **Astronomical experiments.**
V. Icke.
Third edition. Burgess, Minneapolis, Minn., USA. 198 pp. (1983).
From Phys. Today, Vol. 37, No. 12, p. 72 (1984).

003.109 **Astronomy: fundamentals and frontiers.**
R. Jastrow, M. H. Thompson.
Fourth edition. John Wiley and Sons, Chichester - New York - Brisbane - Toronto - Singapore. 564 pp. Price $ 30.95 (1984). ISBN 0-471-89700-0.
Review in Sky Telesc., Vol. 68, No. 1, p. 37 (1984).

003.110 **Schwerkraft - Rätsel des Gewohnten.**
U. Kasper.
Urania-Verlag, Leipzig - Jena - Berlin. 122 pp. Price M 7.80 (1984).
Review in Astron. Raumfahrt, 22. Jahrg., Heft 4, p. 95-96; 1984 (*A. Unkroth, H. Albert*).

003.111 **Speurtocht door het heelal.**
R. Kerrod (Editor).
Elsevier Publication, Amsterdam - Brussel. 160 pp. Price *f* 39.75 (1984). ISBN 90-10-05110-2.
Review in Zenit, 11. Jaarg., No. 10, p. 442; 1984 (*G. Schilling*).

003.112 **Chondrules and their origins.**
E. A. King (Editor).
Lunar and Planetary Institute, Houston, Texas, USA. 377 pp. Price $ 33.00 (1983).
Review in Geochim. Cosmochim. Acta, Vol. 48, No. 7, p. 1578; 1984 (*R. T. Dodd*).

003.113 **Pocket computer programs for astronomy.**
F. Klein.
Klein Publications, Los Altos, Calif., USA. 100 pp. Price $ 12.95 (1983).
Review in Observatory, Vol. 104, No. 1063, p. 281; 1984 (*W. G. Rees*).

003.114 **L'Univers.**
P. Kohler.
Editions Hachette, Paris. 139 pp. Price FF 69.50 (1984).
Review in Ciel Terre, Vol. 100. No. 6, p. 178; 1984 (*R. Dejaiffe*).

003.115 **In search of ancient astronomies.**
E. C. Krupp (Editor).
Penguin Books, London, England. 18+281 pp. Price £ 4.95 paper (1984). ISBN 0-14-00-5481-2. For the hard cover edition see 30.003.089.
From Archaeoastronomy (U.K.), No. 7, p. S152 (1984).

003.116 **Constructing the universe.**
D. Layzer.
Scientific American Library, New York. 14+313 pp. Price US$ 21.95, £ 15.95 (1984).
Reviews in Astron. Express, Vol. 1, Issue 2, p. 88 (1984); Nature, Vol. 312, No. 5991, p. 212-213; 1984 (*J. D. Barrow*).

003.117 **Sterrekunde door de telescoop.**
R. Learner, with an introduction by C. Titulaer.
De Lantaarn, Amsterdam, The Netherlands. 224 pp. Price *f* 49.95, BF 900 (1984). ISBN 90-70485-11-7. For the English original see 31.003.084.
Review in Zenit, 11. Jaarg., No. 12, p. 537 (1984).

003.118 **The hidden sun: solar eclipses and astrophotography.**
J. Lowenthal.
Avon Books, New York. 107 pp. Price $ 5.95 (1984).
From Phys. Today, Vol. 37, No. 12, p. 72 (1984).

003.119 **The Greenwich Meridian.**
S. Malin, C. Stott.
Ordnance Survey, Southampton, England. 48 pp. Price £ 1.95 (1984). ISBN 0-319-00038-9.
Reviews in J. Br. Astron. Assoc., Vol. 95, No. 5, p. 246; 1984 (*C. A. Ronan*); Observatory, Vol. 104, No. 1063, p. 281-282; 1984 (*D. Stickland*).

003.120 **Quantum gravity.**
M. A. Markov, P. C. West (Editors).
Plenum Publishing Corporation, New York - London. 13+548 pp. Price $ 79.50 (1984). ISBN 0-306-41513-5.
From Nature, Vol. 311, No. 5982, p. 185 (1984).

003.121 **La science de l'univers à l'âge du positivisme, étude sur les origines de la cosmologie contemporaine.**
J. Merleau-Ponty.
Collection l'Histoire des Sciences, éditeur Vrin, Paris, France. 367 pp. Price FF 270.00 (1983).
Review in Astronomie, Vol. 98, p. 412 - 413; 1984 (*J.-C. Pecker*).

003.122 **The earth's magnetic field. Its history, origin and planetary perspective.**
R. T. Merrill, M. W. McElhinny.
Academic Press Inc., Orlando - San Diego - San Francisco - New York - London - Toronto - Montreal - Sydney - Tokyo - São Paulo. 401 pp. Price £ 39.50, $ 67.50 cloth; £ 17.50, $ 30.00 paper (1983). ISBN 0-12-491240-0.
Reviews in Nature, Vol. 310, No. 5976, p. 436; 1984 (*D. H. Tarling*); Phys. Abstr., Vol. 87, No. 1238, Entry 73186 (1984).

003.123 **Astronomiewijzer. Gids voor amateur-astronomen.**
P. Moore.
Uitgeverij Het Spectrum, Utrecht - Antwerpen. 144 pp. Price *f* 19.90 (1984). ISBN 90-274-7755-8.
Review in Zenit, 11. Jaarg. No. 12, p. 537 (1984).

003.124 **Patrick Moore's history of astronomy.**
P. Moore.
Macdonald, London. 327 pp. Price £ 14.95 (1983). ISBN 0-356-08607-0.
Review in J. Br. Astron. Assoc., Vol. 95, No. 1, p. 41; 1984 (*R. Baum*).

003.125 **Numerical solution of problems of atmospheric optics.**
Collection of scientific papers. Inst. prikl. mat. Akad. Nauk SSSR. Moskva, 234 pp. (1984). In Russian.
Review in Ref. Zh., 51. Astron., 11.51.38 (1984).

003.126 **Bernhard Schmidt, 1879 - 1935.**
P. Müürsepp, U. Veismann, edited by N. N. Mikhel'son.
Nauka, Leningradskoe Otdelenie, Leningrad. 134 pp. Price 20 Kop. (1984). In Russian.

003.127 **Astronomy with binoculars.**
J. Muirden.
Second edition. Arco Publishing Inc., 219 Park Ave., South, New York, N.Y. 10003, USA. 170 pp. Price $ 7.95 (1983). ISBN 0-668-05832-3. (Available from Sky Publishing Corporation).
Reviews in Sky Telesc., Vol. 68, No. 1, p. 37 (1984); Strolling Astron., Vol. 30, Nos. 9 - 10, p. 207; 1984 (*R. C. Maag*).

003.128 **The Petersburg Astronomical School of the XVIIIth century.**
N. I. Nevskaya.
Nauka, Leningrad. 238 pp. (1984). In Russian.
Review in Ref. Zh., 51. Astron., 9.51.50 (1984).

003.129 **Strichfiguren der Sternbilder.**
K. Oechslin.
Available from Naturforschende Gesellschaft Uri, CH-6460 Altdorf, Switzerland. 79 pp. Price SFr. 12.00.
Review in Orion, 42. Jahrg., Nr. 205, p. 229; 1984 (*A. Tarnutzer*).

003.130 **To know the stars: a simple guide to the night sky.**
G. Ottewell.
Astronomical Calendar Publications, c/o Department of Physics, Furman University, Greenville, S.C. 29613, USA. Price $ 5.00.
Review in Sci. Am., Vol. 251, No. 6, p. 25 (1984).

003.131 **The catastrophic universe. An essay in the philosophy of cosmology.**
A. G. Pacholczyk.
The Philosophy in Science Library, Vol. 2. Pachart Publishing House, 1130 San Lucas Circle, Tucson, Ariz. 85704, USA. 127 pp. Price $ 9.95 (1984). ISBN 0-88126-702-3.
From Science, Vol. 225, No. 4667, p. 1143 (1984).

003.132 **Teacher's guide to astronomy: from the earth to the universe.**
J. M. Pasachoff.
2nd edition. Sanders College Publishing, 383 Madison Avenue, New York, N.Y. 10017, USA. 236 pp. Price $ 19.95 (1983). A set of aids for use with "Astronomy: from the earth to the universe", see 33.003.096.
From Strolling Astron., Vol. 30, Nos. 11 - 12, p. 255 (1984).

003.133 **Effects of LTE-deviation in atmospheres of M giants.**
Ya. V. Pavlenko.
Valgus, Tallin. 98 pp. (1984). In Russian.
Review in Ref. Zh., 51. Astron., 9.51.58 (1984).

003.134 **The Stonehenge Watch, with accompanying instructional booklet entitled "Stonehenge unraveled".**
P. Payack.
The Idea Works, Inc., 7 Wildwood Street, West Milford, N.J. 07480, USA. Price $ 9.95 (1982).
Review in Archaeoastronomy (U.S.A), Vol. 6, Nos. 1 - 4, p. 154 - 155; 1983 (*R. J. C. Atkinson*).

003.135 **Exploring the universe.**
W. M. Protheroe, E. R. Capriotti, G. H. Newsom (Editors).
3rd edition. Charles E. Merrill. 586 pp. (1984). ISBN 0-675-20145-4.
From Nature, Vol. 311, No. 5982, p. 186 (1984).

003.136 **Exploring the Universe. Instructor's manual.**
W. M. Protheroe, E. R. Capriotti, G. H. Newsom (Editors).
3rd edition. Charles E. Merrill. 116 pp. (1984).
From Nature, Vol. 310, No. 5980, p. 805 (1984).

003.137 **Atoms of silence: an exploration of cosmic evolution.**
H. Reeves.
The MIT Press, Cambridge, Mass. - London, England. 12+244 pp. Price US$ 14.95, £ 14.20 (1984). ISBN 0-262-18112-6.
Reviews in Astron. Express, Vol. 1, Issue 2, p. 87 (1984); Sky Telesc., Vol. 68, No. 5, p. 425 (1984).

003.138 **Il sistema solare.**
M. Rigutti.
Corso Editore, Ferrara, Italy. 184 pp. (1984).
Review in G. A.A.B., N. 76, p. 18; 1984 (*L. Baldinelli*).

003.139 **A hundred billion stars.**
M. Rigutti, translated by M. Giacconi.
The MIT Press, Cambridge, Mass. - London, England. 15+285 pp. Price US$ 25.00, £ 23.75 (1984). ISBN 0-262-18111-8.
Reviews in Nature, Vol. 312, No. 5991, p. 212; 1984 (*J. Silk*); Sky Telesc., Vol. 68, No. 2, p. 136 (1984); Spaceflight, Vol. 26, No. 12, p. 476 (1984).

003.140 **Amateur astronomy: a comprehensive and practical survey.**
C. Ronan (Editor), revised and updated by S. Dunlop.
Newnes Books, 84-88 The Centre, Feltham, Middlesex TW13 4BH, England. 256 pp. Price £ 8.95 (1984). ISBN 0-600-35667-1.
Reviews in J. Br. Astron. Assoc., Vol. 94, No. 5, p. 245 - 246; 1984 (*P. Weston*); J. R. Astron. Soc. Can., Vol. 78, No. 6, p. 268 - 269; 1984 (*D. C. Wright*); Observatory, Vol. 104, No. 1062, p. 242; 1984 (*P. Moore*); Space Educ., Vol. 1, No. 8, p. 384 (1984).

003.141 **The golden book of astronomy.**
C. Ronan, S. Dunlop (Editors).
Western Publishing Co., 1220 Mound Ave., Racine, Wis. 53404, USA. 256 pp. Price $ 19.95 (1984). ISBN 0–307–46649–3.
Review in Sky Telesc., Vol. 68, No. 5, p. 425 (1984).

003.142 **Connected components in binary images: the detection problem.**
C. Ronse, P. A. Devijver.
John Wiley & Sons, Chichester – New York – Brisbane – Toronto – Singapore. 165 pp. Price £ 24.00 (1984). ISBN 0–86380–016–5.
From Nature, Vol. 311, No. 5987, p. 685 (1984).

003.143 **Sind wir allein im Universum?**
R. T. Rood, J. S. Trefil.
Birkhäuser Verlag, Basel. 300 pp. Price DM 32.00 (1982). ISBN 3–7643–1245–5. For the English original see 34.003.129.
Review in Sterne Weltraum, 23. Jahrg., Nr. 12, p. 660 – 661; 1984 (*D. Johannsmann*).

003.144 **Copernicus and the scientific revolution.**
E. Rosen.
Krieger, Malabar, Fla., USA. 220 pp. Price $ 6.50 (1984).
From Science, Vol. 225, No. 4667, p. 1144 (1984).

003.145 **Elementary particles and the structure of the universe.**
I. L. Rozental'.
Nauka, Moskva. 113 pp. (1984). In Russian.
Review in Ref. Zh., 51. Astron., 12.51.62 (1984).

003.146 **Megalithic astronomy: a new archaeological and statistical study of 300 western Scottish sites.**
C. L. N. Ruggles.
BAR British Series 123. 320 pp. Price £ 15.00 (1984). ISBN 0–86054–253–X.
From Nature, Vol. 311, No. 5982, p. 186 (1984).

003.147 **Computational optics. Handbook.**
M. M. Rusinov (Editor).
Mashinostroenie, Leningrad. 423 pp. (1984). In Russian.
Review in Ref. Zh., 51. Astron., 9.51.60 (1984).

003.148 **Sonne, Zeit und Ewigkeit. Alte Sonnenuhren.**
H. Sadler.
Reihe "Die bibliophilen Taschenbücher". Harenberg Kommunikation Dortmund, F.R. Germany. 103 pp. Price DM 16.80 (1983).
Review in Sterne Weltraum, 23. Jahrg., Nr. 11, p. 602; 1984 (*G. D. Roth*).

003.149 **Bernard Lovell: a biography.**
D. Saward.
Robert Hale, London. 320 pp. Price £ 12.95 (1984).
Review in Nature, Vol. 312, No. 5991, p. 210 – 211; 1984 (*R. Cockburn*).

003.150 **Die Harmonie der Sphären: Die Geschichte der Idee des Welteneinklangs und der Seeleneinstimmung.**
H. Schavernoch.
Orbis academicus, Sonderband 6. Verlag Karl Alber, Freiburg – München, F.R. Germany. 227 pp. Price DM 58.00 (1981).
Review in J. Hist. Astron., Vol. 15, Part 3, No. 44, p. 215; 1984 (*H. Druba*).

003.151 **From falling bodies to radio waves.**
E. Segrè.
W. H. Freeman and Co., 41 Madison Ave., New York, N.Y. 10010, USA. 298 pp. Price $ 24.95 cloth; $ 13.95 paper (1984). ISBN 0–7167–1481–7 cloth; ISBN 0–7167–1482–5 paper.
Review in Sky Telesc., Vol. 68, No. 6, p. 529 (1984).

003.152 **Fission products of transuranian elements in space.**
Yu. A. Shukolyukov, Dang Vu Minh.
Nauka, Moskva. 120 pp. (1984). In Russian.
Review in Ref. Zh., 51. Astron., 9.51.56 (1984).

003.153 **Interferometry.**
W. H. Steel.
Second edition. Cambridge University Press, Cambridge – London – New York – New Rochelle – Melbourne – Sydney. 15 + 308 pp. Price $ 39.50 (1983). ISBN 0–521–25320–9.
Reviews in Icarus, Vol. 60, No. 3, p. 722 – 723; 1984 (*E. K. Hege*); Phys. Abstr., Vol. 87, No. 1243, Entry 95081 (1984); Sky Telesc., Vol. 68, No. 3, p. 230 (1984).

003.154 **Stjernekort, Stjernehimlen hver dag, og hver time året rundt.**
B. Svanhof, A. G. Bak.
6th edition. G. E. C. Gads–Forlag, København. Price Dkr 92.75 (1983).
Review in Astron. Tidsskr., Årg. 17, Nr. 4, p. 189 – 190; 1984 (*O. J. Knudsen*).

003.155 **Time and the Highland Maya.**
B. Tedlock.
University of New Mexico Press, Albuquerque, USA. 245 pp. Price $ 27.50 (1982).
Review in Archaeoastronomy (U.S.A.), Vol. 6, Nos. 1 – 4, p. 161 – 163; 1983 (*A. F. Aveni*).

003.156 **Subatomic physics: nuclei and particles.**
L. Valentin.
North–Holland Publishing Company, Amsterdam, The Netherlands. Vol. 1: 300 pp., Vol. 2: 300 pp. Price $ 122.00 the set (1981). ISBN 0–444–86117–3.
Review in Astrophys. Lett., Vol. 24, No. 2, p. 119 – 120; 1984 (*S. E. Woosley*).

003.157 **Galileo's early notebooks: the physical questions.**
W. A. Wallace.
University of Notre Dame Press, Notre Dame, Indiana, USA. 16 + 321 pp. Price $ 14.95 (1977).
Review in J. Hist. Astron., Vol. 15, Part 3, No. 44, p. 220 – 221; 1984 (*R. S. Westfall*).

003.158 **The telescope.**
L. Yount.
Walker and Co., 720 Fifth Ave., New York, N.Y. 10019, USA. 64 pp. Price $ 8.95 (1983). ISBN 0–8027–6492–4.
Review in Sky Telesc., Vol. 68, No. 1, p. 37 (1984).

003.159 **Astronomia sferica e teorica.**
F. Zagar.
Zanichelli, Bologna, Italy. 510 pp. Price L. 24,000 (1984).
Review in Coelum, Vol. 52, N. 6, p. 386; 1984 (*F. Mancuso*).

003.160 **Recomendaciones sobre el futuro de la investigacion en el IAC, 1983.**
Instituto de Astrofisica de Canarias, Universidad de La Laguna, Tenerife, Spain. 123 pp.
Review in Sky Telesc., Vol. 68, No. 1, p. 37 (1984).

003.161 **Astrofysik, Kosmos 1983.**
Forskningsrådens Förlagstjänst, Stockholm. 239 pp. Price Sv.kr. 90.00 (1983).
Review in Astron. Tidsskr., Årg. 17, Nr. 4, p. 190 – 191; 1984 (*R. Simonsen*).

003.162 **Friedrich Wilhelm Bessel: Ich habe Euch lieb – aber der Himmel ist mir näher. Eine Autobiographie in Briefen.**
Edited by Stadt Minden and Edith Schlieper, Minden, F.R. Germany. 190 pp. Price DM 24.00 (1984).
Review in Nachr. Olbers–Ges. Bremen, Nr. 131, p. 3 – 5 (1984).

003.163 **Cosmic rays with energies above $10^{17}$eV. Collected papers.**
Institute of cosmophysical investigations and aeronomy, Yakutsk Department of the Special Observatory of the USSR Academy of Sciences, Yakutsk. 154 pp. (1983). In Russian.
Review in Ref. Zh., 51. Astron., 8.51.32 (1984).

# 004 History of Astronomy

004.001 **The Harvard summer school in astronomy.**
D. H. DeVorkin.
Phys. Today, Vol. 37, No. 7, p. 48 – 55 (1984).
The annual conferences that Harlow Shapley organized from 1935 through 1942 fostered the growth of astrophysics in the US, and were the model that inspired similar sessions on other campuses.

004.002 **Année Diderot: «Qui a découvert les satellites de Jupiter?»**
J.-P. Parisot.
Astronomie, Vol. 98, p. 341 – 342 (1984).

004.003 **Due eclissi storiche in Abruzzo.**
S. De Meis.
Coelum, Vol. 52, N. 4, p. 219 – 226 (1984).

004.004 **Aufenthalt in Hveen. Einen Augenblick zurück zu Tychos Zeit.**
G. W. E. Beekman.
Sterne Weltraum, 23. Jahrg., Nr. 8-9, p. 420 – 421 (1984).

004.005 **Exploitation des observations de Delambre des satellites galiléens (1775 – 1802).**
J. E. Arlot, B. Morando, W. Thuillot.
Astrométrie, géodésie, mécanique céleste, p. 49 – 52 (1984). – See Abstr. 012.021.

004.006 **L'obliquité de l'écliptique à l'époque d'Hipparque.**
J. P. Brunet, R. Nadal.
Astrométrie, géodésie, mécanique céleste, p. 101 – 104 (1984). – See Abstr. 012.021.

004.007 **Archives du Bureau des Longitudes. Éclipses de Delambre.**
B. Morando.
Astrométrie, géodésie, mécanique céleste, p. 107 – 114 (1984). – See Abstr. 012.021.

004.008 **The end of Pulkovo Observatory's reign as the "astronomical capital of the world".**
K. Krisciunas.
Q. J. R. Astron. Soc., Vol. 25, No. 3, p. 301 – 305 (1984).

004.009 **Longitude zero 1884 – 1984.**
C. Stott.
Q. J. R. Astron. Soc., Vol. 25, No. 3, p. 306 – 308 (1984).

004.010 **Length of the year in pre-epicyclic planetary theory.**
G. Abraham.
Arch. Hist. Exact Sci., Vol. 30, Nos. 3 – 4, p. 189 – 195 (1984).
An estimate is made of the length of the year implied in the pre-epicyclic planetary theory of the Pañcasiddhāntikā. It is shown that this could be the same as the year-length of the corresponding Babylonian theory.

004.011 **Les instruments anciens de la navigation astronomique.**
J. Hourrière.
Astronomie, Vol. 98, p. 439 – 448 (1984).

004.012 **L. Foucault et H. Fizeau: exploitation d'une information nouvelle.**
P. Costabel.
C. R. Acad. Sci., Sér. Gén., Vie Sci., Tome 1, No. 3, p. 235 – 249 (1984).

004.013 **Early radio astronomy at Wheaton, Illinois.**
G. Reber.
The early years of radio astronomy, p. 43 – 66 (1984). – See Abstr. 003.012.

004.014 **Optical and radio astronomers in the early years.**
J. L. Greenstein.
The early years of radio astronomy, p. 67 – 82 (1984). – See Abstr. 003.012.

004.015 **The origins of radio astronomy in Australia.**
E. G. Bowen.
The early years of radio astronomy, p. 84 – 111 (1984). – See Abstr. 003.012.

004.016 **The first decade of solar radio astronomy in Australia.**
W. N. Christiansen.
The early years of radio astronomy, p. 112 – 131 (1984). – See Abstr. 003.012.

004.017 **Early days in radio and radar astronomy in Australia.**
F. J. Kerr.
The early years of radio astronomy, p. 132 – 145 (1984). – See Abstr. 003.012.

004.018 **Radio sources and the log N–log S controversy.**
B. Y. Mills.
The early years of radio astronomy, p. 146 – 166 (1984). – See Abstr. 003.012.

004.019 **Early work on imaging theory in radio astronomy.**
R. N. Bracewell.
The early years of radio astronomy, p. 167 – 190 (1984). – See Abstr. 003.012.

004.020 **The origins and early history of Jodrell Bank.**
B. Lovell.
The early years of radio astronomy, p. 192 – 211 (1984). – See Abstr. 003.012.

004.021 **Paraboloids, galaxies and stars: memories of Jodrell Bank.**
R. Hanbury Brown.
The early years of radio astronomy, p. 212 – 235 (1984). – See Abstr. 003.012.

004.022 **Early work on radio stars at Cambridge.**
F. G. Smith.
The early years of radio astronomy, p. 236 – 248 (1984). – See Abstr. 003.012.

004.023 **The development of aperture synthesis at Cambridge.**
P. A. G. Scheuer.
The early years of radio astronomy, p. 249 – 265 (1984). – See Abstr. 003.012.

004.024 **The first steps of Soviet radio astronomy.**
A. E. Salomonovich.
The early years of radio astronomy, p. 268 – 288 (1984). – See Abstr. 003.012.

004.025 **Remarks on my work in radio astronomy.**
V. L. Ginzburg.
The early years of radio astronomy, p. 289 – 302 (1984). – See Abstr. 003.012.

004.026 **The early years of radio astronomy in France.**
J. F. Denisse.
The early years of radio astronomy, p. 303 – 315 (1984). – See Abstr. 003.012.

004.027 **Beginnings of solar radio astronomy in Canada.**
A. E. Covington.
The early years of radio astronomy, p. 317 – 334 (1984). – See Abstr. 003.012.

004.028 **Development of solar radio astronomy in Japan up until 1960.**
H. Tanaka.
The early years of radio astronomy, p. 335 – 348 (1984). – See Abstr. 003.012.

004.029 **Styles of research in three early radio astronomy groups.**
D. O. Edge.
The early years of radio astronomy, p. 350 – 364 (1984). – See Abstr. 003.012.

004.030 **The influence of radio astronomy on cosmology.**
W. H. McCrea.
The early years of radio astronomy, p. 365 – 384 (1984). – See Abstr. 003.012.

004.031 **Nanohertz astronomy.**
H. C. van de Hulst.
The early years of radio astronomy, p. 385 – 398 (1984). – See Abstr. 003.012.

The paper is an attempt at what with a sophistication understandable to radio astronomers might be termed nanohertz astronomy: the art of registering the coming and going of astronomical convictions in periods of the order of $10^9$ seconds = 30 years.

004.032 **Radio astronomy and the nature of science.**
O. Gingerich.
The early years of radio astronomy, p. 399 – 407 (1984). – See Abstr. 003.012.

004.033 **Wàs dat Halley wel, in 1910?**
W. Zanstra.
Zenit, 11. Jaarg., No. 10, p. 419 – 421 (1984).

004.034 **De laatste totale zonsverduistering in Nederland.**
W. Carton.
Zenit, 11. Jaarg., No. 10, p. 422 – 423 (1984).

004.035 **Ptolemy's Almagest: fourteen centuries of neglect.**
N. S. Hetherington, C. A. Ronan.
J. Br. Astron. Assoc., Vol. 94, No. 6, p. 256 – 262 (1984).

Ptolemy's Almagest remained the ultimate in astronomical theory for more than a millennium. Yet Ptolemy's system contains internal inconsistencies so evident and disturbing as to raise the question of why the Copernican Revolution was so long delayed. Historical breaks in research in the Ptolemaic tradition, resulting in the interment and then the resurrection of that tradition, help to explain its seemingly long lifetime.

004.036 **Anfänge der Astronomie bei den Griechen.**
Á. Szabó.
Sterne Weltraum, 23. Jahrg., Nr. 10, p. 498 – 501 (1984).

004.037 **Die ersten Entdecker veränderlicher Sterne. Zum 200. Jahrestag der Entdeckung von Delta Cephei.**
R. Lukas.
Sterne Weltraum, 23. Jahrg., Nr. 10, p. 524 – 527 (1984).

004.038 **Prähistorische Kalenderastronomie VI: Weitere Fallstudien.**
H. Hindrichs.
Orion, 42. Jahrg., Nr. 204, p. 179 – 182 (1984).

004.039 **Maria Mitchell's observations of Mira, 1856 – 1859.**
E. P. Belserene.
J. Am. Assoc. Variable Star Obs., Vol. 12, No. 2, p. 49 – 53 (1983).

Maria Mitchell's comparisons of o Ceti with surrounding stars are reduced with modern magnitudes to give the light curve for the maxima in early 1857 and two cycles later in late 1858. The state of astronomical photometry at the time is briefly reviewed.

004.040 **The comet of Charles V.**
D. Hoffleit.
J. Am. Assoc. Variable Star Obs., Vol. 12, No. 2, p. 76 (1983). Abstract. – See Abstr. 010.021.

004.041 **The naming of Pluto.**
P. Moore.
Sky Telesc., Vol. 68, No. 5, p. 400 – 401 (1984).

004.042 **Longitude zero 1884 – 1984.**
J. Dudley.
Observatory, Vol. 104, No. 1062, p. 209 – 211 (1984).

004.043 **Edmond Halley's observations of Halley's comet.**
D. W. Hughes, A. Drummond.
J. Hist. Astron., Vol. 15, Part 3, No. 44, p. 189 – 197 (1984).

004.044 **Galileo, Kepler, and phases of Venus.**
S. Drake.
J. Hist. Astron., Vol. 15, Part 3, No. 44, p. 198 – 208 (1984).

004.045 **Phases of Venus in 1610.**
O. Gingerich.
J. Hist. Astron., Vol. 15, Part 3, No. 44, p. 209 – 210 (1984).

004.046 **The appearances of Venus and Mars in 1610.**
W. T. Peters.
J. Hist. Astron., Vol. 15, Part 3, No. 44, p. 211 – 214 (1984).

004.047 **Hopewellian geometry and astronomy at High Bank.**
R. Hively, R. Horn.
Archaeoastronomy (U.K.), No. 7, p. S85 – S100 (1984).

004.048 **The interpretation of the pecked cross symbols at Teotihuacan: a methodological note.**
C. L. N. Ruggles, N. J. Saunders.
Archaeoastronomy (U.K.), No. 7, p. S101 – S110 (1984). With a comment by A. F. Aveni.

004.049 **A statistical method of evaluating megalithic observatories.**
G. Rosenfeldt.
Archaeoastronomy (U.K.), No. 7, p. S111 – S118 (1984).

004.050 **The astronomical orientation of the Egyptian pyramids.**
S. C. Haack.
Archaeoastronomy (U.K.), No. 7, p. S119 – S125 (1984).

004.051 **The great astrolabe of Jaipur and its sister unit.**
V. N. Sharma.
Archaeoastronomy (U.K.), No. 7, p. S126 – S128 (1984).

004.052 **The two major megalithic observatories in Scotland.**
A. Thom, A. S. Thom.
Archaeoastronomy (U.K.), No. 7, p. S129 – S148 (1984).

004.053 **Astronomical scrapbook: the discovery of the Milky Way's spiral arms.**
O. Gingerich.
Sky Telesc., Vol. 68, No. 1, p. 10 – 12 (1984).

004.054 **The glory of gravity – Halley's Comet 1759.**
R. Wallis.
Ann. Sci., Vol. 41, p. 279 – 286 (1984).

During the eighty years since the publication of the "Principia", Newton's theories had been popularized. Based upon them, the first predicted return of any comet aroused anticipation and then elation. Contemporary accounts in widely-read periodicals show the interest in the vindication of gravitational theory provided by fulfilment of the prediction.

004.055 **The origin of the constellations.**
A. E. Roy.
Vistas Astron., Vol. 27, Part 2, p. 171 – 197 (1984).
The problem of the origin of the stellar constellations familiar to western astronomers from ancient times is discussed in an attempt to answer the classical detective story questions: Who? When? Why? and Where? The available astronomical, literary and archaeological evidence is examined to suggest a possible solution.

004.056 **Space and time as existential forms of matter in the theory of Rudjer Bošković.**
B. Jovanović.
Vasiona, Année 32, No. 3, p. 58 – 61 (1984). In Croatian.

004.057 **Anfänge der Astronomie bei den Griechen. Teil 2.**
A. Szabó.
Sterne Weltraum, 23. Jahrg., Nr. 11, p. 556 – 558 (1984).

004.058 **Astronomical scrapbook: Galileo and the phases of Venus.**
O. Gingerich.
Sky Telesc., Vol. 68, No. 6, p. 520 – 522 (1984).

004.059 **Sunspot–story.**
G. Rüdiger.
Sterne, 60. Band, Heft 5, p. 267 – 277 (1984).

004.060 **Bessels Hypothese der spezifischen Gravitation und das Problem der Uranusbewegung. Ein bisher unbekanntes Manuskript von F. W. Bessel.**
J. Hamel.
Sterne, 60. Band, Heft 5, p. 278 – 283 (1984).

004.061 **Die Kometen im Weltbild der Antike und des Mittelalters.**
J. Hamel.
Sterne, 60. Band, Heft 6, p. 347 – 356 (1984).

004.062 **Anfänge der Astronomie bei den Griechen. Teil 3.**
Á. Szabó.
Sterne Weltraum, 23. Jahrg., Nr. 12, p. 618 – 621 (1984).

004.063 **Draughtsmen of the constellations.**
D. W. Hughes.
Nature, Vol. 312, No. 5996, p. 697 (1984).
This note comments on recent studies of the historic origins of the star constellations.

004.064 **The history of the discovery of the thermonuclear nature of stellar energy.**
A. Masani, D. Cacace.
G. Fis., Vol. 24, No. 4, p. 235 – 244 (1983). In Italian. Abstr. in Phys. Abstr., Vol. 88, No. 1247, Entry 43 (1985).

004.065 **Ancient heliocentrists' unburial & the Almagest crimes' motive.**
D. Rawlins, H. Perry.
Bull. Am. Astron. Soc., Vol. 16, No. 3, Part I, p. 724 (1984). Abstract. – See Abstr. 010.065.

004.066 **Light in the temples.**
E. C. Krupp.
Bull. Am. Astron. Soc., Vol. 16, No. 3, Part I, p. 732 (1984). Abstract. – See Abstr. 010.065.

004.067 **A woodcut comet riddle – possible solution.**
T. A. Nagy.
Bull. Am. Astron. Soc., Vol. 16, No. 3, Part I, p. 732 (1984). Abstract. – See Abstr. 010.065.

004.068 **The discovery of Neptune: essential revisions to the history.**
D. Rawlins.
Bull. Am. Astron. Soc., Vol. 16, No. 3, Part I, p. 734 (1984). Abstract. – See Abstr. 010.065.

004.069 **A Lutheran astrologer: Johannes Kepler.**
J. V. Field.
Arch. Hist. Exact Sci., Vol. 31, No. 3, p. 189 – 272 (1984).

004.070 **Stone–age burials as a hint to prehistoric astronomy.**
T. Schmidt–Kaler, W. Schlosser.
J. R. Astron. Soc. Can., Vol. 78, No. 5, p. 178 – 188 (1984).
The authors describe statistical studies of neolithic burial sites in central Europe and conclude that the alignments of graves and bodies show that stone–age people had abstract concepts of the cardinal directions of the compass, rather than simple notions of the direction of sunrise.

004.071 **Standardization of early surveying instruments.**
J. E. Kennedy.
J. R. Astron. Soc. Can., Vol. 78, No. 5, p. 200 (1984). Abstract. – See Abstr. 010.401.

004.072 **Commemorating the Toronto Observatory of 1840.**
D. A. MacRae.
J. R. Astron. Soc. Can., Vol. 78, No. 5, p. 201 (1984). Abstract. – See Abstr. 010.401.

004.073 **Early Canadian observatories.**
P. M. Millman.
J. R. Astron. Soc. Can., Vol. 78, No. 5, p. 201 (1984). Abstract. – See Abstr. 010.401.

004.074 **Archaeoastronomy and the Andean world.**
A. F. Aveni.
Archaeoastronomy (U.S.A.), Vol. 6, Nos. 1 – 4, p. 10 (1983). Abstr. – See Abstr. 012.056.

004.075 **Nabua: an important celestial observer from ancient Assyria.**
R. Chadwick.
Archaeoastronomy (U.S.A.), Vol. 6, Nos. 1 – 4, p. 10 (1983). Abstr. – See Abstr. 012.056.

004.076 **The sky as an ethnographic resource.**
V. D. Chamberlain.
Archaeoastronomy (U.S.A.), Vol. 6, Nos. 1 – 4, p. 10 – 11 (1983). Abstr. – See Abstr. 012.056.

004.077 **Celestial myths and metaphors in the rock art of Sears Point, Arizona.**
T. Hoskinson.
Archaeoastronomy (U.S.A.), Vol. 6, Nos. 1 – 4, p. 11 (1983). Abstr. – See Abstr. 012.056.

004.078 **Cosmovision in northern Mexico.**
R. Opperman.
Archaeoastronomy (U.S.A.), Vol. 6, Nos. 1 – 4, p. 11 (1983). Abstr. – See Abstr. 012.056.

004.079 **Notes about the archaeoastronomical investigations in the central area of Ingapirca, Ecuador.**
M. S. Ziolkowski, R. M. Sadowski.
Archaeoastronomy (U.S.A.), Vol. 6, Nos. 1 – 4, p. 11 (1983). Abstr. – See Abstr. 012.056.

004.080 **Venus and the Inca calendar.**
R. T. Zuidema.
Archaeoastronomy (U.S.A.), Vol. 6, Nos. 1 – 4, p. 11 (1983). Abstr. – See Abstr. 012.056.

004.081 **Archaeoastronomy in the Southwestern United States: attitudes, concepts, strategies, and altering the course of action.**
A. F. Aveni.
Archaeoastronomy (U.S.A.), Vol. 6, Nos. 1 – 4, p. 13 (1983). Abstr. – See Abstr. 012.057.

004.082 **The sun personified: some preliminary observations regarding celestial cosmology and the ordering of Pueblo Indian society.**
T. S. Jojola.
Archaeoastronomy (U.S.A.), Vol. 6, Nos. 1 – 4, p. 13 (1983). Abstr. – See Abstr. 012.057.

004.083 **Chaco Canyon and the role of archaeoastronomy in the prehistoric Southwest.**
W. J. Judge.
Archaeoastronomy (U.S.A.), Vol. 6, Nos. 1 – 4, p. 13 – 14 (1983). Abstr. – See Abstr. 012.057.

004.084 **The discovery of 19 prehistoric calendric petroglyph sites in Arizona.**
R. A. Preston, A. L. Preston.
Archaeoastronomy (U.S.A.), Vol. 6, Nos. 1 – 4, p. 14 (1983). Abstr. – See Abstr. 012.057.

004.085 **Astronomical markings at three sites on Fajada Butte.**
A. Sofaer, R. Sinclair.
Archaeoastronomy (U.S.A.), Vol. 6, Nos. 1 – 4, p. 14 – 15 (1983). Abstr. – See Abstr. 012.057.

004.086 **Light and shadow, ritual and astronomy in Anasazi structures.**
R. A. Williamson.
Archaeoastronomy (U.S.A.), Vol. 6, Nos. 1 – 4, p. 15 (1983). Abstr. – See Abstr. 012.057.

004.087 **Anticipation in ceremony: the readiness is all.**
M. Zeilik.
Archaeoastronomy (U.S.A.), Vol. 6, Nos. 1 – 4, p. 15 (1983). Abstr. – See Abstr. 012.057.

004.088 **Methodology and validity in California archaeoastronomy.**
K. Hedges.
Archaeoastronomy (U.S.A.), Vol. 6, Nos. 1 – 4, p. 17 (1983). Abstr. – See Abstr. 012.058.

004.089 **Shadows of Kakunupmawa.**
T. Hoskinson.
Archaeoastronomy (U.S.A.), Vol. 6, Nos. 1 – 4, p. 17 (1983). Abstr. – See Abstr. 012.058.

004.090 **Skywatchers of ancient California.**
T. Hudson.
Archaeoastronomy (U.S.A.), Vol. 6, Nos. 1 – 4, p. 17 (1983). Abstr. – See Abstr. 012.058.

004.091 **Mockingbird Canyon's solar and stellar observations.**
V. Hunter, J. Rafter.
Archaeoastronomy (U.S.A.), Vol. 6, Nos. 1 – 4, p. 17 (1983). Abstr. – See Abstr. 012.058.

004.092 **Under California skies.**
E. Krupp.
Archaeoastronomy (U.S.A.), Vol. 6, Nos. 1 – 4, p. 17 (1983). Abstr. – See Abstr. 012.058.

004.093 **The discovery of 19 prehistoric calendric petroglyph sites in Arizona.**
R. A. Preston, A. L. Preston.
Archaeoastronomy (U.S.A.), Vol. 6, Nos. 1 – 4, p. 17 (1983). Abstr. – See Abstr. 012.058.

004.094 **Archaeoastronomy of Counsel Rocks, a ring of pictured stones.**
J. Rafter.
Archaeoastronomy (U.S.A.), Vol. 6, Nos. 1 – 4, p. 17 (1983). Abstr. – See Abstr. 012.058.

004.095 **Archaeoastronomical investigations at Burro Flats: aspects of ceremonialism at a Chumash rock art and habitation site.**
J. Romani, G. Romani, D. Larson.
Archaeoastronomy (U.S.A.), Vol. 6, Nos. 1 – 4, p. 17 (1983). Abstr. – See Abstr. 012.058.

004.096 **Rock art sites as solar observatories: two possible examples from Vandenberg Air Force Base.**
L. Spanne.
Archaeoastronomy (U.S.A.), Vol. 6, Nos. 1 – 4, p. 18 (1983). Abstr. – See Abstr. 012.058.

004.097 **Cosmic symbolism of the Stūpa in India.**
A. Agee.
Archaeoastronomy (U.S.A.), Vol. 6, Nos. 1 – 4, p. 35 (1983). Abstr. – See Abstr. 012.059.

004.098 **Indigenous astronomies of the Indo–Malay Archipelago.**
G. Ammarell.
Archaeoastronomy (U.S.A.), Vol. 6, Nos. 1 – 4, p. 35 (1983). Abstr. – See Abstr. 012.059.

004.099 **The moon and the Venus Table: an example of commensuration in the Maya calendar.**
A. F. Aveni.
Archaeoastronomy (U.S.A.), Vol. 6, Nos. 1 – 4, p. 35 (1983). Abstr. – See Abstr. 012.059.

004.100 **The zero–point of the zodiac of Hindu astronomers in ancient India.**
A. Bandyopadhyay.
Archaeoastronomy (U.S.A.), Vol. 6, Nos. 1 – 4, p. 35 (1983). Abstr. – See Abstr. 012.059.

004.101 **Some speculation about the astronomical identification of Moche deities.**
E. P. Benson.
Archaeoastronomy (U.S.A.), Vol. 6, Nos. 1 – 4, p. 35 (1983). Abstr. – See Abstr. 012.059.

004.102 **African astronomy and art: an introduction and case study.**
S. P. Blier.
Archaeoastronomy (U.S.A.), Vol. 6, Nos. 1 – 4, p. 36 (1983). Abstr. – See Abstr. 012.059.

004.103 **Classic Maya observations of planetary retrograde motion.**
V. R. Bricker.
Archaeoastronomy (U.S.A.), Vol. 6, Nos. 1 – 4, p. 36 (1983). Abstr. – See Abstr. 012.059.

004.104 **Mesoamerican cosmovision: an overview.**
J. B. Carlson.
Archaeoastronomy (U.S.A.), Vol. 6, Nos. 1 – 4, p. 36 (1983). Abstr. – See Abstr. 012.059.

004.105 **Celestial observations in Mesopotamia: three millennia of celestial beliefs and practices in the ancient Near East.**
R. Chadwick.
Archaeoastronomy (U.S.A.), Vol. 6, Nos. 1 – 4, p. 36 (1983). Abstr. – See Abstr. 012.059.

004.106 **A bibliography of native American concepts and traditions of the sky.**
V. D. Chamberlain, A. Meadows, C. Johnson, S. Brown, K. Kirkpatrick.
Archaeoastronomy (U.S.A.), Vol. 6, Nos. 1 – 4, p. 37 (1983).
Abstr. – See Abstr. 012.059.

004.107 **Jean-Sylvain Bailly and "Antediluvian" astronomy.**
S. L. Chapin.
Archaeoastronomy (U.S.A.), Vol. 6, Nos. 1 – 4, p. 37 (1983).
Abstr. – See Abstr. 012.059.

004.108 **South Asian ethnoastronomy: artifacts and paradigms.**
J. R. Conway.
Archaeoastronomy (U.S.A.), Vol. 6, Nos. 1 – 4, p. 37 (1983).
Abstr. – See Abstr. 012.059.

004.109 **The conjurer's lodge: star lore, mythology, pictographs and religious practices found among the Canadian Ojibwa and Temagami Indians.**
T. Conway.
Archaeoastronomy (U.S.A.), Vol. 6, Nos. 1 – 4, p. 37 (1983).
Abstr. – See Abstr. 012.059.

004.110 **Icicle gnomonic sundial among the Antarcticans.**
N. Count, I. M. Cold, W. Freeze.
Archaeoastronomy (U.S.A.), Vol. 6, Nos. 1 – 4, p. 37 (1983).
Abstr. – See Abstr. 012.059.

004.111 **Reflections on Muyuw spatial and temporal notions.**
F. H. Damon.
Archaeoastronomy (U.S.A.), Vol. 6, Nos. 1 – 4, p. 37 (1983).
Abstr. – See Abstr. 012.059.

004.112 **Linguistic representation of astronomical cycles in ancient Egypt.**
V. L. Davis.
Archaeoastronomy (U.S.A.), Vol. 6, Nos. 1 – 4, p. 38 (1983).
Abstr. – See Abstr. 012.059.

004.113 **Identification of the ancient Egyptian constellations.**
V. L. Davis.
Archaeoastronomy (U.S.A.), Vol. 6, Nos. 1 – 4, p. 38 (1983).
Abstr. – See Abstr. 012.059.

004.114 **Astronomical calculations for ethnoastronomy.**
L. E. Doggett, S. M. Hoban.
Archaeoastronomy (U.S.A.), Vol. 6, Nos. 1 – 4, p. 38 (1983).
Abstr. – See Abstr. 012.059.

004.115 **"You're standing on my Indian watch!": Mescalero Apache ceremonial timing.**
C. R. Farrer.
Archaeoastronomy (U.S.A.), Vol. 6, Nos. 1 – 4, p. 38 (1983).
Abstr. – See Abstr. 012.059.

004.116 **Navigation without instruments.**
B. R. Finney.
Archaeoastronomy (U.S.A.), Vol. 6, Nos. 1 – 4, p. 38 (1983).
Abstr. – See Abstr. 012.059.

004.117 **Lakota star knowledge.**
R. Goodman.
Archaeoastronomy (U.S.A.), Vol. 6, Nos. 1 – 4, p. 39 (1983).
Abstr. – See Abstr. 012.059.

004.118 **Something burns in half here: the astronomical orientation of a Mexican village.**
T. L. Grigsby, J. Ostermiller.
Archaeoastronomy (U.S.A.), Vol. 6, Nos. 1 – 4, p. 39 (1983).
Abstr. – See Abstr. 012.059.

004.119 **The use and abuse of ethnoastronomy.**
E. Hadingham.
Archaeoastronomy (U.S.A.), Vol. 6, Nos. 1 – 4, p. 39 (1983).
Abstr. – See Abstr. 012.059.

004.120 **Medicine wheels, sun circles, and the magic of world center shrines.**
R. L. Hall.
Archaeoastronomy (U.S.A.), Vol. 6, Nos. 1 – 4, p. 39 (1983).
Abstr. – See Abstr. 012.059.

004.121 **A new interpretation of pages 65–69 of the Dresden Codex.**
M. P. Hatch.
Archaeoastronomy (U.S.A.), Vol. 6, Nos. 1 – 4, p. 39 (1983).
Abstr. – See Abstr. 012.059.

004.122 **Sparks of the sun.**
T. Hoskinson.
Archaeoastronomy (U.S.A.), Vol. 6, Nos. 1 – 4, p. 39 (1983).
Abstr. – See Abstr. 012.059.

004.123 **Celestial motifs and metaphors in the Gila River region of southwest Arizona.**
T. Hoskinson.
Archaeoastronomy (U.S.A.), Vol. 6, Nos. 1 – 4, p. 40 (1983).
Abstr. – See Abstr. 012.059.

004.124 **Orion and the Pleiades in Easter Island ethnoastronomy.**
T. Hoskinson.
Archaeoastronomy (U.S.A.), Vol. 6, Nos. 1 – 4, p. 40 (1983).
Abstr. – See Abstr. 012.059.

004.125 **Sky signs of the Mother Goddess in the ancient Middle East and Greece.**
H. C. Hostetter.
Archaeoastronomy (U.S.A.), Vol. 6, Nos. 1 – 4, p. 40 (1983).
Abstr. – See Abstr. 012.059.

004.126 **Venus in the east and west.**
S. Iwaniszewski.
Archaeoastronomy (U.S.A.), Vol. 6, Nos. 1 – 4, p. 40 (1983).
Abstr. – See Abstr. 012.059.

004.127 **Arawak constellations.**
F. Jara.
Archaeoastronomy (U.S.A.), Vol. 6, Nos. 1 – 4, p. 40 (1983).
Abstr. – See Abstr. 012.059.

004.128 **Ritual and calendar in ancient Hawaii.**
R. K. Johnson.
Archaeoastronomy (U.S.A.), Vol. 6, Nos. 1 – 4, p. 40 (1983).
Abstr. – See Abstr. 012.059.

004.129 **Astral aspects of the Eddas.**
B. Johnson.
Archaeoastronomy (U.S.A.), Vol. 6, Nos. 1 – 4, p. 41 (1983).
Abstr. – See Abstr. 012.059.

004.130 **Ethnoastronomy on the American Plains.**
A. B. Kehoe.
Archaeoastronomy (U.S.A.), Vol. 6, Nos. 1 – 4, p. 41 (1983).
Abstr. – See Abstr. 012.059.

004.131 **Over the rainbow: the color of cosmic order.**
E. C. Krupp.
Archaeoastronomy (U.S.A.), Vol. 6, Nos. 1 – 4, p. 41 (1983).
Abstr. – See Abstr. 012.059.

004.132 **Tzotzil starlore.**
W. Lamb.
Archaeoastronomy (U.S.A.), Vol. 6, Nos. 1 – 4, p. 41 (1983).
Abstr. – See Abstr. 012.059.

004.133 **Astronomy and lunar ritual among the Ngas of Nigeria.**
D. LaPin.
Archaeoastronomy (U.S.A.), Vol. 6, Nos. 1 – 4, p. 41 (1983). Abstr. – See Abstr. 012.059.

004.134 **The Mije calendrical system.**
F. Lipp.
Archaeoastronomy (U.S.A.), Vol. 6, Nos. 1 – 4, p. 42 (1983). Abstr. – See Abstr. 012.059.

004.135 **Carib tribal astronomy.**
E. Magaña.
Archaeoastronomy (U.S.A.), Vol. 6, Nos. 1 – 4, p. 42 (1983). Abstr. – See Abstr. 012.059.

004.136 **Astronomical discovery and the evolution of symbols in Hindu mythology.**
J. M. Malville.
Archaeoastronomy (U.S.A.), Vol. 6, Nos. 1 – 4, p. 42 (1983). Abstr. – See Abstr. 012.059.

004.137 **Ice age lunar and solar notation.**
A. Marshack.
Archaeoastronomy (U.S.A.), Vol. 6, Nos. 1 – 4, p. 42 (1983). Abstr. – See Abstr. 012.059.

004.138 **Nineteenth–century Siberian and American Indian calendar sticks.**
A. Marshack.
Archaeoastronomy (U.S.A.), Vol. 6, Nos. 1 – 4, p. 42 (1983). Abstr. – See Abstr. 012.059.

004.139 **Indigenous and Greco–Roman astronomy in sixth–century Gaul.**
S. McCluskey.
Archaeoastronomy (U.S.A.), Vol. 6, Nos. 1 – 4, p. 43 (1983). Abstr. – See Abstr. 012.059.

004.140 **The seven sisters.**
N. S. Momaday.
Archaeoastronomy (U.S.A.), Vol. 6, Nos. 1 – 4, p. 43 (1983). Abstr. – See Abstr. 012.059.

004.141 **Time and calendar in Thailand.**
T. E. Morgan.
Archaeoastronomy (U.S.A.), Vol. 6, Nos. 1 – 4, p. 43 (1983). Abstr. – See Abstr. 012.059.

004.142 **The Lakota constellation Tayamni.**
L. Nesper.
Archaeoastronomy (U.S.A.), Vol. 6, Nos. 1 – 4, p. 43 (1983). Abstr. – See Abstr. 012.059.

004.143 **The correlation of dynastic and planetary cycles in ancient China.**
D. Pankenier.
Archaeoastronomy (U.S.A.), Vol. 6, Nos. 1 – 4, p. 43 (1983). Abstr. – See Abstr. 012.059.

004.144 **Documenting Navajo star lore in a planetarium.**
M. B. Peterson.
Archaeoastronomy (U.S.A.), Vol. 6, Nos. 1 – 4, p. 43 (1983). Abstr. – See Abstr. 012.059.

004.145 **Ursa Major in North American Indian mythology.**
D. C. Raffalovich.
Archaeoastronomy (U.S.A.), Vol. 6, Nos. 1 – 4, p. 44 (1983). Abstr. – See Abstr. 012.059.

004.146 **The ethnoastronomy project of the Colombian Orient.**
E. Reichel de von Hildebrand.
Archaeoastronomy (U.S.A.), Vol. 6, Nos. 1 – 4, p. 44 (1983). Abstr. – See Abstr. 012.059.

004.147 **Tabwa cosmology.**
A. F. Roberts.
Archaeoastronomy (U.S.A.), Vol. 6, Nos. 1 – 4, p. 44 (1983). Abstr. – See Abstr. 012.059.

004.148 **Mythic substitution and the stars: aspects of Shipibo and Quechua ethnoastronomy compared.**
P. G. Roe.
Archaeoastronomy (U.S.A.), Vol. 6, Nos. 1 – 4, p. 44 (1983). Abstr. – See Abstr. 012.059.

004.149 **The haphazard astronomy of the Mursi.**
C. Ruggles.
Archaeoastronomy (U.S.A.), Vol. 6, Nos. 1 – 4, p. 44 (1983). Abstr. – See Abstr. 012.059.

004.150 **Ethnoastronomy at Vijayanagar in South India.**
T. S. Smith.
Archaeoastronomy (U.S.A.), Vol. 6, Nos. 1 – 4, p. 45 (1983). Abstr. – See Abstr. 012.059.

004.151 **Humanities and the stars: interpreting the astronomy and mythology of other cultures.**
E. M. Starr.
Archaeoastronomy (U.S.A.), Vol. 6, Nos. 1 – 4, p. 45 (1983). Abstr. – See Abstr. 012.059.

004.152 **Mayan sidereal moon reckoning.**
B. Tedlock.
Archaeoastronomy (U.S.A.), Vol. 6, Nos. 1 – 4, p. 45 (1983). Abstr. – See Abstr. 012.059.

004.153 **Mayan astronomy in the Popol Vuh.**
D. Tedlock.
Archaeoastronomy (U.S.A.), Vol. 6, Nos. 1 – 4, p. 45 (1983). Abstr. – See Abstr. 012.059.

004.154 **Celestial lore of some Australian aboriginal tribes.**
N. B. Tindale.
Archaeoastronomy (U.S.A.), Vol. 6, Nos. 1 – 4, p. 45 (1983). Abstr. – See Abstr. 012.059.

004.155 **Ethnoastronomy, social organization and the ritual calendar in an Andean community.**
G. Urton.
Archaeoastronomy (U.S.A.), Vol. 6, Nos. 1 – 4, p. 45 – 46 (1983). Abstr. – See Abstr. 012.059.

004.156 **The morning star sings like Raven: Northwest Coast Indian ethnoastronomy.**
D. Vogt.
Archaeoastronomy (U.S.A.), Vol. 6, Nos. 1 – 4, p. 46 (1983). Abstr. – See Abstr. 012.059.

004.157 **Medicine wheels, boulder by boulder.**
D. Vogt, M. Ovenden.
Archaeoastronomy (U.S.A.), Vol. 6, Nos. 1 – 4, p. 46 (1983). Abstr. – See Abstr. 012.059.

004.158 **The Navajo hogan as cosmic model.**
R. A. Williamson.
Archaeoastronomy (U.S.A.), Vol. 6, Nos. 1 – 4, p. 46 (1983). Abstr. – See Abstr. 012.059.

004.159 **Amahuaca star lore.**
J. Woodside.
Archaeoastronomy (U.S.A.), Vol. 6, Nos. 1 – 4, p. 47 (1983). Abstr. – See Abstr. 012.059.

004.160 **Astronomy in Pueblo and Navajo world views.**
M. J. Young.
Archaeoastronomy (U.S.A.), Vol. 6, Nos. 1 – 4, p. 47 (1983). Abstr. – See Abstr. 012.059.

004.161 **"La Lune sale" (The dirty moon): ethnoastronomy in Central Africa.**
D. Zahan.
Archaeoastronomy (U.S.A.), Vol. 6, Nos. 1 - 4, p. 47 (1983). Abstr. - See Abstr. 012.059.

004.162 **Historic Puebloan sun and moon watching.**
M. Zeilik.
Archaeoastronomy (U.S.A.), Vol. 6, Nos. 1 - 4, p. 47 (1983). Abstr. - See Abstr. 012.059.

004.163 **Origin and significance of the 28 Chinese lunar mansions or "Xiu".**
H. J. Zhong, Z. Z. Jiu, T. Dan, K. Brecher.
Archaeoastronomy (U.S.A.), Vol. 6, Nos. 1 - 4, p. 47 (1983). Abstr. - See Abstr. 012.059.

004.164 **The Inca solar calendar and its Huari and Tiahuanaco antecedents.**
R. T. Zuidema.
Archaeoastronomy (U.S.A.), Vol. 6, Nos. 1 - 4, p. 47 (1983). Abstr. - See Abstr. 012.059.

004.165 **Navajo constellations in literature, art, artifact and a New Mexico rock art site.**
V. D. Chamberlain.
Archaeoastronomy (U.S.A.), Vol. 6, Nos. 1 - 4, p. 48 - 58 (1983).

004.166 **Sky symbolism in a Navajo rock art site. Chaco Canyon National Historical Park, New Mexico.**
R. A. Williamson.
Archaeoastronomy (U.S.A.), Vol. 6, Nos. 1 - 4, p. 59 - 65 (1983).

004.167 **Wijiji at Chaco Canyon: a winter solstice sunrise and sunset station.**
M. Zeilik, R. Elston.
Archaeoastronomy (U.S.A.), Vol. 6, Nos. 1 - 4, p. 66 - 73 (1983).

004.168 **The reliability of historical maps of earthworks in the Ohio Valley.**
T. H. Robertson.
Archaeoastronomy (U.S.A.), Vol. 6, Nos. 1 - 4, p. 74 - 79 (1983).

004.169 **A possible aboriginal winter solstice viewing site on the southern Oregon coast.**
R. L. Hall, R. D. Hall.
Archaeoastronomy (U.S.A.), Vol. 6, Nos. 1 - 4, p. 86 - 88 (1983).

004.170 **The ancient Mexican astronomical apparatus: an iconographical criticism.**
M. E. R. G. N. Jansen, G. A. Perez.
Archaeoastronomy (U.S.A.), Vol. 6, Nos. 1 - 4, p. 89 - 95 (1983).

004.171 **A painted Venus glyph in the tower at Palenque.**
K. H. Mayer.
Archaeoastronomy (U.S.A.), Vol. 6, Nos. 1 - 4, p. 96 - 98 (1983).

004.172 **Chinese "astronomical" jade disks: the pi.**
B. E. Schaefer.
Archaeoastronomy (U.S.A.), Vol. 6, Nos. 1 - 4, p. 99 - 101 (1983).

004.173 **Invention of the sky.**
E. Magaña, F. Jara.
Archaeoastronomy (U.S.A.), Vol. 6, Nos. 1 - 4, p. 102 - 113 (1983).

004.174 **The Mallia table: kernos or clock?**
C. F. Herberger.
Archaeoastronomy (U.S.A.), Vol. 6, Nos. 1 - 4, p. 114 - 117 (1983).

004.175 **The solstice eclipses of Stonehenge II.**
J. H. Robinson.
Archaeoastronomy (U.S.A.), Vol. 6, Nos. 1 - 4, p. 124 - 131 (1983).

004.176 **Stonehenge: the Aubrey holes and the avenue.**
R. J. C. Atkinson.
Archaeoastronomy (U.S.A.), Vol. 6, Nos. 1 - 4, p. 132 - 133 (1983).

004.177 **Observatory chairs.**
C. Stott.
Vistas Astron., Vol. 27, Part 3, p. 291 - 302 (1984).

Since the early seventeenth century a variety of pieces of furniture has been used to comfort and support the observer in his aim of keeping his eye to the telescope eyepiece. This paper is concerned with a selection of designs which in large were for and by astronomers with private observatories.

004.178 **On the main instrument of the Samarkand Observatory.**
O. S. Tursunov.
Pamyatniki nauk. i tekh., 1982 - 1983. Moskva, p. 12 - 27 (1984). In Russian. Abstr. in Ref. Zh., 51. Astron., 11.51.5 (1984).

004.179 **On the history of construction of the celestial globe of the Museum–Lyceum in Pushkino.**
K. G. Bol'shakova, L. E. Majstrov.
Pamyatniki nauk. i tekh., 1982 - 1983. Moskva, p. 28 - 33 (1984). In Russian. Abstr. in Ref. Zh., 51. Astron., 11.51.6 (1984).

004.180 **Beyond the planets: early nineteenth–century studies of double stars.**
M. Williams.
Br. J. Hist. Sci., Vol. 17, Part 3, No. 57, p. 295 - 309 (1984).

004.181 **L'enigma di Stonehenge.**
A. Curir.
Orione, Vol. 4, N. 4, p. 144 - 149 (1984).

004.182 **Briefe Franz Xaver von Zachs in sein Vaterland.**
P. Brosche, M. Vargha.
Publ. Astron. Dep. Eötvös Univ., No. 7, 116 pp. (1984).

004.183 **The Louvain Lectures (Lectiones Lovanienses) of Bellarmine and the autograph copy of his 1616 declaration to Galileo.**
U. Baldini, G. V. Coyne.
Stud. Galileiani, Vol. 1, No. 2, 48 pp. (1984).

004.184 **Zur Geschichte von Bessel's großem Fraunhofer–Heliometer.**
E. H. Geyer.
Mitt. Astron. Ges., Nr. 62, p. 192 - 194 (1984). - See Abstr. 010.341.

004.185 **Frühe Beobachter veränderlicher Sterne - 200 Jahre Delta Cephei.**
R. Lukas.
Mitt. Astron. Ges., Nr. 62, p. 199 - 200 (1984). - See Abstr. 010.341.

004.186 **På väg mot en ny världsbild.**
A. Sundman.
Astron. Tidsskr., Årg. 17, Nr. 4, p. 151 - 159 (1984).

004.187 **Rings in astronomy and cosmology, 1600 – 1900.**
A. Van Helden.
Planetary rings, p. 12 – 22 (1984). – See Abstr. 012.086.
A brief outline of historical ideas on the nature and origin of planetary rings is presented.

004.188 **The centenary of the prime meridian and of international standard time.**
H. D. Black, G. Gebel, R. R. Newton.
Johns Hopkins APL Tech. Dig., Vol. 5, No. 4, p. 381 – 389 (1984).

**Mathematical astronomy in Copernicus's De revolutionibus. Parts 1 + 2.**
See Abstr. 003.010.

**Leben und Werk des Königsberger Astronomen Friedrich Wilhelm Bessel.**
See Abstr. 003.028.

**James E. Keeler. Pioneer American astrophysicist and the early development of American astrophysics.**
See Abstr. 003.048.

**Prehistoric astronomy and ritual.**
See Abstr. 003.067.

**Die ersten Astronomen. Eine Einführung in die Ursprünge der Astronomie.**
See Abstr. 003.075.

**Time stands still: new light on megalithic science.**
See Abstr. 003.076.

**Stonehenge. An ancient masonic temple.**
See Abstr. 003.100.

**The history of astronomy from Herschel to Hertzsprung.**
See Abstr. 003.101.

**In search of ancient astronomies.**
See Abstr. 003.115.

**The Greenwich Meridian.**
See Abstr. 003.119.

**La science de l'univers à l'âge du positivisme, étude sur les origines de la cosmologie contemporaine.**
See Abstr. 003.121.

**Patrick Moore's history of astronomy.**
See Abstr. 003.124.

**The Stonehenge Watch, with accompanying instructional booklet entitled "Stonehenge unraveled".**
See Abstr. 003.134.

**Copernicus and the scientific revolution.**
See Abstr. 003.144.

**Megalithic astronomy: a new archaeological and statistical study of 300 western Scottish sites.**
See Abstr. 003.146.

**Die Harmonie der Sphären: Die Geschichte der Idee des Welteneinklangs und der Seeleneinstimmung.**
See Abstr. 003.150.

**Time and the Highland Maya.**
See Abstr. 003.155.

**Galileo's early notebooks: the physical questions.**
See Abstr. 003.157.

**Friedrich Wilhelm Bessel (1784 – 1846).**
See Abstr. 005.001.

**Bessel's Arbeiten zur Positionsastronomie.**
See Abstr. 005.024.

**The early history of the Dominion Observatories.**
See Abstr. 009.033.

**The history of British astronomy and geophysics. Summary of the RAS specialist discussion, held 1983 December 9 in the Scientific Societies' Lecture Theatre, Savile Row.**
See Abstr. 010.681.

**Historia de la astronomia en Mexico.**
See Abstr. 012.090.

**Megalithic astronomy: the last five years.**
See Abstr. 013.056.

**On the astronomical orientation of St. Vitus' cathedral and St. George's church in the Castle of Prague.**
See Abstr. 015.007.

**Time computations and Dionysius Exiguus.**
See Abstr. 046.028.

**Die durchschnittliche Dauer der Jahre und Monate im babylonischen und im jüdischen Kalender.**
See Abstr. 046.030.

**An analysis of recently rediscovered eclipse observations of the first satellite of Jupiter made between 1775 and 1802, and collected by Delambre.**
See Abstr. 099.003.

**Saturn through the telescope: a brief historical survey.**
See Abstr. 100.130.

**F. W. Bessels Beobachtungen des Kometen Halley im Jahre 1835. Eine Neureduktion.**
See Abstr. 103.925.

**Supernova progenitors and Be stars: stellar variability from a 21 century perspective.**
See Abstr. 122.197.

# 005 Biography

005.001 **Friedrich Wilhelm Bessel (1784 – 1846).**
G. W. E. Beekman.
Zenit, 11. Jaarg., No. 9, p. 374 – 375 (1984).

005.002 **Der Astronom Friedrich Wilhelm Bessel. Zu seinem 200. Geburtstag am 22. Juli 1984.**
F. Schmeidler.
Nachr. Olbers–Ges. Bremen, Nr. 130, p. 1 – 3 (1984).

005.003 **Bessel und Bremen.**
W. Stein.
Nachr. Olbers–Ges. Bremen, Nr. 130, p. 4 (1984).

005.004 **Sternwarte Lilienthal. Von Johann Hieronymus Schroeter und den Instrumenten seiner Sternwarte.**
D. Gerdes.
Nachr. Olbers–Ges. Bremen, Nr. 130, p. 5 – 9 (1984).

005.005 **Bessel in Lilienthal.**
D. Gerdes.
Nachr. Olbers–Ges. Bremen, Nr. 130, p. 10 (1984).

005.006 **Bessel als Prüfer von Seeschiffsführern.**
K. Walter.
Nachr. Olbers–Ges. Bremen, Nr. 130, p. 12 – 16 (1984).

005.007 **Friedrich Wilhelm Bessel. Eine Biographie zum 200. Geburtstag.**
B. Wöbke.
Sterne Weltraum, 23. Jahrg., Nr. 7, p. 360 – 366 (1984).

005.008 **In search of Henry Fitz.**
J. Lankford.
Sky Telesc., Vol. 68, No. 3, p. 214 – 218 (1984).

005.009 **Karl Jansky and the discovery of extraterrestrial radio waves.**
W. T. Sullivan III.
The early years of radio astronomy, p. 2 – 42 (1984). – See Abstr. 003.012.

005.010 **Kunisuke Kinoshita – his life and astronomy.**
W. Unno.
Astron. Her., Vol. 77, No. 9, p. 224 – 225 (1984). In Japanese.

005.011 **Astronomical scrapbook: Herschel's 1784 autobiography.**
O. Gingerich.
Sky Telesc., Vol. 68, No. 4, p. 317 – 319 (1984).

005.012 **Friedrich Wilhelm Bessel zum 200. Geburtstag.**
A. R. Serban.
Orion, 42. Jahrg., Nr. 204, p. 174 – 175 (1984).

005.013 **Friedrich Wilhelm Bessel – seine Bedeutung in der Vergangenheit und heute.**
K.-G. Steinert.
Astron. Raumfahrt, 22. Jahrg., Heft 5, p. 104 – 110 (1984).

005.014 **The Old Guard.**
R. N. Mayall.
J. Am. Assoc. Variable Star Obs., Vol. 12, No. 2, p. 77 (1983). Abstract. – See Abstr. 010.021.

005.015 **F. Bessel, on the occasion of the 200th anniversary of his birthday.**
V. G. Demin.
Zemlya Vselennaya, No. 4, p. 46 – 52 (1984). In Russian.

005.016 **The rise and fall of Edward S. Holden: Part 2.**
D. E. Osterbrock.
J. Hist. Astron., Vol. 15, Part 3, No. 44, p. 151 – 176 (1984).

005.017 **An astronomical life.**
J. L. Greenstein.
Annu. Rev. Astron. Astrophys., Vol. 22, p. 1 – 35 (1984). – See Abstr. 003.024.

005.018 **P. M. Gorshkov, 1883 – 1975, on the occasion of the 100th anniversary of his birthday.**
V. K. Abalakin.
Zemlya Vselennaya, No. 5, p. 67 – 72 (1984). In Russian.

005.019 **Paul Ledoux's academic career.**
M. Dehousse.
Theoretical problems in stellar stability and oscillations, p. IV – VI (1984). – See Abstr. 012.065.

005.020 **Paul Ledoux's scientific career.**
A. Noels, M. Gabriel.
Theoretical problems in stellar stability and oscillations, p. IX – XVI (1984). – See Abstr. 012.065.

005.021 **Der Galilei–Prozess (12. April – 22. Juni 1633).**
G. Hamann.
Tätigkeitsbericht der Österreichischen Akademie der Wissenschaften 1982/83, Sonderdr., Nr. 2, p. 21 – 45 (1984).

005.022 **William Herschel and the first two asteroids.**
C. Cunningham.
Minor Planet Bull., Vol. 11, No. 1, p. 3 (1984).

005.023 **Al–K̄hwarizmi – a 9th century mathematician and astronomer.**
A. R. Serban.
Madà, Vol. 28, No. 4, p. 181 (1984). In Hebrew.

005.024 **Bessel's Arbeiten zur Positionsastronomie.**
F. Schmeidler.
Mitt. Astron. Ges., Nr. 62, p. 75 – 85 (1984). – See Abstr. 010.341.

After some reflections on the importance of the history of astronomy, the scientific work of F. W. Bessel is considered as a consequence of the astronomy of the 18th century in which celestial mechanics and the improvement of the astrometric instruments were two main subjects of reseach. Bessel has made important contributions to theoretical astronomy; he has also been a successful observer who made use of the new instruments available in his time. Some details about his works are reported. Information on the influence exerted by him on the astronomers of the 19th century and a general characterization of his astronomical work are given.

005.025 **Das Bessel–Portrait von Jensen.**
H. W. Duerbeck, E. H. Geyer.
Mitt. Astron. Ges., Nr. 62, p. 191 (1984). Abstract. – See Abstr. 010.341.

005.026 **Thomas Bugge og hans berømte refraktor.**
P. Darnell.
Astron. Tidsskr., Årg. 17, Nr. 4, p. 145 – 150 (1984).

**Leben und Werk des Königsberger Astronomen Friedrich Wilhelm Bessel.**
See Abstr. 003.028.

**James E. Keeler. Pioneer American astrophysicist and the early development of American astrophysics.**
See Abstr. 003.048.

**Stephen Hawking's Universe.**
See Abstr. 003.063.

**Ascent to orbit: a scientific autobiography.**
See Abstr. 003.073.

**Copernicus and the scientific revolution.**
See Abstr. 003.144.

**Bernard Lovell: a biography.**
See Abstr. 003.149.

**Galileo's early notebooks: the physical questions.**
See Abstr. 003.157.

**Friedrich Wilhelm Bessel: Ich habe Euch lieb – aber der Himmel ist mir näher. Eine Autobiographie in Briefen.**
See Abstr. 003.162.

**A Lutheran astrologer: Johannes Kepler.**
See Abstr. 004.069.

**Paris Observatory in 1784 – Cassini de Thury's legacy to Jean-Dominique Cassini.**
See Abstr. 009.039.

# 006 Personal Notes

006.001 **Les 80 ans d'Emile Antonini/Zum 80. Geburtstag von Emile Antonini.**
W. Maeder.
Orion, 42. Jahrg., Nr. 204, p. 176 (1984).

006.002 **Deborah Byrd received the Klumpke–Roberts Award 1984 of the Astronomical Society of the Pacific.**
S. Wolff, A. Fraknoi.
Mercury, Vol. 13, No. 6, p. 188 (1984).

006.003 **Subrahmanyan Chandrasekhar and William A. Fowler received the 1983 Nobel Prize for Physics.**
J. Elliott.
Ir. Astron. J., Vol. 16, No. 4, p. 287 – 288 (1984).

006.004 **Nobel Prize in Physics 1983: Stellar structure and evolution; origin of the elements in the universe. Awarded to S. Chandrasekhar and W. A. Fowler.**
A. R. Serban.
Madà, Vol. 28, No. 2, p. 74 – 76, 83 (1984). In Hebrew.

006.005 **Subrahmanyan Chandrasekhar – Nobel laureate.**
Patrika, No. 7, p. 1 – 2 (1984).

006.006 **William A. Fowler and Subrahmanyan Chandrasekhar received the 1983 Nobel Prize for Physics.**
J. Elliott.
Ir. Astron. J., Vol. 16, No. 4, p. 287 – 288 (1984).

006.007 **Nobel Prize in Physics 1983: Stellar structure and evolution; origin of the elements in the universe. Awarded to W. A. Fowler and S. Chandrasekhar.**
A. R. Serban.
Madà, Vol. 28, No. 2, p. 74 – 76, 83 (1984). In Hebrew.

006.008 **Russel M. Genet received the Amateur Achievement Award 1984 of the Astronomical Society of the Pacific.**
S. Wolff, A. Fraknoi.
Mercury, Vol. 13, No. 6, p. 168 (1984).

006.009 **George Greenstein received the AIP–US Steel Foundation Science–Writing award in physics.**
Phys. Today, Vol. 37, No. 12, p. 65 (1984).

006.010 **Deidre Hunter received the Robert J. Trumpler Award 1984 of the Astronomical Society of the Pacific.**
S. Wolff, A. Fraknoi.
Mercury, Vol. 13, No. 6, p. 188 (1984).

006.011 **Deidre Hunter received the Robert J. Trumpler Award 1984 of the Astronomical Society of the Pacific.**
Phys. Today, Vol. 37, No. 12, p. 80 (1984).

006.012 **On the occasion of the 75th birthday of E. K. Kharadze.**
Eh. A. Shevardnadze.
Istor. nauk, Tom 6, Tbilisi, p. 142 (1984). In Russian. Abstr. in Ref. Zh., 51. Astron., 10.51.12 (1984).

006.013 **B. Yu. Levin received the Leonard Medal.**
Zemlya Vselennaya, No. 5, p. 105 (1984). In Russian.

006.014 **D. Lynden–Bell received the 1984 Eddington Medal of the Royal Astronomical Society.**
R. Hide.
Q. J. R. Astron. Soc., Vol. 25, No. 3, p. 231 (1984).

006.015 **Ein Leben für die Raumfahrt. Zum 90. Geburtstag von Hermann Oberth.**
F.-E. Rietz.
Astron. Raumfahrt, 22. Jahrg., Heft 4, p. 74 – 77 (1984).

006.016 **Jan Hendrik Oort received the Balzan–Prize 1984 for astrophysics.**
Naturwissenschaften, 71. Jahrg., Nr. 12, p. 646 (1984).

006.017 **Danie Overbeek received the Gill Medal 1984.**
Mon. Notes Astron. Soc. S. Afr., Vol. 43, Nos. 9 – 10, p. 41 (1984).

006.018 **S. K. Runcorn received the 1984 Gold Medal of the Royal Astronomical Society.**
R. Hide.
Q. J. R. Astron. Soc., Vol. 25, No. 3, p. 229 – 230 (1984).

006.019 **V. P. Shcheglov, on the occasion of his 80th birthday on 28 June 1984.**
K. A. Kulikov.
Zemlya Vselennaya, No. 5, p. 63 – 65 (1984). In Russian.

006.020 **Eugene M. Shoemaker received the Kuiper Prize of the American Astronomical Society.**
Phys. Today, Vol. 37, No. 12, p. 79 – 80 (1984).

006.021 **David Stevenson received the Urey Prize of the American Astronomical Society.**
Phys. Today, Vol. 37, No. 12, p. 79 – 80 (1984).

006.022 **William G. Unruh received the Steacie Prize for 1983.**
Phys. Today, Vol. 37, No. 7, p. 82 (1984).

006.023 **Olin C. Wilson received the Catherine Wolfe Bruce Medal 1984 of the Astronomical Society of the Pacific.**
S. Wolff, A. Fraknoi.
Mercury, Vol. 13, No. 6, p. 187 (1984).

006.024 **Olin C. Wilson received the Catherine Wolfe Bruce Medal of the Astronomical Society of the Pacific.**
Phys. Today, Vol. 37, No. 12, p. 80 (1984).

# 007 Obituaries

007.001 **George O. Abell (1927 – 1983).**
D. M. Popper, D. A. Pierce.
Mercury, Vol. 13, No. 4, p. 108 – 112 (1984).

007.002 **Alfio Betti.**
L. Baldinelli.
G. A.A.B., N. 75, p. 4 (1984).

007.003 **C. R. Burch died 1983 July 19.**
D. F. Gibbs.
Q. J. R. Astron. Soc., Vol. 25, No. 4, p. 533 – 534 (1984).

007.004 **In memoriam: Louis A. Capone (1943 July 29 – 1984 April 25).**
J. N. Cuzzi, R. C. Whitten, O. B. Toon.
Icarus, Vol. 59, No. 3, p. 303 (1984).

007.005 **Giuseppe Colombo died 1984 February 20.**
G. Field, I. Shapiro.
Phys. Today, Vol. 37, No. 10, p. 111 – 112 (1984).

007.006 **Ernst A. Dibaj, 1931 – 1983.**
Sov. Astron., Vol. 28, No. 1, p. 121 – 122 (1984). English translation of 37.007.015.

007.007 **Alan Fletcher, 1903 December 6 – 1981 July 4.**
P. J. Message.
Q. J. R. Astron. Soc., Vol. 25, No. 3, p. 372 – 373 (1984).

007.008 **Scott Ellsworth Forbush died 1984 April 4.**
M. A. Pomerantz.
Phys. Today, Vol. 37, No. 10, p. 111 (1984).

007.009 **In memoriam: R. G. Giovanelli died 1984 January 27.**
J. T. Jefferies.
Sol. Phys., Vol. 94, No. 1, p. 1 – 2 (1984).

007.010 **Willy Hartner, 1905 January 22 – 1981 May 16.**
M. Hoskin.
Q. J. R. Astron. Soc., Vol. 25, No. 3, p. 373 (1984).

007.011 **Otto H. L. Heckmann, 1901 June 23 – 1983 May 13.**
W. Fricke.
Q. J. R. Astron. Soc., Vol. 25, No. 3, p. 374 – 376 (1984).

007.012 **In memoriam: Jeanne Hopkins (1923 – 1984).**
Astrophys. J., Vol. 286, No. 1, p. 1 (1984).

007.013 **Abdullah Kizilirmak died 1983 December 5.**
Mitt. Astron. Ges., Nr. 61, p. 5 – 6 (1984).

007.014 **E. L. Krinov, 1906 March 3 – 1984 January 2.**
Meteoritika, Vyp. 43, p. 186 – 187 (1984). In Russian.

007.015 **In memoriam: E. L. Krinov, 1906 March 3 – 1984 January 2.**
Zemlya Vselennaya, No. 5, p. 73 – 75 (1984). In Russian.

007.016 **Sir Harrie Massey, 1908 – 1983 November 27.**
M. J. Seaton.
Q. J. R. Astron. Soc., Vol. 25, No. 4, p. 530 – 532 (1984).

007.017 **Norman Graham Matthew, 1908 May 13 – 1982 August 15.**
D. C. Heggie.
Q. J. R. Astron. Soc., Vol. 25, No. 3, p. 377 – 379 (1984).

007.018 **A. A. Mikhajlov, 1888 April 26 – 1983 September 29.**
Sov. Astron., Vol. 28, No. 2, p. 243 – 244 (1984). English translation of 37.007.022.

007.019 **In memoriam: A. A. Mikhajlov (1888 – 1983).**
M. S. Zverev.
Tsirk. Vses. astron.-geod. o-vo Akad. Nauk SSSR, No. 37, p. 78 – 88 (1984). In Russian. Abstr. in Ref. Zh., 51. Astron., 9.51.16 (1984).

007.020 **A. A. Mikhajlov, 1888 – 1983.**
Vestn. AN SSSR, No. 2, p. 99 (1984). In Russian. Abstr. in Ref. Zh., 51. Astron., 8.51.6 (1984).

007.021 **Alois Purgathofer, 1925 February 18 – 1984 March 13.**
A. Schnell, W. M. Tscharnuter.
Mitt. Astron. Ges., Nr. 61, p. 7 – 8 (1984).

007.022 **Sir Martin Ryle 1918 – 1984.**
Gemini, No. 12, p. 12 (1984).

007.023 **Martin Ryle 1918 – 1984.**
F. G. Smith.
Nature, Vol. 312, No. 5989, p. 18 (1984).

007.024 **Professor Sir Martin Ryle died 1984 October 14.**
G. G. Pooley.
Observatory, Vol. 104, No. 1063, p. 283 – 284 (1984).

007.025 **Charles Donald Shane, 1895 September 6 – 1983 March 19.**
S. Vasilevskis.
Q. J. R. Astron. Soc., Vol. 25, No. 4, p. 532 – 533 (1984).

007.026 **In memoriam: Frano Simović, 1907 – 1984.**
Vasiona, Année 32, No. 2, p. 41 (1984). In Serbo-Croatian.

007.027 **Bohumil Šternberk, 1897 January 21 – 1983 March 24.**
Z. Horský.
Q. J. R. Astron. Soc., Vol. 25, No. 3, p. 379 – 380 (1984).

007.028 **In memoriam: Pol Swings (1906 – 1983).**
P. Ledoux.
Astrophys. Space Sci., Vol. 102, No. 1, p. 1 – 2 (1984).

007.029 **In memoriam: Pol Swings (1906 – 1983).**
P. Ledoux.
Ciel Terre, Vol. 100, No. 4, p. 135 – 138 (1984).

007.030 **Giuseppe Tagliaferri: 1924 – 1984.**
Coelum, Vol. 52, N. 6, p. 325 (1984).

007.031 **V. P. Tsesevich, 1907 October 11 – 1983 October 28.**
Astron. Tsirk., No. 1319, p. 6 – 8 (1984). In Russian.

007.032 **In memoriam: V. P. Tsesevich (1907 – 1983).**
V. A. Smirnov.
Tsirk. Vses. astron.-geod. o-vo Akad. Nauk SSSR, No. 37, p. 89 – 92 (1984). In Russian. Abstr. in Ref. Zh., 51. Astron., 9.51.20 (1984).

007.033 **V. P. Tsesevich, 1907 October 11 – 1983 October 28.**
Zemlya Vselennaya, No. 4, p. 58 – 60 (1984). In Russian.

007.034 **John Marsh Wilcox, 1925 January 31 – 1983 October 14.**
A. B. C. Walker Jr., R. A. Helliwell, P. A. Sturrock.
Q. J. R. Astron. Soc., Vol. 25, No. 4, p. 534 – 536 (1984).

# 008 Publications of Observatories, Institutes

Reports, communications, and publications of observatories and astronomical institutes are recorded in this section; included are numbered series of reprints. Whenever possible, the numbers of the abstracts referring to the publications are given. The places of observatories and institutes are listed in alphabetical order. If only the formal name of an observatory or institute is known, its place can be found in the following index list.

| | |
|---|---|
| Aerospace Corporation | **El Segundo,** Calif. |
| Algonquin Radio Observatory | **Lake Traverse,** Canada |
| Allegheny Observatory | **Pittsburgh,** Pa. |
| Anglo-Australian Observatory | **Epping,** Australia |
| Antares Observatory | **Feira de Santana,** Brazil |
| Applied Research Corporation | **Landover,** Md. |
| Archenhold Observatory | **Berlin,** German Democratic Republic |
| Argentine Institute of Radio Astronomy | **Villa Elisa,** Argentina |
| Argentine Radio Astronomy Institute | **Pereyra,** Argentina |
| Arizona State University | **Tempe,** Ariz. |
| Astronomisches Rechen-Institut | **Heidelberg,** F.R. Germany |
| | |
| Babelsberg Observatory | **Potsdam,** German Democratic Republic |
| Bartol Research Foundation | **Newark,** Del. |
| Battelle Memorial Institute | **Richland,** Wash. |
| Bell Telephone Laboratories | **Holmdel,** N.J. |
| Bergedorf Observatory | **Hamburg,** F.R. Germany |
| Blindern Institute of Theoretical Astrophysics | **Oslo,** Norway |
| Bosscha Observatory | **Lembang,** Indonesia |
| Bouzareah Observatory | **Algiers,** Algeria |
| Boyden Observatory | **Bloemfontein,** South Africa |
| Bureau International de l'Heure | **Paris,** France |
| | |
| C. E. Kenneth Mees Observatory | **Rochester,** N.Y. |
| California Institute of Technology | **Pasadena,** Calif. |
| Cantonal Observatory | **Neuchâtel,** Switzerland |
| Carter Observatory | **Wellington,** New Zealand |
| Center for Astrophysics | **Cambridge,** Mass. |
| Center for High Angular Resolution Astronomy, Georgia State University | **Atlanta,** Ga. |
| Centre de Données Stellaires | **Strasbourg,** France |
| Centro Astronomico Hispano-Aleman | **Calar Alto,** Spain |
| Centro de Investigación de Astronomía (CIDA) | **Mérida,** Venezuela |
| Cerro Calan National Astronomical Observatory | **Santiago,** Chile |
| Cerro Tololo Interamerican Observatory | **La Serena,** Chile |
| Chamberlin Observatory | **Denver,** Colo. |
| Charles University Astronomical Institute | **Prague,** Czechoslovakia |
| Climenhaga Observatory | **Victoria,** Canada |
| Cointe Observatory | **Liège,** Belgium |
| Computer Sciences Corporation | **Silver Spring,** Md. |
| Cornell University | **Ithaca,** N.Y. |
| Crawford Hill Observatory | **Holmdel,** N.J. |
| | |
| David Dunlop Observatory | **Richmond Hill,** Canada |
| Dearborn Observatory | **Evanston,** Ill. |
| Department of Astronomy and Space Sciences, Punjabi University | **Patiala,** India |
| Department of Mathematics, University of Poona | **Poona,** India |
| Deutsches Hydrographisches Institut | **Hamburg,** F.R. Germany |
| Dominion Astrophysical Observatory | **Ottawa,** Canada |
| Dominion Astrophysical Observatory | **Victoria,** Canada |
| Dominion Radio Astrophysical Observatory | **Penticton,** Canada |
| Dudley Observatory | **Schenectady,** N.Y. |
| Dunsink Observatory | **Dublin,** Ireland |
| Dyer Observatory | **Nashville,** Tenn. |
| | |
| Ebro Observatory | **Roquetas,** Spain |
| Effelsberg Radio Observatory | **Bonn,** F.R. Germany |
| Ege University Observatory | **Izmir,** Turkey |
| Electronics Research Laboratory | **Göteborg,** Sweden |
| Engelhardt Observatory | **Kazan,** USSR |
| Erwin W. Fick Observatory | **Ames,** Iowa |
| European Southern Observatory | **Garching,** F.R. Germany |
| European Southern Observatory | **La Silla,** Chile |

| | |
|---|---|
| Fabra Observatory | **Barcelona,** Spain |
| Felix Aguilar Observatory | **San Juan,** Argentina |
| Fernbank Science Center | **Atlanta,** Ga. |
| Figl Observatory | **Vienna,** Austria |
| Five College Astronomy Department | **Amherst,** Mass. |
| Floirac Observatory | **Bordeaux,** France |
| Florida State University Radio Observatory | **Tallahassee,** Fla. |
| Flower and Cook Observatory | **Malvern,** Pa. |
| Fort Skala Station | **Cracow,** Poland |
| Franko State University Astronomical Observatory | **Lvov,** USSR |
| Georgia State University | **Atlanta,** Ga. |
| Goddard Space Flight Center | **Greenbelt,** Md. |
| Goethe Link Observatory | **Bloomington,** Indiana |
| Hale Observatories | **Pasadena,** Calif. |
| Hartebeesthoek Radio Astronomy Observatory | **Johannesburg,** South Africa |
| Haute Provence Observatory | **Saint Michel,** France |
| Haystack Observatory | **Westford,** Mass. |
| Heinrich-Hertz-Institut | **Berlin,** German Democratic Republic |
| Heliophysical Observatory | **Debrecen,** Hungary |
| Herzberg Institute of Astrophysics | **Ottawa,** Canada |
| High Altitude Observatory | **Boulder,** Colo. |
| H.M. Nautical Almanac Office | **Greenwich,** England |
| Hoher List Observatory | **Bonn,** F.R. Germany |
| Hopkins Observatory | **Williamstown,** Mass. |
| Horn d'Arturo Observatory | **Bologna,** Italy |
| Hvar Observatory | **Zagreb,** Yugoslavia |
| IBM Thomas J. Watson Research Center | **Yorktown Heights,** N.Y. |
| Indian Institute of Astrophysics | **Bangalore,** India |
| Infrared Telescope Facility | **Honolulu,** Hawaii |
| Institut Astrofiziki | **Dushanbe,** USSR |
| Instituto Nacíonal de Astrofisica, Optica y Electronica | **Puebla,** Pue., México |
| International Latitude Observatory | **Carloforte-Cagliari,** Italy |
| International Latitude Observatory | **Mizusawa,** Japan |
| IUE Observatory | **Villafranca,** Spain |
| James Mims Observatory | **Baton Rouge,** La. |
| Jet Propulsion Laboratory | **Pasadena,** Calif. |
| Jodrell Bank Radio Observatory | **Manchester,** England |
| Judson B. Coit Observatory | **Boston,** Mass. |
| Kandilli Observatory | **Istanbul,** Turkey |
| Kanzelhöhe Solar Observatory | **Graz,** Austria |
| Kapteyn Astronomical Laboratory | **Groningen,** The Netherlands |
| Karl Schwarzschild Observatory | **Tautenburg,** German Democratic Republic |
| Kiepenheuer Institut | **Freiburg,** F.R. Germany |
| Kiso Observatory | **Tokyo,** Japan |
| Kitt Peak National Observatory | **Tucson,** Ariz. |
| Kodaikanal Observatory | **Bangalore,** India |
| Königstuhl Observatory | **Heidelberg,** F.R. Germany |
| Konkoly Observatory | **Budapest,** Hungary |
| Korean National Astronomical Observatory | **Seoul,** Korea |
| Kwasan and Hida Observatories | **Kyoto,** Japan |
| Las Campanas Observatory | **Pasadena,** Calif. |
| Latitude Station of the Polish Academy of Sciences | **Borowiec,** Poland |
| Lawrence Livermore National Laboratory | **Livermore,** Calif. |
| Leander McCormick Observatory | **Charlottesville,** Va. |
| Lick Observatory | **Santa Cruz,** Calif. |
| Lindheimer Astronomical Research Center | **Evanston,** Ill. |
| Llano del Hato Observatory | **Merida,** Venezuela |
| Lockheed Palo Alto Research Laboratory | **Palo Alto,** Calif. |
| Lockheed Solar Observatory | **Saugus,** Calif. |
| Lohrmann Observatory | **Dresden,** German Democratic Republic |
| Louisiana State University Observatory | **Baton Rouge,** La. |
| Lowell Observatory | **Flagstaff,** Ariz. |
| Lunar and Planetary Laboratory | **Tucson,** Ariz. |
| Main Astronomical Observatory of the USSR Academy of Sciences | **Pulkovo,** USSR |
| Max-Planck-Institut für Astronomie | **Calar Alto,** Spain |

| | |
|---|---|
| Max-Planck-Institut für Astronomie | **Heidelberg,** F.R. Germany |
| Max-Planck-Institut für Kernphysik | **Heidelberg,** F.R. Germany |
| Max-Planck-Institut für Physik und Astrophysik | **Garching,** F.R. Germany |
| Max-Planck-Institut für Radioastronomie | **Bonn,** F. R. Germany |
| McDonald Observatory | **Fort Davis,** Tex. |
| McDonnell Center for the Space Sciences | **St. Louis,** Mo. |
| Meudon Observatory | **Paris,** France |
| Michigan State University Observatory | **East Lansing,** Mich. |
| Millstone Hill Radar Observatory | **Westford,** Mass. |
| Minor Planet Center | **Cambridge,** Mass. |
| Molonglo Radio Observatory | **Sydney,** Australia |
| Monterey Institute for Research in Astronomy | **Carmel Valley,** Calif. |
| Mount Hamilton | **Santa Cruz,** Calif. |
| Mount John University Observatory | **Lake Tekapo,** New Zealand |
| Mount Palomar Observatory | **Pasadena,** Calif. |
| Mount Stromlo Observatory | **Canberra,** Australia |
| Mount Wilson Observatory | **Pasadena,** Calif. |
| Mullard Radio Astronomy Observatory | **Cambridge,** England |
| Mullard Space Science Laboratory | **London,** England |
| Multiple Mirror Telescope Observatory | **Tucson,** Ariz. |
| N. Copernicus Astronomical Center | **Warsaw,** Poland |
| N. Copernicus University Observatory | **Torun,** Poland |
| Narrabri Observatory | **Sydney,** Australia |
| NASA Headquarters | **Washington,** D.C. |
| National Astronomy and Ionosphere Center | **Ithaca,** N.Y. |
| National Bureau of Standards | **Washington,** D.C. |
| National Radio Astronomy Observatory | **Green Bank,** W.Va. |
| National Radio Astronomy Observatory | **Socorro,** N.M. |
| National Solar Observatory | **Sunspot,** N.M. |
| Naval Observatory | **San Fernando** (Cadiz), Spain |
| New Mexico State University Observatory | **Las Cruces,** N.M. |
| Nicholas Copernicus Observatory | **Brno,** Czechoslovakia |
| Nizamiah and Japal-Rangapur Observatories | **Hyderabad,** India |
| Nuffield Radio Astronomy Laboratories | **Manchester,** England |
| Oak Ridge Observatory | **Cambridge,** Mass. |
| Observatorio del Ebro | **Roquetas,** Spain |
| Ole Roemer Observatory | **Aarhus,** Denmark |
| Onsala Space Observatory | **Göteborg,** Sweden |
| Owens Valley Radio Observatory | **Big Pine,** Calif. |
| Pacific Northwest Laboratory | **Richland,** Wash. |
| Pennsylvania State University | **University Park,** Pa. |
| Perkins Observatory | **Delaware,** Ohio |
| Perth Observatory | **Bickley,** Australia |
| Pic du Midi Observatory | **Toulouse,** France |
| Pino Torinese Observatory | **Turin,** Italy |
| Purple Mountain Observatory | **Nanking,** China |
| Raman Research Institute | **Bangalore,** India |
| Rattlesnake Mountain Observatory | **Richland,** Wash. |
| Remeis Observatory | **Bamberg,** F.R. Germany |
| Rensselaer Observatory | **Troy,** N.Y. |
| Ritter Astrophysical Research Center | **Toledo,** Ohio |
| Rosemary Hill Observatory | **Bronson,** Fla. |
| Rothney Astrophysical Observatory | **Calgary,** Canada |
| Royal Observatory | **Edinburgh,** Scotland |
| Royal Observatory | **Greenwich,** England |
| Royal Observatory of Belgium | **Uccle,** Belgium |
| Rutgers University | **Piscataway,** N.J. |
| Sagamore Hill Radio Observatory | **Hanscom,** Mass. |
| Saltsjöbaden Observatory | **Stockholm,** Sweden |
| San Vittore Observatory | **Bologna,** Italy |
| Satelliten-Beobachtungsstation | **Zimmerwald,** Switzerland |
| Shaanxi Astronomical Observatory | **Lintong,** China |
| Smithsonian Astrophysical Observatory | **Cambridge,** Mass. |
| Sommers-Bausch Observatory | **Boulder,** Colo. |
| Sonnenborgh Observatory | **Utrecht,** The Netherlands |
| Sonoma State University | **Rohnert Park,** Calif. |
| South African Astronomical Observatory | **Cape Town,** South Africa |
| Space Telescope Science Institute | **Baltimore,** Md. |
| Special Astrophysical Observatory | **Zelenchukskaya,** USSR |

| | |
|---|---|
| Sproul Observatory | **Swarthmore,** Pa. |
| Stanford Center for Radar Astronomy | **Menlo Park,** Calif. |
| Stellar Data Center | **Strasbourg,** France |
| Sternberg State Astronomical Institute | **Moscow,** USSR |
| Steward Observatory | **Tucson,** Ariz. |
| Stockert Radio Observatory | **Bonn,** F.R. Germany |
| Struve Astrophysical Observatory | **Tartu,** USSR |
| | |
| Tata Institute of Fundamental Research | **Bombay,** India |
| Tôhoku University Observatory | **Sendai,** Japan |
| | |
| United Kingdom Infrared Telescope (UKIRT) | **Hilo,** Hawaii |
| University of Sussex | **Brighton,** England |
| U.S. Naval Observatory | **Washington,** D.C. |
| UT Radio Astronomy Observatory | **Marfa,** Tex. |
| Uttar Pradesh State Observatory | **Naini Tal,** India |
| | |
| Van Vleck Observatory | **Middletown,** Conn. |
| Vatican Observatory | **Castel Gandolfo,** Vatican City |
| | |
| Warner and Swasey Observatory | **Cleveland,** Ohio |
| Washburn Observatory | **Madison,** Wis. |
| Wesleyan Radio Observatory | **Delaware,** Ohio |
| Western Ontario University Observatory | **London,** Canada |
| Whipple Observatory (MMT) | **Tucson,** Ariz. |
| Wilhelm Foerster Observatory | **Berlin,** F.R. Germany |
| | |
| Yale University Astronomical Observatory | **New Haven,** Conn. |
| Yerkes Observatory | **Williams Bay,** Wis. |
| | |
| Zentralinstitut für Physik der Erde | **Potsdam,** German Democratic Republic |

008.039 **Arecibo**

**National Astronomy and Ionosphere Center, Newsletter.**
No. 4 (1984).

008.042 **Armagh**

**Armagh Observatory in 1982/83.**
M. de Groot.
Ir. Astron. J., Vol. 16, No. 4, p. 276 – 286 (1984).
This report covers the period 1st April, 1982 until 31st March, 1983.

**Armagh Observatory Publications.**
List No. 3, 1983–84 (1984): Nos. 1 (34.012.026), 2 (34.122.076), 3 (34.122.088), 4 (34.072.016), 5 (34.117.086), 6 (37.122.084), 7 (34.122.119), 8 (34.114.126), 9 (37.122.010), 10 (34.008.043), 11, 12 (34.073.147), 13 (37.073.012), 14 (33.022.061), 15 (34.076.033), 16 (34.022.046), 17 (34.065.001), 18 (34.114.078), 19 (34.114.052), 20 (33.073.142).

008.054 **Austin, Tex.**

**Department of Astronomy of the University of Texas, Reprints.**
Nos. 1151 (34.155.002), 1152 (33.113.053), 1153 (34.157.042), 1154 (34.117.006), 1155 (33.099.088), 1156 (34.122.005), 1157 (34.126.007), 1158 (34.122.066), 1159 (34.067.025), 1160 (34.126.009), 1161 (34.131.055), 1162 (34.157.033), 1163 (33.117.137), 1164 (34.101.017), 1165 (34.119.030), 1166 (34.131.019), 1167 (34.157.048), 1168 (34.124.321), 1169 (34.117.107), 1170, 1171 (34.099.034), 1172 (34.124.401), 1173 (34.131.169), 1174 (34.155.035), 1175 (34.122.041), 1176 (34.159.069), 1177 (34.114.065), 1178 (34.119.064), 1179 (34.091.016), 1180 (37.022.099), 1181 (34.159.200), 1182 (34.114.117), 1183 (34.042.065), 1184 (34.120.025), 1185 (34.112.045), 1186 (34.096.019), 1187 (34.158.192), 1188 (34.161.416), 1189 (37.120.005), 1190 (37.119.022), 1191 (37.151.011), 1192 (37.131.008), 1193 (37.117.016), 1194 (34.098.134), 1195 (34.098.016), 1196 (37.117.088), 1197 (37.151.032), 1198 (37.159.061), 1199 (37.131.093), 1200 (37.131.045), 1201 (37.131.096), 1202 (37.041.003), 1203 (37.158.014), 1204 (37.121.019), 1205 (37.114.049), 1206 (37.114.050), 1207 (37.131.113), 1208 (37.064.023), 1209 (37.125.060), 1210 (37.126.056), 1211 (37.126.063), 1212 (37.122.078), 1213 (37.101.005), 1214 (37.098.022), 1215 (37.114.096), 1216 (37.131.173), 1217 (37.114.115), 1218 (37.114.086), 1219 (37.114.087), 1220 (37.114.089), 1221 (37.131.231), 1222 (37.121.038), 1223 (37.112.003), 1224 (37.157.197), 1225 (37.118.031), 1226 (37.065.084), 1227 (38.064.001), 1228 (38.122.013), 1229 (37.101.020), 1230 (37.131.120), 1231 (37.008.055), 1232 (37.065.102), 1233 (37.124.152).

008.057 **Baltimore, Md.**

**Space Telescope Science Institute.**
AURA. The first twenty-five years 1957 – 1982, p. 34 – 43 (1983). – See Abstr. 010.351.

**Space Telescope Science Institute, Preprint Series.**
Nos. 18 (38.131.197), 19 (38.157.166), 20 (38.131.198), 21 (38.160.063), 22 (38.126.069), 23 (38.042.075), 24 (38.158.205), 25 (38.132.048), 26 (38.064.055), 27 (38.158.206), 28 (38.112.079), 29 (38.034.180), 30 (38.021.015), 31 (38.117.210), 32 (38.036.153), 33 (38.114.103).

**Space Telescope Science Institute, Newsletter.**
Vol. 1, Nos. 3 – 4 (1984).

008.060 **Bamberg**

**Dr.–Remeis–Sternwarte Bamberg, Astronomisches Institut der Universität Erlangen–Nürnberg. Report for 1983.**
I. Bues.
Mitt. Astron. Ges., Nr. 61, p. 11 – 16 (1984).

008.063 **Bangalore**

**Indian Institute of Astrophysics. Reports for the periods 1981 April 1 – 1982 March 31 and 1982 April 1 – 1983 March 31.**
M. K. V. Bappu.
Bull. Astron. Soc. India, Vol. 12, No. 3, p. 269 – 286, 293 – 309 (1984).

**Raman Research Institute. Reports for the periods 1981 April 1 – 1982 March 31 and 1982 April 1 – 1983 March 31.**
V. Radhakrishnan.
Bull. Astron. Soc. India, Vol. 12, No. 3, p. 287 – 292, 310 – 316 (1984).

008.069 **Basel**

**Astronomisches Institut der Universität Basel. Report for 1983.**
G. A. Tammann.
Mitt. Astron. Ges., Nr. 61, p. 17 – 22 (1984).

008.084 **Berlin (East)**

**Heinrich–Hertz–Institut Solar Data.**
Vol. 34, October – December 1983 (38.072.077).

**Heinrich–Hertz–Institut Supplement Series of Solar Data.**
Vol. 5, No. 4 (38.072.078).

008.087 **Berlin (West)**

**Institut für Astronomie und Astrophysik der Technischen Universität Berlin. Report for 1983.**
R. Wielen.
Mitt. Astron. Ges., Nr. 61, p. 23 – 30 (1984).

**Institut für Astronomie und Astrophysik der Technischen Universität Berlin, Mitteilungen.**
No. 5 (38.008.087).

008.093 **Bloemfontein**

**University of the Orange Free State: Boyden Observatory, Department of Astronomy. Report 1983.**
A. H. Jarrett.
Mon. Notes Astron. Soc. S. Afr., Vol. 43, Nos. 1 – 4, p. 8 – 9 (1984).

**Boyden Observatory, Reprint.**
Nos. 55 (37.132.063), 56 (37.034.071).

008.099 **Bochum**

**Astronomisches Institut der Ruhr–Universität. Report for 1983.**
T. Schmidt–Kaler.
Mitt. Astron. Ges., Nr. 61, p. 31 – 39 (1984).

**Bereich Extraterrestrische Physik in der Abteilung XII der Ruhr–Universität Bochum. Report for 1983.**
R. H. Giese.
Mitt. Astron. Ges., Nr. 61, p. 41 – 42 (1984).

**Institut für Theoretische Physik, Lehrstuhl IV der Ruhr–Universität Bochum.**
K. Schindler.
Mitt. Astron. Ges., Nr. 61, p. 43 – 44 (1984).

008.111 **Bonn**

**Astronomische Institute der Universität Bonn:**
**I. Sternwarte mit Observatorium Hoher List. Report for 1983.**
H. Schmidt.
Mitt. Astron. Ges., Nr. 61, p. 45 – 55 (1984).

**II. Radioastronomisches Institut und Teilprojekt A5 und A7 des Sonderforschungsbereichs 131, Radioastronomie. Report for 1983.**
U. Mebold.
Mitt. Astron. Ges., Nr. 61, p. 57 – 62 (1984).

**III. Institut für Astrophysik und Extraterrestrische Forschung. Report for 1983.**
W. Priester.
Mitt. Astron. Ges., Nr. 61, p. 63 – 70 (1984).

**Veröffentlichungen der Astronomischen Institute Bonn.**
Nr. 96 (38.117.211), 97 (38.117.212), 98 (38.119.089).

**Max–Planck–Institut für Radioastronomie. Report for 1983.**
P. G. Mezger, R. Wielebinski.
Mitt. Astron. Ges., Nr. 61, p. 71 – 101 (1984).

008.117 **Borowiec**

**Circular Time and Latitude Service.**
Nos. 168 – 170 (38.044.046).

008.126 **Brighton**

**Astronomy Centre, University of Sussex. Report for the year ending 1983 September 30.**
R. J. Tayler.
Q. J. R. Astron. Soc., Vol. 25, No. 3, p. 327 – 343 (1984).

008.141 **Budapest**

**Communications from the Konkoly Observatory.**
No. 84 (38.122.195).

**Publications of the Astronomy Department of the Eötvös University.**
No. 7 (38.004.182).

008.156 **Cambridge, England**

**Institute of Astronomy, University of Cambridge. Report for the period 1982 September 1 – 1983 August 31.**
D. Lynden–Bell.
Q. J. R. Astron. Soc., Vol. 25, No. 4, p. 449 – 473 (1984).

008.159 **Cambridge, Mass.**

**Smithsonian Astrophysical Observatory, Special Report.**
No. 394 (38.107.048).

**The Minor Planet Circulars/Minor Planets and Comets.**
Nos. 8827 – 9314 (1984).

**IAU Circulars.**
Nos. 3957 – 4024 (1984).

008.165 **Cape Town**

**South African Astronomical Observatory, Circulars.**
No. 8 (38.122.196, 38.113.049, 38.113.050, 38.114.104).

**SAAO Newsletter.**
No. 4 (1984).

**University of Cape Town: Department of Astronomy. Report 1983.**
B. Warner.
Mon. Notes Astron. Soc. S. Afr., Vol. 43, Nos. 1 – 4, p. 4 – 6 (1984).

**University of Cape Town: Department of Applied Mathematics. Report 1983.**
G. F. R. Ellis.
Mon. Notes Astron. Soc. S. Afr., Vol. 43, Nos. 1 – 4 p. 7 (1984).

008.171 **Cardiff**

**Astrophysics and Relativity, Preprint Series.**
Nos. 106 (38.103.037), 107 (38.015.034), 108 (38.015.035), 109 (38.091.056), 110 (38.015.036), 111 (38.015.037), 112 (38.015.038).

008.180 **Castel Gandolfo**

**Studi Galileiani.**
Vol. 1, No. 2 (38.004.183).

008.183 **Catania**

**Osservatorio Astrofisico di Catania, Pubblicazione.**
N. 169 (38.072.079).

008.208 **Cologne**

**I. Physikalisches Institut der Universität. Report for 1983.**
G. Winnewisser.
Mitt. Astron. Ges., Nr. 61, p. 219 – 222 (1984).

008.237 **Dresden**

**Lohrmann–Observatorium, Technische Universität Dresden, Zirkular.**
Nr. 108 – 109 (38.044.047).

008.242 **Durham**

**Department of Physics, University of Durham. Report for the period 1982 October 1 – 1983 December 31.**
B. H. Bransden.
Q. J. R. Astron. Soc., Vol. 25, No. 4, p. 474 – 481 (1984).

008.249 **Edinburgh**

**Occasional Reports of the Royal Observatory, Edinburgh.**
Nos. 13 (38.012.001), 14 (38.012.067).

008.255 **Epping**

**Anglo–Australian Observatory, Preprints.**
Nos. 196 (38.159.112), 197 (38.117.213), 198 (38.154.064), 199 (38.154.065).

**AAO Newsletter.**
Nos. 30 – 31 (1984).

008.258 **Erlangen**

**Physikalisches Institut, Angewandte Optik. Report for 1983.**
G. Weigelt.
Mitt. Astron. Ges., Nr. 61, p. 103 – 104 (1984).

008.264 **Farnborough**

**Royal Aircraft Establishment, Farnborough, Geophysical Studies in Space Department. Report for the year ending 1984 March 31.**
Q. J. R. Astron. Soc., Vol. 25, No. 4, p. 504 – 506 (1984).

008.276 **Frankfurt**

**Institut für Theoretische Physik (Astrophysik). Report for 1983.**
W. H. Kegel.
Mitt. Astron. Ges., Nr. 61, p. 105 – 106 (1984).

008.279 **Freiburg**

**Kiepenheuer–Institut für Sonnenphysik. Report for 1983.**
E. H. Schröter.
Mitt. Astron. Ges., Nr. 61, p. 107 – 118 (1984).

008.285 **Garching**

**Max–Planck–Institut für Physik und Astrophysik:**
**I. Institut für Astrophysik. Report for 1983.**
R. Kippenhahn.
Mitt. Astron. Ges., Nr. 61, p. 223 – 243 (1984).

**II. Institut für extraterrestrische Physik. Report for 1983.**
G. Haerendel.
Mitt. Astron. Ges., Nr. 61, p. 244 – 258 (1984).

**MPA Reports.**
Nos. 90 (38.012.020), 154, 155 (38.062.151), 156 (38.042.076), 157 (38.075.017), 158 (38.067.131), 159 (38.067.132), 160 (38.065.135, 38.065.136), 161 (38.080.110), 162 (38.151.072), 163 (38.066.251), 164 (38.066.252), 165 (38.151.073), 166 (38.061.122), 167 (38.066.253), 168 (38.065.137), 169 (38.125.012).

**MPE Reports.**
Nos. 183 (38.034.183), 184 (38.012.039), 185 (38.036.154).

**MPE Preprints.**
Nos. 28 (38.125.067, 38.142.058, 38.117.214, 38.117.215, 38.142.059),
29 (38.112.080, 38.124.143, 38.125.068, 38.126.070, 38.117.216 – 38.117.221).

**ESO Scientific Preprint.**
Nos. 331 (38.131.201), 332 (38.151.074), 333 (38.117.222), 334 (38.157.169), 335 (38.160.064), 336 (38.131.202), 337 (38.114.105), 338 (38.065.138), 339 (38.158.208), 340 (38.157.170), 341 (38.158.209), 342 (38.151.075), 343 (38.151.076), 344 (38.159.113), 345 (38.157.171), 346 (38.157.172), 347 (38.154.067), 348 (38.111.024), 349 (38.036.155), 350 (38.157.173), 351 (38.065.139), 352 (38.160.065), 353 (38.160.066), 354 (38.157.174).

**The Messenger – El Mensajero.**
Nos. 37 – 38 (1984).

008.294 **Glasgow**

**Department of Astronomy, University of Glasgow. Report for the period 1982 October to 1983 September.**
A. E. Roy.
Q. J. R. Astron. Soc., Vol. 25, No. 3, p. 344 – 347 (1984).

008.297 **Goeteborg**

**Onsala Space Observatory, Preprint.**
Nos. 84:13 (38.131.199), 84:42 (38.131.200), 84:48 (38.157.167), 84:58 (38.157.168).

008.300 **Göttingen**

**Universitätssternwarte Göttingen und Institut für Sonnenforschung Locarno–Orselina (Tessin). Report for 1983.**
K. J. Fricke.
Mitt. Astron. Ges., Nr. 61, p. 119 – 126 (1984).

008.306 **Graz**

**Institut für Astronomie der Universität Graz: Institut für Astronomie (Universitäts–Sternwarte), Observatorium Lustbühel, Sonnenobservatorium Kanzelhöhe. Report for 1983.**
H. Haupt.
Mitt. Astron Ges., Nr. 61, p. 127 – 132 (1984).

**Institut für Weltraumforschung. Tätigkeitsbericht 1982 – 1983.**
O. M. Burkhard.
Tätigkeitsbericht der Österreichischen Akademie der Wissenschaften 1982 – 1983, p. 126 – 136 (1984).

008.315 **Greenwich**

**The Royal Greenwich Observatory. Report for the period 1980 October 1 to 1982 September 30.**
F. G. Smith, A. Boksenberg.
Q. J. R. Astron. Soc., Vol. 25, No. 3, p. 348 – 370 (1984).

**Royal Greenwich Observatory, Bulletins.**
Nos. 192 (38.002.060), 193 (38.041.035).

**Gemini.**
Nos. 11 – 12 (1984).

008.318 **Groningen**

**Kapteyn Astronomical Institute. Annual report 1983.**
P. C. van der Kruit, R. Gathier (Editors).
Kapteyn Astron. Inst., Annu. Rep., 103 pp. (1984).

008.321 **Hamburg**

**Hamburger Sternwarte. Report for 1983.**
D. Reimers.
Mitt. Astron. Ges., Nr. 61, p. 135 – 140 (1984).

**Deutsches Hydrographisches Institut Hamburg. 38. Jahresbericht 1983.**
G. Zickwolff.
Dtsch. Hydrogr. Inst. Hamb., 114 pp. (1984).

**Deutsches Hydrographisches Institut (DHI). Report for 1983.**
H. Enslin.
Mitt. Astron. Ges., Nr. 61, p. 133 (1984).

**Deutsches Hydrographisches Institut Hamburg, Zeit– und Breitendienst.**
Januar – Dezember 1983 (38.044.049).

008.327 **Hannover**

**Astronomische Station am Institut für Erdmessung der Universität. Report for 1983.**
G. Seeber.
Mitt. Astron. Ges., Nr. 61, p. 141 – 143 (1984).

008.333 **Heidelberg**

**Astronomisches Rechen–Institut. Report for 1983.**
W. Fricke.
Mitt. Astron. Ges., Nr. 61, p. 145 – 155 (1984).

**Astronomisches Rechen–Institut Heidelberg, Mitteilungen, Serie A.**
Nr. 153 (34.155.102), 154 (34.155.113), 155 (34.155.123), 156 (34.041.047), 157 (34.041.048).

**Astronomisches Rechen–Institut Heidelberg, Mitteilungen, Serie B.**
Nr. 126 (34.043.006), 127 (37.046.018), 128 (34.111.019).

**Astronomische Grundlagen für den Kalender 1986.**
(38.046.008).

**Apparent Places of Fundamental Stars 1985.**
(38.046.031).

**Astronomy and Astrophysics Abstracts.**
Vol. 37 (38.002.061).

**Institut für Theoretische Astrophysik. Report for 1983.**
B. Baschek, G. Traving.
Mitt. Astron. Ges., Nr. 61, p. 157 – 161 (1984).

**Landessternwarte Heidelberg–Königstuhl. Report for 1983.**
I. Appenzeller.
Mitt. Astron Ges., Nr. 61, p. 169 – 175 (1984).

**Max–Planck–Institut für Astronomie und Deutsch–Spanisches Astronomisches Zentrum Calar Alto/Almeria. Report for 1983.**
H. Elsässer.
Mitt. Astron. Ges., Nr. 61, p. 177 – 201 (1984).

**Max–Planck–Institut für Astronomie, Heidelberg–Königstuhl, Preprints.**
(38.121.039 – 38.121.041).

**Max–Planck–Institut für Kernphysik. Abteilung Kosmophysik. Report for 1983.**
H. Fechtig, H. Völk.
Mitt. Astron. Ges., Nr. 61, p. 163 – 167 (1984).

008.354 **Innsbruck**

**Institut für Astronomie. Report for 1983.**
J. Pfleiderer.
Mitt. Astron. Ges., Nr. 61, p. 203 – 207 (1984).

008.363 **Istanbul**

**Publications of the Istanbul University Observatory.**
Nos. 121 (38.113.051), 122 (38.072.080), 123 (38.071.035), 124 (38.071.036), 125 (38.113.052), 126 (38.072.081), 127 (38.072.082), 128 (38.072.083), 129 (38.072.084), 130 (38.114.106), 131 (38.071.037), 132 (38.044.048), 133 (38.072.085), 134 (38.072.086), 135 (38.072.087).

008.373 **Johannesburg**

**Hartebeesthoek Radio Astronomy Observatory. Annual report 1983.**
G. D. Nicolson.
Mon. Notes Astron. Soc. S. Afr., Vol. 43, Nos. 9 – 10, p. 37 – 40 (1984).

008.378 **Kiel**

**Institut für Theoretische Physik und Sternwarte der Universität. Report for 1983.**
K. Hunger.
Mitt. Astron. Ges., Nr. 61, p. 209 – 213 (1984).

**Institut für Reine und Angewandte Kernphysik. Arbeitsgruppe Mathematische Physik. Report for 1983.**
K. O. Thielheim.
Mitt. Astron. Ges., Nr. 61, p. 215 – 217 (1984).

008.381 **Kiev**

**Astrometriya i Astrofizika.**
Vyp. 52 (1984).

008.396 **La Serena**

**Cerro Tololo Inter–American Observatory.**
AURA. The first twenty–five years 1957 – 1982, p. 20 – 29 (1983). – See Abstr. 010.351.

008.402 **Lake Tekapo**

**Physics Department and Mt John University Observatory, University of Canterbury. Annual report.**
B. G. Wybourne.
Q. J. R. Astron. Soc., Vol. 25, No. 4, p. 494 – 497 (1984).

**Astronomy at the University of Canterbury, Physics Department and the Mount John University Observatory. Annual report 1983.**
J. B. Hearnshaw.
South. Stars, Vol. 30, Nos. 7 – 8, p. 462 – 466 (1984).

008.417 **Leicester**

**Department of Astronomy, University of Leicester. Report for the period 1982 October 1 – 1983 December 31.**
A. J. Meadows.
Q. J. R. Astron. Soc., Vol. 25, No. 4, p. 498 – 503 (1984).

008.429 **Lintong**

**Time and Frequency Services, Bulletin.**
Nos. 55 – 60 (38.044.050).

008.438 **London, England**

**Atmospheric Physics Group, Imperial College, London. Report for the period 1982 January 1 – 1983 December 31.**
G. E. Hunt.
Q. J. R. Astron. Soc., Vol. 25, No. 4, p. 507 – 512 (1984).

008.447 **Lund**

**Reports from the Observatory of Lund.**
No. 19 (38.082.084).

008.459 **Madrid**

**Seminario de Astronomia y Geodesia, Publicación.**
Nos. 108 (33.021.008), 109 (38.043.005), 110 (34.044.018), 111 (38.044.051), 112 (38.021.016), 113 (38.045.015), 114, 121 (38.021.017).

008.465 **Manchester**

**Astronomical Contributions from the University of Manchester, Series II: Jodrell Bank Reprints.**
Nos. 695 (34.141.017), 696 (37.157.019), 697 (34.008.460), 698 (34.159.035), 699 (33.141.035), 700 (34.112.004), 701 (33.141.085), 702 (34.158.015), 703 (34.141.087), 704 (34.004.028), 705, 706 (37.131.028), 707 (34.141.023), 708 (34.112.008), 709 (37.158.021), 710 (37.112.027), 711 (37.158.083), 712 (34.125.301), 713 (34.159.094), 714 (37.159.131), 715 (37.131.124), 716 (37.112.075), 717 (38.117.025), 718 (37.125.007), 719 (38.157.125), 720 (37.008.460), 721 (38.126.045), 722 (38.158.175), 723 (38.041.014), 724 (38.157.088), 725 (37.159.079), 726, 727 (37.159.133), 728 (37.155.070), 729 (38.131.060), 730 (38.126.003), 731 (38.159.065), 732 (38.158.030).

**Astronomical Contributions from the University of Manchester, Series III.**
Nos. 451 (27.132.009), 452 (27.119.065), 453 (27.117.120), 454 (27.119.064), 455 (27.117.119), 456 (28.117.003), 457 (28.119.010), 458 (27.119.129), 459 (28.159.002), 460 (27.134.009), 461 (28.119.001), 462 (28.053.007), 463 (28.119.032), 464 (27.121.017), 465 (26.119.001), 466 (27.159.003), 467 (27.094.028), 468 (27.154.013), 469 (27.154.014), 470 (27.154.015), 471 (27.132.019), 472 (28.022.158), 473 (28.135.040), 474 (28.121.001), 475 (28.015.008), 476 (28.134.028), 477 (29.134.002), 478 (28.159.009), 479 (28.065.062), 480 (28.119.055), 481 (28.119.047), 482 (29.158.033), 483 (28.132.038), 484 (28.131.284), 485 (29.132.013), 486 (29.002.005), 487 (29.134.007), 488 (30.132.024), 489 (30.135.024), 490 (30.064.003), 491 (29.065.025), 492 (29.119.065), 493 (30.132.003), 494 (30.159.006), 495 (29.117.124), 496 (29.119.108), 497 (29.119.111), 498 (29.117.122), 499 (30.155.052), 500 (30.159.018), 501 (30.132.013), 502 (29.159.022), 503 (30.151.061), 504 (30.131.176), 505 (30.117.101), 506 (31.135.040), 507 (31.134.014), 508 (31.119.023), 509 (31.151.012), 510 (31.141.506), 511 (31.121.003), 512 (31.158.315), 513 (31.158.062), 514 (31.133.005), 515 (31.119.023), 516 (31.159.004), 517 (31.115.031), 518, 519 (32.117.150), 520 (32.117.138), 521 (31.119.134), 522 (31.141.100), 523 (31.134.044), 524 (32.132.036), 525 (32.132.051), 526 (32.021.032), 527 (32.002.074), 528 (32.141.126), 529 (32.117.066), 530 (32.106.026), 531 (32.132.006), 532 (32.131.236), 533 (32.121.028), 534 (32.119.104), 535 (33.099.031), 536 (33.132.013), 537, 538 (34.132.008), 539 (33.141.025), 540 (33.157.217), 541 (32.117.222), 542 (32.117.226), 543 (33.159.052), 544 (33.132.052), 545 (33.134.038), 546 (33.119.043), 547 (33.063.037), 548 (33.131.173), 549 (33.117.107), 550 (34.114.010).

008.468 **Manila**

**Solar Maps and Activity.**
(38.072.088).

008.480 **Middletown, Conn.**

**Contributions from the Van Vleck Observatory.**
No. 2 (38.002.062 – 38.002.065).

008.489 **Mizusawa**

**Publications of the International Latitude Observatory of Mizusawa.**
Vol. 17, No. 1 (38.044.053, 38.036.156).

**Bulletins, Time Service of the Mizusawa Observatory.**
Vol. 27 (38.044.054).

**Monthly Notes of the International Polar Motion Service.**
Nos. 5 – 11 (1984) (38.044.052).

008.498 **Moscow**

**Trudy Gosudarstvennogo Astronomicheskogo Instituta im. P. K. Shternberga.**
Tom 54 (38.114.011, 38.157.024, 38.126.012, 38.022.008, 38.094.007, 38.071.006, 38.009.010, 38.009.011, 38.041.010, 38.042.005).
Tom 55 (38.041.011, 38.031.003, 38.041.012, 38.044.006, 38.041.013, 38.093.003, 38.034.018, 38.158.023, 38.114.012, 38.071.007, 38.094.008, 38.097.004, 38.094.009, 38.021.004, 38.081.007, 38.094.010, 38.081.008, 38.081.009).
Tom 56 (38.052.009, 38.052.010, 38.042.023, 38.094.022 – 38.094.025).

008.501 **Münster**

**Astronomisches Institut der Universität und Außenstation Schalkenmehren. Report for 1983.**
W. C. Seitter.
Mitt. Astron. Ges., Nr. 61, p. 267 – 273 (1984).

**Forschergruppe Erde–Mond–System (Planetologie). Report for 1983.**
W. C. Seitter, D. Stöffler.
Mitt. Astron. Ges., Nr. 61, p. 274 – 278 (1984).

008.504 **Munich**

**Institut für Astronomie und Astrophysik der Universität München. Universitäts–Sternwarte. Report for 1983.**
R. P. Kudritzki.
Mitt. Astron. Ges., Nr. 61, p. 259 – 266 (1984).

**Mitteilungen aus dem Institut für Astronomische und Physikalische Geodäsie der Technischen Hochschule München.**
Nr. 166 (38.009.040), 167 (38.081.034).

008.513 **Nanking**

**Publications of the Purple Mountain Observatory.**
Vol. 3, No. 2 (38.098.099, 38.041.036, 38.044.055, 38.095.006, 38.095.007, 38.034.184, 38.072.089, 38.078.016, 38.042.077, 38.034.185, 38.066.254).

008.540 **Oslo**

**Institute of Theoretical Astrophysics, University of Oslo, Publication Series.**
No. 1 (38.022.145).

008.549 **Oxford**

**Department of Astrophysics, University of Oxford. Report for the period 1982 August 1 – 1983 July 31.**
D. E. Blackwell.
Q. J. R. Astron. Soc., Vol. 25, No. 4, p. 482 – 493 (1984).

008.555 **Paris**

**L'Observatoire de Paris – son histoire (1667 – 1963).**
P. Charvin.
Observatoire de Paris, 61, avenue de l'Observatoire, F–75014 Paris, France. 9 + 68 pp. Price FF 35.00 (1984). ISBN 2–901057–04–7.

**L'Observatoire de Paris aujourd'hui.**
Observatoire de Paris, 61, avenue de l'Observatoire, F–75014 Paris, France. 14 pp. (1984). ISBN 2–901057–06–3.

**Cartes Synoptiques.**
Vol. VI, Fasc. 5 (38.073.089).

**Institut d'Astrophysique de Paris, Pré–Publication.**
Nos. 63 (38.157.175), 64 (38.106.051), 65 (38.074.057), 66 (38.102.055), 67 (38.035.074), 68 (38.157.176), 69 (38.063.070), 70 (38.117.223), 71 (38.158.210), 72 (38.132.049), 73 (38.112.081), 74 (38.103.305), 75 (38.125.069), 76 (38.063.071), 77 (38.063.072), 78 (38.063.073), 79 (38.160.067), 80 (38.159.114), 81 (38.075.018).

**Bureau International de l'Heure (B.I.H.) Circulars.**
A (38.044.058).
D 212 – D 217 (38.044.059).

008.567 **Peking**

**Publications of the Beijing Astronomical Observatory.**
No. 3/1983 (38.077.043).

008.586 **Poona**

**Department of Mathematics, University of Poona, Pune. Report for the period 1982 April 1 – 1984 March 31.**
Bull. Astron. Soc. India, Vol. 12, No. 3, p. 317 – 318 (1984).

008.594 **Potsdam**

**Zeit– und Breitenbestimmungen, Zeitsysteme, Präzisionszeitvergleiche.**
Jahrg. 82, Nr. 3 – 6, Sonderheft (38.044.057).

008.597 **Prague**

**Circular of the Czechoslovak Observatories, Time and Latitude.**
January – September 1983 (38.044.056).

008.613 **Quito**

**Publicacion Especial del Observatorio Astronomico de Quito.**
No. 2 (38.013.075, 38.045.019, 38.036.157, 38.045.020, 38.096.028, 38.102.060, 38.013.076).

008.654 **Santa Cruz, Calif.**

**Contributions from the Lick Observatory.**
Nos. 428 (34.131.158), 429 (34.114.011), 430 (34.117.050), 431 (34.117.098), 432 (37.114.007), 433 (37.158.030).

**Lick Observatory Bulletin.**
Nos. 951 (34.116.006), 952 (34.153.015), 953 (34.158.064), 954 (33.123.011), 955 (34.153.014), 956 (34.159.071), 957 (33.153.034), 958 (34.121.001), 959 (34.122.039), 960 (34.158.113), 961 (34.158.114), 962 (34.121.003), 963 (34.158.052), 964 (34.114.102), 965 (34.122.219), 966 (34.158.111), 967 (37.116.004), 968 (34.153.051), 969 (34.117.131), 970 (38.034.186), 971 (37.122.075), 972 (37.159.025), 973 (34.082.103), 974 (37.160.115), 975 (37.158.131), 976 (37.159.155), 977 (37.065.092), 978 (38.158.003), 979 (37.098.077), 980 (37.114.141), 981 (37.153.072).

008.672 **Sendai**

**Annual report of research activities of the Astronomy Group in 1983/84.**
Sci. Rep. Tôhoku Univ., Eighth Ser., Vol. 5, No. 2, p. 172 – 179 (1984).

**Sendai Astronomiaj Raportoj.**
Nos. 261 (37.065.006), 263 (38.106.005), 264 (34.117.171), 266 (38.131.044), 269 (38.008.672), 270 (38.160.021).

008.678 **Shanghai**

**Annals of Shanghai Observatory, Academia Sinica.**
No. 5/1983 (38.044.023 – 38.044.025, 38.094.051, 38.036.073, 38.081.039, 38.044.026, 38.032.031, 38.032.032, 38.044.027, 38.021.012, 38.021.013, 38.041.027, 38.118.025, 38.036.074, 38.118.026, 38.118.027, 38.036.075, 38.111.015, 38.111.016, 38.066.216, 38.045.008 – 38.045.011, 38.034.097, 38.034.098, 38.044.028, 38.034.099, 38.034.100, 38.032.033, 38.035.056, 38.034.101, 38.033.030, 38.033.031).

**Time Service Annual Report.**
1983 (38.044.074).

008.693 **Sonneberg**

**Veröffentlichungen der Sternwarte in Sonneberg.**
Band 10, Heft 1 (38.121.042).

**Mitteilungen über Veränderliche Sterne.**
Band 10, Heft 3 (38.117.224, 38.119.090, 38.123.030, 38.122.198, 38.117.225, 38.123.031, 38.123.032).

008.696 **St. Andrews**

**Communications from the University Observatory, St. Andrews.**
Nos. 38 (32.119.042), 39 (33.117.027), 40 (34.117.021), 41 (37.117.048), 42 (37.117.125).

**University Observatory, St. Andrews, Reprint.**
Nos. 90 (32.122.025), 91 (32.119.060), 92 (32.122.041), 93 (34.131.023), 94 (34.034.190), 95 (34.113.023), 96 (37.113.033), 97 (37.114.030), 98 (34.042.008), 99 (37.122.057), 100 (37.117.126).

008.705 **Stockholm**

**Stockholms Observatorium, Report.**
Nos. 23 (38.065.140), 24 (38.158.211), 25 (38.155.063), 26 (38.133.018).

008.711 **Strasbourg**

**Bulletin d'Information du Centre de Données Stellaires.**
No. 27 (1984).

**Bulletin d'Information sur les Etoiles Be.**
No. 10 (1984).

**Bulletin sur les Etoiles Tardives à Spectre Particulier.**
No. 1 (1984).

008.714 **Sunspot, N.M.**

**Sacramento Peak Observatory.**
AURA. The first twenty-five years 1957 – 1982, p. 30 – 33 (1983). – See Abstr. 010.351.

008.729 **Tartu**

**W. Struve nimeline Tartu Astrofüüsika Observatoorium Teated.**
Nr. 71 (38.114.019, 38.036.013, 38.036.014), 72 (38.111.011, 38.155.013, 38.111.012), 73 (38.160.012).

008.732 **Tashkent**

**Chronicle.**
Tsirk. Astron. Inst. Tashkent, No. 112, p. 39 – 43 (1983). In Russian.

**Tsirkulyar Astronomicheskogo Instituta Tashkent.**
Nos. 109 (38.044.031, 38.072.065, 38.044.032, 38.032.061, 38.041.028),
110 (38.041.029),
111 (38.041.030),
112 (38.032.062, 38.008.732).

008.744 **Tokyo**

**Annals of the Tokyo Astronomical Observatory, Second Series.**
Vol. 20, No. 1 (38.035.075, 38.117.226),
Vol. 20, No. 2 (38.117.227, 38.113.055).

**Tokyo Astronomical Observatory Report.**
Vol. 20, No. 2 (38.034.049, 38.074.021, 38.118.015, 38.034.050, 38.035.029, 38.034.051, 38.044.016, 38.033.012, 38.033.013, 38.034.052, 38.041.021, 38.034.053).

**Tokyo Astronomical Observatory Reprints.**
Nos. 696 (37.064.026), 697 (38.115.009), 698 (38.117.080), 699 (38.141.008), 700 (38.141.009), 701 (38.157.086), 702 (38.098.021), 703 (38.033.014), 704 (38.033.015), 705 (38.159.041), 706 (38.044.061), 707 (38.155.004), 708 (38.073.020), 709 (38.131.111), 716 (38.153.109).

**Tokyo Astronomical Observatory, Kiso Information Bulletin.**
Vol. 2, No. 1 (38.002.069).

**Tokyo Astronomical Observatory, Time and Latitude Bulletins.**
Vol. 58, Nos. 1 – 3 (38.044.060).

**Quarterly Bulletin on Solar Activity.**
Vol. 23, Part III (38.073.090).

**Contributions from the Department of Astronomy, University of Tokyo.**
Nos. 324 (38.063.074), 325 (37.074.026), 326 (34.076.037), 327 (37.021.013), 328 (37.042.011), 329 (37.042.012), 330 (37.073.154), 331 (37.022.033), 332 (37.062.090), 333 (37.074.060).

008.753 **Torino**

**Contributi dell'Osservatorio Astronomico di Torino.**
Nos. 217 (32.062.103), 218 (34.103.031), 219 (33.112.032), 220 (33.022.069), 221 (33.098.025), 222 (33.098.022), 223 (33.098.024), 224 (33.155.066), 225 (34.098.002), 226 (33.100.097), 227 (34.098.010), 228 (33.122.157), 229 (33.151.087), 230 (33.098.116), 231 (33.098.117), 232 (33.100.102), 233 (33.098.097), 234 (33.036.153), 235 (34.098.039), 236 (34.098.038), 237 (37.098.002), 238, 239 (34.118.049), 240 (34.118.048), 241 (34.118.020), 242 (37.044.003), 243 (37.117.011), 244 (37.098.003), 245, 246 (37.118.003), 247 (37.117.123), 248 (37.064.034), 249 (37.101.002), 250 (37.118.002), 251 (34.159.199, 37.159.016, 37.159.068), 252, 253, 254 (37.117.198), 255 (37.117.218), 256 (37.042.019), 257 (37.100.004), 258, 259 (38.117.081), 260 (31.114.042), 261 (32.114.098), 262 (33.114.093), 263 - 265, 266 (37.036.098), 267 (37.022.045), 268 (38.098.018), 269 (38.098.014), 270 - 277, 278 (34.103.031), 279 (38.098.019), 280 (38.098.066), 281, 282 (38.112.021), 283 (37.098.020), 284 (38.098.002), 285 - 288, 289 (38.157.150), 290 (38.098.024), 291, 292 (38.122.065).

008.765 **Tucson, Ariz.**

**Kitt Peak National Observatory.**
AURA. The first twenty-five years 1957 - 1982, p. 12 - 19 (1983). - See Abstr. 010.351.

**Preprints of the Steward Observatory.**
Nos. 529 (38.121.044), 530 (38.158.228), 531 (38.031.060), 532 (38.156.016), 533 (38.002.071), 534 (38.131.209), 535 (38.031.061), 536 (38.032.067), 537 (38.158.229), 538 (38.154.071), 539 (38.031.062), 540 (38.158.230), 541 (38.158.231), 542 (38.158.232), 543 (38.117.232), 544 (38.157.182), 545 (38.160.071), 546 (38.099.133), 547 (38.082.087), 548 (38.158.233), 549 (38.158.234), 550 (38.126.071), 551 (38.126.072), 552 (38.142.061), 553 (38.158.235), 554 (38.118.043), 555 (38.141.021), 556 (38.126.073), 557 (38.126.074), 558 (38.157.183), 559 (38.113.057), 560 (38.113.058), 561 (38.119.092), 562 (38.002.072).

008.768 **Tübingen**

**Universität Tübingen, Astronomisches Institut. Report for 1983.**
M. Grewing, H. Ruder.
Mitt. Astron. Ges., Nr. 61, p. 279 - 290 (1984).

**Lehrstuhl für Theoretische Astrophysik der Universität Tübingen, Preprints.**
(38.076.021, 38.073.095, 38.076.022, 38.073.096).

008.771 **Turku**

**Turku University Observatory, Informo.**
Nos. 82 (38.160.069), 83 (38.160.070), 84 (38.012.079), 85 (38.158.224), 86 (38.151.077), 87 (38.161.319), 88 (38.158.225), 89 (38.031.059).

**Report Series, Department of Physical Sciences, University of Turku.**
FTL-R63 (38.160.069), FTL-R64 (38.160.070), FTL-R68 (38.012.079), FTL-R69 (38.158.224), FTL-R70 (38.151.077), FTL-R71 (38.161.319), D12 (38.158.225).

008.774 **Uccle**

**Observatoire Royal de Belgique, Communications, Série A.**
Nos. 73, 74 (38.071.004).

**Observatoire Royal de Belgique, Communications, Série B.**
Nos. 130 (37.118.029), 131.

**Bulletin Astronomique.**
Vol. 9, No. 5 (38.098.100, 38.099.131, 38.098.101, 38.099.132, 38.118.039 - 38.118.042).

008.786 **Urbana, Ill.**

**Erratum: "University of Illinois at Urbana-Champaign, Department of Astronomy, Urbana, Illinois 61801-3000. Report for the period 1 September 1982 - 31 August 1983" [Bull. Am. Astron. Soc., Vol 16, No. 1, p. 162 - 168 (1984)].**
C. B. Foltz, E. C. Olson.
Bull. Am. Astron. Soc., Vol. 16, No. 3, Part I, p. 737 - 738 (1984).

008.795 **Victoria**

**Dominion Astrophysical Observatory, Victoria, British Columbia. Report for the year 1983 April 1 - 1984 March 31.**
S. van den Bergh.
Q. J. R. Astron. Soc., Vol. 25, No. 4, p. 513 - 528 (1984).

008.798 **Vienna**

**Institut für Astronomie der Universität Wien. Report for 1983.**
W. M. Tscharnuter.
Mitt. Astron. Ges., Nr. 61, p. 291 - 300 (1984).

008.810 **Vilnius**

**Vilniaus Astronomijos Observatorijos Biuletenis.**
Nos. 66 (38.002.005), 67 (38.131.233, 38.131.234, 38.113.061, 38.103.762), 68 (38.002.075, 38.113.062).

008.813 **Warsaw**

**Latitude Circular.**
Nos. 89 - 91 (38.044.065).

**Publications of the Institute of Geophysics, Polish Academy of Sciences, Series F (Planetary Geodesy).**
F-11 (173).

008.816 **Washington, D.C.**

**Publications of the United States Naval Observatory, Second Series.**
Vol. 24, Part VII (38.002.073).

**U.S. Naval Observatory, Time Service Publications, Series 4.**
Nos. 909 - 934 (38.044.062).

**U.S. Naval Observatory, Time Service Announcement, Series 14.**
No. 38 (38.044.063).

**Earth Orientation Bulletin.**
Vol. 2, Nos. 27 - 52 (38.044.064).

008.819 **Wellington**

**National Observatory of New Zealand. Annual report of the Carter Observatory Board for the year ending 1984 March 31.**
I. W. G. Cochrane.
Astron. Bull., Carter Obs., No. 102, 12 pp. (1984).

**Carter Observatory. Report for the year ended 31 March 1984.**
R. J. Dodd.
South. Stars, Vol. 30, Nos. 7 – 8, p. 457 – 461 (1984).

**Astronomical Bulletin, Carter Observatory.**
Nos. 101 (38.046.045), 102 (38.008.819).

008.837 **Würzburg**

**Institut für Astronomie und Astrophysik, Lehrstuhl Astronomie. Report for 1983.**
F.-L. Deubner.
Mitt. Astron. Ges., Nr. 61, p. 301 – 304 (1984).

008.843 **Zagreb**

**Hvar Observatory Bulletin.**
Vol. 7, No. 1 (38.012.037).

008.846 **Zelenchukskaya**

**Soobshcheniya Spetsial'noj Astrofizicheskoj Observatorii.**
Vyp. 40 (38.012.076), 41 (38.141.018), 42 (38.141.019), 43 (38.131.195, 38.141.020, 38.155.061, 38.132.047, 38.155.062).

**Bulletin of the Special Astrophysical Observatory–North Caucasus.**
Vol. 17 (38.117.163, 38.142.047, 38.036.076, 38.082.070, 38.082.071, 38.036.077, 38.158.181, 38.131.171, 38.132.042, 38.131.172, 38.022.125, 38.033.032, 38.036.078, 38.033.033, 38.033.034, 38.036.079, 38.033.035 – 38.033.037).

008.849 **Zimmerwald**

**Progress report 1984 of the Zimmerwald Satellite Observation Station.**
P. Kloeckler, I. Bauersima, G. Beutler, W. Gurtner, M. Rothacher.
Mitt. Satell.-Beobachtungsstn. Zimmerwald, Nr. 16, 4+41 pp. (1984).

**Mitteilungen der Satelliten–Beobachtungsstation Zimmerwald.**
Nr. 9 (38.021.018), 14 (38.045.016), 15 (38.045.017), 16 (38.008.849), 17 (38.045.018).

008.852 **Zurich**

**Institut für Astronomie. Report for 1983.**
J. O. Stenflo.
Mitt. Astron. Ges., Nr. 61, p. 305 – 311 (1984).

# 009 Notes on Observatories, Planetaria, Exhibitions

009.001 **Een tijd van komen en van gaan.**
G. W. E. Beekman.
Zenit, 11. Jaarg., No. 7/8, p. 332 – 333 (1984).

009.002 **Potchefstroomse Universiteit Departement Wiskunde en Toegepasde Wiskunde.**
J. H. De Klerk.
Mon. Notes Astron. Soc. S. Afr., Vol. 43, Nos. 1 – 4, p. 9 – 10 (1984).

009.003 **Rhodes University: Department of Physics and Electronics. Report 1983.**
Mon. Notes Astron. Soc. S. Afr., Vol. 43, Nos. 1 – 4, p. 10 – 11 (1984).

009.004 **University of South Africa: Department of Mathematics, Applied Mathematics and Astronomy.**
J. Wolterbeek.
Mon. Notes Astron. Soc. S. Afr., Vol. 43, Nos. 1 – 4, p. 11 – 12 (1984).

009.005 **L'ESO (European Southern Observatory).**
S. Cristiani.
Coelum, Vol. 52, N. 4, p. 213 – 217 (1984).

009.006 **Optical astronomy at the Monash Observatory.**
D. W. Coates, K. Thompson, J. L. Innis, T. T. Moon.
Advances in photoelectric photometry, Vol. 2, p. 74 – 118 (1984). – See Abstr. 003.002.

009.007 **Photoelectric photometry at Van Vleck Observatory.**
W. Herbst, D. W. Carlson.
Advances in photoelectric photometry, Vol. 2, p. 134 – 143 (1984). – See Abstr. 003.002.

009.008 **The photoelectric photometry of Yunnan Observatory of the Chinese Academy of Science.**
Q. Liu, Y. Yang.
Advances in photoelectric photometry, Vol. 2, p. 144 – 155 (1984). – See Abstr. 003.002.

009.009 **Archenhold Observatory past and present.**
D. B. Herrmann.
Sky Telesc., Vol. 68, No. 1, p. 5 – 6 (1984).

009.010 **A short historical description of the Kuchino Astrophysical Observatory.**
G. F. Sitnik.
Tr. Gos. Astron. Inst. Shternberg, Tom 54, p. 100 – 129 (1984). In Russian.
A short description of the Kuchino Observatory is given with a more detailed account of its post-war development and research programs.

009.011 **Basic work of the Solar Physics Department beginning with 1958. (A historical review).**
G. F. Sitnik.
Tr. Gos. Astron. Inst. Shternberg, Tom 54, p. 130 – 148 (1984). In Russian.

009.012 **The Rutherford Appleton Laboratory's role in space.**
J. T. Houghton.
J. Br. Interplanet. Soc., Vol. 37, No. 7, p. 306 – 308 (1984).

009.013 **Solar research at RAL (*Rutherford Appleton Laboratory*).**
A. H. Gabriel.
J. Br. Interplanet. Soc., Vol. 37, No. 7, p. 317 – 319 (1984).

009.014 **IRAS mission operations experience at RAL (*Rutherford Appleton Laboratory*).**
J. R. Macdougall, P. H. McPherson, K. E. Mount, G. R. Thomas.
J. Br. Interplanet. Soc., Vol. 37, No. 7, p. 337 – 347 (1984).
The Infra Red Astronomical Satellite (IRAS) was launched in January 1983. During its 300–day lifetime IRAS carried out a complete survey of the sky at infrared wavelengths and also made about 14,000 additional observations of selected sources. This paper presents a summary of the development of the ground operations hardware and software, and discusses the experience gained during IRAS mission operations.

009.015 **Greenwich: where east meets west.**
C. Stott.
Sky Telesc., Vol. 68, No. 4, p. 300 – 303 (1984).

009.016 **20 Jahre Beobachtungsstation Großschwabhausen der Universitäts–Sternwarte Jena.**
W. Pfau.
Jenaer Rundsch., 29. Jahrg., Heft 3, p. 121 – 122 (1984).

009.017 **Automatische Vorführungen – Herausforderung an die moderne Planetariumstechnik.**
V. Schorcht, K. Weniger.
Jenaer Rundsch., 29. Jahrg., Heft 3, p. 132 – 134 (1984).

009.018 **10 Jahre Planetarium Cottbus – 10 Jahre Erfahrungen mit dem Raumflugplanetarium.**
G. Golka.
Jenaer Rundsch., 29. Jahrg., Heft 3, p. 136 – 137, 152 (1984).

009.019 **Das Planetarium Prag.**
A. Rükl.
Jenaer Rundsch., 29. Jahrg., Heft 3, p. 136 – 138, 152 (1984).

009.020 **Das Planetarium und Observatorium Olsztyn.**
K. Schilling.
Jenaer Rundsch., 29. Jahrg., Heft 3, p. 139 (1984).

009.021 **Planetarium und Sternwarte "Nikolaus Kopernikus" Chorzów.**
H. Chrupała.
Jenaer Rundsch., 29. Jahrg., Heft 3, p. 140, 152 (1984).

009.022 **Planetarien aus Jena in Frankreich.**
L. Bolgert.
Jenaer Rundsch., 29. Jahrg., Heft 3, p. 141 – 142 (1984).

009.023 **Das Planetarium Calouste Gulbenkian Lissabon.**
J. H. F. Cyrne de Castro.
Jenaer Rundsch., 29. Jahrg., Heft 3, p. 143 – 144 (1984).

009.024 **Das McLaughlin–Planetarium des Royal Ontario Museum.**
T. R. Clarke.
Jenaer Rundsch., 29. Jahrg., Heft 3, p. 144 – 146 (1984).

009.025 **Australiens erstes modernes Planetarium – das Sir Thomas Brisbane Planetarium.**
J. W. Ryder.
Jenaer Rundsch., 29. Jahrg., Heft 3, p. 146 – 148 (1984).

009.026 **Das Planetarium als Bildungszentrum für Astronomie und Raumfahrt.**
J. J. Salas Vasquez.
Jenaer Rundsch., 29. Jahrg., Heft 3, p. 148 – 149 (1984).

009.027 **The Maria Mitchell Observatory: a jubilee salute.**
B. L. Welther.
J. Am. Assoc. Variable Star Obs., Vol. 12, No. 2, p. 76 (1983). Abstract. – See Abstr. 010.021.

009.028 **Das Fenster zum Weltall. Observatorium auf dem Calar Alto mit 3.5 m Teleskop.**
Umschau, 84. Jahrg., Nr. 22, p. 666 – 668 (1984).

009.029 **Sterrenwacht van Peking in oude glorie hersteld.**
G. W. E. Beekman.
Zenit, 11. Jaarg., No. 11, p. 466 – 468 (1984).

009.030 **The Hurbanovo Observatory.**
J. Francisty.
Vasiona, Année 32, No. 2, p. 36 – 41 (1984). In Croatian.
An account of the development and present activities of the Hurbanovo Observatory in Czechoslovakia is presented.

009.031 **Calar Alto 1984.**
H. Elsässer.
Sterne Weltraum, 23. Jahrg., Nr. 11, p. 564 – 565 (1984).

009.032 **A report from European Southern Observatory.**
M. Iye.
Astron. Her., Vol. 77, No. 12, p. 312 – 316 (1984). In Japanese.

009.033 **The early history of the Dominion Observatories.**
J. H. Hodgson.
J. R. Astron. Soc. Can., Vol. 78, No. 5, p. 200 – 201 (1984). Abstract. – See Abstr. 010.401.

009.034 **The NRAO–AIPS based image–processing station at the University of Toronto.**
P. P. Kronberg.
J. R. Astron. Soc. Can., Vol. 78, No. 5, p. 201 (1984). Abstract. – See Abstr. 010.401.

009.035 **Some non–atlas work at ESO Sky Atlas Laboratory.**
C. Madsen.
Occas. Rep. R. Obs., Edinb., No. 14, p. 143 – 152 (1984). – See Abstr. 012.067.
The ESO Sky Atlas Laboratory (SAL) was set up in 1972 with the aim of producing the ESO Quick Blue Survey and later the joint ESO/SERC Survey of the Southern Sky. With the establishment of a Scientific Group, it became apparent that ESO had additional photographic needs, the fullfilment of which was also entrusted to SAL. Thus, in the course of the years, the "Photographic Section" evolved as a subdivision of the Sky Atlas Laboratory.

009.036 **Processing of photographic observing material at the National Astronomical Observatory Rozhen.**
B. Zh. Kovachev, M. K. Tsvetkov, V. I. Tsintsarov.
Occas. Rep. R. Obs., Edinb., No. 14, p. 165 – 171 (1984). – See Abstr. 012.067.

009.037 **The UK Schmidt Telescope and the facilities in Edinburgh.**
R. D. Cannon.
Occas. Rep. R. Obs., Edinb., No. 14, p. 173 – 176 (1984). – See Abstr. 012.067.
In this note the author concentrates on the UKST facilities available in Edinburgh, especially in the Plate Library.

009.038 **The work of the "Photolabs" at the Royal Observatory, Edinburgh.**
B. W. Hadley.
Occas. Rep. R. Obs., Edinb., No. 14, p. 177 – 184 (1984). – See Abstr. 012.067.

009.039 **Paris Observatory in 1784 – Cassini de Thury's legacy to Jean–Dominique Cassini.**
I. Howard–Duff.
J. Br. Astron. Assoc., Vol. 95, No. 1, p. 26 – 28 (1984).
This short paper commemorates the appointment of Jean–Dominique Cassini (Cassini IV) as Director of the Paris Observatory in 1784 following the death of his father, Cassini de Thury (Cassini III). It also reviews the status of Paris Observatory at that time.

009.040 **The Satellite Observation Station Wettzell of the Sonderforschungsbereich 78 Satellitengeodäsie (SFB 78) as a Geodynamic Fundamental Station.**
R. Sigl.
Allg. Vermess.-Nachr., Int. Suppl., Heft 1, p. 6 – 19 (1984). = Mitt. Inst. Astron. Phys. Geod. Tech. Hochsch. Münch., Nr. 166.

After a brief presentation of the Sonderforschungsbereich 78 Satellitengeodäsie (SFB 78) and its research programme, the scope and state of the Satellite Observation Station Wettzell as a "Geodynamic Fundamental Station" is specified.

**The Greenwich Meridian.**
See Abstr. 003.119.

**Recomendaciones sobre el futuro de la investigacion en el IAC, 1983.**
See Abstr. 003.160.

**The end of Pulkovo Observatory's reign as the "astronomical capital of the world".**
See Abstr. 004.008.

**Le Colloque des Planétariums Européens.**
See Abstr. 011.016.

**Plate measurements and reductions at Münster.**
See Abstr. 034.157.

**The Toruń measuring system for astronomical spectra.**
See Abstr. 034.158.

**Space plasma research at RAL.**
See Abstr. 035.010.

**The remote control run from La Serena, June 10 – 17, 1984.**
See Abstr. 036.027.

**Data saving and banking at La Silla.**
See Abstr. 036.028.

# 010 Societies, Associations, Organizations

### American Association of Variable Star Observers (AAVSO)

010.021 **Abstracts of papers presented at the 72nd annual meeting of the AAVSO held in Nantucket, Massachusetts, October 21 – 23, 1983.**
J. Am. Assoc. Variable Star Obs., Vol. 12, No. 2, p. 76 – 78 (1983).

010.022 **Meetings and activities of the Society, committee reports.**
J. Am. Assoc. Variable Star Obs., Vol. 12, No. 2, p. 79 – 82, 103 – 109 (1983).

010.023 **Annual report of the director for fiscal year 1982 – 1983.**
J. A. Mattei.
J. Am. Assoc. Variable Star Obs., Vol. 12, No. 2, p. 83 – 102 (1983).

010.024 **The Journal of the American Association of Variable Star Observers.**
Vol. 12, No. 2 (1983), Vol. 13, No. 1 (1984).

010.025 **Abstracts of papers presented at the spring meeting of the AAVSO in Ames, Iowa, May 25 – 26, 1984.**
J. Am. Assoc. Variable Star Obs., Vol. 13, No. 1, p. 41 – 45 (1984).

010.026 **Meetings and activities of the Society, committee reports.**
J. Am. Assoc. Variable Star Obs., Vol. 13, No. 1, p. 46 – 55 (1984).

### American Astronomical Society (AAS)

010.061 **The 16th annual meeting of the Division for Planetary Sciences, held 8 – 12 October 1984 at Kona, Hawaii. Abstracts of presented papers.**
Bull. Am. Astron. Soc., Vol. 16, No. 3, Part I, p. 621 – 715 (1984).

010.062 **Historical notes on the DPS/AAS.**
D. P. Cruikshank.
Bull. Am. Astron. Soc., Vol. 16, No. 3, Part I, p. 716 – 717 (1984).

010.063 **Minutes of the Division for Planetary Sciences annual business meeting, 19 October 1983, Sheraton Conference Center, Ithaca, New York.**
D. P. Cruikshank.
Bull. Am. Astron. Soc., Vol. 16, No. 3, Part I, p. 717 – 719 (1984).

010.064 **The 15th meeting of the Division on Dynamical Astronomy, held 14 – 15 June 1984 at Baltimore, Maryland. Abstracts of presented papers.**
Bull. Am. Astron. Soc., Vol. 16, No. 3, Part I, p. 720 – 723 (1984).

010.065 **Late-paper abstracts from the 164th meeting of the American Astronomical Society, held 10 – 13 June 1984 at Baltimore, Maryland.**
Bull. Am. Astron. Soc., Vol. 16, No. 3, Part I, p. 724 – 734 (1984).

010.066 **Bulletin of the American Astronomical Society.**
Vol. 16, No. 3, Parts I, II (1984).

010.067 **AAS Photo-Bulletin.**
No. 36 (1984).

### Association Française des Observateurs d'Etoiles Variables

010.101 **La vie de l'Association.**
E. Schweitzer.
Bull. Assoc. Fr. Obs. Etoiles Variables, No. 29, p. 113, No. 30, p. 155 (1984).

010.102 **Activité de l'A.F.O.E.V. en 1983.**
E. Schweitzer.
Bull. Assoc. Fr. Obs. Etoiles Variables, No. 29, p. 85 – 98 (1984).

010.103 **Bulletin de l'Association Française des Observateurs d'Etoiles Variables.**
Nos. 29 – 30 (1984).

**Association of Lunar and Planetary Observers (A.L.P.O.)**

010.121 **The 1984 Business Meeting of the A.L.P.O.**
J. L. Benton Jr.
Strolling Astron., Vol. 30, Nos. 11 – 12, p. 255 – 260 (1984).

010.122 **The A.L.P.O. at ALCON '84.**
J. L. Benton Jr.
Strolling Astron., Vol. 30, Nos. 11 – 12, p. 256 – 260 (1984).

010.123 **The Minor Planet Bulletin.**
Vol. 11, Nos. 1 – 4 (1984).

010.124 **The Strolling Astronomer. The Journal of the Association of Lunar and Planetary Observers.**
Vol. 30, Nos. 9 – 12 (1984).

**Astronomical Society of Australia (ASA)**

010.161 **Proceedings of the Astronomical Society of Australia.**
Vol. 5, No. 3 (1984).

**Astronomical Society of India**

010.201 **Bulletin of the Astronomical Society of India.**
Vol. 12, Nos. 2 – 3 (1984).

**Astronomical Society of Japan**

010.221 **Publications of the Astronomical Society of Japan.**
Vol. 36, Nos. 3 – 4 (1984).

**Astronomical Society of New York**

010.241 **Abstracts of papers given at the Spring meeting of the Astronomical Society of New York, held April 28, 1984 at Columbia University, New York.**
A. G. D. Philip.
News Lett. Astron. Soc. N.Y., Vol. 2, No. 6, p. 1 – 20 (1984).

010.242 **News Letter of the Astronomical Society of New York.**
Vol. 2, No. 6 (1984).

**Astronomical Society of Southern Africa (ASSA)**

010.261 **Proceedings of the Annual General Meeting, 1984.**
Mon. Notes Astron. Soc. S. Afr., Vol. 43, Nos. 9 – 10, p. 43 – 50 (1984).

010.262 **Centre reports: Bloemfontein Centre, Cape Centre, Harare Centre, Natal Centre, Natal Midlands Centre, Pretoria Centre, Transvaal Centre.**
J. J. van Ellinckhuyzen, M. G. Soltynski, R. Fleet,
J. P. G. Cramb, C. S. Lake, J. R. Starkey, M. A. Gray.
Mon. Notes Astron. Soc. S. Afr., Vol. 43, Nos. 11 – 12, p. 78 – 86 (1984).

010.263 **Section reports.**
Mon. Notes Astron. Soc. S. Afr., Vol. 43, Nos. 9 – 10, p. 50 – 55 (1984).

010.264 **Monthly Notes of the Astronomical Society of Southern Africa.**
Vol. 43, Nos. 1 – 11 (1984).

**Astronomical Society of the Pacific (ASP)**

010.281 **Abstracts of papers presented at the summer scientific meeting of the Astronomical Society of the Pacific at the University of California, Santa Cruz, California, 9 – 12 July 1984.**
Publ. Astron. Soc. Pac., Vol. 96, No. 584, p. 783 – 799 (1984).

010.282 **Publications of the Astronomical Society of the Pacific.**
Vol. 96, Nos. 581 – 586 (1984).

010.283 **Mercury. The Journal of the Astronomical Society of the Pacific.**
Vol. 13, Nos. 4, 6 (1984).

**Astronomical Society of Western Australia (ASWA)**

010.301 **Journal of the Astronomical Society of Western Australia.**
Vol. 34, No. 10, October, Vol. 35, No. 3 (1984).

**Astronomical Society "Rudjer Bošković"**

010.321 **The founding and activity of Belgrade's Astronomical Club "Rudjer Bošković" during 1951 and 1952.**
D. Teleki.
Vasiona, Année 32, No. 3, p. 54 – 58 (1984). In Croatian.

The paper describes the activity of the Astronomical Club "R. Bošković" which existed in Belgrade during 1951 and 1952, and which later grew into the present Astronomical Society "Rudjer Bošković".

**Astronomische Gesellschaft (AG)**

010.341 **Bericht über die Versammlung in Minden, 11. – 14. September 1984.**
Mitt. Astron. Ges., Nr. 62, p. 11 – 376 (1984).

010.342 **Lebt die Astronomische Gesellschaft?**
G. A. Tammann.
Mitt. Astron. Ges., Nr. 62, p. 13 – 16 (1984). – See Abstr. 010.341.

010.343 **Mitteilungen der Astronomischen Gesellschaft.**
Nr. 61 – 62 (1984).

**AURA (The Association of Universities for Research in Astronomy)**

010.351 **AURA. The first twenty–five years, 1957 – 1982.**
H. Abt, V. M. Blanco, G. Burbidge,
F. K. Edmondson, J. W. Evans, M. Fults, R. Giacconi,
D. Hall, M. Hanna, W. A. Hiltner, C. R. O'Dell, P. Osmer,
A. Paulsen, E. Spicer, J. M. Teem, A. B. Weaver, K. Welch,
H. J. Wood, J. B. Zirker.
The Association of Universities for Research in Astronomy, Inc., 1002 North Warren Avenue, Tucson, Ariz. 85719, USA. 56 pp. (1983).

Contents: Introduction. History of AURA, Inc. Kitt Peak National Observatory. Cerro Tololo Inter–American Observatory. Sacramento Peak Observatory. Space Telescope Science Institute. A look to the future. Facts about AURA, NSF, NASA. Observatories' annualized user statistics.

**British Astronomical Association (BAA)**

010.361 **Report of the Council on work done during and accounts for the session 1983 August 1 to 1984 July 31.**
J. Br. Astron. Assoc., Vol. 94, No. 6, p. 288 – 299 (1984).

010.362 **British Astronomical Association Circular.**
Nos. 641 – 646 (1984).

010.363 **Meetings and activities of the Association.**
J. Br. Astron. Assoc., Vol. 94, No. 5, p. 230 – 238, No. 6, p. 273 – 282 (1984).

010.364 **Section reports.**
J. Br. Astron. Assoc., Vol. 94, No. 5, p. 224 – 228, No. 6, p. 269 – 272, Vol. 95, No. 1, p. 33 – 36 (1984).

010.365 **Journal of the British Astronomical Association.**
Vol. 94, Nos. 5 – 6, Vol. 95, No. 1 (1984).

010.366 **The British Astronomical Association Newsletter.**
Nos. 7 – 9 (1984).

**British Interplanetary Society (BIS)**

010.381 **JBIS. Journal of the British Interplanetary Society.**
Vol. 37, Nos. 7 – 12 (1984).

010.382 **Spaceflight. A publication of the British Interplanetary Society.**
Vol. 26, Nos. 7 – 12 (1984).

010.383 **Space Education. A publication of the British Interplanetary Society.**
Vol. 1, No. 8 (1984).

**Canadian Astronomical Society**

010.401 **Fifteenth meeting of the Canadian Astronomical Society held in Ottawa, June 4 – 7, 1984. Abstracts of papers.**
J. R. Astron. Soc. Can., Vol. 78, No. 5, p. 197 – 214 (1984).

**European Space Agency (ESA)**

010.461 **ESA and its programmes: present and future.**
E. Mallet.
ESA Spec. Publ., ESA SP–202, p. 40 – 42 (1984). – See Abstr. 012.004.

010.462 **The ESA technological research programmes.**
M. Trella.
ESA Spec. Publ., ESA SP–202, p. 43 – 46 (1984). – See Abstr. 012.004.

010.463 **ESA IUE Newsletter.**
Nos. 20 – 21 (1984).

010.464 **EXOSAT Express.**
Nos. 6 – 8 (1984).

010.465 **ESA Bulletin.**
Nos. 39 – 40 (1984).

010.466 **ESA Journal.**
Vol. 7, No. 4 (1983), Vol. 8, Nos. 2 – 4 (1984).

**International Amateur–Professional Photoelectric Photometry**

010.481 **Membership directory.**
I.A.P.P.P. Commun., No. 18, 32 pp. (1984).

010.482 **I.A.P.P.P. Communication.**
Nos. 16 – 18 (1984).

**International Astronomical Union (IAU)**

010.521 **IAU Circulars.**
Nos. 3957 – 4024 (1984).

010.522 **Circulaire d'Information.**
No. 94 (1984).

010.523 **Commission 27 of the IAU, Information Bulletin on Variable Stars.**
Nos. 2544 – 2649 (1984).

010.524 **The Minor Planet Circulars/Minor Planets and Comets.**
Nos. 8827 – 9314 (1984).

010.525 **Quarterly Bulletin on Solar Activity.**
Vol. 23, Part III (1981).

010.526 **IAU Colloquia.**
Nos. 75 (012.085), 79 (012.064), 81 (012.089).

**Korean Astronomical Society**

010.581 **The Journal of the Korean Astronomical Society.**
Vol. 17, No. 2 (1984).

**LEST Foundation**

010.591 **LEST Foundation, Technical Report.**
No. 2 (1983), Nos. 3 – 7, 9 (1984).

010.592 **LEST Foundation, Annual Report 1983.**
Ø. Hauge.
LEST Found., Annu. Rep., 35 pp. (1984).

010.593 **LEST Large European Solar Telescope. Executive summary.**
A. A. Wyller.
Separate Print, 32 pp. (1984).

010.594 **LEST project study. Management plan and capital budget.**
T. E. Andersen, K.–I. Hillerud.
LEST Found., Tech. Rep., No. 9, 35 pp. (1984).

**Meteoritical Society**

010.601 **Meteoritics. The Journal of the Meteoritical Society.**
Vol. 19, Nos. 2 – 3 (1984).

**Oriental Astronomical Association**

010.661 **The Heavens.**
Vol. 65, Nos. 6 – 12 (1984). In Japanese.

**Royal Astronomical Society (RAS)**

010.681 **The history of British astronomy and geophysics. Summary of the RAS specialist discussion, held 1983 December 9 in the Scientific Societies' Lecture Theatre, Savile Row.**
Observatory, Vol. 104, No. 1061, p. 181 – 186 (1984).

010.682 **Summaries of papers presented at the RAS Junior Members' Day, held 1984 February 10 in the Scientific Societies' Lecture Theatre, Savile Row.**
Observatory, Vol. 104, No. 1062, p. 211 – 216 (1984).

010.683 **Galaxies: activity and environment. Summary of the RAS specialist discussion, held 1984 April 13 at the University of Leicester.**
Observatory, Vol. 104, No. 1063, p. 254 – 255 (1984).

010.684 **Minor bodies in the solar system. Summary of the RAS specialist discussion held 1984 April 13 at the University of Leicester.**
A. J. Meadows.
Observatory, Vol. 104, No. 1063, p. 256 – 257 (1984).

010.685 **Close binary systems. Summary of the RAS specialist discussion held 1984 April 13 at the University of Leicester.**
Observatory, Vol. 104, No. 1063, p. 257 – 259 (1984).

010.686 **Meetings of the Society.**
Observatory, Vol. 104, No. 1061, p. 173 – 181, No. 1063, p. 245 – 254 (1984).

010.687 **Meetings and activities of the Society.**
Q. J. R. Astron. Soc., Vol. 25, No. 3, p. 381 – 403 (1984).

010.688 **Monthly Notices of the Royal Astronomical Society.**
Vol. 209, Nos. 1 – 4, Vol. 210, Nos. 1 – 4, Vol. 211, Nos. 1 – 4 (1984).

010.689 **Geophysical Journal of the Royal Astronomical Society.**
Vol. 78, Nos. 1 – 3, Vol. 79, Nos. 1 – 3 (1984).

010.690 **The Quarterly Journal of the Royal Astronomical Society.**
Vol. 25, Nos. 3 – 4 (1984).

**Royal Astronomical Society of Canada**

010.701 **Meetings of the Society.**
J. R. Astron. Soc. Can., Vol. 78, No. 5, p. 215 – 222 (1984).

010.702 **The Journal of the Royal Astronomical Society of Canada.**
Vol. 78, Nos. 4 – 6 (1984).

010.703 **National Newsletter. Supplement to the Journal of the Royal Astronomical Society of Canada.**
Vol. 78, Nos. 4 – 6 (1984).

**Royal Astronomical Society of New Zealand**

010.721 **Royal Astronomical Society of New Zealand. President's report 1983 – 1984.**
F. Andrews.
South. Stars, Vol. 30, Nos. 7 – 8, p. 452 – 456 (1984).

010.722 **Southern Stars. Journal of the Royal Astronomical Society of New Zealand.**
Vol. 30, Nos. 7 – 8 (1984).

**Schweizerische Astronomische Gesellschaft (SAG)**

010.741 **BBSAG Bulletin.**
Nos. 73 – 74 (1984).

010.742 **Mitteilungen.**
Orion, 42. Jahrg., Nr. 203, p. 147 – 154, Nr. 204, p. 185 – 188, Nr. 205, p. 219 – 222 (1984).

010.743 **Orion. Zeitschrift der Schweizerischen Astronomischen Gesellschaft. Revue de la Société Astronomique de Suisse.**
Jahrg. 42, Nr. 203 – 205 (1984).

**Società Astronomica Italiana (S.A.It.)**

010.761 **Memorie della Società Astronomica Italiana.**
Vol. 55, Nos. 1 – 2 (1984).

010.762 **Giornale di Astronomia.**
Vol. 10, N. 1 – 2 (1984).

**Société Astronomique de France**

010.781 **La vie de la Société.**
B. Clouet.
Astronomie, Vol. 98, p. 565 – 571 (1984).

010.782 **L'Astronomie et Bulletin de la Société Astronomique de France.**
Vol. 98, juillet – décembre (1984).

**Société Astronomique de Liège**

010.801 **Le Ciel.**
Vol. 46, septembre – décembre, p. 177 – 296 (1984).

**Société Royale Belge d'Astronomie**

010.821 **Ciel et Terre. Bulletin de la Société Royale Belge d'Astronomie, de Météorologie et de Physique du Globe.**
Vol. 100, Nos. 4 – 6 (1984).

**Vereinigung der Sternfreunde e.V. (VdS)**

010.861 **Nachrichten der Vereinigung der Sternfreunde e.V.**
Sterne Weltraum, 23. Jahrg., Nr. 7, p. 396 – 399, Nr. 8 – 9, p. 476 – 478, Nr. 10, p. 533 – 537, Nr. 11, p. 594 – 597, Nr. 12, p. 652 – 657 (1984).

010.862 **Sonne. Mitteilungsblatt der Amateursonnenbeobachter.**
Jahrg. 8, Nr. 31 – 32 (1984).

# 011 Reports on Colloquia, Congresses, Meetings, Symposia, Expeditions

011.001 **Kiel: Plasmaphysik, extraterrestrische Physik, Kurzzeitphysik.**
E. Hintz, M. Hugenschmidt, E. Keppler, J. Richter, G. Wibberenz.
Phys. Bl., 40. Jahrg., Heft 7, p. 224 – 226 (1984).
Report of a conference, held at Kiel, March 19 – 23, 1984.

011.002 **Compte rendu de l'exposition «La mesure du temps».**
J. De Graeve.
Ciel Terre, Vol. 100, No. 4, p. 132 – 134 (1984).

011.003 **Ein gemeinsames amerikanisch–europäisches Programm zur Planetenerforschung.**
H. Fechtig.
Sterne Weltraum, 23. Jahrg., Nr. 8–9, p. 416 – 417 (1984).

011.004 **Planetary dynamics meeting held at Herstmonceux 1983 September 15 – 17.**
Q. J. R. Astron. Soc., Vol. 25, No. 3, p. 319 – 320 (1984).

011.005 **Middle atmosphere MIST meeting.**
H. Rishbeth.
Q. J. R. Astron. Soc., Vol. 25, No. 3, p. 321 – 326 (1984).

011.006 **Sonnen– und Planetentagung '84 in der Volkssternwarte Hof.**
R. Stark.
Sonne, Jahrg. 8, Nr. 31, p. 109 – 122 (1984).

011.007 **Geodesy at the XVIIIth General Assembly of the International Union of Geodesy and Geophysics. Hamburg, 1983, August 15 – 27.**
Yu. D. Bulanzhe, M. T. Prilepin.
Zemlya Vselennaya, No. 4, p. 61 – 64 (1984). In Russian.

011.008 **Symposium in Pamporovo, Bulgaria, 1983, July 18 – 22.**
V. M. Lipunov.
Zemlya Vselennaya, No. 4, p. 65 – 67 (1984). In Russian.
Concerning successes in high–energy astrophysics and cosmology.

011.009 **Problems of investigation of star clusters and associations. Symposium held in Prague, 1983, September 27 – October 1.**
A. S. Rastorguev.
Zemlya Vselennaya, No. 4, p. 67 – 69 (1984). In Russian.

011.010 **Search for extraterrestrial civilizations is continued. Symposium No. 12 of the International Astronautical Federation, held in Budapest, 1983, October 10 – 15.**
V. S. Troitskij.
Zemlya Vselennaya, No. 4, p. 70 – 71 (1984). In Russian.

011.011 **Trends in planetary data analysis. Executive summary of the Planetary Data Workshop, held at Goddard Space Flight Center, 29 November – 1 December 1983.**
N. Evans.
NASA Conf. Publ., NASA CP–2333, 5+40 pp. (1984).
The Planetary Data Workshop was held to define the science and computational requirements for a Planetary Data System. Additionally, the workshop provided systems concepts which take advantage of present and future advances in technology, avoid past mistakes in the archiving and curating of data, and ensure that essential planetary datasets are easily accessible to current and future generations of scientists.

011.012 **Report on the gamma–ray astronomy papers presented at the 18th International Cosmic–Ray Conference, Bangalore, India, September, 1983.**
N. A. Porter.
Ir. Astron. J., Vol. 16, No. 3, p. 202 – 209 (1984). – See Abstr. 012.036.

011.013 **Second seminar–conference on physical investigations of asteroids.**
D. F. Lupishko.
Astron. Vestn., Tom 18, No. 2, p. 158 – 159 (1984). In Russian. Abstr. in Ref. Zh., 51. Astron., 10.51.31 (1984).

011.014 **Soviet–French conference in Cannes, October 1983.**
V. A. Kotel'nikov.
Zemlya Vselennaya, No. 5, p. 76 – 78 (1984). In Russian.

011.015 **Lunar and planetary investigations. XVth Lunar and Planetary Science conference in Houston, Tex., March 1984.**
V. V. Shevchenko.
Zemlya Vselennaya, No. 6, p. 43 – 45 (1984). In Russian.

011.016 **Le Colloque des Planétariums Européens.**
A. Acker, W. Bodenmuller.
Astronomie, Vol. 98, p. 573 – 575 (1984).

011.017 **Die Tagung der Astronomischen Gesellschaft in Minden.**
R. Lukas.
Sterne Weltraum, 23. Jahrg., Nr. 12, p. 617 (1984).

011.018 **Neutrinos and the Sun.**
V. Trimble.
Nature, Vol. 311, No. 5987, p. 608 – 609 (1984).
This note reports on the Homestake Neutrino Conference, held at Lead, South Dakota, 23 – 25 August 1984.

011.019 **Very–long–baseline interferometry takes to the sky.**
P. N. Wilkinson.
Nature, Vol. 312, No. 5989, p. 17 – 18 (1984).
Report of a workshop on space–VLBI projects, held at Gross-enzersdorf, Vienna, Austria, 18 – 22 July 1984.

011.020 **Assembling the cloudy jigsaw.**
A. Lawrence, M. Ward.
Nature, Vol. 312, No. 5990, p. 100 – 101 (1984).
This note reports on a workshop on active galactic nuclei, held in honour of the 60th birthday of Donald E. Osterbrock at Santa Cruz, California, 16 – 27 July 1984.

011.021 **The Abastumani conference on astrophotography.**
O. D. Dokuchaeva.
Sov. Astron., Vol. 28, No. 3, p. 365 – 366 (1984). English translation of 37.011.034.

011.022 **Planetary science experiments aboard Space Station: workshop report.**
R. Greeley.
Bull. Am. Astron. Soc., Vol. 16, No. 3, Part I, p. 651 (1984). Abstract. – See Abstr. 010.061.

011.023 **RAS discussion meeting on the plasmapause and autumn MIST meeting.**
M. Lockwood, P. A. Hadjiry, M. J. Laird.
Q. J. R. Astron. Soc., Vol. 25, No. 4, p. 429 – 434 (1984).
Reports on two meetings held 1983 November 25 in London.

011.024 **Twelfth meeting of the Indian Association of General Relativity and Gravitation.**
T. Padmanabhan, A. Kembhavi.
Bull. Astron. Soc. India, Vol. 12, No. 3, p. 319 – 320 (1984).
Report on the IAGRG XII meeting held at the University of Poona, 1983 November 9 – 12.

011.025 **Cambridge Society – TIFR workshop on galaxy interactions.**
S. P. Bhavsar, D. Bhattacharya.
Bull. Astron. Soc. India, Vol. 12, No. 3, p. 320 – 324 (1984).
Report on the Cambridge Society – TIFR workshop on galaxy interactions, held 1983 December 25 – 30 at Tata Institute of Fundamental Research, Bombay.

011.026 **Second Indo–US workshop on solar–terrestrial physics.**
N. Gopalswamy.
Bull. Astron. Soc. India, Vol. 12, No. 3, p. 324 – 328 (1984).
Report on the second Indo–US workshop on solar–terrestrial physics, held 1984 January 30 – February 3 at National Physical Laboratory, New Delhi.

011.027 **Scientific session of the Department of General Physics and Astronomy and the Department of Nuclear Physics of the USSR Academy of Sciences, 18 – 19 January 1984.**
Usp. fiz. nauk, Tom 43, No. 4, p. 673 – 677 (1984). In Russian. Abstr. in Ref. Zh., 51. Astron., 12.51.22 (1984).

011.028 **Wissenschaftliche Tagung Minden 1984. Bericht über die Tagung.**
W. Seggewiß.
Mitt. Astron. Ges., Nr. 62, p. 11 – 12 (1984). – See Abstr. 010.341.

011.029 **Workshop über astrometrische und geodätisch–geophysikalische Nutzung der Radio–Interferometrie.**
J. Campbell.
Mitt. Astron. Ges., Nr. 62, p. 137 (1984). – See Abstr. 010.341.

011.030 **Workshop über wechselwirkende Galaxien.**
O. E. Gerhard.
Mitt. Astron. Ges., Nr. 62, p. 145 (1984). – See Abstr. 010.341.

011.031 **Workshop über Sternbedeckungen.**
H.-J. Bode.
Mitt. Astron. Ges., Nr. 62, p. 153 (1984). – See Abstr. 010.341.

011.032 **HIPPARCOS workshop.**
U. Bastian, S. Röser.
Mitt. Astron. Ges., Nr. 62, p. 156 – 157 (1984). – See Abstr. 010.341.

011.033 **Workshop über Photographie.**
J. D. Schumann.
Mitt. Astron. Ges., Nr. 62, p. 160 – 161 (1984). – See Abstr. 010.341.

011.034 **Workshop über Sonnenflecken.**
U. Grossmann–Doerth.
Mitt. Astron. Ges., Nr. 62, p. 184 (1984). – See Abstr. 010.341.

011.035 **Conference of young scientists in memory of V. P. Tsesevich, held in Odessa, 21 – 25 May 1984.**
E. V. Menchenkova, I. L. Andronov.
Astron. Zh., Tom 61, Vyp. 6, p. 1238 – 1239 (1984). In Russian. English translation in Sov. Astron., Vol. 28, No. 6.

011.036 **IAU Symposium 110: Very long baseline interferometry and compact radio sources.**
D. B. Shaffer, D. Weistrop.
Astron. Q., Vol. 5, No. 17, p. 39 – 42 (1984).
Report on the IAU Symposium No. 110, held at Bologna, Italy, 1983 June 27 – July 1.

011.037 **International conference on astrometric binaries.**
R. D. Chapman, F. B. Wood.
Comments Astrophys., Vol. 10, No. 4, p. 161 – 165 (1984).
This note reports on a conference on astrometric binaries, held at Bamberg, F.R. Germany, June 1984.

011.038 **Mass loss from red giants – a report on the June 1984 meeting at UCLA.**
V. Trimble.
Comments Astrophys., Vol. 10, No. 4, p. 167 – 177 (1984).

011.039 **Symposium on pulsars and active objects sponsored by the Beijing Astronomy Society was held in Peking University, Aug. 2 – 8, 1983.**
Acta Astrophys. Sin., Vol. 4, No. 1, p. 83 – 85 (1984). In Chinese and English.

011.040 **The Academic Meeting of Chinese Radio Astronomy was held in Beijing, Sept. 2 – 8, 1983.**
Acta Astrophys. Sin., Vol. 4, No. 1, p. 86 – 88 (1984). In Chinese and English.

011.041 **The International Workshop on Solar Physics and Interplanetary Travelling Phenomena was held at Yunnan Observatory in Kunming, Nov. 21 – 25, 1983.**
Acta Astrophys. Sin., Vol. 4, No. 2, p. 170 – 172 (1984). In Chinese.

011.042 **On the All–Union conference on physics and dynamics of comets at the Crimean Astrophysical Observatory of the USSR Academy of Sciences, 1984 October 15 – 20.**
Komet. Tsirk., No. 330 (1984). In Russian.

011.043 **A report on the 96th Annual Meeting of the A.S.P.**
A. Fraknoi.
Mercury, Vol. 13, No. 6, p. 162 – 164, 189 (1984).

011.044 **The 1984 Scientific Meeting of the A.S.P.: a symposium on stellar populations.**
E. Friel.
Mercury, Vol. 13, No. 6, p. 165 – 167, 185 – 186 (1984).

# 012 Proceedings of Colloquia, Congresses, Meetings, Symposia

012.001 **Proceedings of the Workshop on Star Formation, held in Edinburgh, 4 – 6 October 1983.**
R. D. Wolstencroft (Editor), with a summary by A. Webster.
Occas. Rep. R. Obs., Edinb., No. 13, 8+227 pp. (1984). ISBN 0–902553–30–5.
The individual contributions are included in their corresponding subject categories – see abstracts 064.007, 065.010, 116.006, 121.007, 131.027 – 131.036, 155.006, 155.007, 157.027 – 157.030.

012.002 **Proceedings of the Third Marcel Grossmann Meeting on General Relativity, held in Shanghai, China, 30 August – 3 September 1982.**
N. Hu (Editor).
Part A: Session Papers + Part B: Workshop Papers. Science Press, Beijing and North–Holland Publishing Company, Amsterdam – New York – Oxford. Part A: 17+887 pp. Part B: 21+625 pp. Price US$ 150.00, Dfl. 390.00 (1983). ISBN 0–444–86746–5.
The individual contributions are included in their corresponding subject categories – see abstracts 022.017 – 022.019, 034.022 – 034.031, 061.009 – 061.026, 066.023 – 066.121, 067.019 – 067.034, 117.041, 126.017, 159.022 – 159.025, 161.042 – 161.069.

012.003 **Towards an improved International Reference Ionosphere. Proceedings of the URSI/COSPAR Workshop held in Stara Zagora, Bulgaria, 30th August – 3rd September 1983.**
K. Rawer, C. M. Minnis, K. B. Serafimov (Editors), with a report on the opening of the workshop by Y. V. Ramanamurty, a short scientific report on the workshop by I. Kutiev, and final summary and conclusions by K. Rawer.
Adv. Space Res., Vol 4, No. 1, 6+171 pp. (1984). ISBN 0–08–031511–9.
The individual contributions within the subject scope of Astronomy and Astrophysics Abstracts are included in their corresponding categories – see abstracts 083.005 – 083.024.

012.004 **Proceedings of an ESA Workshop on SPLAT – Space Laser Applications and Technology, held at Les Diablerets, Switzerland, 25 – 30 March 1984.**
B. Battrick, E. Rolfe (Editors).
ESA Spec. Publ., ESA SP–202. 6+46+272 pp. Price FF 140.00 (1984).
The individual contributions within the subject scope of Astronomy and Astrophysics Abstracts are included in their corresponding categories – see abstracts 010.461, 010.462, 035.011, 036.011, 044.007, 081.012, 082.025.

012.005 **Technology in the 1990s: the industrialization of space. Proceedings of a Royal Society Meeting held at London, 7 – 8 December 1983.**
G. K. C. Pardoe (Editor).
Philos. Trans. R. Soc. London, Ser. A, Vol. 312, No. 1519, p. 1 – 140 (1984). Available as separate print from: The Royal Society, 6 Carlton House Terrace, London, SW1Y 5AG, England. Price £ 22.00 U.K., £ 23.00 overseas. ISBN 0–85403–233–9.

012.006 **Proceedings of the Fifth Pacific Northwest Conference on Nuclear Track Registration, held at Richland, Wash., USA, 28 – 29 July 1982.**
Nucl. Tracks Radiat. Meas., Vol. 7, No. 1 – 2 (1983).
Review in Phys. Abstr., Vol. 87, No. 1237, Entry 67851 (1984).
See abstracts 034.008, 035.002.

012.007 **Quark matter '83. Proceedings of the Third International Conference on Ultra–Relativistic Nucleus–Nucleus Collisions, held at Upton, Long Island, N. Y., USA, 26 – 30 September 1983.**
Nucl. Phys. A, Vol. A418 (1984).
Review in Phys. Abstr., Vol. 87, No. 1237, Entry 67855 (1984).
See abstracts 061.005, 144.004, 161.013.

012.008 **Proceedings of the 7th International Conference on Mathematical Physics, held at Boulder, Colo., USA, 1 – 10 August 1983.**
Physica A, Vol. 124A, No. 1 – 3 (1984).
Review in Phys. Abstr., Vol. 87, No. 1237, Entry 67856 (1984).
See Abstr. 067.005.

012.009 **Proceedings of the XVIIIth Solvay Conference on Physics, held at Austin, Tex., USA, 8 – 12 November 1982.**
Phys. Rep., Vol. 104, No. 2 – 4 (1984).
Review in Phys. Abstr., Vol. 87, No. 1237, Entry 67857 (1984).
See Abstr. 161.022.

012.010 **Field theory in elementary particles. Proceedings of Orbis Scientiae 1982. Conference held at Coral Gables, Fla., USA, 18 – 21 January 1982.**
A. Perlmutter (Editor).
Plenum Publishing Corporation, New York, USA. 9+469 pp. (1983). ISBN 0–306–41345–0.
Review in Phys. Abstr., Vol. 87, No. 1237, Entry 67869 (1984).
See abstracts 061.004, 117.016, 117.017.

012.011 **ERS–1 radar altimeter data products. Proceedings of an ESA Workshop held at Frascati, Italy, 8 – 11 May, 1984.**
T. D. Guyenne, J. J. Hunt (Editors).
ESA Spec. Publ., ESA SP–221. 6+259 pp. Price FF 140.00 (1984).

012.012 **IGARSS '84. Remote sensing – from research towards operational use. Volumes I, II. International Geoscience and Remote Sensing Symposium held at Strasbourg, 27 – 30 August 1984.**
T. D. Guyenne, J. J. Hunt (Editors).
ESA Spec. Publ., ESA SP–215. 16+557 pp. (Vol. I), 6+319 pp. (Vol. II). Price $ 44.00 (Vol. I), $ 34.00 (Vol. II) (1984).
The individual contributions within the subject scope of Astronomy and Astrophysics Abstracts are included in their corresponding categories – see abstracts 051.018, 082.026 – 082.030.

012.013 **1983 Annual Meeting of the Division of Particles and Fields of the American Physical Society, held at Blacksburg, Va., USA, 15 – 17 September 1983.**
AIP Conf. Proc., No. 112 (1984).
Review in Phys. Abstr., Vol. 87, No. 1238, Entry 73135 (1984).
See abstracts 161.074, 161.075.

012.014 **Deployable optical systems. Conference held at Los Angeles, Calif., USA, 18 – 19 January 1983.**
Proc. SPIE Int. Soc. Opt. Eng., Vol. 383 (1983).
Review in Phys. Abstr., Vol. 87, No. 1238, Entry 73138 (1984).
See abstracts 031.006 – 031.008, 032.006, 033.001.

012.015 **Eighth International Conference on Infrared and Millimeter Waves, held at Miami Beach, Fla., USA, 12 – 17 December 1983. Conference Digest.**
R. J. Temkin (Editor).
IEEE, New York, USA. 22+290 pp. (1984).
Review in Phys. Abstr., Vol. 87 No. 1238, Entry 73154 (1984).
See abstracts 033.002, 033.003, 035.014, 077.008, 101.006.

012.016 **Antarctic earth science. Fourth International Symposium, held at Adelaide, South Australia, 16 – 20 August 1982.**
R. L. Oliver, P. R. James, J. B. Jago (Editors).
Cambridge University Press, Cambridge – London – New York – New Rochelle – Melbourne – Sydney. 22+697 pp. (1983). ISBN 0–521–25836–7.
Review in Phys. Abstr., Vol. 87, No. 1239, Entry 78045 (1984).
See abstracts 105.033 – 105.035.

012.017 **XVI International Conference on Phenomena in Ionized Gases, held at Düsseldorf, F.R. Germany, 29 August – 2 September 1983.**
W. Botticher, H. Wenk, E. Schulz–Gulde (Editors).
University of Düsseldorf, Düsseldorf, F.R. Germany. 9+121 pp. (1983).
Review in Phys. Abstr., Vol. 87, No. 1240, Entry 82598 (1984).
See abstracts 022.022, 022.064, 062.032, 062.035 – 062.038, 067.039, 067.040, 071.009, 073.009, 073.010, 074.009, 126.020.

012.018 **Proceedings of the 7th International Conference on Vacuum Ultraviolet Radiation Physics, held at Jerusalem, Israel, 8 – 12 August 1983.**
Ann. Isr. Phys. Soc., Vol. 6 (1984).
Review in Phys. Abstr., Vol. 87, No. 1240, Entry 82583 (1984).
See abstracts 022.023 – 022.025, 035.016.

012.019 **Structural mechanics of optical systems. Conference held at Cambridge, Mass., USA, 7 – 8 November 1983.**
Proc. SPIE Int. Soc. Opt. Eng., Vol. 450 (1984).
Review in Phys. Abstr., Vol. 87, No. 1241, Entry 86104 (1984).
See abstracts 031.009, 031.010, 032.007, 035.017.

012.020 **Proceedings of the second workshop on "Nuclear astrophysics", held at Schloß Ringberg, Tegernsee, F.R. Germany, June 6 – 10, 1983.**
W. Hillebrandt, E. Müller, W. Ober, F.–K. Thielemann (Editors).
MPA Rep., No. 90, 5+141 pp. (1983).
The individual contributions are included in their corresponding subject categories – see abstracts 061.041 – 061.059, 065.020 – 065.028, 067.054 – 067.057, 114.025,114.026, 126.022, 155.016, 155.017, 161.122.

012.021 **Astrométrie, géodésie, mécanique céleste. 5e colloque du G.S. 5, held at Paris, 25 – 27 octobre 1983.**
Y. Requième (Editor).
Observatoire de l'Université de Bordeaux I, F–33270 Floirac, France. 4+266 pp. (1984).
The individual contributions are included in their corresponding subject categories – see abstracts 002.011, 004.005 – 004.007, 013.015 – 013.018, 034.037 – 034.039, 036.020 – 036.025, 041.015 – 041.017, 042.013 – 042.015, 044.008 – 044.012, 045.002, 046.010, 080.013, 082.032, 094.013, 094.014, 098.017, 099.017, 099.018, 100.010, 101.007, 102.007 – 102.009, 115.007, 118.013.

012.022 **Oscillations as a probe of the Sun's interior. Proceedings of an EPS study conference, held at Catania, Italy, 20 – 24 June 1983.**
G. Belvedere, L. Paternò (Editors).
Mem. Soc. Astron. Ital., Vol. 55, Nos. 1 – 2, p. 1 – 384 (1984).
The individual contributions are included in their corresponding subject categories – see abstracts 051.023, 051.024, 065.030, 073.013, 073.014, 074.012, 080.016 – 080.057, 115.008.

012.023 **Climatic changes on a yearly to millennial basis. Geological, historical and instrumental records. Proceedings of the Second Nordic Symposium on Climatic Changes and Related Problems, held in Stockholm, Sweden, May 16 – 20, 1983.**
N.–A. Mörner, W. Karlén (Editors), with concluding remarks by N.–A. Mörner.
D. Reidel Publishing Company, Dordrecht – Boston – Lancaster. 18+667 pp. Price Dfl. 210.00, $ 79.00 (1984). ISBN 90–277–1779–6.
The individual contributions within the subject scope of Astronomy and Astrophysics Abstracts are included in their corresponding categories – see abstracts 082.033, 082.034, 085.005 – 085.008.

012.024 **Proceedings of the Fourth International Ultraviolet Explorer Conference held in Rome, Italy, 15 – 18 May 1984.**
E. Rolfe, B. Battrick (Editors), with a foreword by R. Viotti, and opening addresses by L. Gratton, P. Benvenuti, Y. Kondo.
ESA Spec. Publ., ESA SP–218. 20+518 pp. Price FF 175.00 (1984).
The individual contributions are included in their corresponding subject categories – see abstracts 002.014, 002.015, 022.043, 035.020, 036.032 – 036.034, 051.025, 051.026, 064.014, 102.010, 103.141, 112.014 – 112.021, 114.032 – 114.046, 116.014, 117.055 – 117.073, 119.030, 119.031, 120.008, 121.015, 121.016, 122.064 – 122.067, 124.002, 125.321, 126.025, 131.075 – 131.084, 132.013, 134.018 – 134.021, 142.018 – 142.020, 154.021, 154.030, 156.006, 157.074 – 157.076, 158.060 – 158.067, 159.034, 159.035.

012.025 **Large–scale structure of the universe, cosmology and fundamental physics. Proceedings of the First ESO–CERN Symposium, held at CERN, Geneva, Switzerland, 21 – 25 November 1983.**
G. Setti, L. Van Hove (Editors).
European Southern Observatory, Karl–Schwarzschild–Str. 2, D–8046 Garching bei München, F. R. Germany. 6+455 pp. Price DM 35.00 (1984).
The individual contributions are included in their corresponding subject categories – see abstracts 035.021, 061.062 – 061.069, 066.153, 157.077 – 157.080, 160.019, 161.131 – 161.148.

012.026 **Observations of comets: radio and otherwise. A collection of extended abstracts of the papers presented at a workshop held at the Max–Planck–Institut für Radioastronomie, Bonn, F. R. Germany, 4 June 1984.**
M. K. Bird (Editor).
Technischer Bericht Nr. 64. Max–Planck–Institut für Radioastronomie, Auf dem Hügel 69, D–5300 Bonn 1, F. R. Germany. 6+72 pp. (1984).
The individual contributions are included in their corresponding subject categories – see abstracts 102.011 – 102.017, 103.005, 103.103, 103.906.

012.027 **Colloquium on atomic spectra and oscillator strengths for astrophysics and fusion research. Proceedings of a meeting held at Lund, Sweden, August 17 – 19, 1983.**
U. Litzén (Editor).
Phys. Scr., Vol. T8, 146 pp. Price SEK 400.00, US$ 64.00 (1984).
The individual contributions are included in their corresponding subject categories – see abstracts 022.046 – 022.061, 051.027, 064.019, 064.020, 071.010, 112.026, 114.048, 114.049.

012.028 **Cosmic ray intensity and cosmogenic isotopes. Papers of the 13th Leningrad seminar on cosmophysics, Leningrad, 19 – 21 November, 1982.**
G. E. Kocharov (Editor).
Leningrad. 202 pp. (1983). In Russian. From Ref. Zh., 51. Astron., 8.51.30 (1984).
The individual contributions are included in their corresponding subject categories – see abstracts 072.011 – 072.013, 073.037, 076.005, 076.006, 078.002, 080.071, 106.012, 144.014.

012.029 **Proceedings of the miniconference on low energy tests of conservation laws in particle physics, held at Blacksburg, Va., USA, 11 – 14 September 1983.**
AIP Conf. Proc., No.114 (1984).
Review in Phys. Abstr., Vol. 87, No. 1242, Entry 90281 (1984).
See Abstr. 061.073.

012.030 **Advanced infrared sensor technology. Conference held at Geneva, Switzerland, 18 – 19 April 1983.**
Proc. SPIE Int. Soc. Opt. Eng., Vol. 395 (1983).
Review in Phys. Abstr., Vol. 87, No. 1243, Entry 95018 (1984).
See abstracts 034.055 – 034.057, 034.081, 035.030, 035.040.

012.031 **The 27th International Geological Congress, held at Moscow, 1984 August 4 – 14.**
Reports Tom 5, Sections 10 – 11; Tom 11, Section S 11 (Geochemistry and Cosmochemistry), Nauka, Moskva (1984).
From Ref. Zh., 62. Issled. Kosm. Prostranstva, 8.62.398 (1984), Ref. Zh., 51. Astron., 9.51.55 (1984).
See abstracts 022.133, 061.080, 082.045, 082.046, 091.019, 091.025, 093.013, 093.023, 094.016, 094.017, 094.028, 094.029,

094.032, 094.036 – 094.038, 102.022, 105.047, 105.056, 105.058 – 105.065, 105.072, 107.007, 107.010 – 107.012, 107.014, 107.044 – 107.046.

012.032 **Mass loss from astronomical objects. Proceedings of a Rutherford Appleton Laboratory workshop on astronomy and astrophysics, held at Chilton, Didcot, Oxon, England, 26 – 28 April 1982.**
P. M. Gondhalekar (Editor).
RL–82–075. Rutherford Appleton Laboratory, Chilton, Didcot, Oxon OX11 0QX, England. 3 + 146 pp. (1984).
The individual contributions are included in their corresponding subject categories – see abstracts 112.031 – 112.034, 125.032, 155.028, 157.057, 159.052 – 159.054, 160.031.

012.033 **Eleventh Texas Symposium on Relativistic Astrophysics. Proceedings of a conference held in Austin, Texas, December 12 – 17, 1982.**
D. S. Evans (Editor).
Ann. N.Y. Acad. Sci., Vol. 422, 11 + 396 pp. Price US$ 79.00 cloth (1984). ISBN 0–89766–234–2 cloth; ISBN 0–89766–235–0 paper.
The individual contributions are included in their corresponding subject categories – see abstracts 031.013, 032.018 – 032.020, 034.059, 035.034, 061.075 – 061.078, 065.044 – 065.047, 066.177 – 066.182, 067.073 – 067.075, 080.068, 117.101, 117.102, 125.033, 125.034, 125.361, 126.36 – 126.040, 141.011, 142.025, 143.015, 143.016, 144.025, 157.098, 158.084 – 158.092, 159.055 – 159.059, 160.032, 161.192 – 161.212.

012.034 **Instabilités hydrodynamiques et applications astrophysiques. Comptes rendus de l'École de Goutelas, France, 5 – 9 avril 1983.**
A. Baglin, M. Auvergne, C. Caseneuve (Editors).
Société Française des Spécialités d'Astronomie, 61, Avenue de l'Observatoire, 75014 Paris, France. 586 pp. (1983).
The individual contributions are included in their corresponding subject categories – see abstracts 062.072 – 062.082, 064.030, 064.031, 065.048 – 065.052, 091.017, 091.018, 117.106, 117.107, 122.071.

012.035 **Workshop on solar and space physics, held at Hilton Head Island, South Carolina, 20 – 23 September 1983.**
ESA J., Vol. 7, No. 4, p. 325 – 429 (1983).
The individual contributions are included in their corresponding subject categories – see abstracts 062.172 – 062.174, 074.073, 080.124, 082.106.

012.036 **Proceedings of the meeting held by the Astronomical Science Group of Ireland at University College Dublin on 29th September, 1983.**
Ir. Astron. J., Vol. 16, No. 3, p. 188 – 242 (1984).
The individual contributions are included in their corresponding subject categories – see abstracts 011.012, 051.044, 073.040, 082.047, 121.027, 134.035, 156.010.

012.037 **Rapid variability of early–type stars. Proceedings of the workshop held at Hvar, Yugoslavia, 19 – 23 September 1983.**
P. Harmanec, K. Pavlovski (Editors).
Hvar Obs. Bull., Vol. 7, No. 1, 13 + 392 pp. (1983).
The individual contributions are included in their corresponding subject categories – see abstracts 064.032, 112.038 – 112.046, 113.023, 114.060, 117.115, 120.013, 122.098 – 122.111.

012.038 **Solar terrestrial physics: present and future. A report based on the Solar Terrestrial Physics Workshop, December 1982 to November 1983.**
D. M. Butler, K. Papadopoulos (Editors), with an executive summary.
NASA Ref. Publ., NASA RP–1120, 22 + 25 + 515 + 23 pp. (1984).
The individual contributions are included in their corresponding subject categories – see abstracts 062.087, 074.031, 074.032, 075.004, 080.074, 082.048, 083.036, 084.029 – 084.031, 085.017, 085.018.

012.039 **X–ray and UV emission from active galactic nuclei. Proceedings of a conference held at the Max–Planck–Institut für extraterrestrische Physik, Garching, F.R. Germany, July 9 – 13, 1984.**
W. Brinkmann, J. Trümper (Editors).
MPE Rep., No. 184. 12 + 280 pp. (1984).
The individual contributions are included in their corresponding subject categories – see abstracts 035.037, 051.045, 051.046, 063.045 – 063.047, 067.080 – 067.089, 142.032 – 142.036, 158.114 – 158.143, 159.071 – 159.076.

012.040 **Relativity, cosmology, topological mass and supergravity. Proceedings of the 4th SILARG Symposium on gravity, gauge theories and supergravity, held at USB Campus, Caracas, Venezuela, 5 – 11 December 1982.**
C. Aragone (Editor).
World Scientific Publishing Co. Pte. Ltd., Singapore. 8 + 296 pp. Price US$ 44.00 (1983). ISBN 9971–950–95–2.
The individual contributions are included in their corresponding subject categories – see abstracts 061.083 – 061.094, 066.192 – 066.194, 067.090, 161.227 – 161.233.

012.041 **Uranus and Neptune. Proceedings of a workshop held in Pasadena, California, February 6 – 8, 1984.**
J. T. Bergstralh (Editor).
NASA Conf. Publ., NASA CP–2330. 7 + 627 pp. (1984).
The individual contributions are included in their corresponding subject categories – see abstracts 074.043, 101.019 – 101.046.

012.042 **Synthetic aperture systems. Conference held at San Diego, Calif., USA, 25 – 26 August 1983.**
Proc. SPIE Int. Soc. Eng., Vol. 440 (1984).
Review in Phys. Abstr., Vol. 87, No. 1243, Entry 95020 (1984).
See Abstr. 031.015, 031.016, 032.026 – 032.028, 035.041.

012.043 **International conference on nonimaging concentrators, held at San Diego, Calif., USA, 25 – 26 August 1983.**
Proc. SPIE Int. Soc. Opt. Eng., Vol. 441 (1984).
Review in Phys. Abstr., Vol. 87, No. 1243, Entry 95021 (1984).
See Abstr. 034.082.

012.044 **Infrared detectors. Conference held at San Diego, Calif., USA, 25 – 26 August 1983.**
Proc. SPIE Int. Soc. Opt. Eng., Vol. 443 (1984).
Review in Phys. Abstr., Vol. 87, No. 1243, Entry 95022 (1984).
See Abstr. 034.077.

012.045 **12th International Conference on Solid State Nuclear Track Detectors, held at Acapulco, Mexico, 4 – 10 September 1983.**
Nucl. Tracks Radiat. Meas., Vol. 8, No. 1 – 4 (1984).
Review in Phys. Abstr., Vol. 87, No. 1243, Entry 95027 (1984).
See abstracts 034.078 – 034.080, 035.038, 035.042, 144.033.

012.046 **Symposium on mantle convection, held at the Second Meeting of the European Union of Geosciences at Strasbourg, France, 29 March 1983.**
Ann. Geophys., Vol. 2, No. 3 (1984).
Review in Phys. Abstr., Vol. 87, No. 1243, Entry 95029 (1984).
See abstracts 081.026, 081.028 – 081.030.

012.047 **Particle induced X–ray emission and its analytical applications. Proceedings of the Third International Conference on PIXE and its Analytical Applications, held at Heidelberg, F. R. Germany, 18 – 22 July 1983.**
Nucl. Instrum. Methods Phys. Res., Sect. B, Vol. 231, No. 1 – 3 (1984).
Review in Phys. Abstr., Vol. 87, No. 1243, Entry 95034 (1984).
See abstracts 094.040, 105.083, 105.084.

012.048 **Workshop on semiclassical methods in nuclear physics, held in Grenoble, France, 5 - 8 March 1984.**
J. Phys. Colloq., Vol. 45, No. C6 (1984).
Review in Phys. Abstr., Vol. 87, No. 1244, Entry 101567 (1984).
See Abstr. 065.068.

012.049 **Contemporary methods of optical manufacturing and testing, Conference held at San Diego, Calif., USA, 24 - 26 August 1983.**
Proc. SPIE Int. Soc. Opt. Eng., Vol. 433 (1983).
Review in Phys. Abstr., Vol. 87, No. 1245, Entry 106104 (1984).
See abstracts 031.017, 035.046.

012.050 **Precision surface metrology. Conference held at San Diego, Calif., USA, 23 - 24 August 1983.**
Proc. SPIE Int. Soc. Opt. Eng., Vol. 429 (1983).
Review in Phys. Abstr., Vol. 87, No. 1245, Entry 106103 (1984).
See Abstr. 035.053.

012.051 **Infrared technology IX. Conference held at San Diego, Calif., USA, 23 - 25 August 1983.**
Proc. SPIE Int. Soc. Opt. Eng., Vol. 430 (1983).
Review in Phys. Abstr., Vol. 87, No. 1246, Entry 110800 (1984).
See abstracts 034.086, 035.047 - 035.052, 084.060.

012.052 **18th General Assembly of the International Union of Geodesy and Geophysics, held at Hamburg, F.R. Germany, 15 - 27 August 1983.**
GeoJournal, Vol. 8, No. 4 (1984).
Review in Phys. Abstr., Vol. 87, No. 1245, Entry 106108 (1984).
See Abstr. 081.034.

012.053 **Problems in unification and supergravity. Record of a workshop, held at La Jolla, Calif., USA, 1983.**
AIP Conf. Proc., No. 116 (1984)
Review in Phys. Abstr., Vol. 87, No. 1246, Entry 110797 (1984).
See abstracts 061.103, 061.104, 161.267.

012.054 **Thermal conductivity 16. Proceedings of an international conference, held at Chicago, Ill., USA, 7 - 9 November 1979.**
D. C. Larson (Editor).
Plenum Publishing Corporation, New York, USA. 20 + 623 pp. (1983). ISBN 0-306-41032-X.
Review in Phys. Abstr., Vol. 88, No. 1247, Entry 29 (1985).
See Abstr. 067.112.

012.055 **Orages solaires de type I. Proceedings of the C.N.R.S. workshop "Solar noise storms", held at Abbaye de Sénanque, Gordes, France, 17 - 21 March 1980.**
J. L. Bougeret, A. O. Benz (Editors), with working group reports by B. Roberts, D. G. Wentzel, P. Zlobec and concluding remarks by D. B. Melrose.
Editions du Centre National de la Recherche Scientifique, 15 quai Anatole-France, F-75700 Paris, France. 218 pp. Price FF 98.00 (1984). ISBN 2-222-03040-4.
The individual contributions are included in their corresponding subject categories - see abstracts 062.101, 062.102, 072.059, 073.064, 074.045, 077.025 - 077.038, 084.062, 091.034, 091.035.

012.056 **Anthropology and astronomy: a marriage made in heaven. Symposium, held in Quebec City, Canada, 16 August 1983.**
Archaeoastronomy (U.S.A.), Vol. 6, Nos. 1 - 4, p. 10 - 11 (1983).
See abstracts 004.074 - 004.080.

012.057 **Astronomy and ceremony in the prehistoric Southwest. Symposium, held in Albuquerque, New Mexico, 12 - 14 October 1983.**
Archaeoastronomy (U.S.A.), Vol. 6, Nos. 1 - 4, p. 11 - 15 (1983).
See abstracts 004.081 - 004.087.

012.058 **First western regional conference on archaeoastronomy. Symposium, held in Northridge, Calif., USA, 1983 November 12.**
Archaeoastronomy, (U.S.A.), Vol. 6, Nos. 1 - 4, p. 16 - 18 (1983).
See abstracts 004.088 - 004.096.

012.059 **First International Conference on Ethnoastronomy: indigenous astronomical and cosmological traditions of the world, held at Washington, D.C., 5 - 9 September 1983.**
Archaeoastronomy (U.S.A.), Vol. 6, Nos. 1 - 4, p. 26 - 47 (1983).
See abstracts 004.097 - 004.164.

012.060 **General relativity and gravitation. Invited papers and discussion reports of the 10th International Conference on General Relativity and Gravitation, held at Padua, Italy, July 3 - 8, 1983.**
B. Bertotti, F. de Felice, A. Pascolini (Editors).
D. Reidel Publishing Company, Dordrecht - Boston - Lancaster. 16 + 517 pp. Price Dfl. 180.00, US$ 69.00, £ 45.75 (1984). ISBN 90-277-1819-9.
The individual contributions are included in their corresponding subject categories - see abstracts 061.118, 066.217 - 066.236, 067.116 - 067.118, 161.291 - 161.296.

012.061 **Problems of collapse and numerical relativity. Proceedings of a NATO Advanced Research Workshop, held at Toulouse, France, November 7 - 11, 1983.**
D. Bancel, M. Signore (Editors), with a conference summary by G. Baym.
NATO Advanced Science Institutes, Series, C, Vol. 134.
D. Reidel Publishing Company, Dordrecht - Boston - Lancaster. 10 + 381 pp. Price Dfl. 145.00, US$ 59.00, £ 36.95 (1984). ISBN 90-277-1816-4.
The individual contributions are included in their corresponding subject categories - see abstracts 062.107, 065.079 - 065.086, 066.237 - 066.242, 125.053 - 125.057, 143.029 - 143.031.

012.062 **Proceedings of the discussion meeting on Stellar and Planetary Magnetic Fields, held in Potsdam, German Democratic Republic, August 29 - September 2, 1983.**
F. Krause (Editor).
Astron. Nachr., Vol. 305, No. 5, p. 213 - 276 (1984).
The individual contributions are included in their corresponding subject categories - see abstracts 062.108, 065.087 - 065.091, 073.074, 075.011 - 075.013, 080.087, 081.040, 091.047.

012.063 **Proceedings of a Workshop on High Angular Resolution Optical Interferometry from Space, held in Baltimore, USA, June 13, 1984.**
P. B. Boyce, R. D. Reasenberg (Editors).
Bull. Am. Astron. Soc., Vol. 16, No. 3, Part II, p. 747 - 837 (1984).
The individual contributions are included in their corresponding subject categories - see abstracts 035.057 - 035.068, 036.083 - 036.086.

012.064 **Very large telescopes, their instrumentation and programs. Proceedings of IAU Colloquium No. 79, held in Garching bei München, F.R. Germany, April 9 - 12, 1984.**
M.-H. Ulrich, K. Kjär (Editors).
European Southern Observatory, Karl-Schwarzschild-Str. 2, D-8046 Garching bei München, F.R. Germany. 12 + 908 pp. Price DM 90.00, US$ 30.00 (1984). ISBN 3-923524-19-6.
The individual contributions are included in their corresponding subject categories - see abstracts 031.020 - 031.033, 032.037 - 032.060, 034.105 - 034.119, 036.087 - 036.099, 082.073 - 082.076.

012.065 **Theoretical problems in stellar stability and oscillations. Proceedings of the 25th Liège International Astrophysical Colloquium, July 10 - 13, 1984.**
Université de Liège, Institut d'Astrophysique, B-4200 Cointe-Ougrée, Belgium. 22 + 478 pp. Price US$ 25.00 (1984).

The individual contributions are included in their corresponding subject categories – see abstracts 005.019, 005.020, 062.121 – 062.124, 064.047, 064.048, 065.097 – 065.126, 066.243, 067.122, 071.033, 080.090 – 080.105, 117.185 – 117.187, 122.148 – 122.155, 151.066.

012.066 **Spectral evolution of galaxies. Proceedings of a Rutherford Appleton Laboratory workshop on astronomy and astrophysics, held at Chilton, Didcot, Oxon, England, 23 – 25 May 1983.**
P. M. Gondhalekar (Editor).
RAL–84–008. Rutherford Appleton Laboratory, Chilton, Didcot, Oxon OX11 0QX, England. 9+210 pp. (1984).
The individual contributions are included in their corresponding subject categories – see abstracts 065.127, 131.187, 157.154 – 157.162, 158.194, 161.310.

012.067 **Astronomical photography 1984. Proceedings of a meeting organised by the IAU Working Group on Photographic Problems, held in Edinburgh, 4 – 6 April 1984.**
M. E. Sim, K. Ishida (Editors), with an introduction by R. M. West, a welcoming address by R. Cannon and a summary by J.-L. Heudier.
Occas. Rep. R. Obs., Edinb. No. 14, 12+322 pp. (1984). ISBN 0–902553–31–3.
The individual contributions are included in their corresponding subject categories – see abstracts 009.035 – 009.038, 034.133 – 034.160, 036.103 – 036.111.

012.068 **Rotation in the solar system. Discussion held 8 and 9 March 1984.**
R. Hide (Editor), with introductory remarks by W. H. McCrea, and concluding remarks by S. K. Runcorn.
Philos. Trans. R. Soc. London, Ser. A, Vol. 313, No. 1524, p. 1 – 186 (1984).
The individual contributions are included in their corresponding subject categories – see abstracts 042.062, 044.034 – 044.037, 080.107, 081.043, 091.051, 094.055, 098.085, 100.071, 101.073, 102.050, 107.041 – 107.043.

012.069 **Classical general relativity.**
W. B. Bonnor, J. N. Islam, M. A. H. MacCallum (Editors).
Cambridge University Press, Cambridge – London – New York – New Rochelle – Melbourne – Sydney. 16+269 pp. Price $ 44.50 (1984).
Review in Astron. Express, Vol. 1, Issue 2, p. 82 (1984).

012.070 **Problems of physics of solar flares. Proceedings of the First Seminar of the working group "Special theoretical and experimental investigations of the solar plasma" held together with the Astronomical Council of the USSR Academy of Sciences. Materials of the yearly seminar of the section "Sun". 1981, December 15 – 18.**
A. A. Korchak (Editor).
Moskva, 220 pp. (1983). In Russian.
Review in Ref. Zh., 51. Astron., 11.51.37 (1984).
The individual contributions are included in their corresponding subject categories – see abstracts 062.154, 072.093 – 072.096, 073.100 – 073.105, 074.058, 075.020 – 075.022, 078.017, 078.018.

012.071 **Fourier transform spectroscopy. Proceedings of the International Conference, held at Durham, England, 5 – 9 September 1983.**
J. R. Birch (Editor).
Infrared Phys., Vol. 24, No. 2/3, 5+289 pp. (1984).
The individual contributions are included in their corresponding subject categories – see abstracts 022.137, 034.165, 036.123, 036.124.

012.072 **Middle atmosphere science. IAMAP/IAGA Symposium during the XVIII General Assembly of the IUGG, held at Hamburg, August 1983.**
A. Ebel, P. C. Simon (Editors).
J. Atmos. Terr. Phys., Vol. 46, No. 9, 133 pp. (1984).
The individual contributions are included in their corresponding subject categories – see abstracts 082.081, 085.031.

012.073 **Space: a developing role for Europe. 18th European Space Symposium held in London, England, 6 – 9 June 1983.**
L. J. Carter, P. M. Bainum (Editors), with a foreword by H. Jacobs.
Science and Technology Series, Vol. 56. A Supplement to Advances in the Astronautical Sciences. Published for the American Astronautical Society by Univelt, Inc., PO Box 28130, San Diego, Calif. 92128, USA. 10+268 pp. Price $ 45.00 cloth, $ 35.00 paper (1984). ISBN 0–87703–193–2 cloth, ISBN 0–87703–194–0 paper.
Review in Sky Telesc., Vol. 68, No. 6, p. 529 (1984).
The individual contributions within the subject scope of Astronomy and Astrophysics Abstracts are included in their corresponding categories – see abstracts 051.062 – 051.065.

012.074 **Kleine Teilchen und Lichtstreuung. Workshop im Rahmen der Frühjahrstagung 1984 der Arbeitsgemeinschaft Extraterrestrische Physik der Deutschen Physikalischen Gesellschaft, Kiel, 19. – 23. März 1984.**
K. O. Thielheim (Editor).
Verhandlungen der Deutschen Physikalischen Gesellschaft, Reihe VI, Band 19, p. 1475 – 1558 (1984).
The individual contributions within the subject scope of Astronomy and Astrophysics Abstracts are included in their corresponding categories – see abstracts 022.141, 022.142, 063.069, 106.037.

012.075 **Particle acceleration processes, shockwaves, nucleosynthesis and cosmic rays. Proceedings of Symposia 6 and 8 and the Joint Sessions 6/8 of the COSPAR Twenty-fifth Plenary Meeting held in Graz, Austria, 25th June – 7th July 1984.**
L. Koch-Miramond, M. A. Lee (Editors).
Adv. Space Res., Vol. 4, No. 2 – 3, 9+532 pp. (1984). ISBN 0–08–032711–7.
The individual contributions are included in their corresponding subject categories – see abstracts 022.144, 062.148, 062.149, 067.127 – 067.129, 073.081, 074.053, 078.008 – 078.015, 084.071 – 084.084, 093.064, 106.038 – 106.049, 116.030, 125.065, 125.066, 144.053 – 144.075, 158.203, 158.204.

012.076 **Reports of the meeting "Astrophotography-82", held at the Special Astrophysical Observatory of the USSR Academy of Sciences, 18 – 20 May 1982.**
K. V. Bychkov (Editor).
Soobshch. Spets. Astrofiz. Obs., Vyp. 40, 88 pp. (1983). In Russian.
The individual contributions are included in their corresponding subject categories – see abstracts 002.049, 034.169 – 034.174, 034.190, 036.128, 036.129, 036.131 – 036.133.

012.077 **The case for Mars. Proceedings of a conference held at Univerity of Colorado, Boulder, Colorado, USA, April 29 – May 2, 1981.**
P. J. Boston (Editor).
Science and technology series, Vol. 57. A Supplement to Advances in the Astronautical Sciences. Published for the American Astronautical Society by Univelt, Inc., PO Box 28130, San Diego, Calif. 92128, USA. 25+322 pp. Price $ 45.00 cloth, $ 25.00 paper (1984). ISBN 0–87703–197–5 cloth, ISBN 0–87703–198–3 paper.
Review in Sky Telesc., Vol. 68, No. 5, p. 426 (1984).
Contents: Introduction. Mission strategy. Spacecraft design. Life support. Surface activities and materials processing. Social and political aspects.

012.078 **Space research prospects in stellar activity and variability. Proceedings of a workshop, held at Observatoire de Paris, Meudon, France, February 29 – March 2, 1984.**
A. Mangeney, F. Praderie (Editors).
Observatoire de Paris – Meudon, France. 385 pp. (1984).
The individual contributions are included in their corresponding subject categories – see abstracts 013.057 – 013.060, 035.073, 036.139 – 036.148, 051.066 – 051.069, 064.052, 064.053, 065.129 – 065.132, 116.031 – 116.033, 121.038, 122.192, 122.193.

012.079 **Proceedings of the Fifth Finnish–Soviet Astronomical Symposium, held at Seili, Finland, 14 – 17 June 1983.**
P. Teerikorpi (Editor).
Rep. Ser., Dep. Phys. Sci., Univ. Turku, No. FTL–R68, 49 pp. (1984). = Turku Univ. Obs., Informo, No. 84. ISBN 951–642–574–7.
The individual contributions are included in their corresponding subject categories – see abstracts 013.074, 053.026, 071.038, 080.113, 080.114, 113.056, 114.119, 114.120, 116.036, 131.208, 158.226, 158.227, 161.320.

012.080 **Modern theoretical and experimental problems of the theory of relativity and gravitation. Proceedings of the 6th Soviet Gravitational Conference, held in Moscow, 3 – 5 July 1984.**
Univ. druzhby narodov, Moskva. 372 pp. (1984). In Russian.
Review in Ref. Zh., 51. Astron., 12.51.64 (1984).
The individual contributions are included in their corresponding subject categories – see abstracts 066.267 – 066.277, 067.134, 161.325.

012.081 **Space investigations made by the USSR in 1982, 1983. 25th COSPAR plenum, held in Graz, 1984.**
Institut kosmicheskikh issledovanij, Moskva. 117 pp. (1984). In Russian.
Review in Ref. Zh., 62. Issled. Kosm. Prostranstva, 12.62.105 (1984).

012.082 **Magnetism, planetary rotation, and convection in the solar system: retrospect and prospect. Conference held in the School of Physics at the University of Newcastle upon Tyne, 6 – 8 April 1983, in honour of Prof. S. K. Runcorn. Part I.**
W. O'Reilly (Editor), with a preface about Runcorn's scientific achievement.
Geophys. Surv., Vol. 7, No. 1, 14 + 123 pp. (1984).
The individual contributions are included in their corresponding subject categories – see abstracts 022.149, 044.067, 081.054, 084.087, 091.063, 094.066.

012.083 **Gravitation and relativistic astrophysics. Proceedings of a workshop held at Ahmedabad, India, 18 – 20 January 1982.**
A. R. Prasanna, J. V. Narlikar, C. V. Vishveshwara (Editors).
Published for the Indian Academy of Sciences, Bangalore, by World Scientific Publishing Co. Pte. Ltd., Singapore. 7 + 151 pp. Price US$ 18.40 (1984). ISBN 9971–966–67–0.
The individual constributions are icluded in their corresponding subject categories – see abstracts 061.127, 061.128, 065.152, 066.279, 067.136 – 067.140, 161.331 – 161.335.

012.084 **Relativistic astrophysics and cosmology. Proceedings of the 14th GIFT International Seminar on Theoretical Physics, held at Sant Feliu de Guixols, Spain, 27 June – 1 July 1983.**
X. Fustero, E. Verdaguer (Editors).
World Scientific Publishing Co. Pte. Ltd., Singapore. 9 + 301 pp. Price US$ 32.00 (1984). ISBN 9971–966–60–3.
The individual contributions are included in their corresponding subject categories – see abstracts 066.280, 066.281, 067.141, 161.336 – 161.338.

012.085 **Planetary rings (Anneaux des planètes). IAU Colloquium No. 75, held at Toulouse, France, 30 August – 2 September 1982.**
A. Brahic (Editor), with concluding remarks by E. C. Stone.
Cepadues–Editions, 111 rue Nicolas–Vauquelin, F–31100 Toulouse, France. 767 pp. Price FF 460.00 (1984). ISBN 2–85428–100–4.
The individual contributions are included in their corresponding subject categories – see abstract 013.077, 013.078, 042.088 – 042.091, 051.073 – 051.075, 091.064 – 091.081, 098.113 – 098.115, 099.139, 100.081 – 100.126, 101.076 – 101.083, 104.035, 107.050 – 107.053.

012.086 **Planetary rings.**
R. Greenberg, A. Brahic (Editors), with the assistance of M. S. Matthews.
The University of Arizona Press, Tucson, Arizona, USA. 11 + 784 pp. Price US$ 46.20, £ 34.50. ISBN 0–8165–0828–3.
The contributions of this volume are based on review papers presented at IAU Colloquium No. 75 ("Planetry rings"), held at Toulouse, France, 30 August – 2 September 1982. (See also Abstr. 012.085).
The individual contributions are included in their corresponding subject categories – see abstracts 004.187, 013.080, 051.078, 091.082 – 091.091, 099.140, 100.127, 100.128, 101.084, 107.055, 107.056.

012.087 **First achievements of MAP. Proceedings of Symposium 5 of the COSPAR Twenty–fifth Plenary Meeting held in Graz, Austria, June 25 – July 7, 1984.**
E. V. Thrane (Editor).
Adv. Space Res., Vol. 4, No. 4, 6 + 191 pp. (1984).
The individual contributions within the subject scope of Astronomy and Astrophysics Abstracts are included in their corresponding categories – see abstracts 082.094 – 082.102, 085.038, 085.039.

012.088 **Clusters and groups of galaxies. International Meeting held in Trieste, Italy, September 13 – 16, 1983.**
F. Mardirossian, G. Giuricin, M. Mezzetti (Editors), with a summary by P. J. E. Peebles.
Astrophysics and Space Science Library, Vol. 111. D. Reidel Publishing Company, Dordrecht – Boston – Lancaster. 22 + 659 pp. Price Dfl. 240.00, $ 89.50 (1984). ISBN 90–277–1772–9.
The individual contributions are included in their corresponding subject categories – see abstracts 051.076, 131.236, 131.237, 151.098 – 151.100, 157.203 – 157.219, 158.249 – 158.252, 159.126, 159.127, 160.078 – 160.137, 161.340 – 161.354.

012.089 **Local interstellar medium. International Astronomical Union Colloquium No. 81, held at the University of Wisconsin, Madison, Wisconsin, June 4 – 6, 1984.**
Y. Kondo, F. C. Bruhweiler, B. D. Savage (Editors).
NASA Conf. Publ. NASA CP–2345, 16 + 356 pp. (1984).
The individual contributions are included in their corresponding subject categories – see abstracts 035.079, 035.080, 051.077, 106.060, 112.091, 126.080, 126.081, 131.240 – 131.282, 142.084 – 142.089, 155.079, 155.080.

012.090 **Historia de la astronomia en Mexico.**
M. A. M. Corral (Editor).
Instituto de Astronomia, Universidad Nacional Autonoma de Mexico, Apartado Postal 70–264, Ciudad Universitaria, Mexico 20, D.F., Mexico. 336 pp. Price $ 12.00 (1983).
Review in Sky Telesc., Vol. 68, No. 5, p. 426 (1984).

012.091 **Cosmology of the early universe.**
L. Z. Fang, R. Ruffini (Editors).
World Scientific Publishing Co. Pte. Ltd., Singapore/John Wiley & Sons, Chichester – New York – Brisbane – Toronto – Singapore. 8 + 305 pp. Price £ 31.35 (1984).

Reviews in Astron. Express, Vol. 1, Issue 2, p. 83 (1984); Astron. Tidsskr., Årg. 17, Nr. 3, p. 141 (1984).

012.092 **Albert Einstein: historical and cultural perspectives. The centennial symposium in Jerusalem.**
G. Holton, Y. Elkana (Editors).
Princeton University Press, Princeton, N.J., USA. 32+439 pp. Price $ 35.00 (1982).
Review in J. Hist. Astron., Vol. 15, Part 3, No. 44, p. 222; 1984 (*O. Gingerich*).

012.093 **Chondrules and their origins. Proceedings of a workshop held at the Lunar and Planetary Institute, Houston, November 1982.**
E. A. King (Editor).
Lunar and Planetary Institute, Houston, Tex., USA. 377 pp. Price $ 33.00 (1984).
Review in Nature, Vol. 311, No. 5981, p. 88; 1984 (*D. W. G. Sears*).

012.094 **Proceedings of the Eighth Symposium on Antarctic Meteorites.**
T. Nagata (Editor).
Memoirs of National Institute of Polar Research, Special Issue No. 30. National Institute of Polar Research, Tokyo, Japan. 457 pp. (1983).
Review in Geochim. Cosmochim. Acta, Vol. 48, No. 12, p. 2773; 1984 (*E. Olsen*).

012.095 **The first 25 years in space: a symposium.**
A. Needell (Editor).
Random House (Smithsonian Inst. Press), New York. 152 pp. Price $ 12.50.
Review in Icarus, Vol. 60, No. 3, p. 718 – 719; 1984 (*W. H. Pickering*).

012.096 **Application de la photométrie bidimensionnelle à l'astronomie.**
Astropresse, Obs. du Pic du Midi et de Toulouse. 272 pp. (1980).
Review in Astrophys. Lett., Vol. 24, No. 2, p. 108 – 109; 1984 (*R. Racine*).

012.097 **Advanced technology optical telescopes II. Conference held at London, England, 5 – 6 September 1983.**
Proc. SPIE Int. Soc. Opt. Eng., Vol. 444 (1984).
Review in Phys. Abstr., Vol. 87, No. 1245, Entry 106098 (1984).

012.098 **Instrumentation in astronomy V. Conference held at London, England, 7 – 9 September 1983.**
Proc. SPIE Int. Soc. Opt. Eng., Vol. 445 (1984).
Review in Phys. Abstr., Vol. 87, No. 1245, Entry 106099 (1984).

012.099 **Joint Meeting of Mineralogical Society and Meteoritical Society on meteoritic research in progress (papers in summary form only received). Conference held at London, England, 24 May 1984.**
Mineral. Soc. Bull., No. 62 (1984).
Review in Phys. Abstr., Vol. 87, No. 1243, Entry 95036 (1984).

# 013 Reports on Astronomy in Various Countries and Particular Fields

013.001 **Nederland in de ruimte.**
H. C. van de Hulst.
Zenit, 11. Jaarg., No. 7/8, p. 292 – 294, 296 – 304 (1984).

013.002 **Report on the research in geomagnetism and aeronomy in Poland for the period 1978 – 1982.**
S. Grzedzielski, J. Jankowski.
Acta Geophys. Pol., Vol. 31, No. 3, p. 317 – 324 (1983). Abstr. in Phys. Abstr., Vol. 87, No. 1237, Entry 72486 (1984).

013.003 **Astronomi come amatori.**
T. R. Williams.
G. A.A.B., N. 75, p. 5 – 7 (1984).

013.004 **Why do we prefer photoelectric photometry?**
B. Szeidl, L. Szabados.
Advances in photoelectric photometry, Vol. 2, p. 1 – 7 (1984). – See Abstr. 003.002.

013.005 **Designing and implementing a photoelectric photometry program: the AAVSO experience.**
J. R. Percy.
Advances in photoelectric photometry, Vol. 2, p. 8 – 14 (1984). – See Abstr. 003.002.

013.006 **Photoelectric photometry in Estonia.**
P. Kalv, U. Veismann.
Advances in photoelectric photometry, Vol. 2, p. 119 – 133 (1984). – See Abstr. 003.002.

013.007 **The Geneva photometry.**
F. Rufener.
Advances in photoelectric photometry, Vol. 2, p. 156 – 185 (1984). – See Abstr. 003.002.

013.008 **La Fondation LEST: vers un grand télescope solaire européen.**
G. Mathys.
Ciel Terre, Vol. 100, No. 4, p. 118 – 120 (1984).
The author presents the LEST Foundation, an organization aiming at the realization of a Large European Solar Telescope. The need for such an instrument is briefly outlined; its requested characteristics and the solutions that have been planned to achieve them are described.

013.009 **Meteorbeobachtung in der Tschechoslowakei.**
H. Nováková.
Sterne Weltraum, 23. Jahrg., Nr. 8–9, p. 418 – 419 (1984).

013.010 **The Southampton Astronomical Society, 1924 – 1984.**
M. R. Hobbs.
J. Br. Astron. Assoc., Vol. 94, No. 5, p. 238 – 239 (1984).

013.011 **Astronomy from Spacelab 1.**
D. Dooling.
Sky Telesc., Vol. 68, No. 1, p. 18 – 19 (1984).

013.012 **The Planetary Observer Program.**
W. H. Blume.
J. Br. Interplanet. Soc., Vol. 37, No. 8, p. 355 – 360 (1984).
This paper presents an overview of NASA plans for a Planetary Observer Program, including the key assumptions intended to constrain the cost of each mission in the programme. Descriptions are given of several candidate missions, their scientific objectives and launch opportunities.

013.013 **TARO Cosmic–ray Observatory Report (I).**
H. Sakuyama, N. Suzuki.
Res. Bull. Meisei Univ., Phys. Sci. Eng., No. 20, p. 23 – 34 (1984). In Japanese.

013.014 **Illuminating incidents in antipodean astronomy: attempted formation of the Australian Astronomical Society in 1892.**
W. Orchiston, R. Bhathal.
Proc. Astron. Soc. Aust., Vol. 5, No. 3, p. 422 – 425 (1984).

013.015 **Observations d'objets du système solaire à l'instrument méridien de Bordeaux.**
M. Rapaport, J. M. Mazurier.
Astrométrie, géodésie, mécanique céleste, p. 69 – 71 (1984). – See Abstr. 012.021.

013.016 **Géodésie, géodynamique et sciences de la Terre.**
F. Barlier.
Astrométrie, géodésie, mécanique céleste, p. 117 – 124 (1984). – See Abstr. 012.021.

013.017 **Trois ans de variation des coordonnées de Paris et de Green Bank.**
S. Débarbat, J.–C. Hascoët.
Astrométrie, géodésie, mécanique céleste, p. 139 – 143 (1984). – See Abstr. 012.021.

013.018 **MEDOC. Mesure Expérimentale par DOppler de la Cinématique terrestre. Motion of the Earth by Doppler Observing Campaign.**
A. Piuzzi.
Astrométrie, géodésie, mécanique céleste, p. 145 – 148 (1984). – See Abstr. 012.021.

013.019 **L'Europe et la France à l'heure du télescope spatial.**
A. Vidal–Madjar.
Astronomie, Vol. 98, p. 423 – 430 (1984).

013.020 **The future of space astronomy.**
G. B. Field.
Mercury, Vol. 13, No. 4, p. 98 – 105 (1984).

013.021 **IRAS ground–based follow–up at ESO.**
T. de Jong.
Messenger, No. 37, p. 26 – 28 (1984).

013.022 **Extraterrestrische Astronomie.**
R. Lüst.
Sitzungsber. Heidelb. Akad. Wiss., Math. Naturwiss. Kl., Jahrg. 1984, 1. Abh., 24 pp. (1984).

013.023 **Development of planetology in the USSR.**
K. P. Florenskij.
O. Yu. Shmidt i sov. geofiz. 80–kh godov. Moskva, p. 141 – 148 (1983). In Russian. From Ref. Zh., 51. Astron., 8.51.137 (1984).

013.024 **Development of meteor investigations at the Odessa Astronomical Observatory.**
V. P. Tsesevich.
Meteor. issled., Moskva, No. 9, p. 14 – 18 (1984). In Russian. Abstr. in Ref. Zh., 51. Astron., 8.51.180 (1984).

013.025 **Meteor investigations in Dushanbe.**
P. B. Babadzhanov, L. N. Rubtsov, V. S. Getman, R. P. Chebotarev, V. M. Kolmakov, R. Sh. Bibarsov.
Meteor. issled., Moskva, No. 9, p. 19 – 42 (1984). In Russian. Abstr. in Ref. Zh., 51. Astron., 8.51.181 (1984).

013.026 **Probleme der extragalaktischen Astrophysik.**
K.–H. Schmidt.
Jenaer Rundsch., 29. Jahrg., Heft 3, p. 108 – 112 (1984).

013.027 **Die indischen Zwillinge – 1–m–Teleskope aus Jena in Indien.**
H. G. Beck.
Jenaer Rundsch., 29. Jahrg., Heft 3, p. 112 – 113, 121 (1984).

013.028 **The optical behavior of SS Cygni for the international campaign.**
J. A. Mattei.
J. Am. Assoc. Variable Star Obs., Vol. 13, No. 1, p. 43 (1984). Abstract. – See Abstr. 010.025.

013.029 **Die "American Association of Variable Star Observers – Solar Division".**
P. O. Taylor.
Sonne, Jahrg. 8, Nr. 31, p. 124 – 125 (1984).

013.030 **International cooperation in studying the middle atmosphere.**
A. D. Danilov.
Zemlya Vselennaya, No. 4, p. 72 – 77 (1984). In Russian.

013.031 **NASA's first 25 years.**
C. Simpson.
Space Educ., Vol. 1, No. 8, p. 346 – 349 (1984).

013.032 **On astronomical telegrams.**
H. Kosai.
Astron. Her., Vol. 77, No. 10, p. 261 – 263 (1984). In Japanese.

013.033 **Nyt nordisk astronomisk samarbejdsprojekt.**
S. Laustsen.
Astron. Tidsskr., Årg. 17, Nr. 3, p. 93 – 100 (1984).

013.034 **Astronomy in Egypt.**
N. S. Awadalla.
Astron. Her., Vol. 77, No. 11, p. 291 – 293 (1984). In Japanese.

013.035 **Work of the Department of General Physics and Astronomy of the USSR Academy of Sciences solving the problems of scientifically–technical progress.**
A. M. Prokhorov.
Vestn. Akad. Nauk SSSR, No. 5, p. 14 – 21 (1984). In Russian. Abstr. in Ref. Zh., 51. Astron., 10.51.38 (1984).

013.036 **Present problems of fundamental investigations in astrophysics.**
L. Filipov.
Spisanie Blg. Akad. Nauk, Vol. 30, No. 2, p. 52 – 57 (1984). In Bulgarian. Abstr. in Ref. Zh., 51. Astron., 10.51.58 (1984).

013.037 **USSR – India: in cosmic orbits.**
S. A. Nikitin.
Zemlya Vselennaya, No. 5, p. 44 – 50 (1984). In Russian.
Concerning the cooperation in exploring the cosmos.

013.038 **L'observation des étoiles variables par les amateurs.**
E. Schweitzer.
Astronomie, Vol. 98, p. 559 – 560 (1984).

013.039 **Die Geminga–Story.**
A. Krabbe.
Sterne Weltraum, 23. Jahrg., Nr. 11, p. 566 – 567 (1984).

013.040 **A celebration of Finnish astronomy.**
H. Oja, T. Markkanen.
Sky Telesc., Vol. 68, No. 6, p. 503 – 505 (1984).

013.041 **Infrared astronomy.**
R. Papoular.
Ciel Espace, No. 200, p. 37 – 41 (1984). In French. Abstr. in Phys. Abstr., Vol. 88, No. 1247, Entry 4629 (1985).

013.042 **Lowell Observatory's International Planetary Patrol: current status.**
L. J. Martin, P. B. James.
Bull. Am. Astron. Soc., Vol. 16, No. 3, Part I, p. 683 (1984). Abstract. – See Abstr. 010.061.

013.043 **The Pilot Planetary Data System: the 3rd annual report: transition phase from concepts to experimentation.**
A. L. Lane.
Bull. Am. Astron. Soc., Vol. 16, No. 3, Part I, p. 687 (1984). Abstract. – See Abstr. 010.061.

013.044 **Planetary astronomy in support of exploration – the 1959 proposals.**
J. N. Tatarewicz.
Bull. Am. Astron. Soc., Vol. 16, No. 3, Part I, p. 687 – 688 (1984). Abstract. – See Abstr. 010.061.

013.045 **The Infrared Astronomical Satellite (IRAS) and the asteroids: a progress report.**
E. F. Tedesco, D. L. Matson, G. J. Veeder, J. W. Fowler, T. Kia.
Bull. Am. Astron. Soc., Vol. 16, No. 3, Part I, p. 689 (1984). Abstract. – See Abstr. 010.061.

013.046 **The Multimission Image Processing Laboratory and the planetary scientist.**
R. J. Bambery.
Bull. Am. Astron. Soc., Vol. 16, No. 3, Part I, p. 703 (1984). Abstract. – See Abstr. 010.061.

013.047 **Uplink radio science.**
R. A. Simpson, G. L. Tyler, V. R. Eshleman.
Bull. Am. Astron. Soc., Vol. 16, No. 3, Part I, p. 703 (1984). Abstract. – See Abstr. 010.061.

013.048 **Dynamical astronomy via optical astrometric interferometry in space.**
R. D. Reasenberg, J. F. Chandler.
Bull. Am. Astron. Soc., Vol. 16, No. 3, Part I, p. 723 (1984). Abstract. – See Abstr. 010.064.

013.049 **The USNO pole–to–pole astrometric program.**
T. J. Rafferty, J. A. Hughes.
Bull. Am. Astron. Soc., Vol. 16, No. 3, Part I, p. 725 (1984). Abstract. – See Abstr. 010.065.

013.050 **Infrared Astronomical Satellite (IRAS) data products.**
N. W. Boggess.
Bull. Am. Astron. Soc., Vol. 16, No. 3, Part I, p. 727 (1984). Abstract. – See Abstr. 010.065.

013.051 **The case for a national facility for astronomical image processing.**
C. T. Bolton.
J. R. Astron. Soc. Can., Vol. 78, No. 5, p. 202 (1984). Abstract. – See Abstr. 010.401.

013.052 **Astronomy as a leisure activity of the future.**
P. A. Wayman.
Ir. Astron. J., Vol. 16, No. 4, p. 251 – 255 (1984).

013.053 **CompuServe for IAPPP announcements.**
J. L. Hopkins.
I.A.P.P.P. Commun., No. 17, p. 11 – 12 (1984).

013.054 **Getting started in photometry.**
F. H. Spickler.
I.A.P.P.P. Commun., No. 17, p. 39 – 43 (1984).

013.055 **The study of variable stars using small telescopes.**
J. R. Percy.
I.A.P.P.P. Commun., No. 17, p. 66 – 67 (1984).

013.056 **Megalithic astronomy: the last five years.**
C. L. N. Ruggles.
Vistas Astron., Vol. 27, Part 3, p. 231 – 289 (1984).
This paper attempts to review developments in megalithic astronomy since 1979 and to examine the topic in the wider contexts of world archaeoastronomy and of archaeology and anthropology in general.

013.057 **Ground–based long–term photoelectric photometry as support for space observations of variables.**
C. Sterken.
Space research prospects in stellar activity and variability, p. 107 – 111 (1984). – See Abstr. 012.078.

013.058 **E.V.E. project (Etude de la Variabilité des Etoiles au Pôle Sud).**
J. M. Le Contel, J. C. Valtier.
Space research prospects in stellar activity and variability, p. 213 – 219 (1984). – See Abstr. 012.078.
The basic idea of this experiment can be resumed as follows: a continuous photometric observation of a variable star with a precision of $10^{-3}$ magnitude, covering a period of at least several times 24 hours without interruption and extending over a frequency spectrum from 200 to 0.1 c/j.

013.059 **Observations of stellar activity: advantages and disadvantages of ground based multisites, polar region observatories, and space observations.**
B. R. Pettersen.
Space research prospects in stellar activity and variability, p. 343 – 358 (1984). – See Abstr. 012.078.
Future problems in stellar activity will require simultaneous observational efforts over a long wavelength baseline. Parallel use of groundbased and space instrumentation will be essential. Some programs are clearly of a long term nature and will require considerable dedication of telescopes and manpower. The author suggests some research areas that are ripe for investigation, and discusses the use of multisite observations, polar region observations, and space platforms to obtain relevant data.

013.060 **On the need to go to space to study stellar variability and activity.**
A. Mangeney, F. Praderie.
Space research prospects in stellar activity and variability, p. 379 – 385 (1984). – See Abstr. 012.078.
The authors consider stellar seismology and stellar activity phenomena, examining what can be reached from ground–based observations and what imperatively requests space–borne observations.

013.061 **A photometric geodesy program for main belt asteroids.**
D. H. Levy, D. R. Davis, S. J. Weidenschilling, C. R. Chapman, R. Greenberg.
Minor Planet Bull., Vol. 11, No. 4, p. 31 – 33 (1984).
This article describes a long–term program to measure the lightcurves of several main belt asteroids, to test whether they have shapes similar to fluid bodies in equilibrium.

013.062 **Relazione sulla attività svolta dal Gruppo di Lavoro per il Telescopio Nazionale Italiano.**
C. Barbieri, A. Braccesi, O. Citterio, P. Salinari, M. Tarenghi.
G. Astron., Vol. 10, N. 2, p. 101 – 122 (1984).

013.063 **The C.D.S. and astrometry.**
G. Westerhout.
Bull. Inf. Cent. Données Stellaires, No. 27, p. 3 – 5 (1984).

013.064 **L'astrométrie au télescope de Schmidt de Calern.**
J. L. Heudier.
Bull. Inf. Cent. Données Stellaires, No. 27, p. 9 – 10 (1984).

013.065 **Catalogues astrométriques et programmes spatiaux.**
J. Kovalevsky.
Bull. Inf. Cent. Données Stellaires, No. 27, p. 11 – 13 (1984).
Three astrometric catalogues will be constructed in the ten years to come, in connection with space missions. A summary is given of the contents of these three catalogues: HIPPARCOS main mission catalogue, TYCHO catalogue and the Space Telescope Guide Star Selection System. The accuracy expected and the number of stars included in the results represent a major advance in the field.

013.066 **Les données astrométriques accessibles au CDS. État actuel et développements prochains.**
D. Egret.
Bull. Inf. Cent. Données Stellaires, No. 27, p. 15 – 16 (1984).
The astrometric data available in the SIMBAD Data Base of the Stellar Data Center (CDS) are presented. The impact of the HIPPARCOS, TYCHO and HST missions is briefly discussed.

013.067 **Le C.D.S. et les programmes d'astrométrie méridienne.**
Y. Requième.
Bull. Inf. Cent. Données Stellaires, No. 27, p. 17 – 19 (1984).

013.068 **Le CDS, pourquoi pas centre de logiciels astrométriques?**
F. Chollet.
Bull. Inf. Cent. Données Stellaires, No. 27, p. 51 – 54 (1984).
Astrometric observations obtained by various techniques in different places may be easily compared if they are homogeneous. Regarding catalogs, theory of Earth motions and constants, the preceding condition is verified. It is however not always true for the mathematical methods used to compute apparent places. To remove this inhomogeneity the author proposes to create a bank of programs. This bank should be used as a reference, and should be included, for example, in the CDS data bank which is very easily accessible from observatories.

013.069 **Conclusions de la Réunion Scientifique du CDS: sur le rôle du CDS en astrométrie.**
J. Kovalevsky.
Bull. Inf. Cent. Données Stellaires, No. 27, p. 55 – 56 (1984).

013.070 **A progress report on a catalogue and atlas of galactic novae.**
H. W. Duerbeck.
Bull. Inf. Cent. Données Stellaires, No. 27, p. 131 – 132 (1984).

013.071 **Catalogue des étoiles cibles utilisées pour l'étude spectroscopique du milieu interstellaire diffus.**
R. Freire Ferrero, R. Ferlet.
Bull. Inf. Cent. Données Stellaires, No. 27, p. 135 – 136 (1984).

013.072 **The Astrographic Catalogue and the Carte du Ciel.**
C. Jaschek.
Bull. Inf. Cent. Données Stellaires, No. 27, p. 197 – 201 (1984).
Circular letter concerning the Astrographic Catalog (*H. Eichhorn, C. Jaschek*) with answers from Oxford (*D. E. Blackwell*), Cape (*J. Churms*), Perth (*I. Nikoloff*), Hamburg (*C. de Vegt*), Uccle (*P. Melchior*), Bordeaux (*M. Rousseau*).

013.073 **Photometric observing campaign. Progress Report No. 7.**
P. Harmanec, J. Horn, P. Koubsky.
Bull. Inf. Etoiles Be, No. 10, p. 64 – 67 (1984).

013.074 **On the investigation of close binary stars in Tallinn.**
P. Kalv.
Rep. Ser., Dep. Phys. Sci., Univ. Turku, No. FTL–R68, p. 15 (1984). – See Abstr. 012.079.

013.075 **Ultimos logros de la astronomía contemporánea.**
H. Dávila, V. Yurevich.
Publ. Espec. Obs. Astron. Quito, No. 2, p. 1 – 17 (1981).

013.076 **Conjunto de investigaciones astronómicas y geofisicas en el Ecuador.**
H. Davila, S. Tatevian (*S. Tatevyan*).
Publ. Espec. Obs. Astron. Quito, No. 2, p. 75 – 85 (1981).

013.077 **Future observations of planetary rings from ground–based observatories and earth–orbiting satellites.**
B. A. Smith.
Planetary rings, p. 725 – 729 (1984). – See Abstr. 012.085 (IAU Colloq. No. 75).

013.078 **Objectives of U.V. observations in the study of the Saturnian rings.**
P. Bruston.
Planetary rings, p. 737 – 738 (1984). – See Abstr. 012.085 (IAU Colloq. No. 75).

013.079 **Important center of educational and scientific work on astrophysics (50 years of the Leningrad University Astrophysics Chair).**
V. G. Nagnibeda.
Astrofizika, Tom 21, Vyp. 1, p. 189 – 205 (1984). In Russian. English translation in Astrophysics, Vol. 21, No. 1.
The educational and scientific work at the Leningrad University Chair of Astrophysics for 50 years since the year of the chair foundation is outlined.

013.080 **Future observations of planetary rings from groundbased observatories and earth–orbiting satellites.**
B. A. Smith.
Planetary rings, p. 704 – 712 (1984). – See Abstr. 012.086.

013.081 **Probing the depths of a star: the study of solar oscillations from space. Report of the NASA Science Working Group on the study of solar oscillations from space.**
R. W. Noyes, E. J. Rhodes Jr. (Editors).
National Aeronautics and Space Administration – Jet Propulsion Laboratory, California Institute of Technology, 4800 Oak Grove Drive, Pasadena, Calif. 91109, USA. 15+76 pp. (1984).
This is the result of an 18–month study of the scientific potential and feasibility of a space helioseismology investigation – that is, the measurement of the spectrum of solar oscillations and the use of that spectrum to probe the structure and dynamics of the solar interior. The conclusions are that, while much useful work can and should be done with ground–based instruments, this should be complemented by imaging Doppler observations from a spacecraft in full or nearly full sunlit orbit.

**Recomendaciones sobre el futuro de la investigacion en el IAC, 1983.**
See Abstr. 003.160.

**Photoelectric photometry at Van Vleck Observatory.**
See Abstr. 009.007.

**The photoelectric photometry of Yunnan Observatory of the Chinese Academy of Science.**
See Abstr. 009.008.

**Sterrenwacht van Peking in oude glorie hersteld.**
See Abstr. 009.029.

**Some non–atlas work at ESO Sky Atlas Laboratory.**
See Abstr. 009.035.

**The first US–Japan VLBI test observation by use of K–3 system at the Radio Research Laboratories.**
See Abstr. 033.004.

**ESA and microgravity research.**
See Abstr. 051.062.

**Die "International Halley Watch".**
See Abstr. 103.901.

# 014 Teaching in Astronomy

014.001 **Simulators for assisting the teaching of observational astronomy. II. Photometry of normal variable stars in total visible light.**
A. A. Mills.
Eur. J. Phys., Vol. 4, No. 4, p. 198 – 204 (1983). Abstr. in Phys. Abstr., Vol. 87, No. 1237, Entry 67940 (1984).

014.002 **Temperaturbestimmungen in der Astronomie. Teil 2.**
O. Zimmermann.
Sterne Weltraum, 23. Jahrg., Nr. 7, p. 373 – 377 (1984).

014.003 **Sonnenuhrenmodelle auf dem Riesenglobus.**
A. Riss.
Orion, 42. Jahrg., Nr. 203, p. 161 – 162 (1984).

014.004 **Laboratory exercises in astronomy – The Wilson–Bappu effect.**
D. J. Bord, V. M. Ogle.
Sky Telesc., Vol. 68, No. 1, p. 20 – 23 (1984).

014.005 **Backyard astronomy – 6. Close–up of an alien world.**
A. MacRobert.
Sky Telesc., Vol. 68, No. 1, p. 29 – 31 (1984).

014.006 **Astronomie von Anfang an. Teil 2.**
H. Baderschneider.
Sterne Weltraum, 23. Jahrg., Nr. 7, p. 390 – 392 (1984).

014.007 **Temperaturbestimmung in der Astronomie. Teil 3.**
O. Zimmermann.
Sterne Weltraum, 23. Jahrg., Nr. 8–9, p. 446 – 449 (1984).

014.008 **Backyard astronomy. 7 A guided tour of the Moon.**
A. Chaikin.
Sky Telesc., Vol. 68, No. 3, p. 211 – 213 (1984).

014.009 **Starmaps from a home computer.**
G. C. Carter.
Space Educ., Vol. 1, No. 8, p. 350 – 353 (1984).

014.010 **Comets: why they are important.**
M. J. Hendrie.
Space Educ., Vol. 1, No. 8, p. 355 – 358 (1984).

014.011 **Symmetry of the spinning Earth.**
J. Sivardiere.
Eur. J. Phys., Vol. 5, No. 1, p. 46 – 48 (1984). Abstr. in Phys. Abstr., Vol. 87, No. 1239, Entry 78084 (1984).

014.012 **Entropy.**
K. Andrew.
Am. J. Phys., Vol. 52, No. 6, p. 492 – 496 (1984). Abstr. in Phys. Abstr., Vol. 87, No. 1240, Entry 82622 (1984).

014.013 **Small black holes: ionization tracks and range.**
G. Greenstein, J. O. Burns.
Am. J. Phys., Vol. 52, No. 6, p. 531 – 534 (1984). Abstr. in Phys. Abstr., Vol. 87, No. 1240, Entry 82631 (1984).

014.014 **Astronomical computing: "Jumping off an asteroid" and other puzzles.**
R. W. Sinnott.
Sky Telesc., Vol. 68, No. 1, p. 62 – 63 (1984).

014.015 **Astronomical computing: 1. Computing under the open sky. 2. Virtues of the haversine.**
B. P. Shumaker, R. W. Sinnott.
Sky Telesc., Vol. 68, No. 2, p. 158 – 159 (1984).

014.016 **Astronomical computing: Counting the sunny hours.**
J. Meeus.
Sky Telesc., Vol. 68, No. 3, p. 254 (1984).

014.017 **Meridiandurchgang und Kulmination.**
G. Groschopf.
Sterne Weltraum, 23. Jahrg., Nr. 10, p. 512 – 513 (1984).

014.018 **Danziger Lehrplan Astronomie war vorbildlich.**
A. Kunert.
Sterne Weltraum, 23. Jahrg., Nr. 10, p. 513 (1984).

014.019 **Astronomie von Anfang an. Teil 3.**
H. Baderschneider.
Sterne Weltraum, 23. Jahrg., Nr. 10, p. 530 – 532 (1984).

014.020 **Astronomical computing: Telescope control comes of age.**
R. W. Sinnott.
Sky Telesc., Vol. 68, No. 4, p. 346 – 347 (1984).

014.021 **TELEMENTOR und TELEMATOR – empfehlenswerte Fernrohre für den Amateur.**
H. Mucke.
Jenaer Rundsch., 29. Jahrg., Heft 3, p. 123 – 126 (1984).

014.022 **Der Einsatz des Schulfernrohrs TELEMENTOR bei der Ausbildung von Physik/Astronomie–Lehrern.**
H.–G. Reimann.
Jenaer Rundsch., 29. Jahrg., Heft 3, p. 126 – 129 (1984).

014.023 **25 Jahre Astronomieunterricht in der DDR.**
V. Schorcht.
Jenaer Rundsch., 29. Jahrg., Heft 3, p. 129 – 130 (1984).

014.024 **Excellence in science: a challenge.**
R. C. Maag.
J. Am. Assoc. Variable Star Obs., Vol. 13, No. 1, p. 43 – 44 (1984). Abstract. – See Abstr. 010.025.

014.025 **Observations of Delta Cephei by 9th grade earth science students.**
J. C. Troeger.
J. Am. Assoc. Variable Star Obs., Vol. 13, No. 1, p. 44 (1984). Abstract. – See Abstr. 010.025.

014.026 **Backyard astronomy – 8. Observing double stars.**
A. MacRobert.
Sky Telesc., Vol. 68, No. 5, p. 417 – 419 (1984).

014.027 **Astronomical computing: Designing an aplanatic doublet.**
J. A. Church.
Sky Telesc., Vol. 68, No. 5, p. 450 – 451 (1984).

014.028 **Las microcomputadoras en la astronomía amateur.**
M. L. Alvarez.
Rev. Astron., Tomo 55, No. 227, p. 17 – 21 (1983).

014.029 **Die Entfernung der Fixsterne.**
T. Schmidt.
Sterne Weltraum, 23. Jahrg., Nr. 11, p. 575 – 577 (1984).

014.030 **Laboratory exercises in astronomy – how far is the galactic center?**
A. Hirshfeld.
Sky Telesc., Vol. 68, No. 6, p. 498 – 502 (1984).

014.031 **Astronomical computing: Viewing our Galaxy form afar.**
R. P. Waldron.
Sky Telesc., Vol. 68, No. 6, p. 555 – 556 (1984).

014.032 **Bestimmung und Konstruktion der Jupiterumlaufbahn anhand von Dias.**
P. Ehbets.
Orion, 42. Jahrg., Nr. 205, p. 232 – 234 (1984).

014.033 **Extragalaktische Entfernungen.**
T. Schmidt.
Sterne Weltraum, 23. Jahrg., Nr. 12, p. 633 – 634 (1984).

014.034 **How to estimate the distance between Earth and Vega.**
M. David, I. Beauvois.
Ciel Espace, No. 197, p. 36 – 40 (1984). In French. Abstr. in Phys. Abstr., Vol. 88, No. 1247, Entry 4722 (1985).

014.035 **Binocoli per l'astronomia.**
M. Boetto.
Orione, Vol. 4, N. 3, p. 109 – 119 (1984).

014.036 **Microcomputers in amateur astronomy.**
R. H. Warden.
J. Br. Astron. Assoc., Vol. 95, No. 1, p. 22 – 25 (1984).
This paper gives an introduction to a number of applications of a home micro in amateur astronomy.

014.037 **Bestimmung der differentiellen Rotation der Sonne.**
R. Zerm.
Astron. Raumfahrt, 22. Jahrg., Heft 6, p. 127 – 130 (1984).

**An introduction to experimental astronomy.**
See Abstr. 003.077.

**The light–hearted astronomer.**
See Abstr. 003.089.

**Astronomical experiments.**
See Abstr. 003.108.

**The Petersburg Astronomical School of the XVIIIth century.**
See Abstr. 003.128.

**Teacher's guide to astronomy: from the earth to the universe.**
See Abstr. 003.132.

**The telescope.**
See Abstr. 003.158.

**Important center of educational and scientific work on astrophysics (50 years of the Leningrad University Astrophysics Chair).**
See Abstr. 013.079.

# 015 Miscellanea (Philosophical Aspects, Extraterrestrial Life, etc.)

015.001 **The number of inhabited planets in the Galaxy.**
T. B. Tang.
J. Br. Interplanet. Soc., Vol. 37, No. 9, p. 410 – 413 (1984).

015.002 **Astrologie – Wissenschaft oder Aberglaube?**
J. Tobler–Mutti.
Orion, 42. Jahrg., Nr. 203, p. 136 – 139 (1984).

015.003 **On the visibility of Sirius in daylight.**
C. Henshaw.
J. Br. Astron. Assoc., Vol. 94, No. 5, p. 221 – 222 (1984).

015.004 **A universe in our own image.**
A. Finkbeiner.
Sky Telesc., Vol. 68, No. 2, p. 106 – 111 (1984).

015.005 **A survey of space biology and space medicine.**
H. Planel, H. Oser.
ESA Brochure, ESA BR–17, 3+25 pp. Price FF 60.00 (1984).

015.006 **Roundtrip interstellar travel using laser–pushed lightsails.**
R. L. Forward.
J. Spacecr. Rockets, Vol. 21, No. 2, p. 187 – 195 (1984). Abstr. in Phys. Abstr., Vol. 87, No. 1239, Entry 82364 (1984).

015.007 **On the astronomical orientation of St. Vitus' cathedral and St. George's church in the Castle of Prague.**
C. Köberl.
Bull. Astron. Inst. Czech., Vol. 35, No. 4, p. 216 – 219 (1984).
For both cases considered in this paper a connection between the alignment of the church's axis and the solar rising azimuth on the days of St. Vitus and St. George could be confirmed.

015.008 **Can mirage phenomena explain some of the strange stellar observations made in olden days up chimneys?**
Y. Öhman.
Q. J. R. Astron. Soc., Vol. 25, No. 3, p. 290 – 292 (1984).

015.009 **Natural selection of stellar civilizations by the limits of growth.**
M. D. Papagiannis.
Q. J. R. Astron. Soc., Vol. 25, No. 3, p. 309 – 318 (1984).

015.010 **Scientific responses to pseudoscience. An annotated bibliography.**
A. Fraknoi.
Mercury, Vol. 13, No. 4, p. 121 – 124 (1984).

015.011 **Der zerteilte Himmel – über die Geometrie der Sternbilder.**
M. Gossler.
Sternenbote, 27. Jahrg., Nr. 339, p. 190 – 194 (1984).

015.012 **La "Spacewatch Camera" ovvero possiamo evitare una fine come quella dei dinosauri?**
V. Zappalà.
Orione, Vol. 4, N. 2, p. 46 – 51 (1984).

015.013 **SETI auf neuen Wegen.**
J. Dorschner, J. Gürtler.
Sterne, 60. Band, Heft 4, p. 195 – 200 (1984).

015.014 **Bioastronomie – Herausforderungen und Gelegenheiten bei der astronomischen Suche nach außerirdischem Leben.**
M. D. Papagiannis.
Sterne, 60. Band, Heft 4, p. 201 – 211 (1984).

015.015 **Die Suche nach elektromagnetischen Signalen von außerirdischen Wesen.**
B. Zuckerman.
Sterne, 60. Band, Heft 4, p. 214 – 223 (1984).

015.016 **Zehn Jahre Suche nach außerirdischen intelligenten Wesen.**
R. S. Dixon.
Sterne, 60. Band, Heft 4, p. 224 – 227 (1984).

015.017 **Wird es eine kosmische Universalsprache geben? – Ein CETI–Modell.**
J. Lehmann.
Sterne, 60. Band, Heft 4, p. 228 – 236 (1984).

015.018 **Das Paradoxon des interstellaren Verkehrs.**
R. L. Forward.
Sterne, 60. Band, Heft 4, p. 237 – 245 (1984).

015.019 **Wir sind etwas Besonderes.**
F. J. Tipler.
Sterne, 60. Band, Heft 4, p. 246 – 252 (1984).

015.020 **Existence of extra–terrestrial life.**
A. J. Dunlop.
J. Br. Interplanet. Soc., Vol. 37, No. 11, p. 513 – 514 (1984).

015.021 **The astronomical visions of Lucien Rudaux.**
R. Miller.
Sky Telesc., Vol. 68, No. 4, p. 293 – 295 (1984).

015.022 **Astrologie – Wissenschaft oder Aberglaube?**
J. Tobler–Mutti.
Orion, 42. Jahrg., Nr. 204, p. 172 – 173 (1984).

015.023 **The Christmas Star, novae, and pulsars.**
D. Hoffleit.
J. Am. Assoc. Variable Star Obs., Vol. 13, No. 1, p. 15 – 20 (1984).
A ceiling painting in the Priscilla catacomb in Rome appears to represent a star map showing the Christmas Star. If so, this might be a portrait of the nova or comet of 5 B. C., or it could be Venus in 6 B. C. It could not represent the conjunction of Jupiter and Saturn in 7 B. C., which occurred in a different part of the sky.

015.024 **SETI: its reasonings and present status.**
H. Hirabayashi, J. Jugaku.
Kagaku, Vol. 54, No. 7, p. 437 – 444 (1984). In Japanese.
The present status of SETI programs is surveyed.

015.025 **Searching for extraterrestrial artifacts.**
R. A. Freitas.
Spaceflight, Vol. 26, No. 12, p. 438 – 441 (1984).
The Fermi Paradox, attributed to a famous question from physicist Enrico Fermi in 1943, asks: if there are intelligent beings elsewhere then, in time, they must achieve the technology of nuclear power and space flight and would explore and colonize the Galaxy. Thus, they should have been able to travel to Earth, but we see no evidence of such visitations. Ergo, they cannot exist. The author, of the Xenology Research Institute in California, discusses this viewpoint and suggests how and where we might be able to detect an alien presence in the Solar System.

015.026 **Komeet Halley was geen Ster van Bethlehem.**
G. Schilling.
Zenit, 11. Jaarg., No. 12, p. 500 – 511 (1984).

015.027 **L'influenza della retina sulle osservazioni di stelle variabili.**
A. Cora.
Orione, Vol. 4, N. 3, p. 103 – 108 (1984).

015.028 **A twinkle in the eye of the Universe.**
E. R. Harrison.
Q. J. R. Astron. Soc., Vol. 25, No. 4, p. 423 – 428 (1984).
In this paper some comments on the origin of astronomical structure in the Universe are mixed with thoughts of a general nature.

015.029 **Bayes' theorem and the real SETI equation.**
T. L. Wilson.
Q. J. R. Astron. Soc., Vol. 25, No. 4, p. 435 – 448 (1984).
The anthropomorphic assumption that because our own society uses electromagnetic radiation (based upon Maxwell's equations) then so do others as advanced as ourselves, is averted. The assumption that extraterrestrial intelligence exists a priori is likewise averted by use of the conditional probability.

015.030 **Recherche observationnelle d'une émission radio fortement polarisée d'une intelligence extraterrestre.**
J. P. Vallée, M. Simard–Normandin.
J. R. Astron. Soc. Can., Vol. 78, No. 5, p. 210 (1984). Abstract. – See Abstr. 010.401.

015.031 **One approach to archaeoastronomy: an astronomer's view.**
M. Zeilik.
Archaeoastronomy (U.S.A.), Vol. 6, Nos. 1 – 4, p. 4 – 7 (1983).

015.032 **A Milky Way search strategy for extraterrestrial intelligence.**
W. T. Sullivan III, K. J. Mighell.
Icarus, Vol. 60, No. 3, p. 675 – 684 (1984).
The authors assume that detectable extraterrestrial civilizations do indeed exist and that microwave frequencies are a good choice for any first large–scale effort. They then focus on the question of optimizing the choice of directions in the sky toward which one should seek signals. They examine whether or not a different mix of search directions might be more profitable. Finally they develop a search strategy.

015.033 **Una stella sulla Murgia.**
A. Tavolaro.
Coelum, Vol. 52, N. 6, p. 327 – 354 (1984).

015.034 **An object within a particle of extraterrestrial origin compared with an object of presumed terrestrial origin.**
F. Hoyle, N. C. Wickramasinghe, H. D. Pflug.
Astrophys. Relativ., Prepr. Ser., No. 107, 3 pp. (1984).

015.035 **The relation of bacterial absorptivity in the 3 to 4 μm waveband to observations of GC–IRS 7 and sundry comments.**
F. Hoyle, N. C. Wickramasinghe.
Astrophys. Relativ., Prepr. Ser., No. 108, 26 pp. (1984).
It is shown that bacterial absorptivity fits very well to observations of GC–IRS 7 over the whole waveband from 2.9 μm to 3.9 μm. Claims which have recently been made to the contrary are without merit.

015.036 **The ultraviolet absorbance of interstellar bacteria and related matters.**
F. Hoyle, N. C. Wickramasinghe, S. Al–Mufti.
Astrophys. Relativ., Prepr. Ser., No. 110, 16 pp. (1984).
It is shown that the well–known 2200 Å peak in the extinction of starlight is explained by microorganisms. A mixed culture of diatoms and bacteria, which previously is found to give excellent fits to astronomical data in the infrared, has a peak absorption slightly shortward of 2200 Å, in extremely close agreement with the absorptions found in directions towards most early–type stars.

015.037 **On a possibly fundamental principle in chemistry as viewed in a cosmogonic context.**
F. Hoyle, N. C. Wickramasinghe.
Astrophys. Relativ., Prepr. Ser., No. 111, 25 pp. (1984).

Subject to the condition that atoms be conserved – i.e. without nuclear transmutations – the authors conjecture that it is impossible to synthesise organic materials in appreciable quantity from inorganic materials without the intervention of biological systems. The restriction is not a consequence of a mystic quality in the laws of physics and chemistry but of the practical disposition of materials on the Earth and in the cosmos generally. It is a corollary that biology is the means by which an approximation to thermodynamic equilibrium is maintained in materials at temperatures $\sim$300K.

015.038 **Interstellar microorganisms.**
F. Hoyle, N. C. Wickramasinghe, S. Al-Mufti.
Astrophys. Relativ., Prepr. Ser., No. 112, 2 pp. (1984).

015.039 **No relations between the four symbolic animals of the lunar mansions and the celestial phenomena at the four seasons.**
S.-l. Wang.
Acta Astron. Sin., Vol. 25, No. 3, p. 304 – 308 (1984). In Chinese.

**Chemical evolution and the origin of life. Bibliography Supplement 1982.**
See Abstr. 002.004.

**Das intelligente Universum. Eine neue Sicht von Entstehung und Evolution.**
See Abstr. 003.041.

**Plurality of worlds: the origins of the extraterrestrial life debate from Democritus to Kant.**
See Abstr. 003.081.

**The origin of the constellations.**
See Abstr. 004.055.

**Search for extraterrestrial civilizations is continued. Symposium No. 12 of the International Astronautical Federation, held in Budapest, 1983, October 10 – 15.**
See Abstr. 011.010.

**Los cometas y el origen de la vida. 2ª Parte.**
See Abstr. 102.052.

# Applied Mathematics, Physics

## 021 Mathematical Papers Related to Astronomy and Astrophysics, Computing

021.001 **A computer program for light curve synthesis of X–ray binary systems.**
N. I. Balog, A. V. Goncharskij, Z. Yu. Metlitskaya, A. M. Cherepashchuk.
Perem. Zvezdy, Tom 21, No. 5, p. 695 – 708 (1982). In Russian.
A computer program in Fortran for light curve synthesis of X–ray binary systems is presented. The ellipticity of the optical star, X–ray heating and eclipses connected with an optically thick accretion disk are considered.

021.002 **Calculs astronomiques pour amateurs.**
J. Meeus.
Astronomie, Vol. 98, p. 348 – 354, 410 – 412, 577 – 582 (1984).

021.003 **Computerized photometry for intelligent, efficient, and flexible observing.**
M. Zeilik, R. Elston.
Advances in photoelectric photometry, Vol. 2, p. 54 – 63 (1984). – See Abstr. 003.002.

021.004 **Formulae for construction of digitized contour maps on IBM computers.**
S. G. Pugacheva.
Tr. Gos. Astron. Inst. Shternberg, Tom 55, p. 149 – 158 (1983). In Russian.
A method of digitized contour maps plotting is discussed, and a program of construction of such maps is given for the "Benson" computer. Maps plotted by BESM–4 and –6 computers are presented.

021.005 **La gestion du variabiliste sur micro–ordinateur.**
J. Gunther.
Bull. Assoc. Fr. Obs. Etoiles Variables, No. 29, p. 99 – 101 (1984).

021.006 **Presentazione di un programma astronomico.**
P. Pognant.
Orione, Vol. 4, N. 2, p. 53 – 61 (1984).

021.007 **A computer program to calculate first order extinction coefficients.**
M. K. Kaitting.
J. Am. Assoc. Variable Star Obs., Vol. 13, No. 1, p. 3 – 10 (1984).
A description is given of a computer program written in Applesoft BASIC on an Apple II Plus. It is designed to calculate the first order extinction K' by least–squares and to display the observations as function of air mass.

021.008 **Variable star/asteroid appulse predicting.**
P. L. Sventek.
J. Am. Assoc. Variable Star Obs., Vol. 13, No. 1, p. 45 (1984). Abstract. – See Abstr. 010.025.

021.009 **The computer as a tool in photoelectric photometry.**
D. M. Slauson, J. C. Troeger.
J. Am. Assoc. Variable Star Obs., Vol. 13, No. 1, p. 45 (1984). Abstract. – See Abstr. 010.025.

021.010 **Algorithms for automated photoelectric photometry.**
L. J. Boyd.
I.A.P.P.P. Commun., No. 17, p. 1 – 10 (1984).

021.011 **A better heliocentric correction algorithm.**
H. J. Landis.
I.A.P.P.P. Commun., No. 17, p. 62 – 63 (1984).

021.012 **Flow chart on the numerical solution to Poisson equation.**
G. Song.
Ann. Shanghai Obs., Acad. Sin., No. 5, p. 77 – 88 (1983). In Chinese.
A flow chart on the numerical solution to Poisson equation is given according to the method suggested by James, who improves the MFT convolution model.

021.013 **The linear multistep methods with back–corrections.**
W. Zhu, Z. Cheng, Z. Teng.
Ann. Shanghai Obs., Acad. Sin., No. 5, p. 89 – 95 (1983). In Chinese.
The linear multistep methods with back–corrections have been proposed for solving the equations of motion of artificial satellites. The authors derived a complete set of formulas of linear multistep methods with back–corrections for prediction and correction, which is available for programming.

021.014 **Computer tracking charts.**
J. Jahn.
Minor Planet Bull., Vol. 11, No. 4, p. 39 – 41 (1984).
To aid in photographing asteroids, a computer program is used to generate finding charts for the selection of guide stars. The data base contains all asteroids with a given minimum magnitude and the stars from the Yale catalogue down to 7th magnitude.

021.015 **Corrections for biases in slope estimation.**
H. L. Marshall.
Space Telesc. Sci. Inst., Prepr. Ser., No. 30, 33 pp. (1984). To appear in Astrophys. J.

021.016 **Programas de cálculo para la determinación de la dirección de la vertical por fotografías cenitales de estrellas.**
M. J. Sevilla.
Semin. Astron. Geod., Univ. Madr., Publ., No. 111, p. 10 – 32 (1980).

021.017 **Programación del método de la cuerda.**
M. J. Sevilla.
Bol. Inf., Serv. Geogr. Ejército, núm. 51, p. 1 – 5 (1981). = Semin. Astron. Geod., Univ. Madr., Publ. No. 121.

021.018 **Programmsystem und Datenorganisation zur Berechnung von Kleinplaneten– und Kometenbahnen.**
M. Rothacher.
Mitt. Satell.–Beobachtungsstn. Zimmerwald, Nr. 9, 96 pp. (1984).

021.019 **A method of computing the complex probability function and other related functions over the whole complex plane.**
S. J. McKenna.
Astrophys. Space Sci., Vol. 107, No. 1, p. 71 – 83 (1984).

Using the symmetry relations of the complex probability function (CPF), the algorithm developed by Humlíček (1979) to compute this function over the upper half plane can be extended to cover the entire complex plane. Using the Humlíček algorithm, the real and imaginary components of the CPF can be computed over the whole complex plane. Because of the relation between the CPF and other interesting mathematical functions, fast and accurate computer programs can be written to compute them. Such functions include the derivatives of the CPF, the complex error function, the complex Fresnel integrals, and the complex Dawson's functions. FORTRAN implementations of these functions are included in an appendix.

021.020 **Laguerre polynomial series for Sobolev functions.**
K. I. Selyakov.
Astrofizika, Tom 21, Vyp. 2, p. 295 – 308 (1984). In Russian. English translation in Astrophysics, Vol. 21, No. 2.

The solution of the integral equation for Sobolev functions is represented by Laguerre polynomial series. The coefficients of these series are at the same time the coefficients of the power series for the Ambartsumyan–Chandrasekhar H–functions. Infinite systems of linear algebraic equations with Toeplitz matrices are derived for the coefficients of these series. Numerical results and approximate formulae are presented for isotropic scattering.

021.021 **Remarks on regression analysis and Malmquist bias.**
P. Notni.
Astrophys. Lett., Vol. 24, No. 3, p. 133 – 138 (1984).

Arguments are given claiming that in regression analysis of two variables which *both* are affected by errors or variances, the mostly forgotten method of orthogonal regression should be more widely used. Further, precepts are given how samples limited by observational constraints or selection must be cut to give an unbiased orthogonal regression line or an unbiased mean.

**Orbits for amateurs with a microcomputer.**
See Abstr. 003.019.

**Pocket computer programs for astronomy.**
See Abstr. 003.113.

**Starmaps from a home computer.**
See Abstr. 014.009.

**Las microcomputadoras en la astronomía amateur.**
See Abstr. 014.028.

**Expansion theory for the elliptic motion of arbitrary eccentricity and semi–major axis. VI. Elliptic expansions in terms of the sectorial variables for the fifth and sixth categories.**
See Abstr. 042.022.

**Calulation of elliptic functions in problems of celestial mechanics.**
See Abstr. 042.049.

**Computation of convective model atmospheres of cool white dwarfs.**
See Abstr. 064.027.

**The solar eclipse of 1984 November 22 – 23.**
See Abstr. 079.002.

**Computer program for precalculations of lunar occultations of stars and its application.**
See Abstr. 096.001.

**Interstellare Blasen im Weltall und im Computer.**
See Abstr. 131.026.

**Simulating spiral galaxies.**
See Abstr. 151.014.

# 022 Physical Papers Related to Astronomy and Astrophysics

022.001 **The standard problem of line formation theory: second–order approximations taking into account continuous absorption.**
V. M. Serbin.
Pis'ma Astron. Zh., Tom 10, No. 7, p. 554 – 560 (1984). In Russian. English translation in Sov. Astron. Lett., Vol. 10.

For the standard problem of line formation (two–level atom, complete frequency redistribution, plane–parallel atmosphere) simple approximate expressions for the line source function which take into account the effect of continuous absorption are given.

022.002 **Si III emission line strengths in low density plasmas.**
P. L. Dufton, F. P. Keenan, A. E. Kingston.
Mon. Not. R. Astron. Soc., Vol. 209, No. 1, p. 1P – 4P (1984).

Theoretical emissivity ratios appropriate to homogeneous optically thin plasmas are presented for the Si III lines at 1883 and 1892 Å. These lines are a useful density diagnostic for $N_e \geqslant 10^4 \mathrm{cm}^{-3}$, their use being illustrated for the planetary nebulae NGC 7662 and V1016 Cygni.

022.003 **Micrometre–sized polymer layers synthesized by MeV ions impinging on frozen methane.**
G. Foti, L. Calcagno, K. L. Sheng, G. Strazzulla.
Nature, Vol. 310, No. 5973, p. 126 – 128 (1984).

There has recently been much experimental and theoretical interest among astrophysicists in the synthesis of complex organic materials (polymers) as a consequence of the energy deposition by external agents on carbon–containing molecular solids such as $CH_4$, $C_2H_6$, $C_3H_8$, $C_6H_6$ and mixtures with $CO_2$, CO, $NH_3$ and $H_2O$. The authors present here the first quantitative results on the rate of production of solid residues, obtained by bombarding thick frozen layers ($T \approx 4$K) of methane with 1.5–MeV protons produced at the Catania University van de Graaff accelerator. This process converts the methane molecules to a polymer–like residue which is stable at room temperature and above.

022.004 **Relativistic $^{238}$U ion tracks in olivine and cosmic–ray track studies.**
C. Perron.
Nature, Vol. 310, No. 5976, p. 397 – 399 (1984).

Fossil nuclear tracks recorded in meteoritic material have been used by Fleischer et al. to identify cosmic ray nuclei heavier than Fe. Attempts to use tracks in meteorites to study the chemical composition of ultraheavy cosmic rays, however, have led to ambiguous results. The authors report here the first results on tracks of $^{238}$U ions of 190 MeV per nucleon in the mineral olivine. They show that the etchable part of a U ion trail is surprisingly long (nearly 3 mm). These results have an important bearing both on the interpretation of cosmic–ray tracks in lunar and meteoritic crystals, and on models of track formation in minerals.

022.005 **Atomic partition functions for iron.**
J. Halenka, B. Grabowski.
Astron. Astrophys., Suppl. Ser., Vol. 57, No. 1, p. 43 – 49 (1984).
Numerical values of the atomic partition functions $U^{(r)}$, for Fe I, Fe II and Fe III are given for the temperature range 2000 – 40000K, and for various lowering of the ionization energy (in the range 0.01 to 2.00 eV).

022.006 **Stark broadening of neutral helium lines of astrophysical interest. Regularities within spectral series.**
M. S. Dimitrijević, S. Sahal–Bréchot.
Astron. Astrophys., Vol. 136, No. 2, p. 289 – 298 (1984).
Following their earlier work, and using a semiclassical–perturbation formalism for the Stark broadening of atomic lines, the authors have calculated electron and proton impact line widths and shifts of 36 neutral He lines. Stark broadening parameters were calculated for lines originating from upper energy levels with $4 \leqslant n \leqslant 10$, and for an electron density of $10^{13}$ cm$^{-3}$. The comprehensive set of results obtained is used for the investigation of line–broadening parameter regularities within the spectral series.

022.007 **Isotopic dependence of infrared transition moments in methane.**
K. Fox.
J. Chem. Phys., Vol. 88, No. 3, p. 531 – 532 (1984). Abstr. in Phys. Abstr., Vol. 87, No. 1237, Entry 69146 (1984).

022.008 **A new method of determination of the gravitational constant using a superconductive suspension system.**
U. Bleyer, R. John, D. E. Liebscher.
Tr. Gos. Astron. Inst. Shternberg, Tom 54, p. 53 – 60 (1984). In Russian.
The authors propose the use of a pendulum with two degrees of freedom to determine the gravitational constant. Such a pendulum may be realized by a superconducting suspension system. The resulting method is compared with the Eötvös method using the oscillations of a torsion pendulum.

022.009 **Studies of proton–irradiated $SO_2$ at low–temperatures: implications for Io.**
M. H. Moore.
Icarus, Vol. 59, No. 1, p. 114 – 128 (1984).
The infrared absorption spectrum from 3.3 to 27 μm (3030 – 370 cm$^{-1}$) of $SO_2$ ice films has been measured at 20 and 88K before and after 1–MeV proton irradiation. The radiation flux was chosen to simulate the estimated flux of Jovian magnetospheric 1–MeV protons incident on Io. Based on the experiments, it is reasonable to propose that small amounts of $SO_3$ may exist on the surface of Io as a result of irradiation synthesis in $SO_2$ frosts.

022.010 **Oscillator strengths and photoionization cross–sections for positive ions in the magnesium isoelectronic sequence.**
K. Butler, C. Mendoza, C. J. Zeippen.
Mon. Not. R. Astron. Soc., Vol. 209, No. 2, p. 343 – 351 (1984). With microfiche MN 209/1.
Photoionization cross–sections of the ground states of $Al^{+}$, $Si^{2+}$ and $S^{4+}$ are calculated in a 7–state close–coupling approximation. Resonances are shown to be important in the region close to the ionization threshold although the background contribution is found to be in good agreement with previous results. Oscillator strengths for the principal series are also presented and compared with other theoretical determinations.

022.011 **Emission of large hydrocarbons from frozen $CH_4$ by keV proton irradiation.**
A. E. de Vries, R. Pedrys, R. A. Haring, A. Haring, F. W. Saris.
Nature, Vol. 311, No. 5981, p. 39 – 40 (1984).
Ever since the first description of the synthesis of molecules by impinging protons on ices, the possibility of formation of various molecules on cold surfaces by particle radiation has been widely discussed. The authors have recently studied sputtering of species by bombarding simple frozen gases such as water, carbon monoxide and ammonia with light ions. With these substances, only rather simple molecules were found. However, when bombarding methane, much larger molecules were detected in the sputtered beam–hydrocarbons containing up to at least 13 atoms. This suggests that bombardment of frozen methane by keV protons may produce large molecules in the Solar System.

022.012 **Vom Erdmeridian zum Lichtzeitmeter.**
F. Rotter.
Österr. Z. Vermessungswes. Photogramm., 72. Jahrg., Heft 1, p. 1 – 10 (1984).
The metric unit of length, the metre, has a long history which begins in the year 1791. At the beginning, the metre was defined as one tenmillionth of the quarter of the meridian of the earth. This definition was often changed corresponding to the increasing precision of length–measurements. In 1983, a new definition was found which is based on the value of velocity of light in vacuo: c = 299792458 m/s.

022.013 **Investigation of shock–induced changes of matter.**
A. K. Lavrukhina, Z. K. Mil'nikova, K. I. Ignatenko, S. A. Stakheeva, S. V. Pershin.
Meteoritika, Vyp. 43, p. 83 – 90, 190 (1984). With 3 plates. In Russian.

022.014 **Ionization equilibrium in rarefied gas. The effect of charge–exchange reactions.**
A. A. Suchkov, Yu. A. Shchekinov.
Astrofizika, Tom 19, Vyp. 3, p. 569 – 574 (1983). In Russian. English translation in Astrophysics, Vol. 19, No. 3.
The ionization of elements H, He, C, N, O, Ne, Mg, Si and S in low–density gas is considered. The ion composition of such a gas at $T < 10^5$K is mainly controlled by charge–exchange reactions. The results are tabulated for the temperature range of $10^4 - 10^5$K.

022.015 **The effect of resonant photoexcitation on fine–structure line intensities.**
S. O. Kastner.
Astrophys. Space Sci., Vol. 103, No. 2, p. 331 – 344 (1984).
Resonant photoexcitation of one atomic species by another often involves a ground configuration level from which an important fine structure line, used for density or temperature determination, may be emitted. It is pointed out that if the photoexcitation rate is sufficiently high the intensity of the latter may be affected. Estimates of the effect are made for several cases of astrophysical interest.

022.016 **The radiation of microwaves and infrared by slender graphite needles.**
F. Hoyle, J. V. Narlikar, N. C. Wickramasinghe.
Astrophys. Space Sci., Vol. 103, No. 2, p. 371 – 377 (1984).
It has become clear from sources observed by IRAS that there are many examples which attain maximum emission per unit (frequency) bandwidth longward of 100 μm. Here, the authors show that a large emissivity at such long wavelengths can be obtained by needle–shaped particles of free carbon. Indeed for needles with a sufficiently high ratio of length to diameter the large emissivity extends from the infrared through the whole microwave region of the spectrum.

022.017 **Progress in precision rotations for matter creation experiment and inertial clock.**
R. C. Ritter, G. R. Jones.
Proceedings of the Third Marcel Grossmann Meeting on General Relativity, Part B, p. 1485 – 1489 (1983). – See Abstr. 012.002.

022.018 **Molecular interferometry: a possible gravitational field measuring technique.**
G. I. Opat.
Proceedings of the Third Marcel Grossmann Meeting on General Relativity, Part B, p. 1491 – 1495 (1983). – See Abstr. 012.002.

022.019 **Experiments with a cryogenic torsion balance.**
R. D. Newman.
Proceedings of the Third Marcel Grossmann Meeting on General Relativity, Part B, p. 1497 – 1500 (1983). – See Abstr. 012.002.

The torsion balance is a remarkably capable instrument for the measurement of slowly varying exceedingly small forces. The author outlines here some of the virtues (and limitations) of the torsion balance, and presents a list of gravitation–related experiments to which it may be applied.

022.020 **Radiative lifetimes of the second negative system of $O_2^+$.**
R. W. Wetmore, J. L. Fox, A. Dalgarno.
Planet. Space Sci., Vol. 32, No. 9, p. 1111 – 1113 (1984).

The dipole moment of the $A^2\Pi_u$–$X^2\Pi_g$ transition of $O_2^+$ is calculated as a function of internuclear distance using ab initio methods. The band absorption oscillator strengths and band transition probabilities of the second negative system are derived and the resulting lifetimes are compared with experimental data. The high–lying v″ levels of the ground state may decay into low–lying v′ levels of the excited state. The corresponding radiative lifetimes are calculated.

022.021 **High resolution absorption cross section measurements of $SO_2$ at 213K in the wavelength region 172 – 240 nm.**
D. E. Freeman, K. Yoshino, J. R. Esmond, W. H. Parkinson.
Planet. Space Sci., Vol. 32, No. 9, p. 1125 – 1134 (1984).

Laboratory measurements at high resolution of the absorption cross section of $SO_2$ at the temperature 213K have been performed in the wavelength region 172 – 240 nm with a 6.65 m scanning spectrometer/spectrograph operated at an instrumental width of 0.002 nm. The measured cross sections are presented graphically in representative wavelength regions and are available throughout the region 172 – 240 nm at wavenumber intervals of 0.4 – 0.1 $cm^{-1}$ as a numerical tabulation stored on magnetic tape. The measured cross sections, which are relevant to the photochemistry of planetary atmospheres, possess significantly more spectroscopic structure, and are more accurate, than previous measurements made at lower resolution.

022.022 **Laboratory terrella experiment at Osaka City University.**
S. Minami, S. Tsutsumi.
XVI International Conference on Phenomena in Ionized Gases, p. 90 (1983). Abstr. in Phys. Abstr., Vol. 87, No. 1239, Entry 79771 (1984). – See Abstr. 012.017.

022.023 **The photodissociation of $H_2O$ at 157 nm: full internal state distributions and alignment of nascent $OH(X^2\Pi)$ radicals.**
P. Andresen, G. S. Ondrey, E. W. Rothe, B. Titze.
Ann. Isr. Phys. Soc., Vol. 6, p. 226 – 228 (1984). Abstr. in Phys. Abstr., Vol. 87, No. 1240, Entry 83439 (1984). – See Abstr. 012.018.

022.024 **Photoabsorption and photodissociation of CO in the 900 – 1200 Å region.**
M. Eidelsberg, F. Launay, F. Rostas, A. Le Floch, J. Breton, B. Thieblemont.
Ann. Isr. Phys. Soc., Vol. 6, p. 240 – 242 (1984). Abstr. in Phys. Abstr., Vol. 87, No. 1240, Entry 83441 (1984). – See Abstr. 012.018.

022.025 **Theoretical studies of photon and electron–impact excitation and ionization in stable gas–phase compounds.**
P. W. Langoff.
Ann. Isr. Phys. Soc., Vol. 6, p. 222 – 224 (1984). Abstr. in Phys. Abstr., Vol. 87, No. 1240, Entry 83481 (1984). – See Abstr. 012.018.

022.026 **Laboratory millimeter and submillimeter spectra of HNO and DNO.**
K. V. L. N. Sastry, P. Helminger, G. M. Plummer, E. Herbst, F. C. De Lucia.
Astrophys. J., Suppl. Ser., Vol. 55, No. 4, p. 563 – 570 (1984).

Millimeter and submillimeter wave spectra of the molecular species HNO and DNO have been measured in the frequency region 150 – 500 GHz. Based on these spectral line measurements, models of HNO and DNO have been constructed which yield the frequencies of all transitions below 500 GHz involving $J \leqslant 10$.

022.027 **The millimeter and submillimeter laboratory spectrum of methyl formate in its ground symmetric torsional state.**
G. M. Plummer, E. Herbst, F. C. De Lucia, G. A. Blake.
Astrophys. J., Suppl. Ser., Vol. 55, No. 4, p. 633 – 656 (1984).

Over 200 rotational lines of methyl formate in its ground ($v_t = 0$), symmetric ($A$) torsional state have been measured in the frequency range 140 – 550 GHz. Analysis of these and lower frequency transitions permits accurate prediction ($\leqslant$0.1 MHz) of over 10,000 transitions at frequencies below 600 GHz with angular momentum $J \leqslant 50$. The measured spectral lines have permitted identification of over 100 new methyl formate lines in Orion.

022.028 **Oscillator strengths and transition probabilities of pseudo–resonance lines.**
A. G. Egikyan.
Astrofizika, Tom 20, Vyp. 2, p. 341 – 350 (1984). In Russian. English translation in Astrophysics, Vol. 20, No. 2.

Computations for the determination of the absolute oscillator strengths and the transition probabilities for pseudo–resonance absorption lines of O I–O III, C II, N I, N II, Al II, Si III, S II and S III have been carried out using analytical one–configuration Hartree–Fock radial orbitals.

022.029 **On stimulated processes in annihilation line radiation of cosmic sources. II.**
V. V. Zheleznyakov, A. A. Litvinchuk.
Astron. Zh., Tom 61, Vyp. 5, p. 860 – 873 (1984). In Russian. English translation in Sov. Astron., Vol. 28, No. 5.

This paper investigates the influence of stimulated annihilation on the steady–radiation characteristics of a double–quantum system with a given power of a positron source. The coherent–radiation conditions and criteria for gamma–laser realization, i. e. population inversion, are determined. The laser functioning as an amplifier of incident radiation is studied. The possibility of population inversion in cosmic sources is discussed.

022.030 **On the theoretical calculation of wavelengths and oscillator strengths for Fe X and similar spectra.**
R. D. Cowan, G. E. Bromage, B. C. Fawcett.
Mon. Not. R. Astron. Soc., Vol. 210, No. 2, p. 439 – 442 (1984).

The authors re–emphasize the importance of both spin–orbit interaction and configuration–mixing effects in the calculation of wavelengths and oscillator stengths for heavy–element moderately–ionized spectra; with particular reference to Fe X resonance lines. They correct several rather misleading statements in a recent paper by R. Glass on Fe X, and dispute the argument that three– and four–electron correlation effects are important. They discuss afresh the (well–established) method of allowing for weak correlations by the scaling of radial integrals, and consider its implications for oscillator stength calculations.

022.031 **Laboratory measurement of the $1_{10} - 1_{11}$ submillimeter line of $H_2D^+$.**
M. Bogey, C. Demuynck, M. Denis, J. L. Destombes, B. Lemoine.
Astron. Astrophys., Vol. 137, No. 2, p. L15 – L16 (1984).

The $1_{10} - 1_{11}$ line of $H_2D^+$, at 372421.34(20) MHz, has been observed in the laboratory by discharging a mixture of $H_2 + D_2 + Ar$. The line frequency is in good agreement with ab initio calculations and with the prediction derived from the infrared spectrum.

022.032 **Millimeter wave spectrum of methyldiacetylene, $CH_3C_4H$.**
M. Bester, K. Yamada, G. Winnewisser, W. Joentgen, H.-J. Altenbach, E. Vogel.
Astron. Astrophys., Vol. 137, No. 2, p. L20 – L22 (1984).
The ground state rotational spectrum of methyldiacetylene, $CH_3C_4H$, has been measured between 8 and 224 GHz yielding precise values for the rotational constant $B_0$ = 2035.74706(8) MHz and the $<P^4>$ and $<P^6>$ centrifugal distortion constants. From the J = 2–1 transition the electric dipole moment has been determined to be 1.2071(10) Debye. For $CH_3C_3N$ the authors obtained 4.75(7) Debye. These data allow an estimate of the dipole moment of $CH_3C_6H$ and of $CH_3C_5N$ to be about 1.5(1) and 5.4(1) Debye, respectively.

022.033 **Laboratory detection of the protonated carbon dioxide by submillimeter wave spectroscopy.**
M. Bogey, C. Demuynck, J. L. Destombes.
Astron. Astrophys., Vol. 138, No. 1, p. L11 – L12 (1984).
Six lines corresponding to the $K = 0$ and $K = 2$ components of the $J = 15 \rightarrow 16$ and $16 \rightarrow 17$ transitions of $HCO_2^+$ have been observed in the laboratory by discharging a mixture of $H_2 + CO_2 + Ar$. These lines have been fitted together with the astrophysical lines attributed to $HCO_2^+$ or HOCN, and the $B$ and $C$ constants as well as centrifugal distortion constants have been determined.

022.034 **Influence of line interference on the vibration–rotation band shapes.**
M. O. Bulanin, A. B. Dokuchaev, M. V. Tonkov, N. N. Filippov.
J. Quant. Spectrosc. Radiat. Transfer, Vol. 31, No. 6, p. 521 – 543 (1984).
The shapes of the CO, $\nu_3$ $CO_2$, and $\nu_3$ $N_2O$ fundamental vibration–rotation bands have been studied at various temperatures and in the presence of several perturbing gases. Also the half-widths of CO vibration–rotation lines have been measured at 78K.

022.035 **On the temperature dependence of self–broadening in the first overtone band of CO.**
J.-P. Bouanich.
J. Quant. Spectrosc. Radiat. Transfer, Vol. 31, No. 6, p. 561 – 567 (1984).
The halfwidths of CO spectral lines have been measured in the 2–0 band at 298, 198, 131, and 85K using a high resolution spectrometer. The temperature dependence of the linewidths has been calculated by applying two semi-classical theories developed within the impact approximation. The theoretical results are compared with experimental data.

022.036 **Line parameter measurements and calculations of CO broadened by $H_2O$ and $CO_2$ at elevated temperatures.**
H. S. Lowry III, C. J. Fisher.
J. Quant. Spectrosc. Radiat. Transfer, Vol. 31, No. 6, p. 575 – 581 (1984).
Measurements of collisional broadening of four fundmental CO lines in atmospheres of $CO_2$ and $H_2O$ at 300 – 600K were made. The Voigt line-shape model was fitted parametrically to the absorption spectra line-shapes. The results are compared with other experimental studies and with theoretical calculations for $CO_2$ broadening.

022.037 **Tables for direct determination of spectral line parameters, from spectrometric data. II. Intermediate Voigt broadening.**
M. L. Claude.
J. Quant. Spectrosc. Radiat. Transfer, Vol. 32, No. 1, p. 17 – 48 (1984).
Tables and curves are presented for the calculation of spectral-line parameters from observed spectra for combined Doppler and Lorentz Broadening (Voigt profile), using a spectrometer of known instrument response function.

022.038 **Oscillator strengths of ultraviolet V(II) lines from measurements in a vanadium–helium plasma.**
A. Goly, S. Weniger.
J. Quant. Spectrosc. Radiat. Transfer, Vol. 32, No. 1, p. 61 – 67 (1984).
A V–He plasma has been produced in a wall-stabilized arc. Oscillator strengths for 66 V(II) lines in the range 2342 – 2617 Å have been determined. Comparisons with theoretical values show large discrepancies for about 25% of the lines, possibly due to errors in calculations and in term classifications.

022.039 **Lifetimes, branching ratios, and transition probabilities in Mo I.**
W. Whaling, P. Hannaford, R. M. Lowe, E. Biémont, N. Grevesse.
J. Quant. Spectrosc. Radiat. Transfer, Vol. 32, No. 1, p. 69 – 80 (1984).
The radiative lifetimes of 56 levels in Mo I have been measured by the method of laser-induced fluorescence from sputtered metal vapor. Emission branching ratios have been measured for these levels to determine transition probabilities for 570 Mo I transitions in the wavelength range 2944 – 9767 Å. The uncertainty in the transition probability of stronger branches is typically less than 5%.

022.040 **Optical properties of an isolated resonance line with high density self–broadening.**
W. Heering.
J. Quant. Spectrosc. Radiat. Transfer, Vol. 32, No. 1, p. 81 – 89 (1984).

022.041 **SPECTRA – a model for *K*-shell spectroscopy.**
R. W. Lee, B. L. Whitten, R. E. Stout II.
J. Quant. Spectrosc. Radiat. Transfer, Vol. 32, No. 1, p. 91 – 101 (1984).
X-ray spectroscopy is a powerful tool for understanding the kinetics of highly ionized plasmas. The authors have developed a computer code for modeling and analyzing plasmas which is highly accurate as well as fast and easy to use. It produces synthetic spectra for hydrogen-like and helium-like ions at arbitrary density and temperature.

022.042 **Relativistic model–potential oscillator strengths and transition probabilities for $4f^n6s$–$4f^n6p$ transitions in Eu(II), Tb(II), and Ho(II) in $J,j$ coupling.**
J. Migdałek.
J. Quant. Spectrosc. Radiat. Transfer, Vol. 32, No. 1, p. 103 – 108 (1984).

022.043 **New Fe II lines in IUE spectra of stars.**
S. Johansson.
ESA Spec. Publ., ESA SP-218, p. 307 – 308 (1984). – See Abstr. 012.024.
The never-ceasing need for more atomic data of Fe II for astrophysical purposes is discussed in terms of present state and future work on the term analysis of this spectrum. More data already exist and an extended analysis of laboratory data has just started. Special excitation conditions in stars allow highly excited states to be populated for which data is needed. Some new presumptive interstellar lines are reported.

022.044 **Collisional pumping of main line OH–maser sources.**
P. Andresen, D. Häusler, H. W. Lülf, W. H. Kegel.
Astron. Astrophys., Vol. 138, No. 2, p. L17 – L18 (1984).
Recently cross-sections for collisions between OH– and $H_2$-molecules have been measured. The autors discuss consequences of the results with respect to the interpretation of type I OH-maser sources. Considerable modifications of previous collisional pumping schemes are necessary.

022.045 **The hydrogenic $2s \rightarrow 1s$ two–photon emission.**
H. Nussbaumer, W. Schmutz.
Astron. Astrophys., Vol. 138, No. 2, p. 495 – 496 (1984).
The authors give an analytical expression for the probability of the hydrogenic two-photon decay. The expression allows a very

accurate description of the two-photon continuum. For the total hydrogen $2s \rightarrow 1s$ two-photon transition probability a value of $A_{2q} = 8.2249\ s^{-1}$ is found.

022.046 **Forbidden lines in hot plasmas.**
B. Edlén.
Phys. Scr., Vol. T8, p. 5 – 9 (1984). – See Abstr. 012.027.

A survey is given of the numerous forbidden lines that have recently been discovered in the spectra of Tokamak plasmas. They represent magnetic-dipole transitions in three different groups of configurations: $2s2p$ and $2s^22p^k$ (k = 1 – 5), $3s3p$ and $3s^23p^k$ (k = 1 – 5), and $3d^k$ (k = 1, 2, 8, 9).

022.047 **Highly-ionized atoms in fusion research plasmas.**
N. J. Peacock, M. F. Stamp, J. D. Silver.
Phys. Scr., Vol. T8, p. 10 – 20 (1984). – See Abstr. 012.027.

Topical problems involving the interpretation of line emission from impurities or diagnostic test ions in Tokamak fusion devices are discussed briefly. The particular use of Tokamaks for precision wavelength spectroscopy of $\Delta n = 0$ transitions in highly ionised atoms with between 2 and 12 electrons is described in more detail. The Tokamak data for ions with $Z \geqslant 3$-electrons lead to improved isoelectronic extrapolations which reveal errors in the previous data and provide transfer standards of wavelength in the XUV and VUV regions.

022.048 **A comparison between the observed and predicted Fe II spectrum in different plasmas.**
S. Johansson.
Phys. Scr., Vol. T8, p. 63 – 69 (1984). – See Abstr. 012.027.

This paper gives a survey of the spectral distribution of emission lines of Fe II, predicted from a single atomic model. The observed differences between the recorded and the predicted spectrum are discussed in terms of deficiencies of the model and interactions within the emitting plasma. The author gives a number of illustrative examples of unexpected features with applications to astrophysics.

022.049 **Measurements of lifetimes and oscillator strengths of neutral and singly ionized atoms. Experimental results of the last five years.**
J. Richter.
Phys. Scr., Vol. T8, p. 70 – 76 (1984). – See Abstr. 012.027.

After a short description of the standard methods to determine $f$-values and atomic lifetimes, the most important development of the last years – the application of pulsed dye lasers – is discussed and some results are compared. The further development in the field of $f$-value measurement is described by some typical activities: emission and absorption measurements with thermal light sources and the application of emission and absorption branching ratio measurements.

022.050 **Lifetime measurements in highly ionised atoms.**
L. J. Curtis.
Phys. Scr., Vol. T8, p. 77 – 83 (1984). – See Abstr. 012.027.

Recent developments in the measurement of lifetimes of excited states in highly ionised atoms using fast ion beam excitation methods are reviewed.

022.051 **Fe I oscillator strengths.**
M. Kock, S. Kroll, S. Schnehage.
Phys. Scr., Vol. T8, p. 84 – 87 (1984). – See Abstr. 012.027.

By a combination of hook and emission measurements oscillator strengths of 122 Fe I lines have been determined without any assumption concerning the plasma state. The relative $f$-values have been converted to an absolute scale by means of lifetime data found in the literature.

022.052 **Measurements of transition probabilities for spin-changing lines of atomic ions used in diagnostics of astrophysical plasmas.**
P. L. Smith, B. C. Johnson, H. S. Kwong, W. H. Parkinson, R. D. Knight.
Phys. Scr., Vol. T8, p. 88 – 94 (1984). – See Abstr. 012.027.

The intensities of ultraviolet, spin-changing, "intersystem" lines of low-$Z$ atomic ions are frequently used in determinations of electron densities and temperatures in astrophysical plasmas as well as in measurements of element abundances in the interstellar gas. A radio-frequency ion trap has been used for the first measurements of transition probabilities for intersystem lines seen in astronomical spectra. The measurement procedure is discussed and results for Si III, O III, N II, and C III are reviewed and compared to calculated values.

022.053 **Branching ratios of the $2s^23p\ ^2P^0$ term of singly-ionized carbon.**
M. C. E. Huber, R. J. Sandeman, G. P. Tozzi.
Phys. Scr., Vol. T8, p. 95 – 99 (1984). – See Abstr. 012.027.

The branching ratios of the $2s^23p\ ^2P^0{}_{1/2,\ 3/2}$ levels of $C^+$ have been measured by use of a Fourier-transform spectrometer and a vacuum ultraviolet spectrometer. The strong multiplets occurring at 658, 284, and 176 nm can be used for determining the interstellar extinction between the red and vacuum ultraviolet spectral regions.

022.054 **Argon branching ratios and absolute transition probabilities in the $4s$–$4p$ transition array.**
K.-P. Nick, V. Helbig.
Phys. Scr., Vol. T8, p. 100 – 103 (1984). – See Abstr. 012.027.

Absolute transition probabilities for five lines in the $4s$–$4p$ transition array of neutral argon were determined. The plasma source was an argon cascade arc in local thermodynamic equilibrium (LTE). The measured line emission coefficients were corrected for line-wing losses and for optically thin emission.

022.055 **Lifetimes of some excited levels in Cu I and Cu II.**
H. Cederquist, S. Mannervik, M. Kisielinski, P. Forsberg, I. Martinson, L. J. Curtis, P. S. Ramanujam.
Phys. Scr., Vol. T8, p. 104 – 106 (1984). – See Abstr. 012.027.

Using the beam-foil method the authors have determined the lifetimes of the $nd^2D$ ($n$ = 4 – 6) and $4s5s^4D$ levels in Cu I and of the $4p^3D$, $^3F$, $4f^3H$ and $5s^3D$ levels in Cu II. Special emphasis was placed on the problems of measuring comparatively long lifetimes in heavy atoms with the beam-foil excitation method.

022.056 **Determination of autoionization rates by means of optical emission spectroscopy.**
S. Mannervik, H. Cederquist, M. Kisielinski.
Phys. Scr., Vol. T8, p. 107 – 111 (1984). – See Abstr. 012.027.

Extraction of accurate values of autoionization rates from broadened emission lines in optical spectra is discussed. Results for autoionizing states in He I and Li I are given and compared with theory.

022.057 **Extended relative efficiency calibration of a grazing incidence monochromator and applications in fast-beam spectroscopy.**
E. Träbert.
Phys. Scr., Vol. T8, p. 112 – 116 (1984). – See Abstr. 012.027.

A relative efficiency calibration curve for a 2.2 m McPherson grazing incidence monochromator has been extended to cover the wavelength range 10 nm $\leqslant \lambda \leqslant$ 80 nm using branching ratios for $\Delta n \neq 0$ transition in Li-like ions. A summary of various measurements applicable to problems of astrophysical and plasma physics interest is given.

022.058 **Fifteen years on – the calculation of atomic transition probabilities revisited.**
R. Crossley.
Phys. Scr., Vol. T8, p. 117 – 128 (1984). – See Abstr. 012.027.

An up-date is presented of a previous review by the author of atomic transition probability calculations published in 1969. Attention is again restricted, in the main, to allowed electric dipole transitions. Attention is in particular paid to the construction of accurate atomic wave-functions, to the random phase approximation and to recent progress in relativistic calculations.

022.059 **Relativistic calculations of atomic structure.**
B. Fricke.
Phys. Scr., Vol. T8, p. 129 – 133 (1984). – See Abstr. 012.027.
A review of relativistic atomic structure calculations is given with an emphasis on the Multiconfigurational–Dirac–Fock method. Its problems and deficiencies are discussed together with the contributions which go beyond the Dirac–Fock procedure.

022.060 **Calculation of atomic data for astrophysics at University College London.**
H. E. Saraph.
Phys. Scr., Vol. T8, p. 134 – 140 (1984). – See Abstr. 012.027.
This paper reports on methods for the calculation of atomic data employed at University College London, giving the essence of the approximations involved and some guidance as to the accuracy that can be expected from the data so produced.

022.061 **Oscillator strengths and the accuracy of atomic wavefunctions.**
Z. B. Rudzikas, M. Szulkin, I. Martinson.
Phys. Scr., Vol. T8, p. 141 – 144 (1984). – See Abstr. 012.027.
General expressions (both relativistic and non–relativistic) for the electric multipole transition operators, having not specified values of the gauge condition of the electromagnetic field potential, are presented. The dependence of oscillator strengths for electric dipole transition on the gauge is studied. Transitions in Be–like ions are considered and the results of the calculations are compared with experimentally determined oscillator strengths.

022.062 **Impact fragmentation experiments of basalts and pyrophyllites.**
Y. Takagi, H. Mizutani, S.–I. Kawakami.
Icarus, Vol. 59, No. 3, p. 462 – 477 (1984).
Impact fragmentation experiments were performed on basalts and pyrophyllite. Comparing the present experimental data with those from previous studies, the authors found that the nondimensional impact stress $P_I$ introduced by Mizutani et al. (1984) is a useful parameter to describe impact fragmentation phenomena. The results show that fragment size and fragment size distributions form a record of impact conditions.

022.063 **Analytical calculation of the magnetic dipole field in a conducting radially expanding and differentially rotating medium.**
A. I. Laptukhov, V. P. Shabanskij.
Geomagn. Aehron., Tom 24, No. 4, p. 635 – 643 (1984). In Russian. English translation in Geomagn. Aeron.

022.064 **Ternary association reaction of $CH_3{}^+$ and $CH(OH)_2{}^+$ with $H_2O$ in He buffer.**
H. Villinger, A. Saxer, H. Bryant, W. Lindinger.
XVI International Conference on Phenomena in Ionized Gases, p. 572 – 573 (1983). Abstr. in Phys. Abstr., Vol. 87, No. 1242, Entry 93525 (1984). – See Abstr. 012.017.

022.065 **Experimental simulation of asteroidal fragmentation by macroscopic hypervelocity impacts against free falling bodies.**
R. Bianchi, F. Capaccioni, P. Cerroni, M. Coradini, E. Flamini, P. Hurren, G. Martelli, P. N. Smith.
Astron. Astrophys., Vol. 139, No. 1, p. 1 – 6 (1984).
Free falling triaxial ellipsoids, having a mass of $\sim$10 kg, were impacted in vacuo with aluminium projectiles moving at $\sim$10 km s$^{-1}$. The impact specific energy lay between 4 and $5 \times 10^7$erg/g. The impact generated fragments have been recovered, measured and weighed. Their mass and shape distributions have been obtained and compared with those obtained by other authors in experiments performed under different conditions. The analysis of the main fracture planes, carried out by reconstructing the targets with the recovered fragments has provided new information on the generation of fractures and their distribution in space.

022.066 **Unified theory calculations of Stark–broadened hydrogen lines in the presence of a magnetic field.**
G. Mathys.
Astron. Astrophys., Vol. 139, No. 1, p. 196 – 210 (1984).
Previously derived expressions of Stark–broadened profiles of hydrogen lines in a hot dense plasma, in the presence of a large scale, uniform magnetic field (whose strength, a few hundreds to a few thousands Gauss, is relevant to a number of astrophysical situations), are discussed in detail.

022.067 **The C II 2325 Å multiplet and the 158 µm transition.**
M. A. Hayes, H. Nussbaumer.
Astron. Astrophys., Vol. 139, No. 1, p. 233 – 236 (1984).
Effective electron–ion collision strengths for the excitation of the fine–structure levels of the terms $2s2p^2\,{}^4P$ and ${}^2D$ in $C^+$ and for transitions among the levels of $2s2p^2\,{}^4P$ are calculated. The authors investigate the temperature and density dependence of relative emissivities of the individual components within the intercombination multiplet at λ2325, including the contributions from proton collisions. They also look at the influence of stellar radiation fields. The ratio of the emissivities of the intercombination multiplet to the infrared 158 µm transition is calculated as a function of temperature and density.

022.068 **Oscillator strengths and collision strengths for S III.**
Y. K. Ho, R. J. W. Henry.
Astrophys. J., Vol. 282, No. 2, p. 816 – 819 (1984).
Collision strengths for electron impact excitation of S III from the ground $3p^2\,{}^3P$ state to excited states $3s3p^3\,{}^3D^0, {}^3P^0, {}^3S^0$, $3d\,{}^3D^0$, ${}^3P^0$, and $4s\,{}^3P^0$ are calculated in a close–coupling approximation for the energy range up to $10^6$K. Also given are oscillator strengths for those transitions and for a number of others which involve $3p^2\,{}^1D$ and ${}^1S$.

022.069 **The $\upsilon = 0 \rightarrow 0$ spectrum of $H_2$.**
D. E. Jennings, S. L. Bragg, J. W. Brault.
Astrophys. J., Lett. Ed., Vol. 282, No. 2, p. L85 – L88 (1984).
Ground–state rotational parameters for molecular hydrogen are evaluated using a combination of laboratory and observational data. High–accuracy $\upsilon = 0 \rightarrow 0$ $S(0)$–$S(5)$ frequencies from infrared laboratory measurements are analyzed simultaneously with available observed frequencies for the $S(8)$, $S(9)$, $S(12)$, $S(13)$, $S(14)$, and $S(15)$ Orion emission lines. The analysis produces the parameters (in cm$^{-1}$): $B_0 = 59.334627(30)$, $D_0 = 0.0456912(30)$, $H_0 = 4.8974(90) \times 10^{-5}$, $L_0 = -6.320(110) \times 10^{-8}$, $M_0 = 7.19(30) \times 10^{-11}$, where a correction is made for the estimated contribution of sixth and higher orders.

022.070 **Laboratory submillimeter transition frequencies of $^7$LiH and $^6$LiH.**
G. M. Plummer, E. Herbst, F. C. De Lucia.
Astrophys. J., Lett. Ed., Vol. 282, No. 2, p. L113 – L114 (1984).
The fundamental ($J = 1 \leftarrow 0$) rotational transition frequencies of $^7$LiH and $^6$LiH in their ground ($\upsilon = 0$) vibrational states and of $^7$LiH in its first excited ($\upsilon = 1$) vibrational state have been measured in the laboratory. Use of these transition frequencies should permit astronomical investigations of LiH abundances in red giant stars of high lithium abundance.

022.071 **Is the metal contribution to the astrophysical opacity incorrect?**
N. H. Magee Jr., A. L. Merts, W. F. Huebner.
Astrophys. J., Vol. 283, No. 1, p. 264 – 272 (1984).
Numerical tests by Simon have indicated that an arbitrary increase in the metal (atomic number $Z > 2$) contribution to the opacity by a factor of 2 to 3 leads to Cepheid models that, in some aspects, are in better agreement with observations than models using the standard opacity. The authors show that such a large increase in opacity is incompatible with atomic physics.

022.072 **Discrepancy in the CNO opacity bump resolved.**
T. R. Carson, W. F. Huebner, N. H. Magee Jr., A. L. Merts.
Astrophys. J., Vol. 283, No. 1, p. 466 – 468 (1984).
The discrepancy between the Los Alamos and the Carson opacities for mixtures containing "metals" at low densities around a temperature of $10^6$K ("the CNO bump") has been investigated and resolved. The bump in the Carson opacities arises from a misapplied correction to the atomic level occupancies. Results from a modified Carson opacity program now show much closer agreement with the Los Alamos opacities. There is no evidence for a CNO bump.

022.073 **Modelling of processes of cyclic evolution type. Synchronization by a random signal.**
V. A. Antonov.
Vestn. Leningr. Univ., Mat. Mekh. Astron., No. 7, p. 67 – 76 (1984). In Russian. Abstr. in Ref. Zh., 51. Astron., 9.51.146 (1984).

022.074 **Optical constants of organic tholins produced in a simulated Titanian atmosphere: from soft X–ray to microwave frequencies.**
B. N. Khare, C. Sagan, E. T. Arakawa, F. Suits, T. A. Callcott, M. W. Williams.
Icarus, Vol. 60, No. 1, p. 127 – 137 (1984).
As part of a continuing series of experiments on the production of dark reddish organic solids, called tholins, the synthesis from a simulated Titanian atmosphere of a tholin with a visible reflection spectrum similar to that of the high altitude aerosols responsible for the albedo and reddish color of Titan has been reported (C. Sagan et al., 1981, 1982, 1984). The determination of the real ($n$) and imaginary ($k$) parts of the complex refractive index of thin films of such tholin prepared by continuous D.C. discharge through a 0.9 $N_2$/0.1 $CH_4$ gas mixture at 0.2 mb are reported. For 250 Å $\leqslant \lambda \leqslant$ 1000 μm, $n$ and $k$ have been determined from a combination of transmittance, specular reflectance, interferometric, Brewster angle, and ellipsometric polarization measurements. Values of $n$ and $k$ in the visible range are consistent with deductions made by ground–based and spacecraft observations of Titan.

022.075 **Microdunes and other aeolian bedforms on Venus: wind tunnel simulations.**
R. Greeley, J. R. Marshall, R. N. Leach.
Icarus, Vol. 60, No. 1, p. 152 – 160 (1984).
Previous studies confirm that sand can be entrained at the wind velocities recorded on Venus. Present results describe bedforms produced in the Venus wind tunnel simulating the average Venusian environment. Even at the low wind speeds measured on Venus, dunelike structures form in fine–grained quartz sands (particles 50 – 200 μm in diameter). The dunelike structures, referred to as microdunes, are considered to be true dunes analogous to those on Earth. It is predicted that relatively short periods of higher winds may destroy microdunes and other small bedforms which could account for the sparsity of definitive aeolian features observed in Venera images. Some apparent cross–bedding observed in Venera images, however, could represent preserved aeolian structures.

022.076 **Collision strengths for λ1199 and λ1729 of S III.**
Y. K. Ho, R. J. W. Henry.
Astrophys. J., Vol. 284, No. 1, p. 435 – 437 (1984).
Collision strengths for electron impact excitation of S III from the ground $3p^2\ ^3P$ state to excited states $3s3p^3\ ^5S^0$ and $3s3p^3\ ^3D^0$ are calculated in a three–state close–coupling approximation for the energy range up to $1.5\times10^5$K. Ratios for the emission of UV radiation at λ1729 and λ1199 are presented as a function of temperature. The calculations are consistent with observed ratios of intensities for the Io torus.

022.077 **Dissociative recombination coefficients for $H_3^+$, $HCO^+$, $N_2H^+$, and $CH_5^+$ at low temperature: interstellar implications.**
D. Smith, N. G. Adams.
Astrophys. J., Lett. Ed., Vol. 284, No. 1, p. L13 – L16 (1984).
The authors summarize the results obtained from a recent laboratory study of the dissociative recombination coefficients, α, at 95 and 300K for the important interstellar ions $H_3^+$, $HCO^+$, $N_2H^+$, and $CH_5^+$. The most important result of the study is that $\alpha(H_3^+)$ is much smaller than previous studies had indicated. However, the values for the other ions are in accordance with previous laboratory data. The smaller $\alpha(H_3^+)$ implies a larger fractional ionization of dense interstellar clouds than previous estimates had indicated. Some possible consequences of a higher ionization density to the physics and chemistry of dense clouds are mentioned.

022.078 **A set of partition functions and equilibrium constants for 300 diatomic molecules of astrophysical interest.**
A. J. Sauval, J. B. Tatum.
Astrophys. J., Suppl. Ser., Vol. 56, No. 2, p. 193 – 209 (1984).
Polynomial developments for partition functions and equilibrium constants of 300 diatomic molecules and of relevant atoms are given for a temperature range from 1000 to 9000K. All molecular spectroscopic data, with a few exceptions, are derived from the book by Huber and Herzberg in order to lead to a homogeneous set of data.

022.079 **Line strengths and positions of the submillimeter magnetic dipole transitions of $O_2$.**
R. T. Boreiko, T. L. Smithson, T. A. Clark, H. Wieser.
J. Quant. Spectrosc. Radiat. Transfer, Vol. 32, No. 2, p. 109 – 117 (1984).
Strengths and positions of lines between 14 and 160 $cm^{-1}$ from pure rotational magnetic dipole transitions of the homonuclear molecule $O_2$ have been measured under carefully controlled conditions over a range of pressures from 52 to 672 torr in a long–path cell using a Fourier transform spectrometer. The data show excellent agreement with the theory of Tinkham and Strandberg over the above range of measurement. The possible implication of this work to the calibration of balloon–borne measurements of the cosmologically significant microwave background is also briefly discussed.

022.080 **Line–by–line and narrow–band statistical model calculations for $H_2O$.**
J. M. Hartmann, R. Levi Di Leon, J. Taine.
J. Quant. Spectrosc. Radiat. Transfer, Vol. 32, No. 2, p. 119 – 127 (1984).
In line–by–line calculations the authors use hot–band lines generated from the AFGL or GEISA data. The parameters of a random statistical model are calculated from these lines for 25 $cm^{-1}$ wide intervals. Comparisons are made between experimental spectra and results of the two theoretical models for the 1595 and 3755 $cm^{-1}$ bands between 900 and 1500K. The validities of the statistical model and of the Curtis–Godson and Lindquist–Simmons approximations for nonisothermal and inhomogeneous media are tested.

022.081 **Theoretical lifetimes, transition energies, fluorescence yields, and nonradiative branching ratios for highly excited states of lithium–like argon.**
C. P. Bhalla, T. W. Tunnell.
J. Quant. Spectrosc. Radiat. Transfer, Vol. 32, No. 2, p. 141 – 158 (1984).
The authors have calculated theoretical transition rates and transition energies for the states of the excited electron configurations, $1s\,2s\,nl$ and $1s\,2p\,nl$, of lithium–like argon. The values of $n$ considered range from $n = 2$ to $n = 4$ with all allowed values of $l$ included, $l = 0$ to $l = n$–1. The authors present numerical results for the theoretical lifetimes, transition energies, fluorescence yields, and nonradiative branching ratios. These quantities are related to the dielectronic recombination rates and cross sections of helium–like argon.

022.082 **Absolute OH absorption cross sections (for lidar measurements).**
T. J. McGee, T. J. McIlrath.
J. Quant. Spectrosc. Radiat. Transfer, Vol. 32, No. 2, p. 179 – 184 (1984).
The authors have used existing experimental data to calculate the effective cross sections for a number of transitions in the $A$–$X$ system of OH, taking into account such factors as temperature and the spectral shape and width of both the absorption line and the excitation source. Details of the calculations are presented to allow easy extension to other transitions.

022.083 **High resolution absorption cross–section measurements of $N_2O$ at 295 – 299K in the wavelength region 170 – 222 nm.**
K. Yoshino, D. E. Freeman, W. H. Parkinson.
Planet. Space Sci., Vol. 32, No. 10, p. 1219 – 1222 (1984).

022.084 **Measured ion–molecule reaction rates for modelling Titan's atmosphere.**
S. Dheandhanoo, R. Johnsen, M. A. Biondi.
Planet. Space Sci., Vol. 32, No. 10, p. 1301 – 1305 (1984).
Rate coefficients for several two– and three–body ion–molecule reactions involving hydrocarbons have been determined at thermal energies and above using drift tube–mass spectrometer techniques. The measured rates for clustering and breakup reactions involving $CH_5^+$ and $C_2H_5^+$ ions in methane are found to be strongly temperature dependent in the range from 80 to 240K. Rate coefficients for two–body reactions of $CH_5^+$, $C_2H_5^+$, $N^+$, $H^+$ and $D^+$ ions with methane and/or ethane have been measured. The results indicate that the product yields of several of the fast ion–molecule reactions depend strongly on ion energy (temperature), and therefore previous room–temperature results may be of limited value for model calculations of Titan's atmosphere.

022.085 **Collisional excitation of HCN–hyperfine transitions.**
T. Monteiro.
Mon. Not. R. Astron. Soc., Vol. 211, No. 2, p. 257 – 266 (1984).
Rate coefficients for the rotational excitation of HCN with the inclusion of nuclear spin, were evaluated for collisions with the helium atom. The coupled states method was used with an interaction potential calculated for the HCN–He system by Green & Thaddeus. Rate coefficients were obtained for all levels up and including $j = 5$.

022.086 **Remark on the correct consideration of motion of electrons in astrophysical problems and their interaction with photons.**
G. T. Ter–Kazaryan.
Dokl. Akad. Nauk SSSR. Ser. Mat. Fiz., Tom 276, No. 1, p. 106 – 110 (1984). In Russian. Abstr. in Ref. Zh., 51. Astron., 10.51.159 (1984).

022.087 **On rays of Alfvén waves.**
A. V. Gul'el'mi, B. I. Klajn, V. N. Repin, A. R. Polyakov.
Issled. Geomagn. Aehron. Fiz. Solntsa, Moskva, No. 66, p. 29 – 37 (1983). In Russian. Abstr. in Ref. Zh., 62. Issled. Kosm. Prostranstva, 10.62.445 (1984).

022.088 **A suggested mechanism for Fe II emission.**
I. Furenlid.
Astron. Astrophys., Vol. 140, No. 1, p. 49 – 51 (1984).
Emission lines of Fe II are seen in a large number of emission line objects. It is suggested that in cases where densities and temperatures are relatively low the Fe II line emission can be induced by Lyman line emission. A Lyman line photoionizes a Fe I atom in the ground state and excites the Fe II ion to the upper level of the observed emission transition. It is argued that the most likely mechanism is autoionization, where a Lyman line photon excites a Fe I atom resonantly to an autoionizing state, which decays to an excited Fe II state. Subsequent downward cascading produces the observed Fe II emission lines.

022.089 **On the excitation of CH microwave lines.**
V. Bujarrabal, I. Gonzalo, F. Salinas.
Astrophys. J., Vol. 285, No. 1, p. 312 – 321 (1984).
Along with the very frequent occurrence of every CH 9 cm line inversion, CH radio observations also show the existence of another almost universal characteristic of this line's excitation: the overexcitation of the satellite $F = 0 \rightarrow 1$ line. It is shown here that the overlap of two components of the $^2\Pi_{1/2}$ $J = 3/2 \rightarrow J = 1/2$ rotational transition leads to these satellite–line anomalies. The numerical results are in reasonable agreement with the observations.

022.090 **Isotope shifts of some ultraviolet transitions of first row elements.**
C. W. Clark.
Astrophys. J., Vol. 285, No. 1, p. 322 – 324 (1984).
Attention is drawn to the existence of unusually large isotope shifts in the spectra of first row elements, and a simple explanation for the phenomenon is given. Calculated values of these shifts are presented, which are believed accurate to within 10 – 20%; they agree well with the available experimental data. Some of these shifts could be employed for isotopic abundance studies of interstellar matter, when sufficient resolution in the ultraviolet range becomes available.

022.091 **Selective Λ–doublet population of OH in inelastic collisions with $H_2$: a possible pump mechanism for the $^2\Pi_{1/2}$ astronomical OH maser.**
P. Andresen, D. Hausler, H. W. Lulf.
J. Chem. Phys., Vol. 81, No. 1, p. 571 – 572 (1984). Abstr. in Phys. Abstr., Vol. 87, No. 1243, Entry 96652 (1984).

022.092 **The geometry of the propynol cation, $HC_3O^+$.**
J. D. Goddard.
Chem. Phys. Lett., Vol. 109, No. 2, p. 170 – 174 (1984). Abstr. in Phys. Abstr., Vol. 87, No. 1244, Entry 102602 (1984).

022.093 **The role of stimulated processes in cosmic annihilation–line sources. I.**
V. V. Zheleznyakov, A. A. Litvinchuk.
Sov. Astron., Vol. 28, No. 2, p. 160 – 165 (1984). English translation of 37.022.044.

022.094 **Heat flow through dusty and porous ices.**
R. Smoluchowski.
Bull. Am. Astron. Soc., Vol. 16, No. 3, Part I, p. 643 (1984). Abstract. – See Abstr. 010.061.

022.095 **Collection and analysis techniques for cometary dust.**
B. C. Clark.
Bull. Am. Astron. Soc., Vol. 16, No. 3, Part I, p. 644 (1984). Abstract. – See Abstr. 010.061.

022.096 **Propane absorption at 200K near 748 $cm^{-1}$; implications for interpretations of spectra of Jupiter and Saturn.**
L. P. Giver, F. P. J. Valero, P. Varanasi.
Bull. Am. Astron. Soc., Vol. 16, No. 3, Part I, p. 647 (1984). Abstract. – See Abstr. 010.061.

022.097 **Hydrocarbon production in simulated lightning discharges.**
T. W. Scattergood, C. P. McKay, W. Borucki, J. Kasting, S. Miller.
Bull. Am. Astron. Soc., Vol. 16, No. 3, Part I, p. 648 (1984). Abstract. – See Abstr. 010.061.

022.098 **Thermodynamics of metallic hydrogen mixed with helium, carbon, and oxygen.**
W. B. Hubbard, J. J. MacFarlane.
Bull. Am. Astron. Soc., Vol. 16, No. 3, Part I, p. 649 (1984). Abstract. – See Abstr. 010.061.

022.099 **Complex refractive index of liquid and solid sulfur.**
E. T. Arakawa, R. Wright, B. N. Khare, C. Sagan.
Bull. Am. Astron. Soc., Vol. 16, No. 3, Part I, p. 653 (1984). Abstract. – See Abstr. 010.061.

022.100 **Spectral reflectance (0.39 – 0.70 μm) of carbon–sulfur mixtures.**
A. Clare, J. Gradie.
Bull. Am. Astron. Soc., Vol. 16, No. 3, Part I, p. 653 – 654 (1984). Abstract. – See Abstr. 010.061.

022.101 **Surface photolysis of $SO_2$ and photophysical changes in sulfur may affect Io's surface UV–vis reflectance.**
D. B. Nash, J. I. Moses.
Bull. Am. Astron. Soc., Vol. 16, No. 3, Part I, p. 654 – 655 (1984). Abstract. – See Abstr. 010.061.

022.102 **Time–evolution of infrared absorption features in an oxygen–free Titan organic synthesis experiment.**
B. N. Khare, W. R. Thompson, C. Sagan, M. L. Cooke, J. DiMauro, B. G. J. P. T. Murray, E. T. Arakawa.
Bull. Am. Astron. Soc., Vol. 16, No. 3, Part I, p. 664 – 665 (1984). Abstract. – See Abstr. 010.061.

022.103 **Titan: multiple light scattering by organic tholins and condensates.**
C. Sagan, W. R. Thompson, B. N. Khare, E. T. Arakawa.
Bull. Am. Astron. Soc., Vol. 16, No. 3, Part I, p. 665 (1984). Abstract. – See Abstr. 010.061.

022.104 **Coloration and darkening of $CH_4$ hydrate and other ices upon charged–particle irradiation.**
C. Sagan, B. N. Khare, W. R. Thompson, B. G. J. P. T. Murray, S. W. Squyres.
Bull. Am. Astron. Soc., Vol. 16, No. 3, Part I, p. 668 (1984). Abstract. – See Abstr. 010.061.

022.105 **Spectroscopic data on planetary atmospheric hydrocarbons at relevant temperatures.**
P. Varanasi, S. Chudamani.
Bull. Am. Astron. Soc., Vol. 16, No. 3, Part I, p. 668 (1984). Abstract. – See Abstr. 010.061.

022.106 **Rotational assignments in the 6200 Å band of methane.**
C. de Bergh, J. Brault, G. J. Scherer, K. K. Lehmann, W. Klemperer, A. S. Pine.
Bull. Am. Astron. Soc., Vol. 16, No. 3, Part I, p. 669 (1984). Abstract. – See Abstr. 010.061.

022.107 **The collision–induced translational/rotational absorption spectra of hydrogen and hydrogen–helium mixtures.**
L. Frommhold, W. Meyer, G. Birnbaum.
Bull. Am. Astron. Soc., Vol. 16, No. 3, Part I, p. 669 (1984). Abstract. – See Abstr. 010.061.

022.108 **Improved molecular constants for HD.**
M. E. Mickelson, B. A. Hare, J. T. Trauger.
Bull. Am. Astron. Soc., Vol. 16, No. 3, Part I, p. 669 (1984). Abstract. – See Abstr. 010.061.

022.109 **Developments in ultra–high pressure diamond cell experiments: applications to planetary interiors.**
K. A. Goettel.
Bull. Am. Astron. Soc., Vol. 16, No. 3, Part I, p. 682 (1984). Abstract. – See Abstr. 010.061.

022.110 **Effects of electron bombardment on the light scattering properties of icy planetary surfaces.**
W. D. Smythe.
Bull. Am. Astron. Soc., Vol. 16, No. 3, Part I, p. 684 (1984). Abstract. – See Abstr. 010.061.

022.111 **Determination of mineral abundance from reflectance spectra.**
R. N. Clark, T. L. Roush.
Bull. Am. Astron. Soc., Vol. 16, No. 3, Part I, p. 685 (1984). Abstract. – See Abstr. 010.061.

022.112 **Rheological measurements of ammonia–water melt.**
W. B. McKinnon, M. Meadows.
Bull. Am. Astron. Soc., Vol. 16, No. 3, Part I, p. 686 (1984). Abstract. – See Abstr. 010.061.

022.113 **The effects of viewing geometry on mineral absorption.**
J. Gradie.
Bull. Am. Astron. Soc., Vol. 16, No. 3, Part I, p. 692 (1984). Abstract. – See Abstr. 010.061.

022.114 **Measurements of the microwave opacity and vapor pressure of gaseous sulfuric acid under simulated Venus conditions.**
P. G. Steffes, P. S. Stellitano, R. C. Lott.
Bull. Am. Astron. Soc., Vol. 16, No. 3, Part I, p. 694 (1984). Abstract. – See Abstr. 010.061.

022.115 **Laboratory simulation of asteroidal catastrophic collisions.**
P. Cerroni, R. Bianchi, F. Capaccioni, M. Coradini, E. Flamini, G. Martelli.
Bull. Am. Astron. Soc., Vol. 16, No. 3, Part I, p. 698 – 699 (1984). Abstract. – See Abstr. 010.061.

022.116 **Laboratory measurements of asteroid models with dusty surfaces.**
M. A. Barucci, M. Fulchignoni, J. Piironen.
Bull. Am. Astron. Soc., Vol. 16, No. 3, Part I, p. 699 (1984). Abstract. – See Abstr. 010.061.

022.117 **The effect of the large crater on the asteroids lightcurve.**
M. Fulchignoni, M. A. Barucci.
Bull. Am. Astron. Soc., Vol. 16, No. 3, Part I, p. 699 – 700 (1984). Abstract. – See Abstr. 010.061.

022.118 **Laboratory studies of interplanetary dust: relationships to comets and interstellar clouds.**
A. Fahey, K. D. McKeegan, S. A. Sandford, R. M. Walker, B. Wopenka, E. Zinner.
Bull. Am. Astron. Soc., Vol. 16, No. 3, Part I, p. 704 (1984). Abstract. – See Abstr. 010.061.

022.119 **Solar photo rates for planetary atmospheres and atmospheric pollutants.**
W. F. Huebner, J. J. Keady, S. P. Lyon.
Bull. Am. Astron. Soc., Vol. 16, No. 3, Part I, p. 705 (1984). Abstract. – See Abstr. 010.061.

022.120 **Band model coefficients of methane from 4000 to 9200 $cm^{-1}$, determined at 112, 188, and 294K.**
L. P. Giver, D. C. Benner, R. W. Boese.
Bull. Am. Astron. Soc., Vol. 16, No. 3, Part I, p. 711 – 712 (1984). Abstract. – See Abstr. 010.061.

022.121 **Wavelengths and radiative transition rates for selected lines of 2–, 3– and 4–electron systems for the elements from Ca to Cu.**
J. Hata, I. P. Grant.
Mon. Not. R. Astron. Soc., Vol. 211, No. 3, p. 549 – 557 (1984).

The MCDF–EAL method has been applied to predict the wavelengths and radiative transition rates for the *w*, *x*, *y*, *z* lines of helium–like ions from Ca to Cu (omitting Sc), and for the stronger satellites of the corresponding lithium–like and beryllium–like ions in the range $\lambda = 1.5 - 3$ Å. Some comparisons are made of new data with other methods of calculation and, to a limited extent, with observations.

022.122 **Line intensity ratios of transitions within the $1s^2 2s^2 2p^2$ ground configuration of Mg VII.**
K. M. Aggarwal.
Sol. Phys., Vol. 94, No. 1, p. 75 – 84 (1984).
Using the *R*–matrix approach new calculations have been made for the electron impact excitation of the fine structure transitions within the $1s^2 2s^2 2p^2$ ground configuration of Mg VII. The computations have been made at a large number of energies in order to account for the contribution of resonances. All partial waves with $L \leqslant 9$ are included in the calculations which are considered to be sufficient for the convergence of collision strengths in the energy range below 65 Ry.

022.123 **Theoretical Ne IX line ratios compared to solar observations.**
F. P. Keenan, S. S. Tayal, A. E. Kingston.
Sol. Phys., Vol. 94, No. 1, p. 85 – 89 (1984).
*R*–matrix calculations of the $1^1S$–$2^3S$ and $1^1S$–$2^3P$ electron excitation rates in He–like C V, O VII, and Mg XI by Kingston and Tayal are used to interpolate results for Ne IX. Adoption of these in emission line strength calculations leads to values for the density–sensitive *R* ratio very similar to those of Pradhan et al. and Wolfson et al., although the temperature–sensitive *G* ratios are approximately 10 to 20% lower than those deduced by these authors. However the present theoretical value of *G* at the temperature of maximum Ne IX emission is in excellent agreement with those observed by the SMM and P78–1 satellites for the 1980, November 5 flare and nonflaring active regions.

022.124 **Mg IX and Si XI line ratios in the Sun.**
F. P. Keenan, A. E. Kingston, P. L. Dufton, J. G. Doyle, K. G. Widing.
Sol. Phys., Vol. 94, No. 1, p. 91 – 98 (1984).
New theoretical emission line ratios for the Be–sequence ions Mg IX and Si XI are presented. A comparison with observational data for two solar flares and an active region loop obtained with the Harvard EUV spectrometer and NRL XUV spectroheliograph aboard Skylab reveals that these plasmas are in ionization equilibrium at coronal temperatures. Unfortunately most of the density diagnostics are not particularly useful under solar plasma conditions, as they vary only slightly over the electron density range $10^8$–$10^{13}$cm$^{-3}$. However the Si XI ratio $I(^3P^\circ_2 - ^3P^\circ_2)/I(^3P^\circ_1 - ^1S^\circ_0)$ is density sensitive in the range $10^8$ to $10^{10}$cm$^{-3}$, which is representative of electron densities found in solar active regions or small flares.

022.125 **The Seaton matrix for vibrational transitions in the hydrogen molecule.**
V. K. Khersonskii (*V. K. Khersonskij*).
Bull. Spec. Astrophys. Obs. – North Caucasus, Vol. 17, p. 72 – 73 (1984). English translation of 34.022.174.

022.126 **Bound–bound radiative transitions in C I.**
H. Nussbaumer, P. J. Storey.
Astron. Astrophys., Vol. 140, No. 2, p. 383 – 389 (1984).
For the six energetically lowest configurations of $C^0$, radiative transition probabilities are calculated by two different methods. For the prominent interstellar C I multiplets longwards of 1260 Å, the authors find considerable differences against oscillator strengths derived from astronomical observations. They further investigate a radiative excitation process for the cometary C I λ1931 emission.

022.127 **N I forbidden lines revisited.**
K. Butler, C. J. Zeippen.
Astron. Astrophys., Vol. 141, No. 1, p. 274 – 278 (1984).
Radiative transition probabilities for the astrophysically important forbidden lines between the $2p^3$ energy levels of neutral nitrogen have been calculated using the computer program Superstructure. The necessity for semiempirical term energy corrections is confirmed and the importance of the relativistic corrections to the magnetic dipole transition operator is reemphasized. The implications of the new data for studies of nebulae is discussed briefly.

022.128 **Einstein *A*–values for purely rotational transitions in the HDO–molecule.**
S. Chandra, W. H. Kegel, D. A. Varshalovich.
Astron. Astrophys., Suppl. Ser., Vol. 58, No. 3, p. 687 – 691 (1984).
The authors report the Einstein *A*–values for the electric dipole transitions in th HDO–molecule between the rotational levels of the ground vibrational state up to 1000 cm$^{-1}$. The Einstein *A*–values are used to compute the mean–radiative–lifetimes of the levels.

022.129 **Laboratory measurement of $H_3O^+$ line positions around 10 μm for astrophysical searches.**
P. B. Davies, P. A. Hamilton, S. A. Johnson.
Astron. Astrophys., Vol. 141, No. 2, p. L9 – L10 (1984).
Several transitions of the $1^- \leftarrow O^+$ infrared band of the $H_3O^+$ ion of interest in astronomy have been measured near the band centre at 10 μm. The measurements are more accurate than earlier results and in very good agreement with the predicted spectrum.

022.130 **Ab initio study of the vibronic structure in the $^1\Delta_g$ state of $NH_2{}^+$.**
M. Perić, R. J. Buenker, S. D. Peyerimhoff.
Astrophys. Lett., Vol. 24, No. 2, p. 69 – 73 (1984).
The vibronic levels of the lowest $^1\Delta_g$ state in $NH_2{}^+$ are calculated on the basis of ab initio potential energy curves for the two Renner–Teller component states $^1A_1$ and $^1B_1$. Energies are given for K = 0, 1 – 4; the intensity distribution for transitions in $NH_2{}^+$, $NHD^+$ and $ND_2{}^+$ is predicted to serve as a guideline for future experimental observation.

022.131 **Use of the recent Oxford data to calibrate the Mn I oscillator strengths measured by Warner and Bowell.**
A. J. Booth, D. E. Blackwell, A. D. Petford, M. J. Shallis.
Observatory, Vol. 104, No. 1063, p. 265 – 266 (1984).
A formula is given for calibrating the 610 Mn I oscillator strengths measured in emission by Warner and Bowell (WB) against the more accurate Oxford data. The standard deviation for log *gf*(WB, corrected) – log *gf*(Oxford) is 0.10 dex (26 per cent), compared with the standard deviation of 0.21 dex (62 per cent) for the uncorrected data. The change in absolute scale, resulting chiefly from the use of recent lifetime measurements in the Oxford calibration, is 0.72 dex (× 5.3).

022.132 **Mendeleev's periodic system and the law of galactic distribution of chemical elements.**
Yu. K. Burkov.
Zap. Vses. mineral. o–va, Tom 113, No. 3, p. 258 – 261 (1984). In Russian. Abstr. in Ref. Zh., 51. Astron., 11.51.161 (1984).

022.133 **Long–lived radioactive isotopes in cosmochemistry.**
M. Honda.
The 27th International Geological Congress, Tom 11, Sect. S 11, p. 12 – 19 (1984). In Russian. Abstr. in Ref. Zh., 51. Astron., 11.51.262 (1984). – See Abstr. 012.031.

022.134 **Stark broadening measurements of some N(II) and N(III) lines.**
S. T. Purcell, A. J. Barnard.
J. Quant. Spectrosc. Radiat. Transfer, Vol. 32, No. 3, p. 205 – 209 (1984).
The Stark widths of seventeen N(II) and ten N(III) spectral lines, emitted from a wall–stabilized arc plasma with electron densities in the range $0.9 - 2.9 \times 10^{17}$cm$^{-3}$, were measured.

022.135 **Wavelength coincidence of allowed and intercombination multiplets in N III and O IV: a density and/or temperature diagnostic.**
S. O. Kastner, A. K. Bhatia.
J. Quant. Spectrosc. Radiat. Transfer, Vol. 32, No. 3, p. 249 – 251 (1984).
Two multiplets of N III with different intensity dependence on source conditions fall in the same wavelength interval near 1750 Å, and the analogous multiplets in O IV also lie close to-

gether. It is suggested that the multiplet pair can be useful for diagnostic purposes. The calculated ratios are consistent with available observations. A preliminary inspection of solar flare spectra shows the presence of the higher–density–indicating allowed multiplet in O IV.

022.136 **Collision–strength dependence of solar emission lines of Si(X) and S(XII).**
S. Chandra.
J. Quant. Spectrosc. Radiat. Transfer, Vol. 32, No. 3, p. 253 – 256 (1984).
The author has calculated the normalized populations of 15 levels in Si(X) and S(XII) ions in an optically thin medium, under steady–state condition, by using (1) constant collision strengths computed at the incident electron energy $kT_e$ and (2) incident electron–energy–dependent collision strengths. To calculate intensity ratios for solar emission lines, one may use constant collision strengths at the electron energy $E'$.

022.137 **High–resolution IR spectra of methyleneimine ($CH_2NH$) in the 10 μm region, the $\nu_8$ band.**
G. Duxbury, M. L. Le Lerre.
Infrared Phys., Vol. 24, No. 2/3, p. 261 – 272 (1984). – See Abstr. 012.071.
The IR spectrum of methyleneimine ($CH_2NH$) has been observed in the gas phase with a resolution of 0.0048 $cm^{-1}$ using the KPNO Solar Fourier transform spectrometer. The short–lived $CH_2NH$ was produced in a flow system by the pyrolysis of $CH_3NH_2$ at ca. 1000°C. The origin of the $\nu_8$ band was determined to be 1126.95(6) $cm^{-1}$. Preliminary calculations of the strong Coriolis interactions between the $\nu_7$, $\nu_8$ and $\nu_9$ bands have been made.

022.138 **Spectral properties of ice–particulate mixtures and implications for remote sensing. 1. Intimate mixtures.**
R. N. Clark, P. G. Lucey.
J. Geophys. Res., Vol. 89, No. B7, p. 6341 – 6348 (1984).
The spectral properties of water ice–particulate mixtures are studied for the purpose of deriving the ice and particulate abundances from remotely obtained spectra (particulates referring to nonicy materials in the form of grains). The ice absorption band depths are related to the mean optical path length of photons in ice through Beers law, Fresnel reflection from the ice–crystal faces on the surface, and ice absorption coefficient as a function of wavelength. Laboratory spectra of many ice–particulate mixtures are studied with high–, medium and low–albedo particulates. From the laboratory spectra the band depth, the continuum reflectance, and the ratio of band depth to continuum reflectance are derived for each particulate albedo as a function of the logarithm of the particulate weight fraction in the sample.

022.139 **Oxygen absorption cross sections in the Herzberg continuum and between 206 and 327K.**
H. S. Johnston, M. Paige, F. Yao.
J. Geophys. Res., Vol. 89, No. D7, p. 11661 – 11665 (1984).
The ultraviolet absorption cross sections for molecular oxygen have been determined for wavelengths between 205 and 225 nm, for temperatures between 206 and 327K, and at a pressure from 100 to 750 torr. No temperature effect on the $O_2$ cross sections was observed in the Herzberg continuum. It was necessary to separate the extremely weak $O_2$ absorption from the stronger collision–induced absorption. These experimental data and similar data by others were interpreted in terms of a theoretical expression for the shape of the Herzberg continuum absorption spectrum. The $O_2$ absorption cross sections are in good agreement with the values measured in the stratosphere by Herman and Mentall (1982).

022.140 **The importance of kinetically excited ions in the synthesis of interstellar molecules.**
N. G. Adams, D. Smith, T. J. Millar.
Mon. Not. R. Astron. Soc., Vol. 211, No. 4, p. 857 – 865 (1984).
The authors have presented laboratory data which show that elevated ion–molecule centre–of–mass kinetic energies, $KE_{cm}$, can overcome activation–energy barriers and small endoergicities in ion–molecule reactions. Thus, $KE_{cm}$ can promote some ion–molecule reactions involving $H_2$ which may be important in interstellar clouds. Kinetically–excited ions result from interstellar MHD shocks and therefore many of the reactions discussed in this paper, which are otherwise slow or energetically inhibited, can proceed rapidly under such conditions. The authors have re–investigated $NH_3$ and $H_2S$ synthesis in interstellar clouds but have only been able to reconcile their model predictions with observations when reactions involving N and S atoms are included in the schemes. Use of their experimental data for the rate coefficient at elevated $KE_{cm}$ for the $C^+$–$H_2$ reaction in a model of $CH^+$ production in shocked interstellar gas leads to good agreement between the model predictions and the observations.

022.141 **Laboruntersuchungen zur Deutung der Infrarotspektren des interstellaren Staubes.**
W. Krätschmer.
Kleine Teilchen und Lichtstreuung, p. 1478 – 1490 (1984). – See Abstr. 012.074.

022.142 **Experimentelle Methoden zur Bestimmung der Streufunktionen unregelmäßiger Staubpartikeln.**
R. H. Zerull, R. H. Giese, K. Weiß–Wrana.
Kleine Teilchen und Lichtstreuung, p. 1516 – 1525 (1984). – See Abstr. 012.074.

022.143 **Electron impact excitation of forbidden transitions in Mg VII.**
K. M. Aggarwal.
Astrophys. J., Suppl. Ser., Vol. 56, No. 3, p. 303 – 314 (1984).
The electron impact excitation of forbidden transitions among the $1s^22s^22p^2$, $1s^22s2p^3$, and $1s^22p^4$ configurations of Mg VII has been studied using the $R$–matrix approach. All partial waves with $L \leqslant 9$ are included, and the results are calculated in the $LS$ coupling scheme at 1192 energies in a wide energy range below 65.0 Ry. The entire collision strength spectrum has been averaged over a Maxwellian distribution of electron velocities to get the so–called effective collision strengths. The results are listed at a series of electron temperatures below $5 \times 10^7$K which are suitable for astrophysical applications.

022.144 **Particle acceleration during reconnection in laboratory plasmas.**
R. L. Stenzel, W. Gekelman.
Adv. Space Res., Vol. 4, No. 2 – 3, p. 459 – 470 (1984). – See Abstr. 012.075.
The basic physics of magnetic field line reconnection near magnetic null points is investigated in a large collisionless laboratory plasma. The observations focus on the close vicinity of the neutral sheet, a region which is difficult to resolve in space. Thus, the high resolution laboratory data are complimentary to the more global space observations.

022.145 **Photo cross–sections for stellar atmosphere calculations – compilation of references and data.**
R. Mathisen.
Inst. Theor. Astrophys., Univ. Oslo, Publ. Ser., No. 1, 63 pp. (1984).
Photo cross–section data for continuum absorption and scattering processes of importance to stellar model atmosphere calculations are reviewed. The author presents the results of an evaluation of the most reliable cross–section data available.

022.146 **Comment on "Scaling laws for hydrogen–like atoms in magnetic fields of arbitrary strength".**
H. Herold, H. Ruder, G. Wunner.
Astrophys. J., Vol. 285, No. 2, p. 870 – 871 (1984).
It is shown that the generalization of the scaling laws for hydrogen–like systems in magnetic fields published in *Astrophys. J.* by Wadehra in 1983 (see Abstr. 34.022.057) is false on account of an incorrect treatment of the two–body problem in a magnetic field.

022.147 **Quantal calculations of charge transfer in collisions between N V and atomic hydrogen.**
C. A. Feickert, R. J. Blint, G. T. Surratt, W. D. Watson.
Astrophys. J., Vol. 286, No. 1, p. 371 – 376 (1984).
Calculations are presented for the cross section and rate coefficient for charge transfer in collisions between N V ions and hydrogen atoms. Kinetic temperatures from 30K to $10^5$K are considered. Molecular orbital and quantum mechanical scattering techniques are utilized. A noteworthy aspect of the N V – H charge transfer is the possibility for competition among final states of N IV which are excited.

022.148 **Low–temperature dissociative recombination of $e+H_3^+$.**
H. H. Michels, R. H. Hobbs.
Astrophys. J., Lett. Ed., Vol. 286, No. 1, p. L27 – L29 (1984).
Accurate ab initio calculations of the ground and excited $H_3$ potential energy surfaces have been carried out. These surfaces have been examined at geometries appropriate for an analysis of the reaction products of $e+H_3^+$ dissociative recombination. It is found that direct recombination of electrons with $H_3^+$ ions in their ground vibrational state is very unlikely for low collisional energies. The authors estimate that the currently accepted values of the $e+H_3^+$ recombination coefficient for interstellar conditions are two orders of magnitude too large.

022.149 **On the grain size dependence of the behaviour of fine magnetic particles in rocks.**
M. Fuller.
Geophys. Surv., Vol. 7, No. 1, p. 75 – 87 (1984). – See Abstr. 012.082.
The grain size dependence of the ratio of saturation remanent magnetization to saturation magnetization ($J_R$:$J_S$), weak field susceptibility ($X_0$), thermoremanent magnetization (TRM) and its stability against AF demagnetization are interpreted in terms of nucleation theory. It is concluded that each of these parameters exhibits grain size dependence due to two effects. The first is the increasing difficulty with which domain walls are nucleated as grain size decreases. The second is an intrinsic grain size dependence of the parameters in multidomain particles.

022.150 **Cross sections for electron–impact excitation of the helium atom.**
Yu. M. Smirnov.
Astron. Zh., Tom 61, Vyp. 6, p. 1087 – 1097 (1984). In Russian. English translation in Sov. Astron., Vol. 28, No. 6.
Helium atom excitation by electron impact was studied in detail. In the spectral region 260 – 800 nm the cross sections of 73 transitions of He I, belonging to the six main spectral series of the helium atom, were measured. Varying the electron energy from the threshold up to 250 eV the energy dependences for all the cross sections were measured. The character of variation of the values of the excitation cross sections depending on the quantum number of the upper level, following a power law, is discussed.

022.151 **Excitation of rotational levels of $H_2O$ by electron impact.**
A. M. Sobolev, S. V. Makarov.
Astron. Zh., Tom 61, Vyp. 6, p. 1228 – 1229 (1984). In Russian. English translation in Sov. Astron., Vol. 28, No. 6.
A simple formula for the calculation of excitation coefficients of rotational levels of $H_2O$ by electron impact is derived. The formula was obtained by averaging quantum–mechanical cross–sections of Itikawa over a Maxwellian velocity distribution.

022.152 **Model–potential oscillator strengths for $4f^n6s^2$–$4f^n6s6p$ transitions in Sm(I), Eu(I), Dy(I), Er(I) and Yb(I).**
J. Migdałek, R. Marcinek.
J. Quant. Spectrosc. Radiat. Transfer, Vol. 32, No. 4, p. 269 – 277 (1984).

022.153 **Prediction and assignment of the FIR spectrum of hydrogen peroxide.**
P. Helminger, J. K. Messer, W. C. Bowman, F. C. De Lucia.
J. Quant. Spectrosc. Radiat. Transfer, Vol. 32, No. 4, p. 325 – 333 (1984).

022.154 **Measurements of Stark broadening of He(II) 4686 Å.**
Y. Arata, S. Miyake, H. Matsuoka.
J. Quant. Spectrosc. Radiat. Transfer, Vol. 32, No. 4, p. 343 – 351 (1984).
Stark broadening of the He(II) 4686 Å line emitted from a high–current arc source preionized by CW microwave radiation is described for a two–dimensional multichannel analyzer. Comparisons of the experimental profiles with applicable theory are discussed.

022.155 **Application of the invariance principle to problems of gamma–radiation transfer in crystals.**
A. R. Mkrtchyan, R. G. Gabrielyan.
Astrofizika, Tom 20, Vyp. 3, p. 607 – 614 (1984). In Russian. English translation in Astrophysics, Vol. 20, No. 3.
The possibility of the application of Ambartsumyan's invariance principle to solve the problem of gamma–radiation transfer in crystals with Mössbauer isotopes is considered in the case when nuclear resonance and Rayleigh scattering take place.

022.156 **log(*gf*) for singly–ionized elements of the iron group.**
L. A. Milone, A. A. E. Milone.
Astrophys. Space Sci., Vol. 107, No. 2, p. 303 – 312 (1984).
Moderately accurate oscillator strengths are obtained by application of systematic corrections to Warner's–log(*gf*) for the once ionized elements of the iron group. Corrections are derived by comparison of Warner's values with those of Kurucz–Peytremann (Sc II, Ti II, V II, Cr II, Mn II, Co II, and Ni II) and Kurucz (Fe II). Using Warner's corrected log(*gf*) for these elements, the authors have shown in a rather qualitative way that abundances consistent with the values accepted at present are found in the solar photosphere.

022.157 **On the time of achievement of ionization equilibrium in a hot low–density gas.**
K. A. Sidorov.
Astrofizika, Tom 21, Vyp. 2, p. 353 – 359 (1984). In Russian. English translation in Astrophysics, Vol. 21, No. 2.
The time of achievement of ionization equilibrium in a hot low–density gas was calculated. Substantial deviations from the ionization equilibrium are expected for solar flares and for supernova remnants. A moving galaxy in clusters of galaxies has to leave the track with non–equilibrium ionization, which exists longer than $10^8$ – $10^9$ years.

022.158 **Oscillator strengths for Al IV to Ar IX and revised energy levels for Ar IX.**
B. C. Fawcett.
Phys. Scr., Vol. 30, No. 5, p. 326 – 334 (1984).
Weighted oscillator strengths are computed for ions in the isoelectronic sequence between Al IV and Ar IX. They were computed with a program–suite which uses the Hartree–Fock–Relativistic and Blume–Watson methods and takes into account configuration interaction. Slater parameters were optimised based on empirical data. Values of energy levels of Ar IX are revised.

022.159 **The EUV spectrum of $H_2O$ by electron impact.**
J. M. Ajello.
Geophys. Res. Lett., Vol. 11, No. 12, p. 1195 – 1198 (1984).
The author presents the vacuum ultraviolet (VUV) spectrum of $H_2O$ produced by electron impact at 200 eV. A total of 25 spectral features is identified at a resolution of 0.5 nm over the wavelength range of 40 to 280 nm. Absolute emission cross sections were measured for each of these features.

022.160 **New in-situ measurements of the absorption cross-sections of $O_2$ in the Herzberg continuum.**
M. Pirre, P. Rigaud, D. Huguenin.
Geophys. Res. Lett., Vol. 11, No. 12, p. 1199 – 1202 (1984).

The absorption cross-sections of both $O_2$, from 200 to 220 nm, and $O_3$ from 220 to 235 nm have been deduced from measurements of the atmospheric absorption performed at night aboard a stratospheric balloon. The $O_2$ and $O_3$ cross sections have been found to meet most of the laboratory measurements. From 200 to 210 nm the results concerning the $O_2$ cross-sections do not confirm the low values measured recently in the stratosphere and in the laboratory.

022.161 **Oscillator strengths of the $C_2$ $A\,^1\Pi_u - X\,^1\Sigma_g^+$ Phillips system.**
S. P. Davis, W. H. Smith, J. W. Brault, R. Pecyner, J. Wagner.
Astrophys. J., Vol. 287, No. 1, p. 455 – 460 (1984).

The authors have made a comparison between oscillator strengths (*f*-values) of lines in the $C_2$ Swan (0 – 0) band and the Phillips (1 – 0) and (0 – 0) bands, by means of absorption measurements in a carbon tube King-type high-temperature furnace. For these bands, the ratio of values $f_{00}$ Swan/$f_{10}$ Phillips is $14.8 \pm 0.3$, and the ratio $f_{00}$ Swan/$f_{00}$ Phillips is $18.2 \pm 0.7$.

022.162 **Infrared spectrum of the laboratory-synthesized quenched carbonaceous composite (QCC): comparison with the infrared unidentified emission bands.**
A. Sakata, S. Wada, T. Tanabé, T. Onaka.
Astrophys. J., Lett. Ed., Vol. 287, No. 1, p. L51 – L54 (1984).

Infrared spectra (2.5 – 25 μm) of the experimentally synthesized quenched carbonaceous composite (QCC), which has been shown to be a likely candidate for the interstellar 220 nm extinction hump, are presented. QCC is shown to have infrared bands at 3.29, 3.42, 3.48, 6.25, 6.94, 7.27, 11.40, 11.96, and 13.24 μm. The spectrum taken after heating the QCC sample to 773K shows that only the 6.25 and 11.40 μm features survive and the other features almost disappear at that temperature. The infrared absorption features in QCC are in good agreement with six of the nine infrared unidentified emission bands observed in celestial objects.

022.163 **Refractive indices for amorphous carbon.**
W. W. Duley.
Astrophys. J., Vol. 287, No. 2, p. 694 – 696 (1984).

Refractive indices *n* and *k* have been measured over the range $1.5 \leqslant \lambda^{-1} \leqslant 8.5\ \mu m^{-1}$ for an amorphous carbon solid formed by direct condensation from a gas containing C, $C_2$, and $C_3$ as well as trace quantities of linear polyyne molecules. Small particles of this material will exhibit a high albedo in the extreme ultraviolet as required by observations of VUV scattering. In hydrogenated form amorphous carbon particles have significant IR absorption at 3.3 – 3.4 μm.

022.164 **On the establishment of internally consistent solar scales of oscillator strengths and abundances of chemical elements. VII. Mn I.**
Eh. A. Gurtovenko, R. I. Kostyk, T. V. Orlova.
Soln. Dannye, Byull., 1984, No. 9, p. 85 – 87 (1984). In Russian.

Oscillator strengths of 30 Mn I lines are determined up to $\pm(0.06 - 0.07)$ dex from the equivalent widths of solar Fraunhofer lines.

**CRC handbook of high resolution infrared laboratory spectra of atmospheric interest.**
See Abstr. 002.081.

**Physics of gravitating systems. I. Equilibrium and stability.**
See Abstr. 003.008.

**Physics of gravitating systems. II. Nonlinear collective processes: nonlinear waves, solitons, collisionless shocks, turbulence. Astrophysical applications.**
See Abstr. 003.009.

**Understanding relativity. Origin and impact of a scientific revolution.**
See Abstr. 003.025.

**Relativity and engineering.**
See Abstr. 003.032.

**Relativity and its roots.**
See Abstr. 003.102.

**Fission products of transuranian elements in space.**
See Abstr. 003.152.

**Subatomic physics: nuclei and particles.**
See Abstr. 003.156.

**Time-resolved FTS of molecular and atomic IR emission.**
See Abstr. 034.165.

**Cerenkov line emission: basic theory.**
See Abstr. 063.058.

**The depletion of pre-interstellar matter and an examination of Ni II oscillator strengths.**
See Abstr. 112.076.

**Selective excitation of Fe II in the laboratory and late-type stellar atmospheres.**
See Abstr. 114.015.

**Remarks on a vibronic origin for the diffuse band spectrum.**
See Abstr. 131.054.

# Astronomical Instruments and Techniques

## 031 Astronomical Optics

031.001 **Investigating atmospheric distortions of a wave front by the Hartmann method.**
V. F. Matyukhin, L. I. Snezhko, A. L. Shapovalov.
Astron. Tsirk., No. 1286, p. 4 – 6 (1983). In Russian.

031.002 **An automatic Hartmann measuring device.**
K. L. Maslennikov.
Astron. Tsirk., No. 1295, p. 6 – 8 (1983). In Russian.

031.003 **Study of an astronomical lens with the photoelectric Hartmann method.**
A. I. Beslik, M. A. Konstantinova, K. V. Kuimov, E. S. Safonova, P. V. Shcheglov.
Tr. Gos. Astron. Inst. Shternberg, Tom 55, p. 21 – 29 (1983). In Russian.

A method of investigating the angular aberration of a lens is described. Monochromatic angular aberration curves for a 20 cm 4–lens system are measured.

031.004 **The modified Couder telescope.**
R. V. Willstrop.
Mon. Not. R. Astron. Soc., Vol. 209, No. 3, p. 587 – 606 (1984).

The optical quality of the Couder telescope has recently been shown to be comparable to that of the Schmidt camera, giving images smaller than 0.5 arcsec in diameter at f/3 over a field 3° in diameter in its original form, and if its design is modified the aperture can be increased to f/2.5 (or even f/2) with images under 0.4 arcsec over 3°.6. The present paper describes the method of designing an f/1.5 modified Couder telescope that gives images smaller than 1 arcsec over a field of 2°. Other details of this and the f/2 and f/2.5 systems are discussed, including the sensitivity of the image quality to misalignment and the properties of the system in terms of Burch's method of plate–diagram analysis.

031.005 **Adaptive optical systems for LEST.**
O. von der Lühe.
LEST Found., Tech. Rep., No. 2, 36 pp. (1983).

Some fundamental properties of adaptive optical systems and some basic design considerations for such a system on LEST have been discussed.

031.006 **Point spread function for a segmented mirror system.**
S. T. Smith.
Proc. SPIE Int. Soc. Opt. Eng., Vol. 383, p. 32 – 35 (1983). Abstr. in Phys. Abstr., Vol. 87, No. 1238, Entry 74433 (1984). – See Abstr. 012.014.

031.007 **Mirror segment alignment using a multiple–order radial–grating shearing interferometer.**
P. D. Henshaw.
Proc. SPIE Int. Soc. Opt. Eng., Vol. 383, p. 38 – 43 (1983). Abstr. in Phys. Abstr., Vol. 87, No. 1238, Entry 74642 (1984). – See Abstr. 012.014.

031.008 **Deployable reflector configurations.**
A. B. Meinel, M. P. Meinel.
Proc. SPIE Int. Soc. Opt. Eng., Vol. 383, p. 2 – 10 (1983). Abstr. in Phys. Abstr., Vol. 87, No. 1238, Entry 77707 (1984). – See Abstr. 012.014.

031.009 **Optical ray tracing in finite element models.**
N. A. Roy, W. B. Haile.
Proc. SPIE Int. Soc. Opt. Eng., Vol. 450, p. 168 – 175 (1984). Abstr. in Phys. Abstr., Vol. 87, No. 1241, Entry 87425 (1984). – See Abstr. 012.019.

031.010 **Beam jitter transmission and correction.**
D. D. Durante, C. G. Hunter.
Proc. SPIE Int. Soc. Opt. Eng., Vol. 450, p. 176 – 181 (1984). Abstr. in Phys. Abstr., Vol. 87, No. 1241, Entry 87426 (1984). – See Abstr. 012.019.

031.011 **Temporal characteristics of an adaptive astronomical system.**
V. V. Vojtsekhovich.
Inst. kosm. issled. Akad. Nauk SSSR. Prepr., No. 873, 24 pp. (1984). In Russian. Abstr. in Ref. Zh., 51. Astron., 8.51.653 (1984).

031.012 **Modal compensation of turbulent atmospheric distortions and isoplanatism.**
Inst. kosm. issled. Akad. Nauk SSSR. Prepr., No. 881, 22 pp. (1984). In Russian. Abstr. in Ref. Zh., 51. Astron., 8.51.654 (1984).

031.013 **Steps toward eight–meter honeycomb mirror blanks. I. Rationale and approach.**
J. R. P. Angel, N. J. Woolf.
Ann. N.Y. Acad. Sci., Vol. 422, p. 163 – 170 (1984). – See Abstr. 012.033.

031.014 **The different concepts of resolution.**
A. Alvim De Rezende Chaves.
Atti Fond. Giorgio Ronchi, Vol. 38, No. 5 – 6, p. 606 – 610 (1983). Abstr. in Phys. Abstr., Vol. 87, No. 1243, Entry 96809 (1984).

031.015 **A synthetic aperture phase measurement system using a metering rod bridge with corner cubes.**
K. Sun, T. Waite, K. Kissell, F. White.
Proc. SPIE Int. Soc. Opt. Eng., Vol. 440, p. 40 – 49 (1984). Abstr. in Phys. Abstr., Vol. 87, No. 1243, Entry 97016 (1984). – See Abstr. 012.042.

031.016 **The testing of large telescope systems using multiple subaperture.**
T. W. Stuhlinger.
Proc. SPIE Int. Soc. Opt. Eng., Vol. 440, p. 91 – 98 (1984). Abstr. in Phys. Abstr., Vol. 87, No. 1243, Entry 97021 (1984). – See Abstr. 012.042.

031.017 **Off–axis parabola fabrication at Kitt Peak National Observatory.**
L. D. Barr, P. E. Gillett, K. L. Shu.
Proc. SPIE Int. Soc. Opt. Eng., Vol. 433, p. 114 – 122 (1983). Abstr. in Phys. Abstr., Vol. 87, No. 1244, Entry 103028 (1984). – See Abstr. 012.049.

031.018 **Assessment of a large aperture mirror by the Hartmann test.**
J. R. Yang, K. H. Hong, J. D. Park.
New Phys., Korean Phys. Soc., Vol. 24, No. 1, p. 54 – 59 (1984.). In Korean. Abstr. in Phys. Abstr., Vol. 87, No. 1246, Entry 115688 (1984).

031.019 **Aspheric plate corrector for prime focus.**
L. Shao.
Astron. Astrophys., Vol. 141, No. 2, p. 401 – 406 (1984).
This paper deals with two kinds of aspheric plate corrector for prime focus. The first is a four–element corrector consisting of three aspheric plates and a spherical lens. The other is a three–element corrector consisting of only three aspheric plates. Compared with a spherical lens corrector, these systems will give a better aberration correction covering a wide field and a broader spectral region.

031.020 **Steps towards 8 m honeycomb mirrors. V. A method for polishing aspheres as fast as f/1.**
J. R. P. Angel.
Very large telescopes, their instrumentation and programs, p. 11 – 21 (1984). – See Abstr. 012.064 (IAU Colloq. No. 79).
A method is proposed for polishing fast aspherics with a lap whose shape is continuously changed under computer control as it moves over a rigid mirror blank. The required changes of radius, astigmatism and coma in a circular lap are made with edge bending levers and tensioning members with screw actuators. This method of bending has been demonstrated in the laboratory.

031.021 **ESO's New Technology Telescope (NTT) metallic primary mirror project. Status report.**
K. N. Mischung.
Very large telescopes, their instrumentation and programs, p. 57 – 66 (1984). – See Abstr. 012.064 (IAU Colloq. No. 79).
ESO has investigated the technological feasibility of a metallic primary mirror for the NTT. A brief status report is given indicating that a nickel coated aluminum or aluminum–alloy mirror blank can be manufactured within 15 – 18 months at costs well below those for a glass or ceramic material of identical shape and dimensions.

031.022 **Optical wavefront analysis of thermally cycled 500 mm metallic mirrors.**
L. Noethe, F. Franza, P. Giordano, R. N. Wilson.
Very large telescopes, their instrumentation and programs, p. 67 – 74 (1984). – See Abstr. 012.064 (IAU Colloq. No. 79).
Warping is a major concern in the use of aluminum as a material for large mirrors. The warping characteristics are influenced by (1) the type of material, i.e. pure aluminum or some alloy, (2) the type of fabrication, i.e. cast, rolled or forged. So far there is little data available on the warping characteristics of aluminum mirrors. Therefore ESO has tested the warping behaviour of 18 aluminum mirrors under thermal cycling conditions with a glass ceramic mirror, which has not been cycled, as a reference mirror.

031.023 **A finite element approach to the design of the support system for the ESO 1 m active optics experiment.**
G. Ballio, R. Contro, C. Poggi, O. Citterio.
Very large telescopes, their instrumentation and programs, p. 75 – 93 (1984). – See Abstr. 012.064 (IAU Colloq. No. 79).

031.024 **Mirrors: fabrication, processing, testing and tolerances.**
A. B. Meinel, M. P. Meinel, R. R. Shannon, R. E. Parks.
Very large telescopes, their instrumentation and programs, p. 109 – 117 (1984). – See Abstr. 012.064 (IAU Colloq. No. 79).

031.025 **The ESO off–line telescope testing technique illustrated with results for the MPIA 2.2 m telescope II.**
R. N. Wilson, F. Franza, L. Noethe, M. Tarenghi.
Very large telescopes, their instrumentation and programs, p. 119 – 130 (1984). – See Abstr. 012.064 (IAU Colloq. No. 79).
The ESO test device is based on the proposal of Shack (1971). The method is derived from the conventional Hartmann technique.

031.026 **Mirror compensator for testing convex secondary.**
J. Pan.
Very large telescopes, their instrumentation and programs, p. 131 – 133 (1984). – See Abstr. 012.064 (IAU Colloq. No. 79).
The author describes a fabrication testing technique for the secondary of China's 2.16 m telescope. The method is similar to the Hindle concept but uses a concave ellipsoid mirror as aberration compensator.

031.027 **How to quantify ripple.**
H. Geib, C. Kühne, E. Morgenbrod.
Very large telescopes, their instrumentation and programs, p. 135 – 146 (1984). – See Abstr. 012.064 (IAU Colloq. No. 79).
The polishing process may not only lead to large–area surface errors, such as astigmatism or spherical aberration, but also to short–period errors. In particular, this applies to aspherical surfaces because they require polishing tools of small dimensions. These errors are not entirely irregular and may therefore not be treated as statistical errors; nor are they sufficiently regular to be described in terms of amplitudes of Zernike polynomials. The description by means of Zernike polynomials is based on the assumption of a two–dimensional regularity which simply does not exist in practice. The authors omitted the two–dimensional aspect altogether when trying to find a method of making these small–area errors – which are called "ripple" – accessible to quantitative description.

031.028 **Optical shop testing of liquid mirrors.**
E. F. Borra, M. Beauchemin, R. Arsenault, R. Lalande.
Very large telescopes, their instrumentation and programs, p. 147 – 156 (1984). – See Abstr. 012.064 (IAU Colloq. No. 79).
Borra (1982) has argued that it is scientifically useful and technically feasible to build very large optical telescopes (diameters over 15 meters) having as primary mirror a rotating container filled with mercury. The authors report here the results of various optical tests on prototype liquid mirrors. The results are very encouraging and seem to indicate that large liquid mirrors are feasible.

031.029 **Adaptive optics.**
N. J. Woolf.
Very large telescopes, their instrumentation and programs, p. 221 – 233 (1984). – See Abstr. 012.064 (IAU Colloq. No. 79).
Adaptive optics is a technique for correcting atmospheric wavefront disturbances to yield diffraction limited imaging. It is a technique whose advantages are most apparent in the 2 – 5 μ spectral region, where wavefront corrections are derived from study of visible objects. Graphs are presented to show performance needs of adaptive optics systems.

031.030 **The Mersenne–Schmidt telescope.**
R. V. Willstrop.
Very large telescopes, their instrumentation and programs, p. 395 – 404 (1984). – See Abstr. 012.064 (IAU Colloq. No. 79).
A three–mirror telescope is described which covers a field 4° in diameter at f/1.6 with images smaller than 0.31 arcsec total spread (0.17 arcsec r.m.s.); an alternative version, also f/1.6, covers a field 2° in diameter with images smaller than 0.049 arcsec (0.023 arcsec r.m.s.). Both forms of this telescope are fully baffled against sky fog contamination, the illumination is uniform over the central 2° field and decreases by about 13 per cent at the edge of the 4° field, and the tube length is little more than the focal length. The insertion of a plane parallel filter in such steeply converging light introduces spherical aberration and coma, but

these aberrations can be corrected adequately by adjustments to the mirror spacings.

031.031 **A modular structural design for a spectrographical MMT.**
R. Freire Ferrero, M. Chevreton, P. Felenbok.
Very large telescopes, their instrumentation and programs, p. 405 – 414 (1984). – See Abstr. 012.064 (IAU Colloq. No. 79).

The authors present some ideas about a MMT for high resolution spectroscopy purposes. The alternative solution is intended both in technical conception and in economical costs. This possible project considers the use of a system of 36 mirrors about 1.5 meter in diameter, collecting light independently of each other and transmitting the light down to the slit spectrograph by optical fibers.

031.032 **Advanced wide–field broad–passband refracting field correctors for large telescopes.**
H. W. Epps, J. R. P. Angel, E. Anderson.
Very large telescopes, their instrumentation and programs, p. 519 – 548 (1984). – See Abstr. 012.064 (IAU Colloq. No. 79).

A preliminary 30–arcmin prime focus (f/2.0) refracting field corrector system for the University of California Ten–Meter Telescope (UC TMT) is presented which features 1/4–arcsec images containing more than 80% of the energy, over limited passbands within the wavelength range λ3300 Å to λ1.0 μ. A compact 40–arcmin internal Cassegrain (f/1.75 hyperbola to f/5.0) broad–passband (λ3300 Å to λ1.0 μ) corrector, suitable for imaging and multi–object spectroscopy at the UC TMT, is presented which features 1/4–arcsec images containing more than 90% of the energy when averaged over field angle and color. Three 60–arcmin external Cassegrain correctors for 300–inch f/1.8 and f/2.0 parabolic primary mirrors are presented which are suitable for a Fifteen–Meter NNTT/MMT. Image quality is comparable to the UC TMT Cassegrain corrector.

031.033 **Refractive vs reflective correctors.**
E. H. Richardson, C. L. Morbey.
Very large telescopes, their instrumentation and programs, p. 549 – 567 (1984). – See Abstr. 012.064 (IAU Colloq. No. 79).

Improved designs of refractive correctors produce excellent images with fast telescopes such as those with an f/1.5 prime focus and f/3.5 secondary focus. The fields are flat and there is compensation for the chromatic effect caused by windows. Reflective correctors can produce diffraction limited images at all wavelengths and the mirrors can be supported across their backs as well as at their edges. The refractive corrector is preferable at fast foci. A good combination is a Ritchey–Chrétien (R–C) telescope with refractive correctors at the fast prime and secondary foci, and a reflective corrector–magnifier for the slow infrared focus.

031.034 **Optimization program by using generalized bending method.**
K.–c. Chen, M.–w. Chang.
J. Opt. Soc. Am. A, Vol. 1, No. 12, p. 1219 (1984). Abstract.

031.035 **Optimization of constrained optical systems.**
J. L. Rayces, L. Lebich.
J. Opt. Soc. Am. A, Vol. 1, No. 12, p. 1219 (1984). Abstract.

031.036 **Caustic merit function for lens design.**
A. M. Kassim, D. L. Shealy.
J. Opt. Soc. Am. A, Vol. 1, No. 12, p. 1219 (1984). Abstract.

031.037 **Method for defining tolerances in a millimeter–wavelength telescope.**
R. E. Cofield, S. T. Smith.
J. Opt. Soc. Am. A, Vol. 1, No. 12, p. 1226 (1984). Abstract.

031.038 **Coma corrector for a classical Cassegrain telescope.**
D. Shafer.
J. Opt. Soc. Am. A, Vol. 1, No. 12, p. 1226 (1984). Abstract.

031.039 **Correcting secondary spectrum in infrared optical systems with materials having normal dispersion characteristics.**
V. Doherty, M. Hatch.
J. Opt. Soc. Am. A, Vol. 1, No. 12, p. 1226 (1984). Abstract.

031.040 **Optical design of a catadioptric lens with aberrations balanced by aspherizing one surface.**
K. M. Bystricky, P. R. Yoder Jr..
J. Opt. Soc. Am. A, Vol. 1, No. 12, p. 1237 (1984). Abstract.

031.041 **Nonstandard representations of aspheric surfaces in optical design.**
J. M. Rodgers.
J. Opt. Soc. Am. A, Vol. 1, No. 12, p. 1237 (1984). Abstract.

031.042 **Large telescope with spherical primary mirror.**
D. Shafer.
J. Opt. Soc. Am. A, Vol. 1, No. 12, p. 1238 (1984). Abstract.

031.043 **Optical engineering aspects of the University of Texas 7.6–meter telescope.**
A. B. Meinel, M. P. Meinel, R. G. Tull.
J. Opt. Soc. Am. A, Vol. 1, No. 12, p. 1249 (1984). Abstract.

031.044 **Spin casting an *f*/1 honeycomb sandwich glass mirror blank of 1.8–meter diameter.**
J. R. P. Angel, L. Goble.
J. Opt. Soc. Am. A, Vol. 1, No. 12, p. 1250 (1984). Abstract.

031.045 **Comparison of various grazing incidence spectrometer designs based on conical diffraction.**
W. Werner.
Appl. Opt., Vol. 23, No. 14, p. 2408 – 2411 (1984).

A spectrometer design covering the complete beam collected by a telescope system of full revolution was compared with designs covering only a sector of 25% out of the same telescope. The author found that the effective area of one sector is practically equal to that of the full beam system. This is due to the difference in polarization effects and to the smaller average value for the grazing angle of the sector systems. He also compared the throughput of a specified full beam system found by the author's calculations with the throughput published by other authors for the same system. Deviations have been observed ranging from a factor of 0.5 to a factor of 10.

031.046 **Hyperbolic primary telescope configuration.**
A. B. Meinel, M. P. Meinel, D.–q. Su.
Appl. Opt., Vol. 23, No. 15, p. 2469 – 2471 (1984).

The trend toward fast primary focal ratios for astronomical telescopes, driven by the reduced cost that results from a reduced telescope length, has accentuated the problem of achieving a wide field of excellent imagery and focal ratios that match detector characteristics. The authors discuss some of the designs considered under the University of Texas 7.6–m telescope studies. The basic focal ratio of the primary is *f*/2.0.

031.047 **Aberrations of an IR chopping secondary.**
A. B. Meinel, M. P. Meinel.
Appl. Opt., Vol. 23, No. 16, p. 2675 – 2676 (1984).

031.048 **Design and analysis of a two–channel three–mirror X–ray telescope.**
A. M. Kassim, D. L. Shealy.
Appl. Opt., Vol. 23, No. 16, p. 2697 – 2701 (1984). With a correction in Appl. Opt., Vol. 23, No. 19, p. 3482 – 3486 (1984).

Alternate configurations for a two–channel three–mirror glancing incidence X–ray telescope have been designed and evaluated. A Wolter–Schwarzschild type I telescope is used as a base system for one channel of operation. The second channel consists of the secondary mirror of the Wolter–Schwarzschild type I telescope used as a primary mirror and a third aspheric mirror as a secondary for the inner channel. Two alternate aspheric mirror desgins for the secondary of the inner channel were produced, one corrected for spherical aberration and the other corrected for coma. A ray–tracing technique was used to calculate the rms blur circle radius on the focal plane for both channels and each configuration.

031.049 **Asymmetric spline surfaces: characteristics and applications.**
J. E. Stacy.
Appl. Opt., Vol. 23, No. 16, p. 2710 – 2714 (1984).
Asymmetric spline surfaces appear useful for the design of high–quality general optical systems (systems without symmetries). A spline influence function defined as the actual surface resulting from a simple perturbation in the spline definition array shows that a subarea is independent of others four or more points away. Optimization methods presented in this paper are used to vary a reflective spline surface near the focal plane of a decentered Schmidt–Cassegrain to reduce rms spot radii by a factor of 3 across the field.

031.050 **Simple technique for obtaining approximate thermal expansion coefficients of low expansion materials: applications to Zerodur.**
H. E. Bennett, J. J. Shaffer, D. K. Burge.
Appl. Opt., Vol. 23, No. 16, p. 2734 – 2737 (1984).

031.051 **Four–mirror spherical–primary submillimeter telescope design.**
A. B. Meinel, M. P. Meinel, D.–q. Su, Y.–n. Wang.
Appl. Opt., Vol. 23, No. 17, p. 3020 – 3023 (1984).
A four–mirror Nasmyth–focus configuration with a 3–min of arc field of diffraction–limited imagery has been developed utilizing an *f*/0.6 spherical–primary mirror. Limitations were found when a traditional polynomial was used to describe the aspheric surfaces. To get the desired imagery, eleven aspheric terms were necessary. Exponential series were explored, but they gave no significant improvement.

031.052 **Glancing incidence EUV telescopes: analysis and comparison.**
P. M. Davila, T. T. Saha.
Appl. Opt., Vol. 23, No. 20, p. 3660 – 3666 (1984).
Steps in the derivation of generalized surface equations describing either a Wolter or Wolter–Schwarzschild type II extreme ultraviolet telescope are outlined. These surface equations are incorporated in an optical design program where first–order design parameters are varied to produce a number of different series of type II telescopes. These series are then ray traced to compare and analyze the geometric image quality for different designs in terms of minimum rms angular blur and field curvature. Results of an analysis illustrating the dependence of effective area on first–order design parameters are also presented. The differences and similarities between equivalent Wolter and Wolter–Schwarzschild designs are discussed.

031.053 **Fringe scanning Ronchi test for aspherical surfaces.**
T. Yatagai.
Appl. Opt., Vol. 23, No. 20, p. 3676 – 3679 (1984).
A method is proposed for measuring the wave front aberration of a fast aspherical surface. The combination of the Ronchi test and the synchronous phase detection method provides a simple but powerful tool for aspherical surface evaluation. In a common setup for the Ronchi test, translating the Ronchi ruling sideways causes the irradiance of the Ronchigram to vary periodically with time. This enables to make synchronous phase detection in the Ronchi test. The procedure for Ronchigram analysis is presented, and experimental results are shown.

031.054 **Grazing incidence telescopes: a new class for soft X–ray and EUV spectroscopy.**
M. C. Hettrick, S. Bowyer.
Appl. Opt., Vol. 23, No. 21, p. 3732 – 3735 (1984).

031.055 **Thermal expansion uniformity of materials for large telescope mirrors.**
S. F. Jacobs, D. Shough, C. Connors.
Appl. Opt., Vol. 23, No. 23, p. 4237 – 4244 (1984).

031.056 **Aspheric concave grating spectrographs.**
W. C. Cash Jr.
Appl. Opt., Vol. 23, No. 24, p. 4518 – 4522 (1984).
Modifications to the surface of ellipsoidal gratings with straight rulings have been investigated. The author finds that with the addition of a term proportional to $wl^2$ the stigmatic Rowland circle geometry can be greatly improved. He also shows that with the addition of a term proportional to $w^3$, comatic aberration can be corrected, and geometries outside the Rowland circle become well behaved.

031.057 **Elementi disperdenti.**
G. Tondello.
G. Astron., Vol. 10, N. 1, p. 5 – 28 (1984).

031.058 **Efficient modification of the compensation method of astronomical mirror control.**
N. N. Fashchevskij.
Astron. Tsirk., No. 1315, p. 5 – 7 (1984). In Russian.

031.059 **Interferometric method for optical testing and wavefront error sensing.**
T. K. Korhonen.
Proc. SPIE Int. Soc. Opt. Eng., Vol. 444, p. 249 – 252 (1984). = Turku Univ. Obs., Informo No. 89.
A new interferometric method for optical testing has been developed. The test arrangement is similar to that of the Hartmann method using a screen with a rectangular grid of holes. The detection takes place close to the focus where the Hartmann images are partly overlapping and interfering. The positions of the interference maxima contain information on the wavefront errors of the optical system.

031.060 **Why NNTT *(National New Technology Telescope)* should be an MMT.**
N. Woolf, J. R. P. Angel.
Prepr. Steward Obs., No. 531, 15+21 pp. (1984).

031.061 **Steps towards 8 m mirrors. V. A method for polishing aspheres as fast as f/1.**
J. R. P. Angel.
Prepr. Steward Obs., No. 535, 11 pp. (1984). To be published in Proceedings of IAU Colloquium No. 79: Very large telescopes, their instrumentation and programs, April 9 – 12, 1984.

031.062 **Adaptive optics.**
N. J. Woolf.
Prepr. Steward Obs., No. 539, 13 pp. (1984). To be published in Proceedings of IAU Colloquium No. 79: Very large telescopes, their instrumentation and programs, April 9 – 12, 1984.

031.063 **Two useful schemes of astronomical mirror control.**
N. N. Fashchevskij.
Astron. Tsirk., No. 1319, p. 3 – 5 (1984). In Russian.

031.064 **The use of adjusting surfaces.**
N. N. Fashchevskij.
Astron. Tsirk., No. 1332, p. 5 – 7 (1984). In Russian.

031.065 **A design scheme of the monochromatic coronagraph.**
T. Li.
Acta Astrophys. Sin., Vol. 4, No. 1, p. 74 – 79 (1984). In Chinese.
A design scheme of a monochromatic coronagraph for solar eclipse observations is described. This telescope can be used to photograph the corona in 5324.9 Å and 6374 Å, and gives the possibility of analysing the linear polarization of the corona. In this instrument, a birefrigent filter which has a half bandwidth of 3 Å or 4 Å in 6374 Å or 5303 Å, respectively, is used. The possi-

bility of using the KDP crystal as birefringent element is considered. The necessary exposure time for the corona at different solar radii is estimated as well.

031.066 **Using a wide grating in a spectrograph having only one camera to provide large dispersion with high light efficiency.**
S.-y. Jiang, S.-d. Lu.
Acta Astron. Sin., Vol. 25, No. 3, p. 300 - 303 (1984). Abstract. In Chinese.

**Computational optics. Handbook.**
See Abstr. 003.147.

**The Mersenne-Schmidt: a three-mirror survey telescope.**
See Abstr. 032.010.

**Le projet de très grand télescope européen (VLT: Very Large Telescope).**
See Abstr. 032.023.

**Sundai's 75 cm alt-az telescope with an E6 honeycomb mirror.**
See Abstr. 032.041.

**Update on the 18 m aperture multi-mirror telescope concept.**
See Abstr. 032.057.

**New applications of X-ray optical techniques.**
See Abstr. 035.004.

**On the importance of active image stabilisation.**
See Abstr. 036.090.

# 032 Astronomical Instruments

032.001 **Investigation of the optics of the 125-cm ZTE telescope by the Hartmann method.**
L. I. Snezhko, V. Yu. Terebizh, Z. U. Shkhagosheva.
Astron. Tsirk., No. 1287, p. 5 - 7 (1983). In Russian.

032.002 **CCD/transit instrument deep photometric and polarimetric survey - a progress report.**
J. T. McGraw, H. S. Stockman, J. R. P. Angel, H. Epps, J. T. Williams.
Opt. Eng., Vol. 23, No. 2, p. 210 - 215 (1984). Abstr. in Phys. Abstr., Vol. 87, No. 1237, Entry 72886 (1984).

032.003 **Drifts in the orientation parameters of the Pulkovo horizontal transit circle.**
T. R. Kir'yan, G. I. Pinigin, G. M. Timashkova.
Sov. Astron. Lett., Vol. 10, No. 1, p. 57 - 59 (1984). English translation of 37.032.001.

032.004 **Small automated photometric telescopes.**
R. M. Genet, L. J. Boyd, D. S. Hall.
Advances in photoelectric photometry, Vol. 2, p. 64 - 73 (1984). - See Abstr. 003.002.

032.005 **Erste Beobachtungen mit dem 2.2-m-Teleskop II.**
T. Neckel.
Sterne Weltraum, 23. Jahrg., Nr. 7, p. 352 - 353 (1984).

032.006 **Alignment and phasing of deployable telescopes.**
N. J. Woolf, B. L. Ulich.
Proc. SPIE Int. Soc. Opt. Eng., Vol. 383, p. 11 - 20 (1983). Abstr. in Phys. Abstr., Vol. 87, No. 1238, Entry 77708 (1984). - See Abstr. 012.014.

032.007 **Analysis of mirror supports for the University of California Ten Meter Telescope.**
M. P. Budiansky, J. E. Nelson.
Proc. SPIE Int. Soc. Opt. Eng., Vol. 450, p. 59 - 67 (1984). Abstr. in Phys. Abstr., Vol. 87, No. 1241, Entry 90090 (1984). - See Abstr. 012.019.

032.008 **Il telescopio di Cassegrain: i tubi paraluce.**
C. Giordanengo.
Orione, Vol. 4, N. 2, p. 70 - 81 (1984).

032.009 **Technischer Fortschritt im Teleskopbau - das neue 1-m-Spiegelteleskop.**
D. Gutcke.
Jenaer Rundsch., 29. Jahrg., Heft 3, p. 116 - 118 (1984).

032.010 **The Mersenne-Schmidt: a three-mirror survey telescope.**
R. V. Willstrop.
Mon. Not. R. Astron. Soc., Vol. 210, No. 3, p. 597 - 609 (1984).

The Schmidt camera with an achromatic corrector is an excellent survey instrument, but is impractical in sizes much larger than the Palomar and UK Schmidts, because of the size of the dome required. Other one- and two-mirror systems also suffer from limitations of focal ratio, field or tube length. Three-mirror systems show more promise, and some previous designs are described. A new design of survey telescope is presented which is more compact, has a larger relative aperture, and has a larger field than most earlier designs. A large central obscuration is its weakest feature.

032.011 **Coelostat and heliostat: theory of alignment.**
M. Demianski, J. M. Pasachoff.
Sol. Phys., Vol. 93, No. 1, p. 211 - 217 (1984).

For perfectly aligned heliostats and coelostats tracking at the solar rate and half the solar rate, respectively, the solar beam has no translational motion. But, particularly in the field at eclipses, it is not possible to align heliostats and coelostats with infinite precision. The authors derive the effect of small misalignments on the translational motion of the beam, and give tables to allow the calculation of the accuracy to which the instruments must be mounted and adjusted to attain a desired accuracy over a given duration. Further, it is shown how to derive the necessary adjustments to improve alignment, given measurements of the tracking error.

032.012 **Investigation of a telescope support.**
N. G. Bochkarev, D. N. Spitsyna, E. V. Sosnovskij.
Calculation of deformations of mechanical knots of a telescope, p. 6 - 41 (1983). In Russian. - See Abstr. 003.014.

The results of a theoretical investigation of a frame-type telescope support are given. The wind load and the observer's weight influence are considered. The wind load acting is represented in form of a random stationary process. The maximum of possible shift and angular displacement of the upper platform is found. Main frequencies of natural vibrations are calculated.

032.013 **Determination of optimum dimensions of the fork of a telescope mounting.**
N. G. Bochkarev, G. G. Telepneva.
Calculation of deformations of mechanical knots of a telescope, p. 42 - 60 (1983). In Russian. - See Abstr. 003.014.

Results of an investigation of a fork deformation for its different positions are given. The own weight of the fork and the weight of the telescope are considered. The influence of fork dimension changes on its deformation values is studied. A new construction of forks is suggested.

032.014 **Investigation of deformations of a telescope tube under static loads.**
N. G. Bochkarev, E. V. Sosnovskij, D. N. Spitsyna.
Calculation of deformations of mechanical knots of a telescope, p. 61 – 89 (1983). In Russian. – See Abstr. 003.014.
Results of a theoretical analysis of telescope Serrurier type tube deformations in conditions of its rotation in space are given. The action of the tube and equipment weights are considered. The maximum reciprocal shift of the main and secondary mirror are found.

032.015 **Determination of natural vibration frequencies of a telescope tube.**
N. G. Bochkarev, D. N. Spitsyna, E. V. Sosnovskij.
Calculation of deformations of mechanical knots of a telescope, p. 90 – 104 (1983). In Russian. – See Abstr. 003.014.
Results of a theoretical determination of natural vibration frequencies of a Serrurier-type telescope tube are given. The tube is considered to be a system of weightless elastic bars. Each knot possesses mass and inertia moments when rotated with respect to the three reciprocally perpendicular axes.

032.016 **Investigation of forced telescope frame vibrations.**
N. G. Bochkarev, D. N. Spitsyna.
Calculation of deformations of mechanical knots of a telescope, p. 105 – 114 (1983). In Russian. – See Abstr. 003.014.
Results of a forced telescope frame vibrations study are given. These vibrations take place, when equipment with disbalanced mass is present. The influence of different factors on vibration amplitudes is considered.

032.017 **Investigation of telescope tube deformations under wind load.**
N. G. Bochkarev, D. N. Spitsyna, E. V. Sosnovskij.
Calculation of deformations of mechanical knots of a telescope, p. 115 – 133 (1983). In Russian. – See Abstr. 003.014.
Results of a theoretical analysis of telescope (Serrurier type) tube deformations under wind load are given. The dynamical influence of wind load is considered, the wind load being taken in the form of a random function which depends on the wind velocity. The maximum of possible reciprocal shift of the main and secondary telescope mirror are found.

032.018 **The 300–inch new Texas telescope.**
H. J. Smith.
Ann. N.Y. Acad. Sci., Vol. 422, p. 126 – 139 (1984). – See Abstr. 012.033.
The basic design considerations and a status report for the 7.6–m telescope project at McDonald Observatory are presented.

032.019 **The University of California ten–meter telescope project.**
J. Nelson.
Ann. N.Y. Acad. Sci., Vol. 422, p. 140 – 162 (1984). – See Abstr. 012.033.

032.020 **Large optical telescopes. New views into space and time.**
S. M. Faber.
Ann. N.Y. Acad. Sci., Vol. 422, p. 171 – 179 (1984). – See Abstr. 012.033.
The scientific opportunities of very large optical–IR telescopes are illustrated via comparison of a proposed national 15–m telescope with existing 4–m–class telescopes. The topics selected include lookback studies of galaxies and galaxy clustering, the distribution of nonluminous matter in the universe, spectroscopic studies of QSOs, and the structure of infrared protostars.

032.021 **The first telescope of the Astronomical Society "Rudjer Bošković".**
A. Tomić.
Vasiona, Année 32, No. 2, p. 32 – 34 (1984). In Serbo–Croatian.

032.022 **A wind tunnel study of the LEST structure.**
O. Engvold, B. Fossum, A. E. Holdø.
LEST Found., Tech. Rep., No. 6, 19 pp. (1984).
This study focuses on the flow pattern and the associated structural generated turbulence in front of the protubing telescope aperture of LEST. The object is to survey and to investigate structural generated turbulence, and to study practical modifications of the telescope aperture structures which could lead to substantial reduction of the turbulence level. A successful structural design will also reduce vibrations resulting from the fluctuating forces and pressures produced by the air turbulence.

032.023 **Le projet de très grand télescope européen (VLT: Very Large Telescope).**
D. Enard.
Astronomie, Vol. 98, p. 549 – 558 (1984).

032.024 **Das grösste optische Teleskop Europas.**
K. Städeli.
Orion, 42. Jahrg., Nr. 205, p. 208 – 210 (1984).

032.025 **Das Kühn–Slevogt–System 300/900.**
M. Liemen.
Sterne, 60. Band, Heft 5, p. 291 – 294 (1984).

032.026 **Multimirror telescope alignment systems which employ fan beams and TI interferometers.**
T. R. O'Meara.
Proc. SPIE Int. Soc. Opt. Eng., Vol. 440, p. 22 – 29 (1984). Abstr. in Phys. Abstr., Vol. 87, No. 1243, Entry 101162 (1984). – See Abstr. 012.042.

032.027 **The use of the Multiple Mirror Telescope as a phased array.**
J. M. Beckers, E. K. Hege, F. J. Low, D. W. McCarthy, P. A. Strittmatter.
Proc. SPIE Int. Soc. Opt. Eng., Vol. 440, p. 136 – 142 (1984). Abstr. in Phys. Abstr., Vol. 87, No. 1245, Entry 110523 (1984). – See Abstr. 012.042.

032.028 **Fiber optic synthetic aperture.**
J. E. Ludman, J. L. Sampson.
Proc. SPIE Int. Soc. Opt. Eng., Vol. 440, p. 161 – 166 (1984). Abstr. in Phys. Abstr., Vol. 87, No. 1245, Entry 110524 (1984). – See Abstr. 012.042.

032.029 **Automatic photoelectric telescopes.**
R. M. Genet, L. J. Boyd.
Publ. Astron. Soc. Pac., Vol. 96, No. 584, p. 789 (1984). Abstract. – See Abstr. 010.281.

032.030 **Measuring electron density in coronal active regions. II. A multichannel coronagraph with a photoelectric spectrograph and a reflex monitor at λ5303 Å.**
J. C. Noëns, J. Pageault, G. Ratier.
Sol. Phys., Vol. 94, No. 1, p. 117 – 131 (1984).
In order to study the electron density at the scale of the most encountered structures in coronal active regions a new multichannel coronagraph associated with a photoelectric spectrograph is now used at the Pic–du–Midi Observatory. In its quasi–routine mode this instrument, which is decribed in this paper, works with a 30" field aperture in a parallel manner with a *K*–polarimeter. On each observed region it obtains maps of intensities of the λλ3388, 10747, and 10798 Å emission lines due to Fe XIII ion. Each measurement point is associated with a quasi–simultaneous image of the emission corona structures viewed in the light of the λ5303 Å line of Fe XIV. Three examples of observations are given and the capabilities are discussed.

032.031 **Correction of the PZT's glass–plate at Tianjin Latitude Station.**
S. Luo, D. Liao, S. You, A. Li, H. Ji.
Ann. Shanghai Obs., Acad. Sin., No. 5, p. 57 – 62 (1983). In Chinese.

032.032 **Analysis of some systematic errors in observations with astrolabe.**
Z. Yang, Y. Chen.
Ann. Shanghai Obs., Acad. Sin., No. 5, p. 63 – 71 (1983). In Chinese.

032.033 **Using the microcomputer system for the 1.56 m astrometric telescope.**
Y. Wan, C. Dai.
Ann. Shanghai Obs., Acad. Sin., No. 5, p. 262 – 272 (1983). In Chinese.

This paper deals with the method of using a microcomputer system to steer the 1.56 m astrometric telescope at Shanghai Observatory. The main computer of this system is used for the general control of the telescope system, and single board computers are used for carrying out its operation instructions and orders. Its software and hardware design is also described in this paper. It is economical and practicable to apply this microcomputer system to the equipment with many auxiliary instruments.

032.034 **Multipurpose "Ukus" camera with an image–tube for astronomical photography. I. General description.**
V. L. Kuznetsov, E. I. Laptev, Yu. I. Malakhov, A. M. Mechetin, M. H. Rodriguez, Ya. O. Romanyuk, O. A. Svyatogorov, M. G. Sosonkin, L. M. Shul'man.
Astrometr. Astrofiz., Vyp. 53, p. 70 – 72 (1984). In Russian.

A multipurpose camera with an image–tube for astronomical photography has been designed at the Main Astronomical Observatory of the Ukrainian SSR Academy of Sciences. The block diagram of the camera is given. First observations with the camera have been made.

032.035 **Multipurpose "Ukus" camera with an image–tube for astronomical photography. II. Control unit.**
E. I. Laptev, Yu. I. Malakhov, V. N. Petukhov, O. A. Svyatogorov, M. G. Sosonkin.
Astrometr. Astrofiz., Vyp. 53, p. 73 – 75 (1984). In Russian.

Technical description of the control unit and its functional capabilities are reported.

032.036 **Multipurpose "Ukus" camera with an image–tube for astronomical photography. III. A stable high–voltage power source.**
O. A. Svyatogorov, M. G. Sosonkin.
Astrometr. Astrofiz., Vyp. 53, p. 75 – 78 (1984). In Russian.

The main requirements are considered for a high–voltage power source of an image–tube used in astronomical investigations. The parameters, description and circuit diagram of the source as well as the analysis of full–scale tests are given.

032.037 **Thoughts about very large telescopes.**
E. J. Wampler.
Very large telescopes, their instrumentation and programs, p. 3 – 8 (1984). – See Abstr. 012.064 (IAU Colloq. No. 79).

032.038 **From passive support systems to the NTT active support.**
R. N. Wilson, F. Franza, L. Noethe.
Very large telescopes, their instrumentation and programs, p. 23 – 40 (1984). – See Abstr. 012.064 (IAU Colloq. No. 79).

ESO has two telescope projects: the NTT (New Technology Telescope) with 3.5 m aperture and the VLT (Very Large Telescope) with 16 m equivalent aperture. The former is already in the engineering phase and should be completed in 1987; the latter is still in the study phase. This paper is concerned with the primary support system of the NTT.

032.039 **Support system for the 1 m ESO active optics experiment.**
O. Citterio, G. Conti, E. Mattaini, F. Franza.
Very large telescopes, their instrumentation and programs, p. 41 – 56 (1984). – See Abstr. 012.064 (IAU Colloq. No. 79).

The aim of this paper is to describe the support system which has been realized for the 1 m ESO active optics experiment. This experiment has been conceived in order to verify on a smaller scale the concept of the active optics, which is foreseen for the ESO NNT 3.5 m telescope.

032.040 **Consideration for the design of the driving and bearing system for large telescopes.**
F. Taddei, R. Tomelleri.
Very large telescopes, their instrumentation and programs, p. 95 – 107 (1984). – See Abstr. 012.064 (IAU Colloq. No. 79).

Within the limits of the feasibility study for the 3.5 m Italian national telescope, the authors have investigated the alternative possible design for the supporting and driving system for the azimuthal and elevation axes in order to verify the possibility and ability to obtain the requested performances.

032.041 **Sundai's 75 cm alt–az telescope with an E6 honeycomb mirror.**
S. Isobe, H. Seo, H. Osawa, M. Konno, N. Shinohara, S. Norizuki.
Very large telescopes, their instrumentation and programs, p. 157 – 163 (1984). – See Abstr. 012.064 (IAU Colloq. No. 79).

A 75 cm alt–az telescope with a primary mirror made from the Ohara E6 glass was completed in December, 1983, at Kitakaruizawa, Japan. The authors describe the surface accuracy of the mirror during its polishing and in a telescope cell, and also discuss some new technical points developed for this telescope.

032.042 **Large telescope wind loading.**
F. F. Forbes.
Very large telescopes, their instrumentation and programs, p. 165 – 175 (1984). – See Abstr. 012.064 (IAU Colloq. No. 79).

The dynamic image degradation caused by wind forces upon various telescope structures and optical elements is discussed and loading reduction techniques are suggested.

032.043 **Pointing errors of large telescopes due to wind.**
B. L. Ulich.
Very large telescopes, their instrumentation and programs, p. 177 – 181 (1984). – See Abstr. 012.064 (IAU Colloq. No. 79).

Scaling laws are presented which show the dependence of the tracking error of large telescopes on structural materials, mechanical properties and dimensions of the mount, and wind speed. Based on direct measurements on the Multiple Mirror Telescope, predictions are made for future very large telescopes.

032.044 **Large European Solar Telescope (LEST) – a solar telescope for high spatial resolution and polarimetric observations.**
O. Engvold.
Very large telescopes, their instrumentation and programs, p. 183 – 187 (1984). – See Abstr. 012.064 (IAU Colloq. No. 79).

032.045 **Pathlength stability of synthetic aperture telescopes. The case of the 25 cm CERGA interferometer.**
J. M. Mariotti, G. P. Di Benedetto.
Very large telescopes, their instrumentation and programs, p. 257 – 265 (1984). – See Abstr. 012.064 (IAU Colloq. No. 79).

Synthetic aperture telescopes are more and more considered as a promising scheme for the next generation of very large telescopes: the basic idea is to combine the beams of two or more independent telescopes. The question immediately arising when considering such a technique is the pathlength stability and the amount of pathlength difference that can be tolerated, assuming a reasonable bandwidth, to preserve coherent addition of the wavefronts and hence, high spatial frequencies observations. The I2T (two telescopes interferometer) in operation at CERGA constitutes a remarkable test–bench for studies of mechanical, optical and atmospheric constraints that are to be faced when dealing with amplitude interferometry between several telescopes.

032.046 **Progress of the large interferometer at CERGA.**
A. Labeyrie, G. Schumacher, C. Thom, M. Dugué, F. Foy, R. Foy, P. Cormier, F. Vakili.
Very large telescopes, their instrumentation and programs, p. 267 – 278 (1984). – See Abstr. 012.064 (IAU Colloq. No. 79).

The coudé beams from a pair of 1.5 meter telescopes are recombined to achieve coherent aperture synthesis. The spherical telescopes move along a 70–meter baseline, expandable to 300 meters. For the VLT, a grid of railway tracks with 2–D carriages meets the requirements for incoherent and coherent observing.

032.047 **The use of the MMT for interferometric imaging.**
J. M. Beckers, E. K. Hege.
Very large telescopes, their instrumentation and programs, p. 279 – 293 (1984). – See Abstr. 012.064 (IAU Colloq. No. 79).

The authors describe recent progress with interferometric imaging using the Multiple Mirror Telescope (MMT). All six telescopes can now be phased over a wide field of view simultaneously resulting in a (u, v) plane coverage corresponding to that of a 686 cm aperture telescope. The authors describe the open–loop phasing control of the MMT for gravitational changes and they describe a concept of an internal cophasing/coalignment system for MMT type telescopes.

032.048 **16 m Large Slit Aperture Telescope for very high angular resolution astronomy.**
R. G. Petrov, C. Aime, J. Borgnino, F. Martin, G. Ricort.
Very large telescopes, their instrumentation and programs, p. 295 – 308 (1984). – See Abstr. 012.064 (IAU Colloq. No. 79).

The authors suggest that a telescope with a long and thin rectangular primary (16 m X .4 m), able to rotate around the optical axis to ensure a full coverage of the frequency plane, would do almost as well as a conventional 16 m aperture telescope for high angular resolution astronomy for a cost substantially reduced. The performances of such a Large Slit Aperture Telescope (LSAT) for classical and differential speckle interferometry are examined and the releases on the optical and mechanical constraints allowed by the dedication of the instrument to speckle techniques are discussed.

032.049 **The Two–Mirror Telescope (2MT).**
R. G. Bingham.
Very large telescopes, their instrumentation and programs, p. 347 – 365 (1984). – See Abstr. 012.064 (IAU Colloq. No. 79).

Two eight–metre mirrors on an alt–azimuth mount are considered as a Future Large Telescope. The combined collecting area is ten times that of a four–metre class telescope, and the main modes of use are for optical spectroscopy and infrared imaging. In one square arc second, at 10 μm wavelength, the telescope with adaptive auxiliary optics will yield 30 to 100 fully resolved image elements (not speckles). The mechanical arrangement proposed to give path length stability is to link the two tube centre sections along a horizontal axis and to use a total of four altitude bearings. Twin prime and Cassegrain foci and horizontal Nasmyth platforms are intended. In addition to the conventional uses of the coudé light path, there is the possibility of using it to produce interferometric links to other telescopes.

032.050 **Cost scaling laws and their origin: design strategy for an optical array telescope.**
C. M. Humphries, V. C. Reddish, D. J. Walshaw.
Very large telescopes, their instrumentation and programs, p. 379 – 394 (1984). – See Abstr. 012.064 (IAU Colloq. No. 79).

Power law relationships between cost and aperture of optical telescopes are shown to be approximations to polynomial expressions. These polynomials, which are derived for telescopes of traditional and cost–reduced design, have implications for the cost–effective design of an optical array telescope.

032.051 **Pointing, tracking, guiding, and dome seeing at the AAT.**
P. R. Gillingham.
Very large telescopes, their instrumentation and programs, p. 415 – 416 (1984). Abstract. – See Abstr. 012.064 (IAU Colloq. No. 79).

032.052 **The 15–meter National New Technology Telescope: an update.**
L. D. Barr.
Very large telescopes, their instrumentation and programs, p. 727 – 744 (1984). – See Abstr. 012.064 (IAU Colloq. No. 79).

The two telescope concepts currently under evaluation for the 15 m National New Technology Telescope (NNTT) are presented. A choice of either the Multiple–Mirror Telescope (MMT) or the Segmented Mirror Telescope (SMT) has to be made. Principal design characteristics are discussed.

032.053 **The 6–m optical telescope of the USSR: instrumentation and observational possibilities.**
I. M. Kopylov, A. F. Fomenko.
Very large telescopes, their instrumentation and programs, p. 745 – 754 (1984). – See Abstr. 012.064 (IAU Colloq. No. 79).

032.054 **The University of California ten meter telescope project: recent developments.**
J. E. Nelson.
Very large telescopes, their instrumentation and programs, p. 755 – 765 (1984). – See Abstr. 012.064 (IAU Colloq. No. 79).

The University of California has been actively designing a ten meter telescope (TMT) for optical and infrared ground–based astronomical observations since 1977. In order to achieve an acceptable cost for the project, the design departs substantially from conventional telescope designs. A light–weight segmented primary mirror was designed. The parabolic primary is a mosaic of 36 hexagonal segments. In addition, the primary focal ratio is f/1.75, thus resulting in a short telescope tube; this allows a very compact dome.

032.055 **ESO 16 metre very large telescope: the linear array concept.**
D. Enard.
Very large telescopes, their instrumentation and programs, p. 767 – 787 (1984). – See Abstr. 012.064 (IAU Colloq. No. 79).

The present concept is based on a linear array of four 8 metre telescopes, and provides the collecting power equivalent to a 16 metre telescope. The telescopes can be used either independently, or can be optically combined at a small field coudé focus, or "electronically" combined (post detection), or combined by pairs for long baseline IR interferometry. The pros and cons of the concept are summarized.

032.056 **The University of Texas 7.6–m telescope.**
R. G. Tull.
Very large telescopes, their instrumentation and programs, p. 789 – 816 (1984). – See Abstr. 012.064 (IAU Colloq. No. 79).

032.057 **Update on the 18 m aperture multi–mirror telescope concept.**
B. Mack, D. J. Harman.
Very large telescopes, their instrumentation and programs, p. 817 – 830 (1984). – See Abstr. 012.064 (IAU Colloq. No. 79).

The paper describes the proposals currently under investigation for the design of the support system of a 7.5 m primary mirror and describes the development to be carried out so that an efficient and accurate finite element model can be constructed to compute both the structural and thermal deflections.

032.058 **Options for a new Arizona observatory: configurations, costs and site.**
N. J. Woolf, J. R. P. Angel, J. T. Williams.
Very large telescopes, their instrumentation and programs, p. 831 – 843 (1984). – See Abstr. 012.064 (IAU Colloq. No. 79).

This paper is concerned with the process of creating a major observatory at minimum cost. Its premises are that cost optimization starts with an appropriately modestly priced building block, a 7.5 m telescope. There are many possible configurations in which this building block can be used, ranging from a single telescope or a set of separate telescopes with at most electronic data addition, to an MMT in which the light of all telescopes can be phased when desired. In addition there are options for using elements in a non–redundant linear array or Michelson interferometer. The authors discuss the selection process which has led to Mt. Graham in Arizona.

032.059 **Japanese National Large Telescope (JNLT).**
S. Isobe.
Very large telescopes, their instrumentation and programs, p. 845 – 849 (1984). – See Abstr. 012.064 (IAU Colloq. No. 79).

The background and description are briefly given of the plan to build a large telescope of 5 m to 7.5 m class as the Japanese National Large Telescope.

032.060 **The Penn State spectroscopic survey telescope.**
L. W. Ramsey, D. W. Weedman.
Very large telescopes, their instrumentation and programs, p. 851 – 860 (1984). – See Abstr. 012.064 (IAU Colloq. No. 79).

To map the distribution of galaxies and quasars in the universe, a telescope is needed large enough to obtain spectra to about 20th magnitude in under 20 minutes integration and which is devoted entirely to such spectroscopic survey projects in dark time. These projects are suited to a telescope with a large fixed primary, assembled from many spherically figured segments. The authors present a design for such a telescope that consists of 73 segments, each of 0.9–m diameter and 26–m radius of curvature. Effective aperture of the telescope exceeds 7–m; the focal plane system can track objects for 40 minutes, and sky coverage of 48 degrees is obtained by using a fixed tilt for the primary and making the entire telescope and dome rotatable.

032.061 **Determination of the error of centering of objectives of the Kitab Zeiss double astrograph.**
Eh. Rakhmatov.
Tsirk. Astron. Inst. Tashkent, No. 109, p. 32 – 34 (1983). In Russian.

032.062 **Investigation of the pivots of a meridian circle with the help of a suspended level and an interferometer.**
M. F. Bykov, N. N. Yuldashev.
Tsirk. Astron. Inst. Tashkent, No. 112, p. 21 – 38 (1983). In Russian.

032.063 **The University of California 10–meter telescope – status report.**
T. S. Mast, J. E. Nelson, G. Gabor, M. P. Budiansky.
J. Opt. Soc. Am. A, Vol. 1, No. 12, p. 1250 (1984). Abstract.

032.064 **Nuove tecnologie per i grandi telescopi al suolo degli anni 80 – 90.**
O. Citterio.
G. Astron., Vol. 10, N. 2, p. 89 – 100 (1984).

032.065 **Study of the limb division errors of some parts of the Repsold meridian circle.**
V. G. Tauber.
Astron. Tsirk., No. 1305, p. 6 – 8 (1984). In Russian.

032.066 **On the method of adjustment of a Cassegrain telescope.**
V. N. Dudinov, S. B. Novikov, V. V. Shul'ga.
Astron. Tsirk., No. 1307, p. 4 – 6 (1984). In Russian.

032.067 **Options for a new Arizona Observatory: configurations, costs and site.**
N. J. Woolf, J. R. P. Angel, J. T. Williams.
Prepr. Steward Obs., No. 536, 13 pp. (1984). To be published in Proceedings of IAU Colloquium No. 79: Very large telescopes, their instrumentation and programs, April 9 – 12, 1984.

032.068 **A three–mirror astrograph and the method of its adjustment.**
N. N. Fashchevskij.
Astron. Tsirk., No. 1316, p. 6 – 8 (1984). In Russian.

032.069 **Focal reducer for a three–mirror telescope.**
N. N. Fashchevskij.
Astron. Tsirk., No. 1338, p. 6 – 8 (1984). In Russian.

032.070 **Resolution of the 1–m Zeiss–telescope on Mt. Majdanak.**
S. B. Novikov.
Astron. Tsirk., No. 1339, p. 7 – 8 (1984). In Russian.

032.071 **Method for calculation of special telescope systems.**
V. V. Tarabukin.
Opt.–mekh. prom–st', No. 6, p. 19 (1984). In Russian. Abstr. in Ref. Zh., 51. Astron., 12.51.1153 (1984).

032.072 **The ESO VLT project – major concept discriminations.**
D. Enard.
Mitt. Astron. Ges., Nr. 62, p. 123 – 135 (1984). – See Abstr. 010.341.

Major concept discriminators between the segmented mirror telescope, the Multi–Mirror–Telescope and an array are discussed and the ESO linear array concept presented.

032.073 **LEST design study.**
T. E. Andersen, R. B. Dunn, O. Engvold.
LEST Found., Tech. Rep., No. 7, 59 pp. (1984).

032.074 **Investigation of the Zeiss double astrograph in Kitab.**
Eh. Rakhmatov, Eh. Mirmakhmudov.
Komet. Tsirk., No. 333 (1984). In Russian.

**Telescopes for the 1980s.**
See Abstr. 003.066.

**Advanced technology optical telescopes II. Conference held at London, England, 5 – 6 September 1983.**
See Abstr. 012.097.

**La Fondation LEST: vers un grand télescope solaire européen.**
See Abstr. 013.008.

**Le C.D.S. et les programmes d'astrométrie méridienne.**
See Abstr. 013.067.

**Steps toward eight–meter honeycomb mirror blanks. I. Rationale and approach.**
See Abstr. 031.013.

**The testing of large telescope systems using multiple subaperture.**
See Abstr. 031.016.

**ESO's New Technology Telescope (NTT) metallic primary mirror project. Status report.**
See Abstr. 031.021.

**A finite element approach to the design of the support system for the ESO 1 m active optics experiment.**
See Abstr. 031.023.

**The ESO off–line telescope testing technique illustrated with results for the MPIA 2.2 m telescope II.**
See Abstr. 031.025.

**A modular structural design for a spectrographical MMT.**
See Abstr. 031.031.

**Optical engineering aspects of the University of Texas 7.6–meter telescope.**
See Abstr. 031.043.

**ESO 16 metre Very Large Telescope. Instrument matching in spectroscopy and direct imaging.**
See Abstr. 034.106.

**First results with the Cassegrain cluster on the INT (*Isaac Newton Telescope*).**
See Abstr. 034.181.

**Photoelectric device for limb readings of meridian instruments based on the PZS matrix.**
See Abstr. 034.187.

**Measuring the sizes of stars.**
See Abstr. 036.031.

**Reduction of instrumental polarization of coelostat telescopes.**
See Abstr. 036.049.

**Transit time determination using a pseudo–noise grid.**
See Abstr. 036.057.

**Interstellar matter with very large telescopes.**
See Abstr. 036.094.

**Star formation.**
See Abstr. 036.095.

**Very large telescopes and stellar astrophysics.**
See Abstr. 036.096.

**Uses of very large telescopes for galaxy research.**
See Abstr. 036.097.

**Polarimetry with the coronograph of the Norikura Observatory. II.**
See Abstr. 074.021.

**Instrumentation for the recording of microthermal and image–quality data.**
See Abstr. 082.084.

# 033 Radio Telescopes and Equipment

033.001 **Design concepts for reflector antenna structures.**
J. M. Hedgepeth.
Proc. SPIE Int. Soc. Opt. Eng., Vol. 383, p. 21 – 31 (1983). Abstr. in Phys. Abstr., Vol. 87, No. 1238, Entry 77709 (1984). – See Abstr. 012.014.

033.002 **Superconducting tunnel junction receiver for 2.6 mm.**
S.–K. Pan, M. J. Feldman, A. R. Kerr, E. S. Palmer, J. A. Grange, P. Timbie.
Eighth International Conference on Infrared and Millimeter Waves, p. M6.2/1–2 (1984). Abstr. in Phys. Abstr., Vol. 87, No. 1238, Entry 77714 (1984). – See Abstr. 012.015.

033.003 **Properties of an open structure mixer at 584 GHz and 803 GHz.**
H. P. Röser, E. J. Durwen, R. Wattenbach, G. V. Schultz.
Eighth International Conference on Infrared and Millimeter Waves, p. M6.5 (1984). Abstr. in Phys. Abstr., Vol. 87, No. 1238, Entry 77715 (1984). – See Abstr. 012.015.

033.004 **The first US–Japan VLBI test observation by use of K–3 system at the Radio Research Laboratories.**
Y. Saburi, K. Yoshimura, S. Kato, M. Imae, H. Morikawa, T. Sato, K. Takahashi, N. Kawajiri, N. Kawano, F. Takahashi, N. Kawaguchi, K. Koike, T. Yoshino, H. Sugimoto, H. Kuroiwa, T. Kondo, S. Hama, H. Kunimori, S. Takahashi, H. Murakami, N. Kurihara, J. Amagai, H. Kiuchi, A. Kaneko, S. Kozono.
J. Radio Res. Lab., Vol. 31, No. 132, p. 31 – 37 (1984). Abstr. in Phys. Abstr., Vol. 87, No. 1239, Entry 78367 (1984).

033.005 **A broad–band ultra–low–noise Schottky diode mixer receiver from 80 to 115 GHz.**
C. R. Predmore, A. V. Raisanen, N. R. Erickson, P. F. Goldsmith, J. L. R. Marrero.
IEEE Trans. Microwave Theory Tech., Vol. MTT–32, No. 5, p. 498 – 507 (1984). Abstr. in Phys. Abstr., Vol. 87, No. 1240, Entry 85875 (1984).

033.006 **Message from the space – recent development in radio astronomy.**
M. Morimoto.
J. Inst. Electron. Commun. Eng. Jpn., Vol. 66, No. 9, p. 912 – 917 (1983). In Japanese. Abstr. in Phys. Abstr., Vol. 87, No. 1240, Entry 85876 (1984).

033.007 **Radiometric correction of atmospheric path length fluctuations in interferometric experiments.**
G. M. Resch, D. E. Hogg, P. J. Napier.
Radio Sci., Vol. 19, No. 1, p. 411 – 422 (1984). Abstr. in Phys. Abstr., Vol. 87, No. 1240, Entry 85877 (1984).

033.008 **Acousto–optical radiospectrometer for radio astronomy.**
N. Kaifu.
Oyo Buturi, Vol. 52, No. 12, p. 1027 – 1031 (1983). In Japanese. Abstr. in Phys. Abstr., Vol. 87, No. 1241, Entry 90092 (1984).

033.009 **Interpretation of absolute line intensities on the NRAO 11 meter and other millimeter wave telescopes.**
M. L. Kutner, L. Mundy, R. J. Howard.
Astrophys. J., Vol. 283, No. 2, p. 890 – 894 (1984).

The auhors have made a series of observations of the Sun, Moon, and planets (Venus, Mars, and Jupiter) for the purpose of determining source coupling and aperture efficiencies on the NRAO 11 m telescope. They present aperture efficiencies as a function of frequency over the full operating range of the telescope, and compare the results with Ruze theory. The authors have mapped the main beam and diffraction sidelobes at several frequencies and have mapped the error pattern at 91 GHz.

033.010 **Laser levelling instrument for surface adjustment of radio telescopes.**
N. A. Esepkina, Yu. K. Zverev, S. A. Ioffe, A. P. Lavrov, P. A. Fridman, O. V. Chukanov.
Pribory i tekh. ehksp., No. 2, p. 175 – 176 (1984). In Russian. Abstr. in Ref. Zh., 51. Astron., 8.51.681 (1984).

033.011 **VLBI–receiver in the meter wave range for astrophysical investigations.**
B. N. Lipanov, A. S. Sizov.
Izv. Vyssh. Uchebn. Zaved., Radiofiz., Tom 27, No. 2, p. 257 – 260 (1984). In Russian. Abstr. in Ref. Zh., 51. Astron., 8.51.683 (1984).

033.012 **The 80 GHz–radiometer for solar burst observations.**
H. Sekiguchi, H. Nakajima, N. Shinohara, S. Kawashima, Y. Shiomi.
Tokyo Astron. Obs. Rep., Vol. 20, No. 2, p. 327 – 341 (1984). In Japanese.

033.013 **The 35 GHz polarimeter for solar burst observations.**
M. Sawa, H. Sekiguchi, H. Nakajima.
Tokyo Astron. Obs. Rep., Vol. 20, No. 2, p. 342 – 351 (1984). In Japanese.

033.014 **The improved λ 8–cm radioheliograph at Toyokawa.**
M. Nishio, C. Torii, S. Enome, K. Shibasaki, Y. Tsukiji, S. Kobayashi, N. Yoshimi, S. Takata, T. Takayanagi.
Publ. Astron. Soc. Jpn., Vol. 36, No. 2, p. 371 – 381 (1984).

033.015 **Improvement in the angular resolution of the 17–GHz solar radio interferometer at Nobeyama.**
H. Nakajima, H. Sekiguchi, T. Kosugi, Y. Shiomi, M. Sawa, S. Kawashima, K. Kai.
Publ. Astron. Soc. Jpn., Vol. 36, No. 2, p. 383 – 386 (1984).
The one–dimensional spatial resolution of the 17–GHz solar radio interferometer at Nobeyama has been improved from $\sim 50''$ to $\sim 25''$ since July 1981. It is demonstrated, as an example, that a burst source which is hardly resolved with the 50″ resolution is clearly decomposed into two separate sources with the 25″ resolution.

033.016 **Complex cross correlators with three–level quantization: design tolerances.**
L. R. D'Addario, A. R. Thompson, F. R. Schwab, J. Granlund.
Radio Sci., Vol. 19, No. 3, p. 931 – 945 (1984). Abstr. in Phys. Abstr., Vol. 87, No. 1242, Entry 94616 (1984).

033.017 **System of coherent signal transformation for astrometric VLBI in the centimeter range.**
B. N. Lipatov, A. S. Sizov.
Izv. Vyssh. Uchebn. Zaved., Radiofiz., Tom 27, No. 2, p. 131 – 136 (1984). In Russian. Abstr. in Ref. Zh., 51. Astron., 9.51.845 (1984).

033.018 **Radio astronomical complex for measurements according to the "black disk" method in the decimeter wavelength range.**
N. M. Tsejtlin, I. V. Mosalov, N. V. Bakharev, B. M. Bekkerman, N. A. Dugin, E. A. Miller, M. E. Miller, S. A. Pelyushenko, A. A. Romanychev.
Gor'k. nauchn.–issled. radiofiz. inst. Prepr., No. 176, 13 pp. (1984). In Russian. Abstr. in Ref. Zh., 51. Astron., 9.51.873 (1984).

033.019 **The 4 m mm–wave telescope at Nagoya University.**
K. Kawabata, H. Ogawa, Y. Fukui, M. Fujimoto.
Astron. Her., Vol. 77, No. 10, p. 244 – 247 (1984). In Japanese.

033.020 **Investigation of the RT–15 FTI radio telescope of the Academy of Sciences of the Turkmen SSR.**
P. V. Beketov, M. I. Grigor'eva, G. Mukhiev, I. A. Rakhimov, V. A. Tarasov, A. Kh. Khanberdiev.
Izv. Akad. Nauk TSSR. Ser. fiz.–tekh., khim. i geol. nauk, No. 1, p. 20 – 24 (1984). In Russian. Abstr. in Ref. Zh., 51. Astron., 10.51.841 (1984).

033.021 **Sensitivity in two–antenna correlation reception in a radiometric system.**
A. A. Bocharov, N. Ya. Shapirovskaya.
Inst. kosm. issled. Akad. Nauk SSSR. Prepr., No. 895, 14 pp. (1984). In Russian. Abstr. in Ref. Zh., 51. Astron., 10.51.842 (1984).

033.022 **Image formation by self–calibration in radio astronomy.**
T. J. Pearson, A. C. S. Readhead.
Annu. Rev. Astron. Astrophys., Vol. 22, p. 97 – 130 (1984). – See Abstr. 003.024.
Contents: (1) Introduction. (2) Closure quantities: Interferometry. Closure phase. Closure amplitude. Interpretation of closure phase and closure amplitude. (3) Aperture synthesis: Basic theory. Image restoration. (4) Internal calibration methods. (5) Hybrid–mapping methods: Method of Baldwin & Warner. Method of Readhead & Wilkinson. Method of Schwab and Cornwell & Wilkinson. Maximum entropy methods. Redundancy. Global fringe–fitting in VLBI. (6) Capabilities and limitations: Uniqueness and *uv*–plane coverage. Noise. Convergence. Closure errors. (7) The future.

033.023 **A coaxial cable delay system for a synthesis radio telescope.**
T. L. Landecker.
IEEE Trans. Instrum. Meas., Vol. IM–33, No. 2, p. 78 – 83 (1984). Abstr. in Phys. Abstr., Vol. 87, No. 1243, Entry 101164 (1984).

033.024 **SERC's submillimeter–wave telescope proceeding on schedule.**
B. Mitchell.
Microwaves RF, Vol. 23, No. 3, p. 43 – 46 (1984). Abstr. in Phys. Abstr., Vol. 87, No. 1244, Entry 105879 (1984).

033.025 **Test of phase–retrieval holography on the Onsala 20 m radiotelescope.**
J. Ellder, L. Lundahl, D. Morris.
Electron. Lett., Vol. 20, No. 17, p. 709 – 710 (1984). Abstr. in Phys. Abstr., Vol. 87, No. 1245, Entry 110527 (1984).

033.026 **A large millimeter wave antenna.**
K. Akabane.
Int. J. Infrared Millimeter Waves, Vol. 4, No. 5, p. 793 – 808 (1983). Abstr. in Phys. Abstr., Vol. 87, No. 1245, Entry 110528 (1984).

033.027 **A low noise SIS array receiver for radioastronomical applications in the 35 – 50 GHz band.**
L. Olsson, S. Rudner, E. Kollberg, C. O. Lindstrom.
Int. J. Infrared Millimeter Waves, Vol. 4, No. 6, p. 847 – 858 (1983). Abstr. in Phys. Abstr., Vol. 87, No. 1245, Entry 110529 (1984).

033.028 **A new maser receiver for 36 to 50 GHz.**
S. H. Zisk.
Bull. Am. Astron. Soc., Vol. 16, No. 3, Part I, p. 725 (1984). Abstract. – See Abstr. 010.065.

033.029 **A 435 MHz comparison radiometer.**
C. Clements.
Ir. Astron. J., Vol. 16, No. 4, p. 256 – 275 (1984).
Detailed information is provided for the construction of a comparison type radio telescope suitable for amateur use. Various receiver systems are examined and observing programmes outlined. The paper includes a list of standard radio sources, and of suppliers of the specialized apparatus used in the installation.

033.030 **The accuracy of determination of the baseline using the Experimental VLBI System at Shanghai Observatory.**
Z. Xue, X. Huang, H. Liang, R. Zhou, S. Li, Z. Qian.
Ann. Shanghai Obs., Acad. Sin., No. 5, p. 286 – 290 (1983). In Chinese.
In order to demonstrate the accuracy of the determination of baseline components, four VLBI experiments were carried out

using the Experimental VLBI System at Shanghai Observatory in January – February 1982. Each experiment observed the four same radio sources and lasted about twenty hours. The Experimental VLBI System has been improved since 1981.

033.031 **An efficient, reliable, high harmonic suppression frequency multiplier using step recovery diode.**
Y. Zhang.
Ann. Shanghai Obs., Acad. Sin., No. 5, p. 291 – 296 (1983). In Chinese.
The purpose of this article is to discuss a multiplier using step recovery diode. The theory of operation of highly efficient harmonic generating circuits is described, with experimental results and practical circuitry in the frequency range: 140 MHz to 1400 MHz.

033.032 **Algorithms for calculation of settings for the elements of the RATAN–600 primary mirror.**
O. N. Shivris, Yu. K. Postoenko, V. V. Trunov.
Bull. Spec. Astrophys. Obs. – North Caucasus, Vol. 17, p. 74 – 81 (1984). English translation of 34.033.040.

033.033 **Signal/noise ratio in a very long baseline radio-interferometer.**
P. A. Fridman.
Bull. Spec. Astrophys. Obs. – North Caucasus, Vol. 17, p. 86 – 91 (1984). English translation of 34.033.041.

033.034 **Calculation of VPA three–mirror periscopic system.**
E. K. Maiorova (*E. K. Majorova*).
Bull. Spec. Astrophys. Obs. – North Caucasus, Vol. 17, p. 92 – 99 (1984). English translation of 34.033.042.

033.035 **Investigation of broadband acousto–optical spectrum analyzers for radio astronomy.**
N. A. Esepkina, G. M. Blaer, Z. V. Dravskikh, A. A. Zhukov, I. I. Saenko.
Bull. Spec. Astrophys. Obs. – North Caucasus, Vol. 17, p. 104 – 111 (1984). English translation of 34.033.043.

033.036 **A multi–wavelength traveling–wave–resonator feed with a single phase center for application in radio astronomy.**
V. M. Bogod, V. N. Dikii (*V. N. Dikij*), D. V. Korol'kov, V. E. Sorel (*V. E. Sorel'*).
Bull. Spec. Astrophys. Obs. – North Caucasus, Vol. 17, p. 112 – 118 (1984). English translation of 34.033.044.

033.037 **An experimental investigation of the directional patterns of the north sector of the RATAN–600 radio telescope.**
A. V. Temirova.
Bull. Spec. Astrophys. Obs. – North Caucasus, Vol. 17, p. 119 – 131 (1984). English translation of 34.033.045.

033.038 **A 45–MHz array for radio astronomy.**
J. May, F. Reyes, J. Aparici, M. Bitran, H. Alvarez, F. Olmos.
Astron. Astrophys., Vol. 140, No. 2, p. 377 – 382 (1984).
The main characteristics of a 45–MHz meridian radiotelecope built at the Maipú Radio Astronomy Observatory are presented. It is a rectangular filled array of 528 full–wavelength dipoles distributed in six groups connected to a Butler matrix. The four central beams formed by this matrix are used for simultaneous observation. The beams are $2^{\circ}.4(\delta) \times 4^{\circ}.6(\alpha)$ and they are separated by $1^{\circ}.8$ in declination. The set of beams can be steered along the meridian approximately within 40° from the zenith. The receiver and the data acquisition system are described.

033.039 **An acousto–optical spectrometer for the RATAN–600 radio telescope.**
N. A. Esepkina, N. F. Ryzhkov, V. G. Grachev, I. S. Kochergina, S. V. Pruss–Zhukovskij, A. I. Shishkin.
Sov. Astron. Lett., Vol. 10, No. 3, p. 196 – 199 (1984). English translation of 37.033.098.

033.040 **L'interferometria come strumento per ottenere elevata risoluzione angolare. Applicazioni all'astronomia ottica e alla radioastronomia.**
R. Fanti.
G. Astron., Vol. 10, N. 1, p. 39 – 56 (1984).

033.041 **A frequency–agile interferometer for solar microwave spectroscopy.**
G. J. Hurford, R. B. Read, H. Zirin.
Sol. Phys., Vol. 94, No. 2, p. 413 – 426 (1984).
A high–resolution microwave spectrometer has been developed by converting the Owens Valley solar interferometer to frequency–agile operation. The system uses 27 m antennas equipped with phase–locked receivers which can change their observing frequency in 25 or 50 ms. Microwave spectra between 1 and 18 GHz are obtained in a few seconds by successive observations at up to 86 discrete frequencies. At each frequency the data are equivalent to the total power from each antenna and the interferometric amplitude and phase. All data are fully calibrated with respect to cosmic sources. Early observations with the system are illustrated by a sequence of flare spectra featuring cases with exceptionally narrow continuum bandwidths.

**Telescopes for the 1980s.**
See Abstr. 003.066.

**Alignment and phasing of deployable telescopes.**
See Abstr. 032.006.

**A helium–3 cooled bolometer system for one millimeter continuum observations.**
See Abstr. 034.084.

**Relaxing the isoplanatism assumption in self–calibration; applications to low–frequency radio interferometry.**
See Abstr. 036.001.

**Enhancements to the deconvolution algorithm "CLEAN".**
See Abstr. 036.019.

# 034 Auxiliary Instrumentation, Photographic Materials, Clocks

034.001 **Investigation of the SD–26 developer with short development time.**
O. D. Dokuchaeva.
Astron. Tsirk., No. 1283, p. 6 – 8 (1983). In Russian.

034.002 **Hydrogen hypersensitization of ORWO plates.**
I. I. Brejdo, O. M. Mikhajlova, R. V. Yudin.
Astron Tsirk., No. 1285, p. 5 – 7 (1983). In Russian.

034.003 **A polarimeter with a new polarizer and a new device for eliminating the sky background polarization.**
K. Metz.
Astron. Astrophys., Vol. 136, No. 1, p. 175 – 177 (1984).
The European Southern Observatory has decided to procure a new two channel polarimeter which is a modification of the well known Serkowski–polarimeter (Frecker and Serkowski, 1976). The modification mainly consists of two parts: First, instead of a stationary superachromatic half–wave plate in front of the polarizing prism a pair of two phase plates with optical axes rotated by 45° against each other is located closely behind a double aperture diaphragm, thus allowing for an automatic elimination of the sky background polarization (Metz, 1977). Secondly instead of a Wollaston prism a new type of polarizer will be used which combines the polarimetric accuracy of a Wollaston prism with the mechanical superiorities of a Foster prism.

034.004 **A wide–slit stellar spectrograph.**
E. S. Kulagin.
Astron. Tsirk., No. 1290, p. 5 – 7 (1983). In Russian.

034.005 **Asservissement du télescope.**
J. Funel.
Astronomie, Vol. 98, p. 333 – 340 (1984).

034.006 **A drift chamber telescope for heavy ion track detection.**
M. Simon, M. Henkel, R. Hundt, K. D. Mathis, G. Schieweck, T. Suck.
Nucl. Instrum. Methods Phys. Res., Sect. A, Vol. 221, No. 2, p. 466 – 471 (1984). Abstr. in Phys. Abstr., Vol. 87, No. 1237, Entry 68898 (1984).

034.007 **Testing of an image intensifier tube for astronomical applications.**
M. Singh, D. B. Jadhav, S. C. Joshi.
J. Opt. (Calcutta), Vol. 12, No. 3, p. 123 – 127 (1983). Abstr. in Phys. Abstr., Vol. 87, No. 1237, Entry 72889 (1984).

034.008 **Design of emulsion chambers for cosmic ray experiments at balloon altitudes.**
S. C. Strausz, R. J. Davisson, J. J. Lord, R. J. Wilkers.
Nucl. Tracks Radiat. Meas., Vol. 7, No. 1 – 2, p. 79 – 83 (1983). Abstr. in Phys. Abstr., Vol. 87, No. 1237, Entry 72890 (1984). – See Abstr. 012.006.

034.009 **Cosmic rays.**
T. A. H. Abbas.
Thesis, Univ. Durham, England. (1984). Abstr. in Phys. Abstr., Vol. 87, No. 1237, Entry 72891 (1984).

034.010 **Input–output control device attached to the electron–ray tube with dark–trace recording for a digital image treatment system.**
V. G. Parusimov, A. V. Polyanskij, P. V. Nevodovskij, N. K. Zaets, A. A. Babichev.
Astrometr. Astrofiz., Vyp. 52, p. 47 – 49 (1984). In Russian.
Description and operating principle of a video–control device for digital image treatment systems are given. The authors substantiate the choice of electron–ray tubes with dark–trace recording of the skiatron type for the control of the input–output operations carried out by slow mechanical devices.

034.011 **Instrumental profile of the fifth order of the solar spectrograph at the Main Astronomical Observatory of the Ukrainian SSR Academy of Sciences.**
S. I. Gandzha.
Astrometr. Astrofiz., Vyp. 52, p. 49 – 52 (1984).
The instrumental profile of the fifth order was studied photometrically using the krypton emission line λ5570 Å. The method of photometric data processing is described. A comparison of instrumental profiles of the 5th and 4th orders shows fairly good coincidence.

034.012 **An infrared photometric and spectrometric system in San Pedro Martir.**
M. Roth, A. Iriarte, M. Tapia, G. Reséndiz.
Rev. Mex. Astron. Astrofis., Vol. 9, No. 1, p. 25 – 29 (1984).
A new infrared photometer–spectrometer has recently been put into operation on the 2.1–m telescope of the Observatorio Astronómico Nacional in San Pedro Mártir, Baja California. The system allows wide–band photometry in the 1 to 20 μm wavelength range and low resolution ($\lambda/\Delta\lambda \cong 100$) spectrometry in the 2.0 to 3.9 μm region using InSb photovoltaic and Ge bolometer detectors. The various components of the system are described and tests carried out on the telescope are presented.

034.013 **PULSAR: a balloon–borne experiment to detect variable low energy gamma–ray emissions.**
N. A. Buivan, I. M. Martin, F. G. Blanco, J. Braga.
Rev. Mex. Astron. Astrofis., Vol. 9, No. 1, p. 85 – 89 (1984).
The main goal of the balloon–borne "PULSAR" experiment is to observe γ–ray photons of variable sources and pulsars in the energy range 0.1 – 5.0 MeV. The geometrical arrangement of the telescope has been designed according to detector sensitivity estimations for the pulsed radiation, which are carried out by empirical and analytical methods.

034.014 **Warm lopen voor diepkoelen.**
K. van der Helm, J. Steenkamp.
Zenit, 11. Jaarg., No. 9, p. 392 – 394 (1984).

034.015 **Photographie extrem schwacher H II–Regionen.**
W. Celnik, P. Riepe.
Sterne Weltraum, 23. Jahrg., Nr. 8–9, p. 458 – 463 (1984).

034.016 **Sternzeituhr hoher Genauigkeit.**
E. Scheucher.
Sterne Weltraum, 23. Jahrg., Nr. 8–9, p. 473 – 474 (1984).

034.017 **Sternbild–Fotografie mit dem neuen 1000–ASA–Dia Farbfilm.**
H. Kaiser.
Orion, 42. Jahrg., Nr. 203, p. 163 (1984).

034.018 **An automatic electronic device for studying anomalous refraction.**
A. O. Mogilin, Yu. I. Prodan, K. B. Shchigolev.
Tr. Gos. Astron. Inst. Shternberg, Tom 55, p. 59 – 63 (1983). In Russian.
An electronic device for automatic measurement is described. This mounting is to be used for studying the anomalous refraction of the atmosphere close to the Earth's surface.

034.019 **A multichannel Pockels cell spectropolarimeter for the Anglo–Australian Telescope.**
I. S. McLean, S. R. Heathcote, M. J. Paterson, J. Fordham, K. Shortridge.
Mon. Not. R. Astron. Soc., Vol. 209, No. 3, p. 655 – 664 (1984).
A powerful new faint–object spectropolarimeter for the 3.9–m Anglo–Australian Telescope is described. It employs an existing spectrograph and Image Photon Counting System (IPCS) together with a custom–built Pockels cell (electro–optic) polarimeter. Quarter–wave plates may be introduced above the cell

and thus both linear and circular polarization may be measured. A very high polarimetric efficiency is obtained over the entire optical spectrum; peak values of 96 per cent are observed. Instrumental polarization is negligible and the measurement precision is limited only by photon counting statistics.

034.020 **A DC photometric system.**
A. J. Hollis.
I.A.P.P.P. Commun., No. 16, p. 17 – 19 (1984).

034.021 **A review of the Lumicon Swan Band Comet Filter.**
C. S. Morris.
Int. Comet Q., Vol. 6, No. 3, p. 62 – 63 (1984).

034.022 **The atomic hydrogen maser. Is it the best clock for gravity and relativity measurements?**
R. F. C. Vessot.
Proceedings of the Third Marcel Grossmann Meeting on General Relativity, Part A, p. 803 – 821 (1983). – See Abstr. 012.002.

The atomic hydrogen maser currently offers the best stability for averaging intervals, $\tau$, between 1 sec and 1 hr. For $\tau$ approximately 1 hr the available fractional frequency stability of $6 \times 10^{16}$ makes it very useful for astrophysics and relativity measurements. An outline of the theory of operation and the major frequency perturbing influences is given. The significance of high stability oscillators to the measurement of angle, velocity, distance and time for gravitation and relativity experiments are described in terms of Allen variance statistics.

034.023 **Practical cryogenic environment for gravitational radiation detector with parametric transducer.**
T. Suzuki.
Proceedings of the Third Marcel Grossmann Meeting on General Relativity, Part B, p. 1419 – 1422 (1983). – See Abstr. 012.002.

034.024 **The mechanical background noise of Weber–type gravitational wave antenna.**
J. Chen, E. Hu, T. Guan, P. Yu.
Proceedings of the Third Marcel Grossmann Meeting on General Relativity, Part B, p. 1437 – 1442 (1983). – See Abstr. 012.002.

034.025 **Results of measurement on a harmonic oscillator using a back–action–evading scheme.**
F. Fuligni, V. Iafolla.
Proceedings of the Third Marcel Grossmann Meeting on General Relativity, Part B, p. 1451 – 1453 (1983). – See Abstr. 012.002.

The authors report preliminary results obtained by using a "Quantum non Demolition Scheme" (QND) to make observations in a classical harmonic oscillator. These tests are related to the design of an extremely sensitive gravitational radiation detector.

034.026 **Flight suspension for the relativity gyro.**
R. A. Van Patten.
Proceedings of the Third Marcel Grossmann Meeting on General Relativity, Part B, p. 1455 – 1461 (1983). – See Abstr. 012.002.

The Stanford relativity gyroscope experiment GP–B (Gravity Probe–B) is a new test of Einstein's General Theory of Relativity based on comparing gyroscopes against the star Rigel using a telescope in Earth orbit. A suspension system for levitation and precision positioning of the niobium coated spherical quartz gyro rotor during orbital flight has been simulated and is described in this paper.

034.027 **Present status of the 9.5 GHz parametric transducer at the University of Tokyo.**
K. Tsubono.
Proceedings of the Third Marcel Grossmann Meeting on General Relativity, Part B, p. 1463 – 1468 (1983). – See Abstr. 012.002.

The authors report here on the present status of the parametric transducer incorporating a microwave cavity developed at the University of Tokyo. The most advantageous feature of this transducer is that the cavity is rigidly supported and non–contacting to the gravitational wave antenna.

034.028 **How to make a back–action evading (BAE) measurement.**
W. C. Oelfke.
Proceedings of the Third Marcel Grossmann Meeting on General Relativity, Part B, p. 1469 – 1472 (1983). – See Abstr. 012.002.

This paper describes some test results for a superconducting cavity transducer and mechanical oscillator used to test back–action evasion coupling schemes for gravitational radiation antennas.

034.029 **Development of transducers for gravitational wave antennas at the University of Rome.**
M. G. Castellano, C. Cosmelli, P. Rapagnani.
Proceedings of the Third Marcel Grossmann Meeting on General Relativity, Part B, p. 1473 – 1476 (1983). – See Abstr. 012.002.

The research now being performed at the University of Rome on transducers concerns two different kinds of resonant transducers: a capacitive transducer coupled to a low noise cooled FET amplifier and an inductive superconducting transducer coupled to a DC SQUID.

034.030 **Parametric upconverter transducer with a mechanical resonator for low frequency continuous gravitational wave detection.**
Y. Yan, R. Qin.
Proceedings of the Third Marcel Grossmann Meeting on General Relativity, Part B, p. 1477 – 1480 (1983). – See Abstr. 012.002.

The authors propose a tuning fork antenna gravitational wave detector for low frequency continuous gravitational wave detection. For this system they have designed a parametric upconverter transducer. A brief description of this transducer is presented.

034.031 **Transducer for a three–mode cryogenic gravitational radiation detector.**
W. Folkner, M. V. Moody, J.–P. Richard.
Proceedings of the Third Marcel Grossmann Meeting on General Relativity, Part B, p. 1481 – 1484 (1983). – See Abstr. 012.002.

The present Maryland transducer program is to instrument a 2400 kg Al 5056 bar as a three–mode antenna–transducer system to be operated at 0.050K and instrumented with a dc SQUID. This paper is devoted to a discussion of the parameters selected for the system to achieve the projected sensitivity and a brief discussion of the present status of development.

034.032 **Window considerations for LEST.**
R. B. Dunn.
LEST Found., Tech. Rep., No. 3, 50 pp. (1984).

This contribution gathers together material that concerns the 250–cm diameter vacuum or helium window required for the Large European Solar Telescope (LEST). The author applies the formulae developed by Mehltretter and Bernet to the larger LEST window and makes recommendation on the material and thickness. He discusses material availability and fabrication problems.

034.033 **Polarimeter package for LEST.**
J. O. Stenflo.
LEST Found., Tech. Rep., No. 4, 18 pp. (1984).

Various polarimeter schemes for LEST are investigated. To eliminate the spherical aberration in the polarimeter, a solution has been found for which the polarimeter package with collimators, in the form of a "black box" of length about 20 cm, may be inserted in the beam without introducing any focus changes or beam deviations. Piezoelastic modulation appears to be superior to other modulation methods. Two categories of schemes are presented, one for use in systems that have a fast frequency response ($\sim 50$ kHz), one for systems with a slow response (CCD or CID–type detectors). All four Stokes parameters can be recorded simultaneously using two modulators, without any beam splitter or moving parts.

034.034 **On the band width of resonant gravitational-wave antennas.**
G. Pizzella.
Lett. Nuovo Cimento, Vol. 40, Ser. 2, No. 8, p. 240 - 242 (1984). Abstr. in Phys. Abstr., Vol. 87, No. 1241, Entry 86320 (1984).

034.035 **Molecular astronomy using heterodyne detection at 691 GHz.**
D. D. Peck, H. R. Fetterman, D. Buhl, G. Chin, S. Petuchowski.
Int. J. Infrared Millimeter Waves, Vol. 5, No. 3, p. 329 - 340 (1984). Abstr. in Phys. Abstr., Vol. 87, No. 1241, Entry 90218 (1984).

034.036 **A dedicated echelle spectrometer for the Anglo-Australian Telescope.**
J. Meaburn, B. Blundell, R. Carling, D. F. Gregory, D. Keir, C. G. Wynne.
Mon. Not. R. Astron. Soc., Vol. 210, No. 2, p. 463 - 477 (1984).
An echelle spectrometer has been manufactured for use combined with the image photon counting system on the Anglo-Australian Telescope. This instrument employs a collimator/camera lens and in its principal configuration uses broad interference filters to isolate separate echelle orders. Very long multi-slits or multi-image masks can cover large areas of extended sources to obtain line profiles from many spatial elements simultaneously. Provision is made for a grism to be inserted to achieve cross-dispersion with a short slit or produce a simple low dispersion spectrum of the source. Furthermore, the insertion of a plane mirror produces a direct image of the field when a clear area is driven into position to replace the multi-slits. All complex operations are remotely controlled with a BBC microcomputer.

034.037 **Utilisation d'une grille pseudo-aléatoire en astrométrie.**
J.-M. Pochet.
Astrométrie, géodésie, mécanique céleste, p. 245 - 250 (1984). - See Abstr. 012.021.

034.038 **Le cercle de déclinaison de l'instrument méridien automatique de Bordeaux. Détermination des corrections de division.**
J. M. Fournier, M. Rapaport, Y. Requième.
Astrométrie, géodésie, mécanique céleste, p. 251 - 256 (1984). - See Abstr. 012.021.

034.039 **La machine automatique à mesurer pour l'astronomie (M.A.M.A.).**
J. Guibert.
Astrométrie, géodésie, mécanique céleste, p. 263 - 266 (1984). - See Abstr. 012.021.

034.040 **Imaging photon detectors for optical astronomy.**
J. R. Allington-Smith, H. E. Schwarz.
Q. J. R. Astron. Soc., Vol. 25, No. 3, p. 267 - 289 (1984).
This paper discusses the application of imaging photon detectors (IPDs) to optical astronomy. The authors review the theory of their operation and suggest ways in which the design can be optimized for these applications.

034.041 **L'horloge parlante de l'Observatoire de Paris.**
P. Parcelier.
Astronomie, Vol. 98, p. 381 - 391 (1984).

034.042 **First QSO spectra with EFOSC (*ESO Faint Object Spectroscopic Camera*).**
H. Dekker, S. D'Odorico.
Messenger, No. 37, p. 7 - 8 (1984).

034.043 **IDS (*Image Dissector Scanner*) spectroscopy of faint emission-line objects.**
I. Lundström, B. Stenholm.
Messenger, No. 37, p. 35 - 36 (1984).

034.044 **Instrument for recording heavy particles of cosmic radiation with dielectric detectors.**
E. V. Gorchakov, N. L. Grigorov, V. A. Iozenas, R. A. Nymmik, Yu. V. Trigubov, G. G. Zenchev.
Pribory i tekh. ehksp., No. 2, p. 35 (1984). In Russian. Abstr. in Ref. Zh., 51. Astron., 8.51.685 (1984).

034.045 **La Technical Pan 2415 Kodak ipersensibilizzata: tecnica di trattamento e risultati ottenibili.**
P. G. Barbero, A. Bertoglio.
Orione, Vol. 4, N. 2, p. 62 - 69 (1984).

034.046 **Zonnewijzers uit bolprojecties.**
J. A. F. de Rijk.
Zenit, 11. Jaarg., No. 10, p. 415 - 418 (1984).

034.047 **Das Photoelektrische Meßsystem (PMS) - eine universell einsetzbare Zusatzeinheit zu astronomischen Beobachtungsgeräten.**
K. Hörning, D. Söll, K.-H. Heinemann.
Jenaer Rundsch., 29. Jahrg., Heft 3, p. 119 - 121 (1984).

034.048 **L'effet Schwarzschild/Der Schwarzschildeffekt.**
W. Maeder.
Orion, 42. Jahrg., Nr. 204, p. 194 - 195 (1984).

034.049 **Spectro-heliograph at the Norikura Observatory.**
I. Shimizu, S. Hamana, S. Ishikawa.
Tokyo Astron. Obs. Rep., Vol. 20, No. 2, p. 225 - 239 (1984). In Japanese.

034.050 **Development procedure for making velocity plate.**
H. Morishita, K. Mizugaki.
Tokyo Astron. Obs. Rep., Vol. 20, No. 2, p. 270 - 275 (1984). In Japanese.

034.051 **On some improvements of the control system of PZT (Part II).**
M. Yoshinari, K. Nakajima, T. Yamazaki, K. Matsuda, H. Ishizaki.
Tokyo Astron. Obs. Rep., Vol. 20, No. 2, p. 293 - 310 (1984). In Japanese.

034.052 **Test observation with a two dimensional Photon-counting Image Acquisition System (PIAS).**
T. Noguchi, W. Tanaka, H. Hamabe, T. Aoki, T. Soyano, S. Okamura, H. Maehara, K. Ishida, Y. Tsuchiya, T. Kurono, E. Inuzuka.
Tokyo Astron. Obs. Rep., Vol. 20, No. 2, p. 352 - 366 (1984). In Japanese.

034.053 **A system for detecting anomalous room refraction.**
T. Kuwabara, S. Suzuki.
Tokyo Astron. Obs. Rep., Vol. 20, No. 2, p. 382 - 401 (1984). In Japanese.

034.054 **Coupling of oscillation modes in an antenna for gravitational radiation.**
K. Iso, K. Kuroda, H. Hirakawa.
Phys. Lett. A, Vol. 103A, No. 5, p. 262 - 264 (1984). Abstr. in Phys. Abstr., Vol. 87, No. 1242, Entry 90479 (1984).

034.055 **Infrared multiplex encoding imager as telescope's pointing system.**
J. Gay, A. Journet, Y. Rabbia.
Proc. SPIE Int. Soc. Opt. Eng., Vol. 395, p. 149 - 153 (1983). Abstr. in Phys. Abstr., Vol. 87, No. 1242, Entry 94613 (1984). - See Abstr. 012.030.

034.056 **The ESO IR photometer/spectrophotometer detectors.**
A. van Dijsseldonk, A. F. M. Moorwood.
Proc. SPIE Int. Soc. Opt. Eng., Vol. 395, p. 173 - 176 (1983). Abstr. in Phys. Abstr., Vol. 87, No. 1242, Entry 94624 (1984). - See Abstr. 012.030.

034.057 **52 element InSb array and associated electronics.**
M. Meyer.
Proc. SPIE Int. Soc. Opt. Eng., Vol. 395, p. 177 – 181 (1983). Abstr. in Phys. Abstr., Vol. 87, No. 1242, Entry 94625 (1984). – See Abstr. 012.030.

034.058 **Heterodyne spectroscopy for astrophysical applications using tunable laser sidebands at 10 microns.**
H. Rothermel, U. Schrey, G. Magerl, W. Schupita.
Int. J. Infrared Millimeter Waves, Vol. 5, No. 2, p. 147 – 158 (1984). Abstr. in Phys. Abstr., Vol. 87, No. 1242, Entry 94637 (1984).

034.059 **A camera for very high energy gamma–ray astronomy.**
D. J. Fegan, D. McLaughlin, J. Clear, M. F. Cawley, N. A. Porter, K. Gibbs, T. C. Weekes, R. C. Lamb, P. Gorham, V. J. Stenger.
Ann. N.Y. Acad. Sci., Vol. 422, p. 345 – 346 (1984). Abstract. – See Abstr. 012.033.

034.060 **Sterrenfoto's: puntgaaf en spatscherp.**
A. Nagel, S. Wadman.
Zenit, 11. Jaarg., No. 11, p. 478 – 485 (1984).

034.061 **Recommendations on post focus instrumentation for LEST.**
H. Wöhl, M. C. E. Huber, P. Mein, L. Smaldone.
LEST Found., Tech. Rep., No. 5, 30 pp. (1984).

034.062 **A microcomputer–based sytstem for data acquisition and instrument control.**
N. M. White, L. H. Wasserman.
Publ. Astron. Soc. Pac., Vol. 96, No. 582, p. 667 – 670 (1984).
A microcomputer–based system is described which can monitor and control a variety of instruments and detectors commonly used with optical telescopes. The system is composed of a DEC LSI–11/02 computer and modular, commercially available, industrial quality interface hardware. The data–acquisition system can be purchased as a complete, tested system or by components and configured in–house.

034.063 **A fast, low–cost photometer for the measurement of stellar magnitudes from photographic plates.**
G. K. G. Moore, P. Ihnat.
Publ. Astron. Soc. Pac., Vol. 96, No. 582, p. 671 – 673 (1984).
A new stellar photometer is described. Its main advantage is that it can be incorporated without modification into any system which provides video images of stars from a photographic plate. It is a completely electronic system and as such is completely transportable.

034.064 **Una pellicola particolare: la Kodak Technical Pan 2415.**
A. Maitan.
G. A.A.B., N. 76, p. 7 – 8 (1984).

034.065 **Fabry–Perot spectroscopy in space science.**
J. N. Desai.
Proc. Indian Acad. Sci., Earth Planet. Sci., Vol. 93, No. 3, p. 189 – 200 (1984).
The paper describes use of Fabry–Perot (FP) systems in spectroscopic studies of astronomy and earth's upper atmosphere. Factors influencing the resolution, luminosity and instrumental line shape of FP spectrometer are discussed and the need for their optimization is emphasised if desired luminosity gain is to be realised in practice. Use of FP as imaging spectrometer, scanning spectrometer and as a narrow band filter are illustrated.

034.066 **An amateur filar micrometer.**
P. B. Byrne, D. E. Beesley, P. Dunsby.
Ir. Astron. J., Vol. 16, No. 3, p. 181 – 187 (1984).

034.067 **The OMA–system of the Rozhen National Observatory.**
M. K. Tsvetkov, H. S. Markov.
Dokl. Bolg. Akad. Nauk, Vol. 37, No. 1, p. 11 – 14 (1984). Abstr. in Ref. Zh., 51. Astron., 10.51.824 (1984).

034.068 **Erratum: "Optimal sampling of charged–coupled devices" [Astron. J., Vol. 85, No. 10, p. 1421 – 1424 (1980)].**
D. J. Hegyi, A. Burrows.
Astron. J., Vol. 89, No. 10, p. 1628 (1984). See Abstr. 28.034.027.

034.069 **An installation for recording the velocity field of prominences.**
I. S. Kim, O En Den, A. I. Stepanov.
Soln. Dannye, Byull., 1984, No. 7, p. 90 – 93 (1984). In Russian.
An installation for the study of the velocity field of prominences is described. A Fabry–Perot interferometer is used for spectral analysis.

034.070 **Wide–field coronograph for observations of coronal transient phenomena.**
G. M. Nikol'skij, S. Koutchmy, M. P. Fat'yanov.
Soln. Dannye, Byull., 1984, No. 8, p. 53 – 56 (1984). In Russian.
A small Lyot coronograph is described. The results of the trial observations are discussed.

034.071 **Comparative hypersensitization of Kodak spectroscopic plate, type 103a–D, with forming gas and nitrogen.**
R. L. Scott, S. H. Newnam.
AAS Photo–Bull., No. 36, p. 3 – 4 (1984).
Baking in an atmosphere of either 5% forming gas or dry nitrogen has been evaluated as a technique for hypersensitizing Kodak spectroscopic plate, type 103a–D. The results suggest that baking in forming gas is the superior treatment.

034.072 **A comparison of CCD's and photographic plates.**
I. Furenlid.
AAS Photo–Bull., No. 36, p. 5 – 8 (1984).
The reduction procedures for data obtained by CCD's and photographic plates are compared and used to discuss basic similarities and differences of these detectors.

034.073 **Astronomical tests of a photon–counting image acquisition system.**
S. Isobe, K. Saito, K. Tomita, T. Nakamura, E. Kaneda, Y. Tsuchiya, Y. Suzuki.
AAS Photo–Bull., No. 36, p. 9 – 11 (1984).
Astronomical tests of a photon–counting image acquisition system (PIAS) were carried out with a 91–cm reflector at the Dodaira Station of the Tokyo Astronomical Observatory on December 20, 1982. It was shown that this system is a powerful astronomical instrument for faint objects in a dark sky where the photon rate in the detector field is less than $10^5$ photons $s^{-1}$.

034.074 **A birefringent filter for measuring solar vector magnetic field and sight–line velocity field.**
G. Ai, Y. Hu, T. Li, F. He, H. Hou, Z. Gu, H. Ni.
Sci. Sin., Ser. A, Vol. 27, No. 10, p. 1086 – 1095 (1984).
A new birefringent filter has been produced for measuring the solar magnetic field. It is a kind of monochromator with a wide field of view that receives a lot of signals at the same time for realtime observations. It can be used in polarization–free telescopes without any additional polarization.

034.075 **Kiso image detection system.**
H. Maehara.
Astron. Her., Vol. 77, No. 12, p. 300 – 305 (1984). In Japanese.

034.076 **Digitale Ablesung von Mikrometerschrauben. Ein Beitrag zur Astrometrie.**
R. Bendel.
Sterne Weltraum, 23. Jahrg., Nr. 12, p. 640 – 641 (1984).

034.077 **Indium antimonide detectors for ground–based astronomy.**
R. R. Joyce.
Proc. SPIE Int. Soc. Opt. Eng., Vol. 443, p. 50 – 58 (1984). Abstr. in Phys. Abstr., Vol. 87, No. 1243, Entry 95517 (1984). – See Abstr. 012.044.

034.078 **Preliminary results on the registration temperature effect for uranium ions in various solid state nuclear track detectors.**
A. Thompson, D. O'Sullivan.
Nucl. Tracks Radiat. Meas., Vol. 8, No. 1 – 4, p. 567 – 570 (1984). Abstr. in Phys. Abstr., Vol. 87, No. 1243, Entry 96387 (1984). – See Abstr. 012.045.

034.079 **Identification of relativistic nuclei with $10 \lesssim Z \lesssim 92$.**
S. P. Ahlen, T. E. Coan, J. Drach, P. B. Price, M. H. Salamon, G. Tarle, M. L. Tincknell, S. Guo.
Nucl. Tracks Radiat. Meas., Vol. 8, No. 1 – 4, p. 571 – 574 (1984). Abstr. in Phys. Abstr., Vol. 87, No. 1243, Entry 96388 (1984). – See Abstr. 012.045.

034.080 **Dinonyl phthalate as a dopant for CR–39 in cosmic ray studies.**
D. O'Sullivan, A. Thompson, J. H. Adams Jr.
Nucl. Tracks Radiat. Meas., Vol. 8, No. 1 – 4, p. 579 – 581 (1984). Abstr. in Phys. Abstr., Vol. 87, No. 1243, Entry 96390 (1984). – See Abstr. 012.045.

034.081 **Two–dimensional imaging in infrared astronomy.**
F. Sibille, P. Lena.
Proc. SPIE Int. Soc. Opt. Eng., Vol. 395, p. 110 – 113 (1983). Abstr. in Phys. Abstr., Vol. 87, No. 1243, Entry 101159 (1984). – See Abstr. 012.030.

034.082 **Focal plane optics in far–infrared and submillimeter astronomy.**
R. H. Hildebrand.
Proc. SPIE Int. Soc. Opt. Eng., Vol. 441, p. 40 – 50 (1984). Abstr. in Phys. Abstr., Vol. 87, No. 1243, Entry 101173 (1984). – See Abstr. 012.043.

034.083 **Screening out obstructive radiation from thermoluminescence sheets ($BaSO_4$:Eu) exposed to cosmic rays.**
I. Yamamoto, T. Tomiyama, H. Miyai, T. Wada, Y. Yamashita.
Nucl. Instrum. Methods Phys. Res., Sect. A, Vol. 224, No. 3, p. 573 – 575 (1984). Abstr. in Phys. Abstr., Vol. 87, No. 1244, Entry 102523 (1984).

034.084 **A helium–3 cooled bolometer system for one millimeter continuum observations.**
T. P. L. Roellig, J. R. Houck.
Int. J. Infrared Millimeter Waves, Vol. 4, No. 3, p. 299 – 309 (1983). Abstr. in Phys. Abstr., Vol. 87, No. 1244, Entry 105882 (1984).

034.085 **Some considerations of a Michelson spatial interferometer in the far IR.**
V. De Cosmo, M. Petrizzelli.
Int. J. Infrared Millimeter Waves, Vol. 4, No. 4, p. 541 – 548 (1983). Abstr. in Phys. Abstr., Vol. 87, No. 1245, Entry 106367 (1984).

034.086 **Spectral radiometer for the (AMOS) 1.6 meter telescope.**
G. W. Ashley.
Proc. SPIE Int. Soc. Opt. Eng., Vol. 430, p. 93 – 98 (1983). Abstr. in Phys. Abstr., Vol. 87, No. 1246, Entry 115691 (1984). – See Abstr. 012.051.

034.087 **Control system of an imaging triple Fabry–Perot filter.**
J. G. Winter.
Opt. Acta, Vol. 31, No. 7, p. 823 – 830 (1984). Abstr. in Phys. Abstr., Vol. 88, No. 1247, Entry 4608 (1985).

034.088 **Multi–anode microchannel arrays: pulse–counting array detectors for imaging and spectroscopy.**
J. G. Timothy.
Publ. Astron. Soc. Pac., Vol. 96, No. 584, p. 796 (1984). Abstract. – See Abstr. 010.281.

034.089 **CCD scanning for asteroids and geostationary objects.**
T. Gehrels.
Bull. Am. Astron. Soc., Vol. 16, No. 3, Part I, p. 700 (1984). Abstract. – See Abstr. 010.061.

034.090 **The MIT snapshot CCD camera.**
R. L. Baron, E. Dunham, J. L. Elliot.
Bull. Am. Astron. Soc., Vol. 16, No. 3, Part I, p. 707 – 708 (1984). Abstract. – See Abstr. 010.061.

034.091 **The ESO Faint Object Spectrograph and Camera (EFOSC).**
B. Buzzoni, B. Delabre, H. Dekker, S. D'Odorico, D. Enard, P. Focardi, B. Gustafsson, W. Nees, J. Paureau, R. Reiss.
Messenger, No. 38, p. 9 – 13 (1984).

034.092 **The RPCS detector.**
P. R. Christensen, E. Hviid, G. Thomsen, O. Ulfbeck.
Messenger, No. 38, p. 38 – 41 (1984).

034.093 **A two–dimensional photon–counting camera for low–flux astronomical applications.**
D. Durant, E. Hardy.
J. R. Astron. Soc. Can., Vol. 78, No. 5, p. 214 (1984). Abstract. – See Abstr. 010.401.

034.094 **The Four–Shooter: two dimensional photometry at Mt. Palomar.**
A. Porter.
I.A.P.P.P. Commun., No. 17, p. 33 – 38 (1984).

034.095 **An accurate sidereal clock using a digital clock module.**
D. L. DuPuy.
I.A.P.P.P. Commun., No. 17, p. 55 – 61 (1984).

034.096 **Possible problems with the EMI Starlight–1 photometer.**
M. K. Kaitting.
I.A.P.P.P. Commun., No. 17, p. 64 – 65 (1984).

034.097 **The improvement of the performance of the hydrogen maser clock at Shanghai Observatory in recent years.**
Ann. Shanghai Obs., Acad. Sin., No. 5, p. 227 – 231 (1983). In Chinese.

034.098 **Discussion on dual–mixer time difference measurement technique for Atomic Time.**
J. Hu, D. Ma, N. Wan.
Ann. Shanghai Obs., Acad. Sin., No. 5, p. 232 – 240 (1983). In Chinese.

034.099 **The application of a 100 MHz phase comparison method.**
J. Hu, D. Ma.
Ann. Shanghai Obs., Acad. Sin., No. 5, p. 249 – 253 (1983). In Chinese.

034.100 **The design of a temperature control system for hydrogen maser resonator.**
C. Lin.
Ann. Shanghai Obs., Acad. Sin., No. 5, p. 254 – 261 (1983). In Chinese.

034.101 **The establishment of the hypersensitizing laboratory and the preliminary results of plate hypersensitization test.**
B. Yao, Z. Yang, J. Jing, Z. Xu, P. Fun.
Ann. Shanghai Obs., Acad. Sin., No. 5, p. 283 – 285 (1983). In Chinese.

A laboratory for plate hypersensitization has been established at Zô–Sè section of Shanghai Observatory. The preliminary hypersensitization test of Kodak 103a–0 plates which were produced in December 1977 and in November 1981 respectively has been carried out.

034.102 **The registration system of an infrared photometer.**
S. I. Prikhod'ko, V. L. Karbovskij.
Astrometr. Astrofiz., Vyp. 53, p. 78 – 84 (1984). In Russian.

A registration system version of an infrared photometer is proposed. The synchronous detection method is used. The functional scheme of the registration system and basic scheme of the most significant units are presented.

034.103 **Investigation of films for astronomical image–converter observations.**
T. P. Lisovskaya, S. V. Marchenko.
Astrometr. Astrofiz., Vyp. 53, p. 84 – 90 (1984). In Russian.

034.104 **Stability of high–sensitive level scales.**
Yu. S. Dobrokhotov.
Astrometr. Astrofiz., Vyp. 53, p. 90 – 93 (1984). In Russian.

The Dresden automatic level comparator was used for examinations of two half–second and two one–second levels in order to estimate stabilities of their scales. Results of these examinations are discussed.

034.105 **Optical and infrared instruments requirements for a VLT.**
H. R. Butcher.
Very large telescopes, their instrumentation and programs, p. 431 – 438 (1984). – See Abstr. 012.064 (IAU Colloq. No. 79).

The author discusses auxiliary instrumentation needs for the use with a VLT. The paper is divided into three parts: a summary of the principal observational advances made possible by a VLT in the 15 – 20 meter class; a list of what is arguably the resulting basic set of needed instruments; and a brief consideration of the telescope design features, technology development programs, and political/financial efforts required to make those instruments a reality.

034.106 **ESO 16 metre Very Large Telescope. Instrument matching in spectroscopy and direct imaging.**
D. Enard.
Very large telescopes, their instrumentation and programs, p. 439 – 464 (1984). – See Abstr. 012.064 (IAU Colloq. No. 79).

Preliminary feasibility studies have been carried out in the framework of the ESO VLT project which have permitted the evaluation of several possible approaches to the instrument matching problem. This paper briefly discusses the detector long term perspective, recalls the main scaling laws and attempts to provide some quantitative analysis of spatial performance of instrumentation in a few typical cases. Field slicing technique is proposed as a potential solution to solve detector matching problems.

034.107 **C.F.H.T. focal reducer: image and multiaperture spectroscopy.**
B. Fort, G. Lelièvre, J. P. Picat, Y. Rio, L. Vigroux.
Very large telescopes, their instrumentation and programs, p. 465 – 468 (1984). – See Abstr. 012.064 (IAU Colloq. No. 79).

The authors present recent development and preliminary results obtained with the focal reducer of the Canada French Hawaii 3.6 meter telescope with new trends for deep imagery and multiaperture spectroscopy.

034.108 **Image–slicers.**
E. H. Richardson, J. M. Fletcher, W. A. Grundmann.
Very large telescopes, their instrumentation and programs, p. 469 – 492 (1984). – See Abstr. 012.064 (IAU Colloq. No. 79).

The diameter of stellar images produced by Earth–based telescopes often exceeds the width of the entrance slit of the associated spectrograph. It is not uncommon for more starlight to be wasted on the jaws of the slit than is transmitted by the slit. The purpose of an image–slicer is to redirect the light from the slit jaws into the slit; to do so, the length of the slit illuminated by starlight must be increased, either by stacking slices of the image along the slit, or by elongating the image and superimposing full–length slices of it. Thus, in order to work, an image–slicer of whatever type must lose spatial resolution of the sky along the slit.

034.109 **Fabry–Perot spectrographs for VLT's.**
G. Monnet.
Very large telescopes, their instrumentation and programs, p. 493 – 496 (1984). – See Abstr. 012.064 (IAU Colloq. No. 79).

Four reasons are advocated to stress the interest of Fabry–Perot spectrographs as potentially useful instruments for the future very large telescopes: (1) reasonably wide spectrum of astronomical uses, (2) good use of telescope area, (3) relative cheapness, (4) versatility.

034.110 **The new coudé échelle spectrograph for the Anglo Australian Telescope.**
D. D. Walker, F. Diego.
Very large telescopes, their instrumentation and programs, p. 499 – 513 (1984). – See Abstr. 012.064 (IAU Colloq. No. 79).

This paper is intended as a general introduction to the design philosophy of the instrument, paying particular attention to the properties of the detectors with which it will be used. Some implications on the requirements for similar instruments at future very large telescopes are also discussed.

034.111 **Focal reducer techniques for direct imaging and field spectroscopy with large telescopes.**
E. H. Geyer, B. Nelles.
Very large telescopes, their instrumentation and programs, p. 575 – 583 (1984). – See Abstr. 012.064 (IAU Colloq. No. 79).

To exploit the full effectiveness of large ground based telescopes, the focal reducer technique can be applied very adequately. The authors review recent Hoher List Observatory developments of observing techniques for direct imaging, field spectroscopy with multi–slit and masking methods for background supression and double–grism arrangements (heliometer principle) for absolute radial velocity determinations with a focal reducer field spectrograph.

034.112 **Two–dimensional infrared detector arrays.**
C. R. McCreight.
Very large telescopes, their instrumentation and programs, p. 585 – 602 (1984). – See Abstr. 012.064 (IAU Colloq. No. 79).

The technology and terminology of integrated infrared detector arrays are discussed. Specific infrared array multiplexer designs include the charge–coupled device (CCD) and charge–injection device (CID). Laboratory data and telescope imagery obtained with three arrays (InSb CCD, Si:In CCD, and Si:Bi CID) are used to illustrate the performance and promise of integrated arrays in astronomical applications.

034.113 **Qualifying a mosaic of detectors for infrared astronomy.**
F. Lacombe, D. Rouan, D. Stefanovitch,
B. Talureau, J. Berezné, G. Epstein, S. Pau, D. Phan Van,
Y. Zéau.
Very large telescopes, their instrumentation and programs, p. 603 – 606 (1984). – See Abstr. 012.064 (IAU Colloq. No. 79).

An 8 × 8 InSb CID array, developed in France for terrestrial applications, has been tested at Meudon Observatory in order to evaluate if its performances at LHe temperature meet the requirements of infrared astronomy. The paper gives: (1) the description of the array and the experimental conditions; (2) the test results; (3) work under development.

034.114 **Recent developments in coatings, gratings, and reflective components.**
R. R. Shannon.
Very large telescopes, their instrumentation and programs, p. 607 – 616 (1984). – See Abstr. 012.064 (IAU Colloq. No. 79).

034.115 **Prospects for fibre optics in future telescope instrumentation.**
G. Lund.
Very large telescopes, their instrumentation and programs, p. 617 – 635 (1984). – See Abstr. 012.064 (IAU Colloq. No. 79).

The complex nature of fibre beam transmission efficiency and its dependance on not only intrinsic fibre parameters but also on instrument design is considered. The respective advantages and drawbacks of fibres, compared with conventional optics, are reviewed. The relative merits of optical fibres and classical optics are compared. A practical example is given of the overall instrumental efficiency which can be achieved with a moderately long fibre link.

034.116 **Considerations in potographic detection for very large telescopes.**
A. G. Millikan.
Very large telescopes, their instrumentation and programs, p. 637 – 638 (1984). Abstract. – See Abstr. 012.064 (IAU Colloq. No. 79).

034.117 **Emploi de la caméra électronique grand champ avec les très grands télescopes. Extrapolation des résultats obtenus avec le télescope C.F.H.**
G. Wlérick, L. Renard, D. Horville, F. Gex, J. M. Munier, G. Lelièvre, A. Llebaria, A. Bijaoui, R. Dubout, M. Petit.
Very large telescopes, their instrumentation and programs, p. 639 – 657 (1984). – See Abstr. 012.064 (IAU Colloq. No. 79).

The large field Lallemand electronographic camera has proved to be an ideal receptor for bidimensional photometry with the C.F.H. 3.6 m telescope. It permits to measure faint stars up to about B = 25. The limits of the receptor with a very large telescope are discussed.

034.118 **Rubicon – a new diode array detector system.**
T. Schmidt–Kaler, R. Rudolph, H. Tüg.
Very large telescopes, their instrumentation and programs, p. 659 – 663 (1984). – See Abstr. 012.064 (IAU Colloq. No. 79).

A photon–counting system with a 512–channel parallel output digital image tube is presented. Electronics developed separately for each detector channel as well as data aquisition are optimized for low power consumption and high counting rates. This detector, characterized by wide dynamic range, very low noise and high photometric accuracy, is especially suitable for spectrophotometry and calibrations.

034.119 **A simultaneous five colour three channel photometer with fiber optics – design and first results.**
H. Barwig, R. Schoembs.
Very large telescopes, their instrumentation and programs, p. 665 – 671 (1984). – See Abstr. 012.064 (IAU Colloq. No. 79).

034.120 **Sensitometric and information characteristics of the GOSNIICHIMPHOTOPROECT films for an electronographic image tube.**
S. B. Dostovalov, P. P. Petrov, D. A. Ptitsyn, T. A. Ryabchikova, V. L. Khokhlova.
Nauchn. Inf., Vyp. 54, p. 4 – 15 (1983). In Russian.

034.121 **On the possibility of speed increasing of Kodak IIIa–J and IN plates by cool developing.**
I. Eglitis.
Nauchn. Inf., Vyp. 54, p. 16 – 18 (1983). In Russian.

034.122 **On the apparatus for hypersensitization of photomaterial in a nitrogen atmosphere.**
V. V. Natriashvili.
Nauchn. Inf., Vyp. 54, p. 19 – 20 (1983). In Russian.

034.123 **Hypersensitization of photographic plates at the Pulkovo Observatory.**
I. I. Brejdo, O. M. Mikhajlova.
Nauchn. Inf., Vyp. 54, p. 21 – 27 (1983). In Russian.

034.124 **Hypersensitization of infrared photomaterials (literature survey).**
I. I. Brejdo.
Nauchn. Inf., Vyp. 54, p. 28 – 34 (1983). In Russian.

034.125 **Numerical method for measuring exposure, information and noise parameters of astronomical photoemulsions.**
M. F. Shabanov.
Nauchn. Inf., Vyp. 54, p. 35 – 44 (1983). In Russian.

034.126 **Organization of astrophotographic works at the Astrophysical Institute of the Kazakh Academy of Sciences.**
D. A. Rozhkovskij, A. A. Semenikin.
Nauchn. Inf., Vyp. 54, p. 50 – 60 (1983). In Russian.

034.127 **Compiled catalogue of spectral sensitivity data for astronomical photographic materials.**
O. D. Dokuchaeva.
Nauchn. Inf., Vyp. 54, p. 61 – 144 (1983). In Russian.

034.128 **On the use of the Kazan films for photographing supernova remnants.**
T. G. Sitnik.
Nauchn. Inf., Vyp. 54, p. 146 – 147 (1983). In Russian.

034.129 **Testing photographic materials at the Pulkovo Observatory.**
I. I. Brejdo, O. M. Mikhajlova.
Nauchn. Inf., Vyp. 54, p. 148 – 150 (1983). In Russian.

034.130 **On storage of the baked films A–500H and A–600H.**
M. D. Mirmil'shtejn, I. A. Baranova.
Nauchn. Inf., Vyp. 54, p. 151 – 159 (1983). In Russian.

034.131 **Experience in hypersensitizing photomaterials in the photolaboratory of the Sternberg State Astronomical Institute.**
O. D. Dokuchaeva, T. A. Birulya, M. S. Toropova.
Nauchn. Inf., Vyp. 54, p. 160 – 167 (1983). In Russian.

034.132 **On the use and study of the astronomical photographic films with improved speed made in Kazan.**
O. D. Dokuchaeva.
Nauchn. Inf., Vyp. 54, p. 175 – 219 (1983). In Russian.

034.133 **Recalibration of the KPNO spot sensitometers.**
W. E. Schoening, C. Nink.
Occas. Rep. R. Obs., Edinb., No. 14, p. 1 – 9 (1984). – See Abstr. 012.067.

The authors discuss the techniques used to recalibrate the spot sensitometers that are used on Kitt Peak. They explain the methods used including those that did not work.

034.134 **FLACS: a new sensitometer for measurement of extreme low intensity reciprocity failure in a controlled environment.**
A. G. Smith, H. W. Schrader.
Occas. Rep. R. Obs., Edinb., No. 14, p. 11 – 15 (1984). – See Abstr. 012.067.

A compact sensitometer has been developed that facilitates exposures as long as a week in a fully lighted laboratory and in a completely controlled atmosphere. The instrument can readily be placed in a refrigerator, freezer, or oven for temperature control.

034.135 **A spectrosensitometer with an absolute calibration.**
O. Zichová, J. Zicha, J. Honsa.
Occas. Rep. R. Obs., Edinb., No. 14, p. 21 – 30 (1984). – See Abstr. 012.067.

A portable prism spectrosensitometer with an absolute calibration has been designed for evaluation of the spectral sensitivity of photographic materials. The spectral region from 380 nm to 880 nm has been achieved and it is possible to place one, two or three exposures on one sample. An easy measurement of reciprocity characteristics and a quick checking of various hypersensitizing procedures is also possible.

034.136 **A new calibration spectrograph at Kitt Peak.**
W. E. Schoening.
Occas. Rep. R. Obs., Edinb., No. 14, p. 31 – 33 (1984). Abstract. – See Abstr. 012.067.

034.137 **Comparison of the FAH (*fine grain compensation hypersensitization*) process with gas hypersensitization techniques.**
B.-a. Yao, J.-y. Jing, Z.-h. Xu, J.-h. Tong.
Occas. Rep. R. Obs., Edinb., No. 14, p. 35 – 43 (1984). – See Abstr. 012.067.
The authors have been interested in the FAH process. As a gas hypersensitizing laboratory was established at Shanghai Observatory in 1982, they have been able to make a series of comparison experiments between the FAH and gas hypersensitization. The results are summarized in this paper.

034.138 **The results of hydrogen hypersensitization of photo-emulsions.**
R. I. Kiladze, V. B. Natriashvili, D. G. Chipashvili.
Occas. Rep. R. Obs., Edinb., No. 14, p. 45 – 52 (1984). – See Abstr. 012.067.
The study of the astronomical RP type films (produced by the Kazan NII Techphotoproject, USSR) is considered to be important in resolutions of a conference held by the Problem Working Group of the Astronomical Council of the USSR Academy of Sciences. From this standpoint different emulsions have been compared at the Abastumani Astrophysical Observatory – both those widely used in the astronomical practice and emulsions less known, even non-astronomical ones.

034.139 **On the storage of hypersensitized plates.**
J. D. Schumann.
Occas. Rep. R. Obs., Edinb., No. 14, p. 53 – 55 (1984). – See Abstr. 012.067.
This poster reports on the author's experiences since 1977 with storing Kodak photographic plates either unsensitized or treated by forming gas (FG) baking or hypered by the so-called $n_{30}$ process.

034.140 **Some remarks on the hypersensitization of astronomical photomaterials.**
O. D. Dokuchaeva.
Occas. Rep. R. Obs., Edinb., No. 14, p. 57 – 58 (1984). – See Abstr. 012.067.

034.141 **Nitrogen flushing during exposure of large Schmidt telescope plates.**
J. A. Dawe, E. Coyte, N. Metcalfe.
Occas. Rep. R. Obs., Edinb., No. 14, p. 59 – 67 (1984). – See Abstr. 012.067.
The authors have compared sets of hypersensitized plates taken on the U.K. Schmidt telescope before and since the use of nitrogen-flushing during exposure. Substantial improvements are found, both in terms of large-scale and point-to-point variations in plate uniformity as a result of nitrogen-flushing. Plate exposure times are slightly reduced, though such reductions are of the same level as variations found between Kodak batches. The difficulties in ensuring the smooth flow of dry nitrogen over plates as large as 14 inches are very great. However, the resulting gains in uniformity and, hence, potential photometric accuracy are beneficial.

034.142 **The most effective astronomical photographic emulsions for recording of faint and low-contrast objects.**
M. F. Shabanov.
Occas. Rep. R. Obs., Edinb., No. 14, p. 69 – 81 (1984). – See Abstr. 012.067.
As a result of exposure, measurement and computer processing of more than 400 samples of astronomical photographic emulsions, nine of the most effective of them have been selected. The averaged data of the characteristic curves, spectral sensitivity, noise and gradation characteristics are presented for them. Brief analysis is made of properties and features of application of these photographic emulsions in astronomical observations.

034.143 **Astronomical applications of new infrared films.**
A. S. Heinman, B. I. Shapiro, A. G. Shcherbakov, N. I. Shakhovskaya.
Occas. Rep. R. Obs., Edinb., No. 14, p. 83 – 88 (1984). – See Abstr. 012.067.
The photographic plates which were used until recently in infrared astronomy have a very low speed and need hypersensitisation prior to exposure. Therefore, on the basis of a new method a series of high speed infrared films have been made in the Scientific Institut Chimphotoproject. These films do not need hypersensitisation prior to exposure. The possibility of astronomical applications of these new infrared films has been studied in the Crimean Astrophysical Observatory.

034.144 **The IIIa-F emulsion and the quasar redshift cut-off.**
A. Savage.
Occas. Rep. R. Obs., Edinb., No. 14, p. 89 – 92 (1984). Summary of a paper by Savage and Peterson (1983). – See Abstr. 012.067.

034.145 **Microspots on spectroscopic plates.**
A. R. Good, G. Gourlay.
Occas. Rep. R. Obs., Edinb., No. 14, p. 93 – 106 (1984). – See Abstr. 012.067.
Extensive inspections of the large UKSTU collection of IIIa-J plates have shown many of these plates to be affected by microspot blemishes. These blemishes destroy both the qualitative and quantitative properties of parts of the photograph. Results from some preliminary tests suggest environmental and processing causes. Contrary to manufacturers recommendations, the introduction of iodide into the processing line may be beneficial rather than detrimental to spot formation.

034.146 **Procedures for improving the image stability of KODAK spectroscopic plates, type IIIa-J.**
W. E. Lee, F. J. Drago.
Occas. Rep. R. Obs., Edinb., No. 14, p. 107 – 108 (1984). Abstract. – See Abstr. 012.067.

034.147 **Some aspects of the preparation and use of the photographic detector in astronomy.**
A. G. Millikan.
Occas. Rep. R. Obs., Edinb., No. 14, p. 109 – 110 (1984). Abstract. – See Abstr. 012.067.

034.148 **Where to next in astronomical photography?**
D. F. Malin.
Occas. Rep. R. Obs., Edinb., No. 14, p. 111 – 118 (1984). – See Abstr. 012.067.
Recent developments in photographic technology, in particular Eastman Kodak's introduction of a 1000 ASA colour film, have made some of the investigators realise that the 20 year-old IIIa J emulsion may be superseded by something even better. If that is true, what form might such an emulsion take and what properties should it have to be most useful? The purpose of this note is to explore these questions and look at the options.

034.149 **What do we want – increased speed, resolution, contrast or DQE?**
R. D. Cannon.
Occas. Rep. R. Obs., Edinb., No. 14, p. 119 – 133 (1984). – See Abstr. 012.067.
In this paper the author attempts to answer the question posed in the title, using mainly an empirical approach based on experience with the UK 1.2 m Schmidt Telescope (UKST), but supported in some aspects by rather crude photon-statistical arguments.

034.150 **Using a Gossen light meter to produce enlarged interpositive film copies from glass plates.**
M. M. Hanna.
Occas. Rep. R. Obs., Edinb., No. 14, p. 153 – 155 (1984). – See Abstr. 012.067.
The Gossen Luna Pro light meter is primarily used by photographers as a hand-held light meter for determining outdoor and studio exposures. By adding the enlargement attachment, the

author has used the Gossens as film plane light meters for making enlarged interpositives.

034.151 **Unsharp masking at the enlarger.**
M. M. Hanna.
Occas. Rep. R. Obs., Edinb., No. 14, p. 157 – 163 (1984). – See Abstr. 012.067.
Unsharp masking enables the user to produce film or print copies with a greater detail and improved tonal range. It is accomplished on enlarged images rather than contact. Exposure and processing steps have more latitude while registration is accomplished with simple everyday tools. This application of unsharp masking has added flexibility in terms of contrast control when making the corrected interpositive.

034.152 **An investigation into the fidelity of reproduction of UKST original objective prism plates on to film.**
M. T. Brück, J. D. Waldron.
Occas. Rep. R. Obs., Edinb., No. 14, p. 185 – 190 (1984). – See Abstr. 012.067.
In this investigation, the performance of machine processed Kodak 4421 film in conjunction with glass positives on process plates has been examined in order to evaluate the extent to which the negative film copies can be used for photometry.

034.153 **Developer/film combinations in photomacrography.**
W. J. Roberton.
Occas. Rep. R. Obs., Edinb., No. 14, p. 191 – 195 (1984). – See Abstr. 012.067.
This paper serves to illustrate the versatility of Technical Pan Film 2415 as a material for photomacrography from astronomical plates to give enlarged views at selected and wide–ranging contrast levels, and to demonstrate even higher contrast developer/emulsion combinations for extraction of very faint signal. The object selected for photography was a spiral galaxy, NGC 300.

034.154 **Extending magnitude standards to faint stars with a sub–beam prism.**
K. Ishida.
Occas. Rep. R. Obs., Edinb., No. 14, p. 197 – 202 (1984). – See Abstr. 012.067.
A method is developed to extrapolate a magnitude scale to faint stars on photographic plates taken with a Schmidt telescope. The design of the sub–beam prism and the derivation procedure of the two numerical constants worked out by T. Noguchi, T. Soyano, and other staff members of the Kiso Observatory, are presented.

034.155 **Sub–beam prism used at the Bosscha Schmidt telescope.**
T. Noguchi, M. Raharto, B. Hidayat.
Occas. Rep. R. Obs., Edinb., No. 14, p. 203 – 205 (1984). – See Abstr. 012.067.
The purpose of this note is to present the result of tests of the new sub–beam prism, recently acquired and used in conjunction with the Bosscha Schmidt telescope (The, 1961).

034.156 **Recent developments on the COSMOS plate–scanning machine.**
H. T. MacGillivray.
Occas. Rep. R. Obs., Edinb., No. 14, p. 237 – 244 (1984). – See Abstr. 012.067.
Modifications have recently been made to COSMOS in order to enhance the performance of the machine and improve the quality of the data it produces. These modifications are described together with a description of the present level of production on the machine, major projects being carried out, research highlights and some planned future developments which are designed to further enhance the machine's performance.

034.157 **Plate measurements and reductions at Münster.**
D. Teuber, H.–J. Tucholke, C. C. Volkmer, R. Budell, R. Duemmler, H. Horstmann, H.–G. Scheuer, W. C. Seitter.
Occas. Rep. R. Obs., Edinb., No. 14, p. 245 – 252 (1984). – See Abstr. 012.067.
The microdensitometer of the Astronomy Department at Münster University together with various tests and investigations performed during the early part of its active period has been described before. This paper concerns the present status of the machine and its programmes with respect to configuration of the system, measuring accuracies, major limitations to its full employment, new logarithmic amplifier, programme organization, application programmes.

034.158 **The Toruń measuring system for astronomical spectra.**
S. Krawczyk, Z. Turlo, B. Wikierski, A. Woszczyk.
Occas. Rep. R. Obs., Edinb., No. 14, p. 253 – 260 (1984). – See Abstr. 012.067.
The authors have undertaken a modernization of the GII Zeiss Rapid microphotometer and the implementation of the digital techniques. The consecutive steps of their efforts have been already reported (Turlo 1974, Turlo et al. 1975, Krawczyk et al. 1976). This note describes the present state of this system.

034.159 **First tests on the astrometric performance of the new PDS1010G at Hamburg Observatory.**
C. de Vegt, T. Kleine.
Occas. Rep. R. Obs., Edinb., No. 14, p. 295 – 300 (1984). – See Abstr. 012.067.
Various repeatability tests have been performed on the new PDS1010G at Hamburg Observatory. From these measurements, extended up to 11 hours, significant drift of coordinates reaching several μm and occasional discrete 1 μm glitches have been found in accordance with similar results from other PDS granite systems. With the present system configuration therefore precise astrometric measurements cannot be performed without substantial instrumental modifications.

034.160 **Astrometric applications of the Mt. Stromlo PDS 1010A.**
P. A. Ianna.
Occas. Rep. R. Obs., Edinb., No. 14, p. 301 – 310 (1984). – See Abstr. 012.067.
Several problems in photographic astrometry to which the Mt. Stromlo PDS 1010A is being applied are described briefly. A few tests of the overall astrometric accuracy and stability of the machine, as well as recent results, show some micron–level peculiarities, but overall a quite satisfactory level of performance for the system.

034.161 **Cryogenic metal mesh bandpass filters for submillimeter astronomy.**
M. Dragovan.
Appl. Opt., Vol. 23, No. 16, p. 2798 – 2802 (1984).
A bandpass filter of the double–halfwave design has been developed for use in submillimeter astronomy. The filter is rugged, easily tunable, easy to construct, and cryogenically coolable without loss of performance. Its bandpass character is superior to the single Fabry–Perot.

034.162 **Coelostat and heliostat: alignment and use for eclipse and other field purposes.**
J. M. Pasachoff, W. Livingston.
Appl. Opt., Vol. 23, No. 16, p. 2803 – 2808 (1984).
The authors discuss solar coelostats and heliostats and their alignment, with special attention to setting up portable heliostats at eclipse and other field sites. They consider the sources of tracking errors and the relation of small misalignments with such tracking errors.

034.163 **Frequency standards stability cooperating with DOG–O2 Doppler receiver.**
P. Fraczyk.
Artif. Satell., Vol. 19, No. 1, p. 41 – 47 (1984).
The paper presents measurement results of the Allan variance of rubidium standards Cz1–50, quartz oscillators HP 10544C and GWM–5 for sampling times ranging from 1s to 5000s. These results may be the basis for designing the systems which require stable frequency.

034.164 **Controller of a photometer for ground–based airglow measurements.**
S. I. Surgoychev.
Dokl. Bolg. Akad. Nauk, Vol. 37, No. 2, p. 161 – 163 (1984). In Russian. Abstr. in Ref. Zh., 51. Astron., 11.51.855 (1984).

034.165 **Time–resolved FTS of molecular and atomic IR emission.**
W. Barowy, H. Sakai.
Infrared Phys., Vol. 24, No. 2/3, p. 251 – 256 (1984). – See Abstr. 012.071.

034.166 **The Queen Mary College/University of Oregon photometer for submillimetre continuum observations.**
P. A. R. Ade, M. J. Griffin, C. T. Cunningham, J. V. Radostitz, S. Predko, I. G. Nolt.
Infrared Phys., Vol. 24, No. 4, p. 403 – 415 (1984).

034.167 **A FIR photometer for ground–based astronomical observations.**
C. Ceccarelli, G. Dall'Oglio, G. Ferri, L. Pietranera, S. Radford, G. Z. Xie.
Infrared Phys., Vol. 24, No. 6, p. 493 – 497 (1984).
A description is given of a photometer which is suitable for astronomical observations through millimetric atmospheric windows. It consists of a Ge–Ga bolometer operating at 1.0K, with a throughput $AW = 0.3\,cm^2sr$ and a $NET = 1\,mK\,Hz^{-1/2}$. Two metallic mesh filters set the windows at about 10 and $5\,cm^{-1}$.

034.168 **A device for applying plastic film antireflection coatings to optical elements.**
G. C. Augason.
Infrared Phys., Vol. 24, No. 6, p. 567 – 569 (1984).
A film–application device (FAD) has been developed to simplify the process of thermally–bonding thin plastic films to optical elements. These coatings may be used for protecting optical elements or to reduce surface reflection of radiation with wavelengths greater than 20 μm. The FAD may also be used without the center plate to stretch plastic films to make pellicles or beam–splitters. A description of the FAD and its use is given.

034.169 **The practice of photographic works with the main stellar spectrograph of the 6–m telescope.**
V. E. Panchuk.
Soobshch. Spets. Astrofiz. Obs., Vyp. 40, p. 13 – 19 (1983). In Russian. – See Abstr. 012.076.
The methods of preliminary preparation and processing of photographic materials used in the observations with the main stellar spectrograph of the 6–m telescope are shortly described.

034.170 **Compiled catalogue of data on spectral sensitivity of photographic materials which are interesting for astronomy (Short information).**
O. D. Dokuchaeva.
Soobshch. Spets. Astrofiz. Obs., Vyp. 40, p. 33 – 37 (1983). In Russian. – See Abstr. 012.076.
Information is presented on the catalogue creation of spectral sensitivity data for 384 emulsion types used in astronomy. Spectral sensitivity curves are given for 22 types of presented astronomical photoemulsions.

034.171 **A new device for hypersensitization of astrophotoemulsions by heating in vacuum, nitrogen atmosphere and hydrogen.**
M. M. Kononov, M. F. Shabanov.
Soobshch. Spets. Astrofiz. Obs., Vyp. 40, p. 53 – 58 (1983). In Russian. – See Abstr. 012.076.

034.172 **Results of hydrogen hypersensitization of astronomical photoemulsions.**
V. V. Natriashvili.
Soobshch. Spets. Astrofiz. Obs., Vyp. 40, p. 69 – 72 (1983). In Russian. – See Abstr. 012.076.

034.173 **Results of measuring the micronoise of the Kodak, ORWO and Kazan astronomical photoemulsions.**
M. F. Shabanov.
Soobshch. Spets. Astrofiz. Obs., Vyp. 40, p. 73 – 79 (1983). In Russian. – See Abstr. 012.076.

034.174 **Results of hydrogen hypersensitization of astronomical films of the Kazan production.**
N. A. Tikhonov.
Soobshch. Spets. Astrofiz. Obs., Vyp. 40, p. 81 – 85 (1983). In Russian. – See Abstr. 012.076.

034.175 **The Spacewatch Camera.**
T. Gehrels, R. P. Binzel.
Minor Planet Bull., Vol. 11, No. 1, p. 1 – 2 (1984).

034.176 **Alcuni metodi di ipersensibilizzazione di emulsioni fotografiche per uso astronomico.**
R. Mazzella.
Orione, Vol. 4, N. 4, p. 171 – 173 (1984).

034.177 **Esperienze fotografiche col Newton da 114 mm.**
G. Cortini.
Orione, Vol. 4, N. 4, p. 174 – 179 (1984).

034.178 **Polarizzatori.**
E. Landi Degl'Innocenti.
G. Astron., Vol. 10, N. 1, p. 29 – 37 (1984).

034.179 **Rivelatori per astronomia.**
F. Bortoletto.
G. Astron., Vol. 10, N. 2, p. 79 – 88 (1984).

034.180 **Laser interferometer enhanced positional precision of a scanning microdensitometer for astrometric plate scanning.**
J. H. Kinsey, R. J. Denman, A. Evzerov.
Space Telesc. Sci. Inst., Prepr. Ser., No. 29, 16 pp. (1984). To be published in Rev. Sci. Instrum.

034.181 **First results with the Cassegrain cluster on the INT (*Isaac Newton Telescope*).**
M. Pettini, R. Ellis.
Gemini, No. 11, p. 1 – 4 (1984).

034.182 **An introduction to the CCD on the INT (*Isaac Newton Telescope*) Cassegrain spectrograph.**
P. Jorden.
Gemini, No. 12, p. 5 (1984).

034.183 **Bau eines Heterodynempfängers für 10 μm Wellenlänge zum Einsatz am Cassegrainfokus astronomischer Teleskope und spektroskopische Untersuchung der Mars– und Venusatmosphäre.**
H. U. Käufl.
MPE Rep., No. 183, 113 pp. (1984).

034.184 **On the preliminary measurements of the Loran–C signals of both BPL and the SH3 chain.**
Y. Zheng, J.–y. Tan.
Publ. Purple Mt. Obs., Vol. 3, No. 2, p. 33 – 37 (1984). In Chinese.

034.185 **Photoelectric automatic guiding of a transit instrument.**
J.–s. Yao, Z.–l. Zou.
Publ. Purple Mt. Obs., Vol. 3, No. 2, p. 61 – 63 (1984). In Chinese.

034.186 **CCD use at Lick Observatory.**
T. R. Lauer, J. S. Miller, C. S. Osborne, L. B. Robinson, R. J. Stover.
Proc. SPIE Int. Soc. Opt. Eng., Vol. 445, p. 132 – 139 (1984).

034.187 **Photoelectric device for limb readings of meridian instruments based on the PZS matrix.**
B. A. Golovko.
Astron. Tsirk., No. 1306, p. 5 – 8 (1984). In Russian.

034.188 **Test marks on astronomical plates and method for setting them.**
Yu. A. Shokin.
Astron. Tsirk., No. 1311, p. 7 – 8 (1984). In Russian.

034.189 **Some results of hypersensitization of photomaterials by hydrogen at the Odessa Astronomical Observatory.**
Yu. N. Bondarenko.
Astron. Tsirk., No. 1314, p. 5 – 7 (1984). In Russian.

034.190 **Hydrogen hypersensitization of ORWO photographic plates at astronomical observations at the Pulkovo Observatory of the USSR Academy of Sciences.**
I. I. Brejdo, R. V. Yudin.
Soobshch. Spets. Astrofiz. Obs., Vyp. 40, p. 45 – 51 (1983). In Russian. – See Abstr. 012.076.

034.191 **Investigation of the non–linearity of the scanning micrometer scale based on observations of stars.**
V. N. Ershov.
Astron. Tsirk., No. 1316, p. 4 – 5 (1984). In Russian.

034.192 **Gillison spectrophotometer for observations in the 0.6 – 1.1 μm range.**
A. P. Ipatov, V. I. Shenavrin.
Astron. Tsirk., No. 1322, p. 1 – 3 (1984). In Russian.

034.193 **Atmospheric seeing compensator.**
Yu. V. Baranov, N. D. Belkin, V. N. Gorobinskij, S. B. Novikov, A. A. Ovchinnikov, V. V. Shul'ga.
Astron. Tsirk., No. 1329, p. 2 – 4 (1984). In Russian.

034.194 **Automatic Ritchey cassette.**
O. I. Mitin, S. B. Novikov, A. A. Ovchinnikov, V. V. Shul'ga.
Astron. Tsirk., No. 1339, p. 2 – 5 (1984). In Russian.

034.195 **Hochgeschwindigkeitsphotometrie für Sternbedeckungen.**
C. Bittner.
Mitt. Astron. Ges., Nr. 62, p. 155 (1984). Abstract. – See Abstr. 010.341.

034.196 **Kurze Einführung in die astronomische Photographie.**
W. C. Seitter, R. Budell.
Mitt. Astron. Ges., Nr. 62, p. 162 – 169 (1984). – See Abstr. 010.341.

034.197 **Bildmaterialien zum photographischen Prozess.**
R. Budell.
Mitt. Astron. Ges., Nr. 62, p. 170 – 174 (1984). – See Abstr. 010.341.

034.198 **Moderne photographische Kopiermethoden in der Astronomie.**
C. Madsen.
Mitt. Astron. Ges., Nr. 62, p. 175 – 179 (1984). – See Abstr. 010.341.

034.199 **Auflösungsvermögen und Kontrastübertragungsfunktionen astronomischer Emulsionen.**
F.–T. Lentes.
Mitt. Astron. Ges., Nr. 62, p. 180 – 181 (1984). – See Abstr. 010.341.

034.200 **Die characteristischen Kurven der Kodak–Platte IIa–O.**
J. D. Schumann.
Mitt. Astron. Ges., Nr. 62, p. 182 – 183 (1984). – See Abstr. 010.341.

034.201 **Astronomical test of a photon–counting detector.**
J. Barnstedt, M. Gutekunst, M. Grewing, L. Bianchi.
Mitt. Astron. Ges., Nr. 62, p. 356 – 359 (1984). – See Abstr. 010.341.

034.202 **Das Bonner Mikrodensitometer, ein Prozessor–Verbund in der Praxis.**
K.–H. Gebler, M. Böhmer.
Mitt. Astron. Ges., Nr. 62, p. 364 – 367 (1984). – See Abstr. 010.341.

034.203 **Ein schneller logarithmischer Verstärker für die PDS 2020GM.**
C. C. Volkmer.
Mitt. Astron. Ges., Nr. 62, p. 368 – 369 (1984). – See Abstr. 010.341.

034.204 **The instrumental profiles of the coudé spectrograph of the 1 m telescope at Yunnan Observatory.**
S.–n. Qin, J.–k. Li.
Acta Astron. Sin., Vol. 25, No. 2, p. 212 – 214 (1984). In Chinese.

034.205 **Double–camera image converter for the 6–m telescope.**
G. I. Bryukhnevich, L. V. Gyavgyanen, Eh. I. Zak, S. V. Lipatov, A. E. Melamid, V. A. Miller, V. S. Rylov, B. M. Stepanov, T. A. Skosyrskaya, E. I. Titkov.
Astrofizika, Tom 21, Vyp. 2, p. 379 – 386 (1984). In Russian. English translation in Astrophysics, Vol. 21, No. 2.
The results of observations on the 6–m telescope (BTA) and laboratory investigations of the double–camera image converter of UMK–91 v type with magnetic focusing with $25 \times 40\ mm^2$ photocathode field and a plane fiber optic output are given.

034.206 **The possibility of a slitless spectrograph on BTA and on the one–meter Zeiss telescope.**
N. A. Tikhonov, M. F. Shabanov.
Astrofizika, Tom 21, Vyp. 2, p. 387 – 391 (1984). In Russian. English translation in Astrophysics, Vol. 21, No. 2.
The authors have prepared a slitless spectrograph on the basis of a transmission grating tested on sample observations on the 6–m telescope BTA and one–meter Zeiss telescope.

034.207 **Fatigue effects in a pulse counting FW 130 photometer.**
F. R. Chromey, W. A. Rosen.
News Lett. Astron. Soc. N.Y., Vol. 2, No. 6, p. 13 (1984). Abstract. – See Abstr. 010.241.

034.208 **A magnetograph scanning along the whole profile of a line.**
G. M. Nikol'skij, N. S. Kim, S. Koutchmy, V. Yu. Klepikov, O En Den, A. I. Stepanov.
Soln. Dannye, Byull., 1984, No. 9, p. 88 – 93 (1984). In Russian.
A magnetograph recording the Zeeman–effect along the whole profile of a line with modulator of a LiNbO crystal and a Fabry–Perot interferometer is described.

**Astrophysical techniques.**
See Abstr. 003.004.

**Séparation des radiations par les filtres optiques.**
See Abstr. 003.087.

**Das Astrolabium und die Uhr.**
See Abstr. 003.107.

**Sonne, Zeit und Ewigkeit. Alte Sonnenuhren.**
See Abstr. 003.148.

**Processing of photographic observing material at the National Astronomical Observatory Rozhen.**
See Abstr. 009.036.

**The UK Schmidt Telescope and the facilities in Edinburgh.**
See Abstr. 009.037.

**The work of the "Photolabs" at the Royal Observatory, Edinburgh.**
See Abstr. 009.038.

**Workshop über Photographie.**
See Abstr. 011.033.

**Instrumentation in astronomy V. Conference held at London, England, 7 – 9 September 1983.**
See Abstr. 012.098.

**TARO Cosmic–ray Observatory Report (I).**
See Abstr. 013.013.

**Computerized photometry for intelligent, efficient, and flexible observing.**
See Abstr. 021.003.

**Comparison of various grazing incidence spectrometer designs based on conical diffraction.**
See Abstr. 031.045.

**Aspheric concave grating spectrographs.**
See Abstr. 031.056.

**Small automated photometric telescopes.**
See Abstr. 032.004.

**The 6–m optical telescope of the USSR: instrumentation and observational possibilities.**
See Abstr. 032.053.

**L'interferometria come strumento per ottenere elevata risoluzione angolare. Applicazioni all'astronomia ottica e alla radioastronomia.**
See Abstr. 033.040.

**Numérisation des comparateurs Zeiss.**
See Abstr. 036.025.

**Data acquisition and recording system for two–dimensional images of the solar corona during balloon–borne observations of the total solar eclipse of June 11, 1983.**
See Abstr. 036.055.

**System requirements for polarization capability.**
See Abstr. 036.093.

**Some aspects of photography with a small Schmidt camera.**
See Abstr. 036.104.

**Astronomers' problems with the calibration of Schmidt plates.**
See Abstr. 036.105.

**Power spectrum analysis of astronomical photographs digitised with small apertures.**
See Abstr. 036.106.

**Method and apparatus for complex measurements af astroclimatic meteorological parameters.**
See Abstr. 036.116.

**The accuracy limit of ground–based stellar photometry.**
See Abstr. 036.145.

**Instrumental approach to multi–site spectral variability study.**
See Abstr. 036.148.

**Para investigar nuestro planeta.**
See Abstr. 045.020.

**Possibility of detecting the magnetic field of a rotating neutron star with SQUID.**
See Abstr. 067.073.

**Polarization measurements at La Silla.**
See Abstr. 116.021.

# 035 Space Instrumentation

035.001 **Spacelab I. Expérience ondes dans la couche émissive OH.**
M. Hersé.
Astronomie, Vol. 98, p. 343 – 347 (1984).

035.002 **Some potential applications of solid state nuclear track detectors in aerospace.**
H. B. Knowles.
Nucl. Tracks Radiat. Meas., Vol. 7, No. 1 – 2, p. 35 – 37 (1983). Abstr. in Phys. Abstr., Vol. 87, No. 1237, Entry 68880 (1984). – See Abstr. 012.006.

035.003 **A large area, high pressure, sealed, xenon–filled imaging proportional counter for applications in X–ray astronomy.**
H. D. Thomas, M. J. L. Turner.
Nucl. Instrum. Methods Phys. Res., Sect. A, Vol. 221, No. 1, p. 175 – 182 (1984). Abstr. in Phys. Abstr., Vol. 87, No. 1237, Entry 68888 (1984). – See Abstr. 37.012.073.

035.004 **New applications of X–ray optical techniques.**
J. L. Culhane, W. Cash, R. C. Catura.
Nucl. Instrum. Methods Phys. Res., Sect. A, Vol. 221, No. 1, p. 251 – 264 (1984). Abstr. in Phys. Abstr., Vol. 87, No. 1237, Entry 72880 (1984). – See Abstr. 37.012.073.

035.005 **Imaging systems for low energy γ–ray astronomy.**
A. J. Dean.
Nucl. Instrum. Methods Phys. Res., Sect. A, Vol. 221, No. 1, p. 265 – 277 (1984). Abstr. in Phys. Abstr., Vol. 87, No. 1237, Entry 72881 (1984). – See Abstr. 37.012.073.

035.006 **A gamma–ray telescope with arc minute resolution.**
T. L. Jenkins, G. M. Frye Jr., A. Owens, J. N. Carter, D. Ramsden.
Nucl. Instrum. Methods Phys. Res., Sect. A, Vol. 221, No. 1, p. 278 – 283 (1984). Abstr. in Phys. Abstr., Vol. 87, No. 1237, Entry 72882 (1984). – See Abstr. 37.012.073.

035.007 **A coded mask telescope for the Spacelab 2 mission.**
A. P. Willmore, G. K. Skinner, C. J. Eyles, B. Ramsey.
Nucl. Instrum. Methods Phys. Res., Sect. A, Vol. 221, No. 1, p. 284 – 287 (1984). Abstr. in Phys. Abstr., Vol. 87, No. 1237, Entry 72883 (1984). – See Abstr. 37.012.073.

035.008 **A rotation modulation collimator for imaging in high energy X–ray astronomy.**
J. Theinhardt, R. Burk, E. Kendziorra, R. Staubert.
Nucl. Instrum. Methods Phys. Res., Sect. A, Vol. 221, No. 1, p. 288 – 292 (1984). Abstr. in Phys. Abstr., Vol. 87, No. 1237, Entry 72884 (1984). – See Abstr. 37.012.073.

035.009 **Deep–space measurements of the microwave background anisotropy: first results of the Relikt experiment.**
I. A. Strukov, D. P. Skulachev.
Sov. Astron. Lett., Vol. 10, No. 1, p. 1 – 4 (1984). English translation of 37.035.001.

035.010 **Space plasma research at RAL.**
D. A. Bryant, R. Bingham, T. Edwards, D. S. Hall, A. K. Ward.
J. Br. Interplanet. Soc., Vol. 37, No. 7, p. 309 – 316 (1984).
The Rutherford Appleton Laboratory (RAL) is contributing instruments and a spacecraft to several imminent and exciting new explorations of the plasma phenomena arising from the interaction between the solar wind and the Earth, and the solar wind and a comet. The projects in which the Laboratory is engaged include the AMPTE mission, which will trace the flow of particles injected into the solar wind, the Giotto encounter with Comet Halley, the Viking exploration of the generation of the aurora, and the CRRES and ISTP missions to clarify the structure and dynamics of the Earth's magnetosphere.

035.011 **Spaceborne lasers in atmospheric research.**
J. E. Blamont.
ESA Spec. Publ., ESA SP–202, p. P25 – P28 (1984). – See Abstr. 012.004.
The paper reviews: (1) General principles of lidar technique. (2) Major results obtained in atmospheric physics and chemistry up to 1983. (3) New lasers and their potential utilization in space.

035.012 **Calibration of a space–borne charged particle telescope using protons in the energy range 0.4 to 20 MeV.**
R. G. Marsden, J. Henrion, T. R. Sanderson, K.–P. Wenzel, N. De Bray, H. P. Blok.
Nucl. Instrum. Methods Phys. Res., Sect. A, Vol. 221, No. 3, p. 619 – 626 (1984). Abstr. in Phys. Abstr., Vol. 87, No. 1238, Entry 77710 (1984).

035.013 **Cryogenic telescope on the Infrared Astronomical Satellite (IRAS).**
F. J. Low.
Opt. Eng., Vol. 23, No. 2, p. 122 – 127 (1984). Abstr. in Phys. Abstr., Vol. 87, No. 1238, Entry 77712 (1984).

035.014 **A helium cooled Michelson interferometer for far infrared astronomy on Spacelab.**
S. Drapatz, R. Hofmann, R. Katterloher.
Eighth International Conference on Infrared and Millimeter Waves, p. TH5.2/1–2 (1984). Abstr. in Phys. Abstr., Vol. 87, No. 1238, Entry 77716 (1984). – See Abstr. 012.015.

035.015 **A microprocessor–based position control system for a telescope secondary mirror.**
K. R. Lorell, R. R. Clappier, W. F. Barrows, G. K. Lee.
Automatica, Vol. 20, No. 3, p. 295 – 308 (1984). Abstr. in Phys. Abstr., Vol. 87, No. 1240, Entry 85873 (1984).

035.016 **Performance of a solar soft X–ray spectrograph–telescope.**
W. A. Brown, M. E. Bruner, L. W. Acton.
Ann. Isr. Phys. Soc., Vol. 6, p. 60 – 62 (1984). Abstr. in Phys. Abstr., Vol. 87, No. 1240, Entry 85874 (1984). – See Abstr. 012.018.

035.017 **The effects of epoxy shrinkage on the advanced X–ray astrophysics facility technology mirror assembly.**
L. M. Cohen.
Proc. SPIE Int. Soc. Opt. Eng., Vol. 450, p. 95 – 103 (1984). Abstr. in Phys. Abstr., Vol. 87, No. 1241, Entry 86512 (1984). – See Abstr. 012.019.

035.018 **A modulation collimator for X–ray astronomy at high energies.**
R. Lieu, K. J. Beurle, A. R. Engel, J. J. Quenby.
J. Phys. E, Vol. 17, No. 7, p. 577 – 580 (1984). Abstr. in Phys. Abstr., Vol. 87, No. 1241, Entry 90050 (1984).

035.019 **The ultraviolet telescope aboard the astrophysical space station Astron.**
A. A. Boyarchuk, R. E. Gershberg, L. V. Granitskij, V. M. Kovtunenko, Yu. M. Khodzhayants, V. I. Pronik, A. B. Severnyj, M. N. Krmoyan, G. M. Tovmasyan, P. Cruvellier, G. Courtès, S. T. Hua.
Sov. Astron. Lett., Vol. 10, No. 2, p. 67 – 72 (1984). English translation of 37.035.022.

035.020 **Instrumentation studies for FUSE/Columbus.**
J. F. Osantowski.
ESA Spec. Publ., ESA SP–218, p. 489 - 493 (1984). - See Abstr. 012.024.

The paper briefly summarizes several presentations given by NASA representatives to the Second ESA/NASA Ultraviolet Workshop on May 14–15, 1984 in Frascati, Italy. Primary topics discussed are telescope and spectrograph concepts for the FUSE/Columbus mission.

035.021 **Space Telescope and cosmology.**
R. Giacconi.
Large–scale structure of the universe, cosmology and fundamental physics, p. 393 - 413 (1984). - See Abstr. 012.025.

This paper is identical to the contribution in Adv. Space Res., Vol. 3, Nos. 10 - 12, p. 351 - 365 (1983). See Abstr. 37.035.054.

035.022 **Instrumentation for measuring solar ultraviolet emission aboard artificial earth satellites and meteorological rockets.**
D. A. Gonyukh, T. V. Kazachevskaya, E. D. Mishchenko.
Inst. prikl. geofiz. Moskva, 9 pp. (1983). In Russian. Abstr. in Ref. Zh., 51. Astron., 8.51.663 (1984).

035.023 **Television systems of the space probe "Venera–Halley".**
Sh. Salai, L. Sabo, I. Reni, M. Zhenei, T. Gombosi, B. Seidl, M. Gardosh, R. Z. Sagdeev, G. A. Avanesov, I. V. Barinov, V. I. Tarnopol'skij, V. I. Kostenko, G. I. Tsukanova.
Nauchn. kosm. priborostr., Moskva, No. 2, p. 47 - 60 (1983). In Russian. Abstr. in Ref. Zh., 62. Issled. Kosm. Prostranstva, 8.62.113 (1984).

035.024 **Characteristics of calculation of the optical systems of the television cameras of the "Venera–Halley" project.**
G. I. Tsukanova, E. V. Krivopustova, V. I. Kostenko, V. I. Tarnopol'skij.
Nauchn. kosm. priborostr., Moskva, No. 2, p. 60 - 64 (1983). In Russian. Abstr. in Ref. Zh., 62. Issled. Kosm. Prostranstva, 8.62.114 (1984).

035.025 **Spectrophotometric complex aboard the automatic interplanetary stations Venera 13 and Venera 14.**
V. I. Gnedykh, Yu. M. Golovin, A. V. Grigor'ev, B. E. Moshkin, S. R. Tabaldyev, A. P. Ehkonomov, I. D. Anisimova, S. A. Ignatenko, Z. A. Grishina.
Nauchn. kosm. priborostr., Moskva, No. 2, p. 67 - 71 (1983). In Russian. Abstr. in Ref. Zh., 62. Issled. Kosm. Prostranstva, 8.62.118 (1984).

035.026 **Balloon spectrometer for atmospheric investigations in the infrared range.**
A. S. Beshmenev, B. V. Dement'ev, V. V. Ivanov, S. G. Kukin, M. N. Markov.
Nauchn. kosm. priborostr., Moskva, No. 2, p. 86 - 90 (1983). In Russian. Abstr. in Ref. Zh., 62. Issled. Kosm. Prostranstva, 8.62.119 (1984).

035.027 **Photometric scanning system EhMO–5 on board the IK–Bolgariya 1300 artificial earth satellite.**
N. P. Petkov, Ts. N. Gogosheva, M. M. Gogoshev, A. K. Kuz'min, V. N. Angarov.
Nauchn. kosm. priborostr., Moskva, No. 2, p. 77 - 83 (1983). In Russian. Abstr. in Ref. Zh., 62. Issled. Kosm. Prostranstva, 8.62.121 (1984).

035.028 **Photometric panorama systems for investigation of emissions of the upper atmosphere.**
M. N. Markov, V. E. Tukish.
Nauchn. kosm. priborostr., Moskva, No. 2, p. 83 - 86 (1983). In Russian. Abstr. in Ref. Zh., 62. Issled. Kosm. Prostranstva, 8.62.122 (1984).

035.029 **Pointing of a solar telescope on a rocket or balloon.**
S. Hamana.
Tokyo Astron. Obs. Rep., Vol. 20, No. 2, p. 276 - 292 (1984). In Japanese.

035.030 **A near–infrared mapping spectrometer for investigation of Jupiter and its satellites.**
I. M. Aptaker.
Proc. SPIE Int. Soc. Opt. Eng., Vol. 395, p. 133 - 148 (1983). Abstr. in Phys. Abstr., Vol. 87, No. 1242, Entry 94612 (1984). - See Abstr. 012.030.

035.031 **Cryogenic infrared radiance instrument for Shuttle (CIRRIS) telescope.**
J. S. Titus, D. Wang, M. Ahmadjian, D. R. Smith.
Opt. Eng., Vol. 23, No. 3, p. 319 - 322 (1984). Abstr. in Phys. Abstr., Vol. 87, No. 1242, Entry 94614 (1984).

035.032 **Charge–coupled device television camera for NASA's Galileo mission to Jupiter.**
K. P. Klaasen, M. C. Clary, J. R. Janesick.
Opt. Eng., Vol. 23, No. 3, p. 334 - 342 (1984). Abstr. in Phys. Abstr., Vol. 87, No. 1242, Entry 94615 (1984).

035.033 **Conditions for the visibility of astronomical objects at the work with a submillimeter space telescope.**
M. Z. Khokhlov, G. B. Sholomitskij.
Inst. kosm. issled. Akad. Nauk SSSR. Prepr., No. 823, 34 pp. (1983). In Russian. Abstr. in Ref. Zh., 51. Astron., 8.51.657 (1984).

035.034 **Gamma–ray and X–ray telescope using variable–metric diffraction crystals.**
R. K. Smither.
Ann. N.Y. Acad. Sci., Vol. 422, p. 384 (1984). Abstract. - See Abstr. 012.033.

035.035 **Romteleskopet. Eit framtidig hovudinstrument i astronomisk forsking.**
O. Kjeldseth Moe.
Astron. Tidsskr., Årg. 17, Nr. 3, p. 101 - 111 (1984).

035.036 **Europa bouwde supercamera voor Space Telescope.**
H. C. van de Hulst, T. M. Kamperman.
Zenit, 11. Jaarg., No. 11, p. 452 - 454, 456 - 463 (1984).

035.037 **The ROSAT XUV wide field camera.**
J. P. Pye, A. Wells, R. Willingale, K. A. Pounds.
MPE Rep., No. 184, p. 261 - 267 (1984). - See Abstr. 012.039.

The XUV wide field camera (WFC) is being developed for the ROSAT mission by a consortium of five UK institutions led by the University of Leicester. The WFC will perform the first all–sky survey in the XUV band, to locate point sources to $\lesssim$ 1 arc min, map extended sources and the XUV background, and by use of selected filters, perform broadband photometry.

035.038 **An extensive solid state nuclear track detector array for the study of ultra heavy cosmic ray nuclei aboard the NASA Long Duration Exposure Facility (LDEF).**
A. Thompson, D. O'Sullivan, K.–P. Wenzel, V. Domingo, C. O'Ceallaigh, J. Daly, A. Smit.
Nucl. Tracks Radiat. Meas., Vol. 8, No. 1 - 4, p. 575 - 578 (1984). Abstr. in Phys. Abstr., Vol. 87, No. 1243, Entry 96389 (1984). - See Abstr. 012.045.

035.039 **Purging flow protection of infrared telescopes.**
E. P. Muntz, M. Hanson.
AIAA J., Vol. 22, No. 5, p. 696 - 704 (1984). Abstr. in Phys. Abstr., Vol. 87, No. 1243, Entry 101158 (1984).

035.040 **High sensitivity Si:P and Ge:Be infrared detectors for low photon background astronomy.**
U. Brunsmann.
Proc. SPIE Int. Soc. Opt. Eng., Vol. 395, p. 127 – 132 (1983). Abstr. in Phys. Abstr., Vol. 87, No. 1243, Entry 101160 (1984). – See Abstr. 012.030.

035.041 **Coherent optical system of modular imaging collectors (COSMIC): an approach for a large aperture high angular resolution telescope in space.**
B. Davis, G. Hunt, D. Korsch, M. Nein.
Proc. SPIE Int. Soc. Opt. Eng., Vol. 440, p. 13 – 21 (1984). Abstr. in Phys. Abstr., Vol. 87, No. 1243, Entry 101161 (1984). – See Abstr. 012.042.

035.042 **ADC(CR–39) detector module for Space Shuttle Spacelab–3 cosmic ray experiment.**
S. Biswas, N. Durgaprasad, P. J. Kajarekar, M. N. Vahia, J. S. Yadav, C. Basu, J. N. Goswami, L. M. Kukreja, D. D. Bhawalkar.
Nucl. Tracks Radiat. Meas., Vol. 8, No. 1 – 4, p. 559 – 562 (1984). Abstr. in Phys. Abstr., Vol. 87, No. 1243, Entry 101180 (1984). – See Abstr. 012.045.

035.043 **Two cosmic–ray detectors with high charge and energy resolution in the GeV/nucleon range.**
R. Dwyer, S. Jordan, P. Meyer.
Nucl. Instrum. Methods Phys. Res., Sect. A, Vol. 224, No. 1 – 2, p. 247 – 271 (1984). Abstr. in Phys. Abstr., Vol. 87, No. 1244, Entry 102502 (1984).

035.044 **SIGMA: high resolution space observatory project for gamma–ray sources.**
G. Riviere, J. Paul, P. Mandrou.
Acta Astronaut., Vol. 11, No. 5, p. 251 – 262 (1984). Abstr. in Phys. Abstr., Vol. 87, No. 1244, Entry 105875 (1984).

035.045 **The capture–cell experiment for the first LDEF mission.**
E. Igenbergs, H. Fechtig, F. Hoerz, H. Kuczera, N. Pailer, E. Schneider, R. Walker, E. Zinner.
Acta Astronaut., Vol. 11, No. 5, p. 263 – 268 (1984). Abstr. in Phys. Abstr., Vol. 87, No. 1244, Entry 105876 (1984).

035.046 **Infrared telescope design: implications from cryogenic tests of fused–silica mirrors.**
R. K. Melugin, J. H. Miller.
Proc. SPIE Int. Soc. Opt. Eng., Vol. 433, p. 171 – 177 (1983). Abstr. in Phys. Abstr., Vol. 87, No. 1244, Entry 105878 (1984). – See Abstr. 012.049.

035.047 **Thermal modeling of the Shuttle Infrared Telescope Facility.**
H. L. Gier, W. D. Taylor.
Proc. SPIE Int. Soc. Opt. Eng., Vol. 430, p. 279 – 286 (1983). Abstr. in Phys. Abstr., Vol. 87, No. 1245, Entry 110488 (1984). – See Abstr. 012.051.

035.048 **System design parameter study for Shuttle Infrared Telescope Facility (SIRTF).**
A. J. Mord, M. Bottema, W. P. Devereux, R. L. Poley, D. W. Strecker, F. Tai.
Proc. SPIE Int. Soc. Opt. Eng., Vol. 430, p. 325 – 334 (1983). Abstr. in Phys. Abstr., Vol. 87, No. 1245, Entry 110489 (1984). – See Abstr. 012.051.

035.049 **Cryogenic telescope on the Infrared Astronomical Satellite (IRAS).**
F. J. Low, C. A. Beichman, F. C. Gillett, J. R. Houck, G. Neugebauer, D. E. Langford, R. G. Walker, R. H. White.
Proc. SPIE Int. Soc. Opt. Eng., Vol. 430, p. 288 – 296 (1983). Abstr. in Phys. Abstr., Vol. 87, No. 1245, Entry 110525 (1984). – See Abstr. 012.051.

035.050 **Near–Infrared Mapping Spectrometer optical subsystem development and testing.**
S. A. Macenka.
Proc. SPIE Int. Soc. Opt. Eng., Vol. 430, p. 313 – 324 (1983). Abstr. in Phys. Abstr., Vol. 87, No. 1245, Entry 110526 (1984). – See Abstr. 012.051.

035.051 **Telescope–critical system performance analysis for the Cryogenic Limb Array Etalon Spectrometer (CLAES) on the Upper Atmospheric Research Satellite.**
A. E. Roche, P. B. Forney, H. E. Morrow, M. Anapol.
Proc. SPIE Int. Soc. Opt. Eng., Vol. 430, p. 106 – 114 (1983). Abstr. in Phys. Abstr., Vol. 87, No. 1246, Entry 115668 (1984). – See Abstr. 012.051.

035.052 **A modular approach to developing a large deployable reflector (*astronomical facility*).**
R. Pittman, C. Leidich, F. Mascy, B. Swenson.
Proc. SPIE Int. Soc. Opt. Eng., Vol. 430, p. 115 – 128 (1983). Abstr. in Phys. Abstr., Vol. 87, No. 1246, Entry 115669 (1984). – See Abstr. 012.051.

035.053 **Contour line measurements on ROSAT X–ray mirrors.**
K. Beckstette, E. Heynacher.
Proc. SPIE Int. Soc. Opt. Eng., Vol. 429, p. 126 – 129 (1983). Abstr. in Phys. Abstr., Vol. 87, No. 1246, Entry 115686 (1984). – See Abstr. 012.050.

035.054 **A scanning Fabry–Perot interferometer and its application to space research. VII. Rocket–borne pure interferometric high resolution spectrometer.**
T. Kohno, N. Yajima.
J. Mech. Eng. Lab., Vol. 38, No. 2, p. 74 – 87 (1984). In Japanese. Abstr. in Phys. Abstr., Vol. 87, No. 1246, Entry 115687 (1984).

035.055 **Design and construction of the "Mars Ball" prototype exploration vehicle.**
D. Hilton, C. Cunningham, E. Eplee, D. Grinspoon, A. Hildebrand, T. Jones, S. Pope, N. Schneider, M. Sykes, D. M. Hunten.
Bull. Am. Astron. Soc., Vol. 16, No. 3, Part I, p. 707 (1984). Abstract. – See Abstr. 010.061.

035.056 **The electric and electronic apparatus for a balloon–borne telescope.**
Z. Wu.
Ann. Shanghai Obs., Acad. Sin., No. 5, p. 273 – 282 (1983). In Chinese.

This paper describes the electric and electronic apparatus of a balloon–borne telescope for far infrared observation of the sun. Its major parts – the operating principle and circuit feature of drivers, the equipment for measurement of telescope position and infrared photometer are introduced. The results of high and low temperature tests and indication measurement as well as the observing data and main engineering parameters (e.g. cell temperature) during the flights are given.

035.057 **UV spectro–interferometry from the shuttle bay.**
N. Woolf, S. Synnott, J. Breckinridge.
Bull. Am. Astron. Soc., Vol. 16, No. 3, Part II, p. 747 – 749 (1984). – See Abstr. 012.063.

The 18m–long shuttle bay offers potential for sub–milliarc second resolution at near–UV wavelengths. The authors explore its potential for spectro–interferometric comparison of the continuum source and the broad emission line regions of the quasar 3C273 and Seyfert galaxy NGC 4151. In their concept, two moderate–size telescopes are linked to each other and a beam combiner by a laser. The direction of the baseline is set by a photoelectric guider on the telescopes. Diffraction and light collection combine to make the coherence time and thus sensitivity a rapid function of telescope diameters.

035.058 **The technology requirements for a small space-based astrometric interferometer.**
M. Shao, M. Colavita, D. Staelin, K. Johnston.
Bull. Am. Astron. Soc., Vol. 16, No. 3, Part II, p. 750 - 757 (1984). - See Abstr. 012.063.

Most of the technology needed for space-based interferometers has already been or will soon be demonstrated on ground-based two-telescope interferometers at Mt. Wilson, including the Mark III 20-meter interferometer under construction. This technology includes active systems for sub-arcsecond alignment of the optics, for equalizing optical path lengths to a fraction of a wavelength of light, for finding and tracking stellar frings at low photon levels, and for stabilizing large, 20-meter, optical interferometers, against thermal changes using multiple laser interferometer sensors.

035.059 **Microarcsecond astrometric interferometry.**
R. D. Reasenberg.
Bull. Am. Astron. Soc., Vol. 16, No. 3, Part II, p. 758 - 766 (1984). - See Abstr. 012.063.

POINTS (Precision Optical INTerferometry in Space) is a design concept for an astrometric interferometer originally conceived as a means of performing the light deflection experiment to second order. A version with 2 m baselines, Mini-POINTS, would fit fully assembled with a support spacecraft in one-third of the Shuttle bay. For a pair of tenth-magnitude stars about 90° apart, Mini-POINTS would yield the separation uncertain by 5 μas after a 15-minute observation. The author considers both the design of the instrument and aspects of a mission.

035.060 **Search for other planetary systems using space astronomy.**
D. C. Black.
Bull. Am. Astron. Soc., Vol. 16, No. 3, Part II, p. 767 - 769 (1984). - See Abstr. 012.063.

The author reviews general aspects of the search for other planetary systems, indicates what constitutes a comprehensive search, discusses some of the techniques that can be employed in a search, and concludes with specific remarks concerning the possible role of high-precision optical interferometry from space in a search effort.

035.061 **An astrometry satellite.**
D. G. York, B. Jones, S. Faber, D. Lin, W. van Altena, P. Demarque, J. Hughes, K. Johnston, A. Bunner.
Bull. Am. Astron. Soc., Vol. 16, No. 3, Part II, p. 775 (1984). Abstract. - See Abstr. 012.063.

035.062 **High resolution observations of galactic nuclei.**
T. R. Lauer.
Bull. Am. Astron. Soc., Vol. 16, No. 3, Part II, p. 776 - 778 (1984). - See Abstr. 012.063.

Although much useful work on galactic nuclei will be done by the Space Telescope, much higher resolution will be required for direct observation of many nuclear structures such as broad line regions and central star clusters in any but the nearest galaxies. Direct images are needed to examine the geometry and morphology of active galactic nuclei. High resolution spectroscopic observations of the nuclear stellar dynamics offer the best method to investigate the occurrence of massive nuclear black holes. High resolution arrays will work best on bright nuclear components, but may have difficulty observing the much fainter nuclear star fields.

035.063 **Point response functions for unfilled optical arrays.**
J. H. Spencer, R. S. Simon.
Bull. Am. Astron. Soc., Vol. 16, No. 3, Part II, p. 779 - 786 (1984). - See Abstr. 012.063.

This paper describes a family of optical arrays whose parameters were evaluated by a computer analysis system. For simplicity all the arrays considered were in the same family (a circular ring of evenly spaced array elements), so that comparisons could be made as the number of array elements was varied. The authors conclude that a spaceborne, unfilled, minimum-redundancy array can be built that has performance similar to a filled aperture significantly reducing the cost of space-based imaging systems.

035.064 **High angular resolution optical interferometry from space: technical challenges.**
P. Silverglate.
Bull. Am. Astron. Soc., Vol. 16, No. 3, Part II, p. 787 - 792 (1984). - See Abstr. 012.063.

A monolithic primary mirror greater than ten times ST's diameter, smoothed to diffraction limited accuracy, would present formidable technological challenges and be extremely costly. Lower cost, technologically achievable alternatives for obtaining the same angular resolution would be a thinned aperture or a fringe interferometer. In a thinned aperture system, light from several separated primary mirrors is coherently combined to form an image with high angular resolution. In a fringe interferometer, light from separated telescopes is combined to produce fringes. The requirements of both types of systems on technologies such as optical path compensation, structures, jitter, pointing, and space transport and assembly appear to indicate both types of systems are achievable. The fringe interferometers generally have less stringent requirements than thinned apertures.

035.065 **A 50-meter Michelson stellar interferometer on a space platform.**
M. Faucherre, M. G. Lacasse, P. Nisenson, R. D. Reasenberg, M. Shao, R. V. Stachnik, W. A. Traub.
Bull. Am. Astron. Soc., Vol. 16, No. 3, Part II, p. 793 - 796 (1984). - See Abstr. 012.063.

The authors discuss a 50-meter Michelson stellar interferometer that could be mounted on a space platform. For this device, the estimated magnitude limit would be near +15 mag and, in the ultraviolet (2000 Angstroms), upper limits on angular diameters may be determined down to about 0.2 milliarcsecond. The authors regard the development of such a relatively simple, low-cost device as an important demonstration of the technical and scientific potential of long baseline optical interferometers in space.

035.066 **COSMIC: a high resolution, large collecting area telescope.**
W. A. Traub, N. P. Carleton.
Bull. Am. Astron. Soc., Vol. 16, No. 3, Part II, p. 805 - 809 (1984). - See Abstr. 012.063.

The key advantages of COSMIC (Coherent Optical System of Modular Imaging Collectors) are high angular resolution and the ability to produce images of complex, low-surface-brightness objects such as distant galaxies. If configured as a 36 m filled linear array, COSMIC can have 15 times better angular resolution and 10 times greater collecting area than ST.

035.067 **Space Station-SAMSI: a Spacecraft Array for Michelson Spatial Interferometry.**
R. V. Stachnik, K. Ashlin, S. Hamilton.
Bull. Am. Astron. Soc., Vol. 16, No. 3, Part II, p. 818 - 827 (1984). - See Abstr. 012.063.

A modified version of SAMSI (Spacecraft Array for Michelson Spatial Interferometry), which makes use of the fuel resupply capability of Space Station, could be a powerful device. The combination of a refueling facility with a technique for obtaining image Fourier transform phase information from spectrally dispersed fringes would permit SAMSI to synthesize images equivalent to those produced by huge apertures in space. The authors discuss synthesis of apertures in the 100 to 500 meter range. Reconstruction can be performed to a visual magnitude of at least 8 for 100 Å passbands in 9 hours.

035.068 **TRIO: a kilometric optical array controlled by solar sails.**
A. Labeyrie, B. Authier, J. L. Boit, T. De Graauw, E. Kibblewhite, L. Koechlin, P. Rabout, G. Weigelt.
Bull. Am. Astron. Soc., Vol. 16, No. 3, Part II, p. 828 - 831 (1984). - See Abstr. 012.063.

Concepts studied for ESA's TRIO project are presented: system control in the vicinity of the L5 Lagrangian point, pointing and alignment procedures, sky coverage optical train, image reconstruction through aperture synthesis, scattered light contamination of the coudé beams, algorithms for stabilizing drifting fringes in computer memory, limiting magnitudes.

035.069 **Novel approach to the design and manufacture of the Advanced X-Ray Astrophysics Facility primary mirror.**
G. M. Sanger, W. R. Sigman.
J. Opt. Soc. Am. A, Vol. 1, No. 12, p. 1305 (1984). Abstract.

035.070 **Design principles of the high-resolution spectrograph for the Space Telescope.**
M. Bottema.
J. Opt. Soc. Am. A, Vol. 1, No. 12, p. 1305 – 1306 (1984). Abstract.

035.071 **El Telescopio Espacial.**
J. N. Bahcall, L. Spitzer Jr.
Rev. Astron., Tomo 56, No. 228/229, p. 9 – 21 (1984).

035.072 **Methods for increasing the threshold sensitivity of airborne photometers.**
V. N. Angarov, L. M. Efremkina, V. A. Gladyshev, A. K. Kuz'min.
Polyar. siyaniya i svechenie noch. neba, Moskva, No. 31, p. 168 – 175 (1984). In Russian. Abstr. in Ref. Zh., 62. Issled. Kosm. Prostranstva, 11.62.92 (1984).

035.073 **A stabilized platform carrying one or two telescopes.**
P. Lemaire.
Space research prospects in stellar activity and variability, p. 177 – 190 (1984). – See Abstr. 012.078.

The study of a mission dedicated to stellar variability and activity leads to two specific instrumentations: a white light photometer and an ultraviolet multichannel spectrometer. Although it is possible to combine the two objectives in one instrumentation, for the optimization of the mission two separate instruments will give the best scientific return. The optical schemes presented here are not intended to provide the ultimate design; but they give the sensitivity and resolution expected from classical instrumentation and are useful to define a preliminary mission.

035.074 **Photometric calibrations of an MCP (*Micro-Channel Plate*) space qualified photographic-camera.**
S. Koutchmy, A. Verlhac.
Inst. Astrophys. Paris, Pré-Publ., No. 67, 25 pp. (1984). To appear in Astron. Astrophys.

035.075 **Image processing software system of the X-ray telescope aboard the HINOTORI spacecraft.**
S. Tsuneta.
Ann. Tokyo Astron. Obs., Second Ser., Vol. 20, No. 1, p. 1 – 50 (1984).

Hard X-ray imaging of solar flares were successfully made with the imaging X-ray telescope (SXT) aboard the Japanese solar flare observation satellite HINOTORI, using the technique of the rotating modulation collimator. The attitude-determination/image processing software systems for the SXT are developed and verified. The method of attitude analysis of the spacecraft, and image synthesis procedure are described in detail. By the accurate determination of the spacecraft aspect, the effect of nutation and time variation of the attitude of the spacecraft on the X-ray data are almost completely corrected. This makes possible the extremely accurate determination of the flare location on the solar disk, and the synthesis of the hard X-ray image with excellent quality.

035.076 **GeV gamma-ray astronomy telescopes with high angular resolution.**
B. McBreen.
Astron. Express, Vol. 1, Issue 3, p. 105 – 109 (1984).

Gamma-ray telescopes flown on satellites have poor angular resolution with typical point source error circles of a few square degrees. It is shown that a major improvement in angular resolution for the detection of gamma-rays in the GeV region can be obtained with a single crystal as converter.

035.077 **X-ray astronomy satellite Tenma.**
Y. Tanaka, M. Fujii, H. Inoue, N. Kawai, K. Koyama, Y. Maejima, F. Makino, K. Makishima, M. Matsuoka, K. Mitsuda, T. Murakami, J. Nishimura, M. Oda, Y. Ogawara, T. Ohashi, N. Shibazaki, K. Suzuki, I. Waki, T. Yamagami, I. Kondo, H. Murakami, S. Hayakawa, T. Hirano, H. Kunieda, K. Masai, F. Nagase, N. Sato, Y. Tawara, S. Kitamoto, S. Miyamoto, H. Tsunemi, K. Yamashita, M. Nakagawa.
Publ. Astron. Soc. Jpn., Vol. 36, No. 4, p. 641 – 658 (1984).

Instruments and performance of the X-ray satellite Tenma are described.

035.078 **Performance verification of the gas scintillation proportional counters on board the Tenma satellite.**
K. Koyama, T. Ikegami, H. Inoue, N. Kawai, K. Makishima, M. Matsuoka, K. Mitsuda, T. Murakami, Y. Ogawara, T. Ohashi, K. Suzuki, Y. Tanaka, I. Waki, E. E. Fenimore.
Publ. Astron. Soc. Jpn., Vol. 36, No. 4, p. 659 – 666 (1984).

Eight gas scintillation proportional counters (GSPC) with a total effective area of 640 $cm^2$ are the major X-ray detectors on board the X-ray astronomy satellite Tenma. The energy resolution at 5.9 keV is 9.5% (FWHM), which is twice as good as conventional proportional counters. Detected X-ray photons are pulse-height analyzed into either 128 or 256 energy channels. The channel energy can be set to an accuracy of less than 50 eV. The in-flight performance of the GSPC was verified by observations of the Crab Nebula.

035.079 **Observations of local ISM emission with the Berkeley EUV/FUV Shuttle Telescope.**
C. Martin, S. Bowyer.
NASA Conf. Publ., NASA CP-2345, p. 344 – 347 (1984). – See Abstr. 012.089 (IAU Colloq. No. 81).

The Berkeley Extreme Ultraviolet/Far Ultraviolet Shuttle Telescope (BEST) will be launched on the Space Shuttle in November, 1984, as part of the NASA UVX project. The Berkeley spectrometer will make observations of the cosmic diffuse background in the 600 – 1900 Å band, with a spectral resolution of 10 Å. The sensitivity and spectral resolution of the instrument make it ideal for the study of components of the interstellar medium in the $10^4 - 10^6$K range.

035.080 **Measuring EUV line emission from the hot interstellar medium.**
S. Labov, C. Martin, S. Bowyer.
NASA Conf. Publ., NASA CP-2345, p. 348 – 352 (1984). – See Abstr. 012.089 (IAU Colloq. No. 81).

In the local interstellar medium, the presence of hot gas ($\log T \sim 5$ to 6) has been inferred from measurements of oxygen VI absorption in the ultraviolet, and from diffuse emission in the extreme ultraviolet (EUV) and soft X-ray regions. Here the authors describe a spectrometer that is very sensitive to gas in this temperature range. The instrument is currently under construction for a sounding rocket flight.

035.081 **The Giotto spacecraft system and subsystem design.**
P. Lo Galbo.
ESA J., Vol. 8, No. 3, p. 215 – 244 (1984).

The Giotto mission is a scientific in-situ exploration of the comet Halley. The paper summarises the design constraints imposed on the spacecraft by the mission profile and gives a concise review of the spacecraft system design together with a more detailed description of the various subsystems.

**Explicando astronáutica. O homem e a conquista do espaço.**
See Abstr. 003.021.

**Gamma ray astronomy.**
See Abstr. 003.026.

**Spacelab. Research in earth orbit.**
See Abstr. 003.049.

**IRAS mission operations experience at RAL (*Rutherford Appleton Laboratory*).**
See Abstr. 009.014.

**Application de la photométrie bidimensionnelle à l'astronomie.**
See Abstr. 012.096.

**Advanced technology optical telescopes II. Conference held at London, England, 5 – 6 September 1983.**
See Abstr. 012.097.

**Instrumentation in astronomy V. Conference held at London, England, 7 – 9 September 1983.**
See Abstr. 012.098.

**L'Europe et la France à l'heure du télescope spatial.**
See Abstr. 013.019.

**Deployable reflector configurations.**
See Abstr. 031.008.

**Design and analysis of a two–channel three–mirror X–ray telescope.**
See Abstr. 031.048.

**Glancing incidence EUV telescopes: analysis and comparison.**
See Abstr. 031.052.

**Grazing incidence telescopes: a new class for soft X–ray and EUV spectroscopy.**
See Abstr. 031.054.

**Alignment and phasing of deployable telescopes.**
See Abstr. 032.006.

**Design concepts for reflector antenna structures.**
See Abstr. 033.001.

**Effects produced by the ionosphere on radio interferometry.**
See Abstr. 036.016.

**Comparison of Space Telescope and 4–meter ground–based telescope: faint galaxy detection and photometry.**
See Abstr. 036.017.

**Scientific expectations from high accuracy astrometry.**
See Abstr. 036.083.

**A computer simulation of faint object image formation by a rotating linear aperture.**
See Abstr. 036.085.

**The uses of a VLT for complementing observations with radio and space telescopes.**
See Abstr. 036.099.

**Data analyis of electron measurements of the Plasma Experiment aboard the Helios probes.**
See Abstr. 036.154.

**European X–ray satellite on a successful course.**
See Abstr. 051.021.

**The ROSAT all–sky survey in X–rays.**
See Abstr. 051.045.

**Stellar activity and variability with UVSAT (Japanese project for UV astronomy).**
See Abstr. 051.068.

**Recent solar observations from the X–ray polychromator on the repaired SMM satellite.**
See Abstr. 076.013.

# 036 Methods of Observation and Reduction, Data Processing

036.001 **Relaxing the isoplanatism assumption in self–calibration; applications to low–frequency radio interferometry.**
F. R. Schwab.
Astron. J., Vol. 89, No. 7, p. 1076 – 1081 (1984).

A fundamental assumption of the self–calibration/hybrid mapping techniques, which are in widespread use in aperture synthesis data reduction, is one of isoplanatism: that over any given element of the array, all incident wave fronts, regardless of their directions of arrival, are subject to the same tropospheric/ionospheric path delays. This is a poor assumption in the case of a low–frequency array (operating at frequencies lower than $\sim$300 MHz), because of the large field of view of such an instrument and because of the wavelength dependence of wave–front perturbation by the ionosphere. The isoplanatism assumption may be relaxed by incorporating an interpolation formula in the solution for antenna/IF phases, and the usual mapping/deconvolution techniques then must be modified in order to cope with space–variant effects.

036.002 **Optimum signal to noise ratios for observations of faint galaxies.**
S. Phillipps.
Astron. Astrophys., Vol. 136, No. 2, p. 371 – 372 (1984).

The aperture size necessary to optimise the signal to noise ratio for observations of faint galaxies is calculated in terms of the galaxies' scale sizes. Angular sizes are calculated for characteristic cases. It is found that the optimum is attainable from the ground for faint spirals, but that the Space Telescope should be able to observe ellipticals more than a magnitude fainter than an identical ground based system.

036.003 **Statistical analysis of observations in gamma–ray astronomy.**
T. Li, Y. Ma.
Kexue Tongbao (Beijing), Vol. 28, No. 5, p. 717 (1983). Abstr. in Phys. Abstr., Vol. 87, No. 1237, Entry 73102 (1984).

036.004 **Variable stars: observer effects on estimates of magnitude.**
N. T. Higgs, J. V. Vincent, R. Fleet.
Mon. Notes Astron. Soc. S. Afr., Vol. 43, Nos. 1 – 4, p. 13 – 19 (1984).

036.005 **Photographic determination of high–precision coordinates of artificial celestial bodies.**
D. P. Duma.
Sov. Astron., Vol. 28, No. 1, p. 108 – 111 (1984). English translation of 37.036.002.

036.006 **Sonnen–, Mond– und Planetenphotographie mit Amateurteleskopen. Teil 2.**
H. Hornung, P. Hückel.
Sterne Weltraum, 23. Jahrg., Nr. 7, p. 390 – 395 (1984).

036.007 **Die visuelle Kometenbeobachtung und ihre Auswertung.**
A. Kammerer, J. Linder.
Sterne Weltraum, 23. Jahrg., Nr. 8–9, p. 466 – 470 (1984).

036.008 **Faint object detection on photographic plates.**
W. J. Couch, R. S. Ellis, E. J. Kibblewhite, D. F. Malin, J. Godwin.
Mon. Not. R. Astron. Soc., Vol. 209, No. 2, p. 307 – 315 (1984).
The authors compare the detection limits achieved by Malin's high–contrast print method and Kibblewhite's plate measuring machine on a test region examined on a high quality prime–focus plate. The high–contrast print is able to reveal images at the completeness limit of the measuring machine data and thus offers a quick way of finding very faint objects whilst avoiding the need for extensive data processing. Some aspects of using this technique for finding distant clusters of galaxies are briefly discussed.

036.009 **IRAS fast mover program.**
B. C. Stewart, J. K. Davies, S. F. Green.
J. Br. Interplanet. Soc., Vol. 37, No. 7, p. 348 – 352 (1984).
During the IRAS mission, quick–look processing of the science data was carried out to look for fast–moving Solar System objects. Rapid identification of such objects allowed immediate, ground based, follow–up observations to confirm their existence and measure their orbital characteristics. Although designed to discover asteroids approaching the Earth's orbit, the analysis unexpectedly detected a number of comets and a previously unobserved tail for comet Tempel 2.

036.010 **IBM PC interface for the Starlight–1 photometer.**
D. McDavid.
I.A.P.P.P. Commun., No. 16, p. 12 – 16 (1984).

036.011 **Future of laser interferometry in space.**
B. Bertotti.
ESA Spec. Publ., ESA SP–202, p. P147 – P150 (1984). – See Abstr. 012.004.
The author discusses two applications of laser interferometry in space: the measurement of the fine structure of the gravity field of the earth by means of low orbiting satellites and low frequency gravitational wave detection in interplanetary space.

036.012 **An astronomical processing problem.**
B. Jones.
Comput. Syst., Vol. 4, No. 5, p. 55 – 57 (1984). Abstr. in Phys. Abstr., Vol. 87, No. 1239, Entry 82421 (1984).

036.013 **IUE data reduction algorithms and programs for ESHI files.**
A. Sapar, L. Sapar.
Tartu Astrofüüs. Obs. Teated, Nr. 71, p. 17 – 57 (1984).
The IUE data, obtained with a high resolution mode of on–board echelle spectrographs, have been reduced by the staffs of the IUE ground observatories. The paper contains both a general description of the processing routines used and listings of the composed programs written in FORTRAN IV. Some general results are given about the IUE echelle spectrum data obtained in the process of data reduction.

036.014 **Low dispersion IUE data reduction program from ESSR files.**
A. Sapar, L. Sapar.
Tartu Astrofüüs. Obs. Teated, Nr. 71, p. 58 – 66 (1984).

036.015 **Long–baseline Michelson interferometry with large ground–based telescopes operating at optical wavelengths. I. General formalism. Interferometry at visible wavelengths.**
F. Roddier, P. Lena.
J. Opt. (Paris), Vol. 15, No. 4, p. 171 – 182 (1984). Abstr. in Phys. Abstr., Vol. 87, No. 1241, Entry 90094 (1984).

036.016 **Effects produced by the ionosphere on radio interferometry.**
T. A. T. Spoelstra, H. Kelder.
Radio Sci., Vol. 19, No. 3, p. 779 – 788 (1984). Abstr. in Phys. Abstr., Vol. 87, No. 1241, Entry 90095 (1984).

036.017 **Comparison of Space Telescope and 4–meter ground–based telescope: faint galaxy detection and photometry.**
J. A. Tyson.
Publ. Astron. Soc. Pac., Vol. 96, No. 581, p. 566 – 573 (1984).
The Space Telescope (ST) wide–field CCD camera is compared with a 4–meter ground–based telescope with existing CCD camera, for the detection, total photometry, surface photometry, and classification of faint galaxy images. A signal–to–noise ratio calculation shows that for equal–length exposures and for the expected angular sizes of faint galaxy images at high redshift, the ground–based telescope should out–perform the space telescope. The ST out–performs ground–based systems for very high resolution imaging of brighter galaxies and in the UV.

036.018 **A bootstrap resampling analysis of galaxy clustering.**
J. D. Barrow, S. P. Bhavsar, D. H. Sonoda.
Mon. Not. R. Astron. Soc., Vol. 210, No. 2, p. 19P – 23P (1984).
The authors illustrate the use of the bootstrap resampling technique for determining standard errors. This is done by applying it to the calculation of the galaxy–galaxy correlation function in the Zwicky catalogue (Zwicky et al. 1961 – 68). Besides giving specific significance levels for the parameters of the calculated two–point correlation function the authors aim to display how this simple technique can be useful in assigning significance levels to single parameters derived from non–repeatable observations.

036.019 **Enhancements to the deconvolution algorithm "CLEAN".**
D. G. Steer, P. E. Dewdney, M. R. Ito.
Astron. Astrophys., Vol. 137, No. 2, p. 159 – 165 (1984).
The authors describe changes to the deconvolution algorithm CLEAN, which are effective in preventing the formation of "stripes" or "corrugations" during the processing of images with extended features. The modifictions also reduce the arithmetic rounding problems that may be encountered when a very large number of subtractions are performed with limited precision. The major modification is to remove components in groups rather than individually.

036.020 **Le problème de la phase dans l'observation des planètes à l'astrolabe.**
F. Chollet.
Astrométrie, géodésie, mécanique céleste, p. 93 – 97 (1984). – See Abstr. 012.021.

036.021 **Préparation de la réduction des données Hipparcos.**
J. Kovalevsky.
Astrométrie, géodésie, mécanique céleste, p. 191 – 196 (1984). – See Abstr. 012.021.

036.022 **Observabilité des étoiles doubles par Hipparcos.**
P. Bacchus.
Astrométrie, géodésie, mécanique céleste, p. 197 – 201 (1984). – See Abstr. 012.021.

036.023 **Précision astrométrique en présence de bruit.**
J. Gay, J.–M. Pochet.
Astrométrie, géodésie, mécanique céleste, p. 233 – 237 (1984). – See Abstr. 012.021.

036.024 **Estimation de la qualité des observations à l'astrolabe photoélectrique.**
G. Billaud.
Astrométrie, géodésie, mécanique céleste, p. 239 – 244 (1984). – See Abstr. 012.021.

036.025 **Numérisation des comparateurs Zeiss.**
J. M. Rousseau.
Astrométrie, géodésie, mécanique céleste, p. 257 – 262 (1984). – See Abstr. 012.021.

036.026 **Le dessin de la chevelure interne d'une comète: généralités et conseils pratiques.**
J.–C. Merlin.
Astronomie, Vol. 98, p. 407 – 409 (1984).

036.027 **The remote control run from La Serena, June 10 – 17, 1984.**
G. Raffi, M. Tarenghi.
Messenger, No. 37, p. 1 – 3 (1984).

036.028 **Data saving and banking at La Silla.**
S. Vidal.
Messenger, No. 37, p. 13 – 14 (1984).

036.029 **On the accuracy of the wavelength calibration and of the flat–fielding of the CCD CASPEC spectra.**
S. D'Odorico, D. Ponz.
Messenger, No. 37, p. 24 – 26 (1984).

036.030 **Examination of a proposed technique for the economical detection and analysis of ultra–high energy cosmic ray showers.**
R. W. Clay, B. R. Dawson.
Aust. J. Phys., Vol. 37, No. 3, p. 309 – 315 (1984).

A study has been made of a recently suggested scheme for the economical detection and analysis of ultra–high energy cosmic ray showers. It is demonstrated that the scheme should be feasible, particularly if detectors are used to sample the γ–ray component of the showers.

036.031 **Measuring the sizes of stars.**
R. Hanbury Brown.
Patrika, No. 7, p. 9 – 10 (1984).

036.032 **Standard stars: automatic line identification of IUE high resolution spectra.**
M. Ramella, F. Castelli, M. L. Malagnini, C. Morossi, F. Pasian.
ESA Spec. Publ., ESA SP–218, p. 469 – 471 (1984). – See Abstr. 012.024.

The authors present an automatic line identification procedure of IUE high resolution spectra. As an example they show its application to the SWP 3582 image of the B2.5 IV star HD 35708.

036.033 **IUE data reduction facilities on Starlink.**
G. E. Bromage.
ESA Spec. Publ., ESA SP–218, p. 473 – 475 (1984). – See Abstr. 012.024.

The Starlink Software Collection is implemented throughout Starlink and on many other Vax (VMS) systems in the U.K. and worldwide, and is freely available to any astronomical institute on request. It includes a large amount of software of relevance to IUE data reduction. These programs and facilities are listed and summarised in the paper. In particular, the versatile and friendly IUEDR facility is briefly described and advertised to the IUE community.

036.034 **The NASA IUE data analysis facilities.**
R. Thompson, N. Remage Evans.
ESA Spec. Publ., ESA SP–218, p. 477 – 479 (1984). – See Abstr. 012.024.

After three years of operation, the staff and users of the NASA IUE Data Analysis Facilities have developed a wide variety of software for the further reduction and analysis of IUE data. Routines are written in the Interactive Data Language and run under the RSX–11M operating system on a PDP 11/44 computer. This presentation highlights the recent developments of the data analysis facilities.

036.035 **Thermoluminescence method for measurement of short-wave solar radiation in space and its instrumental realization.**
D. A. Gonyukh.
Inst. prikl. geofiz. Moskva, 8 pp. (1983). In Russian. Abstr. in Ref. Zh., 51. Astron., 8.51.664 (1984).

036.036 **Construction of confidence intervals in determining photometric distances of stars.**
R. M. Dzigvashvili, T. G. Mdzinarishvili.
Soobshch. Akad. Nauk Gruz. SSR, Tom 112, No. 3, p. 525 – 528 (1983). In Russian. Abstr. in Ref. Zh., 51. Astron., 8.51.667 (1984).

036.037 **Use of the photographic equidensitometric method for measurement of photographic images of the solar disk.**
A. V. Arkhangel'skij.
L'vov. politekh. inst. L'vov, 13 pp. (1984). In Russian. Abstr. in Ref. Zh., 51. Astron., 8.51.703 (1984).

036.038 **S wie Sharpless – Beobachtergruppe hat Erfolg beim Aufsuchen von S–Nebeln.**
K. Buse.
Sterne Weltraum, 23. Jahrg., Nr. 10, p. 522 – 523 (1984).

036.039 **Möglichkeiten der Meteorfotografie.**
J. Rendtel.
Orion, 42. Jahrg., Nr. 204, p. 196 – 198 (1984).

036.040 **Data management at AAVSO Headquarters.**
E. O. Waagen.
J. Am. Assoc. Variable Star Obs., Vol. 13, No. 1, p. 23 – 26 (1984).

The acquisition at AAVSO Headquarters of in–house Ithaca Intersystems microcomputers, as well as graphics hardware, has led to some significant changes in the manner in which incoming observations are processed and prepared for publication. Current data management procedures are discussed.

036.041 **The DDO data network: observing with a remote computer.**
S. W. Mochnacki, W. G. Weller, S. Chew.
Publ. Astron. Soc. Pac., Vol. 96, No. 582, p. 663 – 666 (1984).

The authors have implemented a serial data network linking the terminals and computers at the David Dunlop Observatory with those of the Department of Astronomy, 25 kilometers away. The hardware and communication protocols are described. A remote superminicomputer is used for on–line data acquisition and analysis.

036.042 **Some methods for searching for weak spectra.**
J. I. Katz.
Astron. J., Vol. 89, No. 9, p. 1429 – 1434 (1984).

The author examines some methods for searching for and determining the orbital parameters of weak spectra buried beneath the noise of a series of digitally recorded spectrograms.

036.043 **An intercontinental baseline coincidence search for optical flashes with two Schmidt telescopes.**
B. E. Schaefer, R. Vanderspek, H. V. Bradt, G. R. Ricker.
Astrophys. J., Vol. 283, No. 2, p. 887 – 889 (1984).

The authors have searched for short duration flashes of light from the sky over a wide field of view. No celestial flashes were found and a 3 $\sigma$ upper limit on the flash rate is derived. The search technique consisted of simultaneously photographing an area of sky with two widely separated Schmidt cameras. This arrangement eliminates local "flash" sources. This study can be used to place the probabilities of $9 \times 10^{-5}$, $2 \times 10^{-5}$, and $2 \times 10^{-10}$ that the three flashes of Schaefer et al. were caused by background celestial flashes, local celestial flashes, or photographic defects respectively.

036.044 **On an approach to automation of reduction of astronomical information.**
O. B. Vasyunina.
Vestn. Leningr. Univ., Mat. Mekh. Astron., 13 pp. (1984). In Russian. Abstr. in Ref. Zh., 51. Astron., 9.51.45 (1984).

036.045 **Esperienze fotografiche sulla fotosfera solare.**
L. Piovan, A. Milani.
Astronomia, N. 3, p. 3 – 8 (1984).

036.046 **Influence of errors using an eyepiece micrometer in zenith methods of astronomical determinations.**
V. Z. Khalkhunov.
Sb. nauchn. tr. TsNII geod., aehrosemki i kartogr., No. 234, p. 87 – 99 (1983). In Russian. Abstr. in Ref. Zh., 52. Geod. Aehrosemka, 9.52.57 (1984).

036.047 **Wiener filtering and cleaning in a general image processing context.**
J. H. T. Bates, W. R. Fright, R. H. T. Bates.
Mon. Not. R. Astron. Soc., Vol. 211, No. 1, p. 1 – 14 (1984).

The algorithm CLEAN is known to be effective for removing sidelobes from an image of Fourier transformed incomplete data when the true image consists of isolated point sources. It is shown here that Cornwell's recent modification of CLEAN is closely related to another result (established in this paper): by cleaning with a dirty beam that has been modified in accord with the Wiener approach to deconvolution, clean maps are obtained equivalent to those produced by the conventional two stage procedure of cleaning followed by reconvolving with a clean beam. It is argued both that the standard theory of the Wiener filter is inappropriate in most image processing contexts and that the filter's operation should be interpreted more straightforwardly.

036.048 **Maximum entropy image reconstruction: general algorithm.**
J. Skilling, R. K. Bryan.
Mon. Not. R. Astron. Soc., Vol. 211, No. 1, p. 111 – 124 (1984).

Maximum entropy is an optimal technique of image reconstruction, widely applicable in astronomy and elsewhere. The authors present a general–purpose algorithm, capable of generating maximum entropy images from a wide variety of types of data.

036.049 **Reduction of instrumental polarization of coelostat telescopes.**
G. Bachmann.
Bull. Astron. Inst. Czech., Vol. 35, No. 5, p. 308 – 312 (1984).

A new vector magnetograph now in operation permits measurements of polarization of 0.0005 and thus requires precautions to be taken with respect to the instrumental polarization caused by the mirrors of tower telescopes. The author carried out measurements of the instrumental polarization in the meridian position of the coelostat. In the channels for the transverse magnetic field he obtained signals as strong as from magnetic fields of 2000 G. As a first precaution, the author introduced compensation by means of a tilted glass plate. He took further precautions by searching for optimum azimuth angles of the auxiliary coelostat mirror which provide less instrumental polarization. The results have shown that for each hour angle an auxiliary azimuth exists, where one of the three components of instrumental polarization, caused by intensity crosstalk, is zero. This offers the possibility to diminish the two linear components or to eliminate the circular component.

036.050 **Determination of observational conditions of celestial bodies by means of the DISPO system.**
R. K. Kazakova, A. V. Krivov.
Inst. prikl. mat. Akad. Nauk SSSR. Prepr., No. 73, 37 pp. (1984). In Russian. Abstr. in Ref. Zh., 51. Astron., 10.51.861 (1984).

036.051 **Investigation of statistical characteristics of the brightness of the atmospherical background and of the density of the stellar background in the near–equatorial region of the celestial sphere.**
S. I. Vereshchagin, S. A. Likhomanov.
Izv. vuzov. Geod., aehrofotosemka, No. 2, p. 56 – 60 (1984). In Russian. Abstr. in Ref. Zh., 52. Geod. Aehrosemka, 10.52.94 (1984).

036.052 **Astronomical Fourier transform spectroscopy revisited.**
S. T. Ridgway, J. W. Brault.
Annu. Rev. Astron. Astrophys., Vol. 22, p. 291 – 317 (1984). – See Abstr. 003.024.

The principal objective of this review is to introduce the Fourier transform spectrometer (FTS) and to discuss both its demonstrated capabilities and its limitations for astronomical measurements. A very brief introduction to the optical characteristics of the FTS is followed by a point–by–point consideration of the distinctive features of the FTS that make it attractive or unattractive for particular astronomical applications. To further clarify the possible uses of the technique, the final section of the review describes in some detail numerous successful applications of the FTS.

036.053 **Numerical restoration of astronomical images.**
J. N. Heasley.
Publ. Astron. Soc. Pac., Vol. 96, No. 583, p. 767 – 772 (1984).

The two iterative deconvolution algorithms are examined in the context of restoring optical images for the effects of seeing. Both algorithms are easy to program, computationally efficient, and are well suited to image restoration applications even on minicomputers. Examples are presented showing restorations of computer simulated images, CCD observations of Uranus and Neptune, and of the M87 jet. Both procedures are found to give resolution improvements in the restored image comparable to that obtained by the Maximum Entropy Method.

036.054 **On a displacement of the photometric center from the center of mass in positional observations of comets and minor planets.**
G. Sitarski.
Acta Astron., Vol. 34, No. 2, p. 269 – 280 (1984).

The problem is considered in two aspects: (1) as an effect of phase which could be taken into account in the case of observations of big asteroids, and (2) as a displacement of the photometric center from the center of mass along the radius–vector in the case of observations of comets. In the latter case the displacement of both centers is assumed as $\Delta r = Df(r)$ where $r$ is a distance of the comet from the Sun; a constant value of $D$ should be determined from the observational equations for an individual comet. Investigations based on analyzing of observations of Comet Burnham 1960 II show that one can obtain good results taking $f(r) = r^{-3}\exp(-r^2/2)$. Numerical results concerning the orbit determination of Pallas and of some comets are presented.

036.055 **Data acquisition and recording system for two–dimensional images of the solar corona during balloon–borne observations of the total solar eclipse of June 11, 1983.**
S. Isobe, Y. Koma, S. Miyaki.
AAS Photo–Bull., No. 36, p. 12 – 15 (1984).

036.056 **Investigation of the constancy of scale of astronomical negatives from results of photographic field observations of stars.**
V. M. Kolgunov, M. P. Kulinich.
Geod., kartogr., aehrofotosemka, L'vov, No. 39, p. 27 – 31 (1984). In Russian. Abstr. in Ref. Zh., 52. Geod. Aehrosemka, 10.52.93 (1984).

036.057 **Transit time determination using a pseudo–noise grid.**
J.–M. Pochet.
Astron. Astrophys., Vol. 139, No. 2, p. 517 – 520 (1984). In French.

A new technique of signal modulation fitted to a photoelectric astronomical instrument is described. It consists of a pseudo–

noise grid generated from a Hadamar sequence of 0's and 1's. The transit time of the star is recovered performing the correlation between the recorded signal and a theoretical reference. In the case of the astrolabe, two images of the observed star are present in the focal plane. The instrument can work either with a periodic grid or with a pseudo-noise grid which allows the detection of the position of each image independently of the position of the grid in the focal plane. Finally, an example of a signal obtained with the photoelectric astrolabe of CERGA, and the corresponding correlation function are presented.

036.058 **A revised Gull–Daniell algorithm in the maximum entropy method.**
N.-I. Wu.
Astron. Astrophys., Vol. 139, No. 2, p. 555 – 557 (1984).

The Gull–Daniell maximum entropy method algorithm for data processing of images in radio astronomy is briefly reviewed. Its merits and drawbacks are discussed. Two significant revisions to this algorithm are presented. The performance of the original and revised versions are compared.

036.059 **On autonomous tracking of a minor celestial body.**
A. A. Sukhanov, P. E. Elyasberg (*P. E. Ehl'yasberg*).
Acta Astronaut., Vol. 11, No. 5, p. 247 – 250 (1984). Abstr. in Phys. Abstr., Vol. 87, No. 1244, Entry 105865 (1984).

036.060 **Measurement of stellar diameters by combining the light issuing from two telescopes.**
M. Faucherre.
Ciel Espace, Vol. 195, p. 21 – 30 (1983). In French. Abstr. in Phys. Abstr., Vol. 87, No. 1244, Entry 105888 (1984).

036.061 **Climatology capabilities of a gamma-ray spectrometer at Mars.**
A. E. Metzger.
Bull. Am. Astron. Soc., Vol. 16, No. 3, Part I, p. 678 – 679 (1984). Abstract. – See Abstr. 010.061.

036.062 **Techniques and concepts of near-earth asteroid searches.**
R. S. Dunbar, E. F. Helin, D. L. Matson.
Bull. Am. Astron. Soc., Vol. 16, No. 3, Part I, p. 690 (1984). Abstract. – See Abstr. 010.061.

036.063 **Photometric approaches to detecting other solar systems.**
W. J. Borucki.
Bull. Am. Astron. Soc., Vol. 16, No. 3, Part I, p. 703 (1984). Abstract. – See Abstr. 010.061.

036.064 **A finite-element approach to two-dimensional photoclinometry.**
R. L. Kirk.
Bull. Am. Astron. Soc., Vol. 16, No. 3, Part I, p. 709 (1984). Abstract. – See Abstr. 010.061.

036.065 **Graphics standards in astronomy.**
C. Stoll.
Bull. Am. Astron. Soc., Vol. 16, No. 3, Part I, p. 732 (1984). Abstract. – See Abstr. 010.065.

036.066 **Corrections for biases in slope estimation.**
H. L. Marshall.
Bull. Am. Astron. Soc., Vol. 16, No. 3, Part I, p. 733 (1984). Abstract. – See Abstr. 010.065.

036.067 **Separation of large-scale photospheric Doppler patterns.**
H. B. Snodgrass.
Sol. Phys., Vol. 94, No. 1, p. 13 – 31 (1984).

Mount Wilson solar Doppler data spanning January 1967 to March 1984 are refit with an expanded set of functions representing the line-of-sight components of rotation, limbshift and meridional flow. The long-standing problem of crosstalk among the fit-determined coefficients is eliminated by orthonormalization with respect to the solar disk of the function space representing each motion class. Examination of the new coefficients shows clear evidence for their variation over the solar cycle: for the rotation coefficients, this variation is a low mode torsional oscillation, and for the limbshift, it appears consistent with the suppression of small-scale convection by magnetic activity. The meridional flow is found to be poleward and slightly faster at low latitudes.

036.068 **Ermittlung des Flächenbedeckungsgrades aus Sonnenfinsternisfotos.**
J. Lapsien.
Sonne, Jahrg. 8, Nr. 32, p. 172 – 174 (1984).

036.069 **Digital image analysis for astronomy.**
F. J. Ahern.
J. R. Astron. Soc. Can., Vol. 78, No. 5, p. 200 (1984). Abstract. – See Abstr. 010.401.

036.070 **Enhancements to the deconvolution algorithm "CLEAN".**
P. E. Dewdney, D. G. Steer, M. R. Ito.
J. R. Astron. Soc. Can., Vol. 78, No. 5, p. 201 (1984). Abstract. – See Abstr. 010.401.

036.071 **Progress report on the CFH image processing system.**
C. A. Christian, D. Crabtree.
J. R. Astron. Soc. Can., Vol. 78, No. 5, p. 202 (1984). Abstract. – See Abstr. 010.401.

036.072 **Photoelectric photometric data acquisition.**
J. L. Hopkins.
I.A.P.P.P. Commun., No. 17, p. 13 – 21 (1984).

036.073 **Influence of earth tide and $M_2$ wave of ocean tide on VLBI.**
W. Jin, C. Qin.
Ann. Shanghai Obs., Acad. Sin., No. 5, p. 31 – 40 (1983). In Chinese.

With the rapid development of VLBI used more and more widely in the field of astrometry and geodesy, it has become one of the new techniques that has a promising future. Recently the accuracy of length of baseline determined by VLBI has attained 3 – 5 cm. The influential factors such as the earth and the ocean tides, which are greater than the order of 1 cm should be considered.

036.074 **An iteration algorithm of the stellar trigonometric parallax reduction with the central overlap technique.**
J. Wang.
Ann. Shanghai Obs., Acad. Sin., No. 5, p. 111 – 121 (1983). In Chinese.

036.075 **On precision of measurement of galaxy images at Zô-Sè.**
Z. Chu.
Ann. Shanghai Obs., Acad. Sin., No. 5, p. 132 – 135 (1983). In Chinese.

There are plates with galaxy images at the section Zô-Sè of Shanghai Observatory. They have been photographically obtained with the 400 mm astrograph at Zô-Sè from the 1950's to the beginning of the 1960's. With the purpose of investigating measurement errors of images of galaxies and stars on plates and whether these plates can be used to determine absolute proper motions of stars, the author has estimated the precision of measurement of galaxy and star images on plates. The probable errors of measuring images of galaxies and stars are $\pm 0\farcs21$ and $\pm 0\farcs14$ respectively, being close to those of Pulkovo or Goloseevo, which indicate that these plates can be used to determine the absolute proper motions of stars.

036.076 **A set of programs for random process spectral analysis.**
V. S. Lebedev.
Bull. Spec. Astrophys. Obs. – North Caucasus, Vol. 17, p. 16 – 23 (1984). English translation of 34.036.241.

036.077 **Automated processing of the spectra of celestial objects.**
A. A. Korovyakovskaya, Yu. P. Korovyakovskii (*Yu. P. Korovyakovskij*).
Bull. Spec. Astrophys. Obs. – North Caucasus, Vol. 17, p. 44 – 53 (1984). English translation of 34.036.242.

036.078 **Number of readings as a factor in coordinate measurements on very long baseline interferometers.**
V. S. Itenberg, A. M. Finkel'shtein (*A. M. Finkel'shtejn*).
Bull. Spec. Astrophys. Obs. – North Caucasus, Vol. 17, p. 82 – 85 (1984). English translation of 34.036.243.

036.079 **The role of the frequency standard in radiointerferometric measurements.**
A. F. Dravskikh, A. M. Finkel'shtein (*A. M. Finkel'shtejn*).
Bull. Spec. Astrophys. Obs. – North Caucasus, Vol. 17, p. 100 – 103 (1984). English translation of 34.036.244.

036.080 **Compiling technique of special reference catalogues of star positions and its application to comet Halley watch.**
Yu. A. Shokin.
Astrometr. Astrofiz., Vyp. 53, p. 48 – 53 (1984). In Russian.

For successful solution of some astrometric problems it is worth compiling special photographic catalogues of star positions. The paper describes the compiling technique of special reference catalogues containing from several hundred to several thousand stars. The project of a special reference catalogue of faint stars is proposed for determination of comet Halley's positions with large telescopes.

036.081 **Taking into account the magnitude equation using exposures of different duration. I. A mathematical model for taking into account the magnitude equation.**
G. A. Ivanov, A. B. Onegina, A. I. Yatsenko.
Astrometr. Astrofiz., Vyp. 53, p. 53 – 56 (1984). In Russian.

The mathematical modelling method used eliminates the linear part of the magnitude equation error owing to an increase in both the number of star images and their diameters' range on two exposures of different duration.

036.082 **On a possibility of quantitative estimation of meteorological conditions constancy in observations with zenith telescopes.**
I. V. Dzhun'.
Astrometr. Astrofiz., Vyp. 53, p. 93 – 97 (1984). In Russian.

A simple method is proposed to control the constancy of meteorological conditions in observations with zenith telescopes. The method is based on the construction of confidence intervals for the hypothesis that $E > 0$, where E is the excess of observational error distribution. A table has been compiled to estimate the probability of the hypothesis $E > 0$ while calculating normal latitude values. In the case of positive distribution excess the Jeffreys estimation method is recommended.

036.083 **Scientific expectations from high accuracy astrometry.**
J. L. Russell.
Bull. Am. Astron. Soc., Vol. 16, No. 3, Part II, p. 770 – 774 (1984). – See Abstr. 012.063.

Star constants, i.e. position, proper motion, parallax, perspective acceleration, and orbital motion, are the measured quantities in the reduction of a series of astrometric observations made over time. This discussion follows the effects on the accuracy of the star constants and on the quantities calculated from them (distances, masses, and radial velocities) for positional measurements with the historical level of accuracy (+/–100 mas) to those currently in development (+/–2 mas) to that for proposed space instruments (up to +/–0.001 mas). Also considered is the expanded list of objects for which astrometric information will become important in increasing the understanding of the universe.

036.084 **Stars, atmospheres, and shells: potential for high–resolution imaging.**
A. K. Dupree, S. L. Baliunas, E. F. Guinan.
Bull. Am. Astron. Soc., Vol. 16, No. 3, Part II, p. 797 – 804 (1984). – See Abstr. 012.063.

Substantial progress in stellar physics can result from one– or two–dimensional imaging at a spatial resolution of $10^{-3}$ to $10^{-5}$arc sec. The direct measurement of stellar diameters, surface and atmospheric features, and extended shells can significantly advance the knowledge of stellar structure, activity, and evolution. A dwarf star – the sun – and a supergiant star – Alpha Orionis – are used to illustrate the possibilities of measurements with high spatial resolution. In addition, magnetically sensitive lines can trace the distribution of activity across a stellar surface.

036.085 **A computer simulation of faint object image formation by a rotating linear aperture.**
M. G. Lacasse, W. A. Traub.
Bull. Am. Astron. Soc., Vol. 16, No. 3, Part II, p. 810 – 817 (1984). – See Abstr. 012.063.

The authors are developing numerical methods of image reconstruction for use with an orbiting array of optical telescopes, such as COSMIC, in which a rotating rectangular aperture mimics a large circular aperture. They have performed numerical simulations to test the robustness of the algorithms in the presence of photon noise. Good quality images are readily reconstructed from both numerical and laboratory simulations.

036.086 **Interferometric image reconstruction using the L.D.R. (*large deployable reflector*) in a light bucket mode.**
F. Roddier, J. B. Breckinridge.
Bull. Am. Astron. Soc., Vol. 16, No. 3, Part II, p. 832 – 837 (1984). – See Abstr. 012.063.

The application of a rotational shear interferometer to the large deployable reflector may lead to very high spatial resolution image reconstruction. It is shown that interferometric techniques could extend the imaging capabilities of the reflector down to the visible where complete image reconstruction, with an angular resolution limit of a few milliarcseconds, may still be achieved to stellar magnitudes as faint as 21.

036.087 **Interferometry with large telescopes.**
P. Léna.
Very large telescopes, their instrumentation and programs, p. 245 – 255 (1984). – See Abstr. 012.064 (IAU Colloq. No. 79).

036.088 **Infrared speckle methods.**
A. Chelli.
Very large telescopes, their instrumentation and programs, p. 309 – 336 (1984). – See Abstr. 012.064 (IAU Colloq. No. 79).

036.089 **Speckle interferometry, speckle masking, speckle spectroscopy and speckle frame selection.**
G. Weigelt.
Very large telescopes, their instrumentation and programs, p. 337 – 345 (1984). – See Abstr. 012.064 (IAU Colloq. No. 79).

High–resolution speckle imaging is one of the most fascinating possibilities of a very large telescope (VLT). Various speckle methods can yield a resolution of about 0.01″ with a 8–m to 10–m VLT. As exciting as the resolution is the limiting magnitude of the speckle methods. The limiting magnitude is extremely seeing–dependent. The following limiting magnitudes are possible with speckle interferometry, speckle masking and speckle spectroscopy: 4″ seeing: limiting magnitude 15; 2″ seeing: limiting magnitude 17.5; 1″ seeing: limiting magnitude 20.

036.090 **On the importance of active image stabilisation.**
J. E. Noordam.
Very large telescopes, their instrumentation and programs, p. 367 – 378 (1984). – See Abstr. 012.064 (IAU Colloq. No. 79).

The performance of an optical telescope can be improved considerably by opening the shutter only at moments of good seeing. The resulting instantaneous high–resolution images must be corrected for shift before adding them all together. The optimum aperture diameter for which this technique works well is 1 meter or less, depending on atmospheric conditions. A promising approach would be to build an array of 1-meter telescopes, preferably mounted in a single large frame so that it can be pointed in all directions without extensive pathlength compensation. Such a "Many Mirror Telescope" has many advantages, and is feasible because image stabilisation helps to solve the beam-combining problem.

036.091 **High-resolution Fourier transform spectroscopy with a VLT.**
D. N. B. Hall.
Very large telescopes, their instrumentation and programs, p. 497 (1984). Abstract. – See Abstr. 012.064 (IAU Colloq. No. 79).

036.092 **Imaging spectroscopy on very large telescopes.**
P. Atherton.
Very large telescopes, their instrumentation and programs, p. 515 – 517 (1984). – See Abstr. 012.064 (IAU Colloq. No. 79).

Imaging spectroscopy is a technique in which a spectrum is obtained for each spatial resolution element across a wide field. The data is essentially 3-D, and may be viewed as a series of monochromatic images, or as a two dimensional array of spectra. A device generating such data may be called an imaging spectrometer. In a previous paper (Atherton, 1983) three different imaging spectrometers – based on grating, Fabry-Perot and Fourier Transform devices – were compared in terms of their ability to obtain spectral and spatial information over a wide field and broad band, to the same spectral resolution and S/N ratio, using the same detector array. From such a study it is clear that interferometer based devices are significantly faster than conventional grating spectrographs. Within the next years it is expected that most major observatories will have systems of this kind, using servo stabilized Fabry-Perots and image photon counting systems.

036.093 **System requirements for polarization capability.**
J. Tinbergen.
Very large telescopes, their instrumentation and programs, p. 569 – 573 (1984). – See Abstr. 012.064 (IAU Colloq. No. 79).

The purpose of this paper is to point out that (1) proper treatment of polarization by a telescope is not only important for polarimetry, but also for other observations requiring high signal-to-noise ratio, and (2) this does put a constraint on telescope design, which, however, is not unduly restrictive. It is essential that at least some of the large optical telescopes are capable of observing polarization cleanly. It is even more essential that the largest telescope of a generation has this capability, since, for the required precision, one must collect on average 100 times as many photons as for (spectro) photometry with similar resolution.

036.094 **Interstellar matter with very large telescopes.**
J. Lequeux.
Very large telescopes, their instrumentation and programs, p. 675 – 678 (1984). – See Abstr. 012.064 (IAU Colloq. No. 79).

Interstellar matter is certainly one of the fields where a very large telescope (VLT) will prove to be most fruitful. This includes the study of extended emissions. The author examines the different domains of interest for a VLT.

036.095 **Star formation.**
A. F. M. Moorwood.
Very large telescopes, their instrumentation and programs, p. 679 – 694 (1984). – See Abstr. 012.064 (IAU Colloq. No. 79).

For star formation studies, the main interest in VLT's will centre on their infrared performance advantages and in particular the possibility of achieving a factor of 2 gain in spatial resolution simultaneously with a factor of more than 10 in sensitivity for direct imaging of star formation regions in our own and nearby galaxies. It is thus important to pay attention to minimizing the telescope emissivity during the detailed design phase in order to maximize sensitivity in the thermal infrared ($\sim 2 - 20$ μm).

036.096 **Very large telescopes and stellar astrophysics.**
V. Castellani.
Very large telescopes, their instrumentation and programs, p. 695 – 702 (1984). – See Abstr. 012.064 (IAU Colloq. No. 79).

036.097 **Uses of very large telescopes for galaxy research.**
M.-H. Ulrich.
Very large telescopes, their instrumentation and programs, p. 703 – 711 (1984). – See Abstr. 012.064 (IAU Colloq. No. 79).

Many areas of extragalactic research will greatly benefit from observations with a very large telescope. The author concentrates on four of them as illustrative examples. She briefly discusses the study of the absorption lines in quasar spectra. The author suggests the possibility of doing serendipity large field imaging during high spectral resolution exposures on point sources. Finally, she discusses two topics which have long been recognized as important drivers for the construction of very large telescopes, the determination of the nature of the missing mass, and the study of elliptical galaxies.

036.098 **The nature of galaxies and clusters of galaxies at very large distances.**
J. B. Oke.
Very large telescopes, their instrumentation and programs, p. 713 – 721 (1984). – See Abstr. 012.064 (IAU Colloq. No. 79).

The author has outlined a few of the crucial programs which can be effectively studied with a very large ground-based telescope. In addition to the telescopes one must have: (1) a high efficiency low- and moderate-resolution spectrograph; (2) a multi-slit or multi-fiber capability for spectroscopy; (3) a site with superb seeing; (4) a telescope designed for use for 3200 Å to a few microns.

036.099 **The uses of a VLT for complementing observations with radio and space telescopes.**
D. N. B. Hall.
Very large telescopes, their instrumentation and programs, p. 723 – 724 (1984). – See Abstr. 012.064 (IAU Colloq. No. 79).

Major advances in observational sensitivity in other regions of the electromagnetic spectrum invariably lead to heavy demand for complementary observations with existing large (3- to 5-meter) optical/infrared telescopes. The Einstein X-ray mission, the International Ultraviolet Explorer (IUE), the Very Large Array (VLA) and the recent Infrared Astronomical Satellite (IRAS) are all clear demonstrations of this effect.

036.100 **Preliminary processing of *LAGEOS* laser ranging data acquired in the 1980 MERIT campaign.**
V. V. Nesterov.
Sov. Astron. Lett., Vol. 10, No. 3, p. 167 – 168 (1984). English translation of 37.036.119.

036.101 **Prospects for post-detection enhancement of astronomical images.**
V. S. Tsvetkova, V. G. Chernyj.
Sov. Astron. Lett., Vol. 10, No. 3, p. 194 – 196 (1984). English translation of 37.036.199.

036.102 **A spectrogram reducing system based on the microdensitometer MDM-6 Joyce Loëbl coupled with the mini-computer Nova 3/12".**
N. E. Piskunov, D. A. Ptitsyn, T. A. Ryabchikova, V. L. Khokhlova.
Nauchn. Inf., Vyp. 54, p. 45 – 49 (1983). In Russian.

036.103 **Sensit.for: a program for sensitometric reduction.**
A. Maury, J. Marchal.
Occas. Rep. R. Obs., Edinb., No. 14, p. 17 – 20 (1984). – See Abstr. 012.067.

This program was mainly written to provide a better sensitometric evaluation of what is done when hypering astronomical plates.

036.104 **Some aspects of photography with a small Schmidt camera.**
C. R. Martys.
Occas. Rep. R. Obs., Edinb., No. 14, p. 135 – 141 (1984). – See Abstr. 012.067.

Following the construction of a Schmidt camera in 1976/77 the author has been deeply involved with many aspects of long exposure astronomical photography. The telescope has a clear aperture of 0.35 m and a focal ratio of f2.9. The instrument will accept a 12.5 cm diameter sheet of film at the focal plane and is capable of photographing a 6 x 6 degrees area of sky. This paper demonstrates that interesting and useful work may be undertaken with a minimum of sophisticated equipment and on a very small budget.

036.105 **Astronomers' problems with the calibration of Schmidt plates.**
R. Fong, J. G. Godwin, N. Metcalfe.
Occas. Rep. R. Obs., Edinb., No. 14, p. 207 – 217 (1984). – See Abstr. 012.067.

A problem is found in the comparison of photographic aperture magnitudes from the PDS measurement of a sky–limited UK Schmidt plate, J3721, with calibrated magnitudes for the stars in Hawkins' SGPG2 sequence, in the sense that the photographic magnitudes for faint stars are too bright. This discrepancy is seen to persist when the magnitudes are compared with similar results from the PDS measurement of a sky–limited AAT plate of the same area, J1888. Stars fainter than $B \approx 20$ are consistently estimated brighter when using the Schmidt plate than when using the AAT plate, giving a relative scale error of about 0.1 mag/mag. The problem is most puzzling. However, it remains to be seen whether or not it is intrinsic to Schmidt plates in general.

036.106 **Power spectrum analysis of astronomical photographs digitised with small apertures.**
R. S. Stobie, S. Okamura, A. C. Davenhall, H. T. MacGillivray.
Occas. Rep. R. Obs., Edinb., No. 14, p. 219 – 236 (1984). – See Abstr. 012.067.

A total of 8 UK Schmidt Telescope plates, 2 Anglo–Australian Telescope prime focus plates and 2 Palomar Observatory Sky Survey (POSS) copies were measured with a scanning aperture of 11 µm and a pixel spacing of 16 µm. Power spectrum analysis of the resulting data shows the plate noise (with the exception of the POSS plates) to correspond to almost white noise over the frequency range 0.1 – 60 cycles $mm^{-1}$. The signal–to–noise power spectrum is shown to be a useful tool for measuring the information content of an astronomical photograph. A comparison was made of the information content of a survey quality UKST IIIaJ plate and atlas quality copies on very fine–grained emulsion. Power spectrum analysis of exactly the same area on the original plate and copies showed no significant degradation in either emulsion noise or image content of the copy relative to the original.

036.107 **The Malin–effect in stellar photometry.**
T. Wesselink.
Occas. Rep. R. Obs., Edinb., No. 14, p. 261 – 267 (1984). – See Abstr. 012.067.

The Malin–effect on stellar photometry has been examined on filtered and hypersensitized IIIaJ and IIIaF plates exposed at the United Kingdom Schmidt Telescope in Siding Spring. A comparison with photographic plates exposed in a stream of dry nitrogen shows that the Malin–effect causes a change in stellar response as function of the position on the plate. It is demonstrated that the observed gradients in stellar response can be described in terms of sensitivity variations combined with changes in reciprocity failure. For filtered and hypersensitized photographic plates that are exposed in a stream of dry nitrogen it is possible to define a flat photometric system with an accuracy of $0^{m}.02$.

036.108 **Some applications of signal–to–noise in photographic stellar photometry.**
V. K. Golev, G. R. Ivanov, N. S. Nikolov.
Occas. Rep. R. Obs., Edinb., No. 14, p. 269 – 277 (1984). – See Abstr. 012.067.

The concept of output signal–to–noise $(S/N)_{OUT}$ is used to determine the limiting stellar magnitude $m_{lim}(B)$, the signal–to–noise $(S/N)_{STAR}$ of stellar images, and the precision of the photometry of faint stars on a given plate taken by the 2–m telescope of the Bulgarian National Observatory. A few independent tests of the results obtained show their reasonableness.

036.109 **Stellar photographic photometry using a double wedge technique with a focal reducer system.**
E. H. Geyer.
Occas. Rep. R. Obs., Edinb., No. 14, p. 279 – 284 (1984). – See Abstr. 012.067.

The photometric autocalibration of astronomical plates with a "half–filter" bi–prism arrangement (heliometer principle), as proposed by Lentes (1980), has been tested with the focal reducer system at the 106 cm Cassegrain reflector of the Hoher List Observatory. Photographic stellar magnitude scales free of systematic errors can be obtained by this method in a range of at least $7^{m}$.

036.110 **Spectrophotometry of cool stars from objective prism spectra.**
L. Labhardt.
Occas. Rep. R. Obs., Edinb., No. 14, p. 285 – 293 (1984). – See Abstr. 012.067.

A technique is presented for obtaining low dispersion spectrophotometry of cool stars from objective prism plates. Photographic stellar spectra are calibrated by means of a spot sensitometer and combined with published photoelectric scans. The resulting spectrophotometry covers an extended spectral region with a wavelength spacing increasing from 1.5 Å in the blue to 27 Å in the near infrared.

036.111 **Automated starcounts in nebulous areas.**
C. P. de Vries.
Occas. Rep. R. Obs., Edinb., No. 14, p. 311 – 319 (1984). – See Abstr. 012.067.

The properties of an automatic starcounting program are being described. Special care is taken to model the background as accurately as possible. This background is used, together with stellar density maps, to investigate a complex nebulous region.

036.112 **Advances in optical heterodyne spectroscopy.**
J. L. Hall.
J. Opt. Soc. Am. A, Vol. 1, No. 12, p. 1306 – 1307 (1984). Abstract.

036.113 **Maximum–magnitude estimation of the object's power spectrum in stellar speckle interferometry.**
D. J. Granrath.
Opt. Lett., Vol. 9, No. 11, p. 478 – 480 (1984).

The maximum Fourier magnitude taken over a set of speckle images is proposed as a replacement for the standard power–spectrum average. This new statistic estimates the diffraction–limited object–times–telescope magnitude values in a statistically unbiased and sufficient fashion and has a variance that decreases with sample size faster than the average magnitude's variance. Simultaneous imaging of a nearby point source is unnecessary with this technique, and any object that lies within a single isoplanatic patch of the atmosphere can be estimated. Magnitude–estimation results of the double star Capella are shown to corroborate these claims.

036.114 **Group corrections in zenith distance.**
L. B. F. Clauzet, P. Benevides–Soares.
Rev. Mex. Astron. Astrofis., Vol. 9, No. 2, p. 115 – 117 (1984).

The actual zenith distance of observation in the Danjon astrolabe is a function of the prism angle and of the atmospheric refraction variation during the observations. The zenith distance instabilities preclude the obtention of reliable group corrections and, consequently, of the individual corrections to star positions. In this paper the authors present group corrections in zenith distance for astrolabe data at Valinhos ($\phi = -23°00'$, $\lambda = +3^h07^m$), freed from systematic effects of instrumental origin.

036.115 **The photography of comets.**
H. B. Ridley.
J. Br. Astron. Assoc., Vol. 95, No. 1, p. 8 – 15 (1984).

The photographic observation of comets is an increasingly important activity for amateurs, and covers four main types of work: discovery, confirmation of discoveries, and recovery of periodic comets; astrometry – the determination of precise positions; the recording of structural features and angular dimensions; and the determination of photographic magnitudes. A further form of photographic work is spectrography, a more specialized discipline that will be dealt with in a later paper.

036.116 **Method and apparatus for complex measurements af astroclimatic meteorological parameters.**
O. G. Ovezgel'dyev, M. B. Berkeliev, S. I. Vereshchagin, S. A. Likhomanov, S. Mukhamednazarov, G. A. Nasyrov, Yu. V. Khan.
Izv. Akad. Nauk TSSR. Ser fiz.–tekh., khim. geol. nauk, No. 2, p. 46 – 49 (1984). In Russian. Abstr. in Ref. Zh., 51. Astron., 11.51.766 (1984).

036.117 **Application of the method of photographic reduction of images using equidensitometry for investigating the velocity fields of solar formations.**
A. A. Bessonov.
Chelyab. univ., Chelyabinsk, 43 pp. (1984). In Russian. Abstr. in Ref. Zh., 51. Astron., 11.51.837 (1984).

036.118 **Digital reduction of photographic images of extended objects.**
V. A. Yakovleva, A. B. Shnejvajs, V. B. Il'in, N. V. Yablokova.
Vestn. Leningr. Univ., Mat. Mekh. Astron., No. 3, p. 91 – 99 (1984). In Russian. Abstr. in Ref. Zh., 51. Astron., 11.51.843 (1984).

036.119 **On automation of reduction of results of astronomical observations with the aim of determination of time and right ascensions of stars.**
S. B. Volodin, E. P. Stepanov.
Vestn. Leningr. Univ., Mat. Mekh. Astron., No. 13, p. 86 – 91 (1984). In Russian. Abstr. in Ref. Zh., 51. Astron., 11.51.853 (1984).

036.120 **Calculation of the parameters characterizing the possibility of finding planets in the region of a scanning star camera during cosmic surveying.**
I. G. Zhurkin, V. A. Kuz'minykh.
Izv. vuzov. Geod. aehrofotosemka, No. 3, p. 85 – 89 (1984). In Russian. Abstr. in Ref. Zh., 52. Geod. Aehrosemka, 11.52.209 (1984).

036.121 **Automation of photometric investigation of lunar surface images with the help of computers.**
S. G. Pugacheva.
Obrab. izobrazhenij i distants. issled. Chast' 1. Metod. i algoritmy obrab. izobrazhenij. Tez. dokl. Vses. konf. Novosibirsk, 18 – 20 iyunya, 1984. Novosibirsk, 43 pp. (1984). In Russian. Abstr. in Ref. Zh., 52. Geod. Aehrosemka, 11.52.211 (1984).

036.122 **Dialogue system for measuring graphical photographs of lunar ring structures.**
A. A. Karlov, G. M. Koomov, Zh. F. Rodionova, V. A. Senchenko, T. P. Skobeleva, I. I. Skryl', V. V. Shevchenko, V. N. Shkudenkov.
Obrab. izobrazhenij i distants. issled. Chast' 1. Metod. i algoritmy obrab. izobrazhenij. Tez. dokl. Vses. konf. Novosibirsk, 18 – 20 iyunya, 1984. Novosibirsk, p. 43 – 44 (1984). In Russian. Abstr. in Ref. Zh., 52. Geod. Aehrosemka, 11.52.212 (1984).

036.123 **Early history of Fourier transform spectroscopy.**
P. Connes.
Infrared Phys., Vol. 24, No. 2/3, p. 69 – 93 (1984). – See Abstr. 012.071.

036.124 **Absolute radiometry with Fourier transform spectrometers.**
W. D. Duncan.
Infrared Phys., Vol. 24, No. 2/3, p. 279 – 290 (1984). – See Abstr. 012.071.

A minimum definition, and the essential concepts, of absolute radiometry are given. The spectrometer is assumed to be a Fourier transform instrument. The topics discussed are the differential nature of a measurement, the multiport nature of spectral encoders and modulators, internal signals, calibration and antenna patterns. The performance constraints necessary for absolute radiometry of the cosmic background are used as performance references for the various elements of a radiometer.

036.125 **A simple internal modulation technique for the spectral calibration of circular variable filter spectrometers in the near–infrared.**
T. Chandrasekhar, N. M. Ashok, H. C. Bhatt, K. S. B. Manian.
Infrared Phys., Vol. 24, No. 6, p. 571 – 572 (1984).

A simple, inexpensive, method of spectral calibration of a continuously variable filter spectrometer is outlined, wherein the harmonic of the power line frequency is used to discriminate spectral lines of the discharge tube from the thermal emission of the envelope.

036.126 **Reflectance spectroscopy: quantitative analysis techniques for remote sensing applications.**
R. N. Clark, T. L. Roush.
J. Geophys. Res., Vol. 89, No. B7, p. 6329 – 6340 (1984).

Several methods for the analysis of remotely sensed reflectance data are compared, including empirical methods and scattering theories, both of which are important for solving remote sensing problems in planetary surface studies.

036.127 **Modeling crater topography and albedo from monoscopic Viking Orbiter images. 1. Methodology.**
P. A. Davis, L. A. Soderblom.
J. Geophys. Res., Vol. 89, No. B11, p. 9449 – 9457 (1984).

A new photoclinometric technique for extraction of topographic data from single planetary images is presented that overcomes many previous limitations of photoclinometry.

036.128 **The experience of the use of the digital microphotometer "Spectrum–Kod" and the program complex of Shapovalova–Shcherbanovskij for the reduction of stellar spectrograms of the main stellar spectrograph of the 6–m telescope.**
O. Eh. Aab, E. A. Barsukova, I. Kh. Iliev, I. A. Lebedeva, E. L. Chentsov.
Soobshch. Spets. Astrofiz. Obs., Vyp. 40, p. 21 – 25 (1983). In Russian. – See Abstr. 012.076.

The experience is generalized of the use of automatic reduction of the 6–m telescope main stellar spectrograph spectrograms for stars of O, B, A spectral classes.

036.129 **Automatic processing of negatives on the basis of the "Atlas–2" system and the computer M–220 M.**
A. I. Povtar', D. A. Rozhkovskij, A. A. Semenikin.
Soobshch. Spets. Astrofiz. Obs., Vyp. 40, p. 39 – 43 (1983). In Russian. – See Abstr. 012.076.

The possibility of photometry of negatives of diffuse nebulae and comets with the help of an automatic microphotometer on the basis of the "Atlas–2" system and the computer "M–220 M" is investigated.

036.130 **Automation of stellar spectrophotometry.**
V. A. Bandrin, V. V. Il'in.
Astron. Tsirk., No. 1321, p. 1 – 2 (1984). In Russian.

036.131 **Astrophotography at high-altitude stations.**
D. A. Rozhkovskij.
Soobshch. Spets. Astrofiz. Obs., Vyp. 40, p. 59 – 62 (1983). In Russian. – See Abstr. 012.076.

A brief discussion of the perspectives favourable for photographic observations at the high-altitude station of the Astrophysical Institute is given. The determining criterion for the evaluation of the efficiency of photographic observations is considered.

036.132 **Level contrasts at registration of stars on the sky background and background influence on the detection and resolution of small objects.**
I. I. Brejdo.
Soobshch. Spets. Astrofiz. Obs., Vyp. 40, p. 63 – 68 (1983). In Russian. – See Abstr. 012.076.

036.133 **Experience of the use of the automatic measuring complex Zenit-2 for solving some astrometrical and astrophysical problems.**
A. A. Kiselev, M. V. Kushnir.
Soobshch. Spets. Astrofiz. Obs., Vyp. 40, p. 87 (1983). In Russian. – See Abstr. 012.076.

036.134 **Optimization of the process of selection of faint optical signals in satellite laser ranging.**
G. S. Kurbasova, V. V. Kurbasov.
Nauchn. Inf., Vyp. 55, p. 17 – 26 (1982). In Russian.

The possibility of optimization of the faint signal detection and selection process by the correlate method is discussed. This method rises the signal-to-noise ratio on the small interval of digitization. An example of operative detection of the faint signal with laser ranging of the Lageos satellite is given.

036.135 **Special features of the method of observations of geostationary satellites with the VAU camera.**
F. N. Masumi.
Nauchn. Inf., Vyp. 55, p. 57 – 63 (1982). In Russian.

The method of stationary artificial satellites observations with camera VAU is considered in order to receive exact coordinates and to study visible movements of these objects.

036.136 **On the possibility of extending information obtained from satellite tracking with the AFU-75 camera.**
N. P. Erpylev.
Nauchn. Inf., Vyp. 55, p. 78 – 81 (1982). In Russian.

Information in a single sputnikogram by AFU-75 gives a possibility to determine both location and angular velocity of a satellite. An example of calculation of an osculating satellite orbit with the use of one simultaneous pair of sputnikograms obtained at Zvenigorod and Riga stations is given.

036.137 **On the application of the routine of processing of AFU-75 plates to photos of the VAU camera.**
N. S. Bakhtigaraev, N. D. Kostyuk, V. A. Yurevich.
Nauchn. Inf., Vyp. 55, p. 82 – 88 (1982). In Russian.

A description of the method for astrometrical reduction of VAU camera astroplates and a list of formulas are given as a model of bright satellites photos processing.

036.138 **Some methods of indirect observations adjustment with the covariance matrix of the initial data.**
O. A. Petrova.
Nauchn. Inf., Vyp. 55, p. 127 – 134 (1982). In Russian.

036.139 **Should one measure radial velocity or brightness fluctuations?**
E. Fossat.
Space research prospects in stellar activity and variability, p. 77 – 90 (1984). – See Abstr. 012.078.

Low degree p-modes of the sun have been measured in spatially integrated sunlight both in Doppler shift and intensity fluctuations. The guideline of this paper is to assume that the sun is removed far enough in space to become an ordinary star of magnitude zero to one. Evidently, another star will oscillate with different frequencies and different amplitudes, but one must use some reference to start with. With this scheme, a detailed investigation of the limitations of observational accuracy in the search for global p-modes is made. The sources of noise stand in the sun itself, in the instrumentation, in the observing time duration, in the corpuscular nature of the light and mostly in the earth atmosphere (for ground-based observations). Finally, a detection of five minute p-modes on $\alpha$ Cen are reported.

036.140 **On the detection of solar oscillations through radial velocity fluctuations.**
P. Connes.
Space research prospects in stellar activity and variability, p. 91 – 106 (1984). – See Abstr. 012.078.

A general discussion of the sources of error in present or proposed helioseismology methods is given. First, resonance cells, then magnetographs used by the two-plage technique, are treated; the former are considered distinctly superior. The need for extremely precise integration of the solar beam is stressed; using an optical fiber and an image rotator, this can be actually achieved. Then, the proposed solar accelerometer eliminates all interactions between the enormously greater observer motion and the studied oscillations. Lastly, the perturbations brought in by time-varying atmospheric transparency gradients are cancelled by a compensating system.

036.141 **Correction of atmospheric transparency gradient effect on solar radial velocities.**
T. Appourchaux.
Space research prospects in stellar activity and variability, p. 117 – 123 (1984). – See Abstr. 012.078.

The author has found that the major part of the atmospheric gradient is that computable from sec z and standard atmospheric tables. Thus a reasonable first approximation of the velocity correction can be obtained simply by computation. The autor has demonstrated a simple system for measuring the actual gradient from which the effective spurious velocity can be computed using standard figures for limb darkening and solar rotation.

036.142 **Experiences with high precision ground-based photometry of the sun.**
C. Winkler.
Space research prospects in stellar activity and variability, p. 125 – 131 (1984). – See Abstr. 012.078.

A new method is presented to measure the integrated light of the solar disk. Systematic errors are significantly reduced. The photometric system is highly stable. The method is convenient to study solar intensity fluctuations with small amplitudes. First measurements of the solar intensity in the spectral region B show variations with significant maxima near periods of 5 minutes. The contribution of pressure variations of the earth's atmosphere is studied.

036.143 **A numerical solution of the data-gaps problem.**
J. Connes, P. Connes.
Space research prospects in stellar activity and variability, p. 135 – 149 (1984). – See Abstr. 012.078.

036.144 **Period finding in gapped data and its application to the pulsating $\lambda$ Bootis type star 29 Cygni.**
E. N. Walker, C. D. Pike, K. F. Hartley.
Space research prospects in stellar activity and variability, p. 151 – 156 (1984). – See Abstr. 012.078.

The authors address themselves to a very specific question which is "under what circumstances is it possible to discover the underlying power spectrum of a multiply periodic phenomenon from data containing large gaps?". The motivation for this study was the discovery in 1970 by ENW that the chemically peculiar $\lambda$ Bootis type star, 29 Cygni, showed photometric variations with an apparent period of $\sim$40 minutes and an amplitude which varied from $\sim$0 to $\sim 0^m.06$ in a manner which suggested that if the variations were strictly periodic then many frequencies must be present.

036.145 **The accuracy limit of ground-based stellar photometry.**
J. R. W. Heintze, C. de Jager, W. van der Veen.
Space research prospects in stellar activity and variability, p. 157 - 167 (1984). - See Abstr. 012.078.

The ultimate accuracy that can be reached in ground-based stellar photometry depends primarily on the fluctuations of the atmosphere causing scintillations and variations of transparency, and besides on the instrumental stability and available integration time. Fluctuations in instrumental stability can be made negligibly small by basing the optical system and observational methods strictly on differential techniques. The integration time constitutes, together with variations in the atmosphere, the real limiting factor. Considerations based on available atmospheric and photometric observational data, yield that the present-days practical limiting accuracy of ground-based photometry is of the order of $(3-7)\times 10^{-4}$mag for the observation of a 7th mag star by a one-meter telescope at an altitude of 4000 m and an integration time of $10^3$s.

036.146 **Ground-based solar and stellar intensity measurements: practical limits of accuracy.**
T. Schmidt-Kaler.
Space research prospects in stellar activity and variability, p. 169 - 174 (1984). - See Abstr. 012.078.

The subject of this paper are the practical limitations of bright object photometry, in particular the errors due to positioning, scintillation and extinction fluctuations, and methods to compensate or to correct these errors. Solar observations with a diffraction coronograph suppress scintillation very effectively and allow accurate correction of extinction fluctuations.

036.147 **Direct and indirect methods of measurement of stellar magnetic fields.**
M. S. Giampapa.
Space research prospects in stellar activity and variability, p. 309 - 326 (1984). - See Abstr. 012.078.

Standard polarization methods for measuring stellar magnetic fields are inappropriate for solar-type field topologies. The author is therefore utilizing a "white-light" method that involves the interpretation of spectral line shapes as seen in high spectral resolution data. Through the application of Fourier deconvolution techniques, the field strength and area coverage on a stellar surface can be ascertained. The applicability and results of this method for both solar and stellar magnetic field measurements are discussed. Besides the aforementioned direct method, there potentially exist complementary indirect methods. They are less satisfactory than direct methods. They require a knowledge of the underlying physics that relates on observable radiative emission to the local magnetic field strength or flux.

036.148 **Instrumental approach to multi-site spectral variability study.**
P. Felenbok, J. Guérin.
Space research prospects in stellar activity and variability, p. 359 - 368 (1984). - See Abstr. 012.078.

The pre-main-sequence star AB Aur, that the authors have observed since 1980 shows spectral variability while its photometric behaviour is steady on the same period. In October 1980 the authors recorded Ca II K line variation (Praderie, F. et al. 1982) on a night to night basis. To follow this effect more precisely this data must be recorded on a shorter time interval. This can be done only by three different means: satellite, polar or multi-site observations. In this study the authors deal with ground-based observations only and compare polar and multi-site approaches.

036.149 **Variation in time of the parameters (u + m) and $M_0$ of the Belgrade meridian circle.**
S. Sadžakov, M. Dačić.
Bull. Appl. Math., No. 190, p. 13 - 17 (1983).

036.150 **Minor planet observing programs.**
R. P. Binzel.
Minor Planet Bull., Vol. 11, No. 1, p. 4 - 6 (1984).

Various minor planet observing programs are described for observers having all levels of experience and equipment.

036.151 **Comprehensive star catalogues.**
H. Eichhorn.
Bull. Inf. Cent. Données Stellaires, No. 27, p. 21 - 37 (1984).

The questions pertaining to the construction of comprehensive complete star catalogues out of existing observations are discussed. In particular, it will be necessary to tighten the meaning of the concept of a system as well as the precepts for reducing specific data to a specific system. The existence of computers urges that one should use more adequate reduction models as well as more adequate reduction methods than were used in the past, in particular, that a generalized overlap approach be employed for the reduction of all catalogues to a common system and, indeed, the establishment of the system itself. The paper discusses problems arising in connection with a definitive reduction of the astrographic catalogue, and a possible first measuring of the plates taken originally for the Carte du Ciel, and briefly mentions some contemporary projects of constructing comprehensive position catalogues.

036.152 **Mesure de clichés de la Carte du Ciel.**
M. Rousseau, J. Guibert.
Bull. Inf. Cent. Données Stellaires, No. 27, p. 43 - 48 (1984).

The main objectives of the Astrographic Chart and Catalogue Enterprise are briefly summarised. The authors describe recent measurements and comparisons performed on test plates of the Astrographic Catalogue in Paris and Bordeaux zones. Results are presented as an element for the discussion concerning possible remeasurement of Catalogue and/or measurement of Chart plates.

036.153 **GKS (*Graphical Kernel System*) for imaging.**
C. Stoll.
Comput. Graphics, Vol. 18, No. 3, p. 271 - 274 (1984). = Space Telesc. Sci. Inst., Prepr. Ser., No. 32.

036.154 **Data analyis of electron measurements of the Plasma Experiment aboard the Helios probes.**
W. G. Philipp, H. Miggenrieder, K.-H. Mühlhäuser, H. Rosenbauer, R. Schwenn.
MPE Rep., No. 185, 46 pp. (1984).

The methods to derive distribution functions for solar wind electrons from the measurements of the "Plasma Experiment" aboard the Helios probes are outlined. Electrons are measured in the energy range from 0.5 to 1660 eV with a one-dimensional angular resolution along the ecliptic plane. From these measurements electron distribution functions with high time resolution (typically 18 s) but poor angular resolution in velocity space (45 degrees) and electron distribution functions with high angular resolution (better than 5 to 10 degrees) but poor time resolution (typically several minutes to few hours) have been derived.

036.155 **Galaxy photometry in the infrared.**
A. F. M. Moorwood.
ESO Sci. Prepr., No. 349, 10 pp. (1984). Invited review to appear in the Proceedings of New Aspects of Galaxy Photometry, specialized Colloquium at the Eighth IAU Regional European Meeting, Toulouse, Sept. 1984.

036.156 **Analytical and numerical studies about effects of natural radio source structures on astrometric and geodetic observations with a very long baseline interferometer.**
M. Fujishita.
Publ. Int. Latitude Obs. Mizusawa, Vol. 17, No. 1, p. 13 - 33 (1983).

Some characteristics of effects of natural radio source structures on astrometric and geodetic observations with VLBI are herein clarified on the basis of theorems of Fourier transform. Using the convolution theorem, it is shown that the effects of a structure, of which the components are all symmetric, can be presented as a function of only the positions and flux densities of the components, but not of its shape. The effects of variations of various parameters such as position, flux density ratio and FWHM are calculated numerically for a double Gaussian component model. The effect of the time variation of the real source structure of 3C345 is investigated by using the frequency–independent double Gaussian component models of 1975 and 1980.

036.157 **Elaboración matemática de las observaciones del astrolabio.**
V. S. Gubanov.
Publ. Espec. Obs. Astron. Quito, No. 2, p. 27 – 55 (1981).

036.158 **On the method of photography of two stellar fields on one plate.**
N. N. Matveev.
Astron. Tsirk., No. 1323, p. 7 – 8 (1984). In Russian.

036.159 **Present–day problems of polarimetry of atmosphereless cosmic bodies.**
Yu. G. Shkuratov, L. A. Akimov, V. P. Tishkovets.
Astron. Vestn., Tom 18, No. 3, p. 163 – 178 (1984). In Russian. Abstr. in Ref. Zh., 51. Astron., 12.51.1180 (1984).

036.160 **Automation of photometric investigation of lunar surface images with computers.**
S. G. Pugacheva.
Obrab. izobrazhenij i distants. issled. Ch. 1. Metody i algoritmy obrab. izobrazhenij. Tez. dokl. Vses. konf. Novosibirsk, 18–20 iyunya, 1984. Novosibirsk, 43 pp. (1984). In Russian. Abstr. in Ref. Zh., 51. Astron., 12.51.1199 (1984).

036.161 **Astrometrische Parameter hergeleitet aus HIPPARCOS–Messungen.**
H.–H. Bernstein, H. G. Walter.
Mitt. Astron. Ges., Nr. 62, p. 158 – 159 (1984). – See Abstr. 010.341.

036.162 **Lösung des "Grid Step Ambiguity" Problems des HIPPARCOS.**
U. Bastian.
Mitt. Astron. Ges., Nr. 62, p. 159 (1984). – See Abstr. 010.341.

036.163 **Hamburger Software–Entwicklung zur Reduktion astrometrischer Meßdaten.**
C. de Vegt, J. Fehlberg, T. Kleine, J. Schramm, N. Zacharias.
Mitt. Astron. Ges., Nr. 62, p. 210 (1984). Abstract. – See Abstr. 010.341.

036.164 **Vollautomatische Analyse komplexer P Cygni–Linien.**
R. Dümmler.
Mitt. Astron. Ges., Nr. 62, p. 266 – 267 (1984). – See Abstr. 010.341.

036.165 **Ein Konzept zum Generieren von virtuellen Betriebssystemen in der Bildverarbeitung.**
D. Teuber.
Mitt. Astron. Ges., Nr. 62, p. 360 – 363 (1984). – See Abstr. 010.341.

036.166 **A new method of absolute determination of a transit instrument's azimuth and latitude.**
W. Mao, Z. Li, Y. Fan, X. Hu, M. Du, H. Li.
Sci. Sin., Ser. A, Vol. 27, No. 11, p. 1183 – 1189 (1984).
This article deals with a new method of the absolute determination for the meridian observations in mid and low latitude areas and raises a limitation that the traditional method cannot be used at low latitudes. The new method will play an important role in improving the reference system of the fundamental celestial sphere. Experimental observations with a reconstructed Zeiss transit instrument show that the principle of the new method is right.

036.167 **A high–accuracy timing method for photographic tracking satellite. An improvement in the SBG timing system.**
H. Feng, B. Wang, Z. Qu, X. Cen.
Sci. Sin., Ser. A, Vol. 27, No. 11, p. 1209 – 1216 (1984).
This new timing method is based on the concept not to look for directly the start or stop signal of the plate carriage, but to collect all of the time–displacement informations concerning the movement of the plate carriage during tracking exposure processes, so that the corresponding tracking time and the timing accuracy can be well determined through a microcomputer.

036.168 **Statistical analysis in gamma–ray astronomy: the upper limit.**
R. J. Protheroe.
Astron. Express, Vol. 1, Issue 1, p. 33 – 38 (1984).
The usual procedures for estimating upper limits to the flux of $\gamma$–rays from suspected sources or in spectral lines are inadequate if the background flux is poorly determined. A method of correctly estimating upper limits in this case, and which is universally applicable, is described.

036.169 **Statistical reliability of X–ray and gamma–ray line observations.**
Y. Ma, T. Li.
Acta Astrophys. Sin., Vol. 4, No. 1, p. 65 – 73 (1984). In Chinese.
It is shown that the significances of gamma–ray lines are underestimated by Cherry et al. (1980) systematically. A formula to evaluate the statistical significance of positive line observations is proposed and tested by Monte–Carlo simulations. Applying this method the authors obtain new estimates for the confidence levels of reported gamma–ray lines.

036.170 **A possible method for the determination of the magnitude and inclination of a sunspot magnetic field by a quarter–wave plate and polarizers.**
J.–h. Jin.
Acta Astron. Sin., Vol. 25, No. 3, p. 234 – 240 (1984). In Chinese.
By using a quarter–wave plate with fast axis at X axis and two polarizers with transmission axis at $+45°$ and $-45°$ respectively, a method for the determination of H and $\gamma$ of a sunspot magnetic field is presented.

036.171 **Laboratory simulation of stellar speckle interferometry.**
M.–z. Shen, W.–j. Wang.
Acta Astron. Sin., Vol. 25, No. 3, p. 293 – 299 (1984). In Chinese.
After giving a brief introduction to the principle of speckle interferometry the paper describes a laboratory simulation, including: (1) simulation of atmospheric effects on the recording of short exposure images, (2) data processing and experimental results. Finally, some possible problems in the application of this method are discussed.

036.172 **Algorithm for astronomical, extended source, signal–to–noise ratio calculations.**
R. R. Jayroe.
NASA Tech. Pap., NASA TP–2396, 3 + 19 pp. (1984).
An algorithm was developed to simulate the expected signal–to–noise ratio as a function of observation time in the charge coupled device detector plane of an optical telescope located outside the earth's atmosphere for an extended, uniform astronomical source embedded in a uniform cosmic background. By choosing the appropriate input values, the expected extended source signal–to–noise ratios can be computed for the Hubble Space Telescope using the Wide Field/Planetary Camera science instrument.

036.173 **Algorithm for astronomical, point source, signal to noise ratio calculations.**
R. R. Jayroe, D. J. Schroeder.
NASA Tech. Pap., NASA TP–2397, 3 + 36 pp. (1984).
An algorithm was developed to simulate the expected signal to noise ratios as a function of observation time in the charge coupled device detector plane of an optical telescope located outside the earth's atmosphere for a signal star, and an optical secondary star, embedded in a uniform cosmic background. By choosing the

appropriate input values, the expected point source signal to noise ratio can be computed for the Hubble Space Telescope using the Wide Field/Planetary Camera science instrument.

036.174 **Visual observations of comets. II. Methods and brightness estimates.**
K. I. Churyumov.
Komet. Tsirk., No. 326 (1984). In Russian.

036.175 **Imaging with a single detector: the Wyoming approach.**
G. L. Grasdalen, J. A. Hackwell, R. D. Gehrz.
Publ. Astron. Soc. Pac., Vol. 96, No. 586, p. 1017 – 1023 (1984).
The computerized techniques currently employed at the Wyoming Infrared Observatory to obtain infrared photometric and imaging data are described. The techniques used for bolometers are described in detail. Methods used for other detector systems are sketched. The first-order reduction processes; removing background gradients, aligning and coadding images, and photometric calibration, are described in detail. Some of the more sophisticated means of processing images are outlined.

**Planetary imaging data archive–laser videodisc storage and retrieval.**
See Abstr. 002.031.

**Astrophysical techniques.**
See Abstr. 003.004.

**Colours of the stars.**
See Abstr. 003.011.

**The hidden sun: solar eclipses and astrophotography.**
See Abstr. 003.118.

**Interferometry.**
See Abstr. 003.153.

**Trends in planetary data analysis. Executive summary of the Planetary Data Workshop, held at Goddard Space Flight Center, 29 November – 1 December 1983.**
See Abstr. 011.011.

**Application de la photométrie bidimensionnelle à l'astronomie.**
See Abstr. 012.096.

**The case for a national facility for astronomical image processing.**
See Abstr. 013.051.

**Observations of stellar activity: advantages and disadvantages of ground based multisites, polar region observatories, and space observations.**
See Abstr. 013.059.

**Le CDS, pourquoi pas centre de logiciels astrométriques?**
See Abstr. 013.068.

**Simulators for assisting the teaching of observational astronomy. II. Photometry of normal variable stars in total visible light.**
See Abstr. 014.001.

**The use of the MMT for interferometric imaging.**
See Abstr. 032.047.

**16 m Large Slit Aperture Telescope for very high angular resolution astronomy.**
See Abstr. 032.048.

**The first US–Japan VLBI test observation by use of K–3 system at the Radio Research Laboratories.**
See Abstr. 033.004.

**Image formation by self–calibration in radio astronomy.**
See Abstr. 033.022.

**Cosmic rays.**
See Abstr. 034.009.

**A microcomputer–based sytstem for data acquisition and instrument control.**
See Abstr. 034.062.

**An amateur filar micrometer.**
See Abstr. 034.066.

**Kiso image detection system.**
See Abstr. 034.075.

**The establishment of the hypersensitizing laboratory and the preliminary results of plate hypersensitization test.**
See Abstr. 034.101.

**Focal reducer techniques for direct imaging and field spectroscopy with large telescopes.**
See Abstr. 034.111.

**The IIIa–F emulsion and the quasar redshift cut–off.**
See Abstr. 034.144.

**What do we want – increased speed, resolution, contrast or DQE?**
See Abstr. 034.149.

**Plate measurements and reductions at Münster.**
See Abstr. 034.157.

**Kurze Einführung in die astronomische Photographie.**
See Abstr. 034.196.

**Das Bonner Mikrodensitometer, ein Prozessor–Verbund in der Praxis.**
See Abstr. 034.202.

**Ein schneller logarithmischer Verstärker für die PDS 2020GM.**
See Abstr. 034.203.

**Image processing software system of the X–ray telescope aboard the HINOTORI spacecraft.**
See Abstr. 035.075.

**Tianjin photographic zenith tube observations.**
See Abstr. 041.033.

**Limb effect of solar absorption lines. Observational method and results for Fe I 557.6 nm.**
See Abstr. 071.027.

**Correction for the observed intensity distribution in solar corona.**
See Abstr. 074.028.

**Straylight influence on measurements of solar velocity oscillations.**
See Abstr. 080.054.

**On a method of reduction of observations of 160–min brightness oscillations of the limb of the solar disc.**
See Abstr. 080.112.

**A method to estimate Fried's seeing parameter from a time series of arbitrary resolved structures imaged through the atmosphere.**
See Abstr. 082.074.

**Determination of the coordinates of an observation point on the lunar surface by equal altitudes.**
See Abstr. 094.043.

**Lunar eclipse crater timing programme.**
See Abstr. 095.005.

**First results of Mars observations with a PVC.**
See Abstr. 097.001.

**Convex–profile inversion of asteroid lightcurves: calibration and application of the method.**
See Abstr. 098.058.

**Komet Halley 1909 abermals in Heidelberg entdeckt.**
See Abstr. 103.915.

**Systematic errors in radial velocities measured on image–tube spectra from the CTIO 1–m telescope.**
See Abstr. 111.020.

**Stability and accuracy of the *uvby* system.**
See Abstr. 113.034.

**Analysis of the stellar fluxes at 2740 Å as observed by S2/68 and OAO–2.**
See Abstr. 114.018.

**The spectra of late–type dwarfs and sub–dwarfs in the near ultraviolet. III. An atlas of Mg II h and k profiles.**
See Abstr. 114.094.

**Spectral classification at ultra–low dispersion.**
See Abstr. 114.100.

**New narrow band light curves of the eclipsing binary u Her.**
See Abstr. 119.008.

**Narrow–band imaging and velocity maps of young stellar objects: initial results.**
See Abstr. 121.043.

**Optical polarization as a probe of the local interstellar medium.**
See Abstr. 131.259.

**Automation of the process of optical identification of radio sources at AMD–1.**
See Abstr. 141.023.

**Faint galaxy number–magnitude counts at high galactic latitude. I.**
See Abstr. 157.094.

**Deconvolution of photographic images of quasars from point spread functions.**
See Abstr. 159.037.

**Detection of lines by calibrated algorithm in a close pair of quasars.**
See Abstr. 159.069.

**An objective prism quasar survey with the Schmidt telescope on Calar Alto.**
See Abstr. 159.076.

# Positional Astronomy, Celestial Mechanics

## 041 Astrometry

041.001 **On a possible improvement of the right ascension system of the General Catalogue of the Time Services of the USSR.**
P. M. Afanas'eva, V. L. Gorshkov.
Astron. Tsirk., No. 1284, p. 6 – 8 (1983). In Russian.

041.002 **Optical positions and proper motions of radio stars.**
E. D. Clements, R. W. Argyle.
Mon. Not. R. Astron. Soc., Vol. 209, No. 1, p. 1 – 6 (1984).
Positions relative to AGK3 with an accuracy of about 0.1 arcsec are presented for 69 radio stars observed with the 26– and 13–in refractors at RGO. Independent proper motions of these stars have been obtained wherever possible. Comparison is made with optical and radio positions given by Walter.

041.003 **Radio interferometric determination of source positions utilizing Deep Space Network antennas – 1971 to 1980.**
J. L. Fanselow, O. J. Sovers, J. B. Thomas, G. H. Purcell Jr., E. J. Cohen, D. H. Rogstad, L. J. Skjerve, D. J. Spitzmesser.
Astron. J., Vol. 89, No. 7, p. 987 – 998 (1984).
During the past decade, a series of observing sessions has been conducted by the Jet Propulsion Laboratory with NASA's Deep Space Network stations to develop a celestial radio reference frame based on the positions of compact extragalactic sources. Approximately 2400 observations were made between 1971 August and 1980 February, with more than half of them at dual frequencies, to eliminate the effects of charged particles. The resulting catalog includes 117 sources, distributed fairly uniformly over the celestial sphere north of -40 declination. Most of these sources have formal positional uncertainties between 1 and 5 milliarcsec.

041.004 **Observations of the sun at the CERGA astrolabe in 1982.**
F. Laclare.
Astron. Astrophys., Suppl. Ser., Vol. 57, No. 2, p. 281 – 284 (1984). In French.
The results of a new campaign of solar position observations are reported. The author has obtained the following mean value for right ascension: $\Delta\alpha = -0\overset{s}{.}044 \pm 0\overset{s}{.}007$, and for the semi-diameter reduced to one astronomical unit: $R = 959\overset{''}{.}35 \pm 0\overset{''}{.}03$.

041.005 **Comparison of the SAO and AGK3 catalogues.**
D. D. Polozhentsev, H. I. Potter.
Astron. Tsirk., No. 1288, p. 7 – 8 (1983). In Russian.

041.006 **On the origin of "East–West" differences in declination observations with a vertical circle.**
M. S. Chubej.
Astron. Tsirk., No. 1296, p. 4 – 6 (1983). In Russian.

041.007 **On the accuracy of star catalogues used for the compilation of PFKSZ–2.**
A. D. Polozhentsev.
Astrometr. Astrofiz., Vyp. 52, p. 8 – 14 (1984). In Russian.
The methods of comparison of catalogues are considered. The $\gamma$ criterion has been used for determination of a number of series of development in the method of spherical harmonics. Estimation of the errors of the FKSZ catalogues used in PFKSZ–2 is given in tables.

041.008 **Accurate positions for young objects: NGC 7000 and IC 5070.**
F. Della Prugna, N. Calvet, M. del C. Araque.
Rev. Mex. Astron. Astrofis., Vol. 9, No. 1, p. 31 – 33 (1984).
As a part of an ungoing observational program to determine astrometric positions for objects associated with molecular clouds and/or located in regions of recent star formation the authors report in this work results for Herbig emission–line objects in the regions of NGC 7000 and IC 5070. Herbig's original positions are within 10 arcsec of the astrometrically determined positions. In particular, the astrometric position for V1057 Cyg agrees within 1 arcsec with the position of the OH maser in the region, confirming the physical association between the two objects.

041.009 **Plans for astrometry for late 1984 asteroidal and planetary occultations.**
D. W. Dunham.
Occultation Newsl., Vol. 3, No. 8, p. 160 – 161 (1984).

041.010 **Catalogue of the right ascensions of 200 stars observed with the Moscow PZT during 1973.8 – 1981.0.**
A. A. Tochilina.
Tr. Gos. Astron. Inst. Shternberg, Tom 54, p. 149 – 159 (1984). In Russian.
The right ascensions of 200 stars observed with the Moscow PZT are improved in 1973.8 – 1981.0. The corrections to the right ascensions are calculated by free chain methods. The mean epoch of the observations is 1978.5; the mean square error of one observation is $\pm 0\overset{s}{.}018$.

041.011 **A wide–angle astrometric standard in the Moscow zenith zone.**
Yu. A. Shokin, N. M. Evstigneeva.
Tr. Gos. Astron. Inst. Shternberg, Tom 55, p. 3 – 20 (1983). In Russian.
A method of the determination of 400 accurate star positions in the area 8 × 8° centered at $\alpha = 22^h30^m$, $\delta = +55°42'$ (1950.0) is described. A catalog of positions and proper motions of stars proposed as a standard for wide–angle astrographic study is given.

041.012 **A remark on calculation of the apparent places of near objects.**
K. V. Kuimov.
Tr. Gos. Astron. Inst. Shternberg, Tom 55, p. 30 – 32 (1983). In Russian.
A viewpoint on apparent places calculation is given. The procedure can be considered as an application of the Lorentz transformation. This viewpoint enables to avoid errors in calculating apparent places of near objects (Moon). A program for convenient application of the Lorentz transformation and for calculating coordinates and the Doppler effect is given.

041.013 **A catalogue of declinations and proper motions of Moscow PZT stars.**
A. A. Volchkov, G. A. Gutsalo.
Tr. Gos. Astron. Inst. Shternberg, Tom 55, p. 40 – 46 (1983). In Russian.
A technique of declinations and proper motions correction for Moscow PZT stars is described. A catalogue of 169 stars is given.

041.014 **Astrometry of 59 pulsars: a comparison of interferometric and timing positions.**
E. B. Fomalont, W. M. Goss, A. G. Lyne, R. N. Manchester.
Mon. Not. R. Astron. Soc., Vol. 210, No. 1, p. 113 – 130 (1984).

Using the VLA, positions have been measured for 58 pulsars which also have accurate estimates of position based on pulse arrival times. Estimated errors in the VLA positions are typically 0.2 arcsec and are comparable with the errors in positions obtained from the pulse timing analyses. Agreement between the VLA and timing positions is generally good, although systematic discrepancies at the 0.2 – 0.5 arcsec level are observed. Timing positions based on an MIT ephemeris differ systematically by ~0.2 arcsec from positions based on a JPL ephemeris. A systematic variation of $\Delta\alpha \cos \delta$ with right ascension $\alpha$ between the VLA and both the MIT and JPL timing positions is observed.

041.015 **Compilation du Catalogue d'Entrée. Programmes astrométriques sur Hipparcos.**
C. Turon.
Astrométrie, géodésie, mécanique céleste, p. 183 – 190 (1984). – See Abstr. 012.021.

041.016 **Liaison des systèmes fondamentaux au système radio extragalactique.**
A. Baudry, J. M. Mazurier.
Astrométrie, géodésie, mécanique céleste, p. 203 – 208 (1984). – See Abstr. 012.021.

041.017 **Astrométrie des Pléiades.**
J. M. Rousseau.
Astrométrie, géodésie, mécanique céleste, p. 215 – 218 (1984). – See Abstr. 012.021.

041.018 **The effect of differential color refraction on declinations determined in meridian circle programs.**
R. C. Stone.
Astron. Astrophys., Vol. 138, No. 2, p. 275 – 284 (1984).

This paper presents a study of differential color refraction from the standpoint of both empirical data and synthetic results generated with a computational model. By comparing the northern Washington $W5_{50}$ and southern Washington $WL_{50}$ and Perth 70 meridian catalogs, this paper finds that the three instruments respectively responsible for these programs have very similar differential color curves, amounting to $0''.0087 \pm 0''.0011$ (s.e.) per spectral class in the mean. This slope of the color curve is significantly smaller than that found by van Herk (1952) but is in good agreement with that determined at Pulkovo. The model can closely match all three of these color curves and has further been used to investigate the dependence of the color curve on various atmospheric and instrumental factors.

041.019 **Results of a comparison of the Kharkov catalogue of declinations of 1407 nearpolar stars with the catalogue of Fabritius.**
L. S. Pavlenko.
Khar'k. univ. Khar'kov, 24 pp. (1983). In Russian. Abstr. in Ref. Zh., 51. Astron., 8.51.77 (1984).

041.020 **Catalogue of declinations of 254 stars in the neighbourhood of 72 extragalactic radio sources.**
P. F. Lazorenko.
Glav. Astron. Obs. Akad. Nauk USSR. Kiev, 48 pp. (1984). In Russian. Abstr. in Ref. Zh., 51. Astron., 8.51.78 (1984).

041.021 **A preliminary determination of the accuracy of observations of right ascensions with the photoelectric meridian circle.**
R. Fukaya, S. Suzuki, A. Sakamoto.
Tokyo Astron. Obs. Rep., Vol. 20, No. 2, p. 367 – 381 (1984). In Japanese.

041.022 **Evaluation of phase corrections on planet position measurements.**
F. Chollet.
Astron. Astrophys., Vol. 139, No. 1, p. 215 – 219 (1984). In French.

Since 1963, the capability of astrolabe for planetary observations, has been fully demonstrated. But, the quality of the results was not as good as those obtained for star or, even, solar observations. This shows that the problem does not come from instrumental errors but, in fact, from an insufficient correction of the phase on observations. The author presents a model of distribution of light intensity and, assuming that the point observed is the photocenter, he computes the corresponding phase corretion. Analysis of Mars observations, compared to the Jet Propulsion Laboratory ephemeris, shows that there is a strong "phase effect" in right ascension. The new phase correction applied to these results gives the best residuals. A comparison of the author's model with Jupiter observations is being made at San Fernando Observatory. Studies of Venus observations will begin soon in Sao Paulo Observatory.

041.023 **On the dependence of fluctuations of measured zenith distances on the change of the field of the refractive index.**
S. G. Savchuk.
L'vov. politekh. inst. L'vov, 7 pp. (1984). In Russian. Abstr. in Ref. Zh., 51. Astron., 10.51.151 (1984).

041.024 **Results of an astrometric experiment on the Crimea–Pushchino interferometer.**
V. E. Zharov, L. R. Kogan, R. L. Sorochenko, E. N. Fedoseev.
Sov. Astron., Vol. 28, No. 2, p. 217 – 221 (1984). English translation of 37.041.013.

041.025 **Natural satellite astrometry with the Space Telescope planetary camera.**
G. W. Null.
Bull. Am. Astron. Soc., Vol. 16, No. 3, Part I, p. 723 (1984). Abstract. – See Abstr. 010.064.

041.026 **Astrometric studies in the development of the Space Telescope Guide Star Selection System.**
J. L. Russell.
Bull. Am. Astron. Soc., Vol. 16, No. 3, Part I, p. 723 (1984). Abstract. – See Abstr. 010.064.

041.027 **The treatment of systematic errors in the astrolabe observations and astrolabe catalogue.**
H. Yan, Z. Li.
Ann. Shanghai Obs., Acad. Sin., No. 5, p. 96 – 100 (1983). In Chinese.

For treating systematic errors, in the astrolabe observations and astrolabe catalogue, the astrolabe error equation: $15\cos\phi\sin A_i\Delta u + \cos A_i\Delta\phi - \Delta z + \delta h_i = v_i$ is the only proper fundamental equation to be used.

041.028 **Group corrections of declinations from observations with the ZTL–180 zenith telescope for the period 1967.0 – 1979.0.**
S. K. Ehshonkulov.
Tsirk. Astron. Inst. Tashkent, No. 109, p. 35 – 39 (1983). In Russian.

041.029 **Catalogue of right ascensions of 109 stars in the system of the compiled catalogue of the USSR Time Service (CTS).**
M. Zakirov, Eh. A. Sanakulov.
Tsirk. Astron. Inst. Tashkent, No. 110, p. 24 – 39 (1983). In Russian.

041.030 **Catalogue of right ascensions of 376 stars from the FK4 with declinations from −20° to +90° obtained from observations with the Tashkent meridian circle in 1976 – 1978, and systematic and random differences TMK–FK4 for these stars.**
I. M. Boroditskij.
Tsirk. Astron. Inst. Tashkent, No. 111, 39 pp. (1983). In Russian.

041.031 **Prospects for space stellar astrometry.**
J. Kovalevsky.
Space Sci. Rev., Vol. 39, Nos. 1/2, p. 1 – 63 (1984).
The present achievements of the Earth–based astrometry are described. The astrometric activities such as measurements of star diameters, double star relative positioning or stellar parallaxes, search for invisible companions, photographic plate reduction, visual and photoelectric meridian and astrolabe astrometry are reviewed. Then, the methods used to construct a quasi–inertial celestial reference frame and to materialize it by a fundamental catalogue are presented and discussed. Then, the ESA astrometric satellite HIPPARCOS presently under development is presented. The satellite and the payload are described as well as the observing procedures. The establishment of an absolute reference frame is subsequently discussed. Plans using simultaneously VLBI, HIPPARCOS, and Space Telescope observations are described. Finally, several space astrometry proposals are described: long focus space astrometry and two versions of space interferometry.

041.032 **Systematic errors of Belgrade NPZT programme star catalogue.**
S. Sadžakov, M. Dačić.
Bull. Appl. Math., No. 197, p. 121 – 125 (1983).
Results of a comparison – Belgrade NPZT programme star catalogue to six other catalogues, compiled by use of meridian circles of the same, or similar properties are given. Existence of small systematic erros of $\Delta\alpha_\alpha$ and $\Delta\delta_s{}^p$ type is found.

041.033 **Tianjin photographic zenith tube observations.**
H.–q. Ji, Z.–x. Sun, S.–r. You.
Chin. Astron. Astrophys., Vol. 8, No. 4, p. 328 – 334 (1984). English translation of Acta Astron. Sin., Vol. 25, No. 2, p. 191 – 200 (1984).
The authors analyse the observational data obtained by the Chinese–made PZT in the two periods 1979 February – 1980 May and 1981 December – 1983 March. The internal accuracy for individual stars is found to be $m_u = \pm 13.0$ ms, $m_\phi = \pm 0''.144$ for the first period, and $m_u = \pm 14.6$ ms, $m_\phi = \pm 0''.152$ for the second. The systematic error caused by the sealed window of the evacuated chamber and the temperature effect of the plate scale are investigated. Monthly means of time and latitude are given.

041.034 **Les données astrométriques utilisées en cinématique stellaire.**
J. Delhaye.
Bull. Inf. Cent. Données Stellaires, No. 27, p. 7 – 8 (1984).

041.035 **Herstmonceux observations of the Sun, planets and Moon 1957 – 1982.**
R. H. D. Swifte, P. Gibbs, M. E. Buontempo, P. Eldridge.
R. Greenwich Obs., Bull., No. 193, 40 pp. (1984). With a microfiche.
This bulletin contains the results of observations made at the Royal Greenwich Observatory (Herstmonceux) with the Cooke Transit Circle from 1957 – 1982. The positions are based on the system of the FK4; the results extend and replace the provisional positions published in RO Bulletins Nos. 178, 179 and 180.

041.036 **Calculation of apparent positions of stars according to MERIT standards.**
G.–c. Shi, Z.–x. Li.
Publ. Purple Mt. Obs., Vol. 3, No. 2, p. 15 – 22 (1984). In Chinese.

041.037 **Results of the first observations with a meridian circle under mountain conditions. I. Declination processing.**
B. A. Golovko, A. P. Gulyaev, V. G. Tauber, V. G. Shamaev.
Astron. Tsirk., No. 1304, p. 1 – 3 (1984). In Russian.

041.038 **On the catalogue of declinations of stars around radio sources.**
P. F. Lazorenko.
Astron. Tsirk., No. 1306, p. 4 – 5 (1984). In Russian.

041.039 **A photoelectric catalogue of right ascensions of circumpolar stars.**
V. N. Ershov.
Astron. Tsirk., No. 1339, p. 5 – 7 (1984). In Russian.

041.040 **Declinations of 35 radio stars.**
P. F. Lazorenko.
Astron. Tsirk., No. 1320, p. 7 – 8 (1984). In Russian.

041.041 **Calculation of the error of determination of equatorial coordinates of observed objects.**
D. I. Bertov, A. A. Gavrilov, N. V. Shoferova.
Sistemy upr. i tsifr. obrab. inf. Leningrad, p. 116 – 121 (1983). In Russian. Abstr. in Ref. Zh., 51. Astron., 12.51.165 (1984).

041.042 **The HIPPARCOS space astrometry mission.**
J. Kovalevsky.
Mitt. Astron. Ges., Nr. 62, p. 63 – 74 (1984). – See Abstr. 010.341.
The payload of the HIPPARCOS mission is shortly described and the accuracy expected is given. The instrumental limitations in the optics, the grid, the mechanical structure and the detector are described. They all lie in the few millisecond of arc range. Applications of the expected results are listed for astrometry, celestial mechanics, double stars, variable stars, stellar physics and evolution, clusters, our Galaxy and distance scale. Finally, the respective roles and achievements of ground based and space astrometry are reviewed and compared showing their complementary aspects.

041.043 **Rotationsmatrizen der Fundamentalepoche B 1950.0.**
N. P. Wieth–Knudsen.
Mitt. Astron. Ges., Nr. 62, p. 154 (1984). Abstract. – See Abstr. 010.341.

041.044 **On the variation of the obliquity of the ecliptic.**
A. D. Wittmann.
Mitt. Astron. Ges., Nr. 62, p. 201 – 204 (1984). – See Abstr. 010.341.

041.045 **Zum astrometrischen Nutzen von PZT–Katalogen.**
K. Lorenzen.
Mitt. Astron. Ges., Nr. 62, p. 205 – 206 (1984). – See Abstr. 010.341.

041.046 **Über die Systeme der Positionen und Eigenbewegungen in den Katalogen AGK3, AGK3R (Corbin's Version) und N30.**
H. Schwan.
Mitt. Astron. Ges., Nr. 62, p. 208 (1984). Abstract. – See Abstr. 010.341.

041.047 **CCD–Astrometrie ausgewählter Radioquellen für radio/optisches extragalaktisches Referenzsystem.**
J. Schramm.
Mitt. Astron. Ges., Nr. 62, p. 209 (1984). Abstract. – See Abstr. 010.341.

041.048 **Optical "proper motions" of compact radio sources.**
P. Brosche, M. Geffert.
Mitt. Astron. Ges., Nr. 62, p. 332 (1984). Abstract. – See Abstr. 010.341.

041.049 **The equinox value as deduced from the current observations of the Sun, Venus and Mars.**
A. S. Kharin, N. F. Minyajlo, Yu. I. Safronov.
Pis'ma Astron. Zh., Tom 10, No. 12, p. 931 – 935 (1984). In Russian. English translation in Sov. Astron. Lett., Vol. 10.
The reduction of 24366 observations of the Sun, Venus and Mars, carried out in 1960 – 1977 with meridian instruments, astrographs and astrolabes was made. Differences between the equinox values calculated from the observations of the Sun and Venus and the observations of Mars were revealed. They can be explained by differences between day and night systematic observational errors. The 20-years current observations of the Sun and Venus do not show any trend of the equinox.

041.050 **An analytical method for the study of a system of star positions.**
C.-w. Zhang.
Acta Astron. Sin., Vol. 25, No. 2, p. 177 – 185 (1984). In Chinese.
There are some difficulties in dealing with the analytical method proposed by Brosche et al. for the star catalogs in which the distribution of star positions is nonuniform. For this reason, a proposition of a set of complete and orthogonal functions has been made. Using the proposition, it is easy to construct the set of complete and orthogonal functions that is satisfied by the arbitrary distribution density of star positions. Using a new model of smoothing, it is possible to obtain more details of the systematic errors of the star catalogs as described above. The improvements have been used successfully in the study of the General Catalog of stars observed with photoelectric astrolabes in China (GCPA).

041.051 **Preliminary results of absolute determinations of right ascensions with a photoelectric transit instrument (BAC).**
Z.-g. Yao, S.-h. Jiang.
Acta Astron. Sin., Vol. 25, No. 3, p. 280 – 286 (1984). In Chinese.

041.052 **Astrometric network SOPROG.**
S. P. Major.
Komet. Tsirk., No. 332 (1984). In Russian.

**Les catalogues astrométriques du C.D.S.**
See Abstr. 002.011.

**Catalogue of right ascensions of near-polar stars.**
See Abstr. 002.025.

**The ST Guide Star Catalog.**
See Abstr. 002.032.

**Documentation for the machine-readable version of the *AGK3 Star Catalogue of Positions and Proper Motions North of −2°.5 Declination* (Dieckvoss and Collaborators 1975).**
See Abstr. 002.040.

**Compiled catalogue of positions and proper motions of 4949 geodetic stars from +90° to −90°. Part I.**
See Abstr. 002.043.

**Compiled catalogue of positions and proper motions of 4949 geodetic stars from +90° to −90°. Part II.**
See Abstr. 002.044.

**Bessel's Arbeiten zur Positionsastronomie.**
See Abstr. 005.024.

**Workshop über astrometrische und geodätisch-geophysikalische Nutzung der Radio-Interferometrie.**
See Abstr. 011.029.

**HIPPARCOS workshop.**
See Abstr. 011.032.

**Observations d'objets du système solaire à l'instrument méridien de Bordeaux.**
See Abstr. 013.015.

**Dynamical astronomy via optical astrometric interferometry in space.**
See Abstr. 013.048.

**The USNO pole-to-pole astrometric program.**
See Abstr. 013.049.

**The C.D.S. and astrometry.**
See Abstr. 013.063.

**L'astrométrie au télescope de Schmidt de Calern.**
See Abstr. 013.064.

**Catalogues astrométriques et programmes spatiaux.**
See Abstr. 013.065.

**Les données astrométriques accessibles au CDS. État actuel et développements prochains.**
See Abstr. 013.066.

**Le C.D.S. et les programmes d'astrométrie méridienne.**
See Abstr. 013.067.

**First tests on the astrometric performance of the new PDS1010G at Hamburg Observatory.**
See Abstr. 034.159.

**Astrometric applications of the Mt. Stromlo PDS 1010A.**
See Abstr. 034.160.

**Le problème de la phase dans l'observation des planètes à l'astrolabe.**
See Abstr. 036.020.

**Préparation de la réduction des données Hipparcos.**
See Abstr. 036.021.

**Précision astrométrique en présence de bruit.**
See Abstr. 036.023.

**Estimation de la qualité des observations à l'astrolabe photoélectrique.**
See Abstr. 036.024.

**An iteration algorithm of the stellar trigonometric parallax reduction with the central overlap technique.**
See Abstr. 036.074.

**Variation in time of the parameters (u + m) and $M_0$ of the Belgrade meridian circle.**
See Abstr. 036.149.

**Comprehensive star catalogues.**
See Abstr. 036.151.

**Mesure de clichés de la Carte du Ciel.**
See Abstr. 036.152.

**Analytical and numerical studies about effects of natural radio source structures on astrometric and geodetic observations with a very long baseline interferometer.**
See Abstr. 036.156.

**Elaboración matemática de las observaciones del astrolabio.**
See Abstr. 036.157.

**Astrometrische Parameter hergeleitet aus HIPPARCOS-Messungen.**
See Abstr. 036.161.

**Lösung des "Grid Step Ambiguity" Problems des HIPPARCOS.**
See Abstr. 036.162.

**Hamburger Software–Entwicklung zur Reduktion astrometrischer Meßdaten.**
See Abstr. 036.163.

**Relation between the dynamical fundamental system FK5 and the system of positions of extragalactic sources.**
See Abstr. 043.007.

**Präzisions–Zeitmessung mit dem Wiener Passage–Instrument Askania Ap 70 anhand ausgewählter FK4 Sterne.**
See Abstr. 044.068.

**Contrast sensitivity of the eye and the systematic errors of the coordinates of planets.**
See Abstr. 091.057.

**First results of Mars observations with a PVC.**
See Abstr. 097.001.

**First observations of Mars with the photoelectric meridian circle in Pulkovo.**
See Abstr. 097.050.

**Study into the accuracy of astrometric observations of comets.**
See Abstr. 102.072.

**Photographic astrometry of binary and proper–motion stars.**
See Abstr. 111.001.

**Radial velocities and positions from objective prism plates.**
See Abstr. 111.009.

**Astrometric study of the subdwaf HR 4550.**
See Abstr. 111.014.

**CCD astrometry and photometry of IAU reference frame benchmark radio sources.**
See Abstr. 158.177.

# 042 Celestial Mechanics, Figures of Celestial Bodies

042.001 **Coplanar solutions in the photo–gravitational restricted circular three–body problem.**
L. G. Luk'yanov.
Astron. Zh., Tom 61, Vyp. 4, p. 789 – 794 (1984). In Russian. English translation in Sov. Astron., Vol. 28, No. 4.

The before–known results for the coplanar libration points are supplemented and refined. The possibility of the existence of four coplanar libration points is shown. The regions of the existence of the libration points in the plane of two free parameters are determined.

042.002 **On the structure of absolutely rigid bodies permitting plane motions.**
V. V. Vidyakin.
Astron. Zh., Tom 61, Vyp. 4, p. 795 – 799 (1984). In Russian. English translation in Sov. Astron., Vol. 28, No. 4.

Necessary and sufficient conditions under which the centre of masses of absolutely rigid bodies can move in an invariable plane are found. The conditions found are also given for an approximate expression of the force function of interaction of bodies.

042.003 **On 6π–periodic solutions of the plane restricted elliptic problem of three bodies.**
A. Ikromov.
Astron. Zh., Tom 61, Vyp. 4, p. 800 – 805 (1984). In Russian. English translation in Sov. Astron., Vol. 28, No. 4.

The plane restricted elliptic problem of three bodies is shown to admit a family of 6π–periodic solutions as series of powers of eccentricity near the libration point. Recurrent formulae for the construction of periodic solutions are found.

042.004 **Wandel der Problemstellungen und Lösungsmethoden in der Himmelsmechanik während der letzten 400 Jahre.**
G. Beutler.
Orion, 42. Jahrg., Nr. 203, p. 139 – 141 (1984).

042.005 **Power series in the framework of the averaged hyperbolic restricted three–body problem.**
A. B. Sorokovich.
Tr. Gos. Astron. Inst. Shternberg, Tom 54, p. 160 – 176 (1984). In Russian.

The planar single–averaged hyperbolic restricted three–body problem is considered. The expansion of the perturbation function of the problem has been obtained, and the differential equations of motion have been worked out. The author has found the general solution for the system of differential equations of motion in the form of power series aranged in powers of the true anomaly of the perturbed body for the case when the first term of the perturbation function is taken into account.

042.006 **Rotational dynamics of celestial bodies: generalized rotation and considerations of angular momentum.**
V. S. Geroyannis.
Astrophys. Space Sci., Vol. 102, No. 2, p. 255 – 261 (1984).

The transformation equations under generalized rotation are obtained for an initially defined reduced velocity tensor governing the motion of a deformable finite material continuum. Then angular momentum considerations lead to relations between flow properties of the continuum and properties of a coordinate system introduced to describe generalized rotation of the continuum. Such relations could define preferable coordinate systems perceiving zero angular momentum for the continuum or referring that it moves according to linear laws.

042.007 **The Gaussian curvature of associated manifold of dynamical systems.**
E. A. Evangelidis, J. D. Neethling.
Astrophys. Space Sci., Vol. 103, No. 1, p. 99 – 113 (1984).

A connection is shown to exist between the Gaussian curvature of the associated manifold and the ergodic or non–ergodic behaviour of certain dynamical systems of astronomical and astrophysical importance.

042.008 **The destruction of a continuous ring revolving around a gravitating center.**
A. M. Fridman, A. I. Morozov, V. L. Polyachenko.
Astrophys. Space Sci., Vol. 103, No. 1, p. 137 – 142 (1984).

In an article by Morozov and Fridman (1984) the general equation of motion of an elastic thread in an external gravitational field was formulated and also the instability of an elastic thread rotating around a gravitating center was established. On this basis, in the present article the authors show that the classical model of a hypothetical continuous ice ring around Saturn, proposed by Laplace and Maxwell, was incorrectly interpreted by them. In conclusion, estimates are given which show that the steel ring (thread) revolving the Earth and rotating around it at a distance of 200 km from the surface would be torn before it falls on the Earth. For threads of firm aluminum fusions the opposite picture is to be expected.

042.009 **The Kepler problem spinor regularization and quantization on the manifolds $S^2$.**
M. D. Vivarelli.
Meccanica, Vol. 19, No. 1, p. 11 – 14 (1984). Abstr. in Phys. Abstr., Vol. 87, No. 1238, Entry 77699 (1984).

042.010 **Numerical simulations of collisions in self–gravitating systems.**
J. Lukkari, H. Salo.
Earth, Moon, Planets, Vol. 31, No. 1, p. 1 – 13 (1984).
The influence of partially elastic collisions, gravitational encounters, and different gravitational potentials is studied in terms of computer simulations involving a ring of 200 identical particles. If the masses of the particles are small, their mutual gravitational attraction is found to produce a decrease in the dispersion of velocities below the collisional equilibrium value, but for heavier particles the dispersion again begins to increase. The vertical component of the force which is produced by the self–gravitation of the ring reduces its thickness. The results of the simulations are in good agreement with the predictions of the collisional theory.

042.011 **Second order secular planetary theory for the four major planets.**
O. M. Kamel, A. A. Bakry.
Earth, Moon, Planets, Vol. 31, No. 2, p. 123 – 134 (1984).
The authors generalize their results of a second order Jupiter–Saturn planetary theory to be applicable for the case of the four major planets. They use the von Zeipel method and neglect powers higher than the third with respect to the eccentricities and sines of the inclinations in their expansions. They consider the critical terms as the only periodic terms.

042.012 **Theory of the motion of Jupiter and Saturn over a period of 6000 years. Solution JASON84.**
J. L. Simon, P. Bretagnon.
Astron. Astrophys., Vol. 138, No. 1, p. 169 – 178 (1984). In French.
The JASON84 solution gives the motion of Jupiter and Saturn with a good precision over a period of 6000 years. In the first part the authors describe the method used to compute the mutual perturbations of Jupiter and Saturn with a good precision over 6000 years. The variables used are $a$, $\lambda$, $e$, $\tilde{\omega}$, $\gamma$, $\Omega$ and the equations are Lagrangian equations. In the second part the solution JASON84 is described. It contains the perturbations of Jupiter and Saturn by Uranus and Neptune from TOP82 and VSOP82 (Bretagnon, 1982) and the perturbations by the inner planets from VSOP82. The precision of the geocentric longitudes of the planets, calculated from this solution, is better than 0°.01 over the period –4000, +2000, and better than 0°.003 over the period –1000, +2000.

042.013 **Précision des théories planétaires.**
P. Bretagnon.
Astrométrie, géodésie, mécanique céleste, p. 5 – 8 (1984). – See Abstr. 012.021.

042.014 **Prolongement de théories planétaires à variations séculaires sur plusieurs milliers d'années.**
J. L. Simon.
Astrométrie, géodésie, mécanique céleste, p. 9 – 14 (1984). – See Abstr. 012.021.
Les théories planétaires à variations séculaires construites au Bureau des Longitudes, VSOP82 (Bretagnon, 1982) et TOP82 (Simon, 1983) ne contiennent pas de termes de degré supérieur à 6 dans le développement en puissance du temps des perturbations. Il en résulte que leur précision se dégrade avec le temps pour des intervalles de temps supérieurs à 1000 ans. L'auteur présente ici un travail effectué en collaboration avec P. Bretagnon qui a consisté à construire une solution du mouvement de Jupiter et Saturne gardant une bonne précision sur 6000 ans.

042.015 **Les paramètres dans une théorie générale planétaire.**
L. Duriez.
Astrométrie, géodésie, mécanique céleste, p. 15 – 19 (1984). – See Abstr. 012.021.

042.016 **Intermediate resonance motion of an equatorial satellite in the gravitational field of a rotating nonspherical planet.**
V. A. Kitova.
Vestn. AN KazSSR. Alma–Ata, 11 pp. (1984). In Russian. Abstr. in Ref. Zh., 62. Issled. Kosm. Prostranstva, 8.62.182 (1984).

042.017 **Application of averaging methods to the problem of motion of a satellite in the gravitational field of a rotating planet.**
V. A. Kitova.
Vestn. AN KazSSR. Alma–Ata, 8 pp. (1984). In Russian. Abstr. in Ref. Zh., 62. Issled. Kosm. Prostranstva, 8.62.183 (1984).

042.018 **Deformations of extended elastic bodies in a Newtonian gravitational field.**
A. I. Lomachenko.
Probl. mekh. i upr. dvizheniem. Tr. 6 Obedin. nauchn. chtenij po kosmonavt., posvyashch. pamyati vydayushch. sov. uchenykh–pionerov osvoeniya kosm. prostranstva Moskva, 11–15 yanv., 1982. Moskva, p. 103 – 111 (1983). In Russian. Abstr. in Ref. Zh., 62. Issled. Kosm. Prostranstva, 8.62.191 (1984).

042.019 **Planets in binary star systems.**
M. Schechter.
Sky Telesc., Vol. 68, No. 5, p. 394 – 395 (1984).

042.020 **Resonance rotations of a satellite in a polar orbit in a magneto–gravitational field.**
V. V. Beletskij, A. N. Shlyakhtin.
Kosm. Issled., Tom 22, Vyp. 4, p. 499 – 506 (1984). In Russian. English translation in Cosm. Res.

042.021 **Construction of conditionally–periodic solutions in the problem of translatory–rotational motion of an axisymmetric satellite of a triaxial planet.**
A. A. Zlenko.
Kosm. Issled., Tom 22, Vyp. 4, p. 531 – 536 (1984). In Russian. English translation in Cosm. Res.

042.022 **Expansion theory for the elliptic motion of arbitrary eccentricity and semi–major axis. VI. Elliptic expansions in terms of the sectorial variables for the fifth and sixth categories.**
M. A. Sharaf.
Astrophys. Space Sci., Vol. 104, No. 2, p. 267 – 284 (1984).
In this paper of the series, elliptic expansions in terms of the sectorial variables $\theta_j^{(i)}$ introduced recently in Paper IV (Sharaf, 1982) are established analytically and computationally for the fifth and sixth categories. For each of the elliptic expansions belonging to a category, literal analytical expressions for the coefficients of its trigonometric series representation are established. Moreover, some recurrence formulae satisfied by these coefficients are also established to facilitate their computations; numerical results are included to provide test examples for constructing computational algorithms.

042.023 **The influence of secular perturbations of gravitational nature on the stability of the collinear libration points of the circular restricted three–body problem.**
G. I. Shirmin.
Tr. Gos. Astron. Inst. Shternberg, Tom 56, p. 38 – 85 (1984). In Russian.
Within the range of a model obtained from the twice–restricted circular four–body problem with the help of the Fatou averaging method the influence of a perturbing body attraction on the infinitesimal mass motion is considered.

042.024 **The influence of the effect of rotation on the evolution of triple systems.**
Zh. P. Anosova, D. I. Bertov, V. V. Orlov.
Astrofizika, Tom 20, Vyp. 2, p. 327 – 339 (1984). In Russian. English translation in Astrophysics, Vol. 20, No. 2.
The dynamical evolution of triple systems with negative total energy and other than zero momentum of rotation is investigated by the method of numerical experiment; the masses of bodies are equal. The initial configurations were chosen randomly; the initial quantities of the virial coefficient are $k = 0.1, 0.2, 0.3, 0.4, 0.5$; the value of angular momentum $L \sim \sqrt{(k(1-k))}$. 1000 triple systems for every $k$ are considered.

042.025 **The stability of binary star systems during encounters with a third star.**
J. R. Donnison.
Mon. Not. R. Astron. Soc., Vol. 210, No. 4, p. 915 – 927 (1984).
The interaction of a binary star system and a third star not moving on a closed orbit is discussed in terms of three–body zero velocity surfaces. The ranges of orbital parameters for which the binary system is stable against disruption or collision by a third star moving on a parabolic or hyperbolic orbit are determined. It is found that the regions of stable motion decreased as the mass and eccentricity of the passing star increased.

042.026 **Translational–rotational motion of a rigid body in the restricted problem of three rigid bodies.**
E. N. Eremenko.
Astron. Zh., Tom 61, Vyp. 5, p. 1013 – 1020 (1984). In Russian. English translation in Sov. Astron., Vol. 28, No. 5.
The stationary solutions of averaged differential equations are investigated for the case of the translational–rotational motion of an infinitesimal body of arbitrary structure in the field of attraction of a triaxial body and of a homogeneous sphere assuming the existence of a commensurability between the rotational motion of the triaxial body and the orbital and rotational motion of the infinitesimal body. The stability of the solutions obtained, due to the averaged equations, is considered.

042.027 **A test of escape valid even for very small mutual distances. I. The acceleration and the escape velocities of the third body.**
C. Marchal, J. Yoshida, Y.–s. Sun.
Celest. Mech., Vol. 33, No. 3, p. 193 – 207 (1984).
The authors consider a three–body system with three given point–masses and analyze its bounded solutions with a given angular momentum. In the first part they analyze the motion of the third mass, its acceleration and its escape velocities when, with a suitable scalar k, a condition of the type $r \leqslant kR$ remains forever satisfied (r being the mutual distance of the two masses of the "binary" and R the distance between the center of mass of the binary and the "third mass"). In the second part the authors look for initial conditions under which the inequality $r \leqslant kR$ will remain forever satisfied and develop the corresponding test and its applications.

042.028 **Periodic orbits in a three–dimensional potential. II. Bifurcations.**
P. Cartigny, N. Desolneux, A. Hayli.
Celest. Mech., Vol. 33, No. 3, p. 217 – 227 (1984). In French.
In a previous paper, Hayli et al. (1983), two families of periodic orbits in a given three–dimensional potential were described. A property of the characteristic curves of the two families was found empirically. This property is demonstrated in the present paper by writing explicitely the Poincaré mapping and by giving an approximation directly comparable with the numerical results obtained in Hayli et al. (1983).

042.029 **About an unsuspected integrable problem.**
F. Mignard, M. Henon.
Celest. Mech., Vol. 33, No. 3, p. 239 – 250 (1984).
The motion of a particle orbiting a planet is investigated when both the gravitational attraction of the planet and the radial radiation force due to the Sun are taken into account. The authors establish that the secular problem, which has two degrees of freedom, is completely integrable and that it possesses only periodical solutions. An explicit analytical solution is provided.

042.030 **Two notes on the Copenhagen Problem. I. Asymptotic branches from $L_4$.**
J. M. A. Danby.
Celest. Mech., Vol. 33, No. 3, p. 251 – 260 (1984).
Poincaré's method of consequents is applied to the computation of asymptotic branches from the unstable equilibrium $L_4$ in the restricted problem of three bodies with equal masses. Because of looping in the orbits, the "Poincaré mapping" in this problem is not continuous, and the structure of the branches is more complicated than that for simpler dynamical systems for which computations have appeared.

042.031 **The evolution of periodic orbits close to heteroclinic points.**
J. M. A. Danby.
Celest. Mech., Vol. 33, No. 3, p. 261 – 270 (1984).
Several families of periodic orbits close to heteroclinic points were computed, and their evolution, as a function of the energy, was followed. All appeared spontaneously at the occasion of the formation of the heteroclinic points, and had no genealogical links with other families. This behavior is in contrast to the behavior of periodic orbits close to homoclinic points, some of which have genealogical links with periodic orbits existing in the "quasi–integrable" phase of the system. The orbits found here appeared in regions where stochasticity was well established; so their appearance does not seem to be connected with the onset of stochasticity.

042.032 **Equilibrium solutions of the restricted problem of 2+2 bodies.**
A. L. Whipple.
Celest. Mech., Vol. 33, No. 3, p. 271 – 294 (1984).
Fourteen equilibrium solutions of the restricted problem of 2+2 bodies are shown to exist. Six of these solutions are located about the collinear Lagrangian points of the classical restricted problem of three bodies. Eight solutions are found in the neighborhood of the triangular Lagrangian points. Linear stability analysis reveals that all of the equilibrium solutions are unstable with the exception of four solutions; two in the vicinity of each of the triangular Lagrangian points. These four solutions are found to be stable provided the mass parameter of the primary masses is less than a critical value which depends also on the mass of the minor bodies.

042.033 **Lagrangian formulation for the MacGehee change of coordinates.**
F. Nahon.
C. R. Acad. Sci., Sér. II, Tome 299, No. 14, p. 933 – 936 (1984). In French.
The author gives in Lagrangian formulation the generalization of the system proposed by MacGehee for the rectilinear three–body problem, and he lists some applications.

042.034 **Lösung des Hauptproblems der Theorie der Satellitenbahnen mit Hilfe von Lie–Reihen.**
M. Schneider.
Veröff. Bayer. Komm. Int. Erdmessung Bayer. Akad. Wiss., Astron. –Geod. Arb., Heft Nr. 45, p. 190 – 194 (1984).

042.035 **On initial orbit determination.**
L. G. Taff.
Astron. J., Vol. 89, No. 9, p. 1426 – 1428 (1984).
The classical methods of initial orbit determination are brought together within a larger viewpoint. This new synthesis stresses that all such techniques follow one of three approaches. Either they seek to compute the orbital element set, or its equivalent, by attacking the differential equations of motion (Laplace), the first integrals of the equations of motions (Taff), or the solution itself (Gauss). The efficacy of some algorithms for each approach is discussed briefly.

042.036 **On the stability of motion in the restricted circular three–body problem.**
R. I. Kiladze, V. S. Safronov.
O. Yu. Shmidt i sov. geofiz. 80–kh godov, Moskva, p. 221 – 227 (1983). In Russian. Abstr. in Ref. Zh., 51. Astron., 9.51.89 (1984).

042.037 **The perturbed motion of a distant satellite of a spheroidal planet.**
A. M. Iskakova.
Kaz. univ. Alma–Ata, 11 pp. (1984). In Russian. Abstr. in Ref. Zh., 51. Astron., 9.51.112 (1984).

042.038 **Non–integrable classical dynamical systems.**
H. Yoshida.
Astron. Her., Vol. 77, No. 11, p. 283 – 285 (1984). In Japanese.

042.039 **Excitation of inclinations in ring–satellite systems.**
N. Borderies, P. Goldreich, S. Tremaine.
Astrophys. J., Vol. 284, No. 1, p. 429 – 434 (1984). = Contrib. No. 3930, Div. Geol. Planet. Sci., Calif. Inst. Technol., Pasadena, Calif. 91125.

Resonant gravitational interactions between a ring and a satellite produce secular variations of their orbital inclinations. Interactions at vertical resonances, analogous to Lindblad resonances but involving inclinations instead of eccentricities, excite inclinations. An equatorial ring changes the inclination of a nearby satellite in qualitatively the same way that a satellite in an equatorial orbit changes the inclination of a nearby ring. Viscous dissipation in a ring leads to an equilibrium value of its inclination. These results provide a basis for discussing the origins of the inclinations of planetary rings.

042.040 **Libration–type trajectories in the planar circular restricted three–body problem.**
A. Yu. Kogan.
Probl. mekh. i upr. dvizheniem. Tr. 6 Obedin. nauchn. chtenij po kosmonavt., posvyashch. pamyati vydayushch. sov. uchenykh–pionerov osvoeniya kosm. prostranstva, Moskva, 11 – 15 yanv., 1982. Moskva, p. 39 – 52 (1983). Abstr. in Ref. Zh., 51. Astron., 10.51.94 (1984).

042.041 **The restricted four–body problem.**
U. A. Bajzakov.
Vestn. Akad. Nauk KazSSR, 6 pp. (1984). In Russian. Abstr. in Ref. Zh., 62. Issled. Kosm. Prostranstva, 10.62.172 (1984).

042.042 **Integration of equations of motion of a material point in the gravitational field of an axisymmetrical planet.**
A. R. Golikov.
Kosm. Issled., Tom 22, Vyp. 5, p. 651 – 662 (1984). In Russian. English translation in Cosm. Res.

042.043 **Periodic regimes of rotational motion of a solid body around the $L_4$ libration point in the restricted elliptical three–body problem.**
Yu. V. Barkin, S. N. Lelyavin.
Kosm. Issled., Tom 22, Vyp. 5, p. 663 – 674 (1984). In Russian. English translation in Cosm. Res.

042.044 **Five bodies to infinity.**
I. Stewart.
Nature, Vol. 312, No. 5993, p. 398 – 399 (1984).

This note reviews recent developments in the mathematical theory of the motion of five bodies under Newtonian gravitation.

042.045 **Perihelic shift of planets due to the gravitational field of the charged Sun.**
M. T. Teli, D. Palaskar.
Nuovo Cimento C, Vol. 7C, Ser. 1, No. 1, p. 130 – 136 (1984). Abstr. in Phys. Abstr., Vol. 87, No. 1245, Entry 110507 (1984).

042.046 **Influence of variable tidal lag on orbital evolution in a system of two celestial bodies.**
I. V. Kuznetsova, M. M. Khapaev.
Sov. Astron., Vol. 28, No. 2, p. 215 – 217 (1984). English translation of 37.042.037.

042.047 **Convergence of classic expansions of celestial mechanics.**
K. V. Kholshevnikov.
Sov. Astron., Vol. 28, No. 3, p. 325 – 329 (1984). English translation of 37.042.074.

042.048 **Lagrangian solutions in the photogravitational, restricted, circular three–body problem.**
L. G. Luk'yanov.
Sov. Astron., Vol. 28, No. 3, p. 329 – 333 (1984). English translation of 37.042.075.

042.049 **Calulation of elliptic functions in problems of celestial mechanics.**
I. A. Gerasimov.
Sov. Astron., Vol. 28, No. 3, p. 356 – 357 (1984). English translation of 37.042.076.

042.050 **Hill stability for the sun–earth–moon system.**
H. Ding, T. Huang.
J. Nanjing Univ., No. 2, p. 257 – 266 (1984). In Chinese.

In this paper Hill stability of the sun–earth–moon system is discussed numerically. The numerical results show that the moon is probably stable and that if the distance between the moon and the earth be enlarged to 1.85 times of the actual distance, the moon will escape within 160 years; and if the distance be enlarged to 1.9 times, the moon will escape within 5 years. These results are quite different from those obtained theoretically.

042.051 **On initial orbit determination.**
L. G. Taff.
Bull. Am. Astron. Soc., Vol. 16, No. 3, Part I, p. 721 (1984). Abstract. – See Abstr. 010.064.

042.052 **On the Brown conjecture.**
B. Garfinkel.
Bull. Am. Astron. Soc., Vol. 16, No. 3, Part I, p. 721 (1984). Abstract. – See Abstr. 010.064.

042.053 **On the tidal evolution of small binaries (asteroids?).**
A. W. Harris.
Bull. Am. Astron. Soc., Vol. 16, No. 3, Part I, p. 721 (1984). Abstract. – See Abstr. 010.064.

042.054 **Gravitational encounters between orbiting bodies.**
J. B. Hoffer.
Bull. Am. Astron. Soc., Vol. 16, No. 3, Part I, p. 724 (1984). Abstract. – See Abstr. 010.065.

042.055 **The use of the energy and angular momentum integrals to obtain a stability criterion in the general hierarchical three–body problem.**
A. E. Roy, A. Carusi, G. B. Valsecchi, I. W. Walker.
Astron. Astrophys., Vol. 141, No. 1, p. 25 – 29 (1984).

The integrals of energy and angular momentum in the general hierarchical three–body problem are used to construct a stability criterion which may be used not only to decide if hierarchical stability of the system is guaranteed but will give information, if stability is guaranteed, about the orbital elements of the orbits at closest possible approach of the second and third masses. The criterion is applied to the Sun – Jupiter – Saturn system and certain cosmogonic conclusions regarding the growth of Jupiter and Saturn by accretion in the early days of the Solar System are drawn.

042.056 **A new compact representation of ephemerides: application to the motion of Pluto, the Sun and the Galilean satellites of Jupiter.**
J. Chapront, D. T. Vu.
Astron. Astrophys., Vol. 141, No. 1, p. 131 – 143 (1984).
A new method of approximation is proposed to represent ephemerides in a compact form. As a basis of approximating functions, instead of polynomials as it is the case with Chebychev's approximation, one chooses a basis of mixed functions with secular and sinusoidal terms, in order to take into account the quasi–periodic character of solutions.

042.057 **Evolution of families of collision periodic orbits in the plane restricted three–body problem with mass ratio μ near 1.**
A. D. Pinotsis, M. K. Zikides.
Astron. Astrophys., Vol. 141, No. 1, p. 162 – 170 (1984).
The evolution of symmetric collision periodic orbits with mass ratio μ near 1 in the plane circular restricted three–body problem is described. An infinite number of *j*–type families of periodic orbits is established. Each family of periodic orbits is represented by its collision periodic orbit, that appears for a specific value of the Jacobi costant *C*. Thus the evolution of each family is described by a μ–*C* curve. Collision periodic orbits in the Sun–Jupiter case are given, and their interconnections with other known families of collision periodic orbits are found. Finally a discussion of the periods of the orbits is made.

042.058 **Third–order and tidal effects in the stellar three–body problem.**
S. Söderhjelm.
Astron. Astrophys., Vol. 141, No. 1, p. 232 – 240 (1984).
A stable three–body system is characterized by two orbital periods $P_1$ and $P_2$. For $P_2/P_1$ greater than some 50:1, the point–mass problem has a well–known second–order analytical solution. It is shown how this solution can be modified to take tidal deformation into account. As expected, the motion becomes qualitatively different when the apsidal advance due to deformation becomes larger than that due to the third body. In particular, there is no longer a "critical" relative inclination, and circular close orbits are allowed at all inclinations. For large period–ratios, even a small deformation may have this effect. For small period–ratios, the previously neglected third–order terms become important. The tidal forces have important evolutionary consequences and as a first step towards better understading, a "classification" of triple stars is introduced.

042.059 **The dynamics of the minor bodies in the solar system.**
I. P. Williams.
Vistas Astron., Vol. 27, Part 3, p. 217 – 229 (1984).
The dynamics of those minor bodies in the solar system that move within the region where gravitational perturbation by Jupiter is important are reviewed. These bodies are the short–period comets, the Apollo–Amor asteroids and meteor streams. The dynamics are now well understood and give a hint of an evolutionary sequence from short–period comets to Apollo–Amor–Aten asteroids, meteor streams being generated in the process.

042.060 **Obtaining transition matrixes for different parameters of motion of celestial bodies.**
A. A. Sukhanov.
Kosm. Issled., Tom 22, Vyp. 6, p. 819 – 827 (1984). In Russian. English translation in Cosm. Res.

042.061 **Approximate solution of the Euler–Lambert equation for the case of a few revolutions around a central body.**
L. B. Livanov.
Kosm. Issled., Tom 22, Vyp. 6, p. 942 – 944 (1984). In Russian. English translation in Cosm. Res.

042.062 **Rotation and the internal structures of the major planets and their inner satellites.**
S. F. Dermott.
Philos. Trans. R. Soc. London, Ser. A, Vol. 313, No. 1524, p. 123 – 139 (1984). – See Abstr. 012.068.
Measurements of the rotational periods coupled with those of the masses, the mean radii and the shapes or the gravitational moments ($J_2$ and $J_4$) enable important constraints to be placed on the internal structures of some remote bodies. Values of $J_2$ for Uranus and Neptune have been calculated from the observed precession rates of the narrow eccentric and inclined Uranian rings and of the orbit of Triton, Neptune's massive satellite.

042.063 **Polytropic equilibrium figures of equal mass.**
M. J. Miketinac.
Astrophys. Space Sci., Vol. 106, No. 1, p. 151 – 160 (1984).
A numerical method for the solution of the (astrophysical) potential problem is presented. The problem is formulated as a free boundary problem for a mildly nonlinear elliptic partial differential equation and the method is obtained by combining Newton–Raphson's procedure and two different types of discretization. The performance of the method is discussed.

042.064 **The structure of rotating polytropes (second order).**
G. R. Singh, M. Singh.
Astrophys. Space Sci., Vol. 106, No. 1, p. 161 – 171 (1984).
In order to carry out a numerical study of the structure of rotationally distorted polytropes, the Monaghan and Roxburgh (1965) method is extended to include the second–order term in the perturbation parameter α.

042.065 **On the possibility of application of the extended family of periodic orbits related to the Lagrangian point $L_4$ of the Sun–Jupiter system for investigation of the solar system.**
B. Szczodrowska–Kozar.
Artif. Satell., Vol. 19, No. 1, p. 27 – 40 (1984).
In the paper the class of short–period orbits of type II, which were investigated first by Goodrich, has been extended. This class of orbits gives approaches to the Earth as well as to the Sun, and on the other side reaches distances beyond the Saturn's orbit.

042.066 **On a method for obtaining the third generalized Keplerian law.**
A. A. Petrutskij.
Izv. vuzov. Fiz. Tomsk, 5 pp. (1984). In Russian. Abstr. in Ref. Zh., 51. Astron., 11.51.60 (1984).

042.067 **On the stability of infinite motions in the theory of Newtonian gravitation.**
A. A. Bakhan'kov, A. P. Ryabushko, S. G. Yudin.
Vestsi Akad. Nauk BSSR. Ser. fiz.–mat. nauk, No. 3, p. 85 – 90 (1984). In Russian. Abstr. in Ref. Zh., 51. Astron., 11.51.61 (1984).

042.068 **On the solution of the Kepler equation.**
N. A. Sorokin.
Nauchn. Inf., Vyp. 55, p. 135 – 143 (1982). In Russian.

042.069 **From rotations and inclinations to zero configurational velocity surfaces. I. A natural rotating coordinate system.**
D. G. Saari.
Celest. Mech., Vol. 33, No. 4, p. 299 – 318 (1984).
There are three purposes for this and a companion paper. The first is to derive the conditions which will lead to the sharpest possible 'zero velocity' surfaces for the solutions of both the 3–body and the n–body problems. The second purpose is to determine the 'natural rotating coordinate system' for an n–body system; that is, one where the coordinate system contains all of the rotational motion. The third purpose is to derive an equality which will yield, as a special case, the important and very useful Sundman inequality.

042.070 **Account of additional perturbations in canonical system solutions obtained by Lie transforms.**
I. V. Tupikova.
Celest. Mech., Vol. 33, No. 4, p. 337 – 342 (1984).
The analytical or semi-analytical solution of some canonical system (basic theory) constructed by the perturbation technique based on the Lie transforms is supposed to be known. A method allowing one to obtain the solution of perturbed canonical or non-canonical system using the known basic theory is discussed.

042.071 **The resonance 4:4:1 for Hamiltonian systems with three degrees of freedom.**
I. Stellmacher.
Celest. Mech., Vol. 33, No. 4, p. 343 – 365 (1984). In French.
Periodic solutions for a Hamiltonian system are investigated analytically. The included frequencies $\omega_\alpha$, $\alpha = 1, 2, 3$ are assumed near the ratio 4:4:1.

042.072 **Trojan orbits with mass exchange of the primaries.**
G. P. Horedt.
Celest. Mech., Vol. 33, No. 4, p. 367 – 374 (1984).
The librational motion round the Lagrangian triangular points $L_4$, $L_5$ with mass exchange of the primaries is investigated according to Brown's theory. The results are the same as in the case of isotropic mass variation studied earlier (Horedt, 1974): (1) The extrema of the elongations with respect to the small mass are unaffected by mass exchange. (2) The equations for the extrema of the Trojan's distance from the Sun and for the libration period are formally the same as in the constant mass problem, but with the understanding that the masses are now time dependent quantities.

042.073 **Collision orbits in the oblate planet problem.**
J. M. Alfaro.
Celest. Mech., Vol. 33, No. 4, p. 375 – 383 (1984).
Some of the properties of the oblate planet problem are derived. The author uses the technique of blowing up the singularity to study the collision orbits. He defines some families of them in terms of their asymptotic behavior.

042.074 **Periodic orbits of the first sort in the restricted problem of three bodies with an ellipsoidal primary.**
H. Ding, F. Tong.
Chin. Astron. Astrophys., Vol. 8, No. 3, p. 255 – 259 (1984). English translation of 34.042.100.

042.075 **Millisecond pulsars and the location of the solar system barycenter.**
M. Prószyński.
Space Telesc. Sci. Inst., Prepr. Ser., No. 23, 5 pp. (1984). To appear in the proceedings of Green Bank Workshop on Millisecond Pulsars, Green Bank, 6 – 8 June, 1984.

042.076 **Fluid satellite.**
I. Hachisu, Y. Eriguchi.
MPA Rep., No. 156, 27 pp. (1984). Submitted to Astron. Astrophys.

042.077 **Methods of solution of Kepler's equation.**
M. Qian, L.-y. She.
Publ. Purple Mt. Obs., Vol. 3, No. 2, p. 48 – 53 (1984). In Chinese.

042.078 **Expansion of the main values of the elliptical motion with the help of iterations.**
M. P. Imnadze.
Tr. Inst. vychisl. mat. AN GSSR, Tom 24, No. 1, p. 55 – 69 (1984). In Russian. Abstr. in Ref. Zh., 51. Astron., 12.51.94 (1984).

042.079 **On a mechanical interpretation of the problem of four fixed centres.**
I. S. Kozlov.
Idei K. Eh. Tsiolkovskogo i sovrem. nauchn. probl. Moskva, p. 116 – 121 (1984). In Russian. Abstr. in Ref. Zh., 51. Astron., 12.51.98 (1984).

042.080 **Applying averaging methods to the problem of motion of a satellite in the gravitational field of a rotating planet.**
V. A. Kitova.
Vestn. AN KazSSR. Alma-Ata, 8 pp. (1984). In Russian. Abstr. in Ref. Zh., 51. Astron., 12.51.110 (1984).

042.081 **Intermediate resonance motion of an equatorial satellite in the gravitational field of a rotating nonspherical planet.**
V. A. Kitova.
Vestn. AN KazSSR. Alma-Ata, 11 pp. (1984). In Russian. Abstr. in Ref. Zh., 51. Astron., 12.51.116 (1984).

042.082 **The Roche problem and its place in the theory of figures of planets.**
V. V. Pavlovskaya.
Inst. mat. Akad. Nauk USSR. Prepr., No. 64, p. 3 – 13 (1983). In Russian. Abstr. in Ref. Zh., 51. Astron., 12.51.129 (1984).

042.083 **N-Körper Hybrid Integrations-Methode.**
S. Aarseth, E. Bettwieser.
Mitt. Astron. Ges., Nr. 62, p. 351 (1984). Abstract. – See Abstr. 010.341.

042.084 **On the evolution of rotations of a viscoelastic planet in a circular orbit of a central-force field.**
V. G. Vil'ke, S. A. Kopylov, Yu. G. Markov.
Astron. Zh., Tom 61, Vyp. 6, p. 1198 – 1204 (1984). In Russian. English translation in Sov. Astron., Vol. 28, No. 6.
An averaging method has been applied to obtain approximate equations describing the evolution of rotational motion of a viscoelastic planet in a Newtonian central-force field. The equations are written down in the form of Andoyer's canonical variables. The evolution of the planet's rotation has been studied based on smoothed equations.

042.085 **On the coefficients of the series of the planar restricted three-body problem.**
E. I. Timoshkova, V. B. Titov.
Astron. Zh., Tom 61, Vyp. 6, p. 1211 – 1217 (1984). In Russian. English translation in Sov. Astron., Vol. 28, No. 6.
Some properties of trigonometric polynomials $W_k(\theta)$ are studied, these polynomials being the coefficients of the earlier obtained series, which are the first integrals of the planar restricted three-body problem. The convergence of the series is discussed.

042.086 **Bifurcation points on the one-ring sequence of uniformly rotating and self-gravitating fluid.**
Y. Eriguchi, I. Hachisu.
Publ. Astron. Soc. Jpn., Vol. 36, No. 3, p. 491 – 496 (1984).
The one-ring equilibrium sequence of a uniformly rotating and self-gravitating incompressible fluid has been studied against a nonaxially symmetric perturbation of the surface. Bifurcation points for modes $m = 1,..., 5$ are obtained on the concave hamburger-to-toroid sequence. There are no bifurcations belonging to modes $m = 2$ and 3.

042.087 **Bifurcation points on the Maclaurin sequence.**
I. Hachisu, Y. Eriguchi.
Publ. Astron. Soc. Jpn., Vol. 36, No. 3, p. 497 – 503 (1984).
All bifurcation points on the Maclaurin spheroidal sequence are calculated. Deformation types can be expressed by associated Legendre functions.

042.088 **The gravitational interaction between inclined, elliptical rings.**
C. F. Yoder.
Planetary rings, p. 447 – 455 (1984). – See Abstr. 012.085 (IAU Colloq. No. 75).
The gravitational torque between two wire rings or within a ring of finite width can prevent differential precession caused by planetary oblateness. Goldreich and Tremaine (1979) proposed this model to explain the observed eccentricity and width variations of the Uranian ε ring. A general expression for the potential for two elliptical and inclined rings is derived in this paper. The stationary solutions and stability of this system are briefly examined.

042.089 **Stochastic modeling of planetary ring systems.**
R. W. Hart.
Planetary rings, p. 461 – 469 (1984). – See Abstr. 012.085 (IAU Colloq. No. 75).
This paper models maximum entropy configurations of idealized gravitational ring systems. Such configurations are of interest because systems generally evolve toward an ultimate state of maximum randomness. For simplicity, attention is confined to ultimate states for which interparticle interactions are no longer of first order importance. The planets, in their orbits about the sun, are one example of such a ring system. The extent to which the present approximation yields insight into ring systems such as Saturn's is explored briefly.

042.090 **Applications of three–body resonances to satellites and planetary rings.**
K. Aksnes.
Planetary rings, p. 479 – 487 (1984). – See Abstr. 012.085 (IAU Colloq. No. 75).
Formulas have been derived for the strengths and libration/circulation characteristics of resonances for the motions of any three gravitationally interacting bodies in coplanar orbits. The formulas are applied to the well known Laplacian satellite resonances as well as to other possible 3–body resonances in the satellite and ring systems of Jupiter, Uranus, and Saturn.

042.091 **The adiabatic invariant: a tool for the study of the dynamics of the solar system.**
J. Henrard.
Planetary rings, p. 489 – 497 (1984). – See Abstr. 012.085 (IAU Colloq. No. 75).
A summary of the main results of the theory of the adiabatic invariant is presented. They are applied to a simple one degree of freedom Hamiltonian system which is a good model for many resonance problems. As an illustration, the capture into resonance of the pair of satellites Enceladus – Dione is analysed.

042.092 **The Radzievskij–Tisserand criterion.**
V. P. Tomanov.
Komet. Tsirk., No. 328 (1984). In Russian.

**Grundlagen der Ephemeridenrechnung. Eine Zusammenstellung der wichtigsten Formeln und Daten zur Berechnung der Bahnen von Körpern im Sonnensystem.**
See Abstr. 003.027.

**Modern numerical methods in problems of celestial mechanics.**
See Abstr. 003.062.

**Erratum: "Reduction, Brouwer's Hamiltonian, and the critical inclination" [Celest. Mech., Vol. 31, No. 4, p. 401 – 429 (1983)].**
See Abstr. 052.012.

**A criterion for the fragmentation of a rotating and collapsing gas cloud.**
See Abstr. 062.114.

**Equilibrium models of differentially rotating polytropes and the collapse of rotating stellar cores.**
See Abstr. 065.137.

**On orbits and rays in Schwarzschild– and in Reissner–Nordström space.**
See Abstr. 066.144.

**Resonance locking between Jupiter and Uranus.**
See Abstr. 091.004.

**Angular momenta in the solar system. II.**
See Abstr. 091.009.

**Gravitační pole a tvary planet.**
See Abstr. 091.033.

**On quantization of the solar system and of the stationary structure of megawaves in it.**
See Abstr. 091.058.

**Quantization of the sectorial velocity of the solar system and two groups of planetary orbits.**
See Abstr. 091.059.

**How a shepherd guides a sheep herd?**
See Abstr. 091.067.

**Evolution of collisional systems.**
See Abstr. 091.071.

**Size distribution in collisional systems.**
See Abstr. 091.072.

**Dynamics of neutral small particles.**
See Abstr. 091.076.

**Waves in planetary rings.**
See Abstr. 091.088.

**Ring particle dynamics in resonances.**
See Abstr. 091.089.

**Dynamics of narrow rings.**
See Abstr. 091.090.

**Unsolved problems in planetary ring dynamics.**
See Abstr. 091.091.

**Structure of the 3:1 Jovian resonance: a comparison of numerical methods.**
See Abstr. 098.013.

**Investigation of motion of six minor planets.**
See Abstr. 098.029.

**On the formation of the gaps in the asteroid belt.**
See Abstr. 098.086.

**Investigation of the errors of the parameters of translatory–rotational motion of an asteroid from measurements of the radial velocity of a space vehicle.**
See Abstr. 098.106.

**Perturbations on the orbits of companions of the satellites of Saturn.**
See Abstr. 100.002.

**A simple model of Saturn's rings – revisited.**
See Abstr. 100.109.

**Dense but "collisionless" orbiting rings.**
See Abstr. 100.115.

**The Laplacian resonance amongst Uranian inner satellites.**
See Abstr. 101.018.

**Capture de comètes par Jupiter.**
See Abstr. 102.009.

**Recurrent power series integration of equations of comet's motion including the nongravitational terms in Marsden's form.**
See Abstr. 102.030.

**Averaging equations of orbital motion of a comet with account for nongravitational effects.**
See Abstr. 102.062.

**The disintegration of small stellar systems.**
See Abstr. 151.005.

**Periodic orbits in a resonant dynamical system.**
See Abstr. 151.042.

**Families of periodic orbits in a quartic potential.**
See Abstr. 151.064.

**Numerical simulations of encounters of hard binaries.**
See Abstr. 151.077.

**Sind gas- bzw. thermodynamische Konzepte gültig für Ergebnisse von direkten n-Körper Rechnungen?**
See Abstr. 151.092.

# 043 Astronomical Constants, Reference Systems

043.001 **Geodetic Reference System 1980.**
H. Moritz.
Bull. Géod., Vol. 58, No. 3, p. 388 – 398 (1984). See Abstr. 28.043.006.

043.002 **Using minor planets for coordinate system improvement, an Hipparcos zero point, and Space Telescope scale calibration: physical considerations and error analysis.**
P. D. Hemenway, G. F. Benedict, R. L. Duncombe, W. H. Jefferys, P. J. Shelus.
Bull. Am. Astron. Soc., Vol. 16, No. 3, Part I, p. 722 – 723 (1984). Abstract. – See Abstr. 010.064.

043.003 **Solar observations with the astrolabe at CERGA.**
A. Journet, F. Laclare.
Astron. Astrophys., Vol. 141, No. 1, p. L7 (1984). In French.
First analysis of the effects of the adoption of a new reference system by I.A.U. in 1984. The $\Delta\alpha$ for solar observations are compared in the old and new system. A good agreement is found with the correction of equinox proposed by Fricke (1981).

043.004 **Dimensional fundamental constants and their application in astronomy.**
P. S. Wesson.
Space Sci. Rev., Vol. 39, Nos. 1/2, p. 153 – 161 (1984).
A discussion is given of the role of dimensional fundamental constants in gravitational and particle physics. It is concluded that such constants can most usefully be interpreted as representing asymptotic states. This interpretation is in agreement with the widespread use of dimensional analysis in astronomy, and implies that angular momentum can be expected to vary like the mass squared in the astronomical limit of large masses.

043.005 **El sistema IAU (1976) de constantes astronómicas y su repercusión en la reducción de posiciones de estrellas.**
M. J. Sevilla.
Bol. Inf., Serv. Geogr. Ejército, núm. 48, p. 1 – 42 (1980). = Semin. Astron. Geod., Univ. Madr., Publ. No. 109.

043.006 **Determination of the errors of the precession constant from the Pulkovo Catalogue of stellar proper motions relative to galaxies.**
N. V. Kharchenko.
Astron. Tsirk., No. 1313, p. 6 – 7 (1984). In Russian.

043.007 **Relation between the dynamical fundamental system FK5 and the system of positions of extragalactic sources.**
W. Fricke.
Mitt. Astron. Ges., Nr. 62, p. 138 – 143 (1984). – See Abstr. 010.341.

043.008 **Ein Katalog ausgewählter kompakter Radioquellen für die Konstruktion eines extragalaktischen radio/optischen Referenzsystems. I. Einführung und optische Daten. II. Radiodaten.**
C. de Vegt, A. Witzel.
Mitt. Astron. Ges., Nr. 62, p. 209 (1984). Abstract. – See Abstr. 010.341.

043.009 **Der CPC2 Katalog, ein photographisches Referenzsternsystem hoher Genauigkeit am Südhimmel.**
N. Zacharias.
Mitt. Astron. Ges., Nr. 62, p. 209 (1984). Abstract. – See Abstr. 010.341.

043.010 **Ein photographischer PZT-Referenzsternkatalog für den Deklinationsgürtel +52°5.**
C. de Vegt, T. Kleine.
Mitt. Astron. Ges., Nr. 62, p. 209 (1984). Abstract. – See Abstr. 010.341.

043.011 **Folgerungen aus mehrjährigen radiointerferometrischen Beobachtungen extragalaktischer Objekte.**
H. G. Walter.
Mitt. Astron. Ges., Nr. 62, p. 211 – 214 (1984). – See Abstr. 010.341.

**On executing the MERIT Standards and homogenizing the classical observational series.**
See Abstr. 044.026.

**Reinvestigation of O, B star kinematics and determination of corrections to precession.**
See Abstr. 111.025.

**Galactic rotation, correction to the constant of precession and coordinates of the solar apex according to proper motions of bright stars.**
See Abstr. 155.038.

# 044 Time and Latitude Determination, Earth Rotation, Polar Motion

044.001 **Time and latitude results of observations made with the Astrolabe Danjon at Cagliari before the main campaign of Project MERIT.**
A. Poma, V. Gusai.
Astron. Astrophys., Suppl. Ser., Vol. 57, No. 1, p. 51 - 53 (1984).
Results are given for the observations made with the Astrolabe Danjon during the period August 1982 - August 1983. The results are given in the FK4 system.

044.002 **Results of observations made in Paris with the astrolabe. Time and latitude 1982 and 1983.**
F. Chollet, S. Débarbat, J.-C. Hascoët, S. K. Lam, P. Texier, M. Tomas.
Astron. Astrophys., Suppl. Ser., Vol. 57, No. 2, p. 197 - 204 (1984). In French.
Results are given for the observations made with the astrolabe APP during the years 1982 and 1983. No change has been made in the star catalogue since 1956. The observing technique and computation method also remain unchanged since 1972. The results are in the FK4 system.

044.003 **The ephemeris time obtained from photographic measurements of the positions of lunar craters relative to stars. II.**
N. A. Vasilenko, V. S. Kislyuk, R. L. Semerenko.
Astron. Tsirk., No. 1299, p. 7 - 8 (1983). In Russian.

044.004 **Spectral analysis of the diurnal non-uniformity in the earth's rotation.**
G. P. Pil'nik.
Astron. Zh., Tom 61, Vyp. 4, p. 806 - 811 (1984). In Russian. English translation in Sov. Astron., Vol. 28, No. 4.
Astronomical observations of the International Time Service for 1968 - 1978 were studied in order to search for periodicities in the diurnal rotation of the earth. The spectral analysis of the mean system as well as of the more numerous observations of individual observatories shows that the earth's diurnal rotation is not uniform.

044.005 **Multiple harmonics in the diurnal rotation of the earth.**
G. P. Pil'nik.
Sov. Astron., Vol. 28, No. 1, p. 112 - 114 (1984). English translation of 37.044.002.

044.006 **On causes of the origin of systematic errors in latitude determination with the Moscow PZT.**
A. A. Volchkov, G. A. Gutsalo.
Tr. Gos. Astron. Inst. Shternberg, Tom 55, p. 33 - 39 (1983). In Russian.
Peculiarities of eye response during visual measurements of star positions on photographic plates are considered. It is shown that variations of the plate background density can be a source of systematic errors during latitude determinations with a PZT.

044.007 **Earth's rotation and its geophysical implications.**
F. Barlier, H.-G. Kahle.
ESA Spec. Publ., ESA SP-202, p. P21 - P23 (1984). - See Abstr. 012.004.
The geophysical implications of the Earth's rotation are a broad field of research. Specific recent results based on laser techniques are presented and progress in the determination of Earth's rotation parameters by laser techniques are emphasized.

044.008 **Rotation de la Terre et systèmes de référence terrestres.**
C. Boucher, M. Feissel.
Astrométrie, géodésie, mécanique céleste, p. 125 - 129 (1984). - See Abstr. 012.021.

044.009 **Effets luni-solaires sur la rotation de la Terre.**
N. Capitaine.
Astrométrie, géodésie, mécanique céleste, p. 131 - 134 (1984). - See Abstr. 012.021.

044.010 **Nouvelle réduction des mesures de latitude et de temps universel pour la détermination des paramètres de la rotation de la Terre de 1962 à 1981.**
Z.-X. Li, E. Van Lauwe.
Astrométrie, géodésie, mécanique céleste, p. 135 - 138 (1984). - See Abstr. 012.021.

044.011 **Utilisation des techniques laser satellites pour l'étude de la rotation de la Terre.**
D. Gambis.
Astrométrie, géodésie, mécanique céleste, p. 155 - 156 (1984). - See Abstr. 012.021.

044.012 **Étude des marées océaniques à partir des observations des satellites TRANSIT.**
R. Biancale.
Astrométrie, géodésie, mécanique céleste, p. 157 - 161 (1984). - See Abstr. 012.021.

044.013 **Effects of terrestrial tides in time and latitude observations.**
A. A. Korsun'.
Inst. teor. fiz. Akad. Nauk USSR. Prepr., No. 143 R, 27 pp. (1983). In Russian. Abstr. in Ref. Zh., 51. Astron., 8.51.82 (1984).

044.014 **Velocity variations of the earth's rotation due to zonal tide and its manifestation in the atmospheric pressure field.**
F. I. Rudyaev.
Izv. Vses. geogr. o-va, Tom 116, No. 2, p. 120 - 126 (1984). In Russian. Abstr. in Ref. Zh., 51. Astron., 8.51.83 (1984).

044.015 **On the excitation of the Earth's free wobble and reference frames.**
B. Fong Chao.
Geophys. J. R. Astron. Soc., Vol. 79, No. 2, p. 555 - 563 (1984).
The author studies the excitation of the Earth's polar motion in connection with problems that are associated with the diversity of reference frames involved in observations and theoretical computations.

044.016 **Abnormal changes in the rate of the earth's rotation in the first half of 1983.**
S. Okazaki, T. Sakai.
Tokyo Astron. Obs. Rep., Vol. 20, No. 2, p. 311 - 326 (1984). In Japanese.

044.017 **Irregular variations of gravity and the earth's rotation.**
N. N. Parijskij.
O. Yu. Shmidt i sov. geofiz. 80-kh godov, Moskva, p. 106 - 111 (1983). In Russian. Abstr. in Ref. Zh., 51. Astron., 9.51.129 (1984).

044.018 **Scale of pulsar time.**
V. G. Il'in, Yu. P. Ilyasov, A. D. Kuz'min, S. B. Pushkin, G. N. Palij, T. V. Shabanova, Yu. P. Shitov.
Dokl. Akad. Nauk SSSR. Ser. Mat. Fiz., Tom 275, No. 4, p. 835 - 838 (1984). In Russian. Abstr. in Ref. Zh., 51. Astron., 10.51.134 (1984).

044.019 **New State standard of time and frequency.**
S. B. Pushkin.
Zemlya Vselennaya, No. 5, p. 51 - 55 (1984). In Russian.

044.020 **Scale determination in the uniform reduction of observations of the International Latitude Service.**
R. M. Rasulov.
Sov. Astron., Vol. 28, No. 2, p. 221 – 223 (1984). English translation of 37.044.012.

044.021 **Instability of the astronomical system of terrestrial coordinates.**
A. A. Korsun', R. M. Rasulov.
Sov. Astron., Vol. 28, No. 3, p. 340 – 343 (1984). English translation of 37.044.023.

044.022 **On multi–peak phenomena in Chandler wobble spectra.**
Y. Ding.
J. Nanjing Univ., No. 2, p. 267 – 278 (1984). In Chinese.
In this paper the data of pole coordinates from results of ILS in a homogeneous system are in parts treated with the periodgram and AR–spectral estimation which checks multi–and/or mono–peak phenomena in Chandler wobble spectra.

044.023 **On the work of Universal Time in the Chinese Joint System in 1982.**
Ann. Shanghai Obs., Acad. Sin., No. 5, p. 1 – 7 (1983). In Chinese.

044.024 **Tidal short periodic terms in Chinese Joint UT System.**
H. Huang, S. Luo.
Ann. Shanghai Obs., Acad. Sin., No. 5, p. 8 – 12 (1983). In Chinese.
UT1–UTC in Chinese Joint UT System (CJUS) is computed by using the data of UT0–UTC obtained by 13 instruments during 3 months.

044.025 **Influence of earth tide and M2 wave of ocean tide on lunar laser ranging.**
W. Jin, C. Qian, Z. Yang.
Ann. Shanghai Obs., Acad. Sin., No. 5, p. 13 – 22 (1983). In Chinese.

044.026 **On executing the MERIT Standards and homogenizing the classical observational series.**
M. Zhao.
Ann. Shanghai Obs., Acad. Sin., No. 5, p. 47 – 56 (1983). In Chinese.
The MERIT Standards for reduction of observational data of the earth's rotation deal with the changes of fundamental catalogue system, astronomical constant system, the definition of Universal Time as well as the algorithm for computing apparent places of stars. These changes will introduce some nonuniformities into the observational series. The author discusses the effects of changes in the working catalogue system, the nutation series, the adopted aberration constant, and the correction of gravitational bending of light.

044.027 **Covariance analysis for choosing the adjusted parameters and determining the effect of model errors on adjusted parameters.**
S. Zhu.
Ann. Shanghai Obs., Acad. Sin., No. 5, p. 72 – 76 (1983). In Chinese.

044.028 **Time synchronization ability of Loran C signal of Northwest Pacific chain.**
Q. Zhuang.
Ann. Shanghai Obs., Acad. Sin., No. 5, p. 241 – 248 (1983). In Chinese.
Some time research laboratories in China and Japan receive Loran C signal of Northwest Pacific chain for time comparison, which is the most important link between those with BIH at present. The author recomputed the propagation distance and propagation delay of each laboratory. The geocentric coordinate system of WGS–72, its reference ellipsoid and the geodetic major formula were adopted for the calculation of propagation distance. Timing results and systematic deviations via several methods are given.

044.029 **Results of astrolable observations made at Merate Observatory: time and latitude, 1983.**
L. Buffoni, F. Carta, F. Chlistovsky, A. Manara, F. Mazzoleni.
Astron. Astrophys., Suppl. Ser., Vol. 58, No. 3, p. 499 – 501 (1984).
Results of the observations made with the Astrolabe Danjon OPL no. 32 are given. The results are in the FK4 system.

044.030 **Earth rotation parameters from *LAGEOS* laser ranging in the 1980 MERIT campaign.**
V. V. Nesterov.
Sov. Astron. Lett., Vol. 10, No. 3, p. 193 – 194 (1984). English translation of 37.044.064.

044.031 **Determination of time (UT1–UTC). 1982 April – 1982 October, 1982 December.**
O. S. Tursunov.
Tsirk. Astron. Inst. Tashkent, Nos. 109, 110, 112 (1983). In Russian.

044.032 **Investigation of the mean latitude of Kitab from observations with the ZTL–180 zenith telescope during the period 1967.0 – 1979.0.**
S. K. Ehshonkulov.
Tsirk. Astron. Inst. Tashkent, No. 109, p. 26 – 31 (1983). In Russian.

044.033 **The IRIS Earth orientation parameters.**
IRIS Bull. A, Nos. 5 – 10 (1984).

044.034 **Long–term changes in the rotation of the Earth: 700 B.C. to A.D. 1980.**
F. R. Stephenson, L. V. Morrison.
Philos. Trans. R. Soc. London, Ser. A, Vol. 313, No. 1524, p. 47 – 70 (1984). – See Abstr. 012.068.
Occultations of stars by the Moon, and solar and lunar eclipses are analysed for variations in the Earth's rotation over the past 2700 years. Although tidal braking provides the dominant, long–term torque, it is found that the rate of rotation does not decrease uniformly as would be expected if tidal friction were the only mechanism affecting the Earth's rotation. There are also non–tidal changes present that vary on timescales ranging from decades to millennia. The magnitudinal and temporal behaviour of these non–tidal variations are evaluated in this paper.

044.035 **The determination of the fluctuations in the rotation of the Earth.**
G. A. Wilkins.
Philos. Trans. R. Soc. London, Ser. A, Vol. 313, No. 1524, p. 85 – 94 (1984). – See Abstr. 012.068.
In recent years new space techniques have been used to determine with higher precision the fluctuations in the rate of rotation of the Earth and in the motion of the pole of rotation with respect to the conventional terrestrial reference frame. The international MERIT–COTES programme of observation and analysis is now in progress and is providing a high–quality data set for use in the study of the nature and causes of the fluctuations and in the determination of the relationships between the reference systems of the different techniques. The results and operational experience gained will be reviewed to provide a basis for recommendations about the future international services for Earth–rotation and for the establishment and maintenance of a new terrestrial reference frame.

044.036 **Causes of fluctuations in the rotation of the Earth.**
M. G. Rochester.
Philos. Trans. R. Soc. London, Ser. A, Vol. 313, No. 1524, p. 95 – 105 (1984). – See Abstr. 012.068.
Over the last decade considerable progress has been made in closing the gap between geophysical theory and the observed changes in the Earth's spin rate and polar motion, which are affected by the Earth's internal structure, properties and processes. New observational techniques and data have had a marked impact on understanding the short–term variations in the length of day and pole position, but we can expect between

another 5 and 30 years to pass before they permit improved explanations of longer–term phenomena (18.6 year tidal effects, the Markowitz wobble, decade fluctuations in the length of day, etc.). This review summarizes recent advances, problems newly illuminated by recent Earth rotation data (for example mantle anelasticity, core–mantle coupling) and problems still unresolved.

044.037 **The rotation of the Earth's inner core.**
A. M. K. Szeto, D. E. Smylie.
Philos. Trans. R. Soc. London, Ser. A, Vol. 313, No. 1524, p. 171 – 184 (1984). – See Abstr. 012.068.
The inner core rotates under the influence of inertial, frictional, electromagnetic and gravity torques. When its symmetry axis is inclined by an angle $\theta_i$ to the mantle symmetry axis, calculation shows the gravity restoring torque to be $2.50 \times 10^{24} \cos\theta_i \sin\theta_i$ N m, several orders of magnitude larger than the other torques combined. Investigation of its rotational dynamics under this torque shows that it can have two modes of steady precession.

044.038 **A comparison of the complete and truncated versions of the polar wander equations.**
R. Sabadini, D. A. Yuen, E. Boschi.
J. Geophys. Res., Vol. 89, No. B9, p. 7609 – 7620 (1984).
The authors give a theoretical development of a uniformly valid solution for describing the polar motions of a layered viscoelastic earth. Such solutions require the usage of two classes of eigenspectra to describe the processes of rotational deformation and viscous relaxation. One set consisting of real eigenvalue describes the isostatic relaxation of the mantle by viscous creep. The second one governs the readjustment of the rotational axis of the viscoelastic earth, and the eigenspectrum is complex valued. These solutions are capable of describing rotational phenomena ranging from Chandler wobble excitation to long–term polar drift. A comparative study is conducted between the complete solution for polar wander and one in which a number of the rotational modes has been truncated.

044.039 **Normal modes of the coupled earth and ocean system.**
J. M. Wahr.
J. Geophys. Res., Vol. 89, No. B9, p. 7621 – 7630 (1984).
When two physical systems are dynamically coupled, the eigenmodes for the combined system will differ, in general, from the modes for the two original, uncoupled systems. The paper examines certain normal modes of the coupled earth and ocean system and discusses differences from corresponding results for the two individual systems. Particular attention is paid to those modes which involve wobble of the solid earth. It has been suggested that the coupled system may possess a long–period wobble mode, not part of the spectrum for the solid earth alone, which might be identified as the Markowitz wobble. The results presented in the paper do not support that conclusion.

044.040 **Tidal terms in Universal Time: effects of zonal winds and mantle Q.**
J. B. Merriam.
J. Geophys. Res., Vol. 89, No. B12, p. 10109 – 10114 (1984).
The zonal response coefficient $\varkappa$ is studied, and some improvements in its determination are made.

044.041 **Determination of the motion of the earth's poles with precise laser tracking data.**
A. I. Khadzhijskij.
Nauchn. Inf., Vyp. 55, p. 92 – 97 (1982). In Russian.
The possibility of the components of the vector motion of the earth's pole by laser observation of earth artificial satellites on the Intercosmos stations net is investigated.

044.042 **The role of air temperature in the seasonal variation of the Earth's rotation.**
G.–b. Peng.
Chin. Astron. Astrophys., Vol. 8, No. 3, p. 245 – 250 (1984). English translation of 34.044.096.

044.043 **On the effects of lunar tides in the Tianjin latitude observations.**
T.–q. Han, Y.–m. Qian, S.–g. Li, L.–z. Ma.
Chin. Astron. Astrophys., Vol. 8, No. 4, p. 325 – 327 (1984). English translation of Acta Astron. Sin., Vol. 25, No. 2, p. 186 – 190 (1984).
The authors use the Tianjin latitude series over the seven years, 1960 – 1966 to analyse the effect of the lunar tide in the Tianjin region on the latitude observations and to evaluate the parameter Λ.

044.044 **A derivation of the rate of spin–down of the Earth using ancient Chinese records of central eclipses.**
Y.–b. Han, Z.–s. Li, B.–s. Lin, X.–h. Yang.
Chin. Astron. Astrophys., Vol. 8, No. 4, p. 348 – 353 (1984). English translation of Acta Astrophys. Sin., Vol. 4, No. 2, p. 107 – 114 (1984).
The authors describe a method of studying the variation in the rotation of the Earth in historical times using ancient records of central solar eclipses. Their result shows that the Earth is spinning down at a relative rate of $-(1.88 \pm 0.10) \times 10^{-10}$/yr.

044.045 **Hydrostatic equilibrium and deceleration of the Earth's rotation.**
M. Burša.
Bull. Astron. Inst. Czech., Vol. 35, No. 6, p. 353 – 355 (1984).
It has been demonstrated that the hydrostatic flattening of the Earth computed on the basis of the running hydrostatic theory using recent satellite data and the precession constant, differs from the actual Earth's flattening $\alpha$ in magnitude as $\sim\alpha^2$ which is the actual accuracy of the hydrostatic theory. The discrepancy between the observed deceleration of the Earth and that computed from the deceleration of the Moon's mean motion using the tidal friction theory, can be explained by the decrease of the Earth's principal moment of inertia, i.e. the decrease of its flattening.

044.046 **Time and Latitude Service, 1983 October – 1984 June.**
Circ. Time Latitude Serv., Nos. 168 – 170 (1983/84).

044.047 **Breitenbestimmungen. Mai – Oktober 1984.**
Lohrmann–Obs., Tech. Univ. Dresden, Zirk., Nr. 108 – 109 (1984).

044.048 **The longitude of the Istanbul University Observatory.**
O. Gölbasi.
Rev. Fac. Sci. Univ. Istanbul, Ser. C, Vol. 47, p. 43 – 52 (1982). = Publ. Istanbul Univ. Obs., No. 132 (1984).

044.049 **Zeit– und Breitendienst. Januar bis Dezember 1983.**
Dtsch. Hydrogr. Inst. Hamb., Zeit–Breitendienst, 29 pp. (1984).

044.050 **Time and Frequency Services Bulletin, 1984 April – September.**
Time Freq. Serv., Bull., Nos. 55 – 60, 49 pp. (1984).

044.051 **Determinación de la latitud y la longitud por fotografías cenitales de estrellas.**
M. J. Sevilla.
Semin. Astron. Geod., Univ. Madr., Publ., No. 111, p. 1 – 9 (1980).

044.052 **Monthly Notes of the International Polar Motion Service.**
Mon. Notes Int. Polar Motion Serv., Nos. 5 – 11, p. 75 – 198 (1984).

044.053 **Comments on "Results of the International Latitude Service in a homogeneous system 1899.0 – 1979.0" by Yumi and Yokoyama.**
Y. Wako, I. Okamoto.
Publ. Int. Latitude Obs. Mizusawa, Vol. 17, No. 1, p. 1 – 12 (1983).

044.054 **Bulletins, Time Service of the Mizusawa Observatory in 1982.**
I. Tsubokawa.
Bull., Time Serv. Mizusawa Obs., Vol. 27, 2+38 pp. (1983).

044.055 **Local effect on determining the time system.**
B.-y. Yang, J.-s. Yao.
Publ. Purple Mt. Obs., Vol. 3, No. 2, p. 23 – 25 (1984). In Chinese.

Annual and semi-annual periodic phases of system errors of the instruments in the Joint system and the Joint signal corrections with respect to the BIH are given.

044.056 **Time and Latitude. January – September 1983.**
V. Ptáček, J. Vondrák, R. Weber.
Circ. Czech. Obs., Time Latitude, 26+22+24 pp. (1984).

044.057 **Zeit- und Breitenbestimmungen, Zeitsysteme, Präzisionszeitvergleiche. 1982 Mai – Dezember.**
Zeit-Breitenbestimmungen, Jahrg. 1982, Nr. 3 – 6, Sonderheft, 15+15+15+13+6 pp. (1983/84).

044.058 **UT2–UT1 for 1985.**
B.I.H. Circ., A, 1 p. (1984).

044.059 **Bureau International de l'Heure (B.I.H.). Circular D.**
B.I.H. Circ., D212 – D217 (1984).

Contents: UTC, TAI, UT and coordinates of the pole, rotation of the Earth. Data for 1984 May – October.

044.060 **Time and Latitude Bulletins. 1984 January – September.**
Y. Kozai.
Tokyo Astron. Obs., Time Latitude Bull., Vol. 58, Nos. 1 – 3, p. 1 – 78 (1984).

044.061 **Cumulative rate changes in the rotation of the Earth.**
N. Sekiguchi.
J. Geod. Soc. Jpn., Vol. 30, No. 2, p. 87 – 91 (1984). = Tokyo Astron. Obs. Repr., No. 706.

Brouwer proposed a model for the irregular variations of the Earth's rotation, regarding them as the results of cumulative small rate changes. It is found that this model is valid for time range about from 50 days to 50 years.

044.062 **Daily time differences and relative phase values. 1984 July – December.**
U.S. Nav. Obs., Time Serv. Publ., Ser. 04, Nos. 909 – 934 (1984).

044.063 **Time Service Announcement.**
G. M. R. Winkler.
U.S. Nav. Obs., Time Serv. Publ., Ser. 14, No. 38 (1984).

044.064 **Earth Orientation. 1984 July – December.**
Earth Orientation Bull., Vol. 2, Nos. 27 – 52 (1984).

044.065 **Results of determination of latitude in Józefosław. 1983 October – 1984 June.**
L. Pieczyński.
Latitude Circ., Nos. 89 – 91 (1984).

044.066 **Fluctuations in the velocity of the earth's rotation in 1979 – 1983.**
D. Yu. Belotserkovskij.
Astron. Tsirk., No. 1328, p. 7 – 8 (1984). In Russian.

044.067 **A systematic approach to radiometric and paleomagnetic studies in a mobile orogenic belt: I. The waning phase of activity in the southern Appalachians of South Carolina.**
R. E. Dooley, M. Manzoni, A. E. M. Nairn.
Geophys. Surv., Vol. 7, No. 1, p. 27 – 54 (1984). – See Abstr. 012.082.

The survey of radiometric and paleomagnetic work on the mafic rocks of South Carolina is consistent with, and amplifies the studies on the acidic rocks of the southeast by Ellwood (1982). The paleopoles of the mafic rocks are in agreement with paleopoles on the North American apparent polar wander path at about 300 m.y.

044.068 **Präzisions-Zeitmessung mit dem Wiener Passage-Instrument Askania Ap 70 anhand ausgewählter FK4 Sterne.**
E. Göbel, M. Firneis.
Mitt. Astron. Ges., Nr. 62, p. 207 – 208 (1984). – See Abstr. 010.341.

044.069 **Analysis of the Earth's variable rotation.**
B. F. Chao.
NASA Tech. Memo., NASA TM-86123, p. 2-2 – 2-5 (1984). – See Abstr. 003.039.

This report analyzes the Earth's rotation, and compares the observations with the atmospheric and seismic excitation functions.

044.070 **The observed excitation function of the Chandler wobble.**
R. Gross.
NASA Tech. Memo., NASA TM-86123, p. 2-6 – 2-9 (1984). See Abstr. 003.039.

The objective of this research effort is to gain a better understanding of the mechanism that is keeping the Chandler wobble excited.

044.071 **Extreme cycles of the motion of the pole of the earth.**
L. D. Kostina, V. I. Sakharov.
Astron. Zh., Tom 61, Vyp. 6, p. 1192 – 1197 (1984). In Russian. English translation in Sov. Astron., Vol. 28, No. 6.

Extreme cycles of the Chandler motion of the pole of the earth have been detected. It is found that they consist of the phases of "growth" and "damping". Parameters $(a, \alpha)$ characterizing the resonance mechanism of the excitation of the motion of the pole of the earth are different for different cycles, and the parameters $\beta$ of the phase of damping are practically constant. This suggests the non-stationary character of the mechanism of the excitation and the constant character of the process of the relaxation of the motion of the pole of the earth.

044.072 **On the global nature of the spectrum of nearly diurnal latitude variations.**
L. D. Kovbasyuk.
Astron. Zh., Tom 61, Vyp. 6, p. 1218 – 1225 (1984). In Russian. English translation in Sov. Astron., Vol. 28, No. 6.

Comparing the vector sums of the relative to the mean solar day symmetric nearly diurnal harmonics based on day-and-night latitude observations made in Gorky with transformed harmonics of the latitude variations of the international latitude stations, the author draws a conclusion on the global nature of the spectrum of the nearly diurnal latitude variations.

044.073 **The moon's zonal tidal effect on the variation in the earth's rotation rate.**
Y.-z. Chu.
Acta Astron. Sin., Vol. 25, No. 3, p. 287 – 292 (1984). In Chinese.

Using the time observations obtained by 8 instruments in the Chinese Joint System during the years 1966 – 1980, the author analyses the moon's zonal tidal effect. The results show that the effects of the $M_f$ and $M_m$ waves are obvious. From this, the parameters $K/C$ of the zonal tide are estimated and the weighted averages of the 8 instruments are $(K/C)_{M_f} = 0.909 \pm 0.114$ and $(K/C)_{M_m} = 0.905 \pm 0.083$ respectively.

044.074 **Time Service Annual Report 1983. (Shanghai Observatory).**
Time Serv. Annu. Rep., 3+181 pp. (1984). In Chinese.

**The centenary of the prime meridian and of international standard time.**
See Abstr. 004.188.

**Trois ans de variation des coordonnées de Paris et de Green Bank.**
See Abstr. 013.017.

**Symmetry of the spinning Earth.**
See Abstr. 014.011.

**Discussion on dual-mixer time difference measurement technique for Atomic Time.**
See Abstr. 034.098.

**The application of a 100 MHz phase comparison method.**
See Abstr. 034.099.

**On the preliminary measurements of the Loran-C signals of both BPL and the SH3 chain.**
See Abstr. 034.184.

**A new method of absolute determination of a transit instrument's azimuth and latitude.**
See Abstr. 036.166.

**On a possible improvement of the right ascension system of the General Catalogue of the Time Services of the USSR.**
See Abstr. 041.001.

**Tianjin photographic zenith tube observations.**
See Abstr. 041.033.

**Calculation of apparent positions of stars according to MERIT standards.**
See Abstr. 041.036.

**Doppler satellite tracking. Some geodetic and geophysical applications.**
See Abstr. 045.003.

**Radio interferometric determination of intercontinental baselines and earth orientation utilizing Deep Space Network antennas: 1971 to 1980.**
See Abstr. 045.012.

**Precise orbit determination of the satellite "STARLETTE" with laser ranging data.**
See Abstr. 052.035.

**Tidal friction in the Earth-Moon system.**
See Abstr. 081.043.

**Verification of the accuracy of GEM-L2 in response to criticism by Lambeck and Coleman.**
See Abstr. 081.055.

**Signatures of troposphere-stratosphere momentum coupling: implications for global atmospheric angular momentum and earth rotation budgets.**
See Abstr. 082.096.

**Rotation of the atmospheres of the Earth and planets.**
See Abstr. 091.051.

**Influence of lunar tide on LLR.**
See Abstr. 094.051.

**Millisecond pulsars.**
See Abstr. 126.046.

# 045 Astronomical Geodesy, Satellite Geodesy, Navigation

045.001 **The influence of reference system disparity on navigation and positioning.**
G. A. Robinson, D. E. Miller.
Navigation (Wash.), Vol. 30, No. 3, p. 252 – 260 (1983). Abstr. in Phys. Abstr., Vol. 87, No. 1237, Entry 72484 (1984).

045.002 **Télémétrie laser-satellites. Station de Grasse.**
F. Pierron, J. L. Hatat, M. Laplanche, R. Futaully.
Astrométrie, géodésie, mécanique céleste, p. 149 – 153 (1984). – See Abstr. 012.021.

045.003 **Doppler satellite tracking. Some geodetic and geophysical applications.**
M. Catalan, J. Fernandez.
Rev. Geofis., Vol. 39, No. 1, p. 3 – 22 (1983). In Spanish. Abstr. in Phys. Abstr., Vol. 87, No. 1242, Entry 94273 (1984).

045.004 **Die Arbeiten des Sonderforschungsbereiches 78 Satellitengeodäsie der Technischen Universität München im Jahre 1983.**
M. Schneider.
Veröff. Bayer. Komm. Int. Erdmessung Bayer. Akad. Wiss., Astron.-Geod. Arb., Heft Nr. 45, 271 pp. (1984). ISBN 3-7696-9788-X.

Contents: Entfernungsmessungen, mobiles Laserentfernungsmeßsystem, Aufbau und Inbetriebnahme einer Empfangsanlage für Radiointerferometrie, Mikrowellenmessungen, Ergänzungsmessungen und allgemeine Dienste, Geopotential, Dynamik des Erde-Mond-Systems, Kinematik geodätischer Punktfelder, astrometrische und geodätisch-geophysikalische Nutzung interferometrischer Methoden.

045.005 **Verwendung von Lie-Reihen zur hochauflösenden Gravitationsfeldbestimmung.**
M. Schneider.
Veröff. Bayer. Komm. Int. Erdmessung Bayer. Akad. Wiss., Astron.-Geod. Arb., Heft Nr. 45, p. 157 – 172 (1984).

045.006 **Zur numerischen Quadratur von Integralen im Rahmen der Bestimmung des hochfrequenten Anteils des Gravitationsfeldes.**
K. H. Ilk, U. Klose.
Veröff. Bayer. Komm. Int. Erdmessung Bayer. Akad. Wiss., Astron.-Geod. Arb., Heft Nr. 45, p. 173 – 189 (1984).

045.007 **Compression of LAGEOS laser data.**
D. Gambis.
Bull. Géod., Vol. 58, No. 4, p. 518 – 526 (1984).

045.008 **The use of satellite laser ranging for single point positioning in short-arc analysis.**
M. He, C. Fong.
Ann. Shanghai Obs., Acad. Sin., No. 5, p. 199 – 204 (1983). In Chinese.

The method of single point positioning in short-arc analysis was used to estimate the geodetic coordinates of the transportable laser ranging station: TLRS-6200 (located at McDonald Observatory in the early time of 1980).

045.009 **Comparison of the broadcast ephemeris and the MEDOC precise ephemeris.**
H. Xu, Y. Zhu.
Ann. Shanghai Obs., Acad. Sin., No. 5, p. 205 – 211 (1983). In Chinese.

Using part of the data acquired during the MERIT short campaign, accuracy estimates of the broadcast ephemeris of the "TRANSIT" satellites have been obtained by comparing with the MEDOC the precise ephemeris taken as reference at common two–minute epochs.

045.010 **The on–site prediction of laser ranging satellite.**
S. Huang.
Ann. Shanghai Obs., Acad. Sin., No. 5, p. 212 – 220 (1983). In Chinese.

This paper describes the procedure of generation of a data block, and gives the formulas of partial derivatives of satellite azimuth A and height H to its elements M and Ω, they are necessary to search for satellites in a small sky region.

045.011 **A note on the simplified formula for atmospheric correction in satellite laser ranging.**
Q. Lin, D. Tan, F. Yang.
Ann. Shanghai Obs., Acad. Sin., No. 5, p. 221 – 226 (1983). In Chinese.

The paper is a supplement to a previous paper (1981) in which a simplified formula for atmospheric correction in satellite laser ranging has been given.

045.012 **Radio interferometric determination of intercontinental baselines and earth orientation utilizing Deep Space Network antennas: 1971 to 1980.**
O. J. Sovers, J. B. Thomas, J. L. Fanselow, E. J. Cohen, G. H. Purcell Jr., D. H. Rogstad, L. J. Skjerve, D. J. Spitzmesser.
J. Geophys. Res., Vol. 89, No. B9, p. 7597 – 7607 (1984).

A series of experiments has been conducted during the last decade to develop a radio interferometry system capable of measuring crustal and rotational motions of the earth, as well as source positions for a reference frame based on compact extragalactic radio sources. Approximately 2400 observations of extragalactic radio sources were made between August 1971 and February 1980 during 48 separate sessions. A single multiparameter fit was applied to the observed values of delay and delay rate to extract astrometric and geophysical parameters from this decade–long sequence. The fit produced estimates of 684 parameters, including station locations, radio source positions, polar motion, universal time, the precession constant, and solid earth tides.

045.013 **Preliminary calculations of the Simeiz Station coordinates by the orbital method.**
O. M. Bulygina.
Nauchn. Inf., Vyp. 55, p. 89 – 91 (1982). In Russian.

A preliminary calculation of Simeiz coordinates by photographic observations of Midas–4 has been made.

045.014 **Use of the a priori information about the accuracy of the unknown parameters in the adjustment of space geodetic networks.**
B. M. Klenitskij.
Nauchn. Inf., Vyp. 55, p. 98 – 126 (1982). In Russian.

The methods of utilizing an a priori information about the accuracy of the preliminary values of unknown parameters are considered. The possibility of utilising those methods for adjustment of different types of space geodetic networks is shown.

045.015 **Sobre un método de cálculo para la resolución de los problemas geodésicos directo e inverso.**
M. J. Sevilla.
Bol. Inf., Serv. Geogr. Ejército, núm. 50, p. 1 – 20 (1981). = Semin. Astron. Geod., Univ. Madr., Publ. No. 113.

045.016 **Some theoretical and practical aspects of geodetic positioning using carrier phase difference observations of GPS (*Global Positioning System*) satellites.**
G. Beutler, D. A. Davidson, R. B. Langley, R. Santerre, P. Vaníček, D. E. Wells.
Mitt. Satell.–Beobachtungsstn. Zimmerwald, Nr. 14, 4 + 79 pp. (1984).

Carrier phase difference observations are potentially the most precise measurements that can be made using the Global Positioning System. The authors discuss the development and use of a software package for processing such data.

045.017 **NAVSTAR/Global Positioning System (GPS) (IV). (Ergänzungen zu GPS (I)).**
I. Bauersima.
Mitt. Satell.–Beobachtungsstn. Zimmerwald, Nr. 15, 5 + 34 pp. (1984).

045.018 **NAVSTAR/Global Positioning System (GPS) (V). Satellite radiointerferometry.**
I. Bauersima.
Mitt. Satell.–Beobachtungsstn. Zimmerwald, Nr. 17, 2 + 61 pp. (1984).

045.019 **La organización y utilización de las observaciones ópticas de los satélites para la geodesia cósmica.**
S. Dziamko.
Publ. Espec. Obs. Astron. Quito, No. 2, p. 18 – 26 (1981).

045.020 **Para investigar nuestro planeta.**
Y. Balodis.
Publ. Espec. Obs. Astron. Quito, No. 2, p. 56 – 65 (1981).

045.021 **Geodätische Modelle der Punkt– und Gravitationsfeldbestimmung.**
S. Heitz.
Mitt. Astron. Ges., Nr. 62, p. 45 – 62 (1984). – See Abstr. 010.341.

045.022 **Relativistic effects in geodetic VLBI measurements.**
M. Soffel, H. Ruder, J. Campbell, H. Schuh.
Mitt. Astron. Ges., Nr. 62, p. 144 (1984). – See Abstr. 010.341.

**Geodesy at the XVIIIth General Assembly of the International Union of Geodesy and Geophysics. Hamburg, 1983, August 15 – 27.**
See Abstr. 011.007.

**Workshop über astrometrische und geodätisch–geophysikalische Nutzung der Radio–Interferometrie.**
See Abstr. 011.029.

**Géodésie, géodynamique et sciences de la Terre.**
See Abstr. 013.016.

**Programas de cálculo para la determinación de la dirección de la vertical por fotografías cenitales de estrellas.**
See Abstr. 021.016.

**Programación del método de la cuerda.**
See Abstr. 021.017.

**Analytical and numerical studies about effects of natural radio source structures on astrometric and geodetic observations with a very long baseline interferometer.**
See Abstr. 036.156.

**The contribution of satellite geodesy to the geosciences.**
See Abstr. 081.034.

# 046 Ephemerides, Almanacs, Calendars, Chronology

046.001 **Éphémérides Nautiques pour l'an 1985.**
Ouvrage publié par le Bureau des Longitudes spécialement à l'usage des marins.
Gauthier-Villars, Paris, France. 485 pp. (1984). ISBN 2-04-015581-3.

046.002 **Nautisches Jahrbuch oder Ephemeriden und Tafeln für das Jahr 1985 zur Bestimmung der Zeit, Länge und Breite auf See nach astronomischen Beobachtungen.**
Deutsches Hydrographisches Institut, Hamburg, F.R. Germany. 134. Jahrg., 45+365+30 pp. (1984). ISSN 0077-6211.

046.003 **Himmelskalender 1985. Ein kleines astronomisches Jahrbuch für Österreich.**
H. Mucke.
Astronomisches Büro, Hasenwartgasse 32, 1238 Wien, Austria. 128 pp. (1984).

046.004 **Das Himmelsjahr 1985. Sonne, Mond und Sterne im Jahreslauf.**
Herausgegeben von H.-U. Keller unter Mitarbeit von E. Karkoschka.
Astro-Kosmos, Kosmos, Gesellschaft der Naturfreunde, Franckh'sche Verlagshandlung, Stuttgart, F.R. Germany. 157 pp. Price DM 9.80 (1984). ISBN 3-440-05348-2.

046.005 **1985 Abridged Nautical Almanac. Pub. No. 683.**
Hydrographic Office of Japan, Maritime Safety Agency, Tokyo, Japan. 4+245+8 pp. (1984).

046.006 **Connaissance des Temps. Éphémérides Astronomiques pour l'An 1985.**
Bureau des Longitudes, 77 Avenue Denfert Rochereau, 75014 Paris, France. 54+124 pp. Price FF 85.00 (1984). ISBN 2-11-080436-X, ISSN 0181-3048.

046.007 **1985 Nautical Almanac. Pub. No. 681.**
Hydrographic Office of Japan, Maritime Safety Agency, Tokyo, Japan. 5+469+12 pp. (1984).

046.008 **Astronomische Grundlagen für den Kalender 1986.**
Compiled by T. Lederle, edited by Astronomisches Rechen-Institut, Heidelberg.
Verlag G. Braun, Karl-Friedrich-Straße 14 - 18, D-7500 Karlsruhe 1, F.R. Germany. 88 pp. Price DM 59.00 (1984). ISBN 3-7650-0186-4, ISSN 0067-0014.

046.009 **Kalender für Sternfreunde 1985. Kleines astronomisches Jahrbuch.**
P. Ahnert.
Johann Ambrosius Barth, Leipzig, German Democratic Republic. 172 pp. Price M 5.70 (1984). ISSN 0075-4706.

046.010 **Calcul automatique des nouvelles éphémérides.**
G. Francou.
Astrométrie, géodésie, mécanique céleste, p. 21 - 22 (1984). - See Abstr. 012.021.

046.011 **Erratum: "Le comput ecclésiastique" [Astronomie, Vol. 98, p. 295 - 303 (1984)].**
H. Naudot.
Astronomie, Vol. 98, p. 471 (1984). See Abstr. 37.046.032.

046.012 **The Air Almanac 1985, January - June.**
Air Publication 1602. Issued by Her Majesty's Nautical Almanac Office (London) and the Nautical Almanac Office, United States Naval Observatory (Washington).
UK Edition: Her Majesty's Stationery Office, London, England; US Edition: Superintendent of Documents, US Government Printing Office, Washington, D.C. 20402, USA. p. 1 - 364, A1 - A104, F1 - F4, Price £ 16.95 (1984). ISBN 0-11-772383-5.

046.013 **The Handbook of the British Astronomical Association 1985.**
Prepared by the Computing Section of the Association under the supervision of G. E. Taylor.
Office of the Association, Burlington House, Piccadilly, London, W1V 0NL, England. 108 pp. Price £ 5.00 (1984).

046.014 **Almanaque Nautico, 1985. Con Suplemento para la Navegación Aerea.**
Instituto y Observatorio de Marina, San Fernando (Cádiz), Spain. 414+30+5 pp. (1984). ISBN 84-7469-021-8, ISSN 0210-735X.

046.015 **1985 Polaris Almanac for Azimuth Determination. Pub. No. 685.**
Hydrographic Office of Japan, Maritime Safety Agency, Tokyo, Japan. 16 pp. (1984).

046.016 **Anuarul Astronomic 1985.**
Centrul de Astronomie si Stiinte Spatiale, Bucuresti.
Editura Academiei Republicii Socialiste România, R. 79717, Bucuresti, Calea Victoriei 125, Rumania. 319 pp. Price Lei 17.00 (1984).

046.017 **Ephémérides astronomiques et calendrier des événements célestes pour 1985.**
P. Cugnon, A. Koeckelenbergh, G. Evrard.
Ciel Terre, Vol. 100, No. 5, 73 pp. (1984).

046.018 **Efemérides Astronómicas para o ano de 1984.**
Observatório Astronómico da Universidade de Coimbra, Coimbra, Portugal. 16+252 pp. (1984).

046.019 **Observer's Handbook 1985.**
R. L. Bishop (Editor).
The Royal Astronomical Society of Canada, 136 DuPont Street, Toronto, Ontario M5R 1V2, Canada. 184 pp. Price $ 8.50 (1984). ISSN 0080-4193.

046.020 **The Astronomical Almanac for the year 1985. Data for astronomy, space sciences, geodesy, surveying, navigation and other applications.**
Issued by the Nautical Almanac Office, United States Naval Observatory (Washington) and Her Majesty's Nautical Almanac Office, Royal Greenwich Observatory (London).
US Edition: Superintendent of Documents, US Government Printing Office, Washington, D.C. 20402, USA. UK Edition: Her Majesty's Stationery Office, London, England. 10+529 pp. Price $ 17.00 (1984). ISBN 0-11-886922-1.

046.021 **Almanac for Computers 1985.**
Nautical Almanac Office, United States Naval Observatory, Washington, D.C. 20390, USA. 93 pp. Price $ 4.25 (1984). Available from the Superintendent of Documents, US Government Printing Office, Washington, D.C. 20402, USA.

046.022 **Astronomical Phenomena for the year 1986.**
Prepared by The Nautical Almanac Office, United States Naval Observatory and Her Majesty's Nautical Almanac Office, Royal Greenwich Observatory.
US Edition: Superintendent of Documents, US Government Printing Office, Washington, D.C. 20402, USA; UK Edition: Her Majesty's Stationery Office London, England. 77 pp. (1984). ISBN 0-11-886918-3.

046.023 **Anuário de Astronomia 1984.**
R. R. de Freitas Mourão.
Livraria Francisco Alves Editora S/A, Rua Sete de Setembro, 177 -Centro 20050, Rio de Janeiro, RJ, Brazil. 289 pp. (1984).

046.024 **Dados Astronómicos para os Almanaques de 1985 para Portugal.**
Observatório Astronómico de Lisboa, Lisboa, Portugal. 55 pp. (1984).

046.025 **Satellites Galiléens de Jupiter.**
Phénomènes et configurations pour 1985, suivis d'une méthode permettant de calculer les phénomènes pour 1986. Supplément à la Connaissance des Temps, Bureau des Longitudes, 77 ave. Denfert–Rochereau, F–75014 Paris, France. 70 pp. (1984).

046.026 **Configurations des huit premiers satellites de Saturne.**
Supplément à la Connaissance des Temps. Bureau des Longitudes, 77 ave. Denfert–Rochereau, F–75014 Paris, France. 33 pp. (1984).

046.027 **Efemérides Astronómicas 1985.**
Instituto y Observatorio de Marina, San Fernando (Cádiz), Spain. Vol. 194, 4+459 pp. (1984). ISBN 84–7469–022–6, ISSN 0080–5971.

046.028 **Time computations and Dionysius Exiguus.**
G. Teres.
J. Hist. Astron., Vol. 15, Part 3, No. 44, p. 177 – 188 (1984).

046.029 **Calculation of the ephemerides of the sun taking into account the perturbed motion of the earth.**
V. P. Brusentsov.
Teor. i prakt. mor. navigats. Moskva, p. 21 – 24 (1984). In Russian. Abstr. in Ref. Zh., 51. Astron., 10.51.124 (1984).

046.030 **Die durchschnittliche Dauer der Jahre und Monate im babylonischen und im jüdischen Kalender.**
K. Ferrari d'Occhieppo.
Sternenbote, 27. Jahrg., Nr. 12, p. 209 – 211 (1984).

046.031 **Apparent Places of Fundamental Stars 1985. Containing the 1535 stars in the Fourth Fundamental Catalogue (FK4).**
W. Fricke, T. Lederle, H. Schwan (Editors).
Published for Astronomisches Rechen–Institut, Heidelberg, F.R. Germany by Verlag G. Braun, Karl–Friedrich–Straße 14–18, D–7500 Karlsruhe, F.R. Germany. 46+510 pp. Price DM 62.00 (1984). ISBN 3–7650–0085–1, ISSN 0174–254–X.

046.032 **Ephémérides astronomiques 1985. Calendriers – soleil – lune – planètes – astéroides – satellites – comètes – étoiles.**
Annuaire du Bureau des Longitudes.
Gauthier–Villars, Paris, France. 18+277 pp. (1984). ISBN 2–04–015682–8.

046.033 **Hvězdářská ročenka 1985.**
V. Vanýsek (Editor).
Ročník 61. Academia, nakladatelstvi Československé akademie věd, Praha, Czechoslovakia. 217 pp. Price 30.00 Kčs (1984).

046.034 **Rocznik Astronomiczny Obserwatorium Krakowskiego 1985. International Supplement Nr. 56.**
K. Rudnicki, P. Flin (Editors).
Państwowe Wydawnictwo Naukowe, Warszawa – Kraków, Poland. 6+138 pp. Price zł 240.00 (1984). ISBN 83–01–06067–0, ISSN 0075–7047.
See abstracts 046.035, 119.056, 119.057, 122.033.

046.035 **Geocentric ephemerides for the year 1985 of the libration points $L_4$ and $L_5$ in the Earth–Moon system and in the Sun–Venus system.**
J. Kordylewski, R. Szafraniec.
Rocznik Astronomiczny Obserwatorium Krakowskiego 1985, p. 133 – 137 (1984). – See Abstr. 046.034.

046.036 **Ephémérides des satellites naturels des planètes: Jupiter, Saturne, Uranus pour 1985.**
Supplément à la Connaissance des Temps.
Bureau des Longitudes, 77 ave. Denfert–Rochereau, F–75014 Paris, France. 87 pp. (1984).

046.037 **Planetary phenomena near the end of the 20th century.**
A. Satō.
Heavens, Vol. 65, No. 11, p. 311 – 317 (1984).

046.038 **Limb shortening and the limiting elongation for the lunar crescent's visibility.**
M. Ilyas.
Q. J. R. Astron. Soc., Vol. 25, No. 4, p. 421 – 422 (1984).
The practical Danjon's limiting elongation limit remains close to 10.5° for the crescent's earliest visibility irrespective of the cause of the phenomenon.

046.039 **A computer study of Babylonian almanacs.**
N. A. Roughton.
Archaeoastronomy (U.S.A.), Vol. 6, Nos. 1 – 4, p. 118 – 123 (1983).

046.040 **Ephémérides 1985.**
Supplément à l'Astronomie de Janvier 1985. 18+277 pp. (1984). ISBN 2–04–015682–8.

046.041 **Astronomical Handbook for Southern Africa 1985.**
R. F. Hurly (Editor).
The Astronomical Society of Southern Africa, Cape Town, South Africa. 4+46 pp. (1985). ISSN 0571–7191.

046.042 **Almanacco Astronomico della Rivista Coelum per l'anno 1985.**
P. Battistini, A. Buzzoni.
Coelum, Supplemento al n. 6, 43+38 pp. Price L 8000 (1984). ISSN 0366–7588.

046.043 **Almanacco 1985.**
Astronomia, N. 4, p. 1 – 164 (1984).

046.044 **The Nautical Almanac for the year 1986.**
N.P. 314–86. Issued by Her Majesty's Nautical Almanac Office, London and Nautical Almanac Office, United States Naval Observatory, Washington.
UK Edition: Her Majesty's Stationery Office, London, England.
US Edition: Superintendent of Documents, US Government Printing Office, Washington, D.C. 20402, USA. A4+276+36 pp. Price £ 9.50 (1985). ISBN 0–11–772357–6.

046.045 **Astronomical Handbook for 1984.**
Astron. Bull., Carter Obs., No. 101, 32 pp. (1983).

046.046 **Astronomiskais Kalendārs 1985.**
J. Bikše, I. Daube, M. Dīrikis, J. Francmanis, V. Freijs, J. Miezis.
Latvijas PSR Zinātnu Akadēmija, Radioastrofizikas Observatorija, Vissavienibas Astronomijas un Geodēzijas Biedribas Latvijas Nodala. Zinātne, Riga. 191 pp. Price 60 Kop. (1984).

046.047 **The Air Almanac 1985, July – December.**
Air Publication 1602. Issued by Her Majesty's Nautical Almanac Office, London and Nautical Almanac Office, United States Naval Observatory, Washington.
UK Edition: Her Majesty's Stationery Office, London, England; US Edition: Superintendent of Documents, US Government Printing Office, Washington, D.C. 20402, USA. p. 364 – 732, A1 – A104, F1 – F4, Price £ 16.95 (1985). ISBN 0–11–772384–3.

046.048 **Nautisches Jahrbuch 1985.**
Seehydrographischer Dienst der Deutschen Demokratischen Republik, Rostock, German Democratic Republic. 35. Jahrg., 8721, 45+365 pp. (1984). ISSN 0433–681X.

046.049 **Chinese Astronomical Ephemeris 1985.**
Purple Mountain Observatory, Academia Sinica, Nanking, People's Republic of China. 5+544 pp. (1984). In Chinese.

046.050 **Tartu Tähetorni Kalender 1985. Aastaks.**
Compiled by M. Jõeveer.
Eesti NSV Teaduste Akadeemia, Astrofüüsika ja Atmosfäärifüüsika Instituut, W. Struve nimeline Tartu Astrofüüsika Observatoorium. Publishing House Valgus, Tallinn. 96 pp. Price 35 Kop. (1984). ISSN 0202-2214.

046.051 **To know the stars. (Astronomical calendars).**
G. Ottewell.
Department of Physics, Furman University, Greenville, S.C. 29613, USA. 41 pp. Price $ 5.00 (1984).
Review in Strolling Astron., Vol. 30, Nos. 9 - 10, p. 208; 1984 (*J. R. Smith*).

046.052 **The Indian Astronomical Ephemeris for the year 1985.**
Prepared by the Positional Astronomy Centre, India Meteorological Department, New Alipore, Calcutta-700053, India.
Available from the Controller of Publications, Civil Lines, Delhi-110054, India; The High Commission of India, India House, Aldwych, London W.C.2, England. 6+562 pp. Price Rs. 100.00, £ 11.66, US$ 36.00 (1984).

046.053 **Astronomical Yearbook of the USSR for the year 1986.**
V. K. Abalakin (Editor).
Institut Teoreticheskoj Astronomii Akademii Nauk SSSR. Izdatel'stvo Nauka, Leningradskoe Otdelenie, Leningrad. 692 pp. Price 13 Rbl. 20 Kop. (1984). In Russian.

046.054 **Rocznik Astronomiczny na rok 1985.**
Instytut Geodezji i Kartografii, Państwowe Przedsiebiorstwo Wydawnictw Kartograficznych, Warszawa, Poland. Vol. 40, 160 pp. Price zł 405.00 (1984). ISBN 83-7000-048-7, ISSN 0209-0341.

**Grundlagen der Ephemeridenrechnung. Eine Zusammenstellung der wichtigsten Formeln und Daten zur Berechnung der Bahnen von Körpern im Sonnensystem.**
See Abstr. 003.027.

**A new compact representation of ephemerides: application to the motion of Pluto, the Sun and the Galilean satellites of Jupiter.**
See Abstr. 042.056.

**Millisecond pulsars and the location of the solar system barycenter.**
See Abstr. 042.075.

**Comparison of the broadcast ephemeris and the MEDOC precise ephemeris.**
See Abstr. 045.009.

**Canon der Sonnenfinsternisse, -2003 bis +2526.**
See Abstr. 079.001.

**Canon der Mondfinsternisse, -2002 bis +2526.**
See Abstr. 095.001.

**The Comet Handbook 1985 of the Oriental Astronomical Association, Comet Section.**
See Abstr. 103.036.

# Space Research

## 051 Extraterrestrial Research Related to Astronomy and Astrophysics

051.001 **Global geospace study.**
M. J. Rycroft, A. J. Smith, D. Jones, J. R. Dudeney, S. D. Shawhan.
Nature, Vol. 310, No. 5977, p. 449 (1984).

This note discusses the aims of the International Solar Terrestrial Programme, where a total of 4 spacecraft is intended to investigate the interaction of the solar wind with the Earth plasmasphere and to study in detail a number of physical processes in geospace.

051.002 **The Spacelab experience: a synopsis.**
C. R. Chappell, K. Knott.
Science, Vol. 225, No. 4658, p. 163 – 165 (1984).

The Spacelab 1 mission, a joint venture of the European Space Agency and the National Aeronautics and Space Administration took place during the period 28 November through 8 December 1983. An overview of the first flight of the orbiting laboratory is presented.

051.003 **The conquest of Mars.**
L. W. David.
Spaceflight, Vol. 26, Nos. 7 – 8, p. 290 – 293 (1984).

With the Space Shuttle reaching operational status, manned missions to Mars are technologically achievable, scientifically legitimate, biologically necessary and financially affordable. The author reviews the reasons for humans voyaging to the Red Planet.

051.004 **En route to the giant planet.**
Spaceflight, Vol. 26, Nos. 7 – 8, p. 302 – 303 (1984).

051.005 **Die Raumstation kommt!**
D. Lemke.
Sterne Weltraum, 23. Jahrg., Nr. 8–9, p. 434 – 439 (1984).

051.006 **New lease on life for the Solar Maximum Mission.**
E. L. Chupp.
Nature, Vol. 310, No. 5980, p. 725 – 726 (1984).

051.007 **The cosmos from space.**
D. H. Clark.
J. Br. Interplanet. Soc., Vol. 37, No. 7, p. 320 – 322 (1984).

051.008 **Comet rendezvous.**
R. F. Draper, S. A. Lundy.
J. Br. Interplanet. Soc., Vol. 37, No. 8, p. 361 – 365 (1984).

The first sustained visit to a comet will begin in 1994 when a spacecraft will rendezvous with the Comet Kopff. Arriving near aphelion, a variety of instruments will be able to study the comet nucleus close up in its quiescent state. They will then record the changes as the comet heats up on its approach to the Sun, forms a coma and tail, and passes through perihelion.

051.009 **Titan atmospheric probe.**
B. L. Swenson.
J. Br. Interplanet. Soc., Vol. 37, No. 8, p. 366 – 369 (1984).

Since the Voyager 1 close–pass observations of Titan, scientific interest in the nature and composition of the atmosphere and surface of Titan has increased greatly. This enhanced interest has led to the study of mission options to the Saturnian system with a principal goal of depositing a probe into the atmosphere of Titan. Such a probe could carry instrumentation to perform in situ measurements of the structure and composition of the atmosphere and clouds and of the morphology and composition of the surface. This paper provides an overview of the mission and system concepts for such an atmospheric probe mission to Titan.

051.010 **Mars exploration in the 1990's.**
A. L. Albee.
J. Br. Interplanet. Soc., Vol. 37, No. 8, p. 370 – 374 (1984).

Beginning with the objectives of continued Mars exploration as set forth by the Committee on Planetary Exploration in 1977, this paper discusses the portion of the NASA Core Program that pertains to Mars. Objectives are listed of the Mars Geo–Science–Climatology Observer, the Mars Aeronomy Orbiter, and the Mars Network Mission.

051.011 **Starprobe: to confront the Sun.**
W. I. McLaughlin, J. E. Randolph.
J. Br. Interplanet. Soc., Vol. 37, No. 8, p. 375 – 380 (1984).

Starprobe is an advanced–technology mission that is being planned for the purpose of making a very close approach to the Sun; perhaps as near as four solar radii. The scientific experiments will accurately measure the solar quadrupole moment, conduct relativity experiments, measure the fields and particles environment, and image solar features at high resolution.

051.012 **Titan buoyant station.**
A. L. Friedlander.
J. Br. Interplanet. Soc., Vol. 37, No. 8, p. 381 – 387 (1984).

The proposed mission concept evolves about one or more buoyant stations (balloons and/or airship) operating at varying altitudes in Titan's atmosphere. An orbiter of Titan provides communications link support and accomplishes remote sensing science experiments. The paper describes science objectives and payloads, the reference mission concept and various aspects of mission design analysis.

051.013 **Comet nucleus sample return.**
H. Feingold.
J. Br. Interplanet. Soc., Vol. 37, No. 8, p. 388 – 393 (1984).

A comet nucleus sample return mission has been proposed for implementation near the end of this century. The objective of the mission is to collect a sample of undisturbed material from beneath the surface of an active comet and return it to Earth in a minimally altered state. Potential targets include the short–period comets Encke, Tempel 2 and Wild 2. This article defines such a mission and describes its requirements with regard to science, sampling thermal protection and performance.

051.014 **Rendezvous with Giotto.**
A. Wilson.
Space Educ., Vol. 1, No. 8, p. 359, 362 – 363 (1984).

051.015 **The Venus Radar Mapper mission.**
E. Cutting, J. H. Kwok, S. N. Mohan.
J. Br. Interplanet. Soc., Vol. 37, No. 10, p. 443 – 452 (1984).
The Venus Radar Mapper (VRM) mission is sponsored by NASA to put a single spacecraft in orbit around Venus to map the surface using a synthetic aperture mapping radar. This paper describes the VRM mission at its present state of design. The science objectives and project constraints are described. Key features of the spacecraft system and radar system are discussed. The interplanetary and mapping orbit design are covered.

051.016 **Pioneer to Venus: the multiprobe and orbiter missions.**
R. A. Craig, L. Colin, R. O. Fimmel.
J. Br. Interplanet. Soc., Vol. 37, No. 10, p. 453 – 467 (1984).
This paper outlines the history of the development of the Pioneer Venus Program, discusses its scientific achievements and contrasts them with the important questions about Venus. This paper also presents the expectations for the remainder of the mission, including the exciting observations of Comet Halley.

051.017 **Exploring beyond the planets: the Pioneer 10 and 11 missions.**
P. Dyal, R. O. Fimmel.
J. Br. Interplanet. Soc., Vol. 37, No. 10, p. 469 – 479 (1984).
The Pioneer 10 spacecraft is now further from the Sun than all of the nine planets and is the most distant man–made object in our Solar System. Pioneer 11 is 15 AU from the Sun and travelling (in a direction opposite to that of Pioneer 10) in the "upstream" direction of the heliosphere in our galactic arm. These two Pioneer spacecraft have provided the first large–scale examination of the Sun as a star with its surrounding gas and dust. Since launch, the Pioneers have measured large–scale properties of the heliosphere during one 11–year solar sunspot cycle, and have detected the magnetic field polarity reversal that occurs every 22 years. The charged particle experiments have measured the properties of the expanding solar atmosphere, the transport of cosmic rays into our heliosphere, and the high energy trapped radiation around the planets Jupiter and Saturn.

051.018 **The Venus Radar Mapper: radar system design for highly elliptical orbits.**
W. T. K. Johnson.
ESA Spec. Publ., ESA SP–215, Vol. II, p. 809 – 813 (1984). – See Abstr. 012.012.
The Venus Radar Mapper mission will observe the planet Venus at microwave frequencies in three simultaneous modes: Synthetic Aperture Radar (SAR or "imaging") mode, an altimeter mode, and a radiometer mode. The nominal mission will last one full Venusian rotation or 243 days. This paper describes the science objectives, mission constraints, radar system design, flight hardware, and ground processing methods.

051.019 **Astronomy from the Shuttle.**
M. Engle.
Astronomy, Vol. 12, No. 4, p. 66 – 70 (1984). Abstr. in Phys. Abstr., Vol. 87, No. 1238, Entry 77688 (1984).

051.020 **Preliminary scientific results from the first six months of the Infrared Astronomical Satellite (IRAS).**
B. T. Soifer, C. A. Beichman, J. R. Houck, G. Neugebauer, M. Rowan–Robinson.
Opt. Eng., Vol. 23, No. 2, p. 128 – 130 (1984). Abstr. in Phys. Abstr., Vol. 87, No. 1238, Entry 77692 (1984).

051.021 **European X–ray satellite on a successful course.**
M. J. Schmidt.
Schweiz. Tech. Z., Vol. 81, No. 10, p. 12 – 15 (1984). Abstr. in Phys. Abstr., Vol. 87, No. 1241, Entry 90054 (1984).

051.022 **Plans for future space missions.**
A. Johnstone.
Nature, Vol. 311, No. 5983, p. 209 (1984).
This note comments on plans for a new generation of spacecraft intended for cometary encounters in the post–Halley era.

051.023 **Report on a special session on space observations of stellar oscillations, held at the EPS study conference "Oscillations as a probe of the Sun's interior", Catania, 20 – 24 June 1983.**
Mem. Soc. Astron. Ital., Vol. 55, Nos. 1 – 2, p. 369 – 373 (1984). – See Abstr. 012.022.

051.024 **The $L_1$–Lagrangian point orbit as a suitable site for helioseismologic measurements.**
V. Domingo, D. Wyn–Roberts.
Mem. Soc. Astron. Ital., Vol. 55, Nos. 1 – 2, p. 375 – 381 (1984). – See Abstr. 012.022.
The orbit around the Earth – Sun Lagrangian point $L_1$ has the advantages of providing continuous solar viewing and eliminating the Earth rotation velocity when compared with ground–based observations. ESA has completed a study on a spacecraft (DISCO) which was to have been placed in this orbit, and is now studying a potential solar observatory (SOHO) to be placed in the same orbit. The aims of SOHO are the understanding of the solar wind origin and of the heating and energy balance of the corona and the investigation of the solar interior structure.

051.025 **The International Ultraviolet Explorer (IUE) – 6 years in orbit.**
J. Fälker.
ESA Spec. Publ., ESA SP–218, p. 465 – 468 (1984). – See Abstr. 012.024.
Spacecraft and scientific instrument status of the International Ultraviolet Explorer mission at 6 years in orbit is briefly discussed.

051.026 **The ultraviolet astronomy mission – Columbus.**
R. Wilson.
ESA Spec. Publ., ESA SP–218, p. 483 – 487 (1984). – See Abstr. 012.024.
A possible future mission in ultraviolet astronomy, named Columbus, is currently being studied on both sides of the Atlantic under the auspices of ESA and NASA. Its prime aim is to exploit the relatively unexplored region of the spectrum between 900 and 1200 Å which is extremely rich in containing the Lyman lines of hydrogen and deuterium and the Lyman band of their molecules together with the resonance lines of many important ions.

051.027 **Spectroscopy from space.**
G. F. Gahm.
Phys. Scr., Vol. T8, p. 59 – 62 (1984). – See Abstr. 012.027.
This article gives a brief review of some existing or planned space missions where spectroscopy of celestial objects is involved. Its focus is on projects where European countries participate through the European Space Agency (ESA).

051.028 **Space Report.**
K. T. Wilson.
Spaceflight, Vol. 26, Nos. 7 – 8, p. 296 – 300 (1984).
July/August 1984 review of space news and events.

051.029 **Space Report.**
Spaceflight, Vol. 26, Nos. 9 – 10, p. 348 – 353 (1984).
September/October 1984 review of space news and events.

051.030 **Space at JPL.**
W. I. McLaughlin.
Spaceflight, Vol. 26, Nos. 7 – 8, p. 304 – 307 (1984).
Contents: Voyaging to Neptune. Assembling and testing Galileo. Galileo flight path. An Arctic polesitter.

051.031 **Space at JPL.**
W. I. McLaughlin.
Spaceflight, Vol. 26, Nos. 9 – 10, p. 354 – 356 (1984).
Contents: Touring the Galilean satellites. Space resources. Solar Max. The early laboratory.

051.032 **De Europese Exosat: röntgensterrenkunde op eenzame hoogten.**
G. W. E. Beekman.
Zenit, 11. Jaarg., No. 10, p. 404 – 406, 408 – 409 (1984).

051.033 **Space report.**
Spaceflight, Vol. 26, No. 11, p. 394 – 400 (1984).
November 1984 review of space news and events.

051.034 **Space at JPL.**
W. I. McLaughlin.
Spaceflight, Vol. 26, No. 11, p. 401 – 403 (1984).
Contents: Comets about Vega? The flash of sunlight. Images of Neptune. Photometry of Halley's comet. Systems analysis.

051.035 **Through the tail of a comet.**
W. I. McLaughlin.
Spaceflight, Vol. 26, No. 11, p. 405 – 409 (1984).
The first space mission for cometary exploration will be conducted by the International Cometary Explorer (ICE). The craft will pass through the tail of Comet Giacobini–Zinner on 11 September 1985, some 10,000 km from the nucleus . Its instruments consist of fields and particles experiments (no imaging) which will provide an in situ study of the comet and its interaction with the solar wind.

051.036 **Raumsonde soll den Sonnennordpol überfliegen.**
M. J. Schmidt.
Orion, 42. Jahrg., Nr. 204, p. 177 – 178 (1984).

051.037 **Development of Japanese artificial satellites.**
J. Tabata, M. Kunsanagi.
J. Soc. Instrum. Control Eng., Vol. 23, No. 1, p. 30 – 36 (1984). In Japanese. Abstr. in Phys. Abstr., Vol. 87, No. 1242, Entry 94545 (1984).

051.038 **Lunar exploration by polar orbiter.**
T. Shiho.
J. Soc. Instrum. Control Eng., Vol. 23, No. 1, p. 49 – 55 (1984). In Japanese. Abstr. in Phys. Abstr., Vol. 87, No. 1242, Entry 94547 (1984).

051.039 **Observations with scientific satellites.**
K. Makishima.
J. Soc. Instrum. Control Eng., Vol. 23, No. 1, p. 111 – 116 (1984). In Japanese. Abstr. in Phys. Abstr., Vol. 87, No. 1242, Entry 94549 (1984).

051.040 **The universe in ultraviolet.**
Space Educ., Vol. 1, No. 8, p. 372 – 373 (1984).
In late 1988 a new observatory will be launched into orbit by the Space Shuttle: the Extreme Ultraviolet Explorer. EUVE will observe the universe in a poorly–explored portion of the spectrum.

051.041 **Space report.**
Spaceflight, Vol. 26, No. 12, p. 442 – 445 (1984).
1984 December review of space news and events.

051.042 **Space at JPL.**
W. I. McLaughlin.
Spaceflight, Vol. 26, No. 12, p. 448 – 450 (1984).
Contents: Keeping Voyager current. AMPTE/CCE operations. Wide field/planetary camera. Radio astronomy with the DSN (*Deep Space Network*).

051.043 **The infrared sky.**
H. J. Habing, G. Neugebauer.
Am. Sci., Vol. 251, No. 5, p. 42 – 51 (1984).
An orbiting, liquid–helium–cooled telescope has made panoramic images of the infrared sky, recording the glow from cold, solid matter in the solar system, the galaxy and the universe at large.

051.044 **Infrared astronomy with IRAS.**
R. E. Jennings.
Ir. Astron. J., Vol. 16, No. 3, p. 210 – 217 (1984). – See Abstr. 012.036.

051.045 **The ROSAT all–sky survey in X–rays.**
J. Trümper.
MPE Rep., No. 184, p. 254 – 260 (1984). – See Abstr. 012.039.
The Roentgen satellite ROSAT which is to be launched in 1987 will carry a large X–ray telescope and an XUV telescope. The major objective is to perform the first all–sky survey in the X–ray (6 – 80 Å) and XUV (> 70 Å) bands by means of imaging telescopes. The other general objective is to make detailed observations of selected sources with respect to spatial structures, spectra and time variabilities. This paper concentrates on some scientific and technical aspects of the X–ray sky survey.

051.046 **On the unique capabilities of the Hubble Space Telescope for studies of ROSAT deep–survey X–ray sources.**
R. E. Griffiths.
MPE Rep., No. 184, p. 271 – 274 (1984). – See Abstr. 012.039.

051.047 **AMPTE and the artificial comet.**
M. Washburn.
Sky Telesc., Vol. 68, No. 6, p. 516 (1984).

051.048 **Expériences actives dans un plasma cosmique: le projet AMPTE.**
M. Roth.
Ciel Terre, Vol. 100, No. 6, p. 150 – 152 (1984).

051.049 **IRAS ou le ciel avec des yeux infrarouges.**
A. Heck.
Ciel Terre, Vol. 100, No. 6, p. 163 – 167 (1984).
The Infrared Astronomical Satellite (IRAS) gathered an impressive amount of data in slightly more than nine months of operations. They led to an overall new view of the universe, unknown previously in the infrared wavelength range. This paper offers a brief review of the first results acquired in various astronomical fields.

051.050 **The Voyager encounters with Saturn.**
E. C. Stone.
AIAA J., Vol. 22, No. 4, p. 498 – 503 (1984). Abstr. in Phys. Abstr., Vol. 87, No. 1243, Entry 101112 (1984).

051.051 **First success for a space mission and a comet for Christmas.**
D. J. Southwood, D. A. Bryant.
Nature, Vol. 312, No. 5995, p. 594 (1984).
This note comments on the recent achievements of the Active Magnetospheric Particle Tracers (AMPTE) space mission, a joint project of the United States, Britain and West Germany.

051.052 **Europe X–rays the skies.**
D. Whitehouse, G. Branduardi–Raymont.
New Sci., Vol. 102, No. 1411, p. 34 – 38 (1984). Abstr. in Phys. Abstr., Vol. 87, No. 1244, Entry 105867 (1984).

051.053 **NASA space programs in infrared astronomy.**
N. W. Boggess.
Laser Focus/Electro–Opt., Vol. 20, No. 6, p. 116 – 127 (1984). Abstr. in Phys. Abstr., Vol. 87, No. 1245, Entry 110490 (1984).

051.054 **Interplanetary navigation: an overview.**
J. F. Jordan, L. J. Wood.
J. Astronaut. Sci., Vol. 32, No. 1, p. 17 – 28 (1984). Abstr. in Phys. Abstr., Vol. 87, No. 1245, Entry 110493 (1984).

051.055 **Recent study of a comet nucleus penetrator experiment.**
S. W. Squyres, J. I. Trombka, A. E. Metzger, B. L. Swenson.
Bull. Am. Astron. Soc., Vol. 16, No. 3, Part I, p. 643 (1984). Abstract. – See Abstr. 010.061.

051.056 **The Mars Geoscience/Climatology Observer (MGCO) mission.**
A. L. Albee, F. D. Palluconi.
Bull. Am. Astron. Soc., Vol. 16, No. 3, Part I, p. 652 (1984). Abstract. – See Abstr. 010.061.

051.057 **Studies of potential Mars Rover Sample–Return sites.**
H. Masursky, D. H. Scott, E. C. Morris.
Bull. Am. Astron. Soc., Vol. 16, No. 3, Part I, p. 681 – 682 (1984). Abstract. – See Abstr. 010.061.

051.058 **Venus Radar Mapper: precision positioning of SAR images.**
S. N. Mohan.
Bull. Am. Astron. Soc., Vol. 16, No. 3, Part I, p. 706 (1984). Abstract. – See Abstr. 010.061.

051.059 **Mars sample return mission.**
D. P. Blanchard, J. P. deVries, R. D. Bourke.
Bull. Am. Astron. Soc., Vol. 16, No. 3, Part I, p. 706 (1984). Abstract. – See Abstr. 010.061.

051.060 **An assessment study of a joint NASA/ESA comet coma sample return mission.**
B. L. Swenson, R. Reinhard, P. Tsou, A. Friedlander.
Bull. Am. Astron. Soc., Vol. 16, No. 3, Part I, p. 711 (1984). Abstract. – See Abstr. 010.061.

051.061 **Comet Rendezvous Asteroid Flyby Mission – Mariner Mark II.**
R. F. Draper.
Bull. Am. Astron. Soc., Vol. 16, No. 3, Part I, p. 711 (1984). Abstract. – See Abstr. 010.061.

051.062 **ESA and microgravity research.**
D. J. Shapland.
Space: a developing role for Europe, p. 19 – 34 (1984). – See Abstr. 012.073.

An overview of how the European Space Agency (ESA) is helping to coordinate European activities in the domain of microgravity research is presented. The flight durations under microgravity conditions vary from minutes to days and months. The element of ESA's microgravity programme that provide these conditions are described. These include sounding rockets, Spacelab, EURECA and, eventually, Space Stations. For all these carriers, facilities are being developed and will be made available to experimenters for conducting research in the fascinating environment of near–zero–gravity attainable over long periods only through the use of space vehicles.

051.063 **The Far Ultraviolet Spectroscopic Explorer.**
W. Cash.
Space: a developing role for Europe, p. 107 – 115 (1984). – See Abstr. 012.073.

A new astronomy mission to study stellar spectra in the 912 – 1200 Å band is being seriously studied in both Europe and the USA. It is conceived as a follow–on to IUE, which would extend astronomical observations into a new spectral region containing a wealth of atomic transitions useful as probes of astrophysical objects. The observatory is to be for the use of guest investigators and will make high spectral resolution observations of stars fainter than IUE could reach. It is seen as an observatory which would complement the tremendous power of the Space Telescope by operating in the bandpass just shortward of where ST ceases to operate efficiently.

051.064 **IRAS, the first four months in orbit.**
J. K. Davies, S. F. Green.
Space: a developing role for Europe, p. 117 – 122 (1984). – See Abstr. 012.073.

This paper reviews briefly the first phase of a project which is proving an outstanding example of cooperation, both within Europe and across the Atlantic. This is the mission of the Infrared Astronomical Satellite (IRAS), in its fifth month at the time of writing. The objective of the IRAS mission is to carry out a highly reliable, all–sky survey in four infrared wavelength bands, centred on approximately 12, 25, 60 and 100 microns.

051.065 **The US planetary exploration program opportunities for international cooperation.**
G. A. Briggs.
Space: a developing role for Europe, p. 257 – 259 (1984). – See Abstr. 012.073.

051.066 **A modified satellite proposal for stellar and solar seismology.**
T. Schmidt–Kaler.
Space research prospects in stellar activity and variability, p. 191 – 196 (1984). – See Abstr. 012.078.

The eigenfrequencies of the Sun can be observed by means of precise intensity measurements (Woodard, Hudson 1983; Schmidt–Kaler, Winkler 1983). The SMM–ACRIM experiment is interrupted by technical malfunction, and the DISCO satellite project has been turned down. Solar intensity measurements from space are of fundamental importance, especially if done simultaneously with precise radial velocity observations from the ground. The requirements for such measurements are defined. Three options for a corresponding low cost modification of a Stellar Activity and Variability Space Project are sketched.

051.067 **An All–sky Photometric Explorer.**
H. S. Hudson.
Space research prospects in stellar activity and variability, p. 197 – 211 (1984). – See Abstr. 012.078.

These notes give ideas for an "All–sky Photometric Explorer", which would survey the sky from space for visual photometric variability of as many objects as possible. The observations would be as precise as possible and would have the maximum duty cycle possible for each object. These conflicting requirements present many difficulties for optimization, and of course the basic objective – that of studying variability on scales much finer than those of Earth–based photometry – also presents many difficulties. Several technical routes may be available; the author is basing the discussion upon a particular concept sketched in this paper.

051.068 **Stellar activity and variability with UVSAT (Japanese project for UV astronomy).**
T. Watanabe, K. Kodaira, M. Iye.
Space research prospects in stellar activity and variability, p. 221 – 225 (1984). – See Abstr. 012.078.

The 600 mm aperture Cassegrain telescope (VUVT f/4) of UVSAT is aimed for stellar spectrophotometry and spectroscopic survey of faint objects in the vacuum ultraviolet region $\lambda\lambda$1200 – 3000 Å. Two instruments (low and high resolution spectrographs) attached to VUVT would be useful for the observation of stellar variability and activity. It is hoped that UVSAT will be launched by ISAS/Japan in the early 90's. Intensive studies and experiments for final designs are now underway.

051.069 **Space mission analysis: a review of constraints and a proposition for a logical approach.**
P. Brudieu.
Space research prospects in stellar activity and variability, p. 227 – 240 (1984). – See Abstr. 012.078.

This paper presents a short study performed at CNES about a space mission devoted to stellar activity and variability. The general mission concept is shortly described.

051.070 **Infrared heavens unveiled [by IRAS].**
Y. Unna.
Madà, Vol. 28, No. 2, p. 55 – 56 (1984). In Hebrew.

051.071 **What's new in the Soviet space program?**
A. R. Serban.
Madà, Vol. 28, No. 4, p. 162 (1984). In Hebrew.

051.072 **EXOSAT, European X–Ray Astronomy Satellite.**
EXOSAT Express, Nos. 6 – 8, 48 + 45 + 43 pp. (1984).

051.073 **Future studies of planetary rings by spaceprobes.**
E. C. Stone.
Planetary rings, p. 731 – 735 (1984). – See Abstr. 012.085 (IAU Colloq. No. 75).

In the second half of this decade, Voyager 2 will encounter Uranus and Neptune and Galileo will orbit Jupiter, returning new information on those planetary ring systems. The opportunities and capabilities for ring studies are briefly discussed.

051.074 **A priority of space exploration: observation of Saturn's rings from a Saturn orbiter.**
A. Brahic.
Planetary rings, p. 739 – 742 (1984). – See Abstr. 012.085 (IAU Colloq. No. 75).

051.075 **Rendez–vous with Saturn's rings.**
K. T. Nock.
Planetary rings, p. 743 – 759 (1984). – See Abstr. 012.085 (IAU Colloq. No. 75).

A mission to rendez–vous with the rings of Saturn is studied with regard to science rationale and instrumentation and engineering feasibility and design. Future detailed in situ exploration of the rings of Saturn will require spacecraft systems with enormous propulsive capability. NASA is currently studying the critical technologies for just such a system, called Nuclear Electric Propulsion (NEP). An important aspect of this mission is the ability of the low thrust propulsion system to continuously boost the spacecraft above the ring plane as it spirals in toward Saturn.

051.076 **EXOSAT observations.**
G. Branduardi–Raymont.
Clusters and groups of galaxies, p. 341 – 346 (1984). – See Abstr. 012.088.

A short status report of the EXOSAT mission is given: the in–flight performance of the four experiments onboard is described and a brief comment is made on the problems encountered in orbit and on their resolution. At the time of the meeting only CMA image data of one cluster, Virgo, were available and these are presented, as well as, for demonstration purposes, the only PSD image of an extended X–ray source existing at the time, that of the supernova remnant Cassiopeia A.

051.077 **Future studies of the local interstellar medium with Space Telescope and Columbus.**
B. D. Savage.
NASA Conf. Publ., NASA CP–2345, p. 333 – 343 (1984). – See Abstr. 012.089 (IAU Colloq. No. 81).

The spectrographs aboard Space Telescope and Columbus will provide important new information about the interstellar medium in the immediate vicinity of the Sun. The Space Telescope high resolution spectrograph (HRS) will produce resolutions, $\lambda/\Delta\lambda$, of about 18,000 and 70,000 with high sensitivity between 1200 and 3200 Å and greatly reduced sensitivity between 1060 and 1200 Å. The Columbus mission is in the planning stages. However, it is likely that the spacecraft will contain spectrographs capable of resolutions of $3 \times 10^4$ between 912 and 1200 Å and 500 between 100 and 900 Å. Astronomers using the ST HRS and Columbus spectrographs will be able to study a wide range of problems relating to cold, warm, and hot gas in the local ISM. Some of the most important observing projects are described.

051.078 **Future studies of planetary rings by space probes.**
E. C. Stone.
Planetary rings, p. 687 – 703 (1984). – See Abstr. 012.086.

Recent observations by the Pioneer and Voyager missions have been the basis for major advances in our knowledge and understanding of the Jovian and Saturnian rings. Future space probe observations offer further opportunities for studying the ring systems of giant planets.

**Explicando astronáutica. O homem e a conquista do espaço.**
See Abstr. 003.021.

**Planeten, Monde, Ringsysteme: Kamerasonden erforschen unser Sonnensystem.**
See Abstr. 003.036.

**Spacelab. Research in earth orbit.**
See Abstr. 003.049.

**IRAS mission operations experience at RAL (*Rutherford Appleton Laboratory*).**
See Abstr. 009.014.

**ESA and its programmes: present and future.**
See Abstr. 010.461.

**The ESA technological research programmes.**
See Abstr. 010.462.

**Ein gemeinsames amerikanisch–europäisches Programm zur Planetenerforschung.**
See Abstr. 011.003.

**Very–long–baseline interferometry takes to the sky.**
See Abstr. 011.019.

**Planetary science experiments aboard Space Station: workshop report.**
See Abstr. 011.022.

**The case for Mars. Proceedings of a conference held at Univerity of Colorado, Boulder, Colorado, USA, April 29 – May 2, 1981.**
See Abstr. 012.077.

**The Planetary Observer Program.**
See Abstr. 013.012.

**The future of space astronomy.**
See Abstr. 013.020.

**Probing the depths of a star: the study of solar oscillations from space. Report of the NASA Science Working Group on the study of solar oscillations from space.**
See Abstr. 013.081.

**Space plasma research at RAL.**
See Abstr. 035.010.

**Observations of local ISM emission with the Berkeley EUV/FUV Shuttle Telescope.**
See Abstr. 035.079.

**The Giotto spacecraft system and subsystem design.**
See Abstr. 035.081.

**On autonomous tracking of a minor celestial body.**
See Abstr. 036.059.

**The HIPPARCOS space astrometry mission.**
See Abstr. 041.042.

**Investigating the internal structure and evolution of massive stars by space techniques.**
See Abstr. 065.130.

**Space experiments. Report of workshop C2 (*10th International Conference on General Relativity and Gravitation, Padua, July 3 – 8, 1983*).**
See Abstr. 066.229.

**Questions for the geologic exploration of Venus.**
See Abstr. 093.006.

**Neutron stars, black holes and Exosat.**
See Abstr. 142.039.

# 052 Astrodynamics, Navigation of Space Vehicles

052.001 **Analysis of the orbit of the satellite Cosmos 482 in its last 15 days.**
D. G. King-Hele.
Proc. R. Soc. London, Ser. A, Vol. 393, No. 1805, p. 235 – 236 (1984). Abstr. in Phys. Abstr., Vol. 87, No. 1237, Entry 72849 (1984).

052.002 **Mixed secular perturbations of satellite orbits.**
V. A. Tamarov.
Sov. Astron., Vol. 28, No. 1, p. 104 – 107 (1984). English translation of 37.052.002.

052.003 **Analysis of the orbital inclination of Tansei 3 rocket 1977–12B.**
P. Moore.
Planet. Space Sci., Vol. 32, No. 9, p. 1101 – 1109 (1984).
The orbit of Tansei 3 rocket (1977–12B) has been determined at 47 epochs between 1 October 1977 and 19 March 1979 using over 1700 observations and the RAE orbit refinement program PROP6.

052.004 **Construction of the analytical theory of motion of artificial earth satellites in rectangular coordinates.**
L. A. Isakovich, Yu. A. Pupyshev, L. I. Fedorova.
Vopr. mekh. i protsessov upr., Leningrad, No. 6, p. 134 – 142 (1984). In Russian. Abstr. in Ref. Zh., 62. Issled. Kosm. Prostranstva, 8.62.177 (1984).

052.005 **Numerical model of motion of NAVSTAR–type artificial earth satellites and possibilities of its use.**
T. V. Bordovitsyna, L. E. Bykova, V. A. Tamarov, N. A. Sharkovskij.
Probl. mekh. i upr. dvizheniem. Tr. 6 Obedin. nauchn. chtenij po kosmonavt., posvyasyshch. pamyati vydayushch. sov. uchenykh–pionerov osvoeniya kosm. prostranstva Moskva, 11–15 yanv., 1982. Moskva, p. 180 – 188 (1983). In Russian. Abstr. in Ref. Zh., 62. Issled. Kosm. Prostranstva, 8.62.180 (1984).

052.006 **Using the variation method in differential improvement of orbits of artificial earth satellites.**
I. M. Tsyupak.
L'vov. politekh. inst. L'vov, 15 pp. (1984). In Russian. Abstr. in Ref. Zh., 62. Issled. Kosm. Prostranstva, 8.62.190 (1984).

052.007 **Determination of the orientation of artificial earth satellites using measurements of angular velocities.**
M. L. Pivovarov.
Inst. kosm. issled. AN SSSR. Prepr., No. 859, 11 pp. (1983). In Russian. Abstr. in Ref. Zh., 62. Issled. Kosm. Prostranstva, 8.62.196 (1984).

052.008 **On the optimization of interplanetary trajectories.**
H. Matsuo.
J. Soc. Instrum. Control Eng., Vol. 23, No. 1, p. 100 – 105 (1984). In Japanese. Abstr. in Phys. Abstr., Vol. 87, No. 1242, Entry 94548 (1984).

052.009 **A comparative analysis of perturbations in the motion of artificial earth satellites based on some models of the geopotential.**
N. V. Emel'yanov, V. N. Salyamov.
Tr. Gos. Astron. Inst. Shternberg, Tom 56, p. 16 – 25 (1984). In Russian.
On the basis of a method of calculation of perturbations in the elements of an AES intermediate orbit caused by the planet's aspherity a number of harmonics in the expansion of the potential of attraction is determined which have to be taken into account under a given accuracy of calculation of satellite coordinates. The calculations have been fulfilled for some variants of the given accuracy, the intermediate orbit of the AES and the model of the geopotential.

052.010 **Problems of improvement of the parameters of motion of artificial earth satellites from observations.**
N. V. Emel'yanov.
Tr. Gos. Astron. Inst. Shternberg, Tom 56, p. 26 – 37 (1984). In Russian.
The author gives a detailed formulation of the problem of improvement of elements of the artificial earth satellite intermediate orbit from observations.

052.011 **An expansion of the perturbation function caused by the earth's asphericity.**
N. V. Emel'yanov, L. P. Nasonova.
Astron. Zh., Tom 61, Vyp. 5, p. 1021 – 1028 (1984). In Russian. English translation in Sov. Astron., Vol. 28, No. 5.
An expansion of the perturbation function caused by the asphericity of the terrestrial gravitational field has been obtained, the zero– and the first–order members with respect to the earth's oblateness being taken into consideration.

052.012 **Erratum: "Reduction, Brouwer's Hamiltonian, and the critical inclination" [Celest. Mech., Vol. 31, No. 4, p. 401 – 429 (1983)].**
R. Cushman.
Celest. Mech., Vol. 33, No. 3, p. 297 (1984). See Abstr. 37.052.010.

052.013 **Über die Neigungs– und Exzentrizitätsfunktionen der Theorie der Satellitenbahnen.**
M. Soffel, R. Wirrer, H. Ruder, M. Schneider.
Veröff. Bayer. Komm. Int. Erdmessung Bayer. Akad. Wiss., Astron. –Geod. Arb., Heft Nr. 45, p. 195 – 208 (1984).

052.014 **Nearly circular equatorial orbits of artificial earth satellites taking into account the light pressure.**
E. N. Polyakhova, E. I. Timoshkova.
Vestn. Leningr. Univ., Mat. Mekh. Astron., No. 1, p. 100 – 106 (1984). In Russian. Abstr. in Ref. Zh., 62. Issled. Kosm. Prostranstva, 9.62.112 (1984).

052.015 **On the existence of a solution of a system of arbitrary differential equations of elliptical motion of space vehicles.**
Yu. I. Trofimtsev.
Differ. i integral. uravn. Gor'kij, p. 69 – 72 (1983). In Russian. Abstr. in Ref. Zh., 62. Issled. Kosm. Prostranstva, 9.62.121 (1984).

052.016 **On stable circular orbits of artificial earth satellites with daily period and different inclinations.**
M. L. Lidov, M. A. Vashkov'yak.
Probl. mekh. i upr. dvizheniem. Tr. 6 Obedin. nauchn. chtenij po kosmonavt., posvyashch. pamyati vydayushch. sov. uchenykh–pionerov osvoeniya kosm. prostranstva, Moskva, 11 – 15 yanv., 1982. Moskva, p. 5 – 21 (1983). In Russian. Abstr. in Ref. Zh., 51. Astron., 10.51.114 (1984).

052.017 **Selection of a very simple model of motion of artificial earth satellites with fixed accuracy in a given time interval.**
L. L. Sokolov, E. I. Timoshkova, V. B. Titov, K. V. Kholshevnikov.
Probl. mekh. i upr. dvizheniem. Tr. 6 Obedin. nauchn. chtenij po kosmonavt., posvyashch. pamyati vydayushch. sov. uchenykh–pionerov osvoeniya kosm. prostranstva, Moskva, 11 – 15 yanv., 1982. Moskva, p. 57 – 62 (1983). In Russian. Abstr. in Ref. Zh., 51. Astron., 10.51.115 (1984).

052.018 **Periodic motions of an axisymmetric satellite relative to the mass center under the action of the gravitational momentum.**
V. V. Sazonov, M. E. Sidoryuk.
Izv. Akad. Nauk Mekh. tverd. tela, No. 3, p. 6 – 16 (1984). In Russian. Abstr. in Ref. Zh., 62. Issled. Kosm. Prostranstva, 10.62.179 (1984).

052.019 **The problem of the translatory–rotational motion of satellites taking into account the influence of a third body.**
A. T. Tagaev.
Vestn. Akad. Nauk KazSSR, Alma–Ata, 22 pp. (1984). In Russian. Abstr. in Ref. Zh., 62. Issled. Kosm. Prostranstva, 10.62.180 (1984).

052.020 **A priori guaranteed estimate of the accuracy of determination of a space vehicle's orbit with the method of least squares.**
P. E. Ehl'yasberg.
Kosm. Issled., Tom 22, Vyp. 5, p. 643 – 650 (1984). In Russian. English translation in Cosm. Res.

052.021 **Trajectory control for the Agency's deep–space missions.**
M. Rosengren.
ESA Bull., No. 40, p. 30 – 37 (1984).
A software system has been developed at ESOC for trajectory control for deep–space missions. It will be first used for the Agency's Giotto and Ulysses (formerly ISPM) missions, but can be extended to serve future missions without changing the basic design concept.

052.022 **Galileo Jupiter approach orbit determination.**
J. K. Miller, F. T. Nicholson.
J. Astronaut. Sci., Vol. 32, No. 1, p. 63 – 80 (1984). Abstr. in Phys. Abstr., Vol. 87, No. 1245, Entry 110494 (1984).

052.023 **Optimum two–impulse transfers for preliminary interplanetary trajectory design.**
N. D. Hulkower, C. O. Lau, D. F. Bender.
J. Guid. Control Dyn., Vol. 7, No. 4, p. 458 – 461 (1984). Abstr. in Phys. Abstr., Vol. 87, No. 1246, Entry 115676 (1984).

052.024 **Globally optimal impulsive transfers via Green's theorem.**
G. A. Hazelrigg.
J. Guid. Control Dyn., Vol. 7, No. 4, p. 462 – 470 (1984). Abstr. in Phys. Abstr., Vol. 87, No. 1246, Entry 115677 (1984).

052.025 **An unorthodox method of determining the unperturbed orbits of unknown space objects from partial data of optical observations.**
N. I. Perov.
Sov. Astron., Vol. 28, No. 3, p. 334 – 336 (1984). English translation of 37.052.076.

052.026 **Numerical determination of trajectories of artificial earth satellites by the Adams variable order method. I.**
V. K. Taradij, M. L. Tsesis.
Astrometr. Astrofiz., Vyp. 53, p. 56 – 65 (1984). In Russian.
An algorithm and a computer programme are proposed for the Adams method to solve the second–order differential equations. The efficiency of the numerical integration is provided by introducing variable order and variable integration step as well as by using divided differences.

052.027 **Optimalization of the process of tesseral harmonics perturbations computation.**
E. Wnuk, I. Wytrzyszczak.
Artif. Satell., Vol. 19, No. 1, p. 5 – 16 (1984).
Two methods of optimalization of the computation process of tesseral harmonics perturbations are analysed in this paper. Computation examples illustrate the number of coefficients necessary in perturbations computation for chosen satellites.

052.028 **The unmodelled effects in the motion of artificial satellites.**
W. Popiński.
Artif. Satell., Vol. 19, No. 1, p. 17 – 26 (1984).
A method of tackling satellite motion variations which are caused by the effects not included in the integrated equations of motion is described in the paper. It is particularly important when station positions are determined from the observations of a satellite. Numerical results showing the influence of the unmodelled effects and remarks concerning the application of the method are presented as well.

052.029 **Application of the two–cycle integration method to the investigation of motion of a 12–hour satellite.**
K. P. Pleshkov.
Vestn. Leningr. Univ., Mat. Mekh. Astron., No. 13, p. 75 – 80 (1984). In Russian. Abstr. in Ref. Zh., 51. Astron., 11.51.83 (1984).

052.030 **Perturbed motion of a distant satellite of a spheroidal planet.**
A. M. Iskakova.
Kaz. univ., Alma–Ata, 11 pp. (1984). In Russian. Abstr. in Ref. Zh., 62. Issled. Kosm. Prostranstva, 11.62.175 (1984).

052.031 **Analytical investigations of visible satellite trajectories.**
Yu. Kh. Zhagar.
Nauchn. Inf., Vyp. 55, p. 37 – 50 (1982). In Russian.
It is shown that the coefficients of the equation of satellite visible trajectories form a symmetric tensor. Some investigations of the properties of this tensor are presented as well as descriptions of the coordinate systems based on satellite visible trajectories.

052.032 **Determination of the preliminary orbit of a geostationary satellite using four observations.**
Yu. V. Rusin.
Nauchn. Inf., Vyp. 55, p. 51 – 56 (1982). In Russian.
Method of orbit determination of stationary satellite by 4 observations is described and example showing the opportunity of the utilization of this method for GSS identification purposes is given.

052.033 **Identification of geostationary satellites by observations from one tracking station.**
M. V. Bratijchuk, A. G. Kirichenko.
Nauchn. Inf., Vyp. 55, p. 68 – 73 (1982). In Russian.
The method of geostationary satellites identification is considered. For this aim the elements of satellites are calculated using the methods of Gauss, Kiselev. The evolution of the orbits of two geostationary satellites for three years is given.

052.034 **Optimal configurations for dual–spin satellites subject to gravitational torques.**
T. Li, R. W. Longman.
Celest. Mech., Vol. 33, No. 4, p. 319 – 336 (1984).
Dual–spin or gyrostat satellites subject to gravitational torques can adopt an infinite number of possible equilibria obtained by adjusting the magnitude and direction of the rotor angular momentum within the satellite. This paper seeks to answer the question, which of these equilibria is best – and best is chosen here to mean most stable in the sense that the energy required to perturb the orientation by any prescribed amount is maximized.

052.035 **Precise orbit determination of the satellite "STARLETTE" with laser ranging data.**
M.–f. He.
Chin. Astron. Astrophys., Vol. 8, No. 3, p. 251 – 254 (1984). English translation of 37.052.097.

052.036 **Opportunities of ballistic missions to long-period comets.**
L. Kresák, E. M. Pittich.
Bull. Astron. Inst. Czech., Vol. 35, No. 6, p. 364 - 375 (1984).
The opportunities of ballistic missions to long-period comets are discussed. The possible transfer orbits of zero inclination are mapped as a function of the heliocentric distance of the comet's node, the minimum geocentric hyperbolic excess speed of launching, and the minimum flight time required.

052.037 **Approximation of functions of variation of the orbital elements of artificial earth satellites in the presence of discontinuity points.**
G. S. Kurbasova.
Astron. Tsirk., No. 1311, p. 3 - 5 (1984). In Russian.

052.038 **Approximation of the ephemerides of artificial earth satellites for control of a two-axial mounting of a laser rangefinder.**
G. S. Kurbasova.
Astron. Tsirk., No. 1311, p. 5 - 7 (1984). In Russian.

052.039 **Construction of the analytical theory of motion of artificial earth satellites in rectangular coordinates.**
L. A. Isakovich, Yu. A. Pupyshev, L. I. Fedorova.
Vopr. mekh. i protsessov upr. Leningrad, No. 6, p. 134 - 142 (1984). In Russian. Abstr. in Ref. Zh., 51. Astron., 12.51.113 (1984).

052.040 **On the use of the variation method in differential improvement of the orbits of artificial earth satellites.**
I. M. Tsyupak.
L'vov. politekh. inst. L'vov, 15 pp. (1984). In Russian. Abstr. in Ref. Zh., 51. Astron., 12.51.118 (1984).

052.041 **On the existence of a solution of a system of random differential equations of the elliptical motion of space vehicles.**
Yu. I. Trofimtsev.
Differ. i integral. uravneniya. Gor'kij, p. 69 - 72 (1983). In Russian. Abstr. in Ref. Zh., 51. Astron., 12.51.128 (1984).

052.042 **Lageos orbit and the albedo problem.**
D. P. Rubincam.
NASA Tech. Memo., NASA TM-86123, p. 2-17 - 2-18 (1984).
See Abstr. 003.039.
This report discusses modeling the perturbative force due to radiation pressure which acts on the orbit of Lageos.

052.043 **Lageos orbit and solar eclipses.**
D. P. Rubincam.
NASA Tech. Memo., NASA TM-86123, p. 2-19 - 2-20 (1984).
See Abstr. 003.039.
The diminuation of radiation pressure occurring when the Lageos satellite moves into the moon's shadow is assessed.

052.044 **The orbital variation of a $24^h$ synchronous satellite and its calculation method.**
L. Liu.
Acta Astron. Sin., Vol. 25, No. 3, p. 255 - 264 (1984). In Chinese.
The orbital variation of a $24^h$ synchronous satellite has been researched in the paper. First the characteristics of the synchronous orbit is analysed, and then a calculating method of the perturbation of the synchronous orbit is given. These results provide a theoretical basis for the design and calculation of the orbit of a communication satellite.

**The linear multistep methods with back-corrections.**
See Abstr. 021.013.

**Influence of the model of the geopotential on the accuracy of the determination of orbital elements.**
See Abstr. 081.051.

**On the accuracy of laser observations of the ISAGEX experiment.**
See Abstr. 081.052.

**Determination of the Love number $k_2$ from laser observations of the GEOS-A and GEOS-B satellites.**
See Abstr. 081.053.

**Comparative studies of atmospheric density models used for Earth satellite orbit estimation.**
See Abstr. 082.014.

**Determination of satellite gravity harmonics from Galileo radio tracking data.**
See Abstr. 099.106.

# 053 Artificial Satellites, Space Probes

053.001 **Periods of brightness changes of some artificial satellites.**
M. V. Bratijchuk, I. I. Isak, Ya. M. Motrunich, I. F. Neubauer (*I. F. Najbauer*).
Astron. Tsirk., No. 1282, p. 6 - 8 (1983). In Russian.

053.002 **Observation of specular reflection from a geostationary satellite.**
A. V. Bagrov, E. P. Pavlenko, V. V. Prokof'eva, M. A. Smirnov.
Astron. Tsirk., No. 1286, p. 3 - 4 (1983). In Russian.

053.003 **An advanced concept for the pointing of a spaceborne observatory: Exosat.**
A. Massart, A. Schütz, V. Wood.
ESA Bull., No. 39, p. 90 - 96 (1984).

053.004 **On the flight of the automatic interplanetary stations Venera 13 and Venera 14 to Venus.**
V. L. Barsukov.
Gagarin. nauchn. chteniya po kosmonavt. i aviats., 5 - 9 apr. 1982. Moskva, p. 37 - 48 (1984). In Russian. Abstr. in Ref. Zh., 62. Issled. Kosm. Prostranstva, 8.62.402 (1984).

053.005 **Satellite Digest-175.**
R. D. Christy.
Spaceflight, Vol. 26, Nos. 7 - 8, p. 334 (1984).
July/August 1984 listing of satellite and spacecraft launches.

053.006 **Satellite Digest-176.**
R. D. Christy.
Spaceflight, Vol. 26, Nos. 9 - 10, p. 383 - 384 (1984).
September/October 1984 listing of satellite and spacecraft launches.

053.007 **Satellite Digest–177.**
R. D. Christy.
Spaceflight, Vol. 26, No. 11, p. 428 – 429 (1984).
November 1984 listing of satellite and spacecraft launches.

053.008 **Visual observations, horizontal and equatorial (1950) coordinates of Bolgariya 1300.**
Rezul't. Nablyud. Iskusstv. Sputnikov Zemli, Vyp. 100 (240), 66 pp. (1982). In Russian.
Concerning Bolgariya 1300–1981–75–1. August – December 1981. January – June 1982.

053.009 **Visual observations, horizontal coordinates, equatorial coordinates (1950.0) of artificial earth satellites.**
Rezul't. Nablyud. Iskusstv. Sputnikov Zemli, Vyp. 101 (241), 58 pp. (1982). In Russian.
Concerning the following satellites: Poljot 1–1963–43–1. September – December 1981. January – June 1982. Intercosmos 19–1979–20–1. September – November 1981. Visual observations. Horizontal coordinates. Poljot 1–1963–43–1. April, June, August 1982. Intercosmos 19–1979–20–1. April – September 1982. Equatorial coordinates (1950.0).

053.010 **Experiment "Relikt".**
I. A. Strukov.
Zemlya Vselennaya, No. 4, p. 5 – 12 (1984). In Russian.

053.011 **Satellite Digest – 178.**
R. D. Christy.
Spaceflight, Vol. 26, No. 12, p. 475 (1984).
1984 December listing of satellite and spacecraft launches.

053.012 **Electron beam emission from the EXOS–B (JIKIKEN) satellite as a powerful diagnostic tool in the magnetosphere.**
N. Kawashima, K. Akai, Y. Murasato, S. Sasaki.
Astrophys. Space Sci., Vol. 106, No. 1, p. 117 – 123 (1984).
In an electron beam emission experiment on board the EXOS–B (JIKIKEN) satellite (200 V, 1 mA–maximum), several types of waves are strongly excited by the beam such as plasma frequency, upper hybrid fequency, electron cyclotron frequency, their harmonics and nonlinear coupling of these waves. Measurements of these waves give information on local plasma density and magnetic field strength and it is revealed that the electron beam emission from the spacecraft is a powerful diagnostic tool in the magnetosphere.

053.013 **"Astron" – window to the ultraviolet space.**
A. A. Boyarchuk.
Zemlya Vselennaya, No. 5, p. 36 – 43 (1984). In Russian.

053.014 **Fifteen years on the trajectories of "Interkosmos".**
M. A. Rimsha.
Zemlya Vselennaya, No. 6, p. 26 – 32 (1984). In Russian.

053.015 **Satellite locations – 1984.**
W. L. Morgan.
Proc. IEEE, Vol. 72, No. 11, p. 1434 – 1444 (1984).
In a brief 20 years the number of satellites in the geosynchronous and geostationary orbits have increased by enormous bounds. This paper lists those satellites in geographical order, starting with the longitude of Greenwich and proceeding eastward, and the second table starting at Greenwich and going westward.

053.016 **Ultraviolet observations aboard the Astron station.**
A. A. Boyarchuk.
Vestn. Akad. Nauk SSSR, No. 8, p. 16 – 20 (1984). In Russian. Abstr. in Ref. Zh., 62. Issled. Kosm. Prostranstva, 11.62.145 (1984).

053.017 **X–ray experiment aboard the Astron station.**
V. G. Kurt.
Vestn. Akad. Nauk SSSR, No. 8, p. 20 – 26 (1984). In Russian. Abstr. in Ref. Zh., 62. Issled. Kosm. Prostranstva, 11.62.146 (1984).

053.018 **Visual observations, horizontal coordinates of artificial earth satellites.**
Rezul't. Nablyud. Iskusstv. Sputnikov Zemli, Vyp. 103(243), 67 pp. (1983). In Russian.
Concerning the following satellites: Poljot 1–1963–43–1. August – September 1981. February – April, June 1982. Intercosmos 19–1979–20–1. August 1981. January – December 1982. Intercosmos 21–1981–11–1. August – October 1981. January – March 1982. Visual observations. Horizontal coordinates.

053.019 **Visual observations, equatorial coordinates (1950.0) of artificial earth satellites.**
Rezul't. Nablyud. Iskusstv. Sputnikov Zemli, Vyp. 104(244), 71 pp. (1983). In Russian.
Concerning the following satellites: Intercosmos 19–1979–20–1. October – December 1982. January – April 1983. Arkad 3–1981–94–1. January – April 1983. Visual observations. Equatorial coordinates (1950.0).

053.020 **Modernization of the Intercosmos laser system of the first generation at the Zvenigorod Station.**
D. T. Matveev, B. D. Chepurnov.
Nauchn. Inf., Vyp. 55, p. 3 – 6 (1982). In Russian.
The results of tests of the first generation ranging laser Intercosmos in 1980 – 81 are given. The accuracy of distance measured is increased by 1.5 – 2 times.

053.021 **Calibration of the Intercosmos laser ranging system.**
D. T. Matveev.
Nauchn. Inf., Vyp. 55, p. 7 – 12 (1982). In Russian.
The method of measurement of parameters of the reception system of the Intercosmos laser of the first generation by cosmic objects is described.

053.022 **Experimental measurements of the energy of signals reflected from satellites.**
A. G. Masevich, G. S. Kurbasova, V. A. Mazurov, L. S. Shtirberg.
Nauchn. Inf., Vyp. 55, p. 13 – 16 (1982). In Russian.
Data on measurement of returning signals from GEOS–C, GEOS–A, Starlette, Be–C satellites at the Intercosmos ranging laser are given.

053.023 **Prediction of satellite visibility.**
Yu. Kh. Zhagar.
Nauchn. Inf., Vyp. 55, p. 27 – 36 (1982). In Russian.
The D. W. Scott and K. Arnold satellite visibility prediction methods are discussed, as well as causes of their limited precision. Some serious modifications of these methods are suggested to improve the precision. Numerical examples are presented for IKB–1300, GEOS–A and GEOS–C satellites over the period of 1982.

053.024 **Brightness changes of some geostationary satellites.**
F. N. Masumi.
Nauchn. Inf., Vyp. 55, p. 64 – 67 (1982). In Russian.
The results of photometric determinations of geostationary satellites brightness are discussed.

053.025 **Calculation of ephemerides for geostationary satellite observations using the VAU camera.**
N. P. Erpylev.
Nauchn. Inf., Vyp. 55, p. 74 – 77 (1982). In Russian.

053.026 **The spacecraft "Astron" for ultraviolet and X-ray observations.**
A. A. Boyarchuk.
Rep. Ser., Dep. Phys. Sci., Univ. Turku, No. FTL-R68, p. 1 – 2 (1984). – See Abstr. 012.079.

053.027 **Energy distribution in the spectrum of artificial celestial bodies.**
A. V. Bagrov, M. A. Smirnov.
Astron. Tsirk., No. 1327, p. 7 – 8 (1984). In Russian.

053.028 **Multi-colour electrophotometric observations of a geostationary satellite.**
A. V. Didenko, F. K. Rspaev.
Astron. Tsirk., No. 1333, p. 6 – 8 (1984). In Russian.

**Optimization of the process of selection of faint optical signals in satellite laser ranging.**
See Abstr. 036.134.

# Theoretical Astrophysics

## 061 General Aspects (Nucleosynthesis, Elementary Particles, Neutrino Astronomy, etc.)

061.001 **Antineutrino astronomy and geophysics.**
L. M. Krauss, S. L. Glashow, D. N. Schramm.
Nature, Vol. 310, No. 5974, p. 191 – 198 (1984).

Radioactive decays inside the Earth produce antineutrinos that may be detectable at the surface. Their flux and spectrum contain important geophysical information. New detectors need to be developed, discriminating between sources of antineutrinos, including the cosmic–background. The latter can be related to the frequency of supernovas.

061.002 **On the detection of positrons via the optical lines of positronium.**
J. E. McClintock.
Astrophys. J., Vol. 282, No. 1, p. 291 – 295 (1984).

This paper proposes that in the search for astrophysical positron sources it may prove fruitful to make use of the optical lines emitted when a positron and an electron combine to form positronium. Positronium, an analog of the hydrogen atom, has a recombination–line spectrum with wavelengths which are twice the wavelengths of the corresponding hydrogen lines. On the average, for a wide range of physical conditions, the emission of each 511 keV annihilation–line photon will be preceded by the emission of about one positronium Lyman–α λ2430 photon. The detectability of this line in possible positron sources is examined.

061.003 **Renormalization–group analysis of grand unified theories in curved spacetime.**
L. Parker, D. J. Toms.
Phys. Rev. D, Vol. 29, No. 8, p. 1584 – 1608 (1984). Abstr. in Phys. Abstr., Vol. 87, No. 1237, Entry 68091 (1984).

061.004 **Intermediate vector bosons and neutrino cosmology.**
Z. Parsa.
Field theory in elementary particles, p. 139 – 149 (1983). Abstr. in Phys. Abstr., Vol. 87, No. 1237, Entry 68485 (1984). – See Abstr. 012.010.

061.005 **Is it safe to disturb the vacuum?**
P. Hut.
Nucl. Phys. A, Vol. A418, p. 301 – 311 (1984). Abstr. in Phys. Abstr., Vol. 87, No. 1237, Entry 68714 (1984). – See Abstr. 012.007.

061.006 **Neutrinos in gravitational collapse: analysis of the flux profile.**
S. V. Dhurandhar, C. V. Vishveshwara.
Pramāna, Vol. 22, No. 3 – 4, p. 159 – 171 (1984). Abstr. in Phys. Abstr., Vol. 87, No. 1237, Entry 72869 (1984).

061.007 **Resonant nuclear fusion processes and the gamma rays of SS433.**
R. N. Boyd, G. H. Newsom, G. W. Collins II, M. Wiescher.
Science, Vol. 225, No. 4661, p. 508 – 510 (1984).

Gamma–ray spectral lines have recently been reported coming from the celestial object SS433, which is known to emit high–speed jets in opposite directions. The proposed identification of the lines as coming from fusion reactions on nitrogen nuclei as part of the carbon–nitrogen–oxygen cycle operating in the jets has now received observational support. Predictions of strengths and widths of additional lines which, if seen, would provide valuable new information about conditions giving rise to the jets are presented.

061.008 **On the validity of the local approximation for the *s*–process in the Os region, and implications for the $^{187}$Re–$^{187}$Os cosmochronology.**
M. Arnould, K. Takahashi, K. Yokoi.
Astron. Astrophys., Vol. 137, No. 1, p. 51 – 57 (1984).

The *s*–process number abundance ratio of $^{186}$Os and $^{187}$Os, which is of key importance for the $^{187}$Re–$^{187}$Os cosmochronology, is calculated in two schematic *s*–process models. It is concluded that the relative *s*–process yields of $^{186}$Os and $^{187}$Os are still uncertain enough for preventing the $^{187}$Re–$^{187}$Os pair from being accepted as a reliable cosmochronometer. In particular, it is stressed that some of those predicted yields are incompatible with the results of a study of that pair in the framework of a chemical evolution model of the Galaxy which satisfies various observational constraints in the solar neighborhood, and which predicts an age of the Galaxy in the $11 \lesssim T_G \lesssim 15$ Gyr range (Yokoi et al., 1983).

061.009 **Subquark pregeometry and the possible origin of our Universe.**
H. Terazawa.
Proceedings of the Third Marcel Grossmann Meeting on General Relativity, Part A, p. 239 – 250 (1983). – See Abstr. 012.002.

A fundamental theory of particles and fields which is based on two principles (of relativity and of quantum) and two hypotheses (of the fundamental length and of compositeness) is proposed. It effectively reproduces not only gauge theories for the strong and electroweak forces of the fundamental fermions (leptons and quarks) but general relativity for gravity at low energies. A possible explanation of the big–bang of our Universe is also presented in this supergrand unified theory.

061.010 **Spontaneously broken supergravity and the scale of weak interactions.**
S. Ferrara.
Proceedings of the Third Marcel Grossmann Meeting on General Relativity, Part A, p. 251 – 267 (1983). – See Abstr. 012.002.

061.011 **Supergravity, the seven–sphere, and spontaneous symmetry breaking.**
M. J. Duff.
Proceedings of the Third Marcel Grossmann Meeting on General Relativity, Part A, p. 269 – 293 (1983). – See Abstr. 012.002.

061.012 **$N = 8$ supergravity.**
B. de Wit.
Proceedings of the Third Marcel Grossmann Meeting on General Relativity, Part A, p. 295 – 307 (1983). – See Abstr. 012.002.

061.013 **Gravitational perturbations of the hydrogen atom.**
L. Parker.
Proceedings of the Third Marcel Grossmann Meeting on General Relativity, Part A, p. 343 – 352 (1983). – See Abstr. 012.002.

The strength of a gravitational field is characterized by the Riemann curvature tensor. This paper presents a brief summary of work on the effects of curvature on the spectrum of the hydrogen atom.

061.014 **The anisotropic space–time: a possible arena of subnuclear phenomena.**
G. Preparata.
Proceedings of the Third Marcel Grossmann Meeting on General Relativity, Part A, p. 431 – 446 (1983). – See Abstr. 012.002.

061.015 **Quaternions and $N = 2$ supersymmetry.**
P. Townsend.
Proceedings of the Third Marcel Grossmann Meeting on General Relativity, Part B, p. 1087 – 1090 (1983). – See Abstr. 012.002.

061.016 **On the stability of Kaluza–Klein vacuum.**
I. An, S. Chen.
Proceedings of the Third Marcel Grossmann Meeting on General Relativity, Part B, p. 1091 – 1094 (1983). – See Abstr. 012.002.

061.017 **Discussion about the irreducible representations of superalgebras osp(m/2n).**
Q. Han, H. Sun.
Proceedings of the Third Marcel Grossmann Meeting on General Relativity, Part B, p. 1095 – 1101 (1983). – See Abstr. 012.002.

061.018 **Superspace and the super Killing equation, super algebra.**
L. Zhang.
Proceedings of the Third Marcel Grossmann Meeting on General Relativity, Part B, p. 1103 – 1106 (1983). – See Abstr. 012.002.

061.019 **Torsion constraints in supergravity.**
S. Bedding.
Proceedings of the Third Marcel Grossmann Meeting on General Relativity, Part B, p. 1107 – 1109 (1983). – See Abstr. 012.002.

061.020 **Is supergravity well–posed?**
J. Isenberg, D. Bao, P. B. Yasskin.
Proceedings of the Third Marcel Grossmann Meeting on General Relativity, Part B, p. 1111 – 1120 (1983). – See Abstr. 012.002.

061.021 **Can leptons and quarks achieve superunification at composite level?**
D. Zhang.
Proceedings of the Third Marcel Grossmann Meeting on General Relativity, Part B, p. 1121 – 1128 (1983). – See Abstr. 012.002.

061.022 **An approach to the supergravity theories.**
C. Shao.
Proceedings of the Third Marcel Grossmann Meeting on General Relativity, Part B, p. 1129 – 1133 (1983). – See Abstr. 012.002.

061.023 **Supergravity as a generalized Einstein–Cartan connection.**
B. Morel.
Proceedings of the Third Marcel Grossmann Meeting on General Relativity, Part B, p. 1147 – 1152 (1983). – See Abstr. 012.002.

061.024 **Quantization ambiguity and the Aharanov–Bohm effect.**
G. Kunstatter.
Proceedings of the Third Marcel Grossmann Meeting on General Relativity, Part B, p. 1247 – 1251 (1983). – See Abstr. 012.002.

A brief review is given of the role of quantization ambiguity in both quantum mechanics and quantum field theory. It is pointed out that quantization ambiguity is not relevant to discussions of physical experiments designed to test the Aharanov–Bohm effect.

061.025 **An almost trivial gauge theory in the limit of infinite gauge coupling constant.**
S. Kaptanoglu.
Proceedings of the Third Marcel Grossmann Meeting on General Relativity, Part B, p. 1253 – 1263 (1983). – See Abstr. 012.002.

A local SU(2) gauge theory with one multiplet of scalars in the adjoint representation is considered. In the limit of infinite gauge coupling constant Yang–Mills fields become auxiliary and the action possesses a larger invariance than the usual gauge invariance; hence, the system develops a richer structure of constraints. The constraint analysis is carried out.

061.026 **Gravitational lens effects of degenerate neutrino celestial objects.**
C. Xu, X. Wu, S. Xiang.
Proceedings of the Third Marcel Grossmann Meeting on General Relativity, Part B, p. 1325 – 1328 (1983). – See Abstr. 012.002.

In this paper a neutrino celestial object (NCO) is treated as a transparent gravitational lens. The trajectory of a light ray emitted from the source inside or outside an NCO is obtained by means of geodesic equations. The brightness amplification ratio $K$ inside an NCO is also obtained.

061.027 **Particles do not exist.**
P. C. W. Davies.
Quantum theory of gravity, p. 66 – 77 (1984). – See Abstr. 003.005.

The concept of a particle is purely an idealized model of some utility in flat space quantum field theory. Away from that limited context, however, the concept becomes much less useful and has been the source of much confusion. The study of DeWitt–style particle detectors has exposed the nebulousness of the particle concept and suggests that it should be abandoned completely.

061.028 **Supersymmetry, finite theories and quantum gravity.**
K. S. Stelle.
Quantum theory of gravity, p. 345 – 355 (1984). – See Abstr. 003.005.

061.029 **Properties of extended theories of rigid supersymmetry.**
P. C. West.
Quantum theory of gravity, p. 356 – 373 (1984). – See Abstr. 003.005.

061.030 **Cosmological topological supergravity.**
S. Deser.
Quantum theory of gravity, p. 374 – 381 (1984). – See Abstr. 003.005.

Recently, the locally supersymmetric extension of topologically massive gravity in three space–time dimensions was constructed. It is shown here that, due to an initial conformal–type invariance of the free topological part which is miraculously maintained by the supersymmetric coupling, it is possible to extend the model further to include a supercosmological sector. The latter consists as usual of a cosmological term together with an explicit, non–topological, mass term for the fermion.

061.031 **Future avenues in supergravity.**
J. G. Taylor.
Quantum theory of gravity, p. 382 – 387 (1984). – See Abstr. 003.005.

061.032 **Dynamical applications of the gauge–invariant effective action formalism.**
S. L. Adler.
Quantum theory of gravity, p. 388 – 402 (1984). – See Abstr. 003.005.

This article gives a self–contained introduction to two recent developments in the theory of the effective action functional, within the context of a simple example, consisting of an Abelian gauge field interacting with a light fermion species and with a set of infinitely massive test charges.

061.033 **Towards a quantum theory without "quantization".**
D. Deutsch.
Quantum theory of gravity, p. 421 – 430 (1984). – See Abstr. 003.005.

061.034 **On quantum gravity and the many-worlds interpretation of quantum mechanics.**
L. Smolin.
Quantum theory of gravity, p. 431 – 454 (1984). – See Abstr. 003.005.

061.035 **The positivity of the Jacobi operator on configuration space and phase space.**
C. R. Doering, C. DeWitt-Morette.
Quantum theory of gravity, p. 455 – 462 (1984). – See Abstr. 003.005.

061.036 **Unstable photino mass bound from cosmology.**
J. E. Kim, A. Masiero, D. V. Nanopoulos.
Phys. Lett. B, Vol. 139B, No. 5 – 6, p. 346 – 350 (1984). Abstr. in Phys. Abstr., Vol. 87, No. 1239, Entry 78505 (1984).

061.037 **Time development of Higgs field at finite temperature.**
A. Hosoya, M. Sakagami.
Phys. Rev. D, Vol. 29, No. 10, p. 2228 – 2239 (1984). Abstr. in Phys. Abstr., Vol. 87, No. 1239, Entry 78513 (1984).

061.038 **Short-range and long-range correlations in cosmic ray interactions.**
R. Hasan, M. Q. R. Khan, M. S. Ahmad, M. S. Swami.
Can. J. Phys., Vol. 62, No. 5, p. 442 – 445 (1984). Abstr. in Phys. Abstr., Vol. 87, No. 1239, Entry 78576 (1984).

061.039 **Measurement of cross sections for $^{22}$Na, $^{20-22}$Ne and $^{36-42}$Ar in the spallation of Mg, Al, Si, Ca and Fe. Production ratios of some cosmogenic nuclides in meteorites.**
F. Baros, S. Regnier.
J. Phys., Vol. 45, No. 5, p. 855 – 861 (1984). Abstr. in Phys. Abstr., Vol. 87, No. 1239, Entry 78631 (1984).

061.040 **On a lower limit to the rest mass of ions in the halo of our Galaxy.**
L. Z. Fang, J. G. Gao.
Phys. Lett. B, Vol. 139B, No. 5 – 6, p. 351 – 352 (1984). Abstr. in Phys. Abstr., Vol. 87, No. 1239, Entry 82409 (1984).

061.041 **The impact of nuclear input data on expected signals for the Cl and Ga solar neutrino experiments.**
W. Hampel.
MPA Rep., No. 90, p. 11 – 15 (1983). – See Abstr. 012.020.

061.042 **The Ne-Na cycle and the $^{12}$C + α reaction.**
H. P. Trautvetter, J. Görres, K. U. Kettner, C. Rolfs.
MPA Rep., No. 90, p. 24 – 33 (1983). – See Abstr. 012.020.

061.043 **The $^{187}$Re – $^{187}$Os chronology and chemical evolution of the Galaxy.**
M. Arnould.
MPA Rep., No. 90, p. 69 – 70 (1983). Abstract. – See Abstr. 012.020.

061.044 **The $^{176}$Lu – $^{176}$Hf *s*-process chronometer and planned projects with the 4π BGO detector.**
G. Schatz.
MPA Rep., No. 90, p. 71 – 72 (1983). – See Abstr. 012.020.

061.045 **Actinide *r*-process chronometers and β-delayed fission.**
F.-K. Thielemann.
MPA Rep., No. 90, p. 73 – 77 (1983). – See Abstr. 012.020.
When including new predictions for β-delayed fission and neutron emission rates in an *r*-process calculation, the production ratios for $^{232}$Th/$^{238}$U and $^{244}$Pu/$^{238}$U are drastically lowered. Employing the chronometric pairs $^{232}$Th/$^{238}$U, $^{235}$U/$^{238}$U, $^{244}$Pu/$^{238}$U and $^{129}$I/$^{127}$I within a simple exponential model for chemical evolution of the Galaxy results in an age ranging roughly from 14 to $20 \times 10^9$a. This is much larger than previous age determinations from actinide chronometers.

061.046 **β-decays of heavy nuclei in astrophysics.**
K. Takahashi.
MPA Rep., No. 90, p. 78 – 81 (1983). – See Abstr. 012.020.
The present status of the study of β-decays of heavy nuclei in astrophysical environments is reviewed by way of a few selected examples. In particular, the importance of bound-state $\beta^-$ decays in *s*-process conditions is stressed.

061.047 **The status of the σN-curve and the *s*-process neutron density.**
F. Käppeler.
MPA Rep., No. 90, p. 82 – 85 (1983). – See Abstr. 012.020.
Results of new measurements of neutron capture cross sections for isotopes of Nd and Sm are used to improve the empirical σN-curve for *s*-process elements.

061.048 **On the origin of $^{180}$Ta$^m$, the rarest naturally occurring isotope.**
K. Yokoi, K. Takahashi.
MPA Rep., No. 90, p. 89 – 91 (1983). – See Abstr. 012.020.

061.049 **β-decays of $^{180}$Hf$^m$ and $^{180}$Lu and the origin of $^{180}$Ta.**
E. B. Norman, S. E. Kellogg.
MPA Rep., No. 90, p. 92 – 93 (1983). – See Abstr. 012.020.

061.050 **Calculation of nuclear properties far from the line of β-stability.**
J. Krumlinde, P. Möller, J. R. Nix.
MPA Rep., No. 90, p. 94 – 98 (1983). – See Abstr. 012.020.

061.051 **Hartree-Fock predictions far from stability: comparison with phenomenological models and sensitivity to the interaction.**
F. Tondeur.
MPA Rep., No. 90, p. 99 – 100 (1983). – See Abstr. 012.020.

061.052 **Masses of neutron-rich nuclei, derived from experimental beta-decay Q-values, in comparison with mass formula predictions.**
F. Münnich.
MPA Rep., No. 90, p. 101 – 105 (1983). – See Abstr. 012.020.

061.053 **Beta-decay properties of exotic nuclei.**
K.-L. Kratz.
MPA Rep., No. 90, p. 106 – 110 (1983). – See Abstr. 012.020.

061.054 **Beta-delayed neutron emission as the inverse process to neutron-capture on radioactive isotopes.**
K.-L. Kratz, W. Ziegert, F.-K. Thielemann, W. Hillebrandt.
MPA Rep., No. 90, p. 111 – 112 (1983). – See Abstr. 012.020.

061.055 **Beta-decay far from stability and astrophysics.**
H. V. Klapdor.
MPA Rep., No. 90, p. 113 – 115 (1983). – See Abstr. 012.020.

061.056 **Large basis shell model calculations of β-decay strength functions for applications in nuclear astrophysics.**
G. J. Mathews, S. D. Bloom, R. F. Hausman Jr.
MPA Rep., No. 90, p. 116 – 127 (1983). – See Abstr. 012.020.

061.057 **The role of direct radiative neutron capture rates in the *r*-process, and radioactive ion beam research at Livermore.**
G. J. Mathews.
MPA Rep., No. 90, p. 128 – 136 (1983). – See Abstr. 012.020.

061.058 **Nuclear forces and the systematics of giant dipole resonances.**
M. Rayet.
MPA Rep., No. 90, p. 137 – 139 (1983). – See Abstr. 012.020.

061.059 **Strong interacting systems in astrophysics – a realistic theoretical access?**
E. R. Hilf, L. Polley.
MPA Rep., No. 90, p. 140 – 141 (1983). – See Abstr. 012.020.

061.060 **Neutrino scattering on electrons in a strong magnetic field and the influence of this process on energy loss of neutron stars.**
V. P. Tsvetkov.
Astrofizika, Tom 20, Vyp. 2, p. 359 – 364 (1984). In Russian. English translation in Astrophysics, Vol. 20, No. 2.
Neutrino scattering on electrons in a strong magnetic field is investigated taking into account the polarization of the electron, the chirality of the neutrino. Owing to this effect the change in the energy loss of a neutron star which is caused by neutrinos going out of the star is estimated.

061.061 **Deuterium and helium–3 production from massive neutrino decay.**
R. J. Scherrer.
Mon. Not. R. Astron. Soc., Vol. 210, No. 2, p. 359 – 371 (1984).
A massive ($m_H > 3.6$ MeV) antineutrino can produce free neutrons in the course of its decay and the neutrons can then bind into deuterium and $^3$He. The production of deuterium and $^3$He from the decay of a massive neutrino with mass $m_H \sim 10 - 100$ MeV and lifetime $\tau_H \sim 10^3 - 10^4$ s is investigated over a wide range of values for $\eta$ (the baryon–to–photon ratio). The photofission of the deuterium from the high–energy photons generated by the neutrino decay is also considered. It is found that cosmologically significant mass fractions ($\gtrsim 10^{-5}$) of both deuterium and $^3$He can be produced for $\eta \sim 3\times10^{-9} - 10^{-6}$. Thus, a universe with $\Omega_{B} \sim 1$ can be made consistent with the observed deuterium and $^3$He abundances.

061.062 **Electroweak unification and its experimental status.**
P. Darriulat.
Large–scale structure of the universe, cosmology and fundamental physics, p. 17 – 33 (1984). – See Abstr. 012.025.
A brief account of the most important features of electroweak unification and of supporting experimental facts is given, with particular emphasis on the recent discovery of the electroweak bosons.

061.063 **Unified field theories.**
P. Fayet.
Large–scale structure of the universe, cosmology and fundamental physics, p. 35 – 71 (1984). – See Abstr. 012.025.
This paper presents the main approaches to a unified theory of particles and interactions – namely grand unification and supersymmetry – and their consequences in particle physics and astrophysics.

061.064 **Naturally vanishing cosmological constant and dynamical generation of $m_W$, $m_{3/2}/m_p$ hierarchy.**
C. Kounnas.
Large–scale structure of the universe, cosmology and fundamental physics, p. 72 – 77 (1984). – See Abstr. 012.025.

061.065 **Nucleon stability and neutron–antineutron oscillations.**
E. Fiorini.
Large–scale structure of the universe, cosmology and fundamental physics, p. 81 – 110 (1984). – See Abstr. 012.025.
The various experiments aiming to the detection of nucleon decay and presently running are reviewed, and the results obtained compared and discussed. Future experiments presently being designed to improve the present sensitivity are also considered. The results are used to obtain a "static" limit on neutron–antineutron oscillations, and compared with those obtained in the single "direct" experiment on neutron–antineutron experiments performed so far, and with the expected sensitivities of the planned experiments.

061.066 **Magnetic monopoles.**
G. Giacomelli.
Large–scale structure of the universe, cosmology and fundamental physics, p. 111 – 126 (1984). – See Abstr. 012.025.
This paper reviews the properties of cosmic GUT magnetic monopoles, their history in the early universe and discusses cosmological and astrophysical limits on their present density in the universe. The various experimental methods for detecting a flux of monopoles in cosmic radiation are briefly surveyed.

061.067 **The search for massive neutrinos.**
R. L. Mössbauer.
Large–scale structure of the universe, cosmology and fundamental physics, p. 273 – 288 (1984). – See Abstr. 012.025.
Experimental efforts to determine neutrino masses are reviewed. The subjects discussed include high resolution beta spectroscopy, neutrinoless double beta decay and various attempts to detect neutrino oscillations. At present, there is no confirmed evidence for finite neutrino masses of for lepton mixing, but experiments provide rather severe constraints on possible masses and mixing parameters.

061.068 **The relevance of limits on the neutrino restmass deduced from double beta decay.**
T. Kirsten.
Large–scale structure of the universe, cosmology and fundamental physics, p. 289 – 292 (1984). – See Abstr. 012.025.
Double beta decay (DBD) has the potential to distinguish whether neutrinos are Majorana or Dirac particles. However, neutrinoless DBD (in violation of lepton number conservation) has not yet been observed. From the respective upper limits, limits on the neutrino restmass can be deduced if the neutrino is of Majorana type.

061.069 **Cosmological and experimental constraints on the tau neutrino.**
S. Sarkar, A. M. Cooper.
Large–scale structure of the universe, cosmology and fundamental physics, p. 319 – 323 (1984). – See Abstr. 012.025.
Cosmological arguments concerning the survival of primordially synthesized helium and deuterium, together with recent experimental results, imply that the tau neutrino is stable, and consequently lighter than $\sim$100 eV.

061.070 **Deuterium, helium, and the Big–Bang nucleosynthesis.**
A. Vidal–Madjar, C. Gry.
Astron. Astrophys., Vol. 138, No. 2, p. 285 – 289 (1984).
A review of the helium and deuterium observations in interstellar space is presented. From these observations, through the use of simple galactic evolution models, the corresponding primordial values are estimated. The results obtained seem to fit poorly the prediction of the standard Big–Bang nucleosynthesis model. This could possibly be due to the fact that the presently available data represent a peculiar selection of observations. If this is not the case, it appears that either the standard Big–Bang model has to be revised or the models of galactic evolution are too simple and should include other pregalactic or galactic events able to burn deuterium more quickly.

061.071 **Gravitational polarization and magnetization.**
V. I. Grigor'ev, E. V. Grigor'eva.
Vestn. Mosk. Univ., Ser. 3. Fiz. Astron., Tom 25, No. 3, p. 40 – 45 (1984). In Russian. Abstr. in Ref. Zh., 51. Astron., 8.51.120 (1984).

061.072 **Neutrino oscillations and the atmospheric neutrino fluxes.**
G. V. Dass.
Phys. Rev. D, Vol. 30, No. 1, p. 80 – 88 (1984). Abstr. in Phys. Abstr., Vol. 87, No. 1242, Entry 90775 (1984).

061.073 **Monopole search and neutrino astrophysics with liquid scintillation detectors.**
K. Brown, M. L. Cherry, S. Corbato, I. Davidson, D. Keida, K. Lande, C. K. Lee, R. I. Steinberg.
AIP Conf. Proc., No. 114, p. 136 – 153 (1984). Abstr. in Phys. Abstr., Vol. 87, No. 1242, Entry 91149 (1984). – See Abstr. 012.029.

061.074 **Detection of massive cosmic neutrinos by second–order in *G* coherent scattering.**
R. Opher.
Astrophys. J., Vol. 282, No. 2, p. 398 – 401 (1984).
Coherent scattering for detecting cosmic neutrinos has been suggested by Opher and by Lewis. If the massive cosmic neutrino flux equals the massive cosmic antineutrino flux, however, then only second–order in $G$ coherent scattering is possible. ($G$ is the weak coupling constant.) The author studies the characteristics of inhomogeneous matter, characterized by independent scattering spheres, in order to obtain a maximum value for the momentum transfer due to second–order in $G$ coherent scattering. A neutrino mass $m_\nu \approx 30$ eV is assumed. A superconducting detector of the type discussed by Opher (1982) which is less than 1 $m^3$ in volume is found to be sufficiently sensitive to detect the predicted flux of massive cosmic neutrinos.

061.075 **Supernovae, neutrino rest mass, and the middle–energy neutrino background in the universe.**
G. S. Bisnovatyi–Kogan (*G. S. Bisnovatyj–Kogan*), Z. F. Seidov.
Ann. N.Y. Acad. Sci., Vol. 422, p. 319 – 327 (1984). – See Abstr. 012.033.
The authors have calculated the density and spectrum of neutrinos with energy 3 – 30 MeV due to SN explosions and their influence on the solar neutrino detector. The density of these neutrinos in the universe at present is $2 - 10 \times 10^{-33}$g cm$^{-3}$, that is, larger than the density of the relict microwave radiation. Propagation of a nonzero rest–mass neutrino in the Friedmann universe is considered, and the temporal characteristics of the neutrino background are calculated.

061.076 **Program and results of the Homestake muon, neutrino, and proton decay experiment.**
E. J. Fenyves.
Ann. N.Y. Acad. Sci., Vol. 422, p. 347 – 348 (1984). Abstract. – See Abstr. 012.033.

061.077 **Radiation from test charges in Yang–Mills theories.**
T. A. Morgan.
Ann. N.Y. Acad. Sci., Vol. 422, p. 371 (1984). Abstract. – See Abstr. 012.033.

061.078 **Just how fast are tachyons?**
T. A. Weber.
Ann. N.Y. Acad. Sci., Vol. 422, p. 392 (1984). Abstract. – See Abstr. 012.033.

061.079 **On detecting stellar collapse with neutrinos.**
A. Burrows.
Astrophys. J., Vol. 283, No. 2, p. 848 – 852 (1984).
The author estimates the distances out to which various terrestrial neutrino detectors can see a stellar collapse. With these ranges, neutrino burst detection rates are obtained. The mean time between detections is expected to be rather long. A realistic estimate for this time is $\sim 50 - 100$ years. Whether this is indeed the case depends sensitively on the as yet unknown mean energy of the emitted neutrinos and the galactic collapse rate.

061.080 **On the nucleosynthesis of matter of the solar system in the light of new data on isotopic anomalies in meteorites.**
A. K. Lavrukhina.
The 27th International Geological Congress, Tom 5, Sect. 10 – 11, p. 332 – 333 (1984). In English and Russian. Abstr. in Ref. Zh., 51. Astron., 9.51.190 (1984). – See Abstr. 012.031.

061.081 **Searches for proton decay and superheavy magnetic monopoles.**
B. V. Sreekantan.
J. Astrophys. Astron., Vol. 5, No. 3, p. 251 – 271 (1984).

061.082 **Numerical simulation of the plasma and gravitational dynamics of a Galactic magnetic monopole halo.**
R. Farouki, S. L. Shapiro, I. Wasserman.
Astrophys. J., Vol. 284, No. 1, p. 282 – 298 (1984).
The authors present numerical and analytical calculations designed to assess the viability of a Galactic magnetic monopole halo. The dynamical interplay between magnetic currents and the Galactic magnetic field is treated self–consistently, taking full account of the temporal evolution of currents and fields. The authors find that in general self–consistent bounds on the Galactic magnetic monopole flux based on the survival of the Galactic magnetic field are far less restrictive than the Parker limit. The halo models are consistent with monopole masses $m_M \lesssim 7 \times 10^{19}$GeV, and monopole fluxes in the range $F_M \gtrsim 3 \times 10^{-13}$cm$^{-2}$s$^{-1}$.

061.083 **The background field method.**
L. F. Abbott.
Relativity, cosmology, topological mass and supergravity, p. 2 – 10 (1983). – See Abstr. 012.040.
The background field formalism now exists for calculations in gauge theories to arbitrary numbers of loops. The method can be used for general discussions about the structure and particularly about the divergences of gauge theories, and for specific calculations.

061.084 **An introduction to spontaneously broken supergravity.**
S. Ferrara.
Relativity, cosmology, topological mass and supergravity, p. 11 – 56 (1983). – See Abstr. 012.040.
This paper reviews the construction of locally supersymmetric Lagrangians for particle interactions, discusses the super–Higgs effect, and derives their low–energy limit.

061.085 **Superunification from eleven dimensions.**
M. J. Duff, B. E. W. Nilsson, C. N. Pope.
Relativity, cosmology, topological mass and supergravity, p. 57 – 99 (1983). – See Abstr. 012.040.
The authors show how $N = 8$ supersymmetry can break spontaneously to $N = 1$ at the Planck scale via a Kaluza–Klein compactification of $d = 11$ supergravity on the squashed seven–sphere. They ask whether the resulting $N = 1$ theory can provide a realistic $SU(3) \times SU(2) \times U(1)$ unification and speculate that it might if some of the gauge bosons and fermions are composite. The special properties of the seven–sphere suggest a possible solution of the cosmological constant problem involving fermion condensates.

061.086 **Global solutions of hyperbolic equations of gauge theories.**
Y. Choquet–Bruhat.
Relativity, cosmology, topological mass and supergravity, p. 108 – 134 (1983). – See Abstr. 012.040.
The author reviews different methods which have been used, and results which have been obtained, for the classical equations of gauge theories in hyperbolic signature.

061.087 **Vacua and solitons in extended supergravity.**
G. W. Gibbons.
Relativity, cosmology, topological mass and supergravity, p. 163 – 176 (1983). – See Abstr. 012.040.

061.088 **Topologically massive $N = 1$ supersymmetric Yang–Mills in superspace.**
C. Aragone.
Relativity, cosmology, topological mass and supergravity, p. 178 – 195 (1983). – See Abstr. 012.040.
The supersymmetric formulation of topological massive 3–D Yang–Mills theory is presented in the superspace formalism, emphasizing the properties of the relevant bosonic algebra. The

superfield strength and the corresponding Bianchi's identity is given.

061.089 **The inverse problem of the calculus of variations.**
S. Hojman.
Relativity, cosmology, topological mass and supergravity, p. 196 – 204 (1983). – See Abstr. 012.040.
A review of the inverse problem of the calculus of variations is presented. Conservation–law theorems and ambiguities in quantum theory arising from the multiplicity of Lagrangians for systems of differential equations are discussed. A direct relationship between Lagrangians and their equations of motion is established.

061.090 **Twistor bundles as a formalism for unified gauge theories of gravitation and other Yang–Mills fields.**
M. Rosenbaum.
Relativity, cosmology, topological mass and supergravity, p. 205 – 218 (1983). – See Abstr. 012.040.

061.091 **Geometric quantization of supersymmetry.**
V. Aldaya, J. A. de Azcárraga.
Relativity, cosmology, topological mass and supergravity, p. 234 – 241 (1983). – See Abstr. 012.040.
The authors briefly describe the motivation and some of the results of a programme of geometric quantization entirely based on the consideration of a group manifold, and analyse its application to the case of supersymmetry groups.

061.092 **Classical localized solutions in a two–dimensional supersymmetric model.**
R. Montemayor, L. Urrutia.
Relativity, cosmology, topological mass and supergravity, p. 266 – 271 (1983). – See Abstr. 012.040.
The authors present a classical finite–energy exact solution of a "spin" 0–1/2 two–dimensional supersymmetric model. The solution describes localized bosonic and fermionic excitations and cannot be rotated to the purely bosonic sector by a supersymmetry transformation.

061.093 **A symmetric stress energy tensor for the gravitino field.**
O. Obregón, M. P. Ryan Jr.
Relativity, cosmology, topological mass and supergravity, p. 278 – 284 (1983). – See Abstr. 012.040.
A variational derivation of a symmetric energy–momentum tensor for the spin–3/2 field of supergravity is given by means of an Einstein–Cartan formulation of supergravity. That is, a formulation in which the basic variables are the metric, the torsion and the gravitino (spin–3/2) field.

061.094 **Vector fields quantization in curved space time.**
O. Zandron.
Relativity, cosmology, topological mass and supergravity, p. 291 – 296 (1983). – See Abstr. 012.040.
This paper presents a quantum field formulation for massive and massless vector fields in curved space–time. This is carried out by using the general theory of Green's functions.

061.095 **Neutrinos investigate the earth.**
V. A. Tsarev.
Zemlya Vselennaya, No. 6, p. 46 – 50 (1984). In Russian.

061.096 **The production of very high energy photons and neutrinos from cosmic proton sources.**
V. J. Stenger.
Astrophys. J., Vol. 284, No. 2, p. 810 – 816 (1984).
The emission of neutrinos and photons with energies of the order of $10^{12}$eV from the matter surrounding a source of very high energy cosmic ray protons is computed. The optimum conditions for neutrino and photon emission are established in terms of the matter density and integrated path length seen by the protons as they pass through this matter. The results imply that compact objects are the best candidate sources. It is shown how observations of both γ–rays and neutrinos can be used to otain information about primary proton flux at the source and the matter surrounding it.

061.097 **Sonic search for monopoles, gravitational waves and newtorites.**
C. Bernard, A. De Rujula, B. Lautrup.
Nucl. Phys. B, Part. Phys., Vol. B242, No. 1, p. 93 – 144 (1984). Abstr. in Phys. Abstr., Vol. 87, No. 1243, Entry 95286 (1984).

061.098 **Nuclearites – a novel form of cosmic radiation.**
A. De Rújula, S. L. Glashow.
Nature, Vol. 312, No. 5996, p. 734 – 737 (1984).
E. Witten (personal communication) has raised the intriguing possibility that nuclear matter consisting of aggregates of up, down and strange quarks in roughly equal proportions may be less massive than ordinary nuclear matter of the same quark number consisting of protons and neutrons (triplets of non–strange quarks). These nuggets of strange quark matter may be stable for almost any baryon number ($A$), including values intermediate between those of ordinary nuclei ($A \lesssim 263$) and neutron stars ($A \sim 10^{57}$). The authors use the term "nuclearite" to describe such strange quark nuggets in collision with Earth and suggest experiments to detect these encounters.

061.099 **Does wave–particle duality apply to galaxies?**
M. Dersarkissian.
Lett. Nuovo Cimento, Vol. 40, Ser. 2, No. 13, p. 390 – 394 (1984). Abstr. in Phys. Abstr., Vol. 87, No. 1244, Entry 101697 (1984).

061.100 **Measurement of the β–decay branch in $^{180}$Lu to $^{180m}$Hf for investigation of the *r*–process nucleosynthesis of $^{180m}$Ta.**
W. Eschner, W.–D. Schmidt–Ott, K.–L. Gippert, E. Runte, H. Beer, G. Walter, R. Kirchner, O. Klepper, E. Roeckl, D. Schardt.
Z. Phys., A, Vol. 317, No. 3, p. 281 – 286 (1984). Abstr. in Phys. Abstr., Vol. 87, No. 1244, Entry 102225 (1984).

061.101 **Destruction of $^{26}$Al via the $^{26}$Al(*n*,*p*)$^{26}$Mg–reaction.**
H. P. Trautvetter, F. Käppeler.
Z. Phys., A, Vol. 318, No. 1, p. 121 – 122 (1984). Abstr. in Phys. Abstr., Vol. 87, No. 1244, Entry 102274 (1984).

061.102 **Neutron capture cross sections and solar abundances of $^{160,161}$Dy, $^{170,171}$Yb, $^{175,176}$Lu, and $^{176,177}$Hf for the *s*–process analysis of the radionuclide.**
H. Beer, G. Walter, R. L. Macklin, P. J. Patchett.
Phys. Rev. C, Vol. 30, No. 2, p. 464 – 478 (1984). Abstr. in Phys. Abstr., Vol. 87, No. 1245, Entry 110590 (1984).

061.103 **Magnetic monopoles and the Kaluza–Klein theory.**
M. J. Perry.
AIP Conf. Proc., No. 116, p. 121 – 125 (1984). Abstr. in Phys. Abstr., Vol. 87, No. 1246, Entry 110959 (1984). – See Abstr. 012.053.

061.104 **Light–cone superfields for extended supergravity in ten dimensions and type II superstrings.**
J. H. Schwarz.
AIP Conf. Proc., No. 116, p. 135 – 146 (1984). Abstr. in Phys. Abstr., Vol. 87, No. 1246, Entry 110980 (1984). – See Abstr. 012.053.

061.105 **Conflicts between B–violating phenomena in supergravity models.**
J. E. Kim, A. Masiero, D. V. Nanopoulos.
Phys. Lett. B, Vol. 145B, No. 3 – 4, p. 187 – 191 (1984). Abstr. in Phys. Abstr., Vol. 87, No. 1246, Entry 111002 (1984).

061.106 **A simple linear description of gauge interactions.**
A. Zee.
Phys. Lett. B, Vol. 143B, No. 4 – 6, p. 368 – 372 (1984). Abstr. in Phys. Abstr., Vol. 87, No. 1246, Entry 111284 (1984).

061.107 **Effective Lagrangian for $\lambda\phi^4$ theory in curved spacetime with varying background fields: quasilocal approximation.**
B. L. Hu, D. J. O'Connor.
Phys. Rev. D, Vol. 30, No. 4, p. 743 – 755 (1984). Abstr. in Phys. Abstr., Vol. 87, No. 1246, Entry 111302 (1984).

061.108 **Radiative breaking of cosmologically acceptable grand unified theories.**
B. Gato, J. Leon, M. Quiros.
Phys. Lett. B, Vol. 144B, No. 3 – 4, p. 207 – 211 (1984). Abstr. in Phys. Abstr., Vol. 87, No. 1246, Entry 111354 (1984).

061.109 **Coloured scalars and cosmology.**
V. A. Kuz'min, M. E. Shaposhnikov.
Nuovo Cimento A, Vol. 82A, Ser. 11, No. 1, p. 35 – 50 (1984). Abstr. in Phys. Abstr., Vol. 87, No. 1246, Entry 111371 (1984).

061.110 **Cosmological equations in Kaluza–Klein $d$ = 11 supergravity theory.**
R. G. Moorhouse, J. Nixon.
Phys. Lett. B, Vol. 145B, No. 1 – 2, p. 39 – 44 (1984). Abstr. in Phys. Abstr., Vol. 88, No. 1247, Entry 158 (1985).

061.111 **More dimensions – less entropy.**
E. W. Kolb, D. Lindley, D. Seckel.
Phys. Rev. D, Vol. 30, No. 6, p. 1205 – 1213 (1984). Abstr. in Phys. Abstr., Vol. 88, No. 1247, Entry 161 (1985).

061.112 **Light gauginos: masses and instabilities.**
R. Barbieri, L. Maiani.
Nucl. Phys. B, Part. Phys., Vol. B243, No. 3, p. 429 – 448 (1984). Abstr. in Phys. Abstr., Vol. 88, No. 1247, Entry 175 (1985).

061.113 **Cosmological solutions of $N$–1 supergravity in 11 dimensions.**
E. Alvarez.
Phys. Rev. D, Vol. 30, No. 6, p. 1394 – 1396 (1984). Abstr. in Phys. Abstr., Vol. 88, No. 1247, Entry 182 (1985).

061.114 **Helium plasma: parameter domain for strong screening effect.**
A. E. M. Khairozzaman.
Proc. Natl. Acad. Sci. India, Sect. A, Vol. 53, Part 1, p. 23 – 34 (1983). Abstr. in Phys. Abstr., Vol. 88, No. 1247, Entry 4600 (1985).

061.115 **Gravitational instability of the primordial heavy–neutrino background.**
A. N. Shvarts.
Sov. Astron., Vol. 28, No. 2, p. 132 – 136 (1984). English translation of 37.061.025.

061.116 **Galactic neutrino halos.**
W. Y. Chau, K. Lake, J. Stone.
J. R. Astron. Soc. Can., Vol. 78, No. 5, p. 212 (1984). Abstract. – See Abstr. 010.401.

061.117 **Unconventional $^{12}$C production.**
R. Mitalas.
J. R. Astron. Soc. Can., Vol. 78, No. 5, p. 213 (1984). Abstract. – See Abstr. 010.401.

061.118 **Supergravity and the Kaluza–Klein program. Report of workshop D2 (*10th International Conference on General Relativity and Gravitation, Padua, July 3 – 8, 1983*).**
P. van Nieuwenhuizen.
General relativity and gravitation, p. 471 – 489 (1984). – See Abstr. 012.060.

061.119 **Can cosmic neutrinos be detected with superconductors?**
J. D. Lewin, P. F. Smith.
Astrophys. Lett., Vol. 24, No. 2, p. 59 – 67 (1984).
The development of techniques for the experimental detection of cosmic neutrinos would be of great importance to both astrophysics and particle physics. In this paper the authors discuss ideas based on the coherent response of conducting materials to long–wavelength cosmic or galactic neutrinos, via the weak neutral current interaction.

061.120 **Relativistic electron transport in a quantizing magnetic field.**
L. Hernquist.
Astrophys. J., Suppl. Ser., Vol. 56, No. 3, p. 325 – 367 (1984).
Electron transport in a strong magnetic field is considered. Relativistic effects are included by calculating scattering rates from exact solutions to the Dirac equation, assuming elastic scattering. Expressions for all of the independent components of the transport tensors (electrical conductivity, thermoelectric coefficient, and thermal conductivity) are derived for arbitrary degree of degeneracy and scattering mechanism. Numerical results are presented for the scattering potentials of most importance to physical processes in neutron star crusts ($e$–ion and $e$–phonon collisions).

061.121 **The line profile of a relativistic Keplerian ring in a massive neutrino halo.**
T.–r. Deng, S. p. Xiang.
Chin. Astron. Astrophys., Vol. 8, No. 4, p. 290 – 293 (1984). English translation of Acta Astrophys. Sin., Vol. 4, No. 3, p. 184 – 190 (1984).
If a massive neutrino halo transparent to light surrounds a compact body (galactic nucleus or black hole), then the strong gravitational field inside the halo will radically alter the spectral line profiles of a light–emitting Keplerian ring. This paper gives precise solutions of the spectral profile in a core–halo structure of matter and neutrino, and the result is compared with the no–halo case. The distribution of the neutrino mass is also given under typical circumstances.

061.122 **Yang–Mills particles in a spherically symmetric monopole field.**
A. W. Wipf.
MPA Rep., No. 166, 7 pp. (1984). Submitted to J. Phys. A.

061.123 **Electrical and thermal conductivities of dense matter in the crystalline lattice phase.**
N. Itoh, Y. Kohyama, N. Matsumoto, M. Seki.
Astrophys. J., Vol. 285, No. 2, p. 758 – 765 (1984).
The electrical and thermal conductivities of dense matter in the crystalline lattice–phase are calculated by accurately taking into account the Debye–Waller factor. The inclusion of the Debye–Waller factor results in a reduction of the resistivities (enhancement of the conductivities) typically by a factor of 2 – 4 near the melting temperature. This change will have important consequences that bear on the cooling of neutron stars.

061.124 **Enhancement of thermonuclear reaction rate due to screening by relativistic degenerate electrons: long–range correlation effect.**
S. Ichimaru, K. Utsumi.
Astrophys. J., Vol. 286, No. 1, p. 363 – 365 (1984).
Revised calculations for the screening potential in dense plasmas from the polarization of relativistic degenerate electrons are carried out taking account of the long–range as well as the short–range ion–ion correlations. The computed results of the enhancement factor for the thermonuclear reaction rate are parameterized in analytic formulae applicable over the domain $\Gamma \leqslant 200$, $r_s \leqslant 0.1$, and $1 \leqslant Z \leqslant 26$, where $\Gamma$, $r_s$, and $Z$ refer to the Coulomb coupling constant of the ions, the dimensionless density parameter of the electrons, and the ionic charge number, respectively.

061.125 **Some astrophysical consequences of the dynamical interpretation of gravitation.**
Yu. V. Baryshev, V. V. Sokolov.
Astrofizika, Tom 21, Vyp. 2, p. 361 – 366 (1984). In Russian. English translation in Astrophysics, Vol. 21, No. 2.
It has been shown that the dynamical (relativistic field theory) interpretation of gravitation predicts the existence of scalar gravi-

tational radiation. The energy loss by this radiation is of the order of $Mc^2$ at the stage of relativistic collapse. The collapse will be finished by a powerful impulse (or several impulses) of scalar radiation instead of black hole formation. This gives a new possibility to explain SN explosions and high peculiar velocities of some O–B stars. Powerful scalar radiation is also expected from active galactic nuclei. Superluminal expansion observed in some compact extragalactic radio sources may be caused by scalar radiation.

061.126 **Angular momentum versus mass, re–examined.**
V. Trimble.
Comments Astrophys., Vol. 10, No. 4, p. 127 – 135 (1984).

A plot of angular momentum vs. mass for a wide variety of astronomical objects and systems is well fit by $J \propto M^b$ and $b = 1.9$. Some authors have seen in this the working of a new physical principle that might be used to help understand the origin and evolution of structure in the universe. It is concluded here that the relationship can be fully understood using Newtonian mechanics and other familiar physics.

061.127 **Gauging the Lorentz group and long range interaction between spins.**
T. Pradhan.
Gravitation and relativistic astrophysics, p. 17 – 22 (1984). – See Abstr. 012.083.

There exists and elementary universal interaction between spins of particles which follow from a gauge principle. This force is of long range and weaker than the weak interaction. It can provide stability to the otherwise unstable classical electron and can remove all divergences in quantum electrodynamics.

061.128 **Introduction to supergravity.**
P. Majumdar.
Gravitation and relativistic astrophysics, p. 23 – 37 (1984). – See Abstr. 012.083.

061.129 **Neutron–capture nucleosynthesis of neodymium isotopes and the *s*–process from $A = 130$ to 150.**
G. J. Mathews, F. Käppeler.
Astrophys. J., Vol. 286, No. 2, p. 810 – 821 (1984).

New measurements of neutron capture cross sections for $^{142,143,144}$Nd are reported. These are combined with other recent measurements and applied to a detailed study of the *s*–process and *r*–process systematics for $A = 130 - 150$ nuclei. The influence of these results on the interpretation of isotopic anomalies observed in acid–insoluble residues and inclusions from the Allende meteorite is also examined.

061.130 **The $^{13}N(p,\gamma)^{14}O$ thermonuclear reaction rate and the hot CNO cycle.**
G. J. Mathews, F. S. Dietrich.
Astrophys. J., Vol. 287, No. 2, p. 969 – 976 (1984).

New estimates for the $^{13}N(p,\gamma)^{14}O$ reaction rate are presented. A direct–semidirect radiative capture model with a hydrodynamic core–polarization correction for quenching of the $E1$ transitions is utilized, together with spectroscopic information from measurements and shell–model calculations, to study the systematics of nucleon capture reactions on $A = 12$ and 13 nuclei. The new reaction rate is compared with other estimates and the implications of this rate on the conditions for the hot CNO cycle are discussed.

**Die Physik der Zeitrichtung.**
See Abstr. 003.016.

**Magnetic monopoles.**
See Abstr. 003.069.

**Symposium in Pamporovo, Bulgaria, 1983, July 18 – 22.**
See Abstr. 011.008.

**Gamma–ray line emission, nuclear destruction and neutron production in hot astrophysical plasmas. The deuterium boiler as a gamma–ray source.**
See Abstr. 062.044.

**Nucleosynthesis during core He burning in massive stars.**
See Abstr. 065.023.

**Presupernova yields and their dependence on the $^{12}C(\alpha, \gamma)\,^{16}O$ reaction.**
See Abstr. 065.024.

**On the production of technetium in thermally pulsing red giant stars.**
See Abstr. 065.042.

**La produzione degli elementi chimici nelle stelle. I – L'evoluzione stellare e la nucleosintesi primaria.**
See Abstr. 065.072.

**Very massive stars: evolution with mass loss. II. Nucleosynthesis, mass loss, and related problems.**
See Abstr. 065.128.

**La produzione degli elementi chimici nelle stelle. II – La nucleosintesi secondaria e la formazione degli elementi più pesanti.**
See Abstr. 065.134.

**Do monopoles keep white dwarfs hot?**
See Abstr. 065.146.

**Accreting white dwarf models for Type I supernovae. III. Carbon deflagration supernovae.**
See Abstr. 065.156.

**The quantum effect in general relativity.**
See Abstr. 066.041.

**A possible unified field theory of gravitation with other interactions.**
See Abstr. 066.153.

**Stability properties of gravitational theories and three dimensional gravities.**
See Abstr. 066.192.

**Positive energy and Kaluza–Klein theory.**
See Abstr. 066.225.

**Quantum gravity and quantum field theory in a curved space. Report of workshop D1 (*10th International Conference on General Relativity and Gravitation, Padua, July 3 – 8, 1983*).**
See Abstr. 066.235.

**Neutrino halos around the sun and precession of perihelia of planets.**
See Abstr. 066.245.

**Gauge approach to classical gravity.**
See Abstr. 066.279.

**The inner horizon in the Kerr–Newman metric.**
See Abstr. 067.090.

**Neutrinos in gravitational collapse.**
See Abstr. 067.138.

**Limits from the soft X–ray background on the temperature of old neutron stars and on the flux of superheavy magnetic monopoles.**
See Abstr. 067.145.

**Solar neutrinos and other problems and their relation to energy production in the Sun.**
See Abstr. 080.082.

**Barium stars and normal red giants: evidence for a modified *s*-process.**
See Abstr. 114.026.

**An upper limit to the abundance of $^9$Be in the Population II star HD 76932 from a high resolution spectrum with IUE.**
See Abstr. 114.070.

**Loss of angular momentum in binary system due to collisionless particles as monopoles, gravitinos, or mirror particles: does it exceed the gravitational radiation emission in binary system PSR 1913+16?**
See Abstr. 117.102.

**Nucleosynthesis in type I and type II supernovae.**
See Abstr. 125.065.

**Star formation and extinct radioactivities.**
See Abstr. 131.151.

**Search for the radiative decay of a cosmic relic particle with I.U.E.**
See Abstr. 142.020.

**Stellar nucleosynthesis and the cosmic-ray source composition up to Zn.**
See Abstr. 144.057.

**Isotopic anomalies in cosmic rays and winds from Wolf-Rayet stars: a new model.**
See Abstr. 144.058.

**Clues to the origin of cosmic rays derived from their source composition and energy spectra.**
See Abstr. 144.059.

**On the contribution of stellar winds to the composition of cosmic rays: exotic Ne and Fe from Wolf-Rayet stars.**
See Abstr. 144.060.

**Nuclear fragmentation cross sections and cosmic-ray source abundances.**
See Abstr. 144.061.

**Abundances of ultraheavy nuclei and the acceleration process.**
See Abstr. 144.064.

**Nucleosynthesis in Wolf-Rayet stars and galactic cosmic-ray isotopic composition.**
See Abstr. 144.065.

**Rapporteur paper on nucleosynthesis and cosmic rays.**
See Abstr. 144.073.

**Firm bounds on the neutrino mass from the distribution of dark matter in galaxies.**
See Abstr. 161.009.

**Quark-hadron and chiral transitions and their relation to the early Universe.**
See Abstr. 161.013.

**The role of massive neutrinos in the evolution of gravitational cosmological perturbations.**
See Abstr. 161.025.

**Particle physics and cosmology: the inner space/outer space connection.**
See Abstr. 161.074.

**Neutrino cosmology.**
See Abstr. 161.086.

**Supersymmetric cosmology and running masses and coupling constants.**
See Abstr. 161.089.

**Nonlinear photons in the universe. II. The anisotropic case.**
See Abstr. 161.103.

**Does the Newton's world model revive?**
See Abstr. 161.104.

**Cosmology, galactic astronomy and elementary particle physics.**
See Abstr. 161.114.

**Stellar He production in a cold Universe?**
See Abstr. 161.122.

**Elementary particles and the large-scale structure of the universe.**
See Abstr. 161.139.

**Early nucleosynthesis.**
See Abstr. 161.141.

**Helium production by pregalactic stars.**
See Abstr. 161.142.

**The quark-hadron transition in cosmology.**
See Abstr. 161.143.

**Unified field theories and the early universe.**
See Abstr. 161.145.

**Concluding comments: a cosmologist's view. (First ESO-CERN Symposium on "Large-scale structure of the universe, cosmology and fundamental physics", CERN, Geneva, 21 - 25 November 1983).**
See Abstr. 161.147.

**A brilliant past in front of us.**
See Abstr. 161.148.

**Cosmic separation of phases.**
See Abstr. 161.166.

**Massive neutrinos and cosmology.**
See Abstr. 161.172.

**On the relation of the cosmological constraints on neutrino flavors to the width of the $Z^0$.**
See Abstr. 161.173.

**Cooling pancakes.**
See Abstr. 161.175.

**Physical processes in the universe at the epoch of the quark-hadron transition.**
See Abstr. 161.178.

**Massive magnetic monopoles in cosmology and astrophysics.**
See Abstr. 161.194.

**Helium formation in cosmologies with anisotropic curvature.**
See Abstr. 161.217.

**Formation of structure in a universe with unstable neutrinos.**
See Abstr. 161.220.

**Axion cosmology.**
See Abstr. 161.227.

**Photon mass and cosmic microwave background anisotropy.**
See Abstr. 161.238.

**Cosmological gravitino regeneration and decay.**
See Abstr. 161.265.

**Constraints on the mass of muon neutrinos and their possible role in the galaxy formation.**
See Abstr. 161.266.

**Formation of monopole–proton bound states in the hot Universe.**
See Abstr. 161.271.

**Galaxy formation in a monopole dominated universe.**
See Abstr. 161.272.

**The quantum mechanics of the universe.**
See Abstr. 161.296.

**Big bang cosmology, relic neutrinos, and absorption of neutrino cosmic rays.**
See Abstr. 161.316.

**Quantum fluctuations near the classical space–time singularity.**
See Abstr. 161.334.

**Stationary states in quantum cosmology.**
See Abstr. 161.335.

# 062 Hydrodynamics, Magnetohydrodynamics, Plasma

062.001 **The continuous spectrum in differentially rotating perfect fluids: a model with an analytic solution.**
E. Balbinski.
Mon. Not. R. Astron. Soc., Vol. 209, No. 1, p. 145 – 157 (1984).
The first known analytic solution for the perturbations of a differentially rotating perfect fluid system is derived. This analytic solution is then used to facilitate a discussion of the continuous spectrum of eigenfrequencies which occurs in such systems. The temporal behaviour represented by the continuum is exhibited, and it is demonstrated how the "eigenfunctions" of the continuum can represent a physical perturbation, even though they are singular in form.

062.002 **Steady mildly relativistic thermal plasmas: processes and properties.**
R. Svensson.
Mon. Not. R. Astron. Soc., Vol. 209, No. 1, p. 175 – 208 (1984).
The physical properties of a finite, thermal plasma in pair balance are investigated as a function of the three dimensionless parameters: (1) the temperature $\theta$, (2) the "proton optical depth" $\tau_p$, and (3) the proton density $n_p$. At a dimensionless luminosity $l \gtrsim 1$ Comptonization of bremsstrahlung and double Compton photons dominates the cooling. A determination of the steady pair density in such a Wien equilibrium plasma shows the existence of the two pair density branches previously found for optically thin plasmas. The complicated interplay between photon and pair processes in such a plasma is elucidated. The main result is the existence of a pair–dominated luminosity curve that separates a pair–free region from a forbidden one in $l - \theta$ space. Requiring pair balance limits allowed luminosities at mildly relativistic temperatures to values two orders of magnitude smaller than the Eddington luminosity normally employed in accretion models.

062.003 **Magnetic field generation in celestial bodies.**
A. Z. Dolginov.
Astron. Astrophys., Vol. 136, No. 1, p. 153 – 160 (1984).
The possibility of magnetic field generation in various celestial bodies by "battery" effects of different kinds is analyzed. The effect is caused by the difference in the directions of the electron pressure gradient and the gradient of the electron number density. The possibility that conditions suitable for the "battery mechanism" exist in the Earth, the Sun and stars are considered. It is shown that the observational data are not in contradiction with the hypothesis that the effect considered plays a major or even an essential role in many cases.

062.004 **The thermal instability in a sheared magnetic field: filament condensation with anisotropic heat conduction.**
G. Van Hoven, Y. Mok.
Astrophys. J., Vol. 282, No. 1, p. 267 – 273 (1984).
The condensation–mode growth rate of the thermal instability in an empirically motivated sheared field is shown to depend upon the existence of perpendicular thermal conduction. This typically very small effect increases the spatial–derivative order of the compressible temperature–perturbation equation. The resulting growth rate is less than 1.5 times the controlling constant–density radiation rate, and has a clear maximum at a cross–field length of order $10^2$ times and a width of $\sim 10^{-1}$ the magnetic shear scale for solar conditions. The relation of this result with observable filaments and prominences is discussed.

062.005 **The effect of plasma rotation on axisymmetric magnetic field configuration.**
W. Hu.
Kexue Tongbao (Beijing), Vol. 28, No. 5, p. 657 – 660 (1983). Abstr. in Phys. Abstr., Vol. 87, No. 1237, Entry 69908 (1984).

062.006 **Gravitational instability of rotating gaseous disks with magnetic field. II. Dispersion relation.**
T. Nakamura.
Prog. Theor. Phys., Vol. 71, No. 1, p. 212 – 214 (1984). Abstr. in Phys. Abstr., Vol. 87, No. 1237, Entry 72866 (1984).

062.007 **Hydrodynamic flows in accretion disks.**
V. A. Urpin.
Sov. Astron., Vol. 28, No. 1, p. 50 – 53 (1984). English translation of 37.062.012.

062.008 **Confinement of magnetic fields in field tubes of variable cross section.**
A. A. Solov'ev.
Sov. Astron., Vol. 28, No. 1, p. 54 – 59 (1984). English translation of 37.062.013.

062.009 **Adiabatic theory in rapidly rotating magnetospheres.**
A. F. Cheng.
J. Geophys. Res., Vol. 89, No. A7, p. 5453 – 5459 (1984).
The Jovian magnetosphere is a rapid rotator in the sense that the corotation velocity exceeds the gyrovelocity for most of the low–energy plasma particles. Ordinarily, an electric drift velocity greater than the gyrovelocity will destroy the near periodicity of the bounce motion, so the parallel invariant $J$ will no longer be valid. However, this paper shows that $J$ invariance actually remains valid in rapidly rotating magnetospheres so long as departures from rigid rotation are not too great. Northrop and Birmingham have already proved $J$ invariance for the case of rapid but exactly rigid rotation, and this paper extends their results to nonrigid, rapid rotation with corotation lag and/or convection.

062.010 **A note on Southwood's instability criterion.**
C. Uberoi.
J. Geophys. Res., Vol. 89, No. A7, p. 5652 – 5654 (1984).
It is shown that Southwood's instability criterion for the onset of the Kelvin–Helmholtz instability at the magnetopause can be directly obtained from the marginal instability condition for the

pure Alfvén surface waves propagating along the interface between two incompressible media in the limit when the wave propagation direction is nearly perpendicular to the direction of the largest magnetic field.

062.011 **Fast–time–scale plasma turbulence and the collisionless tearing mode.**
R. O. Dendy, D. ter Haar.
Mon. Not. R. Astron. Soc., Vol. 209, No. 2, p. 335 – 342 (1984).
The authors consider the effect of fast–time–scale turbulence in a magnetized collisionless Langmuir plasma on the growth rate of the tearing mode and show that it may under typical conditions enhance this growth rate.

062.012 **The continuous spectrum in differentially rotating perfect fluids. II. The effect of gravitational radiation reaction.**
E. Balbinski.
Mon. Not. R. Astron. Soc., Vol. 209, No. 4, p. 721 – 727 (1984).
The effect of gravitational radiation reaction on the continuous spectrum of normal modes in a differentially rotating perfect fluid is considered for the first time. This is done by evaluating the criterion for secular stability for a specific case, for which the continuum is found not to give rise to a secular instability. The calculation highlights certain differences between discrete and continuum modes, as well as between the Eulerian and Lagrangian representations of the continuum modes.

062.013 **The hotspots from a time–varying jet.**
M. J. Wilson.
Mon. Not. R. Astron. Soc., Vol. 209, No. 4, p. 923 – 943 (1984).
Numerical simulations are presented for a model of an extragalactic radio source in which the jet speed varies cosinusoidally with time. The distribution of radio emission on the plane of the sky from the hotspot produced by the jet is predicted, and found to change profoundly in both morphology and power as a result of the variations in the jet speed.

062.014 **On the viability of acceleration mechanisms for high–velocity clouds and quasar emission–line clouds.**
A. J. Allen.
Mon. Not. R. Astron. Soc., Vol. 210, No. 1, p. 147 – 158 (1984).
The acceleration of dense features in astrophysical systems to high velocities is discussed in the light of the Rayleigh–Taylor instability, and it is shown that, unless stabilized by viscosity, such features must be formed comoving relative to an ambient hot flux. This principle is then applied to specific mechanisms for the production of high–velocity emission– and absorption–line clouds in active galactic nuclei.

062.015 **Propagation of waves in an atmosphere in the presence of a magnetic field. VII. Magneto–acoustic–gravity modes in an oblique *B*.**
S. J. Schwartz, N. Bel.
Astron. Astrophys., Vol. 137, No. 1, p. 128 – 132 (1984).
The authors investigate the propagation of magneto–acoustic–gravity modes in the solar atmosphere with a magnetic field inclined at an arbitrary angle to the direction of gravity. The linearized equations of motion in this isothermally stratified atmosphere are solved via a Frobenius series method coupled with appropriate boundary conditions. Strong magnetic fields yield essentially acoustic modes propagating along the field lines and therefore suffer similar problems to previous studies. Weak magnetic fields produce a complicated competition amongst the various forces which is inefficient in transporting energy to coronal heights. Only Alfvénic motions seem to remain as a viable option for coronal heating.

062.016 **On geometrization of steady rotating hydromagnetic flows.**
S. N. Singh, H. P. Singh.
Astrophys. Space Sci., Vol. 102, No. 1, p. 3 – 9 (1984).
Employing the anholonomic geometric results, the complex–lamellar motion and Beltrami flow of steady rotating magneto-fluids are discussed and some results of physical importance are derived.

062.017 **Heat and mass transfer in rotating fluid with Hall current. II.**
H. L. Agrawal, P. C. Ram, V. Singh.
Astrophys. Space Sci., Vol. 102, No. 1, p. 41 – 47 (1984).
Effects of Hall current on the hydromagnetic free–convection resulting from the combined effects of thermal and mass diffusion of an electrically conducting liquid past an infinite vertical porous plate in a rotating frame of reference have been analysed. As the mean steady flow has been presented in Part I, only the solutions for the transient primary and secondary velocity profiles, transient temperature profiles in the boundary layer and the transient skin friction, the transient rate of heat transfer on the plate are presented in this work. Different parameters entering into the problem are discussed.

062.018 **An investigation of grain equilibrium potential in a plasma environment.**
M. T. Ahmadian, T. P. Armstrong.
Astrophys. Space Sci., Vol. 102, No. 1, p. 179 – 188 (1984).
It is found that grains with zero initial potential gradually gain a negative potential and approach equilibrium. This potential depends most strongly on the plasma temperature. If secondary and photoemission of electrons from this grain is not considered, this equilibrium potential is about –2.7 times the plasma thermal energy. If secondary and photoemission is included, the grain can approach equilibrium at positive rather than negative potential. In this case the sign and the value of the equilibrium potential depends strongly on the material properties and to lesser extent on plasma temperature.

062.019 **Hall effects on MHD free–convection flow past an accelerated vertical porous plate.**
A. K. Singh.
Astrophys. Space Sci., Vol. 102, No. 2, p. 213 – 221 (1984).
The effect of Hall currents on the hydromagnetic free–convection flow of an electrically conducting and incompressible viscous fluid past a uniformly accelerated infinite vertical porous plate is discussed. The magnetic Reynolds number is assumed to be small so that the induced magnetic field can be neglected. The governing equations of the flow are solved by defining a complex velocity with the help of the Laplace transform method when the Prandtl number is equal to unity. The influence of the various parameters on the unsteady flow field is presented for both the cases, cooling and heating of the porous plate by free–convection currents.

062.020 **Symmetric flow of a viscous and heat conducting gas following a normal shock of variable strength.**
S. K. Srivastava, R. K. Singh.
Astrophys. Space Sci., Vol. 102, No. 2, p. 247 – 254 (1984).
This paper makes a comparative study of viscous and heat conducting one dimensional, spherical and cylindrical flows following a normal shock of variable strength. An exact solution representing progressive waves is applied to the problem of a shock advancing into a viscous and heat conducting region in which the pressure is constant but the density varies according to a simple power law.

062.021 **Self–similar cylindrical shock waves with radiation heat flux.**
J. B. Singh.
Astrophys. Space Sci., Vol. 102, No. 2, p. 263 – 268 (1984).
Similarity solutions describing the flow of a perfect gas behind cylindrical shock waves with radiation heat flux are investigated. The total energy of the expanding waves has been supposed to remain constant. The solution, however, is only applicable to a gaseous medium where the undisturbed pressure falls as the inverse square of the distance from the line of explosion.

062.022 **An explosion with cylindrical symmetry in magnetogasdynamics.**
P. R. Vishwakarma, R. B. Dube.
Astrophys. Space Sci., Vol. 102, No. 2, p. 269 – 276 (1984).
The propagation of magnetogasdynamic cylindrical shock waves in an exponentially increasing medium including the effects of the azimuthal magnetic field, is investigated. The shock wave moves with variable velocity and the total energy of the wave is variable. It is shown that the magnetic field has a significant effect on the pressure flow velocity and the inner expanding vacuum region.

062.023 **Weak MHD discontinuities in a radiation–induced flow field.**
R. Ram, A. Rai, R. P. Upadhyay.
Astrophys. Space Sci., Vol. 102, No. 2, p. 277 – 286 (1984).
The growth of weak MHD discontinuities have been studied in a radiation induced flow field at very high temperature. Growth and decay properties of weak MHD discontinuities have been discussed under the influences of time–dependent gasdynamic field, the radiation field and the magnetic field with finite electrical conductivity. The effects of thermal radiation and conduction on the global behaviour of weak MHD discontinuities have been studied under a quasi–equilibrium and quasi–isotropic hypothesis of the differential approximation to the radiative heat transfer equation.

062.024 **Effect of magnetic resistivity on the stability of a partially ionized compressible Hall plasma.**
N. Ogbonna, P. K. Bhatia.
Astrophys. Space Sci., Vol. 103, No. 1, p. 9 – 19 (1984).
The dynamic stability of a partially ionized, compressible Hall plasma of finite electrical conductivity has been investigated when the plasma is immersed in a uniform, horizontal magnetic field. Based on the variational principle, which is shown to characterize the problem, the solution has been obtained for a semi–infinite plasma confined between two planes and having an exponential density stratification along the vertical. It is found that the effect of neutral gas friction is stabilizing while magnetic resistivity, Hall currents and compressibility all have destabilizing influence.

062.025 **Homothermal magnetogasdynamic shock waves caused by instantaneous monochromatic radiation.**
B. G. Verma, R. C. Srivastava, A. H. Khan.
Astrophys. Space Sci., Vol. 103, No. 1, p. 27 – 31 (1984).
The self–similar motion of a gas heated by an instantaneous isotropic point source of monochromatic radiation has been studied under the influence of magnetic field, within the framework of a homothermal model.

062.026 **Kelvin–Helmholtz discontinuity in two superposed viscous conducting fluids.**
P. K. Bhatia, V. D. Sankhla.
Astrophys. Space Sci., Vol. 103, No. 1, p. 33 – 38 (1984).
The Kelvin–Helmholtz discontinuity in two superposed viscous conducting fluids has been investigated in the presence of a two–dimensional horizontal uniform magnetic field. The streaming motion is also assumed to be two–dimensional. The stability analysis has been carried out for two highly viscous fluids of uniform densities. It is found that the streaming motion has dual influence on the unstable system, destabilizing for low values of streaming velocity and stabilizing for high values of streaming velocity. The effect of viscosity is, however, found to be stabilizing as the growth rate of the unstable configuration decreases on increasing the viscosity.

062.027 **MHD free–convection flow past an accelerated vertical porous plate in a rotating fluid.**
A. K. Singh.
Astrophys. Space Sci., Vol. 103, No. 1, p. 155 – 163 (1984).
The three–dimensional flow of an electrically conducting and incompressible viscous fluid past a uniformly accelerated infinite vertical porous plate is studied in a rotating fluid. The flow is assumed to be at small magnetic Reynolds number so that the induced magnetic field is neglected. An exact solution has been obtained by defining a complex velocity with the help of the Laplace transform method for the Prandtl number equal to unity. The effects of rotation, magnetic and free–convection parameters are discussed for the whole problem. Also, the skin–friction components on the plate are discussed.

062.028 **Magnetic field reconnection in a compressible plasma.**
B. K. Shivamoggi.
Astrophys. Space Sci., Vol. 103, No. 2, p. 219 – 232 (1984).
An extension of Sonnerup's model for the magnetic field–line reconnection for a compressible plasma is given. The plasma is considered to be only slightly compressible so that the leading wave in Sonnerup's model can still be taken to be a thin discontinuity. The flow is assumed to occur under adiabatic conditions, and de Hoffmann–Teller jump conditions are used to connect the state variables across the shocks. The compressibility effects are found to increase the reconnection rate. The signaling problem is finally considered to study the evolution of MHD waves in a compressible, dissipative plasma so as to investigate the conditions under which the assumption of MHD waves in a compressible plasma to be thin discontinuities is valid.

062.029 **The Rayleigh–Taylor instability of superposed partially–ionized plasmas.**
N. Ogbonna, P. K. Bhatia.
Astrophys. Space Sci., Vol. 103, No. 2, p. 233 – 240 (1984).
The Rayleigh–Taylor instability of the plane interface separating two superposed, partially–ionized, viscous plasmas of different densities has been studied to include the effects of finite Larmor radius. The solution of the relevant linearized perturbation equations has been developed by the normal mode technique, taking the prevalent magnetic field to be uniform and horizontal. The potentially unstable case of a dense fluid superimposed on a lighter one has been considered. It is found that neutral gas friction, viscosity as well as finite Larmor radius all have a stabilizing influence.

062.030 **Stokes problem for a porous vertical plate with heat sinks by finite difference method.**
A. K. Singh.
Astrophys. Space Sci., Vol. 103, No. 2, p. 241 – 248 (1984).
The effect of heat sinks on the free–convection flow of an incompressible viscous fluid past an impulsively started infinite vertical porous plate is considered. Finite difference method has been used to obtain the numerical solution of the governing equations. The influence of the various parameters, occurring into the problem, on the velocity field, the temperature field and on their related quantities is extensively discussed with the help of graphs and tables.

062.031 **Modulational instability of finite–amplitude Alfvén waves.**
B. K. Shivamoggi.
Astrophys. Space Sci., Vol. 103, No. 2, p. 391 – 393 (1984).
A finite–amplitude plane–polarised Alfvén wave is shown to undergo a modulational instability consequent to its trapping in density cavities which are created by the ponderomotive force associated with the Alfvén wave. The instability arises when the density cavity travels at subsonic speeds.

062.032 **Particle–plasma coupling through magnetohydrodynamic turbulence.**
A. L. Brinca.
XVI International Conference on Phenomena in Ionized Gases, p. 80 – 81 (1983). Abstr. in Phys. Abstr., Vol. 87, No. 1239, Entry 79770 (1984). – See Abstr. 012.017.

062.033 **Unified theory of whistler mode instability in the presence of a parallel electric field.**
I. M. L. Das, R. P. Singh.
J. Plasma Phys., Vol. 31, Part 2, p. 263 – 274 (1984). Abstr. in Phys. Abstr., Vol. 87, No. 1239, Entry 79789 (1984).

062.034 **Effects of a cold electron beam on Brillouin backscattering in a plasma.**
J. E. Willett, Y. Aktas.
Plasma Phys. Controlled Fusion, Vol. 26, No. 6, p. 837 – 840 (1984). Abstr. in Phys. Abstr., Vol. 87, No. 1239, Entry 79807 (1984).

062.035 **Magnetic field generation via parametric instabilities in relativistic plasma.**
J. G. Lominadze, Eh. G. Tsikarishvili, O. V. Chedia (*O. V. Chediya*).
XVI International Conference on Phenomena in Ionized Gases, p. 106 – 107 (1983). Abstr. in Phys. Abstr., Vol. 87, No. 1239, Entry 79834 (1984). – See Abstr. 012.017.

062.036 **The application of collisionless plasma expansion processes to space plasmas.**
U. Samir, N. H. Stone.
XVI International Conference on Phenomena in Ionized Gases, p. 84 – 85 (1983). Abstr. in Phys. Abstr., Vol. 87, No. 1239, Entry 82377 (1984). – See Abstr. 012.017.

062.037 **Laboratory simulation effects of collisionless interaction in super–Alfvénic space plasma flows.**
V. M. Antonov, V. P. Bashurin, L. B. Gevorkyan, V. A. Zhmailo, Yu. P. Zakharov, V. V. Maximov (*V. V. Maksimov*), A. M. Orishich, A. G. Ponomarenko, V. G. Posukh, V. N. Syntnikov.
XVI International Conference on Phenomena in Ionized Gases, p. 88 – 89 (1983). Abstr. in Phys. Abstr., Vol. 87, No. 1239, Entry 82410 (1984). – See Abstr. 012.017.

062.038 **A numerical approach to equilibrium and stability properties in astrophysical plasmas.**
W. Zwingmann.
XVI International Conference on Phenomena in Ionized Gases, p. 100 – 101 (1983). Abstr. in Phys. Abstr., Vol. 87, No. 1239, Entry 82411 (1984). – See Abstr. 012.017.

062.039 **The summation of Bessel products.**
E. A. Evangelidis.
J. Math. Phys., Vol. 25, No. 7, p. 2151 – 2153 (1984). Abstr. in Phys. Abstr., Vol. 87, No. 1241, Entry 87884 (1984).

062.040 **Inverse Compton reflection from an ionized non–relativistic plasma.**
S. R. Kel'ner, E. S. Shikhovtseva.
Astrofizika, Tom 19, Vyp. 4, p. 803 – 814 (1983). In Russian. English translation in Astrophysics, Vol. 19, No. 4.

The analytical solution of the statistical matrix transfer equation under inverse Compton reflection from a perfectly ionized non–relativistic semi–infinite plasma is obtained. Angular distributions, polarization and spectrum of the reflected photons in the frequency region $\omega \gg kT/\hbar$ are found.

062.041 **Propagation of waves in a one–dimensional relativistic electron–positron plasma.**
L. S. Kuz'menkov, P. A. Polyakov.
Astrofizika, Tom 19, Vyp. 4, p. 815 – 821 (1983). In Russian. English translation in Astrophysics, Vol. 19, No. 4.

Dispersion formulae and damping decrement for waves in a magneto–active relativistic one–dimensional plasma are found. In particular, relativistic cyclotronic waves are discovered.

062.042 **Langmuir soliton propagating in a relativistic plasma at an angle to the external magnetic field.**
G. I. Melikidze, A. D. Pataraya.
Astrofizika, Tom 20, Vyp. 1, p. 157 – 163 (1984). In Russian. English translation in Astrophysics, Vol. 20, No. 1.

The possibility of generation of electromagnetic waves depending on the propagation of nonlinear Langmuir waves at a small angle to an external magnetic field in a relativistic electron–positron plasma is discussed. The appearance of charged bunches the existence of which is caused either by the presence of a small admixture of protons in the plasma or by a slight difference between the undisturbed functions of electrons and positrons is studied.

062.043 **Equilibrium states of a degenerate magnetized electron–nuclear plasma.**
S. S. Lipovetskij, V. S. Sekerzhitskij.
Astrofizika, Tom 20, Vyp. 2, p. 315 – 320 (1984). In Russian. English translation in Astrophysics, Vol. 20, No. 2.

Thermodynamics and nuclear parameters of degenerate strongly magnetized matter with density $\sim 10^7 - 10^{13}$g cm$^{-3}$ stabilized relative to β–processes and picnonuclear reactions are calculated. The correctness of the used independent particle model for similar calculations is discussed.

062.044 **Gamma–ray line emission, nuclear destruction and neutron production in hot astrophysical plasmas. The deuterium boiler as a gamma–ray source.**
F. A. Aharonian (*F. A. Agaronyan*), R. A. Sunyaev (*R. A. Syunyaev*).
Mon. Not. R. Astron. Soc., Vol. 210, No. 2, p. 257 – 277 (1984).

In hot astrophysical plasmas with ion temperatures high enough for nuclear excitation ($\geqslant 10^{11}$K) inelastic spallation reactions can proceed at a higher or similar rate leading to nuclei destruction during the stationary high–temperature plasma formation time. In the quasi–stationary sources of hard radiation a proton–neutron plasma is formed due to nuclear destruction without a noticeable content of heavy nuclei. In the two–temperature plasma ($T_i > T_e$) the nucleons lose their energy mainly through elastic collisions. Besides, the nucleons radiatively cool due to neutron capture by protons (with deuterium production) and bremsstrahlung during the proton–neutron scattering. The possibility of neutron evaporation from the two–temperature accretion discs is discussed.

062.045 **On compressional modes in radiative plasmas.**
M. A. Schweizer.
Mon. Not. R. Astron. Soc., Vol. 210, No. 2, p. 303 – 322 (1984).

Dispersion relations associated with compressional modes in radiative fluids are derived from Israel's relativistic theory of transient thermodynamics. The main focus is on regimes where the photons dominate the thermal energy and pressure of the two–component fluid. A comparison with standard results, derived earlier by Weinberg from conventional theory, shows that transient thermodynamics modifies not only the short–wavelength but also parts of the long–wavelength regime. The question of the survival of protogalaxies during the acoustic phase in an expanding universe is touched upon.

062.046 **Diffusive shock acceleration of energetic eletrons subject to synchrotron losses.**
G. M. Webb, L. O'C. Drury, P. Biermann.
Astron. Astrophys., Vol. 137, No. 2, p. 185 – 201 (1984).

Steady–state solutions of equations describing first order Fermi acceleration of energetic electrons at a plane shock and including synchrotron losses for the electrons are derived. It is shown that the problem in general reduces to solving two linear integral equations at the shock that arise from matching the upstream and downstream solutions. For the special case where the upstream and downstream diffusion coefficients are independent of momentum the two integral equations are solved by a Laplace transform method, and the resulting solution for the distribution function is expressed as an inverse Laplace transform. Numerical examples are obtained. Applications of the results to the acceleration of electrons in supernova remnant shocks and in the shocks of radio–jets are briefly discussed.

062.047 **Variable energy self–similar blast waves.**
R. Bedogni, A. Messina.
Astron. Astrophys., Vol. 137, No. 2, p. 261 – 268 (1984).

The authors discuss the general solutions of adiabatic self–similar blast waves with variable energy $[E(t) = E^* t^s]$ expanding in inhomogeneous media $[\varrho(r) = \varrho^* r^{-\omega}]$ when one has energy deposition only at the front. Assuming either the same physical process (any constant $s$) or the same medium (any constant $\omega$), the solutions can exhibit dramatic differences with the variation of either ω or *s*.

062.048 **On the effect of thermal processes on the resistive tearing instability of a current sheet.**
B. I. Meerson.
Pis'ma Astron. Zh., Tom 10, No. 9, p. 716 – 720 (1984). In Russian. English translation in Sov. Astron. Lett., Vol. 10.
The role of Ohmic heating and radiative cooling of a current sheet with a force–free magnetic field is investigated for the early phase of the evolution of the sheet. Under the approximation of incompressibility it is shown that occurrence of the thermal mode in the system may lead to stabilization of the tearing instability at high conductivity of the plasma.

062.049 **Propagation of an intense relativistic electron beam through a plasma region into vacuum.**
W. W. Destler, P. G. O'Shea, M. Reiser.
Phys. Fluids, Vol. 27, No. 7, p. 1897 – 1903 (1984). Abstr. in Phys. Abstr., Vol. 87, No. 1243, Entry 97808 (1984).

062.050 **Magnetohydrodynamics of a bow shock.**
V. G. Eselevich.
Kosm. Issled., Tom 22, Vyp. 4, p. 599 – 608 (1984). In Russian. English translation in Cosm. Res.

062.051 **The effect of Coriolis force on acceleration covariance in turbulent flow with rotational symmetry.**
N. Kishore, S. R. Singh.
Astrophys. Space Sci., Vol. 104, No. 1, p. 121 – 125 (1984).
The authors have considered turbulent flow of an incompressible, viscous fluid which is nearly isotropic with an extra vector argument (i.e. rotational symmetry) and spatially homogeneous. The expression for acceleration covariance in presence of Coriolis force has been derived and solution has been obtained in terms of defining scalars.

062.052 **Effect of cavitation on spherical blast waves.**
S. Kumar.
Astrophys. Space Sci., Vol. 104, No. 1, p. 127 – 132 (1984).
For spherical blast waves propagating through a self–gravitating gas with an energy input $E_\alpha = E_0 t^\beta$, where $E_\alpha$ is the energy released up to time *t*, $E_0$ is a functional constant, and β is a constant, kinetic, internal heat, and gravitational potential energies have been computed. Taking the parameter $A^2$, which characterises the gravitational field, equal to 2, variations of the percentages of these energies for β = 0, 1/2, 4/3, and 3 with shock strength have been presented. For β = 3, the effect of cavitation on the percentages of kinetic energy and internal heat energies has been explored.

062.053 **The flow of viscoelastic non–Newtonian fluid in the Ekman layer.**
K. R. Choubey.
Astrophys. Space Sci., Vol. 104, No. 1, p. 157 – 161 (1984).
In this paper the unsteady flow in the Ekman layer of a visco–elastic non–Newtonian fluid near a flat plate is discussed. Laplace transform technique has been employed to show the basic differential equations. Expressions for velocity profile, the skin friction have been calculated. It is shown that the time to attain the steady state increases with the elastic parameter. It is shown that normally the ultimate steady state is reached through a decay of inertial oscillations whose frequency decreases with increase in the elastic parameter.

062.054 **Energization of plasma in a weak magnetic field and structure of the geomagnetic tail.**
A. Hruška.
Astrophys. Space Sci., Vol. 104, No. 1, p. 169 – 179 (1984).
A "two–fluid" plasma is described as a single continuum characterised by the generalised tensor of mechanical pressure and generalised vector of flow of mechanical energy. Plasma energization due to the transfer of mechanical energy inside the plasma body is emphasised and the energization of plasma by conversion of the electromagnetic energy into the mechanical energy is discussed. Two kinds of conversion associated with the convection electric field $-(1/c)\mathbf{V}\times\mathbf{B}$ and with the deviation $\mathbf{E}^*$ of the total electric field from $-(1/c)\mathbf{V}\times\mathbf{B}$ are distinguished. The $\mathbf{V}\times\mathbf{B}$–field is related to the work done upon the plasma, while the $\mathbf{E}^*$–field is related to the plasma heating. Plasma motions with scale length larger than the Debye distance, taking place in the central part of the Earth–plasma sheet, are considered.

062.055 **Propagation of spherical shock waves in non–homogeneous medium with idealized magnetic field.**
V. Sharan.
Astrophys. Space Sci., Vol. 104, No. 2, p. 189 – 197 (1984).
By use of the approximate method of Whitham (1958), the propagation of magnetogasdynamic spherical shock waves is considered for adiabatic and isothermal flows in a decreasing density medium. The effect of initial magnetic fields on the shock velocity is discussed; and a comparison made between adiabatic and isothermal cases.

062.056 **Hydromagnetic free–convection flow near a time–varying accelerated vertical plate.**
A. K. Singh.
Astrophys. Space Sci., Vol. 104, No. 2, p. 259 – 265 (1984).
The effect of a uniform transverse magnetic field on the free–convection flow of an electrically conducting fluid past an exponentially accelerated infinite vertical plate is analysed for both cases, when the magnetic lines of force are fixed relative to the fluid and the plate, respectively. The Laplace transform method is used to obtain the expressions for velocity and skin–friction. The effect of a magnetic parameter is to decrease the velocity of water when the magnetic field is fixed to the fluid, while it increases the velocity when the magnetic lines of force are fixed relative to the plate.

062.057 **Some geometric properties of a dusty fluid in MFD flows.**
S. N. Singh, R. Babu.
Astrophys. Space Sci., Vol. 104, No. 2, p. 285 – 292 (1984).
By reformulating the basic equations governing the steady, compressible dusty fluid in magnetofluid flows, certain geometric results of physical importance are obtained. The congruence of the stream lines and magnetic field lines has been taken as the orthogonal coordinate system on the Maxwellian surfaces.

062.058 **Simulations of cosmic–ray particle diffusion.**
I. H. Urch.
Astrophys. Space Sci., Vol. 104, No. 2, p. 357 – 366 (1984).
The diffusion of charged particles in a stochastic magnetic field (strength *B*') which is superimposed on a uniform magnetic field $B_0\mathbf{k}$ is studied. A slab model of the stochastic magnetic field is used. Many particles were released into different realizations of the magnetic field and their subsequent displacements Δ*z* in the direction of the uniform magnetic field were numerically computed. The particle trajectories were calculated over periods of many particles scattering times. The ensemble average $\langle(\Delta z)^2\rangle$ was then used to find the parallel diffusion coefficient. The simulations were performed for several types of stochastic magnetic fields and for a wide range of particle gyro–radius and the parameter $B'/B_0$. The calculations have shown that the theory of charged particle diffusion is a good approximation even when the stochastic magnetic field is of the same strength as the uniform magnetic field.

062.059 **Free–convection flow of water at 4°C past an infinite vertical porous plate with constant heat flux.**
A. K. Singh, A. Raptis.
Astrophys. Space Sci., Vol. 104, No. 2, p. 399 – 404 (1984).
An analysis of the two dimensional flow of water at 4°C past an infinite porous plate is presented, when the plate is subjected to a normal suction velocity and the heat flux at the plate is constant. Approximate solutions are derived for the velocity and temperature fields and the skin–friction. The effects of *G* (Grashof number) and *E* (Eckert number) on the velocity and temperature fields are discussed.

062.060 **Propagation of axisymmetric disturbances on a twisted magnetic flux tube.**
T. J. Bogdan.
Astrophys. J., Vol. 282, No. 2, p. 769 – 775 (1984).
The author considers the propagation of axisymmetric disturbances on a twisted magnetic flux tube in an inviscid, incompressible, unstratified atmosphere of infinite electrical conductivity. Focusing and distortion of the radial wave profile occur on time scales $t \sim 10R/V$ ($R$ is the flux tube radius, $V$ is the Alfvén velocity along the flux tube axis) for disturbances of wavelength $\lambda \sim R$ on highly twisted flux tubes. It is suggested that this dispersion may be an important factor in hydromagnetic wave propagation on magnetic flux tubes in the tenuous solar atmosphere.

062.061 **Efficient particle acceleration in shocks.**
A. F. Heavens.
Mon. Not. R. Astron. Soc., Vol. 210, No. 4, p. 813 – 827 (1984).
A self–consistent non–linear theory of acceleration of particles by shock waves is developed, using an extension of the two–fluid hydrodynamical model by Drury and Völk. The transport of the accelerated particles is governed by a diffusion coefficient which is initially assumed to be independent of particle momentum, to obtain exact solutions for the spectrum. It is found that steady–state shock structures with high acceleration efficiency are only possible for shocks with Mach numbers less than about 12. A more realistic diffusion coefficient is then considered, and this maximum Mach number is reduced to about 6. The efficiency of the acceleration process determines the relative importance of the non–relativistic and relativistic particles in the distribution of accelerated particles, and this determines the effective specific heat ratio.

062.062 **On the geometry of magnetohydrodynamic flows.**
S. N. Singh, H. P. Singh.
Astrophys. Space Sci., Vol. 105, No. 1, p. 33 – 37 (1984).
In this paper it is shown that the motion of magnetofluid is circulation preserving if the magnetic field is in a direction of no variation. Employing the anholonomic geometric results it is proved that for the circulation preserving and complex lamellar motion of magnetofluids, the Lamb surfaces are developable if and only if the magnitude of the vorticity is constant along the streamlines.

062.063 **Hydromagnetic Ekman layer on free convection past an infinite porous flat plate.**
A. Raptis, P. C. Ram, V. Singh, A. K. Agrawal.
Astrophys. Space Sci., Vol. 105, No. 1, p. 47 – 53 (1984).
Free convection in a conducting liquid past an infinite porous vertical flat plate in a rotating frame of reference when the Hall current is present is considered. Exact solutions for the velocity and temperature fields have been derived. The effects of $M$ (Hartmann number), $m$ (Hall parameter), and $E$ (Ekman number) on the velocity field are discussed.

062.064 **Self–similar problem with increasing energy in radiative magneto–gas dynamics.**
J. B. Singh, S. K. Srivastava.
Astrophys. Space Sci., Vol. 105, No. 1, p. 161 – 170 (1984).
In the present paper self–similar solutions have been investigated for the propagation of piston driven, radiative gas–dynamic shocks into an inhomogeneous ideal gas permeated by a current free azimuthal magnetic field for spherical symmetry. The effects of radiation flux and magnetic field together have been seen in the region of interest on the other flow variables. The total energy of the flow between the piston and the shock is taken to be dependent on the shock radius obeying a power law. The radiative pressure and energy have been neglected. This problem is more general than the others done so far. The word piston implies some means to drive plasma radially onwards.

062.065 **On the Kelvin–Helmholtz instabilities of supersonic shear layers.**
S. Roy Choudhury, R. V. E. Lovelace.
Astrophys. J., Vol. 283, No. 1, p. 331 – 342 (1984).
The authors consider the Kelvin–Helmholtz instability of a finite–width supersonic shear layer with "linear" and "sinusoidal" velocity profiles in the transition region. They describe a simple but quite general numerical technique which generates contour maps of the real and imaginary parts of the perturbation frequency in the (reduced Mach number, wavenumber)–plane. A comparison is given between the numerical results and analytical expressions derived for small reduced Mach numbers.

062.066 **Stellar fibril magnetic systems. I. Reduced energy state.**
E. N. Parker.
Astrophys. J., Vol. 283, No. 1, p. 343 – 348 (1984).
The remarkable fibril structure of the magnetic fields at the surface of the Sun (with fibrils compressed to $1-2\times10^3$gauss) lies outside existing statistical theories of magnetohydrodynamic turbulence. The magnetic energy density within a fibril is of the order of $10^2$ times the local kinetic energy density, so that no simple application of equipartition principles is possible. It is pointed out that the total energy of the atmosphere (thermal + gravitational + magnetic) is reduced by the fibril state of the field by avoiding the magnetic inhibition of the convective overturning. It is shown that minimizing the total energy, as well as it can be formulated, predicts the observed compression of the fibrils to $1-2\times10^3$gauss at the visible surface, irrespective of the strength of the mean field.

062.067 **On some properties of force–free magnetic fields in infinite regions of space.**
J. J. Aly.
Astrophys. J., Vol. 283, No. 1, p. 349 – 362 (1984).
This paper discusses and solves, by a very general approach, two boundary value problems for three–dimensional force–free magnetic fields existing in an infinite domain which may be a half–space or the exterior of a bounded "star–shaped" region. These general results are applied to some important classes of two–dimensional configurations and are used to discuss aspects of several specific models presented in the literature. Examples of these domains are the solar corona, the environment of magnetic stars and the outer regions of the magnetospheres of accreting compact objects.

062.068 **Combined effect of potential and nonpotential magnetic fields on equilibrium in stellar atmospheres.**
E. B. Gliner.
Astrophys. J., Vol. 283, No. 1, p. 363 – 372 (1984).
An equilibrium in a plasma atmosphere around a gravitating body is considered with regard to both a magnetic field of electrical currents in the atmosphere and a magnetic field originating inside the star. The relation between the combined non–force–free magnetic field and the thermodynamic parameters of atmospheric plasma is treated analytically. Though restricted by the axisymmetrical situation, the approach allows for multipole structure of a stellar magnetic field and arbitrary radial variation of a toroidal atmospheric magnetic field. Applications are discussed, in particular for the explanation of solar corona asymmetry.

062.069 **Addendum: "On the propagation of acoustic waves in a radiating fluid" [Astrophys. J., Vol. 273, No. 1, p. 355 – 362 (1983)].**
D. Mihalas, B. W. Mihalas.
Astrophys. J., Vol. 283, No. 1, p. 469 (1984). See Abstr. 34.062.104.

062.070 **Perturbation theory for free oscillations of a self–gravitating sphere with phase boundaries.**
S. V. Vorontsov.
Astron. Zh., Tom 61, Vyp. 5, p. 854 – 859 (1984). In Russian. English translation in Sov. Astron., Vol. 28, No. 5.
Standard boundary conditions used at the discontinuity in the material parameter distribution in the problem of free oscil-

lations must be modified in the case when this boundary is a phase one. The perturbation theory is constructed for the estimation of corrections to eigenfrequencies due to the modification of boundary conditions without direct computations with new conditions at the phase boundary. The applicability of the perturbation theory is tested by the comparison with the results of direct computations for a model of Jupiter.

062.071 **On the reverse influence of accelerated particles on the structure of shock wave fronts.**
S. V. Bulanov, I. V. Sokolov.
Astron. Zh., Tom 61, Vyp. 5, p. 882 - 891 (1984). In Russian. English translation in Sov. Astron., Vol. 28, No. 5.
Acceleration of particles in the vicinity of shock wave fronts is considered. Particular attention is paid to the reverse influence of the accelerated particles on the flow of plasma in the vicinity of the front. The effectiveness of particles acceleration by weakly non-linear waves is also discussed. It is demonstrated that the shock waves of the type considered are evolutionary. If the viscosity is taken into account, these waves are continuous by their structure. The results obtained are applied to the problem of acceleration of cosmic rays in supernova remnants.

062.072 **Notions générales d'hydrodynamique.**
B. Leroy.
Instabilités hydrodynamiques et applications astrophysiques, p. 3 - 37 (1983). - See Abstr. 012.034.
Contents: Introduction. Equations fondamentales. Similitude dynamique - analyse dimensionnelle. Approximation de Boussinesq - critères de similitude de la convection. Linéarisation - instabilité.

062.073 **Simulation numérique directe et hydrodynamique.**
A. Pouquet.
Instabilités hydrodynamiques et applications astrophysiques, p. 39 - 64 (1983). - See Abstr. 012.034.
Contents: Introduction. Turbulence: modèles et phénoménologie. Simulations numériques. Résultats de simulations hydrodynamiques et MHD. Conclusion.

062.074 **Instabilités hydrodynamiques ou instabilités numériques?**
M. Farge.
Instabilités hydrodynamiques et applications astrophysiques, p. 65 - 82 (1983). - See Abstr. 012.034.
Contents: L'approche numérique: simulation ou simulacre des phénomènes? Le problème de l'existence et de l'unicité des solutions. Le problème de la consistance et de la précision du schéma de discrétisation. Le problème de la stabilité et de la convergence de l'algorithme numérique. Quelques conseils en guise de conclusion.

062.075 **Instabilité convective et equations d'amplitude.**
S. Fauve.
Instabilités hydrodynamiques et applications astrophysiques, p. 83 - 115 (1983). - See Abstr. 012.034.
Contents: Problème de Rayleigh-Bénard. Equations d'Oberbeck-Boussinesq. Analyse linéaire. Sélection de la structure convective. Effets dépendant du temps au seuil de convection.

062.076 **La convection astrophysique.**
J.-P. Poyet.
Instabilités hydrodynamiques et applications astrophysiques, p. 117 - 147 (1983). - See Abstr. 012.034.
Contents: Les approximations des équations utilisées en astrophysique. Les problèmes linéaires. Quelques notions de structure interne. Calibration de la théorie de la longueur de mélange. Résultats expérimentaux et numériques. Quelques problèmes exotiques.

062.077 **Modélisations numériques de la convection pénétrative.**
J. Latour.
Instabilités hydrodynamiques et applications astrophysiques, p. 149 - 165 (1983). - See Abstr. 012.034.
Contents: Introduction. Equations fondamentales. Modélisations de la convection pénétrative. Prédictions des simulations numériques. Conclusions et perspectives.

062.078 **Le couplage pulsation-convection.**
J.-P. Poyet.
Instabilités hydrodynamiques et applications astrophysiques, p. 187 - 223 (1983). - See Abstr. 012.034.
Contents: Quelques problèmes Boussinesq bien definis. Les théories de couplage pulsation radiale-convection. Quelques pas dans le domaine du couplage des pulsations non radiales avec la convection. Conclusion.

062.079 **Interactions non-linéaires entre instabilités dans les systèmes loin de l'équilibre thermodynamique.**
A. Arneodo.
Instabilités hydrodynamiques et applications astrophysiques, p. 229 - 284 (1983). - See Abstr. 012.034.
Contents: 1. Généralités: Convection Rayleigh-Benard. Double convection ou convection doublement diffusive. Triple convection ou convection triplement diffusive. 2. Formes normales décrivant la dynamique au voisinage du seuil simultane de plusieurs instabilités: Théorie linéaire. Méthode directe et pédagogique de réduction aux formes normales décrivant l'interaction entre instabilités au voisinage de la surface polycritique. 3. Convection thermohaline: Equations. Classification de la dynamique au voisinage de la surface polycritique. 4. Triple convection: Forme normale sans symétrie. Forme normale avec symétrie de réflexion.

062.080 **Un scénario typique de la transition vers le chaos: la cascade de dédoublement de période.**
P. Coullet, C. Tresser.
Instabilités hydrodynamiques et applications astrophysiques, p. 285 - 317 (1983). - See Abstr. 012.034.
De nombreuses expériences, à la fois réelles et numériques, ont mis en évidence une route vers le chaos temporel qui consiste en une cascade d'instabilités paramétriques dans laquelle un signal périodique voit à chaque étape doubler sa période. Les auteurs donnent une analyse élémentaire de ce processus.

062.081 **Eléments de phénoménologie de la turbulence développée. Transport turbulent.**
J.-D. Fournier.
Instabilités hydrodynamiques et applications astrophysiques, p. 333 - 358 (1983). - See Abstr. 012.034.
Contents: 1. Introduction. 2. Caractères de l'état turbulent: Turbulence développée. Influence du nombre de Reynolds. Universalité des propriétés statistiques. Cas du spectre d'énergie. Intermittence. 3. Phénoménologie à la Kolmogorov: Echelles de la turbulence, self-similarité et analyse dimensionnelle. Tourbillons et temps caractéristiques. Remarques sur la consistance, l'inconsistance et les extensions de cette phénoménologie. Turbulence bidimensionnelle. 4. Viscosité turbulente: Introduction. Critique. Extension. Turbulence avec trou spectral. Interactions turbulentes, fermetures et renormalisation.

062.082 **Ondes et turbulence en milieu tournant.**
E. J. Hopfinger.
Instabilités hydrodynamiques et applications astrophysiques, p. 359 - 388 (1983). - See Abstr. 012.034.
Contents: 1. Introduction. 2. Concepts fondamentaux: Ordres de grandeur et paramètres principaux. Ecculement géostrophique. Théorème de Proudman-Taylor. Vorticité. Rigidité rotationelle. Ondes inertielles. 3. Turbulence bidimensionnelle. 4. Turbulence barotrope en rotation rapide: Situation expérimentale et observations. Genèse des tourbillons intenses. Expulsion de la vorticité. Indications d'une dynamique 2D. 5. Conclusions.

062.083 **Scattering of shock waves by a spherical cloud.**
S. Ikeuchi, L. Spitzer Jr.
Astrophys. J., Vol. 283, No. 2, p. 825 – 832 (1984).

A two-dimensional fluid-dynamical computer program is used to examine the acoustic pulses produced when a shock wave, moving into a diffuse medium strikes a spherical cloud of higher density. Initially the cloud and the intercloud gas are taken to be in pressure equilibrium. The resultant flow shows clearly the pressure pulses, positive in some directions, negative in others, which propagate outward. The energy content of these pulses is discussed. It is found that the acoustic energy available for heating the warm clouds is increased by roughly an order of magnitude over earlier very rough estimates.

062.084 **The time development of a blast wave with shock-heated electrons.**
R. J. Edgar, D. P. Cox.
Astrophys. J., Vol. 283, No. 2, p. 833 – 841 (1984).

Accurate approximations are presented for the time development of both edge conditions and internal structures of a blast wave with shock-heated electrons, and equal ion and electron temperatures at the shock. The cases considered evolve in cavities with power law ambient densities (including the uniform ambient density case) and have negligible external pressure. Account is taken of possible saturation of the thermal conduction flux.

062.085 **Linear astrophysical dynamos in rotating spheres: mode transition between steady and oscillatory dynamos as a function of dynamo strength and anisotropic turbulent magnetic diffusivity.**
H. Yoshimura, Z. Wang, F. Wu.
Astrophys. J., Vol. 283, No. 2, p. 870 – 878 (1984).

Linear solutions of the dynamo equation with anisotropic diffusivity in a spherical shell are investigated for a wide range of dynamo numbers as a basis for understanding the behavior of nonlinear dynamos with variable dynamo strength as a function of magnetic field. Implications of the results for understanding solar cycle multiple period nonlinear oscillations with Maunder-minimum-like era and sporadic polarity transitions of the Earth's almost steady magnetic fields through temporary oscillatory phases are discussed.

062.086 **Optically thick ablation fronts.**
A. Königl.
Astrophys. J., Vol. 284, No. 1, p. 303 – 316 (1984).

It is pointed out that, under optically thick conditions, the evaporation of cold clumps of gas that are immersed in a hot medium could proceed very rapidly because of the relatively high efficiency of radiative heat conduction. The purpose of this paper is to study the basic properties of optically thick ablation fronts. The structure of the fronts in both planar and spherical geometries, and the modifications introduced by a gravitational field or by an acceleration of the evaporating clump, are considered. The effects of heat conduction on the development of a Rayleigh-Taylor instability within the front are discussed. Potential astrophysical applications ranging from stellar winds to extragalactic jets are outlined.

062.087 **Particle acceleration.**
R. Rosner, E. L. Chupp, G. Gloeckler, D. J. Gorney, S. M. Krimigis, Y. Mok, R. Ramaty, D. W. Swift, L. Vlahos, E. G. Zweibel.
NASA Ref. Publ., NASA RP-1120, p. 2-1 – 2-54 (1984). – See Abstr. 012.038.

Contents: 1. Introduction. 2. Phenomenology: Solar flares. Acceleration processes in the interplanetary medium. Magnetospheric and ionospheric observations. Particle acceleration outside the solar system. 3. Theoretical particle acceleration mechanisms: Adiabatic compression, magnetic pumping, and diffusion. Acceleration in direct electric fields. Stochastic acceleration. Shock particle acceleration. Coherent wave acceleration. Injection. 4. The remaining questions.

062.088 **Drift wave instabilities in a high $\beta$ multispecies plasma.**
V. L. Patel, P. H. Ng, G. R. Ludlow.
J. Geophys. Res., Vol. 89, No. A10, p. 8851 – 8856 (1984).

The dispersion relation for drift-Alfvén waves in two-component (cold and hot), high $\beta$ inhomogeneous multispecies plasma containing protons, oxygen, and sulphur ions is solved numerically. The magnetic field is assumed to have a gradient in simple slab geometry configuration. The plasma ion composition consists of H, O, S, $SO_2$, and Na ions. The effect of heavy ions and multiple charge states ($Z > 1$) on the growth rates has been numerically estimated for assumed ion compositions. The numerical calculations of growth rates also have been performed by using ion composition and plasma parameters based on Voyager spacecraft in the magnetospheres of Jupiter and Saturn.

062.089 **Some geometric properties of a dusty fluid in MFD flows.**
S. N. Singh, H. S. Tiwari, H. P. Singh.
Astrophys. Space Sci., Vol. 105, No. 2, p. 237 – 244 (1984).

By employing a nonholonomic description of the governing equations, the complex lamellar motion of a viscous, incompressible and perfectly conducting dusty fluid is discussed, and some results of physical importance are derived.

062.090 **The instability of a stratified layer of rotating and conducting self-gravitating fluid.**
P. K. Bhatia, V. D. Sankhla.
Astrophys. Space Sci., Vol. 105, No. 2, p. 287 – 294 (1984).

Instability of a horizontal rotating layer of a self-gravitating electrically conducting fluid has been studied. The prevailing magnetic field is uniform and acts along the vertical direction along which the fluid has a one-dimensional density gradient. The solution has been obtained through the variational methods. The dispersion relation obtained has been solved numerically and it is found that Hall currents as well as magnetic resistivity have a destabilizing influence. Coriolis forces, however, have a stabilizing influence.

062.091 **Unsteady flow of a non-Newtonian visco-elastic fluid between two coaxial circular cylinders.**
K. R. Choubey.
Astrophys. Space Sci., Vol. 105, No. 2, p. 301 – 305 (1984).

The flow of a visco-elastic liquid between two coaxial circular cylinders has been studied when an inner cylinder is moving from rest for a certain period with linearly growing speed and then stops suddenly. The Laplace transform technique has been employed to solve the basic differential equations. The expression for the velocity field is obtained.

062.092 **Oscillatory free-convective flow of an elastico-viscous fluid past an impulsively started infinite vertical porous plate. II.**
A. K. Singh.
Astrophys. Space Sci., Vol. 105, No. 2, p. 307 – 313 (1984).

The author presents the two-dimensional free-convective flow of an elastico-viscous fluid past an infinite vertical porous plate for the Stokes problem when the flow is subjected to a constant suction velocity through the porous plate. As the mean steady flow has been presented in Part I, only the solution for the transient velocity profiles, transient temperature profiles, the amplitude and the phase of the skin-friction and the rate of heat transfer are presented in this work. As in the case of mean steady flow, the influence of various parameters on the unsteady flow field is discussed for both the cases cooling and heating of the porous plate by free-convection currents.

062.093 **The critical layers and other singular regions in ideal hydrodynamics and magnetohydrodynamics.**
J. A. Adam.
Astrophys. Space Sci., Vol. 105, No. 2, p. 401 – 412 (1984).

The governing singular differential equations are stated for certain systems in ideal hydrodynamics and magnetohydrodynamics. The solutions of these equations are studied in the neighborhood of the singular regions, and can be conveniently characterized by the roots of the corresponding indicial equation. The nature of the solution determines the nature of the dispersion relation and of the Green's function for the problem. The differences between the various classes of problems are discussed, and exemplified by considering the initial value problem for three cases: (1) unstratified shear flow, (2) stratified shear flow, and (3) a static magnetofluid.

062.094 **A study of Comptonization of radiation in an electron plasma using the Monte Carlo method.**
A. Górecki, W. Wilczewski.
Acta Astron., Vol. 34, No. 2, p. 141 – 160 (1984).
Comptonization of soft photons in a hot plasma cloud with fixed electron distribution is studied using the Monte Carlo method. A complete description of this method, being a significant improvement compared to that of Pozdnyakov, Sobol', and Sunyaev (1977), is given. The authors make comparisons with earlier analytical and numerical studies for different temperatures, Thomson optical depths, and geometries; for Maxwellian and power law electron distributions; for different spatial distributions of internal photon sources; and for an external photon source.

062.095 **Model calculations of magnetic flux tubes. I. Equations and method.**
W. Deinzer, G. Hensler, M. Schüssler, E. Weisshaar.
Astron. Astrophys., Vol. 139, No. 2, p. 426 – 434 (1984).
The authors give a description of the equations, boundary conditions, geometry, approximations and numerical methods used in a numerical study of the physics of magnetic flux tubes in stellar convection zones and atmospheres. The MHD equations are solved for a compressible medium in two spatial dimensions together with an energy equation which includes radiative and convective energy transport. Material properties like opacity, adiabatic temperature gradient and heat capacity are calculated. Diagnostics of continuum radiation and spectral lines as well as the calculation of radiative equilibrium is carried out with the 2D radiation transfer code provided by Mihalas et al. (1978).

062.096 **Model calculations of magnetic flux tubes. II. Stationary results for solar magnetic elements.**
W. Deinzer, G. Hensler, M. Schüssler, E. Weisshaar.
Astron. Astrophys., Vol. 139, No. 2, p. 435 – 449 (1984).
Based on the methods of a preceding paper (Deinzer et al., 1984) the authors present models of stationary magnetic flux elements which are thought to build the basic structure of the solar photospheric magnetic field.

062.097 **Linear astrophysical dynamos in rotating spheres: solar and stellar cycle north–south hemispheric parity selection mechanism and turbulent magnetic diffusivity.**
H. Yoshimura, F. Wu, Z. Wang.
Astrophys. J., Vol. 285, No. 1, p. 325 – 338 (1984).
A simple but powerful method of linear solutions of the dynamo equation that describes magnetic field generation processes in rotating spheres is formulated. This method is applied to the basic and simplest case of purely radial differential rotation. A new mechanism in the dynamo processes is proposed and investigated with this method to understand why the Sun has odd magnetic parity, viz., why polarities of the solar cycle general (zonal) magnetic fields are opposite in the north and south hemispheres, and whether or not this odd parity dominance is universal among other stars. The mechanism is due to anisotropic turbulent magnetic diffusivity.

062.098 **Nonlinear generation of electromagnetic waves in a magnetoplasma.**
G. Murtaza, P. K. Shukla.
J. Plasma Phys., Vol. 31, Part 3, p. 423 – 436 (1984). Abstr. in Phys. Abstr., Vol. 87, No. 1244, Entry 103481 (1984).

062.099 **Nonlinear convection in a spherical shell.**
N. Raihi.
J. Phys. Soc. Jpn., Vol. 53, No. 8, p. 2506 – 2512 (1984). Abstr. in Phys. Abstr., Vol. 87, No. 1245, Entry 108076 (1984).

062.100 **A current generation by Compton scattering in a relativistic plasma with velocity shear and temperature gradient.**
S. Hinata.
Plasma Phys. Controlled Fusion, Vol. 26, No. 9, p. 1111 – 1118 (1984). Abstr. in Phys. Abstr., Vol. 88, No. 1247, Entry 2232 (1985).

062.101 **Current instabilities: laboratory measurements vs. theory.**
K. Appert.
Orages solaires de type I, p. 41 – 44 (1984). – See Abstr. 012.055.
This brief paper discusses recent theoretical progress in understanding current–driven ion–acoustic turbulence in plasmas.

062.102 **Possible wave–wave interactions for radio emission.**
J. Vaclavik.
Orages solaires de type I, p. 55 – 56 (1984). – See Abstr. 012.055.
Nonlinear wave–conversion processes in a plasma, leading to the generation of high–frequency electromagnetic waves with $\omega \lesssim \omega_{pe}$ (electron plasma frequency), are discussed within the framework of perturbation theory (weak nonlinearities).

062.103 **A novel instability in dense hydrogen–helium fluids.**
L. D. Cloutman.
Bull. Am. Astron. Soc., Vol. 16, No. 3, Part I, p. 704 – 705 (1984). Abstract. – See Abstr. 010.061.

062.104 **Monte Carlo shock–like solutions to the Boltzmann equation with collective scattering.**
D. C. Ellison, D. Eichler.
Bull. Am. Astron. Soc., Vol. 16, No. 3, Part I, p. 733 (1984). Abstract. – See Abstr. 010.065.

062.105 **Cyclotron radio emission in plasma with steep temperature gradient.**
M. Siarkowski.
Sol. Phys., Vol. 94, No. 1, p. 105 – 115 (1984).
In the paper it is investigated how the non–Maxwellian electron velocity distribution function associated with steep temperature gradient affects the cyclotron radio emission at centimeter wavelengths. The results obtained were used to calculate the distribution of the brightness temperature across a sunspot. The presence of the large electron temperature gradient in the transition region may lead to a significant increase of the observed brightness temperature (up to several tens of percent) in comparison with the brightness temperature obtained by assumption of the Maxwellian distribution function.

062.106 **The rapid dissipation of magnetic fields with internal strains.**
E. Parker.
J. R. Astron. Soc. Can., Vol. 78, No. 5, p. 205 (1984). Abstract. – See Abstr. 010.401.

062.107 **Numerical simulations of supersonic turbulent flows.**
J. Léorat, A. Pouquet, J. P. Poyet.
Problems of collapse and numerical relativity, p. 287 – 310 (1984). – See Abstr. 012.061.
Numerical simulations of 2D compressible time dependent flows are presented. It is shown that spectral methods are able to treat shocks. Preliminary results involving turbulence and gravitation are given.

062.108 **Mean–field dynamo with cubic non–linearity.**
N. I. Kleeorin (*N. I. Kliorin*), A. A. Ruzmaikin (*A. A. Ruzmajkin*).
Astron. Nachr., Vol. 305, No. 5, p. 265 – 275 (1984). – See Abstr. 012.062.
The turbulent mean–field dynamo of $\alpha\omega$–type with a mean helicity quadratically dependent on the magnetic field is investigated. A nonlinear system of ordinary differential equations is derived for the amplitudes of the magnetic field expansion over the eigenvectors of the linear problem. In a one–mode approximation the non–linear supercritical solution is stable when

$d\gamma/dD > 0$, where $\gamma$ is the growth rate of the linear solution and $D$ is the dynamo number. Non–linear interaction between two modes of dipole and quadrupole symmetry is considered. The conditions are found for the synchronization and beats of these modes under the assumption that the quadrupole mode is weaker than the dipole one.

062.109 **Simple nonlinear recursions and the mean–field concept.**
F. Krause.
Astron. Nachr., Vol. 305, No. 6, p. 281 – 288 (1984).

Some questions concerning the formation of mean–field theories are analyzed by a consideration of simple nonlinear recursions. The results suggest that a general closure scheme does not exist, rather there is a great variety of closures, each of them valid in only a certain interval of the characteristic parameter. Furthermore, questions connected with the appearance of random behaviour in these deterministic systems and the modelling of random reversals are discussed.

062.110 **On spherical mean–field dynamo models with coupling between axisymmetric and non–axisymmetric field modes.**
K.–H. Rädler.
Astron. Nachr., Vol. 305, No. 6, p. 289 – 294 (1984).

The mean–field dynamo theory, which was developed on the basis of mean–field magnetohydrodynamics, played a crucial part in finding–out the mechanisms responsible for generation and maintenance of the magnetic fields of the Earth and some planets, of the Sun and of other cosmical bodies. There are, however, a number of observational facts which are not satisfactorily comprehended with the models developed so far. This concerns in particular the multimodal structure of the observed fields, that is, the fact that they are composed of several parts showing different symmetries with respect to rotation axis and equatorial plane. The author discusses how mean–field models could cover also this property of the observed fields.

062.111 **Green's functions of the induction equation on regions with boundary.**
H.–J. Bräuer, K.–H. Rädler.
Astron. Nachr., Vol. 305, No. 6, p. 295 – 296 (1984).

Numerous problems of magnetohydrodynamics make it necessary to trace the evolution of a magnetic field permeating an electrically conducting fluid and its non–conducting surroundings. The magnetic field evolves according to a system of equations among which the induction equation plays the central role. The solution to the initial value problem posed by this system can be represented by means of tensorial Green's functions.

062.112 **Secondary α–effect in the nearly symmetric dynamo.**
I. Cupal.
Astron. Nachr., Vol. 305, No. 6, p. 297 – 299 (1984).

An attempt has been made to examine the role of the secondary α–effect in the nearly symmetric dynamo. The tensor form of the effect has been found and it is concluded that this effect only plays a role under special assumption of the disturbance velocity field. The secondary α–effect can amplify the generation of the azimuthal field from the poloidal field in the $\alpha^2$–dynamo rather than in the αω–dynamo.

062.113 **Theory of magnetic stars.**
L. Mestel.
Astron. Nachr., Vol. 305, No. 6, p. 301 – 309 (1984).

The principal observational facts on the early–type magnetic stars are summarized and interpreted in terms of the oblique rotator model. In the theoretical discussion of the interaction between a large–scale magnetic field and the star's thermal, gravitational, and rotation fields, emphasis is laid on effects that are relevant to both fossil and dynamo theories of the origin of the magnetic field.

062.114 **A criterion for the fragmentation of a rotating and collapsing gas cloud.**
I. Hachisu, Y. Eriguchi.
Astron. Astrophys., Vol. 140, No. 2, p. 259 – 264 (1984).

A criterion for the fragmentation of a rotating and collapsing isothermal gas cloud is obtained by the use of hydrostatic equilibrium solution. The authors assume that (1) the initial state of the cloud is a uniformly rotating homogeneous sphere and that (2) during the contraction, the cloud conserves angular momentum locally and has a uniform density distribution. Then the cloud can maintain a Maclaurin spheroidal configuration throughout the collapse. There are three outcomes of the collapse – a stable cloud, a solar type system, and a multiple system.

062.115 **Explosions into a 'vacuum'.**
M. Nepveu.
Astron. Astrophys., Vol. 140, No. 2, p. 314 – 316 (1984).

The early phases of explosions of dense gas into a vacuum are studied. These phases are defined by the fact that shocks have not yet formed. Similarity solutions are studied and applied to recently discovered jet–like features in spiral galaxies.

062.116 **Gyrosynchrotron radiation from a source with spatially varying field and density.**
K.–L. Klein, G. Trottet.
Astron. Astrophys., Vol. 141, No. 1, p. 67 – 76 (1984).

A computation of gyrosynchrotron emission from electrons radiating in an inhomogeneous source is presented. For the purpose of illustration it is applied to a simple model with typical parameters of some types of solar flare associated radio emission from microwaves to the decametre band. The source morphology at discrete frequencies and the spatially integrated spectrum are represented in their dependence on parameters such as the observer's location relative to the source, the magnetic field and the density of the ambient medium.

062.117 **A theory of hydrogen line broadening in the presence of a magnetic field accounting for ion dynamics.**
G. Mathys.
Astron. Astrophys., Vol. 141, No. 1, p. 248 – 254 (1984).

The unified theory for ions and electrons (accounting for ion–radiator motions) is adapted to the derivation of Stark broadened profiles of hydrogen lines in the presence of a magnetic field. The theory is only valid in a restricted range of electron densities, whose lower and upper limits are set respectively by the applicability of the straight line path approximation for the electrons and of the impact approximation for the ions.

062.118 **Relativistic positron stream in the electric field generated by radiation pressure.**
I. Kovner.
Astron. Astrophys., Vol. 141, No. 2, p. 341 – 350 (1984).

Radiation driven relativistic particle streams can be produced by a very luminous source via two related mechanisms: the acceleration of a plasma as a whole by radiation pressure and the acceleration of positrons through a free plasma by radiation pressure and the electric field generated by radiation in the plasma. The second mechanism yields much larger particle energies than the first one. The asymptotic analytic solutions of the equations of motion for both mechanisms are found for the acceleration along the axis of an axisymmetric radiation source in two configurations: an arbitrary single source of finite size and an infinite, thin, Keplerian accretion disk. These solutions provide a simple method for estimating the effectiveness of possible radiation sources as particle accelerators for both mechanisms. This method is applied to the Jaroszinski model of a thick accretion disk. A simple model of an ultrarelativistic positron stream illustrates the possibility for such a stream to be stable.

062.119 **Conical reflections of interstellar shocks. The formation of interstellar bullets.**
G. Tenorio–Tagle, M. Rózyczka.
Astron. Astrophys., Vol. 141, No. 2, p. 351 – 361 (1984).
Two–dimensional models of a conical shock reflection are obtained under the assumption of an adiabatic and a radiative flow. The converging conical shock forms when a strong, plane or spherical shock overtakes an obstacle cloudlet. If the reflection proceeds adiabatically, the passage of the conical shocks is followed by a fast re–expansion of the overtaken gas and the original shock is only slightly distorted. If cooling becomes effective shortly before the reflection is completed, condensation takes place instead of re–expansion, and a bullet able to overtake the original shock forms. The calculations support the ideas of bullet formation put forward in an earlier communication as an alternative explanation for the origin of Herbig–Haro objects.

062.120 **Gravity modes of inviscid plane Couette flow in a stratified fluid.**
P. Veugelen.
Astron. Astrophys., Vol. 141, No. 2, p. 389 – 393 (1984).
Eigenfrequencies of Couette flows or linear velocity fields in stratified fluids are numerically determined for various values of the Froude number. The behavior of the eigenfrequencies as a function of the Froude number makes it possible to investigate the influence of a velocity field on the stability and the normal modes of a static configuration. All modes of Couette flows can be identified as $g^+$– or $g^-$–modes of a static fluid modified by the velocity field.

062.121 **Continuous spectra of oscillation frequencies of an axisymmetric incompressible equilibrium with a poloidal magnetic field.**
M. Goossens, D. Hermans, S. Poedts.
Theoretical problems in stellar stability and oscillations, p. 382 – 386 (1984). – See Abstr. 012.065.
This paper concerns the continuous spectrum of incompressible linear motions in a static self–gravitating equilibrium with a poloidal magnetic field.

062.122 **Linear stability of an incompressible horizontal layer with a nonuniform horizontal magnetic field.**
D. Hermans, M. Goossens, R. Polfliet.
Theoretical problems in stellar stability and oscillations, p. 387 – 392 (1984). – See Abstr. 012.065.
The purpose of this paper is to study the linear stability of an incompressible horizontal layer of fluid in the presence of a non–uniform horizontal magnetic field and to show that the Alfvén continuum of frequencies is fundamental for the understanding of the linear stability.

062.123 **Stability of rotation laws.**
J.–P. Zahn.
Theoretical problems in stellar stability and oscillations, p. 407 – 419 (1984). – See Abstr. 012.065.
This paper reviews recent theoretical work on fluid dynamical instabilities in rotating flows. The discussion concentrates on two families of instabilities: (1) Rayleigh–Taylor type instabilities; (2) shear instabilities.

062.124 **Most effective potentials for stability analysis.**
J. Katz.
Theoretical problems in stellar stability and oscillations, p. 420 – 424 (1984). – See Abstr. 012.065.
The paper investigates steady flows of self gravitating gases with a barotropic equation of state $P(\varrho)$. It discusses stability with respect to non–small, non–dissipative perturbations. Sufficient conditions are obtained by minimizing the energy.

062.125 **Reynolds stresses and differential rotation. III. Zonal momentum fluxes in rapidly rotating stars.**
G. Rüdiger.
Geophys. Astrophys. Fluid Dyn., Vol. 25, No. 3, p. 213 – 233 (1983).

062.126 **Nonlinear convection in an imposed horizontal magnetic field.**
W. Arter.
Geophys. Astrophys. Fluid Dyn., Vol. 25, No. 4, p. 259 – 292 (1983).
Nonlinear two–dimensional magnetoconvection, with a Boussinesq fluid driven across the field–lines, is taken as a model for giant–cell convection in the sun and late–type stars. A series of numerical experiments shows the sensitivity of the horizontal scale of convection to the applied field and to the Rayleigh number $R$. Overstable oscillations occur in cells as broad as they are deep, but increasing $R$ leads to steady motions of much greater wavelength. Purely geometrical effects can cause oscillation: this work implies that strong horizontal field will in general lead to time–dependent convection.

062.127 **Relativistic magnetohydrodynamic winds of finite temperature.**
C. F. Kennel, F. S. Fujimura, I. Okamoto.
Geophys. Astrophys. Fluid Dyn., Vol. 26, Nos. 3 – 4, p. 147 – 222 (1983).

062.128 **Global magnetohydrostatic fields in stellar atmosphere.**
P. K. Browning, E. R. Priest.
Geophys. Astrophys. Fluid Dyn., Vol. 28, No. 2, p. 141 – 160 (1984).
The equilibrium properties of the magnetic field of an axisymmetric star are studied. A family of analytical solutions to the magnetohydrostatic equations is found, which are used to model the slow evolution of the field through a series of equilibria.

062.129 **An energy method for the stability of solar magnetohydrostatic atmospheres.**
A. W. Hood.
Geophys. Astrophys. Fluid Dyn., Vol. 28, Nos. 3 – 4, p. 223 – 241 (1984).
The energy method of Bernstein et al. (1958) was used by Schindler et al. (1983) to derive a useful form of the linearised potential energy for coronal arcades. The basic state magnetic field and plasma are assumed to be in magnetohydrostatic equilibrium and independent of variations along the length of the arcade. A sufficient condition for stability is derived and is applied to two particular magnetic structures. When the axial field component vanishes, this condition becomes necessary and sufficient.

062.130 **Alfvén waves in a thermally stratified fluid.**
E. N. Parker.
Geophys. Astrophys. Fluid Dyn., Vol. 29, No. 1, p. 1 – 12 (1984).
The propagation of Alfvén waves along a uniform horizontal field in a highly conducting incompressible fluid, subject to the convective forces produced by a uniform vertical temperature gradient, is treated in a Boussinesq approximation. It is shown that there are exact solutions with large amplitude but restricted form.

062.131 **On the geometry of streamlines of a dusty fluid in MHD flows.**
J. Singh, S. N. Singh.
Astrophys. Space Sci., Vol. 106, No. 1, p. 35 – 40 (1984).
By employing the nonholonomic description of basic equations governing the steady, incompressible dusty fluid in hydromagnetic flows, certain geometric results of physical importance are obtained. The conditions are derived for which the Maxwellian surfaces can be developed and coincide with surfaces of constant pressure.

062.132 **Flow of elasto–viscous fluid past an accelerated porous plate.**
J. Singh.
Astrophys. Space Sci., Vol. 106, No. 1, p. 41 – 45 (1984).
The flow problem of an elasto–viscous liquid (Walter–B type) due to impulsively accelerated motion of an infinite porous plate is considered. The effect of suction and flow parameters is analysed. It is found that at a fixed time the fluid velocity and skin

friction at any point decrease with the increase in suction and fluid elastic parameters.

062.133 **Isothermal shock waves in uniform atmospheres.**
J. B. Singh, K. S. Singh.
Astrophys. Space Sci., Vol. 106, No. 1, p. 47 – 51 (1984).
Similarity solutions describing the isothermal flow of a perfect gas behind shock waves are investigated for spherical symmetry. The flow is caused by a propelling contact surface (or expanding piston) and its total energy increases as a power of the time. The shock is propagating in a medium at rest with a uniform density.

062.134 **Plane and cylindrical hydromagnetic shock waves.**
S. Kumar.
Astrophys. Space Sci., Vol. 106, No. 1, p. 53 – 59 (1984).
The Chisnell–Chester–Whitham method has been used to investigate the propagation of diverging plane and cylindrical shock waves through an ideal gas in presence of a magnetic field having only constant axial and variable azimuthal components, simultaneously for both weak and strong cases.

062.135 **Comparative effect of viscous dissipation on heat transfer in mixed convection flow: UWT and UHF cases.**
R. Rajasekaran, M. G. Palekar.
Astrophys. Space Sci., Vol. 106, No. 1, p. 73 – 80 (1984).
The effect of viscous dissipation on mixed convection flow about a rotating sphere with a prescribed uniform surface heat flux (UHF) has been investigated analytically. As in the case of uniform wall temperature (UWT), the heat transfer rate decreases with viscous dissipation. It is also observed that for an equivalence viscous dissipation effect, heating by UHF yields larger Nusselt number than heating by UWT.

062.136 **Flow behind an attached shock wave in a radiating gas.**
L. P. Singh, A. Kumar, R. Shyam.
Astrophys. Space Sci., Vol. 106, No. 1, p. 81 – 92 (1984).
A simple method is used to determine the curvature of an attached shock wave and the flow variable gradients behind the shock curve at the tip of a straight–edged wedge placed symmetrically in a supersonic flow of a radiating gas near the optically thin limit. The shock curvature and the flow variable gradients along the wedge at the tip are computed for a wide range of upstream flow Mach numbers and wedge angles. Several interesting results are noted; in particular, it is found that the effect of an increase in the upstream flow Mach number or the radiative flux is to enhance the shock wave curvature which, however, decreases with an increase in the specific heat ratio or the wedge angle.

062.137 **Magneto–atmospheric waves from a localized source.**
J. A. Adam, J. H. Thomas.
Astrophys. Space Sci., Vol. 106, No. 1, p. 125 – 150 (1984).
A technique developed by Lighthill for finding the asymptotic solution of an inhomogeneous wave equation with constant coefficients is applied to the study of wave propagation in magneto–atmospheres. The geometry of the wave number surface plays an important role in determining the generation and propagation of various types of magneto–atmospheric waves from a localized forcing region. Examples of these wavenumber surfaces are exhibited for various magnetic field strengths and wave frequencies. The asymptotic far field is tabulated for a time–harmonic, spatially gaussian, localized forcing term.

062.138 **Note on nonlinear Rayleigh–Taylor instability.**
S. K. Malik, M. Singh.
Astrophys. Space Sci., Vol. 106, No. 1, p. 205 – 206 (1984).

062.139 **Hodograph transformations in steady plane rotating hydromagnetic flows.**
S. N. Singh, H. P. Singh, R. Babu.
Astrophys. Space Sci., Vol. 106, No. 2, p. 231 – 243 (1984).
Steady plane flows of an incompressible rotating viscous fluid with infinite electrical conductivity are studied. Certain results of physical importance have been deduced for orthogonal flows. Lastly, the solutions to vortex and source flow problems are obtained.

062.140 **Effects of rotation on Rayleigh–Taylor instabilities of an accelerating, compressible, perfectly conducting plane layer.**
A. H. Khater, M. H. Obied Allah.
Astrophys. Space Sci., Vol. 106, No. 2, p. 245 – 255 (1984).
The stability of a plane–stratified slab of perfectly conducting, rotating, compressible, inviscid plasma accelerated by a magnetic field is considered. Exact solutions for the growth rates of the unstable modes are determined for a $\gamma$–law gas when the undisturbed equilibrium is an isothermal atmosphere of semi–infinite extent with no frozen–in field. It is shown that the rotation has no effect on the region of unstable modes which has wavelengths long compared with the atmospheric scale height. On the other side, the growth rates in the presence of rotation are less than those in the absence of rotation for unstable modes.

062.141 **Effects of Hall current and rotation.**
A. Raptis, P. C. Ram.
Astrophys. Space Sci., Vol. 106, No. 2, p. 257 – 264 (1984).
Effects of Hall current and rotation on the flow of electrically conducting rarefied gas due to combined buoyant effects of thermal and mass diffusion, past an infinite porous plate in the presence of transverse magnetic fields have been investigated. The equations governing the flow problem have been solved and the profiles are shown on graphs. Effects of $m$ (Hall parameter) and $E$ (Ekman number) on velocity are discussed.

062.142 **Thermal effect on the pulses of the rotating Sun.**
G. C. Das, S. Sarkar, B. Karmakar.
Astrophys. Space Sci., Vol. 106, No. 2, p. 265 – 272 (1984).
A theoretical investigation is undertaken of the group travel–time for the Sun's pulses to travel from the source of the pulse to the solar surface. The expression for the group travel time gives two terms: the term arising from the linear theory which varies inversely to $(2\Omega–\omega)^{1/2}$ and the term introduced by the thermal effect which is inversely proportional to $(2\Omega–\omega)^{7/2}$. The thermal effect variation has been shown and an estimation of the temperature of the medium has been made. Furthermore, the velocity distribution and the amplitude of the magnetic field of the wave, arising from the damping of ionized particles, have been calculated.

062.143 **Numerical solution of MHD free–convection flow in the Stokes problem by the Crank–Nicolson method.**
N. G. Kafousias, J. Daskalakis.
Astrophys. Space Sci., Vol. 106, No. 2, p. 381 – 389 (1984).
A numerical solution of magnetohydrodynamic free–convection flow, in the Stokes' problem, for a porous vertical plate, is obtained, when Prandtl number $P$ is not equal to one. The magnetic lines of force are assumed to be fixed relative to the plate which is started moving impulsively in its own plane or it is uniformly accelerated. The solution is obtained by the Crank–Nicolson method, which is an implicit second–order method, for $P = 0.71$ (air) and $P = 7$ (water).

062.144 **Magnetic curtailment of the shock–induced thermal instability.**
S. S. Hong, B. C. Koo.
J. Korean Astron. Soc., Vol. 17, No. 2, p. 115 – 125 (1984).
Effect of magnetic field on the thermal instability is studied in the radiatively cooling region behind an interstellar shock of moderate propagation velocity ($\sim$10 km/sec). It is shown that the presence of interstellar magnetic field of a few micro gauss is very effective in preventing the thermal instability from building–up density concentration. In the absence of magnetic field, the shock–induced thermal instability amplifies preshock density inhomogeneity by more than an order of magnitude.

062.145 **On the asymptotic theory of gravitational capture of gas.**
A. N. Golubyatnikov.
Dokl. Akad. Nauk SSSR. Ser. Mat. Fiz., Tom 277, No. 1, p. 57 – 61 (1984). In Russian. Abstr. in Ref. Zh., 51. Astron., 11.51.130 (1984).

062.146 **Transport coefficients and collision frequencies for aeronomic plasmas.**
K. Suchy.
Handbuch der Physik. Geophysik III/7, p. 57 – 221 (1984). –See Abstr. 003.035.

062.147 **Strong turbulence of plasma waves.**
M. V. Goldman.
Rev. Mod. Phys., Vol. 56, No. 4, p. 709 – 735 (1984).
The paper reviews recent work related to modulational instability and wave envelope self–focusing in dynamical and statistical systems. After introductory remarks pertinent to nonlinear optics realizations of these effects, the author summarizes the status of the subject in plasma physics, where it has come to be called "strong Langmuir turbulence". The paper treats the historical development of pertinent concepts, analytical theory, numerical simulations, laboratory experiments, and spacecraft observations. The role of self–similar self–focusing Langmuir envelope wave packets is emphasized, both in the Zakharov equation model for the wave dynamics and in a statistical theory based on this dynamical model.

062.148 **Magnetoacoustic instability driven by Alfvén waves excited by cosmic rays.**
J. F. McKenzie, G. M. Webb.
Adv. Space Res., Vol. 4, No. 2 – 3, p. 219 – 223 (1984). – See Abstr. 012.075.
This paper analyzes the question of the stability of a four component system consisting of a magnetic field embedded in a thermal plasma in the presence of cosmic rays and Alfvén waves. Using MHD equations, modified to account self–consistently for the various exchanges of energy and momentum between the four components, the authors present the wave equation governing the propagation of long wavelength compressive perturbations of the system.

062.149 **Temporal evolution of the convective plasma flow during a reconnection process.**
V. S. Semenov, I. V. Kubyshkin, M. F. Heyn, H. K. Biernat.
Adv. Space Res., Vol. 4, No. 2 – 3, p. 471 – 474 (1984). – See Abstr. 012.075.
Following the initial suggestion of Petschek that the efficiency of the energy conversion from magnetic into kinetic energy of the plasma may be essentially enhanced by shock waves, the authors investigate the temporal evolution of these shocks in a reconnection configuration.

062.150 **Dimensionless jump conditions for MHD shock waves.**
F.–s. Wei.
Chin. Astron. Astrophys., Vol. 8, No. 3, p. 209 – 216 (1984). English translation of Chin. J. Space Sci., Vol. 4, No. 1, p. 22 – 33 (1984).
Jump conditions for MHD shock waves are expressed in a simple form in terms of dimensionless variables, and a generalized analytical solution valid for both MHD and gas–dynamical shocks is given. The cases of oblique, perpendicular and parallel MHD shocks are discussed in detail.

062.151 **Magnetoconvection.**
D. J. Galloway.
MPA Rep., No. 155, 9 pp. (1984).
Theoretical attempts to understand the interaction between convection and magnetic fields are summarized, excluding cases where the flow is turbulent or where rotational effects are also important. Theories and computations are grouped into kinematic, Boussinesq, or fully compressible categories, and are contrasted in terms of the amount of physics they contain and the number of dimensions represented.

062.152 **Pair production rates in mildly relativistic, magnetized plasmas.**
M. L. Burns, A. K. Harding.
Astrophys. J., Vol. 285, No. 2, p. 747 – 757 (1984).
Electron–positron pairs may be produced by either one or two photons in the presence of a strong magnetic field. In magnetized plasmas with temperatures $kT \approx mc^2$, both of these processes may be important and could be competitive. The authors calculate the rates of one–photon and two–photon pair production by photons with Maxwellian, thermal bremsstrahlung, thermal synchrotron, and power–law spectra, as a function of temperature or power–law index and field strength. The two rates are compared and their respective significance in astrophysical plasmas is discussed.

062.153 **The equilibrium of polytropic self–gravitating sheetlike masses.**
M. H. Ibañez S., L. di G. Sigalotti.
Astrophys. J., Vol. 285, No. 2, p. 784 – 790 (1984).
The equilibrium distributions of pressure, density, gravitational potential, internal and gravitational energies per unit area, are found for infinite self–gravitating slabs, with pressure and density related by a polytropic equation. A general but approximate series expansion for the plane Lane–Emden solution $\theta_n(\xi)$ is obtained.

062.154 **Current layers and reconnection.**
I. M. Podgornyj.
Problems of physics of solar flares, p. 126 – 148 (1983). In Russian. Abstr. in Ref. Zh., 51. Astron., 12.51.658 (1984). – See Abstr. 012.070.

062.155 **Compressible convection at large initial superadiabatic temperature gradient.**
S. Yamaguchi.
Publ. Astron. Soc. Jpn., Vol. 36, No. 3, p. 613 – 632 (1984).
Compressible convection in polytropic layers with the polytropic index $m = 1$ and 2 is studied, when the initial superadiabatic temperature gradient is large. The maximum Mach number easily exceeds unity even at fairly lower Rayleigh numbers ($R = 2 - 5\ R_c$), when the grade of temperature stratification is large ($\gtrsim 10$). The boundary values of the pressure are calculated from the conditions derived from the equations of motion.

062.156 **Magnetic flux tube in a stratified atmosphere under the influence of the vertical magnetic field.**
V. A. Osherovich.
Sol. Phys., Vol. 94, No. 2, p. 207 – 217 (1984).
The magnetohydrostatic equilibrium of a magnetic flux tube in a homogeneous gravitational and vertical magnetic field is studied. Gas pressure and density are presented explicitly as a function of the external magnetic field. The influence of the external magnetic field on the magnetic and thermodynamic structures is illustrated by two exact solutions. The possible applications to sunspot and facular modeling are discussed.

062.157 **The propagation direction of Alfvénic waves in the inhomogeneous plasma stream.**
C. Shi.
Acta Astrophys. Sin., Vol. 4, No. 1, p. 25 – 32 (1984). In Chinese.
Using the smooth perturbation method the author treats the problem of Alfvénic wave propagation in a double plasma system. He solves the wave equation in the inhomogeneous plasma stream under the condition of axial symmetry. The approximate solution of the first order is obtained. It can be demonstrated that when the stream is inhomogeneous, self–scattering is present. Meanwhile the pure Alfvénic waves can propagate in a direction different from the static magnetic field and the longitudinal component of the magnetic field fluctuation may exist too. Both values in the approximation are small but not zero. Thus the author shows a mechanism to explain the presence of the longitudinal magnetic field in waves in the solar wind.

062.158 **Stepwise ionization in a non–equilibrium, steady–state hydrogen plasma.**
J. A. Kunc.
J. Quant. Spectrosc. Radiat. Transfer, Vol. 32, No. 4, p. 311 – 323 (1984).
The paper considers a non–equilibrium, steady–state hydrogen plasma with $10^{10} \lesssim N_e, \mathrm{cm}^{-3} \lesssim 10^{17}$ and $8000 \lesssim T_e$, $\mathrm{K} \lesssim 64{,}000$. The following two cases are analyzed:

(1) the plasma is optically thin for all atomic lines and (2) the plasma is optically thick towards the Lyman lines and optically thin for all other lines. Analytical expressions have been obtained for populations and ionization frequencies of excited levels. Populations of the excited levels obtained from the analytical formulas are in good agreement with numerical calculations.

062.159 **Plasma potential and grain charges in a dust cloud embedded in a plasma.**
O. Havnes, G. E. Morfill, C. K. Goertz.
J. Geophys. Res., Vol. 89, No. A12, p. 10999 – 11003 (1984).

The authors have calculated the plasma potential and dust potential throughout the border region of semiinfinite dust clouds embedded in a plasma. A relevant ordering parameter is $P = (N_d{}^O/N_O)a_\mu$ (interior dust density/exterior plasma density) × dust radius (in microns). For comparatively high dust densities ($P > 10^{-3}$) the cloud potential is high, screening out electrons, with low interior dust charges as a result. For decreasing $P$ the cloud potential becomes less, with a corresponding increase in the dust charges. At $P < 10^{-5}$ the cloud potential is practically zero and the dust particles have their maximum charge which is identical to that of a single dust particle in a plasma. For diffuse dust border regions with density scale length $H \gg \lambda_D$ (Debye length), the deviations from charge neutrality are small. When $H \lesssim \lambda_D$, large deviations from charge neutrality occur in the border region and also some penetration of electric fields inside the border region.

062.160 **Reverse shock wave in relativistic explosions.**
M. Yokosawa.
Astrophys. Space Sci., Vol. 107, No. 1, p. 109 – 123 (1984).

The dynamical evolution of a relativistic explosion in a homogeneous medium is studied by means of a time-dependent, hydrodynamic code. When the expanding velocity of the shock front reduces to the sound velocity in the relativistic fluid, the reverse shock wave propagating inward through the expanding material is generated. The radius of the turning point of the reverse shock wave is proportional to the explosion energy and hardly depends on the mass of the explosion products. In the case of the non-relativistic explosion, the reverse shock wave is generated just after the free expansion stage. The radius of the turning point of the reverse shock wave is proportional to the mass of the explosion products and little depends on the explosion energy. In both cases of the non-relativistic and relativistic explosion, the reverse shock wave is strong in a spherical explosion and weak in a cylindrical one. The plane symmetric explosion does not generate the reverse shock wave.

062.161 **Oscillatory hydromagnetic free-convection flow in the Stokes problem past an infinite vertical porous plate in a rotating system. I.**
A. K. Singh.
Astrophys. Space Sci., Vol. 107, No. 1, p. 125 – 134 (1984).

Rotation effect on the hydromagnetic free-convection flow of an electrically conducting, viscous, and incompressible fluid past a steadily moving vertical porous plate has been analysed in the presence of a transverse magnetic field. The free-stream velocity oscillates in time about a constant mean, while the suction velocity, normal to the porous plate, is constant. The magnetic Reynolds number of the flow is taken small enough so that the induced magnetic field can be neglected. The plate temperature is constant and the difference between the temperature of the plate and the free stream is moderately large causing the free-convection currents. The flow field is described by nonlinear coupled system of equations. With viscous dissipative heat taken into account, approximate solutions of the problem are obtained for the components of velocity field and temperature field as well as for the skin-friction components and rate of heat transfer.

062.162 **Waves in a homogeneous nonmagnetized electron-nuclear plasma.**
G. P. Alodzhants, A. A. Saaryan.
Astrofizika, Tom 20, Vyp. 3, p. 571 – 584 (1984). In Russian. English translation in Astrophysics, Vol. 20, No. 3.

Solitary, stationary periodic and shock waves are considered in a degenerate electron-nuclear plasma. It is shown that the account of relativism and degeneration of electrons only affects the plasma parameters, but does not significantly alter the character of such waves.

062.163 **Low-frequency radiation from solitary magnetic sound waves in a nonhomogeneous magnetic field.**
R. G. Dzhangiryan, F. A. Kostanyan.
Astrofizika, Tom 20, Vyp. 3, p. 599 – 605 (1984). In Russian. English translation in Astrophysics, Vol. 20, No. 3.

A perturbation theory has been developed for a nonlinear equation of the Boussinesq type. In a low-temperature magnetoactive plasma the solitary magnetic sound wave travelling in the direction transverse to that of the external magnetic field is shown to generate pulses of the magnetic field of the same shape as a soliton when it crosses the jump of the magnetic field. The radiation propagates in both forward and backward directions relative to the direction of solitary wave propagation at Alfvén velocity. When a solitary wave propagates in a nonhomogeneous magnetic field low-frequency radiation is shown to take place which is considered as the transition scattering of the magnetic sound solitary wave.

062.164 **Statistical theory of the decay-process of weak homogeneous hydromagnetic turbulence.**
N. Kishore, S. R. Singh.
Astrophys. Space Sci., Vol. 107, No. 2, p. 241 – 254 (1984).

A statistical model for the spectral analysis of one dimensional hydromagnetic turbulence is presented for the purpose of gaining insight into the nonlinear relaxation in the decay-process. The random fluctuations of velocity and magnetic fields have been considered on the basis of wavenumbers in Fourier-space together with linearized mode approximation. Finally, with the aid of Bogoliubov expansion method, the dynamical equations for the decay of kinetic and magnetic energy have been derived for weak homogeneous hydromagnetic turbulence.

062.165 **Numerical treatment of hydromagnetic thermal boundary-layer flow of an infinite porous limiting surface.**
G. C. Pande, P. Hatzikonstantinou.
Astrophys. Space Sci., Vol. 107, No. 2, p. 313 – 322 (1984).

The finite difference approximation technique using the explicit method is used for solving the unsteady flow of an electrically conducting viscous and incompressible fluid, subjected to a normal homogeneous magnetic field. The flow is confined on one side of a non-magnetic infinite limiting surface (wall) which is initially at rest and then is suddenly accelerated in its own plane with a velocity which is a general function of time. The wall is porous and the authors assume that the Prandtl number of the fluid corresponds to the case of water and that the magnetic Prandtl number is equal to one. Quantitative discussion of the results is presented for the case of uniformly accelerated motion of the wall.

062.166 **Strong electromagnetic waves in a magnetized relativistic electron-positron plasma.**
M. Y. Yu, P. K. Shukla, N. N. Rao.
Astrophys. Space Sci., Vol. 107, No. 2, p. 327 – 332 (1984).

It is shown that in a strongly magnetized relativistic electron-positron plasma, strongly localized large amplitude circularly polarized electromagnetic wave pulses exist. The localization is due to relativistic mass variation as well as ponderomotive force effects. Three types of pulses are found analytically: the sharply spiked pulse in a strongly magnetized cold plasma, the smooth pulse in a weak magnetized warm plasma, and the moderately spiked pulse for a weakly magnetized cold plasma. The physical mechanisms giving rise to these pulses are distinct for each case. Possible implications of the investigation to pulsar radiation are discussed.

062.167 **Magnetic reconnection through the current sheet: quasi-stationary and explosive stages.**
S. Yu. Bogdanov, A. G. Frank, V. S. Markov.
Phys. Scr., Vol. 30, No. 4, p. 279 – 281 (1984).

The magnetic field structure is determined. The magnetic reconnection is revealed to be accompanied by a sharp decrease in plasma conductivity, generation of inductive electric fields, and electron acceleration.

062.168 **Plasma dynamics inside the current sheet.**
S. Yu. Bogdanov, G. V. Dreiden, A. G. Frank, N. P. Kirei, A. Z. Khodzhaev, I. I. Komissarova, V. S. Markov, G. V. Ostrovskaya, Yu. I. Ostrovskij, V. N. Philippov, M. M. Savchenko, E. N. Shedova.
Phys. Scr., Vol. 30, No. 4, p. 282 – 283 (1984).
Plasma dynamics is investigated by the five–frames holographic interferometry and the Doppler broadening of spectral lines. The explosive stage of magnetic reconnection is accompanied by plasma sheet destruction and the appearance of high velocity plasma flows.

062.169 **Caustic of Alfvén waves.**
V. N. Senatorov.
Geomagn. Aehron., Tom 24, No. 6, p. 1036 – 1037 (1984). In Russian. English translation in Geomagn. Aeron.

062.170 **Monte Carlo shock–like solutions to the Boltzmann equation with collective scattering.**
D. C. Ellison, D. Eichler.
Astrophys. J., Vol. 286, No. 2, p. 691 – 701 (1984).
The authors present the results of Monte Carlo simulations of steady state shocks generated by a collision operator that isotropizes the particles by means of elastic scattering in some locally defined frame of reference. The simulations include both the back–reaction of accelerated particles on the inflowing plasma and the free escape of high–energy particles from finite shocks. Energetic particles are found to be naturally extracted out of the background plasma by the shock process with an efficiency in good quantitative agreement with (1) an earlier analytic approximation and (2) observations of the entire particle spectrum at a quasi–parallel interplanetary shock.

062.171 **Helical twisting on an adiabatically expanding jet. I. Propagation and temporal growth.**
P. E. Hardee.
Astrophys. J., Vol. 287, No. 2, p. 523 – 533 (1984).
The effect of adiabatic expansion on the propagation and temporal growth of helical twisting on a supersonic jet is investigated. This is accomplished by deriving the local linearized dispersion relation for perturbations to the surface of a constantly expanding adiabatic jet assumed to be in static pressure balance with an isothermal confining medium. The paper follows the consequences of assuming that a change in the phase velocity of a propagating helical wave along a jet, given by solution of the local dispersion relation, causes the wavelength of a helical wave to change.

062.172 **Collisionless shocks.**
G. Paschmann.
ESA J., Vol. 7, No. 4, p. 357 – 368 (1983). – See Abstr. 012.035.
Recent spacecraft observations pertaining to the structure of collisionless shocks and their role in the generation of energetic particles are reviewed.

062.173 **Recent advances in numerical simulation of space–plasma–physics problems.**
T. J. Birmingham.
ESA J., Vol. 7, No. 4, p. 369 – 383 (1983). – See Abstr. 012.035.
Computer simulations have become an increasingly popular, important and insightful tool for studying space plasmas. The review describes MHD and particle simulations, both of which treat the plasma and the electromagnetic field in which it moves in a self consistent fashion but on drastically different spatial and temporal scales. The complementary roles of simulation, observations and theory are stressed. Several examples of simulations being carried out in the area of magnetospheric plasma physics are described to illustrate the power, potential and limitations of the approach.

062.174 **Magnetic–field–aligned electric fields.**
C.–G. Fälthammar.
ESA J., Vol. 7, No. 4, p. 385 – 404 (1983). – See Abstr. 012.035.

**Cold plasma waves.**
See Abstr. 003.031.

**Magnetic fields in astrophysics.**
See Abstr. 003.045.

**Artificial particle beams in space plasma studies.**
See Abstr. 003.093.

**SPECTRA – a model for *K*–shell spectroscopy.**
See Abstr. 022.041.

**Forbidden lines in hot plasmas.**
See Abstr. 022.046.

**Unified theory calculations of Stark–broadened hydrogen lines in the presence of a magnetic field.**
See Abstr. 022.066.

**Particle acceleration during reconnection in laboratory plasmas.**
See Abstr. 022.144.

**Equilibrium emission of low–density, optically thick plasma.**
See Abstr. 063.018.

**Spectra from pair–equilibrium plasmas.**
See Abstr. 063.037.

**Radiative transfer in a hot plasma.**
See Abstr. 063.048.

**Soft X–ray opacity in hot and photoionized gases.**
See Abstr. 063.075.

**Magnetohydrodynamic critical levels and radiative damping.**
See Abstr. 064.003.

**Molecular viscosity as a mechanism of generation of magnetic field in accretion discs.**
See Abstr. 064.008.

**On the solution of a class of high–order wave equations.**
See Abstr. 064.048.

**On the fission theory of binary stars. IV. Exact solutions in polynomial spaces.**
See Abstr. 065.056.

**The generation of high–frequency gravitational waves in a perfect fluid.**
See Abstr. 066.018.

**A class of spherically symmetric perfect fluid distributions in general relativity.**
See Abstr. 066.141.

**On the gravitational radiation of an oblique rotator.**
See Abstr. 066.157.

**A numerical investigation of the Kelvin–Helmholtz instability in the context of accreting neutron stars.**
See Abstr. 067.068.

**Spherical propagation of large amplitude pulsar waves.**
See Abstr. 067.095.

**The decay of torque free toroidal magnetic fields confined beneath the solar convection zone.**
See Abstr. 075.003.

**Non–linear wave coupling and Type I solar radio bursts theories.**
See Abstr. 077.014.

**Injection of particles into the diffusive shock acceleration process: Monte Carlo simulation of heliospheric shocks.**
See Abstr. 078.010.

**On mass and energy transfer by waves in an atmosphere.**
See Abstr. 080.105.

**Whistler wave amplitude oscillation and frequency modulation in the magnetospheric cavity.**
See Abstr. 084.038.

**The problem of reconnection: energy release in the geomagnetic tail.**
See Abstr. 084.062.

**The MHD Rankine–Hugoniot jump conditions and the terrestrial bow shock: a statistical comparison.**
See Abstr. 084.071.

**On the microphysics of the magnetotail reconnection region.**
See Abstr. 084.072.

**Numerical simulations of double layers and auroral electric fields.**
See Abstr. 084.076.

**Numerical simulation of compressible convection in dense hydrogen–helium fluids. A novel instability.**
See Abstr. 099.026.

**Strong cylindrical magnetogasdynamic shock waves in a rotating interplanetary medium.**
See Abstr. 106.006.

**Computer simulations of collisionless shock waves.**
See Abstr. 106.038.

**The physical mechanisms of subcritical collisionless shock–wave formation.**
See Abstr. 106.039.

**The heating and acceleration of electrons by shocks.**
See Abstr. 106.040.

**Relaxation of plasma at the shock front.**
See Abstr. 106.041.

**Particle acceleration and MHD wave excitation upstream of interplanetary shocks.**
See Abstr. 106.043.

**Temperature relaxation in supernova remnants, revisited.**
See Abstr. 125.070.

**A warm magnetoactive plasma in a large volume of space.**
See Abstr. 131.273.

**Erratum: "Non–linear theory of cosmic ray shocks including self–generated Alfvén waves" [Astron. Astrophys., Vol. 116, p. 191 – 200 (1982)].**
See Abstr. 144.003.

**Importance of transverse diffusion for cosmic–ray propagation.**
See Abstr. 144.032.

**Thermal particle injection into the shock acceleration process.**
See Abstr. 144.066.

**Reaction effects in diffusive shock acceleration.**
See Abstr. 144.067.

**Stochastic Fermi acceleration and the origin of cosmic rays.**
See Abstr. 144.068.

**Numerical studies of time–dependent cosmic–ray shocks.**
See Abstr. 144.069.

**Contribution of turbulent cascades to the luminosity of extragalactic radio–jets and Mach number constraints.**
See Abstr. 158.011.

**MHD Kelvin–Helmholtz instability in extended radio jets.**
See Abstr. 158.097.

**Jet deflection and hairpin bends.**
See Abstr. 158.199.

**A model for the surface brightness of a turbulent low Mach number jet. I. Theoretical development and application to 3C 31.**
See Abstr. 158.239.

# 063 Radiative Transfer, Scattering

063.001 **Escape probability methods.**
G. B. Rybicki.
Methods in radiative transfer, p. 21 – 64 (1984). – See Abstr. 003.001.

Escape probability methods for the approximate treatment of non–LTE line and continuum transfer have recently undergone considerable development. The author reviews these methods, starting from a detailed description of the escape probabilities in static and moving media, and proceeding through the original (first–order) escape probability methods, the core saturation method, the newer (second–order) escape probability methods, and Scharmer's method.

063.002 **Mean escape probabilities and mean numbers of scatterings for resonance lines.**
H. Frisch.
Methods in radiative transfer, p. 65 – 77 (1984). – See Abstr. 003.001.

A mean escape–probability approximation for resonance lines, which encompasses both the effectively thin and the effectively thick limits, is described. Global conservation of photons and the large scale diffusive behavior of the radiation field are reasonably well preserved in this approximation. This approximation is tested on a two–level atom by comparison with accurate numerical results.

063.003 **Probabilistic equations for line source functions: accurate and approximate.**
R. G. Athay.
Methods in radiative transfer, p. 79 – 100 (1984). – See Abstr. 003.001.

An exact expression for line source functions is derived in terms of the local and non–local probabilities for gaining and losing photons in the bandwidth of the line. Examples of exact and approximate solutions are given for Voigt profiles, for constant and variable Planck function, and for constant and variable Doppler width. Two approximate second–order differential equations are derived in terms of purely local probabilities for gaining and losing photons. Examples of solutions using terms in first–order differentials demonstrate that the approximate equations are quite accurate for both constant and variable Planck functions.

063.004 **Probabilistic radiative transfer.**
R. C. Canfield, A. N. McClymont, R. C. Puetter.
Methods in radiative transfer, p. 101 – 129 (1984). – See Abstr. 003.001.

The authors have developed a computationally efficient method for highly nonlinear problems in which radiative transfer is an important aspect of the heating and cooling of the medium. This paper summarizes all essential aspects of the method. The authors derive an approximate probabilistic radiative transfer equation for one–dimensional plane–parallel atmospheres of finite or semi–infinite extent, for both spectral lines and bound–free continua. They also discuss boundary conditions, accuracy, escape probabilities, and practical aspects of complete linearization, which is a key element of the method.

063.005 **Core saturation in a moving medium.**
W. Kalkofen, P. Ulmschneider.
Methods in radiative transfer, p. 131 – 156 (1984). – See Abstr. 003.001.

The authors discuss a method of solving line transfer problems in moving media with plane symmetry. The method is an adaptation of the core saturation method of Rybicki (1972) to a medium with internal structure, such as shocks. Its speed follows from two simplifications: the neglect of detailed transfer in the line core, and the neglect of atmospheric regions that contribute negligibly to the formation of the emergent spectrum of interest. The method has been tested with the calculation of a line source function in a static, isothermal medium and in a model solar atmosphere traversed by multiple shocks.

063.006 **Non–local perturbation techniques in radiative transfer.**
C. J. Cannon.
Methods in radiative transfer, p. 157 – 172 (1984). – See Abstr. 003.001.

063.007 **Accurate solutions to non–LTE problems using approximate Lambda operators.**
G. B. Scharmer.
Methods in radiative transfer, p. 173 – 210 (1984). – See Abstr. 003.001.

A numerical method for the approximate solution of non–LTE problems is described. The essence of this method is the evaluation of an approximate operator, relating the intensity $I_\nu$ to the source function $S_\nu$ along a ray. Using this operator, an approximate matrix equation for the line–source function is derived. The solution of this equation generally gives less than 10 – 30% errors in the line–source function even when there are large gradients in the Doppler width or in the macroscopic velocity. Furthermore, a linearization method for the exact solution on non–LTE problems is presented.

063.008 **Iterative solution of radiative transfer problems with spherical symmetry using a single–ray approximation.**
Å. Nordlund.
Methods in radiative transfer, p. 211 – 233 (1984). – See Abstr. 003.001.

An iterative technique for solving radiative transfer problems with spherical symmetry is described. An approximate relation between changes in a source function and the resulting changes in the mean intensity is obtained by using the radiative transfer equation along a single representative ray, combined with a geometrical factor relating the mean intensity to the intensity along the representative ray. This approximate relation is then used iteratively to improve the estimate of the source function until a prespecified accuracy is obtained. The convergence properties of this technique are very good, with typically 2 – 3 iterations being sufficient for better than 1% accuracy in the source function.

063.009 **Difference equations and linearization methods for radiative transfer.**
L. H. Auer.
Methods in radiative transfer, p. 237 – 279 (1984). – See Abstr. 003.001.

Difference equation methods are developed for the treatment of radiation transport in planar and spherical geometries. Using the Hermite method equations are derived, which give fourth–order accuracy for the spatial discretization, yet are no more difficult to solve than the usual second–order Feautrier system. It is shown that the angular dependence of the radiation field may be handled by the use of "variable Eddington factors", which reduce the system to just one equation per frequency. An efficient iterative method for evaluating these factors is presented and its convergence discussed. Finally, the difference equations are linearized in terms of the local state of the material, and methods are developed for solving the general transfer problem in which there are additional constraints, e.g., hydrostatic, statistical, energy equilibria.

063.010 **Discrete space theory of radiative transfer.**
A. Peraiah.
Methods in radiative transfer, p. 281 – 306 (1984). – See Abstr. 003.001.

This contribution describes a general numerical method for solving the equation of radiative transfer in inhomogeneous media with curved geometries. The method is based on the interaction principle and the star product algorithm. It is applied to line formation in expanding media.

063.011 **Radiative transfer in spherical atmospheres.**
W. Kalkofen, R. Wehrse.
Methods in radiative transfer, p. 307 – 339 (1984). – See Abstr. 003.001.

The authors discuss the construction of extended model atmospheres in radiative–convective and hydrostatic equilibrium for given luminosity and mass of the star and for prescribed net flux, of $T_{eff}$, at a standard monochromatic optical depth. The transfer equation is in the form of a first–order partial integro–differential equation for the specific intensity. For the solution of the corresponding difference equation the authors use the finite–difference matrix to develop "integral operators" in curvilinear coordinates that are analogous to the integral operator of the formal solution of the transfer equation in plane geometry.

063.012 **Line transfer in static and expanding spherical atmospheres.**
E. H. Avrett, R. Loeser.
Methods in radiative transfer, p. 341 – 379 (1984). – See Abstr. 003.001.

The authors show how to solve transfer problems using simultaneous equations based on the integral dependence of the intensity $I$ on the source function $S$. The method is simple in concept, easy to apply, and applicable to a wide range of practical line and continuum transfer problems. Explicit equations with numerical solutions are derived for some examples of static spherical atmospheres and for line transfer in expanding spherical atmospheres with complete frequency redistribution and with partial redistribution.

063.013 **Transfer in spherical media using integral equations.**
J. Schmid–Burgk, M. Scholz.
Methods in radiative transfer, p. 381 – 394 (1984). – See Abstr. 003.001.

An integral method for solving transfer problems in media of spherically–symmetric geometry is presented. Particular attention is paid to aspects of practical importance, like reduction in computer time by maximizing the number of quantities reusable in each iteration step and an optimal representation of the solution in terms of spline functions. The accuracy and speed of the method is discussed, and results of some applications are given.

063.014 **Line blanketing.**
D. F. Carbon.
Methods in radiative transfer, p. 395 – 426 (1984). – See Abstr. 003.001.

This contribution reviews the two most powerful methods currently available for treating extensive atomic and molecular line blanketing in a model atmosphere calculation: (1) the Opacity Distribution Function (ODF), a representation for the collected line opacities, and (2) Monte Carlo sampling of the blanketed radiation field at random frequencies.

063.015 **Operator perturbation methods: a synthesis.**
W. Kalkofen.
Methods in radiative transfer, p. 427 – 450 (1984). – See Abstr. 003.001.

This paper shows the equivalence of the perturbation methods decribed by Cannon, Scharmer, and Nordlund with a method based on the expansion of an inverse operator, discusses the treatment of non–linearities in the transfer equation, sketches an integral equation method for solving partial redistribution problems, and shows how the source function equation in a moving medium is modified. It also considers the impact of operator perturbation on methods for building model atmospheres and discusses the properties and construction of approximate integral operators.

063.016 **Transfer of line radiation: general analysis of approximate solutions.**
V. V. Ivanov, V. M. Serbin.
Astron. Zh., Tom. 61, Vyp. 4., p. 691 – 699 (1984). In Russian. English translation in Sov. Astron., Vol. 28, No. 4.

A critical analysis is given of various approximate solutions of the line transfer equation for static plane atmospheres under the assumption of complete frequency redistribution.

063.017 **Photon scattering coefficients for $H_2O^+$ molecules.**
N. N. Shefov.
Sov. Astron., Vol. 28, No. 1, p. 70 – 72 (1984). English translation of 37.063.002.

063.018 **Equilibrium emission of low–density, optically thick plasma.**
S. R. Kel'ner, E. S. Shikhovtseva.
Sov. Astron. Lett., Vol. 10, No. 1, p. 29 – 31 (1984). English translation of 37.063.001.

063.019 **Bidirectional reflectance spectroscopy. 3. Correction for macroscopic roughness.**
B. Hapke.
Icarus, Vol. 59, No. 1, p. 41 – 59 (1984).

A mathematically rigorous formalism is derived by which an arbitrary photometric function for the bidirectional reflectance of a smooth surface may be corrected to include effects of general macroscopic roughness. The correction is applied to the bidirectional reflectance function of B. Hapke (1981). Expressions for both the differential and integral brightnesses are obtained. Photometric profiles on hypothetical smooth and rough planets of low and high albedo are shown to illustrate the effects of macroscopic roughness. The theory is applied to observations of Mercury and predicts the integral phase function, the apparent polar darkening, and the lack of limb brightness surge on the planet.

063.020 **Experiments to test theoretical models of the polarization of light by rough surfaces.**
J. E. Geake, M. Geake, B. H. Zellner.
Mon. Not. R. Astron. Soc., Vol. 210, No. 1, p. 89 – 112 (1984).

A number of attempts have been made to provide theoretical models of the physical processes involved in the polarization of light scattered by a rough surface, such as the regolith of an atmosphereless planet. The authors describe some laboratory experiments designed to test different aspects of these models. They conclude that double Fresnel reflection is usually the dominant process in producing negative polarization, but that diffraction effects may play a significant part in double events involving small–scale surface features.

063.021 **The *H*–functions of radiative transfer in an active amplifying medium.**
S. R. Das Gupta, S. Das Gupta.
Astrophys. Space Sci., Vol. 103, No. 1, p. 81 – 93 (1984).

The authors obtain two explicit closed form representations of Chandrasekhar's *H*–functions $H(z)$ characterising transfer of radiation in an active amplifying medium corresponding to the dispersion function $T(z)$. Their basic properties are derived and some of their values are approximately computed.

063.022 **The numerical evaluation of the *H*–functions of radiative transfer in active amplifying media.**
S. R. Das Gupta, C. R. Nayak, S. Das Gupta.
Astrophys. Space Sci., Vol. 103, No. 2, p. 271 – 284 (1984).

The present paper contains extensive tables of the values of the *H*–function $H(z,\omega < 0)$ and of the moments of $Q(x)$ (in terms of which the moments of $H(z, \omega)$ can be determined) appropriate for transfer of radiation in active amplifying media in which $\omega < 0$. These values have been computed correct to the 7th decimal place for values of $\omega$ in the range $(-10^{-12}) - (-10^{30})$ and for values of $z \varepsilon [0, 1]$ with the aid of a 48–point Gaussian quadrature formula.

063.023 **On the theory of anisotropically scattered radiation. The explicit expression for the resolvent function.**
Eh. Kh. Danielyan.
Astrofizika, Tom 19, Vyp. 4, p. 711 – 718 (1983). In Russian. English translation in Astrophysics, Vol. 19, No. 4.

The solution of the basic integral equation of anisotropically scattered radiation in a homogeneous plane parallel slab of finite optical thickness is obtained.

063.024 **Integral relations and asymptotic formulae of the theory of radiative transfer.**
V. V. Sobolev.
Astrofizika, Tom 20, Vyp. 1, p. 123 – 132 (1984). In Russian. English translation in Astrophysics, Vol. 20, No. 1.

The problem of radiation transfer in a plane layer with arbitrary phase function and arbitrary value of particle albedo is considered. The integral relations are obtained for the intensities of radiation immediately from the transfer equation. The primary sources of energy may be arbitrary. Using these relations asymptotic formulae may be found for the emergent intensities from an optically thick layer. As examples such formulae are obtained for the following three cases: 1) plane layer illuminated by parallel rays; 2) homogeneous distribution of sources; 3) a radiant plane in the middle of the layer.

063.025 **Green's functions for the equation of radiative transfer in an infinite homogeneous medium with radial distribution of the sources.**
A. K. Kolesov.
Astrofizika, Tom 20, Vyp. 1, p. 133 – 147 (1984). In Russian. English translation in Astrophysics, Vol. 20, No. 1.

The homogeneous equation of radiative transfer in an anisotropically scattering spherical medium is solved. The complete set of singular eigenfunctions of this equation is found. A full–range orthogonality relation for these eigenfunctions is obtained. It is shown that the eigenvalues of the radiative transfer equation for spherical and plane geometries is the same. Green's functions of spherically symmetric problems for a homogeneous infinite spherical medium are constructed.

063.026 **Scattering of space–homogeneous isotropic radiation on cool electron gas.**
D. I. Nagirner.
Astrofizika, Tom 20, Vyp. 1, p. 149 – 156 (1984). In Russian. English translation in Astrophysics, Vol. 20, No. 1.

Multiple Compton scattering on a gas of electrons at rest is considered. The energies of photons are supposed to be much smaller than the rest energy of the electrons. The exact solutions of the Fokker–Planck equation are obtained in linear and non–dispersional approximations. The evolution of the spectrum of radiation is studied in one special case.

063.027 **On a single scattering approximation in the problem of transfer of polarized radiation.**
V. M. Loskutov, J. A. Freimanis.
Astrofizika, Tom 20, Vyp. 2, p. 295 – 302 (1984). In Russian. English translation in Astrophysics, Vol. 20, No. 2.

Approximate and exact values of the degree of polarization for radiation emergent from an atmosphere with nonconservative Rayleigh scattering are compared. Two cases of the source distribution are considered: 1) uniform distribution, 2) illumination by parallel rays. The conditions of applicability for single scattering approximation (both ordinary and modified) are determined.

063.028 **Radiation field in an infinite atmosphere.**
Zh. M. Dlugach, Eh. G. Yanovitskij.
Astrofizika, Tom 20, Vyp. 2, p. 303 – 314 (1984). In Russian. English translation in Astrophysics, Vol. 20, No. 2.

An infinite homogeneous atmosphere with a plane unidirectional source is considered. The reflection coefficient $\varrho_\infty$ for such an atmosphere is studied. A numerical method for the computation of $\varrho_\infty$ is suggested and the efficiency of this method is estimated. Some results of computations for the Henyey–Greenstein phase function with different anisotropy and absorption are given. An asymptotic formula for $\varrho_\infty$ in a nearly conservative case is obtained.

063.029 **Regularized Monte–Carlo calculations of the optical polarization induced by scattering in X–ray binaries.**
J. F. Dolan.
Astron. Astrophys., Vol. 138, No. 1, p. 1 – 9 (1984).

The polarization induced when light from the primary in a close binary scatters off material in the vicinity of the secondary is calculated by a regularized Monte–Carlo approach. The Stokes parameters of light from up to 5000 elements on a primary assumed to fill its Roche lobe, each element weighted in intensity by limb darkening and gravity darkening caused by tidal distortion, are followed through the scattering process. The net polarization observed from the binary system, including the unpolarized contributions from the unresolved primary, is calculated in increments of 30° in orbital longitude and 15° in orbital inclination for a wide variety of scattering sites and orbital and physical parameters of the binary, including elliptical orbits.

063.030 **Adaptive–mesh radiation hydrodynamics – I. The radiation transport equation in a completely adaptive coordinate system.**
K.–H. A. Winkler, M. L. Norman, D. Mihalas.
J. Quant. Spectrosc. Radiat. Transfer, Vol. 31, No. 6, p. 473 – 478 (1984).

The authors formulate the radiation transport equation in a completely adaptive coordinate system, which they define as a system in which the mesh in spacetime, angles and frequency adapts automatically to the dynamical evolution of the radiation field and fluid flow.

063.031 **Adaptive–mesh radiation hydrodynamics – II. The radiation and fluid equations in relativistic flows.**
D. Mihalas, K.–H. A. Winkler, M. L. Norman.
J. Quant. Spectrosc. Radiat. Transfer, Vol. 31, No. 6, p. 479 – 489 (1984).

The authors derive the radiation and fluid equations for relativistic flows in conservative form in a completely adaptive coordinate system.

063.032 **The range of validity of the Rayleigh limit for computing Mie scattering and extinction efficiencies.**
J. C. Ku, J. D. Felske.
J. Quant. Spectrosc. Radiat. Transfer, Vol. 31, No. 6, p. 569 – 574 (1984).

For the small particle limit for scattering by either absorbing or dielectric spheres, values of the size parameter are determined below which various approximate expressions for the scattering and extinction efficiencies are accurate to within 1%. Simple, analytical criteria for ensuring 1% accuracy are given in terms of the size parameter and the complex refractive index. A much broader range of refractive indices is considered than has been studied previously.

063.033 **Formation of emission spectra in moving media. Review.**
V. P. Grinin.
Astrofizika, Tom 20, Vyp. 2, p. 365 – 417 (1984). In Russian. English translation in Astrophysics, Vol. 20, No. 2.

Contents: Introduction. 1) Sobolev's escape–probability method; 2) A modification of the EPM for flows with non–local radiative coupling; 3) The main characteristic lengths and asymptotes of the kernel functions; 4) Formation of spectral lines in an envelope with supersonic motions; 5) Other asymptotic and approximate methods; 6) Numerical methods; 7) The multilevel problems; 8) Formation of maser lines in moving media; 9) The radiative force in the spectral lines; 10) Conclusion.

063.034 **The interaction principle in radiative transfer.**
A. Peraiah.
Astrophys. Space Sci., Vol. 105, No. 1, p. 209 – 212 (1984).

The author describes the interaction principle which is of fundamental importance to the theory of radiative transfer in one–, two–, and three–dimensional geometry. The practical difficulties associated with this principle in these geometries are also described.

063.035 **Line locking and SS 433.**
M. Pekarevich, T. Piran, J. Shaham.
Astrophys. J., Vol. 283, No. 1, p. 295 – 302 (1984).

The authors reconsider the general problem of acceleration by line locking in an optically thin medium. They present analytic solutions to the coupled equations of radiation transfer and matter motion. Using these solutions, the authors derive restrictions on the physical conditions of the accelerated matter. By applying these conditions to SS 433 it is found that if the absorbing ion is light (hydrogen or helium), the matter must be highly clumped, and the acceleration has to begin at $\sim 10^{12}$cm. Line–locking acceleration with a hydrogen–like heavy ion such as iron demands higher than solar ion abundance.

063.036 **Transfer of line radiation: approximate solutions for semi–infinite atmospheres.**
V. V. Ivanov, V. M. Serbin.
Astron. Zh., Tom 61, Vyp. 5, p. 900 – 912 (1984). In Russian. English translation in Sov. Astron., Vol. 28, No. 5.

An atmosphere which scatters radiation with complete frequency redistribution is suggested to be considered as an amplifier predominantly filtering low space frequencies of the primary source distribution. The usual thermalization length is closely related to its transmission bandwidth.

063.037 **Spectra from pair–equilibrium plasmas.**
A. A. Zdziarski.
Astrophys. J., Vol. 283, No. 2, p. 842 – 847 (1984).

Spectra from relativistic Maxwellian plasmas without magnetic fields or external photon sources and in pair equilibrium are presented. Photon emission by bremsstrahlung and pair annihilation processes, multiple Compton scatterings, and photon absorption by free–free and photon–photon pair production processes are modeled numerically using the Monte Carlo technique. Optical depth effects and equilibrium effects are solved for self–consistency by iteration. The results are applied to modelling X–ray spectra from active galactic nuclei.

063.038 **The Monte–Carlo method in radiative transfer problems.**
V. V. Karyukin.
Vestn. Leningr. Univ., Mat. Mekh. Astron., No. 7, p. 93 – 96 (1984). In Russian. Abstr. in Ref. Zh., 51. Astron., 9.51.153 (1984).

063.039 **Half–range moment methods for radiative transfer in spherical geometry. II. Implementation of the method.**
P. G. Martin, C. Rogers, G. B. Rybicki.
Astrophys. J., Vol. 284, No. 1, p. 317 – 326 (1984).

The authors implement and demonstrate the accuracy of a new numerical method for problems of radiative transfer in spherical geometry, based on half–range power moments of the specific intensity. Use of Ricatti transformation allows an accurate, though computationally inefficient, radial integration of the moment equations. The method was checked against both analytic (exact) solutions and results from several other numerical methods, for a variety of problems including those with central sources or with an isotropic external radiation field and including anisotropic and nonconservative scattering. When the radial integration is treated with a fast finite–difference scheme, this half–range moment method offers an efficient solution to a wide range of problems in radiative transfer.

063.040 **Half–range moment methods for radiative transfer in spherical geometry. III. Numerical solution and applications.**
C. Rogers, P. G. Martin.
Astrophys. J., Vol. 284, No. 1, p. 327 – 336 (1984).
The authors describe a fast, accurate numerical technique for solving the half–range moment equations for spherical transfer problems. Models of objects where the source of opacity is dust, as in circumstellar envelopes and interstellar clouds, provide practical demonstrations of the method. The authors calculate the optical visibility function of a circumstellar dust envelope and examine the effects of multiple scattering on extinction. For an interstellar cloud the apparent extinction, the interior radiation field, and the surface brightness profile are computed.

063.041 **Nonthermal synchrotron radiation in a strong magnetic field.**
R. W. Bussard.
Astrophys. J., Vol. 284, No. 1, p. 357 – 363 (1984).
Gamma–ray burst spectra are investigated under the assumption that mildly relativistic electrons are injected into a strong magnetic field ($B \sim 10^{12}$gauss). Such field strengths lead to extremely rapid synchrotron energy losses, and the radiation spectrum is calculated with the losses taken into account self–consistently. The conversion of electron kinetic energy to γ–rays ($\hbar\omega > 30$ keV) is found to be quite efficient, and the resulting spectra are not inconsistent with those observed.

063.042 **Non–coherent scattering in subordinate lines – V. Solutions of the transfer problem.**
I. Hubený, P. Heinzel.
J. Quant. Spectrosc. Radiat. Transfer, Vol. 32, No. 2, p. 159 – 168 (1984).
The authors present several examples of the numerical solution of the radiative transfer in subordinate lines. Using a simplified physical model that yields the line source function analogous to the usual two–level–atom form modified by the presence of the redistribution function $R_V$ in the scattering integral, the authors have solved the transfer problem for isothermal, plane–parallel atmospheres, both finite and semi–infinite.

063.043 **Hilbert transforms of resonance functions.**
S. C. Mehrotra.
J. Quant. Spectrosc. Radiat. Transfer, Vol. 32, No. 2, p. 169 – 171 (1984).
Expressions have been derived for the Hilbert transforms of resonance functions for induction and dispersion forces in spectral line–shape theories. These functions are evaluated numerically and are also represented by power series.

063.044 **Padé approximant in radiative transfer.**
S. A. El Wakil, M. H. Haggag, H. M. Machali, E. A. Saad.
J. Quant. Spectrosc. Radiat. Transfer, Vol. 32, No. 2, p. 173 – 177 (1984).
A functional relation is obtained between radiative transfer in an inhomogeneous medium with internal sources and diffuse reflection. The intensity of the emerging radiation for a linear source is obtained by using the Padé approximation. The single scattering albedo is assumed to decrease exponentially with optical depth. Numerical results are given.

063.045 **Physical processes in relativistic plasmas.**
R. Svensson.
MPE Rep., No. 184, p. 152 – 163 (1984). – See Abstr. 012.039.
This paper first discusses the processes taking place in relativistic plasmas and the standard models for the continuum emission from AGNs. It then summarizes the effects pair production has on these models and the observational implications of the presence of electron–positron pairs.

063.046 **Power law X– and γ–ray emission from relativistic thermal plasmas.**
A. A. Zdziarski.
MPE Rep., No. 184, p. 188 – 192 (1984). – See Abstr. 012.039.
This paper summarizes results of work on pair equilibrium in thermal plasmas emitting power law photon spectra by repeated Compton scatterings of a soft photon source. Model spectra are compared with observational data from active galactic nuclei.

063.047 **Pair production at moderate optical depths.**
S. Stepney, P. W. Guilbert.
MPE Rep., No. 184, p. 193 – 196 (1984). – See Abstr. 012.039.
Electron–positron pair production must be included when considering the hard X–ray spectra of AGN's and γ–ray bursters. The authors investigate thermal plasmas with temperatures $kT \cong m_e c^2$ and optical depths $1 \lesssim \tau \lesssim 5$, including pair processes, Comptonization and bremsstrahlung. Analytic and numerical results are presented for equilibrium and impulsively heated plasmas.

063.048 **Radiative transfer in a hot plasma.**
M. J. Goupil.
Astron. Astrophys., Vol. 139, No. 2, p. 353 – 358 (1984).
The radiative heat flux in a hot (1 MeV), optically thick plasma depends on the opacity of the medium, the Rosseland mean opacity which is defined by an appropriate averaging of the photon mean path $l(\omega)$ over frequency. Small departures from equilibrium give rise to a photon momentum distribution which is assumed to be the Planck function corrected by a small quantity which depends on $l(\omega)$ and positrons and electrons are assumed in thermal balance. It is then possible to linearize the relativistic Boltzmann transport equation for the photons, which yields a linear, integral equation for $l(\omega)$. The relativistic Boltzmann transport equation for the photons involves microscopic processes of interactions for the radiation. For $T$ around 1 MeV, two dominant processes have to be considered, Compton scattering and annihilation–creation of pairs.

063.049 **Boundary layer conditions for the transport of radiation in stars.**
H. Frisch, M. Faurobert.
Astron. Astrophys., Vol. 140, No. 1, p. 57 – 66 (1984).
This paper discusses the diffusion limit of non–grey radiative transport and gives a surface boundary condition for the temperature consistent with this limit. The emphasis is on stationary situations. An asymptotic analysis of the transfer equation coupled to an energy equation is carried out to first–order in an expansion parameter ε, which is essentially the ratio of a typical photon mean free–path, evaluated at atmospheric level, to a stellar characteristic dimension.

063.050 **The asymmetry of resonance line scattering in a velocity gradient.**
L. B. Lucy.
Astron. Astrophys., Vol. 140, No. 1, p. 210 – 212 (1984).
Monte Carlo experiments are reported that investigate the asymmetry of resonance–line scattering in the presence of a velocity gradient. The asymmetry found by Caroff et al. (1972) when individual scatterings are taken to be coherent in the atom's frame is shown to decay at large optical thicknesses. This contradicts a claim by Panagia and Macchetto (1982) and eliminates a possible explanation of the extended, black absorption troughs seen in the circumstellar spectra of hot stars. Multiply nonmonotonic flow appears to be the only convincing explanation of such troughs.

063.051 **Linear polarization of molecular lines at radio frequencies.**
S. Deguchi, W. D. Watson.
Astrophys. J., Vol. 285, No. 1, p. 126 - 133 (1984).

Two-level calculations by others for the linear polarization expected for molecular emission lines from interstellar clouds at radio frequencies are extended by considering a number of coupled rotational states of a diatomic molecule. Higher transitions commonly have strengths comparable with that of the $J = 1 \rightarrow 0$ transition so that depolarization due to coupling between the various $J$-states might be expected to reduce the predicted linear polarization. The authors find that the inclusion of additional states tends to decrease (by about a factor of 2) the maximum polarizations that are calculated, though the calculated polarization is substantially increased under conditions for which "superthermal" excitation occurs.

063.052 **Linear polarization of astronomical masers and magnetic fields.**
L. R. Western, W. D. Watson.
Astrophys. J., Vol. 285, No. 1, p. 158 - 173 (1984).

The ability of magnetic fields to cause high linear polarization in the radiation emitted by astronomical masers is examined. Because substantial linear polarization is a common feature of the circumstellar SiO masers and because their emitted power indicates masing under saturated conditions, the constraints provided by the SiO masers in late-type giants are emphasized. The transfer equations for polarized radiation in a one-dimensional maser are integrated numerically both for $J = 1 \rightarrow 0$ and $J = 2 \rightarrow 1$ transitions. It is shown that for circumstellar SiO masers fractional polarization up to about one-third is readily produced as a result of magnetic fields alone.

063.053 **The cross section for double Compton scattering.**
R. J. Gould.
Astrophys. J., Vol. 285, No. 1, p. 275 - 278 (1984).

Employing elementary methods in nonrelativistic quantum electrodynamics, the cross section for $\gamma_0 + e \rightarrow e + \gamma + \gamma'$ is computed for arbitrary energy in the spectrum of the outgoing photons. The final result is given, differential in the energy of one of these photons, for the case where the incident photon is unpolarized and has energy $E_0 \ll mc^2$, a polarization sum and angular integration being performed for the final-state photons.

063.054 **Minimum entropy production in the steady state and radiative transfer.**
C. Essex.
Astrophys. J., Vol. 285, No. 1, p. 279 - 293 (1984).

This paper attempts to establish a thermodynamic extremum principle for radiative transfer. It grapples with the bilinear formalism of irreversible thermodynamics, and introduces a simple thermodynamic approximation for entropy radiation that is theoretically and practically compatible with the local thermodynamic equilibrium approximation for energy radiation. An example, due to Planck, is generalized to apply to problems with nonequilibrium energy balance solutions. With the local thermodynamic equilibrium approximation for entropy radiation the discussion is broadened to encompass gray absorption.

063.055 **Radiative transfer in horizontally inhomogeneous media.**
K. Lumme, H. Karttunen, W. M. Irvine.
Bull. Am. Astron. Soc., Vol. 16, No. 3, Part I, p. 669 (1984). Abstract. - See Abstr. 010.061.

063.056 **Radiative transfer in a homogeneous sphere illuminated by a parallel beam.**
R.-L. Shia, Y. L. Yung.
Bull. Am. Astron. Soc., Vol. 16, No. 3, Part I, p. 669 - 670 (1984). Abstract. - See Abstr. 010.061.

063.057 **Three-dimensional radiative transfer in a scattering atmosphere using doubling and adding matrix algorithms.**
J. V. Martonchik, D. J. Diner.
Bull. Am. Astron. Soc., Vol. 16, No. 3, Part I, p. 670 (1984). Abstract. - See Abstr. 010.061.

063.058 **Cerenkov line emission: basic theory.**
J. H. You, T. Kiang, F. H. Cheng, F. Z. Cheng.
Mon. Not. R. Astron. Soc., Vol. 211, No. 3, p. 667 - 677 (1984).

Cerenkov radiation in the neighbourhood of an atomic line has been numerically demonstrated (You & Cheng). The basic theory of this 'Cerenkov line emission' is developed in the present paper. It is shown that there will be an emergent flux, that the emission will be redshifted, and that its profile will be broad and asymmetrical. Formulae are given for the profile, the total intensity, the displacement of the emission peak and the total width. The theory predicts a strict relation between any two intensity ratios of Balmer lines. It offers a new way of solving some of the outstanding problems of quasar emission lines, the low Ly$\alpha$/H$\beta$ ratio, their differing widths, their differing redshifts etc.

063.059 **Effects of grain models on UV penetration into clouds.**
G. Chlewicki, J. M. Greenberg.
Mon. Not. R. Astron. Soc., Vol. 211, No. 3, p. 719 - 736 (1984).

A derivation of the average scattering characteristics of interstellar grains in the ultraviolet based mainly on observations of extinction is shown to be more reliable than a determination from direct measures of scattered light. Several representative scattering models are constructed and their consequences for the transfer of UV radiation in clouds are discussed using the spherical harmonic expansion method (Flannery et al., 1980). Because of the presence of large grains ($a \sim 0.15$ μm) in the size distribution, all models are characterized by strongly anisotropic scattering throughout the UV ($g \approx 0.7 - 0.9$) with a relatively low albedo ($\alpha < 0.4$ for $\lambda < 2500$ Å) resulting from the presence of very small grains ($a \leqslant 0.01$ μm). An approximate algorithm replacing the exact solution of the radiative transfer equation by a greatly simplified solution for $g = 1$ is presented.

063.060 **The angle-dependent redistribution functions $R_{III}$ and $R_{IV}$: line absorption probability distribution functions and reemission coefficients.**
S. J. McKenna.
Astrophys. Space Sci., Vol. 106, No. 2, p. 299 - 307 (1984).

The line absorption probability distribution functions and the reemission coefficients are derived for the non-coherent scattering functions $R_{III}$ and $R_{IV}$. The appropriate line profile function for $R_{III}$ is shown to be a simple Voigt function, while for $R_{IV}$, the line absorption probability distribution function is more complex involving a linear combination of two Voigt functions and another more complex probability distribution. The structure of the reemission coefficients for $R_{III}$ and $R_{IV}$ is then discussed.

063.061 **On numerical evaluation of the $H$-functions of transport problems by kernel approximation for the albedo $0 < \omega \leqslant 1$.**
Z. Islam, S. R. Das Gupta.
Astrophys. Space Sci., Vol. 106, No. 2, p. 355 - 369 (1984).

063.062 **Standard problem of the theory of line formation in moving atmospheres.**
V. M. Serbin.
Vestn. Leningr. Univ., Mat. Mekh. Astron., 31 pp. (1984). In Russian. Abstr. in Ref. Zh., 51. Astron., 11.51.126 (1984).

063.063 **Higher order Levermore theory.**
G. C. Pomraning.
J. Quant. Spectrosc. Radiat. Transfer, Vol. 32, No. 3, p. 185 - 190 (1984).

The transport analysis of Levermore which in lowest order leads to a flux-limited diffusion equation, is extended to the next highest order. The resulting approximate description of transport processes is diffusion-like in character but is not flux limited.

063.064 **A generalized emissivity problem.**
G. C. Pomraning.
J. Quant. Spectrosc. Radiat. Transfer, Vol. 32, No. 3, p. 191 - 204 (1984).

The emissivity problem involving specular and diffuse reflection at the surface is considered. A variational principle is derived to estimate both the hemispherical and angular emissivities. Asymptotic and once iterated asymptotic angular distributions are used as trial functions to produce numerical results.

063.065 **The use of the Galerkin method for radiation transfer in an anisotropically scattering slab with reflecting boundaries.**
Y. A. Cengel, M. N. Özisik.
J. Quant. Spectrosc. Radiat. Transfer, Vol. 32, No. 3, p. 225 – 233 (1984).
Radiation transfer in an absorbing, emitting, anisotropically scattering, plane–parallel medium with diffusely reflecting boundaries is solved by application of the Galerkin method. With this approach, the radiation heat flux, angular distribution of radiation intensity, and the divergence of the radiation heat flux anywhere in the medium can be determined highly accurately. For optical thicknesses up to about 10, exact results are also readily obtainable if sufficient number of terms are considered in the expansion. Numerical results are presented for representative cases.

063.066 **A method for performing radiative transfer calculations in media where population inversion may occur.**
D. J. Bond.
J. Quant. Spectrosc. Radiat. Transfer, Vol. 32, No. 3, p. 235 – 240 (1984).

063.067 **Effects of individual line shapes on composite profiles and line series.**
C. A. Iglesias, R. W. Lee, C. F. Hooper, B. L. Whitten.
J. Quant. Spectrosc. Radiat. Transfer, Vol. 32, No. 3, p. 241 – 247 (1984).
The authors consider (1) the use of approximate ion microfield distributions as inputs into spectral line shapes and (2) the effects of individual lines on unresolved transition array (UTA) profiles observed in highly ionized plasmas. Spectral line shapes and line wings have significant effects on mean opacity calculations.

063.068 **Scattering of scalar waves by an impenetrable rough sphere.**
R. Schiffer, K. O. Thielheim.
Opt. Acta, Vol. 31, No. 10, p. 1085 – 1100 (1984).
The authors have calculated the incoherent differential cross–section of an impenetrable, slightly rough sphere of arbitrary size, but taking into account only the lowest–order term in the root–mean–square deviation of the surface from a smooth sphere. They have examined more thoroughly the back–scattering cross–section, and have shown that, in the large sphere limit and under certain conditions, it reduces to a result known in the literature.

063.069 **Theoretische Methoden zur Lichtstreuung an Teilchen mit unregelmäßiger Oberfläche.**
R. Schiffer.
Kleine Teilchen und Lichtstreuung, p. 1502 – 1515 (1984). – See Abstr. 012.074.

063.070 **Redistribution functions for astrophysical problems.**
C. Magnan.
Inst. Astrophys. Paris, Pré–Publ., No. 69, 34 pp. (1984). To appear in Astron. Astrophys.

063.071 **Line formation in the presence of turbulence.**
C. Magnan.
Inst. Astrophys. Paris, Pré–Publ., No. 76, 24 pp. (1984). To appear in Astron. Astrophys.

063.072 **Non local effects on the redistribution of resonant scattered photons.**
E. Simonneau.
Inst. Astrophys. Paris, Pré–Publ., No. 77, 14 pp. (1984). To appear in the Proceedings of Theory Workshop on Spectral Line Formation. Trieste, September 1984.

063.073 **Rapid non–LTE calculations of Balmer lines and hydrogen ionization: the solar case.**
S. Dumont, S. Collin–Souffrin.
Inst. Astrophys. Paris, Pré–Publ., No. 78, 12 pp. (1984). To appear in Astron. Astrophys.

063.074 **Convection and gravity waves in two layer models. II. Overstable convection of large horizontal scales.**
M.–A. Kondo, W. Unno.
Geophys. Astrophys. Fluid Dyn., Vol. 27, p. 229 – 252 (1983). = Contrib. Dep. Astron., Univ. Tokyo, No. 324.
Penetrative convection is studied for idealized two–layer models composed of convectively unstable and stable layers with constant viscosity and conductivity in each layer but different from that in the other layer.

063.075 **Soft X–ray opacity in hot and photoionized gases.**
J. H. Krolik, T. R. Kallman.
Astrophys. J., Vol. 286, No. 1, p. 366 – 370 (1984).
New calculations are presented of the soft X–ray opacity of gas having cosmic elemental abundances, but in a variety of ionization states. The authors consider coronal ionization equilibrium over the range of temperatures $10^4$ to $3\times10^8$K and photoionization equilibrium with power–law and thermal bremsstrahlung spectra over a wide range of ionization parameters. It is shown that the soft X–ray opacity of the gas declines as it is heated and photoionized and becomes significantly smaller than the opacity predicted in the neutral atom limit.

063.076 **Non–LTE Line–Formation mit vereinfachtem Strahlungstransport.**
K. Werner.
Mitt. Astron. Ges., Nr. 62, p. 250 (1984). Abstract. – See Abstr. 010.341.

063.077 **Two–dimensional radiative transfer.**
G. Hensler.
Mitt. Astron. Ges., Nr. 62, p. 280 (1984). Abstract. – See Abstr. 010.341.

063.078 **Atmospheric transfer of radiation above an inhomogeneous non–Lambertian reflective ground. II. Computational considerations and results.**
D. J. Diner, J. V. Martonchik.
J. Quant. Spectrosc. Radiat. Transfer, Vol. 32, No. 4, p. 279 – 304 (1984).
The theoretical foundation for solution of the three–dimensional radiative transfer problem described in the authors' preceding paper is reviewed. Practical considerations involved in implementing the Fourier transform/Gauss–Seidel method on a minicomputer are discussed, along with derivations of symmetry relations and approximations which can be used to enhance the computational efficiency. Model results for a surface whose albedo varies as a step function are presented and compared with published solutions obtained by using the Monte Carlo method.

063.079 **Particle transfer in decelerating regions.**
E. A. Saad, A. Hendi.
Astrophys. Space Sci., Vol. 107, No. 1, p. 1 – 9 (1984).
Equations connecting the transmission and reflection functions of a finite medium to the reflection function of a semi–infinite one are used to get the albedos in the slowing–down region. The transport equation is solved by the modified Eddington method using Fermi's backward–forward scattering model, modulated to allow for different orders of backward–forward and isotropic scattering. Numerical results for the energy albedos are obtained and compared.

063.080 **Energy albedo for multiregion slabs.**
E. A. Saad, M. A. El Gazzar, M. S. Abdel Krim, S. A. El Wakil.
Astrophys. Space Sci., Vol. 107, No. 1, p. 11 – 18 (1984).
An asymptotic solution for the equation of radiative transfer in an inhomogeneous medium was obtained on the basis of the corresponding solutions for homogeneous sub–layers in the slow-

ing down region. Function relations between the reflection and transmission coefficients for the whole slab and those of the sub-layers are given. The invariant embedding concepts are used to get the reflection and transmission coefficients for the sub-layers. The authors assumed different models for the slowing-down kernels. Laplace transform was used to transform the Boltzmann equation to one velocity approximation with re-scaled mean-free path and single-scattering albedo. Numerical results are given for energy albedo as a function of the mass number of the host medium.

063.081 **Radiative transfer equation in spherically-symmetric non-scattering media.**
A. Peraiah, B. A. Varghese.
Astrophys. Space Sci., Vol. 107, No. 1, p. 177 - 190 (1984).
The authors solved the equation of radiative transfer in spherically-symmetric shells with arbitrary internal sources. They integrated the equation of transfer on a discrete grid of angle and radius. Two cases of variation of the Planck function are considered: (1) constant throughout the medium and (2) varying as $1/r^2$. It is found that in the inner shells, the radiation directed toward the centre of the sphere is greater than that directed away from the centre of the sphere. In the outer shells the converse is true.

063.082 **The comoving-frame equation of radiative transfer in a curvilinear coordinate system.**
N. Kaneko, K. Morita, M. Maekawa.
Astrophys. Space Sci., Vol. 107, No. 2, p. 333 - 346 (1984).
The comoving-frame equation of radiative transfer and moment equations are derived in orthogonal, curvilinear coordinates, inclusive of terms of order $v/c$. The equation of radiative transfer, which contains the terms due to the effect of curvature of coordinate lines explicitly as well as those of Doppler shift and aberration, is the generalization of Castor's equation for spherical symmetry and of Buchler's equation for Cartesian coordinates. The moment equations agree with Buchler's.

063.083 **On the asymptotic light regime in the outer layers of a homogeneous sphere of large optical radius.**
A. K. Kolesov.
Astrofizika, Tom 21, Vyp. 2, p. 309 - 322 (1984). In Russian. English translation in Astrophysics, Vol. 21, No. 2.
The radiation field in the outer layers of a homogeneous sphere of large optical radius is investigated in the case of anisotropic scattering. Asymptotic expressions for Green's functions for the sphere with radial distributions of light sources and in particular with a point source in the sphere center are obtained. Corresponding formulae for the intensities of radiation are found.

063.084 **Statistical description of a radiation field on the basis of the invariance principle. I. The mean number of scatterings in a medium illuminated from outside.**
A. G. Nikogosyan.
Astrofizika, Tom 21, Vyp. 2, p. 324 - 341 (1984). In Russian. English translation in Astrophysics, Vol. 21, No. 2.
A new approach to determine the mean number of scatterings based on Ambartsumyan's invariance principle and on the systematic application of the generating function method is suggested. The mean quantities evaluated in the paper concern the case when the medium is illuminated from outside. The photons destroyed as a result of diffusion in the medium and the photons escaping the medium are considered separately. It is shown that the suggested approach enables one to reveal the dependence of the mean number of scatterings on the characteristics of the incident photon and can be used under general assumptions concerning the elementary act of scattering.

063.085 **A remark on the consideration of the motion of electrons in their interaction with photons in astrophysical problems.**
G. T. Ter-Kazaryan.
Astrofizika, Tom 21, Vyp. 2, p. 343 - 352 (1984). In Russian. English translation in Astrophysics, Vol. 21, No. 2.
The formulas which take into account the electron's motion in problems of Compton scattering are improved by means of correct definition of the solid angle in the scattered photon direction. The correct expressions of the scattering function and energy losses are obtained.

063.086 **Half-range moment methods for radiative transfer in spherical geometry. V. Line formation in a static atmosphere.**
C. Rogers.
Astrophys. J., Vol. 286, No. 2, p. 659 - 665 (1984).
The author describes a fast, accurate numerical technique for solving the problem of line formation in a static spherical atmosphere. Two numerical examples provide a comparison of this method with two other independent calculations. The method of solution can accommodate the effects of atmospheric expansion and frequency redistribution.

063.087 **On Bowen enhancement of the N III spectrum under solar and nebular conditions.**
S. O. Kastner, A. K. Bhatia.
Astrophys. J., Vol. 287, No. 2, p. 945 - 951 (1984).
Expected enhancement of the N III emission-line spectrum by the Bowen He II → O III → N III resonant photoexcitation process is computed using a six-configuration model. Preliminary comparisons with available solar and nebular observations indicate that the process may play only a small role, although photoexcitation quenching of the N III and O III far-infrared fine-structure lines is noted as a possible important mechanism. Some features are discussed in connection with the nebular emission at 1750 Å, which is suggested as an indicator of excitation class for planetary nebulae.

**Determination of mineral abundance from reflectance spectra.**
See Abstr. 022.111.

**Steady mildly relativistic thermal plasmas: processes and properties.**
See Abstr. 062.002.

**Inverse Compton reflection from an ionized non-relativistic plasma.**
See Abstr. 062.040.

**Optically thick ablation fronts.**
See Abstr. 062.086.

**A study of Comptonization of radiation in an electron plasma using the Monte Carlo method.**
See Abstr. 062.094.

**Model calculations of magnetic flux tubes. I. Equations and method.**
See Abstr. 062.095.

**Monte Carlo shock-like solutions to the Boltzmann equation with collective scattering.**
See Abstr. 062.104.

**Polarization of continuum radiation in magnetic atmospheres.**
See Abstr. 064.024.

**The $H^-$ equilibrium using coupled rate equations for $H^-$, H, $H^+$, $H_2$, and $H_2^+$.**
See Abstr. 064.029.

**Theory of astrophysical SiO masers.**
See Abstr. 064.037.

**The transfer of polarized radiation in spectral lines: solar-type stellar atmospheres.**
See Abstr. 064.049.

**Numerical solution of the equation of radiative equilibrium.**
See Abstr. 064.063.

**Self-consistent radiation-hydrodynamic equations for stellar oscillations. I. Nonlinear form.**
See Abstr. 065.055.

**Duelling winds: a model for broad-line regions.**
See Abstr. 067.088.

**A note on the variational method of Stokes and DeMarcus for radiative transfer in planetary atmospheres.**
See Abstr. 091.006.

**Application of a radiative transfer model to bright icy satellites.**
See Abstr. 091.043.

**Comet dust – applications of Mie scattering.**
See Abstr. 102.024.

**Radiative processes in cool star chromospheres.**
See Abstr. 112.026.

**The ultraviolet spectrum of stars having optical Fe II emission lines.**
See Abstr. 114.044.

**Self-absorption of high-energy gamma-rays in Cygnus X-3.**
See Abstr. 117.257.

**On the theory of cyclotron lines in the spectra of magnetic white dwarfs.**
See Abstr. 126.032.

**Polarization angle in magnetic white dwarfs.**
See Abstr. 126.041.

**Radiative transfer effect in $H_2O$ masers.**
See Abstr. 131.174.

**A semi-classical description of radiation transport in astrophysical masers.**
See Abstr. 131.192.

**Erratum: "Non-linear theory of cosmic ray shocks including self-generated Alfvén waves" [Astron. Astrophys., Vol. 116, p. 191 – 200 (1982)].**
See Abstr. 144.003.

**A universal spectrum for AGNs and QSOs.**
See Abstr. 158.138.

**An inverse Compton model of the X-ray emission in Seyfert galaxies.**
See Abstr. 158.139.

**Theoretical studies of emission line variability in type 1 Seyfert galaxies.**
See Abstr. 158.140.

**On the stability of broad line clouds in active galactic nuclei.**
See Abstr. 158.141.

**Photon-photon absorption and the uniqueness of the spectra of active galactic nuclei.**
See Abstr. 158.268.

**Photoionization models and diagnosis of physical properties of the broad-line emission gas in quasars and Seyfert nuclei.**
See Abstr. 159.050.

# 064 Stellar Atmospheres, Stellar Envelopes, Mass Loss, Accretion

064.001 **The vertical structure and stability of alpha model accretion disks.**
J. K. Cannizzo, J. C. Wheeler.
Astrophys. J., Suppl. Ser., Vol. 55, No. 3, p. 367 – 388 (1984).

The results of a numerical study of the steady-state vertical structure of thin accretion disks are presented for ranges of $\alpha$, $M$, and $r$ appropriate to compact binary systems. Particular attention is paid to the role of convection and to the location of critical points which represent the onset of thermal instability relevant to outbursts in dwarf novae and certain X-ray transients. The temperature of the bulk of the matter in quiescence is likely to be low, $T \leqslant 2500$K.

064.002 **The infrared and radio spectrum of early-type stars with mass loss. I. The curve of growth method.**
H. J. G. L. M. Lamers, L. B. F. M. Waters.
Astron. Astrophys., Vol. 136, No. 1, p. 37 – 52 (1984).

The theory of free-free and free-bound emission from stellar winds is reformulated and simple but accurate diagnostic methods are developed. For isothermal winds the spectrum of the infrared and radio excess radiation can be expressed in terms of a curve of growth the shape of which depends only on the velocity law. By comparing observed and predicted curves of growth the mass loss rate as well as the velocity law can be derived. The effect of electron scattering on the energy distribution is studied and an easy method is developed to allow for it. The excess radiation can also be expressed in terms of an effective radius or an effective optical depth. This concept of effective radius can be used to find first estimates of the velocity law in the wind. The curve of growth method is demonstrated in an application to the infrared and radio observations of the star $\varepsilon$ Ori (B0 Ia).

064.003 **Magnetohydrodynamic critical levels and radiative damping.**
P. S. Cally.
Astron. Astrophys., Vol. 136, No. 1, p. 121 – 126 (1984).

The full MHD wave equation for radiatively damped wave propagation in a stratified magnetoatmosphere permeated by a horizontal magnetic field is derived. In the adiabatic case a discussion of the various types of singularity (critical level) is given. For an isothermal atmosphere with uniform magnetic field the damped wave equation is solved in terms of hypergeometric functions, and it is concluded that for many cases of interest the erstwhile discontinuity in the wave energy flux is considerably "smoothed out" by the damping. Substantial fluxes ($\sim 10^8$ erg cm$^{-2}$ s$^{-1}$) are found to be carried by some modes which, in the absence of critical levels, would carry no energy.

064.004 **Generation of acoustic energy from convection zones of late type stars.**
H. U. Bohn.
Astron. Astrophys., Vol. 136, No. 2, p. 338 – 350 (1984).

Former computations relating to the acoustic energy generation from convection zones of late type stars need to be revised. By taking into account the effect of absorption by dust and molecules for the opacity tables and improving the calculation of thermodynamical quantities by including molecular dissociation the stellar convection zone models yield larger convective velocities. Moreover, for low temperatures and high surface gravities, the neglect of monopole- and dipole-source terms for the calculation of the sound generation by the Lighthill-Proudman theory causes a large deficiency in the resulting acoustic output. Improved calculations are carried out for stellar parameters in the range $0 \leqslant \log g \leqslant 6$ and $2500\text{K} \leqslant T_{\text{eff}} \leqslant 10{,}500\text{K}$ and show for the latest stars up to five orders of magnitude more acoustic energy than the usual models.

064.005 **Recombination stellar wind model for the coronae of early–type stars.**
W. L. Waldron.
Astrophys. J., Vol. 282, No. 1, p. 256 – 266 (1984).

The wind structures of early–type stars which have base coronal zones are discussed. The calculations give rise to what is called a recombination stellar wind structure because of the transition from a coronally driven wind to a radiatively driven wind. The coronal X–rays are a dominant factor in determining the wind structure because of their effect on the ionization structure, which in turn determines the heating, cooling, and the attenuating properties of the stellar wind. The X–ray spectra are compared with those observed with the IPC instrument on the Einstein Observatory and are found to be in excellent agreement. The comparison suggests that the X–ray IPC spectra reflect the absorption properties of the wind and not the nature of the X–ray source.

064.006 **The infrared and radio spectrum of early type stars with mass loss. II. Tables of theoretical curves of growth for IR and radio excess and gaunt factors.**
L. B. F. M. Waters, H. J. G. L. M. Lamers.
Astron. Astrophys., Suppl. Ser., Vol. 57, No. 3, p. 327 – 352 (1984).

The authors present an extensive grid of theoretical curves of growth for the IR and radio excess of early type stars with mass loss. The curve of growth for IR and radio excess is a relation between the measured flux normalized to the expected photospheric flux, and the wavelength, and has been introduced by Lamers and Waters (1984, Paper I). Comparison of observed curves of growth with theoretical ones yields information on the mass loss rate and shape of the velocity law. Also, the authors present an extensive grid of bound–free and bound–free + free–free gaunt factors for a large range of temperatures ($8000 - 10^6$K) for a gas consisting of H, He and CNO. Finally, the dependence of excess flux on model parameters is illustrated by means of a difference table.

064.007 **Shapes of wind driven circumstellar shells.**
F. Kahn, K. West.
Occas. Rep. R. Obs., Edinb., No. 13, p. 129 – 132 (1984). – See Abstr. 012.001.

064.008 **Molecular viscosity as a mechanism of generation of magnetic field in accretion discs.**
R. Byrka, J. Jaskólska, P. Majer.
Astron. Astrophys., Vol. 137, No. 1, p. 101 – 107 (1984).

A mechanism of generation of magnetic field by molecular viscosity, originally presented by Browne (1968) is reexamined and its application to accretion discs is made. The difference between the viscous retardations of electrons and ions causes electric current which, in turn, generates magnetic field. Estimates of the intensity of the magnetic field component normal to the disc equatorial plane (i.e. $B_z$ component) are performed for very simplified disc models of two types: (1) with diffusing magnetic field, (2) with magnetic field frozen in the matter (infinite conductivity approximation). In both types of discs, however, the magnetic field strength does not exceed $10^{-2}$G. This field can be used as a "seed" field in dynamo action as proposed by Pudritz (1981).

064.009 **Molecular abundances and nucleation in protostellar envelopes.**
S. C. F. Rossi, W. J. Maciel.
Astrophys. Space Sci., Vol. 103, No. 1, p. 143 – 154 (1984).

This work is basically concerned with grain nucleation occurring in protostellar envelopes. On the basis of the dissociation equilibrium theory, molecular and atomic abundances are obtained for massive protostar envelopes. Application of time–independent homogeneous nucleation theory results in the possibility of atomic Fe condensation.

064.010 **Non–critical solution of a magnetic accretion disk.**
R. Horiuchi.
Prog. Theor. Phys., Vol. 71, No. 2, p. 295 – 308 (1984). Abstr. in Phys. Abstr., Vol. 87, No. 1239, Entry 82467 (1984).

064.011 **Mechanisms of heating of the coronae around accretion disks.**
Yu. É. Lyubarskij (*Yu. Eh. Lyubarskij*).
Sov. Astron., Vol. 28, No. 1, p. 59 – 61 (1984). English translation of 37.112.004.

064.012 **On the analytical properties of "alpha–beta" discs.**
R. Dgani, M. Livio.
Mon. Not. R. Astron. Soc., Vol. 210, No. 2, p. 393 – 398 (1984).

The authors study the equations describing standard accretion discs, with the effects of stream penetration included in a parametrized form. Fully analytic solutions can be obtained in several simplified cases. The ones that can relatively easily be analyzed, exhibit some of the properties expected from the competition between the action of the stream, which tends to shrink the disc into an annulus and the action of the viscous stresses which tends to broaden the disc.

064.013 **The infrared and radio spectrum of early type stars with mass loss. III. The effect of coronae, hot blobs and clumping on the IR and radio spectrum.**
H. J. G. L. M. Lamers, L. B. F. M. Waters.
Astron. Astrophys., Vol. 138, No. 1, p. 25 – 35 (1984).

The effect of different temperature structures and of clumping and hot blobs in stellar winds on the infrared and radio excess is studied. It is shown that the presence of a corona can lead to either an increase or a decrease (albeit marginally detectable) of the IR excess, depending on the optical depth of the corona. The authors describe the effect of hot blobs or hot filaments on the IR spectrum. They describe a simple method for calculating the excess flux from an inhomogeneous stellar wind and they calculate the possible errors in the derived mass loss rates when clumping is neglected.

064.014 **Corona models tested with IUE and Einstein observations.**
R. Hammer, J. L. Linsky.
ESA Spec. Publ., ESA SP–218, p. 25 – 32 (1984). – See Abstr. 012.024.

The authors review recent compilations of IUE and Einstein observations which show that the emissions from the various outer layers of cool stars are nonlinearly correlated. This result can be used to test theoretical corona models as well as hypotheses on the mechanism that determines the location of the transition region.

064.015 **Calculations of two–parameter representations of limb darkening laws for classical stars. III. Application to particular darkening laws.**
A. A. Rubashevskij.
Astrometr. Astrofiz., Vyp. 52, p. 18 – 25 (1984). In Russian.

Calculation of two–parameter nonlinear approximations was carried out for the limb darkening laws of classical stars derived by Grygar, Cooper and Jurkevich (1972), Gingerich (1966) and Margrave (1969) from models of atmospheres. In all cases the best possible fitting of surface brightness distribution over the stellar disc was derived provided that the total flux of its radiation is retained.

064.016 **The stability limit of hypergiant photospheres.**
C. De Jager.
Astron. Astrophys., Vol. 138, No. 2, p. 246 – 252 (1984).

The author examines the consequences of the hypothesis that the stability limit of the atmospheres of hypergiants is reached when the effective acceleration of gravity $g_{\text{eff}} = g_{\text{grav}}(1 - \Gamma) - g_{\text{turb}} \approx 0$, where $g_{\text{grav}} = GM/R^2$, $g_{\text{turb}}$ is the outward acceleration due to the gradient of turbulent pressure, and $\Gamma$ is the Eddington ratio $|g_{\text{rad}}/g_{\text{grav}}|$. In stars close to the stability limit dissipation of mechanical energy builds up a field of supersonic turbulent motions in the photosphere which tends to destabilize the atmosphere. The larger the absolute luminosity, the deeper in the photosphere dissipation becomes important. This manifests itself in a rate of mass–loss, increasing with $L$. For near–unstable stars with $T_e \approx 10^4$K the rate of mass–loss is predicted to be nearly by a factor ten larger for a star brighter by one magnitude. This effect greatly changes the evolutionary tracks and explains why brighter stars have not been detected.

064.017 **Evolution of the silicon abundance distribution in the atmosphere of magnetic Bp and Ap stars.**
C. Mégessier.
Astron. Astrophys., Vol. 138, No. 2, p. 267 – 274 (1984).

It has been shown already that diffusion processes in the presence of a magnetic field in the atmosphere of Bp and Ap stars lead to surface inhomogeneities of the chemical abundances. Here the author gives a complete description of the distribution of the silicon abundance in the atmosphere of magnetic stars for two effective temperatures: 12000K and 14000K, and for dipolar magnetic field strengths up to $10^5$Gauss. As usual, diffusion processes are considered during the main sequence stage of the star.

064.018 **Magnetospheres of early–type chemically peculiar magnetic stars.**
O. Havnes, C. K. Goertz.
Astron. Astrophys., Vol. 138, No. 2, p. 421 – 430 (1984).

Magnetospheric plasma densities and temperatures are calculated for early type magnetic stars (chemically peculiar stars) on the assumption that the magnetic field is that of a dipole.

064.019 **The role of oscillator strengths in modelling and analysing stellar spectra.**
B. Baschek.
Phys. Scr., Vol. T8, p. 21 – 24 (1984). – See Abstr. 012.027.

In order to elucidate the needs for atomic data, in particular for oscillator stengths, an overview is given over the role of line absorption in the quantitative analysis of high–resolution stellar spectra.

064.020 **On the need for spectroscopic data in ultraviolet stellar spectral synthesis calculations.**
S. J. Adelman, D. S. Leckrone.
Phys. Scr., Vol. T8, p. 25 – 27 (1984). – See Abstr. 012.027.

The ultraviolet spectra of sharp–lined normal and peculiar stars are rich in information about the abundances of the chemical elements, patterns of nucleosynthesis in our galaxy and the mechanisms responsible for the production of abundance anomalies. The authors illustrate their efforts to theoretically synthesize observed line profiles in the spectra of three sharp–lined, normal, moderately hot stars. These computations are seriously complicated and the results made ambiguous by the low accuracy of the oscillator strengths used (primarily those of Kurucz and Peytremann) and by the incompleteness of finding lists.

064.021 **Radiative forces and the abundance of beryllium, magnesium and barium in HgMn stars.**
J. Borsenberger, G. Michaud, F. Praderie.
Astron. Astrophys., Vol. 139, No. 1, p. 147 – 160 (1984).

Detailed radiative accelerations for Be, Mg, and Ba have been calculated in atmospheres and envelopes of main sequence ($\log g = 4$) stars with $T_{eff} = 10{,}000$, 12,500, and 15,000K i.e. in the temperature range of HgMn stars. For the three elements, the non–LTE radiative acceleration in the atmosphere is able to support the abundances observed in HgMn stars.

064.022 **Stark broadening in non–LTE stellar atmospheres: effects on occupation numbers and line profiles.**
K. Werner.
Astron. Astrophys., Vol. 139, No. 1, p. 237 – 239 (1984).

For stars whose parameters are within $32{,}000\text{K} \leqslant T_{eff} \leqslant 100{,}000\text{K}$ and $4.1 \leqslant \log g \leqslant 8.0$ it is shown, that their atmospheres are described satisfactorily by non–LTE models in which the hydrogen absorption coefficients are approximated by depth independent Doppler profiles. When the depth dependent linear Stark effect is included in the rate equations and the equation of radiative transfer, occupation numbers and line profiles change only marginally.

064.023 **Density inversion in a red giant envelope.**
A. Harpaz.
Mon. Not. R. Astron. Soc., Vol. 210, No. 3, p. 633 – 639 (1984).

The phenomenon of density inversion in the recombination zone is discussed through detailed analysis of a particular envelope of a red giant star. It is found that the completion of the recombination of the hydrogen causes a drop in the efficiency of energy transfer by convection. This drop raises the need for a steep temperature gradient, which requires an opposite density gradient to balance the hydrostatic equation.

064.024 **Polarization of continuum radiation in magnetic atmospheres.**
K. N. Nagendra, A. Peraiah.
Astrophys. Space Sci., Vol. 104, No. 1, p. 61 – 72 (1984).

A numerical solution is presented for the problem of continuum radiative transfer in a magnetoactive medium. The continuum opacities are calculated in the presence of a strong magnetic field ($H = 10^7$G) typical of magnetic white dwarfs. The L.T.E. pure absorption model is assumed for calculating the polarized radiation field emitted by a realistic model atmosphere in the plane parallel approximation. The wavelength dependence of the linear and circular polarizations are calculated for both uniform and dipole field configurations.

064.025 **Winds from rotating, magnetic, hot stars. I. General model results.**
D. B. Friend, K. B. MacGregor.
Astrophys. J., Vol. 282, No. 2, p. 591 – 602 (1984).

The authors present a rotating magnetic wind model for hot stars, which combines the Weber and Davis description of a rotating magnetic solar wind with the Castor, Abbott, and Klein (CAK) theory of a line radiation–driven wind. The wind model is unique for a given set of stellar parameters, and passes through *two* CAK–type critical points. The main effect of an increase in the magnetic field is to increase the Lorentz force at large radii, thereby raising the terminal velocity. The rotational velocity, on the other hand, acts mainly to increase the centrifugal force near the base of the wind, which raises the mass loss rate. The observational consequences of these results for the winds from early–type stars are discussed.

064.026 **On the possibility of resonant electrodynamic coupling in the coronae of red dwarfs.**
D. J. Mullan.
Astrophys. J., Vol. 282, No. 2, p. 603 – 611 (1984).

The authors suggest that, in the context of Ionson's electrodynamic coupling theory of coronal heating, the coupling efficiency may pass through a resonance in lower main–sequence stars. It is proposed that this resonance contributes to the strong observed X–ray emission in early–M dwarfs. In the coolest M dwarfs, the efficiency is expected to decrease rapidly as conditions move away from resonance.

064.027 **Computation of convective model atmospheres of cool white dwarfs.**
A. Galdikas (*A. K. Gal'dikas*).
Astrophys. Space Sci., Vol. 105, No. 1, p. 39 – 45 (1984).

It is shown that, in constructing model atmospheres of cool white dwarfs with strong convection, the computation of the radiative flux can be omitted in a number of iterations. This does not increase the number of iterations needed to obtain convergent model and, consequently, saves computation time.

064.028 **Corotating interaction regions in stellar winds.**
D. J. Mullan.
Astrophys. J., Vol. 283, No. 1, p. 303 – 312 (1984).

In a stellar wind, a corotating interaction region (CIR) is formed when a rotating star emits wind in a non–spherically symmetric manner. The radial distance at which CIRs form (in terms of the stellar radius) is proportional to the ratio of wind speed to the rotational speed at the stellar surface: in the Sun, this ratio is ~200, but it may be much smaller than that in other stars. The properties of CIRs appear to be able to account for certain features of "hybrid stars" (i.e., cool giants with cool winds plus "warm" lines in their UV spectra), for X-ray emitting structures in hot star winds, and for the appearance of extended chromospheres in cool supergiants.

064.029 **The $H^-$ equilibrium using coupled rate equations for $H^-$, H, $H^+$, $H_2$, and $H_2^+$.**
B. W. Lites, D. Mihalas.
Sol. Phys., Vol. 93, No. 1, p. 23 – 35 (1984).
The authors formulate rate equations for the reaction network coupling H, $H^-$, $H^+$, $H_2$, and $H_2^+$. They attempt to systematize the notation, and to write the equations in a form suitable for modern computational methods of handling the coupled rate equations and radiative transfer equations, for both dynamical and static atmospheres. The authors have accounted for more processes than are generally considered in most current work; some of these may have an impact on the equilibrium of $H^-$ (hence its opacity) and on charge conservation (hence the proton density) in the atmospheres of solar-type stars.

064.030 **Disques d'accrétion. Modèles simplifiés.**
M. Friedjung.
Instabilités hydrodynamiques et applications astrophysiques, p. 471 – 478 (1983). – See Abstr. 012.034.
Contents: Structure d'un disque d'accrétion stationnaire: L'équation de conservation de la quantité de mouvement radial. L'équation de conservation du moment angulaire. Energie émise par un disque stationnaire. Tests observationnels.

064.031 **Revue générale de la théorie des disques d'accrétion.**
J.-P. Poyet.
Instabilités hydrodynamiques et applications astrophysiques, p. 497 – 582 (1983). – See Abstr. 012.034.
Contents: Introduction. Historique. La théorie des disques minces stationnaires. Comment fabriquer un disque d'accrétion. Les disques dépendent du temps. L'origine de la viscosité: Les disques convectifs. Les disques en rotation turbulents. Les disques magnétiques. Les disques supercritiques. Les disques épais. Formation de jet. Conclusion.

064.032 **Comments concerning magnetic fields as a possible cause of rapid, irregular variability of early-type stars.**
A. B. Underhill.
Hvar Obs. Bull., Vol. 7, No. 1, p. 345 – 358 (1983). – See Abstr. 012.037.
The spectra of stars are interpreted by noting the agreements and discrepancies between observed spectra and predicted spectra from model atmospheres. The purpose of this paper is to describe a model which allows to understand a possible cause of the irregular spectroscopic changes of early-type stars. The model can help to understand what may be happening in intervals of time from hours to decades.

064.033 **Instabilities in line-driven stellar winds. I. Dependence on perturbation wavelength.**
S. P. Owocki, G. B. Rybicki.
Astrophys. J., Vol. 284, No. 1, p. 337 – 350 (1984).
The authors investigate the stability of absorption-line-driven flows against small-amplitude velocity perturbations. Their work represents a generalization of previous linear stability analyses in that they assume neither that the flow perturbations are optically thin, nor that they satisfy the Sobolev approximation. The assumption that one may neglect the dynamical effect of any scattered radiation field is retained. It is confirmed that absorption-line-driven flows are indeed very unstable to optically-thin perturbations with a wavelength $\lambda$ approximately equal to or less than the Sobolev length $L$, but that they become asymptotically stable to any perturbations with $\lambda \gg L$.

064.034 **Wave amplification on line-driven winds.**
L. B. Lucy.
Astrophys. J., Vol. 284, No. 1, p. 351 – 356 (1984).
The dynamical effect of scattered line photons is investigated in relation to the amplification of radiative-acoustic waves propagating radially in the acceleration zones of line-driven winds. For wavelengths long compared to the Sobolev length, Abbott's analysis is confirmed. For short wavelengths a damping effect (line drag) is found which, for lines formed by coherent scattering, exactly cancels the destabilizing effect found by earlier investigators assuming line formation by pure extinction. Line-driven winds seem therefore not to be as highly unstable as hitherto claimed.

064.035 **Relativistic charge-separated winds.**
F. C. Michel.
Astrophys. J., Vol. 284, No. 1, p. 384 – 388 (1984).
The author presents three-dimensional solutions for the particle and field distribution in a relativistic charge-separated wind, such as expected specifically from pulsars and from other models involving rotating magnetized objects. Particles are accelerated in this wind zone to Lorentz factors of the order of $\sigma^{2/3}$ in contrast to the $\sigma^{1/3}$ estimate given by magnetohydrodynamic theory ($\sigma$ is the Lorentz factor a particle would have if it could cross from the lowest potential field line in the wind to the highest, or vice versa depending on the charge). The quantity $\sigma$ is of the order of $10^7$ (electrons) for typical pulsars.

064.036 **Mass loss and R136a-type stars.**
M. S. Vardya.
Publ. Astron. Soc. Pac., Vol. 96, No. 583, p. 743 – 745 (1984).
Empirical and semiempirical expressions for the rate of mass loss, derived for O and B stars, have been examined vis-a-vis R136a. The best expression has been used to predict the rate of mass loss for some R136a-type objects. The change in predicted values, when these objects are not single stars but are composed of a dense cluster of similar stars has been discussed.

064.037 **Theory of astrophysical SiO masers.**
S. H. Langer, W. D. Watson.
Astrophys. J., Vol. 284, No. 2, p. 751 – 768 (1984).
Circumstellar masing in the vibrationally excited states of SiO has been observed in many sources and is particularly common in oxygen-rich, late-type giants. The authors present model calculations for SiO maser emission in a smooth stellar wind for a variety of system parameters appropriate for a late-type giant. The Sobolev approximation is used to describe the radiative transfer. It is found that models employing a smooth wind yield less maser power than is reported from the neighborhood of late-type giants. The models do, however, exhibit many of the other features of the SiO masers. The accuracy of the Sobolev approximation is examined by numerically integrating the exact equations of radiative transfer.

064.038 **Asymmetries in stellar Mg II *h* and *k* and Ca II H and K line profiles: discrepancies between Mg and Ca asymmetries.**
D. J. Mullan.
Astrophys. J., Vol. 284, No. 2, p. 769 – 773 (1984).
"Discrepant asymmetry" stars are cool giants which show the opposite sense of asymmetry in the emission cores of Ca and Mg (i.e., red peak dominant in Mg, blue peak dominant in Ca). The author suggests that these discrepancies can be understood in terms of corotating interaction regions in the winds.

064.039 **Magnetic starspots.**
K. Jahn, K. Stępień.
Acta Astron., Vol. 34, No. 1, p. 1 – 20 (1984).
Models of large magnetic starspots with an axisymmetric untwisted magnetic field on late type stars are discussed. A unique relation between the stellar mass and the difference of effective temperatures of the spot and the surrounding photosphere is adopted from observations. Recent observations of magnetic fields on G and K type stars are in good agreement with the authors' predictions.

064.040 **On radiative shocks in atomic and molecular stellar atmospheres. II. The precursor structure.**
D. Gillet, J.-P. J. Lafon.
Astron. Astrophys., Vol. 139, No. 2, p. 401 – 410 (1984).

This paper deals with the problem of determining the structure of the radiative shock precursor in dense ($10^{15}$–$10^{21}$ particles per $m^3$), cool (500 – 5000K) atomic and molecular media such as protostars and cool pulsating stars. The medium is assumed to consist of hydrogen only. A one-photon and one-temperature precursor model, taking into account cooling by dissociation and ionization and heating by incident photon flux, is presented. An analytical solution in Lagrangian space is given for chemical and transfer equations. It is shown that the heating clearly dominates the cooling. The variation of the mass density is only of a few per cent whereas the temperature can be multiplied by a factor of a few units.

064.041 **Penetrative cellular convection in a stratified atmosphere.**
J. M. Massaguer, J. Latour, J. Toomre, J.-P. Zahn.
Astron. Astrophys., Vol. 140, No. 1, p. 1 – 16 (1984).

Penetrative convection has been investigated within a simple compressible model consisting of three layers of differing stratification prior to the onset of convection. The middle one is a convectively unstable polytrope bounded above and below by two stably stratified polytropes. Nonlinear anelastic modal equations are used to describe these convective motions. It is shown that: (1) Penetration into the lower stable layer by downward-directed plumes is substantially larger in a stratified medium than in a Boussinesq fluid, and produces an extended region of adiabatic stratification. (2) Overshooting into the upper stable layer can be sharply diminished by a reversal of the buoyancy occurring in the upper part of the unstable zone if the stratification is strong enough. These effects are all related to large pressure perturbations that arise in convective flows.

064.042 **Consequences of a chromospheric temperature gradient on the width of Hα in late-type giants.**
D. M. Zarro.
Astrophys. J., Vol. 285, No. 1, p. 232 – 236 (1984).

An analytic expression for the integrated Hα optical depth profile is derived for a one-dimensional slab geometry model chromosphere, with electron temperature increasing as a power law with height. The formula predicts Hα opacity and profile width to be sensitive functions of the thermal gradient. Application of the model to observation reveals that broad Hα absorption widths in G and K giant stars are consistent with a mean Hα chromospheric optical depth of 50, while narrower widths in M stars indicate slightly lower opacities.

064.043 **CO line emission from the expanding circumstellar envelopes of red giants.**
W. K. Rose.
Astrophys. J., Vol. 285, No. 1, p. 237 – 246 (1984).

This is the first in a series of papers describing calculations of theoretical models for red giant envelopes. It is concerned primarily with CO line emission from the outer regions of these envelopes. The author describes the method of solution, which requires the simultaneous solution of the radiative transfer equation for the CO emission lines, the equations of statistical equilibrium that determine the populations of the rotational levels of the CO molecule, and an energy equation, which determines the kinetic temperature distribution within the envelope. Finally, the calculations are compared to observations of IRC +10216.

064.044 **An efficient method for computing spectral line profiles in stellar envelopes.**
C. Bertout.
Astrophys. J., Vol. 285, No. 1, p. 269 – 274 (1984).

This paper presents a method for computing line profiles in spherically symmetric stellar envelopes. It can efficiently handle arbitrary velocity fields, including $v(r) = 0$, which makes it suitable for studying spectral lines in a variety of astrophysical situations.

064.045 **Effect of scattered radiation on the instability of hot star winds.**
G. B. Rybicki, S. P. Owocki.
Bull. Am. Astron. Soc., Vol. 16, No. 3, Part I, p. 725 (1984).
Abstract. – See Abstr. 010.065.

064.046 **A method for adding opacity distribution functions.**
M. Saxner, B. Gustafsson.
Astron. Astrophys., Vol. 140, No. 2, p. 334 – 340 (1984).

An efficient method for calculating composite opacity distribution functions (ODFs) for several opacity sources using the ODFs for each individual source is presented and discussed.

064.047 **On the stability of accretion disks.**
W. Glatzel.
Theoretical problems in stellar stability and oscillations, p. 341 – 345 (1984). – See Abstr. 012.065.

The problem of viscosity in accretion disks is considered. It is proposed that a nonlinear mechanism for instability analogous to plane Couette flow in hydrodynamic flow may also occur in accretion disks.

064.048 **On the solution of a class of high-order wave equations.**
L. M. B. C. Campos.
Theoretical problems in stellar stability and oscillations, p. 399 – 406 (1984). – See Abstr. 012.065.

Waves in ionized, isothermal atmospheric layers can have constant or exponential wave speeds and/or damping rates, leading to wave equations with exponential coefficients. The author presents a method of exact, analytical solution of linear wave equations, of any order, with exponential coefficients. The method is compared with existing results for second-order acoustic-, Alfvén- and magnetosonic-gravity waves, and is applied to the study of a fourth-order hydromagnetic-gravity wave.

064.049 **The transfer of polarized radiation in spectral lines: solar-type stellar atmospheres.**
S. J. McKenna.
Astrophys. Space Sci., Vol. 106, No. 2, p. 283 – 297 (1984).

The polarization structure in several spectral lines in solar type stars is computed. The line profiles and polarization structures have been computed for several weak solar resonance lines including K I λ7664 Å, Sr I λ4607 Å, Ba II λ4554 Å, for various polar angles along the stellar disk. Both the line profiles and polarization structures as well as the center to limb behavior of the line center polarization agree well with observations.

064.050 **Structure of X-ray irradiated stellar wind.**
K. Masai.
Astrophys. Space Sci., Vol. 106, No. 2, p. 391 – 407 (1984).

The effect of X-ray irradiation on the line-driven stellar wind in X-ray binary systems is studied. The product of the X-ray luminosity $L_x$ and the density $n$ in the wind is a measure to distinguish between the optically thin and thick wind regimes. The formation of fully ionized helium regions depends on $L_x$ and $n$ in a way different from that of the Strömgren sphere. In such an optically thick wind, the region behind the He II ionization boundary is little affected by X-ray irradiation and the trace elements remain to be responsible for wind acceleration. Thus, its location is important for the structure of the wind and the interpretation of various phenomena in objects such as wind-fed X-ray pulsars.

064.051 **On accretion from a medium containing a density gradient.**
N. Soker, M. Livio.
Mon. Not. R. Astron. Soc., Vol. 211, No. 4, p. 927 – 932 (1984).

The problem of accretion from an inhomogeneous medium is examined. A scenario in which angular momentum can be deposited into the gravitating body, from a medium containing a density gradient, is proposed. The obtained equations are solved under certain simplifying assumptions. In a different regime of flow, it is suggested that angular momentum can be accreted via a displaced accretion line, within the accretion cone, in the classical Bondi-Hoyle picture.

064.052 **Relation between stellar activity, rotation and stellar dynamos.**
S. Catalano.
Space research prospects in stellar activity and variability, p. 243 – 259 (1984). – See Abstr. 012.078.

In spite of the difficulties in measuring stellar magnetic fields, magnetic activity of late type stars has been investigated from several spectroscopic indicators. Quantitative correlations between such indicators and the main stellar parameters are being looked for. On the other hand the stellar dynamo theory is rapidly evolving so that positive indications exist for explaining more than qualitatively the observed facts.

064.053 **Bimodal mass loss by magnetically controlled stellar winds.**
L. Paternò.
Space research prospects in stellar activity and variability, p. 327 – 333 (1984). – See Abstr. 012.078.

There is evidence of bimodal distributions of stars that seem to be correlated with bimodal distributions of masses and angular velocities. The approach the author proposes to explain the observed bimodality suggests that this is due to bimodal mass and angular momentum losses by magnetically controlled stellar winds during stellar evolution, owing to different magnetic field configurations.

064.054 **An analytical model of the SiO maser velocity spectrum.**
J. Sun, S.–m. Wu, S.–d. Xie, W.–g. Ge.
Chin. Astron. Astrophys., Vol. 8, No. 4, p. 265 – 269 (1984). English translation of Acta Astrophys. Sin., Vol. 4, No. 2, p. 97 – 106 (1984).

SiO maser sources associated with Mira variables show a wide velocity spread in the spectral line and a variety of profiles. The authors think this velocity spread may be ultimately traced to the star's pulsation. They consider the propagation of a perturbation generated by the pulsation in the circumstellar envelope, regarded as a compressible fluid. The authors' model gives, at the distance of the SiO maser, a velocity spread of $\pm 6$ km/s, in good agreement with the observed widths. The question of a correlated variation of the line width with the optical phase is also discussed.

064.055 **Synchrotron emission from chaotic stellar winds.**
R. L. White.
Space Telesc. Sci. Inst., Prepr. Ser., No. 26, 23 pp. (1984). To be published in Astrophys. J.

064.056 **On the determination of the limb darkening coefficient based on models of stellar atmospheres.**
A. A. Rubashevskij.
Astron. Tsirk., No. 1300, p. 5 – 6 (1983). In Russian.

064.057 **Theoretical values of limb darkening coefficients in the uvby system.**
A. A. Rubashevskij.
Astron. Tsirk., No. 1300, p. 6 – 8 (1983). In Russian.

064.058 **Mass loss of early–type stars.**
Eh. Ya. Vil'koviskij, L. V. Tambovtseva.
Astron. Tsirk., No. 1332, p. 4 – 5 (1984). In Russian.

064.059 **Zur Nichtlinearität des Zusammenhangs zwischen chromosphärischer und koronaler Emission bei kühlen Sternen.**
R. Hammer.
Mitt. Astron. Ges., Nr. 62, p. 220 (1984). Abstract. – See Abstr. 010.341.

064.060 **Koronaschalen und ihre Abhängigkeit vom Heizungsgesetz.**
R. Hammer.
Mitt. Astron. Ges., Nr. 62, p. 220 – 221 (1984). Abstract. – See Abstr. 010.341.

064.061 **Synthetische Spektren von Wolf–Rayet–Sternen.**
W.–R. Hamann.
Mitt. Astron. Ges., Nr. 62, p. 242 (1984). Abstract. – See Abstr. 010.341.

064.062 **Geometry and dynamics of envelopes around Be stars.**
J. Dachs, R. Hanuschik, D. Kaiser.
Mitt. Astron. Ges., Nr. 62, p. 268 – 273 (1984). – See Abstr. 010.341.

A quantitative classification of Be star envelopes is presented using three parameters: the projected rotational velocity, $v \cdot \sin i$, of the central star, the Hα emission line halfwidth, and the Hα emission line equivalent width. Data support the rotating disk model for the envelopes. It is demonstrated that Hα line equivalent widths increase approximately as the square of the radius of the Hα emitting region if the radial run of rotational velocity corresponds to that of a Keplerian disk.

064.063 **Numerical solution of the equation of radiative equilibrium.**
K. Nariai, T. Shigeyama.
Publ. Astron. Soc. Jpn., Vol. 36, No. 3, p. 593 – 601 (1984).

The authors solve numerically the equation of radiative equilibrium in a plane–parallel structure. The unknowns are the temperature at every node, and the condition of radiative equilibrium at every node gives the same number of equations. Therefore, the method is straightforward and needs no complicated explanations.

064.064 **A time–dependent magnetohydrodynamic model of the circumstellar envelopes of Be stars.**
M. Saitō, Y. Saito.
Publ. Astron. Soc. Jpn., Vol. 36, No. 3, p. 603 – 612 (1984).

A model for time variation of Be star envelopes is numerically given from the stellar surface to a few thousand stellar radii in the equatorial plane using the photospheric parameters of Pleione, in which the photospheric density varies by one order of magnitude with a period of 17 yr. The magnetic field adopted at the photosphere has a smaller amount of energy as compared with the kinetic energy of the gas for both the radial and azimuthal directions, respectively.

064.065 **Dimensional approach to mass loss: O– and B–type stars.**
M. S. Vardya.
Astrophys. Space Sci., Vol. 107, No. 1, p. 141 – 153 (1984).

An attempt has been made to obtain an expression for the rate of stellar mass loss using dimensional analysis. The best expression for O and B stars is of the form: $\dot{M} = A' G^{-1/2} c^{-4} L^2 (R/M)^{3/2}$. It is also found that $A'$ increases as one goes from B → O stars and from O → O(f) → O((f)), but is not sensitive to luminosity.

064.066 **A preliminary look at accretion disk oscillations.**
B. W. Carroll, W. Cabot, P. N. McDermott, M. P. Savedoff, H. M. Van Horn.
News Lett. Astron. Soc. N.Y., Vol. 2, No. 6, p. 9 (1984). Abstract. – See Abstr. 010.241.

064.067 **On the overstability of axisymmetric oscillations in thin accretion disks.**
G. R. Blumenthal, L. T. Yang, D. N. C. Lin.
Astrophys. J., Vol. 287, No. 2, p. 774 – 784 (1984). = Lick Obs. Bull., No. 988.

The authors examine the stability of geometrically thin and optically thick and thin accretion disks to axisymmetric perturbations. They show that the standard α–disk models admit not only classical thermal and viscous instabilities, which are generally associated with regions where electron scattering and radiation pressure are dominant, but also radial pulsational instabilities. The authors derive a general dispersion relationship and examine the eigenvectors' response to local axisymmetric perturbations. They speculate that quasi–periodic oscillation found in dwarf nova systems may be a manifestation of these pulsation overstabilities.

064.068 **X–rays from accretion of red giant winds.**
M. Jura, D. J. Helfand.
Astrophys. J., Vol. 287, No. 2, p. 785 – 792 (1984).

Mira and R Aqr are, in all probability, mass–losing red giants with companions whose luminosities are powered by accretion of cool stellar winds. Einstein observations of both stars show marginal detections of coincident, very low luminosity X–ray sources ($\leqslant 10^{-3} L_\odot$). It is usually thought that the companions to Mira and R Aqr are white dwarfs; here the authors propose instead that they are main–sequence stars. The authors also consider the general problem of white dwarf accretion from late–type giant winds. They propose that sources more luminous than $10\ L_\odot$ never occur because grains in the red giant's wind drastically reduce the Eddington limit for the inhibition of accretion by radiation pressure.

064.069 **The interaction of a stellar wind with the surrounding interstellar medium.**
S. Sakashita.
Astrophys. Lett., Vol. 24, No. 3, p. 149 – 151 (1984).

The interaction of a stellar wind with the surrounding interstellar medium is investigated. A simple model of the expanding shell induced by a stellar wind is considered, on the basis of momentum conservation, and the efficiency of the energy conversion is derived and compared with observations.

**Effects of LTE–deviation in atmospheres of M giants.**
See Abstr. 003.133.

**Photo cross–sections for stellar atmosphere calculations – compilation of references and data.**
See Abstr. 022.145.

**Hydrodynamic flows in accretion disks.**
See Abstr. 062.007.

**Propagation of waves in an atmosphere in the presence of a magnetic field. VII. Magneto–acoustic–gravity modes in an oblique *B*.**
See Abstr. 062.015.

**Combined effect of potential and nonpotential magnetic fields on equilibrium in stellar atmospheres.**
See Abstr. 062.068.

**Model calculations of magnetic flux tubes. I. Equations and method.**
See Abstr. 062.095.

**Isothermal shock waves in uniform atmospheres.**
See Abstr. 062.133.

**Compressible convection at large initial superadiabatic temperature gradient.**
See Abstr. 062.155.

**Regularized Monte–Carlo calculations of the optical polarization induced by scattering in X–ray binaries.**
See Abstr. 063.029.

**Half–range moment methods for radiative transfer in spherical geometry. III. Numerical solution and applications.**
See Abstr. 063.040.

**The asymmetry of resonance line scattering in a velocity gradient.**
See Abstr. 063.050.

**Linear polarization of astronomical masers and magnetic fields.**
See Abstr. 063.052.

**The angle–dependent redistribution functions $R_{III}$ and $R_{IV}$: line absorption probability distribution functions and reemission coefficients.**
See Abstr. 063.060.

**Half–range moment methods for radiative transfer in spherical geometry. V. Line formation in a static atmosphere.**
See Abstr. 063.086.

**On the main–sequence evolution of massive stars with mass loss.**
See Abstr. 065.033.

**Turbulent compressible convection in a deep atmosphere. II. Two–dimensional results for main–sequence A5 and F0 type envelopes.**
See Abstr. 065.036.

**Two–dimensional compressible convection extending over multiple scale heights.**
See Abstr. 065.037.

**Convection in the envelopes of red giants.**
See Abstr. 065.038.

**Building of abundance stratifications in stars, due to diffusion processes.**
See Abstr. 065.063.

**Line profiles calculated for the oblique pulsator model of Ap stars.**
See Abstr. 065.113.

**Very massive stars: evolution with mass loss. II. Nucleosynthesis, mass loss, and related problems.**
See Abstr. 065.128.

**Magnetic fields and angular momentum loss.**
See Abstr. 065.132.

**Light and line profile variations due to *r*–mode pulsations with an application to the ZZ Ceti star G117–B15A.**
See Abstr. 065.147.

**A test of the Planck–Larkin partition function.**
See Abstr. 071.033.

**Coronal heating by reconnection in DC current systems. A theory based on Taylor's hypothesis.**
See Abstr. 074.006.

**Effects of the nonlocal character of the mean intensity, deviations from radiative equilibrium, and a nongray atmosphere on oscillations in stellar envelopes.**
See Abstr. 080.081.

**Multiple circumstellar shells and radiation pressure on grains in the outflows from late–type giants.**
See Abstr. 112.004.

**Flux variations of 88 Her through phase transition from far UV to visual.**
See Abstr. 112.016.

**Possible identifying characteristics of Be stars.**
See Abstr. 112.017.

**Shock–driven mass loss of stars.**
See Abstr. 112.033.

**Physics of the atmospheres of early–type stars.**
See Abstr. 112.034.

**High–resolution, far–ultraviolet study of Beta Draconis (G2 Ib–II): transition region structure and energy balance.**
See Abstr. 112.036.

**The Be star phenomena. II. Spectral formation and structure of envelopes.**
See Abstr. 112.047.

**On the extended chromosphere of α Orionis.**
See Abstr. 112.050.

**The cool wind of the B hypergiant Cygnus OB 2 No. 12 in the infrared.**
See Abstr. 112.056.

**Metal abundances in 26 halo stars.**
See Abstr. 114.017.

**Ultraviolet and optical observations of metal deficient red giants and chromospheric models.**
See Abstr. 114.034.

**Iron abundances in G dwarfs.**
See Abstr. 114.054.

**Solar and non–solar activity in stars: the cosmic–ray connection.**
See Abstr. 116.030.

**Spectral variability in early–type binary X–ray systems.**
See Abstr. 117.014.

**Models for the active and quiescent regions on the RS CVn–type system II Pegasi (HD 224085).**
See Abstr. 117.057.

**Comparison of the X–ray observations of the AM Herculis objects to theory.**
See Abstr. 117.142.

**Radio emission from symbiotic stars: a binary model.**
See Abstr. 117.236.

**Helium in the atmospheres of binary systems.**
See Abstr. 119.014.

**The helium spectrum variables: the photospheres and magnetospheres of the magnetic B stars.**
See Abstr. 122.102.

**Non–LTE analysis of subluminous O–stars. VI. Feige 110.**
See Abstr. 126.006.

**The formation of condensed carbon particles in the atmospheres of white dwarfs.**
See Abstr. 126.033.

**The rotational velocities of white dwarfs.**
See Abstr. 126.063.

**Feige 24: wind/flare accretion by a hot DAZ1 degenerate.**
See Abstr. 126.083.

**Interstellar polyynes from the disruption of carbon grains.**
See Abstr. 131.131.

# 065 Stellar Structure and Evolution

065.001 **Transport properties of neutrinos in stellar collapse. II. Shear viscosity, heat conduction, and diffusion.**
L. J. van den Horn, C. G. van Weert.
Astron. Astrophys., Vol. 136, No. 1, p. 74 – 80 (1984).

Within the context of relativistic kinetic theory the authors develop a framework for the calculation of transport properties of neutrinos in stellar collapse. Taking into account both scattering and absorption of the neutrinos, they calculate for arbitrary neutrino degeneracy the transport coefficients characterizing the irreversible heat and diffusion flows as well as the viscous phenomena. Previous results, obtained separately for different physical regimes, are recovered and placed into a single perspective. It is concluded that, depending on degeneracy, heat conduction and diffusion are the dominant dissipative modes. Under favorable conditions, conduction by nondegenerate neutrinos may lead to non–negligible heating on dynamical time scales.

065.002 **Wolf–Rayet stars in clusters: the initial stellar masses and evolutionary connections between subtypes.**
H. Schild, A. Maeder.
Astron. Astrophys., Vol. 136, No. 2, p. 237 – 242 (1984).

The membership of Wolf–Rayet stars to open clusters and OB associations is used to determine the initial mass $M_i$, of stars becoming WR stars. A detailed discussion of the relation between the spectral types of the cluster turnoff and the initial mass of WR stars is made on the basis of stellar models. The authors obtain as lower limits 18 $M_\odot$ for WN stars and 35 $M_\odot$ for WC stars. From the study of WR stars in clusters and associations they also establish the evolutionary connections between WR subtypes. Binary evolution appears to have a negligible effect on the lower limit of initial mass but introduces a systematic deviation into the evolutionary sequences mentioned above.

065.003 **Non–linear stellar oscillations. Multi–mode interactions.**
J. Perdang, S. Blacher.
Astron. Astrophys., Vol. 136, No. 2, p. 263 – 281 (1984).

The numerical investigation of the role of the non–linear adiabatic coupling of pairs of linear radial stellar modes on the character of the oscillations (Perdang and Blacher, 1982) is extended to the non–linear coupling of larger numbers of modes. Besides the qualitative result that chaotic stellar oscillations survive under this more realistic alternative, several new phenomena emerge: (1) As the number $F$ of interacting modes is increased, chaos sets in at progressively lower surface amplitudes. (2) For $F$ large, the power spectrum of the surface displacement seemingly tends towards a universal shape, irrespective of the initial conditions. (3) The spectrum exhibits a non–linear peak at a frequency of about 2/5 times the fundamental frequency.

065.004 **The lithium abundance: constraints on stellar evolution.**
G. Michaud, G. Fontaine, G. Beaudet.
Astrophys. J., Vol. 282, No. 1, p. 206 – 213 (1984).

Because lithium burns at low temperatures and settles gravitationally below a mixed zone, the recent observations by Spite and Spite of Li abundance in halo stars impose strict constraints on evolutionary models for these objects. The current evolutionary models do not explain the constant Li abundance for the range of effective temperatures studied by Spite and Spite. The implications for the various particle transport processes in the interiors of these stars are considered.

065.005 **Two– and three–dimensional numerical simulations of the core helium flash.**
R. G. Deupree.
Astrophys. J., Vol. 282, No. 1, p. 274 – 286 (1984).

The effects arising from the time–dependent interaction of convection and the thermonuclear runaway of the core helium flash found previously with coarsely zoned two-dimensional calculations are studied with much more finely zoned two-dimensional and with three-dimensional calculations. These effects include: (1) a cyclic behavior in the strength of both the thermonuclear runaway and convection caused by the failure of convection to adjust instantaneously to the runaway, (2) extensive convective overshooting across the strong temperature inversion into the inner core, (3) the rapid convective heating of the inner core, (4) large nonspherical temperature variations, and (5) the eventual failure of convection to contain the runaway.

065.006 **Remark on a family of static relativistic stellar models.**
H. A. Buchdahl.
Classical Quantum Gravity, Vol. 1, No. 3, p. 301 – 304 (1984). Abstr. in Phys. Abstr., Vol. 87, No. 1237, Entry 72930 (1984).

065.007 **Pion stars.**
Yu. L. Vartanyan, G. S. Adzhyan, G. B. Alaverdyan.
Astron. Zh., Tom 61, Vyp. 4, p. 677 – 685 (1984). In Russian. English translation in Sov. Astron., Vol. 28, No. 4.

Nuclear matter is studied in the case of strong pion condensates. For a set of nuclear constants for which nuclear plasma can form bound systems the equation of state is calculated on the basis of which models of superdense celestial bodies capable to exist also in absence of self-gravitation are considered.

065.008 **On the value of anisotropy for the collapse of a massive star.**
V. G. Kornilov, V. M. Lipunov.
Astron. Zh., Tom. 61, Vyp. 4., p. 686 – 690 (1984). In Russian. English translation in Sov. Astron., Vol. 28, No. 4.

A statistical value of anisotropy during the collapse of a massive star into a neutron star is estimated using the simulation of the massive-binary evolution. The kick velocity of a neutron star is obtained to be $\Delta V = 70 - 100$ km/s. Physical causes of such an anisotropy are discussed.

065.009 **The isotropic electron antineutrino flux: a clue to the rate of stellar gravitational collapse in the universe.**
G. V. Domogatskij.
Sov. Astron., Vol. 28, No. 1, p. 30 – 31 (1984). English translation of 37.065.001.

065.010 **Spherical accretion onto a protostellar core.**
H. Zinnecker, W. M. Tscharnuter.
Occas. Rep. R. Obs., Edinb., No. 13, p. 83 – 90 (1984). – See Abstr. 012.001.

Firstly, the results of an analytical calculation are presented describing the temporal behaviour of spherical accretion onto a hydrostatic protostellar core formed by nonhomologous collapse of a Jeans-unstable gaseous sphere. Secondly, the results of a numerical calculation are reported dealing with the problem of spherical accretion onto a protostar embedded in a larger gas reservoir, in an attempt to relax the boundary condition that the total mass be constant.

065.011 **Time dependent models of rotating magnetic stars.**
D. Moss.
Mon. Not. R. Astron. Soc., Vol. 209, No. 3, p. 607 – 639 (1984).

The problem of the structure of axisymmetric rotating magnetic stars is tackled again, using a new numerical technique. The solutions found differ from those of previous studies in that a time-dependent version of the magneto-hydrodynamic equation is solved on a two dimensional grid. Many of the restrictions which applied to earlier models seem to disappear. In particular, it is now possible to find relatively rapidly rotating models which retain a comparatively low ratio of internal to surface magnetic flux over ohmic time-scales. A tentative extension to the case of perpendicular magnetic and rotation axes is presented.

065.012 **Nucleosynthesis in barium stars.**
V. Trimble.
Nature, Vol. 310, No. 5979, p. 633 – 634 (1984).

065.013 **On the models of early main sequence evolution of moderately massive stars with growing inhomogeneous convection zone.**
M. Umezu, N. Arimoto, K. Suda.
Sci. Rep. Tôhoku Univ., Eighth Ser., Vol. 4, No. 3, p. 226 – 239 (1983).

Models similar to those proposed by Sakashita and Hayashi to represent the main sequence evolution of massive stars have been studied for the evolution of moderately massive stars. Each of the models consists of the inhomogeneous convection zone developing between the outer radiative envelope of initial chemical composition and the homogeneous convection core of the diminishing hydrogen content. In actual computation, such an inhomogeneous convection zone is approximated by the treatment of Sakashita and Hayashi's semiconvection zone. The models are examined for two cases of simplified opacity law.

065.014 **Stability of bidimensional stellar systems for axisymmetric modes.**
D. Gillon.
Astrophys. Space Sci., Vol. 102, No. 2, p. 367 – 373 (1984).

The author considers bidimensional stellar systems with distribution function $F$ depending both on the energy $\varepsilon$ and on the angular momentum $l$. Using the Bartholomew formalism, he presents in a new form the demonstration of the sufficient condition of stability $\partial F/\partial\varepsilon < 0$ when these systems are submitted to axisymmetric modes.

065.015 **The structure of detonation waves and nucleosynthesis during thermonuclear explosions of carbon-oxygen stellar cores and white dwarfs.**
V. S. Imshennik, A. M. Khokhlov.
Pis'ma Astron. Zh., Tom 10, No. 8, p. 631 – 640 (1984). In Russian. English translation in Sov. Astron. Lett., Vol. 10.

The structure of a steady detonation wave in highly degenerate carbon $C^{12}$ is studied in the density range $10^6 - 10^8$ g/cm$^3$. A nuclear reaction network for 84 nuclides including protons, neutrons and $\alpha$-particles and about 600 nuclear reactions is used. Conditions are investigated for existence of all possible regimes of detonation known in hydrodynamical theory of combustion.

065.016 **Effects of the temperature and mass of iron cores on the mass ejection of neutrino-trapping supernovae.**
M. Takahara, K. Sato.
Prog. Theor. Phys., Vol. 71, No. 3, p. 524 – 534 (1984). Abstr. in Phys. Abstr., Vol. 87, No. 1240, Entry 85972 (1984).

065.017 **Electron capture in stellar collapse.**
J. Cooperstein, J. Wambach.
Nucl. Phys. A, Vol. A420, No. 3, p. 591 – 620 (1984). Abstr. in Phys. Abstr., Vol. 87, No. 1241, Entry 86803 (1984).

065.018 **Single star evolution. I. Massive stars and early evolution of low and intermediate mass stars.**
I. Iben Jr., A. Renzini.
Phys. Rep., Vol. 105, No. 6, p. 329 – 406 (1984). Abstr. in Phys. Abstr., Vol. 87, No. 1241, Entry 90140 (1984).

065.019 **Fitting formula for the critical mass of relativistic white dwarfs with different chemical compositions.**
J. M. Ibanez Cabanell, R. Canal.
An. Fis., Ser. B, Vol. 80, No. 1, p. 63 – 66 (1984). Abstr. in Phys. Abstr., Vol. 87, No. 1241, Entry 90151 (1984).

065.020 **Pre-main sequence evolution.**
W. M. Tscharnuter.
MPA Rep., No. 90, p. 1 – 5 (1983). – See Abstr. 012.020.

This paper presents a brief outline of current problems in theoretical studies of star formation and the evolution of protostars up to the zero-age main sequence state.

065.021 **Stellar evolution from the main sequence to central helium burning.**
H. Ritter.
MPA Rep., No. 90, p. 6 – 10 (1983). – See Abstr. 012.020.

The purpose of this article is to give a brief introduction to some basic characteristics of stellar evolution from the main sequence to central helium burning.

065.022 **Synthetic H–R diagrams as an observational test of stellar evolution theory.**
G. J. Mathews, S. A. Becker, W. M. Brunish.
MPA Rep., No. 90, p. 18 – 23 (1983). – See Abstr. 012.020.

The Magellanic Clouds contain numerous young clusters whose H–R diagrams are well populated in the various evolutionary phases of intermediate–mass ($3\,M_\odot < M < 10\,M_\odot$) stars. The authors construct synthetic H–R diagrams from theoretical evolutionary tracks as a function of mass and composition to directly compare with these cluster H–R diagrams.

065.023 **Nucleosynthesis during core He burning in massive stars.**
N. Prantzos, M. Arnould.
MPA Rep., No. 90, p. 34 – 35 (1983). – See Abstr. 012.020.

Nucleosynthesis during core He burning in massive stars has many important astrophysical implications. Some of those implications are discussed on grounds of simplified stellar models based on Arnett's (1972) calculations, and of the latest available nuclear data.

065.024 **Presupernova yields and their dependence on the $^{12}C(\alpha,\gamma)\,^{16}O$ reaction.**
W. D. Arnett, F.–K. Thielemann.
MPA Rep., No. 90, p. 36 – 38 (1983). – See Abstr. 012.020.

Massive stars ($M \gtrsim 20\,M_\odot$) produce abundance patterns during the high temperature shell burning phases of hydrostatic evolution before the core collapse which are little changed by explosive processing of the ejected envelope in a type II supernova explosion. A calculation using a detailed nucleosynthesis network (254 nuclear species), which adds shell burning products at typical burning conditions (temperatures, density) weighted in mass according to stellar evolution calculations, is presented for a $8\,M_\odot$ He–core.

065.025 **Presupernova stellar evolution.**
K. Nomoto.
MPA Rep., No. 90, p. 39 (1983). – See Abstr. 012.020.

065.026 **Carbon combustion supernovae.**
E. Müller.
MPA Rep., No. 90, p. 40 – 41 (1983). – See Abstr. 012.020.

The author has performed a set of axisymmetric hydrodynamic numerical experiments of the core carbon flash, which incorporate the treatment of two–dimensional convection without the need of a phenomenological theory of convection. The numerical experiments show a variety of possible evolutions of degenerate carbon–oxygen cores ranging from spherical detonations to non–spherical deflagrations.

065.027 **Nucleosynthesis in type I supernovae.**
K. Nomoto.
MPA Rep., No. 90, p. 42 (1983). Abstract. – See Abstr. 012.020.

065.028 **Type II supernovae.**
W. Hillebrandt.
MPA Rep., No. 90, p. 59 – 61 (1983). – See Abstr. 012.020.

065.029 **On the effects of molecular weight gradients on the structure of middle main–sequence magnetic stars.**
D. Moss.
Mon. Not. R. Astron. Soc., Vol. 210, No. 2, p. 489 – 495 (1984).

With the aid of idealized model computations an investigation is made of the efficacy of meridional circulations driven by molecular weight gradients in modifying the structure of magnetic fields in middle main sequence stars. Contrary to some earlier speculation, it is deduced that, even if a 'creeping paralysis' does occur eventually, it will not significantly affect either the structure of the field of magnetic (chemically peculiar) stars, nor the distribution of the angle between the magntic and rotation axis.

065.030 **Influence of rotation and magnetic fields on stellar oscillation eigenfrequencies.**
D. O. Gough, P. P. Taylor.
Mem. Soc. Astron. Ital., Vol. 55, Nos. 1 – 2, p. 215 – 226 (1984). – See Abstr. 012.022.

A formalism to calculate rotational and magnetic splitting of otherwise degenerate high–order oscillation $p$ modes of low degree is outlined. For the purposes of illustration, attention is restricted to axisymmetric toroidal magnetic fields, with axes of symmetry that are permitted to be different from the rotation axis. It is found that advection is the most important contributor to rotational splitting; in general, magnetic perturbations add to the complexity of the frequency spectrum when the oscillations are viewed from an inertial frame. The effect of a field whose axis lies close to the sun's equatorial plane is briefly considered. It is found that such a field is not able to split five–minute dipole modes into observable triplets.

065.031 **Lithium depletion in stars. Pre–main sequence burning and extra–mixing.**
F. D'Antona, I. Mazzitelli.
Astron. Astrophys., Vol. 138, No. 2, p. 431 – 442 (1984).

The authors compute complete sets of pre–main sequence and main sequence evolution for masses from 0.6 to 1.2 $M_\odot$, in which lithium burning is considered in detail. By fitting their models with the halo dwarf abundances (Spite and Spite, 1982) it is concluded that the initial lithium in Population II is $\log N\,(\mathrm{Li}) = 2.2 \pm 0.15$.

065.032 **The role of astrophysical constant and screening effect on $_1H^1(p,\beta^+\nu)_1H^2$.**
A. E. M. Khairozzaman.
Proc. Indian Natl. Sci. Acad., Part A, Vol. 49, No. 4, p. 443 – 447 (1984). Abstr. in Phys. Abstr., Vol. 87, No. 1242, Entry 94742 (1984).

065.033 **On the main–sequence evolution of massive stars with mass loss.**
R. Mitalas, H. J. Falk.
Mon. Not. R. Astron. Soc., Vol. 210, No. 3, p. 641 – 653 (1984).

An analytical model of a massive mass–losing star is presented and compared with evolutionary calculations. A mass–luminosity law for stars with $15 < M/M_\odot < 500$, accurate to within 15 per cent during the MS, is given. Variation of ZAMS luminosity with $X$ and $Z$ is discussed. Power–law form of the ZAMS luminosity law is used to derive the critical mass loss rate for stellar evolution with mass loss to occur at constant luminosity. The relative importance of mass loss and nuclear burning on the average mean molecular weight is discussed.

065.034 **Pulsation periods and critical frequencies in low gravity yellow supergiant models.**
T. Aikawa.
Astrophys. Space Sci., Vol. 104, No. 2, p. 405 – 408 (1984).

It is shown that the coincidence between the pulsation periods and the critical frequencies found by Fadeyev (1982) results from strong effects on non–adiabaticity on the linear pulsation periods in low surface gravity models. If the boundary condition of running waves for super–critical pulsations is applied, it is found that the linear growth–rates are considerably affected by the boundary condition.

065.035 **Stellar acoustics. II. Pulse resonance in giant star models.**
T. Aikawa, C. A. Whitney.
Astrophys. J., Vol. 282, No. 2, p. 527 – 532 (1984).

Radial pulsation and the propagation of acoustic pulses generated at the $He^+$ ionization zone are investigated for radiative stellar envelopes in the linear nonadiabatic approximation. A pulse resonance condition is shown to be satisfied for models near the $P_2{:}P_0$ resonance of the Cepheid instability strip, and the time of arrival of the reflected pulse is shown to be in qualitative agreement with the properties of bump Cepheids of different periods.

065.036 **Turbulent compressible convection in a deep atmosphere. II. Two-dimensional results for main-sequence A5 and F0 type envelopes.**
S. Sofia, K. L. Chan.
Astrophys. J., Vol. 282, No. 2, p. 550 – 556 (1984).

The authors report a two-dimensional numerical study of turbulent compressible convection in two types of main-sequence envelopes – an A5 and an F0. In this computation, the effects of ionization are included in the equation of state of the gas, and radiative transfer is modeled as a diffusive process with tabulated gas opacities. The results show that the thermal effects of ionization do affect the dynamics of the flows significantly. It is also shown that the mean structure self-consistently produced by a hydrodynamical computation may be very different from a mean structure calculated by a mixing length theory.

065.037 **Two-dimensional compressible convection extending over multiple scale heights.**
N. E. Hurlburt, J. Toomre, J. M. Massaguer.
Astrophys. J., Vol. 282, No. 2, p. 557 – 573 (1984).

Two-dimensional simulations are used to study fully compressible thermal convection spanning multiple density scale heights typical of a stellar envelope. The fluid is assumed to be a perfect gas with constant thermal conductivity and dynamic viscosity. The mean density ratio (bottom to top of the layer) ranges from about 1 to 21, with Rayleigh numbers up to about 1000 times critical. The convective motions in these simulations span the full height of the unstable layer, with no tendency to form a succession of rolls in the vertical as has been assumed in mixing-length treatments of convection. The flows display prominent downward-directed plumes surrounded by broader regions of upflow.

065.038 **Convection in the envelopes of red giants.**
H. M. Antia, S. M. Chitre, D. Narasimha.
Astrophys. J., Vol. 282, No. 2, p. 574 – 583 (1984).

The nature of convection in the envelopes of red giants is investigated, and the linear convective modes are computed to demonstrate the consistency of the mixing length theory. The mixing length at a given depth is tentatively identified with an equivalent width of the luminosity profile of the convective eigenmodes. The dominant convective element turns out to be comparable in size to the radius of the star, and this could account for the observed irregular variations in red giants.

065.039 **Diffusion in a thermally pulsating star.**
A. Kovetz, D. Prialnik, G. Shaviv.
Astrophys. J., Vol. 282, No. 2, p. 584 – 590 (1984).

The effect of diffusion on thermal pulses in an intermediate-mass star is investigated. The diffusion equations, including concentration, pressure, and thermal terms, are solved for a mixture of three gases: H, He, and a third component representative of all heavier elements. Evolutionary calculations are carried out for an initial model of 6 $M_\odot$ with $Z = 0.03$. Thermal pulses are computed for three core-mass values: 0.788 $M_\odot$, 0.914 $M_\odot$, and 1.009 $M_\odot$. It is found that the rate of change of composition owing to diffusion is very small compared with nuclear reaction rates at the peak of either burning shell. The question of mixing by diffusion between carbon and hydrogen in the intershell zone during pulse power-down is also addressed.

065.040 **Cooling of low-mass carbon-oxygen dwarfs from the planetary nucleus stage through the crystallization stage.**
I. Iben Jr., A. V. Tutukov.
Astrophys. J., Vol. 282, No. 2, p. 615 – 630 (1984).

The evolution of a carbon-oxygen dwarf of mass $M \approx 0.6\ M_\odot$ has been carried all the way from an initial nuclear burning stage, when it is the central star of a planetary nebula, to the stage of complete internal crystallization, after $10^{10}$ yr of cooling. Shell hydrogen and helium burning, neutrino losses, and the effects of liquification and crystallization have been taken into account. A luminosity function based on the results covers the range $-5 \lesssim \log(L/L_\odot) \lesssim 4$ and agrees reasonably well with the observed luminosity function extending from the brightest planetary nebula nuclei to the dimmest observed white dwarfs, except perhaps for $\log(L/L_\odot) \lesssim -4.5$. Possible reasons for the apparent discrepancy at low luminosity are discussed.

065.041 **Roche harmonics for tidally-distorted stellar models.**
P. C. Chakravarti, V. P. Singh.
Astrophys. Space Sci., Vol. 105, No. 1, p. 131 – 138 (1984).

Kopal's analysis for Roche harmonics associated with the Roche model of a star, distorted by the tidal forces of a companion star, has been extended further to obtain the explicit expressions for Roche harmonics when terms up to second order of smallness in tidal effects are considered.

065.042 **On the production of technetium in thermally pulsing red giant stars.**
K. R. Cosner, K. H. Despain, J. W. Truran.
Astrophys. J., Vol. 283, No. 1, p. 313 – 315 (1984).

The concentration of $^{99}$Tc which may be expected to characterize the *s*-process matter in intermediate-mass red-giant stars at the termination of a thermal pulse is shown to be an extremely sensitive function of the temperature and time scale conditions. The $^{99}$Tc beta-decay rate is found to be $\sim 10^5$ times its terrestrial value, at the prevailing temperatures. Critical nuclear uncertainties are noted, as are strong dependences on the properties of the stellar environment.

065.043 **Critique of the iterative theory of stellar pulsations.**
J. R. Buchler, W. D. Pesnell.
Astrophys. J., Vol. 283, No. 1, p. 316 – 320 (1984).

The iterative theory of stellar pulsations as a mathematical tool for calculating finite amplitude behavior of stellar pulsation models is critically examined. The iterative theory as it appears in the literature is an application of the asymptotic method of harmonic balance. When the method is applied correctly, the deficiencies of the iterative theory, namely (1) infinities at resonance and (2) incompleteness (free, undetermined parameters), are removed.

065.044 **General relativistic collapse of rotating stars.**
T. Nakamura.
Ann. N.Y. Acad. Sci., Vol. 422, p. 56 – 81 (1984). – See Abstr. 012.033.

This paper presents results of 2D numerical calculations of the general relativistic collapse of rotating stars. The simulations include the effects of gravitational radiation and are intended to illustrate the conditions leading to either neutron stars or Kerr black holes.

065.045 **Core-collapse models of type II supernova explosions and neutron star formation.**
W. Hillebrandt.
Ann. N.Y. Acad. Sci., Vol. 422, p. 197 – 214 (1984). – See Abstr. 012.033.

Stellar evolution calculations have been performed for the mass range 8 to 10 $M_\odot$, which lead to both supernova explosions and neutron star formation. The results of these numerical models are described. Arguments are presented why more massive stars face severe problems if they are to eject their envelopes in supernova explosions.

065.046 **Relativistic stars with anisotropic pressure.**
S. S. Bayin.
Ann. N.Y. Acad. Sci., Vol. 422, p. 330 (1984). Abstract. – See Abstr. 012.033.

065.047 **Neutrino transport and shock propagation in the core collapse of a massive star.**
S. W. Bruenn.
Ann. N.Y. Acad. Sci., Vol. 422, p. 333 (1984). Abstract. – See Abstr. 012.033.

065.048 **Les oscillations solaires et stellaires comme test de la théorie de la structure interne.**
J. Provost.
Instabilités hydrodynamiques et applications astrophysiques, p. 167 – 186 (1983). – See Abstr. 012.034.
Contents: Introduction. Quelques propriétés théoriques des oscillations stellaires. Les oscillations solaires et leur interprétation. Perspectives de développement de la sismologie solaire et stellaire.

065.049 **Remarque sur les théories de la pulsation des étoiles convectives.**
G. Gonczi.
Instabilités hydrodynamiques et applications astrophysiques, p. 225 – 228 (1983). – See Abstr. 012.034.

065.050 **Les instabilités dues à la rotation.**
J.-P. Zahn.
Instabilités hydrodynamiques et applications astrophysiques, p. 389 – 407 (1983). – See Abstr. 012.034.
Contents: Instabilité de cisaillement. Instabilité barocline. Instabilités multidiffusives. Comparaison des différentes instabilités. Esquisse d'une solution.

065.051 **Circulation méridienne dans les étoiles.**
J.-P. Zahn.
Instabilités hydrodynamiques et applications astrophysiques, p. 409 – 416 (1983). – See Abstr. 012.034.
Contents: Le déséquilibre radiatif dans une étoile en rotation. Les premiers travaux. La solution des Tassoul.

065.052 **Tests observationnels et phénoménologie de la diffusion turbulente dans les étoiles.**
E. Schatzman.
Instabilités hydrodynamiques et applications astrophysiques, p. 417 – 430 (1983). – See Abstr. 012.034.
Contents: Introduction. Diffusion microscopique et diffusion turbulente. Étoiles A. Le problème du lithium. Mélange diffusif en couches profondes: les isotopes du carbone. Mélange diffusif à l'intérieur du soleil. Le test de sismologie solaire. Conclusion.

065.053 **Rotating, pulsating stars.**
J. P. Cox.
Publ. Astron. Soc. Pac., Vol. 96, No. 582, p. 577 – 582 (1984).
The author presents a fairly simple physical interpretation of "*m* splitting" of the frequencies of nonradial oscillations in slowly rotating stars. This lifting of the degeneracy with respect to the azimuthal "quantum number" *m* by slow rotation is interpreted in terms of azimuthal running waves whose phase velocities are affected by the rotation. Hence, in a slowly rotating star the observed oscillation frequency depends on *m*. Another type of frequency splitting by rotation in stars whose pulsation and rotation axes do not coincide is also described briefly. Finally, the author presents a summary of astronomical bodies in which *m* splitting has been observed.

065.054 **Diffusion-induced hydrogen burning and the evolution of DA white dwarfs.**
G. Michaud, G. Fontaine.
Astrophys. J., Vol. 283, No. 2, p. 787 – 793 (1984).
The evolution of the hydrogen distribution of a 0.6 $M_\odot$ DA white dwarf in the presence of diffusion and nuclear burning is followed. An approximate solution is obtained by decoupling the diffusion calculations from the computations of the model structure. A DA white dwarf starting its life with a hydrogen layer containing $q(\mathrm{H}) \approx 1-3\times10^{-4}$ reaches an effective temperature of 10,000K with only $q(\mathrm{H}) \approx 3\times10^{-7}$ left over. Diffusion-induced hydrogen burning thus explains the apparent conflict between standard evolution theory and the results of recent pulsational studies of white dwarfs. Turbulence is shown to be extremely small in hydrogen-rich white dwarfs.

065.055 **Self-consistent radiation-hydrodynamic equations for stellar oscillations. I. Nonlinear form.**
B. W. Mihalas.
Astrophys. J., Vol. 284, No. 1, p. 299 – 302 (1984).
Using the correct comoving frame, or Lagrangean, equations of radiation hydrodynamics derived by Castor and Buchler, the author examines earlier discussions of radiation hydrodynamics in the nonequilibrium diffusion regime. Comparison of corresponding equations reveals a fundamental conceptual error in previous treatments of the interaction between stellar oscillations and the radiation field.

065.056 **On the fission theory of binary stars. IV. Exact solutions in polynomial spaces.**
N. R. Lebovitz.
Astrophys. J., Vol. 284, No. 1, p. 364 – 380 (1984).
The bifurcation of compressible, nonellipsoidal figures from families of Riemann ellipsoids is governed by a certain linear equation, whose solutions are investigated. They are found to be given by polynomials in the Cartesian coordinates; in particular, the basic linear operator $L$ acts invariantly on certain polynomial spaces $V_N$, and solutions may be sought in $V_N$. Since $V_N$ is finite-dimensional, the underlying system of integro-differential equations is reduced to a finite problem exactly, without the need of any spatial discretization for numerical purposes. The bifurcation problem is reduced by a perturbation expansion to a hierarchy of linear, inhomogeneous equations.

065.057 **Fission of dumbbell equilibrium and binary state of rapidly rotating polytropes.**
I. Hachisu, Y. Eriguchi.
Publ. Astron. Soc. Jpn., Vol. 36, No. 2, p. 239 – 257 (1984).
Equilibrium sequences of self-gravitating polytropic binaries are computed when synchronous rotation and congruent components are assumed. For the incompressible case the sequence reaches the dumbbell sequence just at the contact phase. Together with their previous work, the authors have obtained a continuous sequence from the Maclaurin spheroid to the binary with a large separation by way of the Jacobi ellipsoid.

065.058 **Calculation of the collapse of a rotating gaseous cloud in a Lagrangian net.**
N. V. Ardelyan, G. S. Bisnovatyj-Kogan, Yu. P. Popov, S. V. Chernigovskij.
Inst. prikl. mat. Akad. Nauk SSSR, Prepr., No. 80, 25 pp. (1984). In Russian. Abstr. in Ref. Zh., 51. Astron., 10.51.179 (1984).

065.059 **Convection in pulsating stars. V. Limit cycle characteristics.**
R. F. Stellingwerf.
Astrophys. J., Vol. 284, No. 2, p. 712 – 718 (1984).
An estimate of the effects of convection on the limiting amplitude behavior of RR Lyrae stars has been obtained by direct integration of a convective nonlinear model to limiting amplitude. A detailed analysis of the approach to the limit is made, over a 100 period interval. The presence of convection significantly lowers the final amplitude when compared with fully radiative models, but the actual amplitude obtained is sensitive to the form of the eddy viscosity. An analysis of the stability of the limit cycle is presented, using a new analysis technique.

065.060 **The evolution of low-mass close binaries influenced by the radiation of gravitational waves and by a magnetic stellar wind.**
I. Iben Jr., A. V. Tutukov.
Astrophys. J., Vol. 284, No. 2, p. 719 – 744 (1984).
The evolution of low-mass binaries with 0.79 $M_\odot$ and 1.25 $M_\odot$ mass-donating components filling their Roche lobes is studied numerically, taking into account the angular momentum loss by the radiation of gravitational waves and by a magnetic stellar wind. Numerical results are discussed with reference to Algol systems, cataclysmic binaries, W UMa stars, and low-mass X-ray binaries. The comparison of numerical results with parameters of observed low-mass close binaries having initial semimajor axes smaller than $\sim 12\ R_\odot$ displays the possible influence of

a magnetic stellar wind on the evolution of these binaries. Some applications are made for systems containing neutron stars. In particular, the 50 min orbital period of the X-ray burster MXB 1916-05 and the large spin rate of the millisecond pulsar PSR 1953+29 are discussed.

065.061 **Influence of a hydrogen-rich envelope on the critical mass of helium stars.**
A. Noels, E. Magain.
Astron. Astrophys., Vol. 139, No. 2, p. 341 – 343 (1984).

The influence of a hydrogen-rich envelope on the vibrational stability of helium stars is studied. The linear stability analysis towards radial perturbations has been performed. Such an effect increases the amplitude ratio between the surface and the center of the adiabatic eigenfunction corresponding to the fundamental mode. This enhances the stabilizing role of the flux term in the expression of the damping coefficient and the critical mass increases as the hydrogen rich envelope mass, $M_H$, grows. From a value of 16 $M_\odot$ for pure helium stars, the critical mass becomes 25 $M_\odot$ when $M_H = 10^{-5}$, in fraction of the total mass, and is greater than 80 $M_\odot$ when $M_H = 10^{-4}$.

065.062 **The initial mass function of massive stars.**
D. Vanbeveren.
Astron. Astrophys., Vol. 139, No. 2, p. 545 – 548 (1984).

The initial mass function (IMF) of massive stars (mass larger than 20 $M_\odot$) is reinvestigated including the large number of B type stars which are observed. Special attention is given to the possible difference between the IMF of cluster stars and the IMF of field stars.

065.063 **Building of abundance stratifications in stars, due to diffusion processes.**
G. Alecian, R. Grappin.
Astron. Astrophys., Vol. 140, No. 1, p. 159 – 168 (1984).

The authors study the temporal building of density inhomogeneities due to diffusion of a single ion in CP stars. They show that, in the particular case where the diffusion velocity depends on $x$ only through $N$, every perturbation of concentration generates a wave which steepens while propagating, until it is balanced by linear effects. In the general case, the authors observed numerically that equilibrium solutions were almost always reached in times shorter than the stellar lifetime and appeared to be extremely stable. Preceding the final state, a transient phase is obtained eventually, in which significant time and space fluctuations occur. It is suggested that part of the spread of abundances observed on Main Sequence CP stars belonging to the same category may be due to such transient states.

065.064 **Evolution of star-planet systems.**
G. Bath.
Nature, Vol. 312, No. 5992, p. 307 – 308 (1984).

065.065 **Effects of pulsation and mass loss on stellar evolution.**
L. A. Willson, G. H. Bowen.
Nature, Vol. 312, No. 5993, p. 429 – 431 (1984).

This letter points out that there are many possible consequences, some not previously recognized, of pulsation-related mass loss for the evolutionary behaviour of post-main sequence stars. It advances the hypothesis that most mass loss of evolutionary significance is closely related to stellar pulsation.

065.066 **Use of variational principle in the study of the effects of rotational and tidal distortions on the periods of small oscillations of gaseous spheres.**
C. Mohan, R. M. Saxena.
Indian J. Pure Appl. Math., Vol. 15, No. 5, p. 497 – 508 (1984). Abstr. in Phys. Abstr., Vol. 87, No. 1244, Entry 103339 (1984).

065.067 **White dwarfs. Big things in small packages.**
A. J. Nicastro.
Astronomy, Vol. 12, No. 7, p. 6 – 14 (1984). Abstr. in Phys. Abstr., Vol. 87, No. 1244, Entry 105973 (1984).

065.068 **An extended Thomas-Fermi calculation of supernova matter.**
X. Vinas, M. Barranco, M. Pi, A. Polls, A. Perez-Canyellas.
J. Phys. Colloq., Vol. 45, No. C6, p. 103 – 110 (1984). Abstr. in Phys. Abstr., Vol. 87, No. 1244, Entry 106014 (1984). – See Abstr. 012.048.

065.069 **Gravitational collapse of a rotating magnetized star.**
T. Ohnishi.
Tech. Rep. Inst. At. Energy Kyoto Univ., No. 198, p. 1 – 23 (1983). Abstr. in Phys. Abstr., Vol. 87, No. 1244, Entry 106015 (1984).

065.070 **A stellar energy loss mechanism involving axions.**
L. M. Krauss, J. E. Moody, F. Wilczek.
Phys. Lett. B, Vol. 144B, No. 5 – 6, p. 391 – 394 (1984). Abstr. in Phys. Abstr., Vol. 87, No. 1246, Entry 115748 (1984).

065.071 **Numerical simulations of stellar convective dynamos. I. The model and method.**
G. A. Glatzmaier.
J. Comput. Phys., Vol. 55, No. 3, p. 461 – 484 (1984). Abstr. in Phys. Abstr., Vol. 88, No. 1247, Entry 4724 (1985).

065.072 **La produzione degli elementi chimici nelle stelle. I – L'evoluzione stellare e la nucleosintesi primaria.**
M. Busso.
Orione, Vol. 4, N. 3, p. 94 – 102 (1984).

065.073 **The latest stage in the evolution of a very massive star.**
A. F. J. Moffat.
J. R. Astron. Soc. Can., Vol. 78, No. 5, p. 203 (1984). Abstract. – See Abstr. 010.401.

065.074 **Diffusion-induced hydrogen burning in white dwarfs.**
G. Michaud, G. Fontaine.
J. R. Astron. Soc. Can., Vol. 78, No. 5, p. 204 (1984). Abstract. – See Abstr. 010.401.

065.075 **Microturbulence in main-sequence stars.**
P. H. Andersen, I. D. Cameron.
J. R. Astron. Soc. Can., Vol. 78, No. 5, p. 205 – 206 (1984). Abstract. – See Abstr. 010.401.

065.076 **Inferred properties of stellar granulation.**
D. F. Gray, C. G. Toner.
J. R. Astron. Soc. Can., Vol. 78, No. 5, p. 206 (1984). Abstract. – See Abstr. 010.401.

065.077 **Non-axisymmetric collapse and explosion of rapidly rotating objects.**
H. Beltrami, W. Y. Chau.
J. R. Astron. Soc. Can., Vol. 78, No. 5, p. 212 (1984). Abstract. – See Abstr. 010.401.

065.078 **Stellar evolutionary sequences and isochrones for intermediate-age star clusters.**
D. A. VandenBerg.
J. R. Astron. Soc. Can., Vol. 78, No. 5, p. 213 – 214 (1984). Abstract. – See Abstr. 010.401.

065.079 **Collapse and gravitation = the present situation.**
M. Signore.
Problems of collapse and numerical relativity, p. 1 – 26 (1984). – See Abstr. 012.061.

This paper presents an overview of current theoretical studies of the collapse and the ultimate fate of the iron cores of highly evolved massive stars.

065.080 **The evolution of massive stars.**
C. Doom.
Problems of collapse and numerical relativity, p. 49 – 58 (1984). – See Abstr. 012.061.

Detailed comparison of stellar evolution models with observations show that it is necessary to take into account the effects of mass loss by stellar wind and overshooting from convective cores in the computation of the evolution of massive stars. Very massive stars ($M > 34\ M_{\odot}$) evolve into Wolf–Rayet stars, which will explode in a very particular way at the end of their life.

065.081 **Combined effects of mass loss and extended mixing on the structure and evolution of massive stars.**
M. Cassé.
Problems of collapse and numerical relativity, p. 59 – 68 (1984). – See Abstr. 012.061.

Stellar evolutionary models yield contrasted results for different prescriptions of the internal mixing. This paper explores the structural and evolutionary consequences of adopting the Roxburgh criterion for convection rather than the Schwarzschild criterion.

065.082 **The equation of state of hot dense matter.**
D. Vautherin.
Problems of collapse and numerical relativity, p. 69 – 79 (1984). – See Abstr. 012.061.

The approaches which have been used so far to calculate the equation of state of hot dense matter are briefly reviewed.

065.083 **A Thomas–Fermi equation of state of hot dense matter.**
E. Suraud.
Problems of collapse and numerical relativity, p. 81 – 87 (1984). – See Abstr. 012.061.

The author reports on the results of some Thomas–Fermi calculations of the equation of state of supernova matter.

065.084 **Evolution of accreting white dwarfs, Type I supernova explosion, and white dwarf collapse.**
K. Nomoto.
Problems of collapse and numerical relativity, p. 89 – 116 (1984). – See Abstr. 012.061.

The evolution of accreting white dwarfs in close binary systems is studied from the onset of accretion through the supernova stage. Relatively rapid accretion onto a C + O white dwarf eventually leads to a carbon deflagration supernova. This explosion produces 0.5 – 0.6 $M_{\odot}$ of $^{56}$Ni which is sufficient to power the light curve of Type I supernovae. On the other hand, the detonation supernova explosion triggered by slow accretion is inconsistent with observations of SN I because it produces almost exclusively iron peak elements. Accretion onto an O + Ne + Mg white dwarf leads to the formation of a neutron star as a result of electron captures on $^{24}$Mg and $^{20}$Ne.

065.085 **Collapse and explosion of degenerate stars.**
R. Canal, J. Isern, J. Labay.
Problems of collapse and numerical relativity, p. 117 – 124 (1984). – See Abstr. 012.061.

Arguments which favor the simultaneity of collapse and explosion in a fraction of mass–accreting white dwarfs are reviewed. The phase transition to the solid state induced by cooling is singled out as the key physical process allowing collapse upon thermonuclear ignition. Neutron star formation and Type I supernova outbursts would thus be the endpoints of the evolution of "frozen" carbon–oxygen white dwarfs, accreting matter from low–mass, close binary companion stars.

065.086 **Final evolution of 8 – 10 $M_{\odot}$ stars.**
R. Mochkovitch.
Problems of collapse and numerical relativity, p. 125 – 132 (1984). – See Abstr. 012.061.

In stars in the range 8 – 10 $M_{\odot}$ carbon shell burning leads to the formation of an O + Mg + Ne degenerate core. For a central density $\varrho_c \approx 5 \times 10^9 \mathrm{g/cm^3}$ electron captures on the core material begin. A sharp double diffusive interface is formed at the boundary of the capture region. The author discusses the stability and heat transport at the interface and comments on its effect on core evolution.

065.087 **Theory of stellar magnetic activity: a review.**
G. Belvedere.
Astron. Nachr., Vol. 305, No. 5, p. 237 – 242 (1984). – See Abstr. 012.062.

Dynamo models of stellar activity in the main sequence, although subjected to different assumptions, do converge in predicting that activity should increase with the increasing angular velocity of rotation ω and colour index ($B–V$), in agreement with almost all existing observational evidence. However, even if linear dynamo theory seems to have captured the essentials of the mechanism of stellar activity, the complexity of stellar activity phenomenology and operation modes suggests that the future is in the non–linear approach. Finally, observational and theoretical uncertainties and difficulties, which affect the present status of dynamo theory, are briefly discussed.

065.088 **Pre–main sequence stellar activity.**
G. Belvedere.
Astron. Nachr., Vol. 305, No. 5, p. 243 – 244 (1984). – See Abstr. 012.062.

Models of stellar activity based on the α–ω dynamo have recently been proposed for main sequence and giant late type stars. In the present work the author extends the theoretical analysis to the pre–main sequence fully convective phase of stellar evolution. As a measure of magnetic activity he assumes the intensity of the toroidal magnetic field $B_t$. The results show that $B_t$ decreases with the increasing effective temperature along the evolutionary track of a star of given mass. The crucial dependence of the intensity of the magnetic field on the rotation rate is stressed. Comparison with the observations of pre–main sequence stellar activity in the very young galactic cluster NGC 2264 is briefly discussed.

065.089 **The effect of a weak magnetic field on stellar oscillations.**
P. H. Roberts.
Astron. Nachr., Vol. 305, No. 5, p. 245 – 249 (1984). – See Abstr. 012.062.

The effect of a weak magnetic field on the otherwise radial pulsations of a star are discussed.

065.090 **Magnetic starspots.**
K. Jahn, K. Stępień.
Astron. Nachr., Vol. 305, No. 5, p. 250 (1984). Abstract. – See Abstr. 012.062.

065.091 **Bimodal mass and angular momentum losses by magnetically controlled winds during stellar evolution.**
V. Castellani, L. Paternò.
Astron. Nachr., Vol. 305, No. 5, p. 251 – 256 (1984). – See Abstr. 012.062.

Bimodal distribution of stars in the H–R diagram has reached a striking evidence. These bimodal distributions seem to be correlated with a bimodal distribution of masses and angular velocities. A mechanism is assumed analogous to that which produces solar wind, with magnetic field generated by dynamo working in the convection zone. Different field geometries (dipole or quadrupole), which depend on the mode the dynamo operates, can produce different but discrete mass losses during stellar evolution, thus producing bimodal distributions of masses and angular velocities.

065.092 **Long periodic variations of magnetic stars caused by precession.**
E. Gerth.
Astron. Nachr., Vol. 305, No. 6, p. 329 – 332 (1984).

Long–time observations give evidence of secular variations of the effective magnetic fields for some Ap stars. These secular variations may be interpreted by a precessional motion of the star. The precession of the rotationally flattened star is assumed to be caused either by the orbital motion of a satellite where the plane of the orbit is inclined to the axis of rotation of the star or by rotation of the inner and outer layers of the star around differently inclined axes.

065.093 **Linear radial pulsations in stars in thermal imbalance.**
P. Smeyers, P. Bruggen.
Astron. Astrophys., Vol. 141, No. 2, p. 297 – 303 (1984).

Linear radial pulsations of a spherically symmetric star that evolves on the Helmholtz–Kelvin time scale are studied by means of the two–time–variable expansion method. Particular attention is paid to the derivation of an integral expression for the rate of change of the eigenfrequency and of an equation for the rate of change of the radial displacement of an individual mass element. The integral expression shows that the change of the eigenfrequency is due to the star's evolution. It is transformed into a form that is convenient for numerical applications.

065.094 **The rotation of Ap stars.**
P. North.
Astron. Astrophys., Vol. 141, No. 2, p. 328 – 340 (1984).

Photometric periods of silicon stars belonging to clusters and associations as well as to the field are discussed in relation with theories of magnetic braking taking place on the main sequence. It is shown that Si stars do not lose the bulk of their angular momentum during their main–sequence lifetime, but must have lost it before. Any braking mechanism which predicts an *e*–folding time of the order of, or shorter than, the star's lifetime seems ruled out. The theory of purely hydromagnetic rotational braking is much too efficient to fit the observations, while the accretion theory does not fit them either, although it seems more realistic.

065.095 **Third–order perturbation theory for rotating polytropic stars.**
V. S. Geroyannis.
Astron. Astrophys., Vol. 141, No. 2, p. 425 – 429 (1984).

The author presents numerical results interpreting quantitatively the distortion of the structure of a polytropic body owing to rapid rotation either uniform or differential by implementation of algorithms based upon a third–order perturbation theory developed in a previous investigation. The results obtained for the case of uniform rotation are systematically compared with those of previous investigations based on various methods.

065.096 **Collapse of a low–mass iron stellar core.**
S. I. Blinnikov, M. A. Rudzskij.
Sov. Astron. Lett., Vol. 10, No. 3, p. 152 – 155 (1984). English translation of 37.065.046.

065.097 **Bump Cepheid models.**
T. R. Carson, R. B. Stothers.
Theoretical problems in stellar stability and oscillations, p. 29 – 34 (1984). – See Abstr. 012.065.

Models of pulsating variables are constructed for parameters appropriate to the bump Cepheid region. Fourier decomposition of non–linear models and period ratios of linear models are used as objective criteria to test the models for indicators of the bump phenomenon.

065.098 **Nonlinear radial pulsations in R Coronae Borealis stars.**
H. Saio, J. C. Wheeler.
Theoretical problems in stellar stability and oscillations, p. 35 – 37 (1984). – See Abstr. 012.065.

The authors investigated nonlinear radial pulsations in models for R CrB stars by using a nonlinear hydrodynamic code.

065.099 **A perturbative formalism for nonlinear nonadiabatic stellar pulsations.**
J. R. Buchler, M. J. Goupil, J. Klapp.
Theoretical problems in stellar stability and oscillations, p. 38 – 42 (1984). – See Abstr. 012.065.

With the help of an asymptotic perturbation formalism simple ordinary differential equations (amplitude equations) are derived which govern the temporal, regular or chaotic modulations of the amplitudes of the excited modes. An application to realistic models of bump Cepheids is discussed. The agreement with numerical hydrocodes is very good and is achieved with a great reduction in computing time.

065.100 **A non–local theory of time–dependent convection in the stellar pulsation problem.**
G. Gonczi.
Theoretical problems in stellar stability and oscillations, p. 43 – 47 (1984). – See Abstr. 012.065.

065.101 **On the origin of pulsation.**
A. Grandpierre.
Theoretical problems in stellar stability and oscillations, p. 48 – 54 (1984). – See Abstr. 012.065.

This paper demonstrates that an intimate connection exists between the onset of convection and pulsations in stellar interiors.

065.102 **Influence of a hydrogen rich envelope on the critical mass of helium stars.**
A. Noels, E. Magain.
Theoretical problems in stellar stability and oscillations, p. 55 – 58 (1984). – See Abstr. 012.065.

Replacing the outer layers of a pure helium star by hydrogen rich ones increases the critical mass of such stars. The authors present here a short account of the results.

065.103 **On the vibrational stability of WR stars.**
A. Noels, M. Gabriel.
Theoretical problems in stellar stability and oscillations, p. 59 – 63 (1984). – See Abstr. 012.065.

The authors have evolved stars of initial masses 80 $M_\odot$ and 100 $M_\odot$ during core helium burning, taking mass loss into account. At the minimum of the central condensation, a vibrational instability appears. The authors study the effect of an increase in the mass loss rate, tentatively due to this instability.

065.104 **The general expression of the sufficient condition for dynamical stability of stellar configurations.**
M. Gabriel.
Theoretical problems in stellar stability and oscillations, p. 64 – 67 (1984). – See Abstr. 012.065.

The author deduces the general expression of the sufficient condition for dynamical stability toward radial pulsation. A particular form of this condition is that $\Gamma_1 > 4/3$ everywhere in the star. Another one is that the solution regular at the center (surface) computed for $\sigma = 0$ has no node.

065.105 **Non–radial oscillations.**
P. Smeyers.
Theoretical problems in stellar stability and oscillations, p. 68 – 92 (1984). – See Abstr. 012.065.

This paper presents a review of the current status of the general theory of linear non–radial oscillations of non–rotating, non–magnetic, spherically symmetric stars. The discussion is restricted to free oscillations.

065.106 **Sensitivity of eigenfrequencies to model structure.**
R. Loiacono, R. Ventura, G. Belvedere, L. Paternò.
Theoretical problems in stellar stability and oscillations, p. 93 – 96 (1984). – See Abstr. 012.065.

This note reports an attempt to test the sensitivity of eigenfrequencies to the structure of stellar models. The authors use polytropic models, and compute the eigenfrequencies of modes of low degree and high order, varying the polytropic index and the adiabatic exponent, in order to simulate different model properties. The sensitivity of the regression coefficients of the asymptotic expansion to the structure variations is then studied.

065.107 **Normal mode spectra of a self–gravitating sphere with phase boundaries.**
S. V. Vorontsov.
Theoretical problems in stellar stability and oscillations, p. 97 – 102 (1984). – See Abstr. 012.065.

The standard boundary conditions used at a density discontinuity in linear adiabatic normal–mode analysis must be revised if the boundary is a phase one and the entropy discontinuity is small. A short derivation of new boundary conditions is given and their applicability is discussed. A model of Jupiter is used to test the possible influence of the modification of boundary conditions on the normal mode spectrum.

065.108 **Semiconvection and mixing phenomena in massive stars.**
N. Langer, M. F. El Eid, K. J. Fricke.
Theoretical problems in stellar stability and oscillations, p. 120 – 125 (1984). – See Abstr. 012.065.

In stellar evolutionary calculations the authors use a formalism for the treatment of semiconvection (SC) as a vibrational instability which occurs in layers having a gradient of mean molecular weight. This treatment takes into account the dependence of the time scale of SC mixing on the local physical conditions by solving the time-dependent diffusion equation. The results for Pop I stars of 30 $M_\odot$ show that the post main sequence evolution is strongly influenced by SC mixing.

065.109 **Vibrational stability analysis in relation to WR star models.**
H. Kirbiyik, G. Bertelli, C. Chiosi.
Theoretical problems in stellar stability and oscillations, p. 126 – 132 (1984). – See Abstr. 012.065.

Stellar models with initial masses 20 and 60 $M_\odot$ are considered. The models which are tested for vibrational stability are at the core He-burning phase, while some of them have both core He-burning and shell H-burning. They were tested for stability towards $g^+$ modes of nonradial oscillations. It is found that vibrational instability occurs in most of these models.

065.110 **On the rapid oscillations of Ap stars.**
H. Shibahashi, H. Saio.
Theoretical problems in stellar stability and oscillations, p. 133 – 138 (1984). – See Abstr. 012.065.

The authors interpret the rapid oscillations of Ap stars discovered by Kurtz as a superposition of high overtone $p$-modes with the radial order $n \cong 30 \sim 40$ and low degrees $l$ ($l = 0 - 3$). They have calculated the frequency separation of the consecutive overtones with a fixed value of $l$, for stellar evolutionary models, and found that the observed frequency separation is consistent with the theoretical one for models with $M \cong 2\, M_\odot$ during the main sequence stage. However, the observed frequencies of HD 24712 are found to be higher than the critical frequencies of these models. This contradiction probably results from the significant deviation of the $T$–$\tau$ relation of the Ap stars from the standard one.

065.111 **On the rapidly oscillating Ap stars.**
M. Gabriel, A. Noels, R. Scuflaire, G. Mathys.
Theoretical problems in stellar stability and oscillations, p. 139 – 142 (1984). – See Abstr. 012.065.

Frequency separation $\Delta\nu$ between consecutive modes have been computed for A stars. It is shown that $\Delta\nu$ is a good indicator of the star's position in the main sequence band. Applications to HD 101065, HD 24712 and HD 83368 are discussed.

065.112 **Rapidly oscillating Ap stars: an alternative model.**
G. Mathys.
Theoretical problems in stellar stability and oscillations, p. 143 – 147 (1984). – See Abstr. 012.065.

The influence on the rapid light variations of the Ap stars of the surface inhomogeneities (postulated by the oblique rotator model to explain their mean variations) is studied. The *spotted pulsator model* is introduced: the star is assumed to oscillate in a single nonradial mode, symmetric about the rotation axis. The flux and/or the ratio of the flux variation amplitude to the radius variation amplitude are supposed not to be homogeneous over the stellar surface, which thus appears as "spotted". It is shown that the amplitude of the rapid photometric variations is modulated with a frequency exactly equal to the rotation frequency of the star.

065.113 **Line profiles calculated for the oblique pulsator model of Ap stars.**
A. P. Odell, T. J. Kreidl.
Theoretical problems in stellar stability and oscillations, p. 148 – 154 (1984). – See Abstr. 012.065.

The authors have produced simulated Na I and Fe I line profiles which are expected from the oblique pulsator model, and which incorporate $l$ and $m$ spherical harmonics. It is demonstrated that it should be possible to identify these harmonics spectroscopically in pulsating Ap stars.

065.114 **Qualitative aspects of the convection–pulsation interaction.**
W. D. Pesnell.
Theoretical problems in stellar stability and oscillations, p. 294 – 297 (1984). – See Abstr. 012.065.

A method of including time-dependent convection in a non-radial pulsational stability analysis is outlined. The principal assumptions are a thin convection zone and solenoidal mass flux field. Spherical coordinates are used to facilitate the calculation, but the equations describing the convection can be shown to reduce to the plane-parallel case.

065.115 **Non-radial oscillations of white dwarfs and neutron stars.**
H. M. Van Horn.
Theoretical problems in stellar stability and oscillations, p. 307 – 327 (1984). – See Abstr. 012.065.

Substantial advances in our knowledge of the ZZ Ceti stars (pulsating white dwarfs with hydrogen atmospheres) have occurred within the past three years. The developments culminating in our understanding of these stars as non-radial oscillators in which modes trapped by compositional stratification are excited by the hydrogen and helium partial ionization zones is reviewed. The investigation of neutron star oscillations is still in a primitive state. The complex structure of a neutron star allows the existence of a wide variety of non-radial oscillation modes. Early calculations are summarized, and recent advances in our understanding of the mode spectrum are reviewed.

065.116 **On the calibration of the mixing-length theory in pulsating white dwarfs.**
G. Fontaine, M. Tassoul, F. Wesemael.
Theoretical problems in stellar stability and oscillations, p. 328 – 333 (1984). – See Abstr. 012.065.

This paper discusses stellar evolution calculations for white dwarfs located in the ZZ Ceti instability strip of the HR diagram. In particular, the sensitivity of the blue edge of the instability strip on the convective efficiency of the applied mixing-length theory is studied.

065.117 **Rotationally stabilized stars at densities intermediate to white dwarfs and neutron stars.**
J. E. Tohline, G. Chanmugam.
Theoretical problems in stellar stability and oscillations, p. 334 – 338 (1984). – See Abstr. 012.065.

The collapse of a rotating stellar core of an evolved massive star is considered. It is shown that if the initial ratio of rotational to gravitational energy is sufficiently large, dynamical collapse will be halted at a density intermediate to that of white dwarfs and neutron stars. Further evolution, on non-dynamical time scales, proceeds as a result of gravitational radiation, cooling, and magnetic dipole radiation, with the formation of a neutron star.

065.118 **Vibrational instability of stars with rotation and/or magnetic field.**
W. Dziembowski.
Theoretical problems in stellar stability and oscillations, p. 346 – 355 (1984). – See Abstr. 012.065.

Papers suggesting an active role of rotation and magnetism in driving stellar oscillations are critically reviewed. Destabilizing effect of magnetic torque is proposed as a cause of solar gravity-mode excitation. Possible explanation of the observed pulsation amplitude – rotation velocity anticorrelation is given.

065.119 **Instabilities of differentially rotating self-gravitating cylinders.**
E. Balbinski.
Theoretical problems in stellar stability and oscillations, p. 356 – 360 (1984). – See Abstr. 012.065.

065.120 **The effect of shear on *g*-modes.**
P. Veugelen, P. Smeyers.
Theoretical problems in stellar stability and oscillations, p. 361 – 366 (1984). – See Abstr. 012.065.

The influence of differential rotation on the dynamical stability and the isentropic oscillations of stars is studied by considering normal modes of plane inviscid Couette flows in fluids with an exponential law for the density variation.

065.121 **Rotational and tidal perturbation of high-order non-radial *p*-modes.**
L. Martens, P. Smeyers.
Theoretical problems in stellar stability and oscillations, p. 367 – 371 (1984). – See Abstr. 012.065.

A rotating component of a binary system is considered. It is assumed that the axis of rotation is perpendicular to the orbital plane and that the rotation is uniform and synchronous with the orbital motion. This paper studies the linear and isentropic oscillations of a rotationally and tidally distorted star with respect to a co-rotating frame that has its origin at the center of the star.

065.122 **Non-linear dynamics and stellar oscillations.**
J. Perdang.
Theoretical problems in stellar stability and oscillations, p. 425 – 453 (1984). – See Abstr. 012.065.

The author discusses two aspects of the impact of the theory of non-linear systems on our understanding of stellar oscillations. (1) In the framework of *non-linear* oscillations, he presents an inventory of the different types of variability compatible with the conventional stellar hydrodynamics. It is pointed out that the structure of the latter conceals a variety of physically distinct classes of irregular oscillations of which so far only the mildest species have been traced in numerical studies. (2) In the framework of *linear* non-radial oscillations, the author analyses the potential interest of the stellar counterpart of the notorious Quantum Chaos. He discusses a simple technique capable of detecting this phenomenon in stars.

065.123 **Chaos in adiabatic oscillations of a two-shell star model.**
C. A. Whitney.
Theoretical problems in stellar stability and oscillations, p. 454 – 458 (1984). – See Abstr. 012.065.

As a step toward investigating the possibility that semi-regular variable stars may be showing chaotic motions governed by deterministic but non-linear equations, the author studies the behavior of a simple stellar model with two degrees of freedom. The model consists of a rigid spherical core surrounded by two homogeneous layers of perfect gas separated by thin shells that carry the inertia of the system. The purpose of this paper is to describe the periodic solutions and the influence of resonances on the emergence of chaos.

065.124 **Non linear stellar oscillations: Poincaré maps for two mode coupling.**
S. Blacher.
Theoretical problems in stellar stability and oscillations, p. 466 – 473 (1984). – See Abstr. 012.065.

The author investigates numerically the existence conditions of unstable Kolmogorov motions of radial adiabatic stellar oscillations in the context of the Hamiltonian formalism and the polytropic model $n = 3$. To this end, she integrates the Hamiltonian equations and builds Poincaré maps for all two-mode coupling among the 10 first modes and for different values of the oscillation energy.

065.125 **Nonlinear nonradial adiabatic stellar oscillations: numerical results for many-mode couplings.**
W. Däppen.
Theoretical problems in stellar stability and oscillations, p. 474 (1984). Abstract. – See Abstr. 012.065.

065.126 **On variational principles for thermal evolution.**
H. Kähler.
Theoretical problems in stellar stability and oscillations, p. 475 (1984). Abstract. – See Abstr. 012.065.

065.127 **Evolution of massive stars.**
C. de Loore.
Spectral evolution of galaxies, p. 1 – 27 (1984). – See Abstr. 012.066.

The evolution of stars with masses larger than 15 $M_{\odot}$ is reviewed. These stars have large convective cores and lose a substantial fraction of their matter by stellar wind. The treatment of convection and the parameterisation of the stellar wind mass loss are analysed within the context of existing disagreements between theory and observation. Specific solutions like overshooting extended atmospheres, and successive birth waves in OB associations are discussed. The evolution of massive close binaries and the origin of Wolf-Rayet stars and X-ray binaries is also sketched.

065.128 **Very massive stars: evolution with mass loss. II. Nucleosynthesis, mass loss, and related problems.**
J. Klapp.
Astrophys. Space Sci., Vol. 106, No. 2, p. 215 – 229 (1984).

In a previous paper the author has described the evolution of metal-free mass-losing very massive stars (VMS) in the 500 – 10,000 $M_{\odot}$ range. The present paper concerns the nucleosynthesis and mass loss aspects of the hydrogen- and helium-burning phase. Through radiation driven winds, the star loses 20 – 40% of its mass as helium and ~1% as carbon and oxygen. The results show that VMS cannot produce the cosmological helium abundance without overproducing heavier elements. The high oxygen yield of VMS makes them the best candidates for producing an oxygen overabundance in old Population II stars.

065.129 **What will asteroseismology teach us?**
J. Christensen-Dalsgaard.
Space research prospects in stellar activity and variability, p. 11 – 46 (1984). – See Abstr. 012.078.

In the present paper the author considers the study of stars by means of their oscillation frequencies. Assuming high-order modes, the oscillation frequencies are conveniently discussed in terms of asymptotic theory. This is briefly reviewed. Then the author makes some remarks on observations of stellar oscillations and presents results of frequency calculations for a set of main sequence stars: such calculations are useful in giving an indication of the information that may be obtained from stellar oscillation frequencies.

065.130 **Investigating the internal structure and evolution of massive stars by space techniques.**
C. de Jager.
Space research prospects in stellar activity and variability, p. 47 – 59 (1984). – See Abstr. 012.078.

All massive stars ($\gtrsim 15\,M_{\odot}$) are variable in luminosity ($\Delta m \approx 10^{-2}$mag) and radial velocity ($\Delta v \approx 1$ km s$^{-1}$), and have a large rate of mass loss. A precise determination of the spectrum of the $v_R$- and m-variations with noise levels of ca. $10^{-3}$km s$^{-1}$ and $10^{-5}$mag would allow for a direct determination of the internal structure and (hence) evolutionary status of stars with $M \gtrsim 5\,M_{\odot}$. If the above accuracy can be reached in one hour integration time, then a four-years observational period would yield the following results: a precise spectrum of pulsations for some 20 stars, allowing for an accurate determination of the internal structure and evolutionary phase; general information on the internal structure of about 400 stars. A lifetime of the Observatory of 12 years is suggested, which would yield detailed information on 60 and general data on 1200 stars.

065.131 **The dimension of a variable star.**
M. Auvergne, A. Baglin.
Space research prospects in stellar activity and variability, p. 61 - 68 (1984). - See Abstr. 012.078.

The authors discuss the applicability of recent results, concerning dynamical systems, to variable stars. The dynamics of a star is described by a third order differential equation including dissipative terms (due to radiative losses). For some of those stars Fourier spectra of the light curve are complicated and the origin of this complexity is not yet clear. The authors propose to use the concept of dimension to make a step forward in understanding these phenomena: the calculation of the dimension of the attractor (limit cycle) defined by the temporal behaviour of the stellar luminosity will impose some constraints on any modelization of the dynamics, and help to choose between different interpretations. The example of white dwarfs is chosen to illustrate this purpose, as the shortness of their periods has permitted to obtain Fourier spectra of good quality. It is shown that satellite observations are necessary to apply this new tool.

065.132 **Magnetic fields and angular momentum loss.**
R. S. Gill, I. W. Roxburgh.
Space research prospects in stellar activity and variability, p. 335 - 342 (1984). - See Abstr. 012.078.

Angular momentum loss rates are calculated for stars with dipole and quadrupole magnetic fields, and isothermal coronae. The angular momentum loss rate for dipole fields is found to be considerably higher than that for quadrupole fields. If the dynamo produced magnetic field changes from quadrupole to dipole as the angular velocity decreases, there should be a sudden increase in the rate of angular momentum loss. The resulting decrease in the angular velocity with time is calculated assuming that the stars rotate uniformly throughout and used to determine the predicted distribution of angular velocity amongst stars of the same mass. This distribution has two peaks in qualitative agreement with the observations of Vaughan.

065.133 **The stability of stellar structure in non-equilibrium thermodynamics III. The PPI reaction allowing for mass loss.**
Z.-y. Wang, Q.-j. Peng, K.-l. Huang.
Chin. Astron. Astrophys., Vol. 8, No. 3, p. 223 - 227 (1984). English translation of Acta Astrophys. Sin., Vol. 4, No. 3, p. 191 - 198 (1984).

The authors investigate the stability of a solar-type star the structure of which is determined by the PPI reaction according to the criterion of excess entropy production of non-linear, non-equilibrium thermodynamics, taking into account the mass loss in the reaction. Their result is: the star is certainly stable if the hydrogen-to-helium ratio satisfies the inequality $X/Y < 8.916$. This is to be compared with $X/Y < 9.2$, derived previously, neglecting the mass loss.

065.134 **La produzione degli elementi chimici nelle stelle. II - La nucleosintesi secondaria e la formazione degli elementi più pesanti.**
M. Busso.
Orione, Vol. 4, N. 4, p. 162 - 170 (1984).

065.135 **Hydrostatic nucleosynthesis. I. Core helium and carbon burning.**
W. D. Arnett, F.-K. Thielemann.
MPA Rep., No. 160, p. 1 - 50 (1984). Submitted to Astrophys. J.

065.136 **Hydrostatic nucleosynthesis. II. Core neon to silicon burning and presupernova abundance yields of massive stars.**
F.-K. Thielemann, W. D. Arnett.
MPA Rep., No. 160, p. 51 - 94 (1984). Submitted to Astrophys. J.

065.137 **Equilibrium models of differentially rotating polytropes and the collapse of rotating stellar cores.**
Y. Eriguchi, E. Müller.
MPA Rep., No. 168, 28 pp. (1984). Submitted to Astron. Astrophys.

065.138 **Carbon deflagrating supernovae and the chemical history of the solar neighbourhood.**
F. Matteucci, A. Tornambè.
ESO Sci. Prepr., No. 338, 39 pp. (1984). To appear in Astron. Astrophys.

065.139 **Possible scenarios for the [O/Fe] ratio in metal-poor stars.**
F. Matteucci.
ESO Sci. Prepr., No. 351, 16 pp. (1984). To appear in Mem. Soc. Astron. Ital.

065.140 **CNO abundances in low mass stars evolving up to the red giant branch.**
U. O. Frisk.
Stockholms Obs., Rep., No. 23, 71 pp. (1984).

Using a new theoretical calibration of effective temperatures the CNO abundances of stars in the region from the main sequence up to the base of the red giant branch are determined. Material processed by the CN-cycle is found in stars more luminous than $\log L/L_{\odot} = 0.8$. Stars with low $^{12}C/^{13}C$ ratios ($<15$) are found down to the luminosity $\log L/L_{\odot} = 1.2$. In stars where processed material is observed, C is deficient and N is enhanced. The C/N and $^{12}C/^{13}C$ ratios and their correlation with stellar mass are found to be consistent with a slow mixing during the main sequence.

065.141 **Explosive nucleosynthesis in massive stars: comparison with the Cassiopeia A fast-moving knots.**
M. D. Johnston, A. Yahil.
Astrophys. J., Vol. 285, No. 2, p. 587 - 594 (1984).

If the ejecta of a Type II supernova do not undergo extensive mixing, then, based on the explosion of current presupernova models, only a small fraction $\lesssim 0.1\ M_{\odot}$ of the mantle of a massive star can yield abundances similar to those observed in the fast-moving knots of Cas A. This is shown to be independent of the detailed structure of the mantle and the supernova energy. Lack of mixing in Cas A is indicated by strong upper limits on the abundance ratios Ne/O, Mg/O, and Fe/O. The implications of this result for stellar evolution theories of massive stars are briefly outlined.

065.142 **The collapse of rotating stellar cores: equilibria between white dwarf and neutron star densities.**
J. E. Tohline.
Astrophys. J., Vol. 285, No. 2, p. 721 - 728 (1984).

At the end of its normal nuclear burning stage of evolution, the white dwarf-like core of a nonrotating massive star will collapse dynamically to neutron star densities. If the core is rotating such that the ratio of rotational to gravitational energy in the core is greater than some minimum value, the collapse will not proceed all the way to neutron star densities. Instead, a rotating equilibrium at intermediate density will form. In this paper, a simple analytical model is developed through which properties of the core at the endpoint of its collapse can be estimated without performing a detailed collapse calculation. The model allows to specify the range of physical conditions for which dynamical collapse to nuclear density will be prevented.

065.143 **Magnetorotational iron core collapse.**
E. M. D. Symbalisty.
Astrophys. J., Vol. 285, No. 2, p. 729 - 746 (1984).

The author has modeled the collapse of a magnetic and rotating iron core of an evolved 15 $M_{\odot}$ star, using a two-dimensional magnetohydrodynamics numerical model. Nonmagnetic and nonrotating iron cores are typically the evolved quantities of spherically symmetric models seeking to explain Type II supernovae. Models that have greater than $\sim 4.5\%$ rotational energy do not explode but do allow some mass to be ejected. Extreme magnetic fields (precollapse field strength $\gtrsim 10^{13}$gauss), combined with $\sim$4.5% rotational energy initially, allow for a deeper collapse and a full-scale jetlike explosion outside of a nuclear density core. The matter in this explosion and the escaping matter from the strongly rotating model are likely to be exposed to the proper environment for an $r$-process to occur.

065.144 **Electron capture supernovae: one-zone collapse calculation for stars with masses 10 $M_{\odot}$ and 15 $M_{\odot}$.**
A. Ray, S. M. Chitre, K. Kar.
Astrophys. J., Vol. 285, No. 2, p. 766 – 777 (1984).

The equilibrium values of the lepton fraction, the entropy, and other thermodynamic variables are evaluated in a one-zone collapse calculation based on an analytic equation of state for hot, dense matter. The electron capture rates that take account of the forbidden capture and thermal excitation modes in the neutron-shell blocked region are used, and a diffusion scheme for neutrino transport is adopted. It is found that the results are sensitive to a variation in the input physics such as the efficiency of the neutrino trapping and also to the presupernova configurations for 10 $M_{\odot}$ and 15 $M_{\odot}$ stars.

065.145 **Thermal response of stellar envelopes during nonradial pulsations.**
W. D. Pesnell.
Astrophys. J., Vol. 285, No. 2, p. 778 – 783 (1984).

The response of spherical stellar envelopes in hydrostatic and thermal balance to nonradial thermal distributions is outlined. Because of the nature of the initial model, the governing equations yield an operator that is almost identical to the radial case. The use of propagation velocities to estimate the nonadiabatic effects in a pulsation is demonstrated.

065.146 **Do monopoles keep white dwarfs hot?**
K. Freese.
Astrophys. J., Vol. 286, No. 1, p. 216 – 220 (1984).

Although theoretical cooling curves predict large numbers of faint white dwarfs ($M_{bol} \geqslant 16$), none have been observed singly or even inferred as the missing objects in binary systems. The author suggests magnetic monopole-catalyzed nucleon decay as a mechanism for keeping white dwarfs hot ($M_{bol} < 16$). At the same time the observed luminosities of old white dwarfs ($L = 10^{-4} L_{\odot}$) place an upper limit on the product of monopole flux times cross section for catalysis, $(F/\mathrm{cm}^{-2}\mathrm{s}^{-1}\mathrm{sr}^{-1})(\sigma v/10^{-28}\mathrm{cm}^2) \leqslant 2 \times 10^{-18}$.

065.147 **Light and line profile variations due to $r$-mode pulsations with an application to the ZZ Ceti star G117-B15A.**
S. O. Kepler.
Astrophys. J., Vol. 286, No. 1, p. 314 – 327 (1984).

The author presents a general formulation for the effects of $r$-mode nonradial pulsations on the light curves and line profiles of a slowly rotating star, given a pulsation model, and model atmospheres. He shows that the temperature perturbation, the gravity perturbation, and the light variations are proportional to the spheroidal component of the displacement, while the velocity perturbation is proportional to the toroidal component.The formalism is applied to the pulsating DA white dwarf G117-B15A. The wings of the H$\gamma$ line of G117-B15A were observed photometrically. A comparison with the calculations indicates that the observed variations of G117-B15A are not caused by $r$-mode pulsations, but are consistent with $g$-mode pulsations.

065.148 **Equilibrium, stability and pulsations of stars with phase transition.**
Z. F. Seidov.
Inst. kosm. issled. Akad. Nauk SSSR., No. 889, 37 pp. (1984). In Russian. Abstr. in Ref. Zh., 51. Astron., 12.51.694 (1984).

065.149 **Massive-star evolution with semiconvective diffusion.**
N. Langer, M. F. El Eid, K. J. Fricke.
Mitt. Astron. Ges., Nr. 62, p. 353 – 354 (1984). – See Abstr. 010.341.

065.150 **Explosions and nucleosynthesis in rotating very massive stars.**
W. Glatzel, M. F. El Eid, K. J. Fricke.
Mitt. Astron. Ges., Nr. 62, p. 355 (1984). Abstract. – See Abstr. 010.341.

065.151 **The current status of stellar collapse.**
A. Burrows.
Comments Astrophys., Vol. 10, No. 4, p. 149 – 158 (1984).

065.152 **Equations of state, neutrinos and supernova explosions.**
B. Banerjee.
Gravitation and relativistic astrophysics, p. 41 – 42 (1984). Abstract. – See Abstr. 012.083.

065.153 **Time-dependent convection theory generalized for non-radial oscillations of stars.**
D.-r. Xiong.
Acta Astron. Sin., Vol. 25, No. 3, p. 215 – 223 (1984). In Chinese.

The time-dependent convection theory is generalized for non-radial oscillations of stars.

065.154 **Screened reaction rates and mean lifetimes for the production of helium in stars.**
A. E. M. Khairozzaman.
Astrophys. Space Sci., Vol. 107, No. 2, p. 233 – 240 (1984).

Screened reaction rates and mean lifetimes are studied for the reactions $H_1^1(P\beta^+\nu)H_1^2$ and $He_2^3(He^3 2H)He_2^4$. It is seen that a Main-Sequence star develops its helium core in a very short time after the hydrostatic equilibrium phase. The stellar longevity depends entirely on the reaction $H^1(P\beta^+\nu)H^2$ at the densities $\varrho X_A = 10^2\mathrm{g\,cm}^{-3}$ and $10^3\mathrm{g\,cm}^{-3}$. Again, the reaction $He^3(He^3 2H)He^4$ plays a vital role at $\varrho X_A = 10^2\mathrm{g\,cm}^{-3}$ to retain a star for some million years in its Main-Sequence stage.

065.155 **Models of configurations from an incompressible liquid taking into account the role of gravitational vacuum.**
L. Sh. Grigoryan.
Astrofizika, Tom 21, Vyp. 2, p. 367 – 377 (1984). In Russian. English translation in Astrophysics, Vol. 21, No. 2.

Models of superdense star configurations from an incompressible liquid are calculated based on the idea of the existence of a special gravitational vacuum. The vacuum effects prove to be essential in the case of superdense celestial bodies. The most important result is that there may exist equilibrium superdense celestial bodies with masses many times exceeding the solar mass.

065.156 **Accreting white dwarf models for Type I supernovae. III. Carbon deflagration supernovae.**
K. Nomoto, F.-K. Thielemann, K. Yokoi.
Astrophys. J., Vol. 286, No. 2, p. 644 – 658 (1984).

The carbon deflagration models in accreting C + O white dwarfs are presented as a plausible model for Type I supernovae. The relatively rapid accretion studied here leads to the initiation of the carbon deflagration at the center. Subsequent propagation of the convective carbon deflagration wave and associated explosive nucleosynthesis are calculated for several cases of mixing length in the convection theory. As a result of large nuclear energy release, the star is disrupted completely, leaving no compact star remnant behind. Nucleosynthesis in the deflagration models is discussed based on the reaction network calculation including 205 species: The abundance ratios of these species with respect to $^{56}$Fe normalized to the solar values are shown.

065.157 **Magnetic fields in the radiative interior of stars. I. Thermal shadows and forced convection.**
E. N. Parker.
Astrophys. J., Vol. 286, No. 2, p. 666 – 676 (1984).

It is pointed out that magnetic fields in the radiative interior of a star cause a slight reduction in the gas pressure, by a fraction of the order of $B^2/8\pi p$. The effect is a proportionate reduction in the opacity and an increase in the effective heat transport coefficient, so that magnetic inhomogeneities cast thermal shadows in the general outflow of radiant energy. The shadows involve horizontal temperature gradients, forcing convective circulation in the otherwise stably stratified radiative zone. The associated vertical mixing may have important consequences for the thermonuclear burning of $^7$Li in the outer envelope of the star.

065.158 **Magnetic fields in the radiative interior of stars. II. Forced convection and the $^7$Li abundance.**
E. N. Parker.
Astrophys. J., Vol. 286, No. 2, p. 677 – 680 (1984).
Magnetic fields of various forms, and of strengths ranging up to $5 \times 10^8$gauss, in the radiative core of the Sun have been hypothesized by various authors. This paper points out that the thermal shadows of magnetic inhomogeneities in the radiative core cause vertical mixing of fluid, with possible consequences for the $^7$Li abundance in the Sun. Magnetic fields of the order of $4 \times 10^5$gauss with scales of $3 \times 10^4$km are sufficient to provide the partial depletion of $^7$Li inferred from observation. It is found that the rms field in the outer radiative zone cannot much exceed $4 \times 10^5$gauss.

065.159 **Rotational mode splitting about an inclined axis.**
M. L. Aizenman, C. J. Hansen, J. P. Cox, W. D. Pesnell.
Astrophys. J., Lett. Ed., Vol. 286, No. 2, p. L43 – L45 (1984).
The authors have derived the "$m$–splitting" formula for the frequencies of nonradial oscillations of stars due to slow rotation about an axis which is inclined with respect to the pulsation axis. They obtain the simple result that the usual formula applies, except that only the component of the angular rotation velocity along the pulsation axis, rather than the full angular rotation velocity, is to be used in this formula.

065.160 **Two– and three–dimensional numerical simulations of the hydrodynamic phase of the core helium flash.**
R. G. Deupree.
Astrophys. J., Vol. 287, No. 1, p. 268 – 281 (1984).
Two–dimensional (2D) and three–dimensional (3D) numerical simulations of the core helium flash are compared for times after convection has failed to contain the thermonuclear runaway. For a given set of initial conditions, it is found that in 2D the peak temperature is slightly higher and occurs earlier and that higher expansion velocities are obtained than in 3D. It is shown that the events after convection fails to contain the runaway can be well approximated by a 3D calculation initially in hydrostatic equilibrium with a large temperature perturbation in one zone. A 3D survey of the hydrodynamic phase is performed with this model.

065.161 **On the evolution of an intermediate–mass zero–metal star which does not experience thermal instabilities during the double shell burning phase.**
A. Chieffi, A. Tornambé.
Astrophys. J., Vol. 287, No. 2, p. 745 – 748 (1984).
The evolution of a 5 $M_\odot$ zero–metal star has been followed from the pre–main sequence to the asymptotic giant branch (AGB) phase, special attention being paid to the behavior of the nuclear burning regions during the AGB phase. The authors find that (1) both hydrogen and helium central burning phases occur at the blue side of the H–R diagram; (2) during the AGB phase, double shell burning proceeds at a steady state without being interrupted by thermal instabilities of the kind found in Population I and II stellar models. The models spontaneously follow Iben's 1977 prescriptions for steady–state burning, the ratio between the helium and the hydrogen luminosities being of the order of 0.20.

065.162 **Hydrogen and helium burning in zero–metal asymptotic giant branch stars and the existence of thresholds (in core mass and CNO abundance) for the occurrence of helium shell flashes.**
M. Y. Fujimoto, I. Iben Jr., A. Chieffi, A. Tornambé.
Astrophys. J., Vol. 287, No. 2, p. 749 – 760 (1984).
The purpose of this paper is to study the thermal properties of hydrogen and helium shell burning and to explore the dependence of these properties on the CNO abundance in the envelope as well as on the core mass interior to the hydrogen–burning shell. The authors first examine the nature of shell burning in zero–metal stars to reveal what makes them distinctive from those stars in which the CNO abundance is "normal", and then derive the conditions necessary for triggering helium shell flashes.

**The history of the discovery of the thermonuclear nature of stellar energy.**
See Abstr. 004.064.

**Is the metal contribution to the astrophysical opacity incorrect?**
See Abstr. 022.071.

**Discrepancy in the CNO opacity bump resolved.**
See Abstr. 022.072.

**Rotational dynamics of celestial bodies: generalized rotation and considerations of angular momentum.**
See Abstr. 042.006.

**Bifurcation points on the one–ring sequence of uniformly rotating and self–gravitating fluid.**
See Abstr. 042.086.

**Bifurcation points on the Maclaurin sequence.**
See Abstr. 042.087.

**On detecting stellar collapse with neutrinos.**
See Abstr. 061.079.

**Helium plasma: parameter domain for strong screening effect.**
See Abstr. 061.114.

**Unconventional $^{12}$C production.**
See Abstr. 061.117.

**Enhancement of thermonuclear reaction rate due to screening by relativistic degenerate electrons: long–range correlation effect.**
See Abstr. 061.124.

**The continuous spectrum in differentially rotating perfect fluids: a model with an analytic solution.**
See Abstr. 062.001.

**The continuous spectrum in differentially rotating perfect fluids. II. The effect of gravitational radiation reaction.**
See Abstr. 062.012.

**Stellar fibril magnetic systems. I. Reduced energy state.**
See Abstr. 062.066.

**Notions générales d'hydrodynamique.**
See Abstr. 062.072.

**La convection astrophysique.**
See Abstr. 062.076.

**Modélisations numériques de la convection pénétrative.**
See Abstr. 062.077.

**Le couplage pulsation–convection.**
See Abstr. 062.078.

**Linear astrophysical dynamos in rotating spheres: solar and stellar cycle north–south hemispheric parity selection mechanism and turbulent magnetic diffusivity.**
See Abstr. 062.097.

**Continuous spectra of oscillation frequencies of an axisymmetric incompressible equilibrium with a poloidal magnetic field.**
See Abstr. 062.121.

**Stability of rotation laws.**
See Abstr. 062.123.

**Most effective potentials for stability analysis.**
See Abstr. 062.124.

**Boundary layer conditions for the transport of radiation in stars.**
See Abstr. 063.049.

**Generation of acoustic energy from convection zones of late type stars.**
See Abstr. 064.004.

**The stability limit of hypergiant photospheres.**
See Abstr. 064.016.

**Penetrative cellular convection in a stratified atmosphere.**
See Abstr. 064.041.

**Bimodal mass loss by magnetically controlled stellar winds.**
See Abstr. 064.053.

**The collapse of rotating stellar cores: the amount of gravitational radiation predicted by various numerical models.**
See Abstr. 066.242.

**Physically realizable relativistic stellar structures.**
See Abstr. 067.016.

**On the accuracy of the single–nucleus approximation in the equation of state of hot, dense matter.**
See Abstr. 067.097.

**Neutrino–pair bremsstrahlung in dense stars. IV. Phonon contributions in the crystalline lattice phase.**
See Abstr. 067.098.

**Lateral propagation of the helium shell flash on an accreting neutron star.**
See Abstr. 067.133.

**Neutrinos in gravitational collapse.**
See Abstr. 067.138.

**Les oscillations du Soleil et des étoiles.**
See Abstr. 080.014.

**Beyond mixing–length theory in convection.**
See Abstr. 080.048.

**Meridional circulation in rotating stars. VIII. The solar spin–down problem.**
See Abstr. 080.116.

**Lithium in late–type giants. III. The weak G band giants.**
See Abstr. 114.055.

**Oxygen isotopic abundances in the atmospheres of seven red giant stars.**
See Abstr. 114.113.

**Analogy of rapid oscillations of Ap stars to solar oscillations.**
See Abstr. 115.008.

**Massive stars in the Galaxy and the Magellanic Clouds: clues to stellar evolution.**
See Abstr. 115.014.

**The evolution of a star–"planet" system in the double core phase.**
See Abstr. 117.026.

**Accurate integrations of the Roche model.**
See Abstr. 117.044.

**The combined effect of mass loss and overshooting. IV. Mass transfer in massive close binaries.**
See Abstr. 117.052.

**The post–RLOF (*Roche lobe overflow*) structure of the secondary components in close binary systems, with an application to masses of Wolf–Rayet stars.**
See Abstr. 117.086.

**The lower main sequence and the nature of secondary stars in ultracompact binaries.**
See Abstr. 117.094.

**Are cataclysmic variables the progenitors of Type I supernovae?**
See Abstr. 117.095.

**The rotational and orbital evolution of cataclysmic binaries containing magnetic white dwarfs.**
See Abstr. 117.143.

**Main–sequence components of Algol–type binaries as blue stragglers.**
See Abstr. 117.166.

**Excitation of non–adiabatic quasi–toroidal oscillations in double stars.**
See Abstr. 117.186.

**Instabilities of contact binary systems.**
See Abstr. 117.187.

**Close visual binaries. II. Quantitative test of isochrones.**
See Abstr. 118.021.

**Close visual binaries. III. Parameters and evolutionary status.**
See Abstr. 118.037.

**Helium content of the Population II binary system CM Draconis.**
See Abstr. 119.093.

**Photographic photometry of RR Lyrae variables in the globular cluster M15.**
See Abstr. 122.027.

**The evolutionary status of the pulsating helium star BD + 13°3224 (V652 Her).**
See Abstr. 122.082.

**A search for evidence of the neon–sodium cycle in the cepheid X Cygni.**
See Abstr. 122.092.

**Observations and a new interpretation for the Delta Scuti variable star VZ Cancri.**
See Abstr. 122.112.

**The Population II Cepheids.**
See Abstr. 122.120.

**Radial oscillations and their theoretical interpretations.**
See Abstr. 122.148.

**Convective supernovae.**
See Abstr. 125.043.

**Type II supernovae.**
See Abstr. 125.053.

**Simplifications in supernova theory.**
See Abstr. 125.054.

**The physics of supernova shocks.**
See Abstr. 125.055.

**Successful supernovae, the anatomy of shocks: neutrino emission and the adiabatic index.**
See Abstr. 125.056.

**Core collapse models of Type II supernova explosions.**
See Abstr. 125.057.

**Nucleosynthesis in type I and type II supernovae.**
See Abstr. 125.065.

**Models of type I supernovae and observations of SN 1981b.**
See Abstr. 125.361.

**Observations of the upper initial mass limit of white dwarf progenitors $M_{WD}$.**
See Abstr. 126.022.

**Explosive helium burning in white dwarfs.**
See Abstr. 126.024.

**Spectral analysis of helium poor hot subdwarfs.**
See Abstr. 126.025.

**Oscillations of degenerate stars.**
See Abstr. 126.035.

**Mass loss from red giants: a simple evolutionary model for NGC 7027.**
See Abstr. 134.060.

**Stellar nucleosynthesis and the cosmic–ray source composition up to Zn.**
See Abstr. 144.057.

**Isotopic anomalies in cosmic rays and winds from Wolf–Rayet stars: a new model.**
See Abstr. 144.058.

**Clues to the origin of cosmic rays derived from their source composition and energy spectra.**
See Abstr. 144.059.

**Nucleosynthesis in Wolf–Rayet stars and galactic cosmic–ray isotopic composition.**
See Abstr. 144.065.

**Rapporteur paper on nucleosynthesis and cosmic rays.**
See Abstr. 144.073.

**Red giants in old open clusters. A test for stellar evolution.**
See Abstr. 153.005.

**Yellow giants in young clusters. II. A comparison of observation with theory.**
See Abstr. 153.049.

**Ages of globular clusters from their luminosity functions.**
See Abstr. 154.042.

**An interpretation of the gap on the subgiant branch of several globular clusters.**
See Abstr. 154.043.

**Theoretical foundations of evolutionary population synthesis: a progress report.**
See Abstr. 157.157.

**Stellar He production in a cold Universe?**
See Abstr. 161.122.

**The detectability of population III "Jupiters".**
See Abstr. 161.188.

# 066 Relativistic Astrophysics, Gravitation Theory

066.001 **Scaling behavior of semiclassical gravity .**
B. L. Nelson, P. Panangaden.
Gen. Relativ. Gravitation, Vol. 16, No. 7, p. 625 – 643 (1984).

Using the idea of metric scaling the authors examine the scaling behavior of the stress tensor of a scalar quantum field in curved space–time. The process of renormalizing the stress tensor is viewed as being equivalent to renormalizing the coupling constants in the Lagrangian for gravity (with terms quadratic in the curvature included). The results suggest that quantum gravity is asymptotically scale invariant.

066.002 **The geodesic equation in five–dimensional relativity theory of Kaluza–Klein.**
D. Kovacs.
Gen. Relativ. Gravitation, Vol. 16, No. 7, p. 645 – 655 (1984).

In five–dimensional relativity theory of Kaluza–Klein the projection of the 5–geodesic equation is discussed in detail. A simple classification of 5–geodesics is found and pictured with the help of a deformed 5–cone. Intrinsic mass and intrinsic charge are introduced as new particle parameters. Mass and charge formulas are proposed.

066.003 **The geodesic spray, the vertical projection, and Raychaudhuri's equation.**
M. Crampin, G. E. Prince.
Gen. Relativ. Gravitation, Vol. 16, No. 7, p. 675 – 689 (1984).

The metric connection on a space–time manifold $M$ defines on its tangent bundle $TM$ a distribution of subspaces complementary to the vertical subspaces and therefore called horizontal. The authors give a formula for the Lie derivative with respect to the geodesic spray of the tensor field on $TM$ which defines projection onto the vertical subspace along the horizontal subspace.

066.004 **Lyapunov functional approach to radiative metrics.**
B. Lukács, Z. Perjés, J. Porter, Á. Sebestyén.
Gen. Relativ. Gravitation, Vol. 16, No. 7, p. 691 – 701 (1984).

Lyapunov's second method is applied to the spherical radiative Robinson–Trautman vacuum space–times to prove that they asymptotically settle down to Schwarzschild space–time.

066.005 **Kerr metric in the de Sitter background.**
P. C. Vaidya.
Pramāna, Vol. 22, No. 3 – 4, p. 151 – 158 (1984). Abstr. in Phys. Abstr., Vol. 87, No. 1237, Entry 68072 (1984).

066.006 **Gravitational radiation of an isotropic elastic sphere.**
H. Dehnen, F. Romero, G. Schäfer.
Classical Quantum Gravity, Vol. 1, No. 3, p. 305 – 318 (1984). Abstr. in Phys. Abstr., Vol. 87, No. 1237, Entry 68075 (1984).

066.007 **The nonsymmetric non–abelian Jordan–Thiry theory.**
M. W. Kalinowski.
Nuovo Cimento A, Vol. 80A, Ser. 11, No. 1, p. 47 – 76 (1984). Abstr. in Phys. Abstr., Vol. 87, No. 1237, Entry 68094 (1984).

066.008 **Gravitational gauge fields and the cosmological constant.**
H. R. Pagels.
Phys. Rev. D, Vol. 29, No. 8, p. 1690 – 1698 (1984). Abstr. in Phys. Abstr., Vol. 87, No. 1237, Entry 68098 (1984).

066.009 **Chaotic inflation in supergravity.**
A. S. Goncharov, A. D. Linde.
Phys. Lett. B, Vol. 139B, No. 1 – 2, p. 27 – 30 (1984). Abstr. in Phys. Abstr., Vol. 87, No. 1237, Entry 68111 (1984).

066.010 **Violation of cosmic censorship in the gravitational collapse of a dust cloud.**
D. Christodoulou.
Commun. Math. Phys., Vol. 93, No. 2, p. 171 – 195 (1984). Abstr. in Phys. Abstr., Vol. 87, No. 1237, Entry 72863 (1984).

066.011 **Vacuum $<\varphi^2>$ in Schwarzschild spacetime.**
P. Candelas, K. W. Howard.
Phys. Rev. D, Vol. 29, No. 8, p. 1618 – 1625 (1984). Abstr. in Phys. Abstr., Vol. 87, No. 1237, Entry 72868 (1984).

066.012 **Emission of gravitational waves by gravitating systems in the post–Newtonian approximation.**
V. P. Tsvetkov.
Astron. Zh., Tom 61, Vyp. 4, p. 673 – 676 (1984). In Russian. English translation in Sov. Astron., Vol. 28, No. 4.
A new effective method is suggested which gives post–Newtonian corrections to the intensity of gravitational–wave emission of gravitating systems. The influence of the first–order post–Newtonian correction on the gravitational quadrupole radiation of a rotating asymmetric neutron star caused by magnetic strains is estimated.

066.013 **Le singolarità in relatività generale (1916 – 1970): II.**
R. Bergamini, B. Giorgini.
Coelum, Vol. 52, N. 4, p. 227 – 239 (1984).

066.014 **More on the gravitational constant.**
D. W. Dunham.
Occultation Newsl., Vol. 3, No. 8, p. 158 (1984).

066.015 **Gravitational light deflection in the Solar System.**
S. A. Cowling.
Mon. Not. R. Astron. Soc., Vol. 209, No. 2, p. 415 – 427 (1984).
The problems of defining the bending of a light path in general stationary space–times are discussed. It is shown how in situations of axial symmetry the use of "triangulation lines" leads to a coordinate invariant definition of the bending angle between arbitrary points on a light path. An exact expression is derived for the bending angle in a static, spherically symmetric space–time of arbitrarily strong curvature. This formulation is used to calculate, to second PPN order, the apparent positional shift which is measured when observing, for example, an asteroid or a spacecraft situated close to superior conjunction with the Sun.

066.016 **Relativistic deflection of radio signals in the solar gravitational field measured with VLBI.**
D. S. Robertson, W. E. Carter.
Nature, Vol. 310, No. 5978, p. 572 – 574 (1984).
For an observer on Earth the general theory of relativity (GR) predicts an apparent outward displacement of a star seen at the Sun's limb of 1.75 arc s. A generalized formulation for gravitational deflection of light includes a parameter $\gamma$ which ranges from 1 (GR) to –1 (no deflection). Radio interferometer measurements of deflections of extragalactic objects have all been consistent with GR with increasing accuracy. The authors report the most sensitive measurement of deflection so far achieved, in which very long baseline interferometry (VLBI) observations yield a value of $\gamma$ of 1.008 with a 1 $\sigma$ formal standard error of $\pm 0.005$.

066.017 **Continuing doubt on gravitation.**
J. Maddox.
Nature, Vol. 310, No. 5980, p. 723 (1984).
The note discusses recent attempts to obtain precision measurements of the gravitational constant.

066.018 **The generation of high–frequency gravitational waves in a perfect fluid.**
V. Pirronello.
Astrophys. Space Sci., Vol. 102, No. 1, p. 11 – 19 (1984).
The problem of the generation of a gravitational wave by an acoustic one is studied in the high–frequency domain for a perfect fluid. The negative result which is found may be of some relevance for astrophysical and cosmological applications.

066.019 **Equilibrium ensembles in general relativity.**
H. E. Kandrup.
Astrophys. Space Sci., Vol. 103, No. 2, p. 293 – 300 (1984).
The author considers a collection of particles that interact only via elastic point 'collisions' in a fixed background spacetime in the presence of some specified external fields. Is is shown that if there exists a time–like vector field that Lie derives the metric and the external fields, then it is possible to formulate sensible covariant analogues of the standard equilibrium ensembles from Newtonian statistical mechanics.

066.020 **Comment on the LRS–BDT–Bianchi type–V vacuum solution.**
D. Lorenz–Petzold.
Astrophys. Space Sci., Vol. 103, No. 2, p. 395 – 396 (1984).
The author wishes to point out that the 'anisotropic' locally rotationally–symmetric (LRS) – Brans–Dicke (BDT) – Bianchi type–V vacuum solution recently given by Ram and Singh is wrong. Moreover, the correct solution is nothing but an isotropic Friedmann–Robertson–Walker ($k = -1$) solution given by the author previously.

066.021 **The general non–LRS (*locally rotationally–symmetric*)–BDT (*Brans–Dicke theory*)–Bianchi type–V radiation solution.**
D. Lorenz–Petzold.
Astrophys. Space Sci., Vol. 103, No. 2, p. 397 – 399 (1984).
The author presents the most general triaxial BDT–Bianchi type–V radiation solution. The solution is the simplest generalization of the GRT (*theory of general relativity*) solution first given by Ruban.

066.022 **A class of Szekeres space–times with cosmological constant.**
G. M. Covarrubias.
Astrophys. Space Sci., Vol. 103, No. 2, p. 401 – 404 (1984).
The author presents a generalization of a class of Szekeres space–times. The new solutions satisfy Einstein's equations with cosmological constant and have the same geometrical properties as the corresponding class found by Szekeres. Particular cases leading to known solutions are considered.

066.023 **On coordinates and coordinate transformation in Einstein's theory of gravitation.**
P. Chou.
Proceedings of the Third Marcel Grossmann Meeting on General Relativity, Part A, p. 1 – 20 (1983). – See Abstr. 012.002.
This investigation is a further exposition of the significance of coordinates and their transformation in Einstein's theory of gravitation. The author considers the static axisymmetric field as an example, starts with its metric in the cylindrical coordinates, transforms this metric and the field equations into the Weyl–Levi–Civita system of coordinates, and supplements them with the harmonic condition. The results strongly support the argument that the harmonic condition should be a physical supplement to Einstein's theory of gravitation.

066.024 **Applications of the Riemann–Hilbert problem to general relativity.**
F. J. Ernst.
Proceedings of the Third Marcel Grossmann Meeting on General Relativity, Part A, p. 21 – 28 (1983). – See Abstr. 012.002.
This paper outlines the basic ideas involved in the Hauser–Ernst homogeneous Hilbert problem formulation of the Kinnersley–Chitre transformation theory and the proof of the famous Geroch conjecture.

066.025 **Ernst equation Bäcklund transformations revisited: new approaches and results.**
B. K. Harrison.
Proceedings of the Third Marcel Grossmann Meeting on General Relativity, Part A, p. 29 – 39 (1983). – See Abstr. 012.002.
This paper discusses new results relating to the solution, using Bäcklund transformations, of the Ernst equation for axially symmetric fields in general relativity.

066.026 **The gravitational radiation damping and the change of coordinates.**
N. Hu.
Proceedings of the Third Marcel Grossmann Meeting on General Relativity, Part A, p. 41 – 50 (1983). – See Abstr. 012.002.
In the general theory of relativity, the gravitational waves having temporal and longitudinal components of polarization carry no energy and thus represent only the spurious effect of the coordinates. An expression for the radiation damping force on binary stars is obtained from the field equation by requiring that the out-going gravitational radiation should have only transverse components of polarization.

066.027 **A non–local variable for general relativity.**
C. N. Kozameh, E. T. Newman.
Proceedings of the Third Marcel Grossmann Meeting on General Relativity, Part A, p. 51 – 55 (1983). – See Abstr. 012.002.
The usual description of differential geometry and general relativity is in terms of local fields, e.g. the metric, the curvature tensor, etc, which satisfy local differential equations. It is the purpose of this paper to introduce a new non–local field, denoted by $Z$, from which the local fields can be derived.

066.028 **Global solutions of Yang–Mills and associated field equations.**
Y. Choquet–Bruhat.
Proceedings of the Third Marcel Grossmann Meeting on General Relativity, Part A, p. 57 – 67 (1983). – See Abstr. 012.002.
This paper is a survey of results which have been obtained recently about the global properties of solutions of the classical equations of gauge field theories coupled with scalar or spinor multiplets. These results concern global existence and asymptotic (decay) properties.

066.029 **Topologically massive three–dimensional vector and gravity gauge theories.**
R. Jackiw.
Proceedings of the Third Marcel Grossmann Meeting on General Relativity, Part A, p. 69 – 78 (1983). – See Abstr. 012.002.

066.030 **On spacetimes without maximal surfaces.**
D. R. Brill.
Proceedings of the Third Marcel Grossmann Meeting on General Relativity, Part A, p. 79 – 87 (1983). – See Abstr. 012.002.
Whereas asymptotically flat spacetimes are usually assumed to admit a spacelike maximal Cauchy hypersurface, it is shown here that such hypersurfaces do not always exist, by giving explicit counterexamples.

066.031 **A formulation of nonlinear gauge theory and its applications.**
Q. Lu, H. Guo, K. Wu.
Proceedings of the Third Marcel Grossmann Meeting on General Relativity, Part A, p. 89 – 106 (1983). – See Abstr. 012.002.
The purpose of this paper is to give a formulation of nonlinear connections which is suitable for problems of physics and local differential geometry.

066.032 **Vacuum solutions with double duality properties of the Poincaré gauge field theory. II.**
P. Baekler, F. W. Hehl, H. J. Lenzen.
Proceedings of the Third Marcel Grossmann Meeting on General Relativity, Part A, p. 107 – 128 (1983). – See Abstr. 012.002.
The authors consider the Poincaré gauge field theory with a 10–parameter Lagrangian which is a polynomial of 2nd order in torsion and curvature. A survey of the exact vacuum solutions carrying non–vanishing torsion is given.

066.033 **Possible role of torsion in gravitational theories.**
H. T. Nieh.
Proceedings of the Third Marcel Grossmann Meeting on General Relativity, Part A, p. 129 – 137 (1983). – See Abstr. 012.002.

066.034 **On the motion of a string in a curved space–time.**
C. Gu.
Proceedings of the Third Marcel Grossmann Meeting on General Relativity, Part A, p. 139 – 142 (1983). – See Abstr. 012.002.
The motion of an infinite or closed string in a curved space–time can be determined by harmonic maps from the Minkowski plane to the curved space–time.

066.035 **Prolongation structure, Bäcklund transformation and principal homogeneous Hilbert problem in general relativity.**
H. Guo, K. Wu, S. Wang.
Proceedings of the Third Marcel Grossmann Meeting on General Relativity, Part A, p. 143 – 164 (1983). – See Abstr. 012.002.
This paper reviews the prolongation structure theory for nonlinear systems, such as nonlinear evolution equations and soliton equations, and its applications to the solution generating techniques for gravitational fields with two commuting Killing vectors.

066.036 **General relativistic hydrodynamic wave propagation.**
A. H. Taub.
Proceedings of the Third Marcel Grossmann Meeting on General Relativity, Part A, p. 165 – 187 (1983). – See Abstr. 012.002.
The author discusses the propagation of disturbances in a self–gravitating perfect fluid. The equations governing the motion of the fluid and the gravitational field due to the presence of the fluid are the Einstein field equations, their consequences and either a conservation law for the rest mass of the fluid or an equation of state.

066.037 **Relativistic generalization of the Newtonian force.**
A. Qadir, J. Quamar.
Proceedings of the Third Marcel Grossmann Meeting on General Relativity, Part A, p. 189 – 220 (1983). – See Abstr. 012.002.
Whereas there is no denying the essential contribution of geometrodynamics, it must be admitted that our physical intuition is still firmly based in the Newtonian concept of force. Here the authors extend some earlier work re–introducing the Newtonian force concept into relativity theory. Some fundamentally new insights into the relativistic effects due to charge and rotation are presented.

066.038 **Lagrangians, Bäcklund transformations and a stationary solution of the relativistic two–body problem.**
C. Hoenselaers.
Proceedings of the Third Marcel Grossmann Meeting on General Relativity, Part A, p. 221 – 237 (1983). – See Abstr. 012.002.
This paper describes recent progress in finding exact solutions to the equations describing axisymmetric stationary gravitational fields in General Relativity. It discusses the geometric meaning of Neugebauer's Bäcklund transformation. A metric is presented which is a stationary solution to the relativistic two–body problem. It describes the field of two masses whose gravitational attraction is balanced by repulsion due to their angular momenta.

066.039 **Path integration quantization.**
C. DeWitt–Morette.
Proceedings of the Third Marcel Grossmann Meeting on General Relativity, Part A, p. 309 – 319 (1983). – See Abstr. 012.002.

066.040 **Finite temperature effects of quantum gravity.**
C. Mukku, W. A. Sayed.
Proceedings of the Third Marcel Grossmann Meeting on General Relativity, Part A, p. 321 – 342 (1983). – See Abstr. 012.002.
The authors obtain the temperature dependent contributions to the free energy density of a system of thermal gravitons and

scalars. These terms are qualitatively different in character and, unlike the previously examined cases of scalars, vectors, and fermions, can contribute negatively to the free energy density in the high temperature limit.

066.041 **The quantum effect in general relativity.**
Y. Duan, C. Yu.
Proceedings of the Third Marcel Grossmann Meeting on General Relativity, Part A, p. 353 – 358 (1983). – See Abstr. 012.002.

Using the general covariant Dirac equation, the gravitational correction to the anomalous magnetic moment of the electron has been studied. It is shown that this new quantum effect in general relativity due to the earth's gravity could be measured by means of modern experimental devices.

066.042 **On the perturbations of Kerr–Newman metric.**
V. Bellezza, V. Ferrari.
Proceedings of the Third Marcel Grossmann Meeting on General Relativity, Part A, p. 359 – 365 (1983). – See Abstr. 012.002.

The authors derive a set of differential equations governing the linearized perturbations of Kerr–Newman metric. These equations couple the electromagnetic and the gravitational perturbations; they are invariant under infinitesimal coordinates transformations (gauge invariant) and under infinitesimal tetrad rotations (tetrad invariant).

066.043 **Infinitely conserved currents and hidden symmetry algebra related to the Belinskiĭ–Zakharov formulation of gravity.**
Y. Wu, M. Ge, B. Hou.
Proceedings of the Third Marcel Grossmann Meeting on General Relativity, Part A, p. 367 – 381 (1983). – See Abstr. 012.002.

Based on the Belinskiĭ–Zakharov formulation of the vacuum Einstein equation with the metric $g_a{}^b$ depending only on two coordinates, the related infinitely conserved currents are discussed. It is shown that there appears a set of nonlocal conservation currents which can be regarded as a parametric Noether current which arises from "H–transformation".

066.044 **Gauge freedom in perfect fluid spatially homogeneous spacetimes.**
R. T. Jantzen.
Proceedings of the Third Marcel Grossmann Meeting on General Relativity, Part A, p. 447 – 460 (1983). – See Abstr. 012.002.

The class of reference systems compatible with the symmetry of a spatially homogeneous perfect fluid spacetime is discussed together with the associated class of symmetry adapted comoving ADM frames (or computational frames). The fluid equations of motion are related to the four functions on the space of fluid flow lines discovered by Taub and which characterize an isentropic flow.

066.045 **Space–like infinity.**
M. Alexander, P. G. Bergmann.
Proceedings of the Third Marcel Grossmann Meeting on General Relativity, Part A, p. 575 – 582 (1983). – See Abstr. 012.002.

066.046 **Radiation damping in general relativity.**
T. Damour.
Proceedings of the Third Marcel Grossmann Meeting on General Relativity, Part A, p. 583 – 597 (1983). – See Abstr. 012.002.

The author considers the problem of the gravitational interaction of two compact bodies with special emphasis on the issue of 'radiation damping'. After a brief discussion of the inconclusiveness of previous approaches, he outlines a new method where one matches an 'external' gravitational field to the field near each compact body. Equations of motion are derived from the vacuum field equations. Explicit expressions are given for the cubically non–linear external gravitational field and for the $2^{1/2}$ Post Newtonian equations of orbital motion. The final results agree with the phenomena observed in the binary pulsar PSR 1913+16.

066.047 **Coordinate conditions for numerical relativity.**
T. Piran.
Proceedings of the Third Marcel Grossmann Meeting on General Relativity, Part A, p. 643 – 651 (1983). – See Abstr. 012.002.

066.048 **Quantum theory of the gravitational wave antenna.**
J. Weber.
Proceedings of the Third Marcel Grossmann Meeting on General Relativity, Part A, p. 653 – 654 (1983). Abstract. – See Abstr. 012.002.

066.049 **The gravitational wave experiment of the Rome group.**
E. Amaldi, P. Bonifazi, E. Coccia, S. Frasca, I. Modena, G. V. Pallottino, G. Pizzella, F. Ricci.
Proceedings of the Third Marcel Grossmann Meeting on General Relativity, Part A, p. 655 – 666 (1983). – See Abstr. 012.002.

066.050 **A progress report on the Stanford low temperature gravitational wave detector.**
M. Bassan, W. M. Fairbank, E. Mapoles, M. S. McAshan, P. F. Michelson, B. Moskowitz, K. Ralls, R. C. Taber.
Proceedings of the Third Marcel Grossmann Meeting on General Relativity, Part A, p. 667 – 678 (1983). – See Abstr. 012.002.

The sensitivity of a 4800 kg cryogenic resonant mass gravitational wave detector is discussed. The detector has been operated for an extended period of time with a noise temperature for pulse detection of 20mK corresponding to an rms dimensionless strain of $10^{-18}$. The output noise level has been used to establish an observational upper limit for the flux of 1 kHz gravity wave bursts. Future sensitivity improvements are outlined.

066.051 **Gravitational waves: an introduction and a progress report.**
W. O. Hamilton.
Proceedings of the Third Marcel Grossmann Meeting on General Relativity, Part A, p. 679 – 703 (1983). – See Abstr. 012.002.

066.052 **Laser interferometry and gravitational waves.**
H. Billing.
Proceedings of the Third Marcel Grossmann Meeting on General Relativity, Part A, p. 705 – 706 (1983). Abstract. – See Abstr. 012.002.

066.053 **Gravitational wave experiments in China.**
S. Feng, R. Hu, Y. Liu, Z. Liu, R. Qin, Y. Song, D. Tan, J. Tian, G. Wang, Y. Yan, P. Zhang, J. Chen, T. Guan, E. Hu, M. Tang, P. Yu.
Proceedings of the Third Marcel Grossmann Meeting on General Relativity, Part A, p. 707 – 717 (1983). – See Abstr. 012.002.

Chinese physicists have so far constructed two Weber type antennas at room temperature. They are now developing a tuning fork antenna in Beijing and quadrupole antenna in Guangzhou for detecting low frequency continuous gravitational radiation.

066.054 **Progress in the development of a niobium gravitational radiation antenna.**
D. G. Blair, J. Bryant, M. J. Buckingham, B. Candy, C. Edwards, J. Ferreirinho, R. N. James, F. van Kann, A. G. Mann, L. D. Mann, T. Suzuki, P. J. Veitch, W. Back, I. Castledon.
Proceedings of the Third Marcel Grossmann Meeting on General Relativity, Part A, p. 719 – 738 (1983). – See Abstr. 012.002.

At the University of Western Australia a 1.5 tonne niobium gravitational radiation antenna is being constructed with the aim of detecting g.r. pulses of amplitude $\sim 10^{-20}$, that may be produced by the gravitational collapse of stars in nearby galaxies.

066.055 **Developments in laser interferometer gravitational wave detectors.**
R. W. P. Drever, S. Hoggan, J. Hough, B. J. Meers, A. J. Munley, G. P. Newton, H. Ward, D. Z. Anderson, Y. Gursel, M. Hereld, R. E. Spero, S. E. Whitcomb.
Proceedings of the Third Marcel Grossmann Meeting on General Relativity, Part A, p. 739 – 753 (1983). – See Abstr. 012.002.

Some basic ideas and recent developments in gravitational radiation detectors using laser interferometry are described, with emphasis on optical cavity interferometers and on methods of improving detector sensitivity for particular kinds of signal.

066.056 **Tests of the gravitational inverse square law on a laboratory distance scale.**
R. D. Newman.
Proceedings of the Third Marcel Grossmann Meeting on General Relativity, Part A, p. 755 – 769 (1983). – See Abstr. 012.002.
This paper reviews some of the motivations for inverse square tests, and presents a short review of the experimental situation. Preliminary results of a recent experiment are presented, indicating that the variation of the gravitational force from 5 to 105 cm agrees with the Newtonian prediction to about 7 parts in $10^4$.

066.057 **Cryogenic Equivalence Principle Experiment: discussion and present status.**
P. W. Worden Jr.
Proceedings of the Third Marcel Grossmann Meeting on General Relativity, Part A, p. 771 – 779 (1983). – See Abstr. 012.002.
The Weak Equivalence Principle is the hypothesis that the ratio of passive gravitational mass to inertial mass is the same for all bodies regardless of their composition. This principle has a fundamental place in physics as the experimental basis for Einstein's Strong Equivalence Principle, which is a postulate of General Relativity. This paper describes the present state of the Stanford Orbital Equivalence Principle Experiment.

066.058 **Solar system tests of relativity.**
R. D. Reasenberg.
Proceedings of the Third Marcel Grossmann Meeting on General Relativity, Part A, p. 781 – 796 (1983). – See Abstr. 012.002.
This paper reviews the modern solar-system tests of general relativity. It discusses the status of four such tests: (1) a possible secular variation of the gravitational constant; (2) the metric parameter $\gamma$; (3) the metric parameter $\beta$; and (4) the coefficient $\eta$ of Nordtvedt's hypothesized violation of the principle of equivalence. Furthermore, the Viking relativity experiment and its current status is described and prospects for future experiments with higher accuracy are outlined.

066.059 **A proposed measurement of the deflection of starlight by the sun using Hipparcos.**
J. T. Anderson.
Proceedings of the Third Marcel Grossmann Meeting on General Relativity, Part A, p. 797 – 801 (1983). – See Abstr. 012.002.
The Hipparcos Spacecraft of ESA will produce an astrometric catalog containing approximately $10^5$ stars to an accuracy approaching 1 milliarcsec. The measurements on which the catalog is based must be corrected for the deflection of starlight. If the deflection is treated as a free parameter in the data analysis, the possibility exists of a measurement of $(1+\gamma)/2$ to about 0.2 percent.

066.060 **Redetermination of the Newtonian gravitational constant "$G$".**
G. G. Luther.
Proceedings of the Third Marcel Grossmann Meeting on General Relativity, Part A, p. 827 – 838 (1983). – See Abstr. 012.002.

066.061 **A null test of the inverse square law of gravitation.**
H. J. Paik, H. A. Chan, M. V. Moody.
Proceedings of the Third Marcel Grossmann Meeting on General Relativity, Part A, p. 839 – 853 (1983). – See Abstr. 012.002.
This paper reports on an experimental attempt to improve the empirical limits of validity for Newton's law on the laboratory and geological distance scales. A unique feature in the present approach is the examination of the tensor nature of the local gravitational field as opposed to the measurement of a spatial dependence of the gravitational force as is done in conventional experiments.

066.062 **Junction conditions and the propagation of isometries in general relativity.**
L. Herrera, J. Jiménez.
Proceedings of the Third Marcel Grossmann Meeting on General Relativity, Part B, p. 889 – 892 (1983). – See Abstr. 012.002.

066.063 **Some invariance theorems of the geometry in Einstein non-symmetrical field theory and the classifications of pseudo-tensor field $U_{ik}{}^{l}$.**
S. Liu.
Proceedings of the Third Marcel Grossmann Meeting on General Relativity, Part B, p. 893 – 898 (1983). – See Abstr. 012.002.

066.064 **Post-Post-Newton contribution to the gravitational deflection of light in arbitrary region.**
C. Xu, J. Xu, Z. Huang, L. Yang.
Proceedings of the Third Marcel Grossmann Meeting on General Relativity, Part B, p. 899 – 902 (1983). – See Abstr. 012.002.

066.065 **The problem of motion in the Kaluza theory.**
K. Wu, H. Guo.
Proceedings of the Third Marcel Grossmann Meeting on General Relativity, Part B, p. 903 – 909 (1983). – See Abstr. 012.002.

066.066 **Approximate solutions for the $R + R^2$ theories of gravitation with torsion.**
Y. Zhang.
Proceedings of the Third Marcel Grossmann Meeting on General Relativity, Part B, p. 911 – 914 (1983). – See Abstr. 012.002.

066.067 **Riemannian structure on gauge group and metric compatible Yang-Mills theory.**
Y. Liu.
Proceedings of the Third Marcel Grossmann Meeting on General Relativity, Part B, p. 915 – 919 (1983). – See Abstr. 012.002.

066.068 **Rosen's bi-metric theory of gravity as a harmonic map.**
W. R. Stoeger.
Proceedings of the Third Marcel Grossmann Meeting on General Relativity, Part B, p. 921 – 925 (1983). – See Abstr. 012.002.

066.069 **Group covariance and metrical theory.**
L. Halpern.
Proceedings of the Third Marcel Grossmann Meeting on General Relativity, Part B, p. 927 – 931 (1983). – See Abstr. 012.002.

066.070 **A conformally invariant theory of gravitation and electromagnetism.**
G. Papini.
Proceedings of the Third Marcel Grossmann Meeting on General Relativity, Part B, p. 933 – 936 (1983). – See Abstr. 012.002.

066.071 **Dynamical properties of spinor gauge theory.**
Y. Liu, D. Zhu.
Proceedings of the Third Marcel Grossmann Meeting on General Relativity, Part B, p. 937 – 940 (1983). – See Abstr. 012.002.

066.072 **Newtonian limit of an $R + R^2$ theory of gravity with torsion.**
J. Anandan.
Proceedings of the Third Marcel Grossmann Meeting on General Relativity, Part B, p. 941 – 945 (1983). – See Abstr. 012.002.

066.073 **Theories of gravity alternative to general relativity.**
I. Ciufolini.
Proceedings of the Third Marcel Grossmann Meeting on General Relativity, Part B, p. 947 – 950 (1983). – See Abstr. 012.002.

066.074 **A remark on the problem of gravitational radiation.**
A. Papapetrou.
Proceedings of the Third Marcel Grossmann Meeting on General Relativity, Part B, p. 951 – 954 (1983). – See Abstr. 012.002.

066.075 **Gravitational radiation from two compact bodies: the radiative losses at the post linear approximation of general relativity.**
N. Deruelle.
Proceedings of the Third Marcel Grossmann Meeting on General Relativity, Part B, p. 955 – 958 (1983). – See Abstr. 012.002.

Starting from the second order gravitational field and equations of motion of two compact bodies the author deduces the asymptotic expansion of the metric at the post linear approximation of general relativity.

066.076 **On the types of Einstein equations and globally admissible coordinate systems.**
V. Unt.
Proceedings of the Third Marcel Grossmann Meeting on General Relativity, Part B, p. 959 – 962 (1983). – See Abstr. 012.002.

It is argued that by considering the approximate radiative solutions of the Einstein equations in the infinite domain of space–time, one can make a distinction between admissible and non–admissible coordinate systems. There emerges a natural division of these equations into elliptic and hyperbolic types.

066.077 **Gravitational radiation from perturbed astrophysical sources: progress towards an optimally economical and flexible theory.**
U. H. Gerlach.
Proceedings of the Third Marcel Grossmann Meeting on General Relativity, Part B, p. 963 – 967 (1983). – See Abstr. 012.002.

066.078 **Improved equations for gravitational perturbations of the Kerr geometry.**
M. Sasaki, T. Nakamura.
Proceedings of the Third Marcel Grossmann Meeting on General Relativity, Part B, p. 969 – 972 (1983). – See Abstr. 012.002.

A transformation of the Teukolsky equation to a class of equations which are free of divergence difficulties is presented.

066.079 **Pseudoanalyticity and minimal surface in electromagnetic perturbation of the Kerr geometry.**
W. K. Chan, F. G. Shin.
Proceedings of the Third Marcel Grossmann Meeting on General Relativity, Part B, p. 973 – 979 (1983). – See Abstr. 012.002.

The authors reduce Maxwell's equations in the Kerr geometry to a form which shows that, aside from certain "kinematic factors", the NP components $\phi_0$, $\phi_1$ and $\phi_2$ are pseudoanalytic. They derive a representation for these components which may be useful for numerical investigations.

066.080 **The conservation law of energy–momentum in general relativity.**
Y. Duan, Y. Wang.
Proceedings of the Third Marcel Grossmann Meeting on General Relativity, Part B, p. 981 – 988 (1983). – See Abstr. 012.002.

Using Noether's theorem, a satisfactory conservation law of energy–momentum in general relativity is obtained from a generalized translational transformation. This conservation law is discussed with regard to the covariant property of the energy–momentum tensor.

066.081 **Gravitational fields with symmetries: bundle connections and gauge transformations.**
F. J. Chinea.
Proceedings of the Third Marcel Grossmann Meeting on General Relativity, Part B, p. 989 – 992 (1983). – See Abstr. 012.002.

066.082 **Einstein equations, self–dual Yang–Mills field and non–linear σ models.**
N. Sanchez.
Proceedings of the Third Marcel Grossmann Meeting on General Relativity, Part B, p. 993 – 996 (1983). – See Abstr. 012.002.

066.083 **On transformations which leave invariant the Einstein equations.**
Pham Mau Quan.
Proceedings of the Third Marcel Grossmann Meeting on General Relativity, Part B, p. 997 – 1000 (1983). – See Abstr. 012.002.

066.084 **Field of an arbitrarily accelerating charged point mass.**
Y. Wang, X. Lin.
Proceedings of the Third Marcel Grossmann Meeting on General Relativity, Part B, p. 1001 – 1004 (1983). – See Abstr. 012.002.

The authors have derived a metric containing five arbitrary functions of time. It contains the Kinnersley metric, the Reissner–Nordström metric and the Schwarzschild metric as special cases. This metric describes the field of an arbitrarily accelerating charged point mass.

066.085 **The stationary gravitational field of an infinite rod.**
H. Ding.
Proceedings of the Third Marcel Grossmann Meeting on General Relativity, Part B, p. 1005 – 1008 (1983). – See Abstr. 012.002.

The stationary gravitational field of an infinite rod in Einstein's theory of gravitation with the harmonic condition is given.

066.086 **The complexification of a non–rotating sphere: an extension of the Newman–Janis algorithm.**
L. Herrera, J. Jiménez.
Proceedings of the Third Marcel Grossmann Meeting on General Relativity, Part B, p. 1017 – 1022 (1983). – See Abstr. 012.002.

066.087 **Electrovac generalization of Neugebauer's N = 2 solution and a new nine–parameter electrovac solution of the coupled Einstein–Maxwell field equations.**
D. Guo.
Proceedings of the Third Marcel Grossmann Meeting on General Relativity, Part B, p. 1023 – 1027 (1983). – See Abstr. 012.002.

066.088 **On the harmony of the Ernst equation.**
C. Reina.
Proceedings of the Third Marcel Grossmann Meeting on General Relativity, Part B, p. 1029 – 1032 (1983). – See Abstr. 012.002.

The geometrical interpretation of the Ernst equation as a harmonic problem is briefly reviewed. It is suggested how some results from harmonic maps theory can clarify the structure of the "space of solutions" of the Ernst equation.

066.089 **Geometrical meaning of Bäcklund transformation for Ernst equation.**
B. Hou, B. Hou, P. Wang.
Proceedings of the Third Marcel Grossmann Meeting on General Relativity, Part B, p. 1033 – 1038 (1983). – See Abstr. 012.002.

066.090 **The *N*–fold charged Kerr family solution.**
S. Wang, H. Guo, K. Wu.
Proceedings of the Third Marcel Grossmann Meeting on General Relativity, Part B, p. 1039 – 1044 (1983). – See Abstr. 012.002.

066.091 **On the Kramer–Neugebauer spinning masses solutions.**
M. Yamazaki.
Proceedings of the Third Marcel Grossmann Meeting on General Relativity, Part B, p. 1045 – 1048 (1983). – See Abstr. 012.002.

A metric is presented, in terms of six determinations of degree $N$, in a concise closed form, describing collinear $N$ isolated Kerr masses ($N = 1, 2, 3,...$) balanced by their gravitational spin–spin interaction.

066.092 **Structure of the two Kerr–solution.**
M. Kihara, K.–i. Oohara, H. Sato, A. Tomimatsu.
Proceedings of the Third Marcel Grossmann Meeting on General Relativity, Part B, p. 1049 – 1052 (1983). – See Abstr. 012.002.

066.093 **N rank zero HKX transformations.**
W. Dietz.
Proceedings of the Third Marcel Grossmann Meeting on General Relativity, Part B, p. 1053 – 1057 (1983). – See Abstr. 012.002.

The author derives a new representation of an arbitrary number of rank zero HKX transformations acting on an arbitrary Weyl solution as seed metric and presents the resulting Ernst potential by means of determinants.

066.094 **Exact global radiative solutions to the Bondi initial value problem.**
R. A. Isaacson, J. S. Welling, J. Winicour.
Proceedings of the Third Marcel Grossmann Meeting on General Relativity, Part B, p. 1059 – 1066 (1983). – See Abstr. 012.002.

The authors discuss the Bondi–type initial value problem for axisymmetric space–times, with a dust source. Choosing vanishing shear, pressure and initial velocity, they find a class of density distributions for which all the null hypersurface constraint equations, as well as the dynamical equation for the initial time derivative of the shear, can be integrated in closed form.

066.095 **A new approach to the algebraic structure in stationary axially symmetric gravity.**
Y. Wu, M. Ge.
Proceedings of the Third Marcel Grossmann Meeting on General Relativity, Part B, p. 1067 – 1072 (1983). – See Abstr. 012.002.

A new parametric infinitesimal transformation to preserve the equation of motion and linearization formulation in Kinnersley–Chitre–Hauser–Ernst formulation of gravity is proposed.

066.096 **The geometry of sine–Laplace, sinh–Laplace equations.**
H. Hu.
Proceedings of the Third Marcel Grossmann Meeting on General Relativity, Part B, p. 1073 – 1078 (1983). – See Abstr. 012.002.

066.097 **Temporal gauge in general relativity.**
Y. Choquet–Bruhat, T. Ruggeri.
Proceedings of the Third Marcel Grossmann Meeting on General Relativity, Part B, p. 1079 – 1082 (1983). – See Abstr. 012.002.

066.098 **[(2 + 1) + 1]–dimensional representation of the Einstein equations.**
K.–i. Maeda.
Proceedings of the Third Marcel Grossmann Meeting on General Relativity, Part B, p. 1083 – 1086 (1983). – See Abstr. 012.002.

066.099 **Hermitian gravity and supergravity.**
V. de Sabbata, M. Gasperini.
Proceedings of the Third Marcel Grossmann Meeting on General Relativity, Part B, p. 1135 – 1145 (1983). – See Abstr. 012.002.

This paper presents the formulation of a hermitian theory of gravity, in which the antisymmetric part of the metric tensor is generated by the spin density of matter.

066.100 **Is gravity quantized?**
D. N. Page.
Proceedings of the Third Marcel Grossmann Meeting on General Relativity, Part B, p. 1157 – 1159 (1983). – See Abstr. 012.002.

066.101 **Manifestly covariant canonical formalism of quantum gravity.**
N. Nakanishi.
Proceedings of the Third Marcel Grossmann Meeting on General Relativity, Part B, p. 1161 – 1164 (1983). – See Abstr. 012.002.

066.102 **A gauge invariant resummation of quantum gravity.**
A. Strominger.
Proceedings of the Third Marcel Grossmann Meeting on General Relativity, Part B, p. 1165 – 1170 (1983). – See Abstr. 012.002.

Quantum gravity is expanded in powers of 1/D, where D is the number of dimensions. The extra dimensions are highly compactified. The expansion is gauge invariant and renormalizable. Flat space is found to be unstable under small fluctuations.

066.103 **Kaluza–Klein vacua.**
P. G. O. Freund.
Proceedings of the Third Marcel Grossmann Meeting on General Relativity, Part B, p. 1171 – 1174 (1983). – See Abstr. 012.002.

066.104 **Microcanonical quantum gravity.**
A. E. Strominger.
Proceedings of the Third Marcel Grossmann Meeting on General Relativity, Part B, p. 1175 (1983). Abstract. – See Abstr. 012.002.

066.105 **Quantum mechanical tests of general relativity.**
J. Anandan.
Proceedings of the Third Marcel Grossmann Meeting on General Relativity, Part B, p. 1185 – 1188 (1983). – See Abstr. 012.002.

This paper discusses possible experiments aiming the detection of the Lense–Thirring field of the Earth.

066.106 **Blackbody radiation from light cone in flat space–time.**
U. H. Gerlach.
Proceedings of the Third Marcel Grossmann Meeting on General Relativity, Part B, p. 1201 – 1203 (1983). – See Abstr. 012.002.

066.107 **On defining particles in Robertson–Walker space–times.**
T. Dray.
Proceedings of the Third Marcel Grossmann Meeting on General Relativity, Part B, p. 1205 – 1208 (1983). – See Abstr. 012.002.

The author considers the semi–classical quantization of the Klein–Gordon field on a Robertson–Walker background. He derives an expression for the density of particles created during an arbitrary time interval using an arbitrary particle definition.

066.108 **A geometrical foundation of a unified field theory.**
G. E. Tauber.
Proceedings of the Third Marcel Grossmann Meeting on General Relativity, Part B, p. 1221 – 1235 (1983). – See Abstr. 012.002.

066.109 **Generalised field theory and general relativity.**
A. H. Klotz.
Proceedings of the Third Marcel Grossmann Meeting on General Relativity, Part B, p. 1237 – 1242 (1983). – See Abstr. 012.002.

This paper describes the generalised field theory (GFT) as a macrophysical extension of general relativity (GR) and of the non–symmetric unified field theory (UFT) of Einstein and Straus.

066.110 **Diffraction effect of plane electromagnetic wave by a spherical gravitational field of normal star.**
R. Wang, J. You, S. Xiang.
Proceedings of the Third Marcel Grossmann Meeting on General Relativity, Part B, p. 1329 – 1336 (1983). – See Abstr. 012.002.

This paper gives a rediscussion of gravitational diffraction of an electromagnetic wave by a normal star. Using an iteration method, a simple analytic solution of the four–potential equation with Schwarzschild metric is given.

066.111 **Laser interferometer experiments at Caltech.**
S. E. Whitcomb, D. Z. Anderson, R. W. P. Drever, Y. Gursel, M. Hereld, R. Spero.
Proceedings of the Third Marcel Grossmann Meeting on General Relativity, Part B, p. 1399 – 1405 (1983). – See Abstr. 012.002.

The status of the Caltech 40 m laser gravitational wave detector is reported, emphasizing the suspension system for the test masses. Preliminary noise measurements are also presented.

066.112 **The weak plane gravitational wave and the motion of a test particle in the proper reference frame.**
H. Liu.
Proceedings of the Third Marcel Grossmann Meeting on General Relativity, Part B, p. 1407 – 1412 (1983). – See Abstr. 012.002.

The metric of a weak plane gravitational wave in the proper reference frame is obtained using a coordinate transformation. Thus the geodesic equation can directly be employed to determine the motion of test particles. The oscillations of a test particle are analysed and calculated. A significant result is that the particle oscillates also in the propagation direction of the wave.

066.113 **The effect of gravitational wave on electromagnetic field and the possibility about electromagnetic detection of gravitational wave.**
F. Tao, Z. He.
Proceedings of the Third Marcel Grossmann Meeting on General Relativity, Part B, p. 1413 – 1418 (1983). – See Abstr. 012.002.

066.114 **Approach to tunable quadrupole antenna.**
M. Tang, J. Chen.
Proceedings of the Third Marcel Grossmann Meeting on General Relativity, Part B, p. 1423 – 1428 (1983). – See Abstr. 012.002.
This paper describes the basic design of a tunable quadrupole antenna for low-frequency gravitational radiation.

066.115 **Tentative research on how to generate and detect high frequency gravitational waves.**
J. Lin.
Proceedings of the Third Marcel Grossmann Meeting on General Relativity, Part B, p. 1429 – 1432 (1983). – See Abstr. 012.002.

066.116 **The gravitational radiation project at Maryland.**
W. S. Davis, F. Desrosier, W. Folkner, D. Gretz, K. Krack, M. V. Moody, H. J. Paik, J.-P. Richard, J. Weber, G. Wilmot, J. Clarke, J. Martinis.
Proceedings of the Third Marcel Grossmann Meeting on General Relativity, Part B, p. 1433 – 1436 (1983). – See Abstr. 012.002.
The present Maryland program in gravitational wave detection is to instrument a 2400 kg Al 5056 bar antenna as a three-mode antenna coupled to a dc SQUID amplifier and operate it at 0.050K.

066.117 **Primary research on tuning fork antenna for low frequency gravitational radiation.**
S. Feng, Z. Liu, R. Qin, Y. Song, Y. Yan.
Proceedings of the Third Marcel Grossmann Meeting on General Relativity, Part B, p. 1443 – 1446 (1983). – See Abstr. 012.002.
The authors have designed a tuning fork antenna for detecting continuous gravitational radiation from pulsars. It consists of two collective masses linked by a crooked cross beam which serves as a spring between the two masses.

066.118 **Continuous gravitational wave experiment in Japan.**
M.-K. Fujimoto.
Proceedings of the Third Marcel Grossmann Meeting on General Relativity, Part B, p. 1447 – 1450 (1983). – See Abstr. 012.002.

066.119 **A null experiment of gravitational inverse square law.**
H. Liu, P. Zhang, R. Qin.
Proceedings of the Third Marcel Grossmann Meeting on General Relativity, Part B, p. 1501 – 1504 (1983). – See Abstr. 012.002.
An experiment to test the Newtonian inverse square law of gravity has been made by employing a torsion balance suspended inside a long hollow cylinder.

066.120 **A new gravitational effect of a rotating mass.**
B. Mashhoon.
Proceedings of the Third Marcel Grossmann Meeting on General Relativity, Part B, p. 1505 – 1507 (1983). – See Abstr. 012.002.

066.121 **Some observational results about gravitational abnormal phenomena when three celestial bodies were coaxial.**
J. Chen, T. Guan, E. Hu, X. Chen.
Proceedings of the Third Marcel Grossmann Meeting on General Relativity, Part B, p. 1509 – 1513 (1983). – See Abstr. 012.002.
The authors report on a search for abnormal gravitational effects by experiments with a torsion pendulum during lunar and solar eclipses.

066.122 **The contributions of Bryce DeWitt to classical general relativity.**
L. L. Smarr.
Quantum theory of gravity, p. 1 – 20 (1984). – See Abstr. 003.005.

066.123 **The contributions of Bryce DeWitt to quantum gravity.**
C. J. Isham.
Quantum theory of gravity, p. 21 – 41 (1984). – See Abstr. 003.005.

066.124 **What have we learned from quantum field theory in curved space–time?**
S. A. Fulling.
Quantum theory of gravity, p. 42 – 52 (1984). – See Abstr. 003.005.

066.125 **The world of the Schwinger–DeWitt algorithm and the magical $a_2$ coefficient.**
S. M. Christensen.
Quantum theory of gravity, p. 53 – 65 (1984). – See Abstr. 003.005.

066.126 **Is there a quantum equivalence principle?**
P. Candelas, D. W. Sciama.
Quantum theory of gravity, p. 78 – 88 (1984). – See Abstr. 003.005.
The aim of this article is to point out that the understanding that has been gained in recent years of accelerated vacuum states, motivated in large part by a desire to understand the phenomenon of black hole radiance, sheds new light on the classic issue of the radiation reaction suffered by a uniformly accelerated charge and also on the related questions of the radiation emitted by a charge that is either at rest in or freely falling through a static gravitational field.

066.127 **Aspects of interacting quantum field theory in curved space–time: renormalization and symmetry breaking.**
L. H. Ford.
Quantum theory of gravity, p. 125 – 134 (1984). – See Abstr. 003.005.
This essay discusses two topics of interacting quantum field theories in curved space–time: the renormalization of such theories and the role of gravitation in phase transitions in cosmology.

066.128 **The Gospel according to DeWitt.**
G. Vilkovisky.
Quantum theory of gravity, p. 169 – 209 (1984). – See Abstr. 003.005.
This paper discusses some basic topics of quantum gravity: the Abelianization of gauge theory, the covariant–quantization conjecture and the effective action principle.

066.129 **Is there a quantum theory of gravity?**
A. Strominger.
Quantum theory of gravity, p. 210 – 223 (1984). – See Abstr. 003.005.
This essay reviews attempts to construct a unitary, renormalizable quantum field theory of gravity. Particular attention is paid to "non-perturbative" expansions that probe above the Planck energy. It is concluded that there is a consistent flat space perturbation theory capable of describing quantum gravitational processes below the Planck energy.

066.130 **Quantum gravity: the question of measurement.**
J. A. Wheeler.
Quantum theory of gravity, p. 224 – 233 (1984). – See Abstr. 003.005.
The formalism of quantum gravity, in its best developed form, makes three-geometry a central concept; consequently, the finding of a proper way to measure this three-geometry – against the background of the indeterminancy of the conjugate geometrodynamic field momentum – is a central issue.

066.131 **Steps towards a quantum theory of gravity.**
W. G. Unruh.
Quantum theory of gravity, p. 234 – 242 (1984). – See Abstr. 003.005.

066.132 **Quantum gravity at small distances.**
M. R. Brown.
Quantum theory of gravity, p. 243 – 250 (1984). – See Abstr. 003.005.

066.133 **Renormalization and asymptotic freedom in quantum gravity.**
E. T. Tomboulis.
Quantum theory of gravity, p. 251 – 266 (1984). – See Abstr. 003.005.
This article reviews some recent attempts to construct satisfactory theories of quantum gravity within the framework of local, continuum field theory. In particular, it considers renormalizable theories with actions that involve terms quadratic in the Riemann tensor.

066.134 **Quantization of higher derivative theories of gravity.**
D. G. Boulware.
Quantum theory of gravity, p. 267 – 294 (1984). – See Abstr. 003.005.

066.135 **Donaldson's moduli space: a "model" for quantum gravity?**
R. Penrose.
Quantum theory of gravity, p. 295 – 298 (1984). – See Abstr. 003.005.

066.136 **Quantum geometry.**
C. J. Isham.
Quantum theory of gravity, p. 299 – 314 (1984). – See Abstr. 003.005.
This essay studies the problem of quantum geometry within a canonical, Hilbert–space–based approach. Attention is centered on the unitary representations of the affine commutation relations and of the diffeomorphism group of the compact three–manifold.

066.137 **The role of time in path integral formulations of parametrized theories.**
J. B. Hartle, K. V. Kuchař.
Quantum theory of gravity, p. 315 – 326 (1984). – See Abstr. 003.005.
In the history of the efforts to construct a quantum theory of space–time the contrast between the canonical and the covariant approaches to this goal is striking. This paper investigates how the choice of time, so central to the canonical method, is reflected in the formalism of the covariant version of the theory.

066.138 **The Hamiltonian structure of two–dimensional space–time and its relation with the conformal anomaly.**
C. Teitelboim.
Quantum theory of gravity, p. 327 – 344 (1984). – See Abstr. 003.005.
This article is concerned with three different aspects of field dynamics in two–dimensional space–time and their mutual relationship: the algebra of the Hamiltonian generators of surface deformations and its Schwinger terms, gravitation theory, and the conformal anomaly.

066.139 **Liouville field theory: a two–dimensional model for gravity?**
R. Jackiw.
Quantum theory of gravity, p. 403 – 420 (1984). – See Abstr. 003.005.

066.140 **On complete group covariance without torsion.**
L. Halpern.
Quantum theory of gravity, p. 463 – 474 (1984). – See Abstr. 003.005.
This paper discusses a modification of general relativity based on a Kaluza–Klein model, which is in accord with complete de Sitter covariance.

066.141 **A class of spherically symmetric perfect fluid distributions in general relativity.**
R. Tikekar.
Indian J. Pure Appl. Math., Vol. 15, No. 3, p. 289 – 296 (1984). Abstr. in Phys. Abstr., Vol. 87, No. 1238, Entry 73308 (1984).

066.142 **Solitonic gravitational waves.**
V. O. Rivelles.
Rev. Bras. Fis., Vol. 13, No. 2, p. 369 – 373 (1983). Abstr. in Phys. Abstr., Vol. 87, No. 1239, Entry 78209 (1984).

066.143 **Exact Brans–Dicke–Bianchi type–II solutions.**
D. Lorenz–Petzold.
Prog. Theor. Phys., Vol. 71, No. 2, p. 406 – 408 (1984). Abstr. in Phys. Abstr., Vol. 87, No. 1239, Entry 78217 (1984).

066.144 **On orbits and rays in Schwarzschild– and in Reissner–Nordström space.**
F. Gackstatter.
Ann. Phys. (Leipzig), Vol. 40, No. 6, p. 352 – 374 (1983). In German. Abstr. in Phys. Abstr., Vol. 87, No. 1240, Entry 85865 (1984).

066.145 **Viscous fluid collapse.**
A. A. Coley, B. O. J. Tupper.
Phys. Rev. D, Vol. 29, No. 12, p. 2701 – 2704 (1984). Abstr. in Phys. Abstr., Vol. 87, No. 1241, Entry 86272 (1984).

066.146 **Electromagnetic fields invariant up to a duality rotation under a group of isometries.**
M. Henneaux.
J. Math. Phys., Vol. 25, No. 7, p. 2276 – 2283 (1984). Abstr. in Phys. Abstr., Vol. 87, No. 1241, Entry 86286 (1984).

066.147 **Gravitational redshift of gravitational clocks.**
C. M. Will.
Ann. Phys. (N.Y.), Vol. 155, No. 1, p. 133 – 157 (1984). Abstr. in Phys. Abstr., Vol. 87, No. 1241, Entry 90083 (1984).

066.148 **Segal's mechanism for the redshift and its physical bases.**
J. Tarski.
Int. J. Theor. Phys., Vol. 23, No. 5, p. 425 – 439 (1984). Abstr. in Phys. Abstr., Vol. 87, No. 1241, Entry 90086 (1984).

066.149 **On the gravitational effects of rotating masses: the Thirring–Lense papers.**
B. Mashhoon, F. W. Hehl, D. S. Theiss.
Gen. Relativ. Gravitation, Vol. 16, No. 8, p. 711 – 750 (1984).
The purpose of this work is to provide a critical analysis of the classical papers of H. Thirring (1918, 1921) and J. Lense and H. Thirring (1918) on rotating masses in the relativistic theory of gravitation and to render them accessible to a wider circle of scholars. An English translation of these papers is presented which follows the original German text as closely as possible. This is followed by a concise account of the significance of the results of these papers as well as the possibility of measuring the gravitational effects of rotating masses.

066.150 **Einstein's theory in a three–dimensional space–time.**
S. Giddings, J. Abbott, K. Kuchař.
Gen. Relativ. Gravitation, Vol. 16, No. 8, p. 751 – 775 (1984).
The authors analyze unusual features of Einstein's theory of gravitation in a three–dimensional space–time. In three dimensions, matter curves space–time only locally and the gravitational field has no dynamical degrees of freedom. The standard correspondence of Einstein's theory with Newton's theory breaks down. A dust distribution moves without any geodesic deviation between the particles. The cosmological models and relativistic stars behave in a qualitatively different way from their Newtonian counterparts.

066.151 **Nonstatic gas spheres in general relativity.**
H. Knutsen.
Gen. Relativ. Gravitation, Vol. 16, No. 8, p. 777 – 788 (1984).
A model for a time–dependent relativistic "gas" sphere is given. Both pressure $p$ and density $\varrho$ are positive inside the outer boundary of the sphere, and they both drop to zero at the boundary itself. Their respective gradients are negative. For expanding models the pressure and the density are decreasing, and for contracting models the pressure and the density are increasing.

066.152 **Killing vector fields and the Einstein–Maxwell field equations for null electromagnetic fields.**
H. Goenner.
Gen. Relativ. Gravitation, Vol. 16, No. 8, p. 795 – 796 (1984).

066.153 **A possible unified field theory of gravitation with other interactions.**
G. E. Tauber.
Large–scale structure of the universe, cosmology and fundamental physics, p. 78 – 80 (1984). – See Abstr. 012.025.
Generalizing a work by Einstein and Mayer, it is assumed that at each point of space–time there exists a linear vector space of $N$ dimensions. From geometrical considerations and a variational principle several fields (including gauge fields) and their equations are obtained. In the case $N = 6$ there exist an electromagnetic and a charged vector meson field.

066.154 **Gravitational lenses.**
R. Webster.
Observatory, Vol. 104, No. 1062, p. 215 (1984). Abstract. – See Abstr. 010.682.

066.155 **Einstein–Weyl field equations in a Bianchi type–IX space–time.**
C. A. Kolassis, G. Le Denmat.
Phys. Rev. D, Vol. 30, No. 2, p. 309 – 312 (1984). Abstr. in Phys. Abstr., Vol. 87, No. 1242, Entry 90454 (1984).

066.156 **On the geometry of charged dust distribution.**
K. P. Singh.
Proc. Indian Natl. Sci. Acad., Part A, Vol. 49, No. 4, p. 448 – 451 (1983). Abstr. in Phys. Abstr., Vol. 87, No. 1242, Entry 90457 (1984).

066.157 **On the gravitational radiation of an oblique rotator.**
D. V. Gal'tsov, V. P. Tsvetkov.
Phys. Lett. A, Vol. 103A, No. 4, p. 193 – 196 (1984). Abstr. in Phys. Abstr., Vol. 87, No. 1242, Entry 94586 (1984).

066.158 **Precession of a gyroscope on the ground and the existence of magnetic–type gravitation.**
H. Peng, C. Qin.
Phys. Lett. A, Vol. 103A, No. 4, p. 197 – 199 (1984). Abstr. in Phys. Abstr., Vol. 87, No. 1242, Entry 94587 (1984).

066.159 **Phenomenological theory of gravitational vacuum.**
L. Sh. Grigorian (*L. Sh. Grigoryan*), G. S. Sahakian (*G. S. Saakyan*).
Astrophys. Space Sci., Vol. 104, No. 1, p. 19 – 39 (1984).
An idea is developed that the vacuum in the gravitational field acquires properties of an elastic medium described by a definite tension tensor $\tau_{ik}$. The vacuum is stated to also participate in the formation of the space–time metric, together with the usual matter. So, the matter, vacuum, and metric form a complex unity determined by the solution of the field equations. The vacuum may prove to play an essential role in the extremely strong fields existing in superdense celestial bodies. The problem of vacuum is investigated in the case of the central symmetry static field. A number of properties of the tensor $\tau_{ik}$ is found using the symmetry of the field and comparison with the post–Newton limit. The external and internal problems, as well as the procedure of joining the solutions on the surface of a celestial body, have been formulated.

066.160 **Nonlinear Newtonian theory and the effect of time dilatation.**
D. D. Dionysiou.
Astrophys. Space Sci., Vol. 104, No. 2, p. 389 – 397 (1984).
The fact that energy density $\varrho_g$ of a static spherically symmetric gravitational field acts as a source of gravity, gives a harmonic function $f(\varphi) = \exp(\varphi/c^2)$, which is determined by a nonlinear differential equation presented in the paper. Furthermore, the author formulates the infinitesimal time–interval between a couple of events measured by two different inertial observers, one in a position with potential $\varphi$ – i.e., $dt_\varphi$ and the other in a position with potential $\varphi = 0$ – i.e., $dt_0$, as $dt_\varphi = f dt_0$. When the principle of equivalence is satisfied, the well–known effect of time dilatation is obtained.

066.161 **On the stability of a new relativistic kinetic equation.**
H. E. Kandrup.
Astrophys. J., Vol. 282, No. 2, p. 361 – 369 (1984).
Recently, Israel and Kandrup have formulated a new relativistic kinetic equation appropriate for a self–gravitating system. That equation allows for the combined influences of (1) collective mean fields, which define an "average" spacetime geometry, and (2) "fluctuations" from mean field conditions, which are modeled in a Landau–type approximation. It is shown here that this new kinetic equation admits the relativistic isothermal distribution as an exact stationary solution. Linearized perturbation equations are then derived for small spherically symmetric departures from the isothermal "equilibrium".

066.162 **On test particles in general relativity.**
T. W. Noonan.
Astrophys. J., Vol. 283, No. 1, p. 321 – 330 (1984).
The field–to–particle approach is generalized to include the self–gravity of a test particle. The quantity which is identified as the four–momentum of the particle is found, for a charged particle, to be related to the Lorentz force, and, for a suitably defined free, spherically symmetric particle, to be the product of the particle's four–velocity and a constant scalar. In the latter case, the particle's world–line is a geodesic. In the Newtonian approximation, the apparent mass is the usual combination of proper mass and binding energy. The principal assumption is that the particle is small compared with variations in the background metric.

066.163 **One–sided type–*D* gravitational instantons.**
M. Przanowski, B. Baka.
Gen. Relativ. Gravitation, Vol. 16, No. 9, p. 797 – 803 (1984).
It is shown that for every one–sided type–*D* gravitational instanton, Einstein's vacuum equations can be reduced locally to a single second–order, nonlinear differential equation of second degree of one real function.

066.164 **Is the Riemann tensor derivable from a tensor potential?**
E. Massa, E. Pagani.
Gen. Relativ. Gravitation, Vol. 16, No. 9, p. 805 – 816 (1984).
In recent years it has been conjectured that the Riemann tensor itself admits a linear representation in terms of the covariant derivatives of a suitable "potential" tensor of rank 3. This conjecture is shown to be false, at least for a class of space–time geometries including several physically significant ones.

066.165 **Stationary Riemannian space–times with self–dual curvature.**
J. D. Gegenberg, A. Das.
Gen. Relativ. Gravitation, Vol. 16, No. 9, p. 817 – 829 (1984).
This paper attempts to find all Riemannian metrics with self–dual curvature which, in addition, admit a one–parameter group of isometries, i. e., which admit the existence of at least one Killing vector field.

066.166 **The solution of Dirac's equation in algebraically special space–times.**
G. F. Torres del Castillo.
Gen. Relativ. Gravitation, Vol. 16, No. 9, p. 831 – 834 (1984).
A procedure for obtaining solutions to Dirac's equation in algebraically special space–times which admit a shear-free congruence of null geodesics along the repeated principal null direction of the Weyl tensor, is presented.

066.167 **Approximate solutions for the $R + R^2$ theories of gravitation: torsion components.**
Y. Z. Zhang, K. Y. Guan.
Gen. Relativ. Gravitation, Vol. 16, No. 9, p. 835 – 845 (1984).
The authors search for vacuum solutions of a spherically symmetric and static field for the nine-parameter $R + R^2$ gravitational theories in a weak-field limit. In this paper approximate solutions of torsion are obtained.

066.168 **Inertial structure. I.**
D. P. L. Castrigiano, G. Süssmann.
Gen. Relativ. Gravitation, Vol. 16, No. 9, p. 867 – 875 (1984).
The desirability and possibility to generalize space–time structure to non-Euclidean spaces is investigated. To this end, the two basic notions of inertial motion and relativity are revealed. Several relevant statements about kinematics on spaces of constant curvature are pointed out.

066.169 **Galilean inertial structure. II.**
D. P. L. Castrigiano, G. Süssmann.
Gen. Relativ. Gravitation, Vol. 16, No. 10, p. 893 – 900 (1984).
Galilean space–time is characterized in an axiomatic manner within a certain class of fiber bundles over the time axis. The distinction comes about through an examination of the inertial structure constituted by the entirety of the inertial trajectories. The latter are defined geometrically as a relation on the space–time expressing the situation of relative rest for events.

066.170 **Metrical geometry of Galilean inertial structure. III.**
D. P. L. Castrigiano.
Gen. Relativ. Gravitation, Vol. 16, No. 10, p. 901 – 908 (1984).
An axiomatic treatment of Galilean space–time is carried out in the spirit of metrical geometry. According to it only material concepts enter the postulates. The central notion thereby is that of the Galilean inertial relation introduced in a previous paper.

066.171 **Linearized superhairs on multi–black–hole backgrounds.**
F. Embacher.
Gen. Relativ. Gravitation, Vol. 16, No. 10, p. 909 – 920 (1984).
It is shown that the linearized Rarita–Schwinger equation on a multi–black–hole background of the Majumdar–Papapetrou class admits time–independent and regular nongauge solutions that die off at infinity. The number of freely available real constants is supposed to be four, thus being independent of the number of black holes. As a consequence, the gravitino field seems to be uniquely determined by its supercharge.

066.172 **Projective differential geometry and geodesic conservation laws in general relativity. I: Projective actions.**
G. E. Prince, M. Crampin.
Gen. Relativ. Gravitation, Vol. 16, No. 10, p. 921 – 942 (1984).
The Lagrangian structure of the geodesic equation allows an extension of classical projective geometry to one–parameter projective group actions on $R \times TM$ (where $M$ is the space–time). The authors determine all those projective actions which arise by prolongation of one–parameter group actions on $R \times M$.

066.173 **Cosmic censorship and conformal transformations.**
R. P. A. C. Newman.
Gen. Relativ. Gravitation, Vol. 16, No. 10, p. 943 – 953 (1984).
A new definition of a nakedly singular space–time is proposed. Conformal transformations of general, vacuum space–times are considered for conformal factors which are proper mappings into $(0, \infty)$. A space–time generated in this manner which is null convergent on the future Cauchy development of a partial Cauchy surface is shown to be not nakedly singular relative to that surface in the sense of the chosen definition. A censorship theorem for null convergent, conformally flat space–times is obtained as a corollary to the principal result.

066.174 **The quantization of the relativistic string.**
D. C. Salisbury.
Gen. Relativ. Gravitation, Vol. 16, No. 10, p. 955 – 978 (1984).
The classical field–dependent parametrization covariant Hamiltonian formulation of the open and the closed string is discussed. A conformally covariant formalism is developed for the open string. The constraints and gauge conditions are taken into account in the quantum theory as conditions on physical states. A recursive construction is presented which permits the counting of physical states of any given mass, spin, and parity. The quantization is carried out in four space–time dimensions.

066.175 **On axially symmetric solutions of Einstein's field equations in matter and an example of static gravitational shielding.**
H. Pfister, W. Konrad.
Gen. Relativ. Gravitation, Vol. 16, No. 10, p. 979 – 1001 (1984).
A model system, consisting of a thin spherical shell with radius $R$ and mass $M$ and a point mass $m$ at a distance $s > R$ from the center of the sphere, held fixed by an appropriate strut, is solved to order $mM$. The stresses in the shell are not of the canonical Weyl type, and it is argued that the same is true for more realistic situations, e.g., rotating matter. Owing to the nonlinearity of Einstein's field equations, the field of the point mass is shielded from the interior of the shell.

066.176 **Restoration of the shape of the signal in a gravitational–wave experiment.**
S. I. Babichenko, V. N. Rudenko, A. G. Yagola.
Astron. Zh., Tom 61, Vyp. 5, p. 913 – 924 (1984). In Russian. English translation in Sov. Astron., Vol. 28, No. 5.
A computer experiment has been made to reconstruct the shape and parameters of an external signal disturbing a gravitational detector with a relaxation detecting electrical circuit. A mathematical procedure has been developed and results for some typical gravitational–wave pulses are presented, with an account for the fluctuations of the antenna.

066.177 **Energy in gravitational theories. Definition, positivity theorems, and stability.**
S. Deser.
Ann. N.Y. Acad. Sci., Vol. 422, p. 45 – 55 (1984). – See Abstr. 012.033.
This paper explains and summarizes the positive energy and stability results that have recently been established in gravitational theories. The known positive energy and stability theorems are given for the various models of current interest. It is also shown explicitly how they can be derived from simple supersymmetry ideas. Semiclassical stability, Kaluza–Klein models, and some other theories are also briefly considered.

066.178 **Killing tensors and the geodesic deviation equation.**
P. Dolan, J. L. Safko.
Ann. N.Y. Acad. Sci., Vol. 422, p. 340 (1984). Abstract. – See Abstr. 012.033.

066.179 **Gravitational lenses producing a large number of images.**
C. C. Dyer.
Ann. N.Y. Acad. Sci., Vol. 422, p. 341 (1984). Abstract. – See Abstr. 012.033.

066.180 **Implications of an alternative relativistic transport theory.**
J. J. Dykla, A. E. Hwang.
Ann. N.Y. Acad. Sci., Vol. 422, p. 342 (1984). Abstract. – See Abstr. 012.033.

066.181 **Spinning fluid energy–momentum tensors in the Einstein–Cartan theory.**
J. R. Ray, L. L. Smalley.
Ann. N.Y. Acad. Sci., Vol. 422, p. 378 – 379 (1984). Abstract. – See Abstr. 012.033.

066.182 **Does gravity exhibit a $1/r$ force on the scale of galaxies?**
J. E. Tohline.
Ann. N.Y. Acad. Sci., Vol. 422, p. 390 (1984). Abstract. – See Abstr. 012.033.

066.183 **Relativistische Interpretation astrometrischer und spektraler Meßgrößen.**
J. Schastok, M. Soffel, H. Ruder, M. Schneider.
Veröff. Bayer. Komm. Int. Erdmessung Bayer. Akad. Wiss., Astron.-Geod. Arb., Heft Nr. 45, p. 110 – 137 (1984).

066.184 **Über die Lichtablenkung im Gravitationsfeld der Sonne.**
M. Soffel, J. Schastok, H. Ruder, M. Schneider.
Veröff. Bayer. Komm. Int. Erdmessung Bayer. Akad. Wiss., Astron.-Geod. Arb., Heft Nr. 45, p. 138 – 156 (1984).

066.185 **On clock synchronization in a reference system moving in an isotropic universe.**
V. A. Bruk.
Khar'k. univ., Khar'kov, 6 pp. (1983). In Russian. Abstr. in Ref. Zh., 51. Astron., 9.51.828 (1984).

066.186 **Gravitational lensing by stars in a galaxy halo: theory of combined weak and strong scattering.**
R. Nityananda, J. P. Ostriker.
J. Astrophys. Astron., Vol. 5, No. 3, p. 235 – 250 (1984).
The theory of gravitational lensing of background quasars by stars in the halo of a galaxy is considered. In the limiting case of small "optical depth", only one star is close enough to the beam to cause strong scattering, and the effect of all the other stars is treated as a perturbation with both systematic and random components. The perturbation coming from weak scattering can increase the number of images and the amplification in those cases where the amplification is already high; such events are preferentially selected in flux limited observations. The theory is applicable to the apparent association of background quasars with foreground galaxies. The relevance of these results to gravitational lensing by galaxies as perturbed by random inhomogeneities surrounding the ray path is also briefly discussed.

066.187 **A source for the Reissner–Nordstrøm metric.**
R. Tikekar.
J. Astrophys. Astron., Vol. 5, No. 3, p. 273 – 276 (1984).
An exact solution of the coupled Einstein–Maxwell equations representing the gravitational field in the interior of a sphere of charged incoherent matter in equilibrium is obtained which is a charged analogue of the static perfect fluid sphere solution with spheroidal 3-space obtained by Vaidya & Tikekar.

066.188 **Cosmological solution with an energy flux.**
B. Modak.
J. Astrophys. Astron., Vol. 5, No. 3, p. 317 – 322 (1984).
The paper presents some spherically symmetric cosmological solutions in which the velocity field is shear-free but there is a flux of energy. The solutions are believed to be new and the previous known solutions of this class due to Bergmann and Maiti may be obtained as special cases of the author's metrics.

066.189 **Collapse of anisotropic spheres in general relativity: an analytical model.**
J. M. Ibáñez.
Astrophys. J., Vol. 284, No. 1, p. 381 – 383 (1984).
The author has extended the so-called Adam's method in order to obtain an exact solution of the Einstein's field equations for the collapse of fluid spheres governed by a given anisotropic law.

066.190 **Gravitation in the weak-field limit.**
H. J. Rood.
Astrophys. J., Vol. 284, No. 1, p. 407 – 411 (1984).
A perturbed force is suggested which provides a flat-space description of gravitation in the weak-field limit. The theory leads to values for the observed gravitational redshift of photons emitted from the solar surface, the gravitational deflection of solar-grazing photons, and the excess precession of the perihelion of Mercury's orbit that agree with the observed quantities to within the observational uncertainties. The magnitude of Mercury's angular momentum per unit mass is predicted to change by 2 parts in $10^8$ from a minimum at perihelion to a miximum at aphelion.

066.191 **On superdense configurations in astrophysics.**
Yu. N. Vlasenko.
Mater. 7 Konf. mol. uchenykh Univ. druzhby narodov: mat., fiz., khim., Moskva, 13 – 16 marta, 1984. Chast' 1. Moskva, p. 63 – 66 (1984). In Russian. Abstr. in Ref. Zh., 51. Astron., 10.51.178 (1984).

066.192 **Stability properties of gravitational theories and three dimensional gravities.**
S. Deser.
Relativity, cosmology, topological mass and supergravity, p. 135 – 162 (1983). – See Abstr. 012.040.
Two separate subjects are discussed. The first concerns recent advances in the formulation of energy and proofs of its positivity for gravity and supergravity theories, including those with non-vanishing cosmological constants. The second topic deals with gravity theories in three spacetime dimensions. The author first considers ordinary Einstein theory as a soluble model where topology takes the place of geometry as the carrier of information about the sources. Then he presents the locally supersymmetric extensions of topologically massive gravity, with or without cosmological terms.

066.193 **Strong gravity: an approach to its source.**
J. Chela-Flores, V. Varela.
Relativity, cosmology, topological mass and supergravity, p. 248 – 250 (1983). – See Abstr. 012.040.
Within the context of strong gravity the authors study different ways of incorporating simple models for the hadronic source, putting special emphasis on its possible superfluidity.

066.194 **Cylindrical gravitational waves: a source with counter-rotating particles.**
V. H. Hamity, O. Lovera.
Relativity, cosmology, topological mass and supergravity, p. 258 – 261 (1983). – See Abstr. 012.040.
In this paper the static cylindrically symmetric exterior solutions are briefly discussed and a generalization of the concept of disposable gravitational mass per unit length is given. A source with counter-rotating particles is constructed to study the changes in its characteristic parameters after a gravitational pulse is emitted and the system has returned to a new static regime.

066.195 **Erratum: "Star disturbances in gravitational lens galaxies" [Astron. Astrophys., Vol. 132, No. 1, p. 168 – 178 (1984)].**
K. Chang, S. Refsdal.
Astron. Astrophys., Vol. 139, No. 2, p. 558 (1984). See Abstr. 37.066.018.

066.196 **The amplification caused by gravitational bending of light.**
P. Schneider.
Astron. Astrophys., Vol. 140, No. 1, p. 119 – 124 (1984).
Gravitational bending of light rays may not only lead to multiple imaging (gravitational lens effect), but also changes the apparent luminosity of a source. It is shown that a general mass distribution always leads to an enhanced apparent luminosity relative to that which one would observe in a lumpy universe if the lens were absent, for every relative position of source and

observer. The author then discusses that this does not violate flux conservation.

066.197 **On the calculation of the geodetic distance and associated invariants in relativistic gravitational fields.**
R. W. John.
Ann. Phys. (Leipzig), Vol. 41, No. 1, p. 67 – 80 (1984). Abstr. in Phys. Abstr., Vol. 87, No. 1243, Entry 95250 (1984).

066.198 **Tetrads, broken symmetries, and the gravitational constant.**
K. Cahill.
Z. Phys., C, Vol. 23, No. 4, p. 353 – 356 (1984). Abstr. in Phys. Abstr., Vol. 87, No. 1243, Entry 95278 (1984).

066.199 **The Rochester gravitational wave detector.**
M. F. Bocko, M. W. Cromar, D. H. Douglass, R. Q. Gram, W. W. Johnson, M. Karim, C. C. Lam, D. Macaluso, J. R. Marsden, B. Muhlfelder, L. Narici, M. Zucker.
J. Phys. E, Vol. 17, No. 8, p. 694 – 703 (1984). Abstr. in Phys. Abstr., Vol. 87, No. 1243, Entry 95287 (1984).

066.200 **Are "superluminal" expansions superluminal? (Considerations about the apparent superluminal expansions in astrophysics).**
E. Recami, A. Castellino, G. D. Maccarrone, M. Rodono.
Report PP/761, Ist. Naz. Fis. Nucl., Catania, Italy, 33 pp. (1984). Abstr. in Phys. Abstr., Vol. 87, No. 1244, Entry 101650 (1984).

066.201 **Sensitivity improvement of a two–mode gravitational–radiation detector through the use of a transducer not perfectly tuned to the antenna.**
M. G. Castellano, C. Cosmelli.
Nuovo Cimento C, Vol. 7C, Ser. 1, No. 1, p. 9 – 20 (1984). Abstr. in Phys. Abstr., Vol. 87, No. 1245, Entry 106231 (1984).

066.202 **Lagrangian formalism for resonant capacitive transducers for gravitational–wave antennas.**
Y. Ogawa, P. Rapagnani.
Nuovo Cimento C, Vol. 7C, Ser. 1, No. 1, p. 21 – 34 (1984). Abstr. in Phys. Abstr., Vol. 87, No. 1245, Entry 106232 (1984).

066.203 **Back–action–evading measurement of optical fields with the use of a four–wave mixer.**
B. Yurke, L. R. Corruccini.
Phys. Rev. A, Vol. 30, No. 2, 895 – 900 (1984). Abstr. in Phys. Abstr., Vol. 87, No. 1245, Entry 106233 (1984).

066.204 **Some remarks on the transparent gravitational lenses.**
K. Ma, L. Luo.
Kexue Tongbao (Beijing), Vol. 29, No. 2, p. 182 – 185 (1984). Abstr. in Phys. Abstr., Vol. 87, No. 1245, Entry 110513 (1984).

066.205 **Casimir effect and gravitational collapse.**
I. Brevik, H. Kolbenstvedt.
Nuovo Cimento B, Vol. 82B, Ser. 11, No. 1, p. 71 – 84 (1984). Abstr. in Phys. Abstr., Vol. 87, No. 1246, Entry 110956 (1984).

066.206 **Gravitation in similarity mechanics.**
P. S. Gulati, S. Gulati.
Indian J. Theor. Phys., Vol. 30, No. 2, p. 155 – 172 (1982). Abstr. in Phys. Abstr., Vol. 87, No. 1246, Entry 110961 (1984).

066.207 **Fourth–order gravity as general relativity plus matter.**
B. Whitt.
Phys. Lett. B, Vol. 145B, No. 3 – 4, p. 176 – 178 (1984). Abstr. in Phys. Abstr., Vol. 87, No. 1246, Entry 110974 (1984).

066.208 **On the potential of induced gravity.**
K. G. Akdeniz.
Phys. Lett. B, Vol. 145B, No. 1 – 2, p. 49 – 52 (1984). Abstr. in Phys. Abstr., Vol. 88, No. 1247, Entry 160 (1985).

066.209 **On the identification of gravitation with the massless spin 2 field.**
A. Meszaros.
Report KFKI–1984–62, Hungarian Acad. Sci., Budapest, Hungary, 13 pp. (1984). Abstr. in Phys. Abstr., Vol. 88, No. 1247, Entry 172 (1985).

066.210 **Tests of gravitation and relativity.**
R. F. C. Vessot.
Contemp. Phys., Vol. 25, No. 4, p. 355 – 380 (1984). Abstr. in Phys. Abstr., Vol. 88, No. 1247, Entry 183 (1985).

066.211 **Relativistic theory of astrometric observations. I. Determination of the positions of remote sources from the surfaces of solar–system bodies.**
N. V. Pavlov.
Sov. Astron., Vol. 28, No. 2, p. 223 – 227 (1984). English translation of 37.066.052.

066.212 **On the relativistic theory of astrometric observations. II. Determination of the positions of remote sources from the earth's surface.**
N. V. Pavlov.
Sov. Astron., Vol. 28, No. 3, p. 351 – 355 (1984). English translation of 37.066.073.

066.213 **Limits on metric parameters $\beta$ and $\gamma$.**
R. W. Babcock, J. F. Chandler, R. D. Reasenberg, I. I. Shapiro.
Bull. Am. Astron. Soc., Vol. 16, No. 3, Part I, p. 722 (1984). Abstract. – See Abstr. 010.064.

066.214 **New limit on $\dot{G}$.**
J. F. Chandler, R. W. Babcock, R. D. Reasenberg, I. I. Shapiro.
Bull. Am. Astron. Soc., Vol. 16, No. 3, Part I, p. 722 (1984). Abstract. – See Abstr. 010.064.

066.215 **Gravitational lensing by cosmic strings.**
C. Hogan, R. Narayan.
Mon. Not. R. Astron. Soc., Vol. 211, No. 3, p. 575 – 591 (1984).
The authors discuss the possible role of a cosmological distribution of vacuum strings in the gravitational imaging of quasars. Lensing properties are described for open strings and for large and small closed loops. A simplified model shows that small loops typically produce a pair of magnified merging images. This effect introduces an observational 'amplification bias' which causes magnified quasars to appear frequently in a flux–limited sample even though they are intrinsically rare. The probability of observing lensing events with various properties are estimated in two simple scenarios, and the possibility of searching for them is discussed.

066.216 **Axisymmetric gravitational collapse of an isothermal rotating gas cloud.**
R. Wang, B. Qian, P. Jiang, L. Mi.
Ann. Shanghai Obs., Acad. Sin., No. 5, p. 190 – 198 (1983). In Chinese.
The authors have numerically analyzed the isothermal collpase of a gas cloud. The results show that falling velocities of the matter in a collapsing cloud increase gradually outward. The greatest velocity along the rotating axis reaches $1.3 \times 10^4 \mathrm{cm\ sec}^{-1}$. It is close to the velocity of sound ($2.02 \times 10^4 \mathrm{cm\ sec}^{-1}$), and about two times as fast as that on the equatorial plane. The isodensity contours for three models after about one free falling time are shown. The results are sensitive to the angular momentum. The bigger the angular momentum of the cloud, the faster the contraction of the cloud goes. The gas collapse could collapse to form a ring if some appropriate initial parameters are chosen.

066.217 **Gravitational collapse of a dust cloud and the cosmic censorship conjecture.**
D. Christodoulou.
General relativity and gravitation, p. 27 – 36 (1984). – See Abstr. 012.060.

This paper studies the general inhomogeneous Tolman class of solutions for the relativistic collapse of spherically symmetric dust clouds. The results show that for an open subset of initial density distributions, the first singular event, which occurs at the center of symmetry, is the vertex of an infinity of future null geodesic cones which intersect future null infinity. The frequency of the corresponding light rays is infinitely redshifted.

066.218 **Asymptotic properties of isolated systems: recent developments.**
A. Ashtekar.
General relativity and gravitation, p. 37 – 68 (1984). – See Abstr. 012.060.

066.219 **Exact solutions and singularities. Report of workshop A2 (*10th International Conference on General Relativity and Gravitation, Padua, July 3 – 8, 1983*).**
M. A. H. MacCallum.
General relativity and gravitation, p. 69 – 81 (1984). – See Abstr. 012.060.

066.220 **Initial value problem. Report of workshop A3 (*10th International Conference on General Relativity and Gravitation, Padua, July 3 – 8, 1983*).**
K. V. Kuchař.
General relativity and gravitation, p. 83 – 87 (1984). – See Abstr. 012.060.

066.221 **The quadrupole approximation to gravitational radiation.**
M. Walker.
General relativity and gravitation, p. 107 – 123 (1984). – See Abstr. 012.060.

The purpose of this paper is to comment on the status of the quadrupole formulae for gravitational radiation and radiation reaction.

066.222 **Gravitational radiation and the "quadrupole" formula. Report of workshop A1 (*10th International Conference on General Relativity and Gravitation, Padua, July 3 – 8, 1983*).**
J. Ehlers, M. Walker.
General relativity and gravitation, p. 125 – 137 (1984). – See Abstr. 012.060.

066.223 **Algebraic computing in general relativity.**
I. Cohen, I. Frick, J. E. Åman.
General relativity and gravitation, p. 139 – 162 (1984). – See Abstr. 012.060.

066.224 **Computational relativity: numerical and algebraic approaches. Report of workshop A4 (*10th International Conference on General Relativity and Gravitation, Padua, July 3 – 8, 1983*).**
L. Smarr.
General relativity and gravitation, p. 163 – 183 (1984). – See Abstr. 012.060.

The current use of computers in general relativity is reviewed. Today several groups have working codes which can calculate the coupled general relativity/hydrodynamics equations in two spatial dimensions plus time. These codes are compared and a critical review is made of the areas needing improvement. The potential uses of numerical relativity are discussed.

066.225 **Positive energy and Kaluza–Klein theory.**
E. Witten.
General relativity and gravitation, p. 185 – 197 (1984). – See Abstr. 012.060.

066.226 **Classical and quantum alternatives to gravitational theories. Report of workshop A5 (*10th International Conference on General Relativity and Gravitation, Padua, July 3 – 8, 1983*).**
E. T. Newman, H. Goenner.
General relativity and gravitation, p. 199 – 211 (1984). – See Abstr. 012.060.

This workshop consisted of two parts. The first was dedicated to classical alternative theories of gravity. The second part was concerned with the use of spinors in general relativity.

066.227 **Nonlinear wave propagation in relativistic hydrodynamics and cosmology.**
A. M. Anile.
General relativity and gravitation, p. 313 – 335 (1984). – See Abstr. 012.060.

This paper reports on some recent work on nonlinear wave propagation in relativistic fluid dynamics. It concentrates on the analytical methods and the qualitative results. Shock waves are treated with the method of singular hypersurfaces. The author discusses briefly the structure of shock waves and suggests that the present formulation of dissipative processes in relativistic fluid dynamics is not adequate to cope with these phenomena.

066.228 **Testing relativity with solar system dynamics.**
R. W. Hellings.
General relativity and gravitation, p. 365 – 385 (1984). – See Abstr. 012.060.

It is the purpose of this paper to explain the planetary data which are used in the solutions for the relativistic effects, to explain the Solar System model which is used to fit the data and the process by which such fits are performed. The results are discussed with respect to (1) the large number hypothesis and a time–varying gravitational constant; (2) PPN parameters; (3) Moffat's nonsymmetric gravitation theory.

066.229 **Space experiments. Report of workshop C2 (*10th International Conference on General Relativity and Gravitation, Padua, July 3 – 8, 1983*).**
P. L. Bender.
General relativity and gravitation, p. 387 – 395 (1984). – See Abstr. 012.060.

This workshop was concerned with (1) gravitational physics with Earth satellites and suggested new interplanetay missions; (2) gravitational wave experiments using microwave tracking and pulsar timing.

066.230 **Laser interferometer gravitational wave detectors.**
R. W. P. Drever.
General relativity and gravitation, p. 397 – 412 (1984). – See Abstr. 012.060.

066.231 **High frequency gravitational waves: attempts to laboratory detection. Report of workshop C1 (*10th International Conference on General Relativity and Gravitation, Padua, July 3 – 8, 1983*).**
E. Amaldi, F. Fuligni.
General relativity and gravitation, p. 413 – 422 (1984). – See Abstr. 012.060.

The papers presented at this workshop concerned resonant antennas, laser interferometric detectors and various problems ranging from gravitational radiation sources to methods of detection.

066.232 **Laboratory experiments. Report of workshop C3 (*10th International Conference of General Relativity and Gravitation, Padua, July 3 – 8, 1983*).**
H. Hirakawa.
General relativity and gravitation, p. 423 – 429 (1984). – See Abstr. 012.060.

The test of the laws of gravitational force and the test of the equivalence principle were the main topics of the workshop on laboratory experiments.

066.233 **Free differential algebras and relativistic theories.**
T. Regge.
General relativity and gravitation, p. 433 – 437 (1984). – See Abstr. 012.060.

066.234 **Topics in quantum gravity.**
B. S. DeWitt.
General relativity and gravitation, p. 439 – 451 (1984). – See Abstr. 012.060.

Three separate developments in quantum gravity theory are discussed. The first is a computer search for a solution, representing the Big–Bang era, of the dynamical equations generated by a proposed approximate effective action. The second is a surprising new development in the theory of the effective action due to G. Vilkovisky. The third is a new attempt to confront and resolve the old question whether the notion that space may undergo spontaneous changes in topology can make any sense.

066.235 **Quantum gravity and quantum field theory in a curved space. Report of workshop D1 (*10th International Conference on General Relativity and Gravitation, Padua, July 3 – 8, 1983*).**
A. Ashtekar.
General relativity and gravitation, p. 453 – 470 (1984). – See Abstr. 012.060.

066.236 **Highlights of the conference (*10th International Conference on General Relativity and Gravitation, Padua, July 3 – 8, 1983*).**
J. A. Wheeler.
General relativity and gravitation, p. 501 – 509 (1984). – See Abstr. 012.060.

066.237 **The Cauchy problem for Einstein equations.**
D. Bancel.
Problems of collapse and numerical relativity, p. 195 – 202 (1984). – See Abstr. 012.061.

066.238 **[(2+1)+1]–formalism of general relativity.**
M. Sasaki.
Problems of collapse and numerical relativity, p. 203 – 220 (1984). – See Abstr. 012.061.

For a spacetime having a Killing vector field, one can divide out the Killing direction so that the Einstein equations are reduced to the 3–dimensional "Einstein" equations. Then geometrical degrees of freedom related to the Killing direction are represented by separate fields. Assuming the Killing vector to be spacelike one can further transform the equations into the canonical Hamiltonian form. The resulting form of equations is called the [(2+1)+1]–representation of the Einstein equations, which can play a powerful role in numerical relativity.

066.239 **{2+2} formalism in general relativity.**
R. A. d'Inverno.
Problems of collapse and numerical relativity, p. 221 – 238 (1984). – See Abstr. 012.061.

The importance in general relativity of different types of initial value problems, especially those based on null or characteristic hypersurfaces, is discussed. This leads to the idea of decomposing space–time into two families of space–like hypersurfaces. Such a decomposition suggests that the freely specifiable initial data or gravitational degrees of freedom of the theory may be cast into the so–called conformal 2–structure. The formalism for describing this decomposition, called the covariant orthogonal {2+2} formalism, is outlined.

066.240 **Gravitational radiation from initial data.**
R. A. Isaacson, J. S. Welling, J. Winicour.
Problems of collapse and numerical relativity, p. 239 – 252 (1984). – See Abstr. 012.061.

The choice of free gravitational data for the characteristic initial value problem strongly affects the production of gravitational waves. From model calculations it is shown that the shear of the initial null cone must be properly matched to the matter distribution or a spurious flux of incoming radiation will mask outgoing emission. An algorithm for choosing the initial shear is given which successfully suppresses incoming radiation in numerical studies.

066.241 **General relativistic hydrodynamics and accretion physics: a numerical approach.**
J. F. Hawley, L. L. Smarr.
Problems of collapse and numerical relativity, p. 253 – 269 (1984). – See Abstr. 012.061.

The authors describe a numerical approach to the equations of general relativistic hydrodynamics. A discussion of the development and use of their computer code to study axisymmetric accretion flows around black holes is given. The authors outline the equations to be solved, the numerical techniques used and the code calibration "test bed". An illustrative accretion disk formation is presented.

066.242 **The collapse of rotating stellar cores: the amount of gravitational radiation predicted by various numerical models.**
E. Müller.
Problems of collapse and numerical relativity, p. 271 – 286 (1984). – See Abstr. 012.061.

The amount of gravitational radiation predicted by numerical calculations of the collapse of rotating stellar cores is reviewed. Currently the most realistic models suggest an energy loss of the order $10^{-6} M_{\odot} c^2$. For a source situated at a distance of 10 Mpc ($\approx$ Virgo Cluster) the corresponding (dimensionless) wave amplitude $h$ is smaller than $10^{-22}$, which seems to be even beyond the realm of the third generation gravitational wave detectors. On the other hand non–axisymmetric models predict up to 100 times larger radiative efficiencies. The uncertainties in these calculations are examined.

066.243 **Using solar oscillations to search for gravitational radiation.**
J. R. Kuhn, S. P. Boughn.
Theoretical problems in stellar stability and oscillations, p. 224 – 228 (1984). – See Abstr. 012.065.

Solar oscillations may provide an interesting upper bound to the background flux of gravitational radiation. Based on current observations and solar models a low order quadrupole oscillation mode provides the best gravitational flux limit of $2 \times 10^8 \mathrm{erg\, s^{-1} cm^{-2} Hz^{-1}}$ at frequencies near 400 μHz.

066.244 **Exact non–flat Friedmann–Robertson–Walker solutions in Barker's theory of gravitation.**
D. Lorenz–Petzold.
Astrophys. Space Sci., Vol. 106, No. 2, p. 419 – 421 (1984).

The author derives the complete series of the $k = 0$, 1, –1 Friedmann–Robertson–Walker vacuum solutions in Barker's theory of gravitation. Only the $k = 0$ solution was known in a somewhat different form.

066.245 **Neutrino halos around the sun and precession of perihelia of planets.**
K. Tennakone.
Hadronic J., Vol. 7, p. 728 – 733 (1984).

It is shown that, under the assumption that Einstein's general relativity is valid, the experimental data on the precession of the perihelia of planets may be used to obtain upper limits on the density of neutrino halos around the sun.

066.246 **Experiment versus theory in gravitation.**
Y. Mizobuchi, H. Yilmaz.
Hadronic J., Vol. 7, p. 874 – 896 (1984).

It is shown that all four expansion parameters of the metric can be determined using only (planetary and lunar) orbital information. The first three values ($\alpha = \beta = \gamma = 1$) coincide with those of the usual PPN, whereas, the fourth ($\delta = 1$) is beyond and is determined by the "active" principle of equivalence $m_a = m_p$. It is noted that the 532″ per century Newtonian part of the perihelion advance of Mercury is a consequence of $\delta = 1$ and is determined by the same to an accuracy of $10^{-4}$. It is further noted that gravitational radiation determines this same parameter as $\delta = 1$, although in this case (binary pulsar) the present accuracy is about $1.2 \times 10^{-2}$. In conjunction with the moon ranging experiment the gravity radiation is shown to provide experimental evidence for the universality of the mass–energy relation $E = m \cdot c^2$.

066.247 **Gauge fixing and massless, integral–spin fields in de Sitter space.**
J.–P. Gazeau.
Lett. Math. Phys., Vol. 8, No. 6, p. 507 – 516 (1984).
The classical field theories of massless fields of arbitrary, fixed, integral spins reach their simplest Gupta–Bleuler triplet structure when the gauge–fixing parameter $c$ is put equal to zero ("the Feynman Gauge"). This value of $c$ is peculiar to the flat space theory. In a de Sitter massless field theory with spin $s$, the "best value" of $c$ is shown to be equal to $2/2s+1$.

066.248 **A further generalization of Vaidya's star metric.**
Y.–j. Wang.
Chin. Astron. Astrophys., Vol. 8, No. 4, p. 283 – 285 (1984). English translation of Acta Astrophys. Sin., Vol. 4, No. 2, p. 152 – 157 (1984).
A metric containing five arbitrary functions of time is obtained which describes the gravitational field of an arbitrary accelerated and charged point–mass (star). It is a further generalization of Vaidya's star metric. In particular cases, it reduces to the Reissner–Nordström metric and the Kinnersley metric. Since the various parameters of the source can vary arbitrarily in time, this metric is of wide applicability and may prove useful for various specific astronomical objects.

066.249 **Surface temperature of a rotating charged body.**
F.–s. Chen.
Chin. Astron. Astrophys., Vol. 8, No. 4, p. 286 – 289 (1984). English translation of Acta Astrophys. Sin., Vol. 4, No. 2, p. 158 – 165 (1984).
The surface temperature of a rotating, charged body is found separately under the Kerr–Newman metric and the vector graviton metric. It is found that (1) under the Kerr–Newman metric the surface temperature rises from the poles to the equator when the radius $R$ of the body is greater than a certain critical value, $r_n$. When $R = r_n$, the surface temperature is uniform. When $R < r_n$, the above gradient is reversed. For pulsars, the equatorial temperature is some $3 \times 10^4$K higher than the polar temperature. (2) Under the vector graviton field metric a similar temperature differential exists, but it is much smaller in size.

066.250 **A rigorous method of calculating the deflection of light in the gravitational field of the Sun.**
R.–s. Pan, J.–j. Wang, H.–j. Yan, G.–q. Tang, C. Huang.
Chin. Astron. Astrophys., Vol. 8, No. 4, p. 319 – 324 (1984). English translation of Acta Astron. Sin., Vol. 25, No. 2, p. 168 – 176 (1984).
Starting from the equation for the path of a photon in a gravitational field, the authors give a method of calculating the deflection of starlight in the solar field. The transformation from the non–inertial system at the Sun to the locally inertial system of the observer is considered. Calculation was made under the Schwarzschild metric. The authors' calculated results for the deflection before making coordinate transformation agree well with the results according to the IAU formulae, but when an accuracy of 0.″001 is required, then the coordinate transformation must be included.

066.251 **Mutual coherence of gravitationally lensed images.**
P. Schneider, J. Schmid–Burgk.
MPA Rep., No. 163, 35 pp. (1984). Submitted to Astron. Astrophys.

066.252 **Lagrangians for n point–masses at the second post–Newtonian approximation of general relativity.**
T. Damour, G. Schäfer.
MPA Rep., No. 164, 41 pp. (1984). Submitted to Gen. Relativ. Gravitation.

066.253 **Two mass solution of Einstein's vacuum equations: the double Kerr solution.**
W. Dietz, C. Hoenselaers.
MPA Rep., No. 167, 86 pp. (1984).
The newly developed transformations, which allow to generate solutions of Eintein's vacuum equations from given ones, are meanwhile very powerful tools for constructing new solutions. The authors prove the equivalence of Kramer–Neugebauer and HKX transformations and construct the double Kerr solution by means of four HKX transformations. They analyse the resulting solution in detail and present an analytic continuation of the parameters to pass from underextreme to hyperextreme constituents.

066.254 **An outline of the astrometric problem in the determination of Einstein's light–deflection in the solar gravitational field.**
X.–h. Zhou, J.–y. Jing.
Publ. Purple Mt. Obs., Vol. 3, No. 2, p. 64 – 71 (1984). In Chinese.

066.255 **Regge calculus and observations. II. Further applications.**
R. M. Williams, G. F. R. Ellis.
Gen. Relativ. Gravitation, Vol. 16, No. 11, p. 1003 – 1021 (1984).
The method, developed in an earlier paper, for tracing geodesics of particles and light rays through Regge calculus space–times, is applied to a number of problems in the Schwarzschild geometry. It is possible to obtain accurate predictions of light bending by taking sufficiently small Regge blocks. Calculations of perihelion precession, Thomas precession, and the distortion of a ball of fluid moving on a geodesic can also show good agreement with the analytic solution.

066.256 **Clocks in Weyl space–time.**
R. Chandra.
Gen. Relativ. Gravitation, Vol. 16, No. 11, p. 1023 – 1030 (1984).
This paper deals with the usual methods for the construction of clocks as a first step in the direction of measurements in Weyl space–time. The atomic clocks constructed via atomic transitions, and the gravitational clocks constructed by the method of Marzke and Wheeler measure quantities with Weyl weights zero, whereas the Kundt and Hoffmann method for the construction of a clock was not found to be suitable.

066.257 **Constraints on unified theories from the experimental tests of the equivalence principle.**
M. Gasperini.
Gen. Relativ. Gravitation, Vol. 16, No. 11, p. 1031 – 1037 (1984).
The predictions of a general unified theory for the gravitational, electromagnetic and scalar field are compared with the results of the experimental tests of the equivalence principle. It is shown that the theoretical predictions do not disagree with experimental data provided that the coupling of the scalar to the electromagnetic field is suppressed by a factor $k \sim 10^{-3}$, or, alternatively, the scalar field is massive.

066.258 **A relation between the Szekeres quasispherical gravitational collapse solution and the Robinson–Trautman metrics.**
R. J. Gleiser.
Gen. Relativ. Gravitation, Vol. 16, No. 11, p. 1039 – 1043 (1984).
A form of the Robinson–Trautman metric for pure radiation is obtained by applying a limiting procedure to the quasispherical gravitational collapse solution of Szekeres. The procedure can be interpreted as the one allowing the particles making up the dust in the Szekeres solution to become massless.

066.259 **Chronal isomorphism.**
U. D. Vyas, G. M. Akolia.
Gen. Relativ. Gravitation, Vol. 16, No. 11, p. 1045 – 1051 (1984).
Along with certain properties of a chronal isomorphism, it is proved that to convert a chronal isomorphism into a causal isometry, one needs only the distinguishingness of one of the space–times.

**066.260 Surface layers in general relativity and their relation to surface tensions.**
H.-J. Schmidt.
Gen. Relativ. Gravitation, Vol. 16, No. 11, p. 1053 – 1061 (1984).
For a thin shell, the intrinsic 3–pressure is shown to be analogous to $-A$, where $A$ is the classical surface tension: First, interior and exterior Schwarzschild solutions are matched together such that the surface layer generated at the common boundary has no gravitational mass. Second, after a suitable definition of mean curvature, the general relativistic analog to Kelvin's relation is proven to be contained in the equation of motion of the surface layer.

**066.261 Projective differential geometry and geodesic conservation laws in general relativity, II: Conservation laws.**
G. E. Prince, M. Crampin.
Gen. Relativ. Gravitation, Vol. 16, No. 11, p. 1063 – 1075 (1984).
The authors examine the geodesic conservation laws associated with the projective actions discussed earlier. Using the Cartan formalism, a one–to–one correspondence between a class of these actions and all geodesic conservation laws is possible. In particular there is a natural geometric interpretation of Killing tensors.

**066.262 On the physical interpretation of some simple soliton solutions of Einstein's equations.**
R. J. Gleiser.
Gen. Relativ. Gravitation, Vol. 16, No. 11, p. 1077 – 1094 (1984).
The simple soliton solutions of Einstein's equations obtained by Belinski and Zakharov using the inverse scattering method have been interpreted as gravitational (solitary) shock waves, partly on the basis of an analysis of certain singularities apparently inherent to the method. A closer study reveals, however, that such singularities can be removed by an appropriate extension of the solutions, which is given explicitly. A classification of inequivalent flat space–time metrics appropriate for the applications of the method is derived.

**066.263 Gravitational multisolitons in two space dimensions.**
G. Clément.
Gen. Relativ. Gravitation, Vol. 16, No. 11, p. 1095 – 1102 (1984).
Two classes of regular solutions of five–dimensional general relativity, depending only on two space coordinates, and fulfilling boundary conditions appropriate to gravity theory in two space dimensions, are obtained. The solutions of the first class have a Euclidean spatial topology, while the solutions of the second class have two spatial end points connected by wormholes.

**066.264 The gravitational contribution to the stress–energy tensor of a medium in general relativity.**
T. W. Noonan.
Gen. Relativ. Gravitation, Vol. 16, No. 11, p. 1103 – 1118 (1984).
The macroscopic stress–energy tensor of an astronomical "medium" such as a galaxy of stars is determined by the field equation of general relativity from the small–scale variations in mass and velocity. In the weak–field, slow–motion approximation, in which the gravitational fields of the stars are Newtonian, it is found that the contribution by the small–scale gravitational fields to the macroscopic density and stress are, respectively, the Newtonian gravitational energy density and the Newtonian gravitational stress tensor.

**066.265 Milgrom's revision of Newton's laws: dynamical and cosmological consequences.**
J. E. Felten.
Astrophys. J., Vol. 286, No. 1, p. 3 – 6 (1984).
Milgrom's recent revision of Newtonian dynamics was introduced to eliminate the inference that large quantities of invisible mass exist in galaxies. The author shows by simple examples that a Milgrom acceleration, in the form presented so far, implies other far–reaching changes in dynamics. The momentum of an isolated system is not conserved, and the usual theorem for center–of–mass motion of any system does not hold. The model fails to provide a complete description of particle dynamics and should be thought of as a revision of Kepler's laws rather than Newton's. The Milgrom acceleration also implies fundamental changes in cosmology. It is shown that a new theory of gravitation will be needed before the behavior of a Milgrom universe can be predicted.

**066.266 Does the missing mass problem signal the breakdown of Newtonian gravity?**
J. Bekenstein, M. Milgrom.
Astrophys. J., Vol. 286, No. 1, p. 7 – 14 (1984).
The authors consider a nonrelativistic potential theory for gravity which differs from the Newtonian theory. The theory is built on the basic assumptions of the modified dynamics, which were shown earlier to reproduce dynamical properties of galaxies and galaxy aggregates without having to assume the existence of hidden mass. The theory involves a modification of the Poisson equation and can be derived from a Lagrangian. The total momentum, angular momentum, and (properly defined) energy of an isolated system are conserved. It is shown that the weak equivalence principle is satisfied and that the strong equivalence principle is violated in this theory.

**066.267 On an internal spherically–symmetrical radiation source.**
A. M. Baranov, N. N. Paklin.
Modern theoretical and experimental problems of the theory of relativity and gravitation, p. 126 – 127 (1984). In Russian. Abstr. in Ref. Zh., 51. Astron., 12.51.199 (1984). – See Abstr. 012.080.

**066.268 Evolution of an extended body.**
R. I. Khrapko.
Modern theoretical and experimental problems of the theory of relativity and gravitation, p. 237 – 238 (1984). In Russian. Abstr. in Ref. Zh., 51. Astron., 12.51.200 (1984). – See Abstr. 012.080.

**066.269 Use of reduced moments in the dynamics of extended bodies.**
R. I. Khrapko.
Modern theoretical and experimental problems of the theory of relativity and gravitation, p. 239 – 240 (1984). In Russian. Abstr. in Ref. Zh., 51. Astron., 12.51.201 (1984). – See Abstr. 012.080.

**066.270 Possibility of detecting cosmic gravitational waves.**
M. V. Sazhin.
Modern theoretical and experimental problems of the theory of relativity and gravitation, p. 176 – 177 (1984). In Russian. Abstr. in Ref. Zh., 51. Astron., 12.51.205 (1984). – See Abstr. 012.080.

**066.271 Influence of a periodic gravitational wave on the orbit of a planet.**
R. A. Kochkin.
Modern theoretical and experimental problems of the theory of relativity and gravitation, p. 192 – 193 (1984). In Russian. Abstr. in Ref. Zh., 51. Astron., 12.51.207 (1984). – See Abstr. 012.080.

**066.272 On the energy of a gravitationally–inertial field in the arbitrary metric theory of gravitation.**
V. A. Tverdokhlebova.
Modern theoretical and experimental problems of the theory of relativity and gravitation, p. 282 – 283 (1984). In Russian. Abstr. in Ref. Zh., 51. Astron., 12.51.210 (1984). – See Abstr. 012.080.

**066.273 Complement of the condition of Piran for the Papapetrou equations in a Schwarzschild field.**
R. M. Plyatsko.
Modern theoretical and experimental problems of the theory of relativity and gravitation, p. 216 – 217 (1984). In Russian. Abstr. in Ref. Zh., 51. Astron., 12.51.211 (1984). – See Abstr. 012.080.

**066.274 Dynamics of a system of test bodies in GTR.**
A. N. Aleksandrov.
Modern theoretical and experimental problems of the theory of relativity and gravitation, p. 218 – 219 (1984). In Russian. Abstr. in Ref. Zh., 51. Astron., 12.51.212 (1984). – See Abstr. 012.080.

066.275 **Manifestation of the gravitational ultra–relativistic spin–orbital interaction.**
R. M. Plyatsko, A. L. Vynar, Ya. N. Pelekh.
Modern theoretical and experimental problems of the theory of relativity and gravitation, p. 220 – 221 (1984). In Russian. Abstr. in Ref. Zh., 51. Astron., 12.51.213 (1984). – See Abstr. 012.080.

066.276 **Motion of gravitating bodies in GTR.**
S. M. Kopejkin.
Modern theoretical and experimental problems of the theory of relativity and gravitation, p. 241 – 242 (1984). In Russian. Abstr. in Ref. Zh., 51. Astron., 12.51.214 (1984). – See Abstr. 012.080.

066.277 **Lagrangian systems of two gravitating charged bodies in GTR in the second post–Newtonian approximation.**
R. P. Gajda, V. I. Tretyak.
Modern theoretical and experimental problems of the theory of relativity and gravitation, p. 269 – 270 (1984). In Russian. Abstr. in Ref. Zh., 51. Astron., 12.51.215 (1984). – See Abstr. 012.080.

066.278 **Photon trajectories in the vicinity of a rotating, strongly gravitating mass.**
J. Schastok, M. Soffel, H. Ruder, M. Schneider.
Mitt. Astron. Ges., Nr. 62, p. 370 (1984). Abstract. – See Abstr. 010.341.

066.279 **Gauge approach to classical gravity.**
N. Mukunda.
Gravitation and relativistic astrophysics, p. 3 – 16 (1984). – See Abstr. 012.083.

The purpose of this paper is to give a pedagogical introduction to the treatment of gravitation theory as a gauge theory. The basic idea is to apply to the Poincaré group the same kind of gauge principle that leads, in the case of a non–Abelian internal symmetry, to the well–known Yang–Mills theory.

066.280 **The use of perturbation and approximation methods in general relativity.**
B. F. Schutz.
Relativistic astrophysics and cosmology, p. 35 – 97 (1984). – See Abstr. 012.084.

066.281 **Isolated systems in relativistic gravity.**
M. Walker.
Relativistic astrophysics and cosmology, p. 99 – 134 (1984). – See Abstr. 012.084.

These notes contain four chapters. The first two describe those aspects of the theory which are common to all isolated systems, viz. asymptotic flatness, the description of gravitational radiation, asymptotic symmetries, energy–momentum, and positivity of total energy. The third chapter describes an attempt to relate these asymptotic structures to material sources. Finally the fourth chapter contains some remarks on gravitational collapse.

066.282 **Slow rotation in the bimetric theory of gravitation.**
Eh. V. Chubaryan, A. V. Sarkisyan.
Astrofizika, Tom 20, Vyp. 3, p. 585 – 598 (1984). In Russian. English translation in Astrophysics, Vol. 20, No. 3.

In the frame of the bimetric theory of gravitation the equations of the axial–symmetric gravitational field describing the field of rotating configurations are obtained and examined. The analytical solutions of these equations out of configurations are found. The equations are solved numerically in the first approximation to the angular velocity for the configurations consisting of real baryon gas. The moments of inertia of these configurations in the BTG are calculated.

066.283 **Phenomenological approach to the theory of gravitational vacuum.**
L. Sh. Grigoryan, G. S. Saakyan.
Astrofizika, Tom 20, Vyp. 3, p. 615 – 629 (1984). In Russian. English translation in Astrophysics, Vol. 20, No. 3.

In the present paper the following idea is developed: in the presence of gravitation the vacuum acquires properties of a real medium described by the definite energy–momentum tensor $\tau_{ik}$. It is stated that besides usual matter the vacuum also takes part in the formation of the space–time metric. The problem of the vacuum in the case of a central–symmetric static field is examined. Essential properties of the tensor $\tau_{ik}$ are established using the symmetry of the problem, and also the post–Newtonian and Newtonian limits.

066.284 **The gravitational field of a concentrated mass in the generalized theory of gravitation.**
G. G. Arutyunyan, V. V. Papoyan.
Astrofizika, Tom 21, Vyp. 1, p. 175 – 182 (1984). In Russian. English translation in Astrophysics, Vol. 21, No. 1.

The solution of the problem of the gravitational field of a concentrated mass in the generalized theory of gravitation is found.

066.285 **Some characteristics of a non–static uniform density sphere in general relativity.**
H. Knutsen.
Phys. Scr., Vol. 30, No. 4, p. 228 – 233 (1984).

The characteristics of some particular models for a non–static uniform density sphere is investigated. The model is not conformally flat, and the pressure is positive and the pressure gradient is negative. For contracting models pressure, $p$, and density, $\varrho$, are increasing functions of time, and so are the active mass and the Misner–Sharp–Hernandez mass–function. Two kinds of matter "velocities" are investigated, and it is found that they both are decreasing functions of radial coordinate and they are increasing during the contraction.

066.286 **Long–term post–Newtonian effects in a binary system.**
M. Portilla, R. Lapiedra.
Astrophys. J., Vol. 286, No. 2, p. 633 – 643 (1984).

The authors solve the equations of motion for two gravitating pointlike particles by applying the method of the averaging. They obtain an approximate solution valid in a long interval of time (many times the orbital period). As a consequence the periastron epochs and the periastron positions present an oscillation, of long period, which disappears in the center–of–mass frame.

066.287 **All torsion–free spherical vacuum solutions of the quadratic Poincaré gauge theory of gravity.**
P. Baekler, P. B. Yasskin.
Gen. Relativ. Gravitation, Vol. 16, No. 12, p. 1135 – 1155 (1984).

The authors find all torsion–free, spherically symmetric, vacuum solutions to the theory of gravity recently proposed by Hehl, Ne'eman, Nitsch, and von der Heyde. There are three classes of solutions: (1) the Schwarzschild metrics with arbitrary mass and arbitrary cosmological constant; (2) the Nariai–Bertotti metrics with arbitrary positive cosmological constant; (3) the conformally flat metrics.

066.288 **Light–cone cuts of $I^+$ for charged–Kerr geometry.**
P. S. Joshi, E. T. Newman.
Gen. Relativ. Gravitation, Vol. 16, No. 12, p. 1157 – 1162 (1984).

By integrating the equations for null geodesics the authors obtain the light–cone cuts of null infinity for the charged Kerr geometries.

066.289 **Censorship, strong curvature, and asymptotic causal pathology.**
R. P. A. C. Newman.
Gen. Relativ. Gravitation, Vol. 16, No. 12, p. 1163 – 1176 (1984).

Necessary and sufficient conditions are established for a weakly asymptotically simple and empty, null convergent, generic space–time to be future asymptotically predictable. These conditions require that the causal structure of the space–time is well behaved near spatial infinity and future null infinity, and that there are no singularities of less than a certain finite strength in the future asymptotic limit.

066.290 **Persistent curvature and cosmic censorship.**
R. P. A. C. Newman.
Gen. Relativ. Gravitation, Vol. 16, No. 12, p. 1177 – 1187 (1984).
A definition is given which quantifies the strength of persistent Riemann curvature along a null geodesic. A numerical value thereof is identified which ensures the existence of conjugate points on null geodesics of infinite length. A class of examples show that no lesser value can suffice. One is led to a new theorem of cosmic censorship which identifies an upper bound on the persistent curvature strength with which any space–time may violate weak cosmic censorship.

066.291 **Gravitational radiation from colliding compact stars: hydrodynamical calculations in two dimensions.**
D. L. Gilden, S. L. Shapiro.
Astrophys. J., Vol. 287, No. 2, p. 728 – 744 (1984).
Gravitational radiation from the head–on collision of two identical compact stars is calculated using a two–dimensional, time–dependent, Newtonian hydrodynamical code. The star fluid is treated in an adiabatic approximation, and all relativistic effects are ignored. Gravitational radiation is computed as a small perturbation using linearized theory. The efficiency with which collisions generate radiation, $\Delta E/M$, is typically $\lesssim 0.1\%$ for colliding 1.4 $M_{\odot}$ neutron stars. The results indicate that neutron star collisions might be the most copious emitters of gravitational radiation among promising astrophysical sources.

066.292 **Orbital motion of DI Herculis as a test of a theory of gravitation.**
J. W. Moffat.
Astrophys. J., Lett. Ed., Vol. 287, No. 2, p. L77 – L79 (1984).
The eclipsing binary system DI Her has been observed to have a small rate of motion of the periastron during a period of more than 80 years. This disagrees with the predictions of Newtonian theory and general relativity by a large amount (21 σ). The nonsymmetric gravitational theory can explain this observational result. The observed periastron motions of several other binary systems are also fitted by the theory.

**Physics of gravitating systems. I. Equilibrium and stability.**
See Abstr. 003.008.

**Physics of gravitating systems. II. Nonlinear collective processes: nonlinear waves, solitons, collisionless shocks, turbulence. Astrophysical applications.**
See Abstr. 003.009.

**Understanding relativity. Origin and impact of a scientific revolution.**
See Abstr. 003.025.

**Relativity and engineering.**
See Abstr. 003.032.

**General relativity and relativistic astrophysics.**
See Abstr. 003.034.

**Radiation of gravitational waves by means of electrodynamical systems.**
See Abstr. 003.090.

**Schwerkraft – Rätsel des Gewohnten.**
See Abstr. 003.110.

**Quantum gravity.**
See Abstr. 003.120.

**Twelfth meeting of the Indian Association of General Relativity and Gravitation.**
See Abstr. 011.024.

**A new method of determination of the gravitational constant using a superconductive suspension system.**
See Abstr. 022.008.

**Progress in precision rotations for matter creation experiment and inertial clock.**
See Abstr. 022.017.

**Molecular interferometry: a possible gravitational field measuring technique.**
See Abstr. 022.018.

**Experiments with a cryogenic torsion balance.**
See Abstr. 022.019.

**The atomic hydrogen maser. Is it the best clock for gravity and relativity measurements?**
See Abstr. 034.022.

**Practical cryogenic environment for gravitational radiation detector with parametric transducer.**
See Abstr. 034.023.

**The mechanical background noise of Weber–type gravitational wave antenna.**
See Abstr. 034.024.

**Results of measurement on a harmonic oscillator using a back–action–evading scheme.**
See Abstr. 034.025.

**Flight suspension for the relativity gyro.**
See Abstr. 034.026.

**Present status of the 9.5 GHz parametric transducer at the University of Tokyo.**
See Abstr. 034.027.

**How to make a back–action evading (BAE) measurement.**
See Abstr. 034.028.

**Development of transducers for gravitational wave antennas at the University of Rome.**
See Abstr. 034.029.

**Parametric upconverter transducer with a mechanical resonator for low frequency continuous gravitational wave detection.**
See Abstr. 034.030.

**Transducer for a three–mode cryogenic gravitational radiation detector.**
See Abstr. 034.031.

**On the band width of resonant gravitational–wave antennas.**
See Abstr. 034.034.

**Coupling of oscillation modes in an antenna for gravitational radiation.**
See Abstr. 034.054.

**Gravitational perturbations of the hydrogen atom.**
See Abstr. 061.013.

**The anisotropic space–time: a possible arena of subnuclear phenomena.**
See Abstr. 061.014.

**Quaternions and $N = 2$ supersymmetry.**
See Abstr. 061.015.

**On the stability of Kaluza–Klein vacuum.**
See Abstr. 061.016.

**Discussion about the irreducible representations of superalgebras osp(m/2n).**
See Abstr. 061.017.

**Superspace and the super Killing equation, super algebra.**
See Abstr. 061.018.

**Torsion constraints in supergravity.**
See Abstr. 061.019.

**Is supergravity well–posed?**
See Abstr. 061.020.

**Can leptons and quarks achieve superunification at composite level?**
See Abstr. 061.021.

**An approach to the supergravity theories.**
See Abstr. 061.022.

**Supergravity as a generalized Einstein–Cartan connection.**
See Abstr. 061.023.

**Quantization ambiguity and the Aharanov–Bohm effect.**
See Abstr. 061.024.

**An almost trivial gauge theory in the limit of infinite gauge coupling constant.**
See Abstr. 061.025.

**Gravitational lens effects of degenerate neutrino celestial objects.**
See Abstr. 061.026.

**Particles do not exist.**
See Abstr. 061.027.

**Supersymmetry, finite theories and quantum gravity.**
See Abstr. 061.028.

**Properties of extended theories of rigid supersymmetry.**
See Abstr. 061.029.

**Cosmological topological supergravity.**
See Abstr. 061.030.

**Future avenues in supergravity.**
See Abstr. 061.031.

**Dynamical applications of the gauge–invariant effective action formalism.**
See Abstr. 061.032.

**On quantum gravity and the many–worlds interpretation of quantum mechanics.**
See Abstr. 061.034.

**Naturally vanishing cosmological constant and dynamical generation of $m_W$, $m_{3/2}/m_p$ hierarchy.**
See Abstr. 061.064.

**Just how fast are tachyons?**
See Abstr. 061.078.

**An introduction to spontaneously broken supergravity.**
See Abstr. 061.084.

**Superunification from eleven dimensions.**
See Abstr. 061.085.

**Vacua and solitons in extended supergravity.**
See Abstr. 061.087.

**Twistor bundles as a formalism for unified gauge theories of gravitation and other Yang–Mills fields.**
See Abstr. 061.090.

**A symmetric stress energy tensor for the gravitino field.**
See Abstr. 061.093.

**Vector fields quantization in curved space time.**
See Abstr. 061.094.

**Sonic search for monopoles, gravitational waves and newtorites.**
See Abstr. 061.097.

**Supergravity and the Kaluza–Klein program. Report of workshop D2 (*10th International Conference on General Relativity and Gravitation, Padua, July 3 – 8, 1983*).**
See Abstr. 061.118.

**The line profile of a relativistic Keplerian ring in a massive neutrino halo.**
See Abstr. 061.121.

**Some astrophysical consequences of the dynamical interpretation of gravitation.**
See Abstr. 061.125.

**Introduction to supergravity.**
See Abstr. 061.128.

**The continuous spectrum in differentially rotating perfect fluids. II. The effect of gravitational radiation reaction.**
See Abstr. 062.012.

**Pair production rates in mildly relativistic, magnetized plasmas.**
See Abstr. 062.152.

**Reverse shock wave in relativistic explosions.**
See Abstr. 062.160.

**Adaptive–mesh radiation hydrodynamics – II. The radiation and fluid equations in relativistic flows.**
See Abstr. 063.031.

**General relativistic collapse of rotating stars.**
See Abstr. 065.044.

**Relativistic stars with anisotropic pressure.**
See Abstr. 065.046.

**Collapse and gravitation = the present situation.**
See Abstr. 065.079.

**Models of configurations from an incompressible liquid taking into account the role of gravitational vacuum.**
See Abstr. 065.155.

**On the emission of gravitational radiation from inhomogeneous Jacobi configurations.**
See Abstr. 067.004.

**Gravitational radiation from a particle with orbital angular momentum plunging into a Kerr black hole.**
See Abstr. 067.006.

**Gravitational waves from a particle scattered by a Schwarzschild black hole.**
See Abstr. 067.007.

**Physically realizable relativistic stellar structures.**
See Abstr. 067.016.

**Quantum effects near horizons in a stationary Riemann space–time.**
See Abstr. 067.022.

**Feynman's path–integral method and Hawking evaporation.**
See Abstr. 067.025.

**The geometrical origin of temperature of quantum fields in equilibrium on spacetimes with horizons.**
See Abstr. 067.027.

**Behavior of entropy of self-gravitating radiation in collapse.**
See Abstr. 067.033.

**What happens to the horizon when a black hole radiates?**
See Abstr. 067.035.

**Black holes, singularities and predictability.**
See Abstr. 067.037.

**The motion of compact bodies and gravitational radiation.**
See Abstr. 067.117.

**Dynamics of a developing neutron star in general relativity.**
See Abstr. 067.120.

**Linear momentum and gravitational waves: circular orbits around a Schwarzschild black hole.**
See Abstr. 067.126.

**GTR corrections to the intensity of gravitational radiation of a fast-rotating magnetized neutron star.**
See Abstr. 067.134.

**General relativity and gravitational collapse.**
See Abstr. 067.136.

**Search for relativistic companions in "non-X-ray" binary systems.**
See Abstr. 117.034.

**Gravitational and pressure redshifts in white dwarf stars.**
See Abstr. 126.020.

**Arrival-time analysis for a millisecond pulsar.**
See Abstr. 126.089.

**Anti-gravity and galaxy rotation curves.**
See Abstr. 151.006.

**Multiple imaging of quasars by galaxies and clusters.**
See Abstr. 159.007.

**Gravitational lens model of the double QSO 0957+561 A,B incorporating *VLBI* features.**
See Abstr. 159.018.

**Determination of difference in light travel time for QSO 0957+561A, B.**
See Abstr. 159.036.

**Relativistic star clusters with large central redshifts.**
See Abstr. 159.044.

**Gravitational lens models for the triple radio source MG 2016+112.**
See Abstr. 159.062.

**3C 273: a gravitationally lensed quasar?**
See Abstr. 159.085.

**Image separation statistics for multiply imaged quasars.**
See Abstr. 159.131.

**The milli-arcsecond images of Q0957+561.**
See Abstr. 159.134.

**Information on galaxy clustering from gravitational lenses.**
See Abstr. 160.134.

**Some general results in self-creation cosmologies.**
See Abstr. 161.035.

**The neutrino in the Friedmann universe: non-zero rest-mass effects.**
See Abstr. 161.036.

**Some results in the general scalar tensor theory.**
See Abstr. 161.037.

**Exact Bianchi type-I cosmological model with matter in Brans-Dicke theory.**
See Abstr. 161.038.

**Exact Bianchi type $VI_0$ cosmological solutions with matter in Brans-Dicke theory.**
See Abstr. 161.039.

**On Cosmic Censorship: do compact Cauchy horizons imply symmetry?**
See Abstr. 161.051.

**Is our universe an open system?**
See Abstr. 161.054.

**Some cosmological aspects of quantum gravity.**
See Abstr. 161.070.

**Vacuum energy in a squashed Einstein universe.**
See Abstr. 161.071.

**Primordial cosmological inflation versus local supersymmetry breaking in SUSY GUTs coupled to $N = 1$ supergravity.**
See Abstr. 161.072.

**Non-equilibrium thermodynamics of cosmological event horizons.**
See Abstr. 161.119.

**Gravitational interactions of cosmic strings.**
See Abstr. 161.121.

**An Einstein-Strauss-de Sitter model of the Universe.**
See Abstr. 161.123.

**Variation of the gravitational constant, *G*, and the age of the Sun in the Brans-Dicke cosmology.**
See Abstr. 161.124.

**Inhomogeneous cosmologies with cosmological constant.**
See Abstr. 161.160.

**Bianchi type-$VI_0$ conformally invariant scalar fields.**
See Abstr. 161.177.

**A Bianchi type-III cosmological model in Brans-Dicke theory for vacuum field.**
See Abstr. 161.185.

**Structural stability properties of Friedmann cosmology.**
See Abstr. 161.191.

**A single expression for the general solution of Einstein's vacuum and electrovac field equations with cosmological constant for Petrov type D admitting a nonsingular aligned Maxwell field.**
See Abstr. 161.204.

**A gravitation theory with *G* determined by retarded cosmic potential.**
See Abstr. 161.210.

**The statistics of gravitational lenses: the distributions of image angular separations and lens redshifts.**
See Abstr. 161.218.

**On the causal Green function in a Robertson-Walker universe with linear expansion.**
See Abstr. 161.228.

**A class of Szekeres space-times with cosmological constant.**
See Abstr. 161.229.

**An exact solution for a quantum Taub model.**
See Abstr. 161.231.

**Formation of galaxies in G–variable cosmologies. II. The Hoyle–Narlikar and the Brans–Dicke gravitational theories.**
See Abstr. 161.245.

**Relativistic cosmology: its nature, aims and problems.**
See Abstr. 161.291.

**The quantum mechanics of the universe.**
See Abstr. 161.296.

**Gravitational amplification of light at relativistic distance.**
See Abstr. 161.300.

**Some special non–flat BDT–FRW solutions with a cosmological constant.**
See Abstr. 161.312.

**Rotating systems in cosmological background.**
See Abstr. 161.333.

**Quantum fluctuations near the classical space–time singularity.**
See Abstr. 161.334.

**Stationary states in quantum cosmology.**
See Abstr. 161.335.

**Einstein's equations and some cosmological solutions.**
See Abstr. 161.337.

**Magneto–viscous fluid cosmological model of plane symmetry in general relativity.**
See Abstr. 161.355.

**Relation between physical and metric field in general relativity and cosmology.**
See Abstr. 161.356.

**Behavior of higher modes of gravitational waves and gauge–invariant density perturbations in Bianchi IX cosmological models.**
See Abstr. 161.359.

**Limits on a stochastic gravitational wave background from observations of terrestrial and solar oscillations.**
See Abstr. 161.360.

**Some observable effects in relativistic cosmology with non–perfect fluids.**
See Abstr. 161.361.

**Solitonic gravitational waves in Bianchi II cosmologies. 2. One solitonic perturbations.**
See Abstr. 161.367.

# 067 Astrophysics of Compact Objects (Neutron Stars, Black Holes)

067.001 **Magnetic flare model of γ–ray bursts.**
E. P. Liang, S. K. Antiochos.
Nature, Vol. 310, No. 5973, p. 121 – 122 (1984).

One model of the origin of γ–ray bursts assumes emission from a hot, dense and thin region on the surface of a neutron star. The severe energetics and persistence of such an emitting layer prompted the author to consider a picture in which the emission regions lie at the foot points of reconnecting magnetic loops, powered by downward impinging electromagnetic waves. They examine the structure of the emitting sheet, and the generation, propagation and coupling of the electromagnetic energy fluxes to the surface layer, and show that a flare–like model can account for most of the general features of γ–ray bursts.

067.002 **Radiation from accreting magnetized neutron stars.**
P. Mészáros.
Space Sci. Rev., Vol. 38, Nos. 3/4, p. 325 – 351 (1984).

The author reviews some recent developments in our understanding of accreting magnetized neutron stars. A brief summary of the observations is given, on which current phenomenological models are based. The main part of this paper is a discussion of recent work by several groups on the radiative transfer problem in a strong magnetic field and its application to models of the structure and properties of self–consistent neutron star polar cap emission regions. The assumptions and uncertainties involved are discussed, recent progress is evaluated, and current and future problems are indicated.

067.003 **Nonadiabatic self–consistent spherical accretion as a model for quasars and active galactic nuclei.**
A. Wandel, A. Yahil, M. Milgrom.
Astrophys. J., Vol. 282, No. 1, p. 53 – 60 (1984).

New solutions are presented for spherical accretion onto a black hole, which are self–consistent in the sense that the infalling gas itself emits the radiation which controls its thermal profile. In these solutions, radiative cooling dominates in the outer part of the flow, and Compton heating by X–rays becomes important when the gas is already supersonic. The luminosity in these solutions increases with the accretion rate, while the mean photon energy decreases.

067.004 **On the emission of gravitational radiation from inhomogeneous Jacobi configurations.**
J. R. Ipser, R. A. Managan.
Astrophys. J., Vol. 282, No. 1, p. 287 – 290 (1984).

The effects of inhomogeneity on the emission of gravitational radiation from Jacobi configurations (uniformly rotating nonaxisymmetric configurations) are studied. If the equation of state is approximated by a polytropic relation, a neutron–star core with adiabatic index $\gamma \leqslant 3$ can radiate up to a fraction $\sim 10^{-3}$ to $10^{-4}$ of its mass–energy while passing through a Jacobi phase during its evolution. While some of the results suggest that the fast pulsar PSR 1937+214 is hovering near the Jacobi bifurcation point, it appears more likely that the pulsar is not rotating sufficiently rapidly for this to be the case.

067.005 **Initial data for N black holes.**
J. W. York Jr.
Physica A, Vol. 124A, No. 1 – 3, p. 629 – 637 (1984). Abstr. in Phys. Abstr., Vol. 87, No. 1237, Entry 68070 (1984). – See Abstr. 012.008.

067.006 **Gravitational radiation from a particle with orbital angular momentum plunging into a Kerr black hole.**
Y. Kojima, T. Nakamura.
Prog. Theor. Phys., Vol. 71, No. 1, p. 79 – 90 (1984). Abstr. in Phys. Abstr., Vol. 87, No. 1237, Entry 68076 (1984).

067.007 **Gravitational waves from a particle scattered by a Schwarzschild black hole.**
K. Oohara, T. Nakamura.
Prog. Theor. Phys., Vol. 71, No. 1, p. 91 – 99 (1984). Abstr. in Phys. Abstr., Vol. 87, No. 1237, Entry 68077 (1984).

067.008 **R–S model and radio binary pulsars.**
Z. Wang.
Kexue Tongbao (Beijing), Vol. 28, No. 5, p. 640 – 643 (1983). Abstr. in Phys. Abstr., Vol. 87, No. 1237, Entry 72968 (1984).

067.009 **Time–asymmetric initial data for the *N* black hole problem in general relativity.**
A. D. Kulkarni.
J. Math. Phys., Vol. 25, No. 4, p. 1028 – 1034 (1984). Abstr. in Phys. Abstr., Vol. 87, No. 1237, Entry 72974 (1984).

067.010 **A model for evaporating black holes.**
Y. Kuroda.
Prog. Theor. Phys., Vol. 71, No. 1, p. 100 – 108 (1984). Abstr. in Phys. Abstr., Vol. 87, No. 1237, Entry 72975 (1984).

067.011 **A numerical model of an explosion near the surface of a neutron star and gamma–ray bursts.**
G. S. Bisnovatyi–Kogan (*G. S. Bisnovatyj–Kogan*), S. I. Blinnikov, A. F. Zakharov.
Sov. Astron., Vol. 28, No. 1, p. 62 – 66 (1984). English translation of 37.067.012.

067.012 **Pulsar magnetospheres.**
E. Asseo, D. Beaufils, R. Pellat.
Mon. Not. R. Astron. Soc., Vol. 209, No. 2, p. 285 – 306 (1984).
The authors examine static and dynamic solutions of the aligned rotator model introduced for pulsars by Goldreich & Julian (1969). They first develop configurations with a cold and force–free plasma atmosphere and investigate then the possible existence of charged beams above the pulsar polar caps, propagating up and down along the magnetic field lines. Finally, an alternative configuration is studied, in which particles are exchanged between the neutron–star surface and a force–free magnetospheric plasma.

067.013 **Transport coefficients of dense matter in the liquid metal regime.**
R. Nandkumar, C. J. Pethick.
Mon. Not. R. Astron. Soc., Vol. 209, No. 3, p. 511 – 524 (1984).
The authors have calculated the thermal and electrical conductivities and the shear viscosity of dense matter when the ions form a liquid. Effects of ionic correlations are included by use of results for the liquid structure factor of the one component plasma obtained from numerical simulations, and screening of the electron–ion interaction is taken into account. Electron effective mass and vertex corrections are also estimated.

067.014 **Radio pulsars: a theory of subpulse formation and drift.**
P. B. Jones.
Mon. Not. R. Astron. Soc., Vol. 209, No. 3, p. 569 – 586 (1984).
A theory of drifting–subpulse phenomena in radio pulsars is formulated in terms of travelling–wave modulation of the high–energy positron and ion components of the open magnetic–flux–line current density. The necessary condition for growth of travelling waves is derived for a model system. The theory provides qualitative explanations for the effect of nulling on the subpulse drift rate, the subpulse memory phenomenon, and for the observed correlation of drift direction with period derivative.

067.015 **Frequency shifts and line broadening of radiation from rotating neutron stars.**
R. C. Kapoor, B. Datta.
Mon. Not. R. Astron. Soc., Vol. 209, No. 4, p. 895 – 904 (1984).
The frequency shifts and line broadening of radiation from the surface of rapidly rotating neutron stars are calculated using a rotationally perturbed interior spherical metric and a representative choice of the equation of state for neutron star matter. The effect of rotation on the surface redshift and Doppler broadening are found to be substantial for rotation close to the secular instability limit, relevant for fast pulsars. If gamma–ray line emission originates from the surface of such pulsars, the results presented can be used to constrain the validity of the equation of state of matter at high densities.

067.016 **Physically realizable relativistic stellar structures.**
M. C. Durgapal, A. K. Pande, R. S. Phuloria.
Astrophys. Space Sci., Vol. 102, No. 1, p. 49 – 66 (1984).
Relativistic stellar structures can be obtained both analytically and by computation, but all these models do not stand the tests of physical reality. It is shown that the solution discussed by Durgapal and Rawat (1980) is the only exact solution which in its most generalized form for a given density distribution stands all the tests of physical reality. A neutron star model is proposed by assuming $\varrho \geqslant 2\times10^{14}\ \mathrm{g\ cm^{-3}}$. Two specific cases are considered. The structures are found to be bound and stable under radial perturbations. The models have been studied for slow rotation and the mass of the Crab pulsar has been estimated for different mechanisms.

067.017 **The magnetohydrodynamics of black–hole accretion. II. Stationary and axisymmetric flow in Kerr space–time.**
U. S. Pandey, G. S. Dubey.
Astrophys. Space Sci., Vol. 102, No. 1, p. 71 – 79 (1984).
The solution of basic flow equations of perfect magnetofluid in the fixed background of Kerr space–time has been obtained. The authors assume that magnetofluid (test) flow around a Kerr black hole is stationary and axisymmetric. A particular solution exhibits the property that flow lines lie on circles of constant radius. In addition, it is found that magnetofluid is endowed with "frozen–in" magnetic field – i.e., that it moves with the material.

067.018 **Hydrogen molecular ion in the magnetic field of a neutron star. II. Quadrupole moment tensor.**
V. K. Khersonskij.
Astrophys. Space Sci., Vol. 103, No. 2, p. 357 – 370 (1984).
A method for the calculation of $H_2^+$ quadrupole moment tensor components in a strong magnetic field typical for a neutron star surface is discussed. Numerical computations were carried out for the values of a magnetic field in the interval $B = 10^{11} - 10^{13}$ G. The main attention is concentrated on those regions of internuclear distances and angles between the directions of the magnetic field and the internuclear axis which are most interesting for future calculations of the probabilities of the vibrational–rotational transitions in an $H_2^+$ molecular ion in a strong magnetic field.

067.019 **Quasi–normal modes of a black hole.**
B. Mashhoon.
Proceedings of the Third Marcel Grossmann Meeting on General Relativity, Part A, p. 599 – 608 (1983). – See Abstr. 012.002.
This paper attempts to establish a theoretical framework for interpreting quasi–normal mode (QNM) oscillations of black holes. A method is developed for calculating the frequencies of the dominant QNMs.

067.020 **X–ray bursts: nuclear flash model and its difficulties.**
S. Hayakawa.
Proceedings of the Third Marcel Grossmann Meeting on General Relativity, Part A, p. 869 – 882 (1983). – See Abstr. 012.002.
X–ray bursts are considered to be produced by the nuclear flash taking place in the accretion layer of a neutron star (type I) or by accretion instabilities which feed matter suddenly on the surface of a neutron star. A brief account of the nuclear flash model and the main consequences thereof are given. Effects of general relativity on the nuclear flash and the observation of X–ray bursts are reviewed. A number of burst properties observed mainly by Hakucho can be hardly accounted for in terms of the nuclear flash model.

067.021 **Imaginary coordinate, decrease of the horizon area of Schwarzschild black holes and naked singularities.**
Y. Wang.
Proceedings of the Third Marcel Grossmann Meeting on General Relativity, Part B, p. 1009 – 1016 (1983). – See Abstr. 012.002.

067.022 **Quantum effects near horizons in a stationary Riemann space–time.**
Z. Zhao, Y. Guei.
Proceedings of the Third Marcel Grossmann Meeting on General Relativity, Part B, p. 1177 – 1183 (1983). – See Abstr. 012.002.
In this paper it is proved that both the thermal and nonthermal radiations coming from horizons of black holes do not depend on the detailed form of the metric.

067.023 **Approximate stress tensors near a black hole.**
D. N. Page.
Proceedings of the Third Marcel Grossmann Meeting on General Relativity, Part B, p. 1189 – 1192 (1983). – See Abstr. 012.002.

067.024 **The Hawking radiation of charged Dirac particles in Kerr–Newman spacetime.**
D. Xu.
Proceedings of the Third Marcel Grossmann Meeting on General Relativity, Part B, p. 1193 – 1199 (1983). – See Abstr. 012.002.
The Hawking radiation of charged Dirac particles on the event horizon of a Kerr–Newman black hole is studied.

067.025 **Feynman's path–integral method and Hawking evaporation.**
L. Liu.
Proceedings of the Third Marcel Grossmann Meeting on General Relativity, Part B, p. 1209 – 1215 (1983). – See Abstr. 012.002.
This paper demonstrates that, independent of the explicit forms of space–time metrics, every static or stationary space–time with future horizon must have Hawking evaporation.

067.026 **On the reversible processes in the thermodynamics of black hole.**
L. Chen, M. Zhou.
Proceedings of the Third Marcel Grossmann Meeting on General Relativity, Part B, p. 1217 – 1220 (1983). – See Abstr. 012.002.

067.027 **The geometrical origin of temperature of quantum fields in equilibrium on spacetimes with horizons.**
M. Zhou, L. Chen, H. Guo.
Proceedings of the Third Marcel Grossmann Meeting on General Relativity, Part B, p. 1243 – 1246 (1983). – See Abstr. 012.002.

067.028 **Thermal X–ray emission from isolated, older pulsars: a new heating mechanism.**
J. Huang, Q. Peng, K. Huang, R. E. Lingenfelter.
Proceedings of the Third Marcel Grossmann Meeting on General Relativity, Part B, p. 1269 – 1272 (1983). – See Abstr. 012.002.
The observation of thermal X–ray emission from pulsars implies that some heating source must be at work on the neutron star surface. The authors propose here a new heating mechanism, named magnetic dipole radiation from superfluid neutrons.

067.029 **Transport properties of dense matter.**
N. Itoh, S. Mitake, H. Iyetomi, S. Ichimaru.
Proceedings of the Third Marcel Grossmann Meeting on General Relativity, Part B, p. 1273 – 1275 (1983). – See Abstr. 012.002.
In this paper the authors present improved calculations of the electrical and thermal conductivities of dense matter in the liquid metal phase.

067.030 **Abnormal neutron stars with abnormal protons.**
Q. Qu, Z. Wang, T. Lu, Y. Chu.
Proceedings of the Third Marcel Grossmann Meeting on General Relativity, Part B, p. 1277 – 1278 (1983). – See Abstr. 012.002.

067.031 **Quark – abnormal neutron stars.**
Q. Qu, Z. Wang, J. Zhao.
Proceedings of the Third Marcel Grossmann Meeting on General Relativity, Part B, p. 1279 – 1281 (1983). – See Abstr. 012.002.

067.032 **The distribution of the projected density of stars near the Reissner–Nordström black hole in a globular cluster.**
Y. Shen, Q. Zhang.
Proceedings of the Third Marcel Grossmann Meeting on General Relativity, Part B, p. 1283 – 1288 (1983). – See Abstr. 012.002.

067.033 **Behavior of entropy of self–gravitating radiation in collapse.**
Z. Zhang.
Proceedings of the Third Marcel Grossmann Meeting on General Relativity, Part B, p. 1289 – 1293 (1983). – See Abstr. 012.002.
This paper examines the behavior of the entropy of a spherically symmetric and time symmetric distribution of self–gravitating radiation in dynamical equilibrium in the context of general relativity.

067.034 **Clusters of black holes as models for active galactic nuclei.**
W. R. Stoeger, A. G. Pacholczyk.
Proceedings of the Third Marcel Grossmann Meeting on General Relativity, Part B, p. 1295 – 1299 (1983). – See Abstr. 012.002.
The authors propose to model active galactic nuclei (AGN) using a spherically symmetric cluster of accreting black holes. The cluster is characterized by a mass spectrum, and mass segregation is assumed. Each black hole has an accretion disk – together these provide the dominant luminosity from the AGN. This model is fitted to the luminosities and optical spectra of a number of quasars.

067.035 **What happens to the horizon when a black hole radiates?**
J. W. York Jr.
Quantum theory of gravity, p. 135 – 147 (1984). – See Abstr. 003.005.
This paper studies in detail the changes the horizon undergoes when a black hole emits energy. It is shown that the basic reaction is the formation of a "quantum ergosphere" that results from the coincidence, for spherical holes, of the apparent horizon and quasi–static limit surface and from the fact that these surfaces are outside the event horizon.

067.036 **Black hole fluctuations.**
J. D. Bekenstein.
Quantum theory of gravity, p. 148 – 159 (1984). – See Abstr. 003.005.
This paper discusses the thermal fluctuations of the basic parameters (mass, angular momentum, momentum, charge) of an isolated black hole.

067.037 **Black holes, singularities and predictability.**
R. M. Wald.
Quantum theory of gravity, p. 160 – 168 (1984). – See Abstr. 003.005.

067.038 **Particle creation by a black hole as a consequence of the Casimir effect.**
R. M. Nugayev (*R. M. Nugaev*).
Lett. Nuovo Cimento, Vol. 39, Ser. 2, No. 14, p. 331 – 335 (1984). Abstr. in Phys. Abstr., Vol. 87, No. 1238, Entry 73316 (1984).

067.039 **On the limiting mass of dense stellar matter.**
J. P. Sharma.
XVI International Conference on Phenomena in Ionized Gases, p. 108 – 109 (1983). Abstr. in Phys. Abstr., Vol. 87, No. 1239, Entry 82468 (1984). – See Abstr. 012.017.

067.040 **On the induced scattering in the relativistic electron–positron plasma.**
M. Eh. Gedalin, G. Z. Machabeli, P. G. Mamradze.
XVI International Conference on Phenomena in Ionized Gases, p. 102 – 103 (1983). Abstr. in Phys. Abstr., Vol. 87, No. 1239, Entry 82498 (1984). – See Abstr. 012.017.

067.041 **Shape of nuclei in the crust of a neutron star.**
M. Hashimoto, H. Seki, M. Yamada.
Prog. Theor. Phys., Vol. 71, No. 2, p. 320 – 326 (1984). Abstr. in Phys. Abstr., Vol. 87, No. 1239, Entry 82499 (1984).

067.042 **The hoop conjecture on black holes. II.**
W. B. Bonnor.
Phys. Lett. A, Vol. 102A, No. 8, p. 347 (1984). Abstr. in Phys. Abstr., Vol. 87, No. 1239, Entry 82502 (1984).

067.043 **Condition for equilibrium of two Reissner–Nordström black holes.**
A. Tomimatsu.
Prog. Theor. Phys., Vol. 71, No. 2, p. 409 – 412 (1984). Abstr. in Phys. Abstr., Vol. 87, No. 1239, Entry 82503 (1984).

067.044 **Do quantum effects exclude the possibility of black holes?**
M. D. Roberts.
Lett. Nuovo Cimento, Vol. 40, Ser. 2, No. 6, p. 182 – 184 (1984). Abstr. in Phys. Abstr., Vol. 87, No. 1240, Entry 82705 (1984).

067.045 **Cold Bose stars.**
J. D. Breit, S. Gupta, A. Zaks.
Phys. Lett. B, Vol. 104B, No. 5 – 6, p. 329 – 332 (1984). Abstr. in Phys. Abstr., Vol. 87, No. 1241, Entry 90087 (1984).

067.046 **Influence of strong–coupling corrections on the equilibrium phase for $^3P_2$ superfluid neutron–star matter.**
V. Z. Vulovic, J. A. Sauls.
Phys. Rev. D, Vol. 29, No. 12, p. 2705 – 2711 (1984). Abstr. in Phys. Abstr., Vol. 87, No. 1241, Entry 90174 (1984).

067.047 **Thick discs around black holes as energy sources for quasars and other astrophysical objects.**
M. Calvani.
Cesk. Cas. Fyz., Sekce A, Vol. 34, No. 1, p. 23 – 39 (1984). Abstr. in Phys. Abstr., Vol. 87, No. 1241, Entry 90175 (1984).

067.048 **Absorption cross sections of a Schwarzschild black hole.**
L. Z. Fang, R. C. Wang.
Classical Quantum Gravity, Vol. 1, No. 4, p. 403 – 405 (1984). Abstr. in Phys. Abstr., Vol. 87, No. 1241, Entry 90176 (1984).

067.049 **Energy–extraction processes from a Kerr black hole immersed in a magnetic field. I. Negative–energy states.**
S. V. Dhurandhar, N. Dadhich.
Phys. Rev. D, Vol. 29, No. 12, p. 2712 – 2720 (1984). Abstr. in Phys. Abstr., Vol. 87, No. 1241, Entry 90178 (1984).

067.050 **Conformal invariance under spatial inversion of extreme Reissner–Nordström black holes.**
W. E. Couch, R. J. Torrence.
Gen. Relativ. Gravitation, Vol. 16, No. 8, p. 789 – 792 (1984).
The extreme Reissner–Nordström geometry is shown to be conformally invariant under a spatial inversion. Generalization of this result to other geometries is briefly discussed.

067.051 **Probing the interiors of neutron stars.**
F. K. Lamb.
Essays in theoretical physics, p. 75 – 93 (1984). – See Abstr. 003.006.
This paper reviews the information on neutron stars that has been obtained from observations of pulsars and pulsing X–ray stars, with emphasis on results from timing studies. Included are summaries of the most recent theoretical work on neutron star structure and the current evidence concerning neutron star masses and radii. Also included are discussions of large frequency jumps and small–scale timing irregularities, the description of the latter phenomenon in terms of random noise processes, and the use of both as probes of the internal structure and dynamical properties of neutron stars.

067.052 **Nuclei in dense matter.**
C. J. Pethick.
Essays in theoretical physics, p. 187 – 205 (1984). – See Abstr. 003.006.
A review is given of the application of the liquid drop model and extensions of it to calculate properties of nuclei in high density matter encountered in neutron stars and stellar collapse.

067.053 **On magnetic properties of pion condensate in neutron stars.**
G. S. Saakyan, L. Sh. Grigoryan.
Astrofizika, Tom 20, Vyp. 1, p. 203 – 207 (1984). In Russian. English translation in Astrophysics, Vol. 20, No. 1.
The possible role of pion condensate in the problem of powerful magnetic field generation in neutron stars is discussed.

067.054 **A modified thermal Thomas–Fermi approach to supernova matter.**
D. Hartmann.
MPA Rep., No. 90, p. 43 – 47 (1983). – See Abstr. 012.020.

067.055 **Nuclei in supernova matter.**
R. Ogasawara.
MPA Rep., No. 90, p. 48 – 49 (1983). – See Abstr. 012.020.
This paper gives a brief account of an investigation of the properties of the very hot and dense nuclear matter forming during the collapse of iron cores in the final evolutionary stages of massive stars.

067.056 **Mean field and Thomas–Fermi calculations of the equation of state of hot dense matter.**
E. Suraud, D. Vautherin.
MPA Rep., No. 90, p. 50 – 54 (1983). – See Abstr. 012.020.

067.057 **Towards selfconsistent band structure calculations of properties of supernova matter.**
R. G. Wolff.
MPA Rep., No. 90, p. 55 – 58 (1983). – See Abstr. 012.020.

067.058 **Numerical modelling of gas–dynamical processes in shells of neutron stars considering anomalies in the equation of state.**
B. G. Mukanova.
Astrofizika, Tom 20, Vyp. 2, p. 321 – 326 (1984). In Russian. English translation in Astrophysics, Vol. 20, No. 2.
Some periodic regimes have been obtained numerically for the envelope of neutron stars of 10 g/cm density immersed in a superstrong magnetic field of 10 Oe. The character of these oscillations is due to anomalies of the equation of state.

067.059 **Champions of magnetic fields.**
V. M. Lipunov.
Priroda, No. 8, p. 22 – 30 (1984). In Russian.
Concerning neutron stars, radio pulsars, X–ray pulsars.

067.060 **Pulsar space velocities and neutrino chirality.**
N. N. Chugaj.
Sov. Astron. Lett., Vol. 10, No. 2, p. 87 – 88 (1984). English translation of 37.067.049.

067.061 **Critical luminosity of a magnetic neutron star.**
Y. N. Gnedin (*Yu. N. Gnedin*), W. Nagel.
Astron. Astrophys., Vol. 138, No. 2, p. 356 – 358 (1984).
Cyclotron absorption can drastically increase the force exerted by radiation on electrons in a magnetic field. It has been argued (Mitrofanov and Pavlov, 1982) that this effect lowers the critical luminosity of magnetic neutron stars and degenerate dwarfs by several orders of magnitude. The authors have found, however, that the formation of an absorption line in the spectrum of a magnetic neutron star largely reduces the importance of the cyclotron resonance, and consequently leads to critical luminosities not much below the Eddington limit.

067.062 **The temperature regime of neutron stars heated by nucleon decay.**
I. D. Novikov, T. V. Perevodchikova.
Inst. kosm. issled. Akad. Nauk SSSR. Prepr., No. 861, 10 pp. (1984). In Russian. Abstr. in Ref. Zh., 51. Astron., 8.51.347 (1984).

067.063 **Thermal response of neutron stars to shell flashes.**
T. Hanawa, M. Y. Fujimoto.
Publ. Astron. Soc. Jpn., Vol. 36, No. 2, p. 199 – 214 (1984).
Recurrences of shell flashes have been computed for an accreting neutron star, where the whole internal structure is taken into account. The general relativistic equations that describe the evolution of a spherical star are solved exactly; nuclear reactions are followed by means of an approximation network involving sixteen nuclides. It is found that when the accretion rate is higher than $dM/dt \cong 3\times10^{-10}\,M_{\odot}\mathrm{yr}^{-1}$, thermal interaction between the burning shell and the core is very weak. Therefore, the recurrence of a shell flash depends little on the thermal state of the interior. Implications of the present results are also discussed in relation to observations of X–ray bursts.

067.064 **New appraoch to the quasinormal modes of a black hole.**
V. Ferrari, B. Mashhoon.
Phys. Rev. D, Vol. 30, No. 2, p. 295 – 304 (1984). Abstr. in Phys. Abstr., Vol. 87, No. 1242, Entry 90453 (1984).

067.065 **Excitation of the free oscillations of a Schwarzschild black hole by the gravitational waves from a scattered test particle.**
K. Oohara.
Prog. Theor. Phys., Vol. 71, No. 4, p. 738 – 751 (1984). Abstr. in Phys. Abstr., Vol. 87, No. 1242, Entry 94851 (1984).

067.066 **Formation of a single millisecond pulsar by the coalescence of a neutron star and a massive white dwarf.**
E. P. J. van den Heuvel, P. T. J. Bonsema.
Astron. Astrophys., Vol. 139, No. 1, p. L16 – L18 (1984).
It is shown that a close binary ($P < 16^h$) consisting of a neutron star and a massive white dwarf ($M > 0.7\,M_{\odot}$) inevitably evolves towards coalescence as a consequence of gravitational radiation losses. The reason for coalescence into a single object is: the steep mass–radius relation of massive white dwarfs, which causes runaway mass transfer to ensue as soon as the white dwarf overflows its Roche lobe. The white dwarf is disrupted on its dynamical timescale (a few seconds). Part of its matter ($\geqslant 0.10\,M_{\odot}$) is expected to be accreted by the neutron star through a disk. This provides a viable model for forming a single millisecond pulsar. As its progenitor system may have descended from a moderately massive (B–type) binary system with a low or moderate runaway velocity, such a pulsar may be expected to have a location close to the galactic plane, like the 1.5 msec pulsar.

067.067 **Nonstationary accretion onto neutron stars: some constraints and consequences.**
G. E. Morfill, J. Trümper, P. Bodenheimer, G. Tenorio–Tagle.
Astron. Astrophys., Vol. 139, No. 1, p. 7 – 14 (1984).
The extreme case of highly nonstationary, inhomogeneous accretion onto magnetised neutron stars is discussed. Such a model allows super–Eddington luminosities, it appears justified if nonlinear, possibly self–excited instabilities at the inner edge of the accretion disc can occur, and yields results, the consequences of which may be observationally tested.

067.068 **A numerical investigation of the Kelvin–Helmholtz instability in the context of accreting neutron stars.**
Y.–M. Wang, J. A. Robertson.
Astron. Astrophys., Vol. 139, No. 1, p. 93 – 103 (1984).
A 2–D magnetohydrodynamical code is used to study the nonlinear development of the Kelvin–Helmholtz instability at the boundary between two non–equal density fluids in relative motion, with parameters chosen to represent conditions at the magnetopause of an accreting neutron star.

067.069 **Rapid postglitch spin–up of the superfluid core in pulsars.**
M. A. Alpar, S. A. Langer, J. A. Sauls.
Astrophys. J., Vol. 282, No. 2, p. 533 – 541 (1984).
Vortex lines in the superfluid cores of neutron stars carry flux due to the induced proton charge current which results from the Fermi liquid interaction between neutrons and protons. As a consequence the scattering of charges off these magnetic vortex lines equilibrates the core superfluid to the plasma and the crust on time scales of order 1 second after a glitch. Thus, the core superfluid cannot be responsible for the observed time scales of the Vela and Crab pulsars.

067.070 **Rotating neutron star structure: implications of the millisecond pulsar PSR 1937 + 214.**
A. Ray, B. Datta.
Astrophys. J., Vol. 282, No. 2, p. 542 – 549 (1984).
A semi–Newtonian condition of rotational stability applied to the recently discovered millisecond pulsar PSR 1937 + 214 implies lower bounds on the mass and moment of inertia and an upper bound on the radius of the neutron star. These upper and lower bounds are dependent on the equation of state of high–density neutron matter. The authors construct critically rotating realistic neutron star models, using the prescription of Hartle and Thorne, for six representative equations of state. The lower bounds on mass are found to be substantially higher than previous estimates.

067.071 **Generating fields from data on $H^- \cup H^+$ and $H^- \cup J^-$.**
R. J. Torrence, W. E. Couch.
Gen. Relativ. Gravitation, Vol. 16, No. 9, p. 847 – 866 (1984).
This paper examines whether reflection series solutions to both spatial and temporal characteristic boundary value problems can provide an approximation scheme for the calculation of black hole reflection and transmission coefficients.

067.072 **Temperature regime of neutron stars heated by decay of nucleons.**
I. D. Novikov, T. V. Perevodchikova.
Astron. Zh., Tom 61, Vyp. 5, p. 935 – 938 (1984). In Russian. English translation in Sov. Astron., Vol. 28, No. 5.
A calculation of the temperature regime of a neutron star in the remote future is given. The temperature is determined by the balance of heating of the star due to baryon decay, reconstruction of the star because of the change of the baryon number and the energy loss by runaway neutrinos and by photon radiation from the surface of the star.

067.073 **Possibility of detecting the magnetic field of a rotating neutron star with SQUID.**
J. Grygar, M. Odehnal, V. Petříček, K. Prikner.
Ann. N.Y. Acad. Sci., Vol. 422, p. 355 (1984). Abstract. – See Abstr. 012.033.

067.074 **Dynamics of accretion disks around compact objects.**
A. R. Prasanna, D. K. Chakraborty.
Ann. N.Y. Acad. Sci., Vol. 422, p. 377 (1984). Abstract. – See Abstr. 012.033.

067.075 **Optically thick, time–dependent accretion onto black holes.**
P. Vitello.
Ann. N.Y. Acad. Sci., Vol. 422, p. 391 (1984). Abstract. – See Abstr. 012.033.

067.076 **Black hole accretion disks: coronal stabilization of the Lightman–Eardley instability.**
J. A. Ionson, M. Kuperus.
Astrophys. J., Vol. 284, No. 1, p. 389 – 393 (1984).
Black hole accretion disks are thought to play a key role in converting gravitational energy into high–energy (>1 keV) X–rays. Two different theoretical models have been proposed: the coronal disk model of Liang and Price and the two–temperature–disk model of Shapiro, Lightman, and Eardley. To the extent that the Lightman–Eardley (L–E) instability determines the appropri-

ate model, the authors point out that the existence of a corona can significantly raise the instability's threshold. A modified L–E instability criterion for the coronal disk model is derived. It is argued that the stabilizing effect of heating a corona lends further support for the coronal–disk interpretation of Cyg X–1.

067.077 **Optically thick, time–dependent spherical accretion onto a black hole. I. Equations and numerical methods.**
P. Vitello.
Astrophys. J., Vol. 284, No. 1, p. 394 – 406 (1984).
This paper considers the equations for fully relativistic, optically thick, spherical, time–dependent accretion onto a black hole, and an approach to their numerical solution. Radiation energy transport is treated in the diffusion approximation, and magnetic dissipative heating is included. An implicit algorithm with a fixed Eulerian spatial mesh is presented, allowing accurate treatment of flow for which the time scales vary rapidly with radius. With no magnetic dissipation, the net escaping luminosities are much less than the Eddington luminosity. With dissipation however, net luminosities near the Eddington luminosity can be readily achieved.

067.078 **Numerical models of force–free black–hole magnetospheres.**
D. A. Macdonald.
Mon. Not. R. Astron. Soc., Vol. 211, No. 2, p. 313 – 329 (1984).
This paper develops numerical models of stationary, axisymmetric, force–free black–hole magnetospheres, based on the theory originally developed by Blandford & Znajek and reformulated and extended by Macdonald & Thorne. The structure of such a magnetosphere is determined by a single scalar "stream function" satisfying a non–linear, second–order partial differential "stream equation" on a region bounded by the black–hole horizon, the accretion disc, and an outer boundary beyond which the force–free condition breaks down. The stream equation is solved numerically. For the chosen boundary conditions, it is found that the poloidal field structure does not change greatly as the hole and the field are spun up.

067.079 **Black hole models for active galactic nuclei.**
M. J. Rees.
Annu. Rev. Astron. Astrophys., Vol. 22, p. 471 – 506 (1984). – See Abstr. 003.024.
Contents: (1) Introduction. (2) Inferences insensitive to detailed model: Some fiducial numbers. Processes in ultrahot thermal plasma. Cyclotron/synchrotron and inverse Compton cooling. (3) Radial accretion flows: Spherical accretion. (4) Black holes according to general relativity: The Kerr metric. Three astrophysically important relativistic effects. (5) Accretion flows with angular momentum: Origin of infalling matter. Thin disks. General structure of tori or thick disks. Radiation–supported tori. Ion–supported tori. (6) Jet formation. (7) Some comments on phenomenology: The continuum spectrum. Ordinary QSOs. Radio galaxies. Radio quasars and optically violent variables (OVVs). Hard X–ray and gamma–ray sources. (8) Demography of AGNs. (9) Some concluding comments.

067.080 **Shock acceleration in active galactic nuclei and quasars.**
D. Kazanas, D. C. Ellison.
MPE Rep., No. 184, p. 185 – 187 (1984). – See Abstr. 012.039.
The authors consider a model of spherical accretion on a massive black hole as the energy source for active galactic nuclei (AGNs) and quasars (QSOs). According to the model, the kinetic energy of the infalling material is randomized by a collisionless shock around the black hole. Relativistic protons are produced with high efficiency via first–order Fermi shock acceleration. The model can naturally produce relativistic particles with flat spectra needed to account for the radiation from these sources, including their broad band spectral characteristics.

067.081 **On the hydrodynamical equations of astrophysical jets.**
M. Calvani, L. Nobili, R. Turolla.
MPE Rep., No. 184, p. 197 – 199 (1984). – See Abstr. 012.039.
A brief summary of work on the hydrodynamics of jet acceleration and confinement in thick accretion disks around blackholes is presented.

067.082 **Collimated winds from active galactic nuclei.**
A. Ferrari, R. Rosner, E. Trussoni, K. Tsinganos.
MPE Rep., No. 184, p. 200 – 202 (1984). – See Abstr. 012.039.

067.083 **An alternative to relativistic beaming in AGN.**
C. S. Coleman.
MPE Rep., No. 184, p. 203 – 205 (1984). – See Abstr. 012.039.
An alternative to relativistic beaming in AGN is that the radiating particles have a highly anisotropic phase space distribution. Such a mechanism is described here, and the resulting particle distribution is shown to provide natural explanations for the high brightness temperatures, superluminal velocities and spectral features of various classes of AGN.

067.084 **Models of non–Keplerian accretion disks based on dynamo magnetic field: application to AGN.**
G. Belvedere, D. Molteni, R. M. Pidatella.
MPE Rep., No. 184, p. 206 – 210 (1984). = Mitt. Kiepenheuer Inst., No. 240. – See Abstr. 012.039.
The authors present a model of a non–Keplerian accretion disk in which the stress tensor is expressed in terms of the large scale components of a dynamo magnetic field and X–ray emission is due to magnetic energy dissipation through a coronal mechanism. The analytical solutions depend on the mass of the compact object, the accretion rate, the geometric dimensions and the angular momentum per unit mass at the outer disk boundary. Application to AGN is discussed.

067.085 **The models of annihilation $\gamma$–line and continuous gamma–spectrum radiation in the center of the Galaxy.**
I. D. Novikov.
MPE Rep., No. 184, p. 211 (1984). Abstract. – See Abstr. 012.039.

067.086 **Centrifugally driven MHD–winds and the non–thermal emission regions in AGNs.**
M. Camenzind, T. J.–L. Courvoisier.
MPE Rep., No. 184, p. 212 – 214 (1984). – See Abstr. 012.039.
The authors briefly describe their model of non–thermal energy generation in active galactic nuclei. Rotational energy released by the central massive source is carried away by a strong MHD–wind. At a certain distance from the central object this wind is terminated by a strong magnetosonic shock which dissipates its energy radiatively. This mechanism is able to explain the observed energy spectra of AGNs from the mm–range to the X–ray region.

067.087 **Macroscopic currents and wind perturbations in the frame of the wind and shock model.**
T. J.–L. Courvoisier, M. Camenzind.
MPE Rep., No. 184, p. 215 – 217 (1984). – See Abstr. 012.039.
The formation of macroscopic currents in the framework of the wind and shock model for AGNs by Camenzind and Courvoisier is studied. These currents are important for the stability of the whole structure and for defining a preferred direction in the generation of jets.

067.088 **Duelling winds: a model for broad–line regions.**
M. D. Smith, D. J. Raine.
MPE Rep., No. 184, p. 218 – 220 (1984). – See Abstr. 012.039.
A model for broad–line regions in active galactic nuclei is discussed. When the X–ray flux from a galactic nucleus Compton–heats an accretion disc a large amount of hot gas can be released in the form of a wind. This material becomes entrained into a wind blowing from the nucleus. Provided the ram pressure of the nuclear wind exceeds that of the disc wind, it will collapse into clouds. The nuclear wind accelerates small clouds to higher velocities, corresponding to observed line widths. Large clouds fall inwards. Cloud sizes, line asymmetries and variability characteristics are determined.

067.089 **X-ray spectra of accretion disks.**
P. Meszaros, W. Nagel.
MPE Rep., No. 184, p. 221 – 223 (1984). – See Abstr. 012.039.

Some of the problems encountered in the interpretation of AGN radiation spectra are discussed. The authors present preliminary results of calculations of the X-ray spectrum from thin and thick accretion disks around massive black holes.

067.090 **The inner horizon in the Kerr–Newman metric.**
M. Curé, N. Zamorano.
Relativity, cosmology, topological mass and supergravity, p. 219 – 231 (1983). – See Abstr. 012.040.

The evolution of a massive scalar field inside the Kerr–Newman metric is studied. The initial conditions for this field are set on the event horizon. Only small perturbations are considered. It is found that the local energy density in the vicinity of the inner Cauchy horizon grows unbounded. The authors conclude that the analytical extension of the Kerr–Newman solution inside the event horizon must be reconsidered if any small perturbation acting on this family of black holes is taken into account.

067.091 **Infrared emission from accretion disks: detectability and variability.**
J. H. Beall, F. K. Knight, H. A. Smith, K. S. Wood, M. Lebofsky, G. Rieke.
Astrophys. J., Vol. 284, No. 2, p. 745 – 750 (1984).

The thermal spectrum of the infrared and optical emission from an accretion disk around a compact object has been calculated, assuming an optically thick, finite disk and a power–law dependence of disk temperature on radius. This calculated spectrum is then compared to the spectrum of blackbody radiators whose temperatures and radii are equivalent to B0 and O5 stars. For likely values of central disk temperature and maximum radius, the disk emission can dominate over the emitted spectrum of the companion star in a binary system. The authors discuss the consequences of these results for three black hole candidate systems (Cygnus X–1, Circinus X–1, and GX 339–4).

067.092 **Neutron gas of a superdense magnetic star.**
S. S. Lipovetskij, V. S. Sekerzhitskij.
Izv. vuzov. Fiz. Tomsk, 12 pp. (1984). In Russian. Abstr. in Ref. Zh., 51. Astron., 9.51.148 (1984).

067.093 **The formation of radiation driven winds in bursting neutron stars.**
R. Z. Yahel, W. Brinkmann, A. Braun.
Astron. Astrophys., Vol. 139, No. 2, p. 359 – 367 (1984).

The authors discuss the conditions for the formation of radiation driven wind around accreting neutron stars during a helium flash. Time–dependent dynamical models of the star atmosphere suggest that the total luminosity during the mass loss phase is oscillating around a mean value that is limited to a few times (2 – 3) the Eddington luminosity. The authors also consider effects of a dipole–like superstrong magnetic field ($10^{12}$ – $10^{13}$G) on the formation of a radiation–driven wind. In particular, they obtain modified heat transport coefficients and show that the efficiency of momentum transfer from the radiation field to the gas depends crucially on the ratio of the cyclotron energy to the mean thermal energy.

067.094 **Cyclotron emission near stellar mass black holes.**
K. M. V. Apparao.
Astron. Astrophys., Vol. 139, No. 2, p. 375 – 377 (1984).

Cyclotron emission in the inner regions of an accretion disk around a matter accreting black hole can be appreciable. In the case of the X–ray source Cyg X–1, cyclotron emission may provide the soft photons needed for "Comptonization" to produce high energy X–rays. The inverse correlation between the fluxes of high energy and low energy X–rays during the "high" and "low" states of Cyg X–1, may be understood as a result of the variation of the rate of accretion and the Compton scattering of the cyclotron photons. In the case of the X–ray source GX 339–4, the observed optical flux during the high states does not seem to be due to cyclotron emission, but probably due to reprocessing of high energy X–rays by the outer regions of the disk.

067.095 **Spherical propagation of large amplitude pulsar waves.**
E. Asseo, X. Llobet, R. Pellat.
Astron. Astrophys., Vol. 139, No. 2, p. 417 – 425 (1984).

The authors consider the propagation of a large amplitude, superluminous, linearly polarized electromagnetic wave in an electron positron plasma. Their interest in the propagation of large amplitude electromagnetic waves is related to the very existence of these waves in the pulsar atmosphere.

067.096 **Equation of state for dense stellar matter at high temperatures.**
K. Arai.
Astron. Her., Vol. 77, No. 12, p. 318 – 321 (1984). In Japanese.

067.097 **On the accuracy of the single–nucleus approximation in the equation of state of hot, dense matter.**
A. Burrows, J. M. Lattimer.
Astrophys. J., Vol. 285, No. 1, p. 294 – 303 (1984).

The authors develop an analytical formalism for determining the errors committed in the entropy, free energy, pressure, and abundances of hot, dense matter in stellar collapse when it is assumed that all nuclei can be represented by a single isotope. This single–nucleus approximation is the standard Ansatz. The authors compare the properties of a mixture that contains a realistic spread of nuclei with those that contain a single nucleus and show that despite the wide spread of nuclei possible, thermodynamic functions deviate little from their single–nucleus counterparts.

067.098 **Neutrino–pair bremsstrahlung in dense stars. IV. Phonon contributions in the crystalline lattice phase.**
N. Itoh, Y. Kohyama, N. Matsumoto, M. Seki.
Astrophys. J., Vol. 285, No. 1, p. 304 – 311 (1984).

Phonon contributions to the neutrino–pair bremsstrahlung in the crystalline lattice phase are calculated in the framework of the Weinberg–Salam theory for the dense matter relevant to the interior of white dwarfs and neutron stars. The frequency moment sum rules for the pure Coulomb bcc lattice are used to derive the neutrino energy loss rate. It is found that the phonon contribution to the neutrino–pair bremsstrahlung is comparable to that of the static lattice near the melting temperature.

067.099 **Comment on the absolute maximum temperature of thermal radiation.**
M. Nishioka.
Lett. Nuovo Cimento, Vol. 40, Ser. 2, No. 14, p. 443 – 445 (1984). Abstr. in Phys. Abstr., Vol. 87, No. 1243, Entry 101148 (1984).

067.100 **Distorted rotating black holes.**
A. Tomimatsu.
Phys. Lett. A, Vol. 103A, No. 8, p. 374 – 376 (1984). Abstr. in Phys. Abstr., Vol. 87, No. 1243, Entry 101404 (1984).

067.101 **Models of rapidly rotating neutron stars.**
J. L. Friedman, J. R. Ipser, L. Parker.
Nature, Vol. 312, No. 5991, p. 255 – 257 (1984).

Interest in rapidly rotating neutron stars has been reinforced by the discovery of the fast pulsar PSR 1937+214. The authors report results of the first numerical construction of rapidly rotating relativistic stars based on equations of state (EOS) proposed for neutron–star matter. Of nine EOS considered, none permits uniformly rotating equilibrium configurations for which the ratio $T/W$ of rotational energy to gravitational energy exceeds 0.12.

067.102 **A runaway instability in thick accretion disks?**
D. B. Wilson.
Nature, Vol. 312, No. 5995, p. 620 – 621 (1984).

It has been proposed that thick disks of material accreting around black holes may be subject to a runaway instability. If the mass of the disk exceeds a critical value ($m_{\text{disk}} > 0.01\ M_{\text{hole}}$), the disk drains into the hole on a dynamical time scale. The author reports a study of this effect, using models of non–self–gravitating disks in the Kerr metric, which enables proper allowance to be made for the effects of black–hole rotation and the increase in the spin of the hole as it accretes. It is concluded that there is no runaway instability in these systems.

067.103 **Massive black holes generated by cosmic strings.**
C. J. Hogan.
Phys. Lett. B, Vol. 143B, No. 1 – 3, p. 87 – 91 (1984). Abstr. in Phys. Abstr., Vol. 87, No. 1244, Entry 101727 (1984).

067.104 **Singular behavior of irregular wavefunctions in Schwarzschild electrodynamics.**
M.-d. Dong.
Phys. Lett. A, Vol. 104A, No. 1, p. 15 – 20 (1984). Abstr. in Phys. Abstr., Vol. 87, No. 1244, Entry 106022 (1984).

067.105 **Ultra high energy radiation from a black hole.**
J. Oliensis, C. T. Hill.
Phys. Lett. B, Vol. 143B, No. 1 – 3, p. 92 – 96 (1984). Abstr. in Phys. Abstr., Vol. 87, No. 1244, Entry 106023 (1984).

067.106 **Hawking radiation and the back reaction – a first approach.**
R. Balbinot.
Classical Quantum Gravity, Vol. 1, No. 5, p. 573 – 577 (1984). Abstr. in Phys. Abstr., Vol. 87, No. 1246, Entry 110942 (1984).

067.107 **Two black holes with axisymmetric parallel spins: initial data.**
J. Bowen, J. Rauber, J. W. York Jr.
Classical Quantum Gravity, Vol. 1, No. 5, p. 591 – 610 (1984). Abstr. in Phys. Abstr., Vol. 87, No. 1246, Entry 110943 (1984).

067.108 **Mining the Universe.**
P. C. W. Davies.
Phys. Rev. D, Vol. 30, No. 4, p. 737 – 742 (1984). Abstr. in Phys. Abstr., Vol. 87, No. 1246, Entry 115773 (1984).

067.109 **Nucleation rate for black holes.**
J. I. Kapusta.
Phys. Rev. D, Vol. 30, No. 4, p. 831 – 832 (1984). Abstr. in Phys. Abstr., Vol. 87, No. 1246, Entry 115774 (1984).

067.110 **Black holes in astrophysics.**
D. Dultzin-Hacyan.
Rev. Mex. Fis., Vol. 30, No. 3, p. 483 – 511 (1984). Abstr. in Phys. Abstr., Vol. 87, No. 1246, Entry 115775 (1984).

067.111 **Can stable black-hole dyons be observed?**
D. Lohiya.
Phys. Rev. D, Vol. 30, No. 6, p. 1194 – 1197 (1984). Abstr. in Phys. Abstr., Vol. 88, No. 1247, Entry 152 (1985).

067.112 **Thermal conductivity in two regions of a neutron star.**
A. Ray.
Thermal conductivity 16, p. 163 – 177 (1983). Abstr. in Phys. Abstr., Vol. 88, No. 1247, Entry 4789 (1985). – See Abstr. 012.054.

067.113 **Thermal convection in black hole accretion disks.**
D. G. Lominadze, G. D. Chagelishvili.
Sov. Astron., Vol. 28, No. 2, p. 168 – 172 (1984). English translation of 37.067.052.

067.114 **Stars with neutron cores. The possibility of the existence of objects with a low neutrino luminosity.**
G. S. Bisnovatyi-Kogan (*G. S. Bisnovatyj-Kogan*), S. A. Lamzin.
Sov. Astron., Vol. 28, No. 2, p. 187 – 193 (1984). English translation of 37.067.053.

067.115 **Le disque d'accrétion de LMC X-3.**
S. Pineault.
J. R. Astron. Soc. Can., Vol. 78, No. 5, p. 206 – 207 (1984). Abstract. – See Abstr. 010.401.

067.116 **The mathematical theory of black holes.**
S. Chandrasekhar.
General relativity and gravitation, p. 5 – 26 (1984). – See Abstr. 012.060.

067.117 **The motion of compact bodies and gravitational radiation.**
T. Damour.
General relativity and gravitation, p. 89 – 106 (1984). – See Abstr. 012.060.

This paper considers the problem of the gravitational interaction of two compact bodies (neutron stars or black holes). It outlines a new method where one matches an "external" gravitational field, obtained by iterating a Post-Minkowskian approximation scheme, to the field near each compact body. The results are compared with phenomena observed in the binary pulsar PSR 1913+16.

067.118 **Recent developments in the physics and astrophysics of black holes. Report of workshop B1 (*10th International Conference on General Relativity and Gravitation, Padua, July 3 – 8, 1983*).**
R. M. Wald.
General relativity and gravitation, p. 355 – 361 (1984). – See Abstr. 012.060.

067.119 **An origin for quasi-periodic variations in magnetic accreting compact objects.**
M. Livio.
Astron. Astrophys., Vol. 141, No. 1, p. L4 – L6 (1984).

The author proposes a mechanism, by which quasi-periodic luminosity variations can be obtained in accreting magnetic white dwarfs and neutron stars. The expected periods of these variabilities are estimated and an attempt is made to identify them with observed quasi-periodic oscillations. Predictions are made for quasi periodicities in two sources.

067.120 **Dynamics of a developing neutron star in general relativity.**
Yu. A. Berezin, O. E. Dmitrieva.
Sov. Astron. Lett., Vol. 10, No. 3, p. 175 – 176 (1984). English translation of 37.067.132.

067.121 **Exploding neutron stars in close binaries.**
S. I. Blinnikov, I. D. Novikov, T. V. Perevodchikova, A. G. Polnarev.
Sov. Astron. Lett., Vol. 10, No. 3, p. 177 – 179 (1984). English translation of 37.067.133.

067.122 **X-ray burst sources – from limit cycles to chaos.**
M. Livio, O. Regev.
Theoretical problems in stellar stability and oscillations, p. 459 – 465 (1984). – See Abstr. 012.065.

Accretion onto neutron stars and the resultant helium shell burning is described in terms of a simple two zone model. It is first demonstrated that at a constant accretion rate, limit cycle oscillations can be generated. The authors then introduce a feedback model which represents the impact of the burst luminosity on the accretion rate. The system is shown to undergo a sequence of period doublings. Above a certain accretion rate, a totally chaotic behavior, very similar to the one actually observed in X-ray burst sources, is obtained.

067.123 **Axion emission from neutron stars.**
N. Iwamoto.
Phys. Rev. Lett., Vol. 53, No. 12, p. 1198 – 1201 (1984).

The author shows that axion emission from neutron stars is the dominant energy-loss mechanism for a range of values of the Peccei-Quinn symmetry-breaking scale ($F$) not excluded by previous constraints. This gives the possibility of obtaining a better bound on $F$ from measurements of surface temperature of neutron stars.

067.124 **Accretion spin-up of low-magnetic neutron stars.**
V. M. Lipunov, K. A. Postnov.
Astrophys. Space Sci., Vol. 106, No. 1, p. 103 – 115 (1984).

The paper is concerned with the spin-up of low-magnetic neutron stars by the accretion of matter onto the star. Calculations have been made for the evolution of the rotation of a neutron star and applied to different stellar models. It is shown that the existence of a millisecond pulsar imposes no restriction on any of the equations of state considered. It is demonstrated that in the case of continued accretion onto a star, after it has acquired the critical angular frequency allowing no deviation from axial symmetry, specific accretion disks can be formed with a smooth transition into a star.

067.125 **Generation of gamma rays by accreting neutron stars.**
A. I. Tsygan.
Astrophys. Space Sci., Vol. 106, No. 1, p. 199 – 200 (1984).

The spherically symmetric accretion of matter onto a neutron star with a weak magnetic field is shown to be accompanied by the generation of gamma rays due to the Comptonization of X-rays from the neutron star on the flow of incident plasma.

067.126 **Linear momentum and gravitational waves: circular orbits around a Schwarzschild black hole.**
M. J. Fitchett, S. Detweiler.
Mon. Not. R. Astron. Soc., Vol. 211, No. 4, p. 933 – 942 (1984).

Perturbation theory is used to calculate the linear momentum flux in the gravitational waves emitted by a test particle moving on a circular geodesic in the Schwarzschild geometry. Comparisons with previous quasi-Newtonian calculations show good agreement when the radius of the test particle orbit is large and indicate that the quasi-Newtonian approach underestimates the linear momentum flux for smaller orbits. An estimate is made of the recoil velocity induced by a particle plunging into the black hole from the last stable circular orbit.

067.127 **Particle acceleration in compact binary stars.**
W. Kundt.
Adv. Space Res., Vol. 4, No. 2 – 3, p. 381 – 386 (1984). – See Abstr. 012.075.

There may be as many binary neutron stars in the Galaxy as there are pulsars. Their rotational energy can be converted into beams of cosmic rays if there is enough coupling between the corotating magnetosphere and the impinging plasma. Power-law spectra in energy are obtained from a power-law dependence of the accelerating fields. The observed cosmic-ray spectrum would result from a superposition of the injection by $\lesssim 10^6$ young binary neutron stars.

067.128 **Cosmic-ray acceleration by pulsars.**
F. C. Michel.
Adv. Space Res., Vol. 4, No. 2 – 3, p. 387 – 391 (1984). – See Abstr. 012.075.

With the discovery of the millisecond pulsars, the expectation that pulsars exist that are spinning close to breakup was confirmed. Thus the nominal pulsar parameters for spin rate ($\geqslant 6000$ rad/sec) magnetization ($\geqslant 10^{13}$gauss) and size ($\sim 10$ km) give pole-to-equator electric potentials $\gtrsim 3 \times 10^{20}$V, consistent with the highest observed cosmic ray energies. The author reexamines the possibility that pulsars are themselves the primary source of cosmic rays.

067.129 **Frictional acceleration of high-energy particles under astrophysical conditions.**
E. G. Berezhko.
Adv. Space Res., Vol. 4, No. 2 – 3, p. 405 – 407 (1984). – See Abstr. 012.075.

The process of frictional acceleration of high-energy charged particles caused by the presence of shearing flows in an electron-positron plasma of a pulsar wind is considered. It is shown that this process explains the high-energy part of electromagnetic emission from the Crab nebula.

067.130 **The angular momentum of a Kerr black hole.**
Y.-j. Li.
Chin. Astron. Astrophys., Vol. 8, No. 4, p. 277 – 282 (1984). English translation of Acta Astrophys. Sin., Vol. 4, No. 2, p. 142 – 151 (1984).

When calculating the angular momentum of a Kerr black hole, one should allow for the effect of accretion of scattered electrons in the retrograde direction. By using the radius of the smallest circular orbit of a particle as a measure for the capture cross-section by the hole, the calculation is greatly simplified. Thorne's result, $a/M = 0.9982$, when retrograde photons are considered, is easily obtained. If one allows for retrograde electrons, one gets $a/M = 0.9986$, for both retrograde electrons and photons: $a/M = 0.9968$ and for pair annihilation: $a/M = 0.9961$.

067.131 **Accretion onto magnetized neutron stars.**
J. G. Kirk.
MPA Rep., No. 158, 49 pp. (1984). Invited talk delivered at the 1984 meeting of the Astronomical Society of Australia.

Models of accretion onto neutron stars are discussed concentrating on the dynamics of the flow near the stellar surface. Ultimately, these models should reach the level of sophistication which enables them to predict the spectrum of radiation emitted by an accretion flow as well as the beam shape. Nevertheless, there are still basic questions – such as the existence of a limiting luminosity – which could be addressed by fairly simple models, and it is certainly desirable to have a thorough understanding of these before embarking on the inevitably expensive numerical quest for a solution of the full problem.

067.132 **2-dimensional low luminosity accretion onto neutron stars.**
J. G. Kirk.
MPA Rep., No. 159, 22 pp. (1984). Submitted to Astron. Astrophys.

067.133 **Lateral propagation of the helium shell flash on an accreting neutron star.**
T. Nozakura, S. Ikeuchi, M. Y. Fujimoto.
Astrophys. J., Vol. 286, No. 1, p. 221 – 231 (1984).

The propagation of a thermal wave caused by a helium flash on an accreting neutron star is investigated by a one-zone model. The model consists of equations for the energy balance and the mass balance. The energy equation includes the nuclear reaction, the radial heat flux, the lateral heat flux and the accretion rate in the mass equation. For the lateral heat flux the two limiting cases of (1) no convection at all, and (2) strong convection in lateral heat transfer, are examined.

067.134 **GTR corrections to the intensity of gravitational radiation of a fast-rotating magnetized neutron star.**
V. P. Tsvetkov.
Modern theoretical and experimental problems of the theory of relativity and gravitation, p. 173 (1984). In Russian. Abstr. in Ref. Zh., 51. Astron., 12.51.204 (1984). – See Abstr. 012.080.

067.135 **Interaction of $e^{\pm}$ with photons in magnetospheres of neutron stars.**
N. S. Kardashev, I. G. Mitrofanov, I. D. Novikov.
Astron. Zh., Tom 61, Vyp. 6, p. 1113 – 1124 (1984). In Russian. English translation in Sov. Astron., Vol. 28, No. 6.

The dependence of the force of radiative braking of relativistic electrons and positrons on their energy is studied for different elementary processes of interaction: Thomson, Compton, and cyclotron scattering and birth of $e^{\pm}$ pairs. It is shown that these processes may exert significant influence on the acceleration of electrons and positrons by the electric field in the magnetospheres of active neutron stars. The possibility of generation of hard gamma-ray radiation as a result of the processes considered is discussed and some astrophysical implications are presented.

067.136 **General relativity and gravitational collapse.**
C. V. Vishveshwara.
Gravitation and relativistic astrophysics, p. 43 – 45 (1984). Abstract. – See Abstr. 012.083.

067.137 **Core envelope models of collapsed objects.**
B. R. Iyer.
Gravitation and relativistic astrophysics, p. 47 – 57 (1984). – See Abstr. 012.083.

067.138 **Neutrinos in gravitational collapse.**
S. V. Dhurandhar.
Gravitation and relativistic astrophysics, p. 59 – 75 (1984). – See Abstr. 012.083.

This paper discusses the general relativistic effects on neutrinos in the background geometry relevant to spherical gravitational collapse.

067.139 **Accretion disks around compact objects with self-consistent electromagnetic fields.**
A. R. Prasanna.
Gravitation and relativistic astrophysics, p. 79 – 93 (1984). – See Abstr. 012.083.

This paper discusses the effects of self-generated electromagnetic fields in the dynamics of accretion disks around compact objects.

067.140 **Structure and stability of thick disks around black holes.**
D. K. Chakraborty.
Gravitation and relativistic astrophysics, p. 95 – 110 (1984). – See Abstr. 012.083.

067.141 **Extragalactic gravitational collapse.**
M. J. Rees.
Relativistic astrophysics and cosmology, p. 3 – 34 (1984). – See Abstr. 012.084.

After some introductory "numerology", routes towards black hole formation are briefly reviewed; some properties of black holes relevant to theories for active galactic nuclei are then described. Applications are considered to specific models for energy generation and the production of relativistic beams. The paper concludes with a discussion of extragalactic sources of gravitational waves.

067.142 **Disk accretion onto a weakly magnetized neutron star.**
R. Hoshi.
Publ. Astron. Soc. Jpn., Vol. 36, No. 4, p. 785 – 797 (1984).

The structure of an accretion disk surrounding a weakly magnetized neutron star is investigated in a framework of the conventional $\alpha$-model. The author assumes that the surface magnetic fields are so weak that the accretion disk extends down to the neutron star surface. He has found that the innermost region of the disk is optically thin. X-rays emerge from the inner portion of the optically thick disk which is located outside the optically thin inner region and from the neutron star surface.

067.143 **Effect of limb darkening on the radiation flux emitted from the inner region towards the outer region in accretion discs around black holes.**
C.-s. Zhu.
Acta Astron. Sin., Vol. 25, No. 2, p. 112 – 118 (1984). In Chinese.

067.144 **Influence of electron-positron pair production on a pulsar magnetospheric structure.**
Yu. A. Rylov.
Astrophys. Space Sci., Vol. 107, No. 2, p. 381 – 401 (1984).

A self-consistent pulsar magnetospheric model with electron-positron pair production is considered. The primary particles (electrons) are accelerated towards the neutron star and their curvature radiation generates electron-positron plasma near the neutron star. Inside an outflow channel, the generated plasma flows away from the pulsar magnetosphere. Some of the plasma electrons return and regenerate the plasma by their curvature radiation. It is shown that plasma production near the star causes an appearance of positron and electron equatorial belts. The plasma concentration and the flux of the returning electrons are estimated. The portion of the energy entering into the pulsar magnetosphere and its dependence on pulsar parameters are estimated.

067.145 **Limits from the soft X-ray background on the temperature of old neutron stars and on the flux of superheavy magnetic monopoles.**
E. W. Kolb, M. S. Turner.
Astrophys. J., Vol. 286, No. 2, p. 702 – 710 (1984).

Using recent observations of the diffuse, soft X-ray background, the authors obtain upper limits of 49 eV to the average surface temperature and $10^{32}$ergs $s^{-1}$ to the total photon luminosity of old ($\sim 10^{10}$yr) neutron stars in the Galaxy. If neutron stars are kept hot through monopole-induced nucleon decay, this limit corresponds to a monopole flux limit of $F_M \leqslant 3r \times 10^{-23}\mathrm{cm}^{-2}\mathrm{sr}^{-1}\mathrm{s}^{-1}$, where $r = L_{TOT}/L_\gamma$ is the ratio of the total luminosity to the photon luminosity of a neutron star.

067.146 **Analytical models of neutron star envelopes.**
L. Hernquist, J. H. Applegate.
Astrophys. J., Vol. 287, No. 1, p. 244 – 254 (1984).

The thermal structure of neutron star envelopes is investigated using approximate analytical formulae. In the degenerate layers the thermal structure equation is solved exactly for electron-dominated heat transport. In the nondegenerate layers it is shown that if the opacity is a power-low function of density and temperature, then the $T(\varrho)$ profiles lie along curves of constant thermal conductivity. The two solutions are matched at intermediate densities to give an approximate relation between the heat flux and the core temperature of the neutron star.

067.147 **The spectra of X-ray bursting neutron stars.**
R. A. London, R. E. Taam, W. M. Howard.
Astrophys. J., Lett. Ed., Vol. 287, No. 1, p. L27 – L30 (1984).

The authors present self-consistent atmospheric models for X-ray bursting neutron stars which include the effects of inelastic Compton scattering and free-free and bound-free emission and absorption. They show that the calculated spectra are harder than blackbodies of the same effective temperature. This spectral hardening effect is attributable to the reduction of the radiation source function by electron scattering, and the strong frequency dependence of the free-free opacity. The expected strength of iron absorption features and their use as diagnostics for the gravitational redshift at the neutron star surface are also discussed.

067.148 **The evolution of the inner regions of viscous accretion disks surrounding neutron stars.**
R. E. Taam, D. N. C. Lin.
Astrophys. J., Vol. 287, No. 2, p. 761 – 768 (1984). = Lick Obs. Bull., No. 998.

The structure and evolution of the inner regions of an accretion disk surrounding a neutron star has been investigated subject to variations in the mass flow rate. The authors show that under certain circumstances the role of nonlocal energy transfer in the radial direction can be important in stabilizing regions of the disk which are predicted to be unstable by a local analysis. However, in the case where the viscous stress scales with the total pressure, the global analysis confirms the results inferred from the local analysis, and the instability manifests itself in terms of short time scale ($<10$ s) luminosity fluctuations or bursts. The theoretical results bear suggestive resemblance to the general observational characteristics of the rapid X-ray burster.

**Die Physik der Zeitrichtung.**
See Abstr. 003.016.

**General relativity and relativistic astrophysics.**
See Abstr. 003.034.

**Magnetic fields in astrophysics.**
See Abstr. 003.045.

**Entropy.**
See Abstr. 014.012.

**Small black holes: ionization tracks and range.**
See Abstr. 014.013.

**Neutrinos in gravitational collapse: analysis of the flux profile.**
See Abstr. 061.006.

**Neutrino scattering on electrons in a strong magnetic field and the influence of this process on energy loss of neutron stars.**
See Abstr. 061.060.

**The production of very high energy photons and neutrinos from cosmic proton sources.**
See Abstr. 061.096.

**Effective Lagrangian for $\lambda\phi^4$ theory in curved spacetime with varying background fields: quasilocal approximation.**
See Abstr. 061.107.

**Relativistic electron transport in a quantizing magnetic field.**
See Abstr. 061.120.

**The line profile of a relativistic Keplerian ring in a massive neutrino halo.**
See Abstr. 061.121.

**Electrical and thermal conductivities of dense matter in the crystalline lattice phase.**
See Abstr. 061.123.

**Enhancement of thermonuclear reaction rate due to screening by relativistic degenerate electrons: long–range correlation effect.**
See Abstr. 061.124.

**Pair production rates in mildly relativistic, magnetized plasmas.**
See Abstr. 062.152.

**Nonthermal synchrotron radiation in a strong magnetic field.**
See Abstr. 063.041.

**Relativistic charge–separated winds.**
See Abstr. 064.035.

**On the overstability of axisymmetric oscillations in thin accretion disks.**
See Abstr. 064.067.

**On the value of anisotropy for the collapse of a massive star.**
See Abstr. 065.008.

**General relativistic collapse of rotating stars.**
See Abstr. 065.044.

**Core–collapse models of type II supernova explosions and neutron star formation.**
See Abstr. 065.045.

**Neutrino transport and shock propagation in the core collapse of a massive star.**
See Abstr. 065.047.

**The evolution of low–mass close binaries influenced by the radiation of gravitational waves and by a magnetic stellar wind.**
See Abstr. 065.060.

**Collapse and gravitation = the present situation.**
See Abstr. 065.079.

**The equation of state of hot dense matter.**
See Abstr. 065.082.

**A Thomas–Fermi equation of state of hot dense matter.**
See Abstr. 065.083.

**Non–radial oscillations of white dwarfs and neutron stars.**
See Abstr. 065.115.

**Rotationally stabilized stars at densities intermediate to white dwarfs and neutron stars.**
See Abstr. 065.117.

**The collapse of rotating stellar cores: equilibria between white dwarf and neutron star densities.**
See Abstr. 065.142.

**Magnetorotational iron core collapse.**
See Abstr. 065.143.

**Electron capture supernovae: one–zone collapse calculation for stars with masses 10 $M_\odot$ and 15 $M_\odot$.**
See Abstr. 065.144.

**The world of the Schwinger–DeWitt algorithm and the magical $a_2$ coefficient.**
See Abstr. 066.125.

**Gravitational redshift of gravitational clocks.**
See Abstr. 066.147.

**Linearized superhairs on multi–black–hole backgrounds.**
See Abstr. 066.171.

**Fourth–order gravity as general relativity plus matter.**
See Abstr. 066.207.

**General relativistic hydrodynamics and accretion physics: a numerical approach.**
See Abstr. 066.241.

**A further generalization of Vaidya's star metric.**
See Abstr. 066.248.

**Surface temperature of a rotating charged body.**
See Abstr. 066.249.

**Two mass solution of Einstein's vacuum equations: the double Kerr solution.**
See Abstr. 066.253.

**Gravitational radiation from colliding compact stars: hydrodynamical calculations in two dimensions.**
See Abstr. 066.291.

**Clues from the photonuclear time scale on the nature of particle accelerators in Cygnus X–3 and Vela X–1.**
See Abstr. 117.005.

**Sco X–1: a mini quasar?**
See Abstr. 117.007.

**Binary stars.**
See Abstr. 117.018.

**Search for relativistic companions in "non–X–ray" binary systems.**
See Abstr. 117.034.

**Two X–ray sources model of Cyg X–3.**
See Abstr. 117.035.

**Interpretation of spectroscopic variations in SS433.**
See Abstr. 117.040.

**A geometrical model for the SS433 system.**
See Abstr. 117.048.

**LMC X–3: a black hole in a neighbor galaxy.**
See Abstr. 117.053.

**A0538–66 during the extended 'off' state: evidence for continuing mass motions.**
See Abstr. 117.091.

**Photometry of A0538–66 during an active and subsequent inactive state.**
See Abstr. 117.092.

**New evidence on the nature of the neutron star and accretion flow in Vela X–1 from pulse timing observations.**
See Abstr. 117.114.

**Neutron stars in interacting binary systems.**
See Abstr. 117.125.

**Is Cygnus X–3 a monoenergetic $10^{17}$eV accelerator?**
See Abstr. 117.144.

**Cosmic–ray emission from Cygnus X–3.**
See Abstr. 117.145.

**Comments on "The white dwarf collapse in low mass binary systems".**
See Abstr. 126.016.

**Pulsar statistics and two types of pulsars.**
See Abstr. 126.017.

**The millisecond pulsar.**
See Abstr. 126.037.

**On the magnetization and origin of the millisecond pulsar PSR 1937+214.**
See Abstr. 126.038.

**Pulsar magnetospheres.**
See Abstr. 126.039.

**Neutronensterne und Supernova–Überreste. Neuere Ergebnisse der Pulsarforschung.**
See Abstr. 126.044.

**The magnetic field structure of PSR 0809+74.**
See Abstr. 126.045.

**Millisecond pulsars.**
See Abstr. 126.046.

**Models for the formation of binary and millisecond radio pulsars.**
See Abstr. 126.047.

**Energetics of the aligned pulsar magnetosphere.**
See Abstr. 126.050.

**Super–Eddington luminosities in X–ray bursts.**
See Abstr. 142.017.

**Neutron stars, black holes and Exosat.**
See Abstr. 142.039.

**HZ Her/Her X–1: an alternative model for the $35^d$ cycle?**
See Abstr. 142.048.

**Be/neutron star binaries: a relationship between orbital period and neutron star spin period.**
See Abstr. 142.050.

**Are X–ray bursts really of super–Eddington luminosities?**
See Abstr. 142.066.

**Discovery of absorption lines in X–ray burst spectra from X1636–536.**
See Abstr. 142.078.

**The production of gamma–rays in gamma–ray burst sources.**
See Abstr. 143.004.

**Binary star model of the γ–ray burst with special reference to that on 5 March, 1979 and subsequent bursts on 6 March, 4 April, and 24 April, 1979.**
See Abstr. 143.009.

**Resonant absorption, hot electrons, and cosmic gamma–ray bursts.**
See Abstr. 143.013.

**Physics of gamma–ray bursts.**
See Abstr. 143.015.

**On the nature of cosmic gamma–ray bursts.**
See Abstr. 143.017.

**Is Geminga a very close neutron star binary?**
See Abstr. 143.023.

**The theory of gamma–ray bursts.**
See Abstr. 143.030.

**Gamma–ray bursts: a model.**
See Abstr. 143.031.

**Fluxes and distances of gamma–ray bursts.**
See Abstr. 143.043.

**Generation of γ–bursts by old magnetic neutron stars.**
See Abstr. 143.044.

**Astrophysical boosters.**
See Abstr. 144.025.

**Astrophysical boosters.**
See Abstr. 144.027.

**Black hole remnants in globular clusters.**
See Abstr. 154.027.

**Black hole weighs in at one billion suns.**
See Abstr. 158.040.

**Models of active galactic nuclei.**
See Abstr. 158.085.

**Black hole nuclear cluster model of quasars and active galactic nuclei.**
See Abstr. 158.090.

**Comments on the physics of active galactic nuclei.**
See Abstr. 158.136.

**Seyfert nuclei as short–lived stochastic accretion events.**
See Abstr. 158.159.

**Particle acceleration in active galactic nuclei.**
See Abstr. 158.204.

**Some recent developments in problems of quasar like objects.**
See Abstr. 159.022.

**An accretion disk origin for the ultraviolet excess of luminous quasars: the confrontation of theory with observation.**
See Abstr. 159.057.

**Quark–hadron and chiral transitions and their relation to the early Universe.**
See Abstr. 161.013.

**Black–hole formation in the new inflationary universe.**
See Abstr. 161.162.

**Cosmic separation of phases.**
See Abstr. 161.166.

**Rotating systems in cosmological background.**
See Abstr. 161.333.

# Sun

## 071 Photosphere, Spectrum

071.001 **Analysis of Mn I lines in the solar spectrum using accurate oscillator strengths.**
A. J. Booth, D. E. Blackwell, M. J. Shallis.
Mon. Not. R. Astron. Soc., Vol. 209, No. 1, p. 77 – 81 (1984).
The authors present an analysis of both weak and strong lines of Mn I with well–defined profiles in the solar photosphere, using the accurate oscillator strengths measured at Oxford. In an LTE analysis with the Holweger and Müller model solar atmosphere a manganese abundance of $\log A = 5.39$ at a microturbulence value of $0.85\ \mathrm{km\ s^{-1}}$ is found for the disc centre.

071.002 **Recent improvements of the basic solar radiation data for λλ3300 – 12500 Å.**
H. Neckel.
Space Sci. Rev., Vol. 38, Nos. 3/4, p. 187 – 202 (1984).
This paper reviews shortly the most reliable observations, which are available at present to describe the basic characteristics of the solar radiation: absolute measurements of the disk–center intensity, high resolution spectra of disk–center intensity and irradiance, and scans of the center to limb variation at continuum wavelengths and for the line–averaged radiation. Further it reports recent results obtained by the merging of these observations, namely high precision data for disk–center intensity and irradiance, which concern not only the line–averaged radiation, but also the position of the continuum.

071.003 **Measurement of the solar spectral irradiance from 200 to 3000 nanometers.**
G. Thuillier, J. P. Goutail, P. C. Simon, R. Pastiels, D. Labs, H. Neckel.
Science, Vol. 225, No. 4658, p. 182 – 184 (1984).
The solar spectrum experiment on Spacelab 1 measured 98 percent of the sun's total energy output. It improved the absolute accuracy of solar irradiance data, especially in the ultraviolet and infrared regions. In order to detect any variation in the spectrum on future shuttle flights, the data were obtained in a radiation scale that can be preserved with high precision over many years.

071.004 **The pure rotation spectrum of OH and the solar oxygen abundance.**
A. J. Sauval, N. Grevesse, J. W. Brault, G. M. Stokes, R. Zander.
Astrophys. J., Vol. 282, No. 1, p. 330 – 338 (1984).
High–resolution solar observations of about 100 pure rotation lines of the $\upsilon = 0, 1, 2$, and 3 levels of the $X^2\Pi$ electronic state of OH have yielded a very accurate value of the solar abundance of oxygen, $\log N(\mathrm{O}) = 8.91 \pm 0.01$, in perfect agreement with that deduced from the two [O I] lines. Moreover, the present work strongly supports the recent OH electric dipole moment function of Werner, Rosmus, and Reinsch rather than other theoretical results.

071.005 **On the formation of faint Fraunhofer lines of the solar spectrum.**
B. T. Babij, M. M. Koval'chuk.
Astron. Zh., Tom 61, Vyp. 4, p. 771 – 777 (1984). In Russian. English translation in Sov. Astron., Vol. 28, No. 4.
A theoretical calculation of profiles of a number of faint Fraunhofer lines belonging to ions of Ce and Fe assuming pure absorption, coherent and noncoherent scattering is made. It is shown that individual accounting for various mechanisms of line formation leads to essential difference in theoretical contours. The conclusion is made that for bringing into agreement the theoretical and observed profiles of even faint Fraunhofer lines, it is necessary to take into account simultaneously the action of various mechanisms of line formation as well as the deviation of level populations from the equilibrium state.

071.006 **An additional identification of cyanogen lines in the solar spectrum in the wavelength interval from λ4145 Å to λ4190 Å. Some notes on Rowland's tables of 1928 and 1966.**
G. A. Porfir'eva.
Tr. Gos. Astron. Inst. Shternberg, Tom 54, p. 70 – 99 (1984). In Russian.
In Rowland's table of 1966 for the wavelength interval from λ4141 Å to λ4190 Å CN has been indicated in the identification of about 60 solar lines only. An additional identification of weak cyanogen lines has been fulfilled in the present paper. For 26 solar lines there has been confirmed the identification with CN, indicated in Rowland's table of 1928, but missed in Rowland's table of 1966. In the spectra of high resolution (Liège Solar Atlas was used) there were revealed many weak lines not indicated in Rowland's table of 1966. The wavelengths of these solar lines were approximately evaluated, about 30 have been identified with the CN molecule.

071.007 **An additional identification of cyanogen lines in the solar spectrum for the wavelength interval from λ4190 Å to λ4216 Å. Some notes to Rowland's tables of 1928 and 1966.**
G. A. Porfir'eva.
Tr. Gos. Astron. Inst. Shternberg, Tom 55, p. 102 – 113 (1983). In Russian.
An additional identification of weak CN lines has been made for the solar spectrum from 4190 Å to 4216 Å. Laboratory CN wavelengths, the Liège Atlas of the solar spectrum (1973) and Rowland's tables of the solar spectrum have been used. Theoretical equivalent widths of CN lines have been calculated for the centre of the solar disc, they were used to estimate line strengths in the solar spectrum. For 18 solar lines the CN identification indicated in Rowland's tables (1928) has been confirmed, but missing in Rowland's tables of 1966. 15 new lines seen in high resolution spectra have been identified with CN. All these lines are not indicated in Rowland's table of 1966.

071.008 **$^{24}\mathrm{MgH}^+$ in the solar photospheric spectrum.**
P. D. Singh.
Astrophys. Space Sci., Vol. 102, No. 2, p. 287 – 293 (1984).
The $^{24}\mathrm{MgH}^+$ $(A^1\Sigma^+ - X^1\Sigma^+)$ molecular lines have been identified in the photospheric spectrum. The rotational excitation temperature determined from the analysis of molecular line intensities of $^{24}\mathrm{MgH}^+$ is found to be of the order of 4850K which corresponds to the photospheric temperature of the Sun. The CNDO/2 dipole moments of $^{24}\mathrm{MgH}^+$ for internuclear distance range: (1.3 – 2.1) Å in the $X^1\Sigma^+$ state can be approximated by $M(R) = 4.92 + 1.33R$. Estimations for the spontaneous emission Einstein coefficients and the absorption oscillator strengths for the (1, 0), (2, 0) and (2, 1) transitions in the $X^1\Sigma^+$ state of the $^{24}\mathrm{MgH}^+$ ion are also made.

071.009 **The photosphere as the only possible source of Babcock's sunspots due to the field amplification by the differential rotation.**
E. Woyk.
XVI International Conference on Phenomena in Ionized Gases, p. 98 – 99 (1983). Abstr. in Phys. Abstr., Vol. 87, No. 1239, Entry 82462 (1984). – See Abstr. 012.017.

071.010 **Accurate atomic data and solar photospheric spectroscopy.**
N. Grevesse.
Phys. Scr., Vol. T8, p. 49 – 58 (1984). – See Abstr. 012.027.
This paper reviews the requests for high precision atomic data in studies of the solar atmosphere. A table is presented summarizing the presently accepted solar abundances and comparing these data with meteoritic values.

071.011 **On waves in the solar photosphere. I.**
M. B. Kerimbekov, K. V. Manujlov.
Dokl. Akad. Nauk AzSSR, Tom 9, No. 10, p. 53 – 56 (1983). In Russian. Abstr. in Ref. Zh., 51. Astron., 8.51.226 (1984).

071.012 **A study of solar photospheric limb–darkening variations.**
L. D. Petro, P. V. Foukal, W. A. Rosen, R. L. Kurucz, A. K. Pierce.
Astrophys. J., Vol. 283, No. 1, p. 426 – 438 (1984).
The authors have obtained regular observations of photospheric limb darkening, using the McMath Solar Telescope, to study possible slow changes in the global temperature structure, in $T_{eff}$, and in the ultraviolet continuum flux from the quiet Sun. This paper reports on the analysis of data obtained on 15 days between 1980 September and 1982 December in a continuum window at λ4451. There are no variations of global limb darkening exceeding 0.1% at the 99% confidence level. The implications of these measurements for slow changes in solar luminosity, convection zone structure, and ultraviolet flux are discussed.

071.013 **The displacement of the He I 58.4 nm line in the solar spectrum and the interpretation of observations of the interplanetary UV background.**
M. S. Burgin.
Astron. Zh., Tom 61, Vyp. 5, p. 874 – 881 (1984). In Russian. English translation in Sov. Astron., Vol. 28, No. 5.
The gravitational redshift and the correlation between the vertical velocity and the emissivity of the gas in the transition region between the chromosphere and the corona give rise to a displacement of the He I 58.4 nm line in the solar spectrum with respect to the resonant wavelength. The displacement probably exceeds 0.6 km/s and may be as large as several km/s. If the displacement is taken into account, the temperature and the velocity of the interstellar wind deduced from the observations of the backscattered radiation in the He I 58.4 nm line are lower than the values obtained when the displacement is assumed to be absent.

071.014 **On the Sun's limb brightening in the visible.**
G. H. Elste.
Sol. Phys., Vol. 93, No. 1, p. 15 – 21 (1984).
It is shown that in practice the method by Julius (1906) is incapable to determine the limb intensity drop. The brief intensity reversal near the extreme limb as derived from cinematography of flash spectra can be explained by diffraction.

071.015 **Asymmetries and wavelengths of solar spectral lines and the solar rotation determined from Fourier–transform spectra.**
H. Balthasar.
Sol. Phys., Vol. 93, No. 2, p. 219 – 241 (1984).
143 selected lines from Fourier–transform spectra obtained in 1980 and 1981 at the McMath–telescope of Kitt Peak National Observatory are used to determine wavelength shifts and asymmetries of solar spectral lines and the solar rotation. An equatorial rotation velocity of 1.971 km $s^{-1}$ is found. Using all 143 lines, no depth dependence of the solar rotation within the photosphere remains, in contradiction to a preliminary result of Balthasar (1983).

071.016 **Morphological study of the solar granulation. III. The mesogranulation.**
N. Oda.
Sol. Phys., Vol. 93, No. 2, p. 243 – 255 (1984).
Time–sequential high quality photographs of the photospheric granule on a quiet region of the disc center obtained at Pic–du–Midi Observatory by Kawaguchi are analyzed. The size variation of individual granules in the area of 54″ × 52″ on the photosphere are traced over a period of 4 min. The granules are classified according to their morphological features as follows: (1) Active granules, they repeat the expansion and the fragmentation. (2) Quiet granules, they do not alter the size noticeably during the observed time span. (3) Declining granules, they disappear without further fragmentation or merging.

071.017 **On changes from center to limb of the solar disk of central residual intensities of the Fraunhofer lines.**
B. T. Babij, R. E. Rykalyuk.
Vestn. L'vov. univ. Ser. astron., No. 58, p. 10 – 15 (1984). In Russian. Abstr. in Ref. Zh., 51. Astron., 9.51.310 (1984).

071.018 **On the influence of the selection of a model solar atmosphere on the theoretical profiles of absorption lines.**
M. M. Koval'chuk.
Vestn. L'vov. univ. Ser. astron., No. 58, p. 16 – 19 (1984). In Russian. Abstr. in Ref. Zh., 51. Astron., 9.51.311 (1984).

071.019 **Centre–to–limb behaviour of molecular rotational temperatures. III.**
K. Sinha.
Bull. Astron. Soc. India, Vol. 12, No. 2, p. 172 – 181 (1984).
Utilizing the KPNO atlas (Brault and Testerman, 1972) for the (0–0) band of the Swan system of the $C_2$ molecules, the author determines rotational temperatures $T_{rot}$, for $\mu = 1.0$ and $\mu = 0.2$. When saturation in the observed equivalent widths is accounted for, the model–based result for $T_{rot}$ comes out slightly lower than observations at a near limb position ($\mu = 0.2$). There is no evidence for an increase in temperature towards the solar limb discussed earlier (Sinha, 1979).

071.020 **Turbulence in the photosphere of the sun as a star. III. Micro–macroturbulence.**
V. A. Sheminova.
Soln. Dannye, Byull., 1984, No. 7, p. 70 – 78 (1984). In Russian.
Equalizing computed and observed equivalent widths of Fe I lines in the solar emission flux isotropic microturbulence of the sun as a star was found. Macroturbulence has been calculated from central intensities.

071.021 **Limits on photospheric Doppler signatures for solar giant cells.**
H. B. Snodgrass, R. Howard.
Astrophys. J., Vol. 284, No. 2, p. 848 – 855 (1984).
The authors analyze Mount Wilson solar velocity data taken since improvement of the spectrograph in 1982 May to search for photospheric traces of persistent velocity patterns that are anticipated in recent model predictions. The method involves time–averaged auto– and cross–correlations of the residuals that remain after least squares fits for differential rotation, limb shift, and meridional circulation are extracted from the daily magnetogram velocity arrays. The following limits are currently established through this analysis: (1) there is no sharply peaked power spectrum with amplitude above ~1 m $s^{-1}$ per wavenumber, and (2) there is no broad–band power spectrum for which the total integrated power is greater than ~10 $m^2 s^{-2}$.

071.022 **The solar abundance of erbium.**
E. Biemont, N. Y. Youssef.
Astron. Astrophys., Vol. 140, No. 1, p. 177 – 178 (1984).
A new value of the solar abundance of erbium is obtained from the analysis of 7 lines. The result $A_{Er} = 0.93 \pm 0.06$, is in fairly good agreement with the meteoritic result recently proposed by Anders and Ebihara (1982).

071.023 **Properties of solar magnetic fluxtubes as revealed by Fe I lines.**
S. K. Solanki, J. O. Stenflo.
Astron. Astrophys., Vol. 140, No. 1, p. 185 – 198 (1984).
The information contents in the solar spectrum for modelling of magnetic fluxtubes is explored by analysing the statistical properties of the Stokes *I* and *V* line profiles of 400 unblended Fe I lines. Methods of determining the temperature structure, photospheric magnetic field strength, magnetic filling factor, and microturbulence velocity are presented and used to provide estimates of these quantities in enhanced network and plage regions. Analysis of the magnetic line broadening shows that the magnetic field strength is approximately equal in network and plage regions, consistent with previous results. On the other hand, the temperature structures of the plage fluxtubes and the network elements are observed to be substantially different. The variation of the Stokes *V* line asymmetries with line strength are found to be similar in the different solar regions, indicating similar velocity structures in plages and network elements.

071.024 **Heights of faculae above the photosphere, based on observation of the eclipse of July 31, 1981.**
L. A. Akimov, I. L. Belkina, N. P. Dyatel.
Sov. Astron., Vol. 28, No. 3, p. 310 – 312 (1984). English translation of 37.071.025.

071.025 **Solar line asymmetry variations.**
D. H. Bruning.
Bull. Am. Astron. Soc., Vol. 16, No. 3, Part I, p. 730 (1984). Abstract. – See Abstr. 010.065.

071.026 **Variability of the quiet photospheric network.**
R. Muller, T. Roudier.
Sol. Phys., Vol. 94, No. 1, p. 33 – 47 (1984).
High-resolution photographs of the photospheric network taken in the Ca II K 3933 Å line and at λ4308 Å are analysed in order to study the variation, in latitude and over the sunspot cycle, of its density (the density is defined as the number of network elements – also called facular points – per surface unity). It appears that the density is not distributed uniformly at the surface. The density at the equator is varying in antiphase to the sunspot number. Density variations of the photospheric network express in fact variations of the quiet Sun magnetic flux. It thus results that the quiet Sun magnetic flux probably varies in antiphase to the flux in active regions. Finally, the variation over the solar cycle and the latitude distribution of photospheric network density are compared to those of X-ray bright points and ephemeral active regions.

071.027 **Limb effect of solar absorption lines. Observational method and results for Fe I 557.6 nm.**
B. N. Andersen.
Sol. Phys., Vol. 94, No. 1, p. 49 – 56 (1984).
The author has presented measurements of the solar limb shift of the magnetic insensitive Fe I 557.6 nm line with accuracies higher than previously obtained. Apart from an increase at $\mu = 0.85$ the appearance is a monotonic decrease in blueshift towards the limb. The total line shift at limb, corrected for solar rotation, is found to be equal at the poles and at the equator. The difference between the shapes of the limb shifts along the meridian and the equator is caused by either a modulation of the granular sizes, contrast, or velocity field with latitude or by the existence of a meridional flow from the poles and equator towards middle latitudes.

071.028 **Thermal bifurcation in the upper photosphere inferred from heterodyne spectroscopy of OH rotational lines.**
D. Deming, J. J. Hillman, T. Kostiuk, M. J. Mumma, D. M. Zipoy.
Sol. Phys., Vol. 94, No. 1, p. 57 – 74 (1984).
The authors have obtained low noise (S/N $> 10^3$), high spectral resolution ($\nu/\Delta\nu \approx 10^6$) observations of two pure rotation transitions of OH from the solar photosphere. The observations were obtained using the technique of optically null-balanced infrared heterodyne spectroscopy, and consist of center-to-limb line profiles of a $\upsilon = 1$ and a $\upsilon = 0$ transition near 12 μm. These lines should be formed in local thermodynamic equilibrium, and are diagnostics of the thermal structure of the upper photosphere. The data support a two-dimensional model in which horizontal thermal fluctuations of order $\pm$800K occur in the region $\tau_{5000} \approx 10^{-3} - 10^{-2}$. This thermal bifurcation may be maintained by the presence of magnetic flux tubes, and may be related to the solar limb extensions observed in the 30 – 200 μm region.

071.029 **On the change of the iron solar line profile λ5123.7 Å from centre to limb.**
B. T. Babij, E. K. Kokhan, M. M. Koval'chuk.
Astrometr. Astrofiz., Vyp. 53, p. 10 – 14 (1984). In Russian.
Observational and theoretical line profiles of Fe I λ5123.7 Å were compared for six positions on the solar disk. The profiles were calculated from three models of the solar atmosphere and for two mechanisms of line formation: LTE and coherent scattering. The scattering mechanism shows a better agreement with observations. The influence of the model on the profiles as compared to the influence of the mechanism of line formation is not important.

071.030 **Microturbulent velocity distribution in the solar photosphere.**
V. A. Sheminova.
Astrometr. Astrofiz., Vyp. 53, p. 15 – 19 (1984). In Russian.
Application of a new method for simultaneous calculations of equivalent widths and central depths of Fe I Fraunhofer lines permitted to obtain the distribution of velocities of moving small-size elements in the solar photosphere at altitudes of 150 – 550 km. The amplitude of the microturbulent velocity decreases with heights from 1.4 km $s^{-1}$ to 0.3 km $s^{-1}$ in a range of 150 – 350 km and increases with heights in a range of 350 – 550 km.

071.031 **Effect of deviations from LTE on weak Fe I lines.**
V. A. Sheminova, Yu. B. Matveev.
Astrometr. Astrofiz., Vyp. 53, p. 19 – 21 (1984). In Russian.
The equivalent widths and central depths of weak Fe I lines with low excitation potentials are shown to decrease by 10%, on the average, when deviations from LTE are taken into account in calculations of the line profiles. The iron abundance value obtained from the equivalent widths of these lines increases by 0.07 dex and becomes equal to $7.65 \pm 0.03$.

071.032 **An analysis of vibration–rotation lines of OH in the solar infrared spectrum.**
N. Grevesse, A. J. Sauval, E. F. van Dishoeck.
Astron. Astrophys., Vol. 141, No. 1, p. 10 – 16 (1984).
High resolution solar spectra have permitted the authors to measure equivalent widths of vibration-rotation lines of OH in the $X^2\Pi$ state near λ3 μm. The solar abundance of oxygen is $8.92 \pm 0.035$, as inferred from the analysis of 43 vibration-rotation lines and 81 pure rotation lines of the OH molecule.

071.033 **A test of the Planck–Larkin partition function.**
P. Magain.
Theoretical problems in stellar stability and oscillations, p. 234 – 238 (1984). – See Abstr. 012.065.
By comparing computed hydrogen line profiles with observed ones for the center of the solar disk, the author shows that the Planck–Larkin weight factors – sometimes used in the computation of the partition function – do not represent the actual level populations, at least in the solar photosphere.

071.034 **Centre-to-limb variation of CH line profile parameters.**
L. M. Punetha, G. C. Joshi.
Bull. Astron. Soc. India, Vol. 12, No. 3, p. 249 – 251 (1984).
Four CH line profiles at $\mu = 1.0$ and 0.2 were Fourier transformed. They show (1) turbulent velocity $\xi_t = 4.0$ km $s^{-1}$, and (2) an increase in the Doppler width parameter $\Delta\lambda_D$ by a factor of 1.03 in going from the centre to limb.

071.035 **The experimental solar curve of growth for Fe II with different sets of oscillator strengths.**
K. Avcioğlu, C. Bolcal, D. Koçer.
Rev. Fac. Sci. Univ. Istanbul, Ser. C, Vol. 46, p. 39 – 53 (1981). = Publ. Istanbul Univ. Obs., No. 123 (1984).
Three experimental curves of growth have been constructed for Fe II lines of the solar spectrum with different sets of oscillator strengths. The excitation temperatures derived from these curves are 6050K, 6100K and 5150K, respectively.

071.036 **The experimental solar curve of growth for neutral and single ionized titanium through different oscillator strength sets.**
K. Avcioğlu, C. Bolcal, D. Koçer.
Rev. Fac. Sci. Univ. Istanbul, Ser. C, Vol. 46, p. 135 – 150 (1981). = Publ. Istanbul Univ. Obs., No. 124 (1984).
Seven solar experimental curves of growth have been constructed for Ti I and Ti II, using seven different sets of oscillator strengths. Four of them have been constructed for Ti I and three for Ti II. Different excitation temperatures have been found to exist for both Ti I and Ti II as an influence of oscillator strengths on the same experimental curve of growth.

071.037 **The experimental curve of growth for neutral and single ionized cromium through different sets of oscillator strengths.**
K. Avcioğlu, C. Bolcal, D. Koçer.
Rev. Fac. Sci. Univ. Istanbul, Ser. C, Vol. 47, p. 23 – 30 (1982). = Publ. Istanbul Univ. Obs., No. 131 (1984).
Five experimental curves of growth have been constructed for Cr I and Cr II taking the sets of oscillator strengths from different sources. Two of them have been constructed for Cr I and three for the other. Derived excitation temperatures are different for both Cr I and Cr II as an influence of oscillator strengths on the experimental curve of growth.

071.038 **Oscillator strengths for neutral iron and iron abundance in the solar atmosphere.**
A. A. Boyarchuk, I. S. Savanov.
Rep. Ser., Dep. Phys. Sci., Univ. Turku, No. FTL–R68, p. 3 – 5 (1984). – See Abstr. 012.079.

071.039 **Observations of the He I 10830.38 Å line in the spectrum of the sun as a star.**
Eh. V. Kandrashov, I. I. Sokolova.
Astron. Tsirk., No. 1318, p. 6 – 8 (1984). In Russian.

071.040 **Erratum: "A photometric study of heat flow at the solar photosphere" [Astrophys. J., Vol. 281, No. 1, p. 442 – 454 (1984)].**
P. Foukal, L. Fowler.
Astrophys. J., Vol. 286, No. 1, p. 377 (1984). See Abstr. 037.071.031.

071.041 **Gibt es meridionale Strömungen des solaren Plasmas?**
G. Lustig, H. Wöhl.
Mitt. Astron. Ges., Nr. 62, p. 217 (1984). Abstract. – See Abstr. 010.341.

071.042 **Verlauf der kleinräumigen, vertikalen und horizontalen Geschwindigkeit in der tieferen Photosphäre.**
A. Nesis.
Mitt. Astron. Ges., Nr. 62, p. 218 (1984). Abstract. – See Abstr. 010.341.

071.043 **Numerische Simulation der Sonnengranulation.**
M. Steffen, D. Gigas.
Mitt. Astron. Ges., Nr. 62, p. 221 (1984). Abstract. – See Abstr. 010.341.

**log(*gf*) for singly–ionized elements of the iron group.**
See Abstr. 022.156.

**Separation of large–scale photospheric Doppler patterns.**
See Abstr. 036.067.

**Ground–based solar and stellar intensity measurements: practical limits of accuracy.**
See Abstr. 036.146.

**Model calculations of magnetic flux tubes. II. Stationary results for solar magnetic elements.**
See Abstr. 062.096.

**Rapid non–LTE calculations of Balmer lines and hydrogen ionization: the solar case.**
See Abstr. 063.073.

**On Bowen enhancement of the N III spectrum under solar and nebular conditions.**
See Abstr. 063.087.

**Processes observable in the photosphere during the formation of an active region. 3. Development of an active region at the edge of an older bipolar magnetic field.**
See Abstr. 072.007.

**Computation of solar magnetic fields from photospheric observations.**
See Abstr. 075.024.

**A solar spectral line list between 10 and 200 Å modified for application to high spectral resolution X–ray astronomy.**
See Abstr. 076.011.

**Limb darkening fluctuations: Is the five–minute oscillation also terrestrial?**
See Abstr. 080.030.

**Plasma motion in umbrae and the surrounding photosphere derived from spectroscopic Doppler measurements and tracer measurements of spots.**
See Abstr. 080.063.

**Solar rotation from a number of Fraunhofer lines.**
See Abstr. 080.064.

**Determination of the total amount of hydrogen atoms in a quiescent prominence.**
See Abstr. 080.066.

**Solar rotation rate from stable recurrent sunspot tracings.**
See Abstr. 080.088.

**Intensity oscillations in spectral lines in the Sun. Theory and observation.**
See Abstr. 080.104.

**Apparent solar temperature enhancement due to large–amplitude waves.**
See Abstr. 080.123.

**Comparison of UV spectra from solar–type stars.**
See Abstr. 114.083.

# 072 Sunspots, Faculae, Activity Cycles, Solar Patrol

072.001 **Solar faculae observations with an Opton coudé refractor.**
I. V. Mironova.
Astron. Tsirk., No. 1290, p. 3 – 5 (1983). In Russian.

072.002 **Elementary energetic model of a sunspot.**
A. A. Solov'ev.
Astron. Zh., Tom 61, Vyp. 4, p. 764 – 770 (1984). In Russian. English translation in Sov. Astron., Vol. 28, No. 4.

A simple energetic sunspot model is considered. According to the model the sunspot is a vertical magnetic cylinder embedded in the photosphere. The magnetic and gravitational energy is shown to make the main contribution to the sunspot energy balance. The thermal energy plays a role only at the first stage of the sunspot formation.

072.003 **Modeling of the 22–year cycle of solar activity in the framework of the dynamo theory with account for the primary field.**
M. I. Pudovkin, E. E. Benevolenskaya.
Astron. Zh., Tom 61, Vyp. 4, p. 783 – 788 (1984). In Russian. English translation in Sov. Astron., Vol. 28, No. 4.

The effect of a hypothetic quasi–stationary primary field of the sun on the behaviour of solar activity and background magnetic field during the 22–year solar cycle is considered. It is shown that the main features of the behaviour of the surface magnetic field of the sun as well as of the solar activity during the 22–year cycle may be accounted for by a primary solar field in Leighton's model, which is proved by the direct comparison with experimental results.

072.004 **Statistical analysis of the 11–year and 80–year solar cycles.**
M. Bielekova.
Sov. Astron., Vol. 28, No. 1, p. 96 – 100 (1984). English translation of 37.072.002.

072.005 **Resonance oscillations in sunspots.**
Yu. D. Zhugzhda.
Sov. Astron. Lett., Vol. 10, No. 1, p. 19 – 23 (1984). English translation of 37.072.001.

072.006 **Die Sonnenaktivität 1983.**
V. Gericke, E. Junker, K. Reinsch.
Sterne Weltraum, 23. Jahrg., Nr. 7, p. 382 – 384 (1984).

Bericht der SONNE–Beobachternetze "Relativzahlen", "Fackeln" und "Positionsbestimmung".

072.007 **Processes observable in the photosphere during the formation of an active region. 3. Development of an active region at the edge of an older bipolar magnetic field.**
V. Bumba, J. Suda.
Bull. Astron. Inst. Czech., Vol. 35, No. 4, p. 224 – 232 (1984). With plates 1 – 7.

Processes visible during several impulses of new sunspot formations at the periphery of an older large background magnetic field body, in which one rotation earlier the large June–July 1974 proton–flare region developed, are discussed. The coalescence of separate nuclei into the future umbra and the counter–clockwise rotation of the whole leading region, the change of the rotational sense during its further development, as well as the non–simultaneous development of individual parts of the active region are underlined.

072.008 **Number of spot groups formed and their average lifetime in solar activity cycle No. 20.**
M. Kopecký, F. Kopecká.
Bull. Astron. Inst. Czech., Vol. 35, No. 4, p. 233 – 240 (1984).

The paper represents a supplement to the values, published earlier, of the number of sunspot groups formed and their average lifetime for the 11–year cycle No. 20. It is shown that, in cycle No. 20, the average lifetime of sunspot groups decreased further as compared to cycles Nos. 19 and 18 and that the number of groups formed in cycle No. 20 proves the existence of a systematic increase in the number of sunspot groups formed over the last 100 years; this is probably associated with a solar activity variation of several hundred years and supports the authors' forecast of an abnormally high solar activity for the next century (Kopecký, 1983).

072.009 **Surveillance de l'activité solaire. Nombres relatifs de Wolf.**
M.–J. Martres.
Astronomie, Vol. 98, p. 460 – 462 (1984).

072.010 **A dynamo theory prediction for solar cycle 22: sunspot number, radio flux, exospheric temperature, and total density at 400 km.**
K. H. Schatten, A. E. Hedin.
Geophys. Res. Lett., Vol. 11, No. 9, p. 873 – 876 (1984).

Using the "dynamo theory" method to predict solar activity, a value for the smoothed sunspot number of $109 \pm 20$ is obtained for solar cycle 22. The predicted cycle is expected to peak near December, $1990 \pm 1$ year. Concommitantly, $F_{10.7}$ radio flux is expected to reach a smoothed value of $158 \pm 18$ flux units. Global mean exospheric temperature is expected to reach $1060 \pm 50$K and global total average total thermospheric density at 400 km is expected to reach $4.3 \times 10^{-15}$ gm cm$^{-3}$ $\pm 25\%$.

072.011 **On the cyclicity of solar activity over a long period.**
A. A. Zhdanov, V. M. Ostryakov.
Cosmic ray intensity and cosmogenic isotopes. 13th Leningrad seminar on cosmophysics, p. 68 – 74 (1983). In Russian. Abstr. in Ref. Zh., 51. Astron., 8.51.317 (1984). – See Abstr. 012.028.

072.012 **On the dynamics of solar activity in the epoch of the Maunder minimum.**
V. A. Vasil'ev, G. E. Kocharov.
Cosmic ray intensity and cosmogenic isotopes. 13th Leningrad seminar on cosmophysics, p. 75 – 100 (1983). In Russian. Abstr. in Ref. Zh., 51. Astron., 8.51.318 (1984). – See Abstr. 012.028.

072.013 **On the 11–year solar activity cycle during the last 100 years and in the epoch of the Maunder minimum.**
G. E. Kocharov, Yu. N. Markov, R. Ya. Metskhvarishvili, S. L. Tsereteli.
Cosmic ray intensity and cosmogenic isotopes. 13th Leningrad seminar on cosmophysics, p. 101 – 111 (1983). In Russian. Abstr. in Ref. Zh., 51. Astron., 8.51.319 (1984). – See Abstr. 012.028.

072.014 **Sunspot numbers: April 1984.**
Sky Telesc., Vol. 68, No. 1, p. 91 (1984).

072.015 **Sunspot numbers: May 1984.**
Sky Telesc., Vol. 68, No. 2, p. 187 (1984).

072.016 **Sunspot numbers: June 1984.**
Sky Telesc., Vol. 68, No. 3, p. 283 (1984).

072.017 **Die Sonnenfleckenrelativzahlen des S.I.D.C. für Mai 1984.**
Sterne Weltraum, 23. Jahrg., Nr. 7, p. 406 (1984).

072.018 **Die Sonnenfleckenrelativzahlen des S.I.D.C. für Juni 1984.**
Sterne Weltraum, 23. Jahrg., Nr. 8 – 9, p. 486 (1984).

072.019 **The Sun in 1983.**
B. E. Stonehouse.
South. Stars, Vol. 30, Nos. 7 – 8, p. 449 – 451 (1984).

072.020 **Sunspot numbers: July 1984.**
Sky Telesc., Vol. 68, No. 4, p. 377 (1984).

072.021 **Die Sonnenfleckenrelativzahlen des S.I.D.C. für Juli und August 1984.**
Sterne Weltraum, 23. Jahrg., Nr. 10, p. 544 (1984).

072.022 **The present sunspot cycle, No. 21.**
H. A. Luft.
J. Am. Assoc. Variable Star Obs., Vol. 12, No. 2, p. 78 (1983).
Abstract. – See Abstr. 010.021.

072.023 **Sunspot numbers: August 1984.**
Sky Telesc., Vol. 68, No. 5, p. 476 (1984).

072.024 **Jahresauswertungen 1980 bis 1982 der synoptischen Karten.**
J. Hoell, E. Junker, G. Schwaab.
Sonne, Jahrg. 8, Nr. 31, p. 133 – 136 (1984).

072.025 **Some statistical relationships and a forecast for the 11-year solar cycle.**
Yu. M. Slonim.
Soln. Dannye, Byull., 1984, No. 5, p. 78 – 83 (1984). In Russian.
A close statistical relationship has been established between the Wolf numbers 7 – 9 years prior to the cycle maximum and at the maximum. A close statistical relationship has been found between the interval separating the maxima of two neighbouring cycles and the duration of the descending branch of the first solar cycle. A method for forecasting the solar cycle is proposed on the basis of these two relationships.

072.026 **On the method of the Mountain Kislovodsk Station of the Pulkovo Observatory for the construction of "Synoptic Charts of Active Solar Features".**
V. I. Makarov.
Soln. Dannye, Byull., 1984, No. 6, p. 59 – 61 (1984). In Russian.

072.027 **On the absence of exterior and interior bright rings of sunspots.**
Kh. I. Abdusamatov.
Soln. Dannye, Byull., 1984, No. 6, p. 80 – 84 (1984). In Russian.
Preliminary results are given of two-dimensional photometric studies on the distribution of the intensity in the neighbouring photosphere of a sunspot as given by stratospheric observations. Isophotes on the exterior side of the penumbra have an irregular form and show the presence of bright and dark regions. Neither essential variations of intensity, lifetime or size of large-scale photospheric structures in dependence on the distance from the sunspot have been found.

072.028 **Solar luminosity fluctuations and active region photometry.**
G. A. Chapman, A. D. Herzog, J. K. Lawrence, J. C. Shelton.
Astrophys. J., Lett. Ed., Vol. 282, No. 2, p. L99 – L101 (1984).
The authors present monochromatic observations, obtained with a 512 element diode array, of the irradiance fluctuations of the sunspots and faculae of an active region during its disk transit in 1982 August. The maximum sunspot fluctuation, as a fraction of the quiet-Sun irradiance, is -800 parts per million (ppm). Faculae have a maximum irradiance fluctuation of about +200 ppm near the limbs. It is found that the facular energy excess is more than 50% of the sunspot energy deficit, which is $-5.8 \times 10^{35}$ ergs.

072.029 **Three-dimensional structures of two solar active regions from VLA observations at 2, 6, and 20 centimeter wavelengths.**
R. K. Shevgaonkar, M. R. Kundu.
Astrophys. J., Vol. 283, No. 1, p. 413 – 420 (1984). With plates 7 – 9.
Three-dimensional structures of two active region groups are determined from observations with the VLA at 2, 6, and 20 cm. One of the groups exhibits a single magnetic loop of length $\sim 10^{10}$cm. The 2 cm radiation is mostly thermal bremsstrahlung and originates from the footpoints of the loop. The 6 and 20 cm radiation is dominated by low-harmonic gyroresonance radiation and originates from the upper portion of the legs or the top of the loop.

072.030 **Sunspot atmosphere as derived from X-, EUV-, optical and radio observations.**
J. Staude, F. Fürstenberg, J. Hildebrandt, A. Krüger, J. Jakimiec, V. N. Obridko, M. Siarkowski, B. Sylwester, J. Sylwester.
Astron. Zh., Tom 61, Vyp. 5, p. 956 – 967 (1984). In Russian. English translation in Sov. Astron., Vol. 28, No. 5.
The lower umbral chromosphere was found to be best represented by a model similar to that of Teplitskaya et al. (1978). This model has been extended to higher levels by a steep gradient of the temperature $T$ reaching values of $T \approx 40000$K and electron density $n_e \approx 4 \times 10^{10}$ cm$^{-3}$ at a height of $z \approx 2000$ km above the umbral photosphere; these values are mainly suggested by EUV data from HRTS. At higher levels, at least two components must be assumed: the hot component occupying a fraction $\alpha \sim 0.8$ to 0.9 of the volume involved has a shallow transition layer and coronal values of $T \approx 1.8 \times 10^6$K and $n_e \approx 5 \times 10^8$ already at $z = 3000$ km to 5000 km.

072.031 **Comparative spectral characteristics of stars and sunspots.**
O. G. Badalyan, V. N. Obridko.
Astron. Zh., Tom 61, Vyp. 5, p. 968 – 975 (1984). In Russian. English translation in Sov. Astron., Vol. 28, No. 5.
A comparative analysis of the spectra of dwarf stars and sunspots has been carried out. The spectral class of a sunspot has been shown to depend on the index used: the sunspot has M0 class in the continuum approximately K5 in weak lines and G8 – K0 in strong lines. It infers that the atmosphere of a sunspot cannot be approximated by an average standard atmosphere of a star of any spectral class.

072.032 **On stationarization of monthly fluctuations of heliophysical indices.**
K. S. Vojchishin, M. I. Stodilka.
Astron. Zh., Tom 61, Vyp. 5, p. 976 – 984 (1984). In Russian. English translation in Sov. Astron., Vol. 28, No. 5.
A method of stationarization of a high-frequency fluctuating component, conserving its micro- and macrostructure, is proposed. The results of application of this method to the equalization of the monthly fluctuation dispersion of the heliophysical indices are given.

072.033 **On the correlation of longitudinal and latitudinal motions of sunspots.**
P. A. Gilman, R. Howard.
Sol. Phys., Vol. 93, No. 1, p. 171 – 175 (1984).
Using new measurements of positions of individual sunspots and sunspot groups obtained from 62 years of the Mt. Wilson white-light plate collection, the authors have recomputed the correlation between longitude and latitude motion. Their results for groups are similar to those of Ward (1965) computed from the Greenwich record, but for individual spots the covariance is reduced by a factor of about 3 from the Ward values. The authors conclude that there is a real correlation between longitude and latitude movement of individual spots, implying angular momentum transport toward the equator as inferred by Ward. The two thirds reduction in the covariance for individual spots as opposed to groups is probably due to certain properties of spot groups, as first pointed out by Leighton.

072.034 **Evidence for the 22-year-cycle in the longitudinal distribution of sunspots.**
H. Balthasar, M. Schüssler.
Sol. Phys., Vol. 93, No. 1, p. 177 – 179 (1984).
It was shown in an earlier paper that preferred hemispheres of solar activity alternate with the 22-year magnetic cycle, when analyzed in the 27.0 day Bartels rotation. Using data which cover the time between 1818 and 1983 the authors trace back this result

to 1880 (cycle 12). Before 1880 no significant correlations are found.

072.035 **Sunspot activity according to Greenwich observations.**
M. Kopecký.
Sol. Phys., Vol. 93, No. 1, p. 181 – 187 (1984).
This article presents a survey of studies incorporating methods of evaluation of Greenwich observations. It also describes some extreme sunspot groups (i.e., groups with a very large area, a very long life, and/or in very high heliographic latitudes); data on the total areas of the largest sunspots are given, and a survey of the parameters concerning individual 11–year cycles, resulting from Greenwich Photoheliographic Results, is presented.

072.036 **Long–term fluctuations of solar activity during the last thousand years.**
L. Křivský.
Sol. Phys., Vol. 93, No. 1, p. 189 – 194 (1984).
Long–term variations of solar activity during the last thousand years were derived from historical observations (by the naked eye) of large sunspots and from records of aurorae at latitudes $< 55°$. The results shown indicate very good agreement as regards the occurrence of maxima and minima of all three indices (sunspots, aurorae, $^{14}C$), and even between secondary fluctuations of the $^{14}C$–curve, which were usually neglected in the presentation of the smoothed curve.

072.037 **On the diagnostic of magnetic fields in sunspots through the interpretation of Stokes parameters profiles.**
M. Landolfi, E. Landi Degl'Innocenti, P. Arena.
Sol. Phys., Vol. 93, No. 2, p. 269 – 287 (1984).
The inversion routine proposed by Auer et al. (1977), for the determination of vector magnetic fields from Stokes profiles, has been generalized to include magneto–optical and damping effects. Synthetic profiles have been generated from a sunspot model atmosphere accounting for the depth variation of the relevant physical parameters such as the magnetic field amplitude, inclination angle, etc. Alfvén waves and magnetic inhomogeneities over the field of view have also been considered.

072.038 **Broad–band circular polarimetry of sunspots, 0.4–1.7 microns: spatial scans with a 3.4 arc sec diameter aperture.**
G. D. Henson, J. C. Kemp.
Sol. Phys., Vol. 93, No. 2, p. 289 – 299 (1984).
Spatial scans with a resolution of 3.4 arc sec of the broad–band circular polarization of several sunspots have been made in five filter bands over the wavelength range 0.4–1.7μ with a sensitivity of $1 \times 10^{-6}$ fractional polarization. The scans revealed two important features: (1) The broad–band circularly polarized flux $V$ reverses in sign, or diminishes to near zero, at the center of the umbral region relative to the outer penumbra. (2) In all cases the limbward penumbral region exhibited stronger $V$ values than did the disk–center (inward) side of the spot, at all observed wavelengths. Such previously unreported structure in the magnetic circular polarization of sunspots provides new clues for understanding the 'anomalous' large broad–band polarization at short wavelengths and at the same time imposes new constraints on sunspot models.

072.039 **Sharp edges in solar microwave spectra: neutral current sheets or cyclotron lines?**
E. J. Schmahl, R. K. Shevgaonkar, M. R. Kundu, D. McConnell.
Sol. Phys., Vol. 93, No. 2, p. 305 – 315 (1984).
The authors have mapped two solar active regions using the VLA at three closely spaced frequencies (4496, 4716, and 4996 MHz) in an attempt to determine the origin of the steep spectra (index $\gamma \approx -5$ to $-8$) sometimes observed with large single telescopes. One of the regions observed indeed shows an anomalously large slope ($\gamma \approx -6$) compared to the usual ($\gamma \approx -2$ to $-2.5$). The other region shows a similar slope ($\gamma \approx -5$) but with a larger range of statistical error. Two possible explanations for such steep edges in solar spectra are (1) transmission effects of neutral current sheets, and (2) the appearance of cyclotron lines. The internal evidence of the microwave maps and simultaneous optical observations favor an explanation in terms of cyclotron lines.

072.040 **A possible relationship between spectral bands in sunspot number and the space–time organization of our planetary system.**
H. Schwentek, W. Elling.
Sol. Phys., Vol. 93, No. 2, p. 403 – 413 (1984).
For the particular purpose of this paper, Zürich relative sunspot numbers of the time spans 1749 – 1982, 1749 – 1865, and 1866 – 1982, have been analysed anew by two different methods. It is shown that the spectral bands in the power spectra of sunspot numbers between 1 and 234 years obtained from these analyses can be clearly related to the modified configuration frequencies of the giant planets and their harmonics. In particular, the clearly dominant spectral band in sunspot number, the solar cycle of 10.8 years, is given by the configuration period of Jupiter and Saturn (19.859 yr) times the ratio of their distances from the Sun (0.545).

072.041 **Frequent ultraviolet brightenings observed in a solar active region with Solar Maximum Mission.**
J. G. Porter, J. Toomre, K. B. Gebbie.
Astrophys. J., Vol. 283, No. 2, p. 879 – 886 (1984).
Observations in the ultraviolet of sites of enhanced intensity within an active region on the Sun reveal frequent and rapid brightenings in Si IV and O IV line emission. These transition region lines were observed with 0.08 s sampling in time using the ultraviolet spectrometer and polarimeter instrument on the SMM satellite. The observations show that intermittent heating events of modest amplitude are occurring at many sites within an active region. The results suggest that heating due to magnetic field reconnection within an active region is proceeding almost stochastically; events involving only a modest release of energy occur the most frequently.

072.042 **Typisation of the 11–year cycles of solar activity according to the principle of similarity.**
P. M. Mushenko.
Leningr. gidrometeorol. inst. Leningrad, 8 pp. (1984). In Russian. Abstr. in Ref. Zh., 51. Astron., 9.51.431 (1984).

072.043 **Sunspot pair formation by the photospheric dynamo process.**
S.–I. Akasofu.
Planet. Space Sci., Vol. 32, No. 10, p. 1257 – 1261 (1984).
It is suggested that the formation of a pair of sunspots results from amplification of the weak, large–scale fields in the photosphere, rather than from emergence of a hypothetical magnetic flux "tube" from beneath the photosphere.

072.044 **An essay on sunspots and solar flares.**
S.–I. Akasofu.
Planet. Space Sci., Vol. 32, No. 11, p. 1469 – 1496 (1984).
The author reviews some of the recent findings on large–scale magnetic fields and sunspots. Then, instead of relying on the hypothetical flux tube beneath the photosphere, he considers an amplification process of the observed large–scale fields by a dynamo process on the basis of the observed and possible photospheric shear flows. Thus, the photosphere is considered as an active medium, rather than the passive medium through which the hypothetical flux tube merely penetrates. Specifically, the author considers the dynamo process associated with vortex motions which can supply the power needed for the formation of sunspots from the observed weak field and the power needed for solar flares.

072.045 **Types of sunspot groups occurring in the course of solar cycle No. 20.**
Š. Knoška, L. Křivský.
Bull. Astron. Inst. Czech., Vol. 35, No. 5, p. 271 – 272 (1984).
The occurrence of Zurich–type sunspots is investigated year by year throughout the whole solar cycle No. 20 (1965 – 1976) and

the percentage occurrence of the individual types is determined for the cycle as a whole.

072.046 **On the dynamics of active longitudes of sunspots in the 21st solar cycle.**
Yu. I. Vitinskij, E. V. Miletskij.
Soln. Dannye, Byull., 1984, No. 7, p. 85 – 89 (1984). In Russian.
The cross–correlation of active longitudes of sunspot areas in the 21st solar cycle is considered. No significant correlation between active longitudes of the northern hemisphere was found. In the southern hemisphere a close correlation with a time lag of 4 solar rotations was detected between the active longitudes 280° – 240° and 360° – 320°.

072.047 **Strong fluctuations of the total areas of sunspot umbrae and penumbrae for 1879 – 1976.**
Yu. I. Vitinskij.
Soln. Dannye, Byull., 1984, No. 8, p. 79 – 84 (1984). In Russian.
Lists of strong positive and negative fluctuations of the total areas of sunspot umbrae and penumbrae for 1879 – 1976 as detected from Greenwich data by the method previously proposed by the author (1980) are given. A list of strong fluctuations of the total areas of sunspots for 1964 – 1976 is also presented.

072.048 **Large–scale structure of a sunspot and the surrounding photosphere from stratospheric observations.**
Kh. I. Abdusamatov.
Soln. Dannye, Byull., 1984, No. 8, p. 84 – 89 (1984). In Russian.
Preliminary results of studies of photometric properties of the large–scale structure of a sunspot and the surrounding photosphere and their variations with time are given.

072.049 **On the long–period changes of solar activity.**
J. Mergentaler.
Acta Astron., Vol. 34, No. 2, p. 263 – 267 (1984).
A transformation of the curve of variation of Wolf numbers during the solar activity cycle was done. The obtained curve was described by means of the third power polynomial whose coefficients were calculated for twenty cycles (1755 – 1976). A circa 180 years periodicity of coefficients $a$ and $b$ was found but no 90 years periodicity. A simple attempt to forecast the time of the minimum of the XXIst cycle was undertaken.

072.050 **Die Sonnenfleckenrelativzahlen des S.I.D.C. für September 1984.**
Sterne Weltraum, 23. Jahrg., Nr. 11, p. 606 (1984).

072.051 **Sunspot numbers: September 1984.**
Sky Telesc., Vol. 68, No. 6, p. 582 (1984).

072.052 **Center–to–limb variation of a sunspot–associated microwave source.**
C. E. Alissandrakis, M. R. Kundu.
Astron. Astrophys., Vol. 139, No. 2, p. 271 – 284 (1984).
The authors present the analysis of observations of a simple bipolar active region (McMath 16862) at 6.16 cm during six consecutive days (May 22 – 27, 1980). The radio emission consists of two bright sources associated with the two main spots of the region, as well as some weaker emission. The structure of one sunspot–associated source, as well as its center–to–limb variation is compared with model computations.

072.053 **Description of sunspot cycles by orthogonal functions.**
D. L. Teuber, E. J. Reichmann, R. M. Wilson.
Astron. Astrophys., Vol. 139, No. 2, p. 450 – 454 (1984).
Based on the principal component analysis technique and evidence for a 22–yr double–sunspot cycle periodicity, the authors represent the time series of sunspot numbers as a sum of mutually orthogonal eigenvectors in the time domain. They show that the first two eigenvectors account for about 90% of the cumulative "signal power", and that this is sufficient for reconstruction of the raw data curve. They also note that the second eigenvector behaves as the time derivative of the first, and that a phase–plane plot of these eigenvectors suggests that the Sun's sunspot cycle is driven by an oscillator; the implication is that, embedded within the Sun, a chronometer is at work (e.g. Dicke, 1979).

072.054 **L'activité solaire.**
M.-J. Martres, G. Zlicaric.
Astronomie, Vol. 98, p. 362 – 363, 416 – 417, 472 – 473, 584 – 585 (1984).

072.055 **Rotationsverhalten der wiederkehrenden Sonnenflecken von 1948 bis 1976.**
G. Lustig.
Sterne, 60. Band, Heft 5, p. 295 – 301 (1984).

072.056 **Dynamical phenomena in sunspots. I. Observing procedures and oscillatory phenomena.**
J. H. Thomas, L. E. Cram, A. H. Nye.
Astrophys. J., Vol. 285, No. 1, p. 368 – 380 (1984).
This series of papers describes the results of an observing program at Sacramento Peak Observatory designed to obtain a long (1 hr or more) time sequence of high–resolution spectral observations of dynamical phenomena in a large sunspot. These observations comprise simultaneous photographic spectra in several spectral lines that probe a range of heights in the sunspot atmosphere. The present paper describes the observational method and reports results related to oscillatory phenomena in the sunspots, including umbral oscillations and flashes, penumbral waves, 5 minute oscillations, and a new phenomenon of repeated brightenings in an umbral light bridge.

072.057 **Dynamical phenomena in sunspots. II. A moving magnetic feature.**
A. H. Nye, J. H. Thomas, L. E. Cram.
Astrophys. J., Vol. 285, No. 1, p. 381 – 385 (1984).
High–resolution observations of the magnetic field, vertical velocity, and chromospheric intensity in a typical moving magnetic feature (MMF) are obtained over a period of 1 hr as it moves outward across the moat surrounding a large sunspot. The area of the MMF increases and the magnetic field strength decreases as it moves outward. The MMF contains a strong downdraft with a velocity of 400 m $s^{-1}$ in the photosphere. The lifetime assigned to the MMF depends upon the signature by which it is observed.

072.058 **A spot check. Does the sun change its spots?**
S. A. Haines.
Sci. Dimension, Vol. 16, No. 2, p. 27 – 29 (1984). Abstr. in Phys. Abstr., Vol. 87, No. 1244, Entry 105939 (1984).

072.059 **Solar active regions: observation and theory.**
C. J. Durrant.
Orages solaires de type I, p. 21 – 33 (1984). – See Abstr. 012.055.

072.060 **Preliminary observations on the energy budget of a solar activity complex, July – September 1982.**
G. A. Chapman, A. D. Herzog, J. K. Lawrence, K. Eskenas, C. Mallory, J. C. Shelton.
Bull. Am. Astron. Soc., Vol. 16, No. 3, Part I, p. 729 (1984). Abstract. – See Abstr. 010.065.

072.061 **Umbral oscillations measured in the Stokes–V inversion point.**
H. Balthasar, E. Wiehr.
Sol. Phys., Vol. 94, No. 1, p. 99 – 103 (1984).
The inversion point of the circular Zeeman polarization profile (V–Stokes) parameter is used to observe umbral Doppler oscillations free from disturbing influences of parasitic light. In a second step, purely umbral lines are used to avoid remaining influences from the V–profile of the (oscillating) penumbra. Among a total of nine sunspot umbrae, three exhibit oscillations within the various 1.5 to 2.5 hr samples. The periods differ significantly from 300 s, vary with time, and occur within time intervals of high tranquility thus explaining the lack of oscillations in the remaining sunspots.

072.062 **Dynamic evolution of recurrent mass ejections observed in Hα and C IV lines.**
B. Schmieder, P. Mein, M. J. Martres, E. Tandberg–Hanssen.
Sol. Phys., Vol. 94, No. 1, p. 133 – 150 (1984).

During a coordinated SMY program, the consecutive formation of two new active centers merging together within AR 2646 was observed from 28 August, to 5 September, 1980. The two preceding spots compressed an inverse polarity spot on 1 September 1980, causing recurrent ejecta of matter. The authors focus their analyses on observations in the chromospheric Hα line and the transition region C IV 1548 Å line. They discuss the instruments and the geometric evolution of the active region, then the mass ejection. They study the dynamic evolution of Hα matter during and after the ejection, considering the Hα absorbing structure as a cloud "overlying the chromosphere".

072.063 **The influence of penumbral fine structures on line profiles.**
E. Wiehr, A. Koch, M. Knölker, G. Küveler, G. Stellmacher.
Astron. Astrophys., Vol. 140, No. 2, p. 352 – 356 (1984).

The two magnetically insensitive lines Ti 5222.7 and $Fe^+$5264.8 are investigated from a very highly resolved penumbra spectrum. The highly excited $Fe^+$ line shows a conspicuous relation of the line asymmetry to bright and of the line core shift to dark spectrum streaks. The low excited Ti line occasionally shows a reversed asymmetry in bright streaks, where $Fe^+$ shows right–sense asymmetry. A two component model of sunspot penumbrae seems doubtful. Indication is found that the Evershed asymmetry originates mainly from a fine structure mixing.

072.064 **Time–dependence of umbral oscillations.**
D. Soltau, E. Wiehr.
Astron. Astrophys., Vol. 141, No. 1, p. 159 – 161 (1984).

Doppler shifts of the magnetically insensitive line Fe 7090.4 are observed in two sunspot umbrae on three days over time intervals up to 8.5 h. The temporal power spectra show a large variety of periods whose pattern reoccurs periodically. Neither the 300 s nor the 180 s periods are preferred umbral modes. Conspicuous power near 0.8 mHz closely corresponds to the known ~20 min time scale of umbral dots.

072.065 **Solar activity. 1982 April – 1982 December.**
F. G. Mustaeva.
Tsirk. Astron. Inst. Tashkent, Nos. 109, 110, 112 (1983). In Russian.

072.066 **Provisional sunspot–numbers for June – November 1984.**
Yamamoto Circ., Nos. 2018, 2020, 2022, 2026, 2029, 2030 (1984). In Japanese.

072.067 **Provisional sunspot–numbers for July – December 1984.**
M. Dillig.
Sonne, Suppl., 7 pp. (1984).

072.068 **On the height of formation of H and K Ca II lines in an umbra.**
R. B. Teplitskaya, S. A. Grigor'eva.
Issled. Geomagn. Aehron. Fiz. Solntsa, Moskva, No. 68, p. 12 – 21 (1984). In Russian. Abstr. in Ref. Zh., 51. Astron., 11.51.362 (1984).

072.069 **On the spatial structure of a penumbra.**
R. B. Teplitskaya.
Issled. Geomagn. Aehron. Fiz. Solntsa, Moskva, No. 68, p. 3 – 12 (1984). In Russian. Abstr. in Ref. Zh., 51. Astron., 11.51.364 (1984).

072.070 **Investigation of oscillation processes in an umbra from the H Ca II lines with photoelectric methods.**
I. P. Turova, S. A. Druzhinin.
Issled. Geomagn. Aehron. Fiz. Solntsa, Moskva, No. 68, p. 66 – 72 (1984). In Russian. Abstr. in Ref. Zh., 51. Astron., 11.51.366 (1984).

072.071 **Cellular structures of an activity complex, their influence on the development of large solar flares.**
V. G. Banin.
Issled. Geomagn. Aehron. Fiz. Solntsa, Moskva, No. 68, p. 73 – 83 (1984). In Russian. Abstr. in Ref. Zh., 51. Astron., 11.51.379 (1984).

072.072 **Radial velocity fluctuations in an umbra from measurements in two sections of the 5434.5 Å line profile.**
S. A. Druzhinin.
Issled. Geomagn. Aehron. Fiz. Solntsa, Moskva, No. 68, p. 111 – 118 (1984). In Russian. Abstr. in Ref. Zh., 51. Astron., 11.51.391 (1984).

072.073 **Some results of investigations of the 11– and 22–year cycles of Wolf numbers.**
G. V. Kuklin.
Issled. Geomagn. Aehron. Fiz. Solntsa, Moskva, No. 68, p. 45 – 56 (1984). In Russian. Abstr. in Ref. Zh., 51. Astron., 11.51.434 (1984).

072.074 **The mean Evershed flow.**
W.–r. Hu.
Chin. Astron. Astrophys., Vol. 8, No. 3, p. 194 – 200 (1984). English translation of Acta Astrophys. Sin., Vol. 4, No. 1, p. 33 – 42 (1984).

An explanation for the mean Evershed flow in sunspots is given. Because the inertial force of the Evershed flow is small, one can divide the relevant dynamical process into a base state and a perturbation. The base state solution gives the equilibrium relations between the pressure gradient, Lorentz force and gravity and the law of flow. The flow in the theoretical model agrees with the observed Evershed flow. The perturbation describes the force driving the Evershed flow and a specific case is discussed.

072.075 **Model dependence of the rotational temperature of SiO in sunspots.**
L. M. Punetha, G. C. Joshi.
Bull. Astron. Soc. India, Vol. 12, No. 3, p. 229 – 232 (1984).

Centre–to–limb variation of the rotational temperature of the fundamental band of SiO in different sunspot models has been calculated. The possibility of using these calculations to pick up the true model on the basis of infrared observations is indicated.

072.076 **Sunspot continuum from 1 to 10 μm and the umbral temperature models.**
L. M. Punetha, G. C. Joshi.
Bull. Astron. Soc. India, Vol. 12, No. 3, p. 233 – 236 (1984).

Infrared flux in Zwaan's (1974) and Boyer's (1980) sunspot models has been calculated in the wavelength range 1 to 10 μm, taking into account the molecular line haze opacity (Joshi et al. 1979). The calculations are compared with Sitnik's (1980) observations in the available wavelength region 1 to 2.1 μm and corrections in the corresponding models are suggested.

072.077 **Solare Beobachtungsergebnisse. Solar data. Solar radio emission. October – December 1983.**
HHI Sol. Data, Vol. 34, p. 97 – 123 (1983).

072.078 **Magnetic polarities and maximum field strengths of two selected sunspot groups with time distances of about thirty minutes during the period 1981 September 4 – September 7.**
H. Künzel.
HHI Suppl. Ser. Sol. Data, Vol. 5, No. 4, p. 73 – 99 (1983).

072.079 **Solar observations made at Catania Astrophysical Observatory during 1982.**
C. D'Arrigo, R. A. Zappalà.
Oss. Astrofis. Catania, Pubbl., N. 169, 78 pp. (1984).

072.080 **The relation between sunspot–relative and spot–group numbers for the University Observatory of Istanbul.**
A. Marsoğlu.
Rev. Fac. Sci. Univ. Istanbul, Ser. C, Vol. 46, p. 33 – 37 (1981). = Publ. Istanbul Univ. Obs., No. 122 (1984).

072.081 **Sunspot observations in 1976.**
C. Bolcal.
Rev. Fac. Sci. Univ. Istanbul, Ser. C, Vol. 46, p. 181 – 187 (1981). = Publ. Istanbul Univ. Obs., No. 126 (1984).

072.082 **Sunspot observations made in 1977.**
H. G. Tektunali.
Rev. Fac. Sci. Univ. Istanbul, Ser. C, Vol. 46, p. 189 – 198 (1981). = Publ. Istanbul Univ. Obs., No. 127 (1984).

072.083 **Sunspot observations made in 1979.**
M. T. Özkan.
Rev. Fac. Sci. Univ. Istanbul, Ser. C, Vol. 46, p. 199 – 223 (1981). = Publ. Istanbul Univ. Obs., No. 128 (1984).

072.084 **A new definition for relative sunspot numbers.**
A. Marsoğlu.
Rev. Fac. Sci. Univ. Istanbul, Ser. C, Vol. 47, p. 1 – 4 (1982). = Publ. Istanbul Univ. Obs., No. 129 (1984).

072.085 **Sunspot observations made in 1978.**
H. H. Mentese.
Rev. Fac. Sci. Univ. Istanbul, Ser. C, Vol. 47, p. 53 – 72 (1982). = Publ. Istanbul Univ. Obs., No. 133 (1984).

072.086 **Sunspot observations made in 1980.**
O. Gölbasi.
Rev. Fac. Sci. Univ. Istanbul, Ser. C, Vol. 47, p. 73 – 95 (1982). = Publ. Istanbul Univ. Obs., No. 134 (1984).

072.087 **Sunspot observations made in 1981.**
H. G. Tektunali.
Rev. Fac. Sci. Univ. Istanbul, Ser. C, Vol. 47, p. 97 – 122 (1982). = Publ. Istanbul Univ. Obs., No. 135 (1984).

072.088 **Manila Observatory, Solar Division. Solar maps and activity. 1984 February – 1984 September.**
F. J. Heyden.
Sol. Maps Act. (1984).

072.089 **Spectrum of a peculiar bright dot in a sunspot umbra.**
Z.-y. Wang, X.-z. Chen.
Publ. Purple Mt. Obs., Vol. 3, No. 2, p. 38 – 41 (1984). In Chinese.

072.090 **On the nature of bright sunspot rings.**
Eh. V. Kononovich.
Astron. Tsirk., No. 1324, p. 5 – 6 (1984). In Russian.

072.091 **Restoration of contrasts of solar faculae.**
I. V. Mironova.
Astron. Tsirk., No. 1331, p. 5 – 8 (1984). In Russian.

072.092 **On the nature of facular granules.**
Eh. V. Kononovich.
Astron. Tsirk., No. 1336, p. 1 – 2 (1984). In Russian.

072.093 **Separation of current structures in active regions on the solar surface.**
S. I. Gopasyuk, B. Kalman, V. A. Romanov.
Problems of physics of solar flares, p. 66 – 71 (1983). In Russian. Abstr. in Ref. Zh., 51. Astron., 12.51.632 (1984). – See Abstr. 012.070.

072.094 **On calculations of magnetic fields of active regions in potential approximation.**
O Gym Den, E. A. Kornitskaya, M. M. Molodenskij.
Problems of physics of solar flares, p. 72 – 77 (1983). In Russian. Abstr. in Ref. Zh., 51. Astron., 12.51.633 (1984). – See Abstr. 012.070.

072.095 **The influence of a drift in an inhomogeneous magnetic field of an active region on the temporal profile and energy spectrum of a charged particle stream.**
B. P. Filippov.
Problems of physics of solar flares, p. 107 – 112 (1983). In Russian. Abstr. in Ref. Zh., 51. Astron., 12.51.640 (1984). – See Abstr. 012.070.

072.096 **Evolution of a spot group and origin of large-scale current structures in the solar atmosphere.**
S. I. Gopasyuk, B. Kalman, V. A. Romanov.
Problems of physics of solar flares, p. 78 – 81 (1983). In Russian. Abstr. in Ref. Zh., 51. Astron., 12.51.641 (1984). – See Abstr. 012.070.

072.097 **Neuere Beobachtung von Sonnenflecken.**
A. Koch.
Mitt. Astron. Ges., Nr. 62, p. 185 – 188 (1984). – See Abstr. 010.341.

072.098 **Geschwindigkeitskartierung von Sonnenflecken.**
G. Küveler, E. Wiehr.
Mitt. Astron. Ges., Nr. 62, p. 189 (1984). Abstract. – See Abstr. 010.341.

072.099 **Sonnenflecken in extremen Breiten.**
H. Haupt.
Mitt. Astron. Ges., Nr. 62, p. 222 – 224 (1984). – See Abstr. 010.341.

072.100 **Daily maps of the sun and of magnetic fields of sunspots.**
Soln. Dannye, Byull., 1984, Nos. 5 – 9 (1984). In Russian.

072.101 **A case for submergence of magnetic flux in a solar active region.**
D. Rabin, R. Moore, M. J. Hagyard.
Astrophys. J., Vol. 287, No. 1, p. 404 – 411 (1984).
In NOAA Active Region 2372 (1980 April), $4 \times 10^{20}$maxwells of magnetic flux concentrated in an area 30″ across disappeared overnight. Vector magnetograms show that all components of the magnetic field weakened together. The authors argue that the most likely possibility is that flux *submerged.* The observations suggest that even during the growth phase of active regions, submergence is a strong process comparable in magnitude to emergence.

072.102 **Fluctuations of the total area of the sunspot umbra and penumbra and cycles of solar activity.**
Yu. I. Vitinskij.
Soln. Dannye, Byull., 1984, No. 9, p. 61 – 66 (1984). In Russian.
On the basis of lists of strong positive and negative fluctuations of the total area of the umbra ($S_u$) and penumbra ($S_p$) of sunspots and data on relative "perturbations" and coefficient of fluctuation variations of these indices for 1878 – 1976 manifestations of various cycles of solar activity in their properties are studied. It is shown that strong positive fluctuations $S_u$ and $S_p$ are most frequent on the descending branch of the 11 year solar cycle and in the second year on the ascending branch.

072.103 **Some regularities of the behaviour of the high-frequency region of annual Wolf numbers.**
Yu. R. Rivin.
Soln. Dannye, Byull., 1984, No. 9, p. 74 – 76 (1984). In Russian.
Using a narrow-band filter variations with a period of 5 years were found from annual Wolf numbers for 1750 – 1982. The behaviour of these variations is different before and after the year 1870. After 1870 the variation of their amplitude has a period of 22 years which breaks for solar cycles 20 and 21.

072.104 **Note on the flare activity of the July 1982 sunspot group.**
M. B. Ogir', L. D. Parfinenko, M. N. Stoyanova.
Soln. Dannye, Byull., 1984, No. 9, p. 77 – 85 (1984). In Russian.
The flare activity of groups 228 – 229 (Solnechnye Dannye Bulletin) of July 1982 was studied. The group was considered to

consist of three interacting domains. The conclusion is made that the proton flare was caused by that interaction. A histogram of the distribution of the total number of flares is given. During large flares significant variations of the longitudinal magnetic field were observed for several minutes.

**Workshop über Sonnenflecken.**
See Abstr. 011.034.

**Theoretical Ne IX line ratios compared to solar observations.**
See Abstr. 022.123.

**Mg IX and Si XI line ratios in the Sun.**
See Abstr. 022.124.

**A possible method for the determination of the magnitude and inclination of a sunspot magnetic field by a quarter–wave plate and polarizers.**
See Abstr. 036.170.

**New lease on life for the Solar Maximum Mission.**
See Abstr. 051.006.

**The effect of plasma rotation on axisymmetric magnetic field configuration.**
See Abstr. 062.005.

**Cyclotron radio emission in plasma with steep temperature gradient.**
See Abstr. 062.105.

**Magnetic flux tube in a stratified atmosphere under the influence of the vertical magnetic field.**
See Abstr. 062.156.

**Relation between stellar activity, rotation and stellar dynamos.**
See Abstr. 064.052.

**The photosphere as the only possible source of Babcock's sunspots due to the field amplification by the differential rotation.**
See Abstr. 071.009.

**Heights of faculae above the photosphere, based on observation of the eclipse of July 31, 1981.**
See Abstr. 071.024.

**Variability of the quiet photospheric network.**
See Abstr. 071.026.

**On the observed variation of the period of oscillations of the chromosphere above sunspots.**
See Abstr. 073.018.

**The Sun as a star: three–component analysis of chromospheric variability in the calcium K line.**
See Abstr. 073.019.

**High–resolution photography of the solar chromosphere. XIX. Flow velocities along an active region loop.**
See Abstr. 073.021.

**Oscillations in EUV emission lines during a loop brightening.**
See Abstr. 073.026.

**Suspended spicules associated with the enhanced bright network in an active region.**
See Abstr. 073.029.

**Preflare activity of solar prominences.**
See Abstr. 073.030.

**Eruptive prominences of 1980 April 27 observed during STIP interval–X.**
See Abstr. 073.042.

**Occurrence of flares with type II and IV radio bursts in relation to sunspot group types.**
See Abstr. 073.044.

**Occurrences of flares with type II and IV radio events in interacting sunspot groups in the course of revolutions.**
See Abstr. 073.045.

**Long–decay soft X–ray flares. 1. Region McMath 11926, June 15, 1972.**
See Abstr. 073.046.

**On the nature of bright points in the low sunspot chromosphere.**
See Abstr. 073.047.

**Die Chromosphäre über Sonnenflecken.**
See Abstr. 073.054.

**A 154–day periodicity in the occurrence of hard solar flares?**
See Abstr. 073.059.

**Cartes synoptiques de la chromosphère solaire et catalogues des filaments et des centres d'activité.**
See Abstr. 073.089.

**Ca II K bright points and the solar cycle.**
See Abstr. 073.108.

**Analysis of magnesium XI line profiles from solar active regions.**
See Abstr. 074.023.

**Active longitudes of coronal holes in the minimum and maximum of solar activity.**
See Abstr. 074.036.

**Solar activity cycle variation of the K corona.**
See Abstr. 074.042.

**Change of the distribution of magnetic fields of active regions in the process of their evolution.**
See Abstr. 075.015.

**Latitudinal zonal structure of the solar coronal magnetic field.**
See Abstr. 075.026.

**Characteristic structures of the solar disc observed on rocket UV filtergrams.**
See Abstr. 076.001.

**Active region soft X–ray line profiles.**
See Abstr. 076.015.

**Interplanetary radio storms. I. Extension of solar active regions through the interplanetary medium.**
See Abstr. 077.003.

**The spectrum of shortwave radiation for aeronomical calculations at different levels of solar activity.**
See Abstr. 077.010.

**Investigation of the longitudinal heterogeneity of the solar activity from the radio flux of the sun at 10.7 cm.**
See Abstr. 077.012.

**Noise storm association to non–radio phenomena.**
See Abstr. 077.028.

**Structure and polarization of active region microwave emission.**
See Abstr. 077.046.

**Velocity signal produced by the passage of active regions compared with observations.**
See Abstr. 080.052.

**Influence of active regions on global oscillation measurements.**
See Abstr. 080.053.

**Oscillatory motions in the Sun corresponding to the solar cycle.**
See Abstr. 080.055.

**Rotation of the Sun measured from Mount Wilson white–light images.**
See Abstr. 080.060.

**Variations in solar rotation with the sunspot cycle.**
See Abstr. 080.061.

**Plasma motion in umbrae and the surrounding photosphere derived from spectroscopic Doppler measurements and tracer measurements of spots.**
See Abstr. 080.063.

**Pulsations of microwave emission as a consequence of oscillatory transients in the solar atmosphere.**
See Abstr. 080.065.

**Solar neutrino data and the 11–year solar activity cycle.**
See Abstr. 080.067.

**Linear models of acoustic waves in sunspot umbrae.**
See Abstr. 080.070.

**The longitudinal distribution of solar active regions and solar global oscillation measurements.**
See Abstr. 080.079.

**Convective instability in a solar flux tube. I. Nonlinear calculations for an adiabatic inviscid fluid.**
See Abstr. 080.111.

**Temporal variation of the concentration of cosmogenic isotopes in the earth's atmosphere and solar activity.**
See Abstr. 144.019.

**Variations of galactic cosmic rays and evolution of coronal holes with the cycles of solar activity.**
See Abstr. 144.030.

# 073 Chromosphere, Flares, Prominences

073.001 **Observation of the chromospherical flare of May 13, 1981 at the post–maximum phase.**
G. F. Sitnik, V. S. Prokudina, V. I. Bulavina.
Astron. Tsirk., No. 1282, p. 1 – 3 (1983). In Russian.

073.002 **The unimportance of beam momentum in electron–heated models of solar flares.**
A. N. McClymont, R. C. Canfield.
Astron. Astrophys., Vol. 136, No. 1, p. L1 – L4 (1984).

In a previous letter to this journal Brown and Craig (1984) called attention to the possible importance of the hitherto neglected direct collisional acceleration of flare atmospheres heated by particle beams. For the case commonly believed to be of most physical interest – electron beam heating in closed flare loops – the authors find their arguments incomplete. They conclude that acceleration due to beam pressure is important for only a few seconds after the electron beam is switched on and is unlikely to have a significant influence on the global flare evolution. Brown and Craig's conclusions should, however, be taken into account in models of proton beam heating, because of the much higher momentum–to–energy ratio of proton beams.

073.003 **Dynamics of solar filaments. III. Analysis of steady flows in Hα and C IV lines.**
B. Schmieder, J. M. Malherbe, P. Mein, E. Tandberg–Hanssen.
Astron. Astrophys., Vol. 136, No. 1, p. 81 – 88 (1984).

A statistical analysis of intensities and radial velocities of several solar plage filaments observed at disk centre is presented. Intensity and radial velocity maps were derived from simultaneous 2D measurements of the Hα chromospheric line with the Multichannel Subtractive Double Pass spectrograph operating on the Meudon Solar Tower, and also obtained from 2D recordings of the C IV transition–zone line ($\lambda$ = 1548 Å) with the Ultra Violet Spectrometer and Polarimeter aboard the Solar Maximum Mission satellite. A good correlation is found between intensities in both lines, as well as between velocities. Persistent upflows are measured in both lines at the filament location. The analysis of mass fluxes suggests that C IV upflows occur in the transition region around prominences rather than below, in the chromosphere–corona transition zone.

073.004 **Comparison of optical observations of the flare of May 16, 1981 and radio bursts in the kilometer range.**
V. P. Grigor'eva, V. S. Prokudina.
Astron. Tsirk., No. 1287, p. 3 – 5 (1983). In Russian.

073.005 **The Hα spectral signatures of solar flare nonthermal electrons, conductive flux, and coronal pressure.**
R. C. Canfield, T. A. Gunkler, P. J. Ricchiazzi.
Astrophys. J., Vol. 282, No. 1, p. 296 – 307 (1984).

The authors have calculated Hα line profiles based primarily on the static model flare chromospheres of Ricchiazzi and Canfield, which show the effects of enhanced fluxes of nonthermal electrons and heat, as well as enhanced pressure, on the temperature, density, and ionization structure of the chromosphere. Both hydrostatic and impulsive phase models are examined. Two unique qualitative signatures are identified among the mechanisms examined. First, only high values of the flux of energetic electrons produce Hα profiles with obvious broad (Stark) wings of non–Gaussian form. Second, only high values of coronal pressure produce strong Hα emission profiles that are not centrally reversed.

073.006 **Radio–frequency heating of the coronal plasma during flares.**
D. B. Melrose, G. A. Dulk.
Astrophys. J., Vol. 282, No. 1, p. 308 – 315 (1984).

The authors develop a model for the radio–frequency (RF) heating of soft X–ray emitting plasma in solar flares due to absorption of amplified cyclotron radiation. The radiation, carrying $\sim 10^{27}$ to $\sim 10^{30}$ ergs $s^{-1}$, is generated through maser emission following partial precipitation of electrons in one or more flaring loops. The maser operates in a large number of small regions, each producing an "elementary burst" of short duration. RF heating and evaporation models for the production of soft X–ray emitting plasma are compared.

073.007 **Magnetic reconnection in current sheets: solar flares and laboratory experiments.**
S. V. Bulanov, V. A. Dogel', A. G. Frank.
Sov. Astron. Lett., Vol. 10, No. 1, p. 59 – 61 (1984). English translation of 37.073.014.

073.008 **Variation in observed coronal calcium abundance of X–ray flare plasmas.**
J. Sylwester, J. R. Lemen, R. Mewe.
Nature, Vol. 310, No. 5979, p. 665 – 666 (1984).
The authors present the first spectroscopic evidence for the variation of the coronal calcium abundance in high–temperature solar flare plasmas. The analysed data consist of the high–resolution X–ray flare spectra observed with the Bent Crystal Spectrometer on board the SMM satellite. The observed abundance variation has important consequences for the analysis and interpretation of XUV and X–ray spectra.

073.009 **Flare–initiated oscillations of quiescent prominences.**
J. Galindo–Trejo.
XVI International Conference on Phenomena in Ionized Gases, p. 92 – 93 (1983). Abstr. in Phys. Abstr., Vol. 87, No. 1239, Entry 82459 (1984). – See Abstr. 012.017.

073.010 **Wave propagation in a cylindrical layer of a spicule.**
A. D. Pataraya, B. B. Chargeishvili
(*B. B. Chargejshvili*).
XVI International Conference on Phenomena in Ionized Gases, p. 94 – 95 (1983). Abstr. in Phys. Abstr., Vol. 87, No. 1239, Entry 82460 (1984). – See Abstr. 012.017.

073.011 **The development and cooling of a solar limb–flare.**
N. J. Veck, K. T. Strong, C. Jordan, G. M. Simnett, P. J. Cargill, E. R. Priest.
Mon. Not. R. Astron. Soc., Vol. 210, No. 2, p. 443 – 462 (1984).
Observations of a flare that began in soft X–rays at 20:37 UT on 1980 April 12, at the west limb of the Sun are discussed. The Solar Maximum Mission (SMM) satellite was used to obtain X–ray images before the flare and for a period during the decay phase. Hα photographs and the soft X–ray flux measured by the GOES–3 satellite are available throughout the flare. Some alternative models are discussed but insufficient data are available for a full treatment. Overall, the observations are best fitted qualitatively by a model where pre–existing magnetic loop structures are perturbed by the intrusion of new magnetic flux.

073.012 **Distribution law of integrated intensity of solar optical flares.**
T. K. Das, M. K. Das Gupta.
Bull. Astron. Inst. Czech., Vol. 35, No. 4, p. 241 – 243 (1984).
The occurrence frequency distribution of solar optical flares in different values of integrated intensity has been investigated. The number of flares diminishes as $I_{(i)}{}^{-1.9}$ for flares with integrated intensity, $I_{(i)}$, greater than 2 units. This power law distribution has been compared with that found in other kinds of solar activity.

073.013 **On the heating of the solar chromosphere by hydromagnetic waves with a five–minute period.**
L. M. B. C. Campos.
Mem. Soc. Astron. Ital., Vol. 55, Nos. 1 – 2, p. 267 – 272 (1984). – See Abstr. 012.022.
The author presents a theoretical model of the solar chromosphere based on heating by resistive dissipation of Alfvén–gravity waves, with the energy deposition by Joule effect balanced against losses by thermal radiation. The main input to the model is the exact calculation of the amplitudes and phases of the velocity perturbation, magnetic field and electric current associated with an Alfvén–gravity wave propagating vertically in the chromosphere. The analytical temperature and density profiles calculated from the theory are consistent with empirical data.

073.014 **Wave observations in the solar transition region with the Solar Maximum Mission.**
M. E. Bruner, G. Poletto.
Mem. Soc. Astron. Ital., Vol. 55, Nos. 1 – 2, p. 313 – 318 (1984). – See Abstr. 012.022.
Observations of intensity and velocity oscillations in the N V line at 1238 Å, obtained with the UVSP instrument on the Solar Maximum Mission, are analyzed in order to ascertain their nature. About 2/3 of the data are consistent with the behavior expected from acoustic waves. An estimate of the energy flux carried by the waves still yields a value too low to provide for coronal heating. However, there is unequivocal evidence for an upward directed flux.

073.015 **On hydrogen emission in filamentary prominences.**
V. V. Zharkova.
Inst. teor. fiz. Akad. Nauk USSR. Prepr., No. 141, 20 pp. (1983). In Russian. Abstr. in Ref. Zh., 51. Astron., 8.51.236 (1984).

073.016 **Comparison of results of synchronous observations of prominences in the radio– and optical ranges.**
G. P. Apushkinskij, V. I. Bulavina, A. B. Delone, E. F. Rizov, N. A. Topchilo, A. N. Tsyganov, G. V. Yakunina, B. Ya. Losovskij.
Vestn. Leningr. Univ., Mat. Mekh. Astron., No. 1, p. 107 – 109 (1984). In Russian. Abstr. in Ref. Zh., 51. Astron., 8.51.241 (1984).

073.017 **On a local method of long–term forecast of solar flare activity.**
E. V. Miletskij.
Soln. Dannye, Byull., 1984, No. 6, p. 61 – 68 (1984). In Russian.
On the basis of the local approach to the analysis of solar activity the method of the derivation of empirical regression equations is described. The method enables to forecast averaged over one Carrington rotation values of the flare index in various areas of the solar surface for the following Carrington rotations using the known values of other indices of solar activity for the preceding rotations. Systems of equations for various phases of the 11–year solar cycles during 1957 – 1980 are derived. Results of the forecasts of the flare index derived according to the method are given.

073.018 **On the observed variation of the period of oscillations of the chromosphere above sunspots.**
V. S. Loskutnikov.
Soln. Dannye, Byull., 1984, No. 6, p. 91 – 94 (1984). In Russian.
The brightness oscillations of a sunspot umbra in the center line Hα are considered. It has been found that slow variations of the oscillation period exist as well as fast variations, which are difficult to explain in terms of the model of a single linear atmospheric resonator.

073.019 **The Sun as a star: three–component analysis of chromospheric variability in the calcium K line.**
A. Skumanich, J. L. Lean, O. R. White, W. C. Livingston.
Astrophys. J., Vol. 282, No. 2, p. 776 – 783 (1984). With plate 11.
A three–component model of the solar cycle variability of the Ca II K emission is developed using extant contrast and fractional area parameters for (1) cell, (2) network, and (3) plage components that are resolution–consistent. The authors have been able to fit the quiet–Sun Ca II K emission (at solar minimum) as observed by White and Livingston with cell and network features alone using extant limb–darkening laws. The occurrence of plages during the growth of the solar cycle was found to be insufficient to account for the cycle increase in the K emission and required the introduction of additional network in excess of the quiet–Sun value.

073.020 **Solar flare iron Kα emission associated with a hard X–ray burst.**
K. Tanaka, T. Watanabe, N. Nitta.
Astrophys. J., Vol. 282, No. 2, p. 793 – 798 (1984).
An exceptionally intense iron Kα emission associated with a hard X–ray burst was observed by the Bragg crystal spectrometer aboard Hinotori. High–resolution continuum spectra from 1.5 to 12.5 keV and hard X–ray spectra from 18 to 400 keV of this flare were obtained. They show a remarkable power–law photon distribution extending from below 10 keV to above 100 keV in the early phase of the flare. The authors attribute the intense Kα emission in the early phase to fluorescence of photospheric neutral iron by the power–law X–ray flux that extends down to the K–shell ionization threshold at 7.1 keV. The gradually increasing

$K\alpha$ emisson after the decay of the hard X–ray burst can be explained by fluorescence due to the X–ray flux of thermal origin.

073.021 **High–resolution photography of the solar chromosphere. XIX. Flow velocities along an active region loop.**
R. E. Loughhead, R. J. Bray.
Astrophys. J., Vol. 283, No. 1, p. 392 – 397 (1984).
True flow velocities along the central axis of an active region loop have been determined for the first time. The loop material is found to rise from the surface along one leg with an increasing upward velocity despite the retarding effect of gravity. Descending material in the other leg suffers a retarding force and does not attain the velocity of free fall. The derived velocities are supersonic at all points measured; the Mach number exceeds 5 at one location.

073.022 **Flares in Hale 17760: simultaneous microwave spectra and images of the flare of 1981 July 27.**
A. Kattenberg.
Astrophys. J., Vol. 283, No. 1, p. 398 – 403 (1984).
On 1981 July 27 a class 1B flare occurred in Hale region 17760. The corresponding microwave burst was observed with the VLA at 2 cm wavelength, while at the Owens Valley Radio Observatory simultaneous microwave spectra (1 – 18 GHz) of the burst were recorded. The images and spectra are analyzed, and it is shown that the flare occurred as a series of reconnection events in the current layer between the magnetic field of the active region and the differently oriented field of an emerging flux region.

073.023 **The possibility of steady state nonionization equilibrium conditions in soft X–ray flare plasmas.**
G. A. Doschek.
Astrophys. J., Vol. 283, No. 1, p. 404 – 412 (1984).
The possibility of the existence in soft X–ray flare plasmas of conditions that result in a steady state departure of ion abundances from ionization equilibrium values is considered. The observed flare plasma is assumed to be a result of many small "elementary bursts" that occur on time scales comparable to the ionization and recombination times of highly ionized atoms of iron and calcium. Although the behavior of certain line ratios can be explained in the context of the burst models considered in this paper, the behavior of the set of all the available line ratios cannot be explained in this manner.

073.024 **Wavelength–dependence of EUV continuum absorption.**
T. Nishikawa.
Sol. Phys., Vol. 93, No. 1, p. 37 – 51 (1984).
The weakening of EUV line emission due to continuum absorptions of neutral hydrogen and neutral helium is investigated to examine its wavelength–dependence. After convolving the intensities predicted from multilevel calculations over an instrumental profile, the author found a systematic, linear weakening for lines shortward of 912 Å, which can be attributed to Lyman continuum absorption in the cool chromospheric cloud. The degree of the weakening at the quiet Sun seems to be constant in the temperature range of $4.3 < \log T_{\_} < 5.4$. The author also finds that the lines shortward of 504 Å are somewhat weakened by He I continuum absorption. From the comparison of both weakenings the temperature of the absorber is estimated to be rather low ($T_e \lesssim 7 \times 10^3$K).

073.025 **Permanent changes in filaments near solar flares.**
D. M. Rust.
Sol. Phys., Vol. 93, No. 1, p. 73 – 83 (1984).
High resolution $H\alpha$ images obtained before and after 57 importance 1N or larger flares have been examined for changes in the magnetic fields ($B\perp$) transverse to the line of sight. It was assumed that $H\alpha$ chromospheric structures outline $B\perp$. In 37% of the cases, there was a reconfiguration of segments of filaments or of chromospheric fibrils. Examination of data from 21 non–flare intervals shows such changes in 24% of the cases. When changes of any kind, including total disappearance and length changes, are included, the proportions for flare and non–flare intervals increase to 58% and 52%, respectively. It is concluded that flares do not cause enduring magnetic field changes in the chromosphere.

073.026 **Oscillations in EUV emission lines during a loop brightening.**
E. Antonucci, A. H. Gabriel, B. E. Patchett.
Sol. Phys., Vol. 93, No. 1, p. 85 – 94 (1984).
Oscillations in the emission in the ultraviolet lines of C II, O IV, and Mg X, detected by the Harvard College Observatory EUV spectroheliometer on Skylab are observed on August 7, 1973, during a loop brightening. The intensity of the EUV lines varies with a period of 141 s during the time of enhanced intensity of the coronal loop, lasting 10 min. The periodic oscillation is not only localized in the loop region but extends over a larger area of the active region, maintaining the same phase. The authors suggest that the intensity fluctuation of the EUV lines is caused by small–amplitude waves, propagating in the plasma confined in the magnetic loop and that size of the loop might be important in determining its preferential heating in the active region.

073.027 **Flare loops heated by thermal conduction.**
D. M. Rust, B. V. Somov.
Sol. Phys., Vol. 93, No. 1, p. 95 – 104 (1984).
A flare observed with the Hard X–Ray Imaging Spectrometer was studied during its rise to maximum temperature and X–ray emission rate. Two proximate flare loops, of lengths $2.8 \times 10^9$ cm and $1.1 \times 10^{10}$ cm, rose to temperatures of $21.5 \times 10^6$K and $30 \times 10^6$K, respectively, in 30 s. Assuming equal heat flux $F$ into each loop from a thermal source at the point where they met, the authors derive a simple relationship between temperature $T$ and loop length $L$. It is suggested that conductive heating adequately describes the rise and maximum phase emissions in the loops and that long flare loops reach higher temperatures than short loops during the impulsive phase because of an equipartition of energy between them at their point of interaction.

073.028 **$H\alpha$ red asymmetry of solar flares.**
K. Ichimoto, H. Kurokawa.
Sol. Phys., Vol. 93, No. 1, p. 105 – 121 (1984). = Contrib. Kwasan and Hida Obs., Univ. Kyoto, No. 258.
The evolutional characteristics of the red asymmetry of $H\alpha$ flare line profiles were studied by means of a quantitative analysis of $H\alpha$ flare spectra obtained with the Domeless Solar Telescope at Hida Observatory. Red–shifted emission streaks of $H\alpha$ line are found at the initial phase of almost all flares which occur near the disk center, and are considered to be substantial features of the red asymmetry. Referring to the numerical simulations made by Livshits et al. (1981) and Somov et al. (1982), the authors conclude that the conspicuous red–asymmetry or the red–shifted emission streak of $H\alpha$ line is due to the downward motion of the compressed chromospheric flare region produced by the impulsive heating by energetic electron beam or thermal conduction.

073.029 **Suspended spicules associated with the enhanced bright network in an active region.**
V. Gaizauskas.
Sol. Phys., Vol. 93, No. 2, p. 257 – 268 (1984).
The progressive rotation from the limb onto the disk of a long–lived cluster of coaligned $H\alpha$ spicules was observed at high spatial resolution on the fringe of a large complex of activity. Although individual spicules were steadily changing, the organized cluster appeared consistently suspended above the photospheric limb when viewed in the wings of $H\alpha$. The phenomenon is the counterpart near an active region of the dark band discovered in the quiet low chromosphere by Loughhead (1969). But in the present circumstances the effect is perceived as a weakening of emission, i.e. as a gap rather than an obscuration.

073.030 **Preflare activity of solar prominences.**
G. Simon, N. Mein, P. Mein, L. Gesztelyi.
Sol. Phys., Vol. 93, No. 2, p. 325 – 336 (1984).
The preflare activity of a plage filament is analysed from $H\alpha$ observations made with the Multichannel Subtractive Double Pass Spectrograph of the Meudon Solar Tower. The June 22, 1980 event is studied and interpreted in terms of preflare heating

of a filament, connected to the rise of emerging flux, and the relative approach of pores of different magnetic polarity, prior to the onset of a two–ribbon flare. The authors suggest that most of the observed features may be the consequence of emergence of new magnetic flux and the related reconnection processes.

073.031 **Analysis of the magnetic field configuration of a filament–associated flare from X–ray, UV, and optical observations.**
C.–C. Cheng, R. Pallavicini.
Sol. Phys., Vol. 93, No. 2, p. 337 – 350 (1984).
X–ray and ultraviolet observations from SMM of a filament–associated event on 22 November, 1980 are examined in conjunction with ground–based optical observations, in order to determine the magnetic field configuration involved in the flaring process. The authors find evidence that the flare was produced by gradual energy release in a large sheared magnetic loop which interacted with another smaller loop. Non–thermal processes were produced in the interaction of the two loops. Although this flare shared some of the characteristics of Long Duration (class II) Events the magnetic configuration of the event was more similar to that of a compact (class I) flare, although on a much larger spatial scale and longer time scale.

073.032 **Extension of the reconnection theory of two–ribbon solar flares.**
R. A. Kopp, G. Poletto.
Sol. Phys., Vol. 93, No. 2, p. 351 – 361 (1984).
The magnetic reconnection theory for two–ribbon flares and flare loops hypothesizes that the gradual energy release during the decay phase is a direct result of dissipative relaxation of the open coronal magnetic configuration created by an eruptive prominence/coronal transient precursor. This scenario is here developed quantitatively. The major refinements are (1) to use an analytical description of the reconnecting field geometry specifically chosen to characterize the spatial scale of the active region where a flare occurs, and (2) to take account of the fact that the volume occupied by X–ray–emitting plasma (hot loops) generally increases with time during the flare. As a test of the modified theory a new representation of the Skylab observations of the large two–ribbon flare of 29 July, 1973 has been undertaken.

073.033 **Do prominences migrate equatorwards?**
V. I. Makarov.
Sol. Phys., Vol. 93, No. 2, p. 393 – 396 (1984).
The mean latitude of prominence samples referring to one boundary of the polarity division line of the large–scale magnetic field is calculated on the basis of Hα charts in the period 1955 – 1982. It is shown that the magnetic field in that period had a latitude zonal structure. The boundaries of the latitude zones of the magnetic field had no regular equatorwards migration. They either oscillated near the mean boundaries at 0°, ±20°, ±40°, or migrated polewards.

073.034 **High–energy gamma rays and neutrons in solar flares.**
G. E. Kocharov, N. Z. Mandzhavidze.
Pis'ma Astron. Zh., Tom 10, No. 10, p. 771 – 774 (1984). In Russian. English translation in Sov. Astron. Lett, Vol. 10.
It is shown that the observed localization of solar flares near the limb with high–energy gamma–rays and neutrons emissions may be explained assuming the generation of high–energy neutrons and gamma–rays in *pp*, *pα* and *αp* reactions.

073.035 **Coronal extension of flaring region magnetic fields inferred from high–resolution microwave and type III burst observations.**
P. Lantos, M. Pick, M. R. Kundu.
Astrophys. J., Lett. Ed., Vol. 283, No. 2, p. L71 – L74 (1984).
Observations of three solar radio bursts, obtained with the VLA at 6 cm wavelength, have been combined with meter observations from the Mark III Nançay Radioheliograph. There is a good correlation between solar activity observed at the two wavelength domains. The bursts come from two distinct active regions. With two–dimensional spatial resolution, it is shown that, in this sample, each active region possesses a coronal extension that is separated from that of the neighboring active region.

073.036 **Time–resolved spectral observations of spicule velocities at several heights.**
S. S. Hasan, S. L. Keil.
Astrophys. J., Lett. Ed., Vol. 283, No. 2, p. L75 – L77 (1984).
The vacuum telescope of the National Solar Observatory, Sacramento Peak was used to obtain Hα spectral observations of spicules. A set of spectra corresponding to five slit positions above the solar limb were recorded every 8 s in order to study the temporal variation of spicules at several heights with high space and time resolution. A typical flow event in a spicule was found to last 10–15 minutes.

073.037 **Investigation of the correlation of hard X–ray emission bursts and microwave radio radiation accompanying solar flares in March – April 1979.**
G. E. Kocharov, G. A. Matveev, V. F. Mel'nikov,
T. S. Podstrigach, Yu. E. Charikov.
Cosmic ray intensity and cosmogenic isotopes. 13th Leningrad seminar on cosmophysics, p. 9 – 23 (1983). In Russian. Abstr. in Ref. Zh., 51. Astron., 9.51.358 (1984). – See Abstr. 012.028.

073.038 **Forecast of shock wave velocity and of the flux from a powerful flare on the sun from Hα line broadening measurements.**
A. D. Chertkov, S. K. Ryabchuk.
Magnitosfer. issled. No. 2. Plazma i protsessy v magnitosfere. Moskva, p. 127 – 135 (1983). In Russian. From Ref. Zh., 51. Astron., 9.51.420 (1984).

073.039 **Observations of the sun at the Astronomical Observatory of Lvov University in 1982.**
M. B. Girnyak, M. M. Koval'chuk, I. S. Laba, P. A. Olijnyk.
Vestn. L'vov. univ. Ser. astron., No. 58, p. 20 – 23 (1984). In Russian. Abstr. in Ref. Zh., 51. Astron., 9.51.447 (1984).

073.040 **SMM observations of a compact flare. Interpretation and theoretical implications for flare theory.**
J. G. Doyle, P. B. Byrne.
Ir. Astron. J., Vol. 16, No. 3, p. 226 – 242 (1984). – See Abstr. 012.036.
The authors complete an energy–balance analysis of a double impulsive hard X–ray flare. From spatial observations they deduce both flares probably occurred in the same loop. For the first flare, the energy in the fast electrons (assuming a thick–target model) is comparable to the convective up–flow energy, thus implying that these are related successive modes of energy storage and transfer. The total energy loss through radiation and conduction is comparable to the fast–electron energy. For the second flare, the energy in the fast electrons is more than one order of magnitude greater than the energy of the convective up–flow. Total energy losses are within a factor of two of the calculated fast–electron energy.

073.041 **Recurrent mass ejections associated with flare behind the limb on 1978 December 20.**
R. Jain, A. Bhatnagar, R. N. Shelke.
Bull. Astron. Soc. India, Vol. 12, No. 2, p. 162 – 171 (1984).
A solar flare occurred on 1978 December 20 in the McMath active region No. 15700. In this paper a detailed study of the spray and surge activities associated with this flare is presented. A possible explanation for the observed recurring tendency of surges with a quasi–periodicity of 5 – 30 min is suggested as due to additional small magnetic flux brought out by buoyancy in or near a satellite spot which could provide enough collimating energy to squeeze out the surge material at a quasi–periodic time interval. Another interesting surge phenomenon was observed which showed up and down motion of a small surge material between 11.00 – 11.30 UT, with a velocity of about 20 km $s^{-1}$. To explain this event a qualitative model is presented.

073.042 **Eruptive prominences of 1980 April 27 observed during STIP interval–X.**
R. Jain, A. Bhatnagar, R. N. Shelke.
J. Astrophys. Astron., Vol. 5, No. 3, p. 323 – 330 (1984).
Observations and analyses of two similar eruptive prominences on the north–east limb observed on 1980 April 27 at 0231 and 0517 UT, which are associated with the Boulder active region No. 2416 are presented. Both the eruptive prominences gave rise to white–light coronal transients. Type II and moving type IV radio bursts are reported in association with the first Hα eruptive prominence. The authors estimate a magnetic field in the eruptive prominence of about 100 G and a build–up rate $\sim 10^{26}$erg s$^{-1}$. It is proposed that fast–moving Hα features must have initiated the observed coronal transients. From Hα, type II and coronal–transient observations, a magnetic field of 2.8 G at $1.9 R_{\odot}$ from the disc centre is estimated.

073.043 **Implications of solar flare dynamics for reconnection in magnetospheric substorms.**
R. L. Moore, J. L. Horwitz, J. L. Green.
Planet. Space Sci., Vol. 32, No. 11, p. 1439 – 1454 (1984).
From observations of two–ribbon solar flares, the authors present a new line of evidence that magnetic reconnection is of key importance in magnetospheric substorms. They infer that in substorms reconnection of closed field lines in the near–Earth thinned plasma sheet both initiates and is driven by the overall MHD instability that drives the tailward expulsion of the reconnected closed field (0 loops). The general basis for this inference is the longstanding notion that two–ribbon flares and substorms are essentially similar phenomena, driven by similar processes. The authors give an array of observed similarities that substantiate this view.

073.044 **Occurrence of flares with type II and IV radio bursts in relation to sunspot group types.**
Š. Knoška, L. Křivský.
Bull. Astron. Inst. Czech., Vol. 35, No. 5, p. 261 – 270 (1984).
The occurrence of flares was statistically processed, separately for those with type II and type IV radio bursts, in relation to the sunspot groups in which these flares took place. The period in question is cycle No. 20, 1966 – 1976. Zurich and magnetic typification were taken into account. As compared with flares accompanied by type II bursts, flares with type IV display an enhanced tendency to occur in more complex active regions in which more magnetic energy is accumulated in a small space (type δ). The occurrence of flares with type II and type IV bursts in the course of cycle No. 20 shows an anomalous behaviour, in particular a conspicuous decrease of occurrence from 1970 to 1971.

073.045 **Occurrences of flares with type II and IV radio events in interacting sunspot groups in the course of revolutions.**
J. Klimeš, L. Křivský.
Bull. Astron. Inst. Czech., Vol. 35, No. 5, p. 273 – 275 (1984).
Using data from 11–year solar cycle No. 20, it was found that flares with type II radio bursts are more than twice as frequent and flares with type IV bursts nearly twice as frequent in sunspot groups which developed close to each other or which merged in the course of revolutions as in isolated sunspot groups. With both types the occurrence of these flares is concentrated in the revolution of the so–called sunspot group interaction (their approximation, merging).

073.046 **Long–decay soft X–ray flares. 1. Region McMath 11926, June 15, 1972.**
A. Antalová, M. B. Ogir (*M. B. Ogir'*).
Bull. Astron. Inst. Czech., Vol. 35, No. 5, p. 276 – 294 (1984). With plates 1 – 4.
The time–space occurrence and development of H–alpha flares, which were generated in McMath region 11926 on 15 June 1972, were analysed with regard to the activation of filaments and the development of filament channels. H–alpha films were obtained from the Crimean Astrophysical Observatory. The magnetic field of the studied region had the form of an S–field "island" surrounded by an extensive N–field.

073.047 **On the nature of bright points in the low sunspot chromosphere.**
V. N. Dermendjiev.
Dokl. Bolg. Akad. Nauk, Vol. 37, No. 1, p. 7 – 10 (1984). Abstr. in Ref. Zh., 51. Astron., 10.51.315 (1984).

073.048 **Dynamics of hydromagnetic clouds from powerful solar flares.**
K. G. Ivanov, A. F. Harshiladze (*A. F. Kharshiladze*).
Inst. zem. magn., ionos. i rasprostr. radiovoln. Akad. Nauk SSSR. Prepr., No. 63a/474, 12 pp. (1984). Abstr. in Ref. Zh., 51. Astron., 10.51.329 (1984).

073.049 **Photographs of prominences in the He I 10830 Å line on the I–1060V film.**
N. S. Shilova, L. I. Mikheeva, A. G. Vakar.
Soln. Dannye, Byull., 1984, No. 7, p. 78 – 80 (1984). In Russian.
The results are given of the usage of the new Soviet I 1060V film for frames of prominences in the He I 10830 Å line. The advantages of observations in the He I 10830 Å line over those in Hα for detection of rotational structures are shown.

073.050 **A relationship between flare processes in an active region according to X–ray data.**
A. A. Zhdanov, Yu. E. Charikov.
Soln. Dannye, Byull., 1984, No. 8, p. 62 – 69 (1984). In Russian.
Sequences of flare processes in active regions according to X–ray data have been investigated and a scheme interpreting these phenomena has been proposed.

073.051 **Hard X–ray imaging observations of solar hot thermal flares with the *Hinotori* spacecraft.**
S. Tsuneta, N. Nitta, K. Ohki, T. Takakura, K. Tanaka, K. Makishima, T. Murakami, M. Oda, Y. Ogawara.
Astrophys. J., Vol. 284, No. 2, p. 827 – 832 (1984).
Two solar hard X–ray bursts of a new type (*hot thermal flare*) were observed with the spacecraft *Hinotori*. The flares emit intense hard X–rays (10 – 50 keV) with extremely steep spectra from a small region with size of 10″ – 20″. This source contains a hot thermal plasma of $(3 - 3.5) \times 10^7$K with an emission measure of the order of $10^{49}$cm$^{-3}$. One of the flares occurred just on the limb, and the centroid of the hard X–ray source was located at $(6 \pm 3) \times 10^3$ km above the photosphere. It is concluded that the energy continuously released goes into heating rather than acceleration almost throughout the flare.

073.052 **Physical conditions in the cool parts of spicules.**
D. A. Landman.
Astrophys. J., Vol. 284, No. 2, p. 833 – 838 (1984).
The resonance lines of Na, $Sr^+$, and the Mg triplet system, together with the middle H Balmer lines, are used to determine the plasma parameters in the cool parts of spicules. Results are presented for the heights 2100 and 2500 km according to the chromospheric line emission data reported by Dunn et al. for the 1962 eclipse. In addition to deriving values for the electron density, the present method is unique in yielding direct results for the degree of hydrogen ionization, the hydrogen density, and the line–of–sight thickness of emitting material.

073.053 **The impulsive phase of a solar limb flare.**
G. M. Simnett, K. T. Strong.
Astrophys. J., Vol. 284, No. 2, p. 839 – 847 (1984).
The limb flare of 1980 November 18 at 14:51 UT is studied with the X–ray instruments on board the SMM. In the impulsive phase the hard X–rays emitted during the rapidly varying part of the burst are observed on the limb, at a level consistent with the chromosphere. During the final stages of the hard X–ray burst, when the fast time structure has disappeared, the hard X–rays clearly come from the corona, at which time the spectrum is softer. Throughout the decay phase of the flare the X–rays also originate in the corona.

073.054 **Die Chromosphäre über Sonnenflecken.**
W. Mattig.
Sterne Weltraum, 23. Jahrg., Nr. 11, p. 559 – 563 (1984).

073.055 **Electron precipitation in solar flares: collisionless effects.**
L. Vlahos, H. L. Rowland.
Astron. Astrophys., Vol. 139, No. 2, p. 263 – 270 (1984).

A large fraction of the electrons which are accelerated during the impulsive phase of solar flares stream towards the chromosphere and are unstable to the growth of plasma waves. The linear and non–linear evolution of plasma waves as a function of time is analyzed with the use of a set of rate equations that follow in time the nonlinearly coupled system of plasma waves – ion fluctuations. As an outcome of the fast transfer of wave energy from the beam to the ambient plasma, non–thermal electron tails are formed which can stabilize the anomalous Doppler resonance instability that is responsible for the pitch angle scattering of the beam electrons. The authors estimate the non–collisional losses of the precipitating electrons and discuss the observational implications of their results.

073.056 **Comparative magnetographic observations in different prominence lines.**
G. M. Nikolsky (*G. M. Nikol'skij*), I. S. Kim, S. Koutchmy, G. Stellmacher.
Astron. Astrophys., Vol. 140, No. 1, p. 112 – 118 (1984).

Analysis of the longitudinal field strength in prominences obtained by measurements of Zeeman polarization in HeD3, Hα, and Hβ is presented. The measurements were performed with the spectrally scanning magnetograph installed at the 53 cm coronograph of the Kislovodsk high altitude station. The observations allow to correlate the values of the equivalent widths and those of the half widths with magnetographic data. No systematic dependence of field strengths on line widths is found. High field strengths persist for large Doppler shifts and low emissions. In the histogram showing the distribution of observed field strengths two peaks near $B_{\|} \cong 8$ Gauss and $B_{\|} \cong 20$ Gauss appear. The second peak corresponds to prominences mainly related to active regions for latitudes $< \pm 35°$. A similar distribution is obtained from a set of data deduced from observation of the Hanle effect and corresponding to the same epoch.

073.057 **The material motion in a "post"–flare loop prominence.**
X. Gu, B. Li, Q. Li, Y. Ding, A. Xu, J. Ding, Y. Tang.
Sci. Sin., Ser. A, Vol. 27, No. 9, p. 972 – 983 (1984).

The velocity fields of a "post"–flare loop prominence taken in different times by the spectra–spectroheliograph are measured by a 3CS densitometer. According to the fundamental features of the observed velocity fields, the authors consider that the plasma matter within the loop prominence, which is formed by the magnetic reconnection after flares, would flow down from the top of the loops along the helical magnetic lines under the action of the gravity of the Sun and the magnetic stress in the loop prominence.

073.058 **Heating the sun's lower transition region with fine–scale electric currents.**
D. Rabin, R. Moore.
Astrophys. J., Vol. 285, No. 1, p. 359 – 367 (1984).

This paper discusses the hypothesis that the lower transition region is locally heated by the dissipation of electric currents. It proposes a model based on ohmic heating by filamentary electric currents that flow along the magnetic field. The current filaments must be of fine scale, with a narrow dimension in the range 1 cm to 1 km, and the ambient magnetic field must be greater than about 10 gauss. An ensemble of filamentary currents that agree in sign across the horizontal scale of a photospheric granule can generate enough heat to match observations without the need for anomalous resistivity. Thermal conduction perpendicular to the axis of a current filament produces a distribution of emission measure over temperature that is in good agreement with observations.

073.059 **A 154–day periodicity in the occurrence of hard solar flares?**
E. Rieger, G. H. Share, D. J. Forrest, G. Kanbach, C. Reppin, E. L. Chupp.
Nature, Vol. 312, No. 5995, p. 623 – 625 (1984).

Since the launch of the SMM satellite on 1980 February 14, the Gamma–Ray Spectrometer has been monitoring the Sun in the energy range 0.3 – 100 MeV. So far 139 flares have been monitored. The authors report on an analysis of the temporal distribution of these high–energy events to provide information on solar activity and find that, instead of being randomly distributed in time, these events have a tendency to occur in groups with a mean spacing of ~154 days (75 nHz) over the observing interval.

073.060 **Calculation of spectrum and directivity of hard X–rays from solar flares.**
P. N. Khosa, R. R. Rausaria, S. Peshin, K. L. Moza.
Indian J. Radio Space Phys., Vol. 13, No. 2, p. 56 – 62 (1984).
Abstr. in Phys. Abstr., Vol. 87, No. 1244, Entry 105924 (1984).

073.061 **Nonlinear computations of a solar flare model.**
H. Strauss, G. Van Hoven.
Phys. Fluids, Vol. 27, No. 8, p. 2063 – 2066 (1984). Abstr. in Phys. Abstr., Vol. 87, No. 1246, Entry 115735 (1984).

073.062 **The turbulence in the solar atmosphere and in the interplanetary plasma.**
I. V. Chashej, V. I. Shishov.
Sov. Astron., Vol. 28, No. 3, p. 279 – 284 (1984). English translation of 37.073.114.

073.063 **Dynamics of recurrent surges on the sun.**
Yu. V. Platov, B. P. Filippov.
Sov. Astron., Vol. 28, No. 3, p. 321 – 324 (1984). English translation of 37.073.115.

073.064 **Non–radio observations of fast fluctuations.**
G. E. Brueckner.
Orages solaires de type I, p. 87 – 93 (1984). – See Abstr. 012.055.

High spatial (~1″) resolution spectroscopic observations of the transition zone plasma (~100,000K) reveal high energetic events in the quiet sun which show very fast fluctuations. These phenomena fall into two categories: (1) explosive events, and (2) jets.

073.065 **On the formation of temperature plateaus in the solar transition region.**
S. P. Owocki, A. N. McClymont, R. C. Canfield.
Bull. Am. Astron. Soc., Vol. 16, No. 3, Part I, p. 729 (1984). Abstract. – See Abstr. 010.065.

073.066 **Magnetic structure and heating of the upper and lower transition region.**
J. F. Dowdy Jr., S. T. Wu, R. L. Moore.
Bull. Am. Astron. Soc., Vol. 16, No. 3, Part I, p. 729 (1984). Abstract. – See Abstr. 010.065.

073.067 **Heating of the chromosphere and transition region by fast electrons from the temperature minimum.**
A. V. Byalko, E. H. Avrett.
Bull. Am. Astron. Soc., Vol. 16, No. 3, Part I, p. 730 (1984). Abstract. – See Abstr. 010.065.

073.068 **Recent soft X–ray spectra during a flare decay.**
K. J. H. Phillips, J. R. Lemen, K. T. Strong.
Bull. Am. Astron. Soc., Vol. 16, No. 3, Part I, p. 730 (1984). Abstract. – See Abstr. 010.065.

073.069 **A note on magnetic fields and electric currents in solar prominences.**
J. L. Ballester.
Sol. Phys., Vol. 94, No. 1, p. 151 – 154 (1984).

The author determines magnetic fields and electric currents in three prominences without simplifications used in his previous papers.

073.070 **Longitudinal distribution of the cool solar surges.**
V. K. Verma.
Sol. Phys., Vol. 94, No. 1, p. 155 – 159 (1984).
This investigation shows that solar surges are poorly associated with sudden ionospheric disturbances (X-ray) implying that solar surge material is cool and does not heat the corona. The investigation also shows that solar surges are most prolific at longitudes 80°, 110°, 260°, and 290°.

073.071 **Synchrotron or plasma emission in solar microwave flares?**
A. O. Benz.
Sol. Phys., Vol. 94, No. 1, p. 161 – 163 (1984).
The spectral indices of microwave and hard X-ray emissions of a solar flare are found to correlate. Their observed values are in agreement with the expected relation from synchrotron and bremsstrahlung theory. These results are considered as strong evidence for the synchrotron mechanism in the microwave flare, contrary to recent alternative suggestions.

073.072 **Spectral properties of filamentary physically inhomogeneous prominences. IV. Excitation and ionization of Ca II.**
N. N. Morozhenko.
Astrometr. Astrofiz., Vyp. 53, p. 3 – 10 (1984). In Russian.
The source functions of Ca II K λ8542 lines and continuum have been obtained for models of structurally and physically inhomogeneous prominences. The K line profiles have been calculated. The following items are discussed: Ca II excitation and ionization; the role of different levels in these processes; the effect of filamentness and physical inhomogeneity on the line profiles.

073.073 **Investigation of second-level excitation and ionization of a hydrogen atom in filamentary prominences.**
V. V. Zharkova.
Astrometr. Astrofiz., Vyp. 53, p. 22 – 30 (1984). In Russian.
Filamentary prominence models with the structure coefficients $\gamma = 0.00 - 0.99$ are considered. Equations of radiative transfer in $H_\alpha$, $L_\alpha$ and $L_c$ frequencies are solved together with steady-state equations for levels with $n = 2 - 5$. Distributions of relative second-level populations and emission measures of hydrogen atom over the optical depth as well as Lyman-α line intensities are obtained. It is concluded that the effect of the filamentary structure on the Lyman-α profiles is small.

073.074 **A model of the oscillations in the chromosphere and transition region above sunspot umbrae.**
Y. C. Žugžda (*Yu. D. Zhugzhda*), J. Staude, V. Locāns.
Astron. Nachr., Vol. 305, No. 5, p. 236 (1984). Abstract. – See Abstr. 012.062. See 37.073.116 for a detailed version.

073.075 **The gas pressure in solar prominences.**
J. Hellwig, G. Stellmacher, E. Wiehr.
Astron. Astrophys., Vol. 140, No. 2, p. 449 – 452 (1984).
New measurements of the emission relation of $Ca^+8542$ and Hβ are discussed. The emission ratio $E(Ca^+8542)/E(H\beta) \cong 0.16$ is found to be significantly lower than that ($\approx 0.3$) observed by Landman et al. (1977). Comparison with the model (7500, 0.01) by Heasley and Milkey (1978) shows that a much higher value of the gas pressure ($\approx 0.1$ dyn cm$^{-2}$) has to be assumed to match the observed lower ratio.

073.076 **Erratum: "Microwave, soft and hard X-ray imaging observations of two solar flares" [Astron. Astrophys., Vol. 132, No. 2, p. 241 – 252 (1984)].**
M. R. Kundu, M. E. Machado, F. T. Erskine, M. G. Rovira, E. J. Schmahl.
Astron. Astrophys., Vol. 140, No. 2, p. 476 (1984). See Abstr. 37.073.024.

073.077 **Review of solar observations of the $n = 2$–$2$ transitions in fourth period elements.**
K. D. Lawson, N. J. Peacock.
Astron. Astrophys., Suppl. Ser., Vol. 58, No. 3, p. 475 – 497 (1984).
A review is made of the solar observations identified with the $n = 2$–$2$ transitions of elements in the fourth period of the periodic table. The identifications are mainly with transitions of the solar abundant elements calcium and iron. The majority of observations are of solar flares. Some new identifications are presented. The study illustrates the usefulness of the $n = 2$–$2$ transitions for the measurement of electron densities. It, also, clearly demonstrates the need for observations having the highest possible spectral resolution, especially in the spectral region around a wavelength of 100 Å.

073.078 **Oscillations in the chromospheric network as response to resonance oscillations of the sun.**
V. E. Merkulenko, M. N. Mishina.
Issled. Geomagn. Aehron. Fiz. Solntsa, Moskva, No. 68, p. 56 – 65 (1984). In Russian. Abstr. in Ref. Zh., 51. Astron., 11.51.355 (1984).

073.079 **Review on methods and data of observations of turbulent electric fields in solar flares.**
N. M. Firstova.
Issled. Geomagn. Aehron. Fiz. Solntsa, Moskva, No. 68, p. 31 – 45 (1984). In Russian. Abstr. in Ref. Zh., 51. Astron., 11.51.403 (1984).

073.080 **On a possible appearance of the process of energy accumulation connected with powerful flares.**
V. P. Nefed'ev, N. N. Potapov, G. Ya. Smol'kov.
Issled. Geomagn. Aehron. Fiz. Solntsa, Moskva, No. 68, p. 84 – 90 (1984). In Russian. Abstr. in Ref. Zh., 51. Astron., 11.51.405 (1984).

073.081 **Electron spectra deduced from solar hard X-ray bursts.**
G. Trottet, N. Vilmer.
Adv. Space Res., Vol. 4, No. 2 – 3, p. 153 – 156 (1984). – See Abstr. 012.075.
Information regarding the acceleration of electrons in solar flares may be deduced from the hard X-ray emission they produce through bremsstrahlung. From these observations, the electron energy spectra and temporal evolution can be deduced through models. Here, a model of hard X-ray emission is briefly presented.

073.082 **Illustration of the emergent magnetic flux model: observations of the large flare of 1981 May 16.**
F.-m. Hu, Q.-d. Wu.
Chin. Astron. Astrophys., Vol. 8, No. 3, p. 177 – 184 (1984). English translation of 34.073.092.

073.083 **Positional fix of the loop prominence of 1981 April 27 and comparison with X-ray sources.**
S.-c. Li, T.-j. Cao.
Chin. Astron. Astrophys., Vol. 8, No. 3, p. 190 – 193 (1984). English translation of 37.073.187.

073.084 **A possible mechanism of solar surges.**
J.-p. Ding, A.-a. Xu.
Chin. Astron. Astrophys., Vol. 8, No. 3, p. 217 – 222 (1984). English translation of Chin. J. Space Sci., Vol. 4, No. 1, p. 89 – 95 (1984).
The authors propose the following sequence of events as a possible mechanism for producing surges: pulsed emergence of magnetic flux in the active region – periodic, dynamic pinching of the plasma – rapid field annihilation – ejection of surge. This model gives a satisfactory explanation of the recurrence of surges and some other properties of surges.

073.085 **Rotating motion in solar surges.**
A.–a. Xu, J.–p. Ding, S.–y. Yin.
Chin. Astron. Astrophys., Vol. 8, No. 4, p. 294 – 298 (1984). English translation of Acta Astron. Sin., Vol. 25, No. 2, p. 119 – 126 (1984).

The line–of–sight velocity field of the solar limb surge of 1980 Oct 11, observed at Yunnan Observatory, showed a rotating motion. The authors point out that the steep density gradient between the surge plasma and its surrounding atmosphere causes double–pole diffusion, and the electric field generated by the double–pole diffusion causes electric drift of the surge plasma, hence the rotation.

073.086 **Statistics and investigation of white light solar flares.**
X.–z. Chen, Z.–y. Wang.
Chin. Astron. Astrophys., Vol. 8, No. 4, p. 299 – 305 (1984). English translation of Acta Astron. Sin., Vol. 25, No. 2, p. 127 – 135 (1984).

The authors discuss the properties of white light flares on the basis of the published accounts of these events, together with the associated Hα flares, radio bursts, X–ray bursts, proton events and ionospheric disturbances. In addition, spectral plates taken at Purple Mountain Observatory since 1962 have been examined. It is found that 5% of the spectrograms of solar flares show variable white–light emission.

073.087 **A combined optical and radio study of the solar ejection of 1982 January 22.**
J.–l. Wang, F.–x. Kong.
Chin. Astron. Astrophys., Vol. 8, No. 4, p. 335 – 339 (1984). English translation of Chin. J. Space Sci. Vol. 4, No. 3, p. 177 – 183 (1984).

The limb event 1982 January 22 with Type III + V and Type II bursts and Hα mass ejection is described. The bursts in short dm and cm waves are discussed in terms of the flux rate–time curve. The authors estimate the total energy of this event to be $10^{30}$erg.

073.088 **On a peculiar type of prominence activation.**
P. Pant.
Bull. Astron. Soc. India, Vol. 12, No. 3, p. 252 – 257 (1984).

The author describes time–lapse H–alpha observations of a peculiar type of prominence activation of 1981 October 7 and its morphological behaviour. The prominence material seems to have originated from the photosphere as a huge twisted rope–like structure. It reached a height of $\cong 10^5$km, and later on split into braided closed loop–like features. The splitting of the twisted rope–like structure is explained on the basis of screw–type instability in current–carrying plasma column.

073.089 **Cartes synoptiques de la chromosphère solaire et catalogues des filaments et des centres d'activité.**
M.–J. Martres, G. Zlicaric.
Cartes Synoptiques, Vol. 6, Fasc. 5, 175 pp. (1984).

Rotations Nos. 1717 – 1743 (1982 January 2 – 1983 December 14).

073.090 **Eruption chromosphérique, April – September 1981.**
Q. Bull. Sol. Act., Vol. 23, Part III, p. 43 – 120 (1981).

073.091 **A consistent picture of coronal and chromospheric processes in a well–observed solar flare.**
T. A. Gunkler, R. C. Canfield, L. W. Acton, A. L. Kiplinger.
Astrophys. J., Vol. 285, No. 2, p. 835 – 842 (1984).

The authors have analyzed the solar flare of 15:22 UT 1980 June 24 using simultaneous observations in hard X–rays, soft X–rays, and Hα line profiles. The X–ray observations were made with instruments aboard the SMM satellite, and the Hα profiles were taken with a CCD detector at Sacramento Peak Observatory. The paper discusses various flare phenomena, including heating of the chromosphere by nonthermal electrons, enhanced coronal pressure, enhanced thermal conduction, chromospheric evaporation, and mass motion. It is shown that a consistent picture of the flare can be obtained in terms of a model where the energy release occurs at the site of the interaction of two large loop systems.

073.092 **Alfvénic resonances on solar spicules.**
A. C. Sterling, J. V. Hollweg.
Astrophys. J., Vol. 285, No. 2, p. 843 – 850 (1984).

It is suggested that twisting and heating of solar spicules can be produced by Alfvén waves which enter the spicule from below. The spicule is treated as a region of constant Alfvén speed which is bounded above by a region of much higher Alfvén speed (the corona) and below by a region of exponentially increasing Alfvén speed (the photosphere and chromosphere). It is shown how the spicule can act as a resonant cavity. The transmission of the waves into the cavity is analytically determined to be enhanced at certain resonant frequencies. The authors suggest that the Alfvén waves are dissipated via a turbulent cascade of their energy to higher wavenumbers and that the waves can thereby heat the spicules to the observed temperatures.

073.093 **Peculiarities of radio and optical emissions in the post–maximum period of the proton flare on May 16, 1981.**
V. F. Mel'nikov, T. S. Podstrigach, V. S. Prokudina.
Astron. Tsirk., No. 1302, p. 4 – 6 (1983). In Russian.

073.094 **Appearance of global solar 160 min pulsations in flares of individual active regions.**
V. V. Kasinskij.
Astron. Tsirk., No. 1314, p. 3 – 5 (1984). In Russian.

073.095 **Analysis of the 1980 November 18 limb flare observed by the Hard X–ray Imaging Spectrometer (HXIS).**
E. Haug, G. Elwert, P. Hoyng.
Lehrstuhl Theor. Astrophys. Univ. Tübingen, Prepr., 4 pp. (1984). To appear in Adv. Space Res.

073.096 **Energy and angular distributions of energetic flare electrons and their X–radiation. I. Initially monoenergetic electrons.**
E. Haug, G. Elwert, R. R. Rausaria.
Lehrstuhl Theor. Astrophys. Univ. Tübingen, Prepr., 30 pp. (1984). To appear in Astron. Astrophys.

073.097 **Hα emission and electron beams: its relation in bright points of solar flares.**
M. A. Livshits.
Astron. Tsirk., No. 1327, p. 1 – 4 (1984). In Russian.

073.098 **Observation of chromospheric activity and an extended filament in an active complex in July 1982.**
A. B. Delone, V. I. Bulavina, V. S. Prokudina.
Astron. Tsirk., No. 1337, p. 1 – 4 (1984). In Russian.

073.099 **The relation of electron temperature to emission measure and limits of increase in emission measure in soft X–ray flares.**
R. E. Denton, U. Feldman.
Astrophys. J., Vol. 286, No. 1, p. 359 – 362 (1984).

A sample of 126 soft X–ray flares (ranging in X–ray class from C3 up to X3) is studied in order to look for possible correlations between various parameters of the flare plasma. It is found that the emission measure ($N^2V$) correlates positively with the electron temperature ($T$) determined from the Ca XIX resonance line and nearby satellites. Measured values of the maximum value of $T$ range from 8 to $18 \times 10^6$K. The authors observe increases in $N^2V$ of no more than a factor of 400 from X–ray onset to soft X–ray maximum. The largest increases result from the most intense flares.

073.100 **Measurement of magnetic fields in solar prominences.**
G. M. Nikol'skij, I. S. Kim, S. Koutchmy, A. I. Stepanov.
Problems of physics of solar flares, p. 52 – 59 (1983). In Russian. Abstr. in Ref. Zh., 51. Astron., 12.51.624 (1984). – See Abstr. 012.070.

073.101 **On X-ray spectral diagnostics of epithermal electrons in solar flare plasma.**
A. M. Urnov.
Problems of physics of solar flares, p. 113 – 119 (1983). In Russian. Abstr. in Ref. Zh., 51. Astron., 12.51.647 (1984). – See Abstr. 012.070.

073.102 **On the nature of pulsations of X-ray and microwave radiation of solar flares.**
V. V. Zajtsev, A. V. Stepanov.
Problems of physics of solar flares, p. 173 – 180 (1983). In Russian. Abstr. in Ref. Zh., 51. Astron., 12.51.648 (1984). – See Abstr. 012.070.

073.103 **On the dynamics of the transition layer of solar flares.**
O. P. Shmeleva.
Problems of physics of solar flares, p. 187 – 194 (1983). In Russian. Abstr. in Ref. Zh., 51. Astron., 12.51.649 (1984). – See Abstr. 012.070.

073.104 **Some peculiarities of the spatial composition and mechanism of emission of flares in the optical range.**
V. A. Ostapenko.
Problems of physics of solar flares, p. 87 – 93 (1983). In Russian. Abstr. in Ref. Zh., 51. Astron., 12.51.650 (1984). – See Abstr. 012.070.

073.105 **Problems of physics of solar flares.**
B. V. Somov.
Problems of physics of solar flares, p. 5 – 51 (1983). In Russian. Abstr. in Ref. Zh., 51. Astron., 12.51.657 (1984). – See Abstr. 012.070.

073.106 **Disparition brusque of quiescent prominences.**
G. P. Apushkinskij, N. A. Topchilo.
Astron. Zh., Tom 61, Vyp. 6, p. 1150 – 1157 (1984). In Russian. English translation in Sov. Astron., Vol. 28, No. 6.
The disparitions brusques of quiescent prominences of July 2, 1976 and July 30, 1980 are discussed. Radio and optical data at $\lambda = 1.35$ and 0.8 cm and in the $H_\alpha$ line, respectively, confirm the arch structures of prominences. The broadening of arch as well as the plasma outflow through its base were the causes of their destruction, the heat conductivity was negligible. The differential rotation of the solar atmosphere is found to be the most appropriate cause of quiescent prominences' instability.

073.107 **Determination of the magnetic field in quiescent solar prominences from polarization characteristics of radiated light.**
M. B. Gornyj, D. V. Kupriyanov, B. G. Matisov.
Astron. Zh., Tom 61, Vyp. 6, p. 1158 – 1167 (1984). In Russian. English translation in Sov. Astron., Vol. 28, No. 6.
A method of approximate solution of optical pumping equations describing the Hanle effect in quiescent solar prominences is suggested. The method is based on the property of weak anisotropy of the exciting light. Polarization characteristics of the prominences' radiation depending on the value and direction of the magnetic field are calculated for the helium triplet transitions. The influence of the trapping of resonant radiation upon the polarization of an ensemble of radiating atoms is analysed.

073.108 **Ca II K bright points and the solar cycle.**
K. R. Sivaraman.
Sol. Phys., Vol. 94, No. 2, p. 235 – 238 (1984).
Number counts are made of the bright points that populate the interior of the Ca II network in the solar chromosphere for four solar cycles. The author finds that the number of these fine structures during the solar maximum exceeds that during the minimum phase on an average by 30%. The contribution due to this excess number would need to be taken into account while explaining the variation in the Ca II K emission profiles of the integrated sunlight from the solar minimum to the maximum.

073.109 **Oscillations of the Hα emission in solar prominences.**
E. Wiehr, G. Stellmacher, H. Balthasar.
Sol. Phys., Vol. 94, No. 2, p. 285 – 288 (1984).
The time dependence of Doppler shift and line-center intensity is simultaneously observed for the Hα emission of three solar prominences, each one during about two hours. Doppler oscillations with periods near one hour and amplitudes between 1 and 2 km $s^{-1}$ are conspicuously visible in the recordings of all three prominences. Fourier analysis yields periods of 50, 60, and 64 min, as well as slight indications of short periods near 3 and 5 min. No oscillations are found in the line-center brightness.

073.110 **The oscillating loop prominence of July 17, 1981.**
B. Vršnak.
Sol. Phys., Vol. 94, No. 2, p. 289 – 297 (1984).
Oscillatory motions of a loop prominence observed on July 17, 1981 are analysed. The oscillations were mainly horizontal, with a period of 8 min. Restoring force was a result of magnetic tension, and assuming a simplified magnetic field configuration the expression for frequencies of oscillations is derived and compared with the observations. Taking the observed period, the strength of the magnetic field permeating the prominence is estimated as 45 G. Finally, the stability of the prominence is discussed.

073.111 **Evolution of rising helical prominences in a nonuniform atmosphere.**
G. W. Pneuman.
Sol. Phys., Vol. 94, No. 2, p. 299 – 313 (1984).
Recent coronagraph observations of rising prominences such as in the 14 April and 5 May, 1980 coronal transient events, as well as other older observations, have shown evidence for helical structure in the prominences. The author describes a simple model – that of a helical pinch rising in a low beta atmosphere under the influence of an ambient external magnetic field which declines in strength with radial distance from the solar center. Under the general assumption of an internal uniform, but time-varying, temperature and neglecting gravitational stratification within the prominence, expressions are derived for associated variations in the prominence structure as it rises. The author discusses in some detail particular quantities which are potentially most accessible to observation such as prominence radius, density, and pitch angle as they vary with height during the eruptive process.

073.112 **Energy release in solar flares.**
P. A. Sturrock, P. Kaufmann, R. L. Moore, D. F. Smith.
Sol. Phys., Vol. 94, No. 2, p. 341 – 357 (1984).
The authors examine observational evidence concerning energy release in solar flares. They propose that different processes may be operative on four different time scales: (1) on the sub-second time scale of "sub-bursts" which are a prominent feature of mm-wave microwave records; (2) on the few-seconds time scale of "elementary bursts" which are a prominent feature of hard X-ray records; (3) on the few-minutes time scale of the impulsive phase; and (4) on the tens-of-minutes or longer time scale of the gradual phase. They propose that the concentration of magnetic field into "magnetic knots" at the photosphere has important consequences for the coronal magnetic-field structure such that the magnetic field in this region may be viewed as an array of "elementary flux tubes". The case of two-ribbon flares is also considered.

073.113 **Horizontal and vertical heating in the solar chromosphere.**
M.-t. Song.
Acta Astron. Sin., Vol. 25, No. 3, p. 224 – 233 (1984). In Chinese.
Using the anisotropy of the stress tensor of turbulent waves, the effects of Alfvén turbulence on a single magnetic tube are investigated. Under the assumptions of the magnetic field being force-free and of neglecting thermal conduction and convection the equations governing the one-dimensional steady flow in the tube are derived. And then, in terms of these equations the horizontal and vertical heating in the chromosphere are computed.

073.114 **Solar flares and plasma experiments.**
A. T. Altyntsev, V. I. Krasov, V. M. Tomozov.
Itogi Nauki i Tekhniki. Seriya Astronomiya. Tom 25. Physics of the sun, p. 99 – 191 (1984). In Russian. – See Abstr. 003.053.

073.115 **Numerical simulations of loops heated to solar flare temperatures. III. Asymmetric heating.**
C.–C. Cheng, J. T. Karpen, G. A. Doschek.
Astrophys. J., Vol. 286, No. 2, p. 787 – 803 (1984).
The authors have numerically simulated the flare hydrodynamics of a coronal loop subjected to a large, asymmetrically located energy input. The important physical processes involved are the rapid heating of the plasma to high temperatures ($\sim 10^7$K), the generation and propagation of a thermal conduction front, chromospheric evaporation, and the formation of compressed chromospheric plasma. Using the hydrodynamic results, the evolving X–ray and UV spectroscopic signatures of the flaring loop are calculated. Comparisons of the calculated X–ray and UV spectra with available flare observations show significant differences.

073.116 **The energetics of chromospheric evaporation in solar flares.**
E. Antonucci, A. H. Gabriel, B. R. Dennis.
Astrophys. J., Vol. 287, No. 2, p. 917 – 925 (1984).
The typical impulsive–phase development of a soft X–ray solar flare is derived from observations of a large set of flares with the Bent Crystal Spectrometer and the Hard X–Ray Burst Spectrometer on the SMM spacecraft. An indicator of the impulsive phase in soft X–rays is the presence of high–speed plasma upflows with velocities up to 400 km $s^{-1}$, temperatures of $\sim 10^7$K, and turbulent mass motions of $\sim 100$ km $s^{-1}$. The data support the hypothesis that chromospheric evaporation is effective in solar flares, and is the main mechanism responsible for the formation of the flare coronal plasma.

073.117 **Fast variations in high–energy X–rays from solar flares and their constraints on nonthermal models.**
A. L. Kiplinger, B. R. Dennis, K. J. Frost, L. E. Orwig.
Astrophys. J., Lett. Ed., Vol. 287, No. 2, p. L105 – L108 (1984).
Studies of the temporal properties of over 7000 hard X–ray solar flares as observed with the Hard X–Ray Burst Spectrometer on SMM have revealed hundreds of examples of fast spikes with durations of less than 1 s. This study quantitatively compares the spectra observed in a well–observed fast spike with the predicted spectra from a detailed, time–dependent, nonthermal beam model in order to impose new constraints on electron beam dynamics and path lengths.

**Spectrophotometric investigations of quiescent solar prominences.**
See Abstr. 003.007.

**Solar research at RAL (*Rutherford Appleton Laboratory*).**
See Abstr. 009.013.

**Solar terrestrial physics: present and future. A report based on the Solar Terrestrial Physics Workshop, December 1982 to November 1983.**
See Abstr. 012.038.

**Theoretical Ne IX line ratios compared to solar observations.**
See Abstr. 022.123.

**Mg IX and Si XI line ratios in the Sun.**
See Abstr. 022.124.

**Wavelength coincidence of allowed and intercombination multiplets in N III and O IV: a density and/or temperature diagnostic.**
See Abstr. 022.135.

**On the time of achievement of ionization equilibrium in a hot low–density gas.**
See Abstr. 022.157.

**An installation for recording the velocity field of prominences.**
See Abstr. 034.069.

**Image processing software system of the X–ray telescope aboard the HINOTORI spacecraft.**
See Abstr. 035.075.

**The thermal instability in a sheared magnetic field: filament condensation with anisotropic heat conduction.**
See Abstr. 062.004.

**A numerical approach to equilibrium and stability properties in astrophysical plasmas.**
See Abstr. 062.038.

**Particle acceleration.**
See Abstr. 062.087.

**An essay on sunspots and solar flares.**
See Abstr. 072.044.

**Solar active regions: observation and theory.**
See Abstr. 072.059.

**Dynamic evolution of recurrent mass ejections observed in Hα and C IV lines.**
See Abstr. 072.062.

**Cellular structures of an activity complex, their influence on the development of large solar flares.**
See Abstr. 072.071.

**Note on the flare activity of the July 1982 sunspot group.**
See Abstr. 072.104.

**Coronal heating by reconnection in DC current systems. A theory based on Taylor's hypothesis.**
See Abstr. 074.006.

**The role of Alfvén wave dispersion in the dynamics of energetic protons in the solar corona.**
See Abstr. 074.038.

**Revivals of a coronal arch.**
See Abstr. 074.047.

**Associations between coronal mass ejections and solar energetic proton events.**
See Abstr. 074.052.

**Selective heating of minority ions in the lower solar corona by electrostatic ion cyclotron waves.**
See Abstr. 074.053.

**The origin of spicules and heating of the lower transition region.**
See Abstr. 074.071.

**On the problem of formation of magnetic fields of flare–active regions.**
See Abstr. 075.008.

**Remarks on the magnetic support of quiescent prominences.**
See Abstr. 075.017.

**Characteristic structures of the solar disc observed on rocket UV filtergrams.**
See Abstr. 076.001.

**Analysis of the short-period pulsations in the hard X-ray radiation of the solar flares detected by Venera 13 and Venera 14.**
See Abstr. 076.007.

**Are EUV bursts members of the flare family?**
See Abstr. 076.009.

**High-energy neutral radiations from the Sun.**
See Abstr. 076.012.

**Solar X-ray flare emission and energetic particles according to data of observations aboard the Venera 13, 14 automatic interplanetary stations.**
See Abstr. 076.016.

**Calculation of X-ray polarization during flares.**
See Abstr. 076.021.

**Effect of electron energy and angular dispersion on hard X-ray polarization during solar flares.**
See Abstr. 076.022.

**Temporal variations in the solar submillimeter continuum.**
See Abstr. 077.015.

**Characteristics of flares producing metric type II bursts and coronal mass ejections.**
See Abstr. 077.016.

**Pulsations of type IV radio bursts as an indicator of protonability of solar flares.**
See Abstr. 077.017.

**Recent advances in solar radio physics.**
See Abstr. 077.019.

**Type IV radio radiation and orientation of the magnetic field in a flare region as an index of geoefficiency of a flare stream.**
See Abstr. 077.021.

**Microwave emission and polarization of a flaring loop.**
See Abstr. 077.022.

**Radio pulsations associated with the solar flare of 1982 April 14.**
See Abstr. 077.042.

**Survey of $He^+/He^{2+}$ abundance ratios in energetic particle events.**
See Abstr. 078.001.

**Electron-rich particle events and the production of gamma-rays by solar flares.**
See Abstr. 078.004.

**On the energetic dependence of the length of the free path of flare particles.**
See Abstr. 078.007.

**On the problem of relativistic particles acceleration in processes similar to those in solar flares.**
See Abstr. 078.011.

**Acceleration of particles in shocked magnetic neutral sheets.**
See Abstr. 078.012.

**Ionic charge states of N, Ne, Mg, Si and S in solar energetic particle events.**
See Abstr. 078.013.

**Peculiar points of the solar rotation.**
See Abstr. 080.015.

**Intensity oscillations in the calcium – K line.**
See Abstr. 080.043.

**Determination of the total amount of hydrogen atoms in a quiescent prominence.**
See Abstr. 080.066.

**Apparent solar temperature enhancement due to large-amplitude waves.**
See Abstr. 080.123.

**Flares, 200 MHz radio bursts and magnetic storms.**
See Abstr. 084.010.

**Excitation and ionizatioan in the upper atmosphere during solar flares with account for photoelectrons.**
See Abstr. 085.016.

**Shock acceleration of nuclei and electrons in the heliosphere beyond 24 AU.**
See Abstr. 106.019.

**Magnetohydrodynamic modelling of interplanetary disturbances between the Sun and Earth.**
See Abstr. 106.020.

**On the E-W asymmetry and the generation of ESP (*Energetic Storm Particle*) events.**
See Abstr. 106.021.

**Flare-generated heliospheric and geomagnetic disturbances.**
See Abstr. 106.057.

**The effectiveness of charged particles acceleration by shock waves in supernova remnants and in solar flares.**
See Abstr. 125.014.

**On pulsations of cosmic ray intensity before the increase of the relativistic particle stream from solar flares.**
See Abstr. 144.021.

**Cosmic ray fluctuations on the earth during solar flares.**
See Abstr. 144.079.

# 074 Corona, Solar Wind

074.001 **Interferometric observations of radial velocities in the F-corona.**
L. I. Shestakova, P. V. Shcheglov.
Astron. Tsirk., No. 1286, p. 6 – 8 (1983). In Russian.

074.002 **Resonant heating: an interpretation of coronal loop data.**
J. V. Hollweg, A. C. Sterling.
Astrophys. J., Lett. Ed., Vol. 282, No. 1, p. L31 – L33 (1984).
The authors show that the resonant heating theory of Hollweg can be used to organize the coronal loop data of Golub et al. When combined with a reasonable form for the input power spectrum, the resonant heating theory is fully compatible with the loop data.

074.003 **Detection of a slowly moving coronal transient event.**
R. Fisher, C. Garcia.
Astrophys. J., Lett. Ed., Vol. 282, No. 1, p. L35 – L37 (1984).
Data are presented from the K-III coronameter which show the development of a bright coronal structure over a period of 3 hours. Interpreting the upward motion as movement within the plane of the sky, the velocity inferred for the event is 14 km $s^{-1}$. The amount of material removed from the corona at the site of this event over a period of 24 hours is of the order of $10^{15}$ grams. It is concluded that this event was an extremely slow coronal mass ejection transient.

074.004 **Distribution of solar wind angular momentum between particles and magnetic field: inferences about the Alfvén critical point from Helios observations.**
E. Marsch, A. K. Richter.
J. Geophys. Res., Vol. 89, No. A7, p. 5386 – 5394 (1984).
The authors first discuss theoretically the relative importance and the behavior of the two basic terms adding to the total angular momentum flux, the angular momentum of the particles and of the magnetic field stresses, respectively. Second, they analyze these two quantities with respect to their dependence on heliocentric distance by using the Helios 1 and 2 plasma and interplanetary magnetic field observations between 0.3 and 1 AU. Applying these results as well as various combinations of the constants of motion for the solar wind and their observational constraints, as deduced earlier by Marsch and Richter (1984), the authors finally present various methods (1) to derive the values of several characteristic solar wind plasma and magnetic field parameters at the Alfvén critical points, (2) to estimate their locations above the solar surface, and (3) to obtain the radial slope of the associated solar wind velocity profiles for the three solar wind classes separately.

074.005 **A reexamination of rotational tangential discontinuities in the solar wind.**
M. Neugebauer, D. R. Clay, B. E. Goldstein, B. T. Tsurutani, R. D. Zwickl.
J. Geophys. Res., Vol. 89, No. A7, p. 5395 – 5408 (1984).
ISEE 3 magnetic field and proton data are used to study the properties of rotational and tangential discontinuities in the solar wind. A Sonnerup-Cahill minimum variance analysis of the magnetic field data is used to determine the direction of the normal to each discontinuity. The discontinuities are then classified as rotational (RD), tangential (TD), or either (ED), depending on the relative values of the normal field component and the change of field magnitude across them. This process yields substantially more RD's than TD's, in agreement with earlier studies using this method of classification. Other field and plasma parameters are then examined for each of these three groups.

074.006 **Coronal heating by reconnection in DC current systems. A theory based on Taylor's hypothesis.**
J. Heyvaerts, E. R. Priest.
Astron. Astrophys., Vol. 137, No. 1, p. 63 – 78 (1984).
Photospheric fluid motions permanently build up coronal electric currents, likely to be force free. These DC current systems represent a potential source of free energy on which flares, or more regular coronal heating could tap. How much and how fast are the questions considered here. The presented theory also enlightens the possible relation, up to now ill-understood, between coronal heating and flaring activity.

074.007 **On the role of plasma parameters and the Kelvin-Helmholtz instability in a viscous interaction of solar wind streams.**
N. P. Korzhov, V. V. Mishin, V. M. Tomozov.
Planet. Space Sci., Vol. 32, No. 9, p. 1169 – 1178 (1984).
The authors examine a mechanism for breaking down solar wind speed shears within 1 astronomical unit, initiated by the development of the Kelvin-Helmholtz instability for typical parameters of the plasma and magnetic field in the interplanetary medium.

074.008 **Low energy ions in corotating interaction regions at 1 AU: observations.**
I. G. Richardson, R. D. Zwickl.
Planet. Space Sci., Vol. 32, No. 9, p. 1179 – 1193 (1984).
The spacecraft ISEE-3 was launched in August 1978 and subsequently placed in orbit about the Sun-Earth L1 libration point where it continuously monitored the particles and fields in interplanetary space until mid-1982. The ISEE-3 Energetic Proton Anisotropy Spectrometer makes 3-dimensional intensity measurements of 35-1600 keV, $Z \geqslant 1$ ions. This data is used in conjunction with simultaneous solar wind plasma and magnetic field data from the same spacecraft to study the properties of ions in interaction regions lying at the leading edges of nine corotating high speed solar wind streams observed during October 1978 – July 1979.

074.009 **The Palermo-Harvard numerical code for the dynamics of confined coronal plasma.**
G. Peres, S. Serio, G. S. Vaiana, R. Rosner.
XVI International Conference on Phenomena in Ionized Gases, p. 96 – 97 (1983). Abstr. in Phys. Abstr., Vol. 87, No. 1239, Entry 82461 (1984). – See Abstr. 012.017.

074.010 **Ascending arches in the monochromatic corona at the July 1981 eclipse.**
A. B. Delone, E. A. Makarova, J. Sikora.
Sov. Astron. Lett., Vol. 10, No. 2, p. 129 – 130 (1984). English translation of 37.074.035.

074.011 **Photometric analysis of the June 11, 1983 solar corona.**
S. Koutchmy, C. Nitschelm.
Astron. Astrophys., Vol. 138, No. 1, p. 161 – 163 (1984).
The solar corona was observed using different photocameras during the solar total eclipse on June 11, 1983 under good sky conditions. Pictures were used to perform a detailed morphological analysis of *K*-corona structures and an isophotometric analysis. Accurate values of the flattening index over 8 $R_{\odot}$ are deduced.

074.012 **Oscillations and stationary motions in filaments.**
B. Schmieder, J. M. Malherbe, P. Mein, E. Tandberg-Hanssen.
Mem. Soc. Astron. Ital., Vol. 55, Nos. 1 – 2, p. 325 – 329 (1984). – See Abstr. 012.022.
Observation of oscillations in filaments, which are cold magnetic structures suspended in the corona, are used to test theories of coronal heating by acoustic waves high in the atmosphere. Long lived stationary motions observed in filaments suggest exchanges of material between them and the surrounding corona. A coupling between upflows in filaments and subphotospheric convective motions, dragging the field lines of the prominence magnetic support, is possible.

074.013 **Helios observational constraints on solar wind expansion.**
E. Marsch, A. K. Richter.
J. Geophys. Res., Vol. 89, No. A8, p. 6599 – 6612 (1984).
Helios particle and magnetic field observations between 0.3 and 1 AU are used to determine plasma parameters that characterize the bulk and internal energy state of the solar wind.

074.014 **Resonant acceleration and heating of solar wind ions: anisotropy and dispersion.**
P. A. Isenberg.
J. Geophys. Res., Vol. 89, No. A8, p. 6613 – 6622 (1984).
This paper is an extension of recent work by Isenberg and Hollweg dealing with the preferential acceleration and heating of solar wind heavy ions by a resonant cyclotron interaction with a turbulent spectrum of ion–cyclotron waves. Two major approximations of the previous work are eliminated: that of isotropic ion distributions, and that of nondispersive waves. The wave action flux of finite–amplitude Alfvén waves in a multi–ion, differentially flowing, gyrotropic plasma is derived. This quantity is incorporated into the wave–driven solar wind model of Isenberg and Hollweg, which is then applied to gyrotropic protons and alpha particles.

074.015 **Radial variations of large–scale magnetohydrodynamic fluctuations in the solar wind.**
L. F. Burlaga, M. L. Goldstein.
J. Geophys. Res., Vol. 89, No. A8, p. 6813 – 6817 (1984).
Two time periods are studied for which a relatively complete set of magnetic field and plasma measurements are available at both 1 AU (from IMP 8 and ISEE 3) and near 4 to 5 AU (from Voyager 1). One of these periods is characterized by the predominance of corotating streams, whereas many transient flows were observed in the second period. The evolution of these two flow systems with heliocentric distance is studied by using power spectral techniques.

074.016 **On the mechanism of solar wind formation.**
I. V. Chashej, V. I. Shishov.
Fiz. inst. Akad. Nauk SSSR. Prepr., No. 281, 22 pp. (1983). In Russian. Abstr. in Ref. Zh., 51. Astron., 8.51.271 (1984).

074.017 **On solar sources of recurrent high–energy streams of solar wind plasma.**
B. V. Zhiromskij.
Vologod. politekh. inst. Vologda, 10 pp. (1983). In Russian. Abstr. in Ref. Zh., 51. Astron., 8.51.272 (1984).

074.018 **Recurrent high–velocity streams of solar wind plasma in the circumterrestrial space.**
B. V. Zhiromskij.
Vologod. politekh. inst. Vologda, 10 pp. (1984). In Russian. Abstr. in Ref. Zh., 62. Issled. Kosm. Prostranstva, 8.62.492 (1984).

074.019 **Velocities of recurrent streams near the earth and dimensions of their solar sources.**
B. V. Zhiromskij.
Vologod. politekh. inst. Vologda, 9 pp. (1984). In Russian. Abstr. in Ref. Zh., 62. Issled. Kosm. Prostranstva, 8.62.493 (1984).

074.020 **Transition region of solar wind from observations of scintillations of the water vapour line of maser sources.**
D. F. Blooms, N. A. Lotova, R. L. Sorochenko.
Geomagn. Aehron., Tom 24, No. 4, p. 537 – 540 (1984). In Russian. English translation in Geomagn. Aeron.

074.021 **Polarimetry with the coronograph of the Norikura Observatory. II.**
I. Shimizu, T. Okamoto.
Tokyo Astron. Obs. Rep., Vol. 20, No. 2, p. 240 – 254 (1984). In Japanese.

074.022 **The effect of Alfvén wave pressure on temperature distribution in open regions of the solar corona.**
S. I. Molodykh.
Soln. Dannye, Byull., 1984, No. 6, p. 69 – 74 (1984). In Russian.
Calculations are made of the temperature and energy flux of the waves required for the solar wind to form in open regions of the solar corona both with and without taking into account the wave pressure. The results of the calculations enable to conclude that the wave pressure can be neglected in quiet regions at heliocentric distances $<4R_{\odot}$, but it should be taken into account for superradially diverging regions.

074.023 **Analysis of magnesium XI line profiles from solar active regions.**
R. L. Blake, R. D. Cowan, H. Felthauser, E. E. Fenimore, M. P. Hockaday, F. Bely–Dubau, P. Faucher, L. Steenman–Clark.
Astrophys. J., Vol. 282, No. 2, p. 784 – 792 (1984).
High–resolution solar spectra of the Mg XI $1s^2\ {}^1S_0$–$1s2p\ {}^1P_1$ resonance line at 9.169 Å and the associated nearby satellite lines are presented from two rocket–borne crystal spectrometer measurements. Comparisons with two independent sets of theoretical calculations for the $1s^2nl$–$1s2pnl$ dielectronic satellite lines with $3 \leqslant n \leqslant 7$ indicate electron temperatures of $4-4.5 \times 10^6$K. Measured line widths indicate either that the ion temperature exceeds the electron temperature by about $1 \times 10^6$K or that about 28 km s$^{-1}$ of turbulence is present.

074.024 **Structure of the inner corona derived from observations of the eclipse of 16 February 1980.**
O. G. Badalyan, M. A. Livshits, J. Sýkora.
Sol. Phys., Vol. 93, No. 1, p. 143 – 153 (1984).
The volume emission measure $EM(V)$ of the arch systems of the inner corona has been determined by analyzing the pictures of the green corona. It was found that the $EM(V)$–values of these systems are substantially lower than those obtained from X–ray data for the active regions, and this fact should be taken into account in interpreting extra–atmospheric observations. The combined investigation of data on the radiation of the corona in the green line and in the continuum enables to determine the total extension of the radiating matter, $(0.5–1) \times 10^{10}$ cm, as well as the density in the separate arches, $\approx 1.5 \times 10^9$ cm$^{-3}$. It is assumed that matter exists between the arches with a density of $\leqslant 10^8$ cm$^{-3}$.

074.025 **Interpretation of $^3$He abundance variations in the solar wind.**
M. A. Coplan, K. W. Ogilvie, P. Bochsler, J. Geiss.
Sol. Phys., Vol. 93, No. 2, p. 415 – 434 (1984).
The ion composition instrument on ISEE–3 has observed the isotopes of helium of mass 3 and 4 in the solar wind between August 1978 and July 1982. The long–term average $^4He^{++}/^3He^{++}$ flux ratio $\langle R \rangle$, was $2050 \pm 200$. The $\langle R \rangle$ values for 6–month intervals show statistically significant differences. The highest of these values is 2300 and coincides with the solar maximum of cycle 21. It is also found that $\langle R \rangle$ drops under conditions of low proton flux in the solar wind, and that it is high when solar wind speeds are lowest. At times of sector boundary current sheet crossings, identified with coronal streamers, there is a characteristic rise in the value of $\langle R \rangle$. Autocorrelations have been computed for $^4He^{++}$ and $^3He^{++}$ and indicate correlation times of about 14 and 20 hr, respectively.

074.026 **Gasdynamical models of interaction between the solar wind and cometary ionospheres.**
V. B. Baranov, I. P. Mastikov.
Inst. kosm. issled. Akad. Nauk SSSR. Prepr., No. 874, 33 pp. (1984). In Russian. Abstr. in Ref. Zh., 62. Issled. Kosm. Prostranstva, 9.62.336 (1984).

074.027 **Estimation of the amount of heating for the solar coronal loops and kernels.**
S. Chandra.
Bull. Astron. Soc. India, Vol. 12, No. 2, p. 152 – 161 (1984).

The authors have investigated the amount of heating due to an external source for some coronal loops and kernels in the line–dipole and the point–dipole geometries under nonstatic conditions. Two forms for the external source are considered. It is found that the amount of heating for the line–dipole geometry is larger than that for the point–dipole geometry. Further, the amount of heating is larger when the source is an extended one. The heating for the second form of the source is found larger than that for the first form in the line–dipole geometry, whereas in the point–dipole geometry the reverse is true.

074.028 **Correction for the observed intensity distribution in solar corona.**
K. D. Abhyankar, P. V. Subrahmanyam.
Bull. Astron. Soc. India, Vol. 12, No. 2, p. 190 – 195 (1984).

An estimation of the correction for the broadening of the intensity profile of the solar corona caused by atmospheric turbulence, scattering in the photographic emulsion, focusing errors etc. is made from the coronal observations of the sun during 1980 February 16 total solar eclipse.

074.029 **Theoretical model of the solar corona during sunspot minimum. I. Quasi–static approximation.**
V. A. Osherovich, I. Tzur, E. B. Gliner.
Astrophys. J., Vol. 284, No. 1, p. 412 – 421 (1984).

A model of the solar corona during sunspot minimum is constructed. It is suggested that the pole–equator asymmetry is caused by the interaction between global azimuthal electric currents in the corona and potential magnetic fields originated below. The asymmetry between the two hemispheres, which can be due to the magnetic quadrupole, is considered in detail. It is shown that an additional quadrupole term in the magnetic field representation affects the position of polar coronal hole boundaries differently in the two hemispheres. As a result, the outflow from the two polar regions is found to be different.

074.030 **The effect of line–tying on the radiative MHD stability of coronal plasmas with radial pressure profile.**
C.–H. An.
Astrophys. J., Vol. 284, No. 1, p. 422 – 428 (1984).

This paper studies the effects of photospheric line–tying on the localized radiative magnetohydrodynamic (MHD) modes of compressible coronal plasmas. It is found that the effect of line–tying on radiative MHD stability depends critically on the radial pressure profile. A detailed discussion of this dependence is presented.

074.031 **Coupling of the solar wind to the magnetosphere.**
J. K. Alexander, L. F. Bargatze, J. L. Burch, T. E. Eastman, J. G. Lyon, J. D. Scudder, T. W. Speiser, G. H. Voigt, C.–C. Wu.
NASA Ref. Publ., NASA RP–1120, p. 5–1 – 5–25 (1984). – See Abstr. 012.038.

Contents: 1. Introduction: Magnetospheric plasma regimes and boundaries. Major questions concerning solar wind–magnetosphere coupling. 2. Major elements of solar wind–magnetosphere coupling: The entry efficiency of solar wind plasma. Reconnection of geomagnetic and interplanetary magnetic fields. Control of dayside convection patterns by the east–west IMF component. Magnetic topology of the low–altitude boundary layer. Correlation of geomagnetic activity with southward IMF components. Sunward convection in the polar caps. Flux transfer events at the dayside magnetopause. 3. Solar wind coupling mechanisms: Proposed coupling processes. Momentum and energy transfer. Boundary conditions. 4. Global modeling. 5. Summary and outstanding problems.

074.032 **Coronal transients and their interplanetary effects.**
A. J. Hundhausen, L. F. Burlaga, W. C. Feldman, J. T. Gosling, E. Hildner, L. L. House, R. A. Howard, A. S. Krieger, M. R. Kundu, B. C. Low, N. R. Sheeley Jr., R. S. Steinolfson, R. T. Stewart, R. G. Stone, S. T. Wu.
NASA Ref. Publ., NASA RP–1120, p. 6–1 – 6–32 (1984). – See Abstr. 012.038.

Contents: 1. Introduction. 2. Background material: Ancient history – solar flares and geomagnetic storms. Modern history – interplanetary shock waves. Coronal transients or mass ejections. 3. The present: Theoretical models. New observations of coronal mass ejections. 4. The future: Solar origins. Interplanetary effects.

074.033 **Solar wind instability near streamlined bodies.**
I. S. Veselovskij.
Geomagn. Aehron., Tom 24, No. 5, p. 705 – 708 (1984). In Russian. English translation in Geomagn. Aeron.

074.034 **Coronal mass ejections.**
W. J. Wagner.
Annu. Rev. Astron. Astrophys., Vol. 22, p. 267 – 289 (1984). – See Abstr. 003.024.

Contents: (1) Introduction. (2) Models of CME origin. (3) Observations of the CME: Phenomenological description. Empirical density distributions. Kinematics. CME frequency. Fringes of the CME. CME precursors. Observing the finish of a CME. (4) CMEs related to other solar phenomena: The active Sun and CMEs. CMEs and large–scale magnetic fields. (5) Association with interplanetary phenomena: Prompt proton events. Galactic cosmic–ray modulation. Noncompressive density enhancements. Mass and energy comparisons to interplanetary shocks. CMEs from the nearest star. (6) Perspectives.

074.035 **Observation of the small–scale structure of the solar wind at the front of sudden increase of velocity of a plasma stream.**
L. A. Avanov, G. N. Zastenker, O. L. Vajsberg, Yu. I. Ermolaev.
Kosm. Issled., Tom 22, Vyp. 5, p. 774 – 780 (1984). In Russian. English translation in Cosm. Res.

074.036 **Active longitudes of coronal holes in the minimum and maximum of solar activity.**
G. I. Ol'.
Soln. Dannye, Byull., 1984, No. 7, p. 80 – 85 (1984). In Russian.

Active longitudes of coronal holes are different in the minimum and maximum of solar activity. Active longitudes of geomagnetic disturbance coincide with those of coronal holes in the minimum of solar activity. The coincidence is not observed in the maximum. This may be considered as evidence of a relationship between the geometric disturbances and coronal holes near the minimum of solar activity.

074.037 **Electron densities of coronal condensations from observations of the total solar eclipse of 31 July 1981 in the Fe XIII lines λ10747 and λ10798.**
O. G. Badalyan, N. S. Shilova.
Soln. Dannye, Byull., 1984, No. 8, p. 56 – 61 (1984). In Russian.

During the total solar eclipse of 31 July 1981 the corona images in λ10747 and λ10798 Fe XIII were obtained using narrow–band interference filters. From these data it was found that the electron density in the coronal condensation above sunspot group N 327 is several times greater than the electron density in the condensation above the group N 325 consisting of spots of much larger area.

074.038 **The role of Alfvén wave dispersion in the dynamics of energetic protons in the solar corona.**
V. A. Mazur, A. V. Stepanov.
Astron. Astrophys., Vol. 139, No. 2, p. 467 – 473 (1984).

It is shown that the Alfvén wave dispersion associated with the gyrotropy $\omega/\omega_i \ll 1$, leads to waveguiding of Alfvén waves in plasma density inhomogeneities (ducts) of coronal magnetic loops. In flaring loops isotropization of the accelerated protons by Alfvén waves forces them to escape rapidly into the loss–cone and to precipitate in the solar chromosphere. This explains the deficit of energetic protons in interplanetary space observable in some flares as compared to their number, expected from the gamma–ray emission. In sources of pulsating Type IV radio emission, where $\xi \gtrsim 1$, the pitch–angle diffusion of protons due to Alfvén waves determines the pulse train duration.

074.039 **The relationship between coronal transients, Type II bursts and interplanetary shocks.**
H. V. Cane.
Astron. Astrophys., Vol. 140, No. 1, p. 205 – 209 (1984).

Recent observations have shown that the picture of coronal mass ejections (CME's) driving the shocks which produce coronal (C) Type II radio emission, faces severe difficulties. Meanwhile the correspondence between a large number of interplanetary (IP) shocks and CME's has been established. These results can be understood if there are two types of shocks i.e. that coronal shocks are not the progenitors of IP shocks. However, a relationship between C shocks and IP shocks has been established. This is that C Type II bursts associated with IP shocks which generate IP radio emission, commence at lower frequencies than other C Type II bursts. A possible scenario which can explain this result and yet be consistent with a two shock model, is one in which the C shock commences below the transient, which later drives an IP shock, and radio emission is generated when the C shock catches up to the mass ejection. Within the limitations of available data the model fits some events.

074.040 **The far reaches of the solar wind: Pioneer 10 and Pioneer 11 plasma results.**
S. E. Kayser, A. Barnes, J. D. Mihalov.
Astrophys. J., Vol. 285, No. 1, p. 339 – 346 (1984).

Selected plasma parameters observed by Pioneer 10 and Pioneer 11 between launch (1972 and 1973) and the end of 1979 are used to find the large–scale radial structure of the solar wind. Comparison of data from the two spacecraft is used to separate temporal from spatial variations. The average bulk speed is found to remain constant at about 430 km $s^{-1}$, with stream structure still evident, though of diminished amplitude, at 20.5 AU. Proton density, flux, pressure, and kinetic energy flux are found to have radial profiles consistent with $R^{-2}$.

074.041 **Plasma diagnostics for the outer solar corona: UV and XUV Fe XII lines.**
G. L. Withbroe, J. C. Raymond.
Astrophys. J., Vol. 285, No. 1, p. 347 – 353 (1984).

The formation of the UV and XUV spectral lines of Fe XII in the outer corona is investigated. The results indicate that these lines can be an important tool for probing the physical conditions in the coronal regions where the solar wind is generated. Application of the theoretical calculations to existing UV, EUV, and XUV observations of the low corona are presented.

074.042 **Solar activity cycle variation of the K corona.**
R. Fisher, D. G. Sime.
Astrophys. J., Vol. 285, No. 1, p. 354 – 358 (1984).

The 18 year long record of K coronameter data from the Mauna Loa, Hawaii instruments is used to determine the variation of the monthly averages of the integrated polarized brightness of the electron corona. This same data set is used to estimate the variation of coronal hole area as a function of the solar activity cycle. The amplitudes of monthly averaged polarized integrated brightness shows variation of an inferred coronal mass of a factor of 2 between the maximum and minimum of the activity cycle. A simple model is used to demonstrate that this modulation cannot be accounted for solely by the growth and decay of coronal holes.

074.043 **The solar wind at 20 – 30 AU.**
A. Barnes, P. R. Gazis.
NASA Conf. Publ., NASA CP-2330, p. 527 – 540 (1984). – See Abstr. 012.041.

Pioneer 10 sampled the interplanetary plasma over the range 20 – 30 astronomical units, during the period 1979 – 1983. Comparison with nearly simultaneous observations taken from Pioneer 11 in the vicinity of 10 AU shows the average solar wind flow speed does not vary significantly with increasing heliocentric distance, and the density falls off as $R^{-2}$, as predicted by simple solar wind models. The day–to–day variations in solar wind parameters are smaller at larger distance. The dynamic pressure at 20 AU does vary by a factor 5 or more.

074.044 **Near–infrared observation of the circumsolar dust emission during the 1983 solar eclipse.**
K. Mizutani, T. Maihara, N. Hiromoto, H. Takami.
Nature, Vol. 312, No. 5990, p. 134 – 136 (1984).

A near–infrared excess emission superposed on the coronal continuum at $\sim 4\ R_\odot$ from the Sun was first observed at 2.2 μm by Peterson and MacQueen during the total solar eclipse in 1966. The authors have carried out observations of near–infrared brightness distributions of the solar corona, using a balloon–borne photometer, during the total eclipse on 11 June 1983 in Indonesia. Emissions in excess of the strong coronal background emission were recorded in some of the scans at $\sim 4\ R_\odot$ from the Sun. The spatial distribution of the excess emission implies the existence of a circumsolar ring of dust lying approximately in the ecliptic plane.

074.045 **Non–equilibrium and instability in coronal structures.**
B. Roberts.
Orages solaires de type I, p. 75 – 85 (1984). – See Abstr. 012.055.

074.046 **Quasi–steady mass flow in coronal loops.**
A. N. McClymont, I. J. D. Craig.
Bull. Am. Astron. Soc., Vol. 16, No. 3, Part I, p. 729 (1984). Abstract. – See Abstr. 010.065.

074.047 **Revivals of a coronal arch.**
Z. Švestka.
Sol. Phys., Vol. 94, No. 1, p. 171 – 192 (1984).

The giant post–flare arch of 6 November 1980 revived 11 hr and 25 hr after its formation. Both these revivals were caused by two–ribbon flares with growing systems of loops. During the first revival, at an altitude of 100000 km, temperature in the arch peaked ~1 hr, brightness ~2 hr, and emission measure ~3.5 hr after the onset of the brightening. Thus the arch looked like a magnified flare. At ~100000 km altitude the maximum temperature was $\cong 14 \times 10^6$K, max. $n_e \cong 2.5 \times 10^9 \mathrm{cm}^{-3}$, and max. energy density $\cong 11.2$ erg $\mathrm{cm}^{-3}$. The volume of the whole arch can be estimated to $1.1 \times 10^{30} \mathrm{cm}^3$, total energy $\cong 1.2 \times 10^{31}$erg, and total mass $\cong 4.4 \times 10^{15}$g.

074.048 **Type II radio emission in coronal transients.**
R. S. Steinolfson.
Sol. Phys., Vol. 94, No. 1, p. 193 – 202 (1984).

Recent observations demonstrate that some type II radio bursts (1) occur below the top of coronal white light loops in the early stages and (2) travel faster than white light transients when both data sources are recorded concurrently. These characteristics are examined with numerical simulations of a coronal transient in combination with the suggestion by Holman and Pesses (1983) that shock drift acceleration may be the originating mechanism for type II emission.

074.049 **Interplanetary radio storms. II. Emission levels and solar wind speed in the range 0.05 – 0.8 AU.**
J.–L. Bougeret, J. Fainberg, R. G. Stone.
Astron. Astrophys., Vol. 141, No. 1, p. 17 – 24 (1984).

Storms of interplanetary type III radio bursts (IP storms) are commonly observed in the interplanetary medium by the ISEE–3 radio instrument. The emission levels are determined at a number of radio frequencies for four separate storms. The IP storm radiation is found to occur in regions of enhanced density at levels of 0.05 – 0.8 AU. The density in these enhancements falls off faster than $R^{-2}$. The solar wind speed in the storm region is also measured. The analysis is consistent with steady conditions in the storm region during a few days around the central meridian passage of the storm.

074.050 **Rigorous new limits on magnetic helicity dissipation in the solar corona.**
M. A. Berger.
Geophys. Astrophys. Fluid Dyn., Vol. 30, Nos. 1 – 2, p. 79 – 104 (1984).
The Cauchy–Schwarz inequality is employed to find geometry–independent limits on the magnetic helicity dissipation rate in a resistive plasma. As evaluated in the context of a solar coronal loop, these limits strongly suggest that helicity decay occurs on a diffusion timescale which is far too great to be relevant to most coronal processes. Furthermore, rapid reconnection is likely to approximately conserve magnetic helicity. The difficulties involved in determining the free energy residing in a magnetic structure (given the constraint of magnetic helicity conservation) are discussed.

074.051 **Three–dimensional structure of solar wind velocity from observational data of the polarized brightness of the white corona.**
V. A. Kovalenko, S. I. Molodykh.
Issled. Geomagn. Aehron. Fiz. Solntsa, Moskva, No. 68, p. 131 – 138 (1984). In Russian. Abstr. in Ref. Zh., 51. Astron., 11.51.422 (1984).

074.052 **Associations between coronal mass ejections and solar energetic proton events.**
S. W. Kahler, N. R. Sheeley Jr., R. A. Howard, M. J. Koomen, D. J. Michels, R. E. McGuire, T. T. von Rosenvinge, D. V. Reames.
J. Geophys. Res., Vol. 89, No. A11, p. 9683 – 9693 (1984).
The authors have used data from the Naval Research Laboratory white light coronagraph on the P78–1 spacecraft and energetic ($E > 4$ MeV) proton data from the Goddard Space Flight Center detectors on the IMP 8 and ISEE 3 spacecraft to investigate the association between proton events originating in flares and coronal mass ejections.

074.053 **Selective heating of minority ions in the lower solar corona by electrostatic ion cyclotron waves.**
A. Luhn.
Adv. Space Res., Vol. 4, No. 2 – 3, p. 165 – 168 (1984). – See Abstr. 012.075.
Assuming that monochromatic ion cyclotron wave trains are generated in the corona above active regions on the sun, the influence of these waves on the distribution function of minority ion species is studied using a numerical test particle approach. In certain regions of phase space a sensitive dependence on the initial conditions of the nonlinearly coupled system of waves and gyrating ions produces stochastic behaviour and leads to energy transfer between waves and particles. The results of the calculations are compared with the observed abundances in solar flares.

074.054 **White light and monochromatic observations of the corona during the 1980 February 16 eclipse. I. The quiet parts.**
J.–l. Wang, S.–l. Zhang, J.–z. Liu, J.–m. Wang.
Chin. Astron. Astrophys., Vol. 8, No. 4, p. 270 – 276 (1984). English translation of Acta Astrophys. Sin., Vol. 4, No. 2, p. 115 – 123 (1984).
The following results are presented: the white light isophotes and ellipticities, the positions of coronal holes and coronal condensations, the variations of the monochromatic intensities with the radial distance along several chosen directions, the derived coefficients in the empirical corona formula in these directions and the variation of the electron density with the radial distance and the electron temperature.

074.055 **Numerical solution of the steady–state solar wind.**
D. Odstrčil.
Bull. Astron. Inst. Czech., Vol. 35, No. 6, p. 317 – 326 (1984).
This paper deals with the application of the one–dimensional time–dependent hydrodynamic computer code BOREAS to the solution of the steady–state solar wind.

074.056 **The white–light and emission corona on July 31, 1981.**
V. Rušin, M. Rybanský.
Bull. Astron. Inst. Czech., Vol. 35, No. 6, p. 347 – 353 (1984). With 4 plates.
The purpose of this paper is to determine the structure of the white–light corona and its relation to the emission corona and to photospheric activity.

074.057 **Relations between large scale coronal structure, chromospheric filaments and the solar cycle.**
S. Koutchmy, M. Loucif, O. Koutchmy.
Inst. Astrophys. Paris, Pré–Publ., No. 65, 14 pp. (1984). Paper presented at the Solar Terrestrial Prediction Workshop, Session A2, Meudon Observatory, June 18 – 22, 1984.

074.058 **Dynamical phenomena in the solar corona and changes of the photospheric magnetic field.**
B. P. Filippov.
Problems of physics of solar flares, p. 82 – 86 (1983). In Russian. Abstr. in Ref. Zh., 51. Astron., 12.51.592 (1984). – See Abstr. 012.070.

074.059 **Investigation of the electron component of the solar wind.**
S. A. Romanov, V. V. Temnyj, A. N. Omel'chenko, V. A. Styazhkin, N. A. Ehjsmont.
Solar activity investigations and the Prognoz space system, p. 67 – 73 (1984). In Russian. Abstr. in Ref. Zh., 62. Issled. Kosm. Prostranstva, 12.62.373 (1984). – See Abstr. 003.038.

074.060 **Heavy ion streams in the solar wind and their use for solar corona diagnostics.**
O. L. Vajsberg, Yu. I. Ermolaev, G. N. Zastenker, A. N. Omel'chenko.
Solar activity investigations and the Prognoz space system, p. 73 – 80 (1984). In Russian. Abstr. in Ref. Zh., 62. Issled. Kosm. Prostranstva, 12.62.374 (1984). – See Abstr. 003.038.

074.061 **Collisionless deceleration of stabilized electron beams in the solar corona.**
A. B. Eremin, V. V. Zajtsev.
Astron. Zh., Tom 61, Vyp. 6, p. 1142 – 1149 (1984). In Russian. English translation in Sov. Astron., Vol. 28, No. 6.
The influence of nonlinear effects on the propagation of a stabilized electron beam in the corona is investigated. The observed collisionless deceleration increasing with the increase of the initial flux rate testifies to the essential role of the induced plasma–wave scattering on the trajectory line of the order of $3 \times 10^{11}$cm.

074.062 **Coronal temperature measurements near a helmet structure base at the 1973 solar eclipse.**
R. J. Bessey, D. H. Liebenberg.
Sol. Phys., Vol. 94, No. 2, p. 239 – 247 (1984).
Observations of coronal Fe XIV emission lines from the NE quadrant during the 1973 solar eclipse are reported. Temperatures are deduced from a pure thermal broadening model, and, in the region near an observed white–light enhancement, an alternative interpretation of halfwidth as being in part due to turbulent velocities is suggested.

074.063 **The 'melon–seed' mechanism and coronal transients.**
G. W. Pneuman.
Sol. Phys., Vol. 94, No. 2, p. 387 – 411 (1984).
Adopting the point of view that a coronal transient is a defined magnetic structure, it must be diamagnetic with respect to the external ambient magnetic field, i.e., the external field lines cannot penetrate the structure. If this is so, an integral approach involving only external forces can be very useful for studying the conditions for acceleration and large–scale dynamical behavior of the transient. The author demonstrates the diamagnetic approach to this problem through a particularly simplified model. Necessary conditions for upward acceleration of the transient are discussed in some detail. The author has studied the equilibrium conditions and dynamical behavior of the transient using the

mechanism for two limiting cases – that in which the gas pressure in the structure dominates over the magnetic pressure and that in which the magnetic pressure dominates. Based upon the above results, he suggests that the acceleration of the transient and the two–ribbon flare both may be produced by a common force, namely that provided by the constricting effect of the external magnetic field displaced by the presence of the structure.

074.064 **The formation of neutral current sheets due to the submergence of magnetic flux.**
Y. Hu.
Acta Astrophys. Sin., Vol. 4, No. 1, p. 14 – 24 (1984). In Chinese.
In solar active regions, flare eruptions are associated with various complex features, involved in gradual, continuous evolution in time. Among other signatures, it is frequently observed that local magnetic fluxes can disappear or emerge from below the photosphere. Under the assumption of infinite electrical conductivity, both processes produce neutral current sheets in the corona, which are generally considered to be a possible mechanism of preflare energy storage. For the case of flux emergence, the formation of current sheets was treated by Tur and Priest (1976) through two examples assuming different ambient fields, a bipolar field in one case and a constant field in the other. The present paper is a natural extension of their work, considering the case of current sheet formation due to flux submergence in terms of the same examples.

074.065 **Polarity and north–south asymmetry of coronal holes at middle and low solar latitudes during 1977 – 1978.**
K. Sun, J. Zhao.
Acta Astrophys. Sin., Vol. 4, No. 1, p. 80 – 82 (1984). In Chinese.

074.066 **Temporal evolution of the solar wind and the formation of a standing shock.**
S. R. Habbal, R. Rosner.
J. Geophys. Res., Vol. 89, No. A12, p. 10645 – 10657 (1984).
The temporal evolution of the solar wind from one steady state to another is explored when momentum deposition produces multiple critical points in the flow.

074.067 **Comment on "A comparison of coronal and interplanetary current sheet inclinations" by K. W. Behannon, L. F. Burlaga, and A. J. Hundhausen.**
S. T. Suess.
J. Geophys. Res., Vol. 89, No. A12, p. 11059 – 11060 (1984). See Abstr. 34.074.033.

074.068 **Numerical investigation of fluid models with full electron and proton thermal conduction equations for the quiet solar wind.**
S. Cuperman, I. Tzur, M. Dryer.
Astrophys. J., Vol. 286, No. 2, p. 763 – 771 (1984).
The effect of using full equations (rather than Fourier law expressions) for the electron and proton thermal cunductivities in conjunction with continuity, momentum, and temperature equations for the description of spherically symmetric quiet solar wind states is investigated numerically. For this purpose a time–dependent method is used, and steady state solutions in both subsonic and supersonic regimes between the Sun and 5 AU are obtained and presented.

074.069 **On the large–scale magnetostatic coronal structures and their stability.**
B. C. Low.
Astrophys. J., Vol. 286, No. 2, p. 772 – 786 (1984).
Two analytic calculations are presented to study magnetostatic equilibrium states and their stability. In the first calculation, two classes of axisymmetric equilibrium solutions are constructed, which describe a magnetostatic solar corona having a global magnetic field in a $1/r^2$ gravitational field. In the second calculation, a particular equilibrium solution having a closed global magnetic field is shown to be buoyantly unstable, by the equatorial region of the corona, with the respective effect of outwardly distending or inwardly compressing the global magnetic field. The relevance of the theoretical results to eruptive phenomena in the solar corona is discussed.

074.070 **Detailed studies of the dynamics and energetics of coronal bullets.**
J. T. Karpen, E. S. Oran, J. P. Boris.
Astrophys. J., Vol. 287, No. 1, p. 396 – 403 (1984).
Coronal bullets are small ejecta of cool, dense plasma observed to accelerate through the solar atmosphere from 20 to 450 km $s^{-1}$. The authors have used the NRL Dynamic Flux Tube Model to simulate the evolving physical properties of these dynamic events. They have identified a component of shocked plasma which piles up ahead of the bullet and eventually dominates both the dynamics and heating of the original bullet mass. The observational consequences of this shocked component are discussed in terms of the available HRTS EUV data.

074.071 **The origin of spicules and heating of the lower transition region.**
R. G. Athay.
Astrophys. J., Vol. 287, No. 1, p. 412 – 417 (1984).
A model of the solar corona and transition region is proposed in which the atmosphere is kinematic rather than static. The kinematic character results from an assumed lack of stationarity in the energy input, which is considered to fluctuate both spatially and temporally. Fluctuations in the energy input lead to two types of atmospheric regions: one that is experiencing a diminishing heating rate and another that is experiencing an increasing heating rate. Regions of diminishing heating are associated with decreasing temperatures and downflow, whereas regions of increasing heating are associated with rising temperatures and upflow.

074.072 **Rotational characteristics of the white–light solar corona: 1965 – 1983.**
R. Fisher, D. G. Sime.
Astrophys. J., Vol. 287, No. 2, p. 959 – 968 (1984). With plate 9.
Data from the series of HAO K–coronameters located at Mauna Loa, Hawaii, are analyzed with an autocorrelation technique for the purpose of defining the variation of synodic rotation period and white–light structure persistence as a function of the solar activity cycle. The time–averaged results yield a rotation law of $\omega = 14.21–0.57 \sin^2\phi$ degrees per day, where $\omega$ is the sidereal rotation rate at latitude $\phi$. This is equal to the sunspot rotation rate at the equator but is much more rigid than rotational characteristics of photospheric magnetic fields or sunspots. A variation of the rotation rate is observed, with a definite dependence upon the level of solar activity.

074.073 **Solar wind and coronal structure.**
G. L. Withbroe.
ESA J., Vol. 7, No. 4, p. 341 – 356 (1983). – See Abstr. 012.035.

**Solar research at RAL (*Rutherford Appleton Laboratory*).**
See Abstr. 009.013.

**Solar terrestrial physics: present and future. A report based on the Solar Terrestrial Physics Workshop, December 1982 to November 1983.**
See Abstr. 012.038.

**Analytical calculation of the magnetic dipole field in a conducting radially expanding and differentially rotating medium.**
See Abstr. 022.063.

**Collision–strength dependence of solar emission lines of Si(X) and S(XII).**
See Abstr. 022.136.

**Measuring electron density in coronal active regions. II. A multichannel coronagraph with a photoelectric spectrograph and a reflex monitor at λ5303 Å.**
See Abstr. 032.030.

**Data acquisition and recording system for two–dimensional images of the solar corona during balloon–borne observations of the total solar eclipse of June 11, 1983.**
See Abstr. 036.055.

**Data analyis of electron measurements of the Plasma Experiment aboard the Helios probes.**
See Abstr. 036.154.

**The thermal instability in a sheared magnetic field: filament condensation with anisotropic heat conduction.**
See Abstr. 062.004.

**Propagation of waves in an atmosphere in the presence of a magnetic field. VII. Magneto–acoustic–gravity modes in an oblique *B*.**
See Abstr. 062.015.

**Relativistic magnetohydrodynamic winds of finite temperature.**
See Abstr. 062.127.

**An energy method for the stability of solar magnetohydrostatic atmospheres.**
See Abstr. 062.129.

**The propagation direction of Alfvénic waves in the inhomogeneous plasma stream.**
See Abstr. 062.157.

**The Hα spectral signatures of solar flare nonthermal electrons, conductive flux, and coronal pressure.**
See Abstr. 073.005.

**Radio–frequency heating of the coronal plasma during flares.**
See Abstr. 073.006.

**Variation in observed coronal calcium abundance of X–ray flare plasmas.**
See Abstr. 073.008.

**Wave observations in the solar transition region with the Solar Maximum Mission.**
See Abstr. 073.014.

**Wavelength–dependence of EUV continuum absorption.**
See Abstr. 073.024.

**Oscillations in EUV emission lines during a loop brightening.**
See Abstr. 073.026.

**Coronal extension of flaring region magnetic fields inferred from high–resolution microwave and type III burst observations.**
See Abstr. 073.035.

**Eruptive prominences of 1980 April 27 observed during STIP interval–X.**
See Abstr. 073.042.

**Heating the sun's lower transition region with fine–scale electric currents.**
See Abstr. 073.058.

**The turbulence in the solar atmosphere and in the interplanetary plasma.**
See Abstr. 073.062.

**Non–radio observations of fast fluctuations.**
See Abstr. 073.064.

**On the formation of temperature plateaus in the solar transition region.**
See Abstr. 073.065.

**Magnetic structure and heating of the upper and lower transition region.**
See Abstr. 073.066.

**Heating of the chromosphere and transition region by fast electrons from the temperature minimum.**
See Abstr. 073.067.

**A consistent picture of coronal and chromospheric processes in a well–observed solar flare.**
See Abstr. 073.091.

**Numerical simulations of loops heated to solar flare temperatures. III. Asymmetric heating.**
See Abstr. 073.115.

**Numerical simulation of reconnection in an emerging magnetic flux region.**
See Abstr. 075.025.

**Latitudinal zonal structure of the solar coronal magnetic field.**
See Abstr. 075.026.

**Non–collisional decay of solar type U radio bursts.**
See Abstr. 077.002.

**Characteristics of flares producing metric type II bursts and coronal mass ejections.**
See Abstr. 077.016.

**Recent advances in solar radio physics.**
See Abstr. 077.019.

**Type I storms and their free energy supply.**
See Abstr. 077.026.

**Type III bursts in interplanetary space. Fundamental or harmonic?**
See Abstr. 077.040.

**Study of giant cell motions and their possible relationship with the dynamics of filaments.**
See Abstr. 080.051.

**MHD waves and turbulence in the sun and interplanetary medium.**
See Abstr. 080.074.

**Conformation of the structure of the solar atmosphere in the solar wind.**
See Abstr. 080.115.

**Effect of parallel electric field and geomagnetic field distortion on precipitation of charged particles into the lower ionosphere.**
See Abstr. 084.028.

**A note on the energy transfer from the solar wind to the magnetosphere.**
See Abstr. 084.037.

**The problem of reconnection: energy release in the geomagnetic tail.**
See Abstr. 084.062.

**Long and short term relationships between solar wind velocity and geomagnetic field at low latitudes.**
See Abstr. 084.086.

**On the geoefficiency of recurrent high–velocity solar wind streams.**
See Abstr. 085.012.

**Geophysical response to the dynamics of the transition region of the solar wind.**
See Abstr. 085.022.

**Coordinated observations of a local solar wind disturbance and the auroral H Lyα emission from Uranus.**
See Abstr. 101.057.

**The multiple modes of interaction of the solar wind with a comet as it approaches the sun.**
See Abstr. 102.025.

**Solar composition noble gases in the Washington County iron meteorite.**
See Abstr. 105.093.

**K–coronameter observations and potential field model comparison in 1976 and 1977.**
See Abstr. 106.001.

**Interaction between neutral hydrogen and solar wind: spacecraft measurements of $H^+$ at the earth's orbit.**
See Abstr. 106.008.

**Sector structure of the interplanetary magnetic field and high–velocity solar wind streams.**
See Abstr. 106.013.

**Propagation of solar generated disturbances through the solar wind critical points: one–dimensional analysis.**
See Abstr. 106.018.

**Magnetohydrodynamic modelling of interplanetary disturbances between the Sun and Earth.**
See Abstr. 106.020.

**Interplanetary field enhancements in the solar wind: statistical properties at 0.72 AU.**
See Abstr. 106.023.

**Dynamic spectra of interplanetary scintillations.**
See Abstr. 106.028.

**Are interplanetary magnetic clouds manifestations of coronal mass ejections at 1 AU?**
See Abstr. 106.033.

**Low–energy protons associated with interplanetary shocks as an independent population in the solar wind.**
See Abstr. 106.046.

**Flare–generated heliospheric and geomagnetic disturbances.**
See Abstr. 106.057.

**Energy spectra of protons with E = 23 – 480 keV in the interplanetary medium, transition region and the outer magnetosphere.**
See Abstr. 106.062.

**Peculiarities of the daily variation of cosmic rays in high–speed solar wind streams.**
See Abstr. 144.017.

**Variations of the galactic cosmic ray flux and evolution of coronal holes with the solar activity cycle.**
See Abstr. 144.028.

**Variations of galactic cosmic rays and evolution of coronal holes with the cycles of solar activity.**
See Abstr. 144.030.

# 075 Magnetic Fields

075.001 **Torsional oscillations and diagnostics of the structure of the subphotospheric magnetic field.**
S. I. Gopasyuk.
Sov. Astron., Vol. 28, No. 1, p. 93 – 96 (1984). English translation of 37.075.001.

075.002 **Properties of the "island" structures of the background magnetic field in the 20th solar cycle.**
P. A. Golub.
Soln. Dannye, Byull., 1984, No. 5, p. 88 – 94 (1984). In Russian.

It is shown that the "island" structure of the background magnetic field and sunspot groups has a common zone of localization and a common law of latitude drift with the phase of the 20th solar cycle. An important role of the "island" structure is noted in the formation of the 11 year solar cycle.

075.003 **The decay of torque free toroidal magnetic fields confined beneath the solar convection zone.**
D. Moss.
Astrophys. Space Sci., Vol. 104, No. 2, p. 253 – 258 (1984).

Calculation similar to those of Mestel and Moss (1983) are performed to investigate the decay of a toroidal field through configurations satisfying the torque free condition, imposed by the presence of a poloidal field of dipolar form confined beneath the solar convection zone. It is found that initially stable field configurations diffuse into unstable configurations on time–scales of order a few $\times 10^8$yr. The results are similar to those of Tayler (1982) for a simpler field model without any dynamical constraints.

075.004 **Evolution of solar magnetic flux.**
J. P. Boris, C. R. DeVore, L. Golub, R. F. Howard, B. C. Low, N. R. Sheeley Jr., G. W. Simon, K. C. Tsinganos.
NASA Ref. Publ., NASA RP–1120, p. 3–1 – 3–34 (1984). – See Abstr. 012.038.

Contents: Introduction. Appearance of magnetic flux: models for flux emergence, unexplained observations. Dynamics of surface magnetic flux: magnetic flux transport, magnetic flux structure. Disappearance of magnetic flux: theoretical considerations, observations of flux disappearance. Summary.

075.005 **Three–dimensional model of generation of mean magnetic fields of the sun.**
T. S. Ivanova, A. A. Ruzmajkin.
Inst. prikl. mat. Akad. Nauk SSSR. Prepr., No. 25, 23 pp. (1984). In Russian. Abstr. in Ref. Zh., 51. Astron., 10.51.301 (1984).

075.006 **Structure of the large–scale magnetic field of the sun from July 28 to August 3, 1983.**
M. V. Kushnir, V. I. Makarov, M. P. Fat'yanov.
Soln. Dannye, Byull., 1984, No. 7, p. 58 – 62 (1984). In Russian.

The large–scale solar magnetic field is analysed on the basis of daily H$\alpha$ charts.

075.007 **Resistive diffusion in a force–free magnetic sheet.**
A. A. Solov'ev.
Soln. Dannye, Byull., 1984, No. 7, p. 62 – 70 (1984). In Russian.

One–dimensional resistive diffusion of the force–free magnetic field in a slab normal to the x axis is considered. The problem of the plasma turbulence is discussed.

075.008 **On the problem of formation of magnetic fields of flare–active regions.**
P. A. Golub.
Soln. Dannye, Byull., 1984, No. 8, p. 69 – 79 (1984). In Russian.
The dynamics of large–scale magnetic fields is discussed relative to flare regions of July 1974, August 1972, and November 1960. Processes are shown to exist that are connected with transport of magnetic fields to active regions.

075.009 **Magnetic field transfer in the turbulent solar envelope.**
V. N. Krivodubskii (*V. N. Krivodubskij*).
Sov. Astron., Vol. 28, No. 2, p. 205 – 211 (1984). English translation of 37.075.013.

075.010 **Intensity of sources of magnetic fields of the solar αω dynamo.**
V. N. Krivodubskii (*V. N. Krivodubskij*).
Sov. Astron., Vol. 28, No. 3, p. 316 – 320 (1984). English translation of 37.075.023.

075.011 **Solar magnetism: observation and theory.**
M. Stix.
Astron. Nachr., Vol. 305, No. 5, p. 215 – 223 (1984). = Mitt. Kiepenheuer–Inst., Nr. 236. – See Abstr. 012.062.
This review consists of four parts. The first is a discussion of the concentration of magnetic flux into small elements, the large electrical conductivity and the superadiabatic stratification as the main reasons for these "flux tubes", their relationship to the mean field, and the observational evidence of the latter. The second briefly repeats the approximations and merits of kinematic mean field dynamos. The third part treats the question of flux loss from the convection zone through the buoyancy force, with the conclusion that the dynamo most probably has its main shear region at the bottom of the convection zone, or in a transition layer beneath it. The last part discusses further dynamic aspects of the Sun's magnetism, e.g. models of limit cycles and models with chaotic behaviour, and in particular the question of phase stability of the solar cycle.

075.012 **On the mechanism of topological pumping.**
D. J. Galloway.
Astron. Nachr., Vol. 305, No. 5, p. 224 (1984). Abstract. – See Abstr. 012.062.

075.013 **Oscillatory motions in the sun corresponding to the solar cycle.**
I. Tuominen, H. Virtanen.
Astron. Nachr., Vol. 305, No. 5, p. 225 – 228 (1984). – See Abstr. 012.062.
The mean velocities derived from the proper motions of sunspot groups show a similar behaviour as the torsional oscillation found by Howard and Labonte (1980), and the mean velocities have also a meridional component which changes according to the solar cycle. Calculation of the Lorentz force from a kinematical mean–field dynamo shows that both of these components are a natural consequence of an α–ω dynamo acting in the solar convection zone.

075.014 **The interchange instability of small flux tubes.**
M. Schüssler.
Astron. Astrophys., Vol. 140, No. 2, p. 453 – 458 (1984).
The influence of stationary velocity fields on the interchange (fluting) instability of magnetic flux tubes is investigated. Converging convective cells and internal flows along the fieldlines cannot stabilize but, on the contrary, lead to a slightly destabilizing contribution. A small flux tube (magnetic element) is found to be stable if it is surrounded by a whirl which forms in a narrow intergranular downdraft. Consequences for solar surface magnetic fields are discussed.

075.015 **Change of the distribution of magnetic fields of active regions in the process of their evolution.**
L. V. Ermakova.
Issled. Geomagn. Aehron. Fiz. Solntsa, Moskva, No. 68, p. 99 – 108 (1984). In Russian. Abstr. in Ref. Zh., 51. Astron., 11.51.385 (1984).

075.016 **The magnetohydrodynamic instability of a vertically ejected cylinder from the solar surface.**
M.–t. Song.
Chin. Astron. Astrophys., Vol. 8, No. 3, p. 201 – 208 (1984). English translation of 37.075.039.

075.017 **Remarks on the magnetic support of quiescent prominences.**
U. Anzer, E. Priest.
MPA Rep., No. 157, 10 pp. (1984). Submitted to Sol. Phys.

075.018 **Magnetic and velocity field analysis of a quiet region near the center of the sun.**
H. Dara–Papamargariti, S. Koutchmy.
Inst. Astrophys. Paris, Pré–Publ., No. 81, 10 pp. (1984). Paper presented at the 8th European Regional Meeting of the IAU, Toulouse, 17 – 21 September 1984.

075.019 **Motions of magnetic structures to the solar surface.**
Eh. V. Kononovich.
Astron. Tsirk., No. 1335, p. 1 – 3 (1984). In Russian.

075.020 **Dominating periods of rotation of the general magnetic field of the sun.**
V. A. Kotov, L. S. Levitskij.
Problems of physics of solar flares, p. 94 – 99 (1983). In Russian. Abstr. in Ref. Zh., 51. Astron., 12.51.579 (1984). – See Abstr. 012.070.

075.021 **Role of the transversal and longitudinal magnetic field in current layers on the sun.**
B. V. Somov, V. S. Titov.
Problems of physics of solar flares, p. 149 – 165 (1983). In Russian. Abstr. in Ref. Zh., 51. Astron., 12.51.609 (1984). – See Abstr. 012.070.

075.022 **Structure of the thermal front in flare magnetic loops on the sun.**
V. M. Bardakov.
Problems of physics of solar flares, p. 181 – 186 (1983). In Russian. Abstr. in Ref. Zh., 51. Astron., 12.51.660 (1984). – See Abstr. 012.070.

075.023 **Equilibrium models for thin flux tubes in the solar convection zone.**
V. Anton, M. Schüssler.
Mitt. Astron. Ges., Nr. 62, p. 219 (1984). Abstract. – See Abstr. 010.341.

075.024 **Computation of solar magnetic fields from photospheric observations.**
L. Hannakam, G. A. Gary, D. L. Teuber.
Sol. Phys., Vol. 94, No. 2, p. 219 – 234 (1984).
The observational difficulties of obtaining the magnetic field distribution in the chromosphere and corona of the Sun has led to methods of extending photospheric magnetic measurements into the solar atmosphere by mathematical procedures. A new approach to this problem presented here is that a constant alpha force–free field can be uniquely determined from the tangential components of the measured photospheric flux alone. The vector magnetographs now provide measurements of both the solar photospheric tangential and the longitudinal magnetic field. This paper presents derivations for the computation of the solar magnetic field from these types of measurement.

075.025 **Numerical simulation of reconnection in an emerging magnetic flux region.**
T. G. Forbes, E. R. Priest.
Sol. Phys., Vol. 94, No. 2, p. 315 – 340 (1984).
The resistive MHD equations are numerically solved in two dimensions for an initial–boundary–value problem which simulates reconnection between an emerging magnetic flux region and an overlying coronal magnetic field. The emerging region is modelled by a cylindrical flux tube with a poloidal magnetic field lying in the same plane as the external, coronal field. The plasma betas of the emerging and coronal regions are 1.0 and 0.1, respectively, and the magnetic Reynolds number for the system is $2 \times 10^3$. At the beginning of the simulation the tube starts to emerge through the base of the rectangular computational domain, and, when the tube is halfway into the computational domain, its position is held fixed so that no more flux of plasma enters through the base. Because the time–scale of the emergence is slower than the Alfvén time–scale, but faster than the reconnection time–scale, a region of closed loops forms at the base. These loops are gradually opened and reconnected with the overlying, external magnetic field as time proceeds.

075.026 **Latitudinal zonal structure of the solar coronal magnetic field.**
V. I. Makarov.
Soln. Dannye, Byull., 1984, No. 9, p. 52 – 61 (1984). In Russian.
Properties of the latitudinal zonal structure of the solar magnetic field are studied on the basis of Hα charts for the period 1955 – 1981. It is shown that after the sign inversion of the magnetic field in the minimum of solar activity in each solar hemisphere 3 zones of the field exist: the equatorial, mean–latitudinal and polar zones with the boundaries at $\pm 20°$ and $\pm 40°$. Equatorial and mean–latitudinal zones can be considered as $f$ and $p$ fields of active regions. The lifetime of the zones depends on the character of the sign inversion and is about 13.22 and 35 years. A phenomenological model of the solar cycle is discussed.

075.027 **Interaction of the structures of the large–scale magnetic field of the sun.**
P. A. Golub.
Soln. Dannye, Byull., 1984, No. 9, p. 67 – 74 (1984). In Russian.
On the basis of McIntosh's Hα synoptic chart of the sun characteristic structures are detected in the large–scale magnetic field. Their properties are examined. The chronological sequence of Hα synoptic charts enabled the author to trace interaction of structures which may result in the reversal of the polar magnetic field.

**Solar terrestrial physics: present and future. A report based on the Solar Terrestrial Physics Workshop, December 1982 to November 1983.**
See Abstr. 012.038.

**A birefringent filter for measuring solar vector magnetic field and sight–line velocity field.**
See Abstr. 034.074.

**Direct and indirect methods of measurement of stellar magnetic fields.**
See Abstr. 036.147.

**A possible method for the determination of the magnitude and inclination of a sunspot magnetic field by a quarter–wave plate and polarizers.**
See Abstr. 036.170.

**Magnetic field generation in celestial bodies.**
See Abstr. 062.003.

**The effect of plasma rotation on axisymmetric magnetic field configuration.**
See Abstr. 062.005.

**Confinement of magnetic fields in field tubes of variable cross section.**
See Abstr. 062.008.

**Stellar fibril magnetic systems. I. Reduced energy state.**
See Abstr. 062.066.

**Model calculations of magnetic flux tubes. II. Stationary results for solar magnetic elements.**
See Abstr. 062.096.

**Linear astrophysical dynamos in rotating spheres: solar and stellar cycle north–south hemispheric parity selection mechanism and turbulent magnetic diffusivity.**
See Abstr. 062.097.

**Alfvén waves in a thermally stratified fluid.**
See Abstr. 062.130.

**Properties of solar magnetic fluxtubes as revealed by Fe I lines.**
See Abstr. 071.023.

**Variability of the quiet photospheric network.**
See Abstr. 071.026.

**Modeling of the 22–year cycle of solar activity in the framework of the dynamo theory with account for the primary field.**
See Abstr. 072.003.

**On the diagnostic of magnetic fields in sunspots through the interpretation of Stokes parameters profiles.**
See Abstr. 072.037.

**An essay on sunspots and solar flares.**
See Abstr. 072.044.

**Dynamical phenomena in sunspots. II. A moving magnetic feature.**
See Abstr. 072.057.

**Magnetic polarities and maximum field strengths of two selected sunspot groups with time distances of about thirty minutes during the period 1981 September 4 – September 7.**
See Abstr. 072.078.

**On calculations of magnetic fields of active regions in potential approximation.**
See Abstr. 072.094.

**The influence of a drift in an inhomogeneous magnetic field of an active region on the temporal profile and energy spectrum of a charged particle stream.**
See Abstr. 072.095.

**A case for submergence of magnetic flux in a solar active region.**
See Abstr. 072.101.

**Do prominences migrate equatorwards?**
See Abstr. 073.033.

**Eruptive prominences of 1980 April 27 observed during STIP interval–X.**
See Abstr. 073.042.

**Comparative magnetographic observations in different prominence lines.**
See Abstr. 073.056.

**Measurement of magnetic fields in solar prominences.**
See Abstr. 073.100.

**Determination of the magnetic field in quiescent solar prominences from polarization characteristics of radiated light.**
See Abstr. 073.107.

**Dynamical phenomena in the solar corona and changes of the photospheric magnetic field.**
See Abstr. 074.058.

**The formation of neutral current sheets due to the submergence of magnetic flux.**
See Abstr. 074.064.

**Type IV radio radiation and orientation of the magnetic field in a flare region as an index of geoefficiency of a flare stream.**
See Abstr. 077.021.

**Measurements of solar total irradiance and its variability.**
See Abstr. 080.004.

**Pulsations of microwave emission as a consequence of oscillatory transients in the solar atmosphere.**
See Abstr. 080.065.

**Convective instability in a solar flux tube. I. Nonlinear calculations for an adiabatic inviscid fluid.**
See Abstr. 080.111.

**Reconnection of magnetic fields.**
See Abstr. 084.029.

**K-coronameter observations and potential field model comparison in 1976 and 1977.**
See Abstr. 106.001.

**A phenomenological model of the heliospheric current sheet.**
See Abstr. 106.005.

# 076 UV, X, Gamma Radiation

076.001 **Characteristic structures of the solar disc observed on rocket UV filtergrams.**
B. Foing, R. M. Bonnet.
Astron. Astrophys., Vol. 136, No. 1, p. 133 – 141 (1984).

The high resolution ultraviolet filtergrams obtained with the Transition Region Camera have been statistically analysed, using the Optical and the Numerical Fourier Analysis, and autocorrelation of one-dimensional scans of network and active disc areas. The statistical properties and average dimensions of the structures characteristic of ultraviolet solar filtergrams near temperature minimum are deduced. The dimensions of the ultraviolet cell grains characteristic of the 160 nm pictures are found to be of the order of 0.6 Mm. The network and plage elements are brighter and broader ($\sim$1.0 Mm). An intermediate scale of $\sim$8 Mm is found in the cells which is discussed in terms of the velocity field associated with the 3 min chromospheric oscillations. A scale of $\sim$14 Mm is found in plage areas, which the authors relate to the structure of the network in active regions.

076.002 **On scattering of solar X-ray emission by comets.**
S. I. Ibadov.
Astron. Tsirk., No. 1297, p. 3 – 4 (1983). In Russian.

076.003 **The 160-min pulsation period in the *SMM* solar hard X-ray burst data.**
V. V. Kasinskij, V. A. Kotov.
Sov. Astron. Lett., Vol. 10, No. 2, p. 98 – 100 (1984). English translation of 37.076.008.

076.004 **Line width of the solar EUV He I resonance emissions at 584 and 537 Å.**
H. S. Ogawa, E. Phillips, D. L. Judge.
J. Geophys. Res., Vol. 89, No. A9, p. 7537 – 7540 (1984).

The authors present their latest line width values for the solar He I lines at 584 and 537 Å determined from a curve-of-growth technique. The measurements were conducted with a rocket-borne spectrometer viewing the full disk of the sun on August 10, 1982, at a zenith angle of $\sim 30°$ and along a trajectory for which the peak altitude was $\sim$325 km.

076.005 **Characteristics of bursts of hard X-ray emission of the sun and their connection with optical flares.**
G. E. Kocharov, G. A. Matveev, Yu. E. Charikov.
Cosmic ray intensity and cosmogenic isotopes. 13th Leningrad seminar on cosmophysics, p. 24 – 44 (1983). In Russian. Abstr. in Ref. Zh., 51. Astron., 8.51.257 (1984). – See Abstr. 012.028.

076.006 **Peculiarities of hard X-ray emission and electron acceleration in the solar flare of August 4, 1972.**
P. E. Semukhin, G. A. Koval'tsov.
Cosmic ray intensity and cosmogenic isotopes. 13th Leningrad seminar on cosmophysics, p. 45 – 54 (1983). In Russian. Abstr. in Ref. Zh., 51. Astron., 8.51.259 (1984). – See Abstr. 012.028.

076.007 **Analysis of the short-period pulsations in the hard X-ray radiation of the solar flares detected by Venera 13 and Venera 14.**
S. V. Bogovalov, A. F. Iyudin, Yu. D. Kotov, E. V. Shugal, V. Sh. Dolidze, V. M. Zenchenko, G. Vedrenne, M. Niel, C. Barat, G. Chambon, R. Talon.
Pis'ma Astron. Zh., Tom 10, No. 9, p. 685 – 690 (1984). In Russian. English translation in Sov. Astron. Lett., Vol. 10.

A spectral analysis of hard X-ray solar radiation based on Venera 13 and Venera 14 data was made. Recurrent solar flares from the same active regions were discovered with quasi-periodic pulsations of X-ray intensity.

076.008 **Solar hard X-ray microflares.**
R. P. Lin, R. A. Schwartz, S. R. Kane, R. M. Pelling, K. C. Hurley.
Astrophys. J., Vol. 283, No. 1, p. 421 – 425 (1984).

Using balloon-borne instrumentation of very high sensitivity, the authors have detected $\sim$25 hard X-ray bursts with peak fluxes of above $7 \times 10^{-3} (\mathrm{cm}^2\ \mathrm{s\ keV})^{-1}$ at 20 keV, in 141 minutes of observation of the Sun on 1980 June 27. These hard X-ray microflares last from a few seconds to several tens of seconds and have power-law energy spectra. They are generally accompanied by small soft X-ray bursts, but Hα flares and solar radio bursts are reported for only a few of these hard X-ray bursts.

076.009 **Are EUV bursts members of the flare family?**
R. G. Athay.
Sol. Phys., Vol. 93, No. 1, p. 123 – 131 (1984).

A number of properties of EUV bursts are found to be consistent with the conclusion that they are at the low energy extreme of the family of events including flares. The incidence of EUV bursts exceeds that of flares by over two orders of magnitude, which suggests the possibility that, in total, they may be as important energetically as flares. The faster EUV bursts have light curves resembling those observed in hard X-rays at the times of flares.

076.010 **Catalogue of X–ray bursts on the sun in the period from March to May 1979.**
G. E. Kocharov, Yu. G. Derevitskij, P. B. Dmitriev, A. G. Enikeev, A. A. Zhdanov, V. P. Lazutkov, G. A. Matveev, V. O. Najdenov, M. I. Savchenko, Yu. E. Charikov.
Fiz.–tekh. inst. Akad. Nauk SSSR. Prepr., No. 800, 37 pp. (1983). In Russian. Abstr. in Ref. Zh., 51. Astron., 9.51.445; 62. Issled. Kosm. Prostranstva, 9.62.267 (1984).

076.011 **A solar spectral line list between 10 and 200 Å modified for application to high spectral resolution X–ray astronomy.**
G. A. Doschek, R. D. Cowan.
Astrophys. J., Suppl. Ser., Vol. 56, No. 1, p. 67 – 89 (1984).
A spectral line list for the 10 – 200 Å range is developed from existing solar spectra for application to high spectral resolution measurements of astrophysical plasmas. The effect of the solar emission measure distribution is removed from the line intensities, which results in a set of emission rates for the lines that can be applied to many optically thin, low density high temperature plasmas in ionization equilibrium. In addition to the measured solar lines, ~250 theoretical lines are added to this list. Finally, predicted spectra of the two RS CVn stars, α Aur (Capella) and UX Ari, are presented at 1 and 0.25 Å spectral resolution.

076.012 **High–energy neutral radiations from the Sun.**
E. L. Chupp.
Annu. Rev. Astron. Astrophys., Vol. 22, p. 359 – 387 (1984). – See Abstr. 003.024.
Contents: (1) Introduction. (2) Observational status (solar gamma rays and neutrons): Identification and analysis of gamma–ray events. Comparison of GRS gamma rays with other flare emissions. Temporal characteristics of gamma–ray events. Prompt gamma–ray spectra. Delayed gamma–ray lines (511 keV and 2.223 MeV). The species and energies of accelerated solar particles. (3) Theoretical status: Basic yield calculations (secondary production, ion and electron interactions). Gamma–ray lines and continua (origin and yield estimates). Solar neutrons (yield, photosphere interaction, flux at the earth). (4) Toward an interpretation: A thick target model. Accelerator characteristics. Absolute energetic particle spectra. Interaction geometry, accelerated ion and target abundances. (5) Particle acceleration mechanisms. (6) Outstanding questions.

076.013 **Recent solar observations from the X–ray polychromator on the repaired SMM satellite.**
J. L. R. Saba, G. L. Slater, M. X. Levay, K. L. Smith, K. T. Strong, R. D. Bentley, J. R. Lemen, R. R. Caffey, S. L. Freeland Jr., D. P. Mathur, K. J. H. Phillips, T. A. Waters.
Bull. Am. Astron. Soc., Vol. 16, No. 3, Part I, p. 726 (1984). Abstract. – See Abstr. 010.065.

076.014 **Expressions to determine temperatures and emission measures for solar X–ray events from GOES measurements.**
R. J. Thomas, R. D. Starr, C. J. Crannell.
Bull. Am. Astron. Soc., Vol. 16, No. 3, Part I, p. 726 – 727 (1984). Abstract. – See Abstr. 010.065.

076.015 **Active region soft X–ray line profiles.**
K. T. Strong, J. R. Lemen, K. J. H. Phillips.
Bull. Am. Astron. Soc., Vol. 16, No. 3, Part I, p. 730 (1984). Abstract. – See Abstr. 010.065.

076.016 **Solar X–ray flare emission and energetic particles according to data of observations aboard the Venera 13, 14 automatic interplanetary stations.**
S. A. Belyakov, E. I. Dajbog, A. P. D'yachkov, V. M. Zenchenko, V. G. Kurt, Yu. I. Logachev, A. I. Rutkovskij, S. P. Ryumin, V. G. Stolpovskij, E. V. Yurovitskaya, G. Vedrenne, C. Barat.
Kosm. Issled., Tom 22, Vyp. 6, p. 906 – 921 (1984). In Russian. English translation in Cosm. Res.

076.017 **Absolute measurement of the extreme UV solar flux.**
R. W. Carlson, H. S. Ogawa, E. Phillips, D. L. Judge.
Appl. Opt., Vol. 23, No. 14, p. 2327 – 2332 (1984).
A windowless rare–gas ionization chamber has been developed to measure the absolute value of the solar extreme UV flux in the 50 – 575–Å region. For the day of the flight, 10 August 1982, the solar irradiance (50 – 575 Å), normalized to unit solar distance, was found to be $5.71 \pm 0.42 \times 10^{10}$photons $cm^{-2}sec^{-1}$.

076.018 **Modelling of the solar extreme ultraviolet irradiance for aeronomic applications.**
G. Schmidtke.
Handbuch der Physik. Geophysik III/7, p. 1 – 55 (1984). – See Abstr. 003.035.

076.019 **Solar ultraviolet variation between 1977 and 1983.**
L. A. Hall, G. P. Anderson.
J. Geophys. Res., Vol. 89, No. D5, p. 7322 (1984).
A fifth measurement of the solar ultraviolet irradiance at 40 km in the stratosphere has been added to four earlier ones. It confirms the previously published conclusion that photospheric solar irradiance in the range 2350 – 2870 Å has been essentially constant during the current solar activity cycle. Chromospheric emissions seen in the Mg II line cores increased during rising activity and decreased again as the activity declined.

076.020 **Solar irradiance variation near 2075 Å.**
L. A. Hall, G. P. Anderson.
J. Geophys. Res., Vol. 89, No. D6, p. 9677 – 9678 (1984).
Measurements of solar radiation in the vicinity of the Al I absorption edge at 2075 Å provide a means of comparing the magnitude of the 11–year variation in solar activity with that of the 27–day variation. The ratio of the solar irradiance in the band 2030 – 2070 Å to that in the band 2100 – 2140 Å has been determined from five measurements made in the period 1977 – 1983. This ratio, which is an almost purely relative measurement, avoids the problem of absolute irradiance comparisons over long periods of time. The magnitude of the 11–year effect in this ratio is about ±1%, which is comparable to that in the 27–day effect measured at solar maximum.

076.021 **Calculation of X–ray polarization during flares.**
E. Haug, G. Elwert, R. R. Rausaria.
Lehrstuhl Theor. Astrophys. Univ. Tübingen, Prepr., 4 pp. (1984). To appear in Adv. Space Res.

076.022 **Effect of electron energy and angular dispersion on hard X–ray polarization during solar flares.**
E. Haug, G. Elwert, R. R. Rausaria.
Lehrstuhl Theor. Astrophys. Univ. Tübingen, Prepr., 26 pp. (1984). To appear in Astron. Astrophys.

076.023 **Hard X–ray imaging of a solar gradual hard X–ray burst on April 1, 1981.**
T. Takakura, K. Ohki, T. Sakurai, J. L. Wang, J. Y. Xuan, S. C. Li, R. Y. Zhao.
Sol. Phys., Vol. 94, No. 2, p. 359 – 367 (1984).
Study of the source structure and its time variation of the hard X–ray bursts on April 1, 1981 is reported. The hard X–ray images for the intense gradual hard X–ray burst have revealed a steady main source along the ridge of a large coronal arcade. A tongue like sub–source could be a separate source showing a long coronal loop. Since the global structure of the main source did not change considerably in association with the large relative change in the hot thermal component, the emission sources for the non-thermal and the hot thermal components would be common. It suggests that both electron acceleration and heating of plasma to $3.5 \times 10^7$K are occurring simultaneously in the common source.

076.024 **Timing analysis of hard X–ray emission and 22 GHz flux and polarization in a solar burst.**
J. E. R. Costa, P. Kaufmann, T. Takakura.
Sol. Phys., Vol. 94, No. 2, p. 369 – 378 (1984).
A solar flare occurring on 26 February, 1981 at 19:32 UT was observed simultaneously in hard X–rays and microwaves with a

time resolution of a fraction of a second. The authors find that: (1) all 22 GHz flux structures were delayed by 0.2 – 0.9 s relative to similar structures in hard X–rays throughout the burst duration; (2) different burst structures showed different delays, suggesting that they are independent of each other; (3) the time structures of the degree of polarization at 22 GHz precede the total microwave flux time structures by 0.1 – 0.5 s; (4) the time evolutions of time delays of microwaves with respect to hard X–rays and also the degree of microwave polarization show fluctuations which are not clearly related to any other time structures. If one takes mean values for the 32 s burst duration, one finds that hard X–ray emission precedes the degree of microwave polarization by 450 ms, which in turn precedes the total microwave flux by 110 ms.

**Gamma ray astronomy.**
See Abstr. 003.026.

**Cosmic ray intensity and cosmogenic isotopes. Papers of the 13th Leningrad seminar on cosmophysics, Leningrad, 19 – 21 November, 1982.**
See Abstr. 012.028.

**New lease on life for the Solar Maximum Mission.**
See Abstr. 051.006.

**Frequent ultraviolet brightenings observed in a solar active region with Solar Maximum Mission.**
See Abstr. 072.041.

**Variation in observed coronal calcium abundance of X–ray flare plasmas.**
See Abstr. 073.008.

**The development and cooling of a solar limb–flare.**
See Abstr. 073.011.

**Solar flare iron Kα emission associated with a hard X–ray burst.**
See Abstr. 073.020.

**The possibility of steady state nonionization equilibrium conditions in soft X–ray flare plasmas.**
See Abstr. 073.023.

**Flare loops heated by thermal conduction.**
See Abstr. 073.027.

**Investigation of the correlation of hard X–ray emission bursts and microwave radio radiation accompanying solar flares in March – April 1979.**
See Abstr. 073.037.

**Long–decay soft X–ray flares. 1. Region McMath 11926, June 15, 1972.**
See Abstr. 073.046.

**Hard X–ray imaging observations of solar hot thermal flares with the *Hinotori* spacecraft.**
See Abstr. 073.051.

**The impulsive phase of a solar limb flare.**
See Abstr. 073.053.

**A 154–day periodicity in the occurrence of hard solar flares?**
See Abstr. 073.059.

**Calculation of spectrum and directivity of hard X–rays from solar flares.**
See Abstr. 073.060.

**Synchrotron or plasma emission in solar microwave flares?**
See Abstr. 073.071.

**Electron spectra deduced from solar hard X–ray bursts.**
See Abstr. 073.081.

**Positional fix of the loop prominence of 1981 April 27 and comparison with X–ray sources.**
See Abstr. 073.083.

**Analysis of the 1980 November 18 limb flare observed by the Hard X–ray Imaging Spectrometer (HXIS).**
See Abstr. 073.095.

**Energy and angular distributions of energetic flare electrons and their X–radiation. I. Initially monoenergetic electrons.**
See Abstr. 073.096.

**The relation of electron temperature to emission measure and limits of increase in emission measure in soft X–ray flares.**
See Abstr. 073.099.

**On X–ray spectral diagnostics of epithermal electrons in solar flare plasma.**
See Abstr. 073.101.

**On the nature of pulsations of X–ray and microwave radiation of solar flares.**
See Abstr. 073.102.

**Fast variations in high–energy X–rays from solar flares and their constraints on nonthermal models.**
See Abstr. 073.117.

**Analysis of magnesium XI line profiles from solar active regions.**
See Abstr. 074.023.

**Revivals of a coronal arch.**
See Abstr. 074.047.

**On the spatial and temporal correlation of type I radio bursts and soft X–rays.**
See Abstr. 077.032.

**Global pulsations of the sun in the soft X–ray range (1 – 8 Å) during the period of the project "Solar Maximum Year".**
See Abstr. 080.108.

**Errors in theoretical calculations of the composition of the E and F1 regions of the day–time ionosphere due to uncertainties of the solar ultraviolet radiation stream.**
See Abstr. 083.030.

# 077 Radio, Infrared Radiation

077.001 **Observations of the complex structure of a type II burst in the low–frequency radio range.**
V. P. Grigor'eva, V. I. Shavshina.
Astron. Tsirk., No. 1283, p. 1 – 3 (1983). In Russian.

077.002 **Non–collisional decay of solar type U radio bursts.**
M. Poquérusse, J. L. Bougeret, C. Caroubalos.
Astron. Astrophys., Vol. 136, No. 1, p. 10 – 16 (1984).

The authors study the decay time τ of solar type U radio bursts in the frequency range from 470 MHz to 150 MHz using high time resolution digital records of the Nançay Multichannel Radio Spectrograph. τ appears to be much more beam dependent than medium dependent. This is interpreted as a new, strong indication that at high frequencies the time profile of a type III burst is similar to that of the exciter, thus indicating that plasma waves do not decay slower than the exciter. This extends to the low corona a property previously established in the solar wind. It may be accounted for by various damping mechanism; in the low corona conditions, a possible process is also through plasma wave refraction outside narrow overdense fibers, resulting in their fast Landau damping.

077.003 **Interplanetary radio storms. I. Extension of solar active regions through the interplanetary medium.**
J.–L. Bougeret, J. Fainberg, R. G. Stone.
Astron. Astrophys., Vol. 136, No. 2, p. 255 – 262 (1984).

About 100 storms of type III solar radio bursts have been identified in the ISEE–3 radio experiment data during the 4 yr period around the maximum of the 21st solar cycle. They demonstrate the very frequent presence of streams of suprathermal electrons. Their durations range from 1 to 10 d or more. They are observed up to 100 – 170 solar radii. Their rate of occurrence is 2 to 3 per solar rotation near solar maximum. It is shown that the time variations of the daily radio emission intensities correlate with the sunspot number variations and with the solar activity in general. More specifically, the authors find a very good correlation with the meter wave type III and type I storms, which demonstrates that the suprathermal electrons responsible for the radio emission have been accelerated below 2 solar radii heliocentric. The different lags observed between the sunspot number variations, the *S*–component, the type I and type II storms are discussed.

077.004 **On radio emission of local sources on the sun in the decimeter wavelength region.**
L. P. Ipatova, L. V. Yasnov.
Astron. Tsirk., No. 1290, p. 1 – 3 (1983). In Russian.

077.005 **Change of the polarization sign in a powerful microwave burst.**
V. V. Borisov, O. V. Epifanova, A. V. Khokhlov.
Astron. Tsirk., No. 1294, p. 1 – 3 (1983). In Russian.

077.006 **Type II solar radio bursts, interplanetary shocks, and energetic particle events.**
H. V. Cane, R. G. Stone.
Astrophys. J., Vol. 282, No. 1, p. 339 – 344 (1984).

Using the *ISEE* 3 radio astronomy experiment data the authors have identified 37 interplanetary type II bursts in the period 1978 September to 1981 December. They list these events and the associated phenomena. The events are preceded by intense, soft X–ray events with long decay times and type II or type IV bursts, or both, at meter wavelengths. The majority of the interplanetary type II bursts are associated with energetic particle events. Our results support other studies which indicate that energetic solar particles detected at 1 A.U. are generated by shock acceleration. There appears to be a high correlation of type II bursts with white light coronal transients.

077.007 **The generation of type III solar radio bursts: the role of induced scattering of plasma waves by ions.**
B. N. Levin, A. M. Lerner, V. O. Rapoport.
Sov. Astron. Lett., Vol. 10, No. 1, p. 23 – 25 (1984). English translation of 37.077.001.

077.008 **Far–infrared imaging of the Sun with a balloon–borne 60 cm diameter telescope.**
C. Degiacomi, F. K. Kneubühl.
Eighth International Conference on Infrared and Millimeter Waves, p. M2.1 (1984). Abstr. in Phys. Abstr., Vol. 87, No. 1238, Entry 77810 (1984). – See Abstr. 012.015.

077.009 **Dependence of the line intensity of shortwave radiation of the sun on the activity level.**
A. A. Nusinov.
Geomagn. Aehron., Tom 24, No. 4, p. 529 – 536 (1984). In Russian. English translation in Geomagn. Aeron.

077.010 **The spectrum of shortwave radiation for aeronomical calculations at different levels of solar activity.**
E. A. Bruevich, A. A. Nusinov.
Geomagn. Aehron., Tom 24, No. 4, p. 581 – 585 (1984). In Russian. English translation in Geomagn. Aeron.

077.011 **Unerwarteter Anstieg des solaren Radioflusses.**
C. Monstein.
Orion, 42. Jahrg., Nr. 204, p. 190 – 193 (1984).

077.012 **Investigation of the longitudinal heterogeneity of the solar activity from the radio flux of the sun at 10.7 cm.**
D. V. Erofeev, A. P. Kramynin.
Soln. Dannye, Byull., 1984, No. 5, p. 83 – 87 (1984). In Russian.

The longitudinal heterogeneity of the source of the S–component at 10.7 cm was studied by the complex demodulation method. It is shown that there are two regions in which radio sources are concentrated. The location of these regions corresponds to the rigid rotation with a synodic period of about 28 days. The amplitude of the 28–days fluctuations of the radio flux at 10.7 cm is modulated by the solar cycle and variations with a period of a few months.

077.013 **On fluctuations in the solar radio emission flux density at 2800 MHz.**
Yu. I. Vitinskij.
Soln. Dannye, Byull., 1984, No. 6, p. 85 – 90 (1984). In Russian.

Data of fluctuation indices and absolute and relative "perturbations" and lists of strong positive and negative fluctuations in density of the solar radio emission flux at 2800 MHz are given for 1948 – 1982. A comparison is made of peculiarities of these characteristics and corresponding characteristics for indices of sunspots.

077.014 **Non–linear wave coupling and Type I solar radio bursts theories.**
F. Malara, P. Veltri, G. Zimbardo.
Astron. Astrophys., Vol. 139, No. 1, p. 71 – 78 (1984).

The non–linear coupling between a low frequency ion acoustic turbulence and high frequency (whistler, ordinary and extraordinary modes) waves are studied in detail, using the two fluids approximation for the plasma. The results are discussed in connection with Type I solar radio bursts theory, in order to show that such coupling might explain the polarization of the bursts in the O–mode, as a consequence of the stabilization that the coupling produces on the competing instabilities.

077.015 **Temporal variations in the solar submillimeter continuum.**
C. Lindsey, C. Kaminski.
Astrophys. J., Lett. Ed., Vol. 282, No. 2, p. L103 – L106 (1984).
The authors present observations of local intensity variations in the 300 – 800 μm solar continuum. Two–beam difference observations were used to determine power spectrum statistics of small temporal variations over the frequency range 0.1 – 20 mHz. The authors find a significant enhancement in spectral power (~5K rms) for frequencies between 3 and 7 mHz. These and higher frequency variations are attributed to adiabatic response of the chromospheric medium to compression waves.

077.016 **Characteristics of flares producing metric type II bursts and coronal mass ejections.**
S. Kahler, N. R. Sheeley Jr., R. A. Howard, M. J. Koomen, D. J. Michels.
Sol. Phys., Vol. 93, No. 1, p. 133 – 141 (1984).
The authors study the origin of coronal shocks by comparing several flare characteristics for two groups of flares: those with associated metric type II bursts and coronal mass ejections (CMEs) and those with associated metric type II bursts but no CMEs. Hα flare areas, 1–8 Å X–ray fluxes, and impulsive 3 cm fluxes are all statistically smaller for events with no CMEs than for events with CMEs. It appears that both compact and large mass ejection flares are associated with type II bursts. The events with no CMEs imply that at least many type II shocks are not piston–driven, but the large number of events of both groups with small 3 cm bursts does not support the usual assumption that type II shocks are produced by large energy releases in flare impulsive phases. The poor correlation between 3 cm burst fluxes and the occurrence of type II bursts may be due to large variations in the coronal Alfvén velocity.

077.017 **Pulsations of type IV radio bursts as an indicator of protonability of solar flares.**
V. V. Zaitsev (*V. V. Zajtsev*), A. V. Stepanov, G. P. Chernov.
Sol. Phys., Vol. 93, No. 2, p. 363 – 377 (1984).
The analysis of observational data has shown that the duration of a pulse train in type IV radio bursts decreases with increasing hardness of the spectrum of high–energy protons and increases with decreasing proton fluxes from the Sun. It is shown that such a correlation corresponds to a magnetohydrodynamic model of pulsations and is inexplicable within the framework of a nonlinear periodical regime of plasma instabilities. The pulse train duration is determined by proton pitch–angle diffusion caused by Alfvén waves in coronal magnetic loops. A method of predicting solar proton hardness and proton fluxes using type IV radio burst pulsations is proposed.

077.018 **Irregular paths of exciting electrons and the intensity–time profiles of type III bursts.**
A. D. Fokker.
Sol. Phys., Vol. 93, No. 2, p. 379 – 391 (1984).
The trajectories of individual electrons of a stream that excites a type III burst may well deviate from a smooth path, for one or both of the following reasons: (1) the magnetic field lines along which the electrons are guided have an irregular course; (2) the pitch angle of an electron is liable to variations. Irregularities of the individual trajectories imply different path lengths and consequently have an effect on the intensity–time profile of the burst. The solution of the "gamblers ruin" problem, known from probability theory, is successfully applied to discuss this effect. It yields model profiles that are very similar to the observed profiles.

077.019 **Recent advances in solar radio physics.**
M. R. Kundu.
Proc. Indian Acad. Sci., Earth Planet. Sci., Vol. 93, No. 3, p. 201 – 225 (1984).
The author reviews high spatial resolution microwave observations of solar active regions, coronal loops and flares. Observations of preflare active regions are presented; in particular the author discusses the interpretations of reversal of polarization at the flare site and the role of newly emerging flux in triggering the onset of flares. He discusses the spatial locations of microwave burst emitting regions; loops or arcades of loops appear to be the sites of flare energy release in microwave bursts.

077.020 **On the shift velocity of the effective center of solar radio radiation.**
A. Kurbanov, L. V. Yasnov.
Dokl. Akad. Nauk TadzhSSR, Tom 26, No. 12, p. 764 – 765 (1983). In Russian. Abstr. in Ref. Zh., 51. Astron., 10.51.363 (1984).

077.021 **Type IV radio radiation and orientation of the magnetic field in a flare region as an index of geoefficiency of a flare stream.**
M. I. Pudovkin, A. B. Kharin.
Geomagn. Aehron., Tom 24, No. 5, p. 816 – 818 (1984). In Russian. English translation in Geomagn. Aeron.

077.022 **Microwave emission and polarization of a flaring loop.**
C. E. Alissandrakis, P. Preka–Papadema.
Astron. Astrophys., Vol. 139, No. 2, p. 507 – 511 (1984).
The authors present computations of the microwave emission from a loop filled with accelerated electrons. At short wavelengths the loop is optically thin and the emission comes predominantly from the feet of the loop, while at long wavelengths the emission comes from the entire loop with a maximum near the top. If the loop is not located at the center of the solar disk the observed polarization is greatly affected by propagation effects in the corona which may change the sense of the circular polarization ($V$). For the adopted model and for a loop located 30° away from the center of the disk the sense of $V$ corresponds to the direction of the magnetic field in the region of emission only for wavelengths much shorter than 1.4 cm; for wavelengths much longer than 7.5 cm the sense of $V$ corresponds to the polarity of the field at the diskward foot of the loop.

077.023 **Far–infrared–solar imaging from a balloon–borne platform.**
K. Degiacomi, F. K. Kneubühl, D. Huguenin, E. A. Müller.
Int. J. Infrared Millimeter Waves, Vol. 5, No. 5, p. 643 – 653 (1984). Abstr. in Phys. Abstr., Vol. 87, No. 1245, Entry 110589 (1984).

077.024 **An analysis of the data of the total solar flux observations at λ3.2 cm in max. years.**
S. Jiang, C. Han, Y. Chen.
J. Nanjing Univ., No. 2, p. 279 – 298 (1984). In Chinese.
In this paper data of the total flux observations of the sun at λ3.2 cm obtained in the Nanjing University, Department of Astronomy, during the years 1980 – 1981 have been analysed.

077.025 **Noise storms and their place in the solar radio zoo.**
A. O. Benz.
Orages solaires de type I, p. 35 – 37 (1984). – See Abstr. 012.055.

077.026 **Type I storms and their free energy supply.**
D. S. Spicer.
Orages solaires de type I, p. 67 – 74 (1984). – See Abstr. 012.055.

077.027 **Type I radio emission observations.**
A. Mangeney.
Orages solaires de type I, p. 97 – 106 (1984). In French. – See Abstr. 012.055.

077.028 **Noise storm association to non–radio phenomena.**
P. Zlobec, C. Zanelli.
Orages solaires de type I, p. 107 – 119 (1984). – See Abstr. 012.055.
The correlation of solar radio noise storms with various kinds of non–radio activity is reviewed: solar flares, enhancement of the S–component, sunspots and magnetic fields are considered.

077.029 **Observational and theoretical constraints on theories of type I emission.**
D. B. Melrose.
Orages solaires de type I, p. 123 – 126 (1984). – See Abstr. 012.055.

The author discusses both the polarization and the sizes and brightness temperatures of solar type I radio sources, then he summarizes the properties of three versions of fundamental plasma emission and estimates the expected second harmonic component in the currently most popular theory.

077.030 **Coronal evolution and solar type I radio bursts: an ion–acoustic wave model.**
A. O. Benz, D. G. Wentzel.
Orages solaires de type I, p. 127 – 129 (1984). – See Abstr. 012.055.

077.031 **Cyclotron theories for noise storms.**
P. Veltri.
Orages solaires de type I, p. 131 – 140 (1984). – See Abstr. 012.055.

The cyclotron theories to explain type I solar radio bursts emission are in general based on an electromagnetic instability excited by a beam of accelerated electrons spiralling around the magnetic field lines. This paper presents a critical review of these theories.

077.032 **On the spatial and temporal correlation of type I radio bursts and soft X–rays.**
A. O. Benz, M. Jaeggi, J. M. Mosher, G. J. Nelson.
Orages solaires de type I, p. 145 – 148 (1984). – See Abstr. 012.055.

The authors have compared type I radio burst positions and intensities with spatially resolved soft X–ray observations. The spatial coincidence is generally good, but not exact. There seems to be a time lag of about one day between maximum X–ray brightness and maximum type I activity.

077.033 **On a model of relationship between type I and type III bursts activity.**
M. Karlicky.
Orages solaires de type I, p. 149 – 152 (1984). – See Abstr. 012.055.

077.034 **Constraints on second harmonic emission in type I bursts.**
Ø. Elgarøy.
Orages solaires de type I, p. 153 – 155 (1984). – See Abstr. 012.055.

077.035 **Observations of the harmonic frequency of type I bursts.**
M. Jaeggi.
Orages solaires de type I, p. 157 – 159 (1984). – See Abstr. 012.055.

077.036 **Comments on the high directivity of type I bursts and its effect on the detection of the second harmonic.**
J. L. Bougeret.
Orages solaires de type I, p. 161 – 165 (1984). In French. – See Abstr. 012.055.

077.037 **The polarization of the noise storm continuum and the upper limit on harmonic emission.**
A. O. Benz, P. Zolliker.
Orages solaires de type I, p. 167 – 170 (1984). – See Abstr. 012.055.

077.038 **Time series analysis of the emission times of individual type I bursts.**
J. L. Bougeret.
Orages solaires de type I, p. 171 – 173 (1984). In French. – See Abstr. 012.055.

The preliminary results of a new approach to the study of the periodicity of solar noise storms are presented.

077.039 **Partial reconstruction of the initial conditions for streams of energetic electrons associated with a solar type III burst.**
R. J.-M. Grognard.
Sol. Phys., Vol. 94, No. 1, p. 165 – 170 (1984).

A simple test of the applicability of quasilinear theory is presented to be applied to measured velocity distributions of energetic electrons associated with solar type III bursts or similar particle streams: it consists in taking the time integral of these distributions after removing the background. If quasilinear diffusion of weak plasma turbulence applies to the particle dynamics and if all other losses are neglected, the time integral of the distributions injected at the source of the streaming particles is recovered. In particular, if these distributions are approximated by "hot" Maxwellian ones, direct estimates of the initial temperatures are retrieved.

077.040 **Type III bursts in interplanetary space. Fundamental or harmonic?**
G. A. Dulk, J. L. Steinberg, S. Hoang.
Astron. Astrophys., Vol. 141, No. 1, p. 30 – 38 (1984).

The authors study type III bursts in the solar wind using observations of 120 relatively simple, intense and isolated bursts that were observed on the ISEE–3 spacecraft in the frequency range 30 to 1980 kHz. There are a number of bursts whose trajectories encounter the spacecraft and fast electrons and Langmuir waves are observed. For these, and for radio frequencies near the local plasma frequency, the evidence is strong that radio emission is at the fundamental from burst onset until about the time of burst peak. Further, the fundamental emission begins in time coincidence with the start of the Langmuir waves. This finding should remove any doubt about the conventional belief that Langmuir waves are responsible for type III bursts.

077.041 **High resolution observations of solar Type I sources at 460 MHz.**
X.–z. Liu, T. Han.
Chin. Astron. Astrophys., Vol. 8, No. 3, p. 185 – 189 (1984). English translation of Acta Astrophys. Sin., Vol. 4, No. 1, p. 6 – 13 (1984).

The authors have made a high–resolution study of solar Type–I sources. They divide a Type–I source into 5 morphological classes, multiple–centre, single with sharp edges, single with blurred edges, double–component and sources without continuum. Following analyses of the centre–limb distribution for the different classes and of source counts according to peak separation for the multiple–centre sources, the authors ascertain that the cm wave Type–I sources are essentially multiple. They discuss the differences between the metre wave and cm wave sources.

077.042 **Radio pulsations associated with the solar flare of 1982 April 14.**
P. Pant, K. R. Bondal.
Bull. Astron. Soc. India, Vol. 12, No. 3, p. 263 – 268 (1984).

The authors report the H–alpha observations of a large and complicated flare of optical class 2B which occurred on 1982 April 14. The flare showed significant activity for several minutes prior to its impulsive phase, as evidenced by the dynamic spectrum of the associated radio event taken simultaneously at Culgoora. The dynamic spectrograms showed strong chains of pulsating bursts superimposed upon the type IV continuum in close association with intermittent type III burst storms in metric and decametric bands.

077.043 **Table of solar radio bursts (1979,9 – 1981,6).**
Publ. Beijing Astron. Obs., No. 3, p. 1 – 93 (1983).

077.044 **Fine structure of the quasi–periodic oscillations in the radio emission of the sun.**
S. A. Andrianov, L. P. Ipatova, L. V. Yasnov.
Astron. Tsirk., No. 1308, p. 4 – 7 (1984). In Russian.

077.045 **Research of the intensity gradient of solar radio emission.**
A. R. Abbasov, V. A. Magerramov.
Astron. Tsirk., No. 1335, p. 4 – 7 (1984). In Russian.

077.046 **Structure and polarization of active region microwave emission.**
M. R. Kundu, C. E. Alissandrakis.
Sol. Phys., Vol. 94, No. 2, p. 249 – 283 (1984).

The authors present observations of active region radio emission at 6.16 cm wavelength, obtained with an angular resolution of 3″ by 10″ arc using the Westerbork Synthesis Radio Telescope during the action interval May 20 – 27, 1980 of the Solar Maximum Year. They present maps in both total intensity and circular polarization of three regions (Hale numbers 16850, 16863, and 16864) and provide a detailed comparison of these maps with on– and off–band Hα pictures and with magnetograms. The strongest sources were associated with neutral lines and soft X–ray arcades.

077.047 **Association of type II solar radio bursts with coronal structures above Hα filament channels.**
R. T. Stewart.
Sol. Phys., Vol. 94, No. 2, p. 379 – 385 (1984).

A study of type II solar radio bursts recorded at 160 MHz by the Culgoora radioheliograph during 1980 to 1982 shows that the radio emission occurs above Hα filaments rather than above Hα flares. This suggests that the type II radio emission most probably originates from within a coronal helmet streamer overlying the filament channel.

**The first decade of solar radio astronomy in Australia.**
See Abstr. 004.016.

**Beginnings of solar radio astronomy in Canada.**
See Abstr. 004.027.

**Development of solar radio astronomy in Japan up until 1960.**
See Abstr. 004.028.

**A frequency–agile interferometer for solar microwave spectroscopy.**
See Abstr. 033.041.

**Gyrosynchrotron radiation from a source with spatially varying field and density.**
See Abstr. 062.116.

**An analysis of vibration–rotation lines of OH in the solar infrared spectrum.**
See Abstr. 071.032.

**Three–dimensional structures of two solar active regions from VLA observations at 2, 6, and 20 centimeter wavelengths.**
See Abstr. 072.029.

**Sharp edges in solar microwave spectra: neutral current sheets or cyclotron lines?**
See Abstr. 072.039.

**Center–to–limb variation of a sunspot–associated microwave source.**
See Abstr. 072.052.

**Solare Beobachtungsergebnisse. Solar data. Solar radio emission. October – December 1983.**
See Abstr. 072.077.

**Comparison of optical observations of the flare of May 16, 1981 and radio bursts in the kilometer range.**
See Abstr. 073.004.

**Coronal extension of flaring region magnetic fields inferred from high–resolution microwave and type III burst observations.**
See Abstr. 073.035.

**Investigation of the correlation of hard X–ray emission bursts and microwave radio radiation accompanying solar flares in March – April 1979.**
See Abstr. 073.037.

**Eruptive prominences of 1980 April 27 observed during STIP interval–X.**
See Abstr. 073.042.

**Occurrence of flares with type II and IV radio bursts in relation to sunspot group types.**
See Abstr. 073.044.

**Occurrences of flares with type II and IV radio events in interacting sunspot groups in the course of revolutions.**
See Abstr. 073.045.

**Synchrotron or plasma emission in solar microwave flares?**
See Abstr. 073.071.

**A combined optical and radio study of the solar ejection of 1982 January 22.**
See Abstr. 073.087.

**On the nature of pulsations of X–ray and microwave radiation of solar flares.**
See Abstr. 073.102.

**The relationship between coronal transients, Type II bursts and interplanetary shocks.**
See Abstr. 074.039.

**Type II radio emission in coronal transients.**
See Abstr. 074.048.

**Timing analysis of hard X–ray emission and 22 GHz flux and polarization in a solar burst.**
See Abstr. 076.024.

**Possible evidence of solar oscillations in the multifrequency spectra of solar radio brightness fluctuations.**
See Abstr. 080.044.

**Type III radio bursts in the interplanetary medium: the role of propagation.**
See Abstr. 106.025.

# 078 Cosmic Radiation

078.001 **Survey of $He^{+}/He^{2+}$ abundance ratios in energetic particle events.**
D. Hovestadt, B. Klecker, G. Gloeckler, F. M. Ipavich, M. Scholer.
Astrophys. J., Lett. Ed., Vol. 282, No. 1, p. L39 – L42 (1984).
The authors report first results of a systematic study of the helium ionic charge distribution in the energy range 0.4 – 0.62 MeV per nucleon in energetic particle events. Large fluctuations of the $He^{+}/He^{2+}$ ratios were found especially at low interplanetary particle fluxes. The median of the distribution is $He^{+}/He^{2+} = 0.11$. No obvious association with optical flare events could be observed for the events richest in $He^{+}$. It has been found that generally $He^{+}$–rich events are predominantly low in energetic particle intensity.

078.002 **Mechanism of joint particle acceleration and deceleration on the sun.**
L. G. Kocharov.
Cosmic ray intensity and cosmogenic isotopes. 13th Leningrad seminar on cosmophysics, p. 55 – 59 (1983). In Russian. Abstr. in Ref. Zh., 51. Astron., 8.51.261 (1984). – See Abstr. 012.028.

078.003 **Space–time peculiarities of penetration of solar cosmic rays into the high–latitude region of the earth's magnetosphere according to Intercosmos 19 data.**
G. A. Glukhov, Yu. P. Kratenko, Yu. V. Mineev, E. S. Spir'kova.
Kosm. Issled., Tom 22, Vyp. 4, p. 594 – 598 (1984). In Russian. English translation in Cosm. Res.

078.004 **Electron–rich particle events and the production of gamma–rays by solar flares.**
P. Evenson, P. Meyer, S. Yanagita, D. J. Forrest.
Astrophys. J., Vol. 283, No. 1, p. 439 – 449 (1984).
The authors have determined the abundance of electrons of 5 – 50 MeV energies relative to protons of similar energy in a total of 49 particle events of apparently solar origin during the year 1980. Most events with high electron abundances appear to have been produced by flares which also produced observable fluxes of gamma–rays. Electron–rich particle events have harder electron spectra than more normal events and are never accompanied by strong interplanetary shocks. A model is proposed in which the relative electron richness of the particle events from gamma–ray flares is related to the nature of a shock produced by the solar flares.

078.005 **Determination of energy spectra and integral streams of solar cosmic rays in a source.**
A. B. Bajsakalova, E. V. Kolomeets, V. S. Murzin.
Issled. myuonov i nejtrino v bol'sh. vod. obemakh. Alma–Ata, p. 38 – 44 (1983). In Russian. Abstr. in Ref. Zh., 51. Astron., 9.51.427 (1984).

078.006 **Correlation fluctuations of streams of charged particles and of the magnetic field from data of the Intercosmos 19 satellite.**
G. A. Glukhov, Yu. P. Kratenko, S. N. Kuznetsov, Yu. V. Mineev.
Geomagn. Aehron., Tom 24, No. 5, p. 821 – 823 (1984). In Russian. English translation in Geomagn. Aeron.

078.007 **On the energetic dependence of the length of the free path of flare particles.**
E. I. Dajbog, V. G. Kurt, V. G. Stolpovskij.
Kosm. Issled., Tom 22, Vyp. 5, p. 763 – 773 (1984). In Russian. English translation in Cosm. Res.

078.008 **The energy spectra of solar energetic particles.**
R. E. McGuire, T. T. von Rosenvinge.
Adv. Space Res., Vol. 4, No. 2 – 3, p. 117 – 125 (1984). – See Abstr. 012.075.
The authors present a survey of recent results on the shapes and relative slopes of the spectra of various solar energetic particle populations, with emphasis on the more extensive results currently available for protons, alphas and electrons.

078.009 **Characteristics of solar and heliospheric ion populations observed near Earth.**
G. Gloeckler.
Adv. Space Res., Vol. 4, No. 2 – 3, p. 127 – 137 (1984). – See Abstr. 012.075.
The author describes current observations of the composition and spectra of ions in solar energetic particle events, energetic storm particle increases, diffuse upstream (of the Earth's bow shock) ions and tailward moving plasmoids in the deep geomagnetic tail. A number of features common to several of these heliospheric ion populations appear to stand out, suggesting that acceleration associated with shocks and turbulence is common.

078.010 **Injection of particles into the diffusive shock acceleration process: Monte Carlo simulation of heliospheric shocks.**
D. C. Ellison.
Adv. Space Res., Vol. 4, No. 2 – 3, p. 139 – 147 (1984). – See Abstr. 012.075.
The author compares a Monte Carlo simulation of diffusive shock acceleration to observations of the nearly radial bow shock, an interplanetary shock and solar flare energetic particle events. The simulation assumes that the same scattering operator that gives rise to shock acceleration can also describe a viscous shock governed by hydrodynamic turbulence. Good agreement between the simulation and observations lends support to thermal leakage of downstream, shock heated ions as an important mode of injection.

078.011 **On the problem of relativistic particles acceleration in processes similar to those in solar flares.**
S. V. Bulanov, F. Cap.
Adv. Space Res., Vol. 4, No. 2 – 3, p. 149 – 152 (1984). – See Abstr. 012.075.
Motion and acceleration of charged particles in steady electric and magnetic fields typical of magnetic reconnection, are investigated. Characteristic energies and energy spectra of particles are determined. The application to the problem of particle acceleration in space plasmas is discussed.

078.012 **Acceleration of particles in shocked magnetic neutral sheets.**
D. J. Mullan, J. Perez–Peraza, M. Galvez, M. Alvarez.
Adv. Space Res., Vol. 4, No. 2 – 3, p. 157 – 160 (1984). – See Abstr. 012.075.
The authors report on a study of particle acceleration in a magnetic neutral sheet which has been struck by a passing MHD shock, such as occurs in the sun when a flare occurs near a helmet streamer. It is proposed that long–lived particle events with energies up to 10 – 100 MeV following solar flares may be explained by the mechanism described here.

078.013 **Ionic charge states of N, Ne, Mg, Si and S in solar energetic particle events.**
A. Luhn, B. Klecker, D. Hovestadt, G. Gloeckler, F. M. Ipavich, M. Scholer, C. Y. Fan, L. A. Fisk.
Adv. Space Res., Vol. 4, No. 2 – 3, p. 161 – 164 (1984). – See Abstr. 012.075.
The mean ionic charges of nitrogen, neon, magnesium, silicon, and sulfur accelerated in solar flares have been determined for selected solar energetic particle events in the time period from September 1978 to September 1979. The observations were carried out with the ULEZEQ sensor on the ISEE–3 satellite. The

electron temperature of the plasma as calculated from ionization equilibrium tables range from $\sim 2\times10^6$K to $\sim 7\times10^6$K in a single flare.

078.014 **Determination of the effective charge of solar cosmic-ray nuclei using the Earth's magnetic field.**
S. Fischer, M. Vandas, K. Kudela, S. N. Kuznetsov, V. N. Lutsenko.
Adv. Space Res., Vol. 4, No. 2 – 3, p. 169 – 172 (1984). – See Abstr. 012.075.

Using the Earth magnetic field as an analyzer one is able to determine an upper limit of the charge of solar cosmic ray nuclei. The authors analyse data from the low–altitude Intercosmos 17 satellite obtained after the solar flare of February 13, 1978 in the energy range 5 – 20 MeV/nucleon. The results indicate the existence of nuclei not fully ionised in this energy range.

078.015 **Acceleration phenomena in the solar system.**
M. Scholer.
Adv. Space Res., Vol. 4, No. 2 – 3, p. 419 – 421 (1984). – See Abstr. 012.075.

This rapporteur paper discusses a number of poster papers and review papers which were presented in Symposium 6 of the 25th COSPAR Plenary Meeting. These papers were concerned with energetic particle phenomena in the solar system.

078.016 **Some optical characteristics of proton active regions in the 21st solar cycle.**
Z.–y. Zhu.
Publ. Purple Mt. Obs., Vol. 3, No. 2, p. 42 – 47 (1984). In Chinese.

The correlated relationship between proton flux and optical parameters of proton active regions is statistically analysed for 26 proton events which occurred during July 1977 – June 1983.

078.017 **On the relative proportion of different nuclei in acceleration by shock waves.**
A. A. Korchak.
Problems of physics of solar flares, p. 100 – 106 (1983). In Russian. Abstr. in Ref. Zh., 51. Astron., 12.51.670 (1984). – See Abstr. 012.070.

078.018 **On estimates of energetics of solar cosmic rays.**
L. I. Miroshnichenko.
Problems of physics of solar flares, p. 120 – 125 (1983). In Russian. Abstr. in Ref. Zh., 51. Astron., 12.51.673 (1984). – See Abstr. 012.070.

078.019 **Investigation of solar cosmic rays aboard the Prognoz stations.**
Yu. I. Logachev.
Solar activity investigations and the Prognoz space system, p. 97 – 117 (1984). In Russian. Abstr. in Ref. Zh., 62. Issled. Kosm. Prostranstva, 12.62.367 (1984). – See Abstr. 003.038.

078.020 **The spectrum in the source and energetics of the solar proton event on August 7, 1972.**
L. I. Miroshnichenko.
Geomagn. Aehron., Tom 24, No. 6, p. 865 – 870 (1984). In Russian. English translation in Geomagn. Aeron.

078.021 **On estimates of the energy density of solar cosmic rays near the earth.**
L. I. Miroshnichenko.
Geomagn. Aehron., Tom 24, No. 6, p. 871 – 874 (1984). In Russian. English translation in Geomagn. Aeron.

**Electron spectra deduced from solar hard X–ray bursts.**
See Abstr. 073.081.

**Peculiarities of radio and optical emissions in the post–maximum period of the proton flare on May 16, 1981.**
See Abstr. 073.093.

**Coronal mass ejections.**
See Abstr. 074.034.

**Associations between coronal mass ejections and solar energetic proton events.**
See Abstr. 074.052.

**Selective heating of minority ions in the lower solar corona by electrostatic ion cyclotron waves.**
See Abstr. 074.053.

**High–energy neutral radiations from the Sun.**
See Abstr. 076.012.

**Solar X–ray flare emission and energetic particles according to data of observations aboard the Venera 13, 14 automatic interplanetary stations.**
See Abstr. 076.016.

**Type II solar radio bursts, interplanetary shocks, and energetic particle events.**
See Abstr. 077.006.

**Partial reconstruction of the initial conditions for streams of energetic electrons associated with a solar type III burst.**
See Abstr. 077.039.

**Fluctuations of the interplanetary and geomagnetic fields in connection with proton bursts.**
See Abstr. 106.016.

**Shock acceleration of nuclei and electrons in the heliosphere beyond 24 AU.**
See Abstr. 106.019.

**The interplanetary scattering mean free path: collisionless wave–damping effects.**
See Abstr. 106.026.

**Particle acceleration and MHD wave excitation upstream of interplanetary shocks.**
See Abstr. 106.043.

**ISEE–3 observations of energetic protons associated with interplanetary shocks.**
See Abstr. 106.044.

**Diffusion coefficients of low–energy protons upstream of quasi–parallel interplanetary shocks.**
See Abstr. 106.045.

**Low–energy protons associated with interplanetary shocks as an independent population in the solar wind.**
See Abstr. 106.046.

**The time scales of the scattering of energetic protons in interplanetary space.**
See Abstr. 106.047.

**Time–lag of cosmic–ray intensity.**
See Abstr. 144.050.

# 079 Solar Eclipses

079.001 **Canon der Sonnenfinsternisse, –2003 bis +2526.**
H. Mucke, J. Meeus.
Sternenbote, 27. Jahrg., No. 335, p. 106 – 120 (1984).

079.002 **The solar eclipse of 1984 November 22 – 23.**
G. H. Moller.
South. Stars, Vol. 30, Nos. 7 – 8, p. 444 – 448 (1984).
The prediction of the local circumstances of a solar eclipse is not covered in the present-day literature on positional astronomy by calculator. An amateur who merely wishes to predict the circumstances of an eclipse for his place of residence can do so with fairly straightforward calculations. The eclipse of 1984 November 22 – 23 is used to demonstrate the method.

079.003 **Solar eclipse diary: 1985 – 1995.**
J. Meeus.
Sky Telesc., Vol. 68, No. 4, p. 296 – 298 (1984).

079.004 **Saros: the clockwork of eclipses.**
A. Chaikin.
Sky Telesc., Vol. 68, No. 4, p. 299 (1984).

079.005 **Tjugoen minuter i skuggan – ostraffbart?**
P.-Å. Björklund.
Astron. Tidsskr., Årg. 17, Nr. 3, p. 118 – 125 (1984).

**Solar eclipse 1984 May 30**

079.101 **The 1984 May 30th solar eclipse.**
D. W. Dunham.
Occultation Newsl., Vol. 3, No. 8, p. 163 – 166 (1984).

079.102 **May's eclipse: flirting with totality.**
L. J. Robinson.
Sky Telesc., Vol. 68, No. 2, p. 101 – 105 (1984).

079.103 **The great annular eclipse of 1984.**
F. Espenak.
Astronomy, Vol. 12, No. 5, p. 34 – 39 (1984). Abstr. in Phys. Abstr., Vol. 87, No. 1242, Entry 94562 (1984).

079.104 **L'eclisse anulare di Sole del 30 Maggio 1984.**
L. Ceccaroni.
G. A.A.B., N. 76, p. 5 – 6 (1984).

079.105 **Die Perlschnurring-Finsternis am 30. Mai 1984.**
K. Löchel.
Sterne Weltraum, 23. Jahrg., Nr. 12, p. 642 – 649 (1984).

079.106 **Drawing a bead on the solar eclipse.**
G. Moller.
Photonics Spectra, Vol. 18, No. 5, p. 147 – 149 (1984).
For the amateur astronomer: formulas that can be used to track the 1984 May 30 solar eclipse.

079.107 **Observations of the May 30, 1984 annular solar eclipse.**
R. E. Hill.
Strolling Astron., Vol. 30, Nos. 11 – 12, p. 248 – 252 (1984).

**Solar eclipse 1984 November 22 – 23**

079.121 **Éclipse totale du soleil sur un territoire français d'Outre-mer, La Nouvelle-Calédonie, le 23 novembre 1984.**
C. Nitschelm, S. Koutchmy.
Astronomie, Vol. 98, p. 393 – 399 (1984).

079.122 **Predictions for the total solar eclipse of 1984.**
F. Espenak.
Strolling Astron., Vol. 30, Nos. 9 – 10, p. 191 – 199 (1984).

079.123 **Die totale Sonnenfinsternis am 22. November 1984 über Neuguinea.**
K. Löchel.
Sterne Weltraum, 23. Jahrg., Nr. 10, p. 518 – 521 (1984).

**Solar eclipse 1980 February 16**

079.141 **Cultural implications of the eclipse of 16 February 1980 in Lamu, Kenya.**
P. Seitel.
Archaeoastronomy (U.S.A.), Vol. 6, Nos. 1 – 4, p. 134 – 140 (1983).

079.142 **Observations of shadow bands at the total solar eclipse of 16 February 1980.**
L. A. Marschall, R. Mahon, R. C. Henry.
Appl. Opt., Vol. 23, No. 23, p. 4390 – 4393 (1984).
Photoelectric observations of short term light variations (shadow bands) at the 16 Feb. 1980 total solar eclipse have been made using a set of spatially separated PIN diodes. Light variations in a bandpass of 1 – 500 Hz were detected during the half-minutes preceding and following the total phase. Fourier analysis of the noise spectrum of the variations reveals a sharp drop-off for frequencies above 50 Hz and an overall spectrum quite similar to previously reported power spectra of stellar scintillation. This is consistent with an atmospheric origin for the shadow bands.

**The hidden sun: solar eclipses and astrophotography.**
See Abstr. 003.118.

**Due eclissi storiche in Abruzzo.**
See Abstr. 004.003.

**De laatste totale zonsverduistering in Nederland.**
See Abstr. 004.034.

**Ermittlung des Flächenbedeckungsgrades aus Sonnenfinsternisfotos.**
See Abstr. 036.068.

**A derivation of the rate of spin-down of the Earth using ancient Chinese records of central eclipses.**
See Abstr. 044.044.

**On the Sun's limb brightening in the visible.**
See Abstr. 071.014.

**Photometric analysis of the June 11, 1983 solar corona.**
See Abstr. 074.011.

**Structure of the inner corona derived from observations of the eclipse of 16 February 1980.**
See Abstr. 074.024.

**Correction for the observed intensity distribution in solar corona.**
See Abstr. 074.028.

**Near-infrared observation of the circumsolar dust emission during the 1983 solar eclipse.**
See Abstr. 074.044.

**White light and monochromatic observations of the corona during the 1980 February 16 eclipse. I. The quiet parts.**
See Abstr. 074.054.

**Coronal temperature measurements near a helmet structure base at the 1973 solar eclipse.**
See Abstr. 074.062.

**Ionospheric currents and fields during the solar eclipse.**
See Abstr. 083.004.

**Years with seven eclipses in 1900 – 2500.**
See Abstr. 095.006.

# 080 Atmosphere, Figure, Internal Constitution, Neutrinos, Rotation, etc.

080.001 **Rotational frequency splitting of solar oscillations.**
T. L. Duvall Jr., J. W. Harvey.
Nature, Vol. 310, No. 5972, p. 19 – 22 (1984).

Prograde and retrograde sectoral oscillations of the Sun have been observed so as to determine frequency differences produced by rotation. Oscillations in the frequency range 2.1 – 3.7 mHz and with spherical harmonic degrees from 1 to 100 have been identified. Average frequency shifts due to rotation in a sidereal reference frame are found to range from a high of ~660 nHz at degree 1 to a low of ~423 nHz at degree 6, rising to ~471 nHz at degree 100. These results indicate that most of the Sun's volume rotates at a rate close to that of the surface, but also that the energy–generating core may rotate more rapidly than the surface.

080.002 **Internal rotation of the Sun.**
T. L. Duvall Jr., W. A. Dziembowski, P. R. Goode, D. O. Gough, J. W. Harvey, J. W. Leibacher.
Nature, Vol. 310, No. 5972, p. 22 – 25 (1984).

The frequency difference between prograde and retrograde sectoral solar oscillations is analysed to determine the rotation rate of the solar interior, assuming no latitudinal dependence. Much of the solar interior rotates slightly less rapidly than the surface, while the innermost part apparently rotates more rapidly. The resulting solar gravitational quadrupole moment is $J_2 = (1.7 \pm 0.4) \times 10^{-7}$ and provides a negligible contribution to current planetary tests of Einstein's theory of general relativity.

080.003 **Ground–based observations of solar intensity oscillations.**
T. Schmidt–Kaler, C. Winkler.
Astron. Astrophys., Vol. 136, No. 2, p. 299 – 305 (1984).

Very inexpensive ground–based photometric measurements of the full disk solar light were obtained with a diffraction telescope at ESO – La Silla. First results of the analysis are: (1) Power spectra obtained on 1981, Dec. 16 and 17 in Hβ show significant peaks at $\nu = 3.200$ mHz and 3.334 mHz with relative amplitudes $\Delta I/I \cong 10^{-4}$ ($\sigma_\nu = \pm 10.6$ μHz). (2) The data of 1983 (Hβ) show again the peak at $\nu = 3.333$ mHz, $\Delta I/I \cong 10^{-4}$ without any concomitant peak in the air pressure analysis ($\sigma_\nu = \pm 7.8$ μHz). (3) A number of significant peaks in the *B*–region is identified as non–atmospheric–pressure–related and non–instrumental in the range 2.99 mHz – 3.70 mHz with relative amplitudes $\Delta I/I \cong 4 \times 10^{-5}$. (4) The same feature as in *B* seems to be weakly indicated in the analysis of the *V*–measurements.

080.004 **Measurements of solar total irradiance and its variability.**
R. C. Willson.
Space Sci. Rev., Vol. 38, Nos. 3/4, p. 203 – 242 (1984).

The development of electrically self calibrated cavity pyrheliometric instrumentation that occurred in the early 20th century provided the technological base for experiments to detect variability of the solar total irradiance. Refinements in pyrheliometric technology during the 1960's and 1970's and the accessibility of extended experimental opportunities above the Earth's atmosphere in recent years have provided the first direct observations of solar total irradiance variability and provided the cornerstone observations of a long term database on solar irradiance. A program of solar irradiance monitoring has evolved to sustain the database over at least 22 years, corresponding to a single cycle of solar magnetic activity, and the shortest well identified cycle of climate variation. Direct links between total irradiance variations, solar magnetic activity and the solar global "5 min" oscillation phenomena have been derived from recent space flight observations by the SMM/ACRIM I experiment.

080.005 **The Sun as an inconstant star.**
K. Sakurai.
Space Sci. Rev., Vol. 38, Nos. 3/4, p. 243 – 279 (1984).

In this paper, the problem related to the observed results on the neutrino flux from the Sun will first be reviewed by taking into account the physical processes in the solar interior. The possible mechanism on the variability of the solar luminosity and activity will be considered on the basis of the theory of the energy production and transport inside the Sun. Discussion will be given on the possible connection of the Sun's variations with the Earth's climate. Finally, the cause of the inconstancy of the Sun will be briefly considered based on the variability of the internal structure of the Sun.

080.006 **Solar irradiance observations.**
D. Crommelynck, V. Domingo.
Science, Vol. 225, No. 4658, p. 180 – 181 (1984).

The absolute radiometer on Spacelab 1 was used to obtain solar irradiance observations from space. A number of effects must be taken into account in the data reduction. A provisional value was obtained for the mean solar constant during the observation period (6 to 8 December 1983).

080.007 **The overshoot region at the bottom of the solar convection zone.**
J. H. M. M. Schmitt, R. Rosner, H. U. Bohn.
Astrophys. J., Vol. 282, No. 1, p. 316 – 329 (1984).

The authors have investigated the extent and thermal stratification of the region of convective overshoot underneath the convection zone of the Sun. This study has been motivated by recent suggestions that this overshoot zone may be a suitable place for magnetic flux storage during the course of the solar cycle, and that doubly diffusive instabilities may act in this region as a mechanism for breaking up diffuse magnetic fields into flux ropes. The authors derive a set of differential equations suitable for the study of convective overshoot in an astrophysical context. They present numerical solutions of these equations and obtain simple scaling relations for the ensuing overshoot in terms of the initial velocity and the filling factor.

080.008 **A model of the sun with a neutrally stable core.**
Yu. V. Vandakurov.
Sov. Astron., Vol. 28, No. 1, p. 100 – 103 (1984). English translation of 37.080.005.

080.009 **Does the Davis neutrino experiment show a variable $^{37}$Ar production rate?**
V. N. Gavrin, A. V. Kopylov.
Sov. Astron. Lett., Vol. 10, No. 1, p. 62 – 64 (1984). English translation of 37.080.006.

080.010 **Non–linear helioseismology.**
J. Perdang, S. Blacher.
Mon. Not. R. Astron. Soc., Vol. 209, No. 4, p. 905 – 912 (1984).

The differential equations for non–linearly coupled radial adiabatic oscillation modes are integrated for three detailed solar models. In all experiments the resulting power spectrum of the surface displacement shows a peak at a long period $P_L$ exceeding 2 hr.

080.011 **Solar large scale velocity structures from optical resonance method.**
J. M. Robillot, R. Bocchia, E. Fossat, G. Grec.
Astron. Astrophys., Vol. 137, No. 1, p. 43 – 50 (1984).

Preliminary observations have been made to search for large scale velocity fields by means of sodium optical resonance spectrophotometry. The authors recall that the standard method measures only 2 points of the line, in order to derive the Doppler shift. They point out an ambiguity from line profile distortion, because apparent upflows are seen in the wide area of active regions. In any case, this can explain the long-term modulation observed in full disk Doppler measurements (Claverie et al., 1982) without the need of invoking a rapidly rotating core. In the quiet solar atmosphere, an upper limit of 15 m $s^{-1}$ rms for the large scale (≥50,000 km) velocity distribution is deduced. A relatively simple optimization of the optical resonance technique is suggested in order to progress in this field.

080.012 **Solar neutrinos and comets.**
A. J. R. Prentice.
Essays in theoretical physics, p. 207 - 217 (1984). - See Abstr. 003.006.

It is demonstrated how a new class of chemically inhomogeneous solar models can be constructed which leads to a present-day sun possessing a small burnt-out core consisting of at least 95% of elements heavier than He. Such models with a core mass fraction of 1% have a neutrino flux of order 2 SNU, in line with the observed value. It is outlined how such a segregated solar core may have formed from interstellar grains consisting of rocky silicates and CNO ices in the protosolar nebula. Comets seem to have a very similar composition to that postulated for the solar core. This leads the authors to the conclusion that comets may have formed through gravitational fragmentation of the original cloud of dirty CNO grains which have sedimented to form the sun's core.

080.013 **Mesure du diamètre solaire et des éléments orbitaux de la Terre à l'astrolabe du C.E.R.G.A.**
M. Bougeard, F. Laclare.
Astrométrie, géodésie, mécanique céleste, p. 87 - 91 (1984). - See Abstr. 012.021.

080.014 **Les oscillations du Soleil et des étoiles.**
J.-C. Pecker.
C. R. Acad. Sci., Sér. Gén., Vie Sci., Tome 1, No. 3, p. 199 - 222 (1984).

Des travaux récents ont attiré l'attention du monde astronomique, en mettant en évidence des oscillations dans l'atmosphère du Soleil et des étoiles. Que des étoiles oscillent, cela n'est pas nouveau: on connaît depuis des décennies le caractère oscillant, pulsatoire, des étoiles du type δ Cephei, ou Céphéides. Mais les oscillations découvertes dans le Soleil et dans α Centauri A sont loin d'être aussi simples: elles ne sont pas «radiales». Divers harmoniques se superposent les uns aux autres et prennent plus d'importance que l'oscillation fondamentale; pour chacun d'eux, la surface stellaire est un réseau de lignes nodales, non affectées par le mouvement oscillatoire. Certaines de ces oscillations sont commandées par la force de rappel qu'exerce la compression locale des gaz (modes acoustiques); d'autres par celle qu'exercent les forces de gravité (modes de gravité). La distribution d'énergie entre les différents modes et les différents harmoniques est en effet un reflet fidèle de la structure interne de l'étoile. Il est ainsi possible de sonder des milieux jusqu'à présent tout-à-fait inaccessibles à l'observation.

080.015 **Peculiar points of the solar rotation.**
I. Escaut-Soru, M.-J. Martres, Z. Mouradian.
C. R. Acad. Sci., Sér. II, Tome 299, No. 9, p. 545 - 548 (1984). In French.

The spatial distribution of single flares shows that some of them appear out of active regions and faculae. In this paper an association between the flare location and a quiescent filament observed only in a preceding rotation is pointed out.

080.016 **Towards a solar model.**
D. O. Gough.
Mem. Soc. Astron. Ital., Vol. 55, Nos. 1 - 2, p. 13 - 35 (1984). - See Abstr. 012.022.

This paper reviews the basic physics of solar oscillations, the diagnostic inferences that can be drawn from the solar eigenfrequencies and the constraints on solar models imposed by the presently available data.

080.017 **Comparisons of theoretical and observational eigenfrequencies for low, intermediate, and high degree solar oscillations.**
E. J. Rhodes Jr., R. K. Ulrich, W. M. Brunish.
Mem. Soc. Astron. Ital., Vol. 55, Nos. 1 - 2, p. 37 - 44 (1984). - See Abstr. 012.022.

080.018 **Observations of global solar oscillations: past, present and future.**
G. R. Isaak.
Mem. Soc. Astron. Ital., Vol. 55, Nos. 1 - 2, p. 45 (1984). Abstract. - See Abstr. 012.022.

080.019 **Low degree *p*-modes observed in velocity and brightness. Future in the frame of the space-ground competition.**
E. Fossat.
Mem. Soc. Astron. Ital., Vol. 55, Nos. 1 - 2, p. 47 - 62 (1984). - See Abstr. 012.022.

The main observational results for solar *p*-mode oscillations are summarized and the prospects for improving current measurements are examined.

080.020 **Continuous observation of solar oscillations from two suitably spaced ground stations.**
A. Claverie, G. R. Isaak, C. P. McLeod, H. B. van der Raay, P. L. Palle, T. Roca Cortés.
Mem. Soc. Astron. Ital., Vol. 55, Nos. 1 - 2, p. 63 - 66 (1984). - See Abstr. 012.022.

By operating two observing stations, one at Izana on Tenerife and the other at Haleakala on Maui, up to 22 hours of continuous solar surface velocity data per day are obtained. Correlation of two station data provides confirmation of the solar origin of the signals and the improved window function and long time span of data collection yield clear power spectra of the 5 minute oscillations. Detailed analysis of a prominent $l_1$ mode indicates that coherence exists over the 88 days of observation.

080.021 **Total irradiance observations of low-degree *p*-modes.**
M. Woodard, H. S. Hudson.
Mem. Soc. Astron. Ital., Vol. 55, Nos. 1 - 2, p. 67 - 68 (1984). - See Abstr. 012.022.

A spectral analysis of 10 months of brightness data from the Active Cavity Radiometer Irradiance Monitor on the SMM satellite has revealed new information about the low-degree ($l$ = 0,1, and 2) *p*-mode oscillations of the Sun. The mean frequencies, irradiance amplitudes, and line widths of the largest amplitude modes have been determined with previously unattained accuracy.

080.022 **Anisotropy and amplitude structure of low-*l* 5-min solar oscillations.**
J. R. Kuhn.
Mem. Soc. Astron. Ital., Vol. 55, Nos. 1 - 2, p. 69 - 74 (1984). - See Abstr. 012.022.

Line center Doppler shift data has been obtained from CCD-spectrograph observations. The 5 hour time series yields useful information at spatial wavelengths near one solar radius along both the solar north-south and east-west orientations. At these wavelengths the 5-min oscillation velcocity field is not isotropic and not well described by modes of uniform excitation. The velocity power density per spherical harmonic increases with decreasing $l$ to $1.1 \times 10^3$ $cm^2$ $s^{-2}$ per $3.5 \times 10^{-4}$ Hz angular frequency bandwidth at $l = 4$.

080.023 **Solar gravity modes detection and identification.**
P. Delache.
Mem. Soc. Astron. Ital., Vol. 55, Nos. 1 - 2, p. 75 - 82 (1984). - See Abstr. 012.022.

In contrast with the 5 min pressure modes, the solar gravity modes exhibit a very intricate spectral signature. Analyses of presently available global oscillations data show that *g*-modes of low degree and intermediate order can, however, be detected. They lie in a range of periods, several hours, where the signal/noise ratio demands a careful approach. Theoretical asymptotic guide lines of $g$–modes identification, including effects of rotation, are studied in some detail.

080.024 **Detection of solar gravity mode oscillations.**
P. H. Scherrer.
Mem. Soc. Astron. Ital., Vol. 55, Nos. 1 – 2, p. 83 – 89 (1984). – See Abstr. 012.022.

An analysis of solar velocity data obtained at the Stanford Solar Observatory has shown the existence of solar global oscillations. The oscillations are in the range 45 to 105 μHz (160 to 370 minutes) and are interpreted as internal gravity modes of degree $l = 1$ and $l = 2$.

080.025 **Solar *g* modes.**
G. R. Isaak, H. B. van der Raay, P. L. Palle, T. Roca Cortés, P. Delache.
Mem. Soc. Astron. Ital., Vol. 55, Nos. 1 – 2, p. 91 – 97 (1984). – See Abstr. 012.022.

A statistical analysis of the frequency spectrum of 85 days of data obtained from integral solar velocity signals, has shown that significant signals exist in the frequency range 25 – 150 μHz. The signal structure is shown to have characteristics of constant period spacings indicative of the existence of $g$ modes in the solar oscillation spectrum. Utilising the asymptotic Tassoul relation a total of 58 lines are identified covering the $l_1$, $l_2$, $l_3$ and $l_4$ modes.

080.026 **Solar gravity modes from ACRIM/SMM irradiance data.**
C. Fröhlich, P. Delache.
Mem. Soc. Astron. Ital., Vol. 55, Nos. 1 – 2, p. 99 – 105 (1984). – See Abstr. 012.022.

The record of 280 days of continuous data of the ACRIM radiometer on board the SMM satellite is analysed in the frequency range from 2 to 80 μHz. Gravity modes of degree one and two with orders from about 10 to several hundreds can be localized. A statistical method to determine the fundamental period and the rate of rotation as seen by rotational splitting is described and the results are compared with other determinations.

080.027 **Influence of turbulent diffusion on asymptotic low degree solar gravity modes.**
G. Berthomieu, J. Provost, E. Schatzman.
Mem. Soc. Astron. Ital., Vol. 55, Nos. 1 – 2, p. 107 – 114 (1984). – See Abstr. 012.022.

Asymptotic properties of the periods of low frequency and low degree solar oscillations are studied. These periods are computed for a set of solar models with turbulent diffusion mixing. They are found to be very sensitive to the pseudo–Reynolds number that measures the efficiency of the turbulent diffusion and significantly enhanced by an even moderate mixing.

080.028 **Seismonuclear excitation of solar gravity modes and the solar neutrino puzzle.**
Y. S. Kopysov (*Yu. S. Kopysov*).
Mem. Soc. Astron. Ital., Vol. 55, Nos. 1 – 2, p. 115 (1984). Abstract. – See Abstr. 012.022.

080.029 **160–minute and other long–period oscillations of the Sun: analysis and interpretation of 9–year observations.**
V. A. Kotov, A. B. Severny (*A. B. Severnyj*), T. T. Tsap.
Mem. Soc. Astron. Ital., Vol. 55, Nos. 1 – 2, p. 117 – 122 (1984). – See Abstr. 012.022.

Besides the persistent 160–min oscillation Crimean observations show the presence of 10 or more other long–period oscillations interpreted as gravity $g$–modes.

080.030 **Limb darkening fluctuations: Is the five–minute oscillation also terrestrial?**
M. R. Yerle.
Mem. Soc. Astron. Ital., Vol. 55, Nos. 1 – 2, p. 123 – 128 (1984). – See Abstr. 012.022.

High speed scans of the apparent solar limb darkening show fluctuations of the maximum brightness gradient. One spectrum calculated from four consecutive days corresponding to very good seeing conditions is presented with a 2 μHz frequency resolution. In the five–minute range, two periods also could be linked to terrestrial atmosphere oscillations.

080.031 **Instability of non–radial *g*–mode oscillations of the Sun with low *z* interior.**
A. G. Kosovichev, A. B. Severny (*A. B. Severnyj*).
Mem. Soc. Astron. Ital., Vol. 55, Nos. 1 – 2, p. 129 – 134 (1984). – See Abstr. 012.022.

A solar model is calculated with low initial heavy elements abundance. The non–radial oscillations stability of this model are investigated by the quasi–adiabatic approximation. The $g_1$ and $g_2$–modes belonging to the dipole ($l = 1$) oscillations are found unstable due to the ε–mechanism. Thus from this model one expects the resonance excitation of long–period oscillations with degree $l = 4$ and order $n = 10 – 20$.

080.032 **Phase relations of high degree modes revisited.**
F.–L. Deubner, F. Endler, J. Staiger.
Mem. Soc. Astron. Ital., Vol. 55, Nos. 1 – 2, p. 135 – 146 (1984). – See Abstr. 012.022.

Recently developed numerical methods for de–biassing spectra of short period solar oscillations impaired by terrestrial seeing afford an analysis of phases as a function of frequency much further into the range of high frequencies than previously appeared possible. A simple model is used to extract a few preliminary, but quantitative, results from the data. These results seem to indicate that the layer of reflection of the short–period waves is rather higher than the level commonly ascribed to the transition region. Horizontal magnetic fields must be either too weak or too rare to leave any noticeable effect on the short–period acoustic wave field.

080.033 **Phase–spectra of solar oscillations.**
J. Staiger, B. Schmieder, F.–L. Deubner, W. Mattig.
Mem. Soc. Astron. Ital., Vol. 55, Nos. 1 – 2, p. 147 – 152 (1984). – See Abstr. 012.022.

Diodearray–measurements of intensity– and velocity–oscillations were made at the Coudé–telescope of Capri. From these observations phase–spectra were calculated to identify different types of waves within the solar atmosphere. The results indicate the existence of atmospheric gravity modes in the frequency–range below 2.5 mHz. No running acoustic waves were found in the photosphere for frequencies beyond 7.5 mHz. Vertical propagation seems only to take place for frequencies between 5 mHz and 7.5 mHz. No vertical propagation at all was found in the upper chromosphere.

080.034 **Attempt to measure the solar subsurface velocity.**
F. Hill, D. Gough, J. Toomre.
Mem. Soc. Astron. Ital., Vol. 55, Nos. 1 – 2, p. 153 – 161 (1984). – See Abstr. 012.022.

Five–minute oscillation modes are advected by horizontal velocities below the solar surface, and thus can be used as probes of rotation and large–scale convective flows. Results of inverse theory applied to observations of high–degree modes carried out on six separate days reveal variations in horizontal velocities with depth from day to day that may be the result of giant convection cells.

080.035 **Influence of the equations of state and of the *z* value on the solar oscillations.**
H. Shibahashi, A. Noels, M. Gabriel.
Mem. Soc. Astron. Ital., Vol. 55, Nos. 1 – 2, p. 163 – 168 (1984). – See Abstr. 012.022.

The authors have evolved the standard solar models taking account of the electrostatic correction in the equations of state and the Los Alamos Opacity Library, and computed their eigenfrequencies for $p$–mode and $g$–mode oscillations. The contribution of that correction to the pressure is up to several percent in the convective zone, and the present equilibrium models thus obtained have deeper convective zones than the conventional models. These improvements in the equilibrium model reduce the discrepancies between the theoretical frequencies and the observations concurrently for $p$–mode oscillations with low and high degree $l$. Theoretical frequencies of $g$–modes with low $l$ are significantly higher than the observational ones.

080.036 **Influence of opacities, partition functions and hydrogen diffusion on the 5 min solar oscillations.**
M. Gabriel, A. Noels, R. Scuflaire.
Mem. Soc. Astron. Ital., Vol. 55, Nos. 1 – 2, p. 169 – 174 (1984). – See Abstr. 012.022.
The authors compare their frequency calculations of solar oscillations based on a standard solar model with the results from other researchers. A model with diffusion of H and He has also been studied. Its frequencies are nearly the same as those of the standard model.

080.037 **Overstable solar oscillations of intermediate and high degree.**
H. M. Antia, S. M. Chitre, D. Narasimha.
Mem. Soc. Astron. Ital., Vol. 55, Nos. 1 – 2, p. 175 – 180 (1984). – See Abstr. 012.022.
The stability of linear acoustic modes trapped in the solar envelope is investigated with the thermal and mechanical effects of turbulence on the mean flow incorporated through turbulent conductivity, viscosity and turbulent pressure. A number of these modes turn out to be overstable, and the most rapidly growing acoustic modes are found to occupy a region centred around 3.2 mHz and spread over a wide range of length scales. These numerical results are in reasonable accord with the observed power–spectrum of the five–minute oscillations.

080.038 **Simple asymptotic estimates of the fine structure in the spectrum of solar oscillations due to rotation and magnetism.**
W. Dziembowski, P. R. Goode.
Mem. Soc. Astron. Ital., Vol. 55, Nos. 1 – 2, p. 185 – 213 (1984). – See Abstr. 012.022.
Simple asymptotic formulae are presented and used to predict the effect of rotation and magnetism on the fine structure in the spectrum of solar oscillations. The authors compare the linear effect of rotation on the fine structure to the quadratic effect of rotation. The asymptotic formulae for the magnetically induced fine structure are used to schematically determine the requisite field size to cause an observable change in the structure and the nature of the change.

080.039 **On the identification of normal modes of oscillation from observations of the solar periphery.**
D. O. Gough, J. Latour.
Mem. Soc. Astron. Ital., Vol. 55, Nos. 1 – 2, p. 227 (1984). Abstract. – See Abstr. 012.022.

080.040 **A preliminary report on brightness oscillation observations made at South Pole during the 1982 – 1983 austral summer.**
R. Stebbins.
Mem. Soc. Astron. Ital., Vol. 55, Nos. 1 – 2, p. 229 (1984). Abstract. – See Abstr. 012.022.

080.041 **Ground–based observations of solar intensity oscillations.**
T. Schmidt–Kaler, C. Winkler.
Mem. Soc. Astron. Ital., Vol. 55, Nos. 1 – 2, p. 231 – 236 (1984). – See Abstr. 012.022.
A method is presented to make very inexpensive ground–based photometric measurements of the full disk solar light using a diffraction telescope at ESO – La Silla in the spectral domains Hβ, *V* and *B*. The random error of the photometric device is reduced down to $8 \times 10^{-5}$ (for one minute averages), systematic errors over periods of 20 minutes were not detectable.

080.042 **Wavelength dependence of solar luminosity fluctuations in the five minutes range.**
C. Fröhlich.
Mem. Soc. Astron. Ital., Vol. 55, Nos. 1 – 2, p. 237 – 243 (1984). – See Abstr. 012.022.
During a balloon flight high precision measurements of solar spectral irradiance at three wavelengths (368, 500 and 778 nm) have been made during 4 hours. The high correlation with simultaneous measurements from the SMM satellite demonstrate the solar origin of the registered short term fluctuations. Some peaks in the five minutes range may be interpreted as solar *p*–mode oscillations. The magnitudes of the ratios of the amplitudes at different colours are a further indication of pressure induced fluctuations of the solar photosphere.

080.043 **Intensity oscillations in the calcium – K line.**
P. Gouttebroze, L. Damé, J.–M. Malherbe.
Mem. Soc. Astron. Ital., Vol. 55, Nos. 1 – 2, p. 245 – 250 (1984). – See Abstr. 012.022.
The authors analyze a time sequence of filtergrams of the quiet sun, taken in the core of the K line, to investigate the oscillatory properties of the chromosphere. The authors discuss the physical significance of these intensity variations and their diagnostic capabilities. A diagram of oscillatory power versus frequeny and mean intensity shows the different behaviours of bright (chromospheric network) regions, dominated by low frequency (gravity?) waves and darker regions (cell interiors), dominated by high frequency (probably acoustic) waves.

080.044 **Possible evidence of solar oscillations in the multifrequency spectra of solar radio brightness fluctuations.**
S. K. Alurkar, R. V. Bhonsle, S. S. Degaonkar, O. P. N. Calla, G. Raju.
Mem. Soc. Astron. Ital., Vol. 55, Nos. 1 – 2, p. 251 – 261 (1984). – See Abstr. 012.022.
Changes in solar radio brightness temperature were derived at 2.8, 19.3 and 22.2 GHz from the observations of solar radio flux during the total eclipse on 16 February 1980. High resolution MEM spectra of the brightness temperature fluctuations at the three frequencies showed periodicities ranging from 3.5 min to 128 min. Between 3.5 min and 14.6 min there are a number of significant periodicities common to the three operating frequencies. These periodic variations in brightness temperature may be a result of the spatial and/or temporal variations in the solar radio emissions.

080.045 **Solar oscillations and the effect of a comet impact.**
G. R. Isaak, H. B. van der Raay, P. L. Palle, T. Roca Cortés.
Mem. Soc. Astron. Ital., Vol. 55, Nos. 1 – 2, p. 263 – 265 (1984). – See Abstr. 012.022.
The recently reported impacts of comets on the sun, could have profound effects on the apparent coherence of solar oscillations. One such impact was reported during observations taken in 1981 and an analysis of the data shows effects which could be correlated with this event.

080.046 **Rotational instabilities in the solar interior, turbulent diffusion and the solar neutrino problem.**
I. W. Roxburgh.
Mem. Soc. Astron. Ital., Vol. 55, Nos. 1 – 2, p. 273 – 283 (1984). – See Abstr. 012.022.
The sun is losing angular momentum through the mass loss in the solar wind and must therefore have spun down from higher velocities during its lifetime. This angular momentum loss will produce differential rotation in the solar interior unless magnetic fields are retained in the interior and are able to transport angular momentum by magnetic stresses. In this paper the author assumes magnetic fields are not effective in the bulk of the interior and shows that when the radial angular velocity reaches a critical value, baroclynic overstability sets in. This instability drives the inward diffusion of $^1$H. The increase of the central $^1$H abundance will lead to a reduction of the detectable neutrino flux.

080.047 **Non–grey radiative transfer in solar oscillations.**
J. Christensen–Dalsgaard, S. Frandsen.
Mem. Soc. Astron. Ital., Vol. 55, Nos. 1 – 2, p. 285 – 291 (1984). – See Abstr. 012.022.
Preliminary results are presented of a calculation of linear, radial oscillations in a realistic solar model, taking consistently into account the interaction between the oscillations and the non–grey radiation field. The eigenfrequencies and damping rates of the modes are affected only slightly by the inclusion of non–grey radiation; however the effects on the detailed behaviour of the eigenfunctions in the solar atmosphere are larger.

080.048 **Beyond mixing–length theory in convection.**
N. Dolez, A. Legait, J. P. Poyet.
Mem. Soc. Astron. Ital., Vol. 55, Nos. 1 – 2, p. 293 – 298 (1984). – See Abstr. 012.022.

This paper is an attempt to lay out the basis of a theory of convection, easy to perturb, so that its application to the solar case will yield statistical information about amplitudes and coherence times of the solar oscillations. Viewed as an inverse problem, it will also put constraints on the solar interior.

080.049 **Solar oscillations: the challenge of nonlinearity.**
W. Däppen, J. Perdang.
Mem. Soc. Astron. Ital., Vol. 55, Nos. 1 – 2, p. 299 – 304 (1984). – See Abstr. 012.022.

The authors report on numerical experiments and on analytical arguments which favour the view that the observed long period of the Sun might be a nonlinear phenomenon.

080.050 **Solar oscillations as an algebraic eigenvalue problem.**
M. Knölker, M. Stix.
Mem. Soc. Astron. Ital., Vol. 55, Nos. 1 – 2, p. 305 – 311 (1984). = Mitt. Kiepenheuer–Inst., No. 233. – See Abstr. 012.022.

Frequencies of adiabatic solar oscillations are obtained as eigenvalues of a real symmetric matrix. The authors use the "Cowling–approximation", where the Eulerian perturbation of the gravitational potential is neglected, and compare computed *p*–mode spectra of low and high degree, *l*, with observational results.

080.051 **Study of giant cell motions and their possible relationship with the dynamics of filaments.**
B. Schmieder, E. Ribes, P. Mein, J.–M. Malherbe.
Mem. Soc. Astron. Ital., Vol. 55, Nos. 1 – 2, p. 319 – 323 (1984). – See Abstr. 012.022.

Observations and models of steady flows in filaments suggest the existence of subphotospheric motions below the magnetic structure supporting cold material, which could be interpreted in terms of giant cell motions. Differential rotation of the Sun and properties of giant convective cells are investigated by statistical analysis of magnetic outliners evolutions from Meudon spectroheliograms.

080.052 **Velocity signal produced by the passage of active regions compared with observations.**
A. Herrero, R. Jimenez, T. Roca Cortés.
Mem. Soc. Astron. Ital., Vol. 55, Nos. 1 – 2, p. 331 – 337 (1984). – See Abstr. 012.022.

The passage of active regions across the visible disk of the sun generates a velocity signal which can be compared with the observations made by Claverie et al. (1982). The authors have computed a calibrated numerical simulation for the seasons where there are observations; the result is that the computed signal has an amplitude well below that of the observations.

080.053 **Influence of active regions on global oscillation measurements.**
B. N. Andersen.
Mem. Soc. Astron. Ital., Vol. 55, Nos. 1 – 2, p. 339 – 344 (1984). – See Abstr. 012.022.

The influence of active regions on line of sight velocity measurements is discussed. The results from a model that includes the effects of sunspots and closely reproduces the data of Claverie et al. (1982) are presented. The effect of sunspots on the integrated velocity signal is closely related to the active region/photosphere line contrast.

080.054 **Straylight influence on measurements of solar velocity oscillations.**
F. Albregtsen, B. N. Andersen.
Mem. Soc. Astron. Ital., Vol. 55, Nos. 1 – 2, p. 345 – 350 (1984). – See Abstr. 012.022.

Doppler measurements of solar velocity oscillations with apertures without axial symmetric averaging will be strongly influenced by the long range straylight (scattering) and its time variation. The corresponding error signal is of the same magnitude as the global velocity oscillation signal. For spectral lines with a large limb effect, the straylight will also introduce appreciable errors in observations performed with axisymmetric apertures as well as in full disk measurements.

080.055 **Oscillatory motions in the Sun corresponding to the solar cycle.**
F. Krause, J. Tuominen, H. Virtanen.
Mem. Soc. Astron. Ital., Vol. 55, Nos. 1 – 2, p. 351 (1984). Abstract. – See Abstr. 012.022.

080.056 **Further experimental evidence of the 13 day solar periodicity.**
G. R. Isaak, H. B. van der Raay, P. L. Palle, T. Roca Cortés.
Mem. Soc. Astron. Ital., Vol. 55, Nos. 1 – 2, p. 353 – 356 (1984). – See Abstr. 012.022.

Extensive data obtained over the period May to August 1982 are analysed to show the continued existence of a 13 day velocity signal of 6 $ms^{-1}$ amplitude. The coherence of this signal is maintained over six years of observation.

080.057 **Summary and conclusions for the EPS study conference "Oscillations as a probe of the Sun's interior", Catania, 20 – 24 June 1983.**
P. Ledoux.
Mem. Soc. Astron. Ital., Vol. 55, Nos. 1 – 2, p. 357 – 368 (1984). – See Abstr. 012.022.

080.058 **New ways with solar neutrinos.**
J. Maddox.
Nature, Vol. 311, No. 5984, p. 298 (1984).

This note comments on recent proposals for new improved solar neutrino detectors based on the use of $^{71}$Ga or $^{81}$Br as detector isotopes.

080.059 **On the stability of solar gravity mode oscillations and the structure of the sun.**
A. G. Kosovichev, A. B. Severnyj.
Pis'ma Astron. Zh., Tom 10, No. 9, p. 679 – 684 (1984). In Russian. English translation in Sov. Astron. Lett., Vol. 10.

The *g*–mode oscillations stability for the interior of solar models with low heavy–element abundances and models with turbulent diffusion mixing inside is investigated by a quasi–adiabatic approximation. The models with low *Z* are found to be most unstable. It is pointed out that only the models with low *Z* are in agreement with the solar neutrino experiment and the observed solar gravity mode oscillations.

080.060 **Rotation of the Sun measured from Mount Wilson white–light images.**
R. Howard, P. A. Gilman, P. I. Gilman.
Astrophys. J., Vol. 283, No. 1, p. 373 – 384 (1984).

The Mount Wilson white–light plate collection from 1921 through 1982 has been measured to give positions and areas of individual sunspots. A pattern–recognition process was used to determine the next–day returns of sunspots so that the rotation rates and latitude drifts of individual sunspots could be calculated. The rotation rate for all sunspots for the whole interval is determined to be $\omega = 14.522(\pm 0.004) - 2.840(\pm 0.043) \sin^2 B$ deg $day^{-1}$. Sunspot groups rotate slower than individual spots by about 1%. Larger spots rotate slower than smaller spots by about 2%.

080.061 **Variations in solar rotation with the sunspot cycle.**
P. A. Gilman, R. Howard.
Astrophys. J., Vol. 283, No. 1, p. 385 – 391 (1984).

The authors measure changes in the rotation rate of sunspots through the solar cycle for the period 1921 – 1982 by calculating and averaging residual rotation rates after subtracting out the grand average differential rotation. They find that a dominant feature is a variation with two peaks per sunspot cycle, one near solar minimum, the other (smaller) one near maximum. These peaks appear at both low and high sunspot latitudes, and in both north and south hemispheres. In latitudes poleward of 15°, there appears to be a third peak in the declining phase of the cycle. The magnitude of these peaks varies substantially from cycle to cycle; several are as large as 0°.4 – 0°.5 per day or 3% – 4% of the average equatorial rotation rate. Physical interpretations of these results are discussed.

080.062 **Solar irradiance changes caused by *g*–modes and large scale convection.**
C. L. Wolff.
Sol. Phys., Vol. 93, No. 1, p. 1 – 13 (1984).
Solar irradiance measurements from the ACRIM experiment show a clear response to the rotation periods of *g*–mode oscillations ($l$ = 1, 2, and 3) and their first harmonics. Peaks in the ACRIM spectrum at 16.6, 18.3, 20.7, 36.5 and ≅71 days all lie within about 1% of periods arising from *g*–mode rotation. This means that the *g*–modes are a fundamental cause of irradiance fluctuations. On time scales of months and less they modulate the irradiance by means of transient flows of global scale which they stimulate in the Sun's convective envelope. Dimensional arguments indicate that the flows carry up heat at an average rate $\sim 10^{-3} L_{\odot}$. Five additional tests for *g*–modes and large–scale convection are given. An instability is described which undermines diffusion models of sunspot energy storage.

080.063 **Plasma motion in umbrae and the surrounding photosphere derived from spectroscopic Doppler measurements and tracer measurements of spots.**
A. Koch.
Sol. Phys., Vol. 93, No. 1, p. 53 – 72 (1984).
The rotational velocity of the Sun is determined by sunspot tracings and by spectroscopic measurements of the photospheric plasma using the non–Zeeman–split line Fe I 5576 and absolute iodine reference. Stationary line shifts as limb–effect and longperiodical shifts introduced by supergranulation are discussed. The dependence on solar activity as $Ca^+$ emissivity and magnetic fields is investigated including line asymmetries.

080.064 **Solar rotation from a number of Fraunhofer lines.**
A. K. Pierce, J. C. LoPresto.
Sol. Phys., Vol. 93, No. 1, p. 155 – 170 (1984).
Solar differential rotation for 1982 – 83 obtained at the double pass spectrograph of the McMath telescope at Kitt Peak, is presented for lines of Fe, Mg, Na, K, Si, and O in the wavelength interval λλ3820 – 10827. The results are analyzed by representations in Legendre polynomials and a power series. The differential rotation is very close to the result obtained by Howard and Harvey in 1966 – 68 at Mt. Wilson but with a slightly greater amplitude. The authors find a mean equatorial rate of rotation of 1.977 km $s^{-1}$. Day to day variations are observed in the equatorial zone; a stable rate of rotation is observed outside of the sunspot zone.

080.065 **Pulsations of microwave emission as a consequence of oscillatory transients in the solar atmosphere.**
V. G. Zandanov, A. M. Uralov.
Sol. Phys., Vol. 93, No. 2, p. 301 – 304 (1984).
The authors offer a new interpretation of the origin of oscillations in active regions. Quasi–periodic oscillation trains with periods of 3 min, 5 min, and ≅7 min arise due to slow oscillatory transients which accompany the appearance of plasma flow along the sunpot's magnetic field. In oscillatory transients of the fast type, occurring in current loop systems, the typical period of oscillations shifts towards values of the order of a second.

080.066 **Determination of the total amount of hydrogen atoms in a quiescent prominence.**
M. Sh. Gigolashvili.
Sol. Phys., Vol. 93, No. 2, p. 317 – 324 (1984).
A brief quantitative review of the determination of the chemical composition of the Sun and stars is given. The method of estimation of the total amount of hydrogen atoms in a prominence is considered. The values of relative abundances of some infrequent elements in the solar atmosphere are refined using metal emission lines in the spectrum of a quiescent prominence. The most probable values of relative abundances for yttrium $\cong 1 \times 10^{-9}$, zirconium $\cong 2 \times 10^{-9}$, and scandium $\cong 3.8 \times 10^{-9}$ are derived.

080.067 **Solar neutrino data and the 11–year solar activity cycle.**
P. Raychaudhuri.
Sol. Phys., Vol. 93, No. 2, p. 397 – 401 (1984).
Recent data on solar neutrino flux have been analysed and it is shown that there is a statistically significant variation of solar neutrino flux data with the solar activity cycle. Thus the observation suggests that the solar activity cycle is due to the pulsating characters of the nuclear energy generation in the interior of the Sun.

080.068 **A solar breeder to explain constant luminosity and lack of neutrinos.**
L. H. Levine, L. M. Libby.
Ann. N.Y. Acad. Sci., Vol. 422, p. 365 (1984). Abstract. – See Abstr. 012.033.

080.069 **A conjecture concerning thermodynamic limits to solar atmospheric heating.**
K. H. Schatten, H. G. Mayr.
Astrophys. J., Vol. 283, No. 2, p. 853 – 858 (1984).
Considering the solar atmosphere in its entirety, the authors hypothesize that large–scale nonadiabatic processes (MHD waves, and so on) effectively tend to yield an "equalization of temperature," such that the atmospheric temperature is limited to the base temperature associated with its heat source. This conjecture suggests that (1) the chromospheric temperature is limited by the granulation base temperature ($10^4$K), (2) the spicule temperatures are limited by the base temperature ($10^5$K) where the supergranular cells form, and (3) the quiet coronal temperature is less than or equal to the convection zone base temperature ($2 \times 10^6$K).

080.070 **Linear models of acoustic waves in sunspot umbrae.**
J. B. Gurman, J. W. Leibacher.
Astrophys. J., Vol. 283, No. 2, p. 859 – 869 (1984).
The authors investigate the two–dimensional, linear hydrodynamics of quiet solar and umbral model atmospheres in a plane–parallel, adiabatic approximation. They interpret the 5.5 – 8.5 mHz oscillations observed in umbral chromospheres and transition regions as acoustic waves propagating parallel, or nearly parallel, to the temperature gradient. The suggestion by Zugzda, Locans, and Staude (1983) that umbral oscillations are resonances in the transmissivity of the sunspot atmosphere is compared with the observations of chromospheric and transition region oscillations above umbrae.

080.071 **Deficiency of solar neutrinos and superheavy monopole?**
G. E. Kocharov, A. K. Pavlov.
Cosmic ray intensity and cosmogenic isotopes. 13th Leningrad seminar on cosmophysics, p. 112 – 115 (1983). In Russian. Abstr. in Ref. Zh., 51. Astron., 9.51.154 (1984). – See Abstr. 012.028.

080.072 **Parametric generation of Alfvén and sound waves in the solar atmosphere. Homogeneous medium.**
N. S. Petrukhin, S. M. Fainshtein (*S. M. Fajnshtejn*).
Sov. Astron., Vol. 28, No. 2, p. 202 – 205 (1984). English translation of 37.080.032.

080.073 **Analysis of the variations of solar neutrino fluxes.**
Yu. R. Rivin, E. A. Gavryuseva, V. G. Gavryusev, L. V. Kosheleva.
Geomagn. variatsii, ehlektr. polya i toki, Moskva, p. 153 – 172 (1983). In Russian. Abstr. in Ref. Zh., 51. Astron., 9.51.428 (1984).

080.074 **MHD waves and turbulence in the sun and interplanetary medium.**
A. Barnes, M. Goldstein, J. Hollweg, J. Mariska, W. Matthaeus, C. Smith, E. Smith, R. Stein, G. Withbroe, R. Woo.
NASA Ref. Publ., NASA RP–1120, p. 4–1 – 4–21 (1984). – See Abstr. 012.038.
Contents: Introduction. Global oscillations of the sun. Observations related to waves or turbulence in the solar atmosphere. Local waves in the solar atmosphere: theoretical considerations. Interplanetary hydromagnetic fluctuations. Recent studies of the interplanetary plasma based on turbulence theory. Effects of waves and turbulence of the solar wind.

080.075 **The possibility of excitation of a rapidly growing overstability in the solar core.**
Yu. V. Vandakurov.
Pis'ma Astron. Zh., Tom 10, No. 11, p. 873 – 880 (1984). In Russian. English translation in Sov. Astron. Lett., Vol. 10.
The stability is considered for the sequence of solar models assuming that the initial model has an inhomogeneous convectively neutral core. Slight differences of subsequent models due to hydrogen depletion are roughly taken into account. It is shown that after a time period as short as 10 years a configuration can be formed in which a nonadiabatic nonradial overstability is present with characteristic period and growth time of the order of a year. This instability seems to promote both nearly adiabatic equilibrium within the core and low central temperature of the sun.

080.076 **Solar rotation.**
R. Howard.
Annu. Rev. Astron. Astrophys., Vol. 22, p. 131 – 155 (1984). – See Abstr. 003.024.
Contents: (1) Introduction. (2) Surface rotation: Early measurements. Modern measurements. The rotation rate. Differential rotation. Torsional oscillations. Variation of rotation rate with height in the atmosphere. (3) Rotation from tracers: Sunspots. Rotation of photospheric magnetic fields. (4) Chromosphere and corona. (5) Rotation of the solar interior.

080.077 **Helioseismology: oscillations as a diagnostic of the solar interior.**
F.-L. Deubner, D. Gough.
Annu. Rev. Astron. Astrophys., Vol. 22, p. 593 – 619 (1984). – See Abstr. 003.024.
Contents: (1) Introduction. (2) Resonant cavities in the Sun. (3) Observational methods. (4) High–degree modes: Solar structure. Subphotospheric velocities. (5) Low–degree modes: High–order *p* modes. Gravity modes. Solar structure. (6) Five–minute modes of intermediate degree. (7) Limb observations. (8) Problems for the immediate future. Appendix: Classification of stellar oscillations.

080.078 **Non–equilibrium formation of neutral oxygen lines in the solar atmosphere.**
N. G. Shchukina.
Akad. Nauk Ukrain. SSR, Inst. Teor. Fiz., Kiev, Prepr., No. 84–75R, 31 pp. (1984). In Russian.
Non–LTE effects in the occupation numbers of the levels, source functions, line profiles, central residual intensities, equivalent widths for the solar disk oxygen O I spectrum are evaluated.

080.079 **The longitudinal distribution of solar active regions and solar global oscillation measurements.**
E. H. Schröter.
Astron. Astrophys., Vol. 139, No. 2, p. 538 – 544 (1984). = Mitt. Kiepenheuer–Inst., Nr. 241.
The recently discovered 13.1–day periodicity in solar global oscillation measurements as reported by Claverie et al. (1982) is shown to be produced by the rotation of active regions occurring predominantly in long lived active longitudes 180° apart. They cause an artificial oscillatory Doppler signal for long periods whenever the solar activity exceeds a critical minimum value ($\bar{R}_M \geqslant 40$). It is further shown that the properly calibrated active region–weighted Doppler signal for the period July 10 to July 24, 1981 agrees with the measurements of Claverie et al. within the error limits. It is thus concluded that Claverie's observations contain no direct evidence for a fast rotating solar core as was speculated by the authors and by Dicke (1983).

080.080 **Die Atmosphäre der Sonne.**
H. Habermayr.
Orion, 42. Jahrg., Nr. 205, p. 211 – 213 (1984).

080.081 **Effects of the nonlocal character of the mean intensity, deviations from radiative equilibrium, and a nongray atmosphere on oscillations in stellar envelopes.**
H. A. Hill, J. D. Logan.
Astrophys. J., Vol. 285, No. 1, p. 386 – 399 (1984).
A semiempirical model for the solar atmosphere was used to study the effects of three factors on the properties of the oscillations in the photosphere and the low chromosphere. These factors are the nonlocal character of the radiation field, deviations from radiative equilibrium, and the nongray opacity. For radial long–period oscillations within the photosphere, the fractional change in temperature is $\sim 10^3$ times as large as the fractional change in radius when these effects are included. An immediate consequence of this large temperature perturbation is the resolution of a long–standing problem in the study of solar oscillations: the inconsistency between the interpretation of the SCLERA diameter measurements and the Doppler shift measurements of other observers.

080.082 **Solar neutrinos and other problems and their relation to energy production in the Sun.**
C. L. Joseph.
Nature, Vol. 312, No. 5991, p. 254 – 255 (1984).
Models of the solar interior, based on the usual physical assumptions, predict a neutrino flux several times greater than that observed in the Davis $^{37}Cl$ experiment. The author points out some less well known discrepancies between theoretical predictions of the standard solar model and relevant observations. It is suggested that an additional source of energy production that competes with the proton–proton chain and carbon–nitrogen cycles, can reduce these discrepancies if it has a sufficiently strong temperature dependence.

080.083 **The contribution of cosmogenic radioisotopes to solar neutrino flux measurements.**
K. O'Brien, A. de la Zerda.
Trans. Am. Nucl. Soc., Vol. 46, p. 641 – 643 (1984). Abstr. in Phys. Abstr., Vol. 88, No. 1247, Entry 4573 (1985).

080.084 **Parametric generation in Alfvén and sound waves in the solar atmosphere. II. Isothermal atmosphere.**
N. S. Petrukhin, S. M. Fainshtein (*S. M. Fajnshtejn*).
Sov. Astron., Vol. 28, No. 3, p. 313 – 315 (1984). English translation of 37.080.052.

080.085 **Solar models with a comet–like core and low neutrino flux.**
A. J. R. Prentice.
Bull. Am. Astron. Soc., Vol. 16, No. 3, Part I, p. 728 (1984). Abstract. – See Abstr. 010.065.

080.086 **Is the rotation of the sun normal?**
D. F. Gray.
J. R. Astron. Soc. Can., Vol. 78, No. 5, p. 203 (1984). Abstract. – See Abstr. 010.401.

080.087 **A numerical study on turbulent heat transfer in convective envelopes.**
K. Arlt, G. Rüdiger.
Astron. Nachr., Vol. 305, No. 5, p. 229 – 235 (1984). – See Abstr. 012.062.
Using a two–dimensional temperature equation the authors study the well–known solar heat flux problem. A spherically symmetric heat flux may enter the convection zone at its bottom and leave in accord to the Stefan–Boltzmann law. Meridional

circulation is involved, radiation is still ignored. In order to establish the eddy conductivity tensor for the simplest rotating turbulences a second-order correlation approximation is used. The resulting pole-equator differences vary for various profiles of eddy conductivity and turnover time.

080.088 **Solar rotation rate from stable recurrent sunspot tracings.**
G. Lustig, R. Dvorak.
Astron. Astrophys., Vol. 141, No. 1, p. 105 – 107 (1984).
Analyzing the 1948 – 1976 Kanzelhöhe solar-observatory data for the recurrent *H* and *J* sunspots, the authors have derived the following results: Recurrent sunspots show a deceleration of their rotation velocities during their lifetimes. Different methods of analysis led to quite the same results for the rotation of *H* and *J* sunspots. The authors found good agreement between the rotation rates determined from the central meridian passage data of Greenwich and Kanzelhöhe.

080.089 **The 160-minute solar pulsations are not excited by gravitational waves.**
A. G. Kosovichev.
Sov. Astron. Lett., Vol. 10, No. 3, p. 190 – 193 (1984). English translation of 37.080.068.

080.090 **Solar oscillations.**
J. Christensen-Dalsgaard.
Theoretical problems in stellar stability and oscillations, p. 155 – 208 (1984). – See Abstr. 012.065.
Among all stars observed to pulsate, the Sun has by far the largest number and variety of modes of oscillation. To match the observational accuracy, very precise calculations of oscillation frequencies are needed. Asymptotic methods have proved useful in the analysis and interpretation of the frequencies. The results provide tight constraints on solar models. There are difficulties in reconciling the amplitudes obtained in Doppler velocity with those observed in the apparent position of the solar limb. The excitation of the oscillations is so far not well understood, although it is probable that the interaction between pulsation and convection plays an important role.

080.091 **Frequencies of solar *p*-mode oscillations.**
J. W. Harvey, T. L. Duvall Jr.
Theoretical problems in stellar stability and oscillations, p. 209 – 214 (1984). – See Abstr. 012.065.

080.092 **Solar oscillations: excitation decay and rotational splitting.**
H. B. van der Raay.
Theoretical problems in stellar stability and oscillations, p. 215 – 219 (1984). – See Abstr. 012.065.
The study of solar five minute oscillations from two well separated sites provides conclusive evidence as to their authenticity. Correlation of the two data sets demonstrates the fine structure of individual lines and leads to an estimate of the coherence time of particular modes.

080.093 **Intensity oscillations in solar spectral lines.**
B. N. Andersen.
Theoretical problems in stellar stability and oscillations, p. 220 – 223 (1984). – See Abstr. 012.065.
The variation of the amplitude of the 5 minute oscillatory signal as function of wavelength in several medium strong solar spectral lines has been studied observationally. These observations are compared with the calculations carried out by Frandsen (1984). The theoretical calculations reproduce the observations in considerable detail.

080.094 **Solar oscillations and the equation of state.**
M. Kaisig, M. Knölker, M. Stix.
Theoretical problems in stellar stability and oscillations, p. 239 – 243 (1984). – See Abstr. 012.065.
In the framework of the Debye-Hückel theory the authors discuss the internal partition functions of H I, He I and He II. They include the dependence of the ground state energy on the Debye radius and the total number of terms found numerically by Rogers et al. (1970). The results are briefly discussed in the context of the Sun's eigenfrequencies.

080.095 **Influence of the *Z*-value on the solar five-minute oscillation.**
R. Scuflaire, M. Gabriel, A. Noels.
Theoretical problems in stellar stability and oscillations, p. 244 – 249 (1984). – See Abstr. 012.065.
Two solar models with different heavy element abundances ($Z = 0.018$ and $Z = 0.02$) have been computed. Their theoretical oscillation frequencies for low $l$, with periods around five minutes have been computed and compared to the observed ones. The discrepancy between the observed and computed spectra cannot be resolved by the choice of an appropriate value of $Z$.

080.096 **The influence of convection theory on the frequencies of solar oscillations.**
R. K. Ulrich, E. J. Rhodes Jr.
Theoretical problems in stellar stability and oscillations, p. 250 – 258 (1984). – See Abstr. 012.065.
The authors have studied the influence of convective theory uncertainties on the frequencies of solar oscillations. They find that the turbulent pressure plays an important role in setting the high $l$ frequencies and a lesser role in setting the low and intermediate $l$ frequencies. The representation of the turbulent pressure thus is important to the interpretation of the comparison between theory and observation.

080.097 **Effects of core mixing on solar oscillation frequencies.**
A. N. Cox, R. B. Kidman.
Theoretical problems in stellar stability and oscillations, p. 259 – 263 (1984). – See Abstr. 012.065.
This paper discusses predicted *p*- and *g*-mode periods for two solar models provided by Schatzman and Maeder (1981). One model has no mixing while the other one has turbulent diffusion mixing (allowed for by a Reynolds number of 100). The resulting *g*-mode period spacings allow a very close look at the central solar structure which, like the *p*-mode period splittings, indicate little or no mixing over the lifetime of the sun.

080.098 **On the determination of the helium abundance of the solar convection zone.**
W. Däppen, D. O. Gough.
Theoretical problems in stellar stability and oscillations, p. 264 – 269 (1984). – See Abstr. 012.065.
With an appropriate analysis of data from solar oscillations, the sound speed $c$ in the solar interior becomes an observable quantity [1,2]. Since $c^2 = \gamma p/\varrho$ the ionization zones are manifest in the sound speed, because $\gamma$ undergoes substantial variations. The variation in $\gamma$ due to helium ionization is evidently dependent on the solar helium abundance. Therefore by measuring that variation it ought to be possible to infer the helium abundance in the Sun.

080.099 **On the inversion of solar oscillation data.**
A. I. Denis, C. Denis.
Theoretical problems in stellar stability and oscillations, p. 270 – 277 (1984). – See Abstr. 012.065.
This paper reviews the mathematical inversion theory applied to solar oscillation data in the attempt to derive information on the internal structure of the Sun.

080.100 **On the stability of solar gravity mode oscillations and the structure of the Sun.**
A. G. Kosovichev, A. B. Severny (*A. B. Severnyj*).
Theoretical problems in stellar stability and oscillations, p. 278 – 283 (1984). – See Abstr. 012.065.
The *g*-mode oscillations stability for the interior of solar models with low heavy elements abundance and the models with turbulent diffusion mixing are investigated by quasi-adiabatic approximation. The models with low $Z$ are found most unstable. It is pointed out that only the models with low $Z$ are in agreement with the solar neutrino experiment and the observed solar gravity mode oscillations.

080.101 **Solar g–modes: comparison between observations and theoretical periods for 3 solar models.**
M. Gabriel.
Theoretical problems in stellar stability and oscillations, p. 284 – 289 (1984). – See Abstr. 012.065.

In addition to work published earlier the author reports results obtained with 2 other solar models. One is a standard model with $Z = 0.02$. It has $\Delta P = 35.53$ min. The second one is a model with $Z = 0.018$ taking diffusion of H and He into account. This model has a $\Delta P = 35.14$ min.

080.102 **The base of the solar convection zone.**
A. N. Ellis.
Theoretical problems in stellar stability and oscillations, p. 290 – 293 (1984). – See Abstr. 012.065.

A method is given to deduce the nature of the base of the solar convection zone from observations of gravity modes. The approximate order of the modes necessary to resolve the convective interface is derived from numerical results.

080.103 **Helioseismological determination of solar internal rotation.**
J. Leibacher.
Theoretical problems in stellar stability and oscillations, p. 298 – 302 (1984). – See Abstr. 012.065.

Inversion of the observed splitting of 780 $p$ modes, with angular degree from 1 to 131, yields the solar internal rotation. The bulk of the Sun rotates significantly less rapidly than the surface. Although the deep core appears to rotate more rapidly, the resulting gravitational quadrupole moment $J_2$ does not contribute significantly to the orbital precession of Mercury.

080.104 **Intensity oscillations in spectral lines in the Sun. Theory and observation.**
S. Frandsen.
Theoretical problems in stellar stability and oscillations, p. 303 – 306 (1984). – See Abstr. 012.065.

This paper reports on a study of the variations of solar spectral line profiles caused by solar oscillations. As an example, calculations for the Fe I λ5576 line are presented. Some details of the procedure for calculating eigenfunctions in the solar atmosphere are also given.

080.105 **On mass and energy transfer by waves in an atmosphere.**
L. M. B. C. Campos.
Theoretical problems in stellar stability and oscillations, p. 393 – 398 (1984). – See Abstr. 012.065.

This paper considers the four possible wave modes in an ionized atmosphere and shows that each can play a specific role, and give rise to a physically distinct region. In the solar case: acoustic–gravity waves dominate the global mass balance and appear as spicules; Alfvén–gravity waves heat the chromosphere; coupled magnetosonic–gravity waves produce the transition region and heat the closed regions of the corona; fourth–order mixed hydromagnetic–gravity waves accelerate matter in coronal holes and form the high–speed streams in the solar wind.

080.106 **Feasibility of a $^{81}$Br(ν,$e^-$)$^{81}$Kr solar neutrino experiment.**
G. S. Hurst, C. H. Chen, S. D. Kramer, B. T. Cleveland, R. Davis Jr., R. K. Rowley, F. Gabbard, F. J. Schima.
Phys. Rev. Lett., Vol. 53, No. 11, p. 1116 – 1119 (1984).

A solar neutrino experiment utilizing the interaction of $^{81}$Br(ν,$e^-$)$^{81}$Kr to study the $^7$Be neutrino source in the interior of the sun is shown to be feasible.

080.107 **On the rotation of the Sun.**
D. O. Gough.
Philos. Trans. R. Soc. London, Ser. A, Vol. 313, No. 1524, p. 27 – 38 (1984). – See Abstr. 012.068.

An asymptotic method is developed to estimate the rotational splitting of sectoral five–minute solar oscillations. Integral formulae are obtained which can be inverted to yield the variation with depth of the Sun's angular velocity near the equatorial plane. The result is a functional of smoothed data, and does not rely on a detailed theoretical model of the Sun. The method has been tested with artifical data. The results are encouraging, for they reproduce at least the broadest feature of the somewhat arbitrary angular velocity with which the theoretical model was endowed.

080.108 **Global pulsations of the sun in the soft X–ray range (1 – 8 Å) during the period of the project "Solar Maximum Year".**
V. V. Kasinskij.
Issled. Geomagn. Aehron. Fiz. Solntsa, Moskva, No. 68, p. 90 – 99 (1984). In Russian. Abstr. in Ref. Zh., 51. Astron., 11.51.382 (1984).

080.109 **Surface solitary waves and solitons.**
J. V. Hollweg, B. Roberts.
J. Geophys. Res., Vol. 89, No. A11, p. 9703 – 9710 (1984).

The solar atmosphere and solar wind are magnetically structured. The structuring can include tangential discontinuities, which can support surface waves. Such waves can be dispersive. This means that dispersion and nonlinearity can balance in such a way that solitary waves (or solitons) can result. This general point is illustrated by a two–dimensional nonlinear analysis which explicitly demonstrates the presence of long–wavelength solitary waves propagating on tangential discontinuities. If the waves are only weakly nonlinear, then they obey the Korteweg–de Vries equation and are true solitons.

080.110 **Mixing in the solar interior.**
H. C. Spruit.
MPA Rep., No. 161, 17 pp. (1984). To appear in 'Hydromagnetics of the Sun' (Proceedings of the Fourth European Meeting on Solar Physics). ESA SP–220. 1984.

080.111 **Convective instability in a solar flux tube. I. Nonlinear calculations for an adiabatic inviscid fluid.**
S. S. Hasan.
Astrophys. J., Vol. 285, No. 2, p. 851 – 857 (1984).

The object of this investigation is to study the development of the convective instability in a solar flux tube with a view to seeing whether intense fields can be formed by this process. As a first step in this direction, the ideal MHD equations using an adiabatic energy equation are solved in order to delineate the main physics of convective collapse. It is seen that an initial equipartition field state in thermal equilibrium with the external medium can collapse to produce the observed kilogauss field. The final state is found to exhibit stationary oscillations. The analysis also suggests that no appreciable systematic downflow in flux tubes is likely to be observed.

080.112 **On a method of reduction of observations of 160–min brightness oscillations of the limb of the solar disc.**
Eh. V. Kononovich.
Astron. Tsirk., No. 1315, p. 7 – 8 (1984). In Russian.

080.113 **Solar rotation and meridional motions.**
J. Tuominen, I. Tuominen, J. Kyröläinen.
Rep. Ser., Dep. Phys. Sci., Univ. Turku, No. FTL–R68, p. 31 (1984). Abstract. – See Abstr. 012.079.

080.114 **Solar 160 min oscillation at microwaves.**
S. Urpo.
Rep. Ser., Dep. Phys. Sci., Univ. Turku, No. FTL–R68, p. 33 – 36 (1984). – See Abstr. 012.079.

080.115 **Conformation of the structure of the solar atmosphere in the solar wind.**
G. P. Lyubimov, N. V. Pereslegina.
Astron. Tsirk., No. 1323, p. 2 – 4 (1984). In Russian.

080.116 **Meridional circulation in rotating stars. VIII. The solar spin–down problem.**
M. Tassoul, J.–L. Tassoul.
Astrophys. J., Vol. 286, No. 1, p. 350 – 358 (1984).

A detailed analysis is given of the thermally driven meridional circulation and concomitant differential rotation in the hydrogen–burning core of the Sun, while taking into account the continuous removal of angular momentum from the surface convective layers as the Sun leisurely evolves on the main sequence. In particular, by assuming that the surface rotation rate varies as the inverse square root of the age, the authors calculate the inner redistribution of angular momentum caused by both the large–scale circulatory currents and the small–scale baroclinic eddies, which are an ever–present feature of a stellar radiative zone. It is shown that anisotropic turbulence is the main factor that inexorably changes the Sun's inner rotation rate. The transport of angular momentum by meridional streaming is many orders of magnitude smaller than that caused by turbulent viscous friction.

080.117 **Bestimmung von Veränderungen des Sonnendurchmessers.**
H.–J. Bode.
Mitt. Astron. Ges., Nr. 62, p. 154 (1984). Abstract. – See Abstr. 010.341.

080.118 **On the nature of the 11–year torsional ocillations of the sun.**
N. I. Kliorin, A. A. Ruzmajkin.
Pis'ma Astron. Zh., Tom 10, No. 12, p. 925 – 930 (1984). In Russian. English translation in Sov. Astron. Lett., Vol. 10.

It is shown that the 22–year dynamo wave of the large–scale magnetic field produces torsional oscillations of the sun with an 11–year period. Superposition of the modes of dipole and quadrupole symmetry results in a secular asymmetry of the differential rotation relative to the solar equator.

080.119 **On the identification of normal modes of oscillation from observations of the solar periphery.**
D. O. Gough, J. Latour.
Astron. Express, Vol. 1, Issue 1, p. 9 – 24 (1984).

The decomposition of solar oscillations into their constituent normal modes requires a knowledge of both the spatial and temporal variation of the perturbation to the Sun's surface. The task can be especially difficult when only limited spatial information is available. The authors consider a procedure by which the contributions from quite small groups of modes to spatially well resolved data obtained at any instant can be extracted from the remaining modes. Combining these results with frequency information then permits the modes to be identified.

080.120 **On the megawave nature of the 160 min oscillations of the sun.**
A. M. Chechel'nitskij.
Komet. Tsirk., No. 328 (1984). In Russian.

080.121 **The atmosphere of the quiet sun.**
R. A. Gulyaev.
Itogi Nauki i Tekhniki. Seriya Astronomiya. Tom 25. Physics of the sun, p. 3 – 98 (1984). In Russian. – See Abstr. 003.053.

080.122 **Techniques for detecting giant cells using spatially resolved solar velocity data.**
T. M. Brown, P. A. Gilman.
Astrophys. J., Vol. 286, No. 2, p. 804 – 809 (1984).

The authors present techniques for improving the detectability of giant cells in the solar convection zone. These methods are based on the spatial extent and symmetry properties of giant cells as seen in self–consistent dynamical models of the solar convection zone. Simulations suggest that these techniques allow detection of giant cells with photospheric rms velocities of $2\ \mathrm{m\,s^{-1}}$ ($\sim 0.4\ \mathrm{m\,s^{-1}}$ per longitudinal wavenumber), given observations spanning about one year.

080.123 **Apparent solar temperature enhancement due to large–amplitude waves.**
W. Kalkofen, P. Ulmschneider, F. Schmitz.
Astrophys. J., Vol. 287, No. 2, p. 952 – 958 (1984).

The authors suggest that the discrepancies in the temperature structure between recent empirical solar model atmospheres that are based on observations of the infrared continuum or the ultraviolet resonance lines of Ca II and Mg II are caused, at least in part, by acoustic–type slow–mode magnetohydrodynamic waves.

080.124 **Temporal solar variations.**
D. O. Gough.
ESA J., Vol. 7, No. 4, p. 325 – 339 (1983). – See Abstr. 012.035.

Temporal variations of the Sun on time scales ranging from about 5 min to $10^{10}$yr are discussed. Attention is restricted to only the inner 99.999999996% by mass of the Sun. Variations on the longest time scales relate to the overall evolution of the Sun, and are predicted from the theory of stellar structure. Any evidence that can be gathered to confirm such variations is therefore pertinent not only to the Sun, but to stars in general.

**Secrets of the sun.**
See Abstr. 003.013.

**Neutrinos and the Sun.**
See Abstr. 011.018.

**Die Geminga–Story.**
See Abstr. 013.039.

**Probing the depths of a star: the study of solar oscillations from space. Report of the NASA Science Working Group on the study of solar oscillations from space.**
See Abstr. 013.081.

**Separation of large–scale photospheric Doppler patterns.**
See Abstr. 036.067.

**Should one measure radial velocity or brightness fluctuations?**
See Abstr. 036.139.

**On the detection of solar oscillations through radial velocity fluctuations.**
See Abstr. 036.140.

**Correction of atmospheric transparency gradient effect on solar radial velocities.**
See Abstr. 036.141.

**Experiences with high precision ground–based photometry of the sun.**
See Abstr. 036.142.

**A numerical solution of the data–gaps problem.**
See Abstr. 036.143.

**Observations of the sun at the CERGA astrolabe in 1982.**
See Abstr. 041.004.

**Starprobe: to confront the Sun.**
See Abstr. 051.011.

**The $L_1$–Lagrangian point orbit as a suitable site for helioseismologic measurements.**
See Abstr. 051.024.

**The impact of nuclear input data on expected signals for the Cl and Ga solar neutrino experiments.**
See Abstr. 061.041.

**Monopole search and neutrino astrophysics with liquid scintillation detectors.**
See Abstr. 061.073.

**Neutron capture cross sections and solar abundances of $^{160,161}$Dy, $^{170,171}$Yb, $^{175,176}$Lu, and $^{176,177}$Hf for the *s*–process analysis of the radionuclide.**
See Abstr. 061.102.

**Propagation of waves in an atmosphere in the presence of a magnetic field. VII. Magneto–acoustic–gravity modes in an oblique *B*.**
See Abstr. 062.015.

**Propagation of axisymmetric disturbances on a twisted magnetic flux tube.**
See Abstr. 062.060.

**Linear astrophysical dynamos in rotating spheres: mode transition between steady and oscillatory dynamos as a function of dynamo strength and anisotropic turbulent magnetic diffusivity.**
See Abstr. 062.085.

**Linear astrophysical dynamos in rotating spheres: solar and stellar cycle north–south hemispheric parity selection mechanism and turbulent magnetic diffusivity.**
See Abstr. 062.097.

**Thermal effect on the pulses of the rotating Sun.**
See Abstr. 062.142.

**Current layers and reconnection.**
See Abstr. 062.154.

**Influence of rotation and magnetic fields on stellar oscillation eigenfrequencies.**
See Abstr. 065.030.

**Lithium depletion in stars. Pre–main sequence burning and extra–mixing.**
See Abstr. 065.031.

**Les oscillations solaires et stellaires comme test de la théorie de la structure interne.**
See Abstr. 065.048.

**Tests observationnels et phénoménologie de la diffusion turbulente dans les étoiles.**
See Abstr. 065.052.

**Numerical simulations of stellar convective dynamos. I. The model and method.**
See Abstr. 065.071.

**Qualitative aspects of the convection–pulsation interaction.**
See Abstr. 065.114.

**What will asteroseismology teach us?**
See Abstr. 065.129.

**Magnetic fields in the radiative interior of stars. II. Forced convection and the $^{7}$Li abundance.**
See Abstr. 065.158.

**Gravitational light deflection in the Solar System.**
See Abstr. 066.015.

**Relativistic deflection of radio signals in the solar gravitational field measured with VLBI.**
See Abstr. 066.016.

**Using solar oscillations to search for gravitational radiation.**
See Abstr. 066.243.

**A rigorous method of calculating the deflection of light in the gravitational field of the Sun.**
See Abstr. 066.250.

**Measurement of the solar spectral irradiance from 200 to 3000 nanometers.**
See Abstr. 071.003.

**Asymmetries and wavelengths of solar spectral lines and the solar rotation determined from Fourier–transform spectra.**
See Abstr. 071.015.

**Centre–to–limb behaviour of molecular rotational temperatures. III.**
See Abstr. 071.019.

**Limits on photospheric Doppler signatures for solar giant cells.**
See Abstr. 071.021.

**Solar luminosity fluctuations and active region photometry.**
See Abstr. 072.028.

**Dynamical phenomena in sunspots. I. Observing procedures and oscillatory phenomena.**
See Abstr. 072.056.

**Evolution of a spot group and origin of large–scale current structures in the solar atmosphere.**
See Abstr. 072.096.

**Oscillations in the chromospheric network as response to resonance oscillations of the sun.**
See Abstr. 073.078.

**Appearance of global solar 160 min pulsations in flares of individual active regions.**
See Abstr. 073.094.

**Rotational characteristics of the white–light solar corona: 1965 – 1983.**
See Abstr. 074.072.

**On the mechanism of topological pumping.**
See Abstr. 075.012.

**Temporal variations in the solar submillimeter continuum.**
See Abstr. 077.015.

**Recovery of spectral optical depth and zero–airmass solar spectral irradiance under conditions of temporally varying optical depth.**
See Abstr. 082.030.

**Survival of the Earth and the future evolution of the Sun.**
See Abstr. 107.020.

**The interaction of rotation and magnetic field in the solar system.**
See Abstr. 107.042.

**The early solar system and the rotation of the Sun.**
See Abstr. 107.043.

**Analogy of rapid oscillations of Ap stars to solar oscillations.**
See Abstr. 115.008.

**On the change of the period of cosmic ray variation connected with the rotation of the sun.**
See Abstr. 144.018.

**Limits on a stochastic gravitational wave background from observations of terrestrial and solar oscillations.**
See Abstr. 161.360.

# Earth

## 081 Structure, Figure, Gravity, Orbit, etc.

081.001 **Geopotential harmonics of orders 15 and 30.**
D. G. King-Hele, D. M. C. Walker.
Nature, Vol. 310, No. 5974, p. 218 – 219 (1984).

The Earth's gravitational potential is usually expressed as a double infinite series of spherical harmonics. The most precise technique for determining coefficients of a particular order is by analysis of satellite orbits which experience resonance with the Earth's gravitational field. The authors have re-evaluated individual harmonic coefficients of orders 15 and 30 from analyses of 24 orbits having a wide range of inclinations to the Equator. These coefficients have standard deviations equivalent to an accuracy in geoid height of 1 cm, for degree up to 23.

081.002 **Selective extinctions and terminal Cretaceous events.**
E. Buffetaut.
Nature, Vol. 310, No. 5975, p. 276 (1984).

This note discusses recent research on the hypothesis that mass extinctions of living species may have been caused by impacts of extraterrestrial bodies on the Earth.

081.003 **Physical evidence of impact.**
R. A. F. Grieve.
Nature, Vol. 310, No. 5976, p. 370 (1984).

This note discusses recent developments in the hypothesis that abnormal abundances of iridium and other siderophiles at the Cretaceous-Tertiary boundary represent the geochemical signature of a large impact event.

081.004 **Siderophile-rich magnetic spheroids from the Cretaceous-Tertiary boundary in Umbria, Italy.**
J. Smit, F. T. Kyte.
Nature, Vol. 310, No. 5976, p. 403 – 405 (1984).

The magnetic fraction of some Cretaceous-Tertiary (K-T) boundary sediments sampled at two locations in Italy, contains a spheroidal component which has retained some relict mineral textures and, possibly, mineral chemistry from the K-T event. These spheroids contain higher concentrations of Ir and probably other siderophiles than the bulk sediment and may have been an important carrier for these elements at 65 Myr. This discovery supports evidence for a major accretionary event at the end of the Cretaceous.

081.005 **Dynamo action in a stably stratified core.**
H. A. Singer, P. L. Olson.
Geophys. J. R. Astron. Soc., Vol. 78, No. 2, p. 371 – 387 (1984).

It is demonstrated that a field of propagating inertial gravity waves can act as a dynamo in a rotating, stably stratified, electrically conducting fluid. Individual inertial gravity waves possess helicity due to the action of the Coriolis acceleration and therefore an $\alpha$-effect can result. Thus, a field of inertial gravity waves can generate magnetism through dynamo action of the $\alpha^2$-type.

081.006 **Periodicity of extinctions in the geologic past.**
D. M. Raup, J. J. Sepkoski Jr.
Proc. Natl. Acad. Sci. U.S.A., Vol. 81, No. 3, p. 801 – 805 (1984). Abstr. in Phys. Abstr., Vol. 87, No. 1237, Entry 72596 (1984).

081.007 **The second derivatives of the normal field of the earth's gravity potential.**
Z. N. Levitskaya.
Tr. Gos. Astron. Inst. Shternberg, Tom 55, p. 159 – 174 (1983). In Russian.

Analytical expressions for the elements of the normal tensor of gravity potential's second derivatives at the earth's ellipsoid surface in a coordinate system connected with the normal to them are received. Numerical estimates of the tensor elements with error $1 \times 10^{-4} E$ are given.

081.008 **The improved Gershun method of determination of the gravitational constant.**
Z. S. Kopejkina.
Tr. Gos. Astron. Inst. Shternberg, Tom 55, p. 185 – 192 (1983). In Russian.

An estimate of the errors of the determination of the gravitational constant G caused by the difference between the earth's level and a spherical surface is given. Formulae for a more general case than by Gershun are derived.

081.009 **Historical remarks on marine gravimetry at Moscow University.**
P. A. Stroev, N. P. Grushinskij.
Tr. Gos. Astron. Inst. Shternberg, Tom 55, p. 193 – 208 (1983). In Russian.

081.010 **Mineralogy and composition of the upper mantle.**
D. L. Anderson, J. D. Bass.
Geophys. Res. Lett., Vol. 11, No. 7, p. 637 – 640 (1984).

081.011 **Modelling the global climate response to orbital forcing and atmospheric carbon dioxide changes.**
N. G. Pisias, N. J. Shackleton.
Nature, Vol. 310, No. 5980, p. 757 – 759 (1984). With a correction in Nature, Vol. 312, No. 5996, p. 786 (1984).

At least part of the record of climate change during the past million years may be attributed to forcing by changes in the geometry of the Earth's orbital system. The authors investigate here the operation of a model which is forced simultaneously by changes in the distribution of solar insolation caused by orbital variations and by changes in the atmospheric carbon dioxide ($CO_2$) concentration which are themselves forced by orbital parameters.

081.012 **Space laser applications in geophysics.**
H.-G. Kahle, F. Barlier.
ESA Spec. Publ., ESA SP-202, p. P15 – P20 (1984). – See Abstr. 012.004.

In this paper the actual geophysical problems related to global geodynamics are defined and modern geodetic techniques which are capable of contributing to their solutions are described.

081.013 **The oceans and the physics of the Earth.**
A. H. Cook.
Tectonophysics, Vol. 105, No. 1 – 4, p. 71 – 77 (1984). Abstr. in Phys. Abstr., Vol. 87, No. 1239, Entry 81831 (1984).

081.014 **The large-numbers hypothesis and the Earth's expansion. II.**
S. Yabushita.
Earth, Moon, Planets, Vol. 31, No. 1, p. 43 – 47 (1984).
Careful study by Owen of the displacements of the continents renders it likely that the Earth radius increased by almost 20% during the past 200 million years. Ramsey's proposal that the Earth liquid core is a high-pressure phase of the mantle material is adopted to calculate the Earth radius under variable $G$. It is shown that if $G$ varies according to the large-numbers hypothesis of Dirac, the primordial Earth would have had a radius 700 km less than the modern value. If it is assumed that the Earth began to respond and, hence, expand only $2 \times 10^8$yr B.P., an expansion as large as 700 km is shown to be energetically plausible.

081.015 **A Landsat-assisted study of the aquatic areas of the Lake Kemijärvi region, Northern Finland.**
J. Raitala, H. Jantunen, S. Hellsten.
Earth, Moon, Planets, Vol. 31, No. 2, p. 183 – 216 (1984).
Landsat MSS imagery is employed to assess the aquatic complexes of the regulated Lake Kemijärvi and the adjoining small natural lakes in Northern Finland.

081.016 **The precise computation of geoid undulation differences with comparison to results obtained from the global positioning system.**
T. Engelis, R. H. Rapp, C. C. Tscherning.
Geophys. Res. Lett., Vol. 11, No. 9, p. 821 – 824 (1984).

081.017 **Mantle upflow under North America and plate dynamics.**
D. I. Gough.
Nature, Vol. 311, No. 5985, p. 428 – 433 (1984).
In the development of plate dynamics there are two fundamental questions. First, do the plates slide over a passive asthenosphere, or are they driven by a convecting upper mantle? Second, are all ocean ridges passive, pull-apart plate margins, or do some lie above upcurrents in the mantle? This paper presents evidence favouring an active mantle interpretation, both for the North American plate and for the mantle upflow which lay beneath the East Pacific Rise in Tertiary times and which now lies beneath the western United States.

081.018 **Application of hydrostatic models of planets to the analysis of the chemical composition of the earth's interior.**
A. N. Karapats.
Dag. univ. Makhachkala, 15 pp. (1984). In Russian. Abstr. in Ref. Zh., 51. Astron., 8.51.144 (1984).

081.019 **Iron-water reaction under high pressure and evolution of the earth.**
Y. Fukai.
Solid State Phys., Vol. 19, No. 5, p. 310 – 312 (1984). In Japanese. Abstr. in Phys. Abstr., Vol. 87, No. 1242, Entry 94056 (1984).

081.020 **Die Gauß-Krügersche Abbildung des Erdellipsoides mit direkter Berechnung der elliptischen Integrale durch Landentransformation.**
M. Gerstl.
Dtsch. Geod. Komm. Bayer. Akad. Wiss., Reihe C, Heft Nr. 296, 57 pp. (1984).

081.021 **Spectral properties of the Newtonian potential field and their application in the interpretation of the gravity anomalies.**
M. Zadro.
Geophys. J. R. Astron. Soc., Vol. 79, No. 2, p. 489 – 493 (1984).
Some spectral properties of the Newtonian potential field and of fields related to it by derivation processes are examined. Corresponding spectral techniques are proposed to treat several geophysical problems connected with the study of local gravity perturbation fields.

081.022 **Construction of a system of point masses representing the gravitational field of a planet from satellite observations. I. Derivation of the algorithm.**
S. M. Poleshchikov, K. V. Kholshevnikov.
Vestn. Leningr. Univ., Mat. Mekh. Astron., No. 7, p. 76 – 86 (1984). In Russian. Abstr. in Ref. Zh., 51. Astron., 9.51.114 (1984).

081.023 **On the causes of ice-ages.**
F. Hoyle.
Earth, Moon, Planets, Vol. 31, No. 3, p. 229 – 248 (1984).
Light entering homogeneous ice at normal incidence propagates without scattering until it is absorbed by the ice. Air bubbles serve as scattering centres, and if there are enough of them multiple scatterings turn the light around by a random walk before absorption takes place. Much of the light then reemerges from the ice which shines white in the Sun. A sufficient density of submicron particles of high refractive index in surface waters could produce a similar effect in the sea. The reemergent light would tend to be in the blue to orange region of the visual spectrum. Reflection of sunlight in this spectral region could lower the oceanic temperature sufficiently to cause an ice-age. The origin of a suitably large density of submicron particles of large refractive index is attributed either to bolide impact or to an enhanced density within the zodiacal dust.

081.024 **On the global geodynamic model of the Earth.**
J. Nedoma.
Gerlands Beitr. Geophys., Band 93, Heft 4, p. 241 – 260 (1984).
The paper discusses a geodynamic model of the Earth, which is the initial model for a global mathematical Earth model and for mathematical models of local regions.

081.025 **A new isostatic model of the Earth.**
K. F. Tyapkin.
Geofiz. Koeslemenyek, Vol. 30, No. 1, p. 3 – 10 (1984). Abstr. in Phys. Abstr., Vol. 87, No. 1243, Entry 100428 (1984).

081.026 **Geoid heights and lithospheric stresses for a dynamic Earth.**
Y. Ricard, L. Fleitout, C. Froidevaux.
Ann. Geophys., Vol. 2, No. 3, p. 267 – 286 (1984). Abstr. in Phys. Abstr., Vol. 87, No. 1243, Entry 100495 (1984). – See Abstr. 012.046.

081.027 **The dynamics of the Earth's inner and outer cores.**
D. E. Smylie, A. M. K. Szeto, M. G. Rochester.
Rep. Prog. Phys., Vol. 47, No. 7, p. 855 – 906 (1984). Abstr. in Phys. Abstr., Vol. 87, No. 1243, Entry 100512 (1984).

081.028 **Numerical and dynamical stability of convection cells in the Rayleigh number range $10^3 - 8 \times 10^5$.**
U. Hansen, A. Ebel.
Ann. Geophys., Vol. 2, No. 3, p. 291 – 301 (1984). Abstr. in Phys. Abstr., Vol. 87, No. 1243, Entry 100523 (1984). – See Abstr. 012.046.

081.029 **Large-scale gravity profiles as evidences of a convective circulation.**
M. Rabinowicz, B. Lago.
Ann. Geophys., Vol. 2, No. 3, p. 321 – 331 (1984). Abstr. in Phys. Abstr., Vol. 87, No. 1243, Entry 100526 (1984). – See Abstr. 012.046.

081.030 **Plate motions, the driving mechanism, and the geopotential.**
E. Seidler, W. R. Jacoby.
Ann. Geophys., Vol. 2, No. 3, p. 333 – 342 (1984). Abstr. in Phys. Abstr., Vol. 87, No. 1243, Entry 100527 (1984). – See Abstr. 012.046.

081.031 **Is the lower mantle homogeneous?**
A. L. Hales.
J. Geodyn., Vol. 1, No. 1, p. 3 – 9 (1984). Abstr. in Phys. Abstr., Vol. 87, No. 1244, Entry 105481 (1984).

081.032 **The Earth's thermal profile: is there a mid-mantle thermal boundary layer?**
S. Spiliopoulos, F. D. Stacey.
J. Geodyn., Vol. 1, No. 1, p. 61 – 77 (1984). Abstr. in Phys. Abstr., Vol. 87, No. 1244, Entry 105482 (1984).

081.033 **Geodynamics and state of the Earth's interior.**
H. Stiller, S. Franck, D. Mohlmann.
J. Geodyn., Vol. 1, No. 1, p. 79 – 100 (1984). Abstr. in Phys. Abstr., Vol. 87, No. 1244, Entry 105483 (1984).

081.034 **The contribution of satellite geodesy to the geosciences.**
R. Sigl.
GeoJournal, Vol. 8, No. 4, p. 341 – 362 (1984). = Mitt. Inst. Astron. Phys. Geod. Tech. Hochsch. Münch., Nr. 167. – See Abstr. 012.052.

081.035 **Planetary structures: an analytical approach.**
J. P. Sharma.
Indian J. Pure Appl. Math., Vol. 15, No. 6, p. 709 – 712 (1984).
Using the technique of Pade' (2,2) approximant, the author has discussed some classical aspects of the structural features of the self-gravitating planets Earth and Moon. Approximate analytical results are tabulated. The advantage of Pade' technique over those of previously existing methods has been briefly mentioned.

081.036 **Rhenium enriched Kilauea volcano aerosols: evidence for a volcanogenic component in the K/T boundary clay layer.**
A. R. Hildebrand, W. V. Boynton, W. H. Zoller.
Bull. Am. Astron. Soc., Vol. 16, No. 3, Part I, p. 679 (1984). Abstract. – See Abstr. 010.061.

081.037 **The role of mineral precipitates in moderating Shuttle imaging radar signal penetration and subsurface scattering in the hyperarid eastern Sahara: applications to climatic history of Mars.**
G. G. Schaber, J. F. McCauley, C. S. Breed.
Bull. Am. Astron. Soc., Vol. 16, No. 3, Part I, p. 679 (1984). Abstract. – See Abstr. 010.061.

081.038 **Terrestrial analogs for martian and lunar wrinkle ridges.**
J. B. Plescia, M. P. Golombek.
Bull. Am. Astron. Soc., Vol. 16, No. 3, Part I, p. 681 (1984). Abstract. – See Abstr. 010.061.

081.039 **The harmonic analysis of gravimetric tide of the earth at Kunming station.**
C. Fong.
Ann. Shanghai Obs., Acad. Sin., No. 5, p. 41 – 46 (1983). In Chinese.

081.040 **On the dynamics and energetics of the earth's core.**
S. Franck.
Astron. Nachr., Vol. 305, No. 5, p. 276 (1984). Abstract. – See Abstr. 012.062.

081.041 **Lagging mantle convection, the geoid and mantle structure.**
G. F. Davies.
Earth Planet. Sci. Lett., Vol. 69, No. 1, p. 187 – 194 (1984).
In numerical models of convection incorporating migration of a simulated subduction zone, the main descending flow lags far behind the migrating trench, and a geoid low is associated with the main descending flow. This provides physical plausibility for the suggestion that present very long-wavelength (degree 2 – 4) geoid lows are associated with Mesozoic trench locations, and suggests further that the present long-wavelength geoid, deep mantle structure and hotspot distribution may be straightforward consequences of plate evolution since the Upper Paleozoic.

081.042 **FeO and $H_2O$ and the homogeneous accretion of the Earth.**
M. A. Lange, T. J. Ahrens.
Earth Planet. Sci. Lett., Vol. 71, No. 1, p. 111 – 119 (1984).
The authors present new shock devolatilization recovery data for brucite ($Mg(OH)_2$) shocked to 13 and 23 GPa. These data combined with previous data for serpentine ($Mg_3Si_2O_5(OH)_4$) are used to constrain the minimum size terrestrial planet for which planetesimal infall will result in an impact-generated water atmosphere. Assuming a chondritic abundance of minerals including 3 – 6%, by mass water, in hydrous phyllosilicates, the authors carried out model calculations simulating the interaction of metallic iron with impact-released free water on the surface of the accreting Earth. In order to obtain $\sim 10^{25}$ g of atmospheric water by the end of accretion, slightly heterogeneous accretion with initially 36% by mass iron planetesimals, as compared to a homogeneous value of 34% is required. Such models yield final FeO budgets, which either require a higher FeO content of the mantle (17 wt.%) or oxygen as a light element in the outer core of the Earth.

081.043 **Tidal friction in the Earth-Moon system.**
P. Brosche.
Philos. Trans. R. Soc. London, Ser. A, Vol. 313, No. 1524, p. 71 – 75 (1984). – See Abstr. 012.068.
Modern computers enable us to obtain realistic values for the present tidal torque between the Moon and the oceans; those values agree with the observations. In principle, computations for distant geological epochs are possible as well and have been performed. However, the very complex eigenperiod spectrum of the oceans today precludes a continuous reconstruction of the tidal torque for an essential part of the Earth's history. Hence the original state of the Earth-Moon system is still uncertain.

081.044 **The origin and earliest state of the earth's hydrosphere.**
J. G. Cogley, A. Henderson-Sellers.
Rev. Geophys. Space Phys., Vol. 22, No. 2, p. 131 – 175 (1984).

081.045 **Mapping the lower mantle: determination of lateral heterogeneity in *P* velocity up to degree and order 6.**
A. M. Dziewonski.
J. Geophys. Res., Vol. 89, No. B7, p. 5929 – 5952 (1984).

081.046 **Mapping the upper mantle: three-dimensional modeling of earth structure by inversion of seismic waveforms.**
J. H. Woodhouse, A. M. Dziewonski.
J. Geophys. Res., Vol. 89, No. B7, p. 5953 – 5986 (1984).

081.047 **Geoid anomalies in a dynamic earth.**
M. A. Richards, B. H. Hager.
J. Geophys. Res., Vol. 89, No. B7, p. 5987 – 6002 (1984).

081.048 **Subducted slabs and the geoid: constraints on mantle rheology and flow.**
B. H. Hager.
J. Geophys. Res., Vol. 89, No. B7, p. 6003 – 6015 (1984).

081.049 **Geophysical and isotopic constraints on mantle convection: an interim synthesis.**
G. F. Davies.
J. Geophys. Res., Vol. 89, No. B7, p. 6017 – 6040 (1984).

081.050 **Hugoniot data for pyrrhotite and the earth's core.**
J. M. Brown, T. J. Ahrens, D. L. Shampine.
J. Geophys. Res., Vol. 89, No. B7, p. 6041 – 6048 (1984).

081.051 **Influence of the model of the geopotential on the accuracy of the determination of orbital elements.**
G. V. Romanova.
Astron. Tsirk., No. 1305, p. 5 – 6 (1984). In Russian.

081.052 **On the accuracy of laser observations of the ISAGEX experiment.**
G. V. Romanova.
Astron. Tsirk., No. 1304, p. 3 – 6 (1984). In Russian.

081.053 **Determination of the Love number $k_2$ from laser observations of the GEOS–A and GEOS–B satellites.**
G. V. Romanova.
Astron. Tsirk., No. 1304, p. 6 – 7 (1984). In Russian.

081.054 **Paleomagnetism of gabbros of the early proterozoic Blachford Lake Intrusive Suite and the Easter Island dyke, Great Slave Lake, NWT: possible evidence for the earliest continental drift.**
E. Irving, A. Davidson, J. C. McGlynn.
Geophys. Surv., Vol. 7, No. 1, p. 1 – 25 (1984). = Earth Phys. Branch Contrib., No. 1111. – See Abstr. 012.082.

081.055 **Verification of the accuracy of GEM–L2 in response to criticism by Lambeck and Coleman.**
F. J. Lerch.
NASA Tech. Memo., NASA TM–86123, p. 3–2 – 3–14 (1984). See Abstr. 003.039.
The objective is to evaluate the accuracy of GEM–L2 and its improvement in estimating Lageos orbits in order to provide better baselines for plate tectonics, improved polar motion and earth rotation.

081.056 **Accuracy of mapping the earth's gravity field fine structure with a spaceborne gravity gradiometer mission.**
W. D. Kahn.
NASA Tech. Memo., NASA TM–86123, p. 5–2 – 5–7 (1984). See Abstr. 003.039.

**Reversals of the Earth's magnetic field.**
See Abstr. 003.003.

**Evolution of the expanding earth.**
See Abstr. 003.104.

**ERS–1 radar altimeter data products. Proceedings of an ESA Workshop held at Frascati, Italy, 8 – 11 May, 1984.**
See Abstr. 012.011.

**Future of laser interferometry in space.**
See Abstr. 036.011.

**Earth's rotation and its geophysical implications.**
See Abstr. 044.007.

**Irregular variations of gravity and the earth's rotation.**
See Abstr. 044.017.

**Geodätische Modelle der Punkt– und Gravitationsfeldbestimmung.**
See Abstr. 045.021.

**Antineutrino astronomy and geophysics.**
See Abstr. 061.001.

**Mesure du diamètre solaire et des éléments orbitaux de la Terre à l'astrolabe du C.E.R.G.A.**
See Abstr. 080.013.

**The Earth's magnetic field.**
See Abstr. 084.024.

**Geomagnetic secular variation impulses.**
See Abstr. 084.058.

**Outer–core geostrophic flow and secular variation of Earth's geomagnetic field.**
See Abstr. 084.059.

**On a method of construction of the point masses model for the gravitational field of a planet.**
See Abstr. 091.001.

**Geochemical comparison of meteorites, the moon and the earth: trends of element evolution.**
See Abstr. 105.063.

**Temporal variation of the concentration of cosmogenic isotopes in the earth's atmosphere and solar activity.**
See Abstr. 144.019.

**Limits on a stochastic gravitational wave background from observations of terrestrial and solar oscillations.**
See Abstr. 161.360.

# 082 Atmosphere (Refraction, Scintillation, Extinction, Airglow, Site Testing)

082.001 **Night–time measurements of the structure characteristic of the $C_n^2$ refractive index in the atmospheric boundary layer and estimates of astronomical "seeing".**
A. Eh. Gur'yanov, M. A. Kallistratova, G. A. Karyukin, I. V. Petenko.
Astron. Tsirk., No. 1283, p. 3 – 6 (1983). In Russian.

082.002 **Evidence for haematite particles at 60 km altitude.**
C. F. Bohren, J. J. Olivero.
Nature, Vol. 310, No. 5974, p. 216 – 218 (1984).
Ackerman et al. have reported experimental evidence for the existence of a dust layer at an altitude of 60 km. The authors have analysed these data and suggest that the particles are composed of haematite ($\alpha$–$Fe_2O_3$) and have radii around 200 Å.

082.003 **Solar image properties at a site near Norilsk.**
Ya. Boyarskij, Eh. V. Kononovich, Yu. A. Kupryakov, Eh. Ya. Kononov, O. B. Smirnova, V. N. Obridko.
Astron. Tsirk., No. 1289, p. 3 – 5 (1983). In Russian.

082.004 **Comparison of astronomical site testing instruments on a horizontal path.**
A. Eh. Gur'yanov, S. L. Zubkovskij, A. A. Tokovinin, P. V. Shcheglov, B. B. Khodzhadurdyev, A. A. Makarov.
Astron. Tsirk., No. 1291, p. 1 – 4 (1983). In Russian.

082.005 **Atmospheric water vapour content above some sites in Armenia, Pamir and Chile.**
G. A. Alekseeva, V. D. Galkin, L. A. Kamionko.
Astron. Tsirk., No. 1296, p. 6 – 8 (1983). In Russian.

082.006 **Atmospheric spectral imaging.**
M. R. Torr, D. G. Torr.
Science, Vol. 225, No. 4658, p. 169 – 171 (1984).
An array of imaging spectrometers flown on the Spacelab 1 mission was capable of providing spectra of the atmospheric emissions over a broad wavelength range from 300 to 12,700 Å and acquiring each complete spectrum nearly simultaneously. As a result of the mission a data base has been acquired that is valuable for studies of both the upper atmosphere and the shuttle–Spacelab vehicle environment.

082.007 **Sample performance of the grille spectrometer.**
M.-P. Lemaitre, J. Laurent, J. Besson, A. Girard, C. Lippens, C. Muller, J. Vercheval, M. Ackerman.
Science, Vol. 225, No. 4658, p. 171 – 172 (1984).

The grille spectrometer observed the setting and rising sun 18 times during the Spacelab 1 mission. In addition to solar absorption lines, many of which had not been observed before, atmospheric spectral absorptions due to carbon monoxide and carbon dioxide were observed at heights tangent to the thermosphere (greater than 85 kilometers), and absorptions due to ozone, water, methane, and nitrous oxide were observed in the mesosphere (greater than 50 kilometers).

082.008 **Observations of Lyman–α emissions of hydrogen and deuterium.**
J. L. Bertaux, F. Goutail, G. Kockarts.
Science, Vol. 225, No. 4658, p. 174 – 176 (1984).

A spectrophotometer was flown on Spacelab 1 to study various mechanisms of Lyman–α emission in the upper atmosphere. Preliminary results are presented on three topics: the first optical detection of the deuterium Lyman–α emission at 110 kilometers, with an intensity of 330 rayleighs indicating an eddy diffusion coefficient of $1.3 \times 10^6$ square centimeters per second; auroral proton precipitations seen on both the night and the day side; and an emission located above 250 kilometers of altitude, interpreted as the result of charge exchange of magnetospheric protons with geocoronal atoms.

082.009 **Improved scattering formula for calculations of artificial night–sky illumination.**
R. H. Garstang.
Observatory, Vol. 104, No. 1061, p. 196 – 197 (1984).

082.010 **Inversion of infrared spectra obtained at the "Observatoire du Pic du Midi".**
C. Lippens, C. Muller.
Bull. Cl. Sci., Acad. R. Belg., Vol. 69, No. 6, p. 379 – 387 (1983). Abstr. in Phys. Abstr., Vol. 87, No. 1237, Entry 72697 (1984).

082.011 **On the relationship between clear–sky planetary and surface albedos.**
T. S. Chen, G. Ohring.
J. Atmos. Sci., Vol. 41, No. 1, p. 156 – 158 (1984). Abstr. in Phys. Abstr., Vol. 87, No. 1237, Entry 72703 (1984).

082.012 **Dynamics of the lower thermosphere consistent with satellite observations of 5577 Å airglow. I. Method of analysis.**
J. S. Murphree, R. D. Elphinstone, L. L. Cogger.
Can. J. Phys., Vol. 62, No. 4, p. 370 – 381 (1984). Abstr. in Phys. Abstr., Vol. 87, No. 1237, Entry 72766 (1984).

082.013 **Dynamics of the lower thermosphere consistent with satellite observations of 5577 Å airglow. II. Atomic oxygen, local turbulence, and global circulation results.**
R. D. Elphinstone, J. S. Murphree, L. L. Cogger.
Can. J. Phys., Vol. 62, No. 4, p. 382 – 395 (1984). Abstr. in Phys. Abstr., Vol. 87, No. 1237, Entry 72767 (1984).

082.014 **Comparative studies of atmospheric density models used for Earth satellite orbit estimation.**
R. E. Shanklin Jr., T. Lee, M. Samii, M. K. Mallick, J. O. Cappellari Jr.
J. Guid. Control Dyn., Vol. 7, No. 2, p. 235 – 237 (1984). Abstr. in Phys. Abstr., Vol. 87, No. 1237, Entry 72768 (1984).

082.015 **Thermospheric general circulation with coupled dynamics and composition.**
R. E. Dickinson, E. C. Ridley, R. G. Roble.
J. Atmos. Sci., Vol. 41, No. 2, p. 205 – 219 (1984). Abstr. in Phys. Abstr., Vol. 87, No. 1237, Entry 72769 (1984).

082.016 **On the variability of the refraction constant.**
I. G. Kolchinskij.
Astrometr. Astrofiz., Vyp. 52, p. 38 – 46 (1984). In Russian.

Different definitions of the "refraction constant" are discussed. The author considers critically Schmeidler's result, according to which the total annual change of the refraction constant in Munich amounts to 0.″38. Schmeidler's conclusion on the existence of global changes in the refraction constant with the amplitude of 0.″1 and annual period was also critically considered. On the basis of four different refraction theories (Gylden, Radau, Harzer, Emden) it is shown that seasonal changes in temperature height distribution may be responsible for corresponding annual variations of the refraction constant within the limits of 0.″001 – 0.″03 only, depending on the applied theory.

082.017 **Light pollution and its measurement.**
D. W. Dawson.
Advances in photoelectric photometry, Vol. 2, p. 30 – 53 (1984). – See Abstr. 003.002.

082.018 **Grain abundance in the primordial atmosphere of the Earth.**
H. Mizuno, G. W. Wetherill.
Icarus, Vol. 59, No. 1, p. 74 – 86 (1984).

For models of planetary accumulation in the presence of solar nebular gas, the initial surface temperature of the Earth is controlled by the grain opacity of the atmosphere. The surface temperature in turn controls the quantity of neon dissolved and trapped within the interior of the Earth. To compare accumulation theory with observation calculations have been made of the grain opacity expected to be associated with accumulation in a gaseous nebula.

082.019 **Atmospheric gamma radiation: a comparison of experimental results at 4.5 GV and 11.7 GV.**
J. M. Lavigne, M. Niel, G. Vedrenne, B. Agrinier.
J. Geophys. Res., Vol. 89, No. A7, p. 5636 – 5642 (1984).

082.020 **Spectroscopy of the EUV (350 – 1400 Å) nightglow.**
S. Chakrabarti, R. Kimble, S. Bowyer.
J. Geophys. Res., Vol. 89, No. A7, p. 5660 – 5664 (1984).

The authors present time averaged spectra of the extreme ultraviolet nightglow in the 350 to 1400–Å range at 8–Å resolution. The uplooking spectrum is dominated by geocoronal and interplanetary resonantly scattered solar lines (H I 1216, H I 1026, O II 834, and He I 584 Å). The downlooking spectrum contains O I emission features from the nighttime ionosphere, including several previously unobserved features. In contrast to previously reported observations, no significant $N_2$ Lyman–Birge–Hopfield emissions are evident in the data, and a 2–σ upper limit of 135 R can be placed on this radiation for the time of observations.

082.021 **Sodium clouds in the lower thermosphere.**
V. W. J. H. Kirchhoff, H. Takahashi.
Planet. Space Sci., Vol. 32, No. 7, p. 831 – 836 (1984).

Very large short lived enhancements observed in Na nightglow measurements are described. Only nine such events have been observed in about 8 years of regular nocturnal observations at 23°S. These enhancements are believed to be due to excess sodium deposited in the upper atmosphere by meteoroid ablation. The cases observed are not always correlated with known meteor showers, except perhaps in June with χ Scorpiids and θ Ophiuchids.

082.022 **Polar cap OH airglow rotational temperatures at the mesopause during a stratospheric warming event.**
H. K. Myrabø, C. S. Deehr, B. Lybekk.
Planet. Space Sci., Vol. 32, No. 7, p. 853 – 856 (1984).

082.023 **Tidal and solar cycle effects on the O I 5577 Å, NaD and OH(8,3) airglow emissions observed at 23°S.**
H. Takahashi, Y. Sahai, P. P. Batista.
Planet. Space Sci., Vol. 32, No. 7, p. 897 – 902 (1984).

The upper mesosphere airglow emissions O I 5577, NaD and OH have been observed at Cachoeira Paulista (22.7°S; 45.0°W)

Brazil. Nocturnal variations and their seasonal dependencies in amplitude and phase, and the annual variations of these emissions are presented, analysing the data obtained from 1977 to 1982 during the ascending phase of the last solar cycle. Long-term intensity variations are also presented with the solar sunspot numbers and the 10.7 cm flux.

082.024 **Atmospheric science.**
J. E. Harries.
J. Br. Interplanet. Soc., Vol. 37, No. 7, p. 330 – 336 (1984).
This paper presents a review of the use of satellite remote sounding experiments for studies of the Earth in both the stratosphere and mesosphere.

082.025 **Spaceborne lidar investigations of the atmosphere.**
E. V. Browell, S. Ismail.
ESA Spec. Publ., ESA SP–202, p. P181 – P188 (1984). – See Abstr. 012.004.

082.026 **Passive microwave remote sensing of the atmosphere from satellites.**
D. H. Staelin.
ESA Spec. Publ., ESA SP–215, Vol. I, p. 413 – 415 (1984). – See Abstr. 012.012.

082.027 **Ground–based remote sensing of temperature and water vapor by passive microwave radiometers.**
J. Askne.
ESA Spec. Publ., ESA SP–215, Vol. I, p. 423 – 429 (1984). – See Abstr. 012.012.

082.028 **Combined ground–based and satellite remote sensing of temperature and water vapor.**
E. R. Westwater, Z. Wang.
ESA Spec. Publ., ESA SP–215, Vol. I, p. 431 – 436 (1984). – See Abstr. 012.012.
Results of deriving temperature and moisture from combinations of ground–based microwave and orbiting satellite (NOAA 6/7) radiometric data are presented.

082.029 **Differential emission measurement for atmospheric absorption by a ground–based microwave radiometer.**
T. Ojima.
ESA Spec. Publ., ESA SP–215, Vol. I, p. 443 – 448 (1984). – See Abstr. 012.012.

082.030 **Recovery of spectral optical depth and zero–airmass solar spectral irradiance under conditions of temporally varying optical depth.**
J. A. Reagan, I. C. Scott–Fleming, B. M. Herman, R. M. Schotland.
ESA Spec. Publ., ESA SP–215, Vol. I, p. 455 – 459 (1984). – See Abstr. 012.012.
The effect of temporal variations in atmospheric optical depth on the determination of the zero–airmass voltage via Langley plot fitting of solar radiometer data is analyzed.

082.031 **Prediction of wet air refraction at microwaves (research note).**
A. Greve.
Int. J. Infrared Millimeter Waves, Vol. 5, No. 3, p. 315 – 320 (1984). Abstr. in Phys. Abstr., Vol. 87, No. 1241, Entry 89995 (1984).

082.032 **Effet de la turbulence atmosphérique sur le bilan de liaison des échos laser sur la lune et satellites.**
P. Assus, J. Gay, J.–P. Gay.
Astrométrie, géodésie, mécanique céleste, p. 169 – 173 (1984). – See Abstr. 012.021.

082.033 **Planetary, solar, atmospheric, hydrospheric and endogene processes as origin of climatic changes on the earth.**
N.–A. Mörner.
Climatic changes on a yearly to millennial basis, p. 483 – 507 (1984). – See Abstr. 012.023.

082.034 **Planetary periodicities and terrestrial climate stress.**
R. W. Fairbridge.
Climatic changes on a yearly to millennial basis, p. 509 – 520 (1984). – See Abstr. 012.023.

082.035 **Interpretation of $NO_2$ absorption in twilight sky spectra.**
B. B. McMahon.
J. Quant. Spectrosc. Radiat. Transfer, Vol. 32, No. 1, p. 9 – 15 (1984).
A multiple scattering model has been developed to calculate nitrogen dioxide ($NO_2$) absorption in the light from the zenith sky during twilight. Model studies show that this absorption is not very sensitive to the atmospheric temperature profile or to tropospheric $NO_2$. The model was used to interpret some ground–based measurements of $NO_2$ sky absorption.

082.036 **Analytical yield spectrum approach to photoelectron fluxes in the earth's atmosphere.**
R. P. Singhal, S. A. Haider.
J. Geophys. Res., Vol. 89, No. A8, p. 6847 – 6852 (1984).

082.037 **On the determination of the astronomical refraction in the near–horizontal zone by the photographic method.**
A. V. Arkhangel'skij, S. D. Volzhanin.
L'vov. politekh. inst. L'vov, 5 pp. (1984). In Russian. Abstr. in Ref. Zh., 51. Astron., 8.51.90 (1984).

082.038 **Investigation of the spectral transmittance of the atmosphere above the Astrophysical Institute [of the Academy of Sciences of the Kazakh SSR] and the Sternberg Station carried out during a spectrophotometry of stars.**
L. N. Knyazeva, A. V. Kharitonov.
Astrofiz. inst. Akad. Nauk KazSSR. Alma–Ata, 92 pp. (1984). In Russian. Abstr. in Ref. Zh., 51. Astron., 8.51.652 (1984).

082.039 **Investigations of the dynamics of the mean atmosphere in the USSR (Report of the 18th General Assembly of the International Geodetic and Geophysical Union, August 1983, FRG).**
Eh. S. Kazimirovskij.
Issled. Geomagn., Aehron. Fiz. Solntsa, Moskva, No. 67, p. 3 – 23 (1984). In Russian. Abstr. in Ref. Zh., 62. Issled. Kosm. Prostranstva, 8.62.583 (1984).

082.040 **Seasonal variations of atomic oxygen in the lower thermosphere.**
B. P. Komitov, K. B. Serafimov.
Dokl. Bolg. Akad. Nauk, Vol. 36, No. 11, p. 1395 – 1397 (1983). In Russian. Abstr. in Ref. Zh., 62. Issled. Kosm. Prostranstva, 8.62.587 (1984).

082.041 **Atmosphärische Einflüsse auf die astronomische Beobachtung.**
A. Sudy.
Sternenbote, 27. Jahrg., Nr. 338, p. 174 – 178 (1984).

082.042 **Thunder and electric storms.**
J. Maddox.
Nature, Vol. 311, No. 5986, p. 507 (1984).
A new measurement of the stratospheric electrostatic fields throws light on the link between solar activity and terrestrial climate.

082.043 **Improved positive–ion composition measurements in the upper troposphere and lower stratosphere and the detection of acetone.**
G. Hauck, F. Arnold.
Nature, Vol. 311, No. 5986, p. 547 – 550 (1984).
The authors report improved positive–ion composition measurements in the lower stratosphere and upper troposphere, using an improved aircraft–borne mass spectrometer with higher mass resolution and sensitivity than hitherto achieved.

082.044 **Astronomi och hägringar.**
Y. Öhman.
Astron. Tidsskr., Årg. 17, Nr. 3, p. 112 – 117 (1984).

082.045 **The deficiency of atmospheric xenon.**
F. A. Podosek, T. J. Bernatowicz, M. Honda, F. E. Kramer.
The 27th International Geological Congress, Tom 5, Sect. 10 – 11, p. 377 (1984). Abstr. in Ref. Zh., 51. Astron., 9.51.186 (1984). – See Abstr. 012.031.

082.046 **Evolution of silicate clouds of the earth and atmosphere considering current isotopic data.**
I. Ya. Azbel', I. N. Tolstikhin.
The 27th International Geological Congress, Tom 5, Sect. 10 – 11, p. 221 – 222 (1984). In English and Russian. From Ref. Zh., 51. Astron., 9.51.205 (1984). – See Abstr. 012.031.

082.047 **The role of meteoric particles in noctilucent clouds.**
A. F. Roddy.
Ir. Astron. J., Vol. 16, No. 3, p. 194 – 202 (1984). – See Abstr. 012.036.

082.048 **Assessment of plasma transport and convection at high latitudes.**
R. Schunk, A. R. Barakat, H. Carlson, J. B. Evans, J. Foster, R. Greenwald, M. C. Kelley, T. Potemra, M. H. Rees, A. D. Richmond, R. G. Roble.
NASA Ref. Publ., NASA RP–1120, p. 11–1 – 11–36 (1984). – See Abstr. 012.038.

Contents: Introduction. Currents and conductivities. Thermospheric winds and plasma transport. Plasma convection. Polar wind. Conclusions.

082.049 **Rocket observations of the ultraviolet airglow during morning twilight.**
R. P. Cebula, P. D. Feldman.
J. Geophys. Res., Vol. 89, No. A10, p. 9080 – 9088 (1984).

The nitric oxide γ and δ bands, O II λ2470–, O I λλ1356– and 1304–Å lines, and $N_2$ Lyman–Birge–Hopfield and $O_2$ Herzberg band emissions were observed during morning twilight between 90 and 246 km on September 24, 1979, with two rocket–borne scanning monochromators viewing successively at 82° and 90° to the zenith near the sun–earth–rocket plane. The data also contain a significant zodiacal light contribution. Pure airglow spectra are obtained by subtracting the zodiacal light contribution from the data.

082.050 **$O_2$ ($b^1\Sigma_g^+ - X^3\Sigma_g^-$) atmospheric band night airglow measurements in the northern polar cap region.**
H. K. Myrabø, K. Henriksen, C. S. Deehr, G. J. Romick.
J. Geophys. Res., Vol. 89, No. A10, p. 9148 – 9152 (1984).

The $O_2$ atmospheric (0–1) band at 8645 Å has been observed in the airglow during the 1982/1983 winter solstice from the ground at Longyearbyen, West Spitsbergen (78.4°N latitude, 15°E longitude, geographic).

082.051 **Night airglow OH (8–3) band rotational temperatures at Poker Flat, Alaska.**
H. K. Myrabø, G. J. Romick, G. G. Sivjee, C. S. Deehr.
J. Geophys. Res., Vol. 89, No. A10, p. 9153 – 9156 (1984).

Temperatures of the mesopause region (85 – 90 km) have been deduced from OH (8–3) molecular band night airglow emission measurements made at Poker Flat (65°N), Alaska.

082.052 **Solar and terrestrial parameterizations for radiative–convective models.**
I. M. Vardavas, J. H. Carver.
Planet. Space Sci., Vol. 32, No. 10, p. 1307 – 1325 (1984).

The authors present a radiative–convective modelling technique with parameterizations, for both solar and terrestrial radiation transfer, which allow the rapid computation of the mean vertical temperature profile from the ground to the thermosphere. The method has been specifically designed for modelling the evolution of the Earth's mean vertical temperature structure due to changes in atmospheric composition, variations in the solar flux, surface albedo, cloud cover, water vapour and lapse rate, and changes in the temperature of the thermosphere which is associated with solar activity.

082.053 **On the molecular scattering in the terrestrial atmosphere: an empirical formula for its calculation in the homosphere.**
M. Nicolet.
Planet. Space Sci., Vol. 32, No. 11, p. 1467 – 1468 (1984).

A recent determination by D. R. Bates of the Rayleigh scattering cross section ($\sigma_{RS}$) for air from 0.2 to 1 μm leads to a simple empirical formula (λ in μm) $\sigma_{RS} = 4.02 \times 10^{-28}/\lambda^{4+x}\mathrm{cm}^2$ where $x = 0.389\lambda + 0.09426/\lambda - 0.3228$ for the spectral region 0.2 μm $< \lambda <$ 0.55 μm; the accuracy is within ±0.5%. From the visible at 0.55 μm to the infrared (i.r.) at 1 μm, the same accuracy can be obtained using a constant value, $x = 0.04$. The formula accounts for the degree of depolarization which varies with the wavelength according to the latest determination by Bates.

082.054 **Disturbances in the stratosphere and mesosphere from an ultraviolet radiation source.**
S. I. Kozlov.
Kosm. Issled., Tom 22, Vyp. 5, p. 712 – 719 (1984). In Russian. English translation in Cosm. Res.

082.055 **Observatory–site reconnaissance.**
R. Lynds, J. W. Goad.
Publ. Astron. Soc. Pac., Vol. 96, No. 583, p. 750 – 766 (1984).

A cartographic exploration for potential observatory sites in the southwestern United States has been made. The purpose of the investigation was to supply some answers to questions as to the number of such sites and their characteristics. Criteria used in the selection included: possible seeing–related considerations, low sky brightness in the optical and in the infrared regions of the spectrum, and probability of clear weather. The search yielded a total of 56 peaks meeting the adopted criteria; 24 were in Nevada, with the remainder about equally divided between Utah, New Mexico, Arizona, and California. Rough quantitative characterizations have been provided as an aid to ranking the peaks according to various notions of desirability. According to these, Wheeler Peak, in eastern Nevada, stands out as the most promising site in the region surveyed.

082.056 **Optical properties of aerosols of mixed composition.**
C. S. Sloane.
Atmos. Environ., Vol. 18, No. 4, p. 871 – 878 (1984). Abstr. in Phys. Abstr., Vol. 87, No. 1243, Entry 100726 (1984).

082.057 **Altitude dependence of the vibrational distribution of $O_2(c^1\Sigma_u^-)$ in the nightglow and the possible effects of vibrational excitation in the formation of $O(^1S)$.**
I. C. McDade, E. J. Llewellyn, R. G. H. Greer, G. Witt.
Can. J. Phys., Vol. 62, No. 8, p. 780 – 788 (1984). Abstr. in Phys. Abstr., Vol. 87, No. 1244, Entry 105785 (1984).

082.058 **Observation of the helium 10830 Å airglow emission in midlatitude.**
K. Suzuki.
J. Geomagn. Geoelectr., Vol. 35, No. 9, p. 321 – 330 (1983). Abstr. in Phys. Abstr., Vol. 87, No. 1244, Entry 105786 (1984).

082.059 **Radiation belt particles and O II emissions determined as contaminations of geocoronal helium airglow observations.**
H. J. Fahr, G. Lay.
J. Geophys., Vol. 54, No. 3, p. 219 – 229 (1984). Abstr. in Phys. Abstr., Vol. 87, No. 1244, Entry 105787 (1984).

082.060 **A quasi–continuous record of atmospheric opacity at λ = 1.1 mm over 34 days at Mauna Kea Observatory.**
R. L. de Zafra, A. Parrish, P. M. Solomon, J. W. Barrett.
Int. J. Infrared Millimeter Waves, Vol. 4, No. 5, p. 757 – 765 (1983). Abstr. in Phys. Abstr., Vol. 87, No. 1245, Entry 110250 (1984).

082.061 **Long term variations of the Earth atmosphere transparency derived from the extinction measurements at Skalnaté Pleso and Brno observatories.**
J. Papousek, M. Vetesnik, J. Tremko, K. Juza.
Scr. Fac. Sci. Nat. Univ. Purkynianae Brun., Phys., Vol. 14, No. 1 – 2, p. 3 – 19 (1984). Abstr. in Phys. Abstr., Vol. 87, No. 1245, Entry 110253 (1984).

082.062 **Astronomical image quality and the vertical distribution of turbulent optical interference in the night atmosphere.**
A. É. Gur'yanov (*A. Eh. Gur'yanov*).
Sov. Astron., Vol. 28, No. 3, p. 343 – 350 (1984). English translation of 37.082.047.

082.063 **Light pollution – the problem and solutions to it.**
D. L. Crawford.
Publ. Astron. Soc. Pac., Vol. 96, No. 584, p. 786 (1984). Abstract. – See Abstr. 010.281.

082.064 **Night sky brightness and limiting naked–eye star visibility.**
R. H. Garstang.
Publ. Astron. Soc. Pac., Vol. 96, No. 584, p. 788 (1984). Abstract. – See Abstr. 010.281.

082.065 **Reanalysis of the British NHO site–survey observations at the NW Cone and 88–inch ridge sites.**
M. F. Walker.
Publ. Astron. Soc. Pac., Vol. 96, No. 584, p. 829 – 835 (1984). = Lick Obs. Bull., No. 990.
The observations obtained by the British Northern Hemisphere Observatory Site Survey team in 1974 – 75 at the NW Cone and on the 88–inch ridge on Mauna Kea have been reanalyzed. The new analysis confirms the result of the initial study of these data: better seeing conditions existed at the NW Cone than on the 88–inch ridge during the period covered by the observations. The present, more detailed, analysis indicates that the seeing at both sites became poorer and wind–speed dependent with southerly winds. That deterioration was larger at the 88–inch ridge site than at the NW Cone, indicating (1) that the effect was local in origin and was not due to high–altitude meteorological conditions, and (2) that whatever the local cause, the 88 inch ridge site was more adversely affected than was the NW Cone.

082.066 **Characteristics of water vapor over Kitt Peak as determined from FTS data.**
L. Wallace, J. W. Brault, M. Brown, W. Livingston.
Publ. Astron. Soc. Pac., Vol. 96, No. 584, p. 836 – 838 (1984).
Spectra of three water vapor lines obtained with the Fourier transform spectrometer at the McMath telescope have been analyzed to obtain the atmospheric water–vapor content. The results for 52 days between 1978 and 1983 show a maximum of between 20 and 25 mm precipitable water vapor in July and August and minima between February and June and between October and December. Outside of a secondary maximum due to two days in January 1983 the results are in generally good agreement with those obtained with a grating spectrograph. The pressure–broadened linewidths have been used to make corrections not exceeding 11% to the results obtained with the spectrograph.

082.067 **Light pollution out West concerns optical astronomers.**
W. Sweet.
Phys. Today, Vol. 37, No. 12, p. 63 – 64 (1984).

082.068 **Astronomy's problem with light pollution.**
M. Schmidt.
Phys. Today, Vol. 37, No. 12, p. 144 (1984).

082.069 **Visit of STC (*Scientific Technical Committee*) to possible sites for the VLT in Northern Chile.**
D. Enard.
Messenger, No. 38, p. 1 – 2 (1984).

082.070 **Atmospheric extinction at the Special Astrophysical Observatory of the USSR Academy of Sciences (1976 – 1980).**
S. I. Neizvestnyi (*S. I. Neizvestnyj*).
Bull. Spec. Astrophys. Obs. – North Caucasus, Vol. 17, p. 24 – 36 (1984). English translation of 34.082.114.

082.071 **Estimating clear–sky probability and star–image quality for the BTA.**
V. N. Erokhin, S. P. Plyaskin.
Bull. Spec. Astrophys. Obs. – North Caucasus, Vol. 17, p. 37 – 43 (1984). English translation of 34.082.063.

082.072 **Forecasting and optimizing astronomic image quality using acoustic sounding.**
G. Dubosclard, H. Hecquet, G. Coupinot, H. Sauvageot.
Astron. Astrophys., Vol. 140, No. 2, p. 273 – 276 (1984).
Acoustic soundings, performed at the Pic–du–Midi Observatory in July 1982, have shown that the sound–wave reflectivity profiles, associated with temperature fluctuations in atmospheric layers, are correlated to the image quality obtained with the 2 m telescope. Such an observation suggests that systematic observations, by remote sensing techniques, of the atmospheric turbulent structure and of its evolution may permit the measurement and the forecasting of the quality of the nocturnal atmosphere.

082.073 **Measuring atmospheric seeing.**
F. Roddier.
Very large telescopes, their instrumentation and programs, p. 191 – 201 (1984). – See Abstr. 012.064 (IAU Colloq. No. 79).

082.074 **A method to estimate Fried's seeing parameter from a time series of arbitrary resolved structures imaged through the atmosphere.**
O. von der Lühe.
Very large telescopes, their instrumentation and programs, p. 203 – 220 (1984). – See Abstr. 012.064 (IAU Colloq. No. 79).
A method is presented that allows to estimate the effective MTF from the observation of arbitrary structure with the use of Fried–Korff theory. The ratio of the squared modulus of the average Fourier transform and the average power spectrum serves as an estimator for the Fried parameter $r_0$. To a first approximation, this ratio is independent from the observed object. Additionally, the behaviour of the ratio in regions beyond the seeing limit in the Fourier plane may be analyzed to obtain an estimate of the speckle interferometry signal–to–noise ratio.

082.075 **Astronomical seeing at Mauna Kea and in particular at the CFHT (*Canada–France–Hawaii Telescope*).**
R. Racine.
Very large telescopes, their instrumentation and programs, p. 235 – 243 (1984). – See Abstr. 012.064 (IAU Colloq. No. 79).
The astronomical seeing at the CFHT is excellent. It now averages $<FWHM> = 0.9''$ and is continuously improving as a result of seeing conservation measures in the telescope building. Studies suggest that the average natural seeing on Mauna Kea should be $\sim 0.6''$. Very high ($< 0.2''$) resolution imaging at the CFHT, through adaptive optics, appears feasible.

082.076 **Some possible VLT sites in Chile.**
A. Ardeberg, H. Lindgren.
Very large telescopes, their instrumentation and programs, p. 417 – 428 (1984). – See Abstr. 012.064 (IAU Colloq. No. 79).
A report is given on the site–testing programme for a very large telescope in Chile. Data are now available from seven months of investigations. Comments are made on the places which have so far seemed most attractive. Meteorological data from constant monitoring of a coastal mountain, Paranal, are presented and discussed. These data indicate an average percentage of photometric nights of around 85. Atmospheric integrated water–vapour content seems very low in winter but relatively high in summer. Other meteorological parameters seem favourable for astronomical observations.

082.077 **Inelastic scattering in planetary atmospheres. 2: Polarization of the rotational component.**
T. J. Humphreys, G. W. Kattawar, A. T. Young.
Appl. Opt., Vol. 23, No. 23, p. 4422 – 4426 (1984).

The inelastic part of Rayleigh scattering, which can account for the partially filled line profiles of the blue sky observed at high spectral resolution, also explains the variation of polarization across a line profile. At lower spectral resolution, only the rotational part of the molecular scattering is needed to account for the polarization phenomena; translational (Brillouin) shifts can be neglected. Aerosol scattering, like the ground reflection, reduces the polarizations observed in the zenith sky by diluting the molecular (Rayleigh) component.

082.078 **Modelling of neutral and ionized atmospheres.**
K. Rawer.
Handbuch der Physik. Geophysik III/7, p. 223 – 535 (1984). – See Abstr. 003.035.

082.079 **Global distribution of helium in the upper atmosphere during solar minimum.**
R. P. Cageao, R. B. Kerr.
Planet. Space Sci., Vol. 32, No. 12, p. 1523 – 1529 (1984).

082.080 **Non–equilibrium water species as absorbers of infrared and millimetre waves.**
H. A. Gebbie.
Infrared Phys., Vol. 24, No. 5, p. 429 – 436 (1984).

Examples of anomalous absorption of millimetre waves in the atmosphere are given and evidence supporting an explanation in terms of non–equilibrium species is presented. Emphasis is placed on the need for energy sources to maintain a steady–state of the non–equilibrium species and it is shown that some laboratory experiments have unintentionally included such an energy source.

082.081 **Solar Mesosphere Explorer measurements of polar mesospheric clouds (noctilucent clouds).**
G. E. Thomas.
J. Atmos. Terr. Phys., Vol. 46, No. 9, p. 819 – 824 (1984). – See Abstr. 012.072.

The Ultraviolet Spectrometer experiment on board the Solar Mesosphere Explorer satellite has measured scattering of sunlight from the polar mesospheric cloud layer in the 0.2 – 0.3 μm spectral range. The layer is manifested in the limb–scanning measurements as large increases in radiance at heights near 80 km, at latitudes 60 – 90°N and 60 – 90°S and during the summer season. The data reveal the presence of forward scattering behaviour. This permits the estimation of the mean particle sizes through Mie scattering theory. The relationships of the particle radii, optical depths and equivalent water vapor column mass are described. This is followed by a brief description of the potential of the large SME data base for future studies.

082.082 **Diurnal variations of atmospheric nitric oxide: ground–based infrared spectroscopic measurements and their interpretation with time–dependent photochemical model calculations.**
C. P. Rinsland, R. E. Boughner, J. C. Larsen, G. M. Stokes, J. W. Brault.
J. Geophys. Res., Vol. 89, No. D6, p. 9613 – 9622 (1984).

Total vertical column amounts of NO have been derived from infrared solar absorption spectra recorded near sunrise and sunset with the 0.01–$cm^{-1}$ resolution Fourier transform interferometer at the National Solar Observatory on Kitt Peak on February 23, 1981. The results show an increase in NO concentration in the morning, late afternoon values about 40% higher than in the morning, and a decrease in NO concentration prior to sunset. The measured diurnal changes in the total vertical column amount are compared with values obtained from time–dependent photochemical calculations.

082.083 **Atlas of stratospheric submillimeter lines: 1. The 7 – 20 $cm^{-1}$ interval.**
M. G. Baldecchi, B. Carli, F. Mencaraglia, A. Bonetti, M. Carlotti.
J. Geophys. Res., Vol. 89, No. D7, p. 11689 – 11704 (1984).

The stratospheric emission in the submillimeter spectral region has been measured from 38.2 km altitude with a resolution of 0.0033 $cm^{-1}$. In most cases this resolution makes possible the separation of contributions caused by the individual rotational transitions. An atlas of the observed features is given as a guide to both the measurements of minor stratospheric constituents concentration and the evaluation of stratospheric transparency for the measurement of nontelluric sources.

082.084 **Instrumentation for the recording of microthermal and image–quality data.**
A. Ardeberg, N. Hansson, E. Olsen, S. K. Simpson, T. Wiesel.
Rep. Obs. Lund, No. 19, 10 pp. (1984).

A short description is given of the instruments constructed for site testing for the Nordic Optical Telescope. The telescope will be installed at the Spanish–International facility on La Palma, Observatorio del Roque de los Muchachos. Whereas the general excellence of the observatory is beyond doubt, the aim of the site testing is to select an optimum site within the large area available. Data will be collected to allow a study of the dependence of microthermal variations on height above ground. These data are then to be related to the resulting image quality.

082.085 **Interdependence of the emissions of the upper atmosphere and ionospheric characteristics.**
O. A. Matveeva.
Astron. Tsirk., No. 1303, p. 2 – 3 (1983). In Russian.

082.086 **On the increase of atmospheric extinction in 1982.**
V. G. Moshkalev.
Astron. Tsirk., No. 1309, p. 4 – 6 (1984). In Russian.

082.087 **Mt. Graham: a continental United States site option for the NNTT (*National New Technology Telescope*).**
N. J. Woolf, K. M. Merrill.
Prepr. Steward Obs., No. 547, 25 pp. (1984).

This report briefly discusses the reasons for considering additional sites for testing within the continental United States and outlines the rationale for identifying Mt. Graham as a site suitable for comparative study.

082.088 **On the theory of halos of the sun and the moon.**
G. F. Sitnik.
Astron. Tsirk., No. 1324, p. 1 – 4 (1984). In Russian.

082.089 **Generalization of the seeing histogram at the Gornotaezhnoe Station.**
V. L. Oknyanskij, M. I. Fisenko, S. E. Mosunov, S. G. Mozharovskij, V. A. Ziva, A. V. Kuklin.
Astron. Tsirk., No. 1329, p. 4 – 6 (1984). In Russian.

082.090 **A mechanism of corpuscular–atmospheric relations.**
Eh. R. Mustel'.
Astron. Zh., Tom 61, Vyp. 6, p. 1179 – 1183 (1984). In Russian. English translation in Sov. Astron., Vol. 28, No. 6.

The author paid attention to the results of Brazilian scientists who concluded that in the region of the southern magnetic anomaly a "dumping" of high energy particles from the radiation belts takes place immediately after geomagnetic storms towards the stratosphere.

082.091 **Analysis of the phase terms in the visibility function.**
W. Han.
Acta Astrophys. Sin., Vol. 4, No. 2, p. 124 – 132 (1984). In Chinese.

The integral expression of the visibility function, on which the effect of the atmosphere is considered, is derived in the paper.

082.092 **The pure astro–refraction of visible light in the Shanghai region.**
W.–f. Shen.
Acta Astron. Sin., Vol. 25, No. 2, p. 201 – 211 (1984). In Chinese.
The pure astro–refraction of visible light in Shanghai region has been calculated by means of the meteorological data of balloon sounding of the Shanghai Meteorological Observatory about 3 kilometers to the south of Shanghai Observatory and with the aid of a computer. Thus the Shanghai table of refraction and the statistical relations of refraction have been obtained.

082.093 **Surveying and sieving sites for a solar tower telescope in Kunming.**
T.–l. Qian, Q.–s. Li, Z.–h. Liu, Q. Zhang.
Acta Astron. Sin., Vol. 25, No. 3, p. 309 – 320 (1984). In Chinese.
The results of surveying and sieving sites for a large solar tower telescope at Yunnan Observatory are reported.

082.094 **Winds and waves in the middle atmosphere as observed by ground–based radars.**
R. Rüster.
Adv. Space Res., Vol. 4, No. 4, p. 3 – 18 (1984). – See Abstr. 012.087.

082.095 **Recent advances in understanding stratospheric dynamics and transport processes: application of satellite data to their interpretation.**
W. L. Grose.
Adv. Space Res., Vol. 4, No. 4, p. 19 – 28 (1984). – See Abstr. 012.087.

082.096 **Signatures of troposphere–stratosphere momentum coupling: implications for global atmospheric angular momentum and earth rotation budgets.**
H. A. Taylor, H. G. Mayr, R. E. Hartle, L. Kramer, R. Stirling.
Adv. Space Res., Vol. 4, No. 4, p. 41 – 45 (1984). – See Abstr. 012.087.

082.097 **Tidal components in 5577 Å nightglow.**
N. Ranade, P. B. Choudhary, A. D. Tillu.
Adv. Space Res., Vol. 4, No. 4, p. 97 – 100 (1984). – See Abstr. 012.087.
Intensity variations of 5577 Å nightglow observed in IQSY and MAP periods are analysed for the estimation of tidal components. The analysis involves the use of a photochemical model with an assumed tidal wave for the estimation of the solar tidal component. A straightforward harmonic analysis has also been carried out for the estimation of lunar tidal components.

082.098 **Gravity waves from sodium–dayglow observations.**
D. K. Datarange, A. D. Tillu.
Adv. Space Res., Vol. 4, No. 4, p. 101 – 104 (1984). – See Abstr. 012.087.
The sampling interval of sodium dayglow observations was progressively reduced from one hour to two minutes for investigation of mesospheric dynamics, especially for daytime detection of the periods of gravity waves in the mesosphere and the lower thermosphere. The observations were analysed by usual power spectrum technique. The prominent periods, observed for the present, vary between 5 to 27 minutes in general agreement with results obtained by other techniques.

082.099 **The global distribution and variability of stratospheric constituents measured by LIMS.**
J. M. Russell III.
Adv. Space Res., Vol. 4, No. 4, p. 107 – 116 (1984). – See Abstr. 012.087.
The purpose of the Nimbus 7 LIMS experiment was to sound the composition and structure of the upper atmosphere and provide data for study of photochemistry, radiation, and dynamics processes. Vertical profiles were measured of temperature and ozone over the 10–km to 65–km range and water vapor, nitrogen dioxide, and nitric acid over the 10–km to ~50–km range. Latitude coverage extended from 64°S to 84°N. Several general features of the atmosphere have emerged from data analyses thus far.

082.100 **Derivation of OH concentrations from LIMS measurements.**
J. A. Pyle, A. M. Závody, J. E. Harries, P. H. Moffat.
Adv. Space Res., Vol. 4, No. 4, p. 117 – 120 (1984). – See Abstr. 012.087.

082.101 **Satellite measurements of atmospheric composition: three years' observations of $CH_4$ and $N_2O$.**
R. L. Jones.
Adv. Space Res., Vol. 4, No. 4, p. 121 – 130 (1984). – See Abstr. 012.087.

082.102 **Ring effect and sodium dayglow.**
M. R. Tade, A. D. Tillu.
Adv. Space Res., Vol. 4, No. 4, p. 139 – 142 (1984). – See Abstr. 012.087.
A coordinated programme of simultaneous observations of sodium dayglow and Fraunhofer Filling In (FFI) was organised for obtaining diurnal variation of sodium dayglow, free from effects of FFI or 'Ring effect'. Radiative transfer calculations for the estimation of sodium intensity were extended to include the combined effects of ground reflection and FFI. The two–fold approach shows that the Ring effect could account for 30% change in the observed diurnal variation of sodium dayglow. The residual sodium intensity variation, still indicates a small variation of sodium abundance.

082.103 **Effects of collisional narrowing and line parameters on atmospheric composition measurements.**
Y.–j. Wang.
J. Quant. Spectrosc. Radiat. Transfer, Vol. 32, No. 4, p. 305 – 309 (1984).
The effects of line shape and of line width and its temperature dependence on retrieved stratospheric composition from ground–based, high–resolution solar spectra measurements are discussed. The results show that they depend on the method of measurement. For differential transmission measurements, large effects of line–shape and other line parameters are found. Even for transmission measurements, line parameters can also lead to significant error.

082.104 **EUV airglow during active solar conditions. 2. Emission between 530 and 930 Å.**
E. P. Gentieu, P. D. Feldman, R. W. Eastes, A. B. Christensen.
J. Geophys. Res., Vol. 89, No. A12, p. 11053 – 11058 (1984).
Daytime airglow spectra between 530 and 930 Å were obtained at ~4 Å resolution from a rocket launch at White Sands, New Mexico, June 27, 1980. Portions of the spectrum were observed in second order at ~2 Å resolution. The higher resolution of the present data confirms the authors' previous identification of O II transitions and resolves the identification of O II emission at 537 – 539 Å as due to both a doublet and a quartet component.

082.105 **Characteristics of the spacecraft optical glow.**
J.–H. Yee, V. J. Abreu, A. Dalgarno.
Geophys. Res. Lett., Vol. 11, No. 12, p. 1192 – 1194 (1984).
The optical glow induced by spacecraft–environment interaction under daytime conditions is analyzed using photometric data obtained by the Visible Airglow Experiment on board the Atmosphere Explorer Satellite.

082.106 **Global dynamic models of the Earth's thermosphere and ionosphere.**
R. G. Roble.
ESA J., Vol. 7, No. 4, p. 405 – 429 (1983). – See Abstr. 012.035.
There has been steady progress over the years toward the development of global dynamic models of the Earth's thermosphere and ionosphere. These models have been used primarily to investigate high–latitude dynamic phenomena associated with magnetospheric convection and its interaction with the ionosphere and thermosphere. The results of these studies are reviewed.

**UK National Report for the Quadrennium 1979 – 1983 to the International Association of Geomagnetism and Aeronomy (*Bibliography*).**
See Abstr. 002.021.

**CRC handbook of high resolution infrared laboratory spectra of atmospheric interest.**
See Abstr. 002.081.

**Die farbigen Dämmerungserscheinungen.**
See Abstr. 003.017.

**Atmospheric electrodynamics.**
See Abstr. 003.018.

**Rare halos, mirages, anomalous rainbows and related electro–magnetic phenomena: a catalog of geophysical anomalies.**
See Abstr. 003.074.

**Artificial particle beams in space plasma studies.**
See Abstr. 003.093.

**Remote sounding of atmospheres.**
See Abstr. 003.105.

**Numerical solution of problems of atmospheric optics.**
See Abstr. 003.125.

**Advanced technology optical telescopes II. Conference held at London, England, 5 – 6 September 1983.**
See Abstr. 012.097.

**Report on the research in geomagnetism and aeronomy in Poland for the period 1978 – 1982.**
See Abstr. 013.002.

**A computer program to calculate first order extinction coefficients.**
See Abstr. 021.007.

**Oxygen absorption cross sections in the Herzberg continuum and between 206 and 327K.**
See Abstr. 022.139.

**New in–situ measurements of the absorption cross–sections of $O_2$ in the Herzberg continuum.**
See Abstr. 022.160.

**Investigating atmospheric distortions of a wave front by the Hartmann method.**
See Abstr. 031.001.

**Modal compensation of turbulent atmospheric distortions and isoplanatism.**
See Abstr. 031.012.

**Adaptive optics.**
See Abstr. 031.029.

**Adaptive optics.**
See Abstr. 031.062.

**Options for a new Arizona observatory: configurations, costs and site.**
See Abstr. 032.058.

**Options for a new Arizona Observatory: configurations, costs and site.**
See Abstr. 032.067.

**An automatic electronic device for studying anomalous refraction.**
See Abstr. 034.018.

**A system for detecting anomalous room refraction.**
See Abstr. 034.053.

**Spacelab I. Expérience ondes dans la couche émissive OH.**
See Abstr. 035.001.

**Spaceborne lasers in atmospheric research.**
See Abstr. 035.011.

**Précision astrométrique en présence de bruit.**
See Abstr. 036.023.

**On the importance of active image stabilisation.**
See Abstr. 036.090.

**Correction of atmospheric transparency gradient effect on solar radial velocities.**
See Abstr. 036.141.

**The accuracy limit of ground–based stellar photometry.**
See Abstr. 036.145.

**Ground–based solar and stellar intensity measurements: practical limits of accuracy.**
See Abstr. 036.146.

**Precise orbit determination of the satellite "STARLETTE" with laser ranging data.**
See Abstr. 052.035.

**Neutrino oscillations and the atmospheric neutrino fluxes.**
See Abstr. 061.072.

**Unified theory of whistler mode instability in the presence of a parallel electric field.**
See Abstr. 062.033.

**Limb darkening fluctuations: Is the five–minute oscillation also terrestrial?**
See Abstr. 080.030.

**Modelling the global climate response to orbital forcing and atmospheric carbon dioxide changes.**
See Abstr. 081.011.

**On the causes of ice–ages.**
See Abstr. 081.023.

**Possibilities for mutual comparison of the IRI and the neutral atmosphere models by optical measurements.**
See Abstr. 083.022.

**Excitation and ionizatioan in the upper atmosphere during solar flares with account for photoelectrons.**
See Abstr. 085.016.

**Influence of outer factors on the vertical structure of planetary waves in the troposphere and stratosphere.**
See Abstr. 085.019.

**The photoelectron spectrum in the upper atmosphere of the earth during solar flares.**
See Abstr. 085.026.

**Rotation of the atmospheres of the Earth and planets.**
See Abstr. 091.051.

**Production of high–energy gamma rays by cosmic ray interactions in the atmosphere and lunar surface.**
See Abstr. 144.077.

# 083 Ionosphere

083.001 **De veranderlijke ionosfeer.**
H. Kelder, T. A. T. Spoelstra.
Zenit, 11. Jaarg., No. 9, p. 365 – 369 (1984).

083.002 **An investigation of the consistency of the ionospheric measurements of the photoelectron flux and solar EUV flux.**
P. G. Richards, D. G. Torr.
J. Geophys. Res., Vol. 89, No. A7, p. 5625 – 5635 (1984).
In this paper the authors demonstrate that there is an inconsistency between ionospheric measurements of the photoelectron flux and the solar EUV flux. The problem which occurs for EUV wavelengths near 250 Å or photoelectron energies near 35 eV has serious implications for theoretical photoelectron flux calculations and could help to explain up to 40% of the difference between calculated and measured photoelectron fluxes for electron energies below 35 eV.

083.003 **Electric fields and potential patterns in the high–latitude ionosphere for different situations in interplanetary space.**
Y. I. Feldstein (*Ya. I. Fel'dshtejn*), A. E. Levitin,
D. S. Faermark, R. G. Afonina, B. A. Belov, V. Y. Gaidukov (*V. Yu. Gajdukov*).
Planet. Space Sci., Vol. 32, No. 7, p. 907 – 923 (1984).

083.004 **Ionospheric currents and fields during the solar eclipse.**
M. Takeda, T. Araki.
Planet. Space Sci., Vol. 32, No. 8, p. 1013 – 1019 (1984).
The authors calculate electric fields and currents in the ionosphere during a solar eclipse. Three dimensional anisotropic and inhomogeneous conductivity distribution is taken into consideration. Electrical conductivity is assumed to be decreased by the solar eclipse, but winds are not altered. It is shown that the additional currents and fields due to the eclipse are essentially understood as those generated by the obstruction to *Sq* currents. However, the currents flow preferably in the region nearer to the noon meridian and at the equator and partly escape through the field–aligned currents to the opposite hemisphere, and therefore the effect of the solar eclipse is also seen in the opposite hemisphere in the deformed shape.

083.005 **New description of the electron density profile.**
K. Rawer.
Adv. Space Res., Vol. 4, No. 1, p. 11 – 15 (1984). – See Abstr. 012.003.
Methods are described by which the desired analytical representation of the whole profile might be reached while enforcing the most important observed physical features. An outline of future work in this connection is given.

083.006 **Geometry of the "exponential" middle ionosphere.**
T. L. Gulyaeva.
Adv. Space Res., Vol. 4, No. 1, p. 17 – 23 (1984). – See Abstr. 012.003.
The required numbers of exponential sections fitting an electron density profile between the E– and F–layer peak heights of the ionosphere are considered. The range of the scale height variation and the angles of slope of the F region sub–peak profile are obtained under the different solar–geophysical conditions.

083.007 **Prediction of total electron content using the International Reference Ionosphere.**
L. F. McNamara.
Adv. Space Res., Vol. 4, No. 1, p. 25 – 50 (1984). – See Abstr. 012.003.
The International Reference Ionosphere (IRI) is a model of the ionosphere, based on experimental data, which has been proposed as a standard ionospheric model. One of the ways in which the electron density profile given by the IRI, especially above the peak of the F layer, can be tested is to compare calculated and observed values of total electron content (TEC). The author has therefore studied the discrepancies between calculated and observed values of TEC recorded at 15 stations covering a wide range of longitudes and latitudes, mainly in the northern hemisphere, and mainly for high levels of solar activity.

083.008 **Model representation of mid–latitudinal electron density by means of "Interkosmos–19" data.**
N. P. Benkova (*N. P. Ben'kova*), N. A. Kochenova,
A. D. Legenka (*A. D. Legen'ka*), M. N. Fatkullin, M. D. Fligel (*M. D. Fligel'*).
Adv. Space Res., Vol. 4, No. 1, p. 51 – 58 (1984). – See Abstr. 012.003.

083.009 **N(h) profiles from ground and "Interkosmos–19" satellite data.**
T. Pashova.
Adv. Space Res., Vol. 4, No. 1, p. 59 (1984). Abstract. – See Abstr. 012.003.

083.010 **The state of the art in D–region IRI modelling.**
K. Rawer, Y. V. Ramanamurty.
Adv. Space Res., Vol. 4, No. 1, p. 71 – 78 (1984). – See Abstr. 012.003.
While describing the formulation of IRI–79, in so far as it is applicable to the sub E–peak electron density profile, an attempt is made to compare its predictions with the available experimental evidence. The improvements needed for D–region electron density modelling are summarized. The current D–region IRI modelling effort is illustrated with a typical example.

083.011 **Improvement of the solar–cycle variation of IRI lower ionosphere models by means of radio wave propagation data.**
W. Singer, J. Bremer, J. Taubenheim.
Adv. Space Res., Vol. 4, No. 1, p. 79 – 85 (1984). – See Abstr. 012.003.
Data on day–time and night–time radio wave absorption in the frequency range 50 to 2614 kHz, obtained in long–term observational programmes in Central Europe, are compared with corresponding absorption values calculated from electron density profiles of the International Reference Ionosphere (IRI–1979) using the full–wave method. Discrepancies between calculated and observed absorption values were found for the diurnal and the solar–cycle variation. A modification of the solar–activity dependence of the D–region electron density parameters is derived, which provides an improvement of the solar–cycle variation as well as the diurnal variation of the IRI electron density profiles.

083.012 **Solar cycle effects upon the relationship of $N_e$ and $T_e$ in the F–region.**
L. H. Brace, R. F. Theis.
Adv. Space Res., Vol. 4, No. 1, p. 89 – 91 (1984). – See Abstr. 012.003.
Employing Atmosphere Explorer–C measurements made in 1974, just prior to solar minimum, Brace and Theis (1978) demonstrated that a remarkably consistent inverse relationship existed between the electron density $N_e$ and temperature $T_e$ in the F–region. In this paper the authors use later data from AE–C, taken when solar activity was rising (1975 – 1978), and Dynamics Explorer–2 data taken at solar maximum (1981), to examine how the temperature and density relationship changes with solar activity.

083.013 **Improved analytical representation of electron temperature in the IRI.**
D. Bilitza.
Adv. Space Res., Vol. 4, No. 1, p. 93 – 95 (1984). – See Abstr. 012.003.
In the last decade extensive measurements from incoherent scatter stations and from several satellite missions have consider-

ably improved the knowledge of long- and short-term variations in the ionospheric electron temperature. Comparisons with IRI have revealed some shortcomings of this earlier model. It is obvious that in different altitude regions one has to concentrate the modelling efforts on different parameters. A model representation is proposed that will facilitate approaches (for the different altitude regions) in one analytical form.

083.014 **Comparison of various empirical models of electron temperature with experimental measurements during low solar activity.**
V. K. Pandey, K. K. Mahajan.
Adv. Space Res., Vol. 4, No. 1, p. 97 – 99 (1984). – See Abstr. 012.003.

Comparisons of various available empirical models of electron temperature are made with actual measurements from incoherent scatter radar and rocket and satellite probes, during low solar activity period. The models compared are those of Pandey et al. (1983), Brace and Theis (1978), IRI (1979) and Bilitza (1983). It is found that the authors' model and the Brace and Theis model are closer to actual measurements than the IRI and Bilitza models.

083.015 **Note on a discussion about the ion composition.**
C. R. Philbrick, K. Rawer.
Adv. Space Res., Vol. 4, No. 1, p. 103 – 105 (1984). – See Abstr. 012.003.

083.016 **Comparison between the IRI ion composition and incoherent scatter measurement and theoretical values.**
D. Bilitza.
Adv. Space Res., Vol. 4, No. 1, p. 107 – 109 (1984). – See Abstr. 012.003.

083.017 **Ion composition behaviour in low and mid-latitudes during high solar activity.**
M. K. Goel, B. C. N. Rao.
Adv. Space Res., Vol. 4, No. 1, p. 111 – 117 (1984). – See Abstr. 012.003.

In the present analysis, the mass spectrometer data from the ISS-b satellite, available in the form of contour plots at an average height of 1100 km for every alternate hour, is used. This analysis showed some interesting results in terms of the diurnal variation of the transition heights; at times dominance of $He^+$ over the other ions, and the seasonal variations of different constituents. In the development of ion composition models, it is suggested that this type of result should be taken into consideration.

083.018 **An approximation of the height of the $O^+ - H^+$ transition level for use in IRI.**
I. Kutiev, P. Marinov, K. B. Serafimov.
Adv. Space Res., Vol. 4, No. 1, p. 119 – 121 (1984). – See Abstr. 012.003.

A set of data including over 1400 transition level (TL) encounters of the OGO-6 satellite is used to construct analytical presentations of this altitude upon dip-latitude and longitude. The formula represents the TL variation between $\pm 50°$ dip-latitude and 20 h and 04 h local time intervals. The data were obtained in the high solar activity period in 1969 and 1970.

083.019 **Hybrid quadri-ionic model of the lower ionosphere.**
P. Velinov, N. A. Smirnova, V. A. Vlascov (*V. A. Vlaskov*).
Adv. Space Res., Vol. 4, No. 1, p. 123 – 130 (1984). – See Abstr. 012.003.

An ion model of the lower ionosphere is proposed. It consists of four positive ions: $O_2{}^+$, $NO^+$ and two cluster ions – a simpler $CI_1$ and a more complex $CI_2$. This model well explains the normal component of the winter anomaly in the D-region, which is recorded by absorption measurements on short radiowaves and rocket experiments at middle (40°N) and high (70°) latitudes.

083.020 **Some measurement possibilities for the improvement of IRI.**
K. B. Serafimov.
Adv. Space Res., Vol. 4, No. 1, p. 133 – 138 (1984). – See Abstr. 012.003.

The purpose of the paper is to demonstrate, very briefly, some methodological concepts for the development of various measurements, or for the use of the available ones through which it will be possible to obtain additional data for the systematic development of IRI.

083.021 **The improvement of IRI profiles of $O_2{}^+$, $NO^+$ and $O^+$ by means of airglow measurements.**
K. B. Serafimov.
Adv. Space Res., Vol. 4, No. 1, p. 139 – 142 (1984). – See Abstr. 012.003.

The author shows the possibilities for the determination of the vertical ion distribution of $[O_2{}^+](h)$, $[NO^+](h)$ and $[O^+](h)$ through combined satellite airglow and plasma measurements.

083.022 **Possibilities for mutual comparison of the IRI and the neutral atmosphere models by optical measurements.**
K. B. Serafimov, T. Gogosheva, M. Gogoshev.
Adv. Space Res., Vol. 4, No. 1, p. 143 – 147 (1984). – See Abstr. 012.003.

083.023 **A plan for compilation of experimental data on drifts for IRI.**
E. S. Kazimirovsky (*Eh. S. Kazimirovskij*), E. I. Zhovty.
Adv. Space Res., Vol. 4, No. 1, p. 149 – 151 (1984). – See Abstr. 012.003.

Methods for constructing a dynamical part of the IRI model are described.

083.024 **Highlights of the URSI/COSPAR workshop on IRI.**
Y. V. Ramanamurty.
Adv. Space Res., Vol. 4, No. 1, p. 153 – 163 (1984). – See Abstr. 012.003.

083.025 **Detection of radiation from a heated and modulated equatorial electrojet current system.**
R. J. Lunnen, H. S. Lee, A. J. Ferraro, T. W. Collins, R. F. Woodman.
Nature, Vol. 311, No. 5982, p. 134 – 135 (1984).

In May 1983, ionospheric heating experiments were conducted using the Jicamarca very high frequency radar facility located ~11 km east of Lima, Peru. These experiments resulted in the detection of radiation from a heated and modulated equatorial electrojet current system in the Western Hemisphere.

083.026 **Observation of interplanetary magnetic field and of ionospheric plasma convection in the vicinity of the dayside polar cleft.**
C. R. Clauer, P. M. Banks, A. Q. Smith, T. S. Jorgensen, E. Friis-Christensen, S. Vennerstrom, V. B. Wickwar, J. D. Kelly, J. Doupnik.
Geophys. Res. Lett., Vol. 11, No. 9, p. 891 – 894 (1984).

Dayside ionospheric convection at high latitudes has been examined during a series of experiments using the Sondrestrom radar together with ancillary observations of the interplanetary magnetic field by the IMP-8 spacecraft. The radar experiments obtained a latitude coverage of 67.6°Λ to 81.3°Λ and a temporal resolution of between 14 to 25 minutes. A total of 17 rotations through the dayside cleft region during April, June and July, 1983 have been examined.

083.027 **Quasi steady state multiple-plasma-cloud configurations in the *F* region ionosphere.**
J. Chen, P. Satyanarayana, S. L. Ossakow.
J. Geophys. Res., Vol. 89, No. A8, p. 6787 – 6792 (1984).

The morphology of a finite array of plasma density enhancements (multiple plasma clouds) in the ionosphere is studied in order to determine the condition under which the system undergoes a quasi steady state $\mathbf{E} \times \mathbf{B}$ drift.

083.028 **Change of the parameters of the F1 layer in the solar activity cycle.**
G. P. Kushnarenko.
Issled. Geomagn., Aehron. Fiz. Solntsa, Moskva, No. 67, p. 114 – 118 (1984). In Russian. Abstr. in Ref. Zh., 62. Issled. Kosm. Prostranstva, 8.62.598 (1984).

083.029 **Golven in de ionosfeer.**
H. Kelder, T. A. T. Spoelstra.
Zenit, 11. Jaarg., No. 10, p. 410 – 414 (1984).

083.030 **Errors in theoretical calculations of the composition of the E and F1 regions of the day–time ionosphere due to uncertainties of the solar ultraviolet radiation stream.**
A. V. Pavlov.
Geomagn. Aehron., Tom 24, No. 4, p. 577 – 580 (1984). In Russian. English translation in Geomagn. Aeron.

083.031 **Evolution of strong irregularities of the ionospheric plasma. II.**
V. A. Rozhanskij, L. D. Tsendin.
Geomagn. Aehron., Tom 24, No. 4, p. 598 – 602 (1984). In Russian. English translation in Geomagn. Aeron.

083.032 **Nitric oxide in the D region. I. Experimental data on the [NO] distribution.**
A. D. Danilov, S. Yu. Ledomskaya.
Geomagn. Aehron., Tom 24, No. 4, p. 614 – 619 (1984). In Russian. English translation in Geomagn. Aeron.

083.033 **Band of the fast drift of ions in the sub–auroral ionospheric F–region and its manifestation in the structure of the high–latitude ionosphere.**
V. M. Filippov, L. V. Shestakova, Yu. I. Gal'perin.
Kosm. Issled., Tom 22, Vyp. 4, p. 557 – 564 (1984). In Russian. English translation in Cosm. Res.

083.034 **Ionospheric sporadic–E parameters: long–term trends.**
W. J. Baggaley.
Science, Vol. 255, No. 4664, p. 830 – 833 (1984).
The results of observations over 3 1/2 solar cycles of ionosonde parameters describing sporadic–E patches ($E_s$) are given for two Southern Hemisphere stations. Analysis has revealed systematic changes in the occurrence probabilities of $f_0E_s$ and $f_bE_s$, changes that are independent of time of day or season. It is unlikely that observational effects can be entirely responsible for the trends, which suggest that major long–term changes may be occurring at E region heights.

083.035 **Rocket borne studies of electron density irregularities in equatorial D and E regions.**
S. Prakash, R. Pandey.
Proc. Indian Acad. Sci., Earth Planet. Sci., Vol. 93, No. 3, p. 283 – 308 (1984).
Electron density irregularities in the equatorial D and E regions have been studied using ground–based and rocket–borne experiments. In this paper this subject is reviewed with an emphasis on the rocket–borne measurements of electron density and electric field structures and other parameters relevant for identifying processes responsible for generation of electron density irregularities. These studies have led to the identification of various causative mechanisms, in terms of plasma and neutral dynamical processes for producing irregularities. Based on these mechanisms, the irregularities are classified and discussed.

083.036 **High latitude ionospheric structure.**
S. L. Ossakow, W. Burke, H. C. Carlson, P. Gary, R. Heelis, M. Keskinen, N. Maynard, C. Meng, E. Szuszczewicz, J. Vickrey.
NASA Ref. Publ., NASA RP–1120, p. 12–1 – 12–39 (1984). – See Abstr. 012.038.
Contents: 1. Introduction: Ionospheric structure in general. Equatorial spread–F irregularities – a success story and a guide for high latitudes. Focus on high–latitude structure. 2. Sources and observations of high–latitude structure: Electron precipitation structures. Electric fields. Field–aligned currents. Plasma density structure. 3. Plasma instability theory: Macroinstabilities and high latitude structure. Microinstabilities and high latitude structure. 4. An emerging picture. 5. Future studies: Theoretical thrusts. Experimental emphasis.

083.037 **Statistical connection between ionospheric and geomagnetic disturbances.**
G. K. Solodovnikov, V. M. Russkin.
Geomagn. Aehron., Tom 24, No. 5, p. 830 – 833 (1984). In Russian. English translation in Geomagn. Aeron.

083.038 **On the interpretation of data obtained by vertical sounding of the ionosphere.**
G. I. Gordienko.
Geomagn. Aehron., Tom 24, No. 5, p. 833 – 835 (1984). In Russian. English translation in Geomagn. Aeron.

083.039 **Formation of the plasmapause and its connection with the structure of the polar ionosphere.**
N. K. Osipov, N. M. Maksimova.
Geomagn. Aehron., Tom 24, No. 5, p. 835 – 838 (1984). In Russian. English translation in Geomagn. Aeron.

083.040 **Excitation of electric fields by longitudinal currents conditioned by the northern component of the interplanetary magnetic field.**
V. M. Uvarov.
Geomagn. Aehron., Tom 24, No. 5, p. 840 – 842 (1984). In Russian. English translation in Geomagn. Aeron.

083.041 **A possible effect of magnetospheric convection on ionospheric conductivities.**
K. Schlegel.
Ann. Geophys., Vol. 2, No. 2, p. 207 – 210 (1984). Abstr. in Phys. Abstr., Vol. 87, No. 1243, Entry 100847 (1984).

083.042 **Duration of equatorial spread–*F*.**
J. H. Sastri.
Ann. Geophys., Vol. 2, No. 3, p. 353 – 358 (1984). Abstr. in Phys. Abstr., Vol. 87, No. 1243, Entry 100848 (1984).

083.043 **An improved empirical model for electron temperature in the F2–region.**
V. K. Pandey, K. K. Mahajan, V. C. Jain.
Indian J. Radio Space Phys., Vol. 13, No. 2, p. 42 – 47 (1984). Abstr. in Phys. Abstr., Vol. 87, No. 1244, Entry 105795 (1984).

083.044 **A semi–empirical model of ionospheric electron content at low latitudes for solar minimum.**
P. K. Bhuyan, Tuhi Ram Tyagi.
Indian J. Radio Space Phys., Vol. 13, No. 2, p. 63 – 68 (1984). Abstr. in Phys. Abstr., Vol. 87, No. 1244, Entry 105796 (1984).

083.045 **Quantitative explanation of some electron temperature profiles measured in situ in the high latitude ionospheric *E*–region.**
K. Schlegel, K.–I. Oyama, K. Hirao.
J. Geomagn. Geoelectr., Vol. 35, No. 10, p. 357 – 366 (1983). Abstr. in Phys. Abstr., Vol. 87, No. 1244, Entry 105800 (1984).

083.046 **Local time dependence of the response of the equatorial electrojet to DP2 and SI disturbances.**
I. Papamastorakis, G. Haerendel, W. Baumjohann.
J. Geophys., Vol. 54, No. 3, p. 213 – 218 (1984). Abstr. in Phys. Abstr., Vol. 87, No. 1244, Entry 105802 (1984).

083.047 **On the recurrence of sudden ionospheric disturbances.**
R. N. Shelke, M. C. Pande.
Bull. Astron. Soc. India, Vol. 12, No. 3, p. 223 – 228 (1984).
In this paper, an attempt is made to investigate the recurrence tendency of sudden ionospheric disturbances (SID) and, in turn, of the energetic solar flares through power spectrum analysis of SID time series. Such analysis may throw some light on the flare build–up mechanism and relaxation times for the flares on the

build-up mechanism and relaxation times for the flares on the solar disc.

083.048 **Variations of the parameters of the ionospheric D-region during the 27-daily solar cycle.**
S. V. Pakhomov.
Tr. Tsentr. aehrol. obs., No. 152, p. 100 - 106 (1984). In Russian. Abstr. in Ref. Zh., 51. Astron., 12.51.688 (1984).

083.049 **Distribution of the electron concentration above the geomagnetic equator in the day-time F2 region of the ionosphere.**
T. Yu. Leshchinskaya, A. V. Mikhajlov.
Geomagn. Aehron., Tom 24, No. 6, p. 894 - 901 (1984). In Russian. English translation in Geomagn. Aeron.

083.050 **Formation of small-scale irregularities in the high-latitude upper ionosphere.**
M. G. Gel'berg.
Geomagn. Aehron., Tom 24, No. 6, p. 911 - 915 (1984). In Russian. English translation in Geomagn. Aeron.

083.051 **Diffusion of strong irregularities of an ionospheric plasma.**
V. A. Rozhanskij, L. D. Tsendin.
Geomagn. Aehron., Tom 24, No. 6, p. 916 - 922 (1984). In Russian. English translation in Geomagn. Aeron.

083.052 **Relative contribution of X-ray and ultraviolet radiation to ionization of the near-equatorial E-region.**
A. Kh. Depueva, G. S. Ivanov-Kholodnyj, A. A. Nusinov, I. N. Odintsova.
Geomagn. Aehron., Tom 24, No. 6, p. 1006 - 1008 (1984). In Russian. English translation in Geomagn. Aeron.

083.053 **On a model of planetary distribution of the integral electron content of the ionosphere.**
I. G. Zakharov, O. F. Tyrnov.
Geomagn. Aehron., Tom 24, No. 6, p. 1009 - 1011 (1984). In Russian. English translation in Geomagn. Aeron.

083.054 **Daily variation of conditions of Alfvén waves passing in the ionosphere.**
G. V. Givishvili, L. A. Lobachevskij, M. V. Fiskina, A. B. Shvartsburg.
Geomagn. Aehron., Tom 24, No. 6, p. 1011 - 1013 (1984). In Russian. English translation in Geomagn. Aeron.

**Atmospheric electrodynamics.**
See Abstr. 003.018.

**Solar terrestrial physics: present and future. A report based on the Solar Terrestrial Physics Workshop, December 1982 to November 1983.**
See Abstr. 012.038.

**Effects produced by the ionosphere on radio interferometry.**
See Abstr. 036.016.

**Longitudinal distribution of the cool solar surges.**
See Abstr. 073.070.

**Spectroscopy of the EUV (350 - 1400 Å) nightglow.**
See Abstr. 082.020.

**Assessment of plasma transport and convection at high latitudes.**
See Abstr. 082.048.

**Modelling of neutral and ionized atmospheres.**
See Abstr. 082.078.

**Interdependence of the emissions of the upper atmosphere and ionospheric characteristics.**
See Abstr. 082.085.

**Global dynamic models of the Earth's thermosphere and ionosphere.**
See Abstr. 082.106.

**Effect of parallel electric field and geomagnetic field distortion on precipitation of charged particles into the lower ionosphere.**
See Abstr. 084.028.

**Connection between the magnetosphere and ionosphere.**
See Abstr. 084.030.

# 084 Aurorae, Geomagnetic Field, Magnetosphere

084.001 **A physical model for palaeosecular variation.**
P. L. McFadden, M. W. McElhinny.
Geophys. J. R. Astron. Soc., Vol. 78, No. 3, p. 809 - 830 (1984).

A new model to describe the latitude dependence of the angular dispersion of the palaeomagnetic field (palaeosecular variation) is developed.

084.002 **Associations of geomagnetic activity with plasma sheet thinning and expansion: a statistical study.**
E. W. Hones Jr., T. Pytte, H. I. West Jr.
J. Geophys. Res., Vol. 89, No. A7, p. 5471 - 5478 (1984).

Associations of geomagnetic activity in the auroral zone with thinnings and expansions of the magnetotail plasma sheet are examined statistically in this paper.

084.003 **Energetic oxygen in a mid-latitude aurora.**
M. R. Torr, D. G. Torr.
J. Geophys. Res., Vol. 89, No. A7, p. 5547 - 5553 (1984).

On the night of September 21/22, 1982, ground-based observations were made of an aurora which extended as far equatorward as Logan, Utah (41°N, 111°W). An imaging spectrometer was used to obtain spectral images over the wavelength range 3300 - 5300 Å at a spectral resolution of 2.7 Å. This segment of the spectrum was found to be rich in atomic and molecular features. The vibrational and rotational structures of the molecular nitrogen bands provide extremely valuable information on the velocity of the precipitating particles. From a comparison of the $N_2^+$ vibrational and rotational structures, which are strongly populated at the higher levels, and similar data for the $N_2$ bands the authors conclude that the events indicate strong precipitation of ring current energetic heavy atoms. Because of the absence of observable $H_\beta$ emission and because of the characteristic energy inferred from the velocity the energetic atoms are probably oxygen.

084.004 **Quantitative magnetospheric magnetic field modelling with toroidal and poloidal vector fields.**
J. C. Kosik.
Planet. Space Sci., Vol. 32, No. 8, p. 965 - 974 (1984).

Toroidal and poloidal vector fields allow divergence free magnetic field representations in regions where currents flow. The author derives general magnetospheric magnetic fields using combinations of spherical harmonic expansions of the toroidal and poloidal fields. Adding restrictive conditions like the field line topology symmetry or the magnetic field measurements, more specific magnetospheric magnetic field models can be derived. Two examples of this technique are given: an axisymmetric model with a ring current in the equatorial region and a time-dependent model of the Earth's magnetosphere. The results are compared with the Olson-Pfitzer model.

084.005 **Ring current simulation in connection with interplanetary space conditions.**
Y. I. Feldstein (*Ya. I. Fel'dshtejn*), V. Yu. Pisarsky (*V. Yu. Pisarskij*), N. M. Rudneva, A. Grafe.
Planet. Space Sci., Vol. 32, No. 8, p. 975 – 984 (1984).
A ring current model has been obtained which permits calculations of $D_s^t$ variations on the Earth's surface during magnetic storms.

084.006 **Dawn-dusk (*y*) component of the interplanetary magnetic field and the local time of the Harang discontinuity.**
A. S. Rodger, S. W. H. Cowley, M. J. Brown, M. Pinnock, D. A. Simmons.
Planet. Space Sci., Vol. 32, No. 8, p. 1021 – 1027 (1984).
The authors describe a simple method for determining the time at which the meridian of a sub-auroral magnetic observatory crosses that of the Harang discontinuity – the separation of the eastward and westward electrojets which flow in the evening and morning sectors of the auroral oval. They then consider how this time, determined from examination of magnetograms from sub-auroral observatories, varies with the dawn-dusk (*y*) component of the interplanetary magnetic field.

084.007 **The nightside magnetosphere.**
S. W. H. Cowley.
Nature, Vol. 310, No. 5978, p. 543 (1984).
This note discusses some of the first results obtained by the ISEE-3 spacecraft during its passage through the nightside Earth magnetosphere.

084.008 **A comment on proposed mechanisms for the excitation of $O(^1S)$ in the aurora.**
I. C. McDade, E. J. Llewellyn.
Planet. Space Sci., Vol. 32, No. 9, p. 1195 – 1198 (1984). See Abstr. 37.084.014.

084.009 **Reply to I. C. McDade and E. J. Llewellyn's comments on proposed mechanisms for the excitation of $O(^1S)$ in the aurora.**
M. H. Rees.
Planet. Space Sci., Vol. 32, No. 9, p. 1199 – 1200 (1984). See abstracts 37.084.014, 084.008.

084.010 **Flares, 200 MHz radio bursts and magnetic storms.**
Y. Guo.
Acta Sci. Nat. Univ. Pekin., No. 1, p. 68 – 73 (1984). In Chinese. Abstr. in Phys. Abstr., Vol. 87, No. 1238, Entry 77684 (1984).

084.011 **Auroral kilometric radiation/aurora correlation.**
R. F. Benson, S.-I. Akasofu.
Radio Sci., Vol. 19, No. 2, p. 527 – 541 (1984). Abstr. in Phys. Abstr., Vol. 87, No. 1241, Entry 90039 (1984).

084.012 **Ordinary mode auroral kilometric radiation, with harmonics observed by ISIS 1.**
R. F. Benson.
Radio Sci., Vol. 19, No. 2, p. 543 – 550 (1984). Abstr. in Phys. Abstr., Vol. 87, No. 1241, Entry 90040 (1984).

084.013 **Secular variation of the earth's magnetic field.**
A. M. Karasik.
Priroda, No. 7, p. 116 – 117 (1984). In Russian.

084.014 **Geomagnetic and solar data. February 1984.**
H. E. Coffey.
J. Geophys. Res., Vol. 89, No. A7, p. 5666 (1984).

084.015 **The evanescence of striation parameters along the geomagnetic field.**
J. L. Sperling.
J. Geophys. Res., Vol. 89, No. A8, p. 6793 – 6803 (1984).

084.016 **Geomagnetic and solar data. March, April 1984.**
H. E. Coffey.
J. Geophys. Res., Vol. 89, No. A8, p. 6853 – 6854 (1984).

084.017 **Geomagnetic and solar data. May 1984.**
H. E. Coffey.
J. Geophys. Res., Vol. 89, No. A9, p. 7593 (1984).

084.018 **Aurora seen in daylight.**
A. Johnstone.
Nature, Vol. 311, No. 5985, p. 413 (1984).

084.019 **Using the magnetosphere as an analyser of the form of spectra of radiation belt particles.**
A. S. Kovtyukh.
Geomagn. Aehron., Tom 24, No. 4, p. 566 – 570 (1984). In Russian. English translation in Geomagn. Aeron.

084.020 **Observations of aurorae from board the scientific orbital station Salyut 6 during the strong geomagnetic storm of April 11 – 13, 1981.**
A. I. Lazarev, V. V. Kovalenok, L. S. Evlashin.
Geomagn. Aehron., Tom 24, No. 4, p. 620 – 624 (1984). In Russian. English translation in Geomagn. Aeron.

084.021 **On the magnetosphere's energetics.**
A. Prigantsova.
Geomagn. Aehron., Tom 24, No. 4, p. 644 – 649 (1984). In Russian. English translation in Geomagn. Aeron.

084.022 **Propagation of hydromagnetic waves through an ionospheric plasma and space characteristics of geomagnetic variations.**
L. S. Al'perovich, E. N. Fedorov.
Geomagn. Aehron., Tom 24, No. 4, p. 650 – 657 (1984). In Russian. English translation in Geomagn. Aeron.

084.023 **Response in latitude distributions of auroral phenomena to variations of magnetic activity.**
N. S. Zaretskij, M. G. Gusev.
Geomagn. Aehron., Tom 24, No. 4, p. 674 – 676 (1984). In Russian. English translation in Geomagn. Aeron.

084.024 **The Earth's magnetic field.**
D. Gubbins.
Contemp. Phys., Vol. 25, No. 3, p. 269 – 290 (1984). Abstr. in Phys. Abstr., Vol. 87, No. 1242, Entry 93997 (1984).

084.025 **Polar aurorae – from myths to satellites.**
Ya. I. Fel'dshtejn.
Zemlya Vselennaya, No. 4, p. 27 – 33 (1984). In Russian.

084.026 **Invariant and changing transitional field configurations in a sequence of geomagnetic reversals.**
J.-P. Valet, C. Laj.
Nature, Vol. 311, No. 5986, p. 552 – 555 (1984).
The authors report the results of a palaeomagnetic study of four different geomagnetic reversals, three of which are successive. This sequence represents $\sim$1.5 Myr of geomagnetic field history. The results obtained suggest that the process of reversal from either starting polarity may remain invariant for a long period of time, then change rather abruptly between successive reversals.

084.027 **Observations of the electric fields that accelerate auroral particles.**
M. Temerin, F. S. Mozer.
Proc. Indian Acad. Sci., Earth Planet. Sci., Vol. 93, No. 3, p. 227 – 245 (1984).
In this report the authors are concerned with the observations of electric fields on auroral field lines and the interaction of these electric fields with charged particles – both electrons and ions.

084.028 **Effect of parallel electric field and geomagnetic field distortion on precipitation of charged particles into the lower ionosphere.**
R. N. Singh, R. Prasad.
Proc. Indian Acad. Sci., Earth Planet. Sci., Vol. 93, No. 3, p. 247 – 255 (1984).
Mead axisymmetric distortions in the geomagnetic field and uniform electrostatic field parallel to geomagnetic field have been assumed to derive an expression for the inhomogeneity parameter. Consequent change in the pitch angle diffusion of charged particles has been obtained. Using these parameters, the variations in the precipitating electron influx with varying stand–off distances and parallel electrostatic fields have been computed.

084.029 **Reconnection of magnetic fields.**
B. U. Ö. Sonnerup, P. J. Baum, J. Birn, S. W. H. Cowley, T. G. Forbes, A. B. Hassam, S. W. Kahler, W. H. Matthaeus, W. Park, G. Paschmann, E. R. Priest, C. T. Russell, D. S. Spicer, R. Stenzel.
NASA Ref. Publ., NASA RP–1120, p. 1–1 – 1–71 (1984). – See Abstr. 012.038.
Contents: Introduction. Reconnection at the magnetopause. Reconnection in the geomagnetic tail. Particle acceleration in current sheets. The diffusion region. Laboratory simulation of magnetospheric reconnection. Reconnection and the evolution of solar fields. Reconnection in solar flares. Acceleration of solar–flare electrons by reconnection. MHD simulation and small scale structure. Laboratory observations of neutral sheets. Reconnection in tokamaks. Conclusion.

084.030 **Connection between the magnetosphere and ionosphere.**
Y. T. Chiu, R. Anderson, J. Fennell, L. Frank, R. Hoffman, M. Hudson, L. Lyons, P. Palmadesso, E. Ungstrup, R. Vondrak, D. Williams, R. Wolf.
NASA Ref. Publ., NASA RP–1120, p. 7–1 – 7–48 (1984). – See Abstr. 012.038.
Contents: 1. Introduction. 2. Global scale coupling: Magnetospheric configuration effects. Electrical coupling. Plasma transport. 3. Intermediate scale coupling: Evening potential structure. Electric field signatures. Morningside structures. Westward traveling surge. 4. Microscopic scale effects: Auroral kilometric radiation. Ion conics and beams. 5. Summary and conclusions.

084.031 **Substorms in the magnetosphere.**
D. N. Baker, S.–I. Akasofu, W. Baumjohann, J. W. Bieber, D. H. Fairfield, E. W. Hones Jr., B. Mauk, R. L. McPherron, T. E. Moore.
NASA Ref. Publ., NASA RP–1120, p. 8–1 – 8–55 (1984). – See Abstr. 012.038.
Contents: 1. Introduction. 2. Solar wind–magnetosphere coupling: Interplanetary parameter changes. "Best" substorm predictors in the solar wind. Time delays: solar wind–magnetosphere coupling response time. 3. Driven processes in substorms: Dissipation. Storage. 4. The unloading process in substorms: Onset. Release. Recovery. 5. Summary and future directions.

084.032 **A fast Fermi process: energetic electrons accelerated by a nearly perpendicular bow shock.**
C. S. Wu.
J. Geophys. Res., Vol. 89, No. A10, p. 8857 – 8862 (1984).
The highly localized acceleration of electrons at the foreshock reported by Anderson et al. (1979) is explained in terms of a fast Fermi process. The basic notion is that in the solar wind frame the nearly perpendicular bow shock at the point of tangency of the interplanetary magnetic field acts as a fast–moving magnetic mirror which can reflect electrons with sufficiently large pitch angles. The reflection process can effectively energize electrons and drive them upstream. If the seed electrons have energies of several hundred electron volts, they can attain energies of several keV through the acceleration process.

084.033 **Interaction between an interplanetary shock and the earth's magnetosphere on August 27, 1978: ISEE 1 electric field and ISEE 2 plasma observations.**
K. H. Alfsen, C. Bonifazi, A. Pedersen, P.–A. Lindqvist.
J. Geophys. Res., Vol. 89, No. A10, p. 8863 – 8871 (1984).
The authors study a fast compressional motion of the magnetopause resulting from the interaction between an interplanetary shock and the earth's magnetosphere on August 27, 1978.

084.034 **Thick current sheets in the renovated model of the magnetosphere.**
G. Atkinson.
J. Geophys. Res., Vol. 89, No. A10, p. 8949 – 8955 (1984).
Earlier work using the observed location and direction of field–aligned currents to diagnose plasma distributions and magnetic field configurations in the outer magnetosphere is expanded to include thick adjacent current sheets mapping to the whole plasma sheet.

084.035 **Geomagnetic and solar data. June 1984.**
H. E. Coffey.
J. Geophys. Res., Vol. 89, No. A10, p. 9163 – 9165 (1984).

084.036 **Magnetic field inclination angle at geosynchronous orbit.**
C. S. Lin, J. N. Barfield.
Planet. Space Sci., Vol. 32, No. 10, p. 1283 – 1290 (1984).
Simultaneous magnetic field data from the geostationary satellites GOES 2 and GOES 3 have shown significant differences between the inclination angles measured at these two satellites. The relationship between the two inclination angles is examined for three magnetic activity groups and eight local time sectors by using data from GOES 2 and GOES 3 for the 12–month period from March 1979 to February 1980.

084.037 **A note on the energy transfer from the solar wind to the magnetosphere.**
L. C. Lee, S.–I. Akasofu.
Planet. Space Sci., Vol. 32, No. 11, p. 1423 – 1425 (1984).
The dimensional analysis of the energy transfer from the solar wind to the magnetosphere is formulated to include the effect of finite lifetime of the ring current particles. It is found that the solar wind energy input rate should be re–evaluated by considering the lifetime of the ring current particles.

084.038 **Whistler wave amplitude oscillation and frequency modulation in the magnetospheric cavity.**
U. Motschmann, K. Sauer, K. Baumgaertel.
Astrophys. Space Sci., Vol. 105, No. 2, p. 373 – 377 (1984).
A magnetospheric cavity model is presented to explain self–pulsing and frequency modulation of ducted whistler waves associated with triggered emissions. Partial reflection of these waves at the ends of the duct suggests a magnetospheric cavity analogous to a laser cavity.

084.039 **Seasonal displacement of the auroral zone.**
N. S. Zaretskij.
Issled. Geomagn. Aehron. Fiz. Solntsa, Moskva, No. 66, p. 164 – 169 (1983). In Russian. Abstr. in Ref. Zh., 62. Issled. Kosm. Prostranstva, 10.62.482 (1984).

084.040 **Measurement of the kinetic temperature in polar aurorae.**
Yu. I. Vakulin, G. A. Zherebtsov, Yu. A. Nadubovich, E. A. Ponomarev, L. N. Popov, L. M. Chagarov.
Issled. Geomagn. Aehron. Fiz. Solntsa, Moskva, No. 66, p. 80 – 85 (1983). In Russian. Abstr. in Ref. Zh., 62. Issled. Kosm. Prostranstva, 10.62.498 (1984).

084.041 **On the nature of the latitude variation of the frequency of polar aurorae appearance.**
N. S. Zaretskij.
Issled. Geomagn. Aehron. Fiz. Solntsa, Moskva, No. 66, p. 85 – 89 (1983). In Russian. Abstr. in Ref. Zh., 62. Issled. Kosm. Prostranstva, 10.62.499 (1984).

084.042 **Space structure of 60– and 30–year variations of the geomagnetic field from data of analytical models.**
V. I. Kalugin, N. M. Rotanova, V. P. Golovkov.
Geomagn. Aehron., Tom 24, No. 5, p. 780 – 786 (1984). In Russian. English translation in Geomagn. Aeron.

084.043 **Spectrum of paleosecular variations of the geomagnetic field of the recent 50 thousand years from data of sections of Molodova V and Korman IV (The Ukraine).**
L. S. Kulikova.
Geomagn. Aehron., Tom 24, No. 5, p. 787 – 794 (1984). In Russian. English translation in Geomagn. Aeron.

084.044 **Finding of the components of the main geomagnetic field, their nature and motion.**
A. V. Dolitskij.
Geomagn. Aehron., Tom 24, No. 5, p. 795 – 804 (1984). In Russian. English translation in Geomagn. Aeron.

084.045 **Measurement of the streams of energetic protons and electrons at two oppositely directed spectrometers at the Cosmos 368 artificial earth satellite.**
L. S. Bratolyubova–Tsulukidze, N. N. Volodichev,
O. Yu. Nechaev, I. A. Savenko, A. A. Suslov, P. I. Shavrin.
Geomagn. Aehron., Tom 24, No. 5, p. 812 – 813 (1984). In Russian. English translation in Geomagn. Aeron.

084.046 **Bursts of relativistic electrons in the magnetopause and in the outer radiation belt.**
I. P. Bezrodnykh, E. G. Berezhko, E. I. Morozova,
N. F. Pisarenko, I. Ya. Plotnikov, Yu. G. Shafer.
Geomagn. Aehron., Tom 24, No. 5, p. 818 – 820 (1984). In Russian. English translation in Geomagn. Aeron.

084.047 **The role of the interplanetary magnetic field in generation of electric fields in the magnetospheric tail.**
V. G. Pivovarov.
Geomagn. Aehron., Tom 24, No. 5, p. 838 – 840 (1984). In Russian. English translation in Geomagn. Aeron.

084.048 **Secular variations of the geomagnetic field on Kamchatka**
N. P. Ben'kova, V. P. Nazarets.
Geomagn. Aehron., Tom 24, No. 5, p. 849 – 851 (1984). In Russian. English translation in Geomagn. Aeron.

084.049 **Excitation of atomic oxygen emissions by auroral electron streams.**
G. D. Gefan, G. S. Kudryashov.
Geomagn. Aehron., Tom 24, No. 5, p. 855 – 857 (1984). In Russian. English translation in Geomagn. Aeron.

084.050 **Registration of heavy ions in the auroral region during magnetic storms.**
Yu. V. Gotselyuk, S. N. Kuznetsov, K. Kudela.
Kosm. Issled., Tom 22, Vyp. 5, p. 742 – 748 (1984). In Russian. English translation in Cosm. Res.

084.051 **Formation of proton radiation belts in the energy range of a few MeV.**
M. I. Panasyuk, Eh. N. Sosnovets.
Kosm. Issled., Tom 22, Vyp. 5, p. 756 – 762 (1984). In Russian. English translation in Cosm. Res.

084.052 **Charge composition of the electrons of the earth's radiation belt.**
B. I. Luchkov.
Kosm. Issled., Tom 22, Vyp. 5, p. 809 – 812 (1984). In Russian. English translation in Cosm. Res.

084.053 **Optical Doppler imaging of the Aurora Borealis.**
C. G. Shepherd, W. A. Gault, R. A. Koehler,
J. C. McConnell, K. V. Paulson, E. J. Llewellyn, C. D. Anger,
L. L. Cogger, J. W. Haslett, D. R. Moorcroft, R. L. Gattinger.
Geophys. Res. Lett., Vol. 11, No. 10, p. 1003 – 1006 (1984).
The Fabry–Perot spectrometer method of measuring upper atmospheric winds and temperatures has now been firmly established. The Michelson interferometer offers an alternative approach that has so far received little attention. A brief description is given of a stable achromatic field–widened Michelson interferometer, that uses a CCD detector. Four phase–stepped images are reduced to images of phase (leading to velocity), of fringe visibility (leading to temperature), and intensity. Preliminary results on the 557.7 nm emission in aurora, made during field trials in February, 1984 are presented.

084.054 **Unusual auroral features observed on January 10 – 11, 1983 and their possible relationships to the interplanetary magnetic field.**
S. I. Akasofu, B. Tsurutani.
Geophys. Res. Lett., Vol. 11, No. 10, p. 1086 – 1089 (1984).
The authors report unusual auroral features observed by the DMSP–F6 satellite on January 10 and 11, 1983. These features include the shift of the most active region to the morning sector, the shift of the polar cap center toward the evening sector, a large number of sun–earth aligned arcs and the oval–shaped diffuse glow without bright oval arcs. The simultaneous magnetic field observations by the ISEE–3 and IMP–8 satellites, together with a simple magnetospheric modeling result, suggest that the unusual features result from both the unusually large $B_y$ and a large positive $B_z$ component of the IMF.

084.055 **The wandering path of Virtual Geomagnetic Pole during the last 6000 years.**
Q.–y. Wei, D.–j. Li, G.–y. Cao, W.–x. Zhang, S.–p. Wang.
Acta Geophys. Sin., Vol. 27, No. 6, p. 562 – 572 (1984). In Chinese.
The declinations and inclinations of 17 epochs were determined using the baked earth collected from Shanxi, Hubei, Zhejiang, Jiangxi provinces and Beijing region. According to the geocentric dipole model for the magnetic field of the earth the geographic coordinates of Virtual Geomagnetic Pole (VGP) for all these epochs were calculated and the wandering path of VGP for the last 6000 years was obtained. The preliminary analyses indicate that the motion of VGP during the last several thousands of years is not uniform.

084.056 **La structure interne de la magnétopause.**
M. Roth.
Acad. R. Belg., Mém. Cl. Sci., Tome 44, Fasc. 7, p. 1 – 222 (1984).
The kinetic model developed in this paper aims to give a description of the internal structure of the magnetopause. It takes, among other factors, the interplanetary magnetic field and the thermal motion of the particles into account. These factors have been neglected in previous theories. This model describes the structure of a tangential discontinuity in a multi–component collisionless plasma. Furthermore, from the consideration of the two–stream instabilities a minimum magnetopause thickness can be deduced. The tangential discontinuities observed at the magnetopause are linked to regions where magnetosheath and magnetospheric plasmas are interpenetrated.

084.057 **Low energy enhancement of the O I 6300 Å line and enhancement due to resonance excitation of the $N_2^+$ first negative bands.**
K. Henriksen, C. S. Deehr, G. J. Romick, G. G. Sivjee,
N. I. Fedorova, G. F. Totunova.
Ann. Geophys., Vol. 2, No. 2, p. 191 – 196 (1984). Abstr. in Phys. Abstr., Vol. 87, No. 1243, Entry 100841 (1984).

084.058 **Geomagnetic secular variation impulses.**
V. Courtillot, J. L. Le Mouël.
Nature, Vol. 311, No. 5988, p. 709 – 716 (1984).

The secular variation of the Earth's internal magnetic field changes on time scales of the order of a year. Such impulses occurred in 1969 and probably in 1913; they provide constraints on deep mantle conductivity and on the motions of the top of the fluid core. The impulses can be correlated with extrema in the Earth's rotation rate and interpreted in terms of electromagnetic core–mantle coupling.

084.059 **Outer–core geostrophic flow and secular variation of Earth's geomagnetic field.**
J. L. Le Mouël.
Nature, Vol. 311, No. 5988, p. 734 – 735 (1984).

084.060 **Calculation of auroral infrared Earth limb spatial structure based on all–sky TV 4278 and 6300 Å measurements.**
J. B. Kumer, R. D. Sears, J. E. Evans, S. E. Harris, R. C. Gunton.
Proc. SPIE Int. Soc. Opt. Eng., Vol. 430, p. 342 – 355 (1983). Abstr. in Phys. Abstr., Vol. 87, No. 1245, Entry 110343 (1984). – See Abstr. 012.051.

084.061 **Terrestrial magnetosphere.**
D. C. Pande, D. C. Agarwal.
Indian J. Theor. Phys., Vol. 30, No. 2, p. 111 – 132 (1982). Abstr. in Phys. Abstr., Vol. 87, No. 1246, Entry 115660 (1984).

084.062 **The problem of reconnection: energy release in the geomagnetic tail.**
R. Treumann.
Orages solaires de type I, p. 45 – 53 (1984). – See Abstr. 012.055.

084.063 **Energetics of the magnetosphere.**
D. P. Stern.
Space Sci. Rev., Vol. 39, Nos. 1/2, p. 193 – 213 (1984).

Energy flow in various large–scale processes of the Earth's magnetosphere is examined. This energy comes from the solar wind. Global energy inputs, energy coupling, the ring current energy, Birkeland currents, and other processes are discussed. Recent results on all those processes, and attempts to correlate them with interplanetary conditions, have been described, and estimates of the various powers are listed. Substorms, their mechanism, and the roles which magnetic merging and magnetic energy storage play in them are briefly mentioned.

084.064 **Palaeomagnetic evidence for stationary sources of geomagnetic secular variation.**
M. E. Evans.
Phys. Earth Planet. Inter., Vol. 35, No. 4, p. 223 – 226 (1984).

Asymmetrical curvilinear patterns of secular variation with approximately superposed "outward" and "return" trajectories can be attributed to stationary magnetic sources in the core. Some appropriate palaeomagnetic examples are presented here.

084.065 **Geomagnetic evolution: 400 years of change on planet Earth.**
R. Thompson.
Phys. Earth Planet. Inter., Vol. 36, No. 1, p. 61 – 77 (1984).

Spherical harmonic coefficients of the geomagnetic field, calculated from historical observations of declination, inclination and intensity, and from archaeomagnetic inclination results, have been used to produce a film of geomagnetic change since 1600 A.D.

084.066 **Short–period geomagnetic secular variation.**
S. I. Braginsky (*S. I. Braginskij*).
Geophys. Astrophys. Fluid Dyn., Vol. 30, Nos. 1 – 2, p. 1 – 78 (1984).

In this paper the main problems of the theory of the short–period geomagnetic secular variation are described. An approximate analytical representation of the observational data on both the geomagnetic field variations and the associated variations in the length of the day, the kinematics of the generation of geomagnetic variations in the skin–layer at the core–mantle boundary and the influence of the electrical conductivity of the mantle on the geomagnetic variations are considered. A theory of torsional oscillations in the Earth's core is developed and an attempt is made to evaluate the associated geomagnetic variations. Magnetohydrodynamic effects in a possible thin layer of stably stratified fluid near the core–mantle boundary are investigated.

084.067 **Plate motions and the geomagnetic field. II. Jurassic to Tertiary.**
R. A. Livermore, F. J. Vine, A. G. Smith.
Geophys. J. R. Astron. Soc., Vol. 79, No. 3, p. 939 – 961 (1984).

The analysis of the time–averaged geomagnetic field is extended back to 200 Ma. Palaeomagnetic poles from the major plates have been carefully selected from recent compilations of reliable results for each region. These were returned, with their corresponding sampling sites, to their locations at the estimated dates of magnetization, in a fixed–hotspots framework. The corrected results were then grouped into 20 Ma windows at intervals of 10 Ma representing the past 100 Ma, and 40 Ma windows at 30 Ma intervals for the more sparse 100 – 200 Ma data. Global means and Fisher statistics were calculated for each window having included the axial quadrupole in the calculation.

084.068 **The interplanetary electric field, cleft currents and plasma convection in the polar caps.**
P. M. Banks, T. Araki, C. R. Clauer, J. P. St. Maurice, J. C. Foster.
Planet. Space Sci., Vol. 32, No. 12, p. 1551 – 1557 (1984).

This report investigates the suggestion that the pattern of plasma convection in the polar cleft region is directly determined by the interplanetary electric field.

084.069 **Global compressional oscillations of the terrestrial magnetosphere: the evidence and a model.**
M. G. Kivelson, J. Etcheto, J. G. Trotignon.
J. Geophys. Res., Vol. 89, No. A11, p. 9851 – 9856 (1984).

Compressional oscillations of nearly constant frequency (period $\lesssim$ 8 min) were observed from $L \cong 5$ to $L \cong 10$ near local noon over an interval of almost 3 hours during a dayside radial pass of the ISEE 1 spacecraft on August 12, 1982. The density fluctuations, measured by the electron density experiment, were in phase with the compressional magnetic oscillations measured by the magnetometer. The authors relate the observations to an analytical model of a global compressional wave reflected at the gradient of the magnetic field and plasma density near the plasmapause and standing in the outer magnetosphere.

084.070 **Geomagnetic and solar data. July 1984.**
H. E. Coffey.
J. Geophys. Res., Vol. 89, No. A11, p. 9863 (1984).

084.071 **The MHD Rankine–Hugoniot jump conditions and the terrestrial bow shock: a statistical comparison.**
D. Winterhalter, M. G. Kivelson, R. J. Walker, C. T. Russell.
Adv. Space Res., Vol. 4, No. 2 – 3, p. 287 – 292 (1984). – See Abstr. 012.075.

The authors have examined 204 bow shock crossings observed in the magnetic field and plasma data on ISEE–1. Using the measured upstream field and plasma data they calculated the downstream field magnitudes by using the single–fluid MHD Rankine–Hugoniot jump conditions, and compared the results with the observed downstream field magnitudes.

084.072 **On the microphysics of the magnetotail reconnection region.**
T. Terasawa.
Adv. Space Res., Vol. 4, No. 2 – 3, p. 429 – 438 (1984). – See Abstr. 012.075.

The paper reviews the theories of the tearing mode instability and its implications for the reconnection process in the magnetotail. It is shown that in the thin plasma sheet whose width is comparable to the ion inertial length, the Hall current effect is important.

084.073 **Current sheet particle acceleration: theory and observations for the geomagnetic tail.**
T. W. Speiser.
Adv. Space Res., Vol. 4, No. 2 – 3, p. 439 – 448 (1984). – See Abstr. 012.075.

The current sheet in the geomagnetic tail is a source of plasma and energetic particles for the magnetospheric ring current and radiation belts, a seat for instabilities and magnetospheric substorms, a major transport region of mass, momentum, and energy into the magnetosphere. This paper reviews simple models which allow particle orbits in the current sheet to be calculated analytically.

084.074 **The electric field and plasma energization in computer simulations of magnetotail reconnection.**
J. Birn.
Adv. Space Res., Vol. 4, No. 2 – 3, p. 449 – 458 (1984). – See Abstr. 012.075.

Reconnection in the magnetotail involves structural changes on large spatial and time scales which can, at least up to the present time, be handled only by MHD theory including some kind of resistivity. The author reviews and discusses results from numerical MHD codes on bulk and single particle energization and the electric fields governing possible energization.

084.075 **Distribution of energetic ($E$ > 35 keV) ion anisotropies in the distant geomagnetic tail.**
P. W. Daly, T. R. Sanderson, K.–P. Wenzel.
Adv. Space Res., Vol. 4, No. 2 – 3, p. 475 – 478 (1984). – See Abstr. 012.075.

The direction and magnitude of the anisotropy of energetic ions (35 – 56 keV) have been measured in the plasma sheet during the passage of ISEE–3 through the geomagnetic tail, and sorted according to distance from the earth. The observations support the plasmoid model of magnetotail dynamics, and indicate that the X–line in the magnetotail is normally within 100 $R_e$ of the earth.

084.076 **Numerical simulations of double layers and auroral electric fields.**
N. Singh, R. W. Schunk, H. Thiemann.
Adv. Space Res., Vol. 4, No. 2 – 3, p. 481 – 490 (1984). – See Abstr. 012.075.

Recent numerical simulations on double layers are reviewed and discussed in the light of relevant space observations in the auroral plasma. Some important features, such as the amplitudes and scale lengths of the parallel and perpendicular electric fields, seen in these simulations are qualitatively similar to those observed in space.

084.077 **Origin of auroral electric potential structures.**
Y. T. Chiu.
Adv. Space Res., Vol. 4, No. 2 – 3, p. 491 – 498 (1984). – See Abstr. 012.075.

Careful analyses during the past 5 – 6 years of particle and electric field signatures observed in association with the evening discrete arc have led to major advances in the basic understanding of the structure of the auroral electric potential. The origin of this basic potential structure, which is inherent in the principle of kinetic plasma distributions in globally inhomogeneous magnetic configurations, is reviewed and examined.

084.078 **Potential structures and particle acceleration on auroral field lines.**
D. J. Gorney.
Adv. Space Res., Vol. 4, No. 2 – 3, p. 499 – 508 (1984). – See Abstr. 012.075.

Observations of plasmas and electric fields within the regions of space responsible for auroral particle acceleration have verified the existence of electric fields with components parallel to the magnetic field over large altitude regions. Evidence exists that small–amplitude double layers, spread along the auroral magnetic field lines, provide the mechanism for the maintenance of the auroral potential.

084.079 **Particle acceleration processes on auroral field lines observed by Dynamics Explorers.**
R. A. Hoffman.
Adv. Space Res., Vol. 4, No. 2 – 3, p. 509 (1984). Abstract. – See Abstr. 012.075.

084.080 **Some features of auroral electric fields as seen in 2D numerical simulations.**
H. Thiemann, N. Singh, R. W. Schunk.
Adv. Space Res., Vol. 4, No. 2 – 3, p. 511 – 514 (1984). – See Abstr. 012.075.

Results of 2D plasma simulations, which show the formation of V–shaped double layers with larger perpendicular than parallel electric field amplitudes, the excitation of electrostatic ion–cyclotron waves and the occurrence of return currents, are presented and related to auroral observations.

084.081 **Electron acceleration by lower hybrid waves on auroral field lines.**
D. S. Hall, D. A. Bryant, R. Bingham.
Adv. Space Res., Vol. 4, No. 2 – 3, p. 515 – 518 (1984). – See Abstr. 012.075.

Particle measurements show that electron acceleration on auroral field lines is a statistical, velocity–dependent, process. It is proposed that the process is stochastic acceleration by waves, and demonstrated that lower hybrid waves seen on auroral field lines have the right properties to account for the electron acceleration.

084.082 **Measurements of the angular distribution of precipitating electrons in a breakup aurora.**
C. P. Chaloner, D. A. Bryant, D. S. Hall.
Adv. Space Res., Vol. 4, No. 2 – 3, p. 519 – 522 (1984). – See Abstr. 012.075.

Measurements of the complete pitch–angle distribution of precipitating auroral electrons in the energy range 0.5 to 25 keV in several arcs are presented. The measurements were made with a time resolution of 2 s between altitudes of 400 and 700 km by electrostatic analysers carried on a sounding rocket.

084.083 **Small–scale magnetic disturbances in the cusp and polar cap regions.**
I. Arshinkov, A. Bochev, Y. Matviichuk, D. Teodosiev, I. Rumchev, G. Stanev, P. Nenovsky.
Adv. Space Res., Vol. 4, No. 2 – 3, p. 523 – 526 (1984). – See Abstr. 012.075.

The authors report on an investigation of the small–scale transverse magnetic disturbances in the Earth's magnetosphere based upon magnetic field, electric field and particle flux data acquired by three satellite experiments.

084.084 **On the interpretation of ULF fluctuations near the Earth magnetopause.**
J. Buechner, H. R. Lehmann.
Adv. Space Res., Vol. 4, No. 2 – 3, p. 527 – 530 (1984). – See Abstr. 012.075.

In–situ data of ULF–plasma fluctuations in the outer magnetosphere became available from an experiment on board the Prognoz–8 satellite. The objective of this paper is a theoretical analysis of ULF–current fluctuation spectra and a comparison with the experiment.

084.085 **The geomagnetic storm of 1910 May 18 and the tail of Halley's Comet.**
L.–s. Yan, Z.–y. Li, S.–y. Gu.
Chin. Astron. Astrophys., Vol. 8, No. 3, p. 260 – 264 (1984). English translation of Chin. J. Space Sci., Vol. 4, No. 2, p. 149 – 155 (1984).

The authors argue that the geomagnetic storm observed on 1910 May 18 at the Lu–Kia–Pang Observatory near Shanghai is possibly connected with the tail of Halley's Comet, rather than originating in some solar activity.

084.086 **Long and short term relationships between solar wind velocity and geomagnetic field at low latitudes.**
G. K. Rangarajan.
Proc. Indian Acad. Sci., Earth Planet. Sci., Vol. 93, No. 4, p. 343 – 351 (1984).
The solar wind velocity control of the low latitude geomagnetic field both on long and short term basis is studied. It is shown that semiannual averages of the low latitude field are inversely related to solar wind velocity and that there is a dominant local time dependence of the relationship. When the solar wind velocity exhibits a significant recurrent pattern the low latitude geomagnetic field also shows a strong solar synodic rotation periodicity of 27 days.

084.087 **The geomagnetic dynamo – elementary energetics and thermodynamics.**
F. J. Lowes.
Geophys. Surv., Vol. 7, No. 1, p. 91 – 105 (1984). – See Abstr. 012.082.
This paper is a non-mathematical review, summarising the work in this field.

084.088 **The effects of dipole tilt on the structure of the magnetosphere.**
C. C. Wu.
J. Geophys. Res., Vol. 89, No. A12, p. 11048 – 11052 (1984).
The structure of the magnetosphere as a function of dipole tilt was studied by using a global magnetohydrodynamics (MHD) model. The author shows that the cusp current sheet in the MHD model is related to the tail current sheet and that the calculated shape and positions of the tail current sheet are consistent with observations. In addition, he discusses the difference between the MHD model and the Chapman–Ferraro type model which manifests as a function of tilt angle.

084.089 **Geomagnetic and solar data. August 1984.**
H. E. Coffey.
J. Geophys. Res., Vol. 89, No. A12, p. 11063 (1984).

084.090 **On the theory of auroral kilometric radiation.**
S. Bujarbarua, M. Nambu.
Phys. Scr., Vol. 30, No. 3, p. 201 – 211 (1984).
The induced bremsstrahlung interaction between the auroral electrons and electrostatic ion cyclotron turbulence or double layer is effective for the generation of the auroral kilometric radiation. The theory explains the close correlation between the double layer or the electrostatic shocks and AKR.

084.091 **High–latitude investigations of charged particle streams and longitudinal currents at the Intercosmos – Bolgariya–1300 station.**
A. Bochev, Ts. Dachev, L. N. Zhuzgov, V. I. Lazarev, M. V. Tel'tsov, S. I. Shkol'nikova.
Geomagn. Aehron., Tom 24, No. 6, p. 948 – 952 (1984). In Russian. English translation in Geomagn. Aeron.

084.092 **Oscillations of the decrease period on the morning side of the magnetosphere.**
B. V. Dovbnya, Eh. T. Matveeva, V. A. Troitskaya, F. Z. Fejgin.
Geomagn. Aehron., Tom 24, No. 6, p. 977 – 982 (1984). In Russian. English translation in Geomagn. Aeron.

084.093 **Variations of the direction of the geomagnetic field 23900 – 25300 and 43300 – 44100 years ago from data of Molodova V and Korman IV (Ukraine) cross sections.**
L. S. Kulikova.
Geomagn. Aehron., Tom 24, No. 6, p. 986 – 993 (1984). In Russian. English translation in Geomagn. Aeron.

084.094 **The forbidden region, injection boundary and aurora oval.**
A. Ya. Fel'dshtejn.
Geomagn. Aehron., Tom 24, No. 6, p. 1019 – 1021 (1984). In Russian. English translation in Geomagn. Aeron.

084.095 **On the probability of radio aurora appearance at midlatitudes.**
S. V. Bubennikov, A. P. Tarakanov.
Geomagn. Aehron., Tom 24, No. 6, p. 1021 – 1023 (1984). In Russian. English translation in Geomagn. Aeron.

**UK National Report for the Quadrennium 1979 – 1983 to the International Association of Geomagnetism and Aeronomy (*Bibliography*).**
See Abstr. 002.021.

**Reversals of the Earth's magnetic field.**
See Abstr. 003.003.

**Atmospheric electrodynamics.**
See Abstr. 003.018.

**The earth's magnetic field. Its history, origin and planetary perspective.**
See Abstr. 003.122.

**Solar terrestrial physics: present and future. A report based on the Solar Terrestrial Physics Workshop, December 1982 to November 1983.**
See Abstr. 012.038.

**Report on the research in geomagnetism and aeronomy in Poland for the period 1978 – 1982.**
See Abstr. 013.002.

**Laboratory terrella experiment at Osaka City University.**
See Abstr. 022.022.

**Particle acceleration during reconnection in laboratory plasmas.**
See Abstr. 022.144.

**Global geospace study.**
See Abstr. 051.001.

**First success for a space mission and a comet for Christmas.**
See Abstr. 051.051.

**Electron beam emission from the EXOS–B (JIKIKEN) satellite as a powerful diagnostic tool in the magnetosphere.**
See Abstr. 053.012.

**Magnetic field generation in celestial bodies.**
See Abstr. 062.003.

**A note on Southwood's instability criterion.**
See Abstr. 062.010.

**Particle–plasma coupling through magnetohydrodynamic turbulence.**
See Abstr. 062.032.

**Effects of a cold electron beam on Brillouin backscattering in a plasma.**
See Abstr. 062.034.

**Energization of plasma in a weak magnetic field and structure of the geomagnetic tail.**
See Abstr. 062.054.

**Particle acceleration.**
See Abstr. 062.087.

**Nonlinear generation of electromagnetic waves in a magnetoplasma.**
See Abstr. 062.098.

**Temporal evolution of the convective plasma flow during a reconnection process.**
See Abstr. 062.149.

**Current layers and reconnection.**
See Abstr. 062.154.

**Collisionless shocks.**
See Abstr. 062.172.

**Recent advances in numerical simulation of space–plasma–physics problems.**
See Abstr. 062.173.

**Magnetic–field–aligned electric fields.**
See Abstr. 062.174.

**Long–term fluctuations of solar activity during the last thousand years.**
See Abstr. 072.036.

**Implications of solar flare dynamics for reconnection in magnetospheric substorms.**
See Abstr. 073.043.

**Coupling of the solar wind to the magnetosphere.**
See Abstr. 074.031.

**Correlation fluctuations of streams of charged particles and of the magnetic field from data of the Intercosmos 19 satellite.**
See Abstr. 078.006.

**Characteristics of solar and heliospheric ion populations observed near Earth.**
See Abstr. 078.009.

**On the dynamics and energetics of the earth's core.**
See Abstr. 081.040.

**Observations of Lyman–α emissions of hydrogen and deuterium.**
See Abstr. 082.008.

**Assessment of plasma transport and convection at high latitudes.**
See Abstr. 082.048.

**Radiation belt particles and O II emissions determined as contaminations of geocoronal helium airglow observations.**
See Abstr. 082.059.

**Statistical connection between ionospheric and geomagnetic disturbances.**
See Abstr. 083.037.

**A possible effect of magnetospheric convection on ionospheric conductivities.**
See Abstr. 083.041.

**Local time dependence of the response of the equatorial electrojet to DP2 and SI disturbances.**
See Abstr. 083.046.

**Change of the geomagnetic pulsation regime in the period of the solar eclipse on October 12, 1977.**
See Abstr. 085.015.

**A comparative study of distant magnetotail structure at Venus and Earth.**
See Abstr. 093.026.

**A note on anomalous sounds from meteor fireballs and aurorae.**
See Abstr. 104.014.

**Fluctuations of the interplanetary and geomagnetic fields in connection with proton bursts.**
See Abstr. 106.016.

**Magnetohydrodynamic modelling of interplanetary disturbances between the Sun and Earth.**
See Abstr. 106.020.

**The heating and acceleration of electrons by shocks.**
See Abstr. 106.040.

**Relaxation of plasma at the shock front.**
See Abstr. 106.041.

**Investigation of the isotopic and charge composition of energetic particles aboard the Prognoz 5 and Prognoz 6 stations.**
See Abstr. 106.055.

**Energy spectra of protons with E = 23 – 480 keV in the interplanetary medium, transition region and the outer magnetosphere.**
See Abstr. 106.062.

# 085 Solar-terrestrial Relations

085.001 **Sonnenaktivität und Troposphäre. Beeinflußt der Sonnenfleckenzyklus die Temperatur der Troposphäre und das Wetter?**
H. Schwentek.
Umschau, 84. Jahrg., No. 17, p. 505 – 506 (1984).

085.002 **Manifestation of cyclic helio–geomagnetic activity variations of the zonal pressure gradient changes in the troposphere.**
R. V. Smirnov, Eh. V. Kononovich, S. V. Startsev.
Astron. Zh., Tom 61, Vyp. 4, p. 778 – 782 (1984). In Russian. English translation in Sov. Astron., Vol. 28, No. 4.

Using a statistical investigation of the 1891 – 1979 period data, the existence of 11–year solar activity dependence of the zonal component of the surface atmospheric pressure gradient is found. This dependence has a significantly seasonal character.

085.003 **North–south asymmetries of solar particle events in upper stratospheric ozone.**
K. Maeda, D. F. Heath, T. Aruga.
Planet. Space Sci., Vol. 32, No. 7, p. 857 – 870 (1984).

The paper's main purpose is to describe the asymmetries of SPE–effects on stratospheric ozone between the two hemispheres in terms of their spatial and temporal variations and to discuss possible mechanisms for the appearance of these asymmetries. For this purpose, SPEs in January 1971 and August 1972 were chosen for analysis.

085.004 **Polar aurora effects associated with solar flares.**
B. V. Rezhenov, V. G. Vorobjev (*V. G. Vorob'ev*).
Planet. Space Sci., Vol. 32, No. 8, p. 925 – 934 (1984).

The behaviour of polar auroras in the dark part of the auroral oval during solar flares has been examined. For the analysis 29 solar flares during spring and autumn periods when a part of the polar cap was sunlit were selected.

085.005 **Sunspot cycles and global oscillations.**
D. J. Schove.
Climatic changes on a yearly to millennial basis, p. 257 – 259 (1984). – See Abstr. 012.023.

085.006 **Mechanism for linking solar activity to weather–scale effects, climatic changes and glaciations in the northern hemisphere.**
V. Bucha.
Climatic changes on a yearly to millennial basis, p. 415 – 448 (1984). – See Abstr. 012.023.

085.007 **Solar ultraviolet irradiance variations and the earth's atmosphere.**
J. L. Lean.
Climatic changes on a yearly to millennial basis, p. 449 – 471 (1984). – See Abstr. 012.023.
The interaction of solar UV radiation with the earth's atmosphere is described, solar UV irradiance variability is examined, and possible links between solar UV irradiance variations and atmospheric variability are discussed.

085.008 **Cycles of solar flares and weather.**
T. Landscheidt.
Climatic changes on a yearly to millennial basis, p. 473 – 481 (1984). – See Abstr. 012.023.

085.009 **Thermospheric response to the June 11, 1983, solar eclipse.**
E. C. Ridley, R. E. Dickinson, R. G. Roble, M. H. Rees.
J. Geophys. Res., Vol. 89, No. A9, p. 7583 – 7588 (1984).
The NCAR thermospheric general circulation model that includes coupling of dynamics and composition is used to calculate the time–dependent thermospheric response to the June 11, 1983 total solar eclipse.

085.010 **Manifestation of solar activity in the atmosphere and geomagnetic activity.**
V. V. Mikhnevich, R. V. Smirnov.
Soln.–atmos. svyazi i geomagn. akt. Moskva, p. 3 – 11 (1984). In Russian. Abstr. in Ref. Zh., 51. Astron., 8.51.326 (1984).

085.011 **Some questions of the state and perspectives of investigations on the problem "sun–troposphere".**
S. I. Avdyushin, V. V. Mikhnevich, R. V. Smirnov.
Soln.–atmos. svyazi i geomagn. akt. Moskva, p. 12 – 22 (1984). In Russian. Abstr. in Ref. Zh., 51. Astron., 8.51.329 (1984).

085.012 **On the geoefficiency of recurrent high–velocity solar wind streams.**
B. V. Zhiromskij.
Vologod. politekh. inst. Vologda, 16 pp. (1984). In Russian. Abstr. in Ref. Zh., 62. Issled. Kosm. Prostranstva, 8.62.494 (1984).

085.013 **On a possibility of representing geomagnetic variations in the high–latitude region through solar wind parameters.**
R. G. Afonina, B. A. Belov, V. Yu. Gajdukov, A. E. Levitin, Ya. I. Fel'dshtejn.
Soln.–atmos. svyazi i geomagn. aktivnost', Moskva, p. 121 – 137 (1984). In Russian. Abstr. in Ref. Zh., 62. Issled. Kosm. Prostranstva, 8.62.532 (1984).

085.014 **On a connection of the electromagnetic energy of the solar wind with geomagnetic activity indices.**
N. I. Dvinskikh, B. G. Dolgoarshinnykh, N. M. Rudneva.
Soln.–atmos. svyazi i geomagn. aktivnost', Moskva, p. 137 – 142 (1984). In Russian. Abstr. in Ref. Zh., 62. Issled. Kosm. Prostranstva, 8.62.533 (1984).

085.015 **Change of the geomagnetic pulsation regime in the period of the solar eclipse on October 12, 1977.**
A. D'Costa, G. Perez.
Geomagn. Aehron., Tom 24, No. 4, p. 658 – 661 (1984). In Russian. English translation in Geomagn. Aeron.

085.016 **Excitation and ionizatioan in the upper atmosphere during solar flares with account for photoelectrons.**
S. V. Avakyan, G. S. Kudryashev.
Geomagn. Aehron., Tom 24, No. 4, p. 693 – 695 (1984). In Russian. English translation in Geomagn. Aeron.

085.017 **Impact of flares on the solar terrestrial environment.**
L. A. Fisk, R. L. Arnoldy, L. J. Lanzerotti, R. Lin, E. Oran, J. B. Reagan, M. Schulz, B. T. Tsurutani.
NASA Ref. Publ., NASA RP–1120, p. 9–1 – 9–40 (1984). – See Abstr. 012.038.
Contents: Introduction. The flare process. Impact on the heliosphere. Impacts on the magnetosphere. Impacts on the ionosphere and atmosphere. Impacts on technology. Summary.

085.018 **Collisionless shock waves in the solar terrestrial environment.**
E. Greenstadt, V. Formisano, C. Goodrich, J. T. Gosling, M. Lee, M. Leroy, M. Mellott, K. Quest, A. E. Robson, P. Rodriguez, J. Scudder, J. Slavin, M. Thomsen, D. Winske, C. S. Wu.
NASA Ref. Publ., NASA RP–1120, p. 10–1 – 10–60 (1984). – See Abstr. 012.038.
Contents: 1. Introduction. 2. Natural shocks: Occurrence. Global features. Key questions. 3. Quasi–perpendicular, supercritical processes: Ion dynamics and field structure. Electrons. Instabilities and waves. 4. Quasi–perpendicular subcritical processes. 5. Quasi–parallel processes: Terminology. Ion participation. Mingled processes. Phenomenology. 6. Ion acceleration: The foreshock. Proton injection into the upstream flow. Ion acceleration and hydromagnetic wave excitation. 7. Outstanding problems. 8. Investigate avenues: Laboratoy experiments. Theory and simulation. Spacecraft observations.

085.019 **Influence of outer factors on the vertical structure of planetary waves in the troposphere and stratosphere.**
L. R. Rakipova, N. I. Yakovleva.
Tr. Glav. geofiz. obs., No. 471, p. 3 – 9 (1984). In Russian. Abstr. in Ref. Zh., 51. Astron., 10.51.369 (1984).

085.020 **The connection of the amplitude of daily geomagnetic pulsations with the solar wind velocity in the solar activity cycle.**
T. N. Polyushkina, A. S. Potapov.
Issled. Geomagn. Aehron. Fiz. Solntsa, Moskva, No. 66, p. 54 – 61 (1983). In Russian. Abstr. in Ref. Zh., 62. Issled. Kosm. Prostranstva, 10.62.460 (1984).

085.021 **Decrease of the module of geomagnetic field intensity during the solar eclipse of July 31, 1981.**
L. E. Kolokolov, B. S. Russkikh.
Issled. Geomagn. Aehron. Fiz. Solntsa, Moskva, No. 66, p. 108 – 110 (1983). In Russian. Abstr. in Ref. Zh., 62. Issled. Kosm. Prostranstva, 10.62.461 (1984).

085.022 **Geophysical response to the dynamics of the transition region of the solar wind.**
N. A. Lotova, V. D. Kozlovskij.
Geomagn. Aehron., Tom 24, No. 5, p. 709 – 713 (1984). In Russian. English translation in Geomagn. Aeron.

085.023 **Temporal variations of the $^{10}$Be concentration in ice and solar activity.**
A. V. Blinov, A. N. Konstantinov, G. E. Kocharov, V. A. Levchenko.
Soln. Dannye, Byull., 1984, No. 8, p. 90 – 94 (1984). In Russian.
The influence of the 11–year solar cycle on the $^{10}$Be concentration in polar ice is illustrated by experimental data.

085.024 **Influence of the solar cycle upon the upper mesosphere.**
G. von Cossart.
Gerlands Beitr. Geophys., Band 93, Heft 5, p. 329 – 338 (1984).
From indirect phase height measurements over 19 years at the observatory Kühlungsborn (GDR) clear evidence has been obtained that the height of reflection of LF radio waves in the lower

ionosphere is systematically lower during solar maximum than during solar minimum conditions. On the other hand, from an analysis of 600 rocket measurements of upper mesosphere pressure and temperature over 14 years at the Soviet rocket sounding station Volgograd systematic solar cycle variations of pressure and temperature in the height range 70 – 75 km have become evident, indicating an increase of pressure by about 24% and of temperature by about 5K from solar minimum to solar maximum. By model calculations it can be shown that the results of the phase height and rocket measurements are satisfactorily compatible and thus support each other.

085.025 **Rythmes solaires et terrestres.**
N. Grevesse.
Ciel, Vol. 46, p. 273 – 277 (1984).

085.026 **The photoelectron spectrum in the upper atmosphere of the earth during solar flares.**
S. V. Avakyan, G. S. Kudryashev.
Kosm. Issled., Tom 22, Vyp. 6, p. 889 – 894 (1984). In Russian. English translation in Cosm. Res.

085.027 **Analysis of the connection between the parameters of solar wind and the characteristics of daily geomagnetic pulsations according to data of the Observatories Nad'tsenk (Hungary) and Uzur (USSR).**
F. Veres, A. S. Potapov, T. N. Polyushkina.
Issled. Geomagn. Aehron. Fiz. Solntsa, Moskva, No. 68, p. 138 – 150 (1984). In Russian. Abstr. in Ref. Zh., 51. Astron., 11.51.449 (1984).

085.028 **Influence of interplanetary interaction regions on geomagnetic disturbances and tropospheric circulation.**
H. Lundstedt.
Planet. Space Sci., Vol. 32, No. 12, p. 1541 – 1545 (1984).
Proposed solar wind–magnetosphere energy coupling functions are studied. An empirical formula proposed by Svalgaard (1977) is found to predict the geomagnetic activity quite well. The influence of solar wind interaction regions on the tropospheric circulation, through a suggested cosmic ray mechanism, was investigated. The cosmic ray intensity at earth clearly showed a decrease at the time of passage of an interaction region. It is suggested that the well–known dip in the Vorticity Area Index may be caused by an interaction–modulated decrease in cosmic ray intensity.

085.029 **Very long solar periods and the radiocarbon record.**
C. P. Sonett.
Rev. Geophys. Space Phys., Vol. 22, No. 3, p. 239 – 254 (1984).
The ~200–year periodicity in the time variations of atmospheric radiocarbon is shown to extend over the entire 8500–year La Jolla record and appears to be associated with a longer period between about 1500 and 2000 years via amplitude, frequency, or phase modulation, or some combination of these; but the statistical certainty of the source and form of the modulation is hampered by the low signal/noise ratio of the two periods. If it can be confirmed that the sun is the source, the very long periods suggest as one possibility that the core of the sun is the ultimate source, though multiple convective zone dynamo eigenmodes are an equally conjectural possibility. An alternate source for the longer period is pressure variations of the local arm of the galaxy, but this model is not favored.

085.030 **Correlations between thermospheric density and temperature, solar EUV flux, and 10.7–cm flux variations.**
A. E. Hedin.
J. Geophys. Res., Vol. 89, No. A11, p. 9828 – 9834 (1984).
Simultaneously measured thermospheric $N_2$ densities and solar EUV fluxes obtained by the AE–E satellite are compared with ground–based solar 10.7–cm fluxes and calcium plage measurements. Short–wavelength (coronal) EUV emissions correlate better with thermospheric density than the 10.7–cm flux, although the reduction in density residuals is small. Correlation of density with a calcium plage index was somewhat worse than with 10.7. Although the overall correlation of 10.7–cm flux with EUV fluxes is high, the best fit slopes are different for short– and long–term variations, and there are instances where the short–term behavior of various EUV emissions and that of the 10.7–cm flux are distinctly different. For both density and EUV, a two–factor formula based on daily 10.7–cm flux and a running mean of the 10.7–cm flux provides a better fit than using the daily 10.7–cm flux alone.

085.031 **Possible effects of solar variability on the middle atmosphere.**
K. D. Cole.
J. Atmos. Terr. Phys., Vol. 46, No. 9, p. 721 – 725 (1984). – See Abstr. 012.072.
The paper discusses the possible effects on the middle atmosphere of (1) solar variability in emission in the u.v. and far u.v., (2) solar proton events, (3) relativistic electron precipitation events, (4) corpuscular heating in auroras, (5) joule heating by the auroral electrojet, (6) auroral NO production and (7) gravity wave emission by the auroral electrojet. It is important to establish the maximum depth in the atmosphere to which these effects go, in order to be able to determine whether the reflection of planetary scale waves from the troposphere is substantially altered by them.

085.032 **Numerical investigation of photochemical processes in the upper atmosphere. I. Stochastic simulation of solar radiation interaction with a rarefield multi–component gas.**
G. I. Zmievskaya, M. Ya. Marov, V. I. Shematovich.
Nauchn. Inf., Vyp. 55, p. 144 – 159 (1982). In Russian.
The kinetic distribution of the EUV solar radiation interaction with the rarefied gas of the earth's upper atmosphere is presented.

085.033 **Numerical investigation of photochemical processes in the upper atmosphere. II. A numerical stochastic model of photoionization.**
V. I. Shematovich.
Nauchn. Inf., Vyp. 55, p. 160 – 175 (1982). In Russian.
A numerical stochastic model of the upper atmosphere ionization by the EUV solar radiation is constructed and algorithmically realized.

085.034 **Structure of the relation between helio–geomagnetic activity and the tropospheric pressure gradient.**
R. V. Smirnov, Eh. V. Kononovich, S. V. Startsev.
Astron. Tsirk., No. 1336, p. 6 – 8 (1984). In Russian.

085.035 **Effects of solar activity in changes of tropospheric temperature anomalies.**
R. V. Smirnov, Eh. V. Kononovich, S. V. Startsev.
Astron. Tsirk., No. 1337, p. 4 – 6 (1984). In Russian.

085.036 **Study of the interaction between solar wind and the geomagnetosphere aboard the Prognoz stations.**
O. L. Vajsberg, A. N. Omel'chenko, V. N. Smirnov,
G. N. Zastenker, S. I. Klimov, A. O. Fedorov,
M. N. Nozdrachev, S. P. Savin, Yu. I. Ermolaev, A. V. Lejbov.
Solar activity investigations and the Prognoz space system, p. 10 – 32 (1984). In Russian. Abstr. in Ref. Zh., 62. Issled. Kosm. Prostranstva, 12.62.383 (1984). – See Abstr. 003.038.

085.037 **Spatial regularities of solar activity effects in the troposphere.**
R. V. Smirnov.
Astron. Zh., Tom 61, Vyp. 6, p. 1168 – 1178 (1984). In Russian. English translation in Sov. Astron., Vol. 28, No. 6.
Joint analysis of maps of density variations (or density dispersions) in the troposphere after geomagnetic disturbances and of maps of advection, wind velocity divergence, etc. makes it possible to put forward a concept of solar–induced centres of atmospheric action (SICA). Solar–disturbance transfer and planetary–wave development in the atmosphere are accomplished by means of SICA where the level of baroclinic instability is high. Infrasonic waves are considered as an agent connecting the lower thermosphere and the troposphere.

085.038 **Possible influence of solar radiation variability on the stratospheric temperature.**
J. Cacho, M. Gil, M. J. Sainz de Aja, L. F. Alberca.
Adv. Space Res., Vol. 4, No. 4, p. 37 – 40 (1984). – See Abstr. 012.087.

085.039 **Ionospheric and stratospheric effects of a proton flare during unusual solar activity – 22 November 1977.**
P. Velinov, G. Nestorov, C. Spasov, Ts. Dachev, Y. Tasev.
Adv. Space Res., Vol. 4, No. 4, p. 163 – 166 (1984). – See Abstr. 012.087.

The effect of the proton flare of 22 November 1977 on the various levels in the middle atmosphere and the bottom-side ionosphere is studied in order to compare synchronous phenomena in the middle atmosphere with phenomena in upper regions and to investigate the response of middle atmosphere to the penetration of high energy solar particles and radiation.

085.040 **The time of delay of the location of the equatorial boundary of the aurora oval relative to variations of the solar wind electric field and the AE–index.**
M. A. Shukhtina, V. A. Sergeev, N. E. Alekseeva.
Geomagn. Aehron., Tom 24, No. 6, p. 944 – 947 (1984). In Russian. English translation in Geomagn. Aeron.

**Atmospheric electrodynamics.**
See Abstr. 003.018.

**Second Indo–US workshop on solar–terrestrial physics.**
See Abstr. 011.026.

**Solar terrestrial physics: present and future. A report based on the Solar Terrestrial Physics Workshop, December 1982 to November 1983.**
See Abstr. 012.038.

**Experiences with high precision ground–based photometry of the sun.**
See Abstr. 036.142.

**Global geospace study.**
See Abstr. 051.001.

**A dynamo theory prediction for solar cycle 22: sunspot number, radio flux, exospheric temperature, and total density at 400 km.**
See Abstr. 072.010.

**Modelling of the solar extreme ultraviolet irradiance for aeronomic applications.**
See Abstr. 076.018.

**Type IV radio radiation and orientation of the magnetic field in a flare region as an index of geoefficiency of a flare stream.**
See Abstr. 077.021.

**Measurements of solar total irradiance and its variability.**
See Abstr. 080.004.

**Modelling the global climate response to orbital forcing and atmospheric carbon dioxide changes.**
See Abstr. 081.011.

**Planetary, solar, atmospheric, hydrospheric and endogene processes as origin of climatic changes on the earth.**
See Abstr. 082.033.

**Thunder and electric storms.**
See Abstr. 082.042.

**Improvement of the solar–cycle variation of IRI lower ionosphere models by means of radio wave propagation data.**
See Abstr. 083.011.

**Solar cycle effects upon the relationship of $N_e$ and $T_e$ in the F–region.**
See Abstr. 083.012.

**Comparison of various empirical models of electron temperature with experimental measurements during low solar activity.**
See Abstr. 083.014.

# Planetary System

## 091 Physics of the Planets (Interiors, Atmospheres, Magnetic Fields, etc.)

091.001 **On a method of construction of the point masses model for the gravitational field of a planet.**
A. L. Tserklevich, S. D. Volzhanin.
Pis'ma Astron. Zh., Tom 10, No. 7, p. 549 – 553 (1984). In Russian. English translation in Sov. Astron. Lett., Vol. 10.

A method of construction of the point masses model for the gravity potential of a planet if proposed on the basis of replacement of anomalous masses of spherical pyramids by point masses. A model of 288 point masses approximating the geopotential is given.

091.002 **Planetary shape: internal constraints for small bodies.**
J. A. Bastin.
Geophys. J. R. Astron. Soc., Vol. 78, No. 1, p. 219 – 229 (1984).

This paper describes the attainment of sphericity, or equivalent isopotential forms, for the surface of small Solar System objects (linear dimensions $\sim$300 km) as a consequence of viscous relaxation (creep).

091.003 **Free oscillations of giant planets. Influence of conditions at the phase boundary.**
S. V. Vorontsov.
Astron. Zh., Tom. 61, Vyp. 4., p. 700 – 707 (1984). In Russian. English translation in Sov. Astron., Vol. 28, No. 4.

Standard boundary conditions usually used at the discontinuity in the material parameter distribution in the theory of free oscillations of a self–gravitating sphere must be modified in the case when this boundary is a phase boundary. The derivation of new boundary conditions is presented and their applicability is discussed for a model of arbitrary stratification. The normal mode spectrum is calculated with the use of new boundary conditions and is discussed for a model of Jupiter.

091.004 **Resonance locking between Jupiter and Uranus.**
A. Milani, A. M. Nobili.
Nature, Vol. 310, No. 5980, p. 753 – 755 (1984).

Kinoshita and Nakai have recently performed the longest numerical integration of the outer bodies of the Solar System ever attempted. The positions and velocities of Jupiter, Saturn, Uranus, Neptune and Pluto for the 5 Myr that they computed are analysed here using new methods. Stability is monitored by separating the system into three–body subsystems. It is found that the outer Solar System is made up of two main subsystems, Sun–Jupiter–Saturn and Sun–Uranus–Neptune, exchanging angular momentum over a period of 1.1 Myr. The dynamical mechanism responsible for locking the two subsystems together is a libration of the angle between Jupiter's and Uranus' perihelion around 180° with the same period of 1.1 Myr.

091.005 **The boundary of the Solar System.**
R. Smoluchowski, M. Torbett.
Nature, Vol. 311, No. 5981, p. 38 – 39 (1984).

The shape of the boundary of the Solar System, defined as the surface within which the gravitational attraction of the Sun, rather than that of the rest of the Galaxy, controls the orbital motion of bodies such as planets and comets, has been determined. Outside of this surface, the dominant factors are the radial tides due to the galactic centre and, especially, the vertical tides caused by the galactic disk. Orbits which are direct with respect to the galactic plane, have a boundary which differs from that for retrograde orbits, both being 10 – 20% oblate and both larger than the, usually assumed, present Oort cloud. The surface may have been the boundary of the early cloud of comets which was later reduced by the passages of stars and molecular clouds.

091.006 **A note on the variational method of Stokes and DeMarcus for radiative transfer in planetary atmospheres.**
K. S. K. Chow, A. J. Friedson, Y. L. Yung.
Icarus, Vol. 59, No. 2, p. 177 – 187 (1984). = Contrib. No. 3894, Div. Geol. Planet. Sci., Calif. Inst. Technol., Pasadena, Calif.

A generalized functional which yields the Milne integral equation on variation and whose extremum value is proportional to the reflectivity at arbitrary emergent angle is proposed. A similar functional exists for computing the transmissivity at arbitrary emergent angle.

091.007 **Astrophysical aspects of charged–particle radiation in channeling.**
V. A. Muralev.
Nucl. Instrum. Methods Phys. Res., Sect. B, Vol. 230, No. 1 – 3, p. 97 – 98 (1984). Abstr. in Phys. Abstr., Vol. 87, No. 1238, Entry 77762 (1984).

091.008 **Erratum: "High–resolution infrared spectroscopy of planetary atmospheres" [Publ. Astron. Soc. Pac., Vol. 95, No. 572, p. 691 – 699 (1983)].**
A. T. Tokunaga.
Publ. Astron. Soc. Pac., Vol. 96, No. 581, p. 575 (1984). See Abstr. 34.091.043.

091.009 **Angular momenta in the solar system. II.**
R. J. Quiroga.
Earth, Moon, Planets, Vol. 31, No. 2, p. 157 – 174 (1984).

The statistical analysis of the orbital parameters of the planets and small objects of the solar system (meteors, comets and isolated asteroids outside the asteroid belt) shows that they are well associated to discrete levels $n_i$ in the solar potential. These levels are multiples of a fundamental unit of action per unit mass (potential × time). There are also indications of half–integer levels $n_i \pm 1/2$. The level transitions in comets perturbed by close approaches to Jupiter are analysed. A method of determination of the constant $H_0$ is postulated and the possible physical implications of this in relation with the planets and the sun are discussed.

091.010 **Measurements of planetary magnetic fields with space instrumentation.**
Sh. Sh. Dolginov.
O. Yu. Shmidt i sov. geofiz. 80–kh godov. Moskva, p. 172 – 183 (1983). In Russian. From Ref. Zh., 51. Astron., 8.51.139 (1984).

091.011 **On the origin of the magnetic field of a planet due to the inhomogeneous gravitational contraction of its core.**
P. V. Ponomarev, A. A. Rodionov, K. S. Ponomareva.
Kursk. politekh. inst. Kursk, 6 pp. (1984). In Russian. Abstr. in Ref. Zh., 51. Astron., 8.51.140 (1984).

091.012 **Planetary atmospheres.**
G. Hunt.
Spaceflight, Vol. 26, No. 11, p. 419 – 421 (1984).

091.013 **Atmospheres.**
E. J. Ciaccio.
Astronomy, Vol. 12, No. 5, p. 6 – 22 (1984). Abstr. in Phys. Abstr., Vol. 87, No. 1242, Entry 94657 (1984).

091.014 **Data on satellites of planets of the solar system.**
Zemlya Vselennaya, No. 4, p. 26 (1984). In Russian.

091.015 **La géologie des surfaces lunaire et planétaires.**
P. Masson, P. Thomas.
Astronomie, Vol. 98, p. 503 – 519 (1984).

091.016 **Structure interne des planètes telluriques.**
B. Lago.
Astronomie, Vol. 98, p. 521 – 529 (1984).

091.017 **Dynamique des atmosphères planétaires.**
O. Talagrand.
Instabilités hydrodynamiques et applications astrophysiques, p. 431 – 453 (1983). – See Abstr. 012.034.

Contents: 1. Généralités: Structure thermique et circulation. Mouvements hydrostatiques de grande échelle et mouvements convectifs de petite échelle. Influence de la rotation. Effets diurnes et saisonniers. 2. La circulation terrestre. 3. La circulation martienne. 4. La circulation de l'atmosphère de Vénus.

091.018 **Quelques aspects de la dynamique des atmosphères des planètes géantes.**
P. Drossart, T. Encrenaz.
Instabilités hydrodynamiques et applications astrophysiques, p. 455 – 469 (1983). – See Abstr. 012.034.

Contents: 1. La répartition des vents sur Jupiter et Saturne: Equation du vent thermique. Equation du vent géostrophique. Ondes Rossby. 2. Revue des modèles d'atmosphère de Jupiter: Modèle d'instabilité radiatif/convectif. Modèle convectif axisymétrique. Modèle de cascade turbulente à deux dimensions. Modèle de convection en rouleaux. 3. Analyse du profile en latitude des vents.

091.019 **Rare gases in the atmospheres of terrestrial planets and early stages of evolution of the solar system.**
M. N. Izakov.
The 27th International Geological Congress, Tom 5, Sect. 10 – 11, p. 302 – 303 (1984). In English and Russian. Abstr. in Ref. Zh., 51. Astron., 9.51.181 (1984). – See Abstr. 012.031.

091.020 **Connection between the axial rotation of planets and the existence of protoplanetary swarms of small particles.**
R. I. Kiladze.
O. Yu. Shmidt i sov. geofiz. 80–kh godov, Moskva, p. 205 – 214 (1983). In Russian. Abstr. in Ref. Zh., 51. Astron., 9.51.191 (1984).

091.021 **On the cause of differences in theoretical estimates of the angular momentum of planets.**
R. I. Kiladze.
O. Yu. Shmidt i sov. geofiz. 80–kh godov, Moskva, p. 215 – 220 (1983). In Russian. Abstr. in Ref. Zh., 51. Astron., 9.51.192 (1984).

091.022 **On forming of the rotational motion of planets in a boundary model of the accumulation process.**
T. M. Ehneev, N. N. Kozlov.
O. Yu. Shmidt i sov. geofiz. 80–kh godov, Moskva, p. 228 – 233 (1983). In Russian. Abstr. in Ref. Zh., 51. Astron., 9.51.193 (1984).

091.023 **On the regular and arbitrary components of angular momentum of planets in the process of their forming.**
A. V. Vityazev, G. V. Pechernikova.
O. Yu. Shmidt i sov. geofiz. 80–kh godov, Moskva, p. 244 – 249 (1983). In Russian. Abstr. in Ref. Zh., 51. Astron., 9.51.196 (1984).

091.024 **On the time of appearance of cosmogenic structures on the earth and planets.**
A. I. Savchenko.
Izv. Akad. Nauk ArmSSR. Nauk. o Zemle, Tom 37, No. 2, p. 37 – 39 (1984). In Russian. Abstr. in Ref. Zh., 62. Issled. Kosm. Prostranstva, 9.62.293 (1984).

091.025 **Primitive $^3He/^4He$ ratio in terrestrial diamonds and its implication on the origin of noble gases in terrestrial planets. Ozima Minoru, Zashu Shigeo.**
The 27th International Geological Congress, Tom. 5, Sect. 10 – 11, p. 362 – 363 (1984). Abstr. in Ref. Zh., 62. Issled. Kosm. Prostranstva, 9.62.296 (1984). – See Abstr. 012.031.

091.026 **Does crater "saturation equilibrium" occur in the solar system?**
W. K. Hartmann.
Icarus, Vol. 60, No. 1, p. 56 – 74 (1984).

Similarity is found in crater densities on the most heavily cratered surfaces throughout the solar system. This is hypothesized to result from a steady–state "saturation equilibrium" being approached or achieved by cratering processes. This hypothesis conflicts with some recent interpretations. However, it accounts for (1) a similarity in maximum relative crater density, below certain theoretically predicted values, on all heavily cratered surfaces; (2) a leveling off at this same relative density among 100–m scale (secondary?) craters in populations on lunar maria and other sparsely cratered lunar surfaces; (3) the approximate uniformity of maximum relative densities on Saturn satellites.

091.027 **Collisional theory of non–identical particles in a gravitational field.**
K. A. Hämeen–Anttila.
Earth, Moon, Planets, Vol. 31, No. 3, p. 271 – 299 (1984).

The statistical theory of partially elastic collisions and gravitational encounters is generalized for non–identical particles having different masses, internal densities, radii, elastic parameters or other properties.

091.028 **Electrical and electrothermal conductivity of the ionospheres of planets.**
A. V. Pavlov.
Kosm. Issled., Tom 22, Vyp. 5, p. 781 – 791 (1984). In Russian. English translation in Cosm. Res.

091.029 **Names of satellites of planets.**
A. V. Kozenko, E. P. Levitan.
Zemlya Vselennaya, No. 5, p. 90 – 94 (1984). In Russian.

091.030 **Hierarchy of structures: from dust particles to planets.**
M. A. Sadovskij.
Zemlya Vselennaya, No. 6, p. 4 – 9 (1984). In Russian.

091.031 **Axial rotations of accumulating planets.**
S. I. Ipatov.
O. Yu. Shmidt i sov. geofiz. 80–kh godov, Moskva, p. 239 – 243 (1983). In Russian. Abstr. in Ref. Zh., 51. Astron., 9.51.195 (1984).

091.032 **Ringed planets: still mysterious – I.**
J. N. Cuzzi.
Sky Telesc., Vol. 68, No. 6, p. 511 – 515 (1984).

091.033 **Gravitační pole a tvary planet.**
M. Burša.
Hvězdářská ročenka 1985, p. 179 – 216 (1984). – See Abstr. 046.033.

091.034 **The radio emission of the planets.**
A. Boischot.
Orages solaires de type I, p. 57 – 58 (1984). In French. Abstract. – See Abstr. 012.055.

091.035 **Comments on the theories of planetary radio bursts.**
A. Roux.
Orages solaires de type I, p. 59 – 63 (1984). In French. – See Abstr. 012.055.

Recent theories of planetary radio bursts are reviewed. They illustrate two important points for the theory of solar radio bursts: (1) the non–linear conversion factors strongly depend on the static magnetic field; (2) the non–linear processes of wave generation cannot in general be regarded as coherent.

091.036 **Model atmosphere calculations of the four Jovian planets.**
I. de Pater.
Bull. Am. Astron. Soc., Vol. 16, No. 3, Part I, p. 640 (1984). Abstract. – See Abstr. 010.061.

091.037 **Forcing and dissipation of a quasi–geostrophic vortex.**
K. Haines, J. Marshall, G. E. Hunt.
Bull. Am. Astron. Soc., Vol. 16, No. 3, Part I, p. 641 (1984). Abstract. – See Abstr. 010.061.

091.038 **Episodic heatflow in tidally heated satellites.**
G. W. Ojakangas, D. J. Stevenson.
Bull. Am. Astron. Soc., Vol. 16, No. 3, Part I, p. 661 (1984). Abstract. – See Abstr. 010.061.

091.039 **Shepherding moonlets: detection by adjacent ring radial structure.**
J. N. Cuzzi, M. R. Showalter.
Bull. Am. Astron. Soc., Vol. 16, No. 3, Part I, p. 676 (1984). Abstract. – See Abstr. 010.061.

091.040 **The shepherding of narrow rings revisited.**
N. Borderies, P. Goldreich, S. Tremaine.
Bull. Am. Astron. Soc., Vol. 16, No. 3, Part I, p. 676 – 677 (1984). Abstract. – See Abstr. 010.061.

091.041 **The solar system cratering record and the origin of impacting objects.**
R. G. Strom.
Bull. Am. Astron. Soc., Vol. 16, No. 3, Part I, p. 683 (1984). Abstract. – See Abstr. 010.061.

091.042 **Impact metamorphism of ice in the outer solar system: an update.**
E. S. Gaffney, D. L. Matson.
Bull. Am. Astron. Soc., Vol. 16, No. 3, Part I, p. 683 – 684 (1984). Abstract. – See Abstr. 010.061.

091.043 **Application of a radiative transfer model to bright icy satellites.**
B. J. Buratti.
Bull. Am. Astron. Soc., Vol. 16, No. 3, Part I, p. 685 (1984). Abstract. – See Abstr. 010.061.

091.044 **Tidal stresses due to nonsynchronous rotation: implications for fracturing on Europa.**
E. M. Parmentier, P. Helfenstein.
Bull. Am. Astron. Soc., Vol. 16, No. 3, Part I, p. 686 (1984). Abstract. – See Abstr. 010.061.

091.045 **Planetary nomenclature.**
H. Masursky, M. H. Strobell, K. E. Beer.
Bull. Am. Astron. Soc., Vol. 16, No. 3, Part I, p. 710 – 711 (1984). Abstract. – See Abstr. 010.061.

091.046 **The masses of the outer planets.**
E. M. Standish, J. K. Campbell.
Bull. Am. Astron. Soc., Vol. 16, No. 3, Part I, p. 722 (1984). Abstract. – See Abstr. 010.064.

091.047 **The energy flux number and three types of planetary dynamo.**
D. J. Stevenson.
Astron. Nachr., Vol. 305, No. 5, p. 257 – 264 (1984). = Contrib. No. 4009, Div. Geol. Planet. Sci., Calif. Inst. Technol., Pasadena. – See Abstr. 012.062.

It is proposed that the existence and nature of a planetary dynamo can be characterized by a dimensionless number $\Phi \equiv F_e R/\varrho\lambda^2\Omega$, called the energy flux number, where $F_e$ is the energy flux available for dynamo generation, $R$ is the core radius (or thickness of the dynamo generating region), $\varrho$ is the fluid density, $\lambda$ is the magnetic diffusivity and $\Omega$ is the angular velocity. For $\Phi \lesssim 1$, there is no dynamo. For $1 \lesssim \Phi \lesssim 10^{2.5}$ there is an "energy–limited dynamo", in which $F_e$ is insufficient to enable the dynamo to reach the dynamically desirable state $\Lambda \sim 1$. For $10^{2.5} \lesssim \Phi \lesssim 10^5$, there is a dynamically determined dynamo ($\Lambda \sim 1$) in which the magnetic Reynolds number of turbulent eddies is small. For $\Phi \gtrsim 10^5$, there is a turbulent dynamo. Probable planetary examples of these three dynamo states are Mercury ($\Phi \sim 10^2 - 10^3$), Earth ($\Phi \sim 10^4$) and Jupiter ($\Phi \sim 10^{11}$), respectively.

091.048 **Planetary observations at a wavelength of 1.32 mm.**
B. L. Ulich, J. R. Dickel, I. de Pater.
Icarus, Vol. 60, No. 3, p. 590 – 598 (1984).

Observations at a wavelength of 1.32 mm have been made of the Jovian planets, Ceres, the satellites Callisto and Ganymede, and the H II region DR 21. The observed brightness temperatures are presented. Those of the Jovian planets agree with the values expected from model atmosphere calculations, except that of Jupiter, which is lower than expected. Ceres and the satellites do not have atmospheres so their emission arised in their subsurface layers. The observed brightness temperatures are intermediate between those measured at infrared and centimeter wavelengths.

091.049 **Reorientation of planets with elastic lithospheres.**
R. J. Willemann.
Icarus, Vol. 60, No. 3, p. 701 – 709 (1984). = Contrib. 45 Texas A&M Univ. Geodyn. Res. Program.

The orientation of a planet is controlled by the positions of the principal axes of the inertia tensor relative to the planetary surface. Using the theory for the deflection of thin elastic shells the principal axes are computed after emplacement of an arbitrary axisymmetrical load. The partial compensation of the load and the partial relaxation of rotational flattening are included in the computation. It is found that the amount of reorientation is independent of lithosphere thickness. The parameters controlling the amount of reorientation are the location of the load and the size of the load compared to the rotational flattening. The results indicate that the Tharsis rise has probably reoriented Mars by only 3 to 9° and certainly less than 18°. The position of the Caloris Basin on Mercury indicates that if the surrounding lava sheet controls the planetary orientation then the lava sheet is probably less than 2000 m thick.

091.050 **Cratering rate over the surface of a synchronous satellite.**
G. P. Horedt, G. Neukum.
Icarus, Vol. 60, No. 3, p. 710 – 717 (1984).

Equations have been derived for the asymmetries of crater frequency over the surface of a synchronously rotating satellite, when the orbital velocities of projectiles about the parent planet are always larger than the satellite's circular orbital velocity. If the projectiles orbit the planet in moderately eccentric ellipses, no marked apex–antapex asymmetries of crater frequency distribution are expected. Theoretical values of apex–antapex asymmetries are presented for the Earth, the Moon, and some Jovian and Saturnian satellites, and are compared with available observational and theoretical results.

091.051 **Rotation of the atmospheres of the Earth and planets.**
R. Hide.
Philos. Trans. R. Soc. London, Ser. A, Vol. 313, No. 1524, p. 107 – 121 (1984). – See Abstr. 012.068.

The most striking features of the general circulation of the Earth's atmosphere are its average 'super-rotation' relative to the solid Earth at about 10 m $s^{-1}$, and the concentration of much of the motion in jet streams with speeds of about 30 m $s^{-1}$. Venus's atmosphere 'super-rotates' at an average speed of about ten times that of the underlying planet, which, owing to angular momentum exchange with the atmosphere, should (like the solid Earth) undergo detectable changes in rotation period. Each of the giant planets Jupiter and Saturn exhibits a sharply-bounded equatorial atmospheric jet stream moving at 100 m $s^{-1}$ (Jupiter) or 400 m $s^{-1}$ (Saturn) in a positive (i. e. eastward) direction relative to higher latitude parts of the atmosphere, the average rotation of which is close to that of the hydrogen-metallic fluid interior (as determined from radio-astronomical observations).

091.052 **Refraction halos in the solar system. I. Halos from cubic crystals that may occur in atmospheres in the solar system.**
E. Whalley, G. E. McLaurin.
J. Opt. Soc. Am. A, Vol. 1, No. 12, p. 1166 – 1170 (1984).

It is suggested that halos and other refraction effects caused by crystals of various solids that may occur in their atmospheres might occur on the planets of the solar system and their satellites. The detection of the halos would provide valuable information about the nature and the crystalline form of the solids. The halos caused by the octahedral and cubic crystals of carbon monoxide, carbon dioxide, ice Ic, ammonia, methane, and the structure-I clathrate hydrates of nitrogen, carbon monoxide, carbon dioxide, sulfur dioxide, and methane have been predicted over the range of temperatures that may occur in the atmospheres of the planets.

091.053 **Some statistical characteristics of cratered sections of the surface of planets.**
A. E. Altynov, M. Ya. Marov.
Inst. prikl. mat. Akad. Nauk SSSR. Prepr., No. 60, 22 pp. (1984). In Russian. Abstr. in Ref. Zh., 51. Astron., 11.51.178 (1984).

091.054 **Radio observations of planets and satellites.**
V. S. Troitskij.
Gor'k nauchn.-issled. radiofiz. inst. Prepr., No. 169, 63 pp. (1983). In Russian. Abstr. in Ref. Zh., 52. Geod. Aehrosemka, 11.52.197 (1984).

091.055 **Calderas: a planetary perspective.**
C. A. Wood.
J. Geophys. Res., Vol. 89, No. B10, p. 8391 – 8406 (1984).

In the review the author considers calderas in a comparative planetology context to provide perspective for students who investigate only terrestrial calderas and to search for broadscale relationships between calderas and planetary characteristics.

091.056 **The structure of the terrestrial planets and tidal friction theory.**
R. A. Lyttleton.
Astrophys. Relativ., Prepr. Ser., No. 109, 42 pp. (1984).

091.057 **Contrast sensitivity of the eye and the systematic errors of the coordinates of planets.**
B. S. Vozdvizhenskij.
Astron. Tsirk., No. 1314, p. 7 – 8 (1984). In Russian.

091.058 **On quantization of the solar system and of the stationary structure of megawaves in it.**
A. M. Chechel'nitskij.
Astron. Tsirk., No. 1334, p. 3 – 8 (1984). In Russian.

091.059 **Quantization of the sectorial velocity of the solar system and two groups of planetary orbits.**
A. M. Chechel'nitskij.
Astron. Tsirk., No. 1336, p. 2 – 6 (1984). In Russian.

091.060 **On the geospheres of the earth, the moon and terrestrial planets.**
S. A. Bektashi, R. N. Mamedzade.
Inst. geol. Akad. Nauk AzSSR. Baku, 12 pp. (1984). In Russian. Abstr. in Ref. Zh., 51. Astron., 12.51.233 (1984).

091.061 **Interpretation of gravitational anomalies of Mars, Venus and the earth.**
Yu. A. Tarakanov, N. Sh. Kambarov, V. A. Prikhod'ko, D. R. Bondarenko.
Izv. Akad. Nauk SSSR. Fiz. Zemli, No. 12, p. 3 – 15 (1983). In Russian. Abstr. in Ref. Zh., 52. Geod. Aehrosemka, 12.52.166 (1984).

091.062 **On radio physical investigations of planets with space vehicles.**
A. P. Zhukovskij.
Probl. povysh. ehffekt radiotekh. sistem i ustrojstv. Moskva, p. 28 – 33 (1984). In Russian. Abstr. in Ref. Zh., 62. Issled. Kosm. Prostranstva 12.62.120 (1984).

091.063 **The contribution of laboratory dynamo experiments to our understanding of the mechanism of generation of planetary magnetic fields.**
I. Wilkinson.
Geophys. Surv., Vol. 7, No. 1, p. 107 – 122 (1984). – See Abstr. 012.082.

The magnetic fields of the Earth, Jupiter, and Saturn are now accepted as originating in a dynamo mechanism in an electrically conducting fluid region of those planets. In the laboratory it is possible to construct homogeneous dynamos which, while they have a number of important differences from the mechanism which exists inside planets, nevertheless are considerably closer to reality than any theoretical model which can be shown to generate a magnetic field. Observations of the behaviour of such laboratory homogeneous self-exciting dynamos have, over the past twenty years in the Geophysics Department at Newcastle University, together with theoretical predictions on one hand and palaeomagnetic observations on the other, helped towards the development of a consistent picture of both how the geomagnetic field is generated and of its morphology. This review will attempt to show the part played by experimental homogeneous dynamos in the development of the subject.

091.064 **Particle sizes in planetary rings from forward-scattering and spectral-extinction observations.**
E. A. Marouf, G. L. Tyler, V. R. Eshleman.
Planetary rings, p. 281 (1984). Abstract. – See Abstr. 012.085 (IAU Colloq. No. 75).

091.065 **The W-shaped occultation signatures of planetary ring systems: inference of size-sorted particles in charged, tilted ringlets.**
V. R. Eshleman, G. L. Tyler, E. A. Marouf.
Planetary rings, p. 283 – 286 (1984). Abstract. – See Abstr. 012.085 (IAU Colloq. No. 75).

091.066 **Solved and unsolved problems in planetary ring dynamics.**
N. Borderies, P. Goldreich, S. Tremaine.
Planetary rings, p. 327 – 329 (1984). – See Abstr. 012.085 (IAU Colloq. No. 75).

The authors discuss a number of new ideas related to planetary ring dynamics and apply them to Saturn's rings. For simplicity, the ring particles are taken to be identical, smooth, spheres composed of dense and chemically pure ice.

091.067 **How a shepherd guides a sheep herd?**
R. Greenberg.
Planetary rings, p. 331 – 337 (1984). – See Abstr. 012.085 (IAU Colloq. No. 75).

The mechanism by which a shepherd satellite exerts a confining torque on a planetary ring is considered from the point of view of a single ring particle. It is still not clear how one might most meaningfully include damping effects and other collisional processes into this type of approach to the problem.

091.068 **Perturbed particle disks.**
N. Borderies, P. Goldreich, S. Tremaine.
Planetary rings, p. 339 (1984). Abstract. – See Abstr. 012.085 (IAU Colloq. No. 75).

091.069 **Dynamics of eccentric and inclined ringlets.**
N. Borderies, P. Goldreich, S. Tremaine.
Planetary rings, p. 341 – 343 (1984). Abstract. – See Abstr. 012.085 (IAU Colloq. No. 75).

091.070 **Sharp edges of planetary rings.**
N. Borderies, P. Goldreich, S. Tremaine.
Planetary rings, p. 345 – 348 (1984). – See Abstr. 012.085 (IAU Colloq. No. 75).

The ring systems of Saturn and Uranus exhibit a number of sharp edges across which the optical depth drops from order unity to near zero. At least two and perhaps all of these features are associated with the location of orbital resonances between a satellite and a ring particle. The authors illustrate with qualitative argument the results of a numerical simulation concerning the outer edge of the B ring of Saturn.

091.071 **Evolution of collisional systems.**
A. Brahic.
Planetary rings, p. 361 – 362 (1984). – See Abstr. 012.085 (IAU Colloq. No. 75).

The dynamical evolution of planetary disks in the gravitational field of an oblate planet and a satellite is numerically simulated.

091.072 **Size distribution in collisional systems.**
S. Clairemidi.
Planetary rings, p. 397 – 402 (1984). – See Abstr. 012.085 (IAU Colloq. No. 75).

The dynamical evolution of a three dimensional system of particles of different masses and sizes, orbiting in the gravitational field of a central body, and interacting through inelastic collisions is studied.

091.073 **Comments on collision mechanics in ring systems.**
W. K. Hartmann.
Planetary rings, p. 407 – 422 (1984). – See Abstr. 012.085 (IAU Colloq. No. 75).

The nature of collisions within ring systems is reviewed with emphasis on Saturn's rings. The particles may have coherent icy cores and less coherent granular or frosty surface layers, consistent with thermal eclipse observations. Present–day collisions of such ring particles do not cause catastrophic fragmentation of the particles. A plausible ring particle model involves solid ice cores with granular surface layers that exchange material by slow erosion and efficient reaccretion; the granular layers protect the cores from rapid erosion.

091.074 **On the radial structure of planetary rings.**
A. W. Harris, W. R. Ward.
Planetary rings, p. 431 – 437 (1984). – See Abstr. 012.085 (IAU Colloq. No. 75).

A ring of particles in orbit about a planet experiences a viscous shear stress due to the radial gradient of orbital velocity. According to the "shepherd satellite" model of narrow ring confinement, the above viscous shear stress is opposed by tidal torques from neighboring "shepherd" satellites, so that radial spreading does not occur. By requiring that the gradient of the viscous shear stress be equal to the "shepherd" satellite torque density at all radii in the ring, the authors have constructed equilibrium radial density profiles of a confined ring.

091.075 **Diffusion instability in a bi–modal disc.**
W. R. Ward, A. W. Harris.
Planetary rings, p. 439 – 445 (1984). – See Abstr. 012.085 (IAU Colloq. No. 75).

A simplified model of radial diffusion in a particle disc with a bi–modal size distribution is presented. The large particle component excites the small particles' dispersion velocity via gravitational scattering. The resulting small particle viscous stress promotes the development of high and low optical depth zones. The resulting ringlet appearance is due primarily to the distribution of small particles, which produce most of the optical depth but may constitute only a minor portion of the system's mass.

091.076 **Dynamics of neutral small particles.**
F. Mignard.
Planetary rings, p. 513 – 534 (1984). – See Abstr. 012.085 (IAU Colloq. No. 75).

The dynamics of small grains is reviewed when the radiation pressure forces are added to the gravitational attraction. First Mie scattering theories for small particles are reviewed. Then the motions of grains in heliocentric orbits are examined. Emphasis is given to the effects of the Poynting–Robertson drag and the conditions for the ejection of particles from the solar system are analysed. Finally, the motions of grains around a planet are considered and the implications for the dynamics of planetary rings are outlined.

091.077 **Dynamics of charged small particles.**
E. Grün.
Planetary rings, p. 535 – 540 (1984). – See Abstr. 012.085 (IAU Colloq. No. 75).

The interaction of charged dust particles with plasma in planetary magnetospheres is reviewed and some observed effects of the dust–magnetosphere interaction in the ring systems of Jupiter and Saturn are discussed.

091.078 **Diffusion of dust around a central body.**
P. Barge, J. Millet, R. Pellat.
Planetary rings, p. 541 – 548 (1984). – See Abstr. 012.085 (IAU Colloq. No. 75).

The dynamics of particles moving in an external field of force and submitted to a stochastic perturbative force is relevant to a number of astrophysical problems. The Fokker–Planck equation describing the evolution of the distribution function of particles in phase space can be reduced to a diffusion equation. The diffusion coefficients are shown to have a resonant behaviour. This aspect of the diffusion and the simplifications due to the formalism could be interesting in the study of planetary rings.

091.079 **On planetary rings as sources and sinks of magnetospheric plasmas.**
W.–H. Ip.
Planetary rings, p. 575 – 596 (1984). – See Abstr. 012.085 (IAU Colloq. No. 75).

The presence of satellite and ring systems in the magnetospheres of Jupiter, Saturn and Uranus introduces interesting new physical effects affecting the general dynamics and compositions of these magnetospheres. Besides acting as a sink for the magnetospheric particles via solid–body absorption, the satellites and rings could also emit neutral and ionized gas into the magnetospheres. In the case of the Saturnian rings, because of their especially broad structure and close proximity to the planet, strong exospheric and ionospheric couplings can also take place. Several observed features are discussed.

091.080 **On the electric charge carried by dust grains excitation of violent discharge.**
J.–P. J. Lafon, J. M. Millet.
Planetary rings, p. 627 – 630 (1984). – See Abstr. 012.085 (IAU Colloq. No. 75).

Field emission can, when excited by electric influence, destabilize the equilibrium charge carried by bodies in space plasmas and trigger violent discharge. The application to conditions in Saturn's rings is briefly discussed.

091.081 **The gravito–electrodynamics of charged dust in planetary ring systems.**
D. A. Mendis, J. R. Hill, T. G. Northrop.
Planetary rings, p. 631 – 643 (1984). – See Abstr. 012.085 (IAU Colloq. No. 75).

Solid bodies within planetary magnetospheres could be charged, under a variety of circumstances, to sizable electrostatic potentials. This charging can have both physical (e.g., electrostatic erosion and disruption) and dynamical effects on the smaller (micron and sub–micron sized) dust grains. This paper discusses the dynamics of such small grains moving under the combined influence of the planet's gravitational and the electromagnetic forces within the co–rotating regions of planetary magnetospheres, where the observed ring systems are located.

091.082 **Planetary rings: a dynamic, evolving subject.**
R. Greenberg, A. Brahic.
Planetary rings, p. 3 – 11 (1984). – See Abstr. 012.086.

091.083 **Dust–magnetosphere interactions.**
E. Grün, G. E. Morfill, D. A. Mendis.
Planetary rings, p. 275 – 332 (1984). – See Abstr. 012.086.
Mircon–sized dust grains were identified by their light scattering characteristics in most rings of the outer planets. A multitude of interactions between the magnetospheric particles, fields, and dust grains has been proposed. The authors review the major effects and indicate the pertinent observations. Electromagnetic effects are discussed that lead to the halo of Jupiter's ring, the dust distribution in Saturn's E Ring and to levitated dust in the B Ring (spokes) as well as on the Moon.

091.084 **Effects of radiation forces on dust particles in planetary rings.**
F. Mignard.
Planetary rings, p. 333 – 366 (1984). – See Abstr. 012.086.
The effects of radiation forces on a grain in planetary orbit are examined with particular emphasis on implications for planetary rings. The origin of radiation forces is briefly considered. The secular changes of orbital elements are derived and the effects of the planetary shadow assessed. The periodical evolution of the other orbital elements is investigated, both for a spherical and for an oblate planet. This study demonstrates the existence of a forced eccentricity that is strongly enhanced at a particular critical distance where resonance occurs.

091.085 **Ring particles: collisional interactions and physical nature.**
S. J. Weidenschilling, C. R. Chapman, D. R. Davis, R. Greenberg.
Planetary rings, p. 367 – 415 (1984). – See Abstr. 012.086.
Saturn's rings are composed of a myriad of individual particles. Most treatments, however, consider the rings as a continuum, characterized by fluid–like properties such as viscosity. This paper concentrates on the properties of, and dynamical processes affecting, individual ring particles. In particular, accretionary processes within the rings and the counterbalancing tidal disruption of larger aggregates are studied and the resulting particle size distribution is discussed.

091.086 **Transport effects due to particle erosion mechanisms.**
R. H. Durisen.
Planetary rings, p. 416 – 446 (1984). – See Abstr. 012.086.
Various processes can erode the surfaces of planetary ring particles. Erosion rates for the surfaces of icy ring particles due to photosputtering and meteoroid impacts are extremely high. Both these processes produce ejecta that will arc across the rings along Keplerian orbits and will eventually be reabsorbed into the ring system or lost from the ring system by escape, by falling into the planet, or by destruction. Between ejection and reabsorption or loss, atoms, molecules, and chips on ballistic trajectories will exist in steady–state concentrations around a ring system. Exchanges of ejecta between ring regions can lead to significant net transport of mass and angular momentum.

091.087 **Collision–induced transport processes in planetary rings.**
G. R. Stewart, D. N. C. Lin, P. Bodenheimer.
Planetary rings, p. 447 – 512 (1984). – See Abstr. 012.086.
The physics of collision–dominated particle disks around planets is analyzed from both the analytical and numerical point of view. The existence of Saturn's ringlet structure is analyzed in terms of a radial instability induced by viscous diffusion. The detailed model described is based on the assumption of uniform particle size. Modifications caused by finite particle size, gravitational effects, and a distribution of particle sizes are discussed.

091.088 **Waves in planetary rings.**
F. H. Shu.
Planetary rings, p. 513 – 561 (1984). – See Abstr. 012.086.
The Voyager spacecraft revealed the rings of Saturn to have an unexpected richness of structure. Many of the observed features have now been identified as collective effects arising from the self–gravity of the ring material. These effects include spiral density waves and spiral bending waves, the main topics of this review paper. After developing the theory of free and forced waves of both types, the author discusses how the observed waves can be used as diagnostics to obtain crucial parameters that characterize the physical state of the rings.

091.089 **Ring particle dynamics in resonances.**
F. Franklin, M. Lecar.
Planetary rings, p. 562 – 588 (1984). – See Abstr. 012.086.
This paper analyses in a general way the characteristics of gravitational perturbations that arise when a resonance exists between the orbital frequencies of a satellite and a ring particle of negligible mass. Development of a time–averaged Hamiltonian provides an evaluation of the range in semimajor axis where resonant orbits occur, their forced eccentricities and characteristic times for development. The results are used to show that sufficiently strong resonances can produce gaps in the radial distribution of ring particles.

091.090 **Dynamics of narrow rings.**
S. F. Dermott.
Planetary rings, p. 589 – 637 (1984). – See Abstr. 012.086.
The ring models described here were developed to account for the dynamical problems posed by the narrow rings of Uranus. Some of these rings are now known to be eccentric, inclined, nonuniform in width, optically thick, and narrow, with very sharp edges. The theory of shepherding satellites successfully accounts for most of these features and can also account for some features of the narrow Saturnian rings, in particular, waves, kinks, and periodic variations in brightness. The horseshoe–orbit model, although viable, probably has more application to the narrow rings from which the Saturnian coorbital satellites formed.

091.091 **Unsolved problems in planetary ring dynamics.**
N. Borderies, P. Goldreich, S. Tremaine.
Planetary rings, p. 713 – 734 (1984). – See Abstr. 012.086.
The authors discuss a number of unsolved problems in planetary ring dynamics and offer some thoughts concerning their solution. The discussion is specialized to the rings of Saturn. The authors relate the physics of particle collisions to the local velocity dispersion and the particle size distribution. They discuss mechanisms which structure the rings and affect the orbital evolution of satellites which interact with them.

**Bibliography.**
See Abstr. 002.009.

**Planetary imaging data archive–laser videodisc storage and retrieval.**
See Abstr. 002.031.

**A bibliography of planetary geology and geophysics principal investigators and their associates, 1983 – 1984.**
See Abstr. 002.046.

**Cambridge Fotoatlas der Planeten.**
See Abstr. 002.078.

**Physics of planetary interiors.**
See Abstr. 003.015.

**Planeten, Monde, Ringsysteme: Kamerasonden erforschen unser Sonnensystem.**
See Abstr. 003.036.

**Das Sonnensystem.**
See Abstr. 003.040.

**Planetary interiors.**
See Abstr. 003.043.

**Planets and their atmospheres. Origin and evolution.**
See Abstr. 003.046.

**Planeten – Geschwister der Erde?**
See Abstr. 003.083.

**Remote sounding of atmospheres.**
See Abstr. 003.105.

**The earth's magnetic field. Its history, origin and planetary perspective.**
See Abstr. 003.122.

**Il sistema solare.**
See Abstr. 003.138.

**Trends in planetary data analysis. Executive summary of the Planetary Data Workshop, held at Goddard Space Flight Center, 29 November – 1 December 1983.**
See Abstr. 011.011.

**Planetary science experiments aboard Space Station: workshop report.**
See Abstr. 011.022.

**Lowell Observatory's International Planetary Patrol: current status.**
See Abstr. 013.042.

**Dynamical astronomy via optical astrometric interferometry in space.**
See Abstr. 013.048.

**Future observations of planetary rings from ground–based observatories and earth–orbiting satellites.**
See Abstr. 013.077.

**Future observations of planetary rings from groundbased observatories and earth–orbiting satellites.**
See Abstr. 013.080.

**Emission of large hydrocarbons from frozen $CH_4$ by keV proton irradiation.**
See Abstr. 022.011.

**Hydrocarbon production in simulated lightning discharges.**
See Abstr. 022.097.

**Coloration and darkening of $CH_4$ hydrate and other ices upon charged–particle irradiation.**
See Abstr. 022.104.

**Spectroscopic data on planetary atmospheric hydrocarbons at relevant temperatures.**
See Abstr. 022.105.

**Rotational assignments in the 6200 Å band of methane.**
See Abstr. 022.106.

**Improved molecular constants for HD.**
See Abstr. 022.108.

**Developments in ultra–high pressure diamond cell experiments: applications to planetary interiors.**
See Abstr. 022.109.

**Effects of electron bombardment on the light scattering properties of icy planetary surfaces.**
See Abstr. 022.110.

**Determination of mineral abundance from reflectance spectra.**
See Abstr. 022.111.

**Rheological measurements of ammonia–water melt.**
See Abstr. 022.112.

**Solar photo rates for planetary atmospheres and atmospheric pollutants.**
See Abstr. 022.119.

**Band model coefficients of methane from 4000 to 9200 $cm^{-1}$, determined at 112, 188, and 294K.**
See Abstr. 022.120.

**Spectral properties of ice–particulate mixtures and implications for remote sensing. 1. Intimate mixtures.**
See Abstr. 022.138.

**Reflectance spectroscopy: quantitative analysis techniques for remote sensing applications.**
See Abstr. 036.126.

**Excitation of inclinations in ring–satellite systems.**
See Abstr. 042.039.

**Perihelic shift of planets due to the gravitational field of the charged Sun.**
See Abstr. 042.045.

**Rotation and the internal structures of the major planets and their inner satellites.**
See Abstr. 042.062.

**Millisecond pulsars and the location of the solar system barycenter.**
See Abstr. 042.075.

**On the evolution of rotations of a viscoelastic planet in a circular orbit of a central–force field.**
See Abstr. 042.084.

**The gravitational interaction between inclined, elliptical rings.**
See Abstr. 042.088.

**Stochastic modeling of planetary ring systems.**
See Abstr. 042.089.

**Applications of three–body resonances to satellites and planetary rings.**
See Abstr. 042.090.

**The adiabatic invariant: a tool for the study of the dynamics of the solar system.**
See Abstr. 042.091.

**Exploring beyond the planets: the Pioneer 10 and 11 missions.**
See Abstr. 051.017.

**Future studies of planetary rings by spaceprobes.**
See Abstr. 051.073.

**Future studies of planetary rings by space probes.**
See Abstr. 051.078.

**Transport coefficients and collision frequencies for aeronomic plasmas.**
See Abstr. 062.146.

**Radiative transfer in a homogeneous sphere illuminated by a parallel beam.**
See Abstr. 063.056.

**Three–dimensional radiative transfer in a scattering atmosphere using doubling and adding matrix algorithms.**
See Abstr. 063.057.

**Limits on metric parameters β and γ.**
See Abstr. 066.213.

**Testing relativity with solar system dynamics.**
See Abstr. 066.228.

**Neutrino halos around the sun and precession of perihelia of planets.**
See Abstr. 066.245.

**Experiment versus theory in gravitation.**
See Abstr. 066.246.

**A possible relationship between spectral bands in sunspot number and the space–time organization of our planetary system.**
See Abstr. 072.040.

**Grain abundance in the primordial atmosphere of the Earth.**
See Abstr. 082.018.

**Comparison of six crater–scaling laws.**
See Abstr. 094.034.

**An empirical phase relation for atmosphereless bodies.**
See Abstr. 098.036.

**Numerical simulation of compressible convection in dense hydrogen–helium fluids. A novel instability.**
See Abstr. 099.026.

**Equilibrium theory for Saturn's ring with self–gravitation and finite thickness.**
See Abstr. 100.013.

**Predictions for Uranus from a radiometric Bode's law.**
See Abstr. 101.004.

**From icy planetesimals to outer planets and comets.**
See Abstr. 107.001.

**Numerical investigations of the evolution of spatial rings of gravitational bodies corresponding to the feeding zones of giant planets.**
See Abstr. 107.015.

**Origin and history of the outer planets: theoretical models and observational constraints.**
See Abstr. 107.023.

# 092 Mercury

092.001 **The dependence of reflectance spectra of Mercury on surface terrain.**
F. Vilas, M. A. Leake, W. W. Mendell.
Icarus, Vol. 59, No. 1, p. 60 – 68 (1984).

Reflectance spectra of Mercury, covering the spectral range of ~0.3 – 1.1 μm obtained during 1963 – 1976, were examined for any correlations with surface terrain. Mercury's 6.1385°/day rotational rate, the phases of the planet around maximum elongations, and bidirectional reflectance spectroscopy theory were used to identify the surface area associated with each spectrum. The absorption features seen in spectra of Mercury are generally weaker than features seen in lunar spectra.

092.002 **The precession of Mercury's perihelion via perturbation theory.**
M. A. Rosales, J. L. Castro–Quilantan.
Rev. Mex. Fis., Vol. 30, No. 2, p. 341 – 346 (1984). Abstr. in Phys. Abstr., Vol. 87, No. 1245, Entry 110552 (1984).

092.003 **A Mercurian time–stratigraphic classification.**
P. D. Spudis.
Bull. Am. Astron. Soc., Vol. 16, No. 3, Part I, p. 667 – 668 (1984). Abstract. – See Abstr. 010.061.

092.004 **Using new topographic profiles to interpret Mercury's surface structure.**
P. E. Clark, R. F. Jurgens.
Bull. Am. Astron. Soc., Vol. 16, No. 3, Part I, p. 668 (1984). Abstract. – See Abstr. 010.061.

092.005 **Geophysical bounds on the density and composition of Mercury.**
K. A. Goettel.
Bull. Am. Astron. Soc., Vol. 16, No. 3, Part I, p. 668 (1984). Abstract. – See Abstr. 010.061.

092.006 **On the statistical description of fragments of the Mercury surface.**
A. E. Altynov.
Izv. vuzov. Geod. aehrofotosemka, No. 3, p. 78 – 84 (1984). In Russian. Abstr. in Ref. Zh., 52. Geod. Aehrosemka, 11.52.217 (1984).

092.007 **On the statistical description of fragments of the Mercury surface.**
A. E. Altynov.
Izv. vuzov. Geod., aehrofotosemka, No. 3, p. 78 – 84 (1984). In Russian. Abstr. in Ref. Zh., 51. Astron., 12.51.236 (1984).

**Bidirectional reflectance spectroscopy. 3. Correction for macroscopic roughness.**
See Abstr. 063.019.

**Gravitation in the weak–field limit.**
See Abstr. 066.190.

**Reorientation of planets with elastic lithospheres.**
See Abstr. 091.049.

**Secular Love numbers and hydrostatic equilibrium of planets.**
See Abstr. 093.009.

**Filling–in Fraunhofer lines in the spectrum of the moon and Mercury.**
See Abstr. 094.049.

**Geology and tectonics of the Argyre area on Mars: comparisons with other basins in the solar system.**
See Abstr. 097.005.

# 093 Venus

093.001 **Non-equilibrium population of the $CO_2(00^\circ1)$ and CO(1) vibrational states in the upper atmospheres of Venus and Mars.**
G. I. Stepanova, G. M. Shved.
Astron. Tsirk., No. 1294, p. 3 – 4 (1983). In Russian.

093.002 **Heating of the Venusian and Martian upper atmospheres by infrared solar radiation.**
G. I. Stepanova, G. M. Shved.
Astron. Tsirk., No. 1294, p. 5 – 6 (1983). In Russian.

093.003 **Venus observations at the State Astronomical Sternberg Institute in 1975 – 1977.**
B. S. Vozdvizhenskij.
Tr. Gos. Astron. Inst. Shternberg, Tom 55, p. 47 – 58 (1983). In Russian.
A catalogue of 510 Venus positions is given.

093.004 **Photochemistry of fluorine in the mesosphere of Venus.**
J.-P. Parisot, G. Moreels.
Icarus, Vol. 59, No. 1, p. 69 – 73 (1984).
In the laboratory, reactions with fluorine species proceed rapidly with high rates but under mesospheric conditions the efficiency of these compounds is low due to the rapid formation of HF and to the lack of reactivity of this species. Even if diffusion processes are included, the results of calculations lead to fluorine concentrations typically less than 20 $cm^{-3}$. The low photodissociation coefficient of HF leads to the expectation of a scale height of HF greater than or equal to the mean scale height. If the troposphere appears to be a sink for hydrofluoric acid, the maximum value of fluorine is obtained with a downward flux of $1.3 \times 10^8$HF atoms $cm^{-2}s^{-1}$ at the level of the clouds.

093.005 **Comparison of induced magnetospheres at Venus and Titan.**
M. I. Verigin, K. I. Gringauz, N. F. Ness.
J. Geophys. Res., Vol. 89, No. A7, p. 5461 – 5470 (1984).
The paper discusses plasma and magnetic field data from the Venera 9 and 10 spacecraft at Venus and from Voyager 1 at Titan which lead to new conclusions regarding the magnetic tail structure of their induced magnetospheres.

093.006 **Questions for the geologic exploration of Venus.**
R. S. Saunders.
J. Br. Interplanet. Soc., Vol. 37, No. 10, p. 435 – 441 (1984).
Venus is the last terrestrial planet to be mapped in the Solar System. Venus is the most important planet to study for improving our understanding of global terrestrial processes such as plate tectonics. The next NASA mission to Venus, in 1988, will map the planet at a resolution comparable to that of the Mars Mariner 9 mission. The objective is 70% coverage at 1 km per line pair or better. This is the minimum resolution that will allow analysis of tectonic, volcanic, aeolian, meteorite impact and other features. An altimeter will provide a global map of topography that will considerably improve our data base and, together with gravity field measurements, provide better interior geophysical models.

093.007 **Etude de la physico-chimie de la haute atmosphère de Vénus.**
J. P. Parisot, G. Moreels, P. Rigaud, J.-M. Zucconi.
Ann. Sci. Univ. Besançon, Phys., 4ème Sér., Fasc. 2, p. 3 – 34 (1982).

093.008 **Volcanism on Venus: a connecting link?**
L. V. Ksanfomaliti.
Pis'ma Astron. Zh., Tom 10, No. 8, p. 620 – 630 (1984). In Russian. English translation in Sov. Astron. Lett., Vol. 10.
On the basis of data obtained during recent investigations of Venus the suggestion is made that the presence of electric discharges in the atmosphere close to the surface, the variable density of the submicron haze and the peculiarity of cloud microphysics can be jointly explained in the frame of the hypothesis about volcanic eruptions on the surface of Venus.

093.009 **Secular Love numbers and hydrostatic equilibrium of planets.**
M. Burša.
Earth, Moon, Planets, Vol. 31, No. 2, p. 135 – 140 (1984).
The secular Love numbers of planets and of the Moon have been computed by use of the recent satellite data. It has been demonstrated that Venus, Mercury, and the Moon are far from the ideal hydrostatic equilibrium; the angular velocities (rotation periods) which correspond to the hydrostatic equilibrium have been computed.

093.010 **Tectonics of the southern escarpment of Ishtar Terra on Venus from observations of morphology and gravity.**
P. Janle, D. Jannsen.
Earth, Moon, Planets, Vol. 31, No. 2, p. 141 – 155 (1984). = Contrib. No. 273, Inst. Geophys. Univ. Kiel.
It has been shown in this paper that there are some arguments for plate collision and subduction at the southern border of Ishtar Terra. However, an extensive plate recycling on Venus cannot be extrapolated from this result. Former radar observations have revealed several circular features on Venus of unknown nature (volcanic, impact; McGill et al., 1983). First radar pictures of Venera 15 and 16 strongly favour an impact origin of at least a part of these circular structures (Beatty, 1984). This implies the existence of parts of very old surfaces and is an argument against extensive plate recycling. Nevertheless, there are arguments that beside heat transfer by lithospheric conduction and hot spot volcanism plate recycling is or has formerly been active on Venus.

093.011 **Venera 15 and Venera 16 infrared spectrometry: first results.**
D. Oertel, J. Nopirakowsky, H. Becker-Ross, W. Berwald, W. Döhler, H. Drischer, G. Fellberg, H. Jahn, K. Schäfer, R. Schuster, W. Skrbek, D. Spänkuch, W. Stadhaus, H. Studemund, M. Ulich, V. I. Moroz, V. I. Linkin, R. S. Kremnev, A. V. D'yachkov, V. V. Kerzhanovich, A. N. Lipatov, I. A. Matsigorin (*I. A. Matsygorin*), A. A. Shurupov, I. D. Tserenin, E. A. Ustinov, L. V. Zasova, I. A. Zelenov.
Sov. Astron. Lett., Vol. 10, No. 2, p. 101 – 105 (1984). English translation of 37.093.031.

093.012 **On the proper Mach number and ratio of specific heats for modeling the Venus bow shock.**
M. Tatrallyay, C. T. Russell, J. G. Luhmann, A. Barnes, J. D. Mihalov.
J. Geophys. Res., Vol. 89, No. A9, p. 7381 – 7392 (1984).
When considering local solutions for the jump in plasma conditions across planetary bow shocks, a magnetohydrodynamic (MHD) formalism of the Rankine-Hugoniot relations is used. However, when planetary shocks are modeled on a global scale, gas dynamic (GD) treatments are frequently employed which neglect the magnetic field effect on the shock structure and the flowing plasma. These treatments require the specification of a free-stream Mach number and a ratio of specific heats. Observations of the Venus bow shock are used to illustrate how closely one can approximate the observed magnetic field jump with MHD and GD treatments if one chooses a particular Mach number and ratio of specific heats.

093.013 **Physico-chemical modelling of the mineralogical composition of Venus surface rocks.**
V. P. Volkov, I. L. Khodakovskij.
The 27th International Geological Congress, Tom 5, Sect. 10 – 11, p. 431 – 432 (1984). In Russian. Abstr. in Ref. Zh., 62. Issled. Kosm. Prostranstva, 8.62.403 (1984). – See Abstr. 012.031.

093.014 **Radar mapping of the Venus surface with the automatic interplanetary stations Venera 15 and 16.**
A. F. Bogomolov, Yu. S. Tyuflin.
Geod. Kartogr., No. 2, p. 5 – 9 (1984). In Russian. Abstr. in Ref. Zh., 52. Geod. Aehrosemka, 8.52.195 (1984).

093.015 **Buffering of diurnal temperature variations on Venus.**
B. M. Jakosky.
Icarus, Vol. 59, No. 3, p. 478 – 480 (1984).
Previous calculations of the role of degassing of $CO_2$ from calcite in buffering the surface temperature of Venus ignored the efficacy with which energy could be conducted into the subsurface. Conduction into the subsurface plays a minor role in the energy balance, however, such that less than 10% of the insolation will conduct into the subsurface and go into degassing $CO_2$, with the remainder heating up the surface and atmosphere; negligible buffering of the surface temperature will occur.

093.016 **On the mechanism of formation of low temperatures in the nightside thermosphere of Venus.**
B. F. Gordiets, Yu. N. Kulikov.
Pis'ma Astron. Zh., Tom 10, No. 9, p. 696 – 701 (1984). In Russian. English translation in Sov. Astron. Lett., Vol. 10.
A mechanism of cooling of the nightside Venusian thermosphere at heights $\geqslant 100$ km owing to IR radiation in the rotational bands of CO and $H_2O$ is suggested and analysed. The calculations show that while the role of CO is not important, radiation of $H_2O$ can modify the thermospheric temperature profiles for water vapor mixing ratio.

093.017 **Interpretation of the Venus gravitational anomalies.**
Yu. A. Tarakanov, N. Sh. Kambarov, V. A. Prikhod'ko, D. R. Bondarenko.
Kosm. Issled., Tom 22, Vyp. 4, p. 617 – 621 (1984). In Russian. English translation in Cosm. Res.

093.018 **Is Venus younger than Earth?**
R. E. Newell.
Speculations Sci. Technol., Vol. 7, No. 1, p. 51 – 57 (1984). Abstr. in Phys. Abstr., Vol. 87, No. 1242, Entry 94662 (1984).

093.019 **Venus sky polarization.**
V. M. Loskutov, I. N. Minin, K. I. Selyakov.
Pis'ma Astron. Zh., Tom 10, No. 10, p. 780 – 784 (1984). In Russian. English translation in Sov. Astron. Lett, Vol. 10.
Computations of polarization degree and intensity of radiation near the Venus surface are described. It is shown that the radiation field near the surface is practically determined only by the amount of absorbing components in the lower atmospheric layers.

093.020 **Vénus.**
G. Israël.
Astronomie, Vol. 98, p. 493 – 502 (1984).

093.021 **A new interpretation of the Venera 11 spectra of Venus. I. The 0.94–μm water band.**
L. D. G. Young, A. T. Young, L. V. Zasova.
Icarus, Vol. 60, No. 1, p. 138 – 151 (1984).
Band models for $CO_2$ and $H_2O$ absorption are described, and used to model the Venera 11 spectra near 1 μm. An effective–path approximation is used to allow for scattering in the clouds. The model has 10 layers and uses 211 $CO_2$ and 15 $H_2O$ vibrational transitions, at 5 $cm^{-1}$ resolution. Within a factor of 2, a maximum absorption in the 0.94–μm $H_2O$ band just below the clouds, corresponding to 200 ppm by volume, in agreement with V. I. Moroz et al. (1979) is found. More accurate band strengths are needed to model the bottom scale height accurately. The possibility that the 0.94–μm feature is blended with a band of some other molecule has been examined. Ten possible chemical species were examined, with negative results.

093.022 **Hydrogen on Venus: exospheric distribution and escape.**
J. M. Rodriguez, M. J. Prather, M. B. McElroy.
Planet. Space Sci., Vol. 32, No. 10, p. 1235 – 1255 (1984).
Charge exchange between H and $H^+$ and momentum transfer between fast O and H provide comparable sources for suprathermal H atoms in Venus' exosphere. The fast O atoms are produced by dissociative recombination of $O_2^+$. The spatial distribution of suprathermal H was calculated using an approximate numerical solution of the time–independent Boltzmann equation. Sources of suprathermal H atoms were specified on the basis of measurements by Pioneer Venus.

093.023 **A multiple–source model for the volatiles of Venus.**
T. Owen.
The 27th International Geological Congress, Tom 5, Sect. 10 – 11, p. 326 (1984). Abstr. in Ref. Zh., 51. Astron., 10.51.206 (1984). – See Abstr. 012.031.

093.024 **On the possibility of spectroscopic detection of oxygen in the lower Venus atmosphere.**
V. A. Krasnopol'skij.
Kosm. Issled., Tom 22, Vyp. 5, p. 812 – 815 (1984). In Russian. English translation in Cosm. Res.

093.025 **Nightward ion flow in the Venus ionosphere: implications of momentum balance.**
R. C. Elphic, H. G. Mayr, R. F. Theis, L. H. Brace, K. L. Miller, W. C. Knudsen.
Geophys. Res. Lett., Vol. 11, No. 10, p. 1007 – 1010 (1984).
Using global empirical models of Venus ionospheric conditions, the authors solve for the plasma flow field consistent with the horizontal momentum equation in both viscid and inviscid forms.

093.026 **A comparative study of distant magnetotail structure at Venus and Earth.**
J. A. Slavin, E. J. Smith, D. S. Intriligator.
Geophys. Res. Lett., Vol. 11, No. 10, p. 1074 – 1077 (1984).
The authors investigate the Venus and Earth magnetotails for the purpose of better understanding the interaction of the solar wind with these planets.

093.027 **Some problems of Venus' atmospheric dynamics.**
G. S. Golitsyn.
Icarus, Vol. 60, No. 2, p. 289 – 306 (1984).
The author discusses here certain problems in Venus' atmospheric dynamics which concern the intensity and structure of the general circulation, as well as the nature of the near–surface layer.

093.028 **Water loss on Venus: the role of carbon monoxide.**
S. M. Richardson, J. B. Pollack, R. T. Reynolds.
Icarus, Vol. 60, No. 2, p. 307 – 316 (1984).
The abundance of water on Venus is four to five orders of magnitude lower than on the Earth. This difference may reflect initial differences in the bulk volatile contents of the two planets, or may be the result of processes for massive water loss on Venus. A series of thermodynamic calculations are performed on the heterogeneous system C–O–H–N–S in order to evaluate the extent to which either of these possibilities may account for the low water content on Venus. It is concluded that together with the lower initial water abundance that follows from a higher C/H, water loss involving CO may have been a major factor in determining the present low abundance of water on Venus.

093.029 **Dynamics of the Venus ionosphere: a two–dimensional model study.**
R. C. Whitten, P. T. McCormick, D. Merritt, K. W. Thompson, R. R. Brynsvold, C. J. Eich, W. C. Knudsen, K. L. Miller.
Icarus, Vol. 60, No. 2, p. 317 – 326 (1984).
A two–dimensional model of the ionosphere of Venus which simulates ionospheric dynamics by self–consistently solving the plasma equations of motion, including the inertial term, in finite difference form has been constructed. The model simulates the measured horizontal velocity field quite satisfactorily. The ion

density field is somewhat overestimated on the dayside and underestimated on the nightside. It is concluded that solar photoionization on the dayside and ion recombination on the nightside are the processes mainly responsible for accelerating the plasma to the observed velocities. The plasma flow appears to be sufficient to maintain the nightside ionosphere at or near the observed median level of ion densities.

093.030 **Venusdurchgänge.**
W. Wepner.
Sterne Weltraum, 23. Jahrg., Nr. 12, p. 624 – 625 (1984).

093.031 **Observations and theory of the solar semidiurnal tide in the mesosphere of Venus.**
S. B. Fels, J. T. Schofield, D. Crisp.
Nature, Vol. 312, No. 5993, p. 431 – 434 (1984).
The Orbiter Infrared Radiometer experiment, aboard the 1978 Pioneer Venus mission, produced strong evidence for the existence of a semidiurnal solar tide in the mesosphere of Venus. The structure of tidal fields depends sensitively on that of the background zonal mean wind velocity. It is shown here that a physically reasonable mean flow can be found that leads to theoretically predicted tides that are in excellent agreement with those observed. The zonal mean flow shows a maximum velocity of $\sim 130\ \mathrm{ms}^{-1}$ at 70 km, decreasing dramatically to $\sim 0$ at 88 km, and a jet–like meridional structure that deviates strongly from solid–body rotation.

093.032 **Propagation of small–scale acoustic–gravity waves in the Venus atmosphere.**
G. Schubert, R. L. Walterscheid.
J. Atmos. Sci., Vol. 41, No. 7, p. 1202 – 1213 (1984). Abstr. in Phys. Abstr., Vol. 87, No. 1246, Entry 115701 (1984).

093.033 **Venus: journey to the end of hell.**
O. de Goursac.
Ciel Espace, No. 199, p. 22 – 27 (1984). In French. Abstr. in Phys. Abstr., Vol. 88, No. 1247, Entry 4636 (1985).

093.034 **The interpretation of the gravity field of Venus and Mars.**
S. K. Runcorn.
Bull. Am. Astron. Soc., Vol. 16, No. 3, Part I, p. 682 (1984). Abstract. – See Abstr. 010.061.

093.035 **Spectral admittance of Venus.**
R. D. Reasenberg, Z. M. Goldberg.
Bull. Am. Astron. Soc., Vol. 16, No. 3, Part I, p. 693 (1984). Abstract. – See Abstr. 010.061.

093.036 **Venus radio emission: ground–based observations.**
B. D. Chapman, G. H. Pettengill.
Bull. Am. Astron. Soc., Vol. 16, No. 3, Part I, p. 693 (1984). Abstract. – See Abstr. 010.061.

093.037 **High dispersion spectroscopic observations of Venus during 1973.**
L. G. Young, A. Woszczyk.
Bull. Am. Astron. Soc., Vol. 16, No. 3, Part I, p. 693 (1984). Abstract. – See Abstr. 010.061.

093.038 **Theoretical calculations of the Venus semidiurnal tide.**
S. B. Fels, D. Crisp, J. T. Schofield.
Bull. Am. Astron. Soc., Vol. 16, No. 3, Part I, p. 693 – 694 (1984). Abstract. – See Abstr. 010.061.

093.039 **Radiative forcing of the Venus mesosphere.**
D. Crisp.
Bull. Am. Astron. Soc., Vol. 16, No. 3, Part I, p. 694 (1984). Abstract. – See Abstr. 010.061.

093.040 **Tides in the mesosphere of Venus.**
J. T. Schofield, S. B. Fels, D. Crisp.
Bull. Am. Astron. Soc., Vol. 16, No. 3, Part I, p. 694 (1984). Abstract. – See Abstr. 010.061.

093.041 **The radiative drive for the deep atmospheric circulation of Venus.**
H. E. Revercomb, L. A. Sromovsky.
Bull. Am. Astron. Soc., Vol. 16, No. 3, Part I, p. 694 (1984). Abstract. – See Abstr. 010.061.

093.042 **Zonal circulation and angular momentum balance of Venus' atmosphere above the clouds.**
R. L. Walterscheid, G. Schubert.
Bull. Am. Astron. Soc., Vol. 16, No. 3, Part I, p. 694 – 695 (1984). Abstract. – See Abstr. 010.061.

093.043 **A new method of predicting atmospheric waves with applications to Venus.**
A. R. Dobrovolskis, D. J. Diner.
Bull. Am. Astron. Soc., Vol. 16, No. 3, Part I, p. 695 (1984). Abstract. – See Abstr. 010.061.

093.044 **A simple barotropic wave–mean flow model for the cloud level atmosphere of Venus.**
M. Newman, C. B. Leovy.
Bull. Am. Astron. Soc., Vol. 16, No. 3, Part I, p. 695 (1984). Abstract. – See Abstr. 010.061.

093.045 **Pioneer Venus Orbiter Ultraviolet Spectrometer (PVOUVS) Lyman–alpha data analysis: periapsis.**
D. E. Anderson Jr., L. J. Paxton, A. I. F. Stewart.
Bull. Am. Astron. Soc., Vol. 16, No. 3, Part I, p. 695 (1984). Abstract. – See Abstr. 010.061.

093.046 **Temperature contrasts in the lower atmosphere of Venus: further results from the Pioneer Venus entry probes.**
L. A. Sromovsky, H. E. Revercomb.
Bull. Am. Astron. Soc., Vol. 16, No. 3, Part I, p. 695 (1984). Abstract. – See Abstr. 010.061.

093.047 **Mean thermal structure of the Venus atmosphere from Pioneer Venus radio occultation data.**
A. J. Kliore, S. S. Limaye, H. E. Revercomb.
Bull. Am. Astron. Soc., Vol. 16, No. 3, Part I, p. 695 – 696 (1984). Abstract. – See Abstr. 010.061.

093.048 **Long and short–term variations in $SO_2$ on Venus.**
L. W. Esposito, M. Copley, L. J. Gates, H. Worden.
Bull. Am. Astron. Soc., Vol. 16, No. 3, Part I, p. 696 (1984). Abstract. – See Abstr. 010.061.

093.049 **Volcanoes and the clouds of Venus.**
R. G. Prinn.
Bull. Am. Astron. Soc., Vol. 16, No. 3, Part I, p. 696 (1984). Abstract. – See Abstr. 010.061.

093.050 **Topographic variability of the lowlands of Venus.**
P. L. Blake, P. J. Mouginis–Mark.
Bull. Am. Astron. Soc., Vol. 16, No. 3, Part I, p. 696 (1984). Abstract. – See Abstr. 010.061.

093.051 **Possible carbonatite volcanism on Venus.**
G. T. Sill.
Bull. Am. Astron. Soc., Vol. 16, No. 3, Part I, p. 696 (1984). Abstract. – See Abstr. 010.061.

093.052 **Venus radar reflectivity – a re–analysis.**
P. G. Ford, G. H. Pettengill.
Bull. Am. Astron. Soc., Vol. 16, No. 3, Part I, p. 697 (1984). Abstract. – See Abstr. 010.061.

093.053 **A climatology of planetary scale waves in the Venus clouds.**
A. D. Del Genio.
Bull. Am. Astron. Soc., Vol. 16, No. 3, Part I, p. 706 (1984). Abstract. – See Abstr. 010.061.

093.054 **Mean thermal structure and balanced zonal circulation in the middle atmosphere of Venus.**
S. S. Limaye.
Bull. Am. Astron. Soc., Vol. 16, No. 3, Part I, p. 706 – 707 (1984). Abstract. – See Abstr. 010.061.

093.055 **Venus cloud–top circulation.**
W. B. Rossow, E. Kinsella.
Bull. Am. Astron. Soc., Vol. 16, No. 3, Part I, p. 707 (1984). Abstract. – See Abstr. 010.061.

093.056 **New data on Venus' volcanism and tectonism.**
H. Masursky.
Bull. Am. Astron. Soc., Vol. 16, No. 3, Part I, p. 711 (1984). Abstract. – See Abstr. 010.061.

093.057 **Thermal evolution of Venus.**
J. Arkani–Hamed, M. N. Toksöz.
Phys. Earth Planet. Inter., Vol. 34, No. 4, p. 232 – 250 (1984).
The theory of three–dimensional and finite–amplitude convection in a viscous spherical shell with temperature and pressure dependent physical parameters is developed on the basis of a modified Boussinesq fluid assumption. The lateral dependences of the variables are resolved through their spherical harmonic representations, whereas their radial and time dependences are determined by numerical procedures. The theory is then applied to produce thermal evolution models for Venus.

093.058 **The dayside Venus ionosphere. II. Combined numerical model of ion and neutral composition above 120 km.**
T. Shimazaki, R. C. Whitten, H. T. Woodward,
W. C. Knudsen, K. L. Miller.
Icarus, Vol. 60, No. 3, p. 654 – 674 (1984).
A coupled neutral–ionic photochemical model has been used to interpret the ionic composition of the Venusian dayside ionosphere measured by the orbiter retarding potential analyzer (ORPA) experiment on board the NASA Pioneer–Venus orbiter spacecraft. The electron and ion temperatures also measured by the ORPA are used for calculating the plasma–diffusion coefficients and scale heights for ions. The neutral temperature profiles and the densities of neutral constituents, particularly $CO_2$ and O, play key roles in the determination of the height profiles of the ionic constituents. All these quantities vary substantially in the Venusian thermosphere near the terminator; the models presented are representatives of the solar zenith angle $\sim 65°$.

093.059 **Results of an investigation of five regions of the Venus surface with the bistatic radar method.**
A. I. Kucheryavenkov, O. E. Milekhin, A. G. Pavel'ev,
O. I. Yakovlev.
Kosm. Issled., Tom 22, Vyp. 6, p. 895 – 905 (1984). In Russian. English translation in Cosm. Res.

093.060 **Hypsometric peculiarities of Venus.**
Zh. F. Rodionova.
Astron. Vestn., Tom 18, No. 2, p. 117 – 127 (1984). In Russian. Abstr. in Ref. Zh., 51. Astron., 11.51.182 (1984).

093.061 **Preliminary photomap of the Venus surface in the Maxwell Mountains and adjacent surroundings.**
A. F. Bogomolov, Yu. S. Tyuflin.
Geod. kartogr., No. 8, p. 11 – 17 (1984). In Russian. Abstr. in Ref. Zh., 52. Geod. Aehrosemka, 11.52.214 (1984).

093.062 **Two–layer model of formation of the night ionosphere of Venus according to data of radio occultation experiments.**
I. K. Osmolovskij, N. A. Savich, L. N. Samoznaev.
Dokl. Akad. Nauk SSSR. Ser. Mat. Fiz., Tom 276, No. 2, p. 325 – 328 (1984). In Russian. Abstr. in Ref. Zh., 62. Issled. Kosm. Prostranstva, 11.62.403 (1984).

093.063 **Venus banded terrain: tectonic models for band formation and their relationship to lithospheric thermal structure.**
S. C. Solomon, J. W. Head.
J. Geophys. Res., Vol. 89, No. B8, p. 6885 – 6897 (1984).
Recent radar images of the mountains of Ishtar Terra, Venus, at approximately 3–km radar resolution show a series of linear bands of alternatively greater and lesser backscatter; the bands are generally aligned with the topographic strike of the mountain ranges. The authors test the hypothesis that these band features are tectonic in origin, the product either of folding during lithospheric compression or of block faulting during lithospheric extension.

093.064 **Magnetic field overshoots in the Venus bow shock.**
M. Tatrallyay, J. G. Luhmann, C. T. Russell.
Adv. Space Res., Vol. 4, No. 2 – 3, p. 283 – 286 (1984). – See Abstr. 012.075.
An examination of Pioneer Venus Orbiter fluxgate magnetometer data has shown that magnetic field overshoots occur not only behind quasi–perpendicular bow shocks but also behind quasi–parallel shocks. Overshoots are associated only with supercritical shocks. Their amplitudes increase with increasing fast Mach number. The solar wind $\beta$ has a lesser effect.

093.065 **A guide to visual observations of Venus.**
J. L. Benton Jr.
Strolling Astron., Vol. 30, Nos. 11 – 12, p. 239 – 245 (1984).

093.066 **Preliminary data on a radiometric research of the surface layer of the Venus northern hemisphere by the orbital stations Venera 15 and Venera 16.**
A. F. Bogomolov, V. A. Nazarkin, V. I. Eremeev,
N. N. Krupenio, V. V. Kutyashov, S. S. Moseevskij,
N. M. Fejzulla.
Astron. Tsirk., No. 1333, p. 3 – 4 (1984). In Russian.

093.067 **Study of the Maxwell Montes region on Venus by Venera 15 and Venera 16 spacecraft.**
V. A. Kotel'nikov, Eh. L. Akim, Yu. N. Aleksandrov,
N. A. Armand, A. T. Bazilevskij, A. F. Bogomolov,
A. S. Vyshlov, V. M. Dubrovin, N. V. Zherikhin,
A. I. Zakharov, V. E. Zimov, V. I. Kaevitser,
V. M. Kovtunenko, R. S. Kremnev, A. P. Krivtsov,
G. A. Krylov, A. A. Krymov, I. L. Kucheryavenkova,
E. P. Molotov, G. M. Petrov, O. N. Rzhiga, A. S. Selivanov,
A. I. Sidorenko, V. P. Sinilo, A. V. Sknarya, G. A. Sokolov,
V. P. Sorokin, K. G. Sukhanov, V. F. Tikhonov,
Yu. S. Tyuflin, B. Ya. Fel'dman, A. M. Shakhovskoj,
V. A. Shubin.
Pis'ma Astron. Zh., Tom 10, No. 12, p. 883 – 889 (1984). In Russian. English translation in Sov. Astron. Lett., Vol. 10.
The data on the geological structure of Maxwell Montes are obtained by means of the radar system of the Venera 15 and 16 spacecraft. It is found that Maxwell Montes and a significant part of the locality are formed by horizontal tectonic deformations of compression origin. The Patera Cleopatra depression is similar to double–ring craters of impact origin rather than volcanic calderas.

093.068 **Interannual and short–term variations of the Venus nighttime hydrogen bulge.**
H. A. Taylor, H. C. Brinton, H. B. Niemann, H. G. Mayr,
R. E. Hartle, A. Barnes, J. Larson.
J. Geophys. Res., Vol. 89, No. A12, p. 10669 – 10675 (1984).
The predawn bulge in the diurnal distribution of neutral hydrogen at Venus is found to persist over the first three Venus years of the Pioneer Venus mission. Concentrations of H, derived from in situ mass spectrometer measurements of $O^+$, $H^+$, O, and $CO_2$ show a diurnal bulge with peak levels near $2 - 5 \times 10^7/cm^3$ and a night to day concentration ratio of about 200/1. The presence of short–term fluctuations of as much as a factor of three from day to day reflects the variability in the ion and neutral concentrations often detected on the nightside. Allowing for this superimposed variability, relatively little difference in the magni-

tude of $n(H)$ between 1979 and 1980 is indicated. The influence of solar EUV and solar wind are discussed.

093.069 **Growth and maintenance of large-scale magnetic fields in the dayside Venus ionosphere.**
J. L. Phillips, J. G. Luhmann, C. T. Russell.
J. Geophys. Res., Vol. 89, No. A12, p. 10676 – 10684 (1984).
Observations from the Pioneer Venus orbiter magnetometer reveal the presence of large-scale magnetic fields in the dayside ionosphere during or after periods of high solar wind dynamic pressure. Various hypotheses have been proposed concerning the spatial and temporal evolution of these field structures. Here the authors examine the hypothesis that the field is produced by a diffusion/convection process rather than by currents driven by electric fields resulting from the solar wind interaction.

**Galileo, Kepler, and phases of Venus.**
See Abstr. 004.044.

**Phases of Venus in 1610.**
See Abstr. 004.045.

**The appearances of Venus and Mars in 1610.**
See Abstr. 004.046.

**Astronomical scrapbook: Galileo and the phases of Venus.**
See Abstr. 004.058.

**Microdunes and other aeolian bedforms on Venus: wind tunnel simulations.**
See Abstr. 022.075.

**Complex refractive index of liquid and solid sulfur.**
See Abstr. 022.099.

**Measurements of the microwave opacity and vapor pressure of gaseous sulfuric acid under simulated Venus conditions.**
See Abstr. 022.114.

**Molecular astronomy using heterodyne detection at 691 GHz.**
See Abstr. 034.035.

**Bau eines Heterodynempfängers für 10 µm Wellenlänge zum Einsatz am Cassegrainfokus astronomischer Teleskope und spektroskopische Untersuchung der Mars- und Venusatmosphäre.**
See Abstr. 034.183.

**The Venus Radar Mapper mission.**
See Abstr. 051.015.

**Pioneer to Venus: the multiprobe and orbiter missions.**
See Abstr. 051.016.

**The Venus Radar Mapper: radar system design for highly elliptical orbits.**
See Abstr. 051.018.

**Venus Radar Mapper: precision positioning of SAR images.**
See Abstr. 051.058.

**On the flight of the automatic interplanetary stations Venera 13 and Venera 14 to Venus.**
See Abstr. 053.004.

**Temperature effect on the photodissociation rates in the atmospheres of Mars and Venus.**
See Abstr. 097.019.

**The violet absorber on Io (and Venus?): disulfur monoxide derivatives.**
See Abstr. 099.068.

# 094 Moon

094.001 **Source of the oldest lunar basalt.**
S. R. Taylor.
Nature, Vol. 310, No. 5973, p. 98 – 99 (1984).

094.002 **Observations of transient phenomena on the moon.**
V. S. Filonenko.
Astron. Tsirk., No. 1290, p. 7 – 8 (1983). In Russian.

094.003 **The negative branch of polarization curves of lunar regions.**
O. I. Kvaratskheliya.
Astron. Tsirk., No. 1291, p. 5 – 7 (1983). In Russian.

094.004 **Constants of lunar libration obtained with the use of Morrison's corrections.**
Yu. A. Nefed'ev, G. M. Stolyarov.
Astron. Tsirk., No. 1293, p. 7 – 8 (1983). In Russian.

094.005 **Luna Incognita: de laatste uitdaging.**
L. Aerts.
Zenit, 11. Jaarg., No. 7/8, p. 326 – 328 (1984).

094.006 **Orientation of the moon's axes of inertia from earth-based measurements.**
N. G. Rizvanov, E. M. Shchukin.
Sov. Astron. Lett., Vol. 10, No. 1, p. 55 – 56 (1984). English translation of 37.094.007.

094.007 **On determination of the coordinates of lunar features obtained with the "Zond" series space apparatuses.**
S. G. Valeev, Yu. N. Lipskij.
Tr. Gos. Astron. Inst. Shternberg, Tom 54, p. 61 – 69 (1984). In Russian.
Algorithms of the computed treatment of lunar pictures obtained from the spacecraft "Zond" are described; formulae for the precision estimation of the spherical coordinates are deduced; preliminary results of the algorithms application are given.

094.008 **Lunar surface properties in the Mare Orientale region based on data of remote investigations.**
V. V. Shevchenko, S. G. Pugacheva, M. Yu. Shevchenko.
Tr. Gos. Astron. Inst. Shternberg, Tom 55, p. 114 – 129 (1983). In Russian.
A description of Mare Orientale photometric investigation and structure is given. The results of a statistical analysis of selective reflectivity of the lunar surface in the infrared (10 – 12 microns) region are given. Infrared temperatures of lunar features of Mare Orientale are calculated and an a priori isothermal contour map of this region is constructed. A catalogue of brightness temperatures and albedo for 800 lunar regions is given.

094.009 **Classification of lunar regions by optical parameters of reflected light.**
V. V. Novikov.
Tr. Gos. Astron. Inst. Shternberg, Tom 55, p. 137 – 148 (1983). In Russian.
Different optical types of lunar rocks were found by analysing albedo-porosity and albedo-color indexes. The classification re-

flects the correlation of optical characteristics with the physico–mineralogical features of lunar soil.

094.010 **Model of the continuous density distribution in the interior of the moon.**
L. A. Savrov, Kh. G. Tadzhidinov.
Tr. Gos. Astron. Inst. Shternberg, Tom 55, p. 175 – 184 (1983). In Russian.
The continuous distribution of the density in the interior of the spherical moon with the mean radius of 1738 km is considered in dependence of the seismic data, the heat conduction and electrical conductivity of the moon's interior, the inhomogeneous chemical composition and different phase state of the moon's matter. A formula of the gradient of the density is found, diagrams of the change of the density from the surface to the centre of the moon are drawn.

094.011 **Rb–Sr and U–Pb isotopic systematics of lunar soil.**
L. K. Levskij, G. V. Ovchinnikova.
Meteoritika, Vyp. 43, p. 147 – 152 (1984). In Russian.

094.012 **Terra scarps indicating youngest terra faults on the Moon.**
J. Raitala.
Earth, Moon, Planets, Vol. 31, No. 1, p. 63 – 74 (1984).
Lunar crustal shortening does not seem to be restricted to the lava–filled basins alone; but there are some young scarp–like terra ridges in places around mare areas. This crustal shortening has not reached the same intensity as in the case of the lobate scarp overthrusts on Mercury. Young lunar terra ridges indicate that crustal shortening with an areal extent also took place slightly around mare basins. Thus they link tensional rille tectonics with compressional mare ridge tectonics and indicate that areal heating/bending/extension – cooling/shortening/compression may describe an important explaining factor in lunar mare– and near–mare tectonics in addition to the volcanic extrusions.

094.013 **La solution ELP 2000–82 pour le mouvement orbital de la Lune.**
M. Chapront–Touzé, J. Chapront.
Astrométrie, géodésie, mécanique céleste, p. 23 – 28 (1984). – See Abstr. 012.021.

094.014 **La station de télémétrie laser–Lune du C.E.R.G.A. État technique actuel et résultats obtenus.**
J. F. Mangin.
Astrométrie, géodésie, mécanique céleste, p. 163 – 167 (1984). – See Abstr. 012.021.

094.015 **Luna Incognita technical report I: Earthbased selenodetic positions.**
J. E. Westfall.
Strolling Astron., Vol. 30, Nos. 9 – 10, p. 199 – 205 (1984).

094.016 **Structures of disintegration of solid solutions in pyroxenes of the moon and meteorites as indicators of petrological processes.**
N. R. Khisina.
The 27th International Geological Congress, Tom 5, Sect. 10 – 11, p. 309 – 311 (1984). In Russian. Abstr. in Ref. Zh., 62. Issled. Kosm. Prostranstva, 8.62.398 (1984). – See Abstr. 012.031.

094.017 **Geochemical structure of lunar magmatic rocks and their evolution.**
A. A. Yaroshevskij.
The 27th International Geological Congress, Tom 5, Sect. 10 -11, p. 442 – 443 (1984). In Russian. Abstr. in Ref. Zh., 62. Issled. Kosm. Prostranstva, 8.62.400 (1984). – See Abstr. 012.031.

094.018 **Principles and methods of constructing a series of maps of selenodetic reference points.**
L. O. Karas'ov, M. M. Rozanov.
Visn. Kiev univ. Geogr., No. 25, p. 63 – 66 (1983). In Russian. Abstr. in Ref. Zh., 52. Geod. Aehrosemka, 8.52.194 (1984).

094.019 **Improving our knowledge of the lunar radius.**
P. D. Maley.
Mon. Notes Astron. Soc. S. Afr., Vol. 43, Nos. 5 – 8, p. 26 – 30 (1984).

094.020 **Whence came the moon?**
A. E. Rubin.
Sky Telesc., Vol. 68, No. 5, p. 389 – 393 (1984).

094.021 **Some new results of photometric measurements according to panorama pictures obtained from board Lunokhod 2.**
V. V. Busarev, V. V. Zasetskij, Yu. A. Kantarovskij,
V. I. Titov, M. Yu. Shevchenko.
Kosm. Issled., Tom 22, Vyp. 4, p. 609 – 616 (1984). In Russian. English translation in Cosm. Res.

094.022 **On the evolution of lunar craters.**
G. A. Lejkin, A. N. Sanovich.
Tr. Gos. Astron. Inst. Shternberg, Tom 56, p. 94 – 112 (1984). In Russian.
The evolution of the crater form after the completion of its formation, i. e. after the proceeding of rapid slope processes is considered. The role of viscous relaxation and isostatic compensation for impact basins and large craters is noted. A simple model of the evolution of the crater form in the process of slow denudation is constructed.

094.023 **On the mechanism of interconnection of physico–mineralogical and optical properties of finely dispersed soils.**
V. V. Novikov, M. V. Goryachev.
Tr. Gos. Astron. Inst. Shternberg, Tom 56, p. 113 – 125 (1984). In Russian.
Some aspects of albedo dependence on the polarization degree are considered. Some optical characteristics of samples of earth ejective rocks and observations of Moon and Mars are interpreted. Practical recommendations on optical division into districts of the lunar surface are given.

094.024 **Selenodetic investigations of the Mare Orientale region.**
V. I. Chikmachev.
Tr. Gos. Astron. Inst. Shternberg, Tom 56, p. 126 – 144 (1984). In Russian.
Available new selenodetic nets in the Mare Orientale region are given. A relief map and profiles of the Mare Orientale region are made by using this material.

094.025 **Numerical model of infrared temperature distribution of the lunar surface on full moon.**
S. G. Pugacheva.
Tr. Gos. Astron. Inst. Shternberg, Tom 56, p. 145 – 159 (1984). In Russian.
A description of a numerical model representing the brightness temperature of the lunar surface is given. An extrapolated isothermal contour map of the surface of the lunar sphere on full moon is calculated and drawn. The data for brightness temperatures and albedo of the selected sites of the moon are given in a catalogue.

094.026 **On numerical modelling of the moon's physical libration.**
A. A. Shiryaev.
Astron. Zh., Tom 61, Vyp. 5, p. 1029 – 1032 (1984). In Russian. English translation in Sov. Astron., Vol. 28, No. 5.
Some problems concerned with the construction of a numerical theory of the physical libration of the moon are stated. The results of the numerical integration of the equations of the moon's physical libration obtained on the time–span of 6000 days have been compared with Eckhardt's analytical theory (1981).

094.027 **Low–frequency electromagnetic induction in the Moon: linearized inverse theory and lunar core calculations.**
B. A. Hobbs, L. L. Hood, F. Herbert, C. P. Sonett.
Geophys. J. R. Astron. Soc., Vol. 79, No. 2, p. 691 – 696 (1984).
Day–side lunar electromagnetic transfer function estimates of Hood, Herbert & Sonett are analysed using the linearized inverse theory of Backus & Gilbert. A smooth conductivity distribution exponentially increasing with depth in the range 260 – 1200 km is found to fit the data. The conclusions are compatible with the forward model calculations of Hood et al. Further calculations explore the compatibility of a silicate or metallic lunar core with the data. A metallic lunar core of radius 438 km is allowed but not required.

094.028 **Volatile/refractory element fractionation and the lunar fission hypothesis.**
S. R. Taylor.
The 27th International Geological Congress, Tom 5, Sect. 10 – 11, p. 413 (1984). Abstr. in Ref. Zh., 51. Astron., 9.51.240 (1984). – See Abstr. 012.031.

094.029 **Phase convection in regions of particle melting and differentiation of mantle matter in the process of magma formation.**
M. J. Frenkel.
The 27th International Geological Congress, Tom 5, Sect. 10 – 11, p. 262 – 263 (1984). In English and Russian. Abstr. in Ref. Zh., 51. Astron., 9.51.242 (1984). – See Abstr. 012.031.

094.030 **On the nonmonotonic variation of the electrostatic potential in the photoelectron layer above the lunar surface in the period of mean solar activity.**
A. S. Manujlov, E. K. Kolesnikov.
Vestn. Leningr. Univ., Mat. Mekh. Astron., No. 7, p. 101 – 104 (1984). In Russian. Abstr. in Ref. Zh., 51. Astron., 9.51.247 (1984).

094.031 **Contemporary problems of the lunar toponymy.**
V. V. Shevchenko.
Astron. Vestn., Tom 18, No. 2, p. 100 – 106 (1984). In Russian. Abstr. in Ref. Zh., 62. Issled. Kosm. Prostranstva, 9.62.278 (1984).

094.032 **Petrology of lunar matter: rocks of continents, maria and regolith.**
J. J. Papike, S. B. Simon.
The 27th International Geological Congress, Tom 11, Sect. S 11, p. 38 – 58 (1984). In Russian. Abstr. in Ref. Zh., 62. Issled. Kosm. Prostranstva, 9.62.290 (1984). – See Abstr. 012.031.

094.033 **Particle track measurements in lunar regolith breccias.**
G. E. Blanford.
Earth, Moon, Planets, Vol. 31, No. 3, p. 221 – 227 (1984).
Particle track measurements have been reported for 25 (5%) of the regolith breccias in the collection; they have been reported for 16 breccias (30%) in the reference suite. The most frequently reported measurement for these 25 breccias is the maximum surface exposure age of the compacted rock (48% of the published breccia measurements). Information on the nature of the precompaction regolith is given for 9 rocks (36%) and on the nature of the compaction event for 6 rocks (24%). Most of the breccias appear to have simple post–compaction surface exposure histories (89%). It appears that most of these breccias are amenable to studies which separate the contemporary surface exposure age from information about the precompaction regolith.

094.034 **Comparison of six crater–scaling laws.**
G. P. Horedt, G. Neukum.
Earth, Moon, Planets, Vol. 31, No. 3, p. 265 – 269 (1984).
The authors present a compilation of crater–scaling laws suitable for planetological applications, connecting crater diameter to impact energy.

094.035 **Cosmogenic $^{22}$Na and $^{26}$Al and track studies in Luna 16, 20 and 24 samples.**
P. Emrich, V. Jurina, P. Povinec.
Bull. Astron. Inst. Czech., Vol. 35, No. 5, p. 253 – 261 (1984).
Samples from three unmanned Soviet lunar missions, Luna 16, 20 and 24, have been investigated for studying the concentration of cosmogenic radionuclides $^{22}$Na and $^{26}$Al as well as tracks formed by nuclei of the iron group of cosmic rays in mineral grains (feldspar). The depth profiles of $^{22}$Na and $^{26}$Al are compared with theoretical predictions obtained by using three independent methods of calculation of the production rates of cosmogenic radionuclides. The best fit was obtained for an intensity of galactic cosmic rays of 0.24 protons $cm^{-2}s^{-1}sr^{-1}$ with $R > 0.5$ GV for the last million years.

094.036 **Petrology of lunar materials: highland, mare, and regolith.**
J. Papike, S. Steven.
The 27th International Geological Congress, Tom 5, Sect. 10 – 11, p. 367 (1984). In Russian. Abstr. in Ref. Zh., 62. Issled. Kosm. Prostranstva, 10.62.369 (1984). – See Abstr. 012.031.

094.037 **Origin of the lunar highlands: petrographic and geochemical evidence.**
J. Shervais, L. Taylor.
The 27th International Geological Congress, Tom 5, Sect. 10 – 11, p. 393 – 394 (1984). In Russian. Abstr. in Ref. Zh., 62. Issled. Kosm. Prostranstva, 10.62.370 (1984). – See Abstr. 012.031.

094.038 **Pristine nonmare rocks and the diversity of the lunar crust.**
P. H. Warren, J. T. Wasson.
The 27th International Geological Congress, Tom 5, Sect. 10 – 11, p. 433 – 434 (1984). In Russian. Abstr. in Ref. Zh., 62. Issled. Kosm. Prostranstva, 10.62.371 (1984). – See Abstr. 012.031.

094.039 **The mystery of the far side of the moon.**
V. V. Shevchenko.
Zemlya Vselennaya, No. 5, p. 83 – 89 (1984). In Russian.

094.040 **Trace element determination in lunar rocks – quantitative PIXE analyses on thick samples.**
H. Blank, A. El Goresy, J. Janicke, R. Nobiling, K. Traxel.
Nucl. Instrum. Methods Phys. Res., Sect. B, Vol. 231, No. 1 – 3, p. 681 – 687 (1984). Abstr. in Phys. Abstr., Vol. 87, No. 1243, Entry 101210 (1984). – See Abstr. 012.047.

094.041 **Error analysis for the use of presently available lunar radio flux data in broadbeam antenna–system measurements.**
W. C. Daywitt.
Report NBS–TN–1073, Natl. Bur. Stand., Washington, DC, USA, 30 pp. (1984). Abstr. in Phys. Abstr., Vol. 87, No. 1245, Entry 110549 (1984).

094.042 **Division of the lunar surface into regions by albedo and color in the spectral range of 0.62 – 0.95 μm.**
N. N. Evsyukov.
Sov. Astron., Vol. 28, No. 2, p. 212 – 214 (1984). English translation of 37.094.016.

094.043 **Determination of the coordinates of an observation point on the lunar surface by equal altitudes.**
S. S. Peruanskii (*S. S. Peruanskij*).
Sov. Astron., Vol. 28, No. 3, p. 337 – 339 (1984). English translation of 37.094.056.

094.044 **Effects of shock–loading on geologic materials as measured by spectral reflectance.**
T. V. V. King, F. Horz.
Bull. Am. Astron. Soc., Vol. 16, No. 3, Part I, p. 666 (1984). Abstract. – See Abstr. 010.061.

094.045 **Spectral studies of the Imbrium region of the moon.**
P. G. Lucey, B. R. Hawke.
Bull. Am. Astron. Soc., Vol. 16, No. 3, Part I, p. 666 (1984). Abstract. – See Abstr. 010.061.

094.046 **Spectral reflectance studies of Orientale basin deposits.**
B. R. Hawke, P. G. Lucey, J. F. Bell, P. D. Spudis.
Bull. Am. Astron. Soc., Vol. 16, No. 3, Part I, p. 666 – 667 (1984). Abstract. – See Abstr. 010.061.

094.047 **How thick are crater rays?**
C. M. Pieters, P. J. Mouginis–Mark.
Bull. Am. Astron. Soc., Vol. 16, No. 3, Part I, p. 667 (1984). Abstract. – See Abstr. 010.061.

094.048 **Millimeter radio observations of the moon.**
D. A. Swade, F. P. Schloerb.
Bull. Am. Astron. Soc., Vol. 16, No. 3, Part I, p. 667 (1984). Abstract. – See Abstr. 010.061.

094.049 **Filling–in Fraunhofer lines in the spectrum of the moon and Mercury.**
A. E. Potter, T. H. Morgan.
Bull. Am. Astron. Soc., Vol. 16, No. 3, Part I, p. 667 (1984). Abstract. – See Abstr. 010.061.

094.050 **The composition of localized lunar pyroclastic deposits.**
L. R. Gaddis, P. G. Lucey, J. F. Bell, B. R. Hawke.
Bull. Am. Astron. Soc., Vol. 16, No. 3, Part I, p. 710 (1984). Abstract. – See Abstr. 010.061.

094.051 **Influence of lunar tide on LLR.**
W. Jin, Z. Yang.
Ann. Shanghai Obs., Acad. Sin., No. 5, p. 23 – 30 (1983). In Chinese.

Not only the influences of the solid earth tides and the ocean tides which are induced by the tidal forces of both the moon and the sun should be considered on lunar laser ranging (LLR), but also the lunar tides should be taken into account for computing the distance between the earth and the moon because the retroreflector is located on the moon's surface.

094.052 **$^{10}$Be and $^{36}$Cl depth profiles in an Apollo 15 drill core.**
K. Nishiizumi, D. Elmore, X. Z. Ma, J. R. Arnold.
Earth Planet. Sci. Lett., Vol. 70, No. 2, p. 157 – 163 (1984).

094.053 **$^{26}$Al depth in Apollo 15 drill core.**
K. Nishiizumi, J. Klein, R. Middleton, J. R. Arnold.
Earth Planet. Sci. Lett., Vol. 70, No. 2, p. 164 – 168 (1984).

094.054 **Interpretation of the lunar microwave brightness temperature spectrum: feasibility of orbital heat flow mapping.**
S. J. Keihm.
Icarus, Vol. 60, No. 3, p. 568 – 589 (1984).

A detailed model of the lunar regolith is analyzed to examine the feasibility of an orbital mapping of heat flow using microwave radiometers. For the Moon, the most serious spurious effect appears to be emissivity variations due to the near–surface density gradient. A brightness temperature decrease of 10K is predicted from centimeter to decameter wavelengths. If the transition from porous surface fines to compacted regolith soil occurs rapidly (within the upper 3 – 5 cm), most of the emissivity decrease will occur in the 5– to 30–cm wavelength band.

094.055 **The primeval axis of rotation of the Moon.**
S. K. Runcorn.
Philos. Trans. R. Soc. London, Ser. A, Vol. 313, No. 1524, p. 77 – 83 (1984). – See Abstr. 012.068.

The palaeopoles of the Moon have been calculated from the palaeomagnetic directions of lunar crustal strata, determined by Coleman, Russell and Hood from Apollo 15 and 16 sub–satellite data. Because of the dominant role of the Coriolis force in the core dynamo hydromagnetics, the lunar dipole magnetic field would have been aligned along the axis of rotation. As the palaeopoles lie in bipolar groupings along 3 different axes, far from the present axis (the magnetizations are dated at 4.2 Ga, 4.0 Ga and 3.85 Ga ago), it is concluded that the Moon was reoriented at least three times. These polar displacements are attributed to the creation of multi–ring basins at these times by the collision of fragments of at least three lunar satellites, the orbits of which decayed by tidal friction. The actual paths of the pole are explained by Euler's theorem applied to a moon in which the interior can flow by solid state creep.

094.056 **Lunar spectroradiometric measurements.**
W. R. Mallory, H. L. Task.
J. Opt. Soc. Am. A, Vol. 1, No. 12, p. 1334 (1984). Abstract.

094.057 **Lunar dark–haloed impact craters: origin and implications for early mare volcanism.**
J. F. Bell, B. R. Hawke.
J. Geophys. Res., Vol. 89, No. B8, p. 6899 – 6910 (1984).

Because an understanding of the origin of dark–haloed impact craters is important in dealing with serveral major problems in lunar geology, the authors have undertaken a study of these craters using recently developed infrared spectroscopic techniques. The purposes of the investigation include the following: (1) to search for dark halo deposits containing carbonaceous projectile material, (2) to test the presence of basalt in those dark–haloed craters identified by Schulz and Spudis (1979) as having excavated buried basaltic units, (3) to investigate the composition of the supposed ancient basalts, and (4) to determine the extent of the proposed early volcanic activity on the lunar nearside.

094.058 **Effects of thermal cycling on magnetic properties of lunar analogs.**
A. M. Barron, P. N. Shive.
J. Geophys. Res., Vol. 89, No. B13, p. 11573 – 11579 (1984).

The objective of the study is to determine whether stresses associated with thermal cycling cracks can affect the coercivity of remanence carried by iron in lunar samples.

094.059 **Mineralogical comparison and cooling history of lunar and chondritic vesicular melt breccias.**
M. Miyamoto, H. Takeda, T. Ishii.
J. Geophys. Res., Vol. 89, No. B13, p. 11581 – 11588 (1984).

Lunar vesicular melt breccia (77135) was mineralogically compared with LL chondritic analog Yamato 790964 to elucidate impact melting processes. Microprobe analyses were made of pyroxenes, plagioclases, and olivines and of bulk compositions of the glasses which appear to be solidified from impact melts. Notable similarities between the lunar and chondritic melt breccias are (1) abundant vesicles, (2) similar pyroxene chemical zoning trends, (3) presence of variable amounts of clastic material, and (4) similar chemical compositions except for K and Na contents of the glass and mesostasis. Some constraints on the cooling history are estimated from Mg–Fe diffusion profiles in olivine and pyroxene.

094.060 **Luna Incognita: 1983/84 report and 1985 observing schedule.**
J. E. Westfall.
Strolling Astron., Vol. 30, Nos. 11 – 12, p. 221 – 226 (1984).

094.061 **1985 lunar physical ephemeris.**
J. E. Westfall.
Strolling Astron., Vol. 30, Nos. 11 – 12, p. 226 – 229 (1984).

094.062 **Photometric investigation of the ray system of crater Giordano Bruno.**
S. A. Mosina.
Astron. Tsirk., No. 1321, p. 4 – 5 (1984). In Russian.

094.063 **The relationship between the albedo and polarization properties of the moon. Discrete polarimetry data.**
Yu. G. Shkuratov, N. V. Opanasenko.
Astron. Tsirk., No. 1330, p. 6 – 8 (1984). In Russian.

094.064 **Properties of the lunar surface from data of extra-atmospheric IR-astronomy and laboratory IR-spectroscopy of lunar soil.**
M. N. Markov.
Kosm. fiz. Tr. 6 Nauchn. chtenij po kosmonavt., posvyashch. pamyati vydayushch. sov. uchenykh-pionerov osvoeniya kosm. prostranstva, Moskva, 11-15 yanv., 1982. Sekts. Kosm. issled. Moskva, p. 138 – 147 (1982). In Russian. Abstr. in Ref. Zh., 51. Astron., 12.51.305 (1984).

094.065 **Granulometric composition of regolith samples obtained by Luna 24.**
O. D. Rodeh, A. V. Ivanov.
Astron. Vestn., Tom 18, No. 1, p. 3 – 7 (1984). In Russian. Abstr. in Ref. Zh., 51. Astron., 12.51.326 (1984).

094.066 **Past and present magnetism of the Moon.**
D. W. Collinson.
Geophys. Surv., Vol. 7, No. 1, p. 57 – 73 (1984). – See Abstr. 012.082.
The characteristics of the remanent magnetism of lunar samples suggests that it was acquired in a magnetic field on the Moon. The most likely origin of the field is a dynamo process in a molten, electrically-conducting core, but generation of a transient magnetic field during large meteorite impacts cannot be entirely ruled out. Evidence from absolute and relative determinations of the ancient field strength from the sample magnetizations strongly favours a global lunar field. Contributory evidence for a dynamo field is provided by measurements of present lunar surface fields, the present very small dipole moment of the Moon and accumulating evidence of variation of the axis of the lunar field with time.

094.067 **Methods of approximate cluster analysis and the three-dimensional diagram of optical parameters of the lunar surface.**
N. N. Evsyukov.
Astron. Zh., Tom 61, Vyp. 6, p. 1184 – 1191 (1984). In Russian. English translation in Sov. Astron., Vol. 28, No. 6.
A method of approximate multidimensional cluster analysis, which consists in consecutive insertion of new parameters into the analysis and exposition of two-dimensional clusters at each stage is given. Three-dimensional clusters are picked out and a three-parameter diagram of optical parameters: albedo (0.62 μm), colour index $C_1$ (0.62/0.38 μm), colour index $C_2$ (0.95/0.62 μm) of the lunar surface is composed.

**Bibliography.**
See Abstr. 002.009.

**Fotografischer Mondatlas.**
See Abstr. 002.026.

**Automation of photometric investigation of lunar surface images with the help of computers.**
See Abstr. 036.121.

**Dialogue system for measuring graphical photographs of lunar ring structures.**
See Abstr. 036.122.

**Automation of photometric investigation of lunar surface images with computers.**
See Abstr. 036.160.

**The ephemeris time obtained from photographic measurements of the positions of lunar craters relative to stars. II.**
See Abstr. 044.003.

**Lunar exploration by polar orbiter.**
See Abstr. 051.038.

**The oceans and the physics of the Earth.**
See Abstr. 081.013.

**Planetary structures: an analytical approach.**
See Abstr. 081.035.

**Terrestrial analogs for martian and lunar wrinkle ridges.**
See Abstr. 081.038.

**Tidal friction in the Earth-Moon system.**
See Abstr. 081.043.

**Does crater "saturation equilibrium" occur in the solar system?**
See Abstr. 091.026.

**On the geospheres of the earth, the moon and terrestrial planets.**
See Abstr. 091.060.

**Secular Love numbers and hydrostatic equilibrium of planets.**
See Abstr. 093.009.

**Geology and tectonics of the Argyre area on Mars: comparisons with other basins in the solar system.**
See Abstr. 097.005.

**Morphology and origin of ridges in low-latitude areas of Mars.**
See Abstr. 097.006.

**Lunar meteorites: three and counting.**
See Abstr. 105.006.

**Geochemical aspect of impact cratering.**
See Abstr. 105.047.

**Ultrabasic meteorites and the problem of primary hyperbasic magma.**
See Abstr. 105.057.

**Geochemical comparison of meteorites, the moon and the earth: trends of element evolution.**
See Abstr. 105.063.

**Production of high-energy gamma rays by cosmic ray interactions in the atmosphere and lunar surface.**
See Abstr. 144.077.

# 095 Lunar Eclipses

095.001 **Canon der Mondfinsternisse, –2002 bis + 2526.**
J. Meeus, H. Mucke.
Sternenbote, 27. Jahrg., Nr. 9, p. 150 – 162 (1984).

095.002 **Three–color photometry: penumbral lunar eclipse, May 15, 1984 U.T.**
J. E. Westfall.
Strolling Astron., Vol. 30, Nos. 9 – 10, p. 209 – 211 (1984).

095.003 **The cause of dark lunar eclipses.**
A. D. I. Maclean.
J. Br. Astron. Assoc., Vol. 94, No. 6, p. 263 – 265 (1984).
This paper presents a survey of dark lunar eclipses and also of the problems which they create in assigning their possible causes in either astronomical or meteorological terms.

095.004 **Observation of the total lunar eclipse of January 9, 1982 at 3.2 mm wavelength with the RT–7.5 radiotelescope.**
V. N. Ivanov, T. S. Lebedyuk, B. A. Rozanov, G. N. Solov'ev.
Izv. Vyssh. Uchebn. Zaved., Radiofiz., Tom 27, No. 4, p. 403 – 409 (1984). In Russian. Abstr. in Ref. Zh., 51. Astron., 10.51.255 (1984).

095.005 **Lunar eclipse crater timing programme.**
B. W. Soulsby.
J. Br. Astron. Assoc., Vol. 95, No. 1, p. 16 – 21 (1984).
A large number of observations made by amateur astronomers in the Southern Hemisphere for ten lunar eclipses since 1974 are summarized and a method of analysis to determine the size and shape of the umbra is described. The analysis of 1,554 crater timings contributed provide values of umbra enlargement and oblateness which are used to assess observer variability and associated physiological factors.

095.006 **Years with seven eclipses in 1900 – 2500.**
B.–l. Liu.
Publ. Purple Mt. Obs., Vol. 3, No. 2, p. 26 – 30 (1984). In Chinese.

095.007 **Additional notes to the Canon of Lunar Eclipses from 1000 B.C. to A.D. 3000.**
B.–l. Liu.
Publ. Purple Mt. Obs., Vol. 3, No. 2, p. 31 – 32 (1984). In Chinese.

**Occultations during total lunar eclipses.**
See Abstr. 096.013.

# 096 Lunar and Planetary Occultations

096.001 **Computer program for precalculations of lunar occultations of stars and its application.**
E. M. Trunkovskij.
Astron. Tsirk., No. 1288, p. 1 – 5 (1983). In Russian.

096.002 **Photoelectric observations of lunar occultations of stars. The angular diameter of the star 61 $\delta^1$ Tauri.**
V. G. Kornilov, A. V. Mironov, E. M. Trunkovskij, Kh. F. Khaliullin, A. M. Cherepashchuk.
Astron. Zh., Tom 61, Vyp. 4, p. 739 – 749 (1984). In Russian. English translation in Sov. Astron., Vol. 28, No. 4.
Technique and methods of photoelectric observations of lunar occultations of stars are described. A method of restoration of the angular diameter of the star from the analysis of the diffraction pattern observed during the occultation is presented. The angular diameter of the star 61 $\delta^1$ Tau is determined: $d = 0''.0028 \pm 0''.0007$. The linear radius of 61 $\delta^1$ Tau is $R \cong 14\, R_{\odot}$, its effective temperature is $T_{eff} \cong 4700$K. These results are in agreement with the spectral class and luminosity of this star (K0 III CN0.5).

096.003 **An occultation by Pallas.**
V. B. Kapkov.
Sov. Astron. Lett., Vol. 10, No. 1, p. 26 – 27 (1984). English translation of 37.096.001.

096.004 **Decameter observations of a lunar occultation of the Crab Nebula.**
V. P. Bovkun, I. N. Zhuk.
Sov. Astron. Lett., Vol. 10, No. 1, p. 46 – 48 (1984). English translation of 37.096.003.

096.005 **Lunar occultations of planets.**
Occultation Newsl., Vol. 3, No. 8, p. 169 – 170 (1984).

096.006 **Planetary occultations during 1984.**
D. W. Dunham, M. Sôma.
Occultation Newsl., Vol. 3, No. 8, p. 170 – 178 (1984).

096.007 **The occultation of MKE 31 by Neptune on September 12, 1983.**
A. K. Pandey, H. S. Mahra, V. Mohan.
Earth, Moon, Planets, Vol. 31, No. 2, p. 217 – 219 (1984).
The predicted occultation of the star No. 31 (Mink et al., 1981) by Neptune on September 12, 1983 was observed photoelectrically. Four secondary occultations were recorded before the immersion event.

096.008 **Occultations rasantes en France en 1985.**
J. Meeus.
Astronomie, Vol. 98, p. 456 – 459 (1984).

096.009 **Occultations d'étoiles par des petites planètes en 1985.**
Astronomie, Vol. 98, p. 466 – 467 (1984).

096.010 **Upcoming planetary occultations.**
D. W. Dunham.
Occultation Newsl., Vol. 3, No. 9, p. 181 (1984).

096.011 **Proposed changes for asteroidal occultation predictions for 1985.**
D. W. Dunham.
Occultation Newsl., Vol. 3, No. 9, p. 181 – 182 (1984).

096.012 **The September 16th occultation of SAO 146599 by (47) Aglaja.**
D. W. Dunham.
Occultation Newsl., Vol. 3, No. 9, p. 182 – 183 (1984).

096.013 **Occultations during total lunar eclipses.**
D. Herald, D. W. Dunham, P. Maley.
Occultation Newsl., Vol. 3, No. 9, p. 183 – 185 (1984).

096.014 **Observations of asteroidal appulses and occultations.**
J. Stamm.
Occultation Newsl., Vol. 3, No. 9, p. 185 – 188 (1984).

096.015 **IOTA grazing occultation maps.**
D. W. Dunham.
Occultation Newsl., Vol. 3, No. 9, p. 188 – 189 (1984).

096.016 **Format error in USNO total occultation predictions.**
D. W. Dunham.
Occultation Newsl., Vol. 3, No. 9, p. 189 – 190 (1984).

096.017 **Observed profiles of grazing occultations.**
G. Blow.
Occultation Newsl., Vol. 3, No. 9, p. 190 – 191 (1984).

096.018 **Planetary occultations during 1984 – 5.**
Occultation Newsl., Vol. 3, No. 9, p. 191 – 200 (1984).

096.019 **KPNO lunar occultation summary. I.**
P. C. Schmidtke, J. L. Africano.
Astron. J., Vol. 89, No. 9, p. 1371 – 1378 (1984).
The results of a program to photoelectrically record lunar occultations at KPNO are described. The program emphasizes the observation of known or suspected double/multiple stars as well as the observation of stars with resolvable angular diameters.

096.020 **An occultation of a star by the rings of Uranus in 1985.**
G. E. Taylor.
Astron. Astrophys., Vol. 140, No. 1, p. 55 – 56 (1984).
Preliminary predictions of an occultation of the star SAO 184819 by the rings of Uranus are given. It will occur on 1985 June 25 and is likely to be visible from Europe, Asia, Africa, and South America.

096.021 **The 22 April 1982 occultation of star KME–14 by Uranus and its rings.**
L. H. Wasserman, R. L. Millis.
Bull. Am. Astron. Soc., Vol. 16, No. 3, Part I, p. 660 (1984).
Abstract. – See Abstr. 010.061.

096.022 **Predicted occultations by Uranus and its rings: 1984 – 1990.**
D. J. Mink, A. Klemola.
Bull. Am. Astron. Soc., Vol. 16, No. 3, Part I, p. 660 (1984).
Abstract. – See Abstr. 010.061.

096.023 **Grazing occultations.**
D. W. Dunham.
Occultation Newsl., Vol. 3, No. 10, p. 204 – 208 (1984).

096.024 **Planetary occultations during 1985.**
D. W. Dunham.
Occultation Newsl., Vol. 3, No. 10, p. 208 – 222 (1984).

096.025 **Appulse of SAO 186001 to Neptune.**
IAU Circ., No. 3962 (1984).

096.026 **The recent lunar occultation of planet Saturn.**
N. MacDonald, J. V. Vincent.
Mon. Notes Astron. Soc. S. Afr., Vol. 43, Nos. 11 – 12, p. 87 – 88 (1984).

096.027 **Observations of lunar occultations of stars with the 6–m telescope.**
M. B. Bogdanov, G. M. Beskin, S. I. Neizvestnyj, A. A. Pimonov, V. L. Plakhotnichenko, V. F. Shvartsman.
Astron. Tsirk., No. 1301, p. 5 – 7 (1983). In Russian.

096.028 **Las ocultaciones de las estrellas.**
V. Yurevich.
Publ. Espec. Obs. Astron. Quito, No. 2, p. 66 – 68 (1981).

096.029 **Results of radio astronomical observations of the lunar occultation of the Crab nebula on January 26, 1983.**
M. I. Agafonov, A. M. Aslanyan, A. P. Barabanov, I. T. Bubukin, A. G. Gulyan, V. P. Ivanov, R. M. Martirosyan, I. A. Malyshev, K. S. Stankevich, S. P. Stolyarov.
Astrofizika, Tom 21, Vyp. 2, p. 283 – 288 (1984). In Russian. English translation in Astrophysics, Vol. 21, No. 2.
The results are given for the observations of the occultation of the Crab nebula at 960, 750, 178, 128 MHz. Coordinates have been obtained of the gravity centers as well as angular dimensions of the nebula in the direction of small and large axes. Details of the source structure have been investigated at 960 and 750 MHz.

**Catalogue of observations of total occultations of stars by the Moon for the years 1972 to 1980 and of grazing occultations for the years 1963 to 1980.**
See Abstr. 002.060.

**Workshop über Sternbedeckungen.**
See Abstr. 011.031.

**Hochgeschwindigkeitsphotometrie für Sternbedeckungen.**
See Abstr. 034.195.

**Plans for astrometry for late 1984 asteroidal and planetary occultations.**
See Abstr. 041.009.

**Results from the occultation of 14 Piscium by (51) Nemausa.**
See Abstr. 098.026.

**Occultations by Vesta.**
See Abstr. 098.053.

**Occultations of SAO stars by minor planets during 1985.**
See Abstr. 098.103.

**Predictions of the mutual phenomena of the Galilean Satellites of Jupiter for 1985 – 1986.**
See Abstr. 099.016.

**A search for hot spots during a lunar occultation of Io.**
See Abstr. 099.108.

**The occultation of Hyd –20°51695 by Uranus on 1982 May 1.**
See Abstr. 101.074.

# 097 Mars, Mars Satellites

097.001 **First results of Mars observations with a PVC.**
A. V. Devyatkin.
Astron. Tsirk., No. 1282, p. 5 – 6 (1983). In Russian.

097.002 **Investigation of the Martian atmosphere by 10 micron heterodyne spectroscopy.**
H. U. Käufl, H. Rothermel, S. Drapatz.
Astron. Astrophys., Vol. 136, No. 2, p. 319 – 325 (1984).
During Mars opposition 1982 the planetary $CO_2$ atmosphere has been investigated using a 10 μm–heterodyne spectrometer (spectral resolution $\lambda/\Delta\lambda = 6 \times 10^6$) in the Cassegrain focus of a 2.2 m telescope. The spatial resolution on Mars is 1000 km. For the altitude range 50 – 90 km kinetic and rotational temperatures are derived ($T \sim 100$K), much lower than values in the recommended reference atmosphere ($T \sim 140$K). The measurements indicate annual variability of the atmospheric temperature. For the non–LTE emission lines no increase of the intensity from center to limb is observed, thus placing constraints on models for the emission process.

097.003 **The opposition of Mars, 1980.**
R. McKim.
J. Br. Astron. Assoc., Vol. 94, No. 5, p. 197 – 210 (1984).

097.004 **Symmetric and asymmetric hemispheres of Mars.**
Zh. F. Rodionova, K. I. Dekhtyareva.
Tr. Gos. Astron. Inst. Shternberg, Tom 55, p. 130 – 136 (1983). In Russian.
Peculiarities of plains, dark regions, crater density, heights of relief and thickness of the crust have shown that the Martian plain and continental hemispheres are more asymmetric than the northern and southern ones.

097.005 **Geology and tectonics of the Argyre area on Mars: comparisons with other basins in the solar system.**
P. G. Thomas, P. Masson.
Earth, Moon, Planets, Vol. 31, No. 1, p. 25 – 42 (1984).
Despite evident similarities, the Argyre basin exhibits important differences with regard to its lunar counterparts, as the Orientale basin. The Argyre basin is not surrounded by ejecta with radial facies, but by an annulus of structurally uplifted and faulted preimpact basement. The lunar basins exhibit a large annulus of radial facies but only a narrow ring of uplifted terrains. The Argyre basin is surrounded by five or more outer discontinuous rings extending far away from the basis edge. The lunar basins are surrounded by only one, continuous and closer ring. These differences could be partially explained by the external conditions, but mainly by differences in the crustal properties and lithospheres thickness which would have been thinner on Mars than on the Moon.

097.006 **Morphology and origin of ridges in low–latitude areas of Mars.**
R. Battistini.
Earth, Moon, Planets, Vol. 31, No. 1, p. 49 – 61 (1984).
The Martian ridges present a larger diversity of forms and patterns than the lunar ridges. The main explanation is the following: unlike the lunar–ridges, the low latitude ridges on Mars are not related to volcanism, but are in close connection with hydro–lithosphere. Their morphologic similarities with the lunar ridges are probably due to a similar dynamic origin – i.e., compression of a thin crust lying unconformably over the substratum (on Mars, porous sediments saturated with water). It is suggested that in some areas Martian ridges are mainly of extrusive origin (material extruded from the hydro–lithosphere), unlike the lunar ridges.

097.007 **Interpretation of grazing impacts on Mars by the axial rotation theory.**
N. A. Barricelli.
Earth, Moon, Planets, Vol. 31, No. 1, p. 75 – 78 (1984).
This paper calls attention to the fact that the distribution of grazing impacts on Mars, as described by Schultz and Lutz–Garihan (1982), agrees with the distribution predictable by the theory of planetary axial rotation based on the transmission of mass and angular momentum from a satellite system to its primary (Barricelli, 1972). A few additional implications of the theory are pointed out.

097.008 **A guide to observing Mars.**
C. F. Capen.
Strolling Astron., Vol. 30, Nos. 9 – 10, p. 186 – 190 (1984).

097.009 **A major Martian dust storm in 1984.**
J. D. Beish, D. C. Parker, C. F. Capen.
Strolling Astron., Vol. 30, Nos. 9 – 10, p. 211 – 217 (1984).

097.010 **The spatial and seasonal distribution of Martian clouds and some meteorological implications.**
R. Kahn.
J. Geophys. Res., Vol. 89, No. A8, p. 6671 – 6688 (1984).
The spatial and seasonal distributions of Martian atmospheric clouds, organized according to simple morphologic criteria, exhibit distinct patterns. When interpreted self–consistently using analogies to terrestrial cloud forms and Martian temperature, wind direction, and water vapor measurements, patterns of cloud occurrence provide some loose constraints on global–scale near–surface wind speed, static stability, and humidity. The author has examined all Mariner 9 and Viking Orbiter images for this study; he concentrates primarily on the larger Viking data set.

097.011 **Application of the method of mean square collocation for interpolation of the heights of the Martian relief in sampling points.**
V. E. Zinger, L. L. Polyakovskaya.
L'vov. politekh. inst. L'vov, 11 pp. (1984). In Russian. Abstr. in Ref. Zh., 52. Geod. Aehrosemka, 8.52.192 (1984).

097.012 **Mars 1984.**
H. Dierks.
Nachr. Olbers–Ges. Bremen, Nr. 131, p. 7 – 10 (1984).

097.013 **Martian north polar cap and haze 1981 – 1982.**
K. Iwasaki, Y. Saito, T. Akabane.
Publ. Astron. Soc. Jpn., Vol. 36, No. 2, p. 347 – 356 (1984).
Results of the photographic observations of Mars at the Kwasan and Hida Observatories in the retreat and remnant phases of the north polar cap during the 1981 – 1982 apparition are reported. The regression of the north polar cap is examined with the measurements of the photographs, and compared with previous works. The behavior of the haze that covers the north polar region before the northern summer solstice is examined.

097.014 **Modelling the turbulence of mass transfer in the Martian atmosphere.**
Yu. N. Kulikov.
Inst. prikl. mat. Akad. Nauk SSSR, Prepr., No. 40, 19 pp. (1984). In Russian. Abstr. in Ref. Zh., 51. Astron., 9.51.207 (1984).

097.015 **"Dust" streaks on Mars.**
P. Thomas, J. Veverka, D. Gineris, L. Wong.
Icarus, Vol. 60, No. 1, p. 161 – 179 (1984).
Global mapping and photometry of selected areas on Mars are used to investigate the nature of bright and dark wind streaks that extend from topographic obstacles. Data on the colors, albedos, and phase behavior of streaks are consistent with models of bright streaks as mosaics of plains material and brighter, redder dust. Dark streaks can be interpreted as erosional win-

dows in a patchy dust cover. The authors' model of dust deposition in optically thick patches is sedimentologically different from scenarios involving the deposition of ubiquitous, optically thin layers. It has the advantage that large amounts of dust can be deposited without affecting regional albedos.

097.016 **Some observations of Martian cloud systems.**
G. E. Hunt, A. O. Pickersgill.
Vistas Astron., Vol. 27, Part 2, p. 131 – 148 (1984).

The authors discuss the types of cloud phenomena which occur in the main region of the planet – equatorial, mid-latitude and polar – which are strongly influenced by the local orography and surface properties. These phenomena are further discussed in relation to the analogous features seen in the terrestrial atmosphere in order to provide a more detailed understanding of the observed cloud systems.

097.017 **Structure and origin of the Martian satellites.**
V. N. Zharkov, A. V. Kozenko, S. V. Maeva.
Astron. Vestn., Tom 18, No. 2, p. 83 – 99 (1984). In Russian. Abstr. in Ref. Zh., 51. Astron., 10.51.223 (1984).

097.018 **The magnetic field of Mars from data of synchronous measurements in the planet's magnetosphere and in the solar wind.**
Sh. Sh. Dolginov, L. N. Zhuzgov, S. I. Shkol'nikova.
Kosm. Issled., Tom 22, Vyp. 5, p. 792 – 803 (1984). In Russian. English translation in Cosm. Res.

097.019 **Temperature effect on the photodissociation rates in the atmospheres of Mars and Venus.**
J.-P. Parisot, J.-M. Zucconi.
Icarus, Vol. 60, No. 2, p. 327 – 331 (1984).

The available solar flux at a given altitude in the atmospheres of Mars and Venus is attenuated mainly by $CO_2$ with an extra contribution due to $SO_2$ on Venus. The dissociation cross section of $CO_2$ depends on temperature. At temperatures appropriate for these atmospheres ($\sim$250K), the cross sections are about 15% lower than those at room conditions (Y. L. Yung et al., 1982). It is shown that this temperature effect cannot be neglected in the evaluation of photolysis rates. Calculations of the photodissociation coefficients of $CO_2$, $SO_2$, HCl, and $H_2O$ are presented.

097.020 **Threshold wind velocity for particle entrainment at sub-atmospheric pressures as on the planet Mars.**
M. Phillips.
Atmos. Environ., Vol. 18, No. 4, p. 831 – 835 (1984). Abstr. in Phys. Abstr., Vol. 87, No. 1243, Entry 101222 (1984).

097.021 **One hypothesis concerning the evolution of $CO_2$ on Mars.**
R. Kahn.
Bull. Am. Astron. Soc., Vol. 16, No. 3, Part I, p. 671 (1984). Abstract. – See Abstr. 010.061.

097.022 **Vertical structure of hazes on Mars.**
F. Jaquin, R. Kahn.
Bull. Am. Astron. Soc., Vol. 16, No. 3, Part I, p. 671 (1984). Abstract. – See Abstr. 010.061.

097.023 **Determining Martian atmospheric opacity by multiple view angle mapping from orbit.**
D. J. Diner, J. V. Martonchik.
Bull. Am. Astron. Soc., Vol. 16, No. 3, Part I, p. 671 (1984). Abstract. – See Abstr. 010.061.

097.024 **Variations in the temperature of the Martian mesosphere.**
D. Deming, F. Espenak, T. Kostiuk, M. Mumma, D. Zipoy.
Bull. Am. Astron. Soc., Vol. 16, No. 3, Part I, p. 671 (1984). Abstract. – See Abstr. 010.061.

097.025 **The atmospheric circulation on Mars at high and low obliquity: implications for volatile transport.**
R. M. Haberle.
Bull. Am. Astron. Soc., Vol. 16, No. 3, Part I, p. 672 (1984). Abstract. – See Abstr. 010.061.

097.026 **Albedo features in the Solis Planum/Sinai Planum region of Mars: seasonal and secular changes.**
S. W. Lee.
Bull. Am. Astron. Soc., Vol. 16, No. 3, Part I, p. 672 (1984). Abstract. – See Abstr. 010.061.

097.027 **Temporal changes at the Mutch Memorial Station on Mars during the automatic mission.**
J. B. Garvin, O. de Goursac.
Bull. Am. Astron. Soc., Vol. 16, No. 3, Part I, p. 672 (1984). Abstract. – See Abstr. 010.061.

097.028 **Dustiness on Mars as determined from surface meteorological measurements: variability over three years.**
J. E. Tillman, C. B. Leovy.
Bull. Am. Astron. Soc., Vol. 16, No. 3, Part I, p. 672 (1984). Abstract. – See Abstr. 010.061.

097.029 **Accelerated expansion of Martian great dust storms due to atmospheric tides.**
R. W. Zurek.
Bull. Am. Astron. Soc., Vol. 16, No. 3, Part I, p. 672 – 673 (1984). Abstract. – See Abstr. 010.061.

097.030 **Mars: north polar cap regressions, 1975 – 1982.**
P. B. James, M. Pierce, L. J. Martin.
Bull. Am. Astron. Soc., Vol. 16, No. 3, Part I, p. 673 (1984). Abstract. – See Abstr. 010.061.

097.031 **Mars 1984: the transition from polar hood to surface cap.**
L. J. Martin, P. B. James.
Bull. Am. Astron. Soc., Vol. 16, No. 3, Part I, p. 673 (1984). Abstract. – See Abstr. 010.061.

097.032 **Mars at high obliquity: possible precipitation of ice at low latitudes.**
B. M. Jakosky, M. H. Carr.
Bull. Am. Astron. Soc., Vol. 16, No. 3, Part I, p. 673 (1984). Abstract. – See Abstr. 010.061.

097.033 **Polar radiative balance measurements from MGCO.**
D. J. McCleese, D. A. Paige, A. P. Ingersoll.
Bull. Am. Astron. Soc., Vol. 16, No. 3, Part I, p. 673 (1984). Abstract. – See Abstr. 010.061.

097.034 **Mars: astronomical effects on regolith $H_2O$ state.**
F. P. Fanale, J. R. Salvail, A. P. Zent, S. Postawko.
Bull. Am. Astron. Soc., Vol. 16, No. 3, Part I, p. 674 (1984). Abstract. – See Abstr. 010.061.

097.035 **Seasonal cycles of diffusion and phase partitioning of water within the Martian regolith.**
A. P. Zent, F. P. Fanale, J. R. Salvail.
Bull. Am. Astron. Soc., Vol. 16, No. 3, Part I, p. 674 (1984). Abstract. – See Abstr. 010.061.

097.036 **Analysis of Martian surface compositional properties from IRTM and IRIS spectral observations.**
P. R. Christensen, H. H. Kieffer.
Bull. Am. Astron. Soc., Vol. 16, No. 3, Part I, p. 679 (1984). Abstract. – See Abstr. 010.061.

097.037 **First direct detection of clay minerals on Mars.**
R. B. Singer, P. D. Owensby, R. N. Clark.
Bull. Am. Astron. Soc., Vol. 16, No. 3, Part I, p. 679 – 680 (1984). Abstract. – See Abstr. 010.061.

097.038 **Martian terrain ages.**
N. G. Barlow, R. G. Strom.
Bull. Am. Astron. Soc., Vol. 16, No. 3, Part I, p. 680 (1984). Abstract. – See Abstr. 010.061.

097.039 **Mars: eolian erosion of thick regional deposits.**
P. Thomas.
Bull. Am. Astron. Soc., Vol. 16, No. 3, Part I, p. 680 (1984). Abstract. – See Abstr. 010.061.

097.040 **Volcanic rocks in the Valles Marineris.**
B. K. Lucchitta.
Bull. Am. Astron. Soc., Vol. 16, No. 3, Part I, p. 680 – 681 (1984). Abstract. – See Abstr. 010.061.

097.041 **Topographic effects on the visual reflectance values of Ascraeus Mons, Mars.**
J. R. Zimbelman, R. Greeley.
Bull. Am. Astron. Soc., Vol. 16, No. 3, Part I, p. 681 (1984). Abstract. – See Abstr. 010.061.

097.042 **Emplacement of long lava flows on Alba (Mars) and Mauna Loa (Hawaii).**
P. J. Mouginis–Mark, G. P. L. Walker, L. Wilson.
Bull. Am. Astron. Soc., Vol. 16, No. 3, Part I, p. 681 (1984). Abstract. – See Abstr. 010.061.

097.043 **Mapping of the 2 and 6 cm thermal emission from Mars using the Very Large Array.**
D. J. Rudy, D. O. Muhleman, G. L. Berge.
Bull. Am. Astron. Soc., Vol. 16, No. 3, Part I, p. 682 (1984). Abstract. – See Abstr. 010.061.

097.044 **The Martian magnetic field.**
R. S. Wolff, J. A. Slavin, H. A. Taylor.
Bull. Am. Astron. Soc., Vol. 16, No. 3, Part I, p. 682 (1984). Abstract. – See Abstr. 010.061.

097.045 **Zeolites on Mars: prospects for remote sensing.**
E. S. Gaffney, R. B. Singer, T. D. Kunkle.
Bull. Am. Astron. Soc., Vol. 16, No. 3, Part I, p. 710 (1984). Abstract. – See Abstr. 010.061.

097.046 **Indication for $H_2O$ in the upper atmosphere of Mars.**
H. U. Käufl, H. Rothermel, S. Drapatz.
Astron. Astrophys., Vol. 141, No. 2, p. 430 – 432 (1984).

Ground–based observations of non–LTE emission lines of Martian $CO_2$ have been made with adequate spatial and spectral resolution to investigate processes at altitudes between 50 and 90 km above the surface. Measurements of the band integrated steradiancy at various locations on the planetary disc during the opposition 1982 show an increase of the emission over the north polar cap, which cannot be explained by standard models assuming a pure $CO_2$ atmosphere. The results indicate both a substantial influence of $H_2O$ on the emission process, and a somewhat supersaturated $H_2O$ vapour in the emitting atmospheric layer.

097.047 **Aeronomy of Mars. Catalytic cycles and critical chemical reactions.**
Yu. N. Kulikov, M. Ya. Marov.
Inst. prikl. mat. Akad. Nauk SSSR. Prepr., No. 42, 24 pp. (1984). In Russian. Abstr. in Ref. Zh., 51. Astron., 11.51.191 (1984).

097.048 **On the distribution of gravity and deviations of the vertical on the Martian surface.**
T. G. Maksimova, N. A. Chujkova.
Vestn. Mosk. Univ., Ser. 3. Fiz. Astron., Tom 25, No. 1, p. 64 – 69 (1984). In Russian. Abstr. in Ref. Zh., 52. Geod. Aehrosemka, 11.52.206 (1984).

097.049 **Viking bistatic radar experiment: summary of results in near–equatorial regions.**
R. A. Simpson, G. L. Tyler, G. G. Schaber.
J. Geophys. Res., Vol. 89, No. B12, p. 10385 – 10404 (1984).

Viking bistatic radar data have been processed using Hagfors' scattering function to obtain estimates of rms surface roughness $\sigma_B$ from ground tracks primarily in Mars' equatorial region.

097.050 **First observations of Mars with the photoelectric meridian circle in Pulkovo.**
A. V. Devyatkin, V. N. Ershov.
Astron. Tsirk., No. 1313, p. 3 – 4 (1984). In Russian.

**Phobos: close encounter imaging from the Viking Orbiters.**
See Abstr. 002.020.

**Viking orbiter mosaic status report.**
See Abstr. 002.030.

**The surface of Mars.**
See Abstr. 003.068.

**The appearances of Venus and Mars in 1610.**
See Abstr. 004.046.

**The case for Mars. Proceedings of a conference held at Univerity of Colorado, Boulder, Colorado, USA, April 29 – May 2, 1981.**
See Abstr. 012.077.

**Bau eines Heterodynempfängers für 10 µm Wellenlänge zum Einsatz am Cassegrainfokus astronomischer Teleskope und spektroskopische Untersuchung der Mars– und Venusatmosphäre.**
See Abstr. 034.183.

**Climatology capabilities of a gamma–ray spectrometer at Mars.**
See Abstr. 036.061.

**Modeling crater topography and albedo from monoscopic Viking Orbiter images. 1. Methodology.**
See Abstr. 036.127.

**Evaluation of phase corrections on planet position measurements.**
See Abstr. 041.022.

**The conquest of Mars.**
See Abstr. 051.003.

**Mars exploration in the 1990's.**
See Abstr. 051.010.

**The Mars Geoscience/Climatology Observer (MGCO) mission.**
See Abstr. 051.056.

**Studies of potential Mars Rover Sample–Return sites.**
See Abstr. 051.057.

**Mars sample return mission.**
See Abstr. 051.059.

**The role of mineral precipitates in moderating Shuttle imaging radar signal penetration and subsurface scattering in the hyperarid eastern Sahara: applications to climatic history of Mars.**
See Abstr. 081.037.

**Terrestrial analogs for martian and lunar wrinkle ridges.**
See Abstr. 081.038.

**Reorientation of planets with elastic lithospheres.**
See Abstr. 091.049.

**Non-equilibrium population of the $CO_2(00^\circ1)$ and CO(1) vibrational states in the upper atmospheres of Venus and Mars.**
See Abstr. 093.001.

**Heating of the Venusian and Martian upper atmospheres by infrared solar radiation.**
See Abstr. 093.002.

**The interpretation of the gravity field of Venus and Mars.**
See Abstr. 093.034.

**On the mechanism of interconnection of physico-mineralogical and optical properties of finely dispersed soils.**
See Abstr. 094.023.

**Observations photographiques de Mars, de Jupiter et de ses satellites, effectuées à l'Astrographe Double de 40 cm en 1980.**
See Abstr. 099.131.

**Meteorites from Mars.**
See Abstr. 105.039.

**Meteorites from Mars.**
See Abstr. 105.046.

**The case for a martian origin of the shergottites: nitrogen and noble gases in EETA 79001.**
See Abstr. 105.092.

# 098 Minor Planets

098.001 **Minor planet positions obtained at Cerro El Roble astronomical station.**
H. Wroblewski, J. Maza, C. Torres, E. Costa, M. Wischnjewsky, L. E. González.
Astron. Astrophys., Suppl. Ser., Vol. 57, No. 1, p. 37 - 42 (1984).

101 positions of 19 minor planets observed during the period 1979 - 1980 with the Maksutov Astrograph at Cerro El Roble are presented. Data on comparison stars, including dependences, are presented.

098.002 **Rotational rates of very small asteroids: 123 Brunhild, 376 Geometria, 437 Rhodia and 1224 Fantasia.**
M. A. Barucci, M. Di Martino.
Astron. Astrophys., Suppl. Ser., Vol. 57, No. 1, p. 103 - 106 (1984).

This paper presents observations of four small main belt asteroids ($D < 50$ km) obtained during October 1983 at the Observatoire de Haute-Provence (France). For two of them, 123 Brunhild and 376 Geometria, an accurate rotational period was determined. For the other two asteroids, 437 Rhodia and 1224 Fantasia, only tentative periods are suggested.

098.003 **Observacão de asteróides na Estación de Altura "El Leoncito", Observatorio Astronômico "Felix Aguilar", San Juan, Argentina Julho-1983.**
J. F. Caldeira, L. E. Machado, E. R. Netto, G. G. Vieira, H. Debehogne.
Rev. Mex. Astron. Astrofis., Vol. 9, No. 1, p. 91 - 93 (1984).

098.004 **Asteroidal occultations of uncataloged stars, and some Nemausa A.C. events, during 1984.**
R. L. Millis, L. H. Wasserman, O. G. Franz, E. Bowell, A. Klemola, D. W. Dunham.
Occultation Newsl., Vol. 3, No. 8, p. 158 - 160 (1984).

098.005 **Asteroidal occultation news.**
D. W. Dunham.
Occultation Newsl., Vol. 3, No. 8, p. 161 - 162 (1984).

098.006 **Observations of asteroidal appulses and occultations.**
D. W. Dunham.
Occultation Newsl., Vol. 3, No. 8, p. 166 - 169 (1984).

098.007 **The Palomar-Leiden Survey of faint minor planets: conclusion.**
C. J. van Houten, P. Herget, B. G. Marsden.
Icarus, Vol. 59, No. 1, p. 1 - 19 (1984).

The results are given of a revision and small extension of the Palomar-Leiden Survey of minor planets. The majority of the class 4 orbits in the original Survey have been rederived using positions measured from the second plate of each blink pair, and in some cases the orbits could be raised to class 1 quality by the identification of observations in the second month. By using the concept of "*e*-assumed orbits", meaningful - if not always accurate - orbits are given for the cases that previously had to be rejected. The extension to the Survey consists of 170 new objects found in the field used for photometric calibration purposes. The total number of orbits in the Survey is 2403.

098.008 **Diameters and albedos of thirty-six asteroids.**
R. H. Brown, D. Morrison.
Icarus, Vol. 59, No. 1, p. 20 - 24 (1984).

Radiometric diameters and albedos of 36 asteroids, most previously unmeasured by this technique, are reported. These objects were selected primarily to resolve taxonomic ambiguities resulting from the lack of albedo information. Of the sample of 36, most are of the common types C, S, and M, but also represented are types A, D, F, and P. One object, 214 Aschera, is of the rare E type (only the fourth of its kind known); 87 Sylvia is the largest known member of the P class; and albedo combined with Arizona eight-color data from Tholen (1983) indicate that 336 Lacadiera is of type D, one of the few members of this class found in the main asteroid belt, and a prime candidate for future space missions.

098.009 **Mineralogical-petrological characterization of near-Earth asteroids.**
L. A. McFadden, M. J. Gaffey, T. B. McCord.
Icarus, Vol. 59, No. 1, p. 25 - 40 (1984).

The mineralogy and petrology of 17 near-Earth asteroids are characterized using reflectance spectroscopy with ground-based telescopes as one method to address their major issues. With one possible exception, the surface composition of near-Earth asteroids consists of common rock-forming minerals such as olivine, pyroxene, and phyllosilicates. Opaque components are present but cannot be mineralogically identified with existing experimental data. The predominance of mafic silicate absorption features in spectra of near-Earth asteroids compared to the majority of main-belt asteroids may be a primary compositional feature or may be the signature of relatively fresher asteroid material.

098.010 **Erratum: "The discovery of 1982 DB, the most accessible asteroid known" [Icarus, Vol. 57, No. 1, p. 42 - 47 (1984)].**
E. F. Helin, N. D. Hulkower, D. F. Bender.
Icarus, Vol. 59, No. 1, p. 132 (1984). See Abstr. 37.098.009.

098.011 **8 Flora – a minor planet observing project.**
A. J. Hollis.
I.A.P.P.P. Commun., No. 16, p. 20 – 21 (1984).

098.012 **Can gravitational compaction explain the spin–diameter relation for the asteroids?**
R. Halling.
Astrophys. Space Sci., Vol. 103, No. 2, p. 379 – 383 (1984).

Dermott and Murray (1982) have found certain relations between mean spin frequency and mean diameter for asteroids. By assuming that matter gets compacted during the evolution of an asteroid, these relations may be explained by means of a spin formula for accreted bodies given by Alfvén and Arrhenius. The author presents a rough model for gravitational compaction of porous matter in asteroids, combines it with the spin formula, and fits the result to Dermott's and Murray's spin–diameter curves. The values obtained for the parameters (mass densities and material strengths) appear to be of a quite realistic order of magnitude.

098.013 **Structure of the 3:1 Jovian resonance: a comparison of numerical methods.**
C. D. Murray, K. Fox.
Icarus, Vol. 59, No. 2, p. 221 – 233 (1984).

The motion of asteroids near the 3:1 Jovian resonance in the restricted planar case is studied using three numerical methods: (1) integrating the full equations of motion, (2) integrating the averaged equations of motion, and (3) using an algebraic mapping recently developed by Wisdom (1982). The relative merits of each method are investigated. It is concluded that in the regular regions of the phase space, methods (2) and (3) give excellent agreement with each other. In chaotic regions of the phase space there is little agreement between the orbital elements at any given time calculated by each method. However, all methods reflect the qualitative behavior of the chaotic trajectories and give good agreement on the bounds of the motion.

098.014 **Collisional origin of the asteroid families: mass and velocity distributions.**
V. Zappalà, P. Farinella, Z. Knežević, P. Paolicchi.
Icarus, Vol. 59, No. 2, p. 261 – 285 (1984).

The origin of asteroid families in terms of collisional breakup is analyzed using the data by Williams (1979). The distributions of mass and relative velocity of the minor family members with respect to the largest body are derived. These distributions are then compared with the outcomes of catastrophic impacts, predicted from theoretical arguments and observed from laboratory experiments. The general features of the mass distributions can be interpreted in terms of collisional disruption of a parent body followed by self–gravitational reaccumulation on the largest remnant.

098.015 **B and V lightcurves and pole positions of three S–class asteroids.**
R. S. McCheyne, N. Eaton, S. F. Green, A. J. Meadows.
Icarus, Vol. 59, No. 2, p. 286 – 295 (1984).

B and V lightcurves have been obtained for three S–class asteroids – 29 Amphitrite, 39 Laetitia, and 43 Ariadne. Pole positions and shapes for each of these asteroids are derived, and comments on small features present in their lightcurves are made. A small color variation for 39 Laetitia is also found.

098.016 **Minor planet 1983 TB: a dead comet?**
A. L. Cochran, E. S. Barker.
Icarus, Vol. 59, No. 2, p. 296 – 300 (1984).

IDS (*Intensified Dissector Scanner*) spectra of the Apollo asteroid 1983 TB were obtained at McDonald Observatory on 2 December 1983 UT. These spectra were examined for cometary–type emissions and none were found. The reflectance of 1983 TB was found to be similar to an S–type asteroid.

098.017 **Les satellites d'asteroides.**
J. E. Arlot, J. Lecacheux, W. Thuillot.
Astrométrie, géodésie, mécanique céleste, p. 63 – 68 (1984). – See Abstr. 012.021.

Le travail présenté ici aborde le problème de la détection d'éventuels satellites par l'observation. Les résultats obtenus sont également présentés.

098.018 **Analysis of the shape distribution of asteroids.**
V. Catullo, V. Zappalà, P. Farinella, P. Paolicchi.
Astron. Astrophys., Vol. 138, No. 2, p. 464 – 468 (1984).

The shape distribution of asteroids, as indicated by light curve amplitudes, is compared to a similar distribution obtained for fragments generated in high–velocity impact experiments. By dividing the asteroids in three different classes (small objects, less than 100 km in diameter; asteroids larger than 200 km; Apollo–Amor objects), the authors show that only the first class agrees satisfactorily with the fragment distribution; large objects are considerably less elongated, suggesting that they have relaxed to symmetric equilibrium figures, while in the Apollo–Amor group a higher number of very elongated objects is present. These results are briefly discussed in the framework of the current ideas on the collisional evolution of asteroids.

098.019 **Rotation axes of asteroids: results for 14 objects.**
V. Zappalà, Z. Knežević.
Icarus, Vol. 59, No. 3, p. 436 – 455 (1984).

In this paper results of determinations of the rotation axis direction (as given by ecliptic coordinates of its intersection with geocentric celestial sphere) for 14 asteroids are presented for which a sufficient number of good observations are available. The calculation is performed by using the improved amplitude–magnitude method developed by V. Zappalà, M. Di Martino, P. Farinella, and P. Paolicchi (1983). Additionally, an improvement of the method is described regarding a weighting procedure. Whenever possible, the results are compared with already existing data, coming from other methods (mainly from photometric astrometry).

098.020 **2 Pallas: 1982 and 1983 lightcurves and a new pole solution.**
R. P. Binzel.
Icarus, Vol. 59, No. 3, p. 456 – 461 (1984).

Photoelectric lightcurves of asteroid 2 Pallas obtained in March 1982 and May 1983 display amplitudes of 0.04 and 0.10 magnitude respectively. A least–squares best fit to an amplitude–aspect relation for all available lightcurve observations of Pallas between 1951 and 1983 yields two solutions for its pole position.

098.021 **Size distribution of asteroids.**
K. Ishida, T. Mikami, H. Kosai.
Publ. Astron. Soc. Jpn., Vol. 36, No. 2, p. 357 – 370 (1984).

A survey observation of asteroids performed with the 105/150/330 cm Kiso Schmidt telescope yielded about 2700 asteroidal images. The cumulative number function of asteroids of the apparent magnitude up to $m = 19$ and their surface distribution on the sky referred to the solar opposition point are presented. The observed distributions are simulated with a model including the size distribution, space distribution, and optical properties of asteroids.

098.022 **The distribution of asteroid diameters.**
J. R. Donnison, R. A. Sugden.
Mon. Not. R. Astron. Soc., Vol. 210, No. 3, p. 673 – 682 (1984).

The distribution of asteroids with diameters greater than 130 km is analysed using a maximum likelihood method. The mass distribution index of these asteroids for various diameter ranges, taxonomic classes and distance zones is estimated together with the large sample errors and confidence intervals.

098.023 **A note on the determination of the distance between two asteroids during their regular proximity.**
J. L. Simovljević.
Zb. Rad. Mat. Inst., Nova Ser., Tom 4, No. 12, p. 201 – 206 (1984). In Serbo–Croatian.

An approximate method is developed to determine the perturbative action of one asteroid upon his distance to the other one, during the regular proximity of these celestial bodies. The final effect appears to be not sensible during a time interval of «dynamical proximity» inferior to one day, if the mass of the perturbing asteroid is of the order of $10^{-13} M_{\odot}$ and the proximity distance is of the order of 0.000005 AU.

098.024 **Photoelectric photometry of 14 asteroids.**
M. Di Martino, S. Cacciatori.
Icarus, Vol. 60, No. 1, p. 75 – 82 (1984).

Results of observations of 14 asteroids are reported. *V* photoelectric lightcurves were obtained from September 1982 to June 1983 at the Astronomical Observatory of Torino and at the Astrophysical Observatory of Catania. Part of this program aims to obtain complete lightcurves and, when possible, phasecurve information and to determine amplitudes and *V* magnitudes at different longitudes for a selected group of asteroids, in order to enlarge the set of known rotational axis directions.

098.025 **Rotational spectral variations of asteroid (8) Flora: implications for the nature of the S–type asteroids and for the parent bodies of the ordinary chondrites.**
M. J. Gaffey.
Icarus, Vol. 60, No. 1, p. 83 – 114 (1984). = PGD Publ. 402.

Two new interpretive calibrations (olivine–pyroxene abundance, spectral signature of chondritic metal grains) have been applied to visible and near–infrared spectral reflectance data for the typical S–type asteroid, (8) Flora. Chondritic metal grains do not exhibit the steeply reddened spectral signature previously associated with a nickel–iron (NiFe) metal phase. The conditions required to produce a S–type spectral curve from a chondritic material are much more restrictive and improbable than previously supposed. Five mineralogic and petrologic parameters have been derived for the surface material of Flora from the analysis of the spectral reflectance and rotational spectral variations of this object. Flora is most probably the residual core of an intensely heated, thermally evolved, and magmatically differentiated planetesimal which was subsequently disrupted.

098.026 **Results from the occultation of 14 Piscium by (51) Nemausa.**
E. W. Dunham, R. L. Baron, S. Conner, D. W. Dunham, J. B. Dunham, G. Schneider, H. L. Cohen, V. T. Helms III, M. Croom, J. Safko.
Astron. J., Vol. 89, No. 11, p. 1755 – 1758 (1984).

The 11 September 1983 occultation of 14 Psc by (51) Nemausa was observed photoelectrically at six sites across the occultation track. The observations are well represented by an elliptical limb profile having a semimajor axis of $84.9 \pm 2.0$ km and oblateness $(1\text{–}b'/a')$ $0.20 \pm 0.05$. The geocentric position of the asteroid was $0^{s}.18843 \pm 0^{s}.00022$ E and $3''.5696 \pm 0''.0030$ N of 14 Psc at 7:45:00 UTC on 11 September 1983. The quoted uncertainties are all formal errors from the elliptical least–squares fit. No secondary occultations were observed.

098.027 **Radar and photoelectric observations of asteroid 2100 Ra–Shalom.**
S. J. Ostro, A. W. Harris, D. B. Campbell, I. I. Shapiro, J. W. Young.
Icarus, Vol. 60, No. 2, p. 391 – 403 (1984).

Results of 13–cm wavelength radar observations and V–filter photoelectric observations of Ra–Shalom during its 1981 August – September apparition are reported. Most, but certainly not all, of the backscattering is due to single reflections from surface elements that are fairly smooth at decimeter scales. The photoelectric data suggest a value, $P_{syn} = 19.79$ hr, for the synodic rotation period, and yield a composite lightcurve with two pairs of extrema.

098.028 **1984 AB – a unique Mars–crossing asteroid.**
E. F. Helin, R. S. Dunbar.
Icarus, Vol. 60, No. 2, p. 404 – 408 (1984).

Asteroid 1984 AB, discovered in January 1984, proved to be a unique object with a close dynamical relationship to Mars. A brief history of the discovery and subsequent "evolution" of the orbit as it was refined is presented. The preliminary orbit of 1984 AB indicated that it might be a Mars Trojan, and an extended discussion of this interesting possibility is presented, but this hypothesis had to be dismissed after further observations had refined the orbit. The semimajor axis and orbital eccentricity are very similar to that of Mars. No other known Mars–crossing asteroid exists with an orbit as closely associated to Mars.

098.029 **Investigation of motion of six minor planets.**
K. Ziołkowski.
Acta Astron., Vol. 34, No. 1, p. 65 – 78 (1984).

On the basis of all positional observations which were published till the end of 1982 the new osculating orbits of the following minor planets: (457) Alleghenia, (649) Josefa, (1038) Tuckia, (1161) Thessalia, (1162) Larissa and (1297) Quadea, have been obtained. A comparatively well determinable secular deceleration in the motion of (1161) Thessalia has been discovered.

098.030 **A possible determination of asteroid precessions by means of photometric observations.**
T. Michałowski.
Acta Astron., Vol. 34, No. 1, p. 79 – 87 (1984).

By means of numerical studies, it is shown that asteroid lightcurves – in particular their amplitudes – are strongly dependent upon the precessional motion of a minor planet. Such a dependence could be used for the determination of the precession of asteroids from the analysis of observed lightcurves.

098.031 **Positions of minor planets and comets obtained at the Chorzów Observatory.**
I. Włodarczyk.
Acta Astron., Vol. 34, No. 1, p. 89 – 91 (1984).

The paper contains results of photographic observations of minor planets, comet Panther (1980u) and comet Churyumov–Gerasimenko (1982f) made in the years 1977 – 1982.

098.032 **Positions of Comet Bowell (1980b), Asteroids 1982 DV and 51 Nemausa, March 1982, with the GPO, ESO–Chile.**
H. Debehogne, R. R. Mourao, G. Vieira.
Acta Astron., Vol. 34, No. 1, p. 129 – 134 (1984).

22 positions of 1982 DV, 14 of 51 Nemausa and 21 of Comet Bowell, obtained at the Grand Prisme Objectif (GPO, $f = 4$ m, $D = 40$ cm), are presented. The reductions were made by dependences and by least–squares method using 5 reference stars.

098.033 **Minor planet positions in 1981 and 1982.**
H. Debehogne, U. Bezerra, L. Houziaux.
Acta Astron., Vol. 34, No. 1, p. 135 – 138 (1984).

1950.0 positions of asteroids 35, 89, 101, 333, 459, 556, 613, 651, 826, 874, 925, 1043, 1286, 1305, 1434, 1491, 1571, 2094, 2147, 1981 $QG_2$, 1981 SW are given for a period extending from September 1981 to February 1982. The authors also mention 1950.0 positions of asteroids 1981 $ST_2$, 1981 $SU_2$, 1981 $SV_2$, 1981 $WZ_3$, 1981 $WA_4$, 1981 $WB_4$, 1981 $XB_2$, 1981 $XC_2$, 1981 $XD_2$, 1981 $XE_2$, discovered on the same plates. The data are deduced from photographic observations carried with the Liège–CNRS Schmidt telescope at Observatoire de Haute–Provence.

098.034 **Astrometric positions of minor planets in March and April 1982.**
H. Debehogne, L. Houziaux.
Acta Astron., Vol. 34, No. 1, p. 139 – 140 (1984).

28 minor planets have been observed with the 40 cm refractor at the European Southern Observatory. 1950.0 positions in March and April 1982 are given.

098.035 **Asteroids.**
A. Brahic.
Ciel Espace, No. 197, p. 21 – 30 (1984). In French. Abstr. in Phys. Abstr., Vol. 88, No. 1247, Entry 4639 (1985).

098.036 **An empirical phase relation for atmosphereless bodies.**
K. Lumme, E. Bowell, A. W. Harris.
Bull. Am. Astron. Soc., Vol. 16, No. 3, Part I, p. 684 (1984). Abstract. – See Abstr. 010.061.

098.037 **IRAS observations of asteroids.**
H. H. Aumann, R. G. Walker.
Bull. Am. Astron. Soc., Vol. 16, No. 3, Part I, p. 689 (1984). Abstract. – See Abstr. 010.061.

098.038 **Ellipsoidal geometry in the standard radiometric model.**
R. H. Brown.
Bull. Am. Astron. Soc., Vol. 16, No. 3, Part I, p. 689 (1984). Abstract. – See Abstr. 010.061.

098.039 **The discovery circumstances of earth–approaching asteroids.**
L. G. Taff, P. M. S. Randall.
Bull. Am. Astron. Soc., Vol. 16, No. 3, Part I, p. 690 (1984). Abstract. – See Abstr. 010.061.

098.040 **Photometry of earth–approaching asteroids.**
D. J. Tholen.
Bull. Am. Astron. Soc., Vol. 16, No. 3, Part I, p. 690 – 691 (1984). Abstract. – See Abstr. 010.061.

098.041 **Mars–crossing asteroids discovered in the UK–Caltech Asteroid Survey.**
E. M. Shoemaker, S. J. Bus, R. S. Dunbar, E. F. Helin, J. Dawe, J. Barrow, M. Hartley, D. Morgan, K. Russell, A. Savage.
Bull. Am. Astron. Soc., Vol. 16, No. 3, Part I, p. 691 (1984). Abstract. – See Abstr. 010.061.

098.042 **2201 Oljato: an asteroid, a comet, or both?**
L. A. McFadden, S. J. Ostro, E. S. Barker, A. L. Cochran, D. P. Cruikshank, W. K. Hartmann, B. T. Soifer, G. J. Veeder.
Bull. Am. Astron. Soc., Vol. 16, No. 3, Part I, p. 691 (1984). Abstract. – See Abstr. 010.061.

098.043 **Asteroid 1984 BC – an extinct comet?**
E. F. Helin, R. S. Dunbar, J. Gradie, L. A. Lebofsky, D. J. Tholen.
Bull. Am. Astron. Soc., Vol. 16, No. 3, Part I, p. 691 (1984). Abstract. – See Abstr. 010.061.

098.044 **Photoelectric photometry of asteroids 9 Metis, 18 Melpomene, 60 Echo, 116 Sirona, 230 Athamantis, 694 Ekard, and 1984 KD.**
K. W. Zeigler, W. B. Florence.
Bull. Am. Astron. Soc., Vol. 16, No. 3, Part I, p. 691 – 692 (1984). Abstract. – See Abstr. 010.061.

098.045 **Radar detection of 9 Metis, 80 Sappho, 356 Liguria, 694 Ekard, and 2101 Adonis.**
S. J. Ostro, D. B. Campbell, I. I. Shapiro.
Bull. Am. Astron. Soc., Vol. 16, No. 3, Part I, p. 692 (1984). Abstract. – See Abstr. 010.061.

098.046 **Characterization of outer–belt asteroids from narrow-band reflectance spectra ($\sim 0.5 - 1.0$ μm).**
F. Vilas, B. A. Smith.
Bull. Am. Astron. Soc., Vol. 16, No. 3, Part I, p. 692 (1984). Abstract. – See Abstr. 010.061.

098.047 **The 52–color asteroid survey: preliminary results.**
J. F. Bell, B. R. Hawke, M. J. Gaffey, P. D. Owensby.
Bull. Am. Astron. Soc., Vol. 16, No. 3, Part I, p. 692 (1984). Abstract. – See Abstr. 010.061.

098.048 **The nature of C–type asteroids from 3 micron spectrophotometry.**
M. A. Feierberg, L. A. Lebofsky, D. J. Tholen.
Bull. Am. Astron. Soc., Vol. 16, No. 3, Part I, p. 692 – 693 (1984). Abstract. – See Abstr. 010.061.

098.049 **Absolute magnitude statistics of faint asteroids.**
E. Bowell, S. J. Bus.
Bull. Am. Astron. Soc., Vol. 16, No. 3, Part I, p. 693 (1984). Abstract. – See Abstr. 010.061.

098.050 **Speckle interferometry results for 511 Davida and 2 Pallas.**
J. D. Drummond, E. K. Hege.
Bull. Am. Astron. Soc., Vol. 16, No. 3, Part I, p. 697 (1984). Abstract. – See Abstr. 010.061.

098.051 **The Ca–Al rich components in asteroidal regoliths.**
R. S. Rajan, G. J. Veeder, D. L. Matson, E. F. Tedesco, M. J. Gaffey.
Bull. Am. Astron. Soc., Vol. 16, No. 3, Part I, p. 697 (1984). Abstract. – See Abstr. 010.061.

098.052 **Asteroid collision history: evidence from Vesta and Hirayama families.**
D. R. Davis, C. R. Chapman, S. J. Weidenschilling, R. Greenberg.
Bull. Am. Astron. Soc., Vol. 16, No. 3, Part I, p. 697 (1984). Abstract. – See Abstr. 010.061.

098.053 **Occultations by Vesta.**
R. L. Millis, L. H. Wasserman.
Bull. Am. Astron. Soc., Vol. 16, No. 3, Part I, p. 697 – 698 (1984). Abstract. – See Abstr. 010.061.

098.054 **1220 Crocus: a precessing, binary asteroid?**
R. P. Binzel.
Bull. Am. Astron. Soc., Vol. 16, No. 3, Part I, p. 698 (1984). Abstract. – See Abstr. 010.061.

098.055 **Thermal properties of the regolith of asteroid 1 Ceres.**
L. A. Lebofsky, M. V. Sykes, E. F. Tedesco, G. J. Veeder, D. L. Matson, I. G. Nolt, J. V. Radostitz, P. A. R. Ade, W. K. Gear, M. J. Griffin, E. I. Robson.
Bull. Am. Astron. Soc., Vol. 16, No. 3, Part I, p. 698 (1984). Abstract. – See Abstr. 010.061.

098.056 **Observational tests of the differentiated nature of the Flora family parent body.**
M. J. Gaffey, J. F. Bell, B. R. Hawke.
Bull. Am. Astron. Soc., Vol. 16, No. 3, Part I, p. 698 (1984). Abstract. – See Abstr. 010.061.

098.057 **Inversion of asteroid lightcurves: the problem of non-uniqueness and a possible remedy.**
H. Karttunen, K. Lumme, E. Bowell.
Bull. Am. Astron. Soc., Vol. 16, No. 3, Part I, p. 699 (1984). Abstract. – See Abstr. 010.061.

098.058 **Convex–profile inversion of asteroid lightcurves: calibration and application of the method.**
S. J. Ostro, M. D. Dorogi, R. Connelly.
Bull. Am. Astron. Soc., Vol. 16, No. 3, Part I, p. 699 (1984). Abstract. – See Abstr. 010.061.

098.059 **On the estimate of the shape of regular asteroids.**
M. A. Barucci, M. Fulchignoni, V. Zappalà.
Bull. Am. Astron. Soc., Vol. 16, No. 3, Part I, p. 700 (1984). Abstract. – See Abstr. 010.061.

098.060 **Improvement of asteroid pole determinations by scattering corrections.**
V. Zappalà, M. A. Barucci, M. Fulchignoni.
Bull. Am. Astron. Soc., Vol. 16, No. 3, Part I, p. 700 (1984). Abstract. – See Abstr. 010.061.

098.061 **The pole orientation of asteroid 16 Psyche via two independent methods.**
E. F. Tedesco, R. C. Taylor.
Bull. Am. Astron. Soc., Vol. 16, No. 3, Part I, p. 700 (1984).
Abstract. – See Abstr. 010.061.

098.062 **Photometric geodesy of large, rapid rotating asteroids: progress report.**
C. R. Chapman, D. R. Davis, R. Greenberg, S. J. Weidenschilling, S. J. Ostro, J. Goguen.
Bull. Am. Astron. Soc., Vol. 16, No. 3, Part I, p. 708 (1984).
Abstract. – See Abstr. 010.061.

098.063 **Three Amor asteroids: CCD reflectance spectra.**
L. McFadden, F. Vilas.
Bull. Am. Astron. Soc., Vol. 16, No. 3, Part I, p. 708 (1984).
Abstract. – See Abstr. 010.061.

098.064 **Modelling of temperature effects on spectral reflectance of asteroid surfaces.**
T. L. Roush, R. B. Singer.
Bull. Am. Astron. Soc., Vol. 16, No. 3, Part I, p. 710 (1984).
Abstract. – See Abstr. 010.061.

098.065 **Photometry of asteroids. III.**
C. J. Cunningham.
I.A.P.P.P. Commun., No. 17, p. 50 – 54 (1984).

098.066 **Equilibrium models of binary asteroids.**
G. Leone, P. Farinella, P. Paolicchi, V. Zappalà.
Astron. Astrophys., Vol. 140, No. 2, p. 265 – 272 (1984).

Models of binary asteroids based on spherical components are inadequate to analyze lightcurve data, since if the shape is determined by self–gravitation, significant tidal distortions are present. On the other hand, equilibrium binary models proposed so far (e.g. for 624 Hektor and 216 Kleopatra) are based on the arbitrary assumption of equal–sized components. The authors present sequences of equilibrium Roche binaries for values of the mass ratio between 0.01 and 1, and discuss the application of these models to real asteroids. Two items of particular interest are: (1) for asteroids of known rotational properties, the equilibrium models allow to constrain the admissible values of the mass ratio and density; (2) the models allow to describe qualitatively the expected light curve morphology.

098.067 **Orbital elements of numbered minor planets.**
Minor Planet Circ., Nos. 8827 – 9314 (1984).

The minor planets are listed according to their definitive number. Newly numbered objects are indicated by an asterisk. The names of the orbit computers are given behind the respective M.P.C. numbers:
(101) 9065, (154) 8892, (241) 9203, (268), (365) 9065, (386) 9020, (416), (508) 9065, (596), (702) 9066, (717), (738) 9020, (747) 8892, (765) 9020, (766), (788) 9021 W. Landgraf; (994) 9204 L. D. Schmadel; (1141) 9066, (1153) 8892, (1177) 9066, (1221) 9021, (1224), (1797) 8892 W. Landgraf; (1872) 9021, (1873) 9022 C. M. Bardwell; (2101) 8893, 9021, (2169) 8893, (2197) 9066 W. Landgraf; (2234) 8896 L. D. Schmadel; (2550), (2756) 8893, (2775) 8893–8894, (2795), (2970) 8894, (3067)*–(3068)* 8894–8895 W. Landgraf; (3069)* 8896–8897 L. D. Schmadel; (3070)*–(3079)* 8897–8900 C. M. Bardwell; (3080)*–(3092)* 8902–8906 B. G. Marsden; (3093)*–(3098)* 9022–9024 C. M. Bardwell; (3099)*–(3104)* 9026–9028 B. G. Marsden; (3105)* 9066–9067 S. Nakano; (3106)*–(3107)* 9070–9071 B. G. Marsden; (3108)*–(3117)* 9073–9076 C. M. Bardwell; (3118)* 9153 L. D. Schmadel; (3119)*–(3125)* 9154–9157 B. G. Marsden; (3126)*–(3131)* 9158–9160 C. M. Bardwell; (3132)* 9203 W. Landgraf; (3133)*–(3137)* 9204–9205 S. Nakano; (3138)*–(3141)* 9208–9210 C. M. Bardwell; (3142)*–(3143)* 9212 B. G. Marsden; (3144)*–(3155)* 9287–9291 C. M. Bardwell; (3156)*–(3164)* 9293–9296 B. G. Marsden; (3165)* 9303 T. Urata; (3166)* 9305 S. Nakano.

098.068 **Orbital elements of unnumbered minor planets.**
Minor Planet Circ., Nos. 8827 – 9314 (1984).

The unnumbered minor planets are sorted by number. The names of the orbit computers are given behind the respective M.P.C. numbers:
[A916 PC] 9076 C. M. Bardwell; [A922 WB] 9161–9162, [A924 EG] 9305 S. Nakano; [1928 SL] 8906 B. G. Marsden; [1929 PB] 9205–9206, [1931 $TJ_1$] 9305 S. Nakano; [1931 VP] 9212–9213 B. G. Marsden; [1932 CQ], [1932 WB] 9206, [1933 SJ] 9305–9306 S. Nakano; [1934 CU] 9068 K. Hurukawa; [1934 CX] 9206 S. Nakano; [1934 RP] 9069 K. Hurukawa; [1937 UE] 8900 C. M. Bardwell; [1938 AD] 9306, [1938 DZ] 9206–9207, [1940 EF], [1941 UG] 9207, [1949 DA] 9306 S. Nakano; [1952 SG] 8906–8907, [1955 RS] 9071 B. G. Marsden; [1964 UQ] 9160 C. M. Bardwell; [1967 UT] 9031 S. Nakano; [1969 $TT_1$] 9291 C. M. Bardwell; [1969 $TB_2$] 8910–8911 S. Nakano; [1971 QP] 8907, [1971 SC] 9157, [1971 $SP_3$] 9071 B. G. Marsden; [1972 $RU_2$], [1973 DT] 9077 C. M. Bardwell; [1973 $QZ_1$] 9157 B. G. Marsden; [1973 $SW_4$] 9162 K. Hurukawa; [1973 $UU_4$] 9077 C. M. Bardwell; [1973 $UL_5$] 9030–9031, [1974 $QO_2$] 9213 D. W. E. Green; [1974 SL] 9078, [1975 QO] 9291, [1975 $TS_2$], [1975 $TV_2$] 9024 C. M. Bardwell; [1975 $TZ_2$] 8907 B. G. Marsden; [1975 $VN_1$], [1976 $GR_6$] 9078 C. M. Bardwell; [1976 $SU_2$] 9071 B. G. Marsden; [1976 $SZ_5$] 9069 K. Hurukawa; [1977 $QD_2$] 9213 D. W. E. Green; [1977 $RL_6$] 9031 S. Nakano; [1977 $TO_3$] 9160–9161, [1978 $EA_3$] 9210 C. M. Bardwell; [1978 PA] 9157 B. G. Marsden; [1978 QX] 8910 D. W. E. Green; [1978 $QJ_2$] 9291 C. M. Bardwell; [1978 RX] 9296, [1978 UF] 8907, [1979 $FU_2$] 8908 B. G. Marsden; [1979 $MM_5$] 9160, [1979 $SY_9$] 9161 C. M. Bardwell; [1979 $SZ_9$] 9072, [1979 $SN_{11}$] 9028 B. G. Marsden; [1980 DG] 8901, [1980 $EE_2$] 9210 C. M. Bardwell; [1980 FA] 8911 S. Nakano; [1980 JE] 9028–9029 B. G. Marsden; [1980 PH] 9210–9211, [1980 $RJ_2$] 9161 C. M. Bardwell; [1980 SG] 9296 B. G. Marsden; [1980 VO] 9292 C. M. Bardwell; [1981 CK] 8895–8896 M. Kretlow; [1981 $ET_{38}$], [1981 PB] 8908, [1981 PM] 9072, [1981 RF] 8908–8909 B. G. Marsden; [1981 $VW_1$] 8895 W. Landgraf; [1981 WU] 9072, [1982 DA] 9296–9297, [1982 DV], [1982 RA] 9029 B. G. Marsden; [1982 $SA_4$] 9067, [1982 TQ], [1982 $TL_1$] 9032 S. Nakano; [1982 $TC_2$] 9029 B. G. Marsden; [1982 $UJ_2$] 8901 C. M. Bardwell; [1982 $UT_6$], [1982 $UJ_8$] 9032 S. Nakano; [1982 $VZ_4$] 9069 K. Hurukawa; [1983 BN] 9072 B. G. Marsden; [1983 NJ] 8911 S. Nakano; [1983 TB] 9158, [1984 AP], [1984 AQ] 9030 B. G. Marsden; [1984 AZ] 8901 C. M. Bardwell; [1984 $AB_1$] 9020 L. D. Schmadel; [1984 $AC_1$] 9030 B. G. Marsden; [1984 $AF_1$] 9161 C. M. Bardwell; [1984 CQ] 8895 W. Landgraf; [1984 CR] 9024 C. M. Bardwell; [1984 DU] 9067–9068 S. Nakano; [1984 EB] 8896 M. Kretlow; [1984 EV] 9025 C. M. Bardwell; [1984 EZ] 9030 B. G. Marsden; [1984 $EO_1$] 9207, [1984 $ES_1$] 9068 S. Nakano; [1984 FO] 9211 C. M. Bardwell; [1984 GA] 9019 M. Kretlow; [1984 GF] 8901, [1984 $HA_1$] 9078, [1984 JZ] 9211 C. M. Bardwell; [1984 KB] 8909, 9030, [1984 KD] 8909, 9297 B. G. Marsden; [1984 KF] 9025 C. M. Bardwell; [1984 QA] 9073, 9158, 9297 B. G. Marsden; [1984 QC] 9208 S. Nakano; [1984 QE] 9208 T. Urata; [1984 $QJ_1$], [1984 $SF_1$] 9292 C. M. Bardwell; [2055 P–L] 9297, [2091 P–L] 9297–9298, [2093 P–L], [2103 P–L] 9298, [2535 P–L] 9069, [2594 P–L] 9298, [2631 P–L] 9298–9299, [2765 P–L], [2808 P–L] 9033, [3040 P–L], [3524 P–L], [4016 P–L] 9299 K. Hurukawa; [4031 P–L], [4063 P–L] 8909 B. G. Marsden; [4069 P–L] 9299–9300, [4122 P–L], [4237 P–L] 9300, [4260 P–L] 9070, [4601 P–L] 9300, [4650 P–L] 9300–9301, [4657 P–L], [5557 P–L], [6092 P–L] 9301 K. Hurukawa; [6299 P–L] 8910 B. G. Marsden; [6519 P–L], [6536 P–L], [6543 P–L], [6563 P–L] 9302, [6565 P–L] 9033, [6624 P–L], [6787 P–L] 9303 K. Hurukawa.

098.069 **Observations of minor planets.**
Minor Planet Circ., Nos. 8832 – 8887, 8965 – 9018, 9043 – 9062, 9125 – 9152, 9178 – 9202, 9257 – 9286 (1984).

Observations made at the following stations are published: Brescia, Brorfelde, Bulgarian Natl. Obs., Catania, Caussols, Cavriana, Centro Astron. Yebes, Crimean Astrophys. Obs., Eastfield, ESO, Falkensee, Geisei, Goethe Link Obs., Hoher List, IRAS, JCPM Oi Stn., Karasuyama, Kitt Peak, Kleť, Kvistaberg,

Lancaster, Lick Obs., Lincoln Lab., Linz, Lowell Obs., Lowell Obs. Anderson Mesa Stn., Maria Mitchell Obs., Mauna Kea, Mt. John Obs., Mt. Palomar, Mt. Stromlo, Oak Ridge Obs., Ojai, Oss. Chaonis, Perth Obs., Piszkéstetö, Purple Mountain Obs., Reintal, S. Vittore (Bologna), SAAO Cape Town, Siding Spring, Skalnaté Pleso, Sonneberg, Tautenburg, Tokyo Obs. Kiso Stn., Traunstein, Turku, Uccle, Victoria, Yatsuyatake Obs.

098.070 **Identifications and identification changes of minor planets.**
Minor Planet Circ., Nos. 8952 – 8961, 9040 – 9041, 9173 – 9174, 9253 (1984).

098.071 **Ephemerides of minor planets and comets.**
Minor Planet Circ., Nos. 8914 – 8946, 9033 – 9036, 9082 – 9120, 9162 – 9172, 9217 – 9252, 9306 – 9314 (1984).

098.072 **Orbital elements of one–opposition minor planets.**
Minor Planet Circ., Nos. 8887 – 8892, 9018, 9019, 9062 – 9065, 9152 – 9153, 9202 – 9203, 9286 – 9287 (1984).

098.073 **New names of minor planets.**
Minor Planet Circ., Nos. 8911 – 8914, 9079 – 9082, 9214 – 9217 (1984).

098.074 **Index to orbital elements of comets and minor planets.**
Minor Planet Circ., Nos. 9000 – 9006 (1984).
Index concerning all elements published in Minor Planet Circ., Nos. 8000 – 8999.

098.075 **1984 KD.**
IAU Circ., No. 3959 (1984).

098.076 **1982 RA.**
IAU Circ., No. 3966 (1984).

098.077 **1984 QA.**
IAU Circ., Nos. 3980, 3982, 3991 (1984).

098.078 **1983 TB.**
IAU Circ., Nos. 3994, 4022 (1984).

098.079 **Occultation of SAO 146599 by (47) Aglaja.**
IAU Circ., No. 4002 (1984).

098.080 **1984 QA.**
Yamamoto Circ., Nos. 2021 – 2023 (1984). In Japanese.

098.081 **1983 TB.**
Yamamoto Circ., No. 2027 (1984). In Japanese.

098.082 **2060 Chiron: visual and thermal infrared observations.**
L. A. Lebofsky, D. J. Tholen, G. H. Rieke, M. J. Lebofsky.
Icarus, Vol. 60, No. 3, p. 532 – 537 (1984).
Five–color ($\lambda\lambda$ = 0.36 – 0.85 μm) and thermal infrared ($\lambda$ = 22.5 μm) photometric observations of the unusual asteroid 2060 Chiron were made. Between 0.36 and 0.85 μm, Chiron's reflectance spectrum is similar to those of C–class asteroids as well as Saturn's satellite Phoebe. However, the thermal IR measurements imply an albedo $\geqslant$0.05 (i.e., a diameter $\leqslant$250 km at the 2 $\sigma$ level) that is probably higher than those of C–class asteroids or Phoebe.

098.083 **Detection of $\lambda$ = 2 cm emission from minor planet 15 Eunomia.**
W. J. Webster Jr., R. W. Hobbs, P. D. Lowman Jr.
Icarus, Vol. 60, No. 3, p. 538 – 540 (1984).
Emission (2 cm) from 15 Eunomia was detected on March 27, 1983, using the VLA. At this time, 15 Eunomia was 2.0 AU distant from Earth. A flux density of 687 ± 70 μJy was measured at 14.96 GHz (50–MHz bandwidth). If 246 km is adopted for the diameter, a disk temperature of 184 ± 20K results. This is consistent with a rapidly rotating, black sphere with 15 Eunomia's diameter and distance (171K).

098.084 **Physical study of asteroids: lightcurves and rotational periods of six asteroids.**
M. Di Martino.
Icarus, Vol. 60, No. 3, p. 541 – 546 (1984).
V band photoelectric lightcurves and rotational periods are presented for six asteroids: 150 Nuwa, 203 Pompeja, 336 Lacaderia, 545 Messalina, 984 Gretia, and 1240 Fantasia. Except for 984 Gretia, none of these asteroids has been previously observed. The observations were obtained during September 1983 at the Astrophysical Observatory of Catania. For 336 Lacaderia and 984 Gretia the magnitude–phase relations, in terms of $Q_v$ and $\beta_v$, were also obtained.

098.085 **Distribution and evolution of asteroid rotation rates.**
S. F. Dermott, C. D. Murray.
Philos. Trans. R. Soc. London, Ser. A, Vol. 313, No. 1524, p. 157 – 164 (1984). – See Abstr. 012.068.
Data on the rotational characteristics of more than 300 asteroids are currently available and it is now clear that the distribution of the rotation rates is non–random. A plot of rotation rate against asteroid diameter shows large dispersion but is distinctly V–shaped. The minimum of this curve at ca. 120 km may separate primordial asteroids from their collision products. There is also evidence that rotation rate depends on type classification, and weak evidence that it may also depend on family membership. Recent bias–free observations suggest that the marked rise of rotation rate with decreasing diameter $D$ for those asteroids with $D$ < 120 km cannot be completely accounted for by observational selection effects. A significantly large subset of the small asteroids have exceptionally long rotation periods suggestive of either a different nature and origin, or a peculiar history. Models that have been proposed to account for these results are discussed.

098.086 **On the formation of the gaps in the asteroid belt.**
R. Sergysels, I. Wettendorff.
Bull. Cl. Sci., Acad. R. Belg., 5e Sér., Tome 70, No. 1, p. 25 – 32 (1984).
In the framework of the restricted three body problem, if one assumes that Moser's work implies that circular orbits at resonances of order $\leqslant$ 4 evolve towards larger eccentricities, thus towards lesser semi–major axes, all gaps inside the main asteroid belt can be explained. Numerical results confirm this idea. In this model, only the formation of Hilda's group at the 3:2 resonance remains unexplained. The authors think that an improvement of the model would require the full use of statistical mechanics in order to be able to make long time predictions about the asteroids' dynamics.

098.087 **Photoelectric photometry opportunities March – June 1984.**
A. W. Harris, V. Zappalà.
Minor Planet Bull., Vol. 11, No. 1, p. 7 – 8 (1984).

098.088 **General report of position observations by the ALPO Minor Planets Section for the year 1983.**
F. Pilcher.
Minor Planet Bull., Vol. 11, No. 2, p. 11 – 18 (1984).

098.089 **Earth–crossing asteroids: an update.**
E. F. Helin.
Minor Planet Bull., Vol. 11, No. 3, p. 19 – 21 (1984).
Earth–crossing asteroids continue to be discovered at an ever increasing rate. Nearly 60 have been found to date (April 1984). At least 8 Earth–crossers have orbits which are indicative of a cometary origin, though most are probably from the asteroid belt.

098.090 **Photoelectric photometry of asteroid 694 Ekard.**
K. W. Zeigler, D. Tomlinson.
Minor Planet Bull., Vol. 11, No. 3, p. 21 – 22 (1984).
A synodic rotational period for Ekard of 5.925±0.01 hours was obtained. The observed amplitude of the lightcurve was 0.50±0.02 magnitudes.

098.091 **Photoelectric photometry opportunities June – October 1984.**
A. W. Harris, V. Zappalà.
Minor Planet Bull., Vol. 11, No. 3, p. 22 – 24 (1984).

098.092 **Minor planet appulses.**
D. Welch, M. Kaitting, C. Cunningham.
Minor Planet Bull., Vol. 11, No. 3, p. 24 – 26 (1984).

098.093 **Low phase angle asteroids.**
D. J. Tholen.
Minor Planet Bull., Vol. 11, No. 3, p. 26 – 27 (1984).

098.094 **Statistics of asteroids with known periods or amplitudes.**
C. Cunningham.
Minor Planet Bull., Vol. 11, No. 3, p. 27 – 30 (1984).
A statistical analysis is presented of 350 asteroids whose rotational periods or amplitudes are known. Shown is a plot of mean amplitude versus B(1, 0); a histogram of the number of asteroids with known periods up to 350 km diameter; and tables of the number of asteroids with known periods or amplitudes as a function of B(1,0), orbital class, and compositional type.

098.095 **Photoelectric photometry of asteroid 60 Echo.**
W. B. Florence, K. W. Zeigler.
Minor Planet Bull., Vol. 11, No. 4, p. 33 – 34 (1984).
A synodic rotational period for Echo of 25.208±0.048 hours was determined. The observed amplitude of the lightcurve was 0.22±0.02 magnitudes.

098.096 **The phase coefficient of 18 Melpomene as determined by visual observations.**
R. Panther, R. P. Binzel.
Minor Planet Bull., Vol. 11, No. 4, p. 34 – 35 (1984).

098.097 **Photoelectric photometry opportunities November 1984 – January 1985.**
A. W. Harris, V. Zappalà.
Minor Planet Bull., Vol. 11, No. 4, p. 37 – 38 (1984).

098.098 **Low phase angle asteroids.**
D. J. Tholen.
Minor Planet Bull., Vol. 11, No. 4, p. 38 – 39 (1984).

098.099 **Secular perturbations of orbits of some asteroids.**
J.-x. Zhang, G.-y. Li.
Publ. Purple Mt. Obs., Vol. 3, No. 2, p. 1 – 14 (1984). In Chinese.

098.100 **Observations photographiques de petites planètes effectuées à l'Astrographe Double de 40 cm en 1980 et 1981.**
H. Debehogne.
Bull. Astron., Vol. 9, No. 5, p. 224 – 229 (1983).

098.101 **Observations photographiques de petites planètes effectuées en 1979 à l'équatorial GPO de 40 cm de l'Observatoire Austral Européen (ESO) à La Silla (Chili).**
H. Debehogne.
Bull. Astron., Vol. 9, No. 5, p. 236 – 241 (1983).

098.102 **Observations of the asteroid 1983 SA.**
V. P. Goranskij.
Astron. Tsirk., No. 1304, p. 8 (1984). In Russian.
Coordinates of the fast moving object Wild 1983 SA were measured on 40-cm astrograph plates.

098.103 **Occultations of SAO stars by minor planets during 1985.**
E. Goffin.
Separate print, 50 pp. (1984). Available from Edwin Goffin, Aartselaarstraat 14, B2710 Hoboken, Belgium.
Details of predicted occultations, visible in Europe, of stars from the SAO catalogue by a selected number of minor planets are presented. The search was restricted to all numbered minor planets having a diameter of at least 20 km.

098.104 **On the correct account of the rotation effect in calculations of the phase function of asteroids.**
F. A. Tupieva.
Astron. Tsirk., No. 1321, p. 6 (1984). In Russian.

098.105 **Initial state of the asteroids system.**
N. E. Kurochkin.
Astron. Tsirk., No. 1326, p. 2 – 5 (1984). In Russian.

098.106 **Investigation of the errors of the parameters of translatory-rotational motion of an asteroid from measurements of the radial velocity of a space vehicle.**
L. F. Bragar'.
Idei K. Eh. Tsiolkovskogo i sovrem. nauchn. probl. Moskva, p. 110 – 116 (1984). In Russian. Abstr. in Ref. Zh., 51. Astron., 12.51.100 (1984).

098.107 **On revealing genetically related objects in the vicinity of the earth's orbit.**
A. N. Simonenko, A. K. Terent'eva.
Astron. Vestn., Tom 18, No. 3, p. 215 – 224 (1984). In Russian. Abstr. in Ref. Zh., 51. Astron., 12.51.258 (1984).

098.108 **On the Lommel-Seeliger law criteria applied to light reflection by asteroids in opposition.**
V. M. Grigorevskij, S. Ya. Kolesnik.
Astron. Vestn., Tom 18, No. 3, p. 199 – 207 (1984). In Russian. Abstr. in Ref. Zh., 51. Astron., 12.51.261 (1984).

098.109 **Bestimmung von Durchmesser und Form von Planetoiden.**
H.-J. Bode.
Mitt. Astron. Ges., Nr. 62, p. 155 (1984). Abstract. – See Abstr. 010.341.

098.110 **The photoelectric campaign for 51 Nemausa.**
P. Gammelgaard, L. K. Kristensen, H. J. Schober.
Mitt. Astron. Ges., Nr. 62, p. 216 (1984). Abstract. – See Abstr. 010.341.

098.111 **Ist 423 Diotima ein Doppelasteroid?**
H. J. Schober.
Mitt. Astron. Ges., Nr. 62, p. 216 (1984). Abstract. – See Abstr. 010.341.

098.112 **The theory of motion of (126) Velleda.**
G. T. Arazov, S. A. Gabibov.
Astron. Zh., Tom 61, Vyp. 6, p. 1205 – 1210 (1984). In Russian. English translation in Sov. Astron., Vol. 28, No. 6.
Taking (126) Velleda as an example, on the basis of the orbits of the internal variant of the generalized problem of three fixed centres, an intermediate orbit for non-resonance asteroids satisfying the condition $2.15 \leqslant a \leqslant 4.26$ has been constructed. The comparison of the theoretical results with those of the numerical integration and observations during the period of 1905 – 1980 showed that for the investigation of the above-mentioned asteroids it would be reasonable to choose as intermediate orbits the orbits of the internal variant of the generalized problem of three fixed centres.

098.113 **The asteroid ring.**
S. F. Dermott, C. D. Murray.
Planetary rings, p. 695 – 701 (1984). – See Abstr. 012.085 (IAU Colloq. No. 75).
From a statistical analysis of the asteroid orbital period data, it is shown that the present distribution of asteroids is strongly

correlated with the present orbital period of Jupiter (to 1 part in 5000). Furthermore, by analyzing the distribution of orbital eccentricities and inclinations it is shown that the resonant structure of the belt was formed after the asteroids had dispersed from the near-coplanar disk in which they accreted.

098.114 **Asteroid compositional "rings": clues to the compositions of primordial planetesimals in the middle solar system.**
E. F. Tedesco.
Planetary rings, p. 703 – 712 (1984). – See Abstr. 012.085 (IAU Colloq. No. 75).

It has recently been established that the distribution of asteroid taxonomic types at distances between 2.1 and 5.3 astronomical units is highly structured. There are four major, overlapping but nevertheless compositionally distinct, "rings" of asteroids present within this range of heliocentric distance. This structure is consistent with these objects having been formed directly from the solar nebula at, or near, the heliocentric distances at which we find them today. Once the mineralogy of these taxonomic types is firmly established, they may be used as probes of physical conditions in the early solar system.

098.115 **Collisional outcomes in the asteroid belt.**
P. Farinella, P. Paolicchi, V. Zappala.
Planetary rings, p. 713 – 715 (1984). – See Abstr. 012.085 (IAU Colloq. No. 75).

The importance of catastrophic collisions as an evolutionary process determining the main physical properties of asteroids is briefly reviewed.

098.116 **Photographic observations of the minor planet H–501 at the Crimean Astrophysical Observatory.**
L. G. Karachkina.
Komet. Tsirk., No. 328 (1984). In Russian.

**Minor bodies in the solar system. Summary of the RAS specialist discussion held 1984 April 13 at the University of Leicester.**
See Abstr. 010.684.

**Second seminar–conference on physical investigations of asteroids.**
See Abstr. 011.013.

**The Infrared Astronomical Satellite (IRAS) and the asteroids: a progress report.**
See Abstr. 013.045.

**A photometric geodesy program for main belt asteroids.**
See Abstr. 013.061.

**Computer tracking charts.**
See Abstr. 021.014.

**Programmsystem und Datenorganisation zur Berechnung von Kleinplaneten– und Kometenbahnen.**
See Abstr. 021.018.

**Experimental simulation of asteroidal fragmentation by macroscopic hypervelocity impacts against free falling bodies.**
See Abstr. 022.065.

**Laboratory simulation of asteroidal catastrophic collisions.**
See Abstr. 022.115.

**Laboratory measurements of asteroid models with dusty surfaces.**
See Abstr. 022.116.

**The effect of the large crater on the asteroids lightcurve.**
See Abstr. 022.117.

**CCD scanning for asteroids and geostationary objects.**
See Abstr. 034.089.

**On a displacement of the photometric center from the center of mass in positional observations of comets and minor planets.**
See Abstr. 036.054.

**Techniques and concepts of near–earth asteroid searches.**
See Abstr. 036.062.

**Minor planet observing programs.**
See Abstr. 036.150.

**Plans for astrometry for late 1984 asteroidal and planetary occultations.**
See Abstr. 041.009.

**On initial orbit determination.**
See Abstr. 042.035.

**On the tidal evolution of small binaries (asteroids?).**
See Abstr. 042.053.

**The dynamics of the minor bodies in the solar system.**
See Abstr. 042.059.

**Using minor planets for coordinate system improvement, an Hipparcos zero point, and Space Telescope scale calibration: physical considerations and error analysis.**
See Abstr. 043.002.

**New limit on $\dot{G}$.**
See Abstr. 066.214.

**An occultation by Pallas.**
See Abstr. 096.003.

**Proposed changes for asteroidal occultation predictions for 1985.**
See Abstr. 096.011.

**Observations of asteroidal appulses and occultations.**
See Abstr. 096.014.

**Dark ray craters on Ganymede: evidence for impact of D–type asteroids?**
See Abstr. 099.099.

**Continued comparisons of comets and outer solar system small bodies.**
See Abstr. 102.035.

**Identification of the parent bodies of pallasites.**
See Abstr. 105.090.

**A dynamical model for the IRAS solar system dust bands.**
See Abstr. 106.030.

**Formation of the zodiacal dust bands in the asteroid belt.**
See Abstr. 106.031.

**The occultation of 1 Vulpeculae by Pallas.**
See Abstr. 120.005.

# 099 Jupiter, Jupiter Satellites

099.001 **On the soliton model of the Jovian Great Red Spot and other big vortices in planetary atmospheres.**
M. V. Nezlin.
Pis'ma Astron. Zh., Tom 10, No. 7, p. 530 – 541 (1984). In Russian. English translation in Sov. Astron. Lett., Vol. 10.
Results of laboratory experiments supporting a soliton model according to which the biggest anticyclonic vortices upon large planets (first of all – the Great Red Spot of Jupiter) are Rossby solitons are presented. A 3D–soliton model of the Jovian Great Red Spot is suggested.

099.002 **Addendum: "Mutual phenomena of the Galilean and Saturnian satellites in 1973 and 1979/1980" [Astron. J., Vol. 89, No. 2, p. 280 – 288 (1984)].**
K. Aksnes, F. Franklin, R. Millis, P. Birch, C. Blanco, S. Catalano, J. Piironen.
Astron. J., Vol. 89, No. 7, p. 1081 (1984). See Abstr. 37.099.008.

099.003 **An analysis of recently rediscovered eclipse observations of the first satellite of Jupiter made between 1775 and 1802, and collected by Delambre.**
J.–E. Arlot, B. Morando, W. Thuillot.
Astron. Astrophys., Vol. 136, No. 1, p. 142 – 152 (1984).
An important set of old observations of the Galilean satellites have been rediscovered. It represents part of the eclipse observations collected by Delambre and concerns 845 phenomena of Io between 1775 and 1802. The authors give the list of these eclipses and show the importance of such observations for the adjustment of theories of the movements of the Galilean satellites, particularly for the correction of the mean motions.

099.004 **On the observation of a bright formation on Jupiter.**
E. P. Vashchenko, V. N. Zakhlyupanyj.
Astron. Tsirk., No. 1287, p. 7 – 8 (1983). In Russian.

099.005 **Comparative geology of the satellites of the giant planets.**
P. Masson.
Space Sci. Rev., Vol. 38, Nos. 3/4, p. 281 – 324 (1984).
The geologic evolution of the Jovian and Saturnian satellites is reviewed with regard to the major discoveries of the Voyager 1 and 2 encounters with Jupiter and Saturn. Nearly forty satellites are now identified in the Jovian and Saturnian systems. They all have experienced different geologic evolutions related to their composition, and to their location in the two systems. The emphasis is given upon the cratering record of their surfaces, and on their thermal evolution, with regard to the tidal effects produced by the giant planets. The small satellites are also presented with no attempt to review their geologic record since they are almost only known from their orbital properties.

099.006 **La planète Jupiter en 1981 – 1982.**
R. Néel.
Astronomie, Vol. 98, p. 315 – 332 (1984).

099.007 **High–resolution maps of the 1.5 GHz emission from Jupiter's disk and radiation belt.**
J. A. Roberts, G. L. Berge, R. C. Bignell.
Astrophys. J., Vol. 282, No. 1, p. 345 – 358 (1984).
VLA maps of four different faces of Jupiter made with a resolution $\sim 0.3$ Jovian radius show new features of the radiation belt emission. A synchrotron model which reproduces these features serves to define the major characteristics of the relativistic electrons in the radiation belt. The observations provide the best determination to date of the atmospheric emission at 1.5 GHz and yield a disk brightness temperature of $425 \pm 100$K.

099.008 **Experimental test for Jupiter's satellites visible to the naked eye.**
Z. Yang, S. Jiang, X. Hao.
Kexue Tongbao (Beijing), Vol. 28, No. 7, p. 927 – 929 (1983). Abstr. in Phys. Abstr., Vol. 87, No. 1237, Entry 72907 (1984).

099.009 **Optical parameters of the atmosphere at the equatorial belt of Jupiter from observations of 1980.**
A. N. Aksenov, V. D. Vdovichenko, K. Yu. Ibragimov.
Astron. Zh., Tom 61, Vyp. 4, p. 755 – 763 (1984). In Russian. English translation in Sov. Astron., Vol. 28, No. 4.
The intensities of weak, moderate and strong methane absorption bands at $\lambda 0.5 - 1.1$ μm are studied as well as the brightness distribution along the equatorial belt at some wavelengths within the strong absorption band of $CH_4$ 8860 Å. The estimates of the $CH_4$ abundance in the atmosphere above the clouds and the characteristics of the cloud layer are derived from these measurements in terms of two classes of the scattering function.

099.010 **The dynamics of a high–speed Jovian jet.**
T. Maxworthy.
Planet. Space Sci., Vol. 32, No. 8, p. 1053 – 1058 (1984).
The author presents new measurements of the velocity field in the neighborhood of the high–speed jet located at approx. 24°N latitude in the Jovian atmosphere. The maximum zonal velocity is found to be $182 \pm 10$ m s$^{-1}$ located at $23.7 \pm 0.2$°N and representing the largest velocity measured on the planet. The distinctive cloud markings found close to this latitude are discussed and possible dynamical consequences presented.

099.011 **An upper limit to the abundance of lightning–produced amino acids in the Jovian water clouds.**
A. Bar–Nun, N. Noy, M. Podolak.
Icarus, Vol. 59, No. 2, p. 162 – 168 (1984).
The effect of excess hydrogen on the synthesis of amino acids, by high–temperature shock waves in a hydrogen/methane/ammonia/water vapor mixture, was studied experimentally. The energy efficiency results, together with the best estimate of the lightning energy dissipation rate on Jupiter, from the Voyager data, were used to calculate an upper limit to the rate of amino acid production by lightning in the Jovian water clouds.

099.012 **Meridional energy balance of Jupiter.**
J. A. Pirraglia.
Icarus, Vol. 59, No. 2, p. 169 – 176 (1984).
The meridional energy balance of Jupiter is calculated from high spatial resolution observations by the Voyager 1 infrared spectrometer and radiometer. On a hemispheric scale Jupiter radiates thermal energy to space approximately uniform with latitude while solar energy absorption varies approximately as the solar angle. This implies internal adjustment to the solar energy with a larger contribution poleward of $\pm 45°$ than in the equatorial zone. Models have been proposed that would explain the pole–to–equator variation in the thermal emission and it is suggested that the smaller scale variations may be the result of forced convective circulation.

099.013 **Mass pickup in sub–Alfvénic plasma flow: a case study for Io.**
D. J. Southwood, M. W. Dunlop.
Planet. Space Sci., Vol. 32, No. 9, p. 1079 – 1086 (1984).
Mass pickup may be an important process in the vicinity of the Jovian satellite Io and could be largely responsible for the slowing and deviation of the magnetospheric plasma flow in Io's vicinity. Theoretical treatments hitherto have treated a flow dominated by mass pickup near Io in a manner exactly analogous to the case where Io's immediate vicinity is treated as a simple electrical conductor. The analogy although instructive is not exact for it does not take account of the increased mass density in the flow downstream of Io. The authors point out that when correctly treated, mass pickup gives rise to a distinct wake structure behind Io.

099.014 **The Io decametric emission cone.**
J. L. Green.
Radio Sci., Vol. 19, No. 2, p. 556– 570 (1984). Abstr. in Phys. Abstr., Vol. 87, No. 1241, Entry 90108 (1984).

099.015 **A reference model for the upper atmosphere of Jupiter.** B. Petropoulos, C. Banos.
Astron. Astrophys., Suppl. Ser., Vol. 58, No. 1, p. 145 – 161 (1984).
The authors have reviewed the last measurements of Voyager I and II and computed the following physical parameters for the upper atmosphere of Jupiter: temperature, pressure, density, speed of sound, density scale, number density, viscosity, mean pressure scale, mean particle velocity, mean collisional frequency, columnar mass, mean free path. A reference model is proposed for the Jovian atmosphere.

099.016 **Predictions of the mutual phenomena of the Galilean Satellites of Jupiter for 1985 – 1986.**
J.-E. Arlot.
Astron. Astrophys., Vol. 138, No. 1, p. 113 – 120 (1984).
In 1985 and 1986, the Earth and the Sun will be in the equatorial plane of Jupiter, so that mutual phenomena (occultations and eclipses) will occur between the Galilean Satellites. The present paper gives the calculated dates, the durations and the magnitudes of each observable phenomenon for three different ephemerides of the Galilean Satellites. The instrumental possibilities of observations are given and the significance of those observations is explained.

099.017 **Solution d'ordre 2 de la théorie des satellites galiléens de Jupiter.**
W. Thuillot, D. T. Vu.
Astrométrie, géodésie, mécanique céleste, p. 53 – 57 (1984). – See Abstr. 012.021.

099.018 **Les phénomènes mutuels des satellites galiléens de Jupiter en 1985 – 86.**
J. E. Arlot.
Astrométrie, géodésie, mécanique céleste, p. 59 – 62 (1984). – See Abstr. 012.021.

099.019 **Presidential address: the giant planets; Galileo Galilei to Project Galileo.**
R. Hide.
Q. J. R. Astron. Soc., Vol. 25, No. 3, p. 232 – 247 (1984).

099.020 **On the blue cloud features of Jupiter's $NEB_s$–$EZ_n$ region.**
J. Olivarez.
Strolling Astron., Vol. 30, Nos. 9 – 10, p. 181 – 186 (1984).
Twelve vivid blue cloud features were observed in Jupiter's $NEB_s$–$EZ_n$ by the author in 1983 on eleven nights of good seeing. Although remarkable, these blue clouds are not new. Similar blue features in the $NEB_s$–$EZ_n$ were detected visually by F. J. Hargreaves as early as 1928, and they have been photographed by the professional observatories as well as by Voyagers 1 and 2.

099.021 **Strong electron heat flux modes in Jupiter's foreshock.** S. L. Moses, F. V. Coroniti, C. F. Kennel, F. L. Scarf.
Geophys. Res. Lett., Vol. 11, No. 9, p. 869 – 872 (1984).
Analysis of Voyager plasma wave data from the Jovian foreshock has revealed the existence of a new oscillation with center frequency below $f_p$ and typically lying above the upper limit Doppler shift frequency for ion acoustic waves. The properties of these waves are consistent with modes generated by a strong electron heat flux in an unmagnetized plasma. The occasional simultaneous appearance of more than one frequency band is a possible indication of multiple magnetic connection with the shock.

099.022 **Modeling Jupiter's magnetospheric currents using Pioneer data: evidence for a low–latitude cusp.**
B. T. Thomas, D. E. Jones.
J. Geophys. Res., Vol. 89, No. A8, p. 6663 – 6669 (1984).
The Jovian magnetospheric field measured by Pioneer 10 and 11 can be well modeled by a combination of current systems composing an azimuthally symmetric current disc, a dusk–dawn current sheet in both the dayside and the nightside magnetosphere, and an image dipole to represent the effects of currents on the magnetopause. The inclusion of a dusk–dawn current sheet in the dayside magnetosphere allows observations obtained both inbound and outbound to be simultaneously fit by an azimuthally symmetric current disc (i.e., without the need for local time dependent current densities). Similar disc current intensities are found to describe both Pioneer 10 and Pioneer 11 encounters. The most striking feature of the models is that they suggest that the Jovian cusp is at much lower latitudes than is the case with the earth's magnetosphere.

099.023 **Direct evidence for solar wind control of Jupiter's hectometer–wavelength radio emission.**
M. D. Desch, C. H. Barrow.
J. Geophys. Res., Vol. 89, No. A8, p. 6819 – 6823 (1984).
Observations of the solar wind close to Jupiter by the Voyager 1 and Voyager 2 spacecraft in 1978 and 1979 are compared with the hectometer–wavelength radio emission from the planet. A significant positive correlation is found between variations in the solar wind plasma density at Jupiter and the level of Jovian radio emission output. During the 173–day interval studied for the Voyager 2 data the radio emission displayed a long–term periodicity of about 13 days, identical to that shown by the solar wind density at Jupiter and consistent with the magnetic sector structure association already proposed for ground–based observations of the decameter–wavelength emission.

099.024 **Jovian magnetospheric neutral wind and auroral precipitation flux.**
A. Eviatar, D. D. Barbosa.
J. Geophys. Res., Vol. 89, No. A9, p. 7393 – 7398 (1984).
The authors propose a mechanism for the transport of energy from the rotation of Jupiter into the heavy ions whose precipitation into the atmosphere is responsible for the intense ultraviolet aurora. The process involves charge exchange of heavy ions in the Io torus, their flight to the outer magnetosphere as neutrals, and their reionization in a region where the pickup magnetic moment is large.

099.025 **Sulfur and oxygen escape from Io and a lower limit to atmospheric $SO_2$ at Voyager 1 encounter.**
S. Kumar.
J. Geophys. Res., Vol. 89, No. A9, p. 7399 – 7406 (1984).
There is strong evidence that Io is the source of sulfur and oxygen ions and neutrals, and sodium atoms observed in the associated plasma torus and neutral cloud, and significant improvements have been made recently in the estimates of the required supply rates for various constituents. The purpose of this paper is to use these estimates to address the following questions regarding the Io atmosphere: (1) Is an atmosphere necessary in order to supply the material to the torus? (2) If it is, what is the minimum atmosphere that would provide the necessary supply rates of sulfur and oxygen observed in the torus at the Voyager 1 encounter?

099.026 **Numerical simulation of compressible convection in dense hydrogen–helium fluids. A novel instability.**
L. D. Cloutman.
Astron. Astrophys., Vol. 138, No. 2, p. 231 – 236 (1984).
Real gas effects can profoundly influence the stability properties of dense material such as that residing in Jovian planetary interiors. The author presents a simplified stability analysis and multidimensional numerical simulation of an isothermal model of a Jovian planet that demonstrates convective instability caused by real gas effects at high densities in a system that is normally considered to be stable. This instability is driven by the significant increase of the specific heat with density that occurs for densities greater than about 0.1 $g\,cm^{-3}$. The author also reaches some definite conclusions about the methods and procedures required to produce accurate numerical simulations of compressible convection.

099.027 **Radiative–convective equilibrium models of Jupiter and Saturn.**
J. F. Appleby, J. S. Hogan.
Icarus, Vol. 59, No. 3, p. 336 – 366 (1984).

Radiative–convective equilibrium models for Jupiter and Saturn have been produced in a study centered primarily on the stratospheric energy balance and the possible role of aerosol heating. These models are compared directly to the thermal structure profiles obtained from Voyager radio occultation measurements. The balance between absorbed and emitted energies is computed iteratively at each level in the atmosphere, assuming local thermodynamic equilibrium and employing a standard treatment of opacities.

099.028 **Voyager disk–integrated photometry of Io.**
D. P. Simonelli, J. Veverka.
Icarus, Vol. 59, No. 3, p. 406 – 425 (1984).

Voyager full–disk images of Io, available at solar phase angles of $\alpha = 2 - 29°$ and $101 - 159°$, allow comparisons of the satellite's near–opposition photometric behavior with Earth–based results and the determination of the phase curve out to very high phase angles. The near–opposition data were reduced iteratively for self–consistent phase and rotation curves in each Voyager filter; the resulting phase coefficients, geometric albedos, and rotational lightcurves are consistent with Earth–based findings, except for a previously noted tendency for Voyager to yield somewhat redder spectral information.

099.029 **On a possible period of the Jovian activity. Are there seasons?**
A. P. Vid'machenko, A. F. Steklov, N. F. Minyajlo.
Pis'ma Astron. Zh., Tom 10, No. 9, p. 691 – 695 (1984). In Russian. English translation in Sov. Astron. Lett., Vol. 10.

A new index is proposed to characterize the activity of the processes in the Jovian atmosphere. For the time period 1953 through 1983 the 12–year period variations of this index was found. One can assume the variations of reflectivity of the features of the Jovian disk to be caused by seasonal changes in the planet's atmosphere.

099.030 **Geostrophic regimes, intermediate solitary vortices and Jovian eddies.**
G. P. Williams, T. Yamagata.
J. Atmos. Sci., Vol. 41, No. 4, p. 453 – 478 (1984). Abstr. in Phys. Abstr., Vol. 87, No. 1242, Entry 94672 (1984).

099.031 **On the structure of Jovian clouds.**
A. V. Morozhenko.
Pis'ma Astron. Zh., Tom 10, No. 10, p. 775 – 779 (1984). In Russian. English translation in Sov. Astron. Lett, Vol. 10.

The vertical structure of the Jovian cloud layer is studied based on spectrophotometric data. It is found that in the equatorial region two cloud layers are distinguishable.

099.032 **Hydrogen dimer structures in the far–infrared spectra of Jupiter and Saturn.**
L. Frommhold, R. Samuelson, G. Birnbaum.
Astrophys. J., Lett. Ed., Vol. 283, No. 2, p. L79 – L82 (1984).

On the basis of a spectral line shape computation and radiative transfer calculations, the authors show that the unexplained, diminutive structures seen in the Voyager IRIS Jovian spectra near the hydrogen $S_0(0)$ and $S_0(1)$ rotational frequencies are due to bound–free transitions involving hydrogen dimers. The absorption intensities of these transitions, as well as of the collision–induced background, are given.

099.033 **Gravito–electrodynamics and the structure of planetary ring systems.**
D. A. Mendis.
Proc. Indian Acad. Sci., Earth Planet. Sci., Vol. 93, No. 3, p. 177 – 188 (1984).

Recent spacecraft observations of the Saturnian and Jovian ring systems have highlighted a plethora of interesting new phenomena associated with those regions containing fine (micron and sub–micron sized) dust. Recognizing that these dust grains, by virtue of being immersed within the planetary magnetospheres, are electrostatically charged to the point that they experience comparable gravitational and electric forces, a new "gravito–electrodynamic" theory has been developed to decribe their dynamics. This theory has been successful in explaining all these phenomena in a systematic way.

099.034 **Predictions of mutual phenomena of the Galilean satellites in 1985 – 1986.**
K. Aksnes, F. Franklin.
Icarus, Vol. 60, No. 1, p. 180 – 188 (1984).

A thorough search covering more than 3 years shows that nearly 300 observable mutual eclipses and occultations of the Galilean satellites occur between May 1985 and April 1986, marking this apparition as a very favorable one. This paper tabulates quantities needed to obtain light curves of all events, excluding only those taking place either too close to Jupiter or with the planet too near the Sun. Since observations are relatively short and easy to incorporate into photometric programs, the authors urge an active campaign so as to provide the accurate astrometric data required to improve ephemerides and to look for tidal effects in the motion of Io.

099.035 **The Mauna Loa sulfur flow as an analog to secondary sulfur flows (?) on Io.**
R. Greeley, E. Theilig, P. Christensen.
Icarus, Vol. 60, No. 1, p. 189 – 199 (1984).

Independent evidence suggests that both sulfur and silicate materials exist on the surface of Io. Spectral data indicate the presence of sulfur compounds, some of which are suggested to be of fumarolic origin. Morphological evidence and inferences of the physical properties of some landforms suggest that silicate volcanism has occurred, which would involve temperatures $\geqslant 650°C$. Because the liquidus of sulfur is only $\sim 115°C$, it is likely that sulfur in close proximity to "hot spots" or to active silicate volcanic areas on Io would be melted and mobilized as flows. The Mauna Loa sulfur may serve as an analog for such flows.

099.036 **A geometric analysis of surface deformation: implications for the tectonic evolution of Ganymede.**
M. T. Zuber, E. M. Parmentier.
Icarus, Vol. 60, No. 1, p. 200 – 210 (1984).

The visual nonalignment of the furrows and the circularity of impact craters are used to study surface deformation on Ganymede. The furrow system is examined to test the hypothesis that lateral motion has taken place between areas of dark terrain. Results show that while lateral motion cannot be ruled out, it may not be required to explain the geometry of the system. Initial nonconcentricity of the furrows or an early period of penetrative deformation shortly after furrow formation could also account for the present configuration.

099.037 **Jovian decametric arcs: an estimate of the required wave normal angles from three–dimensional ray tracing.**
J. D. Menietti, J. L. Green, S. Gulkis, N. F. Six.
J. Geophys. Res., Vol. 89, No. A10, p. 9089 – 9094 (1984).

A three–dimensional ray tracing code which incorporates the O–4 magnetic field model (Acuna and Ness, 1976) and a realistic plasma model has been used to model high–curvature decametric arcs observed by the Voyager planetary radio astronomy instrument.

099.038 **Correction to "Turbulence analysis of the Jovian upstream "wave" phenomenon" [J. Geophys. Res., Vol. 88, No. A7, p. 5581 – 5593 (1983)].**
C. W. Smith, M. L. Goldstein, W. H. Matthaeus, A. F. Viñas.
J. Geophys. Res., Vol. 89, No. A10, p. 9159 – 9160 (1984). See Abstr. 34.099.005.

099.039 **Correction to "Large amplitude MHD waves upstream of the Jovian bow shock" [J. Geophys. Res., Vol. 88, No. A12, p. 9989 – 9999 (1983)].**
M. L. Goldstein, C. W. Smith, W. H. Matthaeus.
J. Geophys. Res., Vol. 89, No. A10, p. 9161 (1984). See Abstr. 34.099.105.

099.040 **Modelling the stratified cloudiness in the Jovian atmosphere in the presence of two condensates.**
K. Yu. Ibragimov, G. A. Kirienko, A. A. Solodovnik.
Astrofiz. inst. Akad. Nauk KazSSR. Alma-Ata, 45 pp. (1984). In Russian. Abstr. in Ref. Zh., 51. Astron., 10.51.238 (1984).

099.041 **On the formation of multistage stratified cloud structures under the conditions of the Jovian atmosphere.**
K. Yu. Ibragimov, G. A. Kirienko, A. A. Solodovnik.
Astrofiz. inst. Akad. Nauk KazSSR. Alma-Ata, 27 pp. (1984). In Russian. Abstr. in Ref. Zh., 51. Astron., 10.51.239 (1984).

099.042 **Microwave observations of Jupiter and Saturn.**
E. Lellouch, F. Combes, T. Encrenaz.
Astron. Astrophys., Vol. 140, No. 1, p. 216 – 219 (1984).

From microwave observations of the HCN $J = 3 \rightarrow 2$ line at 265.886 GHz on Jupiter and of the CO $J = 2 \rightarrow 1$ line at 230.538 GHz on Saturn, upper limits of the HCN mixing ratio in Jupiter's stratosphere ($4 \times 10^{-9}$) and of the CO mixing ratio in Saturn's stratosphere ($6 \times 10^{-7}$) are derived. Moreover, it is shown that microwave observations support the evidence of a strong $PH_3$ depletion below Jupiter's tropopause.

099.043 **Possible identification of sharp features in the Voyager far-infrared spectra of Jupiter and Saturn.**
A. R. W. McKellar.
Can. J. Phys., Vol. 62, No. 8, p. 760 – 763 (1984). Abstr. in Phys. Abstr., Vol. 87, No. 1244, Entry 105901 (1984).

099.044 **Cloud opacity in Jovian hot spot regions.**
R. A. West, P. N. Kupferman, H. Hart.
Bull. Am. Astron. Soc., Vol. 16, No. 3, Part I, p. 639 (1984). Abstract. – See Abstr. 010.061.

099.045 **The position of cyclonic-anticyclonic Jovian cloud systems relative to the average zonal wind.**
R. F. Beebe.
Bull. Am. Astron. Soc., Vol. 16, No. 3, Part I, p. 639 (1984). Abstract. – See Abstr. 010.061.

099.046 **Jupiter's high altitude cloud features.**
P. H. Smith.
Bull. Am. Astron. Soc., Vol. 16, No. 3, Part I, p. 639 (1984). Abstract. – See Abstr. 010.061.

099.047 **Spatial and temporal monitoring of Jupiter's atmosphere.**
C. C. Cunningham, D. M. Hunten.
Bull. Am. Astron. Soc., Vol. 16, No. 3, Part I, p. 639 – 640 (1984). Abstract. – See Abstr. 010.061.

099.048 **A numerical model of convection in Jupiter's North Temperate Belt.**
D. H. Hathaway, R. C. J. Somerville.
Bull. Am. Astron. Soc., Vol. 16, No. 3, Part I, p. 640 (1984). Abstract. – See Abstr. 010.061.

099.049 **Mesoscale waves as a probe of Jupiter's deep atmosphere.**
F. M. Flasar, P. J. Gierasch.
Bull. Am. Astron. Soc., Vol. 16, No. 3, Part I, p. 640 (1984). Abstract. – See Abstr. 010.061.

099.050 **Hadley circulation on Jupiter.**
J. Magalhães, P. J. Gierasch.
Bull. Am. Astron. Soc., Vol. 16, No. 3, Part I, p. 640 (1984). Abstract. – See Abstr. 010.061.

099.051 **Jovian atmospheric dynamics: shear stability control of diabatically forced, weakly stratified zonal flow.**
M. Allison, P. J. Gierasch.
Bull. Am. Astron. Soc., Vol. 16, No. 3, Part I, p. 640 – 641 (1984). Abstract. – See Abstr. 010.061.

099.052 **1.4- to 33-μm maps of the Jovian disk in 1984.**
G. S. Orton, K. H. Baines, J. T. Bergstralh, T. Z. Martin, R. J. Terrile, R. A. West, J. Caldwell.
Bull. Am. Astron. Soc., Vol. 16, No. 3, Part I, p. 641 (1984). Abstract. – See Abstr. 010.061.

099.053 **Relative temperature fields in the Jovian atmosphere in 1984.**
G. S. Orton, K. H. Baines, J. T. Bergstralh, T. Z. Martin, R. A. West, J. Caldwell.
Bull. Am. Astron. Soc., Vol. 16, No. 3, Part I, p. 641 (1984). Abstract. – See Abstr. 010.061.

099.054 **Inverse problems for near infrared spectra of the Jovian atmosphere.**
S. B. Calcutt, F. W. Taylor.
Bull. Am. Astron. Soc., Vol. 16, No. 3, Part I, p. 641 – 642 (1984). Abstract. – See Abstr. 010.061.

099.055 **The bulk composition of Jupiter's atmosphere from a reexamination of Voyager RSS and IRIS data.**
J. F. Appleby, G. S. Orton.
Bull. Am. Astron. Soc., Vol. 16, No. 3, Part I, p. 642 (1984). Abstract. – See Abstr. 010.061.

099.056 **The $^{14}N/^{15}N$ ratio in the atmosphere of Jupiter.**
P. Drossart, T. Encrenaz, M. Combes.
Bull. Am. Astron. Soc., Vol. 16, No. 3, Part I, p. 642 (1984). Abstract. – See Abstr. 010.061.

099.057 **The hydrogen dimer in Jupiter's atmosphere.**
R. Samuelson, L. Frommhold, G. Birnbaum.
Bull. Am. Astron. Soc., Vol. 16, No. 3, Part I, p. 646 – 647 (1984). Abstract. – See Abstr. 010.061.

099.058 **Constraints on the properties of Jupiter's stratospheric aerosols from ultraviolet photometry.**
A. E. Karkoshka, M. G. Tomasko.
Bull. Am. Astron. Soc., Vol. 16, No. 3, Part I, p. 647 (1984). Abstract. – See Abstr. 010.061.

099.059 **Model studies of the latitudinal variability of Jovian hydrocarbon species.**
M. Allen, Y. L. Yung, G. R. Gladstone, W. Maguire.
Bull. Am. Astron. Soc., Vol. 16, No. 3, Part I, p. 647 (1984). Abstract. – See Abstr. 010.061.

099.060 **Latitudinal variation of acetylene and ethane in the Jovian atmosphere from Voyager IRIS observations.**
W. C. Maguire, R. E. Samuelson, R. A. Hanel, V. G. Kunde.
Bull. Am. Astron. Soc., Vol. 16, No. 3, Part I, p. 647 (1984). Abstract. – See Abstr. 010.061.

099.061 **Colored sulfur species in the atmosphere of Jupiter.**
W. T. Huntress Jr., V. G. Anicich.
Bull. Am. Astron. Soc., Vol. 16, No. 3, Part I, p. 648 (1984). Abstract. – See Abstr. 010.061.

099.062 **Vibrationally excited molecular hydrogen in the upper atmosphere of Jupiter.**
T. E. Cravens.
Bull. Am. Astron. Soc., Vol. 16, No. 3, Part I, p. 648 (1984). Abstract. – See Abstr. 010.061.

099.063 **Spatial dependence of the Jovian north and south pole auroral emissions.**
T. E. Skinner, H. W. Moos.
Bull. Am. Astron. Soc., Vol. 16, No. 3, Part I, p. 648 (1984). Abstract. – See Abstr. 010.061.

099.064 **Spatially resolved ultraviolet spectra of the North Polar Region and the Great Red Spot of Jupiter.**
R. Wagener, J. Caldwell.
Bull. Am. Astron. Soc., Vol. 16, No. 3, Part I, p. 648 – 649 (1984). Abstract. – See Abstr. 010.061.

099.065 **A search for vibrational emission from ammonia in the stratosphere of Jupiter.**
A. Betz, D. Goldhaber.
Bull. Am. Astron. Soc., Vol. 16, No. 3, Part I, p. 649 (1984). Abstract. – See Abstr. 010.061.

099.066 **Energy sources, nutrients, and environmental limits to growth of contaminating terrestrial microorganisms on Jupiter.**
C. R. Stoker, P. J. Boston.
Bull. Am. Astron. Soc., Vol. 16, No. 3, Part I, p. 649 (1984). Abstract. – See Abstr. 010.061.

099.067 **Microwave observations of Jupiter and Saturn.**
E. Lellouch, F. Combes, T. Encrenaz.
Bull. Am. Astron. Soc., Vol. 16, No. 3, Part I, p. 649 (1984). Abstract. – See Abstr. 010.061.

099.068 **The violet absorber on Io (and Venus?): disulfur monoxide derivatives.**
B. W. Hapke, F. G. Graham.
Bull. Am. Astron. Soc., Vol. 16, No. 3, Part I, p. 653 (1984). Abstract. – See Abstr. 010.061.

099.069 **Black is hot on Io.**
A. S. McEwen, D. L. Matson, T. V. Johnson, L. A. Soderblom.
Bull. Am. Astron. Soc., Vol. 16, No. 3, Part I, p. 653 (1984). Abstract. – See Abstr. 010.061.

099.070 **Sulfur lakes on Io.**
J. I. Lunine, D. J. Stevenson.
Bull. Am. Astron. Soc., Vol. 16, No. 3, Part I, p. 654 (1984). Abstract. – See Abstr. 010.061.

099.071 **Accurate albedos of the brightest regions on Io.**
D. Simonelli, J. Veverka.
Bull. Am. Astron. Soc., Vol. 16, No. 3, Part I, p. 654 (1984). Abstract. – See Abstr. 010.061.

099.072 **Spatial variations in the surface composition of Io based on Voyager infrared data.**
J. Pearl.
Bull. Am. Astron. Soc., Vol. 16, No. 3, Part I, p. 654 (1984). Abstract. – See Abstr. 010.061.

099.073 **Io: precise comparison of Voyager spectral curves with laboratory measurements.**
J. Veverka, D. Simonelli.
Bull. Am. Astron. Soc., Vol. 16, No. 3, Part I, p. 654 (1984). Abstract. – See Abstr. 010.061.

099.074 **FTS spectra of Io from 2 – 4 μm.**
U. Fink, H. P. Larson, G. T. Sill.
Bull. Am. Astron. Soc., Vol. 16, No. 3, Part I, p. 655 (1984). Abstract. – See Abstr. 010.061.

099.075 **Volcanic distribution on Io: constraints from IRTF data and Voyager imaging.**
T. V. Johnson, D. L. Matson, G. J. Veeder.
Bull. Am. Astron. Soc., Vol. 16, No. 3, Part I, p. 655 (1984). Abstract. – See Abstr. 010.061.

099.076 **Topographic evidence for shield volcanism on Io.**
J. M. Moore, E. F. Albin, R. Greeley.
Bull. Am. Astron. Soc., Vol. 16, No. 3, Part I, p. 655 (1984). Abstract. – See Abstr. 010.061.

099.077 **Io: a two–layer model of the silicate lithosphere.**
M. H. Carr.
Bull. Am. Astron. Soc., Vol. 16, No. 3, Part I, p. 655 – 656 (1984). Abstract. – See Abstr. 010.061.

099.078 **The determination of the latitude and longitude of an Io hot spot from polarization measurements.**
J. Goguen, W. M. Sinton.
Bull. Am. Astron. Soc., Vol. 16, No. 3, Part I, p. 656 (1984). Abstract. – See Abstr. 010.061.

099.079 **Polarization measurements of the Loki hot spot.**
W. M. Sinton, J. Goguen.
Bull. Am. Astron. Soc., Vol. 16, No. 3, Part I, p. 656 (1984). Abstract. – See Abstr. 010.061.

099.080 **Infrared speckle imaging of Io and its volcanic hot spots.**
R. R. Howell, M. T. McGinn.
Bull. Am. Astron. Soc., Vol. 16, No. 3, Part I, p. 656 (1984). Abstract. – See Abstr. 010.061.

099.081 **Orbital phase coverage of Io at 4.8 microns.**
G. J. Veeder, D. L. Matson, T. V. Johnson, M. D. Morrison, R. H. Brown.
Bull. Am. Astron. Soc., Vol. 16, No. 3, Part I, p. 656 – 657 (1984). Abstract. – See Abstr. 010.061.

099.082 **Shape of Io.**
S. P. Synnott, J. C. Callahan, J. E. Riedel, A. J. Donegan.
Bull. Am. Astron. Soc., Vol. 16, No. 3, Part I, p. 657 (1984). Abstract. – See Abstr. 010.061.

099.083 **Tidally forced viscous heating in a partially molten Io.**
M. N. Ross, G. Schubert.
Bull. Am. Astron. Soc., Vol. 16, No. 3, Part I, p. 661 (1984). Abstract. – See Abstr. 010.061.

099.084 **Physical properties of Io's atmosphere.**
M. E. Summers, Y. L. Yung.
Bull. Am. Astron. Soc., Vol. 16, No. 3, Part I, p. 661 (1984). Abstract. – See Abstr. 010.061.

099.085 **The $SO_2$ atmosphere and ionosphere of Io and their interaction with the plasma torus.**
S. Kumar.
Bull. Am. Astron. Soc., Vol. 16, No. 3, Part I, p. 662 (1984). Abstract. – See Abstr. 010.061.

099.086 **Supersonic meteorology with application to Io's atmosphere.**
A. P. Ingersoll, M. E. Summers.
Bull. Am. Astron. Soc., Vol. 16, No. 3, Part I, p. 662 (1984). Abstract. – See Abstr. 010.061.

099.087 **Long term stability of the Io high–temperature plasma torus.**
H. W. Moos, T. E. Skinner, S. T. Durrance, P. D. Feldman, M. C. Festou, J. L. Bertaux.
Bull. Am. Astron. Soc., Vol. 16, No. 3, Part I, p. 662 (1984). Abstract. – See Abstr. 010.061.

099.088 **An imaging search for optical east–west asymmetries in the Io torus.**
J. S. Morgan, J. H. Fertel.
Bull. Am. Astron. Soc., Vol. 16, No. 3, Part I, p. 662 (1984). Abstract. – See Abstr. 010.061.

099.089 **Magnetic longitude variations and plasma production in the Io torus.**
C. B. Pilcher, J. S. Morgan.
Bull. Am. Astron. Soc., Vol. 16, No. 3, Part I, p. 662 (1984). Abstract. – See Abstr. 010.061.

099.090 **Directional features of the Io sodium cloud.**
B. A. Goldberg, W. H. Smyth.
Bull. Am. Astron. Soc., Vol. 16, No. 3, Part I, p. 662 – 663 (1984). Abstract. – See Abstr. 010.061.

099.091 **Io's sodium cloud: an Io/magnetosphere probe.**
B. A. Goldberg.
Bull. Am. Astron. Soc., Vol. 16, No. 3, Part I, p. 663 (1984). Abstract. – See Abstr. 010.061.

099.092 **Probing Io's ionosphere by modeling observations of its hydromagnetic wake.**
F. L. Herbert.
Bull. Am. Astron. Soc., Vol. 16, No. 3, Part I, p. 663 (1984). Abstract. – See Abstr. 010.061.

099.093 **New evidence for high–density inclusions in the Io plasma torus.**
N. M. Schneider, D. M. Hunten, R. A. Brown.
Bull. Am. Astron. Soc., Vol. 16, No. 3, Part I, p. 663 (1984). Abstract. – See Abstr. 010.061.

099.094 **Understanding the escape of material from Io and its role in the planetary magnetosphere.**
W. H. Smyth, M. R. Combi.
Bull. Am. Astron. Soc., Vol. 16, No. 3, Part I, p. 663 – 664 (1984). Abstract. – See Abstr. 010.061.

099.095 **Fine structure in Jupiter's main ring.**
M. R. Showalter, J. A. Burns, J. N. Cuzzi.
Bull. Am. Astron. Soc., Vol. 16, No. 3, Part I, p. 674 (1984). Abstract. – See Abstr. 010.061.

099.096 **Jupiter's "gossamer" ring.**
M. R. Showalter, J. A. Burns, J. N. Cuzzi.
Bull. Am. Astron. Soc., Vol. 16, No. 3, Part I, p. 674 – 675 (1984). Abstract. – See Abstr. 010.061.

099.097 **Particle size and properties in Jupiter's main ring.**
M. R. Showalter, J. A. Burns, J. N. Cuzzi.
Bull. Am. Astron. Soc., Vol. 16, No. 3, Part I, p. 675 (1984). Abstract. – See Abstr. 010.061.

099.098 **Numerology or not? Jovian ring features are near resonances.**
J. A. Burns, L. E. Schaffer, R. J. Greenberg, M. R. Showalter.
Bull. Am. Astron. Soc., Vol. 16, No. 3, Part I, p. 675 (1984). Abstract. – See Abstr. 010.061.

099.099 **Dark ray craters on Ganymede: evidence for impact of D–type asteroids?**
P. M. Schenk, W. B. McKinnon.
Bull. Am. Astron. Soc., Vol. 16, No. 3, Part I, p. 683 (1984). Abstract. – See Abstr. 010.061.

099.100 **Comparison of spectral properties of Europa and Ganymede surfaces from Voyager Image Data.**
T. B. McCord, M. L. Nelson, J. Hayashi.
Bull. Am. Astron. Soc., Vol. 16, No. 3, Part I, p. 684 (1984). Abstract. – See Abstr. 010.061.

099.101 **Exogenic modification of Europa and Ganymede.**
M. L. Nelson, T. V. Johnson, D. L. Matson.
Bull. Am. Astron. Soc., Vol. 16, No. 3, Part I, p. 684 (1984). Abstract. – See Abstr. 010.061.

099.102 **Ultraviolet phase coefficients of the Galilean satellites from IUE.**
R. M. Nelson, A. L. Lane, B. J. Buratti, D. L. Matson, G. J. Veeder, E. F. Tedesco, T. V. Johnson, M. D. Morrison.
Bull. Am. Astron. Soc., Vol. 16, No. 3, Part I, p. 684 – 685 (1984). Abstract. – See Abstr. 010.061.

099.103 **An ice–poor interpretation of Callisto's reflectance spectrum.**
J. R. Spencer.
Bull. Am. Astron. Soc., Vol. 16, No. 3, Part I, p. 685 (1984). Abstract. – See Abstr. 010.061.

099.104 **Three–layer generic Ganymedes.**
S. W. Mueller, W. B. McKinnon.
Bull. Am. Astron. Soc., Vol. 16, No. 3, Part I, p. 685 – 686 (1984). Abstract. – See Abstr. 010.061.

099.105 **Stability of the Laplace relation in deep resonance.**
R. Greenberg.
Bull. Am. Astron. Soc., Vol. 16, No. 3, Part I, p. 687 (1984). Abstract. – See Abstr. 010.061.

099.106 **Determination of satellite gravity harmonics from Galileo radio tracking data.**
J. K. Campbell.
Bull. Am. Astron. Soc., Vol. 16, No. 3, Part I, p. 705 (1984). Abstract. – See Abstr. 010.061.

099.107 **Search for minor atmospheric species in the microwave spectra of Jupiter and Saturn.**
J. L. Destombes, E. Lellouch.
Bull. Am. Astron. Soc., Vol. 16, No. 3, Part I, p. 705 (1984). Abstract. – See Abstr. 010.061.

099.108 **A search for hot spots during a lunar occultation of Io.**
H. B. Hammel, W. M. Sinton, J. D. Goguen, E. E. Becklin, I. Gatley, S. Ridgway, K. Matthews.
Bull. Am. Astron. Soc., Vol. 16, No. 3, Part I, p. 708 (1984). Abstract. – See Abstr. 010.061.

099.109 **On the detectability of desequilibrium species in the upper atmospheres of Jupiter and Saturn from the submillimeter spectrum of these planets.**
D. Gautier, B. Bézard, A. Marten.
Bull. Am. Astron. Soc., Vol. 16, No. 3, Part I, p. 709 (1984). Abstract. – See Abstr. 010.061.

099.110 **Photometric images of sulfur ions and sodium neutrals in the Jupiter/Io torus.**
J. T. Trauger.
Bull. Am. Astron. Soc., Vol. 16, No. 3, Part I, p. 712 (1984). Abstract. – See Abstr. 010.061.

099.111 **1984 sodium, potassium, and sulfur images of the Io torus.**
J. S. Morgan, J. H. Fertel, C. B. Pilcher.
Bull. Am. Astron. Soc., Vol. 16, No. 3, Part I, p. 712 (1984). Abstract. – See Abstr. 010.061.

099.112 **Modeling the neutral clouds of Io.**
W. H. Smyth, M. R. Combi.
Bull. Am. Astron. Soc., Vol. 16, No. 3, Part I, p. 712 (1984). Abstract. – See Abstr. 010.061.

099.113 **Galilean satellites studies on a microcomputer.**
B. C. Brown.
Bull. Am. Astron. Soc., Vol. 16, No. 3, Part I, p. 721 (1984). Abstract. – See Abstr. 010.064.

099.114 **The detectability of minor atmospheric species in the far infrared spectra of Jupiter and Saturn.**
E. Lellouch, T. Encrenaz, M. Combes.
Astron. Astrophys., Vol. 140, No. 2, p. 405 – 413 (1984).

The authors have calculated the synthetic spectra of Jupiter and Saturn between 10 and 90 $cm^{-1}$ (110 μ – 1 mm). Calculations show that $NH_3$, $PH_3$ and HCN should be easily observable on Jupiter and Saturn so that their vertical profiles could be derived, while CO and $CH_3D$ could be marginally detectable on Saturn. $H_2O$ and $^{15}NH_3$ features are expected to be too weak to be observable, even on Jupiter. Observations of the far infrared spectra of Jupiter and Saturn could be performed from an aircraft or a spacecraft with a good S/N ratio.

099.115 **Visibility of the mutual phenomena of the Galilean Satellites of Jupiter in 1985 and 1986.**
J. E. Arlot, P. Rocher.
Astron. Astrophys., Suppl. Ser., Vol. 58, No. 3, p. 453 – 473 (1984).

099.116 **Vertical distribution of disequilibrium species in Jupiter's troposphere.**
J. S. Lewis, M. B. Fegley Jr.
Space Sci. Rev., Vol. 39, Nos. 1/2, p. 163 – 192 (1984).
Sources of organic matter and inorganic tracers on Jupiter, including solar UV photolysis, lightning discharges, and convective quenching of hot gases from the lower atmosphere, are reviewed in light of Earth–based and Voyager data with the purpose of predicting the tropospheric steady–state abundances and vertical distributions of HCN, $CH_2O$, and other species.

099.117 **Jupiter I (Io).**
IAU Circ., No. 3962 (1984).

099.118 **Jupiter's and Saturn's fine–scale magnetic fields.**
J. W. Warwick, D. R. Evans.
Phys. Earth Planet. Inter., Vol. 36, No. 1, p. 85 – 89 (1984).
Jupiter's field is strongly dipolar but with relatively large high order moments compared to the Earth's. Saturn's radio emissions show complexities similar to Jupiter's. They are strongly asymmetric about the rotational axis, although Saturn's field is nearly axisymmetric. Their strong asymmetry suggests strong longitudinal variations in the magnetic field a few thousand kilometers from the cloud tops, in conflict with the field measured aboard Pioneer 11. An arguable case can be made that the giant planets exhibit complexity of magnetic structure similar to the Sun.

099.119 **Variability of phosphine on Jupiter from 5–µm spectroscopy.**
P. Drossart, T. Encrenaz, A. T. Tokunaga.
Icarus, Vol. 60, No. 3, p. 613 – 620 (1984).
Spectra of Jupiter recorded in the 1900– to 2300–$cm^{-1}$ range at the IRTF in Hawaii, July 1982, provide tentative evidence for variability of the Jovian atmosphere between zones and belts. It is concluded from analysis of the $\nu_1$ and $\nu_3$ bands of $PH_3$ that there is a possible enhancement of the $PH_3/H_2$ ratio in the belts when compared to the zones. There is an apparent reduction of the $PH_3$ abundance between the IRIS Voyager 1 determinations and these spectra, implying temporal or spatial variability of $PH_3$ on Jupiter. Interpretation of this variability in the troposphere could involve both dynamical and thermochemical processes.

099.120 **The Jovian atmospheric window at 2.7 µm: a search for $H_2S$.**
H. P. Larson, D. S. Davis, R. Hofmann, G. L. Bjoraker.
Icarus, Vol. 60, No. 3, p. 621 – 639 (1984).
The atmospheric transmission window at 2.7 µm in Jupiter's atmosphere was observed at a spectral resolution of 0.1 $cm^{-1}$ from the Kuiper Airborne Observatory. From analysis of the $CH_4$ abundance (~80 m–am) and the $H_2O$ abundance (<0.0125 cm–am) it was determined that the penetration depth of solar flux at 2.7 µm is near the base of the $NH_3$ cloud layer. The search for $H_2S$ in Jupiter's atmosphere yielded an upper limit of ~0.1 cm–am. The corresponding limit to the elemental abundance ratio [S]/[H] was $\sim 1.7 \times 10^{-8}$, about $10^{-3}$ times the solar value. Upon modeling the abundance and distribution of $H_2S$ in Jupiter's atmosphere it was concluded that, contrary to expectations, sulfur–bearing chromophores are not present in significant amounts in Jupiter's visible clouds. Rather, it appears that most of Jupiter's sulfur is locked up as $NH_4SH$ in a lower cloud layer. Alternatively, the global abundance of sulfur in Jupiter may be significantly depleted.

099.121 **The zonal distribution of hydrogen in the Jovian atmosphere.**
R. M. Killen, J. W. Chamberlain.
Icarus, Vol. 60, No. 3, p. 640 – 653 (1984).
The authors have used the Lyman alpha intensities measured by Voyagers 1 and 2 and by IUE to deduce the distribution of atomic hydrogen in the Jovian atmosphere, assuming that the sources of the dayside Lyman alpha include resonance scattering of solar Lyman alpha, resonance scattering of the interplanetary Lyman alpha radiation, and direct excitation by charged particles. They assume that 1 kR of the dayside emission is due to resonance scattering of an interstellar wind source, and 2.3 to 7 kR of the dayside Lyman alpha brightness are due to direct excitation by charged particles. Finally, they used the distribution of freely scattering hydrogen with longitude, along with nightside Lyman alpha intensities measured by the Voyager spectrograph, to estimate the longitudinal distribution of particle flux on the nightside.

099.122 **Sulfur flows of Ra Patera, Io.**
D. C. Pieri, S. M. Baloga, R. M. Nelson, C. Sagan.
Icarus, Vol. 60, No. 3, p. 685 – 700 (1984).
Voyager 1 imaging data have been used to investigate the color and morphology of several radial flow–like features at Ra Patera, a broad volcanic structure at approximately 8° latitude and 325° longitude on the Galilean satellite Io (J1). It was found that downstream progressions of flow color and morphology are consistent with lava of a predominately sulfur composition cooling radiatively and erupting in the range of 470 to 520K at effusion rates at $10^{10}$ to $10^{11} cm^3/sec$. This implies global resurfacing rates by volcanic flows on Io of the order of 1 cm/year. Calculated energy content and effusion rates for flows at Ra Patera, using the physical parameters of sulfur, are of the order of the largest known terrestrial basaltic eruptions and are consistent with calculations of globally available energy.

099.123 **Observations photographiques de Jupiter et ses satellites effectuées en 1980 au GPO de l'European Southern Observatory (ESO) La Silla. Considérations sur des théories plus modernes de la réduction des clichés astrographiques.**
H. Debehogne, R. R. de Freitas Mourão, M. R. Nunes, O. C. Tavares.
Rev. Mex. Astron. Astrofis., Vol. 9, No. 2, p. 139 – 140 (1984).

099.124 **On the dynamical structure of the wind field of Jupiter's Great Red Spot.**
Z.–y. Yue, B. Zhang.
Appl. Math. Mech., Vol. 5, No. 2, p. 1163 – 1172 (1984).
Based on the geographic approximation the two–dimensional dynamical structure of the wind fields of Jupiter's Great Red Spot and White Oval BC is obtained. The results of the calculation are in good agreement with the observations. Thus, an explanation of the observed dispersion of the velocities along the horizontal streamline is given. The distributions of the vorticities of the GRS and White Oval BC are also calculated.

099.125 **Scale height of the Jovian current sheet.**
J. L. Parish.
Planet. Space Sci., Vol. 32, No. 12, p. 1531 – 1532 (1984).
The author discusses the parameter *D* in the Goertz et al. (1979) magnetic field model as it relates to the scale height of the current sheet.

099.126 **Comparison of the Jovian north and south pole aurorae using the IUE observatory.**
T. E. Skinner, H. W. Moos.
Geophys. Res. Lett., Vol. 11, No. 11, p. 1107 – 1110 (1984).
New results on the spatial and temporal variability of the auroral emissions from Jupiter have been obtained from three IUE observations of the south pole made during the period July 1983 to March 1984. The current observations, together with previous IUE studies of the north pole aurora, provide convincing evidence for persistent longitudinal asymmetries in the Jovian auroral emissions. The strongest emissions appear to originate from regions centered near $\lambda_{III} \sim 0°$ at the south pole and

$\lambda_{III} \sim 185°$ at the north pole. Possible implications for models of auroral production are discussed.

099.127 **Planetocentric versus heliocentric impacts in the Jovian and Saturnian satellite system.**
G. P. Horedt, G. Neukum.
J. Geophys. Res., Vol. 89, No. B12, p. 10405 – 10410 (1984).
Readily applicable equations are derived for calculating the impact velocities, collision intervals, and cumulative crater frequencies in a satellite system for planetocentric and heliocentric impactors. Observed cumulative crater frequencies in the Saturnian satellite system and a sometimes observed lack of apex–antapex asymmetry of crater frequencies favor crater–producing projectiles orbiting initially in elliptic orbits round the parent planet (planetocentric impactors). On the other hand, a sometimes observed asymmetry of crater size distribution and short collision intervals ($<10^4$ years) of planetocentric projectiles favor heliocentric impactors. The authors conclude that cratering on the Jovian and Saturnian satellites is produced by heliocentric projectiles as well as by captured planetocentric impactors. Small satellites could be crushed several times by major impacts and reaccrete from planetocentric orbits.

099.128 **Is Jupiter getting brighter?**
J.–y. Liu.
Chin. Astron. Astrophys., Vol. 8, No. 4, p. 354 – 356 (1984). English translation of 33.099.054.

099.129 **Statistical studies of latitude and motion of major Jovian features, 1972 – 1983.**
P. C. Sherrod.
Strolling Astron., Vol. 30, Nos. 11 – 12, p. 226, 230 – 239 (1984).

099.130 **Some recent 1984 observations of Jupiter.**
J. Olivarez.
Strolling Astron., Vol. 30, Nos. 11 – 12, p. 260 – 262 (1984).

099.131 **Observations photographiques de Mars, de Jupiter et de ses satellites, effectuées à l'Astrographe Double de 40 cm en 1980.**
H. Debehogne.
Bull. Astron., Vol. 9, No. 5, p. 230 – 235 (1983).

099.132 **Observations photographiques de Jupiter et de ses satellites, effectuées en 1979 à l'équatorial GPO de 40 cm de l'Observatoire Austral Européen (ESO) à La Silla (Chili).**
H. Debehogne.
Bull. Astron., Vol. 9, No. 5, p. 242 – 250 (1983).

099.133 **The Jovian atmospheric window at 2.7 μm: a search for $H_2S$.**
H. P. Larson, D. S. Davis, R. Hofmann, G. L. Bjoraker.
Prepr. Steward Obs., No. 546, 36 pp. (1984).
The authors have observed the atmospheric transmission window at 2.7 μm in Jupiter's atmosphere at a spectral resolution of 0.1 $cm^{-1}$ from the Kuiper Airborne Observatory. From their analysis of the $CH_4$ abundance ($\sim 80$ m–am) and the $H_2O$ abundance ($<0.0125$ cm–am) the authors determined that the penetration depth of solar flux at 2.7 μm is near the base of the $NH_3$ cloud layer. The search for $H_2S$ in Jupiter's atmosphere yielded an upper limit of $\sim 0.1$ cm–am. The corresponding limit to the elemental abundance ratio [S]/[H] was $\sim 1.7 \times 10^{-8}$, about $10^{-3}$ times the solar value. Upon modeling the abundance and distribution of $H_2S$ in Jupiter's atmosphere the authors concluded that, contrary to expectations, sulfur–bearing chromophores are not present in significant amounts in Jupiter's visible clouds.

099.134 **Electropolarimetric observations of Ganymede and Callisto.**
R. A. Chigladze.
Astron. Tsirk., No. 1320, p. 3 – 4 (1984). In Russian.

099.135 **Electropolarimetric observations of Io and Europa.**
R. A. Chigladze.
Astron. Tsirk., No. 1320, p. 4 – 5 (1984). In Russian.

099.136 **Polarization peculiarities of Jovian polar regions.**
O. M. Starodubtseva, V. G. Tejfel'.
Astron. Tsirk., No. 1327, p. 5 – 7 (1984). In Russian.

099.137 **Polarization of light in the polar regions of Jupiter.**
O. M. Starodubtseva, V. G. Tejfel'.
Astron. Vestn., Tom 18, No. 3, p. 179 – 190 (1984). In Russian.
Abstr. in Ref. Zh., 51. Astron., 12.51.271 (1984).

099.138 **Sulfur volcanoes on Io?**
R. Greeley, J. H. Fink.
Astron. Express, Vol. 1, Issue 1, p. 25 – 31 (1984).
The unusual rheological properties of molten sulfur, in which viscosity decreases ~four orders of magnitude as it cools from 170° to 120°C, may result in distinctive volcanic flow morphologies that allow sulfur flows and volcanoes to be identified on Io. Search of high resolution Voyager images reveals three features – Atar Patera, Daedalus Patera, and Kibero Patera – considered to be possible sulfur volcanoes based on their morphology. All three average 250 km in diameter and are distinguished by circular–to–oval central masses surrounded by irregular, widespread flows. Geometric relations indicate that the flows were emplaced after the central zone and appear to have emanated from their margins. The central zones are interpreted to be domes representing the high–temperature stage of sulfur formed initially upon eruption.

099.139 **The Jovian ring.**
J. A. Burns.
Planetary rings, p. 203 – 209 (1984). – See Abstr. 012.085 (IAU Colloq. No. 75).
Lying in Jupiter's equatorial plane is a diaphanous ring having little substructure within its three components (main band, faint disk, and halo). Micron–sized grains account for much of the visible ring, but particles of centimeter sizes and larger must also be present to absorb charged particles. Since dynamical evolution times and survival life times are quite short ($\lesssim 10^{2-3}$yr) for small grains, the Jovian ring is being continually replenished; probably most of the visible ring is generated by micrometeoroids colliding into unseen parent bodies that reside in the main band.

099.140 **The ethereal rings of Jupiter and Saturn.**
J. A. Burns, M. R. Showalter, G. E. Morfill.
Planetary rings, p. 200 – 272 (1984). – See Abstr. 012.086.
The Jovian ring is tenuous, with little known substructure. Its brighter main band lies in Jupiter's equatorial plane at about 1.7 – 1.8 $R_J$, well within the Roche limit. Micron–sized grains account for most of the visible ring's brightness, but particles with sizes of centimeters and larger must also be present to absorb charged particles. The E, F, and G Rings of Saturn share many properties with Jupiter's ring: they have low optical depths, are immersed in a magnetospheric plasma, and contain a significant complement of micron–sized particles. All of these rings are distinguished in two fundamental ways from the rings of Uranus and the major Saturnian rings: single–particle dynamics, rather than collective effects, most likely govern their form, and the majority of ring particles have quite limited lifetimes.

099.141 **Mobility of water ice on Callisto: evidence and implications.**
J. R. Spencer, P. R. Maloney.
Geophys. Res. Lett., Vol. 11, No. 12, p. 1223 – 1226 (1984).
Voyager images of Callisto reveal evidence for the accumulation of a bright volatile, presumably water ice, on north–facing slopes in the north polar region. The geometry of these accumulations suggests that ice migration in Callistoan high latitudes is dominated by thermal sublimation, controlled by insolation–dependent surface temperature contrasts. Such sublimation–dominated migration may result in the cold–trapping of most of the surface ice in discrete high–albedo regions.

099.142 **Jupiter: lord of the planets.**
W. Kaufmann.
Mercury, Vol. 13, No. 6, p. 169 – 184 (1984).

099.143 **Ejection of atoms and molecules from Io by plasma–ion impact.**
E. M. Sieveka, R. E. Johnson.
Astrophys. J., Vol. 287, No. 1, p. 418 – 426 (1984).
The authors examine models for ejection of atoms from the surface of Io or from an atmosphere on Io by plasma ion bombardment. They consider direct collisional ejection from an exosphere, surface sputtering, and atmospheric sputtering produced by this bombardment, as well as Jeans escape. Each of the ejection processes considered produces a distinctive directional and energy distribution for the ejected neutral flux and different ejection rates. These distributions and rates are compared to the source distributions suggested by Pilcher *et al.* for the observed neutral Na clouds associated with Io.

099.144 **Io's sodium directional features: evidence for a magnetospheric–wind–driven gas escape mechanism.**
C. B. Pilcher, W. H. Smyth, M. R. Combi, J. H. Fertel.
Astrophys. J., Vol. 287, No. 1, p. 427 – 444 (1984).
Elongated features in Io's sodium cloud, directed away from Jupiter and inclined both to the north and to the south of the satellite's orbital plane, have been observed. The north/south directions of the features are correlated with Io's magnetic longitude, suggesting a formation mechanism involving the oscillating plasma torus. The authors show by means of a model analysis that the features can result from a source of high–velocity ($\sim 20$ km $s^{-1}$) sodium combined with the oscillating neutral sodium sink provided by the plasma. The initial directions and speeds of sodium atoms escaping Io to form the directional features can be understood in terms of a magnetospheric–wind–driven escape mechanism.

099.145 **Radio emissions from the planets earth, Jupiter, and Saturn.**
M. L. Kaiser, M. D. Desch.
Rev. Geophys. Space Phys., Vol. 22, No. 4, p. 373 – 384 (1984).
The authors review the major observed and inferred features of the three currently known nonthermal radio planets, the earth, Jupiter, and Saturn. Several of the properties of the three planets are quite similar, suggesting that the same processes take place at all three. In addition, Jupiter has two radio components in the low–frequency range which may not have analogs at the earth or Saturn. The authors speculate on the expected addition of Uranus to this list of radio planets.

**Année Diderot: «Qui a découvert les satellites de Jupiter?»**
See Abstr. 004.002.

**Exploitation des observations de Delambre des satellites galiléens (1775 – 1802).**
See Abstr. 004.005.

**Isotopic dependence of infrared transition moments in methane.**
See Abstr. 022.007.

**Studies of proton–irradiated $SO_2$ at low–temperatures: implications for Io.**
See Abstr. 022.009.

**Propane absorption at 200K near 748 $cm^{-1}$; implications for interpretations of spectra of Jupiter and Saturn.**
See Abstr. 022.096.

**Thermodynamics of metallic hydrogen mixed with helium, carbon, and oxygen.**
See Abstr. 022.098.

**Complex refractive index of liquid and solid sulfur.**
See Abstr. 022.099.

**Spectral reflectance (0.39 – 0.70 μm) of carbon–sulfur mixtures.**
See Abstr. 022.100.

**Surface photolysis of $SO_2$ and photophysical changes in sulfur may affect Io's surface UV–vis reflectance.**
See Abstr. 022.101.

**Evaluation of phase corrections on planet position measurements.**
See Abstr. 041.022.

**Second order secular planetary theory for the four major planets.**
See Abstr. 042.011.

**Theory of the motion of Jupiter and Saturn over a period of 6000 years. Solution JASON84.**
See Abstr. 042.012.

**Satellites Galiléens de Jupiter.**
See Abstr. 046.025.

**Ephémérides des satellites naturels des planètes: Jupiter, Saturne, Uranus pour 1985.**
See Abstr. 046.036.

**En route to the giant planet.**
See Abstr. 051.004.

**Galileo Jupiter approach orbit determination.**
See Abstr. 052.022.

**Adiabatic theory in rapidly rotating magnetospheres.**
See Abstr. 062.009.

**Unified theory of whistler mode instability in the presence of a parallel electric field.**
See Abstr. 062.033.

**Perturbation theory for free oscillations of a self–gravitating sphere with phase boundaries.**
See Abstr. 062.070.

**Drift wave instabilities in a high β multispecies plasma.**
See Abstr. 062.088.

**A novel instability in dense hydrogen–helium fluids.**
See Abstr. 062.103.

**Normal mode spectra of a self–gravitating sphere with phase boundaries.**
See Abstr. 065.107.

**Free oscillations of giant planets. Influence of conditions at the phase boundary.**
See Abstr. 091.003.

**Resonance locking between Jupiter and Uranus.**
See Abstr. 091.004.

**Quelques aspects de la dynamique des atmosphères des planètes géantes.**
See Abstr. 091.018.

**Episodic heatflow in tidally heated satellites.**
See Abstr. 091.038.

**Tidal stresses due to nonsynchronous rotation: implications for fracturing on Europa.**
See Abstr. 091.044.

**Geology and tectonics of the Argyre area on Mars: comparisons with other basins in the solar system.**
See Abstr. 097.005.

**Observations et théories des satellites de Saturne et de Jupiter.**
See Abstr. 100.010.

**Microwave emission from Titan and the Galilean satellites.**
See Abstr. 100.059.

# 100 Saturn, Saturn Satellites

100.001 **Photochemistry of the atmosphere of Titan: comparison between model and observations.**
Y. L. Yung, M. Allen, J. P. Pinto.
Astrophys. J., Suppl. Ser., Vol. 55, No. 3, p. 465 – 506 (1984). = Contrib. No. 3813 Div. Geol. Planet. Sci., Calif. Inst. Technol., Pasadena.

A post–Voyager synthesis of our knowledge of the photochemistry of Titan's atmosphere is presented. A comprehensive set of chemical reactions relevant to Titan and the other solar system has been compiled. Using this reaction set and invoking interactions with ultraviolet sunlight, energetic electrons, and cosmic rays, it is possible to explain the observed minor species in the atmosphere of Titan as photochemical products of the parent molecules $N_2$, $CH_4$, and $H_2O$. Detailed model results are discussed.

100.002 **Perturbations on the orbits of companions of the satellites of Saturn.**
A. T. Sinclair.
Astron. Astrophys., Vol. 136, No. 1, p. 161 – 166 (1984).

The effects of orbital resonances on satellites in tadpole or horseshoe orbits relative to Mimas, Enceladus, Tethys or Dione are examined. A tadpole companion of Enceladus would be the most highly perturbed, and although the perturbations would not cause instability they may prevent the initial formation of a satellite in such an orbit. Tadpole companions of Mimas would have a large forced eccentricity, which would probably prevent the simultaneous existence of horseshoe orbits. The effects of tidal forces are also examined, and are found to cause a small displacement of the L4 and L5 equilibrium points.

100.003 **Investigation of the ammonia absorption band near 6450 Å in the spectrum of Saturn. II. Quantitative analysis of observations.**
V. V. Avramchuk, A. I. Karmelyuk.
Astrometr. Astrofiz., Vyp. 52, p. 30 – 37 (1984). In Russian.

Results of investigation of the Saturn spectra published earlier are analyzed. The rotation temperature ($T_{rot} = 129 \pm 8$K) has been determined from the comparison of the intensities of ammonia lines λλ6444.6 Å ($J = 3$, $K = 1$), 6445.6 Å ($J = 3$, $K = 2$) and 6452.6 Å ($J = 2$, $K = 1$) in terms of the homogeneous half–infinite atmosphere model. The pressure is equal to $1.7 \pm 0.1$ atm at the level of the ammonia band λ6450 Å formation. The volume concentration of gaseous ammonia in the cloud layer of Saturn is $(6.7 \pm 1.2) \times 10^{12} cm^{-3}$. The ratio of the volume concentration of ammonia and methane is $(0.76 \pm 0.13) \times 10^{-2}$.

100.004 **Saturn 1943 – 1981: a visual photometric study – II.**
R. J. McKim, K. W. Blaxall.
J. Br. Astron. Assoc., Vol. 94, No. 5, p. 211 – 220 (1984).

Visual intensity estimates for the belts and zones of Saturn during the 1943 – 81 period are discussed in this paper. Spot activity in the atmosphere of Saturn may lead to short–term intensity changes and longer term seasonal effects may arise by virtue of the periodic change in the solar altitude at a particular belt or zone. Thus, while the meteorology of the planet may well be governed by internal heating, it is apparent that infalling solar radiation has a considerable effect too.

100.005 **Application of a polarimetric model to the surface microstructure of particles in the B–ring of Saturn.**
G. A. Steigmann.
Mon. Not. R. Astron. Soc., Vol. 209, No. 2, p. 359 – 371 (1984).

Scattering of light from the surface of particles in the B–ring of Saturn is examined using a polarimetric model. Application of the model allows evaluation of the individual parameters involved in the scattering process and the manner in which they inter–relate to produce the observed polarization curve. Theoretical polarization curves based on a simulated rough–ice surface are compared with published observational data. The model suggests that the surfaces of the particles in the B–ring of Saturn are very fine–grained, being covered with micron sized pits rather than larger, more complex, structures. The results are consistent with laboratory simulations which suggest that the surface of the ring particles consists of a fine–grained water ice.

100.006 **Production and condensation of organic gases in the atmosphere of Titan.**
C. Sagan, W. R. Thompson.
Icarus, Vol. 59, No. 2, p. 133 – 161 (1984).

The rates and altitudes for the dissociation of atmospheric constituents of Titan are calculated for solar UV, solar wind protons, interplanetary electrons, Saturn magnetospheric particles, and cosmic rays. The production rate curves show no localized maxima corresponding to observed altitudes of Titan's hazes and clouds. For simple to moderately complex organic gases in the Titanian atmosphere, condensation occurs below the top of the main cloud deck at 2825 km. Such condensates comprise the principal cloud mass, with molecules of greater complexity condensing at higher altitudes. Visible condensation hazes of more complex organic compounds may occur at altitudes up to ~3060 km provided only that the abundance of organic products declines with molecular mass no faster than laboratory experiments indicate.

100.007 **Spokes in Saturn's rings: dynamical and reflectance properties.**
R. E. Eplee Jr., B. A. Smith.
Icarus, Vol. 59, No. 2, p. 188 – 198 (1984).

Orbital velocities and relative reflectances of a preliminary sample of 15 Saturnian spokes recorded in the Voyager 2 low–resolution ring movie have been examined. While 13 spokes exhibit the expected Keplerian velocities, 2 anomalous spokes deviate from this motion. For approximately 2 hr after their formation these spokes exhibit corotational motion and, only then, accelerate to Keplerian speeds. The atypical dynamics of these 2 spokes suggest that they are generated by plasma clouds of unusually high charge density, while the contrast of these features appears to depend more on ring particulate concentration than on plasma cloud density.

100.008 **The control network of Iapetus.**
M. E. Davies, F. Y. Katayama.
Icarus, Vol. 59, No. 2, p. 199 – 204 (1984).

A control network of the Saturnian satellite Iapetus has been established photogrammetrically from pictures taken by the two Voyager spacecraft. Coordinates of 62 control points have been computed and listed; pixel measurements of these points were made on 14 Voyager 1 and 66 Voyager 2 pictures. Some of these points are identified on the preliminary U.S. Geological Survey map of Iapetus and many are identified by name. The mean radius of Iapetus has been determined as $718 \pm 8$ km.

100.009 **The tectonic and volcanic history of Dione.**
J. M. Moore.
Icarus, Vol. 59, No. 2, p. 205 – 220 (1984). – See Abstr. 37.012.044.

The tectonic and volcanic modifications of Dione are described and interpreted. It is proposed that after the formation of a brittle outer shell, but before the end of heavy meteoritic bombardment, global expansion due to radionuclide heating (and perhaps a loss of oblateness due to tidal despinning and orbital recession) produced a global system of lineaments. An $NH_3 \cdot H_2O$ melt was produced and "erupted" on the surface to form plains units. Cooling of the interior (or a phase change) led to horizontal compression in the surface.

100.010 **Observations et théories des satellites de Saturne et de Jupiter.**
G. Dourneau.
Astrométrie, géodésie, mécanique céleste, p. 35 – 48 (1984). – See Abstr. 012.021.

100.011 **Saturn in 1982 – 83: some European observations.**
G. Adamoli.
Strolling Astron., Vol. 30, Nos. 9 – 10, p. 177 – 181 (1984).
Visual and photographic observations concerning the 1982 – 83 apparition of the planet are presented. Dark and bright spots were recorded on the Equatorial Zone and North Equatorial Belt. Filter visual and photographic observations suggest a redder equatorial region and bluer high latitudes and rings.

100.012 **The $Z_3$ model of Saturn's magnetic field and the Pioneer 11 vector helium magnetometer observations.**
J. E. P. Connerney, M. H. Acuña, N. F. Ness.
J. Geophys. Res., Vol. 89, No. A9, p. 7541 – 7544 (1984).
The authors carefully examine the Pioneer 11 vector helium magnetometer observations as a further evaluation of the $Z_3$ model and to search for evidence of any departure from axisymmetry of Saturn's magnetic field. Comparison of the Pioneer 11 high–field flux gate observations with either the $Z_3$ model or the vector helium magnetometer observations is not fruitful because of the relatively large quantization step size of the Pioneer 11 flux gate magnetometer.

100.013 **Equilibrium theory for Saturn's ring with self–gravitation and finite thickness.**
W. Seboldt, K. Schindler.
Astron. Astrophys., Vol. 138, No. 2, p. 290 – 296 (1984).
The authors discuss the equilibrium of Saturn's ring within kinetic theory taking the self–gravitation and the finite thickness of the ring into account. The present model is a first step towards a future quantitative theory of the evolution of planetary rings. It may also serve as a starting point for stability investigations. It is shown that for typical values of the relevant physical parameters self–gravitation effects are expected to be small but not fully negligible. Two explicit equilibrium examples are given, one with a monotonic density variation and one with oscillatory structure.

100.014 **Phosphine photochemistry in the atmosphere of Saturn.**
J. A. Kaye, D. F. Strobel.
Icarus, Vol. 59, No. 3, p. 314 – 335 (1984).
A model is presented for the photochemistry of $PH_3$ in the upper troposphere and lower stratosphere of Saturn that includes the effects of coupling with $NH_3$ and hydrocarbon photochemistry, specifically the $C_2H_2$ catalyzed photodissociation of $CH_4$.

100.015 **Structure of the satellitary accretion disk of Saturn.**
A. Coradini, G. Magni.
Icarus, Vol. 59, No. 3, p. 376 – 391 (1984).
A detailed computation on the equilibrium structure of an accretion disk around Saturn from which the regular satellites presumably originated is reported. Such a disk is the predecessor of the self–dissipating disk that is formed when the mass infall stops. When determining the disk structure local energy balance was assumed. Convection was taken into account by introducing local energy dissipation and, in an approximate manner, sonic convection. Changes in the disk structure were investigated by varying the free parameters, i.e., the external flux from both the protosun and the protoplanet, the abundance of dust and the strength of turbulence. It has been verified that the external energy flux does not play an important role in the evolution of the disk structure. All the studied models present a temperature distribution that allows the condensation of iron, silicate, and, in the outer regions, ice grains.

100.016 **Voyager disk resolved photometry of the Saturnian satellites.**
B. J. Buratti.
Icarus, Vol. 59, No. 3, p. 392 – 405 (1984).
Photometric analysis of Voyager images of the medium–sized icy satellites of Saturn shows that their surface exhibit a wide range of scattering properties. At low phase angles, Rhea and Dione closely follow lunar behavior with almost no limb darkening. Mimas, Tethys, and especially Enceladus show significant limb darkening at low phase angles, which suggests multiple scattering is important for their surfaces.

100.017 **Voyager photometry of Iapetus.**
S. W. Squyres, B. J. Buratti, J. Veverka, C. Sagan.
Icarus, Vol. 59, No. 3, p. 426 – 435 (1984).
Voyager images of Iapetus ranging in phase angle from 8 to 90° were used to define the satellite's photometric properties and construct an albedo map of its surface. The images confirm that the albedo distribution has a roughly hemispheric asymmetry, as had been inferred from earlier analyses of the disk–integrated lightcurve.

100.018 **Saturn 1943 – 1981: a visual photometric study – III.**
R. J. McKim, K. W. Blaxall.
J. Br. Astron. Assoc., Vol. 94, No. 6, p. 249 – 255 (1984).
The BAA intensity data for the Saturnian ring features are discussed in this final instalment of the authors' visual photometric study. It is found that both ring A and ring C darken as they open to the Earth and Sun, but ring B brightens. These and other results are discussed in terms of the structure of the rings.

100.019 **The helium abundance of Saturn from *Voyager* measurements.**
B. J. Conrath, D. Gautier, R. A. Hanel, J. S. Hornstein.
Astrophys. J., Vol. 282, No. 2, p. 807 – 815 (1984).
The helium abundance in the atmosphere of Saturn is obtained from *Voyager* measurements by two methods. The first method combines infrared spectra and vertical profiles derived from radio occultation measurements and yields a hydrogen mole fraction of $q = 0.963 \pm 0.024$ corresponding to a helium mass fraction of $Y = 0.06 \pm 0.05$. The second method is based on the direct inversion of infrared spectra and yields values consistent with those from the first method. Comparison of the helium abundance of Saturn with that of Jupiter and the Sun suggests that helium precipitation is significant in Saturn but may not have begun in Jupiter.

100.020 **Formation of the Titan plasma tail due to critical ionization velocity phenomenon.**
A. A. Galeev, I. Kh. Khabibrakhmanov.
Astron. Zh., Tom 61, Vyp. 5, p. 892 – 899 (1984). In Russian. English translation in Sov. Astron., Vol. 28, No. 5.
A critical ionization velocity phenomenon is proposed to account for anomalous interaction of Titan's atmosphere with the magnetospheric plasma corotating with Saturn. The limiting plasma density, the electron temperatures as well as the frequency spectrum of the excited waves are determined and compared with the measurements.

100.021 **On collisional dynamics of particles in Saturn's rings.**
I. G. Shukhman.
Astron. Zh., Tom 61, Vyp. 5, p. 985 – 1004 (1984). In Russian. English translation in Sov. Astron., Vol. 28, No. 5.
The problem of the quasi–stationary evolution of Saturn's rings is considered. The rings' dynamics is assumed to be determined by inelastic collisions of solid particles. The kinetic equation is obtained, which contains a collisional term accounting for both "frontal" and "tangential" inelasticity, the possibility of particles twisting (spin) after the collision, as well as finite dimensions of the colliding bodies. On the basis of this kinetic equation the system of moment equations is derived, and therefrom the conditions of heat balance in the rings are analyzed.

100.022 **Further investigation into a recent model for toroidal rings of Saturn.**
J. V. Breakwell, V. R. Eshleman.
Celest. Mech., Vol. 33, No. 3, p. 229 – 238 (1984).
A previous paper described a toroidal ringlet consisting of a large number of charged particles distributed uniformly along slightly inclined, slightly eccentric, orbits around Saturn. The orbits have identical semimajor–axis, identical eccentricity and identical inclination, but their ascending nodes are distributed uniformly around Saturn. Assuming an inclination (in radians) much smaller than the eccentricity, it was shown that the toroidal cross–section could rock slowly back and forth as the inclination fluctuates. In this paper a similar result is obtained for the more

general case when inclinaton is comparable with eccentricity and both fluctuate.

100.023 **The eccentric Saturnian ringlets at 1.29 $R_s$ and 1.45 $R_s$.**
C. Porco, P. D. Nicholson, N. Borderies,
G. E. Danielson, P. Goldreich, J. B. Holberg, A. L. Lane.
Icarus, Vol. 60, No. 1, p. 1 – 16 (1984). = Contrib. No. 3960 Div. Geol. Planet. Sci., Calif. Inst. Technol., Pasadena, Calif. 91125.

The shapes and kinematics of the two major eccentric ringlets in Saturn's C ring are studied in data acquired by four Voyager experiments: imaging science, radio science, ultraviolet spectrometer, and photopolarimeter. It is found that the ringlets have mean widths of ~25 km (Titan, 1.29 $R_s$) and ~64 km (Maxwell, 1.45 $R_s$), eccentricities of order $10^{-4}$, sharp edges on a scale of ~1 km, normal optical depths $\tau \sim 1-2$, and are embedded in essentially empty gaps ($\tau < 0.05$). In addition, they exhibit positive linear width–radius relations. Masses, mean surface mass densities, and mass extinction coefficients have been calculated.

100.024 **Saturn's nonaxisymmetric ring edges at 1.95 $R_s$ and 2.27 $R_s$.**
C. Porco, G. E. Danielson, P. Goldreich, J. B. Holberg, A. L. Lane.
Icarus, Vol. 60, No. 1, p. 17 – 28 (1984). = Contrib. No. 3961 Div. Geol. Planet. Sci., Calif. Inst. Technol., Pasadena, Calif. 91125.

The outer edges of Saturn's A and B rings, at 2.27 $R_s$ and 1.95 $R_s$, have been examined using data acquired by four Voyager experiments. The shapes and kinematics of these features are influenced by their proximity to strong low–order Lindblad resonances. The data for the A–ring edge are consistent with a seven–lobed radial distortion of amplitude $6.7 \pm 1.5$ km which rotates with the mass–weighted mean angular velocity of the coorbital satellite system. The B–ring edge has essentially, a double–lobed figure of radial amplitude $74 \pm 9$ km which rotates with the mean motion of Mimas, though there is an indication that it is not completely described with a simple Saturn–centered ellipse. An upper limit of 10 m has been placed on the vertical thickness in the unperturbed region of the B ring.

100.025 **The ring atmosphere of Saturn: Monte Carlo simulation of ring source models.**
W.–H. Ip.
J. Geophys. Res., Vol. 89, No. A10, p. 8843 – 8849, 9035 (1984).

The spatial distributions of hydrogen atoms emitted from the rings of Saturn with a range of emission speeds are simulated by a Monte Carlo technique. Basically two issues, one concerning the source strength and the other concerning the geometrical configuration (which is specifically discussed in the present work), have to be dealt with. The present study indicates that the loss rate of neutral particles via charge exchange could vary between $4 \times 10^{27} s^{-1}$ and $7 \times 10^{28} s^{-1}$.

100.026 **Correction to "Implications of Voyager data for energetic ion erosion of the icy satellites of Saturn" [J. Geophys. Res., Vol. 88, No. A11, p. 8765 – 8770 (1983)].**
L. J. Lanzerotti, C. G. Maclennan, W. L. Brown,
R. E. Johnson, L. A. Barton, C. T. Reimann, J. W. Garrett,
J. W. Boring.
J. Geophys. Res., Vol. 89, No. A10, p. 9157 (1984). See Abstr. 34.100.045.

100.027 **Absolute spectrophotometry of Titan, Uranus, and Neptune: 3500 – 10,500 Å.**
J. S. Neff, D. C. Humm, J. T. Bergstralh, A. L. Cochran,
W. D. Cochran, E. S. Barker, R. G. Tull.
Icarus, Vol. 60, No. 2, p. 221 – 235 (1984).

Absolute measurements of the geometric albedo spectra of Titan, Uranus, and Neptune from 3500 to 10,500 Å are reported. The measurements have spectral resolution of about 7 Å and high signal–to–noise ratio owing to the superb efficiency of the spectrograph and Reticon detector used to acquire them. Agreement is excellent with albedo measurements of Uranus by G. W. Lockwood et al. (1983). The high precision and spectral resolution of the data make possible quantitative measurements of the effects of Raman scattering by $H_2$ in the atmospheres of Uranus and Neptune.

100.028 **Titan: far–infrared and microwave remote sensing of methane clouds and organic haze.**
W. R. Thompson, C. Sagan.
Icarus, Vol. 60, No. 2, p. 236 – 259 (1984).

It is shown that Titan's surface and plausible atmospheric thermal opacity sources – gaseous $N_2$, $CH_4$, and $H_2$, $CH_4$ cloud, and organic haze – are sufficient to match available Earth–based and Voyager observations of Titan's thermal emission spectrum. Matching computed spectra to the observed Voyager IRIS spectra at 7.3 and 52.7° emission angles yields the following abundances and locations of opacity sources: $CH_4$ clouds: 0.1 g cm$^{-2}$ at a planetocentric radius of 2610 – 2625 km, 0.3 g cm$^{-2}$ at 2590 – 2610 km, total $0.4 \pm 0.1$ g cm$^{-2}$ above 2590 km; organic haze: $4 \pm 2 \times 10^{-6}$ g cm$^{-2}$ above 2750 km; tropospheric $H_2$: $0.3 \pm 0.1$ mol%.

100.029 **Possible production by lightning of aerosols and trace gases in Titan's atmosphere.**
W. J. Borucki, C. P. McKay, R. C. Whitten.
Icarus, Vol. 60, No. 2, p. 260 – 273 (1984).

Although lightning has not yet been observed in Titan's atmosphere, the authors use the results of measurements for lightning activity in the terrestrial and Jovian atmospheres to estimate the energy–dissipation rate of Titan lightning, and its contribution to the composition of the Titan atmosphere relative to that of galactic cosmic rays, electron precipitation, and photochemical sources. It is found that Titan lightning could produce significant quantities of HCN, $C_2N_2$, and soot particles in the lower atmosphere.

100.030 **A seasonal model of the Saturnian upper troposphere: comparison with Voyager infrared measurements.**
B. Bézard, D. Gautier, B. Conrath.
Icarus, Vol. 60, No. 2, p. 274 – 288 (1984).

An extension of the seasonal climate model of R. D. Cess et al. (1979) to Saturn's upper troposphere is presented. The ring–modulated latitudinal dependence of the insolation, the ring thermal emission, the oblateness of the planet, the orbit eccentricity, and the latitudinal variation of the internal heat flux are taken into account. Calculations agree closely with the temperature–latitude profiles retrieved from Voyager IRIS measurements at atmospheric levels located above the 0.2–bar pressure level. The model fails to predict the retrieved temperature–latitude profiles below the 0.3–bar level. The poor correlation between infrared measurements and visible images is discussed and a possible model of Saturn's cloud structure is proposed.

100.031 **Evidence for material between Saturn's A and F rings from the Voyager 2 Photopolarimeter Experiment.**
A. L. Graps, A. L. Lane, L. J. Horn, K. E. Simmons.
Icarus, Vol. 60, No. 2, p. 409 – 415 (1984).

The region in the Saturn system between the F ring and the outer edge of the A ring is an area that appears, in images from the imaging experiment, to be virtually devoid of material except for three small satellites. Near the orbit of 1980S28, Atlas – the innermost satellite – the Voyager Photopolarimeter Stellar Occultation data show a discontinuity in count rate which marks a boundary between the tenuous material near the outer edge of the A ring and the orbit of Atlas. The data pertaining to this region have been examined with the aid of statistics and models generated from other similar ring structures. It is concluded that the discontinuity is real, implying the existence of tenuous material of normal optical depth of 0.01 to 0.006 in this region.

100.032 **Chaotic spinning of Hyperion?**
C. D. Murray.
Nature, Vol. 311, No. 5988, p. 705 (1984).

100.033 **New model of Saturn's ionosphere with an influx of water from the rings.**
J. E. P. Connerney, J. H. Waite.
Nature, Vol. 312, No. 5990, p. 136 – 138 (1984).

The authors propose a model of Saturn's ionosphere in which water plays a major role as a minor constituent present by downward diffusion from an external source. This model Saturn ionosphere is a classical "F2" type layer resulting from the photodissociative production of $H^+$ from $H_2$ and rapid chemical loss by a series of charge exchange reactions with water. A planet-wide influx of $\sim 4 \times 10^7$ molecules $cm^{-2}s^{-1}$ of water from the rings is consistent with the observed ionospheric electron densities. An enhanced influx of water occurs at latitudes (−38°, +44°) connected magnetically to the inner edge of Saturn's B ring.

100.034 **Production of ammonia-depleted surface layers on the Saturnian satellites by ion sputtering.**
L. J. Lanzerotti, W. L. Brown, K. J. Marcantonio, R. E. Johnson.
Nature, Vol. 312, No. 5990, p. 139 – 140 (1984).

Theories of surface processes on the Saturnian satellites Enceladus, Tethys, Dione, Rhea predict that the peritectic melt $2H_2O \cdot NH_3$ should be abundantly present on their surfaces. Spectra, however, have not shown any evidence of ammonia on the surfaces of these satellites. On the basis of laboratory-derived data on the ion erosion of water and ammonia ices, the authors suggest that the satellite surfaces could have become enriched with water ice over a reasonable geological time period ($10^6 - 10^9$ yr) owing to preferential erosion of the ammonia by magnetosphere ion bombardment.

100.035 **Possible detection of a ring system in Saturn's magnetosphere.**
R. Vasundhara, J. C. Bhattacharyya, P. Santhanam, A. K. Pande, V. Mohan, H. S. Mahra.
Nature, Vol. 312, No. 5995, p. 621 – 623 (1984).

The region of the magnetosphere of Saturn around 14 radii and 19 radii has been of special interest since the detection of anomalous dips in the low-energy plasma ion-density in these regions by the Voyager 1, Voyager 2 and Pioneer 11 spacecrafts. The authors have observed the occultation of the star SAO158913 from two stations in India following predictions by Mink. This star showed symmetric dips in the light curves on 24 March 1984 (immersion event) and on 25 March 1984 (emersion event), suggesting the possible presence of particulate matter.

100.036 **The composition and structure of the Saturnian stratosphere as determined from UV observations.**
V. Moore, G. E. Hunt.
Bull. Am. Astron. Soc., Vol. 16, No. 3, Part I, p. 650 (1984). Abstract. – See Abstr. 010.061.

100.037 **Structure of the midlatitude ribbon feature on Saturn from Voyager images.**
D. A. Godfrey, G. E. Hunt.
Bull. Am. Astron. Soc., Vol. 16, No. 3, Part I, p. 650 (1984). Abstract. – See Abstr. 010.061.

100.038 **Scattering properties and vertical distribution of Saturn stratospheric hazes.**
K. A. Rages, J. B. Pollack.
Bull. Am. Astron. Soc., Vol. 16, No. 3, Part I, p. 650 (1984). Abstract. – See Abstr. 010.061.

100.039 **Saturn's equatorial radius and oblateness: atmospheric data acquired with spacecraft occultation experiments.**
G. Lindal, D. Sweetnam, V. Eshleman.
Bull. Am. Astron. Soc., Vol. 16, No. 3, Part I, p. 650 (1984). Abstract. – See Abstr. 010.061.

100.040 **A seasonal climate model of Saturn's stratosphere.**
B. Bezard, D. Gautier.
Bull. Am. Astron. Soc., Vol. 16, No. 3, Part I, p. 650 (1984). Abstract. – See Abstr. 010.061.

100.041 **Sputtering of the icy satellites of Saturn: a source of a neutral torus and magnetospheric plasma.**
R. E. Johnson, E. M. Sieveksa, G. Taylor, L. J. Lanzerotti.
Bull. Am. Astron. Soc., Vol. 16, No. 3, Part I, p. 661 (1984). Abstract. – See Abstr. 010.061.

100.042 **Voyager imaging observations of Titan's atmosphere: I. Disk-resolved photometric properties.**
S. W. Squyres, W. R. Thompson, C. Sagan.
Bull. Am. Astron. Soc., Vol. 16, No. 3, Part I, p. 664 (1984). Abstract. – See Abstr. 010.061.

100.043 **Voyager imaging observations of Titan's atmosphere: II. Variation of particle properties with altitude and latitude.**
W. R. Thompson, S. W. Squyres, C. Sagan.
Bull. Am. Astron. Soc., Vol. 16, No. 3, Part I, p. 664 (1984). Abstract. – See Abstr. 010.061.

100.044 **Vertical and latitudinal distributions of HCN, $HC_3N$ and $C_2N_2$ in Titan's atmosphere.**
S. J. Kim, V. G. Kunde, R. E. Samuelson, D. E. Jennings, Y. L. Yung, M. Allen.
Bull. Am. Astron. Soc., Vol. 16, No. 3, Part I, p. 664 (1984). Abstract. – See Abstr. 010.061.

100.045 **Bolometric geometric albedos of Titan, Uranus, and Neptune.**
J. S. Neff, T. A. Ellis, J. Apt, J. T. Bergstralh.
Bull. Am. Astron. Soc., Vol. 16, No. 3, Part I, p. 665 (1984). Abstract. – See Abstr. 010.061.

100.046 **Temporal variability of the albedo of Titan.**
G. W. Lockwood, B. L. Lutz, D. T. Thompson.
Bull. Am. Astron. Soc., Vol. 16, No. 3, Part I, p. 665 (1984). Abstract. – See Abstr. 010.061.

100.047 **The role of scattering in the infra-red on the heat budget of Titan's lower atmosphere.**
C. P. McKay, J. B. Pollack.
Bull. Am. Astron. Soc., Vol. 16, No. 3, Part I, p. 665 (1984). Abstract. – See Abstr. 010.061.

100.048 **A partially collisional model of the Titan hydrogen torus.**
D. A. Hilton, D. M. Hunten.
Bull. Am. Astron. Soc., Vol. 16, No. 3, Part I, p. 665 – 666 (1984). Abstract. – See Abstr. 010.061.

100.049 **Oceans in the outer solar system.**
M. L. Delitsky.
Bull. Am. Astron. Soc., Vol. 16, No. 3, Part I, p. 666 (1984). Abstract. – See Abstr. 010.061.

100.050 **Classification of Saturn's ring features from radio occultation measurements.**
G. L. Tyler, P. A. Rosen, E. A. Marouf.
Bull. Am. Astron. Soc., Vol. 16, No. 3, Part I, p. 675 (1984). Abstract. – See Abstr. 010.061.

100.051 **Constraints on composition of Saturn ring particles from knowledge of particle size distribution functions.**
H. A. Zebker, G. L. Tyler.
Bull. Am. Astron. Soc., Vol. 16, No. 3, Part I, p. 675 – 676 (1984). Abstract. – See Abstr. 010.061.

100.052 **Photometry of spokes in Saturn's rings.**
R. E. Eplee Jr., B. A. Smith.
Bull. Am. Astron. Soc., Vol. 16, No. 3, Part I, p. 676 (1984). Abstract. – See Abstr. 010.061.

100.053 **New measurements of spoke activity.**
E. Grün, G. W. Garneau, R. J. Terrile, T. V. Johnson, G. E. Morfill.
Bull. Am. Astron. Soc., Vol. 16, No. 3, Part I, p. 676 (1984). Abstract. – See Abstr. 010.061.

100.054 **On the maximum size of a satellite at Saturn's G ring.**
J. B. Blake, W. Fillius.
Bull. Am. Astron. Soc., Vol. 16, No. 3, Part I, p. 676 (1984). Abstract. – See Abstr. 010.061.

100.055 **Evidence from the Voyager 2 photopolarimeter for material outside of the Saturn F ring at 2.33 Rs.**
A. L. Graps, A. L. Lane, K. E. Simmons.
Bull. Am. Astron. Soc., Vol. 16, No. 3, Part I, p. 677 (1984). Abstract. – See Abstr. 010.061.

100.056 **Low optical depth features in Saturn's D ring.**
L. J. Horn, A. L. Lane, J. N. Cuzzi.
Bull. Am. Astron. Soc., Vol. 16, No. 3, Part I, p. 677 (1984). Abstract. – See Abstr. 010.061.

100.057 **Bending waves and the structure of Saturn's B ring.**
J. J. Lissauer.
Bull. Am. Astron. Soc., Vol. 16, No. 3, Part I, p. 677 (1984). Abstract. – See Abstr. 010.061.

100.058 **Features in Saturn's rings.**
L. W. Esposito, C. Harris.
Bull. Am. Astron. Soc., Vol. 16, No. 3, Part I, p. 677 (1984). Abstract. – See Abstr. 010.061.

100.059 **Microwave emission from Titan and the Galilean satellites.**
D. O. Muhleman, G. L. Berge, D. J. Rudy.
Bull. Am. Astron. Soc., Vol. 16, No. 3, Part I, p. 686 (1984). Abstract. – See Abstr. 010.061.

100.060 **Do current observations support the hypothesis of chaotic rotation for Hyperion?**
S. J. Peale, J. Wisdom.
Bull. Am. Astron. Soc., Vol. 16, No. 3, Part I, p. 686 – 687 (1984). Abstract. – See Abstr. 010.061.

100.061 **Masses of Janus and Epimetheus.**
C. F. Yoder, S. P. Synnott.
Bull. Am. Astron. Soc., Vol. 16, No. 3, Part I, p. 687 (1984). Abstract. – See Abstr. 010.061.

100.062 **Photometry of Titan, Uranus, and Neptune: lightcurves through solar maximum.**
D. T. Thompson, G. W. Lockwood.
Bull. Am. Astron. Soc., Vol. 16, No. 3, Part I, p. 705 – 706 (1984). Abstract. – See Abstr. 010.061.

100.063 **An identification of $N_2$–$H_2$ dimers in the atmosphere of Titan.**
K. Fox, S. Kim.
Bull. Am. Astron. Soc., Vol. 16, No. 3, Part I, p. 706 (1984). Abstract. – See Abstr. 010.061.

100.064 **The light curve of Hyperion.**
J. Wisdom, S. J. Peale.
Bull. Am. Astron. Soc., Vol. 16, No. 3, Part I, p. 707 (1984). Abstract. – See Abstr. 010.061.

100.065 **Rings on parade – a Saturn ring atlas preview.**
J. Diner, S. A. Collins, G. W. Garneau, A. L. Lane, E. D. Miner, S. P. Synnott, R. J. Terrile, J. Holberg, B. A. Smith, G. L. Tyler.
Bull. Am. Astron. Soc., Vol. 16, No. 3, Part I, p. 709 – 710 (1984). Abstract. – See Abstr. 010.061.

100.066 **The distribution of atomic hydrogen in Saturn's magnetosphere.**
D. E. Shemansky, G. R. Smith, W. H. Smyth, M. R. Combi.
Bull. Am. Astron. Soc., Vol. 16, No. 3, Part I, p. 712 (1984). Abstract. – See Abstr. 010.061.

100.067 **Numerical simulation of light curves of a chaotically rotating Hyperion.**
S. J. Peale, J. Wisdom.
Bull. Am. Astron. Soc., Vol. 16, No. 3, Part I, p. 722 (1984). Abstract. – See Abstr. 010.064.

100.068 **A comparison of the theory of the motion of Hyperion with observations made during 1967 – 1982.**
D. B. Taylor.
Astron. Astrophys., Vol. 141, No. 1, p. 151 – 158 (1984).
A model of the orbital motion of Hyperion is developed from Woltjer's theory. The orbital model includes the short-period perturbations due to the action of Titan which is assumed to move in a circular orbit. The orbital elements are referred to a standard reference frame B1950.0. The model is fitted to photographic observations from 1967 to 1982, and improved orbital elements are derived. In addition, a new determination is made of the amplitudes of the libration term in the mean longitude, and of the long-period terms in the mean longitude, longitude of pericentre and eccentricity. The theories of Rhea, Dione, Titan and Iapetus are also fitted to observations over the same period, and give a considerably better fit than is found for Hyperion. Hence it is concluded Woltjer's theory of Hyperion is not completely adequate. A brief discussion of possible improvements to the model is given.

100.069 **Ring torque of Saturn from interplanetary meteoroid impact.**
W.-H. Ip.
Icarus, Vol. 60, No. 3, p. 547 – 552 (1984).
Reevaluation of the interplanetary meteoroid mass flux at 10 AU obtains a value of $M \approx 6 \times 10^4 \mathrm{g\ sec}^{-1}$ for the meteoroid mass loading rate to the rings of Saturn. This meteoroid impact flux suggests that a large change to the configuration of the ring system could occur in a relatively short time ($\lesssim 10^9$ years). This new element thus should be taken into consideration in discussion of the dynamical evolution of the rings.

100.070 **The millimeter spectrum of Titan: detectability of HCN, $HC_3N$, and $CH_3CN$ and the CO abundance.**
G. Paubert, D. Gautier, R. Courtin.
Icarus, Vol. 60, No. 3, p. 599 – 612 (1984).
The flux emitted by Titan's disk in millimeter lines of HCN, $HC_3N$, $CH_3CN$, and CO is calculated by means of a radiative transfer formulation which takes into account the sphericity of the atmosphere. It is demonstrated that the plane-parallel approximation for radiative transfer is no longer valid, especially in the core of emission lines, when Titan is not spatially resolved. The observability of (1–0) and (2–1) transitions of CO, of (1–0), (2–1), and (3–2) transitions of HCN is investigated.

100.071 **The rotation of Hyperion.**
S. J. Peale.
Philos. Trans. R. Soc. London, Ser. A, Vol. 313, No. 1524, p. 147 – 156 (1984). – See Abstr. 012.068.
For almost the entire range of dimensions of Hyperion allowed by uncertainties in the observations, the satellite cannot librate stably about a rotation rate which is synchronous with its orbital mean motion. Rather, the large gravitational torques on the asymmetric satellite coupled with the large eccentricity forced by the orbital resonance with Titan cause Hyperion to tumble in a random manner. Large changes in orientation of body axes relative to inertial space and in the instantaneous spin rate occur on timescales of the order of the orbit period. Numerical evaluation of the exponential divergence of nearby trajectories in the phase space of the motion verifies that the tumbling is truly chaotic. This newly defined state of chaotic rotation for Hyperion is likely to be the only example of confined, continuously observable chaotic motion in the solar system.

100.072 **Saturn 1980 – 1981.**
A. W. Heath.
J. Br. Astron. Assoc., Vol. 95, No. 1, p. 1 – 7 (1984).

100.073 **Reflection capactiy of the Saturn satellites Dione and Rhea.**
V. V. Avramchuk, L. R. Lisina.
Astron. Vestn., Tom 18, No. 2, p. 128 – 136 (1984). In Russian. Abstr. in Ref. Zh., 51. Astron., 11.51.216 (1984).

100.074 **The E ring of Saturn and satellite Enceladus.**
K. D. Pang, C. C. Voge, J. W. Rhoads, J. M. Ajello.
J. Geophys. Res., Vol. 89, No. B11, p. 9459 – 9470 (1984).
Saturn's outermost (E) ring appears to be closely related to its satellite Enceladus. The radial optical profile of the E ring and systematic trends in the albedos of the seven Saturnian satellites residing within the E ring suggest that Enceladus bears a generative relationship to the E ring. The authors compared the relationship between Enceladus and the E ring with that between Io and Io's torus and discovered many parallels, which are useful in understanding outer solar system volcanism and ring formation.

100.075 **Saturn central meridian ephemeris: 1985.**
J. E. Westfall.
Strolling Astron., Vol. 30, Nos. 11 – 12, p. 247 – 249 (1984).

100.076 **The resonance argument of the Mimas–Tethys system.**
I. G. Chugunov, G. V. Stolyarov, V. V. Stolyarov.
Astron. Tsirk., No. 1326, p. 5 – 6 (1984). In Russian.

100.077 **Seasonal variation of the methane absorption in the Saturn atmosphere.**
A. P. Vid'machenko.
Astron. Tsirk., No. 1331, p. 1 – 2 (1984). In Russian.

100.078 **On the question of the resonance structure of the VI, VII Saturn satellites.**
I. G. Chugunov, G. V. Stolyarov.
Astron. Tsirk., No. 1333, p. 5 – 6 (1984). In Russian.

100.079 **The albedo of the southern equatorial region of Saturn in 1977 – 1981.**
A. P. Vid'machenko.
Astron. Vestn., Tom 18, No. 3, p. 191 – 198 (1984). In Russian. Abstr. in Ref. Zh., 51. Astron., 12.51.285 (1984).

100.080 **Formation of the multistage strato–shaped cloudiness in the Saturn atmosphere.**
K. Yu. Ibragimov, G. A. Kirienko, A. A. Solodovnik.
Astrofiz. inst. Akad. Nauk KazSSR Alma–Ata, 33 pp. (1984). In Russian. Abstr. in Ref. Zh., 51. Astron., 12.51.287 (1984).

100.081 **A review of radar observations of Saturn's rings.**
S. J. Ostro, G. H. Pettengill.
Planetary rings, p. 49 – 55 (1984). – See Abstr. 012.085 (IAU Colloq. No. 75).
The authors review radar observations of Saturn's rings, obtained during the past decade by using the λ3.5–cm and λ12.6–cm systems at the JPL's Goldstone Tracking Station and the λ12.6–cm system at the National Astronomy and Ionosphere Center's Arecibo Observatory.

100.082 **Microwave emission from Saturn's rings.**
D. O. Muhleman, G. L. Berge.
Planetary rings, p. 57 – 70 (1984). – See Abstr. 012.085 (IAU Colloq. No. 75).
Passive radio measurements of Saturn's rings are reviewed and new interferometric measurements at $\lambda = 2.7$ mm are presented. The brightness temperatures of the A plus B rings at a ring angle of $B \cong 10°$ were found to be 17K and 38K in two separate experiments, the latter being the more reliable. These results are interpreted in terms of ring particle emission and scattering of the planet's disk emission by the rings.

100.083 **Saturn's rings resolved by the VLA.**
I. de Pater, J. R. Dickel.
Planetary rings, p. 71 – 74 (1984). – See Abstr. 012.085 (IAU Colloq. No. 75).
High resolution radio data of Saturn have been obtained at 1.3, 2, 6 and 21 cm, at different inclination angles of the ring plane. Preliminary results on optical depth measurements in the rings are described.

100.084 **Apparent thickness of Saturn's rings from the observations of the ring plane crossing.**
B. Sicardy, P. Laques, J. Lecacheux, A. Brahic, G. Wlerick.
Planetary rings, p. 75 – 86 (1984). – See Abstr. 012.085 (IAU Colloq. No. 75).
The reductions of electronographic plates taken during ground–based observations of Saturn at the times of the transits of the Sun and the Earth through the ring plane (March 1980) are interpreted. From plates taken near Earth transit an equivalent thickness of 1.1 km has been determined for the rings.

100.085 **Azimuthal variations in Saturn's A ring.**
K. Lumme, W. M. Irvine.
Planetary rings, p. 87 – 91 (1984). – See Abstr. 012.085 (IAU Colloq. No. 75).
The observations defining the azimuthal brightness variations in ring A are reviewed. This phenomenon is entirely separate from the "spokes" which are present in ring B. An explanation in terms of scattering by "density wakes" produced by large ring particles seems most reasonable.

100.086 **A model containing "blobs" for the A ring of Saturn.**
W. Thompson, D. van Blerkom, W. Irvine.
Planetary rings, p. 93 – 98 (1984). – See Abstr. 012.085 (IAU Colloq. No. 75).
Observations of the brightness distribution of Saturn's ring A from four separate years, including specifically the azimuthal variations, are calibrated and interpreted through the use of a scattering model. In this model the A ring is composed of elongated density enhancements ("blobs") embedded in a ring of uniform density.

100.087 **Saturn's E–ring in the red spectral region.**
W. A. Feibelman.
Planetary rings, p. 99 – 102 (1984). – See Abstr. 012.085 (IAU Colloq. No. 75).
Observations of the 1966 edge–on presentation of Saturn's E–ring in the red spectral region are presented. Techniques to suppress the glare from the planet are briefly described. The E–ring profile did not exhibit a maximum near the orbit of Enceladus in 1966.

100.088 **Saturn's E ring.**
W. A. Baum, T. J. Kreidl, L. H. Wasserman.
Planetary rings, p. 103 – 109 (1984). – See Abstr. 012.085 (IAU Colloq. No. 75).
The spatial distribution of material in Saturn's E ring has been derived from ground–based CCD observations obtained at the time of the 1980 edge–on presentation. The ring commences abruptly near 3 $R_S$, peaks at the orbit of Enceladus (3.94 $R_S$), and has tenuous outskirts to more than 8 $R_S$. On reasonable assumptions, the authors calculate the attenuation factor to be $5 \times 10^{-11}$ km$^{-1}$ in the equatorial plane near Enceladus, and the physical thickness of the ring there to be 7500 km. If the E–ring particles are predominantly icy and small, the total mass of the ring is probably less than $10^6$tons.

100.089 **Summary of optical groundbased E ring observations at the University of Arizona.**
S. Larson.
Planetary rings, p. 111 – 113 (1984). – See Abstr. 012.085 (IAU Colloq. No. 75).
The application and refinements of coronagraphic imaging techniques and new detectors have permitted observations of the E ring of Saturn with the University of Arizona 154 cm telescope since the ring plane crossing in 1979. The measurements, briefly summarized here, are derived from photographic and CCD imaging data obtained over 6 nights in 1980, and 7 nights in 1981.

100.090 **Photometric study of Saturn's E ring.**
P. Laques, B. Sicardy, A. Brahic, J. Lecacheux.
Planetary rings, p. 115 – 120 (1984). – See Abstr. 012.085 (IAU Colloq. No. 75).

Observations of Saturn's E ring at the Pic du Midi Observatory with an electronographic camera are presented. Faint material has been observed up to Dione's orbit with a maximum of brightness near the Enceladus orbit.

100.091 **The Saturn ring particles from optical reflectance polarimetry.**
A. Dollfus.
Planetary rings, p. 121 – 145 (1984). – See Abstr. 012.085 (IAU Colloq. No. 75).

The light reflected by the Saturn rings is polarized; the degree and azimuth of the polarization varies with the phase angle, the wavelength and the position on the ring. This paper presents an extensive review of the polarization data for rings A and B and examines the methods applied in deriving physical parameters for the scattering particles.

100.092 **Saturn rings: divisions, gaps and ringlets from ground-based telescopes.**
A. Dollfus.
Planetary rings, p. 147 – 154 (1984). – See Abstr. 012.085 (IAU Colloq. No. 75).

High magnification visual telescopic observations of Saturn's rings exhibit divisions, gaps and bright sub-rings. Some gaps and concentric bright or dark sub-rings are phase angle dependent; the steep luminance peaks of their light curves around zero phase angle are volume-density dependent (opposition effect).

100.093 **Physical properties of Saturn's rings.**
J. N. Cuzzi.
Planetary rings, p. 211 – 217 (1984). – See Abstr. 012.085 (IAU Colloq. No. 75).

A brief review of the important features of Saturn's ring system as revealed by the Voyager encounters is presented.

100.094 **The kinematics of spokes.**
C. C. Porco, G. E. Danielson.
Planetary rings, p. 219 – 222 (1984). – See Abstr. 012.085 (IAU Colloq. No. 75).

A short summary of a study of the long term behavior of spokes in Saturn's rings is presented. The results are based on Voyager 2 data covering an interval of 90 Saturnian rotations.

100.095 **Dynamics of spokes in Saturn's B ring.**
R. E. Eplee Jr., B. A. Smith.
Planetary rings, p. 223 – 224 (1984). Abstract. – See Abstr. 012.085 (IAU Colloq. No. 75).

100.096 **Unusual imaging observations of Saturn's rings.**
R. J. Terrile.
Planetary rings, p. 225 – 227 (1984). Abstract. – See Abstr. 012.085 (IAU Colloq. No. 75).

100.097 **Structure and dynamics of Saturn's rings as seen by the Voyager stellar occultation.**
L. W. Esposito.
Planetary rings, p. 229 – 250 (1984). – See Abstr. 012.085 (IAU Colloq. No. 75).

The stellar occultation of δ Scorpii by Saturn's rings shows radial structure in Saturn's rings down to the resolution element of ~100 m. Density waves, embedded moonlets, and other proposed mechanisms are insufficient to explain the observed structure. A limit of less than 100 density waves are likely to be found in the rings in the current data. Unresolved waves do not seem to play a major role, either. A detailed search and histogram of optical depth in the outer B ring show no evidence of embedded moonlets. The mass of Saturn's rings is $5(\pm 3) \times 10^{-8}$ of Saturn's mass. The vertical thickness of the rings is probably about 50 m.

100.098 **The C ring of Saturn: a high resolution view of some of its structure.**
A. L. Lane, A. L. Graps, K. E. Simmons.
Planetary rings, p. 251 – 258 (1984). – See Abstr. 012.085 (IAU Colloq. No. 75).

The high resolution examination of selected features in Saturn's C ring has demonstrated that they are far more complex than previously considered. This brief paper illustrates several classes of the phenomena observed by the Voyager photopolarimeter star occultation experiment.

100.099 **The eccentric ringlet at 1.29 $R_S$.**
C. Porco, N. Borderies, G. E. Danielson, P. Goldreich, J. B. Holberg, A. L. Lane, P. D. Nicholson.
Planetary rings, p. 259 – 262 (1984). – See Abstr. 012.085 (IAU Colloq. No. 75).

The authors present preliminary results of an examination of the eccentric ringlet near 1.29 $R_s$ in Saturn's inner C ring. Situated near the Titan 1:0 apsidal resonance, this feature provides an opportunity to study the behavior of a ring under the influence of both the gravity field of a major planet and the potential of an external satellite.

100.100 **Voyager UVS observations of Saturn's rings and the ring atmosphere.**
J. B. Holberg.
Planetary rings, p. 263 (1984). Abstract. – See Abstr. 012.085 (IAU Colloq. No. 75).

100.101 **Voyager UVS observations of stellar occultations by the rings of Saturn.**
J. B. Holberg, W. T. Forrester.
Planetary rings, p. 265 – 277 (1984). – See Abstr. 012.085 (IAU Colloq. No. 75).

During the Voyager 1 and 2 Saturn encounters the ultraviolet spectrometers observed three separate stellar occultations by Saturn's rings. These observations can be used to establish a highly accurate distance scale allowing the identification of numerous ring features associated with resonances due to exterior satellites. Three separate observations of an eccentric ringlet near the location of the Titan apsidal resonance are discussed along with other ringlet-resonance associations occurring in the C ring. Density waves occurring in the A and B rings are reviewed and a detailed discussion of one of these features is presented.

100.102 **Structure of Saturn's rings from Voyager radio occultation measurements.**
G. L. Tyler, E. A. Marouf, V. R. Eshleman.
Planetary rings, p. 279 (1984). Abstract. – See Abstr. 012.085 (IAU Colloq. No. 75).

100.103 **Micron-size particles detected near Saturn by the Voyager plasma wave instrument.**
D. A. Gurnett, E. Grün, D. Gallagher, W. S. Kurth, F. L. Scarf.
Planetary rings, p. 287 – 288 (1984). Abstract. – See Abstr. 012.085 (IAU Colloq. No. 75).

100.104 **Voyager planetary radio astronomy: grain and particle impacts shot noise in Saturn's ring plane.**
M. G. Aubier, N. Meyer-Vernet, B. M. Pedersen.
Planetary rings, p. 289 – 297 (1984). – See Abstr. 012.085 (IAU Colloq. No. 75).

Radio emissions were observed by Voyager 1 and Voyager 2 planetary radio astronomy experiments during Saturnian ring plane crossings. Several mechanisms are proposed and tested. For the strong noise in the ring plane, observed by Voyager 2 at ~2.9 $R_S$, one must take into account shot noise to grain impacts. A quantitative evaluation of this process gives an estimate of the dust size ~1.1 μm close to the G ring.

100.105 **Saturn electrostatic discharges.**
J. W. Warwick, J. H. Romig, D. R. Evans.
Planetary rings, p. 299 – 306 (1984). – See Abstr. 012.085 (IAU Colloq. No. 75).

Electrostatic discharges were recorded by the Planetary Radio Astronomy instrument aboard both Voyager 1 and Voyager 2 at Saturn closest approach. This paper discusses possible locations and mechanisms of these discharges. An origin in the planet's ring system and involving small satellites is favored.

100.106 **Statistical study of Saturnian electrostatic discharges observed by the Voyager Planetary Radio Astronomy experiment.**
P. Zarka, B. M. Pedersen.
Planetary rings, p. 307 – 317 (1984). – See Abstr. 012.085 (IAU Colloq. No. 75).

During the two Voyager–Saturn missions, the Planetary Radio Astronomy experiment observed sporadic broadband (20 kHz to at least 40 MHz) emissions during a few days around the closest approaches. These emissions are interpreted to arise from short-lived electrostatic discharges in the vicinity of Saturn. Their characteristics have been studied for the experiment's high band (>1.2 MHz).

100.107 **An Earth–based attempt to detect S.E.D. (*Saturn Electric Discharges*) at 21 centimeters.**
A. Lecacheux, F. Biraud.
Planetary rings, p. 319 – 324 (1984). – See Abstr. 012.085 (IAU Colloq. No. 75).

The authors tried to observe Saturn Electrostatic Discharges at 1420 MHz using the radiotelescope at Nançay, first during Saturn 1981 opposition, then during Voyager 2 fly–by. The obtained flux upper limit indicates a spectrum steeper than $f^{-1}$.

100.108 **Resonance identifications in Saturn's rings.**
W. E. Wiesel, F. A. Franklin.
Planetary rings, p. 349 – 359 (1984). – See Abstr. 012.085 (IAU Colloq. No. 75).

When Saturn's oblateness perturbations are included, a single classical resonance splits into a resonance band structure. The authors have derived expressions for resonance location, libration region width, and the maximum librator range. These predictions are compared to areas of the C and B rings, the Cassini division, and the F ring.

100.109 **A simple model of Saturn's rings – revisited.**
M. Henon.
Planetary rings, p. 363 – 384 (1984). – See Abstr. 012.085 (IAU Colloq. No. 75).

The author describes a crude model of the rings based on a power–law distribution for particle size. The parameters of the model are determined from apparent thickness and radar–radio observations. A number of quantities are derived: mass of the rings, mean surface density, size of largest particle, time scale for dynamical evolution, minimal size of gap–forming particles, number and size distribution of the gaps, kinematic viscosity. Comparisons with observations meet with mixed success. The most serious problem lies in the derived time scale for the radial spreading of the disk, which is less than 10 million years.

100.110 **Viscosity in Saturn's rings.**
J. J. Lissauer, F. H. Shu, J. N. Cuzzi.
Planetary rings, p. 385 – 392 (1984). – See Abstr. 012.085 (IAU Colloq. No. 75).

The authors discuss the technique of estimating the viscosity in Saturn's rings from the damping rate of waves observed to be propagating within the rings. They present a new method that considers the damping of spiral bending waves, a diagnostic which is largely free of major complications. The authors deduce a kinematic viscosity of 260 (+150, –100) $cm^2/s$ for the middle of the A ring where bending waves have been excited by the 5:3 vertical resonance with Mimas.

100.111 **Structure and dynamical processes in Saturn's rings.**
R. J. Terrile.
Planetary rings, p. 393 – 395 (1984). Abstract. – See Abstr. 012.085 (IAU Colloq. No. 75).

100.112 **Physical and geological nature of Saturn's larger ring particles.**
C. R. Chapman, R. Greenberg, S. Weidenschilling, D. R. Davis.
Planetary rings, p. 403 – 406 (1984). – See Abstr. 012.085 (IAU Colloq. No. 75).

The evolution and behavior of Saturn's rings may be influenced by relatively large ring particles. The authors describe how an important mass fraction of the rings may reside in particles with dimensions of tens of meters to kilometers, which are too few to be seen by radio occultation and too small to be detected in Voyager pictures. The possible physical and geological structure of such bodies are considered.

100.113 **Effects of shepherd conjunctions on Saturn's F–ring.**
M. R. Showalter.
Planetary rings, p. 423 – 429 (1984). – See Abstr. 012.085 (IAU Colloq. No. 75).

Synnott (1982) has reported that the F–ring braids are often found near the ring segment which was between the shepherds at their last conjunction. The author studies the possible perturbing effects of such conjunctions on the ring, and finds that they are much too small to directly account for the braids.

100.114 **Orbital elements of Saturn's F ring and its shepherding satellites.**
S. P. Synnott.
Planetary rings, p. 459 – 460 (1984). Abstract. – See Abstr. 012.085 (IAU Colloq. No. 75).

100.115 **Dense but "collisionless" orbiting rings.**
F. C. Michel.
Planetary rings, p. 471 – 478 (1984). – See Abstr. 012.085 (IAU Colloq. No. 75).

The discovery by Voyager, that the rings of Saturn are substructured into a multitude of narrow ringlets, has posed some severe theoretical difficulties. These rings have substantial optical depths, implying high collision rates and therefore rapid diffusive spreading, which would lead to the ringlet structure being eradicated. The author proposes that convective motion, a helical rolling motion of a ringlet as it orbits, provides a stabilizing influence. For ringlets of negligible mass orbiting a point mass, the ringlets can be arbitrarily dense without collisions acting to dissipate them.

100.116 **Instabilities and the fine structure of Saturn's rings.**
M. S. Bobrov, A. M. Fridman, V. L. Polyachenko.
Planetary rings, p. 499 – 510 (1984). – See Abstr. 012.085 (IAU Colloq. No. 75).

100.117 **Spoke electrodynamics.**
A. Acterberg, R. Blandford, P. Goldreich.
Planetary rings, p. 549 – 550 (1984). Abstract. – See Abstr. 012.085 (IAU Colloq. No. 75).

100.118 **The formation of spokes in Saturn's rings.**
G. E. Morfill.
Planetary rings, p. 551 – 568 (1984). – See Abstr. 012.085 (IAU Colloq. No. 75).

An electromagnetic model for the origin of spokes in Saturn's rings is presented. The model is based on the evolution of small (~100 km) plasma clouds, interacting both with the magnetosphere and small dust particles in the rings.

100.119 **On the role of mother–planet magnetic field in spiral structure explanation of formation of spokes.**
G. X. Song.
Planetary rings, p. 569 – 574 (1984). – See Abstr. 012.085 (IAU Colloq. No. 75).

A mechanism for the formation of spokes in Saturn's B ring is proposed, which is based on the existence of a spiral sector structure in the planet's equatorial magnetic field.

100.120 **Saturn's E, G and F rings: modulated by the plasma sheet.**
E. Grün, G. E. Morfill, T. V. Johnson, G. H. Schwehm.
Planetary rings, p. 597 (1984). Abstract. – See Abstr. 012.085 (IAU Colloq. No. 75).

100.121 **The E–ring: interactions with plasma and implications for its evolution.**
T. V. Johnson, G. E. Morfill, E. Grün.
Planetary rings, p. 599 – 602 (1984). Abstract. – See Abstr. 012.085 (IAU Colloq. No. 75).

100.122 **The E–ring of Saturn.**
R. Smoluchowski.
Planetary rings, p. 603 – 606 (1984). – See Abstr. 012.085 (IAU Colloq. No. 75).

It is proposed that the E–ring of Saturn is formed as a consequence of micrometeoritic and meteoritic bombardment of Enceladus. It is shown that the mass reaching the ring replaces that lost by erosion, drag and other mechanisms. The effect of the presence of amorphous ice and the possible opposition effect is indicated.

100.123 **Macrostructure and microphysics of Saturn's E–ring.**
K. D. Pang, C. C. Voge, J. W. Rhoads.
Planetary rings, p. 607 – 613 (1984). – See Abstr. 012.085 (IAU Colloq. No. 75).

All observed optical and infrared properties of Saturn's E–ring can be explained in terms of Mie scattering by a narrow size distribution of ice spheres of 2 – 2.5 micron diameter. The spherical shape of the ring particles and their narrow size distribution imply a molten (possibly volcanic) origin on Enceladus. The E–ring consists of many layers, possibly stratified by electrostatic levitation.

100.124 **Multivalued electrostatic potentials; "chance or necessity" for the charge of solid particles near Saturn?**
N. Meyer–Vernet.
Planetary rings, p. 615 – 625 (1984). – See Abstr. 012.085 (IAU Colloq. No. 75).

It is shown that secondary emission plays a critical role in the electrostatic potential of solid grains in the environments of the outer planets, and particularly in Saturn's ring system. The paper investigates the charge's response to a slight variation of ambient parameters near Saturn (flux of suprathermal electrons; passage from shadow to sunlit regions). The importance of the process is evaluated for different Saturnian environments.

100.125 **On the heterogeneous appearance of Saturn's rings at high optical resolution.**
A. F. Cook, F. A. Franklin, E. M. Shoemaker, R. J. Terrile.
Planetary rings, p. 691 (1984). Abstract. – See Abstr. 012.085 (IAU Colloq. No. 75).

100.126 **Resonance structure in the early evolution of the Saturn system and the location of the inner satellites.**
M. Torbett, R. Smoluchowski.
Planetary rings, p. 693 (1984). Abstract. – See Abstr. 012.085 (IAU Colloq. No. 75).

100.127 **Saturn's rings: properties and processes.**
J. N. Cuzzi, J. J. Lissauer, L. W. Esposito, J. B. Holberg, E. A. Marouf, G. L. Tyler, A. Boischot.
Planetary rings, p. 73 – 199 (1984). – See Abstr. 012.086.

The structural and particle properties of Saturn's rings are described, with particular emphasis on spacecraft observations. Properties are discussed generically rather than regionally, and the authors attempt in all cases to relate observed properties to favored causative processes. The ring particles are primarily icy. Most of the particles are in the 1 cm to 5 m radius range, but there is reason to suspect the existence of some particles of all sizes up to 10 km in radius. Ring structure is, in general, dominated by gravitational and collisional dynamics. Orbital resonances with various satellites drive spiral density and bending waves, and define the sharp outer edges of the A and B Rings. Electromagnetic processes are evident in the spokes, which are radially elongated regions enhanced in fine dust relative to their surroundings.

100.128 **Atlas of Saturn's rings.**
S. A. Collins, J. Diner, G. W. Garneau, A. L. Lane, E. D. Miner, S. P. Synnott, R. J. Terrile, J. B. Holberg, B. A. Smith, G. L. Tyler.
Planetary rings, p. 737 – 743 (1984). – See Abstr. 012.086.

An effort is underway to prepare an atlas of Voyager Saturn ring data. A preliminary and greatly–abridged version of this compilation is presented here. Specifically, radial profiles of ring opacity (defined here as normal optical thickness) are presented at a resolution of approximately 50 km for wavelengths of $2.6 \times 10^{-5}$cm, 3.6 cm and 13 cm. Ring reflectance (at $\lambda = 5 \times 10^{-5}$cm) is also presented for two geometries of ring viewing (illuminated and unilluminated faces).

100.129 **The Saturn system.**
E. C. Stone, T. C. Owen.
Saturn, p. 3 – 22 (1984). – See Abstr. 003.052.

The introductory chapter provides an overview of the results of recent studies of the Saturn system.

100.130 **Saturn through the telescope: a brief historical survey.**
A. Van Helden.
Saturn, p. 23 – 43 (1984). – See Abstr. 003.052.

This chapter is a survey of telescopic studies of Saturn, its ring system, and its satellites, from 1610 to about 1900. It covers early observations of the Saturn system and the changing beliefs concerning the constitution of the rings. It shows that what an observer sees in the heavens depends not only on the quality of his instruments, but also to a surprising degree on theories and expectations.

100.131 **Interior structure of Saturn.**
W. B. Hubbard, D. J. Stevenson.
Saturn, p. 47 – 87 (1984). – See Abstr. 003.052.

The authors summarize the principal observational data that constrain interior models of Saturn, and explain why they are relevant. They discuss the behavior of hydrogen, Saturn's major constituent, at pressures on the order of 0.1 to 10 Mbar and temperatures on the order of $10^4$K. Possible behavior and distributions of minor constituents are also considered, along with processes for their transport. The authors interpret Saturn's external gravitational and magnetic fields in terms of interior structure, and discuss the relationship between atmospheric zonal flows and the deep interior. The constraint imposed by tidal evolution considerations is evaluated. Calculations for the thermal evolution of Saturn are presented, both with and without consideration of possible gravitational unmixing. Possible scenarios for Saturn's mode of origin and their implications for presently observed atmospheric abundances are discussed.

100.132 **Composition and chemistry of Saturn's atmosphere.**
R. G. Prinn, H. P. Larson, J. J. Caldwell, D. Gautier.
Saturn, p. 88 – 149 (1984). – See Abstr. 003.052.

The authors provide a comprehensive discussion and review of the chemistry and composition of Saturn as determined by Earth–based (infrared, visible), Earth–orbital (UV), and Voyager 1 and 2 (infrared, UV) spectroscopic observations. The observations imply that there are important differences between the actual composition of Saturn's atmosphere and that predicted for a homogeneous solar–composition planet. Thermochemical equilibrium models and nonequilibrium phenomena are discussed: both equilibrium and nonequilibrium chemical processes are important on Saturn.

100.133 **Clouds and aerosols in Saturn's atmosphere.**
M. G. Tomasko, R. A. West, G. S. Orton, V. G. Tejfel'.
Saturn, p. 150 – 194 (1984). – See Abstr. 003.052.

The connections between the observable radiation scattered or emitted by Saturn and the distribution and optical properties of cloud and aerosol particles are outlined. The authors review the observations available for constraining the properties of the aerosols (from the ultraviolet to the near infrared). They discuss the implications of the observational data for the vertical and horizontal distributions, single–scattering optical properties and, so far as possible, the composition of the various aerosol and cloud particles.

100.134 **Structure and dynamics of Saturn's atmosphere.**
A. P. Ingersoll, R. F. Beebe, B. J. Conrath, G. E. Hunt.
Saturn, p. 195 – 238 (1984). – See Abstr. 003.052.

This paper treats the large–scale structure and dynamics of Saturn's atmosphere, as revealed in the visible markings, the wind patterns, and the horizontal variations of temperature. Examples of this structure include the cloud bands, the zonal jets, the long–lived oval spots, and the active convective regions, all of which are large compared to the atmospheric scale height (of order 60 km) but small compared to the planetary radius ($\sim$60,000 km). The equator–to–pole temperature contrast and interhemispheric seasonal contrasts are examples of planetary–scale structure. All of these give evidence of a complicated and interesting dynamics.

100.135 **Theory, measurements, and models of the upper atmosphere and ionosphere of Saturn.**
S. K. Atreya, J. H. Waite Jr., T. M. Donahue, A. F. Nagy, J. C. McConnell.
Saturn, p. 239 – 277 (1984). – See Abstr. 003.052.

The structure and composition of the thermosphere, exosphere and ionosphere of Saturn have been determined from observations at optical and radio wavelengths, principally by instruments aboard Voyager spacecraft. Section I is devoted to the discussion of temperature and density distributions and eddy diffusion in the neutral upper atmosphere. Section II deals with the ionospheric measurements, characteristics, modeling, heat sources, and the coupling with the neutral atmosphere.

100.136 **Energetic particles in the inner magnetosphere of Saturn.**
J. A. Van Allen.
Saturn, p. 281 – 317 (1984). – See Abstr. 003.052.

A number of salient features of the energetic particle population of the inner magnetosphere of Saturn can be listed: (1) The rings of particulate matter and the many large and small satellites inside $r = 10\ R_S$ reduce the population of particles $E > 0.5$ MeV to values of the order of $10^3$ times less than would otherwise be present. (2) All trapped particles lie outside the magnetic shell through the outer edge of Ring A. (3) The radial distribution of very energetic protons $E_p >$ tens of MeV exhibits three major peaks at $r = 3.37$, 2.68, and 2.44 $R_S$. (4) A second component of the proton population $E_p \sim 1$ MeV lies outside the orbit of Enceladus and is apparently diffused into the inner magnetosphere from the magnetopause. (5) A third component of the proton population, in the energy range $E_p < 0.25$ MeV, apparently has an internal origin associated with Dione, Tethys, Enceladus, Ring E, Mimas, and Ring G. (6) The distribution of electrons $E_e > 0.040$ MeV extends throughout the magnetosphere and exhibits spectral features that are indicative of relatively unimpeded resonant diffusion across the orbits of satellites. The source of these electrons is thermalized solar wind at the magnetopause.

100.137 **Measurements of plasma, plasma waves, and suprathermal charged particles in Saturn's inner magnetosphere.**
F. L. Scarf, L. A. Frank, D. A. Gurnett, L. J. Lanzerotti, A. Lazarus, E. C. Sittler Jr.
Saturn, p. 318 – 353, 869 – 871 (1984). – See Abstr. 003.052.

The Pioneer 11 and Voyager 1, 2 traversals of Saturn's inner magnetosphere provided direct information on the complex and highly structured distributions of plasma and suprathermal charged particles present in this region. The Voyager wave instruments also yielded absolute electron density measurements in certain inner magnetosphere locations; the wave data were used to evaluate the magnitudes of several wave–particle interactions. The authors summarize the plasma and wave measurements for 24–hr periods centered around closest approach, evaluate pitch–angle scattering effects possibly associated with measured whistler mode turbulence, and speculate on the effects of wave–particle interactions associated with electrostatic waves.

100.138 **Magnetic field models.**
J. E. P. Connerney, L. Davis Jr., D. L. Chenette.
Saturn, p. 354 – 377 (1984). – See Abstr. 003.052.

The planetary magnetic field of Saturn, as measured by the Pioneer 11, Voyager 1, and Voyager 2 spacecraft, is primarily that of a dipole of moment 0.21 Gauss–$R_S{}^3$ (where Saturn's radius $R_S = 60{,}330$ km). In spite of many similarities Saturn is distinguished from Earth and Jupiter by the symmetry of its internal field about its axis of rotation (spin symmetry). The axial symmetry of Saturn's main field makes possible additional analyses of the dipole position and orientation based on the energetic charged particle absorption signatures. An equatorial ring current of $\sim 10^7$A, confined to a 5–$R_S$–thick annulus with inner and outer radii of $\sim$ 8 and 16 $R_S$, has a major effect on the geometry of Saturn's outer magnetosphere. A well–developed magnetic tail results from the interaction of Saturn's magnetic field with the solar wind.

100.139 **Saturn as a radio source.**
M. L. Kaiser, M. D. Desch, W. S. Kurth, A. Lecacheux, F. Genova, B. M. Pedersen, D. R. Evans.
Saturn, p. 378 – 415 (1984). – See Abstr. 003.052.

Saturn's radio radiation takes place in the frequency range from as low as 3 kHz to $\sim$ 1200 kHz, with peak intensity near 200 kHz in the kilometer wavelength band. The intense storms of km–wavelength radiation display periodicities. A strong correlation exists between the solar wind and the intensity of Saturn's radio emission. The observations are interpreted in terms of radiation near the electron gyrofrequency escaping in the extraordinary mode from two source regions, one in the northern hemisphere and one in the southern hemisphere. In addition to the intense component of Saturn's magnetospheric radio spectrum, there are weak narrowband tones observable at very low frequencies ($<$ 10 kHz) within Saturn's magnetosphere. These two magnetospheric radio components are complemented by a third component, Saturn electrostatic discharges, which appeared as very brief, intense, broadband bursts in episodes separated by $\sim$ 10 hr 10 min.

100.140 **The outer magnetosphere.**
A. W. Schardt, K. W. Behannon, R. P. Lepping, J. F. Carbary, A. Eviatar, G. L. Siscoe.
Saturn, p. 416 – 459 (1984). – See Abstr. 003.052.

The magnetic field in Saturn's outer magnetosphere requires a planetary dipole field, a ring–current field, plus contributions from magnetopause and tail currents. Saturn, like Earth, has a fully developed magnetic tail, 80 to 100 $R_S$ in diameter. One major difference between the two outer magnetospheres is the hydrogen and nitrogen torus produced by Titan which maintains the plasma density between $10^{-2}$ and $5 \times 10^{-1}$ions cm$^{-3}$ and a temperature of $\sim 10^6$K. A large increase in the proton and electron population below $\sim$500 keV is found at the magnetopause. Interplanetary protons and ions above 2 MeV have free access to the outer magnetosphere. As a result, the proton energy spectrum has two components. Proton pitch–angle distributions are generally pancake (perpendicular to the field), while electron pitch–angle distributions may be either dumbell (field–aligned) or pancake. Acceleration of particles into the several–hundred–keV range must be taking place in the tail.

100.141 **Saturn's rings: structure, dynamics, and particle properties.**
L. W. Esposito, J. N. Cuzzi, J. B. Holberg, E. A. Marouf, G. L. Tyler, C. C. Porco.
Saturn, p. 463 – 545 (1984). – See Abstr. 003.052.

Spacecraft measurements have a resolution as good as 100 m on the rings, revealing unexpected microstructure down to the limit of resolution. Density waves and bending waves propagate through the rings. Analysis of these waves gives the mass, viscosity, and thickness of the rings. The mass of the rings is $\sim 5 \times 10^{-8} M_S$, about the same as Mimas'. The rings are thinner than 100 m, but thicker than a single particle. The ring particles have a broad size distribution, ranging from 1 cm to 5 m, with some areas containing particles as small as several microns. Moonlets (particles as large as 10 km) are quite rare. The ring particles are bright and strongly backscattering, consistent with rough, icy blocks. Recently observed features in the rings include dark, nearly radial lines (spokes) in the B Ring and noncircular ringlets and ring edges. Several regions in the rings contain fine dust, which allows a coupling between the magnetosphere and the observable rings.

100.142 **Electrodynamic processes in the ring system of Saturn.**
D. A. Mendis, J. R. Hill, W.-H. Ip, C. K. Goertz, E. Grün.
Saturn, p. 546 – 589 (1984). – See Abstr. 003.052.

The authors discuss a number of recently observed ring phenomena including Voyager 1 and 2 observations of the rotating near-radial spokes in the B Ring; waves and braids of the F Ring; and discrete episodic bursts of broadband radio emission, claimed by some authors to originate in the ring. In addition, several other phenomena are discussed, including the origin and evolution of the diffuse E Ring and G Ring (which appear to be composed of fine dust), as well as the existence of a number of sharp discontinuities in the main ring system, within the context of gravitoelectrodynamics of charged dust in the magnetosphere. The possible role of electromagnetic processes in the formation of the ring system is considered.

100.143 **Orbital resonances among Saturn's satellites.**
R. Greenberg.
Saturn, p. 593 – 608 (1984). – See Abstr. 003.052.

This article reviews the kinetics of the various classes of resonances represented among the classical and newly discovered satellites and describes how resonances have played a fundamental role in shaping the Saturn system as it is seen today.

100.144 **Satellites of Saturn: geological perspective.**
D. Morrison, T. V. Johnson, E. M. Shoemaker, L. A. Soderblom, P. Thomas, J. Veverka, B. A. Smith.
Saturn, p. 609 – 639 (1984). – See Abstr. 003.052.

The satellites of Saturn are surprisingly heterogeneous, in spite of their apparently common composition of more than half water ice. All are more or less heavily cratered, but most also show clear evidence of substantial endogenic modification and resurfacing during the first few hundred million years of their existence. In the case of Enceladus, this internal activity has been most dramatic and may have persisted into the past billion years of solar system history. Iapetus also remains a major mystery with its still unexplained hemispheric dichotomy, for which both external and internal causes have been suggested. The authors review the current knowledge of the satellites from a geologic perspective, using primarily Voyager imaging data. Emphasis is placed on the six larger bodies (except Titan): Rhea, Iapetus, Dione, Tethys, Enceladus, and Mimas.

100.145 **Satellites of Saturn: optical properties.**
D. P. Cruikshank, J. Veverka, L. A. Lebofsky.
Saturn, p. 640 – 667 (1984). – See Abstr. 003.052.

This review concentrates on composition and microstructures derived from photometric, spectrophotometric, and thermal observations from spacecraft and from groundbased telescopes. Photometric phase integrals of Mimas, Enceladus, Tethys, Dione, and Rhea were determined from Voyager photometry over a wide range in phase angles, and similar values were derived from Voyager infrared observations. General agreement is found for most objects with the exception of Rhea. A lunar-like law of surface optical scattering is found for some satellites, but others are somewhat more complex, involving a significant component of Lambert-like scattering. The high albedos of the small, innermost satellites suggest a water-ice composition. Low mean densities of the larger satellites (including Titan) indicate a bulk composition dominated by water, but the surface exposures of water ice are contaminated in varying degree by dust or other material fairly neutral or reddish in color.

100.146 **Titan.**
D. M. Hunten, M. G. Tomasko, F. M. Flasar, R. E. Samuelson, D. F. Strobel, D. J. Stevenson.
Saturn, p. 671 – 759 (1984). – See Abstr. 003.052.

Titan is the only satellite with a substantial atmosphere, and this atmosphere is denser than that of any terrestrial planet except Venus. Methane clouds are probably present. The upper atmosphere is remarkably close to isothermal all the way from the stratosphere to the exosphere, and surface temperatures are globally uniform within a few degrees. The dark-orange or brown stratospheric aerosol particles are slightly $>0.1$ µm in radius, and there may be larger particles at greater depths. The aerosols are thought to be the end product of methane photochemistry. Reasonable photochemical paths can be set out for production of the observed hydrocarbons and organic molecules. A puzzling aspect of the upper atmosphere is the ultraviolet emissions confined to the day side but too bright to be excited by the available solar energy. Hydrogen escapes rapidly to populate the observed torus, accompanied by some nitrogen ejected by electron impact dissociation of $N_2$. A global wind system can be constructed on the basis of the observed temperature field. The interior is made up of roughly equal parts of rocky and icy material. The methane supply must have been adequate to replenish the atmospheric amount many times over.

100.147 **Titan's magnetospheric interaction.**
F. M. Neubauer, D. A. Gurnett, J. D. Scudder, R. E. Hartle.
Saturn, p. 760 – 787, 872 (1984). – See Abstr. 003.052.

The authors review the plasma, magnetic field and plasma wave observations and analyses from the plasma dynamist's point of view. Properties of incident plasma flow during Titan encounter are discussed as well as the question of an internal magnetic field of Titan, and the induced magnetosphere of Titan (mass loading and field-line draping, electron density variation in Titan's wake along the Voyager 1 encounter trajectory, magnetopause and neutral sheet boundaries of the induced magnetosphere and associated density structure, flow asymmetry and the picture of four tail lobes, dynamical phenomena, and relevant future investigations).

100.148 **Organic matter in the Saturn system.**
C. Sagan, B. N. Khare, J. S. Lewis.
Saturn, p. 788 – 807 (1984). – See Abstr. 003.052.

The albedo pattern of Iapetus is suggestive of organic matter produced by photolysis of ices and distributed by collisional volatilization of ices and ballistic diffusion. On Titan, the Voyager IRIS experiment has identified nine organic species heavier than methane. Their detection confirms earlier deductions that synthesis of complex organic molecules is occurring in the atmosphere of Titan. The observed atmospheric composition is roughly consistent with expectations from kinetic calculations and with laboratory experiments on irradiated mixtures of methane and nitrogen. Contemporary production rates on Titan imply, over the age of the solar system, the deposition of a surface layer of condensed organic matter at least 100 m thick.

100.149 **Origin and evolution of the Saturn system.**
J. B. Pollack, G. Consolmagno.
Saturn, p. 811 – 866 (1984). – See Abstr. 003.052.

As was the case for Jupiter, Saturn formed either as a result of a gas instability within the solar nebula or the accretion of a solid core that induced an instability within the surrounding solar nebula. In either case, the protoplanet's history can be divided into three major stages: early, quasi-hydrostatic evolution (stage 1); very rapid contraction (stage 2); and late, quasi-hydrostatic contraction (stage 3). The irregular satellite Phoebe was most likely captured by gas drag experienced in its passage through a bloated Saturn, just prior to the onset of stage 2. The nebular disk served as the birthplace of Saturn's regular satellites and probably the ring material. As Saturn's excess luminosity declined during the early portion of stage 3, water was able to condense at progressively closer distances to the center of the system and water clathrates and hydrates were able to form. Despite the comparatively small size of the inner satellites of

Saturn, a number (especially Dione and Rhea) may have experienced significant expansion and melting during the first $\sim 1 \times 10^9$ yr. During the early history of the Saturn system, giant impact events may have catastrophically disrupted most of the original satellites of Saturn. This may be responsible, in part, for the occurrence of Trojans and coorbital moons in the Saturn system. Titan's atmosphere formed from hydrates and clathrates. Over the age of the solar system, a nontrivial amount of atmospheric nitrogen and much more methane than is presently in the atmosphere have been lost.

100.150 **The composition of Saturn's atmosphere at northern temperate latitudes from Voyager IRIS spectra: $NH_3$, $PH_3$, $C_2H_2$, $C_2H_6$, $CH_3D$, $CH_4$, and the Saturnian D/H isotopic ratio.**
R. Courtin, D. Gautier, A. Marten, B. Bézard, R. Hanel.
Astrophys. J., Vol. 287, No. 2, p. 899 – 916 (1984).

The abundances of minor atmospheric constituents at northern Saturnian latitudes have been inferred from infrared emission spectra recorded by Voyager. The $NH_3$, $PH_3$, $C_2H_2$, and $C_2H_6$ vertical distributions have been determined from spectra selected in a cloud–free region centered approximately at the latitude of the Voyager 2 radio–occultation point (36°.5 N). The D/H value resulting from the $CH_3D/CH_4$ ratio is $1.6(+1.3, -1.2) \times 10^{-5}$, significantly lower than previous determinations obtained from HD lines; it is also less than half of the IRIS–inferred Jovian value.

**Voyager 1 and 2 atlas of six Saturnian satellites.**
See Abstr. 002.024.

**Rings in astronomy and cosmology, 1600 – 1900.**
See Abstr. 004.187.

**Objectives of U.V. observations in the study of the Saturnian rings.**
See Abstr. 013.078.

**Isotopic dependence of infrared transition moments in methane.**
See Abstr. 022.007.

**Optical constants of organic tholins produced in a simulated Titanian atmosphere: from soft X–ray to microwave frequencies.**
See Abstr. 022.074.

**Measured ion–molecule reaction rates for modelling Titan's atmosphere.**
See Abstr. 022.084.

**Propane absorption at 200K near 748 cm$^{-1}$; implications for interpretations of spectra of Jupiter and Saturn.**
See Abstr. 022.096.

**Thermodynamics of metallic hydrogen mixed with helium, carbon, and oxygen.**
See Abstr. 022.098.

**Time–evolution of infrared absorption features in an oxygen–free Titan organic synthesis experiment.**
See Abstr. 022.102.

**Titan: multiple light scattering by organic tholins and condensates.**
See Abstr. 022.103.

**The destruction of a continuous ring revolving around a gravitating center.**
See Abstr. 042.008.

**Second order secular planetary theory for the four major planets.**
See Abstr. 042.011.

**Theory of the motion of Jupiter and Saturn over a period of 6000 years. Solution JASON84.**
See Abstr. 042.012.

**Configurations des huit premiers satellites de Saturne.**
See Abstr. 046.026.

**Ephémérides des satellites naturels des planètes: Jupiter, Saturne, Uranus pour 1985.**
See Abstr. 046.036.

**Titan buoyant station.**
See Abstr. 051.012.

**The Voyager encounters with Saturn.**
See Abstr. 051.050.

**A priority of space exploration: observation of Saturn's rings from a Saturn orbiter.**
See Abstr. 051.074.

**Rendez–vous with Saturn's rings.**
See Abstr. 051.075.

**Drift wave instabilities in a high $\beta$ multispecies plasma.**
See Abstr. 062.088.

**Quelques aspects de la dynamique des atmosphères des planètes géantes.**
See Abstr. 091.018.

**Does crater "saturation equilibrium" occur in the solar system?**
See Abstr. 091.026.

**Shepherding moonlets: detection by adjacent ring radial structure.**
See Abstr. 091.039.

**Particle sizes in planetary rings from forward–scattering and spectral–extinction observations.**
See Abstr. 091.064.

**The W–shaped occultation signatures of planetary ring systems: inference of size–sorted particles in charged, tilted ringlets.**
See Abstr. 091.065.

**Solved and unsolved problems in planetary ring dynamics.**
See Abstr. 091.066.

**Dynamics of eccentric and inclined ringlets.**
See Abstr. 091.069.

**Sharp edges of planetary rings.**
See Abstr. 091.070.

**Comments on collision mechanics in ring systems.**
See Abstr. 091.073.

**On the radial structure of planetary rings.**
See Abstr. 091.074.

**Diffusion instability in a bi–modal disc.**
See Abstr. 091.075.

**On planetary rings as sources and sinks of magnetospheric plasmas.**
See Abstr. 091.079.

**On the electric charge carried by dust grains excitation of violent discharge.**
See Abstr. 091.080.

**Ring particles: collisional interactions and physical nature.**
See Abstr. 091.085.

**Transport effects due to particle erosion mechanisms.**
See Abstr. 091.086.

**Collision–induced transport processes in planetary rings.**
See Abstr. 091.087.

**Dynamics of narrow rings.**
See Abstr. 091.090.

**Unsolved problems in planetary ring dynamics.**
See Abstr. 091.091.

**Comparison of induced magnetospheres at Venus and Titan.**
See Abstr. 093.005.

**The recent lunar occultation of planet Saturn.**
See Abstr. 096.026.

**Comparative geology of the satellites of the giant planets.**
See Abstr. 099.005.

**Presidential address: the giant planets; Galileo Galilei to Project Galileo.**
See Abstr. 099.019.

**Radiative–convective equilibrium models of Jupiter and Saturn.**
See Abstr. 099.027.

**Hydrogen dimer structures in the far–infrared spectra of Jupiter and Saturn.**
See Abstr. 099.032.

**Gravito–electrodynamics and the structure of planetary ring systems.**
See Abstr. 099.033.

**Microwave observations of Jupiter and Saturn.**
See Abstr. 099.042.

**Possible identification of sharp features in the Voyager far–infrared spectra of Jupiter and Saturn.**
See Abstr. 099.043.

**Microwave observations of Jupiter and Saturn.**
See Abstr. 099.067.

**Search for minor atmospheric species in the microwave spectra of Jupiter and Saturn.**
See Abstr. 099.107.

**On the detectability of desequilibrium species in the upper atmospheres of Jupiter and Saturn from the submillimeter spectrum of these planets.**
See Abstr. 099.109.

**The detectability of minor atmospheric species in the far infrared spectra of Jupiter and Saturn.**
See Abstr. 099.114.

**Jupiter's and Saturn's fine–scale magnetic fields.**
See Abstr. 099.118.

**Planetocentric versus heliocentric impacts in the Jovian and Saturnian satellite system.**
See Abstr. 099.127.

**The ethereal rings of Jupiter and Saturn.**
See Abstr. 099.140.

**Radio emissions from the planets earth, Jupiter, and Saturn.**
See Abstr. 099.145.

**Magnetosphere, rings, and moons of Uranus.**
See Abstr. 101.041.

# 101 Uranus, Neptune, Pluto, Transplutonian Planets

101.001 **A photometric study of Pluto near perihelion. II. Refined rotation period. Color indices.**
V. M. Lyutyj, V. P. Tarashchuk.
Pis'ma Astron. Zh., Tom 10, No. 7, p. 542 – 548 (1984). In Russian. English translation in Sov. Astron. Lett., Vol. 10.

Results of *UBV* photometry of Pluto during 1982 – 1983 are reported. Using all photoelectric observations of 1953 – 1983 a more accurate period is calculated. Significant reddening at maximum light (phases 0.6 – 0.7) is revealed. The secular variations of the mean brightness of Pluto does not contradict a 124–year periodicity.

101.002 **Pluto.**
P. Moore.
J. Br. Astron. Assoc., Vol. 94, No. 5, p. 223 – 224 (1984).

101.003 **The positions of Pluto obtained during the period May–June 1980: observations carried out at the European Southern Observatory (ESO), La Silla.**
H. Debehogne, R. R. De Freitas Mourão, M. R. Nunes, O. C. Tavares.
Astron. J., Vol. 89, No. 8, p. 1262 – 1263 (1984).

Observations of Pluto were made in May and June 1980, using the ESO Astrograph at La Silla. Measurement and reduction of the plates were performed at the Observatório do Valongo with the Ascorecord measuring machine with a positional accuracy of 0.1 μ. The position of Pluto was determined relative to the position of five reference stars.

101.004 **Predictions for Uranus from a radiometric Bode's law.**
M. D. Desch, M. L. Kaiser.
Nature, Vol. 310, No. 5980, p. 755 – 757 (1984).

The magnetospheres of three planets, Earth, Jupiter and Saturn, are sources of intense, nonthermal radio bursts. An analysis of the overall efficiency of the solar wind in stimulating these emissions leads the authors to postulate a "radiometric Bode's law" from which a planet's magnetic moment can be estimated from its radio emission output. Applying the radiometric scaling law to Uranus, the authors estimate the low–frequency radio power likely to be measured by the Voyager 2 spacecraft as it approaches this planet.

101.005 **Uranus: microwave images.**
W. J. Jaffe, G. L. Berge, T. Owen, J. Caldwell.
Science, Vol. 225, No. 4662, p. 619 – 621 (1984).

Observations of Uranus at wavelengths of 2 and 6 centimeters with the Very Large Array were made in 1980 and 1981. The resulting maps of brightness temperature show a subsolar symmetry at 2 centimeters but a near–polar symmetry at 6 centimeters. The 6–centimeter maps show an increase in temperature from equator to pole with some evidence for a warm "ring" surrounding the north pole. The variations in brightness temperature probably reflect variations in ammonia abundance in the planet's atmosphere, but the mechanism driving these variations is still unclear.

101.006 **Sub-mm and millimeter observations of the outer planets.**
I. G. Nolt, J. V. Radostitz, G. S. Orton, M. J. Griffin, C. T. Cunningham, P. A. R. Ade, I. A. Robson.
Eighth International Conference on Infrared and Millimeter Waves, p. M2.2 (1984). Abstr. in Phys. Abstr., Vol. 87, No. 1238, Entry 77764 (1984). – See Abstr. 012.015.

101.007 **Observation et mouvement des satellites d'Uranus.**
C. Veillet.
Astrométrie, géodésie, mécanique céleste, p. 29 – 33 (1984). – See Abstr. 012.021.

101.008 **On the origin of Triton and Pluto.**
W. B. McKinnon.
Nature, Vol. 311, No. 5984, p. 355 – 385 (1984).

Lyttleton hypothesized long ago that Triton and Pluto originated as adjacent prograde satellites of Neptune. This letter shows that with the presently accepted masses of Triton and Pluto–Charon, however, the momentum and energy exchange that would be required to set Triton on a retrograde trajectory is impossible. The simplest hypothesis is that Triton and Pluto are independent representatives of large outer Solar System planetesimals. Triton is simply captured, with potentially spectacular consequences that include runaway melting of interior ices and release to the surface of clathrated $CH_4$, CO and $N_2$. Condensed remnants of this proto-atmosphere could account for features in Triton's unique spectrum.

101.009 **The seasonal variation of the thermal structure of the atmosphere of Neptune.**
L. Wallace.
Icarus, Vol. 59, No. 3, p. 367 – 375 (1984).

Previous work on the atmosphere of Uranus is extended to Neptune. The variation of effective temperature with latitude and season is evaluated within the approximations that the redistribution of internal heat in the interior results in the temperature at fixed pressure near the top of the convective region being independent of latitude and time, and that the transport of heat in the atmosphere is by means of radiation and convection. Meridional heat transport in the atmosphere is neglected. It is found that as the absorbed solar flux varies with season the flux of internal heat varies in the opposite sense such that the variation in the sum of the two is much smaller than the variation in either.

101.010 **Das Ringsystem des Planeten Uranus – eine bemerkenswerte Entdeckung im Sonnensystem.**
J. C. Bhattacharyya.
Jenaer Rundsch., 29. Jahrg., Heft 3, p. 114 – 115 (1984).

101.011 **Pluto – distant and puzzling.**
V. A. Bronshtehn.
Zemlya Vselennaya, No. 4, p. 20 – 25 (1984). In Russian.

101.012 **The Tisserand criterion as indicator of the source of perturbation and program of search for transplutonian planets.**
V. V. Radzievskij.
Gor'k. gos. ped. inst. Gor'kij, 10 pp. (1984). In Russian. Abstr. in Ref. Zh., 51. Astron., 10.51.109 (1984).

101.013 **Estimate of the parameters of a transplutonian planet.**
N. A. Shirokov.
VNII yader. geofiz. i geokhim., Moskva, 13 pp. (1984). In Russian. Abstr. in Ref. Zh., 51. Astron., 10.51.200 (1984).

101.014 **Structure of the Uranian rings. I. Square-well model and particle-size constraints.**
J. L. Elliot, R. G. French, K. J. Meech, J. H. Elias.
Astron. J., Vol. 89, No. 10, p. 1587 – 1603 (1984).

A square-well model for occultation profiles produced by the Uranian rings is described. The model includes diffraction effects and treats the occulting ring segment as a uniformly transmitting gray screen with abrupt edges. Up to ten parameters of the model can be fit by least squares to a ring profile; those describing the profiles of the ring are: (1) the midtime of the profile, which specifies the position of the occulting ring segment; (2) the duration of the ring occultation, which is proportional to the radial width of the ring; and (3) the fractional transmission of the ring, which is simply related to its optical depth. The diameter of the occulted star can also be fit as a free parameter. For their data with the highest signal-to-noise ratio, the authors can achieve a precision of tens of meters in the placement of a ring segment, a precision of one-hundred meters in the width of a segment, and a precision of a few hundredths in its fractional transmission. This model has been used to fit occultation profiles obtained simultaneously at more than one wavelength in order to constrain the distribution of particle sizes within the rings.

101.015 **Photometry of Pluto during the 1983 opposition: a new determination of the phase coefficient.**
R. P. Binzel, J. D. Mulholland.
Astron. J., Vol. 89, No. 11, p. 1759 – 1761 (1984).

Photoelectric observations of Pluto's light curve were made on 19 nights during the 1983 opposition. High time-resolution measurements of both syzygy phases made on a total of six nights in April, May, and June showed no evidence for eclipses between Pluto and its satellite. These syzygy observations occurred over an illumination phase-angle range of >1 deg, and overlapping rotational phases were used to compute a phase coefficient of $0.041 \pm 0.003$ mag/deg.

101.016 **The Uranus system.**
E. L. Ruskol.
Zemlya Vselennaya, No. 6, p. 16 – 20 (1984). In Russian.

101.017 **Photometrische Strukturen des Neptun?**
H. Kohl.
Sterne Weltraum, 23. Jahrg., Nr. 11, p. 585 – 587 (1984).

101.018 **The Laplacian resonance amongst Uranian inner satellites.**
D. Lazzaro, S. Ferraz-Mello, C. Veillet.
Astron. Astrophys., Vol. 140, No. 1, p. 33 – 38 (1984).

Computation of the coefficients of the differential equation that gives the inequalities of period 12.5 years in the longitude of Miranda. All contributing terms of second order in the satellites masses are considered and the resulting inequalities are compared to results from a numerical integration of the equations of motion. The results allow to suspect that the masses of the satellites are smaller than those generally given.

101.019 **The origin and evolution of Uranus and Neptune.**
J. S. Lewis.
NASA Conf. Publ., NASA CP-2330, p. 3 – 24 (1984). – See Abstr. 012.041.

The present state of our understanding of the origin and evolution of the planets Uranus and Neptune is briefly reviewed. Emphasis is placed on the need for improved physical and compositional data on both planets, particularly those data needed to resolve certain major inconsistencies in the present data set.

101.020 **Atmospheres of Uranus and Neptune.**
D. M. Hunten.
NASA Conf. Publ., NASA CP-2330, p. 27 – 54 (1984). – See Abstr. 012.041.

Current knowledge of the atmospheric compositions and structures are surveyed. Although the two planets are similar in some ways, they are strikingly different in a number of others. The similarities include mass, size, and effective temperature. There are differences in heat flow, stratospheric thermal structure and composition, and spectral appearance. Presumably these are traceable to some combination of Uranus' unusual obliquity and smaller distance from the sun, but most of the details are obscure.

101.021 **Aeronomy.**
S. K. Atreya.
NASA Conf. Publ., NASA CP-2330, p. 55 – 88 (1984). – See Abstr. 012.041.

From the known composition ($H_2$, $CH_4$, $C_2H_2$(?), $NH_3$(?) at Uranus, and $H_2$, $CH_4$, $C_2H_6$ at Neptune) and the inversion and photolysis region temperatures, reasonable theoretical models for the upper atmospheric distribution of the neutral and ionospheric species can be reconstructed on the basis of the expected physical and chemical processes. The models indicate that $C_2H_2$ would condense over an extensive height range on Uranus. Ionospheric calculations indicate peak electron concentrations in the 3000 – 5000 $cm^{-3}$ range, with an upper limit of $2 \times 10^4 cm^{-3}$ at Uranus. The maximum atomic hydrogen abundance is calculated to be $2 \times 10^{16} cm^{-2}$, including flux from the ionosphere.

101.022 **Temperature structures and infrared-derived properties of the atmospheres of Uranus and Neptune.**
G. S. Orton, J. F. Appleby.
NASA Conf. Publ., NASA CP-2330, p. 89 – 155 (1984). – See Abstr. 012.041.

Temperature structures of Uranus and Neptune are currently derivable from a combination of stellar occultation and thermal radiance observations. The globally-averaged temperatures of Uranus and Neptune appear quite similar in their convective regions, reaching temperatures near 75K at 1 bar and 150K near 10 bar. Temperature minima near 200 mbar are also similar, near 53 – 55K. The temperature in the stratosphere of Neptune rises much more steeply with altitude than for Uranus in the range of 0.1 to 100 mbar. Above this level, near 1 μbar, both atmospheres reach temperatures of 120 – 160K. The bolometric radiant energy output is equivalent to the output of a blackbody at $58.3 \pm 2.0$K for Uranus and $60.3 \pm 2.0$ for Neptune. These values imply approximate equilibrium with absorbed sunlight for Uranus but an internal energy source 1.9 to 2.6 times the absorbed sunlight for Neptune. The implication of the thermal spectra is that while the bulk composition may be consistent with a solar mixture of $H_2$ and He, the mixing ratio of $CH_4$ is probably about 2% in the deep atmospheres of both planets. Observations near 10 μm are also consistent with clouds at the $CH_4$ condensation level in Uranus. Stratospheric clouds (of $CH_4$ and other hydrocarbons) are implied for Neptune, with $CH_4$ transported (convectively) through the cold trap. Such vigorous convection is consistent with the variability of the stratospheric cloud reflectivity, the large internal heat source, and the strongly inhomogeneous visual appearance of the planet.

101.023 **Ultraviolet observations of Uranus and Neptune below 3000 Å.**
J. Caldwell, R. Wagener, T. Owen, M. Combes, T. Encrenaz.
NASA Conf. Publ., NASA CP-2330, p. 157 – 177 (1984). – See Abstr. 012.041.

From 2000 to 3000 Å, both Uranus and Neptune have albedos that are about two times higher than Jupiter or Saturn's, implying that the outer giants have stratospheres that are relatively free of aerosol absorption. Uncertainties in the absolute calibration procedure allow discrepancies of order 15% between conservative models and the observations. A small amount of aerosol absorption is therefore possible. Below 2000 Å the derived albedo is highly dependent on the solar spectrum source used in the data reduction. The most recent result for Uranus, first reported here, is consistent with a secular change in $C_2H_2$ mixing ratio from $\sim 3 \times 10^{-8}$ in 1980 to $\leqslant 10^{-9}$ in 1983.

101.024 **Properties of the upper tropospheres of Uranus and Neptune derived from observations at "visible" to near-infrared wavelengths.**
J. T. Bergstralh, K. H. Baines.
NASA Conf. Publ., NASA CP-2330, p. 179 – 212 (1984). – See Abstr. 012.041.

Recent observations of Uranus and Neptune at wavelengths between 0.3 and 2.5 μm are reviewed and compared with predictions based on models of the atmospheric structures. Significant results for Uranus include the presence of an opaque lower boundary to the visible atmosphere very near the level corresponding to 2 bars pressure, and consequently a methane/hydrogen ratio no less than 3 percent.

101.025 **Near infrared imaging of Uranus and Neptune.**
B. A. Smith.
NASA Conf. Publ., NASA CP-2330, p. 213 – 223 (1984). – See Abstr. 012.041.

Imaging of Uranus and Neptune in the deep methane absorption band at 890 nm has been used to detect high-altitude atmospheric hazes and to search for possible undiscovered close-in satellites. The appearances of Uranus and Neptune are very different from one another and Uranus seems to be changing with time. The Neptune images show rotation in the direct sense. No new satellites have been found to date.

101.026 **A review of the millimeter and centimeter observations of Uranus.**
S. Gulkis, I. de Pater.
NASA Conf. Publ., NASA CP-2330, p. 225 – 262 (1984). – See Abstr. 012.041.

Millimeter and centimeter wave astronomical data of Uranus are presented and discussed. Some recent observations of the brightness distribution across the disk at 2 cm and 6 cm wavelength are reported. Data which relate to the variability of the microwave spectrum are reanalysed taking into account the strong variation of temperature with wavelength, and including up-to-date data. The results indicate that the brightness temperatures of Uranus have probably reached their maximum values. The evidence for the depletion of ammonia in the upper atmosphere is discussed.

101.027 **Atmospheric dynamics of Uranus and Neptune: theoretical considerations.**
A. P. Ingersoll.
NASA Conf. Publ., NASA CP-2330, p. 263 – 269 (1984). – See Abstr. 012.041.

Uranus, because of its pole-on orientation and low internal heat source, is in a dynamically different atmospheric regime from Jupiter and Saturn. Neptune resembles Jupiter and Saturn in orientation and internal heating, but its extremely long radiative time constant puts Neptune in a different class. Voyager observations of seasonal temperature gradients, equator-to-pole temperature gradients, infrared emission, Bond albedo, possible cloud structures (bands, spots, eddies), and cloud motions should greatly improve our ability to classify planetary atmospheres according to their dynamical regimes.

101.028 **The possibility of detection and measurement of cloud structure in the atmosphere of Uranus.**
R. F. Beebe.
NASA Conf. Publ., NASA CP-2330, p. 271 – 278 (1984). – See Abstr. 012.041.

Historical data concerning variability of surface brightness of the visible cloud deck are reviewed. The data do not rule out low contrast small cloud structures of the type measured in the Saturnian Voyager wind measurements. Convective models in which methane can condense suggest similar structure may be present at the level of the visible cloud deck in Uranus' atmosphere. Voyager 2 may obtain significant dynamical information about the Uranian atmosphere.

101.029 **Variability of Neptune.**
D. P. Cruikshank.
NASA Conf. Publ., NASA CP-2330, p. 279 – 287 (1984). – See Abstr. 012.041.

Earth-based observers of Neptune have found that the planet varies in brightness at various wavelengths in ways that suggest that changes occur in the planet's atmosphere on several different time-scales. Global inhomogeneities in high-altitude haze distribution that are stable for several days permit measurements of the planet's rotation period (about 18 hours), but this stability sometimes breaks down, obscuring the diurnal lightcurve. In addition, there is an apparent long-term variability of the brightness of Neptune in anticorrelation with the cycle of solar activity. This slow variability of low amplitude may be punctuated by outbursts of high-altitude condensation of particles in the atmosphere, whose decay time is several months.

101.030 **Interior structure of Uranus.**
W. B. Hubbard.
NASA Conf. Publ., NASA CP–2330, p. 291 – 325 (1984). – See Abstr. 012.041.

This paper discusses key measurements which are diagnostic of Uranus' interior structure and evolutionary history, and reviews their present status. The author compares Uranus' interior with Neptune's in a number of ways, considering heat flow, degree of internal differentiation, and possible magnetic field.

101.031 **Rotational properties of Uranus and Neptune.**
M. J. S. Belton, R. Terrile.
NASA Conf. Publ., NASA CP–2330, p. 327 – 347 (1984). – See Abstr. 012.041.

An assessment of recent research bearing on the rotational properties of Uranus and Neptune is given: Direction of pole, period, and sense of spin. There is a clear signature of large–scale, zonal, atmospheric flows in observations of the atmosphere of Neptune. Wind velocities are at least as great as 109 m $sec^{-1}$. There is no evidence pro or con for atmospheric motions in the Uranus case.

101.032 **Oblateness of Uranus and Neptune.**
R. G. French.
NASA Conf. Publ., NASA CP–2330, p. 349 – 355 (1984). – See Abstr. 012.041.

Analysis of Stratoscope II images of Uranus yields $\varepsilon = 0.022 \pm 0.001$, and stellar occultation observations yield $\varepsilon = 0.024 \pm 0.003$. Neptune's oblateness has been determined from stellar occultation observations made in 1968 and 1983 and yield an oblateness of $0.021 \pm 0.004$.

101.033 **Physical properties of Neptune and Triton inferred from the orbit of Triton.**
A. W. Harris.
NASA Conf. Publ., NASA CP–2330, p. 357 – 373 (1984). – See Abstr. 012.041.

The orbital motion of Triton was redetermined from photographic data spanning the interval 1899 – 1981. The resulting ephemeris should provide Triton positions with respect to Neptune with an accuracy of $\pm 500$ km through the end of the century. Physical results following from the orbit solution are given.

101.034 **Origin and evolution of the Uranian and Neptunian satellites: some dynamical considerations.**
S. F. Dermott.
NASA Conf. Publ., NASA CP–2330, p. 377 – 404 (1984). – See Abstr. 012.041.

The satellite system of Neptune is so irregular that some formation mechanism or subsequent dynamical evolution that sets it apart from the other, more regular, satellite systems is obviously indicated. McKinnon has argued that satellite capture is the most likely possibility and has shown that tidal circularization of Triton's presumably highly eccentric initial orbit probably resulted in melting of the satellite's interior. The satellite system of Uranus, although somewhat bland, also has a number of special features that indicate an interesting dynamical history. These include the anomalously high orbital inclination of Miranda and the probable coexistence of small satellites and narrow rings inside the planet's Roche limit. The author discusses the possibility that orbital evolution due to tidal dissipation is involved in both of these phenomena. Other topics discussed are: the origin of rings; the formation of coorbital satellites; the lack of stable orbit–orbit resonances in the Uranian satellite system; and chaos, tidal heating and the shapes of Miranda and Ariel.

101.035 **Composition, structure and evolution of Uranian and Neptunian satellites.**
D. J. Stevenson.
NASA Conf. Publ., NASA CP–2330, p. 405 – 423 (1984). = Contrib. No. 4042, Div. Geol. Planet. Sci., Calif. Inst. Technol., Pasadena, Calif. 91125. – See Abstr. 012.041.

Large uncertainties in the current estimated densities of all of these satellites prevent detailed modeling or predictions. Nevertheless, current evidence suggests that at least Titania and Oberon might have "anomalously" high densities ($2 - 3$ g $cm^{-3}$), possibly requiring almost ice–free hydrated silicates or formation in a CO–rich environment, implying presence of CO–clathrate and a small ice/rock ratio. Triton and the four largest satellites of Uranus are massive enough to have undergone significant accretional heating and early differentiation; $NH_3$–$H_2O$ volcanism; partial outgassing of CO, $N_2$, $CH_4$; formation of dark surficial deposits of carbon–rich material obtained by UV irradiation of outgassed material; and, at least in the cases of Ariel and Triton, a possibility of weak ongoing icy volcanic activity. Triton may be the largest captured body in the solar system, with an unusual history and composition, including the possibility of substantial liquid or solid nitrogen obtained from either primordial $NH_3$ photolysis or clathrate decomposition.

101.036 **Physical properties of the satellites of Neptune.**
D. P. Cruikshank.
NASA Conf. Publ., NASA CP–2330, p. 425 – 436 (1984). – See Abstr. 012.041.

Spectrophotometric studies of Triton suggest a surface covered in part by solid methane. Nitrogen is suspected. The radius of Triton has been determined by indirect means to be 1750 km. There is no information on physical properties of Nereid. A third suspected satellite has not been confirmed.

101.037 **Physical properties of the Uranian satellites.**
R. H. Brown.
NASA Conf. Publ., NASA CP–2330, p. 437 – 461 (1984). – See Abstr. 012.041.

Radiometric measurements have shown that the Ariel, Umbriel, Titania and Oberon have diameters which range from 1630 to 1110 km and albedos which range from 0.30 to 0.18. Spectrophotometric observations of Miranda suggest that it may have the highest albedo of the known Uranian satellites and a diameter of about 500 km. Near–infrared measurements show that Ariel, Titania and Oberon have the largest known opposition surges. All five known satellites of Uranus have surfaces which are composed of water ice contaminated with small amounts of dark material. The dark material on the surfaces of Ariel, Umbriel, Titania and Oberon is spectrally bland and has spectral similarities to carbon black, charcoal, carbonaceous chondritic material and other dark, spectrally neutral materials. Recent density determinations suggest that there may be large density differences among Ariel, Umbriel, Titania and Oberon, with density increasing with distance from Uranus.

101.038 **Satellite masses in the Uranus and Neptune systems.**
R. Greenberg.
NASA Conf. Publ., NASA CP–2330, p. 463 – 480 (1984). – See Abstr. 012.041.

Our knowledge of satellite masses, and how they are derived, is reviewed with emphasis on implications for bulk densities and albedos. A more severe constraint than previously published is here placed on Miranda's mass, and hence on its density and albedo. The recent radiometric value for Triton's diameter, combined with now–rather–old mass determinations, yields a density greater than 4 $gm/cm^3$, but systematic errors are possible in both mass and diameter.

101.039 **Seasonal variations in Triton's atmospheric mass and composition.**
L. M. Trafton.
NASA Conf. Publ., NASA CP–2330, p. 481 – 493 (1984). – See Abstr. 012.041.

101.040 **Magnetospheric structures: Uranus and Neptune.**
T. W. Hill.
NASA Conf. Publ., NASA CP–2330, p. 497 – 525 (1984). – See Abstr. 012.041.

The purpose of this paper is to describe magnetospheric structures that Voyager might reasonably be expected to encounter at Uranus (1986) and Neptune (1989). Although the existence of magnetospheres at Uranus and Neptune has not been established, the available statistics indicate a sufficiently high probability. The reported observation of auroral emission from Uranus has motivated the development of a model in which solar–wind motion is coupled directly to the rotation of the ionosphere to establish a dynamo circuit which generates Birkeland currents and polar–cap aurora. This model predicts the strength and configuration of the aurora as functions of the magnitude and polarity, respectively, of the planetary magnetic moment.

101.041 **Magnetosphere, rings, and moons of Uranus.**
A. F. Cheng.
NASA Conf. Publ., NASA CP–2330, p. 541 – 556 (1984). – See Abstr. 012.041.

The observation of an ultraviolet aurora on Uranus implies the existence of a magnetosphere. It is suggested here that the magnetospheres of Uranus and Saturn may be very similar.

101.042 **Do the satellites of Uranus control its magnetosphere?**
A. F. Cheng, T. W. Hill.
NASA Conf. Publ., NASA CP–2330, p. 557 – 558 (1984). – See Abstr. 012.041.

101.043 **Lyman–alpha aurora.**
S. T. Durrance, J. T. Clarke.
NASA Conf. Publ., NASA CP–2330, p. 559 – 572 (1984). – See Abstr. 012.041.

The existence of intense and variable H Ly $\alpha$ emission from Uranus is demonstrated utilizing the monochromatic imaging capabilities of the International Ultraviolet Explorer satellite. A series of 14 observations, using the IUE short wavelength spectrograph in low dispersion and covering the period from 3 March 1982 through 2 September 1983, shows the disk averaged Ly $\alpha$ brightness of Uranus to vary between 690 and 2230 Rayleighs. Model calculations indicate that $<400$ R of this emission can be attributed to resonant scattering of solar Ly $\alpha$ radiation. In addition to being unexpectedly strong, the Uranian Ly $\alpha$ emission has been observed to vary by a factor of two in one 24 hr period and by about 50% in one 5 hr period. These data thus offer the first strong evidence for the presence of aurora and therefore a magnetic field on Uranus.

101.044 **The structure of the Uranian rings and the search for rings around Neptune.**
J. L. Elliot.
NASA Conf. Publ., NASA CP–2330, p. 575 – 588 (1984). – See Abstr. 012.041.

The nine narrow rings of Uranus, presently the only confirmed features of this system, have been observed with the diffraction–limited resolution (3.5 km) of ground–based occultations since their discovery in 1977. These data have been used to establish an orbit model, from which the five Keplerian orbit parameters for each ring, the pole of the mean ring plane, and the gravitational harmonic coefficients $J_2$ and $J_4$ have been determined. The rings are typically a few kilometers wide with eccentricities of about $10^{-3}$ and inclinations of a few hundredths of a degree, although a few have no measurable eccentricity or inclination. For Neptune, occultation searches have revealed no rings to a limit of a few hundredths optical depth, within a few hundred kilometers from the top of the planet's atmosphere (for equatorial rings).

101.045 **Infrared images of the Uranian rings.**
P. D. Nicholson.
NASA Conf. Publ., NASA CP–2330, p. 589 – 608 (1984). – See Abstr. 012.041.

Images of the Uranian rings at a wavelength of 2.2 μm, chosen to reduce the problem of scattered light from the planet, have been acquired on several occasions spanning the period 1978 – 1983. Although individual rings are not resolved in these images, the integrated brightness of the system is observed to vary with azimuth. The phase and amplitude of this variation is found to be consistent with the precession rate and variable width of the ε ring, as deduced from stellar occultation data. Preliminary absolute calibration of the 1983 data yields an average ring geometric albedo, at 2.2 μm, of $0.022 \pm 0.002$, consistent with previous estimates.

101.046 **Dynamics of the Uranian rings.**
S. F. Dermott.
NASA Conf. Publ., NASA CP–2330, p. 609 – 627 (1984). – See Abstr. 012.041.

The author describes some of the problems of the shepherding satellite model of Goldreich and Tremaine. The following questions are discussed: Why are the optical depths of all the observed narrow rings close to unity? Why are the satellite and ring separation timescales so short? Why are the ring edges sharp? How important is shock formation in narrow rings? Do we have any evidence for the existence of small satellites near the Uranian rings? What are the probable sizes of these satellites? How are the apse and node alignments of the eccentric and inclined rings maintained?

101.047 **Methane frost on Pluto: model implications from spectrophotometry.**
M. W. Buie, U. Fink.
Bull. Am. Astron. Soc., Vol. 16, No. 3, Part I, p. 651 (1984). Abstract. – See Abstr. 010.061.

101.048 **A two spot model for the surface of Pluto.**
R. L. Marcialis.
Bull. Am. Astron. Soc., Vol. 16, No. 3, Part I, p. 651 (1984). Abstract. – See Abstr. 010.061.

101.049 **A constraint on Pluto's origin.**
T. Owen.
Bull. Am. Astron. Soc., Vol. 16, No. 3, Part I, p. 651 (1984). Abstract. – See Abstr. 010.061.

101.050 **Cloud features on Uranus from CCD imaging.**
R. J. Terrile, B. A. Smith.
Bull. Am. Astron. Soc., Vol. 16, No. 3, Part I, p. 657 (1984). Abstract. – See Abstr. 010.061.

101.051 **A determination of the bolometric albedo of Uranus and Neptune.**
J. B. Pollack, K. A. Rages, D. D. Wenkert, G. E. Danielson, J. T. Bergstralh, K. H. Baines, J. S. Neff.
Bull. Am. Astron. Soc., Vol. 16, No. 3, Part I, p. 657 (1984). Abstract. – See Abstr. 010.061.

101.052 **Solar heating and the internal heat flow of Uranus.**
A. J. Friedson, A. P. Ingersoll.
Bull. Am. Astron. Soc., Vol. 16, No. 3, Part I, p. 657 – 658 (1984). Abstract. – See Abstr. 010.061.

101.053 **Recent observations of the brightness temperature of Uranus at 3.56 cm wavelength.**
M. J. Klein, S. Gulkis, A. Rius.
Bull. Am. Astron. Soc., Vol. 16, No. 3, Part I, p. 658 (1984). Abstract. – See Abstr. 010.061.

101.054 **The deep atmosphere of Uranus.**
D. J. Stevenson.
Bull. Am. Astron. Soc., Vol. 16, No. 3, Part I, p. 658 (1984). Abstract. – See Abstr. 010.061.

101.055 **Raman scattering in the atmosphere of Uranus: solar line pofiles.**
W. D. Cochran, L. M. Trafton.
Bull. Am. Astron. Soc., Vol. 16, No. 3, Part I, p. 658 (1984). Abstract. – See Abstr. 010.061.

101.056 **The structure of the Uranus atmosphere: constraints from $H_2$ 4–0 quadrupole line shapes and the visible geometric albedo spectrum.**
K. H. Baines, J. T. Bergstralh, J. T. Trauger.
Bull. Am. Astron. Soc., Vol. 16, No. 3, Part I, p. 658 (1984). Abstract. – See Abstr. 010.061.

101.057 **Coordinated observations of a local solar wind disturbance and the auroral H Lyα emission from Uranus.**
J. T. Clarke, S. T. Durrance, A. Barnes, J. Mihalov, J. Belcher.
Bull. Am. Astron. Soc., Vol. 16, No. 3, Part I, p. 659 (1984). Abstract. – See Abstr. 010.061.

101.058 **The HD/$H_2$ ratio in the atmosphere of Uranus.**
J. T. Trauger.
Bull. Am. Astron. Soc., Vol. 16, No. 3, Part I, p. 659 (1984). Abstract. – See Abstr. 010.061.

101.059 **Analysis of Uranus' (3–0) quadrupole $H_2$ shapes at 0.1 Å resolution.**
L. M. Trafton.
Bull. Am. Astron. Soc., Vol. 16, No. 3, Part I, p. 659 (1984). Abstract. – See Abstr. 010.061.

101.060 **Imaging of Uranus and Neptune from Voyager 1 and 2 and implications for their internal heat sources.**
D. D. Wenkert, G. E. Danielson, J. B. Pollack.
Bull. Am. Astron. Soc., Vol. 16, No. 3, Part I, p. 659 (1984). Abstract. – See Abstr. 010.061.

101.061 **Observations of Uranus and Neptune with the IUE satellite.**
J. Caldwell, R. Wagener, K. Fricke.
Bull. Am. Astron. Soc., Vol. 16, No. 3, Part I, p. 659 (1984). Abstract. – See Abstr. 010.061.

101.062 **Aerosols in the atmospheres of Uranus and Neptune: comparison of their physical properties and distribution.**
J. T. Bergstralh, K. H. Baines, J. S. Neff.
Bull. Am. Astron. Soc., Vol. 16, No. 3, Part I, p. 660 (1984). Abstract. – See Abstr. 010.061.

101.063 **Photochemistry of methane in the atmosphere of Neptune.**
P. N. Romani, S. K. Atreya.
Bull. Am. Astron. Soc., Vol. 16, No. 3, Part I, p. 660 (1984). Abstract. – See Abstr. 010.061.

101.064 **Hydrogen blow–off from Uranus: an extrapolation from Jupiter and Saturn.**
D. E. Shemansky, G. R. Smith.
Bull. Am. Astron. Soc., Vol. 16, No. 3, Part I, p. 660 (1984). Abstract. – See Abstr. 010.061.

101.065 **The widths, radii, and optical depths of the rings of Uranus.**
R. G. French, J. L. Elliot.
Bull. Am. Astron. Soc., Vol. 16, No. 3, Part I, p. 677 – 678 (1984). Abstract. – See Abstr. 010.061.

101.066 **Possible detection of a Uranian ring shepherding satellite.**
W. M. Sinton, E. E. Becklin.
Bull. Am. Astron. Soc., Vol. 16, No. 3, Part I, p. 678 (1984). Abstract. – See Abstr. 010.061.

101.067 **Uranian rings: a square–well model for occultation profiles.**
J. L. Elliot, R. G. French.
Bull. Am. Astron. Soc., Vol. 16, No. 3, Part I, p. 678 (1984). Abstract. – See Abstr. 010.061.

101.068 **Composition, particle sizes and age of Uranian rings.**
K. D. Pang, P. D. Nicholson.
Bull. Am. Astron. Soc., Vol. 16, No. 3, Part I, p. 678 (1984). Abstract. – See Abstr. 010.061.

101.069 **Build up of carbonaceous material by fast protons on Pluto and Triton.**
G. Strazzulla, L. Calcagno, G. Foti.
Astron. Astrophys., Vol. 140, No. 2, p. 441 – 444 (1984).
The authors summarize their experimental results on the cross section for polymerization induced by fast protons ($E \approx 100$ keV – 2 MeV) on substrates ($T = 10 - 300$K) surrounded by hydrocarbon atmospheres or covered with frozen hydrocarbon layers. On the basis of these results they show that this process, activated by solar cosmic rays, is capable of building up substantial carbonaceous material on Pluto and Triton, both of which are probably covered by methane frost and/or surrounded by methane–rich atmospheres.

101.070 **1984 N1.**
IAU Circ., No. 3968 (1984).

101.071 **Pluto.**
IAU Circ., No. 3986 (1984).

101.072 **Probable ring of Neptune (1984 N1, 1981 N1).**
IAU Circ., No. 4022 (1984).

101.073 **The rotation of the Uranian system.**
M. Podolak.
Philos. Trans. R. Soc. London, Ser. A, Vol. 313, No. 1524, p. 141 – 146 (1984). – See Abstr. 012.068.
The rotation of Uranus is examined for clues as to the origin of the solar system. Both theories based on the formation of planets through the accretion of small planetesimals, and theories based on the formation of giant gaseous protoplanets through a gravitational instability in the primitive solar nebula allow for qualitative explanations of the large tilt of Uranus's equator to the orbital plane, and the fact that its satellites lie in the equatorial plane. Models of the planetary interior show that the mass ratio of ice–forming materials to rock in Uranus's interior must be more than about three if the rotation period is about 16 h. Such a large ratio seems to exclude those accretional theories that require most of the nebular gas to be heated to relatively high temperatures before being accreted into the planet.

101.074 **The occultation of Hyd –20°51695 by Uranus on 1982 May 1.**
A. K. Pandey, H. S. Mahra, V. Mohan.
Bull. Astron. Soc. India, Vol. 12, No. 3, p. 258 – 262 (1984).
The observed emersion event of the occultation of the star Hyd –20°51695 by Uranus on 1982 May 1 has been used to evaluate the temperature of the Uranian upper atmosphere. Assuming an isothermal atmosphere the scale height comes out to be $48 \pm 7$ km which corresponds to an isothermal temperature of $105 \pm 16$K for the Uranian upper atmosphere containing 90% hydrogen and 10% helium by number. From the inverted temperature profile, the mean temperature of the Uranian upper atmosphere comes out to be $100 \pm 25$K for the altitude range 20 km above to 100 km below the half–light level.

101.075 **Electropolarimetric study of Uranus.**
R. A. Chigladze.
Astron. Tsirk., No. 1320, p. 5 – 6 (1984). In Russian.

101.076 **The rings of Uranus.**
J. L. Elliot.
Planetary rings, p. 155 – 164 (1984). – See Abstr. 012.085 (IAU Colloq. No. 75).
The Uranian ring system consists of nine, narrow rings that have an average geometric albedo of 0.02. Since their discovery in 1977, the rings have been regularly observed with the technique of stellar occultations. The occultation data are well fitted by a kinematic model that describes the rings as ellipses, which are inclined to the equatorial plane of Uranus and precess due to harmonic terms in the Uranian potential field.

101.077 **Inclinations of the Uranian rings.**
R. G. French, J. L. Elliot.
Planetary rings, p. 165 – 168 (1984). – See Abstr. 012.085 (IAU Colloq. No. 75).
A kinematic model of the orbits of the nine known Uranian rings was fitted to available stellar occultation data, including a new set of times from the 26 April 1981 occultation obtained from the Anglo–Australian Telescope. The results of the fit reveal that seven of the nine rings have significant inclinations with

respect to the equatorial plane of Uranus. The inclinations are typically a few hundredths of a degree.

101.078 **Infrared observations of the Uranian rings.**
P. D. Nicholson, T. J. Jones, K. Matthews.
Planetary rings, p. 169 – 179 (1984). – See Abstr. 012.085 (IAU Colloq. No. 75).

The authors present maps of the Uranian rings at an effective wavelength of 2.2 μm, and spectrophotometric observations covering the ranges 2.0 – 2.4 μm and 2.9 – 3.9 μm. The maps, constructed from data obtained in May 1978 and July 1979, reveal a 2:1 azimuthal brightness variation whose amplitude and phase are consistent with the known width variation and apsidal precession of the eccentric ε ring. From the spectrophotometric data, the spectral geometric albedos of both Uranus and the integrated system of rings are calculated.

101.079 **Les occultations du 15 Août 1980 et du 22 Avril 1982 par les anneaux d'Uranus.**
B. Sicardy, P. Bouchet, C. Perrier, A. Brahic, J. Lecacheux, P. Laques.
Planetary rings, p. 181 – 191 (1984). – See Abstr. 012.085 (IAU Colloq. No. 75).

This paper reports the results of stellar occultation observations of the Uranus ring system on August 15, 1980 and April 22, 1982 obtained at the methane absorption bands at 0.89 μm and 2.2 μm.

101.080 **Indications of twist in the epsilon ring of Uranus.**
R. Vasundhara, J. C. Bhattacharyya.
Planetary rings, p. 193 – 195 (1984). – See Abstr. 012.085 (IAU Colloq. No. 75).

A comparison of light curves during stellar occultations by the ε ring of Uranus indicates that this ring assembly possesses stable fine structures. The authors discuss the possibility that the ring contains a twisted bundle of narrow lanes of condensed material similar to that of Saturn's F ring.

101.081 **Rings around Neptune?**
J. L. Elliot.
Planetary rings, p. 197 – 199 (1984). – See Abstr. 012.085 (IAU Colloq. No. 75).

Observation of two stellar occultations by Neptune in 1981 led to an upper limit of 0.07 on the optical depth of any equatorial rings with radii greater than 31,400 km.

101.082 **Weak dynamical effects in the Uranian ring system.**
A. P. Freedman, S. Tremaine, J. L. Elliot.
Planetary rings, p. 457 – 458 (1984). Abstract. – See Abstr. 012.085 (IAU Colloq. No. 75).

101.083 **Tidal breakup and origin of the Uranian ring system.**
P. Farinella, A. Milani, A. M. Nobili, P. Paolicchi.
Planetary rings, p. 687 – 689 (1984). – See Abstr. 012.085 (IAU Colloq. No. 75).

101.084 **The rings of Uranus.**
J. L. Elliot, P. D. Nicholson.
Planetary rings, p. 25 – 72 (1984). – See Abstr. 012.086.

Uranus is encircled by nine narrow rings, denoted 6, 5, 4, α, β, η, γ, δ and ε, in order of increasing distance from the planet. The rings have been observed at high spatial resolution (3.5 km) during 13 stellar occultations from 1977 to 1983. This paper presents an extensive review of the current (December 1983) knowledge of the Uranus ring system.

**The naming of Pluto.**
See Abstr. 004.041.

**The discovery of Neptune: essential revisions to the history.**
See Abstr. 004.068.

**Second order secular planetary theory for the four major planets.**
See Abstr. 042.011.

**Ephémérides des satellites naturels des planètes: Jupiter, Saturne, Uranus pour 1985.**
See Abstr. 046.036.

**Titan atmospheric probe.**
See Abstr. 051.009.

**Resonance locking between Jupiter and Uranus.**
See Abstr. 091.004.

**Dynamics of eccentric and inclined ringlets.**
See Abstr. 091.069.

**Sharp edges of planetary rings.**
See Abstr. 091.070.

**Dynamics of narrow rings.**
See Abstr. 091.090.

**An occultation of a star by the rings of Uranus in 1985.**
See Abstr. 096.020.

**The 22 April 1982 occultation of star KME–14 by Uranus and its rings.**
See Abstr. 096.021.

**Predicted occultations by Uranus and its rings: 1984 – 1990.**
See Abstr. 096.022.

**Appulse of SAO 186001 to Neptune.**
See Abstr. 096.025.

**Absolute spectrophotometry of Titan, Uranus, and Neptune: 3500 – 10,500 Å.**
See Abstr. 100.027.

**Bolometric geometric albedos of Titan, Uranus, and Neptune.**
See Abstr. 100.045.

**Oceans in the outer solar system.**
See Abstr. 100.049.

**Photometry of Titan, Uranus, and Neptune: lightcurves through solar maximum.**
See Abstr. 100.062.

**From icy planetesimals to outer planets and comets.**
See Abstr. 107.001.

# 102 Comets (Origin, Structure, Atmospheres, Dynamics)

102.001 **Explanation of the short-period comets' distributions according to the angular elements of orbits.**
A. M. Kazantsev, L. M. Sherbaum.
Astron. Zh., Tom 61, Vyp. 4, p. 750 – 754 (1984). In Russian. English translation in Sov. Astron., Vol. 28, No. 4.

It is shown with the help of numerical model calculations that the short-period comets' distributions on the argument of the perihelion and on the longitude of the perihelion are simply explained by the evolution of their orbits under planetary influence. It is found by comparison of some calculated and observed dependences that these comets, as celestial mechanical bodies, must must have existed hundreds and thousands of years before the moments of their discovery.

102.002 **On the magnetic field penetration in a cometary ionosphere.**
Z. M. Ioffe.
Astron. Zh., Tom 61, Vyp. 4, p. 812 – 814 (1984). In Russian. English translation in Sov. Astron., Vol. 28, No. 4.

A possible mechanism of magnetic field penetration in a cometary ionosphere is considered. It is shown that the velocity of the field motion, if the field is rope-like, may achieve $10^5$ cm $s^{-1}$.

102.003 **Generation of a magnetic barrier around a type I cometary tail.**
Z. M. Ioffe.
Sov. Astron., Vol. 28, No. 1, p. 72 – 73 (1984). English translation of 37.102.002.

102.004 **Die Oort'sche Kometenwolke. Teil 2.**
T. Herczeg.
Sterne Weltraum, 23. Jahrg., Nr. 7, p. 367 – 371 (1984).

102.005 **Constraints on cometary origins from isotropy of the microwave background and other measurements.**
M. E. Bailey, B. McBreen, T. P. Ray.
Mon. Not. R. Astron. Soc., Vol. 209, No. 4, p. 881 – 888 (1984). With a correction in Vol. 211, No. 1, p. 255 (1984).

Results are presented for four ways of limiting the mass of a dense inner core of comets that may surround the planetary system. Deep far-infrared measurements probably provide the most useful constraints, while future measurements with the Cosmic Background Explorer Satellite (COBE) will do even better. Theories of cometary origins which predict a mass of the inner core in excess of $\cong 100\ M_{\delta}$ should take these observations into account. The possibility of detecting comet clouds around nearby stars is also discussed.

102.006 **Cosmic dust and the comet connection.**
R. Knacke.
Sky Telesc., Vol. 68, No. 3, p. 206 – 210 (1984).

102.007 **Éphémérides des comètes.**
B. Morando.
Astrométrie, géodésie, mécanique céleste, p. 73 – 76 (1984). – See Abstr. 012.021.

102.008 **Dynamique du nuage de Oort.**
F. Rémy, F. Mignard.
Astrométrie, géodésie, mécanique céleste, p. 77 – 80 (1984). – See Abstr. 012.021.

102.009 **Capture de comètes par Jupiter.**
D. Benest, R. Bien, H. Rickman.
Astrométrie, géodésie, mécanique céleste, p. 81 – 84 (1984). – See Abstr. 012.021.

Les auteurs montrent que le mouvement de la comète P/Boethin est celui d'un satellite très lointain de Jupiter, et ils l'ont comparé avec le mouvement de la comète P/Slaughter–Burnham, qui est elle aussi en résonance 1/1 avec Jupiter, avec toutefois un comportement actuel un peu différent.

102.010 **Search for "exotic" molecules in bright comets.**
C. B. Cosmovici, L. Biermann, R. Lüst.
ESA Spec. Publ., ESA SP-218, p. 455 – 457 (1984). – See Abstr. 012.024.

The authors propose to take particular care of special molecular features in new bright comets during future IUE-observations. The most interesting molecules are tabulated.

102.011 **Highlights of cometary research.**
P. Stumpff.
Observations of comets: radio and otherwise, p. 1 – 5 (1984). Abstract. – See Abstr. 012.026.

102.012 **Radio line observations of comets.**
L. E. Snyder.
Observations of comets: radio and otherwise, p. 7 – 13 (1984). Abstract. – See Abstr. 012.026.

102.013 **The radio continuum of comets.**
W. J. Altenhoff.
Observations of comets: radio and otherwise, p. 15 – 19 (1984). Abstract. – See Abstr. 012.026.

102.014 **Comets in the UV.**
H. U. Keller.
Observations of comets: radio and otherwise, p. 33 – 43 (1984). Abstract. – See Abstr. 012.026.

102.015 **Theoretical aspects of cometary radio line excitation: OH, $H_2O$ and others.**
J. Crovisier.
Observations of comets: radio and otherwise, p. 45 – 51 (1984). Abstract. – See Abstr. 012.026.

102.016 **Cometary radar and radio science.**
M. K. Bird.
Observations of comets: radio and otherwise, p. 59 – 66 (1984). Abstract. – See Abstr. 012.026.

102.017 **Cometary radio occultation experiments.**
W.-H. Ip.
Observations of comets: radio and otherwise, p. 67 – 72 (1984). Abstract. – See Abstr. 012.026.

102.018 **Distribution of nearly parabolic comets according to Tisserand's constant and its consequences.**
V. V. Radzievskij, A. V. Artem'ev, M. A. Shishkina.
Gor'k. gos. ped. inst. Gor'kij, 11 pp. (1984). In Russian. Abstr. in Ref. Zh., 51. Astron., 8.51.166 (1984).

102.019 **The dissociation of OH and OD in comets by solar radiation.**
E. F. van Dishoeck, A. Dalgarno.
Icarus, Vol. 59, No. 3, p. 305 – 313 (1984).

The radiative lifetimes of cometary OH are calculated as a function of the heliocentric velocity of the comet and the velocity distributions of the product atoms are determined. At a distance of 1 AU from the Sun, the lifetimes vary between $1.2 \times 10^5$ and $1.9 \times 10^5$ sec at solar minimum and between $1.0 \times 10^5$ and $1.4 \times 10^5$ sec at solar maximum, depending upon velocity. The calculated lifetimes are generally consistent with the lifetimes inferred from observations, but suggest that some elaboration of the models is necessary.

102.020 **Investigation of the vectorial model of cometary atmospheres.**
V. A. Dranevich.
Pis'ma Astron. Zh., Tom 10, No. 9, p. 710 – 715 (1984). In Russian. English translation in Sov. Astron. Lett., Vol. 10.

General integral equations both for radical concentration and surface brightness are obtained in the frame of the vectorial model of a neutral cometary atmosphere. It is shown that the vectorial model brightness profiles and Hazer's model profiles practically coincide under the condition that certain dependences between their parameters do exist. These dependences which enable to transform Hazer's model to the vectorial one are shown in graphic form.

102.021 **Wind instability and helical structures in cometary tails.**
S. G. Gestrin, V. M. Kontorovich.
Pis'ma Astron. Zh., Tom 10, No. 10, p. 790 – 796 (1984). In Russian. English translation in Sov. Astron. Lett, Vol. 10.

Development of magnetic hydrodynamic analogue of wind instability at the boundary of two compressible media can cause appearance of helical perturbations with typical wavelength and azimuthal number in cometary tails.

102.022 **Coagulated interstellar dust as the basic ingredient of comets and carbonaceous chondrites.**
J. Greenberg.
The 27th International Geological Congress, Tom 5, Sect. 10 – 11, p. 279 (1984). Abstr. in Ref. Zh., 51. Astron., 9.51.248 (1984). – See Abstr. 012.031.

102.023 **On the problem of determining the element composition of cometary dust particles by a mass–analysis of shock plasma.**
V. B. Leonas, Yu. G. Malama.
Inst. kosm. issled. Akad. Nauk SSSR. Prepr., No. 885, 23 pp. (1984). In Russian. Abstr. in Ref. Zh., 62. Issled. Kosm. Prostranstva, 9.62.45 (1984).

102.024 **Comet dust – applications of Mie scattering.**
N. Eaton.
Vistas Astron., Vol. 27, Part 2, p. 111 – 129 (1984).

The physical properties of cometary dust are described in relation to the interplanetary Brownlee particles found in the upper atmosphere. The application of Mie scattering by smooth spheres is examined for the irregular particles found in the interplanetary medium. The Mie scattering formulae are quoted and discussed in relation to problems in their computation. Finally the formulae are used to investigate the temperature of cometary grains in the solar system.

102.025 **The multiple modes of interaction of the solar wind with a comet as it approaches the sun.**
D. A. Mendis, K. R. Flammer.
Earth, Moon, Planets, Vol. 31, No. 3, p. 301 – 311 (1984).

A quasi-steady 1-D hydrodynamic model, with mass addition, has been used to study the various modes of interaction of the solar wind with a medium-bright, $H_2O$-dominated comet (such as P/Halley) approaching the Sun.

102.026 **The steady–state $1/a$–distribution and the problem of cometary fading.**
M. E. Bailey.
Mon. Not. R. Astron. Soc., Vol. 211, No. 2, p. 347 – 368 (1984).

The problem of the $1/a$–distribution including stellar perturbations and allowing for arbitrary fading per revolution is investigated. It is emphasized that conclusions about comet origins based on the $1/a$–distribution depend entirely on the assumptions made about fading. Assuming the steady-state model to apply the author has solved the concerning one–dimensional integral equation and compared the predicted $1/a$–distribution with observations in order to constrain the fading function. A physical model based on thermal effects associated with a long-period comet's perihelion passage is tentatively proposed. It is emphasized that until a viable explanation for fading has been found, the validity of the primordial hypothesis for comet origins remains unresolved.

102.027 **Ballistic aspects of investigation of comets.**
Eh. L. Akim, N. A. Belyaev, R. K. Kazakova, O. V. Papkov, A. K. Platonov, V. V. Savchenko, K. G. Sukhanov.
Dokl. Akad. Nauk SSSR. Ser. Mat. Fiz., Tom 275, No. 2, p. 323 – 327 (1984). In Russian. Abstr. in Ref. Zh., 51. Astron., 10.51.117 (1984).

102.028 **The method of "bursts" in the program of search for families of long–period comets.**
V. V. Radzievskij, A. V. Artem'ev, V. V. Belova, D. F. Ziganshina.
Gor'k. gos. ped. inst. Gor'kij, 6 pp. (1984). In Russian. Abstr. in Ref. Zh., 51. Astron., 10.51.261 (1984).

102.029 **A possible observation of a cometary bow shock.**
C. T. Russell, J. L. Phillips, M. R. Arghavani, J. D. Mihalov, W. C. Knudsen, K. Miller.
Geophys. Res. Lett., Vol. 11, No. 10, p. 1022 – 1025 (1984).

Three discontinuities within a large disturbance of the interplanetary field, observed at Venus and previously suggested to be due to the possible passage of a comet, are examined to see if they could possibly represent the first observation of a cometary shock. The changes in number density, temperature and velocity as well as the magnetic structure of the discontinuities are consistent with this hypothesis.

102.030 **Recurrent power series integration of equations of comet's motion including the nongravitational terms in Marsden's form.**
G. Sitarski.
Acta Astron., Vol. 34, No. 1, p. 53 – 63 (1984).

Recurrent relations for computing the coefficients in time power expansion of functions $f(r)$ and $g(r)$ are derived which makes possible to integrate the equation of comet's motion when nongravitational forces in the form $A_i f(r)$ or $A_i g(r)$ are taken into account. Differential equations are also given to compute the coefficients in observational equations for determination of values of $A_i$ by the least squares method. Results of practical computations are discussed.

102.031 **CH in comets: a tracer of activity?**
A. L. Cochran.
Bull. Am. Astron. Soc., Vol. 16, No. 3, Part I, p. 638 (1984). Abstract. – See Abstr. 010.061.

102.032 **On the shapes of cometary nuclei.**
Z. Sekanina.
Bull. Am. Astron. Soc., Vol. 16, No. 3, Part I, p. 642 – 643 (1984). Abstract. – See Abstr. 010.061.

102.033 **Morphology of cometary nuclei.**
M. C. Malin, G. E. Danielson.
Bull. Am. Astron. Soc., Vol. 16, No. 3, Part I, p. 643 (1984). Abstract. – See Abstr. 010.061.

102.034 **Sublimation rates of grains in cometary comae.**
W. A. Baum, T. J. Kreidl.
Bull. Am. Astron. Soc., Vol. 16, No. 3, Part I, p. 643 (1984). Abstract. – See Abstr. 010.061.

102.035 **Continued comparisons of comets and outer solar system small bodies.**
W. K. Hartmann, D. P. Cruikshank, D. J. Tholen.
Bull. Am. Astron. Soc., Vol. 16, No. 3, Part I, p. 644 (1984). Abstract. – See Abstr. 010.061.

102.036 **Time dependent dusty hydrodynamical flow near cometary nuclei.**
A. F. Nagy, T. I. Gombosi, T. E. Cravens.
Bull. Am. Astron. Soc., Vol. 16, No. 3, Part I, p. 645 (1984). Abstract. – See Abstr. 010.061.

102.037 **Numerical integration of short–period comet orbits over eight centuries.**
A. Carusi, L. Kresák, E. Perozzi, G. B. Valsecchi.
Bull. Am. Astron. Soc., Vol. 16, No. 3, Part I, p. 645 – 646 (1984). Abstract. – See Abstr. 010.061.

102.038 **Changes in comet characteristics with dynamical evolution.**
C. A. Wood, J. R. Bollinger.
Bull. Am. Astron. Soc., Vol. 16, No. 3, Part I, p. 646 (1984). Abstract. – See Abstr. 010.061.

102.039 **The physical and orbital characteristics of comets with plasma tail disconnections.**
J. R. Bollinger, C. A. Wood.
Bull. Am. Astron. Soc., Vol. 16, No. 3, Part I, p. 646 (1984). Abstract. – See Abstr. 010.061.

102.040 **Peculiar orbital histories of some short–period comets.**
A. Carusi, L. Kresák, E. Perozzi, G. B. Valsecchi.
Bull. Am. Astron. Soc., Vol. 16, No. 3, Part I, p. 704 (1984). Abstract. – See Abstr. 010.061.

102.041 **Emission by OD in comets.**
D. G. Schleicher, M. F. A'Hearn, R. A. West.
Bull. Am. Astron. Soc., Vol. 16, No. 3, Part I, p. 707 (1984). Abstract. – See Abstr. 010.061.

102.042 **The origin of CS in comets.**
P. S. Butterworth, W. M. Jackson, D. Ballard.
Bull. Am. Astron. Soc., Vol. 16, No. 3, Part I, p. 709 (1984). Abstract. – See Abstr. 010.061.

102.043 **Implications of observed distributions of very long period comet orbits.**
R. S. Harrington.
Bull. Am. Astron. Soc., Vol. 16, No. 3, Part I, p. 721 (1984). Abstract. – See Abstr. 010.064.

102.044 **The distribution of the aphelion directions of long–period comets.**
R. Lüst.
Astron. Astrophys., Vol. 141, No. 1, p. 94 – 100 (1984).

The distribution of the aphelia of 223 long–period comets is analysed with respect to the equatorial, the galactic and the solar apex coordinate system. A systematic north–south asymmetry can at least partly be explained by observational biasing due to a better detection probability for comets in the northern sky. A division of the sample into 3 subgroups according to the semimajor axes shows a different behavior of the more primordial "new" and "intermediate" comets on one hand and the "old" comets with smaller semimajor axes on the other hand. No indication is found for an accumulation of aphelia directly around the antapex.

102.045 **Systematic and random deformations of long–period comet orbits.**
V. A. Antonov, Z. P. Todriya.
Sov. Astron. Lett., Vol. 10, No. 3, p. 166 (1984). English translation of 37.102.048.

102.046 **Some radiation–chemical aspects of chemistry in cometary nuclei.**
I. G. Draganić, Z. D. Draganić, S. Vujošević.
Icarus, Vol. 60, No. 3, p. 464 – 475 (1984).

The present contribution analyzes various achievements in radiation chemistry that are of interest to the chemistry of cometary nuclei. This is particularly the case with the radiation chemistry of water and aqueous systems, both liquid and frozen. The findings suggest that cometary material could have been significantly altered by cosmic rays and the imbedded radionuclides, and encourage more simulation experiments with ionizing radiation as an energy source.

102.047 **An idealized short–period comet model: surface insolation, $H_2O$ flux, dust flux, and mantle evolution.**
F. P. Fanale, J. R. Salvail.
Icarus, Vol. 60, No. 3, p. 476 – 511 (1984). = PGD Publ. No. 365, Planet. Geosci. Div., Hawaii Inst. Geophys., Univ. Hawaii, Manoa, Honolulu, Hawaii 96822.

A model of cometary activity is developed which integrates the feedback processes involving heat, gas, and dust transport, and dust mantle development. The model includes the effects of latitude, rotation, and spin axis orientation. Results are obtained for various grain size distributions, dust–to–ice ratios, and spin axis orientations. Attention is focused on the development, change of structure and distribution of dust mantles and their mutual interaction with ice surface temperature and gas and dust production.

102.048 **Temperature profiles and thermal stresses in cometary nuclei.**
E. Kührt.
Icarus, Vol. 60, No. 3, p. 512 – 521 (1984).

Assuming a spherical nucleus of water ice with an isothermal surface, temperature profiles are computed for several heliocentric distances of Halley's comet. Sublimation of ice and the temperature dependence of the material properties are taken into account. The resulting strongly nonlinear heat diffusion problem is solved numerically. With some simplifications an analytical solution is derived. The heat conduction causes a "thermal hysteresis" of the surface temperature and a slow increase of the inner temperature. The complete thermal equilibrium is reached, however, only after 100 or more revolutions in the inner solar system. The calculated temperature profiles are used to estimate the thermal stresses in the nucleus. It is shown that thermal stresses can give a plausible explanation for cometary outbursts and splits.

102.049 **The role of the intrinsic energy source in the physical behaviour of a multiple cometary nucleus.**
V. D. Davydov.
Kosm. Issled., Tom 22, Vyp. 6, p. 922 – 941 (1984). In Russian. English translation in Cosm. Res.

102.050 **Rotation of cometary nuclei.**
M. K. Wallis.
Philos. Trans. R. Soc. London, Ser. A, Vol. 313, No. 1524, p. 165 – 170 (1984). – See Abstr. 012.068.

Asymmetric comas and repetitive appearances of structures in the heads and tails of comets are used to infer nuclei rotation periods. However, periodic behaviour of optically–thick expanding comas or of ion plasma production may contribute spurious results. The spin periods of comets are longer than those of asteroids, ranging generally over 10 – 100 h and above probable limits for gravitational escape. The periods show a flatter distribution, which may reflect an accretional rather than collisional–fragmentation history. Arguments for spin–up with age due to sublimating gases are weak; the converse is possible and spin–down due to preferential escape of particles from equatorial regions appears likely.

102.051 **Comets.**
H. B. Ridley.
J. Br. Astron. Assoc., Vol. 95, No. 1, p. 29 – 32 (1984).

102.052 **Los cometas y el origen de la vida. 2ª Parte.**
A. H. Delsemme.
Rev. Astron., Tomo 56, No. 228/229, p. 22 – 30 (1984).

102.053 **On the statistical verification of the hypothesis of the interstellar origin of comets.**
M. A. Mamedov, V. V. Radzievskij.
Dokl. Akad. Nauk AzSSR, Tom 40, No. 1, p. 17 – 21 (1984). In Russian. Abstr. in Ref. Zh., 51. Astron., 11.51.78 (1984).

102.054 **Prediction of deuterium abundance in comets.**
V. Vanýsek.
Bull. Astron. Inst. Czech., Vol. 35, No. 6, p. 361 – 364 (1984).

Using a chemical scheme based on ion–molecule reactions in cool interstellar clouds, the deuterium enrichment of volatile material in comets is discussed. It is assumed that the hydrogen containing molecules are deuterated before accretion and condensation on the core–mantle dust particles from which the cometesimals are formed. The estimated total ratio in a cometary nucleus is $D/H \sim 2 \times 10^{-4}$.

102.055 **Aeronomical processes in cometary atmospheres: the carbon compounds' puzzle.**
M. C. Festou.
Inst. Astrophys. Paris, Pré-Publ., No. 66, 11 pp. (1984). To appear in Adv. Space Res.

102.056 **Vibrational and rotational excitation of CO in comets: nonequilibrium calculations.**
G. Chin, H. A. Weaver.
Astrophys. J., Vol. 285, No. 2, p. 858 - 869 (1984).
The vibrational and rotational excitation of the CO molecule in cometary comae has been investigated using a model which includes IR vibrational pumping by the solar flux, vibrational and rotational radiative decay, and collisional coupling among rotational states. Steady state is not assumed in solving the rate equations. The evolution of a shell of CO gas is followed as it expands from the nucleus into the outer coma. Collisional effects are treated using a kinetic temperature profile derived from recent theoretical work on the coma energy balance. From these calculations the conditions are derived under which the CO millimeter and IR transitions should be observable.

102.057 **Erratum: "Infrared molecular emissions from comets" [Astrophys. J., Vol. 276, No. 2, p. 782 - 797 (1984)].**
H. A. Weaver, M. J. Mumma.
Astrophys. J., Vol. 285, No. 2, p. 872 - 873 (1984). See Abstr. 37.102.013.

102.058 **The distribution of orbits in the hypothesis of eruption of comets.**
M. V. Nikolaeva, V. P. Tomanov.
Astron. Tsirk., No. 1306, p. 1 - 4 (1984). In Russian.

102.059 **Comets and the Galaxy.**
F. A. Tsitsin, V. M. Chepurova, A. S. Rastorguev.
Astron. Tsirk., No. 1310, p. 5 - 6 (1984). In Russian.

102.060 **Las observaciones y descubrimiento de los cometas.**
V. Yurevich.
Publ. Espec. Obs. Astron. Quito, No. 2, p. 69 - 74 (1981).

102.061 **On the commensurability of the axial rotation periods of cometary nuclei with the 160-min period.**
V. A. Kotov, V. P. Tarashchuk.
Astron. Tsirk., No. 1318, p. 1 - 4 (1984). In Russian.

102.062 **Averaging equations of orbital motion of a comet with account for nongravitational effects.**
E. N. Polyakhova.
Astron. Vestn., Tom 18, No. 3, p. 235 - 239 (1984). In Russian. Abstr. in Ref. Zh., 51. Astron., 12.51.369 (1984).

102.063 **On the non-reality of the "seasonal effect" in cometary brightness flares.**
K. I. Churyumov.
Astron. Vestn., Tom 18, No. 3, p. 252 - 254 (1984). In Russian. Abstr. in Ref. Zh., 51. Astron., 12.51.373 (1984).

102.064 **On the temperature minimum in the interior coma of a comet.**
D. V. Bisikalo, V. S. Stel'nitskij.
Komet. Tsirk., No. 330 (1984). In Russian.

102.065 **Matrices of ultra-low density and strength on the surface of the ice nucleus of a comet.**
Kh. I. Ibadinov, S. A. Aliev, A. A. Rakhmonov.
Komet. Tsirk., No. 330 (1984). In Russian.

102.066 **Close encounters of comets with Jupiter.**
N. Yu. Emel'yanenko.
Komet. Tsirk., No. 331 (1984). In Russian.

102.067 **Interconnection of comets with direct and retrograde motions.**
A. M. Kazantsev.
Komet. Tsirk., No. 331 (1984). In Russian.

102.068 **On the variation of minimum distances between the orbit of Jupiter and the orbits of short-period comets.**
Yu. G. Babenko, V. P. Konopleva.
Komet. Tsirk., No. 333 (1984). In Russian.

102.069 **On an asymmetry in the distribution of perihelion longitudes of the orbits of the periodic comets of the Jupiter family.**
A. S. Guliev.
Komet. Tsirk., No. 333 (1984). In Russian.

102.070 **Photochemistry of Na-containing compounds and their abundance in cometary nuclei.**
I. N. Matveev.
Komet. Tsirk., No. 333 (1984). In Russian.

102.071 **The effects of the diffuse radiation fields due to multiple scattering and thermal reradiation by dust on the dynamics and thermodynamics of a dusty cometary atmosphere.**
M. L. Marconi, D. A. Mendis.
Astrophys. J., Vol. 287, No. 1, p. 445 - 454 (1984).
In an earlier paper the authors presented a self-consistent, multifluid model of the dynamical and thermal structure of a dust/gas atmosphere of a dirty-clathrate cometary nucleus. The main deficiency of that model was the neglect of the diffuse radiation fields due to the circumnuclear dust. The authors have remedied that in the present investigation by introducing a three-stream approximation based on the shadowing geometry of the nucleus for the radiation that is multiply scattered as well as thermally reradiated by the circumnuclear dust. The calculation is carried out for a 3 km radius comet, believed to represent P/Halley, at a heliocentric distance of 0.89 AU, which corresponds to its heliocentric distance during its 1986 March encounter by the European *GIOTTO* spacecraft.

102.072 **Study into the accuracy of astrometric observations of comets.**
S. Röser.
European Space Agency Contractor Report No. 5759/83/D/IM(SC). European Space Agency, Scientific and Technical Publications Branch, ESTEC, Postbus 299, 2200 AG Noordwijk, The Netherlands. 81 pp. (1984).
This study deals with the accuracy with which astrometric observations of comets have been made in the past 200 years. Special emphasis is given to the case of comet Halley at its 1759, 1835 and 1910 apparitions. The following topics are investigated in this report: (1) the influence of the positions of reference stars, (2) the accuracy with which relative measurements between comets and stars can be performed with various techniques, and (3) an optimum equipment for future observations of comets.

**Orbits for amateurs with a microcomputer.**
See Abstr. 003.019.

**Comets: why they are important.**
See Abstr. 014.010.

**Programmsystem und Datenorganisation zur Berechnung von Kleinplaneten- und Kometenbahnen.**
See Abstr. 021.018.

**Heat flow through dusty and porous ices.**
See Abstr. 022.094.

**Collection and analysis techniques for cometary dust.**
See Abstr. 022.095.

**Die visuelle Kometenbeobachtung und ihre Auswertung.**
See Abstr. 036.007.

**On a displacement of the photometric center from the center of mass in positional observations of comets and minor planets.**
See Abstr. 036.054.

**The photography of comets.**
See Abstr. 036.115.

**Visual observations of comets. II. Methods and brightness estimates.**
See Abstr. 036.174.

**Astrometric network SOPROG.**
See Abstr. 041.052.

**The dynamics of the minor bodies in the solar system.**
See Abstr. 042.059.

**The Radzievskij–Tisserand criterion.**
See Abstr. 042.092.

**Plans for future space missions.**
See Abstr. 051.022.

**Recent study of a comet nucleus penetrator experiment.**
See Abstr. 051.055.

**An assessment study of a joint NASA/ESA comet coma sample return mission.**
See Abstr. 051.060.

**Opportunities of ballistic missions to long-period comets.**
See Abstr. 052.036.

**Photon scattering coefficients for $H_2O^+$ molecules.**
See Abstr. 063.017.

**Gasdynamical models of interaction between the solar wind and cometary ionospheres.**
See Abstr. 074.026.

**On scattering of solar X-ray emission by comets.**
See Abstr. 076.002.

**Solar oscillations and the effect of a comet impact.**
See Abstr. 080.045.

**On a possible genetic relation between insufficiently studied meteor streams and comets.**
See Abstr. 104.034.

**From icy planetesimals to outer planets and comets.**
See Abstr. 107.001.

**Do comet showers kill the flowers?**
See Abstr. 107.031.

**Orbital stability at the limits of the solar system: comets and solar companions.**
See Abstr. 107.032.

**The microstructure of terrestrial catastrophism.**
See Abstr. 107.047.

# 103 Comets (Individual Objects)

103.001 **Tabulation of comet observations.**
Int. Comet Q., Vol. 6, No. 3, p. 63 – 71 (1984).
Concerning comets: 1974 II P/Schwassmann–Wachmann 1, 1980 XI P/Encke, 1982 VII P/d'Arrest (=1982e), 1982 VIII P/Churyumov–Gerasimenko (=1982f), 1982d P/Tempel 2, 1982k P/Kopff, 1983d IRAS–Araki–Alcock, 1983l Černis, 1983n P/Crommelin, 1983o IRAS, 1983s P/Wild 2, 1983v P/Hartley–IRAS, 1983w P/Clark, 1984h P/Faye, 1984i Austin, 1984j P/Takamizawa.

103.002 **Recent news and research concerning comets.**
D. W. E. Green.
Int. Comet Q., Vol. 6, No. 3, p. 71 – 74 (1984).
Concerning comets: 1972 IV P/Neujmin 3, 1975 VII P/Smirnova–Chernykh, 1977 VII P/Gehrels 3, 1978 XXIV P/Van Biesbroeck, 1980 XI P/Encke, 1983j P/IRAS, 1983o IRAS, 1983r P/Harrington–Abell, 1983s P/Wild 2, 1983u P/Taylor, 1983v P/Hartley–IRAS, 1983w P/Clark, 1984a P/Bradfield, 1984c P/Neujmin 1, 1984d P/Russell 4, 1984f Shoemaker, 1984g P/Wolf–Harrington, 1984h P/Faye, 1984i Austin, 1984j Takamizawa, 1984k P/Arend–Rigaux.

103.003 **A tale of two comets.**
J. Davies.
New Sci., Vol. 101, No. 1403, p. 16 – 19 (1984). Abstr. in Phys. Abstr., Vol. 87, No. 1239, Entry 82455 (1984).

103.004 **Les comètes de l'année 1982. Désignations définitives.**
A. C. Levasseur-Regourd.
Astronomie, Vol. 98, p. 455 (1984).

103.005 **IRAS observations of comets.**
J. K. Davies.
Observations of comets: radio and otherwise, p. 21 – 25 (1984). Abstract. – See Abstr. 012.026.

103.006 **Comet Digest.**
J. E. Bortle.
Sky Telesc., Vol. 68, No. 1, p. 92 (1984).
Concerning comets: 1982i P/Halley, 1983w P/Clark, 1984e P/Giacobini–Zinner, 1984? P/Faye.

103.007 **Comet Digest.**
J. E. Bortle.
Sky Telesc., Vol. 68, No. 2, p. 188 (1984).
Concerning comets: 1982i P/Halley, 1983v Hartley–IRAS, 1983w P/Clark, 1984h P/Faye.

103.008 **Comet Digest.**
J. E. Bortle.
Sky Telesc., Vol. 68, No. 3, p. 284 (1984).
Concerning comets: 1983n P/Crommelin, 1984h P/Faye, 1984i Austin.

103.009 **Comet Digest.**
J. E. Bortle.
Sky Telesc., Vol. 68, No. 4, p. 378 (1984).
Concerning comets: 1948 XI, 1982i P/Halley, 1984h P/Faye.

103.010 **Comet Digest.**
J. E. Bortle.
Sky Telesc., Vol. 68, No. 5, p. 482 (1984).
Concerning comets: 1982i P/Halley, 1984i Austin, 1984j P/Takamizawa.

103.011 **Infrared photometry of comet Bowell and other comets.**
M. F. A'Hearn, E. Dwek, A. T. Tokunaga.
Astrophys. J., Vol. 282, No. 2, p. 803 – 806 (1984).
The authors present broad–band and intermediate–band infrared photometry of comets Bowell, Elias, and Gunn. The data for comet Bowell show that the strongly depressed albedo near 3 μm observed by Campins and colleagues is very broad and extends at least to 4 μm. Scattering calculations indicate that existing infrared data do not provide direct, observational evidence for the presence of $H_2O$ ice.

103.012 **Radioastronomical observations of comets IRAS–Araki–Alcock (1983d) and Sugano–Saigusa–Fujikawa (1983e).**
W. M. Irvine, Z. Abraham, M. A'Hearn, W. Altenhoff, C. Andersson, J. Bally, W. Batrla, A. Baudry, D. Bockelée-Morvan, G. Chin, J. Crovisier, I. de Pater, D. Despois, L. Ekelund, E. Gerard, T. Hasegawa, C. Heiles, J. M. Hollis, W. Huchtmeier, N. Kaifu, R. Levreault, C. R. Masson, P. Palmer, M. Perault, L. J. Rickard, A. I. Sargent, E. Scalise, F. P. Schloerb, J. Schmidt, A. A. Stark, M. Stevens, P. Stumpff, E. C. Sutton, D. Swade, M. Sykes, B. Turner, C. Wade, M. Walmsley, J. Webber, A. Winnberg, A. Wootten.
Icarus, Vol. 60, No. 1, p. 215 – 220 (1984).
Detections and upper limits to the continuum emission ($1 \leqslant \lambda \leqslant 6$ cm) and spectral line emission (OH, CO, CS, HCN, $HCO^+$, CN, $CH_3CN$, $CH_3C_2H$, $NH_3$, $H_2O$, $HC_3N$, $CH_3CH_2CN$) are reported from radio observations of comets 1983d and 1983e. Comparison is made with observations of CN at optical wavelengths. These results may be useful in planning future cometary observations.

103.013 **Luminosity and spatial distribution of the [O I] 6300 Å emission in comets.**
U. Fink, J. R. Johnson.
Astron. J., Vol. 89, No. 10, p. 1565 – 1572 (1984).
The authors have obtained CCD observations of the forbidden oxygen lines at 6300 and 6364 Å for the comets Tuttle, Stephan–Oterma, and Brooks 2. Their high–quality observations have allowed the authors to perform good cancellation of the night sky [O I] lines and to determine the spatial profiles and the absolute luminosities of the cometary [O I] lines. Analysis of both the spatial emission profiles and the total [O I] flux demonstrates that the source of the observed [O I] photons is direct photodissociation of water vapor. Production of $O^1D$ by dissociation of OH is of minor importance in the inner coma, but becomes dominant at larger distances from the nucleus. Spatial profiles for comets Tuttle and Stephan–Oterma agree well with the model calculations of Festou and Feldman (1981), as well as with a more simple Haser model having a scale length at 1 AU of $8.2 \times 10^4$km for the parent molecule water vapor.

103.014 **Charge coupled device (CCD) spectroscopy of comets: Tuttle, Stephan–Oterma, Brooks 2, and Bowell.**
J. R. Johnson, U. Fink, S. M. Larson.
Icarus, Vol. 60, No. 2, p. 351 – 372 (1984).
Spectra of the four comets, Tuttle, Stephan–Oterma, Brooks 2, and Bowell, were taken with a prototype space telescope charge coupled device (CCD) camera using a $500 \times 500$ Texas Instruments chip. The spectra extended from 5600 to 10,400 Å at a resolution of $\sim 25$ Å. Both absolute flux scales and spectral albedos were determined with the data reduction procedure which included flat fielding and sky subtraction.

103.015 **Stellar occultations by two comets: IRAS–Araki–Alcock (1983d) and P/IRAS (1983j).**
J. Lecacheux, W. Thuillot, T. Encrenaz, P. Laques, D. Rouan, R. Despiau.
Icarus, Vol. 60, No. 2, p. 386 – 390 (1984).
A stellar appulse of comet 1983d was observed on 1983 May 08.956 using a video photometry technique. Although the miss distance did not exceed $420 \pm 50$ km no significant absorption was detected. Another close appulse, of comet 1983j this time, was measured with a photoelectric photometer on 1983 September 14.036. A small feature of the lightcurve, only 1.8% deep, could be attributed to dust surrounding the nucleus within a 350 km radius. The deduced dust ejection rate is $6 \times 10^6$g/sec.

103.016 **Erratum: "Improved orbits of three periodic comets: Reinmuth 1, Slaughter–Burnham and Kearns–Kwee" [Acta Astron., Vol. 33, No. 1, p. 119 – 126 (1983)].**
B. Todorovic–Juchniewicz.
Acta Astron., Vol. 34, No. 2, p. 314 (1984). See Abstr. 34.103.006.

103.017 **Comet Digest.**
J. E. Bortle.
Sky Telesc., Vol. 68, No. 6, p. 583 (1984).
Concerning comets: 1984i Austin, 1984o Meier, P/Encke.

103.018 **Drawings of recently–observed comets.**
J.-C. Merlin, S. J. O'Meara.
Int. Comet Q., Vol. 6, No. 4, p. 80 – 86 (1984).

103.019 **Tabulation of comet observations.**
Int. Comet Q., Vol. 6, No. 4, p. 90 – 95 (1984).
Concerning comets: 1974 II P/Schwassmann–Wachmann 1, 1983n P/Crommelin, 1983s P/Wild 2, 1983w P/Clark, 1984c P/Neujmin 1, 1984d P/Russell 4, 1984f Shoemaker, 1984g P/Wolf–Harrington, 1984h P/Faye, 1984i Austin, 1984j P/Takamizawa, 1984o Meier.

103.020 **Recent news and research concerning comets.**
D. W. E. Green.
Int. Comet Q., Vol. 6, No. 4, p. 95 – 99 (1984).
Concerning comets: 1984c P/Neujmin 1, 1984i Austin, 1984j P/Takamizawa, 1984k P/Arend–Rigaux, 1984m P/Schaumasse, 1984n P/Kowal–Mrkos, 1984o Meier, 1984p P/Tsuchinshan 1.

103.021 **Two close encounters.**
M. Festou.
Ciel Espace, No. 196, p. 8 – 10 (1983). In French. Abstr. in Phys. Abstr., Vol. 87, No. 1246, Entry 115713 (1984).

103.022 **Spectrophotometric and spatially resolved comparison of comets IRAS–Araki–Alcock, SSF, and P/Crommelin.**
H. Spinrad.
Bull. Am. Astron. Soc., Vol. 16, No. 3, Part I, p. 637 (1984). Abstract. – See Abstr. 010.061.

103.023 **Observed $H_2O^+/CO^+$ ratios in comets.**
S. Wyckoff, P. Wehinger, E. Leibowitz, T. Pitman.
Bull. Am. Astron. Soc., Vol. 16, No. 3, Part I, p. 639 (1984). Abstract. – See Abstr. 010.061.

103.024 **Gas–to–dust production ratios for sixteen comets.**
P. D. Feldman, M. F. A'Hearn.
Bull. Am. Astron. Soc., Vol. 16, No. 3, Part I, p. 642 (1984). Abstract. – See Abstr. 010.061.

103.025 **Studies of the spectral continua of seventeen comets.**
R. L. Newburn Jr., H. Spinrad.
Bull. Am. Astron. Soc., Vol. 16, No. 3, Part I, p. 645 (1984). Abstract. – See Abstr. 010.061.

103.026 **Continuum spectra of comets.**
R. Remillard, D. Jewitt.
Bull. Am. Astron. Soc., Vol. 16, No. 3, Part I, p. 645 (1984). Abstract. – See Abstr. 010.061.

103.027 **Comets near the $H_2O$ sublimation threshold zone.**
R. M. Wagner, S. Wyckoff, P. A. Wehinger, D. G. Schleicher, M. Belton, B. L. Lutz, H. Spinrad.
Bull. Am. Astron. Soc., Vol. 16, No. 3, Part I, p. 704 (1984). Abstract. – See Abstr. 010.061.

103.028 **The lost periodic comets of one apparition.**
C. E. Spratt.
J. R. Astron. Soc. Can., Vol. 78, No. 5, p. 189 – 195 (1984).

Of the 48 periodic comets that have passed perihelion only once, 18 have failed to return for a second time, as predicted. This paper surveys those 18 comets.

103.029 **The 1983 – 84 Halley's comet observing season and the comet Crommelin trial run.**
S. J. Edberg.
I.A.P.P.P. Commun., No. 17, p. 31 – 32 (1984).

103.030 **Observations of comets.**
Minor Planet Circ., Nos. 8827 – 8832, 8961 – 8964, 9041 – 9043, 9122 – 9125, 9174 – 9178, 9253 – 9257 (1984).

Observations made at the following stations are published: Abastumani, Alma-Ata, Athens, Bulgarian Natl. Obs., Byurakan, Catania, Caussols, Cerro Tololo, Chamberlin Obs., Crimean Astrophys. Obs., East Chinnock, El Leoncito, ESO, Geisei, Gissar, Hoher List, Kambah (Canberra), Kazan Zelenchukskaya Stn., Kitab, Kitt Peak, Kleť, Kourovskaya, Lowell Obs. Anderson Mesa Stn., Maria Mitchell Obs., Mauna Kea, Mt. John Obs., Mt. Palomar, Nihondaira Obs., Oak Ridge Obs., Oss. Chaonis, Perth Obs., Purple Mountain Obs., S. Vittore (Bologna), SAAO Cape Town, Sapporo, Sendai, Siding Spring, Skalnaté Pleso, Stakenbridge, Steward Obs. Kitt Peak Stn., Tokyo Obs. Dodaira Stn., Tokyo Obs. Kiso Stn., Uzhgorod, Victoria, Woolston Obs., Zimmerwald, Zvenigorod.
Observations of the following comets are published: 1972 VIII Heck-Sause, 1973 II Kojima, 1974 II P/Schwassmann-Wachmann 1, 1974 III Bradfield, 1975 I P/Boethin, 1975 VII P/Smirnova-Chernykh, 1975 IX Kobayashi-Berger-Milon, 1975 X Suzuki-Saigusa-Mori, 1975 XII Mori-Sato-Fujikawa, 1976 XV P/Schaumasse, 1980 XI P/Encke, 1981 XVIII Slaughter-Burnham, 1982 I Bowell, 1982 VI Austin, 1982 VIII P/Churyumov-Gerasimenko, 1982 X P/Gunn, 1982d P/Tempel 2, 1982i P/Halley, 1982j P/Tempel 1, 1982k P/Kopff, 1983d IRAS-Araki-Alcock, 1983e Sugano-Saigusa-Fujikawa, 1983h P/Johnson, 1983j P/IRAS, 1983k IRAS, 1983*l* Černis, 1983n P/Crommelin, 1983o IRAS, 1983p Shoemaker, 1983r P/Harrington-Abell, 1983s P/Wild 2, 1983u P/Taylor, 1983v P/Hartley-IRAS, 1983w P/Clark, 1984a P/Bradfield, 1984c P/Neujmin 1, 1984d P/Russell 4, 1984e P/Giacobini-Zinner, 1984f Shoemaker, 1984g P/Wolf-Harrington, 1984h P/Faye, 1984i Austin, 1984j P/Takamizawa, 1984k P/Arend-Rigaux, 1984*l* P/Gehrels 3, 1984m P/Schaumasse, 1984n P/Kowal-Mrkos, 1984o Meier, 1984p P/Tsuchinshan 1, 1984q P/Shoemaker 1, 1984r Shoemaker, 1984s Shoemaker, 1984t Levy-Rudenko, 1984u P/Shoemaker 2, 1984v Hartley.

103.031 **Orbital elements of comets.**
Minor Planet Circ., Nos. 8827 – 9314 (1984).

The comets are listed according to their Roman numeral designation or preliminary designation. The names of the authors are given behind the respective M.P.C. numbers.
1982 IX P/Russell 3 9304 S. Nakano; 1982i P/Halley 9214, 1983n P/Crommelin 9213 D. K. Yeomans; 1983o IRAS, 1983t Kowal-Vávrová, 1983v P/Hartley-IRAS 9304 S. Nakano; 1984a P/Bradfield 8910 D. W. E. Green, 9025 B. G. Marsden; 1984d P/Russell 4 8902 B. G. Marsden, 9304 S. Nakano; 1984f Shoemaker 8902, 9026, 9154, 1984i Austin 9026, 9154, 1984j P/Takamizawa 9026, 9070, 9221, 1984n P/Kowal-Mrkos 9154, 9211, 1984o Meier 9154, 9212, 1984q P/Shoemaker 1 9212, 9293 B. G. Marsden; 1984r Shoemaker 9204 S. Nakano, 9292 B. G. Marsden; 1984s Shoemaker 9204, 9305 S. Nakano; 1984t Levy-Rudenko 9293 B. G. Marsden; 1984u P/Shoemaker 2 9304 S. Nakano; 1984v Hartley 9293 B. G. Marsden.

103.032 **Ephemerides of comets.**
Minor Planet Circ., Nos. 8914 – 8915, 9034, 9082, 9163 – 9169, 9217 – 9219, 9306 – 9308 (1984).

Concerning ephemerides of the following comets:
1974 II P/Schwassmann-Wachmann 1, 1975 VII P/Smirnova-Chernykh, 1978 XIV P/Ashbrook-Jackson, 1978 XVI Tsuchinshan 2, 1983*l* Černis, 1983p Shoemaker, 1984a P/Bradfield, 1984e P/Giacobini-Zinner, 1984f Shoemaker, 1984i Austin, 1984j P/Takamizawa, 1984o Meier, 1984q P/Shoemaker 1, 1984r Shoemaker, 1984s Shoemaker, 1984t Levy-Rudenko, 1984u P/Shoemaker 2, 1984v Hartley.

103.033 **Identifications of minor planets with comets.**
Minor Planet Circ., No. 8961 (1984).

103.034 **Recoveries of periodic comets.**
Br. Astron. Assoc. Circ., No. 641 (1984).

Concerning comets: 1984e P/Giacobini-Zinner, 1984g P/Wolf-Harrington, 1984h P/Faye.

103.035 **Recoveries of periodic comets.**
Br. Astron. Assoc. Circ., No. 645 (1984).

Concerning comets: 1984k P/Arend-Rigaux, 1984*l* P/Gehrels 3, 1984m P/Schaumasse, 1984q P/Tsuchinshan 1.

103.036 **The Comet Handbook 1985 of the Oriental Astronomical Association, Comet Section.**
S. Nakano.
S. Nakano, PO Box 32, Sumoto Post Office, Hyogo-ken, 656-91, Japan. 84 pp. (1984).

103.037 **On the nature of dust grains in the comae of comets Černis and Bowell.**
F. Hoyle, N. C. Wickramasinghe, M. K. Wallis.
Astrophys. Relativ., Prepr. Ser., No. 106, 10+5 pp. (1984).

The currently available infrared data on scattered light from the dust comae of Černis and Bowell set stringent upper limits to the contribution of icy grains. For comet Černis the data is consistent with only a 10% mass fraction of water ice included within silicate-organic-carbon grains of scale radius 15 microns, while for comet Bowell there is no evidence for any ice component.

103.038 **Photometric parameters and brightness outbursts of comet Stephan-Oterma (1980g) and comet Meier (1980q).**
V. S. Filonenko.
Komet. Tsirk., No. 327 (1984). In Russian.

103.039 **Physical characteristics of the comets of 1979.**
D. A. Andrienko, A. V. Karpenko.
Komet. Tsirk., No. 327 (1984). In Russian.

103.040 **Physical characteristics of the comets of 1980.**
D. A. Andrienko, A. V. Karpenko.
Komet. Tsirk., No. 329 (1984). In Russian.

### Comet 1983 VII IRAS-Araki-Alcock

103.101 **Detection of new molecules in the visible spectrum of comet IRAS-Araki-Alcock (1983d).**
C. B. Cosmovici, S. Ortolani.
Nature, Vol. 310, No. 5973, p. 122 – 124 (1984).

Photographic and spectroscopic plates of comet IRAS-Araki-Alcock were obtained on 9.9 UT May 1983 using the 90-cm Schmidt and the 182-cm Copernicus telescopes at the Asiago Observatory. On the spectroscopic plates taken with the grating spectrograph in the spectral range 3,800 – 8,500 Å with a resolution of 6 Å, 450 lines can be distinguished, although 90 lines are still unidentified. The authors have identified the molecules HCO and $H_2S^+$ which have never before been observed in the visible part of the spectrum of any astronomical object. They suspect that $H_2CO$, DCO, $S_2$ and $NH_4$ are also present.

103.102 **Comet 1983d: a contrast between data from IRAS and data from the ground.**
M. F. A'Hearn, R. L. Millis.
Astrophys. J., Lett. Ed., Vol. 282, No. 1, p. L43 – L45 (1984).

Optical photometry of comet IRAS-Araki-Alcock 1983d provides values for the product of albedo and average optical depth in the coma. These turn out to be at least a factor of 20 lower than expected from the measurements with IRAS as reported by Walker et al.

103.103 **Recent new detections of cometary molecules in the optical.**
C. B. Cosmovici.
Observations of comets: radio and otherwise, p. 27 – 31 (1984).
Abstract. – See Abstr. 012.026.

103.104 **Temporal and spatial behavior of the ultraviolet emissions of comet IRAS–Araki–Alcock 1983d.**
P. D. Feldman, M. F. A'Hearn, R. L. Millis.
Astrophys. J., Vol. 282, No. 2, p. 799 – 802 (1984).
The spatial distribution and temporal variation of the ultraviolet emissions from comet IRAS–Araki–Alcock 1983d, observed by *IUE* near closest approach of the comet to the Earth, suggest a model for this comet of a rotating cometary nucleus with a nonuniform distribution of volatile ices. The short lifetime of $S_2(\sim 500$ s), a species whose presence was first identified in this comet and is presumed to be present generally in the cometary ice, makes $S_2$ a valuable indicator of short–term nuclear activity.

103.105 **A radar study of comet IRAS–Araki–Alcock 1983d.**
R. M. Goldstein, R. F. Jurgens, Z. Sekanina.
Astron. J., Vol. 89, No. 11, p. 1745 – 1754 (1984).
Radar echoes from comet IRAS–Araki–Alcock at wavelengths 3.54 and 12.9 cm indicate that the comet's nucleus is very rough on a scale larger than the radar wavelengths; however, the low polarization ratio (25% at 3.54 cm) indicates that the scattering is not dominated by multiple reflections, internal reflections, or large abundances of sharp edges, cracks, and pits. The shape of the nucleus probably departs greatly from a sphere with average radii near 3 – 4 km. The nucleus does not appear to look significantly different from a number of Apollo and Amor asteroids. Other considerations lead the authors to believe that the pole was at least 45° away from the line of sight on two days of observation, and that the rotation period is approximately 1 – 2 days.

103.106 **Discovery of excited OH radicals produced by photodissociation of water vapor in the head of comet 1983d.**
M. C. Festou.
Bull. Am. Astron. Soc., Vol. 16, No. 3, Part I, p. 637 (1984).
Abstract. – See Abstr. 010.061.

103.107 **High spatial resolution spectra of comet IRAS–Araki–Alcock (1983d).**
M. C. Festou, T. Encrenaz, H. Pedersen, M. Tarenghi, C. Boisson.
Bull. Am. Astron. Soc., Vol. 16, No. 3, Part I, p. 637 (1984).
Abstract. – See Abstr. 010.061.

103.108 **Short–term temporal variability of comet IRAS–Araki–Alcock (1983d).**
B. L. Lutz, R. M. Wagner, S. Wyckoff.
Bull. Am. Astron. Soc., Vol. 16, No. 3, Part I, p. 637 (1984).
Abstract. – See Abstr. 010.061.

103.109 **Limits on the production of $NH_3$ in comet IRAS–Araki–Alcock.**
M. J. Mumma, T. Kostiuk, H. A. Weaver.
Bull. Am. Astron. Soc., Vol. 16, No. 3, Part I, p. 638 (1984).
Abstract. – See Abstr. 010.061.

103.110 **Hydrogen cyanide in comets: excitation conditions and radio observations of comet IRAS–Araki–Alcock 1983d.**
D. Bockelée–Morvan, J. Crovisier, A. Baudry, D. Despois, M. Perault, W. M. Irvine, F. P. Schloerb, D. Swade.
Astron. Astrophys., Vol. 141, No. 2, p. 411 – 418 (1984).
The authors evaluate the emission of the pure rotation and rotation/vibration lines of the HCN molecule in cometary atmospheres. Resonant excitation of the $\nu_2$, $2\nu_2$, and $\nu_3$ vibrational bands by the solar infrared field is the prevailing excitation mechanism. For the inner coma, a dynamical excitation model including collisions is presented. The authors have unsuccessfully searched for the $J = 1$–$0$ HCN radio lines in comet 1983d. The close approach of comet 1983d to the Earth allows to put a very low upper limit to the HCN production rate, smaller than the observed CN production rate. The authors conclude that HCN might not be the only parent of cometary CN.

103.111 **8– to 13–μm spectrophotometry of comet IRAS–Araki–Alcock.**
M. A. Feierberg, F. C. Witteborn, J. R. Johnson, H. Campins.
Icarus, Vol. 60, No. 3, p. 449 – 454 (1984).
The authors present spectrophotometry between 8.0 and 13.0 microns at 2% spectral resolution for areas in and near the nuclear condensation of comet IRAS–Araki–Alcock (1983d) on May 11 and 12, 1983. All the spectra can be fit very well by blackbody curves, and no 10 micron silicate emissions are seen. The temperature structure of the coma suggests the presence of small (radii < 5 microns) dust particles within 150 km of the nucleus and larger ones further out. The change in the spatial distribution of the infrared flux between the two nights suggests that an outburst may have occurred sometime on May 11.

103.112 **Observations of comet IRAS–Araki–Alcock (1983d).**
Komet. Tsirk., Nos. 330, 333 (1984). In Russian.

### Periodic comet Crommelin

103.121 **The 1983 – 84 apparition of periodic comet Crommelin.**
D. W. E. Green, C. S. Morris.
Int. Comet Q., Vol. 6, No. 3, p. 55 – 58 (1984).

103.122 **Some thoughts on the light curve of periodic comet Crommelin.**
C. S. Morris.
Int. Comet Q., Vol. 6, No. 3, p. 58 – 61 (1984).

103.123 **Observations of comet P/Crommelin at ESO in the near infrared range.**
T. Encrenaz, D. Engels, J. Krautter.
Messenger, No. 37, p. 29 – 30 (1984).

103.124 **Near infrared photometry of comet P/Crommelin.**
T. Encrenaz, D. Engels, J. Krautter.
Astron. Astrophys., Vol. 140, No. 1, p. L13 – L15 (1984).
Near IR photometry of comet P/Crommelin has been recorded on March 19, 1984 at ESO (La Silla) in the J, H, and K filters. The (J–H) and (H–K) values are compatible with a silicate composition of micron–size particles; there is no evidence for absorbing particles. The comparison with other observations performed with different diaphragm sizes could indicate a marginal departure from the $r^{-2}$ density distribution (r distance to the nucleus).

103.125 **Infrared photometry of comet Crommelin.**
M. S. Hanner, R. F. Knacke, A. T. Tokunaga.
Bull. Am. Astron. Soc., Vol. 16, No. 3, Part I, p. 644 (1984).
Abstract. – See Abstr. 010.061.

103.126 **Radio observations of OH in P/Crommelin.**
F. P. Schloerb, C. A. McGunigle, A. R. Molloy.
Bull. Am. Astron. Soc., Vol. 16, No. 3, Part I, p. 645 (1984).
Abstract. – See Abstr. 010.061.

103.127 **Development of the atmosphere of P/Crommelin.**
P. A. Wehinger, S. Wyckoff, H. Spinrad, M. Belton, W. Romanishin.
Bull. Am. Astron. Soc., Vol. 16, No. 3, Part I, p. 704 (1984).
Abstract. – See Abstr. 010.061.

103.128 **Periodic comet Crommelin 1983n.**
Komet. Tsirk., Nos. 327, 329, 333 (1984). In Russian.

**Comet 1982 VI Austin**

103.141 **Analysis of hydroxyl data from comet Austin 1982g.**
M. K. Wallis, W. C. Carey, D. W. Hughes, W. M. Burton.
ESA Spec. Publ., ESA SP-218, p. 459 - 462 (1984). - See Abstr. 012.024.

103.142 **A radio source occultation experiment with comet Austin 1982g, with unusual results.**
I. de Pater, W.-H. Ip.
Astrophys. J., Vol. 283, No. 2, p. 895 - 904 (1984).
A surprising result of an occultation of the radio source 1242+41 at 1.5 GHz by comet Austin 1982g is described. While the radio source was at a distance of $2.2 \times 10^5$km from the center of the plasma tail, a change in its position of ~1".3 was observed. If this apparent change is due to refraction by the cometary plasma rather than the Earth's ionosphere, the electron density of the cometary plasma is of the order of $10^4 cm^{-3}$. Several hours later, when the source was at $2.3 \times 10^5$km at the other side of the tail, a gradual change of ~140° in the position angle of the electric vector was observed in a time span of 2 hr. If it is due to Faraday rotation in the cometary plasma, the cometary coma must have an electron density of $\sim 10^4 cm^{-3}$ and a magnetic field strength of ~0.5 gauss.

**Periodic comet Russell 4**

103.151 **Periodic comet Russell 4 1984d.**
S. Nakano.
Nakano wa Kangaeru noda, NK 468, 2 pp. (1984).

103.152 **Observations of comet Russell 1984d.**
Komet. Tsirk., Nos. 328, 332 (1984). In Russian.

**Comet 1983 V Sugano-Saigusa-Fujikawa**

103.161 **A search for millimeter wave spectral lines from Comet Sugano-Saigusa-Fujikawa (1983e).**
T. Hasegawa, M. Ohishi, M. Morimoto, H. Suzuki, N. Kaifu.
Icarus, Vol. 60, No. 1, p. 211 - 214 (1984).
The search for radio spectral lines from Comet Sugano-Saigusa-Fujikawa (1983e) was conducted using the 45-m telescope of Nobeyama Radio Observatory. The frequency ranges of 44.0 - 46.0 and 47.5 - 49.5 GHz were surveyed with a beam size of ~35 arc sec. Upper limits have been established for spectral lines of atomic hydrogen, CS, OCS, $SO_2$, $H_2CO$, $CH_3OH$, HCCCCCN, $HCOOCH_3$, $CH_3OCH_3$, and $CH_3CH_2CN$. The $J = 5-4$ line from HCCCN in the vibrational ground state possibly has been detected but not confirmed. The suggested total amount of HCCCN in the coma is consistent with the possible picture that HCCCN is the main parent molecule of CN.

**Periodic comet Taylor**

103.171 **Periodic comet Taylor 1977 II = 1983u.**
S. Nakano.
Nakano wa Kangaeru noda, NK 471, 2 pp. (1984).

**Periodic comet Tempel 2**

103.181 **The anomalous dust tail of comet P/Tempel 2.**
N. Eaton, J. K. Davies, S. F. Green.
Mon. Not. R. Astron. Soc., Vol. 211, No. 1, p. 15P - 19P (1984).
The narrow dust tail of comet P/Tempel 2, observed by IRAS, is the result of low-velocity emissions of large particles. Dynamical analysis of the dust particles under the influence of solar radiation pressure indicates that some of these emissions must have occurred at least 1500 days prior to the IRAS observations.

**Periodic comet Wolf**

103.191 **Short-period comet Wolf 1983m.**
Komet. Tsirk., No. 326 (1984). In Russian.

**Periodic comet Tempel-Tuttle**

103.201 **On the possibility that comet Tempel-Tuttle will be captured by Jupiter.**
I. N. Murav'eva, E. A. Reznikov.
Chelyab. politekh. inst. Chelyabinsk, 14 pp. (1984). In Russian. Abstr. in Ref. Zh., 51. Astron., 10.51.108 (1984).

**Periodic comet Ashbrook-Jackson**

103.211 **Short-period comet Ashbrook-Jackson.**
Komet. Tsirk., No. 331 (1984). In Russian.

**Periodic comet Kopff**

103.221 **Precession model for the nucleus of periodic comet Kopff.**
Z. Sekanina.
Astron. J., Vol. 89, No. 10, p. 1573 - 1586 (1984).
Information on periodic comet Kopff's light curve, "nuclear" brightness, coma asymmetry, and, especially, the nongravitational perturbation of its orbital motion (caused by direct outgassing) is combined to investigate the nucleus properties of this prime target candidate for NASA's planned rendezvous mission in the 1990's.

103.222 **Spectrophotometry of P/Kopff during its 1983 apparition.**
E. S. Barker, A. L. Cochran.
Bull. Am. Astron. Soc., Vol. 16, No. 3, Part I, p. 644 (1984). Abstract. - See Abstr. 010.061.

103.223 **Observations of comet Kopff.**
R. V. Abdulova, N. V. Sinyaeva.
Astron. Tsirk., No. 1321, p. 7 (1984). In Russian.

103.224 **Short-period comet Kopff (1982k).**
Komet. Tsirk., Nos. 327, 331 (1984). In Russian.

**Comet 1982 I Bowell**

103.241 **Coma expansion and photometry of comet Bowell (1980b).**
D. Jewitt.
Icarus, Vol. 60, No. 2, p. 373 - 385 (1984).
Optical and infrared observations of comet Bowell are presented. The optical observations indicate that the solid grain coma is expanding at only $0.9 \pm 0.2$ m $sec^{-1}$. This is two orders of magnitude slower than the local gas sound speed and may suggest that gas drag is not responsible for stripping the grains from the nucleus. It is concluded that the grain coma may be largely a relic of activity occurring on the nucleus at $R \cong 10$ AU while reported high OH production rates may indicate sublimation from the nucleus near perihelion and from coma grains near $R \cong 4.6$ AU.

103.242 **JHK photometry of comet Bowell (1982 I).**
M. S. Hanner, G. J. Veeder.
Icarus, Vol. 60, No. 3, p. 445 - 448 (1984).
Comet Bowell (1982 I) was observed at JHK (1.25, 1.68, 2.25 μm) with the NASA Infrared Telescope Facility on 10 March 1981, 3 - 5 June 1981, and 22 February 1982. The inner coma was no brighter near perihelion in February 1982 than it was in 1981. Total area × albedo of the grains within $10^4$km of the nucleus was $\sim 5 \times 10^{12} cm^2$.

103.243 **Comet Bowell 1982 I = 1980b.**
S. Nakano.
Nakano wa Kangaeru noda, NK 462, 5 pp. (1984).

### Comet 1970 II Bennett

103.251 **Determination of the scale lengths of parent molecules and daughter cyanogen in the head of comet Bennett (1970 II).**
V. A. Dranevich, I. N. Matveev.
Komet. Tsirk., No. 332 (1984). In Russian.

### Comet 1981 VII González

103.261 **Positions of Comet González (1981g) in September 1981.**
H. Debehogne, R. R. de Freitas Mourao, O. C. Tavares, M. R. Nunes.
Acta Astron., Vol. 34, No. 1, p. 125 – 127 (1984).
Twelve astronomical positions of Comet González are given as obtained from photographic observations made at ESO, La Silla, during the period 1981 September 5 and 6. Reductions are made with 5 reference stars by dependences and least squares method.

### Periodic comet Tempel 1

103.271 **Observations of periodic comet Tempel 1 (1982j).**
Komet. Tsirk., No. 332 (1984). In Russian.

### Periodic comet Giacobini–Zinner

103.281 **The intercept of comet Giacobini–Zinner (G/Z) by the International Cometary Explorer (ICE).**
J. C. Brandt, R. W. Farquhar, M. B. Niedner, T. von Rosenvinge.
Publ. Astron. Soc. Pac., Vol. 96, No. 584, p. 784 – 785 (1984).
Abstract. – See Abstr. 010.281.

103.282 **Periodic comet Giacobini–Zinner (1984e).**
IAU Circ., No. 3979 (1984).

103.283 **Periodic comet Giacobini–Zinner 1972 VI = 1979 III = 1984e.**
S. Nakano.
Nakano wa Kangaeru noda, NK 459, 5 pp. (1984).

103.284 **Non–gravitational effects in the motion of comet Giacobini–Zinner.**
Komet. Tsirk., No. 327 (1984). In Russian.

### Periodic comet Tuttle

103.291 **Comet Tuttle 1980h.**
Komet. Tsirk., No. 332 (1984). In Russian.

### Periodic comet Encke

103.301 **The symmetry of P/Encke about perihelion.**
M. F. A'Hearn, P. D. Feldman, R. L. Millis, A. I. F. Stewart.
Bull. Am. Astron. Soc., Vol. 16, No. 3, Part I, p. 638 (1984).
Abstract. – See Abstr. 010.061.

103.302 **An analysis of Pioneer Venus hydrogen Lyman–α observations of comet P/Encke.**
M. R. Combi, A. I. F. Stewart, W. H. Smyth.
Bull. Am. Astron. Soc., Vol. 16, No. 3, Part I, p. 638 (1984).
Abstract. – See Abstr. 010.061.

103.303 **The infrared light curve of comet Encke.**
H. Campins, M. F. A'Hearn.
Bull. Am. Astron. Soc., Vol. 16, No. 3, Part I, p. 638 (1984).
Abstract. – See Abstr. 010.061.

103.304 **The ultraviolet spectrum of periodic comet Encke (1980 XI).**
P. D. Feldman, H. A. Weaver, M. C. Festou.
Icarus, Vol. 60, No. 3, p. 455 – 463 (1984).
A comparison of the ultraviolet spectrum of periodic comet Encke (1980 XI), recorded by the IUE between 24 October and 5 November 1980 with similar spectra of short– and long–period comets shows the gaseous composition of P/Encke to be nearly identical to that of the other comets observed by the IUE. The OH brightness distribution shows a spatial variation similar to the visible fan–shaped image of the comet, suggestive of a non-uniform distribution of volatile ices on the surface of the nucleus. The total derived water production rate appears to be a factor of 5 higher than that derived from H I Lyman–α observations made during the 1970 apparition and shows a variation with heliocentric distance ($r$) as $r^{-3.3}$ over the range 0.81 to 1.02 AU.

103.305 **The ultraviolet spectrum of periodic comet Encke (1980 XI).**
P. D. Feldman, H. A. Weaver, M. C. Festou.
Inst. Astrophys. Paris, Pré–Publ., No. 74, 29 pp. (1984). To be appear in Icarus.

103.306 **Short–period comet Encke.**
Komet. Tsirk., No. 329 (1984). In Russian.

### Comet 1984i Austin

103.321 **Comet Austin (1984i).**
IAU Circ., Nos. 3957, 3958, 3960, 3964, 3967, 3972, 3975, 3978, 3981, 3984, 3990, 3999, 4006, 4017 (1984).

103.322 **Comet Austin (1984i).**
Yamamoto Circ., Nos. 2018, 2020 – 2026 (1984). In Japanese.

103.323 **Comet Austin 1984i.**
Br. Astron. Assoc. Circ., Nos. 641, 643, 644 (1984).

103.324 **Comet Austin 1984i.**
S. Nakano.
Nakano wa Kangaeru noda, NK 469, 2 pp. (1984).

103.325 **Comet Austin 1984i.**
Komet. Tsirk., Nos. 328 – 333 (1984). In Russian.

### Comet 1984f Shoemaker

103.341 **Comet Shoemaker (1984f).**
IAU Circ., No. 3961 (1984).

103.342 **New comet Shoemaker 1984f.**
Br. Astron. Assoc. Circ., No. 641 (1984).

103.343 **Comet Shoemaker 1984f.**
Komet. Tsirk., No. 326 (1984). In Russian.

### Periodic comet Wild 2

103.361 **Periodic comet Wild 2 (1983s).**
IAU Circ., No. 3961 (1984).

**Periodic comet Takamizawa**

103.381 **Periodic comet Takamizawa (1984j).**
IAU Circ., Nos. 3964, 3965, 3970, 3971, 3973, 3974, 3977, 3985, 4002 (1984).

103.382 **Periodic comet Takamizawa (1984j).**
Yamamoto Circ., Nos. 2019 – 2021, 2027 (1984). In Japanese.

103.383 **Comet P/Takamizawa 1984j.**
Br. Astron. Assoc. Circ., Nos. 643, 644 (1984).

103.384 **New comet Takamizawa (1984j).**
Komet. Tsirk., No. 329 (1984). In Russian.

**Periodic comet Faye**

103.401 **Periodic comet Faye (1984h).**
IAU Circ., Nos. 3969, 3985 (1984).

103.402 **Periodic comet Faye (1984h).**
Yamamoto Circ., No. 2018 (1984). In Japanese.

103.403 **Rediscovery of short–period comet Faye (1984h).**
Komet. Tsirk., No. 327 (1984). In Russian.

**Periodic comet Arend–Rigaux**

103.421 **Periodic comet Arend–Rigaux (1984k).**
IAU Circ., Nos. 3972, 3974, 3978, 4016 (1984).

103.422 **P/Arend–Rigaux (1984k).**
Yamamoto Circ., Nos. 2021, 2030 (1984). In Japanese.

103.423 **Rediscovery of short–period comet Arend–Rigaux (1984k).**
Komet. Tsirk., No. 330 (1984). In Russian.

**Periodic comet Gehrels 3**

103.441 **Periodic comet Gehrels 3 (1984*l*).**
IAU Circ., No. 3977 (1984).

103.442 **Periodic comet Gehrels 3 (1984*l*).**
Yamamoto Circ., No. 2022 (1984). In Japanese.

103.443 **Periodic comet Gehrels 3 1977 VII = 1984*l*.**
S. Nakano.
Nakano wa Kangaeru noda, NK 458, 2 pp. (1984).

103.444 **Rediscovery of short–period comet Gehrels 3 (1984*l*).**
Komet. Tsirk., No. 331 (1984). In Russian.

**Periodic comet Wolf–Harrington**

103.461 **Periodic comet Wolf–Harrington (1984g).**
IAU Circ., No. 3983 (1984).

103.462 **Periodic comet Wolf–Harrington (1984g).**
Yamamoto Circ., Nos. 2018, 2023 (1984). In Japanese.

103.463 **Rediscovery of short–period comet Wolf–Harrington (1984g).**
Komet. Tsirk., No. 326 (1984). In Russian.

**Periodic comet Schaumasse**

103.481 **Periodic comet Schaumasse (1976 XV = 1984m).**
IAU Circ., Nos. 3986, 3987, 4006, 4011, 4016 (1984).

103.482 **P/Schaumasse (1984m).**
Yamamoto Circ., Nos. 2023, 2024, 2030 (1984). In Japanese.

103.483 **Rediscovery of short–period comet Schaumasse (1984m).**
Komet. Tsirk., No. 331 (1984). In Russian.

**Periodic comet Neujmin 1**

103.501 **Periodic comet Neujmin 1 (1984c).**
IAU Circ., Nos. 3987, 3997 (1984).

103.502 **P/Neujmin 1 (1984c).**
Yamamoto Circ., No. 2023 (1984). In Japanese.

**Periodic comet Kowal–Mrkos**

103.521 **Periodic comet Kowal–Mrkos (1984n).**
IAU Circ., Nos. 3988, 4001 (1984).

103.522 **Periodic comet Kowal–Mrkos (1984n).**
Yamamoto Circ., Nos. 2023, 2027 (1984). In Japanese.

103.523 **New comet P/Kowal–Mrkos 1984n.**
Br. Astron. Assoc. Circ., No. 644 (1984).

103.524 **New short–period comet Kowal–Mrkos (1984n).**
Komet. Tsirk., No. 332 (1984). In Russian.

**Comet 1984o Meier**

103.541 **Comet Meier (1984o).**
IAU Circ., Nos. 3991, 3992, 3995, 3999 (1984).

103.542 **Comet Meier (1984o).**
Yamamoto Circ., Nos. 2024 – 2026 (1984). In Japanese.

103.543 **New comet Meier 1984o.**
Br. Astron. Assoc. Circ., No. 644 (1984).

103.544 **New comet Meier (1984o).**
Komet. Tsirk., No. 332 (1984). In Russian.

**Periodic comet Tsuchinshan 1**

103.561 **Periodic comet Tsuchinshan 1 (1984p).**
IAU Circ., No. 3993 (1984).

103.562 **Periodic comet Tsuchinshan 1 (1984p).**
Yamamoto Circ., No. 2025 (1984). In Japanese.

103.563 **Rediscovery of short–period comet Tsuchinshan 1 (1984p).**
Komet. Tsirk., No. 333 (1984). In Russian.

**Periodic comet Shoemaker 1**

103.581 **Periodic comet Shoemaker 1 (1984q).**
IAU Circ., Nos. 3997, 3998, 4000, 4002, 4010, 4017 (1984).

103.582 **Periodic comet Shoemaker 1 (1984q).**
Yamamoto Circ., Nos. 2025 – 2027, 2031 (1984). In Japanese.

103.583 **New comet Shoemaker 1984q.**
Br. Astron. Assoc. Circ., No. 645 (1984).

103.584 **New comet Shoemaker (1984q).**
Komet. Tsirk., No. 333 (1984). In Russian.

**Comet 1984r Shoemaker**

103.601 **Comet Shoemaker (1984r).**
IAU Circ., Nos. 4003, 4004, 4006, 4007, 4011 (1984).

103.602 **Comet Shoemaker (1984r).**
Yamamoto Circ., Nos. 2027, 2028 (1984). In Japanese.

103.603 **New comet Shoemaker 1984r.**
Br. Astron. Assoc. Circ., No. 645 (1984).

**Comet 1984s Shoemaker**

103.621 **Comet Shoemaker (1984s).**
IAU Circ., Nos. 4004, 4005, 4010, 4012, 4021 (1984).

103.622 **Comet Shoemaker (1984s).**
Yamamoto Circ., Nos. 2027 – 2029, 2031 (1984). In Japanese.

103.623 **Comet Shoemaker 1984s.**
Br. Astron. Assoc. Circ., Nos. 645, 646 (1984).

**Comet 1984t Levy–Rudenko**

103.641 **Comet Levy–Rudenko (1984t).**
IAU Circ., Nos. 4007, 4009 – 4011, 4013, 4014, 4016, 4018, 4023 (1984).

103.642 **Comet Levy–Rudenko (1984t).**
Yamamoto Circ., Nos. 2028 – 2031 (1984). In Japanese.

103.643 **Comet Levy–Rudenko 1984t.**
Br. Astron. Assoc. Circ., Nos. 645, 646 (1984).

**Periodic comet Shoemaker 2**

103.661 **Periodic comet Shoemaker 2 (1984u).**
IAU Circ., Nos. 4014, 4015 (1984).

103.662 **Periodic comet Shoemaker 2 (1984u).**
Yamamoto Circ., Nos. 2029, 2031 (1984). In Japanese.

**Comet 1984v Hartley**

103.681 **Comet Hartley (1984v).**
IAU Circ., Nos. 4015, 4016, 4019, 4020 (1984).

103.682 **Comet Hartley (1984v).**
Yamamoto Circ., Nos. 2029 – 2031 (1984). In Japanese.

**Periodic comet Holmes**

103.701 **Comet P/Holmes, 1892 III: a case of duplicity?**
F. L. Whipple.
Icarus, Vol. 60, No. 3, p. 522 – 531 (1984).
The observations of comet P/Holmes 1892 III, exhibiting two 8- to 10-mag bursts, have been carefully analyzed. The phenomena are consistent with the grazing encounter of a small satellite with the nucleus on November 4.6, 1892, and the final encounter on January 16.3, 1893.

**Periodic comet Biela**

103.711 **Periodic comet Biela 1852 III – an update.**
C. E. Spratt.
J. R. Astron. Soc. Can., Vol. 78, No. 6, p. 251 – 254 (1984).
A brief account of this famous comet since it was last observed in 1852 is given, along with a search ephemeris for a possible 1985 perihelion passage.

**Periodic comet Boethin**

103.721 **Periodic comet Boethin 1975 I = 1975a.**
S. Nakano.
Nakano wa Kangaeru noda, NK 456, 2 pp. (1984).

**Periodic comet Whipple**

103.741 **Periodic comet Whipple 1963 II = 1970 XIV = 1978 VIII.**
S. Nakano.
Nakano wa Kangaeru noda, NK 457, 3 pp. (1984).

**Comet 1983 XII Černis**

103.761 **Comet Černis 1983*l*.**
S. Nakano.
Nakano wa Kangaeru noda, NK 460, 5 pp. (1984).

103.762 **Comet Černis 1983*l*.**
K. Černis.
Vilniaus Astron. Obs. Biul., Nr. 67, p. 43 – 47 (1984). In Russian.

103.763 **Comet Černis (1983*l*).**
Komet. Tsirk., No. 327 (1984). In Russian.

**Comet 1983 XV Shoemaker**

103.781 **Comet Shoemaker 1983p.**
S. Nakano.
Nakano wa Kangaeru noda, NK 461, 3 pp. (1984).

**Periodic comet Russell 3**

103.801 **Periodic comet Russell 3 1982 IX = 1983i.**
S. Nakano.
Nakano wa Kangaeru noda, NK 463, 2 pp. (1984).

**Comet 1983 XVI IRAS**

103.821 **Comet IRAS 1983o.**
S. Nakano.
Nakano wa Kangaeru noda, NK 464, 2 pp. (1984).

**Periodic comet Kowal–Vávrová**

103.841 **Periodic comet Kowal–Vávrová 1983t.**
S. Nakano.
Nakano wa Kangaeru noda, NK 465, 1 p. (1984).

**Periodic comet Hartley–IRAS**

103.861 **Periodic comet Hartley–IRAS 1983v.**
S. Nakano.
Nakano wa Kangaeru noda, NK 466, 2 pp. (1984).

103.862 **Comet Hartley–IRAS 1983v.**
Komet. Tsirk., Nos. 328, 330, 331 (1984). In Russian.

**Periodic comet Bradfield**

103.881 **Periodic comet Bradfield 1984a.**
S. Nakano.
Nakano wa Kangaeru noda, NK 467, 2 pp. (1984).

**Periodic comet Halley**

103.901 **Die "International Halley Watch".**
H. Drechsel, R. L. Newburn Jr., J. Rahe.
Sterne Weltraum, 23. Jahrg., Nr. 8–9, p. 440 – 445 (1984).

103.902 **Exploration of Halley's comet from the ground: the International Halley Watch (IHW) and its observing nets.**
ESA Bull., No. 39, p. 37 – 50 (1984).

103.903 **Rotation of comet P/Halley: recurrent brightening observed at the heliocentric distance of 8 AU.**
O. Le Fevre, J. Lecacheux, G. Mathez, G. Lelièvre, J. Baudrand, J. P. Lemonnier.
Astron. Astrophys., Vol. 138, No. 1, p. L1 – L4 (1984).

Rapid recurring increases of comet P/Halley brightness were observed with the 3.60 meter CFH–Telescope during consecutive nights of February 1984, while the mean heliocentric distance of the comet was $r_H = 7.95$ AU. The measured amplitude in blue light was 1.7 magnitude. This phenomenon seemed to occur at constant time intervals of $24.3 \pm 0.3$ hours. Such brightness variations probably reflect the rotation of the nucleus. Some alternative interpretations of the observations are also suggested.

103.904 **Variability of P/Halley.**
R. M. West, H. Pedersen.
Astron. Astrophys., Vol. 138, No. 1, p. L9 – L10 (1984).

21 photometric magnitudes of P/Halley (1982i) have been measured on 1984 Jan. 27 – 30 with the Danish 1.5 m telescope at La Silla and a CCD. They have r.m.s. errors of $\pm 0.2$ mag and show night–to–night variability. The available material does not allow an unambiguous determination of a period of rotation, but short periods ($\leqslant 10$ hours) are considered less likely.

103.905 **Le point sur les récentes observations de la comète de Halley.**
R. Boyer, E. Neyvoz, E. Stram.
Astronomie, Vol. 98, p. 431 – 438 (1984).

103.906 **A proposal for artificial stimulation of the OH maser in comet Halley.**
J. E. B. Ponsonby.
Observations of comets: radio and otherwise, p. 53 – 58 (1984). Abstract. – See Abstr. 012.026.

103.907 **Where to observe Halley's comet.**
E. M. Brooks.
Sky Telesc., Vol. 68, No. 4, p. 304 – 305 (1984).

103.908 **Flop of spektakel? Komeet Halley in 1986.**
E. P. Bus.
Zenit, 11. Jaarg., No. 11, p. 464 – 465 (1984).

103.909 **Coma morphology and dust–emission pattern of periodic comet Halley. II. Nucleus spin vector and modeling of major dust features in 1910.**
Z. Sekanina, S. M. Larson.
Astron. J., Vol. 89, No. 9, p. 1408 – 1425 (1984). With plates 80 – 81.

To explain the evolution of spiral jets that unwind from the nucleus condensation into envelopes or halos in the head of comet Halley, the authors develop a concept of continuous ejection of dust from discrete emission sources on the rotating nucleus. Application of this model to the digitally processed images of three features, each seen for three days on the Mount Wilson plates taken during the 1910 apparition, makes it possible to fit their motions and thereby to determine the nucleus spin vector and, for each emission source, its cometocentric coordinates and a function that relates particle ejection velocity $v$ to solar radiation pressure $\beta$ exerted on the ejecta. The authors also obtain information on the activity cycle of the emission areas.

103.910 **Comet Halley: expecting the encounter.**
Ya. S. Yatskiv, K. I. Churyumov.
Nauk. v SSSR, No. 6, p. 62 – 73 (1983). In Russian. Abstr. in Ref. Zh., 62. Issled. Kosm. Prostranstva, 9.62.56 (1984).

103.911 **Halley's comet and plans for its observations during its return in 1986.**
J. Rahe, J. C. Brandt, L. D. Friedman, R. L. Newburn.
Vasiona, Année 32, No. 2, p. 25 – 31 (1984). In Croatian.

103.912 **La comète de Halley à son retour de 1910.**
R. Boyer, E. Neyvoz, E. Stram.
Astronomie, Vol. 98, p. 535 – 548 (1984).

103.913 **En attendant Halley (I).**
W. Maeder.
Orion, 42. Jahrg., Nr. 205, p. 217 (1984).

103.914 **Halleys Komet 1985/1986.**
E. Hügli.
Orion, 42. Jahrg., Nr. 205, p. 227 – 228 (1984).

103.915 **Komet Halley 1909 abermals in Heidelberg entdeckt.**
S. Röser.
Sterne Weltraum, 23. Jahrg., Nr. 12, p. 612 (1984).

103.916 **Photometry of comet Halley at R = 11.0 and R = 8.2 AU.**
D. Jewitt, G. E. Danielson.
Bull. Am. Astron. Soc., Vol. 16, No. 3, Part I, p. 636 (1984). Abstract. – See Abstr. 010.061.

103.917 **The reflection spectrum of P/Halley.**
M. J. S. Belton, H. Spinrad, P. Wehinger, D. K. Yeomans.
Bull. Am. Astron. Soc., Vol. 16, No. 3, Part I, p. 636 (1984). Abstract. – See Abstr. 010.061.

103.918 **Light variations of comet Halley at large solar distances.**
Z. Sekanina.
Bull. Am. Astron. Soc., Vol. 16, No. 3, Part I, p. 636 (1984). Abstract. – See Abstr. 010.061.

103.919 **The mass of comet Halley.**
A. H. Delsemme, D. K. Yeomans.
Bull. Am. Astron. Soc., Vol. 16, No. 3, Part I, p. 636 – 637 (1984). Abstract. – See Abstr. 010.061.

103.920 **Coma structure of comet Halley (1910 II).**
S. M. Larson, Z. Sekanina.
Bull. Am. Astron. Soc., Vol. 16, No. 3, Part I, p. 637 (1984). Abstract. – See Abstr. 010.061.

103.921 **Periodic comet Halley (1982i).**
IAU Circ., Nos. 3957, 3996 (1984).

103.922 **Periodic comet Halley (1982i).**
Yamamoto Circ., No. 2025 (1984). In Japanese.

103.923 **Charge–coupled device photometry of comet P/Halley.**
D. Jewitt, G. E. Danielson.
Icarus, Vol. 60, No. 3, p. 435 – 444 (1984).

Comet P/Halley has been observed during its approach to perihelion at heliocentric distances $R = 11.0$ AU and $R = 8.2$ AU. No extended coma is seen and limits can be placed on the fraction of the total light contributed by coma. The brightness of the comet varies on a short time scale. The variations may be due to transient activity or to rotation of the irregular nucleus.

103.924 **Positions of comet Halley in 1910 from observations in Tashkent.**
L. I. Batashova, A. A. Latypov.
Astron. Tsirk., No. 1335, p. 7 – 8 (1984). In Russian.

103.925 **F. W. Bessels Beobachtungen des Kometen Halley im Jahre 1835. Eine Neureduktion.**
S. Röser.
Mitt. Astron. Ges., Nr. 62, p. 195 – 196 (1984). – See Abstr. 010.341.

103.926 **Beobachtungen des Kometen Halley im Jahre 1909 vor seiner Entdeckung.**
S. Röser, U. Bastian, G. Klare, G. Schwehm.
Mitt. Astron. Ges., Nr. 62, p. 197 – 198 (1984). – See Abstr. 010.341.

**Wàs dat Halley wel, in 1910?**
See Abstr. 004.033.

**The comet of Charles V.**
See Abstr. 004.040.

**Edmond Halley's observations of Halley's comet.**
See Abstr. 004.043.

**The Giotto spacecraft system and subsystem design.**
See Abstr. 035.081.

**IRAS fast mover program.**
See Abstr. 036.009.

**Le dessin de la chevelure interne d'une comète: généralités et conseils pratiques.**
See Abstr. 036.026.

**On a displacement of the photometric center from the center of mass in positional observations of comets and minor planets.**
See Abstr. 036.054.

**On autonomous tracking of a minor celestial body.**
See Abstr. 036.059.

**Compiling technique of special reference catalogues of star positions and its application to comet Halley watch.**
See Abstr. 036.080.

**Comet rendezvous.**
See Abstr. 051.008.

**Comet nucleus sample return.**
See Abstr. 051.013.

**Rendezvous with Giotto.**
See Abstr. 051.014.

**Pioneer to Venus: the multiprobe and orbiter missions.**
See Abstr. 051.016.

**Through the tail of a comet.**
See Abstr. 051.035.

**Comet Rendezvous Asteroid Flyby Mission – Mariner Mark II.**
See Abstr. 051.061.

**The geomagnetic storm of 1910 May 18 and the tail of Halley's Comet.**
See Abstr. 084.085.

**Positions of minor planets and comets obtained at the Chorzów Observatory.**
See Abstr. 098.031.

**Positions of Comet Bowell (1980b), Asteroids 1982 DV and 51 Nemausa, March 1982, with the GPO, ESO–Chile.**
See Abstr. 098.032.

**2201 Oljato: an asteroid, a comet, or both?**
See Abstr. 098.042.

**Asteroid 1984 BC – an extinct comet?**
See Abstr. 098.043.

**Ephemerides of minor planets and comets.**
See Abstr. 098.071.

**Index to orbital elements of comets and minor planets.**
See Abstr. 098.074.

**Search for a meteor shower associated with the Sugano–Saigusa–Fujikawa comet (1983e).**
See Abstr. 104.030.

# 104 Meteors, Meteor Streams

104.001 **Meteoren waarnemen: een prachtige hobby!**
F. Janssen, J. van der Laan.
Zenit, 11. Jaarg., No. 7/8, p. 334 – 336 (1984).

104.002 **Meteoren boven de Benelux, 1983.**
C. ter Kuile, P. Roggemans.
Zenit, 11. Jaarg., No. 7/8, p. 337 – 339 (1984).

104.003 **The fragmentation of meteoroids. I. Quasi–continuous fragmentation.**
G. G. Novikov, V. N. Lebedinets, A. V. Blokhin.
Sov. Astron. Lett., Vol. 10, No. 1, p. 27 – 29 (1984). English translation of 37.104.001.

104.004 **Meteor astronomy. 1: Theory, methods and practice.**
R. A. Mackenzie.
Space Educ., Vol. 1, No. 8, p. 377 – 380 (1984).

104.005 **An observational study of the range drift of meteor echoes.**
B. A. Lindblad, A. Hajduk.
Bull. Astron. Inst. Czech., Vol. 35, No. 4, p. 244 – 252 (1984). With plate 8.

A study is made of 338 meteor radar echoes in which the echo signal showed a change in range. The sample includes all echoes with observed range drifts greater than about 1.5 km in a radar–visual sample of 2600 meteors. The mean absolute value of all radial velocities is 1.8 km $s^{-1}$. If the radial velocities are related to the orientation of the meteor trail, the range drift in general corresponds to a downward motion of the echoing point along the ionization trail. For 262 echoes associated with Perseid meteors the trail geometry was fully determined and drift velocities along the ionization trail have been computed. Mean and median absolute values of the velocities along the trail are 5.9 km $s^{-1}$ and 1.4 km $s^{-1}$, respectively. The distribution of radial and drift velocities was studied for echo signals of different durations. The analysis shows that the drift velocity is inversely proportional to the echo duration.

104.006 **Meteor–stream density evolution by planetary perturbations.**
V. V. Emel'yanenko.
Sov. Astron. Lett., Vol. 10, No. 2, p. 131 – 132 (1984). English translation of 37.104.021.

104.007 **Observing meteors: I.**
D. H. Levy.
Strolling Astron., Vol. 30, Nos. 9 – 10, p. 208 – 209 (1984).

104.008 **Radio meteor investigations made according to international geophysical projects.**
V. A. Bronshtehn, B. L. Kashcheev.
Meteor. issled., Moskva, No. 9, p. 5 – 13 (1984). In Russian. Abstr. in Ref. Zh., 51. Astron., 8.51.179 (1984).

104.009 **Determination of the parameters of meteor trains according to results of radar observations at a two–frequency coherent radar station.**
Yu. V. Chumak, R. I. Mojsya.
Meteor. issled., Moskva, No. 9, p. 69 – 78 (1984). In Russian. Abstr. in Ref. Zh., 51. Astron., 8.51.183 (1984).

104.010 **Meteors.**
J. W. Mason.
J. Br. Astron. Assoc., Vol. 94, No. 6, p. 266 – 269 (1984).

104.011 **Orioniden 1982. Ergebnisse einer neuen Auswertungsmethode visueller Beobachtungen.**
H.–G. Schmidt, L. Weidinger, S. Deiries.
Sterne Weltraum, 23. Jahrg., Nr. 10, p. 527 – 528 (1984).

104.012 **Features of evolution of the Geminid and Quadrantid meteor streams.**
P. B. Babadzhanov, Yu. V. Obrubov.
Astron. Zh., Tom 61, Vyp. 5, p. 1005 – 1012 (1984). In Russian. English translation in Sov. Astron., Vol. 28, No. 5.

It is shown that a number of features of meteor showers such as the correlation of orbital semimajor axes and longitudes of the ascending nodes with meteor magnitudes, the restriction of visibility periods and the displacement of maximum activity dates is explained by the joint influence of planetary perturbations, of the Poynting–Robertson effect and its corpuscular analogue, by the light pressure and various velocities of the ejection of different mass meteoroids from cometary nuclei.

104.013 **Determination of the parameters of quasi–continuous meteoroid fragmentation from meteor observational data.**
G. G. Novikov, V. N. Lebedinets, A. V. Blokhin.
Pis'ma Astron. Zh., Tom 10, No. 10, p. 785 – 789 (1984). In Russian. English translation in Sov. Astron. Lett, Vol. 10.

The theory of quasi–continuous meteoroid fragmentation developed earlier is applied to the evaluation of meteoroid densities and fragment masses from data on photographic and phototelevision meteor heights.

104.014 **A note on anomalous sounds from meteor fireballs and aurorae.**
D. Y. Wang, T. F. Tuan, S. M. Silverman.
J. R. Astron. Soc. Can., Vol. 78, No. 4, p. 145 – 150 (1984).

104.015 **Catalogo dei bolidi osservati in Italia. FIRECAT 1983.**
R. Balestrieri, F. Fontanelli.
Astronomia, N. 3, p. 9 – 12 (1984).

The authors present the first experimental edition of the fireballs catalogue, concerning the period 1949 – 1982. It collects 257 observations.

104.016 **A rigorous method for radiant determination.**
C. Steyaert.
Bull. Astron. Inst. Czech., Vol. 35, No. 5, p. 312 – 314 (1984).

The problem of finding the mean radiant position of more than two meteor trails has been treated analytically. It leads to a problem of optimization with a constraint. An interpretation of the radius or dispersion of the radiant position is found as well. On the practical side, this method allows a straightforward calculation.

104.017 **On the effective temperature and cross–sections of atomic excitation of silicon in a meteor coma.**
K. Kh. Saidov.
Dokl. Akad. Nauk TadzhSSR, Tom 26, No. 11, p. 701 – 704 (1983). In Russian. Abstr. in Ref. Zh., 51. Astron., 10.51.270 (1984).

104.018 **Possible mechanism of formation of short–period meteor swarms.**
E. N. Kramer, I. S. Shestaka.
Astron. Vestn., Tom 18, No. 2, p. 137 – 144 (1984). In Russian. Abstr. in Ref. Zh., 51. Astron., 10.51.272 (1984).

104.019 **On the activity of the Quadrantid meteor shower.**
A. F. Zausaev, A. N. Pushkarev.
Astron. Vestn., Tom 18, No. 2, p. 145 – 150 (1984). In Russian. Abstr. in Ref. Zh., 51. Astron., 10.51.273 (1984).

104.020 **Peculiarities of interaction with the atmosphere of cosmic bodies forming bolides and their mass scale.**
V. V. Kalenichenko.
Astron. Vestn., Tom 18, No. 2, p. 151 – 157 (1984). In Russian. Abstr. in Ref. Zh., 51. Astron., 10.51.275 (1984).

104.021 **The Saskatchewan fireball of 1922 and a possible link with the Wynyard chondrite.**
D. J. Gendzwill, M. R. Stauffer.
Meteoritics, Vol. 19, No. 3, p. 145 – 152 (1984).

104.022 **An estimate of the capacity of the meteor burst channel.**
J. A. Weitzen, W. P. Birkemeier, M. D. Grossi.
IEEE Trans. Commun., Vol. COM–32, No. 8, p. 972 – 974 (1984). Abstr. in Phys. Abstr., Vol. 88, No. 1247, Entry 4666 (1985).

104.023 **The structure of meteoroids.**
M. Beech.
Mon. Not. R. Astron. Soc., Vol. 211, No. 3, p. 617 – 620 (1984).
The available observational data for the beginning and end heights of meteors in the Perseid, Southern Taurid and Geminid meteor streams is compared to the quantitative theory of dustball meteor ablation due to Hawkes & Jones. A reasonable agreement between model predictions and observations is found. Following the analysis of Hapgood, Rothwell & Royrvik the amount of energy required to fragment unit mass of meteoroid material is calculated, with the result that meteors are possibly more fragile than assumed in theory.

104.024 **Fireball of 8 January 1984.**
P. H. Andersen, I. D. Cameron, D. Weinrich.
J. R. Astron. Soc. Can., Vol. 78, No. 5, p. 205 (1984). Abstract. – See Abstr. 010.401.

104.025 **1984 Orionids.**
Br. Astron. Assoc. Circ., No. 644 (1984).

104.026 **On the mechanism of formation of meteor streams in orbits with small dimensions and small perihelion distances.**
V. N. Lebedinets.
Tr. Inst. ehksp. meteorol. Goskomgidrometa, No. 14/110, p. 3 – 9 (1984). In Russian. Abstr. in Ref. Zh., 51. Astron., 11.51.259 (1984).

104.027 **Estimate of the influence of meteor streams on the apparent structure of the zodiacal light.**
V. N. Lebedinets.
Tr. Inst. ehksp. meteorol. Goskomgidrometa, No. 14/110, p. 9 – 16 (1984). In Russian. Abstr. in Ref. Zh., 51. Astron., 11.51.260 (1984).

104.028 **The seasonal structure of ionosonde $E_s$ parameters and meteoroid deposition rates.**
W. J. Baggaley, D. I. Steel.
Planet. Space Sci., Vol. 32, No. 12, p. 1533 – 1539 (1984).
A comparison covering more than three decades is made between the seasonal variation of radar meteor influx and seasonal variations in the occurrence probabilities of ionosonde sporadic–$E$ parameters $f_0E_s$, and $f_bE_s$ for different diurnal intervals at two Southern Hemisphere stations. The analyses show that for medium intensity $3E_m$ type sporadic–$E$ no clear correlation with major Southern Hemisphere meteor shower activity exists.

104.029 **Observing meteors: II.**
D. H. Levy.
Strolling Astron., Vol. 30, Nos. 11 – 12, p. 246 – 247 (1984).

104.030 **Search for a meteor shower associated with the Sugano–Saigusa–Fujikawa comet (1983e).**
M. Šimek, P. Pecina.
Bull. Astron. Inst. Czech., Vol. 35, No. 6, p. 375 – 379 (1984).
Though there are some indications of the existence of a meteor shower connected with the Sugano–Saigusa–Fujikawa comet, its existence cannot be proved beyond doubt.

104.031 **On the dependence of the coefficient of ionization on the velocity of meteors.**
K. Kh. Saidov.
Astron. Tsirk., No. 1303, p. 6 – 7 (1983). In Russian.

104.032 **Taking into account the influence of the astronomical selection on structural characteristics of meteor streams.**
G. V. Andreev.
Astron. Vestn., Tom 18, No. 3, p. 240 – 248 (1984). In Russian. Abstr. in Ref. Zh., 51. Astron., 12.51.402 (1984).

104.033 **Large–scale structure of the Perseid meteor stream.**
A. I. Grishchenyuk, M. N. Bidnichenko.
Astron. Vestn., Tom 18, No. 3, p. 249 – 251 (1984). In Russian. Abstr. in Ref. Zh., 51. Astron., 12.51.403 (1984).

104.034 **On a possible genetic relation between insufficiently studied meteor streams and comets.**
N. V. Kulikova.
Astron. Vestn., Tom 18, No. 3, p. 225 – 234 (1984). In Russian. Abstr. in Ref. Zh., 51. Astron., 12.51.405 (1984).

104.035 **The evolution of meteor streams.**
K. Fox, I. P. Williams.
Planetary rings, p. 717 – 721 (1984). – See Abstr. 012.085 (IAU Colloq. No. 75).

**Meteoritic and meteor investigations.**
See Abstr. 003.037.

**Meteorbeobachtung in der Tschechoslowakei.**
See Abstr. 013.009.

**Development of meteor investigations at the Odessa Astronomical Observatory.**
See Abstr. 013.024.

**Meteor investigations in Dushanbe.**
See Abstr. 013.025.

**Möglichkeiten der Meteorfotografie.**
See Abstr. 036.039.

**The dynamics of the minor bodies in the solar system.**
See Abstr. 042.059.

**Sodium clouds in the lower thermosphere.**
See Abstr. 082.021.

# 105 Meteorites, Meteorite Craters

105.001 **Titanium isotopic anomalies in meteorites.**
S. Niemeyer, G. W. Lugmair.
Geochim. Cosmochim. Acta, Vol. 48, No. 7, p. 1401 – 1416 (1984).

High-precision analyses of Ti are reported for samples from a variety of meteorite classes. The expanded data base for Allende inclusions still shows Ti isotopic anomalies in every inclusion. All the coarse-grained inclusions give quite similar patterns, but fine-grained inclusions show more variable, and sometimes larger, anomalies. Whether or not Allende "FUN" inclusions are considered, at least 4 distinct isotopic components of Ti are required to account for the observed isotopic variations. The Ti data cannot be plausibly explained in terms of an early solar-system particle irradiation; instead, neutron-rich hydrostatic burning within a star is probably responsible for the dominant $^{50}$Ti anomalies, while s-process mechanisms are viable sources for some of the more subtle Ti variations.

105.002 **Intrinsic oxygen fugacities of diogenites and mesosiderite clasts.**
R. H. Hewins, G. C. Ulmer.
Geochim. Cosmochim. Acta, Vol. 48, No. 7, p. 1555 – 1560 (1984).

This paper presents previously undetermined oxygen fugacity data for clasts from mesosiderites and diogenites obtained using a modified (Elliott et al., 1982) double $ZrO_2$ electrolyte cell technique (after Sato, 1971) and discusses the implications for the number of parent bodies.

105.003 **Nederland bekogeld vanuit de ruimte.**
K. Stevens.
Zenit, 11. Jaarg., No. 7/8, p. 314 – 319 (1984).

105.004 **The mineral chemistry and origin of inclusion matrix and meteorite matrix in the Allende CV3 chondrite.**
A. S. Kornacki, J. A. Wood.
Geochim. Cosmochim. Acta, Vol. 48, No. 8, p. 1663 – 1676 (1984).

This paper describes the mineralogy and mineral chemistry of inclusion matrix and meteorite matrix in the Allende CV3 chondrite, and evaluates the physical and chemical parameters of the equilibrium condensation model to determine how well that model explains their chemistry, mineralogy, texture, and isotopic systematics. The authors develop an alternative model of the origin of the mafic constituent of Allende inclusions that is based on a new model of chondrule petrogenesis and the physical evolution of the primitive solar nebula.

105.005 **Mass spectrometric isotope dilution analyses of palladium, silver, cadmium, and tellurium in carbonaceous chondrites.**
R. D. Loss, K. J. R. Rosman, J. R. De Laeter.
Geochim. Cosmochim. Acta, Vol. 48, No. 8, p. 1677 – 1681 (1984).

105.006 **Lunar meteorites: three and counting.**
J. K. Beatty.
Sky Telesc., Vol. 68, No. 3, p. 224 (1984).

105.007 **The scattering of the Tsarev stony meteorite shower.**
I. T. Zotkin, V. I. Tsvetkov.
Meteoritika, Vyp. 43, p. 3 – 8 (1984). In Russian.

105.008 **Conditions of the fall of the Chinge iron meteorite.**
E. L. Krinov.
Meteoritika, Vyp. 43, p. 9 – 15 (1984). In Russian.

105.009 **Is the Frost rule true?**
V. I. Tsvetkov.
Meteoritika, Vyp. 43, p. 16 – 18 (1984). In Russian.

105.010 **The morphology of the stony meteorites Preobrazhenka, Krutikha, Divnoe and Kamyshla.**
E. L. Krinov.
Meteoritika, Vyp. 43, p. 19 – 25 (1984). In Russian.

105.011 **The Anyujskij iron meteorite.**
A. A. Plyashkevich, G. F. Pavlov.
Meteoritika, Vyp. 43, p. 26 – 29, 188 (1984). With 2 plates. In Russian.

105.012 **The chemical and mineralogical composition and structure of the Zhigansk, Bilibino and Anyujskij iron meteorites.**
N. I. Zaslavskaya, G. M. Kolesov, L. D. Barsukova, L. F. Migdisova.
Meteoritika, Vyp. 43, p. 30 – 35, 188 (1984). With 2 plates. In Russian.

105.013 **The Doroninsk chondrite – a monomict breccia H5–7.**
G. V. Baryshnikova, S. A. Stakheeva.
Meteoritika, Vyp. 43, p. 36 – 43, 188 – 189 (1984). With 4 plates. In Russian.

105.014 **The structural and mineralogical characteristic of the Aleksandrovskij Khutor meteorite.**
V. P. Semenenko, L. G. Samojlovich, B. V. Tertychnaya.
Meteoritika, Vyp. 43, p. 44 – 48, 189 (1984). With 3 plates. In Russian.

105.015 **Mineralogy, petrography and chemical composition of Ca- and Al-rich inclusions of the Efremovka meteorite.**
M. A. Nazarov, M. I. Korina, A. A. Ul'yanov, G. M. Kolesov, E. Ya. Shcherbovskij.
Meteoritika, Vyp. 43, p. 49 – 66, 189 (1984). With 2 plates. In Russian.

105.016 **Shock effects in meteorites. Observations and experiment.**
A. K. Lavrukhina, Z. K. Mil'nikova.
Meteoritika, Vyp. 43, p. 67 – 82, 189 – 190 (1984). With 1 plate. In Russian.

105.017 **Chemical composition of chromite and ilmenite of the shock-metamorphosed Tsarev chondrite.**
N. I. Zaslavskaya, L. F. Migdisova, E. Ya. Shcherbovskij.
Meteoritika, Vyp. 43, p. 91 – 98, 190 (1984). With 2 plates. In Russian.

105.018 **On the normative composition of iron-magnesium silicates in equilibrated ordinary chondrites.**
A. A. Yavnel', L. D. Barsukova.
Meteoritika, Vyp. 43, p. 99 – 101 (1984). In Russian.

105.019 **On the chemical classification of the Krymka chondrite.**
A. A. Yavnel'.
Meteoritika, Vyp. 43, p. 102 – 105 (1984). In Russian.

105.020 **Investigation of the chemical composition of meteorites by the activation method using microtron.**
G. M. Kolesov, A. Hernandez.
Meteoritika, Vyp. 43, p. 106 – 113, 190 (1984). With 1 plate. In Russian.

105.021 **Investigation of the chemical composition of mineral fractions of the Tsarev chondrite.**
L. F. Semenova, A. V. Fisenko, V. G. Kashkarova, L. N. Mel'nikova, E. V. Bezrogova, V. A. Pomytkina, A. K. Lavrukhina.
Meteoritika, Vyp. 43, p. 114 – 122 (1984). In Russian.

105.022 **The Verkhnedneprovsk meteorite from the collection of the Geological Institute of the Estonian SSR Academy of Sciences.**
G. M. Kolesov, N. I. Zaslavskaya, R. T. Tijrmaa.
Meteoritika, Vyp. 43, p. 123 – 124 (1984). In Russian.

105.023 **Results of an experiment of vaporization of the Krymka chondrite.**
O. I. Yakovlev, O. M. Markova, G. A. Semenov, A. N. Belov.
Meteoritika, Vyp. 43, p. 125 – 133 (1984). In Russian.

105.024 **Iron phases under joint heating of the chondrites Indarch EI (EH) and Murchison CM.**
T. V. Malysheva, K. I. Tobelko, A. V. Polosin, E. P. Smirnova.
Meteoritika, Vyp. 43, p. 134 – 139 (1984). In Russian.

105.025 **$^6$Li and $^7$Li/$^6$Li in iron meteorites.**
P. S. Goel, S. V. S. Murty, E. S. Bogomolov, V. G. Zaslavskij, L. K. Levskij.
Meteoritika, Vyp. 43, p. 140 – 141 (1984). In Russian.

105.026 **Radioactivity of the Malakal and Kutais chondrites.**
A. K. Lavrukhina, V. D. Gorin, G. K. Ustinova.
Meteoritika, Vyp. 43, p. 142 – 146 (1984). In Russian.

105.027 **Tracks of heavy cosmic nuclei and effects of channeling in olivine crystals from meteorites.**
G. I. Dolivo–Dobrovol'skaya, V. D. Kolomenskij, V. P. Perelygin, S. G. Stetsenko.
Meteoritika, Vyp. 43, p. 153 – 154, 190 (1984). With 2 plates. In Russian.

105.028 **Magnetism of achondrites.**
E. G. Gus'kova.
Meteoritika, Vyp. 43, p. 155 – 160 (1984). In Russian.

105.029 **Division of the USSR territory into regions on the basis of a relative estimate of density of presumably preserved astroblemes.**
V. L. Masajtis, M. S. Mashchak.
Meteoritika, Vyp. 43, p. 161 – 172 (1984). In Russian.

105.030 **Impactites from the basin of the Big Kuonamka river (Anabar massif).**
E. P. Gurov, E. P. Gurova.
Meteoritika, Vyp. 43, p. 173 – 177, 191 (1984). With 2 plates. In Russian.

105.031 **First Appendix to the Catalogue of Meteorites of the USSR up to January 1983.**
E. L. Krinov.
Meteoritika, Vyp. 43, p. 183 – 185 (1984). In Russian.

105.032 **Impact ejection, spallation, and the origin of meteorites.**
H. J. Melosh.
Icarus, Vol. 59, No. 2, p. 234 – 260 (1984).
This paper presents a model for ejection of material from an impact crater. The model links ejection velocity, fragment size, and shock pressure through a simplified stress–wave propagation and reflection scheme. The approximations used are valid near the free surface, becoming progressively worse as depth below the surface increases. However, since the largest and least shocked fragments are found to originate near the surface, the model may answer some of the most important questions about meteorite origins.

105.033 **Measurements of the triangulation network at the Allan Hills meteorite icefield.**
J. O. Annexstad.
Antarctic earth science, p. 617 – 619 (1983). Abstr. in Phys. Abstr., Vol. 87, No. 1239, Entry 82145 (1984). – See Abstr. 012.016.

105.034 **Results of meteorite search and recovery activities in the vicinity of the Allan Hills, Antarctica, Dec. 1981 – Jan. 1982.**
J. W. Schutt, W. A. Cassidy, G. Crozaz, R. F. Fudali, U. B. Marvin.
Antarctic earth science, p. 617 – 619 (1983). Abstr. in Phys. Abstr., Vol. 87, No. 1239, Entry 82146 (1984). – See Abstr. 012.016.

105.035 **The remarkably low surface density of meteorites at Allan Hills and implications in this for climate change.**
W. A. Cassidy.
Antarctic earth science, p. 623 – 625 (1983). Abstr. in Phys. Abstr., Vol. 87, No. 1239, Entry 82230 (1984). – See Abstr. 012.016.

105.036 **Bismuth in interplanetary dust.**
I. D. R. Mackinnon, F. J. M. Rietmeijer.
Nature, Vol. 311, No. 5982, p. 135 – 138 (1984).
Samples of a large ($\sim$60 μm) chondritic porous (CP) aggregate collected from the stratosphere have been analysed in detail by analytical electron microscopy. Previous studies of CP aggregates have shown that they are extraterrestrial in origin and may be related to cometary debris. The authors report a high concentration of $Bi_2O_3$ grains within the large CP aggregate designated W7029*A ($\sim$60 μm) and suggest they formed by rapid heating ($\sim$300°C) of elemental Bi grains within the aggregate during atmospheric entry.

105.037 **Occurrence of oxidized components in Qingzhen enstatite chondrite.**
E. R. Rambaldi, R. M. Housley, R. S. Rajan.
Nature, Vol. 311, No. 5982, p. 138 – 140 (1984).
Clues to the early history of the Solar System can be acquired by studying the properties of highly unequilibrated chondrites. The authors show that the most unequilibrated enstatite chondrite, Qingzhen, contains a population of enstatite grains which appear to have formed under more oxidizing nebular conditions than the bulk of the meteorite, which is highly reduced. The discovery of this once–oxidized material in Qingzhen poses significant constraints on the existing models of formation of reduced matter in the Solar System.

105.038 **Presence of a superparamagnetic component in the Orgueil meteorite.**
T. J. Wdowiak, D. G. Agresti.
Nature, Vol. 311, No. 5982, p. 140 – 142 (1984).
The Orgueil CI carbonaceous chondrite is of particular interest because of the presence of an organic component which included amino acids. Using Mössbauer spectroscopy, the authors have determined that the Orgueil meteorite contains an appreciable fraction of its iron in the superparamagnetic state. This result implies that the superparamagnetic iron component was formed during an aqueous phase in the history of the parent body, and that the superparamagnetic component may have served as a catalyst for formation of the organic component.

105.039 **Meteorites from Mars.**
L. D. Majorova.
Priroda, No. 7, p. 104 – 105 (1984). In Russian.

105.040 **Distribution of rare earth elements and uranium in various components of ordinary chondrites.**
M. Ebihara, M. Honda.
Meteoritics, Vol. 19, No. 2, p. 69 – 77 (1984).
Rare earth elements (REE) and uranium (U) were studied for their distributions in various component phases of four ordinary chondrites. The present study revealed that both REE and U are highly enriched in the Ca–phosphate minerals with different enrichment factors, implying chemical fractionation between them. The phosphates seem to be responsible for more than 80% of the light REE in all chondrites.

105.041 **Minor elements in Marjalahti olivine.**
G. Ryder.
Meteoritics, Vol. 19, No. 2, p. 79 – 83 (1984).

Precise microprobe determinations of minor elements in olivine from Marjalahti show averages of 0.0267% CaO; 0.0211% $Cr_2O_3$; less than 0.0045% $TiO_2$; 0.288% MnO; and 30 ppm Ni. The calcium is as high as in some terrestrial plutonic olivines but lower than in terrestrial nodule olivines, consistent with very slow cooling to low temperatures.

105.042 **The Machinga, Malawi, meteorite: a recent L6 fall.**
A. L. Graham, M. J. Crow, J. C. Chatupa, A. T. Mndala.
Meteoritics, Vol. 19, No. 2, p. 85 – 88 (1984).

105.043 **The Grayton Beach, Florida meteorite.**
H. Povenmire.
Meteoritics, Vol. 19, No. 2, p. 89 – 90 (1984).

105.044 **Discovery of two australites in lithfied rocks.**
W. H. Cleverly, S. Kirsch.
Meteoritics, Vol. 19, No. 2, p. 91 – 93 (1984).

105.045 **Interobject structure of ungrouped iron meteorites as revealed by advanced clustering: methods and chemical features.**
K. H. Esbensen, L. Kaufman, D. L. Massart.
Meteoritics, Vol. 19, No. 2, p. 95 – 109 (1984).

Using the MASLOC clustering strategy, 61 ungrouped iron meteorites are probed for physico–chemical clusters. The existing classification of 13 resolved chemical groups is used as an internal coherence reference. Results include tentative establishment of a new group: ID (7 members), with chemical features similar to those of the non–magnetic groups IAB and IIICD.

105.046 **Meteorites from Mars.**
D. W. Hughes.
Nature, Vol. 311, No. 5985, p. 411 – 412 (1984).

This note comments on recently renewed speculations that certain classes of meteorites (shergottites, chassignites, nakhlites) may be of Martian origin.

105.047 **Geochemical aspect of impact cratering.**
A. T. Bazilevskij, O. I. Yakovlev.
The 27th International Geological Congress, Tom 5, Sect. 10 – 11, p. 232 – 233 (1984). In Russian. Abstr. in Ref. Zh., 62. Issled. Kosm. Prostranstva, 8.62.399 (1984). – See Abstr. 012.031.

105.048 **Methods for determining the depths of meteorite craters of spherical form.**
V. P. Ivanov, G. V. Aksenov.
Geod. Kartogr., No. 3, p. 26 – 28 (1984). In Russian. Abstr. in Ref. Zh., 52. Geod. Aehrosemka, 8.52.196 (1984).

105.049 **Noble gas contents of shergottites and implications for the Martian origin of SNC meteorites.**
D. D. Bogard, L. E. Nyquist, P. Johnson.
Geochim. Cosmochim. Acta, Vol. 48, No. 9, p. 1723 – 1739 (1984).

The purpose of this paper is to present the additional data on noble gases in EETA 79001 to present the cosmic ray exposure ages, and to discuss various aspects of the data in terms of a possible Martian origin. The authors include noble gas data subsequently measured in two other shergottites, Zagami and Allan Hills (ALH) A77005.

105.050 **Matrix material in type 3 chondrites – occurrence, heterogeneity and relationship with chondrules.**
E. R. D. Scott, A. E. Rubin, G. J. Taylor, K. Keil.
Geochim. Cosmochim. Acta, Vol. 48, No. 9, p. 1741 – 1757 (1984).

The authors present petrologic and electron–microprobe data on some occurrences of matrix material in type 3 chondrites. They conclude that matrix material with diverse compositions existed in the solar nebula before chondrules formed, and suggest that some chondrules formed from matrix–like material.

105.051 **Coarse–grained chondrule rims in type 3 chondrites.**
A. E. Rubin.
Geochim. Cosmochim. Acta, Vol. 48, No. 9, p. 1779 – 1789 (1984).

Relatively coarse–grained rims occur around all types of chondrules in type 3 carbonaceous and ordinary chondrites. Those in H–L–LL3 chondrites are composed primarily of olivine and low–Ca pyroxene; those in CV3 chondrites contain much less low–Ca pyroxene. CV chondrites may have formed in a region where the chondrule formation mechanism was less efficient, probably at greater solar distances than the ordinary chondrites. The lesser efficiency of heating could be responsible for the greater abundance of coarse–grained rims around CV chondrules. Alternatively, CV chondrules may have suffered fewer particle collisions prior to agglomeration.

105.052 **Beryllium–10 contents of core samples from the St. Severin meteorite.**
C. Tuniz, C. M. Smith, R. K. Moniot, T. H. Kruse, W. Savin, D. K. Pal, G. F. Herzog, R. C. Reedy.
Geochim. Cosmochim. Acta, Vol. 48, No. 9, p. 1867 – 1872 (1984).

Eleven samples taken from along the 35–cm core AIII of the LL–chondrite St. Severin have $^{10}Be$ contents ranging from 20 to 28 dpm/kg. Calculations based on cross sections for proton–induced reactions that make $^{10}Be$ disagree seriously with the measured values. Relatively large cross sections were constructed for neutron–induced reactions below 1 GeV. They give calculated $^{10}Be$ production rates that agree with the measured values to within 10%. Cosmogenic $^{10}Be$ in stony meteorites is better described as a medium–energy product than as a high–energy one.

105.053 **Microvesicles in meteorites, a model of pre–biotic evolution.**
H. D. Pflug.
Naturwissenschaften, 71. Jahrg., Heft 10, p. 531 – 533 (1984).

105.054 **Relic interstellar grains in Murchison meteorite.**
J. Yang, S. Epstein.
Nature, Vol. 311, No. 5986, p. 544 – 547 (1984).

The authors show that, in the Murchison meteorite, the D/H ratios of hydrogen are unusually high in several separates and in one case up to 30 times the cosmic value of $2 \times 10^{-5}$. Many phases show high $^{13}C/^{12}C$ ratios, up to 2.5 times the terrestrial value of 0.011. These $^{13}C$–rich and D–rich components of the two chemical elements are not correlated and are heterogeneously distributed. The D–rich hydrogen in the meteorite is probably due to molecules formed by ion–molecule reactions in interstellar clouds while the tiny amount of $^{13}C$–rich carbon is probably due to nucleosynthesis in red giant stars.

105.055 **A negative polarization does not prove the presence of dust on the surface of an atmosphereless cosmic body.**
Yu. G. Shkuratov, L. A. Akimov, V. P. Tishkovets.
Pis'ma Astron. Zh., Tom 10, No. 10, p. 797 – 799 (1984). In Russian. English translation in Sov. Astron. Lett, Vol. 10.

Laboratory polarimetric investigation of meteoritic samples has shown that powder and rock surfaces of C3 type carbonaceous chondrites have similar polarization curves at small phase angles. This is an indication that negative light polarization cannot serve as a proof of dust presence on the surface of an atmosphereless cosmic body.

105.056 **Investigation of cosmic bodies with cosmogenic isotopes.**
V. A. Alekseev, A. K. Lavrukhina, G. K. Ustinova.
The 27th International Geological Congress, Tom 5, Sect. 10 – 11, p. 215 – 217 (1984). In English and Russian. Abstr. in Ref. Zh., 51. Astron., 9.51.189 (1984). – See Abstr. 012.031.

105.057 **Ultrabasic meteorites and the problem of primary hyperbasic magma.**
A. A. Marakushev.
U'ltraosnov. magmy i ikh metallogeniya. Tez. dokl. Vses. Simpoz. Vladivostok, 20–22 sent., 1983. Vladivostok, p. 54 – 56 (1983). In Russian. Abstr. in Ref. Zh., 51. Astron., 9.51.241 (1984).

105.058 **Some questions of classification of meteorites.**
A. A. Yavnel'.
The 27th International Geological Congress, Tom 5, Sect. 10 – 11, p. 443 – 445 (1984). In English and Russian. Abstr. in Ref. Zh., 51. Astron., 9.51.266 (1984). – See Abstr. 012.031.

105.059 **Carbonaceous chondrite Kaidun: intensive mixing of matter during formation of meteoritic parent bodies.**
A. V. Ivanov, A. A. Ul'yanov, A. Ya. Skripnik, M. A. Nazarov, N. N. Konokova.
The 27th International Geological Congress, Tom 5, Sect. 10 – 11, p. 296 – 297 (1984). In English and Russian. Abstr. in Ref. Zh., 51. Astron., 9.51.271 (1984). – See Abstr. 012.031.

105.060 **Possible photochemical production of meteoritic oxygen isotopic anomalies.**
M. H. Thiemens.
The 27th International Geological Congress, Tom 5, Sect. 10 – 11, p. 416 (1984). Abstr. in Ref. Zh., 51. Astron., 9.51.276 (1984). – See Abstr. 012.031.

105.061 **The stable isotopic composition of carbon and stony meteorites: implications and interrelations.**
M. M. Grady, C. T. Pillinger.
The 27th International Geological Congress, Tom 5, Sect. 10 – 11, p. 278 – 279 (1984). Abstr. in Ref. Zh., 51. Astron., 9.51.278 (1984). – See Abstr. 012.031.

105.062 **Long lived radioactivities in cosmochemistry.**
M. Honda.
The 27th International Geological Congress, Tom 5, Sect. 10 – 11, p. 293 (1984). Abstr. in Ref. Zh., 51. Astron., 9.51.284 (1984). – See Abstr. 012.031.

105.063 **Geochemical comparison of meteorites, the moon and the earth: trends of element evolution.**
K. Hanke, J. Harff.
The 27th International Geological Congress, Tom 5, Sect. 10 – 11, p. 288 – 289 (1984). Abstr. in Ref. Zh., 51. Astron., 9.51.285 (1984). – See Abstr. 012.031.

105.064 **Geochemistry of chondrules: fractionation processes in the early solar system.**
G. Kurat.
The 27th International Geological Congress, Tom 5, Sect. 10 – 11, p. 325 – 326 (1984). Abstr. in Ref. Zh., 51. Astron., 9.51.288 (1984). – See Abstr. 012.031.

105.065 **Ca–, Al–rich chondrules in ordinary chondrites.**
F. Wlotzka.
The 27th International Geological Congress, Tom 5, Sect. 10 – 11, p. 438 (1984). Abstr. in Ref. Zh., 51. Astron., 9.51.289 (1984). – See Abstr. 012.031.

105.066 **A group of young meteoritic craters in the west of the Aldansk anteclise.**
E. P. Gurov, E. P. Gurova.
Dokl. Akad. Nauk USSR, No. 2, B, p. 9 – 12 (1984). In Russian. Abstr. in Ref. Zh., 51. Astron., 9.51.295 (1984).

105.067 **Shock–metamorphed rocks and impactites of the Terny astrobleme (North Krivorozhye).**
A. P. Nikol'skij, V. P. Naumov, M. S. Mashchak, V. L. Masajtis.
Tr. Vses. nauchn.–issled. geol. inst., Tom 238, p. 132 – 142 (1982). In Russian. Abstr. in Ref. Zh., 51. Astron., 9.51.296 (1984).

105.068 **Origin of impact bombs from the Ilinetsk astrobleme.**
E. P. Gurov, E. P. Gurova.
Dokl. Akad. Nauk SSSR. Ser. Mat. Fiz., Tom 275, No. 4, p. 958 – 961 (1984). In Russian. Abstr. in Ref. Zh., 51. Astron., 9.51.297 (1984).

105.069 **Thallium and lead in the Allende C3V carbonaceous chondrite. A study of the matrix phase.**
J. W. Arden, G. Cressey.
Geochim. Cosmochim. Acta, Vol. 48, No. 10, p. 1899 – 1912 (1984).
In this study the authors attempt to gain insight into the distribution and siting of Tl and Pb in phases of Allende and in particular the fine–grained volatile–rich matrix. Various fractions of the matrix were prepared by sieving, sedimentation, magnetic separation and acid–etching. However, although only partial mineral separation was achieved due to the sticking properties of the organically coated grains, XRD and analytical SEM provided information on the siting of Tl and Pb from the relative abundances of the phases present, their compositions, and grain sizes.

105.070 **Chemical compositions of refractory inclusions in the Murchison C2 chondrite.**
V. Ekambaram, I. Kawabe, T. Tanaka, A. M. Davis, L. Grossman.
Geochim. Cosmochim. Acta, Vol. 48, No. 10, p. 2089 – 2105 (1984).
The purpose of this study is to explore the range of trace element patterns present in a large number of refractory objects from Murchison, to determine the physico–chemical similarities and differences between the formation sites of refractory inclusions in C2 and C3 chondrites, to ascertain if any relationship exists between chemical and mineralogical compositions of Murchison inclusions and to see how common it is for these objects to have higher equilibration temperatures than Allende coarse–grained inclusions.

105.071 **Infrared transmission spectra from 2.5 to 25 μm of various meteorite classes.**
S. A. Sandford.
Icarus, Vol. 60, No. 1, p. 115 – 126 (1984).
Infrared transmission spectra from 53 meteorites in the spectral range from 2.5 to 25 μm were measured to permit comparisons with data of astronomical objects that are potential meteorite sources. Data were taken for 14 carbonaceous chondrites, 5 LL ordinary chondrites, 6 L ordinary chondrites, 10 H ordinary chondrites, 1 enstatite chondrite, 4 aubrites, 3 eucrites, 4 howardites, 1 diogenite, 1 mesosiderite, 2 nakhlites, 1 shergottite, and the anomalous achondrite Angra dos Reis.

105.072 **Mineralogy of some achondrites and the early evolution of their parent bodies.**
H. Takeda, H. Mori.
The 27th International Geological Congress, Tom 11, Sect. S 11, p. 58 – 64 (1984). In Russian. Abstr. in Ref. Zh., 51. Astron., 10.51.285 (1984). – See Abstr. 012.031.

105.073 **New data on the heterogeneous nature of metal in the Bakhmut meteorite.**
V. P. Semenenko, L. G. Samojlovich, B. V. Tertychnaya.
Dokl. Akad. Nauk USSR, B, No. 5, p. 24 – 28 (1984). In Russian. Abstr. in Ref. Zh., 51. Astron., 10.51.286 (1984).

105.074 **Composition and structure of the Mar'inka iron meteorite.**
V. P. Semenenko, O. P. Sharkin, B. V. Tertychnaya, V. A. Chebotarev.
Mineral. zh., Tom 6, No. 2, p. 83 – 87 (1984). In Russian. Abstr. in Ref. Zh., 51. Astron., 10.51.287 (1984).

105.075 **On evidence of cosmogenic explosions on the Aktogaj and Ajdarly copper–porphyry deposits (South Kazakhstan).**
B. S. Zejlik, S. A. Litvintsev, Eh. Yu. Sejtmuratova.
Dokl. Akad. Nauk SSSR. Ser. Mat. Fiz., Tom 276, No. 1, p. 187 – 191 (1984). In Russian. Abstr. in Ref. Zh., 51. Astron., 10.51.295 (1984).

105.076 **Unusual meteorite.**
Priroda, No. 11, p. 103 – 104 (1984). In Russian.
Concerning meteorite Kajdun, fallen at Kajdun (Yemen) in December 1980.

105.077 **Chromium isotopic anomalies in Allende refractory inclusions.**
J.-L. Birck, C. J. Allègre.
Geophys. Res. Lett., Vol. 11, No. 10, p. 943 – 946 (1984).
Measurements of both $^{53}Cr/^{52}Cr$ and $^{54}Cr/^{52}Cr$ ratios with a precision of $10^{-4}$ in Allende inclusions display widespread chromium isotopic anomalies which are in the range of $5 \times 10^{-4}$. Instrumental mass fractionation precludes specific identification of the affected isotopes, but there are strong arguments favoring excesses in $^{54}Cr$ and a hint of a very small deficit in $^{53}Cr$ ($\cong 10^{-4}$). These samples also display excesses in $^{50}Ti$ and no significant effect in nickel isotopes.

105.078 **Achondrite ALHA77005: alteration of chromite and olivine.**
J. V. Smith, I. M. Steele.
Meteoritics, Vol. 19, No. 3, p. 121 – 133 (1984).

105.079 **Size–distributions of chondrule types in the Inman and Allan Hills A77011 L3 chondrites.**
A. E. Rubin, K. Keil.
Meteoritics, Vol. 19, No. 3, p. 135 – 143 (1984).
A petrographic study of 9 thin sections of Inman (L3) and 18 thin sections of ALHA77011 (L3) served to determine the size–distributions of different chondrule textural types. Inman chondrules are significantly larger than those in ALHA77011, but in each chondrite, there is no statistically significant difference between the size–distributions of barred olivine and radial pyroxene plus cryptocrystalline chondrules. In ALHA77011, barred olivine chondrules outnumber radial pyroxene plus cryptocrystalline chondrules, whereas in Inman, the reverse is true.

105.080 **The Brownell and Ness County (1894) L6 chondrites: further sorting–out of Ness County meteorites.**
A. E. Rubin, W. F. Read.
Meteoritics, Vol. 19, No. 3, p. 153 – 160 (1984).
Brownell is a new, moderately shocked, L6 chondrite from Ness County, Kansas that is petrologically distinct from the other L6 chondrites from Ness County. These latter meteorites, Wellmanville, Franklinville and Ness County (1894), are very similar in their olivine and low–Ca pyroxene compositional distributions, kamacite Co contents, modal abundances of metallic Fe, Ni and presence of martensitic metallic Fe, Ni ($\sim$14 wt.% Ni). However, their silicates indicate that they have been shocked to different extents.

105.081 **Inventory of the Vatican Meteorite Collection.**
R. Salvatori, A. Maras, E. A. King.
Meteoritics, Vol. 19, No. 3, p. 161 – 172 (1984).

105.082 **Thermoluminescence and the shock and reheating history of meteorites – II: Annealing studies of the Kernouve meteorite.**
D. W. G. Sears, N. Bakhtiar, B. D. Keck, K. S. Weeks.
Geochim. Cosmochim. Acta, Vol. 48, No. 11, p. 2265 – 2272 (1984).
Samples of the unshocked, equilibrated chondrite, Kernouve (H6), have been annealed for 1 – 100 hours at 500 – 1200°C, their thermoluminescence sensitivity measured and Na, K, Mn, Ca and Sc determined by instrumental neutron activation analysis.

105.083 **Microprobe analyses of rare earth element fractionation in meteoritic minerals.**
T. M. Benjamin, C. J. Duffy, C. J. Maggiore, P. S. Z. Rogers, D. S. Woolum, D. S. Burnett, M. T. Murrell.
Nucl. Instrum. Methods Phys. Res., Sect. B, Vol. 231, No. 1 – 3, p. 677 – 680 (1984). Abstr. in Phys. Abstr., Vol. 87, No. 1243, Entry 101276 (1984). – See Abstr. 012.047.

105.084 **Determination of magnesium isotopic ratios with a proton microprobe in chondrules of the Allende meteorite.**
C. C. A. H. van der Stap, A. Hogenbirk, A. J. J. Bos, R. D. Vis, H. Verheul.
Nucl. Instrum. Methods Phys. Res., Sect. B, Vol. 231, No. 1 – 3, p. 695 – 699 (1984). Abstr. in Phys. Abstr., Vol. 87, No. 1243, Entry 101277 (1984). – See Abstr. 012.047.

105.085 **The origin and significance of reverse zoning in melilite from Allende Type–B inclusions.**
G. J. MacPherson, J. M. Paque, E. Stolper, L. Grossman.
J. Geol., Vol. 92, No. 3, p. 289 – 305 (1984). Abstr. in Phys. Abstr., Vol. 87, No. 1243, Entry 101278 (1984).

105.086 **Origins of stony–iron meteorites.**
E. R. D. Scott.
Nature, Vol. 311, No. 5988, p. 708 (1984).

105.087 **Shatter cones: epilogue to a dynamic story.**
C. Thompson.
GEOS, Vol. 13, No. 2, p. 1 – 4 (1984). Abstr. in Phys. Abstr., Vol. 87, No. 1246, Entry 115405 (1984).

105.088 **Chemical studies on the Antarctic iron meteorites, Yamato–790724, ALH–77263 and ALH–77289.**
T. Nagata, I. Taguchi, A. Masuda.
JEOL News Electron Opt. Instrum., Vol. 22E, No. 2, p. 25 – 30 (1984). Abstr. in Phys. Abstr., Vol. 88, No. 1247, Entry 4669 (1985).

105.089 **Antarctic meteorites: scientific results and potential.**
M. E. Lipschutz.
Bull. Am. Astron. Soc., Vol. 16, No. 3, Part I, p. 689 (1984). Abstract. – See Abstr. 010.061.

105.090 **Identification of the parent bodies of pallasites.**
J. F. Bell, M. J. Gaffey, B. R. Hawke.
Bull. Am. Astron. Soc., Vol. 16, No. 3, Part I, p. 708 – 709 (1984). Abstract. – See Abstr. 010.061.

105.091 **$^{39}Ar$ recoil and the apparent persistence of the presolar age of an Allende inclusion.**
E. K. Jessberger.
Earth Planet. Sci. Lett., Vol. 69, No. 1, p. 1 – 12 (1984).

105.092 **The case for a martian origin of the shergottites: nitrogen and noble gases in EETA 79001.**
R. H. Becker, R. O. Pepin.
Earth Planet. Sci. Lett., Vol. 69, No. 2, p. 225 – 242 (1984).
Nitrogen and noble gases were measured in samples of a glass inclusion and the surrounding basaltic matrix from the antarctic shergottite EETA 79001. Trapped noble gases in the glass reproduce, within error, the elemental and isotopic compositions measured in Mars' atmosphere by Viking, and are in general agreement with previous measurements except for much lower abundances of neutron–generated krypton and xenon isotopes. The most reasonable explanation at the present time for the noble gas pattern and the isotopically heavy nitrogen is that a sample of martian atmosphere has been trapped in the EETA 79001 glass, and that this meteorite, and thus the shergottites and probably the nakhlites and chassignites as well, originated on Mars.

105.093 **Solar composition noble gases in the Washington County iron meteorite.**
R. H. Becker, R. O. Pepin.
Earth Planet. Sci. Lett., Vol. 70, No. 1, p. 1 – 10 (1984).
The authors have analyzed a sample of the Washington County iron meteorite for its light noble gases. Unless the Washington County iron formed by some unique process in a solar-wind-irradiated regolith, the fact that the trapped gas is of solar composition has implications with regard to the compositional history of the solar wind, to conditions in the early solar nebula during grain formation or accretion, and to primordial helium in the earth.

105.094 **On the formation of meteoritic chondrules by aerodynamic drag heating in the solar nebula.**
J. A. Wood.
Earth Planet. Sci. Lett., Vol. 70, No. 1, p. 11 – 26 (1984).
During the formation of the solar nebula interstellar grains were falling into the nebula with velocities of the order of 10 km/s at the radial distance where the meteorites were to form. This kinetic energy is 20 times the amount of thermal energy needed to melt the grains. The grains were decelerated by aerodynamic drag in the nebula. Where grain-rich parcels of interstellar material fell into the nebula, heat generated by drag could not be radiated away because of the opacity imparted to the system by the grains, and high temperatures were reached. In this situation presolar aggregations of grains would melt to form chondrules.

105.095 **Beardsley vs. Faith: physical diversity among H5 chondrites.**
M. Żbik, B. Lang.
Earth Planet. Sci. Lett., Vol. 70, No. 2, p. 169 – 174 (1984).
The authors examined some physical properties of a pair of ordinary chondrites of the same type, with the objective of determining if the similarity of these properties reflects the identity of their chemical-petrologic classification. Beardsley and Faith were compared, both are H5 chondrites, not suspected of incorrect classification. However, with regard to the microstructure as disclosd by scanning electron microscopy they observed some important differences.

105.096 **Investigations of cosmic-ray-produced nuclides in iron meteorites. 6. The Signer-Nier model and the history of the cosmic radiation.**
H. Voshage.
Earth Planet. Sci. Lett., Vol. 71, No. 1, p. 181 – 194 (1984).

105.097 **The formation age of the Brachina meteorite.**
G. Crozaz, P. Pellas.
Earth Planet. Sci. Lett., Vol. 71, No. 1, p. 195 – 199 (1984).

105.098 **On the chondrite-achondrite transition: mineralogy and chemistry of Yamato 74160(LL7).**
H. Takeda, T. J. Huston, M. E. Lipschutz.
Earth Planet. Sci. Lett., Vol. 71, No. 2, p. 329 – 339 (1984).

105.099 **Matrices of type 3 ordinary chondrites – primitive nebular records.**
H. Nagahara.
Geochim. Cosmochim. Acta, Vol. 48, No. 12, p. 2581 – 2595 (1984).
This paper presents the results of detailed petrological studies of the matrices of type 3 ordinary chondrites, especially those of low petrologic subtype, having low thermoluminescence sensitivities. On the basis of the observations, the origin of matrix-forming materials, differences between matrix materials in chondrites of different chemical groups, and the relationship between chondrules and matrices are discussed.

105.100 **Ca isotope variations in Allende.**
M. H. A. Jungck, T. Shimamura, G. W. Lugmair.
Geochim. Cosmochim. Acta, Vol. 48, No. 12, p. 2651 – 2658 (1984).
The authors report the results from several coarse- and fine-grained CAIs, which show that non-linear isotope effects on Ca may not be as rare as previous data appear to indicate. These effects are at present only resolvable on $^{48}$Ca and are below the 1 per mill level.

105.101 **Light element stable isotopes in meteorites – from grams to picograms.**
C. T. Pillinger.
Geochim. Cosmochim. Acta, Vol. 48, No. 12, p. 2739 – 2766 (1984).
Fifty years work in the study of stable isotopes of hydrogen, carbon, nitrogen, oxygen and sulphur in meteorites is reviewed to mark the anniversary of the first paper dealing with the subject (Manian et al., 1934). Papers up to and including those presented at the 15th Lunar and Planetary Science Conference (March, 1984) are included.

105.102 **New data on the Bragin pallasite.**
A. M. Dymkin, V. A. Vilisov, G. I. Emel'yanov, I. A. Yudin, V. N. Loginov.
Dokl. Akad. Nauk BSSR, Tom 28, No. 5, p. 445 – 447 (1984). In Russian. Abstr. in Ref. Zh., 51. Astron., 12.51.420 (1984).

105.103 **Isotope composition of xenon and crypton of micro- and macro-inclusions in iron meteorites.**
Yu. A. Shukolyukov, Dang Vu Minh, P. S. Goel, N. I. Zaslavskaya.
Geokhimiya, No. 6, p. 771 – 780 (1984). In Russian. Abstr. in Ref. Zh., 51. Astron., 12.51.425 (1984).

105.104 **Anomalous isotopic abundances of $^{128}$Xe in some meteorites.**
Yu. A. Shukolyukov, Dang Vu Minh.
Geokhimiya, No. 7, p. 931 – 948 (1984). In Russian. Abstr. in Ref. Zh., 51. Astron., 12.51.426 (1984).

105.105 **On iridium distribution in the rocks of terrestrial impact craters.**
A. T. Bazilevskij, V. I. Fel'dman, I. G. Kapustkina, G. M. Kolesov.
Geokhimiya, No. 6, p. 781 – 790 (1984). In Russian. Abstr. in Ref. Zh., 51. Astron., 12.51.429 (1984).

105.106 **On radiation effects under the conditions of the nonhomogeneous atmosphere at the Tunguska event.**
V. P. Korobejnikov, B. V. Putyatin, P. I. Chushkin, L. V. Shurshalov.
Meteoritic and meteor investigations, p. 5 – 24 (1983). In Russian. Abstr. in Ref. Zh., 51. Astron., 12.51.430 (1984). – See Abstr. 003.037.

105.107 **Small spheres [30 μm – 50 μm] of (Ca, RE)$F_2$ and CaS in the meteorites Zagami and Murchison.**
V. W. A. Vieira, T. V. V. Costa, J. M. Knudsen, G. B. Jensen.
Phys. Scr., Vol. 30, No. 5, p. 384 – 386 (1984). With a correction in No. 6, p. 447 – 448.
In a sample from the Zagami meteorite (classified as a shergottite) eight grains of perfect spherical shape and with diameters from 30 to 50 μm have been found. Two of the spheres consist of (Ca, RE)$F_2$ (RE = rare earths), one is a mixture of (Ca, RE)$F_2$ and CaS, and another is almost pure CaS. In the Murchison meteorite (classified as a carbonaceous chondrite II) one sphere consisting of (Ca, RE)$F_2$ has been found. The mineral (Ca, RE)$F_2$ has not previously been identified in these meteorites.

**Meteoritic and meteor investigations.**
See Abstr. 003.037.

**Heroic efforts at meteor crater, Arizona: selected correspondence between Daniel Moreau Barringer and Elihu Thomson.**
See Abstr. 003.055.

**Meteorite – Boten aus dem Weltall.**
See Abstr. 003.056.

**Chondrules and their origins.**
See Abstr. 003.112.

**Chondrules and their origins. Proceedings of a workshop held at the Lunar and Planetary Institute, Houston, November 1982.**
See Abstr. 012.093.

**Proceedings of the Eighth Symposium on Antarctic Meteorites.**
See Abstr. 012.094.

**Joint Meeting of Mineralogical Society and Meteoritical Society on meteoritic research in progress (papers in summary form only received). Conference held at London, England, 24 May 1984.**
See Abstr. 012.099.

**An object within a particle of extraterrestrial origin compared with an object of presumed terrestrial origin.**
See Abstr. 015.034.

**Relativistic $^{238}$U ion tracks in olivine and cosmic–ray track studies.**
See Abstr. 022.004.

**Measurement of cross sections for $^{22}$Na, $^{20-22}$Ne and $^{36-42}$Ar in the spallation of Mg, Al, Si, Ca and Fe. Production ratios of some cosmogenic nuclides in meteorites.**
See Abstr. 061.039.

**On the nucleosynthesis of matter of the solar system in the light of new data on isotopic anomalies in meteorites.**
See Abstr. 061.080.

**Neutron–capture nucleosynthesis of neodymium isotopes and the *s*–process from $A = 130$ to 150.**
See Abstr. 061.129.

**Siderophile–rich magnetic spheroids from the Cretaceous–Tertiary boundary in Umbria, Italy.**
See Abstr. 081.004.

**The deficiency of atmospheric xenon.**
See Abstr. 082.045.

**Structures of disintegration of solid solutions in pyroxenes of the moon and meteorites as indicators of petrological processes.**
See Abstr. 094.016.

**Mineralogical comparison and cooling history of lunar and chondritic vesicular melt breccias.**
See Abstr. 094.059.

**Rotational spectral variations of asteroid (8) Flora: implications for the nature of the S–type asteroids and for the parent bodies of the ordinary chondrites.**
See Abstr. 098.025.

**Coagulated interstellar dust as the basic ingredient of comets and carbonaceous chondrites.**
See Abstr. 102.022.

**The Saskatchewan fireball of 1922 and a possible link with the Wynyard chondrite.**
See Abstr. 104.021.

**Accelerator mass spectrometry of $^{10}$Be to study the fossil record of cosmic radiation.**
See Abstr. 144.012.

# 106 Interplanetary Matter, Interplanetary Magnetic Field, Zodiacal Light

106.001 **K–coronameter observations and potential field model comparison in 1976 and 1977.**
R. Bruno, L. F. Burlaga, A. J. Hundhausen.
J. Geophys. Res., Vol. 89, No. A7, p. 5381 – 5385 (1984).
Results of determining the shape and location of the heliospheric current sheet from a potential field model and from K–coronameter observations are compared. Interplanetary magnetic field polarities as observed by IMP 8, Helios 1 and 2, Voyager 2 spacecraft were used to test the two methods over the period May 1976 to August 1977 throughout 18 Carrington rotations. The computed heliospheric current sheets from both methods had a quasi–stationary four–sector structure and very similar shapes. Agreement between interplanetary magnetic field polarity and the results from the potential field model was found on 79% of the days, while agreement between the interplanetary field polarity and the polarities derived from the K–coronameter data was found on 87% of the days.

106.002 **Suprathermal ions upstream from interplanetary shocks.**
J. T. Gosling, S. J. Bame, W. C. Feldman, G. Paschmann, N. Sckopke, C. T. Russell.
J. Geophys. Res., Vol. 89, No. A7, p. 5409 – 5418 (1984).
Suprathermal ions with energies between solar wind thermal energies and ~29 keV are occasionally observed ahead of outward propagating interplanetary shocks with the Los Alamos/Garching fast plasma experiment on ISEE 1 and 2.

106.003 **Plasma and energetic particle structure upstream of a quasi–parallel interplanetary shock.**
C. F. Kennel, F. L. Scarf, F. V. Coroniti, C. T. Russell, K.–P. Wenzel, T. R. Sanderson, P. Van Nes, W. C. Feldman, G. K. Parks, E. J. Smith, B. T. Tsurutani, F. S. Mozer, M. Temerin, R. R. Anderson, J. D. Scudder, M. Scholer.
J. Geophys. Res., Vol. 89, No. A7, p. 5419 – 5435 (1984).
This paper assembles ISEE 1, 2, and 3 observations of the interplanetary magnetic and electric fields, plasma, magnetohydrodynamic waves, electromagnetic and electrostatic waves, 1– to 6–keV protons and electrons, and >30–keV/$Q$ ions for the interplanetary shock of November 12, 1978. The authors study the period from 1800 UT on November 11 to 2400 UT on November 12, which includes the passage over the three spacecraft of a large region of upstream interplanetary disturbance, an energetic ion acceleration front, the shock, at least one neutral sheet downstream, and, finally, the shock piston. The authors collect enough information to begin to examine, for the first time, a quasi–parallel shock from both the collisionless shock structure and particle acceleration points of view. For this shock it can be seen that the two approaches cannot be separated.

106.004 **Structure of the November 12, 1978, quasi–parallel interplanetary shock.**
C. F. Kennel, J. P. Edmiston, F. L. Scarf, F. V. Coroniti, C. T. Russell, E. J. Smith, B. T. Tsurutani, J. D. Scudder, W. C. Feldman, R. R. Anderson, F. S. Mozer, M. Temerin.
J. Geophys. Res., Vol. 89, No. A7, p. 5436 – 5452 (1984).
An objective of this paper is to determine the jump in plasma parameters across the November 12, 1978, interplanetary shock.

The authors use ISEE 1 and 3 measurements of the magnetic field and electron and proton densities, temperatures, and bulk velocities, as well as ISEE 3 alpha particle measurements, and confirm the ISEE 1 electron densities using plasma wave measurements. They solve for the shock normal using four independent methods and show that the upstream and downstream flow parameters are consistent to better than 10% with $\gamma = 5/3$ Rankine–Hugoniot jump conditions. The authors study the complex train of events that followed the November 12 shock, in order to define the downstream region to which shock acceleration theory applies.

106.005 **A phenomenological model of the heliospheric current sheet.**
O. Kaburaki, T. Imai.
Sci. Rep. Tôhoku Univ., Eighth Ser., Vol. 4, No. 3, p. 240 – 254 (1983).
Structure of the heliospheric current sheet is inferred through a phenomenological model, the 'wave model', proposed by Kaburaki and Yoshii. In contrast with another such model, Schetten's 'source surface model' the information about magnetic distribution on the solar surface is propagated by an MHD spherical wave which is assumed to travel isotropically with a constant velocity of the order of the solar wind velocity.

106.006 **Strong cylindrical magnetogasdynamic shock waves in a rotating interplanetary medium.**
B. G. Verma, R. C. Srivastava, V. K. Singh.
Astrophys. Space Sci., Vol. 103, No. 1, p. 95 – 98 (1984).
Strong cylindrical magnetogasdynamic shock waves in rotating interplanetary medium have been studied and an analytic solution for their propagation has been obtained. Using characteristic method and considering the effect of Coriolis force, the authors have shown that magnetic field has significant effect on the velocity of the shock wave.

106.007 **Photographier la lumière zodiacale.**
A.-C. Levasseur-Regourd.
Astronomie, Vol. 98, p. 405 – 406 (1984).

106.008 **Interaction between neutral hydrogen and solar wind: spacecraft measurements of $H^+$ at the earth's orbit.**
A. Egidi, P. Francia, U. Villante.
Geophys. Res. Lett., Vol. 11, No. 8, p. 709 – 912 (1984).
Measurements taken by the solar wind experiment aboard the ISEE 2 satellite show evidence for $H^+$ ions, originating from neutral hydrogen moving with a speed much lower than that of the solar wind. Positive ion fluxes are measured as a function of energy and flow direction and accumulated over a time of few hours, in which the solar wind and magnetic field parameters remain approximately constant. A model given by Vasyliunas and Siscoe (1976) is used to predict the expected location of these ions in velocity space.

106.009 **Interplanetary flow systems associated with cosmic ray modulation in 1977 – 1980.**
L. F. Burlaga, F. B. McDonald, N. F. Ness, R. Schwenn, A. J. Lazarus, F. Mariani.
J. Geophys. Res., Vol. 89, No. A8, p. 6579 – 6587 (1984).
The hydromagnetic flow configurations associated with the cosmic ray modulation in 1977 – 1980 were determined using solar wind plasma and magnetic field data from Voyager 1 and 2 and Helios 1. The modulation was related to two types of large-scale systems of flows: one containing a number of transients such as shocks and postshock flows, the other consisting primarily of a series of quasi-stationary flows following interaction regions containing a stream interface and often bounded by a forward-reverse shock pair. Each of three major episodes of cosmic ray modulation was associated with the passage of a system of transient flows. Plateaus in the cosmic ray intensity-time profile were associated with the passage of systems of corotating streams.

106.010 **Heliospheric current sheet displacements during the solar cycle evolution.**
V. P. Tritakis.
J. Geophys. Res., Vol. 89, No. A8, p. 6588 – 6598 (1984).
The purpose of the present study is to confirm the view of a heliospheric current sheet moving with respect to the solar equator, in relation to the solar cycle phases. For this reason, the author has analyzed both observed and high-accuracy inferred daily interplanetary magnetic field polarities between the maxima of solar cycles 18 and 21 (1947 – 1980).

106.011 **The forward-reverse shock pair at large heliocentric distances.**
Y. C. Whang.
J. Geophys. Res., Vol. 89, No. A9, p. 7367 – 7379 (1984).
An unsteady one-dimensional MHD model is developed to study (1) the essential physical processes involved in the development of the forward-reverse shock pair at large heliocentric distances, (2) the interaction of the shock pair with the rarefaction regions of the stream structure, and (3) the merging of two forward or reverse shocks.

106.012 **$^{26}$Al and capture of interplanetary dust by the earth.**
A. V. Blinov.
Cosmic ray intensity and cosmogenic isotopes. 13th Leningrad seminar on cosmophysics, p. 185 – 194 (1983). In Russian. Abstr. in Ref. Zh., 51. Astron., 8.51.193 (1984). – See Abstr. 012.028.

106.013 **Sector structure of the interplanetary magnetic field and high-velocity solar wind streams.**
B. V. Zhiromskij.
Vologod. politekh. inst. Vologda, 13 pp. (1984). In Russian. Abstr. in Ref. Zh., 62. Issled. Kosm. Prostranstva, 8.62.497 (1984).

106.014 **Catalogue of determinations of the polarity of sectors of the interplanetary magnetic field in the 1981 – 1982 period.**
L. G. Mansurova, G. S. Mansurov, L. S. Okulova, V. O. Papitashvili, V. A. Popov.
Antarktika, Moskva, No. 23, p. 176 – 183 (1984). In Russian. Abstr. in Ref. Zh., 62. Issled. Kosm. Prostranstva, 8.62.498 (1984).

106.015 **Structure of propagating disturbances of the interplanetary plasma.**
V. I. Vlasov, V. I. Shishov, T. D. Shishova.
Geomagn. Aehron., Tom 24, No. 4, p. 541 – 545 (1984). In Russian. English translation in Geomagn. Aeron.

106.016 **Fluctuations of the interplanetary and geomagnetic fields in connection with proton bursts.**
B. D. Shel'ting, T. E. Val'chuk, Ya. I. Fel'dshtejn, V. N. Obridko.
Geomagn. Aehron., Tom 24, No. 4, p. 557 – 565 (1984). In Russian. English translation in Geomagn. Aeron.

106.017 **Variations of the sector structure of the interplanetary magnetic field.**
P. V. Sumaruk.
Soln. Dannye, Byull., 1984, No. 6, p. 74 – 80 (1984). In Russian.
Three types of dynamic processes have been detected that lead to variations in the sector structure of the interplanetary magnetic field. These are as follows: the gradual build-up of a new sector, a gradual decrease of the life-time of the sector and its complete disappearance and the sector border displacement. The above phenomena may be caused by the effects of high-velocity quasi-stationary and flare fluxes of the solar wind on the equatorial current layer of the sun, the latter being responsible for the build-up of the sector structure of the interplanetary magnetic field.

106.018 **Propagation of solar generated disturbances through the solar wind critical points: one-dimensional analysis.**
R. S. Steinolfson, M. Dryer.
Astrophys. Space Sci., Vol. 104, No. 1, p. 111 – 120 (1984).

The authors investigate the proper method for mathematically simulating the formation of an interplanetary disturbance (IPD) in the subsonic, sub-Alfvénic region near the solar surface within the constraints of one-dimensional hydrodynamic and magnetohydrodynamic analyses. They then numerically simulate the subsequent propagation of the IPD through the solar wind critical points in the equatorial plane to the outer corona. They conclude that physically realistic studies of the propagation of IPD which are assumed to originate near the solar surface must (1) initiate the IPD inside the critical points and (2) include the magnetic field. Although this conclusion is based on a one-dimensional analysis, the authors speculate that it would be equally valid in multi-dimensions.

106.019 **Shock acceleration of nuclei and electrons in the heliosphere beyond 24 AU.**
K. R. Pyle, J. A. Simpson, A. Barnes, J. D. Mihalov.
Astrophys. J., Lett. Ed., Vol. 282, No. 2, p. L107 – L111 (1984).

Radially propagating shocks of solar flare origin are analyzed using plasma and charged particle measurements on Pioneer 10 to show that protons and helium nuclei are continuously accelerated to energies of up to roughly 70 MeV per nucleon. The discovery that electrons are simultaneously accelerated to relativistic energies places new constraints on models for acceleration by these quasi-perpendicular shocks in the outer heliosphere.

106.020 **Magnetohydrodynamic modelling of interplanetary disturbances between the Sun and Earth.**
M. Dryer, S. T. Wu, G. Gislason, S. M. Han, Z. K. Smith, J. F. Wang, D. F. Smart, M. A. Shea.
Astrophys. Space Sci., Vol. 105, No. 1, p. 187 – 208 (1984).

A time-dependent, nonplanar, two-dimensional magnetohydrodynamic computer model is used to simulate a series, separately examined, of solar flare-generated shock waves and their subsequent disturbances in interplanetary space between the Sun and the Earth's magnetosphere. Particular attention is addressed to the MHD model's ability to address fundamental operational questions vis-à-vis the long-range forecasting of geomagnetic disturbances. These questions are: (1) will a disturbance (such as the present canonical series of solar flare shock waves) produce a magnetospheric and ionospheric disturbance, and, if so, (2) when will it start, (3) how severe will it be, and (4) how long will it last? The model's output is used to compute various solar wind indices of current interest as a demonstration of the model's potential for providing 'answers' to these questions.

106.021 **On the E-W asymmetry and the generation of ESP (*Energetic Storm Particle*) events.**
E. T. Sarris, G. C. Anagnostopoulos, P. C. Trochoutsos.
Sol. Phys., Vol. 93, No. 1, p. 195 – 210 (1984).

Observations of energetic-ion intensity enhancements ($E \geqslant 290$ keV) associated with solar flare generated shock waves (solar flare ESP events), obtained during nearly a decade by the APL/JHU instruments on board the Earth orbiters IMP-7 and 8, are incorporated in this work in order to examine the role of the heliolongitude depended large scale shock morphology with relation to the upstream interplanetary magnetic field in the formation of these ESP events.

106.022 **On the erosion of small meteoroids.**
I. Kapišinský.
Bull. Astron. Inst. Czech., Vol. 35, No. 5, p. 300 – 308 (1984).

One of the important dissipative processes, continuously but not randomly affecting smaller meteoroids in interplanetary space has been shown to be impact erosion mainly caused by their mutual collisions with zodiacal dust particles. The efficiency of impact erosion in dependence on heliocentric distance is discussed, taking into account the particle concentration of the eroding medium, the collision frequency proportional to the eroding particle flux and the particle energy. Numerical analysis shows that the decay of dust particles in the Solar System is determined not only by their falling into the solar corona but also by total destruction and/or total hyperbolic escape due to impact erosion, regardless of other stronger effects such as catastrophic collisions or corpuscular solar wind sputtering.

106.023 **Interplanetary field enhancements in the solar wind: statistical properties at 0.72 AU.**
C. T. Russell, M. R. Arghavani, J. G. Luhmann.
Icarus, Vol. 60, No. 2, p. 332 – 350 (1984). = Inst. Geophys. Planet. Phys. Rep. No. 2505.

A new class of disturbance in the interplanetary magnetic field has been discovered. This disturbance consists of an enhancement in the magnetic field strength lasting tens of minutes to hours. These events occur randomly with respect to the position of the spacecraft relative to Venus but not randomly with respect to Venus solar ecliptic longitude. There is a significant tendency for these events to cluster near certain ecliptic longitudes. The field distortion is often greater in the direction perpendicular to the solar wind flow rather than along it. These characteristics suggest that the source of the disturbances are weakly outgassing objects, possibly dispersed along their orbits such as in meteor streams.

106.024 **The motion of interplanetary dust particles. I. Radial velocity measurements on Fraunhofer line profiles in the zodiacal light spectrum.**
I. R. East, N. K. Reay.
Astron. Astrophys., Vol. 139, No. 2, p. 512 – 516 (1984).

Radial velocity measurements on the solar Mg I 5183.6 Å Fraunhofer absorption line in the zodiacal light spectrum have been made with a microprocessor-controlled servo-stabilised Fabry-Perot interferometer. Observations were made at 5° and 10° intervals in the ecliptic plane between morning and evening elongation of 25° from the Sun, and these new data are of greater precision and coverage than any previous observations. Obtained over a two year period, the observations show a consistent evening/morning asymmetry in the radial velocity curve with the Gegenschein receeding from the Earth at a velocity of $\sim 2.5$ km s$^{-1}$. These data do not support the hypothesis that the majority of interplanetary dust grains are in hyperbolic trajectories.

106.025 **Type III radio bursts in the interplanetary medium: the role of propagation.**
J. L. Steinberg, G. A. Dulk, S. Hoang, A. Lecacheux, M. G. Aubier.
Astron. Astrophys., Vol. 140, No. 1, p. 39 – 48 (1984).

The authors analyze new observations of interplanetary type III radio bursts in order to determine the role played by propagation effects between the true source and the observer. The observations provide strong evidence that propagation effects, group delays, ducting and/or scattering, greatly affect the observed heights, sizes and brightness temperatures of interplanetary type III bursts and also affect the arrival times of the radiation to a greater or lesser extent depending on the path from the source to the observer.

106.026 **The interplanetary scattering mean free path: collisionless wave-damping effects.**
J. M. Davila, J. S. Scott.
Astrophys. J., Vol. 285, No. 1, p. 400 – 410 (1984).

Waves with small wavelengths are responsible for the scattering of particles in interplanetary space. However, these waves are strongly damped in a thermal plasma by collisionless cyclotron damping. This damping allows a large mean free path for some particles. A specific model for the observed interplanetary magnetic turbulence spectrum is chosen which is consistent with current observational constraints and also reflects the effects of collisionless wave damping. The consequences of these assumptions for the calculation of the interplanetary mean free path are explored. Specific predictions of the theory regarding the propagation of protons are presented.

106.027 **Detection of a radio emission at 3 kHz in the outer heliosphere.**
W. S. Kurth, D. A. Gurnett, F. L. Scarf, R. L. Poynter.
Nature, Vol. 312, No. 5989, p. 27 – 31 (1984).

A radio source in the outer heliosphere has been detected by the plasma wave receivers on Voyagers 1 and 2. The radio emission is observed in the frequency range 2 – 3 kHz, and is above the local solar wind electron plasma frequency whenever supporting plasma density data are available. The maximum spectral density of the emission recorded is $\sim 10^{-14} V^2 m^{-2} Hz^{-1}$. Possible sources include continuum radiation from Jupiter's distant magnetotail and radiation at the second harmonic of the plasma frequency at the heliopause. If the latter interpretation is correct, these data represent the first remote observations of the heliopause.

106.028 **Dynamic spectra of interplanetary scintillations.**
W. A. Coles, J. P. Filice.
Nature, Vol. 312, No. 5991, p. 251 – 254 (1984).

Interplanetary scintillations (IPS) are fluctuations in the apparent intensity of radio sources caused by fluctuations in the electron density of the solar wind. The authors compare the observations with a numerical simulation of the diffraction of plane waves by a two-dimensional random phase screen. The frequency–time structure of the simulated IPS shows good qualitative agreement with the observations when the turbulence spectrum in the phase screen is a simple power law with the Kolmogorov exponent.

106.029 **Origin of the Solar System dust bands discovered by IRAS.**
S. F. Dermott, P. D. Nicholson, J. A. Burns, J. R. Houck.
Nature, Vol. 312, No. 5994, p. 505 – 509 (1984).

Measurements of longitudinal variations in the brightness and the latitude of the Solar System dust bands observed by the Infrared Astronomical Satellite determine the orbital elements of the particles involved. They may also allow discrimination between cometary and asteroidal models of the origin of these bands and even perhaps the zodiacal dust complex.

106.030 **A dynamical model for the IRAS solar system dust bands.**
P. D. Nicholson, S. F. Dermott, J. A. Burns.
Bull. Am. Astron. Soc., Vol. 16, No. 3, Part I, p. 689 – 690 (1984). Abstract. – See Abstr. 010.061.

106.031 **Formation of the zodiacal dust bands in the asteroid belt.**
M. V. Sykes, R. Greenberg, D. M. Hunten.
Bull. Am. Astron. Soc., Vol. 16, No. 3, Part I, p. 690 (1984). Abstract. – See Abstr. 010.061.

106.032 **Survival lifetimes for interplanetary dust.**
D. E. Brownlee, J. P. Bradley, G. M. Raisbeck.
Bull. Am. Astron. Soc., Vol. 16, No. 3, Part I, p. 702 (1984). Abstract. – See Abstr. 010.061.

106.033 **Are interplanetary magnetic clouds manifestations of coronal mass ejections at 1 AU?**
E. Hildner, R. M. Wilson.
Bull. Am. Astron. Soc., Vol. 16, No. 3, Part I, p. 729 (1984). Abstract. – See Abstr. 010.065.

106.034 **Observed perturbations of the velocity distribution of interstellar hydrogen atoms in the solar system with Prognoz Lyman–alpha measurements.**
R. Lallement, J. L. Bertaux, V. G. Kurt, E. N. Mironova.
Astron. Astrophys., Vol. 140, No. 2, p. 243 – 250 (1984).

Observations of the interplanetary Lyman α emission made at five different places in the solar system with high apogee satellites Prognoz 5 and Prognoz 6 are reported. This Lyα emission is the result of resonance scattering of solar photons by H atoms of the local interstellar medium (*LISM*) flowing in the solar system at the velocity – $\mathbf{V}_w$ and the spectral profile is an image of the velocity distribution of the atoms. A hydrogen absorption cell was used to analyze the spectral profile, the variation of Doppler shift with looking direction providing its spectral scanning. A first estimate of the *LISM* temperature is derived, $T \cong 8000$K, and also an approximate location of $\mathbf{V}_w$ in fair agreement with previous measurements. The observed variation of the heating effect is of 3000K, and is independent of the existence of a potential galactic background in the vicinity of the Lyα wavelength. An upper limit of this background is found to be 15 Rayleigh.

106.035 **Transmission electron microscopy of the "low–Ca" hydrated interplanetary dust particle.**
K. Tomeoka, P. R. Buseck.
Earth Planet. Sci. Lett., Vol. 69, No. 2, p. 243 – 254 (1984).

Transmission electron microscopy of a hydrated interplanetary dust particle indicates that it consists largely of a poorly crystalline phyllosilicate containing Fe, Mg and Al with an interlayer spacing of 10 to 12 Å and so is distinct from the major phyllosilicate in CI and CM carbonaceous chondrites.

106.036 **The power spectrum of interplanetary Alfvénic fluctuations: derivation of the governing equation and its solution.**
C.–y. Tu, Z.–y. Pu, F.–s. Wei.
J. Geophys. Res., Vol. 89, No. A11, p. 9695 – 9702 (1984).

A model is presented to explain the radial evolution of the power spectra of interplanetary Aflvénic fluctuations found by Bavassano et al. (1982) based on the magnetic field data of Helios 1 and 2.

106.037 **Staub variabler Porositäten (Dichten) im Sonnensystem.**
H. Fechtig, E. Grün, T. Mukai.
Kleine Teilchen und Lichtstreuung, p. 1491 – 1501 (1984). – See Abstr. 012.074.

106.038 **Computer simulations of collisionless shock waves.**
M. M. Leroy.
Adv. Space Res., Vol. 4, No. 2 – 3, p. 231 – 243 (1984). – See Abstr. 012.075.

A review of the contributions of particle computer simulations to the understanding of the physics of magnetic shock waves in collisionless plasmas is presented. The emphasis is on the relation between the computer simulation results, spacecraft observations of shocks in space, and related theories. It is shown that much has been learned from the comparison of ISEE spacecraft observations of the terrestrial bow shock and particle computer simulations concerning the quasi–perpendicular, supercritical shock.

106.039 **The physical mechanisms of subcritical collisionless shock–wave formation.**
M. M. Mellott.
Adv. Space Res., Vol. 4, No. 2 – 3, p. 245 – 253 (1984). – See Abstr. 012.075.

This paper reviews theoretical descriptions of low Mach number collisionless shock waves from an historical perspective. It discusses weak shocks where shock steepening is limited by dispersion and/or anomalous resistivity. The character of such shocks is determined mainly by the dispersive properties of the ambient plasma. The theoretical results are compared with recent data for bow shocks obtained with the ISEE dual spacecraft pair.

106.040 **The heating and acceleration of electrons by shocks.**
A. A. Galeev.
Adv. Space Res., Vol. 4, No. 2 – 3, p. 255 – 263 (1984). – See Abstr. 012.075.

Two fundamental mechanisms of electron heating by collisionless shocks are considered in detail. The first one is due to the ion–sound turbulence excited by electron current in the shock front. The second mechanism is associated with a magnetosonic turbulence excited by a beam of reflected ions ahead of the shock front. The available experimental data for the Earth bow shock concerning these two processes are compared with theoretical predictions.

106.041 **Relaxation of plasma at the shock front.**
O. Vaisberg (*O. Vajsberg*), S. Klimov, G. Zastenker, M. Nozdratchev (*M. Nozdrachev*), A. Sokolov, V. Smirnov, S. Savin, L. Avanov.
Adv. Space Res., Vol. 4, No. 2 – 3, p. 265 – 275 (1984). – See Abstr. 012.075.
A brief overview of previous studies of ion thermalization at the shock transition is given. One non–quite typical Prognoz–8 quasi–perpendicular bow shock crossing on 11 February 1981 is considered for which high–time resolution data on plasma and ELF electric field fluctuations are available.

106.042 **Multiple spacecraft observations of interplanetary shocks: shock–normal oscillations and their effects.**
C. T. Russell, C. J. Alexander.
Adv. Space Res., Vol. 4, No. 2 – 3, p. 277 – 282 (1984). – See Abstr. 012.075.
Observations of interplanetary shocks with multiple spacecraft and multiple instruments has permitted the determination of their average shock normals with unprecedented accuracy. Nevertheless there are still local deviations from the best–fit normal. These deviations in general appear to be less than 5° but on occasion can be of the order of 20° or more. These fluctuating normals and the fluctuating upstream field can cause variations in the downstream field strength even when the upstream field strength is constant.

106.043 **Particle acceleration and MHD wave excitation upstream of interplanetary shocks.**
M. A. Lee.
Adv. Space Res., Vol. 4, No. 2 – 3, p. 295 – 304 (1984). – See Abstr. 012.075.
The theory of diffusive shock acceleration and its application in interplanetary space are reviewed. Special emphasis is placed on the distinction between diffusive and shock–drift shock acceleration, on ion–excitation of MHD waves upstream of the shock and the resulting self–consistent configuration of waves and accelerated ions. A comparison of predicted and observed proton anisotropies for the 12 November 1978 event is presented.

106.044 **ISEE–3 observations of energetic protons associated with interplanetary shocks.**
T. R. Sanderson.
Adv. Space Res., Vol. 4, No. 2 – 3, p. 305 – 313 (1984). – See Abstr. 012.075.
A considerable number of travelling interplanetary shocks has been observed at ISEE–3 since launch in 1978. Proton events associated with these shocks range from large ESP events with intensities enhanced for several hours to small shock spikes lasting only a few minutes, as well as events where no intensity increases are observable. In this paper, a summary of results of statistical surveys of proton intensity profiles, spectra and 3–dimensional anisotropies in the energy range 35 to 1600 keV is presented.

106.045 **Diffusion coefficients of low–energy protons upstream of quasi–parallel interplanetary shocks.**
P. van Nes, E. C. Roelof, R. Reinhard, T. R. Sanderson, K.–P. Wenzel.
Adv. Space Res., Vol. 4, No. 2 – 3, p. 315 – 318 (1984). – See Abstr. 012.075.
The authors present results of a study of the diffusion coefficient for low–energy protons upstream of three interplanetary shocks observed on ISEE–3. The intensity increase as the shock approaches can be fitted by a running exponential, the e–folding time $t^*$ of which is assumed to be a direct measure of the diffusion coefficient. The energy dependence of $t^*$ close to the shock is derived from the four energy channels covering the range 35 – 238 keV.

106.046 **Low–energy protons associated with interplanetary shocks as an independent population in the solar wind.**
B. Sanahuja, V. Domingo.
Adv. Space Res., Vol. 4, No. 2 – 3, p. 319 – 322 (1984). – See Abstr. 012.075.
The authors investigate the large scale flow pattern of 36 – 1600 keV protons observed in association with several large interplanetary shocks on ISEE–3 between August 1978 and April 1980. The distribution function in the solar wind frame and its temporal evolution indicates that these particles can behave as a coherent population in the solar wind.

106.047 **The time scales of the scattering of energetic protons in interplanetary space.**
P. Hick, G. Stevens.
Adv. Space Res., Vol. 4, No. 2 – 3, p. 323 – 326 (1984). – See Abstr. 012.075.
Observations with the directional spectrometer DFH aboard ISEE 3 have been used to obtain results on the scattering time scales of energetic protons. Depending on the duration of the scattering process the particle distribution will be subjected to either phase scattering or full scattering. Analysis of some representative events shows that full scattering is applicable to shock–associated events.

106.048 **Turbulence associated with interplanetary shock waves within 1 AU.**
A. Luttrell, A. K. Richter, H. Rosenbauer, F. M. Neubauer.
Adv. Space Res., Vol. 4, No. 2 – 3, p. 327 – 330 (1984). – See Abstr. 012.075.
The nature and strength of low–frequency MHD turbulence in association with fast–forward, oblique shock waves is investigated using Helios 1 magnetic field and plasma data. Power and coherence spectra were computed and compared for the pre– and post–shock regions for the magnetic field, velocity and density fluctuations in the frequency range $3.9 \times 10^{-4}$Hz to $1.2 \times 10^{-2}$Hz.

106.049 **The magnetic field turbulence near interplanetary shock fronts.**
E. I. Morozova, N. F. Pissarenko (*N. F. Pisarenko*), W. Riedler, K. Schwingenschuh, G. Schelch, I. P. Bezrodnykh, Ye. G. Yeroshenko (*E. G. Eroshenko*).
Adv. Space Res., Vol. 4, No. 2 – 3, p. 331 – 334 (1984). – See Abstr. 012.075.
The power spectra (frequency range $10^{-3}$ to $10^{-2}$Hz) of the magnetic field fluctuations near interplanetary shock fronts (IPS) are analysed using Venera 13 and Venera 14 data. Ten events in the period from November 1981 to February 1982 are evaluated. The event on February 2, 1982 is considered in detail. A sharp increase of the magnetic turbulence downstream the shock front is common to all detected IPS.

106.050 **Numerical solution of the interplanetary shock waves.**
D. Odstrčil.
Bull. Astron. Inst. Czech., Vol. 35, No. 6, p. 326 – 335 (1984).
This paper gives a detailed analysis of the formation and propagation of the shock wave together with the results obtained under different time durations of the pulse at the inner boundary.

106.051 **Doppler shifts measurements of the zodiacal light at the Pic du Midi Observatory.**
R. Robley, A. Bücher, S. Koutchmy, P. Lamy.
Inst. Astrophys. Paris, Pré–Publ., No. 64, 9 pp. (1984). To be published in the Proceedings of the IAU Colloq. No. 85, Properties and interaction of interplanetary dust.

106.052 **Photographic colour photometry of the inner zodiacal light.**
G. M. Nikol'skij, S. Koutchmy, I. A. Nesmyanovich.
Astron. Tsirk., No. 1310, p. 3 – 4 (1984). In Russian.

106.053 **New specifications of a phenomenological direct data based model of a flare–generated interplanetary stream: slow wave and boundary layer.**
K. G. Ivanov.
Inst. zemn. magn., ionos., rasprostr. radiovoln Akad. Nauk SSSR. Prepr., No. 61a/472, 15 pp. (1983). Abstr. in Ref. Zh., 51. Astron., 12.51.669 (1984).

106.054 **Some results of investigations made in the interplanetary plasma with wide–angle plasma detectors aboard Prognoz 6.**
M. I. Verigin.
Solar activity investigations and the Prognoz space system, p. 81 – 90 (1984). In Russian. Abstr. in Ref. Zh., 62. Issled. Kosm. Prostranstva, 12.62.369 (1984). – See Abstr. 003.038.

106.055 **Investigation of the isotopic and charge composition of energetic particles aboard the Prognoz 5 and Prognoz 6 stations.**
V. N. Lutsenko, N. S. Nikolaeva, S. Fisher.
Solar activity investigations and the Prognoz space system, p. 147 – 155 (1984). In Russian. Abstr. in Ref. Zh., 62. Issled. Kosm. Prostranstva, 12.62.375 (1984). – See Abstr. 003.038.

106.056 **Characteristics and propagation of interplanetary shock waves.**
G. N. Zastenker, O. L. Vajsberg.
Solar activity investigations and the Prognoz space system, p. 55 – 67 (1984). In Russian. Abstr. in Ref. Zh., 62. Issled. Kosm. Prostranstva, 12.62.379 (1984). – See Abstr. 003.038.

106.057 **Flare–generated heliospheric and geomagnetic disturbances.**
G. Zhang, S. Xue, C. Lu.
Sci. Sin., Ser. A, Vol. 27, No. 11, p. 1172 – 1182 (1984).
The authors' purpose is to study the solar wind and IMF disturbance caused by solar flares. They make a statistical study on the variations of various interplanetary medium parameters in the interval from two days before through six days after the occurrence of flares by means of the superposed–epoch method.

106.058 **Configuration of heliospheric current sheets and its cause.**
X. Zhao.
Acta Astrophys. Sin., Vol. 4, No. 2, p. 166 – 169 (1984). In Chinese.
The average configuration of heliospheric current sheets in the years 1974 and 1976 has been figured out. The morphology of the heliospheric current sheets for the individual rotations in January – April 1976 has also been analyzed.

106.059 **Large–scale interplanetary magnetic fields: Voyager 1 and 2 observations between 1 AU and 9.5 AU.**
L. F. Burlaga, L. W. Klein, R. P. Lepping, K. W. Behannon.
J. Geophys. Res., Vol. 89, No. A12, p. 10659 – 10668 (1984).
The purpose of the paper is to describe the large–scale radial and temporal variations of the interplanetary magnetic field observed by Voyagers 1 and 2 between 1 AU and 9.5 AU.

106.060 **Broadening of the interplanetary helium cone structure due to elastic collisions of LISM helium atoms with solar wind ions.**
H. J. Fahr, H. U. Nass, D. Rucinski.
NASA Conf. Publ., NASA CP–2345, p. 32 – 36 (1984). – See Abstr. 012.089 (IAU Colloq. No. 81).
Neutral interstellar particles penetrating into the heliosphere suffer elastic collisions with keV–solar wind ions. The momentum transfer to the neutrals connected with these collisions leads to a loss of angular momentum with respect to the sun and to a fractional compensation of the effective solar gravity. The dynamical particle trajectories hence are changed into non–Keplerians leading to density and temperature distributions differing from those calculated in the past. This is found from a solution of the Boltzmann equation that linearizes the effect of this additional force. It is shown here that the He I–584 Å resonance glow of the heliospheric helium cone lead to substantially lower interstellar helium temperatures if re–interpreted on the basis of this revised theory. These temperatures now seem to be in accordance with the derived temperatures for interstellar hydrogen.

106.061 **The properties of large particles in the zodiacal cloud and in the interstellar medium, and their relation to recent IRAS observations.**
F. Hoyle, N. C. Wickramasinghe.
Astrophys. Space Sci., Vol. 107, No. 2, p. 223 – 232 (1984).
To explain the scattering of sunlight observed from the *F*–corona and from the zodiac, the scattering particles must have radii of order 15 μm, and must have an imaginary component of the refractive index that requires the presence (from 5 to 10%) of free carbon. Such particles absorb sunlight, the absorbed solar energy being reradiated in the infrared with a close approximation to black–body emission, even as far into the infrared as 100 μm, a deduction in good agreement with recently published observations from the IRAS satellite. The IRAS observations at high ecliptic latitudes require similar particles to be present in large quantity in the interstellar medium, $\sim 10^6$ solar masses or more of them.

106.062 **Energy spectra of protons with E = 23 – 480 keV in the interplanetary medium, transition region and the outer magnetosphere.**
R. N. Basilova, I. A. Kurillo, S. P. Ryumin, K. N. Sharvina.
Geomagn. Aehron., Tom 24, No. 6, p. 1000 – 1002 (1984). In Russian. English translation in Geomagn. Aeron.

**Solar terrestrial physics: present and future. A report based on the Solar Terrestrial Physics Workshop, December 1982 to November 1983.**
See Abstr. 012.038.

**An object within a particle of extraterrestrial origin compared with an object of presumed terrestrial origin.**
See Abstr. 015.034.

**Laboratory studies of interplanetary dust: relationships to comets and interstellar clouds.**
See Abstr. 022.118.

**The application of collisionless plasma expansion processes to space plasmas.**
See Abstr. 062.036.

**Particle acceleration.**
See Abstr. 062.087.

**Collisionless shocks.**
See Abstr. 062.172.

**Low energy ions in corotating interaction regions at 1 AU: observations.**
See Abstr. 074.008.

**Recurrent high–velocity streams of solar wind plasma in the circumterrestrial space.**
See Abstr. 074.018.

**Coronal transients and their interplanetary effects.**
See Abstr. 074.032.

**Coronal mass ejections.**
See Abstr. 074.034.

**The relationship between coronal transients, Type II bursts and interplanetary shocks.**
See Abstr. 074.039.

**The far reaches of the solar wind: Pioneer 10 and Pioneer 11 plasma results.**
See Abstr. 074.040.

**Interplanetary radio storms. II. Emission levels and solar wind speed in the range 0.05 – 0.8 AU.**
See Abstr. 074.049.

**Comment on "A comparison of coronal and interplanetary current sheet inclinations" by K. W. Behannon, L. F. Burlaga, and A. J. Hundhausen.**
See Abstr. 074.067.

**Interplanetary radio storms. I. Extension of solar active regions through the interplanetary medium.**
See Abstr. 077.003.

**Type II solar radio bursts, interplanetary shocks, and energetic particle events.**
See Abstr. 077.006.

**Partial reconstruction of the initial conditions for streams of energetic electrons associated with a solar type III burst.**
See Abstr. 077.039.

**Survey of $He^{+}/He^{2+}$ abundance ratios in energetic particle events.**
See Abstr. 078.001.

**Characteristics of solar and heliospheric ion populations observed near Earth.**
See Abstr. 078.009.

**Injection of particles into the diffusive shock acceleration process: Monte Carlo simulation of heliospheric shocks.**
See Abstr. 078.010.

**Acceleration phenomena in the solar system.**
See Abstr. 078.015.

**MHD waves and turbulence in the sun and interplanetary medium.**
See Abstr. 080.074.

**Electric fields and potential patterns in the high–latitude ionosphere for different situations in interplanetary space.**
See Abstr. 083.003.

**Observation of interplanetary magnetic field and of ionospheric plasma convection in the vicinity of the dayside polar cleft.**
See Abstr. 083.026.

**Excitation of electric fields by longitudinal currents conditioned by the northern component of the interplanetary magnetic field.**
See Abstr. 083.040.

**Dawn–dusk (*y*) component of the interplanetary magnetic field and the local time of the Harang discontinuity.**
See Abstr. 084.006.

**Interaction between an interplanetary shock and the earth's magnetosphere on August 27, 1978: ISEE 1 electric field and ISEE 2 plasma observations.**
See Abstr. 084.033.

**The role of the interplanetary magnetic field in generation of electric fields in the magnetospheric tail.**
See Abstr. 084.047.

**Unusual auroral features observed on January 10 – 11, 1983 and their possible relationships to the interplanetary magnetic field.**
See Abstr. 084.054.

**The interplanetary electric field, cleft currents and plasma convection in the polar caps.**
See Abstr. 084.068.

**The MHD Rankine–Hugoniot jump conditions and the terrestrial bow shock: a statistical comparison.**
See Abstr. 084.071.

**Dynamics of neutral small particles.**
See Abstr. 091.076.

**Magnetic field overshoots in the Venus bow shock.**
See Abstr. 093.064.

**Cosmic dust and the comet connection.**
See Abstr. 102.006.

**Comet dust – applications of Mie scattering.**
See Abstr. 102.024.

**Bismuth in interplanetary dust.**
See Abstr. 105.036.

**IUE high resolution spectrophotometry of H Ly α emission from the local interstellar medium.**
See Abstr. 131.155.

**The physics of the heliospheric interface and its implications for LISM (*local interstellar matter*) diagnostics.**
See Abstr. 131.156.

**Analysis of interplanetary Lyman–alpha line profile with a hydrogen absorption cell: theory of the Doppler angular spectral scanning method.**
See Abstr. 131.173.

**Helium and hydrogen of the local interstellar medium observed in the vicinity of the sun.**
See Abstr. 131.240.

**Pioneer 10 and Voyager observations of the interstellar medium in scattered emission of the He 584 Å and H Lyα 1216 Å lines.**
See Abstr. 131.241.

**Changes of LISM characteristics in the heliospheric interface.**
See Abstr. 131.242.

# 107 Cosmogony

107.001 **From icy planetesimals to outer planets and comets.**
R. Greenberg, S. J. Weidenschilling, C. R. Chapman, D. R. Davis.
Icarus, Vol. 59, No. 1, p. 87 – 113 (1984).
Numerical simulations of planet growth in the outer solar system show that growth of Uranus and Neptune occurs in reasonably short time, well below the actual age of the system, without the need for ad hoc assumptions about excess mass or artificially low relative velocities among the icy planetesimals. The authors' model shows that gravitational clumping occurred in successive portions of the material that reached the midplane, producing "initial" planetesimals probably of comet-like sizes. Models of subsequent collisional evolution show that such an initial size distribution, similar to known comets, would have been required in order to have an adequate comet-like size distribution available to feed the Oort cloud as the other planets reach full size. Comets are probably unaltered remnants of the initial population of planetesimals in the outer solar system, not fragments of larger bodies.

107.002 **Possible conditions for formation of satellite from protoearth by ejection of material.**
H. F. Petersons.
Earth, Moon, Planets, Vol. 31, No. 1, p. 15 – 24 (1984).
Conditions for the formation of a large gaseous protoearth and the formation of a satellite are considered. The model is compatible with geochemical, tidal and angular momentum data.

107.003 **Contraction of the solar nebula.**
J. J. Rawal.
Earth, Moon, Planets, Vol. 31, No. 2, p. 175 – 182 (1984).
The concept of Roche limit is applied to the Laplacian theory of the origin of the solar system to study the contraction of a spherical gas cloud (solar nebula). In the process of contraction of the solar nebula, it is assumed that the phenomenon of supersonic turbulent convection described by Prentice (1978) is operative. It is found that the radius of the contracting solar nebula follows Titius–Bode law $R_p = R_\odot a^p$, where $R_\odot$ is the radius of the present Sun and $a = 1.442$. The consequences of the relation are also discussed. The aim, here, is an attempt to explain, on the basis of the concept of Roche limit, the distribution of planets in the solar system and try to understand the physics underlying it.

107.004 **Comments on the physics of planetary equilibrium.**
G. H. A. Cole.
Q. J. R. Astron. Soc., Vol. 25, No. 3, p. 248 – 258 (1984).
It has been deduced from a wide range of evidence that the Solar System formed as a unit some $4.6 \times 10^9$yr ago by an accretion process (probably from the cold) involving material having the cosmic abundance or something close to it. The general properties to be expected of the System can be deduced from various physical arguments and a comparison is then possible with the objects actually observed.

107.005 **Development of the Soviet planetary cosmogony. Review.**
V. S. Safronov.
O. Yu. Shmidt i sov. geofiz. 80-kh godov. Moskva, p. 41 – 57 (1983). In Russian. From Ref. Zh., 51. Astron., 8.51.134 (1984).

107.006 **On regularities of formation of planetary systems.**
B. V. Fedotov, V. M. Zelenev.
Voronezh. gos. ped. inst. Voronezh, 28 pp. (1984). In Russian. Abstr. in Ref. Zh., 51. Astron., 8.51.136 (1984).

107.007 **Hypothesis of the accretion of stratified planets of various composition in the light of the discoveries of Venera 13 and 14 Voyager 1 and 2.**
G. G. Kochemasov.
The 27th International Geological Congress, Tom 5, Sect. 10 – 11, p. 315 – 316 (1984). In Russian. Abstr. in Ref. Zh., 62. Issled. Kosm. Prostranstva, 8.62.404 (1984). – See Abstr. 012.031.

107.008 **On the dynamics of circumplanetary swarms.**
G. V. Pechernikova, S. V. Maeva, A. V. Vityazev.
Pis'ma Astron. Zh., Tom 10, No. 9, p. 702 – 709 (1984). In Russian. English translation in Sov. Astron. Lett., Vol. 10.
The evolution model of a presatellite circumplanetary swarm is investigated. The equations describing the density redistribution in a presatellite swarm due to accretion of matter from the feeding zone of the planet are obtained. Calculations showed the model to be able to explain the origin of systems of satellites with total mass $10^{-4} - 10^{-5}$ of the planet's mass.

107.009 **La formation du système solaire.**
A. Brahic.
Astronomie, Vol. 98, p. 479 – 492 (1984).

107.010 **Physico-chemical model of the evolution of the protoplanetary cloud.**
V. A. Dorofeeva.
The 27th International Geological Congress, Tom 5, Sect. 10 – 11, p. 251 – 253 (1984). In English and Russian. Abstr. in Ref. Zh., 51. Astron., 9.51.178 (1984). – See Abstr. 012.031.

107.011 **Physics and chemistry of the primitive solar nebula.**
G. E. Morfill.
The 27th International Geological Congress, Tom 5, Sect. 10 – 11, p. 352 (1984). Abstr. in Ref. Zh., 51. Astron., 9.51.179 (1984). – See Abstr. 012.031.

107.012 **Isotopes of rare gases and the origin of the earth and other planets.**
L. K. Levskij.
The 27th International Geological Congress, Tom 5, Sect. 10 – 11, p. 336 – 337 (1984). In English and Russian. Abstr. in Ref. Zh., 51. Astron., 9.51.182 (1984). – See Abstr. 012.031.

107.013 **Cosmic model of the origin of the earth.**
Eh. V. Sobotovich.
Geol. zh., Tom 44, No. 2, p. 112 – 123 (1984). In Russian. Abstr. in Ref. Zh., 51. Astron., 9.51.183 (1984).

107.014 **Xenon isotopes and the origin of volatiles of the earth.**
Yu. A. Shukolyukov.
The 27th International Geological Congress, Tom 11, Sect. S 11, p. 94 – 103 (1984). In Russian. Abstr. in Ref. Zh., 51. Astron., 9.51.185 (1984). – See Abstr. 012.031.

107.015 **Numerical investigations of the evolution of spatial rings of gravitational bodies corresponding to the feeding zones of giant planets.**
S. I. Ipatov.
Inst. prikl. mat. Akad. Nauk SSSR, Prepr., No. 1, 28 pp. (1984). In Russian. Abstr. in Ref. Zh., 51. Astron., 9.51.187 (1984).

107.016 **The Ehneev–Kozlov cosmogonic theory and evolution of rotations and accumulation of celestial bodies.**
V. V. Beletskij.
O. Yu. Shmidt i sov. geofiz. 80-kh godov, Moskva, p. 234 – 238 (1983). In Russian. Abstr. in Ref. Zh., 51. Astron., 9.51.194 (1984).

107.017 **On the dynamics of a protosatellite swarm.**
A. V. Vityazev, G. V. Pechernikova.
O. Yu. Shmidt i sov. geofiz. 80-kh godov, Moskva, p. 250 – 257 (1983). In Russian. Abstr. in Ref. Zh., 51. Astron., 9.51.197 (1984).

107.018 **Radial migration of preplanetary material: implications for the accretion time scale problem.**
K. Hourigan, W. R. Ward.
Icarus, Vol. 60, No. 1, p. 29 – 39 (1984).
Radial drift of planetesimals due to density wave interaction with the solar nebula is considered. The mechanism is most effective for large masses and provides mobility over a size range where aerodynamic drag is unimportant. The process could shorten accretion time scales to $\sim O(10^5 - 10^6$ years) throughout the solar system. Accumulation stalls down when growing objects are massive enough to open gaps in the gas disk. Implications of this process for current cosmogonic models are discussed.

107.019 **The range of validity of the two–body approximation in models of terrestrial planet accumulation. I. Gravitational perturbations.**
G. W. Wetherill, L. P. Cox.
Icarus, Vol. 60, No. 1, p. 40 – 55 (1984).
Gravitational perturbations in semimajor axis, eccentricity, and inclination resulting from close planetesimal encounters (near 1 AU) out to 10 Tisserand sphere of influence radii were calculated by two– and three–dimensional numerical integration. These are compared with the results of treating the encounter as a two–body problem, as is customary in Monte Carlo calculations of orbital evolution and in numerical and analytical studies of planetary accumulation. It is found that for values of $(V/V_e) \lesssim 0.35$ ($V$ = relative velocity, $V_e$ = escape velocity of largest body), the two–body approximation fails to describe the outcome of individual encounters. In spite of the differences, average perturbations given by the two–body approximation are valid within a factor of 2 when $V/V_e > 0.07$.

107.020 **Survival of the Earth and the future evolution of the Sun.**
S. C. Vila.
Earth, Moon, Planets, Vol. 31, No. 3, p. 313 – 315 (1984).
The question of the survival of the planet Earth as the Sun becomes a red giant and possibly a planetary nebula depends on the maximum extent of the Sun's envelope. This study shows that if the Earth is reached by this envelope, it will be destroyed; but if it is not, it will survive. Present stellar evolution calculations imply that the Earth will be destroyed.

107.021 **Stochastic model of the tidal evolution of the earth–moon system.**
B. A. Kagan, N. B. Maslova.
Dokl. Akad. Nauk SSSR. Ser. Mat. Fiz., Tom 276, No. 1, p. 88 – 91 (1984). In Russian. Abstr. in Ref. Zh., 51. Astron., 10.51.104 (1984).

107.022 **On a modified vortical model of the origin of the solar system.**
G. I. Kuz'menko.
Tikhookean. geol., No. 2, p. 110 – 116 (1984). In Russian. Abstr. in Ref. Zh., 51. Astron., 10.51.201 (1984).

107.023 **Origin and history of the outer planets: theoretical models and observational constraints.**
J. B. Pollack.
Annu. Rev. Astron. Astrophys., Vol. 22, p. 389 – 424 (1984). – See Abstr. 003.024.
This paper reviews our current understanding of the origin and subsequent development of the Jovian planets and their associated satellite and ring systems within the context of plausible models of the early solar system and modern astrophysical concepts. Stress is given not only to delineating alternative hypotheses but also to comparing their predictions against relevant observations.

107.024 **Nemesis for Nemesis?**
M. E. Bailey.
Nature, Vol. 311, No. 5987, p. 602 – 603 (1984).
This note comments on the recent discussion of Nemesis, a hypothetical companion star of the Sun, which could have triggered the periodic sequence of extinction events found in fossil records.

107.025 **Terrestrial catastrophism – Nemesis or Galaxy?**
S. V. M. Clube, W. M. Napier.
Nature, Vol. 311, No. 5987, p. 635 – 636 (1984).
The ~30 Myr periodicity associated with the Sun's motion through the central plane has been linked to geomagnetic reversals, biological extinctions and crater ages. This periodicity is consistent with galactic theories of terrestrial catastrophism. It has been suggested, however, that the periodicity is controlled by a hypothetical stellar companion of the Sun ("Nemesis") in a highly eccentric orbit of arbitrarily chosen period which periodically upsets the Oort cloud. This letter points out some of the difficulties encountered by this latter hypothesis.

107.026 **Dynamical constraints on the mass and perihelion distance of Nemesis and the stability of its orbit.**
J. G. Hills.
Nature, Vol. 311, No. 5987, p. 636 – 638 (1984).
It has been suggested that the observed periodic extinction of species at intervals of 26 Myr may be catalysed by a hypothetical stellar companion of the Sun, Nemesis, with an orbital period of 26 Myr. This letter investigates the effect of Nemesis on the comets in the inner comet cloud. This is used to determine the minimum required mass of Nemesis and its minimum required perihelion distance. The effect of passing stars on the stability of the orbit of Nemesis is investigated, and the probability of its having passed within the planetary system estimated.

107.027 **How stable is an astronomical clock that can trigger mass extinctions on Earth?**
P. Hut.
Nature, Vol. 311, No. 5987, p. 638 – 641 (1984).
The periodicity in mass extinctions observed in the fossil record may be driven by an astronomical clock consisting of a companion star to the Sun. This letter reports orbital calculations for the hypothetical companion including the influence of the galactic tidal field and the perturbations of passing stars. From these calculations the irregularity in the period of revolution over the past 250 Myr is expected to be at least 10%, and more likely around 20%.

107.028 **Orbital stability of the unseen solar companion linked to periodic extinction events.**
M. V. Torbett, R. Smoluchowski.
Nature, Vol. 311, No. 5987, p. 641 – 642 (1984).
Recent suggestions of astronomical causes for the closely periodic extinction events in the geological record and the possible periodicities in the cratering record have assumed that a low–mass companion to the Sun (Nemesis) in an eccentric orbit ($e \sim 0.7$) may have triggered these events. The authors present here evidence from three–dimensional numerical modelling that only orbits with a limited range in inclination with respect to the galactic plane are formally stable for the required length of time. Perturbations bypassing stars and molecular clouds may make even these orbits unstable.

107.029 **Magnetic field amplification in the solar nebula through interaction with the T Tauri wind.**
W.–H. Ip.
Nature, Vol. 312, No. 5995, p. 625 – 626 (1984).
The remnant magnetization of carbonaceous chondrites can be used to estimate the intensity of the magnetic field in the primordial solar nebula. Estimates vary between 0.2 and 1 G, with the possibility that the field may have been as high as 2 – 3 G. This letter presents a possibility for the origin of a strong magnetic field in the early solar system: magnetic field enhancement at the point of stagnation between the solar nebula and the intense solar outflows. This mechanism, which involves the interaction of the T Tauri wind with the solar nebula, is straightforward and is supported by results from recent space research.

107.030 **A new hypothesis of the origin of the Earth and the Earth–like planets.**
Z. Liu.
J. Zhejiang Univ., Vol. 18, No. 2, p. 35 – 40 (1984). In Chinese. Abstr. in Phys. Abstr., Vol. 88, No. 1247, Entry 4630 (1985).

107.031 **Do comet showers kill the flowers?**
P. R. Weissman.
Bull. Am. Astron. Soc., Vol. 16, No. 3, Part I, p. 644 (1984). Abstract. – See Abstr. 010.061.

107.032 **Orbital stability at the limits of the solar system: comets and solar companions.**
M. V. Torbett, R. Smoluchowski.
Bull. Am. Astron. Soc., Vol. 16, No. 3, Part I, p. 646 (1984). Abstract. – See Abstr. 010.061.

107.033 **3D models of the formation of the presolar nebula.**
A. P. Boss.
Bull. Am. Astron. Soc., Vol. 16, No. 3, Part I, p. 701 (1984). Abstract. – See Abstr. 010.061.

107.034 **Deuterium fractionation in the pre–solar nebula: the role of surface catalysis.**
D. H. Grinspoon, J. S. Lewis.
Bull. Am. Astron. Soc., Vol. 16, No. 3, Part I, p. 701 (1984). Abstract. – See Abstr. 010.061.

107.035 **Turbulent accretion disks in the solar system.**
A. Coradini, G. Magni.
Bull. Am. Astron. Soc., Vol. 16, No. 3, Part I, p. 701 (1984). Abstract. – See Abstr. 010.061.

107.036 **Accretion of planets in the solar nebula: effects of drag and resonances.**
S. J. Weidenschilling, D. R. Davis.
Bull. Am. Astron. Soc., Vol. 16, No. 3, Part I, p. 702 (1984). Abstract. – See Abstr. 010.061.

107.037 **Calculations of the origin of the giant planets.**
P. Bodenheimer, J. B. Pollack.
Bull. Am. Astron. Soc., Vol. 16, No. 3, Part I, p. 702 (1984). Abstract. – See Abstr. 010.061.

107.038 **Pre–accretional thermal processing of planetary materials.**
J. A. Wood.
Bull. Am. Astron. Soc., Vol. 16, No. 3, Part I, p. 702 (1984). Abstract. – See Abstr. 010.061.

107.039 **Cosmogony as an extrapolation of magnetospheric research.**
H. Alfvén.
Space Sci. Rev., Vol. 39, Nos. 1/2, p. 65 – 90 (1984).

A theory of the origin and evolution of the solar system (Alfvén and Arrhenius, 1975, 1976) which considered electromagnetic forces and plasma effects is revised in the light of new information supplied by space research. The results can be extrapolated both outwards in space, to interstellar clouds, and backwards in time, to the formation of the solar system. The first extrapolation leads to a revision of some cloud properties which are essential for the early phases in the formation of stars and solar nebulae. The latter extrapolation makes possible to approach the cosmogonic processes by extrapolation of (rather) well–known magnetospheric phenomena. Pioneer–Voyager observations of the Saturnian rings indicate that essential parts of their structure are "fossils" from cosmogonic times. By using detailed information from these space missions, it seems possible to reconstruct certain events 4 – 5 billion years ago with an accuracy of a few percent. This will cause a change in our views of the evolution of the solar system.

107.040 **Evolution of grains in a turbulent solar nebula.**
S. J. Weidenschilling.
Icarus, Vol. 60, No. 3, p. 553 – 567 (1984).

In accretion disk models of the solar nebula, turbulence is driven by convective instability. This mechanism requires high opacity, which must be provided by solid grains. Evolution of the grain size distribution in a turbulent disk is computed numerically, using realistic collisional outcomes and strengths of grain aggregates, rather than arbitrary "sticking efficiency". The presence of turbulence greatly increases the rate of grain collisions. Aggregates quickly reach sizes $\sim 0.1 - 1$ cm, but erosion and breakup in collisions prevent growth of larger bodies for plausible aggregate impact strengths. The opacity of the disk is reduced by aggregate formation; some combinations of opacity law and surface density produce an optically thin disk, cutting off turbulent convection. The disk may experience alternating periods of turbulence and quiescence. Planetesimals can form only during the quiescent intervals; it is argued that such episodes were rare during the lifetime of the accretion disk.

107.041 **The evolution of rotation in the early history of the solar system.**
M. M. Woolfson.
Philos. Trans. R. Soc. London, Ser. A, Vol. 313, No. 1524, p. 5 – 18 (1984). – See Abstr. 012.068.

Theories that require the co–genetic formation of the Sun and planets have difficulty in explaining the slow rotation of the Sun. An analysis is made of various mechanisms for slowing down the core of an evolving nebula. Two of these involve a high magnetic dipole moment for the early Sun. A mechanical process for transferring angular momentum outward involving dissipation in a solar–nebula disc is incapable of giving what is required. Two processes of star formation in a turbulent cloud are discussed. Various models for producing planets are examined in relation to the spin they would produce. The various tilts of their spin axes could be explained by interactions between protoplanets in the early solar system.

107.042 **The interaction of rotation and magnetic field in the solar system.**
L. Mestel.
Philos. Trans. R. Soc. London, Ser. A, Vol. 313, No. 1524, p. 19 – 25 (1984). – See Abstr. 012.068.

During the early phases of star formation, torsional Alfvén waves propagating along the locally distorted galactic magnetic field can transport away the bulk of the initial angular momentum of a condensation, so enabling it to contract to solar nebula dimensions. Enough primeval magnetic flux may be retained for magnetic redistribution of angular momentum to continue until the proto–sun reaches the pre–main sequence Hayashi phase. A rotating star with a convective envelope has a dynamo–maintained field which brakes the star through coupling to the stellar wind. Observational evidence from young star clusters and from T Tauri stars suggests that the zero–age main sequence sun had about ten times its present angular momentum. The same braking process, scaled up because of the much more powerful T Tauri winds, can explain why the zero–age sun had lost at least nine–tenths of the centrifugal upper limit, and is more acceptable than the suggestion that the 'missing' solar angular momentum has been magnetically fed into the planetary system.

107.043 **The early solar system and the rotation of the Sun.**
T. Gold.
Philos. Trans. R. Soc. London, Ser. A, Vol. 313, No. 1524, p. 39 – 45 (1984). – See Abstr. 012.068.

107.044 **Some short–lived radioactive nuclei at early stages of the Solar System.**
G. J. Wasserburg.
The 27th International Geological Congress, Tom 11, Sect. S 11, p. 26 – 31 (1984). In Russian. Abstr. in Ref. Zh., 51. Astron., 11.51.168 (1984). – See Abstr. 012.031.

107.045 **Accretion of the earth and inner planets.**
G. Dreibus, H. Wänke.
The 27th International Geological Congress, Tom 11, Sect. S 11, p. 3 – 11 (1984). In Russian. Abstr. in Ref. Zh., 51. Astron., 11.51.169 (1984). – See Abstr. 012.031.

107.046 **The $^3He/^4He$ primary ratio in terrestrial diamonds and the problem of origin of rare gases in terrestrial planets.**
M. Ozima, Sh. Zashu.
The 27th International Geological Congress, Tom 11, Sect. S 11, p. 87 – 94 (1984). In Russian. Abstr. in Ref. Zh., 51. Astron., 11.51.170 (1984). – See Abstr. 012.031.

107.047 **The microstructure of terrestrial catastrophism.**
S. V. M. Clube, W. M. Napier.
Mon. Not. R. Astron. Soc., Vol. 211, No. 4, p. 953 – 968 (1984).
The theory of evolution involving episodic terrestrial catastrophism predicts that the Oort cloud is disturbed by close encounters with massive nebulae. Each disturbance generates bombardment pulses of a few million years duration, the pulse frequencies being determined by the Sun's passage through the spiral arms and central plane of the Galaxy where nebulae concentrate. The structure within a pulse is shown here to be dominated by a series of "spikes" of $\sim$0.01 – 0.1 Myr duration separated by $\sim$0.1 – 1.0 Myr, each caused by the arrival in circumterrestrial space of the largest comets followed by their disintegration into short-lived Apollo asteroids. Evidence is presented that a bombardment pulse was induced 3 – 5 Myr ago and that a "spike" in the form of debris from a Chiron-like progenitor of Encke's comet has dominated the terrestrial environment for the last 0.02 Myr. Among the predicted consequences are $^{14}C$ modulations, climatic variations including the last major glaciation and observable zodiacal light phenomena.

107.048 **Calculation of infall heating at the surfaces of model solar nebulae.**
J. A. Wood.
Smithson. Astrophys. Obs., Spec. Rep., No. 394, 43 pp. (1984).
This work attempts to estimate the temperature rise of infalling interstellar gas and dust as it accreted to the surfaces of the primitive solar nebula, where its velocity was checked and its kinetic energy was dissipated as heat. Summations are made of the distribution of mass from all stages of infall in several nebula models with different initial parameters. These are compared with the present distribution of planetary mass in the solar system.

107.049 **On the tidal interaction between protoplanets and the primordial solar nebula. I. Linear calculation of the role of angular momentum exchange.**
J. Papaloizou, D. N. C. Lin.
Astrophys. J., Vol. 285, No. 2, p. 818 – 834 (1984). = Lick Obs. Bull., No. 982.
The authors study the tidal interaction between a protoplanet and a gaseous protoplanetary disk with moderate thickness. They consider the situation when the surface density of the disk changes on a scale length comparable to its thickness in a gap region, of comparable dimensions, containing the protoplanet. In this case it is necessary to solve for the tidal perturbations taking the surface density structure fully into account. The authors develop expressions for the angular momentum exchange rate between protoplanet and disk that require simple numerical calculations only. The methods are used to calculate the strength of the tidal torque in the limit that the protoplanet is sufficiently massive to induce a gap in the nebula.

107.050 **Formation of the solar system.**
V. S. Safronov.
Planetary rings, p. 647 – 659 (1984). – See Abstr. 012.085 (IAU Colloq. No. 75).

107.051 **The formation of the terrestrial planets from lunar sized planetesimals.**
S. J. Aarseth, M. Lecar.
Planetary rings, p. 661 – 674 (1984). – See Abstr. 012.085 (IAU Colloq. No. 75).
The authors have performed a numerical simulation of the formation of the terrestrial planets from lunar sized planetesimals. The simulation was confined to a plane and included the gravitational forces between planetesimals as well as the Sun. When two planetesimals physically collided, they were merged into one. Initially, two hundred moons were distributed with uniform surface density, from 0.5 to 1.5 AU. After 65,000 years one is left with five bodies, the two largest of which each had 3/4 of an Earth mass.

107.052 **Numerical experiment applicable to planet formation.**
A. Cazenave.
Planetary rings, p. 675 – 676 (1984). Abstract. – See Abstr. 012.085 (IAU Colloq. No. 75).

107.053 **Statistical theory of planet growth.**
P. Hornung, R. Pellat.
Planetary rings, p. 677 – 685 (1984). – See Abstr. 012.085 (IAU Colloq. No. 75).
This paper treats the collisional dynamics of a differentially rotating disk of planetesimals. It takes into account simultaneously the inelastic hard sphere collisions (including gravitational focusing) and the long distance self gravitational interaction between the planetesimals.

107.054 **Fission in a graduated form is an apparent characteristic of the nebulous ring formation.**
S. Xue.
Acta Astrophys. Sin., Vol. 4, No. 2, p. 133 – 141 (1984). In Chinese.
According to the law of conservation of angular momentum, it is proved quantitatively, that planets and satellites are formed through the nebulous ring discussed in this paper. On this foundation, the author has gone further into the correspondent relation between the distributed distance of the nebulous ring and the masses of planets and satellites. So he obtains an apparent characteristic of the nebulous ring formation-fission in a graduated form.

107.055 **The origin and evolution of planetary rings.**
A. W. Harris.
Planetary rings, p. 641 – 659 (1984). – See Abstr. 012.086.
The primary issue debated in theories of the origin of planetary rings is whether they represent a failure of the innermost portion of a circumplanetary disk to accumulate into satellites, or whether they are the result of the disruption of preexisting satellites. This paper reviews a number of dynamical processes which act on a gas/solid disk in Keplerian motion: viscous spreading, gas drag, coagulation of particulates, and the effect of further infall of matter from heliocentric orbit onto the planet/disk. It is concluded that the survival of a remnant ring is unlikely and that is appears more plausible that the rings were created by the disruption of sizeable satellites.

107.056 **The solar nebula and the planetesimal disk.**
W. R. Ward.
Planetary rings, p. 660 – 684 (1984). – See Abstr. 012.086.
Two popular theories of solar system formation are briefly reviewed, then used as background in an examination of several new developments related to planetary ring dynamics. Most important are the incorporation of accretion disk and density wave theories into cosmogonic theory. A successful integration of these mechanisms may significantly constrain evolutionary models of the early solar system.

107.057 **Transport of dust and vapor and chemical fractionation in the early protosolar cloud.**
G. E. Morfill, H. J. Völk.
Astrophys. J., Vol. 287, No. 1, p. 371 – 395 (1984).
In this paper, the mathematical framework for discussing the transport of dust, trace gases, and vapors in a turbulent protostellar accretion disk is formulated, and analytical solutions are obtained for these equations in the limit of "small" particle sizes. The solutions give a rough idea of the gross chemical structure and solid-matter distribution in the planetary region of the early protosolar cloud. The results appear to be in basic agreement with a number of cosmochemical measurements.

**Planets and their atmospheres. Origin and evolution.**
See Abstr. 003.046.

**A criterion for the fragmentation of a rotating and collapsing gas cloud.**
See Abstr. 062.114.

**Solar neutrinos and comets.**
See Abstr. 080.012.

**On the rotation of the Sun.**
See Abstr. 080.107.

**FeO and $H_2O$ and the homogeneous accretion of the Earth.**
See Abstr. 081.042.

**Tidal friction in the Earth–Moon system.**
See Abstr. 081.043.

**The origin and earliest state of the earth's hydrosphere.**
See Abstr. 081.044.

**Grain abundance in the primordial atmosphere of the Earth.**
See Abstr. 082.018.

**Rare gases in the atmospheres of terrestrial planets and early stages of evolution of the solar system.**
See Abstr. 091.019.

**On the regular and arbitrary components of angular momentum of planets in the process of their forming.**
See Abstr. 091.023.

**Collisional theory of non–identical particles in a gravitational field.**
See Abstr. 091.027.

**Axial rotations of accumulating planets.**
See Abstr. 091.031.

**The primeval axis of rotation of the Moon.**
See Abstr. 094.055.

**The asteroid ring.**
See Abstr. 098.113.

**Asteroid compositional "rings": clues to the compositions of primordial planetesimals in the middle solar system.**
See Abstr. 098.114.

**Collisional outcomes in the asteroid belt.**
See Abstr. 098.115.

**Resonance structure in the early evolution of the Saturn system and the location of the inner satellites.**
See Abstr. 100.126.

**Origin and evolution of the Saturn system.**
See Abstr. 100.149.

**On the origin of Triton and Pluto.**
See Abstr. 101.008.

**The rotation of the Uranian system.**
See Abstr. 101.073.

**Tidal breakup and origin of the Uranian ring system.**
See Abstr. 101.083.

**Dynamique du nuage de Oort.**
See Abstr. 102.008.

**Solar composition noble gases in the Washington County iron meteorite.**
See Abstr. 105.093.

**On the formation of meteoritic chondrules by aerodynamic drag heating in the solar nebula.**
See Abstr. 105.094.

**Star formation and extinct radioactivities.**
See Abstr. 131.151.

**Resonances in a low–mass disc.**
See Abstr. 151.054.

# Stars

## 111 Parallaxes, Proper Motions, Radial Velocities, Space Motions, Distances

111.001 **Photographic astrometry of binary and proper–motion stars.**
W. D. Heintz, E. R. Borgman.
Astron. J., Vol. 89, No. 7, p. 1068 – 1072 (1984).
Fifty–eight trigonometric parallaxes and 14 binary–star mass ratios are listed and discussed.

111.002 **Erratum: "Non–supergiant high–velocity stars near the Magellanic Clouds" [Astron. Astrophys., Suppl. Ser., Vol. 16, p. 277 – 287 (1974)].**
N. Carozzi.
Astron. Astrophys., Suppl. Ser., Vol. 57, No. 1, p. 169 (1984). In French. See Abstr. 12.112.003.

111.003 **Radial velocities of southern stars obtained with the photoelectric scanner CORAVEL. II. Faint southern potential radial–velocity standards.**
E. Maurice, M. Mayor, J. Andersen, A. Ardeberg, W. Benz, H. Lindgren, M. Imbert, N. Martin, B. Nordström, L. Prévot.
Astron. Astrophys., Suppl. Ser., Vol. 57, No. 2, p. 275 – 279 (1984).
The authors present radial–velocity data for the 14 stars listed by Evans (1968) as a useful southern extension of the existing faint radial–velocity standard stars. The stars have been measured with the photoelectric scanner CORAVEL at La Silla from January, 1981, through August, 1983.

111.004 **The radial velocity of HR 4550 (Groombridge 1830) in 1974 – 1984.**
R. F. Griffin.
Observatory, Vol. 104, No. 1061, p. 192 – 193 (1984).
The radial velocity of HR 4550 has shown no significant departure from the IAU–adopted value during the past decade.

111.005 **Magnitude equation of stellar proper motions.**
N. V. Kharchenko.
Astrometr. Astrofiz., Vyp. 52, p. 3 – 8 (1984). In Russian.
A new method is proposed for correction of the magnitude equation in stellar proper motions using catalogue data only. The method enables to obtain an independent catalogue with taking into account the local stellar motions in individual sky areas.

111.006 **Moving–cluster membership of solar–neighborhood stars.**
T. A. Agekyan, V. V. Orlov.
Sov. Astron., Vol. 28, No. 1, p. 36 – 39 (1984). English translation of 37.111.001.

111.007 **O–star kinematics.**
D. K. Karimova, E. D. Pavlovskaya.
Sov. Astron., Vol. 28, No. 1, p. 40 – 45 (1984). English translation of 37.111.002.

111.008 **Peculiar stars in the Ursa Major stream.**
V. G. Klochkova, I. M. Kopylov.
Sov. Astron., Vol. 28, No. 1, p. 80 – 84 (1984). English translation of 37.111.003.

111.009 **Radial velocities and positions from objective prism plates.**
J. Stock.
Rev. Mex. Astron. Astrofis., Vol. 9, No. 1, p. 77 – 83 (1984).
Objective prism plates are used to determine positions, approximate radial velocities, magnitudes, and spectral types for 10053 stars in a selected southern zone. The catalogue produced is the first part of an extensive survey which serves as a finding list for possible high velocity stars, and which may contribute to the study of the kinematic characteristics of the Milky Way Galaxy.

111.010 **Correction of spuriously large proper–motion determinations for the S–type stars T Cam and S UMa.**
C. B. Stephenson, S. J. Hulbert.
Astron. J., Vol. 89, No. 8, p. 1261 (1984).
Only about 30 S–type stars have published proper motions, and two of these – T Cam and S UMa – are given such large values in the AGK3 as to be almost certainly spurious. The authors have redetermined these two stars' proper motions, comparing new photographic position determinations with late 19th–century first–epoch positions. The new proper motions are much smaller than the AGK3 values, although based mainly upon AGK3 reference stars.

111.011 **Determination of stellar proper motions related to galaxies. Current state and planned work at Pulkovo.**
N. M. Bronnikova, O. N. Orlova, N. A. Shacht.
Tartu Astrofüüs. Obs. Teated, Nr. 72, p. 5 – 11 (1984).
A survey is given of the works on the determination of stellar proper motions related to galaxies. Secular parallaxes were calculated for the star groups $13^{m}_{.}8$ and $15^{m}_{.}7$. The accuracy of the UBV magnitudes is characterized by a table.

111.012 **Comments on the treatment of reductions of proper motions relative to galaxies.**
H. Eelsalu.
Tartu Astrofüüs. Obs. Teated, Nr. 72, p. 20 – 26 (1984).
Expressions for the proper motions of the centroids of stellar motions in terms of measurements relative to reference galaxies are presented and discussed with the aim of elaborating a new approach to reduction equations.

111.013 **Radial velocities of CH Cygni in 1981.**
T. Tomov, L. Luud.
Astrofizika, Tom 20, Vyp. 1, p. 99 – 105 (1984). In Russian. English translation in Astrophysics, Vol. 20, No. 1.
The mean radial velocities from all the spectrograms of CH Cygni taken during JD 2444604 – 2444956 are given for the absorption lines of ionized metals, for Fe II and [Fe II] emission lines and for the absorption lines of neutral metals. It has been found that the absorption lines of ionized metals measured in the ultraviolet and visible spectra correspond to the motion of the hot component and the absorption lines of neutral metals measured in the infrared spectra correspond to the motion of the cold component. The ratio between the masses of the cold and the hot components is about 1:2.

111.014 **Astrometric study of the subdwaf HR 4550.**
W. D. Heintz.
Publ. Astron. Soc. Pac., Vol. 96, No. 581, p. 557 – 558 (1984).

A parallax of 0".122 is measured for the high-velocity star, while evidence for orbital effects of the unconfirmed visual companion is not found in the observations.

111.015 **Determination of relative proper motions and the membership of the open cluster NGC 1817.**
K. Tian, M. Ying, J. Jin, Z. Xu.
Ann. Shanghai Obs., Acad. Sin., No. 5, p. 136 – 160 (1983). In Chinese.

NGC 1817 ($l$ = 186°.1, $b$ = −13°.1) is an open cluster with moderate age of about $6.3 \times 10^8$ years. The purpose of this paper is to find the member stars of NGC 1817 from plates taken with one and the same telescope by means of the improved maximum-likelihood method. From three pairs of photographic plates taken with the 40 cm astrograph at Zô–Sè Section of Shanghai Observatory relative proper motions of 752 stars in the region of NGC 1817 are determined.

111.016 **Absolute proper motions of 774 stars with respect to 10 galaxies in 4 areas of the southern sky.**
G. Zhu, X. Wang.
Ann. Shanghai Obs., Acad. Sin., No. 5, p. 161 – 189 (1983). In Chinese.

The catalogue of absolute proper motions of 774 stars with respect to 10 galaxies in 4 areas of the southern sky is given. Observations are made with the 40 cm astrograph with focal length of 6.9 m and field of view of 2° × 2°.5 at Zô–Sè. The difference of epochs is equal to 19.6 years on the average. The proper motions are derived from 8 plate pairs. The probable error of absolute proper motions is equal to ±0".0066 on the average. The probable error of a reduction to absolute proper motions of a star is equal to ±0".0040 on the average. The mean value of the probable error for proper motions of reference stars is equal to ±0".0054.

111.017 **On the accuracy of relative stellar proper motions determined with the long–focus astrograph of the Main Astronomical Observatory of the Ukrainian SSR Academy of Sciences.**
S. P. Rybka.
Astrometr. Astrofiz., Vyp. 53, p. 37 – 42 (1984). In Russian.

Stellar proper motions accuracy was investigated as dependent on the following factors: epoch differences, measurement errors, proper motion dispersion of reference stars and their number. The results are used to select optimum conditions for determination of stellar proper motions.

111.018 **On the effect of cosmic errors on determinations of stellar proper motions.**
A. I. Yatsenko.
Astrometr. Astrofiz., Vyp. 53, p. 42 – 48 (1984). In Russian.

The main causes affecting the value of a cosmic error in stellar proper motions are considered. A formula is proposed for preliminary estimation of a possible cosmic error. The systematic character of this error is shown to be more pronounced in the areas where there are systematic motions of reference stars.

111.019 **A wave–like effect in the motions of nearby young stars.**
P. Brosche, H. Schwan.
Astron. Astrophys., Vol. 140, No. 2, p. L25 – L26 (1984).

The motions of young stars from astrometric fundamental catalogues show a breaking wave in the velocity field along the local spiral arm as a function of distance along the same direction: wave–length ~1 kpc, total velocity range ~7 km/s. Some consequences of this finding are indicated.

111.020 **Systematic errors in radial velocities measured on image–tube spectra from the CTIO 1–m telescope.**
L. P. Bassino, J. C. Muzzio.
Rev. Mex. Astron. Astrofis., Vol. 9, No. 2, p. 165 – 169 (1984).

The authors investigate the systematic errors of radial velocities measured on spectra obtained with the image–tube spectrograph of the 1–m (Yale) telescope at Cerro Tololo Inter-American Observatory. Data from other authors and the authors' own suggest that those errors are caused by the dispersion of the stellar light in the atmosphere of the Earth, which has been invoked to explain similar errors obtained with other instruments.

111.021 **Southern stars of high radial velocity.**
J. Stock, D. J. MacConnell, W. Osborn, H. Alvarez.
Astron. J., Vol. 89, No. 12, p. 1897 – 1907 (1984).

First, it has been shown that the Stock Velocity Survey is a useful source of velocity information on 8–13th mag southern stars. The particular use has been to identify stars of very large velocity. Thackeray (1975) has tabulated those galactic objects with heliocentric radial velocities over 250 km/s; his list contains 69 objects, of which 59 are stellar. The authors have found twelve more stars. Second, the slit spectra of a sample of high–velocity stars in the Stock Survey revealed several extremely metal–weak stars. Abundance analysis for these objects would be worthwhile. Finally, among the high–velocity objects are a surprising number of late B and early A stars. In most instances, these objects appear to be spectroscopically normal, main–sequence stars, but three are definite A + G spectrum binaries. High–velocity binaries are rare.

111.022 **Problèmes posés par l'emploi du catalogue photographique et de la Carte du Ciel.**
P. Lacroute.
Bull. Inf. Cent. Données Stellaires, No. 27, p. 39 – 42 (1984).

In this paper are discussed the uncertainties of the proper motions which can be obtained from the Astrographic Catalogues put into machine-readable form at Strasbourg. Also are examined the problems posed by the use of the plates from other zones. A comparison is made between the results to be expected with those from modern inventories.

111.023 **Sur l'efficacité des mesures systématiques de mouvements propres.**
A. Robin, M. Crézé.
Bull. Inf. Cent. Données Stellaires, No. 27, p. 49 – 50 (1984).

The authors use a model of stellar populations in the Milky Way for predicting the fraction of significant proper motions which are to be expected at various levels of accuracy.

111.024 **Radial velocities of southern stars obtained with the photoelectric scanner CORAVEL. III. 790 late–type bright stars.**
J. Andersen, B. Nordström, A. Ardeberg, W. Benz, M. Imbert, H. Lindgren, N. Martin, E. Maurice, M. Mayor, L. Prévot.
ESO Sci. Prepr., No. 348, 31 pp. (1984). To appear in Astron. Astrophys., Suppl. Ser.

111.025 **Reinvestigation of O, B star kinematics and determination of corrections to precession.**
J.–l. Zhao.
Sci. Sin., Ser. A, Vol. 27, No. 12, p. 1298 – 1306 (1984).

The purpose of the present investigation is to analyze in detail the kinematics of O, B stars based upon a fairly rigorous reduction procedure by fully utilizing data available at present in order to obtain relatively more appropriate results and to ascertain corrections to the precession constant as well.

111.026 **On two WR stars in the Large Magellanic Cloud: the high velocity of BR 52 and a newly discovered WN star.**
A. P. Cowley, D. Crampton, J. B. Hutchings, I. B. Thompson.
Publ. Astron. Soc. Pac., Vol. 96, No. 586, p. 968 – 972 (1984).

Measurements of BR 52 (= FDA 47, = HV 5947) over three years show it to have a constant, high radial velocity, in spite of the expectation that it might be a spectroscopic binary of large radial–velocity amplitude. Hydrogen absorption–line components, which have been thought to be due to a hot companion, appear rather to be formed in a variable stellar wind from the WR star. The high velocity (~ +470 km $sec^{-1}$) of BR 52 with respect to the LMC (~ +270 km $sec^{-1}$) is noted. A new WN5 star, designated W0536–67, has been found near the X–ray source CAL 60, although it does not appear to be the optical identification for this source. Some details about its spectrum and velocity are given.

**Radial velocities and positions from objective prism plates. The catalogue.**
See Abstr. 002.019.

**A new edition of the Yale Parallax Catalogue.**
See Abstr. 002.029.

**Radial velocities of 232 stars belonging to two fields of 4° × 4°. Third list.**
See Abstr. 002.035.

**An iteration algorithm of the stellar trigonometric parallax reduction with the central overlap technique.**
See Abstr. 036.074.

**On precision of measurement of galaxy images at Zô-Sè.**
See Abstr. 036.075.

**Mesure de clichés de la Carte du Ciel.**
See Abstr. 036.152.

**Optical positions and proper motions of radio stars.**
See Abstr. 041.002.

**A catalogue of declinations and proper motions of Moscow PZT stars.**
See Abstr. 041.013.

**Les données astrométriques utilisées en cinématique stellaire.**
See Abstr. 041.034.

**A photometric study of six nearby common-proper-motion pairs.**
See Abstr. 113.001.

**Intermediate band, Hβ and *RI* photometry of a large sample of stars unbiased with respect to their motion. I. the F-type stars.**
See Abstr. 113.028.

**The A0 stars.**
See Abstr. 114.020.

**New spectrophotometry of the extremely cool proper motion star LHS 2924.**
See Abstr. 114.053.

**Spectroscopic and photometric observations of the suspected binary star HD 123058.**
See Abstr. 114.102.

**Determination of the dynamical parallaxes of some visual binaries by the method of apparent motion parameters.**
See Abstr. 118.044.

**Proper motions of Herbig-Haro objects. V. Southern HH objects.**
See Abstr. 121.031.

**On proper motions and distribution of flare stars in the Pleiades.**
See Abstr. 153.010.

**A radial-velocity study of open clusters containing blue stragglers.**
See Abstr. 153.029.

**Galactic rotation, correction to the constant of precession and coordinates of the solar apex according to proper motions of bright stars.**
See Abstr. 155.038.

**Kinematics of field K giants in the outer galactic halo.**
See Abstr. 155.048.

**Velocity dispersion and metallicity distribution of K giants in the galactic nuclear bulge.**
See Abstr. 155.049.

**Kinematics of nearby gas and stars.**
See Abstr. 155.079.

# 112 Stellar Environments (Chromospheres, Coronae, Stellar Winds, Shells, Masers, etc.)

112.001 **Mass loss rates for northern OB stars.**
U. Peppel.
Astron. Astrophys., Suppl. Ser., Vol. 57, No. 1, p. 107 - 119 (1984).

A mass loss survey of 44 O- and early B-type stars in northern OB associations is presented. For 22 stars mass loss rates were derived from Hα equivalent widths using the method of Klein and Castor (1978). For 15 stars upper limits of the mass loss rates were estimated. For the remaining 7 stars Hα and Hβ equivalent width values are given but no quantitative evaluation of $\dot{M}$ was possible. The results are compared with present ideas concerning the relation between mass loss and other basic stellar parameters.

112.002 **The spectrum of the Be star HD 110432.**
S. J. Codina, J. A. de Freitas Pacheco, D. F. Lopes, D. Gilra.
Astron. Astrophys., Suppl. Ser., Vol. 57, No. 2, p. 239 - 247 (1984).

The UV spectrum of HD 110432, a B0.5IIIe star, is described. The IUE spectra present broad absorption photospheric lines, very narrow absorption lines of interstellar origin and narrow absorption lines of highly ionized species (N V, C IV, and Si IV) displaced to the short wavelengths at velocities of about $-1350$ km s$^{-1}$. The authors have estimated a rotational velocity of 360 km s$^{-1}$ and the distance was calculated to be 430 pc. The asymmetric profiles of the resonance lines of Si IV and C IV suggest an expanding envelope corresponding to a mass-loss of about $3 \times 10^{-9}$ $M_{\odot}$/y.

112.003 **Grain temperature, radiation pressure and electric potential in the vicinity of main sequence and white dwarf stars.**
J. Leiknes, O. Havnes.
Astron. Astrophys., Suppl. Ser., Vol. 57, No. 2, p. 263 - 274 (1984).

The authors present results of calculations of the grain physical parameters temperature, lifetime against evaporation, radiation pressure and electric potential for spherical grains near main sequence stars, hydrogen type (DA) white dwarfs and helium type (DB) white dwarfs. These parameters are essential in determining the behaviour of grains near such stars. The grain temperature as a function of stellar distance is calculated for grains of sizes 0.1 and 1 μ for grain materials of silicate (obsidian), iron and graphite. The lifetime due to thermal evaporation as a function of grain temperature of these materials is also given. The radiation pressure is given for grain sizes from 0.01 to 10 μ for the same three grain materials.

112.004 **Multiple circumstellar shells and radiation pressure on grains in the outflows from late–type giants.**
M. Jura.
Astrophys. J., Vol. 282, No. 1, p. 200 – 205 (1984).
Some mass–losing red giants display multiple P Cygni lines in their spectra. The author examines the possibility that multiple shells can be produced by radiation pressure on grains. He finds that several shells can be produced by episodes of grain growth, but if there is a continuous outflow of matter from the central star, the P Cygni lines should appear as broad absorption troughs between the initial and final velocity rather than distinct, sharp lines. Broad troughs are observed in IRC +10216, but not in some other less dusty stars with multiple shells such as α Ori or Mira. The stars with distinct shells apparently experience impulsive, discontinuous mass loss, and in fact, reanalysis of existing data toward Mira shows that there is actual infall of circumstellar matter into this star.

112.005 **On the properties of stellar winds in early–type stars.**
I. N. Portnova, N. A. Sakhibullin.
Astrofizika, Tom 19, Vyp. 4, p. 697 – 709 (1983). In Russian. English translation in Astrophysics, Vol. 19, No. 4.
An attempt to determine the relation between the physical characteristics of stellar winds and temperatures and luminosities of early–type stars, based on the study of ultraviolet asymmetric lines in 14 OB stars, has been undertaken. The "warm" wind and "corona + cool wind" models have been used. It has been shown that there are two drops in the temperatures of the wind material at $T_{eff}$ near 13000K and $T_{eff}$ = 20000–25000K.

112.006 **OH 32.8–0.3. A second source with absorption features of pure water ice.**
P. F. Roche, D. K. Aitken.
Mon. Not. R. Astron. Soc., Vol. 209, No. 3, p. 33P – 36P (1984).
The authors present infrared spectra of the type II OH/IR source OH 32.8–0.3 showing evidence for absorption bands near 12 and 3.1 μm. The 12 μm absorption feature is similar to that seen in OH 231.8 + 4.2 and is attributed to relatively pure water ice formed in the outflow from an evolved star.

112.007 **Transient components in the Lyman lines of hot stars. Evidence for the ejection of shells or puffs on very short timescales.**
C. Gry, H. J. G. L. M. Lamers, A. Vidal–Madjar.
Astron. Astrophys., Vol. 137, No. 1, p. 29 – 42 (1984).
The existence of H I transient components in the stellar wind of ε Per was recently discovered. The authors investigate here systematically the existence of such components toward 8 early type stars observed by the Copernicus Satellite. They show that such transient components are indeed present in the spectra of 5 of these stars and have characteristics very similar to the one found in the case of ε Per. They show that such components can be interpreted in terms of coherent "puffs" or shells of matter ejected by the stars. It is suggested that such events must be due to instabilities in these stars which are all slightly evolved stars ranging in spectral types between O9 and B1. The instabilities are probably similar to those in β Cep stars, i.e. non–radial pulsations.

112.008 **Spectroscopy of the shell episode of η Car (1981 – 1983).**
R. Zanella, B. Wolf, O. Stahl.
Astron. Astrophys., Vol. 137, No. 1, p. 79 – 84 (1984).
The authors compared high–dispersion spectra (12 Å $mm^{-1}$) of η Car obtained in March 1974, Dec. 1981, and Feb. 1983. The spectrum of Dec. 1981 shows conspicuous deviations from the usual spectral appearance of η Car; all lines originating from high ionization and/or excitation levels have dramatically faded. The P Cygni characteristics of the Balmer lines are more pronounced in the spectra of 1981 and 1983. The spectroscopic variations are explained as a result of changing ionization/excitation conditions due to an episodic ejection of a dense shell with a maximum velocity of about 600 km $s^{-1}$.

112.009 **The shell spectrum of CH Cygni in 1981.**
M. H. Rodríguez.
Astrophys. Space Sci., Vol. 102, No. 1, p. 123 – 129 (1984).
The shell spectrum of CH Cyg observed in 1981 has been analyzed by the curve of growth method and using the equivalent widths of Balmer lines. The following parameters for the region of the shell where the absorption spectrum is formed have been obtained: excitation temperature $T_{ex}$ = 9000K, microturbulent velocity $v_t$ = 16 km $s^{-1}$, electron density $n_e = 6.3 \times 10^{12}$, number of active hydrogen atoms in the second excited state $N_{0.2} = 5.8 \times 10^{15}$, and the effective height of the absorption spectrum formation region $H = 10^5$ km.

112.010 **Ammonia and cyanotriacetylene in the envelopes of CRL 2688 and IRC + 10216.**
Nguyen–Q–Rieu, D. Graham, V. Bujarrabal.
Astron. Astrophys., Vol. 138, No. 1, p. L5 – L8 (1984).
The authors report the first detection of the $NH_3$ (1,1) and (2,2) inversion lines, and the rotational transition $j$ = 21 – 20 of $HC_7N$ in the Egg nebula (CRL 2688). They have also reobserved these lines in the envelope of IRC +10216 and confirm, in particular, the detection of the $NH_3$ (2,2) line which is centred approximately at the same velocity as the $NH_3$ (1,1) line. The data have been interpreted in terms of a radiative transfer model, from which the source size and the molecular abundance have been derived.

112.011 **A study of the expanding envelope around the pole–on Be star HR 5223.**
J. Dachs, R. Hanuschik.
Astron. Astrophys., Vol. 138, No. 1, p. 140 – 160 (1984).
Photometric and spectrometric data for the southern Be–type star HR 5223 (= HD 120991, B2 IIIep, $v \sin i$ = 70 km $s^{-1}$) covering the wavelength range from 1150 Å to 5 microns and obtained between 1978 and 1982 are combined in an effort to develop an ad hoc model for the envelope surrounding the star and to find out the nature of its variations. From the combination of the data it is concluded that the circumstellar envelope consists of a dense, cool disk rotating around the equatorial bulge of the star and a strong, hot stellar wind at higher stellar latitudes. Disk formation is considered to be a natural consequence of energy dissipation and angular momentum transport in the Be star envelope formed by self–accretion from regions of the stellar wind for which the terminal velocity does not reach escape velocity. Possible mechanisms of interaction between the hot stellar wind regions and the cool equatorial disk are considered.

112.012 **The transition layer of eleven late–type stars.**
M. J. Fernandez–Figueroa, B. Montesinos, M. Rego.
Astron. Astrophys., Vol. 138, No. 1, p. 164 – 168 (1984).
Far UV–spectra of eleven F, G, and K stars of different luminosity class have been studied. Si IV, C IV, and N V emission line fluxes have been used to obtain parameters for the transition region. The stars of the sample have been classified according to the activity level. Using two abundance criteria the gaseous pressure upper limits have been computed. They show a correlation with the line surface fluxes.

112.013 **New SiO maser sources: red–supergiants in the vicinity of molecular clouds.**
N. Ukita, P. F. Goldsmith.
Astron. Astrophys., Vol. 138, No. 1, p. 194 – 200 (1984).
A search for SiO maser emission has been made toward $H_2O$/infrared objects which are found in the vicinity of star–forming regions, but which are not coincident with radio continuum sources. SiO emission was detected in GL 2103, GL 2139, GL 2999, and GL 1074. The maser spectra together with the infrared suggest that these objects are red supergiants. They have radial velocities similar to those of the molecular cloud complexes in their vicinities. The luminosities of GL 2103 and GL 2139, based on their kinematic distances, are $3 \times 10^5$ and $4 \times 10^5$ $L_{\odot}$, respectively, which are comparable to that of VY CMa. No SiO masers similar to the Orion SiO source were found.

112.014 **The stellar wind enigma in early-type stars: the high-velocity, narrow absorption lines in the UV.**
H. Henrichs.
ESA Spec. Publ., ESA SP-218, p. 43 - 54 (1984). - See Abstr. 012.024.

Recent work on the topic "narrow, high-velocity absorption components in UV spectra of early-type stars" is reviewed. The distribution of stars showing this phenomenon is presented in an HR diagram.

112.015 **The wind and chromosphere of the Herbig Ae star AB Aur as seen by IUE.**
C. Catala.
ESA Spec. Publ., ESA SP-218, p. 227 - 231 (1984). - See Abstr. 012.024.

The Herbig Ae variable AB Aur (spectral type A0) exhibits a line spectrum characteristic of a stellar wind. The results of a detailed quantitative interpretation of the Mg II resonance lines (λλ 2796, 2802 Å) are summarized and preliminary results of a quantitative interpretation of the C IV resonance line at 1548 Å are presented.

112.016 **Flux variations of 88 Her through phase transition from far UV to visual.**
M. Barylak, V. Doazan.
ESA Spec. Publ., ESA SP-218, p. 291 - 295 (1984). - See Abstr. 012.024.

The energy distribution and light variations of 88 Her at different phases - quasi normal B and Be-shell - from far UV to visual are analyzed.

112.017 **Possible identifying characteristics of Be stars.**
V. Doazan, R. N. Thomas.
ESA Spec. Publ., ESA SP-218, p. 297 - 300 (1984). - See Abstr. 012.024.

The authors search for physical parameters whose size and behaviour may identify a Be star independently of the presence (or non-presence) of Hα-emission.

112.018 **An IUE study of Be stars with Fe II emissions in the visible.**
P. L. Selvelli, F. X. De Araújo.
ESA Spec. Publ., ESA SP-218, p. 301 - 305 (1984). - See Abstr. 012.024.

The IUE spectra of 24 Be stars showing emission lines of Fe II in the optical range have been studied. In this subclass of Be stars, represented by earlier objects, the outflow phenomenon seems to be enhanced with respect to the Be main class. The outflow velocities are not related to the spectral type or to the luminosity class. Unlike in previous studies of Be stars, there seems to be a definite trend associating small outflow velocities with low v sin i values and vice versa.

112.019 **Shell ejection phenomena in the hypergiant star P Cygni.**
H. J. G. L. M. Lamers, P. Korevaar, A. Cassatella.
ESA Spec. Publ., ESA SP-218, p. 315 - 317 (1984). - See Abstr. 012.024.

The authors analyse the time evolution (1978 - 1983) of the shell lines in the hypergiant star P Cygni (B1Ia$^+$). The wind of P Cygni is characterized by a stable shell formed at large distance from the star, only seen in low excitation lines (Fe II) at 206±1.5 km/s, and by a variable shell at lower velocities, seen also in Fe III in addition to Fe II. The variable shell is formed closer to the star and accelerates outwards as time increases. The analysis shows that a new variable shell is formed every 1 - 2 years, so indicating that the shell ejection in P Cygni is a recurrent phenomenon.

112.020 **Ultraviolet and optical observations of the Cygnus OB2 association: the star 8A.**
F. Giovannelli, C. Bartolini, C. Cacciari, A. Guarnieri, A. Piccioni.
ESA Spec. Publ., ESA SP-218, p. 323 - 324 (1984). - See Abstr. 012.024.

Even if the high reddening of the Cygnus OB2 association makes the observations of its members in the ultraviolet region very difficult, few quite good IUE low resolution spectra of N° 8A star were taken in a couple of years. The most exciting feature of the UV spectra is the very strong C IV resonance line, which shows a clear P Cygni profile. A terminal velocity of the order of 5000 km/s can be estimated. From optical spectra a central depth $A_c$ = 12.9% of the interstellar 4430 Å feature is derived.

112.021 **Optical and ultraviolet observations of FK Comae stars.**
L. Bianchi, M. Grewing, N. Kappelmann, A. Cassatella.
ESA Spec. Publ., ESA SP-218, p. 355 - 358 (1984). - See Abstr. 012.024.

The authors observed with IUE FK Comae itself and HD 32918, a suggested member of the FK Comae class in the southern sky, around one minimum and one maximum of the light curve, to search for a possible dependence of the chromospheric and transition zone lines on the (orbital?) phase.

112.022 **Coronal activity in F-, G-, and K-type stars. II. Coronal structure and rotation.**
C. J. Schrijver, R. Mewe, F. M. Walter.
Astron. Astrophys., Vol. 138, No. 2, p. 258 - 266 (1984).

Soft X-ray spectra, obtained with the imaging proportional counter on board the Einstein observatory, are analysed for a sample of 34 late-type stars. Coronal temperatures $T$ and emission measures $\zeta$ per unit area at the stellar surface are determined. The data are discussed in terms of a model for static, magnetic loops. Coronal activity and structure are clearly related to the stellar rotation rate.

112.023 **Stellar OH masers with polarized features at intermediate velocities.**
N. Ukita, A. M. Le Squeren.
Astron. Astrophys., Vol. 138, No. 2, p. 343 - 352 (1984).

Peculiar profiles were discovered in EY And. The 1612-MHz emission has a typical, double-peak structure, but the main-line masers show polarized features at the central velocity of the 1612-MHz emission. The line shape of the SiO maser resembles that of the 1665-MHz emission. Maser sources with polarized OH emission at intermediate velocities are discussed in the context of two kinematic models. Expanding envelopes with a smooth velocity field would require some of the polarized OH features to arise in the innermost part of the envelopes. Distortion of a smooth outflow by interaction with the ambient gas cannot explain the peculiar OH emission. The authors interpret it as a fluctuation of the velocity field of the expanding envelope. The polarized features in both main-line spectra have the same sense of circular polarization when they appear at nearly the same velocity. This evidence is consistent with the mechanism proposed by Cook and Shklovskij.

112.024 **A geometrical model of VY Canis Majoris for SiO maser lines.**
Z.-p. Zhou, N. Kaifu.
Astron. Astrophys., Vol. 138, No. 2, p. 359 - 364 (1984).

A new geometrical model of VY CMa is proposed to explain the three-peaked spectra of transitions $\upsilon$ = 1,2 $J$ = 1-0 of SiO maser emission. In this model the circumstellar envelope of VY CMa is a rotating disk of gas and dust seen nearly edge-on. The disk consists of two regions: a decelerated steady stream near the photosphere of the star and an accelerated one further away. Other geometries are discussed and eliminated. Calculated profiles of SiO maser lines fit well the observations. It is possible to explain the three-peaked profiles of SiO maser lines emitted by NML Cyg, RR Aql, NML Tau, etc.

112.025 **The emission nebula surrounding CPD -59°2661 in the Carina nebula (NGC 3372).**
J. R. Walsh.
Astron. Astrophys., Vol. 138, No. 2, p. 380 - 384 (1984).

An ionised knot with a filament extending from it has been found on an [S II] plate of the Carina nebula (NGC 3372), and high and low dispersion spectra of this region are presented. The knot surrounds the early-type star CPD -59°2661 here classified as O9.5 V, and the filament is most probably an ionisation front.

The Hα emission line shows a velocity splitting of $\sim 25 \text{ km s}^{-1}$ over the star. Positive and negative velocity wings to $\sim 150 \text{ km s}^{-1}$ are found, which are spatially extended on scales up to at least $\sim 2$ pc. These high velocity features are probably related to a stellar wind bubble or bubbles.

112.026 **Radiative processes in cool star chromospheres.**
C. Jordan, P. Judge.
Phys. Scr., Vol. T8, p. 43 – 48 (1984). – See Abstr. 012.027.

Observations made with the IUE satellite between 1200 Å and 3000 Å have shown that the spectra of giant and supergiant stars later than about K0 are dominated by emission from neutral atoms and singly charged ions. The cool star spectra show many examples of photo–excitation, line–leaking and fluorescent processes, some of which are discussed here. There is evidence that the H Ly α and H Ly β lines play an important role in the production of emission from O I, S I and Fe II. Some lines which remain without satisfactory explanations in both the stellar and solar spectra are discussed.

112.027 **Be– und Hüllensterne.**
H. Lehmann.
Astron. Raumfahrt, 22. Jahrg., Heft 5, p. 99 – 104 (1984).

112.028 **Short–period variability of the Be star λ Eri.**
E. Antonello, L. Mantegazza, L. Pastori.
Astrophys. Space Sci., Vol. 104, No. 2, p. 245 – 251 (1984).

Photoelectric and spectrographic observations of the Be star λ Eri made during five nights indicate that the luminosity variations have a different periodicity from that of radial velocity variations. This result seems to show that, at least during the authors' observing time, the observed variability has not to be ascribed to a pulsation of the star.

112.029 **SiO flux measurements of variable stars.**
F. O. Clark, T. H. Troland, G. H. Pepper, D. R. Johnson.
Astrophys. J., Vol. 283, No. 1, p. 174 – 178 (1984).

The authors report measurements of total flux for six circumstellar SiO maser sources. Both polarizations were measured simultaneously for these highly polarized sources. SiO curves are compared with the infrared flux, especially the infrared minima, which are observed to coincide in phase with the optical minima. The comparisons of SiO and infrared flux at light minimum are the most straightforward and indicate a characteristic phase lag for several of the sources. When interpreted in terms of observed physical motions associated with these stars, the SiO–infrared phase lag of 63 – 129 days can be used to infer a scale size on the order of $10^{13}$cm.

112.030 **Line profile variation in δ Orionis A, ι Orionis A, and 15 Monocerotis.**
C. A. Grady, T. P. Snow, W. C. Cash.
Astrophys. J., Vol. 283, No. 1, p. 218 – 225 (1984).

The results of a monitoring program with *IUE* and *Einstein* are presented for three stars, δ Ori A, ι Ori A, and 15 Mon. Line profile variability is observed in the UV profiles accessible to *IUE* and the relation between the variation in the different ions suggests that the ionization level is varying in the winds of these stars. This is consistent with *Einstein* observations of soft X–ray variability for two of the stars.

112.031 **Mass loss from Wolf–Rayet stars.**
A. J. Willis.
Mass loss from astronomical objects, p. 1 – 44 (1984). – See Abstr. 012.032.

Recent results relating to the stellar winds and mass loss rates of the WR stars are reviewed, emphasising new data and their interpretation acquired at UV, IR and radio wavelengths. The mass loss rates derived for 38 galactic WR stars from radio and/or IR data yield a mean value of $3.8 \pm 2.0 \times 10^{-5} M_{\odot}\text{yr}^{-1}$. No correlation is found in $\dot{M}$ with between WN and WC stars, among the WN and WC subclasses or between single and binary WN or WC stars. Acceleration mechanisms and terminal velocities for the winds are examined. Finally, a review of the properties of wind–blown ring nebulae around WR stars is presented.

112.032 **Mass loss from early–type stars.**
P. S. Conti.
Mass loss from astronomical objects, p. 45 – 58 (1984). – See Abstr. 012.032.

This paper reviews the observational data for stellar winds in early–type stars. The determination of mass loss rates is emphasized. It then considers the numbers of early–type stars in the solar neighborhood and discusses the shape of the initial mass function for these stars. Some consequences of mass loss from early–type stars on a galactic scale are also considered: the energy input into interstellar matter via stellar winds, the origin of giant H II complexes in galaxies, and the possible role of stellar winds as the source of mass input to black holes.

112.033 **Shock–driven mass loss of stars.**
C. de Jager, G. H. J. van den Oord.
Mass loss from astronomical objects, p. 59 – 66 (1984). – See Abstr. 012.032.

The authors suggest that shocks, fed by multi–mode pulsational motion of the photospheres of supergiants are the basic mechanism for the mass loss of cool and medium type supergiants. They review the present theoretical situation, and describe the implementations of the observed transient mass loss component of the pulsating Beta Cephei stars BW Vul and σ Sco.

112.034 **Physics of the atmospheres of early–type stars.**
H. J. G. L. M. Lamers.
Mass loss from astronomical objects, p. 67 – 93 (1984). – See Abstr. 012.032.

This paper reviews the observed characteristics of the stellar winds of early–type stars and the theories which have been proposed for their explanation. The following subjects are discussed: the acceleration of the stellar wind; the narrow components of the UV line profiles; the super–ionization of the wind; the X–ray luminosity and the X–ray spectrum.

112.035 **Spektroskopische Untersuchungen der Hüllenphase (1981 – 1983) des eruptiven Veränderlichen Eta Carinae.**
R. Zanella.
Diplomarbeit, Landessternwarte Heidelberg – Königstuhl, Heidelberg, F.R. Germany, 56 pp. (1984).

By comparing high–dispersion spectrograms of Eta Carinae of March 1974, December 1981, and February 1983 it was found that the spectrum of December 1981 shows conspicuous deviations from the usual appearance: All emission lines of the highly ionized/excited ions have greatly reduced their intensities or have disappeared completely. Furthermore, the blueshifted absorption components of the hydrogen lines were stronger than at other times. To explain these observations, the ejection of a thick shell from the central object is proposed, which has a screening effect to UV–photons. On the blue side of some forbidden emission lines, weak emission structures were found and identified as blueshifted components of these lines. A collimated mass loss from the central object is proposed to explain these features.

112.036 **High–resolution, far–ultraviolet study of Beta Draconis (G2 Ib–II): transition region structure and energy balance.**
A. Brown, C. Jordan, R. E. Stencel, J. L. Linsky, T. R. Ayres.
Astrophys. J., Vol. 283, No. 2, p. 731 – 744 (1984).

High resolution far–ultraviolet spectra of the star β Draconis have been obtained with the IUE satellite. The observed emission–line fluxes have been used to derive the mean emission measure distribution, which is used to construct models of the density and temperature variation with height as a function of the transition region pressure. The extension of the models to coronal regions with log $T_e > 5.3$ using the observed X–ray flux is discussed. The energy balance in the outer atmosphere is investigated and possible modes of energy transport and deposition are discussed. It is shown that simple acoustic waves cannot carry a flux to match the radiation losses.

112.037 **Identification of the SiCC radical toward IRC +10216: the first molecular ring in an astronomical source.**
P. Thaddeus, S. E. Cummins, R. A. Linke.
Astrophys. J., Lett. Ed., Vol. 283, No. 2, p. L45 – L48 (1984).
Nine unassigned lines in the millimeter–wave spectrum of IRC +10216 have been identified as the SiCC radical, following a recent laser experiment in a supersonic molecular beam which showed that this molecule is a compact symmetric ring, not linear as previously thought. The lines attributed to SiCC are identical in shape and width as required, and their relative intensities are those expected for a prolate asymmetric top like SiCC under collisional excitation. The mean column density of SiCC over a 90″ beam derived from the rotational analysis is fairly large with respect to molecules of comparable size: $1.5\times10^{14}\mathrm{cm}^{-2}$.

112.038 **Review of observational facts about Be stars.**
P. Harmanec.
Hvar Obs. Bull., Vol. 7, No. 1, p. 55 – 95 (1983). – See Abstr. 012.037.

112.039 **Variable mass–flux in Be stars.**
V. Doazan, R. N. Thomas.
Hvar Obs. Bull., Vol. 7, No. 1, p. 97 – 123 (1983). – See Abstr. 012.037.

112.040 **Photometric variations of some early Be stars.**
C. Waelkens, F. Rufener, M. Burnet.
Hvar Obs. Bull., Vol. 7, No. 1, p. 125 – 131 (1983). – See Abstr. 012.037.
The authors present the results of a detailed investigation of the photometric behaviour of six Be stars, that, according to their colours in the Geneva Photometric System, fall close to or even inside the instability strip defined by the Beta Cephei stars.

112.041 **Photometric variability of Be stars.**
K. Pavlovski.
Hvar Obs. Bull., Vol. 7, No. 1, p. 133 – 142 (1983). – See Abstr. 012.037.
The author demonstrates the complexity of the short–term variations of Be stars on the example of three Be stars, quite extensively observed for several seasons at the Hvar Observatory.

112.042 **The southern Be star photometric campaign: a first report.**
C. Stagg.
Hvar Obs. Bull., Vol. 7, No. 1, p. 143 – 147 (1983). – See Abstr. 012.037.

112.043 **The upcoming photometric campaign on o And, KX And, KY And, LQ And, and EW Lac.**
C. Stagg.
Hvar Obs. Bull., Vol. 7, No. 1, p. 149 – 153 (1983). – See Abstr. 012.037.

112.044 **A note on the variability of λ Eri and the pulsation in Be stars.**
E. Antonello, L. Mantegazza, L. Pastori.
Hvar Obs. Bull., Vol. 7, No. 1, p. 155 – 158 (1983). – See Abstr. 012.037.

112.045 **Rapid spectroscopic variations in γ Cas and ζ Oph.**
Z. Ninkov, S. Yang, G. Walker.
Hvar Obs. Bull., Vol. 7, No. 1, p. 167 – 187 (1983). – See Abstr. 012.037.

112.046 **Variability of Ae and A–shell stars.**
Y. Andrillat, M. Jaschek, C. Jaschek.
Hvar Obs. Bull., Vol. 7, No. 1, p. 193 – 205 (1983). – See Abstr. 012.037.
The authors have started in 1983 a systematic study of Ae and A–shell stars with the aim to compare their behaviour with the one of Be stars specially for the Hα profiles, the shell lines and temporal changes of the spectrum.

112.047 **The Be star phenomena. II. Spectral formation and structure of envelopes.**
R. Hirata, T. Kogure.
Bull. Astron. Soc. India, Vol. 12, No. 2, p. 109 – 151 (1984).
In succession to the general considerations in Paper I, a review of the spectral formation and the structure of envelopes are presented.

112.048 **Ca II K emission in Canopus.**
M. K. V. Bappu, M. V. Mekkaden,
N. Kameswara Rao.
Bull. Astron. Soc. India, Vol. 12, No. 2, p. 196 – 201 (1984).
The changes in the Ca II K line profile of the F0 supergiant Canopus have been studied from the high dispersion coude spectrograms obtained since 1975. The emission in the core of the line is variable in a timescale of few days. Similar activity seems to be occurring in Mg II h and k lines also. This is attributed to the variability in the appearance of plage areas on the stellar surface.

112.049 **SiS in circumstellar shells.**
R. Sahai, A. Wootten, R. E. S. Clegg.
Astrophys. J., Vol. 284, No. 1, p. 144 – 156 (1984).
The authors have observed the spectrum of SiS toward the Mira variable IRC +10216, including nine rotational transitions from $J=1$–0 to $J=16$–15. A detailed model is described which has been used to match these observations. The variation of the line profile with rotational level, infrared flux, and position can sensitively constrain models of the shell. The authors obtain, in the carbon–rich shell of IRC +10216, $[\mathrm{SiS}]/[\mathrm{H_2}]=2.4\times10^{-7}$. A survey in the $J=7$–6 and $J=6$–5 lines of SiS was also conducted, and results are reported toward carbon–rich and oxygen–rich circumstellar shells.

112.050 **On the extended chromosphere of α Orionis.**
L. Hartmann, E. H. Avrett.
Astrophys. J., Vol. 284, No. 1, p. 238 – 249 (1984).
The authors develop preliminary models for the extended chromosphere of α Orionis. A flux of Alfvén waves sufficient to drive an outflow of $\sim10^{-6}M_{\odot}\mathrm{yr}^{-1}$ is assumed. The momentum and energy equations for the wave–driven wind are solved assuming steady, radial, spherically symmetric flow. The mechanical energy fluxes needed to heat the extended chromosphere are of sufficient magnitude that momentum deposition by waves can support an extended chromosphere. The model accounts reasonably well for the observed radio continuum emission, the observed surface brightness distributions, and the emission fluxes in a variety of spectral lines. The detailed kinematics of the solution is less satisfactory in reproducing observed line profiles. It is suggested that the observed, nearly symmetric line profiles can be understood best in terms of a "fountain" model, in which most of the ejected material falls back rather than escaping.

112.051 **Infrared photometric studies of Be stars.**
N. M. Ashok, H. C. Bhatt, P. V. Kulkarni,
S. C. Joshi.
Mon. Not. R. Astron. Soc., Vol. 211, No. 2, p. 471 – 484 (1984).
*JHK* magnitudes for 55 Be stars are presented along with *L* magnitudes for 15 of them. For 20 Be stars the observations reported in the present work are new. By geometric considerations it is shown that the energy input to the circumstellar (CS) envelope, by the Lyman continuum flux $L_L{}^0$ from the central star, cannot account for the observed continuum luminosity, $L_{\mathrm{IR}}$, of the CS envelope. The bolometric luminosity of the central star, $L_*$, is found to be correlated to $L_{\mathrm{IR}}$. It is also shown that $L_{\mathrm{H\alpha}}$, the luminosity of the Hα emission line, is proportional to $L_{\mathrm{IR}}$.

112.052 **"Astrosols" – new object of physical and astronomical investigations.**
I. V. Petryanov–Sokolov, A. G. Sutugin.
Priroda, No. 11, p. 11 – 18 (1984). In Russian.

112.053 **Mass loss in O–type stars: parameters which affect it.**
C. D. Garmany, P. S. Conti.
Astrophys. J., Vol. 284, No. 2, p. 705 – 711 (1984).

The authors present newly determined mass loss rates for sixteen O–type stars in three open clusters. Combining the data with that already in the literature, they find no evidence that the rates are different in clusters with differing galactocentric distances and compositions, at least near the Sun. There is still appreciable dispersion in the relationship between the mass loss rate and the stellar luminosity. It may be that the mass loss depends additionally on the stellar mass and/or radius. The authors find evidence that a stellar wind increases as a massive star evolves from the zero–age main sequence.

112.054 **The origin of low–velocity absorption components in the Mg II resonance lines of hybrid–chromosphere stars.**
S. A. Drake, A. Brown, J. L. Linsky.
Astrophys. J., Vol. 284, No. 2, p. 774 – 783 (1984).

The authors argue that the low–velocity absorption features seen in the Mg II resonance lines of seven confirmed and three probable hybrid–chromosphere stars are interstellar rather than circumstellar in origin. From a comparison of radial velocities based on all available spectra in the IUE archives with estimates of the interstellar velocity along each line of sight, the authors find a good correlation between the observed position of the low–velocity component and the predicted interstellar feature. They also show that previous arguments in favor of the circumstellar origin of the low–velocity absorption features are either incorrect or implausible.

112.055 **Rapid spectroscopic variability of the Be star HR 9070: evidence for double periodicity?**
D. Baade, Y. Bellas, W. Eichendorf, T. Tomov.
Astron. Astrophys., Vol. 139, No. 2, p. 521 – 528 (1984).

Fast spectroscopy of HR 9070 during two consecutive nights revealed rapid variations of the emission line ratios in Hβ and Hγ, the radial velocities, and the visibility of the helium lines and Mg II λ4481. There are indications of a 0.12–day period in addition to the photometric 0.31–day period (Percy, 1983). The situation is further complicated by individual, but not random, differences between various lines or groups of lines. The observations are discussed in terms of the oblique magnetic–rotator model advocated for HR 9070 by Harmanec (1984) and nonradial pulsation suggested by Percy (1983). The new spectroscopic observations of this paper favour a nonradial–pulsation model with possibly two modes.

112.056 **The cool wind of the B hypergiant Cygnus OB 2 No. 12 in the infrared.**
C. Leitherer, C. Bertout, O. Stahl, B. Wolf.
Astron. Astrophys., Vol. 140, No. 1, p. 199 – 204 (1984).

The authors present numerical computations of the infrared flux distribution of the extremely luminous star Cygnus OB 2 No. 12 (spectral type B5Ia$^+$). A mass–loss rate of $\dot{M} = 4 \times 10^{-5} M_\odot \mathrm{yr}^{-1}$ and a cool wind with $T_e$ = 5000K have been derived from radio observations by White and Becker (1983). It is shown that a cool wind produces a flux density less than predicted by model atmospheres for this star in the infrared (3 – 10 μ). The presence of a cool wind removes the apparent IR–flux deficit found in earlier investigations assuming a wind temperature equal to the stellar effective temperature. The authors discuss the temperature– and the velocity–field of the stellar wind of Cygnus OB 2 No. 12 in terms of optical spectroscopy, visual and infrared photometry, and radio–flux measurements.

112.057 **Infrared pumping for $HC_5N$ in the circumstellar envelope of IRC + 10216.**
S. Deguchi, M. Uyemura.
Astrophys. J., Vol. 285, No. 1, p. 153 – 157 (1984).

Infrared excitation is investigated for the rotational transitions of $HC_5N$ with special emphasis on the circumstellar envelope of IRC +10216. The modeling of the circumstellar envelope shows that the rate for infrared pumping through the $\nu_7$ and $\nu_8$ bands exceeds the collisional excitation rate by a factor of 3. From the calculated emission–line intensities of high–$J$ and low–$J$ transitions, the abundance of $HC_5N$ is found to be $2.5 \times 10^{-7}$ per hydrogen in the circumstellar envelope of IRC +10216.

112.058 **Characteristics of the Fe II and C II emission in high–resolution *IUE* spectra (2300–3000 Å) of Alpha Orionis.**
K. G. Carpenter.
Astrophys. J., Vol. 285, No. 1, p. 181 – 189 (1984).

The author presents a study of Fe II and C II emission features in the 2300–3000 Å region of four high–resolution *IUE* spectra of α Ori obtained during the period 1978 April – 1982 November. He has identified the major Fe II features in these spectra and obtained measurements of their velocities, integrated fluxes, and asymmetries, and has searched for correlations of these data with various atomic parameters, including excitation energies and intrinsic line strength. The analysis of the C II lines yields estimates of the electron density in the chromosphere in the range $3.2 \times 10^7$ to $1.3 \times 10^8 \mathrm{cm}^{-3}$ and indicates that the chromosphere is quite extended, with the regions emitting C II reaching out to one–tenth of a photospheric radius.

112.059 **Be stars in binaries.**
H. A. Abt, O. Cardona.
Astrophys. J., Vol. 285, No. 1, p. 190 – 194 (1984).

The authors considered the known companions to 80 Be stars and 355 B stars listed in the Bright Star Catalogue in the range B1–B7 III–V and north of δ= -30°. They confirm the known near–absence of Be binaries with periods less than $10^{-1}$yr. For longer periods up to the limit of $10^4$AU of this survey, the Be and B stars do not differ in binary frequencies. This result implies that during pre–main–sequence contraction, the tidal braking in binaries wider than 0.5 AU was inadequate to prevent the formation of stars with nearly the break–up rotational velocities. The fraction of Be and B stars that have companions is higher in clusters and associations (38%) than among field stars (25%).

112.060 **Time variations of the Hα line profile from the core of Eta Carinae.**
M. T. Ruiz, J. Melnick, P. Ortiz.
Astrophys. J., Lett. Ed., Vol. 285, No. 1, p. L19 – L21 (1984).

Observations with high spectral resolution of the Hα line profile from the core of η Carinae show variations of the shape of the profile in time scales less than a year. A Gaussian analysis of the observed line profiles suggests the presence of several knots of material with high velocities with respect to a central object. The time variation of the profile can be interpreted as due to physical changes in the central object of η Carinae.

112.061 **The Hα emission envelope of EW Lacertae.**
W. J. Spiesman.
Publ. Astron. Soc. Pac., Vol. 96, No. 584, p. 795 (1984). Abstract. – See Abstr. 010.281.

112.062 **Optical searches for protoplanetary disks.**
B. A. Smith, R. J. Terrile.
Bull. Am. Astron. Soc., Vol. 16, No. 3, Part I, p. 702 – 703 (1984). Abstract. – See Abstr. 010.061.

112.063 **Short timescale periodicity in Hα emission from the main–sequence star HZ1883.**
G. W. Marcy, D. K. Duncan, R. D. Cohen.
Bull. Am. Astron. Soc., Vol. 16, No. 3, Part I, p. 726 (1984). Abstract. – See Abstr. 010.065.

112.064 **The remarkable kinematics of the WR wind–blown bubble RCW 58.**
L. J. Smith, M. Pettini, J. E. Dyson, T. W. Hartquist.
Mon. Not. R. Astron. Soc., Vol. 211, No. 3, p. 679 – 693 (1984).

The authors present high resolution, high signal–to–noise ratio, ultraviolet and optical observations of HD 96548, a WN8 star surrounded by the nebula RCW 58 which consists of stellar ejecta enclosed in a wind–blown bubble. The spectra exhibit blueshifted absorption components in a number of interstellar lines arising from a variety of ions. The velocities of the components are well correlated with their ionization potentials. The authors discuss the likely origin of such a correlation and argue that the absorption components arise behind the inwards–facing shock within the wind–blown bubble. Significant modifications to the standard bubble model are, however, required by the results.

112.065 **Un oggetto straordinario: Eta Carinae.**
M. Cerruti Sola.
Orione, Vol. 4, N. 3, p. 128 – 132 (1984).

112.066 **Chromospheric modelling in late–type dwarfs: 1. Quiescent objects.**
J. Beckman, L. Crivellari, B. Foing.
Messenger, No. 38, p. 24 – 28 (1984).

112.067 **Chromospheres of early–type stars and antiflare stars.**
A. F. Pugach.
Astrometr. Astrofiz., Vyp. 53, p. 31 – 34 (1984). In Russian.

It is shown that antiflare stars (i.e. rapid irregular variables with unpredictable Algol–like minima) are the best objects to observe chromospheres of early–type stars. At minimum light the photospheric radiation of antiflare stars is inhibited and indication of chromospheric features is registered in the continuum.

112.068 **μ Cephei.**
IAU Circ., No. 4008 (1984).

112.069 **Photoelectric observations of o And.**
B. Szeidl.
Inf. Bull. Variable Stars, No. 2575, 2 pp. (1984).

112.070 **o And. The overall behaviour of the shell.**
P. Koubský.
Inf. Bull. Variable Stars, No. 2584, 4 pp. (1984).

112.071 **Photometric observations of active chromosphere stars: II Peg, HD 175742, HD 199178.**
J. Kałuzny.
Inf. Bull. Variable Stars, No. 2627, 5 pp. (1984).

112.072 **Variations in the shell spectrum of the Be star EW Lac.**
Y.–l. Guo, H.–l. Cao.
Inf. Bull. Variable Stars, No. 2643, 2 pp. (1984).

112.073 **Structure of the expanding atmosphere of P Cygni.**
K. Avcioğlu.
Astrophys. Space Sci., Vol. 106, No. 2, p. 341 – 353 (1984).

With the aid of spectra taken in the years 1959 – 1968, a physical analysis of the atmosphere of P Cygni has been carried out and the motions of the atmosphere have been studied. The variations of radial velocities, the velocity progressions of Balmer and He I lines, the high rate of mass loss ($2 \times 10^{-5} M_\odot \mathrm{yr}^{-1}$), the features of the observed line profiles, especially that of H – K lines of Ca II and $D_1 - D_2$ lines of Na I confirm the conclusion of Van Blerkom (1978), concerning the assumption of an accelerating atmosphere for P Cygni. An atmospheric model based on the assumption of three envelopes, two of which accelerate gradually with two different velocity laws (up to 11.2 $r_c$), and the third of which accelerates rapidly with a standard velocity law (beyond 11.2 $r_c$) has been developed.

112.074 **On the wind structure of the Wolf–Rayet star in V444 Cyg.**
G. Koenigsberger.
Rev. Mex. Astron. Astrofis., Vol. 9, No. 2, p. 159 – 163 (1984).

The velocity law in the wind of the WN5 star in V444 Cyg is deduced from the photometric observations of Kron and Gordon (1943) and the light curve analysis of Kopal and Shapley (1946). A constant velocity wind region in the inner portions of the flow ($1.6 \lesssim r/R_{WR} \lesssim 3.1$), with a subsequent acceleration ($3.1 \lesssim r/R_{WR} \lesssim 7.7$), is implied. This could be consistent with a multiply non–monotonic wind model.

112.075 **Coronas estelares.**
J. Sahade.
Rev. Astron., Tomo 56, No. 228/229, p. 2 – 8 (1984).

112.076 **The depletion of pre–interstellar matter and an examination of Ni II oscillator strengths.**
J. A. Cardelli.
Astron. J., Vol. 89, No. 12, p. 1825 – 1832 (1984).

IUE observations of the B2.5 V star α Sco B have been used to analyze the absorption lines that arise in the circumstellar shell around the M1.5 Iab star α Sco A. In addition to resonant lines of Fe II, Si II, Mg II, Mn II, and S II, most of the resonant lines of Ni II are also present at strengths above 40 mÅ. A fit of these Ni II lines to a curve of growth, established from the Fe II, Si II, and Mn II lines, shows that the oscillator strengths of Kurucz and Peytremann (1975) are good to within relative errors of ±50%. An analysis of the elemental abundances in the expanding preinterstellar material around α Sco A provides an opportunity to examine the depletions that arise from dust formation as this material expands and cools. The formation of circumstellar dust is considered one of the primary sources for the population of dust in the interstellar medium (Woolf 1973; Field 1974).

112.077 **Photoelectric monitoring of a shell star in Perseus.**
Y.–h. Chu, Z.–w. Zhao, B. Wang.
Chin. Astron. Astrophys., Vol. 8, No. 4, p. 306 – 309 (1984). English translation of Acta Astron. Sin., Vol. 25, No. 2, p. 136 – 141 (1984).

The authors report on the 3–colour photoelectric monitoring of the shell star in Perseus, discovered in 1978. They suppose that the shell activity is now at a stable phase for a few years.

112.078 **Spatial distribution and luminosity function of OH/IR maser sources.**
Y. Tong, J. Sun, S.–d. Xie, X.–x. Yang.
Chin. Astron. Astrophys., Vol. 8, No. 4, p. 343 – 347 (1984). English translation of Acta Astrophys. Sin., Vol. 4, No. 2, p. 89 – 96 (1984).

The authors study the galactic distribution and luminosity function of OH/IR maser sources. All the selected OH/IR sources have optical or infrared identification. Most of them are associated with late–type Mira variables. Their derived density distribution shows a steep peak at a galactocentric distance of $R_0 \cong 7.5$ kpc and decreases rapidly at smaller and larger $R_0$. The authors also derive the luminosity function of the identified OH/IR maser sources from their distances, their detection probabilities, and their corrected OH radio peak flux densities. Some differences and relations between identified and unidentified OH/IR maser sources are discussed.

112.079 **Extreme UV emission from cool star outer atmospheres.**
M. Landini, B. C. Monsignori Fossi, F. Paresce,
R. A. Stern.
Space Telesc. Sci. Inst., Prepr. Ser., No. 28, 43 pp. (1984). To be published in Astrophys. J.

112.080 **X–ray coronae of late type stars: theoretical implications.**
J. H. M. M. Schmitt.
MPE Prepr., No. 29, p. 1 – 6 (1984). Paper presented at the Symposium on X–Ray Astronomy '84, Bologna, Italy, June 26 – 30, 1984.

112.081 **Infrared studies of early–type stars in OB associations. II. A discussion of density and temperature distributions in stellar winds.**
C. Bertout, C. Leitherer, O. Stahl, B. Wolf.
Inst. Astrophys. Paris, Pré–Publ., No. 73, 54 pp. (1984). To be published in Astron. Astrophys.

112.082 **Luminosities and mass loss rates of OH/$H_2O$ maser stars.**
P. F. Bowers, W. Hagen.
Astrophys. J., Vol. 285, No. 2, p. 637 – 647 (1984).

The authors discuss the results of a high–sensitivity ($\lesssim 1$ Jy) search for $H_2O$ maser emission from late–type stars. The survey has resulted in the discovery of 21 circumstellar $H_2O$ masers, all of which are associated with Mira variables. These data and published data are used to examine the range of $H_2O$ maser luminosities. A correlation is found between the $H_2O$ luminosity and the stellar mass loss rate, whereby the most luminous $H_2O$ masers are associated with Mira variables and supergiant stars which have high mass loss rates. OH and $H_2O$ maser luminosities are then compared for late-type stars. It is found that the ratio of $H_2O$ to OH luminosities also depends on the mass loss rate and has a range of at least three orders of magnitude, being systematically larger for nearby Mira variables with low mass loss rates.

112.083 **Circumstellar material around rapidly rotating B Stars. II. On the nature of ultraviolet shell lines in the spectra of Be and shell stars.**
W. R. Oegerle, R. S. Polidan.
Astrophys. J., Vol. 285, No. 2, p. 648 – 655 (1984).

Using high-resolution (FWHM ≈ 0.05 Å) ultraviolet spectral data obtained with Copernicus, the authors have investigated the nature of shell lines in Be and shell stars. Scans were obtained for various excited fine-structure lines of Fe III, Si II, and O I. It is found that Be stars, in general, do not display shell lines in the ultraviolet, contrary to the findings of previous work at lower resolution. The only Be stars that display strong UV shell lines are the "classical" shell stars, i.e., stars that also display shell lines in the optical region. The results presented here argue strongly in favor of a flattened disk structure for the shell line forming region.

112.084 **The distribution of HCN in the circumstellar envelope of IRC + 10216.**
J. H. Bieging, B. Chapman, W. J. Welch.
Astrophys. J., Vol. 285, No. 2, p. 656 – 667 (1984).

The authors present results of aperture synthesis mapping of the 88.632 GHz ($J = 1$–$0$) emission of HCN from the circumstellar envelope of the carbon star IRC +10216. The observations were made with the Hat Creek interferometer. The angular resolution of the maps is 9″, and the spectral resolution is 4.06 km $s^{-1}$. The authors compare the observations with model calculations for HCN excitation in the envelope. The model can reproduce the general features of the maps if one allows for deviations from a simple $r^{-2}$ density distribution in the envelope. The best fit density distribution implies departure from a steady mass outflow and is consistent with photodissociation of the HCN in the outer layers of the envelope.

112.085 **On the nature of the material surrounding Vega.**
D. A. Harper, R. F. Loewenstein, J. A. Davidson.
Astrophys. J., Vol. 285, No. 2, p. 808 – 812 (1984).

Observations of Vega at 193 μm indicate that the far-infrared emission from the circumstellar material discovered by *IRAS* may decline more rapidly than a Planck spectrum at wavelengths greater than 100 μm. This suggests that the emitting particles may be smaller than the millimeter-sized objects proposed by Aumann et al. (1984). Small grains would be driven from the stellar system by radiation pressure and a continuous supply of new particles would be required. The authors hypothesize that the small grains are ejected by sublimation of volatile material from larger comet-like bodies in partially coalesced preplanetary disk.

112.086 **Thermal and maser molecular lines from the envelope of IRC + 10216.**
Nguyen-Q-Rieu, V. Bujarrabal, H. Olofsson, L. E. B. Johansson, B. E. Turner.
Astrophys. J., Vol. 286, No. 1, p. 276 – 283 (1984).

The authors derive the physical conditions in the circumstellar shell of IRC +10216, using high-sensitivity spectra of the diatomic molecules SiS, SiO, $^{12}CO$ and $^{13}CO$, together with a model that takes into account the variation of the excitation temperature throughout the shell. The asymmetry of the observed line profiles (blueshifted wing systematically weaker than the redshifted one) is explained quantitatively in terms of absorption of the emission from the warm internal part of the shell by the cold external layers. Most of the molecular emission regions are confined within a radius smaller than $\sim 2 \times 10^{17}$cm. Detailed radiative transfer calculations for the SiS $\upsilon = 0, J = 1 \to 0$ maser emission are included.

112.087 **The hybrid spectrum of the LMC hypergiant R 126.**
F.-J. Zickgraf, B. Wolf, O. Stahl, C. Leitherer, G. Klare.
Mitt. Astron. Ges., Nr. 62, p. 246 (1984). Abstract. – See Abstr. 010.341.

112.088 **Partial redistribution in the winds of red giants.**
K. Hempe.
Mitt. Astron. Ges., Nr. 62, p. 247 (1984). Abstract. – See Abstr. 010.341.

112.089 **OH/$H_2O$ Maser in unidentifizierten IRAS-Quellen.**
D. Engels, H. J. Habing, F. M. Olnon, J. Schmid-Burgk, C. M. Walmsley.
Mitt. Astron. Ges., Nr. 62, p. 298 (1984). Abstract. – See Abstr. 010.341.

112.090 **Metallic line analysis of Pleione. II. 1977 – 1979.**
J.-i. Katahira, R. Hirata.
Publ. Astron. Soc. Jpn., Vol. 36, No. 3, p. 577 – 592 (1984).

The metallic shell lines of Pleione in the intermediate shell phase in 1977 – 1979 are studied. The shell spectrum has rapidly developed since late 1977. The profiles showed an asymmetry with a widened blue side and a steep red side. The strong metallic shell lines showed an abrupt increase in equivalent widths and central depths. The half-width turned to increase in the Fe II, Cr II, and Ti II ions. The shell began to expand in its outer part, maintaining an ionization structure similar to that in 1973 – 1976.

112.091 **Detection of narrow C IV and Si IV absorption features in spectra of stars within 200 pc of the sun.**
P. Molaro, J. E. Beckman, M. Franco, C. Morossi, M. Ramella.
NASA Conf. Publ., NASA CP-2345, p. 185 – 188 (1984). – See Abstr. 012.089 (IAU Colloq. No. 81).

Detection of narrow ($\delta\lambda < 0.5$ Å) absorption features in C IV at λλ1548 and 1550 have been made in the spectra of 4 late B dwarfs within 200 pc of the sun; the Si IV doublet at λλ1393 and 1403 shows up in two of them. The authors argue that it is difficult to account for the strengths, widths, shapes, and C IV/Si IV ratios in terms consistent with a circumstellar origin except possibly for an asymmetric C IV component in one star (HD 185037). The most probable source is "semi-torrid" gas (cf. Bruhweiler et al. 1980) in the 50,000K range forming the interfaces between cooler H I clouds and the ambient medium at coronal temperatures. The technique, using late B rapid rotators, is useful for LISM probing of this kind.

112.092 **On the mechanism of matter outflow from Wolf-Rayet stars.**
I. F. Malov, V. I. Shishov.
Astrofizika, Tom 21, Vyp. 1, p. 137 – 146 (1984). In Russian. English translation in Astrophysics, Vol. 21, No. 1.

A mechanism of an acceleration of matter in Wolf-Rayet stars due to radiation is considered. It allows a smooth transition from subsonic velocities to supersonic ones. On the basis of some assumptions luminosities of Wolf-Rayet stars and distances to them can be estimated from observed values of masses.

112.093 **Variability of the Hα emission in φ Per.**
P. S. Goraya.
Astrophys. Space Sci., Vol. 107, No. 2, p. 323 – 325 (1984).

Photoelectric spectrum scans of the Be shell star φ Per in the Hα region have been taken to investigate the variations of Hα emission lines. Definite variations with time-scales of a few minutes have been detected. It is hoped that hydromagnetic instability within the envelope is responsible for such variations.

112.094 **JHK photometry for some stellar maser sources.**
K.–p. Zhou, Y.–x. Hao, P.–s. Chen, Y. Zhang, H. Gao.
Astrophys. Space Sci., Vol. 107, No. 2, p. 373 – 380 (1984).

The infrared photometry in J–, H–, and K–bands for 24 stellar maser sources is given in this paper. Compared with normal M giants, some distribution characteristics of the sources are discussed in the (J–H)–(H–K)–diagram. Correlation between infrared colors and periods for the sources are investigated as well.

112.095 **The abrupt onset of a major ω Orionis mass loss episode.**
E. F. Guinan, D. P. Hayes.
Astrophys. J., Lett. Ed., Vol. 287, No. 1, p. L39 – L42 (1984).

Intensive visible wavelength photometric and polarimetric coverage of a major mass loss episode of the Be star ω Orionis is reported. Particularly striking is the sharp onset of pronounced continua light changes. The broad–band polarimetry shows pronounced changes which have a somewhat different temporal development.

112.096 **The temperature of C II emission–line formation regions in cool stars.**
A. Brown, K. G. Carpenter.
Astrophys. J., Lett. Ed., Vol. 287, No. 1, p. L43 – L46 (1984).

The authors have investigated the temperature of C II emission–line formation regions in the outer atmospheres of late–type giant and supergiant stars. A distinct dichotomy is seen in the C II λ2325/λ1335 ratio between coronal and noncoronal stars. It is found that C II emission from noncoronal giant and supergiant stars comes from regions with temperatures of 7000 – 9000K, with the mean temperature being ~8500K, whereas the C II emission from coronal stars likely comes from hotter regions.

112.097 **Near–infrared speckle interferometry of evolved stars and bipolar nebulae.**
H. M. Dyck, B. Zuckerman, C. Leinert, S. Beckwith.
Astrophys. J., Vol. 287, No. 2, p. 801 – 813 (1984).

The authors present near–infrared (1.65 – 4.8 μm) spatial visibility data for 16 late–type stars and bipolar nebulae. The majority of the sample is partially resolved and has linear radii averaging about $10^{15}$cm at 3.8 μm. The measured angular diameters at 3.8 μm are in good agreement with predictions of dust–shell boundaries in the models of Rowan–Robinson and Harris. The authors interpret these observations as general confirmation of the assumption that there is a dust–free zone which extends out to a few stellar radii in the circumstellar envelope. From the limited data available, it is found that the infrared angular diameters lie between those observed for the SiO and $H_2O$ masing regions but are a factor ~50 smaller than those observed for OH.

**Stellar chromospheres or the 2800 Å Mg II doublet in astrophysics.**
See Abstr. 003.095.

**Mass loss from red giants – a report on the June 1984 meeting at UCLA.**
See Abstr. 011.038.

**Stars, atmospheres, and shells: potential for high–resolution imaging.**
See Abstr. 036.084.

**Linear polarization of molecular lines at radio frequencies.**
See Abstr. 063.051.

**Linear polarization of astronomical masers and magnetic fields.**
See Abstr. 063.052.

**The infrared and radio spectrum of early–type stars with mass loss. I. The curve of growth method.**
See Abstr. 064.002.

**Recombination stellar wind model for the coronae of early–type stars.**
See Abstr. 064.005.

**The infrared and radio spectrum of early type stars with mass loss. II. Tables of theoretical curves of growth for IR and radio excess and gaunt factors.**
See Abstr. 064.006.

**The infrared and radio spectrum of early type stars with mass loss. III. The effect of coronae, hot blobs and clumping on the IR and radio spectrum.**
See Abstr. 064.013.

**Corona models tested with IUE and Einstein observations.**
See Abstr. 064.014.

**Radiative forces and the abundance of beryllium, magnesium and barium in HgMn stars.**
See Abstr. 064.021.

**Density inversion in a red giant envelope.**
See Abstr. 064.023.

**On the possibility of resonant electrodynamic coupling in the coronae of red dwarfs.**
See Abstr. 064.026.

**Corotating interaction regions in stellar winds.**
See Abstr. 064.028.

**Theory of astrophysical SiO masers.**
See Abstr. 064.037.

**Asymmetries in stellar Mg II *h* and *k* and Ca II H and K line profiles: discrepancies between Mg and Ca asymmetries.**
See Abstr. 064.038.

**Consequences of a chromospheric temperature gradient on the width of Hα in late–type giants.**
See Abstr. 064.042.

**CO line emission from the expanding circumstellar envelopes of red giants.**
See Abstr. 064.043.

**Effect of scattered radiation on the instability of hot star winds.**
See Abstr. 064.045.

**An analytical model of the SiO maser velocity spectrum.**
See Abstr. 064.054.

**Zur Nichtlinearität des Zusammenhangs zwischen chromosphärischer und koronaler Emission bei kühlen Sternen.**
See Abstr. 064.059.

**Koronaschalen und ihre Abhängigkeit vom Heizungsgesetz.**
See Abstr. 064.060.

**A time–dependent magnetohydrodynamic model of the circumstellar envelopes of Be stars.**
See Abstr. 064.064.

**Dimensional approach to mass loss: O– and B–type stars.**
See Abstr. 064.065.

**X–rays from accretion of red giant winds.**
See Abstr. 064.068.

**The interaction of a stellar wind with the surrounding interstellar medium.**
See Abstr. 064.069.

**10–micron observations of Magellanic Cloud supergiants.**
See Abstr. 113.015.

**Continuum energy distributions of some bright Be stars.**
See Abstr. 113.020.

**Be and shell stars observed with the thirteen–color photometric system. II.**
See Abstr. 113.046.

**The mass loss rate of the galactic hypergiant HD 152236 ($\zeta^1$ Scorpii) from optical and near infrared observations.**
See Abstr. 114.010.

**P Cygni and related profiles in the ultraviolet spectra of O–stars.**
See Abstr. 114.028.

**Ultraviolet and optical observations of metal deficient red giants and chromospheric models.**
See Abstr. 114.034.

**Search for chromospheric indicators in A dwarf stars.**
See Abstr. 114.037.

**The Mg II lines in A dwarf stars.**
See Abstr. 114.038.

**The search for superionization lines in A dwarf stars.**
See Abstr. 114.039.

**Observations of extensive variability in the ultraviolet spectrum of ξ Per.**
See Abstr. 114.045.

**Ultraviolet spectral morphology of the O stars. II. The main sequence.**
See Abstr. 114.134.

**The masses of early–type stars in the galactic halo determined from ultraviolet resonance line profiles.**
See Abstr. 115.010.

**X–ray properties of hot stars. I. On a dichotomy among O–type stars.**
See Abstr. 115.013.

**Microwave radiation from magnetic stars.**
See Abstr. 116.008.

**The appearance of polarization in radiation from hot stars due to the Faraday rotation effect as a possible method of determining stellar magnetic fields.**
See Abstr. 116.009.

**Age–rotation relationship for late–type main–sequence stars.**
See Abstr. 116.020.

**Chromospheric emission and rotation of the Hyades lower main–sequence.**
See Abstr. 116.023.

**Evidence for global oscillations in the K2 dwarf Epsilon Eridani.**
See Abstr. 116.024.

**Solar and non–solar activity in stars: the cosmic–ray connection.**
See Abstr. 116.030.

**The relation between stellar activity, rotation, convection, and implications for the combined study of stellar activity and oscillations from space.**
See Abstr. 116.031.

**Progress on rotational modulation studies of continuum and line fluxes.**
See Abstr. 116.032.

**Chromospheric, transition–region and coronal diagnostics for stellar magnetic activity.**
See Abstr. 116.033.

**The appearance of magnetic flux on the surfaces of the early main–sequence F stars.**
See Abstr. 116.037.

**Einstein observations of selected close binaries and shell stars.**
See Abstr. 117.089.

**An analysis of V861 Sco. II. The stellar wind.**
See Abstr. 117.118.

**High–resolution observations of bipolar mass flow from the symbiotic star HM Sagittae.**
See Abstr. 117.132.

**Dust shells in the peculiar objects CH Cygni and TX Canum Venaticorum.**
See Abstr. 117.149.

**The light and colour curves of the very active southern RS CVn–system HD 127535 in 1984.**
See Abstr. 117.197.

**A study of ultraviolet spectra of ζ Aur/VV Cep systems. VI. Excitation temperature within the wind of the K supergiant 32 Cyg.**
See Abstr. 119.032.

**The stellar wind in the B0 V–type eclipsing binary Y Cygni.**
See Abstr. 119.060.

**The many faces of Capella: a search for rotational modulations and a study of systematic velocities of emission lines in the ultraviolet.**
See Abstr. 120.015.

**Five colour photometry of the chromospherically active southern star HD 155555.**
See Abstr. 120.020.

**A new investigation of the T Tauri star RU Lupi. III. The wind model.**
See Abstr. 121.013.

**Stellar winds and molecular clouds: Herbig Be and Ae type stars.**
See Abstr. 121.021.

**High spatial resolution IR observations of young stellar objects: a possible disk surrounding HL Tauri.**
See Abstr. 121.026.

**Spectroscopic and photometric observations of the Herbig Be star Z Canis Majoris.**
See Abstr. 121.037.

**Characteristic time scales for variability in pre–main sequence stars.**
See Abstr. 121.038.

**Zum Charakter und Verhalten zirkumstellarer Hüllen von T–Tauri–Sternen.**
See Abstr. 121.042.

**Discovery of solar system–size halos around young stars.**
See Abstr. 121.052.

**The photometric behaviour of FG Sagittae in 1975 – 80: pulsations with increasing period.**
See Abstr. 122.001.

**Long–period variables in Baade's window: 1 – 12 μm photometry.**
See Abstr. 122.015.

**Chromospheric lines in red dwarf flare stars. II. EV Lacertae, EQ Pegasi A, and V1054 Ophiuchi.**
See Abstr. 122.020.

**Discovery and preliminary identification of two retrograde non-radial pulsation modes in the Be star μ Centauri.**
See Abstr. 122.098.

**Variable circumstellar obscuration of the carbon star R Fornacis.**
See Abstr. 122.117.

**The photometric behaviour of the young disk carbon star TW Horologii: determination of its physical characteristics.**
See Abstr. 122.127.

**The presence of $Si^{3+}$ and $C^{3+}$ in the wind of AB Aur.**
See Abstr. 122.142.

**Supernova progenitors and Be stars: stellar variability from a 21 century perspective.**
See Abstr. 122.197.

**A search for 183 GHz emission from water in late–type stars.**
See Abstr. 122.211.

**OH maser emission from warm variable stars.**
See Abstr. 122.220.

**Polarization properties and time variations of the SiO maser emission of Omicron Ceti (Mira).**
See Abstr. 122.221.

**Grain growth and destruction in novae.**
See Abstr. 124.001.

**Spectral variation and the expanding envelope of PU Vulpeculae.**
See Abstr. 124.101.

**The infra–red emission from a composite circumstellar dust shell heated by a supernova.**
See Abstr. 125.003.

**The interaction of supernovae with a circumstellar medium.**
See Abstr. 125.033.

**On the theory of cyclotron lines in the spectra of magnetic white dwarfs.**
See Abstr. 126.032.

**Winds from low–mass stars. Self–regulated star formation in Taurus, Ophiuchus, NGC 2264 and Orion.**
See Abstr. 131.050.

**Arc–second IR CCD observations of OH0739–14.**
See Abstr. 131.298.

**High velocities within extensive regions of the 30 Doradus nebula.**
See Abstr. 132.035.

**Nitrogen and helium enrichment in four Wolf–Rayet ring nebulae.**
See Abstr. 132.064.

**A search for OH and $H_2O$ maser emission from unidentified IRAS sources.**
See Abstr. 133.014.

**Dust around deeply embedded infrared sources.**
See Abstr. 133.015.

**Mass loss from the central star of NGC 3242.**
See Abstr. 134.020.

**Mass loss from the central star of NGC 3242.**
See Abstr. 134.037.

**The kinematics of the Wolf–Rayet ring nebula RCW 58.**
See Abstr. 134.048.

**Stellar coronas, X–rays, and Einstein.**
See Abstr. 142.011.

**Color anomalies and starspots in Hyades dwarfs.**
See Abstr. 153.018.

**Distribution of absolute energy of bright Be stars in Pleiades.**
See Abstr. 153.024.

**Main–sequence B stars with strong winds in the core of NGC 6231.**
See Abstr. 153.054.

**A curious, high velocity, component within a small shell in the Large Magellanic Cloud.**
See Abstr. 156.003.

# 113 Photometric Properties

113.001 **A photometric study of six nearby common–proper–motion pairs.**
J. A. R. Caldwell, J. H. Spencer Jones, J. W. Menzies.
Mon. Not. R. Astron. Soc., Vol. 209, No. 1, p. 51 – 58 (1984).

Six nearby common–proper–motion pairs are studied with new photometry. The constituents are found to be approximately one–half normal dwarfs and one–half subdwarfs, with an "ultrasubdwarf" (3.8 mag underluminous) and a degenerate star in the sample as well.

113.002 **CCD observations of photoelectric standard stars.**
A. R. Walker.
Mon. Not. R. Astron. Soc., Vol. 209, No. 1, p. 83 – 91 (1984).

Accurately timed short exposures of bright photoelectric standard stars with a CCD camera show that a photometric accuracy approaching 0.1 per cent is possible. The results obtained are indistinguishable from those achieved using conventional photoelectric photometry on the same stars. Observations of two faint photoelectric sequences are presented and the factors limiting the accuracy of the CCD photometry are discussed.

113.003 **Four *UBVRI* sequences near bright Virgo galaxies.**
D. Kilkenny, G. Malcolm.
Mon. Not. R. Astron. Soc., Vol. 209, No. 1, p. 169 – 174 (1984).

Four photoelectric sequences have been established around bright galaxies in the Virgo cluster. *UBVRI* observations of 37 stars, mostly in the range $9 < V < 15$, are given and serious discrepancies between the present data and earlier *UBV* work are described.

113.004 ***RGU*–photometry of a starfield in the direction to the Praesepe cluster.**
S. Karaali.
Astron. Astrophys., Suppl. Ser., Vol. 57, No. 1, p. 23 – 30 (1984).

A photographic three–colour photometry in the *RGU* system has been carried out in a field of 3.56 square degrees, containing a large fraction of the Praesepe cluster ($l = 205°9$, $b = +32°4$), with 1500 stars down to the limiting magnitude $G = 16^m2$. Space density functions have been determined for late–type giants and for main–sequence stars of the photometric populations I and II. The population I densities are compared with the ones for the solar neighbourhood at corresponding $z$–distances.

113.005 ***RGU* three–colour photometry of a starfield in anticentre direction (A5).**
R. Fenkart, S. Karaali.
Astron. Astrophys., Suppl. Ser., Vol. 57, No. 2, p. 225 – 232 (1984).
A starfield of ~0.3 square degrees in anticentre direction (~0°.5 north of NGC 2141) has been measured on *RGU* plates taken at the 48" Palomar Schmidt telescope with the purpose to determine the reddening, the density, and the luminosity functions in this direction for the ~1700 stars down to the limiting magnitude $G = 18^m$.

113.006 **Theoretical values of limb darkening coefficients in the RIJ system.**
A. A. Rubashevskij.
Astron. Tsirk., No. 1297, p. 6 – 8 (1983). In Russian.

113.007 **Photoelectric V ≠ $m_v$.**
R. H. Stanton.
J. Am. Assoc. Variable Star Obs., Vol. 12, No. 2, p. 78 (1983).
Abstract. – See Abstr. 010.021.

113.008 **Brightness distribution across the disk of μ Geminorum from an analysis of the data of its lunar occultation observations.**
M. B. Bogdanov, A. M. Cherepashchuk.
Astron. Zh., Tom 61, Vyp. 5, p. 944 – 955 (1984). In Russian. English translation in Sov. Astron., Vol. 28, No. 5.
The influence of the brightness distribution across the disk of a star on its lunar occultation curve is investigated. A method is described which allows to find the radial brightness distribution across a stellar disk directly from the occultation curve. From an analysis of the average occultation curve the brightness distribution across the disk of the red giant μ Gem is derived. The results are compared with the calculations of extended atmosphere models for late–type stars.

113.009 **Thirteen–color photometry of B–type stars.**
W. J. Schuster.
Rev. Mex. Astron. Astrofis., Vol. 9, No. 1, p. 53 – 64 (1984).
The author presents data fundamental for the study of B–type stars. Interstellar reddening ratios, intrinsic colors, comparison stars, reddening–free indices, and calibrations for MK spectral type, effective temperature, and absolute visual magnitude are given. In particular the author discusses briefly the use of 13C photometry for measuring stellar distances and for studying intrinsic excesses of Be and shell stars.

113.010 **Near infrared photometry and free–free fluxes of 12 southern OB supergiants.**
J. A. López, J. R. Walsh.
Rev. Mex. Astron. Astrofis., Vol. 9, No. 1, p. 95 – 97 (1984).
Near infrared photometry is presented for 12 OB supergiant stars in the southern hemisphere. The presence of extended atmospheres is investigated by means of an analysis of possible color–excess in the *K* and *L* bands, expected to be caused by free–free radiation in the envelope. Most of the studied objects show significant color–excess.

113.011 ***UBV* photometry of stars whose positions are accurately known.**
T. Oja.
Astron. Astrophys., Suppl. Ser., Vol. 57, No. 3, p. 357 – 359 (1984).
*UBV* photometry is presented for stars of the AGK3R and NPZT catalogues north of declination +55°.

113.012 ***RGU*–photometry of a starfield close to M67 and space density behaviour in the anticentre–northern galactic meridian.**
R. P. Fenkart, S. Karaali.
Astron. Astrophys., Suppl. Ser., Vol. 57, No. 3, p. 419 – 425 (1984).
A photographic three–colour photometry in the *RGU* system in a starfield near M67 of 1.70 square degrees containing 759 stars down to the limiting magnitude of $G = 16^m.5$, with no interstellar absorption, led to logarithmic space density functions for the photometrically separable populations of late–type giants and main–sequence stars of populations I and II. They are all compared with the corresponding functions in Karaali's (1984) Praesepe field. The present field was investigated with the purpose to check the isodensity pattern in the anticentre–northern quarter of the galactic meridian.

113.013 **Four–colour and Hβ photometry of B stars in Fehrenbach and Burnage radial velocity fields at *b* ~ –30°.**
W. Tobin.
Astron. Astrophys., Suppl. Ser., Vol. 57, No. 3, p. 427 – 434 (1984).
*uvbyβ* observations are presented for 39 stars classified as B–type in seven 4° × 4° fields studied by Fehrenbach and Burnage at *b* ~ –30°. About half the stars are photometrically little–evolved B stars; most of the rest are intermediate A stars. Intrinsic colour and absolute magnitude calibrations are applied. Maps of reddening along lines–of–sight indicate that little reddening occurs within 100 pc of the Sun, that essentially all reddening has occurred by $|z|$ ~ 150 – 200 pc, and that the total reddening is essentially consistent with values given by the Burstein and Heiles H I/GC method.

113.014 **Preliminary *uvby* calibrations for G and K type dwarf stars.**
E. H. Olsen.
Astron. Astrophys., Suppl. Ser., Vol. 57, No. 3, p. 443 – 466 (1984).
Four–colour *uvby* photometry for several hundred late type dwarf stars of types G, K and M is discussed. Mean values of photometric indices are given for MK spectral types between F8/G0V and M2V. Preliminary standard relations between the four–colour indices are derived. Based on parallaxes and results from high–dispersion spectroscopic analyses calibrations of the observed indices in terms of $M_v$, log $T_e$ and [Fe/H] have been derived. The calibrations are valid for class V stars of all populations between G0 and M2. The possible contribution of a "fourth–parameter" variation to the mean errors of the calibrations is briefly discussed. This fourth parameter could be the intrinsic helium abundance of the stars.

113.015 **10–micron observations of Magellanic Cloud supergiants.**
I. S. Glass.
Mon. Not. R. Astron. Soc., Vol. 209, No. 4, p. 759 – 763 (1984).
10 μm broad–band (*N*) observations have been made of several Magellanic Cloud supergiants. The brightest F/G types show strong excesses in 2 out of 5 cases. The S Dor type variable R 71, previously interpreted in the infrared as a free–free emitter, has strong 10 μm radiation probably from a cool dust shell. S Dor itself does not have such an excess. Those B supergiants known already to have dust shells are easily detected. The most luminous cloud member at *B* and *V*, R76, was not detected, nor was R81, a Be star with a high mass–loss rate. R84 (Wolf–Rayet with late–type companion) has a strong excess. R94, the brightest LMC star at *K*, was also not detected.

113.016 **On the non–variability of Capella.**
K. Krisciunas.
I.A.P.P.P. Commun., No. 16, p. 10 – 11 (1984).

113.017 **Faint spectrophotometric standard stars for large optical telescopes. I.**
A. V. Filippenko, J. L. Greenstein.
Publ. Astron. Soc. Pac., Vol. 96, No. 581, p. 530 – 536 (1984).
The authors present spectral energy distributions over the range λλ3300 – 10000 for a set of five spectrophotometric standards four of which are white dwarfs. These stars are faint ($V$ ~ $15^m.5$), equatorial, and exhibit nearly featureless continua, so they can be observed without neutral density filters to calibrate the response of photon–counting spectrographs on large telescopes in both hemispheres.

113.018 **Is 109 Virginis a variable star.**
A. G. D. Philip, D. S. Hayes.
Publ. Astron. Soc. Pac., Vol. 96, No. 581, p. 546 – 548 (1984). = Van Vleck Obs. Contrib. No. 209.
109 Vir (HR 5511) is a standard star in the four-color system and a spectrophotometric standard. Recently Taylor has raised questions concerning the stability of 109 Vir. The authors present photometric and spectrophotometric data indicating that over the time period 1969 – 80 no evidence can be found for any variations greater than the rms error of their observations.

113.019 ***RGU* photometry of a field in the Carina region near the cluster IC 2581.**
E. J. Alfaro, J. M. García-Pelayo.
Astron. Astrophys., Suppl. Ser., Vol. 58, No. 1, p. 121 – 130 (1984).
A field of 0.21 sq. degrees containing 2099 stars in the Carina region near the cluster IC 2581, has been measured on 14 plates in the *RGU* system. The reddening is caused by a thin absorbing cloud at 770 pc with $E(G-R) = 0^m23$, as well as an interstellar absorption free zone followed by a moderate gradient, which increases with distance. The density gradients corresponding to late-type giants and main sequence stars of the same absolute magnitude do not display the same behaviour. The results seem to corroborate the existence of external spiral features between 12 and 13 kpc. The distribution of F0 – F5 stars seems to confirm predictions of the theory of density waves.

113.020 **Continuum energy distributions of some bright Be stars.**
P. S. Goraya.
Astron. Astrophys., Vol. 138, No. 1, p. 19 – 24 (1984).
The absolute flux distributions of seven bright Be stars, in the wavelength range λλ350 – 750 nm is presented. The observed energy distribution curves are compared with Kurucz (1979) model atmospheres to estimate their effective temperatures. Some peculiarities in the energy distributions and the evolutionary stages of the Be stars are discussed.

113.021 **VBLUW photometry of some halo field giants.**
B. Nelles, T. Richtler, W. Seggewiß.
Astron. Astrophys., Vol. 139, No. 1, p. 220 – 226 (1984).
Walraven photometry has been obtained for a sample of 43 metal-deficient halo giants and 35 stars with known values of $T_{eff}$, log *g* and [Fe/H]. A calibration of Walraven photometric indices in terms of $\theta = 5040\text{K}/T_{eff}$ and [Fe/H] is established. The metallicities derived for the sample of halo giants are in good agreement with those of Kraft et al. (1982). In contrast to the metallicities derived by Bond (1980), the authors find a considerably lower number of stars with [Fe/H] < -2.7 dex. The calibration yields a metallicity of -2.9 dex for CoD-38°245, which is in contrast to the extremely low value of -4.6 dex based on a spectroscopic abundance determination by Bessell and Norris (1981). The importance of this star for the calibration of photometric metallicities in the domain of extreme metal deficiency is discussed.

113.022 **Photometric standard stars for *L'* and *M* filter bands.**
W. M. Sinton, W. C. Tittemore.
Astron. J., Vol. 89, No. 9, p. 1366 – 1370 (1984).
The authors present *L'* (3.8 μm) and *M* (4.8 μm) photometry of 20 primary and 21 secondary standard stars that are sufficiently faint for large telescope observation. The estimated uncertainties of the magnitudes of the primary standards is 0.01 at *L'*, 0.02 at *M*, and 0.015 for the *L'-M* color. A color transformation between the authors' *L'* magnitudes and the *L* magnitude of Elias et al. (1982) is given.

113.023 **The photometric variability of B stars: a general approach.**
C. Waelkens, F. Rufener.
Hvar Obs. Bull., Vol. 7, No. 1, p. 29 – 53 (1983). – See Abstr. 012.037.
In this paper, the authors report the results of the new observations of variable B-stars that were performed using the Geneva Photometric System.

113.024 **Correlation of the visual surface brightness parameter with *uvby* β photometry.**
T. T. Moon.
Mon. Not. R. Astron. Soc., Vol. 211, No. 1, p. 21P – 24P (1984).
The Barnes-Evans relation between a visual surface brightness parameter, $F_V$, and colour index is defined for the $(b-y)_0$ index of *uvby* β photometry, the relation being applicable for spectral types ranging from about B0 to about G0.

113.025 **The photometric variability of solar-type stars. IV. Detection of rotational modulation among Hyades stars.**
G. W. Lockwood, D. T. Thompson, R. R. Radick, W. H. Osborn, W. E. Baggett, D. K. Duncan, L. W. Hartmann.
Publ. Astron. Soc. Pac., Vol. 96, No. 583, p. 714 – 722 (1984).
High-precision differential photometric observations of 32 Hyades stars, spectral type F4 V to K1 V, were obtained at Lowell Observatory over a five-month season during late 1982 and early 1983. Contemporaneous Ca II H+K emission flux measurements for 30 of these stars were made at Mount Wilson Observatory. Fifteen (possibly 18) of these stars, all later than type F7 V, were found to be photometrically variable. Highly significant correlations between continuum and Ca II emission variations were found in several cases. The photometric variations for nine of the stars were periodic. The authors have interpreted this as rotational modulation, since the measured periods agreed well with rotation periods predicted from measurements of the mean chromospheric emission fluxes for these stars.

113.026 **Luminosity classification with the Washington system.**
D. Geisler.
Publ. Astron. Soc. Pac., Vol. 96, No. 583, p. 723 – 733 (1984).
The DDO 51 filter has been added to the Washington photometric system. The strong surface gravity sensitivity of the Mg I "b" triplet and MgH bands which it monitors allows the system to easily discriminate between dwarfs and giants of late G and K spectral type. The system is especially suited as a membership criterion for abundance studies of distant cluster giants. The Mg index is insensitive to surface gravity variations among G giants. The metallicity sensitivity among giants is also relatively weak. Population I and II giants can be distinguished but no further differentiation is evident.

113.027 **Photometry of K giants in the nuclear bulge of the Galaxy.**
J. A. Frogel, A. E. Whitford, R. M. Rich.
Astron. J., Vol. 89, No. 10, p. 1536 – 1542 (1984).
For 52 K and early-M giants in Baade's Window at a galactic latitude of -3°9, the authors present *RIJHK* photometry. The location of these galactic bulge giants on various color-color plots is markedly different from both globular cluster and solar neighborhood giants. However, their locations in infrared color-magnitude diagrams are not obviously consistent with the interpretation that they are old and metal rich. The authors tentatively conclude that blanketing by an as yet unidentified source (or sources) has a significant effect on the energy distributions of these stars which spectroscopically have been shown to be metal rich.

113.028 **Intermediate band, Hβ and *RI* photometry of a large sample of stars unbiased with respect to their motion. I. the F-type stars.**
O. J. Eggen.
Astron. J., Vol. 89, No. 10, p. 1606 – 1626 (1984).
The intermediate band and Hβ observations for the stars earlier than about type G2 in a sample of 1000 stars, selected without bias, with respect to their motion, are presented and briefly discussed on the basis of their luminosity, metal abundance, and space motion. The sample is divided between stars within 10° of the South Galactic Pole and stars with photographic magnitude 8.5 – 8.6, inclusive, in the *Henry Draper Catalogue*. The methods used to determine reddening, luminosity, and metal abundance are outlined and the space motions are discussed in some detail.

113.029 **Limb darkening laws in photometric systems.**
A. A. Rubashevskij.
Akad. Nauk Ukrain. SSR, Glav. Astron. Obs., Inst. Probl. Materialovedeniya, Kiev, Prepr., No. 20, 65 pp. (1984). In Russian.

113.030 **A photoelectric sequence in the region of the globular cluster NGC 5986.**
G. Alcaino.
Astron. Astrophys., Vol. 139, No. 2, p. 549 – 550 (1984).
A *UBV* photoelectric sequence of 22 stars in the magnitude range $8.6 < V < 16.6$ is presented in the region of the galactic globular cluster NGC 5986. This sequence supplements that of 6 bright stars within $8.6 < V < 11.4$ given by White (1971) and that of 11 stars within $11.3 < V < 16.4$ from Harris et al. (1976).

113.031 **Verification of the calibration of Harold L. Johnson's 13–color photometry.**
G. C. Augason.
Publ. Astron. Soc. Pac., Vol. 96, No. 584, p. 783 – 784 (1984). Abstract. – See Abstr. 010.281.

113.032 **Short term photometric variability in southern Be stars.**
C. Stagg.
J. R. Astron. Soc. Can., Vol. 78, No. 5, p. 207 (1984). Abstract. – See Abstr. 010.401.

113.033 **An additional note on the IR–excess of the helium–variable stars.**
A. P. Odell, M. Lebofsky.
Astron. Astrophys., Vol. 140, No. 2, p. 468 – 469 (1984).
The infrared fluxes of two helium–variable stars in Orion ($\sigma$ Ori E = HD 37479 and HD 37017) have been reobserved through 4.8 µm on 16 January 1982. The *M*–band excesses previously reported are not confirmed by these observations.

113.034 **Stability and accuracy of the *uvby* system.**
J. Manfroid.
Astron. Astrophys., Vol. 141, No. 1, p. 101 – 104 (1984).
In order to evaluate the transformation errors for stars of various types and classes, the author has simulated observations made with standard and non–standard equipments. Large deviations which cannot be corrected are observed mainly for objects outside the range $0.0 < b\text{–}y < 0.4$ and in the *u* filter. The rectangular passbands have many advantages however. They can be very stable and are accurately reproducible. Hence the problem of nontransformable outputs may not exist between such systems. Since many *uvby* observations are to be collected with such equipments, it could be advantageous to define a new standard "rectangular" *uvby* system along with the conventional one.

113.035 **The Böhm–Vitense gap in the Geneva photometric system.**
G. Jasniewicz.
Astron. Astrophys., Vol. 141, No. 1, p. 116 – 126 (1984).
A gap among A – F stars is searched on the main sequence of different stellar samples in the photometric space *M* defined by the four colour indexes U–B1, B1–B2, B2–V1 and V1–G of the Geneva photometric system. A gap is noticed for the field and the eight young open clusters investigated when selected stars are not visual or spectroscopic binaries, and not chemically peculiar. It is shown that this gap is induced by the existence of two differently oriented star branches in *M*, the so–called radiative and convective branches. The gap is suggested not to be only a mere artefact of having previously excluded Am and Ap stars but partly due to an abrupt onset of convection in late A stars as proposed by Böhm–Vitense (1970).

113.036 **Relations between the β index and Geneva photometry. The B–type stars.**
N. Cramer.
Astron. Astrophys., Vol. 141, No. 1, p. 215 – 222 (1984).
The relations linking Geneva photometry with Strömgren's β index have been established for the O, B and early A–type stars via the reddening–free *X,Y,Z* parameter space of the Geneva system. The $\beta(X,Y)$ estimator, when compared with the line–defined β index, creates a new type of colour index which is highly sensitive to departures from a normal $H_\beta$ line width versus spectral energy distribution relationship. As a first application, the author examines the consistency of rotating stellar models with observations, the internal consistency of model atmospheres by means of synthetic photometry and the effects of emission in the hydrogen lines on such a composite index.

113.037 ***JHK* photometry of Be stars.**
H. C. Bhatt, N. M. Ashok, T. Chandrasekhar, P. S. Goraya.
Astron. Astrophys., Suppl. Ser., Vol. 58, No. 3, p. 685 – 686 (1984).
*JHK* photometry of 23 Be stars selected from the list of Harmanec et al. (1983) are reported. For 14 Be stars the observations reported in the present work are new.

113.038 **Evidence for the variability of HD 5210: a comparison star for HD 5303.**
J. L. Innis, D. W. Coates, K. Thompson.
Inf. Bull. Variable Stars, No. 2544, 3 pp. (1984).

113.039 **On the photometric variations of HD 28843 and HD 29009.**
J. Manfroid, G. Mathys.
Inf. Bull. Variable Stars, No. 2551, 4 pp. (1984).

113.040 **Photographic observations of AT Cnc in 1983/84.**
M. Huruhata.
Inf. Bull. Variable Stars, No. 2556, 2 pp. (1984).

113.041 **On the variability of HD 205328.**
I. Lelatko.
Inf. Bull. Variable Stars, No. 2568, 2 pp. (1984).

113.042 **HD 126200 – a suspected eclipsing binary?**
J. B. Srivastava, C. D. Kandpal.
Inf. Bull. Variable Stars, No. 2574, 2 pp. (1984).

113.043 **Additional photometry of the Of–type variable HD 167971.**
D. Forbes.
Inf. Bull. Variable Stars, No. 2605, 5 pp. (1984).

113.044 **A search for low–harmonic pulsation in the Ap star HD 184905.**
T. J. Kreidl.
Inf. Bull. Variable Stars, No. 2607, 5 pp. (1984).

113.045 **RR Ari: not a variable star.**
E. Poretti.
Astrophys. Space Sci., Vol. 106, No. 1, p. 201 – 203 (1984).
New *V* standard measures of RR Ari are presented and the star is found not to be variable. The light of the star $\lambda$ Ari B ≡ CSV 5954 seems also to be constant.

113.046 **Be and shell stars observed with the thirteen–color photometric system. II.**
W. J. Schuster, J. Guichard.
Rev. Mex. Astron. Astrofis., Vol. 9, No. 2, p. 141 – 151 (1984).
Individual 13–color photometric measurements for 89 Be and shell stars observed at the observatory at San Pedro Mártir are presented. Observing procedures, reduction techniques and transformation problems for several colors are discussed. 22 stars show definite photometric variability using this new data.

113.047 **Weak–lined stars in the Michigan spectral survey. I. A–, F–, and G–type stars.**
O. J. Eggen.
Astron. J., Vol. 89, No. 12, p. 1878 – 1886 (1984).
Intermediate–band and Hβ photometry of 89 stars of type earlier than about G2 and classified as weak–lined objects from spectra with the Curtis–Schmidt telescope yields 55 disk and 34 halo stars. The halo stars include 21 objects near the main–sequence turnoff, nine horizontal–branch stars, all but two of

which are on the red side of the branch with one suprahorizontal branch star, and four blue stragglers. The procedures for luminosity and abundance determinations are tested on the members of Kapteyn's star group which contains four very short period Cepheids (VSPC), X Ari, U Lep, SU Dra, and ST Leo, and two blue stragglers, +25°1981 and SX Phe. The group members near the main-sequence turnoff and the VSPC both give $P$[Fe/H] = –1.7 for the group, and the mean luminosity of the VSPC is +0.75 mag.

113.048 **Photometry of faint blue stars. VI. Some "very definitely blue" PHL stars.**
D. Kilkenny.
Mon. Not. R. Astron. Soc., Vol. 211, No. 4, p. 969 – 972 (1984).
Photometry on the *uvby* system is presented for 56 faint blue stars, mostly from the Palomar–Haro–Luyten "very definitely blue" list. The majority of these stars are hot subdwarfs.

113.049 **Standardisation of broad band photometry of equatorial standards.**
A. W. J. Cousins.
S.Afr. Astron. Obs., Circ., No. 8, p. 59 – 67 (1984).
The published photometry of the fainter equatorial standards for broad band photometry is not homogeneous. The U–B systems differ from one another. V from Crawford et al. (1971) and B–V from Moffett and Barnes (1979) and from Landolt (1983) need colour corrections to reduce them to the Johnson system. The R and I photometry is in three different systems and there are zero point errors present. The present note is mainly concerned with B and V. The external s.e. of V and B–V are $\pm 0^m.004$ or $\pm 0^m.005$ for the better series.

113.050 **UBV photometry of equatorial stars.**
A. W. J. Cousins.
S.Afr. Astron. Obs., Circ., No. 8, p. 69 – 84 (1984).
260 stars, nearly all between ±10° declination, have been observed with the photometer used for measuring E region standard stars. The brighter stars have been used to compare the E region system with Johnson's UBV system, showing that there is good agreement for V and B–V but a systematic difference $((U-B)_J-(U-B)_E = 0.015(B-V)-0.003)$ for U–B. The internal s.e. is near $\pm 0^m.002$ for both colours and magnitudes.

113.051 **Revised RGU photometry of an anticentre field around NGC 2158.**
L. Topaktas.
Rev. Fac. Sci. Univ. Istanbul, Ser. C, Vol. 46, p. 1 – 15 (1981). = Publ. Istanbul Univ. Obs., No. 121 (1984).
A galactic anticentre field of 0.4 square degrees with 1506 stars around globular cluster NGC 2158 discussed by Fenkart (1964) has been reduced to the present RGU system and rediscussed. The interstellar reddening, the density functions for different groups of stars and the luminosity functions have been redetermined.

113.052 **RGU photometry of a field in Vela II.**
A. Marsoğlu.
Rev. Fac. Sci. Univ. Istanbul, Ser. C, Vol. 46, p. 151 – 163 (1981). = Publ. Istanbul Univ. Obs., No. 125 (1984).
2199 stars down to a limiting magnitude of G = 16.78 have been measured in the RGU system in a field of 0.309 square degrees in Vela II. The interstellar reddening function, the density functions of different groups of stars and the luminosity functions have been determined.

113.053 **Photometric data for Be stars.**
B. Hauck, M. Mermilliod.
Bull. Inf. Etoiles Be, No. 10, p. 7 – 10 (1984).

113.054 **UBV observations of EW Lac during 1982 and 1983.**
L. H. Iliev, B. Z. Kovachev, N. A. Tomov.
Bull. Inf. Etoiles Be, No. 10, p. 11 – 28 (1984).

113.055 **A search for ultraviolet-excess objects. II.**
M. Kondo, T. Noguchi, H. Maehara.
Ann. Tokyo Astron. Obs., Second Ser., Vol. 20, No. 2, p. 130 – 189 (1984).
The second catalogue of stellar UV-excess objects (KUV) searched with the Kiso Schmidt Telescope is presented. It consists of celestial positions, brightnesses, color indices, and identifications of 598 objects detected in this search. Finding charts of the KUV's are given, including those listed in the first catalogue. A few statistical properties are presented concerning their color indices and brightnesses.

113.056 **Narrow-band photometry of the magnetic star 53 Cam.**
N. S. Polosukhina, V. I. Burnashev, V. P. Malanushenko.
Rep. Ser., Dep. Phys. Sci., Univ. Turku, No. FTL-R68, p. 19 – 20 (1984). – See Abstr. 012.079.

113.057 **An absolute photometric system at 10 and 20 μm.**
G. H. Rieke, M. J. Lebofsky, F. J. Low.
Prepr. Steward Obs., No. 559, 27 pp. (1984).
Two new direct calibrations at 10 and 20 μm are presented, in which terrestrial flux standards are referred to infrared standard stars. These measurements give both good agreement and higher accuracy when compared with previous direct calibrations. As a result, the absolute calibrations at 10 μm and 20 μm have now been determined with accuracies of 3% and 8% respectively. A variety of absolute calibrations based on extrapolation of stellar spectra from the visible to 10 μm are reviewed. The photometric system at 10 and 20 μm is updated to reflect the new absolute calibration, to base its zero point directly on the colors of A0 stars, and to improve the accuracy in the intercomparison of the standard stars.

113.058 **Absolute calibration of photometry at 1 through 5 μm.**
H. C. Campins, G. H. Rieke, M. J. Lebofsky.
Prepr. Steward Obs., No. 560, 16 pp. (1984).
The solar analog method is used to determine the absolute calibration of photometry in the J, H, K, L, and M bands to an accuracy of 3% (5% at M). This calibration agrees well with the direct calibration obtained by Blackwell et al. (1983); this agreement gives confidence that the calibration is known to an accuracy of 2 to 3%. The agreement also implies that the behavior of solar-type stars is well understood in the infrared and that such stars make reliable standards for calibration of non-standard photometric bands and for planetary reflectance studies. Both methods indicate that Vega is slightly brighter than indicated by atmospheric models.

113.059 **Photometrie und Spektrophotometrie schwacher blauer sonnennaher Sterne.**
I. Bues, G. Rupprecht.
Mitt. Astron. Ges., Nr. 62, p. 263 – 264 (1984). – See Abstr. 010.341.

113.060 **Die Verwendung photometrischer Systeme intermediärer Bandbreite zur Bestimmung physikalischer Parameter von Sternen.**
B. Nelles.
Mitt. Astron. Ges., Nr. 62, p. 289 – 290 (1984). – See Abstr. 010.341.

113.061 **Stellar magnitude – photon flux relationship in the Vilnius photometric system.**
D. A. Ralys.
Vilniaus Astron. Obs. Biul., Nr. 67, p. 28 – 42 (1984). In Russian.
The absolute energy distribution curve in the spectrum of Vega is constructed in the wavelength region of 130 – 1250 nm. The photon flux of Vega is estimated for bandpasses of the Vilnius photometric system using spectral transmission curves of the atmosphere and of other light absorbing components of the photoelectric photometer. The photon fluxes for luminosity V stars of different spectral types and for the night sky are calculated.

113.062 **Photoelectric photometry of stars near the north galactic pole in the Vilnius system. I. Kiev Area 112.**
S. Bartašiūtė.
Vilniaus Astron. Obs. Biul., Nr. 68, p. 33 – 41 (1984). In Russian.

Seven–color photoelectric photometry in the Vilnius system is presented for 78 stars in Kiev Area 112 near the north galactic pole. The survey area of 1°20′ diameter is centered at $\alpha_{1950} = 13^h40^m.2$, $\delta_{1950} = +36°27'$. It represents one of the 21 northern areas along the main meridional section of the Galaxy for which absolute proper motions relative to galaxies are measured in Kiev. The photoelectric catalog includes all the stars of the area down to $V = 12.8$.

113.063 **Possible light variation of comparison star for BY Dra: BD +51°2410.**
K. Oláh, M. A. Soliman.
Inf. Bull. Variable Stars, No. 2648, 4 pp. (1984).

**Catalogue of stellar magnitudes, colour indices, spectral classes of stars in the direction to the cluster NGC 6633.**
See Abstr. 002.016.

**Documentation for the machine–readable version of the *ANS Ultraviolet Photometry Catalogue of Point Sources* (Wesselius et al. 1982).**
See Abstr. 002.041.

**An up–to–date catalogue of the Vilnius UPXYZVTS photometry.**
See Abstr. 002.052.

**Colours of the stars.**
See Abstr. 003.011.

**The Geneva photometry.**
See Abstr. 013.007.

**Ground–based long–term photoelectric photometry as support for space observations of variables.**
See Abstr. 013.057.

**Photometric observing campaign. Progress Report No. 7.**
See Abstr. 013.073.

**An infrared photometric and spectrometric system in San Pedro Martir.**
See Abstr. 034.012.

**Extending magnitude standards to faint stars with a sub–beam prism.**
See Abstr. 034.154.

**Astronomers' problems with the calibration of Schmidt plates.**
See Abstr. 036.105.

**The Malin–effect in stellar photometry.**
See Abstr. 036.107.

**Some applications of signal–to–noise in photographic stellar photometry.**
See Abstr. 036.108.

**Stellar photographic photometry using a double wedge technique with a focal reducer system.**
See Abstr. 036.109.

**Spectrophotometry of cool stars from objective prism spectra.**
See Abstr. 036.110.

**Should one measure radial velocity or brightness fluctuations?**
See Abstr. 036.139.

**The accuracy limit of ground–based stellar photometry.**
See Abstr. 036.145.

**Ground–based solar and stellar intensity measurements: practical limits of accuracy.**
See Abstr. 036.146.

**Imaging with a single detector: the Wyoming approach.**
See Abstr. 036.175.

**Short–period variability of the Be star λ Eri.**
See Abstr. 112.028.

**Infrared photometric studies of Be stars.**
See Abstr. 112.051.

**The mass loss rate of the galactic hypergiant HD 152236 ($\zeta^1$ Scorpii) from optical and near infrared observations.**
See Abstr. 114.010.

**The A0 stars.**
See Abstr. 114.020.

**Classification of Ap stars HR 830 and 21 CVn.**
See Abstr. 114.063.

**Spectroscopic tests of photoelectric stellar classification of abnormal stars.**
See Abstr. 114.071.

**Spectroscopic and photometric observations of the suspected binary star HD 123058.**
See Abstr. 114.102.

**The ultraviolet variability of early–type supergiants.**
See Abstr. 114.112.

**Studies of late–type dwarfs. VI. Identification of Population II main–sequence stars at $M_v = +14$.**
See Abstr. 114.121.

**Carbon monoxide band intensities in M giants.**
See Abstr. 114.127.

**The distribution of absorbing matter and of OB stars in the region of the association Per OB1.**
See Abstr. 114.130.

**A redetermination of the Barnes–Evans relation for surface flux in the V band.**
See Abstr. 115.012.

**Studies in Be–star variability. I. A remarkable similarity of the rapid periodic light variations of EM Cep, σ Ori E, and possibly LQ And.**
See Abstr. 116.012.

**Subdwarf companions for HD 36629 and HD 81307.**
See Abstr. 117.083.

**On a possibility of finding nonlinearity in limb darkening from the light curves of classical eclipsing systems.**
See Abstr. 119.013.

**Spectroscopic and photometric observations of the Herbig Be star Z Canis Majoris.**
See Abstr. 121.037.

**Photographic colorimetry of stellar flares in the Pleiades and Orion aggregates. II.**
See Abstr. 122.053.

**IR observations of OH/IR stars.**
See Abstr. 122.081.

**A frequency analysis of the 6.8–min oscillations in the cool magnetic Ap star α Cir.**
See Abstr. 122.088.

**The photometric behaviour of the young disk carbon star TW Horologii: determination of its physical characteristics.**
See Abstr. 122.127.

**Microvariability in the HR diagram and two examples of very low amplitude stars.**
See Abstr. 122.192.

**Photoelectric B and V band photometry of comparison stars for eight Mira–type variable star fields.**
See Abstr. 122.194.

**The halo population G and K subdwarfs.**
See Abstr. 126.027.

**Infrared extinction in the Small Magellanic Cloud.**
See Abstr. 131.193.

**Interstellar reddening in the direction of the globular cluster M71. II.**
See Abstr. 131.234.

**Local interstellar extinction with an emphasis on uvbyβ results.**
See Abstr. 131.258.

**Polarimetric and new photometric observations in the NGC 6334/6357 region.**
See Abstr. 132.009.

**Survey à 2.2 μm du plan galactique austral.**
See Abstr. 133.019.

**Negative detection of the four–hour periodicity of the central star of M1–2 (VV8).**
See Abstr. 134.027.

**Photoelectric search for CP2–stars in open clusters. V. NGC 6281, NGC 6405, IC 4665.**
See Abstr. 153.015.

**Photometry of open cluster King 12.**
See Abstr. 153.023.

**A photometric metallicity scale for globular clusters based on stars with known curve of growth chemical abundances.**
See Abstr. 154.024.

**Photoelectric sequences near nine galactic globular clusters.**
See Abstr. 154.038.

**Photoelectric *UBVRI* sequences in the region of the galactic globular clusters NGC 1261, NGC 3201, NGC 4590, and NGC 5139 (ω Cen).**
See Abstr. 154.040.

**Photometry of faint stars in globular clusters. VII. On some common features of photometric characteristics of faint stars in M10 and M12.**
See Abstr. 154.069.

**A deep optical survey of a small region in Aquarius. I. Stellar photometry and galactic structure.**
See Abstr. 155.008.

**The galactic rotation outside the solar circle.**
See Abstr. 155.022.

**A *B,V* sequence in Baade's window.**
See Abstr. 155.029.

**The stellar distribution in Puppis. I. Distribution of B5 to M5 stars to $V \cong 12.5$.**
See Abstr. 155.035.

**Comparison between *UBV*– and *RGU*–photometrically determined density–functions for the (photometric) disc– and halo–populations in SA 82.**
See Abstr. 155.058.

**Photoelectric comparison sequences in the fields of B2 1308+326 and 1418+54.**
See Abstr. 158.024.

# 114 Spectra, Temperatures, Chemical Composition, etc.

114.001 **Bp and Ap stars in the moving Scorpio–Centaurus cluster.**
V. G. Klochkova, I. M. Kopylov.
Pis'ma Astron. Zh., Tom 10, No. 7, p. 508 – 515 (1984). In Russian. English translation in Sov. Astron. Lett., Vol. 10.

The degree and type of peculiarity as well as quantitative spectral types and rotation velocities are defined for 10 peculiar stars and for 4 stars of Vb luminosity class from the Scorpio–Centaurus cluster. Effective temperatures ($T_e$) and surface gravities (lg $g$) are obtained for 29 peculiar stars from this cluster. It is shown that lg $g$ values for the stars of Vb luminosity class are by 0.2 – 0.4 larger than those for the stars on the zero–age main sequence.

114.002 **IUE low–dispersion spectra flux catalogue. Part 1. Normal stars (magnetic tape).**
A. Heck, D. Egret, M. Jaschek, C. Jaschek.
Astron. Astrophys., Suppl. Ser., Vol. 57, No. 2, p. 213 – 216 (1984).

This catalogue on magnetic tape contains the fluxes (2 Å–step) of 229 selected IUE low–dispersion stellar spectra constituting spectral reference sequences.

114.003 **The lithium isotope ratio in F and G stars.**
J. Andersen, B. Gustafsson, D. L. Lambert.
Astron. Astrophys., Vol. 136, No. 1, p. 65 – 73 (1984).

The lithium doublet at 6707 Å has been observed at high spectral resolution in the spectra of 8 solar–type dwarfs, known to have strong Li lines, and in the Population II star ν Ind. The wavelength shifts of the lines and the line profiles were analyzed in attempts to determine the $^6$Li/$^7$Li ratios. An upper limit of about 0.10 was found, in conflict with earlier claims that some of these stars are rich in $^6$Li. However, it is suggested that some of them have a $^6$Li/$^7$Li ratio significantly different from zero and that the isotopic ratio possibly is correlated with v sin *i*.

114.004 **Metallicity distribution of metal–deficient field giants.**
A. Bartkevičius.
Astron. Tsirk., No. 1289, p. 5 – 7 (1983). In Russian.

114.005 **The metallicity distribution in red giants.**
V. A. Marsakov, A. A. Suchkov.
Astron. Tsirk., No. 1297, p. 1 – 3 (1983). In Russian.

114.006 **Spectral types of some Ap stars.**
R. Radkov.
Astron. Tsirk., No. 1298, p. 7 – 8 (1983). In Russian.

114.007 **Aluminum abundance in metal–deficient stars.**
P. François.
C. R. Acad. Sci., Sér. II, Tome 299, No. 5, p. 195 – 198 (1984). In French.

The explosive carbon burning theory studied by Arnett predicts an overdeficiency of Al relative to Mg in metal–poor stars. Analyses of such stars by Peterson have shown a good agreement with Arnett's theory. But other measurements of Al abundance on six cool metal–poor stars have not shown any significant deficiency. The author has obtained high quality spectra (high resolution, good S/N ratio) in the region of the weak lines of aluminum at 6696 and 6698 Å for six metal–poor stars. The abundances of Al derived from these lines are very reliable; they do not show Al overdeficiency. According to Arnett's theory, the synthesis of light metals by carbon burning has not taken place in explosive conditions.

114.008 **The lithium–rich K giant 9 Boötis.**
L. Hänni.
Sov. Astron. Lett., Vol. 10, No. 1, p. 51 – 53 (1984). English translation of 37.114.019.

114.009 **Abundances of some *s*–process elements on the M giant β Pegasi.**
M. Ya. Orlov, A. V. Shavrina.
Sov. Astron. Lett., Vol. 10, No. 1, p. 53 – 54 (1984). English translation of 37.114.020.

114.010 **The mass loss rate of the galactic hypergiant HD 152236 ($\zeta^1$ Scorpii) from optical and near infrared observations.**
J. A. López, J. R. Walsh.
Rev. Mex. Astron. Astrofis., Vol. 9, No. 1, p. 3 – 8 (1984).

Intermediate dispersion spectra obtained with a photon–counting detector in the blue and red spectral regions of HD 152236, as well as near infrared photometry (*J, H, K, L*) are presented. Equivalent widths of the main lines are tabulated and infrared free–free fluxes derived. Mass loss rates are determined from the free–free excess at 2.2 and 3.5 μm and by means of the Hα emission strength; a consistent value of $\sim 9\times10^{-6} - 2\times10^{-5} M_{\odot} yr^{-1}$ is obtained with both methods. Spectral variability is discussed in connection with previous observations.

114.011 **Energy distribution in the spectra of 72 stars in the region λλ3200 – 7600 Å.**
I. N. Glushneva, I. B. Voloshina, V. T. Doroshenko, L. V. Mossakovskaya, S. L. Ovchinnikov, T. S. Khruzina.
Tr. Gos. Astron. Inst. Shternberg, Tom 54, p. 3 – 32 (1984). In Russian.

Data on energy distribution in the spectra of 72 stars in the region λλ3200 – 7600 Å are presented. The data obtained are in good agreement with other investigation results for common stars.

114.012 **The energy distribution for 50 stars in the λλ3200 – 10800 Å spectral region.**
I. B. Voloshina, I. N. Glushneva, V. I. Shenavrin.
Tr. Gos. Astron. Inst. Shternberg, Tom 55, p. 84 – 101 (1983). In Russian.

The values of stellar energy distribution for 50 stars in the wide spectral region λλ3200 – 10800 Å are reproduced. The observations were made by means of two photoelectric spectrophotometers in the visual region (λλ3200 – 7600 Å) and the near infrared (λλ6300 – 10800 Å).

114.013 **An objective–prism survey for late M dwarf stars.**
T. H. Robertson.
Astron. J., Vol. 89, No. 8, p. 1229 – 1237 (1984).

An objective–prism survey in the red spectral region has been completed which covers approximately two percent of the sky at intermediate south galactic latitudes. Catalogues of 142 stars classified as M2 and later are presented which contain 73 dwarf stars, 41 giant stars, and 28 stars of uncertain luminosity class. Data obtained from photoelectric *BVI* photometry and slit spectrograms are presented for several program stars. Surveys of this type are very effective in the detection of motion unbiased samples of late M dwarf stars. The value of such surveys in studies of the kinematic properties and space density of these stars is seriously jeopardized by systematic errors in the assignment of spectral types for stars near the plate limit. This problem is compounded by the rapid change of limiting magnitude with advancing spectral type.

114.014 **Spectrophotometry of early type supergiants in the Large Magellanic Cloud.**
K. Nandy, G. I. Thompson, D. H. Morgan, L. Houziaux.
Mon. Not. R. Astron. Soc., Vol. 210, No. 1, p. 131 – 139 (1984).

Measures of the equivalent widths of major stellar features in the visible and UV spectral region of a sample of LMC supergiants whose types range from O5 to A1 have been derived. Their low–resolution UV spectra were obtained with the IUE, and complementary visible spectra with the ESO 3.6–m telescope and the SAAO 1.95–m telescope. The methods used are entirely numerical and as far as possible free from subjective estimates. The results for the LMC members are compared with those obtained for their galactic counterparts.

114.015 **Selective excitation of Fe II in the laboratory and late–type stellar atmospheres.**
S. Johansson, C. Jordan.
Mon. Not. R. Astron. Soc., Vol. 210, No. 1, p. 239 – 256 (1984).

High resolution IUE spectra of RR Tel and β Gru (M3 II) show lines which can be definitely attributed to Fe II. The presence of particular transitions and absence of others suggests that the source of excitation for levels with excitation energy around 10 eV is emission in the H Ly α line which is strong and broad in the cool giants and supergiants. New laboratory spectra confirm the excitation route for Ly α pumping to a further $^4G^0$ level at ~13 eV, which decays to give the strong feature at 1870 Å observed at low spectral resolution in several cool stars.

114.016 **Macrostructure of the ultraviolet continuous spectra of hot stars according to OAO–2 data.**
S. S. Rustambekova.
Astrofizika, Tom 19, Vyp. 3, p. 533 – 545 (1983). In Russian. English translation in Astrophysics, Vol. 19, No. 3.

The presence of a depression in ultraviolet continuous spectra is discovered in 22 hot stars among 134 already analyzed ones according to OAO–2 data. The depression lies generally in the region 2200 – 2400 Å.

114.017 **Metal abundances in 26 halo stars.**
R. G. Gratton, S. Ortolani.
Astron. Astrophys., Vol. 137, No. 1, p. 6 – 16 (1984).

Metal abundances are derived for 26 field halo giants by means of high dispersion spectra and a model atmosphere analysis. The stars cover a wide range in metal abundance from slightly less than solar ([Fe/H] = –0.2) to very metal poor ([Fe/H] = –2.4). In this range the relative abundances of most of the studied elements are approximately solar. A deficiency of Ba in very metal poor stars ([Fe/H] < –1.5) such as suggested by Spite and Spite (1979) is confirmed by the present observations.

114.018 **Analysis of the stellar fluxes at 2740 Å as observed by S2/68 and OAO–2.**
R. Faraggiana, M. L. Malagnini.
Astron. Astrophys., Vol. 137, No. 1, p. 149 – 151 (1984).

In previous works by Malagnini et al. (1982, 1983) significant discrepancies between observed and predicted stellar fluxes at 2740 Å were found. A critical analysis of the S2/68 channel centered at this wavelength is presented. The observed stellar fluxes prove to be underestimated by a considerable amount for O and B stars. A dependence on the level of the 2740 Å flux seems to be present.

114.019 **General data on IUE observations of O – F stars within the ES2AS program.**
A. Sapar.
Tartu Astrofüüs. Obs. Teated, Nr. 71, p. 5 – 16 (1984).
In the present introductory paper the author summarizes in the form of tables the main characteristics of the exposures of 48 early–type stars observed at the Goddard Space Flight Center during two weeks in November 1978 in the framework of the ES2AS observing program. 48 stars of the spectral classes O – F were observed. 117 high–dispersion spectra of 39 stars and 41 low–dispersion spectra of 14 stars were obtained.

114.020 **The A0 stars.**
O. J. Eggen.
Astrophys. J., Suppl. Ser., Vol. 55, No. 4, p. 597 – 632 (1984). With a correction in Astrophys. J., Suppl. Ser., Vol. 56, No. 4, p. 650 – 652 (1984).
A photometric grid, based on intermediate–band and Hβ photometry and standardized on the MK spectral standards, has been used to make comparisons with the published spectral types of the B8–A1 stars in the *Bright Star Catalogue*. The spectral type of B9.5 may represent an otherwise unrecognized spectral peculiarity. A different A0/A1 boundary is apparent in two major spectroscopic surveys, and an appreciable percentage of stars called dwarfs spectroscopically are of higher photometric luminosity. The distribution of space motion vectors is also discussed.

114.021 **Non–LTE effect on the determination of microturbulence in the atmospheres of F type supergiants.**
L. S. Lyubimkov, A. A. Boyarchuk.
Astrofizika, Tom 19, Vyp. 4, p. 683 – 696 (1983). In Russian. English translation in Astrophysics, Vol. 19, No. 4.
It has been shown that the analysis of Fe I lines in the spectra of several F type supergiants leads to lower values of microturbulent velocity $\xi_t$ in comparison with lines of ions Fe II and Ti II. Probable causes of the disparity in microturbulence determination – effect of errors in oscillator strengths and in damping, effect of helium overabundance, and effect of departures from LTE are discussed.

114.022 **The atmospheres of the southern supergiants α Lep and ι Car.**
A. A. Boyarchuk, L. S. Lyubimkov.
Astrofizika, Tom 20, Vyp. 1, p. 85 – 98 (1984). In Russian. English translation in Astrophysics, Vol. 20, No. 1.
On the basis of model atmospheres of Kurucz (1979) using high dispersion spectroscopic data of Castley and Watson (1980) the authors have analysed the southern supergiants α Lep (F0 Ib) and ι Car (F0 Iab). The following values of effective temperature and surface gravity were obtained: $T_{eff} = 7300 \pm 150K$, $\log g = 1.75 \pm 0.2$ for α Lep; $T_{eff} = 7300 \pm 200K$, $\log g = 1.4 \pm 0.2$ for ι Car. The distribution of microturbulent velocity was studied. The authors have investigated the chemical composition of α Lep and ι Car using relatively weak lines.

114.023 **Spectrophotometric study of OB stars in the region of the association Cep OB2.**
A. T. Garibdzhanyan.
Astrofizika, Tom 20, Vyp. 1, p. 107 – 114 (1984). In Russian. English translation in Astrophysics, Vol. 20, No. 1.
The results of a spectrophotometric investigation of 39 OB stars in the region of the association Cep OB2 are presented. The blue (4000 – 4800 Å) absolute spectrophotometric gradients, colour excesses and distances of these stars are determined. The analysis of the obtained results suggests that the absorbing matter is concentrated inside the association Cep OB2.

114.024 **Spectral classifications of emisson–line stars.**
G. Welin.
Astrofizika, Tom 20, Vyp. 1, p. 193 – 194 (1984).

114.025 **Abundances in very hot stars.**
R. P. Kudritzki.
MPA Rep., No. 90, p. 17 (1983). Abstract. – See Abstr. 012.020.

114.026 **Barium stars and normal red giants: evidence for a modified *s*–process.**
H. Holweger.
MPA Rep., No. 90, p. 86 – 88 (1983). – See Abstr. 012.020.

114.027 **Spectrophotometry of circumpolar stars.**
L. N. Knyazeva, V. M. Tereshchenko,
A. V. Kharitonov.
Astrofiz. inst. Akad. Nauk KazSSR. Alma–Ata, 5 pp. (1984). In Russian. Abstr. in Ref. Zh., 51. Astron., 10.51.415 (1984).

114.028 **P Cygni and related profiles in the ultraviolet spectra of O–stars.**
R. Costero, R. Stalio.
Astron. Astrophys., Suppl. Ser., Vol. 58, No. 1, p. 95 – 115 (1984).
P Cygni and related profiles of the resonance lines of N V, Si IV and C IV ions are studied over a sample of 40 O–stars observed with IUE in the high resolution mode. The sample includes subdwarf stars, high luminosity population II objects, stars which are not recognized members of open clusters and/or associations, and stars which are members. The profiles give information on the ionization structure of the expanding stellar atmosphere and on the velocity fields. The large variety of these profiles and the wide range of observed velocities allow to take a look at the largely unexplored field of stellar individuality.

114.029 **The Wolf–Rayet spectrum of MS 4 = WR 29a.**
A. F. J. Moffat, W. Seggewiss.
Astron. Astrophys., Suppl. Ser., Vol. 58, No. 1, p. 117 – 119 (1984).
The authors present a fluxed IDS spectrum (4000 – 7000 Å) of a heretofore inadequately classified Wolf–Rayet star number 4 in the catalogue of MacConnell and Sanduleak (1970). MS 4, = WR 29a in the catalogue of van der Hucht et al. (1981), is classified WC4 + O4.

114.030 **Oxygen abundances in metal poor stars.**
R. G. Gratton, S. Ortolani.
Messenger, No. 37, p. 3 – 5 (1984).

114.031 **The Sun and α Cen A.**
I. Furenlid, T. Meylan.
Messenger, No. 37, p. 10 – 11 (1984).

114.032 **Spectral analysis of (very) hot stars: methods and results.**
R. P. Kudritzki.
ESA Spec. Publ., ESA SP–218, p. 33 – 41 (1984). – See Abstr. 012.024.
The methods of combined UV and visual quantitative spectral analysis of hot stars are summarized. The application of these methods to massive O–stars, subluminous OB–stars, and central stars of planetary nebulae is presented. The results are discussed with respect to the evolutionary status of these objects.

114.033 **Ultraviolet fluxes of blue straggler stars.**
M. Gerbaldi, N. Morguleff.
ESA Spec. Publ., ESA SP–218, p. 187 – 190 (1984). – See Abstr. 012.024.
The authors present low–resolution IUE spectra of blue straggler stars and discuss their appearance in comparison with normal stars and stellar atmosphere models.

114.034 **Ultraviolet and optical observations of metal deficient red giants and chromospheric models.**
A. K. Dupree, E. H. Avrett, L. Hartmann, G. Smith.
ESA Spec. Publ., ESA SP–218, p. 191 – 195 (1984). – See Abstr. 012.024.
Three metal deficient field stars have been observed in the ultraviolet and optical spectral regions: HD 165195, HD 110281, and HD 232078. High dispersion spectra near Hα were obtained as well as low dispersion, long wavelength IUE spectra. Semi–empirical atmospheric models suggest that the characteristic emission in the wings of the Hα line can arise within static chromospheres. Wind models with extended atmospheres suggest

mass loss rates $\ll 2 \times 10^{-9} M_{\odot} yr^{-1}$. Thus Hα provides no evidence that steady mass loss is substantial enough to significantly affect the evolution of stars on the red giant branch of globular clusters.

114.035 **An upper limit to the beryllium abundance in the population II star HD 140283 from a high resolution IUE spectrum.**
P. Molaro, J. E. Beckman, F. Castelli.
ESA Spec. Publ., ESA SP-218, p. 197 - 201 (1984). - See Abstr. 012.024.

An IUE archive spectrum of HD 140283 has been analyzed in the region of the 3131 Å $^{9}$Be II resonance doublet. Close comparison of the observed and synthetic spectra yields no measurable Be, and a conservative upper limit of $3 \times 10^{-12}$ for $^{9}$Be, which is one fourth solar, can be set. With this star, and with a previous population II star HD 76932 detailed arguments rule out significant $^{9}$Be depletion. The authors can use these [$^{9}$Be] limits to set upper limits to the fraction of non-primordial $^{7}$Li. The results support Spite and Spite's conclusion that most, (if not all) of their observed $^{7}$Li is primordial and tend to validate the utility of their [$^{7}$Li] numbers in deriving the universal baryon/photon ratio using standard Big Bang model.

114.036 **Emission lines in the UV spectrum of α Tau.**
S. Catalano, C. Blanco, L. Bruca, E. Marilli.
ESA Spec. Publ., ESA SP-218, p. 203 - 206 (1984). - See Abstr. 012.024.

Based on high dispersion IUE spectra, identifications of ultraviolet emission lines of α Tau (K5 III) are presented. In the long-wavelength range the authors have identified Fe I and Fe II emission lines and a possible Al II one in addition to the strong chromospheric Mg II h and k resonance lines. In the short-wavelength range only the prominent lines of O I, Si II and S I and the Ly α have been identified.

114.037 **Search for chromospheric indicators in A dwarf stars.**
R. Freire Ferrero, A. Talavera.
ESA Spec. Publ., ESA SP-218, p. 207 - 210 (1984). - See Abstr. 012.024.

An IUE observational program, using the high resolution mode, with the aim to search for chromospheric indicators in A normal stars was started in 1982. The classical chromospheric indicators are core reversals in collisional dominated lines as C II (1335 Å), Si II (1308 Å) and Mg II (2800 Å) lines. The analysis of the sample of 16 stars observed (from A0 V to A8 IV) do not show any line core reversal on the UV high resolution spectra. The preliminary result that emerges with this small sample of A stars, supported by other results in the literature, is that chromospheres are probably absent in A type IV - V stars.

114.038 **The Mg II lines in A dwarf stars.**
R. Freire Ferrero, A. Talavera, P. Gouttebroze.
ESA Spec. Publ., ESA SP-218, p. 211 - 215 (1984). - See Abstr. 012.024.

High resolution IUE spectra, in the spectral region of the Mg II resonance lines, of 16 A stars, were obtained at the ESA IUE Observatory. To interpret these observations, the authors study the formation of Mg II lines in an A type stellar atmosphere for different values of Mg abundance and microturbulence within the framework of the NLTE theory with complete and partial redistribution on the Mg II line profiles. Finally, the authors compare the observed with the computed line profiles and establish preliminary results for abundance values.

114.039 **The search for superionization lines in A dwarf stars.**
R. Freire Ferrero, A. Talavera.
ESA Spec. Publ., ESA SP-218, p. 217 - 221 (1984). - See Abstr. 012.024.

The authors analyze a sample of 7 stars, observed with IUE at high resolution, belonging to spectral subtypes A0 to A4, searching for the resonance lines of Si IV (1400 Å) and C IV (1550 Å) and also for the doublet of He II (1640 Å). Even if these lines may be present in an A type atmosphere as very weak lines, it seems that the major spectral features in the corresponding spectral regions are blends of metallic lines. This brings more evidence for the absence of transition regions in A type IV - V stars, as also stated by other papers in the literature.

114.040 **IUE spectra of three A0 V stars rotating close to their break-up velocity: HD 119921, HD 129685, HD 181296.**
P. Molaro, C. Morossi, M. Ramella, M. Franco.
ESA Spec. Publ., ESA SP-218, p. 223 - 225 (1984). - See Abstr. 012.024.

The IUE spectra of the three A0 V stars with the highest rotational velocity reported in the literature are analyzed. Superionization lines are surely present only in the spectrum of HD 119921 (Z Cen) and they are found to be variable. Using a visible spectrum, a new value of the rotational velocity of 220 km/sec is given for Z Cen; it is considerably far from its break-up velocity. No clear evidence of superionization is found in the other two stars.

114.041 **Statistical classification of IUE stellar spectra by the variable procrustean bed (VPB) approach.**
A. Heck, D. Egret, P. Nobelis, J. C. Turlot.
ESA Spec. Publ., ESA SP-218, p. 257 - 261 (1984). - See Abstr. 012.024.

Various statistical algorithms have been applied to a set of stellar low-dispersion IUE spectra in order to derive objective spectral-classification schemes in the UV. In what the authors call the variable procrustean bed (VPB) approach, the shape of the continuum, together with the most important emission and absorption features are taken into account to define distances in an adequate multivariate space. Principal-component and cluster analysis algorithms are then applied to construct groups of stars which, together with the statistically most significant spectral features, are compared respectively with the existing classification frames and rules.

114.042 **IUE-high resolution study of ON V-stars.**
D. Schönberner, K. P. Simon, R. P. Kudritzki.
ESA Spec. Publ., ESA SP-218, p. 267 - 271 (1984). - See Abstr. 012.024.

The authors report on a comparison of IUE high resolution spectrograms of six peculiar, early type main sequence stars classified as ON (or BN) with those of normal stars covering the same spectral class range. A differential curve-of-growth analysis using the N III-lines 1748, 1752 Å and the C III-line 1247 Å revealed for these peculiar stars nitrogen enrichments of about a factor of five and carbon depletions of about a factor of ten compared with the normal stars.

114.043 **An analysis of the helium-weak star HD 144334.**
H. J. Walker, A. E. Lynas-Gray.
ESA Spec. Publ., ESA SP-218, p. 279 - 283 (1984). - See Abstr. 012.024.

IUE low resolution spectra of the helium-weak star HD 144334 are combined with published data from other sources, namely S2/68 fluxes, optical and infrared photometry. These data, using the infrared and total flux method, give $T_{eff} = 16000K \pm 100K$. From published Balmer line profiles the surface gravity is redetermined as $\log g = 4.10 \pm 0.05$. Metallic ions in the high resolution IUE SWP spectra are identified.

114.044 **The ultraviolet spectrum of stars having optical Fe II emission lines.**
G. Muratorio, M. Friedjung, R. Viotti.
ESA Spec. Publ., ESA SP-218, p. 309 - 313 (1984). - See Abstr. 012.024.

Emission lines of singly ionized iron occur in the optical spectra of many different kinds of objects. Emission lines can also be present in the UV, though often they appear to be replaced by absorption lines. The physics of the formation of these lines is still not very well understood. The authors briefly review previous study, and describe their present work in progress.

114.045 **Observations of extensive variability in the ultraviolet spectrum of ξ Per.**
R. K. Prinja, H. F. Henrichs, I. D. Howarth, M. van der Klis.
ESA Spec. Publ., ESA SP–218, p. 319 – 322 (1984). – See Abstr. 012.024.

The authors present high resolution IUE observations of the O7.5 III star ξ Per, taken between 1978 and 1984. The spectra reveal extensive variability in the Si IV resonance line doublet. Profile fitting is used to estimate central velocities and column densities of these varying features and the plausibility of the "shell ejection" model is discussed.

114.046 **The nitrogen abundance in WC stars.**
W. Schmutz, C. Morossi, M. Ramella.
ESA Spec. Publ., ESA SP–218, p. 325 – 327 (1984). – See Abstr. 012.024.

Willis (1982) identifies tentatively an absorption at 1235 Å in the IUE spectrum of the WC star HD 192103 as blue shifted P Cygni absorption of N V λ1240. The presence of nitrogen in a WC star is of interest because this observation would disagree with the predictions of theoretical stellar evolution calculations. The authors reanalysed the IUE image of this Wolf–Rayet star. On the grounds of the reprocessed spectrum and of comparisons with the same wavelength region in other Wolf–Rayet spectra, the authors conclude that most probably nitrogen is not observed in the WC star HD 192103.

114.047 **Spectroscopy of five Wolf–Rayet star candidates including a Wolf–Rayet star in the galactic bulge.**
I. Lundström, B. Stenholm.
Astron. Astrophys., Vol. 138, No. 2, p. 311 – 316 (1984).

Spectrograms are presented for the remaining five WR star candidates in the HCLS catalogue, not previously observed in this manner. The spectra are discussed. One candidate is found to be a planetary nebula or an H II region. One WR star showing very small interstellar reddening is suggested as lying within the galactic bulge.

114.048 **Lanthanide rare earths in stellar spectra with emphasis on chemically peculiar stars.**
C. R. Cowley.
Phys. Scr., Vol. T8, p. 28 – 38 (1984). – See Abstr. 012.027.

This paper deals with stellar abundance patterns of the lanthanides. First the author investigates the accuracy with which partition functions may be calculated from known energy levels. Corrections for "missing levels" are made, and current "best estimates" are tabulated. Secondly, the systematics of the behavior of La II through Tm II are reviewed one–by–one. The emphasis is on CP stars, but late–type giants and dwarfs are also discussed.

114.049 **The spectrum of Pt II in chemically peculiar stars.**
M. M. Dworetsky, P. J. Storey, J. M. Jacobs.
Phys. Scr., Vol. T8, p. 39 – 42 (1984). – See Abstr. 012.027.

The authors have obtained a reliable estimate of the platinum abundance in HgMn stars for the first time. It is as high as $\log A(\mathrm{Pt}) = 6$, on the scale $\log A(\mathrm{H}) = 12$; the Pt abundance often matches the remarkably high abundances of Hg already known in these stars. The results are based on calculations of theoretical oscillator strengths which were applied to high resolution spectra taken with the IUE satellite.

114.050 **Bolometric corrections for peculiar B–type stars.**
T. Lanz.
Astron. Astrophys., Vol. 139, No. 1, p. 161 – 170 (1984).

Semi–empirical bolometric corrections are estimated for 25 peculiar stars and 13 comparison stars from IUE low dispersion spectrum and from theoretical fluxes best fitting the Geneva colour indices. When corrected for reddening, the relation BC vs. $(\mathrm{B2{-}G})_0$ for main–sequence stars matches fairly well the theoretical relation computed by Kurucz. The difference on BC between a peculiar star and a comparison star with the same Paschen slope is less than 0.3 mag. A quantitative estimation of the UV deficiency of CP stars is also given.

114.051 **The identification of high–level transitions of Fe II in stellar spectra.**
S. Johansson, C. R. Cowley.
Astron. Astrophys., Vol. 139, No. 1, p. 243 – 247 (1984).

Positive identifications are made in four B and A–type stellar spectra of lines from $3d^6(^5D)4d$–$3d^6(^5D)4f$ transitions in Fe II. The lowest levels of the $4d$ terms have excitation energies of 10.3 eV, while those of the $4f$ terms are at 12.8 eV. Numerous features that were previously unidentified in B and A stars may now be explained. The obvious presence of such transitions makes it probable that many other high–level transitions are present in stellar spectra. Some of these transitions have not yet been identified in laboratory studies analyses, but they could be of considerable importance in the chemical analyses of stellar atmospheres.

114.052 **Spectroscopic studies of Wolf–Rayet stars with absorption lines. VIII. HD 193793.**
P. S. Conti, D. Roussel–Dupré, P. Massey, M. Rensing.
Astrophys. J., Vol. 282, No. 2, p. 693 – 698 (1984).

The authors present absorption–line velocities for the O type star spanning over 16 years and emission–line velocities for the WC star covering 10 years. They find no periodicities in either of these sets of data. In particular, they are unable to confirm the claim of Lamontagne, Moffat, and Seggewiss that the two stars are in orbit about one another. Rather, it seems that a generic relationship between the two components has not been established and one is dealing with a situation in which two stars are in the same line of sight.

114.053 **New spectrophotometry of the extremely cool proper motion star LHS 2924.**
J. Liebert, T. A. Boroson, M. S. Giampapa.
Astrophys. J., Vol. 282, No. 2, p. 758 – 762 (1984).

Improved spectrophotometric scans are presented for this very faint, red, large–proper motion star which has a peculiar spectrum. The Palomar Observatory 5 meter CCD spectrum shows the clear presence of VO λ7400 and λ7900. However, the TiO and hydride bands and the Na I and K I lines appear relatively weaker over the spectrum of LHS 2924 than they do in the benchmark low–luminosity star VB 10. A high–resolution observation with the MMT echelle spectrograph shows very weak Hα emission, at best. These and other recent results eliminate all previously considered explanations of the nature of LHS 2924 other than a very cool, main–sequence star or sub–stellar–mass "brown dwarf." The weakness of Hα emission tends to indicate that the star is an old, main–sequence object.

114.054 **Iron abundances in G dwarfs.**
M. Cornide, M. Rego.
Astrophys. Space Sci., Vol. 105, No. 1, p. 55 – 65 (1984).

High resolution coudé spectra in the blue and violet spectral regions have been obtained for five G dwarfs. Model atmospheres and spectrum synthesis of selected iron lines have been performed. The *gf* values as well as other atomic parameters used in this analysis were previously tested by means of a solar model and the Utrecht Solar Atlas. Revised iron abundances have been derived. Some tentative evidence is presented that these abundances are related with the Wilson–Bappu width of Ca II emissions.

114.055 **Lithium in late–type giants. III. The weak G band giants.**
D. L. Lambert, S. R. Sawyer.
Astrophys. J., Vol. 283, No. 1, p. 192 – 199 (1984).

A new survey of nine southern giants with published results for another 12 stars shows that the Li abundance spans a broad range with the cosmic abundance apparently providing the upper bound. Both $^6$Li and $^7$Li are shown to be present in one star: $^6\mathrm{Li}/^7\mathrm{Li} \approx 0.1$ for HR 1299. It is argued that the upper bound and the presence of $^6$Li rules out all explanations of the weak G band phenomenon calling for the production of Li in parallel or in series with the operation of the CN–cycle. Large–scale diffusion within the pre–main–sequence star is proposed as the key to the weak G band giants.

114.056 **The infrared Ca II triplet: a luminosity indicator for stellar population synthesis.**
J. E. Jones, D. M. Alloin, B. J. T. Jones.
Astrophys. J., Vol. 283, No. 1, p. 457 – 465 (1984).

The authors study the properties of the Ca II infrared triplet (Ca T) from a sample of Reticon spectra of 62 stars. The stars range over spectral types B though mid–M, and the sample spans some four orders of magnitude in gravity $g$, and almost a factor of 10 in metallicity. It is found that, over all spectral types, from F to mid–M, and over this large gravity interval, the equivalent width of Ca T correlates strongly with $\log g$. This is an almost one–valued relation, being relatively insensitive to metallicity. Thus, since most of the light from galaxies is thought to come from G and K stars, the Ca T feature should be a useful tool for constraining the dwarf: giant ratio of the light–dominant stellar population. Existing measurements of Ca T are applied in an analysis of the stellar content in the nucleus of M31.

114.057 **A search for duplicity in the weak G–band stars.**
J. Tomkin, C. Sneden, P. L. Cottrell.
Publ. Astron. Soc. Pac., Vol. 96, No. 582, p. 609 – 612 (1984).

The radial velocities of seven weak G–band stars have been monitored for a period of over two years. Velocity variations are seen in the known spectroscopic binary HR 1023, from which the authors redetermine the orbital solution. Small–amplitude variations are present in the velocities of 37 Com and HR 6791, and the other four stars have constant velocities. These results indicate that, unlike the Ba II stars which have been discovered recently to be all binaries, the weak G–band stars do not have an unusually high incidence of duplicity. The origin of the weak G–band phenomenon must be sought after in a scenario of single–star evolution.

114.058 **Is 89 Herculis a binary star?**
A. Arellano Ferro.
Publ. Astron. Soc. Pac., Vol. 96, No. 582, p. 641 – 645 (1984).

A periodicity of about 285 days has been found in the 1977 – 81 radial velocities of the yellow supergiant variable 89 Her (HR 6685) which has a pulsational period of about 63.5 days. This is interpreted as being due to orbital motion around an unseen companion. The orbital parameters are computed and seem to favor the high mass case ($13 < M/M_\odot < 24$) for 89 Her. Several possibilities for the nature of the presumed companion are considered.

114.059 **Line identifications in the ultraviolet (2590 – 3230 Å) spectrum of the hydrogen–deficient carbon star HD 182040.**
H. R. Johnson, M. M. Ameen, J. A. Eaton.
Astrophys. J., Vol. 283, No. 2, p. 760 – 764 (1984). Summ. Indiana Astron., No. 1.

The authors present identifications for absorption features detected in the ultraviolet (2590 – 3230 Å) spectrum of the hydrogen–deficient supergiant carbon star HD 182040 observed at high resolution with the *IUE* satellite. The spectrum in this range is filled with blended lines of the iron–peak elements, with Fe I and Cr II dominating the spectrum. In this range the spectrum resembles that of a middle–G supergiant.

114.060 **Observational knowledge about the physical properties of O stars.**
A. B. Underhill.
Hvar Obs. Bull., Vol. 7, No. 1, p. 1 – 28 (1983). – See Abstr. 012.037.

Information about the effective temperatures, radii, and masses of O–type stars is presented. It is argued that rapid variations in the amount of light from O stars and the spectral distribution are a result chiefly of changes which occur in the envelope of the star. The stability of the photospheric layers of O stars against convection is reviewed and it is noted that late O stars and early B stars have a convection zone in the deeper parts of the photosphere.

114.061 **Intermediate resolution spectroscopy of the 3.53–μm dust feature in Elias 1.**
D. C. B. Whittet, A. D. McFadzean, T. R. Geballe.
Mon. Not. R. Astron. Soc., Vol. 211, No. 1, p. 29P – 31P (1984).

Spectroscopy at a resolution of 0.01 μm is presented for the pre–main–sequence Ae star Elias 1 in a spectral range which includes the unidentified emission features at 3.28, 3.43 and 3.53 μm. The central wavelengths and profile shapes of the 3.43– and 3.53–μm features in Elias 1 and in the prototype object HD 97048 are identical to within observational errors. The marked asymmetry of the 3.53–μm feature supports an identification with formaldehyde in cold grain mantles.

114.062 **Carbon, nitrogen, and oxygen abundances in Betelgeuse.**
D. L. Lambert, J. A. Brown, K. H. Hinkle, H. R. Johnson.
Astrophys. J., Vol. 284, No. 1, p. 223 – 237 (1984).

Vibration–rotation bands of CO, NH, and OH and infrared lines of the CN red system in the spectrum of the M supergiant Betelgeuse (α Orionis) are analyzed to obtain the C, N, and O abundances. A model atmosphere with $T_{eff} = 3800$K and $\log g = 0.0$ is adopted. Results are: $\log \varepsilon(C) = 8.4$, $\log \varepsilon(N) = 8.6$, and $\log \varepsilon(O) = 8.8$. These abundances suggest that CN–cycle processed material has been mixed to the surface as a result of the first dredge–up phase. This analysis does not confirm recent reports that C is severely underabundant.

114.063 **Classification of Ap stars HR 830 and 21 CVn.**
J. Zverko.
Bull. Astron. Inst. Czech., Vol. 35, No. 5, p. 294 – 299 (1984).

Two ambiguously classified Ap stars, HR 830 and 21 CVn, are studied. The observational data are compared with the data for normal stars α Del and α Lyr. The star HR 830 is classified as Ap of type Si based on the enhanced absorption in the Si lines and photometric variability in *UBV*. 21 CVn is classified as Ap Si based on weaker absorptions of helium, the variability of helium lines, photometric periodic variability and the properties of the UV spectrum. It is concluded that the photometric periodic variability is a good indicator of Ap–properties of rapidly rotating A and late B stars.

114.064 **Hydrogen lines near the Lyman limit in spectra of hot stars.**
N. A. Sakhibullin, V. Ya. Shabert.
Kazan. univ. Kazan', 32 pp. (1984). In Russian. Abstr. in Ref. Zh., 51. Astron., 10.51.441 (1984).

114.065 **On technetium abundances in the atmospheres of red giants.**
T. A. Kipper, M. A. Kipper.
Pis'ma Astron. Zh., Tom 10, No. 11, p. 868 – 872 (1984). In Russian. English translation in Sov. Astron. Lett., Vol. 10.

Technetium abundances are estimated for a sample of 8 red giants by means of matching the synthetic spectra of short regions containing Tc I lines with the observed ones. In the spectra of 7 stars the Tc I lines are present and the estimated Tc abundances turn out to be fairly large.

114.066 **A search for rapid variations in the spectrum of the Ap star Theta Aurigae.**
J. B. Rice, W. H. Wehlau, V. L. Khokhlova.
Publ. Astron. Soc. Pac., Vol. 96, No. 583, p. 734 – 736 (1984).

Spectra of θ Aur taken in rapid succession at several phases showed no real variations, with an upper limit of 0.5% of the continuum. Variations above the noise level which were found for four spectra at one epoch are explained as the result of a spot on the Reticon detector.

114.067 **Short–term spectral variability of stars.**
S. V. Marchenko.
Akad. Nauk Ukrain. SSR, Inst. Teor. Fiz., Kiev, Prepr., No. 84–80R, 27 pp. (1984). In Russian.

Information is given on different type stars which have short–time (from minutes to tens of minutes) variations of absorption and emission spectral features. Practical procedures of data processing to detect and estimate the parameters of the characteristics of fast spectral variations are considered.

114.068 **The $H^-$ flux peak and the hydrogen abundance in N type carbon stars.**
J. H. Goebel, H. R. Johnson.
Astrophys. J., Lett. Ed., Vol. 284, No. 2, p. L39 – L42 (1984).

For galactic N and late R type irregularly variable carbon stars, the authors make a comparison of airborne infrared spectrophotometry to spectral flux distributions derived from model atmospheres. Although the agreement is good in several respects, the $H^-$ emission peak at 1.6 μm, so prominent in the models, is significantly weakened in the stellar spectrophotometry. It seems unlikely that molecular bands in the stars can completely obliterate this peak. A plausible explanation is that hydrogen is deficient in these stars, most probably from an episode of mass loss.

114.069 **IUE observations of UV bright stars in the globular clusters M15 and ω Cen.**
C. Cacciari, V. Caloi, V. Castellani, F. Fusi Pecci.
Astron. Astrophys., Vol. 139, No. 2, p. 285 – 288 (1984).

IUE spectra in the low dispersion mode are reported for three UV–bright stars. Two of them (stars K 559 and K 996) are members of the very metal poor cluster M15 and one (star ROA 5701) is a member of the populous cluster ω Cen. The authors find $E(B\text{–}V) = 0.11$ as a lower limit for the reddening of ω Cen whereas the reddening of M15 should not be larger than $E(B\text{–}V) \cong 0.08$. Effective temperatures and bolometric magnitudes are evaluated for all the stars. Comparison with previous estimates shows that such temperatures tend to be generally lower than previously expected on a $(B\text{–}V)$ basis. This difference is of the order of 2000K for ROA 5701, but it reaches more than 7000K for the two other stars. Evolutionary implications are discussed briefly.

114.070 **An upper limit to the abundance of $^9$Be in the Population II star HD 76932 from a high resolution spectrum with IUE.**
P. Molaro, J. Beckman.
Astron. Astrophys., Vol. 139, No. 2, p. 394 – 400 (1984).

A high resolution spectrum ($\Delta\lambda = 0.2$ Å) of the Population II star HD 76932 from the archive of the IUE satellite has been analysed in a search for measurable resonance absorption lines due to $^9$Be. Matching the observations by a synthesized model of the wavelength region around the doublet at 3130 Å gives a conservative upper limit for the abundance $^9$Be/H in the star of $3 \times 10^{-12}$. When all alternatives for the production modes of $^9$Be and $^7$Li have been considered, this result strongly supports a primordial origin for $^7$Li. The primordial $^7$Li is the best method now available for estimation of the universal baryon/photon ratio η; the range of η derived appears to set a severe constraint on the number of two–component neutrino types permitted within the context of standard Big Bang model.

114.071 **Spectroscopic tests of photoelectric stellar classification of abnormal stars.**
H. A. Abt.
Astrophys. J., Vol. 285, No. 1, p. 247 – 251 (1984).

The author obtained spectroscopic classification for 169 northern A5–G0 stars predicted by Olsen to have abnormal spectra on the basis of Strömgren four–color photometry. The success in identifying reddened early–type stars was nearly 100%, for Am and early–type weak–lined stars ~75%, for stars above the main sequence ~50%, for composite spectra ~25%, and for Ap and λ Boo stars 0%. Thus photoelectric photometry is a successful first step in discovering stars of the more extreme spectroscopic abnormalities.

114.072 **Effective temperatures of stars of spectral classes A–G. Comparison with theoretical models.**
I. N. Glushneva.
Sov. Astron., Vol. 28, No. 2, p. 193 – 199 (1984). English translation of 37.114.065.

114.073 **Metallicity of solar–neighborhood F stars.**
V. A. Marsakov, A. A. Suchkov, Yu. G. Shevelev.
Sov. Astron., Vol. 28, No. 3, p. 284 – 288 (1984). English translation of 37.114.108.

114.074 **Determination of the electron temperatures of the envelopes of the supergiants α Cyg and ν Cep of class A2 Ia.**
I. N. Glushneva.
Sov. Astron., Vol. 28, No. 3, p. 305 – 310 (1984). English translation of 37.114.109.

114.075 **Lithium in population I early–F dwarfs.**
A. M. Boesgaard, M. J. Tripicco.
Publ. Astron. Soc. Pac., Vol. 96, No. 584, p. 784 (1984). Abstract. – See Abstr. 010.281.

114.076 **Carbon, nitrogen, and oxygen abundances in M67 giants.**
J. A. Brown.
Publ. Astron. Soc. Pac., Vol. 96, No. 584, p. 785 (1984). Abstract. – See Abstr. 010.281.

114.077 **Carbon and nitrogen abundances in a sample of 70 metal–poor dwarfs.**
D. F. Carbon, B. Barbuy, R. P. Kraft, N. B. Suntzeff, E. D. Friel.
Publ. Astron. Soc. Pac., Vol. 96, No. 584, p. 786 (1984). Abstract. – See Abstr. 010.281.

114.078 **Carbon and nitrogen abundances in CN–strong and CN–weak giants in M5.**
G. E. Langer, R. P. Kraft, E. Friel.
Publ. Astron. Soc. Pac., Vol. 96, No. 584, p. 790 (1984). Abstract. – See Abstr. 010.281.

114.079 **Short–time–scale periodicity in Hα emission from the main–sequence star HZ 1883.**
G. W. Marcy, D. K. Duncan, R. D. Cohen.
Publ. Astron. Soc. Pac., Vol. 96, No. 584, p. 790 (1984). Abstract. – See Abstr. 010.281.

114.080 **Light element abundances in disk and halo dwarfs.**
J. Tomkin, D. L. Lambert, S. Balachandran.
Publ. Astron. Soc. Pac., Vol. 96, No. 584, p. 796 (1984). Abstract. – See Abstr. 010.281.

114.081 **Two bright peculiar galactic emission–line stars.**
R. A. Downes.
Publ. Astron. Soc. Pac., Vol. 96, No. 584, p. 807 – 810 (1984).

In the course of a survey for galactic–plane, ultraviolet–excess objects, two peculiar emission–lined stars have been detected. One is a newly discovered WC Wolf–Rayet star, designated KPD 0721+0042, which appears to be located in the galactic halo. This peculiar location, as well as the relatively narrow width of the emission lines, argues that the object is a "Population II" Wolf–Rayet star. The other object, MWC 84 = KPD 0415+5552, is a previously cataloged early–type star with strong Balmer and He I emission lines. The spectrum is dominated, however, by emissions of Fe II, and is similar to two previously–discussed bright stars with prominent Fe emission, although the strength of the He I lines in this object indicates that it is substantially hotter.

114.082 **Magnesium abundances in field dwarf stars.**
J. B. Laird.
Bull. Am. Astron. Soc., Vol. 16, No. 3, Part I, p. 729 (1984). Abstract. – See Abstr. 010.065.

114.083 **Comparison of UV spectra from solar–type stars.**
A. Greve, W. Wamsteker.
Sol. Phys., Vol. 94, No. 1, p. 3 – 11 (1984).

The authors compare high–resolution spectra from the Sun and the four solar–type stars 16 Cyg A, 16 Cyg B, HD 32008, HD 34411 obtained with IUE in the wavelength range 2650–2930 Å. The comparison is made for peak intensities between absorption lines. At the level of accuracy of the IUE observations, the stars 16 Cyg A, 16 Cyg B, HD 34411 are indistinguishable from the Sun, in particular the Mg II resonance line profiles are identical. HD 32008 is not a solar analog but is evidently of late G to early K spectral type.

114.084 **CASPEC (*Cassegrain Echelle Spectrograph*) and IUE: a perfect match.**
I. Appenzeller, G. Klare, O. Stahl, B. Wolf, F.-J. Zickgraf.
Messenger, No. 38, p. 28 – 31 (1984).

114.085 **Catching carbon stars in the Baade's windows.**
M. Azzopardi, J. Lequeux, E. Rebeirot.
Messenger, No. 38, p. 31 – 32 (1984).

114.086 **Light element abundances in F stars and the chemical evolution of the galactic disk.**
B. Edvardsson, B. Gustafsson, P. E. Nissen.
Messenger, No. 38, p. 33 – 38 (1984).

114.087 **Spectral classification of faint O–B stars around the cluster IC 1805.**
V. I. Kuznetsov.
Astrometr. Astrofiz., Vyp. 53, p. 34 – 36 (1984). In Russian.
Spectral classification of 89 faint O–B stars around the cluster IC 1805 was carried out using the method developed for unbroadened stellar spectra.

114.088 **Photometry and spectrophotometry of magnetic stars.**
K. Stepień.
Astron. Nachr., Vol. 305, No. 6, p. 311 – 316 (1984).
A review of selected photometric and spectrophotometric properties of magnetic stars is given. It is shown that unreddened B–V of magnetic stars around B9 can be transformed to log $T_e$ using the calibration for normal stars. Mechanisms of light variability are discussed. It is concluded that different mechanisms must be operating in different stars. Recently obtained photometric observations of short-period variability by Kurtz and Hβ variability by Madej et al. are discussed. An information about internal structure and the atmospheric structure of these stars can be derived from these observations.

114.089 **High resolution spectroscopy of Alpha Centauri. I. Lithium depletion near one solar mass.**
D. R. Soderblom, D. Dravins.
Astron. Astrophys., Vol. 140, No. 2, p. 427 – 430 (1984).
The authors have measured the lithium abundance of α Centauri A, and have set an upper limit to Li in α Cen B. For α Cen A, log $N$(Li) = 1.28 (on a scale where log $N$(H) = 12.00), about twice the solar Li abundance, while log $N$(Li) < 0.7 for α Cen B. These abundances are consistent with the probable evolutionary age of α Cen (6 Gyr), given the mass of α Cen A (1.1 $M_\odot$).

114.090 **Variable blue supergiants in the Large Magellanic Cloud: R 84, R 85, and R 99.**
O. Stahl, B. Wolf, C. Leitherer, F.-J. Zickgraf, J. Krautter, M. de Groot.
Astron. Astrophys., Vol. 140, No. 2, p. 459 – 467 (1984).
During a search for luminous blue variables in the LMC the three emission-line stars R 84 (OIafpe), R 85 (B5Iae), and R 99 (OBf:pe) have been found to show significant brightness and colour variations. Photometric and spectroscopic data from the UV to the IR have been used to discuss the nature of these variables. It is argued that they are closely related to the S Dor variables. The implications of this finding for the interpretation of the Ofpe/WN9 stars and the S Dor variables are discussed. Photometric data for 11 additional suspected variables are given.

114.091 **Abundance of lithium in another sample of halo dwarfs, and in the spectroscopic binary BD –0°4234.**
M. Spite, J. P. Maillard, F. Spite.
Astron. Astrophys., Vol. 141, No. 1, p. 56 – 60 (1984).
The observation of the lithium resonance doublet in 13 other metal poor dwarfs (halo dwarfs) confirms the stability of the lithium abundance in the range 5500° < $T_{eff}$ < 6300°. The value of the lithium abundance of this plateau: log $N_{Li}$ = 2.05, is interpreted as the abundance of this element at the beginning of the life of the Galaxy and thus constrains the cosmological models of the primordial Universe. Lithium is also observed in a cold dwarf, BD –0°4234. The authors suggest the possibility that the observed lithium is formed in the star itself.

114.092 **The chemical composition of the He–w Bp star HR 6000.**
F. Castelli, M. Cornachin, M. Hack, C. Morossi.
Astron. Astrophys., Vol. 141, No. 1, p. 223 – 226 (1984).
The abundance anomalies of the CP star HR 6000 (He–w, $T_e$ = 14,000K, log $g$ = 4) are compared with the optical values and with those of other peculiar stars. The slight overabundance of Ga, the strong underabundance of Sc, the doubtful presence of Hg and the very peculiar behaviour of Si make it difficult to classify this star in one of the typical CP groups. The underabundance of Si and the presence of circumstellar (or interstellar) components on the short wavelength side of the resonance lines of C II, O I, Mg II, Si II, and Fe II are discussed.

114.093 **Optical region elemental abundance analyses of B and A stars. III. The mercury–manganese stars Upsilon Herculis and 38 Draconis.**
S. J. Adelman.
Astron. Astrophys., Suppl. Ser., Vol. 58, No. 3, p. 585 – 594 (1984).
Abundance analyses using optical region data and fully line blanketed model atmospheres have been performed for the HgMn stars υ Her and 38 Dra. Comparison with the abundances of other peculiar stars derived in the second paper of this series indicates that the hot Am stars and the HgMn stars might form a temperature sequence. Line blocking coefficients were measured for υ Her in an attempt to explain the slight anomalies in its optical energy distribution.

114.094 **The spectra of late–type dwarfs and sub–dwarfs in the near ultraviolet. III. An atlas of Mg II h and k profiles.**
M. L. Franco, L. Crivellari, P. Molaro, G. Vladilo, M. Ramella, C. Morossi, C. Allocchio, J. E. Beckman.
Astron. Astrophys., Suppl. Ser., Vol. 58, No. 3, p. 693 – 704 (1984).
Using a radically new suite of data extraction routines, IUEARM, high resolution IUE spectra of six F and G stars, taken at regular intervals since 1978 are analysed. The salient features of the new routines are: reduction of interorder scattering via two–dimensional deconvolution of the echelle spectrograph image plane, with high frequency filtering, to reduce the effects of the instrument function; enhanced spectral resolution by data extraction using a small pseudo–slit; improved "ripple correction" yielding mutually consistent inter–order overlap regions; and an improved absolute calibration. Rigorous limits, twice as precise as any previously available, are set on weak spectral features, and on possible variability of the h and k chromospheric emission cores in the G stars. The authors show how the interstellar Mg II components of the stars can be abstracted and used to examine a sector of the local interstellar medium.

114.095 **Evidence for a supersonic turbulent velocity gradient in the outer photosphere of the supergiant α Cygni (A2Ia).**
C. de Jager, P. S. Mulder, Y. Kondo.
Astron. Astrophys., Vol. 141, No. 2, p. 304 – 308 (1984).
Two high–resolution near–ultraviolet spectra of the supergiant α Cyg have been analyzed with the aim of determining the average microturbulent line–of–sight velocity component which was found to be $15.0 \pm 0.5$ km s$^{-1}$. This value is close to the local sound velocity (13.7 km s$^{-1}$). This result, together with previous determinations in the visual spectral region, is consistent with the picture that shock waves are the dominant structure in the outer photosphere of Deneb. The macroturbulent velocity component is ≈10 km s$^{-1}$, a factor two to four smaller than earlier communicated values for the line of sight components of the stochastic macroturbulent velocities.

114.096 **A study of twenty-one cool Ap stars. II. A reassessment of HD 8441.**
S. J. Adelman.
Astron. Astrophys., Vol. 141, No. 2, p. 362 – 368 (1984).
The techniques used to derive the stellar temperatures and the elemental abundances of the cool Ap stars in the author's thesis are reassessed by using HD 8441 as an example. For this star, $T_{eff}$ = 9325K and log $g$ = 3.40 with iron being about three times the solar value. Of the elements with identified lines, chromium and heavier elements tend to be overabundant while carbon through vanadium tend to be underabundant.

114.097 **Maps of the iron-group elements on the magnetic Ap star $\alpha^2$ Canum Venaticorum.**
V. L. Khokhlova, V. M. Pavlova.
Sov. Astron. Lett., Vol. 10, No. 3, p. 158 – 163 (1984). English translation of 37.114.107.

114.098 **HD 107346, a red subgiant with a strong ultraviolet excess.**
V. Straižys.
Sov. Astron. Lett., Vol. 10, No. 3, p. 181 – 183 (1984). English translation of 37.114.151.

114.099 **HD 91948 is not a Be star.**
C. T. Bolton, R. W. Lyons.
Inf. Bull. Variable Stars, No. 2580, 2 pp. (1984).

114.100 **Spectral classification at ultra-low dispersion.**
R. Rajamohan.
Astrophys. Space Sci., Vol. 106, No. 1, p. 173 – 179 (1984).
The author demonstrates the possibility of using very low dispersion (10000 Å $mm^{-1}$) slitless spectroscopy of stars for a fairly accurate spectral classification. The technique is shown to be very useful in studies of star clusters.

114.101 **On the effective temperatures for early type stars from integrated fluxes.**
J. O. Woo.
J. Korean Astron. Soc., Vol. 17, No. 2, p. 104 – 114 (1984).
The effective temperatures, $T_{eff}$ (flux) of 52 early type stars are derived from de-reddened monochromatic and integrated fluxes obtained by absolute spectrophotometry, using the method of graphical analysis pioneered by Blackwell and Shallis (1977). The author also estimates the effective temperatures, $T_{eff}$ (comp) of the same stars by comparing their de-reddened energy distributions with those of the LTE model atmospheres by Kurucz (1979). The effective tempertures derived from these two methods are found to be in good agreement, confirming that they are effective for the estimation of effective temperatures of early type stars.

114.102 **Spectroscopic and photometric observations of the suspected binary star HD 123058.**
I. Skillen, M. de Groot.
Mon. Not. R. Astron. Soc., Vol. 211, No. 4, p. 793 – 798 (1984).
Spectroscopic and photometric observations of the F4 star HD 123058 do not support the hypothesis that it is a spectroscopic binary. Broad absorption lines in its spectrum are attributed to a rather high rotation rate with $v_e \sin i = 61 \pm 9$ km $s^{-1}$ but no evidence is found for chromospheric activity. The mean radial velocity derived from 23 spectrograms is $v_r = -5.2 \pm 4.3$ km $s^{-1}$. The photometric data combined with published data are consistent with a spectral classification of F4V, and yield an effective temperature of 6670K. The metal abundance is [Fe/H] = 0.02 which indicates a probable age of less than $4 \times 10^9$yr, while the small value derived for the colour difference $\delta c_1$ suggests that the star lies near the zero-age main sequence. The distance of HD 123058 is shown to be $70 \pm 8$ pc.

114.103 **Ultraviolet spectral morphology of the O stars. III. The ON and OC stars.**
N. R. Walborn, R. J. Panek.
Space Telesc. Sci. Inst., Prepr. Ser., No. 33, 23 pp. (1984). To appear in Astrophys. J.

114.104 **Early-type stars detected on UK Schmidt objective-prism plates. I. UKST Area 345.**
B. D. Kelly, D. Kilkenny.
S.Afr. Astron. Obs., Circ., No. 8, p. 85 – 99 (1984).
Finding charts and approximate co-ordinates are given for 141 stars, apparently of spectral type F0 or earlier, detected on an objective-prism plate of UKST Area 345 (RA = $22^h30^m$, Dec = –40°).

114.105 **A high dispersion analysis of a giant star in 47 Tucanae.**
S. D'Odorico, R. G. Gratton, D. Ponz.
ESO Sci. Prepr., No. 337, 23 pp. (1984). Submitted to Astron. Astrophys.

114.106 **Spectroscopic study of the star β Canum Venaticorum (G0).**
H. H. Mentese.
Rev. Fac. Sci. Univ. Istanbul, Ser. C, Vol. 47, p. 15 – 21 (1982). = Publ. Istanbul Univ. Obs., No. 130 (1984).
The effective temperature and surface gravity of the star β CVn were determined by means of theoretical profiles calculated by De Jager and Neven (1967 – 1968). The values of the excitation temperatures for Ti, Cr, Mn and Fe were calculated and found to be very close to each other. The abundances of the elements were obtained by the curve of growth method and found to be normal.

114.107 **An absence of Balmer emission from Vega.**
G. A. H. Walker, Z. Ninkov, S. Yang.
Bull. Inf. Etoiles Be, No. 10, p. 3 – 6 (1984).

114.108 **On the behaviour of the Hα and Hβ emission in the spectra of three Be stars: γ Cas, ζ Tau and 66 Oph.**
T. S. Galkina.
Bull. Inf. Etoiles Be, No. 10, p. 29 – 47 (1984).

114.109 **On the reality of the ultra-rapid V/R variations of Be stars. (Comments on a paper by T. S. Galkina).**
P. Harmanec.
Bull. Inf. Etoiles Be, No. 10, p. 48 – 51 (1984). See also Abstr. 114.108.

114.110 **HD 91948 is not a Be star.**
C. T. Bolton, R. W. Lyons.
Bull. Inf. Etoiles Be, No. 10, p. 52 – 53 (1984).

114.111 **The ultra-metal-deficient (Population III?) red giant CD -38°245.**
M. S. Bessell, J. Norris.
Astrophys. J., Vol. 285, No. 2, p. 622 – 636 (1984).
The authors present photometric and high-resolution spectroscopic data on the ultra-metal-deficient red giant CD -38°245. A model atmosphere analysis yields the parameters effective temperature, surface gravity and the element abundances. As a check the same procedures were applied to the well studied Population II red giant HD 2796. CD -38°245 has [Fe/H] = -4.5, and is the most metal-deficient astronomical object currently known. The authors suggest that CD -38°245 is an extreme Population II object, as opposed to its being a member of a new population.

114.112 **The ultraviolet variability of early-type supergiants.**
A. B. Underhill.
Astrophys. J., Vol. 285, No. 2, p. 668 – 673 (1984).
Four early-type supergiants – HD 79186, HD 96919, HD 105056, and HD 148379 – have been observed with *IUE* on 14 days. The behavior of the ultraviolet fluxes with time is studied. The light from all four stars seems to vary. Typically the dispersion about the mean magnitude at any wavelength corresponds to ±0.085, ±0.080, ±0.101, and ±0.106 mag, respectively. It is suggested that the rotation of a mottled disk on which the pattern of mottling is continually changing may be the cause of the light changes.

114.113 **Oxygen isotopic abundances in the atmospheres of seven red giant stars.**
M. J. Harris, D. L. Lambert.
Astrophys. J., Vol. 285, No. 2, p. 674 – 682 (1984).

Abundance ratios of the oxygen isotopes have been measured in α Tau, β And, μ Gem, α Her, β Peg, γ Dra, and α Boo. In all the stars the $^{16}O/^{18}O$ ratios are similar; the mean value is 475, which is consistent with the solar system value 490. The $^{16}O/^{17}O$ ratios range from ~1000 for β Peg and α Boo to 160 for β And. A detailed comparison of these data with predictions from stellar evolution models for stars near the end of the main-sequence evolution is presented and it is demonstrated that standard models must be modified in order to account for the observations.

114.114 **New results for HD 161796.**
J. D. Fernie, R. F. Garrison.
Astrophys. J., Vol. 285, No. 2, p. 698 – 701 (1984). With plates 12 – 13.

The authors have redetermined the MK classification of the high-latitude supergiant HD 161796 by proper MK methods and criteria, confirming Bidelman's original classification of F3 Ib, despite the conclusion of others using different means that it must be more luminous and cooler. New four-color photometry together with existing Geneva photometry leads to a photometrically determined gravity and effective temperature of $\log g = 0.1 \pm 0.5$ and $T_e = 6400 \pm 200$K.

114.115 **Investigation of the linearity of the Alma-Ata spectrophotometric star catalogue.**
E. A. Glushkova, V. M. Tereshchenko, A. V. Kharitonov.
Astron. Tsirk., No. 1310, p. 1 – 2 (1984). In Russian.

114.116 **Observations of spectrophotometric standard stars in 1981 – 1982. I. The energy distribution $E^{83}_{CT}(\lambda)$.**
L. N. Knyazeva, V. M. Tereshchenko, A. V. Kharitonov.
Astron. Tsirk., No. 1311, p. 1 – 2 (1984). In Russian.

114.117 **Observations of spectrophotometric standard stars in 1981 – 1982. II. Comparison with the results of other observations.**
L. N. Knyazeva, V. M. Tereshchenko, A. V. Kharitonov.
Astron. Tsirk., No. 1312, p. 1 – 3 (1984). In Russian.

114.118 **Observations of spectrophotometric standard stars in 1981 – 1982. III. On the uniformity of the system $E^{83}_{CT}(\lambda)$.**
L. N. Knyazeva, V. M. Tereshchenko, A. V. Kharitonov.
Astron. Tsirk., No. 1312, p. 3 – 5 (1984). In Russian.

114.119 **On variations of 53 Cam studied with spectra with high time resolution.**
N. S. Polosukhina, V. P. Malanushenko, H. Karttunen, I. Tuominen, H. Virtanen.
Rep. Ser., Dep. Phys. Sci., Univ. Turku, No. FTL-R68, p. 21 – 28 (1984). – See Abstr. 012.079.

114.120 **On the hydrogen-alpha profile of 56 Peg (HD 218356).**
H. Virtanen.
Rep. Ser., Dep. Phys. Sci., Univ. Turku, No. FTL-R68, p. 39 – 49 (1984). – See Abstr. 012.079.

Variable asymmetry and a core shift of the H-alpha line have been detected from spectra of the peculiar giant 56 Peg (K0IIp+WD). 56 Peg is a mild barium star and has a white dwarf companion. One explanation for the changing asymmetry is a small blueshifted emission caused by a stellar wind and accretion onto the companion star.

114.121 **Studies of late-type dwarfs. VI. Identification of Population II main-sequence stars at $M_v = +14$.**
F. D. A. Hartwick, A. P. Cowley, J. R. Mould.
Astrophys. J., Vol. 286, No. 1, p. 269 – 275 (1984).

Spectroscopic and photometric observations of a sample of 65 faint cool stars from the LHS catalog have been obtained. From these observations broad-band *V, R, I* colors and narrow-band CaH and TiO indices have been determined as well as *B, V, R, I*, and *JHK* photoelectric photometry for some stars. Radial velocities were determined for 24 stars. The authors are able to unambiguously separate out a true halo population of M stars. The observations are in very good agreement with recent theoretical model computations. The authors have obtained an upper limit to the relative denstiy of halo to disk stars at $M_v \approx 14$ and argue that the faint halo M stars are unlikely candidates for the solution of the outstanding "missing mass" problems in the Galaxy.

114.122 **Der supermassive Stern R 136a.**
T. Schmidt-Kaler.
Mitt. Astron. Ges., Nr. 62, p. 245 (1984). Abstract. – See Abstr. 010.341.

114.123 **Spektralanalyse von Horizontalaststernen im Kugelhaufen NGC 6752.**
U. Heber, R. P. Kudritzki, V. Caloi, V. Castellani.
Mitt. Astron. Ges., Nr. 62, p. 249 (1984). Abstract. – See Abstr. 010.341.

114.124 **Iron in the spectra of central stars and very hot subdwarfs.**
J. S. Drilling, D. Schönberner.
Mitt. Astron. Ges., Nr. 62, p. 250 (1984). Abstract. – See Abstr. 010.341.

114.125 **Stickstoff und Kohlenstoff in OBN-Sternen.**
D. Schönberner, K. P. Simon, R. P. Kudritzki.
Mitt. Astron. Ges., Nr. 62, p. 251 (1984). Abstract. – See Abstr. 010.341.

114.126 **Identification of atomic lines in the spectra of magnetic degenerates.**
G. Wunner, H. Herold, W. Rösner, H. Ruder.
Mitt. Astron. Ges., Nr. 62, p. 265 (1984). – See Abstr. 010.341.

114.127 **Carbon monoxide band intensities in M giants.**
A. McWilliam, D. L. Lambert.
Publ. Astron. Soc. Pac., Vol. 96, No. 585, p. 882 – 890 (1984).

The strength of the CO 2.3 μm bands was measured photometrically in an attempt to identify extremely carbon poor M giants. Magnitudes of about 200 bright M stars were obtained through a *J* filter, and narrow filters centered on 2.17 μm, and 2.40 μm. No M giants with CO indices indicative of extremely low carbon abundances were positively identified. The tight correlation of CO index to effective temperature does not extend to the extremely late and variable M giants. The CO index in M giants defined by the 2.17 μm and 2.40 μm magnitudes was calibrated by a spectrum synthesis technique. The dependence of CO index upon carbon abundance, the $^{12}C/^{13}C$ ratio, surface gravity, effective temperature, and microturbulent velocity was investigated. Predicted and observed CO indices are in excellent agreement for stars with a spectroscopically determined carbon abundance.

114.128 **Ultraviolet analysis of the peculiar F supergiant HD 112374 = HR 4912.**
E. Böhm-Vitense, C. Proffitt.
Publ. Astron. Soc. Pac., Vol. 96, No. 585, p. 897 – 902 (1984).

The authors have studied the ultraviolet energy distribution of the metal-poor supergiant HD 112374. They need a temperature $T_{eff} = 5500 + 100$K, $\log g = -0.3 \pm 0.3$ for a deficiency of Mg by a factor of 5 in order to find agreement between theoretical and observed energy distributions. A reddening $E(B-V) \sim 0.1$ is then required, consistent with its galactic latitude of +36°.

114.129 **On the parameters of atmospheres of the chemically peculiar stars α And and $\alpha^2$CVn.**
V. V. Leushin, L. I. Snezhko, V. V. Sokolov.
Astrofizika, Tom 20, Vyp. 3, p. 429 – 435 (1984). In Russian. English translation in Astrophysics, Vol. 20, No. 3.

The hydrogen lines $P_{13} - P_{17}$ and $H_\gamma$ in the spectra of the Ap stars α And and $\alpha^2$CVn have been investigated. For both stars a value of $\log g \cong 3.5$ has been found by hydrogen lines. It has been shown that the values of $T_{eff}$ obtained for both of them by an infrared flux method are consistent with the actual value.

114.130 **The distribution of absorbing matter and of OB stars in the region of the association Per OB1.**
A. T. Garibdzhanyan.
Astrofizika, Tom 20, Vyp. 3, p. 437 – 445 (1984). In Russian. English translation in Astrophysics, Vol. 20, No. 3.
The results of a spectrophotometric study of 86 OB stars in the region of Per OB1 association are presented. The absolute spectrophotometric blue (4000 – 4800 Å) gradients, the colour excesses and the distances of these stars are obtained. The analysis of these results permits to suppose that: 1) The difference of radial distances between two centers of the OB stars concentration h and χ Per seems to be about 1000 pc. 2) The absorbing matter in the direction of Per OB1 association is concentrated in the volumes occupied by OB stars.

114.131 **Balmer lines in the spectrum of P Cygni. Dilemma of the interpretation of variability.**
N. Markova, I. Kolka.
Astrofizika, Tom 20, Vyp. 3, p. 465 – 473 (1984). In Russian. English translation in Astrophysics, Vol. 20, No. 3.
Measurements of 32 spectra of P Cygni have shown that the fine structure variability of the absorption components of hydrogen lines can be interpreted in two ways – either the fine structure components are being shifted one after another from the line centrum towards its violet wing or the components' shifts are periodic, forming 4 curves which are oscillating at opposite phases.

114.132 **Evolutionary increase of nitrogen abundance in the atmospheres of early B stars.**
L. S. Lyubimkov.
Astrofizika, Tom 20, Vyp. 3, p. 475 – 493 (1984). In Russian. English translation in Astrophysics, Vol. 20, No. 3.
On the basis of high dispersion spectroscopic data the atmospheric nitrogen abundance log ε(N) has been determined for 36 stars of spectral types B0 – B3 and luminosity classes V – III. The non–LTE approach was applied. Masses and ages of the investigated stars were found with the aid of evolutionary calculations. It is shown that the abundance log ε(N) increases with the age, moreover this increase is going on gradually from values 7.4 – 7.6 intrinsic for interstellar matter in the solar neighbourhood and to values 8.4 – 8.7 inherent in the same stars when they evolve to F – K supergiants.

114.133 **The MK classification and continuous spectra of early–type stars.**
E. V. Ruban.
Astrofizika, Tom 21, Vyp. 1, p. 111 – 124 (1984). In Russian. English translation in Astrophysics, Vol. 21, No. 1.
The correspondence of the spectral sub–classes in the MK classification to observed parameters of continuous spectra (Balmer discontinuities, spectrophotometric gradients and reddening free $Q_\lambda$ curves) for early–type stars is considered. None of these is shown to agree strictly with the spectral sub–class. The nature of this disparity is discussed.

114.134 **Ultraviolet spectral morphology of the O stars. II. The main sequence.**
N. R. Walborn, R. J. Panek.
Astrophys. J., Vol. 286, No. 2, p. 718 – 724 (1984).
A survey of the 1200 – 1900 Å region in 49 spectra of types O3 – B1 V, based upon *IUE* high–resolution data, is presented. Standard sequences of prominent spectral features are illustrated, and UV classification criteria are introduced. For the majority of the stars, there is an excellent correlation between the behavior of the UV features, both stellar–wind and photospheric, and the optical spectral types. Several cases of enhanced winds are found; the most pronounced of these is the frequently observed "standard", τ Scorpii.

114.135 **The resonance lines of Hg II in *IUE* spectra of chemically peculiar stars.**
D. S. Leckrone.
Astrophys. J., Vol. 286, No. 2, p. 725 – 735 (1984).
High–dispersion *IUE* observations of the two resonance lines of Hg II, at 1650 and 1942 Å, are presented for two normal stars, ν Cap and π Cet; for the HgMn stars κ Cnc, μ Lep, 46 Dra, ι CrB, HR 4072, and χ Lup; and for the magnetic variable $\alpha^2$CVn. These peculiar stars have all been classified as Hg–rich on the basis of their strong 3984 Å features. The six HgMn stars all possess strong Hg II resonance lines, which confirms the reality of the Hg anomaly. The magnetic star $\alpha^2$CVn possesses moderately strong Hg II resonance lines. Quantitative spectrum syntheses of the 1942 Å line in these stars are described. Detailed curve–of–growth analyses of the 3984 Å feature in the HgMn stars, based on published equivalent widths, are also presented. It is shown that departures from LTE are ruled out as source of the Hg anomaly. The primordial abundance of Hg is discussed and possible explanations of the Hg anomaly are considered.

114.136 **An analysis by wavelength coincidence statistics of the ultraviolet spectrum of Kappa Cancri.**
G. T. Chjonacki, C. R. Cowley, D. J. Bord.
Astrophys. J., Vol. 286, No. 2, p. 736 – 740 (1984).
*IUE* wavelengths in the mercury–manganese star κ Cnc have been analyzed using the method of wavelength coincidence statistics. The method has been adapted to include line intensities for the second and third spectra of iron and manganese, which were calculated for a temperature of 12,600K with the help of the Kurucz–Peytremann transition probabilities. The authors have investigated the level of ionization of iron and manganese from single to doubly ionized stages, and find agreement with LTE predictions. It is concluded that the large manganese excesses for κ Cnc (∼2 dex) cannot be attributed wholly to non–LTE or to the concentration of manganese into a cool, thin layer.

114.137 **Optical spectra of R136.**
Y.–H. Chu, N. A. Daod.
Publ. Astron. Soc. Pac., Vol. 96, No. 586, p. 999 – 1006 (1984).
The authors have obtained spectra for the components in R136. There are at least two components showing Wolf–Rayet type spectral features: $R136a_1$–WN4.5 and R136c–WN7. The OB population in R136a has an integrated spectral type of about O6 or earlier. These new and improved data of R136a do not support the previously alleged periodic variation in the He II λ4686 line velocity or the progressive redshifts in the higher Balmer absorption lines.

114.138 **Similar photospheres, different mantles: a study of four O stars.**
A. B. Underhill.
Astrophys. J., Vol. 287, No. 2, p. 874 – 891 (1984).
The shapes and strengths of the continuous and line spectra of four O stars (10 Lac, S Mon, HD 46223, and ζ Pup) having similar effective temperatures (35,000K) are reviewed. It is shown that these stars have similar photospheres but different mantles. It is suggested that the leading members of the series of H, He I, and He II lines seen in absorption in each spectrum are formed in the mantle, and that the strengths and shapes of these lines are chiefly a result of the physical conditions in the mantle. These conditions are reviewed. It is concluded that taking account of the probable presence of bipolar and unipolar magnetic regions on the surfaces of OB supergiants and O and Wolf–Rayet stars will generate the different mantles above similar photospheres which are observed.

**Metal–deficient giants in the galactic field. Catalogue and some physical parameters.**
See Abstr. 002.005.

**An IUE high–resolution atlas of O–type spectra.**
See Abstr. 002.014.

**IUE low–dispersion spectra reference atlas.**
See Abstr. 002.015.

**Catalogue of stellar magnitudes, colour indices, spectral classes of stars in the direction to the cluster NGC 6633.**
See Abstr. 002.016.

**A library of stellar spectra.**
See Abstr. 002.023.

**Documentation for the machine-readable version of *A Library of Stellar Spectra* (Jacoby, Hunter and Christian 1984).**
See Abstr. 002.042.

**Catalogue of Metal-Deficient F – M stars. Part 1. Stars Classified Spectroscopically. Supplement 1.**
See Abstr. 002.075.

**An infrared photometric and spectrometric system in San Pedro Martir.**
See Abstr. 034.012.

**Standard stars: automatic line identification of IUE high resolution spectra.**
See Abstr. 036.032.

**Spectrophotometry of cool stars from objective prism spectra.**
See Abstr. 036.110.

**Automation of stellar spectrophotometry.**
See Abstr. 036.130.

**Should one measure radial velocity or brightness fluctuations?**
See Abstr. 036.139.

**Instrumental approach to multi-site spectral variability study.**
See Abstr. 036.148.

**Vollautomatische Analyse komplexer P Cygni-Linien.**
See Abstr. 036.164.

**The role of oscillator strengths in modelling and analysing stellar spectra.**
See Abstr. 064.019.

**On the need for spectroscopic data in ultraviolet stellar spectral synthesis calculations.**
See Abstr. 064.020.

**Radiative forces and the abundance of beryllium, magnesium and barium in HgMn stars.**
See Abstr. 064.021.

**The lithium abundance: constraints on stellar evolution.**
See Abstr. 065.004.

**Lithium depletion in stars. Pre-main sequence burning and extra-mixing.**
See Abstr. 065.031.

**CNO abundances in low mass stars evolving up to the red giant branch.**
See Abstr. 065.140.

**Comparative spectral characteristics of stars and sunspots.**
See Abstr. 072.031.

**Radial velocities and positions from objective prism plates.**
See Abstr. 111.009.

**Southern stars of high radial velocity.**
See Abstr. 111.021.

**On two WR stars in the Large Magellanic Cloud: the high velocity of BR 52 and a newly discovered WN star.**
See Abstr. 111.026.

**The spectrum of the Be star HD 110432.**
See Abstr. 112.002.

**Spectroscopy of the shell episode of η Car (1981 – 1983).**
See Abstr. 112.008.

**The transition layer of eleven late-type stars.**
See Abstr. 112.012.

**Short-period variability of the Be star λ Eri.**
See Abstr. 112.028.

**Line profile variation in δ Orionis A, ι Orionis A, and 15 Monocerotis.**
See Abstr. 112.030.

**High-resolution, far-ultraviolet study of Beta Draconis (G2 Ib-II): transition region structure and energy balance.**
See Abstr. 112.036.

**The Be star phenomena. II. Spectral formation and structure of envelopes.**
See Abstr. 112.047.

**Ca II K emission in Canopus.**
See Abstr. 112.048.

**Rapid spectroscopic variability of the Be star HR 9070: evidence for double periodicity?**
See Abstr. 112.055.

**Characteristics of the Fe II and C II emission in high-resolution *IUE* spectra (2300-3000 Å) of Alpha Orionis.**
See Abstr. 112.058.

**Be stars in binaries.**
See Abstr. 112.059.

**Structure of the expanding atmosphere of P Cygni.**
See Abstr. 112.073.

**Partial redistribution in the winds of red giants.**
See Abstr. 112.088.

**Variability of the Hα emission in φ Per.**
See Abstr. 112.093.

**Faint spectrophotometric standard stars for large optical telescopes. I.**
See Abstr. 113.017.

**Continuum energy distributions of some bright Be stars.**
See Abstr. 113.020.

**The masses of early-type stars in the galactic halo determined from ultraviolet resonance line profiles.**
See Abstr. 115.010.

**A redetermination of the Barnes-Evans relation for surface flux in the V band.**
See Abstr. 115.012.

**Calibration of effective temperature, bolometric correction and mass for early-type stars using Strömgren photometric indices.**
See Abstr. 115.016.

**General study of the apparently average magnetic Ap star Theta Aurigae. I. Data and method.**
See Abstr. 116.001.

**Polarimetric observations of carbon stars.**
See Abstr. 116.011.

**Spectroscopic evidence for starspots in the G dwarf HD 1835.**
See Abstr. 116.018.

**Polarization measurements of Be stars at Hβ and Ca II *H* and *K*.**
See Abstr. 116.025.

**Rotational velocities of later B type and A type stars as determined from ultraviolet versus visual line profiles.**
See Abstr. 116.039.

**Close visual binaries. I. MK classifications.**
See Abstr. 118.008.

**HD 138168 and HD 138181, a hard binary?**
See Abstr. 118.031.

**Limb–darkening and metal abundances in the sdOB primary of the eclipsing binary LB 3459 (AA Dor).**
See Abstr. 119.030.

**Helium content of the Population II binary system CM Draconis.**
See Abstr. 119.093.

**Spectroscopic and photometric observations of the Herbig Be star Z Canis Majoris.**
See Abstr. 121.037.

**The pure absorption spectrum of FG Sagittae in 1980.**
See Abstr. 122.059.

**The photometric behaviour of the young disk carbon star TW Horologii: determination of its physical characteristics.**
See Abstr. 122.127.

**The study of stellar global oscillations by Ca II H and K variations.**
See Abstr. 122.193.

**Hγ and Hδ emission in the spectrum of Mira Ceti during its 1975 maximum.**
See Abstr. 122.209.

**Non–LTE analysis of subluminous O–stars. VI. Feige 110.**
See Abstr. 126.006.

**The halo population G and K subdwarfs.**
See Abstr. 126.027.

**Inflow of halo gas from the direction of the galactic North Pole.**
See Abstr. 131.010.

**Interstellar lines in high resolution IUE spectra of runaway B type stars.**
See Abstr. 131.082.

**Interstellar Ca II lines in SMC stars.**
See Abstr. 131.190.

**Observations of local interstellar Mg I and Mg II.**
See Abstr. 131.246.

**Mg II spectra of late type stars used to probe the LISM.**
See Abstr. 131.247.

**Fe II and Mg II in the nearby interstellar medium.**
See Abstr. 131.248.

**A type stars as probes of the local interstellar medium.**
See Abstr. 131.249.

**IUE observations of the gaseous component of the local interstellar medium.**
See Abstr. 131.252.

**A high–resolution study of local interstellar sodium.**
See Abstr. 131.255.

**Optical observations of nearby interstellar gas.**
See Abstr. 131.256.

**Distances of local clouds from optical line observations.**
See Abstr. 131.257.

**Observations of highly–ionized interstellar iron.**
See Abstr. 131.264.

**Optical study of the galactic nebula NGC 6357 and of its stellar content, Pis 24 and the WC 6 star HD 157504.**
See Abstr. 132.031.

**The interstellar lines and the energetics of the inner 30 Doradus nebula from the ultraviolet spectrum of R 136a.**
See Abstr. 132.046.

**Hot central stars in planetary nebulae.**
See Abstr. 134.019.

**Negative detection of the four–hour periodicity of the central star of M1–2 (VV8).**
See Abstr. 134.027.

**The central star of the planetary nebula Abell 78.**
See Abstr. 134.028.

**Spectral anomalies in the Hyades and Pleiades and in field stars with active chromospheres.**
See Abstr. 153.009.

**The lithium abundance of Hyades main–sequence stars.**
See Abstr. 153.017.

**Distribution of absolute energy of bright Be stars in Pleiades.**
See Abstr. 153.024.

**Lithium abundances of solar–type stars in Praesepe.**
See Abstr. 153.025.

**UV–bright stars in galactic globular clusters, their far–UV spectra and their contribution to the globular cluster luminosity.**
See Abstr. 154.021.

**A search for cyanogen variations on the lower giant branch of M3.**
See Abstr. 154.077.

**Frequency distribution of metal abundance for faint stars in the direction of the north galactic pole.**
See Abstr. 155.004.

**The galactic rotation outside the solar circle.**
See Abstr. 155.022.

**Distribution of OB stars and absorbing matter in the region around P Cyg.**
See Abstr. 155.026.

**The stellar distribution in Puppis. I. Distribution of B5 to M5 stars to $V \cong 12.5$.**
See Abstr. 155.035.

**Kinematics of field K giants in the outer galactic halo.**
See Abstr. 155.048.

**Velocity dispersion and metallicity distribution of K giants in the galactic nuclear bulge.**
See Abstr. 155.049.

**Luminous late–type stars in NGC 205.**
See Abstr. 157.135.

**New carbon stars in spheroidal galaxies: I. Sculptor, Carina, Leo I and Leo II systems.**
See Abstr. 157.171.

**OB stars in the M33 galaxy.**
See Abstr. 157.187.

**Identifications of blue stellar objects from the Case Low–Dispersion Northern Sky Survey.**
See Abstr. 159.095.

# 115 Luminosities, Masses, Diameters, HR and other Diagrams

115.001 **Improvement of the visual surface brightness scale for the B5 – F5 main–sequence stars.**
L. Pastori, G. Malaspina.
Astron. Astrophys., Suppl. Ser., Vol. 57, No. 2, p. 219 – 223 (1984).

Stellar angular diameters of 593 B5 – F5 main–sequence stars listed in "CADARS" have been analyzed in order to improve the precision of the visual surface brightness $S_v$. The new relationships between this quantity and the color index $(B-V)_0$ and log $T_{eff}$ are examined; in particular the $S_v - (B-V)_0$ relations turn out to be in close agreement with those found with the interferometric method. The possible influence of surface gravity effects on the $S_v - (B-V)_0$ diagram is considered.

115.002 **A superluminous object in the Large Cloud of Magellan.**
J. S. Mathis, B. D. Savage, J. P. Cassinelli.
Sci. Am., Vol. 251, No. 2, p. 42 – 50 (1984).

A giant nebula in this small galaxy close to our own holds an object that is 50 million times brighter than the sun. If it is one body, it is far more massive than any known star.

115.003 **The mass–spectrum relation for main–sequence stars.**
M. A. Svechnikov, T. A. Taidakova (*T. A. Tajdakova*).
Sov. Astron., Vol. 28, No. 1, p. 84 – 89 (1984). English translation of 37.115.002.

115.004 **Atmospheric abundances, masses, radii and other parameters of four Am stars.**
L. S. Lyubimkov, I. S. Savanov.
Astrofizika, Tom 19, Vyp. 3, p. 505 – 511 (1983). In Russian. English translation in Astrophysics, Vol. 19, No. 3.

On the basis of high dispersion spectra a model atmosphere analysis of the Hyades cluster Am stars 68 Tau, 81 Tau, 16 Ori and the field Am star 15 Vul has been fulfilled. Effective temperatures, surface gravities, microturbulent velocities and chemical composition of the atmospheres were determined. Masses, radii, luminosities and ages of the stars were found using evolutionary calculations.

115.005 **Periods, amplitudes and luminosities of red supergiants.**
G. V. Abramyan.
Astrofizika, Tom 20, Vyp. 2, p. 239 – 243 (1984). In Russian. English translation in Astrophysics, Vol. 20, No. 2.

On the basis of photoelectric observations the periods of 22 red supergiants have been determined. For these stars the relations "period–luminosity–spectrum" and "amplitude–luminosity–spectrum" are found. The masses and luminosities of red supergiants have been estimated.

115.006 **The width of the spectroscopically defined main–sequence.**
C. Jaschek, J. C. Mermilliod.
Astron. Astrophys., Vol. 137, No. 2, p. 358 – 359 (1984).

Based upon a study of open clusters having both MK spectral classifications, *UBV* photometry and known distance moduli, it is shown that spectroscopically defined dwarf stars occupy an interval of about two magnitudes above the MS in the range of late B to early F–type stars. Such a large width of the MS implies conversely that absolute magnitudes of individual A–type dwarfs obtained from spectral classification alone are rather uncertain.

115.007 **Détermination des diamètres stellaires par occultations.**
M. Froeschlé, G. Helmer, C. Meyer.
Astrométrie, géodésie, mécanique céleste, p. 223 – 230 (1984). – See Abstr. 012.021.

115.008 **Analogy of rapid oscillations of Ap stars to solar oscillations.**
H. Shibahashi.
Mem. Soc. Astron. Ital., Vol. 55, Nos. 1 – 2, p. 181 – 183 (1984). – See Abstr. 012.022.

It is shown that the rapid oscillations of Ap stars discovered by Kurtz can be interpreted as a superposition of high overtone *p*–mode oscillations ($n \cong 40$) with low spherical harmonic index $l$ ($l = 0, 1, 2$), quite similar to the solar oscillations. A fine structure of triplets in the amplitude spectrum is evidence of the existence of dipole ($l = 1$) modes, while a slight, systematic departure from equal spacing of frequencies gives a strong indication for radial ($l = 0$) modes.

115.009 **On the upper limit of the angular diameter of Gamma Cassiopeiae with the two–telescope interferometer at CERGA.**
F. Vakili, P. Granes, D. Bonneau, M. Noguchi, R. Hirata.
Publ. Astron. Soc. Jpn., Vol. 36, No. 2, p. 231 – 237 (1984).

The authors observed the Be star γ Cas with the stellar interferometer at CERGA in 1982 September. The upper limit of the angular diameter in the north–south direction was 0\".0009, which corresponds to about 46 solar radii for the 230–pc distance. Though the diameter of the equatorial disk is expected to exceed this upper limit at the authors' observation epoch, its contribution to their measurement in the visible continuum is probably small. Thus, their observation gives an upper limit of the stellar diameter of γ Cas.

115.010 **The masses of early–type stars in the galactic halo determined from ultraviolet resonance line profiles.**
F. P. Keenan, P. L. Dufton.
Astron. Astrophys., Vol. 139, No. 1, p. 227 – 229 (1984).

Terminal velocities for stellar winds in intermediate and high galactic latitude OB stars have been determined from high resolution IUE ultraviolet spectra. These velocities, together with previous estimates, have been used to deduce stellar masses which are found to be similar to those derived from conservative and non–conservative evolutionary tracks for Population I stars. It is therefore concluded that the stars are not subluminous, nearby objects but are normal and at the large *z*–distances calculated by previous authors.

115.011 **A re–calibration of the luminosities of early–type stars: its effect on the Cepheid luminosity scale.**
L. A. Balona, R. R. Shobbrook.
Mon. Not. R. Astron. Soc., Vol. 211, No. 2, p. 375 – 390 (1984).

The authors have used the most recently published data on 421 stars in 13 open clusters to recalibrate the Strömgren β index as a luminosity indicator for early type stars. Since all the clusters independently tie up with the Pleiades, previous problems involving the use of associations are avoided. The final calibration is presented as an analytical expression in (β, $c_0$). Analysis of evolutionary diagrams for the calibrating clusters show small trends which are best explained as due to differences in zero–point of the β index. The authors present a calibration for the zero–age main sequence and compare it with results by various authors. The effect of the calibration on the zero–point of the Cepheid luminosity scale is investigated in an appendix.

115.012 **A redetermination of the Barnes–Evans relation for surface flux in the V band.**
J. A. Eaton, C. H. Poe.
Acta Astron., Vol. 34, No. 1, p. 97 – 116 (1984).

Paying special attention to the errors, the authors have redetermined the relation between visual flux at the star and color for stars with measured angular diameters. The results agree moderately well with those of Code et al. (1976), on which they are primarily based, for the hotter stars. However, they imply significant revisions of both the temperature and bolometric–correction scales for cool stars.

115.013 **X-ray properties of hot stars. I. On a dichotomy among O-type stars.**
T. Chlebowski.
Acta Astron., Vol. 34, No. 2, p. 191 – 206 (1984).

All published EINSTEIN X-ray observations of O-type stars are collected. On the basis of the analysed X-ray luminosities of 47 objects it is shown that runaway stars are relatively fainter in the X-ray band compared to normal stars. This dichotomy has serious consequences in the theory of X-ray emission of hot stars. From the most plausible interpretation, based on Lucy's model, it is suggested that a large fraction of X rays in normal stars originate in the shocked winds where the blobs interact with the interstellar matter.

115.014 **Massive stars in the Galaxy and the Magellanic Clouds: clues to stellar evolution.**
C. D. Garmany.
Publ. Astron. Soc. Pac., Vol. 96, No. 584, p. 779 – 782 (1984). Invited paper presented at the Symposium on Stellar Populations, at the 95th Annual Scientific Meeting of the Astronomical Society of the Pacific, University of California, Santa Cruz, July 1984.

A great deal of work has been done on the most massive main-sequence stars, the O stars, and their evolutionary descendants, the Wolf-Rayet stars. Not only do these stars influence their surroundings through mass loss and heating, the O stars are the only main-sequence objects that can be observed beyond the Magellanic Clouds. Variations in either their formation rate or in their initial mass function have important implications for the appearance of later evolutionary stages which include not only the Wolf-Rayet stars but also Hubble-Sandage variables, red supergiants, and supernovae.

115.015 **UV-bright stars in globular clusters: *IUE* observations.**
B. M. Altner, T. M. Matilsky.
Publ. Astron. Soc. Pac., Vol. 96, No. 584, p. 783 (1984). Abstract. – See Abstr. 010.281.

115.016 **Calibration of effective temperature, bolometric correction and mass for early-type stars using Strömgren photometric indices.**
L. A. Balona.
Mon. Not. R. Astron. Soc., Vol. 211, No. 4, p. 973 – 979 (1984).

Using the Kurucz model atmospheres as a guide, the author calibrates effective temperatures and bolometric corrections for early-type stars in terms of the Strömgren $c_0$ and $\beta$ indices. These are placed on an empirical scale by adjusting the zero points for best agreement with measurements by Code et al. The form of the mass calibration is determined from evolutionary calculations and the zero point adjusted for best agreement with empirical determinations. A mass discrepancy problem arises when these masses are compared with those determined from model atmospheres.

115.017 **Some problems in the determination of fundamental stellar parameters from binary stars.**
D. M. Popper.
Mitt. Astron. Ges., Nr. 62, p. 19 – 28 (1984). – See Abstr. 010.341.

115.018 **99 δ-Cepheiden-Geschwindigkeiten im HRD.**
R. Schröder.
Mitt. Astron. Ges., Nr. 62, p. 274 (1984). – See Abstr. 010.341.

115.019 **Die absoluten Helligkeiten der roten Zwergsterne.**
W. Gliese, H. Jahreiß.
Mitt. Astron. Ges., Nr. 62, p. 286 – 288 (1984). – See Abstr. 010.341.

115.020 **R136: multiple or monster?**
A. R. Walker, D. E. O'Donoghue.
Astron. Express, Vol. 1, Issue 2, p. 45 – 57 (1984).

CCD images of R136 taken through V and I filters in excellent seeing have been sharpened using maximum entropy methods to provide spatial detail with a resolution of about 0.4 arc sec. The brightest component, R136$a_1$, has $M_V = -8.3$ which is fainter than previously measured due to a more accurate estimate of the contribution of close stars and the background. These observations lend weight to the view that R136 is the partially resolved core of NGC 2070 and as such, consists of massive and luminous, but normal, early type stars.

**Metal-deficient giants in the galactic field. Catalogue and some physical parameters.**
See Abstr. 002.005.

**Measuring the sizes of stars.**
See Abstr. 036.031.

**Measurement of stellar diameters by combining the light issuing from two telescopes.**
See Abstr. 036.060.

**Report on a special session on space observations of stellar oscillations, held at the EPS study conference "Oscillations as a probe of the Sun's interior", Catania, 20 – 24 June 1983.**
See Abstr. 051.023.

**Mass loss and R136a-type stars.**
See Abstr. 064.036.

**Dimensional approach to mass loss: O- and B-type stars.**
See Abstr. 064.065.

**Wolf-Rayet stars in clusters: the initial stellar masses and evolutionary connections between subtypes.**
See Abstr. 065.002.

**Synthetic H-R diagrams as an observational test of stellar evolution theory.**
See Abstr. 065.022.

**The initial mass function of massive stars.**
See Abstr. 065.062.

**Photoelectric observations of lunar occultations of stars. The angular diameter of the star 61 $\delta^1$ Tauri.**
See Abstr. 096.002.

**Time variations of the Hα line profile from the core of Eta Carinae.**
See Abstr. 112.060.

**Luminosity classification with the Washington system.**
See Abstr. 113.026.

**The A0 stars.**
See Abstr. 114.020.

**The Sun and α Cen A.**
See Abstr. 114.031.

**Spectral analysis of (very) hot stars: methods and results.**
See Abstr. 114.032.

**The infrared Ca II triplet: a luminosity indicator for stellar population synthesis.**
See Abstr. 114.056.

**Observational knowledge about the physical properties of O stars.**
See Abstr. 114.060.

**R31: a massive O + W-R star binary in the Small Magellanic Cloud.**
See Abstr. 120.016.

**Arythmies stellaires.**
See Abstr. 122.071.

**Cepheid radii from infrared photometry.**
See Abstr. 122.140.

**Lyman and Balmer continuum ionization in zero-age main-sequence stars: applications to the line excess phenomenon.**
See Abstr. 132.020.

**Properties of R136a as derived from its optical light distribution.**
See Abstr. 132.022.

**On the nature of the Wolf-Rayet component in the core of the massive galactic H II region NGC 3603.**
See Abstr. 153.026.

**The initial mass function for massive stars in the Galaxy and the Magellanic Clouds.**
See Abstr. 155.041.

**The evolution of asymptotic giant branch stars in the Large Magellanic Cloud.**
See Abstr. 156.011.

**The luminosities of the M supergiants in M33.**
See Abstr. 157.025.

# 116 Rotation, Magnetic Fields, Activity, Polarization, Radio Radiation

116.001 **General study of the apparently average magnetic Ap star Theta Aurigae. I. Data and method.**
S. J. Adelman, H. Hensberge, W. Van Rensbergen.
Astron. Astrophys., Suppl. Ser., Vol. 57, No. 1, p. 121 – 137 (1984).

The authors report spectrophotometric and spectroscopic observations of θ Aurigae, obtained for the major part from 1977 on. The spectrophotometry covers the wavelength region from 3300 Å to 7850 Å, for 19 phases distributed over 80 percent of the rotation cycle. Spectra were taken at 10 phases, 8 of which are concentrated in 60 percent of the rotation cycle. The spectrophotometry is used to refine the rotation period. The spectra show rather broad lines that vary in strength and profile, although the changes are much less obvious than in the classical magnetic spectrum variables.

116.002 **General study of the apparently average magnetic Ap star Theta Aurigae. II. Model.**
W. Van Rensbergen, H. Hensberge, S. J. Adelman.
Astron. Astrophys., Vol. 136, No. 1, p. 31 – 36 (1984).

New spectrophotometric and spectroscopic data for θ Aur are used together with data from the literature to derive $t \cong 2\times10^8$ yr, $M \cong 3.5\,M_\odot$, $R \cong 4.5\,R_\odot$, $\log g \cong 3.30$, $T_{eff} \cong 10{,}500$K, overabundances of Si( × 8), Fe( × 13), Cr( × 65) and an underabundance of He( × 1/35). Thus θ Aur is near the end of its main-sequence life. The authors propose that it is an oblique rotator whose rotation axis is seen under an angle of $\cong 50°$. The polar axis of the magnetic field, whose strength is of the order of 1 kG, is inclined over $\cong 75°$ to the rotation axis. Metals are concentrated towards the magnetic equator (chromium, possibly silicon) or towards negative polarities (iron).

116.003 **On the determination of the upper limit of the rotation velocity of hot stars.**
V. P. Merezhin.
Astron. Tsirk., No. 1287, p. 1 – 3 (1983). In Russian.

116.004 **A note about the comparison between the 'old' and 'new' Slettebak systems of axial rotational velocities.**
B. E. García, H. Levato.
Rev. Mex. Astron. Astrofis., Vol. 9, No. 1, p. 9 – 12 (1984).

The authors redetermined projected-axial-rotational velocities for 207 stars in the 'new' Slettebak system. They discussed the relation between the 'old' and the 'new' systems obtaining regression lines to convert values from one system to the other.

116.005 **Variable polarization observations of the early-type supergiants ϰ Cas and α Cam.**
D. P. Hayes.
Astron. J., Vol. 89, No. 8, p. 1219 – 1228 (1984).

The optical continuum (B-band) linear polarization of the early-type supergiants ϰ Cas and α Cam was intensively monitored over the last two quarters of 1978 and, to a lesser extent, in 1976. Ordered but nonperiodic polarization changes were detected in both stars. These changes are manifestation of variable asymmetric mass loss in the stars' extended atmospheres. They may originate in aperiodic nonradial stellar pulsations.

116.006 **The rotation of B-type stars and processes of star formation.**
B. N. G. Guthrie.
Occas. Rep. R. Obs., Edinb., No. 13, p. 147 – 149 (1984). – See Abstr. 012.001.

116.007 **The rotation of early B-type stars and the problem of star formation.**
B. N. G. Guthrie.
Mon. Not. R. Astron. Soc., Vol. 210, No. 1, p. 159 – 171 (1984).

Distributions of rotational velocities are derived for samples of early B-type stars in young clusters and associations and field stars. Particular attention is given to the youngest portions of associations. The results are discussed under the aspect of star formation.

116.008 **Microwave radiation from magnetic stars.**
V. V. Zheleznyakov, Yu. V. Tikhomirov.
Astrophys. Space Sci., Vol. 102, No. 1, p. 189 – 209 (1984).

Cyclotron microwave emission from magnetic stars is considered, assuming that they have coronae with the temperature $T \lesssim 10^7$K and the emission measure $EM \lesssim 10^{54}\,\mathrm{cm}^{-3}$. It is shown that the cyclotron radiation from a star with a dipole magnetic field has a specific spectrum with a maximum in the frequency range $s\nu_0/2 < \nu < s\nu_0/\sqrt{2}$ ($s$ being the number of cyclotron harmonic, and $\nu_0$ the gyrofrequency corresponding to the polar magnetic field) and radiation flux decreasing towards lower frequencies as $\nu^{4/3}$. The radiation is partially circularly-polarized in the sense of an extraordinary mode. The degree of polarization is almost constant at frequencies $\nu < s\nu_0/2$ and increases with frequency at $\nu > s\nu_0/2$. The estimation of cyclotron radio fluxes of the nearest magnetic stars shows that they are observable in microwaves by means of modern radio astronomy.

116.009 **The appearance of polarization in radiation from hot stars due to the Faraday rotation effect as a possible method of determining stellar magnetic fields.**
Yu. N. Gnedin, N. A. Silant'ev.
Astrophys. Space Sci., Vol. 102, No. 2, p. 375 – 395 (1984).

The effect of Faraday rotation is shown to lead to the appearance of linear polarization of stellar radiation scattered in an optically-thin circumstellar electron-magnetized shell. The spectral dependence of the polarization degree is evaluated for scattering in (1) a spherically-symmetric magnetized shell with a power-law radial dependence of the electron density, and (2) a non-spherical ellipsoidal uniform envelope. The position of maximum in the polarization spectrum permits to determine the magnetic field magnitude on a star surface. If the rotational and magnetic axes do not coincide, a periodic variability of the polarization will be observed with the period of stellar rotation. Some Be and T Tauri stars are proposed as candidates to be investigated for magnetic fields. The method may also be applied to supernovae shells.

116.010 **Polarimetric investigation of background stars in the area of T and RY Tau.**
R. A. Vardanyan, M. A. Eritsyan.
Astrofizika, Tom 19, Vyp. 4, p. 719 - 724 (1983). In Russian. English translation in Astrophysics, Vol. 19, No. 4.
The results of electropolarimetric observations of stars located in the area between the stars T and RY Tau are presented. The authors could prove the earlier made supposition that the curvature of cometary nebulae cannot be caused only by the distortion of lines of forces of the local magnetic field.

116.011 **Polarimetric observations of carbon stars.**
M. A. Eritsyan, S. E. Nersisyan.
Astrofizika, Tom 20, Vyp. 2, p. 355 - 359 (1984). In Russian. English translation in Astrophysics, Vol. 20, No. 2.
Results of polarimetric observations of carbon stars in a filter system close to the B, V, R photometric system are presented. For 10 stars a polarization of 0.5% and higher is detected. For six of them the polarization is considered to be intrinsic.

116.012 **Studies in Be-star variability. I. A remarkable similarity of the rapid periodic light variations of EM Cep, σ Ori E, and possibly LQ And.**
P. Harmanec.
Bull. Astron. Inst. Czech., Vol. 35, No. 4, p. 193 - 202 (1984).
The aim of the present paper is to call attention to a distinct type of truly periodic rapid light variations observed for two - or possibly three Be stars. The variation is characterized by a light curve with two similar minima and maxima, rather short periods (1.191, 0.806, and 0.623 days for σ Ori E, EM Cep, and LQ And, respectively) and a complete lack of the radial-velocity variations associated with these periods. Definite spectral variations, correlated with the photometric period, are known for σ Ori E and can be suspected for the other two stars, too. Very probably, these variations are associated with the rotational periods of the stars. The magnetic oblique rotator model, suggested for He variables by Shore and Bolton (1976, 1982) is worth considering as a general model for this type of stars.

116.013 **Photoelectric rotational velocities of late-type dwarfs.**
W. Benz, M. Mayor.
Astron. Astrophys., Vol. 138, No. 1, p. 183 - 188 (1984).
The photoelectric determination of rotational velocities by means of the correlation technique yields accurate $v_e \sin i$ values for F to M dwarfs, even for slow rotators, down to 2 km s$^{-1}$. The root mean square dispersion between other existing accurate determinations and CORAVEL rotational velocities is about 1 km s$^{-1}$. An additional line-broadening mechanism exists with increasing equatorial velocity and spectral type. This line broadening probably results from an increase of turbulence and/or magnetic field. CORAVEL rotational velocities are given for 95 stars.

116.014 **Magnetic activity in lower main sequence stars.**
O. Vilhu.
ESA Spec. Publ., ESA SP-218, p. 239 - 242 (1984). - See Abstr. 012.024.
The IUE satellite has provided a good opportunity to study transition region plasma ($10^5$K, coronal loop foot points) in main sequence F-M stars. In the present study the author discusses the radiation from this plasma and its dependence on the crucial stellar parameters: rotation and spectral type.

116.015 **A note on the relation between magnetic fields and the λ5200 feature in helium-weak stars.**
H. M. Maitzen.
Astron. Astrophys., Vol. 138, No. 2, p. 493 - 494 (1984).
Magnetic field strengths and the λ5200 feature appear to be correlated in He-weak stars. This is a further evidence that these stars form the hot extension of the upper main sequence chemically peculiar stars. To take account of this situation an extension of Preston's (1974) nomenclature of CP-stars is suggested.

116.016 **The magnetic-field variations of HD 32633.**
P. Renson.
Astron. Astrophys., Vol. 139, No. 1, p. 131 - 134 (1984). In French.
The magnetic-field measurements of HD 32633 by Borra and Landstreet (1980) are compared with the values of $H_e$ measured by Babcock (mostly unpublished) and those published by Preston and Stepień (1968), in order to improve the value obtained for the period. The result is $P = 6^d.42998 \pm 0^d.00012$. As found previously (Renson, 1969) from Babcock's observations alone, the variation is highly anharmonic and asymmetric, the ascending part of the curve (from negative to positive values of $H_e$) being much steeper than the descending part. Moreover, in the descending part, there is at least one bump, which is also very difficult to explain within the framework of the models proposed up to now.

116.017 **Rapid rotation and Hα emission anomalies among some members of the population of low-mass stars.**
A. Young, A. Skumanich, E. Harlan.
Astrophys. J., Vol. 282, No. 2, p. 683 - 687 (1984). With plate 10.
The authors report on some important anomalous phenomena which were discovered during a systematic investigation of the properties of low-mass (M dwarf) stars. They find that the single M dwarfs GLS 410 and GLS 179 have a diluted Hα absorption line. It is suggested that this is due to weak plage activity. The dMe star GLS 890 is a rapidly rotating single M dwarf with $v \sin i \approx 70$ km s$^{-1}$. The authors find that the Joy and Abt "dMe" assignment is in error not only for GLS 179 but also for GLS 176, GLS 611.3, and GLS 908.

116.018 **Spectroscopic evidence for starspots in the G dwarf HD 1835.**
B. Campbell, R. Cayrel.
Astrophys. J., Lett. Ed., Vol. 283, No. 1, p. L17 - L20 (1984).
High signal-to-noise ratio spectra of HD 1835 (G2 V) reveal very weak lines of titanium oxide and calcium hydride. The authors show that these are unlikely to come from an M dwarf companion. The molecular lines probably originate in starspot umbrae, which cover about 3% of the stellar surface.

116.019 **The magnetic activity of sunlike stars.**
A. H. Vaughan.
Science, Vol. 255, No. 4664, p. 793 - 800 (1984).
Centuries of observation have enabled solar scientists to recognize that the Sun's magnetism exists and varies in a globally organized pattern that is somehow coupled to the Sun's rotation. Within the past decade O. C. Wilson demonstrated that analogs of solar activity exist and can be studied in many other dwarf stars. From the continuing study, knowledge of the precise rates of rotation of the stars under investigation is being gained for the first time. The results are expected to increase our understanding of the origin of solar activity and stellar activity in general.

116.020 **Age-rotation relationship for late-type main-sequence stars.**
T. N. Rengarajan.
Astrophys. J., Lett. Ed., Vol. 283, No. 2, p. L63 - L65 (1984).
The Mount Wilson spectrophotometric data on the Ca II H and K line fluxes are used in conjunction with the 1984 analysis of Noyes et al. relating the stellar rotation period to the H - K line flux, to obtain a plot of $B - V$ versus rotation period. From this it is concluded that the rotation period is a single-valued function of mass and age of main-sequence stars. For a dependence of the form $P \propto \tau^\alpha$, τ being the age of the star, the exponent α is a constant equal to $0.46 \pm 0.10$ in the $B - V$ range 0.5 - 1.0.

116.021 **Polarization measurements at La Silla.**
T. Korhonen, V. Piirola, A. Reiz.
Messenger, No. 38, p. 20 - 24 (1984).

116.022 **The magnetic fields of the late-type stars.**
E. F. Borra, G. Edwards, M. Mayor.
Astrophys. J., Vol. 284, No. 1, p. 211 - 222 (1984).

The authors have assembled a powerful stellar magnetometer capable of measuring longitudinal magnetic fields as small as a few gauss. They have observed a large number of late-type stars, and obtained mostly nulls or detections that are suspect. They do, however, find definite detections for the Ca II emission dwarf ξ Boo A and probable detections in the RS CVn star UX Ari. Observations of the transverse Zeeman effect do not show evidence of strong transverse fields. The authors make an approximate estimate of the magnetic geometries of active dwarfs, concluding that the most likely geometry is complex and composed of several hundred patches of opposite polarities.

116.023 **Chromospheric emission and rotation of the Hyades lower main-sequence.**
D. K. Duncan, S. L. Baliunas, R. W. Noyes, A. H. Vaughan, J. Frazer, H. H. Lanning.
Publ. Astron. Soc. Pac., Vol. 96, No. 583, p. 707 - 713 (1984).

The identity of chromospheric Ca II H- and K-line emission has been monitored for two seasons in a large sample of Hyades F and G dwarfs. Mean emission levels vary smoothly with spectral type, suggesting that the rotation rate at a given spectral type varies relatively little, and that the rotation rate decreases smoothly as a function of mass. In this case the mechanism which controls pre-main-sequence angular momentum loss in late-type stars must be self-regulating to a high degree.

116.024 **Evidence for global oscillations in the K2 dwarf Epsilon Eridani.**
R. W. Noyes, S. L. Baliunas, E. Belserene, D. K. Duncan, J. Horne, L. Widrow.
Astrophys. J., Lett. Ed., Vol. 285, No. 1, p. L23 - L26 (1984).

The authors have obtained evidence of global *p*-mode oscillations in the K2 V star ε Eri (HD 22049), based on observed time variations in the Ca II H and K emission lines. Power spectra of time series of Ca II intensity measures reveal a number of peaks, spaced at about 86 and 172 μHz. The 172 μHz spacing is significantly larger than the corresponding spacing (136 μHz) observed for solar *p*-mode oscillations but is in excellent agreement with predictions for a star of the known radius of ε Eri. The amplitude of the oscillations seen in the H and K flux of this chromospherically active star is much larger than in the Sun.

116.025 **Polarization measurements of Be stars at Hβ and Ca II *H* and *K*.**
D. Clarke, A. Brooks.
Mon. Not. R. Astron. Soc., Vol. 211, No. 3, p. 737 - 752 (1984).

Eight of the brightest Be stars have been measured for linear polarization across the spectral lines of Hβ, Ca II *H* and Ca II *K*. Dips in the polarization at Hβ have been confirmed in the stars which have been measured previously; a dip was discovered for ω Ori, while no dip could be detected for φ And. Polarization dips at Ca II *H* have been discovered for γ Cas, φ Per and possibly for ψ Per and ζ Tau. Enhancements of polarization at Ca II *K* have been discovered for ζ Tau, ϰ Dra and possibly for γ Cas and φ Per. There appears to be a tendency for the polarizations to dip in the hydrogen lines and to be enhanced in the calcium lines. An earlier hypothesis that the polarization dips in the Balmer lines could be ascribed to unpolarized emission flux is shown not to have general validity.

116.026 **Some observational results on long-time variability of stars with magnetic fields.**
G. Scholz.
Astron. Nachr., Vol. 305, No. 6, p. 325 - 328 (1984).

The intention of this paper is a summary of long-time phenomena of magnetic stars which were obtained at the Potsdam Observatory. The considered magnetic stars are γ Equ, HD 9996, 53 Cam, 52 Her, and ν Cep. Many results have been published already; here, they are compiled and completed by new details.

116.027 **Evidence of decay of the magnetic fields of Ap stars.**
P. North, N. Cramer.
Astron. Astrophys., Suppl. Ser., Vol. 58, No. 2, p. 387 - 403 (1984).

The authors provide, on the basis of 708 field and cluster Ap stars, observational evidence that these stars undergo decay of their magnetic field on an evolutionary timescale. Justifications are given for the application of a photometric gravity calibration to peculiar stars. The dependence of the photometrically estimated surface magnetic field on gravity is found to differ markedly from available theoretical calculations. HgMn stars are found to show the same trend, strengthening the impression that they might be slightly magnetic. He-weak stars do not.

116.028 **Evidence for a radial magnetic field gradient on the Ap star $\alpha^2$ Canum Venaticorum.**
I. I. Romanyuk.
Sov. Astron. Lett., Vol. 10, No. 3, p. 184 - 186 (1984). English translation of 37.116.039.

116.029 **Linear-polarization line profiles as an indicator of the magnetic field geometry of Ap stars.**
N. E. Piskunov, V. L. Khokhlova.
Sov. Astron. Lett., Vol. 10, No. 3, p. 187 - 189 (1984). English translation of 37.116.040.

116.030 **Solar and non-solar activity in stars: the cosmic-ray connection.**
T. Montmerle.
Adv. Space Res., Vol. 4, No. 2 - 3, p. 357 - 366 (1984). - See Abstr. 012.075.

This paper reviews recent evidence that some components of cosmic rays may originate in stellar atmospheres. Arguments are presented which indicate that low-energy heavy nuclei may result from solar type activity in late-type stars. On the other hand, stellar winds from hot early-type stars may be a source for a fraction of cosmic-ray protons and alpha particles.

116.031 **The relation between stellar activity, rotation, convection, and implications for the combined study of stellar activity and oscillations from space.**
R. W. Noyes.
Space research prospects in stellar activity and variability, p. 261 - 263 (1984). Abstract. - See Abstr. 012.078.

116.032 **Progress on rotational modulation studies of continuum and line fluxes.**
M. Rodonò.
Space research prospects in stellar activity and variability, p. 277 - 289 (1984). - See Abstr. 012.078.

The methodology commonly adopted in studying quasi-periodic flux variations, which are produced by surface inhomogeneities on a rotating star - the so-called rotational modulation study - is briefly reviewed with particular emphasis on recent results concerning the physical characteristics and topology of stellar active areas at various atmospheric levels, such as photospheric spots, chromospheric and transition-region plages and coronal features. The present observational limits are discussed and the need of highly accurate and dedicated photometry is stressed as a necessary and, maybe, sufficient condition for a break-through in the study of stellar activity.

116.033 **Chromospheric, transition-region and coronal diagnostics for stellar magnetic activity.**
C. J. Schrijver, C. Zwaan.
Space research prospects in stellar activity and variability, p. 291 - 308 (1984). - See Abstr. 012.078.

Using various diagnostics magnetic activity has been inferred for the vast majority of stars with convective envelopes. The authors discuss relations between emissions from different temperature regimes, and their dependence on various stellar parameters. They suggest several practical indicators for stellar magnetic structure and acitivity, and outline a few topics for which space observations would be useful.

116.034 **Rotation of 53 Be stars.**
W. Gao, H. Cao.
Bull. Inf. Etoiles Be, No. 10, p. 54 – 59 (1984).

116.035 **Change of light from a spherical star with two hot precessing spots.**
D. I. Kyurkchieva, V. G. Shkodrov.
Astron. Tsirk., No. 1308, p. 3 – 4 (1984). In Russian.

116.036 **Magnetic cycles in solar type stars.**
I. Tuominen.
Rep. Ser., Dep. Phys. Sci., Univ. Turku, No. FTL–R68, p. 29 – 30 (1984). – See Abstr. 012.079.

116.037 **The appearance of magnetic flux on the surfaces of the early main–sequence F stars.**
M. S. Giampapa, R. Rosner.
Astrophys. J., Lett. Ed., Vol. 286, No. 1, p. L19 – L22 (1984).
The authors examine available chromospheric, transition region, and coronal observations of the early main–sequence F stars to find that while these objects exhibit enhanced levels of magnetic–field–related radiative emissions, significant inhomogeneities in surface activity are not present. The authors discuss this phenomenon within the context of the calculations published by Schmitt and Rosner in 1983 for the production of flux ropes of various spatial scales at a given rotation rate at the bottom of a stellar convection zone.

116.038 **Emergent polarization structure in spectral lines from rotating stars of spectral type B.**
S. J. McKenna.
Astrophys. Space Sci., Vol. 107, No. 1, p. 61 – 69 (1984).
The emergent polarization structure within spectral lines from rapidly rotating stars of spectral type B5V has been computed for the Mg II 4481 Å doublet and Hγ for several angles of inclination and for rotational speeds of 0.5 and 0.9 of the critical rotational speed. It is found in this study that the emergent polarization within the Mg II 4481 Å doublet is currently unobservable with modern polarimeters. However, the polarization within Hγ should be observable.

116.039 **Rotational velocities of later B type and A type stars as determined from ultraviolet versus visual line profiles.**
K. G. Carpenter, A. Slettebak, G. Sonneborn.
Astrophys. J., Vol. 286, No. 2, p. 741 – 746 (1984).
Theoretical line profiles of the Si III 1299 and Fe II 2756 lines are computed for rotating B5 – A7 model stars, and compared with observed profiles from *IUE* spectra to derive rotational velocities. The comparison gives rotational velocities that are in good agreement with visually determined $v \sin i$ values for the same stars, which suggests that the shape–distorted, gravity–darkened models are reasonable.

116.040 **The space orientation of stars.**
L. R. Doyle, T. J. Wilcox, J. J. Lorre.
Astrophys. J., Vol. 287, No. 1, p. 307 – 314 (1984). With plate 7.
In order to determine the space orientation of a star, one must first find a way to isolate the rotational–velocity component from the stellar inclination to the line of sight, as well as determine the clocklike, on–the–plane–of–the–sky orientation. The authors have found that these components may be separable by looking at the stars in the Ca II H and K line emission associated with starspot activity. The application of this method to current searches for extrasolar planetary systems is discussed.

116.041 **Ultraviolet observations of the Be star Mu Centauri during its recent active phase.**
G. J. Peters.
Publ. Astron. Soc. Pac., Vol. 96, No. 586, p. 960 – 967 (1984).
*IUE* and *Copernicus* ultraviolet observations and ground–based image–tube spectrograms of the "pole–on" Be star μ Cen have been combined to investigate the nature of this star's activity during its Balmer emission–line episode of 1980 – 81. Within the uncertainties of the observations, the UV data show *no* evidence for a (1) wind, (2) a classical type shell ejection, or (3) a region of superionization, either before, during, or after the active phase. Evidence is presented that the activity most likely resides in or near the photosphere.

116.042 **The relation between stellar rotation rate and activity cycle periods.**
R. W. Noyes, N. O. Weiss, A. H. Vaughan.
Astrophys. J., Vol. 287, No. 2, p. 769 – 773 (1984).
The empirical relation between rotation period, spectral type, and activity cycle period is investigated for a sample of 13 slowly rotating lower main–sequence stars, including the Sun, all of which show long–term chromospheric variability like that of the solar cycle. It is found that for slowly rotating stars of similar spectral type, the cycle period $P_{cyc}$ and rotation period $P_{rot}$ are related by $P_{cyc} \propto P^{n}_{rot}$, where $n \approx 1.25$. When the stars are considered as a group, their cycle periods are found to be consistent with the relation $P_{cyc} \approx (P_{rot}/\tau_c)^n$, where $\tau_c$ is the convective turnover time near the bottom of the convection zone. These relations are interpreted in terms of simple nonlinear dynamo models.

**Observations of stellar activity: advantages and disadvantages of ground based multisites, polar region observatories, and space observations.**
See Abstr. 013.059.

**On the need to go to space to study stellar variability and activity.**
See Abstr. 013.060.

**Direct and indirect methods of measurement of stellar magnetic fields.**
See Abstr. 036.147.

**Declinations of 35 radio stars.**
See Abstr. 041.040.

**Magnetic field generation in celestial bodies.**
See Abstr. 062.003.

**Stellar fibril magnetic systems. I. Reduced energy state.**
See Abstr. 062.066.

**Theory of magnetic stars.**
See Abstr. 062.113.

**Corotating interaction regions in stellar winds.**
See Abstr. 064.028.

**Comments concerning magnetic fields as a possible cause of rapid, irregular variability of early–type stars.**
See Abstr. 064.032.

**Magnetic starspots.**
See Abstr. 064.039.

**Relation between stellar activity, rotation and stellar dynamos.**
See Abstr. 064.052.

**Magnetic starspots.**
See Abstr. 065.090.

**Long periodic variations of magnetic stars caused by precession.**
See Abstr. 065.092.

**The rotation of Ap stars.**
See Abstr. 065.094.

**The photometric variability of solar–type stars. IV. Detection of rotational modulation among Hyades stars.**
See Abstr. 113.025.

**IUE spectra of three A0 V stars rotating close to their break–up velocity: HD 119921, HD 129685, HD 181296.**
See Abstr. 114.040.

**Maps of the iron–group elements on the magnetic Ap star $\alpha^2$ Canum Venaticorum.**
See Abstr. 114.097.

**Spectroscopic and photometric observations of the suspected binary star HD 123058.**
See Abstr. 114.102.

**Similar photospheres, different mantles: a study of four O stars.**
See Abstr. 114.138.

**Detection of a sub–milliarcsecond radio component in the RS CVn system HR 1099.**
See Abstr. 117.015.

**Rotational modulation of spots and plages on RS CVn stars.**
See Abstr. 117.056.

**Ultraviolet flare on Lambda Andromedae.**
See Abstr. 117.090.

**The FK Comae candidate UZ Librae.**
See Abstr. 117.140.

**Starspot cycle of RS CVn.**
See Abstr. 117.151.

**Ca II H and K, and Hα emission variability in AR Lacertae.**
See Abstr. 117.195.

**Time–resolved spectroscopy of long–period DQ Her stars.**
See Abstr. 117.232.

**Flare of HR 1099 in the radio range.**
See Abstr. 117.235.

**Measurement of the radial velocity of 32 Cygni during its 1981 eclipse.**
See Abstr. 119.087.

**The synchronous rotation of Iota Pegasi.**
See Abstr. 120.006.

**The many faces of Capella: a search for rotational modulations and a study of systematic velocities of emission lines in the ultraviolet.**
See Abstr. 120.015.

**On the photometric activity of R Coronae Australis.**
See Abstr. 121.001.

**Activity and latent periodicity of BN Orionis.**
See Abstr. 122.002.

**Polarimetric observations of μ Cep.**
See Abstr. 122.056.

**IUE observations of BY Draconis.**
See Abstr. 122.064.

**Pulsation of CP2 stars.**
See Abstr. 122.103.

**Variability of the lambda–5200 feature in CP2 stars.**
See Abstr. 122.104.

**Discovery of a possibly rotation–induced resonant coupling of two nonradial pulsation modes in γ Arae (B1 Ib).**
See Abstr. 122.129.

**A further search for the δ–Sct–like variability of the Ap star HD 10088.**
See Abstr. 122.169.

**Determination of the magnetic field in the star V474 Mon.**
See Abstr. 122.187.

**Model of the phenomena responsible for variations of polarization and brightness of μ Cep.**
See Abstr. 122.217.

**The rotational velocities of white dwarfs.**
See Abstr. 126.063.

**Polarimetric and new photometric observations in the NGC 6334/6357 region.**
See Abstr. 132.009.

**Spectral anomalies in the Hyades and Pleiades and in field stars with active chromospheres.**
See Abstr. 153.009.

**The Pleiades cluster. I. Polarization and reddening of the brighter stars.**
See Abstr. 153.011.

**Rotational studies of lower main sequence stars in nearby open galactic clusters. I. Velocity distributions and age dependence.**
See Abstr. 153.014.

**Color anomalies and starspots in Hyades dwarfs.**
See Abstr. 153.018.

# 117 Close Binaries (Observations, Theory)

117.001 **The change of 4.8 hour period of Cygnus X–3.**
Yu. I. Neshpor, Yu. L. Zyskin.
Perem. Zvezdy, Tom 21, No. 5, p. 709 – 713 (1982). In Russian.
The data of observations of gamma–ray quanta from Cyg X–3 with energy $>2\times10^{12}$eV have been analysed to search for possible variations of the 4.8 hr period. It is shown that the 4.8 hr period of Cyg X–3 is increasing at a rate $\dot{P} = (3.0\pm1.4)\times10^{-9}$s/s.

117.002 **Light curves of close binaries with common scattering envelopes.**
I. B. Pustyl'nik, L. Einasto.
Pis'ma Astron. Zh., Tom 10, No. 7, p. 516 – 522 (1984). In Russian. English translation in Sov. Astron. Lett., Vol. 10.
A method for calculating light curves of a close binary system consisting of two spherical components imbedded in a common, spherically symmetric, isotropically scattering envelope with radial opacity gradient is proposed. It is shown that the observed light curves of close binaries can be interpreted in the framework of the adopted model with accounting for stellar eclipses combined with the effect of periodically variable screening of stellar radiation by the envelope. The model curves are compared with the light curves of AO Cas, RY Gem, VW Cep, the orbital elements and the parameters of the envelope are determined for these systems. The conclusion is drawn that AO Cas is a spectroscopic, rather than an eclipsing binary imbedded in a common scattering envelope of free electrons.

117.003 **On the parameters of the Cygnus X–1 system.**
V. P. Merezhin.
Astron. Tsirk., No. 1284, p. 3 – 4 (1983). In Russian.

117.004 **Mass transfer and the period gap of cataclysmic variables.**
F. Verbunt.
Mon. Not. R. Astron. Soc., Vol. 209, No. 1, p. 227 – 240 (1984).
Three different explanations for the period gap of cataclysmic variables are investigated in some detail, and compared with the observations. The static picture is ruled out; strong continued magnetic braking is shown to be unlikely; disrupted magnetic braking is shown to provide a good explanation. A simple derivation is given for the magnetic braking of a star as a function of the magnetic–field strength and the wind mass flux. A field strength of $\gtrsim 100$ gauss and a wind of $10^{-10}M_{\odot}\text{yr}^{-1}$ are needed for the secondary of a cataclysmic variable to explain the braking.

117.005 **Clues from the photonuclear time scale on the nature of particle accelerators in Cygnus X–3 and Vela X–1.**
R. J. Protheroe.
Nature, Vol. 310, No. 5975, p. 296 – 298 (1984).
Ultrahigh–energy $\gamma$ rays ($\gtrsim 10^{15}$ eV) have recently been observed from the binary X–ray sources Cygnus X–3 and Vela X–1. The source of these $\gamma$ rays remains unknown but it seems likely that protons or nuclei are accelerated, the $\gamma$ rays resulting from nuclear interactions with matter or radiation. The author has calculated the attenuation times of protons and heavier nuclei in the radiation fields of the stellar companions to the Cyg X–3 and Vela X–1 compact objects. The maximum observed $\gamma$–ray energy is used to put an upper limit on the acceleration time scale of protons at $3\times10^{16}$ eV of $10^4 - 3\times10^5$ s for Cyg X–3 and $10^5$ s for Vela X–1. This result can be used to put some constraints on the nature of the particle accelerator.

117.006 **Mass transfer in dwarf binaries: a compilation.**
F. Verbunt, R. A. Wade.
Astron. Astrophys., Suppl. Ser., Vol. 57, No. 2, p. 193 – 196 (1984).
A compilation is presented of mass transfer rates in dwarf binaries as derived from observations and published in the literature. The problems in determining accurate values for the mass transfer rate, and in comparing these values with theory, are enumerated.

117.007 **Sco X–1: a mini quasar?**
W. Kundt, Gopal–Krishna.
Astron. Astrophys., Vol. 136, No. 1, p. 167 – 170 (1984).
A model is proposed for the Galactic triple radio source and X–ray binary Sco X–1 that envisages its origin inside an H I cloud: the high–pressure ambient medium postulated for confining the bright radio lobes is due to the supernova explosion which occurred some $3\times10^3$ yr ago and which gave birth to the (now accreting) neutron star. The supernova shell has already faded away, similar to what is likely to happen to Tycho's shell whose expansion speed has decreased by a factor of 2.6 after no more than 400 yr. The feeding jets are relativistic.

117.008 **Orbital period changes in cataclysmic variables?**
K. Beuermann, M. W. Pakull.
Astron. Astrophys., Vol. 136, No. 2, p. 250 – 254 (1984).
The authors report observations of times of minima for four eclipsing cataclysmic variables: U Gem, EM Cyg, T Aur, and V2051 Oph. For all systems, no secular changes of the orbital period $P$ were detected. The authors derived $3\sigma$ lower limits to the time constant $\tau = P/\dot{P}$ for such changes of $1.4\times10^8$ years, $7.5\times10^6$ yr, $4.4\times10^6$ yr and $7.6\times10^5$ yr for U Gem, EM Cyg, T Aur, and V2051 Oph, respectively. The corresponding upper limits on the mean mass–transfer rate $\dot{M}$ derived on the assumption of mass and angular momentum being conserved are still significantly above independent estimates of $\dot{M}$, using accretion disk models. For U Gem the existence of quasiperiodic O–C variations with a time scale of $\sim 18$ yr is confirmed.

117.009 **Apsidal motion and the pulsating transient X–ray source A0538–66.**
C. B. Boyle, H. F. Henrichs.
Astron. Astrophys., Vol. 136, No. 2, p. 326 – 330 (1984).
It is pointed out that current models which are proposed for the pulsating transient X–ray binary system A0538–66 imply an apsidal–motion period which, according to the "classical" formula, may be as low as about 150 yr. Such a short period should show up as a small but significant difference between the sidereal binary period and the accurately known outburst cycle of $16.6515\pm0.0005$ d, as the latter is attributed to periastron passage. Therefore, after a single accurate determination of the sidereal binary period (which is, in principle, possible because of the presence of the fast pulsar), the apsidal–motion period can be immediately calculated. This would give important information on the internal structure of the massive, post mass–gaining B star in this system. A discussion of the evolutionary stage of A0538–66 is given. It is concluded that the system is rather old in the sense that the X–ray pulsar is spun back up to its present very short rotation period.

117.010 **Cygnus X–1 revisited.**
E. P. Liang, P. L. Nolan.
Space Sci. Rev., Vol. 38, Nos. 3/4, p. 353 – 384 (1984).
The authors review the X– and gamma–ray observations of Cygnus X–1 and their theoretical interpretations, with emphasis on new developments since the mid–1970's. The overall data base at present is most consistent with the inverse Compton model by hot thermal electrons of $T_e \sim 10^9$K, for the hard X–ray luminosity (10 – 200 keV). However, the origin of the soft X–rays ($\leqslant 10$ keV) in high states and gamma rays ($>200$ keV) remain unsettled.

117.011 **Can activity of the secondary components explain the emission–line spectra of cataclysmic binaries?**
S. M. Rucinski.
Observatory, Vol. 104, No. 1061, p. 186 – 188 (1984).
It is shown that the emission lines of C IV observed in cataclysmic dwarf–nova binaries in quiescence cannot be due to the

magnetic activity of the secondary components; these stars are much too faint to provide the necessary emission–line fluxes.

117.012 **Symbiotic stars.**
M. Kafatos, A. G. Michalitsianos.
Sci. Am., Vol. 251, No. 1, p. 70 – 80 (1984).

Satellite observations have revealed that certain celestial objects with a peculiar spectrum consist of a red–giant star surrounded by a small dense nebula heated by a compact hot companion star.

117.013 **IR Geminorum: indications of a massive white dwarf and a heated secondary in this new SU Ursae Majoris cataclysmic variable.**
P. Szkody, A. W. Shafter, A. P. Cowley.
Astrophys. J., Vol. 282, No. 1, p. 236 – 244 (1984).

The authors report the discovery of a new SU UMa type cataclysmic variable through confirmation of a 102 minute superhump period during a supermaximum. Spectroscopic and photometric studies during quiescence show a 98 – 101 minute orbital period with evidence for heating of the secondary star. The flux distribution from 0.1 to 1.25 μm fits with a 20,000K blackbody. The low semiamplitude of the radial velocity curve implies a high mass white dwarf and possibly a degenerate secondary, but may be complicated by an unresolved component in the emission lines.

117.014 **Spectral variability in early–type binary X–ray systems.**
R. McCray, T. R. Kallman, J. I. Castor, G. L. Olson.
Astrophys. J., Vol. 282, No. 1, p. 245 – 255 (1984).

The authors present theoretical models for the ionization of trace elements in a strong stellar wind by a compact binary X–ray source and for the resulting orbital phase dependence of the emergent soft X–ray spectra and the profiles of ultraviolet resonance lines. Model results agree qualitatively with the X–ray and ultraviolet spectra of the system 4U 0900–40/HD 77581 and explain the suppression of the absorption profiles of the Si IV λ1394 and C IV λ1548 lines when the X–ray source is in front of the star. The models predict phase–dependent linear polarization in the resonance line profiles.

117.015 **Detection of a sub–milliarcsecond radio component in the RS CVn system HR 1099.**
J.–F. Lestrade, R. L. Mutel, R. B. Phillips, J. C. Webber, A. E. Niell, R. A. Preston.
Astrophys. J., Lett. Ed., Vol. 282, No. 1, p. L23 – L26 (1984).

The RS CVn system HR 1099 was observed with a five station VLBI array at a frequency of 8.4 GHz during a strong radio outburst of approximately 400 mJy. The data are consistent with a circular Gaussian source of $0.8 \pm 0.12$ milli–arcsec, corresponding to a linear size of $4 \pm 0.6 \times 10^{11}$cm. This is comparable to the distance between the surfaces of the two stars, or to 75% of the diameter of the chromospherically active K star. The high equivalent brightness temperature, approximately $10^{10}$K, is consistent with gyrosynchrotron emission from electrons in a magnetic field of strength $B \sim 30$ gauss.

117.016 **Precessional period and modulation of relativistic radial velocities in SS433.**
A. Mammano, F. Ciatti, R. Turolla.
Field theory in elementary particles, p. 151 – 161 (1983). Abstr. in Phys. Abstr., Vol. 87, No. 1237, Entry 72959 (1984). – See Abstr. 012.010.

117.017 **Periodic and secular changes in SS433.**
G. W. Collins II, G. H. Newsom.
Field theory in elementary particles, p. 163 – 191 (1983). Abstr. in Phys. Abstr., Vol. 87, No. 1237, Entry 72960 (1984). – See Abstr. 012.010.

117.018 **Binary stars.**
B. Paczyński.
Science, Vol. 225, No. 4659, p. 275 – 280 (1984).

Most stars in the solar neighborhood are either double or multiple systems. They provide a unique opportunity to measure stellar masses and radii and to study many interesting and important phenomena. The best candidates for black holes are compact massive components of two X–ray binaries: Cygnus X–1 and LMC X–3. The binary radio pulsar PSR 1913+16 provides the best available evidence for gravitational radiation. Accretion disks and jets observed in close binaries offer a very good testing ground for models of active galactic nuclei and quasars.

117.019 **An interpretation of the SS433 light curve.**
S. N. Fabrika.
Sov. Astron. Lett., Vol. 10, No. 1, p. 16 – 19 (1984). English translation of 37.117.001.

117.020 ***UBVRI* catalogue of dwarf novae.**
J. Echevarría.
Rev. Mex. Astron. Astrofis., Vol. 9, No. 1, p. 99 – 105 (1984).

*UBVRI* photometry of seventy–five dwarf novae compiled from the literature is presented in this catalogue. The catalogue lists observations for each object per night per observer for statistical purposes.

117.021 **Der Lichtwechsel von Lambda Andromedae.**
D. Böhme.
Sterne Weltraum, 23. Jahrg., Nr. 7, p. 386 – 387 (1984).

117.022 **A new photoelectric investigation of the W UMa system U Pegasi.**
D. Zhai, K.–C. Leung, R. Zhang.
Astron. Astrophys., Suppl. Ser., Vol. 57, No. 3, p. 487 – 494 (1984).

Over 400 photoelectric observations of U Peg in *B* and *V* were secured with a 0.6 m reflector at Beijing Observatory in 1978. The system was found to have a secular period decrease and a short term (17 years) sinusoidal oscillation. It is suggested that the oscillating term is caused by the light–time effect of an unseen third body. The third body may be an M6 main sequence star with a mass of 0.16 $M_{\odot}$. The secular change may be associated with a slow mass transfer. It is suggested that the system U Peg has an overcontact configuration of about 9%. It has the characteristic of a W–type W UMa system. The photometric mass ratio is between 3 and 2.5. The absolute dimensions are of 0.6 and 1.8 $M_{\odot}$ and of 0.8 and 1.4 $R_{\odot}$, for components 1 and 2 respectively.

117.023 **IUE spectroscopy of the degenerate components in the Hyades close binaries V471 Tauri and HZ9.**
E. F. Guinan, E. M. Sion.
Astron. J., Vol. 89, No. 8, p. 1252 – 1255 (1984).

The authors have obtained low–resolution International Ultraviolet Explorer (IUE) spectra of the Hyades close binaries V471 Tauri and HZ9. Phase–dependent spectra of the eclipsing, spectroscopic binary V471 Tauri are described. Pure hydrogen, high gravity, LTE model atmospheres are used to obtain effective temperatures and gravities.

117.024 **Contact and near–contact binary systems. II. RR Cen, EZ Hya, V502 Oph and RS Sct.**
D. J. King, R. W. Hilditch.
Mon. Not. R. Astron. Soc., Vol. 209, No. 3, p. 645 – 653 (1984).

Reticon spectroscopy of four binary systems displaying *EW* or *EB* light curves are presented. Orbital parameters and minimum masses are derived for the first time for RR Cen, EZ Hya and RS Sct. From the five systems investigated in this survey to date, it is clear that contact systems and near–contact systems populate the same region of the total mass–total angular momentum plane.

117.025 **Flux density and structure variations in SS433.**
R. E. Spencer.
Mon. Not. R. Astron. Soc., Vol. 209, No. 4, p. 869 – 879 (1984).

Single baseline interferometers and the MERLIN instrument at Jodrell Bank have been used to investigate the flux density and structure variations in SS433. The interferometers enabled the flux density variations and spectra of an unresolved (< 0.3 arcsec) core and an arcsec halo or jet to be measured in

1979 August. Subsequent observations with MERLIN show structures confirming the twin–jet models for the optical emission. The angular velocity of the moving knots of radio emission can be combined with the linear velocity of the optical clouds to give a distance of 4.9±0.2 kpc. The velocity of the knots is constant and they appear to move along a channel formed by continuous jets from the core of SS433. The brightness of the knots decays with a time constant of 35 days, or alternatively as $(\mathrm{time})^{-1.8}$, a flatter power law than expected for adiabatic expansion.

117.026 **The evolution of a star–"planet" system in the double core phase.**
N. Soker, A. Harpaz, M. Livio.
Mon. Not. R. Astron. Soc., Vol. 210, No. 1, p. 189 – 195 (1984).
The spiralling–in of a black dwarf ("planet") inside the envelope of a red giant is studied. The calculation involves both the interaction of the black dwarf with the envelope and the evolution of the giant. The structural changes induced in the giant by an energy deposition and mass ablation are followed using a fully evolutionary code. The authors find that a black dwarf of an initial mass of 0.02 $M_\odot$ is transformed into a 0.14 $M_\odot$ (stellar) binary companion to the red giant's core. This process could be important for the formation of low mass cataclysmic binaries.

117.027 **Dwarf novae in outburst: simultaneous ultraviolet and optical observations of UZ Serpentis, RX Andromedae and AH Herculis.**
F. Verbunt, J. E. Pringle, R. A. Wade, J. Echevarria,
D. H. P. Jones, R. W. Argyle, A. Schwarzenberg–Czerny,
C. la Dous, R. Schoembs.
Mon. Not. R. Astron. Soc., Vol. 210, No. 1, p. 197 – 221 (1984).
The authors present spectroscopic and photometric observations in the optical and ultraviolet wavelength regions made during outbursts of the three dwarf novae UZ Serpentis, RX Andromedae and AH Herculis. The observations do not cover the rapid initial rise to outburst, but rather the peak and the subsequent decline. The overall changes of the continuum distributions are different in the three objects, and for UZ Ser the authors present evidence that the distribution repeats from outburst to outburst, although the line strengths do not. The decline from outburst is characterized by a drop in flux at all wavelengths, the drop being faster at shorter wavelengths. The spectrum of RX And observed at the decline is similar to that taken by Klare et al. at the same optical brightness but during a prolonged standstill.

117.028 **A visible and infrared study of the eclipsing dwarf nova OY Carinae. II. The infrared light curves.**
G. Berriman.
Mon. Not. R. Astron. Soc., Vol. 210, No. 1, p. 223 – 237 (1984).
The paper presents three simultaneous visible ($V$) and infrared ($J, H, K$) light curves of the eclipsing dwarf nova binary system OY Carinae in quiescence. The infrared light curves show a secondary minimum, not seen in the visible, which is the ellipsoidal variation of the red dwarf and its eclipse by the accretion disc surrounding the white dwarf companion. The red star, an M dwarf, supplies between 30 and 60 per cent of the total light at $J$, $H$ and $K$. This requires that the system is between 100 and 300 pc away. The infrared continuum of the accretion disc around the white dwarf companion comes largely from the optically thin gas giving rise to the emission lines seen in the visible and ultraviolet.

117.029 **Simultaneous infrared and optical observations of H2215–086.**
M. R. Sherrington, R. F. Jameson, J. Bailey.
Mon. Not. R. Astron. Soc., Vol. 210, No. 1, p. 1P – 6P (1984).
Simultaneous optical and infrared light curves of the intermediate polar H2215–086 are presented. The observations confirm the 21 min period found by Patterson and Steiner. A 2 μm excess, probably due to cyclotron emission, is found. The system distance is estimated to be $\gtrsim$400 pc.

117.030 **EXOSAT observations of H2215–086: detection of the X–ray pulse period.**
M. C. Cook, M. G. Watson, I. M. McHardy.
Mon. Not. R. Astron. Soc., Vol. 210, No. 1, p. 7P – 10P (1984).
A 2.8 hr observation of the intermediate polar H2215–086 using the EXOSAT Medium Energy experiment has revealed a strong, quasi–sinusoidal modulation of the hard X–ray flux with a period (20.65±0.32 min) consistent with, and in phase with, that of the large amplitude optical variations. The detection of this hard X–ray modulation confirms that the optical pulse period represents the white dwarf rotation period in this system. At an assumed distance of 500 pc the 2 – 10 keV X–ray luminosity of H2215–086 is $\sim 3.0\times10^{32}\,\mathrm{erg\,s^{-1}}$.

117.031 **Optical polarization from magnetized cataclysmic variables.**
P. E. Barrett, G. Chanmugam.
Mon. Not. R. Astron. Soc., Vol. 210, No. 1, p. 15P – 18P (1984).
The absence of optically polarized radiation from certain cataclysmic variables has been interpreted, by a number of authors, to imply that the white dwarfs' magnetic field is weaker ($B \lesssim 10^7$G) than in the AM Her binaries. The authors show that this interpretation is, in general, incorrect.

117.032 **Low–level radio flares from Cygnus X–3.**
L. A. Molnar, M. J. Reid, J. E. Grindlay.
Nature, Vol. 310, No. 5979, p. 662 – 665 (1984).
The authors report observations with the VLA of Cyg X–3 at several radio wavelengths which show that the radio emission in its low state may be interpreted as arising entirely from the superposition of a series of flares that are wavelength dependent with an apparent period near the 4.8–h X–ray period. A model of the source, in which each flare represents the injection of relativistic particles into a small volume which subsequently expands at constant velocity, is qualitatively consistent with the data.

117.033 **The V1010 Ophiuchi binaries.**
J. S. Shaw.
I.A.P.P.P. Commun., No. 16, p. 7 (1984).

117.034 **Search for relativistic companions in "non–X–ray" binary systems.**
A. M. Cherepashchuk, A. A. Aslanov.
Astrophys. Space Sci., Vol. 102, No. 1, p. 97 – 122 (1984).
Consideration is given to a search for relativistic objects in massive close binary systems without strong X–ray emission ($L_x < 10^{34}\,\mathrm{erg\,s^{-1}}$). It is pointed out that, according to the present–day theory on the evolution of massive close binaries, the number of neutron stars and black holes in "non–X–ray" binary systems must be 100 times the number of the known X–ray binaries comprising OB supergiant stars. Criteria are formulated for the relativistic nature of companions in the binary systems. Data are collected on Wolf–Rayet stars with low–mass companions which can be relativistic objects. Presented are new findings in respect of spectral examination of the "runaway" OB stars, eight OB stars can represent binary systems with relativistic companions. A list of 28 OB stars which offer a good chance for finding relativistic companions is given.

117.035 **Two X–ray sources model of Cyg X–3.**
L. M. Singh, H. L. Duorah.
Astrophys. Space Sci., Vol. 103, No. 1, p. 1 – 7 (1984).
A model of Cyg X–3, as a binary cocooned star system with two sources of X–rays, one above the polar caps of the neutron star and the other around the equatorial plane of the magneto–bounding surface formed due to the interaction of the infalling plasma and the magnetic field of the neutron star, is made. The X–ray, γ–ray, and IR radiation light curves are considered from the shadow effect. An upper limit on the mass of the neutron star is estimated. A comparison is made with the results of Elsner et al. (1980) and Ghosh et al. (1981).

117.036 **Loss of orbital angular momentum and period changes in Algols.**
U. S. Chaubey.
Astrophys. Space Sci., Vol. 103, No. 1, p. 55 – 59 (1984).
The observational data of Algols have been examined in order to clarify the implication of the assumption concerning the orbital angular momentum loss on their orbital period change. It is found that the agreement between theory and observational data of Algols is much better when a non–conservative approach of evolution is adopted.

117.037 **Evolution of low–mass close binaries and the problem of orbital period gap in cataclysmic variables.**
A. V. Fedorova, L. R. Yungelson (*L. R. Yungel'son*).
Astrophys. Space Sci., Vol. 103, No. 1, p. 125 – 135 (1984). With a correction in Vol. 107, No. 1, p. 207 (1984).
Evolution of close binary composed of a white dwarf primary and a main–sequence secondary has been calculated. Angular–momentum loss via gravitational radiation and magnetic stellar wind have been taken into account. It has been found that magnetic stellar wind with a rate greater than $(10^{-10}–10^{-9})M_{\odot}\ yr^{-1}$ is able to drive the evolution with mass exchange. The location and width of the gap in the orbital periods are comparable to those observed.

117.038 **V758 Centauri: *UBV* photometry.**
S. L. Lipari, R. F. Sistero.
Astrophys. Space Sci., Vol. 103, No. 2, p. 285 – 292 (1984).
The W UMa system V758 Centauri has been observed photoelectrically in the *UBV* system at Bosque Alegre Station, and the comparison and check stars at Cerro Tololo Observatory. Full light curves were constructed from 1043 differential *UBV* observations. A study of 11 photoelectric times of minimum light and of earlier photographic data shows that the period remained constant in the last century; V758 Cen shows a total occultation at secondary minimum, thus being an A–type W UMa system. Asymmetries are observed at primary minima whereas maxima are of different heights. Thermal decoupling of the components is suggested by the depth ratio of the minima; V758 Cen may be at the broken phase (or marginal contact) of a thermal relaxation oscillation.

117.039 **Evolutionary processes in close binary stars.**
U. S. Chaubey.
Astrophys. Space Sci., Vol. 103, No. 2, p. 385 – 389 (1984).
She evolutionary processes in close binary stars have been examined. It is found that there is a strong deficit of binaries evolving in Case A in the whole mass range. Probably the binary stars having binary separation corresponding to Case A evolution cannot form.

117.040 **Interpretation of spectroscopic variations in SS433.**
G. H. Newsom, G. W. Collins II.
Proceedings of the Third Marcel Grossmann Meeting on General Relativity, Part A, p. 625 – 636 (1983). – See Abstr. 012.002.
The peculiar celestial object SS433 emits three sets of strong emission lines, two of which move back and forth across the spectrum with a variable period of about 163 days. The object appears to be a binary system with a 13.1–day period. Models proposed to explain the 'moving' lines involve relativistic jets or non–thermal line emission in an accretion ring about a black hole. The 163–day period presumably results from precessional motion in the binary system, but no consensus yet exists as to the dynamics in this system.

117.041 **Nutation–like effects in the relativistic radial velocities of SS433.**
A. Mammano, F. Ciatti.
Proceedings of the Third Marcel Grossmann Meeting on General Relativity, Part A, p. 637 – 642 (1983). – See Abstr. 012.002.

117.042 **SVS 2310 is a W UMa–type variable.**
V. P. Goranskij.
Perem. Zvezdy, Prilozh., Tom 4, No. 20, p. 169 – 170 (1982). In Russian.

117.043 **W Ursae Majoris in 1977 – 1978.**
C. Pampaloni.
Perem. Zvezdy, Prilozh., Tom 4, No. 20, p. 271 – 274 (1982). In Russian.

117.044 **Accurate integrations of the Roche model.**
S. W. Mochnacki.
Astrophys. J., Suppl. Ser., Vol. 55, No. 4, p. 551 – 561 (1984).
The Roche model for close binary systems in synchronous rotation is accurately integrated to give volume radii, surface areas, mean gravities, and mean inverse gravities in normalized units. The range of fill–out is $0.2 \leqslant F \leqslant 2.0$, and for smaller fill–out factors a simple analytical approximation is adequate. The mass ratio range is $0.04 \leqslant q \leqslant 1.0$. A table of contact potentials is given for $0.01 \leqslant q \leqslant 1.0$.

117.045 **On the influence of magnetic dipole orientation onto the accretion rate in close binary systems.**
I. L. Andronov.
Astrofizika, Tom 20, Vyp. 1, p. 165 – 177 (1984). In Russian. English translation in Astrophysics, Vol. 20, No. 1.
A model for plasma ejection from a nondegenerate star is investigated. The main assumption is that matter while moving along magnetic field lines must penetrate through a potential barrier, the altitude of which is determined by the field orientation. It is supposed that convection velocities are pointed mainly perpendicular to the equipotential Roche surface. An analytical expression for the accretion rate depending on physical parameters of the system, dipole orientation and velocity distribution function of convective elements in the companion's envelope is derived.

117.046 **An orbital period for the dwarf nova AH Herculis.**
A. F. J. Moffat, M. M. Shara.
Publ. Astron. Soc. Pac., Vol. 96, No. 581, p. 552 – 556 (1984).
AH Her is a dwarf nova which displays rapid coherent oscillations at maximum light. Eleven nights of photometric observations with the 1–meter Wise Observatory telescope yield a period $P = 0.247 \pm 0.008$ day, in excellent agreement with that predicted from the decline rate of the outburst. The source of flickering in the system (presumably the hot spot) is shown to fluctuate with the same amplitude whether AH Her is in outburst or not.

117.047 **The orbit of VY Sculptoris determined near minimum light.**
J. B. Hutchings, A. P. Cowley.
Publ. Astron. Soc. Pac., Vol. 96, No. 581, p. 559 – 562 (1984).
The authors present spectroscopic data for the cataclysmic variable VY Scl taken during the rise from the low state of 1983 when the spectrum still showed strong emission lines. The period is not definitely established but is probably 0.1662 days. No strong phase–related spectral changes are seen, but there are night–to–night changes in the line intensities. The system appears to be viewed at a low inclination, with a white dwarf mass of $0.8\ M_{\odot}$ or more.

117.048 **A geometrical model for the SS433 system.**
E. M. Leibowitz.
Mon. Not. R. Astron. Soc., Vol. 210, No. 2, p. 279 – 301 (1984).
A geometrical model is suggested for the SS433 stellar system, consisting of a compact object and a star revolving around each other. The system includes a precessing accretion disc around the compact object and a hotspot on the disc surface. With these elements, most of the photometric characteristics of the system described in Leibowitz et al. can be accounted for, as well as a few other observations of this star. The parameters of the model yield a mass ratio $m(\text{compact})/M_{*} \geqslant 0.8$. Combined with results by Crampton and Hutchings, the model parameters imply that the compact object is probably a black hole, the star is an early B, an O or a super–massive star.

117.049 **Search for the long–term light variations of AO Psc = H2252–035.**
R. Hudec.
Bull. Astron. Inst. Czech., Vol. 35, No. 4, p. 203 – 204 (1984).
The photographic magnitude of AO Psc, the optical counterpart of the X–ray source H2252–035, has been measured on about 700 plates taken in the years 1928 – 1981. It was found that neither light variations, nor brightenings, nor inactive periods with amplitudes greater than 0.4 mag were present during these years.

117.050 **The linear polarization of SS433 = V1343 Aql.**
Y. S. Efimov (*Yu. S. Efimov*), V. Piirola, N. M. Shakhovskoy (*N. M. Shakhovskoj*).
Astron. Astrophys., Vol. 138, No. 1, p. 62 – 70 (1984).
Mean curves of polarization variations over the precession period $P = 164d$ are derived from new observations in 1979 – 1982 and from published data. According to the analysis of the curves, the first harmonic is dominating. The new *UBVRI* polarimetric data suggest a value of $P_{is} \sim 4.7\%$ for the interstellar polarization of SS433. The average intrinsic polarization is approximately orthogonal to the interstellar polarization and parallel to the direction of the radio jets ($\theta \sim 100°$). Such a polarization could arise from scattering by matter concentrated approximately to the orbital plane.

117.051 **Optical photometry of massive X–ray binaries: SMC X–1/Sk 160.**
J. van Paradijs, L. Kuiper.
Astron. Astrophys., Vol. 138, No. 1, p. 71 – 76 (1984).
The authors present the results of photometric observations (Walraven system) of Sk 160, the optical counterpart of the massive X–ray binary SMC X–1, obtained between 1979 and 1982. Combining these with previously published data the authors derive an orbital period of $3.89233 \pm 0.00008$ days, in agreement with the period obtained from X–ray data. The double–wave optical light curve of Sk 160 is the result of ellipsoidal variations and X–ray heating of the primary. The data support the previous conclusion that SMC X–1 contains a luminous accretion disk.

117.052 **The combined effect of mass loss and overshooting. IV. Mass transfer in massive close binaries.**
C. Doom.
Astron. Astrophys., Vol. 138, No. 1, p. 101 – 112 (1984).
The author determines the bifurcation periods between different cases of mass transfer in massive close binaries, taking into account mass loss and overshooting. Evolutionary computations are presented for 12 – 20 $M_\odot$ primaries, evolving through Case B of mass transfer. The evolution of 20 and 40 $M_\odot$ primaries through Case A of mass transfer is presented. After core hydrogen exhaustion, the remnants after Case A of the 20 $M_\odot$ models evolve through Case B of mass transfer.

117.053 **LMC X–3: a black hole in a neighbor galaxy.**
J. B. Hutchings, D. Crampton, A. P. Cowley.
Mercury, Vol. 13, No. 4, p. 106 – 107 (1984).

117.054 **Results of a search for super–rapid optical variability of the object SS433 according to 6–meter telescope observations.**
G. M. Beskin, S. I. Neizvestnyj, A. A. Pimonov, V. L. Plakhotnichenko, S. N. Fabrika, V. F. Shvartsman.
Astron. Tsirk., No. 1298, p. 1 – 4 (1983). In Russian.
Limits of the variable component of the optical flux from SS433 on time–scales of $3 \times 10^{-7}$ to 10 s are obtained. The observations were carried out at the moments corresponding to the different phases of precession and orbital periods of SS433. The absence of super–rapid variability indicates the great optical depth of the envelope surrounding the relativistic companion of SS433. The conclusion is drawn that the size of the funnel formed by the relativistic jet is smaller than 120°.

117.055 **IUE impact on the studies of the Zeta Aur, Algol and W Ser interacting binary systems.**
M. Hack.
ESA Spec. Publ., ESA SP–218, p. 55 – 61 (1984). – See Abstr. 012.024.
A review of the main results obtained with IUE for the Zeta Aur systems (including Epsilon Aur), the Algol systems and the W Ser systems (including Beta Lyr) is given. In spite of the different length of the periods, the interaction of the wind from the star that is losing mass with its companion or with its surroundings produces the same spectroscopically observable effects in all these systems: high temperature plasma and a clumpy medium of relativley high density.

117.056 **Rotational modulation of spots and plages on RS CVn stars.**
P. B. Byrne, J. G. Doyle, A. D. Andrews, C. J. Butler, N. Marstad, J. L. Linsky, T. Simon, M. Rodono, S. Catalano, C. Blanco, E. Marilli, V. Pazzani.
ESA Spec. Publ., ESA SP–218, p. 343 – 349 (1984). – See Abstr. 012.024.
Observations of three RS CVn stars made with the IUE satellite are presented. Emission line fluxes are found to vary in anti–phase with the stars' optical variations. The authors interpret these correlations in terms of large–scale spots in the stellar photospheres with overlying magnetic loops, giving rise to non–thermal heating of the layers above the spots. Evidence of non–thermal gas motions is also presented which appear to be associated with the most active regions of the stars.

117.057 **Models for the active and quiescent regions on the RS CVn–type system II Pegasi (HD 224085).**
J. L. Linsky, A. Brown, N. C. Marstad, M. Rodono, A. D. Andrews, C. J. Butler, P. B. Byrne.
ESA Spec. Publ., ESA SP–218, p. 351 – 354 (1984). – See Abstr. 012.024.
IUE observations of the RS CVn binary system II Pegasi obtained during the period 1981 October 1 – 7 have been used by Marstad to deduce the presence of a compact bright active region covering less than 10% of the primary's visible hemisphere. These same observations have now been used to derive the emission measure distributions and to calculate transition region models of the quiescent and plage (active) regions of II Peg.

117.058 **Dwarf novae through the outburst cycle.**
J. E. Pringle, F. Verbunt.
ESA Spec. Publ., ESA SP–218, p. 377 – 383 (1984). – See Abstr. 012.024.
Spectroscopic and photometric observations at optical and ultraviolet wavelengths have been obtained of a number of dwarf novae through their outburst cycles. The authors discuss the implications of their results for models of accretion discs and of dwarf nova outbursts.

117.059 **The outbursts of dwarf novae.**
B. J. M. Hassall.
ESA Spec. Publ., ESA SP–218, p. 385 – 386 (1984). – See Abstr. 012.024.
The dwarf nova VW Hydri has been observed simultaneously in the UV and optical at several stages of outburst. An attempt is made to model the spectrum using a steady–state model disc that radiates locally as a Kurucz atmosphere, together with an optically thin component. The spectrum on the rise is inconsistent with a steady disc, and may potentially be used to discriminate between possible outburst mechanisms.

117.060 **Constraints on stellar winds of cataclysmic variables derived from ultraviolet resonance lines.**
J. E. Drew, F. Verbunt.
ESA Spec. Publ., ESA SP–218, p. 387 – 389 (1984). – See Abstr. 012.024.
To investigate the structure of winds in cataclysmic variables the authors study the ultraviolet spectra of two such systems during eclipse.

117.061 **SS Cygni: accretion disk morphology.**
F. Giovannelli, C. Bartolini, P. Candy, S. Gaudenzi, A. Guarnieri, R. Lombardi, A. Piccioni.
ESA Spec. Publ., ESA SP–218, p. 391 – 393 (1984). – See Abstr. 012.024.
The authors present an extensive study of IUE spectra of SS Cygni in quiescence together with optical medium dispersion spectra. The ultraviolet spectrum shows a power law distribution $F_\lambda \propto \lambda^\alpha$. The values of $\alpha$ best–fitting the spectra, are roughly –4 and –1 shortward and longward $\sim$1450 Å, respectively. The optical spectra show double peaked Balmer lines occasionally masked by a low velocity emission.

117.062 **UV and optical observations of the novalike object V3885 Sgr.**
K. Haug, H. Drechsel, J. Rahe.
ESA Spec. Publ., ESA SP–218, p. 395 – 397 (1984). – See Abstr. 012.024.
The authors present the results of an UV– and optical line and continuum analysis of V3885 Sgr, together with some conclusions about the physical structure of the system.

117.063 **Two density components in symbiotic stars?**
W. Deuel, H. Nussbaumer.
ESA Spec. Publ., ESA SP–218, p. 399 – 402 (1984). – See Abstr. 012.024.
From IUE spectra the authors derive evidence that the nebular emission of V1016 Cyg originates from at least two emission regions of widely different particle densities. The implications for models are discussed.

117.064 **A new outburst of Z Andromedae.**
R. Viotti, A. Cassatella, A. Altamore.
ESA Spec. Publ., ESA SP–218, p. 403 – 406 (1984). – See Abstr. 012.024.
The symbiotic star Z And, after a long period of quiescence underwent in March 1984 a luminosity brightening which could be the beginning of a new active phase of the star. From a preliminary analysis of the IUE observations made on April 6 the authors found a large decrease of the continuum temperature. The high ionization lines of He II, C IV and O III appear stonger, while N V remained nearly constant and, contrary to the expectation, O I weakened. This behaviour during outburst appears different from the UV variability observed during the preoutburst quiescent phase.

117.065 **The spectrum of the symbiotic star CH Cygni in January 1984.**
C. Boehm, M. Hack, M. Persic.
ESA Spec. Publ., ESA SP–218, p. 407 – 408 (1984). – See Abstr. 012.024.
The continuum energy distribution and the emission spectrum indicate that the activity phase started in 1977 is still in progress. Significant variations in the appearance of the emission spectrum have been found with respect to the previous spectra, while the continuum seems to have remained nearly unchanged.

117.066 **Optically–thick, highly–variable plasma in interacting binaries.**
Y. Kondo, G. E. McCluskey, S. B. Parsons.
ESA Spec. Publ., ESA SP–218, p. 409 – 410 (1984). – See Abstr. 012.024.
The ultraviolet spectrum of HD 207739 undergoes complex variations in both the lines and the continuum. Considerable circumstellar matter is present. R Arae exhibits Si IV, C IV and even N V absorption features. Both the lines and continuum are variable. An unusual secondary eclipse which is deeper in the ultraviolet than in the visible is present near phase 0.6. The C II line has a P Cygni profile with emission present at all observed phases. These systems are probably in a relatively short–lived phase of evolution which is not often observed.

117.067 **UV and optical spectroscopy of W Serpentis objects.**
W. Strupat, H. Drechsel, J. Rahe.
ESA Spec. Publ., ESA SP–218, p. 411 – 413 (1984). – See Abstr. 012.024.
UV and optical spectroscopic observations of W Serpentis objects have been analyzed, indicating the presence of circumstellar plasma with high degree of ionization. These objects are believed to be near contact or even contact systems, being presently in an evolutionary stage, where a common envelope is formed.

117.068 **A study of mass accretion in Algol from high resolution IUE spectra.**
H. Cugier, P. Molaro.
ESA Spec. Publ., ESA SP–218, p. 415 – 418 (1984). – See Abstr. 012.024.
Several high resolution IUE spectra of Algol taken in 1978, 1979 and 1983 are analysed. Variable additional absorption components of Mg II, Al III, C II and Si IV resonance lines are detected. The presence of Si IV lines is a clear indication of a high temperature region in the system. The observations are interpreted in terms of a variable mass–accretion process.

117.069 **Preliminary results of coordinated optical, UV and X–ray observations of magnetic white dwarfs in binaries.**
L. Maraschi, K. Beuermann, J. M. Bonnet–Bidaud, P. A. Charles, L. Chiappetti, G. Hammerschlag, I. Howarth, C. Motch, M. Mouchet, J. Osborne, L. Stella, E. G. Tanzi, A. Treves, J. Van Paradijs, A. J. Willis, R. Wilson.
ESA Spec. Publ., ESA SP–218, p. 427 – 430 (1984). – See Abstr. 012.024.

117.070 **Ultraviolet orbital variability of the intermediate polars.**
M. Mouchet, J. M. Bonnet–Bidaud.
ESA Spec. Publ., ESA SP–218, p. 431 – 434 (1984). – See Abstr. 012.024.
An analysis of the variability with orbital phase of the ultraviolet spectrum of the intermediate polars 2A0526–328 (TV Col), 3A0729+103, 4U1849–31 (V1223 Sgr), H2215–086 and H2252–035 (AO Psc) is presented. The presence of an ultraviolet orbital modulation of the continuum and emission lines is investigated and discussed in terms of an X–ray heating. Other variable contributions are also present and affect the detection of the orbital modulation.

117.071 **Evidence for variations in the phase–dependent ionization structure of the stellar wind in Vela X–1.**
M. van der Klis, G. Hammerschlag–Hensberge.
ESA Spec. Publ., ESA SP–218, p. 443 – 447 (1984). – See Abstr. 012.024.
Results are reported of high–resolution IUE observations of Vela X–1 between Feb. 16 and 22, 1984. The UV spectra were obtained exactly simultaneously with EXOSAT observations of the source. Some preliminary X–ray data are presented. The ionization structure of the stellar wind, as diagnosticized by the shape of the Si IV resonance line profiles near binary phase 0.5 shows a clear difference with respect to previous observations. Several possibilities to explain this difference are briefly discussed.

117.072 **IUE observations of the X–ray binary A0538–66: recent results on a remarkable recurrent.**
I. D. Howarth.
ESA Spec. Publ., ESA SP–218, p. 449 – 451 (1984). – See Abstr. 012.024.
Four characteristic states are identified for A0538–66; all observed UV spectra can be classified according to this scheme. Model fits show that the off state (type I) luminosity is significantly greater than the interoutburst on state (type III) value. Previously aired doubts as to the validity of key observations by John Raymond are withdrawn.

117.073 **Phase-dependent stellar wind absorptions in interacting Wolf-Rayet binaries.**
G. E. Bromage.
ESA Spec. Publ., ESA SP-218, p. 511 - 513 (1984). - See Abstr. 012.024.

A zone of apparent depletion of moderately ionised wind absorbers has been detected by IUE around the secondary in some single-lined Wolf-Rayet spectroscopic binaries. There is a superficial analogy with the Hatchett-McCray effect predicted and observed in the spectra of massive X-ray binaries. For some WR SB1's such data may be circumstantial evidence in favour of compact secondaries.

117.074 **Accretion disk limit cycle mechanism in twin-degenerate interacting binaries.**
J. K. Cannizzo.
Nature, Vol. 311, No. 5985, p. 443 - 444 (1984).

The twin-degenerate interacting binary stars are thought to consist of two pure helium white dwarfs in close orbit. The candidates are G61-29 (also GP Com), HZ 29 (also AM CVn) and, tentatively, PG1346+082. Large variations in the optical light of PG1346+082 suggest variable accretion onto the star gaining mass. Such behaviour is thought to be caused by either episodic mass overflow from the mass-losing star or the instability in the accretion disk. The authors investigate the latter mechanism, and explore the non-thermal equilibrium limit cycle behaviour of the accretion disk expected for a twin-degenerate interacting binary system.

117.075 **Scattering/absorption dips in 4U1755-33. Evidence for disc oscillations?**
J. Frank, M. Sztajno.
Astron. Astrophys., Vol. 138, No. 2, p. L15 - L16 (1984).

The authors suggest that the internal structure, variable depth and duration of the dips observed in 4U1755-33 are due to partial occultation of an extended source by the oscillating bulge of the accretion disc.

117.076 **"Eclipses" of superhumps in dwarf novae.**
R. Whitehurst.
Observatory, Vol. 104, No. 1062, p. 214 (1984). Abstract. - See Abstr. 010.682.

117.077 **The binary system HD 5303.**
E. Budding.
South. Stars, Vol. 30, Nos. 7 - 8, p. 435 - 443 (1984).

Photometric observations of the RS CVn binary HD 5303 from various southern locations are reviewed. Previously reported "wave-like" variations appear to have largely subsided in 1983.

117.078 **A field guide to close binary stars.**
V. Trimble.
Sky Telesc., Vol. 68, No. 4, p. 306 - 311 (1984).

117.079 **Binary fluid star.**
I. Hachisu, Y. Eriguchi.
Publ. Astron. Soc. Jpn., Vol. 36, No. 2, p. 259 - 276 (1984).

Equilibrium sequences of self-gravitating, synchronously rotating and incompressible fluid binaries are computed.

117.080 **Short-period noncontact close binary systems. III. RU Eridani.**
Y. Nakamura, A. Yamasaki, M. Kitamura.
Publ. Astron. Soc. Jpn., Vol. 36, No. 2, p. 277 - 286 (1984).

The close binary system RU Eri is studied on the basis of spectroscopic observations. Its spectral type is classified to be F3-4 IV-V and its orbital elements are deduced to be $K_1$ = 109 km s$^{-1}$, $f(M)$ = 0.0853 $M_\odot$, and $a_1 \sin i$ = 1.36 $R_\odot$. With the spectroscopic mass ratio $q$ = 0.55 determined in the present work, the light curves of Sarma and Sanwal (1981; AAA 29.119.024) are analyzed with Yamasaki's (1981; AAA 29.117.041) synthesis method. The system is found to consist of a slightly evolved F3-4 primary and an early K secondary. Both components almost fill their critical Roche lobes but are still inside of them.

117.081 **Photometric changes and spot motion in λ Andromedae.**
F. Scaltriti, M. Busso, A. Cellino, D. S. Hall, C. H. Poe, L. P. Lovell, H. J. Landis, R. E. Fried, T. R. Renner, P. Nielsen, H. J. Stelzer, K. Krisciunas, O. H. Brettman, A. J. Stokes, H. Louth, D. M. Slauson, C. W. Rogers, R. Miles, B. E. Powell.
Astron. Astrophys., Vol. 139, No. 1, p. 25 - 29 (1984).

The authors present $V$-band photometry of the long-period RS CVn-type binary λ Andromedae ($P$(phot.) = 53.952 days) carried out in the observing seasons 1981 - 82 and 1982 - 83. They covered one complete cycle in 1981 - 82 and more than three contiguous cycles in 1982 - 83. Maximum amplitudes (full range) are about $0^m.22$ and $0^m.19$, respectively. An analysis of the existing epochs of maximum and minimum light confirms the photometric period given by Boyd et al. (1983) but discloses the existence of a quasi-periodic trend in the ($O$-$C$) curve suggesting cyclic changes on a time scale of about 9 years. The fourier analysis of the 1982 - 83 record shows a certain spread in frequency around the fundamental periodicity; this can be interpreted as due to the presence of dark spots highly spread in latitude, moving at different velocities owing to the differential rotation of the star.

117.082 **Identification of 3A 0729+103 with an intermediate polar-type cataclysmic variable.**
I. M. McHardy, J. P. Pye, A. P. Fairall, B. Warner, M. Cropper, S. Allen.
Mon. Not. R. Astron. Soc., Vol. 210, No. 3, p. 663 - 672 (1984).

On the basis of the Einstein Observatory IPC position the authors identify the X-ray source 3A 0729+103 with a newly-discovered intermediate polar-type cataclysmic variable having a hard X-ray spectrum and showing evidence of low energy absorption. Detailed optical photometry and time-resolved spectroscopy, and X-ray variability studies, reveal the system to be similar to H2252-035. It has a spin-related period of 15.2 min and an orbital period of 3 hr 14.1 min in the optical photometry but there is no evidence of a beat period between these two.

117.083 **Subdwarf companions for HD 36629 and HD 81307.**
B. R. Wade.
Mon. Not. R. Astron. Soc., Vol. 210, No. 3, p. 33P - 36P (1984).

Two bright late B stars observed by the S2/68 package on TD1A have been observed with $uvby$Hβ photometry. They appear to have hot faint companions that produce a steeply increasing flux in the far UV. Separation of the flux distributions indicates that the companions have temperatures of about 100000K, radii of $(0.04\pm0.01)R_\odot$ and $(0.4\pm0.2)R_\odot$, and luminosities of $(2.2\pm0.2)L_\odot$ and $(3.9\pm0.6)L_\odot$ respectively. These luminosities and temperatures lie between values typical of planetary nebula nuclei and white dwarfs. The stars fall close to the evolutionary path in the $L$, $T_e$ diagram predicted by Seaton.

117.084 **Circular spectropolarimetry of VV Puppis.**
D. T. Wickramasinghe, I. N. Reid, M. S. Bessell.
Mon. Not. R. Astron. Soc., Vol. 210, No. 3, p. 37P - 41P (1984).

The authors present circular spectropolarimetric observations of VV Puppis obtained when the system was in a high state. Although the intensity spectrum shows no evidence of cyclotron harmonic features, the circular polarization spectrum shows a broad minimum centred at 5040 Å which could be interpreted as cyclotron harmonic number $n$ = 7 in a magnetic field of $3.2\times10^7$G. The observations are interpreted in terms of current models of cyclotron emission from an accretion shock.

117.085 **Evaluation of the photometric perturbations in close binary systems. II. Rotational distortion.**
H. Rovithis-Livaniou.
Astrophys. Space Sci., Vol. 104, No. 1, p. 1 - 17 (1984).

In this study the photometric perturbations arising from the axial rotation of the two components in close binaries are calculated and discussed. The evaluation is made for occultation eclipses terminating in totality – for both stars together as well as for each component separately.

117.086 **The post-RLOF (*Roche lobe overflow*) structure of the secondary components in close binary systems, with an application to masses of Wolf–Rayet stars.**
P. Hellings.
Astrophys. Space Sci., Vol. 104, No. 1, p. 83 – 109 (1984).
The post-RLOF structure of the secondary after relaxation towards thermal equilibrium is calculated for a large grid of massive close binaries evolving through an early case *B* of mass transfer. The initial primary masses range between 15 and 30 $M_0$, the initial mass ratio between 0.3 and 0.9. It is found that all systems starting from initial conditions in the grid specified above evolve sequentially, i.e. the primary evolves into a supernova before the end of core H burning of the secondary. No WR + WR systems are encountered. The results are used to determine the masses of ten double lined spectroscopic WR + OB binaries.

117.087 **Case *A* evolution of massive close binary systems.**
M. Nakamura, Y. Nakamura.
Astrophys. Space Sci., Vol. 104, No. 2, p. 367 – 387 (1984).
The case *A* evolution of close binary systems with total mass of 20.4 $M_\odot$ is investigated by following the evolution of both components simultaneously. The evolution is followed up to the stage at which the system overflows the outer critical surface or evolves into the phase of mode *Br* mass-transfer. It is found that the evolution of the systems can be classified into six types. The evolutionary features of each evolution type are described in detail.

117.088 **Conservative mass transfer. II. The effective perturbing force: equations of motion for the eccentricity and orbital period.**
J. J. Matese, D. P. Whitmire.
Astrophys. J., Vol. 282, No. 2, p. 522 – 526 (1984).
The dynamical effects of conservative mass transfer on the eccentricity and orbital period are obtained by deriving the effective perturbing force associated with mass flux. In the context of a model in which mass transfer occurs primarily at periastron, it is found that mass transfer alone can rapidly circularize an initially eccentric orbit on a quasi-linear time scale, in contrast to the quasi-exponential time scale associated with uniform mass flow and viscous forces. Application of this result to Her X-1 indicates that the transfer of $\sim 0.1\ M_\odot$ is sufficient to reduce the eccentricity by $\sim 0.1$.

117.089 **Einstein observations of selected close binaries and shell stars.**
E. F. Guinan, R. H. Koch, M. J. Plavec.
Astrophys. J., Vol. 282, No. 2, p. 667 – 674 (1984).
Several evolved close binaries and shell stars were observed with the IPC aboard the HEAO 2 Einstein Observatory. No eclipsing target was detected, and only two of the shell binaries were detected. It is argued that there is no substantial difference in $L_X$ for eclipsing and non-eclipsing binaries. The close binary and shell star CX Dra was detected as a moderately strong source, and the best interpretation is that the X-ray flux arises primarily from the corona of the cool member of the binary. HD 51480 was detected as a weak source. Substantial background information from IUE and ground scanner measurements are given for this binary.

117.090 **Ultraviolet flare on Lambda Andromedae.**
S. L. Baliunas, E. F. Guinan, A. K. Dupree.
Astrophys. J., Vol. 282, No. 2, p. 733 – 740 (1984).
On 1982 November 5/6, a luminous, flarelike brightening of the ultraviolet emissions was observed with *IUE* from the active RS CVn type star λ And during the phase of rotation period corresponding to maximum area coverage of the visible hemisphere by starspots and active regions. Enhancements during the flare in the ultraviolet emission lines as large as factors of several and in the ultraviolet continuum up to 80% persisted for over 5 hours. The energy radiated by the flare alone is in excess of $10^{35}$ ergs just in the ultraviolet region. This is the most energetic stellar flare ever recorded in the ultraviolet.

117.091 **A0538–66 during the extended 'off' state: evidence for continuing mass motions.**
A. P. Smale, P. A. Charles, R. H. Densham, G. T. Bath, J. van Paradijs, J. W. Menzies, G. K. Skinner.
Mon. Not. R. Astron. Soc., Vol. 210, No. 4, p. 855 – 862 (1984).
The authors present spectroscopic and photometric observations of A0538–66 during an extended period of inactivity, with particularly good coverage around phase zero of cycle $N = 116$ (1983 January). The effects of the 16.6 day outburst cycle are no longer visible and there are no periodic flares in brightness or pronounced emission line variations. However, in contrast with previous 'off' states, the Balmer Hα line is observed in emission, and the Hβ and some Hγ absorption line profiles display weak blue-shifted emission components which suggest that some form of mass motions with related low-intensity emission activity is still occurring irregularly, with little or no phase correlation and without associated optical brightness variations.

117.092 **Photometry of A0538–66 during an active and subsequent inactive state.**
J. van Paradijs, S. van Amerongen, M. de Kool, M. Pakull, E. R. Deul, J. Lub, A. Greve.
Mon. Not. R. Astron. Soc., Vol. 210, No. 4, p. 863 – 871 (1984).
The authors present the results of photometric observations of the optical counterpart of the recurrent X-ray transient A0538–66. Three outbursts were observed in 1982 April and May. During the period 1982 October – 1983 February the source did not show outbursts. Combining their data with those of Densham et al. the authors find evidence for a secular decrease of the size of the outbursts throughout a period of activity. The interoutburst *UBV* properties of A0538–66 likewise show a progressive change. During the off-state A0538–66 shows small-amplitude variability, similar to that observed in some 'normal' Be stars. In the (*V*, *B–V*) and (*U–B*, *B–V*) diagrams both these variations and the secular variations of the previous inter-outburst data lie along tracks which appear to be connected.

117.093 ***IUE* observations of the high-velocity symbiotic star AG Draconis. II. The spectral variations during 1979 – 1983.**
R. Viotti, A. Altamore, G. B. Baratta, A. Cassatella, M. Friedjung.
Astrophys. J., Vol. 283, No. 1, p. 226 – 231 (1984).
After a long period of quiescence the high-velocity symbiotic star AG Dra underwent a large outburst in the visual in 1980 November, followed by a second light maximum and by a gradual fading to a new minimum phase in 1983. The authors analyze here a complete set of *IUE* spectra covering the period 1979 June to 1983 June. The results seem to support a binary model of AG Dra, consisting of a nonvariable K giant with a strong wind, and a dwarf companion accreting matter from the cool star wind.

117.094 **The lower main sequence and the nature of secondary stars in ultracompact binaries.**
S. Rappaport, P. C. Joss.
Astrophys. J., Vol. 283, No. 1, p. 232 – 240 (1984).
The authors systematically investigate the possible nature of the secondary stars in ultracompact binary stellar systems (with orbital periods $\lesssim 1$ hr). Using a simplified stellar evolution code, which assumes isentropic stellar models, they generate $\sim 3000$ separate models for hydrogen-burning main-sequence stars with masses less than 0.3 $M_\odot$. The authors explore in detail the effects of the (homogeneous) chemical composition on the properties of such stars and on the minimum main-sequence mass. They find that this minimum mass is a sensitive function of the hydrogen content and can be as small as 0.035 $M_\odot$. The available observational information on the three known ultracompact binary systems (4U 1626–67, G61–29, and 4U 1916–05) is summarized, and the results of the stellar model calculations are combined with the empirical results to place constraints on the properties of the secondary stars.

117.095 **Are cataclysmic variables the progenitors of Type I supernovae?**
J. MacDonald.
Astrophys. J., Vol. 283, No. 1, p. 241 – 248 (1984).

The long term evolution of carbon–oxygen white dwarfs accreting solar composition material at fixed rate is investigated in order to determine whether short–period cataclysmic binary systems are likely to be Type I supernova progenitors. In addition to mass loss from the white dwarf in response to hydrogen shell flashes in the accreted material, a number of previously neglected physical processes are considered, including mass loss due to helium flashes, dredge–up of core material by turbulence induced by shears due to accretion of angular momentum, and mass loss by stellar winds during the hot remnant phase. It is found that the white dwarf can explode as a Type I supernova only if it is initially massive ($\gtrsim 0.9\ M_{\odot}$) and the accretion rate is high. For the Galaxy, the frequency of supernovae from cataclysmic variables is estimated to be $\ll 3 \times 10^{-3} \mathrm{yr}^{-1}$.

117.096 **Ultraviolet high–resolution spectroscopy of the X–ray binary Sk 160/SMC X–1.**
G. Hammerschlag–Hensberge, T. R. Kallman, I. D. Howarth.
Astrophys. J., Vol. 283, No. 1, p. 249 – 253 (1984).

High–resolution *IUE* spectra of Sk 160, the optical counterpart of the X–ray source SMC X–1, show substantial variations with orbital phase of the resonance P Cygni profiles. These variations are ascribed to the changing ionization state in the stellar wind caused by the X–rays emitted by the companion, as suggested by Hatchett and McCray and first established for the galactic X–ray binary Vela X–1.

117.097 **Two–frequency radio spectra during an outburst of the periodic radio star LSI +61°303.**
A. R. Taylor, P. C. Gregory.
Astrophys. J., Vol. 283, No. 1, p. 273 – 278 (1984).

The authors present the results of simultaneous observations of the periodic source LSI +61°303 at 5.0 and 10.5 GHz during the radio outburst of 1981 August – September. The source is optically thin during both the latter portion of the rise and the subsequent decay of the outburst. From analysis of the data the authors derive a source dimension of $\sim 10^{13}$cm and a magnetic field strength of $\sim 1$ gauss. Adiabatic expansion is the dominant energy loss mechanism for the radiating electrons. Two mechanisms are considered for the source of the radio–emitting particles: a young pulsar wind and supercritical accretion. The authors calculate an improved value of the period for LSI +61°303 of $26.496 \pm 0.008$ days.

117.098 **High–resolution soft X–ray spectra of Scorpius X–1: the structure of circumsource accreting material.**
S. M. Kahn, F. D. Seward, T. Chlebowski.
Astrophys. J., Vol. 283, No. 1, p. 286 – 294 (1984).

Four observations of Scorpius X–1 with the Objective Grating Spectrometer of the Einstein Observatory have provided high–resolution spectra ($\lambda/\Delta\lambda \sim$ 20 – 50) in the wavelength range 7 – 46 Å. The spectra reveal the presence of absorption structure due to oxygen, nitrogen, and iron, and variable emission structure associated with ionized iron and nitrogen. The strengths of these features suggest that the N/O abundance ratio in the absorbing and line emitting gas is anomalously high. Possible models for the origin and nature of this circumsource medium are discussed.

117.099 **Nuclear processes in the jets of SS 433.**
R. Ramaty, B. Kozlovsky, R. E. Lingenfelter.
Astrophys. J., Lett. Ed., Vol. 283, No. 1, p. L13 – L16 (1984).

The authors propose that the very narrow (FWHM $\lesssim$ 10 keV) $\gamma$–ray lines observed at 1.495 and 6.695 MeV from SS 433 are blueshifted 1.369 and 6.129 emissions from deexcitations of $^{24}\mathrm{Mg}^{*}$ and $^{16}\mathrm{O}^{*}$ in grains moving with the jets and inelastically excited by interactions with the ambient medium. The presence of grains in the jets resolves hitherto discussed difficulties of inelastic excitation models for $\gamma$–ray production in SS 433, namely the very narrow widths of the observed lines and the absence of other strong lines expected from abundant elements.

117.100 **On the precession of the optical star in the Cyg X–1 system.**
I. M. Kopylov, V. V. Sokolov.
Pis'ma Astron. Zh., Tom 10, No. 10, p. 756 – 764 (1984). In Russian. English translation in Sov. Astron. Lett, Vol. 10.

Some results are analysed of a previous spectroscopic investigation of the supergiant HDE 226868 (based on six–year observations at the 6 m telescope) with the purpose of searching for possible variations in the spectrum connected with the precession of the rotation axis of the star's upper layers. It is noted that the spectral type and the He I λ4471 line halfwidth show a coordinated behaviour with the phase of the 39–day period in the frame of the precession hypothesis.

117.101 **Dwarf nova instabilities.**
J. K. Cannizzo, J. C. Wheeler.
Ann. N.Y. Acad. Sci., Vol. 422, p. 337 (1984). Abstract. – See Abstr. 012.033.

117.102 **Loss of angular momentum in binary system due to collisionless particles as monopoles, gravitinos, or mirror particles: does it exceed the gravitational radiation emission in binary system PSR 1913+16?**
D. Fargion.
Ann. N.Y. Acad. Sci., Vol. 422, p. 343 – 344 (1984). Abstract. – See Abstr. 012.033.

117.103 **Search for relativistic companions of runaway OB stars.**
A. A. Aslanov, L. N. Kornilova, A. M. Cherepashchuk.
Pis'ma Astron. Zh., Tom 10, No. 9, p. 666 – 672 (1984). In Russian. English translation in Sov. Astron. Lett., Vol. 10.

Spectroscopic observations of the runaway stars HD 188001 (9 Sge), HD 201345, HD 214930, HD 218915 allow to suggest the presence of periodic variations in their radial velocities. This periodicity implies that these runaway stars may be close binary systems containing relativistic companions.

117.104 **Rapid flickering of CH Cygni.**
V. P. Reshetnikov, T. N. Khudyakova.
Pis'ma Astron. Zh., Tom 10, No. 9, p. 673 – 678 (1984). In Russian. English translation in Sov. Astron. Lett., Vol. 10.

The observations of rapid flickering activity of CH Cygni made quasi–simultaneously in the *U, B, V, R, I, H, K* spectral bands during the active phase in 1983 are discussed. The conclusion is made that this variability of the star is irregular and similar to flickering of cataclysmic variables. However, the mechanism of flickering in its present form is not suitable to explain the rapid variability of CH Cygni.

117.105 **La natura fisica delle stelle simbiotiche: I.**
L. Del Giudice.
Coelum, Vol. 52, N. 5, p. 261 – 270 (1984).

117.106 **L'observation des disques d'accrétion dans les sources X binaires à haute luminosité.**
C. Motch.
Instabilités hydrodynamiques et applications astrophysiques, p. 479 – 487 (1983). – See Abstr. 012.034.

Contents: Introduction. Comment se manifeste le disque d'accrétion observationnellement: en optique, en rayon X. Le cas des sources X dites "non compactes". GX 339–4.

117.107 **Variabilité rapide des systèmes binaires cataclysmiques.**
M. Mouchet.
Instabilités hydrodynamiques et applications astrophysiques, p. 489 – 496 (1983). – See Abstr. 012.034.

Contents: Qu'est–ce qu'une variable cataclysmique? Classification et temps caractéristiques. Pulsations. Oscillations cohérentes. Oscillations quasipériodiques. Flickering. Conclusion.

117.108 **A study of photoelectric observations of SW Lacertae.**
K.-C. Leung, D. Zhai, R. Zhang.
Publ. Astron. Soc. Pac., Vol. 96, No. 582, p. 634 – 640 (1984).

SW Lac is a well-known late-type close binary system with large variation in its light curve. New photoelectric observations in *B* and *V* were obtained in 1978 at the Beijing Observatory. The authors attempted to derive a consistent photometric solution by analyzing individually three separate sets of light curves (1953, 1961, and 1978). All the solutions suggest that the system is overcontact and belongs to the W-type W UMa systems. The 1961 and 1978 light curves have essentially identical photometric parameters. The derived mass ratio is identical with the spectroscopic value. However, the 1953 light curves have distinctly different parameters from those of the later years. Two different interpretations are discussed.

117.109 **Synthetic parameters of five W-type contact binaries.**
L. Binnendijk.
Publ. Astron. Soc. Pac., Vol. 96, No. 582, p. 646 – 656 (1984).

Synthetic parameters have been determined from observations defining symmetrical light curves of the five W-type contact binary systems SW Lac, U Peg, RZ Com, AM Leo, and AH Vir. In addition, a second observed light curve of each system was studied. The difference in the shape of the light curves for a given system can be explained by one or two subluminuous regions on the larger component. Such a region may be caused by circumstellar material or by a starspot area. The position in longitude is well determined. In every case the absolute value of the latitude can be found but only for SW Lac can the sign of the latitude be obtained without ambiguity.

117.110 **SMC X-1 variability observed from *HEAO 1*.**
D. E. Gruber, R. E. Rothschild.
Astrophys. J., Vol. 283, No. 2, p. 546 – 551 (1984).

The 13 – 70 keV emission from the 0.7 s X-ray pulsar SMC X-1 was monitored by the UCSD/MIT instrument aboard *HEAO 1* during three ~80 day epochs in 1977 and 1978. The X-ray flux was seen to wander slowly between a maximum value and unobservably small levels with a rough time scale of 60 days. This variability is found to be consistent with a random walk in intensity and could be the result of unsteady mass transfer from the primary. A possible alternate explanation involves regular shadowing of the X-ray emitting neutron star by a tilted, precessing accretion disk. The time-averaged spectrum shows a gradual exponential cutoff with a characteristic energy of 10 – 20 keV. There is no evidence for linelike features.

117.111 **X-ray emissions from symbiotic novae.**
S. Kwok, D. A. Leahy.
Astrophys. J., Vol. 283, No. 2, p. 675 – 678 (1984).

X-ray observations of three symbiotic novae made with the Einstein Observatory by Allen have been further analyzed. The observed spectra are consistent with thermal bremsstrahlung from circumstellar hot gas of $\sim 10^7$K. The authors propose that this hot region is the shocked gas resulting from the fast wind from the white dwarf component of the symbiotic nova colliding with the slow wind from the M giant component.

117.112 **Cygnus X-2: neutron star or degenerate dwarf?**
J. E. McClintock, L. D. Petro,
G. Hammerschlag-Hensberge, C. R. Proffitt, R. A. Remillard.
Astrophys. J., Vol. 283, No. 2, p. 794 – 800 (1984). With plate 18.

Two conflicting models have been proposed for Cyg X-2: a degenerate dwarf model which predicts a distance of $250 \pm 50$ pc and a neutron star model which implies a distance of about 8000 pc. Based on a reddening study, the authors find that the distance to Cyg X-2 is greater than 1100 pc, which rules strongly against the degenerate dwarf model. This conclusion is based on observations of the 2200 Å feature in the spectrum of Cyg X-2 made with the IUE, and *UBV* and spectroscopic observations of 38 field stars.

117.113 **On the bizarre gamma-ray spectrum of SS 433.**
H. L. Helfer, M. P. Savedoff.
Astrophys. J., Lett. Ed., Vol. 283, No. 2, p. L49 – L52 (1984).

The observed 1.37 MeV γ-ray line emission of $^{24}$Mg in the jets of SS 433 is analyzed in an attempt to derive physical constraints on the line-emitting region. The line flux implies the mass loss rate is greater than $10^{-5} M_\odot \mathrm{yr}^{-1}$. The γ-emitting region is probably $\sim 10^{12}$cm in extent. The same beam parameters can be used to characterize the γ-emitting region and the Hα-emitting region. An "aspirator" model in which a "magnesium" jet from the neutron star interacts with hydrogen from the accretion disk is proposed.

117.114 **New evidence on the nature of the neutron star and accretion flow in Vela X-1 from pulse timing observations.**
P. E. Boynton, J. E. Deeter, F. K. Lamb, G. Zylstra,
S. H. Pravdo, N. E. White, K. S. Wood, D. J. Yentis.
Astrophys. J., Lett. Ed., Vol. 283, No. 2, p. L53 – L56 (1984).

The authors report Vela X-1 pulse timing results based on data spanning the period 1975 – 1982. The frequency variations in this source are consistent with the occurrence of a succession of temporally unresolved events that can be characterized mathematically as second-order red noise in the pulse phase (random walk in pulse frequency). Several of the important constraints on the properties of the neutron star and the temporal and velocity structure of the accretion flow implied by the present results are outlined.

117.115 **Some episodes of rapid variability of HDE 245770 = A 0535+26.**
C. Bartolini, G. Bianco, A. Guarnieri, A. Piccioni,
F. Giovannelli.
Hvar Obs. Bull., Vol. 7, No. 1, p. 159 – 166 (1983). – See Abstr. 012.037.

The X-ray binary HDE 245770, a Be star, presented in February 1976 and December 1977 two unusual phenomena of light variations connected with X-ray outbursts. Probably in both cases the authors observed the tail of the optical phenomenon that for the event of December 1977 preceded the X-ray flare up. No optical light variations were observed during the decay of other X-ray outbursts.

117.116 **Spin evolution of magnetic white dwarfs in AM Herculis binaries.**
C. G. Campbell.
Mon. Not. R. Astron. Soc., Vol. 211, No. 1, p. 69 – 74 (1984).

The spin evolution of the magnetic primary stars in the 10 known AM Herculis binaries is investigated, using the theory formulated in a previous paper. It is found that, in all cases, magnetic dissipation in the convective secondary can synchronize the primary if it operated before mass transfer began, provided that the initial degree of asynchronism is not excessively large. This dissipation mechanism can also explain a small degree of asynchronism of the primary if accretion began before synchronization was achieved, or if accretion broke a previously attained synchronous state. This slightly asynchronous state exists provided that the mass of the primary is not appreciably greater than 0.8 $M_\odot$. However, this state may exist for higher primary masses if the accreting magnetic pole lies out of the orbital plane.

117.117 **Mass transfer in strongly magnetic binaries.**
C. G. Campbell.
Mon. Not. R. Astron. Soc., Vol. 211, No. 1, p. 83 – 95 (1984).

Magnetically controlled mass transfer is considered in an AM Herculis-type binary. An expression is derived for the accretion torque on the primary star for the case in which its magnetic moment lies in the orbital plane. Stable orientations are found at which this torque vanishes. Other orientations are found which may restrict mass transfer, and a possible connection with the low states observed in these systems is discussed. An expression can be derived for the accretion torque on an asynchronous primary as a function of time, provided that the degree of asynchronism is not too large. The average of this torque over a synodic rotation period of the primary gives the same expression as that derived by assuming that the specific angular momentum of material is conserved in an inertial frame centred on the star, despite the fact that this is not the case for the flow along an instantaneous path.

117.118 **An analysis of V861 Sco. II. The stellar wind.**
I. D. Howarth.
Mon. Not. R. Astron. Soc., Vol 211, No. 1, p. 167 – 194 (1984).
The stellar wind from the B0 Ia primary in the V861 Sco eclipsing binary system is investigated using *IUE* data. The observed P Cygni profiles show a correlation between maximum velocity and ionization potential. There is a further correlation between the range of observed blue-edge velocities and the mean value for a given ion. No significant phase-dependent effects are found, however. Calculations of line profiles are made using the Sobolev approximation. The results are combined with a model for the ionization balance in the wind to derive a mass-loss rate $\dot{M} = 3 \times 10^{-6} M_{\odot} yr^{-1}$. An expression for the escape velocity in a binary system is presented. The ratio $v_{esc}/v_{\infty}$ for the V861 Sco primary is similar to that found for single stars.

117.119 **Mass transfer and accretion in close binaries: a review.**
C. B. Boyle.
Vistas Astron., Vol. 27, Part 2, p. 149 – 169 (1984).
The discovery of the strong galactic X-ray sources (Giacconi et al., 1962; Giacconi, 1974; Gursky and Schreier, 1975) and the suggestion that these sources may be powered by accretion onto a compact object orbiting a "normal" stellar companion (Hayakawa and Matsouko, 1964; Zel'dovich and Guseynov, 1965) has, over the past twenty years or so, led to renewed interest in the processes of mass transfer and accretion in close binary stars.

117.120 **A radio survey of symbiotic stars.**
E. R. Seaquist, A. R. Taylor, S. Button.
Astrophys. J., Vol. 284, No. 1, p. 202 – 210 (1984).
The authors present the results of a VLA survey of 59 symbiotic stars at 4.9 GHz. Seventeen of these were detected as radio sources, and eight of the detected stars were also observed at 1.5 GHz. The radio emission is consistent with optically thick thermal bremsstrahlung from an ionized gas in all cases. A model is discussed in which the hot companion in a binary system ionizes a portion of a stellar wind from the late-type giant.

117.121 **Close binary stars and old stellar populations: the blue straggler problem revisited.**
A. C. Collier, C. R. Jenkins.
Mon. Not. R. Astron. Soc., Vol. 211, No. 2, p. 391 – 419 (1984).
The authors present the results of a series of Monte Carlo simulations of the spectral evolution of synthetic star clusters containing a proportion of close binaries. They find that the numbers of "blue stragglers" produced by the mass-transfer process are consistent with the numbers observed in old disc clusters of similar ages to the models. The synthetic straggler orbital period, velocity amplitude and mass distributions are consistent with the available observational data. The effects of binary evolution on the integrated UBVRI colours of the cluster models are discussed.

117.122 **Light curves of gas-eclipsed binary systems.**
I. B. Pustylnik (*I. B. Pustyl'nik*), L. Einasto.
Astrophys. Space Sci., Vol. 105, No. 2, p. 259 – 285 (1984).
Light curves have been calculated for the model of a binary system consisting of two spherical components, imbedded in a common spherically-symmetric, isotropically-scattering envelope with radially-varying opacity. The periodically-varying screening of stellar radiation by the envelope gives rise to regular light changes, which are similar to stellar eclipses during a superior conjunction of the components and in maxima resemble the light changes due to the reflection effect for Algol-type binaries, whereas for systems with nearly equal luminosities of the components they match light variations, caused by the gravitational distortions of stellar figures. The observed light curves of AO Cas, RY Gem, RS CVn, VW Cep are compared with the model curves. The model can be regarded as a certain optical counterpart of the Roche model.

117.123 **On a model of CH Cygni.**
L. Luud, T. Tomov.
Pis'ma Astron. Zh., Tom 10, No. 11, p. 860 – 867 (1984). In Russian. English translation in Sov. Astron. Lett., Vol. 10.
An analysis of photometric and spectroscopic observations of the peculiar variable star CH Cygni is given. CH Cygni is shown to be a binary star with visible spectra of both components. The masses of the components are evaluated from dynamical and evolutional considerations.

117.124 **Observations of SS 433.**
B. Margon.
Annu. Rev. Astron. Astrophys., Vol. 22, p. 507 – 536 (1984). – See Abstr. 003.024.
Contents: (1) Introduction: Basic data. (2) Optical and infrared observations: Doppler-shifted spectral lines. Deviations from the kinematic model. The stationary spectrum and "normal" star. Interstellar lines and extinction. Optical and infrared photometry. (3) Radio observations: The central source. The extended components. (4) X- and gamma-ray observations. (5) W50. (6) Some major uncertainties: Jet energetics, collimation, and acceleration. The disk and precession mechanism. The nature of the compact star. The extragalactic analogy. (7) Analogous galactic objects.

117.125 **Neutron stars in interacting binary systems.**
P. C. Joss, S. A. Rappaport.
Annu. Rev. Astron. Astrophys., Vol. 22, p. 537 – 592 (1984). – See Abstr. 003.024.
Contents: (1) Introduction. (2) Binary X-ray pulsars: Pulse profiles and spectra. Pulse period variations. Orbital determinations and the evaluation of binary system parameters. (3) X-ray burst sources: Observational properties. Burst sources as binary stellar systems. The thermonuclear flash model. Theory versus observation. (4) Summary.

117.126 **Infrared spectrum of Cygnus X-3.**
J. I. Katz, E. L. Wright, C. R. Lawrence.
Astron. J., Vol. 89, No. 10, p. 1604 – 1605 (1984).
The authors obtained infrared spectrophotometry of Cyg X-3 between 1.94 and 2.50 μ with a nominal bandpass of 0.07 μ. No evidence of the Brackett γ line was seen, setting an upper bound on its equivalent width of about 50 Å. These data impose weak constraints on theoretical models of this enigmatic object.

117.127 **Eruptive binaries. XI. Disk-radius variations in U Gem.**
J. Smak.
Acta Astron., Vol. 34, No. 1, p. 93 – 96 (1984).
A new set of geometrical elements is determined from an improved analysis of eclipses of the hot spot. The radius of the disk, which expands during an outburst, shrinks exponentially from about $r_d = 0.39$ immediately after outburst to about $r_d = 0.29$ just before the next outburst. The expansion of the disk begins at the very onset of an outburst.

117.128 **Accretion in cataclysmic binaries. IV. Accretion disks in dwarf novae.**
J. Smak.
Acta Astron., Vol. 34, No. 2, p. 161 – 189 (1984).
Time-dependent disk models are constructed, including the effects of thermal instability due to the ionization of hydrogen. The results are very sensitive to the assumptions concerning the viscosity. Two types of outbursts are predicted by the models: Type A, with the onset of instability in the outer parts of the disk, and Type B, with the onset of instability in the inner parts. The two types are tentatively identified with observed examples of dwarf nova outbursts.

117.129 **VY Lac – a possible semidetached system of early type.**
I. Semeniuk, J. Kałużny.
Acta Astron., Vol. 34, No. 2, p. 207 – 216 (1984).
483 V and 476 B photoelectric observations of VY Lac have been obtained and analysed with the Wilson and Devinney method. The photometric solution suggests that VY Lac is one of few semidetached systems in the slow stage of case A mass exchange.

117.130 **Determination of parameters of W UMa–type systems: V757 Cen, GW Cep, BX Peg, AH Vir.**
J. Kałużny.
Acta Astron., Vol. 34, No. 2, p. 217 – 224 (1984).
Photoelectric light curves of four W UMa–type systems (V757 Cen, GW Cep, BX Peg, AH Vir) have been solved using the Ruciński code. The solutions were obtained for two assumed values of the gravity darkening exponent ($\beta = 0.00$ and $\beta = 0.08$). It is concluded that it is hard to distinguish between those two values basing on the wide band photometry in the visual wavelength domain.

117.131 **Variable polarization in X–ray binaries: HD 153919 (= 4U1700–37) and HD 152667 (= OAO1653–40?).**
J. F. Dolan, S. Tapia.
Astron. Astrophys., Vol. 139, No. 2, p. 249 – 257 (1984).
Variable linear polarization has been observed in the optical radiation from two presumed X–ray binary systems, HD 153919 (=4U1700–37) and HD 152667 (probably not OAO1653–40). This intrinsic polarization appears correlated with the orbital phase of the binary, but the polarimetric light curve shows variations from year to year. The Stokes parameter $q$, $u$ light curves are consistent with the polarization being caused by the Rayleigh scattering of light from the primary by gas streams in the vicinity of the secondary. Several orbital and physical parameters of the binary systems are derived from the polarimetric observations.

117.132 **High–resolution observations of bipolar mass flow from the symbiotic star HM Sagittae.**
J. Solf.
Astron. Astrophys., Vol. 139, No. 2, p. 296 – 304 (1984).
Coudé observations of high spectral and high spatial resolution of the [N II] and [O III] emission lines from HM Sge obtained at various position angles of the spectrograph entrance slit are presented. They reveal bipolar mass flow ($\approx 200$ km s$^{-1}$) collimated into two narrow lobes separated by 1.″5, two low–velocity features of 0.″2 separation within the equatorial plane compatible with slowly expanding ($\approx 25$ km s$^{-1}$) blobs or a rotating ($\approx 10$ km s$^{-1}$) ring, and a shell of about 0.″5 diameter expanding at an intermediate velocity ($\approx 60$ km s$^{-1}$). A distance of $\approx 400$ pc is inferred from the expansion rate of the nebula. The geometric and kinematical structure of HM Sge seems less compatible with a single–star model for symbiotic stars, but is found in excellent agreement with the model of a binary consisting of a mass–losing M–type giant and a compact companion accreting part of the mass from the giant through an accretion disk.

117.133 **The accretion disk spectrum of LMC X–3.**
S. Pineault.
Astron. Astrophys., Vol. 139, No. 2, p. 313 – 316 (1984).
Recent observations of the optical counterpart of LMC X–3, by Cowley et al. (1983), have led to the suggestion that the optical continuum contains a component associated with the X–ray source. Under the assumption that standard accretion disk theory applies, it is shown that the above suggestion is quantitatively consistent with the observed level of X–ray emission and leads to the prediction that excess infrared and UV radiation could be distinguishable. Expected flux values are calculated, including a discussion of the effect of the outer disk radius on the shape of the postulated IR continuum.

117.134 **The origin of the high and low luminosity states in magnetic cataclysmic variables.**
A. R. King, J. P. Lasota.
Astron. Astrophys., Vol. 140, No. 1, p. L16 – L18 (1984).
The authors show that the high and low states of magnetic cataclysmic variables are probably due to X–ray heating of the companion star by the white dwarf. Shielding by the mass transfer stream stabilizes the accretion rate at a value only slightly above the average rate due to angular momentum losses from the binary, resulting in long–lived high states alternating with short lived low states in which the mass transfer reservoir is refilled. The authors speculate that the complex behaviour of GK Per, recently discovered to be magnetized, may also be explicable in a similar way.

117.135 **The symbiotic eclipsing binary CI Cyg: an Algol symbiotic system.**
D. Chochol, A. Vittone, L. Milano, L. Rusconi.
Astron. Astrophys., Vol. 140, No. 1, p. 91 – 104 (1984).
Spectroscopic and photometric observations of the semidetached symbiotic eclipsing binary CI Cyg are analysed. A model of gaseous streams in the system is proposed. The difference between an outburst and an eruption is proposed and explained. While the outburst is connected with the expansion of the photosphere of the mass accreting star, the eruptions are caused by the outflow of hot matter from a turbulent region of the accretion disk after a deep penetration of the disk by the inflowing stream. The turbulent region is responsible for the rapid photometric variations observed in the system. The masses, radii and distance between the components are derived. According to the classification proposed by Plavec (1982), CI Cyg belongs to the class of Algol symbiotics.

117.136 **Accretion of mass in Algol.**
H. Cugier, P. Molaro.
Astron. Astrophys., Vol. 140, No. 1, p. 105 – 111 (1984).
Several high resolution IUE spectra of Algol taken in 1978 and 1979 are analysed. Variable additional absorption components for the ultraviolet resonance lines of Mg II, Al III, and Si IV are detected. The additional components show eclipsing effects indicating that they originate in a region near the primary star, Algol A. Moreover, their behaviour is variable with the time of observation. The observations are interpreted in terms of a mass–accretion process, during which high temperature plasma is produced by dissipation of the kinetic energy of the gas falling onto the surface of the primary star. Moreover, the analysed material indicates that a region with high rotational velocity is created near the accreting star. The time–dependence of this region suggests that this takes place during an increase in the mass–transfer rate.

117.137 **Hard X–ray temporal and spectral observation of an active state of Sco X–1.**
A. Jain, G. Hasinger, W. Pietsch, R. Proctor, C. Reppin, J. Trümper, W. Voges, E. Kendziorra, R. Staubert.
Astron. Astrophys., Vol. 140, No. 1, p. 179 – 184 (1984).
Sco X–1 was observed on September 20, 1977 and September 28, 1981 in the 18 – 250 keV energy range by the MPI/AIT balloon borne phoswich detectors and also by cooled Ge solid state detectors in the 1981 observation (HEXE experiment), flown from Palestine, Texas/USA. The paper presents details of the 1981 observation along with the first reported observation of Sco X–1 with a Ge spectrometer and of the 1977 observation. The implications of these results in terms of the source phenomenology are briefly discussed. The repetitive flux modulations observed in the 1981 observation are discussed in terms of quasi–periodic oscillations detected in the optical flux of Sco X–1 and in some of the dwarf–novae occurring after their peak eruptions.

117.138 **L'histoire de CH Cygni.**
J.–P. Vaidis.
Bull. Assoc. Fr. Obs. Etoiles Variables, No. 30, p. 145 – 152 (1984).

117.139 **Contact binaries.**
M. Nakamura.
Astron. Her., Vol. 77, No. 12, p. 306 – 311 (1984). In Japanese.

117.140 **The FK Comae candidate UZ Librae.**
B. W. Bopp, B. D. Goodrich, J. L. Africano, P. V. Noah, R. J. Meredith, L. H. Palmer, R. J. Quigley.
Astrophys. J., Vol. 285, No. 1, p. 202 – 207 (1984).
New optical spectroscopic and photometric data are presented for the active chromosphere FK Com candidate UZ Lib. The photometric and spectroscopic variability of UZ Lib is shown to be explicable in terms of cool starspots on a rapidly rotating K giant. The period of rotation is established from optical photometry as 4.75 days. New radial velocity measures are combined with published data to demonstrate that UZ Lib is a member of

a binary system, in synchronous rotation with a secondary of mass $\sim 0.5\ M_{\odot}$.

117.141 **Coordinated UV and optical observations of the AM Herculis object E1405–451 in the high and low states.**
L. Maraschi, A. Treves, E. G. Tanzi, M. Mouchet, A. Lauberts, C. Motch, J. M. Bonnet Bidaud, M. M. Phillips.
Astrophys. J., Vol. 285, No. 1, p. 214 – 222 (1984).

Simultaneous UV and optical observations of the magnetic white dwarf binary E1405–451 taken in 1983 February, March, and April are reported. In the low state the ultraviolet spectrum of E1405–451 can be interpreted as the emission from a hot white dwarf ($T$ = 26,500K), and the $R$ and $I$ magnitudes can be accounted for by a Roche lobe filling secondary, of spectral type between M3 V and M8 V at a distance between 50 and 200 pc. In the high state two additional components seem to appear in the composite spectrum. One could be associated with a hot polar cap on the white dwarf, the other could be due to cyclotron radiation.

117.142 **Comparison of the X–ray observations of the AM Herculis objects to theory.**
J. N. Imamura.
Astrophys. J., Vol. 285, No. 1, p. 223 – 231 (1984).

The author reexamines the X–ray observations of the original AM Her objects (AM Her, VV Pup, EF Eri, and AN UMa). This work differs from that carried out previously in that it compares the observed X–ray spectral properties to theoretical models calculated using a dipolar accretion geometry rather than a spherically symmetric one. The observed X–ray light curves are compared to theoretical light curves calculated assuming that the X–ray sources are cataclysmic variable–like binaries which contain a strongly magnetic white dwarf. It is found that the X–ray observations of AM Her can be self–consistently explained by accretion powered radiative shock models.

117.143 **The rotational and orbital evolution of cataclysmic binaries containing magnetic white dwarfs.**
G. Chanmugam, A. Ray.
Astrophys. J., Vol. 285, No. 1, p. 252 – 257 (1984).

This paper considers the evolution of a binary containing a magnetic white dwarf rotating with an initial period $P_i$ in a binary with $P_{orb} \sim 5 - 10$ hr. The orbital angular momentum losses are assumed to be due to gravitational radiation and to magnetic braking caused by a stellar wind from the tidally locked secondary. It is shown that the systems appear first as DQ Her binaries. As the orbital evolution proceeds, the decrease in the mass transfer rate and the orbital separation leads to white dwarfs, with sufficiently strong fields, becoming synchronized as in the AM Her binaries. A drastic reduction in the mass transfer rate when $P_{orb} < 2$ hr enhances formation of AM Her binaries in the model, in agreement with observations.

117.144 **Is Cygnus X–3 a monoenergetic $10^{17}$eV accelerator?**
A. M. Hillas.
Nature, Vol. 312, No. 5989, p. 50 – 51 (1984).

Cygnus X–3 seems to be a binary system that includes a compact object, but it is unusual in being enshrouded in gas, in displaying occasional huge outbursts and in emitting much more energy above $10^{11}$eV than in the keV X–ray region. If the compact object is a pulsar, the ultra–high–energy photons can be products of an accelerated beam of protons striking the companion's atmosphere. It is noted here that a roughly monoenergetic spray of $\sim 10^{17}$eV protons, generating an electron–photon cascade in the environs of the companion star, best accounts for all photons above a few GeV, and might generate the gas shroud, as beam energy deposition inside the star could lift off a thick layer of matter to generate a fountain under the pulsar.

117.145 **Cosmic–ray emission from Cygnus X–3.**
N. A. Porter.
Nature, Vol. 312, No. 5992, p. 347 – 348 (1984).

Recent observations of the X–ray binary source Cygnus X–3 have revealed a flux of ultra–high energy $\gamma$ rays. Other observations and subsequent models suggest that Cyg X–3 may be a dominant source of galactic high–energy cosmic rays. This letter investigates whether it is possible to escape this conclusion with a model in which most of the energy of accelerated particles at the source is emitted as curvature radiation. It is concluded that such a model is not consistent with the observations.

117.146 **A disk origin for superhumps in Ursae Majoris variables.**
K. Horne.
Nature, Vol. 312, No. 5992, p. 348 – 351 (1984).

Light curve synthesis results for the superhump phenomenon in the SU Ursae Majoris binary Z Cha are presented. It is shown that the observations are consistent with the superhumps originating in the accretion disk around the white dwarf. Models invoking starspots on the red dwarf companion require extreme radiation geometries, which seem rather implausible.

117.147 **Contact binary stars.**
O. Demircan.
Doga Bilim Dergisi, Ser. A1, Vol. 8, No. 2, p. 137 – 147 (1984). In Turkish. Abstr. in Phys. Abstr., Vol. 87, No. 1246, Entry 115782 (1984).

117.148 **Structure of the accretion disk in the dwarf nova Z Cha.**
E. S. Dmitrienko, A. N. Matvienko,
A. M. Cherepashchuk, A. G. Yagola.
Sov. Astron., Vol. 28, No. 2, p. 180 – 187 (1984). English translation of 37.117.202.

117.149 **Dust shells in the peculiar objects CH Cygni and TX Canum Venaticorum.**
O. G. Taranova, B. F. Yudin.
Sov. Astron., Vol. 28, No. 3, p. 299 – 301 (1984). English translation of 37.117.217.

117.150 **Observations of eight close binaries with the *Einstein Observatory*.**
G. E. McCluskey Jr., Y. Kondo.
Publ. Astron. Soc. Pac., Vol. 96, No. 584, p. 817 – 820 (1984).

The binaries R Ara, $\zeta$ Aur, RZ Cas, 31 Cyg, $\delta$ Lib, U Sge, RZ Sct, and RW Tau all show strong indications of active mass flow in their ground–based and/or ultraviolet spectra. The systems R Ara, RZ Cas, 31 Cyg, $\delta$ Lib, and U Sge were detected as X–ray sources while only upper limits on the X–ray luminosities of $\zeta$ Aur, RZ Sct, and RW Tau could be determined. Non-detection of the latter three is probably attributable to their distance. With the exception 31 Cyg which has an X–ray luminosity of about $10^{32}$erg s$^{-1}$, the detected sources have X–ray luminosities in the range $10^{30-31}$erg s$^{-1}$. These luminosities are approximately equal to those observed for single stars of similar spectral types. It appears that the X–ray luminosity of 31 Cyg is at least 60 times higher than that of $\zeta$ Aur.

117.151 **Starspot cycle of RS CVn.**
Y. W. Kang, R. E. Wilson.
Bull. Am. Astron. Soc., Vol. 16, No. 3, Part I, p. 726 (1984). Abstract. – See Abstr. 010.065.

117.152 **The evolution of the optical spectrum of the dwarf nova SS Cygni over one complete outburst cycle.**
J. T. Clarke, D. Capel, S. Bowyer.
Bull. Am. Astron. Soc., Vol. 16, No. 3, Part I, p. 733 (1984). Abstract. – See Abstr. 010.065.

117.153 **Photometry of CPD –48°1577.**
G. A. Williams, W. A. Hiltner.
Mon. Not. R. Astron. Soc., Vol. 211, No. 3, p. 629 – 636 (1984).

$U$, $B$, and $V$ photometry is presented for CPD –48°1577, the newly discovered cataclysmic variable with the greatest known apparent brightness. The light curves are dominated by flickering but do suggest an orbital period of nearly 4 hr. Power spectra of 5 nights of $U$–band observations failed to reveal the presence of any periodic or quasi–periodic oscillations. Autocorrelation functions suggest that two time–scales exist in the flickering activity, one of 77 s and the other 500 s. The implications for a shot noise model of flickering are discussed.

117.154 **Enge Doppelsterne als Indikatoren der Sternentwicklung.**
H. Mauder.
BAV Rundbrief, 33. Jahrg., Nr. 4, p. 145 – 150 (1984).

117.155 **Revidierte Lichtwechselelemente für die W UMa-Sterne V417 Aql und SS Com.**
B.-C. Kämper.
BAV Rundbrief, 33. Jahrg., Nr. 4, p. 152 – 159 (1984).
The author presents period studies for V417 Aql and SS Com, using all available material. Recently published revised period for V417 Aql (Faulkner, 1983) is confirmed and further improved. There is evidence for period changes in the past, but not exceeding $10^{-5}$ P. In the A-type system SS Com, since 1937 a marked period increase of almost $2^s$ has taken place.

117.156 **Elementenkorrektur für den W UMa-Stern GW Cep.**
B.-C. Kämper.
BAV Rundbrief, 33. Jahrg., Nr. 4, p. 160 – 162 (1984).
Because of an error in the epoch counts of old relative to new minima the period quoted for GW Cep in the SAC 55 has to be corrected. Revised elements are given as Min I = HJD 2438651.545 + $0\overset{d}{.}3188311 \cdot E$. New observations indicate a possible period change in this system.

117.157 **Ein photographisches Minimum von SS Com.**
B.-C. Kämper.
BAV Rundbrief, 33. Jahrg., Nr. 4, p. 163 – 172 (1984).

117.158 **The theory of contact binaries.**
R. C. Smith.
Q. J. R. Astron. Soc., Vol. 25, No. 4, p. 405 – 420 (1984).
Contact binary stars have been the subject of intense research for the last 15 years but a completely satisfactory theoretical model has yet to emerge. This article reviews what has been done and argues that evolution with angular momentum loss is the most promising area for future work.

117.159 **A radio survey of symbiotic stars.**
E. R. Seaquist, A. R. Taylor.
J. R. Astron. Soc. Can., Vol. 78, No. 5, p. 207 (1984). Abstract. – See Abstr. 010.401.

117.160 **Radio emission from symbiotic stars: a binary model.**
A. R. Taylor, E. R. Seaquist.
J. R. Astron. Soc. Can., Vol. 78, No. 5, p. 207 (1984). Abstract. – See Abstr. 010.401.

117.161 **Evolution of a distorted low-mass close binary system driven by gravitational radiation losses: implications for the minimum period of cataclysmic variables.**
L. A. Nelson, W. Y. Chau.
J. R. Astron. Soc. Can., Vol. 78, No. 5, p. 213 (1984). Abstract. – See Abstr. 010.401.

117.162 **Near-infrared and optical light variations of symbiotic stars.**
M. H. Slovak.
I.A.P.P.P. Commun., No. 17, p. 44 – 49 (1984).

117.163 **Optical spectroscopy of the source LSI + 61°303.**
O. E. Aab (*O. Eh. Aab*), L. V. Bychkova, I. M. Kopylov.
Bull. Spec. Astrophys. Obs. – North Caucasus, Vol. 17, p. 1 – 3 (1984). English translation of 34.117.163.

117.164 **The UV variations of the symbiotic star CH Cygni from 1978 to 1981.**
M. Persic, M. Hack, P. L. Selvelli.
Astron. Astrophys., Vol. 140, No. 2, p. 317 – 324 (1984).
The spectroscopic changes of CH Cygni from early 1978 to late 1981 are discussed by means of several ultraviolet spectra obtained at the Villafranca Satellite Tracking Station. Correlations have been found to exist between the fluxes of the continuum and the fluxes of the forbidden and semiforbidden lines. Three new forbidden-line identifications are tentatively reported, and the physical mechanisms responsible for several emissions, including Fe II-lines, are discussed. The radial velocities of the circumstellar lines are discussed, together with their variations in time.

117.165 **Spectrophotometry of the outburst and quiescent states of the dwarf novae RX And and KT Per.**
J. T. Clarke, S. Bowyer.
Astron. Astrophys., Vol. 140, No. 2, p. 345 – 351 (1984).
Outburst and quiescent spectra of moderate (3.5 – 10 Å) resolution are presented for the dwarf novae RX And and KT Per. The changing physical nature of the accretion disks of these objects was investigated by analyzing the spectral line strengths and profiles and their evolution over the course of the outburst cycle. In the outburst spectra of both systems there appears a pronounced narrowing of the emission lines and a steepening of the Balmer decrement, indicating a substantial reduction in the strength of the line emission from the inner part of the disk during outburst. The absorption line shapes in the outburst spectra are better fit by modeled profiles of accretion disk absorption than by pressure broadening in a white dwarf type atmosphere.

117.166 **Main-sequence components of Algol-type binaries as blue stragglers.**
M. A. Giannuzzi.
Astron. Astrophys., Vol. 140, No. 2, p. 373 – 376 (1984).
The hot components of Algol-type binaries which are main-sequence stars generally have positions in the H-R diagram which correspond to their masses. Consequently they look like old main-sequence stars. However the real age of a system in which transfer and loss of matter has occurred may be greater than the apparent age of the most massive component. Some evolved binaries have been selected which, under reasonable hypotheses, are estimated to be in this situation. The relation to the blue-straggler problem is examined.

117.167 **Photoelectric light curves and elements of VW Cephei.**
P. G. Niarchos.
Astron. Astrophys., Suppl. Ser., Vol. 58, No. 2, p. 261 – 271 (1984).
The W UMa-type system VW Cephei has been observed in two wavelengths ($V$ and $B$) during the years 1979 – 1980. The light curves of the system are analysed using frequency domain techniques and new geometric and photometric elements are given.

117.168 **Determination of parameters of W UMa systems. V. V757 Cen, ER Cep, AH Cnc.**
C. Maceroni, L. Milano, G. Russo.
Astron. Astrophys., Suppl. Ser., Vol. 58, No. 2, p. 405 – 410 (1984).
From photoelectric and spectroscopic data available in the literature, the authors have determined the elements of three W Ursae Majoris systems, by means of the Roche-model-based method of Wilson and Devinney. According to their solutions, the three systems are all of W-type, that is, primary minima are caused by occultation of the smaller (and less massive) component.

117.169 **Synchronization in late-type binary stars.**
G. Giuricin, F. Mardirossian, M. Mezzetti.
Astron. Astrophys., Vol. 141, No. 1, p. 227 – 231 (1984).
The authors have rediscussed the synchronism between rotation and revolution in late-type close binaries by an inspection of the published rotation velocities and rotation (photometric) periods of ~50 binary components. They have found that, as a rule, dwarfs having fractional radii $r > 0.05$ appear to be in synchronous rotation and those with $r < 0.03$ are in asynchronous rotation. The observed degree of synchronism appears to be in substantial agreement with current theoretical views (Zahn, 1977) on tidal friction in late-type binaries.

117.170 **Two very similar closely detached binary systems: AT and AZ Camelopardalis.**
D.-s. Zhai, K.-c. Leung, R.-x. Zhang, J.-t. Zhang.
Astron. Astrophys., Vol. 141, No. 1, p. 266 – 270 (1984).
Photoelectric observations, in *B* and *V*, of AT and AZ Cam have been analyzed with the Wilson–Devinney method. For AT Cam the eclipse at the primary is found to be total, even though the inclination is only 83 degrees. For AZ Cam, it is a partial occultation, with an inclination of 81 degrees. Both systems are found to be detached, but each of them is only separated by less than the radius of the large component. In both cases, the low mass components almost fill their Roche lobes.

117.171 **Photoelectric observations and minima times for RZ Comae Berenices.**
P. Rovithis, H. Rovithis-Livaniou.
Astron. Astrophys., Suppl. Ser., Vol. 58, No. 3, p. 679 – 684 (1984).
*B* and *V* photoelectric observations for RZ Comae Berenices are presented. The resulting minima times are also given.

117.172 **An exploratory eccentric orbit "Roche lobe" overflow model for recurrent X-ray transients.**
J. C. Brown, C. B. Boyle.
Astron. Astrophys., Vol. 141, No. 2, p. 369 – 375 (1984).
An explanatory estimate is made of the variations in mass transfer rate $\dot{M}$ which could occur with orbital phase in an eccentric semi-detached binary star system. The calculation is based on the assumption that the rate can be instantaneously described in terms of overflow through a "Roche lobe" nozzle in the neighbourhood of the inner Lagrange point, analogously to the usual analyses for circular orbits. A correction factor is also estimated for the effect on $\dot{M}$ of the instantaneous motion of the effective Roche surface. This factor makes the peak in $\dot{M}$ precede periastron. The results are compared to available data on the transient X-ray sources GX301-2 and A0538-66.

117.173 **Wolf-Rayet stars with possible relativistic companions: optical eclipses in the HD 191765 system.**
I. I. Antokhin, A. M. Cherepashchuk.
Sov. Astron. Lett., Vol. 10, No. 3, p. 155 – 158 (1984). English translation of 37.117.226.

117.174 **The precession angle of the optical component of the Cygnus X-1 system.**
V. V. Sokolov, V. V. Tsymbal.
Sov. Astron. Lett., Vol. 10, No. 3, p. 172 – 174 (1984). English translation of 37.117.272.

117.175 **Optical variability of HZ Herculis, 1982 – 1983: normal state and anomalous 35-day behavior.**
A. V. Mironov, V. G. Moshkalev, E. M. Trunkovskij, A. M. Cherepashchuk.
Sov. Astron. Lett., Vol. 10, No. 3, p. 180 – 181 (1984). English translation of 37.117.273.

117.176 **The SU Ursae Majoris stars. II. RZ Sagittae.**
N. F. Vojkhanskaya, I. I. Nazarenko.
Sov. Astron. Lett., Vol. 10, No. 3, p. 183 – 184 (1984). English translation of 37.117.298.

117.177 **AM Herculis.**
IAU Circ., No. 3967 (1984).

117.178 **SS Cygni.**
IAU Circ., Nos. 3971, 3977 (1984).

117.179 **PG 1550+191.**
IAU Circ., No. 3979 (1984).

117.180 **TV Columbae.**
IAU Circ., No. 3980 (1984).

117.181 **Z Andromedae.**
IAU Circ., No. 3985 (1984).

117.182 **CH Ursae Majoris.**
IAU Circ., No. 3989 (1984).

117.183 **VZ Sculptoris.**
IAU Circ., No. 3996 (1984).

117.184 **PG 1346+082.**
IAU Circ., No. 4021 (1984).

117.185 **A model for superoutbursts and superhumps of SU UMa stars.**
Y. Osaki.
Theoretical problems in stellar stability and oscillations, p. 339 – 340 (1984). Abstract. – See Abstr. 012.065.

117.186 **Excitation of non-adiabatic quasi-toroidal oscillations in double stars.**
J. Provost, G. Berthomieu, A. Rocca.
Theoretical problems in stellar stability and oscillations, p. 372 – 376 (1984). – See Abstr. 012.065.
In a close binary system, the stars are subject to the time dependent tidal potential caused by their companion. Forced oscillations are induced in each component and constitute the dynamical tide. In the present paper, the authors take into account the rotation of a component assuming a small uniform rotation and they study numerically the excitation of non-adiabatic quasi-toroidal modes by the perturbing tidal potential.

117.187 **Instabilities of contact binary systems.**
H. Kähler, A. Weigert.
Theoretical problems in stellar stability and oscillations, p. 377 – 381 (1984). – See Abstr. 012.065.

117.188 **Photoelectric light curves and period of HD 199497.**
C. Sezer, Ö. Gülmen, N. Güdür.
Inf. Bull. Variable Stars, No. 2553, 2 pp. (1984).

117.189 **The investigation of the binary system TX Cas.**
J. M. Kreiner, J. Tremko.
Inf. Bull. Variable Stars, No. 2554, 2 pp. (1984).

117.190 **Amplitude-wavelength dependence for W UMa-systems.**
G. V. Zhukov.
Inf. Bull. Variable Stars, No. 2560, 4 pp. (1984).

117.191 **Period variations of SS Ari.**
J. Kałuzny, G. Pojmański.
Inf. Bull. Variable Stars, No. 2564, 2 pp. (1984).

117.192 **Period and light-curve of the close eclipsing binary FZ Orionis.**
J. F. Le Borgne, A. Figer, M. Dumont.
Inf. Bull. Variable Stars, No. 2566, 2 pp. (1984).

117.193 **FO Vir, a close eclipsing binary, not an RR Lyr-variable.**
E. Antonello, L. Mantegazza, E. Poretti.
Inf. Bull. Variable Stars, No. 2567, 2 pp. (1984).

117.194 **A new period change of 44i Bootis in 1979.**
J. Bergeat, M. Lunel, R. Garnier.
Inf. Bull. Variable Stars, No. 2570, 3 pp. (1984).

117.195 **Ca II H and K, and Hα emission variability in AR Lacertae.**
P. S. Goraya, R. K. Srivastava.
Inf. Bull. Variable Stars, No. 2579, 4 pp. (1984).

117.196 **UBV photometry for BD +37°2356.**
X. Liu, H. Tan.
Inf. Bull. Variable Stars, No. 2590, 2 pp. (1984).

117.197 **The light and colour curves of the very active southern RS CVn-system HD 127535 in 1984.**
A. Udalski, E. H. Geyer.
Inf. Bull. Variable Stars, No. 2594, 5 pp. (1984).

117.198 **Hα emission in RS CVn stars: HD 8357, HD 175742 and HR 7428.**
X. Liu, H. Tan.
Inf. Bull. Variable Stars, No. 2606, 4 pp. (1984).

117.199 **Recent changes in the spectrum of CH Cyg.**
T. Tomov.
Inf. Bull. Variable Stars, No. 2610, 2 pp. (1984).

117.200 **Outbursts of the star Case 1.**
S. Yu. Shugarov.
Inf. Bull. Variable Stars, No. 2612, 2 pp. (1984).

117.201 **V566 Ophiuchi – period update.**
H. D. Kennedy.
Inf. Bull. Variable Stars, No. 2613, 4 pp. (1984).

117.202 **AM Cas – a dwarf nova with a very short cycle-length.**
P. Notni, G. A. Richter.
Inf. Bull. Variable Stars, No. 2634, 1 p. (1984).

117.203 **B and V photometry of VZ Psc.**
T. J. Davidge, E. F. Milone.
Inf. Bull. Variable Stars, No. 2639, 3 pp. (1984). = Publ. Rothney Astrophys. Obs., Ser. B, No. 11.

117.204 **Can the activity of the secondary components explain the X-ray emission from cataclysmic variables?**
S. M. Rucinski.
Observatory, Vol. 104, No. 1063, p. 259 – 264 (1984).

The accretion-powered emission of the primary components of cataclysmic binaries seems to dominate the observed X-ray fluxes; but in some dwarf-nova systems with relatively large secondary components this emission is close to, or only marginally exceeds, that expected from the secondary components on the assumption that for them $L_X/L_{bol} = 10^{-2}$, as is actually observed in rapidly rotating active M dwarfs.

117.205 **Did the binary stars in globular clusters come to an early end by being too eccentric?**
J. G. Hills.
Astron. J., Vol. 89, No. 12, p. 1811 – 1815 (1984).

The orbits of hard binaries in globular clusters shrink and develop high eccentricities in encounters with other stars. The most eccentric binaries suffer the physical coalescence of their two components. Slightly less eccentric binaries shrink down to very tightly bound binaries as a result of dissipation (tidal, gravitational radiation) at periastron. Visual binaries in dense globular clusters have either been dissociated by stellar encounters or they have shrunk by these mechanisms to form a combination of coalesced stars and very close binaries. These mechanisms may contribute significantly to the observed scarcity of binaries among globular cluster giants, but at the same time, they form and allow the retention of the close binaries that are needed to produce the X-ray sources in globular clusters.

117.206 **A study of the early-type semi-detached binary system u Herculis.**
R. W. Hilditch.
Mon. Not. R. Astron. Soc., Vol. 211, No. 4, p. 943 – 951 (1984).

Radial velocities derived from 67 spectrograms obtained over a period of 37 yr at the DAO, Victoria, are presented. It is shown that the orbit is circular and of constant period and that the primary component is rotating synchronously. A mass ratio of $0.39 \pm 0.02$ is derived and analysis of recently published light-curves using the synthesis code LIGHT confirms the semi-detached nature of the system. The final values of mass, radius and luminosity together with published *uvbyβ* photometry then indicate that the primary component might be expected to display β Cephei-type variability and therefore provide an explanation for the well-known asymmetries in the light-curves. The binary system belongs to a small group of short-period early-type semi-detached systems which have significantly different characteristics from classical Algol systems and are presumably the result of case A mass-exchange processes.

117.207 **Observations of the O'Connell effect system VY Crucis.**
E. F. Milone, B. J. Hrivnak.
Astrophys. J., Suppl. Ser., Vol. 56, No. 3, p. 295 – 301 (1984). = Publ. Rothney Astrophys. Obs., No. 18.

*UBV* and spectroscopic observations of the $0^d.696$ period binary VY Cru were made in Chile in 1977. The *B* and *V* light curves bear strong resemblance to the photographic light curve of Bruna, which is the only previously published light curve. The system has a sizable O'Connell effect (+0.11 mag in *V* and +0.14 mag in *B*). Despite the large O'Connell effect, the properties of the system appear more characteristic of the A-type W UMa systems. The spectral type of the system is ~A8.

117.208 **La natura fisica delle stelle simbiotiche. II.**
L. Del Giudice.
Coelum, Vol. 52, N. 6, p. 355 – 364 (1984).

117.209 **A simple kinetic model of a multiparticle cloud. II. Gaseous streams in Algols.**
P. Hadrava.
Bull. Astron. Inst. Czech., Vol. 35, No. 6, p. 335 – 343 (1984).

A method described in a previous paper is used for computation of a three-dimensional gas dynamical model of streams in binaries. The results confirm the inapplicability of the assumption of hydrostatic equilibrium for vertical structure of streams. The influence of adiabatic or isothermal expansion, or radiative equilibrium in a stream on its structure, and the change of ionization are also investigated.

117.210 **TT Arietis: the low state.**
A. W. Shafter, P. Szkody, J. Liebert, W. R. Penning, H. E. Bond, A. D. Grauer.
Space Telesc. Sci. Inst., Prepr. Ser., No. 31, 62 pp. (1984).

The authors present photometric and spectroscopic observations extending from the ultraviolet through the infrared of the novalike variable TT Ari, which were obtained during its recent (1982) low state (V ≈ 16.5). The observations indicate that mass transfer did not cease completely, but appeared to be occurring sporadically during the low state. The spectroscopic observations reveal strong, narrow (~150 km/s FWHM) Balmer and He I emission superimposed on a blue continuum when mass transfer is diminished. In addition to the narrow emission lines, the authors have detected broad Balmer absorptions at Hβ, Hγ, Hδ and He II λ4686 which are most likely produced in the photosphere of the white dwarf with $T_{eff} \gtrsim 50{,}000$K. The radial velocity variations of the low-state emission lines are 180 degrees out of phase. The most plausible interpretation of this phasing is that the low-state emission arises in the chromosphere of the secondary star. The authors also consider the possibility that the emission lines arise from a residual accretion disk in the system.

117.211 **Photoelectric observations of contact binary stars.**
M. Hoffmann.
Veröff. Astron. Inst. Bonn, Nr. 96, 9 pp. (1984). In German. With 2 microfiches.

This publication presents photoelectric measures of contact binaries obtained in the years 1976 through 1983 at the European Southern Observatory and at the Hoher List Observatory of the University Observatory of Bonn.

117.212 **Photoelectric observations of the dwarf nova VW Hydri during its November 1977 supermaximum.**
H. W. Duerbeck, N. Vogt.
Veröff. Astron. Inst. Bonn, Nr. 97, 11 pp. (1984). With a microfiche.

117.213 **An optical and near infrared study of the AM Herculis type binary CW 1103+254.**
J. Bailey, D. J. Watts, M. R. Sherrington, D. J. Axon, A. B. Giles, D. A. Hanes, S. R. Heathcote, J. H. Hough, S. Hughes, R. F. Jameson, I. McLean.
Anglo–Aust. Obs., Prepr., No. 197, 41 pp. (1984). Submitted to Mon. Not. R. Astron. Soc.

117.214 **Optical and X–ray observations of 4U2129+47/V1727 Cyg in a quiescent state.**
W. Pietsch, H. Steinle, M. Gottwald.
MPE Prepr., No. 28, p. 9 – 12 (1984). Paper presented at the XXVth COSPAR Meeting on X–Ray Astronomy, Graz, Austria, July 3 – 7, 1984.

117.215 **Simultaneous EXOSAT and optical observations of the pulsing X–ray binary H2252–035.**
W. Pietsch, M. Pakull, S. Tjemkes, W. Voges, E. Kendziorra, J. van Paradijs.
MPE Prepr., No. 28, p. 13 (1984). Abstract. – Paper presented at the XXVth COSPAR Meeting on X–Ray Astronomy, Graz, Austria, July 3 – 7, 1984.

117.216 **EXOSAT observations of the 35$^d$ intensity variations of Her X–1.**
H. Ögelman, P. Kahabka, W. Pietsch, J. Trümper, W. Voges, E. Kendziorra, R. Staubert.
MPE Prepr., No. 29, p. 16 – 19 (1984). Paper presented at the Symposium on X–Ray Astronomy '84, Bologna, Italy, June 26 – 30, 1984.

117.217 **Pulse phase spectroscopy of Her X–1 with EXOSAT.**
P. Kahabka, W. Pietsch, J. Trümper, W. Voges, E. Kendziorra, R. Staubert.
MPE Prepr., No. 29, p. 20 – 23 (1984). Paper presented at the Symposium on X–Ray Astronomy '84, Bologna, Italy, June 26 – 30, 1984.

117.218 **Fine structure in the absorption dips of Her X–1 as observed by EXOSAT.**
W. Voges, P. Kahabka, H. Ögelman, W. Pietsch, J. Trümper.
MPE Prepr., No. 29, p. 24 (1984). Abstract. Paper presented at the Symposium on X–Ray Astronomy '84, Bologna, Italy, June 26 – 30, 1984.

117.219 **Simultaneous EXOSAT and optical observations of the pulsing X–ray binary H2252–035/AO Psc.**
W. Pietsch, M. Pakull, S. Tjemkes, W. Voges, E. Kendziorra, J. van Paradijs.
MPE Prepr., No. 29, p. 25 – 30 (1984). Paper presented at the Symposium on X–Ray Astronomy '84, Bologna, Italy, June 26 – 30, 1984.

117.220 **Spectral studies of low–mass X–ray binaries observed by EXOSAT.**
M. Sztajno, J. Trümper, A. Langmeier.
MPE Prepr., No. 29, p. 31 – 34 (1984). Paper presented at the Symposium on X–Ray Astronomy '84, Bologna, Italy, June 26 – 30, 1984.

117.221 **EXOSAT observations of 4U1705–44.**
A. Langmeier, M. Sztajno, J. Trümper.
MPE Prepr., No. 29, p. 35 – 38 (1984). Paper presented at the Symposium on X–Ray Astronomy '84, Bologna, Italy, June 26 – 30, 1984.

117.222 **Coordinated UV and optical observations of the AM–Her object E1405–451 in high and low state.**
L. Maraschi, A. Treves, E. G. Tanzi, M. Mouchet, A. Lauberts, C. Motch, J. M. Bonnet Bidaud, M. M. Phillips.
ESO Sci. Prepr., No. 333, 34 pp. (1984). Submitted to Astrophys. J.

117.223 **High resolution problems in astrometry and spectroscopy of binaries and their significance for the progress of stellar physics.**
F. Van't Veer.
Inst. Astrophys. Paris, Pré–Publ., No. 70, 23 pp. (1984). Review paper presented at the International Bessel Conference, Bamberg, 13 – 15 June 1984.

117.224 **Eruptions of the cataclysmic variable T Leonis.**
W. Wenzel.
Mitt. Veränderliche Sterne, Band 10, Heft 3, p. 51 – 54 (1984).

Observations on Sonneberg Sky Patrol plates yield 14 maxima. These and further eruptions taken from literature are listed and statistical estimations of the cycle length (400 to 500 d) are given.

117.225 **Zum Helligkeitsverhalten des Polars AM Her im Jahre 1983.**
W. Götz.
Mitt. Veränderliche Sterne, Band 10, Heft 3, p. 69 – 74 (1984).

In supplementing and completing the long time light curve of AM Her and the list of observations the single estimations of the star on blue–sensitive Schmidt camera plates of Sonneberg Observatory obtained from 39 nights on the year 1983 are given. A flare of the star was observed and new elements of the occultation changes could be determined.

117.226 **A detailed reflection model for close binary systems with equal components.**
M. Kitamura, A. Yamasaki.
Ann. Tokyo Astron. Obs., Second Ser., Vol. 20, No. 1, p. 51 – 74 (1984).

A reflection model for the synthesis of theoretical light curves of close binary systems is developed, considering multiple reflections and Roche geometry. The amount of each of multiple reflections is calculated for more than seventy cases of model close binary systems with equal components, which are characterized by different values of the fractional radius, limb–darkening, gravity–darkening and albedo. The effect of the secondary reflection attains, in some cases, to more than twenty percent of that of the primary reflection.

117.227 **A spectroscopic study of the symbiotic star CH Cygni.**
K. H. Yoo.
Ann. Tokyo Astron. Obs., Second Ser., Vol. 20, No. 2, p. 75 – 129 (1984).

High–dispersion spectra of the low–excitation symbiotic star CH Cygni observed in 1981 and 1982 at Okayama were analyzed for both the emission and absorption components of metals. In spite of complexity of their line profiles, the method of curve of growth is used as a promising one for the approach of analysis. The physical and geometrical parameters thus obtained will be interpreted by a binary model.

117.228 **A radial velocity study of XY Leonis.**
B. J. Hrivnak, E. F. Milone, G. Hill, W. A. Fisher.
Astrophys. J., Vol. 285, No. 2, p. 683 – 687 (1984). = Publ. Rothney Astrophys. Obs., No. 24.

Spectra have been obtained over the past three seasons for the short–period ($0^d.28$) contact binary XY Leo. By cross correlating the blended stellar spectra of this system with spectra of velocity standards of similar spectral type, accurate component velocities were measured. Corrections for tidal and eclipse effects were then investigated. With these corrections, the following stellar masses were determined: $M_1 = 0.87 \pm 0.04\,M_\odot$ and $M_2 = 0.44 \pm 0.02\,M_\odot$.

117.229 **On the origin of X–ray variability of SS 433.**
D. L. Band, J. E. Grindlay.
Astrophys. J., Vol. 285, No. 2, p. 702 – 711 (1984).

The X–ray flares observed from the central source in SS 433 by the Einstein telescope are attributed to surges in the mass transfer rate due to changes in the critical Roche volume of the companion. Analysis of the Roche potential for a primary with spin misaligned with the orbital axis, as required by the slaved disk model, predicts that the critical Roche volume will contract twice per orbit if the orbit is circular. A critical Roche volume fractional change of 1% – 2% is found by applying this potential to SS 433. Aspects of this work strengthen the evidence that the compact object might be a black hole.

117.230 **Short-term X-ray variability of GX 339-4.**
Y. Maejima, K. Makishima, M. Matsuoka, Y. Ogawara, M. Oda, Y. Tawara, K. Doi.
Astrophys. J., Vol. 285, No. 2, p. 712 – 720 (1984).

The black hole candidate GX 339-4 (4U 1659-48) was observed in 1981 and 1982 with the X-ray satellite Hakucho for more than 50 days. The authors encountered three distinct intensity states. They witnessed a source transition from a hard-spectrum (low) state into a much brighter state with a very soft spectrum. During the low state, a random, persistent X-ray variability with ~1 s characteristic time scale was observed. This variability disappeared after the transition. The similarity between GX 339-4 and Cyg X-1 is discussed.

117.231 **Light curves of close binary systems with common scattering envelopes.**
I. B. Pustyl'nik, L. A. Einasto.
Astron. Tsirk., No. 1305, p. 1 – 4 (1984). In Russian.

117.232 **Time-resolved spectroscopy of long-period DQ Her stars.**
W. R. Penning.
Prepr. Steward Obs., No. 543, 24 pp. (1984).

The author presents high-dispersion time-resolved spectroscopic observations of four long-period DQ Her stars (intermediate polars). In two of these objects he is able to clearly detect orbital modulations, permitting an investigation of dynamic parameters. The source of He II line emission appears to have a physically different location from that of hydrogen. In all four objects short-period coherent variations in wavelength were discovered. The periods in each case turned out to be close to, and within error the same as, the published photometric white dwarf coherent periods. This lends the strongest kinematic evidence to date in support of an oblique magnetic rotator model.

117.233 **The next minimum of Z And in 1983.**
O. G. Taranova, B. F. Yudin.
Astron. Tsirk., No. 1325, p. 8 (1984). In Russian.

117.234 **Photometry of SS433.**
D. E. Mkrtichyan, A. A. Rajkov.
Astron. Tsirk., No. 1321, p. 7 – 8 (1984). In Russian.

117.235 **Flare of HR 1099 in the radio range.**
A. G. Gorshkov, V. K. Konnikova, M. G. Larionov.
Astron. Tsirk., No. 1326, p. 1 (1984). In Russian.

117.236 **Radio emission from symbiotic stars: a binary model.**
A. R. Taylor, E. R. Seaquist.
Astrophys. J., Vol. 286, No. 1, p. 263 – 268 (1984).

The authors examine a binary model for symbiotic stars to account for their radio properties. The system is comprised of a cool, mass-losing star and a hot companion. Radio emission arises in the portion of the stellar wind photoionized by the hot star. Model fits to radio spectra allow the binary separation, wind density, and ionizing photon luminosity to be calculated. The model is applied to the symbiotic star H1-36.

117.237 **The magnetic field of the AM Herculis object H0139-68.**
D. T. Wickramasinghe, N. Visvanathan, I. R. Tuohy.
Astrophys. J., Vol. 286, No. 1, p. 328 – 331 (1984).

The authors present time-resolved spectrophotometric observations of the AM Herculis type variable H0139-68 obtained when the system was in its recent low state ($V \approx 17.4$ mag). The spectra show Zeeman absorption components due to hydrogen from the underlying magnetic white dwarf. The observations are consistent with the assumption that the spectra arise from a hydrogen-rich white dwarf atmosphere ($T_e$ = 20,000K) with a centered dipole field distribution of polar field strength $B_p = 3 \times 10^7$gauss viewed at an angle $\theta \approx 45° - 90°$ to the magnetic axis.

117.238 **Photographic observations of AM Herculis.**
I. L. Andronov, S. V. Vasil'eva, V. P. Tsesevich.
Odes. univ. Odessa, 29 pp. (1984). In Russian. Abstr. in Ref. Zh., 51. Astron., 12.51.849 (1984).

117.239 **W UMa systems – the outside story.**
H. W. Duerbeck.
Mitt. Astron. Ges., Nr. 62, p. 87 – 100 (1984). – See Abstr. 010.341.

The present state of research on the global properties of W UMa type systems is reviewed: classification, space distribution, photometric properties and analyses, spectroscopic data and reduction methods. Finally, the evolutionary status of W UMa systems and related binaries is discussed, and some evolutionary sequences are outlined.

117.240 **Spektroskopischer Nachweis von bipolarem Massenfluß bei symbiotischen Sternen.**
J. Solf.
Mitt. Astron. Ges., Nr. 62, p. 239 (1984). Abstract. – See Abstr. 010.341.

117.241 **Spektroskopische und photometrische Untersuchungen von W Serpentis-Objekten.**
W. Strupat.
Mitt. Astron. Ges., Nr. 62, p. 275 – 278 (1984). – See Abstr. 010.341.

117.242 **The spectral energy distribution of cataclysmic variables.**
A. Burkert, G. Hensler.
Mitt. Astron. Ges., Nr. 62, p. 279 – 280 (1984). – See Abstr. 010.341.

117.243 **Eccentric disks in binary systems.**
G. Hensler.
Mitt. Astron. Ges., Nr. 62, p. 280 (1984). Abstract. – See Abstr. 010.341.

117.244 **Spektroskopische Untersuchung des novaähnlichen Systems V3885 Sgr.**
K. Haug, H. Drechsel.
Mitt. Astron. Ges., Nr. 62, p. 281 – 282 (1984). – See Abstr. 010.341.

117.245 **Ein Atlas und Katalog nördlicher Zwergnovae.**
A. Bruch, F.-J. Fischer, U. Wilmsen.
Mitt. Astron. Ges., Nr. 62, p. 283 – 284 (1984). Abstract. – See Abstr. 010.341.

117.246 **On the ratio of X-ray and γ-ray emission of SS433.**
S. V. Bogovalov, Yu. D. Kotov.
Pis'ma Astron. Zh., Tom 10, No. 12, p. 903 – 907 (1984). In Russian. English translation in Sov. Astron. Lett., Vol. 10.

The ratio of the fluxes of X-ray emission generated by heavy particles in the interaction with ambient electrons and of γ-ray line emission generated by the same particles is obtained. A probable model of interaction of the matter ejected from SS433 with its environment is distinguished from comparison of the calculations with observations of the SS433 object. The abundance of nuclei emitting γ-quanta in the ejected matter is obtained to be no less than 1/50 with respect to hydrogen.

117.247 **The orbit of HDE 245770.**
O. Eh. Aab.
Pis'ma Astron. Zh., Tom 10, No. 12, p. 918 – 924 (1984). In Russian. English translation in Sov. Astron. Lett., Vol. 10.

Radial velocities of absorption and emission lines in the spectrum of HDE 245770, the optical counterpart of the transient X-ray source A0535+26, have been measured. The orbital parameters $P = 35^d$, K = 35 km/s, $V_0 = -5$ km/s, e = 0.2, $f(M) = 0.15\, M_\odot$ have been calculated from absorptions. The orbits calculated from emission lines have a considerable eccentricity: e = 0.6 (Hβ) and e = 0.7 (Hα) at $P = 35^d$ and, apparently, are not connected directly with the star.

117.248 **Radial velocity components and their variations in the circumstellar spectrum of CH Cygni.**
K. H. Yoo, Y. Yamashita.
Publ. Astron. Soc. Jpn., Vol. 36, No. 3, p. 567 – 575 (1984).
Several high–dispersion plates (5 Å $mm^{-1}$ in the blue region) of the low–excitation symbiotic star CH Cygni were obtained at Okayama in June and August, 1982. The splitting of metallic absorption lines into several velocity components was found. Radial velocities were measured for both emission and absorption components. As was noted previously, they are markedly different from element to element. A large amount of variation of the absorption profile was found during the observed period. From the velocity data now available, the velocity curves were tentatively made for both emission and absorption components.

117.249 **The optical behaviour of the polar AM Her in 1984.**
W. Götz.
Inf. Bull. Variable Stars, No. 2649, 4 pp. (1984).

117.250 **Energy spectra and time variations of the X–ray source Cygnus X–2.**
T. Hirano, S. Hayakawa, H. Kunieda, F. Makino, K. Masai, F. Nagase, K. Yamashita.
Publ. Astron. Soc. Jpn., Vol. 36, No. 4, p. 769 – 784 (1984).
The energy spectra and time variations of Cyg X–2 observed by Hakucho are presented. The energy spectrum was obtained over a wide energy range 0.3 – 20 keV. This cannot be fitted with a thermal bremsstrahlung spectrum but with a combination of blackbody spectra from the stellar surface and the accretion disk, both being Comptonized by a hot central corona and an accretion disk corona, respectively. Time variations are rather complex; the intensity varies by a factor of four, while the spectrum changes little. These properties are considered to be common for weakly magnetized neutron stars with accretion disks.

117.251 **The variations of the light curve and the orbital period of the close binary 44i Boo.**
X. Liu, G. Wang, Z. Guo, G. Jia.
Acta Astrophys. Sin., Vol. 4, No. 1, p. 1 – 5 (1984). In Chinese.
The authors collected a lot of data about the time of minimum determined in the last 65 years. The values of (O–C) are calculated from the ephemeris MinI = $JD_{\odot}242113.2588 + 0.26780832 \cdot E$. The (O–C) curve can be expressed as follows: (O–C) = $0.0238 + 0.0000102 \cdot X + 0.349\cos[0.00449(X + 8221.798)]$. The difference Δ(O–C) has a periodic variation, its period is 12000 days and the amplitude is 0.005 day. This variation is possibly caused by the material transfer between the star and the common envelope. Some jumps in the observed period are considered to be responsible to some material jets.

117.252 **Effect of perturbed potentials on the positions of Lagrangian collinear points and forms of the Roche lobes in close binary systems.**
H.–n. Zhou, T.–w. Lu, Y.–j. Yu.
Acta Astron. Sin., Vol. 25, No. 3, p. 265 – 279 (1984). In Chinese.
Four different types of models for close binary systems are presented in this paper, where perturbed potentials are given as a function of the oblateness factors A1 and A2 of the two components and the radiation factor Q. On the basis of the models, the effect of perturbed potentials on the positions of the Lagrangian collinear points and forms of the Roche lobes in close binary systems are discussed. Applying the analytical formulae to five real binary systems, the authors have performed numerical calculations for different values of the parameters and obtained positions of collinear points and the radii and volumes of the Roche lobes, respectively. The results have been compared with those for the cases with non–perturbation in potentials.

117.253 **The shell spectrum of CH Cygni since brightening in the fall of 1981.**
L. Luud, T. Tomov.
Astrofizika, Tom 20, Vyp. 3, p. 419 – 428 (1984). In Russian. English translation in Astrophysics, Vol. 20, No. 3.
CH Cygni has had a well–developed shell spectrum since the brightening in the fall of 1981. Using two spectrograms with the dispersion 4.2 Å/mm, the shell of the star has been studied. It has been shown that the shell lines are formed around the hot component. The shell ionization temperature 10,000K, electron density $10^{12}cm^{-3}$ and the dilution coefficient $10^{-1}$–$10^{-3}$ have been evaluated from the curve of growth. It has been pointed out that the physical conditions in the accretive shell are close to the conditions in the atmospheres of extremely bright B8 – A2 supergiants.

117.254 **A photometric study of the contact binary XY Boo.**
N. S. Awadalla, A. Yamasaki.
Astrophys. Space Sci., Vol. 107, No. 2, p. 347 – 356 (1984).
The contact binary XY Boo is observed photoelectrically, and variations of the orbital period and the light curves are indicated. New *BV* light curves and those of Binnendijk (1971) and Winkler (1977) are analysed by use of the light curve synthesis method. Combining the derived photometric elements with the spectroscopic elements of McLean and Hilditch (1983), the authors obtain values of the masses and the radii of the components of the system.

117.255 **Time–resolved spectroscopy of SS Cygni at minimum and maximum light.**
F. V. Hessman, E. L. Robinson, R. E. Nather, E.–H. Zhang.
Astrophys. J., Vol. 286, No. 2, p. 747 – 759 (1984).
Time–resolved spectroscopic observations of the cataclysmic variable SS Cygni at minimum light and during an eruption are presented. The authors find $K_{wd} = 92 \pm 2$ km $s^{-1}$ for the white dwarf and $K_{late} = 155 \pm 2$ km $s^{-1}$ for the late–type star during minimum light. At the height of the outburst, the apparent *K* velocity of the late–type star increases to roughly 195 km $s^{-1}$. With the fading of the system the velocity gradually returns to its quiescent value. This phenomenon is explained as the result of the heating of the late–type star by the accretion disk.

117.256 **Hot subdwarfs in detached binary systems and thick–disk cataclysmic variables from the Palomar–Green survey.**
D. H. Ferguson, R. F. Green, J. Liebert.
Astrophys. J., Vol. 287, No. 1, p. 320 – 333 (1984).
The authors discuss 19 stars from the Palomar Green (PG) survey which were found to show composite spectra, generally with a hot subdwarf primary and a cool main–sequence secondary star. The primary components have $T_{eff}$ values between 27,000 and 38,000K, while the secondaries generally show spectral types G8 to K4.5; however, these ranges are undoubtedly limited by selection effects. Approximately half of all the PG subdwarfs could be in undetected binaries. Only two thick–disk cataclysmic variables with predominantly absorption–line spectra were found in the PG survey, which provides strong evidence that this kind of cataclysmic variable is luminous and confined to the galactic plane.

117.257 **Self–absorption of high–energy gamma–rays in Cygnus X–3.**
K. M. V. Apparao.
Astrophys. J., Vol. 287, No. 1, p. 338 – 340 (1984).
Absorption of high–energy γ–rays from Cyg X–3 in its own photon field by the reaction $\gamma + \gamma \rightarrow e^+ + e^-$ is considered. It is found that γ–rays with an energy around $10^{12}$eV are attenuated and the observed flux has to be corrected. Gamma–rays with energy greater than $10^{15}$eV are not affected by the photon field.

117.258 **Analysis of the AAVSO light curves of 21 dwarf novae.**
P. Szkody, J. A. Mattei.
Publ. Astron. Soc. Pac., Vol. 96, No. 586, p. 988 – 995 (1984).
The authors report the measurement of all observed outbursts of 21 dwarf novae during a three–year interval (JD2442300 –

JD2443300), using the data records of the AAVSO. Relationships among the rise time, outburst length, decline time, quiescent interval, maximum $V$ magnitude, and physical-system parameters such as orbital period and masses of the component stars were investigated. The results confirm some, but not all, of the correlations previously found for the best-studied system SS Cyg and other, less consistent, surveys. The autors use the long- and short-outburst separation characteristic of the SU UMa class to identify several prime suspects for this group.

117.259 **Dramatic variability in the X-ray flux of Hercules X-1 preceding the binary eclipse.**
S. D. Vrtilek.
News Lett. Astron. Soc. N.Y., Vol. 2, No. 6, p. 16 – 17 (1984).
Abstract. - See Abstr. 010.241.

117.260 **Pseudo white dwarfs.**
J. Patterson.
News Lett. Astron. Soc. N.Y., Vol. 2, No. 6, p. 20 (1984). Abstract. - See Abstr. 010.241.

117.261 **The evolution of the optical spectrum of the dwarf nova SS Cygni over one complete outburst cycle.**
J. T. Clarke, D. Capel, S. Bowyer.
Astrophys. J., Vol. 287, No. 2, p. 845 – 855 (1984).

Visible spectra covering 3900 – 5000 Å with a spectral resolution of ~6% of the full width of the quiescent emission lines have been obtained each night during one full outburst cycle of the dwarf nova SS Cygni. These spectra cover all phases of the outburst centered near 1983 August 1 and allow the authors to tabulate for the first time the sequence of pronounced spectral changes in the H and He lines and their phasing with respect to the continuum brightening.

117.262 **The 41.5 day binary X-ray pulsar 4U 1223-62 (GX 301-2).**
N. E. White, J. H. Swank.
Astrophys. J., Vol. 287, No. 2, p. 856 – 867 (1984).

The orbital period of the 700 s X-ray pulsar 4U 1223-62 (GX 301-2) has been determined to be 41.5 days from regular X-ray outbursts reported by Watson, Warwick, and Corbet in 1982. The authors reexamine Ariel 5 and SAS 3 X-ray pulse timing data to deduce the orbital elements of this system assuming a 41.5 day orbital period. Spectral observations by HEAO 1 and OSO 8 show a highly variable absorbed X-ray pulsar spectrum. The authors show that the global X-ray properties can be explained in terms of variable stellar wind accretion caused by the moderately elliptical $e \approx 0.47$ orbit of the pulsar about the B2 supergiant star WRA 977, provided that the mass loss properties of WRA 977 are extreme compared to single stars of a similar spectral type.

117.263 **On the spectral types of cool components of U Gem-type stars.**
G. G. Tovmasyan.
Astrofizika, Tom 21, Vyp. 2, p. 289 – 294 (1984). In Russian. English translation in Astrophysics, Vol. 21, No. 2.

The dependence of spectral type of the red components of U Gem-type binary systems on their orbital period is revealed. The obtained dependence permits to determine spectral types of the secondaries. From the dependences of the orbital periods on the mass of the secondaries on the one hand and on their spectral types on the other it follows that the secondaries are main-sequence stars. The latter permits to estimate the luminosities of the secondaries by their periods.

**Catalogue of cataclysmic binaries, low-mass X-ray binaries and related objects. Third edition.**
See Abstr. 002.003.

**A catalogue of symbiotic stars.**
See Abstr. 002.010.

**Veränderliche Sterne.**
See Abstr. 003.023.

**Close binary systems. Summary of the RAS specialist discussion held 1984 April 13 at the University of Leicester.**
See Abstr. 010.685.

**The optical behavior of SS Cygni for the international campaign.**
See Abstr. 013.028.

**On the investigation of close binary stars in Tallinn.**
See Abstr. 013.074.

**A computer program for light curve synthesis of X-ray binary systems.**
See Abstr. 021.001.

**IDS (*Image Dissector Scanner*) spectroscopy of faint emission-line objects.**
See Abstr. 034.043.

**The stability of binary star systems during encounters with a third star.**
See Abstr. 042.025.

**Resonant nuclear fusion processes and the gamma rays of SS433.**
See Abstr. 061.007.

**Regularized Monte-Carlo calculations of the optical polarization induced by scattering in X-ray binaries.**
See Abstr. 063.029.

**Line locking and SS 433.**
See Abstr. 063.035.

**The vertical structure and stability of alpha model accretion disks.**
See Abstr. 064.001.

**Disques d'accrétion. Modèles simplifiés.**
See Abstr. 064.030.

**Structure of X-ray irradiated stellar wind.**
See Abstr. 064.050.

**X-rays from accretion of red giant winds.**
See Abstr. 064.068.

**On the fission theory of binary stars. IV. Exact solutions in polynomial spaces.**
See Abstr. 065.056.

**Fission of dumbbell equilibrium and binary state of rapidly rotating polytropes.**
See Abstr. 065.057.

**The evolution of low-mass close binaries influenced by the radiation of gravitational waves and by a magnetic stellar wind.**
See Abstr. 065.060.

**Evolution of star-planet systems.**
See Abstr. 065.064.

**Evolution of accreting white dwarfs, Type I supernova explosion, and white dwarf collapse.**
See Abstr. 065.084.

**Collapse and explosion of degenerate stars.**
See Abstr. 065.085.

**Rotational and tidal perturbation of high-order non-radial *p*-modes.**
See Abstr. 065.121.

**Long-term post-Newtonian effects in a binary system.**
See Abstr. 066.286.

**X-ray bursts: nuclear flash model and its difficulties.**
See Abstr. 067.020.

**Formation of a single millisecond pulsar by the coalescence of a neutron star and a massive white dwarf.**
See Abstr. 067.066.

**Infrared emission from accretion disks: detectability and variability.**
See Abstr. 067.091.

**Cyclotron emission near stellar mass black holes.**
See Abstr. 067.094.

**Exploding neutron stars in close binaries.**
See Abstr. 067.121.

**X–ray burst sources – from limit cycles to chaos.**
See Abstr. 067.122.

**Particle acceleration in compact binary stars.**
See Abstr. 067.127.

**A solar spectral line list between 10 and 200 Å modified for application to high spectral resolution X–ray astronomy.**
See Abstr. 076.011.

**The shell spectrum of CH Cygni in 1981.**
See Abstr. 112.009.

**Be stars in binaries.**
See Abstr. 112.059.

**Photometric observations of active chromosphere stars: II Peg, HD 175742, HD 199178.**
See Abstr. 112.071.

**The depletion of pre–interstellar matter and an examination of Ni II oscillator strengths.**
See Abstr. 112.076.

**Is 89 Herculis a binary star?**
See Abstr. 114.058.

**Some problems in the determination of fundamental stellar parameters from binary stars.**
See Abstr. 115.017.

**Light time effect in TW Draconis.**
See Abstr. 118.017.

**Fourier analysis of the light changes of four Algol–type eclipsing variables.**
See Abstr. 119.018.

**RR Trianguli Australis: a new analysis of a semi–detached system of unusually small dimensions.**
See Abstr. 119.037.

**Light–curve analysis of Z Herculis.**
See Abstr. 119.038.

**Two–colour photoelectric investigation of SZ Piscium.**
See Abstr. 119.041.

**New BV photoelectric observations of BX Andromedae.**
See Abstr. 119.043.

**Evidence for a high–temperature accretion region in Algol–type binary systems.**
See Abstr. 119.045.

**Light curve analyses for the early–type eclipsing binaries TT Aur and XZ Cep.**
See Abstr. 119.046.

**Simultaneous differential photometry with the St. Andrews twin photometric telescope. I. The eclipsing binary TT Aurigae.**
See Abstr. 119.047.

**Photometric study of the eclipsing binary system AR Lacertae.**
See Abstr. 119.053.

**The new eclipsing magnetic binary system E1114 + 182.**
See Abstr. 119.092.

**Egress spectra of the new eclipsing supergiant 22 Vulpeculae.**
See Abstr. 120.008.

**Variable stars V830, V1095 Ophiuchi and CSV 7725.**
See Abstr. 122.048.

**Coordinated IUE and ground–based observations of active stars: flare events on YZ CMi, V1005 Ori, AD Leo and AR Lac.**
See Abstr. 122.065.

**Period changes and a possible second pulsation mode in BW Vulpeculae.**
See Abstr. 122.094.

**Intense X–ray flares from active stellar systems: EV Lacertae and HD 8357.**
See Abstr. 122.113.

**Supernova progenitors and Be stars: stellar variability from a 21 century perspective.**
See Abstr. 122.197.

**WY Sagittae (Nova 1783): a transition object between classical and dwarf novae?**
See Abstr. 124.181.

**Nova Aquilae 1982 – a short report.**
See Abstr. 124.201.

**Binary progenitors of supernovae.**
See Abstr. 125.083.

**Orbital inclination and mass of the binary pulsar PSR 0655 + 64.**
See Abstr. 126.003.

**Discovery of a white dwarf companion of the "hybrid" K giant HD 81817.**
See Abstr. 126.004.

**Comments on "The white dwarf collapse in low mass binary systems".**
See Abstr. 126.016.

**The millisecond pulsar.**
See Abstr. 126.037.

**Millisecond pulsars.**
See Abstr. 126.046.

**The hard X–ray light–curves of accreting magnetized white dwarfs.**
See Abstr. 126.068.

**Feige 24: wind/flare accretion by a hot DAZ1 degenerate.**
See Abstr. 126.083.

**Angular momentum transfer by gravitational torques and the evolution of binary protostars.**
See Abstr. 131.037.

**Infrared light curves of the central object of NGC 2346: the shape of the obscuring cloud.**
See Abstr. 134.014.

**Variability of SS433 and extragalactic compact radio sources.**
See Abstr. 141.011.

**Electrophotometric observations of the X-ray source Cygnus X-1 in 1982 according to a coordinated programme.**
See Abstr. 142.002.

**A tentative interpretation of the cyclotron line energy variation in Her X-1.**
See Abstr. 142.004.

**Determination of the mass of globular cluster X-ray sources.**
See Abstr. 142.009.

**Infrared studies of X-ray sources.**
See Abstr. 142.014.

**A possible optical counterpart for the M15 globular cluster X-ray source.**
See Abstr. 142.021.

**A study of the spectra and pulse profiles of Centaurus X-3 from Hakucho.**
See Abstr. 142.022.

**Evidence for 4.4 hour periodic dips in the X-ray flux from 4U 1755-33.**
See Abstr. 142.024.

**Discovery of a 7.1 hour period and eclipses from MXB 1659-29.**
See Abstr. 142.027.

**Delays of optical bursts in simultaneous optical and X-ray observations of MXB 1636-53.**
See Abstr. 142.028.

**Discovery of a 176 day period in 4U 1820-30.**
See Abstr. 142.030.

**Characteristics of X-ray spectra from galactic and extra-galactic X-ray sources.**
See Abstr. 142.033.

**New radial velocities of the supergiant in the Cygnus X-1 system.**
See Abstr. 142.047.

**HZ Her/Her X-1: an alternative model for the $35^d$ cycle?**
See Abstr. 142.048.

**Multispectral analysis in the UV, optical and IR of HDE 245770 = A0535+26.**
See Abstr. 142.051.

**Search for millisecond rotational periods in some low-mass X-ray binaries observed by EXOSAT.**
See Abstr. 142.058.

**High resolution spectroscopy of the optical candidate for the X-ray transient X0331+53.**
See Abstr. 142.061.

**Discovery of a 437.5-s X-ray pulsation from 4U 1907+09.**
See Abstr. 142.068.

**A possible 67-s X-ray pulsation period from 4U 1700-37.**
See Abstr. 142.069.

**Variation of the pulse profile of Hercules X-1.**
See Abstr. 142.072.

**Energy spectra of low-mass binary X-ray sources observed from Tenma.**
See Abstr. 142.074.

**Detection of iron K-emission lines from two low-mass binary X-ray sources: Scorpius X-1 and 4U 1608-52.**
See Abstr. 142.075.

**Evidence for an increasing 59-s period in the X-ray emission from Geminga.**
See Abstr. 143.002.

**Binary star model of the γ-ray burst with special reference to that on 5 March, 1979 and subsequent bursts on 6 March, 4 April, and 24 April, 1979.**
See Abstr. 143.009.

**Is Geminga a very close neutron star binary?**
See Abstr. 143.023.

**Evidence for a binary system at the 5 March 1979 gamma-ray burst position.**
See Abstr. 143.027.

**Absorption of high-energy gamma rays by photon-photon interactions near a star.**
See Abstr. 143.028.

**Astrophysical boosters.**
See Abstr. 144.027.

**Effects of encounters with field stars on the evolution of low-mass semidetached binaries.**
See Abstr. 151.043.

**Wolf-Rayet stars in open clusters and associations.**
See Abstr. 153.012.

**Young stars in the old galactic cluster NGC 188?**
See Abstr. 153.028.

**The appearance, number, and history of highly compact binary systems in globular clusters.**
See Abstr. 154.025.

**The evolution of highly compact binary stellar systems in globular clusters.**
See Abstr. 154.026.

**X-ray sources in globular clusters.**
See Abstr. 154.028.

**Quasars as supermassive binaries.**
See Abstr. 159.056.

# 118 Visual Binaries, Multiple Stars, Astrometric Binaries

118.001 **Orbits of 16 visual binaries.**
W. D. Heintz.
Astron. J., Vol. 89, No. 7, p. 1063 – 1067 (1984).
Orbital elements, dynamical parallaxes, and masses have been computed for six Northern and ten Southern pairs from recent reobservations. Mass anomalies are noted for ADS 3959, Fin 357, and EQ Peg.

118.002 **Speckle interferometric measurements of binary stars. II.**
Y. Balega (*Yu. Yu. Balega*), D. Bonneau, R. Foy.
Astron. Astrophys., Suppl. Ser., Vol. 57, No. 1, p. 31 – 36 (1984).
The authors report speckle interferometric measurements of 87 binaries or suspected binaries observed with the 6 m SAO, 3.60 m CFH and ESO, and 1.93 m OHP telescopes. Four binaries are first resolved. Preliminary orbital parameters are proposed for two systems.

118.003 **Erratum: "Photometric and astrometric observations of close visual binaries" [Astron. Astrophys., Suppl. Ser., Vol. 47, p. 221 – 235 (1982)].**
K. D. Rakos, R. Albrecht, H. Jenkner, T. Kreidl, R. Michalke, D. Oberlerchner, E. Santos, A. Schermann, A. Schnell, W. Weiss.
Astron. Astrophys., Suppl. Ser., Vol. 57, No. 1, p. 170 (1984). See Abstr. 31.118.009.

118.004 **Orbits of six visual double stars.**
P. Couteau.
Astron. Astrophys., Suppl. Ser., Vol. 57, No. 2, p. 171 – 175 (1984). In French.
Elements of the pairs ADS 746, 1786, 9692, 10085, 16665 and Cou 542 have been determined with recent observations. Comments are given on dynamical parallaxes, absolute magnitudes and masses.

118.005 **Revised orbital elements of the visual binary star ADS 10235–STF 2107.**
M. Scardia.
Astron. Astrophys., Suppl. Ser., Vol. 57, No. 2, p. 257 – 262 (1984). In French.
Using a sample of 414 observations the author has computed the revised orbital elements of the visual binary star ADS 10235–STF 2107. He has calculated the dynamical parallax and the total mass of the system.

118.006 **On the properties of planets of other stars. Considerations of similarity.**
K. A. Burlov–Vasil'ev, A. F. Steklov.
Astrometr. Astrofiz., Vyp. 52, p. 26 – 30 (1984). In Russian.
A method is suggested for obtaining information on planetary properties of other stars using the minimum known parameters. Considerations of similarity and some consequences of the gas dissipation theory by Jeans are applied. The method is employed for estimation of properties of supposed planets of Barnard's star.

118.007 **Double star measurements made at Nice.**
P. Muller.
Astron. Astrophys., Suppl. Ser., Vol. 57, No. 3, p. 467 – 486 (1984). In French.
The author presents observations of double stars taken from two lists: A (General program), mainly neglected pairs, and B (Mlr pairs), made with the 50 cm refractor of Nice Observatory.

118.008 **Close visual binaries. I. MK classifications.**
C. J. Corbally.
Astrophys. J., Suppl. Ser., Vol. 55, No. 4, p. 657 – 677 (1984).
Each component of 170 close visual binaries has been classified with new procedures for controlling contamination problems. These classifications are presented and are shown to be on the MK system. Two sources of area scanner *UBV* photometry were compared in order to establish homogeneous photometric as well as spectroscopic data. Absolute magnitudes for each binary are derived from the $V$ magnitude differences between the components via a tested MK – $M_v$ map.

118.009 **Astronomy of binaries and multiple stars.**
A. A. Tokovinin.
Zemlya Vselennaya, No. 6, p. 10 – 15 (1984). In Russian.

118.010 **Micrometric measurements of binary stars (second list).**
M. Scardia.
Astron. Astrophys., Suppl. Ser., Vol. 58, No. 1, p. 89 – 94 (1984). In French.
The results of 286 micrometric measurements of 88 binaries are given. The measurements have been made at Brera–Merate observatory during the period March 1983 – January 1984 with a 23 cm refractor.

118.011 **Digital speckle interferometry of binary stars.**
Yu. Yu. Balega, V. P. Ryadchenko.
Sov. Astron. Lett., Vol. 10, No. 2, p. 95 – 98 (1984). English translation of 37.118.015.

118.012 **Interferometer orbits for seven binaries.**
A. A. Tokovinin.
Sov. Astron. Lett., Vol. 10, No. 2, p. 121 – 123 (1984). English translation of 37.118.026.

118.013 **Étoiles doubles visuelles et binaires fictives.**
D. Benest, M. Fulconis.
Astrométrie, géodésie, mécanique céleste, p. 209 – 212 (1984). – See Abstr. 012.021.

118.014 **Alcune stelle binarie.**
S. De Meis, S. Baroni.
Coelum, Vol. 52, N. 5, p. 271 – 292 (1984).

118.015 **Stellar speckle interferometry using the Dodaira 91 cm telescope.**
J. Ohtsubo, T. Eiju, T. Kohno, K. Tomita, M. Noguchi.
Tokyo Astron. Obs. Rep., Vol. 20, No. 2, p. 255 – 269 (1984). In Japanese.

118.016 **Photographic observations of the star ADS 11632 with possible unseen companion.**
N. A. Shakht.
Pis'ma Astron. Zh., Tom 10, No. 10, p. 765 – 770 (1984). In Russian. English translation in Sov. Astron. Lett, Vol. 10.
Relative positions of the components of the double star ADS 11632 are obtained from observations during 1961 – 1982. The results are compared with the Washington, Dearborn and Flagstaff series. Perturbations from an unseen component with a period of 5 years and amplitude of about 0.″01 can be assumed.

118.017 **Light time effect in TW Draconis.**
K. D. Abhyankar, T. Panchatsaram.
Mon. Not. R. Astron. Soc., Vol. 211, No. 1, p. 75 – 82 (1984).
The run of the photoelectric minimum time residuals of TW Draconis is interpreted in terms of light time effect caused by the presence of additional components. While the ephemeris based on the period of 2.806847 day due to Tremko & Kreiner would indicate a quintet system, a new ephemeris based on a period of 2.80685833 day gives a simple triple system. The former is found to be unstable according to the empirical stability criteria for hierarchical systems. It is therefore concluded that TW Dra is most probably a triple system with a third component of mass between 0.71 and 0.84 $M_{\odot}$ for $i > 60°$.

118.018 **Period study of the Algol–type eclipsing binary – R Canis Majoris.**
K. R. Radhakrishnan, K. D. Abhyankar, M. B. K. Sarma.
Bull. Astron. Soc. India, Vol. 12, No. 2, p. 182 – 189 (1984).

Photoelectric observations of the Algol–type eclipsing binary R Canis Majoris during three consecutive observing seasons of 1979 – 82 gave a total of ten primary minima in each of the $V$ and $B$ passbands and nine in the $U$ passband. The $(O-C)$ curve obtained by using these minima and those available in the literature indicates that the period changes noticed in R CMa are due to light–time effect. The elements of the light–time orbit are determined. The mass of the third body is derived to be $\approx 0.5\ M_{\odot}$ for $i$ values ranging from 60° to 90°. It is suggested that the third body may be an M dwarf or a white dwarf.

118.019 **The triple star Zeta Aquarii.**
W. D. Heintz.
Astrophys. J., Vol. 284, No. 2, p. 806 – 809 (1984).

Visual, photographic, and parallax observations are used in a discussion of the orbits of the system: $a = 4''.51$ and $P = 760$ yr are found for the visual pair; the unresolved companion associated with star A has an eccentric orbit and a mass near $0.4\ M_{\odot}$.

118.020 **Planeten bei Beta Pictoris?**
A. M. Quetsch.
Sterne Weltraum, 23. Jahrg., Nr. 12, p. 627 (1984).

118.021 **Close visual binaries. II. Quantitative test of isochrones.**
C. J. Corbally.
Astrophys. J., Vol. 285, No. 1, p. 195 – 201 (1984).

Homogeneous and accurate color–magnitude data are available from Paper I for each component of 56 close visual binaries, 30 with evolved components. The 21 binaries with main–sequence primaries are compared with new Yale isochrones to determine the appropriate isochrone composition: $Y = 0.25$, $Z = 0.04$. In general, the mean apparent age difference between binary components ($0.0 \pm 0.1$ Gyr) shows stellar evolution to be well modeled by the new Yale isochrones.

118.022 **The peculiar emission–line object T21.**
J. H. Lutz.
Publ. Astron. Soc. Pac., Vol. 96, No. 584, p. 790 (1984). Abstract. – See Abstr. 010.281.

118.023 **The binary frequency of CN stars.**
G. C. L. Aikman, R. D. McClure.
J. R. Astron. Soc. Can., Vol. 78, No. 5, p. 205 (1984). Abstract. – See Abstr. 010.401.

118.024 **The precision radial–velocity search for low–mass companions to solar–like stars.**
B. Campbell, G. A. H. Walker, J. Amor.
J. R. Astron. Soc. Can., Vol. 78, No. 5, p. 206 (1984). Abstract. – See Abstr. 010.401.

118.025 **Establishment and improvement of the initial orbit of the visual binary star ADS 8630.**
D. Pan, L. Yan.
Ann. Shanghai Obs., Acad. Sin., No. 5, p. 101 – 110 (1983). In Chinese.

The elements of the true orbit of the visual binary ADS 8630 have been determined from 131 observations.

118.026 **On the dynamical method of the orbit determination of visual binary stars.**
Z. Cheng.
Ann. Shanghai Obs., Acad. Sin., No. 5, p. 122 – 127 (1983). In Chinese.

The dynamical methods of orbit determination for ellipse, hyperbola and parabola types are established. The parabola orbit of the visual binary star ADS 8094 has been determined as an example.

118.027 **Photographic measurement of 40 visual double stars.**
G. Zhu, X. Wang, L. Yan.
Ann. Shanghai Obs., Acad. Sin., No. 5, p. 128 – 131 (1983). In Chinese.

The position angles and angular separations between components of 40 double stars are given, which are referred to the epoch of 1950. RMS errors on the average of position angles and angular separations are $\pm 0°.67$ and $\pm 0''.11$ respectively. The accuracy of orienting measured plates is about 0°.03.

118.028 **The control of calculations of orbital elements of binary systems.**
V. K. Drofa.
Astrometr. Astrofiz., Vyp. 53, p. 66 – 69 (1984). In Russian.

The orbital elements and masses of binary system components as well as the parallaxes for visual systems being available, it is possible to estimate the value of the gravity constant k according to Kepler's third law. On the assumption of k being constant in space, deviation of the estimated value from the generally accepted one indicates that the orbital elements are erroneous. The method has been tested on visual and spectroscopic binary systems.

118.029 **The binary nature of the magnetic star β CrB.**
L. Oetken, R. Orwert.
Astron. Nachr., Vol. 305, No. 6, p. 317 – 323 (1984).

The radial velocity of the binary star β CrB was reinvestigated to look for a hypothetical third body. The eccentricity and the angle of periastron passage of the visual orbit, derived from published speckle interferometric measurements, agree very well with the corresponding elements of the spectroscopic orbit. For the masses of the components those of giant resp. subgiant stars of type A8 and F5 are found. The geometry of the binary β CrB with a magnetic star as the primary component is demonstrated. The probability for the existence of a third component in the system is reduced.

118.030 **Micrometer measurements of visual double stars.**
J. A. Docobo, J. M. Costa, J. F. Ling.
Astron. Astrophys., Suppl. Ser., Vol. 58, No. 2, p. 287 – 289 (1984).

The authors present 190 micrometer measurements of 36 visual double stars made at the Santiago de Compostela Observatory in 1983.

118.031 **HD 138168 and HD 138181, a hard binary?**
L. O. Lodén.
Astron. Astrophys., Suppl. Ser., Vol. 58, No. 2, p. 431 – 434 (1984).

Relevant physical data for the two F3 stars HD 138168 and HD 138181 have been mutually compared with respect to possible existence of significant individual properties. It has not been possible to detect any palpable deviation, however.

118.032 **Observations of binary stars by speckle interferometry at the Pic du Midi T2m.**
D. Bonneau, J. M. Carquillat, J. L. Vidal.
Astron. Astrophys., Suppl. Ser., Vol. 58, No. 3, p. 729 – 733 (1984). In French.

The authors present the results of the first observations of binary stars by speckle interferometry at the Pic du Midi 2 m telescope. Out of 53 observed systems, 31 couples have been measured. The star with composite spectrum HD 52822/3 has been resolved for the first time.

118.033 **Orbites nouvelles.**
Circ. Inf., No. 94 (1984).

118.034 **Etoiles doubles nouvelles.**
P. Couteau.
Circ. Inf., No. 94 (1984).

118.035 **G208–44.**
IAU Circ., No. 3989 (1984).

118.036 **HR 1321 = ADS 3085 B: a new variable star.**
K. Ziegler, M. Shirley, B. Francom, W. Florence, D. S. Hall.
Inf. Bull. Variable Stars, No. 2619, 3 pp. (1984).

118.037 **Close visual binaries. III. Parameters and evolutionary status.**
C. J. Corbally.
Astron. J., Vol. 89, No. 12, p. 1887 – 1896 (1984).
New Yale isochrones, which have been tested for accuracy (Paper II), provide the means to investigate interesting visual binaries, especially those whose classifications and photometry do not match well (Paper I). Various parameters are deduced for those binaries which fitted the isochrones (e.g., ages, metal abundances, luminosities of peculiar stars); various solutions are systematically developed for those which did not fit, and a likely status of evolution proposed (e.g., duplicity of the components, pre–main–sequence, blue straggler, horizontal branch, optical pair, data inaccuracies). Evolution around the helium flash and diffusion theory are briefly considered.

118.038 **Other planetary systems?**
A. R. Serban.
Madà, Vol. 28, No. 2, p. 103 (1984). In Hebrew.

118.039 **Orbite pour le couple visuel IDS 01170 S 2439 = SEE 13 = ADS 1102 = BDS 721.**
L. Zimmermann.
Bull. Astron., Vol. 9, No. 5, p. 251 – 253 (1983).

118.040 **New orbital elements for the visual binary IDS 14527 S 2715 = ADS 9453.**
E. Van Dessel.
Bull. Astron., Vol. 9, No. 5, p. 254 – 255 (1983).

118.041 **Orbit of the visual binary IDS 15154 S 5858 = h 4757 = γ Circini.**
O. Nys.
Bull. Astron., Vol. 9, No. 5, p. 256 – 261 (1983).

118.042 **Orbite pour le couple visuel IDS 23058 S 4331 = B594 = HD 218812.**
A. Norro.
Bull. Astron., Vol. 9, No. 5, p. 262 – 264 (1983).

118.043 **Infrared detection of a close cool companion to Van Biesbroeck 8.**
D. W. McCarthy Jr., R. G. Probst, F. J. Low.
Prepr. Steward Obs., No. 554, 10 pp. (1984).
The authors have detected, via infrared speckle interferometry, a cool object one arcsecond from the very low luminosity star VB 8, three magnitudes fainter than VB 8 at 2.2 microns. Measurements at 1.6 and 2.2 μm give $T_e = 1360K$ and (assuming a physical association) $R = 0.09\ R_\odot$, $L = 3 \times 10^{-5} L_\odot$, consistent with a substellar brown dwarf. These observations may constitute the first direct detection of an extra–solar planet.

118.044 **Determination of the dynamical parallaxes of some visual binaries by the method of apparent motion parameters.**
O. V. Kiyaeva.
Astron. Tsirk., No. 1322, p. 6 – 8 (1984). In Russian.

118.045 **Orbits of three visual binaries.**
G. A. Starikova.
Astron. Tsirk., No. 1325, p. 4 – 5 (1984). In Russian.

118.046 **Quasi–stationary distribution of binary semiaxes.**
V. N. Sementsov, F. A. Tsitsin.
Astron. Tsirk., No. 1328, p. 1 – 3 (1984). In Russian.

118.047 **The evolution of a massive binary in a star field.**
A. S. Baranov.
Astron. Zh., Tom 61, Vyp. 6, p. 1098 – 1107 (1984). In Russian. English translation in Sov. Astron., Vol. 28, No. 6.
The paper deals with the evolution of the orbit of a massive binary system interacting with surrounding single stars, the phase density of field stars being supposed homogeneous in the configuration space. It is assumed that: 1) the distribution function of velocities is isotropic up to a certain limit value and 2) the velocity of a typical field star is greater as compared to the orbital velocity of the pair members. The formula describing the transfer of energy from the binary to field stars has been derived. As a result of the above process the formulae for changes with time of the semimajor axis and the eccentricity have been obtained. These formulae have been applied to estimate the rate of the orbital evolution of a central binary system of black holes in the nucleus of the Galaxy.

118.048 **The parallax and mass of 70 Ophiuchi.**
A. H. Batten, J. M. Fletcher, B. Campbell.
Publ. Astron. Soc. Pac., Vol. 96, No. 585, p. 903 – 909 (1984).
Radial velocities measured for both components of 70 Oph during the recent passage of the stars through the nodes of their orbits enable the parallax and mass of the system to be determined independently of absolute astrometric measurements. The absolute parallax derived form radial velocities alone is $0''.194 \pm 0''.004$. Combining this value with the best astrometric determination yields an absolute parallax of $0''.199$ and individual masses for the stars of $0.81\ M_\odot \pm 0.06\ M_\odot$ and $0.72\ M_\odot \pm 0.09\ M_\odot$. It is also shown that the new radial–velocity observations agree with the best micrometer measures in providing no evidence for a perturbation in the orbital motion of the brighter star.

**Composantes de systèmes doubles ou multiples de l'Index (1976.5) ayant un même No. BD mais figurant séparément dans l'AGK2/3.**
See Abstr. 002.050.

**Fourth Catalog of Orbits of Visual Binary Stars.**
See Abstr. 002.073.

**International conference on astrometric binaries.**
See Abstr. 011.037.

**Observabilité des étoiles doubles par Hipparcos.**
See Abstr. 036.022.

**Photometric approaches to detecting other solar systems.**
See Abstr. 036.063.

**Planets in binary star systems.**
See Abstr. 042.019.

**The influence of the effect of rotation on the evolution of triple systems.**
See Abstr. 042.024.

**The stability of binary star systems during encounters with a third star.**
See Abstr. 042.025.

**Third–order and tidal effects in the stellar three–body problem.**
See Abstr. 042.058.

**Evolution of star–planet systems.**
See Abstr. 065.064.

**On regularities of formation of planetary systems.**
See Abstr. 107.006.

**Photographic astrometry of binary and proper–motion stars.**
See Abstr. 111.001.

**Be stars in binaries.**
See Abstr. 112.059.

**A photometric study of six nearby common–proper–motion pairs.**
See Abstr. 113.001.

**Some problems in the determination of fundamental stellar parameters from binary stars.**
See Abstr. 115.017.

**Binary stars.**
See Abstr. 117.018.

**A new photoelectric investigation of the W UMa system U Pegasi.**
See Abstr. 117.022.

**High resolution problems in astrometry and spectroscopy of binaries and their significance for the progress of stellar physics.**
See Abstr. 117.223.

**Non–radial pulsator in the binary system Y Cam.**
See Abstr. 119.033.

**Periods in the light variation of σ Scorpii.**
See Abstr. 122.116.

# 119 Eclipsing Binaries

119.001 **BM Orionis: the unique eclipsing binary in Trapezium. The components and extended matter around the cool star. I. Photoelectric observations.**
M. M. Zakirov, V. S. Shevchenko.
Perem. Zvezdy, Tom 21, No. 5, p. 629 – 637 (1982). In Russian.
Results of photoelectric observations of the eclipsing binary BM Ori are given. From 150 to 200 measurements in each filter were obtained. The light curve of BM Ori is distorted out of eclipse by oscillations of $0^{m}.1$. The amplitude of the primary minimum of BM Ori is linearly dependent on wavelength.

119.002 **BM Orionis: the unique eclipsing binary in Trapezium. The components and extended matter around the cool star. II. Spectroscopic and polarimetric investigation. Discussion of the results.**
M. M. Zakirov, V. S. Shevchenko.
Perem. Zvezdy, Tom 21, No. 5, p. 638 – 653 (1982). In Russian.
Results of a spectroscopic and polarimetric investigation of BM Ori made in 1971 – 1977 are given. The radial velocity curve for the hot component was derived from 15 spectrograms. Radial velocities of the secondary component were estimated from the weak absorption K Ca II. The spectral class of the hot star B1–B1.5 was determined by three independent methods. A change of the Balmer discontinuity with phase was not found. A possible variability of the hydrogen line profiles was noted. The polarization of BM Ori from 0.8%(U) to 0.2%(V) does not considerably change with phase.

119.003 **On the binarity of RY Tauri.**
U. A. Nurmanova.
Perem. Zvezdy, Tom 21, No. 5, p. 655 – 664 (1982). In Russian.
An examination of published data of multicolour photometry and polarimetry of RY Tau shows that the star is an eclipsing binary. Two working models are supposed, both founded on the assumption of duplicity and allowing with different degree of completeness to explain the properties of RY Tau. The amounts of extinction toward the star and interstellar polarization are evaluated.

119.004 **On the period of IP Persei.**
V. I. Kardopolov, G. K. Filip'ev.
Perem. Zvezdy, Tom 21, No. 5, p. 688 – 690 (1982). In Russian.
Results of UBV photoelectric measurements of the irregular variable star IP Per in 1978 – 1980 are given. The photoelectric data are in very good accordance with the elements: JD $2441932^{d}.74 + 1^{d}.94672$ E. So IP Per may be classified as an eclipsing variable star.

119.005 **Emission components of the $H_{\gamma}$ line in the spectrum of V367 Cyg.**
E. V. Menchenkova.
Astron. Tsirk., No. 1284, p. 5 – 6 (1983). In Russian.

119.006 **Rediscussion of eclipsing binaries. XIV. The bright Am system V624 Herculis.**
D. M. Popper.
Astron. J., Vol. 89, No. 7, p. 1057 – 1062 (1984).
V624 Her is a 6th magnitude $3^{d}.6$ double–lined eclipsing metallic–lined binary of middle A type with eclipses about 0.18 mag deep. Analysis of a new set of 6.7 Å $mm^{-1}$ spectrograms, combined with a reanalysis of Zissell's curve on the basis of the spectrophotometric light ratio, yields well–determined properties of the components. The hotter component is the most massive and the largest of all Am stars with well–determined properties.

119.007 **Differential *UBV* photometry of the short–period eclipsing binaries RW Dor, EZ Hya and V701 Cen.**
G. Russo, A. A. Vittone, L. Milano.
Astron. Astrophys., Suppl. Ser., Vol. 57, No. 1, p. 69 – 73 (1984).
Photoelectric *UBV* differential observations of three short–period eclipsing binaries (RW Dor, EZ Hya and V701 Cen) are presented; new times of minimum light and improved linear ephemerides have been computed.

119.008 **New narrow band light curves of the eclipsing binary u Her.**
W. E. C. J. van der Veen.
Astron. Astrophys., Suppl. Ser., Vol. 57, No. 1, p. 139 – 153 (1984).
This paper presents light curves in the Utrecht Photometric System (UPS) for the eclipsing binary u Her which have been derived from photoelectric observations obtained at the Dutch observatory at Ausserbinn (Switzerland). For the reduction of the raw data an improved and rapid process was developed. It yields iteratively the normal points in a semi–automatic way.

119.009 **UBVR photometry of the primary minimum of RZ Cas.**
N. N. Nosova.
Astron. Tsirk., No. 1288, p. 6 – 7 (1983). In Russian.

119.010 **New data on the apsidal motion of RU Monocerotis.**
D. Ya. Martynov.
Astron. Tsirk., No. 1289, p. 1 – 3 (1983). In Russian.

119.011 **Determination of the masses of the components and spectral classification of the companion of the eclipsing binary CQ Cep.**
T. A. Kartasheva, L. I. Snezhko.
Astron. Tsirk., No. 1293, p. 5 – 7 (1983). In Russian.

119.012 **Outflow of matter in the system of the eclipsing binary star V367 Cyg.**
V. G. Karetnikov, E. V. Menchenkova.
Astron. Tsirk., No. 1299, p. 3 – 5 (1983). In Russian.

119.013 **On a possibility of finding nonlinearity in limb darkening from the light curves of classical eclipsing systems.**
A. A. Rubashevskij.
Astron. Tsirk., No. 1299, p. 5 – 7 (1983). In Russian.

119.014 **Helium in the atmospheres of binary systems.**
V. V. Leushin.
Astron. Zh., Tom. 61, Vyp. 4., p. 733 – 738 (1984). In Russian. English translation in Sov. Astron., Vol. 28, No. 4.

The helium abundances were obtained for 25 brighter components of binary stars by model atmosphere analysis. The helium abundances for the binary stars that lie on the main sequence are larger in average than in single normal stars. The stars with enhanced helium abundance lie at a larger distance from ZAMS than those with normal helium abundance.

119.015 **Characteristics of the object SS 433 (V1343 Aql) as a massive eclipsing binary system.**
A. V. Goncharskii (*A. V. Goncharskij*), Z. Yu. Metlitskaya, A. M. Cherepashchuk.
Sov. Astron., Vol. 28, No. 1, p. 74 – 80 (1984). English translation of 37.119.003.

119.016 **Algol binaries: variable photometric disturbances and evolution.**
E. C. Olson.
Advances in photoelectric photometry, Vol. 2, p. 15 – 29 (1984). – See Abstr. 003.002.

119.017 **Photometric elements of the eclipsing binary EK Cephei.**
G. Hill, E. G. Ebbighausen.
Astron. J., Vol. 89, No. 8, p. 1256 – 1260 (1984).

Photometric elements are presented, based on new $V$ and $R$ observations, of the astrophysically important binary EK Cephei. The light–curve analysis supports the recent conclusion of Khaliullin (1983), who measured a rate of advance of periastron of 0.076°/yr. The authors' rate is 0.082°/yr. A study of the system's evolutionary state reveals that both components are zero–age main–sequence objects of chemical composition ($X$,$Z$; 0.66, 0.037).

119.018 **Fourier analysis of the light changes of four Algol–type eclipsing variables.**
H. M. K. Al–Naimiy, A. O. Al–Sikab.
Astrophys. Space Sci., Vol. 103, No. 1, p. 115 – 124 (1984).

The aim of this paper is to present the analysis of the light curves of four totally–eclipsing systems (BM Ori, VW Cyg, TW Dra, and RW Mon), using the automated Fourier techniques in the frequency–domain, developed recently by Kopal and his school at Manchester University. New physical and geometrical elements for the above systems are derived.

119.019 **106th list of minima of eclipsing binaries.**
BBSAG Bull., No. 73, p. 1 – 5 (1984).

119.020 **V442 Cygni: the duration of totality.**
R. Diethelm.
BBSAG Bull., No. 73, p. 6 (1984).

119.021 **V456 Ophiuchi: improved period.**
R. Diethelm.
BBSAG Bull., No. 73, p. 6 (1984).

119.022 **Increase of the period of the eclipsing binary V444 Cyg and mass loss by the Wolf–Rayet star.**
Kh. F. Khaliullin, A. I. Khaliullina, A. M. Cherepashchuk.
Pis'ma Astron. Zh., Tom 10, No. 8, p. 600 – 603 (1984). In Russian. English translation in Sov. Astron. Lett., Vol. 10.

New photoelectric observations of V444 Cyg have confirmed the increase of the period found earlier on giving $\dot{P} = 0.202 \pm 0.018$ s/yr. Mass loss by the MN5 star of $\dot{M} = -(1.02 + 0.20) \times 10^{-5}\ M_{\odot}$/yr is obtained.

119.023 **More precise elements of V442 Cassiopeiae.**
N. E. Kurochkin.
Perem. Zvezdy, Prilozh., Tom 4, No. 20, p. 166 – 168 (1982). In Russian.

119.024 **MZ Lyrae.**
E. N. Makarenko.
Perem. Zvezdy, Prilozh., Tom 4, No. 20, p. 171 – 173 (1982). In Russian.

119.025 **CSV 6884 = V585 Centauri.**
V. P. Goranskij, V. P. Fedorovich.
Perem. Zvezdy, Prilozh., Tom 4, No. 20, p. 210 (1982). In Russian.

119.026 **Investigation of variable stars SVS 2336 and SVS 2337.**
S. M. Kopejkin.
Perem. Zvezdy, Prilozh., Tom 4, No. 20, p. 221 – 224 (1982). In Russian.

119.027 **A study of the O'Connell effect in the light curves of eclipsing binaries.**
T. J. Davidge, E. F. Milone.
Astrophys. J., Suppl. Ser., Vol. 55, No. 4, p. 571 – 584 (1984). = Publ. Rothney Astrophys. Obs., No. 22.

O'Connell's classic study of light curve asymmetries at maximum light has been redone with *UBV* photoelectric data. An attempt has been made to improve the data base by considering only constant asymmetry systems. The correlations originally found by O'Connell were reexamined, and new correlations explored with parametric and nonparametric statistical methods. The findings suggest that more than one mechanism is responsible for the O'Connell effect.

119.028 **Absolute dimensions of eclipsing binaries. VI. The F–type system DM Virginis.**
J. Andersen, J. V. Clausen, B. Nordström.
Astron. Astrophys., Vol. 137, No. 2, p. 281 – 286 (1984).

New, accurate, and complete *uvby* light curves (Andersen et al., 1983) of the late F–type, double–lined eclipsing binary DM Virginis have been analysed to determine the absolute dimensions, chemical composition, and evolutionary stage of the components. The favourable geometry of the system leads to very well–determined photometric elements, including the limb darkening coefficients. Above the Balmer discontinuity, the latter are 0.10 – 0.15 smaller than the theoretical values. Combination with Popper's (1971) spectroscopic elements yields very precise absolute masses and radii. The standard *uvbyβ* indices show DM Vir to have Hyades metal abundance. A low helium abundance ($Y = 0.18$) is then required to fit Hejlesen's (1980) theoretical models, with a resulting age of $4 \times 10^9$yr.

119.029 **Zustandsgrößen von DI Pegasi.**
T. Dümichen.
BAV Rundbrief, 33. Jahrg., Nr. 3, p. 118 – 121 (1984).

119.030 **Limb–darkening and metal abundances in the sdOB primary of the eclipsing binary LB 3459 (AA Dor).**
A. E. Lynas–Gray, U. Heber, R. P. Kudritzki, K. P. Simon.
ESA Spec. Publ., ESA SP–218, p. 285 – 289 (1984). – See Abstr. 012.024.

The paper reports IUE low and high resolution observations of LB 3459 with the SWP camera. Low resolution SWP spectra were obtained before and during primary eclipse, enabling an improved determination of $R_2/R_1$ to be made. The two high resolution SWP spectra obtained are used, with non–LTE line formation calculations, to determine carbon and silicon abundances in the sdOB primary.

119.031 **Observed peculiarities during the eclipse of Epsilon Aurigae.**
S. Ferluga, M. Hack.
ESA Spec. Publ., ESA SP–218, p. 419 - 422 (1984). – See Abstr. 012.024.
Low and high resolution IUE spectra have been used, in order to study the totality phase of the eclipse of Epsilon Aurigae. In the far UV, the peculiar intrinsic variability, produced by a hot companion of the F star, is studied in function of wavelength and time. In the mid UV, the phase–dependence of lines is examined at high resolution, putting in evidence the existence of normal and peculiar features, Finally, the unexpected mid–eclipse brightening, discovered photometrically in the visible, is detected also in the UV continuum.

119.032 **A study of ultraviolet spectra of ζ Aur/VV Cep systems. VI. Excitation temperature within the wind of the K supergiant 32 Cyg.**
A. Che–Bohnenstengel.
Astron. Astrophys., Vol. 138, No. 2, p. 333 - 336 (1984).
This paper is an attempt to determine the wind temperature of cool supergiants empirically, independent of wind models. Fe II P–Cygni lines from the first excitation levels have been detected in the line spectra of Zeta Aurigae and VV Cephei binary systems. These lines are formed in the winds of the cool supergiants. The Zeta Aurigae system 32 Cyg has been chosen for this investigation due to the high quality of spectra at various characteristic phases. Radiative transfer analysis of the lines enables to determine the population densities of the lowest excitation levels, which are then compared with existing model calculations to deduce the excitation temperature. The lower limit of the excitation temperature is found to be 4200K (LTE population) in the Fe II line–formation zone. When less than 1% of the circumstellar matter is ionized $T_e \sim O(10^4\mathrm{K})$ is to be expected.

119.033 **Non–radial pulsator in the binary system Y Cam.**
P. Broglia, P. Conconi.
Astron. Astrophys., Vol. 138, No. 2, p. 443 - 450 (1984).
On the basis of new *B* and *V* observations the large variation in the orbital period of the eclipsing system Y Cam is studied along an interval over ninety years. An explanation of the minimum time residuals in terms of a third or a fourth body is discussed. Photometric solutions are calculated according to the Roche model. Multiperiod analysis of the Delta Scuti–type primary component of the binary system gives a period ratio of 0.97 consistent with non–radial pulsations. This ratio moreover seems to conform to the linear relation between rotational velocity and period ratio proposed for non–radial pulsators. The oscillation with larger amplitude has a period constant over at least an interval of 120,000 cycles.

119.034 **Epsilon Aurigae.**
M. Hack.
Sci. Am., Vol. 251, No. 4, p. 92 - 99 (1984).
Infrared and ultraviolet observations made during the latest eclipse of this binary star system suggest the eclipsing object is a hot, young star surrounded by a large cloud of dust and gas that gave it birth.

119.035 **Period changes in V726 Cygni: a linear solution.**
L. M. Cook.
J. Am. Assoc. Variable Star Obs., Vol. 13, No. 1, p. 1 - 2 (1984).
The changes in the period of V726 Cygni are discussed. Three instances where the period changed abruptly are noted. A new ephemeris is presented.

119.036 **Studies of early–type variable stars. II. The orbit and masses of HR 7551.**
G. Hill, W. A. Fisher.
Astron. Astrophys., Vol. 139, No. 1, p. 123 - 130 (1984).
Masses of the B0.5 Ib and B2 V components of the system HR 7551 have been determined. Contrary to previous results no apsidal motion has been detected with certainty at the present time, although a new orbit might confirm it.

119.037 **RR Trianguli Australis: a new analysis of a semi–detached system of unusually small dimensions.**
C. R. Chambliss.
Astrophys. Space Sci., Vol. 104, No. 2, p. 205 - 217 (1984).
RR TrA is an Algol–type eclipsing binary with a period of $0^{\mathrm{d}}.7131$. The components are believed to be of spectral types B8 V and F0 IV. The analysis presented is based on previously unreduced observations made by Somerville in 1961 - 63. Orbital elements are obtained using the Wood model, and these indicate that the system is semi–detached, and that the eclipses are marginally complete. The components appear to be of unusually small size for their spectral classes. The solution presented is compared with earlier ones for this system obtained by Kwee and Giuricin and Mardirossian, and similarities and differences between these solutions are noted.

119.038 **Light–curve analysis of Z Herculis.**
O. Tümer, C. Ibanoğlu, Z. Tunca, S. Evren.
Astrophys. Space Sci., Vol. 104, No. 2, p. 225 - 244 (1984).
The blue and yellow light curves of the RS CVn–type eclipsing binary Z Her were analyzed by using Wood's and Nelson, Davis and Etzel's methods. The results obtained by two different approaches are in good agreement. The absolute parameters of the components were also calculated using the parameters derived by Popper from the radial velocities. The physical parameters of the components indicate that the cooler one has separated or has been departing from the Main Sequence.

119.039 **Near–infrared observations of the peculiar binary system Epsilon Aurigae.**
H. C. Bhatt, T. Chandrasekhar, N. M. Ashok, J. N. Desai.
Astrophys. Space Sci., Vol. 104, No. 2, p. 293 - 296 (1984).
Near–infrared photometric and low–resolution spectroscopic observations of ε Aurigae at two phases during the current eclipse are presented. The eclipse depths are found to be wavelength–independent in the infrared right up to 2.5 μm. There is no infrared excess at wavelengths shorter than 2.5 μm. The light absorbing particles in the eclipsing body must be larger than 10 μm in size.

119.040 **Observations, analysis, and absolute parameters of the evolved binary AI Phoenicis.**
B. J. Hrivnak, E. F. Milone.
Astrophys. J., Vol. 282, No. 2, p. 748 - 757 (1984). = Publ. Rothney Astrophys. Obs., No. 21.
*UBVRI* light curves have been obtained of the $24^{\mathrm{d}}.6$ totally eclipsing, eccentric–orbit binary AI Phoenicis. These have been analyzed using the eccentric–orbit synthetic light curve program of Wilson. Masses and radii are $1.12 \pm 0.03\,M_\odot$ and $1.77 \pm 0.03\,R_\odot$ for star 1 and $1.16 \pm 0.03\,M_\odot$ and $2.85 \pm 0.03\,R_\odot$ for star 2, respectively. Both stars have evolved from the ZAMS, with the more massive component already evolved to the subgiant region. A comparison with published evolutionary tracks indicates an age for the system of about 4 billion years.

119.041 **Two–colour photoelectric investigation of SZ Piscium.**
Z. Tunca.
Astrophys. Space Sci., Vol. 105, No. 1, p. 23 - 31 (1984).
The RS CVn–type eclipsing binary SZ Psc was observed in two colours *B* and *V*, in 1981. The rate of change of the period was calculated using two different least–squares procedures. The amplitudes of the light variations at outside eclipses are $0^{\mathrm{m}}.181$ and $0^{\mathrm{m}}.213$ in *B* and *V* colours, respectively. The light curves of the system have been solved using a new approximation. The radius of the cooler component is only 82% of its Roche–radius and the system is detached.

119.042 **The light curve of the eclipsing binary HR 4560.**
A. Giménez, J. M. Garcia, J. A. Quesada.
Astrophys. Space Sci., Vol. 105, No. 1, p. 151 - 160 (1984).
In this paper, 704 photoelectric observations in the Johnson B–band of the eclipsing binary recently discovered in the multiple system HR 4560, are presented. All measurements were made at the Observatory of Mojón del Trigo in Granada (Spain) during

the years 1979, 1980, and 1981. The light curve derived from these observations is also given.

119.043 **New BV photoelectric observations of BX Andromedae.**
P. Rovithis, H. Rovithis–Livaniou.
Astrophys. Space Sci., Vol. 105, No. 1, p. 171 – 186 (1984).
Photoelectric observations of BX Andromedae made with the 48 inch Cassegrain reflector at the Kryonerion Astronomical Station of the National Observatory of Athens during 1981 and 1982 are presented.

119.044 **The curious case of the "eclipsing" WC9 star HD 164270.**
P. Massey, I. Lundström, B. Stenholm.
Publ. Astron. Soc. Pac., Vol. 96, No. 582, p. 618 – 624 (1984).
The WC9 star HD 164270 faded by $\gtrsim 1$ mag in June 1980 and September 1909. The photoelectric observations of the 1980 event, however, show that this remarkable occurrence was accompanied by no discernible change in the colors of the star. Null results of other Harvard patrol plates limit the allowed periods of these events to 70.8 yr, 35.4 yr, or 17.7 yr, and the authors give times of predicted eclipse in the hopes that others may confirm or eliminate one of these possibilities. If these events are eclipses, then any companion must be large and cool. The authors discuss limitations of the physical parameters of this alleged companion, and discuss alternate interpretations.

119.045 **Evidence for a high–temperature accretion region in Algol–type binary systems.**
G. J. Peters, R. S. Polidan.
Astrophys. J., Vol. 283, No. 2, p. 745 – 759 (1984).
The recent discovery of strong *absorption* lines of N V, C IV, and Si IV in the UV spectra of Algol primaries, whose photospheric temperatures are much too cool to form such ions, has suggested the presence of a high–temperature component in the circumstellar matter in these systems. The authors have employed timed, high–resolution *IUE* observations to determine the location and distribution of this gas and the type of system in which it is found. The observed radial velocity behavior and phase dependence in the strengths of the lines provide compelling evidence that the high–temperature matter is associated with the primary, and the source of the heating is the shock from the impact of the gas stream onto the stellar surface. It is postulated that the accretion region is formed from recently transferred matter from the original core of the mass–losing secondary.

119.046 **Light curve analyses for the early–type eclipsing binaries TT Aur and XZ Cep.**
G. Giuricin, F. Mardirossian, M. Mezzetti.
Mon. Not. R. Astron. Soc., Vol. 211, No. 1, p. 39 – 46 (1984).
Using Wood's model the authors have analysed the photoelectric light curves of the early–type eclipsing binaries TT Aur and XZ Cep. The computations indicate that TT Aur cannot be a normal main–sequence system, but it is likely to be a post–mass–exchange binary consisting of a B3 dwarf accompanied by a less massive and smaller B6 secondary, which appears to be over–luminous and oversized for its spectral type. XZ Cep may be tentatively regarded as a semi–detached binary with a Roche lobe filling primary; this component may be transferring mass – via Roche lobe overflow – towards its smaller detached secondary.

119.047 **Simultaneous differential photometry with the St. Andrews twin photometric telescope. I. The eclipsing binary TT Aurigae.**
S. A. Bell, R. W. Hilditch.
Mon. Not. R. Astron. Soc., Vol 211, No. 1, p. 229 – 234 (1984). With microfiche MN 211/1.
The refurbished twin photometric telescope at St. Andrews has been used to obtain a complete *B* light curve ($\sim$1150 observations) of the B–type interacting binary system TT Aurigae. Analysis of the light curve by the synthesis code LIGHT (Hill 1979) results in an rms scatter of the individual observations about the theoretical model fit of $\pm 0.005$ mag. The system is shown to be in a semi–detached state but with too large a mass ratio to be considered as a classical Algol system.

119.048 ***UBV* photometry of the 1981 eclipse of RZ Ophiuchi.**
D. Forbes, C. D. Scarfe.
Publ. Astron. Soc. Pac., Vol. 96, No. 583, p. 737 – 742 (1984).
Three–color photometry of the 1981 eclipse of the long–period binary RZ Oph is presented. The data are combined with those of Baldwin (1978) to improve the determination of the times of contacts. A solution assuming $i = 90°$, similar to Baldwin's, leads to the conclusion that the cooler star does not fill its Roche lobe. A solution similar to Smak's (1981), which assumes that this star does fill its lobe, is also possible, but leads to a value of $k = 0.022 \pm 0.001$, which is hard to reconcile with the stars' spectral types, relative visual luminosities, and bolometric corrections. This leaves unanswered the questions of the origin of the material in the disk surrounding the hotter star, and of the evolutionary state of the system.

119.049 **A hot companion to MU Sagittarii: an opportunity to sound the atmosphere of a B8 Ia supergiant.**
R. S. Polidan, M. J. Plavec.
Astron. J., Vol. 89, No. 11, p. 1721 – 1727 (1984).
It is argued that the bright supergiant star μ Sagittarii is accompanied by a smaller and hotter star, of spectral type approximately B1.5 V. The single–line radial–velocity curve of the B8 star leads to a fairly large mass function, $f(m) = 2.64\, m_{\odot}$, implying that the companion should have at least 50% of the mass of the visible star. Older optical observations indicated the presence of a shallow eclipse at the time of the conjunction with the supergiant behind the companion. Since the Copernicus, IUE, and Voyager observations show that the companion is the hotter component, that eclipse must have been the secondary eclipse (if it was an eclipse at all). A deeper, primary eclipse was observed as a marked decrease of the far–ultraviolet flux from the system both with the Copernicus and the IUE satellites. The presence of a hotter but smaller component in μ Sagittarii offers a unique opportunity to study the outer atmospheric layers of a supergiant which is of a much earlier spectral type than the supergiants in the ζ Aurigae systems.

119.050 **Infrared observations of the eclipse of Epsilon Aurigae: direct measurement of the 500K secondary at 5, 10, and 20 microns.**
D. E. Backman, E. E. Becklin, D. P. Cruikshank, R. R. Joyce, T. Simon, A. Tokunaga.
Astrophys. J., Vol. 284, No. 2, p. 799 – 805 (1984).
The secondary object in the eclipsing binary system ε Aur has been directly detected at 5, 10, and 20 μm. During eclipse, an opacity that is wavelength–independent from 0.4 to 20 μm removes 48% of the F0 Ia primary's light. The flux from the cool secondary remains visible as the eclipse proceeds. The flux from the secondary corresponds to a blackbody at 500K with size $8 \times 10^{-16}$sr. The secondary probably has the structure of a compact mass of stellar dimensions surrounded by a comparatively large, optically thick cloud; however, the observed bolometric luminosity of the secondary is less than $10^{-2}$ of the luminosity expected for a stellar object with the secondary's dynamical mass ($>14\, M_{\odot}$).

119.051 **Light–curve synthesis and element determination for TV Cassiopeiae.**
K. Włodarczyk.
Acta Astron., Vol. 34, No. 1, p. 47 – 51 (1984).
A new solution for the two–color light curve of the eclipsing binary TV Cas by the application of Wilson–Devinney method is presented. Masses, dimensions, color indices and absolute magnitudes of both components are given. Also, the interstellar absorption in *V* and *B* is determined.

119.052 **Photometric elements of the eclipsing binary AE Cas.**
J. B. Srivastava, C. D. Kandpal.
Acta Astron., Vol. 34, No. 2, p. 281 – 290 (1984).
The light curves and *U, B,* and *V* observations of the system AE Cas have been presented. An improved period of $0\overset{d}{.}7591215$

is given on the basis of present observations. Photometric elements of the system have been obtained by employing Kopal's method of Fourier analysis of light curves. The spectral types for the primary and secondary components have been found to be G6 – 7 and G9 – K4, respectively.

119.053 **Photometric study of the eclipsing binary system AR Lacertae.**
R. K. Srivastava.
Acta Astron., Vol. 34, No. 2, p. 291 – 302 (1984).

The geometrical elements of the eclipsing binary system AR Lacertae have been obtained by Fourier analysis of the light changes in the frequency–domain which was developed by Kopal (1975) and compared with the author's earlier (Srivastava, 1981) results obtained by employing Russell and Merrill's (1952) method. The absolute dimensions have been obtained using the spectroscopic elements given by Sanford (1951), and the new geometrical elements. The primary component lies fairly on the main sequence while the secondary component falls above the main sequence. The Roche constants indicate that the system seems to be a detached one apparently. However, evolutionary discussion shows that the possibility of AR Lac being in a semi–detached state of evolution can not be ruled out.

119.054 **S Cancri: a semidetached system with a very low–mass secondary.**
D. M. Popper, J. Tomkin.
Astrophys. J., Vol. 285, No. 1, p. 208 – 213 (1984).

Spectroscopic observations of the $9^{d}_{.}5$ Algol system S Cnc give velocity amplitudes of 10 and 127 km $s^{-1}$ for the A0 main–sequence and late G subgiant components, respectively, and masses of 2.3 and 0.18 $M_{\odot}$. The radii, from the Crawford and Olson photometry, are 2.2 and 4.8 $R_{\odot}$. The Roche lobe–filling subgiant of S Cnc has the smallest known stellar mass established directly from spectrographic observations.

119.055 **107th list of minima of eclipsing binaries.**
BBSAG Bull., No. 74, p. 1 – 6 (1984).

119.056 **Ephemerides of eclipsing binaries for the year 1985.**
M. Kurpińska–Winiarska, I. Lelatko.
Rocznik Astronomiczny Obserwatorium Krakowskiego 1985, p. 1 – 108 (1984). – See Abstr. 046.034.

119.057 **Ephemerides of eclipsing binaries among cataclysmic variables for the year 1985.**
J. M. Kreiner.
Rocznik Astronomiczny Obserwatorium Krakowskiego 1985, p. 109 – 113 (1984). – See Abstr. 046.034.

119.058 **Iterative method of differential corrections for the analysis of light curves of eclipsing binary stars.**
A. I. Khaliullina, Kh. F. Khaliullin.
Sov. Astron., Vol. 28, No. 2, p. 228 – 234 (1984). English translation of 37.119.045.

119.059 **The binaries–in–cluster program at the RAO (*Rothney Astrophysical Observatory*).**
E. F. Milone, S. J. Schiller.
Publ. Astron. Soc. Pac., Vol. 96, No. 584, p. 791 (1984). Abstract. – See Abstr. 010.281.

119.060 **The stellar wind in the B0 V–type eclipsing binary Y Cygni.**
N. D. Morrison, C. D. Garmany.
Publ. Astron. Soc. Pac., Vol. 96, No. 584, p. 791 – 792 (1984). Abstract. – See Abstr. 010.281.

119.061 **Upcoming eclipses of 22 Vulpeculae: needed photometry.**
S. B. Parsons, T. B. Ake.
Bull. Am. Astron. Soc., Vol. 16, No. 3, Part I, p. 726 (1984). Abstract. – See Abstr. 010.065.

119.062 **Epsilon Aurigae–Minimum 1982 – 1984.**
K.-P. Timm.
BAV Rundbrief, 33. Jahrg., Nr. 4, p. 180 – 185 (1984).

119.063 **Update on Epsilon Aurigae (May 1984).**
J. L. Hopkins.
I.A.P.P.P. Commun., No. 17, p. 22 – 30 (1984).

119.064 **Spectrophotometry of RZ Cassiopeiae, October 1977.**
É. F. Kuznetsova (*Eh. F. Kuznetsova*),
M. A. Svechnikov.
Sov. Astron. Lett., Vol. 10, No. 3, p. 163 – 165 (1984). English translation of 37.119.068.

119.065 **LFT 349.**
IAU Circ., No. 4014 (1984).

119.066 **Times of minima of eclipsing variables – III.**
C. D. Scarfe, D. W. Forbes, P. A. Delaney, J. Gagne.
Inf. Bull. Variable Stars, No. 2545, 4 pp. (1984).

119.067 **HR 3337: a bright new eclipsing binary.**
L. J. Boyd, R. M. Genet, D. S. Hall.
Inf. Bull. Variable Stars, No. 2547, 3 pp. (1984).

119.068 **Spectral types of eclipsing binaries.**
E. M. Halbedel.
Inf. Bull. Variable Stars, No. 2549, 2 pp. (1984).

119.069 **Spectral types of eclipsing binaries in optical coincidence with clusters and associations.**
E. M. Halbedel.
Inf. Bull. Variable Stars, No. 2550, 2 pp. (1984).

119.070 **Preliminary orbital elements of HD 184035.**
A. Gaspani.
Inf. Bull. Variable Stars, No. 2552, 2 pp. (1984).

119.071 **Une autre étoile Ap à éclipses?**
P. Renson.
Inf. Bull. Variable Stars, No. 2559, 2 pp. (1984).

119.072 **A photometric anomaly around third contact in Epsilon Aurigae.**
L. J. Boyd, R. M. Genet, D. S. Hall.
Inf. Bull. Variable Stars, No. 2562, 3 pp. (1984).

119.073 **Period variation in the white dwarf eclipsing binary V471 Tauri.**
C. Ibanoğlu, S. Evren.
Inf. Bull. Variable Stars, No. 2573, 3 pp. (1984).

119.074 **Photometry of the eclipsing star W Crucis.**
B. F. Marino, W. S. G. Walker, G. Herdman.
Inf. Bull. Variable Stars, No. 2582, 2 pp. (1984).

119.075 **New photometry of the eclipsing binary 5 Ceti.**
R. D. Lines, L. J. Boyd, R. M. Genet, D. S. Hall.
Inf. Bull. Variable Stars, No. 2589, 4 pp. (1984).

119.076 **The period and photoelectric light curves of DM Del.**
N. Güdür, C. Sezer, Ö. Gülmen.
Inf. Bull. Variable Stars, No. 2591, 3 pp. (1984).

119.077 **Photoelectric minima times of ER Orionis.**
P. Rovithis, H. Rovithis–Livaniou.
Inf. Bull. Variable Stars, No. 2595, 1 p. (1984).

119.078 **Photoelectric observations of the recent eclipse of 22 Vulpeculae.**
D. R. Faulkner.
Inf. Bull. Variable Stars, No. 2599, 2 pp. (1984).

119.079 **The bright eclipsing binary HD 208496.**
J. Manfroid, G. Mathys.
Inf. Bull. Variable Stars, No. 2616, 2 pp. (1984).

119.080 **Primary minimum of W Crucis.**
J. W. Menzies, J. Spencer Jones.
Inf. Bull. Variable Stars, No. 2623, 3 pp. (1984).

119.081 **Photometry showing HD 219989 is an eclipsing binary.**
R. C. Crawford, S. I. Ingvarsson, L. J. Boyd, R. M. Genet, W. S. Barksdale Jr., W. T. Persinger, H. D. Powell, W. J. Bisard, H. J. Stelzer, D. B. Hoff, A. M. Heiser, D. S. Hall, F. C. Fekel Jr.
Inf. Bull. Variable Stars, No. 2624, 5 pp. (1984).

119.082 **HD 208496 (HR 8369): a double-lined eclipsing binary.**
J. Andersen, K. S. Jensen, B. Nordström.
Inf. Bull. Variable Stars, No. 2642, 2 pp. (1984).

119.083 **On the apsidal motion of AG Persei.**
N. Güdür, C. Sezer, O. Gülmen.
Inf. Bull. Variable Stars, No. 2644, 2 pp. (1984).

119.084 **Apsidal motion in the eclipsing binary system V451 Ophiuchi.**
Kh. F. Khaliullin, V. S. Kozyreva.
Astrophys. Space Sci., Vol. 106, No. 1, p. 93 – 101 (1984).

Multicolour *WBVR* photoelectric observations of the eclipsing binary V451 Oph were carried out, and a highly accurate light curve was obtained. The angular velocity of the orbital rotation, $\dot{\omega} = 2.1$ deg yr$^{-1}$, and the apsidal motion constant $k_2 = 0.0045$ are given.

119.085 **Photometric elements of V364 Cassiopeiae.**
U. S. Chaubey.
Astrophys. Space Sci., Vol. 106, No. 2, p. 273 – 282 (1984).

The *UBV* photometric observations of the eclipsing binary V364 Cas have been discussed. Two sets of photometric elements, one based on the Russell and Merrill methods and the other based on Kopal's new method of the analysis of the light curves of eclipsing binary stars in the frequency-domain, have been derived. An examination of all the available times of minimum light of V364 Cas does not reveal any change in its orbital period.

119.086 **Photoelectric photometry of the binary system AZ Cam.**
R.-x. Zhang, J.-t. Zhang, D.-s. Zhai.
Chin. Astron. Astrophys., Vol. 8, No. 3, p. 228 – 234 (1984). English translation of 37.119.106.

119.087 **Measurement of the radial velocity of 32 Cygni during its 1981 eclipse.**
H.-s. Tan, S.-q. Peng, X.-f. Liu.
Chin. Astron. Astrophys., Vol. 8, No. 3, p. 235 – 237 (1984). English translation of 37.119.107.

119.088 **Revised elements and evolutionary status of six eclipsing binary stars.**
U. S. Chaubey.
Bull. Astron. Soc. India, Vol. 12, No. 3, p. 237 – 248 (1984).

Twelve photoelectric light curves of six eclipsing binary stars namely GG Cas, WX Cep, UZ Cyg, VW Cyg, AQ Peg and ST Per in two different filters (*B* and *V* for the first five binaries and *R* and *I* for ST Per) have been analysed by Kopal's new method of analysis of the light curves of eclipsing system in the frequency domain. The elements of the systems have been evaluated and their evolutionary status discussed.

119.089 **Photoelectric observations of the eclipsing binary V836 Cygni.**
H. W. Duerbeck, R. A. Breinhorst.
Veröff. Astron. Inst. Bonn, Nr. 98, 16 pp. (1984).

119.090 **Neue Periodenbestimmungen der Bedeckungsveränderlichen DS And, VY Lac und VZ Cep.**
S. Rößiger.
Mitt. Veränderliche Sterne, Band 10, Heft 3, p. 54 – 56 (1984).

A primary minimum could be determined for each of the three eclipsing binaries DS And, VY Lac and VZ Cep by observation with the Sonneberg photoelectric 60 cm telescope II. This enables new computation of their periods.

119.091 **The Algol variable KR Cas.**
A. V. Kusakin, A. V. Mironov, M. S. Frolov.
Astron. Tsirk., No. 1309, p. 7 – 8 (1984). In Russian.

119.092 **The new eclipsing magnetic binary system E1114+182.**
P. Biermann, G. D. Schmidt, J. Liebert, H. S. Stockman, S. Tapia, H. Kühr, P. A. Strittmatter, S. West, D. Q. Lamb.
Prepr. Steward Obs., No. 561, 61 pp. (1984). Submitted to Astrophys. J.

119.093 **Helium content of the Population II binary system CM Draconis.**
B. Paczyński, R. Sienkiewicz.
Astrophys. J., Vol. 286, No. 1, p. 332 – 336 (1984).

CM Draconis is an eclipsing binary with a space velocity of 163 km s$^{-1}$. Observations by Lacy in 1977 established masses, radii, and luminosities of the two components. As the masses are small, only 0.2 $M_{\odot}$, the stars are fully convective. The authors show that a comparison with fully convective stellar models can be used to determine the helium abundance of the stars and that this method is independent of opacities. They estimate the helium abundance (by mass fraction) to be $0.25 \pm 0.14$ and $0.38 \pm 0.12$ for the components A an B, respectively. Since CM Dra is almost certainly a very old, Population II system, its helium abundance is of considerable cosmological importance.

119.094 **The nova-like binary system RW Trianguli.**
N. F. Vojkhanskaya.
Astron. Zh., Tom 61, Vyp. 6, p. 1137 – 1141 (1984). In Russian. English translation in Sov. Astron., Vol. 28, No. 6.

Spectral observations have been carried out and the spectrum variations during the orbital period have been studied. It is shown that outside eclipses the equivalent widths of the emission lines are constant, and during the eclipse moments they increase. The radial velocity curves from H and He II lines are strongly distorted sinusoids. This distortion is explained by the escape of matter from the system which indicates a supercritical-accretion regime. A model of the system is suggested.

119.095 **Radial velocity of HD 77581 (Vela X-1) in the period of 1972 – 1983.**
K. Sadakane, J. Jugaku, M. Matsuoka.
Publ. Astron. Soc. Jpn., Vol. 36, No. 4, p. 715 – 718 (1984).

Radial velocities of HD 77581 (Vela X-1 = 4U 0900-40) were measured from photospheric Fe IV and Fe III lines observed with the International Ultraviolet Explorer in the period 1978 – 1983. No long-term variation of the system radial velocity was found within an accuracy of $|V_r| = 30$ km s$^{-1}$.

119.096 **The Epsilon Aurigae secondary: a binary embedded within a disk?**
J. J. Lissauer, D. E. Backman.
Astrophys. J., Lett. Ed., Vol. 286, No. 2, p. L39 – L41 (1984).

Epsilon Aurigae is an F supergiant in a spectroscopic binary system that undergoes a partial eclipse of 2 years duration every 27 years. Eclipse characteristics indicate that the secondary is an elongated object ~10 AU by >1 AU; orbital properties indicate the secondary's mass is ~16 $M_{\odot}$. Recent infrared observations show that the secondary has a color temperature $T_c \approx 500$K and a luminosity apparently <1% of a 16 $M_{\odot}$ star. The authors propose that the secondary consists of a close binary embedded within an optically thick disk, which we view edge-on. Two 8 $M_{\odot}$ stars are only ~10% as luminous as one 16 $M_{\odot}$ star, and it is reasonable to expect as much as 90% of the energy emitted in such a configuration would escape through the poles of the disk, resulting in the observed low luminosity of the secondary.

**A catalogue of classical (evolved) Algol-type binary candidate stars.**
See Abstr. 002.051.

**Veränderliche Sterne.**
See Abstr. 003.023.

**Orbital motion of DI Herculis as a test of a theory of gravitation.**
See Abstr. 066.292.

**Photoelectric observations of o And.**
See Abstr. 112.069.

**o And. The overall behaviour of the shell.**
See Abstr. 112.070.

**HD 126200 – a suspected eclipsing binary?**
See Abstr. 113.042.

**RR Ari: not a variable star.**
See Abstr. 113.045.

**Some problems in the determination of fundamental stellar parameters from binary stars.**
See Abstr. 115.017.

**Orbital period changes in cataclysmic variables?**
See Abstr. 117.008.

**A visible and infrared study of the eclipsing dwarf nova OY Carinae. II. The infrared light curves.**
See Abstr. 117.028.

**V758 Centauri: *UBV* photometry.**
See Abstr. 117.038.

**Short-period noncontact close binary systems. III. RU Eridani.**
See Abstr. 117.080.

**Einstein observations of selected close binaries and shell stars.**
See Abstr. 117.089.

**A study of photoelectric observations of SW Lacertae.**
See Abstr. 117.108.

**An analysis of V861 Sco. II. The stellar wind.**
See Abstr. 117.118.

**Light curves of gas-eclipsed binary systems.**
See Abstr. 117.122.

**The symbiotic eclipsing binary CI Cyg: an Algol symbiotic system.**
See Abstr. 117.135.

**Accretion of mass in Algol.**
See Abstr. 117.136.

**Two very similar closely detached binary systems: AT and AZ Camelopardalis.**
See Abstr. 117.170.

**Photoelectric observations and minima times for RZ Comae Berenices.**
See Abstr. 117.171.

**A simple kinetic model of a multiparticle cloud. II. Gaseous streams in Algols.**
See Abstr. 117.209.

**Light time effect in TW Draconis.**
See Abstr. 118.017.

**Period study of the Algol-type eclipsing binary – R Canis Majoris.**
See Abstr. 118.018.

**Simultaneous light and velocity curve solutions for AS Eridani and V1143 Cygni.**
See Abstr. 120.017.

**Studies of early-type variable stars. III. The orbit and physical dimensions for V380 Cygni.**
See Abstr. 120.018.

**Investigation of two variable stars in Ursa Minor.**
See Abstr. 122.044.

**The variable stars V962 and V987 Ophiuchi.**
See Abstr. 122.045.

**The β Cephei stars 12 and 16 Lacertae in 1979.**
See Abstr. 122.122.

**The W Serpentis stars: an observational program.**
See Abstr. 123.002.

**WY Sagittae (Nova 1783): a transition object between classical and dwarf novae?**
See Abstr. 124.181.

**Properties of the iron line from Vela X-1.**
See Abstr. 142.070.

**Sudden disappearence of Vela X-1 pulses.**
See Abstr. 142.071.

# 120 Spectroscopic Binaries

120.001 **Radial velocities of northern mercury stars.**
D. J. Stickland, J. Weatherby.
Astron. Astrophys., Suppl. Ser., Vol. 57, No. 1, p. 55 – 67 (1984).
About 200 radial velocities of 96 bright northern Hg–Mn or candidate Hg–Mn stars are presented. Past and present data have been reexamined for periodic variability in cases that were neither previously known to be binaries with well determined orbits nor were considered to have constant velocity.

120.002 **Radial velocity studies of spectroscopic binaries with the objective prism. IV. Spectroscopic orbits of the new binaries HD 102010 and HD 102465.**
F. Gieseking.
Astron. Astrophys., Suppl. Ser., Vol. 57, No. 1, p. 99 – 102 (1984).
HD 102010 and HD 102465 have been detected to be spectroscopic binaries. 33 objective prism measurements of the radial velocities of these new binaries permit a first determination of their orbital elements.

120.003 **Complicated structure of the λ3933 Ca II line in the spectrum of HD 187399 (V1507 Cyg).**
N. L. Ivanova.
Astron. Tsirk., No. 1291, p. 7 – 8 (1983). In Russian.

120.004 **Spectroscopic binary orbits from photoelectric radial velocities. Paper 57: HD 222018.**
R. F. Griffin.
Observatory, Vol. 104, No. 1061, p. 189 – 191 (1984).

120.005 **The occultation of 1 Vulpeculae by Pallas.**
W. J. Schuster, M. A. Moreno, J. Guichard, G. R. Sánchez.
Rev. Mex. Astron. Astrofis., Vol. 9, No. 1, p. 21 – 24 (1984).
Observing procedures and timings of the occultation of 1 Vulpeculae by the asteroid Pallas, observed with the 150–cm and 84–cm telescopes of the Observatorio Astronómico Nacional at San Pedro Mártir, are given and discussed. Measured quantities for the spectroscopic binary 1 Vul are also presented.

120.006 **The synchronous rotation of Iota Pegasi.**
D. F. Gray.
Publ. Astron. Soc. Pac., Vol. 96, No. 581, p. 537 – 538 (1984).
Measured rotation rates of the spectroscopic binary stars ι Peg A ($6.5 \pm 0.3$ km $s^{-1}$) and B($5 \pm 1$ km $s^{-1}$) are found to be equal to the orbital synchronization values. The question of whether synchronization has actually occurred is discussed and the evidence is found to favor the hypothesis.

120.007 **The WC5 + O7 binary system HD 63099.**
V. S. Niemela, P. Massey, P. S. Conti.
Publ. Astron. Soc. Pac., Vol. 96, No. 581, p. 549 – 551 (1984).
The authors present an orbit for the double–lined spectroscopic binary HD 63099 based on 77 spectra. The mass ratio is typical of other known WC5 binary systems; likewise the inferred mass of the Wolf–Rayet component is relatively small.

120.008 **Egress spectra of the new eclipsing supergiant 22 Vulpeculae.**
T. B. Ake III, S. B. Parsons, Y. Kondo.
ESA Spec. Publ., ESA SP–218, p. 423 – 425 (1984). – See Abstr. 012.024.
The G–type primary in the single–lined spectroscopic binary 22 Vul was found with IUE in April 1983 to eclipse its recently identified B9 main sequence companion. Outside of eclipse, a wind from the system with terminal velocity of 300 km $s^{-1}$ is conspicuous in the Mg II lines. Enhanced Fe II absorption seen well away from geometrical eclipse implies an extensive envelope around the G star. There is a close correspondence between the 22 Vul spectra and those of interacting binaries suggesting common properties due to relatively dense, warm plasmas in these systems.

120.009 **Spectroscopic binary orbits from photoelectric radial velocities. Paper 58: HR 1105.**
R. F. Griffin.
Observatory, Vol. 104, No. 1062, p. 224 – 231 (1984).
The nature and causes of S–type spectra are reviewed. The S star HR 1105 is shown to be a spectroscopic binary with a period of 596 days. It is the first binary of type S to have its orbit determined, notwithstanding that it is arguably the brightest S star and that the variability of its velocity has been known since 1922, the same year as type S was recognized.

120.010 **Photoelectric radial velocities, Paper XI. The orbits of four more spectroscopic binaries in the Clube Selected Areas.**
R. F. Griffin.
Mon. Not. R. Astron. Soc., Vol. 210, No. 3, p. 745 – 751 (1984).
Orbits are presented for four ninth–magnitude stars about which the only spectroscopic datum previously known is the HD type, K0 in each case. The most interesting of the four stars is perhaps HD 145677, whose lack of proper motion suggests a high luminosity, which is hardly compatible with the shortness (4.4 days) of its orbital period.

120.011 **The angular momentum–vs–mass relation for spectroscopic binaries.**
V. Trimble.
Astrophys. Space Sci., Vol. 104, No. 1, p. 133 – 143 (1984).
The spectroscopic binaries fall roughly on the "universal" power–law of index 1.8 for angular momentum vs total mass, as defined by planets, spiral galaxies, and numerous other objects. But the individual systems in the Seventh Catalogue of the Orbital Elements of Spectroscopic Binary Systems define a curve of rather shallower slope, $1.63 \pm 0.07$, over more than two orders of magnitude in mass and four in angular momentum. Various subsets (long and short periods; single and double line systems; known and unknown orbital orientations) all yield slopes from 1.48 to 1.77. These values, as well as the slightly larger one found for eclipsing systems by Sisteró and Marton, are very much what one would expect, given the form of Kepler's Third Law and the stellar mass–radius relation.

120.012 **The spectroscopic orbit of HR 1878.**
R. F. Griffin.
J. R. Astron. Soc. Can., Vol. 78, No. 4, p. 151 – 154 (1984).
HR 1878, whose radial velocity was found to be variable by Young at the David Dunlap Observatory, is shown to be a spectroscopic binary with an eccentricity of 0.51 and a period of 382 days.

120.013 **The line profile variations of Spica.**
G. Fraser, S. Yang, G. A. H. Walker.
Hvar Obs. Bull., Vol. 7, No. 1, p. 231 – 239 (1983). – See Abstr. 012.037.
The authors present high quality data of the Si III triplet at λλ4553, 4568, 4575.

120.014 **Spectroscopy and photometry of the binary system Sigma Coronae Borealis.**
G. A. Bakos.
Astron. J., Vol. 89, No. 11, p. 1740 – 1744 (1984).
From 18 spectrograms, the orbital elements of this double–line spectroscopic binary were redetermined. In addition, from photoelectric observations, a light curve was derived. A general increase of brightness during one season accompanied by small amplitude and short time–scale variations were observed. The radial velocities of the visual companion have also been measured.

120.015 **The many faces of Capella: a search for rotational modulations and a study of systematic velocities of emission lines in the ultraviolet.**
T. R. Ayres.
Astrophys. J., Vol. 284, No. 2, p. 784 – 798 (1984).

The ultraviolet emission of the spectroscopic binary Capella has been monitored with IUE over the course of half an orbital cycle (2 months). The first objective was to detect changes in the fluxes or shapes of ultraviolet emission lines in response to the evolution of large–scale active regions or their rotation onto and off the visible hemisphere of the active F–type secondary star. Capella proved to be surprisingly steady in its ultraviolet emissions. The second objective was to investigate systematic velocities of the low–excitation and high–excitation emission lines. It is found that the latter are significantly redshifted with respect to the former. The author concludes that the redshifts indicate true downflows of $10^5$K plasma in the stellar "transition zone", rather than the operation of a "P Cygni" effect in an optically thick outflow.

120.016 **R31: a massive O + W–R star binary in the Small Magellanic Cloud.**
J. B. Hutchings, D. Crampton, A. P. Cowley, I. B. Thompson.
Publ. Astron. Soc. Pac., Vol. 96, No. 584, p. 811 – 816 (1984).

The authors present spectroscopic data on the O6 + WN3 binary R31 in the SMC, which improve the orbital determination of Moffat (1982). The period is found to be 6.538 days. From several considerations they consider an eccentricity of ~0.2 to be possible. From the absence of known eclipses and lack of evidence that the primary fills its Roche lobe, they infer that the orbital inclination is ~60° or less. The stellar masses are most probably 65 $M_\odot$ and 12 $M_\odot$. This makes the O star one of the most massive yet measured. Velocity differences between absorption lines of different ions are thought to be due to stellar wind effects, and may be phase dependent. The unusually large mass ratio is discussed.

120.017 **Simultaneous light and velocity curve solutions for AS Eridani and V1143 Cygni.**
W. Van Hamme, R. E. Wilson.
Astron. Astrophys., Vol. 141, No. 1, p. 1 – 9 (1984). = Contrib. Dep. Astron., Univ. Florida, No. 73.

The authors have carried out computational examples of a previously developed procedure for making simultaneous solutions of radial velocity curves and multi–color light curves. Because of a reduction in the number of free parameters, such simultaneous solutions strengthen overall results. For purposes of comparison, the authors have also carried out separate velocity and light curve solutions. For AS Eri, the mass ratio and reference phase are found jointly from both sources of information. For V1143 Cyg the main joint parameters are the orbital eccentricity, longitude of periastron, and reference phase, although there seems to be some small information about the mass ratio in the photometry. In their examples the authors have found the joint photometric–spectroscopic parameters to be determined mainly by the photometric observations in the simultaneous solutions. They find a definite indication of third light in the AS Eri system, so that the solution for this binary is somewhat less well determined than was previously believed.

120.018 **Studies of early–type variable stars. III. The orbit and physical dimensions for V380 Cygni.**
G. Hill, A. H. Batten.
Astron. Astrophys., Vol. 141, No. 1, p. 39 – 48 (1984).

The spectroscopic orbits for the primary and secondary components of V380 Cygni have been redetermined by cross–correlation techniques. Relative radii, temperatures and orbital inclination have resulted from a solution of three independent light curves. When combined with the spectroscopic data they yield physical dimensions for the component stars, a chemical composition of $(X, Y)$, (0.66, 0.30) and a system age of $\sim 8 \times 10^6$yr. In $\sim 5 \times 10^5$yr the primary will reach the inner Lagrangian surface and the system will begin the mass–transfer phase of its evolution. Theoretical and observational apsidal periods near 1500 yr are in good accord.

120.019 **Radial velocity variations in the Ca II emission star HD 36705.**
J. L. Innis, D. W. Coates, K. Thompson.
Inf. Bull. Variable Stars, No. 2571, 4 pp. (1984).

120.020 **Five colour photometry of the chromospherically active southern star HD 155555.**
A. Udalski, E. H. Geyer.
Inf. Bull. Variable Stars, No. 2593, 4 pp. (1984).

120.021 **UBV light variation of the WR star HDE 311884.**
M. A. Cerruti.
Inf. Bull. Variable Stars, No. 2637, 4 pp. (1984).

120.022 **Spectroscopic binary orbits from photoelectric radial velocities. Paper 59: 18 Monocerotis.**
R. F. Griffin.
Observatory, Vol. 104, No. 1063, p. 267 – 270 (1984).

120.023 **Copernicus observations of Iota Herculis velocity variations.**
J. B. Rogerson Jr.
Astron. J., Vol. 89, No. 12, p. 1876 – 1877 (1984).

Observations taken in May 1979 with the Copernicus spectrophotometer show a velocity variation that is well represented by the sum of two sinusoids with frequencies of 0.660 and 0.618 cycles per day and amplitudes of 9.18 km/s and 8.11 km/s, respectively.

120.024 **Analysis of peculiarities of the magnetically variable star β CrB with the method of quantitative spectral classification.**
Z. N. Fenina.
Astron. Tsirk., No. 1302, p. 6 – 8 (1983). In Russian.

120.025 **Feinstruktur in den Atmosphären von WC–Sternen.**
S. Lührs.
Mitt. Astron. Ges., Nr. 62, p. 243 – 244 (1984). – See Abstr. 010.341.

**Spectroscopic binaries. 14th Complementary Catalogue.**
See Abstr. 002.037.

**Automatic photoelectric telescopes.**
See Abstr. 032.029.

**Some methods for searching for weak spectra.**
See Abstr. 036.042.

**Spectroscopic studies of Wolf–Rayet stars with absorption lines. VIII. HD 193793.**
See Abstr. 114.052.

**A search for duplicity in the weak G–band stars.**
See Abstr. 114.057.

**Spectroscopic and photometric observations of the suspected binary star HD 123058.**
See Abstr. 114.102.

**Some problems in the determination of fundamental stellar parameters from binary stars.**
See Abstr. 115.017.

**The post–RLOF (*Roche lobe overflow*) structure of the secondary components in close binary systems, with an application to masses of Wolf–Rayet stars.**
See Abstr. 117.086.

**Period study of the Algol–type eclipsing binary – R Canis Majoris.**
See Abstr. 118.018.

**The control of calculations of orbital elements of binary systems.**
See Abstr. 118.028.

**Orbital motion of the dwarf cepheid SZ Lyn.**
See Abstr. 122.018.

**The light variations of the Delta Scuti star $\vartheta^2$Tauri.**
See Abstr. 122.110.

**Periods in the light variation of σ Scorpii.**
See Abstr. 122.116.

**Light variability of σ Scorpii.**
See Abstr. 122.141.

**Determination of radii of cepheids. III. Radial velocities, spectroscopic orbit and dimension of SU Cyg.**
See Abstr. 122.145.

**On the nature of the Wolf–Rayet component in the core of the massive galactic H II region NGC 3603.**
See Abstr. 153.026.

# 121 Early-stage Stars (T Tauri Stars, Herbig-Haro Objects, etc.)

121.001 **On the photometric activity of R Coronae Australis.**
V. I. Kardopolov.
Perem. Zvezdy, Tom 21, No. 5, p. 677 – 681 (1982). In Russian.
Results of BV photoelectric observations of the R CrA irregular variable star in 1977 – 79 are given.

121.002 **Long–time changes of the mean brightness of WW Vul.**
L. V. Timoshenko.
Astron. Tsirk., No. 1285, p. 4 – 5 (1983). In Russian.

121.003 ***UBVRI* monitoring of five late–type pre–main–sequence stars.**
A. E. Rydgren, D. S. Zak, F. J. Vrba, P. F. Chugainov (*P. F. Chugajnov*), G. V. Zajtseva.
Astron. J., Vol. 89, No. 7, p. 1015 – 1021 (1984).
Two T Tauri stars (DF Tau and DK Tau) and three pre–main–sequence K stars with weak line emission (WK X–Ray 1, WK X–Ray 2, and FK X–Ray 3) were monitored with *UBVRI* photometry over a 14–night period during October 1982. Neither T Tauri star shows an obvious rotational modulation in the light curve, but DF Tau did experience a multi–day flare–like brightening. The light variations of the three weak–emission stars are well described by sine curves with periods of 5.6, 2.8, and 7.0 days, respectively; these values presumably represent the axial rotation periods of these stars.

121.004 **Line profiles and radial velocities of Herbig Ae/Be stars.**
U. Finkenzeller, I. Jankovics.
Astron. Astrophys., Suppl. Ser., Vol. 57, No. 2, p. 285 – 326 (1984).
Spectrograms of intermediate and high resolution (4.3 to 0.2 Å FWHM) are presented for 27 Herbig Ae/Be stars in the wavelength range λλ3620 – 4960 Å. The spectra are characterized by complex Balmer lines and an obviously normal appearance of the other photospheric lines. The Balmer line profiles can be grouped into 3 categories: "single–peak", "double–peak", and P Cyg profiles and retain their shape within the Balmer series of a particular star. Higher numbered Balmer lines are less affected by the non–photospheric emission and absorption components. The material presented provides a kinematic argument for the Herbig Ae/Be stars being PMS objects physically associated with their parent molecular clouds. Obviously, the stars are surrounded by extended envelopes in which the hydrogen emission is originating.

121.005 **Energy sources of T Tauri stars.**
N. Calvet, J. Albarrán.
Rev. Mex. Astron. Astrofis., Vol. 9, No. 1, p. 35 – 47 (1984).
The authors have empirically estimated the total energy loss from the atmospheric regions above the photosphere in T Tauri stars. They have also estimated the flux input into the atmosphere by magnetohydrodynamic waves produced in the subphotospheric convection zone. Within the uncertainties of both theory and observations, this flux seems to represent the basic energy input into the atmosphere provided that a large surface coverage of magnetic regions exists. In addition to this basic energy input from the convective zone the T Tauri atmospheres must have other energy sources, originating in the stellar surface. Among those one can include the flux of energy carried by Alfvén waves resulting from the action of surface material motions on magnetic flux tubes, as well as dissipation and annihilation of magnetic fields in flare events.

121.006 **Comments on the He I ionization in T Tauri stars.**
N. Calvet.
Rev. Mex. Astron. Astrofis., Vol. 9, No. 1, p. 49 – 52 (1984).
The predictions of an approximate model for XUV ionization of He I in the atmospheres of T Tauri stars are compared with observations of the luminosity in He I λ5876 Å and in soft X–rays. The comparison shows that XUV ionization cannot account for the observations. Collisional processes are then indicated to be the dominant mechanism determining the excitation and ionization of He in T Tauri atmospheres.

121.007 **Molecular hydrogen emission associated with high velocity flows. A tracer for low velocity shocks.**
W. J. Zealey, P. M. Williams, J. Storey, K. Taylor, G. Sandell.
Occas. Rep. R. Obs., Edinb., No. 13, p. 109 – 115 (1984). – See Abstr. 012.001.

121.008 **Results of photoelectric observations of V1057 Cygni.**
A. V. Oskanyan Jr.
Astrofizika, Tom 19, Vyp. 3, p. 525 – 532 (1983). In Russian. English translation in Astrophysics, Vol. 19, No. 3.
The results of photoelectric *UBV* and $H_\alpha$ observations of V1057 Cyg accomplished during 1978 – 1981 are presented. It is shown that the brightness of the star in *B* and *V* bands diminishes in waves, having a characteristic time of more than 200 days. The mean rate of brightness decrease during 1979 – 1981 has increased in comparison to that during 1976 – 1979. An irregular change of brightness is observed in the *U* band and $H_\alpha$ line.

121.009 **The magnetic field of T Tauri.**
Yu. N. Gnedin, N. P. Red'kina.
Pis'ma Astron. Zh., Tom 10, No. 8, p. 613 – 619 (1984). In Russian. English translation in Sov. Astron. Lett., Vol. 10.
From the linear polarization spectrum the magnetic field of T Tau is determined. The magnitude of the field is estimated to be 250 – 1800 gauss for various data and various geometrical models of the envelope. The increasing of the degree of intrinsic polarization of T Tau at $\lambda = 2.2\ \mu$ is explained by reflection of the radiation of the infrared component of T Tau by a circumstellar gas–dust shell.

121.010 **Some questions concerning the formation and structure of Herbig–Haro objects.**
A. L. Gyul'budagyan.
Astrofizika, Tom 20, Vyp. 1, p. 115 – 122 (1984). In Russian. English translation in Astrophysics, Vol. 20, No. 1.
On the basis of data concerning the tangential velocities of condensations, the groups of which compose the objects HH 1, HH 2, HH 39, a conclusion about the expansion of these condensations in the centre of the inertia system has been made. Hence a suggestion on the formation of these groups by a decay of a more massive body is inferred.

121.011 **An analysis of light curves of Fuor V1057 Cyg.**
E. N. Kopatskaya.
Astrofizika, Tom 20, Vyp. 2, p. 263 – 275 (1984). In Russian. English translation in Astrophysics, Vol. 20, No. 2.
From observations obtained by the author and collected from literature the reduced light curves of V1057 Cyg up to 1982 were constructed in *UBVR* bands. Circumstellar reddening, radius and effective temperature of the star before and after an outburst were found. During the outburst the star increased its radius from $3.5R_\odot$ to $17R_\odot$, its temperature from 4600K to 9000K, and ejected a dust shell. After the outburst up to 1982 the dust shell remained constant, various peculiarities in the star were weakened.

121.012 **On the connection between Herbig–Haro objects and flare stars in the solar neighbourhood.**
A. L. Gyul'budagyan.
Astrofizika, Tom 20, Vyp. 2, p. 277 – 281 (1984). In Russian. English translation in Astrophysics, Vol. 20, No. 2.
The following hypothesis is suggested: Herbig–Haro objects, which are supposed to be composed of a dense nucleus surrounded by a diffuse envelope, after escaping from the dark clouds, lose their envelopes and become UV Ceti–type flare stars. Some facts in favour of this hypothesis are suggested.

121.013 **A new investigation of the T Tauri star RU Lupi. III. The wind model.**
M. T. V. T. Lago.
Mon. Not. R. Astron. Soc., Vol. 210, No. 2, p. 323 – 340 (1984).
The observational constraints on the stellar wind of the T Tauri star RU Lupi are used to develop a model of the dynamics of the line–emitting region. A magnetically driven wind model, in which Alfvén waves provide the main driving mechanism, is shown to be in good agreement with the optical observations: it is able to reproduce the observed velocity–distance relationship, the density requirements and seems also able to reproduce the observed line profiles. Furthermore, it suggests a possible explanation for the non–uniformity of the temperature in the line–emitting region and for the observed stratification in the line widths.

121.014 **The continuum from the Herbig–Haro object M16–HH1 from 1200 Å to 2.2 μm.**
J. Meaburn, J. R. Walsh.
Astron. Astrophys., Vol. 138, No. 1, p. 36 – 40 (1984).
The atomic continuum emission of the Herbig–Haro object M16–HH1 in the near infra–red has been measured. This is found to be comparable with the atomic recombination continuum expected from a radiatively ionized plasma satisfying Menzel's Case B. The line spectrum and continuum between 3400 Å and 6850 Å for M16–HH1, and a nearby area of M16, have also been obtained. The ultraviolet spectrum is shown to be dominated by scattered starlight within M16.

121.015 **A UV glimpse of T Tauri stars.**
M. T. V. T. Lago, M. V. Penston, R. Johnstone.
ESA Spec. Publ., ESA SP–218, p. 233 – 237 (1984). – See Abstr. 012.024.
After a brief review of the most relevant IUE observational results in the field of T Tauri stars and related objects the impact of such results is discussed. Topics highlighted include: the presence around T Tauri stars of high temperature emitting regions where mechanical rather than radiative heating dominates, high resolution line profiles and their indication of the presence of strong winds, the origin of the continuum and molecular emission, variability, the UV flux and the early solar system.

121.016 **Coordinated X–ray and ultraviolet observations of RU Lupi.**
F. Giovannelli, G. S. Bisnovatyi–Kogan (*G. S. Bisnovatyj–Kogan*), I. M. Golynskaya, V. G. Kurt, S. A. Lamzin, T. A. Mizyakina, E. Yu. Shafer, V. M. Shamolin, E. K. Sheffer, A. S. Smirnoff (*A. S. Smirnov*), A. A. Vittone.
ESA Spec. Publ., ESA SP–218, p. 359 – 361 (1984). – See Abstr. 012.024.
Within the framework of a large campaign of observations of RU Lupi in X–Ray, UV, Optical, and IR, the authors present partial results from X–ray and UV observations. A flux of $3\times10^{-11}$erg/s was detected in the range 2 – 6 keV. Nothing was detected at higher energies. The UV spectrum of RU Lupi shows strong similarities with those of other T Tauri–type stars, with the exception of the presence of He II (λ1640 Å) emission line at a level of $\sim6\times10^{-14}$erg/cm²s. A general description of the UV spectrum is given together with some theoretical considerations on this object.

121.017 **Jets von jungen Sternen.**
R. Mundt.
Sterne Weltraum, 23. Jahrg., Nr. 10, p. 502 – 505 (1984).

121.018 **T–Tauri–Sterne.**
B. Wolf.
Sterne Weltraum, 23. Jahrg., Nr. 10, p. 506 – 511 (1984).

121.019 **New Herbig–Haro objects.**
T. Yu. Magakyan.
Pis'ma Astron. Zh., Tom 10, No. 9, p. 661 – 665 (1984). In Russian. English translation in Sov. Astron. Lett., Vol. 10.
Results of further searches for Herbig–Haro objects are presented. One more Herbig–Haro object in the vicinity of the NGC 7129 nebula and an isolated group consisting of probably a T Tauri–type star, a Herbig–Haro object and a reflection nebula are discovered. These data once more confirm the close connection between T Tauri stars and Herbig–Haro objects.

121.020 **A study of the Herbig–Haro object HH 120 and the associated cometary globule CG 30.**
B. Pettersson.
Astron. Astrophys., Vol. 139, No. 1, p. 135 – 146 (1984).
Infrared and spectroscopic observations are presented of a new Herbig–Haro object, here named HH 120, in the cometary globule CG 30. The emission line spectrum indicates a very low degree of excitation of the same order as that found for HH 47, HH 7, and HH 11. From several spectra a radial velocity of $-42\ \mathrm{km}^{-1}\pm12\ \mathrm{km\ s}^{-1}$ is deduced and the electron temperature, $T_e$, and electron number density, $N_e$, are found to be $9100\mathrm{K}\pm400\mathrm{K}$ and $1700\ \mathrm{cm}^{-3}\pm600\ \mathrm{cm}^{-3}$, respectively. Infrared scans of CG 30 revealed five sources, two of which are very close to HH 120. It is argued that one of them, CG 30–IRS4, is the source of energy producing HH 120 and an associated reflection nebula.

121.021 **Stellar winds and molecular clouds: Herbig Be and Ae type stars.**
J. Cantó, L. F. Rodríguez, N. Calvet, R. M. Levreault.
Astrophys. J., Vol. 282, No. 2, p. 631 – 640 (1984).
The authors searched for high–velocity carbon monoxide emission in the environments of 11 Herbig Be– and Ae–type stars with the purpose of determining if young stars of intermediate luminosity are commonly associated with molecular outflows, as is the case for both high– and low–luminosity objects. Outflows were detected in LkHα 198 and MWC 1080. The outflow associated with LkHα 198 appears to be bipolar and is of moderate velocity. MWC 1080 exhibits a powerful, apparently isotropic outflow. Analysis of optical and infrared data suggests a correlation between abundant circumstellar material and detectable molecular outflows.

121.022 **A paradoxical gap in the relative ages of T Tauri stars.**
W. B. Weaver.
Astrophys. J., Vol. 282, No. 2, p. 688 – 692 (1984).
The frequency distribution of T Tauri stars of different relative age shows a pronounced gap at 5% of their time to the zero–age main sequence. This gap, which occurs in all of the four major T Tauri associations, is too large to be filled by unclassifiable veiled stars. It is nearly vertical on the Hertzsprung–Russell diagram, is centered near spectral class K5, and lies close to the transition between the convective and radiative tracks of the pre–main–sequence stars.

121.023 **VLA observations of T Tauri stars. II. A luminosity–limited survey of Taurus–Auriga.**
J. H. Bieging, M. Cohen, P. R. Schwartz.
Astrophys. J., Vol. 282, No. 2, p. 699 – 708 (1984).
The authors have surveyed all T Tauri stars in the Taurus–Auriga dark clouds with $\log L \geqslant 0.2$ for 4.885 GHz continuum radio emission. No new T Tauri stars were detected. However, two infrared sources associated with Herbig-Haro objects (Haro 6–10 in Taurus, HH 24 IRS in Orion) have radio emission. The star DG Tau has been resolved at 4.885 GHz into an elongated radio structure aligned with an associated HH object. Z CMa is found to be a double radio source. Radio spectral indices are derived for 10 objects. Based on radio data for 46 stars in Taurus–Auriga, it is concluded that T Tauri stars which excite HH objects are much more likely to have detectable radio emission than those which are not associated with HH objects.

121.024 **Velocity–resolved infrared spectroscopy of LkHα 101.**
M. Simon, L. Cassar.
Astrophys. J., Vol. 283, No. 1, p. 179 – 187 (1984).
The authors present new high–resolution spectra of the 2 – 2.4 μm and Brackett–α emission of LkHα 101. Transitions of H I, He I, Fe II, [Fe II], Mg II, and possibly [Cr II] are detected in the 2.4 μm spectrum. The authors associate the line emission region and the core radio component with a flowing circumstellar envelope around LkHα 101. They draw a distinction, on the basis of ionization mechanisms and excitation conditions, between the ionized gas in the circumstellar envelope (ICE) and the more extended (~1′) H II region around LkHα 101. It is suggested that the simultaneous presence of the H II region and the ICE region indicates that the envelope must be asymmetrically distributed around the central star of LkHα 101.

121.025 **The proper motions of Herbig–Haro objects. IV. RNO 40 and 43.**
B. F. Jones, M. Cohen, M. Sirk, R. Jarrett.
Astron. J., Vol. 89, No. 9, p. 1404 – 1407 (1984). = Lick Obs. Bull. No. 984. With plates 78 – 79.
The authors have obtained proper motions and spectra of the Herbig–Haro objects RNO 40 and 43. The proper–motion vectors point away from far–infrared sources. The authors have used the spectra to infer shock velocities. These shock velocities agree with the tangential motion in the case of RNO 40, but are smaller than the tangential motion in the case of RNO 43. RNO 43 appears to be associated with an entire family of co–moving knots.

121.026 **High spatial resolution IR observations of young stellar objects: a possible disk surrounding HL Tauri.**
G. L. Grasdalen, S. E. Strom, K. M. Strom, R. W. Capps, D. Thompson, M. Castelaz.
Astrophys. J., Lett. Ed., Vol. 283, No. 2, p. L57 – L61 (1984). With plates L3 – L4.
High spatial resolution images of the T Tauri star HL Tau were obtained at 1.6 μm and 2.2 μm. The original images as well as maximum entropy image reconstructions reveal a circumstellar envelope structure, similar at both wavelengths, and extended along P.A. = 112°; the 10% width of the structure is 1″.9 (300 AU at 160 pc). The extended structure is interpreted as light scattered toward Earth by dust in a disk surrounding this young stellar object. Polarization measurements made at 2.2 μm support this hypothesis.

121.027 **Herbig–Haro objects and star formation.**
J. A. Graham.
Ir. Astron. J., Vol. 16, No. 3, p. 218 – 219 (1984). – See Abstr. 012.036.

121.028 **X–ray photometry and spectroscopy of T Tauri stars.**
F. M. Walter, L. V. Kuhi.
Astrophys. J., Vol. 284, No. 1, p. 194 – 201 (1984).
The authors present follow–up Einstein X–ray observations of four fields containing pre–main–sequence stars. They have searched for evidence of variability on time scales of hours and of 1 year. A large, slow flare in AS 205 and the X–ray disappearance of AA Tau are reported. Otherwise, T Tauri stars appear no more variable in X–rays than do active solar–like stars. The importance of these observations for models of T Tauri stars and their coronae is considered.

121.029 **High–resolution spectroscopy of the T Tauri star RU Lupi.**
A. M. Boesgaard.
Astron. J., Vol. 89, No. 11, p. 1635 – 1649 (1984). With plates 83 – 84.
A program to monitor the spectra of the strong emission–line star, RU Lup, of the T Tauri class has been conducted. Echelle spectrograms were obtained over five nights with the 4–m telescope at CTIO to examine changes in the emission features on time scales of hours and days, as well as to look for rotational modulation. A rotation period of 3.7 days has been associated with RU Lup from photometric variations. Sets of spectra were taken in the blue from 3700 – 4600 Å and in the red from Hβ to Hα at a resolution of 0.1 – 0.15 Å. Clear evidence of changes in the details of the line–profile shapes, in the velocities and intensities of the components, has been found.

121.030 **Photometric variations of Orion population stars. III. RY Tau, T Ori, NV Ori, and HH Aur.**
W. Herbst, P. C. Stine.
Astron. J., Vol. 89, No. 11, p. 1716 – 1720 (1984).
*UBVRI* and Hα photometry of four Orion population stars has been obtained during the interval October 1983 to March 1984. RY Tau underwent a dramatic increase in brightness during this period, reaching a level ($V = 9.43$) not seen in more than two decades. At the same time, the equivalent width of its Hα emission line decreased and, in the longer–wavelength color indices, has begun to show the "bluer when brighter" syndrome characteristic of other weak emission–line Orion population stars. T Ori and NV Ori also showed substantial variability, and their color and Hα characteristics are discussed. HH Aur showed little or no variation during our observations.

121.031 **Proper motions of Herbig–Haro objects. V. Southern HH objects.**
R. Schwartz, B. F. Jones, M. Sirk.
Astron. J., Vol. 89, No. 11, p. 1735 – 1739 (1984). With plates 90 – 94. = Lick Obs. Bull., No. 989.
The authors have measured proper motions for the southern HH objects HH–46 to HH–50, HH–52 to HH–54, HH–56, –57, –99, and –101, and knots near R CrA using CTIO 4–m plates. Several of these objects show motions away from their probable exciting stars. HH–49 and –50 may have undergone repeated acceleration.

121.032 **Jets from young stars. III. The case of Haro 6–5B, HH33/40, HH19, and 1548C27.**
R. Mundt, T. Bührke, J. W. Fried, T. Neckel, M. Sarcander, J. Stocke.
Astron. Astrophys., Vol. 140, No. 1, p. 17 – 23 (1984).
The authors report the discovery of 3 jetlike emission–line nebulosities associated with Haro 6–5B, HH33/40, and HH19 and confirm the 1548C27–jet. These jets are 0.05 – 0.2 pc long and have opening angles of about 5°. CCD images and spectroscopic observations of these new and of previously discovered jets strongly suggest that they are all representations of well–collimated, high–velocity (50 – 400 km $s^{-1}$) flows emanating

from young stars. The possibility of powering many HH objects by such jets is discussed.

121.033 **Simultaneous photometric and polarimetric observations of four T Tauri stars.**
P. Bastien, A. Le Van Suu.
J. R. Astron. Soc. Can., Vol. 78, No. 5, p. 206 (1984). Abstract. – See Abstr. 010.401.

121.034 **Infrared mapping of the Herbig–Haro complex HH7–11 in the S(1) emission line of molecular hydrogen.**
W. J. Zealey, P. M. Williams, G. Sandell.
Astron. Astrophys., Vol. 140, No. 2, p. L31 – L34 (1984).
The authors have mapped the $\upsilon = 1\text{–}0$ S(1), 2.12 μm emission line of molecular hydrogen in the region of HH7–11. The emission is present throughout the strikingly aligned complex, but ceases abruptly to the south and east of HH7. The emission peaks near HH7, with a secondary maximum in intensity near HH8. The observations are compatible with current shock models of Herbig–Haro objects.

121.035 **Forbidden–line profiles of T Tauri stars.**
I. Appenzeller, I. Jankovics, R. Östreicher.
Astron. Astrophys., Vol. 141, No. 1, p. 108 – 115 (1984).
Profiles of forbidden emission lines were derived from high resolution ($1.5 \lesssim R \times 10^{-4} \lesssim 3$) spectrograms of 12 T Tauri stars. These observational results are compared to theoretical line profiles, derived from simple geometric models of T Tauri winds, assuming the presence of circumstellar dust or obscuring protostellar disks. Most observed profiles are not compatible with spherically symmetric radial winds but agree qualitatively with profiles predicted for axially symmetric focussed velocity fields and filled or hollow cone emission patterns.

121.036 **Photographic photometry of V1515 Cygni in the years 1983 – 1984.**
K. P. Tsvetkova.
Inf. Bull. Variable Stars, No. 2617, 2 pp. (1984).

121.037 **Spectroscopic and photometric observations of the Herbig Be star Z Canis Majoris.**
E. Covino, L. Terranegra, A. A. Vittone, G. Russo.
Astron. J., Vol. 89, No. 12, p. 1868 – 1875 (1984).
Spectroscopic and photometric (*UBV*) observations of Z CMa were carried out at ESO, La Silla during February 1983. Photometric variations of the order of a few tenths of magnitude were detected in contrast with its past more variable behavior. The spectrum of Z CMa appears to have a complex structure showing at the same time characteristics of both early (B5 – B8) and late (middle F) spectral types. A rather peculiar and variable profile with three different absorption components was observed at the Hβ line. From a comparison between the present and the past behaviour it seems that the star is now in a relatively quiescent phase.

121.038 **Characteristic time scales for variability in pre–main sequence stars.**
F. Praderie, C. Catala, J. Czarny, P. Felenbok.
Space research prospects in stellar activity and variability, p. 265 – 276 (1984). – See Abstr. 012.078.
The authors conducted a program to observe both short and long term spectral variability in two Herbig stars. They present the results for one of them, AB Aur. Finally, they describe the observational requirements for a program that would elucidate the kind of variability mechanisms present in the pre–main–sequence stars.

121.039 **Highly collimated mass outflows from young stars.**
R. Mundt.
Max–Planck–Inst. Astron., Prepr., 42 pp. (1984). Review paper presented at the second Protostars and Planets Meeting, Tucson, Arizona, January 4 – 8, 1984. To appear in Protostars & Planets II.

121.040 **Jets from young stars. III. The case of Haro 6–5B, HH 33/40, HH 19 and 1548C27.**
R. Mundt, T. Bührke, J. W. Fried, T. Neckel, M. Sarcander, J. Stocke.
Max–Planck–Inst. Astron., Prepr., 22 pp. (1984). To appear in Astron. Astrophys.

121.041 **Jets from young stars. IV. The case of R Monocerotis.**
E. W. Brugel, R. Mundt, T. Bührke.
Max–Planck–Inst. Astron., Prepr., 14 pp. (1984). Submitted to Astrophys. J., Lett. Ed.

121.042 **Zum Charakter und Verhalten zirkumstellarer Hüllen von T–Tauri–Sternen.**
W. Götz.
Veröff. Sternw. Sonneberg, Band 10, Heft 1, 168 pp. (1984).
The investigations, which concern the character and the behaviour of the circumstellar shells at T Tauri stars and which lead to the construction of a shell model on the base of observational facts, rest upon the causal connection between the shell phenomenon and the mass loss of the stars. The author tried to construct the curves of the total flux from the continuous radiation of the stars and their shells in the optical and infrared regions, to separate the shares of radiation from the stars and the shells, to estimate the character of the shell medium and its radiation, to point out possible physical mechanisms and processes which produce this character and, to give statements about the re–distribution of energy and radiation in the shells.

121.043 **Narrow–band imaging and velocity maps of young stellar objects: initial results.**
J. S. Morgan, S. C. Wolff, S. E. Strom, K. M. Strom.
Astrophys. J., Lett. Ed., Vol. 285, No. 2, p. L71 – L73 (1984). With plates L5 – L7.
The first trials of a new technique, designed to map low–excitation ionized gas surrounding young stellar objects, are reported. The region surrounding the T Tau stars HL Tau and XZ Tau, that near HH 101, and that near IRS 5 in L1551 have been imaged through a narrow–band (4.7 Å FWHM) [S II] filter; a Texas Instruments three–phase CCD chip was used as the detector. By tilting the narrow–band filter, it is possible to vary the wavelength of peak transmission and thus to detect high–velocity radial flows as well as map the morphology of the excited gas near these young stars.

121.044 **Jets from young stars. III. The case of Haro 6–5B, HH 33/40, HH 19 and 1548C27.**
R. Mundt, T. Bührke, J. W. Fried, T. Neckel, M. Sarcander, J. Stocke.
Prepr. Steward Obs., No. 529, 22 pp. (1984). To appear in Astron. Astrophys.

121.045 **Brackett–alpha line profiles of young stellar objects.**
S. E. Persson, T. R. Geballe, P. J. McGregor, S. Edwards, C. J. Lonsdale.
Astrophys. J., Vol. 286, No. 1, p. 289 – 301 (1984).
Profiles of the Brα line of H I at a velocity resolution of 45 km $s^{-1}$ are presented for the compact embedded infrared objects BN, S106/IRS 3, GL 490, GL 961, GL 989, Mon R2/IRS 2, and for the visible objects LkHα 101, T Tau, and R Mon. A proportionality between Brα luminosity and bolometric luminosity, analogous to that found by Thompson for Brγ luminosities, is shown to extend over three orders of magnitude for these young stellar objects. The Brα line strengths are compared to radio continuum flux densities in the context of stellar wind models. Fair agreement is found for the Simon et al. formulation, in nearly all cases. A comparison is made between the Hα and Brα profiles for T Tau. The Brα line probably arises in the envelope of T Tau itself and not the binary companion T Tau(S).

121.046 **The nature of AFGL 2591 and its associated molecular outflow: infrared and millimeter–wave observations.**
C. J. Lada, H. A. Thronson Jr., H. A. Smith, P. R. Schwartz, W. Glaccum.
Astrophys. J., Vol. 286, No. 1, p. 302 – 309 (1984).
The results of infrared photometry from 2 μm to 160 μm of AFGL 2591 and $^{12}CO$ observations of its associated molecular cloud and high–velocity molecular outflow are presented and discussed. Observations between 60 and 160 μm provide the first detailed far–infrared spectrum of this source. The observed luminosity between 2 and 160 μm is $6.7 \times 10^4 L_\odot$ at a distance of 2 kpc. The 2 – 160 μm spectrum of AFGL 2591 is interpreted in the context of a model in which a single embedded object is the dominant source of the infrared luminosity. This object may also be responsible for ionizing the compact H II regions near AFGL 2591. The $^{12}CO$ observations show the high–velocity molecular flow near AFGL 2591 to be extended, bipolar, and roughly centered on the infrared emission.

121.047 **Hinweis für eine kollimierte Hochgeschwindigkeitsströmung von T Tauri.**
T. Bührke, R. Mundt.
Mitt. Astron. Ges., Nr. 62, p. 234 – 236 (1984). – See Abstr. 010.341.

121.048 **Jets von jungen Sternen.**
R. Mundt, T. Bührke.
Mitt. Astron. Ges., Nr. 62, p. 237 – 238 (1984). – See Abstr. 010.341.

121.049 **Speckle–Interferometrie von T–Tauri–Sternen und verwandten Objekten.**
G. Baier, E. Keller, G. Weigelt, U. Bastian, R. Mundt.
Mitt. Astron. Ges., Nr. 62, p. 241 (1984). Abstract. – See Abstr. 010.341.

121.050 **Forbidden emission line profiles as indicators of non–isotropic mass loss.**
I. Appenzeller, R. Östreicher, I. Jankovics.
Mitt. Astron. Ges., Nr. 62, p. 247 (1984). Abstract. – See Abstr. 010.341.

121.051 **On objects looking like Herbig–Haro objects.**
A. L. Gyul'budagyan.
Astrofizika, Tom 20, Vyp. 3, p. 631 – 634 (1984). In Russian. English translation in Astrophysics, Vol. 20, No. 3.
A comparison between the objects looking like HH objects and the genuine HH objects is made. A list of 12 new objects looking like HH objects found on the survey of PSS prints is given.

121.052 **Discovery of solar system–size halos around young stars.**
S. Beckwith, B. Zuckerman, M. F. Skrutskie, H. M. Dyck.
Astrophys. J., Vol. 287, No. 2, p. 793 – 800 (1984).
Near–infrared speckle interferometric observations of five pre–main–sequence stars reveal a core–halo structure around two of these stars: HL Tau and R Mon. The halo light distribution is shown to arise from scattered light from small circumstellar particles. Halo sizes of 320 × 200 AU (α × δ) and 1300 × 1300 AU are deduced for HL Tau and R Mon, respectively, and the halo light is substantially bluer than the stellar light. The minimum mass of small particles in the scattering regions is comparable to the Earth's mass in HL Tau and ten times greater in R Mon. Mass loss from the stars is almost certainly insufficient to produce the halo matter.

121.053 **Jets from young stars: the case of R Monocerotis.**
E. W. Brugel, R. Mundt, T. Bührke.
Astrophys. J., Lett. Ed., Vol. 287, No. 2, p. L73 – L75 (1984). With plates L9 – L10.
The authors report the discovery of both components of a highly collimated bipolar outflow from R Mon. The [S II] 6716 Å, 6731 Å line emission regions tracing this outflow are situated in narrow extensions on the opposite sides of R Mon. These emission regions are located at a distance of about 10″ from R Mon and lie along a line joining R Mon and HH 39. It is argued that the flow collimation occurs close to the star.

**Pre–main sequence evolution.**
See Abstr. 065.020.

**The wind and chromosphere of the Herbig Ae star AB Aur as seen by IUE.**
See Abstr. 112.015.

**On the effective temperatures for early type stars from integrated fluxes.**
See Abstr. 114.101.

**Winds from low–mass stars. Self–regulated star formation in Taurus, Ophiuchus, NGC 2264 and Orion.**
See Abstr. 131.050.

**Molecular hydrogen jets from the Orion nebula.**
See Abstr. 131.055.

**Velocity gradients in the bipolar CO gas–flow around L1551 IRS5.**
See Abstr. 131.062.

**Optical and infrared observations of Cep–A/GGD37.**
See Abstr. 131.063.

**Bullets: "interstellar plops and plunks".**
See Abstr. 131.066.

**Magnetic field structure in the Taurus dark cloud.**
See Abstr. 131.098.

**Observations of diffuse interstellar bands toward T Tauri stars.**
See Abstr. 131.109.

**High–velocity molecular jets.**
See Abstr. 131.137.

**The reflection nebula NGC 1999.**
See Abstr. 131.203.

**Bipolar outflows in dark clouds.**
See Abstr. 131.290.

**Absorption features in the 5 – 8 micron spectra of protostars.**
See Abstr. 131.304.

**Internal reddening of H II regions and HH–objects, with special reference to NGC 896(W3).**
See Abstr. 132.017.

**Infrared polarization of sources with bipolar mass outflow.**
See Abstr. 133.016.

**Spectrophotometry of compact embedded infrared sources in the 0.6 – 1.0 micron wavelength region.**
See Abstr. 133.023.

**The Coronet, an obscured cluster adjacent to R Corona Austrina.**
See Abstr. 152.001.

**A deep photometric study of IC 5146.**
See Abstr. 153.022.

**The nature of the embedded population in the Rho Ophiuchi dark cloud: mid–infrared observations.**
See Abstr. 153.053.

# 122 Intrinsic Variables (Pulsating Variables, Spectrum Variables, etc.)

122.001 **The photometric behaviour of FG Sagittae in 1975 – 80: pulsations with increasing period.**
V. P. Arkhipova.
Perem. Zvezdy, Tom 21, No. 5, p. 617 – 627 (1982). In Russian.
The results of photoelectric UBV observations of FG Sge in 1975 – 80 are given. In 1975 – 79 FG Sge showed periodic variations of brightness with continually growing period: its rate is about 2 – 4 days per one period which is unique for a variable star.

122.002 **Activity and latent periodicity of BN Orionis.**
I. M. Ishchenko.
Perem. Zvezdy, Tom 21, No. 5, p. 665 – 676 (1982). In Russian.
A diagram for BN Ori is given, showing the mean annual brightness and its mean variation. A latent periodicity of the star for high activity seasons has been discovered. An attempt was made to explain the periodicity by the axial rotation of the star. The period increase is explained by motion of activity sources along an untwisted spiral and probable mass loss.

122.003 **Preliminary discussion of the light curve of VV Serpentis.**
V. I. Kardopolov, G. K. Filip'ev, V. P. Kuleshov.
Perem. Zvezdy, Tom 21, No. 5, p. 682 – 687 (1982). In Russian.
Results of BVR photoelectric measurements of the rapid irregular variable star VV Ser are given. The observed light curve peculiarities of VV Ser are interpreted to be caused by several different origins.

122.004 **On progressive period changes of classical cepheids.**
G. E. Erleksova, B. N. Irkaev.
Perem. Zvezdy, Tom 21, No. 5, p. 715 – 731 (1982). In Russian.
24 classical cepheids were found to show progressive period changes (secular increase or decrease) according to the O–C diagrams. Assuming linear period changes with time in the first approximation the authors calculated the rates $(\delta P/P)_{100 \text{ years}}$ of these cepheids.

122.005 **Evolution and pulsation masses of classical cepheids.**
Ts. Tsvetkov.
Perem. Zvezdy, Tom 21, No. 5, p. 733 – 740 (1982). In Russian.
Both evolution and pulsation masses of 156 classical cepheids of our Galaxy are calculated. If the radii derived by Wesselink's method are used, then the pulsation masses $M_Q$ are not systematically smaller than the evolution masses $M_e$.

122.006 **Luminosities and ages of classical cepheids.**
Ts. Tsvetkov.
Perem. Zvezdy, Tom 21, No. 5, p. 741 – 746 (1982). In Russian.
The luminosities and ages of 156 classical cepheids of our Galaxy are calculated. The luminosities derived in different ways are compared.

122.007 **Evolution and pulsation masses of Delta Scuti stars.**
Ts. Tsvetkov.
Perem. Zvezdy, Tom 21, No. 5, p. 747 – 749 (1982). In Russian.
Both evolution $M_e$ and pulsation $M_Q$ masses of 53 Delta Scuti stars are calculated using Breger and Bregman's (1975) observational data. It is shown that the two kinds of masses are consistent taking into account the determination errors.

122.008 **On a possible mass "gap" between classical cepheids and Delta Scuti stars.**
Ts. Tsvetkov.
Perem. Zvezdy, Tom 21, No. 5, p. 750 – 752 (1982). In Russian.
The problem of a possible mass "gap" between the two types of Population I pulsating stars – classical cepheids and Delta Scuti stars – is studied. For this purpose the results of evolution and pulsation mass calculations for 156 classical cepheids and 53 Delta Scuti stars are used. The evolution masses indicate a "gap" at (2.5 – 4) $M_\odot$. The estimates of the pulsation masses, however, do not give evidence for the existence of such a "gap".

122.009 **Influence of the chemical composition on the theoretical blue edge location of the instability strip of Delta Scuti stars.**
Ts. Tsvetkov.
Perem. Zvezdy, Tom 21, No. 5, p. 753 – 761 (1982). In Russian.
Radial pulsations of Delta Scuti star envelope models in the four lowest modes are computed using a linear, non–adiabatic approximation. It is shown that the theoretical blue edge location of the instability strip in the H–R diagram depend strongly on variations of the envelope helium content Y. Variations of the heavy element content Z have a small effect. A non–linear dependence of the theoretical blue edge location on Y is derived.

122.010 **A statistical analysis of the light curve of the variable star W Persei.**
I. A. Klyus.
Perem. Zvezdy, Tom 21, No. 5, p. 763 – 765 (1982). In Russian.
Narrow–band occasional noises (cycles) with average durations of 485 and 2667 days, have been found on the light curve. Peaks in the power spectrum exceed the 90% confidence interval. For the first time a cycle with an average duration of 2667 days has been found.

122.011 **A statistical analysis of the light curves of the variable stars R Pictoris and Z Ursae Majoris.**
I. A. Klyus.
Perem. Zvezdy, Tom 21, No. 5, p. 766 – 770 (1982). In Russian.
It is shown that the process of light variations of the investigated stars is a stochastic process with nonstationarity connected with the presence of oscillations more prolonged than observational time intervals in both objects. The cycle with an average duration of 170 days has been found on the light curve of R Pic and 190 days for Z UMa. These oscillations are not harmonic. The light variations of these stars are very much like the light variations of Mira Ceti type variable stars.

122.012 **New variable star in Coma Berenices.**
N. V. Metlova.
Astron Tsirk., No. 1285, p. 8 (1983). In Russian.

122.013 **Observational studies of Cepheids. II. *BVRI* photometry of 112 Cepheids.**
T. J. Moffett, T. G. Barnes III.
Astrophys. J., Suppl. Ser., Vol. 55, No. 3, p. 389 – 432 (1984).
Over 4000 differentially determined photoelectric *BVRI* observations and the resulting light curves are presented for 112 Cepheids accessible from northern hemisphere observatories. The internal precision of these data is better than $\pm 0.01$ mag, and the accuracy of transformation to the Johnson *BVRI* system is nearly as good.

122.014 **Comparison of variations in the visible and ultraviolet spectra of a Centauri.**
R. P. Fahey.
Astrophys. J., Suppl. Ser., Vol. 55, No. 3, p. 507 – 516 (1984).
Equivalent–width and central depth measurements of some ultraviolet absorption lines from the spectrum of the He I spectrum variable star a Cen (HD 125823) are presented. The measurements are from 53 short–wavelength and 52 long–wavelength, high–resolution spectrographs taken with the *IUE* satellite over a $25^d$ period during 1978 July and August. The *IUE* observations are consistent with a period of $8^d\!.8163 \pm 0^d\!.0003$ and with a previously unreported period of $9^d\!.2532 \pm 0^d\!.0003$.

122.015 **Long–period variables in Baade's window: 1 – 12 μm photometry.**
T. J. Jones, A. R. Hyland, G. Robinson.
Astron. J., Vol. 89, No. 7, p. 999 – 1004 (1984).

Infrared photometry of long–period variables in the direction of Baade's window ($l = 0, b = -4$) from 1 to 12 μm is presented. Dust–shell models, blackbody fits, and the [8.7] – [11.6] silicate color index are used to investigate the extent to which the circumstellar shell affects the *JHK* colors in comparison to local Miras. The observed silicate color index is within the range found in local Miras with similar periods. The dust–shell models indicate that on the whole, the *JHK* colors are good indicators of the core–color temperature.

122.016 **Erratum: "*BV* photometry of RR Lyrae variables in the globular cluster NGC 3201" [Astron. J., Vol. 89, No. 2, p. 231 – 262 (1984)].**
C. Cacciari.
Astron. J., Vol. 89, No. 7, p. 1082 (1984). See Abstr. 37.122.058.

122.017 **Photoelectric photometry of the Delta Scuti star BD +43°1894.**
V. Costa, R. Garrido, P. Lopez de Coca, R. Peniche, J. H. Peña, J. M. Quintana, A. Rolland.
Astron. Astrophys., Suppl. Ser., Vol. 57, No. 2, p. 233 – 238 (1984).

Consecutive differential photoelectric photometry of the Delta Scuti star BD +43°1894 has been carried out at three different observatories to avoid the problem of aliasing. Analysis of the data by the classical O–C method and the Fourier Transform converged to establish BD +43°1894 as a monoperiodic pulsator with a frequency of pulsation of 10.175611 c/d.

122.018 **Orbital motion of the dwarf cepheid SZ Lyn.**
C. Bardin, M. Imbert.
Astron. Astrophys., Suppl. Ser., Vol. 57, No. 2, p. 249 – 252 (1984). In French.

The dwarf cepheid SZ Lyn, whose binarity, first suspected in 1967, was established in 1981, was observed from 1979 November 27 to 1983 November 21 with the CORAVEL photoelectric spectrometer. Measurements secured over 30 pulsational cycles yield a sequence of 21 mean radial velocity values. Orbital elements for the system were derived from these data.

122.019 **On variations of (B–V) and (U–B) colours in brightness minima of RZ Piscium.**
G. V. Zajtseva.
Astron. Tsirk., No. 1299, p. 1 – 3 (1983). In Russian.

122.020 **Chromospheric lines in red dwarf flare stars. II. EV Lacertae, EQ Pegasi A, and V1054 Ophiuchi.**
B. R. Pettersen, D. S. Evans, L. A. Coleman.
Astrophys. J., Vol. 282, No. 1, p. 214 – 225 (1984).

Line profile observations with a spectral resolution of 0.45 Å are presented for three similar solar neighborhood dM4e stars with luminosity 1/60 of the Sun. EV Lac is a spotted flare star, EQ Peg A is the brighter member of a flaring visual binary, and V1054 Oph consists of two flare stars in a sextuple. The lines observed were selected because they are useful diagnostics of the chromospheric structure of these stars. Results are presented for Hα (λ6563), the Na D lines (λλ5890, 5896), He I (λ5876), and the Ca II infrared triplet lines (λλ8498, 8542, 8662). To avoid contamination from flares the spectroscopic observations were obtained simultaneously with high–speed photometry. Hα is found in emission in all three stars. The observed profiles of the Hα emission line are compared with recent theoretical results.

122.021 **Stable nonradial pulsations in 53 Persei from 1977 to 1983.**
M. A. Smith, W. S. Fitch, J. L. Africano, B. D. Goodrich, W. Halbedel, L. H. Palmer, G. W. Henry.
Astrophys. J., Vol. 282, No. 1, p. 226 – 235 (1984).

New photometric and line profile data are combined with old in a reanalysis of the nonradial pulsation properties of the middle–B star 53 Persei, the prototype of its class. Seventeen nights of *uvby* photometry in late 1981 have been combined with 1977 and 1978 photometry. An analysis of the data suggests the presence of stable frequencies of nearly equal amplitude at 0.464 and 0.595 cycles per day. Results of fits to Si II λ4130 line profile observations lead to the conclusion that 53 Persei changed its pulsational characteristics to excite the present pair of dominant frequencies sometime during 1977. The new photometric and line profile variations are consistent with the identifications of the frequencies as nonradial modes of high overtone and $l = 3$, $m = -3$ and $-2$.

122.022 **On the energy balance of graphite grains in the atmospheres of R CrB stars.**
B. E. Zhilyaev, V. G. Zubko.
Astrometr. Astrofiz., Vyp. 52, p. 15 – 18 (1984). In Russian.

The problem of graphite grain energy balance in the atmospheres of R CrB stars is reconsidered. For 8 types of graphite materials the equilibrium temperature of graphite grains is shown to be much less than the temperatures of the stars.

122.023 ***UBV* photometry of the short period pulsating variable V1719 Cyg.**
E. Poretti.
Astron. Astrophys., Suppl. Ser., Vol. 57, No. 3, p. 435 – 442 (1984).

*UBV* observations of V1719 Cyg were performed at Merate Observatory during August 1983. The star is a short period pulsating variable showing variations in shape and amplitude of the light curve. Photometric characteristics are discussed and V1719 Cyg is found to be like other pulsating variables classified as RRc, but its nature is not completely determined. Other photometric and spectroscopic observations are discussed.

122.024 **An RR Lyrae survey with the Lick Astrograph. VI. New photometric observations and ephemerides for the variables in the field RR I (MWF 361A).**
T. D. Kinman, B. Wong–Swanson, M. Wenz, E. A. Harlan.
Astron. J., Vol. 89, No. 8, p. 1200 – 1218 (1984).

New ephemerides are given for 57 RR Lyrae variables in the field RR I (MFW 361A) of the Lick Survey. Photoelectric (*B,V*) observations of 29 of these stars show that the systematic errors in the photographic magnitudes are $\lesssim 0.1$ mag over the range $15 < m_{pg} < 17$ which includes 74% of the variables of type *ab* in the field. A zero–point correction of +0.05 mag to the $E(B-V)$ contours of Burstein and Heiles is derived from p.e. observations of the variables. The random observational errors of the new photographic magnitudes ($\pm 0.06$ mag at $m_{pg} = 16$) are about one–half the size of those in the original survey. Of the 42 type *ab* variables in this sample, 7 definitely and 5 probably have more scatter in their light curves than can be accounted for by observational error. This frequency is comparable with the known rate of occurrence of the Blazhko effect in these objects.

122.025 **The pulsating extreme helium star BD +13°3224. II. Ultraviolet fluxes and effective temperature variations.**
A. E. Lynas–Gray, D. Schönberner, P. W. Hill, U. Heber.
Mon. Not. R. Astron. Soc., Vol. 209, No. 2, p. 387 – 400 (1984).

Ultraviolet flux variations are reported for the pulsating extreme helium star BD +13°3224 (V652 Her). Effective temperature and angular radius variations over a cycle are determined from static plane–parallel LTE model atmospheres. When compared with radius changes derived from ground–based spectroscopy, the angular radius variations indicate radial pulsations and correspond to a distance of $1.5 \pm 0.1$ kpc. BD +13°3224 is thought to be a helium–burning star of 0.7 $M_\odot$, which has an envelope contracting as the helium–burning core grows.

122.026 **Physical parameters and pulsational mode of short–period small–amplitude Cepheids.**
A. Arellano Ferro.
Mon. Not. R. Astron. Soc., Vol. 209, No. 3, p. 481 – 502 (1984). With microfiche MN 209/2.

Nine galactic short–period ($P < 8$ days) small–amplitude ($\Delta V < 0.3$ mag) Cepheids, have been the subject of a study to determine their physical parameters and pulsational properties. *UBVRI* photometry and radial velocities were obtained in the same observing seasons. The wings of Hα were used to determine effective temperatures. Empirical radii, luminosities and masses were determined. The short-period stars lie inside and seem to be on the red side of the observational Cepheid instability strip. It was found that these stars are likely pulsating in the radial fundamental mode except HR 8084 (DT Cyg) which is suspected to be a first overtone pulsator.

122.027 **Photographic photometry of RR Lyrae variables in the globular cluster M15.**
E. A. Bingham, C. Cacciari, R. J. Dickens, F. Fusi Pecci.
Mon. Not. R. Astron. Soc., Vol. 209, No. 4, p. 765 – 824 (1984).

Light curves in *B* and *V* are presented for 56 RR Lyrae variables in the Oosterhoff Group II globular cluster M15. The measures are combined with original measures published by Sandage, Katem & Sandage to yield data of improved precision for a total of 62 RR Lyraes. Correlations between the various light curve parameters are obtained, and their significance is discussed. The observations are compared in detail with those of RR Lyraes in the Oosterhoff I cluster M3 in a discussion of the Oosterhoff problem.

122.028 **Two new rapidly oscillating Ap stars: HD 60435 and HD 203932.**
D. W. Kurtz.
Mon. Not. R. Astron. Soc., Vol. 209, No. 4, p. 841 – 850 (1984).

The discovery of two new rapidly oscillating Ap stars is announced. HD 60435 is the largest amplitude rapidly oscillating Ap star yet dicovered with a peak-to-peak variability of $\Delta B \leqslant 15.7$ mmag; HD 203932 is the smallest with $\Delta B \leqslant 2.4$ mmag. A qualitative description of the complex frequency pattern in HD 60435 is given. A frequency analysis of HD 203932 shows the presence of one frequency, $f = 10.09724 \pm 0.00012\ \mathrm{hr}^{-1}$ with a semi-amplitude of 0.65 mmag. A list of the nine known rapidly oscillating Ap stars with references is given.

122.029 **Photoelectric and spectral observations of WW Vul.**
L. V. Timoshenko, G. K. Filip'ev.
Astrofizika, Tom 19, Vyp. 3, p. 513 – 524 (1983). In Russian. English translation in Astrophysics, Vol. 19, No. 3.

The results of photoelectric UBVR observations of the non-periodic variable star WW Vul obtained during the seasons of 1976 – 1981 and those of spectral observations in 1981 are given.

122.030 **The UU Her-type stars: possible tracers of recent star formation at high galactic latitudes.**
D. D. Sasselov.
Astrophys. Space Sci., Vol. 102, No. 1, p. 161 – 178 (1984).

A general review of the group of UU Her-type stars proposed recently as a possible new type of variables is given. The available observations reveal them as luminous Population I F-type supergiants of normal composition. This fact contradicts their great distances above the galactic plane. As there is no reason so far to distrust the results from the spectroscopic analyses and photometry, the galactic plane should be excluded as a possible birthplace of the UU Her stars. In this respect they might probably indicate that recent star formation was possible at high galactic latitudes. Moreover, normal main-sequence B stars seem now also be found there. The type of pulsations of the UU Her stars remains conjectural, but simple radial pulsations should obviously be ruled out.

122.031 **The period-luminosity relation of Mira stars.**
S. van den Bergh.
Astrophys. Space Sci., Vol. 102, No. 2, p. 295 – 300 (1984).

Published photometry in Sgr I is used to derive the period-luminosity relation for Mira variables in yellow and in blue light. The observed dispersion about the adopted relation is comparable to that of RR Lyrae stars around their mean luminosity. This suggests that the Mira star period-luminosity relation has a relatively small dispersion and is, therefore, a powerful tool for astronomical distance determination. The calibration of the Mira period-luminosity relation indicates that Glass and Feast may have overestimated the absorption in the direction of Sgr I and NGC 6522.

122.032 **Les aventures de Mira Ceti.**
D. Proust.
Bull. Assoc. Fr. Obs. Etoiles Variables, No. 29, p. 103 – 112 (1984).

122.033 **Ephemerides of RR Lyrae-type variables for the year 1985.**
W. P. Zessewitsch (*V. P. Tsesevich*), B. N. Firmaniuk (*B. N. Firmanyuk*), J. M. Kreiner.
Rocznik Astronomiczny Obserwatorium Krakowskiego 1985, p. 115 – 132 (1984). – See Abstr. 046.034.

122.034 **On the semiregular variable TW Aquilae.**
E. N. Makarenko.
Perem. Zvezdy, Prilozh., Tom 4, No. 20, p. 175 – 178 (1982). In Russian.

The results of photographic observations of TW Aql within the period JD 2436049–42280 are given. TW Aql is a semiregular SRd-type variable star. Cyclic light variations are characteristic of it, the average cycles duration and light variation amplitude are ranging widely.

122.035 **Photographic and photovisual photometry of TZ Cassiopeiae.**
M. Yu. Skul'skij.
Perem. Zvezdy, Prilozh., Tom 4, No. 20, p. 179 – 183 (1982). In Russian.

122.036 **Observations of nine semiregular variables.**
L. I. Shakun.
Perem. Zvezdy, Prilozh., Tom 4, No. 20, p. 185 – 202 (1982). In Russian.

The results of photographic observations of EF, EH And, BX, GZ, HH, IM, IY, KS, V365 Cas and investigation of their periods are given.

122.037 **Photographic observations of six RR Lyrae-type variables.**
V. P. Smykov, L. I. Shakun.
Perem. Zvezdy, Prilozh., Tom 4, No. 20, p. 203 – 204 (1982). In Russian.

The results of photographic observations of RX, RZ Cet, DL Her, SS Psc, AR and DV Vir are presented. New elements of light variations were determined for DV Vir.

122.038 **On the period and light curve of ET Vulpeculae.**
G. E. Erleksova, O. G. Suyarkova.
Perem. Zvezdy, Prilozh., Tom 4, No. 20, p. 205 – 207 (1982). In Russian.

Photographic observations of ET Vul, a W Vir type variable, were carried out. The mean period was improved. The detected period change was large. One can allow that there are real variations of the light curve from cycle to cycle in the limits of $0^{m}\!.25$ pg. However, the mean light curve is sufficiently stable.

122.039 **Investigation of the variable star V844 Cygni based on photographic plates of 1972 – 1978.**
A. O. Yuferov.
Perem. Zvezdy, Prilozh., Tom 4, No. 20, p. 215 – 216 (1982). In Russian.

122.040 **New variable star SVS 2338 in Cygnus.**
M. B. Ivanij.
Perem. Zvezdy, Prilozh., Tom 4, No. 20, p. 217 – 219 (1982). In Russian.

122.041 **Investigation of variable stars SY Serpentis and CSV 3326.**
E. A. Zabrodina.
Perem. Zvezdy, Prilozh., Tom 4, No. 20, p. 225 – 226 (1982). In Russian.

122.042 **Variable stars UY and TZ Serpentis.**
D. A. Mitrofanov.
Perem. Zvezdy, Prilozh., Tom 4, No. 20, p. 227 – 229 (1982). In Russian.

122.043 **RR Lyrae–type variable stars V1013 and V2033 Ophiuchi.**
S. F. Yarikov.
Perem. Zvezdy, Prilozh., Tom 4, No. 20, p. 236 – 238 (1982). In Russian.

122.044 **Investigation of two variable stars in Ursa Minor.**
V. S. Veremchuk.
Perem. Zvezdy, Prilozh., Tom 4, No. 20, p. 239 – 240 (1982). In Russian.

The variable stars SVS 2508 and 2509 were studied. SVS 2508 is an RR Lyrae variable with the period of $0^d.5317771$. Limits of brightness variations are $14^m.0 - 15^m.8$ pg. SVS 2509 is an eclipsing variable. Two light minima were observed: JD 2443190.281 ($17^m.14$ pg) and 44076.376 ($>17^m.4$ pg). Brightness in maximum is $16^m.0$ pg. Period was not determined.

122.045 **The variable stars V962 and V987 Ophiuchi.**
S. A. Gavryushov.
Perem. Zvezdy, Prilozh., Tom 4, No. 20, p. 241 – 244 (1982). In Russian.

122.046 **Observations of the variable stars V570 and V1096 Ophiuchi.**
A. Yu. Pogosyants.
Perem. Zvezdy, Prilozh., Tom 4, No. 20, p. 245 – 248 (1982). In Russian.

On the basis of 160 photographic observations new light elements for V570 Oph have been obtained: Max hel $= 2442873.598 + 0^d.574851 \times E$. The Mira type variable V1096 Oph has the following elements: Max $= 2443254.5 + 281^d.5 \times E$.

122.047 **New Mira–type variables SVS 2449 and V832 Ophiuchi.**
A. Yu. Pogosyants.
Perem. Zvezdy, Prilozh., Tom 4, No. 20, p. 249 – 252 (1982). In Russian.

On 177 plates light elements of the long–period variable SVS 2449 have been determined: Max $= 2443406 + 331^d.0 \times E$. A new period for a red semiregular variable V832 Oph is $105^d$. Photographic light curves are given.

122.048 **Variable stars V830, V1095 Ophiuchi and CSV 7725.**
O. G. Surikov.
Perem. Zvezdy, Prilozh., Tom 4, No. 20, p. 253 – 256 (1982). In Russian.

122.049 **RR Lyrae variable stars V933 and V960 Ophiuchi.**
O. G. Surikov.
Perem. Zvezdy, Prilozh., Tom 4, No. 20, p. 257 – 259 (1982). In Russian.

122.050 **Investigation of variable stars ZZ Virginis and SVS 2507.**
I. V. Khatuntsev.
Perem. Zvezdy, Prilozh., Tom 4, No. 20, p. 261 – 263 (1982). In Russian.

122.051 **The variable stars RX and RY Serpentis.**
A. M. Chernenko.
Perem. Zvezdy, Prilozh., Tom 4, No. 20, p. 265 – 266 (1982). In Russian.

122.052 **On the variability of UU Herculis.**
D. D. Seselov.
Perem. Zvezdy, Prilozh., Tom 4, No. 20, p. 267 – 270 (1982). In Russian.

Visual observations of the semiregular variable UU Her in 1978 – 1979 are presented. A period of 80.1 days is determined from the analysis of the light curve by the method of Jurkevich (1971). A light curve has some RV Tauri features, but is as a whole peculiar. UU Her can hardly be classified as a cepheid or a typical SRd–variable if all available facts are considered.

122.053 **Photographic colorimetry of stellar flares in the Pleiades and Orion aggregates. II.**
L. V. Mirzoyan, O. S. Chavushyan, N. D. Melikyan, R. Sh. Natsvlishvili, V. V. Ambaryan, G. A. Brutyan.
Astrofizika, Tom 19, Vyp. 4, p. 725 – 737 (1983). In Russian. English translation in Astrophysics, Vol. 19, No. 4.

The results of a new series of synchronous photographic observations of stellar flares in the Pleiades and Orion regions in the *UBV* system are presented. During 86 hours of observations 32 flares were detected – 25 in the Pleiades and 7 in the Orion regions. 8 new flare stars were found, 4 in each aggregate. The colour indices (*U–B*) and (*B–V*) of flare radiation were determined in all points of flare light curves where they were available.

122.054 **On the interpretation of multicolour observations of rapid irregular variables with nonperiodic light fadings.**
A. F. Pugach.
Astrofizika, Tom 19, Vyp. 4, p. 739 – 745 (1983). In Russian. English translation in Astrophysics, Vol. 19, No. 4.

Multicolour observations of VX Cas, UX Ori and WW Vul are interpreted on the ground of the hypothesis of light veiling of the photospheric radiation by a variable dust layer. Optical parameters of the layer and relative intensities of photospheric and chromospheric radiation are obtained. Calculated "magnitude – colour–index" relations agree with *UBV* observations.

122.055 **Synchronous UBV observations of UV Ceti flares.**
N. D. Melikyan, N. N. Kilyachkov, V. N. Igolkin.
Astrofizika, Tom 19, Vyp. 4, p. 835 – 838 (1983). In Russian. English translation in Astrophysics, Vol. 19, No. 4.

The results of synchronous UBV observations of UV Ceti flares made in September 1982 are presented. During the 23.5 hours 28 flares were detected, some of them in all U,B,V.

122.056 **Polarimetric observations of μ Cep.**
M. A. Eritsyan.
Astrofizika, Tom 20, Vyp. 1, p. 195 – 199 (1984). In Russian. English translation in Astrophysics, Vol. 20, No. 1.

The correlation between the brightness of the star μ Cep and the degree of polarization of its light is investigated. It is shown that a negative correlation exists for observations in the yellow spectral region and a positive one for those in the blue spectral region.

122.057 **New variable stars in Cygnus.**
S. V. Ignatov.
Perem. Zvezdy, Prilozh., Tom 4, No. 20, p. 211 – 214 (1982). In Russian.

Two variable stars in Cygnus were investigated. SVS 2315 ($21^h17^m.2 + 36°52'$, 1950) is a Mira Ceti type star or a semiregular variable with the period $103^d.5$ and the limits of variability $15^m.0$–($17^m.7$(B). SVS 2381 ($21^h50^m.5 + 39°15'$, 1950) is a SRa–type variable with the period $289^d.3$ and the limits $15^m.7$–$17^m.5$(B).

122.058 **Metal abundances of RR Lyrae stars: results from delta–*S* spectroscopy.**
H. A. Smith.
Publ. Astron. Soc. Pac., Vol. 96, No. 581, p. 505 – 517 (1984).

The $\Delta S$ method is a means of determining the metal abundances of RR Lyrae stars and related variable stars from low–resolution spectroscopy. This approach has been applied to the measurement of metallicities of field and cluster stars in the Galaxy and in the Magellanic Clouds. This paper reviews the $\Delta S$ method, discusses its results and comments on some prospects for future work.

122.059 **The pure absorption spectrum of FG Sagittae in 1980.**
T. A. Kipper.
Sov. Astron. Lett., Vol. 10, No. 2, p. 91 - 92 (1984). English translation of 37.122.079.

122.060 **The extended atmosphere of VZ Cancri, a δ Scuti variable.**
G. A. Garbuzov, A. S. Mitskevich.
Sov. Astron. Lett., Vol. 10, No. 2, p. 92 - 95 (1984). English translation of 37.122.080.

122.061 **Is the flare star EV Lacertae an eclipsing binary?**
G. Sh. Rojzman.
Sov. Astron. Lett., Vol. 10, No. 2, p. 116 - 118 (1984). English translation of 37.122.086.

122.062 **GY Cygni – Klassifikation des Lichtwechsels.**
D. Böhme.
BAV Rundbrief, 33. Jahrg., Nr. 3, p. 105 - 107 (1984).

GY Cyg in GCVS 1969 classified as Is? is now SRb with a period of $130^d$ and mag. (vis) $9^m.5 - 10^m.8$.

122.063 **T Vul – ein Beispiel zur Physik von Cepheiden.**
W. Braune.
BAV Rundbrief, 33. Jahrg., Nr. 3, p. 114 - 117 (1984).

122.064 **IUE observations of BY Draconis.**
C. J. Butler, J. G. Doyle, A. D. Andrews, P. B. Byrne, J. L. Linsky, T. Simon, N. Marstad, M. Rodonò, V. Pazzani, G. Cutispoto.
ESA Spec. Publ., ESA SP–218, p. 243 - 246 (1984). – See Abstr. 012.024.

Phased IUE observations of BY Draconis show no significant modulation of the prominent SWP emission lines over one rotation period. However, a marginally significant anticorrelation of the Mg II flux, and the flux in the LWR 'continuum', with the V light curve is observed, and is interpreted as due to 'plage' type areas over the photospheric spots. Two SWP spectra show sporadic enhancements of the emission lines as has been seen in other IUE spectra of flares.

122.065 **Coordinated IUE and ground–based observations of active stars: flare events on YZ CMi, V1005 Ori, AD Leo and AR Lac.**
M. Rodonò, G. Cutispoto, S. Catalano, J. L. Linsky, D. M. Gibson, A. Brown, B. M. Haisch, C. J. Butler, P. B. Byrne, A. D. Andrews, J. G. Doyle, D. E. Gary, G. W. Henry, G. Russo, A. Vittone, F. Scaltriti, B. Foing.
ESA Spec. Publ., ESA SP–218, p. 247 - 252 (1984). – See Abstr. 012.024.

The authors present a preliminary report on coordinated observations of stellar flares obtained with IUE and several ground–based facilities. The observations include time–resolved IUE and optical spectroscopy, narrow and wide–band optical photometry, IR photometry and microwave observations. Among the major results is the first detection of IR flux decrease, or negative flare, in coincidence with flux increase at all of the other wavelengths.

122.066 **The hydrogen–deficient irregular variable star V348 Sgr.**
N. Kameswara Rao, K. Nandy.
ESA Spec. Publ., ESA SP–218, p. 363 - 366 (1984). – See Abstr. 012.024.

The distribution of the reddening material around V348 Sgr has been studied using the IUE low resolution spectra.

122.067 **An estimate of V348 Sgr effective temperature.**
U. Heber, A. Heck, L. Houziaux, J. Manfroid, D. Schönberner.
ESA Spec. Publ., ESA SP–218, p. 367 - 370 (1984). – See Abstr. 012.024.

IUE low–dispersion spectra of the irregular variable V348 Sgr taken close to maximum light have been compared, on one hand, with those of the B–type stars HD 53138 and HD 58350 and, on the other hand, with those of the extreme helium stars HD 124448 and LS II +33°005. In this paper, the authors concentrate on deriving an effective temperature for the object and study its energy distribution.

122.068 **Results of observations of the variable stars V346 Ori and V351 Ori.**
G. U. Koval'chuk.
Glav. Astron. Obs. Akad. Nauk USSR. Kiev, 46 pp. (1984). In Russian. Abstr. in Ref. Zh., 51. Astron., 8.51.428 (1984).

122.069 **SY Fornacis and the Mira Ceti B phenomenon, II.**
M. W. Feast, P. A. Whitelock, R. M. Catchpole, B. S. Carter.
Observatory, Vol. 104, No. 1062, p. 217 - 220 (1984).

122.070 **The spectra of seven variable stars.**
T. Lloyd Evans.
Observatory, Vol. 104, No. 1062, p. 221 - 223 (1984).

The spectra of seven stars considered to be possible Type II Cepheids or RV Tauri stars show that they belong to other groups. Their properties are unusual in several cases, and attention is drawn to GR Nor which has outstandingly strong Ca II emission.

122.071 **Arythmies stellaires.**
M. Auvergne.
Instabilités hydrodynamiques et applications astrophysiques, p. 319 - 332 (1983). – See Abstr. 012.034.

Contents: Introduction. Les étoiles variables dans un diagramme de Hertzsprung–Russell. Les étoiles normales. Quelques objets bizarres. Conclusion.

122.072 **VZ Aquilae: a study in the evolution of a BL Herculis star.**
J. Christianson.
J. Am. Assoc. Variable Star Obs., Vol. 12, No. 2, p. 54 - 57 (1983).

VZ Aquilae is presented as a BL Herculis star. An increasing period is determined and used to suggest evolutionary changes.

122.073 **A spectral catalogue of southern hemisphere Mira variable stars.**
R. A. Crowe.
J. Am. Assoc. Variable Star Obs., Vol. 12, No. 2, p. 58 - 65 (1983).

The catalogue of Keenan, Garrison, and Deutsch (1974) has been extended by the addition of 483 blue spectrograms of 72 southern hemisphere Mira variables. About 190 direct and image–tube plates at a dispersion of 120 Å/mm were obtained in Chile between 1977 and 1982. Along with the spectral types and photographic magnitudes, emission ratios Hδ λ4101/Hγ λ4340/Hβ λ4861 and eye–estimated intensities of the absorption lines Ca I λ4226, Cr I λ4254, and Sr II λ4077 have been tabulated.

122.074 **Invisible variable stars.**
L. A. Willson.
J. Am. Assoc. Variable Star Obs., Vol. 12, No. 2, p. 77 (1983). Abstract. – See Abstr. 010.021.

122.075 **Revised elements for V1510 Cygni.**
L. Hoecker.
J. Am. Assoc. Variable Star Obs., Vol. 13, No. 1, p. 11 - 14 (1984).

V 1510 Cygni is an RR Lyrae star whose published period, epoch, and amplitude have been modified in the light of more observational evidence.

122.076 **Preliminary results of observations of EU Delphini.**
H. J. Landis.
J. Am. Assoc. Variable Star Obs., Vol. 13, No. 1, p. 21 - 22 (1984).

The first results of photoelectric observations in V light of the small–range SRb star EU Delphini are given. A light curve is shown, representing about two cycles of the star's activity.

122.077 **PZ Aquilae: period changes in a population II Cepheid.**
D. Wallentinsen.
J. Am. Assoc. Variable Star Obs., Vol. 13, No. 1, p. 27 – 30 (1984).
The photometric behavior of the galactic variable PZ Aquilae is examined during the fifty years 1934 – 1983. Using photographic observations from the Maria Mitchell Observatory, two intervals of constant period were found. The current period is 8.7546 days.

122.078 **Semiregular variables: are they chaotic or ringing?**
C. A. Whitney.
J. Am. Assoc. Variable Star Obs., Vol. 13, No. 1, p. 31 – 37 (1984).
Two types of irregular behavior in semiregular variables, labeled randomness and chaos, are described and the results of recent theoretical explorations are illustrated. The need for precise long–term observations to distinguish between multi–periodic and irregular pulsations in semiregular variables is emphasized.

122.079 **Radial velocities of northern hemisphere Miras.**
R. A. Crowe.
J. Am. Assoc. Variable Star Obs., Vol. 13, No. 1, p. 41 (1984).
Abstract. – See Abstr. 010.025.

122.080 **The application of visual observations to the study of a small–amplitude variable star: Rho Cassiopeiae.**
J. R. Percy, D. Keith.
J. Am. Assoc. Variable Star Obs., Vol. 13, No. 1, p. 42 (1984).
Abstract. – See Abstr. 010.025.

122.081 **IR observations of OH/IR stars.**
J. Herman, R. Isaacman, A. Sargent, H. J. Habing.
Astron. Astrophys., Vol. 139, No. 1, p. 171 – 180 (1984).
Results of broad band photometry in eight bands between $\lambda = 3.8$ μm and 20 μm for a number of OH/IR stars are presented. Combination with other available infrared and radio data yields good estimates of the mean luminosities for these (very) long–period variables. Only a small fraction have large luminosities, in excess of $10^4 L_{\odot}$, but most OH/IR stars have luminosities comparable to those of the classical Mira variables. The depth of the 9.7 μm silicate feature, the infrared colour temperature, and the mass loss rate are indicative of the strength of the OH maser. A few objects seem radically different in character from the usual OH/IR stars. They are not variable and appear to be nearby low mass stars with a very large mass loss rate.

122.082 **The evolutionary status of the pulsating helium star BD + 13°3224 (V652 Her).**
C. S. Jeffery.
Mon. Not. R. Astron. Soc., Vol. 210, No. 3, p. 731 – 743 (1984).
Zero–age models and evolutionary sequences for horizontal–branch stars with very low envelope hydrogen abundances ($X < 0.01$) have been constructed. Models with a helium core mass 0.475 $M_{\odot}$ reproduce the observed luminosity and surface temperature of the pulsating helium star BD +13°3224 (V652 Her) very well. The rate of period decrease of BD +13°3224 can also be reproduced by considering the contraction of the models as they evolve from their zero–age position towards the Helium Main Sequence.

122.083 **A survey of Cepheid sizes.**
J. D. Fernie.
Astrophys. J., Vol. 282, No. 2, p. 641 – 649 (1984).
The literature prior to 1982 May has been surveyed for radii of classical Cepheids determined by one or other variation of the Baade–Wesselink method. Of 594 such determinations, 394 for 55 stars are used to define a period–radius relation. This relation is compared with (1) a theoretical *P–R* relation; (2) a *P–R* relation based on 27 classical Cepheids in clusters and associations; (3) a *P–R* relation based on radii determined from mixed–mode pulsations or bumps on velocity curves. These relations are found to be significantly discrepant. It is suggested that the results from Cepheids in clusters/associations are likely the most accurate, but no good reason for the difference from theoretical results has been found.

122.084 **Surface brightness radii, distances, and absolute magnitudes of classical Cepheids.**
W. Gieren.
Astrophys. J., Vol. 282, No. 2, p. 650 – 654 (1984).
High–quality *V, R* photometric data and radial velocity curves of 14 short–period classical Cepheids have been used to obtain their mean radii, distances, and absolute magnitudes by the surface brightness method of Barnes et al. It is shown that the surface brightness technique is now essentially able to yield Cepheid radii in agreement with those found from other methods, in the period range from $4^d$ to $7^d$. For shorter periods, the present results indicate the possibility of surface brightness radii which are too small. It is further shown that the distances derived from the surface brightness method are in agreement with the *P–L–C* scale.

122.085 **Abundances of distant classical Cepheid variables.**
H. C. Harris, C. A. Pilachowski.
Astrophys. J., Vol. 282, No. 2, p. 655 – 666 (1984).
A high–dispersion abundance analysis has been carried out for eight Cepheid variables spanning 8 kpc in galactocentric radius. A differential analysis relative to the comparison supergiant μ Per was used to derive the temperature, microturbulence, and abundances for each star. The spectroscopic abundances are in good agreement with photometric abundances found previously from Washington–system colors and provide support for the abundance gradient in iron among young disk stars of $-0.07$ kpc$^{-1}$.

122.086 **Ultraviolet and visible flare observations of EQ Pegasi B.**
S. L. Baliunas, J. C. Raymond.
Astrophys. J., Vol. 282, No. 2, p. 728 – 732 (1984).
EQ Peg AB (dM4e+dM5.5e) was monitored in the visible at the Whipple Observatory and ultraviolet with *IUE* on 1981 September 2. In the visible spectrophotometry of EQ Peg B the Hβ emission strengthened by a factor of 2 relative to the nearby stellar continuum within a few minutes and decayed over an hour. This flare in EQ Peg B was coincident with the enhancement of ultraviolet emission lines of C IV λ1550, He II λ1640, and C II λ1335 in the combined light of EQ Peg AB. The ultraviolet fluxes during the flare can be interpreted as similar to those either in the thermal phases of large two–ribbon solar flares or in gas cooling quickly from X–ray emitting temperatures.

122.087 **The remarkable multiple mode δ Scuti star BDS 1269A.**
B. J. McNamara, S. J. Horan.
Astrophys. J., Vol. 282, No. 2, p. 741 – 747 (1984).
Over 1600 differential photoelectric Strömgren *b* measurements on BDS 1269A obtained during a 6 month period in 1982 – 1983 have been analyzed using periodiograms and convergent least squares. Seven frequencies are identified in the data set. This frequency set, when combined with other frequencies found in data obtained by Rucinski in 1976, suggests that the main pulsation mode of BDS 1269A is nonradial. The complete frequency representation also includes lower amplitude radial modes.

122.088 **A frequency analysis of the 6.8–min oscillations in the cool magnetic Ap star α Cir.**
D. W. Kurtz, L. A. Balona.
Mon. Not. R. Astron. Soc., Vol. 210, No. 4, p. 779 – 789 (1984).
High speed photometric observations of α Cir have been obtained during 138 hr on 38 nights in 1981 and 1983. A frequency analysis reveals the presence of only two frequencies in the light variations of α Cir down to a noise level of 0.2 mmag. The principal frequency is $f_1 = 8.791347 \pm 0.000008$ hr$^{-1}$ ($P = 409.4936 \pm 0.0004$ s); the secondary frequency is $f_2 = 8.7821 \pm 0.0007$ hr$^{-1}$. A comparison of these two frequencies with appropriately scaled solar models indicates a radial overtone of $k \approx 25$. It is suggested that these two frequencies are associated with pulsation in modes of overtone and order $(k, l)$

and ($k$-1, $l$+2). A phase shift for $f_1$ of $\Delta\phi(V, B-V) = 13° \pm 6°$ is found, the interpretation of which is problematical.

122.089 **The $^{12}C/^{13}C$ ratio of the carbon star V460 Cygni revisited.**
I. R. Little-Marenin, S. J. Little.
Astrophys. J., Vol. 283, No. 1, p. 188 - 191 (1984).
In order to find the reasons for the disagreement between the $^{12}C/^{13}C$ ratios of carbon stars determined by Climenhaga et al. (CHHS) and other authors (Dominy et al.; Olson and Richer; and Johnson et al.), V460 Cygni was reanalyzed using the original data and methods (Minnaert formula) of CHHS. The authors now obtain a $^{12}C/^{13}C$ ratio between 22 and 29 for this star using their original photographic data and methods as compared to the CHHS value of 9. The reasons for disagreement in the isotopic ratio determinations are analyzed.

122.090 **A spectroscopic study of the peculiar giant FK Comae. I. The radial velocity variation and its implications.**
J. K. McCarthy, L. W. Ramsey.
Astrophys. J., Vol. 283, No. 1, p. 200 - 204 (1984).
The authors present observations of the radial velocity behavior of the peculiar, rapidly rotating late-type giant FK Comae. Observations of the Na D lines in FK Comae and several well-studied radial velocity standards were obtained on 15 nights during late winter and spring 1982. Using cross-correlation techniques to search for variations against the standards, the authors set an upper limit of 5 km $s^{-1}$. The implications of this limit on models for FK Comae extant in the literature are discussed in detail, with the conclusion that models requiring accretion from a binary companion are inconsistent with the current data.

122.091 **Determination of the interstellar component of polarized radiation of cepheids.**
T. A. Polyakova.
Pis'ma Astron. Zh., Tom 10, No. 10, p. 749 - 755 (1984). In Russian. English translation in Sov. Astron. Lett, Vol. 10.
The normalized Stokes parameters of polarized radiation for ten observed cepheids are shown to occupy on the $p_x p_y$-plane rosette-shaped regions having three jets. The interstellar polarization radius vector for cepheids with interstellar polarization derived from observations of neighbourhood stars (Z Lac, VX Per) points close to the base of the "rosette". In the case of absence of interstellar polarization (W Vir, RR Lyr) the base of the "rosette" is located near the zero point.

122.092 **A search for evidence of the neon-sodium cycle in the cepheid X Cygni.**
G. Wallerstein, C. A. Pilachowski, H. C. Harris.
Publ. Astron. Soc. Pac., Vol. 96, No. 582, p. 613 - 617 (1984).
The authors have determined abundances of sodium, aluminum, silicon, calcium, and iron from 16 red spectra of 10 Å $mm^{-1}$ of X Cyg to search for a sodium enhancement that might be expected from partial operation of the neon-sodium hydrogen burning cycle. Such burning is possible if the low oxygen and very high nitrogen abundances in X Cyg and other supergiants reported by Luck and Lambert (1981) are correct. The authors find no significant enhancement of sodium indicating that insufficient material that may have burned hydrogen at temperatures in the 30 - 50 × $10^6$K range has reached the surface.

122.093 **Additional radial velocities of the cepheid binary W Sagittarii.**
T. S. Jacobsen, G. Wallerstein, H. A. Abt.
Publ. Astron. Soc. Pac., Vol. 96, No. 582, p. 630 - 633 (1984).
Radial-velocities of W Sgr from 28 coudé spectrograms taken during 1980 - 83 are presented along with velocities of certain standard stars. A mean velocity of $-28.0 \pm 0.7$ km $s^{-1}$ for a mean epoch of 1981.96 is found. The 1980 - 83 data indicate that the orbital period is longer than the 80 years previously suggested and that the spectroscopic orbit cannot be obtained at this time.

122.094 **Period changes and a possible second pulsation mode in BW Vulpeculae.**
A. P. Odell.
Publ. Astron. Soc. Pac., Vol. 96, No. 582, p. 657 - 660 (1984).
BW Vul, the largest-amplitude β Cephei pulsator, has been observed through eight photometric maxima during 1980 - 81, and a new quadratic ephemeris has been derived which includes all known data through 1980. An examination of all historic maxima shows a cyclic variation of the residuals (O-C) with a period around 25 years and an amplitude of 17 minutes. Two hypotheses can be made to explain this phenomenon: light travel time across a binary orbit, or a heating between two pulsation modes of nearly identical period.

122.095 **Erratum: "Emission lines in the spectra of the RV Tauri stars U Monocerotis and AC Herculis" [Publ. Astron. Soc. Pac., Vol. 96, No. 580, p. 432 - 436 (1984)].**
B. W. Bopp.
Publ. Astron. Soc. Pac., Vol. 96, No. 582, p. 674 (1984). See Abstr. 37.122.192.

122.096 **Stellar distances of Mira variables derived from near-infrared photometry parameters.**
L. Celis S.
Astron. J., Vol. 89, No. 9, p. 1343 - 1349 (1984).
The visual absolute magnitude of red variable stars Me spectra depends on the color indices $(V-R)_0$ and $(R-I)_0$ with the relationship $M_V = 1.90(V-R)_0$-4.67 and $M_V = 0.27(R-I)_0{}^3$-2.87, with a $\sigma(M_V) = 0.4$ in both cases. The final expression is $M_V = 0.95(V-R)_0 + 0.135(R-I)_0{}^3 - 3.77$. The color indices-absolute magnitude relation is applicable in different points of light curves for individual stars to obtain their distances with an error of 20%. The interstellar absorption $a_V$, the distance modulus, and the distances of 40 Mira variables were estimated.

122.097 **A search for distant halo RR Lyrae stars.**
A. Saha.
Astrophys. J., Vol. 283, No. 2, p. 580 - 597 (1984). With plates 14 - 16.
The 48 inch (1.2 m) Schmidt telescope on Palomar mountain has been used to search for faint RR Lyrae stars in selected regions of the galactic halo. Results are presented here for three fields centered at $l = 180°$, $b = 24°$; $l = 180°$, $b = 30°$; and $l = 110°$, $b = -30°$. Photoelectric photometry has been used to obtain accurate light curves and ephemerides for all the survey RR Lyrae stars. The survey has also identified a large number of faint eclipsing stars, mainly of the W UMa type.

122.098 **Discovery and preliminary identification of two retrograde nonradial pulsation modes in the Be star μ Centauri.**
D. Baade.
Hvar Obs. Bull., Vol. 7, No. 1, p. 189 - 192 (1983). - See Abstr. 012.037.

122.099 **Observational properties of the β CMa stars.**
J. M. Le Contel.
Hvar Obs. Bull., Vol. 7, No. 1, p. 207 - 213 (1983). - See Abstr. 012.037.

122.100 **Light variations of σ Sco.**
M. Goossens, P. Lampens, D. De Maerschalck, M. Schrooten.
Hvar Obs. Bull., Vol. 7, No. 1, p. 215 - 221 (1983). - See Abstr. 012.037.
The light variability of σ Sco can be represented by two mathematically equivalent but physically different models: (1) two intrinsic pulsations with constant periods and amplitudes; (2) one intrinsic pulsation of which the amplitude and the shape of the mean light curve are modulated by tidal action.

122.101 **Photoelectric photometry of the new β Cephei-variable HD 129557.**
D. vander Linden.
Hvar Obs. Bull., Vol. 7, No. 1, p. 223 - 230 (1983). - See Abstr. 012.037.

122.102 **The helium spectrum variables: the photospheres and magnetospheres of the magnetic B stars.**
C. T. Bolton.
Hvar Obs. Bull., Vol. 7, No. 1, p. 241 – 262 (1983). – See Abstr. 012.037.

The author attempts to describe the general properties of the complex variations that are observed in helium spectrum variables and shows how these can be interpreted with an oblique rotator model.

122.103 **Pulsation of CP2 stars.**
W. W. Weiss.
Hvar Obs. Bull., Vol. 7, No. 1, p. 263 – 273 (1983). – See Abstr. 012.037.

The author discusses some of the known and questionable pulsating CP2 stars.

122.104 **Variability of the lambda–5200 feature in CP2 stars.**
H. Hensberge, H. M. Maitzen.
Hvar Obs. Bull., Vol. 7, No. 1, p. 275 – 281 (1983). – See Abstr. 012.037.

The authors summarize the knowledge about the strength and variability of the λ5200 feature, and argue that magnetic intensification of spectral lines may be responsible for most of this depression. If true, the variation of the λ5200 feature during the rotation cycle in any CP2 star contains information on the magnetic field supplementary to that found by measuring longitudinal fields.

122.105 **Delta Scuti and related stars: review.**
M. Breger, H. Stockenhuber.
Hvar Obs. Bull., Vol. 7, No. 1, p. 283 – 300 (1983). – See Abstr. 012.037.

122.106 **Light curves of four previously unknown Delta Scuti stars.**
C. Waelkens, F. Rufener.
Hvar Obs. Bull., Vol. 7, No. 1, p. 301 – 308 (1983). – See Abstr. 012.037.

The authors have identified four new southern Delta Scuti stars from the Bright Star Catalogue. In this paper they present light curves of these stars obtained with the Geneva Photometer attached to the Swiss Telescope, La Silla, Chile. The authors have performed a preliminary period analysis, using Deeming's (1975) method.

122.107 **Spectroscopic variations of the δ Scuti variables 20 CVn, ϱ Pup, and 38 Eri.**
S. Yang, G. Walker.
Hvar Obs. Bull., Vol. 7, No. 1, p. 309 – 322 (1983). – See Abstr. 012.037.

The HF absorption cell technique has proven to be valuable in determining the small radial velocity variations of δ Scuti variables. It was most essential in obtaining the result which suggests that ϱ Pup is multiperiodic. The small velocity differences between different spectral lines can be used as probes for the structure of pulsating stellar atmospheres.

122.108 **HR 4684: a possible example of resonance in Delta Scuti stars.**
E. Antonello, M. Bossi, G. Guerrero, L. Mantegazza, M. Scardia.
Hvar Obs. Bull., Vol. 7, No. 1, p. 323 – 327 (1983). – See Abstr. 012.037.

The light curve analysis of HR 4684, a Delta Scuti star observed in 1976 at the Brera–Merate Observatory, is presented. The results show three significant periodicities probably bound by a simple relation: the highest frequency seems to be the sum of the other ones. This fact represents an observational indication for the actuality of the parametric resonance mechanism suggested by Dziembowski for the pulsating dwarf stars.

122.109 **The pulsational amplitudes of Delta Scuti stars.**
E. Antonello.
Hvar Obs. Bull., Vol. 7, No. 1, p. 329 – 334 (1983). – See Abstr. 012.037.

The author applied the multiple regression analysis and other methods of the multivariate statistical analysis (factor analysis, discriminant analysis) to the study of δ Scuti and other variable stars. In particular, the author is interested in deriving a relation among the amplitude of light variations and other physical parameters of low amplitude δ Scuti stars. It is well known that the amplitude increases roughly with the period of these stars, and the author has tried to define quantitatively this dependence.

122.110 **The light variations of the Delta Scuti star $\vartheta^2$Tauri.**
E. Antonello, L. Mantegazza.
Hvar Obs. Bull., Vol. 7, No. 1, p. 335 – 338 (1983). – See Abstr. 012.037.

122.111 **Two color diagram for Delta Scuti stars and HR 1287 in UBV.**
J. K. Badalia, H. S. Gurm.
Hvar Obs. Bull., Vol. 7, No. 1, p. 339 – 343 (1983). – See Abstr. 012.037.

122.112 **Observations and a new interpretation for the Delta Scuti variable star VZ Cancri.**
A. N. Cox, B. J. McNamara, W. Ryan.
Astrophys. J., Vol. 284, No. 1, p. 250 – 256 (1984).

Extensive numbers of new photoelectric data have been collected and analyzed to investigate the modal behavior of VZ Cnc. Its previously observed periods yield a period ratio of 0.80 implying pulsation in the first and second radial overtones. The aim of this investigation is to redetermine the pulsation characteristics of VZ Cnc in an attempt to establish its modes of pulsation, their stability, and whether this star demonstrates helium settling within the envelope of the star.

122.113 **Intense X–ray flares from active stellar systems: EV Lacertae and HD 8357.**
C. Ambruster, W. A. Snyder, K. S. Wood.
Astrophys. J., Vol. 284, No. 1, p. 270 – 277 (1984).

X–ray light curves (0.5 – 20 keV), each approximately 1 week in duration, are presented for the dMe flare star, EV Lac, and the RS CVn binary system, HD 8357, based on data obtained with the A–1 Sky Survey experiment on HEAO 1. Four flares in all were seen, two from EV Lac and two from HD 8357. Variable low–level quiescent emission was also observed from both objects. These light curves represent the first unambiguous observations of more than one flare plus intervening quiescent X–ray emission from active stellar systems. Estimates of time averaged X–ray flaring luminosities show that for at least some epochs, up to an order of magnitude more energy is expended in X–ray flaring than in optical flaring.

122.114 **Phase lags and pulsation modes of classical Cepheids.**
N. R. Simon.
Astrophys. J., Vol. 284, No. 1, p. 278 – 281 (1984).

The first–order phase lag between light and velocity is calculated from Fourier decompositions of the variations of 12 classical Cepheids. The phase lags have a narrow range, with maximum expansion velocity always retarded with respect to maximum light. The average value of the first–order phase lag is found to be approximately –0.3, close to the theoretical value obtained from the models of Simon and Davis. The stars SU Cas and AZ Cen appear exceptional. Detailed analysis of AZ Cen confirms the identification of this star as an overtone pulsator and demonstrates the possible role of the first–order phase lag as a discriminator of mode.

122.115 **Time series infrared spectroscopy of Mira variables. II. CO Δυ = 3 in eight Mira variables and one SRa variable.**
K. H. Hinkle, W. W. G. Scharlach, D. N. B. Hall.
Astrophys. J., Suppl. Ser., Vol. 56, No. 1, p. 1 – 17 (1984).

CO Δυ = 3 lines have been analyzed in time series spectra of nine long–period variables. Velocity, temperature, and column density derived from the lines are presented for each star. The program stars have periods of 293 to 445 days and visual amplitudes of 2 to 8 mag. The CO results for all the stars show that the spectroscopic and photometric variability results from an outwardly propagating wave driven by a general stellar oscillation. The wave traverses the photosphere as a shock.

122.116 **Periods in the light variation of σ Scorpii.**
M. Jerzykiewicz, C. Sterken.
Mon. Not. R. Astron. Soc., Vol. 211, No. 2, p. 297 – 312 (1984).

Analysis is carried out of an extensive series of differential *uvby* observations of this β Cephei variable. Shortward of 0.5 day four periodic components are found. In order of decreasing light-range their periods are equal to 0.246836, 0.23967, 0.16745 and 0.15595 day. Longward of 0.5 day, the results are somewhat less definite.

122.117 **Variable circumstellar obscuration of the carbon star R Fornacis.**
M. W. Feast, P. A. Whitelock, R. M. Catchpole, G. Roberts, M. D. Overbeek.
Mon. Not. R. Astron. Soc., Vol. 211, No. 2, p. 331 – 337 (1984).

In 1983 the carbon Mira-type variable R For became unusually faint in the visible and infrared. This is interpreted as a change in circumstellar obscuration. For absorption by graphite spheres the required particle radius is 0.15 μm.

122.118 **FM Com: has this δ Scuti–star variable frequency spectra?**
M. Paparó, G. Kovács.
Astrophys. Space Sci., Vol. 105, No. 2, p. 357 – 367 (1984).

Component frequencies of FM Com were analysed using 43 hr *V* band photoelectric observations. Both the usual Fourier and the autoregressive extrapolation techniques applied to the closely spaced nights of observation yielded variable frequency patterns. Comparison of the observed and synthetic light curves supports the suggestion of the nonstationary light variation of the star.

122.119 **Wesselink radii of 33 classical cepheids.**
G. R. Ivanov.
Astrophys. Space Sci., Vol. 105, No. 2, p. 369 – 372 (1984).

The radii of 33 classical cepheids are determined. The author has used the simultaneous radial velocity and photoelectric observation of 15 cepheids of Gieren (1981, 1982). He has suggested that the best phase alignment between the light and the radial velocity curves in different epochs of 18 cepheids appears when the points of the phases at equal radius on the $\Delta V$, $\Delta(B-V)$ diagram are transformed from a loop into a straight line.

122.120 **The Population II Cepheids.**
G. Wallerstein, A. N. Cox.
Publ. Astron. Soc. Pac., Vol. 96, No. 583, p. 677 – 691 (1984).

The authors describe the Population II Cepheids in terms of their period distribution, light curves, velocity curves, chemical composition, galactic kinematics, atmospheric shock waves, and location in the color–magnitude diagram. The relationship of normal Population II Cepheids to similar variables such as the RV Tauri stars and anomalous Cepheids is also described.

122.121 **Spectra of seven long–period Cepheids in the Small Magellanic Cloud.**
G. Wallerstein.
Astron. J., Vol. 89, No. 11, p. 1705 – 1706 (1984).

Spectra with a dispersion of 40 Å/mm of seven Cepheids in the Small Magellanic Cloud have been classified on the MK system, and their types compared with their $T_1$–$T_2$ colors. Their spectral types are found to be 4.1 ± 1.3 or 2.6 ± 1.3 subtypes too early for their colors for reddenings of 0.02 and 0.06 in $B$–$V$, respectively. Transforming this difference into a metal deficiency yields [A/H] = –0.65 or –0.56 for each reddening value. Radial velocities of the seven Cepheids have been measured and compared with the velocity of neutral hydrogen in the same line of sight. No significant difference between the Cepheid and H I velocities was found.

122.122 **The β Cephei stars 12 and 16 Lacertae in 1979.**
M. Jerzykiewicz, K. J. Borkowski, B. Musielok.
Acta Astron., Vol. 34, No. 1, p. 21 – 39 (1984).

About 650 blue filter photoelectric observations of 12 and 16 Lacertae are presented. An analysis of these data is then carried out. In the case of 12 Lacertae no substantial change in the frequency spectrum is detected relative to the earlier results of Jarzebowski et al. (1980). As far as 16 Lacertae is concerned, it is found that the decrease of all three pulsation amplitudes, noted in 1977, has probably continued through 1979. The observations of 16 Lacertae include fragments of three successive eclipses of the star by its invisible companion. These new data confirm the ephemeris derived previously from the 1965 and 1977 observations.

122.123 **Simultaneous spectroscopic and photoelectric observations of β Cephei stars. III. HR 6684 (V2052 Oph).**
M. Kubiak, W. Seggewiss.
Acta Astron., Vol. 34, No. 1, p. 41 – 46 (1984).

Simultaneous radial velocity and *uvby* photoelectric observations show that the minimum of radius occurs 0.016 day after the maximum of brightness and 0.009 day after the maximum of $c_1$–index. Radial velocity and brightness variations do not show evidences of perturbations in simple oscillatory motion, so the observed phase–shift seems to be related directly to the character of oscillations.

122.124 **Observational determination of pulsation modes and photometric masses of Cepheid variables.**
A. Opolski.
Acta Astron., Vol. 34, No. 2, p. 225 – 262 (1984).

On the basis of Wesselink radii and dereddened colour indices, $\langle B-V\rangle_0$, the $P_F$–$R_{Max}$–$\langle B-V\rangle_0$ relationship for fundamental pulsators is established. It is then possible to single out the following five Cepheids as first overtone pulsators: BG Cru, BF Oph, V482 Sco, Y Sgr, and U Aql. By means of photometric gravity the "photometric masses" are determined for Cepheids investigated by Pel (1978). A discussion of these quantities enables to divide the stars according to their instability strip crossing. The mass–luminosity–temperature relations for different crossings are then fixed and the following problems related to Cepheids considered: (1) the differences of pulsation constants, (2) period changes, (3) space distribution, and (4) light amplitudes.

122.125 **Photometry of the pulsating variable HD 37819.**
T. D. Padalia, S. K. Gupta.
Acta Astron., Vol. 34, No. 2, p. 303 – 313 (1984).

Comprehensive light curves of the newly discovered δ Scuti star HD 37819 in *U, B*, and *V* filters have been presented and discussed. An improved period of $0\overset{d}{.}1892666$ has been obtained. The colours and physical parameters have been determined.

122.126 **Cepheids: cosmic yardsticks, celestial mysteries?**
J. R. Percy.
Sky Telesc., Vol. 68, No. 6, p. 517 – 519 (1984).

122.127 **The photometric behaviour of the young disk carbon star TW Horologii: determination of its physical characteristics.**
P. Bouchet.
Astron. Astrophys., Vol. 139, No. 2, p. 344 – 352 (1984).

Broad–band photometry from the *U* band up to 30 μm, carried out over a 4–year period, and spectrophotometry from 1 μm to 13 μm are presented for the carbon star TW Hor. It is found that the star is an $SR_b$ variable with a period of 157 days. The observed spectral energy distribution and the known distance of the star are used to determine its effective temperature, $T_{eff}$ = 3250K, its bolometric luminosity, $L = 9\times10^3 L_\odot$, its radius, $R = 3\times10^2 R_\odot$ and their variations. The initial mass at the beginning of the AGB phase of the star is deduced from its membership to the cluster NGC 1252. The results are in agreement with the models of pulsation in the nonadiabatic approximation. A raw comparison of the data with theories of stellar evolution is made.

122.128 **Studies of Cepheid type variability. II. Fourier description of RR Lyrae variables in the globular cluster ω Centauri.**
J. O. Petersen.
Astron. Astrophys., Vol. 139, No. 2, p. 496 – 506 (1984).

Photographic mean light curves for 75 RRab variables and 55 RRc variables in the globular cluster ω Centauri are Fourier analysed. A discussion of the low order amplitude ratios and phase differences show good agreement with similar results for field RR Lyrae stars. The first overtone RRc variables in ω Cen show clear systematic changes in Fourier parameters with period. It is suggested that these changes can be interpreted as a bump progression sequence probably of a new type. It is noted that the long period RRab variables in ω Cen do not share the unique pulsation properties of the small-amplitude variable XZ Ceti recently discussed by Simon and Teays. Evidence is shown to exist for a Cepheid-like progression in the fundamental mode variables in ω Cen from periods about $0^{d}.5$ through the RRab region towards the resonance expected for periods about $1^{d}.5$. A quantitative comparison of the progression sequences in Type I and II Cepheids is attempted with some success.

122.129 **Discovery of a possibly rotation-induced resonant coupling of two nonradial pulsation modes in γ Arae (B1 Ib).**
D. Baade, R. Ferlet.
Astron. Astrophys., Vol. 140, No. 1, p. 72 – 76 (1984).

The first discovery that is based on line profile variations, of stable nonradial pulsations in an early type supergiant is reported. Two periods, $P_1 = 0.87$ d and $P_2 = 0.17$ d have been found in γ Ara and identified with two sectorial modes of orders $l_1 = 2$ and $l_2 = 10$. The two pulsation modes of γ Ara are resonantly coupled, and this second discovery of the "magic" number 5 as the ratio of the two mode orders supports the previous conclusion (Baade, 1984) that for this ratio the mode coupling is the strongest. The episodic character of the line profile variability and the absence of stable nonradial pulsations in the very similar star ϱ Leo (Smith and Ebbets, 1981) are hence reinterpreted as a lack of stabilization through resonant coupling. The unusually rapid rotation of γ Ara and the normal $v \sin i$ of ϱ Leo are in accord with the earlier suggestion (Baade, 1984) that resonant mode coupling of nonradial pulsations is favoured by rapid rotation.

122.130 **Photométrie photographique des Miras.**
M. O. Mennessier, J. M. Roques.
Bull. Assoc. Fr. Obs. Etoiles Variables, No. 30, p. 153 (1984).

122.131 **Zu Beziehungen der Helligkeits- und Polarisationsschwankungen bei My Cephei.**
D. Böhme.
Sterne, 60. Band, Heft 5, p. 315 – 318 (1984).

122.132 **A model of the R Coronae Borealis phenomenon.**
A. F. Pugach.
Sov. Astron., Vol. 28, No. 3, p. 288 – 293 (1984). English translation of 37.122.132.

122.133 **EV Lacertae: relationship of the flares to the nonflare variability.**
G. Sh. Rojzman.
Sov. Astron., Vol. 28, No. 3, p. 293 – 298 (1984). English translation of 37.122.133.

122.134 **Constraining dMe flare characteristics with Ca I/Ba II strengths.**
W. B. Weaver.
Publ. Astron. Soc. Pac., Vol. 96, No. 584, p. 798 (1984). Abstract. – See Abstr. 010.281.

122.135 **Ultraviolet spectrophotometry of the fading of R Coronae Borealis.**
A. V. Holm, J. Hecht, C.-C. Wu, B. Donn.
Bull. Am. Astron. Soc., Vol. 16, No. 3, Part I, p. 733 (1984). Abstract. – See Abstr. 010.065.

122.136 **Optical photometry and UV spectroscopy of the flare star Gl 735 (= V1285 Aql).**
P. B. Byrne, J. G. Doyle, C. J. Butler, A. D. Andrews.
Mon. Not. R. Astron. Soc., Vol. 211, No. 3, p. 607 – 616 (1984).

The authors present optical photometry of flares and possible spot variations on the star Gl 735. Consideration of the energetics of the optical flares suggests that the star may be underactive when compared to other flare stars of the same spectral type. This conclusion is not supported however by an estimate of the radiative loss rate from the lower chromosphere due to Mg II *h* and *k*. Radiative losses from the upper chromosphere, transition zone and corona are also normal for a flare star of its spectral type.

122.137 **The period-luminosity relation for Mira variables and the distance of the Large Magellanic Cloud.**
M. W. Feast.
Mon. Not. R. Astron. Soc., Vol. 211, No. 3, p. 51P – 55P (1984).

New data by Wood, Bessell & Paltoglou are shown to provide strong confirmation of the LMC Mira *P–L* relation discovered by Glass & Lloyd Evans. At *K*, both carbon and non-carbon Miras fit the same relation. Results from Cepheids, OB stars, main-sequence fitting, RR Lyrae and Mira variables are consistent with a true LMC distance modulus of $18.50 \pm 0.10$ if $\langle M_v(\mathrm{RR}) \rangle = +0.6$ at all metallicities including the metal-rich globular clusters containing the Miras studied by Menzies & Whitelock. However the results would also be consistent with the metal-rich RR Lyrae variables being ~0.15 mag fainter than metal-poor ones. There is some suggestion of such a difference in ω Cen.

122.138 **Confirmation of the decreased pulsational amplitude of Polaris.**
K. W. Kamper, N. R. Evans, R. W. Lyons.
J. R. Astron. Soc. Can., Vol. 78, No. 5, p. 173 – 177 (1984).

The decrease in the pulsational amplitude of Polaris recently reported by Arellano Ferro (1983) has been confirmed with new radial-velocity observations. The authors have also made additional analyses of previously published Lick velocities to clarify the past pulsational history of this Cepheid.

122.139 **The quasi-cepheid nature of ϱ Cassiopeiae.**
J. R. Percy, D. Keith.
J. R. Astron. Soc. Can., Vol. 78, No. 5, p. 206 (1984). Abstract. – See Abstr. 010.401.

122.140 **Cepheid radii from infrared photometry.**
D. L. Welch, N. R. Evans.
J. R. Astron. Soc. Can., Vol. 78, No. 5, p. 207 (1984). Abstract. – See Abstr. 010.401.

122.141 **Light variability of σ Scorpii.**
M. Goossens, P. Lampens, D. De Maerschalck, M. Schrooten.
Astron. Astrophys., Vol. 140, No. 2, p. 223 – 229 (1984).

Van Hoof's light observations are used to study the light variability of the spectroscopic binary and β Cephei star σ Scorpii. Period analyses of the light observations in individual years reveal two oscillations with periods and mean light curves that do not show any detectable year to year variation. The light variability of σ Sco can be modelled by two mathematically equivalent but physically different models: (1) two intrinsic pulsations with constant periods and amplitudes, (2) one intrinsic pulsation of which the amplitude and the shape of the mean light curve are modulated by tidal action.

122.142 **The presence of $Si^{3+}$ and $C^{3+}$ in the wind of AB Aur.**
C. Catala, A. Talavera.
Astron. Astrophys., Vol. 140, No. 2, p. 421 – 426 (1984).

The Herbig Ae variable AB Aur (spectral type A0) exhibits a line spectrum characteristic of a stellar wind. IUE observations of the Si IV and C IV resonance lines in AB Aur are reported. These lines are strong and in absorption. Two models can explain the presence of $Si^{3+}$ and $C^{3+}$ in the wind of AB Aur: a "remote chromosphere" model, with a temperature rise up to $10^5$K, occurring beyond the point where the ionizing radiation begins to depart from the local Planck function, and a "deep chromosphere" model, with a temperature rise up to 15,000 – 18,000K, located just above the stellar photosphere. Several additional observational arguments are in favor of the "deep chromosphere" model.

122.143 **New evidence for mass loss in classical cepheids.**
M. J. Stift.
Astron. Astrophys., Vol. 140, No. 2, p. 445 – 448 (1984). In French.

The author discusses the cepheid mass anomalies in the light of mass loss. Interpreting observational evidence that pulsational masses are significantly smaller than evolutionary masses at least in the higher mass range as a consequence of mass loss in previous stages of evolution, he tries to evaluate the effect of mass loss on the observed cepheid properties. It turns out that cepheids having suffered mass loss exhibit longer periods at a given luminosity than canonical cepheids. Comparing these predictions with the observed frequency–period distribution of cepheids in our Galaxy, the LMC, and the SMC, the author finds it no longer necessary to postulate a two-component birthrate function. Finally the author discusses alternative scenarios, especially convective overshooting and turbulent diffusion, which lead to similar effects on the cepheid frequency–period distribution as mass loss but fail to predict the observed relations between cepheid, red supergiant and Wolf–Rayet luminosities.

122.144 **Mode identification for the Delta Scuti variable HR 7222.**
P. López de Coca, R. Garrido, A. Rolland.
Astron. Astrophys., Suppl. Ser., Vol. 58, No. 2, p. 441 – 445 (1984).

*B*-band photoelectric photometry for the δ Scuti type variable HR 7222 has been carried out. A synthetic light curve with two periods with a ratio of 0.85 reproduces all the data. It appears that HR 7222 pulsates in the first and second overtones.

122.145 **Determination of radii of cepheids. III. Radial velocities, spectroscopic orbit and dimension of SU Cyg.**
M. Imbert.
Astron. Astrophys., Suppl. Ser., Vol. 58, No. 3, p. 529 – 536 (1984). In French.

The classical cepheid SU Cyg, known to be a double system, was observed from 1979 May to 1982 June with the Coravel photoelectric radial velocity spectrometer. The radial velocities plus the *UBV* measurements of Szabados, Fernie and Moffett and Barnes have also allowed to calculate the radius of the cepheid by the method of Wesselink. A radius of 43.9 $R_{\odot}$ for the cepheid is derived.

122.146 ***uvby* photometry of Sigma Scorpii.**
C. Sterken.
Astron. Astrophys., Suppl. Ser., Vol. 58, No. 3, p. 657 – 663 (1984).

Approximately 1900 *uvby* magnitude differences σ minus τ Scorpii, derived from photoelectric observations on 24 nights between 29 April 1972 and 10 May 1974 are listed.

122.147 **TT Arietis.**
IAU Circ., Nos. 3992, 4004, 4010, 4011, 4018 (1984).

122.148 **Radial oscillations and their theoretical interpretations.**
A. N. Cox.
Theoretical problems in stellar stability and oscillations, p. 1 – 20 (1984). – See Abstr. 012.065.

This paper reviews the basic properties and our current theoretical understanding of radially pulsating variable stars. Observationally, it is the position in the H–R diagram that determines whether the oscillations are radial or nonradial. Stars on the cool and luminous side of this diagram select radial modes; the smaller radii stars are nonradial pulsators. Except for a few supergiant stars, all stars with a surface effective temperature of less than about 8000K and more luminous than the sun are radial pulsators, and they are driven by hydrogen and helium ionization effects.

122.149 **A bump progression sequence in the first overtone RR Lyrae variables in ω Centauri.**
J. O. Petersen.
Theoretical problems in stellar stability and oscillations, p. 21 – 25 (1984). – See Abstr. 012.065.

Fourier decomposition parameters are derived for 55 first overtone RR Lyrae variables in the globular cluster ω Centauri from mean light curves taken from Martin's 1938 paper. The low order phase differences show clear systematic changes with period. It is noted that these changes can be interpreted as a bump progression sequence probably of a new type. A preliminary discussion indicates that the Christy wave may be the primary bump mechanism.

122.150 **Three interesting stars.**
N. R. Simon.
Theoretical problems in stellar stability and oscillations, p. 26 – 28 (1984). – See Abstr. 012.065.

This note briefly presents some results of a recent study of the population I Cepheids CO Aur, HR 7308 and of the probable RR Lyrae variable XZ Ceti.

122.151 **The international photometric observing campaign on BW Vul: first results.**
C. Sterken, M. Snowden, A. Young, T. Margrave, M. Alvarez, P. Antonelli, F. Catalano, M. Chahbenderian, C. Chavarria, H. L. Cohen, V. Costa, E. de Lara, A. J. Delgado, D. Ducatel, R. Fried, H. H. Fu, R. Garrido, S. Gonzalez, H. Hensberge, S. W. Lee, J. M. Le Contel, J. Manfroid, S. Naftilan, R. Peniche, J. H. Pena, S. Ratajczyk, A. Rolland, W. Sandmann, J. P. Sareyan, W. Schlosser, E. Szuszkiewicz, Z. Tunca, J. C. Valtier, D. Vander Linden.
Theoretical problems in stellar stability and oscillations, p. 103 – 107 (1984). – See Abstr. 012.065.

122.152 **Optical spectroscopy of upper and lower photospheric regions of BW Vulpeculae.**
A. Young, I. Furenlid, C. Sterken.
Theoretical problems in stellar stability and oscillations, p. 108 – 109 (1984). – See Abstr. 012.065.

122.153 **Critical observations of β Cephei pulsation.**
I. Furenlid, A. Young, C. Sterken.
Theoretical problems in stellar stability and oscillations, p. 110 – 114 (1984). – See Abstr. 012.065.

This paper presents an observational approach to β Cephei pulsation, which is essentially confined to studies of the continuum emission of these stars. Reference to specific observations concerns BW Vul, the β Cep star of largest known amplitude.

122.154 **Photospheric velocity fields of early-type supergiants. A preliminary observational report.**
D. Baade.
Theoretical problems in stellar stability and oscillations, p. 115 – 119 (1984). – See Abstr. 012.065.

In a first step to identify the nature of the semi-regular variability of most supergiants 90 high signal-to-noise Reticon spectra were obtained mainly of photospheric lines in 12 early-type supergiants. Results are: (1) Like γ Ara, also the second very rapidly rotating star of the sample, HR 3090, is pulsating in two modes of fairly different nonradial orders. (2) All narrower lined stars with more than two observations display variable line cores. This can easily be reproduced by model calculations of low-amplitude, low-order nonradial pulsations. (3) Apparently without being correlated with any other quantity, the line profiles of 50% of the stars show an additional, persistent (blue or red) overall asymmetry which, however, is variable.

122.155 **Five-minute *p* modes detected in Doppler shift measurement on Alpha Centauri.**
E. Fossat, B. Gelly, G. Grec.
Theoretical problems in stellar stability and oscillations, p. 229 – 233 (1984). – See Abstr. 012.065.

122.156 **Detection of an ultraviolet continuum in T Lyn.**
N. Sanduleak.
Inf. Bull. Variable Stars, No. 2548, 1 p. (1984).

122.157 **FK Ophiuchi = V2 (NGC 6273).**
N. N. Samus'.
Inf. Bull. Variable Stars, No. 2555, 3 pp. (1984).

122.158 **Flare activity of HD 282773 (G5).**
G. U. Kovalchuk (*G. U. Koval'chuk*), A. F. Pugach.
Inf. Bull. Variable Stars, No. 2557, 3 pp. (1984).

122.159 **Electrophotometry of DH Peg and AV Peg.**
I. P. Alania (*I. F. Alaniya*).
Inf. Bull. Variable Stars, No. 2558, 3 pp. (1984).

122.160 **1983 – 1984 photometry of three semi–regular variables.**
L. J. Boyd, R. M. Genet, D. S. Hall.
Inf. Bull. Variable Stars, No. 2563, 3 pp. (1984).

122.161 **Ultraviolet ANS photometry for the five suspected long period Ap stars HD 89822, HD 137389, HD 187474, HD 204411 and HD 221760.**
W. Schöneich.
Inf. Bull. Variable Stars, No. 2569, 4 pp. (1984).

122.162 **Unusual light variations of the carbon star AFGL 2881.**
A. Alksnis, A. Rudzinskis.
Inf. Bull. Variable Stars, No. 2572, 3 pp. (1984).

122.163 **V 1050 Aql not of U Geminorum but of Mira type.**
H. Gessner.
Inf. Bull. Variable Stars, No. 2577, 1 p. (1984).

122.164 **New flare star.**
A. R. Good.
Inf. Bull. Variable Stars, No. 2581, 1 p. (1984).

122.165 **The photometric variability of Epsilon Persei.**
J. R. Percy, M. Bietenholz, A. Fullerton.
Inf. Bull. Variable Stars, No. 2588, 3 pp. (1984).

122.166 **Times of maximum light of the β CMa star BW Vulpeculae.**
M. Rios Herrera, M. Rios Berumen, J. H. Pena, R. Peniche.
Inf. Bull. Variable Stars, No. 2597, 2 pp. (1984).

122.167 **On the supposed periodicity of IP Persei.**
W. Wenzel.
Inf. Bull. Variable Stars, No. 2598, 2 pp. (1984).

122.168 **Long period variable carbon stars at galactic longitude 86°.**
A. Alksnis, Z. Alksne.
Inf. Bull. Variable Stars, No. 2600, 3 pp. (1984).

122.169 **A further search for the δ–Sct–like variability of the Ap star HD 10088.**
T. J. Kreidl.
Inf. Bull. Variable Stars, No. 2602, 4 pp. (1984).

122.170 **HD 91816: a new BY Draconis star.**
B. W. Bopp, D. S. Hall, J. L. Africano, B. D. Goodrich, G. W. Henry, W. S. Barksdale Jr.
Inf. Bull. Variable Stars, No. 2604, 5 pp. (1984).

122.171 **Period and light curve of the RRc star VZ Virginis.**
R. Luthardt.
Inf. Bull. Variable Stars, No. 2608, 2 pp. (1984).

122.172 **The absence of interchanges in AC Herculis.**
G. E. Erleksova.
Inf. Bull. Variable Stars, No. 2614, 2 pp. (1984).

122.173 **Photoelectric observations of AD Leonis.**
M. K. Tsvetkov, A. P. Antov, L. N. Mavridis, G. A. Asteriadis.
Inf. Bull. Variable Stars, No. 2618, 3 pp. (1984).

122.174 **New flare stars in the Mon I association.**
L. Rosino, E. S. Parsamian (*Eh. S. Parsamyan*), H. S. Chavushian (*O. S. Chavushyan*).
Inf. Bull. Variable Stars, No. 2620, 2 pp. (1984).

122.175 **Flare stars in Orion.**
N. D. Melikian (*N. D. Melikyan*).
Inf. Bull. Variable Stars, No. 2621, 4 pp. (1984).

122.176 **Unusual "red" flare on the star T 231 = AZ Ori.**
N. D. Melikian (*N. D. Melikyan*), H. S. Chavushian (*O. S. Chavushyan*), R. Sh. Natsvlishvili.
Inf. Bull. Variable Stars, No. 2622, 3 pp. (1984).

122.177 **On the He weak variable HD 28843.**
J. Manfroid, G. Mathys, A. W. J. Cousins.
Inf. Bull. Variable Stars, No. 2625, 2 pp. (1984).

122.178 **The P–L–C relation of galactic field Cepheids and the P–L and P–C relations of open cluster and association Cepheids.**
T. Ciurla, A. Opolski.
Inf. Bull. Variable Stars, No. 2629, 2 pp. (1984).

122.179 **UV Ceti: synchronous UBV flare observations.**
N. D. Melikian (*N. D. Melikyan*).
Inf. Bull. Variable Stars, No. 2630, 6 pp. (1984).

122.180 **Near–infrared light elements of 13 very cool Mira variables.**
G. W. Lockwood.
Inf. Bull. Variable Stars, No. 2631, 2 pp. (1984).

122.181 **Discovery that HR 454 is a variable star.**
W. S. Barksdale Jr., L. J. Boyd, R. M. Genet, S. Chang, D. S. Hall, S. I. Ingvarsson, W. T. Persinger, H. D. Powell, H. J. Stelzer, N. F. Wasson, K. W. Ziegler.
Inf. Bull. Variable Stars, No. 2632, 4 pp. (1984).

122.182 **A faint flare star in Vela.**
D. J. MacConnell, J.–C. Mermilliod.
Inf. Bull. Variable Stars, No. 2633, 2 pp. (1984).

122.183 **Stellar flares in Taurus.**
A. S. Hojaev (*A. S. Khodzhaev*).
Inf. Bull. Variable Stars, No. 2635, 4 pp. (1984).

122.184 **New flare stars in Taurus.**
A. S. Hojaev (*A. S. Khodzhaev*).
Inf. Bull. Variable Stars, No. 2636, 4 pp. (1984).

122.185 **On the time distribution of flares on UV Ceti.**
N. D. Melikian (*N. D. Melikyan*), A. Grandpierre.
Inf. Bull. Variable Stars, No. 2638, 3 pp. (1984).

122.186 **A new pulsating variable in Vela.**
L. Kohoutek.
Inf. Bull. Variable Stars, No. 2640, 2 pp. (1984).

122.187 **Determination of the magnetic field in the star V474 Mon.**
Yu. S. Romanov, S. N. Udovichenko, M. S. Frolov.
Inf. Bull. Variable Stars, No. 2645, 2 pp. (1984).

122.188 **Photoelectric observations of R CrB.**
D. Böhme.
Inf. Bull. Variable Stars, No. 2646, 1 p. (1984).

122.189 **Photoelectric monitoring of Butler's flare star.**
G. J. Malcolm.
Inf. Bull. Variable Stars, No. 2647, 2 pp. (1984).

122.190 **Flare stars: postflare behaviour of emission lines.**
G. A. Gurzadyan.
Astrophys. Space Sci., Vol. 106, No. 1, p. 1 – 34 (1984).
The problem of the postflare behaviour of chromospheric emission lines in UV Cet–type flare stars is considered. It is shown that the postflare behaviours of different emission lines differ essentially from one another. Theoretical postflare light curves are derived for the most important chromospheric lines: helium, hydrogen, ionized magnesium and calcium. The conditions of the intensification of emission lines during the flare are examined, as is the possibility of the existence of an empirical relationship between chromospheric concentration and flare frequency.

122.191 **RR Lyrae variables in Baade's Window.**
B. M. Blanco.
Astron. J., Vol. 89, No. 12, p. 1836 – 1867 (1984). With plates 98 – 105.
A blink survey for Galactic bulge RR Lyraes in Baade's Window has been carried out in a 777–arcmin$^2$ area that includes part of the area surveyed by Baade. Data are presented for 77 RR Lyraes, including 58 that had been previously detected by Baade, and for a few eclipsing or other short–period variables. Light curves and identification charts for all of these stars are given. The period–frequency and period–amplitude relationships, as well as estimated $\Delta S$ values, suggest that although the Galactic bulge RR Lyraes show an appreciable range in metallicity, the majority are relatively metal rich as compared to the globular cluster M3.

122.192 **Microvariability in the HR diagram and two examples of very low amplitude stars.**
G. Burki.
Space research prospects in stellar activity and variability, p. 69 – 76 (1984). – See Abstr. 012.078.
At the present time, the Geneva catalogue contains 1168 stars brighter than $m_V = 10$ and with at least 10 good measurements. From these well measured stars, three samples have been chosen: (1) Stars with $\sigma_V \leqslant 0^m.005$. These 306 stars can be considered as "stable" stars. (2) Stars with $0^m.012 \leqslant \sigma_V \leqslant 0^m.025$. These 173 stars are microvariable stars. About one third of them belongs to the list of new variable stars by Rufener and Bartholdi (1982). (3) Stars with $\sigma_V > 0^m.025$. Almost all these 131 stars are listed in Kukarkin's catalogue or appear in the list by Rufener and Bartholdi. Finally, two cases of low amplitude multi–periodic variable stars are presented.

122.193 **The study of stellar global oscillations by Ca II H and K variations.**
R. W. Noyes.
Space research prospects in stellar activity and variability, p. 113 – 115 (1984). Abstract. – See Abstr. 012.078.

122.194 **Photoelectric B and V band photometry of comparison stars for eight Mira–type variable star fields.**
M. Dumont.
GEOS Circ., SR 5, 10 pp. (1984).
The comparison star sequences of 8 Mira–type stars have been observed in two colours, B and V in the Geneva photometric system at the Junfraujoch Observatory. The stars involved were R Cas, V Cas, T Cep, R Aql, W Peg, T Her, Z Cyg and W Lyr.

122.195 **Period changes in dwarf cepheids, III. SZ Lyncis.**
B. Szeidl.
Commun. Konkoly Obs., No. 84, p. 251 – 287 (1983).
2371 photoelectric UBV observations obtained between 1962 and 1979 and twelve radial velocity measurements made in 1971 are reported. The long–periodic variation in time of maximum light discovered by van Genderen has been confirmed and the values of its period ($P_1$ = 1170 days = 3.20 yr) and semiamplitude (A = 0.00540 day = 467 sec) were improved. A secular change in the pulsation period ($\beta = 10^{-12}$ day · cycle$^{-1}$) has been found. The observations are compatible with the hypothesis that SZ Lyn is a single–lined spectroscopic binary.

122.196 **BVRI photoelectric photometry of Cepheids in the Small Magellanic Cloud.**
J. A. R. Caldwell, I. M. Coulson.
S.Afr. Astron. Obs., Circ., No. 8, p. 1 – 58 (1984).
1060 new photoelectric BVRI (Cousins system) observations of 53 SMC Cepheids are presented. Older photoelectric data from the literature are combined with the new data to produce light curves for a total of 81 SMC Cepheids with P > 8.4d.

122.197 **Supernova progenitors and Be stars: stellar variability from a 21 century perspective.**
K. P. Hertzog.
Mon. Not. R. Astron. Soc., Vol. 209, No. 3, p. 533 – 541 (1984).
The possible value, for contemporary stellar investigations, of the 21–century–old observations in the *Almagest* star catalogue is discussed. No progenitors of historic supernovae were recorded, which strongly suggests that SN 1006 was not type II and casts doubt on the high–mass interpretation of its remnant's X–ray spectrum. One noteworthy feature of the catalogue is the excessive number of Be stars among those having observations consistent with a variability hypothesis. It is concluded that the catalogue observations may not be as worthlessly unreliable as seems to have been assumed, and that the unique potential afforded by the 21 century baseline makes the catalogue worthy of serious consideration.

122.198 **Photoelektrische Beobachtung von Theta and Eta Lyrae.**
D. Böhme.
Mitt. Veränderliche Sterne, Band 10, Heft 3, p. 68 (1984).

122.199 **The luminosity scale of Cepheid variable stars: a revision.**
E. G. Schmidt.
Astrophys. J., Vol. 285, No. 2, p. 501 – 514 (1984).
The period–luminosity–color relation for classical Cepheids is discussed in terms of a recent redetermination of the distance moduli of open clusters with Cepheid members. The distance moduli of eight clusters with well established Cepheid members were determined from Stroemgren four–color and H$\beta$ photometry of the B stars. Zero points are derived from these data for the period–luminosity–color relation on the *UBV* system and the period–luminosity relations for near–infrared magnitudes. The present relations yield absolute magnitudes which are fainter by 0.4–0.6 mag than those now commonly used. The source of this discrepancy is discussed. The implications of this revision of the Cepheid luminosity scale for the cosmic distance scale are discussed.

122.200 **Spectroscopy and spectrophotometry of distant halo RR Lyrae stars.**
A. Saha, J. B. Oke.
Astrophys. J., Vol. 285, No. 2, p. 688 – 697 (1984).
Spectra of the faint RR Lyrae stars found in a recent survey (Saha 1984) have been used to derive chemical abundances and line–of–sight velocities to individual objects. Spectrophotometry at the minimum light phase has also been obtained. By comparison with suitable model atmospheres, the interstellar reddening and extinction to some of the ab–type RR Lyrae stars have been derived individually. The results are seen to be consistent with other ways of estimating extinction. Reliable distance estimates are thus possible.

122.201 **New Mira–type variable star.**
T. M. Seregina, S. Yu. Shugarov.
Astron. Tsirk., No. 1301, p. 8 (1983). In Russian.

122.202 **Supplement to the catalogue of RR Lyrae stars.**
V. G. Derevyagin, L. E. Lysova, B. N. Firmanyuk.
Astron. Tsirk., No. 1308, p. 7 – 8 (1984). In Russian.

122.203 **ZZ Scorpii.**
M. G. Shirokova.
Astron. Tsirk., No. 1313, p. 8 (1984). In Russian.

122.204 **Flare of the star BD +26°730.**
A. V. Kurchakov, F. K. Rspaev.
Astron. Tsirk., No. 1323, p. 1 – 2 (1984). In Russian.

122.205 **The peculiarities of pulsations of SW And.**
A. N. Rudenko.
Astron. Tsirk., No. 1323, p. 4 – 6 (1984). In Russian.

122.206 **On the character of the light variability of SY Sgr.**
V. I. Kardopolov, G. K. Filip'ev.
Astron. Tsirk., No. 1325, p. 3 – 4 (1984). In Russian.

122.207 **New eruptive stars.**
N. E. Kurochkin.
Astron. Tsirk., No. 1325, p. 5 – 7 (1984). In Russian.

122.208 **Photometric variability of the flare star HII 1306 in the Pleiades.**
G. Sh. Rojzman.
Astron. Tsirk., No. 1328, p. 4 – 7 (1984). In Russian.

122.209 **Hγ and Hδ emission in the spectrum of Mira Ceti during its 1975 maximum.**
E. A. Depenchuk, V. F. Karamysh, R. I. Chuprina.
Astron. Tsirk., No. 1330, p. 4 – 5 (1984). In Russian.

122.210 **Photoelectric observations of Pleione (BU Tau) from September 1980 to March 1984.**
A. S. Sharov, V. M. Lyutyj.
Astron. Tsirk., No. 1333, p. 1 – 2 (1984). In Russian.

122.211 **A search for 183 GHz emission from water in late–type stars.**
T. B. H. Kuiper, P. N. Swanson, D. F. Dickinson,
E. N. Rodriguez Kuiper, P. Zimmermann.
Astrophys. J., Vol. 286, No. 1, p. 310 – 313 (1984).
A search was made for 183 GHz line emission from water vapor in the direction of 12 Mira and two semiregular variables. Upper limits to the emission are in the range of 2000 to 5000 Jy. It is estimated that thermal emission from the inner regions of late–type stellar envelopes will be on the order of 10 Jy. Maser emission, according to one model, would be an order of magnitude stronger.

122.212 **Shock waves in Mira variables. I. Emission–line spectra.**
M. W. Fox, P. R. Wood, M. A. Dopita.
Astrophys. J., Vol. 286, No. 1, p. 337 – 349 (1984).
A catalog of Balmer emission–line profiles (Hγ, Hδ, Hζ, and Hη) is presented for a sample of nine Mira variables ranging in period from 150 to 394 days. These were obtained at very high dispersion with phase coverage of about half a cycle centered around maximum. Observations of emission lines of Mg, Si, and Fe are also described. Variations with phase and cycle number of line profile shape and flux are discussed in relation to a simple shock model in which a spherically symmetric shock wave is propagating outward through the atmosphere of a pulsating Mira variable.

122.213 **Die hellsten blauen Veränderlichen in der Großen Magellanschen Wolke.**
O. Stahl, B. Wolf, C. Leitherer, F.–J. Zickgraf, J. Krautter,
M. de Groot.
Mitt. Astron. Ges., Nr. 62, p. 246 (1984). Abstract. – See Abstr. 010.341.

122.214 **ZZ Canis Minoris as a symbiotic star.**
B. W. Bopp.
Publ. Astron. Soc. Pac., Vol. 96, No. 585, p. 894 – 896 (1984).
The M6 giant ZZ CMi, listed as a semiregular variable in the GCVS, is shown to have spectroscopic and photometric characteristics similar to the symbiotic star EG And.

122.215 **On the period–luminosity–colour relation for galactic Cepheids.**
B. Cester, C. Marsi.
Astrophys. Space Sci., Vol. 107, No. 1, p. 167 – 176 (1984).
A period–luminosity–colour relation is derived from the data of 29 galactic Cepheids with known distances collected from the literature, and is then compared with the simpler period–luminosity relation. The insignificant reduction in the dispersion seems to argue against the introduction of the colour term. It is, however, demonstrated that this evidence may be fictitious and due to the use of Cepheids, whose membership in a star aggregate is still doubtful. The exclusion of these stars leads to a PLC relation which shows a significantly smaller dispersion than that given by the corresponding PL relation.

122.216 **Light variations of WW Vul and their interpretation.**
G. V. Zajtseva, P. F. Chugajnov.
Astrofizika, Tom 20, Vyp. 3, p. 447 – 463 (1984). In Russian. English translation in Astrophysics, Vol. 20, No. 3.
Results of the photoelectric measurements of the brightness of the irregular variable WW Vul carried out in 1967 – 1982 in the *UBV* system are presented.

122.217 **Model of the phenomena responsible for variations of polarization and brightness of μ Cep.**
T. A. Polyakova.
Astrofizika, Tom 21, Vyp. 1, p. 125 – 135 (1984). In Russian. English translation in Astrophysics, Vol. 21, No. 1.
A model is proposed explaining variations of brightness and intrinsic polarization of μ Cep by two active antipodal regions rotating with the star. The polarization is caused by scattering of starlight on dust streams ejected from spots in active regions. The dust particles are oriented when moving through gas along the magnetic field lines above the bipolar active regions. The parameters of polarization sources were chosen within known values to obtain maximum observed polarization. On the basis of this model brightness and polarization variations of μ Cep are forecasted for the period of the next five or six years. Also the results of observations of μ Cep for VI.1981 – VI.1983 are given.

122.218 **On the determination of the distribution function of stellar flare frequency.**
G. A. Arutyunyan.
Astrofizika, Tom 21, Vyp. 1, p. 163 – 173 (1984). In Russian. English translation in Astrophysics, Vol. 21, No. 1.
A possibility of generalization of the method suggested by Ambartsumyan (1978) for the determination of the star distribution function depending on mean flare frequency is discussed. Formulae are given for using the information of chronology of high–order flares for this purpose. To estimate the accuracy of the proposed method a numerical experiment has been realized, in the process of which an ensemble of flare stars has been modelled with known frequency distribution. The results of the mentioned experiments and following treatments are shown on diagrams.

122.219 **Space reddenings for six Cepheid variables.**
D. G. Turner.
J. R. Astron. Soc. Can., Vol. 78, No. 6, p. 229 – 240 (1984).
Colour excesses are derived for six galactic Cepheids from reddenings found for neighbouring stars which seem likely to be physically associated with each variable. Their derived reddenings are in good agreement with other, independent, estimates from the literature.

122.220 **OH maser emission from warm variable stars.**
J. D. Fix, M. J. Claussen.
Astrophys. J., Lett. Ed., Vol. 287, No. 1, p. L35 – L37 (1984).
The authors have discovered main line (1665 MHz/1667 MHz) OH emission from three warm variable stars: TW Aql, TX Per, and RV Tau. These stars are spectral class G when they are warmest and are considerably hotter than nearly all known stellar masers.

122.221 **Polarization properties and time variations of the SiO maser emission of Omicron Ceti (Mira).**
J. S. Miller, F. O. Clark, T. H. Troland.
Astrophys. J., Vol. 287, No. 2, p. 892 – 898 (1984).

The authors have measured the polarization properties of the $\upsilon = 1$, $J = 2$–1, SiO circumstellar maser emission from Mira over a period of 3.5 years. As in previous reports, spectra are presented of Stokes parameter $I$ and linear polarization. The line profiles from Mira are made up of a small number of narrow features which may be discerned in the profile, each having a well–defined position angle of maximum linear polarization. Individual features seem to persist for an optical period, then disappear at or near optical maximum. Flux variations in the majority of individual profile features appear to follow the optical cycle, as does the total integrated flux.

**Veränderliche Sterne.**
See Abstr. 003.023.

**Maria Mitchell's observations of Mira, 1856 – 1859.**
See Abstr. 004.039.

**Why do we prefer photoelectric photometry?**
See Abstr. 013.004.

**Ground–based long–term photoelectric photometry as support for space observations of variables.**
See Abstr. 013.057.

**E.V.E. project (Etude de la Variabilité des Etoiles au Pôle Sud).**
See Abstr. 013.058.

**On the need to go to space to study stellar variability and activity.**
See Abstr. 013.060.

**Is the metal contribution to the astrophysical opacity incorrect?**
See Abstr. 022.071.

**Period finding in gapped data and its application to the pulsating λ Bootis type star 29 Cygni.**
See Abstr. 036.144.

**An analytical model of the SiO maser velocity spectrum.**
See Abstr. 064.054.

**Pulsation periods and critical frequencies in low gravity yellow supergiant models.**
See Abstr. 065.034.

**Stellar acoustics. II. Pulse resonance in giant star models.**
See Abstr. 065.035.

**Rotating, pulsating stars.**
See Abstr. 065.053.

**Convection in pulsating stars. V. Limit cycle characteristics.**
See Abstr. 065.059.

**Bump Cepheid models.**
See Abstr. 065.097.

**Nonlinear radial pulsations in R Coronae Borealis stars.**
See Abstr. 065.098.

**On the rapid oscillations of Ap stars.**
See Abstr. 065.110.

**On the rapidly oscillating Ap stars.**
See Abstr. 065.111.

**Rapidly oscillating Ap stars: an alternative model.**
See Abstr. 065.112.

**Line profiles calculated for the oblique pulsator model of Ap stars.**
See Abstr. 065.113.

**What will asteroseismology teach us?**
See Abstr. 065.129.

**The dimension of a variable star.**
See Abstr. 065.131.

**Light and line profile variations due to r–mode pulsations with an application to the ZZ Ceti star G117–B15A.**
See Abstr. 065.147.

**Radial velocities of CH Cygni in 1981.**
See Abstr. 111.013.

**Transient components in the Lyman lines of hot stars. Evidence for the ejection of shells or puffs on very short timescales.**
See Abstr. 112.007.

**SiO flux measurements of variable stars.**
See Abstr. 112.029.

**Photometric variations of some early Be stars.**
See Abstr. 112.040.

**Photometric variability of Be stars.**
See Abstr. 112.041.

**The southern Be star photometric campaign: a first report.**
See Abstr. 112.042.

**A note on the variability of λ Eri and the pulsation in Be stars.**
See Abstr. 112.044.

**Rapid spectroscopic variations in γ Cas and ζ Oph.**
See Abstr. 112.045.

**SiS in circumstellar shells.**
See Abstr. 112.049.

**Rapid spectroscopic variability of the Be star HR 9070: evidence for double periodicity?**
See Abstr. 112.055.

**Photometric observations of active chromosphere stars: II Peg, HD 175742, HD 199178.**
See Abstr. 112.071.

**Luminosities and mass loss rates of OH/$H_2O$ maser stars.**
See Abstr. 112.082.

**The photometric variability of B stars: a general approach.**
See Abstr. 113.023.

**Weak–lined stars in the Michigan spectral survey. I. A–, F–, and G–type stars.**
See Abstr. 113.047.

**Possible light variation of comparison star for BY Dra: BD + 51°2410.**
See Abstr. 113.063.

**CASPEC (*Cassegrain Echelle Spectrograph*) and IUE: a perfect match.**
See Abstr. 114.084.

**Variable blue supergiants in the Large Magellanic Cloud: R 84, R 85, and R 99.**
See Abstr. 114.090.

**New results for HD 161796.**
See Abstr. 114.114.

**A re–calibration of the luminosities of early–type stars: its effect on the Cepheid luminosity scale.**
See Abstr. 115.011.

**99 δ–Cepheiden–Geschwindigkeiten im HRD.**
See Abstr. 115.018.

**Progress on rotational modulation studies of continuum and line fluxes.**
See Abstr. 116.032.

**Variabilité rapide des systèmes binaires cataclysmiques.**
See Abstr. 117.107.

**Eruptive binaries. XI. Disk–radius variations in U Gem.**
See Abstr. 117.127.

**The UV variations of the symbiotic star CH Cygni from 1978 to 1981.**
See Abstr. 117.164.

**Spectrophotometry of the outburst and quiescent states of the dwarf novae RX And and KT Per.**
See Abstr. 117.165.

**A study of the early–type semi–detached binary system u Herculis.**
See Abstr. 117.206.

**Non–radial pulsator in the binary system Y Cam.**
See Abstr. 119.033.

**On the connection between Herbig–Haro objects and flare stars in the solar neighbourhood.**
See Abstr. 121.012.

**Observations of variable stars in the Scorpius T1 association. II. Irregular variables.**
See Abstr. 123.007.

**G238–53: a new pulsating DA white dwarf.**
See Abstr. 126.054.

**Discovery of oxygen in the PG1159 degenerate stars: a direct evolutionary link to O VI planetary nebula nuclei and confirmation of pulsation theory.**
See Abstr. 126.073.

**The effective temperature of the white–dwarf star and ZZ Ceti candidate Wolf 485A.**
See Abstr. 126.086.

**An investigation of the stellar association containing the 1.95 day Cepheid SU Cassiopeiae.**
See Abstr. 152.005.

**On proper motions and distribution of flare stars in the Pleiades.**
See Abstr. 153.010.

**Search for variable stars in the young open cluster NGC 6871.**
See Abstr. 153.036.

**Search for B–type variable stars in open clusters.**
See Abstr. 153.038.

**Yellow giants in young clusters. II. A comparison of observation with theory.**
See Abstr. 153.049.

**Variable stars in the globular cluster Messier 28.**
See Abstr. 154.001.

**A possible P–L relation for RR–Lyrae variables in globular clusters.**
See Abstr. 154.014.

**A new investigation of the variable stars in the globular cluster Messier 9.**
See Abstr. 154.039.

**The period–luminosity relation in the Magellanic Clouds.**
See Abstr. 156.010.

**The distance to NGC 2403 based on near–infrared observations of Cepheids.**
See Abstr. 157.015.

**The Cepheid variables in NGC 300.**
See Abstr. 157.104.

# 123 Variable Stars (Surveys, Lists of Observations, Charts, etc.)

123.001 **New variable stars.**
D. A. Stepanyan.
Perem. Zvezdy, Tom 21, No. 5, p. 691 – 694 (1982). In Russian.
In the course of a search for galaxies with ultraviolet continuum the author has discovered five new variable stars.

123.002 **The W Serpentis stars: an observational program.**
R. E. Wilson, J. B. Rafert, N. L. Markworth.
I.A.P.P.P. Commun., No. 16, p. 1 – 6 (1984).

123.003 **RZ Eridani needs a light curve.**
D. S. Hall.
I.A.P.P.P. Commun., No. 16, p. 8 – 9 (1984).

123.004 **Tableaux des observations reçues à l'AFOEV en avril, mai et juin 1984.**
Bull. Assoc. Fr. Obs. Etoiles Variables, No. 29, p. 114 – 144 (1984).

123.005 **10th report on visual survey of NSV stars suspected to be eclipsing.**
K. Locher.
BBSAG Bull., No. 73, p. 6 (1984).

123.006 **Observations of variable stars in the association of Scorpius–Ophiuchus. I. Regular variable stars.**
V. Satyvoldiev.
Perem. Zvezdy, Prilozh., Tom 4, No. 20, p. 109 – 126 (1982). In Russian.
Results of photometric investigations of 22 variable stars are given. For 17 of them the elements of the light curves are determined. For 5 variables the elements are improved.

123.007 **Observations of variable stars in the Scorpius T1 association. II. Irregular variables.**
V. Satyvoldiev.
Perem. Zvezdy, Prilozh., Tom 4, No. 20, p. 127 – 153 (1982). In Russian.
The results of investigation of light curves of 64 irregular variables based on photographic observations are presented.

123.008 **On the variability of seven stars in the Cygnus T4 association region.**
V. Satyvoldiev.
Perem. Zvezdy, Prilozh., Tom 4, No. 20, p. 154 – 161 (1982). In Russian.
172 photographic observations of the Cyg T4 association region were carried out. The variability of seven previously suspected variables CSV 5360, 5370, 5428, 8659, V837 Cyg, V1239 Cyg, V1338 Cyg is confirmed. The light elements for V833 Cyg and V1239 Cyg are checked.

123.009 **New results of SA 9 region variable stars SVS 1139–1154.**
N. E. Kurochkin.
Perem. Zvezdy, Prilozh., Tom 4, No. 20, p. 163 – 165 (1982). In Russian.

123.010 **New variable star SVS 2451 in Cygnus.**
I. A. Dubyago.
Perem. Zvezdy, Prilozh., Tom 4, No. 20, p. 209 (1982). In Russian.

123.011 **Investigation of five variable stars in Virgo.**
S. F. Yarikov, V. P. Goranskij.
Perem. Zvezdy, Prilozh., Tom 4, No. 20, p. 231 – 235 (1982). In Russian.

123.012 **Results of observations of variable stars sent by amateur astronomers.**
Perem. Zvezdy, Prilozh., Tom 4, No. 20, p. 275 – 282 (1982). In Russian.

123.013 **HD 207739. Der neue Veränderliche.**
M. Fernandes.
BAV Rundbrief, 33. Jahrg., Nr. 3, p. 108 – 110 (1984).
Further observations in addition to those reported in IBVS 2536 revealed fluctuations of HD 207739 with the following elements: min = JD 2444482.92 + $69^{d}.97 \cdot E$. The period must be doubled if variability is assumed to come from the binary nature of the star.

123.014 **The Orion variables: a symphony of delicacy and brilliance.**
D. H. Levy.
J. Am. Assoc. Variable Star Obs., Vol. 12, No. 2, p. 66 – 68 (1983).

123.015 **Osservazioni visuali di sei stelle variabili.**
L. Talamoni, A. Bertocco, G. Gambato.
Coelum, Vol. 52, N. 5, p. 293 – 303 (1984).

123.016 **Nuove variabili nei Gemelli.**
R. Di Luca.
G. A.A.B., N. 76, p. 9 – 10 (1984).

123.017 **Tableaux des observations reçues à l'AFOEV pendant les mois de juillet, août et septembre 1984.**
Bull. Assoc. Fr. Obs. Etoiles Variables, No. 30, p. 156 – 202 (1984).

123.018 **11th report on visual survey of NSV stars suspected to be eclipsing.**
K. Locher.
BBSAG Bull., No. 74, p. 6 (1984).

123.019 **Beobachtungsergebnisse der Berliner Arbeitsgemeinschaft für Veränderliche Sterne e.V. (BAV).**
J. Hübscher, E. Mundry.
BAV Mitt., Nr. 38, 12 pp. (1984).
This 17th compilation of BAV results contains 661 observed minima and maxima on 217 variable stars including 61 photoelectric results.

123.020 **NSV 2229.**
IAU Circ., No. 3998 (1984).

123.021 **V1343 Aquilae.**
IAU Circ., No. 3963 (1984).

123.022 **Five bright new variable stars.**
L. J. Boyd, R. M. Genet, D. S. Hall.
Inf. Bull. Variable Stars, No. 2546, 5 pp. (1984).

123.023 **Automatic Photoelectric Telescope: first quarter 1984 observations.**
L. J. Boyd, R. M. Genet, D. S. Hall.
Inf. Bull. Variable Stars, No. 2561, 3 pp. (1984).

123.024 **New non-stable stars in Orion.**
R. Sh. Natsvlishvili.
Inf. Bull. Variable Stars, No. 2565, 4 pp. (1984).

123.025 **Two new variable stars near Chi Ursae Majoris.**
G. Romano.
Inf. Bull. Variable Stars, No. 2592, 1 p. (1984).

123.026 **A possible BL Herculis variable in Cetus.**
J. F. Waugh.
Inf. Bull. Variable Stars, No. 2601, 2 pp. (1984).

123.027 **A new variable star (No. 218) near the globular cluster NGC 6401.**
A. Terzan.
Inf. Bull. Variable Stars, No. 2609, 4 pp. (1984).

123.028 **A possible Delta Scuti variable in NGC 6405.**
H. Schneider.
Inf. Bull. Variable Stars, No. 2626, 2 pp. (1984).

123.029 **IRC + 10371, a new variable star.**
W. P. Bidelman.
Inf. Bull. Variable Stars, No. 2628, 1 p. (1984).

123.030 **Veränderliche im Feld β Andromedae.**
L. Meinunger.
Mitt. Veränderliche Sterne, Band 10, Heft 3, p. 56 – 68 (1984).
On plates of the field β And taken with the 40/160 and 40/195 cm astrographs of Sonneberg, 14 new variable stars have been found. For these objects and some other variable stars the type of variability, light curves, coordinates, and charts are given.

123.031 **Ist LD 23 ein Bedeckungsveränderlicher?**
D. Böhme.
Mitt. Veränderliche Sterne, Band 10, Heft 3, p. 75 (1984).

123.032 **Visuelle Beobachtung junger Veränderlicher mit frühem Spektraltyp in den Jahren 1977 bis 83.**
D. Böhme.
Mitt. Veränderliche Sterne, Band 10, Heft 3, p. 76 – 78 (1984).

123.033 **List of variable stars discovered in the USSR and preliminary SVS designations.**
Astron. Tsirk., No. 1303, p. 7 – 8 (1983). In Russian.

123.034 **CD Herculis.**
N. N. Samus'.
Astron. Tsirk., No. 1315, p. 3 – 4 (1984). In Russian.

123.035 **Information on photoelectric observations of variable stars deposited at the Odessa Astronomical Observatory.**
E. N. Makarenko.
Astron. Tsirk., No. 1329, p. 6 – 7 (1984). In Russian.

123.036 **Observations of three variable stars in Cygnus.**
S. Yu. Shugarov.
Astron. Tsirk., No. 1329, p. 8 (1984). In Russian.

123.037 **List of variable stars discovered in the USSR and preliminary SVS designations.**
Astron. Tsirk., No. 1332, p. 8 (1984). In Russian.

123.038 **Five suspected β Cephei stars revisited.**
J. R. Percy.
J. R. Astron. Soc. Can., Vol. 78, No. 6, p. 241 – 250 (1984).
The following suspected β Cephei stars have been observed photometrically or rediscussed: HDE 232999, HD 34626, 43818, 43837 and 190467. The period and amplitude of the photometric variability of these five stars can be understood in the general context of stellar variability, even though none of them appears to be a classical β Cephei star.

**Charts for southern variables. Series No. 17.**
See Abstr. 002.018.

**106th list of minima of eclipsing binaries.**
See Abstr. 119.019.

**107th list of minima of eclipsing binaries.**
See Abstr. 119.055.

**New variables in M80.**
See Abstr. 154.059.

**Eight new variable stars in the globular cluster M 15.**
See Abstr. 154.070.

# 124 Novae

124.001 **Grain growth and destruction in novae.**
R. M. Mitchell, A. Evans.
Mon. Not. R. Astron. Soc., Vol. 209, No. 4, p. 945 – 954 (1984).
The authors investigate the effects of carbon and hydrogen ionization on grain growth and destruction in novae. Grain charge effects are shown to play a significant role in terminating grain growth and initiating destruction concurrently with the ionization of hydrogen. Grain surface reactions may be an important destructive process, particularly in regions of high density. Hydrogen atoms and hydrocarbon surface groups on carbon grains should lead to infrared features, the observation of which would confirm the presence of hydrogenated carbon grains in circumnova outflows.

124.002 **UV and optical spectroscopy of classical novae during outburst.**
H. Drechsel, J. Rahe.
ESA Spec. Publ., ESA SP–218, p. 371 – 375 (1984). – See Abstr. 012.024.
High and low dispersion UV spectra (1200 – 3200 Å) of three recent classical novae (N Sgr 1982, N Mus 1983, and N Ser 1983) have been obtained during early stages of decline. High dispersion IUE spectra of N Mus 1983 were obtained a few weeks and more than a year after the outburst in the nebular phase. Simultaneous optical spectroscopy was carried out for N Sgr 1982 and for N Mus 1983. The line spectra and continua are discussed. The electron density and temperature of the dilute envelope of N Sgr 1982 were derived. Some conclusions about CNO overabundances, shell geometry, and speed classes of the different objects are drawn.

124.003 **Novae in M87.**
IAU Circ., No. 3974 (1984).

124.004 **RS Ophiuchi.**
IAU Circ., No. 4010 (1984).

124.005 **Information on six novae in Sagittarius.**
A. Sarajedini.
Inf. Bull. Variable Stars, No. 2587, 5 pp. (1984).

124.006 **Ein Katalog und Atlas galaktischer Novae.**
H. W. Duerbeck.
Mitt. Astron. Ges., Nr. 62, p. 285 (1984). Abstract. – See Abstr. 010.341.

124.007 **Novae and galactic chemical evolution.**
M. Peimbert, A. Sarmiento.
Astron. Express, Vol. 1, Issue 3, p. 97 – 103 (1984).
It is estimated that the ejected shell masses of novae are about one order of magnitude smaller than previously thought. The implications of this result for galactic chemical evolution are analyzed. It is found that novae may be important sources for the interstellar enrichment of $^{15}N$ but not of $^{14}N$.

**Nova Vulpeculae 1979 = PU Vulpeculae**

124.101 **Spectral variation and the expanding envelope of PU Vulpeculae.**
T. Iijima, S. Ortolani.
Astron. Astrophys., Vol. 136, No. 1, p. 1 – 9 (1984).
The peculiar variable star PU Vul showed significant spectral changes in 1982. The mean radial velocity of absorption lines changed from $+30$ km s$^{-1}$ in May to $-50$ km s$^{-1}$ in November. Strong emission lines with P Cygni profile of Hα, Fe I, Fe II, Ca I, Ba II, Sc II, and La II were detected on a high dispersion spectrum taken in November 1982. The mean radial velocity of the emission lines in the spectrum is about $+30$ km s$^{-1}$. The blue–shift of the absorption lines and the emergence of the emission lines suggest the presence of an expanding envelope. This object probably entered an active period in 1982. Some points of similarity between PU Vul and the peculiar F type supergiant ϱ Cas are pointed out.

124.102 **Spectroscopy of PU Vulpeculae, the novalike object Kuwano, in 1981 – 1983.**
E. A. Kolotilov.
Sov. Astron. Lett., Vol. 10, No. 2, p. 118 – 121 (1984). English translation of 37.124.162.

124.103 **Optical variability of PU Vulpeculae (nova–like Kuwano's object 1979) in 1983.**
E. A. Kolotilov.
Pis'ma Astron. Zh., Tom 10, No. 8, p. 609 – 612 (1984). In Russian. English translation in Sov. Astron. Lett., Vol. 10.
Results of photoelectric *UBV* observations of PU Vul made in March – December 1983 are presented. The star brightness in *U*, *B*, *V* did not show significant variations during the first half of the year. From July to October 1983 the UV flux has been increasing by $\Delta U \approx 0^{m}.5$, that was followed by a $\approx 0^{m}.15$ drop to the end of the year. In the same time a small brightness decrease $\Delta V \approx 0^{m}.1$ and appearance of a slight instability of *B* and *V* brightness, probably of a flare type, were observed.

124.104 **Spectroscopic observations of PU Vul.**
Z. Liu.
Inf. Bull. Variable Stars, No. 2576, 2 pp. (1984).

### Nova Vulpeculae 1984 No. 1

124.121 **Nova Vulpeculae 1984.**
E. Schweitzer.
Bull. Assoc. Fr. Obs. Etoiles Variables, No. 29, p. 102 (1984).

124.122 **Nova Vulpeculae 1984.**
R. Lukas.
BAV Rundbrief, 33. Jahrg., Nr. 3, p. 111 – 112 (1984).

124.123 **Ergänzungen zur Nova Vul 1984.**
M. Fernandes.
BAV Rundbrief, 33. Jahrg., Nr. 3, p. 113 (1984).

124.124 **Nova Vulpecula 1984 No. 1.**
IAU Circ., Nos. 3963, 3964, 3968, 3969, 3971, 3976 – 3979, 3985, 3993, 3999, 4004, 4008, 4023 (1984).

124.125 **Nova Vulpeculae 1984 (Wakuda Object).**
Yamamoto Circ., Nos. 2019 – 2022 (1984). In Japanese.

124.126 **Nova Vulpeculae 1984.**
Br. Astron. Assoc. Circ., No. 642 (1984).

124.127 **Photographic and photoelectric observations of Nova Vulpeculae 1984 near maximum.**
R. H. Schult.
Inf. Bull. Variable Stars, No. 2578, 2 pp. (1984).

124.128 **Position and distance of Nova Vulpeculae 1984.**
H. W. Duerbeck, M. Geffert, B. Nelles, R. Dümmler, M. Nolte.
Inf. Bull. Variable Stars, No. 2641, 2 pp. (1984).

### Nova Vulpeculae 1984 No. 2

124.131 **Nova Vulpeculae 1984 No. 2.**
IAU Circ., Nos. 4023, 4024 (1984).

124.132 **Nova Vulpeculae (Collins).**
Yamamoto Circ., No. 2032 (1984). In Japanese.

124.133 **Nova Vulpeculae (2) 1984.**
Br. Astron. Assoc. Circ., No. 646 (1984).

### Nova Muscae 1983

124.141 **Observations of Nova Muscae 1983 from 1200 Å – 10 μm during its early decline stage.**
J. Krautter, K. Beuermann, C. Leitherer, E. Oliva, A. F. M. Moorwood, E. Deul, W. Wargau, G. Klare, L. Kohoutek, J. van Paradijs, B. Wolf.
Astron. Astrophys., Vol. 137, No. 2, p. 307 – 326 (1984).

Observations of Nova Muscae 1983 over a wide spectral range from 1200 Å to 10 μm in the time interval January 21, 1983 to March 4, 1983 are presented. Two evolutionary phases were covered: diffuse enhanced and Orion–phase. Spectroscopic observations were carried out in the visual spectral range at medium and high resolution, in the IR range at low resolution, and in the UV range at low and high resolution with the IUE. Broad band photometry was carried out in both the visual (*UBV*, Walraven) and IR (*JHKLMN*) spectral ranges.

124.142 **Infrared and optical observations of Nova Mus 1983.**
P. A. Whitelock, B. S. Carter, M. W. Feast, I. S. Glass, D. Laney, J. W. Menzies, J. Walsh, P. M. Williams.
Mon. Not. R. Astron. Soc., Vol. 211, No. 2, p. 421 – 432 (1984).

Extensive optical (*UBVRI*) and infrared (*JHKL*) photometry of Nova Mus 1983 obtained over a period of 300 days is tabulated. Infrared and optical spectra are described. Although by classical definition this was a fast nova its later development was slower than for typical objects of this class. Surprisingly the development of infrared thermal dust emission, which has been predicted for an object with a decay rate $\dot{m}_V \sim 0.06$ mag day$^{-1}$, did not occur. Throughout the period covered, the infrared emission was characteristic of a bound–free plus free–free plasma continuum with emission lines. The absolute magnitude at maximum was $M_{vis} \sim -7.6$ and the distance of Nova Mus is about 5 kpc.

124.143 **Observation of soft X–rays from Nova Muscae 1983 with the EXOSAT satellite.**
J. Krautter, K. Beuermann, H. Ögelman.
MPE Prepr., No. 29, p. 7 (1984). Abstract. Paper presented at the Symposium on X–Ray Astronomy '84, Bologna, Italy, June 26 – 30, 1984.

124.144 **The detection of X–rays from Nova Muscae 1983 with the EXOSAT satellite.**
H. Ögelman, K. Beuermann, J. Krautter.
Astrophys. J., Lett. Ed., Vol. 287, No. 1, p. L31 – L34 (1984).

Nova Muscae 1983 was observed with the EXOSAT satellite on 1984 April 20, about $4 \times 10^7$s after the maximum. In the medium energy X–ray detectors no significant flux was observed. The low–energy telescope (0.04 – 2 keV) operated with 3000 Å Lexan and Parlene–N–Aluminium filters detected the source with 4.5 $\sigma$ significance yielding count rates of $(4.3 \pm 1.3) \times 10^{-3}$ and $(3.7 \pm 1.2) \times 10^{-3}$counts s$^{-1}$ for the respective filters. Possible mechanisms for the X–ray emission are briefly discussed.

### Nova Coronae Austrinae 1981

124.161 **Spectrophotometry of Nova Coronae Austrinae 1981.**
B. S. Shylaja.
Astrophys. Space Sci., Vol. 104, No. 1, p. 163 – 168 (1984).

Spectrophotometric observations of Nova Coronae Austrinae 1981, during its nebular phase are reported. The various emission lines are identified. The electron densities and the mass of the ionized hydrogen in the shell have been calculated. The excess flux in the red continuum probably indicates the appearance of the dust component.

### Nova Sagittae 1783 = WY Sagittae

124.181 **WY Sagittae (Nova 1783): a transition object between classical and dwarf novae?**
M. M. Shara, A. F. J. Moffat, J. T. McGraw, D. S. Dearborn, H. E. Bond, E. Kemper, R. Lamontagne.
Astrophys. J., Vol. 282, No. 2, p. 763 – 768 (1984).

WY Sge, the 19th magnitude remnant of Nova WY Sagittae 1783, has been found to be an eclipsing binary with a period of 3 hr 41 minutes. Radial velocity observations show a low velocity amplitude for the emission lines originating from the vicinity of the primary star, possibly suggesting that the primary is a fairly massive ($>1\ M_\odot$) degenerate object. The secondary, if it obeys the main–sequence mass–radius relation, is an M dwarf of 0.44 $M_\odot$. On one night, WY Sge was observed to be 1.6 mag brighter than normal. This outburst, along with a spectrum in quiescence more typical of dwarf novae than of classical novae at minimum, suggests that WY Sge is an object intermediate (in mass–transfer rate from the secondary star) between classical and dwarf novae.

### Nova Aquilae 1982

124.201 **Nova Aquilae 1982 – a short report.**
M. A. J. Snijders, T. J. Batt, M. J. Seaton, J. C. Blades, D. C. Morton.
Mon. Not. R. Astron. Soc., Vol. 211, No. 1, p. 7P – 13P (1984).

Observations of Nova Aquilae 1982, obtained using the *IUE* satellite and the Anglo–Australian telescope, are reported. The nova was unusual in having very high expansion velocities (up to 10000 km s$^{-1}$), high abundances of heavier elements and large fluxes (reported by other observers) at IR and radio wavelengths. The ejecta had comparable masses in gas and dust, and grain formation may have led to gas–phase element depletions. The hydrogen mass fraction for the ejecta is estimated to be about 0.03. The progenitor may have been a close–binary system.

124.202 **The development of the 8 – 13 μm spectrum of Nova Aquilae 1982.**
P. F. Roche, D. K. Aitken, B. Whitmore.
Mon. Not. R. Astron. Soc., Vol. 211, No. 3, p. 535 – 542 (1984).
Four 8 – 13 μm spectra of the dust emission from Nova Aquilae 1982 are presented. The data, obtained between 140 and 280 days after outburst, show a prominent silicate emission feature superposed on a smooth warm continuum. The presence of the silicate grain signature, which is similar to laboratory emissivity curves of amorphous olivine smoke, indicates that these grains originated in an oxygen–rich environment where graphite or other carbon–rich grains are unlikely to have formed.

**Nova Delphini 1967 = HR Delphini**

124.221 **Nova Delphini (1967) – fünfzehn Jahre später.**
J. Solf.
Sterne Weltraum, 23. Jahrg., Nr. 12, p. 628 – 632 (1984).

**Nova Sagittarii 1984**

124.241 **Nova Sagittarii 1984.**
IAU Circ., Nos. 3995, 3997, 3998, 4002, 4008 (1984).

124.242 **Nova Sagittarii 1984.**
Yamamoto Circ., No. 2025 (1984). In Japanese.

**Nova Aquilae 1984**

124.261 **Nova Aquilae 1984.**
IAU Circ., Nos. 4020, 4023 (1984).

124.262 **Aquila probable nova (Honda).**
Yamamoto Circ., No. 2030 (1984). In Japanese.

**Nova Ophiuchi = V906 Ophiuchi**

124.281 **Nova V906 Ophiuchi rediscussed.**
W. Wenzel.
Inf. Bull. Variable Stars, No. 2585, 1 p. (1984).

**Nova Herculis 1934 = DQ Herculis**

124.301 **UBVR photometry of DQ Her.**
G. A. Voitenko (*G. A. Vojtenko*), G. U. Kovalchuck (*G. U. Koval'chuk*).
Inf. Bull. Variable Stars, No. 2596, 3 pp. (1984).

**Nova Herculis 1963 = V533 Herculis**

124.321 **Photographic observations of Nova Her 1963 (= V533 Herculis).**
X.-l. Hao, B. Mei.
Inf. Bull. Variable Stars, No. 2611, 2 pp. (1984).

**Veränderliche Sterne.**
See Abstr. 003.023.

**A progress report on a catalogue and atlas of galactic novae.**
See Abstr. 013.070.

# 125 Supernovae, Supernova Remnants

125.001 **Radio lines of heavy elements from the hot gas in supernova remnants and clusters of galaxies.**
R. A. Syunyaev, E. M. Churazov.
Pis'ma Astron. Zh., Tom 10, No. 7, p. 483 – 493 (1984). In Russian. English translation in Sov. Astron. Lett., Vol. 10.
The transitions between the hyperfine sublevels in the hydrogen–like and lithium–like ions in isotopes of nitrogen, magnesium, sodium, aluminum, manganese and a number of other elements, such as silicon and iron, are calculated; they lie in the millimetre and submillimetre region of the spectrum. The estimation of the radiation fluxes in lines of nitrogen $\lambda = 5.64$ mm and iron $\lambda = 3.06$ mm from a number of SNRs and the central parts of clusters allow to hope to observe them.

125.002 **G357.7 + 0.3 and G359.1–0.5 – two shell–type supernova remnants in the galactic centre region.**
W. Reich, E. Fürst.
Astron. Astrophys., Suppl. Ser., Vol. 57, No. 1, p. 165 – 167 (1984).
A new shell–type supernova remnant (SNR), G357.7 + 0.3, has been identified from radio continuum observations at 11 cm and 6 cm wavelength. G359.1–0.5 was previously suggested to be a SNR by Downes et al. (1978) from its morphology. A study of spectral index and linear polarization confirm the object to be a SNR.

125.003 **The infra–red emission from a composite circumstellar dust shell heated by a supernova.**
G. Pearce, A. Evans.
Astron. Astrophys., Vol. 136, No. 2, p. 306 – 312 (1984).
The infra–red emission from dust grains in a composite circumstellar dust shell heated by a supernova is investigated. The grain population in the dust shell is composed of both core–mantle (corundum–silicate) and homogeneous (corundum) grains. It is found that specific features in the infrared spectrum are characteristic of such core–mantle grains only. Suitable observations could therefore provide a means of determining the nature of the grains that condense in an oxygen rich environment around supernovae and of determining the period of time over which grains condense.

125.004 **Photometric observations of supernovae 1983 in NGC 3044 and 3106.**
N. V. Metlova, V. G. Metlov.
Astron. Tsirk., No. 1292, p. 6 – 8 (1983). In Russian.

125.005 **Spectrophotometry of young supernova remnants.**
M. A. Dopita, I. R. Tuohy.
Astrophys. J., Vol. 282, No. 1, p. 135 – 141 (1984).
Spectrophotometric data are presented for four young oxygen–rich supernova remnants: G292.0 + 1.8 in the Galaxy, N132D and 0540–69.3 in the Large Magellanic Cloud, and

1E 0102.2–7219 in the Small Magellanic Cloud. These data show that a common physical process is responsible for the excitation in all four remnants. Sensitive upper limits are derived on emission from elements other than oxygen and neon from global spectra of N132D and 1E 0102.2–7219 of very high signal-to-noise ratio.

125.006 **Radiative shock wave theory. III. The nature of the optical emission in young supernova remnants.**
M. A. Dopita, L. Binette, I. R. Tuohy.
Astrophys. J., Vol. 282, No. 1, p. 142 – 153 (1984).
The mechanism which excites the optical emission in the knots of fast-moving oxygen-rich ejecta found in young supernova remnants is generally assumed to be the propagation of a slow, radiative, reverse shock into them. This paper investigates the structure and emission spectra of such shocks, extends the results of Itoh by the inclusion of elements other than oxygen, and compares the results with observations. The fit between theory and observation is poor, and alternative interpretations for modeling are suggested.

125.007 **An optical investigation of the peculiar supernova remnant CTB 80.**
W. P. Blair, R. P. Kirshner, R. A. Fesen, T. R. Gull.
Astrophys. J., Vol. 282, No. 1, p. 161 – 171 (1984). With plates 7 – 8.
Interference-filter photographs and optical spectrophotometry are used to investigate the optical emission associated with the peculiar supernova remnant CTB 80. Optically, the remnant consists of outlying diffuse and filamentary emission surrounding a small, central ring of filaments which are coincident with the central radio emission peak. Spectra of two filaments in the southwestern part of the remnant indicate that the filaments are shock heated. The central filaments form a well-defined shell structure in the light of [O III] $\lambda 5007$ but are irregular in $H\alpha$. These filaments are probably not shock-heated. The authors consider the optical observations in relation to radio and X-ray data on CTB 80 and suggest an age of at least several thousand years for the remnant.

125.008 **Prospects for correlated observations of galactic supernovae and young remnants.**
V. S. Berezinskij, V. L. Ginzburg, O. F. Prilutskij.
Sov. Astron. Lett., Vol. 10, No. 1, p. 38 – 41 (1984). English translation of 37.125.005.

125.009 **A 21-cm study of the Crab Nebula region.**
L. I. Matveenko.
Sov. Astron. Lett., Vol. 10, No. 1, p. 44 – 45 (1984). English translation of 37.125.006.

125.010 **On the mass ejected by supernova explosions.**
J. Bohigas.
Rev. Mex. Astron. Astrofis., Vol. 9, No. 1, p. 13 – 20 (1984).
A simple model is developed in order to calculate the mass ejected by supernovae. The author finds that the 185, 1006, 1572 and 1604 AD events, all of them classified as either probable or possibly type I supernovae, ejected between 0.1 and 0.4 solar masses with an expansion velocity of roughly 10000 km $s^{-1}$. This range of masses suggests that a collapsed object is at the center of the remnants produced by these supernovae if the precursor was a white dwarf whose mass was close to the Chandrasekhar limit. For the Crab the author obtains an ejected mass of 0.45 $M_\odot$ and points out that this value is not in contradiction with a proposal in which the moderate helium stars are good candidates for producing this kind of supernovae.

125.011 **Statistical studies of supernova remnants.**
D. A. Green.
Mon. Not. R. Astron. Soc., Vol. 209, No. 2, p. 449 – 478 (1984). With microfiche MN 209/1.
This paper reviews critically the distance estimates available for Galactic supernova remnants (SNRs). The remnants with reliable distances have a much wider spread of intrinsic properties than has been evident before, which implies that the radio surface brightness-linear diameter relation is of little use for determining distances to Galactic SNRs, and that the cumulative number-diameter relation cannot be applied with any confidence to present studies of Galactic SNRs. It is shown that selection effects are responsible for the apparent free expansion of remnants to large diameters (> 30 pc) in both the Galaxy and the Magellanic Clouds. A detailed catalogue of 145 Galactic SNRs is presented.

125.012 **Accreting white dwarf models for type I supernovae. IV. The optical spectrum of a carbon-deflagration supernova.**
D. Branch, J. B. Doggett, K. Nomoto, F.-K. Thielemann.
MPA Rep., No. 169, 23 pp. (1984). Submitted to Astrophys. J.

125.013 **Erratum: "The evolution of supernova remnants as radio sources" [Mon. Not. R. Astron. Soc., Vol. 207, No. 4, p. 745 – 775 (1984)].**
R. Cowsik, S. Sarkar.
Mon. Not. R. Astron. Soc., Vol. 209, No. 3, p. 719 (1984). See Abstr. 37.125.038.

125.014 **The effectiveness of charged particles acceleration by shock waves in supernova remnants and in solar flares.**
S. V. Bulanov, I. V. Sokolov.
Pis'ma Astron. Zh., Tom 10, No. 8, p. 594 – 599 (1984). In Russian. English translation in Sov. Astron. Lett., Vol. 10.
An assumption on cosmic rays acceleration by shock waves in supernova shells is shown to be able to explain some characteristics of radio emission of old supernova remnants. The role of accelerated particles pressure in shell dynamics seems not to be great. A level of turbulence is evaluated which is necessary for this mechanism of acceleration taking place in solar flares.

125.015 **Infrared emission from silicate grains in the dust shells of two supernovae.**
G. Pearce.
Astron. Astrophys., Vol. 137, No. 2, p. 255 – 260 (1984).
The extent to which silicate grains in a circumstellar dust shell can account for observed infra-red emission from two supernovae is investigated. The two supernovae used in the modelling are SN1979c and SN1980k both of which are known to show a variable infra-red excess. The amorphous silicate used in the modelling is a composite silicate. Predictions are made regarding the spectral features which should be present if this type of silicate exists in such dust shells. It is concluded that the amorphous silicate used in the modelling may not be entirely appropriate to that which exists in the dust shells of the Type II supernovae discussed. The usefulness of this type of modelling in providing estimates of masses and sizes of such circumstellar dust shells is also examined.

125.016 **Magnetic field amplification and particle acceleration in radio supernovae.**
V. N. Fedorenko.
Sov. Astron. Lett., Vol. 10, No. 2, p. 89 – 91 (1984). English translation of 37.125.037.

125.017 **A new look at radio supernova remnants.**
B. Y. Mills, A. J. Turtle, A. G. Little, J. M. Durdin.
Aust. J. Phys., Vol. 37, No. 3, p. 321 – 357 (1984).
Catalogues are presented of radio supernova remnants in both Large (LMC) and Small (SMC) Magellanic Clouds together with maps of the LMC remnants prepared at a frequency of 843 MHz and with a resolution of 43 × 45 arcsec. The 38 confirmed remnants are believed to represent a reasonably complete sample for diameters less than about 40 pc although the LMC catalogue is probably very incomplete for large diameters. The supernova remnants (SNRs) are compared with the Galactic remnants in the catalogue of Clark and Caswell (1976) and they are found to be similar in many ways. In particular a distance scale derived from the Cloud SNRs reproduces well the mean distances of directly measured Galactic calibrators. The usual model of a SNR expanding adiabatically from a small diameter is not supported by the statistics of the number-diameter relation which suggest that rapid expansion up to quite large diameters is common.

125.018 **X-ray haloes around supernova remnants.**
G. E. Morfill, L. O'C. Drury, B. Aschenbach.
Nature, Vol. 311, No. 5984, p. 358 – 359 (1984).

Recent observations of the Cas A supernova remnant have shown X-ray emissions not only from the interior, but also from a fainter "halo" extending beyond what is normally regarded as the outer boundary, or shock front. The authors suggest that this may be due to the diffusion of energetic, charged particles out of the remnant giving rise to precursor structure of the type predicted by the theory of diffusive shock acceleration. If this is the case we are seeing thermal emission from ambient gas heated by compression and wave dissipation.

125.019 **Non-spherical supernova remnants. I. Calculations of the interaction of two remnants.**
P. Bodenheimer, H. W. Yorke, G. Tenorio-Tagle.
Astron. Astrophys., Vol. 138, No. 2, p. 215 – 224 (1984).

Two-dimensional numerical hydrodynamical calculations are presented which show the interaction of the remnants of two supernovae that explode in the low-density (1 particle $cm^{-3}$) interstellar medium. Four cases are considered: (1) the supernovae explode in the same place but at different times, (2) the second supernova explodes outside the expanding shell of the first, (3) the second explosion occurs outside the expanding shell of the first but at a greater distance than in case (2), and (4) the second supernova explodes precisely at the expanding shell of the first.

125.020 **High resolution observations of extended radio sources near $l$ = 94°.**
W. M. Goss, F. Mantovani, C. J. Salter, P. Tomasi, T. Velusamy.
Astron. Astrophys., Vol. 138, No. 2, p. 469 – 476 (1984).

Radio continuum observations of a low galactic latitude region near $l$ = 94° are presented at 608.5 MHz and 4.75 GHz. At 4.75 GHz both total power and linear polarization measurements were made. The non-thermal source 3C434.1 is shown to be a shell-type SNR with a non-uniform distribution of emission. The nearby source NRAO 655 has a flat spectrum above 600 MHz and is unpolarized. A thermal nature seems more likely, although it cannot yet be ruled out that the source is an unusually flat spectrum SNR. Three other probable thermal sources are identified, two of which are candidate compact H II regions. Linear polarization has been detected from the extended extragalactic source 4C50.55.

125.021 **Statistical studies of supernova remnants.**
D. A. Green.
Observatory, Vol. 104, No. 1062, p. 213 (1984). Abstract. – See Abstr. 010.682.

125.022 **Exosat observations of the supernova remnant W49B.**
L. R. Jones.
Observatory, Vol. 104, No. 1062, p. 213 – 214 (1984). Abstract. – See Abstr. 010.682.

125.023 **A radio study of the flat spectrum component in CTB 80.**
R. G. Strom, P. E. Angerhofer, J. R. Dickel.
Astron. Astrophys., Vol. 139, No. 1, p. 43 – 49 (1984).

High resolution radio observations of the compact flat spectrum core of CTB 80 are presented and discussed. From the average rotation measure the authors derive the magnetic field direction projected on the plane of the sky. This reveals a component running tangentially along the southern and western periphery of the radio plateau. In addition, there may be a field component directed east-west, roughly parallel to the major axis of the radio structure. The thermal density suggested by the rate of depolarization is sufficient, if the plasma has a temperature of several million degrees, to account for X-ray emission found by the Einstein Observatory. In addition to an extended component, the X-ray map reveals a small scale feature near the edge of the radio plateau which agrees in position with the brightest radio peak. The possible nature of this small scale feature is discussed.

125.024 **Some properties of the filamentary nebula at 1723–46.**
D. K. Bedford, K. H. Elliott, B. Ramsey, J. Meaburn.
Mon. Not. R. Astron. Soc., Vol. 210, No. 3, p. 693 – 699 (1984).

High resolution spectra of the interstellar Na I $D_2$ absorption line have been obtained for seven B-type stars aligned with an extensive filamentary nebula. Very fine filaments have now been found on a high contrast print of a deep Hα + [N II] plate, making identification as a fossilized SNR probable but uncertain. The inferred age is $\cong 2 \times 10^4$yr, the distance $< 200$ pc, the diameter $< 17.5$ pc, but the initial energy is only $\cong 2 \times 10^{47}$erg.

125.025 **Are supernovae round? I. The case for spectropolarimetry.**
M. L. McCall.
Mon. Not. R. Astron. Soc., Vol. 210, No. 4, p. 829 – 837 (1984). With a correction in Vol. 211, No. 4, p. 991 (1984).

Lines with P Cygni profiles produced by resonance scattering in an expanding atmosphere with non-circular isophotes should show a linear polarization in excess of that in the continuum. Thus, spectropolarimetry offers a means of testing the spherical symmetry of supernovae and assessing the reliability of Baade method distances without invoking assumptions about the polarization introduced by intervening dust.

125.026 **Observations of the remnants of three historical(?) supernovae.**
A. Downes.
Mon. Not. R. Astron. Soc., Vol. 210, No. 4, p. 845 – 853 (1984).

Radio maps are presented of the supernova remnants G11.2–0.3, G348.5 + 0.1 and G348.7 + 0.3, all three of which have been suggested as candidates for the remnants of supernovae in the 4th century AD. An HRI X-ray image of G11.2–0.3 is also shown. G11.2–0.3 is a bright, symmetric shell remnant which looks very like Tycho's or Kepler's SNRs. Its parameters are consistent with an (undetected) supernova explosion 300 – 500 yr ago, but an alternative interpretation, that it is the remnant of the explosion of a massive star in AD 386, cannot be excluded. G348.5 + 0.1 and G348.7 + 0.3 have weaker, more ragged shells than does G11.2–0.3 and are probably much older. This would be consistent with one of these two remnants being the relic of the AD 393 explosion.

125.027 **Extended radio observations of the SNR CTB 109.**
V. A. Hughes, R. H. Harten, C. H. Costain, L. A. Nelson, M. R. Viner.
Astrophys. J., Vol. 283, No. 1, p. 147 – 153 (1984).

Observations have been made of the SNR CTB 109 using the Westerbork Synthesis Radio Telescope at λ49 cm and λ21 cm, the Synthesis Telescope of the Dominion Radio Astrophysical Observatory at λ21 cm, and the 46 m telescope of the Algonquin Radio Observatory at λ4.6 cm. The spectral index of the integrated flux density of the SNR is given by $\alpha = 0.50 \pm 0.04$; various maps are compared and show that, over the range of λ4.6 – λ49 cm, α does not vary across the SNR. The SNR appears to consist of a shell with an X-ray pulsar displaced by 3′.6 from the center of curvature, but there is no indication, so far, that the pulsar is associated with the SNR.

125.028 **New radio observations of the composite supernova remnant G29.7–0.3.**
R. H. Becker, D. J. Helfand.
Astrophys. J., Vol. 283, No. 1, p. 154 – 157 (1984). With plate 3.

The galactic supernova remnant G29.7–0.3, recently identified as a new member of the class of composite remnants, has been imaged with the VLA at three wavelengths: 2, 6, and 20 cm. Two spectrally distinct components, a flat-spectrum core surrounded by a shell with $\alpha \sim -0.7$, suggested by early observations, have been confirmed. Neutral hydrogen absorption measurements of these two components confirm their physical association and place the remnant at a distance of ~21 kpc.

125.029 **Erratum: "Explosive nucleosynthesis and type I supernovae" [Astrophys. J., Lett. Ed., Vol. 279, No. 1, p. L23 – L26 (1984)].**
K. Nomoto, F.-K. Thielemann, J. C. Wheeler.
Astrophys. J., Lett. Ed., Vol. 283, No. 1, p. L25 (1984). See Abstr. 37.125.060.

125.030 **Why type II supernovae do not explode in irregular galaxies?**
I. S. Shklovskij.
Pis'ma Astron. Zh., Tom 10, No. 10, p. 723 – 725 (1984). In Russian. English translation in Sov. Astron. Lett., Vol. 10.
The conclusion is drawn that the reason for an absence of type II supernovae explosions in irregular galaxies is their peculiar chemical composition. The observed lack of stellar wind from massive hot giants is due to relatively low heavy element abundance. For this reason evolving massive stars do not form an extended dense envelope which is a necessary condition for the type II supernova phenomenon.

125.031 **Results of radio astronomical observations of a lunar occultation of the Crab nebula on December 2, 1982.**
M. I. Agafonov, A. M. Aslanyan, A. P. Barabanov, I. T. Bubukin, A. G. Gulyan, V. P. Ivanov, R. M. Martirosyan, I. A. Malyshev, K. S. Stankevich, S. P. Stolyarov.
Pis'ma Astron. Zh., Tom 10, No. 10, p. 730 – 736 (1984). In Russian. English translation in Sov. Astron. Lett., Vol. 10.
Results of observations are given of the Crab nebula occultation by the moon at 1640, 960, 750, 178, 140 MHz. Strip distributions outside the central part of the nebula show similar angular sizes. Coordinates of the gravity center are estimated.

125.032 **The energetics and statistics of supernovae.**
D. H. Clark.
Mass loss from astronomical objects, p. 94 – 100 (1984). – See Abstr. 012.032.
This paper reviews recent work on the birthrate of supernovae in galaxies, on the energetics of supernova explosions and on their probable progenitor stars.

125.033 **The interaction of supernovae with a circumstellar medium.**
R. A. Chevalier.
Ann. N.Y. Acad. Sci., Vol. 422, p. 215 – 232 (1984). – See Abstr. 012.033.
There have been recent observations of type II supernovae over the whole wavelength range from radio to X-rays that are not accounted for by the standard light curve models. These observations point to the importance of interaction with a circumstellar medium and are discussed here. Arguments are presented which show that there is a close correlation between the explosion event and the presupernova mass loss evolution.

125.034 **Interaction of supernova ejecta with interstellar and circumstellar matter.**
E. P. Liang, R. A. Chevalier.
Ann. N.Y. Acad. Sci., Vol. 422, p. 233 – 236 (1984). – See Abstr. 012.033.
This paper summarizes the results of a series of one-dimensional numerical simulations of supernova ejecta interacting with interstellar and circumstellar matter.

125.035 **Confinement of the Crab pulsar's wind by its supernova remnant.**
C. F. Kennel, F. V. Coroniti.
Astrophys. J., Vol. 283, No. 2, p. 694 – 709 (1984).
The authors construct a steady state, spherically symmetric, magnetohydrodynamic model of the Crab nebula. A highly relativistic, positronic pulsar wind is terminated by a strong MHD shock that decelerates the flow and increases its pressure to match boundary conditions imposed by the recently discovered supernova remnant that surrounds the nebula. If the magnetic luminosity of the pulsar wind upstream of the shock is about 0.3% of its particle luminosity, the pressure and velocity boundary conditions imposed by the remnant place the shock where it is inferred to be: near the outer boundary of an underluminous region observed to surround the pulsar.

125.036 **Magnetohydrodynamic model of Crab nebula radiation.**
C. F. Kennel, F. V. Coroniti.
Astrophys. J., Vol. 283, No. 2, p. 710 – 730 (1984).
The authors calculate the properties of the synchrotron continuum expected on the basis of a steady spherically symmetric MHD model of the flow in the Crab nebula. The model accounts for the spectral distribution of the continuum from the optical to γ-rays, and semiquantitatively for its observed spatial distribution. The model does not account for the radio portion of the Crab nebula's spectrum.

125.037 **Kinetics of nuclear reactions and the structure of detonation waves in supernova explosions.**
V. S. Imshennik, A. M. Khokhlov.
Inst. teor. i ehksp. fiz. Moskva, Prepr., No. 177, 76 pp. (1983). In Russian. Abstr. in Ref. Zh., 51. Astron., 9.51.477 (1984).

125.038 **Particle acceleration and Tycho's supernova remnant.**
A. F. Heavens.
Mon. Not. R. Astron. Soc., Vol 211, No. 1, p. 195 – 213 (1984).
Particle acceleration in shock waves may be very efficient at converting bulk kinetic energy into energy density in high-energy particles. The author investigates the implications for the dynamics and evolution of a Sedov-type blast wave, with particular reference to determining the energy of the supernova event associated with Tycho's supernova remnant. He finds that X-ray and radio observations favour a low efficiency of conversion of kinetic energy to high-energy particles, a supernova explosion energy of about $8 \times 10^{43}$J, and an ambient density of about 4 atom $cm^{-3}$.

125.039 **Observations of the western part of the Cygnus Loop at 408 MHz: the 5C8 survey.**
D. A. Green.
Mon. Not. R. Astron. Soc., Vol. 211, No. 2, p. 433 – 441 (1984).
The western part of the Cygnus Loop supernova remnant has been observed with the Cambridge One-Mile telescope at 408 MHz with a resolution of 80 × 160 $arcsec^2$ (RA × Dec). These observations reveal the radio emission from the remnant in great detail, including a remarkable radio filament that stretches over much of the length of the remnant. The nature of this feature, and others, are discussed in terms of the interaction of the remnant with differing regions of the interstellar medium.

125.040 **Observations of supernova remnants.**
J. C. Raymond.
Annu. Rev. Astron. Astrophys., Vol. 22, p. 75 – 95 (1984). – See Abstr. 003.024.
Contents: (1) Introduction. (2) Young remnants: Metal-rich SNRs: Cas A. Type I remnants: Tycho's SNR. The Crab. (3) Middle-aged remnants: The Cygnus Loop. (4) Old supernova remnants: The North Polar Spur. (5) Transitional supernova remnants. (6) Supernova remnants in other galaxies.

125.041 **X-ray emission from supernova remnants. II. The effect of remnant structure on nonequilibrium X-ray spectra.**
A. J. S. Hamilton, C. L. Sarazin.
Astrophys. J., Vol. 284, No. 2, p. 601 – 611 (1984).
A technique is presented for characterizing the ionization structure and consequent thermal X-ray emission of a supernova remnant (SNR) when nonequilibrium ionization effects are important. The technique allows different theoretical SNR models to be compared and contrasted rapidly in advance of detailed computations. The technique can be used whether or not electron and ion temperatures are equilibrated at the shock front. In particular it is shown that the spectrum of a Sedov remnant is probably a good approximation for other SNR models in which emission is dominated by gas undergoing quasi-steady state ionization close to a shock front.

125.042 **X-ray emission from supernova remnants MSH 14-63 and RCW 103.**
J. J. Nugent, S. H. Pravdo, G. P. Garmire, R. H. Becker, I. R. Tuohy, P. F. Winkler.
Astrophys. J., Vol. 284, No. 2, p. 612 – 630 (1984).
A nonequilibrium ionization computer model has been developed to predict the X-ray spectra from the hot ($10^{6-8}$K), shock-heated plasmas that are found in the remnants of supernovae. The model accounts for the lack of collisional ionization equilibrium and for the possible lack of thermal equilibrium between the electrons and ions behind the shock fronts. Both a spectral component arising from the supernova ejecta and a component arising from the shocked interstellar medium (ISM) surrounding the supernova are calculated. The model has been fitted to the spectral data from two young SNRs, MSH 14-63 and RCW 103.

125.043 **Convective supernovae.**
R. Bandiera.
Astron. Astrophys., Vol. 139, No. 2, p. 368 – 374 (1984).
The author studies the onset of instabilities during the first phases of supernova explosions, when the material is still optically thick and radiation dominated. Convection appears to be much more effective than the Rayleigh-Taylor instability. The author uses a class of self-similar analytical solutions for the explosion in a medium with a power law density distribution to evaluate the effectiveness of convection. It is found, for a slope of the original density profile close to -3, that convection not only saturates, but also noticeably affects the dynamics of the explosion. A slope close to -3 is in fact in general agreement with models of high mass, radiation dominated stars, which are commonly accepted as Type II supernova progenitors.

125.044 **Association of type II supernovae with giant H II regions: the case of M83 and M101.**
O.-G. Richter, M. Rosa.
Astron. Astrophys., Vol. 140, No. 1, p. L1 – L4 (1984).
The authors report first results of a statistical study of the association of type II supernovae with H II regions. Using the most accurate positions for the supernovae in the galaxies M83 and M101 they find a strong correlation between SNe II and giant H II regions. The association with H II regions implies very high initial masses (50 to 120 $M_{\odot}$) for the progenitors of these type II supernovae. The evidence from extragalactic supernova events suggests that most of the supernova remnants seen in H II regions in our Galaxy, the LMC and M33 originated from type II supernovae.

125.045 **The evolution of nonthermal supernova remnants. II. Can radio supernovae become plerions?**
R. Bandiera, F. Pacini, M. Salvati.
Astrophys. J., Vol. 285, No. 1, p. 134 – 140 (1984).
The authors discuss the evolution of pulsar-driven supernova remnants. It is shown that the early stages closely resemble the phenomenon of radio supernovae and that one can establish a direct evolutionary link between the two classes of objects. Light curves are obtained in the radio and the X-ray ranges under simple assumptions about the injected spectrum. The derived relations between surface brightness and diameter, both at low and at high frequencies, are compared with the observational data.

125.046 **Two new young galactic supernova remnants.**
D. A. Green, S. F. Gull.
Nature, Vol. 312, No. 5994, p. 527 – 529 (1984).
The authors have used the VLA and the Cambridge 5-km telescope to observe 32 small-diameter objects selected from galactic plane radio surveys. These observations reveal both a new young shell remnant (which, at only ~1 arc min in extent, is the smallest angular size galactic remnant yet discovered), and a very peculiar radio source, probably a supernova remnant only a few hundred years old, which is still in the free expansion phase.

125.047 **The radio emission of galaxies and their supernova rate.**
M. A. Smirnov, F. Kh. Sakhibov.
Sov. Astron., Vol. 28, No. 2, p. 137 – 140 (1984). English translation of 37.125.045.

125.048 **Possible detection of an old supernova remnant associated with HD 50896.**
J. N. Heckathorn, R. A. Fesen.
Bull. Am. Astron. Soc., Vol. 16, No. 3, Part I, p. 725 (1984). Abstract. – See Abstr. 010.065.

125.049 **Convective supernovae.**
R. Bandiera.
Bull. Am. Astron. Soc., Vol. 16, No. 3, Part I, p. 728 (1984). Abstract. – See Abstr. 010.065.

125.050 **A theoretical formulation of the radio surface-brightness – diameter relation for SNR.**
N. Duric, E. R. Seaquist.
J. R. Astron. Soc. Can., Vol. 78, No. 5, p. 209 (1984). Abstract. – See Abstr. 010.401.

125.051 **High-resolution radio observations of the supernova remnants CTB 104A (G94.3-0.3) and 3C 434.1 (G94.0+1.0).**
T. L. Landecker, L. A. Higgs, R. S. Roger.
J. R. Astron. Soc. Can., Vol. 78, No. 5, p. 210 (1984). Abstract. – See Abstr. 010.401.

125.052 **X-ray observations of the supernova remnant HB3.**
D. A. Leahy.
J. R. Astron. Soc. Can., Vol. 78, No. 5, p. 210 (1984). Abstract. – See Abstr. 010.401.

125.053 **Type II supernovae.**
S. E. Woosley, T. A. Weaver.
Problems of collapse and numerical relativity, p. 27 – 48 (1984). – See Abstr. 012.061.
Evidence is briefly reviewed for believing that the core bounce mechanism actually gives rise to Type II supernovae in massive stars. Strongest evidence comes from the historical observation of the Crab Nebula light curve, the existence of pulsars, and from the nucleosynthesis of a few very neutron-rich isotopes such as $^{48}$Ca. A new pre-supernova model for a 25 $M_{\odot}$ star is presented. The evolution of a Population III ($Z = 0$) 25 $M_{\odot}$ star (currently evolved up to the point of silicon shell burning) is also briefly reviewed and implications for nucleosynthesis discussed.

125.054 **Simplifications in supernova theory.**
J. Cooperstein.
Problems of collapse and numerical relativity, p. 133 – 146 (1984). – See Abstr. 012.061.
Numerical calculations of Type II supernovae are considered, concentrating on the entropy and how it changes. Calculations of electron capture rates during the gravitational collapse phase indicate only a small change in the entropy and lepton number. The equation of state of hot dense matter is considered and the transition density to uniform nuclear matter is calculated. An analysis of the post-bounce hydrostatic stage, contrasting successful and unsuccessful models, provides criteria for success.

125.055 **The physics of supernova shocks.**
J. M. Lattimer, A. Burrows.
Problems of collapse and numerical relativity, p. 147 – 162 (1984). – See Abstr. 012.061.
The authors focus on the bounce-shock mechanism for Type II supernovae. They show semi-analytically that the initial shock energy is equal (with minor qualifications) to the binding energy of the hydrostatic remnant. They present analytical formulae for the structure and binding energy of the remnant, and determine the remnant mass by the self-similar arguments. Unless the initial presupernova core is sufficiently small, both analytical and numerical calculations show the bounce shock mechanism is incapable of powering a *prompt* (<10 ms) explosion.

125.056 **Successful supernovae, the anatomy of shocks: neutrino emission and the adiabatic index.**
S. Kahana, E. Baron, J. Cooperstein.
Problems of collapse and numerical relativity, p. 163 – 182 (1984). – See Abstr. 012.061.

Hydrodynamic calculations of stellar collapse in Type II supernovae are described using a variable nuclear adiabatic index and compressibility for the nuclear equation of state at high density. Initial models employing a relatively small mass core with low central entropy are necessary to achieve viable shocks; near success the models are sensitive to both neutrino emission and the high density equation of state. The treatment of neutrino production and transport is sketched and recent results reported.

125.057 **Core collapse models of Type II supernova explosions.**
W. Hillebrandt, E. Müller.
Problems of collapse and numerical relativity, p. 183 – 193 (1984). – See Abstr. 012.061.

Core–collapse models of Type II supernova explosions are discussed. It is shown that stellar models with about 9 $M_\odot$ on the main sequence do indeed lead to very energetic explosions, whereas more massive stars ($M \cong 20\ M_\odot$) so far do not explode. It is found, however, that the latter models are very close to mass ejection, and therefore future improvements of stellar models, input physics and numerical treatment may change these rather disappointing conclusions.

125.058 **The velocity structure of the supernova remnant N157B.**
R. Gilmozzi, P. Murdin, D. H. Clark.
Astron. Astrophys., Vol. 140, No. 2, p. 390 – 392 (1984).

The supernova remnant N157B has been likened to the Crab Nebula on the basis of its X–ray and radio spectral characteristics. Velocity data are presented suggesting that, like the Crab, N157B is expanding as a prolate spheroid.

125.059 **Supernova rate and the number of supernova remnants in M31, M33 and the Galaxy.**
E. M. Berkhuijsen.
Astron. Astrophys., Vol. 140, No. 2, p. 431 – 440 (1984).

The SN rates in M31, M33 and the Galaxy have been estimated from the upper part of the present day mass functions scaled with $\dot{N}_{\rm Lyc}$ derived from the integrated thermal radio emission. The expected rates are in agreement with statistical and historical rates if all stars of mass $\gtrsim 5\ M_\odot$ become SN II. Each of the three galaxies may contain about $10^3$ to $10^4$SNRs depending on the remnants' lifetime. Hence, only about 10 to 1% of all SNRs and about 50 to 5% of the SNRs of diameter $<50$ pc has been detected.

125.060 **Supernova in anonymous galaxy.**
IAU Circ., No. 4011 (1984).

125.061 **Possible supernova in NGC 3675.**
IAU Circ., No. 4021 (1984).

125.062 **Supernovae.**
IAU Circ., No. 4024 (1984).

125.063 **Supernovae discovered by Evans.**
Yamamoto Circ., No. 2019 (1984). In Japanese.

125.064 **Possible supernova in NGC 3675.**
Yamamoto Circ., No. 2031 (1984). In Japanese.

125.065 **Nucleosynthesis in type I and type II supernovae.**
F.-K. Thielemann.
Adv. Space Res., Vol. 4, No. 2 – 3, p. 67 – 78 (1984). – See Abstr. 012.075.

Nucleosynthesis in type I and type II supernovae is reviewed in the framework of existing models and observational constraints. A type II supernova is interpreted by the core bounce mechanism after the onset of a gravitational collapse following core silicon burning of a massive star, leaving a neutron star and ejecting the surrounding envelope. A type I supernova is explained by explosive ignition under degenerate conditions in an accreting white dwarf, situated in a binary system. This mechanism causes the complete disruption of the star.

125.066 **The supernova envelope shock origin of cosmic rays – a review.**
S. A. Colgate.
Adv. Space Res., Vol. 4, No. 2 – 3, p. 367 – 379 (1984). – See Abstr. 012.075.

The hydrodynamic shock origin of cosmic rays in the envelope of a Type I presupernova star is reviewed. The spectrum produced by the relativistic hydrodynamic shock is one power of E steeper than observed and so is unlikely to be the primary source of cosmic rays. On the other hand the possibility of accelerating ultrahigh energy particles to $\gtrsim 10^{18}$eV is unique to the shock mechanism and currently no other suggested galactic or extragalactic site is likely.

125.067 **The X–ray structure of the Crab nebula.**
W. Brinkmann, B. Aschenbach, A. Langmeier, G. Hasinger, T. Bork.
MPE Prepr., No. 28, p. 1 – 4 (1984). Paper presented at the XXVth COSPAR Meeting on X–Ray Astronomy, Graz, Austria, July 3 – 7, 1984.

125.068 **The X–ray structure of the Crab nebula.**
B. Aschenbach, W. Brinkmann, A. Langmeier, G. Hasinger, T. Bork.
MPE Prepr., No. 29, p. 8 – 11 (1984). Paper presented at the Symposium on X–Ray Astronomy '84, Bologna, Italy, June 26 – 30, 1984.

125.069 **The Crab nebula's jet at high spatial resolution.**
S. Koutchmy, C. Lebecq, F. Couturier, P. Lamy, A. Bücher, G. Coupinot, J. Hecquet.
Inst. Astrophys. Paris, Pré–Publ., No. 75, 10 pp. (1984). In French. Paper presented at the 8th European Assembly of the IAU, Toulouse, France, September 17 – 21, 1984.

125.070 **Temperature relaxation in supernova remnants, revisited.**
H. Itoh.
Astrophys. J., Vol. 285, No. 2, p. 601 – 606 (1984).

Some supernova remnants are expanding into a partially neutral medium. The neutral atoms which are engulfed by the fast blast shock are collisionally ionized to eject low–energy secondary electrons. The author calculates the temperature relaxation through Coulomb collisions among the secondary electrons, the shocked electrons, and the ions, assuming that the three species have Maxwellian velocity distributions. The results are applied to a self–similar blast wave. Some observational consequences for the X–ray continua of young supernova remnants are indicated.

125.071 **An X–ray investigation of the unusual supernova remnant CTB 80.**
Z. R. Wang, F. D. Seward.
Astrophys. J., Vol. 285, No. 2, p. 607 – 612 (1984).

The X–ray properties of SNR CTB 80 (G68.8+2.8) are discussed based on both low– and high–resolution images from the Einstein satellite. The X–ray maps show a point source coinciding with the region of maximum radio emission. Diffuse X–ray emission is evident mainly along the radio lobe extending ~8′east of the point source and aligned with the projected magnetic field lines. The unusual radio and X–ray morphologies are interpreted as a result of relativistic jets energized by the central object, and the possible association of CTB 80 with SN 1408 as recorded by Chinese observers is discussed.

125.072 **The kinematics of the Crab Nebula's jet.**
P. Shull Jr., U. Carsenty, M. Sarcander, T. Neckel.
Astrophys. J., Lett. Ed., Vol. 285, No. 2, p. L75 – L78 (1984). With plates L8 – L10.

The optical emission lines in spectra of the northern jet of the Crab Nebula form ringlike velocity ellipses. The spectrograph slit was oriented perpendicularly to the long axis of the jet and centered near the jet's midpoint and base. At both locations, the major and minor axes of the velocity ellipses measure 45″ (=jet diameter) × 720 km $s^{-1}$.The authors propose that the jet is an accelerated beam of hot plasma and discuss its dynamics.

125.073 **Observations of supernovae in NGC 4185 and NGC 5485.**
D. Yu. Tsvetkov.
Astron. Tsirk., No. 1308, p. 1 – 2 (1984). In Russian.

125.074 **On the mass of supernova progenitors.**
M. A. Smirnov, D. Yu. Tsvetkov.
Astron. Tsirk., No. 1315, p. 1 – 2 (1984). In Russian.

125.075 **On the behaviour of photospheres of supernovae.**
O. S. Bartunov, D. Yu. Tsvetkov.
Astron. Tsirk., No. 1325, p. 1 – 3 (1984). In Russian.

125.076 **Der nordpolare Sporn nahe der galaktischen Ebene.**
P. Müller.
Mitt. Astron. Ges., Nr. 62, p. 318 – 321 (1984). – See Abstr. 010.341.

125.077 **Photometric classification and basic characteristics of type I supernovae.**
Yu. P. Pskovskij.
Astron. Zh., Tom 61, Vyp. 6, p. 1125 – 1136 (1984). In Russian. English translation in Sov. Astron., Vol. 28, No. 6.

Uniform estimates of photometric classes are compared with basic characteristics of type I supernovae. Spectrophotometric temperatures and their relations with colour indices of type I supernovae are obtained.

125.078 **On the frequency of supernovae in Sb galaxies.**
R. G. Mnatsakanyan.
Astrofizika, Tom 21, Vyp. 1, p. 81 – 86 (1984). In Russian. English translation in Astrophysics, Vol. 21, No. 1.

The mean frequency of SN in Sb galaxies is calculated. For a galaxy of Sb type, brighter than $12^m.0$, the mean time between two successive bursts of SN is equal to 122 years. Not all bright Sb galaxies ($B \leqslant 10^m.5$) have the same capability to produce SN. The mean frequency of SN in Sb galaxies is more than two times less than in Sc galaxies. The mean frequency of SN depends more on the morphological type of the galaxy than on the amount of stellar population in it.

125.079 **A supernova in Zw1 16.7 + 1.57.**
R. K. Shakhbazyan, F. Börngen.
Astrofizika, Tom 21, Vyp. 2, p. 393 – 395 (1984). In Russian. English translation in Astrophysics, Vol. 21, No. 2.

A supernova near the galaxy Zw1 16.7 + 1.57, has been found on the maps of the Palomar Observatory. The eye estimation of photographic and red magnitudes gives: $m_{pg} = 18.3$, $m_r = 18.6$. The blue colour and the supposed luminosity ($M_{pg} = -17.2$) of the object suggest that it is of type II near the maximum.

125.080 **Heating and cooling in reverse shocks into pure heavy–element supernova ejecta.**
A. J. S. Hamilton, C. L. Sarazin.
Astrophys. J., Vol. 287, No. 1, p. 282 – 294 (1984).

This paper examines the question of electron heating and cooling in reverse shocks into pure heavy–element supernova ejecta, with shock velocities in the range of a few hundred to a few thousand km $s^{-1}$. Collisionless heating of electrons at a shock front in pure heavy–element ejecta tends to have a smaller, though still appreciable, effect than is the case for a cosmic–abundance plasma. The results of numerical computations are presented for some particular cases of shocks.

125.081 **A detailed X–ray and radio comparison of Kepler's supernova remnant.**
Y. Matsui, K. S. Long, J. R. Dickel, E. W. Greisen.
Astrophys. J., Vol. 287, No. 1, p. 295 – 306 (1984). With plates 4 – 6.

High–resolution radio images of Kepler's SNR (SN 1604) at 6 and 21 cm have been obtained using the VLA. The orientation of linear polarization in the SNR implies a substantial nonrandom radial component for the magnetic field of the remnant. Based on a new X–ray picture of the SNR, obtained with the *Einstein Observatory*, a detailed comparison of the X–ray and radio morphology is presented. The total magnetic field in the SNR shell is $7 \times 10^{-5}$gauss. The strong correlation between radio emissivity and X–ray emissivity found over the SNR can be ascribed to acceleration of thermal electrons at the shock wave and insignificant diffusion of resultant relativistic electrons.

125.082 **An optical synchrotron nebula around the X–ray pulsar 0540–693 in the Large Magellanic Cloud.**
G. A. Chanan, D. J. Helfand, S. P. Reynolds.
Astrophys. J., Lett. Ed., Vol. 287, No. 1, p. L23 – L26 (1984). With plate L1.

The authors report the discovery of extended optical continuum emission around the recently discovered 50 ms X–ray pulsar in the supernova remnant 0540–693. Exposures in blue and red broad–band filters made with the CTIO 4 m telescope and prime focus CCD show a center–brightened but clearly extended nebula about 4″ in diameter (FWHM), while an image in an [O III] filter shows an 8″ diameter shell which encloses the continuum source. The emission from $10^{14.5}$Hz to $10^{18}$Hz is probably synchrotron radiation from a single population of particles.

125.083 **Binary progenitors of supernovae.**
V. Trimble.
J. Astrophys. Astron., Vol. 5, No. 4, p. 389 – 402 (1984).

Among the massive stars expected to produce Type II (hydrogen–rich) supernovae, the presence of a close companion seems to be largely negative. Among the less massive stars thought to make Type I (hydrogen–free) supernovae, a close companion could be an innocent bystander to carbon detonation/deflagration in the primary, or it can be a vital participant, transferring material to a white dwarf primary and so driving it to explosive conditions. A widely discussed recent scenario allows both stars first to become degenerate dwarfs, which then spiral together, giving rise to a hydrogen–poor explosion arbitrarily long after the star formation event, even for quite massive binaries.

125.084 **On the supernova remnants produced by pulsars.**
G. Srinivasan, D. Bhattacharya, K. S. Dwarakanath.
J. Astrophys. Astron., Vol. 5, No. 4, p. 403 – 423 (1984).

The authors conclude that pulsar–driven supernova remnants (SNRs) are extremely rare objects. Indeed an analysis of the known sample of plerions suggests a very low birthrate ~1 in 240 years. Long–lived and bright plerions like the Crab nebula are likely to be produced only when the pulsar has an initial period ~10 – 20 milliseconds and a field ~$10^{12}$G. Such pulsars inside rapidly expanding shell remnants should also produce detectable plerions. The extreme rarity of SNRs with such hybrid morphology leads the authors to conclude that these pulsars must have been born with an initial period larger than ~35 – 70 milliseconds.

125.085 **H I absorption observations of two radio sources near the supernova remnant G 127.1 + 0.5.**
W. M. Goss, J. H. van Gorkom.
J. Astrophys. Astron., Vol. 5, No. 4, p. 425 – 428 (1984).

The compact source 0125 + 628 in the centre of the galactic supernova remnant G 127.1 + 0.5 has been re–observed in H I absorption using the Westerbork Synthesis Radio Telescope. The outer arm H I absorption at $V = -95$ km $s^{-1}$ has been confirmed. The absorption spectrum is similar to that of the nearby extragalactic source 0123 + 633. The authors discuss the arguments concerning an extragalactic origin of 0125 + 628 and conclude that it is most likely extragalactic and not an SS 433 type object.

**Supernova in MCG 6–36–55**

125.301 **Light curve of a supernova in MCG 6–36–55. Preliminary data.**
N. V. Metlova.
Astron. Tsirk., No. 1289, p. 8 (1983). In Russian.

**Supernova in NGC 5236 (1983n)**

125.321 **Supernovae.**
N. Panagia.
ESA Spec. Publ., ESA SP-218, p. 21 – 24 (1984). – See Abstr. 012.024.

After a summary of the kowledge on supernovae as for 1982 the author briefly presents and discusses the results of observations of SN 1983n in NGC 5236 which have offered precious insights and opened new horizons on the properties and the nature of Type I supernovae.

125.322 **Radio emission from a Type I supernova: SN 1983.51 in NGC 5236.**
R. A. Sramek, N. Panagia, K. W. Weiler.
Astrophys. J., Lett. Ed., Vol. 285, No. 2, p. L59 – L62 (1984).

For the first time, radio emission has been detected from a Type I supernova. SN 1983.51 (SN 1983n) was discovered optically on 1983 July 3 and reached maximum light on approximately 1983 July 17. Radio observation of the position of the optical supernova detected a source of flux density $2.0 \pm 0.5$ mJy at 6 cm wavelength already on 1983 July 6. The radio emission is almost certainly nonthermal in origin with an optically thin spectral index between 20 and 6 cm of $\alpha = -1.0 \pm 0.1$. A detailed radio "light curve" has been determined at 6 cm with additional data also available at 20 cm and 2 cm wavelengths.

125.323 **The circumstellar interaction model for the radio emission from a Type I supernova.**
R. A. Chevalier.
Astrophys. J., Lett. Ed., Vol. 285, No. 2, p. L63 – L66 (1984).

The radio emission observed from the Type I supernova Evans in M83 is interpreted in terms of the circumstellar interaction model developed for the radio emission from Type II supernovae. The model agrees with the observations in that at late times the flux has a power-law dependence on time since the explosion and the power-law index is that expected for an exploding white dwarf in a circumstellar medium. The relatively rapid evolution compared to that of Type II supernovae is attributed to a flat supernova density profile and a steep relativistic electron spectrum. The implications of this model for observations at other wavelengths and for the progenitors of Type I supernovae are discussed.

**Supernova in NGC 4753 (1983g)**

125.341 **Are supernovae round? II. Spectropolarimetry of SN 1983g in NGC 4753.**
M. L. McCall, N. Reid, M. S. Bessell, D. Wickramasinghe.
Mon. Not. R. Astron. Soc., Vol. 210, No. 4, p. 839 – 843 (1984).

Spectropolarimetric data have been obtained near maximum light for the Type I supernova SN 1983g in NGC 4753. No change in polarization through the P Cygni lines is observed, indicating that the apparent axis ratio of the expanding atmosphere was greater than 0.5. The Baade method would overestimate the distance by at most a factor of 1.4 if spherical symmetry were assumed.

**Supernova in NGC 4536 (1981b)**

125.361 **Models of type I supernovae and observations of SN 1981b.**
D. Branch.
Ann. N.Y. Acad. Sci., Vol. 422, p. 186 – 196 (1984). – See Abstr. 012.033.

The analysis of the photometry and spectroscopy of SN 1981b is generally consistent with the carbon deflagration of an accreting white dwarf. The assumption of blackbody emissivity at maximum light leads to a high luminosity, as expected for any model that synthesizes on the order of a solar mass of $Ni^{56}$. The spectra establish that intermediate mass elements were present with roughly solar relative abundances in the outer ($\gtrsim$8,000 km/sec) parts of the ejecta, and suggest that freshly synthesized iron was present in the inner parts.

**Supernova in NGC 3169**

125.381 **The 1984 supernova in NGC 3169 – a new kind of type II supernova?**
C. M. Gaskell.
Publ. Astron. Soc. Pac., Vol. 96, No. 584, p. 789 (1984). Abstract. – See Abstr. 010.281.

125.382 **The 1984 supernova in NGC 3169: evidence for a super-wind.**
M. A. Dopita, R. Evans, M. Cohen, R. D. Schwartz.
Astrophys. J., Lett. Ed., Vol. 287, No. 2, p. L69 – L71 (1984).

Optical spectra of the Type II supernova in NGC 3169 obtained near maximum light on 1984 April 1.484 UT and April 2.563 UT show a double P Cygni structure in the Balmer lines. The high-velocity component suggests a velocity of expansion of the photosphere of 15,500 km $s^{-1}$ whereas the helium lines give a somewhat lower velocity, 13,500 km $s^{-1}$. The narrow P Cygni profiles seen in Hα to Hδ suggest that the precursor star underwent a strong episode of mass loss immediately prior to the supernova event and that this wind material was outflowing at a velocity close to 3,000 km $s^{-1}$ at the time of observation.

**Supernova in NGC 1559**

125.401 **Supernova in NGC 1559.**
IAU Circ., Nos. 3963, 3967, 3980, 3989, 3997 (1984).

**Supernova in NGC 6850**

125.421 **Supernova in NGC 6850.**
IAU Circ., No. 3971 (1984).

**Supernova in NGC 991 (1984*l*)**

125.441 **Supernova 1984*l* in NGC 991.**
IAU Circ., Nos. 3979, 3981, 3983, 3985, 3989, 4001, 4003, 4008, 4014 (1984).

125.442 **Supernova in NGC 991.**
Yamamoto Circ., No. 2023 (1984). In Japanese.

**Supernova in IC 121**

125.461 **Supernova in IC 121.**
IAU Circ., Nos. 3981, 3985, 3989 (1984).

**Supernova in IC 4839**

125.481 **Supernova in IC 4839.**
IAU Circ., Nos. 4002, 4007 (1984).

125.482 **Supernova in IC 4839.**
Yamamoto Circ., No. 2026 (1984). In Japanese.

**Supernova in NGC 7184**

125.501 **Supernova in NGC 7184.**
IAU Circ., Nos. 3962, 3994, 4013, 4019 (1984).

**Supernova in NGC 4419**

125.521 **Supernova in NGC 4419.**
G. N. Kimeridze, R. Ya. Inasaridze.
Astron. Tsirk., No. 1321, p. 2 – 3 (1984). In Russian.

**A revised supernova catalogue.**
See Abstr. 002.038.

**Check a possible supernova (*Forty photographs of galaxies*).**
See Abstr. 002.082.

**On the time of achievement of ionization equilibrium in a hot low–density gas.**
See Abstr. 022.157.

**Supernovae, neutrino rest mass, and the middle–energy neutrino background in the universe.**
See Abstr. 061.075.

**On detecting stellar collapse with neutrinos.**
See Abstr. 061.079.

**On the reverse influence of accelerated particles on the structure of shock wave fronts.**
See Abstr. 062.071.

**A current generation by Compton scattering in a relativistic plasma with velocity shear and temperature gradient.**
See Abstr. 062.100.

**Effects of the temperature and mass of iron cores on the mass ejection of neutrino–trapping supernovae.**
See Abstr. 065.016.

**Electron capture in stellar collapse.**
See Abstr. 065.017.

**Presupernova yields and their dependence on the $^{12}C(\alpha, \gamma)\,^{16}O$ reaction.**
See Abstr. 065.024.

**Presupernova stellar evolution.**
See Abstr. 065.025.

**Carbon combustion supernovae.**
See Abstr. 065.026.

**Nucleosynthesis in type I supernovae.**
See Abstr. 065.027.

**Type II supernovae.**
See Abstr. 065.028.

**Core–collapse models of type II supernova explosions and neutron star formation.**
See Abstr. 065.045.

**An extended Thomas–Fermi calculation of supernova matter.**
See Abstr. 065.068.

**Gravitational collapse of a rotating magnetized star.**
See Abstr. 065.069.

**Collapse and gravitation = the present situation.**
See Abstr. 065.079.

**Evolution of accreting white dwarfs, Type I supernova explosion, and white dwarf collapse.**
See Abstr. 065.084.

**Collapse and explosion of degenerate stars.**
See Abstr. 065.085.

**Final evolution of 8 – 10 $M_\odot$ stars.**
See Abstr. 065.086.

**Carbon deflagrating supernovae and the chemical history of the solar neighbourhood.**
See Abstr. 065.138.

**Explosive nucleosynthesis in massive stars: comparison with the Cassiopeia A fast–moving knots.**
See Abstr. 065.141.

**Magnetorotational iron core collapse.**
See Abstr. 065.143.

**Electron capture supernovae: one–zone collapse calculation for stars with masses 10 $M_\odot$ and 15 $M_\odot$.**
See Abstr. 065.144.

**The current status of stellar collapse.**
See Abstr. 065.151.

**Equations of state, neutrinos and supernova explosions.**
See Abstr. 065.152.

**Accreting white dwarf models for Type I supernovae. III. Carbon deflagration supernovae.**
See Abstr. 065.156.

**A modified thermal Thomas–Fermi approach to supernova matter.**
See Abstr. 067.054.

**Nuclei in supernova matter.**
See Abstr. 067.055.

**Towards selfconsistent band structure calculations of properties of supernova matter.**
See Abstr. 067.057.

**Decameter observations of a lunar occultation of the Crab Nebula.**
See Abstr. 096.004.

**Results of radio astronomical observations of the lunar occultation of the Crab nebula on January 26, 1983.**
See Abstr. 096.029.

**Sco X–1: a mini quasar?**
See Abstr. 117.007.

**Are cataclysmic variables the progenitors of Type I supernovae?**
See Abstr. 117.095.

**Supernova progenitors and Be stars: stellar variability from a 21 century perspective.**
See Abstr. 122.197.

**Magnetically ordered jets from pulsars.**
See Abstr. 126.007.

**Comments on "The white dwarf collapse in low mass binary systems".**
See Abstr. 126.016.

**Einstein observations of the Crab nebula pulsar.**
See Abstr. 126.034.

**Neutronensterne und Supernova–Überreste. Neuere Ergebnisse der Pulsarforschung.**
See Abstr. 126.044.

**Are we in an old superbubble?**
See Abstr. 131.002.

**Ammonia clouds in absorption against Cas A.**
See Abstr. 131.012.

**Further studies of nearby high velocity interstellar gas.**
See Abstr. 131.079.

**Gas–cloud/supernova interactions and the properties of the interstellar medium.**
See Abstr. 131.132.

**Westerbork observations of 6 cm $H_2CO$ in the direction of Cas A.**
See Abstr. 131.153.

**The molecular cloud associated with W50 and SS433.**
See Abstr. 131.169.

**Hydrogen recombination lines towards the SNR IC 443 and the H II region S 249.**
See Abstr. 131.176.

**Review of radio line observations with the RATAN–600 radio telescope.**
See Abstr. 131.195.

**Observations of absorption lines from highly ionized atoms.**
See Abstr. 131.262.

**Modelling the local interstellar medium as a supernova remnant in a multiphase gas.**
See Abstr. 131.279.

**Non equilibrium ionization in the local hot bubble.**
See Abstr. 131.280.

**One–millimeter continuum observations of galactic and extragalactic sources.**
See Abstr. 132.010.

**Radio observations of the CMa OB1 H II regions.**
See Abstr. 132.028.

**Galactic H II regions and supernova remnants near $l$ = 55°.**
See Abstr. 132.036.

**A catalog of coordinates, angular dimensions, and flux densities of the components of galactic radio sources at 21 cm.**
See Abstr. 132.042.

**A 10–GHz radio continuum survey of the galactic plane region. I. A complex region at $l$ = 21° – 26°.**
See Abstr. 141.008.

**Very high resolution observations of compact sources in the direction of supernova remnants: G74.87 + 1.22.**
See Abstr. 141.024.

**Astronomical observations with the FAUST telescope.**
See Abstr. 142.007.

**On the theory of X–ray emission from the extended galactic Monogem Ring source in the Monoceros constellation.**
See Abstr. 142.013.

**Discovery of a 50 millisecond pulsar in the Large Magellanic Cloud.**
See Abstr. 142.090.

**Balloon observations of the 5 – 100 MeV gamma rays from the Crab nebula.**
See Abstr. 143.007.

**Generation of γ–bursts by old magnetic neutron stars.**
See Abstr. 143.044.

**Numerical studies of time–dependent cosmic–ray shocks.**
See Abstr. 144.069.

**H I shells, supershells, shell–like objects, and "worms".**
See Abstr. 155.014.

**The correlation between the stars and the continuous radio emission intensity distributions in the north polar spur region.**
See Abstr. 155.073.

**Young supernovae in the starburst galaxy M82.**
See Abstr. 158.201.

**Energetic particles in radio sources.**
See Abstr. 158.203.

**Endpoints of stellar evolution: X–ray surveys of the Local Group.**
See Abstr. 160.146.

# 126 Degenerate Stars, White Dwarfs, Pulsars

126.001 **The absorption near 1400 Å in the spectra of additional DA white dwarfs.**
G. Wegner.
Astron. J., Vol. 89, No. 7, p. 1050 – 1056 (1984).

Ultraviolet spectroscopic observations of five white dwarfs having published spectral types of DA made with the IUE satellite are reported. The three coolest objects: GD279, W485A, and G231–40 have easily detected absorptions near λ1400. Of the two remaining stars, PG0039 + 045 may not be a true DA and GD140 is apparently too hot to show any feature near λ1400. It is argued that the λ1400 feature is produced by the λ1420 photoionization edge of Ca II.

126.002 **γ rays of 0.3 – 30 MeV from PSR 0833–45.**
O. T. Tümer, B. Dayton, J. Long, T. O'Neill, A. Zych, R. S. White.
Nature, Vol. 310, No. 5974, p. 214 – 216 (1984).

Pulsed γ rays from the Vela pulsar PSR 0833–45 are reported here for the first time at the medium energies, $E$, of 0.3 – 30 MeV. They were observed with a double scatter γ–ray telescope flown on a balloon on 10 November 1981, 31 days after a large glitch in the pulsar period. The first and second pulses were detected. The energy distribution with six points from 0.3 to 30 MeV appears to bend away from the COS B power law at the lower energies.

126.003 **Orbital inclination and mass of the binary pulsar PSR 0655 + 64.**
A. G. Lyne.
Nature, Vol. 310, No. 5975, p. 300 – 302 (1984).

The author reports here variations in the rate of interstellar scintillation that have been measured as the pulsar PSR 0655 + 64 moves around its orbit, allowing a direct determination of the inclination of the orbit, the transverse velocity of the whole binary system through space and also of the linear scale of the interstellar scintillation pattern. The observations also provide an upper limit to the diameter of the companion star and the electron density within the orbit.

126.004 **Discovery of a white dwarf companion of the "hybrid" K giant HD 81817.**
D. Reimers.
Astron. Astrophys., Vol. 136, No. 1, p. L5 – L6 (1984).

IUE observations show that the bright K giant HD 81817 has a faint blue companion which is identified as a white dwarf. A fit of theoretical DA model atmosphere fluxes to the observed energy distribution in the wavelength range λλ1250 Å to 1600 Å yields an effective temperature $T_e = 20000 \pm 2000$K for the WD companion and places the WD at the same distance as the red giant. On the basis of an estimate of the giant's mass, it is argued that the white dwarf must have had a progenitor with an initial mass above $\sim 4\,M_\odot$. Finally, it is shown by comparison with θ Her that HD 81817 is a member of the 'hybrid star' class.

126.005 **Interpretation of pulsar long-term intensity variations.**
N. Y. Shapirovskaya (*N. Ya. Shapirovskaya*), W. Sieber.
Astron. Astrophys., Vol. 136, No. 1, p. 171 – 174 (1984).

Pulsar long-term intensity variations can be understood as scintillation patterns produced by a thin phase screen located near to the observer. If one assumes a screen velocity of the order of the earth's velocity, and a typical scale size of the inhomogeneities in the interstellar medium of about $10^{13}$ cm, pulsar long-term intensity variations as well as low-frequency variations of extragalactic sources may be explained by the same screen. Observational tests for the validity of the model are proposed.

126.006 **Non-LTE analysis of subluminous O-stars. VI. Feige 110.**
U. Heber, W.-R. Hamann, K. Hunger, R. P. Kudritzki, K. P. Simon, R. H. Méndez.
Astron. Astrophys., Vol. 136, No. 2, p. 331 – 337 (1984).

The ultraviolet (λ1150 – 2000 Å) and the blue spectrum (λ4000 – 5100 Å) of the subdwarf O-star Feige 110 is analyzed. Effective temperature, gravity and helium abundance of Feige 110 are determined by means of NLTE model atmospheres. It is suggested that the strange abundance pattern of Feige 110 might have been affected by diffusion. It must, hence, be concluded that the photospheric abundance of helium, carbon and silicon may have been much larger prior to the onset of diffusion. It even cannot be excluded that also the nitrogen abundance has been altered by diffusion.

126.007 **Magnetically ordered jets from pulsars.**
G. Benford.
Astrophys. J., Vol. 282, No. 1, p. 154 – 160 (1984).

It is proposed that a small fraction of the magnetic field energy expelled by a pulsar may leave in an ordered, propagating form. These plasma-loaded field lines leave close to the spin axis and form a high-pressure beam. If the beam encounters no significant opposing pressure it will form a cone, limb brightened in synchrotron emission, until it strikes a supernova shell. Filled supernova remnants will slow the beam, winding up the toroidal field and producing a well-collimated, nearly force-free equilibrium. Return currents to the pulsar are formed naturally through induction. These types of ordered forms are compared with the jets in the Crab nebula and MSH 15-52.

126.008 **Pulsars: nature's most precise clocks.**
J. H. Taylor.
Proceedings of the 37th Annual Frequency Control Symposium 1983, held at Philadelphia, Pa., USA. IEEE, New York, USA. (1983). p. 6. Abstr. in Phys. Abstr., Vol. 87, No. 1237, Entry 72970 (1984).

126.009 **Space distribution of pulsars.**
O. Kh. Gusejnov, I. M. Yusifov.
Astron. Zh., Tom. 61, Vyp. 4., p. 708 – 726 (1984). In Russian. English translation in Sov. Astron., Vol. 28, No. 4.

A method of determining the electron concentration described in an earlier work by Gusejnov, Kasumov, and Yusifov for the Galaxy volume occupied by the known 331 pulsars is developed. A large amount of new data on the interstellar medium on objects affecting the $n_e$ increase, and on pulsars in general is considered. The distribution of $n_e$ against galactic longitude and the distance from the sun is determined. Distances to the pulsars are calculated and their new parameters are found. The space distribution of pulsars is investigated.

126.010 **Observations of interstellar scattering of the radio emission of PSR 0329+54 at 102.5 MHz.**
M. V. Popov, V. A. Soglasnov.
Astron. Zh., Tom. 61, Vyp. 4., p. 727 – 732 (1984). In Russian. English translation in Sov. Astron., Vol. 28, No. 4.

The parameters of interstellar scattering of PSR 0329+54 at the frequency 102.5 MHz were obtained using a new method. The characteristic decorrelation time is 40 s; two scales of decorrelation bandwidth were detected, 100 and 750 Hz. The frequency dependences of scintillation time and decorrelation bandwidth for PSR 0329+54 follow those expected for a power law spectrum of electron density inhomogeneities.

126.011 **Two meter-emission modes for PSR 0943+10.**
S. A. Sulejmanova, V. A. Izvekova.
Sov. Astron., Vol. 28, No. 1, p. 32 – 35 (1984). English translation of 37.126.012.

126.012 **Determination of the transversal velocities of pulsars from scintillations of the interstellar medium.**
F. Kh. Sakhibov.
Tr. Gos. Astron. Inst. Shternberg, Tom 54, p. 46 – 52 (1984). In Russian.

Estimates of the transversal velocities for 15 pulsars have been carried out. The results of these estimates have been compared with the results of other authors. From the obtained data follows that the probability of accidental coincidence of the pulsars with supernova remnants is large.

126.013 **Search for pulsed optical emission from the millisecond pulsar PSR 1937+214.**
R. N. Manchester, B. A. Peterson, P. T. Wallace.
Nature, Vol. 310, No. 5978, p. 569 – 572 (1984).

A search has been made for optical pulses from the 1.6-ms pulsar PSR 1937+214, using the 3.9-m AAT. Despite previously reported indications of pulsed emission at this period from the red star discussed by Djorgovski, the authors believe the results of the search to be negative with an upper limit on the *R* magnitude of the pulsed emission of ~25. Furthermore, astrometric observations show that the red star is displaced about 2.4 arc s north of the pulsar position and, hence, that it is likely to be a chance association.

126.014 **Spin-down of pulsars by the current: comparison of theory with observations.**
V. S. Beskin, A. V. Gurevich, Ya. N. Istomin.
Astrophys. Space Sci., Vol. 102, No. 2, p. 301 – 326 (1984).

Observational data are analysed on the basis of pulsar magnetosphere theory and the Ruderman-Sutherland plasma generation mechanism. The results of the theory and experiments are shown to be in good agreement. The existence of a special group of long-lived, weakly-radiating fast (the period $P \cong 0.1 - 0.4$ s) and superfast ($P \lesssim 0.01$) pulsars is predicted.

126.015 **Propagation of γ-radiation in strong magnetic fields of pulsars.**
A. E. Shabad, V. V. Usov.
Astrophys. Space Sci., Vol. 102, No. 2, p. 327 – 358 (1984).

The propagation, decay and absorption of electromagnetic waves in magnetospheres of pulsars are considered with emphasis on the resonant effects in the vacuum polarization.

126.016 **Comments on "The white dwarf collapse in low mass binary systems".**
B. S. Shylaja, R. K. Kochhar.
Astrophys. Space Sci., Vol. 103, No. 1, p. 199 – 200 (1984). See Abstr. 34.126.038.

White dwarfs have been shown to possess high-angular momenta, in low-mass binary systems, in the context of formation of "spinars". Whether the collapse to a neutron star occurs spontaneously or is triggered by mass transfer from the companion, is an unresolved question. Many models to type I SN (Whelan and Iben, 1973) refer to the latter case. However, in his recent paper Lipunov (1983) did not consider the formation of a neutron star from a SN explosion in greater detail. The occurrence of pulsars in binary systems also poses serious evolutionary problems. It is possible that low mass binary systems with massive and weak magnetic field WDs as primaries form neutron stars associated with high-space velocities.

126.017 **Pulsar statistics and two types of pulsars.**
J. Huang, K. Huang, Q. Peng.
Proceedings of the Third Marcel Grossmann Meeting on General Relativity, Part B, p. 1265 – 1268 (1983). – See Abstr. 012.002.
The observed period statistics of pulsars is analyzed in an attempt to find empirical evidence for a recently postulated new mechanism of pulsar spindown.

126.018 **On space velocities of radio pulsars.**
A. V. Tutukov, N. N. Chugaj, L. R. Yungel'son.
Pis'ma Astron. Zh., Tom 10, No. 8, p. 586 – 593 (1984). In Russian. English translation in Sov. Astron. Lett., Vol. 10.
Taking into account selection effects it has been shown that about half of all radio pulsars have low space velocities ($v < 50$ km/s). The theoretical spectrum of space velocities of neutron stars which are products of the massive close binaries evolution has been found.

126.019 **Stars too small to burn.**
K. Croswell.
Astronomy, Vol. 12, No. 4, p. 16 – 22 (1984). Abstr. in Phys. Abstr., Vol. 87, No. 1238, Entry 77823 (1984).

126.020 **Gravitational and pressure redshifts in white dwarf stars.**
B. Grabowski, J. Halenka, J. Madej.
XVI International Conference on Phenomena in Ionized Gases, p. 104 – 105 (1983). Abstr. in Phys. Abstr., Vol. 87, No. 1239, Entry 82474 (1984). – See Abstr. 012.017.

126.021 **On plasma beam instabilities in the magnetospheres of pulsars.**
V. D. Egorenkov, D. G. Lominadze, P. G. Mamradze.
Astrofizika, Tom 19, Vyp. 4, p. 753 – 759 (1983). In Russian. English translation in Astrophysics. Vol. 19, No. 4.
It has been shown that for the recently accepted models of relativistic electron–positron plasmas of the lower layer in the magnetosphere of pulsars only the development of "hydrodynamic" beam instabilities is possible; kinetic instabilities of longitudinal waves in such plasmas do not exist.

126.022 **Observations of the upper initial mass limit of white dwarf progenitors $M_{WD}$.**
D. Reimers.
MPA Rep., No. 90, p. 16 (1983). – See Abstr. 012.020.

126.023 **Electromagnetic instabilities of a relativistic neutral beam with inhomogeneous velocity profile in the magnetospheres of pulsars.**
T. D. Kaladze, P. G. Mamradze.
Astrofizika, Tom 20, Vyp. 2, p. 289 – 294 (1984). In Russian. English translation in Astrophysics, Vol. 20, No. 2.
The instability of electromagnetic Alfvén type waves in the magnetospheres of pulsars is investigated. The instability is due to the inhomogeneous velocity profile of the relativistic beam which outflows from the pulsar.

126.024 **Explosive helium burning in white dwarfs.**
A. M. Khokhlov.
Sov. Astron. Lett., Vol. 10, No. 2, p. 123 – 126 (1984). English translation of 37.126.048.

126.025 **Spectral analysis of helium poor hot subdwarfs.**
U. Heber, K. Hunger.
ESA Spec. Publ., ESA SP–218, p. 273 – 277 (1984). – See Abstr. 012.024.
Ultraviolet and visual spectra of 19 helium poor hot subdwarfs (sdB's and sdOB's) are analyzed for effective temperature, gravity and helium abundance by means of model atmospheres. Three helium poor subdwarfs are discovered to be binaries.

126.026 **Spectrum and polarization of the gravitational radiation of pulsars.**
D. V. Gal'tsov, V. P. Tsvetkov, A. N. Tsirulev.
Zh. Ehksp. Teor. Fiz., Tom 86, No. 3, p. 809 – 818 (1984). In Russian. Abstr. in Ref. Zh., 51. Astron., 8.51.128 (1984).

126.027 **The halo population G and K subdwarfs.**
V. Straižys, K. Zdanavičius, G. Tautvaišiene, K. Černis.
Astrophys. Space Sci., Vol. 104, No. 2, p. 219 – 223 (1984).
The photometry of G and K subdwarfs in the Vilnius photometric system permits to recognize them photometrically and to determine their metallicity, despite the presence of interstellar reddening. The weakening of the strong metallic lines in their spectra can also be used to estimate their metallicity.

126.028 **Limits on the $^{12}C$ to $^{13}C$ isotopic ratio in two white dwarfs.**
G. Wegner.
Astrophys. Space Sci., Vol. 104, No. 2, p. 347 – 355 (1984).
A search for the molecular $^{12}C^{13}C$ isotopic bands in two white dwarfs is described. Spectroscopic observations of the two carbon band white dwarfs, BPM 27606 and L 879–14 have been obtained. The isotopic bands have not been detected. However, upper limits to the $^{12}C$:$^{13}C$ abundance ratio derived from the data yield $^{12}C$:$^{13}C > 40$ for BPM 27606 and $^{12}C$:$^{13}C > 8$ for L 879–14. If these ratios are representative of the material in the interiors of the carbon band white dwarfs and taking the low upper limits on the N and O abundances relative to C, this is consistent with the carbon having been formed by the 3α process and not having undergone any subsequent mixing with H–rich material. In addition, the $C_2$ vibrational bandheads are blue-shifted, most probably by pressure shifts.

126.029 **Implications of the absolute magnitude distribution functions of DA and non–DA white dwarfs.**
E. M. Sion.
Astrophys. J., Vol. 282, No. 2, p. 612 – 614 (1984).
The enlarged sample of proper motion selected spectroscopic degenerates is analyzed. The frequency distribution of degenerate spectral types is presented. Color–luminosity calibrations are used to derive the number–absolute magnitude distribution functions of DA and non–DA white dwarfs. The ratio of DA to non–DA white dwarfs is significantly lower (2:1) at the cool end of the distribution than at the hot end (6:1). Evidence for the evolution of some DA white dwarfs into DB white dwarfs in the range $11.25 \leqslant M_v \leqslant 11.75$ is presented and discussed.

126.030 **On the magnetic field in the white dwarf Grw +70°8247.**
R. J. W. Henry, R. F. O'Connell.
Astrophys. J., Lett. Ed., Vol. 282, No. 2, p. L97 – L98 (1984).
Greenstein has observed a sharp absorption feature near 1345 Å in the ultraviolet spectrum of the white dwarf Grw +70°8247. He interprets this as a component of hydrogen Lyα in a strong magnetic field, which he estimates to be in the region of $3.5 \times 10^8$ gauss. Using a very accurate multiparameter variational approach the authors deduce that the maximum Lyα wavelength obtainable is 1342.6 Å (or slightly greater if one includes a gravitational redshift as well as a Stark–induced redshift), at which wavelength the magnetic field is $5.6 \times 10^8$ gauss, all other magnetic field values giving rise to smaller wavelengths.

126.031 **Electrodynamics of radio pulsars.**
V. S. Beskin.
Zemlya Vselennaya, No. 4, p. 13 – 19 (1984). In Russian.

126.032 **On the theory of cyclotron lines in the spectra of magnetic white dwarfs.**
V. V. Zheleznyakov, A. A. Litvinchuk.
Astrophys. Space Sci., Vol. 105, No. 1, p. 73 – 84 (1984).
The radiation transfer at the gyrofrequency in the coronae of magnetic white dwarfs is considered. The electron distribution over Landau levels, taking both radiative and collisional transitions into account, is obtained. The emissivity and absorption coefficients of extraordinary radiation at the gyrofrequency are

calculated. The ranges of parameters where cyclotron lines are observed in emission or absorption are found. The upper limit on coronal plasma density ($2 \times 10^{11}$ cm$^{-3}$) for isolated magnetic white dwarfs with absorption lines in the spectrum is specified.

126.033 **The formation of condensed carbon particles in the atmospheres of white dwarfs.**
B. E. Zhilyaev, V. G. Zubko.
Astrophys. Space Sci., Vol. 105, No. 1, p. 99 – 107 (1984).
The existence of condensed carbon in the form of liquid droplets and graphite grains is found in white dwarf atmospheres with parameters $g = 10^8$cm s$^{-2}$, H/He $\leqslant 10^{-5}$, C/He $= 10^{-3}$, and $T_{eff} \leqslant 6000$K on the basis of model atmospheres techniques. It is shown that the condensation layers are dynamically stable and, consequently, that white dwarfs cannot supply the condensed particles to the interstellar medium. Possible observable effects are considered.

126.034 **Einstein observations of the Crab nebula pulsar.**
F. R. Harnden Jr., F. D. Seward.
Astrophys. J., Vol. 283, No. 1, p. 279 – 285 (1984). With plates 5 – 6.
The authors present Einstein Observatory high–resolution ($\sim 4''$) images of the Crab nebula which clearly resolve the pulsar from the diffuse X–ray emission. The pulsar contributes $3.9 \pm 0.2$% of the observed 0.1 – 4.5 keV flux and shows no evidence (other than the 33 ms pulsations) for variations on time scales up to $\sim 2.3$ days. The pulsar flux decreases to zero over about one–eighth of the cycle, with the upper limit to point source emission during this interval implying a blackbody surface temperature of less than $2.5 \times 10^6$K for the neutron star.

126.035 **Oscillations of degenerate stars.**
A. N. Cox.
Nature, Vol. 311, No. 5986, p. 509 (1984).

126.036 **The 1.5–millisecond pulsar.**
D. C. Backer.
Ann. N.Y. Acad. Sci., Vol. 422, p. 180 – 181 (1984). – See Abstr. 012.033.
A brief summary of the basic properties of the millisecond pulsar PSR 1937+214 is presented.

126.037 **The millisecond pulsar.**
A. F. Cheng.
Ann. N.Y. Acad. Sci., Vol. 422, p. 182 – 185 (1984). – See Abstr. 012.033.
A possible scenario for the formation of a millisecond pulsar is outlined. It is suggested that these pulsars can be formed in binary systems with accretion disks, in which the pulsar progenitor was spun up by accretion torques.

126.038 **On the magnetization and origin of the millisecond pulsar PSR 1937+214.**
J. Arons.
Ann. N.Y. Acad. Sci., Vol. 422, p. 328 (1984). Abstract. – See Abstr. 012.033.

126.039 **Pulsar magnetospheres.**
E. Asséo, D. Beaufils, R. Pellat.
Ann. N.Y. Acad. Sci., Vol. 422, p. 329 (1984). Abstract. – See Abstr. 012.033.

126.040 **High–energy gamma rays from pulsars.**
S. K. Gupta, P. V. Ramana Murthy,
B. V. Sreekantan, S. C. Tonwar, P. R. Viswanath.
Ann. N.Y. Acad. Sci., Vol. 422, p. 356 (1984). Abstract. – See Abstr. 012.033.

126.041 **Polarization angle in magnetic white dwarfs.**
B. Martin, D. T. Wickramasinghe.
Astrophys. J., Vol. 283, No. 2, p. 782 – 786 (1984).
In magnetic white dwarfs, the angle of linear polarization is observed in some stars to be roughly constant except for a change by 90° at a particular wavelength. One proposed explanation for this phenomenon, the presence of cyclotron absorption, is examined and rejected. Another explanation relying on shifted Balmer jumps is proposed, and applied to Grw +70°8247.

126.042 **Far–ultraviolet spectrophotometry of two very hot O type subdwarfs.**
J. S. Drilling, J. B. Holberg, D. Schönberner.
Astrophys. J., Lett. Ed., Vol. 283, No. 2, p. L67 – L70 (1984). = Contrib. Louisiana State Univ. Obs. No. 189.
The authors have observed the very hot O type subdwarfs LSE 21 and LS IV +10°9 with the Voyager 1 ultraviolet spectrometer over the range in wavelength from 540 to 1700 Å. Combination of the Voyager 1 data with published *IUE* spectrophotometry and *UBV* photometry indicates an effective temperature of 65,000K for LS IV +10°9, and an effective temperature in excess of 100,000K for LSE 21, making it one of the hottest known stars.

126.043 **Pulsar electrodynamics.**
S. Shibata.
Astron. Her., Vol. 77, No. 11, p. 272 – 276 (1984). In Japanese.

126.044 **Neutronensterne und Supernova–Überreste. Neuere Ergebnisse der Pulsarforschung.**
W. Sieber.
Naturwissenschaften, 71. Jahrg., Heft 11, p. 560 – 566 (1984).
Recent results on the relationship between radio pulsars and supernova remnants are presented. No clear general picture has as yet emerged from the observations, although some interrelationships between pulsars or neutron stars and supernova remnants are now well documented and understood. The most interesting cases are presented and discussed.

126.045 **The magnetic field structure of PSR 0809+74.**
J. G. Davies, A. G. Lyne, F. G. Smith,
V. A. Izvekova, A. D. Kuzmin (*A. D. Kuz'min*), Yu. P. Shitov.
Mon. Not. R. Astron. Soc., Vol. 211, No. 1, p. 57 – 68 (1984).
The pulsar PSR 0809+74 has been observed simultaneously at frequencies 102, 406 and 1412 MHz. Additional observations were made at 102, 92, 62 and 39 MHz to obtain the alignment of the integrated pulse profiles. In the generally accepted polar cap model, pulsar emission occurs in a region defined by open dipolar magnetic–field lines, lower radio frequencies being emitted at larger distances from the neutron star surface. Although the observations generally support this model, the more detailed characteristics of pulse arrival times, pulse drifting and the cross–correlation of intensity fluctuations at different frequencies require a more complex model of the magnetic field at the polar cap. It seems that the magnetic field is not purely dipolar but includes a multipole component and contra–rotation twisting at large radii.

126.046 **Millisecond pulsars.**
D. C. Backer.
J. Astrophys. Astron., Vol. 5, No. 3, p. 187 – 207 (1984).
In 1982 a pulsar was discovered with the phenomenal rotation rate of 642 Hz, 20 times faster than the spin rate of the Crab pulsar. The absence of supernova debris in the vicinity of the pulsar at any wavelength indicates an age of the neutron star greater than $10^5$yr. The miniscule spindown rate of $1.1 \times 10^{-19}$ confirms the old age and indicates a surface magnetic field of $10^9$G. A second millisecond pulsar was discovered by Boriakoff et al. (1983) in a 120–day orbit. These fast pulsars may have been spun–up by mass transfer in a close binary evolutionary stage. Arrival–time observations of the 642–Hz pulsar display remarkably low residuals over the first 14 months. The stability implied by these observations, $3 \times 10^{-14}$, suggests that millisecond pulsars will provide the most accurate basis for terrestrial dynamical time.

126.047 **Models for the formation of binary and millisecond radio pulsars.**
E. P. J. van den Heuvel.
J. Astrophys. Astron., Vol. 5, No. 3, p. 209 – 233 (1984).
The peculiar combination of a relatively short pulse period and a relatively weak surface dipole magnetic field strength of binary

radio pulsars finds a consistent explanation in terms of (1) decay of the surface dipole component of neutron–star magnetic fields on a timescale of $(2-5) \times 10^6$yr, in combination with (2) spin–up of the rotation of the neutron star during a subsequent mass–transfer phase. The four known binary radio pulsars appear to fall into two different categories. Two of them, PSR 0655+64 and PSR 1913+16, have short orbital periods (<25 h) and high mass functions, indicating companion masses $\geqslant 0.7\ M_{\odot}$. The other two, PSR 0820+02 and PSR 1953+29, have long orbital periods ($\geqslant$117 d), nearly circular orbits, and low mass functions, suggesting companion masses of about $0.3\ M_{\odot}$. Among the binary models proposed for the formation of the 1.5–millisecond pulsar, the only ones that appear to be viable are those in which the companion disappeared by coalescence with the neutron star.

126.048 **Braking index diagnostics of pulsars. I. Alignment, counteralignment and slowing–down noise.**
P. Ghosh.
J. Astrophys. Astron., Vol. 5, No. 3, p. 307 – 316 (1984).

The author shows that the strong correlation observed between the braking indices ($n$) and the slowing–down ages ($\tau$) of pulsars is inconsistent with counteralignment between their rotation and magnetic axes, but that the data on pulsars with positive braking indices is consistent with alignment. Alternatively, slowing–down noise can quantitatively account of the data on all pulsars except the Crab and the Vela, and so for the apparent $|n| \sim \tau^2$ correlation observed for the older pulsars.

126.049 **Analyses of the visual spectra of white dwarfs containing carbon.**
G. Wegner, F. H. Yackovich.
Astrophys. J., Vol. 284, No. 1, p. 257 – 269 (1984).

Spectroscopic observations of white dwarfs showing bands of the $C_2$ molecule in their visual spectra were obtained using the Kitt Peak 2.1 m telescope and IIDS detector system. These data have signal–to–noise ratios high enough to determine atmospheric carbon abundances and to search for faint features of other elements, particularly the H$\alpha$ line and CN and CO bands to derive upper limits for the H, N, and O abundances. Model atmospheres were constructed using a modified Kurucz ATLAS(5) program and were employed to fit the observed $C_2$ Swan bands and investigate the visibility of other spectral features. The implications of the results for the evolution of white dwarfs and the possibility of the accretion–diffusion mechanism in these stars are discussed.

126.050 **Energetics of the aligned pulsar magnetosphere.**
R. F. Martin Jr.
Astrophys. Space Sci., Vol. 105, No. 2, p. 339 – 356 (1984).

An energy analysis is performed on two explicit models, due to Jackson, of a pulsar with aligned magnetic and rotational axes. The unknown parameters of these models are determined by calculating the minimum total energy states of the models. It is found that the minimum energy analysis favors states with extended, dynamically active magnetospheres with a high degree of corotation. It is determined that the minimum energy states are the only states where a power balance can be achieved. Consideration of a local power balance condition and dissipative flows in the magnetosphere shows that neither model is completely self–consistent, but one is considerably better than the other. Properties of both models and implications for other models are discussed.

126.051 **Periodic solutions in a model of pulsar rotation.**
O. I. Bogoyavlenskij.
Dokl. Akad. Nauk SSSR. Ser. Mat. Fiz., Tom 276, No. 2, p. 343 – 347 (1984). In Russian. Abstr. in Ref. Zh., 51. Astron., 10.51.547 (1984).

126.052 **Orientation of the magnetic axis of pulsars and its evolution.**
A. D. Kuz'min, I. M. Dagkesamanskaya, V. D. Pugachev.
Pis'ma Astron. Zh., Tom 10, No. 11, p. 854 – 859 (1984). In Russian. English translation in Sov. Astron. Lett., Vol. 10.

The inclination of the magnetic axis of a pulsar relative to the rotation axis $\beta$ and to the line–of–sight are determined. The secular decrease of $\beta$ has been confirmed. Some geometric characteristics of subpulses for drifting pulsars are obtained.

126.053 **An upper limit to the mass of the white dwarf in UX Ursae Majoris.**
A. W. Shafter.
Astron. J., Vol. 89, No. 10, p. 1555 – 1558 (1984).

Time–resolved spectroscopic observations of the novalike variable UX UMa are presented. The radial–velocity variations of the H$\alpha$ emission lines are found to have a semiamplitude of $157 \pm 6$ km/s. This result, in conjunction with the results of a radial velocity undertaken by Schlegel, Honeycutt, and Kaitchuck, imply that the mass of the white dwarf in this system must be less than $0.7\ M_{\odot}$.

126.054 **G238–53: a new pulsating DA white dwarf.**
G. Fontaine, F. Wesemael.
Astron. J., Vol. 89, No. 11, p. 1728 – 1731 (1984).

High–speed photometric observations of the DA white dwarf G238–53 show that it is another star of the ZZ Ceti type. Its colors place it within the instability strip. Fourier analysis of three independent light curves of G238–53 reveals that it is a very low–amplitude variable ($\Delta m \cong 0.009$ mag) with a quasiperiod of about 206 s. Only one peak with significant power is seen in the Fourier spectrum, but the data suggest that this peak could be made up of at least two unresolved frequency components. G238–53 is likely to be a stable, linear pulsator.

126.055 **A search for radial pulsations in DA white dwarfs.**
E. L. Robinson.
Astron. J., Vol. 89, No. 11, p. 1732 – 1734 (1984).

Following several recent theoretical calculations that suggested DA white dwarfs should pulsate in radial modes, the author conducted a photometric search for these pulsations. The shortest periods examined were 0.1 s and the survey was sensitive to variations as small as $\Delta L/L = 5 \times 10^{-4}$. None of the stars observed showed radial pulsations. If the author includes observations made in other surveys, 19 DA white dwarfs have now been examined for radial pulsations. The failure of any of these stars to pulsate suggests that modifications are required in either pulsation theory or the white dwarf models.

126.056 **The discontinuity near 1600 Å in the spectra of DA white dwarfs.**
G. Wegner.
Astrophys. J., Lett. Ed., Vol. 284, No. 2, p. L43 – L46 (1984).

Ultraviolet spectroscopic observations of two relatively cool DA white dwarfs, L481–60 (=WD 1544–37) and BPM 1266 (=WD 2105–82), with the IUE satellite show a strong drop in their spectral energy distributions below $\lambda$1600. Published model atmospheres and their visual spectra suggest that these two stars have effective temperatures in the vicinity of 9,000 – 10,000K, and it is proposed that the $\lambda$1600 feature could be due to the $3s^2\ {}^1S$ photoionization edge of Mg I.

126.057 **Recent arguments supporting a suggestion to observe pulsars via their direct mode of magnetic–dipole radiation.**
B. Z. Kaplan.
Lett. Nuovo Cimento, Vol. 40, Ser. 2, No. 14, p. 446 – 448 (1984). Abstr. in Phys. Abstr., Vol. 87, No. 1243, Entry 101388 (1984).

126.058 **Those amazing white dwarfs.**
B. Parker.
Astronomy, Vol. 12, No. 7, p. 15 – 22 (1984). Abstr. in Phys. Abstr., Vol. 87, No. 1244, Entry 105982 (1984).

126.059 **Search for periodic low–energy gamma–ray emission from pulsar PSR 0531+21.**
K. K. C. Yau.
Thesis, Univ. Durham, England, 1 p. (1984). Abstr. in Phys. Abstr., Vol. 87, No. 1244, Entry 106019 (1984).

126.060 **Observations of pulsar PSR 0809+74 at a frequency of 25 MHz with a time resolution of 100 μsec.**
A. Yu. Novikov, M. V. Popov, V. A. Soglasnov, Yu. M. Bruk, B. Yu. Ustimenko.
Sov. Astron., Vol. 28, No. 2, p. 199 – 201 (1984). English translation of 37.126.046.

126.061 **Determination of the distance to pulsar 1937+214.**
V. S. Artyukh, A. V. Pynzar', T. V. Smirnova, V. A. Udal'tsov.
Sov. Astron., Vol. 28, No. 3, p. 302 – 304 (1984). English translation of 37.126.074.

126.062 **Kinematics of anti–center subdwarfs.**
J. R. Pier, C. A. Pilachowski.
Publ. Astron. Soc. Pac., Vol. 96, No. 584, p. 793 (1984). Abstract. – See Abstr. 010.281.

126.063 **The rotational velocities of white dwarfs.**
C. A. Pilachowski, R. W. Milkey.
Publ. Astron. Soc. Pac., Vol. 96, No. 584, p. 821 – 828 (1984).
High–resolution profiles of the sharp core of Hα have been observed in 7 DA white dwarfs. Theoretical NLTE Hα profiles have been calculated using blanketed, LTE model atmospheres. Profiles synthesized using a $T_{eff}$ = 20,000K model closely match the observed profiles for stars in the range 19,000K to 23,000K, but profiles synthesized from lower–temperature model atmospheres are too weak to match the Hα cores in stars cooler than 17,000K. The theoretical profiles have been rotationally broadened to compare with the observed profiles to derive rotational velocities ($v \sin i$) for the hotter stars. For the cooler stars, the authors have used the rotationally broadened profile of 40 Eri B to provide limits on $v \sin i$.

126.064 **Discovery of a new, high field, magnetic white dwarf, KUV813–14.**
E. M. Sion, J. Liebert, G. Schmidt, S. G. Starrfield.
Bull. Am. Astron. Soc., Vol. 16, No. 3, Part I, p. 725 (1984). Abstract. – See Abstr. 010.065.

126.065 **A two frequency study of the subpulse drift patterns in PSR 0809+74 and PSR 1919+21.**
A. Wolszczan, M. Prószyński, D. A. Graham, G. A. E. Wright.
Bull. Am. Astron. Soc., Vol. 16, No. 3, Part I, p. 728 (1984). Abstract. – See Abstr. 010.065.

126.066 **The short–term and long–term decay components of the Vela pulsar in the starquake model.**
J. Lodenquai.
Astrophys. Lett., Vol. 24, No. 2, p. 91 – 94 (1984).
The latest glitch frequency observed in the Vela pulsar PSR 0833–45 contains a short–term as well as a long–term decay component. Both components together are incompatible with the generally successful two–component starquake theory of pulsar glitches. A suggestion based on the two–component model is proposed to explain the observations and would require the coupling between the solid core of the pulsar and its crust to be weaker than previously assumed, at least immediately after a corequake.

126.067 **Pulsar in Large Magellanic Cloud.**
IAU Circ., No. 3984 (1984).

126.068 **The hard X–ray light–curves of accreting magnetized white dwarfs.**
A. R. King, G. Shaviv.
Mon. Not. R. Astron. Soc., Vol. 211, No. 4, p. 883 – 893 (1984).
The authors point out that the hard ($\gtrsim 3$ keV) X–ray light–curves of accreting magnetic white dwarfs can be understood as due solely to occultations of the emission region by the white–dwarf body. The prevalence of quasi–sinusoidal hard X–ray light–curves amongst the intermediate polars probably implies that the emission regions occupy a large fraction ($2f \gtrsim 0.5$) of the white–dwarf surface; for the AM Her stars the authors tentatively conclude that $f \gtrsim 10^{-3}$. They discuss the likely behaviour of the light–curve amplitude as the X–ray brightness changes, and show that optical emission directly from the polecaps may dominate the pulsed light in some intermediate polars.

126.069 **Pulsar statistics: a study of pulsar luminosities.**
M. Prószyński, D. Przybycień.
Space Telesc. Sci. Inst., Prepr. Ser., No. 22, 7 pp. (1984). To appear in the proceedings of Green Bank Workshop on Millisecond Pulsars, Green Bank, 6 – 8 June, 1984.

126.070 **Recent results on the Crab–pulsar X–ray light curve.**
G. Hasinger.
MPE Prepr., No. 29, p. 12 – 15 (1984). Paper presented at the Symposium on X–Ray Astronomy '84, Bologna, Italy, June 26 – 30, 1984.

126.071 **Two new color–selected magnetic DA white dwarfs.**
J. Liebert, G. D. Schmidt, E. M. Sion, S. G. Starrfield, R. F. Green, T. A. Boroson.
Prepr. Steward Obs., No. 550, 23 pp. (1984). To appear in Publ. Astron. Soc. Pac.

126.072 **The optical spectrum of hydrogen at 160 to 350 million Gauss in the white dwarf GRW+70°8247.**
J. R. P. Angel, J. Liebert, H. S. Stockman.
Prepr. Steward Obs., No. 551, 25 pp. (1984). To appear in Astrophys. J.

126.073 **Discovery of oxygen in the PG1159 degenerate stars: a direct evolutionary link to O VI planetary nebula nuclei and confirmation of pulsation theory.**
E. M. Sion, J. Liebert, S. G. Starrfield.
Prepr. Steward Obs., No. 556, 22 pp. (1984).
The authors report the discovery of strong O VI absorption/emission lines in five members of the group of very hot degenerate stars whose prototype is the pulsating star PG1159–035. The optical spectra are dominated by C IV, O VI and He II, but with little evidence for nitrogen features. The authors confirm that the central star of the planetary nebula K–16, found to be a pulsating star by Grauer and Bond, is a member of the spectroscopic group.

126.074 **Detection and analysis of photospheric CNO features in the ultraviolet spectrum of the hot DO white dwarf PG 1034+001.**
E. M. Sion, J. Liebert, F. Wesemael.
Prepr. Steward Obs., No. 557, 28 pp. (1984).
High–resolution ultraviolet and optical spectra of the hot DO white dwarf PG 1034+001 are presented. The ultraviolet observations revealed the presence of absorption features due to highly–ionized species, N V (λλ1238, 1242), C IV (λλ1548, 1551), and O V (λ1371), at large redshifts (>35 km s$^{-1}$) with respect to the interstellar lines. On the basis of these shifts, the authors identify these features as being of photospheric origin. The optical observations of the He II λ4686 line reveal a sharp central emission reversal superposed on the broad Stark absorption wings. A preliminary abundance analysis of the ultraviolet metal features is made. The nitrogen abundance is a factor 7 above the solar value, the first occurrence of such a high abundance in a white dwarf star.

126.075 **On interpretation of the optical spectra of magnetic white dwarfs.**
Yu. V. Tikhomirov.
Astron. Tsirk., No. 1317, p. 3 – 8 (1984). In Russian.

126.076 **Spektralanalyse von Helium–armen heißen Subdwarfs.**
U. Heber, K. Hunger.
Mitt. Astron. Ges., Nr. 62, p. 248 (1984). Abstract. – See Abstr. 010.341.

126.077 **CASPEC–Beobachtungen des sdO's ROB 162 in NGC 6397.**
U. Heber, R. P. Kudritzki.
Mitt. Astron. Ges., Nr. 62, p. 249 (1984). Abstract. – See Abstr. 010.341.

126.078 **Braking index of pulsars.**
H.–r. Hang, D.–y. Wang.
Acta Astron. Sin., Vol. 25, No. 2, p. 107 – 111 (1984). In Chinese.
The statistical analysis of relation between log $|\dot{\omega}|$ and log $|\omega|$ for more than two hundred pulsars is made, from which it can be obtained apparently that the braking index of pulsars n ≅ 3. The reason why the statistical analysis of the log $|\dot{\omega}| \sim$ log $|\omega|$ is better than that of the log $|\dot{p}| \sim$ log $|p|$ is analysed. The authors also discuss a variety of energy loss mechanisms of the pulsar and several effects, which influence the braking index.

126.079 **Van Biesbroeck 3: a low–luminosity white dwarf, not an M dwarf.**
W. E. Kunkel, J. Liebert, T. A. Boroson.
Publ. Astron. Soc. Pac., Vol. 96, No. 585, p. 891 – 893 (1984).
Photometry and spectrophotometry of VB3, the distant companion to HR 3018, show it to be a white dwarf of low luminosity ($M_v \cong +15.6$) and not an M dwarf. The pair provides the best available example of a very old white dwarf with a main–sequence star of sufficiently early type that a useful empirical limit may eventually be derivable for the ages of the oldest degenerate stars.

126.080 **The parallax of pulsar 0950 + 08 and the local free electron density.**
C. R. Gwinn, J. H. Taylor, J. M. Weisberg, L. A. Rawley.
NASA Conf. Publ., NASA CP–2345, p. 281 – 283 (1984). – See Abstr. 012.089 (IAU Colloq. No. 81).
The authors report a parallax of $7.9 \pm 0.8$ mas for PSR 0950+08, corresponding to a distance of $130 \pm 15$ pc. The measured pulse dispersion of this pulsar implies an average free electron density of $0.023 \pm 0.002$ cm$^{-3}$ along the line of sight. This parallax measurement is subject to systematic errors and questions of interpretation which have not yet been fully explored.

126.081 **Strömgren trails of hot white dwarfs.**
J. C. Raymond.
NASA Conf. Publ., NASA CP–2345, p. 311 – 314 (1984). – See Abstr. 012.089 (IAU Colloq. No. 81).
High resolution IUE spectra of Feige 24 reveal sharp C IV absorption features at both the stellar velocity and the velocity of the interstellar lines. Dupree and Raymond have shown that the ionization zone created by Feige 24 itself accounts for the interstellar components of C IV and Si IV, and that the density of hot white dwarfs in the galaxy implies that similar ionized zones around hot white dwarfs make a significant contribution to the C IV column densities observed toward distant O and B stars. Typical white dwarf velocities of around 30 km/s and C IV ionization zone diameters less than 1 pc imply that the crossing time is comparable to the ionization time, so that non–equilibrium ionization models are required. Models of the resulting elongated trails are presented.

126.082 **On the nonlinear theory of generation of radiation in a pulsar plasma.**
A. B. Mikhajlovskij, G. I. Suramlishvili.
Astrofizika, Tom 20, Vyp. 3, p. 565 – 570 (1984). In Russian. English translation in Astrophysics, Vol. 20, No. 3.
Taking into account the electron–positron structure of pulsar plasma, the nonlinear mechanisms of generation of its radiation have been considered. Estimates are presented of a growth rate of emission energy on condition that the cyclotron frequency is less than the Langmuir frequency. This condition could be realized near the light cylinder. The considered mechanisms lead to radio emission for old pulsars and radio and infrared emission for young pulsars.

126.083 **Feige 24: wind/flare accretion by a hot DAZ1 degenerate.**
E. M. Sion, S. G. Starrfield.
Astrophys. J., Vol. 286, No. 2, p. 760 – 762 (1984).
The authors have constructed model stellar envelopes in order to represent the outer layers of the hot DAZ white dwarf component of the spectroscopic binary Feige 24. The results show that the metals, discovered by Dupree and Raymond, should diffuse rapidly into the interior unless a continuous source of these elements is present in the system. The authors propose that this source is a stellar wind from the red dwarf component of this detached close binary, and they provide diffusion and accretion calculations to support the hypothesis.

126.084 **Evidence for pulsed 1000 GeV gamma–rays from the Crab pulsar.**
J. C. Dowthwaite, A. B. Harrison, I. W. Kirkman,
H. J. Macrae, T. J. L. McComb, K. J. Orford, K. E. Turver,
M. Walmsley.
Astrophys. J., Lett. Ed., Vol. 286, No. 2, p. L35 – L38 (1984).
The authors have detected a persistent weak periodic emission of $\gtrsim 1000$ GeV γ–rays during observations of PSR 0531 in 1982 – 1983. The light curve for steady VHE γ–ray emission shows a narrow peak which is coincident with the main pulse at other wavelengths. The narrowness of the main pulse of the VHE γ–ray light curve and its coincidence with the radio and optical main pulse is indicative of a close coupling between the VHE γ–ray and the radio/optical emissions.

126.085 **Far and extreme–ultraviolet spectrophotometry of hot white dwarfs: prospects for the determination of the He/H ratio in DA stars.**
F. Wesemael, J. Liebert, R. F. Green.
Publ. Astron. Soc. Pac., Vol. 96, No. 586, p. 981 – 987 (1984).
The various spectroscopic methods currently available to determine the He/H ratio in hot DA white dwarfs are compared and contrasted. The observation of the optical He II λ4686 line, as well as that of the λ1640 line with the low–dispersion IUE camera, can only be used to determine the helium abundance for log(He/H) $\gtrsim$ -3. Low–resolution ($\lambda/\Delta\lambda \sim 20$) rocket–borne extreme ultraviolet spectrophotometry, such as that performed on HZ 43, is more sensitive to the abundance ratio, but only through the strength of the He II edge at 227 Å. The authors contrast these results with those expected from the Faint Object Spectrograph on the Space Telescope and from the Far Ultraviolet Spectroscopic Explorer.

126.086 **The effective temperature of the white–dwarf star and ZZ Ceti candidate Wolf 485A.**
S. W. Digel, H. L. Shipman.
Publ. Astron. Soc. Pac., Vol. 96, No. 586, p. 996 – 998 (1984).
The authors use high–resolution spectra of W 485A, and improved hydrogen–line broadening routines in their model–atmospheres program to find the temperature of W 485A; the authors' best estimate of $T_{eff}$, most consistent with the other data on the star, is 14,600K, outside the instability strip.

126.087 **Whatever happened to the millisecond pulsar's companion? – Or "too close for comfort".**
J. Shabam.
News Lett. Astron. Soc. N.Y., Vol. 2, No. 6, p. 5 – 6 (1984). Abstract. – See Abstr. 010.241.

126.088 **A new Crab–like pulsar in the Large Magellanic Cloud.**
D. J. Helfand, F. D. Seward, F. R. Harnden.
News Lett. Astron. Soc. N.Y., Vol. 2, No. 6, p. 7 – 8 (1984). Abstract. – See Abstr. 010.241.

126.089 **Arrival–time analysis for a millisecond pulsar.**
R. Blandford, R. Narayan, R. W. Romani.
J. Astrophys. Astron., Vol. 5, No. 4, p. 369 – 388 (1984).
The authors develop an analysis of pulsar arrival times so as to estimate the sensitivity of fast pulsars as detectors of gravitational radiation and dispersion–measure fluctuations under the assumption that they remain as good clocks as is indicated by

present observations. They explore the limits to the use of accurate arrival times to measure pulsar spin–down, position, proper motion and parallax distance, in the presence of a particular noise spectrum.

126.090 **Short–term intrinsic–intensity variations of pulsars.**
S. Krishnamohan, V. Balasubramanian.
J. Astrophys. Astron., Vol. 5, No. 4, p. 483 – 493 (1984).

The authors present a method to separate the autocorrelation function of the short–term broadband intensity variations of pulsars from that of the interstellar scintillations. The method assumes that the intrinsic variations are correlated over a bandwidth much larger than the decorrelation bandwidth for scintillations. Hence the ratio of the power in the variations due to the two causes depends on the bandwidth used. By applying the method to the intensity variations of 24 pulsars, the authors show that the presence of short–term intrinsic variations is very common in the radiation of pulsars. Quasi–periodicities were detected in the intensity variations of many pulsars, but their origin is not clear.

126.091 **Metal abundances in the hot DA white dwarfs Wolf 1346 and Feige 24.**
F. Wesemael, R. B. C. Henry, H. L. Shipman.
Astrophys. J., Vol. 287, No. 2, p. 868 – 873 (1984).

Photospheric metal lines observed in high–dispersion *IUE* spectra of the hot DA white dwarfs Wolf 1346 and Feige 24 are analyzed with detailed model synthetic spectra in order to derive C, N and Si abundances in these two stars. The implications of these abundance determinations for theories of element diffusion in, and accretion onto, the envelopes of hot DA stars are discussed.

**PULSAR: a balloon–borne experiment to detect variable low energy gamma–ray emissions.**
See Abstr. 034.013.

**Astrometry of 59 pulsars: a comparison of interferometric and timing positions.**
See Abstr. 041.014.

**Millisecond pulsars and the location of the solar system barycenter.**
See Abstr. 042.075.

**On the detection of positrons via the optical lines of positronium.**
See Abstr. 061.002.

**Strong electromagnetic waves in a magnetized relativistic electron–positron plasma.**
See Abstr. 062.166.

**Computation of convective model atmospheres of cool white dwarfs.**
See Abstr. 064.027.

**X–rays from accretion of red giant winds.**
See Abstr. 064.068.

**The structure of detonation waves and nucleosynthesis during thermonuclear explosions of carbon–oxygen stellar cores and white dwarfs.**
See Abstr. 065.015.

**Fitting formula for the critical mass of relativistic white dwarfs with different chemical compositions.**
See Abstr. 065.019.

**Cooling of low–mass carbon–oxygen dwarfs from the planetary nucleus stage through the crystallization stage.**
See Abstr. 065.040.

**Diffusion–induced hydrogen burning and the evolution of DA white dwarfs.**
See Abstr. 065.054.

**White dwarfs. Big things in small packages.**
See Abstr. 065.067.

**Diffusion–induced hydrogen burning in white dwarfs.**
See Abstr. 065.074.

**Evolution of accreting white dwarfs, Type I supernova explosion, and white dwarf collapse.**
See Abstr. 065.084.

**Collapse and explosion of degenerate stars.**
See Abstr. 065.085.

**Non–radial oscillations of white dwarfs and neutron stars.**
See Abstr. 065.115.

**On the calibration of the mixing–length theory in pulsating white dwarfs.**
See Abstr. 065.116.

**The dimension of a variable star.**
See Abstr. 065.131.

**Do monopoles keep white dwarfs hot?**
See Abstr. 065.146.

**Light and line profile variations due to *r*–mode pulsations with an application to the ZZ Ceti star G117–B15A.**
See Abstr. 065.147.

**Radiation damping in general relativity.**
See Abstr. 066.046.

**R–S model and radio binary pulsars.**
See Abstr. 067.008.

**Pulsar magnetospheres.**
See Abstr. 067.012.

**Radio pulsars: a theory of subpulse formation and drift.**
See Abstr. 067.014.

**Physically realizable relativistic stellar structures.**
See Abstr. 067.016.

**Thermal X–ray emission from isolated, older pulsars: a new heating mechanism.**
See Abstr. 067.028.

**On the induced scattering in the relativistic electron–positron plasma.**
See Abstr. 067.040.

**Probing the interiors of neutron stars.**
See Abstr. 067.051.

**Champions of magnetic fields.**
See Abstr. 067.059.

**Pulsar space velocities and neutrino chirality.**
See Abstr. 067.060.

**Formation of a single millisecond pulsar by the coalescence of a neutron star and a massive white dwarf.**
See Abstr. 067.066.

**Rapid postglitch spin–up of the superfluid core in pulsars.**
See Abstr. 067.069.

**Rotating neutron star structure: implications of the millisecond pulsar PSR 1937 + 214.**
See Abstr. 067.070.

**Spherical propagation of large amplitude pulsar waves.**
See Abstr. 067.095.

**Cosmic–ray acceleration by pulsars.**
See Abstr. 067.128.

**Frictional acceleration of high–energy particles under astrophysical conditions.**
See Abstr. 067.129.

**Influence of electron–positron pair production on a pulsar magnetospheric structure.**
See Abstr. 067.144.

**Astrometric study of the subdwaf HR 4550.**
See Abstr. 111.014.

**Grain temperature, radiation pressure and electric potential in the vicinity of main sequence and white dwarf stars.**
See Abstr. 112.003.

**Faint spectrophotometric standard stars for large optical telescopes. I.**
See Abstr. 113.017.

**Photometry of faint blue stars. VI. Some "very definitely blue" PHL stars.**
See Abstr. 113.048.

**Iron in the spectra of central stars and very hot subdwarfs.**
See Abstr. 114.124.

**Identification of atomic lines in the spectra of magnetic degenerates.**
See Abstr. 114.126.

**Apsidal motion and the pulsating transient X–ray source A0538–66.**
See Abstr. 117.009.

**The evolution of a star–"planet" system in the double core phase.**
See Abstr. 117.026.

**Optical polarization from magnetized cataclysmic variables.**
See Abstr. 117.031.

**Evolution of low–mass close binaries and the problem of orbital period gap in cataclysmic variables.**
See Abstr. 117.037.

**Preliminary results of coordinated optical, UV and X–ray observations of magnetic white dwarfs in binaries.**
See Abstr. 117.069.

**Accretion disk limit cycle mechanism in twin–degenerate interacting binaries.**
See Abstr. 117.074.

**Subdwarf companions for HD 36629 and HD 81307.**
See Abstr. 117.083.

**Are cataclysmic variables the progenitors of Type I supernovae?**
See Abstr. 117.095.

**Two–frequency radio spectra during an outburst of the periodic radio star LSI +61°303.**
See Abstr. 117.097.

**Loss of angular momentum in binary system due to collisionless particles as monopoles, gravitinos, or mirror particles: does it exceed the gravitational radiation emission in binary system PSR 1913+16?**
See Abstr. 117.102.

**Spin evolution of magnetic white dwarfs in AM Herculis binaries.**
See Abstr. 117.116.

**The origin of the high and low luminosity states in magnetic cataclysmic variables.**
See Abstr. 117.134.

**Coordinated UV and optical observations of the AM Herculis object E1405–451 in the high and low states.**
See Abstr. 117.141.

**Comparison of the X–ray observations of the AM Herculis objects to theory.**
See Abstr. 117.142.

**The rotational and orbital evolution of cataclysmic binaries containing magnetic white dwarfs.**
See Abstr. 117.143.

**The magnetic field of the AM Herculis object H0139–68.**
See Abstr. 117.237.

**Hot subdwarfs in detached binary systems and thick–disk cataclysmic variables from the Palomar–Green survey.**
See Abstr. 117.256.

**Pseudo white dwarfs.**
See Abstr. 117.260.

**Period variation in the white dwarf eclipsing binary V471 Tauri.**
See Abstr. 119.073.

**Confinement of the Crab pulsar's wind by its supernova remnant.**
See Abstr. 125.035.

**Magnetohydrodynamic model of Crab nebula radiation.**
See Abstr. 125.036.

**The X–ray structure of the Crab nebula.**
See Abstr. 125.067.

**The X–ray structure of the Crab nebula.**
See Abstr. 125.068.

**An optical synchrotron nebula around the X–ray pulsar 0540–693 in the Large Magellanic Cloud.**
See Abstr. 125.082.

**On the supernova remnants produced by pulsars.**
See Abstr. 125.084.

**Voyager EUV and FUV observations.**
See Abstr. 131.253.

**Statistical survey of planetary nebulae: distances, masses, and distribution in the Galaxy.**
See Abstr. 134.058.

**Variations of the pulse profile and the X–ray intensity of GX301–2 (4U1223–62).**
See Abstr. 142.015.

**Discovery of a 50 millisecond pulsar in the Large Magellanic Cloud.**
See Abstr. 142.090.

**Search for $10^{15}$eV gamma–rays from the Crab pulsar and surrounding regions.**
See Abstr. 143.020.

**Pulsars in the Hyades Supercluster.**
See Abstr. 153.044.

**Ultraviolet radiation from white dwarfs and the carbon ionization in the galactic halo.**
See Abstr. 155.082.

**The anisotropic Compton effect in quasars and pulsars.**
See Abstr. 159.015.

# Interstellar Matter, Nebulae

## 131 Interstellar Matter (Molecular Clouds, Reflection Nebulae, etc.), Star Formation

131.001 **Interstellar extinction in the direction of the globular cluster M71.**
R. Janulis.
Astron. Tsirk., No. 1286, p. 1 – 2 (1983). In Russian.

131.002 **Are we in an old superbubble?**
D. E. Innes, T. W. Hartquist.
Mon. Not. R. Astron. Soc., Vol. 209, No. 1, p. 7 – 13 (1984).
Using a simple model for the thermal properties of very old supernova remnants and superbubbles, it is found that the observed background soft X-ray and extreme ultraviolet spectra can be produced by a superbubble formed by the injection of about $10^{52}$ erg into the interstellar medium surrounding the Earth roughly $4 \times 10^6$ yr ago.

131.003 **On the ortho–$H_2$/para–$H_2$ ratio in molecular clouds.**
D. R. Flower, G. D. Watt.
Mon. Not. R. Astron. Soc., Vol. 209, No. 1, p. 25 – 31 (1984).
The value of the ortho–$H_2$/para–$H_2$ ratio is relevant to discussions of collisional excitation processes in interstellar clouds. The authors consider the temporal evolution of the relative abundance of the ortho– and para–modifications of $H_2$ in interstellar clouds with $H_2$ densities ranging from $10^2$ to $10^5\ cm^{-3}$.

131.004 **An investigation of diffuse cloud chemistry – I. Observational tests for surface chemistry on grains.**
A. P. C. Mann, D. A. Williams.
Mon. Not. R. Astron. Soc., Vol. 209, No. 1, p. 33 – 49 (1984).
Theoretical modelling of the chemistry in diffuse clouds, both including and excluding a contribution from chemical reactions on dust grain surfaces, is compared with observational data. The authors illustrate the possible ambiguities arising when interstellar parameters are inferred from such models by reference to OH and CO. They show that the molecule NH should be very close to detection if grains do contribute to the chemistry, and a deeper search is recommended. The CH and CN observations are discussed in detail.

131.005 **The mass distribution of dark clouds.**
H. C. Bhatt, D. P. Rowse, I. P. Williams.
Mon. Not. R. Astron. Soc., Vol. 209, No. 1, p. 69 – 75 (1984).
Using the Lynds catalogue of dark clouds the authors have, by making plausible assumptions regarding their distances, obtained a number of mass distributions for such clouds in various regions. The distributions of cloud masses in the Orion and ϱ Oph complexes, where the distances are known, indicate a mass function of the form $N(M) \propto M^{-s}$ with the index $s$ near 1. In the field and in the Taurus complex, by contrast, the value of $s$ is near 1.5.

131.006 **Optical and ultraviolet study of five stars in the Pleiades open cluster.**
K. F. Younan, P. L. Dufton.
Mon. Not. R. Astron. Soc., Vol. 209, No. 1, p. 123 – 137 (1984).
Optical and ultraviolet observations of five low–reddened stars in the Pleiades open cluster have been obtained with an echelle spectrograph and the *IUE* satellite. The observed interstellar absorption lines have been analysed with different cloud models to estimate interstellar gas abundances for each line–of–sight. These results have been discussed in the light of current models for diatomic formation.

131.007 **On the distribution of interstellar matter around the Sun.**
F. Paresce.
Astron. J., Vol. 89, No. 7, p. 1022 – 1037 (1984).
Available data on interstellar line or EUV continuum absorption attributable to the local ($r \leqslant 100$ pc) interstellar medium are presented and analyzed as a whole for clues leading to a deeper understanding of the spatial distribution and physical properties of this material. By using these data in conjunction with upper limits provided by polarization and color excess surveys and relating everything to the neutral–hydrogen column density, it is shown that the data are consistent with a simple model that assumes the relatively pervasive presence of a tenuous ($n_H = 0.07\ cm^{-3}$), possibly warm ($T = 10^3 - 10^4$K) gas extending uniformly out to considerable distances from the Sun in most directions.

131.008 **A new interstellar molecule: tricarbon monoxide.**
H. E. Matthews, W. M. Irvine, P. Friberg,
R. D. Brown, P. D. Godfrey.
Nature, Vol. 310, No. 5973, p. 125 – 126 (1984).
The authors report the detection in TMC–1 of a new molecule, tricarbon monoxide ($C_3O$), whose pure rotational spectrum has only very recently been studied in the laboratory. As $C_3O$ is the first known interstellar carbon chain molecule to contain oxygen, its existence places an important new constraint on chemical schemes for cold interstellar clouds.

131.009 **A survey of the 21 cm line of neutral hydrogen in the galactic region $-3° \leqslant l \leqslant 21°$ and $-4° \leqslant b \leqslant +3°$.**
E. Braunsfurth, K. Rohlfs.
Astron. Astrophys., Suppl. Ser., Vol. 57, No. 2, p. 189 – 192 (1984). With 2 microfiches.
The 100 m telescope at Effelsberg has been used to survey the 21 cm line radiation of neutral hydrogen in the galactic region $-3° \leqslant l \leqslant 21°$, $-4° \leqslant b \leqslant 3°$ and the velocity range $-300\ km\ s^{-1} \leqslant v_{LSR} \leqslant +300\ km\ s^{-1}$ with sampling intervals 0°.1 and $2\ km\ s^{-1}$. The sensitivity of the survey is about 0.5K.

131.010 **Inflow of halo gas from the direction of the galactic North Pole.**
K. S. de Boer, B. D. Savage.
Astron. Astrophys., Vol. 136, No. 1, p. L7 – L8 (1984).
A far–UV echelle spectrum has been obtained of the UV–bright star vZ1128 in the globular cluster M3. This cluster lies in a direction 11° off the galactic North Pole at a distance of about 10 kpc. Only the usually strong interstellar lines are recognizable in the faint spectrum, and they show absorption at velocities from near $+40\ km\ s^{-1}$ to $-100\ km\ s^{-1}$. Since in the direction observed the line of sight component of the Milky Way rotation is negligible, the detected velocities demonstrate for this direction the motion of gas from the Milky Way halo toward the disk. Gas returning to the disk in a galactic fountain like flow may explain the observations.

131.011 **Star formation and the galactic magnetic field.**
L. Mestel, R. B. Paris.
Astron. Astrophys., Vol. 136, No. 1, p. 98 – 120 (1984).

The paper reviews the gross features of the interactions between the thermal, gravitational, centrifugal, and magnetic fields of a cool, massive gas cloud, condensing from a warm low-density inter-cloud medium permeated by the local galactic magnetic field. Special attention is paid to the consequences of spontaneous flattening of the cloud along the field, both for interpretation of observations and for the dynamics of break-up into sub-condensations. Magnetic braking is treated by approximate methods that are far simpler than the more rigorous earlier studies for spherical clouds, but which nevertheless yield the essential qualitative results.

131.012 **Ammonia clouds in absorption against Cas A.**
W. Batrla, C. M. Walmsley, T. L. Wilson.
Astron. Astrophys., Vol. 136, No. 1, p. 127 – 132 (1984).

Observations of the (1,1) and (2,2) inversion lines of ammonia in absorption against the supernova remnant Cas A are presented. A line at $\sim 0$ km $s^{-1}$ is present over the whole area of the continuum source; the features at –40 km $s^{-1}$ and –47 km $s^{-1}$ were only found towards the south of the supernova remnant. The rotational temperatures, derived from the relative populations of the (1,1) and (2,2) lines are 18K. This should correspond to a kinetic temperature of 22K. The lines are subthermally excited and $H_2$ densities of $5 \times 10^3$ to $10^4$ $cm^{-3}$ are consistent with the lines seen in absorption against the continuum source. In one position the authors find that this cloud is clumped and has an extent of 0.12 pc (8"). The relative abundances $x(NH_3) = (2.9 \pm 1.3) \times 10^{-9}$ and $x(H_2CO) = (2.2 \pm 0.3) \times 10^{-9}$ are lower than the values found in dark clouds and globules.

131.013 **Detection of $^{13}$CN in three galactic sources.**
M. Gerin, F. Combes, P. Encrenaz, R. Linke,
J. L. Destombes, C. Demuynck.
Astron. Astrophys., Vol. 136, No. 2, p. L17 – L20 (1984).

The authors report the detection of the isotopic radical $^{13}$CN in Sagittarius B2, Orion A and IRC + 10216. $^{13}$CN column densities are derived and compared with previous observations of CN in the same sources. Isotopic ratios $^{12}$CN/$^{13}$CN are obtained and discussed, in agreement with other determinations.

131.014 **Temperatures in the Orion molecular cloud as estimated from methyl cyanide observations.**
M. Andersson, J. Askne, Å. Hjalmarson.
Astron. Astrophys., Vol. 136, No. 2, p. 243 – 249 (1984).

The $J = 6 \to 5$ ($K = 0$–5) and $J = 4 \to 3$ ($K = 0$–3) transitions of methyl cyanide ($CH_3CN$) have been observed in Orion A. Because of the high dipole moment of $CH_3CN$ a thermal population distribution among its energy states does not usually exist in molecular clouds and the authors have found that neither a simple rotation diagram analysis nor the method of Hollis et al. (1981) is appropriate for determining kinetic temperatures. This has been verified by means of a statistical equilibrium analysis including 91 energy levels up to 200K. Using the statistical equilibrium analysis the authors estimate the kinetic temperatures in the Orion A hot core and compact ridge clouds to be $120 \pm 15$K and $\gtrsim 110$K, respectively. The Gaussian half power sizes of both emission regions appear to be $\lesssim 35''$. The statistical equilibrium analysis also gives the column density and an estimate of the density of the regions.

131.015 **G 35.2–0.74: a molecular cloud with bipolar and monopolar outflow?**
N. Matthews, L. T. Little, L.-A. Nyman, G. H. Macdonald.
Astron. Astrophys., Vol. 136, No. 2, p. 282 – 288 (1984).

Observations are described of the $J = 1 \to 0$ and $J = 3 \to 2$ $HCO^+$ emission from the molecular cloud G 35.2–0.74. Near the northern water maser, three components are evident in the spectra: low velocity, intermediate velocity material and high velocity outflow. Splitting of the $HCO^+$ lines near the southern nebulosity is similar to that observed in all other molecular transitions to date. Evidence is also presented for the existence of regions in both sources having molecular hydrogen densities of $10^5 - 10^6$ $cm^{-3}$. It is suggested that in the northern source, outflow from a central object embedded in a flattened distribution of ambient material is channelled in directions of largest density gradient, leading to bipolarity. In the southern source, the ambient cloud possesses greater spherical symmetry and prevents a similar outflow from escaping, leading instead to line splitting.

131.016 **Kinematics of local interstellar clouds.**
P. Belfort, J. Crovisier.
Astron. Astrophys., Vol. 136, No. 2, p. 368 – 370 (1984).

The authors have performed a statistical analysis of the radial velocities of H I clouds observed in recent 21–cm absorption surveys at Arecibo, Effelsberg and Green Bank. The study is analogous to that performed by Crovisier (1978) on the Nançay 21–cm survey and leads to similar results. The solar motion is found to coincide with the Standard Solar Motion within $\approx 1$ km $s^{-1}$. The nodal deviation $l_0$ and expansion velocity term $K$ are not significantly different from 0. The mean distance of H I clouds to the galactic plane is $<|z|> \approx 100$ pc and the cloud velocity dispersion is $<V_t^2>^{1/2} \approx 6.9$ km $s^{-1}$. The latter does not seem to be different for thick ($\tau_{21\ cm} > 0.1$) and thin clouds. The velocity dispersion of local molecular clouds, derived from $H_2CO$ and $^{13}$CO observations, appears to be notably smaller ($\approx 3.5$ km $s^{-1}$).

131.017 **Photodissociation rates of molecules by the interstellar radiation field.**
L. C. Lee.
Astrophys. J., Vol. 282, No. 1, p. 172 – 177 (1984).

The photodissociation rates of CO, NO, $H_2O$, HCN, $NO_2$, $SO_2$, $CS_2$, OCS, $NH_3$, $CH_4$, $H_2O_2$, and $C_2H_2$, dissociated by the interstellar ultraviolet background in the solar neighborhood are calculated for the 106 – 200 nm wavelength region. The processes for photodissociation of various interstellar molecules into fragments are discussed.

131.018 **Radiatively driven dust-bounded implosions: formation and stability of dense globules.**
M. T. Sandford II, R. W. Whitaker, R. I. Klein.
Astrophys. J., Vol. 282, No. 1, p. 178 – 190 (1984).

Two-phase radiation hydrodynamic calculations of the collapse of dusty globules illuminated by an external radiation source are presented. Anisotropic scattering by the dust grains and momentum coupling between the dust and gas phases are included. Two models are presented, an optically thin case and a 1 $M_\odot$ optically thick globule. Radiation pressure acts on the dust, which moves with subsonic velocity, and in both cases the ionization front eventually becomes dust bounded. The presence of dust lengthens the time scale for globule compression by a factor of 3 for the optically thin case. Results for the optically thick model indicate that the globule survives long enough for self-gravity to act, resulting in the formation of low mass stars or long-lived, stable dark globules.

131.019 **A measurement of the hydrogen recombination rate in the diffuse interstellar medium.**
R. J. Reynolds.
Astrophys. J., Vol. 282, No. 1, p. 191 – 196 (1984).

The intensity of the galactic Hα background at high galactic latitudes implies an average hydrogen recombination rate of $4 \times 10^6$ $s^{-1}$ per $cm^2$ of galactic disk for the region within 2 – 3 kpc of the Sun. This rate places an important constraint on possible sources of the ionization for the diffuse interstellar medium. O stars and perhaps planetary nebula nuclei are the only known sources of ionizing radiation with Lyman continuum fluxes equal to or greater than the hydrogen recombination rate.

131.020 **Detection of U45.379: an intense, peculiar unidentified line.**
H. Suzuki, N. Kaifu, T. Miyaji, M. Morimoto, M. Ohishi,
S. Saito.
Astrophys. J., Vol. 282, No. 1, p. 197 – 199 (1984).

The authors have detected a new interstellar line at 45.379 GHz from Sgr B2, TMC1, TMC1($NH_3$), and TMC2. Particularly in TMC1, the line is remarkably intense ($T_r \sim 4.2$K).

TMC1 has been surveyed for a wide range of frequency, but no hfs, harmonically related lines, nor fine structures near U45.379 have been detected.

131.021 **High-latitude molecular clouds.**
L. Blitz, L. Magnani, L. Mundy.
Astrophys. J., Lett. Ed., Vol. 282, No. 1, p. L9 – L12 (1984).
The authors report the detection of a large number of molecular clouds at high galactic latitude, a few of which were previously cataloged as dark nebulae. A large fraction of the clouds detected at $|b| > 30°$ has been mapped in order to determine the properties of this class of objects. The clouds are nearby ($\sim$100 pc) with typical diameters and masses of about 2 pc and roughly 65 $M_\odot$. The total number of these clouds, their surface density projected on the galactic plane and their various physical properties are estimated.

131.022 **How astronomers watch the birth of stars.**
L. Little.
New Sci., Vol. 101, No. 1399, p. 12 – 15 (1984). Abstr. in Phys. Abstr., Vol. 87, No. 1237, Entry 72931 (1984).

131.023 **Compaction of dust clumps in a protostellar cloud.**
A. Z. Dolginov, V. S. Kessel'man.
Sov. Astron., Vol. 28, No. 1, p. 45 – 50 (1984). English translation of 37.131.029.

131.024 **The rate of star formation in a galaxy derived from its integrated spectrum.**
O. K. Sil'chenko.
Sov. Astron. Lett., Vol. 10, No. 1, p. 7 – 10 (1984). English translation of 37.131.001.

131.025 **The ion composition of cooling interstellar gas.**
A. A. Suchkov, Yu. A. Shchekinov.
Sov. Astron. Lett., Vol. 10, No. 1, p. 13 – 15 (1984). English translation of 37.131.002.

131.026 **Interstellare Blasen im Weltall und im Computer.**
M. Różyczka.
Sterne Weltraum, 23. Jahrg., Nr. 8–9, p. 425 (1984).

131.027 **Compact infrared sources – protostars before IRAS.**
P. Brand.
Occas. Rep. R. Obs., Edinb., No. 13, p. 27 – 29 (1984). – See Abstr. 012.001.

131.028 **IRAS observations of galactic star formation.**
S. Harris, B. Baud, C. A. Beichman, J. P. Emerson.
Occas. Rep. R. Obs., Edinb., No. 13, p. 31 – 36 (1984). – See Abstr. 012.001.

131.029 **Far-infrared observations of some regions of star formation.**
W. M. Glencross, J. F. Lightfoot.
Occas. Rep. R. Obs., Edinb., No. 13, p. 37 – 40 (1984). – See Abstr. 012.001.

131.030 **Open cluster formation.**
M. Margulis, C. J. Lada.
Occas. Rep. R. Obs., Edinb., No. 13, p. 41 – 62 (1984). – See Abstr. 012.001.

131.031 **Numerical models of the spectral appearance of protostellar envelopes.**
H. W. Yorke.
Occas. Rep. R. Obs., Edinb., No. 13, p. 63 – 82 (1984). – See Abstr. 012.001.

131.032 **Relation of core structure to high velocity outflow in G35.2–0.74 and other molecular clouds.**
L. T. Little.
Occas. Rep. R. Obs., Edinb., No. 13, p. 91 – 107 (1984). – See Abstr. 012.001.

131.033 **Masers associated with regions of recent star formation.**
R. J. Cohen, P. R. Rowland.
Occas. Rep. R. Obs., Edinb., No. 13, p. 117 – 128 (1984). – See Abstr. 012.001.
The authors have recently carried out observations of $H_2O$ masers which promise to shed new light on the two related questions: how do the masers work, and what is the structure of the masing regions?

131.034 **Time-dependent chemistry of molecular clouds.**
G. D. Watt.
Occas. Rep. R. Obs., Edinb., No. 13, p. 133 – 146 (1984). – See Abstr. 012.001.

131.035 **The significance of comets for star formation.**
S. V. M. Clube.
Occas. Rep. R. Obs., Edinb., No. 13, p. 151 – 158 (1984). – See Abstr. 012.001.

131.036 **Future theoretical needs in studies of star formation.**
H. Zinnecker.
Occas. Rep. R. Obs., Edinb., No. 13, p. 211 – 220 (1984). – See Abstr. 012.001.

131.037 **Angular momentum transfer by gravitational torques and the evolution of binary protostars.**
A. P. Boss.
Mon. Not. R. Astron. Soc., Vol. 209, No. 3, p. 543 – 567 (1984).
The efficiency of angular momentum transport by gravitational torques is investigated semi-analytically for two idealized models. The first model, a rotating ellipsoid embedded within another ellipsoid, is compared with numerical results for the fission instability of a rapidly-rotating polytrope. The second model is a binary system embedded within a rotating ellipsoid. The binary orbital angular momentum decreases significantly when the phase angle with the ellipsoid is constant; the binary separation may then decrease by a factor of $10^2$ within about an orbital period. For a variable phase angle, little secular loss of orbital angular momentum occurs.

131.038 **The 3 μm ice band in Taurus: implications for interstellar chemistry.**
A. P. Jones, D. A. Williams.
Mon. Not. R. Astron. Soc., Vol. 209, No. 4, p. 955 – 960 (1984).
Observations of the ice band in the Taurus molecular cloud are shown to require the formation of ice by surface reactions between oxygen atoms and hydrogen. Reactions with a variety of species lead to a mantle with $H_2O : NH_3$ composition which is consistent with that deduced from profile fitting the observed feature. Mantle loss from warm grains or by sputtering in shocks makes a substantial contribution to interstellar molecular abundances, confirming the importance for interstellar chemistry of surface reactions in addition to gas phase networks.

131.039 **Rotational excitation of molecular ions in interstellar clouds: $HCO^+$ and $HCS^+$.**
T. Monteiro.
Mon. Not. R. Astron. Soc., Vol. 210, No. 1, p. 1 – 11 (1984).
Potential energy surfaces for the $HCO^+$/He and $HCS^+$/He systems have been computed. These surfaces have been used to evaluate collisional cross sections and rate coefficients for the rotational excitation of $HCO^+$ and $HCS^+$ by collision with helium. The astrophysical implications of the results are discussed.

131.040 **Observations of the 14.5-GHz transition of interstellar $H_2CO$ towards continuum sources and dark clouds.**
F. F. Gardner, J. B. Whiteoak.
Mon. Not. R. Astron. Soc., Vol. 210, No. 1, p. 23 – 41 (1984).
Observations of the $2_{11} - 2_{12}$ transition of ortho-$H_2CO$ at 14.5 GHz have been made with the Parkes radio telescope towards about 100 continuum sources and dark clouds and compared with observations of the $1_{10} - 1_{11}$ transition. The observed linewidths were similar for both transitions, with median values of 4.8 km $s^{-1}$ for the clouds towards continuum sources, and 1.7 km $s^{-1}$ for dark clouds. Absorbing clouds near H II regions

have higher excitation temperature (about 7K) and median optical depths (0.045 and 0.14 at 14.5 and 4.8 GHz) than the values (3.5K, 0.01, 0.08) for clouds well-separated from the H II regions. Comparison of the observations with an LVG model based on collisional excitation is discussed.

131.041 **Star formation from hierarchical cloud fragmentation: a statistical theory of the log-normal Initial Mass Function.**
H. Zinnecker.
Mon. Not. R. Astron. Soc., Vol. 210, No. 1, p. 43 – 56 (1984).
The dispersion and the shape of the observed Initial Mass Function (IMF) of field stars at birth is explained in terms of a theoretical model in which star formation is due to a hierarchical cloud fragmentation process, much like a family tree with a number of levels ($\leqslant 5$). The model is a Monte Carlo version of Bodenheimer's (1978) scheme of hierarchical fragmentation which is based on successive stages of ring formation and ring fragmentation of a rotating interstellar cloud. The predictive power of the model includes the black dwarf IMF and furthermore the frequency distribution of the orbital periods of binary stars.

131.042 **On $C^0$ and CO in dense interstellar clouds – evidence that cloud material is frequently shocked.**
D. A. Williams, T. W. Hartquist.
Mon. Not. R. Astron. Soc., Vol. 210, No. 1, p. 141 – 145 (1984).
A model in which low velocity shocks in molecular clouds periodically return grain mantle material to the gas provides an understanding of the observed high C:CO ratio observed in a variety of objects; of the low fractional abundance of molecules observed in some quiescent dark clouds; and of the process limiting the accretion of molecules on to grains.

131.043 **The size and mass distribution of Galactic molecular clouds.**
S. Drapatz, H. Zinnecker.
Mon. Not. R. Astron. Soc., Vol. 210, No. 1, p. 11P – 14P (1984).
By combining observational results on small and large molecular clouds, the spectra of average cloud sizes and masses are obtained, resembling log-normal distributions. The shape of these distributions should place constraints on the mechanism of cloud formation.

131.044 **Thermal evolution of primordial gas clouds and mass range of zero-metal stars.**
Y. Sabano, Y. Yoshii.
Sci. Rep. Tôhoku Univ., Eighth Ser., Vol. 5, No. 1, p. 51 – 86 (1984).
The authors investigate the evolution and fragmentation of a gas cloud with the primordial chemical composition. The present paper primarily deals with the fragmentation due to gravitational and thermal instabilities to see what mass range of density fluctuations is actually excited during the evolution of the primordial gas cloud.

131.045 **Dynamics of interstellar clouds sweeping from a rotating galaxy during its motion in the intracluster medium.**
A. G. Kritsuk.
Astrofizika, Tom 19, Vyp. 3, p. 471 – 482 (1983). In Russian. English translation in Astrophysics, Vol. 19, No. 3.
The dynamics of interaction between interstellar clouds of a disk-like galaxy and the intracluster medium is considered for arbitrary inclination $i$ of the rotation axis of the galaxy to its velocity. Numerical calculations have shown the time of sweeping of clouds from the disk.

131.046 **On a particuliarity of cometary nebulae.**
R. A. Vardanyan.
Astrofizika, Tom 19, Vyp. 3, p. 579 – 583 (1983). In Russian. English translation in Astrophysics, Vol. 19, No. 3.
The aim of this paper is to determine the directions of inclinations of cometary nebulae (in or against the clock direction) and to find out whether the inclinations in different parts of the Galaxy are randomly distributed. It is shown that the inclinations of cometary nebulae have definite directions in different parts of the Galaxy. This can be explained by the influence of the local magnetic field and not by the influence of randomly oriented stellar magnetic fields (in different parts of the Galaxy) on the matter thrown out from the stars.

131.047 **Star formation in Bok globules and low-mass clouds. II. A collimated flow in the Horsehead.**
B. Reipurth, P. Bouchet.
Astron. Astrophys., Vol. 137, No. 1, p. L1 – L4 (1984).
The Horsehead has been studied at optical and near-infrared wavelengths, and has been found to be an active site of low-mass star formation. In particular, deep CCD images have revealed a shock-excited collimated flow-region emanating from an infrared source. Several other nebulous regions are discussed. The Horsehead is a cloud core, which is presently emerging from its parental cloud due to the destructive influence of the nearby OB-system σ Orionis. The onset of low-mass star formation is discussed in this context.

131.048 **Identification of the "unidentified" IR emission features of interstellar dust?**
A. Leger, J. L. Puget.
Astron. Astrophys., Vol. 137, No. 1, p. L5 – L8 (1984).
The authors propose a consistent explanation of the so-called unidentified IR emission features of interstellar dust. They are one of the longest-standing puzzles in IR astronomy. Two mechanisms have been proposed: IR fluorescence of molecules frozen on grains (Allamandola et al. 1979) and equilibrium thermal emission of small grains ($a \sim 100$ Å) by Dwek et al. (1980). In this letter it is shown that the transient heating of very small grains ($a \sim 5$ Å) suggested by Sellgren (1984) can give a much more satisfactory explanation.

131.049 **Ammonia observations and star counts in the Taurus dark cloud complex.**
M. Gaida, H. Ungerechts, G. Winnewisser.
Astron. Astrophys., Vol. 137, No. 1, p. 17 – 25 (1984).
The $NH_3$(1,1) and (2,2) lines have been observed towards the cores of the clouds L1489, Taurus 3C, TMC1, and L1517 in the Taurus-Auriga dark cloud complex. From these measurements the authors estimate the gas kinetic temperature to be close to 10K and the gas density to be in the range $10^4 cm^{-3}$– $10^5 cm^{-3}$. For the central Taurus region and the isolated filament L1517 the visual extinction $A_v$ has been determined from star counts in 4'.5 × 4'.5 fields. The measurements show that $A_v = 6^m$ is a lower limit for the visual extinction in the densest parts of the optical condensations. The cloud sizes derived from the extinction maps suggest that the elongated clouds tend to break into smaller cloudlets by fragmentation along their major axis.

131.050 **Winds from low-mass stars. Self-regulated star formation in Taurus, Ophiuchus, NGC 2264 and Orion.**
J. Franco.
Astron. Astrophys., Vol. 137, No. 1, p. 85 – 91 (1984).
Using the self-regulated star formation model proposed by Franco and Cox, the effects of winds from low-mass stars are analyzed. Two types of winds are considered; hypothetical protostellar winds and the observed T Tauri winds. It is found that the radial momentum input, in both cases, could regulate the evolution of the star-forming regions. The birthrates due to protostellar rotationally driven winds and T Tauri winds are very similar, and are proportional to the square root of the local density in the star-forming region. The predicted rates for Taurus-Auriga, ϱ Ophiuchi, NGC 2264, and the Trapezium region in Orion are in agreement with observational results.

131.051 **Ammonia in B335.**
K. M. Menten, C. M. Walmsley, E. Krügel, H. Ungerechts.
Astron. Astrophys., Vol. 137, No. 1, p. 108 – 112 (1984).
A map of the $NH_3$ (1,1) line towards the nearby isolated dark globule B335 is presented. The core of the globule with $H_2$ density of order $10^4$ $cm^{-3}$ is extended over a region 0.04 × 0.09 pc. The authors put an upper limit of $2 \times 10^{-14} s^{-1}$ upon the angular velocity of rotation of this gas. From measurements at 5 positions

of the $NH_3$ (2,2) line, they find that the temperature of the gas is 10 – 12K and is uniform over the core region. They interpret the data in the light of the discovery by Keene et al. (1983) of a compact far infrared object close to the peak of the ammonia emission. Model fits which have been made to the spectrum of this object are consistent with the presence of a low mass star surrounded by a circumstellar shell with density of order $10^6$ $cm^{-3}$ and diameter roughly 0.03 pc. However, it is suggested that ammonia is underabundant in the high density region. A search for $H_2O$ masers towards the core of the globule proved negative.

131.052 **The fluctuation spectrum of molecular clouds.**
L. G. Stenholm.
Astron. Astrophys., Vol. 137, No. 1, p. 133 – 137 (1984).

New ways to study the fluctuations (turbulence) in molecular clouds have been explored. The author studied the fluctuations within a single cloud. Further, he has analysed fluctuations in intensity and velocity in addition to the line width commonly used. Additionally, he made a strict separation between different molecules (isotopes) thus analysing each line parameter from each molecule separately. Ambiguities in the spatial definition were eliminated by working in Fourier transform space. The following tentative results were derived: The spectrum of fluctuations is similar to that predicted by the Kolmogorov relation. It is found, contrary to previous results, that the column density increases with increasing size. The smallest thermal fluctuations may be controlled by the mean free path of the $^{12}CO$ photons.

131.053 **The availability of phosphorus in the bacterial model of the interstellar grains.**
F. Hoyle, N. C. Wickramasinghe.
Astrophys. Space Sci., Vol. 103, No. 1, p. 189 – 193 (1984).

It is shown that phosphorus availability is in close balance with the phosphorus requirement of the bacterial grain model. This correspondence, which would be fortuitous if the interstellar grains were of inorganic origin, points to the correctness of the biological model.

131.054 **Remarks on a vibronic origin for the diffuse band spectrum.**
J. A. Nuth, B. Donn.
Astrophys. Space Sci., Vol. 103, No. 2, p. 353 – 356 (1984).

The authors have demonstrated that the diffuse band spectrum could not originate as a series of vibronic transitions from atomic impurities in interstellar MgO grains as has recently been proposed since laboratory spectra of such materials show no evidence for such complex structure. Furthermore, recent observational evidence indicates that the extinction feature at 160 nm in the standard interstellar extinction curve upon which the identification of interstellar MgO was founded is actually the result of a calibration error in the original observational data.

131.055 **Molecular hydrogen jets from the Orion nebula.**
K. N. R. Taylor, J. W. V. Storey, G. Sandell, P. M. Williams, W. J. Zealey.
Nature, Vol. 311, No. 5983, p. 236 – 237 (1984).

In an attempt to understand the relationship of the recently discovered complex of Herbig Haro objects in Orion to the IR sources in this region, the authors have carried out a survey of the molecular hydrogen *S*(1) line distribution. These observations have led to the discovery of a previously unsuspected structure of finger–like filaments of $H_2$ emission extending radially outwards from a common centre at IRC9.

131.056 **Determination of distances to interstellar clouds from OH absorption features.**
I. G. Kolesnik, L. V. Yurevich.
Astrofizika, Tom 19, Vyp. 4, p. 761 – 777 (1983). In Russian. English translation in Astrophysics, Vol. 19, No. 4.

The linewidths $\Delta V$ and depths $T_A$ of OH absorption features are shown to yield a value $\delta = \sqrt{(\Delta V/T_A)}$ which linearly increases with the distance $R$ to the absorbing cloud. The slope of the $\delta(R)$ relation depends on the antenna beam size. An interpretation of the relation obtained is proposed. Some examples for the determination of distances and kinematic properties of interstellar clouds are given.

131.057 **A flare of the $H_2O$ radio line emission in W49.**
L. Eh. Abramyan, A. P. Venger, I. V. Gosachinskij, R. A. Kandalyan, R. M. Martirosyan, V. A. Sanamyan, N. A. Yudaeva.
Astrofizika, Tom 19, Vyp. 4, p. 830 – 834 (1983). In Russian. English translation in Astrophysics, Vol. 19, No. 4.

The profiles of powerful $H_2O$ radio line emission at 1.35 cm wavelength obtained in May 1981 – December 1982 are given.

131.058 **Bipolar flow objects, NGC 2071 and GL 490.**
R. Kawabe, T. Takano, Y. Fukui.
Astron. Her., Vol. 77, No. 11, p. 277 – 281 (1984). In Japanese.

131.059 **Three new $H_2O$ masers in the vicinity of NGC 6357 and NGC 6334.**
S. Sakellis, M. I. Taylor, K. N. R. Taylor, R. A. Vaile, T. D. Han.
Publ. Astron. Soc. Pac., Vol. 96, No. 581, p. 543 – 545 (1984).

A search for water–vapour masers in the vicinity of the star–formation regions NGC 6357 and NGC 6334 has led to the discovery of three new sources. Their location in relation to infrared objects and the continuum emission is considered. Possible differences are observed between the two regions.

131.060 **The source of the bipolar outflow in Cepheus A.**
R. J. Cohen, P. R. Rowland, M. M. Blair.
Mon. Not. R. Astron. Soc., Vol. 210, No. 2, p. 425 – 438 (1984).

High–resolution MERLIN maps are presented of the OH and $H_2O$ masers in Cepheus A. These masers occur in four well–defined groups associated with four of the ultracompact H II regions in Cep A. From the present measurements it appears that the most compact of the H II regions is the source of the large–scale bipolar outflow which has been observed in CO, and that the outflow is collimated within $\sim 10^{15}$ cm of the central young star. The bipolar outflow is aligned in position angle with the interstellar magnetic field lines, and the collapsing molecular cloud is oriented at right angles to the field lines. A similar configuration is found in many other bipolar sources. This suggests that the interstellar magnetic field has played an important role in the evolution of these sources.

131.061 **Interstellar grain composition: a model based on elemental depletions.**
D. C. B. Whittet.
Mon. Not. R. Astron. Soc., Vol. 210, No. 2, p. 479 – 487 (1984).

Recent results on interstellar atomic depletions are reviewed. The total density of depleted material is $\sim 1.8 \times 10^{-26}$g $cm^{-3}$, consistent with that required in grains to explain the rate of interstellar extinction in the galactic plane. Carbon appears to be less important as a grain constituent than has previously been assumed, and models which attribute the optical extinction to bacteria are excluded. The metallic elements constitute approximately a third of the grain mass: this can be reconciled with dielectric optical properties only if the dominant chemical configurations are silicates and oxides, as in collected interplanetary dust. Problems involved in attributing the bulk of the optical extinction to silicates are briefly discussed.

131.062 **Velocity gradients in the bipolar CO gas–flow around L1551 IRS5.**
C. V. M. Fridlund, Å. Sandqvist, H. L. Nordh, G. Olofsson.
Astron. Astrophys., Vol. 137, No. 2, p. L17 – L19 (1984).

The authors report on high spatial resolution $^{12}CO$, J = 1–0, observations of the bipolar gas–flow surrounding L1551 IRS5. The data indicate an apparent acceleration of the gas and abrupt velocity changes within the flow. In at least one case, such a velocity change is spatially coincident with an optical nebulosity. This supports the theory that some nebulosities in molecular clouds are the result of the interaction between a wind from a pre–main–sequence star and the surrounding medium.

131.063 **Optical and infrared observations of Cep–A/GGD37.**
R. Lenzen, K.–W. Hodapp, J. Solf.
Astron. Astrophys., Vol. 137, No. 2, p. 202 – 210 (1984).

Photometric, spectrophotometric and polarimetric data of the near infrared emission region in Cep–A are presented. This region is shown to be an infrared reflection nebula illuminated by a central source near the position of the $H_2O$–maser spots. The supposed Herbig–Haro object GGD37 and the extended infrared source near this nebula are argued to be physically associated to this complex. A large–slit coudé–spectrum of GGD37 obtained in the red reveals high velocity dispersions of at least 300 km $s^{-1}$, indicating powerful dynamical processes going on in this region. The relative strength of the [S II] 6717, 6731 lines compared to Hα and [N II] suggests shock–excitation.

131.064 **Measurement of the formaldehyde ortho to para ratio in three molecular clouds.**
C. Kahane, M. A. Frerking, W. D. Langer, P. Encrenaz, R. Lucas.
Astron. Astrophys., Vol. 137, No. 2, p. 211 – 222 (1984).

The authors have measured the $H_2CO$ ortho to para ratio in two types of molecular clouds (a warm cloud, Orion A, and two dark clouds, TMC1 and L183) using the mm rotation lines of formaldehyde. This ratio provides information about the physical conditions for the formation of $H_2CO$ in these sources.

131.065 **Large–scale motions within the Orion Nebula (M42).**
C. Goudis, H. Hippelein, J. Meaburn, R. Songsathaporn.
Astron. Astrophys., Vol. 137, No. 2, p. 245 – 249 (1984).

Two separate sets of profiles of some visible emission lines over the whole 1° area of M42 have been obtained with Fabry–Pérot monochromators on the 1.9–m Kottamia and 1.23–m Calar Alto telescopes. Many large–scale, separate velocity components have been found and related to the underlying molecular cloud and filamentary structure of the ionized gas. It is difficult to explain these within a simple model of a radially expanding shell. Various alternatives are considered.

131.066 **Bullets: "interstellar plops and plunks".**
G. Tenorio–Tagle, M. Rozyczka.
Astron. Astrophys., Vol. 137, No. 2, p. 276 – 280 (1984).

A new model is proposed for the generation of "bullets" in interstellar space. The authors have investigated the phenomena that occur as a strong shock overtakes a high density condensation. If allowances for cooling are made and the shock speed is $\sim$100 km $s^{-1}$ (typical of a T Tauri star) a large degree of compression is reached and the material overtaken by the conical shock condenses into a fast moving bullet. The propagation of such a supersonic bullet when confined by ram pressure of the surrounding gas would generate a bow shock capable of explaining the hard spectrum of Herbig–Haro objects (HH). Because of the velocity difference between the bullet and the incident shock the authors' model predicts the detachment of HH objects from fast molecular flows.

131.067 **A large–scale dark cloud system in the northern sky.**
W. Schlosser, V. Görnandt.
Astron. Astrophys., Vol. 137, No. 2, p. 287 – 290 (1984). In German.

Wide–angle photographs of the northern Milky Way reveal a complex system of dark clouds covering the galactic longitudes from $l$ = 90° to about 210°. This Northern Dark Cloud System (NDS) is a coherent structure of individual clouds of preferentially cellular appearance. The Northern Dark Cloud System has to be assigned to the local spiral arm. This is substantiated by the large extent of the NDS in galactic latitudes and by a $b(l)$–behaviour similar to that of well established tracers of the local spiral arm. With a descending node of $l$ = 145° and an inclination of $i$ = 17° against the galactic equator it agrees well with other young objects belonging to the local arm. A dominant contribution of +I(=Perseus–arm) dust can be excluded since that arm is characterized by a descending node of $l$ = 100° (Kolesnik and Vedenicheva, 1979).

131.068 **The large–scale bubble structure of the interstellar medium, and the properties of the local spiral arm.**
N. G. Bochkarev.
Sov. Astron. Lett., Vol. 10, No. 2, p. 76 – 78 (1984). English translation of 37.131.106.

131.069 **Rotation measure of the $H_2O$ maser flare region in Orion–KL.**
L. I. Matveenko.
Sov. Astron. Lett., Vol. 10, No. 2, p. 82 – 83 (1984). English translation of 37.131.107.

131.070 **$HCO^+$ in NGC 6334.**
R. A. Batchelor, B. J. Robinson, M. G. McCulloch.
Proc. Astron. Soc. Aust., Vol. 5, No. 3, p. 363 – 367 (1984).

The authors present maps showing the distribution of $HCO^+$ ($J = 1 \rightarrow 0$ transition at 89.18855 GHz) in the direction of NGC 6334 as a function of radial velocity. The observations were carried out in May – June at the Epping 4–m millimetre–wave radio telescope.

131.071 **Ion–molecule synthesis of $C_3O$.**
E. Herbst, D. Smith, N. G. Adams.
Astron. Astrophys., Vol. 138, No. 1, p. L13 – L14 (1984).

The fractional abundance of the newly detected interstellar radical $C_3O$ in TMC–1 can be accounted for by ion–molecule, gas phase reactions.

131.072 **On the statistics of galactic H I clouds.**
K. R. Anantharamaiah, V. Radhakrishnan, P. A. Shaver.
Astron. Astrophys., Vol. 138, No. 1, p. 131 – 139 (1984).

The number density and random motions of interstellar H I clouds have been studied using an entirely novel method involving the comparison of the terminal velocities of H I absorption spectra in the direction of H II regions with their recombination line velocities. The authors confirm the conclusions of an earlier study attributing these velocity differences mainly to the random motions of the clouds. They also show that these motions are better described in terms of two populations with quite different velocity dispersions than by a single population with an exponential distribution of velocities.

131.073 **A circumstellar disk in Cep A.**
R. Güsten, R. Chini, T. Neckel.
Astron. Astrophys., Vol. 138, No. 1, p. 205 – 211 (1984).

$NH_3$ observations towards the Cep A molecular cloud reveal a circumstellar torus to be the relic of a recent star forming process. The authors derive its characteristics by fitting synthetic line profiles to the data. The line shapes are consistent with Keplerian orbits around a central source and contraction (or expansion) of the torus. The disk is warm (kinetic temperature $\sim$50K) and massive ($\sim$15 $M_{\odot}$). It fits into the evolutionary scheme of a collapsing interstellar cloud, possessing sufficient angular momentum to form a large–scale rotating ring in its equatorial plane.

131.074 **Two Bok globules with active star formation.**
B. Reipurth.
Messenger, No. 37, p. 14 – 16 (1984).

131.075 **UV studies of the interstellar medium.**
J. C. Blades.
ESA Spec. Publ., ESA SP–218, p. 11 – 20 (1984). – See Abstr. 012.024.

Investigation of the chemical composition and physical condition of interstellar gas and dust in our Galaxy, and in external systems, has continued to be pursued vigorously, making use of much new data from the *IUE* satellite. This review sets out to describe some of the advances that have been made over the last few years in several research areas, including the local interstellar environment, the dust, and the distribution of gas away from the plane of our Galaxy.

131.076 **On the distribution of interstellar matter around the sun.**
F. Paresce.
ESA Spec. Publ., ESA SP–218, p. 131 (1984). Abstract. – See Abstr. 012.024.

131.077 **The interstellar absorption in the direction of some nearby A stars.**
R. Freire Ferrero.
ESA Spec. Publ., ESA SP–218, p. 133 – 137 (1984). – See Abstr. 012.024.

The author investigates the interstellar medium in the direction of 4 A stars that are fast rotators, in the range 1250 to 1800 Å, and 7 fast rotators and 2 low rotators in the range 2000 to 3000 Å. Interstellar Mg II absorption is the most striking observational result appearing centered or not in the stellar line cores. Other interstellar lines are most probably present like Fe II, Al II, Si II, C II and S II ones.

131.078 **Use of high resolution IUE spectra of Mg II in cool stars to probe the local interstellar medium.**
P. Molaro, J. E. Beckman, L. Crivellari, M. L. Franco, G. Vladilo.
ESA Spec. Publ., ESA SP–218, p. 139 – 143 (1984). – See Abstr. 012.024.

The authors demonstrate that, given suitable radial velocity values for the stars and the local interstellar medium, the interstellar absorption can be distinguished from chromospheric self–absorption within the Mg II emissions from late–type stars.

131.079 **Further studies of nearby high velocity interstellar gas.**
B. Bates, W. Brown–Kerr, D. L. Giaretta, F. P. Keenan.
ESA Spec. Publ., ESA SP–218, p. 145 – 149 (1984). – See Abstr. 012.024.

Red–shifted high velocity gas components (HVC) have been detected in spectra of 9 out of a current survey of 47 stars. The majority of the survey stars whose spectra show HVC lie close to the Galactic plane near $l \sim 88$ to 110°. A new distance estimate places the HVC material within $\cong 190$ pc. The improved mapping of the sky and the distance estimate to the HVC material are discussed in terms of an association with nearby Radio Loops. In particular, the possibility that the material may be produced by two interacting SNR (Loops II and III) is considered.

131.080 **A general survey of the correlation of interstellar element depletion with line–of–sight density indicators.**
A. W. Harris, G. E. Bromage, C. Gry.
ESA Spec. Publ., ESA SP–218, p. 151 – 155 (1984). – See Abstr. 012.024.

The authors have investigated the density dependence of depletion for 14 elements using published gas–phase abundance data, including IUE data, and two different indicators of gas density in a sight–line: the column density of hydrogen, N(H), and the mean space density of hydrogen, $\bar{n}$(H). The authors find that the depletion estimates of many elements correlate much better with $\bar{n}$(H) than with N(H), whilst for Zn, O, N and D there is virtually no correlation with $\bar{n}$(H). The implications of the results for the distribution of depleted gas in the interstellar medium are discussed.

131.081 **Interstellar chlorine abundance: results and implications of a large–scale survey.**
A. W. Harris, G. E. Bromage.
ESA Spec. Publ., ESA SP–218, p. 157 – 160 (1984). – See Abstr. 012.024.

Gas–phase abundances of interstellar chlorine have been estimated for over 40 Galactic sight–lines using published Copernicus data and newly derived equivalent widths from IUE spectra. The major result of this survey is that Cl appears to be significantly depleted: the derived abundance values range over more than an order of magnitude and correlate very well with the mean space density of hydrogen, $\bar{n}$(H).

131.082 **Interstellar lines in high resolution IUE spectra of runaway B type stars.**
J. Berger, R. Freire Ferrero, A. M. Fringant, M. Gerbaldi, N. Morguleff.
ESA Spec. Publ., ESA SP–218, p. 161 – 166 (1984). – See Abstr. 012.024.

About 60 interstellar lines have been detected in the spectra of 3 runaway stars situated at large distance above the galactic plane. The lines identified come from (1) lower ionised species like C I, N I, O I, Mg I, S I and like C II, Si II, S II, Mg II, Fe II; (2) higher ionised species as: Al III, Si III, C IV, Si IV and probably N V. Some of these are very saturated ones like the Mg II lines and suggest the presence of Mg II away from the galactic disk. Provisional equivalent widths are given. The authors discuss about the possibility of a circumstellar origin for some of the narrow absorption lines detected.

131.083 **The interstellar medium in the Scorpius–Ophiuchus region.**
T. H. Pwa, S. R. Pottasch.
ESA Spec. Publ., ESA SP–218, p. 167 – 173 (1984). – See Abstr. 012.024.

From a sample of stars in the Sco–Oph complex the authors present a detailed analysis of seven lines–of–sight. The data used are high resolution IUE spectra with additional information from the visual and 21 cm radio observations. Abundances are compared to solar values. Using ionization equilibria electron densities are computed.

131.084 **A comprehensive ultraviolet extinction study: grain model and grain evolution implications.**
G. Chlewicki, J. M. Greenberg, S. Aiello, B. Barsella, P. Patriarchi, M. Perinotto.
ESA Spec. Publ., ESA SP–218, p. 507 – 510 (1984). – See Abstr. 012.024.

Reliable ultraviolet extinction curves for about 110 stars have been derived. At least 90% of these exhibit a uniform shape in the FUV ($\lambda < 1800$ Å) confirming earlier results. Strong variations in the amount of FUV extinction and, in some cases, in the hump occur systematically in some young OB associations. In the diffuse medium the correlation with visual extinction is much stronger for the hump than for the FUV. The data give further support to the tri–modal grain model.

131.085 **Spatial relations between neutral atomic hydrogen, carbon monoxide and hydroxyl radical in diffuse clouds.**
J. Crovisier, I. Kazès, J. Brillet.
Astron. Astrophys., Vol. 138, No. 2, p. 237 – 245 (1984).

The authors have observed the $^{12}$CO and $^{13}$CO 2.6–mm lines in the direction of a dozen diffuse clouds. In a selection of 5 of these regions they have also observed the OH 18–cm lines and partially mapped the $^{12}$CO and H I 21–cm lines. The $^{12}$CO, $^{13}$CO, and OH lines are well correlated. The $^{13}$CO/$^{12}$CO line ratio appears to be $\sim$4 times smaller in these clouds than in dense molecular clouds; the resulting $^{13}$CO/$^{12}$CO abundance ratio is of the order of 1/60. The relation between the 21–cm line and molecular lines is not clear. Some molecular lines have 21–cm self–absorption counterparts, others have 21–cm emission counterparts. In many cases the 21–cm profile does not show any narrow feature at the molecular velocity. This may be attributed to saturation of the 21–cm line.

131.086 **Interstellar lithium and the $^{7}$Li/$^{6}$Li ratio in diffuse clouds.**
R. Ferlet, M. Dennefeld.
Astron. Astrophys., Vol. 138, No. 2, p. 303 – 310 (1984).

The $\lambda$6708 doublet of interstellar Li I has been observed at high spectral resolution (3 km s$^{-1}$) toward $\delta$ Sco and $\zeta$ Oph with a Reticon detector yielding an extremely high signal to noise ratio around 4000. This first detection toward $\delta$ Sco gives a measured equivalent width of $0.25 \pm 0.03$ mÅ. For the first time in the interstellar medium, the authors derive a $^{7}$Li/$^{6}$Li ratio of 38 for a diffuse cloud in front of $\zeta$ Oph. The lower limit of the error bar (25) is clearly incompatible with the ratio measured in meteorites, and is not yet explained by recent models of galactic evolution.

It is suggested that a local inhomogeneity possibly due to an enrichment by material processed in an earlier generation of stars is observed. For both stars, the total lithium column-densities are calculated and compared with those of potassium, sodium, rubidium and hydrogen. It is concluded that all the alkalis, potassium, rubidium, and lithium are depleted by the same amount.

131.087 **TMC1-like cloudlets in HCL2.**
J. Cernicharo, M. Guélin, J. Askne.
Astron. Astrophys., Vol. 138, No. 2, p. 371 – 379 (1984).

The authors have studied, in the lines of $NH_3$, $HC_3N$, $HC_5N$, $HC_7N$, $C_3N$, and $C_4H$, the three most conspicuous gas condensations of Heiles' Cloud 2 (HCL2) besides TMC1 (Taurus molecular cloud one). One of these condensations appears to be the strongest source of cyanopolyyne ($HC_{2n+1}N$) emission in the sky, after TMC1, and is the fourth source in which $HC_7N$ and $C_3N$ have been detected. Another condensation shows a strong gradient in the $NH_3$/cyanopolyyne abundance ratio over a distance of 0.2 pc.

131.088 **Sub-millimetre observations of the Bok globule B335.**
G. Gee.
Observatory, Vol. 104, No. 1062, p. 211 – 212 (1984). Abstract. – See Abstr. 010.682.

131.089 **A mathematical model for interaction network of interstellar gas and interstellar diffusive energy.**
M. Fujimoto, S. Ikeuchi.
Publ. Astron. Soc. Jpn., Vol. 36, No. 2, p. 319 – 332 (1984).

Interstellar gas condenses and protostars are born therefrom. Diffusive energy is released from newborn stars and supernova explosions in the form of cosmic rays, magnetic fields, turbulence, and gaseous thermal motion. The authors assume a model network that the diffusive energy enhances or suppresses the condensation of interstellar gas and subsequent stellar formation depending on its magnitude. A linear analysis is made and shows that a wavy pattern arises in the interstellar matter and propagates as a decaying wave. A stochastic "phase transition" is examined by superposing fluctuations of small amplitude on the model network.

131.090 **Large-scale HCN hyperfine anomalies.**
J. Cernicharo, A. Castets, G. Duvert, S. Guilloteau.
Astron. Astrophys., Vol. 139, No. 1, p. L13 – L15 (1984).

The authors report observations of the J = 1–0 line of HCN in Heiles Cloud 2 (HCL2). About half a square degree was mapped in the emission of this line. In each observed position the two ratios $R_{02} = I(F=0–1)/I(F=2–1)$ and $R_{12} = I(F=1–1)/I(F=2–1)$ are far from their LTE values, and they show a complicated spatial distribution. The authors suggest that absorption and diffusion of the radiation emitted from the core in a low density envelope is responsible for the observed behaviour of the HCN emission. There is some evidence that the HCN abundance changes from source to source.

131.091 **CO (J = 2→1) observations of molecular clouds associated with H II regions from the southern hemisphere.**
J. Brand, M. D. P. van der Bij, C. P. de Vries, F. P. Israel, T. de Graauw, H. van de Stadt, J. G. A. Wouterloot, A. Leene, H. J. Habing.
Astron. Astrophys., Vol. 139, No. 1, p. 181 – 195 (1984).

The authors have searched for CO ($J$ = 2→1) emission at 47 positions in the direction of southern H II regions, mainly taken from the RCW-catalogue. The observations are compared with CO results obtained by other workers and briefly discussed. There seems to be no significant difference between the $T_A{}^*(J = 2\rightarrow1)$ and $T_A{}^*(J = 1\rightarrow0)$ transitions. In general no systematic velocity difference between the molecular and ionized gas is found. Three objects were observed in more detail: G265.1+1.5 (RCW 36), G327.3-0.6, and G35.2-1.8 (W 48). The source G327.3-0.6, associated with RCW 97, is a relatively small CO cloud, showing the characteristics of bipolar outflow. The CO emission associated with W 48 is complex; there is evidence for self absorption. In the direction of RCW 36 there are at least two clouds or cloud components along the line of sight.

131.092 **The mobility of species adsorbed on the surfaces of interstellar grains: consequences for the formation of interstellar molecules.**
M. A. Leitch-Devlin, D. A. Williams.
Mon. Not. R. Astron. Soc., Vol. 210, No. 3, p. 577 – 587 (1984).

A fundamental parameter in determining the efficiency of surface reactions on interstellar grains is the mobility with which adsorbed atoms and molecules move on the surface to locate either chemically active sites or another adsorbed species. This surface mobility is calculated here in a fully quantum mechanical treatment for a general atom-surface interaction. The results are expressed as a contour plot in terms of a scaled grain temperature and a scaled height of the energy barrier to motion across the surface.

131.093 **Dynamical evolution of fragmenting gas clouds.**
S. Mikkola, S. Saarinen, M. Valtonen.
Astrophys. Space Sci., Vol. 104, No. 2, p. 297 – 321 (1984).

Previous numerical simulations have shown that under certain conditions rotating gas clouds break up in two equal parts. The authors explore the consequences, if this process continues and a planar, equal mass four-body system forms. The results of the four-body calculations are used in Monte Carlo simulation of fragmentation of a rotating gas cloud in a galactic nucleus. Then it appears possible that clusters of large numbers of objects form; the dynamical evolution of these clusters has been calculated. The results were applied to ejection of supermassive objects from galactic nuclei. The authors find that several statistical properties of double radio sources may be understood as resulting from the fragmentation process.

131.094 **Interstellar extinction and polarization by spheroidal dust grains.**
D. B. Vaidya, H. C. Bhatt, J. N. Desai.
Astrophys. Space Sci., Vol. 104, No. 2, p. 323 – 336 (1984).

Using the recently available exact computations of the scattering efficiencies of spheroidal particles numerical calculations of the extinction and polarization curves have been made for a distribution of particle sizes, shapes and orientations. The results are presented and compared with the observed interstellar extinction and polarization. Possible models for interstellar dust with nonspherical grains have been discussed.

131.095 **Interstellar magnesium abundances.**
M. J. Murray, P. L. Dufton, A. Hibbert, D. G. York.
Astrophys. J., Vol. 282, No. 2, p. 481 – 484 (1984).

An improved evaluation of the Mg II 1240 Å doublet oscillator strength is used in conjunction with recently published Copernicus observations to derive accurate Mg II column densities toward 74 stars. These imply an average of 40% of interstellar magnesium is in the gaseous phase. Magnesium depletion is examined as a function of various interstellar extinction and density parameters.

131.096 **Observations of narrow diffuse interstellar bands toward stars with low reddening.**
S. R. Federman, C. K. Kumar, P. A. Vanden Bout.
Astrophys. J., Vol. 282, No. 2, p. 485 – 490 (1984).

Measurements of the narrow diffuse interstellar bands at λλ5780, 5797, and 6613 are presented for 22 stars with $E(B-V) \lesssim 0.1$. For this low amount of reddening the lines of sight have molecular column densities that are largely either greater than $10^{18} cm^{-2}$ or less than $10^{17} cm^{-2}$. The bands are observed to be stronger where $N(H_2)$ is large. The correlation with $H_2$ content suggests a molecular origin for the bands, but abundance considerations argue against a gas-phase carrier. An interaction between impurity sites and molecular complexes on grain surfaces is suggested as an origin of the diffuse interstellar bands.

131.097 **CO line emission from shock waves in molecular clouds.**
B. T. Draine, W. G. Roberge.
Astrophys. J., Vol. 282, No. 2, p. 491 – 507 (1984).

Calculations are presented of the emission spectrum of rotationally and vibrationally excited carbon monoxide in shocked interstellar molecular clouds. The calculations are based upon hydrodynamical shock models that include the effects of magnetically driven ion–neutral streaming. They incorporate a variety of collision processes that produce excited CO molecules, including $H_2$ – CO collisions. Intensities are given for lines emitted from levels with $J \leqslant 60$, for $C$-type shocks with speeds $v_s$ from 5 to 50 km s$^{-1}$, in clouds with densities $n_H$ between $10^2$ and $10^6$ H nuclei per cm$^3$.

131.098 **Magnetic field structure in the Taurus dark cloud.**
A. Moneti, J. L. Pipher, H. L. Helfer, R. S. McMillan, M. L. Perry.
Astrophys. J., Vol. 282, No. 2, p. 508 – 515 (1984).

The authors have obtained optical and infrared polarimetry of sources in the direction of the Taurus cloud in order to study the magnetic field in this cloud and its possible role in the cloud's evolution. Most of the stars are background giant stars whose light shines through the cloud and is polarized by the cloud material. The transverse component of the magnetic field, as delineated by the polarization vectors, is generally perpendicular to the galactic plane, and the stratified structure of the cloud could be due to the effect of the magnetic field during the early stages of collapse. Three of the 13 embedded stars are strongly polarized with position angles nearly perpendicular to those of nearby field stars.

131.099 **Cyanide and isocyanide abundances in the cold, dark cloud TMC–1.**
W. M. Irvine, F. P. Schloerb.
Astrophys. J., Vol. 282, No. 2, p. 516 – 521 (1984).

The authors have observed emission from HCN, $H^{13}CN$, $HC^{15}N$, $HN^{13}C$, $H^{15}NC$, $HC_3N$, $CH_3CN$, and possibly $CH_3NC$, and determined and upper limit for $NH_2CN$, toward the cold, dark cloud TMC–1. The abundance ratio [HNC]/[HCN] = $1.55 \pm 0.16$ is at least a factor ~4 and ~100 greater than that observed toward the giant molecular clouds DR 21(OH) and Orion KL respectively.

131.100 **The inner core of a Bok globule.**
T. J. Jones, A. R. Hyland, J. Bailey.
Astrophys. J., Vol. 282, No. 2, p. 675 – 682 (1984).

Near–infrared photometry, spectroscopy, and polarimetry of field stars behind Bok globule 2 in the Southern Coalsack are presented. An upper limit on the far–infrared flux from the globule is given also. The projected reddening distribution shows no signs of a central density enhancement. Both the gas and the dust are very cold ($\lesssim$ 10K) and probably similar in value. The globule is not collapsing, but is supported by turbulence or possibly a magnetic field. There are definitely aligned grains in the globule itself. There is weak evidence for structure in the polarization map indicative of a mildly compressed magnetic field.

131.101 **High angular resolution CS ($J$ = 1–0) observations of the bipolar flow source near NGC 2071: can the CS compact cloud collimate the flow?**
T. Takano, Y. Fukui, H. Ogawa, H. Takaba, R. Kawabe, Y. Fujimoto, K. Sugitani, M. Fujimoto.
Astrophys. J., Lett. Ed., Vol. 282, No. 2, p. L69 – L71 (1984).

The authors have observed the CO bipolar flow source near NGC 2071 in the $J$ = 1–0 line of carbon monosulfide with 30″ resolution provided by the new 45 m Nobeyama telescope. They find a compact cloud with a size of ~0.3 pc × 0.15 pc embedded in an extended cloud. This compact molecular cloud is elongated orthogonally to the bipolar flow, and shows strong signs of rotation. An estimate of its mass (~10 $M_\odot$) indicates that it is probably not responsible for collimating the bipolar flow.

131.102 **High angular resolution CS observations of the CO bipolar flow source GL 490 with the 45 meter telescope.**
R. Kawabe, H. Ogawa, Y. Fukui, T. Takano, H. Takaba, Y. Fujimoto, K. Sugitani, M. Fujimoto.
Astrophys. J., Lett. Ed., Vol. 282, No. 2, p. L73 – L75 (1984).

Using the Nobeyama 45 m millimeter–wave telescope the authors have observed the CO bipolar flow source associated with an infrared source, GL 490, in the $J$ = 1–0 line of carbon monosulfide. The high angular resolution of the telescope, 30″, has allowed to find a compact CS cloud about 0.3 pc × 0.13 pc in extent around GL 490, which is elongated orthogonally to the direction of the CO bipolar flow. Part of the cloud shows signs of expansion at velocities of $\lesssim$ 4 km s$^{-1}$.

131.103 **High spatial resolution observations of Orion bright bar: CS ($J$ = 1–0) and H51α emission.**
T. Omodaka, M. Hayashi, T. Hasegawa.
Astrophys. J., Lett. Ed., Vol. 282, No. 2, p. L77 – L79 (1984).

High–resolution strip maps of CS ($J$ = 1–0) and H51α line emission across the Orion bright bar are presented. The intensity of CS emission is weaker in the H II region with a rapid increase at the optical bar, while the H51α line was detected only on the side of the H II region. The variation of the CS intensity can be reproduced by a beam–convolved step function with the position of the step coincident with the ionization front. The molecular cloud associated with the bright bar is redshifted relative to the ambient molecular cloud by about 2 km s$^{-1}$.

131.104 **The detection of interstellar methyldiacetylene ($CH_3C_4H$).**
J. M. MacLeod, L. W. Avery, N. W. Broten.
Astrophys. J., Lett. Ed., Vol. 282, No. 2, p. L89 – L92 (1984).

A new interstellar molecule methyldiacetylene ($CH_3C_4H$) has been detected in the molecular cloud TMC–1. The $J = 6 \rightarrow 5$ and $J = 5 \rightarrow 4$ transitions have been observed in their $K = 0$ and $K = 1$ components. The total column density of $CH_3C_4H$ lies in the range $2.0 \times 10^{13}$cm$^{-2} \leqslant [CH_3C_4H] \leqslant 3.1 \times 10^{13}$cm$^{-2}$ for rotational temperatures from 4K to 6K.

131.105 **Ammonia absorption toward NGC 7538 IRS 1: 2 arc second observations in the (3, 3) line.**
C. Henkel, T. L. Wilson, K. J. Johnston.
Astrophys. J., Lett. Ed., Vol. 282, No. 2, p. L93 – L96 (1984).

Absorption from the (3, 3) line of $NH_3$ has been detected at –60.0 km s$^{-1}$ toward the ultracompact H II region IRS 1 in NGC 7538 with the VLA. In single–dish spectra, the line is blended with an emission component at –56.5 km s$^{-1}$. With the 2″.5 × 2″.2 beam, however, line emission is resolved. This corresponds to a spatial extent of more than 10″. The molecular cloud covers at least 60% of the continuum source; this is approximately 1″, corresponding to 0.017 pc. From the line shape, the cloud may consist of several clumps of size less than 0.01 pc, each with a different radial velocity and smaller line width. In addition to the spectral line measurements, a 23.87 GHz continuum map of IRS 1, 2, 3 is presented.

131.106 **General constraints on the average scattering characteristics of interstellar grains in the ultraviolet.**
G. Chlewicki, J. M. Greenberg.
Mon. Not. R. Astron. Soc., Vol. 210, No. 4, p. 791 – 801 (1984).

The sizes and optical properties of interstellar grains broadly determined by the full interstellar extinction law extending from infrared to ultraviolet wavelengths lead to rather strict constraints on the average scattering characteristics of grains in the ultraviolet. The presence of relatively large (radii ~0.1 – 0.2 μm) grains implied by the visual extinction is the key to determining the maximum and minimum values of the average albedo $\bar{\alpha}$ and the scattering asymmetry $\bar{g}$ in the ultraviolet. Additional restrictions on the values of $\bar{\alpha}$ and $\bar{g}$ are derived from the most basic physical properties of the small particle population. This procedure produces a range of possible pairs of $\bar{\alpha}$ and $\bar{g}$ at each UV wavelength, which in turn provides a basis for critical evaluation of grain models and results derived from direct measures of scattered light.

131.107 **Modelling the ice band towards HD 29647.**
C. E. P. M. van de Bult, J. M. Greenberg.
Mon. Not. R. Astron. Soc., Vol. 210, No. 4, p. 803 – 811 (1984).

Calculations have been made on a core–mantle–mantle grain model whose outer mantle consists of dirty ice material accreted in a molecular cloud. The theoretical predictions of both the strength and shape of the $H_2O$ band are consistent with infrared observations of HD 29647. It is shown that the absence of a 2200 Å hump cannot be caused by mantles on small graphite particles and therefore must be due to a real depletion of the hump carrier.

131.108 **A submillimetre map of the W51 region.**
C. T. Cunningham, M. J. Griffin, G. Gee, P. A. R. Ade, I. G. Nolt.
Mon. Not. R. Astron. Soc., Vol. 210, No. 4, p. 891 – 897 (1984).

Submillimetre mapping of a section of the molecular cloud/H II region complex W51 with a resolution of 40 arcsec reveals an extended distribution of dust encompassing two separate density enhancements. The total mass of the cloud is up to one thousand times that of previously mapped objects such as OMC1, NGC 6334 and W3 and, like these other sources, it exhibits star formation sites which are coincident with the density peaks.

131.109 **Observations of diffuse interstellar bands toward T Tauri stars.**
D. M. Meyer, R. K. Ulrich.
Astrophys. J., Vol. 283, No. 1, p. 98 – 105 (1984).

The authors present observations of the diffuse interstellar bands at 5778 Å, 5780 Å, and 5797 Å toward several T Tauri stars. Equivalent widths of these bands are presented for Z CMa, DR Tau and DF Tau. The diffuse features in these objects appear to be much weaker than normal with respect to the extinction estimates of Cohen and Kuhi. The most likely explanation for this result is that the dust grains responsible for the diffuse bands are modified by some mechanism within the dense clouds enveloping the T Tauri stars. The impact of this modification on the wavelength dependence of T Tauri extinction is unclear beyond the likelihood that the value of *R* is slightly larger than normal.

131.110 **183 GHz water line variation: an energetic outburst in Orion KL.**
T. B. H. Kuiper, E. N. Rodriguez Kuiper, P. N. Swanson, D. F. Dickinson, M. J. Klein, P. Zimmermann.
Astrophys. J., Vol. 283, No. 1, p. 106 – 116 (1984).

Observations of the $3_{13}$–$2_{20}$ transition of water vapor in the direction of Ori MC1 in 1980 February show a 50% flux increase and an apparent additional redshift of approximately 2 km $s^{-1}$ relative to the line observed in 1977 December. The increase is attributed to the appearance of a new component at a velocity of 12 km $s^{-1}$ with respect to the local standard of rest. The authors have carried out statistical equilibrium calculations using the large–velocity–gradient formalism to develop a model for the emission and to identify the physical conditions to which the excitation of the $3_{13}$–$2_{02}$ transition is sensitive. The temporal coincidence between the 183 GHz increase and the 22 GHz water maser outburst suggests a common, impulsive cause, which has heated the gas in a part of the high–velocity source, enhancing the emission in both transitions.

131.111 **CS around Orion–KL: a large rotating disk.**
T. Hasegawa, N. Kaifu, J. Inatani, M. Morimoto, Y. Chikada, H. Hirabayashi, H. Iwashita, K.–i. Morita, A. Tojo, K. Akabane.
Astrophys. J., Vol. 283, No. 1, p. 117 – 122 (1984).

$J = 1$–$0$ emission of CS and $C^{34}S$ around Orion–KL has been mapped with an angular resolution of 33″. The region of large CS integrated intensity has a single maximum near the KL nebula and is elongated NE–SW. The peak velocity shows a systematic shift along the axis of the elongated feature. This is consistent with an existence of a large rotating disk around the KL nebula. The disk is $4 \times 10^4$AU (0.2 pc) in diameter and is rotating with a period of $5 \times 10^5$yr × sin *i*, where *i* is the inclination.

131.112 **The ratio of total–to–selective extinction in the Chamaeleon T1 and R Coronae Australis dark clouds.**
F. J. Vrba, A. E. Rydgren.
Astrophys. J., Vol. 283, No. 1, p. 123 – 128 (1984).

The ratio of total–to–selective extinction [$R = A(V)/E(B–V)$] in the Cha T1 and R CrA dark clouds is investigated via observations of background field stars seen through these clouds. The observations consist of *UBVRI* and *JHKL* photometry as well as classification spectrograms and yield individual *R* values by the color difference method. *R* is found to be essentially normal in the outer parts of the Cha T1 dark cloud, but several stars near the opaque core of this cloud show evidence of anomalous extinction with $R > 4$. In the R CrA dark cloud, *R* increases with depth into the cloud and reaches clearly anomalous values of about 5 for lines of sight with $A(V) = 3$ mag.

131.113 **Large–scale structure of molecular gas in Heiles Cloud 2: a remarkable rotating ring.**
F. P. Schloerb, R. L. Snell.
Astrophys. J., Vol. 283, No. 1, p. 129 – 139 (1984).

The authors have mapped the $J = 1 \rightarrow 0$ transition of $^{13}CO$ over a 3 $deg^2$ region in Heiles Cloud 2 using the Five College Radio Astronomy Observatory 14 m telescope. The map suggests that the structure of Heiles Cloud 2 is dominated by a quasi equilibrium rotating ring similar to those found in numerical calculations of the gravitational collapse of a rotating cloud. Within this ring, several dense condensations have been identified. These subcondensations, among which is the dense filament TMC–1, probably result from the instability of the ring to fragmentation and have masses on the order of the Jeans mass.

131.114 **Structure of the L1535 dark cloud and the velocity field in the Taurus molecular complex.**
P. F. Goldsmith, M. J. Sernyak Jr.
Astrophys. J., Vol. 283, No. 1, p. 140 – 146 (1984).

The dark cloud L1535 has been studied in the $J = 1 \rightarrow 0$ transitions of $^{12}CO$ and $^{13}CO$ with the Five College Radio Astronomy Observatory's 14 m antenna. This region, a condensation located at the eastern end of an extended filament in the Taurus complex, is found to have a size of 0.7 × 0.35 pc, and a mass of 20 $M_{\odot}$. The H$\alpha$ emission star Taurus–Auriga 3 and surrounding nebulosity are found to be intimately connected to the molecular cloud. The velocity field in L1535 contains a significant systematic component suggestive of rotation with a velocity gradient 1.3 km $s^{-1}$ $pc^{-1}$.

131.115 **Observations of the peculiar OH maser source G24.3+0.1.**
M. I. Pashchenko.
Pis'ma Astron. Zh., Tom 10, No. 10, p. 726 – 729 (1984). In Russian. English translation in Sov. Astron. Lett., Vol. 10.

18 cm OH observations and accurate position measurements of the peculiar maser source G24.3+0.1 were carried out. The profile of the 1667 MHz OH line consists of many variable emission features in the large velocity range from +60 to +120 km/s. An OH emission source of the type IIc has been detected near the maser. The G24.3+0.1 source seems to be a type Ia circumstellar maser in a star–formation region.

131.116 **Bursts of $H_2O$ maser emission in the source W75N.**
E. E. Lekht, R. L. Sorochenko.
Pis'ma Astron. Zh., Tom 10, No. 10, p. 737 – 740 (1984). In Russian. English translation in Sov. Astron. Lett., Vol. 10.

Results of observations of $H_2O$ maser emission from the source W75N during the period from October 1982 to December 1983 are presented. Two bursts of maser emission quite different in the character of evolution were detected. The nature of these phenomena is discussed.

131.117 **Spatial distribution of interstellar clouds toward the Rosette nebula.**
N. G. Guseva, I. G. Kolesnik, S. G. Kravchuk.
Pis'ma Astron. Zh., Tom 10, No. 10, p. 741 – 748 (1984). In Russian. English translation in Sov. Astron. Lett., Vol. 10.

In a 5° × 5° region the distribution of absorbing matter and O–B2 stars are investigated. The presence of a dust clouds complex at 0.8 kpc, which provides the main part of the observed CO emission is revealed. The infrared source CRL 961 is shown to belong to this complex and to be forming a B2 star of the age of $\sim 3 \times 10^6$ years. The distance of the supernova remnant Mon Loop is also 0.8 kpc. All these objects seem to belong to a star formation region related to NGC 2264. The Rosette nebula and the associated cluster NGC 2244 belong to a star formation region in the Perseus spiral arm at 1.7 kpc distance. Clouds collision is probably responsible for massive star clusters formation.

131.118 **Probing molecular clouds.**
M. L. Kutner.
Fundam. Cosmic Phys., Vol. 9, No. 3, p. 233 – 316 (1984).
Contents: (1) Introduction. (2) Molecular spectra: Diatomic molecules. Symmetric top molecules. Asymmetric top molecules. Line strengths. Fine structure and hyperfine structure. Molecules with internal motions. Isotopic substitutions. (3) Probing physical conditions: Molecular excitation. Radiative transfer. Combining radiative transfer and excitation. (4) Results for molecular clouds: Dust clouds. Star formation in dark clouds. Giant molecular clouds. Dense cloud cores. Maps of our galaxy and other galaxies. Interstellar chemistry and isotopic abundances. (5) Summary.

131.119 **Moléculas en el espacio interestelar.**
W. Pöppel.
Rev. Astron., Tomo 55, No. 227, p. 2 – 8 (1983).

131.120 **A model to determinate the slope of the initial mass function in a cluster of young stars.**
K. De Gioia–Eastwood.
Publ. Astron. Soc. Pac., Vol. 96, No. 582, p. 625 – 629 (1984).
The author presents a simple model aimed at determining the slope of the IMF in a cluster of young stars embedded in a cloud of hydrogen. In the model she adds the stellar and gaseous fluxes for this cluster and calculates the ratio of line to continuum fluxes and continuum colors as a function of the slope of a power–law IMF and the age of the cluster. Predictions at far ultraviolet wavelengths are included. The author finds that a determination of the slope of the IMF is possible if the cluster is older than about $2 \times 10^7$years or if the IMF is much steeper than in the solar neighborhood.

131.121 **Rapid variation of a narrow feature in the ground state $\upsilon = 0$, $J = 1 \rightarrow 0$ SiO emission from Orion A.**
A. D. Haschick, S. H. Zisk.
Astron. J., Vol. 89, No. 9, p. 1387 – 1390 (1984).
The letter reports the appearance of a narrow feature on the characteristically broad ground–state $\upsilon = 0$, $J = 1 \rightarrow 0$ spectrum of $^{28}$SiO in Orion–KL and its subsequent rapid evolution and disappearance over a seven–day period.

131.122 **High–angular–resolution observations of the OH masers in W51 (main).**
J. M. Benson, R. L. Mutel, R. A. Gaume.
Astron. J., Vol. 89, No. 9, p. 1391 – 1397 (1984).
The authors have mapped the W51 ground–state OH masers using a VLBI array. The masers are clustered in two separate regions, one of which is coincident with an ultracompact H II region. Both maser clusters are nearly $3 \times 10^{16}$cm in diameter. Two pairs of Zeeman split masers are identified in the 1720–MHz transition and indicate magnetic field strengths of 6 and 10 mG at two locations 500 AU apart in the compact H II region G49.5–0.4($e_2$).

131.123 **The interstellar bubble around θ Muscae.**
C. Cappa de Nicolau, V. S. Niemela.
Astron. J., Vol. 89, No. 9, p. 1398 – 1403 (1984).
H I observations of the interstellar medium surrounding θ Mus are presented. They show the existence of a hole 100 pc in diameter in the neutral hydrogen distribution due to the action of the strong stellar wind of the WR star which produces an expanding bubble of swept–up ambient material. Using the observational parameters, the authors discuss kinematical models for the H I shell around the bubble, and find that only the momentum conserving model applies.

131.124 **A low density molecular cloud in the vicinity of the Pleiades.**
S. R. Federman, R. F. Willson.
Astrophys. J., Vol. 283, No. 2, p. 626 – 631 (1984).
The central region of a small, low density molecular cloud, which lies to the south of the Pleiades cluster, has been studied through the use of molecular line observations. Column densities for CH, OH, $^{12}$CO, and $^{13}$CO are derived form the radio data. Analysis of the carbon monoxide data leads to a relatively high kinetic temperature of at least 20K, a low total gas density of $\sim 300 - 500$ cm$^{-3}$, and a column density of $\sim 4 \times 10^{17}$cm$^{-2}$ for $^{12}$CO.

131.125 **Interaction of the outflow and quiescent gas in Orion: HCO$^+$ aperture synthesis maps.**
S. N. Vogel, M. C. H. Wright, R. L. Plambeck, W. J. Welch.
Astrophys. J., Vol. 283, No. 2, p. 655 – 667 (1984).
The authors present high–resolution (17″) aperture synthesis maps of HCO$^+$ $J = 1$–0 emission at 89 GHz in the Orion molecular cloud obtained with the Hat Creek millimeter interferometer. The low spatial frequencies of the brightness distribution, which are not sampled by the interferometer, were derived from single–dish data taken with the Onsala 20 m telescope. The HCO$^+$ maps probe the interaction of an outflow from IRc2 with quiescent gas. Four components, with different distributions, kinematics, and chemistry, can be identfied in the HCO$^+$ maps. (1) The quiescent molecular ridge, which is located just outside the expanding ring, or "doughnut", of gas inferred from SO observations. (2) A compact component centered near IRc2. (3) A high–velocity bipolar outflow. It is argued that IRc2 is the source of the outflow. (4) An extended outflow, which consists largely of ambient gas which has been swept up by the high–velocity outflow.

131.126 **CN in dark interstellar clouds.**
R. M. Crutcher, E. Churchwell, L. M. Ziurys.
Astrophys. J., Vol. 283, No. 2, p. 668 – 674 (1984).
Observations have been made at 113 GHz of the $N = 1 \rightarrow 0$ transition of CN toward seven positions in six dark clouds. Five clouds were detected. Millimeter CN lines were detected for the first time toward dark clouds outside Taurus. The six detected positions were also observed at 226 GHz for $N = 2 \rightarrow 1$ transitions of CN; detections were made at two positions. The weakness of CN lines is explained by the low level of excitation rather than by low column densities.

131.127 **Thermal instability in molecular clouds.**
D. L. Gilden.
Astrophys. J., Vol. 283, No. 2, p. 679 – 686 (1984).
The cooling function for molecular gas is calculated, and it is found that in environments where CO cooling dominates, molecular gas may be thermally unstable. Two regimes of the condensation mode are isolated; isochorically cooling gas with temperature $10 < T < 200$K and density $n(H_2) \lesssim 10^5$cm$^{-3}$, and thermally equilibrated gas with temperature $35 < T$75K and density near 1000 cm$^{-3}$. It is suggested that thermal instability may be an important source of small–scale structure in fully molecular clouds.

131.128 **Interferometric observations of CO in Orion: hot core and plateau.**
C. R. Masson, G. L. Berge, M. J. Claussen, G. M. Heiligman, R. B. Leighton, K. Y. Lo, A. T. Moffet, T. G. Phillips, A. I. Sargent, S. L. Scott, P. G. Wannier, D. P. Woody.
Astrophys. J., Lett. Ed., Vol. 283, No. 2, p. L37 – L40 (1984).
The authors have observed $^{12}$CO and $^{13}$CO emission from Orion A, using the Owens Valley Millimeter Interferometer. They have detected CO emission from a source at 5 km s$^{-1}$ $V_{LSR}$ in the vicinity of the $NH_3$ "hot core" source. A mass of 5 $M_\odot$ is estimated for this source with a size of $\sim 12''$. The E–W size of the plateau source is measured to be $\sim 17''$, implying a high density ($\sim 5 \times 10^6$cm$^{-3}$) for the gas in the outflow.

131.129 **On the interpretation of the broad–band millimeter–wave flux from Orion.**
E. C. Sutton, G. A. Blake, C. R. Masson, T. G. Phillips.
Astrophys. J., Lett. Ed., Vol. 283, No. 2, p. L41 – L44 (1984).
Spectral observations of the core of Orion A at wavelengths around 1.3 mm show a high density of strong, broad emission lines. The combined flux in lines with peak antenna temperatures stronger than 0.2K accounts for approximately 40% of the

broad-band millimeter-wave flux from the region. Thus the broad-band flux from Orion A is in large part due to sources other than dust emission.

131.130 **The origin of the infrared luminosity in violent star-formation regions.**
A. W. Campbell, R. Terlevich.
Mon. Not. R. Astron. Soc., Vol. 211, No. 1, p. 15 – 30 (1984).
The authors present near-infrared photometry of a sample of 14 violent star-formation regions. Their colours show that most of the 2 µm flux in these objects is produced by ionized gas and cool evolved stars. The detection of a deep CO index ($\lambda$ = 2.3 µm) in two objects reveals that the cool evolved stars detected in them are red supergiants. The distribution of the violent star-formation regions in the infrared two-colour diagram is consistent with a sequence in ages ranging from about 1 – 7 million yr determined from optical spectrophotometry.

131.131 **Interstellar polyynes from the disruption of carbon grains.**
W. W. Duley, D. A. Williams.
Mon. Not. R. Astron. Soc., Vol. 211, No. 1, p. 97 – 103 (1984).
Carbon grains disrupted by shocks in diffuse interstellar clouds release substantial abundances of large carbon-chain molecules into the gas, where they are destroyed by photodissociation and gas-phase reaction. The column density arising from a single shock for molecules containing ~10 atoms is calculated to be $\sim 10^{11} cm^{-2}$. This is a value unlikely to be obtained by gas-phase reactions in diffuse clouds. The chemistry of the large carbon-chain molecules should reflect the conditions appropriate to the cool atmospheres of carbon-rich stars, rather than the situation in an interstellar cloud.

131.132 **Gas-cloud/supernova interactions and the properties of the interstellar medium.**
P. Clifford.
Mon. Not. R. Astron. Soc., Vol. 211, No. 1, p. 125 – 148 (1984).
A theoretical model of the interstellar medium is derived, in which the neutral hydrogen clouds and supernova explosions have a controlling influence. A review of observations shows that the quantity of warm gas ($T \sim 8000$K) in the Galaxy can be generated only if there is a cycling of mass from clouds which is much greater than evaporative losses alone. Mass loss during cloud-cloud collisions may provide an adequate source. If such a mass input into the diffuse medium is assumed, a stable model of the interstellar medium can be constructed which reconciles the observed supernova rate with the presence of warm gas.

131.133 **A search for high-velocity carbon monosulfide emission in star-forming regions.**
H. A. Thronson Jr., C. J. Lada.
Astrophys. J., Vol. 284, No. 1, p. 135 – 140 (1984).
High signal-to-noise $J = 2 \rightarrow 1$ millimeter-wave spectra of $^{12}C^{32}S$ were obtained toward nine active star-forming regions in an attempt to determine whether or not high-velocity flows were also characterized by high densities. Three of the objects (Ori A, S140, and Cep A) showed definite broad wings. One other (AFGL 490) may have such wings. However, on the basis of the results presented here, it appears that easily detectable, high-velocity CS emission is not widespread.

131.134 **The role of the galactic magnetic field in the evolution of a dark globular filament in Cygnus.**
D. McDavid.
Astrophys. J., Vol. 284, No. 1, p. 141 – 143 (1984). With plates 9 – 11.
Optical polarimetry is presented for background stars and adjacent field stars in the direction of the dark globular filament GF 7 in Cygnus. Assuming that the observed polarization is due to the alignment of dust grains by paramagnetic relaxation, the magnetic field associated with the filament is mapped. The direction of the magnetic field is found to coincide with the long axis of the filament, as would be expected if the filament were a cylindrical fragment of a perturbed layer containing an ordered magnetic field. This observation provides strong evidence that the magnetic field of the Galaxy has been a significant factor in the formation and collapse of at least one dark globular filament.

131.135 **The correlation of interstellar element depletions with mean gas density.**
A. W. Harris, C. Gry, G. E. Bromage.
Astrophys. J., Vol. 284, No. 1, p. 157 – 160 (1984).
The dependence of interstellar element depletion on line-of-sight mean hydrogen space density $\bar{n}$(H) is investigated for 14 elements using mainly published gas-phase abundances. For six elements the correlation with $\bar{n}$(H) is clearly stronger than with column density, $N$(H), whereas for Zn, O, N, and D there is virtually no correlation with $\bar{n}$(H). The results clearly demonstrate that depletion is mainly governed by density-dependent processes and emphasize that $\bar{n}$(H) is a more useful sight-line parameter than $N$(H).

131.136 **Infrared and millimeter-wave observations of the Sharpless 156 molecular cloud.**
M. Joy, N. J. Evans II, P. M. Harvey, B. A. Wilking.
Astrophys. J., Vol. 284, No. 1, p. 161 – 175 (1984).
Comprehensive near-infrared, far-infrared, and CO measurements of the Sharpless 156 region reveal five distinct sources embedded in a giant molecular cloud. These sources include a prestellar condensation, massive stars with associated reflection and emission nebulosity, a bright compact H II region, and possibly even a luminous, post-main-sequence M star. The majority of the cloud has been mapped in $^{12}CO$ and $^{13}CO$ and has been surveyed at 2.2 µm. The authors also present 40 – 160 µm far-infrared maps and 1 – 20 µm near-infrared photometry of the active regions along with 6″resolution 10 µm and 20 µm maps of the central nebula.

131.137 **High-velocity molecular jets.**
R. L. Snell, N. Z. Scoville, D. B. Sanders, N. R. Erickson.
Astrophys. J., Vol. 284, No. 1, p. 176 – 193 (1984).
The high-velocity CO emissions in NGC 2071, GL 490, S140, and Orion have been mapped at 23″ or 48″ resolution. The outflow in NGC 2071 was clearly resolved into two highly collimated opposing jets. The outflows in GL 490 and Orion do not appear as collimated, but are clearly bipolar. Only S140 shows no strong evidence of bipolar outflow. Additional observations were made in all four clouds of $^{13}CO$ and in NGC 2071 of HCN and CS to determine the physical conditions in the flow and to map the dense gas in the ambient cloud. An analysis of the data indicates that the high-velocity CO emission arises in numerous dense clumps of material that have been swept up and incorporated into a slower moving shell at the end and edges of a stellar wind-evacuated cavity.

131.138 **Synthesis of complex molecules in dense interstellar clouds via gas-phase chemistry: a pseudo time-dependent calculation.**
C. M. Leung, E. Herbst, W. F. Huebner.
Astrophys. J., Suppl. Ser., Vol. 56, No. 2, p. 231 – 256 (1984).
A pseudo-time-dependent model for the gas-phase chemistry of dense interstellar clouds is presented which includes complex molecular species. The model involves 200 chemical species and over 1800 gas-phase reactions. The results indicate that gas-phase chemistry can produce sufficient abundances of most complex molecules to explain observations but only at times before the steady-state condition is reached when the atomic carbon abundance is still relatively high.

131.139 **Star formation efficiency in giant molecular clouds.**
V. G. Surdin.
Astron. Tsirk., No. 1318, p. 5 – 6 (1984). In Russian.

131.140 **Atoms of the interstellar medium in the solar system.**
V. G. Kurt.
Priroda, No. 10, p. 50 – 57 (1984). In Russian.

131.141 **The excited low–frequency carbon radio lines toward Cassiopeia A. Observations at frequencies 42, 57, and 84 MHz.**
A. A. Ershov, Yu. P. Ilyasov, E. E. Lekht, G. T. Smirnov, V. T. Solodkov, R. L. Sorochenko.
Pis'ma Astron. Zh., Tom 10, No. 11, p. 833 – 845 (1984). In Russian. English translation in Sov. Astron. Lett., Vol. 10.

The carbon recombination lines C538α, C486α, C612β, and C427α have been detected toward Cas A in absorption at frequencies 42, 57 and 84 MHz. The Doppler width of the lines is estimated to be $17.5 \pm 2.7$ km/s. The models with hydrogen–like and thermal dielectronic–like level populations are shown to be in best agreement with the observed lines.

131.142 **Observations of carbon recombination lines at decametric wavelengths in the direction of Cassiopeia A.**
A. A. Konovalenko.
Pis'ma Astron. Zh., Tom 10, No. 11, p. 846 – 853 (1984). In Russian. English translation in Sov. Astron. Lett., Vol. 10.

A number of carbon recombination lines are detected in the 16.7 – 30 MHz range toward Cassiopeia A. Principal parameters of the C603α – C732α lines are measured. Physical conditions are analyzed of the region where observed absorption lines have been formed. The low–temperature dielectronic recombination mechanism is shown to be needed to explain the observational data.

131.143 **Erratum: "The geometry of 'The Infrared Nebula' in Cha–1" [Astron. J., Vol. 89, No. 2, p. 277 – 279 (1984)].**
M. Cohen, R. D. Schwartz.
Astron. J., Vol. 89, No. 10, p. 1627 (1984). See Abstr. 37.131.048.

131.144 **A reanalysis of the interstellar CH abundance.**
D. J. Lien.
Astrophys. J., Vol. 284, No. 2, p. 578 – 588 (1984).

A detailed investigation of the term structure of the $X^2\Pi$, $A^2\Delta$, $B^2\Sigma^-$, and $C^2\Sigma^+$ systems of CH is presented, as well as a discussion of a number of errors that have been found with respect to the historical analysis of the interstellar CH abundance. The primary cause for these errors is the neglect of the Λ doubling of the ground state. The effects of including this splitting are considered in detail. A significant result from this analysis is that the optical data can be used to determine the column densities of each half of the ground–state Λ doublet. This is done for the CH along the lines of sight towards the stars ζ Per, ζ Oph, X Per and χ Oph. The results are compared with published radio observations of the 9 cm $F = 1$–1 hyperfine transition.

131.145 **W3 North: far–infrared and radio molecular observations.**
H. A. Thronson Jr., P. R. Schwartz, H. A. Smith, C. J. Lada, W. Glaccum, D. A. Harper.
Astrophys. J., Vol. 284, No. 2, p. 597 – 600 (1984).

The authors report the results of far–infrared and radio molecular CO observations of the W3 North star–forming region. The object shows extended dust and gas emission, which is interpreted as the result of fairly advanced disruption of a molecular cloud. This idea is supported by an estimate for the age of the embedded H II region. It is possible that W3 North is the oldest object among the W3 complex of sources.

131.146 **Far–infrared and submillimeter observations of the multiple cores in S255, W3, and OMC–1: evidence for fragmentation?**
D. T. Jaffe, J. A. Davidson, M. Dragovan, R. H. Hildebrand.
Astrophys. J., Vol. 284, No. 2, p. 637 – 642 (1984).

The authors present far–IR and sub–mm maps of S255 which reveal the presence of a previously undiscovered cloud core 1′ north of S255–IR. This second core has a mass greater than the core in S255–IR and a temperature ~10K lower. The authors also present 40 μm – 400 μm photometry of these two cores and of the previously observed double cloud cores in W3 and OMC–1 and use the mapping and photometry results to derive temperatures, densities, masses, and visual extinctions for the cores. Molecular line observations indicate velocity differences of 1.4 – 4 km $s^{-1}$ between the two cores in a given cloud. If the multiple cores at different velocities result from large–scale fragmentation of a rotating parent cloud, approximately 90% of the initial angular momentum is now in the orbital motion of the cores.

131.147 **Interstellar matter near the Pleiades. I. Optical absorption–line observations and foreground atomic gas.**
R. E. White.
Astrophys. J., Vol. 284, No. 2, p. 685 – 694 (1984). = Contrib. No. 549 Five Coll. Obs.

High–resolution spectroscopy of 15 Pleiades members in the spectral region λλ3874 – 4300 provides the basis for a new examination of the interstellar gas toward the cluster. The markedly different patterns exhibited by the Ca II and $CH^+$ absorption lines across the cluster, together with kinematic information from radio studies of H I and CO, lead to the conclusion that the Ca II absorption occurs in mostly atomic gas at $v_{LSR} \sim +7$ km $s^{-}1$ in front of the Pleiades and that the $CH^+$ absorption is associated with mostly molecular gas at $v_{LSR} \sim +10$ km $s^{-1}$ within the cluster itself. The foreground gas has a sheetlike structure, with thickness $\lesssim 0.3$ pc.

131.148 **Interstellar matter near the Pleiades. II. $CH^+$ formation.**
R. E. White.
Astrophys. J., Vol. 284, No. 2, p. 695 – 704 (1984). = Contrib. No. 550 Five Coll. Obs.

$CH^+$ can form near stars in the Pleiades via the endothermic reaction of $C^+$ with $H_2$ in photoelectrically heated gas flowing through the cluster. This conclusion follows from either of two models, one in which gas and dust remain well mixed, the other in which radiation pressure separates dust from gas in a layer well outside the $H_2$ dissociation front near each star. The models differ in their ability to account for the bimodal distribution of $N(CH^+)$ with $E(B-V)$. This seems to be a natural consequence of emergence of stars from the molecular gas if radiation pressure creates a dust layer, but not if gas and dust remain well mixed.

131.149 **On interstellar [Fe X] absorption toward Cepheus OB2.**
L. M. Hobbs.
Astrophys. J., Lett. Ed., Vol. 284, No. 2, p. L47 – L50 (1984).

A weak interstellar absorption line which can be identified as [Fe X] λ6375 is observed in the spectrum of the A2 Ia star ν Cep, which is a member of the Cepheus OB2 association. The equivalent width, width, and radial velocity of the line differ indistinguishably from those of similar absorption previously detected uniquely toward the O6f star λ Cep, another association member located at a projected distance of about 50 pc from ν Cep. The absorption may reveal a particularly dense, low–latitude concentration of hot, foreground interstellar gas of the kind generally responsible for the observed soft X–ray background.

131.150 **Detection of submillimeter polarization in the Orion Nebula.**
R. H. Hildebrand, M. Dragovan, G. Novak.
Astrophys. J., Lett. Ed., Vol. 284, No. 2, p. L51 – L54 (1984).

The authors have observed linear polarization of the submillimeter (270 μm) continuum radiation from two regions of Orion: one centered on the Kleinmann–Low nebula and one centered on the 400 μm peak 1.′5 south of the nebula. The polarizations measured for these regions are $P = (1.7 \pm 0.4)\%$ at $\phi = 23° \pm 7°$ and $P = (1.7 \pm 0.5)\%$ at $\phi = 27° \pm 7°$, respectively. A 2σ upper limit, $P \leqslant 1.6\%$, is found for the nebula W3(OH).

131.151 **Star formation and extinct radioactivities.**
A. G. W. Cameron.
Icarus, Vol. 60, No. 2, p. 416 – 427 (1984).

A supernova trigger for star formation lacks observational confirmation. Observations of star formation currently point toward nearly isolated individual events taking place in the interiors of dense molecular clouds, in which cores are formed in the clouds and collapse to form stars. If the formation of the solar system proceeded in this manner, then a viable theory of cloud core formation must provide circumstances in which it is not uncommon for the cloud cores to be formed with a significant supply of fresh short-lived radioactivities. A review is made of the evidence for the existence of now extinct radioactivities in primitive solar system material and an examination is made of the implications for the early stages of formation of the Sun and solar system. The characteristics of possible disturbances in dense molecular clouds which can initiate the formation of cloud cores is discussed.

131.152 **$NH_3$ hyperfine intensity anomalies.**
J. Stutzki, J. M. Jackson, M. Olberg, A. H. Barrett, G. Winnewisser.
Astron. Astrophys., Vol. 139, No. 2, p. 258 – 262 (1984).
A search for intensity anomalies within the hyperfine structure of the $NH_3(J,K) = (1,1)$ transition was performed. Out of 23 selected sources (associated with $H_2O$–maser and IR–emission) in the rotational temperature regime between $T_R = 15$ and 30K, 13 showed non LTE behaviour. Thus, nonthermal population of the (1,1) hyperfine states seems to be common in intermediate warm clouds, although no obvious correlation with other known properties of the clouds has emerged.

131.153 **Westerbork observations of 6 cm $H_2CO$ in the direction of Cas A.**
W. M. Goss, P. M. W. Kalberla, H. R. Dickel.
Astron. Astrophys., Vol. 139, No. 2, p. 317 – 329 (1984).
The 6 cm $1_{10} - 1_{11}$ $H_2CO$ absorption in the Perseus arm feature in the direction of Cas A has been mapped with the Westerbork Synthesis Radio Telescope (WSRT). Only a small part of Cas A is absorbed. In agreement with the 2 cm $2_{11}$–$2_{12}$ results of Batrla et al. (1983), there is no $H_2CO$ absorption in the northern half of Cas A. The low optical depths measured with single dish telescopes ($\leqslant 0.06$) are due to small filling factors in a clumpy medium. With higher angular resolution optical depths of $\sim 1$ are observed. These clumps have $H_2$ densities of $\sim 10^4 cm^{-3}$, sizes in the range 0.1 – 0.4 pc and total masses in the range 0.2 – 66 $M_{\odot}$. A comparison with CO and H I absorption observations has been carried out. Twelve of the $H_2CO$ clumps have possible H I counterparts.

131.154 **3D models of rotating magnetic gas clouds. I. Time evolution, mass spectrum and angular momentum.**
W. Benz.
Astron. Astrophys., Vol. 139, No. 2, p. 378 – 388 (1984).
The author presents results of 3D numerical calculations on the collapse of rotating magnetized isothermal gas clouds. The magnetic field is assumed to remain "frozen in" to the gas. In order to solve the MHD equations the author uses the "smoothed particles" hydrodynamics approach, using 500 particles. He finds that a magnetic field strongly prevents the cloud from collapsing and that in the case of the perpendicular rotator differential rotating nonhomologous bar–like structures are formed. In the cases without magnetic field merged particles are sufficiently numerous to allow the determination of a mass spectrum, angular momentum as a function of mass and the distribution of the orientation of individual angular momenta. The dependence on initial conditions of these quantities is shown to be very small.

131.155 **IUE high resolution spectrophotometry of H Ly α emission from the local interstellar medium.**
J. T. Clarke, S. Bowyer, H. J. Fahr, G. Lay.
Astron. Astrophys., Vol. 139, No. 2, p. 389 – 393 (1984).
The authors present IUE high dispersion spectra obtained in April 1981 and in April 1983 of resonantly scattered solar Ly α emission from H moving into the solar system with the local interstellar wind. They have observed a heliocentric velocity of $-29 \pm 5$ km $s^{-1}$ from the ISW component and a surface brightness which decreased from roughly 1000 Rayleighs to 800 R over the two year interval. A preliminary derivation of the velocity of the ISM at large distances from the sun yields $v_{\infty} = 25.6 \pm 5$ km $s^{-1}$.

131.156 **The physics of the heliospheric interface and its implications for LISM (*local interstellar matter*) diagnostics.**
H. J. Fahr, H. W. Ripken.
Astron. Astrophys., Vol. 139, No. 2, p. 551 – 554 (1984).
The effect of the plasma interface ahead of the heliosphere on neutral interstellar gases traversing this region of a perturbed interstellar plasma flow has recently been subject to some discussion. It is therefore considered necessary to briefly review the intentions and the conclusions of earlier work concerning this effect (Ripken and Fahr, 1983) and then to present some new results concerning (1) the specific extinctions of hydrogen, helium, and oxygen atoms in the interface region and (2) the interface–dependent interpretation of the Ly–alpha resonance glow data. A correct modelling of the interface effects is an indispensable prerequisite for reliable derivations of interstellar gas properties from heliospheric observations of UV/EUV resonance luminescence patterns.

131.157 **Infrared observations of recent star formation regions in the Large Magellanic Cloud: N 160 A and N 105 A.**
N. Epchtein, M. A. Braz, F. Sèvre.
Astron. Astrophys., Vol. 140, No. 1, p. 67 – 71 (1984).
The authors report on the detection of 2 new probable protostellar objects in the Large Magellanic Cloud associated with the radio optical nebulae N 160 A and N 105 A. Both sources were found near $H_2O$ masers. 1.25 – 20 μm photometric data were obtained on N 160 A IR. It is suggested that the infrared emission is due to the combination of an extended free–free component seen at short wavelengths and in the Brackett γ line and a compact source which consists of a hot (450K) core and a dense optically thick cold envelope (65K). This object is shown to be very similar to galactic protostellar objects like W 51–IRS 2. The near infrared data obtained on N 105 A are consistent with a similar interpretation.

131.158 **Ammonia in the W3 region.**
Q. Zeng, W. Hermsen, T. L. Wilson, W. Batrla.
Astron. Astrophys., Vol. 140, No. 1, p. 169 – 173 (1984).
Maps of the $(J,K) = (1,1)$ and (2,2) lines of $NH_3$ over the region measured in the 400 μm emission by Jaffe et al. (1983) show that $NH_3$ emission is strongest toward the western 400 μm peak, near IRS4 and the H II region W3C. The rotational temperature there is $\sim 30$K, and the deconvolved angular size of the $NH_3$ region is $57'' \times 47''$ (FWHP). Toward the eastern 400 μm peak, centered on IRS5, the $NH_3$ is hotter, with $T_R \sim 50$K. The abundance is one–half of that in the western peak. Comparisons with the distribution of other molecules mapped in the core of W3 are made. Between the clouds associated with W3 and W3(OH), there is no detectable $NH_3$, although $NH_3$ is present in a $\sim 1'$ region around W3(OH) itself.

131.159 **Optical properties of interstellar graphite and silicate grains.**
B. T. Draine, H. M. Lee.
Astrophys. J., Vol. 285, No. 1, p. 89 – 108 (1984).
Available laboratory and astronomical data are used to construct dielectric functions for graphite and "astronomical silicate" material. The interaction of small particles with radiation is reviewed, and it is noted that the "magnetic dipole" contribution to absorption in the infrared can be important for conducting particles such as graphite. Absorption cross sections are evaluated for graphite and silicate particles with sizes in the range 0.003 – 1.0 μm, for wavelengths from 300 Å – 1000 μm; Planck–averaged absorption cross sections are also given. Polarization cross sections are computed for both prolate and oblate spheroids which reproduce the observed emissivity of interstellar and circumstellar silicate material. Extinction curves are calculated for the Mathis–Rumpl–Nordsieck graphite–silicate grain mixture and compared to observations.

131.160 **Density fluctuations in the interstellar medium: evidence for anisotropic magnetogasdynamic turbulence. I. Model and astrophysical sites.**
J. C. Higdon.
Astrophys. J., Vol. 285, No. 1, p. 109 – 123 (1984).

A model of anisotropic magnetogasdynamic turbulence is developed in order to investigate quantitatively a turbulent fluid origin for the small spatial scale density fluctuations observed in 1981 by Armstrong, Cordes, and Rickett in the interstellar medium. The anisotropy at these small spatial scales results from the presence of a large-scale approximately uniform magnetic field that is a sum of the steady-state galactic magnetic field and the large-scale turbulent magnetic field fluctuations. The observed density fluctuations are interpreted to be two-dimensional isobaric entropy variations with oppositely directed gradients in temperature and density projected transverse to the local approximately uniform magnetic field. Three possible sites – cloud shell H II regions, diffuse H II regions produced by O stars, and the tenuous intercloud medium – are investigated as possible locations for turbulent flows.

131.161 **On "The ambipolar diffusion time scale and the location of star formation in magnetic interstellar clouds": setting the record straight.**
T. C. Mouschovias.
Astrophys. J., Vol. 285, No. 1, p. 124 – 125 (1984).

This paper replies to critical comments by E. H. Scott (see Abstr. 34.131.297) on an earlier paper by the author (see Abstr. 25.062.035).

131.162 **Interstellar scattering in the inner parts of the Galaxy.**
A. Pramesh Rao, S. Ananthakrishnan.
Nature, Vol. 312, No. 5996, p. 707 – 711 (1984).

A new survey of the galactic plane for sources with size less than 1 arc s at 327 MHz shows that towards the inner parts of the Galaxy for galactic latitudes less than 5°, interstellar scattering is much larger than expected from data at higher latitudes. The enhanced scattering varies both with galactic latitude and longitude. A two-component model for the distribution of scattering matter in the Galaxy is proposed to interpret the observations.

131.163 **Structure of two dark interstellar clouds as determined by microwave ammonia observations.**
P. J. Benson, P. C. Myers.
Bull. Am. Astron. Soc., Vol. 16, No. 3, Part I, p. 724 (1984). Abstract. – See Abstr. 010.065.

131.164 **The bipolar molecular outflows in NGC 7023.**
S.-U. Choe.
Bull. Am. Astron. Soc., Vol. 16, No. 3, Part I, p. 727 (1984). Abstract. – See Abstr. 010.065.

131.165 **Distant molecular clouds toward the Cygnus-X region.**
Y.-L. Huang, T. M. Dame, S.-K. Pan, P. Thaddeus.
Bull. Am. Astron. Soc., Vol. 16, No. 3, Part I, p. 727 (1984). Abstract. – See Abstr. 010.065.

131.166 **Far-infrared photometry of the ϱ Ophiuchi dark cloud.**
W. Cudlip, J. P. Emerson, I. Furniss,
W. M. Glencross, R. E. Jennings, K. J. King, J. F. Lightfoot,
W. A. Towlson.
Mon. Not. R. Astron. Soc., Vol. 211, No. 3, p. 563 – 573 (1984).

The region of star formation in the ϱ Ophiuchi dark cloud has been mapped with a three-band far-infrared photometer flown on a balloon-borne telescope. The maps provide further detail on the spatial and temperature structure of known sources, in addition to detecting several others. Surveys carried out at shorter wavelengths have detected numerous stars embedded in this cloud. However, it has been difficult to differentiate between main-sequence and pre-main-sequence objects in some cases where the absorption of the cloud is uncertain. A number of these problems are resolved with measurements of source-luminosities in the far-infrared. It is concluded that quiescent conditions in the cloud favour the development of low-mass stars, whereas the formation of massive objects to the west has been triggered by some local disturbance.

131.167 **Absorption lines of interstellar $C_2$ and CN molecules.**
E. F. van Dishoeck, J. H. Black.
Messenger, No. 38, p. 16 – 19 (1984).

131.168 **The spectrum of the "IRN" nebula in the Chamaeleon T association.**
A. P. Fairall, K. G. Henize.
Mon. Notes Astron. Soc. S. Afr., Vol. 43, Nos. 9 – 10, p. 68 (1984).

Attempted spectra of the tip and central body of the IRN nebula fail to reveal any emission lines.

131.169 **The molecular cloud associated with W50 and SS433.**
D. Forbes.
J. R. Astron. Soc. Can., Vol. 78, No. 5, p. 210 (1984). Abstract. – See Abstr. 010.401.

131.170 **Sulphur chemistry in a shocked interstellar cloud.**
G. F. Mitchell.
J. R. Astron. Soc. Can., Vol. 78, No. 5, p. 210 – 211 (1984). Abstract. – See Abstr. 010.401.

131.171 **Plotting sky maps from profiles of the H I radio line.**
N. V. Bystrova.
Bull. Spec. Astrophys. Obs. – North Caucasus, Vol. 17, p. 58 – 61 (1984). English translation of 34.131.301.

131.172 **Calculation of self-absorption effect in profiles of the radio line of neutral hydrogen.**
Z. A. Alferova, I. V. Gosachinskii (*I. V. Gosachinskij*).
Bull. Spec. Astrophys. Obs. – North Caucasus, Vol. 17, p. 68 – 71 (1984). English translation of 34.131.302.

131.173 **Analysis of interplanetary Lyman-alpha line profile with a hydrogen absorption cell: theory of the Doppler angular spectral scanning method.**
J. L. Bertaux, R. Lallement.
Astron. Astrophys., Vol. 140, No. 2, p. 230 – 242 (1984).

The use of a hydrogen absorption cell as a means of diagnostic for interstellar hydrogen is described. Through resonance scattering of solar Lyα photons, the interstellar atoms penetrating into the solar system present a Lyα emission the profile of which is an image of their velocity distribution. From the study of hydrogen Lyα profiles the temperature $T$ of the interstellar hydrogen and the three components of the relative velocity vector $\mathbf{V}_w$ can be determined. The case in which the velocity distribution of H atoms in the interplanetary medium is the same Maxwell-Boltzmann distribution as at infinity (the uniform gaussian case) is completely described.

131.174 **Radiative transfer effect in $H_2O$ masers.**
S. Chandra, W. H. Kegel, D. A. Varshalovich,
M. A. Albrecht.
Astron. Astrophys., Vol. 140, No. 2, p. 295 – 302 (1984).

The influence of different physical effects on the efficiency of $H_2O$-masers has been studied by solving, numerically, the coupled set of statistical and radiative transfer equations in an on-the-spot-approximation. The authors used a physical model in which the excitation is basically due to collisions. The results indicate that the combined influence of resonant collisions and the absorption of IR-photons by cold dust can lead to luminosities in the 1.35 cm-line comparable to those observed.

131.175 **The local interstellar electron spectrum: constraints from gamma ray data.**
F. L. Gualandris, A. W. Strong.
Astron. Astrophys., Vol. 140, No. 2, p. 357 – 362 (1984).

The authors present a simple but accurate formulation of the bremsstrahlung emission from cosmic ray electrons in the interstellar medium, simplifying the direct solution of the inverse problem, the calculation of the spectrum corresponding to a given gamma ray emissivity spectrum. Using the emissivity of the local interstellar medium determined in recent COS–B studies, the range of possible electron spectra is derived. The resulting electron spectra are consistent with gamma–ray and radio data only if either (1) the optical depth to radio at 1 MHz is larger than expected, or (2) the electron flux at 1 GeV in the nearby interstellar medium is a factor 2 – 3 times larger than the solar demodulated value given by Rockstroh and Webber (1978).

131.176 **Hydrogen recombination lines towards the SNR IC 443 and the H II region S 249.**
A. Donati–Falchi, G. Tofani.
Astron. Astrophys., Vol. 140, No. 2, p. 395 – 398 (1984).
The SNR IC 443 (S 248) and the adjacent low density H II region S 249 were observed in the H 166α and H 200α lines. The complex seemed to be a good candidate to search for line emission differences in the direction of the SNR and of the H II region. The authors obtained a positive detection of the H 166α line in S 249 above the (3σ) r.m.s. limit, while no detection was obtained in the SNR direction. This indicates that S 249 is not an extended foreground source in the direction of the SNR. The radial velocity with respect to LSR of S 249, as deduced from the line center, is -5 km $s^{-1}$. This velocity is in good agreement with that of the CO cloud associated with this region and with that of the "preshock" molecular cloud associated to the SNR.

131.177 **Molecular observations of B 1: a dense globule in Perseus.**
R. Bachiller, J. Cernicharo.
Astron. Astrophys., Vol. 140, No. 2, p. 414 – 420 (1984).
The authors report CO, $^{13}CO$, $C^{18}O$, $HCO^+$, $H^{13}CO^+$, and $NH_3$ observations and star counts of B 1, a strong molecular source in the ridge connecting B 5, IC 348 and NGC 1333 in Perseus. The size of B 1 is 2 pc and its mass 400 $M_\odot$ (at the adopted distance of 200 pc), its kinetic temperature is 12K. The core, where the $NH_3$ (1,1) line was detected, has a size of 0.5 pc, a temperature of 12K, a mass of 60 $M_\odot$, the visual extinction is $\geqslant 10$ mag and the $H_2$–density $\approx 3 \times 10^4 cm^{-3}$. The chemical abundances of the observed molecules are estimated. Rotation, if it occurs, is too slow to prevent the collapse of the cloud.

131.178 **The Cygnus X region. XV. A 4.8 GHz continuum survey with the 100 m–telescope.**
H. J. Wendker.
Astron. Astrophys., Suppl. Ser., Vol. 58, No. 2, p. 291 – 316 (1984).
Results are given of a radio continuum survey with a resolution of 2ʹ.6. A bewildering amount of details and structures of sources is present. Around 800 local maxima are counted. The overall results and several individual features are commented on.

131.179 **A high density molecular fragment in the ϱ Oph cloud.**
Q. Zeng, W. Batrla, T. L. Wilson.
Astron. Astrophys., Vol. 141, No. 1, p. 127 – 130 (1984).
Ammonia observations of the high density ($n(H_2) \geqslant 10^6 cm^{-3}$) fragment ϱ Oph B and the $HCO^+$ and ϱ Oph A peaks are presented. In ϱ Oph B, the ammonia emission appears to be formed in the high density region, since the linewidth, radial velocity and spatial distribution agree with that of the dense gas. At the intensity peak, $T_K$ is $\sim$18K; toward the edges of the $NH_3$ cloud, $T_K = 19$K. There is no hint of collapse motions from the spatial variations of linewidths or velocities. This suggests that no high mass stars have yet formed in this region.

131.180 **Detection of CS and $C_2H$ in absorption.**
L.–Å. Nyman.
Astron. Astrophys., Vol. 141, No. 2, p. 323 – 327 (1984).
CS and $C_2H$ were detected in absorption against W 49 and $C_2H$ against DR 21. The fractional abundances (relative to $H_2$) towards W 49 are estimated to be $\sim 5 \times 10^{-9}$ for CS, $5 \times 10^{-8}$ for $C_2H$, and $< 5 \times 10^{-10}$ for $N_2H^+$. Somewhat lower values are obtained as upper limits towards Cas A. The detections of absorption features of these and other molecules imply that many molecular species are abundant also in clouds where the density is so low that the energy levels of the molecules are not appreciably excited above the cosmic background radiation. These observations will be compared with earlier observations of $HCO^+$ and HCN towards the same sources.

131.181 **Observations of the high latitude far ultraviolet background in the 1040 – 1080 Å band.**
J. Bixler, S. Bowyer, M. Grewing.
Astron. Astrophys., Vol. 141, No. 2, p. 422 – 424 (1984).
Observations of the far ultraviolet background at high galactic latitudes in a band from 1040 – 1080 Å were carried out with a rocket–borne 1 meter telescope and a 0.23 square degree field of view photometer. The photometer had a sharp cutoff shortward of 1040 Å and a sensitivity at hydrogen Lyman α of a factor of $10^{-4}$ below its sensitivity at 1060 Å. The hydrogen Lyman α flux contributed only 17% of the total counts detected in the cosmic background detector. The authors obtain an upper limit to the emission in the 1040 – 1080 Å band of 9700 photons $cm^{-2}s^{-1}sr^{-1}Å^{-1}$. This upper limit is compared with previous measurements and theoretical predictions.

131.182 **Studies of the Orion–KL water maser flare region.**
A. M. Romanov, L. I. Matveenko.
Sov. Astron. Lett., Vol. 10, No. 3, p. 145 – 147 (1984). English translation of 37.131.200.

131.183 **Water masers observed with a 1008–channel acousto–optical radio spectrometer.**
V. I. Vasil'kov, S. V. Kalenskij, V. I. Slysh, G. T. Smirnov.
Sov. Astron. Lett., Vol. 10, No. 3, p. 148 – 150 (1984). English translation of 37.131.201.

131.184 **Water maser in NGC 3079.**
IAU Circ., No. 3983 (1984).

131.185 **OH megamasers.**
IAU Circ., No. 3993 (1984).

131.186 **Positron annihilation in a simulated low–density galactic environment.**
B. L. Brown, M. Leventhal, A. P. Mills Jr., D. W. Gidley.
Phys. Rev. Lett., Vol. 53, No. 24, p. 2347 – 2350 (1984).
A low–energy positron beam is used to study the behavior of positrons in low–density ($\approx 10^{-3}$ Torr) $H_2$ and He. Positronium is formed in flight and the collision–free Doppler–broadened singlet annihilation linewidth in $H_2$ is found to be $6.4 \pm 0.1$ keV in agreement with calculations. The measured fraction of positrons surviving below the positronium threshold is larger than predicted. The authors conclude that the neutral interstellar environment cannot be ruled out as a galactic annihilation site.

131.187 **Interstellar dust.**
K. Nandy.
Spectral evolution of galaxies, p. 62 – 66 (1984). – See Abstr. 012.066.
This paper presents a brief comparison of the mean interstellar curves in both Magellanic Clouds with the mean interstellar extinction curve in the Galaxy.

131.188 **The interpretation of flows in molecular clouds.**
J. E. Dyson.
Astrophys. Space Sci., Vol. 106, No. 1, p. 181 – 197 (1984).
Stellar winds interacting with gas in dense molecular clouds produce flows which may be initially energy or momentum driven. A criterion for this is derived which depends sensitively on the wind velocity. In general, the flows observed in CO associated with infrared point sources seem to be in the energy driven regime. By combining CO observations with radio continuum flux measurements, wind parameters are derived for several of these sources. There is some evidence from the derived parameters that high luminosity sources have radiatively–driven winds.

131.189 **Analytical form for the interstellar extinction law.**
R. W. Nicholls.
J. Opt. Soc. Am. A, Vol. 1, No. 12, p. 1237 (1984). Abstract.

131.190 **Interstellar Ca II lines in SMC stars.**
J. G. Cohen.
Astron. J., Vol. 89, No. 12, p. 1779 – 1783 (1984).
Interstellar Ca II lines of reddened early–type stars in the SMC were observed. In addition to the component from the Milky Way, strong lines were detected at radial velocities corresponding to gas in the SMC. The interstellar Ca II lines are abnormally strong, given the small color excess of the stars and the low metallicity of the SMC gas, which suggests that the grains contain a much smaller fraction of the interstellar calcium than they do in our galaxy. The radial velocities do not conflict with the two–galaxy model for the SMC of Matthewson and Ford (1983) if the higher–velocity system is behind the lower–velocity system.

131.191 **A search for SiO masers in Orion–like regions.**
R. Barvainis, D. P. Clemens.
Astron. J., Vol. 89, No. 12, p. 1833 – 1835 (1984).
A sensitive search for SiO maser emission toward objects possessing high–velocity flows and/or disk geometries has revealed no Orion–like sources. The ratios of the luminosity limits reported here to the Orion SiO maser luminosity range from 0.04 to 0.0002, indicating that the Orion masers are associated with a unique set of physical conditions and not any widespread phenomena, even at a weak level. The authors report detection of a new SiO maser toward the reflection nebula OH231.8+4.2, exhibiting characteristics typical of the SiO masers seen toward late–type stars.

131.192 **A semi–classical description of radiation transport in astrophysical masers.**
D. Field, I. M. Richardson.
Mon. Not. R. Astron. Soc., Vol. 211, No. 4, p. 799 – 811 (1984).
A semi–classical theory has been developed to describe amplification of radiation in a 2–level astrophysical maser. A simple expression is derived which relates the degree of saturation of the maser population inversion to the intensity of the radiation. On the basis of the relatively well–known physical conditions in W3(OH), the observed intensities of OH masers in that region can be readily reproduced using semi–classical theory.

131.193 **Infrared extinction in the Small Magellanic Cloud.**
K. Nandy, D. H. Morgan, L. Houziaux.
Mon. Not. R. Astron. Soc., Vol. 211, No. 4, p. 895 – 900 (1984).
*JHK* photometry of bright reddened and unreddened early–type SMC members is presented, and the extinction law in those wavebands is studied. *VRI* photometry of several early–type SMC members is also presented and combined with existing observations to study the extinction at *R* and *I* also. It is found that, within the errors, the extinction law from *V* to *K* is the same for the SMC as for the Galaxy and the LMC, despite differences previously found in the ultraviolet.

131.194 **Optical interstellar absorption lines toward 29 stars.**
L. M. Hobbs.
Astrophys. J., Suppl. Ser., Vol. 56, No. 3, p. 315 – 323 (1984).
Observations of optical interstellar absorption lines of Ti II, Ca II, and/or Na I acquired at an instrumental resolution typically of 5.6 km $s^{-1}$ are reported for 29 stars. The data were recorded primarily with a Digicon detector and achieve 2 $\sigma$ detection limits generally near an equivalent width of 1 mÅ. New high–velocity optical interstellar lines at $|v_{LSR}| \geqslant 25$ km $s^{-1}$ are found toward four of the 29 stars.

131.195 **Review of radio line observations with the RATAN–600 radio telescope.**
I. V. Gosachinskij.
Soobshch. Spets. Astrofiz. Obs., Vyp. 43, p. 3 – 32 (1984).
The brief review of the results of H I, OH, $H_2O$ and $H_2CO$ observations made with RATAN–600 during some last years is presented.

131.196 **Formation and evolution of rings in disk–shaped protostellar clouds.**
B.–l. Liu.
Chin. Astron. Astrophys., Vol. 8, No. 4, p. 310 – 318 (1984). English translation of Acta Astron. Sin., Vol. 25, No. 2, p. 156 – 167 (1984).
Numerical simulation of the evolution of a disk–like system of protostellar clouds is made by means of N gravitating bodies. It is shown that the ring structure is physically present in the course of the gravitational collapse, that the ring structure represents only one stage of the cloud's evolution, and that it will eventually break up into new structures.

131.197 **UV studies of the interstellar medium.**
J. C. Blades.
Space Telesc. Sci. Inst., Prepr. Ser., No. 18, 10 pp. (1984). Paper presented at the Fourth European IUE Conference, Rome, May 1984.

131.198 **Absorption and emission of EUV radiation by the local ISM.**
F. Paresce.
Space Telesc. Sci. Inst., Prepr. Ser., No. 20, 26 pp. (1984). To appear in Proc. IAU Colloq. No. 81.

131.199 **Detection of CS and $C_2H$ in absorption.**
L.–Å. Nyman.
Onsala Space Obs., Prepr., No. 84:13, 21 pp. (1984). To appear in Astron. Astrophys.

131.200 **Radio observations of interstellar molecules, of their behaviour, and of their physics.**
O. E. H. Rydbeck, Å. Hjalmarson.
Onsala Space Obs., Prepr., No. 84:42, 133 pp. (1984).
Scans towards Ori A and the evolved, carbon–rich star IRC+10216, through the range 72 – 91 GHz are made. Results of these scans will be presented. Quite generally it can be said that many new molecular transitions are found but no unexpected molecules and radicals, except perhaps for $C_3H$.

131.201 **Star formation in Bok globules and low–mass clouds. II. A collimated flow in the Horsehead.**
B. Reipurth, P. Bouchet.
ESO Sci. Prepr., No. 331, 16 pp. (1984). Submitted to Astron. Astrophys.

131.202 **The CN radical in diffuse interstellar clouds.**
S. R. Federman, A. C. Danks, D. L. Lambert.
ESO Sci. Prepr., No. 336, 35 pp. (1984). Submitted to Astrophys. J.

131.203 **The reflection nebula NGC 1999.**
J. A. Cardelli, K. H. Böhm.
Astrophys. J., Vol. 285, No. 2, p. 613 – 617 (1984).
The authors have studied the ultraviolet properties of the reflection nebula NGC 1999, which lies near the well–known Herbig–Haro objects HH 1, HH 2, and HH 3. For a position 15″ east from the central star V380 Ori the ratio of nebular intensity to stellar flux has been determined in the wavelength range 1300 Å to 3000 Å. The ultraviolet extinction curve of V380 Ori has also been derived. The data are compatible with the (tentative) hypothesis that dust properties in NGC 1999 are very similar or identical to those which have been determined for NGC 7023.

131.204 **Theoretical reinvestigation of hydrocarbon and cyanoacetylene abundances in TMC–1.**
E. Herbst, N. G. Adams, D. Smith.
Astrophys. J., Vol. 285, No. 2, p. 618 – 621 (1984).
The products of the reaction between $C_3H^+$ and $H_2$ at low temperatures were previously reported incorrectly by the authors to be $C_3H_2{}^+$ + H. The only product at 80K (and presumably at lower temperatures) is $C_3H_3{}^+$, formed via a saturated three–body association reaction. Use of the analogous radiative association reaction with a rate coefficient of $10^{-11}$cm$^3$s$^{-1}$ as well as

a newly measured rate coefficient for the radiative association reaction between $CH_3^+$ and $H_2$ in Herbst's "semidetailed" model of the dense interstellar cloud TMC–1 results in improved agreement between theoretical and observed hydrocarbon and cyanoacetylene abundances.

131.205 **Observations of interstellar hydrogen and deuterium toward Alpha Centauri A.**
W. B. Landsman, R. C. Henry, H. W. Moos, J. L. Linsky.
Astrophys. J., Vol. 285, No. 2, p. 801 – 807 (1984).
The authors present a composite profile of the Ly$\alpha$ emission line of $\alpha$ Cen A, obtained from 10 individual spectra with the high–resolution spectrograph aboard the satellite. There is excellent overall agreement with two previous Copernicus observations. Interstellar deuterium is detected, and a lower limit is set on the deuterium to hydrogen ratio of $n_{DI}/n_{HI} > 8\times10^{-6}$. In addition, the deuterium bulk velocity appears blueshifted by $8 \pm 2$ km s$^{-1}$ with respect to interstellar hydrogen, suggesting a nonuniform medium along the line of sight.

131.206 **Aperture synthesis observations of CO emission from the W3 molecular cloud core.**
M. J. Claussen, G. L. Berge, G. M. Heiligman, R. B. Leighton, K. Y. Lo, C. R. Masson, A. T. Moffet, T. G. Phillips, A. I. Sargent, S. L. Scott, P. G. Wannier, D. P. Woody.
Astrophys. J., Lett. Ed., Vol. 285, No. 2, p. L79 – L83 (1984).
The authors present the first aperture synthesis maps of $^{12}$CO ($J$ = 1–0) emission from a galactic star–forming region. They have mapped the line and continuum emission from a 1′ field centered on W3 IRS 5 with the Owens Valley millimeter–wave interferometer. Although the authors did not detect any radio continuum emission from IRS 5 to a 5 $\sigma$ limit of 130 mJy, they have found, associated with IRS 5, compact ($\sim$8″) sources of high velocity red– and blueshifted CO emission with $T_B \approx$ 20K. The outflow is similar to other known sources with high–velocity emission and has a short dynamic lifetime such as found in Orion.

131.207 **$2_{11} - 2_{12}$ formaldehyde emission from DR 21 (OH).**
K. J. Johnston, C. Henkel, T. L. Wilson.
Astrophys. J., Lett. Ed., Vol. 285, No. 2, p. L85 – L88 (1984).
The $2_{11} - 2_{12}$ K–doublet line of $H_2CO$ at 14.5 GHz has been measured toward DR 21 and DR 21(OH) with an angular resolution of 9″. The emission near DR 21(OH) is consistent with quasi–thermal excitation. It displays two peaks with brightness temperatures ≥2K separated by 9″. The overall deconvolved FWHP size of the region is 18″ × ≤ 6″. Absorption with an optical depth of 0.15 of size 10″ was found near the continuum component C of DR 21.

131.208 **The variations of interstellar $H_2O$ masers.**
K. Mattila.
Rep. Ser., Dep. Phys. Sci., Univ. Turku, No. FTL–R68, p. 17 (1984). Abstract. – See Abstr. 012.079.

131.209 **Visible and ultraviolet studies of interstellar molecules.**
J. H. Black.
Prepr. Steward Obs., No. 534, 22 pp. (1984). Invited paper, to be published in the Proceedings of a NATO Advanced Research Workshop on Molecular Astrophysics – State of the Art and Future Directions, 8 – 14 July 1984, Bad Windsheim, West Germany.

131.210 **Spectrophotometric studies of galactic nebulae. XXIII. 1. S 191 – compact red reflection nebula. 2. Anon $3^h03^m$, 58°19′ – really a nebula associated with T Tauri stars?**
Yu. I. Glushkov, Eh. K. Denisyuk, Z. V. Karyagina, L. A. Usol'tseva.
Astron. Tsirk., No. 1339, p. 1 – 2 (1984). In Russian.

131.211 **The methyl cyanide hot and warm cores in Orion: statistical equilibrium excitation models of a symmetric–top molecule.**
R. B. Loren, L. G. Mundy.
Astrophys. J., Vol. 286, No. 1, p. 232 – 251 (1984).
More than 50 lines of the symmetric–top molecule $CH_3CN$, from $J$ = 6–5 up to J = 16–15 and components from $K$ = 0 to $K$ = 7, have been observed toward OMC–1. High–velocity resolution line profiles reveal two kinematically and thermally distinct regions; broad emission ($\Delta V \sim$ 12 km s$^{-1}$) at 5.1 km s$^{-1}$ arising from an unresolved hot core and narrow emission ($\Delta V \sim$ 4 km s$^{-1}$) at 8.7 km s$^{-1}$ arising from dense warm clumps of the quiescent cloud. Statistical equilibrium models yield kinetic temperatures for the regions emitting the broad and narrow components of $275 \pm 25$ K and $95 \pm 8$ K, respectively. The excitation of the statistical equilibrium (SE) models is compared to that predicted by the rotational temperature equilibrium (RTE) analytic method. In addition to the lines of $CH_3CN$, 34 other lines were observed, including the H35$\alpha$ hydrogen recombination line and the first detection of deuterated formaldehyde (HDCO) toward OMC–1.

131.212 **The abundance of interstellar lithium.**
L. M. Hobbs.
Astrophys. J., Vol. 286, No. 1, p. 252 – 254 (1984).
The interstellar Li I $\lambda$6707 absorption line is detected toward $\varrho$ Oph A at an equivalent width $W_\lambda = 3.1 \pm 1.9$ mÅ. Upper limits near 1 mÅ also are found toward seven other stars, including 55 Cyg, toward which relatively strong Li I absorption was previously reported. The four positive detections and the eight upper limits now available for the $\lambda$6707 line indicate that Li is not significantly depleted from the interstellar gas.

131.213 **Large–scale structure of the Taurus molecular complex. I. Density fluctuations – a fossil jeans length?**
S. C. Kleiner, R. L. Dickman.
Astrophys. J., Vol. 286, No. 1, p. 255 – 262 (1984).
The authors present a study of the large–scale density structure of the Taurus molecular cloud complex. Two–dimensional autocorrelation functions of molecular hydrogen column density within a 8° × 14° region are calculated using $^{13}$CO emission as well as visual extinctions from the Lynds catalog of dark clouds as tracers of $N_{H_2}$. A statistically significant length scale characterizing the separation of condensations within the complex is found. The authors argue that this scale may be regarded as the signature of an instability frozen into the complex at the time of its formation.

131.214 **The detection of interstellar methyl–diacetylene.**
R. B. Loren, A. Wootten, L. G. Mundy.
Astrophys. J., Lett. Ed., Vol. 286, No. 1, p. L23 – L26 (1984).
Four transitions of the molecule methyl–diacetylene ($CH_3C_4H$) have been detected toward the cold molecular cloud TMC–1. The presence of this molecule shows that long carbon chain molecules in the interstellar medium can be built with the methyl end group. The column density of $CH_3C_4H$ toward TMC–1 is $2.8\times10^{13}$cm$^{-2}$ at $T_k$ = 15K or $2.0\times10^{13}$cm$^{-2}$ at $T_k$ = 10K, a factor of 4 larger than that of the isoelectronic species $CH_3C_3N$ ($5\times10^{12}$cm$^{-2}$) which has recently been detected in this same cloud by Broton and colleagues.

131.215 **Spektroskopische Beobachtungen des bipolaren Massenflußobjektes L1551–IRS 5.**
M. Sarcander, T. Neckel, H. Elsässer.
Mitt. Astron. Ges., Nr. 62, p. 225 – 228 (1984). – See Abstr. 010.341.

131.216 **Binary mass ratios and star formation.**
H. Zinnecker.
Mitt. Astron. Ges., Nr. 62, p. 239 (1984). Abstract. – See Abstr. 010.341.

131.217 **Sternentstehung in der Dunkelwolke Lynds 810.**
T. Neckel, R. Chini, R. Güsten.
Mitt. Astron. Ges., Nr. 62, p. 240 (1984). Abstract. – See Abstr. 010.341.

131.218 **$H_2$ Emission bei 1.064 μm im Orion Nebel.**
H. Hippelein, G. Münch, E. Pitz.
Mitt. Astron. Ges., Nr. 62, p. 299 – 301 (1984). – See Abstr. 010.341.

131.219 **Morphologie der Dunkelwolke 'Kohlensack'.**
K. J. Seidensticker.
Mitt. Astron. Ges., Nr. 62, p. 304 – 305 (1984). – See Abstr. 010.341.

131.220 **Dunkelwolken und kosmische Gamma–Strahlung.**
J. V. Feitzinger, J. A. Stüwe.
Mitt. Astron. Ges., Nr. 62, p. 307 (1984). Abstract. – See Abstr. 010.341.

131.221 **Interstellar element abundances in the direction of τ Tauri.**
R. Skuppin, M. Grewing.
Mitt. Astron. Ges., Nr. 62, p. 308 – 310 (1984). – See Abstr. 010.341.
From two high–resolution IUE spectra equivalent widths of the interstellar lines seen towards τ Tau (d = 110 pc) are derived. An empirical curve–of–growth based on Fe II– Si II–lines is used to derive column densities. On comparing the relative element abundances to solar values one finds Si, Fe, Mg, and Mn to be depleted whereas others like S or Zn seem almost undepleted.

131.222 **Kalte Kerne und heiße Hüllen im H I Gas.**
P. M. W. Kalberla.
Mitt. Astron. Ges., Nr. 62, p. 311 – 314 (1984). – See Abstr. 010.341.

131.223 **Morphologie eines H I–Jets in hohen galaktischen Breiten.**
U. Herbstmeier, P. M. W. Kalberla, U. Mebold.
Mitt. Astron. Ges., Nr. 62, p. 322 – 325 (1984). – See Abstr. 010.341.

131.224 **H I Shells in hohen galaktischen Breiten und ihre Gegenstücke im Radiokontinuum.**
A. Heithausen.
Mitt. Astron. Ges., Nr. 62, p. 326 – 328 (1984). – See Abstr. 010.341.

131.225 **Parameter für Hochgeschwindigkeitswolken.**
A. Kaelble, K. S. de Boer, M. Grewing.
Mitt. Astron. Ges., Nr. 62, p. 329 – 330 (1984). – See Abstr. 010.341.

131.226 **Wave interpretation of velocity measurements of interstellar gas motion made by Mars 7 and Venera 9 probes.**
E. M. Grivnev.
Pis'ma Astron. Zh., Tom 10, No. 12, p. 908 – 911 (1984). In Russian. English translation in Sov. Astron. Lett., Vol. 10.
It is proposed that the motion of interstellar gas in the neighborhood of the solar system detected by Mars 7 and Venera 9 probes consists of chaotic motion and regular gas flow in a spiral density wave. The angular velocity of density waves of ~(12 – 16)km/s kpc and the value of disturbance of the basic potential of the Galaxy of ~(5 – 10)% are obtained under the assumption that the chaotic velocity of the interstellar gas is smaller than the local sound velocity.

131.227 **Detection of excited carbon lines at decametric waves in some galactic objects.**
A. A. Konovalenko.
Pis'ma Astron. Zh., Tom 10, No. 12, p. 912 – 917 (1984). In Russian. English translation in Sov. Astron. Lett., Vol. 10.
The absorption lines C638α – C641α are detected at 25 MHz in the region with galactic coordinates $l^{II} = 75°$; $b^{II} = 0°$ and near the NGC 2024 nebula. The observed parameters of lines can be only explained in the frame of the low–temperature dielectronic recombination mechanism. The advantages of decametric observations of carbon recombination lines in various components of the interstellar medium are pointed out.

131.228 **Radial distribution of giant molecular clouds. II. Comparison of the viscous cloud system with observations.**
M. Fukunaga.
Publ. Astron. Soc. Jpn., Vol. 36, No. 3, p. 417 – 432 (1984).
The dynamics of the viscous giant molecular cloud (GMC) system in various types of galaxies shows that the distribution of GMCs is characterized by the rotation curve of individual galaxies. When the rotation curve of a galaxy is rigid in the inner region and differential in the outer region, GMCs accumulate in the "transition region" of rotation, and in the outer region they show a special distribution in which the viscous torque is nearly zero. If the rotation curve is flat, such a density distribution is proportional to $r^{-1.5}$, where $r$ is the galactocentric radius. If the rotation curve has two maxima like those of our Galaxy, the GMCs accumulate in the two transition regions of rotation. These results explain the observed CO distributions in galaxies.

131.229 **On the random motion of giant molecular clouds.**
M. Fukunaga.
Publ. Astron. Soc. Jpn., Vol. 36, No. 3, p. 433 – 456 (1984).
The author proposes that the energy source of the random motion of giant molecular clouds (GMCs) is the mechanical energy of the galactic rotation. If one regards the system of GMCs as a viscous fluid, the mechanical energy of rotation of the system dissipates into the energy of random motion of the GMCs due to viscosity. In this paper the author estimates the steady–state velocity dispersion of a differentially rotating uniform GMC system based on the equation of the viscous stress tensor, in which the collision term includes both the gravitational encounter and the inelastic direct collision.

131.230 **Coagulation of interstellar clouds in spiral gravitational potential and formation of giant molecular clouds.**
K. Tomisaka.
Publ. Astron. Soc. Jpn., Vol. 36, No. 3, p. 457 – 475 (1984).
The formation process of giant molecular clouds is investigated from the standpoint of the coagulation theory of molecular clouds. Small clouds collide with one another and grow to massive ones. Ultimately they form giant molecular clouds with a finite lifetime. The occurrence of star formation in a giant cloud destroys it and small clouds are formed again consequently. The author studies the time evolution of the clouds which move through a spiral gravitational potential by an *N*–body simulation. It is shown that giant molecular clouds are formed in the spiral arm region by mutual collisions between clouds.

131.231 **CO observations of the W51 region. II.**
M. Ohishi, T. Nakamura, S. Kodaira, K. Ishii, J. Inatani.
Publ. Astron. Soc. Jpn., Vol. 36, No. 3, p. 505 – 515 (1984).
The $^{13}CO$ ($J = 1 \rightarrow 0$) line was observed in the W51 molecular cloud complex with the 1.5–m telescope at the Kisarazu Technical College. A region of $1° \times 0.7°$ around W51A and B was mapped and some small nearby clouds were also observed. The spatial and kinematical structure of the W51 molecular cloud complex is studied by comparing the $^{13}CO$ results with the $^{12}CO$ data which were previously taken by the same telescope. The main part of the W51 molecular cloud complex is composed of two massive clouds.

131.232 **Two large dark cloud complexes in $l = 351°$ to $1°$ and $b = -2°$ to $+2°$.**
T. Ichikawa.
Publ. Astron. Soc. Jpn., Vol. 36, No. 3, p. 525 – 538 (1984).
Interstellar extinction within 4 kpc from the sun is investigated in the field of $351° \leqslant l \leqslant 1°$ through $l = 0°$ and $|b| \leqslant 2°$ by means of the star–count method on the basis of the observed cumulative number and a space distribution model of late M giants. The general interstellar extinction in visual wavelengths is

about 2.5 mag $kpc^{-1}$ in the galactic plane and the scale height is about 100 pc in a distance range from 1.5 to 4 kpc from the sun for the distance, 10 kpc, of the sun from the galactic center. The space distribution analysis reveals two large dark cloud complexes associated with the spiral arms.

131.233 **On the distance of the Scorpius–Ophiuchus dark cloud.**
V. Straižys.
Vilniaus Astron. Obs. Biul., Nr. 67, p. 3 – 17 (1984). In Russian.

The dependence of interstellar extinction on distance in the direction of the Scorpius–Ophiuchus dark cloud is determined. It is based on photoelectric photometry of 44 stars in the Vilnius photometric system and on the published *UBV* photometry and MK spectral types of 38 other stars. Interstellar extinction $A_V$ is plotted against distance. It is concluded that the front side of the Scorpius–Ophiuchus dark cloud is at $120 \pm 30$ pc distance and that the aggregate of nebulae and young stars extends at least up to 200 pc. The mean distance of B–type stars is 140 pc.

131.234 **Interstellar reddening in the direction of the globular cluster M71. II.**
R. Janulis.
Vilniaus Astron. Obs. Biul., Nr. 67, p. 18 – 27 (1984). In Russian.

The results of photoelectric photometry of 29 stars in the direction of the globular cluster M71 in the Vilnius photometric system are given. Several stars with peculiar energy distribution are discovered.

131.235 **Optical study of the star–forming region Sharpless 252 (NGC 2175).**
L. K. Haikala.
Diss. Naturwiss.–Math. Gesamtfak. Ruprecht–Karls–Univ., Heidelberg, F. R. Germany, 6+98+A8+B12 pp. (1984).

The H II region–molecular cloud complex Sharpless 252 and the stars associated with it have been studied by means of Fabry–Perot interferometry, photography, photoelectric and photographic UBVRI potometry and polarimetry. A three dimensional model has been constructed for the molecular cloud and S 252A, B, C, E and F combining photographic and Fabry–Perot observations and data from literature. The blister and/or champagne model can well explain the observed features of these sources.

131.236 **Dust far infrared emission at high galactic latitudes.**
C. Ceccarelli, G. Dall'Oglio, P. de Bernardis, S. Masi, B. Melchiorri, F. Melchiorri, G. Moreno, L. Pietranera.
Clusters and groups of galaxies, p. 279 – 280 (1984). – See Abstr. 012.088.

131.237 **Dust distribution nearby the solar system and its possible implications on cosmic radiation background anisotropy.**
C. Ceccarelli, G. Dall'Oglio, P. de Bernardis, S. Masi, B. Melchiorri, F. Melchiorri, G. Moreno, L. Pietranera.
Clusters and groups of galaxies, p. 281 – 282 (1984). – See Abstr. 012.088.

131.238 **The properties of molecular clouds with high redshifts ($z = 2 - 3$).**
Yu. A. Shchekinov, M. B. Entel (*M. B. Ehnteh'l'*), A. I. Dudenko.
Astrophys. Space Sci., Vol. 107, No. 1, p. 51 – 59 (1984).

This paper deals with molecular clouds discovered in the absorption spectra ($z = 2 - 3$) of distant quasars. It is argued that these clouds belong to the gaseous subsystems of young galaxies. The authors estimate the gas concentration to be $n < 10^4 cm^{-3}$ in the cloud observed in the direction of the quasar PHL 957. It is shown that this cloud is exposed to ultraviolet radiation. The UV–energy flux does not exceed the value typical for our Galaxy by an order of magnitude ($F \cong 2 \times 10^{-6}$ergs $cm^{-2}s^{-1}Å^{-1}$ at $\lambda = 2000$ Å). The mechanisms maintaining the thermal balance in this cloud are discussed.

131.239 **On the possible equation of state of a giant molecular cloud.**
A. Mitra.
Astrophys. Space Sci., Vol. 107, No. 1, p. 135 – 140 (1984).

The author intends to probe into the nature of the thermodynamical equilibrium of an idealized isothermal, spherical, self–gravitating, giant molecular cloud. The necessity of invoking a frozen–in magnetic field has been pointed out for clouds with uniform mass distribution. An equation of state corresponding to the macroscopic statistical equilibrium of the cloud, steeped into a weak homogeneous magnetic field, has also been derived.

131.240 **Helium and hydrogen of the local interstellar medium observed in the vicinity of the sun.**
J.–L. Bertaux.
NASA Conf. Publ., NASA CP–2345, p. 3 – 23 (1984). – See Abstr. 012.089 (IAU Colloq. No. 81).

From the intensity distribution of the glow in the solar system, one can derive the density of H and He in the local interstellar medium (LISM) and the direction of the relative motion $\vec{V}_w$ between the sun and the LISM in the very vicinity of the sun. A summary of results is presented, together with a discussion of the methods of interpretation and their difficulties. The vector $\vec{V}_w$ is found to be $20 \pm 1$ km $s^{-1}$ in the direction $\alpha = 254 \pm 3°$, $\delta = -17 \pm 3°$, quite different from the Apex direction. This means that the LISM is moving also in respect to the local frame of reference giving rise to the so–called Interstellar Wind. The temperature of the LISM is T = 8,000±1,000K, the density n(H) ≅ 0.04 to 0.06 $cm^{-3}$, and the helium density n(He) ≅ 0.015 to 0.020. The high helium/hydrogen ratio, in respect to the cosmological ratio, would imply that a substantial part of the hydrogen is ionized.

131.241 **Pioneer 10 and Voyager observations of the interstellar medium in scattered emission of the He 584 Å and H Lyα 1216 Å lines.**
D. E. Shemansky, D. L. Judge, J. M. Jessen.
NASA Conf. Publ., NASA CP–2345, p. 24 – 27 (1984). – See Abstr. 012.089 (IAU Colloq. No. 81).

The combination of Pioneer photometric and Voyager spectrometric observations of EUV interstellar–interplanetary emissions in the region beyond 5 AU have been applied to a determination of atomic hydrogen and helium densities. The spacecraft data have been combined with daily full sun averages of the H Lyα 1216 Å line obtained by the Solar Mesospheric Explorer satellite, to obtain a measure of atomic hydrogen density independent of instrument absolute calibration. The density estimates from preliminary work on the observations are [H] = 0.12 $cm^{-3}$ and [He] = .016 $cm^{-3}$.

131.242 **Changes of LISM characteristics in the heliospheric interface.**
H. W. Ripken, H. J. Fahr.
NASA Conf. Publ., NASA CP–2345, p. 28 – 31 (1984). – See Abstr. 012.089 (IAU Colloq. No. 81).

It is possible to deduce LISM properties from observations of interstellar neutral gases in the inner solar system. The subsonic LISM plasma interface ahead of the heliopause causes profound changes in the properties of the neutral LISM gas traversing this region. Mainly p–H charge exchange processes give rise to the destruction of primary hydrogen and the production of secondary hydrogen atoms, the net effect being a depletion of the neutral hydrogen component of the LISM by about 50%. Details on the depletion mechanisms, the hydrogen and oxygen extinctions, and the consequences for the Ly–alpha resonance luminescence intensity interpretations are presented.

131.243 **Absorption line studies and the distribution of neutral gas in the local interstellar medium.**
F. C. Bruhweiler.
NASA Conf. Publ., NASA CP–2345, p. 39 – 50 (1984). – See Abstr. 012.089 (IAU Colloq. No. 81).

Previous published absorption line studies performed at ultraviolet and visual wavelengths are combined with new ultraviolet data in order to map out the distribution of H I within 150 pc of the Sun.

131.244 **Synthesis of data on the local interstellar medium.**
D. G. York, P. C. Frisch.
NASA Conf. Publ., NASA CP–2345, p. 51 – 59 (1984). – See Abstr. 012.089 (IAU Colloq. No. 81).

While no completely definite picture of the local interstellar medium can be presented yet, some general conclusions can be drawn. The authors attempt to draw together the most relevant data. Some new UV results are collated with published results to help state the physical properties of the gas. A morphological view of the local medium is given. The authors confine themselves to the region within 100 pc with the caveat that the distance of much of the material under discussion is poorly known.

131.245 **Observations of interstellar H I toward nearby late–type stars.**
W. B. Landsman, R. C. Henry, H. W. Moos, J. L. Linsky.
NASA Conf. Publ., NASA CP–2345, p. 60 – 63 (1984). – See Abstr. 012.089 (IAU Colloq. No. 81).

High–dispersion Copernicus and IUE observations of chromospheric Ly α emission are used to study the distribution of H I in the local interstellar medium. Interstellar parameters are derived toward 3 stars within 5 pc of the sun, and upper limits are given for the Ly α flux from 9 other stars within 10 pc.

131.246 **Observations of local interstellar Mg I and Mg II.**
F. Bruhweiler, W. Oegerle, E. Weiler, R. Stencel, Y. Kondo.
NASA Conf. Publ., NASA CP–2345, p. 64 – 66 (1984). – See Abstr. 012.089 (IAU Colloq. No. 81).

The authors have combined *Copernicus* and *IUE* observations of 5 stars within 50 pc of the Sun to study the ionization of magnesium in the local interstellar medium (LISM). The high resolution *Copernicus* spectrometer was used to detect interstellar Mg I 2852 in the spectra of α Gru, α Eri, and α Lyr, while placing upper limits on Mg I in the spectra of α CMa and α PsA. Observations of Mg II 2795, 2802 for these stars were also obtained with *IUE* and *Copernicus*. The column densities of Mg I and Mg II are used to place constraints on the temperature of the LISM.

131.247 **Mg II spectra of late type stars used to probe the LISM.**
J. E. Beckman, L. Crivellari, M. L. Franco, P. Molaro, G. Vladilo.
NASA Conf. Publ., NASA CP–2345, p. 67 – 70 (1984). – See Abstr. 012.089 (IAU Colloq. No. 81).

IUE spectra of Mg II h and k in late type dwarfs and giants have been used to detect and measure absorption components due to the LISM. This technique gives a method of probing the awkward range from d = 3 pc to d = 80 pc from the sun. The authors can plot the H I component of the LISM well enough to confirm it as a cloud some 20 – 30 pc in extent, moving towards the sun at 28 km/sec.

131.248 **Fe II and Mg II in the nearby interstellar medium.**
K. S. de Boer, H. Lenhart.
NASA Conf. Publ., NASA CP–2345, p. 71 – 74 (1984). – See Abstr. 012.089 (IAU Colloq. No. 81).

UV spectra with 15 km $sec^{-1}$ resolution of bright stars have been searched for Fe II and Mg II interstellar lines. The authors present equivalent widths of absorption for 8 stars and find average line–of–sight densities n(Fe II) = $4 \times 10^{-8}$ and n(Mg II) = $2 \times 10^{-7} cm^{-3}$. This represents approximately n(H) = 0.03 $cm^{-3}$.

131.249 **A type stars as probes of the local interstellar medium.**
R. Freire Ferrero, R. Ferlet, A. Vidal–Madjar.
NASA Conf. Publ., NASA CP–2345, p. 75 – 79 (1984). – See Abstr. 012.089 (IAU Colloq. No. 81).

With the aim to sample well the local interstellar medium (LISM), the authors propose to use A stars as targets. The Mg II UV lines seem to be the best interstellar absorption candidates. Several hundreds of A stars can be reached within 100 pc. First preliminary results (20 lines of sight) are presented, based on previous Copernicus and actual IUE observations.

131.250 **Probe of the nearby interstellar medium by the vacant line of sight to β CMa.**
C. Gry, D. G. York, A. Vidal–Madjar.
NASA Conf. Publ., NASA CP–2345, p. 80 – 83 (1984). – See Abstr. 012.089 (IAU Colloq. No. 81).

The line of sight to β CMa has been probed by Copernicus observations. This particular line of sight is remarkable for the low mean densities. The authors find $\bar{n}_H \sim .002\ cm^{-3}$. They can distinguish two separate regions: (1) A local nearby H I region extends over a few parsecs from the sun with a density of the order of 0.1 $cm^{-3}$ and a temperature of 11000 to 12500K. (2) An H II region lies somewhere beyond the H I region and is spread over about 60 pc. Its total hydrogen mean density is of the order of $\sim 0.1\ cm^{-3}$ and it contains only elements in low ionization state. All the data are coherent with the picture of a cloud in ionization equilibrium at T ~ 23000K.

131.251 **Deuterium abundance in the local interstellar medium.**
R. Ferlet, C. Gry, A. Vidal–Madjar.
NASA Conf. Publ., NASA CP–2345, p. 84 – 88 (1984). – See Abstr. 012.089 (IAU Colloq. No. 81).

The authors underline the present situation of deuterium abundance evaluation in interstellar space and show that it should be $\leqslant 10^{-5}$. Studying in more detail the λ Sco line of sight and having observed two Na I interstellar components toward this star, they can show that the D/H evaluation made toward λ Sco is in fact related to the local interstellar medium (less than 10 pc from the sun). Because this evaluation is also $\leqslant 10^{-5}$, it is in striking contrast with the one made toward α Aur (D/H $\geqslant 1.8 \times 10^{-5}$) confirming the fact that the deuterium abundance in the local interstellar medium varies by at least a factor of two over few parsecs.

131.252 **IUE observations of the gaseous component of the local interstellar medium.**
D. P. Gilra.
NASA Conf. Publ., NASA CP–2345, p. 89 – 90 (1984). – See Abstr. 012.089 (IAU Colloq. No. 81).

The discovery of interstellar Ge, Ga, and Kr is reported. Several intercombination lines of Fe II are detected. The depletion is most pronounced in Ca, Ti, and V. The highly ionized gas C IV and Si IV is not co–extensive with the Si II and C II gas nor with the O VI gas. The molecular gas (CO) shows very small velocity dispersion (b ~ 1 km/sec).

131.253 **Voyager EUV and FUV observations.**
J. B. Holberg.
NASA Conf. Publ., NASA CP–2345, p. 91 – 94 (1984). – See Abstr. 012.089 (IAU Colloq. No. 81).

The Voyager 1 and 2 ultraviolet spectrometers are sensitive over the wavelength range 500 to 1700 Å. In the EUV, at wavelengths shortward of the Lyman limit (912 Å), Voyager observations have detected emission from three out of a sample of 11 nearby hot DA white dwarfs. These observations imply very low H I column densities in the direction of the three stars detected. In the FUV, at wavelengths between 912 and 1200 Å, Voyager observations of O and B stars can be used to study interstellar reddening at the shortest wavelengths and to provide useful estimates of interstellar $H_2$ column densities.

131.254 **Optical emission line studies and the warm, ionized component of the local interstellar medium.**
R. J. Reynolds.
NASA Conf. Publ., NASA CP–2345, p. 97 – 108 (1984). – See Abstr. 012.089 (IAU Colloq. No. 81).

Observations of diffuse, galactic Hα, [N II] λ6583, and [S II] λ6716 emission lines provide evidence for a warm ($\sim 10^4$K), primarily ionized component of the interstellar medium distributed throughout the galactic disk. This component of the interstellar gas is an important constituent of the interstellar medium in the solar neighborhood. The intensity of the Hα background at high galactic latitudes implies that this component is maintained by an average hydrogen ionization rate in the vicinity of the sun of $(2-4) \times 10^6 s^{-1}$ per $cm^2$ of galactic disk. The sources of this ionization have not yet been identified.

131.255 **A high-resolution study of local interstellar sodium.**
A. Ardeberg, H. Lindgren, E. Maurice.
NASA Conf. Publ., NASA CP-2345, p. 109 – 112 (1984). – See Abstr. 012.089 (IAU Colloq. No. 81).

A high-resolution spectroscopic investigation has been made of interstellar lines of sodium. From identifications of $D_1$ and $D_2$ line components concentrations of interstellar sodium gas have been studied. Some preliminary data are provided on the spatial distribution of stronger sodium concentrations.

131.256 **Optical observations of nearby interstellar gas.**
P. C. Frisch, D. G. York.
NASA Conf. Publ., NASA CP-2345, p. 113 – 116 (1984). – See Abstr. 012.089 (IAU Colloq. No. 81).

The authors have chosen high-resolution observations of the yellow sodium lines as a means of locating and mapping nearby low density interstellar clouds.

131.257 **Distances of local clouds from optical line observations.**
R. M. Crutcher, D. J. Lien.
NASA Conf. Publ., NASA CP-2345, p. 117 – 120 (1984). – See Abstr. 012.089 (IAU Colloq. No. 81).

The authors observed the interstellar D-lines of Na I toward 49 stars in order to determine the distance ($125 \pm 25$ pc) to a cold H I cloud. This sheet of gas may be part of the back side of the shell formed by the Loop I supernova.

131.258 **Local interstellar extinction with an emphasis on uvbyβ results.**
J. Knude.
NASA Conf. Publ., NASA CP-2345, p. 123 – 144 (1984). – See Abstr. 012.089 (IAU Colloq. No. 81).

Three-dimensional mapping of the local dust has been attempted. Two interpretations of the color excess observations have been suggested. Either is the solar vicinity virtually free from dust within 75 – 100 pc or $4.4 \times 10^{-4}$clouds/pc$^3$ filling about 2% of the volume are distributed randomly in a low density medium.

131.259 **Optical polarization as a probe of the local interstellar medium.**
J. Tinbergen.
NASA Conf. Publ., NASA CP-2345, p. 145 – 148 (1984). – See Abstr. 012.089 (IAU Colloq. No. 81).

The use of interstellar polarization as a measurement tool for dust or magnetic field presents practical difficulties (dust and magnetic field configurations inextricably mixed up; limited number of suitable stars). The precision with which polarimetry can nowadays be performed at a number of observatories is about $1 \times 10^{-4}$. The most immediate ($<30$ pc) local region may be thought of as mostly intercloud medium. The precision quoted above corresponds to a distance of 40 pc. It is therefore unrealistic to expect to use polarimetry to see detailed structure in this intercloud medium. At larger distances the precision quoted corresponds to perhaps 5 pc, so that more spatial detail may be seen. This detail should, however, be interpreted in terms of cloud statistics, rather than a continuous medium.

131.260 **Do the arching H I filaments show in local reddening data?**
J. Knude.
NASA Conf. Publ., NASA CP-2345, p. 149 – 151 (1984). – See Abstr. 012.089 (IAU Colloq. No. 81).

Probable dust counterparts of arching structures are looked for in two cases. (1) The Eridanus expanding shell, Heiles (1976) and (2) the extended Sco-Cen feature, Weaver (1977). Apparently the color excess data may be used to derive distance and dimension estimates of these structures.

131.261 **A large-scale dark cloud system in the northern sky.**
W. Schlosser.
NASA Conf. Publ., NASA CP-2345, p. 152 (1984). Abstract. – See Abstr. 012.089 (IAU Colloq. No. 81).

131.262 **Observations of absorption lines from highly ionized atoms.**
E. B. Jenkins.
NASA Conf. Publ., NASA CP-2345, p. 155 – 168 (1984). – See Abstr. 012.089 (IAU Colloq. No. 81).

The pervasiveness of O VI absorption lines in the spectra of hot stars coupled with complementary observations of a diffuse background in soft X-rays and EUV radiation, shows that there is an extensive network of low density gas existing at "coronal" temperatures. Thus, shocks created by supernova explosions or mass loss from early-type stars can propagate freely through space and transfer a large amount of energy to the medium. To create the coronal temperatures, the shocks must have velocities in excess of 150 km sec$^{-1}$; shocks at somewhat lower velocity ($v \lesssim 100$ km sec$^{-1}$) can be directly observed in the lines of Si III. Observations of other lines in the ultraviolet, such as Si IV and C IV, may highlight the widespread presence of energetic UV radiation from very hot, dwarf stars.

131.263 **Absorption and emission of EUV radiation by the local ISM.**
F. Paresce.
NASA Conf. Publ., NASA CP-2345, p. 169 – 184 (1984). – See Abstr. 012.089 (IAU Colloq. No. 81).

The objective of this review is to assemble and critically analyze the available information on the local ISM derived from the limited EUV observations carried out so far. These include measurements of the spectra of bright EUV sources that reveal clear evidence of H photo absorption at $\lambda > 400$ Å and of the He ionization edge at 228 Å. The EUV diffuse background is found to convey interesting information on the density, temperature and possible location of the hot ISM component that can be profitably compared with similar data on far UV absorption lines, principally O VI.

131.264 **Observations of highly-ionized interstellar iron.**
L. M. Hobbs.
NASA Conf. Publ., NASA CP-2345, p. 189 – 191 (1984). – See Abstr. 012.089 (IAU Colloq. No. 81).

The spectra of 24 stars have been observed in the regions of the coronal [Fe X] λ6375 and [Fe XIV] λ5303 lines at detection limits near an equivalent width of 1 mÅ in the best cases. In general agreement with predictions based on a multi-phase model of the interstellar medium, no absorption which can be attributed to Fe X or Fe XIV ions in hot interstellar gas emitting the soft X-ray background is seen in any of these spectra, except for two. Toward λ Cephei an absorption line near λ6375 is measured with an equivalent width of $8.1 \pm 2$ mÅ, a width corresponding to $20 \pm 5$ km s$^{-1}$ or a temperature $T \leqslant (0.5 \pm 0.25) \times 10^6$K, and, if it is caused by Fe X ions, a radial velocity of $-355$ km s$^{-1}$. On that hypothesis, the hot interstellar gas constitutes at least 63% of the column density of gas along this light path.

131.265 **The nature of the soft X-ray emitting region in the direction of the North Polar Spur.**
J. P. Morrison, W. T. Sanders.
NASA Conf. Publ., NASA CP-2345, p. 211 – 214 (1984). – See Abstr. 012.089 (IAU Colloq. No. 81).

The spatial structure of the X-ray sky in the direction of the North Polar Spur was examined in two energy bands, the B band (0.10 – 0.18 keV) and the C band (0.15 – 0.28 keV). A model with two emitting regions, one local with unabsorbed emission, and the other more distant with emission partially absorbed by spatially varying amounts was investigated.

131.266 **Inverse relations between 0.25 keV counts and local interstellar dust.**
J. Knude, P. Jakobsen, S. Labov, S. Bowyer.
NASA Conf. Publ., NASA CP–2345, p. 226 – 228 (1984). – See Abstr. 012.089 (IAU Colloq. No. 81).

A preliminary comparison of color excesses and HEAO–1, A2 0.25 keV counts have been performed on two angular levels in order to search for indications of local photoelectric absorption.

131.267 **Molecular clouds within 100 pc.**
L. Blitz, L. Magnani, L. Mundy.
NASA Conf. Publ., NASA CP–2345, p. 231 – 234 (1984). – See Abstr. 012.089 (IAU Colloq. No. 81).

Observations at the 2.6 mm line of CO reveal the presence of a large number of molecular clouds at high galactic latitude. If the velocity dispersion of the clouds is a measure of their scale height, the mean distance of the ensemble the authors have detected is 100 pc. The clouds are unusual in that either they are not gravitationally bound or they are very deficient in CO relative to molecular hydrogen. These clouds represent a heretofore unrecognized component of the local interstellar medium. If they are pervasive in the Milky Way, they probably represent the small molecular cloud component of the interstellar medium.

131.268 **A wide latitude CO survey of molecular clouds in the northern Milky Way.**
T. M. Dame, P. Thaddeus.
NASA Conf. Publ., NASA CP–2345, p. 235 – 238 (1984). – See Abstr. 012.089 (IAU Colloq. No. 81).

The authors have used the Goddard–Columbia 1.2–meter telescope to carry out a wide latitude, low angular resolution survey of CO along most of the first galactic quadrant and a small part of the second. The survey is uniform and fully sampled in galactic longitude from 12° to 100°, and in latitude from –5° to +6°, with extensions to as high as +11° to include specific dark clouds; the total area surveyed, 1128 $deg^2$, is much larger than any region previously studied in CO or any other interstellar molecule.

131.269 **Dark cloud and globule distribution for galactic longitudes 230 to 360 degrees.**
J. V. Feitzinger, J. A. Stuewe.
NASA Conf. Publ., NASA CP–2345, p. 239 – 242 (1984). – See Abstr. 012.089 (IAU Colloq. No. 81).

A catalogue of dark nebulae and globules has been compiled from a study of the ESO–B and SRC–J sky atlas for galactic longitudes $230° < l < 360°$. This catalogue closes the great southern gap open since the work of Lynds (1962). The authors have listed 489 dark nebulae and 311 globules. The catalogue contains positions, sizes, opacities and the van den Bergh classification (1972) on the filamentary morphology of dark clouds. The authors present statistics concerning the northern and southern distributions and sizes of the nebulae.

131.270 **A neutral gas jet in a low velocity shock front at the boundary of the Draco Nebula.**
P. M. W. Kalberla, U. Herbstmeier, U. Mebold.
NASA Conf. Publ., NASA CP–2345, p. 243 – 247 (1984). – See Abstr. 012.089 (IAU Colloq. No. 81).

21–cm line observations with the Westerbork Synthesis Radio Telescope of a dust and molecular filament at the boundary of the Draco Nebula ($l \approx 91°$, $b \approx 38°$) reveal a jet–like neutral hydrogen feature funneling through an outlet in the low velocity shock front at the interface between the Draco Nebula and the surrounding gas. The jet–like feature is apparently connected with a high velocity filament at VLSR = –180 km $s^{-1}$. The authors suggest that the soft X–ray emission observed in the area is thermal bremsstrahlung produced by the deceleration of high velocity gas in galactic gas.

131.271 **The Draco Nebula, a molecular cloud associated with a high velocity cloud?**
U. Mebold, P. M. W. Kalberla.
NASA Conf. Publ., NASA CP–2345, p. 248 – 252 (1984). – See Abstr. 012.089 (IAU Colloq. No. 81).

An extended and faint bright nebula at $l \approx 91°$, $b \approx 38°$ exhibits H I λ21–cm, $^{12}CO$ λ2.6–mm and $^{13}CO$ λ2.7–mm emission lines in detailed positional agreement with the brightness distribution at the Palomar Observatory Sky Survey. Three colour UBV photometry of ~60 stars in the field of the bright nebula indicates a distance of ~800 pc. With this distance the total mass of two molecular clumps in the nebula is 27 and 219 $M_\odot$ and they appear to be gravitationally bound systems. A possible astrophysical relation to the high velocity H I–cloud phenomenon is pointed out.

131.272 **The smallest sizes of diffuse interstellar clouds.**
J. M. Dickey, J. Crovisier, I. Kazes.
NASA Conf. Publ., NASA CP–2345, p. 258 – 262 (1984). – See Abstr. 012.089 (IAU Colloq. No. 81).

By observing the difference in optical depths between absorption spectra toward the two components of double sources the authors have measured the variations in opacity over lengths of less than 0.1 up to 10 pc inside diffuse interstellar clouds. Significant variations are detected on scales larger than about 0.2 pc, but not less. This may represent the minimum size for diffuse cloud structure. By comparing the variations of Gaussian fitted line parameters the authors find that variations in the internal velocity field of diffuse clouds explain the data rather better than tiny independent "cloudlets".

131.273 **A warm magnetoactive plasma in a large volume of space.**
C. Heiles.
NASA Conf. Publ., NASA CP–2345, p. 263 – 267 (1984). – See Abstr. 012.089 (IAU Colloq. No. 81).

This paper shows that a diffuse ionized warm gas fills a large volume of space in the general direction of Radio Loop II. There are three types of observational evidence: Faraday rotation measures of extragalactic sources; emission measures derived from the Hα emission line in the diffuse interstellar medium; and magnetic field strengths in H I clouds derived from Zeeman splitting observations.

131.274 **An investigation of small scale H I structure at high galactic latitude.**
K. Jahoda, J. M. Dickey, F. J. Lockman, D. McCammon.
NASA Conf. Publ., NASA CP–2345, p. 268 (1984). Abstract. – See Abstr. 012.089 (IAU Colloq. No. 81).

131.275 **The z–dependence of the spin temperature of H I.**
S. R. Kulkarni, C. Heiles, J. H. van Gorkom, J. M. Dickey.
NASA Conf. Publ., NASA CP–2345, p. 269 – 273 (1984). – See Abstr. 012.089 (IAU Colloq. No. 81).

In 1981 the authors conducted an extensive, low–latitude, H I 21–cm absorption survey from the VLA. The goal of the survey was to answer the following questions: (1) What is the vertical distribution of the cold H I clouds? (2) Do H I clouds get systematically warmer at high |z|? (3) Does the fraction of warm intercloud medium H I increase with high |z|? (4) How much H I has been missed in previous H I emission surveys because of optical depth effects? (5) What is the galactic distribution of cool H I clouds? The authors present preliminary results which shed some light on the first four questions.

131.276 **Polarization of radio molecular lines and mapping of magnetic field direction.**
S. Deguchi, W. D. Watson.
NASA Conf. Publ., NASA CP–2345, p. 274 – 275 (1984). – See Abstr. 012.089 (IAU Colloq. No. 81).

Two level calculations by others for the linear polarizations expected for molecular emission lines from interstellar clouds are extended by considering a number of coupled rotational states of a diatomic molecule. Higher transitions commonly have strengths comparable to that of the $J = 1 \rightarrow 0$ transition so that depolarization due to coupling between the various J–states might be expected to reduce the predicted linear polarization. The significance of the null observational results obtained recently by Wannier, Scoville and Barvainis is assessed. The inclusion of additional states tends to decrease the maximum polarizations, though the calculated polarization is increased under conditions for which "superthermal" excitation occurs.

131.277 **Nearby molecular hydrogen.**
F. Lebrun.
NASA Conf. Publ., NASA CP-2345, p. 276 – 279 (1984). – See Abstr. 012.089 (IAU Colloq. No. 81).

If the gas-to-dust ratio is sufficiently uniform throughout the local interstellar medium, galaxy counts may provide a useful probe of the large scale structure of the interstellar gas. This idea substantiated by gamma-ray observations has led to the discovery of nearby molecular cloud complexes. The reddening studies indicate that one of them lies between 80 and 140 pc from the Sun. From CO observations, its molecular mass is estimated to be a few $10^3 M_{\odot}$.

131.278 **Radio and UV observations of high latitude H I.**
F. J. Lockman, L. M. Hobbs, J. M. Shull.
NASA Conf. Publ., NASA CP-2345, p. 280 (1984). Abstract. – See Abstr. 012.089 (IAU Colloq. No. 81).

131.279 **Modelling the local interstellar medium as a supernova remnant in a multiphase gas.**
L. L. Cowie.
NASA Conf. Publ., NASA CP-2345, p. 287 – 296 (1984). – See Abstr. 012.089 (IAU Colloq. No. 81).

A multiphase SNR does seem to provide a plausible description of the local ISM. The local region could quite likely have been deficient (by about a factor of 2 – 3) in cool gas prior to the supernova both on theoretical and observational grounds. However, there are no totally compelling arguments for this as yet. The Sun may lie in a region towards the outside of the supernova remnant which has not yet come into pressure equilibrium with the hot gas. Evaporative models give a much more satisfactory description of the O VI observations than Sedov solutions.

131.280 **Non equilibrium ionization in the local hot bubble.**
M. Arnaud, R. Rothenflug, R. Rocchia.
NASA Conf. Publ., NASA CP-2345, p. 301 – 304 (1984). – See Abstr. 012.089 (IAU Colloq. No. 81).

Soft X-ray surveys proved the existence of a local hot bubble surrounding the solar system (McCammon et al. 1983). Recently two experiments measured the spectrum of this hot interstellar medium with solid state detectors during rocket flights between 300 and 1000 eV (Rocchia et al. 1984). The good resolution of the experiment permitted to measure C V, C VI and O VII line emissivities with good accuracy. The authors use these data to put some constraints on a model where the emission comes from a supernova remnant expanding in a finite pressure medium (Cox and Anderson 1982). They show the importance of non equilibrium ionization phenomena.

131.281 **Large-scale bubble structure of the interstellar medium (ISM) and properties of the local spiral arm (LSA).**
N. G. Bochkarev.
NASA Conf. Publ., NASA CP-2345, p. 305 – 310 (1984). – See Abstr. 012.089 (IAU Colloq. No. 81).

Radio, optical, infrared, and X-ray observations of the Cyg superbubble (CSB) region of the sky show that the CSB is not a single bubble object. 50 to 75 percent of its X-ray emission can be ascribed to discrete sources. The other 25 to 50% of X-ray emission probably originate from bubbles around 8 OB associations of the region. All bubbles located within the spiral structure of the Galaxy, M31, and M33 have diameters $\lesssim 300$ pc. The large distance of stellar association from the galactic plane combined with a picture of the gas distribution within the LSA shows that a Rayleigh-Taylor instability in the LSA can develop and give use to the formation of compact stellar clusters, such as the Cyg OB2 association.

131.282 **Magnetic alignment theory and the interpretation of polarization.**
P. Cugnon.
NASA Conf. Publ., NASA CP-2345, p. 315 – 318 (1984). – See Abstr. 012.089 (IAU Colloq. No. 81).

This paper presents some reflexions on the theory of magnetic spinning alignment. It is shown that a classical enhanced thermal Davis and Greenstein mechanism may be considered as a limiting case of Purcell's suprathermal spinning alignment theory when the number of active sites on a given grain becomes very important and/or for very short-lived suprathermal sites. This result is applied to the interstellar medium and some problems in which it may be important are briefly discussed, with special attention to local implications.

131.283 **Colorimetry of the diffuse nebulae S 156, S 157A, S 158, NGC 7635.**
Eh. S. Parsamyan, V. M. Petrosyan.
Astrofizika, Tom 20, Vyp. 3, p. 495 – 502 (1984). In Russian. English translation in Astrophysics, Vol. 20, No. 3.

The results of a colorimetric investigation of the diffuse nebulae S 156, S 157A, S 158, NGC 7635 excited by O type stars are presented. The *UBV* magnitudes and colour indices *U–B*, *B–V* are determined. Isodensitometric pictures of the nebulae are presented. The colorimetric data show that they are similar in their physical properties.

131.284 **The structure of massive molecular clouds.**
L. N. Arshutkin, I. G. Kolesnik.
Astrofizika, Tom 21, Vyp. 1, p. 147 – 161 (1984). In Russian. English translation in Astrophysics, Vol. 21, No. 1.

The structure of massive molecular clouds supported by turbulent pressure is considered. On the basis of model calculations for spherical-symmetric hydrostatic clouds it has been shown that the necessary condition for the existence of the dense cores of molecular clouds is the turbulent velocity decrease towards the cloud's center. The best consistency with observed data have the models in which turbulent and thermal energies are comparable. According to this condition the turbulent velocities have to be moderately supersonic. The obtained structure of a massive ($\sim 10^5 M_{\odot}$) cloud is in quality consistency with the observed parameters of giant molecular clouds.

131.285 **New cometary nebulae and tight Trapezium-like systems.**
A. L. Gyul'budagyan.
Astrofizika, Tom 21, Vyp. 1, p. 185 – 188 (1984). In Russian. English translation in Astrophysics, Vol. 21, No. 1.

Eight new cometary nebulae and four tight Trapezium-like star systems were found on the Palomar Sky Survey prints.

131.286 **The collapse of the cores of slowly rotating isothermal clouds.**
S. Terebey, F. H. Shu, P. Cassen.
Astrophys. J., Vol. 286, No. 2, p. 529 – 551 (1984).

The authors present here a semianalytic perturbational solution for the collapse of a slowly rotating cloud core. The initial equilibrium state is exact and corresponds to the uniformly rotating analogue of the singular isothermal sphere. Star (and disk) formation proceeds because the equilibrium is unstable to core collapse. The evolution in time can be followed by performing a perturbational analysis on the known similarity solution for the nonrotating case. The hydrodynamic equations, including self-gravity, can be reduced to a set of linear ordinary differential equations, which is solved by the method of matched asymptotic expansions. These calculations provide a self-consistent description of the dynamical collapse of rotating molecular cloud cores and a framework for the study of the formation of stars and nebular disks.

131.287 **High–resolution far–infrared observations of the extended W51 complex.**
T. N. Rengarajan, L. H. Cheung, G. G. Fazio, K. Shivanandan, B. McBreen.
Astrophys. J., Vol. 286, No. 2, p. 573 – 577 (1984).

A 1 $deg^2$ area around the H II region W51 has been scanned at far–infrared wavelengths with $\sim 1'$ resolution. Four main emissions centers were identified on an extended ($\sim 1°$) background. The distribution of the far–infrared emission is similar to that of the radio continuum emission, suggesting that both types of sources have a common origin. A comparison of the infrared and radio data indicates that the embedded energy sources are stellar clusters consisting of early O type stars.

131.288 ***HEAO 3* discovery of $^{26}$Al in the interstellar medium.**
W. A. Mahoney, J. C. Ling, W. A. Wheaton, A. S. Jacobson.
Astrophys. J., Vol. 286, No. 2, p. 578 – 585 (1984).

The *HEAO 3* gamma–ray spectroscopy experiment has detected gamma–ray emission from the galactic equatorial plane in a narrow line at 1809 keV. These gamma–rays are interpreted as resulting from the decay of approximately 3 $M_\odot$ of $^{26}$Al in the interstellar medium. The line has an energy of $1808.49 \pm 0.41$ keV, consistent with the decay of $^{26}$Al at rest, and an intrinsic width of $\leqslant 3.0$ keV (1 $\sigma$), consistent with the Doppler broadening expected from differential galactic rotation and random motions of the interstellar medium. The implied mass fraction ratio of $^{26}Al/^{27}Al$ in the present interstellar medium is $\sim 1.0 \times 10^{-5}$. Possible sources of interstellar $^{26}$Al are discussed.

131.289 **$^{13}CH_3OH$ in OMC–1.**
G. A. Blake, E. C. Sutton, C. R. Masson, T. G. Phillips, E. Herbst, G. M. Plummer, F. C. De Lucia.
Astrophys. J., Vol. 286, No. 2, p. 586 – 590 (1984).

The authors report the identification of several previously unidentified or misidentified lines toward OMC–1 as components of the $J = 5 \rightarrow 4$ $a$–type band of $^{13}CH_3OH$, deduced by combining accurate laboratory spectroscopy with sensitive broad–band astronomical line searches. An LTE fit to the data yields a rotational temperature of $T_{rot} \sim 120$K and a $^{13}CH_3OH$ column density of $\sim 1.3 \times 10^{15} cm^{-2}$, averaged over a 30″ beam. A $^{12}C/^{13}C$ ratio of $\sim 30$ is derived.

131.290 **Bipolar outflows in dark clouds.**
P. F. Goldsmith, R. L. Snell, M. Hemeon–Heyer, W. D. Langer.
Astrophys. J., Vol. 286, No. 2, p. 599 – 608 (1984).

The authors have studied four dark clouds thought to contain bipolar molecular outflows by mapping the $J = 1 \rightarrow 0$ transition of $^{12}$CO and observing the $J = 1 \rightarrow 0$ transition of $^{13}$CO at selected positions. The clouds B335 and L723 each contain an outflow of gas which is distinctly bipolar, well collimated, and centered on low–luminosity far–infrared objects. Observations of L1455 suggest that there are two individual well–collimated molecular outflows, each with its own central source. The high–velocity molecular material is attributed to the acceleration of the ambient molecular gas by stellar winds from the embedded objects. The energies of the molecular outflows and the stellar mass loss rates are calculated.

131.291 **The heating of cloud–hot phase interface regions by the dissipation of hydromagnetic waves.**
T. W. Hartquist, G. E. Morfill.
Astrophys. J., Vol. 287, No. 1, p. 194 – 199 (1984).

The authors show that the dissipation of Alfvén waves in the interface region between clouds and gas in the hot phase could heat $10^4$K gas to temperatures exceeding $10^5$K. Wave dissipation may thus drive mass loss from clouds.

131.292 **Star cloud turbulence.**
R. N. Henriksen, B. E. Turner.
Astrophys. J., Vol. 287, No. 1, p. 200 – 207 (1984).

For the ensemble of galactic molecular clouds, Larson (1984) and Myers (1983) have shown that a power law $v_l \propto l^n$ relates internal velocity dispersions $v_l$ and cloud sizes $l$. They interpret this relation in terms of the Kolmogorov law for incompressible turbulence. The authors argue that Kolmogorov incompressible turbulence is inapplicable in molecular clouds because of the importance of gravitational collapse. They analyze the ensemble of clouds as elements in a self–similar regime of compressible turbulence, in particular, of gravitationally driven, angular–momentum–limited, turbulence. Scaling laws for this "star–cloud turbulence" are derived. Several predictions are made about the time dependence of the initial stellar mass function.

131.293 **The CN radical in diffuse interstellar clouds.**
S. R. Federman, A. C. Danks, D. L. Lambert.
Astrophys. J., Vol. 287, No. 1, p. 219 – 227 (1984).

A survey of 15 lines of sight for the CN $B\,^2\Sigma^+ - X\,^2\Sigma^+$ interstellar absorption lines shows that the CN column density in diffuse interstellar clouds follows the relation $\log N(CN) \propto m \log N(H_2)$, where $m \approx 3$. This result is reproduced by a reaction network in which CN is produced primarily from $C_2$ by the neutral–neutral reaction $C_2 + N \rightarrow CN + C$, and photodissociation is the main destruction pathway for the neutral molecules CH, $C_2$, and CN.

131.294 **Extinction in inhomogeneous clouds.**
A. Natta, N. Panagia.
Astrophys. J., Vol. 287, No. 1, p. 228 – 237 (1984).

Model calculations of the attenuation produced by a thick, inhomogeneous dust layer in front of an extended source show that the apparent optical depth is lower than the average value. Correspondingly, the apparent extinction curve as a function of wavelength is flatter than the intrinsic one. In particular, an inhomogeneous dust layer of high optical depth tends to produce a gray extinction whose value depends only on the geometry of the layer itself. A method to determine the dust parameters from observational data is discussed and applied to the case of the heavily obscured, compact H II region K3–50.

131.295 **Interstellar abundances and depletions inferred from observations of neutral atoms.**
T. P. Snow.
Astrophys. J., Vol. 287, No. 1, p. 238 – 243 (1984).

Data on neutral atomic species are analyzed for the purpose of inferring relative elemental abundances and depletions in diffuse cloud cores, where it is assumed that densities are enhanced in comparison with mean densities over integrated lines of sight. Column densities of neutral atoms are compared to yield relative column densities of singly ionized species, which are assumed dominant in cloud cores. No systematic enhancement in the depletions of calcium or iron in cloud cores is found, except for $\zeta$ Ophiuchi. It is concluded either that in general all elements are depleted together in dense regions so that their relative abundances remain constant, or that typical diffuse clouds do not have significant cores, but instead are reasonably homogeneous.

131.296 **Rocket observations of the Orion reflection nebulosity in the range 130 – 200 nanometers and scattering properties of dust grains in the far–ultraviolet.**
T. Onaka, M. Sawamura, W. Tanaka, T. Watanabe, K. Kodaira.
Astrophys. J., Vol. 287, No. 1, p. 359 – 370 (1984).

Vacuum ultraviolet surface brightness maps of the southern part of the Orion region at the wavelengths 195.1, 179.6, 164.0, 148.4, and 132.8 nm were obtained by a rocket experiment. Large–scale bright nebulosity and a ridge structure in the western circumference of the Orion regions are clearly seen in these contour maps. Simple analysis shows that the observations can be explained by a front–dust–layer model, in which the diffuse light arises from the starlight scattered by dust grains in front of the stars, as well as by a back–dust–layer model, where dust grains behind the stars scatter the starlight to produce the diffuse nebulosity. A detailed discussion of the far–ultraviolet scattering properties of the dust grains in the observed region is included.

131.297 **Deuterium fractionation mechanisms in interstellar clouds.**
A. Dalgarno, S. Lepp.
Astrophys. J., Lett. Ed., Vol. 287, No. 1, p. L47 – L50 (1984).

The theory of the fractionation of deuterated molecules is extended to include reactions with atomic deuterium. With the recognition that dissociative recombination of $H_3^+$ is not rapid, observational data can be used in conjunction with the theory to derive upper and lower bounds to the cosmic deuterium–hydrogen abundance ratio. The authors find that [D]/[H] is at least $3.4 \times 10^{-6}$ and at most $4.0 \times 10^{-5}$ with a probable value of $1 \times 10^{-5}$.

131.298 **Arc-second IR CCD observations of OH0739–14.**
C. E. Woodward, W. J. Forrest, A. Moneti, J. L. Pipher.
News Lett. Astron. Soc. N.Y., Vol. 2, No. 6, p. 14 (1984). Abstract. – See Abstr. 010.241.

131.299 **IR CCD observations of the BN–KL region.**
J. L. Pipher, W. J. Forrest, A. Moneti, C. E. Woodward.
News Lett. Astron. Soc. N.Y., Vol. 2, No. 6, p. 15 (1984). Abstract. – See Abstr. 010.241.

131.300 **Effects of shocks on the sulfur chemistry of a dense interstellar cloud.**
G. F. Mitchell.
Astrophys. J., Vol. 287, No. 2, p. 665 – 670 (1984).

Time-dependent kinetic calculations have been carried out to follow the postshock abundances of sulfur molecules after the passage of shocks with speeds in the range $5 \leqslant v_s \leqslant 15$ km $s^{-1}$. The initial gas density is $10^4 cm^{-3}$. Abundances are followed for $10^5$yr after the shock. Abundance profiles behind shocks of speeds 5, 10, and 15 km $s^{-1}$ are presented for those sulfur species which show large abundance changes, namely, S, $S_2$, HS, $H_2S$, $H_2CS$, CS and SO. Postshock column abundances are presented for these same species.

131.301 **An estimate of star formation efficiency in molecular clouds.**
T. N. Rengarajan.
Astrophys. J., Vol. 287, No. 2, p. 671 – 675 (1984).

From existing data it is shown that there is a linear correlation between cloud mass derived from CO observations and associated luminosity obtained from far-IR observations. Over a 4 dex range of luminosity ($10^4 - 10^8 L_\odot$), the luminosity per unit mass is constant to within a factor of $\sim 2$ and is equal to $\sim 6\ L_\odot/M_\odot$. Based on the observed correlation and Miller–Scalo IMF, it is found that the star formation efficiency over the cloud life is 0.2 – 0.4, and the present-day star formation rate is 6 – 11 $M_\odot yr^{-1}$.

131.302 **Reevaluation of the rate constants for the reactions $C^+ + H_2 \rightarrow CH^+ + H$ and $CH^+ + H \rightarrow C^+ + H_2$ at temperatures characteristic of ambient and shocked diffuse interstellar clouds.**
W. J. Chesnavich, V. E. Akin, D. A. Webb.
Astrophys. J., Vol. 287, No. 2, p. 676 – 680 (1984).

The most recent experimental data on the cross section for the reaction $C^+(H_2, H)CH^+$ were used to determine the rate constants for the reactions $C^+(H_2, H)CH^+$ and $CH^+(H, H_2)C^+$ at temperatures characteristic of ambient and shocked diffuse interstellar clouds. The results of this paper indicate that the phase-space theory results of Herbst and Knudson should be used when explicitly modeling nonequilibrium effects in shocked clouds.

131.303 **The relative abundances of cyanogenated molecules.**
E. Churchwell, A. G. Nash, C. M. Walmsley.
Astrophys. J., Vol. 287, No. 2, p. 681 – 693 (1984).

The authors have observed with approximately the same spatial resolution HCN ($J$ = 1–0), HNC ($J$ = 1–0), their $^{13}C$ isotopic species, and $HC_3N$ ($J$ = 2–1 and $J$ = 10–9) toward six locations in the Taurus dark cloud complex, one dark cloud (L183) outside Taurus, and two warm clouds with associated H II regions. The column densities of HCN, HNC, and $HC_3N$ are determined and compared with those of $NH_3$ and CN taken from the literature. The measured abundances are used to estimate the importance of chemical shuffling reactions which convert HNC to HCN and vice versa.

131.304 **Absorption features in the 5 – 8 micron spectra of protostars.**
A. G. G. M. Tielens, L. J. Allamandola, J. Bregman, J. Goebel, L. B. d'Hendecourt, F. C. Witteborn.
Astrophys. J., Vol. 287, No. 2, p. 697 – 706 (1984).

High signal-to-noise ratio 5 – 8 μm spectra of four sources embedded in molecular clouds are presented. All four sources show evidence for the presence of absorption features. These range from two relatively narrow bands at 6.0 and 6.8 μm in W33A to a broad, shallow feature, which extends from about 5.2 to 7.8 μm and shows some structure, in Mon R2–IRS2, BN, and NGC 2264. The 2 – 13 μm spectra of simple molecules and mixtures of simple molecules have been measured at low temperatures in the laboratory. Comparison of the observed spectra with the laboratory spectra leads to the conclusion that the rather narrow 6.0 and 6.8 μm absorption features in W33A are largely due to the OH bending mode in $H_2O$ and the CH deformation modes in saturated hydrocarbons, respectively. The implications of the observed differences in the 6.0 and 6.8 μm absorption features for interstellar grain chemistry are discussed.

131.305 **High-velocity $HCO^+$ in Cepheus A: ionization levels in high-density clumps within a molecular flow.**
R. B. Loren, A. Wootten, Aa. Sandqvist, P. Friberg, Å. Hjalmarson.
Astrophys. J., Vol. 287, No. 2, p. 707 – 722 (1984).

High-velocity $HCO^+$ wings have been detected in both the $J$ = 1–0 and $J$ = 3–2 lines toward Cep A. Modeling the intensity ratio in the wings of these two lines indicates that this emission arises in high-density ($10^6 cm^{-3}$) clumps within a high-velocity outflow. The highest velocity $HCO^+$ emission is confined to an unresolved region at the location of the IR cluster and $H_2O$ masers. Both the $HCO^+$ abundance ($3 - 6 \times 10^{-9}$) and the electron abundance ($< 10^{-8}$) in the high-velocity flow are similar to more quiescent cold molecular clouds.

131.306 **The gas density gradient for three dark interstellar clouds.**
S. A. Fulkerson, F. O. Clark.
Astrophys. J., Vol. 287, No. 2, p. 723 – 727 (1984).

A grid of models has been constructed of the surface brightness of selected transitions of interstellar formaldehyde for three interstellar clouds. The grid included radial gas density gradients over a considerable range from uniform to very steep. The model results were then compared to observations of the interstellar dark clouds B361, L183, and L134. In all three cases, the comparison indicates that the gas is centrally condensed and follows a gradient in density which closely approximates an inverse square law.

131.307 **Discovery of a young stellar object near the water masers in W3(OH).**
J. L. Turner, W. J. Welch.
Astrophys. J., Lett. Ed., Vol. 287, No. 2, p. L81 – L84 (1984).

The authors present a high-resolution (1″.2) aperture synthesis map of the ultracompact H II region W3(OH) in the 88.6 GHz transition of HCN. They detected an HCN emission feature near the position of the $H_2O$ masers, $\sim 7''$ east of the continuum source. The velocity of the HCN emission indicates a physical association with the masers. The luminosity and high brightness temperature of the emission and the broad HCN line width indicate the presence of an obscured young star, possibly a protostar, within the clump. A dust optical depth of more than 50 at 10 μm is inferred toward the source, corresponding to a visual extinction of greater than 900 mag.

131.308 **The detection of the $F^1\Pi_u - X^1\Sigma_g^+$ system of diatomic carbon toward X Persei.**
D. J. Lien.
Astrophys. J., Lett. Ed., Vol. 287, No. 2, p. L95 – L98 (1984).

The 0 – 0 vibrational band of the $F^1\Pi_u - X^1\Sigma_g^+$ electronic system of $C_2$ has been detected with the *IUE* in the interstellar medium along the line of sight toward X Per. The column density derived from profile fitting is $3(\pm 1) \times 10^{12}/f_{00}cm^{-2}$; a lower limit on the 0 – 0 band oscillator strength has been determined from comparison with the upper limits of observations of the $A - X$ and $D - X$ systems: $f_{00} > 0.1$. The excitation temperature derived from profile fitting is 40K.

**A catalogue of visual extinction in Taurus and Perseus.**
See Abstr. 002.033.

**Catalogue of dark nebulae and globules for galactic longitudes 240 to 360 degrees.**
See Abstr. 002.034.

**A catalogue of equivalent widths of the interstellar 2200 Å band.**
See Abstr. 002.053.

**Interstellar chemistry.**
See Abstr. 003.033.

**Magnetic fields in astrophysics.**
See Abstr. 003.045.

**Physics of thermal gaseous nebulae (physical processes in gaseous nebulae).**
See Abstr. 003.047.

**Catalogue des étoiles cibles utilisées pour l'étude spectroscopique du milieu interstellaire diffus.**
See Abstr. 013.071.

**Micrometre–sized polymer layers synthesized by MeV ions impinging on frozen methane.**
See Abstr. 022.003.

**Emission of large hydrocarbons from frozen $CH_4$ by keV proton irradiation.**
See Abstr. 022.011.

**The radiation of microwaves and infrared by slender graphite needles.**
See Abstr. 022.016.

**The photodissociation of $H_2O$ at 157 nm: full internal state distributions and alignment of nascent $OH(X^2\Pi)$ radicals.**
See Abstr. 022.023.

**Photoabsorption and photodissociation of CO in the 900 – 1200 Å region.**
See Abstr. 022.024.

**Laboratory millimeter and submillimeter spectra of HNO and DNO.**
See Abstr. 022.026.

**The millimeter and submillimeter laboratory spectrum of methyl formate in its ground symmetric torsional state.**
See Abstr. 022.027.

**Laboratory detection of the protonated carbon dioxide by submillimeter wave spectroscopy.**
See Abstr. 022.033.

**New Fe II lines in IUE spectra of stars.**
See Abstr. 022.043.

**Collisional pumping of main line OH–maser sources.**
See Abstr. 022.044.

**Ternary association reaction of $CH_3^+$ and $CH(OH)_2^+$ with $H_2O$ in He buffer.**
See Abstr. 022.064.

**Dissociative recombination coefficients for $H_3^+$, $HCO^+$, $N_2H^+$, and $CH_5^+$ at low temperature: interstellar implications.**
See Abstr. 022.077.

**On the excitation of CH microwave lines.**
See Abstr. 022.089.

**Isotope shifts of some ultraviolet transitions of first row elements.**
See Abstr. 022.090.

**Selective Λ–doublet population of OH in inelastic collisions with $H_2$: a possible pump mechanism for the $^2\Pi_{1/2}$ astronomical OH maser.**
See Abstr. 022.091.

**The geometry of the propynol cation, $HC_3O^+$.**
See Abstr. 022.092.

**Einstein *A*–values for purely rotational transitions in the HDO–molecule.**
See Abstr. 022.128.

**Laboratory measurement of $H_3O^+$ line positions around 10 μm for astrophysical searches.**
See Abstr. 022.129.

**The importance of kinetically excited ions in the synthesis of interstellar molecules.**
See Abstr. 022.140.

**Laboruntersuchungen zur Deutung der Infrarotspektren des interstellaren Staubes.**
See Abstr. 022.141.

**Low–temperature dissociative recombination of $e + H_3^+$.**
See Abstr. 022.148.

**Infrared spectrum of the laboratory–synthesized quenched carbonaceous composite (QCC): comparison with the infrared unidentified emission bands.**
See Abstr. 022.162.

**Refractive indices for amorphous carbon.**
See Abstr. 022.163.

**Molecular astronomy using heterodyne detection at 691 GHz.**
See Abstr. 034.035.

**Interstellar matter with very large telescopes.**
See Abstr. 036.094.

**Star formation.**
See Abstr. 036.095.

**Automated starcounts in nebulous areas.**
See Abstr. 036.111.

**Future studies of the local interstellar medium with Space Telescope and Columbus.**
See Abstr. 051.077.

**Scattering of shock waves by a spherical cloud.**
See Abstr. 062.083.

**Optically thick ablation fronts.**
See Abstr. 062.086.

**Conical reflections of interstellar shocks. The formation of interstellar bullets.**
See Abstr. 062.119.

**Magnetic curtailment of the shock–induced thermal instability.**
See Abstr. 062.144.

**Plasma potential and grain charges in a dust cloud embedded in a plasma.**
See Abstr. 062.159.

**Half–range moment methods for radiative transfer in spherical geometry. III. Numerical solution and applications.**
See Abstr. 063.040.

**Linear polarization of molecular lines at radio frequencies.**
See Abstr. 063.051.

**Effects of grain models on UV penetration into clouds.**
See Abstr. 063.059.

**The interaction of a stellar wind with the surrounding interstellar medium.**
See Abstr. 064.069.

**Spherical accretion onto a protostellar core.**
See Abstr. 065.010.

**Pre–main sequence evolution.**
See Abstr. 065.020.

**Relic interstellar grains in Murchison meteorite.**
See Abstr. 105.054.

**Observed perturbations of the velocity distribution of interstellar hydrogen atoms in the solar system with Prognoz Lyman–alpha measurements.**
See Abstr. 106.034.

**Broadening of the interplanetary helium cone structure due to elastic collisions of LISM helium atoms with solar wind ions.**
See Abstr. 106.060.

**On the tidal interaction between protoplanets and the primordial solar nebula. I. Linear calculation of the role of angular momentum exchange.**
See Abstr. 107.049.

**Transport of dust and vapor and chemical fractionation in the early protosolar cloud.**
See Abstr. 107.057.

**Grain temperature, radiation pressure and electric potential in the vicinity of main sequence and white dwarf stars.**
See Abstr. 112.003.

**Ammonia and cyanotriacetylene in the envelopes of CRL 2688 and IRC + 10216.**
See Abstr. 112.010.

**New SiO maser sources: red–supergiants in the vicinity of molecular clouds.**
See Abstr. 112.013.

**The origin of low–velocity absorption components in the Mg II resonance lines of hybrid–chromosphere stars.**
See Abstr. 112.054.

**OH/$H_2O$ Maser in unidentifizierten IRAS–Quellen.**
See Abstr. 112.089.

**Detection of narrow C IV and Si IV absorption features in spectra of stars within 200 pc of the sun.**
See Abstr. 112.091.

**Four–colour and Hβ photometry of B stars in Fehrenbach and Burnage radial velocity fields at $b \sim -30°$.**
See Abstr. 113.013.

***RGU* photometry of a field in the Carina region near the cluster IC 2581.**
See Abstr. 113.019.

**Revised RGU photometry of an anticentre field around NGC 2158.**
See Abstr. 113.051.

**RGU photometry of a field in Vela II.**
See Abstr. 113.052.

**The rotation of B–type stars and processes of star formation.**
See Abstr. 116.006.

**The rotation of early B–type stars and the problem of star formation.**
See Abstr. 116.007.

**Light–curve synthesis and element determination for TV Cassiopeiae.**
See Abstr. 119.051.

**The continuum from the Herbig–Haro object M16–HH1 from 1200 Å to 2.2 μm.**
See Abstr. 121.014.

**A study of the Herbig–Haro object HH 120 and the associated cometary globule CG 30.**
See Abstr. 121.020.

**Stellar winds and molecular clouds: Herbig Be and Ae type stars.**
See Abstr. 121.021.

**Herbig–Haro objects and star formation.**
See Abstr. 121.027.

**The nature of AFGL 2591 and its associated molecular outflow: infrared and millimeter–wave observations.**
See Abstr. 121.046.

**The UU Her–type stars: possible tracers of recent star formation at high galactic latitudes.**
See Abstr. 122.030.

**Supernova progenitors and Be stars: stellar variability from a 21 century perspective.**
See Abstr. 122.197.

**X–ray haloes around supernova remnants.**
See Abstr. 125.018.

**Some properties of the filamentary nebula at 1723–46.**
See Abstr. 125.024.

**Interaction of supernova ejecta with interstellar and circumstellar matter.**
See Abstr. 125.034.

**X–ray emission from supernova remnants MSH 14–63 and RCW 103.**
See Abstr. 125.042.

**Possible detection of an old supernova remnant associated with HD 50896.**
See Abstr. 125.048.

**The parallax of pulsar 0950 + 08 and the local free electron density.**
See Abstr. 126.080.

**Strömgren trails of hot white dwarfs.**
See Abstr. 126.081.

**VLA observations of collimated outflow at NGC 7538 IRS 1.**
See Abstr. 132.003.

**Compact H II regions in the molecular cloud G35.2–0.74.**
See Abstr. 132.007.

**Polarimetric and new photometric observations in the NGC 6334/6357 region.**
See Abstr. 132.009.

**One–millimeter continuum observations of galactic and extragalactic sources.**
See Abstr. 132.010.

**Internal reddening of H II regions and HH–objects, with special reference to NGC 896(W3).**
See Abstr. 132.017.

**Detection of shocked atomic gas in the Kleinmann–Low nebula.**
See Abstr. 132.018.

**Far–infrared and CO observations of the W33 complex.**
See Abstr. 132.023.

**Kinematics and dynamics of the H II region Sharpless 142. I. The velocity field of the ionized hydrogen.**
See Abstr. 132.025.

**A search for $O^{18}H$ maser emission from G351.8–0.5.**
See Abstr. 132.026.

**Radio observations of the CMa OB1 H II regions.**
See Abstr. 132.028.

**OH maser emission at 4765 MHz in NGC 7538 (IRS1).**
See Abstr. 132.030.

**Study of star formation regions with high angular resolution and high sensitivity.**
See Abstr. 132.047.

**Neutral hydrogen in the vicinity of galactic radio sources. I. H I envelopes around H II regions.**
See Abstr. 132.059.

**Neutral hydrogen in the vicinity of galactic radio sources. II. Model of the interaction of H II regions with an interstellar medium.**
See Abstr. 132.060.

**The molecular disk in S106.**
See Abstr. 132.061.

**Nitrogen and helium enrichment in four Wolf–Rayet ring nebulae.**
See Abstr. 132.064.

**The H51α emission associated with the shocked gas in Orion–KL.**
See Abstr. 132.065.

**IRAS 1827–145 P01: a possible bipolar nebula.**
See Abstr. 133.011.

**Infrared polarimetric maps of protostellar infrared sources.**
See Abstr. 133.013.

**Dust around deeply embedded infrared sources.**
See Abstr. 133.015.

**Infrared polarization of sources with bipolar mass outflow.**
See Abstr. 133.016.

**Near–infrared observations of S 106–IRS 9.**
See Abstr. 133.017.

**A study of embedded objects by means of near– and far–infrared observations.**
See Abstr. 133.018.

**Galactic far infrared astronomy.**
See Abstr. 133.021.

**Infrarot–Polarimetrie an Quellen mit bipolarem Massenfluß.**
See Abstr. 133.022.

**Spectrophotometry of compact embedded infrared sources in the 0.6 – 1.0 micron wavelength region.**
See Abstr. 133.023.

**Interstellar and nebular absorption lines in planetary nebulae.**
See Abstr. 134.018.

**Nebular and interstellar absorption lines in planetary nebulae: the case of NGC 6543.**
See Abstr. 134.023.

**Rocket observations of the diffuse ultraviolet background.**
See Abstr. 142.040.

**The soft X–ray diffuse background: implications for the nature of the local interstellar medium.**
See Abstr. 142.084.

**Non–local origin of a substantial portion of the soft X–ray background.**
See Abstr. 142.085.

**Local contributions to the 0.6 keV diffuse X–ray background.**
See Abstr. 142.086.

**Ultrasoft X–ray background observations of the local interstellar medium.**
See Abstr. 142.088.

**SAS 3 survey of the soft X–ray background.**
See Abstr. 142.091.

**Gamma rays from atomic and molecular gas in the large complex of clouds in Orion and Monoceros.**
See Abstr. 143.012.

**On the nature of the galactic gamma–ray sources.**
See Abstr. 143.019.

**Energy dependence of the correlation of cosmic gamma–ray intensity with atomic and molecular gas in the ISM.**
See Abstr. 143.025.

**The isotopic composition of the anomalous low–energy cosmic rays.**
See Abstr. 144.024.

**The galactic distribution of cosmic–ray particles.**
See Abstr. 144.072.

**Spiral instabilities provoked by accretion and star formation.**
See Abstr. 151.007.

**Spiral structure and star formation. II. Stellar lifetimes and cloud kinematics.**
See Abstr. 151.008.

**Some evolutionary consequences of shock–induced star formation.**
See Abstr. 151.023.

**Bursts of star formation in computer simulations of dwarf galaxies.**
See Abstr. 151.038.

**Can the ballistic particle model explain the Hubble sequence?**
See Abstr. 151.047.

**The formation and early dynamical evolution of bound stellar systems.**
See Abstr. 151.048.

**The Coronet, an obscured cluster adjacent to R Corona Austrina.**
See Abstr. 152.001.

**Optical observations of the bubble nebula N 157C (= 0536.6–6914) and the tight clusters in the association LH 90 in LMC.**
See Abstr. 152.006.

**The Pleiades cluster. I. Polarization and reddening of the brighter stars.**
See Abstr. 153.011.

**The nature of the embedded population in the Rho Ophiuchi dark cloud: mid–infrared observations.**
See Abstr. 153.053.

**Distribution and properties of molecular clouds in the Galaxy.**
See Abstr. 155.006.

**The number and distribution of molecular clouds in the inner Galaxy.**
See Abstr. 155.007.

**Far–infrared survey of the galactic disc in the southern hemisphere. I. The large–scale structure.**
See Abstr. 155.012.

**H I shells, supershells, shell–like objects, and "worms".**
See Abstr. 155.014.

**On the statistical similarity of high–velocity clouds and stars of the spherical component of the Galaxy.**
See Abstr. 155.015.

**Interstellar nitrogen isotope ratios.**
See Abstr. 155.016.

**Shocked molecular hydrogen emission from the centre of the Galaxy.**
See Abstr. 155.023.

**Distribution of OB stars and absorbing matter in the region around P Cyg.**
See Abstr. 155.026.

**The H I halo in the inner Galaxy.**
See Abstr. 155.027.

**Far–infrared and submillimeter survey of the galactic plane from $l = 11\overset{\circ}{.}5$ to $l = 17\overset{\circ}{.}5$.**
See Abstr. 155.030.

**The distribution of CO in the Galaxy for longitudes 294° to 86°.**
See Abstr. 155.032.

**The stellar distribution in Puppis. I. Distribution of B5 to M5 stars to $V \cong 12.5$.**
See Abstr. 155.035.

**Small–scale variations in the galactic magnetic field: the rotation measure structure function and birefringence in interstellar scintillations.**
See Abstr. 155.037.

**Sagittarius A and its environment.**
See Abstr. 155.039.

**The distribution of near–infrared obscuration in the innermost region of the Galaxy.**
See Abstr. 155.044.

**Submillimeter wavelength survey of the galactic plane from $l = -5°$ to $l = +62°$: structure and energetics of the inner disk.**
See Abstr. 155.046.

**$^{13}CO$ in the inner galactic plane.**
See Abstr. 155.055.

**A survey of the 21 cm line of neutral hydrogen in the Cas–Per region $120° \leqslant l \leqslant 142°$ and $-5° \leqslant b \leqslant +5°$ with the 100 m telescope at Effelsberg.**
See Abstr. 155.059.

**The Galaxy as a self–regulated star–forming system: the case of the OB associations.**
See Abstr. 155.064.

**Determinations of the molecular mass of the Galaxy from observations of CO.**
See Abstr. 155.075.

**Kinematics of nearby gas and stars.**
See Abstr. 155.079.

**Dark matter near the Sun: simulated star counts and the Oort limit.**
See Abstr. 155.080.

**Limits on soft X–ray flux from distant emission regions.**
See Abstr. 155.081.

**Ultraviolet radiation from white dwarfs and the carbon ionization in the galactic halo.**
See Abstr. 155.082.

**The diffuse interstellar medium in the Magellanic Clouds.**
See Abstr. 156.006.

**Interstellar matter and star formation in the Large Magellanic Cloud.**
See Abstr. 156.008.

**Masers in the nuclei of galaxies.**
See Abstr. 157.006.

**Water–vapour maser emission from galactic nuclei.**
See Abstr. 157.007.

**A galactic disk is not a true exponential.**
See Abstr. 157.014.

**Star formation bursts in the central regions of Sc–galaxies.**
See Abstr. 157.019.

**Star formation in interacting galaxies.**
See Abstr. 157.028.

**Infrared mapping of galaxian centers: star forming regions with kiloparsec scale sizes.**
See Abstr. 157.029.

**Molecules in external galaxies.**
See Abstr. 157.030.

**Comparison of M33 and NGC 7793: stochastic models of spiral galaxies modulated by density waves.**
See Abstr. 157.060.

**Past star formation rates in spiral galaxies.**
See Abstr. 157.063.

**Shock–induced star formation in IC 2153. A colliding pair of galaxies.**
See Abstr. 157.064.

**Neutral hydrogen in M31. V. The formation rate of H II regions and a correction factor for the star formation rate.**
See Abstr. 157.086.

**Star formation in a dwarf elliptical of the Virgo cluster.**
See Abstr. 157.087.

**On the correlation of CO line and radio continuum emission in nearby galaxies and the efficiency of OB star formation.**
See Abstr. 157.095.

**Far-infrared observations of galactic nuclei.**
See Abstr. 157.114.

**CO observations of M31 with 100 pc resolution.**
See Abstr. 157.124.

**Gas dynamics of flow past galaxies: radiative cooling.**
See Abstr. 157.137.

**Ultraviolet luminosities and colors of spiral and irregular galaxies. The present-day star formation rate.**
See Abstr. 157.145.

**Star formation in late type galaxies.**
See Abstr. 157.154.

**An empirical study of star formation in galaxies.**
See Abstr. 157.158.

**Warps, spiral structure and star bursts caused by perturbed flattened haloes in disc galaxies.**
See Abstr. 157.165.

**Star formation in early-type galaxies.**
See Abstr. 157.170.

**Infrared observations of interacting/merging galaxies.**
See Abstr. 157.232.

**CO emission from *IRAS* galaxies.**
See Abstr. 157.243.

**Dust in the nuclei of the Seyfert galaxies Markarian 231 and NGC 4151.**
See Abstr. 158.144.

**Strong $H_2O$ maser emission from the peculiar galaxy NGC 3079.**
See Abstr. 158.186.

**$10^{12}$ $L_\odot$ starbursts and shocked molecular hydrogen in the colliding galaxies Arp 220 (= IC 4553) and NGC 6240.**
See Abstr. 158.231.

**Molecular clouds in M82.**
See Abstr. 158.269.

**Gravitational collapse of cosmic gas clouds and formation of star clusters.**
See Abstr. 161.186.

# 132 H II Regions, Emission Nebulae

132.001 **Observations of an ionization ridge containing an embedded O-B cluster in the G10.2-0.3 H II region of W31.**
C. E. Woodward, H. L. Helfer, J. L. Pipher.
Mon. Not. R. Astron. Soc., Vol. 209, No. 1, p. 209 - 214 (1984).

VLA radio maps at 5-GHz of the G10.2-0.3 H II region in the W31 complex reveal an extended ionization ridge of dimensions $5.7 \times 2.9$ pc$^2$ at an assumed distance of 7 kpc. In the ridge the authors have found 21 dense clumps with characteristic emission measures of $\sim 10^6$pc cm$^{-6}$ and electron densities $n_e > 10^3$cm$^{-3}$, suggesting embedded stars. From estimates of the number of Lyman continuum photons required to support the radio structure, the authors find spectral types in the ridge which range from B0.5 to O6. The observed radio structure cannot be explained by excitation from a single embedded source, or by expansion of shock-heated gas into the interstellar medium.

132.002 **A high resolution study of M17 at 1.3, 2, 6, and 21 cm.**
M. Felli, E. Churchwell, M. Massi.
Astron. Astrophys., Vol. 136, No. 1, p. 53 - 64 (1984).

The authors present maps of M17 at wavelengths of 21, 6, and 1.3 cm with spatial resolutions ranging from $\sim 10''$ to $\sim 0''.1$. The overall structure of M17 is consistent with that found from other lower resolution studies at radio and infrared wavelengths but the small scale radio structure is far more extensive than previously realized. The authors have identified and derived properties of 4 physically different types of small scale structures in M17.

132.003 **VLA observations of collimated outflow at NGC 7538 IRS 1.**
B. Campbell.
Astrophys. J., Lett. Ed., Vol. 282, No. 1, p. L27 - L30 (1984).

VLA observations of NGC 7538 IRS 1 at 5 and 15 GHz using the A configuration are presented, showing evidence for a disk collimator of the ionized gas on scales less than 100 AU. The link between these observations and the previously mapped high-velocity molecular gas is briefly discussed.

132.004 **Observations of the filamentary nebula Simeiz 22.**
T. A. Lozinskaya, T. G. Sitnik, M. S. Toropova, A. Yu. Klement'eva.
Sov. Astron. Lett., Vol. 10, No. 1, p. 48 - 51 (1984). English translation of 37.132.004.

132.005 **H II-Regionen, Radiosterne und galaktisches Radiokontinuum. Beobachtungen am Effelsberger 100-m-Teleskop.**
W. J. Altenhoff, W. Reich.
Sterne Weltraum, 23. Jahrg., Nr. 8-9, p. 426 - 431 (1984).

132.006 **IUE UV spectra of extragalactic H II regions. I. The catalogue and the atlas.**
M. Rosa, M. Joubert, P. Benvenuti.
Astron. Astrophys., Suppl. Ser., Vol. 57, No. 3, p. 361 - 384 (1984).

More than 150 spectrograms of giant H II complexes in spiral, blue compact and irregular galaxies, excluding the Magellanic Clouds, obtained within the period March, 1 1978 and April, 1 1982 with the International Ultraviolet Explorer (IUE) in the wavelength range 1150 Å to 3200 Å are catalogued. A comparison of repeated observations for two objects in M33 and M101 is made to investigate the reliability of single observations. General considerations on the UV spectra of giant extragalactic H II regions are given and a comparison of the spectra of H II regions at various distances from the centres of M33 and M101 is presented.

132.007 **Compact H II regions in the molecular cloud G35.2-0.74.**
W. R. F. Dent, L. T. Little, G. J. White.
Mon. Not. R. Astron. Soc., Vol. 210, No. 1, p. 173 - 181 (1984).

The authors present high-resolution 6-cm continuum observations of the molecular cloud G35.2-0.74. Two sources associated with the cloud were detected, one of which (G35.2S) is a near-symmetric H II region with a diameter of $\sim 0.37$ pc. It is thought that a stellar wind, rather than pressure expansion of the

H II region, could be responsible for the unusual 3-km $s^{-1}$ molecular line-splitting observed around this object. The second continuum source (G35.2N) has a double-peaked structure, with an orientation very close to that of the surrounding dense elongated molecular cloud. A model which might explain both the structure of G35.2N and the detection of bipolar outflow in $HCO^+$ is that of a disc-constrained outflow.

132.008 **Shell-like H II zones.**
E. L. Logvinenko, Yu. A. Shchekinov.
Astrofizika, Tom 19, Vyp. 3, p. 559 – 567 (1983). In Russian. English translation in Astrophysics, Vol. 19, No. 3.

The structure of an H II zone near a star with strong stellar wind is considered. At initial stages, when the stellar wind pressure is large, ionized gas is contracted in a thin shell. After the time $t \sim 4 \times 10^5$ years, the wind pressure decreases and the shell-like structure of the H II gas practically disappears.

132.009 **Polarimetric and new photometric observations in the NGC 6334/6357 region.**
T. Neckel.
Astron. Astrophys., Vol. 137, No. 1, p. 58 – 62 (1984).

*UBV* data for faint stars in NGC 6357 and in the dark cloud between NGC 6357 and NGC 6334 and polarization data for the brighter stars in NGC 6334 and NGC 6357 are given. 4 new O stars are identified in NGC 6357. Together with the already known O stars they can explain the excitation of the total H II region NGC 6357. The alignment of dust grains in the local spiral arm differs from the usual alignment parallel to the galactic equator by +45° and in the Sagittarius spiral arm by -45°.

132.010 **One-millimeter continuum observations of galactic and extragalactic sources.**
R. Chini, E. Kreysa, P. G. Mezger, H.-P. Gemünd.
Astron. Astrophys., Vol. 137, No. 1, p. 117 – 127 (1984).

Continuum observations of 32 galactic and 8 extragalactic objects have been made at a wavelength of 1 mm. The sources detected include: (1) H II regions, where most of the dust emission appears to come from dust within the H II region. (2) Warm clouds, where most of the dust heating is due to embedded B and A stars. (3) Cold clouds, heated by the general interstellar radiation field. (4) Dusty spiral galaxies and (5) miscellaneous sources, especially the SNRs Tau A and Cas A and the planetary nebula NGC 7027. Colour temperatures of ~50 – 60K for H II regions, 40 – 50K for warm clouds and ~20K for the integrated emission of normal spiral galaxies are obtained.

132.011 **Observations of the Rosette Nebula using the decameter wave radio telescope at Gauribidanur.**
A. A. Deshpande, R. K. Shevgaonkar, C. V. Sastry.
Astrophys. Space Sci., Vol. 102, No. 1, p. 21 – 27 (1984).

A map of the Rosette Nebula in continuum absorption made at 34.5 MHz using the decameter wave radio telescope at Gauribidanur, India, with a resolution of 26′ × 40′, is presented. These observations are combined with the 2700 MHz measurements of Graham et al. (1982) to derive the electron temperature distribution across the nebula. It is found that the temperatures in the southeastern parts of the nebula are around 5000K and increase up to 8000K towards the northwestern regions. It is suggested that the lower electron temperatures in the southeastern regions are due to the presence of more dust there compared to other regions in the nebula.

132.012 **Line profile shapes in optical H II regions.**
H. W. Yorke.
Messenger, No. 37, p. 30 – 34 (1984).

132.013 **Ultraviolet study of the stellar population of the giant extragalactic H II region NGC 604.**
L. Sanz Fernández de Córdoba, P. Benvenuti.
ESA Spec. Publ., ESA SP-218, p. 73 – 75 (1984). – See Abstr. 012.024.

The brightest part of the H II region NGC 604 has been repeatedly observed in the ultraviolet range, in low resolution, with IUE. The ultraviolet spectra of NGC 604 are fitted with synthetic spectra, a mixture of O and B early type stars and WR stars. The observed spectra of NGC 604 can be satisfactorily fitted to the synthetic spectra not only in the continuum but also in the stellar emission and absorption features.

132.014 **Radio continuum and recombination line observations of Orion A.**
T. L. Wilson, T. Pauls.
Astron. Astrophys., Vol. 138, No. 2, p. 225 – 230 (1984).

A 42″ resolution continuum map of Orion A, at 23 GHz, and H110α line data toward, and at 4′ from the continuum peak are presented, as well as measurements of the H39α, 41α, and 42α lines. Toward the peak, the electron temperature, averaged over 2′, is 8300K to 8600K, assuming electron densities in the range 5000 $cm^{-3}$ to 7400 $cm^{-3}$. If the H110α lines at 4′ from the peak are formed in LTE, $T_e$ there is 7600±200K. Comparisons are made with previous observations and models of the Orion H II region. Molecular line results are combined with the continuum data to produce a simplified picture of the H II molecular complex.

132.015 **Line formation in H II regions during the champagne phase.**
H. W. Yorke, G. Tenorio-Tagle, P. Bodenheimer.
Astron. Astrophys., Vol. 138, No. 2, p. 325 – 332 (1984).

The line transfer problem is solved along many lines-of-sight in order to calculate profiles for optical recombination lines of hydrogen and forbidden lines of oxygen in H II regions during the champagne phase ("disruption" of molecular clouds by ionizing radiation). The density, velocity and ionization structure of axisymmetric hydrodynamical models at various evolutionary times were used to calculate the source term. Some observational consequences are discussed.

132.016 **Observations of the motions of the neutral and ionized gas in M8 (NGC 6523).**
K. H. Elliott, C. Goudis, H. Hippelein, J. Meaburn.
Astron. Astrophys., Vol. 138, No. 2, p. 451 – 456 (1984).

Profiles of the interstellar absorption Na I lines and [O III] and [N II] emission lines have been obtained over a large area of the galactic H II, H I, molecular complex of M8. Many separate velocity components are found in each set of profiles. These are compared with each other and with the structural features of M8 that are revealed by filtering low spatial frequencies in enlargements from a Schmidt photograph through an [S II] filter. Separate velocity components, in the emission lines over the ionization fronts at the edge of M8, suggest that a neutral and ionized expanding bubble, driven by stellar winds, supernova explosions or radiation pressure cannot be their explanation if the ambient medium is very uniform. However, dense neutral concentrations at the edge of an expanding shell could produce this splitting. On the other hand the large scale motions in the neutral clouds, prior to the formation of M8 could still be present in M8 to give the same effect without the presence of an expanding shell.

132.017 **Internal reddening of H II regions and HH-objects, with special reference to NGC 896(W3).**
P. Gómez Garrido, G. Münch.
Astron. Astrophys., Vol. 139, No. 1, p. 30 – 36 (1984).

The effect of scattering by dust internal to H II regions on their spectral characteristics has been studied on the basis of simple analytical models, for assumed universal optical parameters of the dust. A procedure to determine the external interstellar extinction and the internal gas to dust abundance ratio of H II regions has been developed and applied as a test to the Orion Nebula. Newly acquired observations of the H II region NGC 896 have been analyzed to identify the effects of internal dust on its spectrum and to establish its relationship to the background radio source W3 (cont.). Finally, the effects that the probable existence of scattering dust internal to HH-objects might have on the color of their spectrum has been included.

132.018 **Detection of shocked atomic gas in the Kleinmann–Low nebula.**
M. W. Werner, M. K. Crawford, R. Genzel, D. J. Hollenbach, C. H. Townes, D. M. Watson.
Astrophys. J., Lett. Ed., Vol. 282, No. 2, p. L81 – L84 (1984).

The authors have mapped the 63 μm $^3P_1 \rightarrow {}^3P_2$ fine–structure line emission of neutral atomic oxygen at the center of the Orion nebula with a resolution of 30″. There are three main emission peaks. One is associated with the region of strongest thermal radio continuum radiation close to the Trapezium cluster and probably arises at the interface between the H II region and the dense Orion molecular cloud. The other two [O I] line emission peaks, associated with the Kleinmann–Low nebula, are similar in both distribution and velocity to those of the 2 μm $S(1)$ line of molecular hydrogen and of the high–velocity wings of rotational CO emission.

132.019 **Internal structure of the diffuse nebula Sh 152.**
R. A. Sarkisyan.
Astrofizika, Tom 20, Vyp. 1, p. 199 – 203 (1984). In Russian. English translation in Astrophysics, Vol. 20, No. 1.

The results of a structural study of the galactic nebula Sh 152 are presented. The comparison of data obtained by different methods (optical, radio) confirms that Sh 152 has a compound structure. It probably is a reflection of active processes which took place during the formation of the nebula. The integral stellar magnitude and isophotes in the wavelength range 6100 – 7000 Å of Sh 152 are given.

132.020 **Lyman and Balmer continuum ionization in zero–age main–sequence stars: applications to the line excess phenomenon.**
R. I. Thompson.
Astrophys. J., Vol. 283, No. 1, p. 165 – 168 (1984).

New calculations of the ionization from zero–age main–sequence (ZAMS) stars are presented. Ionization–bounded calculations for photoionization, both from the $n = 1$ and the $n = 2$ level of hydrogen, are presented. The $n = 2$ calculations are used to show that the suggestion of Balmer continuum ionization proposed by Simon et al. can account for the observed hydrogen–line excess in obscured ZAMS objects.

132.021 **Some comments on models of photoionized nebulae.**
H. Netzer, G. J. Ferland.
Publ. Astron. Soc. Pac., Vol. 96, No. 582, p. 593 – 597 (1984).

The authors compare emission–line predictions of several independent photoionization codes and show that details of the techniques and assumptions made in calculating such models do not strongly affect the results. Such predictions are both noncontroversial and reliable. The authors outline the basic processes which must be included in modern models, and identify the He II $L\alpha$ transfer problem as one of the major sources of uncertainty in models today. Given the present atomic data base, accuracies better than 10% are easily attainable even for the relatively complicated conditions in quasars' clouds. A comparison with models computed nearly a decade ago shows that, even with dramatic changes in many atomic rate coefficients, strong lines are predicted with accuracies better than 30%.

132.022 **Properties of R136a as derived from its optical light distribution.**
Y.–h. Chu, J. P. Cassinelli, M. G. Wolfire.
Astrophys. J., Vol. 283, No. 2, p. 560 – 565 (1984).

The authors have used short exposure 4 m prime focus plates taken with interference filters centered on blue continuum 4765 Å, He II 4686 line, red continuum 6485 Å, and Hα line to study the light distribution within R136a. The brightest component, unresolved under sub–arcsec seeing condition, contributes about 37% of the total light from a 3″ diameter aperture. Combining the optical and UV information, the authors find this brightest component R136a$_1$ may be a single star with a mass of $\sim 750\ M_\odot$ with a brightness of six HD 93129A or 20 O3 V stars, or it could be a cluster of such stars. In either case, R136a$_1$ supplies no more than one–half of the ionization of the 30 Doradus nebula.

132.023 **Far–infrared and CO observations of the W33 complex.**
M. T. Stier, D. T. Jaffe, T. N. Rengarajan, G. G. Fazio, C. W. Maxson, B. McBreen, L. Loughran, S. Serio, S. Sciortino.
Astrophys. J., Vol. 283, No. 2, p. 573 – 579 (1984).

The authors present multi–band far–infrared (20 – 250 μm) and CO observations of the giant H II region W33. At a resolution of $\sim 1'$, they resolve the complex into four distinct far–infrared (far–IR) sources, derive their color temperatures and optical depths, and examine their relationship to the molecular and radio continuum properties. Most of the W33 radio continuum features observed at 3.5 cm have far–IR counterparts. The northern source (W33 A) has a high far–IR luminosity ($\sim 10^5 L_\odot$) but is undetectable in the radio continuum. This source is either a normal ZAMS O7 star with an optically thick H II region of density $10^5 - 10^6 \mathrm{cm}^{-3}$ or a 30 – 50 $M_\odot$ protostar in its gravitational contraction stage.

132.024 **Ultracompact structure in the H II region W49N.**
J. W. Dreher, K. J. Johnston, W. J. Welch, R. C. Walker.
Astrophys. J., Vol. 283, No. 2, p. 632 – 639 (1984).

The component of W49N that coincides with the strongest $H_2O$ masers is resolved into seven distinct ultracompact H II regions. These range in size from less than 0.04 to 0.3 pc and are distributed over an area of 0.8 pc diameter. Central optical depths range from 0.1 to 1.2 at 15 GHz. If each source is ionized by a single star, seven stars with ZAMS types from O9 to O4 are needed, and the total inferred luminosity is $2 \times 10^6 L_\odot$. The $H_2O$ masers coincide with intensity maxima in the largest component, which has the appearance of a limb–brightened shell.

132.025 **Kinematics and dynamics of the H II region Sharpless 142. I. The velocity field of the ionized hydrogen.**
G. Joncas, J.–R. Roy.
Astrophys. J., Vol. 283, No. 2, p. 640 – 652 (1984).

Fabry–Perot interferometry at Hα and efficient image processing techniques have allowed the measurement of nearly 41,000 radial velocities across the H II region S142. Maps of radial velocities were constructed by taking the average value of all velocities measured in each $1' \times 1'$ window across the H II region. S142 and most of its associated neutral complex seem to be seen mostly edge–on. The locus of the most negative Hα velocities ($\sim -50\ \mathrm{km\ s^{-1}}$) coincides with the position of the "hot" component of the CO molecular cloud east of the nebula. The observed velocity field is readily explained as a systematic expansion of the ionized gas away from the molecular cloud and from the observer.

132.026 **A search for $O^{18}H$ maser emission from G351.8–0.5.**
J. D. Fix, R. L. Mutel.
Astrophys. J., Vol. 283, No. 2, p. 653 – 654 (1984).

The authors report an unsuccessful search for 1637 MHz maser emission from $O^{18}H$ molecules in the H II/OH maser source G351.8–0.5. Their detection limit of 25 mJy for 1637 MHz emission sets a lower limit of about 20,000:1 for the ratio of $O^{16}H/O^{18}H$ emission.

132.027 **Detection of the H142α line from the Barnard Loop.**
M. J. Gaylard.
Mon. Not. R. Astron. Soc., Vol. 211, No. 1, p. 149 – 153 (1984).

The 13–cm H142α recombination line has been observed at three positions on the Barnard Loop. The H II region parameters are derived and compared to those obtained from published optical line measurements. Optical and radio radial velocity data suggest that rotation of the Loop is occurring in the same sense as that of the associated neutral cloud, in addition to expansion.

132.028 **Radio observations of the CMa OB1 H II regions.**
M. J. Gaylard, A. J. Kemball.
Mon. Not. R. Astron. Soc., Vol 211, No. 1, p. 155 – 165 (1984).

A sensitive $10° \times 15°$ 13–cm map made of the CMa OB1 H II regions' radio emission shows a strong similarity to Hα emission photographs. Sharpless 296 is shown to consist of a prominent central and western arc completed by a weaker southern loop, and with a faint northern bar. The emission is thermal, superimposed over a predominantly non-thermal background. The H142α recombination line has been detected at eight positions in S296, and in S292 and S297. The average electron temperature in S296 is 6900±1300K. The UV fluxes from the CMa OB1 stars account for the observed emission measures of the H II regions. The H142α spectral line velocities indicate that the ionized material is in contact with the molecular clouds.

132.029 **The H II regions ionized by σ and τ Sco.**
M. J. Gaylard.
Mon. Not. R. Astron. Soc., Vol. 211, No. 2, p. 339 – 345 (1984).
The H142α line has been detected in Sharpless 9, which is ionized by σ Sco, and in RCW 129, ionized by τ Sco. The electron temperatures in the two H II regions are 5700±340K and 4200±600K respectively. The thermal radio emission from S9 is asymmetric with respect to the stellar position, and the emission peak coincides with the position of the optical red emission features to the north and west of the star. There is no evidence for collisional excitation. S9 is a density-bounded H II region in the champagne phase, the bright rims and radio peak marking the ionization front.

132.030 **OH maser emission at 4765 MHz in NGC 7538 (IRS1).**
P. Palmer, F. F. Gardner, J. B. Whiteoak.
Mon. Not. R. Astron. Soc., Vol. 211, No. 2, p. 41P – 44P (1984).
VLA observations with an angular resolution of 1.25 arcsec have shown that most of the 4765–MHz OH emission originates in front of the compact H II region associated with IRS1. The emission positions agree in general with those of ground–state (1720 and 1665 MHz) masers. The absence of 1612–MHz emission imposes restrictions on possible OH cloud models. There appears to be no association of OH emission with $H_2CO$ or $H_2O$ masers.

132.031 **Optical study of the galactic nebula NGC 6357 and of its stellar content, Pis 24 and the WC 6 star HD 157504.**
M. C. Lortet, G. Testor, V. Niemela.
Astron. Astrophys., Vol. 140, No. 1, p. 24 – 32 (1984).
The authors have studied the detailed optical and radio morphology, stellar content and physical conditions in different parts of the galactic nebula NGC 6357, from new stellar and nebular observations. The stars in this H II region are found to be much hotter than indicated by previous photometric studies. The WR star HD 157504 shows absorption lines of an O7–9 companion star. Close to the open cluster Pis 24, the core of the nebula G353.2+0.9 is of outstandingly high electron density and high excitation. The overall ring morphology is found to be ionization–bound, several other types of structures are studied, including high excitation filaments. The hypothesis of successive events including recent supernova explosions near the present position of Pis 24 is studied. The WC 6 star and its progenitor are not wholly responsible for the formation and expansion of the large outer shell.

132.032 **Abundances in galactic H II regions. III. G25.4–0.2, G45.5+0.06, M8, S159, and DR 22.**
J. L. Pipher, H. L. Helfer, T. Herter, D. A. Briotta Jr., J. R. Houck, S. P. Willner, B. Jones.
Astrophys. J., Vol. 285, No. 1, p. 174 – 180 (1984).
Measurements of the [Ar II](6.99 μm), [Ar III](8.99 μm), [Ne II](12.81 μm), [S III](18.71 μm), and [S IV](10.51 μm) lines are presented for five compact H II regions along with continuum spectroscopy. From these data and radio data the authors deduce lower limits to the elemental abundances of Ar, S, and Ne.

132.033 **On the hydrogen recombination radio lines in H II regions of high emission measure.**
Z. V. Dravskikh.
Sov. Astron., Vol. 28, No. 2, p. 166 – 168 (1984). English translation of 37.132.041.

132.034 **On the composition of H II regions in southern galaxies – III. NGC 2997 and 7793.**
M. G. Edmunds, B. E. J. Pagel.
Mon. Not. R. Astron. Soc., Vol. 211, No. 3, p. 507 – 519 (1984).
AAT/IPCS spectra have been obtained for a part of the nuclear region and for four H II regions in the SABc galaxy NGC 2997, and for the nucleus and five H II regions in the prototype Sd Sculptor group galaxy NGC 7793. Additional observations of an H II region in the Scd galaxy NGC 300 extend previously reported results. Abundances are deduced from emission–line intensities. The abundance gradient found in NGC 7793 is very similar to that of NGC 300 and characteristic of late–type spirals, while that of NGC 2997 is milder. The relationship between abundance gradients and other characteristics of spiral galaxies is discussed.

132.035 **High velocities within extensive regions of the 30 Doradus nebula.**
J. Meaburn.
Mon. Not. R. Astron. Soc., Vol. 211, No. 3, p. 521 – 533 (1984).
A dedicated echelle spectrograph has been used on the Anglo–Australian telescope to obtain profiles of the [O III] emission line along three parallel lines each 238 pc long over the face of the 30 Doradus nebula. Many sheets, 10 – 50 pc across, of ionized material have been found each having a separate radial velocity. Superimposed on these motions are those produced by several shells, 15 pc in diameter, expanding at 200 km $s^{-1}$. If these are pressure–driven bubbles they must be $\lesssim 2\times 10^4$yr old and could be a consequence of stellar winds from individual Wolf–Rayet stars in the halo of 30 Doradus. Also, a large number of small shells, ~2.5 pc in diameter, are present which are expanding at ~25 km $s^{-1}$. These again must be young, $4.8\times 10^4$yr old, but most likely formed by the low power winds from minor O and B stars. Very extensive, systematic trends in separate velocity components are also found in the halo of 30 Doradus. These could be the manifestations of shells of $\lesssim$100 pc in diameter, expanding radially at 35 km $s^{-1}$.

132.036 **Galactic H II regions and supernova remnants near $l = 55°$.**
J. L. Caswell.
J. R: Astron. Soc. Can., Vol. 78, No. 5, p. 202 (1984). Abstract. – See Abstr. 010.401.

132.037 **Velocity dispersion in giant extragalactic H II regions.**
R. Arsenault, J.–R. Roy.
J. R. Astron. Soc. Can., Vol. 78, No. 5, p. 208 (1984). Abstract. – See Abstr. 010.401.

132.038 **Observations of the H II region, S142, in H I–line and continuum.**
P. E. Dewdney, L. A. Higgs, G. Joncas, J.–R. Roy.
J. R. Astron. Soc. Can., Vol. 78, No. 5, p. 208 – 209 (1984). Abstract. – See Abstr. 010.401.

132.039 **Kinematics and dynamics of the H II region S142: the velocity field of the ionized hydrogen.**
G. Joncas, J.–R. Roy.
J. R. Astron. Soc. Can., Vol. 78, No. 5, p. 209 (1984). Abstract. – See Abstr. 010.401.

132.040 **Spectrophotometry of emission nebulae in the 5000 – 6000 Å spectral region.**
C. Pritchet, C. Grillmair.
J. R. Astron. Soc. Can., Vol. 78, No. 5, p. 209 (1984). Abstract. – See Abstr. 010.401.

132.041 **Far infrared maps of M42 – deconvolution by maximum entropy.**
D. A. Naylor.
J. R. Astron. Soc. Can., Vol. 78, No. 5, p. 211 (1984). Abstract. – See Abstr. 010.401.

132.042 **A catalog of coordinates, angular dimensions, and flux densities of the components of galactic radio sources at 21 cm.**
Z. A. Alferova, A. P. Venger, I. V. Gosachinskii (*I. V. Gosachinskij*), E. N. Kurochkina, N. P. Komar, V. G. Mogileva, V. K. Khersonskii (*V. K. Khersonskij*).
Bull. Spec. Astrophys. Obs. – North Caucasus, Vol. 17, p. 62 – 67 (1984). English translation of 34.132.081.

132.043 **Turbulent gas motions in giant H II regions. I. The case of NGC 604.**
H. Hippelein, J. W. Fried.
Astron. Astrophys., Vol. 141, No. 1, p. 49 – 55 (1984).
CCD–imagery of NGC 604 in Hα, obtained through a tunable filter with 22 km $s^{-1}$ passband is presented. Radial velocity and dispersion maps therefrom synthesized provide a global characterization of the nebular kinematics in terms of four components: (1) a large scale rotation–like or rolling motion, (2) local expansion of wind–driven shells with a scale of 60 pc, (3) relative motions of these shells, and (4) an overall expansion of the nebular complex. The velocity dispersion at the core is 25 km $s^{-1}$. The overall dispersion, 17.8 km $s^{-1}$, cannot be solely interpreted in terms of the virial theorem. The radial dependence of the nebular brightness, nevertheless, resembles that of a gravitationally bound system.

132.044 **The structure and dynamics of the Car II ("Keyhole") region in the Carina nebula.**
J. A. López, J. Meaburn.
Rev. Mex. Astron. Astrofis., Vol. 9, No. 2, p. 119 – 125 (1984). With 2 plates.
A set of intermediate and low dispersion spectra have been obtained of the Car II ("Keyhole") nebula, located to the north of η Carinae in NGC 3372 (= RCW 53), in order to search for clues on the origin of the high velocity components superimposed on the extended wings of the Hα and Hβ lines reported by Elliott (1979) and their possible association with a supernova remnant. No apparent evidence of such a SNR is found in the spectra which closely resemble those of classical H II regions of moderate excitation.

132.045 **1300 + 361 A,B is not a gravitational lens.**
J. P. Halpern, H. L. Marshall, J. B. Oke.
Astron. J., Vol. 89, No. 12, p. 1802 – 1807 (1984). With plate 95.
A pair of 20th–mag emission–line objects separated by 10".6 has been found at $z = 0.107$. The spectra are nearly identical, and the velocity difference is less than 80 km $s^{-1}$. The line ratios and widths are characteristic of metal–poor extragalactic H II regions. This is not likely to be a gravitational–lens system in view of the large separation, low redshift, and absence of a plausible lensing object. In addition, there is a small but significant difference in the [O III] λ5007/Hβ ratio. A likely explanation involves star–forming regions in a strongly interacting galaxy system. Objects of this type may be fairly common at $z < 1.2$ and could be difficult to distinguish from true narrow–line gravitational lenses.

132.046 **The interstellar lines and the energetics of the inner 30 Doradus nebula from the ultraviolet spectrum of R 136a.**
J. V. Feitzinger, R. W. Hanuschik, T. Schmidt–Kaler.
Mon. Not. R. Astron. Soc., Vol. 211, No. 4, p. 867 – 882 (1984).
An *IUE* short and long wavelength absorption line spectrum (1150 – 3000 Å) of the central superluminous object in the 30 Doradus nebula, R 136a, is analysed. Radial velocities and equivalent widths of the Magellanic interstellar components are given. The Magellanic interstellar components fall into four well–separated groups. The same groups show up in the velocity components of all interstellar LMC emission and absorption lines so far observed. Their average velocities are given. They can be explained by a model of the inner part of 30 Dor consisting of shells and bubbles, and warp material above the plane of the LMC. The power of R 136a alone is sufficient to energize the multitude of the 30 Dor shells and to structure the whole nebula.

132.047 **Study of star formation regions with high angular resolution and high sensitivity.**
N. M. Lipovka.
Soobshch. Spets. Astrofiz. Obs., Vyp. 43, p. 56 – 62 (1984).
The results of observations of the galactic radio emission using RATAN–600 at the wavelengths 3.9 cm, 7.6 cm, 8.2 cm and 31 cm are presented. More than 200 galactic sources were studied. It is shown that the majority of the objects have complex structures and about 20% of the sources have a core–halo structure. It is derived that their distribution in the galactic plane is extremely uniform, and depends on the population of the sources observed.

132.048 **Detection of molecular hydrogen in the SMC H II region N 81.**
J. Koornneef, F. P. Israel.
Space Telesc. Sci. Inst., Prepr. Ser., No. 25, 18 pp. (1984). To appear in Astrophys. J.

132.049 **Gas and star content and spatial distribution in the giant extragalactic H II region Tol 89.**
F. Durret, J. Bergeron, A. Boksenberg.
Inst. Astrophys. Paris, Pré–Publ., No. 72, 38 pp. (1984). To appear in Astron. Astrophys.

132.050 **Spectrophotometric studies of galactic nebulae. XX. Electron density of Ne(S II) and Ne(Cl III) in the central part of the Orion nebula (NGC 1976).**
Yu. I. Glushkov, Eh. K. Denisyuk, Z. V. Karyagina, V. R. Shutenkov.
Astron. Tsirk., No. 1301, p. 1 – 3 (1983). In Russian.

132.051 **Spectrophotometric studies of galactic nebulae. XXI. The Sh 2–288 nebula and its ionizing star.**
Yu. I. Glushkov, Eh. K. Denisyuk, Z. V. Karyagina, I. V. Nosov.
Astron. Tsirk., No. 1301, p. 3 – 5 (1983). In Russian.

132.052 **GM 3–13 – a new distant H II region.**
T. Yu. Magakyan.
Astron. Tsirk., No. 1307, p. 7 – 8 (1984). In Russian.

132.053 **Insect–eye, Fabry–Perot observations of the large–scale motions within the Carina nebula (NGC 3372, RCW 53).**
J. Meaburn, J. A. Lopez, D. Keir.
Mon. Not. R. Astron. Soc., Vol. 211, No. 2, p. 267 – 276 (1984).
Profiles of the [O III] 5007 Å line have been obtained with an insect–eye Fabry–Perot with high signal–to–noise ratios, a high spectral resolution (≅50000), but low spatial resolution (≅3 arcmin) along two lines 34 and 74 pc long respectively, over the Carina nebula. Very large–scale, separate velocity components are found over a range of 84 km $s^{-1}$. There is some evidence that a radially expanding shell about 70 pc in diameter is present. However, it is possible that some of these large–scale motions are those which were present in the neutral clouds before the ionizing stars were formed. Evidence of the merger of two adjacent spiral features is also considered.

132.054 **Observations of the [S II] doublet λ 6717, 6731 Å in the Orion nebula.**
K. G. Dzhakusheva.
Astron. Tsirk., No. 1320, p. 1 – 3 (1984). In Russian.

132.055 **Spectrophotometric studies of galactic nebulae. XXII. New compact H II region – the Trinity nebula.**
Yu. I. Glushkov, Eh. K. Denisyuk, Z. V. Karyagina.
Astron. Tsirk., No. 1338, p. 3 – 5 (1984). In Russian.

132.056 **Neue Beobachtungen an einem interessanten bipolaren Nebel.**
H. J. Staude, T. Neckel.
Mitt. Astron. Ges., Nr. 62, p. 229 (1984). Abstract. – See Abstr. 010.341.

132.057 **Die Kinematik optischer Emissionslinien im Orionnebel (M42).**
A. Hänel.
Mitt. Astron. Ges., Nr. 62, p. 302 – 303 (1984). – See Abstr. 010.341.
Complete radial velocity maps for the Hα, [N II] and [S II] emission lines have been measured over about 30′ in the Orion Nebula. Some preliminary results and their influence on existing models of the H II region are discussed.

132.058 **GM 3–6: an enigmatic object.**
T. Yu. Magakyan.
Astrofizika, Tom 21, Vyp. 1, p. 183 – 185 (1984). In Russian. English translation in Astrophysics, Vol. 21, No. 1.
The object GM 3–6 is located in the dark cloud L1555 and consists of two late-type dwarf stars (K2V and M2V) with a nebulous $H_\alpha$ emission bar ~27″long. Presently the origin of this bar is not clear.

132.059 **Neutral hydrogen in the vicinity of galactic radio sources. I. H I envelopes around H II regions.**
A. P. Venger, I. V. Gosachinskij, V. G. Grachev, T. M. Egorova, N. F. Ryzhkov, V. K. Khersonskij.
Astrophys. Space Sci., Vol. 107, No. 2, p. 271 – 287 (1984).
Results are reported from investigations of neutral hydrogen in the vicinity of a number of H II regions. The observations were performed on a RATAN-600 radio-telescope with angular and velocity resolutions of 2′ × 130′ and 6.3 km $s^{-1}$, respectively. It is demonstrated that the distribution of the neutral gas in the vicinity of 17 H II regions is similar to that of the expanding envelopes. Dimensions and expansion velocities of the envelopes are determined and their masses and kinetic energies estimated.

132.060 **Neutral hydrogen in the vicinity of galactic radio sources. II. Model of the interaction of H II regions with an interstellar medium.**
I. V. Gosachinskij, V. K. Khersonskij.
Astrophys. Space Sci., Vol. 107, No. 2, p. 289 – 302 (1984).
In the preceding paper by Venger et al. (1984) the results of observations of neutral gas in the vicinity of some galactic H II regions were considered. This paper describes a model of interaction between the H II regions and the surrounding interstellar medium constructed on the basis of these results. It is assumed that the main dynamic factor in the envelope formation mechanism is the total stellar wind from the stars which ionizes the H II regions. The employment of the observations of the line at $\lambda = 21$ cm, of the radio continuum and IR dust continuum made it possible to determine the differential mass spectra of the stars exciting the H II zones and calculate some characteristics of the stellar population.

132.061 **The molecular disk in S106.**
J. H. Bieging.
Astrophys. J., Vol. 286, No. 2, p. 591 – 598 (1984).
Aperture synthesis maps of HCN emission ($J = 1 \rightarrow 0$, 88.631 GHz) toward the bipolar nebula S106 have been obtained with the Berkeley millimeter-wavelength interferometer. Spectral and spatial resolution were 4 km $s^{-1}$ and 13″. The HCN emission has a linear structure coincident with the dust lane seen in optical and infrared images of the nebula. The molecular gas bisects the two lobes of radio continuum emission from the ionized gas and is centered on the position of the infrared point source which is presumed to be the exciting star. These results support the idea that a disk of high-density neutral gas surrounds the exciting star and has directed the ionized gas into the observed bipolar shape. Estimates of the dimensions, the total mass and the angular momentum per unit mass of this disk are presented.

132.062 **New radio sources at AFGL 2591: young cluster or single star?**
B. Campbell.
Astrophys. J., Vol. 287, No. 1, p. 334 – 337 (1984).
New sensitive VLA observations in the region about the galactic infrared source AFGL 2591 are reported. In addition to the two sources already known, two new sources at $S$(5 GHz) < 1 mJy are identified. None of the four sources is coincident with the infrared position of the nearby $H_2O$ maser source. Two broad explanations for this disparity are discussed: (1) the radio sources are "leaks" from an otherwise highly obscured ionizing source at the infrared position, or (2) AFGL 2591 is a cluster of at least five B0 or earlier stars powering individual H II regions.

132.063 **Line emission from H II blister models.**
R. H. Rubin.
Astrophys. J., Vol. 287, No. 2, p. 653 – 664 (1984).
The author develops the techniques to calculate numerical models for H II regions that have a density distribution that is a function of two dimensions, with rotational symmetry specifying the shape in 3D. The thermal and ionization equilibrium equations are solved with all the generality of spherical modeling except for the treatment of the diffuse ionizing radiation. The geometry/density distribution of the specific models is meant to simulate an H II "blister" near the edge of a molecular cloud, and is represented by an exciting source embedded in an exponential density gradient normal to a plane-parallel configuration. In order to make comparison with observations possible, global properties (independent of viewing aspect) are derived from the model, while a separate code generates surface brightness for any inclination to the line of sight.

132.064 **Nitrogen and helium enrichment in four Wolf-Rayet ring nebulae.**
K. B. Kwitter.
Astrophys. J., Vol. 287, No. 2, p. 840 – 844 (1984).
The author presents spectrophotometric data and derives chemical compositions for four ring nebulae around Population I Wolf-Rayet stars: S308 (HD 50896), NGC 3199 (HD 89358), Anon (HD 191765 = MR 100), and RCW 58 (HD 96548). The nitrogen to helium abundance ratio in each nebula is combined with corresponding data for the central star and the general interstellar medium to estimate the fractional contribution that the nitrogen- and helium-rich stellar wind has made to the mass of the nebula. Independently, this contribution is estimated from the dynamical age of each nebula and the mass loss rate of its central star.

132.065 **The H51α emission associated with the shocked gas in Orion-KL.**
T. Hasegawa, K. Akabane.
Astrophys. J., Lett. Ed., Vol. 287, No. 2, p. L91 – L94 (1984).
A possible detection of the radio recombination line emission associated with the shocked gas in Orion-KL is reported. Superposed on the stronger foreground H51α emission from the H II region M42, an excess of emission is observed coincident with the infrared $H_2$ line emission. The emission covers a velocity extent of $-30$ km $s^{-1}$ < $V_{LSR}$ < 30 km $s^{-1}$ and does not show a clear sign of the velocity structure expected for a bipolar jet. A temperature of the shocked gas of $\sim 3.3 \times 10^3$K is derived, which is significantly lower than the temperature in the foreground H II region.

**Physics of thermal gaseous nebulae (physical processes in gaseous nebulae).**
See Abstr. 003.047.

**The $\upsilon = 0 \rightarrow 0$ spectrum of $H_2$.**
See Abstr. 022.069.

**Photographie extrem schwacher H II-Regionen.**
See Abstr. 034.015.

**S wie Sharpless – Beobachtergruppe hat Erfolg beim Aufsuchen von S-Nebeln.**
See Abstr. 036.038.

**On Bowen enhancement of the N III spectrum under solar and nebular conditions.**
See Abstr. 063.087.

**The emission nebula surrounding CPD −59°2661 in the Carina nebula (NGC 3372).**
See Abstr. 112.025.

**Time variations of the Hα line profile from the core of Eta Carinae.**
See Abstr. 112.060.

**The remarkable kinematics of the WR wind–blown bubble RCW 58.**
See Abstr. 112.064.

**Near–infrared speckle interferometry of evolved stars and bipolar nebulae.**
See Abstr. 112.097.

**A superluminous object in the Large Cloud of Magellan.**
See Abstr. 115.002.

**R136: multiple or monster?**
See Abstr. 115.020.

**High resolution observations of extended radio sources near $l = 94°$.**
See Abstr. 125.020.

**Association of type II supernovae with giant H II regions: the case of M83 and M101.**
See Abstr. 125.044.

**Molecular hydrogen jets from the Orion nebula.**
See Abstr. 131.055.

**Three new $H_2O$ masers in the vicinity of NGC 6357 and NGC 6334.**
See Abstr. 131.059.

**Large–scale motions within the Orion Nebula (M42).**
See Abstr. 131.065.

**$HCO^+$ in NGC 6334.**
See Abstr. 131.070.

**CO ($J = 2 \rightarrow 1$) observations of molecular clouds associated with H II regions from the southern hemisphere.**
See Abstr. 131.091.

**High spatial resolution observations of Orion bright bar: CS ($J = 1$–$0$) and H51α emission.**
See Abstr. 131.103.

**Ammonia absorption toward NGC 7538 IRS 1: 2 arc second observations in the (3, 3) line.**
See Abstr. 131.105.

**A submillimetre map of the W51 region.**
See Abstr. 131.108.

**183 GHz water line variation: an energetic outburst in Orion KL.**
See Abstr. 131.110.

**CS around Orion–KL: a large rotating disk.**
See Abstr. 131.111.

**High–angular–resolution observations of the OH masers in W51 (main).**
See Abstr. 131.122.

**Interaction of the outflow and quiescent gas in Orion: $HCO^+$ aperture synthesis maps.**
See Abstr. 131.125.

**Interferometric observations of CO in Orion: hot core and plateau.**
See Abstr. 131.128.

**On the interpretation of the broad–band millimeter–wave flux from Orion.**
See Abstr. 131.129.

**The origin of the infrared luminosity in violent star–formation regions.**
See Abstr. 131.130.

**Infrared and millimeter–wave observations of the Sharpless 156 molecular cloud.**
See Abstr. 131.136.

**High–velocity molecular jets.**
See Abstr. 131.137.

**W3 North: far–infrared and radio molecular observations.**
See Abstr. 131.145.

**Far–infrared and submillimeter observations of the multiple cores in S255, W3, and OMC–1: evidence for fragmentation?**
See Abstr. 131.146.

**Detection of submillimeter polarization in the Orion Nebula.**
See Abstr. 131.150.

**Ammonia in the W3 region.**
See Abstr. 131.158.

**Density fluctuations in the interstellar medium: evidence for anisotropic magnetogasdynamic turbulence. I. Model and astrophysical sites.**
See Abstr. 131.160.

**Hydrogen recombination lines towards the SNR IC 443 and the H II region S 249.**
See Abstr. 131.176.

**Review of radio line observations with the RATAN–600 radio telescope.**
See Abstr. 131.195.

**Aperture synthesis observations of CO emission from the W3 molecular cloud core.**
See Abstr. 131.206.

**Spectrophotometric studies of galactic nebulae. XXIII. 1. S 191 – compact red reflection nebula. 2. Anon $3^h03^m$, 58°19′ – really a nebula associated with T Tauri stars?**
See Abstr. 131.210.

**The methyl cyanide hot and warm cores in Orion: statistical equilibrium excitation models of a symmetric–top molecule.**
See Abstr. 131.211.

**Optical study of the star–forming region Sharpless 252 (NGC 2175).**
See Abstr. 131.235.

**Probe of the nearby interstellar medium by the vacant line of sight to β CMa.**
See Abstr. 131.250.

**New cometary nebulae and tight Trapezium–like systems.**
See Abstr. 131.285.

**High–resolution far–infrared observations of the extended W51 complex.**
See Abstr. 131.287.

**Extinction in inhomogeneous clouds.**
See Abstr. 131.294.

**Rocket observations of the Orion reflection nebulosity in the range 130 – 200 nanometers and scattering properties of dust grains in the far–ultraviolet.**
See Abstr. 131.296.

**Discovery of a young stellar object near the water masers in W3(OH).**
See Abstr. 131.307.

**Spectrophotometry of compact embedded infrared sources in the 0.6 – 1.0 micron wavelength region.**
See Abstr. 133.023.

**Monochromatic observations of three nebular objects in the direction of NGC 1714.**
See Abstr. 134.011.

**[O II] studies of galactic planetary nebulae and extragalactic H II complexes.**
See Abstr. 134.030.

**A 10–GHz radio continuum survey of the galactic plane region. I. A complex region at $l = 21° - 26°$.**
See Abstr. 141.008.

**Discovery of a remarkable cometary–shaped radio source.**
See Abstr. 141.012.

**On the nature of the Wolf–Rayet component in the core of the massive galactic H II region NGC 3603.**
See Abstr. 153.026.

**Large, highly organized radio structures near the galactic centre.**
See Abstr. 155.009.

**A radio lobe over the galactic centre.**
See Abstr. 155.010.

**Far–infrared survey of the galactic disc in the southern hemisphere. I. The large–scale structure.**
See Abstr. 155.012.

**Galactic oxygen gradient: airplane measurements of $O^{++}$.**
See Abstr. 155.017.

**The galactic rotation outside the solar circle.**
See Abstr. 155.022.

**Far–infrared and submillimeter survey of the galactic plane from $l = 11\overset{\circ}{.}5$ to $l = 17\overset{\circ}{.}5$.**
See Abstr. 155.030.

**A far–infrared survey of the galactic center.**
See Abstr. 155.031.

**Sagittarius A and its environment.**
See Abstr. 155.039.

**Where massive stars form: a correlative study of molecular and ionized gas emission in the Milky Way.**
See Abstr. 155.051.

**Compact emission–line nebulae in the Small Magellanic Cloud.**
See Abstr. 156.002.

**A curious, high velocity, component within a small shell in the Large Magellanic Cloud.**
See Abstr. 156.003.

**Complex motions within the giant LMC shell N59A (DEM 241).**
See Abstr. 156.005.

**The spatial distribution of young objects in the Large Magellanic Cloud: a problem of pattern recognition.**
See Abstr. 156.007.

**The nature of blue compact galaxies.**
See Abstr. 157.002.

**Wolf–Rayet stars in "lazy" galaxies.**
See Abstr. 157.072.

**Metallicity estimates for H II regions in M81.**
See Abstr. 157.117.

**Abundances in LINERs.**
See Abstr. 157.140.

**Chemical evolution of galaxies with H II regions.**
See Abstr. 157.155.

**Nuclear H II regions in galaxies with emission lines.**
See Abstr. 157.193.

**Structural properties of giant H II regions in nearby galaxies.**
See Abstr. 157.234.

**Do giant H II regions preferentially form in the outer disks of galaxies?**
See Abstr. 157.239.

**The nature of the IRAS "warm" galaxies.**
See Abstr. 158.246.

# 133 Infrared Sources

133.001 **IRAS circular 13.**
Nature, Vol. 310, No. 5976, p. 370 (1984).

133.002 **IRAS circulars 14 and 15.**
Nature, Vol. 310, No. 5978, p. 548 (1984).

Circular 14 contains 24 IRAS sources in a southern field between right ascension 10 and 14 h and between declination -40 and -20 deg. All sources shown are associated with galaxies on the ESO/SRC skyplates. Circular 15 contains a selection of infrared bright northern galaxies with radial velocities larger than 1000 km $s^{-1}$.

133.003 **Filling in the IRAS blanks.**
Sky Telesc., Vol. 68, No. 2, p. 116 – 117 (1984).

133.004 **IRAS circular No. 13.**
Astron. Astrophys., Vol. 137, No. 1, p. C1 (1984).

133.005 **Search for the near infrared counterparts to the IRAS sources.**
H. C. Bhatt, T. Chandrasekhar, N. M. Ashok.
Astrophys. Space Sci., Vol. 103, No. 2, p. 405 – 407 (1984).

As a part of a program to search for the near infrared counterparts to the far infrared sources discovered by the Infrared Astronomical Satellite (IRAS) the authors have scanned two areas including the IRAS error boxes for the sources IRAS 0453+444 and IRAS 1717–087 at 2.2 μm down to the flux density limit of ~1 Jy. Results on the IRAS sources and the detection of two new bright near infrared sources are reported and discussed.

133.006 ***JHKL* observations of *IRAS* sources. I.**
P. Whitelock, M. Feast.
Mon. Not. R. Astron. Soc., Vol. 210, No. 2, p. 25P – 31P (1984).

Thirty *IRAS* sources have been observed. Seven are galaxies, of which several show a non–stellar emission component. Fourteen lie close to the blackbody lines in *JHKL* colours and, where more than one observation has been made, these objects are variable and may be OH/IR stars or carbon–rich Mira variables. One object is a T Tauri star. Three objects show the combined effect of a star with a cool dust shell. In these cases the *JHKL* and *IRAS* colours demonstrate that a range of temperatures must be present in the circumstellar material. One object is unusual in having a very flat energy distribution from 1.2 – 25 μm. Four objects (if correctly identified with *IRAS* sources) are probably normal M stars with very cool dust shells.

133.007 **IRAS circular No. 14.**
Astron. Astrophys., Vol. 137, No. 2, p. C3 (1984).

133.008 **IRAS circular No. 15.**
Astron. Astrophys., Vol. 137, No. 2, p. C4 (1984).

133.009 **GL961: an infrared double source.**
R. Lenzen, K.-W. Hodapp, T. Reddmann.
Astron. Astrophys., Vol. 137, No. 2, p. 365 – 367 (1984).

High resolution near infrared maps (2.2 μm, 3.6 μm) are presented which demonstrate that GL961 is an infrared double source composed of two pointlike components. Optical (0.9 μm) identifications (Eiroa, 1981) refer to the western component for which the authors deduce a colour temperature of about 1050K at 3 μm. The VLA–position, in contrast, coincides with the eastern much redder component ($T_E \approx 650$K). To explain the 3.1 μm–absorption feature observed for the whole complex (Willner et al., 1982), an absorption depth of at least 0.6 mag is required even for the bluer component. The infrared polarization is parallel to the direction of separation of these two components, the main direction of CO–mass outflow is nearly perpendicular to it. Within the conical nebula GGD18 near GL961 another pointlike IR–source has been found which seems to be the exciting source of this nebula.

133.010 **A survey of faint, near–infrared sources towards the centre of the Galaxy.**
A. W. Jones, M. J. Selby, M. Prieto Muñoz, C. Sánchez Magro.
Astron. Astrophys., Vol. 138, No. 2, p. 297 – 302 (1984).

The authors present the results of a survey made at 1.65 μm (*H*) and 2.2 μm (*K*) in 15 small areas on the galactic equator between $l = 0°$ and $l = 10°$. 775 distinct objects were counted in a total area of 0.17 $deg^2$ to limiting magnitudes $H = 10.5$ and $K = 9.5$ (equivalent to detecting M giants at 6 – 7 kpc). Source confusion was the limiting factor in these crowded fields. The sources account for 40 – 50% of the 2.4 μm surface brightness of the galactic disk. The central bulge has not been clearly detected. The authors find evidence for a population (~1000 $deg^{-2}$) of luminous, reddened objects, with a space density of $2 \times 10^{-4}$ $pc^{-3}$, which are probably supergiants within 0°.5 (90 pc) of the galactic nucleus. The optical counterpart of IRC –20416 is found to be a faint red star, and not the bright B star listed in the Infrared Catalogue.

133.011 **IRAS 1827–145 P01: a possible bipolar nebula.**
T. Le Bertre, N. Epchtein, Nguyen–Q–Rieu.
Astron. Astrophys., Vol. 138, No. 2, p. 353 – 355 (1984).

The authors present *JHKLMN* photometry of IRAS 1827–145 P01 which is the infrared counterpart of OH 17.7–2.0. The 1 – 100 μm energy distribution cannot be easily explained in terms of a spherically symmetric dust shell. The infrared spectrum of this source is reminiscent of a bipolar nebula (BPN). This similarity and further consideration based on OH emission data, suggest that IRAS 1827–145 P01 is likely to be an OH/IR–BPN seen almost edge–on.

133.012 **Nature of the infrared sources in IRAS Circulars 2 – 4.**
K. V. K. Iyengar, R. P. Verma.
Astron. Astrophys., Vol. 139, No. 1, p. 64 – 70 (1984).

The authors optically identify 22 of the 57 unidentified sources in the Infrared Astronomical Satellite (IRAS) Circulars 2 – 4 with galaxies and give their positions. A comparison of their far infrared flux density ratios with those of active galaxies indicates that they may be some type of active galaxies. Fourteen unidentified sources have similar flux density ratios as the optically identified IRAS galaxies. These may also be active galaxies.

133.013 **Infrared polarimetric maps of protostellar infrared sources.**
P. A. Heckert, M. Zeilik.
Astron. J., Vol. 89, No. 9, p. 1379 – 1386 (1984).

The authors present high–resolution polarization maps at 2.2 μm of the suspected protostellar objects S158 IRS 4, NGC 2071 IR, NGC 2264, and W3 IRS 1 and 2 and also of the cluster of infrared sources in W3. The authors evaluate the spatial structure of the maps in terms of scattering and/or grain alignment polarization mechanisms and the physical environments in these star–forming regions.

133.014 **A search for OH and $H_2O$ maser emission from unidentified IRAS sources.**
D. Engels, H. J. Habing, F. M. Olnon, J. Schmid–Burgk, C. M. Walmsley.
Astron. Astrophys., Vol. 140, No. 1, p. L9 – L12 (1984).

OH and $H_2O$ maser emission has been searched for in 33 IRAS sources having far infrared colors approximating those of OH/IR stars. Eleven OH masers and six $H_2O$ masers were found, with two sources common to both groups. Nine of the OH spectra are characteristic of OH/IR stars. One of the OH masers is circularly polarized and is possibly not a late–type star. One of the $H_2O$ masers shows two groups of velocity components separated by 110 km $s^{-1}$.

133.015 **Dust around deeply embedded infrared sources.**
T. N. Rengarajan.
Astron. Astrophys., Vol. 140, No. 1, p. 213 – 215 (1984).

Using existing data in literature, an attempt is made to relate $\tau_{9.7}$, the silicate absorption optical depth to $\tau_{400}$ and $\tau_{1000}$ the 400 µm and 1 millimetre emission optical depths of deeply embedded infrared sources. It is found that the mean values of $\tau_{9.7}/\tau_{400}$ and $\tau_{9.7}/\tau_{1000}$ are 110 and 380 respectively. These ratios are in good agreement with expectations based on the laboratory measurements for amorphous silicate grains with icy mantles. The relationship between the hydrogen column density and $\tau_{400}$ implied by the derived ratio is $N(\mathrm{H}) = 6.8 \times 10^{24} \tau_{400}$ atoms $\mathrm{cm}^{-2}$ with an uncertainty of about $\pm 60\%$.

133.016 **Infrared polarization of sources with bipolar mass outflow.**
K.-W. Hodapp.
Astron. Astrophys., Vol. 141, No. 1, p. 255 – 262 (1984).

For the infrared sources L 1551 IRS 5, HH 7–11 SVS 13, GL 490, GL 2591, NGC 2071 IRS, GL 961, and Cep A, all associated with bipolar CO outflows, and for the T Tauri stars HL Tau and DG Tau, showing optical jets, polarimetric observations are presented and discussed. The infrared polarization tends to be perpendicular to the axis of the bipolar CO outflows. These results give strong evidence that the polarization in this class of objects is produced by scattering in bipolar structures and that the central sources are heavily obscured by equatorial dust disks or tori, seen nearly edgewise. The CO outflows tend to have their axis parallel to the interstellar magnetic field.

133.017 **Near-infrared observations of S 106–IRS 9.**
C. Eiroa.
Astron. Astrophys., Vol. 141, No. 1, p. 263 – 265 (1984).

1 – 5 µm photometry and 2 – 4 µm spectral photometry of the infrared source S 106–IRS 9 are reported. The observed energy distribution together with the absorption bands of CO and $H_2O$ in the infrared spectrum show that IRS 9 is a very reddened late type M giant star, which suffers high extinction produced by the S 106 molecular cloud. The distance to IRS 9 is greater than 1 kpc and thus, it is physically unrelated to the star formation region S 106.

133.018 **A study of embedded objects by means of near- and far-infrared observations.**
C. V. M. Fridlund, H. L. Nordh, J. W. G. Aalders.
Stockholms Obs., Rep., No. 26, 28 pp. (1984).

Six embedded pre-main-sequence objects with known luminosities have been repeatedly observed at 1 – 3.6 microns. All of these objects display variations on the order of several tenths of a magnitude over a few years' time. Three of the objects, NGC 2264–IRS, NGC 2071–IR and AFGL 961, have masses estimated to be in excess of 3 $M_\odot$, and the authors' observations are the first ones to demonstrate variability in embedded high mass objects. Some correlation is found between variability, embeddedness and age. The luminosities have been derived from balloon measurements at far infrared wavelenghts; for the star-forming regions associated with NGC 1999 and AFGL 961, some details about the observations and the analysis are given, since they have not been previously reported.

133.019 **Survey à 2.2 µm du plan galactique austral.**
N. Epchtein.
Bull. Etoiles Tardives Spectre Particulier, No. 1, p. 3 – 4 (1984).

133.020 **Preliminary results of IR observations of He2–442A and He2–442B.**
O. G. Taranova, B. F. Yudin.
Astron. Tsirk., No. 1322, p. 8 (1984). In Russian.

133.021 **Galactic far infrared astronomy.**
S. Drapatz.
Mitt. Astron. Ges., Nr. 62, p. 29 – 43 (1984). – See Abstr. 010.341.

After a short introduction to instrumental techniques the potential of far infrared observations is illustrated by examples of current research dealing with star formation and young stars, evolved stars, galactic structure, and the Galactic Center.

133.022 **Infrarot–Polarimetrie an Quellen mit bipolarem Massenfluß.**
K.-W. Hodapp.
Mitt. Astron. Ges., Nr. 62, p. 230 – 232 (1984). – See Abstr. 010.341.

133.023 **Spectrophotometry of compact embedded infrared sources in the 0.6 – 1.0 micron wavelength region.**
P. J. McGregor, S. E. Persson, J. G. Cohen.
Astrophys. J., Vol. 286, No. 2, p. 609 – 629 (1984).

Spectrophotometry at 20 Å resolution in the 0.6 – 1.0 µm wavelength range is presented for the compact embedded infrared sources GL 490, GL 437S, IC 1848A, GL 961, S106/IRS 3, M17/IRS 1, S235B, and M8E, for the compact H II regions GL 437W and NGC 7538/IRS 2, and for the visible objects LkHα 101, MWC 297, MWC 349, V645 Cyg, and the fan nebula associated with GL 989. The heavily obscured objects display extremely red continua, emission lines due to the Paschen series of hydrogen, and in many cases O I λ8446 and the Ca II infrared triplet lines at λλ8498, 8542, and 8662. These and other newly identified spectral characteristics of the embedded sources show similarities to those of Be stars, T Tauri stars, and to a class of emission-line stars with infrared excesses.

**Documentation for the machine-readable version of *The Revised AFGL Infrared Sky Survey Catalogue* (Price and Murdock 1983).**
See Abstr. 002.039.

**IRAS ground-based follow-up at ESO.**
See Abstr. 013.021.

**Infrared astronomy.**
See Abstr. 013.041.

**Infrared spectrum of the laboratory-synthesized quenched carbonaceous composite (QCC): comparison with the infrared unidentified emission bands.**
See Abstr. 022.162.

**The infrared sky.**
See Abstr. 051.043.

**Infrared astronomy with IRAS.**
See Abstr. 051.044.

**IRAS ou le ciel avec des yeux infrarouges.**
See Abstr. 051.049.

**OH 32.8–0.3. A second source with absorption features of pure water ice.**
See Abstr. 112.006.

**New SiO maser sources: red-supergiants in the vicinity of molecular clouds.**
See Abstr. 112.013.

**Infrared observations of the eclipse of Epsilon Aurigae: direct measurement of the 500K secondary at 5, 10, and 20 microns.**
See Abstr. 119.050.

**A study of the Herbig–Haro object HH 120 and the associated cometary globule CG 30.**
See Abstr. 121.020.

**Brackett-alpha line profiles of young stellar objects.**
See Abstr. 121.045.

**The nature of AFGL 2591 and its associated molecular outflow: infrared and millimeter-wave observations.**
See Abstr. 121.046.

**Unusual light variations of the carbon star AFGL 2881.**
See Abstr. 122.162.

**IRAS observations of galactic star formation.**
See Abstr. 131.028.

**Ammonia in B335.**
See Abstr. 131.051.

**Interaction of the outflow and quiescent gas in Orion: $HCO^+$ aperture synthesis maps.**
See Abstr. 131.125.

**Infrared and millimeter–wave observations of the Sharpless 156 molecular cloud.**
See Abstr. 131.136.

**Far–infrared photometry of the ϱ Ophiuchi dark cloud.**
See Abstr. 131.166.

**Absorption features in the 5 – 8 micron spectra of protostars.**
See Abstr. 131.304.

**VLA observations of collimated outflow at NGC 7538 IRS 1.**
See Abstr. 132.003.

**Far–infrared and CO observations of the W33 complex.**
See Abstr. 132.023.

**New radio sources at AFGL 2591: young cluster or single star?**
See Abstr. 132.062.

**Far–infrared and submillimeter survey of the galactic plane from $l$ = 11°.5 to $l$ = 17°.5.**
See Abstr. 155.030.

**A far–infrared survey of the galactic center.**
See Abstr. 155.031.

**Observations of broad helium and hydrogen lines in the very center of the Galaxy.**
See Abstr. 155.036.

**Submillimeter wavelength survey of the galactic plane from $l$ = –5° to $l$ = +62°: structure and energetics of the inner disk.**
See Abstr. 155.046.

**The location of infrared sources in the galactic center from a deep 1 micron CCD image.**
See Abstr. 155.047.

**IRAS exposes a remarkable infrared galaxy.**
See Abstr. 158.017.

# 134 Planetary Nebulae

134.001 **Spectrophotometric observations of some southern planetary nebulae.**
R. Louise, G. Pascoli.
Astron. Astrophys., Suppl. Ser., Vol. 57, No. 1, p. 155 – 164 (1984).

The authors present spectrophotometric observations of four planetary nebulae (NGC 1535, NGC 2452, NGC 1743, NGC 1722) and one H II region (NGC 1714). The IDS (Image Dissector Scanner) of the ESO 1.52 m telescope coupled with the Boller and Chivens spectrograph is used throughout, and about 300 spectra are obtained. Contour maps of Hα, Hβ and [O III] λ5007 Å intensities are presented.

134.002 **The interpretation of internal motions in planetary nebulae.**
F. Sabbadin, R. G. Gratton, A. Bianchini, S. Ortolani.
Astron. Astrophys., Vol. 136, No. 2, p. 181 – 192 (1984).

The observational data on the expansion velocity in planetary nebulae are discussed. The velocities are proportional to the nebular radius $R_n$ up to $R_n \cong 0.15$ pc and decrease slowly for larger radii. Two models are proposed to interpret the observational trend: (1) the planetary nebula is the region of interaction of a fast wind emitted by the nucleus with a slow wind emitted by the progenitor, accelerated by the transfer of momentum from the radiation field to the nebular material; (2) the planetary nebula is a Strömgren sphere evolving in an expanding nebula which results from a sudden ejection by an AGB star. Only model (2) agrees satisfactorily with all the observational data.

134.003 **The planetary nebulae NGC 1535 and NGC 2022.**
F. Sabbadin, A. Bianchini, E. Hamzaoglu.
Astron. Astrophys., Vol. 136, No. 2, p. 193 – 199 (1984).

The [O III] and $H_\alpha$ expansion velocity fields within the planetary nebulae NGC 1535 and NGC 2022 have been obtained from Echelle spectrograms. NGC 1535 consists of two coaxial prolate spheroids of moderate ellipticity; NGC 2022 is a prolate spheroid surrounded by an almost spherical, fainter emitting region. The observed parameters for the two nebulae (nebular radius, expansion velocity, ionized mass, temperature and luminosity of the exciting star) can be explained in terms of the evolutionary model suggested by Sabbadin et al. (1984). Both objects are planetary nebulae in a rather early evolutionary phase, with NGC 2022 somewhat more evolved than NGC 1535.

134.004 **Internal motions of fourteen planetary nebulae.**
F. Sabbadin, A. Bianchini, E. Hamzaoglu.
Astron. Astrophys., Vol. 136, No. 2, p. 200 – 205 (1984).

High dispersion spectra allowed the authors to obtain the expansion velocity, $H_\beta$ flux and angular dimension of fourteen planetary nebulae. All the nebulae of this sample are in a rather early evolutionary phase (nebular radius <0.07 pc) and present expansion velocities smaller than 15 km $s^{-1}$. The observational parameters, i.e. nebular radius, expansion velocity, nebular mass, temperature, radius and luminosity of the central star (for two objects) can be explained in terms of the "two–phase" evolutionary model suggested by Sabbadin et al. (1984).

134.005 **New observations of the C IV and N V resonance doublets in the planetary nebula NGC 3918.**
S. Torres–Peimbert, M. Peña.
Rev. Mex. Astron. Astrofis., Vol. 9, No. 1, p. 107 – 109 (1984).

The authors present new short wavelength high dispersion *IUE* measurements of the resonance doublets λλ1238/1243 of N V and λλ1548/1551 of C IV. The observed intensity ratios are of 1.87±0.15 for N V and 1.83±0.11 for C IV; compatible, within the observational uncertainties with the expected ratio of 2.

134.006 **Explanatory note to the paper "Photoelectric, absolute Hβ fluxes for 55 planetary nebulae".**
L. Carrasco, A. Serrano, R. Costero.
Rev. Mex. Astron. Astrofis., Vol. 9, No. 1, p. 111 (1984). See Abstr. 37.134.026.

134.007 **High dispersion spectra of compact planetary nebulae.**
F. Sabbadin.
Mon. Not. R. Astron. Soc., Vol. 209, No. 4, p. 889 – 894 (1984).

The [O III] expansion velocity in 14 compact planetary nebulae was obtained from high dispersion spectra. All these nebulae are in a rather early evolutionary phase (nebular radius $<0.07$ pc) with expansion velocities smaller than 15 km $s^{-1}$. The only exception is BD $+30°3639$ which shows double peaked emissions with $\Delta V = 46$ km $s^{-1}$.

134.008 **Kinematics and morphology of the planetary nebula NGC 650–651.**
E. Recillas-Cruz, P. Pismis.
Mon. Not. R. Astron. Soc., Vol. 210, No. 1, p. 57 – 67 (1984).

Radial velocities are obtained at 160 points over the face of the planetary nebula NGC 650–651 using the photographic Fabry–Pérot technique in the emission lines Hα and [N II] λ6584. The splitting of the interference rings at 24 points on the PN from four interferograms indicates an overall radial expansion of 42 – 44 km $s^{-1}$. The velocity data are consistent with those previously obtained by other authors; however, the model the authors propose differs essentially from the models suggested so far. They invoke magnetic phenomena to explain the velocity data known to date as well as the symmetrical morphology of NGC 650–651.

134.009 **The dynamics, evolution and formation rate of planetary nebulae.**
J. P. Phillips.
Astron. Astrophys., Vol. 137, No. 1, p. 92 – 100 (1984).

The author investigates the variation of planetary nebula expansion velocity $V_{exp}$ with radius $R_0$, and finds most nebulae to conform with the relation $V_{exp} \propto R_0^{-0.22}$. Using this expression he determines a planetary nebula formation rate similar to that for white dwarfs. The revised nebular expansion timescale implies central star masses of order $\sim 0.75\, M_\odot$, with an overall range $0.6 \rightarrow 1.0\, M_\odot$. This in turn suggests corresponding masses $0.7 \rightarrow 1.5\, M_\odot$ for the central star progenitors. A clear relation is found between central star spectral type and mass, with O and WC stars located predominantly in the range $M < 0.7\, M_\odot$, and WR + O stars representing the higher mass bracket $M > 0.8\, M_\odot$.

134.010 **The magnitudes and temperatures of central stars of planetary nebulae.**
N. K. Reay, S. R. Pottasch, P. D. Atherton, K. Taylor.
Astron. Astrophys., Vol. 137, No. 1, p. 113 – 116 (1984).

In order to determine the Zanstra temperature of the central star of a planetary nebula, it is necessary to know the visual continuum flux emitted by the star. When the central star is very hot, great care is required to distinguish the weak stellar flux from the sometimes relatively strong nebular continuum. The authors have adopted a technique which involves imaging the nebulae and find central star temperatures between $10^5$K and $2 \times 10^5$K for nine planetary nebulae. For the central star of NGC 2440 the temperature is a lower limit since no star can be detected against the nebular background.

134.011 **Monochromatic observations of three nebular objects in the direction of NGC 1714.**
R. Louise, E. Maurice.
Astrophys. Space Sci., Vol. 103, No. 1, p. 195 – 197 (1984).

Using a CCD as detector, the authors have observed three nebulosities not yet listed in catalogues; they all present a planetary nebula morphology and are situated in the immediate neighbourhood (less than 3 arc min to the East) of NGC 1714. Observations have been made through interference filters centered on the main emission lines of these objects.

134.012 **Spectral and photometric observations of the peculiar objects He2–442A and He2–442B.**
V. P. Arkhipova, V. F. Esipov, B. F. Yudin.
Pis'ma Astron. Zh., Tom 10, No. 8, p. 604 – 608 (1984). In Russian. English translation in Sov. Astron. Lett., Vol. 10.

Spectrograms of the peculiar objects He2–442A and He2–442B obtained in autumn 1983 are presented. The [Fe VII] λ6087 line seems to be detected in the spectrum of He2–442A. Photometric *UBV* observations of He2–442A and He2–442B carried out on 1983 September 9 and 11 have revealed the variation in the brightness of He2–442A of about $0^m.25$ in the V band. The possible nature of He2–442 (He2–442A + He2–442B) is discussed.

134.013 **Spatiokinematical models of five planetary nebulae.**
F. Sabbadin.
Mon. Not. R. Astron. Soc., Vol. 210, No. 2, p. 341 – 358 (1984).

The [O III] and Hα expansion velocity fields in the planetary nebulae NGC 6058 and 6804 and the [O III], Hα and [N II] expansion velocity fields in NGC 6309, 6751 and 6818, were obtained from high dispersion spectra. The author has derived spatiokinematical models of the nebulae assuming an expansion velocity of the gas proportional to the distance from the central star and using the expansion velocity–radius correlation given by Sabbadin et al. (1984). The observational parameters of the nebulae and of the exciting stars closely fit the evolutionary model suggested by Sabbadin et al. for this class of objects.

134.014 **Infrared light curves of the central object of NGC 2346: the shape of the obscuring cloud.**
M. Roth, J. Echevarría, M. Tapia, L. Carrasco, R. Costero, L. F. Rodríguez.
Astron. Astrophys., Vol. 137, No. 2, p. L9 – L11 (1984).

Infrared light curves of the central object of NGC 2346 show that the observed flux originates from the central A-type star and a warm circumstellar dust cloud. Two eclipses are seen in the J and H bands instead of the single eclipse previously reported from visual measurements. No light modulation is present at longer wavelengths. A simple model in terms of the passage of an elongated cloud can account for most of the existing observations.

134.015 **The planetary nebula M 1–7.**
F. Sabbadin, R. Falomo, S. Ortolani.
Astron. Astrophys., Vol. 137, No. 2, p. 177 – 184 (1984).

Direct photographs and low and high dispersion spectra of the planetary nebula M 1–7 (189 + 7°1) were taken at the Astrophysical Observatory of Asiago. It is a rather compact (radius $\cong 0.067$ pc) object located about 2.5 kpc from the Sun. The electron density, the electron temperature and the abundances of helium, oxygen, nitrogen, neon, sulphur and argon are given. A spatiokinematical model of the nebula consists of a prolate spheroid of moderate ellipticity. The central star has $m_V = 18.5\,(\pm 1.0)$, $T_* = 130{,}000\,(\pm 20{,}000)$K, $R_* = 0.037\,(\pm 0.015)\, R_\odot$ and $\log L_*/L_\odot = 2.54\,(\pm 0.4)$.

134.016 **Observations of the emission nebulae S188 and S274 at 2.7 and 5 GHz.**
C. J. Salter, A. Greve, K. W. Weiler, K. Birkle, M. Dennefeld.
Astron. Astrophys., Vol. 137, No. 2, p. 291 – 297 (1984).

The two morphologically similar nebulae S188 and S274 have been mapped in the radio continuum at 2.7 and 5 GHz. The new data effectively rule out that either or both are shell-type supernova remnants. It is demonstrated that a number of distance-independent parameters of the nebulae are closely similar and all present evidence indicates that both are old planetary nebulae. On this hypothesis, the distances of the nebulae are estimated from the radio data. A number of other parameters of the objects are considered in the light of these distances.

134.017 **IRAS measurements of planetary nebulae.**
S. R. Pottasch, B. Baud, D. Beintema, J. Emerson, H. J. Habing, S. Harris, J. Houck, R. Jennings, P. Marsden.
Astron. Astrophys., Vol. 138, No. 1, p. 10 – 18 (1984).

Far-infrared IRAS measurements of 46 of the brighter planetary nebulae are presented. Dust temperatures are computed. Higher dust temperatures are found in the younger, denser nebulae. Lyman α is shown to be a sufficient source of energy to heat the dust for the larger, older nebulae, but another energy source (probably the exciting star) is required for the younger nebulae.

134.018 **Interstellar and nebular absorption lines in planetary nebulae.**
T. H. Pwa, J. E. Mo, S. R. Pottasch.
ESA Spec. Publ., ESA SP–218, p. 175 – 178 (1984). – See Abstr. 012.024.
Using high resolution IUE spectra the authors can separate the narrow absorption lines due to the interstellar material and the nebular matter. For NGC 6543 the column densities of ions in both regions are determined and the ionic and elemental abundances derived.

134.019 **Hot central stars in planetary nebulae.**
J. Mo, T. H. Pwa, S. R. Pottasch.
ESA Spec. Publ., ESA SP–218, p. 329 – 331 (1984). – See Abstr. 012.024.
The Zanstra temperature is determined for five hot stars using IUE spectra. The ultraviolet part of the spectrum is much more favourable than the visual region for such a determination because the stellar continuum flux density is much stronger relative to the nebular continuum in the ultraviolet.

134.020 **Mass loss from the central star of NGC 3242.**
W.–R. Hamann, R.–P. Kudritzki, R. H. Méndez, S. R. Pottasch.
ESA Spec. Publ., ESA SP–218, p. 333 – 336 (1984). – See Abstr. 012.024.
The central star of NGC 3242 has been observed with IUE in high resolution. The N V resonance doublet exhibits a weak P Cygni profile with photospheric components. Theoretical profiles are obtained from comoving–frame calculations, and the nitrogen ionization balance is computed. The mass loss rate and the final velocity are derived. The authors compare the wind parameters with the parameters of the central star and find that the usual relations established for winds from early–type stars are fulfilled.

134.021 **A model of the planetary nebula NGC 3918 based on IUE observations.**
R. E. S. Clegg, J. P. Harrington, M. J. Barlow.
ESA Spec. Publ., ESA SP–218, p. 337 – 340 (1984). – See Abstr. 012.024.
A photo–ionization model of NGC 3918 has been constructed using high– and low–resolution IUE spectra, images of the nebula in H and O III lines, together with optical, infra–red and radio fluxes. NLTE model atmospheres have been calculated for the central star.

134.022 **Planetary nebulae with massive nuclei. II. Discussion of observed candidates.**
R. Tylenda.
Astron. Astrophys., Vol. 138, No. 2, p. 317 – 324 (1984).
Several planetary nebulae suspected of possessing massive nuclei are discussed and compared with the theoretical findings of Paper I. Discussion of the behaviour of the Zanstra luminosities during the photoionization epoch leads to the conclusion that there is a maximum value, $L_{max}$, which the Zanstra luminosity cannot surpass. $L_{max}$, which is a function of the electron density only, can be used as a useful test when applying the Zanstra method to planetaries with luminous nuclei. In particular, it has been shown that high luminosities, and thus high masses, proposed by Stecher et al. for three planetaries in the Magellanic Clouds must be overestimated.

134.023 **Nebular and interstellar absorption lines in planetary nebulae: the case of NGC 6543.**
T. H. Pwa, J. E. Mo, S. R. Pottasch.
Astron. Astrophys., Vol. 139, No. 1, p. L1 – L4 (1984).
Using high resolution IUE spectra it is possible to separate the narrow absorption lines due to the interstellar material and the nebular matter. The analysis of nebular absorption lines has three advantages. First the column densities of ions different from those observed in emission can be determined. The abundance of the elements are then readily derived without applying ionization correction. Second the population ratio of the fine structure levels and the ground level of certain ions and the ionization balance can yield an independent estimate of the electron density. Third using Lyman α absorption and interstellar measurements of S and Zn in the foreground extinction can be estimated. The authors use the above method for NGC 6543. They derive the abundance of C, Si, S and Al.

134.024 **Detection of a nebula surrounding the O5 star RWT 152.**
C. Pritchet.
Astron. Astrophys., Vol. 139, No. 1, p. 230 – 232 (1984).
A small, low surface brightness planetary nebula has been found surrounding the O5 star RWT 152. Application of the Shklovsky distance determination technique yields a distance $d \leqslant 6.5$ kpc. The derived distance favours a scenario in which RWT 152 is a low–mass post–AGB star, but disagrees with the hypothesis that this star is a massive main sequence star at $R \cong 50$ kpc in the galactic halo.

134.025 **CCD spectroscopy and modelling of the planetary nebula NGC 7009.**
S. L. Morris, M. J. Ward.
Mon. Not. R. Astron. Soc., Vol. 210, No. 3, p. 655 – 662 (1984).
The authors present CCD spectroscopy from 6000 Å to 1 μm of the Saturn nebula, NGC 7009. They compare these results with the predictions of Case B recombination theory, obtaining an estimate of the electron temperature of the emitting plasma.

134.026 **The C/O abundance ratio in the planetary nebula IC 2501.**
A. Goharji, S. Adams.
Mon. Not. R. Astron. Soc., Vol. 210, No. 3, p. 683 – 691 (1984).
The detection of a silicon carbide feature in the infrared spectrum of the planetary nebula IC 2501 suggests that the nebular envelope is carbon rich (Aitken and Roche 1982). The authors have observed IC 2501 with the *IUE* satellite and have combined these observations with optical measurements in order to analyse the physical conditions and abundances in the nebula. The derived value of the C/O abundance ratio is 1.2, a result which is insensitive to the adopted value of the electron temperature in the nebula.

134.027 **Negative detection of the four–hour periodicity of the central star of M1–2 (VV8).**
A. Yamasaki, A. Okazaki, M. Nakagiri, S. Suzuki.
Astrophys. Space Sci., Vol. 104, No. 1, p. 181 – 185 (1984).
Four–hour (4.0002 hr) binary periodicity of the emission–line object M1–2 (= VV8), claimed by Drummond (1980), is examined on the basis of photometric and spectroscopic observations. However, neither positive variation of the light nor that of the radial velocity associated with that period was found.

134.028 **The central star of the planetary nebula Abell 78.**
J. B. Kaler, W. A. Feibelman.
Astrophys. J., Vol. 282, No. 2, p. 719 – 727 (1984).
The authors have studied the ultraviolet spectrum of the nucleus of Abell 78, one of the two planetaries known to contain zones of nearly pure helium. Abell 78 is strikingly similar in many respects to Abell 30, the other planetary with embedded helium knots. The ultraviolet–optical color temperature, which the authors tentatively identify with the effective temperature, almost certainly lies between 67,000K and 130,000K. The limitations are set primarily by the error in the interstellar extinction. The best two choices of extinction now available yield 77,000 and 84,000 K. From the current evolutionary tracks and distance, these temperatures imply a mass of 0.56–0.58 $M_{\odot}$. The O V and N V P Cygni lines have terminal velocities similar to the value of 4200 km $s^{-1}$ measured earlier for C IV. A new possible P Cygni line is also seen at λ1310, which is tentatively identified as Si II.

134.029 **Monochromatic observations of planetary nebulae.**
R. Louise, C. T. Hua.
Astrophys. Space Sci., Vol. 105, No. 1, p. 139 – 150 (1984).
Monochromatic images in Hα, Hβ, [N II] λ6584, [S II] λ6717 and [O III] λ5007 lines are presented for a morphological study of planetary nebulae. Three faint objects of the Abell's list of old

planetary nebulae have been observed through narrow band filters, by means of an image tube (A33, A36 in Chile) or the image photon counting device (A79 at the Haute Provence Observatory). Following the Hα/[N II] intensity ratio, a discussion is given about the distance previously derived with some assumptions concerning the measured red fluxes.

134.030 **[O II] studies of galactic planetary nebulae and extragalactic H II complexes.**
C. R. O'Dell, H. O. Castañeda.
Astrophys. J., Vol. 283, No. 1, p. 158 – 164 (1984).
The λ3727 doublet ratio of [O II] was observed in five planetary nebulae and nine extragalactic groupings of H II regions (H II Complexes). The theoretical relation between this doublet ratio and nebular density was rederived using the most up–to–date atomic parameters, permitting columnar densities to be determined for all objects. The structure of extragalactic H II Complexes is discussed, and a preferred model advanced. A new method of distance determination for extragalactic systems is proposed.

134.031 **The relation between radius and expansion velocity in planetary nebulae.**
Y.–H. Chu, K. B. Kwitter, J. B. Kaler, G. H. Jacoby.
Publ. Astron. Soc. Pac., Vol. 96, No. 582, p. 598 – 602 (1984).
The authors have reinvestigated the expansion velocity–radius (*R–V*) relation for planetary nebulae using the existing measurements of expansion velocities and recent calculations of radii. Some of the previously alleged *R–V* relations for PN are not convincingly established. The scatter in the *R–V* plots may be due largely to stratification of ions in individual nebulae and to heterogeneity in the planetary nebula population. In addition, from their new echelle/CCD observations of planetary nebulae, the authors find that spatial information is essential in deriving the internal kinematic properties.

134.032 **The dust content of the planetary nebula IC 3568.**
M. Cohen, J. P. Harrington, R. Hess.
Astrophys. J., Vol. 283, No. 2, p. 687 – 693 (1984).
Far–infrared photometry of IC 3568 shows a substantial excess that is not suggested by 10 μm data. A recently constructed model for the nebula permits the temperature profile for dust to be computed. The calculated infrared spectrum reproduces the observations well and requires that the grains have absorption properties like those of amorphous carbon. The dust coexists in the ionized gas but is heated directly by stellar radiation; most of the Lyα radiation escapes from the nebula.

134.033 **Long expected cosmic wonder?**
B. I. Fesenko.
Gor'k. gos. ped. inst., Gor'kij, 8 pp. (1984). In Russian. Abstr. in Ref. Zh., 51. Astron., 9.51.3 (1984).

134.034 **On the "cosmic wonder" in the constellation Draconis.**
V. V. Radzievskij.
Gor'k. gos. ped. inst., Gor'kij, 4 pp. (1984). In Russian. Abstr. in Ref. Zh., 51. Astron., 9.51.4 (1984).

134.035 **The determination of abundances in the interstellar medium from infrared fine–structure lines.**
C. T. Johnson.
Ir. Astron. J., Vol. 16, No. 3, p. 219 – 226 (1984). – See Abstr. 012.036.
The author has re–interpreted the observations of the Ne V 24.28 μm line from three planetary nebulae by Shure et al. using the recent atomic data of Aggarwal. For NGC 7354 and NGC 7662 the derived abundances are a factor of four smaller than those obtained by Shure et al. For NGC 7027, a range of electron densities have been previously derived with the author's results more compatible with the lower electron–density estimates derived from O II.

134.036 **The ionization structure of planetary nebulae. IV. NGC 6853.**
T. Barker.
Astrophys. J., Vol. 284, No. 2, p. 589 – 596 (1984).
Spectrophotometric observations of emission line intensities have been made in seven positions in the planetary nebula NGC 6853; for five of the positions, coverage is across the entire spectral range 1400 Å to 9600 Å. Standard equations used to correct for the existence of elements in other than the optically observable ionization stages give results over a wide range of ionization that are generally consistent and in agreement with abundances calculated using ultraviolet lines. The logarithmic abundances (relative to H = 12.00) are: He = 11.04, O = 8.92, N = 8.48, Ne = 8.43, C = 8.88, Ar = 6.52, and S = 6.77. These values are compared with results from other planetary nebulae.

134.037 **Mass loss from the central star of NGC 3242.**
W.–R. Hamann, R.–P. Kudritzki, R. H. Méndez, S. R. Pottasch.
Astron. Astrophys., Vol. 139, No. 2, p. 459 – 463 (1984).
The central star of NGC 3242, which was classified to be of "continuous type" before Kudritzki et al. (1981) discovered and analysed photospheric absorptions, has been observed with IUE in high resolution. The N V resonance doublet exhibits a weak P Cygni profile with photospheric components. Theoretical profiles are obtained from comoving–frame calculations, and the nitrogen ionization balance is computed. A mass loss rate of $\log\{\dot{M}/(M_\odot \mathrm{yr}^{-1})\} = -10.1...-8.1$ is derived, while the final wind velocity is $v_\infty = 2200$ km s$^{-1}$. The authors compare these wind parameters with the parameters of the central star, which is probably the hottest mass–losing star studied so far, and find that the usual relations established for winds from early–type stars are fulfilled.

134.038 **Low and high resolution spectroscopic observations of the planetary nebula IC 418.**
R. Louise, G. Pascoli.
Astron. Astrophys., Vol. 139, No. 2, p. 529 – 537 (1984). In French.
The ring planetary nebula IC 418 was observed spectroscopically using both low and high spectral resolving powers. From the observations physical properties of the nebula can be studied in two ways: (1) Parametrization of Hα and [N II] profiles. The authors pointed out that the profiles of the [N II] components are non–symmetrical. The "thickness" of the $N^+$ shell is discussed. (2) The space distribution of $H^+$ and $N^+$ is discussed, in order to construct a three–dimensional model of IC 418.

134.039 **Constraints upon visual dust opacities in planetary nebulae.**
J. P. Phillips.
Astron. Astrophys., Vol. 140, No. 1, p. 141 – 147 (1984).
The author investigates the mean level of visual extinction in planetary nebulae as a function of observed radius $R_0$, and finds the intrinsic variability in $E_{B-V}$ to be less than 0.2 mag over the range $R_0 < 0.53$ pc. This is shown to be consistent with models of nebulae in which ionised mass is variable, and the H I halos are appreciably extended compared with observed H II shells. Substantial departures from this scenario (cf. compact halos and/or invariant H II mass) are found to require greatly depressed values of $A_v/(n_{H_2}+0.5n_H)$ compared to the interstellar medium. Finally, the author briefly discusses four nebulae (NGC 7027, NGC 6302, M2–9, and NGC 6537) which appear to break this trend, and show clear evidence for high levels of intrinsic extinction. For these cases, the high column densities of absorbing material may reflect the large masses ($3 \rightarrow 5\ M_\odot$) shared by these and other type I planetaries.

134.040 **High–speed photometry of central stars of planetary nebulae.**
A. D. Grauer.
Publ. Astron. Soc. Pac., Vol. 96, No. 584, p. 789 (1984). Abstract. – See Abstr. 010.281.

134.041 **Infrared observations of the planetary nebula NGC 2346.**
M. Roth, J. Echevarria, M. Tapia, L. Carrasco, R. Costero, L. F. Rodriguez.
Publ. Astron. Soc. Pac., Vol. 96, No. 584, p. 794 (1984). Abstract. – See Abstr. 010.281.

134.042 **High–dispersion *IUE* observations of the planetary nebula NGC 3918.**
S. Torres–Peimbert, M. Peña.
Publ. Astron. Soc. Pac., Vol. 96, No. 584, p. 796 – 797 (1984). Abstract. – See Abstr. 010.281.

134.043 **A catalogue of expansion velocities in planetary nebulae.**
F. Sabbadin.
Astron. Astrophys., Suppl. Ser., Vol. 58, No. 2, p. 273 – 285 (1984).

A catalogue of expansion motions in 165 planetary nebulae is presented. The results strengthen the suggestion by Grieg (1971, 1972) indicating that planetary nebulae (PN) can be divided into two wide sub–classes. The first group (labeled C) includes PN ejected and excited by post asymptotic giant branch (AGB) stars having a final mass close to 0.60 $M_{\odot}$. These nebulae have a nebular mass of about 0.20 $M_{\odot}$ and spend most of their life in a thin status. The second group (labeled B) includes PN whose central stars are more massive than 0.60 $M_{\odot}$ (up to about 0.80 $M_{\odot}$); B nebulae are more massive than C nebulae (nebular masses from about 0.3 $M_{\odot}$ up to about 1.0 $M_{\odot}$) and spend most of their life in a thick status. The two sub–classes have different morphological and kinematical properties, different mean chemical composition and, very likely, represent the final evolution of stars of quite different initial mass.

134.044 **NGC 2346.**
IAU Circ., No. 4024 (1984).

134.045 **Photographic observations of NGC 2346 during 1983–84.**
B. F. Marino, H. O. Williams.
Inf. Bull. Variable Stars, No. 2583, 3 pp. (1984).

134.046 **On the abundance of atomic and molecular hydrogen in the outer parts of young planetary nebulae.**
L. F. Rodríguez, J. A. García–Barreto.
Rev. Mex. Astron. Astrofis., Vol. 9, No. 2, p. 153 – 157 (1984).

Using the VLA the authors searched for H I in absorption against the continuum of seven compact planetary nebulae. With the exception of the previously reported feature in NGC 6302, all observed absorption lines can be attributed to line of sight clouds unrelated to the planetary nebulae.

134.047 **Spatial spectroscopic diagnostics of planetary nebulae. IV. Regularization by the method of best determined terms.**
J. Hekela, J. Neuberg.
Bull. Astron. Inst. Czech., Vol. 35, No. 6, p. 356 – 360 (1984).

The present paper describes and discusses a method for the three–dimensional determination of an upper level population of an optically thin line. Preliminary knowledge of the kinematical model of a planetary nebula is assumed. The main difficulties are connected with the inversion of the Fredholm integral equation of the first kind and the problems of physical uniqueness, accuracy and resolution of the obtained solutions. Regularization by the method of best determined terms is examined.

134.048 **The kinematics of the Wolf–Rayet ring nebula RCW 58.**
L. Smith, M. Pettini.
Gemini, No. 11, p. 4 – 5 (1984).

Recent ultraviolet and optical observations of high velocity gas in the immediate environment of the Wolf–Rayet star HD 96548 provide new information on the physical processes by which these stars blow bubbles in the interstellar medium.

134.049 **On a method of comparison of computed and observed parameters of planetary nebulae.**
Eh. Ya. Vil'koviskij, L. N. Kondrat'eva, L. V. Tambovtseva.
Astron. Tsirk., No. 1332, p. 1 – 4 (1984). In Russian.

134.050 **Kollimierter bipolarer Materiestrom hoher Geschwindigkeit im "Eskimonebel" NGC 2392.**
I. Becker, J. Solf.
Mitt. Astron. Ges., Nr. 62, p. 233 (1984). Abstract. – See Abstr. 010.341.

134.051 **Zur Entwicklung der Zentralsterne Planetarischer Nebel.**
D. Schönberner.
Mitt. Astron. Ges., Nr. 62, p. 248 (1984). Abstract. – See Abstr. 010.341.

134.052 **LSS 2018 – a new binary central star.**
J. S. Drilling.
Mitt. Astron. Ges., Nr. 62, p. 251 (1984). Abstract. – See Abstr. 010.341.

134.053 **High resolution spectroscopy of the CPN LSS 1362.**
U. Heber, J. S. Drilling.
Mitt. Astron. Ges., Nr. 62, p. 252 (1984). Abstract. – See Abstr. 010.341.

134.054 **Kinematische Modelle der Planetarischen Nebel IC 5210, NGC 6210 und NGC 6826.**
I. Becker, F. Gieseking, J. Solf.
Mitt. Astron. Ges., Nr. 62, p. 253 (1984). Abstract. – See Abstr. 010.341.

134.055 **Über die Interpretation der zeitlichen Entwicklung der Expansionsgeschwindigkeiten Planetarischer Nebel.**
F. Gieseking.
Mitt. Astron. Ges., Nr. 62, p. 254 – 256 (1984). – See Abstr. 010.341.

134.056 **Model calculations of the evolution of the infrared and radio spectra of planetary nebulae.**
V. A. Okorokov, B. M. Shustov, A. V. Tutukov, H. W. Yorke.
Mitt. Astron. Ges., Nr. 62, p. 257 (1984). Abstract. – See Abstr. 010.341.

134.057 **Über die Nachweisbarkeit von Doppelsternen unter den Zentralsternen Planetarischer Nebel.**
F. Gieseking.
Mitt. Astron. Ges., Nr. 62, p. 258 – 262 (1984). – See Abstr. 010.341.

134.058 **Statistical survey of planetary nebulae: distances, masses, and distribution in the Galaxy.**
P. R. Amnuel (*P. R. Amnuehl'*), O. H. Guseinov (*O. Kh. Gusejnov*), H. I. Novruzova (*Kh. I. Novruzova*), Yu. S. Rustamov.
Astrophys. Space Sci., Vol. 107, No. 1, p. 19 – 50 (1984).

The authors offer a new empirical scale of distances to the PNs based on the interdependence between the PN diameter and the value of surface brightness in the radio range, determine the dependence of the mean electron density in PNs on the diameter, estimate the masses of the ionized matter of PNs, investigate the change of PN properties dependent on their belonging to a type of stellar population, evaluate the frequency of PN formation in the Galaxy and compare it with that of white dwarfs formation, define the dependence of the PN expansion velocity on its linear size.

134.059 **Carbon abundances in planetary nebulae.**
A. F. Kholtygin.
Astrofizika, Tom 20, Vyp. 3, p. 503 – 511 (1984). In Russian. English translation in Astrophysics, Vol. 20, No. 3.

The intensities of C III recombination lines are calculated. Using the intensities of observed lines of C II, C III, C IV ions the abundances of C III, C IV, C V ions and total carbon abundances for forty-six planetary nebulae are found.

134.060 **Mass loss from red giants: a simple evolutionary model for NGC 7027.**
M. Jura.
Astrophys. J., Vol. 286, No. 2, p. 630 – 632 (1984).
NGC 7027 is a young planetary nebula with the remnants of a red giant circumstellar envelope surrounding the central, ionized region. By comparing the outer molecular envelope with the inner ionized material, the author argues that the mass loss rate has decreased by at least a factor of 3, and more probably by about a factor of 10, during the past $10^3$years. It is further argued that the luminosity of the central star has also decreased substantially during the same time, consistent with models for the rapid evolution of stars just after they evolve off the asymptotic giant branch. In this picture, the distance to NGC 7027 is less than 1300 pc.

134.061 **The unique planetary nebula NGC 2818.**
R. J. Dufour.
Astrophys. J., Vol. 287, No. 1, p. 341 – 352 (1984).
The planetary nebula NGC 2818, or PK 261 + 8°1, is of special interest due to its apparent association with a Population I open cluster (of the same designation). This paper presents the results of new observations of the nebula, including interference filter imagery, photoelectric photometry, and ground-based and *IUE* spectroscopy. Analysis of its composition suggests that NGC 2818 is a prototype of the He- and N-rich Type I group. Analysis of kinematical and morphological data suggests that the nebula is probably associated with the star cluster (revised distance = 3.5 kpc).

134.062 ***IUE* observations of the "Butterfly" nebula M2–9.**
W. A. Feibelman.
Astrophys. J., Vol. 287, No. 1, p. 353 – 358 (1984).
*IUE* observations of the peculiar "Butterfly" nebula M2–9 indicate that it is not a normal planetary nebula. The ultraviolet spectrum is characterized by few emission lines and a weak continuum. Mg II λ2800 is the strongest emission line present and may be indicative of a binary nucleus. Lines of N V, O I, N III, N IV, Si III, and C III are seen, but C IV and O III are conspicuous by their absence. M2–9 may be an object in the early stages of becoming a planetary nebula.

**Physics of thermal gaseous nebulae (physical processes in gaseous nebulae).**
See Abstr. 003.047.

**Si III emission line strengths in low density plasmas.**
See Abstr. 022.002.

**IDS (*Image Dissector Scanner*) spectroscopy of faint emission–line objects.**
See Abstr. 034.043.

**On Bowen enhancement of the N III spectrum under solar and nebular conditions.**
See Abstr. 063.087.

**Iron in the spectra of central stars and very hot subdwarfs.**
See Abstr. 114.124.

**Discovery of oxygen in the PG1159 degenerate stars: a direct evolutionary link to O VI planetary nebula nuclei and confirmation of pulsation theory.**
See Abstr. 126.073.

**Observations of the filamentary nebula Simeiz 22.**
See Abstr. 132.004.

**Compact emission–line nebulae in the Small Magellanic Cloud.**
See Abstr. 156.002.

# Radio Sources, X-ray Sources, Cosmic Radiation

## 141 Radio Sources (Surveys, etc.)

141.001 **On radio source scintillations caused by plasma inhomogeneities behind a shock wave.**
S. F. Pimenov.
Pis'ma Astron. Zh., Tom 10, No. 7, p. 523 – 529 (1984). In Russian. English translation in Sov. Astron. Lett., Vol. 10.
The turbulence in the interplanetary and interstellar medium is shown to become anisotropic and statistically inhomogeneous after passing a shock wave. Scintillation intensity spectra of radio sources are estimated. The possibilities to derive the inhomogeneity spectra and source brightness distribution from scintillation changes are discussed.

141.002 **A deep 6–centimeter radio source survey.**
E. B. Fomalont, K. I. Kellermann, J. V. Wall, D. Weistrop.
Science, Vol. 225, No. 4657, p. 23 – 28 (1984).
The Very Large Array has been used to survey a small region of sky at a wavelength of 6 centimeters down to a completeness level of 60 microjanskys – about 100 times weaker than the faintest radio sources than have been detected with other instruments. The observed source count at flux densities below 100 millijanskys converges in a manner similar to the lower frequency counts, although there is some evidence for an excess of sources weaker than 100 microjanskys. The sources in the survey are preferentially identified with faint galaxies.

141.003 **High–resolution observations of galactic–plane radio variables.**
A. R. Taylor, E. R. Seaquist, P. C. Gregory.
Astron. J., Vol. 89, No. 8, p. 1180 – 1189 (1984).
The authors report single–epoch VLA observations at 20 and 6 cm for 12 variable and possibly variable sources having variability time scales of a few days, that were discovered during a survey of the galactic plane for variable emission. The sources exhibit a wide range of arcsecond scale structure, including single–point sources, multiple–component emission and asymmetric jet–like structure. The authors are not able to distinguish between a galactic or extragalactic origin based on these results alone. The structures seen are found among both extragalactic sources and some galactic variables such as Sco X–1 and SS 433. No optical counterparts were found to the plate limit of the Palomar Sky Survey. Thus, in all cases, these sources are probably at distances greater than a few kpc.

141.004 **A deep Westerbork survey of areas with multicolor Mayall 4 m plates. I. The 1412 MHz catalogue, source counts and angular size statistics.**
R. A. Windhorst, G. M. van Heerde, P. Katgert.
Astron. Astrophys., Suppl. Ser., Vol. 58, No. 1, p. 1 – 37 (1984).
A deep 21 cm Westerbork radio survey is presented of nine fields, selected in four high latitude areas (SA57, SA68, SA28 and an area in Hercules) for which deep multicolor Kitt Peak 4 m plates are available. The observations have been done with the 3 km array at 1412 MHz. Careful calibration of the data yielded high dynamic range, noise limited maps with absolute uncertainties in position of order 0".4 and in flux of order 3 – 5%. In total 471 sources are tabulated. The 1412 MHz source counts are consistent at different cut–off levels. They are consistent with previously observed 1.4 GHz counts.

141.005 **A deep Westerbork survey of areas with multicolor Mayall 4 m plates. II. Optical identifications.**
R. A. Windhorst, R. G. Kron, D. C. Koo.
Astron. Astrophys., Suppl. Ser., Vol. 58, No. 1, p. 39 – 87 (1984).
A deep multicolor optical identification program is presented for a complete sample of 302 radio sources that were observed with the Westerbork Synthesis Radio Telescope within 5.52 $deg^2$ down to $S_{1.4\ GHz} \gtrsim 0.6$ mJy (5 σ). The astrometry has systematic errors smaller than 0".2 – 0".3, while the random errors are of order 0".4. The agreement between the radio and optical coordinate frames is on average better than 0".25. About 20% of the identifications are stellar objects, the remaining have extended images or are too faint to be classified. About 15% of the identifications appear in possible clusters. The magnitude distributions for galaxies plus objects of unknown type generally increase towards fainter magnitudes; the quasar magnitude distributions are also increasing but not as steeply as for the radio galaxies.

141.006 **RATAN–600 counts of weak galactic 7.6–cm sources.**
T. B. Pyatunina.
Sov. Astron. Lett., Vol. 10, No. 2, p. 79 – 82 (1984). English translation of 37.141.013.

141.007 **A confusion limited radio survey from the Molonglo Cross.**
G. L. White.
Proc. Astron. Soc. Aust., Vol. 5, No. 3, p. 290 – 340 (1984).
The first Molonglo deep survey of extragalactic radio sources at 408 MHz resulted from observations made in 1973. This paper presents a second deep survey, prepared from observations made with the Molonglo cross telescope in 1978. At this time the telescope was working at its maximum sensitivity and the observations reported here take advantage of increased sensitivity resulting in a predicted reduction in noise by a factor of ~2 relative to the MDS1. The increased sensitivity gives better source positions and a lower limit of 60 mJy has been set by combined uncertainties in noise and confusion. The area of deep survey analysed is now approximately doubled. Combining the two surveys, a continuous strip of sky has been observed for all RAs from $01^h$ to $23^h$ excluding areas of low galactic latitude.

141.008 **A 10–GHz radio continuum survey of the galactic plane region. I. A complex region at $l = 21° - 26°$.**
Y. Sofue, H. Hirabayashi, K. Akabane, M. Inoue, T. Handa, N. Nakai.
Publ. Astron. Soc. Jpn., Vol. 36, No. 2, p. 287 – 295 (1984).
The authors present a map of the 10–GHz total intensity distribution of the galactic region at $l = 21°$ to 26°, which contains complexes of H II regions. They compare the results with the Bonn 5–GHz survey and discuss the nature of sources found in the region.

141.009 **Observations of weak radio sources with flat or inverted spectra at 10 GHz.**
H. Tabara, T. Kato, M. Inoue, K. Aizu.
Publ. Astron. Soc. Jpn., Vol. 36, No. 2, p. 297 – 304 (1984).
The flux densities of 135 radio sources with flat or inverted spectra between 1.4 and 5 GHz have been measured at 10 GHz using the 45–m telescope of the Nobeyama Radio Observatory. These sources are selected from surveys by Ledden et al. (1980;

AAA 28.141.001) and Owen et al. (1983; AAA 33.141.029) for the study of peculiar spectra and of statistics of quasar evolution. The flux densities of 112 sources are detected at or above 3σ levels. Details of observations and reductions are presented.

141.010 **A deep survey of the 5C2 area at 4850 MHz.**
J. Maslowski, I. I. K. Pauliny–Toth, A. Witzel, H. Kühr.
Astron. Astrophys., Vol. 139, No. 1, p. 85 – 92 (1984).

The results of a survey, at a frequency of 4850 MHz, of an area of $6.60 \times 10^{-3}$ sterad overlapping the 5C2 region are presented. The survey includes 106 sources above a limiting flux density of 17 mJy. The two–point spectral index distributions for sources selected at three different frequencies are discussed. It is suggested that the dependence of the 408 MHz flux density on radial distance, which is present in the original 5C2 survey, affects mostly stronger sources [S(408) > 230 mJy] and that the 408 MHz flux density scale is overestimated by at least 16%. The source counts are presented, and in combination with the counts from the 5C6 area, are compared with those obtained from all other published surveys.

141.011 **Variability of SS433 and extragalactic compact radio sources.**
B. J. Geldzahler, J. H. Spencer, K. J. Johnston, F. B. Waltman, W. J. Klepczynski, P. E. Angerhofer, D. R. Florkowski, D. D. McCarthy, D. N. Matsakis.
Ann. N.Y. Acad. Sci., Vol. 422, p. 352 (1984). Abstract. – See Abstr. 012.033.

141.012 **Discovery of a remarkable cometary–shaped radio source.**
L. A. Higgs, J. P. Vallée.
J. R. Astron. Soc. Can., Vol. 78, No. 4, p. 155 – 162 (1984).

Radio observations of a 2° square field centred on the optical H II region S121, made with the D.R.A.O. synthesis telescope at a wavelength of 21 cm, have revealed the presence of an unusual radio source 0°.5 to the north–east of the H II region. Its morphology is best described as "cometary", consisting of a "head" plus a long curved "tail".

141.013 **A VLBI survey of peaked spectrum radio sources.**
M. W. Hodges, R. L. Mutel, R. B. Phillips.
Astron. J., Vol. 89, No. 9, p. 1327 – 1331 (1984).

VLBI array observations of nine previously unmapped compact radio sources were made at 1.6 GHz. The sources were chosen on the basis of their spectra, which are peaked between 1.0 and 3.0 GHz and are steep at high frequencies. The measured angular sizes are consistent with the spectral turnovers being due to synchrotron self–absorption. None of the sources had widely spaced symmetrical double structure. This is surprising, since nearly all known compact double sources have similar spectral properties. The authors conclude that sources with a compact double morphology constitute a relatively small fraction of all steep–spectrum compact sources with peaked spectra.

141.014 **The polarization properties of compact radio sources.**
R. E. Rusk.
J. R. Astron. Soc. Can., Vol. 78, No. 5, p. 204 – 205 (1984). Abstract. – See Abstr. 010.401.

141.015 **The nature of BG2107 + 49 – galactic or extragalactic?**
L. A. Higgs, J. P. Vallée, J. S. Albinson.
J. R. Astron. Soc. Can., Vol. 78, No. 5, p. 209 (1984). Abstract. – See Abstr. 010.401.

141.016 **Precise optical positions for radio/optical astrometric sources south of δ ∼ + 12°.**
C. Torres, H. Wroblewski, E. Costa.
Astron. Astrophys., Suppl. Ser., Vol. 58, No. 2, p. 193 – 196 (1984).

Precise positions, relative to either the AGK3 Catalogue or the Perth 70 Catalogue, were obtained for the optical counterparts of 23 sources south of δ ∼ +12°, with accuracy levels of ∼0″.2 in both coordinates. Nineteen of these objects are included in the IAU Commission 24 list of radio/optical astrometric sources for the establishment of an inertial reference system.

141.017 **Optical counterpart candidates of radio sources 1320–446, 1622–297, 1908–202, and 2106–413.**
C. Torres, H. Wroblewski.
Astron. Astrophys., Vol. 141, No. 1, p. 271 – 273 (1984).

141.018 **Experiment Kholod. The first deep sky survey made with the RATAN–600 telescope. Part 1.**
A. B. Berlin, L. G. Gassanov, V. Ya. Gol'nev, D. V. Korol'kov, Yu. N. Parijskij.
Soobshch. Spets. Astrofiz. Obs., Vyp. 41, 84 pp. (1984). In Russian.

The authors describe the details of the first very deep survey made with the help of RATAN–600. The strip $0^h < \alpha_{1950,0} < 24^h$, $\delta_{1950.0} = \delta_{SS433} \pm 10'$ was chosen for the deep survey. This strip crosses twice the galactic plane and goes through the coldest part of the sky. The main tasks of the deep cut of the sky were the following: (1) Source statistics down to ∼1 mJy level. (2) The compilation of a catalogue of all the weak sources in the strip by observations in two different azimuths (positions, structure and spectrum for stronger sources, variability). (3) Search for anisotropy clustering in the distribution of the sources. (4) Search for anisotropy of the 3K emission intended to reach the predicted fluctuation in the background temperature due to protoclusters, protogalaxies, protovoids etc. (5) Construction of the dynamical spectrum of SS433 (and the spectrum of W50 as well). (6) A look deeper than before at the Galaxy radio emission.

141.019 **Experiment Kholod. The first deep sky survey made with the RATAN–600 telescope. Part 2.**
A. B. Berlin, L. G. Gassanov, V. Ya. Gol'nev, D. V. Korol'kov, Yu. N. Parijskij.
Soobshch. Spets. Astrofiz. Obs., Vyp. 42, 73 pp. (1984). In Russian.

141.020 **One–dimensional image of Sgr A at 1.38 cm wavelength.**
A. B. Berlin, Yu. N. Parijskij, N. S. Soboleva.
Soobshch. Spets. Astrofiz. Obs., Vyp. 43, p. 33 – 42 (1984).

New results of observations of the radio source Sgr A made with the help of RATAN–600 at 1.38 cm wavelength with resolution of about 10″ are given. In the vicinity of Sgr A West and East some small details being compact H II regions were detected. The non–thermal emission which dominated at this wavelength may be connected with relativistic electrons emitting in the intensive optical emission from galactic nucleus stars.

141.021 **Infrared study and classification of optically faint steep spectrum radio sources.**
D. Walsh, M. J. Lebofsky, G. H. Rieke, D. Shone, R. Elston.
Prepr. Steward Obs., No. 555, 29 pp. (1984). To appear in Mon. Not. R. Astron. Soc.

141.022 **New radio sources with inverted spectra from the Zelenchuk survey.**
V. R. Amirkhanyan, A. G. Gorshkov, V. K. Konnikova, M. G. Larionov, V. N. Sidorenkov.
Astron. Tsirk., No. 1328, p. 3 – 4 (1984). In Russian.

141.023 **Automation of the process of optical identification of radio sources at AMD–1.**
L. S. Ugol'kova.
Astron. Tsirk., No. 1331, p. 3 – 5 (1984). In Russian.

141.024 **Very high resolution observations of compact sources in the direction of supernova remnants: G74.87 + 1.22.**
B. J. Geldzahler, D. B. Shaffer, H. Kühr.
Astrophys. J., Vol. 286, No. 1, p. 284 – 288 (1984).

The authors present the results of very long baseline interferometer observations at 10.65 GHz and 10 months of flux density measurements at 2.7 and 8.1 GHz of the compact radio source G74.87+1.22. It has been suggested that the compact radio source is physically associated with the nearby supernova remnant G74.9 + 1.2. However, in view of H I absorption studies, radio and optical source morphology, flux density variability, and the lack of any structural alignment with the SNR, the authors believe G74.87 + 1.22 to be an extragalactic radio source with no connection to the SNR.

141.025 **One-dimensional image of Sagittarius A at 1.38 cm wavelength.**
A. B. Berlin, Yu. N. Parijskij, N. S. Soboleva.
Astron. Zh., Tom 61, Vyp. 6, p. 1041 - 1047 (1984). In Russian. English translation in Sov. Astron., Vol. 28, No. 6.
The results of new observations of the radio source Sagittarius A made at the RATAN-600 at the wavelength 1.38 cm with 10″ resolution are presented.

**Early radio astronomy at Wheaton, Illinois.**
See Abstr. 004.013.

**Optical and radio astronomers in the early years.**
See Abstr. 004.014.

**The origins of radio astronomy in Australia.**
See Abstr. 004.015.

**Early days in radio and radar astronomy in Australia.**
See Abstr. 004.017.

**Radio sources and the log N-log S controversy.**
See Abstr. 004.018.

**Early work on imaging theory in radio astronomy.**
See Abstr. 004.019.

**The origins and early history of Jodrell Bank.**
See Abstr. 004.020.

**Paraboloids, galaxies and stars: memories of Jodrell Bank.**
See Abstr. 004.021.

**Early work on radio stars at Cambridge.**
See Abstr. 004.022.

**The development of aperture synthesis at Cambridge.**
See Abstr. 004.023.

**The first steps of Soviet radio astronomy.**
See Abstr. 004.024.

**Remarks on my work in radio astronomy.**
See Abstr. 004.025.

**The early years of radio astronomy in France.**
See Abstr. 004.026.

**Styles of research in three early radio astronomy groups.**
See Abstr. 004.029.

**The influence of radio astronomy on cosmology.**
See Abstr. 004.030.

**Radio astronomy and the nature of science.**
See Abstr. 004.032.

**Karl Jansky and the discovery of extraterrestrial radio waves.**
See Abstr. 005.009.

**The accuracy of determination of the baseline using the Experimental VLBI System at Shanghai Observatory.**
See Abstr. 033.030.

**Two new young galactic supernova remnants.**
See Abstr. 125.046.

**H I absorption observations of two radio sources near the supernova remnant G 127.1 + 0.5.**
See Abstr. 125.085.

**A flare of the $H_2O$ radio line emission in W49.**
See Abstr. 131.057.

**Observations of the Rosette Nebula using the decameter wave radio telescope at Gauribidanur.**
See Abstr. 132.011.

**Very deep cross-cut of the Galaxy at 7.6 cm.**
See Abstr. 155.061.

**The outer Galaxy study at 7.6 cm wavelength.**
See Abstr. 155.062.

**A deep radio/optical survey near the North Galactic Pole. III. A 4.85-GHz survey of the 5C12 area.**
See Abstr. 158.033.

**Galactic jets: two exotic cases.**
See Abstr. 158.072.

# 142 UV Sources, X-ray Sources, X-ray Background

142.001 **New mode of activity of MXB 1730–335.**
H. Kunieda, Y. Tawara, S. Hayakawa, K. Masai, F. Nagase, H. Inoue, K. Koyama, F. Makino, K. Makishima, M. Matsuoka, T. Murakami, M. Oda, Y. Ogawara, T. Ohashi, N. Shibazaki, Y. Tanaka, I. Kondo, S. Miyamoto, H. Tsunemi, K. Yamashita.
Publ. Astron. Soc. Jpn., Vol. 36, No. 2, p. 215 – 230 (1984).

A new mode of activity of MXB 1730–335 (rapid burster) was observed from the X-ray astronomy satellite Hakucho during August 8–22, 1979. The activity began with rapidly repetitive bursts which were similar to those observed earlier. Then, the energy per burst quickly increased and evolved to exhibit a long flat top or of roughly trapezoidal shape. In the last phase, burst sizes became smaller and the activity returned to the normal type II burst mode.

142.002 **Electrophotometric observations of the X-ray source Cygnus X-1 in 1982 according to a coordinated programme.**
Ya. T. Blagodir, V. S. Kaserkevich, I. V. Shpychka.
Astron. Tsirk., No. 1282, p. 3 – 4 (1983). In Russian.

142.003 **The radio and optical identification of Circinus X-1.**
A. N. Argue, D. L. Jauncey, D. D. Morabito, R. A. Preston.
Mon. Not. R. Astron. Soc., Vol. 209, No. 1, p. 11P – 14P (1984).

On a red CCD frame, Circinus X-1 appears multiple with the southern component coinciding with the *VLBI* flare position; on a blue frame this component is no longer visible.

142.004 **A tentative interpretation of the cyclotron line energy variation in Her X-1.**
N. Prantzos, P. Durouchoux.
Astron. Astrophys., Vol. 136, No. 2, p. 363 – 367 (1984).

The hard X-ray spectrum of Her X-1 has been extensively investigated, since the discovery by Trümper et al. (1977) of a feature around 60 keV, interpreted as a cyclotron line. A recent observation with high energy resolution detectors ($\Delta E \sim 1$ keV) exhibits a line at a much smaller energy than previous measurements. It seems that this line energy variation is correlated with the intensity of the line as well as the continuum luminosity emitted by Her X-1. The authors try to explain such a correlation by using the Bonazzola et al. (1979) model which considers the cooling by Comptonization on a cold plasma of a blackbody spectrum, injected at the bottom of the accretion column. It is shown that a temperature variation of the injected spectrum could produce the line energy shift as well as its correlation with the line intensity and the continuum luminosity. The model fits satisfactorily most of the observed characteristics of the hard X-ray spectrum of Her X-1.

142.005 **X-ray gas scintillation spectrometer experiment.**
R. D. Andresen, G. Boella, B. Falconi, P. Lamb, G. Manzo, J. Raymont, S. Re, M. R. Sims, G. Villa.
Science, Vol. 225, No. 4658, p. 177 – 179 (1984).

The payload complement on Spacelab 1 included a spectrometer for observations of the brighter cosmic X-ray sources. The primary scientific objective was to study the detailed spectral features of cosmic X-ray sources and their associated temporal variations over a wide energy range from about 2 to 80 kiloelectron volts. The results presented here show new results from two galactic binary X-ray sources, Cygnus X-3 and Centaurus X-3, and from the Perseus cluster of galaxies. The excellent energy resolution of the instrument permits line features to be identified in these sources with unprecedented quality.

142.006 **Very-wide-field ultraviolet sky survey.**
G. Courtès, M. Viton, J. P. Sivan, R. Decher, A. Gary.
Science, Vol. 225, No. 4658, p. 179 – 180 (1984).

Very-wide-field photographs of the sky were taken on Spacelab 1 at 1650, 1930, and 2530 angstroms with a limiting magnitude of 9.3 at 1930 angstroms. A 1.2 by 2.4 kiloparsec ultraviolet extension of the Shapley wing of the Small Magellanic Cloud is seen in some of the photographs.

142.007 **Astronomical observations with the FAUST telescope.**
J. Bixler, S. Bowyer, J. M. Deharveng, G. Courtès, R. Malina, C. Martin, M. Lampton.
Science, Vol. 225, No. 4658, p. 184 – 185 (1984).

The far-ultraviolet space telescope (FAUST) was flown on Spacelab 1 to provide wide-field imaging in the wavelength range 1300 to 1800 angstroms. Most of the developed film showed high levels of background exposure. Frames with a lower background included exposures of the Cygnus Loop supernova remnant and an exposure in the direction of the galaxy cluster Abell 2634.

142.008 **Infrared observations of galactic bulge X-ray sources.**
P. Hertz, J. E. Grindlay.
Astrophys. J., Vol. 282, No. 1, p. 118 – 124 (1984).

Nine unidentified galactic bulge X-ray sources, the recently identified (with an obscured globular cluster) X-ray burster MXB 1728–34, and two optically identified sources (Sco X-1 and MXB 1735–44) were observed with the NASA 3 m IRTF. The data constrain both the presence of diffuse infrared sources near the X-ray positions and the flux of possible infrared counterparts. None of the nine unidentified sources lies within obscured globular clusters. No infrared counterparts were detected for unidentified sources; the limits derived are consistent with all of the sources observed being similar to the low mass X-ray binary Sco X-1.

142.009 **Determination of the mass of globular cluster X-ray sources.**
J. E. Grindlay, P. Hertz, J. E. Steiner, S. S. Murray, A. P. Lightman.
Astrophys. J., Lett. Ed., Vol. 282, No. 1, p. L13 – L16 (1984). With plates L1 – L2.

The precise ($\sim 1''$) positions of the luminous X-ray sources in eight globular clusters have been measured with the Einstein X-Ray Observatory. When combined with similarly precise measurements of the dynamical centers and core radii of the globular clusters, the distribution of the X-ray source mass is determined to be in the range 0.9 – 1.9 $M_\odot$ (90% confidence interval). The available data indicate the X-ray sources are compact binaries and are probably formed by tidal capture.

142.010 **Physical parameters of the visible component of Cygnus X-1.**
O. É. Aab (*O. Eh. Aab*), L. V. Bychkova, I. M. Kopylov, R. N. Kumaigorodskaya (*R. N. Kumajgorodskaya*).
Sov. Astron., Vol. 28, No. 1, p. 90 – 92 (1984). English translation of 37.142.003.

142.011 **Stellar coronas, X-rays, and Einstein.**
R. A. Stern.
Sky Telesc., Vol. 68, No. 1, p. 24 – 28 (1984).

142.012 **The X-ray spectra of galactic-bulge sources.**
E. N. Ercan, A. M. Cruise.
Mon. Not. R. Astron. Soc., Vol. 209, No. 2, p. 271 – 284 (1984). With microfiche MN 209/1.

Eight galactic-bulge sources were observed by the experiment C on Ariel-V discontinuously over the period 1975 February – 1978 April. The spectra of six of these sources X1636–53, X1642–45, X1702–36, X1735–44, X1758–20 and 1837+04 are reasonably well fitted by single thermal bremsstrahlung or power-law models. For X1813–14 and X1728–16 however, no single spectral model yields an acceptable fit to the data. The sources X1702–36, X1642–45 and X1636–53 are found to be the most active sources in the sample.

142.013 **On the theory of X–ray emission from the extended galactic Monogem Ring source in the Monoceros constellation.**
V. A. Krol'.
Astrofizika, Tom 19, Vyp. 3, p. 483 – 494 (1983). In Russian. English translation in Astrophysics, Vol. 19, No. 3.
X–ray emission from the Monogem Ring source which is explained by thermal bremsstrahlung of the hot optically thin gas filling the inside volume of the shock wave under a supernova burst is studied. An analytical expression for the X–ray spectrum is obtained.

142.014 **Infrared studies of X–ray sources.**
A. G. Ananth, U. R. Rao.
Astrophys. Space Sci., Vol. 102, No. 2, p. 359 – 366 (1984).
A detailed study of the infrared radiation from galactic X–ray sources indicates that the galactic bulge sources and X–ray binary sources have different infrared emission characteristics. The galactic bulge sources seem to show a power law dependence between the X–ray flux and the infrared flux emitted by the X–ray source. The results presented suggest that the infrared radiation in the galactic bulge sources is dominated by free–free radiation and, in the case of eclipsing binary sources, the black–body emission from the early–type companion star contributes significantly to the infrared radiation.

142.015 **Variations of the pulse profile and the X–ray intensity of GX301–2 (4U1223–62).**
K. Mitani, M. Matsuoka, K. Makishima, H. Inoue.
Astrophys. Space Sci., Vol. 103, No. 2, p. 345 – 352 (1984).
Two successive X–ray outbursts of the recurrent X–ray pulsar GX301–2 were observed in April and May, 1982 from Hakucho. Apart from general increase during flares lasting for several days, the X–ray intensity was highly variable on a time–scale of the order of an hour. The outstanding feature is that the amplitude of 700 s pulsation of GX301–2 changes rapidly in a very wide range. When the pulsation is distinct, the depth of modulation is as large as 60%; the pulse amplitude occasionally diminishes almost to an undetectable level so that the observed X–rays become unpulsed. These properties suggest that the X–ray emission from GX301–2 consists of a pulsating component with variable amplitude and a non–pulsating component of less variable intensity.

142.016 **Development of X–ray astronomy.**
M. Oda.
J. Inst. Electron. Commun. Eng. Jpn., Vol. 66, No. 9, p. 876 – 883 (1983). In Japanese. Abstr. in Phys. Abstr., Vol. 87, No. 1240, Entry 85980 (1984).

142.017 **Super–Eddington luminosities in X–ray bursts.**
J. van Paradijs, G. M. Stollman.
Astron. Astrophys., Vol. 137, No. 2, p. L12 – L14 (1984).
According to (quasi–) static model calculations of the outer layers of a neutron star, subject to a thermonuclear flash, the maximum luminosity cannot exceed the Eddington limit by more than a few percent. Yet, maximum burst luminosities substantially in excess of the Eddington limit have been observed for many X–ray bursts. In this paper the authors suggest a possible solution to this "Eddington limit problem", related to the transformation properties of the radiation field, as measured in the rest frame and in a comoving frame.

142.018 **Periodicity searches with IUE – studies of pulsating galactic X–ray sources.**
M. J. Coe, B. J. Payne, F. Giovannelli.
ESA Spec. Publ., ESA SP–218, p. 435 – 438 (1984). – See Abstr. 012.024.
The IUE spacecraft has been used in a special trailing mode to obtain time–resolved spectra of regularly pulsating sources. With this technique, multiple synchronised trails are carried out so that the resulting image represents one complete pulse period across the trail. The data have been subsequently analysed using the Maximum Entropy technique to search for significant variations in the flux from the source. The problems associated with the process are discussed, together with the results from observations of the galactic X–ray sources Vela X–1 and A0535+26.

142.019 **Coordinated X–ray, ultraviolet, optical, and infrared observations of the transient source A0535+26.**
F. Giovannelli, M. Ferrari–Toniolo, P. Persi, C. Bartolini, A. Guarnieri, A. Piccioni, E. L. van Dessel, M. Burger, C. de Loore, G. S. Bisnovatyi–Kogan (*G. S. Bisnovatyj–Kogan*), I. M. Golynskaya, V. G. Kurt, T. A. Mizyakina, E. Yu. Shafer, V. M. Shamolin, E. K. Sheffer, A. S. Smirnoff (*A. S. Smirnov*).
ESA Spec. Publ., ESA SP–218, p. 439 – 442 (1984). – See Abstr. 012.024.
The authors present data of A0535+26/HDE 245770 from various wavelengths (IR, optical, UV and X) before and after one of the strongest X–ray outbursts. Optical spectra taken immediately before the X–ray outburst show Hγ in absorption and Hα in weakened emission. In September and in January 1984 before the X–ray outburst the star presents diminished IR excess. These measures bring new elements for a model of the system and they are briefly discussed looking at the problem of the mass loss.

142.020 **Search for the radiative decay of a cosmic relic particle with I.U.E.**
F. Bochicchio, G. Auriemma, A. Frenkel, F. Massa.
ESA Spec. Publ., ESA SP–218, p. 503 – 505 (1984). – See Abstr. 012.024.
The authors present a new determination of U.V. background at $|b^{II}| \geqslant 45°$ obtained from low resolution large aperture IUE spectra. Aside from the identification of hot interstellar gas emission lines, the authors have fitted to their spectrum a component of the continuum with a power law distribution, as expected from the decay of a primordial relic light particle. From this fit they obtain a candidate mass for the neutrino of $14.8 \pm 0.1$ eV and a lifetime of $10^{16}$ years. The probability that statistical fluctuations could mimic this signal is estimated to be less than 2%. The authors discuss these results in the framework of cosmology and particle physics.

142.021 **A possible optical counterpart for the M15 globular cluster X–ray source.**
M. Aurière, O. Le Fèvre, A. Terzan.
Astron. Astrophys., Vol. 138, No. 2, p. 415 – 420 (1984).
The authors report on a high spatial resolution photographic study of stars resolved in the error box of the M15 X–ray source. Their observations were obtained at the 3.6 m Canada–France–Hawaii telescope and at the 2 m and 1 m Pic–du–Midi Observatory telescopes. They show that star No. 211 of the Aurière–Cordoni (1981) catalogue is probably the optical counterpart of the X–ray source since: (1) it is a "UV star", (2) it is a variable star. Its optical properties are consistent with those of some low (or intermediate) mass X–ray binaries (including Her X–1) scaled to the M15 X–ray flux rather than with those of classical globular cluster variable stars.

142.022 **A study of the spectra and pulse profiles of Centaurus X–3 from Hakucho.**
R. Lieu, D. Venkatesan, K. Mitani.
Astrophys. J., Vol. 282, No. 2, p. 709 – 712 (1984).
The binary X–ray pulsar Cen X–3 was observed by the X–ray satellite Hakucho during 1982 March 16 – 30. The data from the proportional counter FMC 2 have been analyzed to produce the high state spectrum in the energy range 2 – 20 keV. This cannot be fitted with a simple model spectrum. A comparison of the spectra taken at pulse peak and pulse minimum reveals that the X–rays associated with the former are harder. Finally, the pulse profiles constructed for four energy bins reveal that the radiation is more modulated at high energy.

142.023 **Short–term X–ray variability of the globular cluster source 4U 1820–30 (NGC 6624).**
L. Stella, S. M. Kahn, J. E. Grindlay.
Astrophys. J., Vol. 282, No. 2, p. 713 – 718 (1984).
Results of an analysis of a large number of observations of 4U 1820–30 in the globular cluster NGC 6624 are presented. The data were obtained with the MPC on board the Einstein Obser-

vatory. At the epoch of the observations the source was in a "high" state, and no burst activity was detected. Timing analysis performed using correlation and power spectrum techniques reveals the presence of shot noise–like variations on a time scale of $\sim 17$ s. The data suggest that the spectrum of the "shots" is significantly harder than that of the steady component from the source.

142.024 **Evidence for 4.4 hour periodic dips in the X–ray flux from 4U 1755–33.**
N. E. White, A. N. Parmar, M. Sztajno, H. U. Zimmermann, K. O. Mason, S. M. Kahn.
Astrophys. J., Lett. Ed., Vol. 283, No. 1, p. L9 – L12 (1984).
The authors report the discovery by the *EXOSAT* observatory of three equally spaced dips in the flux from 4U 1755–33 that suggest the binary period of this system to be 4.4 hr. The duration of each dip is of order 30 minutes with considerable variability within each event on time scales between 30 and 600 s. The variations are spectrally independent with a maximum reduction in flux of 40%. It is suggested that the dips are caused by the regular occultation of the X–ray source by a thickened region of an accretion disk at the point where it is fed by the gas stream from a binary companion.

142.025 **History of X–ray transients as seen by Vela, 1969 – 1979.**
J. Terrell, W. C. Priedhorsky, R. D. Belian, J. P. Conner, W. D. Evans.
Ann. N.Y. Acad. Sci., Vol. 422, p. 282 – 283 (1984). – See Abstr. 012.033.
A brief summary of X–ray transient observations by the Vela spacecraft 5A and 5B is given.

142.026 **The Medium Sensitivity Survey: a new sample of X–ray sources with optical identifications and the revised extragalactic log *N*–log *S*.**
I. M. Gioia, T. Maccacaro, R. E. Schild, J. T. Stocke, J. W. Liebert, I. J. Danziger, D. Kunth, J. Lub.
Astrophys. J., Vol. 283, No. 2, p. 495 – 511 (1984). With plates 10 – 13.
The Einstein Observatory Medium Sensitivity Survey has been extended. 49 new, high galactic latitude X–ray sources have been detected in $\sim 150$ IPC fields ($\sim 40$ deg$^2$) with sensitivities in the range $7 \times 10^{-14}$ to $2 \times 10^{-12}$ergs cm$^{-2}$s$^{-1}$. The optical program undertaken to identify these new sources has yielded a virtual 100% identification rate. These sources have been combined with those previously selected resulting in a sample of 112 serendipitous X–ray sources. Of these, 75% are associated with extragalactic objects, and 25% are identified with galactic stars. Active galactic nuclei (AGNs–quasars and Seyfert galaxies) constitute the large majority of the extragalactic population (68%). The log *N*–log *S* relations for the two classes of objects for which enough statistics are present, AGNs and clusters of galaxies, have been computed.

142.027 **Discovery of a 7.1 hour period and eclipses from MXB 1659–29.**
L. R. Cominsky, K. S. Wood.
Astrophys. J., Vol. 283, No. 2, p. 765 – 773 (1984).
The authors have discovered a 7.1 hr period in the X–ray emission from the "transient" X–ray burst source MXB 1659–29. Erratic variations in X–ray flux are seen during approximately 25% of the 7.1 hr cycle; at the end of this period, a stable dip in the flux occurs. This dip is, in all probability, an eclipse by the companion star in the low–mass binary system. The short ($\sim 15$ minute) eclipse duration is used to refine the period determination ($7.114104 \pm 0.000168$ hr) and to constrain the companion star mass and binary system parameters.

142.028 **Delays of optical bursts in simultaneous optical and X–ray observations of MXB 1636–53.**
M. Matsuoka, K. Mitsuda, T. Ohashi, H. Inoue, K. Koyama, F. Makino, K. Makishima, T. Murakami, M. Oda, Y. Ogawara, N. Shibazaki, Y. Tanaka, K. Tsuno, S. Miyamoto, H. Tsunemi, K. Yamashita, S. Hayakawa, H. Kunieda, K. Masai, F. Nagase, Y. Tawara, I. Kondo, L. Cominsky, J. G. Jernigan, A. Lawrence, W. H. G. Lewin, H. Pedersen, C. Motch, J. van Paradijs.
Astrophys. J., Vol. 283, No. 2, p. 774 – 781 (1984).
Observations of simultaneous optical and X–ray bursts from 4U/MXB 1636–53 were made using the Hakucho burst monitor system and optical telescopes at ESO during 1979 and 1980. The six best cases among the 10 coinciding observations are analyzed in terms of a model in which the optical emission is the result of reprocessing of X–rays (through blackbody heating). From this analysis the authors obtain the temperature (spatially averaged) and size of a reprocessor, and the smearing and delay of the optical bursts. They discuss the possible constraints which the results put on a low–mass binary model of this burst source.

142.029 **Maximum luminosities and distribution of X–ray burst sources.**
T. Ebisuzaki.
Astron. Her., Vol. 77, No. 11, p. 286 – 289 (1984). In Japanese.

142.030 **Discovery of a 176 day period in 4U 1820–30.**
W. Priedhorsky, J. Terrell.
Astrophys. J., Lett. Ed., Vol. 284, No. 1, p. L17 – L20 (1984).
Long–term X–ray observations of the X–ray burster 4U 1820–30 (in NGC 6624) reveal a significant $176.4 \pm 1.3$ day period. These data were obtained by a 3 – 12 keV detector on the Vela 5B satellite during the years 1969 – 1976. The X–ray flux varies by a factor of 2 over the cycle; the light curve, which has a sawtooth shape with fast rise and slow decline, is quite similar to the 199 day light curve of the burster 4U 1915–05.

142.031 **Correlation between X–ray luminosity and the deviations from the main–sequence colour–diagram for stellar X–ray sources.**
P. K. Kunte, A. R. Rao, M. N. Vahia.
Astrophys. Space Sci., Vol. 105, No. 2, p. 295 – 300 (1984).
The authors have investigated the variations in the X–ray luminosity with deviation from main–sequence colour diagram for stellar X–ray sources. It is observed that the early– and late–type stars follow two different patterns. The authors believe that the anomalies seen in this pattern are related to the interaction of the stars with their neighbourhood and propose that such regions should contribute to the composition of interstellar particles and to low energy cosmic rays.

142.032 **The extragalactic nature of H0323+022.**
B. Margon, G. H. Jacoby.
MPE Rep., No. 184, p. 47 – 50 (1984). – See Abstr. 012.039.
The authors report several imaging and spectroscopic observations of the unusual high–latitude X–ray source H0323+022. It is concluded that this source is probably a relatively normal BL Lacertae object.

142.033 **Characteristics of X–ray spectra from galactic and extra–galactic X–ray sources.**
M. Matsuoka.
MPE Rep., No. 184, p. 96 – 98 (1984). – See Abstr. 012.039.
An overview of X–ray spectra obtained with TENMA is presented. Most X–ray pulsars produce strong iron emission lines with an energy of $\sim 6.4$ keV. The low mass X–ray binaries Sco X–1, Cyg X–2 and 4U 1608–52 produce narrow $\sim 6.7$ keV iron emission lines. The active galactic nuclei NGC 4151 and Cen A, produce strong $\sim 6.4$ keV iron emission lines, while no appreciable iron emission lines were detected in the spectra of the BL Lac objects; PKS 2155–304 and Mkn 421.

142.034 **EXO 1102.8+2539, an X–ray variable AGN.**
K. Beuermann, L. Stella, W. Sieber.
MPE Rep., No. 184, p. 111 – 113 (1984). – See Abstr. 012.039.
The authors have discovered a serendipitous variable X–ray source using the LE experiment of the EXOSAT satellite. It was identified with a starlike object of $\sim 18^m$ and 1950 coordinates $\alpha = 11^h02^m47^s$, $\delta = +25°23'.2$. An IDS spectrum obtained with the ESO 3.6m telescope shows H$\alpha$ and H$\beta$ redshifted with

$z = 0.095$. The FWHM of the lines is $\sim 700 \text{ km s}^{-1}$. No forbidden–line emission was seen above the noise level.

142.035 **The HEAO–1 diffuse component spectrum from 15 to above 500 keV.**
D. E. Gruber, R. E. Rothschild, J. L. Matteson, R. L. Kinzer.
MPE Rep., No. 184, p. 129 – 134 (1984). – See Abstr. 012.039.
The spectrum of the diffuse X– and gamma–ray background has been measured over a band of more than 2 decades with the scintillator detectors of the UCSD/MIT instrument aboard the HEAO–1 satellite. The observed spectrum is presented and compared with other measurements. The possible contribution to the diffuse component above 100 keV from active galaxies is discussed.

142.036 **The origin of the X–ray background.**
A. C. Fabian.
MPE Rep., No. 184, p. 232 – 242 (1984). – See Abstr. 012.039.

142.037 **Cosmic X–ray background radiation.**
B. Durišić.
Vasiona, Année 32, No. 3, p. 48 – 53 (1984). In Serbo–Croatian.
A review of the present–day knowledge about X–ray background emission is presented.

142.038 **Optical counterparts for X–ray sources in M31.**
D. Crampton, A. P. Cowley, J. B. Hutchings, D. J. Schade, L. P. van Speybroeck.
Astrophys. J., Vol. 284, No. 2, p. 663 – 669 (1984).
Einstein Observatory X–ray position for 117 sources around M31 have been examined with $1000 \text{ Å mm}^{-1}$ slitless spectral plates obtained at the prime focus of the Canada–France–Hawaii Telescope. A variety of possible counterparts is found within the error circles of the source positions, including globular clusters, foreground stars, faint blue stars, and a compact galaxy. For about half of the sources, no candidate object sufficiently near the X–ray position was visible on the plates which reach to $m_B \approx 21$. Crowding in the nucleus of M31 made it impossible to examine the source positions $< 2' - 3'$ from the center of the galaxy.

142.039 **Neutron stars, black holes and Exosat.**
A. Peacock.
ESA Bull., No. 40, p. 38 – 42 (1984).

142.040 **Rocket observations of the diffuse ultraviolet background.**
P. Jakobsen, S. Bowyer, R. Kimble, P. Jelinsky, M. Grewing, G. Krämer, C. Wulf–Mathies.
Astron. Astrophys., Vol. 139, No. 2, p. 481 – 495 (1984).
The authors present observations of the diffuse ultraviolet background obtained with three ultraviolet broad band photometers placed in the focal plane of a 95 cm, f/2.8 normal incidence telescope flown onboard an Aries sounding rocket. The hypothesis of a uniform background flux is clearly refuted by the data. The spatial variations exhibited by the data correlate with line of sight neutral hydrogen column density as determined from 21 cm radio observations. The results demonstrate that there is a galactic component to the diffuse ultraviolet background.

142.041 ***HEAO 1* high–energy X–ray observations of three bright transient X–ray sources H1705–250 (Nova Ophiuchi), H1743–322, and H1833–077 (Scutum X–1).**
B. A. Cooke, A. M. Levine, F. L. Lang, F. A. Primini, W. H. G. Lewin.
Astrophys. J., Vol. 285, No. 1, p. 258 – 263 (1984).
The authors have observed the bright X–ray transients H1705–250 (Nova Ophiuchi) and H1743–322 in the energy range 12–180 keV and have revealed the presence of high–energy spectral components. On the basis of X–ray spectra in the 1–10 keV range transients are classified as "soft" or "hard"; and both H1705–250 and H1743–322 should be in the "soft" category. The authors have reexamined data for the archetypal "soft" transient A0600–00 taken by the *SAS 3* satellite, but they find no evidence there for a high–energy spectral component. These results make the "hard"/"soft" distinction between X–ray transients much less clear.

142.042 **The 1973 X–ray transient V0332+53.**
J. Terrell, W. C. Priedhorsky.
Astrophys. J., Lett. Ed., Vol. 285, No. 1, p. L15 – L18 (1984).
A bright transient X–ray source, V0332+53, was observed by Vela 5B for 3 months in 1973. It was first clearly detected in the 10 day period centered on 1973 June 1, rose to an intensity of ~1.4 Crab in July, and was last definitely observed around 1973 August 30. The slow rise and symmetric fall of intensity are unusual for a transient X–ray source. Rapid fluctuations took the source to peak intensities above 2 Crab. The spectrum of this source was among the hardest ever observed by Vela 5B. The location of V0332+53, in Camelopardalis, was $l = 146°.11$, $b = -2°.16$, with an uncertainty of $\sim 0°.1$.

142.043 **X–ray astronomy.**
R. Rocchia.
Ciel Espace, No. 196, p. 19 – 25 (1983). In French. Abstr. in Phys. Abstr., Vol. 87, No. 1246, Entry 115835 (1984).

142.044 **The nature of LMC X–3.**
T. S. Khruzina, A. M. Cherepashchuk.
Sov. Astron., Vol. 28, No. 2, p. 173 – 179 (1984). English translation of 37.142.032.

142.045 **H1504+65: another HZ43?**
J. Nousek.
Bull. Am. Astron. Soc., Vol. 16, No. 3, Part I, p. 731 (1984). Abstract. – See Abstr. 010.065.

142.046 **Far UV background observations in the direction of the Coma cluster.**
J. B. Holberg, H. B. Barber.
Bull. Am. Astron. Soc., Vol. 16, No. 3, Part I, p. 731 (1984). Abstract. – See Abstr. 010.065.

142.047 **New radial velocities of the supergiant in the Cygnus X–1 system.**
O. E. Aab (*O. Eh. Aab*).
Bull. Spec. Astrophys. Obs. – North Caucasus, Vol. 17, p. 4 – 15 (1984). English translation of 34.142.074.

142.048 **HZ Her/Her X–1: an alternative model for the $35^d$ cycle?**
F. Meyer, E. Meyer–Hofmeister.
Astron. Astrophys., Vol. 140, No. 2, p. L35 – L38 (1984).
Self–excited non–linear mass flow oscillations in an irradiated accretion disk provide a physical cause for the $35^d$ cycle.

142.049 **LMC X–4: the optical 30–day cycle and its implications.**
S. A. Ilovaisky, C. Chevalier, C. Motch, M. Pakull, J. van Paradijs, J. Lub.
Astron. Astrophys., Vol. 140, No. 2, p. 251 – 258 (1984).
Analysis of an extensive set of optical photometry of the massive X–ray binary LMC X–4 obtained at ESO from 1976 through 1983 shows the presence of the X–ray 30–day cycle. The periodogram exhibits peaks at the sums of the orbital and half–orbital frequencies with the 30–day frequency and at the 30–day frequency itself. During X–ray OFF states, the amplitude of the 1.408–day orbital period double–wave light curve is 0.07 mag, while during ON states it increases to 0.20 mag. The authors explain the observed effects in terms of an X–ray illuminated, tilted, counter–"precessing" accretion disk. The excess light comes both from the accretion disk itself and from the illuminated hemisphere of the primary, the latter experiencing variable X–ray shadowing due to the changing aspect of the disk.

142.050 **Be/neutron star binaries: a relationship between orbital period and neutron star spin period.**
R. H. D. Corbet.
Astron. Astrophys., Vol. 141, No. 1, p. 91 – 93 (1984).
A strong correlation is shown to exist between the orbital periods of Be/neutron star binaries and the spin period of the

neutron star component. This relationship is interpreted as showing that the neutron stars are each in a state of quasi-equilibrium in which the corotation and Alfvén radii are equal. The correlation enables the orbital periods of the X-ray binaries 2S1417-62 and A1118-61 to be approximately predicted.

142.051 **Multispectral analysis in the UV, optical and IR of HDE 245770 = A0535+26.**
C. de Loore, F. Giovannelli, E. L. van Dessel, C. Bartolini, M. Burger, M. Ferrari-Toniolo, A. Giangrande, A. Guarnieri, P. Hellings, H. Hensberge, P. Persi, A. Piccioni, H. Van Diest.
Astron. Astrophys., Vol. 141, No. 2, p. 279 - 296 (1984).

The characteristics of the Be/X-ray binary system HDE 245770/A0535+26 are examined from UV, visual, IR and X-ray data. A sequence of observations in the optical preceding an X-ray outburst is interpreted as the ejection of a shell by the Be primary. An evolutionary scenario is presented that produces a system like HDE 245770. An effective temperature of 26000±2000K is derived for the Be star. The mass loss rate through stellar wind from the primary is determined. The Be star is not filling its Roche lobe and there is no accretion disk around the compact companion. The system configuration can be simulated with an evolution code allowing for the simultaneous evolution of two components - a 20 $M_{\odot}$ primary and a 15 $M_{\odot}$ secondary - producing Be/X-ray binaries of this kind, with systems like φ Per as progenitor.

142.052 **On the periodic gamma-ray emission of Cygnus X-3.**
A. M. Gal'per, B. I. Luchkov.
Sov. Astron. Lett., Vol. 10, No. 3, p. 150 - 152 (1984). English translation of 37.142.053.

142.053 **MXB 1730-335.**
IAU Circ., No. 3957 (1984).

142.054 **H0850-42.**
IAU Circ., No. 3959 (1984).

142.055 **4U 1323-62.**
IAU Circ., No. 3961 (1984).

142.056 **2S 0142+61.**
IAU Circ., Nos. 3996, 4001 (1984).

142.057 **The anisotropy of the diffuse cosmic X-ray background.**
C.-x. Xu.
Chin. Astron. Astrophys., Vol. 8, No. 4, p. 340 - 342 (1984). English translation of Chin. J. Space Sci., Vol. 4, No. 3, p. 253 - 256 (1984).

From the new data of the 2 - 60 keV diffuse X-ray background from HEAO-1 A2, the model galactic component is subtracted to give the cosmic component. The greater intensity in the northern galactic hemisphere is shown to be probably due to the motion of the Sun relative to the background. The derived degree of anisotropy is $\delta = (0.45 \pm 0.18)\%$ and the velocity is $V = (397 \pm 159)$ km/s. These values are consistent with the known anisotropy of the microwave background.

142.058 **Search for millisecond rotational periods in some low-mass X-ray binaries observed by EXOSAT.**
A. Langmeier, M. Sztajno, J. Trümper.
MPE Prepr., No. 28, p. 5 - 8 (1984). Paper presented at the XXVth COSPAR Meeting on X-Ray Astronomy, Graz, Austria, July 3 - 7, 1984.

142.059 **EXOSAT observations of the galactic bulge X-ray source GX17+2.**
M. Sztajno, J. Trümper, H. U. Zimmermann, A. Langmeier.
MPE Prepr., No. 28, p. 14 - 17 (1984). Paper presented at the XXVth COSPAR Meeting on X-Ray Astronomy, Graz, Austria, July 3 - 7, 1984.

142.060 **A hard X-ray observation of the 1979 outburst of Centaurus X-4 with the Franco-Soviet Signe 2 MP (Prognoz 7) satellite experiment.**
P. Bouchacourt, G. Chambon, M. Niel, A. Refloch, I. V. Estulin (*I. V. Ehstulin*), A. V. Kuznetsov, A. S. Melioransky (*A. S. Melioranskij*).
Astrophys. J., Lett. Ed., Vol. 285, No. 2, p. L67 - L70 (1984).

In 1979 May, a collimated detector in the Signe 2 MP experiment aboard the Prognoz 7 satellite detected intense 13 - 163 keV X-ray emission from the source Centaurus X-4. This was the first time that high-energy X-ray emission was observed from this object, which up to now has been categorized as a type I soft X-ray burster. The light curve of the outburst is presented, as well as the hard X-ray spectrum at various times during the outburst.

142.061 **High resolution spectroscopy of the optical candidate for the X-ray transient X0331+53.**
J. Stocke, D. Silva, J. H. Black, K. Kodaira.
Prepr. Steward Obs., No. 552, 16 pp. (1984).

High resolution spectra (0.1 - 1 Å) of the optical candidate for the X-ray transient X0331+53 in January 1984 show broad hydrogen and helium emission (FWZI = 335 km $s^{-1}$) and strong interstellar absorption corresponding to approximately 7 magnitudes of visual extinction. No photospheric absorption and no radial velocity variations in the emission lines greater than 0.5 km $s^{-1}$ over a three-day period were found. This leads to the conclusion that the optical continuum does not arise at or near the primary star.

142.062 **Photoelectric UBVR photometry of the transient X-ray source A0535+26.**
G. U. Koval'chuk, R. I. Goncharova.
Astron. Tsirk., No. 1317, p. 1 - 3 (1984). In Russian.

142.063 **Variations in the brightness level of H 2215-086.**
N. E. Kurochkin.
Astron. Tsirk., No. 1322, p. 3 - 5 (1984). In Russian.

142.064 **Optical counterparts of the Large Magellanic Cloud X-ray point sources.**
A. P. Cowley, D. Crampton, J. B. Hutchings, D. J. Helfand, T. T. Hamilton, J. R. Thorstensen, P. A. Charles.
Astrophys. J., Vol. 286, No. 1, p. 196 - 208 (1984).

The results of an optical identification program for X-ray point sources in the direction of the Large Magellanic Cloud are presented. With the addition of the new identifications, 22 of the 24 point sources with accurate (~3″) positions obtained with the Einstein Observatory HRI are now identified. These include 13 foreground stars, three background active galaxies, and six LMC members. Of the 24 sources with IPC (~30″) positions two are probably active galaxies. Promising candidates have been found for half of the remainder. Twenty-eight of the certain identifications are LMC members, including 19 SNR and 9 X-ray binaries or probable binaries. On the basis of the new identifications, comparisons are made between the overall stellar X-ray population in the LMC and in the Galaxy.

142.065 **First results from the X-ray satellite EXOSAT.**
P. Biermann.
Mitt. Astron. Ges., Nr. 62, p. 101 - 121 (1984). - See Abstr. 010.341.

Based on a large number of preprints the author reports on the first set of scientific results based on observations with the European X-ray satellite EXOSAT.

142.066 **Are X-ray bursts really of super-Eddington luminosities?**
T. Ebisuzaki, T. Hanawa, D. Sugimoto.
Publ. Astron. Soc. Jpn., Vol. 36, No. 3, p. 551 - 566 (1984).

Some X-ray bursts show high color temperature of about 3 keV. This temperature was so far regarded to be equal to the effective temperature. In the present paper the authors show that this is not the case for X-ray bursting neutron stars whose luminosity is close to the Eddington luminosity.

142.067 **Pulse–period changes of X–ray pulsars measured with Hakucho and Tenma.**
F. Nagase, S. Hayakawa, T. Kii, N. Sato, T. Ikegami, N. Kawai, K. Makishima, M. Matsuoka, K. Mitani, T. Murakami, M. Oda, T. Ohashi, Y. Tanaka, S. Kitamoto.
Publ. Astron. Soc. Jpn., Vol. 36, No. 4, p. 667 – 678 (1984).

Eight X–ray pulsars, Her X–1, Cen X–3, 4U 1626–67, OAO 1657–415, A0535+26, Vela X–1, 4U 1538–52, and GX 301–2, have been observed from Hakucho and Tenma. The pulse periods of all these X–ray pulsars measured with Hakucho and Tenma between March 1979 and October 1983 are summarized together with those of 4U 1700–37 and 4U 1907+09, the pulsations of which were newly discovered by Tenma observations. Several instances of remarkable behavior in pulse–period change are revealed both for the wind–fed and disk–fed pulsars.

142.068 **Discovery of a 437.5–s X–ray pulsation from 4U 1907+09.**
K. Makishima, N. Kawai, K. Koyama, N. Shibazaki, F. Nagase, M. Nakagawa.
Publ. Astron. Soc. Jpn., Vol. 36, No. 4, p. 679 – 689 (1984).

An X–ray pulsation of 437.5–s period was discovered from the X–ray source 4U 1907+09 during the Tenma observation in 1983. Thus 4U 1907+09 has been established as an X–ray pulsar. The pulse–arrival–time analysis revealed that the 8.38–d periodic X–ray modulation in this source discovered by the Ariel–5 observation can be regarded as the binary period.

142.069 **A possible 67–s X–ray pulsation period from 4U 1700–37.**
T. Murakami, N. Kawai, K. Makishima, K. Mitani, S. Hayakawa, F. Nagase, Y. Tawara, H. Kunieda.
Publ. Astron. Soc. Jpn., Vol. 36, No. 4, p. 691 – 698 (1984).

Possible X–ray pulsations with a period of 67.4 s were discovered by Tenma from the massive X–ray source 4U 1700–37. Due to the similarity of characteristics with other massive X–ray pulsars, pulsations of 4U 1700–37 have long been suspected. The pulsations were observed only during the brightest flare of 4U 1700–37, while no indication of pulsation was found during the quiescent state.

142.070 **Properties of the iron line from Vela X–1.**
T. Ohashi, H. Inoue, K. Koyama, F. Makino, M. Matsuoka, K. Suzuki, Y. Tanaka, S. Hayakawa, H. Tsunemi, K. Yamashita.
Publ. Astron. Soc. Jpn., Vol. 36, No. 4, p. 699 – 707 (1984).

Properties of the iron emission line from Vela X–1 are studied with the gas scintillation proportional counters on board Tenma. The energy and the width of the iron line are determined to be $6.43 \pm 0.05$ keV and less than 0.5 keV (FWHM), respectively. The fluorescence efficiency of the line is $\sim 1.5\%$ even when the absorbing column is less than $5 \times 10^{22} cm^{-2}$. The iron line intensity during eclipse suggests the size of the line emitting region to be less than $5 \times 10^{11}$cm from the neutron star.

142.071 **Sudden disappearence of Vela X–1 pulses.**
H. Inoue, Y. Ogawara, T. Ohashi, I. Waki, S. Hayakawa, H. Kunieda, F. Nagase, H. Tsunemi.
Publ. Astron. Soc. Jpn., Vol. 36, No. 4, p. 709 – 713 (1984).

An unusual event associated with Vela X–1 was recorded by Tenma, in which the X–ray intensity suddenly dropped to less than one tenth of the preceding level in a few tens of second and remained at this low level for between 18 min and 78 min. The spectrum during the low intensity period showed an overall decrease at all energies, and was not due to increased circumstellar absorption.

142.072 **Variation of the pulse profile of Hercules X–1.**
T. Ohashi, H. Inoue, N. Kawai, K. Koyama, M. Matsuoka, K. Mitani, Y. Tanaka, F. Nagase, M. Nakagawa, Y. Kondo.
Publ. Astron. Soc. Jpn., Vol. 36, No. 4, p. 719 – 729 (1984).

Her X–1 was observed during an on–state of the 35–d cycle in May 1983 with the gas scintillation proportional counters on board Tenma. The declining phase of the on–state is characterized by two progressive features: (1) decrease of the main pulse amplitude, and at the same time (2) increase of absorption column of relatively cool matter. However, the photoelectric absorption in this cool medium does not explain the feature of the main pulse decrease which occurs at all energies. The spectral shape after the correction for photoelectric absorption is found to remain essentially the same through the main–pulse decrease. This may be interpreted as due to an attenuation of X–ray beam by electron scatterings in a highly ionized plasma at an estimated distance of $\sim 3 \times 10^8$cm from the neutron star.

142.073 **Transient dips of Cygnus X–1 observed from Tenma.**
S. Kitamoto, S. Miyamoto, Y. Tanaka, T. Ohashi, Y. Kondo, Y. Tawara, M. Nakagawa.
Publ. Astron. Soc. Jpn., Vol. 36, No. 4, p. 731 – 740 (1984).

Transient intensity dips of Cyg X–1 were observed with the gas scintillation proportional counters on board Tenma. Several dips occurred during 15 hr before the time of superior conjunction. The duration of the observed dips ranges from several seconds to more than 10 min. Within the dips, there exist complex structures with time scales of several seconds and, in at least one case, shorter than 2 s. For one dip lasted longer than 10 min, a good–quality spectrum was obtained, which reveals the K–absorption edge of iron at $7.18 \pm 0.18$ keV. This provides clear evidence for the dips to be due to absorption by intervening clouds.

142.074 **Energy spectra of low–mass binary X–ray sources observed from Tenma.**
K. Mitsuda, H. Inoue, K. Koyama, K. Makishima, M. Matsuoka, Y. Ogawara, N. Shibazaki, K. Suzuki, Y. Tanaka, T. Hirano.
Publ. Astron. Soc. Jpn., Vol. 36, No. 4, p. 741 – 759 (1984).

Observations of X–ray spectra for four low–mass binary X–ray sources, Sco X–1, 4U 1608–522, GX5–1 and GX349+2 were performed with the gas scintillation proportional counters on board Tenma. Common to the four sources, the spectrum hardens with increasing intensity, and the difference of the spectra before and after an intensity increase shows invariably a blackbody spectrum of $kT \cong 2$ keV. Every observed spectrum can be expressed by a sum of two spectral components: a hard component and a soft component, whose spectral shapes are fixed for each individual source.

142.075 **Detection of iron K–emission lines from two low–mass binary X–ray sources: Scorpius X–1 and 4U 1608–52.**
K. Suzuki, M. Matsuoka, H. Inoue, K. Mitsuda, T. Ohashi, Y. Tanaka, T. Hirano, S. Miyamoto.
Publ. Astron. Soc. Jpn., Vol. 36, No. 4, p. 761 – 767 (1984).

Iron K–emission line from two low–mass binary X–ray sources, Sco X–1 and 4U 1608–52 was observed with the gas scintillation proportional counters on board Tenma. For both sources, the line features are similar to each other and represent a single emission line at about 6.7 keV with an intrinsic line width less than 1.7 keV (FWHM). The line center energy implies that the observed lines are mostly due to helium–like iron (Fe XXV).

142.076 **The transient X–ray source 4U 1543–47 observed from Tenma.**
S. Kitamoto, S. Miyamoto, H. Tsunemi, K. Makishima, M. Nakagawa.
Publ. Astron. Soc. Jpn., Vol. 36, No. 4, p. 799 – 806 (1984).

The strong transient X–ray source, 4U 1543–47, was observed between August 17 and August 24, 1983, with the fan–beam scanning counters on board the Tenma satellite. This is the second observation of an outburst from this source. On August 20, its intensity rose to about one half that of Sco X–1 in the 1.5 – 3.75 keV energy range. During the rise to maximum brightness, a hardening of the spectrum was observed. After reaching the maximum brightness, the intensity was roughly constant and the spectrum was consistent with either a thermal bremsstrahlung spectrum with $kT = 1.6$ keV or a power–law spectrum with a photon index of 5.3, but was inconsistent with a blackbody spectrum.

142.077 **The X–ray activity of the rapid burster in 1983.**
H. Kunieda, Y. Tawara, S. Hayakawa, F. Nagase, H. Inoue, N. Kawai, F. Makino, K. Makishima, M. Matsuoka, T. Murakami, M. Oda, Y. Ogawara, T. Ohashi, Y. Tanaka, I. Waki.
Publ. Astron. Soc. Jpn., Vol. 36, No. 4, p. 807 – 817 (1984).

The rapid burster became active on August 5, 1983 after a long quiescence of four years. The source started to emit a train of bursts which apparently resemble type I bursts, with quasi-periodic occurrence (74 – 90 min) superposed on a strong persistent component of ~50 min Crab. From August 17 to 19 there appeared eight long bursts of trapezoidal profiles similar to those observed in August 1979, and seven exotic long bursts, each being preceded by a gradual intensity increase in a few minutes. From August 20 to the termination of the activity on August 31 were observed about 3000 rapidly repetitive short bursts characteristic to this source in its discovery in 1976. The burst interval changed day to day from 15 to 150 s.

142.078 **Discovery of absorption lines in X–ray burst spectra from X1636–536.**
I. Waki, H. Inoue, K. Koyama, M. Matsuoka, T. Murakami, Y. Ogawara, T. Ohashi, Y. Tanaka, S. Hayakawa, Y. Tawara, S. Miyamoto, H. Tsunemi, I. Kondo.
Publ. Astron. Soc. Jpn., Vol. 36, No. 4, p. 819 – 830 (1984).

Thirteen X–ray bursts were observed from X1636–536 with gas scintillation proportional counters on board Tenma. Significant dips in the X–ray spectra were detected in four different bursts and these dips are best described by absorption lines from X–ray bursts. The observed absorption lines are considered as due to helium–like ions of a heavy element, which are redshifted most probably by the general relativistic effect near the surface of the neutron star. If the element responsible for the absorption were iron, the consequent gravitational redshift factor would imply that the neutron star radius is only about 1.6 times the Schwarzschild radius.

142.079 **Peak luminosity at the Eddington limit for the X–ray bursts from X1636–536.**
H. Inoue, I. Waki, K. Koyama, M. Matsuoka, T. Ohashi, Y. Tanaka, H. Tsunemi.
Publ. Astron. Soc. Jpn., Vol. 36, No. 4, p. 831 – 838 (1984).

X–ray bursts from X1636–536 were observed with the gas scintillation proportional counters on board Tenma. Time resolved spectra of twelve bursts are well fitted by the Planck spectra except for fine structures present in some bursts. The three bursts, with the largest peak flux, exhibited a fast increase of the apparent blackbody radius near the flux peak followed by a quick decrease of the apparent radius during a few seconds while the flux was at the maxima. This peak flux level is the same for these three bursts and can be consistently interpreted as the Eddington limit for this burst source.

142.080 **An indication for ejection of hydrogen–rich envelope from the X–ray burster MXB 1636–53.**
D. Sugimoto, T. Ebisuzaki, T. Hanawa.
Publ. Astron. Soc. Jpn., Vol. 36, No. 4, p. 839 – 844 (1984).

X–ray bursts from MXB 1636–53 observed by the Tenma satellite are shown to consist of two distinct classes. The bright class shows a remarkable expansion in the radius well before the stage of peak luminosity. This is interpreted as an ejection of the hydrogen–rich envelope during a helium shell–flash. The faint class does not show any indication of mass ejection.

142.081 **Interpretation of the X–ray burst with precursor in terms of expanding envelope.**
Y. Tawara, S. Hayakawa, T. Kii.
Publ. Astron. Soc. Jpn., Vol. 36, No. 4, p. 845 – 853 (1984).

A very long X–ray burst with a precursor from XB 1715–321 was observed in four energy bands. The burst consisted of a short precursor and a very long main burst with a gap of 4 s in between. The gap is interpreted as due to a decrease in the temperature which is too low to emit detectable X–rays. The gradual rise is attributed to an increase in temperature, while the total luminosity remains constant. The whole burst profile can be explained by a rapid expansion of the envelope by radiation pressure, a gradual contraction thereof and cooling in the final stage.

142.082 **Two successive X–ray bursts within eight minutes from XB 1745–25.**
H. Inoue, K. Koyama, F. Makino, K. Makishima, M. Matsuoka, T. Murakami, M. Oda, Y. Ogawara, T. Ohashi, N. Shibazaki, Y. Tanaka, S. Hayakawa, H. Kunieda, K. Masai, F. Nagase, Y. Tawara, S. Miyamoto, H. Tsunemi, K. Yamashita, I. Kondo.
Publ. Astron. Soc. Jpn., Vol. 36, No. 4, p. 855 – 859 (1984).

Two successive X–ray bursts separated by only eight minutes were observed from XB 1745–25 (Terzan 5 source). The two bursts were similar to each other in intensity and profile. The blackbody radii during the decay portions of these two bursts agreed within statistical uncertainties, with those determined for ten other bursts from XB 1745–25. It is certain that almost the entire surface of the neutron star was responsible for both bursts.

142.083 **Long X–ray burst from GX 17+2.**
Y. Tawara, T. Hirano, T. Kii, M. Matsuoka, T. Murakami.
Publ. Astron. Soc. Jpn., Vol. 36, No. 4, p. 861 – 867 (1984).

Four long X–ray bursts were observed from GX 17+2 in 1981 and 1982. Their profiles are characterized by the rise times of about 1.5 s, durations of 3 – 15 min, burst peak flux values somewhat smaller than the persistent flux values, and spectral softening in the tails. No other bursts were observed from Hakucho pointings on this source in 27 d in 1981 – 1982.

142.084 **The soft X–ray diffuse background: implications for the nature of the local interstellar medium.**
D. McCammon.
NASA Conf. Publ., NASA CP–2345, p. 195 – 203 (1984). – See Abstr. 012.089 (IAU Colloq. No. 81).

A rather large body of data now exists which pertains to the nature and distribution of interstellar material within ~100 pc of the Sun. Models are available which satisfactorily explain various subsets of these data, but it is not obvious that they can all be combined in any physically reasonable way. Rather than giving a model to explain the X–ray data which would contain oversimplifications and misinterpretations making it impossible to reconcile them with other kinds of observations, the author tries to present the constraints introduced by the X–ray observations which must be satisfied by any general model of the local ISM. Two simplified models illustrating how some of these might be met are given near the end.

142.085 **Non–local origin of a substantial portion of the soft X–ray background.**
G. W. Clark.
NASA Conf. Publ., NASA CP–2345, p. 204 – 210 (1984). – See Abstr. 012.089 (IAU Colloq. No. 81).

Comparison of the SAS–3 soft X–ray sky survey (F. Marshall and G. Clark 1984) with the 21–cm neutral hydrogen survey of Stark et al. (1984) confirms the well–known anticorrelation between the counting rates in the C–band (0.10 – 0.28 keV) and the column density of neutral hydrogen. The question arises as to whether a two–component model can be constructed which fits both the anticorrelation and fluctuation properties of the survey data. The numerical analysis of correlations and fluctuations in the X–ray and radio data were carried out. For analysis of the fit to the two–component model the data were divided into eight sets corresponding to octants of the sky. The best fit is in the "second" octant defined by $0° < b < 90°$, $90° < l < 180°$.

142.086 **Local contributions to the 0.6 keV diffuse X–ray background.**
D. N. Burrows.
NASA Conf. Publ., NASA CP–2345, p. 215 – 218 (1984). – See Abstr. 012.089 (IAU Colloq. No. 81).

The intensity of the X–ray background between 0.5 and 1.0 keV has surprisingly little dependence on galactic latitude. Possible mechanisms for the production of these X–rays include extragalactic emission and emission from dM stars, both of which should be strongly dependent on galactic latitude, and diffuse emission from hot gas ($T \cong 3 \times 10^6$K) surrounding the Sun. These mechanisms can be distinguished by the presence or absence of absorption by gas within a few hundred parsecs of the Sun. The author uses X-ray data from the HEAO-1 LED detectors and H I data from the recent Crawford Hill 21 cm survey to place limits on the 0.6 keV intensity originating within 300 pc of the Sun in the general direction of $(l,b) = (150°,-30°)$.

142.087 **HEAO-1 diffuse soft X-ray sky maps.**
J. A. Nousek, G. P. Garmire, G. Weaver.
NASA Conf. Publ., NASA CP-2345, p. 219 – 221 (1984). – See Abstr. 012.089 (IAU Colloq. No. 81).

Maps of the diffuse soft X-ray background intensity are presented, spanning four energy intervals. The lowest energy interval (0.18 – 0.56 keV) is dominated by local emission, while the next two intervals (0.56 – 1.0 keV and 1.0 – 1.4 keV) reveal more distant and more sharply defined structures. Enlarged maps of several of these structures are presented, including the North Polar Spur, the Galactic Center region and the Eridanus Loop.

142.088 **Ultrasoft X-ray background observations of the local interstellar medium.**
W. T. Sanders, S. L. Snowden, J. J. Bloch, M. Juda, K. M. Jahoda, D. McCammon.
NASA Conf. Publ., NASA CP-2345, p. 222 – 225 (1984). – See Abstr. 012.089 (IAU Colloq. No. 81).

Preliminary results from a May 8, 1984 sounding rocket survey of the soft X-ray background are presented in three soft X-ray bandpasses: 80 – 110 eV, 90 – 188 eV, and 284 – 532 eV. The lowest energy X-rays in this range provide information about the local interstellar medium. The count rate in the 80 – 110 eV energy band (the Be band) tracks the 90 – 188 eV band (the B band) very well, indicating that the same ~1 million degree gas that is responsible for the B band emission may be responsible for the bulk of the Be band X-rays as well. The authors estimate for the flux in the Be band ~1 photon $cm^{-2}s^{-1}sr^{-1}eV^{-1}$, about a factor of four lower than that found by Stern and Bowyer (1979) and Paresce and Stern (1981).

142.089 **A model of the soft X-ray background as a blast wave viewed from inside.**
R. J. Edgar, D. P. Cox.
NASA Conf. Publ., NASA CP-2345, p. 297 – 300 (1984). – See Abstr. 012.089 (IAU Colloq. No. 81).

The suggestion that the soft X-ray background arises in part from the Sun being inside a large (R ~ 100 pc) supernova blastwave is examined by producing models of spherical blastwaves. Such models can produce quantitative fits to both surface brightness and energy band ratios (for the lowest energy bands) when $t \sim 10^5$yr, $E_0 = 5 \times 10^{50}$ergs, and $n_0 \cong 0.004\ cm^{-3}$. Such models can be generalized by varying the relative importance of such factors as thermal conduction, Coulomb heating of electrons, and external pressure; by allowing the explosions to occur in pre-existing cavities with steep density gradients, or by examining the effects of large obstructions or other anisotropies in the ambient medium.

142.090 **Discovery of a 50 millisecond pulsar in the Large Magellanic Cloud.**
F. D. Seward, F. R. Harnden Jr., D. J. Helfand.
Astrophys. J., Lett. Ed., Vol. 287, No. 1, p. L19 – L22 (1984).

The authors have discovered a 50 ms X-ray pulsar in the Crab-like LMC supernova remnant 0540-693. The pulsar is spinning down with $\dot{P} = 4.79 \times 10^{-13} s\ s^{-1}$, implying a characteristic age of 1660 yr and an energy loss rate of $\dot{E} = 1.5 \times 10^{38}$ergs $s^{-1}$, one-third that of the Crab pulsar. The X-ray luminosity of the surrounding nebula is $L_X \approx 10^{37}$ergs $s^{-1}$. Unlike the Crab, the X-ray pulse shape is approximately sinusoidal with one broad maximum per period, and the ratio of pulsed luminosity (in the range 0.2 – 0.4 keV) to $\dot{E}$ is an order of magnitude greater than that of the Crab pulsar.

142.091 **SAS 3 survey of the soft X-ray background.**
F. J. Marshall, G. W. Clark.
Astrophys. J., Vol. 287, No. 2, p. 633 – 652 (1984). With plate 8.

The authors report the results of a survey of the soft X-ray sky in the *C* band (0.10 – 0.28 keV) which provides evidence for a nonlocal origin of a substantial portion of the low-energy X-ray background. The survey, compiled from data obtained with the SAS 3 X-ray observatory, covers 80% of the sky with an angular resolution of 2°9 FWHM and a total exposure of $2.2 \times 10^4 cm^2$s sr. It is found that *C* band counting rates are generally anticorrelated with the column density of neutral hydrogen on all angular scales. A two-component model of the X-ray background is applied in analyzing the data. The local component appears to originate in a hot plasma region surrounding the solar system (a local "hot bubble"). The nonlocal component is probably emission from a galactic corona.

**Catalogue of cataclysmic binaries, low-mass X-ray binaries and related objects. Third edition.**
See Abstr. 002.003.

**The HEAO A-1 X-ray source catalog.**
See Abstr. 002.048.

**Gamma ray astronomy.**
See Abstr. 003.026.

**Extraterrestrische Astronomie.**
See Abstr. 013.022.

**The ROSAT XUV wide field camera.**
See Abstr. 035.037.

**X-ray astronomy satellite Tenma.**
See Abstr. 035.077.

**European X-ray satellite on a successful course.**
See Abstr. 051.021.

**De Europese Exosat: röntgensterrenkunde op eenzame hoogten.**
See Abstr. 051.032.

**The ROSAT all-sky survey in X-rays.**
See Abstr. 051.045.

**On the unique capabilities of the Hubble Space Telescope for studies of ROSAT deep-survey X-ray sources.**
See Abstr. 051.046.

**Europe X-rays the skies.**
See Abstr. 051.052.

**EXOSAT, European X-Ray Astronomy Satellite.**
See Abstr. 051.072.

**EXOSAT observations.**
See Abstr. 051.076.

**Equilibrium emission of low-density, optically thick plasma.**
See Abstr. 063.018.

**Recombination stellar wind model for the coronae of early-type stars.**
See Abstr. 064.005.

**Structure of X-ray irradiated stellar wind.**
See Abstr. 064.050.

**X-ray bursts: nuclear flash model and its difficulties.**
See Abstr. 067.020.

**Probing the interiors of neutron stars.**
See Abstr. 067.051.

**Cyclotron emission near stellar mass black holes.**
See Abstr. 067.094.

**X–ray burst sources – from limit cycles to chaos.**
See Abstr. 067.122.

**Limits from the soft X–ray background on the temperature of old neutron stars and on the flux of superheavy magnetic monopoles.**
See Abstr. 067.145.

**The spectra of X–ray bursting neutron stars.**
See Abstr. 067.147.

**X–ray properties of hot stars. I. On a dichotomy among O–type stars.**
See Abstr. 115.013.

**The change of 4.8 hour period of Cygnus X–3.**
See Abstr. 117.001.

**Clues from the photonuclear time scale on the nature of particle accelerators in Cygnus X–3 and Vela X–1.**
See Abstr. 117.005.

**Sco X–1: a mini quasar?**
See Abstr. 117.007.

**Apsidal motion and the pulsating transient X–ray source A0538–66.**
See Abstr. 117.009.

**Cygnus X–1 revisited.**
See Abstr. 117.010.

**Spectral variability in early–type binary X–ray systems.**
See Abstr. 117.014.

**EXOSAT observations of H2215–086: detection of the X–ray pulse period.**
See Abstr. 117.030.

**Low–level radio flares from Cygnus X–3.**
See Abstr. 117.032.

**Two X–ray sources model of Cyg X–3.**
See Abstr. 117.035.

**Search for the long–term light variations of AO Psc = H2252–035.**
See Abstr. 117.049.

**Optical photometry of massive X–ray binaries: SMC X–1/Sk 160.**
See Abstr. 117.051.

**Evidence for variations in the phase–dependent ionization structure of the stellar wind in Vela X–1.**
See Abstr. 117.071.

**IUE observations of the X–ray binary A0538–66: recent results on a remarkable recurrent.**
See Abstr. 117.072.

**Scattering/absorption dips in 4U1755–33. Evidence for disc oscillations?**
See Abstr. 117.075.

**Identification of 3A 0729 + 103 with an intermediate polar–type cataclysmic variable.**
See Abstr. 117.082.

**Einstein observations of selected close binaries and shell stars.**
See Abstr. 117.089.

**A0538–66 during the extended 'off' state: evidence for continuing mass motions.**
See Abstr. 117.091.

**Photometry of A0538–66 during an active and subsequent inactive state.**
See Abstr. 117.092.

**Ultraviolet high–resolution spectroscopy of the X–ray binary Sk 160/SMC X–1.**
See Abstr. 117.096.

**High–resolution soft X–ray spectra of Scorpius X–1: the structure of circumsource accreting material.**
See Abstr. 117.098.

**L'observation des disques d'accrétion dans les sources X binaires à haute luminosité.**
See Abstr. 117.106.

**SMC X–1 variability observed from *HEAO 1*.**
See Abstr. 117.110.

**New evidence on the nature of the neutron star and accretion flow in Vela X–1 from pulse timing observations.**
See Abstr. 117.114.

**Neutron stars in interacting binary systems.**
See Abstr. 117.125.

**Variable polarization in X–ray binaries: HD 153919 (= 4U1700–37) and HD 152667 (= OAO1653–40?).**
See Abstr. 117.131.

**The accretion disk spectrum of LMC X–3.**
See Abstr. 117.133.

**Hard X–ray temporal and spectral observation of an active state of Sco X–1.**
See Abstr. 117.137.

**Is Cygnus X–3 a monoenergetic $10^{17}$eV accelerator?**
See Abstr. 117.144.

**Cosmic–ray emission from Cygnus X–3.**
See Abstr. 117.145.

**An exploratory eccentric orbit "Roche lobe" overflow model for recurrent X–ray transients.**
See Abstr. 117.172.

**The precession angle of the optical component of the Cygnus X–1 system.**
See Abstr. 117.174.

**Optical and X–ray observations of 4U2129 + 47/V1727 Cyg in a quiescent state.**
See Abstr. 117.214.

**Simultaneous EXOSAT and optical observations of the pulsing X–ray binary H2252–035.**
See Abstr. 117.215.

**EXOSAT observations of the $35^d$ intensity variations of Her X–1.**
See Abstr. 117.216.

**Pulse phase spectroscopy of Her X–1 with EXOSAT.**
See Abstr. 117.217.

**Fine structure in the absorption dips of Her X–1 as observed by EXOSAT.**
See Abstr. 117.218.

**Simultaneous EXOSAT and optical observations of the pulsing X–ray binary H2252–035/AO Psc.**
See Abstr. 117.219.

**Spectral studies of low–mass X–ray binaries observed by EXOSAT.**
See Abstr. 117.220.

**EXOSAT observations of 4U1705–44.**
See Abstr. 117.221.

**On the origin of X–ray variability of SS 433.**
See Abstr. 117.229.

**Short–term X–ray variability of GX 339–4.**
See Abstr. 117.230.

**Energy spectra and time variations of the X–ray source Cygnus X–2.**
See Abstr. 117.250.

**Dramatic variability in the X–ray flux of Hercules X–1 preceding the binary eclipse.**
See Abstr. 117.259.

**The 41.5 day binary X–ray pulsar 4U 1223–62 (GX 301–2).**
See Abstr. 117.262.

**Radial velocity of HD 77581 (Vela X–1) in the period of 1972 – 1983.**
See Abstr. 119.095.

**X–ray emission from supernova remnants. II. The effect of remnant structure on nonequilibrium X–ray spectra.**
See Abstr. 125.041.

**X–ray emission from supernova remnants MSH 14–63 and RCW 103.**
See Abstr. 125.042.

**An X–ray investigation of the unusual supernova remnant CTB 80.**
See Abstr. 125.071.

**An optical synchrotron nebula around the X–ray pulsar 0540–693 in the Large Magellanic Cloud.**
See Abstr. 125.082.

**Einstein observations of the Crab nebula pulsar.**
See Abstr. 126.034.

**The hard X–ray light–curves of accreting magnetized white dwarfs.**
See Abstr. 126.068.

**A new Crab–like pulsar in the Large Magellanic Cloud.**
See Abstr. 126.088.

**On interstellar [Fe X] absorption toward Cepheus OB2.**
See Abstr. 131.149.

**The nature of the soft X–ray emitting region in the direction of the North Polar Spur.**
See Abstr. 131.265.

**Inverse relations between 0.25 keV counts and local interstellar dust.**
See Abstr. 131.266.

**A neutral gas jet in a low velocity shock front at the boundary of the Draco Nebula.**
See Abstr. 131.270.

**Non equilibrium ionization in the local hot bubble.**
See Abstr. 131.280.

**Where cosmic rays come from?**
See Abstr. 143.001.

**Evidence for an increasing 59–s period in the X–ray emission from Geminga.**
See Abstr. 143.002.

**Exosat observation of the candidate X–ray counterpart of Geminga.**
See Abstr. 143.003.

**Binary star model of the γ–ray burst with special reference to that on 5 March, 1979 and subsequent bursts on 6 March, 4 April, and 24 April, 1979.**
See Abstr. 143.009.

**Gammastrahlung extrem hoher Energie von Cygnus X–3.**
See Abstr. 143.011.

**Implications of the observation of ultra–high energy γ–rays. I. The production of diffuse X–rays.**
See Abstr. 143.032.

**3 keV to 2 MeV observations of four gamma–ray bursts.**
See Abstr. 143.045.

**The Einstein X–ray survey of the Pleiades revisited.**
See Abstr. 153.052.

**X–ray sources in globular clusters.**
See Abstr. 154.028.

**Observations of the stellar radiation field at 975 Å.**
See Abstr. 155.025.

**Vergleich zwischen neuen X–ray Messungen und radio-astronomischen Beobachtungen an prominenten galaktischen Strukturen.**
See Abstr. 155.070.

**Limits on soft X–ray flux from distant emission regions.**
See Abstr. 155.081.

**The complex X–ray emission of NGC 253.**
See Abstr. 157.229.

**Very low luminosity active galaxies and the X–ray background.**
See Abstr. 158.094.

**X–ray selected quasars and Seyfert galaxies: cosmological evolution, luminosity function, and contribution to the X–ray background.**
See Abstr. 158.095.

**Hard X–ray variability of three active galactic nuclei.**
See Abstr. 158.107.

**Optical studies of X–ray selected AGN in the direction of the LMC.**
See Abstr. 158.120.

**X–ray emission from active galactic nuclei.**
See Abstr. 158.127.

**EXOSAT observations of active galactic nuclei.**
See Abstr. 158.128.

**Soft X–ray emission from BL Lac objects.**
See Abstr. 158.129.

**X–ray emission from PKS 2155–304.**
See Abstr. 158.130.

**X–ray emission from Centaurus A.**
See Abstr. 158.131.

**X-ray emission from Centaurus A observed by TENMA.**
See Abstr. 158.132.

**The X-ray spectra of NGC 4151.**
See Abstr. 158.133.

**On the surface density of X-ray selected BL Lacertae objects.**
See Abstr. 158.135.

**On the surface density of X-ray selected BL Lacertae objects.**
See Abstr. 158.155.

**X-ray emission from M82.**
See Abstr. 158.240.

**The extragalactic nature of H0323+022.**
See Abstr. 158.266.

**The high energy X-ray spectrum of 3C 273.**
See Abstr. 159.011.

**The properties of X-ray emission of quasars.**
See Abstr. 159.051.

**X-ray surveys of quasars with the Einstein Observatory.**
See Abstr. 159.073.

**One to eight month X-ray light curves for 4 active extragalactic objects.**
See Abstr. 159.074.

**The high energy X-ray spectrum of 3C 273.**
See Abstr. 159.075.

**An X-ray survey of variable radio bright quasars.**
See Abstr. 159.081.

**The X-ray to optical luminosity ratio of optically and X-ray selected quasars.**
See Abstr. 159.098.

**High resolution radio and X-ray observations of A 115.**
See Abstr. 160.023.

**X-ray emission from Stephan's Quintet and other compact groups.**
See Abstr. 160.043.

**Einstein images of clusters of galaxies: galaxy haloes, the intracluster medium, and the intercluster gas.**
See Abstr. 160.119.

**Cluster X-ray luminosity and first ranked galaxy size.**
See Abstr. 160.120.

**The radial X-ray brightness profile of the Coma cluster.**
See Abstr. 160.145.

**Endpoints of stellar evolution: X-ray surveys of the Local Group.**
See Abstr. 160.146.

# 143 Gamma-ray Sources, Gamma-ray Background

143.001 **Where cosmic rays come from?**
J. Maddox.
Nature, Vol. 310, No. 5977, p. 447 (1984).

This note discusses recent suggestions that gamma-ray stars like Geminga, which are also X-ray emitters, may be copious sources of cosmic ray particles.

143.002 **Evidence for an increasing 59-s period in the X-ray emission from Geminga.**
G. F. Bignami, P. A. Caraveo, J. A. Paul.
Nature, Vol. 310, No. 5977, p. 464 - 469 (1984).

Einstein Observatory and Exosat X-ray data from 1E0630+178 reveal a $\sim 50\%$ periodic emission at $\geqslant 59$ s. This confirms the earlier SAS 2 and COS B findings on 2CG195+4 at $E > 50$ MeV, and provides the temporal signature for the identification of Geminga. The observations, spanning over a decade (1973 - 83), also suggest a very high positive period derivative, $\dot{P}$, which has become still higher between 1979 and 1983. A simple source scenario is proposed, based on a binary system containing a compact object responsible for the high energy emission, and also taking into account recent ultra high energy $\gamma$-ray measurements.

143.003 **Exosat observation of the candidate X-ray counterpart of Geminga.**
P. A. Caraveo, G. F. Bignami, P. Giommi, S. Mereghetti, J. A. Paul.
Nature, Vol. 310, No. 5977, p. 481 - 483 (1984).

During a search for X-ray counterparts of the COS B $\geqslant 100$ MeV $\gamma$-ray sources, observations using the Einstein Observatory revealed a new X-ray source, 1E0630+178, inside the $\sim 1/2$ square degree error box of the $\gamma$-ray source 2CG195+4, 'Geminga'. This source has been proposed as the X-ray counterpart of Geminga. The region of Geminga was observed with the ESA Exosat satellite on 9 - 10 September 1983. This letter presents time profiles of the X-ray emission observed with the Channel Multiplier Arrays.

143.004 **The production of gamma-rays in gamma-ray burst sources.**
S. Bonazzola, J. M. Hameury, J. Heyvaerts, J. P. Lasota.
Astron. Astrophys., Vol. 136, No. 1, p. 89 - 97 (1984).

The authors propose, as a model for gamma-ray bursts, that the energy released by the thermonuclear runaway at the surface of a magnetized neutron star is converted into gamma-ray photons by small scale magnetic field reconnection. Most of the energy flux is transported up to the atmosphere by Alfvén waves, driven by oscillatory magnetoconvection. The electric field created by reconnection accelerates electrons to very high energies. The excitation of Landau levels by collisions is followed by gamma-ray synchrotron emission. The time scales, energies, and luminosities involved agree with the observations.

143.005 **De gammahemel geeft enkele geheimen prijs.**
J. B. G. M. Bloemen.
Zenit, 11. Jaarg., No. 7/8, p. 306 - 313 (1984).

143.006 **Balloon measurements of the diffuse gamma-ray component above 5 MeV.**
A. F. Iyudin, V. G. Kirillov-Ugryumov, Yu. D. Kotov, Yu. V. Smirnov, V. G. Tyshkevich, V. N. Yurov, L. V. Kurnosova, M. I. Fradkin.
Sov. Astron. Lett., Vol. 10, No. 1, p. 10 - 13 (1984). English translation of 37.143.001.

143.007 **Balloon observations of the 5 – 100 MeV gamma rays from the Crab nebula.**
A. F. Iyudin, V. G. Kirillov–Ugryumov, Yu. D. Kotov, Yu. V. Smirnov, V. N. Yurov, L. V. Kurnosova, M. I. Fradkin, S. V. Damle, G. S. Gokhale, P. K. Kute, B. V. Sreekantan.
Sov. Astron. Lett., Vol. 10, No. 1, p. 41 – 43 (1984). English translation of 37.143.014.

143.008 **De gammahemel geeft enkele geheimen prijs (2).**
J. B. G. M. Bloemen.
Zenit, 11. Jaarg., No. 9, p. 356 – 358, 360 – 364 (1984).

143.009 **Binary star model of the γ–ray burst with special reference to that on 5 March, 1979 and subsequent bursts on 6 March, 4 April, and 24 April, 1979.**
L. M. Singh, H. L. Duorah.
Astrophys. Space Sci., Vol. 102, No. 2, p. 235 – 245 (1984).
The peculiar γ–ray burst phenomenon of 5 March, 1979, and the other subsequent bursts on 6 March, 4 April, and 24 April, 1979, are studied, using the physically more realistic exponentially increasing accretion rate on a highly magnetized neutron star from its companion, and the conclusions that pycnonuclear reaction flash for the first and thermonuclear flashes for the subsequent bursts as the most probable model for this series of bursts, are made. The authors further conclude that a huge γ–ray burst is a sequel to rapid X–ray transient or type–I X–ray bursts. A rough estimate gives that the next burst will occur within ~0.5 yr since 24 April, 1979.

143.010 **Spectral feature of 31 December 1981 γ–ray burst not confirmed.**
P. L. Nolan, G. H. Share, E. L. Chupp, D. J. Forrest, S. M. Matz.
Nature, Vol. 311, No. 5984, p. 360 – 362 (1984).
The emission lines reported at 400 – 500 keV in the spectra of γ–ray bursts have been interpreted as redshifted positron annihilation lines, and, together with the lower–energy cyclotron lines, are the most direct evidence that bursts occur on the surface of neutron stars. So far, the most significant detections of emission features have been reported from the Konus experiments on Veneras 11–14. Here the authors compare measurements of a γ–ray burst at 01:37 UT on 31 December 1981, using the SMM γ–ray spectrometer with those made by the Konus instruments. The SMM spectra exhibit no evidence for the presence of emission features reported by the Konus group.

143.011 **Gammastrahlung extrem hoher Energie von Cygnus X–3.**
M. Samorski, W. Stamm.
Phys. Bl., 40. Jahrg., Heft 10, p. 329 – 331 (1984).

143.012 **Gamma rays from atomic and molecular gas in the large complex of clouds in Orion and Monoceros.**
J. B. G. M. Bloemen, P. A. Caraveo, W. Hermsen, F. Lebrun, R. J. Maddalena, A. W. Strong, P. Thaddeus.
Astron. Astrophys., Vol. 139, No. 1, p. 37 – 42 (1984).
COS–B gamma–ray observations of the large complex of interstellar clouds in Orion and Monoceros are compared with the Columbia CO survey and the Berkeley H I surveys of this region. A good correlation between the gamma–ray emission and the total gas distribution is found. The observed gamma–ray emission can be explained in terms of interactions of cosmic rays distributed uniformly in this region, with the interstellar gas. This correlation is used to calibrate the ratio between the $H_2$ column density and the integrated CO line intensity: $N(H_2)/W_{CO} = (2.6 \pm 1.2) \times 10^{20}$ mol cm$^{-2}$K$^{-1}$km$^{-1}$s, consistent with the value derived from a similar analysis for the inner Galaxy.

143.013 **Resonant absorption, hot electrons, and cosmic gamma–ray bursts.**
E. P. Liang.
Astrophys. J., Lett. Ed., Vol. 283, No. 1, p. L21 – L23 (1984).
The author proposes that resonant absorption of intense electromagnetic waves impinging onto a neutron star surface during a magnetic flare is the source of the hot electrons whose synchrotron radiation leads to γ–ray bursts. Parameters of the emission layer derived from first principles generally agree with empirical values.The luminosity–temperature scaling law observed by the KONUS experiment, if confirmed, may also be explained by this picture.

143.014 **Rapid spectral variability of cosmic gamma–ray bursts.**
I. G. Mitrofanov, V. Sh. Dolidze, C. Barat, G. Vedrenne, M. Niel, K. Hurley.
Astron. Zh., Tom 61, Vyp. 5, p. 939 – 943 (1984). In Russian. English translation in Sov. Astron., Vol. 28, No. 5.
On the basis of the Soviet–French experiment SIGNE–2M carried out at the space probes Venera 11 and 12 it is established that the continuum of cosmic gamma–ray bursts varies at a time–scale of 250 ms, and the annihilation radiation is generated in the form of short strong pulses. The obtained spectral data allow to draw important conclusions on the nature of the sources of gamma–ray bursts.

143.015 **Physics of gamma–ray bursts.**
D. Q. Lamb.
Ann. N.Y. Acad. Sci., Vol. 422, p. 237 – 281 (1984). – See Abstr. 012.033.
It is now widely believed that gamma–ray bursts come from strongly magnetic neutron stars. This review describes current models of the bursts involving neutron stars. It discusses the physical properties of the emission region and the radiation processes expected in strong magnetic fields. It is emphasized, in particular, that the observed burst spectra require the emission region to be optically thin. It is shown that synchrotron emission is favored as the gamma–ray production mechanism, since for moderate or strong magnetic fields it is far more efficient than bremsstrahlung, inverse Compton scattering, and pair annihilation, and it is the only mechanism capable of satisfying the constraint on optical thinness while producing the observed luminosity. The paper concludes by mentioning several important issues that any attempt to explain gamma–ray bursts must face.

143.016 **Active galaxies and the diffuse gamma–ray background.**
D. Kazanas, R. J. Protheroe.
Ann. N.Y. Acad. Sci., Vol. 422, p. 358 (1984). Abstract. – See Abstr. 012.033.

143.017 **On the nature of cosmic gamma–ray bursts.**
I. G. Mitrofanov.
Astrophys. Space Sci., Vol. 105, No. 2, p. 245 – 251 (1984).
It is shown that cosmic gamma–ray bursts may be generated when the magnetospheres of single neutron stars are excited by strong starquakes. Important new observational data are explained, and several verifiable predictions are proposed on the basis of this approach.

143.018 **Gamma–ray bursts observed from the Hinotori satellite.**
M. Yoshimori, K. Okudaira, Y. Hirasima, I. Kondo.
Astrophys. Space Sci., Vol. 105, No. 2, p. 379 – 392 (1984).
During February, 1981 and June, 1982 the gamma–ray and the hard X–ray spectrometers on the Hinotori satellite observed four gamma–ray bursts. These gamma–ray bursts were simultaneously observed by other satellites. The time histories and energy spectra are shown for these gamma–ray bursts, and the burst sizes (erg cm$^{-2}$) are estimated. Two possible source locations for the burst of 21 July, 1981 are roughly determined from arrival time delays between two pairs of satellites, PVO–Hinotori and ISEE–3–Hinotori. The weak gamma–ray line peak structure around 1.8 MeV was observed for the burst of 13 March, 1982. The line could be interpreted in terms of gravitationally redshifted neutron capture line at 2.22 MeV.

143.019 **On the nature of the galactic gamma–ray sources.**
G. E. Morfill, M. Forman, G. Bignami.
Astrophys. J., Vol. 284, No. 2, p. 856 – 868 (1984).
Available data constrain possible models of the gamma–ray point (or pointlike) sources found by COS B. Apart from the pulsars (Crab, Vela) and possibly the quasar 3C 273, which are

"compact" objects, and which have been identified at other photon energies, there are some 20–odd other sources with no accepted or acceptable counterpart, as yet. This class of "gamma-ray objects" may be due to cosmic ray interaction with aggregates of diffuse interstellar matter – i.e., clouds. The constraints on this hypothesis, both observationally and theoretically, are discussed for various scenarios.

143.020 **Search for $10^{15}$eV gamma–rays from the Crab pulsar and surrounding regions.**
J. Boone, R. Cady, G. L. Cassiday, J. W. Elbert, E. C. Loh, P. Sokolsky, D. Steck, S. Wasserbaech.
Astrophys. J., Vol. 285, No. 1, p. 264 – 268 (1984).

During 1980 December and 1981 February, atmospheric Cerenkov light flashes were oberved in a broad region of the sky including the Crab pulsar. The 1981 February data gave a flux upper limit of $5.3 \times 10^{-13}cm^{-2}s^{-1}$ for $10^{15}$eV γ–rays within 3°.5 of the Crab pulsar. Comparable limits were set over a declination band $\delta = 0°–70°$ and a right–ascension interval from approximately $0^h$ to $13^h$. A weak (3.1 σ) excess from the Crab vicinity was detected on 1980 December 9.

143.021 **The identification of an inverse Compton interaction contribution to the intensity of cosmic gamma rays.**
P. A. Riley, A. W. Wolfendale.
J. Phys. G, Vol. 10, No. 8, p. 1149 – 1161 (1984). Abstr. in Phys. Abstr., Vol. 87, No. 1243, Entry 101088 (1984).

143.022 **Detection of possible optical flashes from the γ–ray burst source GBS0526–66.**
H. Pedersen, J. Danziger, K. Hurley, G. Pizzichini, C. Motch, S. Ilovaisky, N. Gradmann, W. Brinkmann, G. Kanbach, E. Rieger, C. Reppin, W. Trumper, N. Lund.
Nature, Vol. 312, No. 5989, p. 46 – 48 (1984).

The unidentified source of the γ–ray burst of 5 March 1979, GBS0526–66, may be located in the supernova remnant N49 in the Large Magellanic Cloud. The authors report that GBS0526–66 was monitored for ~910 h, using a high speed photoelectric photometer, attached to a 50 cm telescope at ESO/La Silla. Three short optical flashes were found, which may possibly be related to the γ–ray burst phenomenon. The most intense of these occurred on 1984 February 8. It reached a maximum brightness corresponding to $m$ (visual) = 8.7. The shape of the light curve bears remarkable similarity to the initial 200 ms of the 1979 March γ–ray burst.

143.023 **Is Geminga a very close neutron star binary?**
P. E. J. Nulsen, A. C. Fabian.
Nature, Vol. 312, No. 5989, p. 48 – 50 (1984).

Geminga (2CG195+4) is one of the brightest sources in the sky at photon energies around 1 GeV. It has been tentatively identified with an unusual soft X–ray and optical object for which the γ–ray–to–X–ray–to–optical luminosity ratios are $\sim 10^6:10^3:1$. This letter demonstrates that the observed properties of Geminga could be explained by a close binary model containing a neutron star if the system is located at a distance of 10 pc.

143.024 **A 164–day period in γ–ray bursts from GBS0526–66?**
R. E. Rothschild, R. E. Lingenfelter.
Nature, Vol. 312, No. 5996, p. 737 – 740 (1984).

The γ–ray burst history of GBS0526–66 is unique among observed burst sources being that of the brightest burst on record (5 March 1979) and one of only two from which recurrent bursts (at least 15 of similar temporal and spectral pattern) have been observed. The authors have found evidence for a pattern in the recurrence times of these bursts, showing an apparent period of 164 day, which they suggest reflects periodic accretion to a neutron star from an eccentric ($e \sim 0.9$) binary companion. It is located within the supernova remnant N49 in the LMC.

143.025 **Energy dependence of the correlation of cosmic gamma–ray intensity with atomic and molecular gas in the ISM.**
P. A. Riley, A. W. Wolfendale.
J. Phys. G, Vol. 10, No. 9, p. 1269 – 1281 (1984). Abstr. in Phys. Abstr., Vol. 87, No. 1245, Entry 110409 (1984).

143.026 **Gamma astronomy.**
J. Paul.
Ciel Espace, No. 198, p. 27 – 30 (1984). In French. Abstr. in Phys. Abstr., Vol. 87, No. 1246, Entry 115836 (1984).

143.027 **Evidence for a binary system at the 5 March 1979 gamma–ray burst position.**
R. E. Rothschild, R. E. Lingenfelter.
Bull. Am. Astron. Soc., Vol. 16, No. 3, Part I, p. 731 (1984). Abstract. – See Abstr. 010.065.

143.028 **Absorption of high–energy gamma rays by photon–photon interactions near a star.**
L. H. Ford.
Mon. Not. R. Astron. Soc., Vol. 211, No. 3, p. 559 – 561 (1984).

The absorption of high–energy γ rays in the $10^{12}$eV energy range due to photon–photon interactions ($\gamma\gamma \rightarrow e^+e^-$) near a star is discussed. It is found that such γ rays passing near the Sun will have an absorption probability of a few per cent. In the case of hotter stars, there will be a region around the star up to $10^2$ stellar radii across which is optically thick to the passage of high–energy gamma rays. The possible application of these results to high–energy gamma–ray sources in binary systems, such as Cygnus X–3, is discussed.

143.029 **Gamma ray bursts: an observational overview.**
G. Vedrenne.
Problems of collapse and numerical relativity, p. 311 – 323 (1984). – See Abstr. 012.061.

New observational results on γ–ray bursts are discussed. The paper examines in particular their classification, evidence for spectral variability of both the continuum and spectral features, and their $\ln N(>S) - \ln S$ distribution.

143.030 **The theory of gamma–ray bursts.**
S. E. Woosley.
Problems of collapse and numerical relativity, p. 325 – 343 (1984). – See Abstr. 012.061.

A γ–ray burst occurs when a strongly magnetic neutron star experiences either a thermonuclear explosion in degenerate material accumulated over a long period of time or the sudden, greatly super–Eddington accretion of matter. Certain aspects of the thermonuclear model are briefly reviewed and special attention is devoted to mechanisms for producing the hard spectrum of γ–ray bursts. A new model for the production of optical flashes accompanying γ–ray bursts is also developed.

143.031 **Gamma–ray bursts: a model.**
J. M. Hameury, S. Bonazzola, J. Heyvaerts, J. P. Lasota.
Problems of collapse and numerical relativity, p. 345 – 360 (1984). – See Abstr. 012.061.

The authors propose, as a model for gamma–ray bursters, that the energy released by a thermonuclear runaway under the surface of a slowly accreting, magnetized, neutron star, is transported up to the atmosphere by Alfvén waves, driven by overstable magnetoconvection. In the atmosphere, the short scale magnetic field lines reconnection generates an electric field. The electrons accelerated by this process produce gamma–rays by inverse Compton scatterings with blackbody photons of a few keV.

143.032 **Implications of the observation of ultra–high energy γ–rays. I. The production of diffuse X–rays.**
N. C. Rana, M. Sadzinska, J. Wdowczyk, A. W. Wolfendale.
Astron. Astrophys., Vol. 141, No. 2, p. 394 – 400 (1984).

Evidence has recently appeared for the existence of sources of cosmic γ–rays in the Galaxy from which energies as high as $10^{16}$eV are detected, of the sources Cygnus X–3 being the most definite. The authors determine the order of magnitude of the expected intensity of X-rays at high galactic latitudes produced by electron synchrotron radiation in the galactic magnetic field, the electrons arising from the collisions of cosmic γ-rays above about $10^{15}$eV with the microwave background. It is found that for even just one source of the strength of the contemporary Cygnus X-3 source emitting continuously in the Galaxy the X-ray flux at $b = 90°$ is significant for X-ray energies above about 100 keV.

143.033 **Coordinated observations of GBS 0526-66.**
IAU Circ., No. 4008 (1984).

143.034 **On possible optical counterparts of the gamma burst sources GB 790325 and GBS 1703+01.**
M. Scholz.
Inf. Bull. Variable Stars, No. 2615, 1 p. (1984).

143.035 **On the background of hard gamma radiation from a coding collimator of a gamma-ray telescope under cosmic experiment conditions.**
A. P. Aleksandrov, A. N. Berezovoj, A. M. Gal'per, V. M. Grachev, V. V. Dmitrenko, V. G. Kirillov-Ugryumov, V. V. Lebedev, V. A. Lyakhov, A. A. Moiseev, S. E. Ulin, N. I. Shchvets.
Kosm. Issled., Tom 22, Vyp. 6, p. 950 - 951 (1984). In Russian. English translation in Cosm. Res.

143.036 **L'astronomia dei raggi gamma: nuova finestra sull'universo.**
A. W. Wolfendale.
Orione, Vol. 4, N. 4, p. 150 - 161 (1984).

143.037 **On the morphology and spectra of the short gamma-ray bursts.**
C. Barat, R. I. Hayles, K. Hurley, M. Niel, G. Vedrenne, I. V. Estulin (*I. V. Ehstulin*), V. M. Zenchenko.
Astrophys. J., Vol. 285, No. 2, p. 791 - 800 (1984).

The rise and decay time characteristics and the energy spectra of 20 short, single-peaked γ-ray bursts detected by the Franco-Soviet Signe instruments are analyzed in detail. It is found that the time histories are well described by simple exponential functions, with *e*-folding rise times ≈*e*-folding decay times in general. Both the rise and decay times extend over a range of several orders of magnitude, suggesting that there is no "standard" burst profile, and that there is probably a wide variety of physical conditions present in different bursters.

143.038 **Two probable optical flashes from gamma-ray bursters.**
B. E. Schaefer, H. V. Bradt, C. Barat, K. Hurley, M. Niel, G. Vedrenne, T. L. Cline, U. Desai, B. J. Teegarden, W. D. Evans, E. E. Fenimore, R. W. Klebesadel, J. G. Laros, I. V. Estulin (*I. V. Ehstulin*), A. V. Kuznetsov.
Astrophys. J., Lett. Ed., Vol. 286, No. 1, p. L1 - L4 (1984). With plate L1.

The authors have located two images on archival photographic plates which are most likely records of optical flashes from γ-ray bursters (GRBs). One of these images appears on a 1901 plate in the field of the 1979 November 5 GRB, while the other is in the field of the 1979 January 13 GRB on a plate exposed in 1944. Independent limits on the optical flash rate from the sky yield a probability of less than $10^{-4}$ that any one of the optical transients is due to a background flash. A total exposure of 2.7 years has been examined for GRB flashes at known GRB locations on the Harvard plates, and three GRB flashes have been seen which implies that the average recurrence time scale for optical flashes is roughly 1 year.

143.039 **1979 January 13: an intense gamma-ray burst with a possible associated optical transient.**
C. Barat, K. Hurley, M. Niel, G. Vedrenne, T. Cline, U. Desai, B. Schaefer, B. Teegarden, W. D. Evans, E. E. Fenimore, R. Klebesadel, J. G. Laros, I. V. Estulin (*I. V. Ehstulin*), V. M. Zenchenko, A. V. Kuznetsov, V. G. Kurt, S. Ilovaisky, C. Motch.
Astrophys. J., Lett. Ed., Vol. 286, No. 1, p. L5 - L9 (1984). With plate L2.

The time history, energy spectra, and precise localization of the intense 1979 January 13 γ-ray burst, observed by five spacecraft in the interplanetary network, are presented. The time history displays a pulse-afterpulse structure suggestive of periodicity. Fine time resolution spectral analysis shows that a wide variety of models can be used to fit individual spectra, while spectra integrated over longer periods are well fitted by a thermal synchrotron law. The precise localization may be consistent with an archival optical transient having no quiescent counterpart down to 22d mag on recent plates.

143.040 **Possible short annihilation flashes in the 1978 November 4 gamma-ray burst.**
C. Barat, K. Hurley, M. Niel, G. Vedrenne, I. G. Mitrofanov, I. V. Estulin (*I. V. Ehstulin*), V. M. Zenchenko, V. Sh. Dolidze.
Astrophys. J., Lett. Ed., Vol. 286, No. 1, p. L11 - L13 (1984).

The Franco-Soviet Signe experiments on the Venera 11 and Venera 12 spacecraft allow a spectral analysis of γ-ray bursts with a time resolution of 250 ms. Several short (<250 ms) and intense (up to 28% of the continuum energy) spectral features appear in the spectra of the 1978 November 4 event around 400 keV. The authors interpret these features as possible $e^-/e^+$ pair annihilation flashes.

143.041 **Three precise gamma-ray burst source locations.**
T. L. Cline, U. D. Desai, B. J. Teegarden, C. Barat, K. Hurley, M. Niel, G. Vedrenne, W. D. Evans, R. W. Klebesadel, J. G. Laros, I. V. Estulin (*I. V. Ehstulin*), A. V. Kuznetsov, V. M. Zenchenko, V. G. Kurt, B. E. Schaefer.
Astrophys. J., Lett. Ed., Vol. 286, No. 1, p. L15 - L18 (1984).

The precise source regions of three moderately intense γ-ray bursts are derived. These events were observed with the first interplanetary burst sensor network on 1978 November 24 and 1979 November 5 and 16. The optimum locations of the detectors, widely separated throughout the inner solar system, allowed for high-accuracy, overdetermined source fields of size 0.7 - 7 $\text{arcmin}^2$. All three locations are at fairly high galactic latitude in regions of low source confusion; none can be identified with a steady source object.

143.042 **Investigation of gamma-ray bursts with the Prognoz automatic stations.**
I. V. Ehstulin.
Solar activity investigations and the Prognoz space system, p. 185 - 196 (1984). In Russian. Abstr. in Ref. Zh., 62. Issled. Kosm. Prostranstva, 12.62.329 (1984). - See Abstr. 003.038.

143.043 **Fluxes and distances of gamma-ray bursts.**
B. J. Carrigan, J. I. Katz.
Astron. Express, Vol. 1, Issue 3, p. 89 - 95 (1984).

The authors calculate γ-γ pair production and annihilation in γ-ray burst source regions. These processes do not produce effective collimation of the flux above pair production threshold, but when repeated $\sim\alpha^{-1}$ times higher order processes soften the spectrum and permit escape of intense fluxes.

143.044 **Generation of γ-bursts by old magnetic neutron stars.**
V. V. Usov.
Astrophys. Space Sci., Vol. 107, No. 1, p. 191 - 197 (1984).

A model of γ-bursts is considered that treats the flares of neutron stars as a result of convective-oscillation instability associated with the stars having strong internal magnetic fields ($\sim 10^{13}$ to $10^{14}$G). In the context of this model only sufficiently old ($10^4$ to $10^7$yr), drastically cooled-down neutron stars may be sources of γ-bursts. The paper shows that major characteristics of a γ-burster in the supernova N 49 remnant (energy release during burst up to $\sim 10^{44}$erg, age $\sim 10^4$yr, burst-to-burst interval $\sim(1$ to $3)\times 10^6$s; rotation period P = 8 s) may be explained under the assumption that the mass of the neutron star is about 0.14 $M_\odot$ while its mean magnetic field strength is $\sim 1.5\times 10^{14}$G and $\sim 10^{13}$G within the star and on its surface, respectively. The observational tests of the model discussed conclude the paper.

143.045 **3 keV to 2 MeV observations of four gamma–ray bursts.** J. G. Laros, W. D. Evans, E. E. Fenimore, R. W. Klebesadel, S. Shulman, G. Fritz.
Astrophys. J., Vol. 286, No. 2, p. 681 - 690 (1984).

Four cataloged γ–ray bursts that occurred between 1979 March 7 and 1979 July 31 have been observed over the 3 - 10 keV range by a joint Naval Research Laboratory/Los Alamos experiment on the Air Force *P78–1* satellite. The bursts also were well observed above ~100 keV by members of the interplanetary network. Additionally, the 1979 July 31 event (GB 790731) was observed from 25 keV to 2 MeV by an experiment on the *ISEE 3*. In this paper the authors present hardness ratios, X–ray/γ–ray luminosity ratios, time histories, and the GB 790731 spectrum.

**Report on the gamma–ray astronomy papers presented at the 18th International Cosmic–Ray Conference, Bangalore, India, September, 1983.**
See Abstr. 011.012.

**Extraterrestrische Astronomie.**
See Abstr. 013.022.

**Die Geminga–Story.**
See Abstr. 013.039.

**PULSAR: a balloon–borne experiment to detect variable low energy gamma–ray emissions.**
See Abstr. 034.013.

**A camera for very high energy gamma–ray astronomy.**
See Abstr. 034.059.

**Statistical analysis of observations in gamma–ray astronomy.**
See Abstr. 036.003.

**An intercontinental baseline coincidence search for optical flashes with two Schmidt telescopes.**
See Abstr. 036.043.

**Statistical analysis in gamma–ray astronomy: the upper limit.**
See Abstr. 036.168.

**Statistical reliability of X–ray and gamma–ray line observations.**
See Abstr. 036.169.

**Resonant nuclear fusion processes and the gamma rays of SS433.**
See Abstr. 061.007.

**The production of very high energy photons and neutrinos from cosmic proton sources.**
See Abstr. 061.096.

**Gamma–ray line emission, nuclear destruction and neutron production in hot astrophysical plasmas. The deuterium boiler as a gamma–ray source.**
See Abstr. 062.044.

**Magnetic flare model of γ–ray bursts.**
See Abstr. 067.001.

**A numerical model of an explosion near the surface of a neutron star and gamma–ray bursts.**
See Abstr. 067.011.

**The change of 4.8 hour period of Cygnus X–3.**
See Abstr. 117.001.

**Clues from the photonuclear time scale on the nature of particle accelerators in Cygnus X–3 and Vela X–1.**
See Abstr. 117.005.

**Cygnus X–1 revisited.**
See Abstr. 117.010.

**Nuclear processes in the jets of SS 433.**
See Abstr. 117.099.

**On the bizarre gamma–ray spectrum of SS 433.**
See Abstr. 117.113.

**Self–absorption of high–energy gamma–rays in Cygnus X–3.**
See Abstr. 117.257.

**γ rays of 0.3 - 30 MeV from PSR 0833–45.**
See Abstr. 126.002.

**High–energy gamma rays from pulsars.**
See Abstr. 126.040.

**Search for periodic low–energy gamma–ray emission from pulsar PSR 0531 + 21.**
See Abstr. 126.059.

**Evidence for pulsed 1000 GeV gamma–rays from the Crab pulsar.**
See Abstr. 126.084.

**The local interstellar electron spectrum: constraints from gamma ray data.**
See Abstr. 131.175.

**Dunkelwolken und kosmische Gamma–Strahlung.**
See Abstr. 131.220.

***HEAO 3* discovery of $^{26}$Al in the interstellar medium.**
See Abstr. 131.288.

**The HEAO–1 diffuse component spectrum from 15 to above 500 keV.**
See Abstr. 142.035.

**On the periodic gamma–ray emission of Cygnus X–3.**
See Abstr. 142.052.

**Cosmic rays and cosmology.**
See Abstr. 144.048.

**Ultra high energy cosmic rays and ultra high energy γ–rays.**
See Abstr. 144.056.

**The galactic distribution of cosmic–ray particles.**
See Abstr. 144.072.

**M31: an extragalactic source of 1000 GeV gamma rays.**
See Abstr. 157.008.

**Cosmic gamma rays from galaxy clusters.**
See Abstr. 160.011.

# 144 Cosmic Radiation

144.001 **A 30,000–year record of the cosmic–ray intensity.**
A. N. Konstantinov, G. E. Kocharov.
Sov. Astron. Lett., Vol. 10, No. 1, p. 37 – 38 (1984). English translation of 37.144.008.

144.002 **Remote sensing, eigenanalysis, and information content of cosmic variations.**
M. El–Raey, I. S. Sabbah.
J. Geophys. Res., Vol. 89, No. A7, p. 5643 – 5651 (1984).

Remote sensing of the primary component of cosmic 11–year and diurnal variations over a wide rigidity range is carried out by nonlinear inversion of the Fredholm integral equation relating secondary variations to the primary spectra. Data from three and four stations are simultaneously analyzed, using recently developed coupling functions based on Feynman scaling hypothesis. The range of rigidity over which the retrieval of the primary spectra is possible is investigated. Information content of measured intensities is studied by eigenanalysis of the covariance matrix of kernels of the integral equation. It is concluded that a redundancy of existing stations in the low rigidity range and a lack of stations in the high rigidity range already exist. Implications, suggestions, errors, and possible extension are discussed.

144.003 **Erratum: "Non–linear theory of cosmic ray shocks including self–generated Alfvén waves" [Astron. Astrophys., Vol. 116, p. 191 – 200 (1982)].**
J. F. McKenzie, H. J. Völk.
Astron. Astrophys., Vol. 136, No. 2, p. 378 (1984). See Abstr. 32.143.058.

144.004 **Very high energy cosmic ray events.**
W. V. Jones.
Nucl. Phys. A, Vol. A418, p. 139 – 160 (1984). Abstr. in Phys. Abstr., Vol. 87, No. 1237, Entry 72842 (1984). – See Abstr. 012.007.

144.005 **The ultra–high energy nuclear interaction event producing multiple large transverse momentum components.**
J.–r. Ren, H.–h. Kuang, A.–x. Huo, S.–l. Lu, S. Su, D.–c. Wang, W.–j. Hu, H.–k. Fan, C.–r. Wang, P.–y. Cao, J.–y. Li, G.–j. Li, J.–g. Lin.
Phys. Energ. Fortis Phys. Nucl., Vol. 8, No. 2, p. 136 – 141 (1984). In Chinese. Abstr. in Phys. Abstr., Vol. 87, No. 1238, Entry 77685 (1984).

144.006 **Muon sea–level intensity and primary cosmic–ray nucleon spectrum in the (1 – 100) TeV energy range from the Mt. Blanc underground experiment.**
L. Bergamasco, A. Castellina, B. D'Ettorre Piazzoli, G. Mannocchi, P. Picchi, S. Vernetto, H. Bilokon.
Nuovo Cimento C, Vol. 6C, Ser. 1, No. 6, p. 569 – 594 (1983). Abstr. in Phys. Abstr., Vol. 87, No. 1239, Entry 82356 (1984).

144.007 **The lateral distribution of electrons of extensive air shower observed at Akeno (920 g/cm$^2$).**
M. Nagano, Y. Hatano, T. Hara, N. Hayashida, S. Kawaguchi, K. Kamata, T. Kifune, G. Tanahashi.
J. Phys. Soc. Jpn., Vol. 53, No. 5, p. 1667 – 1681 (1984). Abstr. in Phys. Abstr., Vol. 87, No. 1239, Entry 82358 (1984).

144.008 **Interpretation of Cerenkov density spectra.**
R. J. Protheroe, J. R. Patterson.
J. Phys. G, Vol. 10, No. 6, p. 841 – 855 (1984). Abstr. in Phys. Abstr., Vol. 87, No. 1239, Entry 82359 (1984).

144.009 **An interpretation of high–energy hadrons in EAS.**
H. Sakuyama, K. Watanabe.
Lett. Nuovo Cimento, Vol. 40, Ser. 2, No. 3, p. 89 – 93 (1984). Abstr. in Phys. Abstr., Vol. 87, No. 1239, Entry 82360 (1984).

144.010 **Inelastic scattering of cosmic ray muons on iron nuclei and the virtual photon shadowing.**
A. Okada, K. Mitsui, T. Kitamura, Y. Ohashi, Y. Muraki, T. Suda, I. Nakamura, Y. Kawashima, S. Matsuno, F. Kajino, T. Aoki, S. Higashi, S. Ozaki, T. Takahashi, K. Honda, K. Kobayakawa, H. Inazawa, Y. Minorikawa, M. Shibata, H. Shibata, Y. Kamiya, Y. Teramoto.
Fortschr. Phys., Vol. 32, No. 4, p. 135 – 173 (1984). Abstr. in Phys. Abstr., Vol. 87, No. 1240, Entry 85852 (1984).

144.011 **Cosmic rays of very high energy.**
A. W. Wolfendale.
Rep. Prog. Phys., Vol. 47, No. 6, p. 655 – 706 (1984). Abstr. in Phys. Abstr., Vol. 87, No. 1241, Entry 90045 (1984).

144.012 **Accelerator mass spectrometry of $^{10}$Be to study the fossil record of cosmic radiation.**
C. Tuniz, D. K. Pal, R. K. Moniot, W. Savin, T. Kruse, G. F. Herzog.
Nuovo Cimento A, Vol. 81A, Ser. 11, No. 1, p. 201 – 207 (1984). Abstr. in Phys. Abstr., Vol. 87, No. 1241, Entry 90047 (1984).

144.013 **Origin and propagation of cosmic rays.**
S. Ramadurai.
Essays in theoretical physics, p. 219 – 246 (1984). – See Abstr. 003.006.

Current studies of cosmic rays concentrate on delineating the composition, energy spectrum, energy requirements and acceleration and propagation processes of this radiation. These in turn yield information about the nature of the sources of these particles. The astrophysical investigations on the nucleosynthetic processes as well as the various radiative and dynamical processes taking place in stars, interstellar medium, galaxies and beyond, help us to identify possible sources of cosmic rays. This article presents a brief overview of these investigations.

144.014 **Long–term cosmic ray variations and possibilities of the radio carbon method.**
L. I. Dorman.
Cosmic ray intensity and cosmogenic isotopes. 13th Leningrad seminar on cosmophysics, p. 60 – 67 (1983). In Russian. Abstr. in Ref. Zh., 51. Astron., 8.51.315 (1984). – See Abstr. 012.028.

144.015 **Anisotropy of the intensity of super–high energy cosmic rays.**
D. D. Krasil'nikov, A. D. Krasil'nikov, K. N. Makarov, F. K. Shamsutdinova.
Kosm. luchi ehnerg. vyshe $10^{17}$ eV. Yakutsk, p. 82 – 108 (1983). In Russian. Abstr. in Ref. Zh., 51. Astron., 8.51.529 (1984).

144.016 **Cosmic rays in the Galaxy.**
A. A. Mikhajlov.
Kosm. luchi ehnerg. vyshe $10^{17}$ eV. Yakutsk, p. 108 – 117 (1983). In Russian. Abstr. in Ref. Zh., 51. Astron., 8.51.531 (1984).

144.017 **Peculiarities of the daily variation of cosmic rays in high–speed solar wind streams.**
L. I. Dorman, N. S. Kaminer, A. E. Kuz'micheva, N. V. Mymrina.
Geomagn. Aehron., Tom 24, No. 4, p. 546 – 551 (1984). In Russian. English translation in Geomagn. Aeron.

144.018 **On the change of the period of cosmic ray variation connected with the rotation of the sun.**
G. A. Bazilevskaya, E. S. Vernova, M. S. Grigoryan, A. I. Sladkova, M. I. Tyasto, T. N. Charakhch'yan.
Geomagn. Aehron., Tom 24, No. 4, p. 552 – 556 (1984). In Russian. English translation in Geomagn. Aeron.

144.019 **Temporal variation of the concentration of cosmogenic isotopes in the earth's atmosphere and solar activity.**
N. I. Akatova, G. E. Kocharov.
Geomagn. Aehron., Tom 24, No. 4, p. 571 – 576 (1984). In Russian. English translation in Geomagn. Aeron.

144.020 **Quasi–two–year variations of galactic cosmic rays.**
T. V. Dzhapiashvili, O. G. Rogava, L. Kh. Shatashvili, G. V. Shafer.
Geomagn. Aehron., Tom 24, No. 4, p. 680 – 682 (1984). In Russian. English translation in Geomagn. Aeron.

144.021 **On pulsations of cosmic ray intensity before the increase of the relativistic particle stream from solar flares.**
S. A. Starodubtsev, A. T. Filippov.
Geomagn. Aehron., Tom 24, No. 4, p. 682 – 683 (1984). In Russian. English translation in Geomagn. Aeron.

144.022 **Monitoring observations of cosmic radiation aboard the Salyut 4 orbital station.**
V. I. Lyagushin, N. A. Mamontova, B. M. Makhmudov, M. A. Saraeva, P. I. Shavrin, Yu. Yu. Yushkov.
Kosm. Issled., Tom 22, Vyp. 4, p. 637 – 639 (1984). In Russian. English translation in Cosm. Res.

144.023 **Search for long–lived massive particles and time structure of electrons in extensive air showers.**
M. Yoshida, Y. Toyoda, T. Maeda.
J. Phys. Soc. Jpn., Vol. 53, No. 6, p. 1983 – 1994 (1984). Abstr. in Phys. Abstr., Vol. 87, No. 1242, Entry 94533 (1984).

144.024 **The isotopic composition of the anomalous low–energy cosmic rays.**
R. A. Mewaldt, J. D. Spalding, E. C. Stone.
Astrophys. J., Vol. 283, No. 1, p. 450 – 456 (1984).
The authors present new observations of the isotopic composition of low–energy He, C, N, O, and Ne nuclei. The observations cover the energy interval from ~5 to 30 MeV per nucleon, the energy range where the "anomalous cosmic ray" fluxes of He, N, O, and Ne have been observed. The isotopic composition of the anomalous cosmic rays is compared with that of higher–energy galactic cosmic rays in an effort to understand the origin of these two components. Although limited by their statistical accuracy, the present low–energy isotope results are generally consistent with the proposed interstellar–neutral origin for the anomalous cosmic rays.

144.025 **Astrophysical boosters.**
W. Kundt.
Ann. N.Y. Acad. Sci., Vol. 422, p. 359 – 364 (1984). – See Abstr. 012.033.
This paper reviews possible sources and mechanisms capable of accelerating protons and ions to very high energies. In particular, magnetized neutron stars and active galactic nuclei are considered.

144.026 **Temporal averaged energy spectra of galactic cosmic–ray nuclei with Z = 50.**
A. B. Akopova, L. V. Melkumyan, V. P. Perelygin, S. G. Stetsenko.
Erevan. fiz. inst. Prepr., No. 650/40, 20 pp. (1983). In Russian. Abstr. in Ref. Zh., 51. Astron., 9.51.678 (1984).

144.027 **Astrophysical boosters.**
W. Kundt.
J. Astrophys. Astron., Vol. 5, No. 3, p. 277 – 283 (1984).
Constraints are derived on the acceleration of charges in shocks to highly relativistic energies. When applied to the extended extragalactic radio sources and to the cosmic rays, they cast doubt on the mechanism of "in–situ acceleration", both for energy, entropy and statistical mechanics reasons.

144.028 **Variations of the galactic cosmic ray flux and evolution of coronal holes with the solar activity cycle.**
I. S. Samsonov, Z. N. Samsonova, N. P. Chirkov.
Nestatsionar. potoki zaryazh. chastits. v okolozem. kosm. prostranstve. Yakutsk, p. 25 – 35 (1983). In Russian. From Ref. Zh., 51. Astron., 10.51.362 (1984).

144.029 **Local conditions of cosmic ray propagation in the Galaxy.**
A. Ghosh, L. I. Dorman, V. S. Ptuskin.
Pis'ma Astron. Zh., Tom 10, No. 11, p. 827 – 832 (1984). In Russian. English translation in Sov. Astron. Lett., Vol. 10.
Cosmic ray propagation in the vicinity of 1 kpc from the sun is considered. The data of the anisotropy of cosmic rays and on the spectrum of high–energy electrons as well as on temporal cosmic ray variations are analysed.

144.030 **Variations of galactic cosmic rays and evolution of coronal holes with the cycles of solar activity.**
I. S. Samsonov, Z. N. Samsonova, N. P. Chirkov.
Geomagn. Aehron., Tom 24, No. 5, p. 813 – 816 (1984). In Russian. English translation in Geomagn. Aeron.

144.031 **The origin of ultra–high–energy cosmic rays.**
A. M. Hillas.
Annu. Rev. Astron. Astrophys., Vol. 22, p. 425 – 444 (1984). – See Abstr. 003.024.
Contents: (1) Why bother with ultra–high–energy cosmic rays? (2) Observational data: Energy spectrum and composition of high–energy particles. Observed anisotropy of high–energy cosmic rays. Specific identified sources of cosmic rays. (3) Acceleration mechanisms: Problems associated with statistical acceleration. Direct acceleration. Final comments. (4) Propagation of cosmic rays: Propagation from extragalactic sources. Propagation from galactic sources. (5) Conclusions.

144.032 **Importance of transverse diffusion for cosmic–ray propagation.**
P. Barge, J. Millet, R. Pellat.
Astrophys. J., Vol. 284, No. 2, p. 817 – 826 (1984).
The importance of transverse diffusion for cosmic ray propagation is reinvestigated with a description of the turbulent magnetic field based on toroidal plasma theory. A coupling between transverse diffusion and parallel diffusion is found to occur on a galactic scale in a number of cases. The radial diffusion coefficient perpendicular to the average magnetic field is found to be smaller than previously thought by a large factor. So a better confinement of the cosmic rays is expected in the Galaxy.

144.033 **The ultra heavy nuclei of the galactic cosmic radiation.**
D. O'Sullivan, A. Thompson.
Nucl. Tracks Radiat. Meas., Vol. 8, No. 1 – 4, p. 545 – 554 (1984). Abstr. in Phys. Abstr., Vol. 87, No. 1243, Entry 101070 (1984). – See Abstr. 012.045.

144.034 **Search for (4/3)e charged leptons in cosmic rays.**
T. Wada, Y. Yamashita, I. Yamamoto.
Lett. Nuovo Cimento, Vol. 40, Ser. 2, No. 11, p. 329 – 335 (1984). Abstr. in Phys. Abstr., Vol. 87, No. 1243, Entry 101071 (1984).

144.035 **Sub–luminal pulses from cosmic–ray air showers.**
J. Linsley.
J. Phys. G, Vol. 10, No. 8, p. L191 – L195 (1984). Abstr. in Phys. Abstr., Vol. 87, No. 1243, Entry 101080 (1984).

144.036 **Lateral distribution of Cerenkov light in large cosmic ray showers as a measure of longitudinal development.**
M. A. B. Craig.
Thesis, Univ. Durham, England, 2 pp. (1984). Abstr. in Phys. Abstr., Vol. 87, No. 1243, Entry 101081 (1984).

144.037 **SU(2) × U(1) vacuum and the Centauro events.**
D. Kazanas, V. K. Balasubrahmanyan,
R. E. Streitmatter.
Phys. Lett. B, Vol. 142B, No. 1 – 2, p. 23 – 27 (1984). Abstr. in Phys. Abstr., Vol. 87, No. 1243, Entry 101089 (1984).

144.038 **Burgers' equation and cosmic ray shocks.**
G. M. Webb, J. F. McKenzie.
J. Plasma Phys., Vol. 31, Part 3, p. 337 – 368 (1984). Abstr. in Phys. Abstr., Vol. 87, No. 1244, Entry 105815 (1984).

144.039 **A new source of cosmic rays?**
N. Duric.
Astronomy, Vol. 12, No. 7, p. 66 – 70 (1984). Abstr. in Phys. Abstr., Vol. 87, No. 1244, Entry 105816 (1984).

144.040 **Energy spectrum of primary cosmic rays between $10^{14.5}$ and $10^{18}$eV.**
M. Nagano, T. Hara, Y. Hatano, N. Hayashida, S. Kawaguchi, K. Kamata, T. Kifune, Y. Mizumoto.
J. Phys. G, Vol. 10, No. 9, p. 1295 – 1310 (1984). Abstr. in Phys. Abstr., Vol. 87, No. 1245, Entry 110410 (1984).

144.041 **On the mechanism of anomalous cosmic–ray nuclear interactions with photoemulsion nuclei.**
S. Sh. Ameev, V. L. Shmonin.
J. Phys. G, Vol. 10, No. 9, p. 1311 – 1317 (1984). Abstr. in Phys. Abstr., Vol. 87, No. 1245, Entry 110429 (1984).

144.042 **A study of energy spectrum of cosmic rays ranging from $6 \times 10^{14}$ to $5 \times 10^{16}$ eV.**
L.–R. Sun.
Phys. Energ. Fortis Phys. Nucl., Vol. 8, No. 4, p. 502 – 507 (1984). In Chinese. Abstr. Phys. in Abstr., Vol. 87, No. 1246, Entry 115662 (1984).

144.043 **Energy spectrum and angular distribution of prompt cosmic–ray muons.**
C. Castagnoli, P. Picchi, A. Castellina, B. D'Ettorre Piazzoli, G. Mannocchi, S. Vernetto.
Nuovo Cimento A, Vol. 82A, Ser. 11, No. 1, p. 78 – 92 (1984). Abstr. in Phys. Abstr., Vol. 87, No. 1246, Entry 115663 (1984).

144.044 **Anisotropy of cosmic rays at energies of $5 \times 10^{11} - 10^{14}$eV in the diffusional approximation.**
V. S. Ptuskin.
Sov. Astron., Vol. 28, No. 3, p. 357 – 358 (1984). English translation of 37.144.039.

144.045 **Cosmic ray gradients in the Galaxy.**
C. L. Bhat, C. J. Mayer, A. W. Wolfendale.
Astron. Astrophys., Vol. 140, No. 2, p. 284 – 287 (1984).
The analysis of cosmic γ–ray data from the SAS II experiments by Dodds et al. (1975) strongly suggested that there was a gradient of cosmic ray intensity in the outer Galaxy, shared by both electrons and protons and that this indicated that both types of particle were derived largely from galactic sources at the energies in question ($E_e \lesssim 0.5$ GeV, $E_p$: 1–10 GeV). Other workers have come to the opposite conclusion insofar as the protons are concerned. The present work comprises a number of studies and these all indicate that the original conclusion may indeed be correct, i.e. that there is a gradient of protons.

144.046 **Energy dependence of the $\bar{P}/P$ ratio in cosmic rays.**
R. L. Golden, B. G. Mauger, S. Nunn, S. Horan.
Astrophys. Lett., Vol. 24, No. 2, p. 75 – 83 (1984).
The antiproton/proton ($\bar{P}/P$) ratio is presented for seven kinetic energy intervals from 4.4 – 13.4 GeV. The $\bar{P}$ were observed in 1979 by the New Mexico State University balloon–borne magnet spectrometer. Results are presented for fits based on the closed–galaxy and other models.

144.047 **Cosmic–ray scintillations and dynamic processes in space.**
L. I. Dorman, I. Ya. Libin.
Space Sci. Rev., Vol. 39, Nos. 1/2, p. 91 – 152 (1984).
The cosmic–ray scintillations established by different authors from the data of ground–base and high–altitude devices for quiet and disturbed periods, as well as the theoretical calculations of different models and mechanisms of the origin and development of cosmic–ray scintillations are analyzed. High–frequency scintillations of $f \geqslant 10^{-5}$Hz are shown to be precursors of an approaching shock wave, scintillations with periods of the order of 10 – 20 and 40 – 50 min being most sensitive to disturbances of interplanetary medium near the Earth. Since cosmic rays of different energies are sensitive to different processes in interplanetary space at different distances from the Earth, one can sound the conditions in interplanetary medium up to $10^{15}$cm from the Earth by measuring particle fluxes at different energy ranges.

144.048 **Cosmic rays and cosmology.**
N. C. Rana, A. W. Wolfendale.
Vistas Astron., Vol. 27, Part 3, p. 199 – 216 (1984).
A brief discussion is given of a number of topics related to the interface of cosmic rays and astronomy, notably a number of possible explanations for the diffuse extragalactic γ–ray background and the origin of the most energetic cosmic ray particles.

144.049 **Relative abundance of the helium isotopes $^3$He and $^4$He in the cosmic radiation.**
S. P. Jordan, P. Meyer.
Phys. Rev. Lett., Vol. 53, No. 5, p. 505 – 508 (1984).
The geomagnetic field near the equator is used to measure the abundance ratio of $^3$He/$^4$He near 6 GeV/nucleon. This ratio is found to be $0.24 \pm 0.05$. If all $^3$He is of secondary origin this implies an interstellar mean escape path length of $\approx 15$ g/cm$^2$, a value considerably larger than that determined from observations of heavier elements, but compatible with measurements of the helium isotopic ratio at lower energies.

144.050 **Time–lag of cosmic–ray intensity.**
H. Mavromichalaki, B. Petropoulos.
Astrophys. Space Sci., Vol. 106, No. 1, p. 61 – 71 (1984).
The best correlation coefficient between the monthly cosmic–ray intensity of the Inuvik Station and various kinds of solar, interplanetary, and geophysical parameters has been found. It is calculated for different time–lags of cosmic–ray intensity with respect to these parameters. The maximum of these coefficients lead to a useful empirical model for the 11–year cosmic–ray modulation.

144.051 **Composition of cosmic radiation at energies of ~ $10^{15}$eV and higher.**
S. I. Nikol'skij, J. N. Stamenov, S. Z. Ushev.
Zh. Ehksp. Teor. Fiz., Tom 87, No. 1, p. 18 – 36 (1984). In Russian. Abstr. in Ref. Zh., 51. Astron., 11.51.633 (1984).

144.052 **Propagation of heavy cosmic–ray nuclei.**
J. R. Letaw, R. Silberberg, C. H. Tsao.
Astrophys. J., Suppl. Ser., Vol. 56, No. 3, p. 369 – 391 (1984).
Techniques for modeling the propagation of heavy cosmic–ray nuclei, and the required atomic and nuclear data, are assembled in this paper. Emphasis is on understanding nuclear composition in the charge range $3 < Z < 83$. Details of the application of "matrix methods" above a few hundred MeV per nucleon, a new treatment of electron capture decay, and a new table of cosmic–ray – stable isotopes are presented. Computation of nuclear fragmentation cross sections, stopping power, and electron stripping and attachment are briefly reviewed.

144.053 **Cosmic–ray elemental abundances.**
N. Lund.
Adv. Space Res., Vol. 4, No. 2 – 3, p. 5 – 14 (1984). – See Abstr. 012.075.
The elemental composition of the cosmic radiation as function of particle energy is one of the major sources of information

concerning the origin and life history of cosmic rays. The Danish–French spectrometer on the HEAO–3 satellite has recently provided new and detailed data on the composition and energy spectra of the cosmic rays for the elements between beryllium and germanium in the energy range between 0.7 and 25 GeV/nucleon. These data appear to be consistent with a source composition similar to that of the Solar System.

144.054 **The isotopic composition of cosmic rays.**
M. E. Wiedenbeck.
Adv. Space Res., Vol. 4, No. 2 – 3, p. 15 – 24 (1984). – See Abstr. 012.075.

Observations of the isotopic composition of the elements carbon, nitrogen, oxygen, neon, magnesium, silicon, sulfur, calcium, iron and nickel are reviewed and "best" values of selected isotope ratios are derived from the collected data. A simple model of cosmic ray propagation is employed to deduce cosmic ray source abundance ratios both from these best values and from the observations made in a recent high resolution satellite experiment.

144.055 **Elemental abundances of ultraheavy cosmic rays.**
W. R. Binns, D. J. Fixsen, T. L. Garrard, M. H. Israel, J. Klarmann, E. C. Stone, C. J. Waddington.
Adv. Space Res., Vol. 4, No. 2 – 3, p. 25 – 34 (1984). – See Abstr. 012.075.

The elemental composition of the cosmic–ray source is different from that which has been generally taken as the composition of the solar system. No general enrichment of products of either *r*–process or *s*–process nucleosynthesis accounts for the differences over the entire range of ultraheavy ($Z > 30$) elements. Comparison between the cosmic–ray source abundances and the abundances of C1 and C2 chondritic meteorites suggests the possibility that differences between the cosmic–ray source and the "standard (C1) solar system" may not be due to acceleration fractionation of the cosmic rays, but may be due instead to a fractionation of the C1 abundances with respect to the interstellar abundances.

144.056 **Ultra high energy cosmic rays and ultra high energy γ–rays.**
A. A. Watson.
Adv. Space Res., Vol. 4, No. 2 – 3, p. 35 – 44 (1984). – See Abstr. 012.075.

Evidence on the energy spectrum, mass composition and anisotropy of cosmic rays above $10^{14}$eV is briefly surveyed and it is shown that the data are consistent with all cosmic rays below $10^{19}$eV being of galactic origin. Results from the new field of γ–ray astronomy above $10^{15}$eV are described and implications for the origin of cosmic rays are discussed.

144.057 **Stellar nucleosynthesis and the cosmic–ray source composition up to Zn.**
M. Arnould.
Adv. Space Res., Vol. 4, No. 2 – 3, p. 45 – 54 (1984). – See Abstr. 012.075.

The author presents a critical review of the predictions of various nucleosynthetic models which may have some relevance to the elemental and isotopic cosmic–ray source composition. In particular, abundances expected from various supernova models and from Wolf–Rayet star nucleosynthesis are examined.

144.058 **Isotopic anomalies in cosmic rays and winds from Wolf–Rayet stars: a new model.**
A. Maeder.
Adv. Space Res., Vol. 4, No. 2 – 3, p. 55 – 65 (1984). – See Abstr. 012.075.

The insufficiencies of the metal–rich supernova (SN) model and of the Wolf–Rayet (WR) star model for explaining the isotopic anomalies in GCRS (galactic cosmic ray source) are examined. A new model is proposed. It assumes that galactic cosmic rays, which may propagate over significant galactic distances, originate from two main sources (SN and WR). The properties (e.g. chemistry and relative distribution) of these two sources are changing with galactocentric distance.

144.059 **Clues to the origin of cosmic rays derived from their source composition and energy spectra.**
L. Koch–Miramond, J. P. Meyer.
Adv. Space Res., Vol. 4, No. 2 – 3, p. 79 – 88 (1984). – See Abstr. 012.075.

Cosmic ray composition and energy spectra at source have been recently obtained with good precision. The bulk of the observations suggests that cosmic rays originate in unevolved star surface material. Most cosmic rays should be stellar energetic particles first accelerated by flares out of the coronae of F to M stars, that got later on reaccelerated to high energy by strong interstellar shock waves. The limitations of this two–step scenario and of the other plausible stellar and interstellar models are pointed out.

144.060 **On the contribution of stellar winds to the composition of cosmic rays: exotic Ne and Fe from Wolf–Rayet stars.**
J. B. Blake, D. S. P. Dearborn.
Adv. Space Res., Vol. 4, No. 2 – 3, p. 89 – 92 (1984). – See Abstr. 012.075.

Wolf–Rayet stars can significantly contribute to the galactic cosmic–ray source by synthesizing anomalous material and then dispersing it into the interstellar medium by means of their stellar winds. In this note the cases of $^{26}$Al and the heavy isotopes of Fe are discussed.

144.061 **Nuclear fragmentation cross sections and cosmic–ray source abundances.**
T. G. Guzik, J. P. Wefel.
Adv. Space Res., Vol. 4, No. 2 – 3, p. 93 – 96 (1984). – See Abstr. 012.075.

Current experimental data provides constraints upon the nuclear excitation functions for $^{16}O + p \rightarrow {}^{14}N$, $^{15}N$, $^{13}C$ and their progenitors. These cross sections, combined with the elemental and isotopic ratios measured at Earth, are used in a complete model of cosmic ray propagation to study the source composition.

144.062 **Abundances of the sub–Fe group of nuclei and their implications.**
S. Ramadurai, M. N. Vahia, J. S. Yadav, S. Biswas, N. Durgaprasad.
Adv. Space Res., Vol. 4, No. 2 – 3, p. 97 – 99 (1984). – See Abstr. 012.075.

Recent evidence has been presented for a large flux of antiprotons in low energy cosmic rays. It is found that this implies a large amount of matter traversal by low energy cosmic rays. There is already evidence for the large amount of matter traversal in the form of a large ratio of Fe secondaries to Fe nuclei at energies below 100 MeV/nucleon. These two facts are combined into a new model of acceleration and propagation of cosmic rays.

144.063 **Spectra of cosmic–ray hadrons in the range 0.1 – 1000 GeV.**
M. Nieminen, J. J. Torsti, E. Valtonen, H. Arvela, M. Lumme, J. Peltonen, E. Vainikka.
Adv. Space Res., Vol. 4, No. 2 – 3, p. 101 – 104 (1984). – See Abstr. 012.075.

The energy spectra of cosmic–ray neutrons, protons, and pions were determined at sea level in the range 0.1 – 1000 GeV. The results are based on the measurements of the neutron–multiplicity distributions produced by cosmic hadrons in the Turku spectrometer.

144.064 **Abundances of ultraheavy nuclei and the acceleration process.**
S. Ramadurai.
Adv. Space Res., Vol. 4, No. 2 – 3, p. 105 – 107 (1984). – See Abstr. 012.075.

Using a combination of gaussian distribution of path length with $\bar{x}$ of 1.5 g cm$^{-2}$ and exponential distribution with $\bar{x}$ of 7 g cm$^{-2}$, the abundances at the earth of ultraheavy nuclei are derived. It is shown that the source abundances obtained from the current observations indicate an enrichment of *r*–process elements.

144.065 **Nucleosynthesis in Wolf–Rayet stars and galactic cosmic–ray isotopic composition.**
N. Prantzos.
Adv. Space Res., Vol. 4, No. 2 – 3, p. 109 – 114 (1984). – See Abstr. 012.075.

An explanation of the isotopic composition of galactic cosmic rays could provide some clues to the mystery of their origin. It seems now that the strong stellar winds of Wolf–Rayet stars could account for most of the isotopic anomalies that have been observed in cosmic rays. Some results are presented, obtained by detailed nucleosynthesis computations.

144.066 **Thermal particle injection into the shock acceleration process.**
G. F. Krymsky (*G. F. Krymskij*).
Adv. Space Res., Vol. 4, No. 2 – 3, p. 175 – 184 (1984). – See Abstr. 012.075.

The efficiency of the shock acceleration mechanism for cosmic ray particles is investigated.

144.067 **Reaction effects in diffusive shock acceleration.**
L. O'C. Drury.
Adv. Space Res., Vol. 4, No. 2 – 3, p. 185 – 191 (1984). – See Abstr. 012.075.

The importance of the reaction of the accelerated particles for any efficient acceleration process is emphasized. The effects on the large–scale structure of the shock, on the particle spectrum, on the scattering mechanism and on the small–scale flow structure are considered.

144.068 **Stochastic Fermi acceleration and the origin of cosmic rays.**
A. Achterberg.
Adv. Space Res., Vol. 4, No. 2 – 3, p. 193 – 204 (1984). – See Abstr. 012.075.

The role of stochastic acceleration of charged particles by turbulence or plasma waves is reviewed, with special emphasis on the acceleration of cosmic rays in the interstellar medium or supernova remnants. Observational constraints are discussed. A comparison is made between stochastic Fermi acceleration, and direct acceleration of cosmic rays by shock waves.

144.069 **Numerical studies of time–dependent cosmic–ray shocks.**
E. Dorfi.
Adv. Space Res., Vol. 4, No. 2 – 3, p. 205 – 213 (1984). – See Abstr. 012.075.

The paper shows the present status of numerical solutions for flows including cosmic rays. The cosmic rays are described in the hydrodynamical limit. The transition of a gas dominated shock to a cosmic ray dominated shock is investigated in linear geometry. A simple model for the early phases of a supernova–explosion hints at a different remnant–evolution if the backreaction of cosmic rays is taken into account.

144.070 **The cosmic–ray pathlength distribution at low energy: a new probe of the source/acceleration regions.**
T. G. Guzik, J. P. Wefel.
Adv. Space Res., Vol. 4, No. 2 – 3, p. 215 – 218 (1984). – See Abstr. 012.075.

Compiled measurements of secondary to primary ratios covering the charge range $Z = 3 - 26$ and the energy range 0.05 – 50 GeV/nucleon are analyzed in energy dependent galactic propagation plus solar modulation calculations. The cosmic ray pathlength distribution is shown to consist of two energy dependent components interpreted as representing confinement in the Galaxy and confinement in the "source" regions.

144.071 **Importance of transverse diffusion for galactic cosmic rays.**
P. Barge, J. Millet, R. Pellat.
Adv. Space Res., Vol. 4, No. 2 – 3, p. 225 – 227 (1984). – See Abstr. 012.075.

The importance of transverse diffusion for cosmic rays propagation is reinvestigated with a description of the turbulent magnetic field based on toroidal plasma theory. A coupling between transverse diffusion and parallel diffusion is found to occur on a galactic scale in a number of cases.

144.072 **The galactic distribution of cosmic–ray particles.**
W. Hermsen, J. B. G. M. Bloemen.
Adv. Space Res., Vol. 4, No. 2 – 3, p. 393 – 404 (1984). – See Abstr. 012.075.

COS–B gamma–ray observations are compared with large–scale H I and CO surveys in different regions of the sky to derive information on the cosmic–ray density distribution throughout the Galaxy and/or a calibration of the ratio between $H_2$ column density and integrated CO line intensity $[N(H_2)/W_{CO}]$. In the region of the large complex of interstellar clouds in Orion and Monoceros the observed gamma–ray emission can be explained in terms of interactions between cosmic rays distributed uniformly with the local density in this region, and the interstellar gas.

144.073 **Rapporteur paper on nucleosynthesis and cosmic rays.**
M. Cassé.
Adv. Space Res., Vol. 4, No. 2 – 3, p. 411 – 416 (1984). – See Abstr. 012.075.

This paper summarizes some of the major topics related to cosmic ray abundances and nucleosynthesis as presented during Symposium 6 of the 25th COSPAR Plenary Meeting.

144.074 **Galactic cosmic rays.**
S. Hayakawa.
Adv. Space Res., Vol. 4, No. 2 – 3, p. 417 – 418 (1984). – See Abstr. 012.075.

This rapporteur paper, presented at Symposium 6 of the 25th COSPAR Plenary Meeting, discusses some poster papers and invited papers dealing with the distribution and the propagation of cosmic rays in the Galaxy.

144.075 **Cosmic–ray interaction with shocks: particle spectrum and injection, time dependence.**
H. J. Völk.
Adv. Space Res., Vol. 4, No. 2 – 3, p. 423 – 426 (1984). – See Abstr. 012.075.

This rapporteur paper discusses a number of poster papers which were presented together with several reviews in Symposium 6 of the 25th COSPAR Plenary Meeting. These papers were concerned with diffusive shock acceleration of cosmic ray particles.

144.076 **Results of an investigation of low–energy cosmic rays in the AYAKS experiment aboard Prognoz 4 and Prognoz 6.**
A. A. Kolchin, V. V. Lebedev, V. F. Levchenko, A. I. Repin, G. P. Skrebtsov, V. L. Shubin.
Solar activity investigations and the Prognoz space system, p. 168 – 185 (1984). In Russian. Abstr. in Ref. Zh., 62. Issled. Kosm. Prostranstva, 12.62.368 (1984). – See Abstr. 003.038.

144.077 **Production of high–energy gamma rays by cosmic ray interactions in the atmosphere and lunar surface.**
D. J. Morris.
J. Geophys. Res., Vol. 89, No. A12, p. 10685 – 10696 (1984).

The production of gamma rays above 10 MeV by the interaction of cosmic rays in the atmosphere and the lunar surface is simulated using Monte Carlo methods. The calculation incorporates a new model of high–energy nucleon–nucleus interactions based on empirical fits to inclusive cross sections for the production of pions and nucleons. The atmospheric gamma ray flux is calculated as a function of direction, energy, and atmospheric depth. These calculations are compared with observations from balloons and from the SAS 2 satellite. Estimates of the flux of earth albedo electrons produced by cosmic ray interactions are presented. The lunar gamma ray albedo is calculated and compared with an upper limit based on SAS 2 measurements.

144.078 **On the formation of the superhigh–energy spectrum of cosmic rays in the nuclei of active galaxies.**
F. A. Agaronyan, A. S. Ambartsumyan.
Astrofizika, Tom 21, Vyp. 2, p. 275 – 281 (1984). In Russian. English translation in Astrophysics, Vol. 21, No. 2.
The energy loss of the proton–nuclei component of superhigh cosmic rays due to the interaction with the X–ray photon field in nuclei of active galaxies is investigated.

144.079 **Cosmic ray fluctuations on the earth during solar flares.**
Ya. L. Blokh, I. Ya. Libin, K. F. Yudakhin.
Geomagn. Aehron., Tom 24, No. 6, p. 875 – 878 (1984). In Russian. English translation in Geomagn. Aeron.

144.080 **Cosmic ray variations in November – December 1982 at earth level.**
A. V. Belov, Ya. L. Blokh, R. T. Gushchina, E. A. Eroshenko, L. I. Dorman, V. N. Ishkov, N. S. Kaminer, I. V. Novitskaya.
Geomagn. Aehron., Tom 24, No. 6, p. 879 – 884 (1984). In Russian. English translation in Geomagn. Aeron.

144.081 **A measurement of the absolute flux of cosmic–ray electrons.**
R. L. Golden, B. G. Mauger, G. D. Badhwar, R. R. Daniel, J. L. Lacy, S. A. Stephens, J. E. Zipse.
Astrophys. J., Vol. 287, No. 2, p. 622 – 632 (1984).
The absolute flux of cosmic–ray electrons has been measured using a balloon–borne superconducting magnet spectrometer. The flux of electrons in the interstellar medium was determined by correcting the observed flux for energy loss in the atmosphere and the payload and for solar modulation. The flux in the interstellar medium was found to be $367\,E^{-(3.15\pm0.2)}\,e^{-}\mathrm{m}^{-2}\mathrm{sr}^{-1}\mathrm{s}^{-1}\mathrm{GeV}^{-1}$ in the energy interval 4.5 – 63.5 GeV. The uncertainty in the absolute flux is 10%.

144.082 **Evidence that the anomalous cosmic–ray component is singly ionized.**
A. C. Cummings, E. C. Stone, W. R. Webber.
Astrophys. J., Lett. Ed., Vol. 287, No. 2, p. L99 – L103 (1984).
The authors have used data from the Cosmic Ray Subsystem on the *Voyager* spacecraft to measure the spectra of He, C, N, O, and Ne with 4 – 124 MeV per nucleon (for O) during solar minimum conditions near the end of 1977. By subtracting both a low–energy interplanetary component and the high–energy galactic component they have determined the energy spectra of the "anomalous" cosmic–ray species He, N, O, and Ne. If the source spectra at the boundary of the modulation region have similar power–law energy dependences, then the observed energy spectra indicate that the anomalous particles are singly ionized.

**Cosmic rays with energies above $10^{17}$eV. Collected papers.**
See Abstr. 003.163.

**Cosmic ray intensity and cosmogenic isotopes. Papers of the 13th Leningrad seminar on cosmophysics, Leningrad, 19 – 21 November, 1982.**
See Abstr. 012.028.

**TARO Cosmic–ray Observatory Report (I).**
See Abstr. 013.013.

**Relativistic $^{238}$U ion tracks in olivine and cosmic–ray track studies.**
See Abstr. 022.004.

**A drift chamber telescope for heavy ion track detection.**
See Abstr. 034.006.

**Design of emulsion chambers for cosmic ray experiments at balloon altitudes.**
See Abstr. 034.008.

**Cosmic rays.**
See Abstr. 034.009.

**Preliminary results on the registration temperature effect for uranium ions in various solid state nuclear track detectors.**
See Abstr. 034.078.

**Dinonyl phthalate as a dopant for CR–39 in cosmic ray studies.**
See Abstr. 034.080.

**Screening out obstructive radiation from thermoluminescence sheets ($BaSO_4$:Eu) exposed to cosmic rays.**
See Abstr. 034.083.

**An extensive solid state nuclear track detector array for the study of ultra heavy cosmic ray nuclei aboard the NASA Long Duration Exposure Facility (LDEF).**
See Abstr. 035.038.

**ADC(CR–39) detector module for Space Shuttle Spacelab–3 cosmic ray experiment.**
See Abstr. 035.042.

**Two cosmic–ray detectors with high charge and energy resolution in the GeV/nucleon range.**
See Abstr. 035.043.

**Examination of a proposed technique for the economical detection and analysis of ultra–high energy cosmic ray showers.**
See Abstr. 036.030.

**Short–range and long–range correlations in cosmic ray interactions.**
See Abstr. 061.038.

**The production of very high energy photons and neutrinos from cosmic proton sources.**
See Abstr. 061.096.

**Nuclearites – a novel form of cosmic radiation.**
See Abstr. 061.098.

**Diffusive shock acceleration of energetic eletrons subject to synchrotron losses.**
See Abstr. 062.046.

**Simulations of cosmic–ray particle diffusion.**
See Abstr. 062.058.

**Magnetoacoustic instability driven by Alfvén waves excited by cosmic rays.**
See Abstr. 062.148.

**Monte Carlo shock–like solutions to the Boltzmann equation with collective scattering.**
See Abstr. 062.170.

**Particle acceleration in compact binary stars.**
See Abstr. 067.127.

**Cosmic–ray acceleration by pulsars.**
See Abstr. 067.128.

**The contribution of cosmogenic radioisotopes to solar neutrino flux measurements.**
See Abstr. 080.083.

**Atmospheric gamma radiation: a comparison of experimental results at 4.5 GV and 11.7 GV.**
See Abstr. 082.019.

**Cosmogenic $^{22}$Na and $^{26}$Al and track studies in Luna 16, 20 and 24 samples.**
See Abstr. 094.035.

**Tracks of heavy cosmic nuclei and effects of channeling in olivine crystals from meteorites.**
See Abstr. 105.027.

**Interplanetary flow systems associated with cosmic ray modulation in 1977 – 1980.**
See Abstr. 106.009.

**The interplanetary scattering mean free path: collisionless wave–damping effects.**
See Abstr. 106.026.

**Solar and non–solar activity in stars: the cosmic–ray connection.**
See Abstr. 116.030.

**Cosmic–ray emission from Cygnus X–3.**
See Abstr. 117.145.

**The supernova envelope shock origin of cosmic rays – a review.**
See Abstr. 125.066.

**The local interstellar electron spectrum: constraints from gamma ray data.**
See Abstr. 131.175.

**Correlation between X–ray luminosity and the deviations from the main–sequence colour–diagram for stellar X–ray sources.**
See Abstr. 142.031.

**Where cosmic rays come from?**
See Abstr. 143.001.

**Gamma rays from atomic and molecular gas in the large complex of clouds in Orion and Monoceros.**
See Abstr. 143.012.

**On the nature of the galactic gamma–ray sources.**
See Abstr. 143.019.

**Can the ultrahigh–energy cosmic rays stem from the galactic center?**
See Abstr. 155.018.

**Extragalactic cosmic rays from active galaxies.**
See Abstr. 158.125.

**Energetic particles in radio sources.**
See Abstr. 158.203.

**Particle acceleration in active galactic nuclei.**
See Abstr. 158.204.

**Cosmic–ray antiprotons as a probe of a photino–dominated universe.**
See Abstr. 161.304.

**Big bang cosmology, relic neutrinos, and absorption of neutrino cosmic rays.**
See Abstr. 161.316.

# Stellar Systems, Galaxy, Extragalactic Objects, Cosmology

## 151 Stellar Systems (Kinematics, Dynamics)

151.001 **Modelling of the integral B–V colour index for metal-poor stellar systems.**
O. K. Sil'chenko.
Pis'ma Astron. Zh., Tom 10, No. 7, p. 501 – 507 (1984). In Russian. English translation in Sov. Astron. Lett., Vol. 10.

The B–V colour is calculated for stellar systems without star formation having metallicities $Z = 10^{-3}$ and $10^{-4}$ and ages $T = 7$ and 18 billion years. Results of model calculations are consistent with the new metallicity scale of globular clusters. The ages of relatively metal-rich globular clusters ([Fe/H] > –1.2) and M92 are roughly estimated.

151.002 **Dynamical constraints during violent relaxation and their effects on the final state.**
A. May, T. S. van Albada.
Mon. Not. R. Astron. Soc., Vol. 209, No. 1, p. 15 – 23 (1984).

A previous paper (van Albada) described a series of $N$-body models of the collapse of collisionless stellar systems from roughly spherical initial conditions. In the present paper, these models are used to illustrate how certain properties of the final states may be related to dynamical constraints during the phase of violent relaxation.

151.003 **How bar strength and pattern speed affect galactic spiral structure.**
M. P. Schwarz.
Mon. Not. R. Astron. Soc., Vol. 209, No. 1, p. 93 – 109 (1984).

The response of a gaseous disc to rotating stellar bars of various strengths and pattern speeds is studied. The arm–to–interarm density contrast and the pitch angle of the induced spiral arms both increase with increasing bar strength. Furthermore, along the sequence there is a widening of a region in which stable periodic orbits are difficult to find until it encompasses the whole annulus between corotation (CR) and the outer Lindblad resonance. Stars moving in quasi–stochastic orbits originating in this region form an exponential tail in the surface density profile of the stellar disc. As the pattern speed is lowered more structure appears inside CR.

151.004 **Orbital characteristics of polar rings of galaxies.**
N. Katz, D. O. Richstone.
Astron. J., Vol. 89, No. 7, p. 975 – 978 (1984).

Recently, S0 and elliptical galaxies have been observed with polar rings of luminous material. Orbital characteristics are computed for these rings of material under the assumption that they are test particles in near closed orbits. Two potentials for the central galaxy are investigated. The vertical extension and velocity as a function of angle is set by the axial ratio of the potential. These calculated values, used in conjunction with observed values, can give insight into the presence and mass distribution of a possible massive halo.

151.005 **The disintegration of small stellar systems.**
L. J. Tomley.
Astron. J., Vol. 89, No. 7, p. 1073 – 1075 (1984).

Numerical studies of the disintegration of slightly perturbed equilibrium configurations of three and four bodies reveal that the angular momentum vectors of the final subsystems are substantially randomized for four–body (and greater) systems but are not for three–body systems. The cause of this difference and some possible applications of these findings are discussed.

151.006 **Anti–gravity and galaxy rotation curves.**
R. H. Sanders.
Astron. Astrophys., Vol. 136, No. 2, p. L21 – L23 (1984).

A modification of Newtonian gravitational attraction which arises in the context of modern attempts to unify gravity with the other forces in nature can produce rotation curves for spiral galaxies which are nearly flat from 10 to 100 kpc, bind clusters of galaxies, and close the universe with the density of baryonic matter consistent with primordial nucleosynthesis. This is possible if a very low mass vector boson carries an effective anti–gravity force which on scales smaller than that of galaxies almost balances the normal attractive gravity force.

151.007 **Spiral instabilities provoked by accretion and star formation.**
J. A. Sellwood, R. G. Carlberg.
Astrophys. J., Vol. 282, No. 1, p. 61 – 74 (1984).

The authors present a description of spiral structure that appears to account for both the imperfect grand–design nature of the patterns and the importance of gas. They argue that most visible spirals are short–lived features organized by gravitational forces. In the absence of gas, which is continuously cooled, recurrent transient patterns would steadily heat the disk until it became stable. Any method of cooling will ensure that the disk will never completely stabilize and will, consequently, exhibit recurrent spiral structure. Both dissipation in the gas and accretion will cool the disk and provoke instabilities. The authors observe transient spiral patterns in computer simulations which recur repeatedly only when processes designed to mimic accretion and star formation are included.

151.008 **Spiral structure and star formation. II. Stellar lifetimes and cloud kinematics.**
M. A. Hausman, W. W. Roberts Jr.
Astrophys. J., Vol. 282, No. 1, p. 106 – 117 (1984).

The authors present further results of their model (see Abstr. 37.151.034) of star formation and star–gas interactions in the cloud–dominated ISMs of spiral density wave galaxies. The global density distribution and velocity field of the gas clouds are virtually independent of stellar parameters and even of mean free path for the wide range of values studied. Increasing the average life span of "spiral tracer" stellar associations beyond about 20 Myr washes out the spiral pattern which younger associations show. The paper also examines the kinematics of clouds as they orbit under the influence of galactic gravity, collisions, and SNR impulses.

151.009 **The influence of nonlinearity and dispersion on the dynamics of packets of spiral density waves.**
V. I. Korchagin, P. I. Korchagin.
Astron. Zh., Tom 61, Vyp. 4, p. 814 – 817 (1984). In Russian. English translation in Sov. Astron., Vol. 28, No. 4.

The influence of nonlinearity, dispersion and inhomogeneity on the evolution of spiral density waves is considered. It is shown that under these effects the evolution timescale of a wave packet may be essentially smaller than the time of its radial propagation on the disk of the galaxy and must be accounted for in the interpretation of the observed spiral structure.

151.010 **On the triple interaction of spiral density waves.**
V. I. Korchagin, P. I. Korchagin.
Sov. Astron. Lett., Vol. 10, No. 1, p. 5 – 6 (1984). English translation of 37.151.001.

151.011 **N–body problem of clusters of galaxies.**
S. Yabushita.
Astron. Her., Vol. 77, No. 8, p. 192 – 196 (1984). In Japanese.

151.012 **Dynamical friction in spherical systems.**
S. Tremaine, M. D. Weinberg.
Mon. Not. R. Astron. Soc., Vol. 209, No. 4, p. 729 – 757 (1984).
The authors investigate dynamical friction on a test object (such as a bar or satellite) which rotates or revolves through a spherical stellar system. They find that frictional effects arise entirely from near–resonant stars and they derive an analog to Chandrasekhar's dynamical friction formula which applies to spherical systems. The authors show that a formula of this type is valid so long as the angular speed of the test object changes sufficiently rapidly. If the angular speed is slowly changing two new effects appear: a reversible dynamical feedback which can stabilize or destabilize the rotation speed, and permanent capture of near–resonant stars into librating orbits. The authors discuss orbital decay of satellites in the light of these results.

151.013 **Dynamics of globular clusters.**
L. Spitzer Jr.
Science, Vol. 225, No. 4661, p. 465 – 472 (1984).
In their attempt to reach kinetic equilibrium, through gravitational encounters between separate stars, globular clusters are driven to destruction, with their cores collapsing and their outer regions expanding. The effects of core collapse, which apparently produces X–ray sources, are not yet fully understood, but white dwarfs and neutron stars, probably in binary systems, are thought to be involved, and possibly black holes as well.

151.014 **Simulating spiral galaxies.**
A. L. Peratt.
Sky Telesc., Vol. 68, No. 2, p. 118 – 122 (1984).

151.015 **Simple stellar dynamic model of the ring structure and its stability.**
V. A. Antonov, S. N. Nuritdinov.
Astrofizika, Tom 19, Vyp. 3, p. 547 – 558 (1983). In Russian. English translation in Astrophysics, Vol. 19, No. 3.
A certain phase model of the ring structure in the outer field is constructed. Its stability is studied relative to large–scale perturbations sustaining the rotation symmetry.

151.016 **Stellar dynamical models of elliptical systems.**
G. Bertin, M. Stiavelli.
Astron. Astrophys., Vol. 137, No. 1, p. 26 – 28 (1984).
The authors discuss the properties of a family of distribution functions which describe axisymmetric self–gravitating non–rotating stellar systems. These models can be treated self–consistently for the cases where the departure from spherical symmetry is small. One self–consistent equilibrium of the family is calculated in detail. It is found to develop pressure anisotropy, velocity dispersion, and energy distribution well in agreement with results obtained in recent *N*–body simulations. In addition, it satisfies the de Vaucouleurs law for the projected density distribution in a wide radial range.

151.017 **Observable properties of non–axisymmetric galaxies. II. Photometric parameters.**
G. Galletta.
Astrophys. Space Sci., Vol. 103, No. 1, p. 39 – 54 (1984).
The analysis of triaxial, coaxial ellipsoids with different intrinsic axial ratios initiated in a previous paper is extended here, introducing an intrinsic distribution of light for each shell. The general properties of this model are studied, considering the analytical solutions for the projection along a line–of–sight in three theoretical cases: (1) dust–free systems; (2) dusty systems; (3) luminous gaseous shells. The first case is then extended in order to predict some observable consequences. The observed properties of both an exponential and a power–law luminosity profile are compared with those of triaxial and axisymmetric systems. In addition, the variation of the central surface brightness and of the isophotal flattening at a fixed level with respect to the inclination of the galaxy are analyzed.

151.018 **Stellar dynamics with non–classical integrals.**
J. Binney.
Essays in theoretical physics, p. 1 – 18 (1984). – See Abstr. 003.006.
Effective integrals of motion that do not derive from a manifest Lagrangian symmetry appear to play a vital role in structuring galaxies. Recent work suggests that these effective integrals may sometimes be understood in terms of potentials that lead to separable Hamilton–Jacobi equations. However, studies of the action integrals of numerically integrated orbits in given rotating potentials suggest that many effective integrals of astronomical interest cannot be understood in this way. A hybrid approach to stellar dynamics in which the angle–action representations of orbits in model potentials are obtained by numerical integration and are then analyzed by the techniques of Hamiltonian perturbation theory may lead to a more satisfactory understanding of the relationship between potentials and the orbits that they support.

151.019 **On equilibrium states of a collisionless gravitating sphere and disk.**
Z. Ya. Turakulov.
Astrofizika, Tom 19, Vyp. 4, p. 791 – 801 (1983). In Russian. English translation in Astrophysics, Vol. 19, No. 4.
Self–consistent solutions of the kinetic equation in the proper gravitational field are obtained which depend on the integrals of motion: energy and square of angular momentum. For the sphere and disk the uniparametric series of solutions are obtained for gravitational potentials $\Psi(r)$ satisfying the following two conditions: $\Psi(r)$ is a monotonous function, and the function $r(\Psi)$, inverse to $\Psi(r)$ is known.

151.020 **Stability of a finite disc under the influence of a spherical halo.**
A. Ambastha.
J. Astrophys. Astron., Vol. 5, No. 3, p. 285 – 306 (1984).
The author has studied the stability of finite gaseous discs, against large–scale perturbations, under the influence of spherical, massive halos. A surface–density distribution consistent with the observed spiral–tracer profiles in disc galaxies is considered for the disc. It is found that growing eigenmodes with both "trailing" and "leading" spirals exist in "cold" discs for a wide range of values of the halo mass and its radius. The amplification rates of the unstable modes reduce as the ratio of the mass of the halo to the mass of the disc is increased. A uniform halo is not very effective towards stabilizing the disc against these modes.

151.021 **Galactic spiral shocks in a corrugated gas disc.**
A. H. Nelson, T. Johns, T. Matsuda.
Mon. Not. R. Astron. Soc., Vol. 210, No. 2, p. 381 – 392 (1984).
The interaction between a stellar density wave driven spiral shock and a corrugation wave in a galactic disc is calculated numerically under the tightly wound approximation. The results show that the detailed structure of the galactic shock can be significantly modified by even a small amplitude corrugation wave.

151.022 **Nonlinear perturbations of a rotating, self–gravitating gaseous disk.**
M. G. Abramyan, S. V. Arutyunyan.
Sov. Astron. Lett., Vol. 10, No. 2, p. 126 – 129 (1984). English translation of 37.151.041.

**151.023 Some evolutionary consequences of shock–induced star formation.**
M. Nepveu, M. Robnik.
Astron. Astrophys., Vol. 138, No. 1, p. 201 – 204 (1984).
Evolutionary consequences of shock induced star formation are calculated for galaxy gas contents on the basis of a simple two–component model. It is shown that a state where the mass density in stars is stationary is reached within less than $10^{9.5}$ yr and that the majority of systems evolve into a particular state.

**151.024 Equipartition and mass segregation in 1–dimensional gravitational systems.**
G. Severne, M. Luwel, P. J. Rousseeuw.
Astron. Astrophys., Vol. 138, No. 2, p. 365 – 370 (1984).
Numerical experiments are reported on the relaxation process in a series of 1–dimensional systems with a bimodal mass distribution and containing from 100 to 600 particles. There appears a clear trend towards equipartition, as measured by the ratio of the average kinetic energies for the light and heavy particles, or equivalently by the decreasing separation between the energy distribution functions for the two species. The variation of the half–mass radii exhibits a corresponding trend towards mass segregation in the systems. The relaxation time for equipartition $t_{eq}$ is of order $Nt_{C_0}$, $t_{C_0}$ being a characteristic dynamical time, rather than $N^2t_{C_0}$ as predicted by existing theory. The results indicate that the relaxation is to be attributed to the effect of binary encounters.

**151.025 Stability of star clusters as galactic satellites. III. Preliminary applications to the system of the globular clusters.**
L. Angeletti, P. Giannone.
Astron. Astrophys., Vol. 138, No. 2, p. 396 – 403 (1984).
A preliminary application of predictions of the linear stability analysis to the system of galactic globular clusters is tried in order to obtain constraints on the space distribution of the galactic mass. A comparison of the representative points of 102 globular clusters with the theoretical expectations turned out to be of interest. Two main conclusions can be drawn:(1) a point–mass model appears to be unacceptable for our galaxy; (2) the present dynamical situation of the galactic satellites is not inconsistent with a galactic mass density $\varrho(R) \propto R^{-2}$, as a function of galactocentric distance $R$. A galactic mass of the order of $10^{12} M_{\odot}$ (and a galactic radius as large as 100 kpc) seems to be consistent with present observational data.

**151.026 Stability of star clusters as galactic satellites. IV. Logarithmic galactic potential.**
L. Angeletti, R. Capuzzo–Dolcetta, P. Giannone.
Astron. Astrophys., Vol. 138, No. 2, p. 404 – 407 (1984).
The authors present the results of an investigation on the linear stability of star clusters orbiting a galaxy with a logarithmic gravitational potential. This galactic model gives rise to a flat rotation curve, which appears to approximate satisfactorily the rotation curve of our Galaxy (as well as other spiral galaxies). The present model favours the cluster stability more than the point–mass model, as expected. A qualitative agreement of the results with King's (1962) and Keenan's (1981) approaches, as discussed in a previous paper (Angeletti and Giannone, 1984) is confirmed.

**151.027 New approach to the dynamics of stellar systems.**
L. E. Lebedev.
Dnepropetr. univ. Dnepropetrovsk, 35 pp. (1984). In Russian.
Abstr. in Ref. Zh., 51. Astron., 8.51.535 (1984).

**151.028 A check on the non–linear behaviour of galaxy simulations.**
S. Inagaki, M. T. Nishida, J. A. Sellwood.
Mon. Not. R. Astron. Soc., Vol. 210, No. 3, p. 589 – 596 (1984).
The authors have obtained solutions from an $N$–body simulation and a collisionless Boltzmann simulation which predict the evolution of the same model galaxy. The principal interest was to compare the non–linear development of a bar instability. The authors find excellent agreement; in particular, the limiting amplitude of the bar differs by merely 2 per cent between the two calculations.

**151.029 Constructing distribution functions for spherical galaxies.**
A. J. Newton, J. Binney.
Mon. Not. R. Astron. Soc., Vol. 210, No. 3, p. 711 – 730 (1984).
Given a complete set of photometric and spectroscopic observations of a spherical galaxy, and an adopted form of the potential $\Phi(r)$, Binney and Mamon (1982) showed how to calculate stellar density $\nu(r)$ and radial velocity dispersion $\sigma_r(r)$ profiles. However, their work left open the question of whether these profiles can be derived from a non–negative distribution function, $f$. Here the authors have developed an algorithm for constructing a non–negative distribution function $f(\varepsilon, J)$ that generates given runs of $\nu$ and $\sigma_r$, in the adopted $\Phi$. Tests of the algorithm demonstrate its ability to reproduce a distribution function which yields specified $\nu$ and $\sigma_r$ profiles, and illustrate the behaviour of the algorithm when $\nu$ and $\sigma_r$ have been derived from a partly negative distribution function. The authors have applied their algorithm to the $\nu$ and $\sigma_r$ profiles of M87 obtained by Binney and Mamon.

**151.030 Galaxy clustering in two dimensions: some analytic results.**
H. E. Kandrup.
Astrophys. Space Sci., Vol. 104, No. 1, p. 41 – 59 (1984).
The key results of Fall, Saslaw, and Kandrup regarding the early stages of galaxy clustering are reformulated here for the model problem of two–dimensional gravity with an $r^{-1}$–interaction force. This should lead to new insights into the various physical processes involved and, more importantly, may prove of some use in interpreting numerical results obtained from two–dimensional $N$–body simulations. The calculations suggest that, for a two–, rather than three–dimensional system, the correlational energy density should grow more rapidly with time, but should exhibit only a weak logarithmic dependence upon the typical particle speed. Various other measures of the clustering are independent of the dimensionality.

**151.031 Numerical study of the relaxation of one–dimensional gravitational systems: a reply.**
B. N. Miller, H. L. Wright, W. E. Stein.
Astrophys. Space Sci., Vol. 104, No. 1, p. 145 – 149 (1984).
This paper is a response to "Numerical Study of the Relaxation of One–Dimensional Gravitational Systems", by Luwel et al. (1984) hereafter referred to as (LSR). The subject of their work is a topic studied earlier by the authors (Wright et al., 1982) hereafter referred to as (WMS). Each group has studied the dynamical evolution of this system by applying the Kolmogorov–Smirnov goodness of fit test to distributions of the density and velocity. The methods employed and conclusions reached by LSR concerning relaxation in this system are questioned. The earlier work of WMS is clarified.

**151.032 Galactic detonation waves: numerical models illustrating the transition from deterministic to stochastic.**
W. L. Freedman, B. F. Madore, S. Mehta.
Astrophys. J., Vol. 282, No. 2, p. 412 – 426 (1984).
Using numerical techniques, the authors have followed the time evolution of blast waves detonated in differentially rotating disks. Predictions from an analytic treatment are tested and extended beyond the first annihilation of opposing wave fronts. Solitary wave solutions are confirmed for galaxy–wide deterministic blast waves propagating in one direction. While azimuthal components of these detonation waves do intersect, resulting in the decay of transient cusplike solutions, it is the finite size of the disk and the coherence of the blast wave, not the collision of wave fronts, that results in the final death of the propagation. When the blast wave is less coherent either because of shorter quiescent times or because of lower efficiencies of propagation, then the detonation is very long–lived and eventually fills the galaxy with randomly propagating fragments.

151.033 **Flat rotation curves according to the fragmentation/shear flow model of galaxy formation.**
W. K. Brown.
Astrophys. Space Sci., Vol. 105, No. 1, p. 109 – 130 (1984).
Initial conditions are derived from the fragmentation/shear flow model of galaxy formation and are used as input to the viscous action presumed to begin as soon as a galactic disk forms. A simple differential equation is found to describe the turbulent viscous evolution of a flat disk. Solutions to this equation produce rotation curves that closely resemble those observed in spiral and elliptical galaxies. For spirals, by using the mass distributions derived from the rotation curves and from Seiden's theory of star formation, exponential luminosity profiles are produced.

151.034 **A unified *N*–body and statistical treatment of stellar dynamics. I. The hybrid code.**
S. L. W. McMillan, A. P. Lightman.
Astrophys. J., Vol. 283, No. 2, p. 801 – 812 (1984).
The authors develop a new, unified method for treating simultaneously both "small–*N*" and "large–*N*" effects in stellar dynamics. The numerical method consists of a "hybrid" computer code, in which all particles in the inner spatial region are followed exactly, via direct *N*–body integrations (including a full treatment of binary interactions), and particles in the outer spatial region are treated statistically, via a distribution function and Fokker–Planck methods. An intermediate transition region, with features of both, allows exchange of particles and energy between the inner and outer regions. Several tests of the hybrid code are described.

151.035 **A unified *N*–body and statistical treatment of stellar dynamics. II. Applications to globular cluster cores.**
S. L. W. McMillan, A. P. Lightman.
Astrophys. J., Vol. 283, No. 2, p. 813 – 824 (1984).
The auhors apply their new "hybrid" stellar dynamical computer code to the gravitational collapse and post collapse evolution of a globular cluster, in the approximation of point particles of initially equal mass and for a time equal to twice the time to collapse. Among the results is the formation of a hard central binary system, reversal of core collapse and expansion due to the heat input from this binary, ejection of the binary from the core, and recollapse of the core.

151.036 **How strong can a simple bar model be?**
M. Petrou.
Mon. Not. R. Astron. Soc., Vol 211, No. 1, p. 1P – 5P (1984).
It is shown that the simple two–dimensional bar models often used in the literature are not suitable for the study of galactic bars, but only appropriate for the study of oval distortions.

151.037 **Violent relaxation and dissipationless collapse.**
J. V. Villumsen.
Astrophys. J., Vol. 284, No. 1, p. 75 – 89 (1984).
A set of *N*–body simulations of dissipationless gravitational collapses of protogalactic systems was performed. The initial systems were either spheres, disks, or bars. All the initial systems were uniform density with solid body rotation. For the spheres, disks, and bars there is a sequence of models with different amounts of random kinetic energy and angular momentum. The spherical collapses are regular, with no significant substructure forming. They lead to systems that do not resemble elliptical galaxies. The nonspherical collapses are chaotic, with significant growth of substructure. The final systems resemble elliptical galaxies.

151.038 **Bursts of star formation in computer simulations of dwarf galaxies.**
N. F. Comins.
Astrophys. J., Vol. 284, No. 1, p. 90 – 97 (1984).
A three–dimensional stochastic self–propagating star formation model of compact galaxies is presented. Two phases of gas, active and inactive, are present, and permanent depletion of gas in the form of long lived, low mass stars and remnants occurs. Similarly, global infall of gas from a galactic halo or through galactic cannibalism is permitted. The parameters are based on the observed properties of the compact blue galaxy I Zw 36. The results suggest that the blue compact galaxies are probably undergoing bursts rather than continuous, nonbursting low–level star formation activity.

151.039 **The response density of irregular orbits in barred potentials.**
M. Petrou.
Mon. Not. R. Astron. Soc., Vol. 211, No. 2, p. 283 – 295 (1984).
Several irregular orbits in a planar bar are followed for up to a Hubble time. Their contribution to the density of the bar is found as well as their evolution on the surface of section. The existence of cantori (partial barriers) in the surface of section is shown and its significance discussed. The trapped orbits along the bar play a very important role in the existence of the bar. They reinforce it so strongly that they may counter–balance the destructive effect of up to 45 per cent of untrapped orbits.

151.040 **The instability of a flat gaseous layer with shear motions and fragmentation of a protocluster.**
P. I. Kolykhalov, S. F. Shandarin.
Pis'ma Astron. Zh., Tom 10, No. 11, p. 820 – 826 (1984). In Russian. English translation in Sov. Astron. Lett., Vol. 10.
Dynamics of flat extensive gaseous layers with strong gradients of shear velocity formed as a result of the gravitational instability in the expanding Universe is investigated. Hydrodynamical instability leads to fragmentation of the layers and makes favourable conditions for galaxy formation. The instability of supersonic flows is considered.

151.041 **Close encounters between a star–planet system and a stellar intruder.**
J. G. Hills.
Astron. J., Vol. 89, No. 10, p. 1559 – 1564 (1984).
The paper reports the results of nearly 40,000 computer simulations of close encounters between a planet–star (P–S) system and a stellar intruder. If the closest approach of the intruder is 2 – 3 times the semimajor axis of the orbit or less, the encounter tends to increase the semimajor axis of the P–S system or even dissociate it. More distant encounters tend to produce a mild shrinking of the P–S orbit. Averaged over all impact parameters, the encounters are disruptive. Close encounters which do not lead to disruption usually lead to the stellar intruder capturing the planet. The average orbital eccentricity of a captured planet is greater than $e = 0.7$, but some close encounters between a solar–mass intruder and an Earth–Sun system would result in the Earth being captured by the stellar intruder in an orbit with an eccentricity of less than $e = 0.10$.

151.042 **Periodic orbits in a resonant dynamical system.**
N. Caranicolas.
Celest. Mech., Vol. 33, No. 3, p. 209 – 215 (1984).
The author studies the periodic orbits in a two dimensional dynamical system, symmetric with respect to both axes, with two equal or nearly equal frequencies. It is shown that the periodic orbits can be found directly from the equations of motion. The form of these orbits depends on the value of the coupling parameter $\alpha$. The author verifies the theoretical results by numerical calculations.

151.043 **Effects of encounters with field stars on the evolution of low–mass semidetached binaries.**
P. Hut, B. Paczyński.
Astrophys. J., Vol. 284, No. 2, p. 675 – 684 (1984).
Consequences of close encounters on the binary parameters and on the mass transfer rate and its stability are discussed. Thousands of numerical experiments have been performed in order to obtain accurate cross sections for inducing eccentricity in binaries with initially circular orbits. Analytic formulae are presented relating the radius excess of the secondary component over its Roche lobe to the mass transfer rate, for two extreme models of mass outflow: adiabatic and isothermal. Mass transfer is found to be stable even if there is no angular momentum transferred from the disk around the primary component back

into binary orbit. It is not likely that close encounters significantly affect the evolution of bright globular cluster X–ray sources, unless central densities in the cores of those clusters are much higher than commonly believed.

151.044 **The tensor virial theorem for subsystems. II.**
R. Caimmi, L. Secco, P. Brosche.
Astron. Astrophys., Vol. 139, No. 2, p. 411 – 416 (1984).

According to an extension of the tensor virial theorem for subsystems formulated in a previous paper, the authors evaluate the new terms for the more general case of two concentric and coaxial homogeneous spheroids, one lying completely within the other. Possible applications to elliptical galaxies and spiral bulges are given.

151.045 **Linear oscillations of isotropic stellar systems. I Basic theoretical considerations.**
Y. Sobouti.
Astron. Astrophys., Vol. 140, No. 1, p. 82 – 90 (1984).

Linear perturbations of stellar systems with step–like distributions of the form $F(E)H(-E)$ are studied, where $E$ is the energy integral, and $H$ is Heaviside's step–function. If (1) $dF/dE > 0$ and (2) $F|dF/dE|^{-1/2} = 0$ at $E = 0$, then the operator generating the self gravitation term in Antonov's equation is positive. Then the system is stable against all infinitesimal perturbations of the Boltzmann–Liouville equation. Polytropes with index $0.5 < n < 1.5$ satisfy conditions (1) and (2) and are stable. If (1) is satisfied and (2) is not satisfied, instability is possible. Polytropes with $n < 0.5$ are examples of such systems. If $dF/dE < 0$, the operator generating the self gravitation term is negative and contributes negatively to the stability of the system.

151.046 **Wie bewegen sich die Sterne in einem Kugelsternhaufen?**
R. Stettler.
Orion, 42. Jahrg., Nr. 205, p. 214 – 217 (1984).

151.047 **Can the ballistic particle model explain the Hubble sequence?**
D. Leisawitz, F. Bash.
Astrophys. J., Vol. 285, No. 1, p. 25 – 43 (1984).

A model is considered in which giant molecular clouds (GMCs) are supported against spontaneous gravitational collapse by the pressure of their internal turbulence fed by collisions with small interstellar clouds whose motions are described by two–armed spiral–shock (TASS) wave hydrodynamics. By varying a model galaxy's total mass, its mass distribution, and its spiral pattern speed, galaxies of different spiral types are simulated. By adopting a picture in which unsupported GMCs collapse to form H II regions whereas those that receive too much energy are disrupted, the authors predict for each model galaxy the H II region and GMC radial distributions, the spiral arm width as a function of galactocentric radius, the overall appearance of the spiral structure, and the kinematics of the H II regions and GMCs. The model galaxies are compared with real spiral galaxies of various morphological types.

151.048 **The formation and early dynamical evolution of bound stellar systems.**
C. J. Lada, M. Margulis, D. Dearborn.
Astrophys. J., Vol. 285, No. 1, p. 141 – 152 (1984).

The authors present the results of numerical $N$–body calculations which simulate the dynamical evolution of young clusters as they emerge from molecular clouds. They follow the evolution of initially virialized stellar systems of 50 and, in some cases, 100 stars from the point in time immediately after the stars have formed in a cloud until a time long after all the residual star–forming gas has been dispersed from the system. It is found that clusters which remain bound after gas dispersal undergo significant expansion as they emerge from a cloud and in the process may lose 10% – 80% of their stars, depending on the star formation efficiency at the time of gas removal and duration of gas removal from the system. By scaling the models to match the observed parameters of real clusters the authors constrain the range of initial mass densities, sizes, and velocity dispersions of protocluster molecular clouds. Arguments are presented which suggest that even if a newly formed stellar system was not initially virialized, it would most likely reach such an equilibrium state before significant gas removal occurred.

151.049 **A new approach to the relaxation process in gravitational systems.**
G. Severne, M. Luwel.
Phys. Lett. A, Vol. 104A, No. 3, p. 127 – 130 (1984). Abstr. in Phys. Abstr., Vol. 87, No. 1246, Entry 110957 (1984).

151.050 **The evolution of rotating star clusters at the inelastic–collision stage. I. Contraction of the stellar component.**
M. M. Romanova.
Sov. Astron., Vol. 28, No. 2, p. 147 – 153 (1984). English translation of 37.151.038.

151.051 **Hydrodynamic equations for a disk of stars.**
S. V. Chalov.
Sov. Astron., Vol. 28, No. 2, p. 154 – 158 (1984). English translation of 37.151.039.

151.052 **Differential stellar space–velocity parameters for motions perturbed by the spiral density wave.**
A. N. Balakirev.
Sov. Astron., Vol. 28, No. 2, p. 158 – 160 (1984). English translation of 37.151.040.

151.053 **Stellar encounter probability and multiplicity in a plane gravitating medium.**
I. V. Petrovskaya, Z. N. Chumak, O. V. Chumak.
Sov. Astron., Vol. 28, No. 3, p. 275 – 278 (1984). English translation of 37.151.081.

151.054 **Resonances in a low–mass disc.**
W. R. Ward.
Bull. Am. Astron. Soc., Vol. 16, No. 3, Part I, p. 701 (1984). Abstract. – See Abstr. 010.061.

151.055 **Discrete methods for the construction of model galaxies.**
P. O. Vandervoort.
Bull. Am. Astron. Soc., Vol. 16, No. 3, Part I, p. 720 (1984). Abstract. – See Abstr. 010.064.

151.056 **An approximate numerical model for a nonrotating galactic bar.**
M. Schwarzschild.
Bull. Am. Astron. Soc., Vol. 16, No. 3, Part I, p. 720 (1984). Abstract. – See Abstr. 010.064.

151.057 **Self–consistency for elliptical Kuzmin–Toomre disks.**
C. Hunter.
Bull. Am. Astron. Soc., Vol. 16, No. 3, Part I, p. 720 (1984). Abstract. – See Abstr. 010.064.

151.058 **Evolution of a stellar system using the tensor–virial theorem.**
P. Pismis, E. Moreno.
Bull. Am. Astron. Soc., Vol. 16, No. 3, Part I, p. 720 (1984). Abstract. – See Abstr. 010.064.

151.059 **Stellar and gas kinematics in disk galaxies.**
B. C. Whitmore, V. C. Rubin, W. K. Ford.
Bull. Am. Astron. Soc., Vol. 16, No. 3, Part I, p. 724 – 725 (1984). Abstract. – See Abstr. 010.065.

151.060 **Tidal encounters of elliptical galaxies.**
L. A. Aguilar, S. D. M. White.
Bull. Am. Astron. Soc., Vol. 16, No. 3, Part I, p. 728 (1984). Abstract. – See Abstr. 010.065.

151.061 **Simulations of galactic cannibalism and multiple mergers.**
M. J. Duncan.
J. R. Astron. Soc. Can., Vol. 78, No. 5, p. 212 (1984). Abstract. – See Abstr. 010.401.

151.062 **Observable parameters using N–body models of Kalnajs Ω discs.**
J. J. E. Hayes, N. F. Comins.
J. R. Astron. Soc. Can., Vol. 78, No. 5, p. 212 (1984). Abstract. – See Abstr. 010.401.

151.063 **The velocity fields of barred galaxies.**
D. Pfenniger.
Astron. Astrophys., Vol. 141, No. 1, p. 171 – 188 (1984).
Schwarzschild's method is used for constructing self–consistent *2D* models of a well pronounced and fast rotating barred galaxy. In particular the basic elements of the distribution function, the invariant tori, are found by integrating orbits until the fluctuations of orbital density stay below a fixed threshold, except for the semi–ergodic orbits which show no diminution of the fluctuations. For finding a self–consistent solution, a non–negative least squares technique is used, which promises interesting applications to the interpretation of real barred galaxies, for which, beside the mass distribution, only some radial velocities and the orientation are known. By extremizing various quantities like the total energy or angular momentum, the set of possible self–consistent solutions is explored.

151.064 **Families of periodic orbits in a quartic potential.**
N. Caranicolas, H. Varvoglis.
Astron. Astrophys., Vol. 141, No. 2, p. 383 – 388 (1984).
The characteristic curves of several families of periodic orbits in a two–dimensional quartic potential field symmetric with respect to both axes $x$ and $y$ are explored. Two types of families are found: (a) regular families which are connected with the families of the unperturber system and (b) irregular families not connected with the above. It is shown that some features of the regular families can be explained by means of a second formal integral of motion. The present results are compared with the results of an earlier study of families of periodic orbits in a cubic potential.

151.065 **Spiral structure generated by infalling matter drifting in the corotation resonance zone of a galaxy.**
Yu. N. Eliseev, K. N. Stepanov.
Sov. Astron. Lett., Vol. 10, No. 3, p. 141 – 144 (1984). English translation of 37.151.078.

151.066 **A theorem due to Jörg Pfleiderer concerning the stability of globular clusters.**
S. C. Temesváry.
Theoretical problems in stellar stability and oscillations, p. 476 – 478 (1984). – See Abstr. 012.065.

151.067 **The formation of ring galaxies.**
T. K. Chatterjee.
Astrophys. Space Sci., Vol. 106, No. 2, p. 309 – 339 (1984).
The conditions under which a head–on collision between a disk galaxy and a spherical galaxy can lead to ring formation are investigated, using the impulsive approximation. The spherical galaxy is modeled as a polytrope of index $n = 4$ and radius $R_S$ and the disk galaxy as an exponential disk. Calculations are made for the case $R_S = R_D$. Since practically the entire mass of the spherical galaxy, for the chosen model, lies within 1/3 of its radius, the radius of the spherical galaxy is effectively 1/3 that of the disk galaxy. It is found that as a result of the collision, the innermost and the outer parts of the disk galaxy are not much affected, but the intermediate region expands and gets evacuated, leading to the crowding of stars in a preferential region forming a ring structure. The rings are best formed for a normal, on–axis collision.

151.068 **The tensor virial theorem applied to an inhomogeneous spheroidal stellar system.**
P. Pismis, E. Moreno.
Rev. Mex. Astron. Astrofis., Vol. 9, No. 2, p. 171 – 177 (1984).
The authors have applied the tensor virial theorem to an initially oblate stellar system without collisions in order to study the variation of its configuration in time. They find that the system undergoes an oscillation between an oblate and a prolate form.

151.069 **The response of a spheroid to a disc field or were bulges ever ellipticals?**
J. Barnes, S. D. M. White.
Mon. Not. R. Astron. Soc., Vol. 211, No. 4, p. 753 – 765 (1984).
A collisionless N–body code is used to study the response of a proto–bulge to the field of an exponential disc. In the limit where the time–scale of the disc's growth is long compared to the dynamical time–scale of the bulge, the final state of the bulge will depend on the final state of the disc, but not on the details of how the disc matter was assembled. The disc field acts to increase the surface brightness, velocity dispersion, rotation velocity and flattening of the bulge, but does not (directly) account for either the "thick disc" or "box–shaped" bulge phenomena. The results suggest that the close similarity between bulges and ellipticals cannot result from disc accretion on to a pre–existing normal elliptical unless discs are considerably less massive than is currently believed.

151.070 **Making warps from bars and haloes.**
L. S. Sparke.
Mon. Not. R. Astron. Soc., Vol. 211, No. 4, p. 911 – 926 (1984).
It is known that some orbits in the principal planes of a triaxial potential well are unstable to oscillations perpendicular to those planes. This fact suggests that if a galaxy has a central bar or a triaxial unseen halo, the disc may become warped. Here, the response of a massive galactic disc to a variety of imposed triaxial forcing potentials is investigated, using both numerical experiment and analytic methods. Galactic bars are unlikely to be responsible for warps. A halo of constant shape, flat rotation curve and reasonable mass, with a figure rotating retrograde to the disc, can amplify a small disturbance to an observable level within reasonable times, if it is strongly flattened towards the disc, and weakly barred.

151.071 **The stochastic behaviour of a galactic model dynamical system.**
B. Barbanis.
Celest. Mech., Vol. 33, No. 4, p. 385 – 395 (1984).
The author studies the stochastic behaviour of a time–independent dynamical system that has closed zero velocity curves for arbitrarily large energies. Thus no escapes appear in such a system and it is possible to study the stochasticity of the system for any value of the energy. The numerical results show that the degree of stochasticity grows as the energy increases from zero, but then it reaches a maximum and for still larger energies it decreases slowly.

151.072 **Stellar dynamics of galaxy interactions.**
O. E. Gerhard.
MPA Rep., No. 162, 5 pp. (1984). To appear in Mitt. Astron. Ges., Nr. 62.

151.073 **N–body simulations and galaxy photometry.**
O. E. Gerhard.
MPA Rep., No. 165, 8 pp. (1984).
N–body models are now a standard tool to improve our understanding of galaxies. Since photometric and kinematic observations of a galaxy do not in general very strongly constrain a dynamical model, the main use of N–body simulations in this respect is to answer qualitative questions about the nature of possible equilibria. This is illustrated by presenting a model with strong radial ellipticity gradients and one with intrinsic principal axes twists.

151.074 **The stochastic behaviour of a galactic model dynamical system.**
B. Barbanis.
ESO Sci. Prepr., No. 332, 22 pp. (1984). Submitted to Celest. Mech.

151.075 **Nonlinear problems in stellar dynamics.**
G. Contopoulos.
ESO Sci. Prepr., No. 342, 18 pp. (1984). To be published in the Proceedings of ELAF 84.

151.076 **Flyby: numerical experiments on a galaxy orbiting within a galaxy cluster.**
R. H. Miller.
ESO Sci. Prepr., No. 343, 81 pp. (1984). Submitted to Astron. Astrophys.

151.077 **Numerical simulations of encounters of hard binaries.**
S. Mikkola.
Rep. Ser., Dep. Phys. Sci., Univ. Turku, No. FTL–R70, 9 pp. (1984). = Turku Univ. Obs., Informo No. 86. Presented as a poster in IAU Symp. No. 113, 29 May – 1 June, 1984, Princeton, USA.

151.078 **Thermodynamics and stellar systems.**
F. A. Tsitsin, V. N. Sementsov.
Astron. Tsirk., No. 1326, p. 6 – 8 (1984). In Russian.

151.079 **A general method for constructing spherical galaxy models.**
D. O. Richstone, S. Tremaine.
Astrophys. J., Vol. 286, No. 1, p. 27 – 37 (1984).
The authors have used Schwarzschild's linear programming method to construct self–consistent dynamical models of spherically symmetric galaxies. The models have mass–to–light ratios which are independent of radius, and surface brightness profiles obeying de Vaucouleurs or King laws. Schwarzschild's method yields the most general solutions of the Vlasov and Poisson equations satisfying these constraints. The authors give special attention to models with the minimum and maximum line–of–sight velocity dispersion at various distances from the galaxy center, in order to investigate the uncertainties associated with the use of a single velocity dispersion measurement to determine the mass–to–light ratio of a galaxy.

151.080 **Angular momentum growth in protogalaxies.**
S. D. M. White.
Astrophys. J., Vol. 286, No. 1, p. 38 – 41 (1984).
The author expands a 1970 analysis by Doroshkevich which shows that the angular momentum of galaxies grew to first order (in proportion to $t$) during the linear phases of protogalactic evolution. This result is confirmed in $N$–body simulations of the formation of structure. The well–known 1969 study of Peebles found growth at second order only (in proportion to $t^{5/3}$), because its analysis was restricted to spherical regions. These considerations do not affect estimates of the *total* angular momentum acquired by a galaxy in the gravitational instability picture.

151.081 **Preferred orbit planes in triaxial galaxies. II. Tumbling about a nonprincipal axis.**
L. P. David, R. H. Durisen, T. Y. Steiman–Cameron.
Astrophys. J., Vol. 286, No. 1, p. 53 – 61 (1984).
The authors use a method from celestial mechanics to determine the preferred orbit planes into which gas will settle in the external potential of both spheroidal and triaxial galaxies tumbling about nonprincipal axes. These preferred planes are directly related to the locations of periodic orbits. The number and orientation of the preferred planes depend on the shape of the galaxy, the orbit radius, the tumble axis orientation, and the tumble rate. One of the major results of this work is that, for galaxies tumbling about a nonprincipal axis, the preferred planes exhibit warps and twists with increasing radial distance from the galaxy center.

151.082 **Gravitational spurs and resonances: effects of small mass disturbers in spiral galaxy disks.**
G. G. Byrd, B. F. Smith, R. H. Miller.
Astrophys. J., Vol. 286, No. 1, p. 62 – 67 (1984).
It has been suggested that the short spurs in both two–armed and chaotic spiral galaxies are localized gravitational disturbances (density wavelets) caused by orbiting clouds of gas and stars in the disk. New $N$–body computer simulations investigate the effects of a disturber on a galaxy's stellar disk. Disturber parameters of the NGC 206 cloud in M31 ($8 \times 10^7\ M_\odot$, 8 kpc) were assumed as a realistic example. These simulations confirm the less general calculations of Julian and Toomre in the small region around an NGC 206 cloud mass orbiting disturber. A simulation with a more massive disturber shows a better defined pattern that clarifies the lower mass runs. The main structure is a spiral arc of density peaks rotating at the same angular rate as the disturber. The density peak radii correspond to theoretical positions of epicyclic resonances with the main spur around the orbiting disturber.

151.083 **Drift motion of particles in an axially–nonsymmetrical gravitational field of a rotating galaxy.**
Yu. N. Eliseev, K. N. Stepanov.
Probl. yader. fiz. i kosm. luchej, Khar'kov, No. 20, p. 57 – 78 (1984). In Russian. Abstr. in Ref. Zh., 51. Astron., 12.51.980 (1984).

151.084 **Stellar dynamics of galaxy interactions.**
O. E. Gerhard.
Mitt. Astron. Ges., Nr. 62, p. 146 – 150 (1984). – See Abstr. 010.341.
In this brief review the author first discusses typical parameters of encounters in different galaxy environments. He then describes stellar–dynamical processes and techniques relevant to galaxy interactions in three astrophysical contexts: Tidal heating of cluster galaxies, shell formation around ellipticals, and formation of tidal tails and merging in strong collisions.

151.085 **Auswirkungen eines Wechselwirkungsterms für inelastische Galaxienbegegnungen auf die Bildung und Entwicklung von Galaxienhaufen unter besonderer Berücksichtigung der Mergerbildung.**
P. Schwekendiek, R. Wielen.
Mitt. Astron. Ges., Nr. 62, p. 152 (1984). – See Abstr. 010.341.

151.086 **Model calculations of the evolution of gas in elliptical galaxies.**
R. Kunze, H. W. Yorke, H.–H. Loose.
Mitt. Astron. Ges., Nr. 62, p. 257 (1984). Abstract. – See Abstr. 010.341.

151.087 **Spiralstruktur in Galaxien: Stochastik und Grenzzyklus.**
J. V. Feitzinger.
Mitt. Astron. Ges., Nr. 62, p. 338 (1984). Abstract. – See Abstr. 010.341.

151.088 **Globale Stabilitätseigenschaften einer speziellen Klasse von Gleichgewichtslösungen für Scheibengalaxien und der Einfluß der Dissipation.**
W. Seboldt.
Mitt. Astron. Ges., Nr. 62, p. 345 – 349 (1984). – See Abstr. 010.341.

151.089 **Dichtewellen–induzierte Entropieänderungen und Entropieströme in Spiralgalaxien.**
B. Fuchs.
Mitt. Astron. Ges., Nr. 62, p. 349 (1984). Abstract. – See Abstr. 010.341.

151.090 **Warum sind Sternsysteme anisotrop?**
E. Bettwieser, R. Spurzem, K. J. Fricke.
Mitt. Astron. Ges., Nr. 62, p. 350 (1984). Abstract. – See Abstr. 010.341.

151.091 **Gasdynamische Simulation der stellardynamischen Entwicklung von Multi–Massen Sternsystemen.**
E. Bettwieser, S. Inagaki.
Mitt. Astron. Ges., Nr. 62, p. 351 (1984). Abstract. – See Abstr. 010.341.

151.092 **Sind gas– bzw. thermodynamische Konzepte gültig für Ergebnisse von direkten n–Körper Rechnungen?**
E. Bettwieser, D. Sugimoto.
Mitt. Astron. Ges., Nr. 62, p. 352 (1984). Abstract. – See Abstr. 010.341.

151.093 **On n–bodies experiments on disk galaxies' evolution.**
Yu. N. Mishurov.
Astron. Zh., Tom 61, Vyp. 6, p. 1074 – 1078 (1984). In Russian. English translation in Sov. Astron., Vol. 28, No. 6.

On the basis of an analysis of computer modelling of disk galaxies' evolution the conclusion is drawn that flat systems used in earlier papers are essentially collisional, unlike the real galaxies. Both systems' heating and structure formation arising due to instability taking place in rotating gravitating systems in the presence of dissipation are connected with this.

151.094 **On equipartition of kinetic energies in two–component star clusters.**
S. Inagaki, P. Wiyanto.
Publ. Astron. Soc. Jpn., Vol. 36, No. 3, p. 391 – 402 (1984).

The dynamical evolution of two–component star clusters is studied by solving the isotropized orbit–averaged Fokker–Planck equation. The simulations are started from Plummer's model with the same velocity dispersion between the components. It is found that the global equipartition is never achieved even if the fraction of the more massive stars is very small. The temperature difference in the core is very small if the fraction of the more massive stars is small. The larger the fraction of the more massive stars is, the larger is the termperature difference in the core. The late stage of the evolution is governed by the more massive component. The more massive component collapses independently of the less massive component and the less massive component follows the collapse of the more massive stars, owing to the strong gravitational field of the more massive component.

151.095 **The approach to self–similar structure during post–collapse evolution of stellar systems.**
E. Bettwieser, U. Fritze.
Publ. Astron. Soc. Jpn., Vol. 36, No. 3, p. 403 – 416 (1984).

According to classical pre–collapse calculations the cores of dense stellar systems collapse to infinite density within a finite time. This paper is concerned with the post–collapse evolution of stellar systems, such as globular clusters or galactic nuclei. Superscattering processes on tight binaries supply the core with additional energy, a process which is taken into account by an artificial heat source. Evolution is found to be directed towards self–similarity except for the very center and for spatial fluctuations in the outer regions.

151.096 **The distribution model of the velocity dispersion of stars within the Galaxy.**
X.–t. Zheng, Y. Tong, S.–g. Wu, Q.–h. Peng.
Acta Astron. Sin., Vol. 25, No. 2, p. 142 – 147 (1984). In Chinese.

The variation of velocity dispersion in $z$–axis with $r$ and $z$, namely the analytical expression of $C_z = C_z(r, z)$ is obtained by direct integration. Besides, the variation of radial velocity dispersion with $r$ and $z$ is also obtained through the stability of the density wave in three–dimensional spiral galaxies and the equation of dispersion. The results show that theoretical values of $C_r$ and $C_z$ are in agreement with the observed values near the sun and also approach the values obtained by Huang et al. (1979).

151.097 **The evolution of some kinds of flat galaxies.**
J.–q. Zheng, H.–n. Zhou.
Acta Astron. Sin., Vol. 25, No. 2, p. 148 – 155 (1984). In Chinese.

Some models are constructed. Using the method proposed in a previous paper, the spiral structure and the dynamical evolution are obtained. The results are roughly consistent with the main observational characteristics of actual galaxies.

151.098 **Computer modelling of the gas dynamics in small groups of interacting galaxies.**
P. A. Foster.
Clusters and groups of galaxies, p. 381 – 384 (1984). – See Abstr. 012.088.

151.099 **Can gas and stars be separated in a tidal interaction?**
P. N. Appleton.
Clusters and groups of galaxies, p. 385 – 387 (1984). – See Abstr. 012.088.

151.100 **Forces between overlapping Plummer–model galaxies.**
F. Ahmad.
Clusters and groups of galaxies, p. 515 – 516 (1984). – See Abstr. 012.088.

151.101 **The stability of elliptical stellar disks. I. Balanced disks.**
G. S. Bisnovatyj–Kogan.
Astrofizika, Tom 20, Vyp. 3, p. 547 – 563 (1984). In Russian. English translation in Astrophysics, Vol. 20, No. 3.

The stability of balanced elliptical disks with quadratic gravitational potential relative to bar–like perturbations for binary systems and in the presence of a spheroidal halo is investigated. It is shown that in binary systems elongated disks and compressed disks close to the circular ones are aperiodically unstable. Strongly compressed disks are stable for large ratio of masses: $M/M_2$ or $M_2/M$ is greater $\sim 30$; oscillating instability takes place for intermediate mass ratio. The disks in the uniform spheroidal halo are stabilized relative to the bar–like perturbations if the gravitational potential of the halo exceeds one half of the gravitational potential of the disk.

151.102 **Integral characteristics of binary galaxies with tidal forces.**
G. S. Bisnovatyj–Kogan.
Astrofizika, Tom 21, Vyp. 1, p. 87 – 92 (1984). In Russian. English translation in Astrophysics, Vol. 21, No. 1.

The expressions for kinetic and potential energies as well as angular momentum are obtained for binary galaxies, described in tidal approximation by a self–consistent solution with quadratic gravitational potential. The corrections to the Keplerian velocity of orbital rotation are calculated and a check of the virial theorem is made.

151.103 **Dissipative formation of an elliptical galaxy.**
R. G. Carlberg.
Astrophys. J., Vol. 286, No. 2, p. 403 – 415 (1984).

Dissipative collapse and star formation within a protogalaxy are modeled using a simple scheme of gas cloud collisions within an $N$–body code. The gas physics model incorporates pressure, dissipation, and star formation, with the associated ejection of metal–enriched gas back into the cloud medium. The gas and stars collapse under their self–gravity alone; adding a static potential corresponding to a massive dark halo suppresses both velocity anisotropies and mass loss in low–mass ellipticals. A set of models in general accord with the kinematic and chemical properties of individual ellipticals is found.

151.104 **Dissipative models for the sequence of elliptical galaxies.**
R. G. Carlberg.
Astrophys. J., Vol. 286, No. 2, p. 416 – 421 (1984).

Protogalaxies with a range of masses and sizes are numerically collapsed with a dissipative model for the formation of elliptical galaxies discussed in a companion paper. The rates of dissipation and star formation in the model are calculated with a simple cloud description of the gas physics. The correlations between the final mass, metal abundance, and kinematics in the models are compared with the empirical correlations of these quantities in an attempt to constrain the size–mass relation of the protogalaxies. These models can be placed in a cosmological context with the results that the ratio of the initial to final half–mass radii is near 10 for the favored models, suggesting formation at a redshift near 10 from sheets of gas gathered up with respect to the dark material.

151.105 **Stochastic models for the evolution of open stellar systems.**
F. Ferrini, F. Marchesoni.
Astrophys. J., Vol. 287, No. 1, p. 17 – 25 (1984).

The authors explore the effects of stochastic processes in a one–zone model galaxy by introducing a noise term in two– and three–phase models (gas, clouds, stars). They show that in the scheme both deterministic and noise–induced phase–transition phenomena arise. The equilibrium states and the dynamical evolution of each phase are influenced greatly. The authors solve some simplified cases analytically and check the results through numerical simulations.

151.106 **On Schwarzschild's method for the construction of model galaxies.**
P. O. Vandervoort.
Astrophys. J., Vol. 287, No. 2, p. 475 – 486 (1984).
The method developed recently by Schwarzschild for the construction of model galaxies is formulated explicitly as a numerical method for the solution of the equations of galactic dynamics governing the equilibrium configurations of a stellar system. In this formulation, the method applies to systems in which the stellar orbits admit of three isolating integrals of the motion, and it provides numerical solutions for the distribution of stars in the phase space of a single star. The present version of Schwarzschild's method is illustrated with the construction of model galaxies with spherically symmetric mass distributions.

151.107 **Interacting binary galaxies. I. A numerical model and preliminary results.**
K. D. Borne.
Astrophys. J., Vol. 287, No. 2, p. 503 – 522 (1984).
A numerical code has been constructed for the study of the evolution of interacting binary galaxies. This "multiple three–body algorithm" (MTBA) essentially involves $N$ concurrent three–body integrations; it is the sum of the interactions of the perturber with each of the $N$ particles comprising the primary "galaxy" that determines the binary orbital evolution. MTBA incorporates a violent relaxation phase that allows the particles to redistribute themselves in the gravitational field of the perturber prior to the full binary orbital evolution calculation. A specific interacting binary simulation is described in detail.

151.108 **Isothermal spheres in the modified dynamics.**
M. Milgrom.
Astrophys. J., Vol. 287, No. 2, p. 571 – 576 (1984).
The author considers self–gravitating (isotropic and anisotropic) isothermal spheres (ISs) using dynamics that differ from the Newtonian in the limit of small accelerations. ISs serve as reference models for various astrophysical objects such as globular and open clusters, galactic bulges, and clusters of galaxies. The modified dynamics have been suggested as an alternative to the hidden mass hypothesis. The author studies spheres in which both the radial and the tangential velocity dispersions are independent of the radius.

151.109 **Recognizing merger remnants among normal elliptical galaxies: NGC 5813.**
J. Kormendy.
Astrophys. J., Vol. 287, No. 2, p. 577 – 585 (1984).
This paper suggests that departures from isothermal structure in the core brightness profiles of elliptical galaxies can be a signature of formation by mergers. The cores of small ellipticals and of bulges of disk galaxies are shown to survive a merger with a giant elliptical provided that the initial core radii differed by a factor of at least 6 – 10. A core–within–a–core structure should result. The observable properties of this structure are derived. The normal giant elliptical NGC 5813 may be such a galaxy. It shows two nested cores with the properties of ellipticals of $M_B \approx -18.9$ and $-21.4$, respectively. Nonisothermal cores can result from mergers or from nuclear mass concentrations such as black holes. Tests for distinguishing between these alternatives are described.

151.110 **Phase–mixing of newly born stars in the galactic disk.**
B. Fuchs.
Astrophys. Lett., Vol. 24, No. 3, p. 153 – 160 (1984).
The evolution of the velocity distribution of stars born in the galactic disk with isotropic peculiar velocities is studied. By following stellar orbits in the epicyclic approximation, it is shown that the coarse–grained distribution function approaches an ellipsoidal one with the correct axial ratios. If the original isotropic distribution is Maxwellian, the resulting function resembles the Schwarzschild distribution.

**Physics of gravitating systems. I. Equilibrium and stability.**
See Abstr. 003.008.

**Physics of gravitating systems. II. Nonlinear collective processes: nonlinear waves, solitons, collisionless shocks, turbulence. Astrophysical applications.**
See Abstr. 003.009.

**Numerical simulations of collisions in self–gravitating systems.**
See Abstr. 042.010.

**Periodic orbits in a three–dimensional potential. II. Bifurcations.**
See Abstr. 042.028.

**Gravitational encounters between orbiting bodies.**
See Abstr. 042.054.

**Third–order and tidal effects in the stellar three–body problem.**
See Abstr. 042.058.

**N–Körper Hybrid Integrations–Methode.**
See Abstr. 042.083.

**Angular momentum versus mass, re–examined.**
See Abstr. 061.126.

**The summation of Bessel products.**
See Abstr. 062.039.

**Explosions into a 'vacuum'.**
See Abstr. 062.115.

**Stability of bidimensional stellar systems for axisymmetric modes.**
See Abstr. 065.014.

**On the stability of a new relativistic kinetic equation.**
See Abstr. 066.161.

**Does gravity exhibit a $1/r$ force on the scale of galaxies?**
See Abstr. 066.182.

**Milgrom's revision of Newton's laws: dynamical and cosmological consequences.**
See Abstr. 066.265.

**Does the missing mass problem signal the breakdown of Newtonian gravity?**
See Abstr. 066.266.

**The distribution of the projected density of stars near the Reissner–Nordström black hole in a globular cluster.**
See Abstr. 067.032.

**The boundary of the Solar System.**
See Abstr. 091.005.

**How stable is an astronomical clock that can trigger mass extinctions on Earth?**
See Abstr. 107.027.

**Orbital stability of the unseen solar companion linked to periodic extinction events.**
See Abstr. 107.028.

**Dynamics of interstellar clouds sweeping from a rotating galaxy during its motion in the intracluster medium.**
See Abstr. 131.045.

**Dynamical evolution of fragmenting gas clouds.**
See Abstr. 131.093.

**Formation and evolution of rings in disk–shaped protostellar clouds.**
See Abstr. 131.196.

**The Galaxy as a self–regulated star–forming system: the case of the OB associations.**
See Abstr. 155.064.

**A galactic disk is not a true exponential.**
See Abstr. 157.014.

**A kinematic method for measuring the pattern speed of barred galaxies.**
See Abstr. 157.017.

**Kinematical tests for the intrinsic shapes of galaxies.**
See Abstr. 157.031.

**Flat rotation curves and mass distribution for spiral galaxies with a disk model.**
See Abstr. 157.042.

**Velocity dispersions of disk stars in spiral galaxies.**
See Abstr. 157.120.

**Neutral hydrogen observations of the interacting galaxies NGC 4725 and NGC 4747.**
See Abstr. 157.125.

**Warps, spiral structure and star bursts caused by perturbed flattened haloes in disc galaxies.**
See Abstr. 157.165.

**Stellar and gas kinematics in disk galaxies.**
See Abstr. 157.166.

**Three–dimensional mass distribution in discoid galaxies with both bulge and disk.**
See Abstr. 157.202.

**Unusual properties of bar–within–ring galaxies and the projection effect model for barred spiral galaxies.**
See Abstr. 157.224.

**Double galaxy redshifts and dynamical analyses.**
See Abstr. 157.227.

**Stellar and gas kinematics in disk galaxies.**
See Abstr. 157.231.

**On the structure of the nucleus of M31.**
See Abstr. 157.233.

**Distribution and kinematics of CO in the galaxy M82.**
See Abstr. 158.008.

**Mass accretion by the nuclei of disk galaxies.**
See Abstr. 158.259.

**Tidal torques and the orientations of galaxies in clusters.**
See Abstr. 160.020.

**The virial masses of clusters of galaxies and the effects of contamination.**
See Abstr. 160.027.

**Monte–Carlo simulations of galaxy systems. VI. The angular correlation function for faint galaxy samples.**
See Abstr. 160.036.

**Globular cluster swapping in clusters of galaxies.**
See Abstr. 160.046.

**On the characteristics of clusters of galaxies with unseen mass.**
See Abstr. 160.047.

**Gravitational collapse of cosmic gas clouds and formation of star clusters.**
See Abstr. 161.186.

**Gravitational clustering for a multicomponent system with a distribution of temperatures.**
See Abstr. 161.321.

# 152 Stellar Associations

152.001 **The Coronet, an obscured cluster adjacent to R Corona Austrina.**
K. N. R. Taylor, J. W. V. Storey.
Mon. Not. R. Astron. Soc., Vol. 209, No. 1, p. 5P – 10P (1984).
A small cluster of obscured, young stars has been discovered around R Corona Austrina. The infrared colours of these stars are consistent with their being recently–formed pre–main–sequence stars. This discovery removes a long–standing anomaly whereby the R CrA region, despite having all the other features of a star formation region, appeared to contain very few young stars. The authors suggest the name "Coronet" for this new cluster.

152.002 **On grouping of B stars in Puppis.**
A. L. Gyul'budagyan.
Astrofizika, Tom 19, Vyp. 3, p. 578 – 579 (1983). In Russian. English translation in Astrophysics, Vol. 19, No. 3.
The results of a search for groupings of B stars in Puppis are given. 13 such groups were obtained.

152.003 **Trapezium–like large systems.**
A. L. Gyul'budagyan.
Astrofizika, Tom 19, Vyp. 4, p. 747 – 752 (1983). In Russian. English translation in Astrophysics, Vol. 19, No. 4.
Data concerning 11 large Trapezium–like systems consisting of B type stars situated in Puppis and 8 systems connected with reflection nebulae are given. A list of 8 narrow Trapezium–like systems consisting of infrared and compact radio sources is also given.

152.004 **Roaming in the Sco OB1 association.**
A. Heske, H. J. Wendker.
Messenger, No. 37, p. 11 – 12 (1984).

152.005 **An investigation of the stellar association containing the 1.95 day Cepheid SU Cassiopeiae.**
D. G. Turner, N. R. Evans.
Astrophys. J., Vol. 283, No. 1, p. 254 – 263 (1984). With plate 4.
Evidence is presented for the membership of the $1^d.95$ Cepheid SU Cas in a poorly populated association of B and A type stars, some of whose members, like SU Cas itself, illuminate reflection nebulosity from dust clouds lying in their immediate vicinity. Core group members of this association appear to share common space velocities, lie at a common distance and have ages of about $1.2 \times 10^8$yr, as predicted for SU Cas from its period. A determination of the luminosity and radius of SU Cas indicates that it is pulsating in the fundamental mode.

152.006 **Optical observations of the bubble nebula N 157C (= 0536.6–6914) and the tight clusters in the association LH 90 in LMC.**
M. C. Lortet, G. Testor.
Astron. Astrophys., Vol. 139, No. 2, p. 330 – 340 (1984).
Monochromatic photographs (Hβ, [O III], Hα) have been obtained of a field 10′ in diameter, centered on the stellar association LH 90 (Lucke and Hodge, 1970), which seems to be the core of N 157C, a bubble of ionized gas. The authors discuss the nebular morphology and excitation and compare with radio structures at 843 MHz (Mills and Turtle, 1984). The photographs allow to distinguish numerous stars (up to 20 – 30) in each of several tight clusters. The stellar content of LH 90 is shown to be in striking contrast with that of the associations LH 85 and 89, from which it is separated by a conspicuous dust lane. The WN3 star Brey 66 is found to be surrounded by a [O III]-rich ring nebula.

152.007 ***VBLUW*** **photometry of the association Sco OB1 (containing the open cluster NGC 6231). A discussion on the evolutionary status of the hypergiant $\zeta^1$ Sco (B1Ia$^+$).**
A. M. van Genderen, W. Bijleveld, E. van Groningen.
Astron. Astrophys., Suppl. Ser., Vol. 58, No. 3, p. 537 – 548 (1984).
*VBLUW* photometry of nearly 190 stars in the very young association Sco OB1, including the cluster NGC 6231, is presented and discussed. Temperature and surface gravity are determined with the aid of theoretical colours. Individual reddenings are also obtained. Practically all reddening is caused by dust in the local spiral arm up to a distance of at most 700 pc from the Sun. The Sagittarius–Carina arm where Sco OB1 is situated is practically transparent over a distance of ~0.5 kpc. The evolutionary status of the hypergiant $\zeta^1$ Sco (B1Ia$^+$) is discussed.

152.008 **New stellar associations in M33 galaxy.**
P. Z. Kunchev, G. R. Ivanov.
Astrophys. Space Sci., Vol. 106, No. 2, p. 371 – 379 (1984).
54 new associations are identified in the galaxy M33 by comparing the plates in the *U* and *B* system. The larger part of the associations are located in the spiral arms. The young associations are prevailing in proximity to the nucleus. A large part of the newly identified associations are connected with H II regions.

152.009 **Radio continuum observations of the Canis Major R1 region.**
M. Nakano, S. Yoshida, T. Kogure.
Publ. Astron. Soc. Jpn., Vol. 36, No. 3, p. 517 – 524 (1984).
Observations of the CMa R1 region in the radio continuum at 10 GHz were made using the 45-m radio telescope at the Nobeyama Radio Observatory. The beam size was 2′.7 in HPBW. An area of 2° × 3° containing three H II regions S292, S296, and S297 was scanned and a calibrated map was obtained. This area has rather uniform foreground extinction and contains no hidden compact H II region. The emission ring (S296) shows a thermal spectrum, confirming the ionization balance between OB stars and ionized gases in the CMa OB1 region.

152.010 **Trapezium-type multiple systems. Review.**
L. V. Mirzoyan, G. N. Salukvadze.
Astrofizika, Tom 21, Vyp. 2, p. 399 – 417 (1984). In Russian. English translation in Astrophysics, Vol. 21, No. 2.
A review of the studies on the Trapezium-type multiple systems and their discussion is given. It is shown that the existing observational and theoretical studies confirm the fundamental feature of these systems i. e. their dynamical instability.

**Problems of investigation of star clusters and associations. Symposium held in Prague, 1983, September 27 – October 1.**
See Abstr. 011.009.

**Wolf–Rayet stars in clusters: the initial stellar masses and evolutionary connections between subtypes.**
See Abstr. 065.002.

**Ultraviolet and optical observations of the Cygnus OB2 association: the star 8A.**
See Abstr. 112.020.

**Infrared studies of early-type stars in OB associations. II. A discussion of density and temperature distributions in stellar winds.**
See Abstr. 112.081.

**The distribution of absorbing matter and of OB stars in the region of the association Per OB1.**
See Abstr. 114.130.

**The rotation of early B-type stars and the problem of star formation.**
See Abstr. 116.007.

**New flare stars in the Mon I association.**
See Abstr. 122.174.

**Molecular observations of B 1: a dense globule in Perseus.**
See Abstr. 131.177.

**Large-scale bubble structure of the interstellar medium (ISM) and properties of the local spiral arm (LSA).**
See Abstr. 131.281.

**Observations of an ionization ridge containing an embedded O–B cluster in the G10.2–0.3 H II region of W31.**
See Abstr. 132.001.

**Radio observations of the CMa OB1 H II regions.**
See Abstr. 132.028.

***UBV*** **photometry of Tr 24 and its relation to Sco OB1.**
See Abstr. 153.003.

**Wolf–Rayet stars in open clusters and associations.**
See Abstr. 153.012.

**The distribution of luminous stars at $(l,b)$ = (245°,0°) in Puppis.**
See Abstr. 155.034.

**The Galaxy as a self-regulated star-forming system: the case of the OB associations.**
See Abstr. 155.064.

**The spatial distribution of young objects in the Large Magellanic Cloud: a problem of pattern recognition.**
See Abstr. 156.007.

# 153 Open Clusters

153.001 **Yellow giants in young clusters. I. Photometric observations.**
E. G. Schmidt.
Astrophys. J., Suppl. Ser., Vol. 55, No. 3, p. 455 – 463 (1984).

A survey has been made to locate yellow giants and supergiants in open clusters. Photometry on the DDO, *uvby*, and *VRI* systems is given for most of the 84 stars in the sample, and the membership of these stars in the clusters is assessed on the basis of this photometry and existing data from the literature. Twenty-two of the stars are likely cluster members, and 48 are likely to be field stars.

153.002 **The radial velocity of the Hyades cluster.**
H. L. Detweiler, K. M. Yoss, R. R. Radick, S. A. Becker.
Astron. J., Vol. 89, No. 7, p. 1038 – 1049 (1984).

Radial velocities for thirty-nine members or suspected members of the Hyades cluster are determined through direct comparison with solar system objects. Using these new data for 17 nonvariable members, plus additional published radial velocities, a new cluster radial velocity ($+39.1 \pm 0.2$ km $s^{-1}$) is derived. A radial-velocity convergent point also is obtained, which is in agreement with those derived through proper-motion data. No significant correction to the standard Hyades distance is required by these new velocity data.

153.003 ***UBV* photometry of Tr 24 and its relation to Sco OB1.**
A. Heske, H. J. Wendker.
Astron. Astrophys., Suppl. Ser., Vol. 57, No. 2, p. 205 – 212 (1984).

In an area of $1\overset{\circ}{.}3 \times 0\overset{\circ}{.}3$ centred on Tr 24 *UBV* photometry down to about $14^m$ has been obtained. From colour-magnitude and colour-colour diagrams a distance of 2300 pc, a foreground reddening of $A_v = 1\overset{m}{.}10$ and a cluster age of $< 10^7$ y is derived. Tr 24 and Sco OB1 form one large cluster containing many pre-main-sequence stars.

153.004 **Spectral classification and photometry of selected Pleiades stars.**
M. Breger.
Astron. Astrophys., Suppl. Ser., Vol. 57, No. 2, p. 217 – 218 (1984).

New spectral classifications and *UBV* photometry of faint stars in the Pleiades cluster are presented. These show that the relatively high reddening of HD 23512 is shared by the other cluster members in the southwest of the cluster. For the cluster members studied in other parts of the cluster, reddening values are normal.

153.005 **Red giants in old open clusters. A test for stellar evolution.**
G. Barbaro, L. Pigatto.
Astron. Astrophys., Vol. 136, No. 2, p. 355 – 362 (1984).

A group of 38 among the oldest open clusters has been analysed using theoretical isochrones derived by Ciardullo and Demarque (1977) and evolutionary tracks for central He-burning of Sweigart and Gross (1976). The following conclusions have been reached: clusters older than 2 or 3 billion years generally agree with theoretical description and allow the determination of reliable ages from the distribution of red giants; for clusters younger than 2 or 3 billion years, the theory fails in the interpretation of their red giant stars and the statistical analysis rejects all the tested distributions. In fact, their behaviour is similar to that of even younger clusters, in which the evolved stars, during shell H-burning, have non-degenerate cores and therefore do not undergo the helium flash. It is suggested that such behaviour could be attributed to overshooting from the convective core during central H-burning in the mass range 1.3 – 2.2 $M_{\odot}$.

153.006 **On reddening and distance moduli of the double open cluster NGC 6882/5.**
U. K. Dzērvitis.
Astron. Tsirk., No. 1294, p. 6 – 8 (1983). In Russian.

153.007 **The abundance of lithium in the old galactic cluster NGC 7789.**
C. A. Pilachowski, J. R. Mould, M. J. Siegel.
Astrophys. J., Lett. Ed., Vol. 282, No. 1, p. L17 – L21 (1984).

The λ6707 doublet of lithium is found to be unusually strong in several red giant stars in the old galactic cluster NGC 7789. In some stars, the lithium abundance is near the cosmic value. The presence of enhanced lithium in a cluster with a well-defined color-magnitude diagram provides specific information on the behavior of lithium in low-mass, evolved stars.

153.008 **Near-infrared and visual photometry of h and χ Persei.**
M. Tapia, M. Roth, R. Costero, S. Navarro.
Rev. Mex. Astron. Astrofis., Vol. 9, No. 1, p. 65 – 75 (1984).

Broad-band near-infrared photometry in the *JHK* (and sometimes *L'* and *M*) bands of 82 stars in h and χ Persei is presented. The stars observed cover the upper main sequence of the central parts of the clusters as well as the giants and supergiants of the extended region. New *UBV* photoelectric photometry of 24 stars for which only photographic photometry was previously available is also reported.

153.009 **Spectral anomalies in the Hyades and Pleiades and in field stars with active chromospheres.**
J. A. Rose.
Astron. J., Vol. 89, No. 8, p. 1238 – 1251 (1984).

Widened photographic image-tube spectra at 50 Å mm$^{-1}$ dispersion have been obtained for a large number of late-type field dwarfs and giants with well-determined atmospheric parameters, and for 35 Hyades dwarfs and 31 Pleiades dwarfs. The spectra have a resolution of 2.5 Å and cover the wavelength range λλ3400 – 4500 Å. A new quantitative three-dimensional spectral classification system is derived for late-type stars, and is calibrated using the sample of field dwarfs and giants with known atmospheric parameters. Diagnostic indices are defined by comparing the counts in the bottoms of two neighboring absorption lines or by comparing the counts in two neighboring pseudocontinuum peaks. It is found that the Hyades and Pleiades dwarfs and field dwarfs with strong Ca II H- and K-emission reversals exhibit well-defined anomalies in their diagnostic indices when compared with normal field dwarfs. It is proposed that active regions on stars with strong Ca II emission reversals have seriously affected many of the principal spectral lines.

153.010 **On proper motions and distribution of flare stars in the Pleiades.**
A. L. Mirzoyan.
Astrofizika, Tom 19, Vyp. 3, p. 588 – 592 (1983). In Russian. English translation in Astrophysics, Vol. 19, No. 3.

It has been shown that the surface and space densities of the Pleiades flare stars decrease with the increase of the distance from the aggregate centre not depending on the probability as to be or not to be a member of the aggregate computed by proper motions. The mean proper motions of flare stars increase with the distance from the aggregate centre.

153.011 **The Pleiades cluster. I. Polarization and reddening of the brighter stars.**
M. Breger.
Astron. Astrophys., Vol. 137, No. 1, p. 145 – 148 (1984).

New linear polarization values have been measured for the bright Pleiades stars at an accuracy level of $10^{-4}$. Suspicions in the literature concerning a high incidence of intrinsic polarization are not confirmed. Multiple measurements show no (or very little) evidence for polarization variability. The critical polarization/extinction ratio is compatible with an interstellar origin. The average ratio of blue to visual polarization is also consistent with an interstellar origin of polarization. Despite the small size of the polarization values, an excellent agreement with the average values measured by Markkanen (1977) is found.

153.012 **Wolf–Rayet stars in open clusters and associations.**
I. Lundström, B. Stenholm.
Astron. Astrophys., Suppl. Ser., Vol. 58, No. 1, p. 163 – 192 (1984).

Existing data on galactic Wolf–Rayet (WR) stars in open clusters and OB associations have been summarized and evaluated. All WR stars within three cluster radii from known clusters have been studied. It was found that 10 – 30% of all WR stars are cluster members. It is also estimated that at least 50% of the WR stars are members of OB associations (including cluster members). These frequencies are in good agreement with the corresponding figures found for O stars. Mean absolute magnitudes for different groups of WR stars have been determined. There is a good correlation between mass and absolute magnitude for binary WR stars. Implications of this on different evolutionary scenarios for WR stars are briefly discussed. The distributions of WR stars in the solar neighbourhood and in LMC are compared and commented.

153.013 **New candidate members of the open cluster NGC 7092 (M39).**
I. K. Platais.
Sov. Astron. Lett., Vol. 10, No. 2, p. 84 – 87 (1984). English translation of 37.153.019.

153.014 **Rotational studies of lower main sequence stars in nearby open galactic clusters. I. Velocity distributions and age dependence.**
W. Benz, M. Mayor, J. C. Mermilliod.
Astron. Astrophys., Vol. 138, No. 1, p. 93 – 100 (1984).

More than 300 new $v \sin i$ values are determined by means of the spectrometer CORAVEL in the following four nearby open galactic clusters: Pleiades, Hyades, Praesepe, and Coma Ber. After excluding all variable stars and known binaries the sample still contains about 200 stars. A determination of the velocity distribution in each cluster is presented. By comparing the mean rotational velocity of the different clusters it is shown that there is equal time dependence of rotation for stars of spectral type F8 V to K3 V and that the $t^{-1/2}$ law for the rotational decay does not fit well the data.

153.015 **Photoelectric search for CP2–stars in open clusters. V. NGC 6281, NGC 6405, IC 4665.**
H. M. Maitzen, H. Schneider.
Astron. Astrophys., Vol. 138, No. 1, p. 189 – 193 (1984).

130 stars in the medium age clusters NGC 6281, NGC 6405, and IC 4665 were observed by photoelectric photometry in the $\Delta a$–system in order to detect CP2–peculiarity by the presence of the $\lambda$5200 depression feature. No CP2–star was found in IC 4665, one in NGC 6281, and 3 in NGC 6405. Star No. 77 of NGC 6405 exhibits the strongest $\lambda$5200 depression ever measured by this type of photometry.

153.016 **Some characteristics of star clusters in M33.**
A. S. Sharov, V. M. Lyutyj.
Pis'ma Astron. Zh., Tom 10, No. 9, p. 654 – 660 (1984). In Russian. English translation in Sov. Astron. Lett., Vol. 10.

Results of photoelectric UBV observations are reported for 11 star clusters in M33. The clusters of M33 are located in the same region of the two–color diagram $(U–B) - (B–V)$ as those of the Large Magellanic Cloud. Ages of the clusters should be estimated from $(0.03 - 0.1) \times 10^9$ to $(8 - 14) \times 10^9$ years. The process of cluster formation in M33 was continuous.

153.017 **The lithium abundance of Hyades main–sequence stars.**
R. Cayrel, G. Cayrel de Strobel, B. Campbell, W. Däppen.
Astrophys. J., Vol. 283, No. 1, p. 205 – 208 (1984).

From high–quality Reticon spectra of Hyades G and K dwarfs, their lithium abundance is found to decline much more rapidly with decreasing temperature than heretofore realized. Previous observations of the $\lambda$6707 Li I doublet for the cooler dwarfs have probably been contaminated by blends. However, the new lithium–temperature relation still does not agree with simple models of lithium burning in the convective zone, either with or without overshooting.

153.018 **Color anomalies and starspots in Hyades dwarfs.**
B. Campbell.
Astrophys. J., Vol. 283, No. 1, p. 209 – 217 (1984).

Hyades dwarfs later than ~F7 show color anomalies which are found to correlate well with various indicators of chromospheric activity. The author proposes that the color anomalies are due to starspots, which apparently also cause the low level photometric variations seen in some stars. That some stars with small color anomaly are variable, plus the finding that Hyades dwarfs show a mean color anomaly relative to field stars, implies that virtually all late–type Hyades are spotted. Some observational implications of starspots on Hyades dwarfs are outlined.

153.019 **The astrometric and kinematic parameters of the Sirius and Hyades superclusters.**
O. J. Eggen.
Astron. J., Vol. 89, No. 9, p. 1350 – 1357 (1984).

Convergent points of $(A, D) = (20^h33^m.7, -38°8')$ and $(6^h24^m.2, +6°30')$, respectively, have been derived for some 30 FK4 stars in each of the Sirius and Hyades superclusters and distributed over the entire sky. The mean motions of the superclusters are $(U, V, W) = (-14.7, 1.5, -10.0) \pm (0.7, 1.3, ,1.2)\ (\sigma)$ km/s for the Sirius Supercluster and $(+41.4, -18.5, -1.9) \pm (0.4, 0.9, 1.1)\ (\sigma)$ km/s for the Hyades Supercluster. The available intermediate band photometry gives photometric moduli that differ, in the mean, from the astrometric results by $0.00 \pm 0.30\ (\sigma)$ mag for the Sirius and $0.08 \pm 0.25\ (\sigma)$ for the Hyades stars.

153.020 **A systematic search for members of the Hyades Supercluster. III. Near–main–sequence G and K stars with very active chromospheres.**
O. J. Eggen.
Astron. J., Vol. 89, No. 9, p. 1358 – 1365 (1984).

Ten probable and four possible Hyades Supercluster members for which there is some indication of very active chromospheres (VAC) are discussed. The VAC stars in the supercluster, HD 18134, V471 Tau B, +26°730 A, DH Leo, Z Her B and, possibly, HD 36705 lie a magnitude or more above the main sequence. Although there is evidence in one or two cases that this elevation may result from long–term light variations, the majority of the displacements can be understood by equating the VAC stars to subgiant (such as Z Her B) remnants of interacting binaries. In at least one (V471 Tau), and probably two, (DH Leo) cases, the defunct primary component is now a white dwarf. Supercluster stars with lower chromospheric activity, indicated mainly by H and K emission which may be periodically variable, are on the main sequence.

153.021 **The clusters NGC 419 and NGC 416 in the Wing of the Small Magellanic Cloud.**
D. Durand, E. Hardy, J. Melnick.
Astrophys. J., Vol. 283, No. 2, p. 552 – 559 (1984).

The authors have constructed deep color–magnitude diagrams, based on du Pont plates, for the rich clusters NGC 419 and NGC 416 located in the Wing of the Small Magellanic Cloud. Integrated photometry in the Washington system provides a lower limit to their metallicity of [Fe/H] = –1.0. Fits of the evolved main sequences to the Yale isochrones, which are dependent on the uncertain distance and reddening scales, indicate ages of $1.2 \pm 0.5$ Gyr for NGC 419 and $2.5 \pm 0.7$ Gyr for NGC 416. The presence of an age difference is consistent with published statistics on the number of bright red giants and carbon stars.

153.022 **A deep photometric study of IC 5146.**
J. C. Forte, A. M. Orsatti.
Astrophys. J., Suppl. Ser., Vol. 56, No. 2, p. 211 – 229 (1984). With plates 3 – 5.

Image processing techniques have been used to obtain *UBVRI* photographic photometry of some 1000 stars brighter than $V = 20.5$ mag in the neighborhood of the young complex IC 5146. The stars were measured on PDS tracings of prime-focus plates taken with the Mayall 4 m telescope. The estimated stellar population of the cluster is $110 \pm 20$ stars. A sample of stars embedded in the darkest nebular region shows that most of them may be contracting objects with maximum ages of $\sim 1 - 3 \times 10^6$yr and spectral type K. From the amount of nebular material in the cluster the star-forming efficiency is estimated as 10%.

153.023 **Photometry of open cluster King 12.**
V. Mohan, A. K. Pandey.
Astrophys. Space Sci., Vol. 105, No. 2, p. 315 – 321 (1984).

Photoelectric magnitudes of 30 stars in the field of King 12 have been obtained. The reddening is variable across the cluster. A distance of 2.49 kiloparsecs has been estimated for the cluster. It is concluded that the age of the cluster lies between NGC 2362 and NGC 884 groups.

153.024 **Distribution of absolute energy of bright Be stars in Pleiades.**
P. S. Goraya.
Astrophys. Space Sci., Vol. 105, No. 2, p. 323 – 330 (1984).

The authors present newly measured energy distributions from λλ3200 – 7500 Å of five late-type Be stars in the Pleiades cluster for search of peculiarities of Be stars continuum energy distributions. Empirical effective temperatures of Be stars have been derived by comparing observed and computed fluxes in the visible region. The variation of the flux in the Balmer continuum region of Pleione (28 Tau) has been discussed.

153.025 **Lithium abundances of solar-type stars in Praesepe.**
D. R. Soderblom, J. R. Stauffer.
Astron. J., Vol. 89, No. 10, p. 1543 – 1544 (1984).

High-resolution spectra have been obtained for seven solar-type stars in Praesepe. Previous observations suggested that Praesepe stars have much more lithium than Hyads of the same color, despite these two clusters having the same age. The present observations show that Praesepe stars and Hyads have essentially the same Li equivalent widths, consistent with the view that stars of the same mass deplete Li at the same rate.

153.026 **On the nature of the Wolf-Rayet component in the core of the massive galactic H II region NGC 3603.**
A. F. J. Moffat, V. S. Niemela.
Astrophys. J., Vol. 284, No. 2, p. 631 – 636 (1984).

The dense, luminous central core (HD 97950) of the massive star cluster NGC 3603 in the galactic giant H II region G291.6-0.5, exhibits a composite WN6 + O5 spectrum. The WN6 emission lines yield a periodic radial velocity variation of 72 km $s^{-1}$ amplitude with $P = 3.7720$ days. This is interpreted in terms of binary motion for one of at least two, and probably three, W-R stars present in the spectroscopically unresolved core. The actual companion is not evident in the absorption spectrum. The alternate explanation for the periodic velocity variations as due to radial oscillations of a very massive star, like that claimed currently for R136, appears improbable.

153.027 **The structure and dynamics of the open cluster M11.**
R. D. Mathieu.
Astrophys. J., Vol. 284, No. 2, p. 643 – 662 (1984).

A complete contamination-free sample of M11 cluster members has been obtained by combining proper-motion data of McNamara, Pratt, and Sanders with magnitude and color information obtained from deep KPNO 4 m plates. The luminosity function of all observed cluster members is in good agreement with the field initial luminosity function for stars with masses greater than 1.6 $M_\odot$ but is somewhat deficient in lower mass stars. Correcting for mass-segregation effects through dynamical models substantially decreases this deficiency. The paper then examines in detail the spatial structure of the cluster and begins an analysis of the cluster dynamics through application of the dynamical theory developed for large-$N$ systems. Multi-mass equipartition King models fit the radial surface-density profiles very well.

153.028 **Young stars in the old galactic cluster NGC 188?**
F. van 't Veer.
Astron. Astrophys., Vol. 139, No. 2, p. 477 – 480 (1984).

The author first briefly discusses the age of the oldest known galactic cluster, according to recently published determinations. The now definitely established membership of four W UMa type contact binaries in this cluster is difficult to understand if the age of these stars is that of the cluster. It appears therefore that these binaries are much younger and that several episodes of star formation took place in NGC 188. This conclusion is reached after a new study of the mean density of the four contact binaries and a critical discussion of the chemical composition and the mixing length parameter.

153.029 **A radial-velocity study of open clusters containing blue stragglers.**
M. Bolte, M. Mateo.
Publ. Astron. Soc. Pac., Vol. 96, No. 584, p. 784 (1984). Abstract. – See Abstr. 010.281.

153.030 **The Einstein soft X-ray survey of the Pleiades.**
J.-P. Caillault, D. J. Helfand.
Publ. Astron. Soc. Pac., Vol. 96, No. 584, p. 785 – 786 (1984). Abstract. – See Abstr. 010.281.

153.031 **A photographic color-magnitude diagram for the remote LMC cluster NGC 1777.**
M. Mateo, P. Hodge.
Publ. Astron. Soc. Pac., Vol. 96, No. 584, p. 791 (1984). Abstract. – See Abstr. 010.281.

153.032 **The abundance of lithium in the old galactic cluster NGC 7789.**
C. A. Pilachowski.
Publ. Astron. Soc. Pac., Vol. 96, No. 584, p. 793 (1984). Abstract. – See Abstr. 010.281.

153.033 **Stellar evolutionary sequences and isochrones for intermediate-age clusters.**
D. A. VandenBerg.
Publ. Astron. Soc. Pac., Vol. 96, No. 584, p. 797 (1984). Abstract. – See Abstr. 010.281.

153.034 ***uvby* β photometry of southern clusters – V. NGC 2422.**
R. R. Shobbrook.
Mon. Not. R. Astron. Soc., Vol. 211, No. 3, p. 659 – 666 (1984).

Strömgren four-colour and Hβ photometry has been obtained for 28 B and A0 members of the young cluster NGC 2422. A comparison of $V_0/c_0$ diagrams clearly shows that the cluster is very similar in age to the Pleiades. Distance estimates for the cluster from three luminosity calibrations of the β index and from a direct comparison of the $V_0/c_0$ diagram with that of the Pleiades are all in good agreement. If the distance modulus of the Pleiades is taken to be 5.50, the adopted modulus for NGC 2422 is 8.0, corresponding to a distance of 400 pc with an estimated uncertainty of 10 pc.

153.035 ***UBV* photometry of the southern galactic cluster NGC 4755 = κ Crucis.**
J. Dachs, D. Kaiser.
Astron. Astrophys., Suppl. Ser., Vol. 58, No. 2, p. 411 – 429 (1984). With a correction in Astron. Astrophys., Suppl. Ser., Vol. 59, No. 2, p. 368 (1985).

*UBV* photoelectric magnitudes and colours are presented for 86 stars together with photographic magnitudes and colours for 553 stars in the vicinity of the southern young open cluster NGC 4755. A true distance modulus of $11^m.82$ is derived corresponding to a linear distance of 2.3 kpc. The evolutionary age is determined to be between 8 and $11.5 \times 10^6$ years, depending on the assumed chemical composition of the cluster stars. The mean interstellar reddening $E(B-V)$ across the cluster increases with galactic latitude and angular distance from the Coalsack from about $0^m.40$ to $0^m.48$ with an average value of $0^m.44$.

153.036 **Search for variable stars in the young open cluster NGC 6871.**
A. J. Delgado, E. J. Alfaro, J. M. García-Pelayo, R. Garrido, S. Vidal.
Astron. Astrophys., Suppl. Ser., Vol. 58, No. 3, p. 447 – 451 (1984).
Photoelectric *uvby* and Hβ observations of the brightest stars in the field of the open cluster NGC 6871 have been performed in order to detect possible light variability of the B-type stars, members of the cluster. Two new variables, a δ Scuti and a β Cephei star, have been discovered.

153.037 **A study of the open cluster NGC 6475 (M7) (and two other stellar clusterings).**
L. O. Lodén.
Astron. Astrophys., Suppl. Ser., Vol. 58, No. 3, p. 595 – 599 (1984).
The open cluster NGC 6475 has been investigated with respect to possible physical properties common to its members and hence useful for discrimination purposes. As comparison objects the author used two loose clusterings of A-type stars. The investigation includes studies of both intermediate-dispersion spectra (20 Å/mm) and *UBV* photometry. The general impression is that the high frequency of multiple stars, visual as well as spectroscopic, will constitute a very serious obstacle as far as physical discrimination of cluster members is concerned and to a large extent also in the case of other studies. The CP stars appearing in NGC 6475 have also been subject to some discussion. If continued investigations should prove multiplicity among cluster stars to be considerably less frequent than in the present cases, the attractive idea to find a membership criterion might be fruitful. Otherwise it probably has to be abandoned.

153.038 **Search for B-type variable stars in open clusters.**
A. J. Delgado, E. J. Alfaro, R. Garrido, J. M. Garcia-Pelayo.
Inf. Bull. Variable Stars, No. 2603, 3 pp. (1984).

153.039 **Positions, proper motions, and magnitudes in the area of the open cluster Tr 10.**
J. Stock.
Rev. Mex. Astron. Astrofis., Vol. 9, No. 2, p. 127 – 137 (1984).
Positions for 979 stars in the area of Tr 10 are derived from fifteen plates taken recently with the CIDA refractor. For the determination of proper motions nine plates of the Cape Astrographic Catalogue are used as first epoch. The results make it appear doubtful that the object is a single cluster.

153.040 **Luminosity function of the integrated magnitudes of open clusters.**
S. van den Bergh, A. Lafontaine.
Astron. J., Vol. 89, No. 12, p. 1822 – 1824 (1984).
The luminosity function of open clusters with integrated magnitudes $-9.0 < M_V \leqslant -2.0$ is determined in the solar vicinity. Between $M_V = -7.5$ and $M_V = -2.5$ the cluster luminosity function increases by a factor of $\sim 10$. The local surface density of clusters with $M_V \leqslant -2.0$ is $\sim 30\ \mathrm{kpc}^{-2}$.

153.041 **Photometry of the open cluster King 21.**
V. Mohan, A. K. Pandey.
Bull. Astron. Soc. India, Vol. 12, No. 3, p. 217 – 222 (1984).
Photoelectric magnitudes of 26 stars in the field of King 21 have been obtained. The reddening is variable across the cluster. A distance of 1.91 kpc has been estimated for the cluster. It is concluded that the age of the cluster lies between that of the NGC 2362 and of the NGC 457 groups.

153.042 **Method of evaluation of star cluster distances.**
A. V. Loktin.
Astron. Tsirk., No. 1337, p. 6 – 8 (1984). In Russian.

153.043 **Feldspektroskopische Radialgeschwindigkeitsbestimmungen an offenen Sternhaufen.**
E. H. Geyer, B. Nelles.
Mitt. Astron. Ges., Nr. 62, p. 294 – 297 (1984). – See Abstr. 010.341.

153.044 **Pulsars in the Hyades Supercluster.**
O. J. Eggen.
Astron. Express, Vol. 1, Issue 2, p. 59 – 60 (1984).

153.045 **Washington photometry of the old open cluster Melotte 66.**
D. P. Geisler, V. V. Smith.
Publ. Astron. Soc. Pac., Vol. 96, No. 585, p. 871 – 873 (1984).
Abundances for a sample of giants in the old open cluster Melotte 66 have been determined from Washington photometry. The mean cluster [Fe/H] = $-0.5 \pm 0.2$, and CN appears to be enhanced. The implications of a comparison to previous abundance determinations are discussed.

153.046 **A search for F and G supergiants in galactic open clusters.**
S. van den Bergh, A. Lafontaine.
Publ. Astron. Soc. Pac., Vol. 96, No. 585, p. 880 – 881 (1984).
Intercomparison of the Michigan Catalogue of Two-Dimensional Spectral Types with a uniform survey of southern open clusters by van den Bergh and Hagen shows that ~6% of F- and G-type supergiants of luminosity classes Ia, Iab, and Ib are cluster members. The most-luminous supergiants are more likely to be cluster members than are less-luminous supergiants.

153.047 **Abundance estimates for the distant anticenter clusters King 8 and Be 19.**
C. A. Christian.
Astrophys. J., Vol. 286, No. 2, p. 552 – 562 (1984).
Intensified Image Dissector Scanner data were used to analyze the spectra of suspected members of the distant anticenter clusters King 8 and Be 19. Two early-type cluster members were identified in King 8. Their spectral types, ⩽A5, are consistent with the stars' photometric positions in the cluster color-magnitude diagram, providing a verification of the photometrically derived age of the cluster. Spectrophotometric data for suspected cluster giants suggest that both clusters may be slightly metal-poor, i.e., with abundances similar to NGC 2158 and NGC 2420 ([Fe/H] = -0.5).

153.048 **Photographic photometry of the open cluster NGC 7790.**
M. Pedreros, B. F. Madore, W. L. Freedman.
Astrophys. J., Vol. 286, No. 2, p. 563 – 572 (1984).
A *UBV* photometric study of the open cluster NGC 7790 containing three Cepheids is presented using new photographic data obtained at the CFH Telescope and calibrated to $V \approx 19$ mag. Using fitting procedures which account for binaries, field contamination, and individual reddenings, a true distance modulus of $12.3 \pm 0.2$ mag is derived for a mean $E(B-V) = 0.64 \pm 0.05$ mag.

153.049 **Yellow giants in young clusters. II. A comparison of observation with theory.**
E. G. Schmidt.
Astrophys. J., Vol. 287, No. 1, p. 261 – 267 (1984).
Luminosities and effective temperatures are calculated for 17 nonvariable yellow giants and supergiants and nine Cepheids which were previously identified as members of open clusters. These stars are identified as being in the helium core burning stage of evolution. A comparison with the predictions of pulsation theory and stellar evolution calculations is presented.

153.050 **DDO and *UBV* photometry of red giant stars in NGC 6791.**
K. A. Janes.
Publ. Astron. Soc. Pac., Vol. 96, No. 586, p. 977 – 980 (1984).
Photoelectric DDO and *UBV* observations of red giant stars in the old open cluster NGC 6791 were used to help derive the reddening to the cluster and to find its distance and metallicity. A comparison of the DDO photometry with previous results leads to a reddening estimate of $E(B-V) = 0^{m}.10 \pm 0^{m}.03$. With this reddening, the DDO metallicity index, $\delta CN = 0.026 \pm 0.015$, corresponding approximately to $[Fe/H] = -0.08 \pm 0.07$. There is no evidence for a spread in composition on the cluster giant branch.

153.051 **A photometric study of a high–latitude cluster: Sculptoris (Blanco).**
A. E. de Epstein, I. Epstein.
News Lett. Astron. Soc. N.Y., Vol. 2, No. 6, p. 12 (1984). Abstract. – See Abstr. 010.241.

153.052 **The Einstein X–ray survey of the Pleiades revisited.**
J.-P. Caillault, D. J. Helfand.
News Lett. Astron. Soc. N.Y., Vol. 2, No. 6, p. 18 – 19 (1984). Abstract. – See Abstr. 010.241.

153.053 **The nature of the embedded population in the Rho Ophiuchi dark cloud: mid–infrared observations.**
C. J. Lada, B. A. Wilking.
Astrophys. J., Vol. 287, No. 2, p. 610 – 621 (1984).
The authors present results of mid–infrared (primarily 10 and 20 μm wavelength) observations of previously identified members of the embedded populatin of the ϱ Ophiuchi dark cloud. These observations enable determination of broad–band energy distributions for 32 of the 44 sources known to be embedded in the cloud. The majority of the sources emit the bulk of their luminosity between 1 μm and 20 μm. These sources apparently are surrounded by dust shells and are probably pre–main–sequence in nature. It is found that the embedded cluster is predominantly composed of low–luminosity objects with luminosities in the range $0.1 \leqslant L/L_{\odot} \leqslant 25$. When corrected for observational selection effects, the resulting luminosity function appears to be deficient in intermediate–luminosity stars compared with the luminosity function for field stars.

153.054 **Main–sequence B stars with strong winds in the core of NGC 6231.**
D. Massa, B. D. Savage, J. P. Cassinelli.
Astrophys. J., Vol. 287, No. 2, p. 814 – 824 (1984).
Low–dispersion *IUE* observations of 13 main sequence B0 – 2 stars in the Sco OB1 association are presented. Eleven of these are in the rich open cluster NGC 6231. Most of the NGC 6231 stars have strong UV resonance lines which are inconsistent with either their spectral types or the rest of their UV spectra. A high–dispersion *IUE* spectrum of one of the peculiar stars, CPD –41°7719, reveals an abnormally strong wind ($\dot{M} > 4 \times 10^{-9} M_{\odot} \mathrm{yr}^{-1}$) with an unusual velocity law. The authors examine some possible causes of such a wind and conclude that an overabundance of carbon is the most likely origin. They discuss how the postulated abundance anomaly could originate and consider some implications of these results.

153.055 **The ultraviolet spectra of the O and B stars in the young galactic cluster NGC 6530.**
E. Böhm–Vitense, P. Hodge, D. Boggs.
Astrophys. J., Vol. 287, No. 2, p. 825 – 839 (1984).
The authors have studied the UV spectra, $1200\,Å < \lambda < 3000\,Å$, of the O and B stars in NGC 6530 and in the surrounding association. The determination of $T_{eff}$ and $L$ shows that these stars are all still on the main sequence. From a comparison of theoretical evolutionary tracks an age of $\sim 5 \pm 2 \times 10^{6}$ years is inferred. The P Cygni line profiles due to the stellar wind were studied and mass loss rates were estimated.

153.056 **Axial rotation in NGC 2287.**
H. Levato, B. Garcia.
Astrophys. Lett., Vol. 24, No. 3, p. 161 – 164 (1984).
The authors have measured axial–rotational velocities of the members of the open cluster NGC 2287. They compared the average rotation of the cluster with the average rotation of field stars of the same types and found that NGC 2287 rotates at a slower rate than field stars.

**Preliminary catalogue of photographic UBV photometry for stars in open clusters.**
See Abstr. 002.054.

**Problems of investigation of star clusters and associations. Symposium held in Prague, 1983, September 27 – October 1.**
See Abstr. 011.009.

**Astrométrie des Pléiades.**
See Abstr. 041.017.

**Wolf–Rayet stars in clusters: the initial stellar masses and evolutionary connections between subtypes.**
See Abstr. 065.002.

**Synthetic H–R diagrams as an observational test of stellar evolution theory.**
See Abstr. 065.022.

**Lithium depletion in stars. Pre–main sequence burning and extra–mixing.**
See Abstr. 065.031.

**Stellar evolutionary sequences and isochrones for intermediate–age star clusters.**
See Abstr. 065.078.

**Determination of relative proper motions and the membership of the open cluster NGC 1817.**
See Abstr. 111.015.

**Mass loss in O–type stars: parameters which affect it.**
See Abstr. 112.053.

**The photometric variability of solar–type stars. IV. Detection of rotational modulation among Hyades stars.**
See Abstr. 113.025.

**The Böhm–Vitense gap in the Geneva photometric system.**
See Abstr. 113.035.

**Ultraviolet fluxes of blue straggler stars.**
See Abstr. 114.033.

**Carbon, nitrogen, and oxygen abundances in M67 giants.**
See Abstr. 114.076.

**Spectral classification of faint O–B stars around the cluster IC 1805.**
See Abstr. 114.087.

**The width of the spectroscopically defined main–sequence.**
See Abstr. 115.006.

**A re–calibration of the luminosities of early–type stars: its effect on the Cepheid luminosity scale.**
See Abstr. 115.011.

**Chromospheric emission and rotation of the Hyades lower main–sequence.**
See Abstr. 116.023.

**IUE spectroscopy of the degenerate components in the Hyades close binaries V471 Tauri and HZ9.**
See Abstr. 117.023.

**Close binary stars and old stellar populations: the blue straggler problem revisited.**
See Abstr. 117.121.

**HD 138168 and HD 138181, a hard binary?**
See Abstr. 118.031.

**The binaries–in–cluster program at the RAO (*Rothney Astrophysical Observatory*).**
See Abstr. 119.059.

**Photographic colorimetry of stellar flares in the Pleiades and Orion aggregates. II.**
See Abstr. 122.053.

**A possible Delta Scuti variable in NGC 6405.**
See Abstr. 123.028.

**Optical and ultraviolet study of five stars in the Pleiades open cluster.**
See Abstr. 131.006.

**Open cluster formation.**
See Abstr. 131.030.

**A model to determinate the slope of the initial mass function in a cluster of young stars.**
See Abstr. 131.120.

**A low density molecular cloud in the vicinity of the Pleiades.**
See Abstr. 131.124.

**Interstellar matter near the Pleiades. I. Optical absorption–line observations and foreground atomic gas.**
See Abstr. 131.147.

**Interstellar matter near the Pleiades. II. $CH^+$ formation.**
See Abstr. 131.148.

**High–resolution far–infrared observations of the extended W51 complex.**
See Abstr. 131.287.

**Properties of R136a as derived from its optical light distribution.**
See Abstr. 132.022.

**Optical study of the galactic nebula NGC 6357 and of its stellar content, Pis 24 and the WC 6 star HD 157504.**
See Abstr. 132.031.

**New radio sources at AFGL 2591: young cluster or single star?**
See Abstr. 132.062.

**The unique planetary nebula NGC 2818.**
See Abstr. 134.061.

**Astronomical observations with the FAUST telescope.**
See Abstr. 142.007.

**The formation and early dynamical evolution of bound stellar systems.**
See Abstr. 151.048.

***VBLUW* photometry of the association Sco OB1 (containing the open cluster NGC 6231). A discussion on the evolutionary status of the hypergiant $\zeta^1$ Sco (B1Ia$^+$).**
See Abstr. 152.007.

**Does the ellipticity of clusters in the LMC correlate with age or luminosity?**
See Abstr. 154.033.

**Magnesium, silicon, and iron abundances and the cluster metallicity scale.**
See Abstr. 154.082.

**Photometry of the active nuclei of galaxies – photographic photometry of the quasar 3C 273.**
See Abstr. 159.124.

# 154 Globular Clusters

154.001 **Variable stars in the globular cluster Messier 28.**
A. Wehlau, H. Sawyer Hogg.
Astron. J., Vol. 89, No. 7, p. 1005 – 1014 (1984). With plates 64 – 65.

Magnitudes for 24 variables have been measured on plates of M28 taken from 1974 through 1983. The variables include 16 RR Lyrae stars, two W Virginis stars, one RV Tauri star, four red irregular variables and one U Geminorum star whose cluster membership has yet to be determined. Light curves have been constructed for the 19 periodic variables. The average length of period for the nine RRab's which are believed to be cluster members is 0.586 days which appears to be intermediate between the average values for Oosterhoff type I and type II clusters.

154.002 **The color–magnitude diagram of NGC 288.**
R. Buonanno, C. E. Corsi, F. Fusi Pecci, G. Alcaino, W. Liller.
Astron. Astrophys., Suppl. Ser., Vol. 57, No. 1, p. 75 – 90 (1984).

The color–magnitude diagram for 2350 stars in the field of the southern globular cluster NGC 288 has been obtained. Metallicity indicators deduced from the CMD suggest a value of [Fe/H] $\sim -1.4$. An age $t = (16 \pm 3) \times 10^9$yrs is derived from $\Delta m$(TO–HB) and from isochrone–fitting techniques. It is consistent with that derived for other well studied clusters.

154.003 **Half–light integrated *UBVRI* photometry of southern globular clusters.**
M. Hamuy.
Astron. Astrophys., Suppl. Ser., Vol. 57, No. 1, p. 91 – 97 (1984).

The author presents integrated *UBVRI* photometry through apertures specially built to accept 50% of the light of globular clusters. Accurate half–light photometry is presented for 72 clusters, each observed on at least two different nights.

154.004 **Deep photometry of globular clusters. II. The remote cluster Pal 3.**
R. G. Gratton, S. Ortolani.
Astron. Astrophys., Suppl. Ser., Vol. 57, No. 2, p. 177 – 188 (1984).

CCD photometry to $V \sim 23.2$ for the remote globular cluster Pal 3 is presented. The C–M diagram closely resembles the Pal 12 one, thus suggesting that Pal 3 be a moderately metal rich cluster ([Fe/H] $= -1.4 \pm 0.4$). The red horizontal branch is found at $V = 20.55 \pm 0.04$. Assuming a reddening $E(B-V) = 0.03$, a distance modulus $(m-M)_0 = 19.8 \pm 0.2$ and a distance of $91 \pm 9$ kpc are derived. An approximate estimate of the cluster diameter and of concentration parameters gives $d = 440''$ and $c = 0.9$.

154.005 **On the two-component structure of a system of galactic globular clusters.**
A. A. Suchkov, M. B. Ehntehl'.
Astron. Tsirk., No. 1292, p. 4 – 6 (1983). In Russian.

154.006 **Investigation of the brightness distribution in globular clusters.**
A. S. Rastorguev, A. V. Mironov, E. V. Naumova.
Astron. Tsirk., No. 1293, p. 1 – 4 (1983). In Russian.

154.007 **Evolution of globular clusters after gravothermal catastrophe.**
S. Inagaki.
Astron. Her., Vol. 77, No. 7, p. 164 – 168 (1984). In Japanese.

154.008 **The main-sequence turnoff of the metal-rich globular cluster NGC 6171.**
G. S. Da Costa, J. R. Mould, S. Ortolani.
Astrophys. J., Vol. 282, No. 1, p. 125 – 134 (1984). With plates 5 – 6.

The KPNO Video Camera system has been used to determine main-sequence color-magnitude diagrams for the metal-rich globular cluster NGC 6171 and the metal-poor cluster M3. Comparisons with theoretical isochrones indicate that the ages of these clusters are identical to within the uncertainties of the determinations. If $(m-M)_v = 14.80$, $[Fe/H] = -0.9$, and $E(B-V) = 0.38$, then the age of NGC 6171 is $16 \pm 3$ Gyr for a helium abundance $Y = 0.24$.

154.009 **Das Alter der Kugelsternhaufen.**
G. Börner.
Sterne Weltraum, 23. Jahrg., Nr. 8-9, p. 422 – 424 (1984).

154.010 **Main-sequence photometry and the age of the metal-rich globular cluster NGC 6171.**
A. Sandage, P. Roques.
Astron. J., Vol. 89, No. 8, p. 1166 – 1179 (1984). With plate 69.

Main-sequence photometry is presented for the metal-rich globular cluster NGC 6171, based on five Palomar Hale reflector plates in each of two colors. The photometry extends to $V = 21$, $B = 21.5$. The main-sequence termination point has been found at $V = 19.35$, which is $\Delta V = 3.60 \pm 0.2$ mag fainter than the zero-age horizontal branch. The reddening has been determined by analysis of the two-color diagram for field stars and for cluster stars separately with the result that $E(B-V) = 0.32$. The effect of low surface gravity on the emergent flux and hence on the colors of horizontal-branch stars is seen in the two-color diagram. The size of the effect relative to main-sequence stars is in agreement with theory. The age determined by three methods is $17 \times 10^9$yr, with an error of at least 20%.

154.011 **Are there S stars in galactic globular clusters?**
T. Lloyd Evans.
Mon. Not. R. Astron. Soc., Vol. 209, No. 4, p. 825 – 839 (1984).

Two very red giant stars in NGC 6723 show enhanced ZrO bands. Comparison with 47 Tuc giants of similar near infrared colour shows that the TiO bands are abnormally weak, which accounts at least in part for the unusually red $B-V$ and $U-B$ as well as strong sodium $D$ lines. Analogy with stars in the Magellanic Clouds suggests that the two NGC 6723 stars have mixed nuclear reaction products into the envelope. Their bolometric mag is $M_{bol} = -3.3$, nearly one mag fainter than the faintest Magellanic Cloud star showing enhanced ZrO. Two red variables in 47 Tuc which have smaller ultraviolet deficiences show no spectroscopic abnormalities, and a report of Ba II stars in M22 is shown to be unfounded.

154.012 **Probing 47 Tucanae.**
W. E. Harris, J. E. Hesser, D. A. VandenBerg.
Sky Telesc., Vol. 68, No. 3, p. 219 – 220 (1984).

154.013 **The rotation of the globular cluster system.**
J. Colin.
Astrophys. Space Sci., Vol. 103, No. 1, p. 71 – 80 (1984).

The author studies the possibility that all globular clusters rotate in the direct sense relatively to the galactic disk motion. He shows with a simple model that this is possible if the projection of the cluster orbits on the galactic plane are elongated. The validity of this hypothesis is discussed.

154.014 **A possible P-L relation for RR-Lyrae variables in globular clusters.**
K. K. Scaria.
Astrophys. Space Sci., Vol. 103, No. 2, p. 207 – 218 (1984).

A study of the mean photographic magnitudes of RR Lyrae variables in galactic globular clusters has shown that the luminosity of RR-ab variables decreases with period. It is found that RR-ab variables, in the cluster ω Centauri, whose distances are greater than 8 arc min from the centre of the cluster, are less luminous than such variables seen in the inner region. The possibility of two transition periods for RR-ab variables, one near $0^d\!.4$ and the other near $0^d\!.9$, is suggested to explain the sharp boundaries of the instability gap and of the periods of RR Lyrae variables. It is also shown that the inner RR Lyrae variables in ω Centauri are more massive than the outer ones.

154.015 **Photoelectric photometry of globular clusters in the Andromeda nebula. X.**
A. S. Sharov, V. M. Lyutyj, V. F. Esipov.
Pis'ma Astron. Zh., Tom 10, No. 8, p. 583 – 585 (1984). In Russian. English translation in Sov. Astron. Lett., Vol. 10.

Photoelectric UBV observations of 21 globular clusters in the Andromeda nebula (M31) are reported.

154.016 **Abundance determinations from low-dispersion spectra of globular-cluster stars.**
R. A. Bell.
Publ. Astron. Soc. Pac., Vol. 96, No. 581, p. 518 – 529 (1984).

A number of synthetic spectra have been computed for comparison with Canterna, Harris, and Ferrall's low-dispersion spectra of stars in the globular clusters 47 Tuc, NGC 288, NGC 362, M15, M2, and Pal 12. Spectral indices have been deduced from the synthetic spectra and compared with the observed indices. The mean abundances of these six clusters are deduced.

154.017 **Mass determination of globular clusters on the basis of proper motions.**
S. Ninkovich.
Astrofizika, Tom 20, Vyp. 2, p. 283 – 288 (1984). In Russian. English translation in Astrophysics, Vol. 20, No. 2.

A determination of globular cluster masses for M15, M92 and M13 by use of the methods proposed by Naumova and Ogorodnikov and by applying the virial theorem is made. A discussion of the obtained results is presented.

154.018 **Deep photometry of globular clusters. I. The outer halo cluster AM-1.**
S. Ortolani.
Astron. Astrophys., Vol. 137, No. 2, p. 269 – 275 (1984).

A color-magnitude diagram of the far globular cluster AM-1 has been obtained from three CCD images to the limiting magnitude of $m_v = 23$ ($M_v = +2.5$). The cluster is the most distant known: assuming $M_v = 0.6$ for the RR Lyrae stars, the derived distance is $(m-M)_0 = 20.4 \pm 0.06$ or 120 kpc. The HR diagram reveals evidence of the "second parameter effect". The diameter of the cluster is unusually large ($D = 168$ pc) when compared with that of classic inner halo clusters. A very approximate mass estimate of $14 \times 10^3 M_\odot$ and a perigalactic distance of about 40 kpc have been obtained.

154.019 **Ellipticity of five globular clusters in M31.**
N. M. Spasova, A. V. Staneva.
Sov. Astron. Lett., Vol. 10, No. 2, p. 114 – 116 (1984). English translation of 37.154.037.

154.020 **$UBV$ photometry for 31 galactic globular clusters.**
D. A. Hanes.
Proc. Astron. Soc. Aust., Vol. 5, No. 3, p. 360 – 363 (1984).

The author presents a homogeneous body of photometry through a single aperture and in the *UBV* system for 31 globular clusters south of declination +12°.

154.021 **UV–bright stars in galactic globular clusters, their far–UV spectra and their contribution to the globular cluster luminosity.**
K. S. de Boer.
ESA Spec. Publ., ESA SP–218, p. 185 – 186 (1984). – See Abstr. 012.024.

Ultraviolet bright stars in 7 galactic globular clusters have been observed with the IUE in the low dispersion mode between 1150 and 3000 Å. The spectral energy distributions are used to derive effective temperatures and stellar luminosities. Such stars produce somewhat less than 1% of the cluster luminosity. Their contribution to the cluster UV intensity varies from about 3% at 3300 Å to over 50% at 1200 Å. In one cluster about 60% of the total 1550 Å intensity comes from a single sdO star.

154.022 **A self–consistent approach to the age of globular cluster M15.**
F. Caputo, V. Castellani, M. L. Quarta.
Astron. Astrophys., Vol. 138, No. 2, p. 457 – 463 (1984).

Adopting for RR Lyrae pulsators in M15 the mass value deduced by Cox et al. (1983) from double mode pulsators and assuming a reddening $E(B-V) = 0.08$, an HB luminosity $\log L/L_{\odot} = 1.75 \pm 0.02$ is found. Adopting Cohen's (1979) metallicity for the cluster and taking into account the increase in the relative amount of CNO elements needed to allow the quoted masses to populate the instability strip, one derives the age $t = 13 \times 10^9$ yr. A reddening $E(B-V) = 0.08 \pm 0.01$ is confirmed for the cluster, both from UV observations and from the properties of RR Lyrae pulsators. Comparisons with previous estimates of the cluster's age are discussed.

154.023 **Far–ultraviolet spectra of the nuclei of globular clusters M30, M54, M70.**
V. Caloi, V. Castellani, D. Galluccio, W. Wamsteker.
Astron. Astrophys., Vol. 138, No. 2, p. 485 – 492 (1984).

IUE spectra in the low dispersion mode have been collected from the nuclei of three galactic globular clusters, M70, M54, and M30. The authors find high UV fluxes from M54 found earlier to be a real characteristic of the cluster population, suggesting a possible connection of this cluster with EB clusters as defined by van Albada et al. (1981). Although the B and EB clusters anticorrelate as groups with metallicity, the flux level in M70 suggests that within the EB group UV colours and metallicity are again correlated. A point–like UV source has been detected at the center of M30. It is shown to be a star in a post–HB phase crossing the HR diagram towards the asymptotic giant branch.

154.024 **A photometric metallicity scale for globular clusters based on stars with known curve of growth chemical abundances.**
B. Nelles, W. Seggewiß.
Astron. Astrophys., Vol. 139, No. 1, p. 79 – 84 (1984).

Intermediate–band Walraven photometry has been obtained for 53 giants in six galactic globular clusters. Using a previously established calibration of *VBLUW* colour indices in terms of [Fe/H], the metallicities of the individual giants are determined. The reliability of the method of individual giant photometry, calibrated by stars with curve–of–growth abundances, is discussed with respect to other methods. The photometry of individual giants leads to the same ranking of globular cluster abundances as from parameters related to the morphology of the colour–magnitude diagram, but it leads to severe discrepancies with metallicities derived by integrated properties of the clusters. Furthermore, while the calibration of photometric indices by curve–of–growth abundances (catalogue of Cayrel de Strobel et al., 1980) is possible with a high degree of correlation, the agreement with the metallicities from spectrum synthesis of the key objects 47 Tuc, NGC 288 and NGC 362 is poor.

154.025 **The appearance, number, and history of highly compact binary systems in globular clusters.**
J. H. Krolik.
Astrophys. J., Vol. 282, No. 2, p. 452 – 465 (1984).

The author analyzes the various evolutionary paths possible for highly compact binaries in globular clusters, where they come under the simultaneous influence of gravitational radiation and gravitational encounters with field stars. The observational appearance of those stages in which mass transfer occurs is discussed, with particular attention given to a new class of objects characterized by massive, nonequilibrium accretion disks created by catastrophic gravitational encounters. The author then constructs an analytic model to predict the number of highly compact binaries in each stage of evolution existing at present in the Galaxy's globular cluster system. Both steady transfer binaries, and systems in which a collapsed stellar remnant acquires a massive, nonequilibrium accretion disk, are shown to be present in significant numbers.

154.026 **The evolution of highly compact binary stellar systems in globular clusters.**
J. H. Krolik, A. Meiksin, P. C. Joss.
Astrophys. J., Vol. 282, No. 2, p. 466 – 480 (1984).

The authors have calculated the secular evolution of a highly compact binary stellar system, composed of a collapsed object and a low–mass secondary star, in the core of a globular cluster. The binary evolves under the combined influences of (1) gravitational radiation losses from the system, (2) the evolution of the secondary star, (3) the resultant gradual mass transfer, if any, from the secondary to the collapsed object, and (4) occasional encounters with passing field stars. The times of encounters with field stars, and the initial parameters specifying those encounters, were chosen by use of a Monte Carlo technique; the subsequent gravitational interactions were calculated utilizing a three–body integrator. It is found that occasional close encounters with field stars strongly dominate the evolution of highly compact binaries in dense globular cluster cores. The authors display the distributions of fractional changes in semimajor axis and changes in eccentricity per encounter, and they discuss the frequency both of catastrophic encounters and of encounters which expel binaries from the cluster by superelastic recoil.

154.027 **Black hole remnants in globular clusters.**
R. B. Larson.
Mon. Not. R. Astron. Soc., Vol. 210, No. 4, p. 763 – 777 (1984).

If a globular cluster contains remnants much more massive than the visible stars, these remnants will dominate the cluster core, and will rapidly attain a nearly singular density distribution. Simple two–component models show that the observed light profiles and velocity dispersions of typical globular cluster cores are reproduced if the density of heavy remnants varies as $r^{-2.5}$ at small radii, and if a representative mass for the heavy remnants is $\sim 2.5\ M_{\odot}$. The cusped light profile of M15 can be explained with a somewhat smaller representative mass of $\sim 1.7\ M_{\odot}$ for the heavy remnants. A possible, but not unique, set of assumptions that can account for the remnant properties in a typical case is that the initial mass function is the same as that of the solar neighbourhood, and that the massive stars leave remnants whose mass is $\sim 15$ per cent of the initial stellar mass. The extremes of small and large fractional content of heavy remnants are represented by 47 Tuc and ω Cen.

154.028 **X–ray sources in globular clusters.**
F. Verbunt, J. van Paradijs, R. Elson.
Mon. Not. R. Astron. Soc., Vol. 210, No. 4, p. 899 – 914 (1984).

The authors show that the bright ($L_x \gtrsim 10^{36}$ erg s$^{-1}$) X–ray sources in globular clusters of our Galaxy are limited to luminosities $L_x \lesssim 10^{37.3}$ erg s$^{-1}$, whereas the sources in the galactic disc range to $\sim 10^{38.4}$ erg s$^{-1}$. The latter limit also applies to the globular–cluster sources of M31. The authors investigate two possible explanations for this difference, viz. a dependence of the mass–transfer rate $\dot{M}$ on the secondary mass in low–mass X–ray binaries, and a dependence of $\dot{M}$ on metallicity.

154.029 **Kinematics of galactic globular clusters.**
A. W. Rodgers, G. Paltoglou.
Astrophys. J., Lett. Ed., Vol. 283, No. 1, p. L5 – L7 (1984).

An analysis of the radial velocities of galactic globular clusters to derive a galactic rotation parameter as a function of metallicity isolates not only the well–established high rotational velocities of the metal–rich clusters but also systematic retrograde motions among a chemically, and also in terms of horizontal–branch morphology, homogeneous group of clusters. This result may indicate that the halo of the Galaxy formed from the coalescence of a small number of galaxies which were parental to halo galactic globular clusters.

154.030 **IUE integrated spectra of galactic globular clusters.**
V. Caloi, V. Castellani, D. Galluccio.
ESA Spec. Publ., ESA SP–218, p. 497 – 501 (1984). – See Abstr. 012.024.

IUE spectra of the central region of galactic globular clusters are presented. The influence on the observed energy distribution on UV–bright objects is taken into account on the basis of a spatial distribution analysis. Comparison with previous observations by the ANS satellite is given.

154.031 **On the relationship between giant–branch color and metallicity for globular clusters.**
A. Michel, H. A. Smith.
Publ. Astron. Soc. Pac., Vol. 96, No. 582, p. 588 – 592 (1984).

The [Fe/H], $(B\text{–}V)_{0,g}$ relationship for globular clusters is reevaluated. When this relationship is empirically derived from iron abundances determined by the $\Delta S$ method, there is good agreement between the empirical relation, and that predicted by stellar evolution theory. This empirical relation implies [Fe/H] = $-0.6 \pm 0.2$ for M71 and 47 Tuc. This result is compared with results from high–resolution spectroscopy of red giants in globular clusters, and the possible effects of nonsolar abundance ratios on the [Fe/H], $(B\text{–}V)_{0,g}$ calibration is discussed.

154.032 **Erratum: "The ages of the Large Magellanic Cloud blue globular clusters NGC 2133 and NGC 2134" [Publ. Astron. Soc. Pac., Vol. 96, No. 575, p. 28 – 37 (1984)].**
P. W. Hodge, R. A. Schommer.
Publ. Astron. Soc. Pac., Vol. 96, No. 582, p. 674 (1984). See Abstr. 37.154.029.

154.033 **Does the ellipticity of clusters in the LMC correlate with age or luminosity?**
S. van den Bergh, C. L. Morbey.
Astrophys. J., Vol. 283, No. 2, p. 598 – 600 (1984).

It is shown that the apparent flattening of heavily reddened Galactic globular clusters is significantly affected by patchy foreground absorption. After eliminating clusters with $A_V > 1.0$ it is found that LMC clusters of all ages are more highly flattened than Galactic globulars. By using statistical "bootstrap" techniques it is shown that LMC cluster ellipticity is significantly correlated with cluster luminosity but *not* with cluster age.

154.034 **Spectroscopy over a range of 5 magnitudes in NGC 6752.**
R. A. Bell, J. E. Hesser, R. D. Cannon.
Astrophys. J., Vol. 283, No. 2, p. 615 – 625 (1984).

SIT vidicon spectra with 3 – 4 Å resolution have been obtained with the CTIO 4 meter telescope for 26 stars in the southern globular cluster NGC 6752. The stars observed range in brightness from the main–sequence turnoff to the tip of the giant branch and are the first spectra of upper–main–sequence stars obtained in a low–metallicity globular cluster. These spectra are compared with an extensive grid of synthetic spectra. The overall cluster abundance is found to be [M/H] $\sim -1.5$. Observed and computed spectral indices due to Ca, CN, CH, and Fe agree well at all $M_v$ after allowing for zero point shifts.

154.035 **Color–magnitude diagram photometry of the LMC red cluster NGC 1978.**
E. W. Olszewski.
Astrophys. J., Vol. 284, No. 1, p. 108 – 117 (1984). With plates 1 – 8.

Photographic photometry has been carried out in the red LMC cluster NGC 1978 using the CTIO 4 meter telescope at prime focus, photographic plates, the Pickering–Racine prism, and a newly derived photoelectric sequence. Color–magnitude diagrams of the cluster and of four adjacent fields are given; statistical subtraction of the fields is needed to derive an age for NGC 1978. The age is $2 \times 10^9$yr. The abundance of NGC 1978 is estimated to be $[Z/Z_\odot] = -0.5$, in agreement with recent spectroscopic work.

154.036 ***BVRI* main–sequence photometry of the globular cluster M4.**
G. Alcaino, W. Liller.
Astrophys. J., Suppl. Ser., Vol. 56, No. 1, p. 19 – 65 (1984).

The authors present *BV* and *RI* photographic photometry of 1421 and 189 stars, respectively, in the intermediate metallicity globular cluster M4 (NGC 6121), demonstrating the potentialities of improved isochrone fittings using longer wavelengths and wider color baselines. Fitting VandenBerg's isochrones to the *BV* photometry, the authors deduce an age of $13 \pm 2$ Gyr for M4 and find, as in other globular clusters, a pronounced gap of 0.6 mag in the subgiant branch. The photoelectrically calibrated photographic photometry agrees excellently with the CCD results of Richer and Fahlman.

154.037 **CN variations among bright NGC 362 giants.**
G. H. Smith.
Astron. J., Vol. 89, No. 10, p. 1545 – 1550 (1984).

Spectroscopic λ3883 CN band strength measurements are presented for bright giants in the globular cluster NGC 362. A bimodal cyanogen distribution appears to be present, with a predominance of CN–rich stars. The results are consistent with previous surveys of the λ4216 CN band among the bright giants, and the λ3883 CN band among the subgiants. An anticorrelation exists between the λ3883 CN band strength and the infrared CO color. The behavior of CN versus CO in NGC 362 is compared with that seen in other globular clusters.

154.038 **Photoelectric sequences near nine galactic globular clusters.**
M. L. Hazen–Liller.
Astron. J., Vol. 89, No. 10, p. 1551 – 1554 (1984).

Photoelectric *V* and *B–V* are presented for sequences of stars in the neighborhoods of the galactic globular clusters NGC 6139, 6355, 6453, 6496, 6558, 6569, 6642, 6652, and 6717. In each case, the faintest stars measured are at $V \sim 14.5 - 15.5$ mag.

154.039 **A new investigation of the variable stars in the globular cluster Messier 9.**
C. M. Clement, P. Ip, N. Robert.
Astron. J., Vol. 89, No. 11, p. 1707 – 1711 (1984). With plate 85.

New observations from the southern hemisphere have been used to check the periods for the 11 RR Lyrae variables in and around the globular cluster Messier 9. As a result, the mean period of the RRc variables has been revised from $0\overset{d}{.}279$ to $0\overset{d}{.}341$, the mean period of the RRab variables has been revised from $0\overset{d}{.}614$ to $0\overset{d}{.}621$, and a star previously believed to be an RR Lyrae has been reclassified as a Population II Cepheid. One of the RRc variables may be a double–mode pulsator.

154.040 **Photoelectric *UBVRI* sequences in the region of the galactic globular clusters NGC 1261, NGC 3201, NGC 4590, and NGC 5139 (ω Cen).**
G. Alcaino, W. Liller.
Astron. J., Vol. 89, No. 11, p. 1712 – 1715 (1984). With plates 86 – 89.

The authors present *UBVRI* photoelectric sequences in the galactic globular clusters NGC 1261, NGC 3201, NGC 4590, and NGC 5139 (ω Cen). For the first three clusters, most of the stars were previously observed in *UBV* by Alcaino (1979, 1976, 1977), and for ω Cen by Cannon and Stobie (1973).

154.041 **Globular clusters, old and young ones.**
N. N. Samus'.
Zemlya Vselennaya, No. 6, p. 20 – 25 (1984). In Russian.

154.042 **Ages of globular clusters from their luminosity functions.**
B. Paczyński.
Astrophys. J., Vol. 284, No. 2, p. 670 – 674 (1984).

The luminosity function near the turnoff point depends on the age and composition of stars in a cluster. It has a number of advantages over the H–R diagram for comparing stellar models with observations. The luminosity function is not affected by the uncertainties in mixing length parameter and complicated low–temperature opacities. It should be possible to obtain bolometric luminosities from multicolor observations without any reference to model atmospheres. A number of theoretical luminosity functions are presented to illustrate their dependence on age and chemical composition. It is proposed that not only H–R diagrams but also luminosity functions should be used to estimate ages of globular clusters and their uncertainty.

154.043 **An interpretation of the gap on the subgiant branch of several globular clusters.**
T. E. Armandroff, P. Demarque.
Astron. Astrophys., Vol. 139, No. 2, p. 305 – 308 (1984).

Four globular clusters whose color–magnitude diagrams show a gap on the subgiant branch are discussed. Standard stellar evolution theory does not predict the existence of a rapid phase of evolution at this point during a star's lifetime. A mechanism is outlined where this rapid evolutionary phase is attributed to the hydrogen–burning shell passing through a discontinuity in chemical composition. This discontinuity is set up by an ad–hoc mixing process. Evolutionary tracks constructed result in a gap crossing time a factor of two smaller than the standard model and almost identical evolution outside the gap. Another possibility, that the gap is caused by a rapid change in atmopsheric structure, is investigated by varying the mixing length in the convection zone. Tracks constructed in this manner are not consistent with the observed gaps.

154.044 **The manifold of globular clusters.**
P. Brosche, F.-T. Lentes.
Astron. Astrophys., Vol. 139, No. 2, p. 474 – 476 (1984).

The observationally significant number of dimensions of the manifold is three. The relation between the observed "tidal radius" and the galactocentric distance is consistent with its interpretation as a tidal radius.

154.045 **Photometric studies of the globular clusters in M31. III. Metallicity gradient.**
A. S. Sharov, V. M. Lyutyj.
Sov. Astron., Vol. 28, No. 2, p. 143 – 146 (1984). English translation of 37.154.036.

154.046 **Color–magnitude diagrams for the clusters and field of the Fornax dwarf spheroidal galaxy.**
R. Buonanno, C. Corsi, F. Fusi Pecci, E. Hardy, R. Zinn.
Publ. Astron. Soc. Pac., Vol. 96, No. 584, p. 785 (1984). Abstract. – See Abstr. 010.281.

154.047 **The color–magnitude diagram of Palomar 11.**
K. Cudworth, R. Schommer.
Publ. Astron. Soc. Pac., Vol. 96, No. 584, p. 786 – 787 (1984). Abstract. – See Abstr. 010.281.

154.048 **CCD photometry to $M_V \sim +11$ in the globular cluster 47 Tucanae.**
J. E. Hesser, W. E. Harris.
Publ. Astron. Soc. Pac., Vol. 96, No. 584, p. 789 – 790 (1984). Abstract. – See Abstr. 010.281.

154.049 **Abundances of globular clusters in the remote halo.**
R. C. Peterson.
Publ. Astron. Soc. Pac., Vol. 96, No. 584, p. 793 (1984). Abstract. – See Abstr. 010.281.

154.050 **The infrared color–luminosity relation for globular clusters in M31.**
M. L. Sitko.
Publ. Astron. Soc. Pac., Vol. 96, No. 584, p. 794 (1984). Abstract. – See Abstr. 010.281.

154.051 **Predicted *BVRI* and Strömgren colors for globular cluster isochrones.**
D. A. VandenBerg, R. A. Bell.
Publ. Astron. Soc. Pac., Vol. 96, No. 584, p. 797 (1984). Abstract. – See Abstr. 010.281.

154.052 **TiO band strengths in metal–rich globular clusters. VI. NGC 5927 and NGC 6352.**
J. R. Mould, M. S. Bessell.
Publ. Astron. Soc. Pac., Vol. 96, No. 584, p. 800 – 803 (1984).

Narrow–band photometry is reported for two clusters of the inner Galactic halo. The authors obtain [M/H] = $-0.4 \pm 0.2$ for NGC 5927 and [M/H] = $-0.5 \pm 0.2$ for NGC 6352 with $E(B-V) = 0^m.49$ and $0^m.29$, respectively. These reddening values lead to consistency with the infrared (giant branch) metallicity scale.

154.053 **Flattening and luminosity profile of the globular cluster Mayall II in M31.**
C. Pritchet, S. van den Bergh.
Publ. Astron. Soc. Pac., Vol. 96, No. 584, p. 804 – 806 (1984).

The cluster Mayall II, which is the brightest ($M_v = -10.9$) globular cluster in M31, has been observed with a CCD detector on the CFH telescope. These observations show that Mayall II is quite flattened and has $\varepsilon = 0.22$. The M31 globular–cluster system therefore resembles that surrounding the Galaxy in which ω Cen is both the most luminous ($M_v = -10.3$) and (intrinsically) the most flattened cluster.

154.054 **The age of the globular cluster Omega Centauri.**
R. A. Bell, R. D. Cannon, G. L. H. Harris, J. E. Hesser.
Bull. Am. Astron. Soc., Vol. 16, No. 3, Part I, p. 734 (1984). Abstract. – See Abstr. 010.065.

154.055 **Colour–magnitude diagrams of star clusters in the Magellanic Clouds from wide–field electronography – I. Hodge 11.**
J. Andersen, A. Blecha, M. F. Walker.
Mon. Not. R. Astron. Soc., Vol. 211, No. 3, p. 695 – 705 (1984).

The colour–magnitude diagram of Hodge 11 is found to resemble that of M92, indicating that the cluster is an old, very metal–poor globular. The CM diagram is thus in agreement with the classification of the cluster derived from the integral photometric observations of Danziger and of Searle et al. Comparing with M92, and assuming an average reddening for the LMC of $E(B-V) = 0.06$, a distance of $(m-M)_0 = 18.1$ is derived for Hodge 11, adopting the distance modulus for M92 given by Sandage (1982), or $(m-M)_0 = 18.4$ if $M_V$(HB) = +0.6 (Harris 1976).

154.056 **A CCD colour–magnitude diagram to V = 25 for the metal–rich globular cluster 47 Tuc.**
W. E. Harris, J. E. Hesser.
J. R. Astron. Soc. Can., Vol. 78, No. 5, p. 201 (1984). Abstract. – See Abstr. 010.401.

154.057 **sdO stars in globular clusters.**
J. Glaspey.
J. R. Astron. Soc. Can., Vol. 78, No. 5, p. 208 (1984). Abstract. – See Abstr. 010.401.

154.058 **Magnitudes and colours for stars in the central part of the globular cluster NGC 7099 (M30).**
J.-P. Cordoni, M. Aurière.
Astron. Astrophys., Suppl. Ser., Vol. 58, No. 3, p. 559 – 564 (1984).

The authors present a colour-magnitude diagram for 184 stars ($V < 17.$) in the central part of M30 ($0 < r < 30''$). $B$ and $V$ radial variations of surface brightness are also presented. A slight colour gradient is detectable in the investigated part of this cluster. An observed central brightness excess is confirmed by computing sampling errors.

154.059 **New variables in M80.**
A. Wehlau, P. Gannon, P. Rice, H. Sawyer Hogg.
Inf. Bull. Variable Stars, No. 2586, 2 pp. (1984).

154.060 **The evolution of the galactic globular clusters. I. Metal abundance calibrations.**
S.-W. Lee, N.-K. Park.
J. Korean Astron. Soc., Vol. 17, No. 2, p. 69 – 103 (1984).

It is attempted to show the difference in five fundamental calibrations of Z-abundance of globular clusters and to examine their effect on the galactic evolution. The fundamental calibrations of Z-abundance are examined and the correlations of these calibration scales with ranking parameters are presented. The spatial distribution of globular clusters and its implication for the galactic evolution are discussed.

154.061 **A colour magnitude diagram for NGC 2210.**
A. R. Walker.
Mon. Notes Astron. Soc. S. Afr., Vol. 43, Nos. 11 – 12, p. 89 – 96 (1984).

A colour magnitude diagram for the LMC globular cluster NGC 2210, obtained with the UCL-SAAO CCD camera, is presented. The photometry extends to V = 21.5 and shows a well-populated blue horizontal branch.

154.062 **The distant globular clusters Palomar 14 and Palomar 15.**
W. E. Harris, S. van den Bergh.
Astron. J., Vol. 89, No. 12, p. 1816 – 1821 (1984). With plates 96, 97.

The authors have obtained new star counts and color-magnitude diagrams for the sparse, distant halo clusters Palomar 14 and 15, from photographic plates obtained with the Hale reflector. The CMD for Pal 14 agrees substantially with that of Da Costa et al. (1982); at $R_{gc} = 70$ kpc, it has core and tidal radii of 15 and 93 pc. Pal 15 is also found to be an outer-halo object ($R_{gc} \cong 60$ kpc), but with a completely blue horizontal branch, unlike all the other outer-halo clusters except NGC 2419. With $r_c \cong 25$ pc and $r_t \cong 100$ pc, Pal 15 also has the largest core radius and lowest central concentration of any known globular cluster. Finally, the authors note a significant correction to the published coordinates of Pal 15.

154.063 **A study of the globular cluster ω Cen.**
S. Ninković.
Bull. Appl. Math., No. 193, p. 55 – 58 (1983).

Examinations of the globular cluster ω Cen on the basis of the virial theorem and a tidal radius determination show that a large mass is possible for this cluster. An empirical model of the cluster, indicating that the largest fraction of the total mass is present near the centre, is constructed.

154.064 **Metallicity determinations for globular clusters through spectrophotometry of their integrated light.**
J. P. Brodie, D. A. Hanes.
Anglo-Aust. Obs., Prepr., No. 198, 64 pp. (1984). Submitted to Astrophys. J.

154.065 **The metallicity of globular clusters associated with the giant elliptical galaxy NGC 4486 (M87).**
D. A. Hanes, J. P. Brodie.
Anglo-Aust. Obs., Prepr., No. 199, 42 pp. (1984). Submitted to Astrophys. J.

154.066 **CCD photometry of the globular cluster NGC 6752.**
A. J. Penny.
Gemini, No. 12, p. 1 – 2 (1984).

154.067 **Reddening of globular clusters in M31.**
M. Iye, O.-G. Richter.
ESO Sci. Prepr., No. 347, 37 pp. (1984). To appear in Astron. Astrophys.

154.068 ***IUE* observations of the clusters of the Magellanic Clouds.**
J. G. Cohen, R. M. Rich, S. E. Persson.
Astrophys. J., Vol. 285, No. 2, p. 595 – 600 (1984).

*IUE* observations of 17 clusters in the Large Magellanic Cloud are presented. The derived ultraviolet colors become progressively redder as the cluster's classification in the scheme of Searle, Wilkinson, and Bagnuolo (SWB) increases. With the aid of theoretical models of the integrated light of coeval clusters of varying ages, it is shown that the SWB classification is one of increasing age accompanied by decreasing metallicity toward later classes.

154.069 **Photometry of faint stars in globular clusters. VII. On some common features of photometric characteristics of faint stars in M10 and M12.**
A. V. Mironov, N. N. Samus', S. Yu. Shugarov, A. O. Yuferov.
Astron. Tsirk., No. 1313, p. 1 – 2 (1984). In Russian.

154.070 **Eight new variable stars in the globular cluster M 15.**
Z. I. Kadla, A. N. Gerashchenko, N. V. Yablokova, N. Spasova.
Astron. Tsirk., No. 1314, p. 1 – 2 (1984). In Russian.

154.071 **The extended giant branches of intermediate age globular clusters in the Magellanic Clouds. IV.**
M. Aaronson, J. Mould.
Prepr. Steward Obs., No. 538, 36 pp. (1984).

A complete survey is now available for asymptotic giant branch stars in the rich star clusters of the Magellanic Clouds. Although data on the main sequence turnoffs of these clusters is still incomplete, some systematic properties of these stars emerge, when grouped by cluster age. Clusters younger than approximately 8 billion years have carbon stars at the tip of the giant branch, produced by the third dredge up mechanism. Clusters younger than approximately 0.8 billion years have giant branches populated by M stars.

154.072 **Infrared photometry of globular clusters in M31.**
M. L. Sitko.
Astrophys. J., Vol. 286, No. 1, p. 209 – 215 (1984).

Photometry of the integrated infrared light of globular clusters in the Andromeda galaxy (M31), including many with projected distances from the galaxy nucleus greater than 10 kpc, are presented. Additional data on globular clusters in NGC 205, NGC 147, and NGC 185 are included. The data obtained indicate a well-defined relation between $M_K$ and ($J$-$K$) for globular clusters in the Local Group. The use of cluster $K$, ($J$-$K$) as a distance indicator is discussed.

154.073 **Astrometrische Untersuchungen am Kugelhaufen M12.**
H.-J. Tucholke, M. Geffert.
Mitt. Astron. Ges., Nr. 62, p. 215 (1984). Abstract. – See Abstr. 010.341.

154.074 **Abschätzung von Kohlenstoff- und Stickstoffhäufigkeiten in extrem metallarmen Halofeldriesen aufgrund von IDS-Spektren geringer Dispersion.**
C. F. Trefzger, M. Grenon.
Mitt. Astron. Ges., Nr. 62, p. 291 – 293 (1984). – See Abstr. 010.341.

154.075 **Radii of globular clusters.**
S. van den Bergh, C. L. Morbey.
Astron. Express, Vol. 1, Issue 1, p. 1 – 7 (1984).

For Galactic globulars, the correlation between cluster radius and Galactocentric distance is found to be marginally stronger than that between cluster metallicity and Galactocentric distance. At a given Galactocentric distance cluster radius does not depend significantly on metallicity. Faint halo globulars tend to be larger than bright halo clusters but for objects within 10 kpc of the Galactic centre the cluster radius is independent of luminosity.

154.076 **Main-sequence photometry of the SMC globular cluster Kron 3.**
R. M. Rich, G. S. Da Costa, J. R. Mould.
Astrophys. J., Vol. 286, No. 2, p. 517 – 528 (1984). With plates 20 – 22.

A color-magnitude (C-M) diagram for the SMC cluster Kron 3 is presented and discussed. The cluster main-sequence turnoff is revealed for the first time, and comparison with theoretical isochrones indicates that the age of Kron 3 is in the range of 5 – 8 billion years. A similar C-M diagram for a SMC background field was also determined. It indicates that, unlike the galactic halo, the SMC halo contains stars with a large range of ages, from 1 billion years upwards.

154.077 **A search for cyanogen variations on the lower giant branch of M3.**
J. Norris, G. H. Smith.
Astrophys. J., Vol. 287, No. 1, p. 255 – 260 (1984).

Spectra have been obtained for 16 giants in the globular cluster M3. These show that variations in the strength of the 3883 Å cyanogen band persist among stars down to the magnitude limit of the sample, which is $M_v = +1.2$ ($V = 16.2$). This is fainter by ~1.6 mag than the luminosity at which the meridional circulation model of Sweigart and Mengel predicts that CN variations should originate. The present data also show that, in the mean, the 4300 Å G band is weaker in the spectra of the CN-rich stars than it is in those of the CN-poor stars.

154.078 **The chemical composition of the globular cluster M55.**
C. A. Pilachowski, C. Sneden, E. M. Green.
Publ. Astron. Soc. Pac., Vol. 96, No. 586, p. 932 – 943 (1984).

The chemical composition of five giant stars in the globular cluster M55 has been determined from the analysis of high-resolution echelle spectrograms in both the red and the blue spectral regions. The overall iron-peak abundances are $[m/H] = -1.3 \pm 0.1$. The abundances of the very heavy elements follow a pattern similar to the solar composition. Carbon in these giants is deficient relative to the iron-peak elements by about a factor of 4.

154.079 **The Large Magellanic Cloud intermediate-age globular cluster NGC 1831.**
P. W. Hodge.
Publ. Astron. Soc. Pac., Vol. 96, No. 586, p. 947 – 959 (1984).

NGC 1831 is one of the largest, most populous "blue globular clusters" of the LMC. A C-M diagram to $V = 21.5$ shows a main sequence extending up to $V = 18.5$ and a well-developed giant branch. Fitting to evolutionary models leads to an age estimate of 300 million years and the giants' colors suggest that $[M/H] \sim -1.0$. Spectra of three cluster stars are shown. One is for one of the "anomalous giants" and it indicates that it is probably a composite image. The spectrum of the cluster shows the dominance of the hot main sequence in the integrated light.

154.080 **The halo and disk systems of globular clusters.**
R. Zinn.
News Lett. Astron. Soc. N.Y., Vol. 2, No. 6, p. 10 – 11 (1984). Abstract. - See Abstr. 010.241.

154.081 **Old stellar populations. I. A spectroscopic comparison of galactic globular clusters, M31 globular clusters, and elliptical galaxies.**
D. Burstein, S. M. Faber, C. M. Gaskell, N. Krumm.
Astrophys. J., Vol. 287, No. 2, p. 586 – 609 (1984). = Lick Obs. Bull., No. 986.

Integrated spectra at 9 Å resolution in the region 4000 – 6300 Å are presented for 17 galactic globular clusters and 19 M31 globular clusters. Indices for 11 molecular bands and lines are measured: the CN 4215 Å band, the G band, Hβ, Mg *b* and MgH, two Fe I blends, Na D, and two TiO bands. The index $Mg_2$ (measuring Mg *b* and MgH) appears to be a good relative metallicity indicator for galactic globular clusters, comparable to Zinn's $Q_{39}$ index. These indices and several broad-band colors are compared between the M31 and galactic cluster systems, and also with the nuclei of 170 E galaxies. The observations are analyzed using detailed stellar population models.

154.082 **Magnesium, silicon, and iron abundances and the cluster metallicity scale.**
D. Geisler.
Astrophys. J., Lett. Ed., Vol. 287, No. 2, p. L85 – L89 (1984).

Abundances have been determined for giants in a number of open and globular clusters from Washington photometry and echelle spectroscopy. Fe abundances derived from these two very different techniques agree to within the uncertainties over the entire range of cluster abundances. The classical metal-rich globular clusters NGC 6352, M69, and M71 are found to have [Fe/H] values $\lesssim -1$. The metal-poor open clusters NGC 2141, 2158, 2506, and Melotte 66 have [Fe/H] values ranging from -0.5 to -1.2. A much better correlation is found between the color of the cluster giant branch and the mean (Mg + Si + Fe) abundance than with the Fe abundance alone.

**Catalog of cluster C-M diagrams.**
See Abstr. 002.071.

**Triennial Report concerning the galactic globular star clusters.**
See Abstr. 002.072.

**The distribution of the projected density of stars near the Reissner-Nordström black hole in a globular cluster.**
See Abstr. 067.032.

**A photoelectric sequence in the region of the globular cluster NGC 5986.**
See Abstr. 113.030.

**IUE observations of UV bright stars in the globular clusters M15 and ω Cen.**
See Abstr. 114.069.

**Carbon and nitrogen abundances in CN-strong and CN-weak giants in M5.**
See Abstr. 114.078.

**A high dispersion analysis of a giant star in 47 Tucanae.**
See Abstr. 114.105.

**Spektralanalyse von Horizontalaststernen im Kugelhaufen NGC 6752.**
See Abstr. 114.123.

**UV-bright stars in globular clusters: *IUE* observations.**
See Abstr. 115.015.

**Close binary stars and old stellar populations: the blue straggler problem revisited.**
See Abstr. 117.121.

**Did the binary stars in globular clusters come to an early end by being too eccentric?**
See Abstr. 117.205.

**Photographic photometry of RR Lyrae variables in the globular cluster M15.**
See Abstr. 122.027.

**The Population II Cepheids.**
See Abstr. 122.120.

**Studies of Cepheid type variability. II. Fourier description of RR Lyrae variables in the globular cluster ω Centauri.**
See Abstr. 122.128.

**A bump progression sequence in the first overtone RR Lyrae variables in ω Centauri.**
See Abstr. 122.149.

**FK Ophiuchi = V2 (NGC 6273).**
See Abstr. 122.157.

**A new variable star (No. 218) near the globular cluster NGC 6401.**
See Abstr. 123.027.

**CASPEC–Beobachtungen des sdO's ROB 162 in NGC 6397.**
See Abstr. 126.077.

**Interstellar extinction in the direction of the globular cluster M71.**
See Abstr. 131.001.

**Inflow of halo gas from the direction of the galactic North Pole.**
See Abstr. 131.010.

**Infrared observations of galactic bulge X–ray sources.**
See Abstr. 142.008.

**Determination of the mass of globular cluster X–ray sources.**
See Abstr. 142.009.

**A possible optical counterpart for the M15 globular cluster X–ray source.**
See Abstr. 142.021.

**Short–term X–ray variability of the globular cluster source 4U 1820–30 (NGC 6624).**
See Abstr. 142.023.

**Modelling of the integral B–V colour index for metal–poor stellar systems.**
See Abstr. 151.001.

**Dynamics of globular clusters.**
See Abstr. 151.013.

**Stability of star clusters as galactic satellites. III. Preliminary applications to the system of the globular clusters.**
See Abstr. 151.025.

**Stability of star clusters as galactic satellites. IV. Logarithmic galactic potential.**
See Abstr. 151.026.

**A unified *N*–body and statistical treatment of stellar dynamics. II. Applications to globular cluster cores.**
See Abstr. 151.035.

**Effects of encounters with field stars on the evolution of low–mass semidetached binaries.**
See Abstr. 151.043.

**Wie bewegen sich die Sterne in einem Kugelsternhaufen?**
See Abstr. 151.046.

**A theorem due to Jörg Pfleiderer concerning the stability of globular clusters.**
See Abstr. 151.066.

**On equipartition of kinetic energies in two–component star clusters.**
See Abstr. 151.094.

**Interaction of the Large Magellanic Cloud with the halo population and the companions of the Galaxy.**
See Abstr. 156.018.

**Chemical evolution of bulges and halos.**
See Abstr. 157.129.

**Kugelsternhaufen in nahen irregulären Galaxien.**
See Abstr. 157.195.

**The NGC 5128 globular cluster system: star counts in *U*, *V*, and *R*.**
See Abstr. 158.270.

**The NGC 5128 globular cluster system: a complete analysis of images using COSMOS.**
See Abstr. 158.271.

**The variability of blue objects in the field of M3.**
See Abstr. 159.084.

**Globular cluster swapping in clusters of galaxies.**
See Abstr. 160.046.

**Globular clusters and galaxy formation in clusters of galaxies.**
See Abstr. 160.094.

# 155 Galaxy

155.001 **Effects of interstellar magnetic bubbles and of galactic tides on galactic magnetic fields.**
J. P. Vallée.
Astron. Astrophys., Vol. 136, No. 2, p. 373 – 377 (1984).
For our Milky Way galaxy, a quantitative estimate is made of the apparent effects of four large interstellar magnetic bubbles near the Sun on the rotation measures of distant radio sources. For other spiral galaxies, a quantitative estimate is made of the apparent correlation found earlier by Vallée (1983) between observed magnetic field distributions and nearby companions. In Spi–Spi galaxies (e.g., with spiral stellar arms and a spiral magnetic field distribution), the tidal radius averages about 50 kpc, so large tidal effects from a companion galaxy could possibly create material arms and align stars, gas and magnetic fields in spiral arms. In Spi–Cir galaxies (e.g., with spiral stellar arms and a circumferential magnetic field distribution), the tidal radius is larger than about 100 kpc, so small tidal effects from a companion galaxy could only possibly excite density–wave arms, i.e. stars in spiral arms but gas and magnetic fields in circumferential orbits.

155.002 **The spiral structure of the Galaxy: a statistical analysis of the space distribution of bright stars and open clusters.**
E. D. Pavlovskaya, A. A. Suchkov.
Astron. Zh., Tom 61, Vyp. 4, p. 665 – 672 (1984). In Russian. English translation in Sov. Astron., Vol. 28, No. 4.
A statistical analysis of the space distribution of bright stars and open clusters in the fourth quadrant of the Galaxy is performed. It is shown that, with a probability equal practically to 100%, the Sagittarius–Carina spiral feature cannot be a chance density fluctuation in an actually homogeneous distribution of stars. The geometrical parameters of this feature are estimated.

155.003 **The radio lobe above the galactic centre.**
Y. Sofue.
Astron. Her., Vol. 77, No. 8, p. 188 – 190 (1984). In Japanese.

155.004 **Frequency distribution of metal abundance for faint stars in the direction of the north galactic pole.**
Y. Yoshii.
Astron. J., Vol. 89, No. 8, p. 1190 – 1199 (1984).
Using the published three–color photometric data of faint stars in the direction of the north galactic pole, the author separated G dwarfs from subgiants by means of a statistical method which utilizes the fact that the star counts are related to both the luminosity function and the density function. Based on the sample of G dwarfs thus separated, the author derived the cumulative metallicity distributions for three intervals of height above the galactic plane, $1.0\ \mathrm{kpc} \leqslant z < 1.5\ \mathrm{kpc}$, $0.2\ \mathrm{kpc} \leqslant z < 1.5\ \mathrm{kpc}$, and $0.0\ \mathrm{kpc} \leqslant z < 1.0\ \mathrm{kpc}$, and found that the gradient in the mean metallicity exists at least up to $z \sim 1.5 - 2.0$ kpc. The derived results are interpreted by a two–zone model of chemical evolution which describes the collapse of gas toward the galactic plane.

155.005 **Erratum: "The galactic extinction of extragalactic objects. I. The csc *b* law and the extinction coefficient" [Astron. J., Vol. 88, No. 7, p. 939 – 961 (1983)].**
G. de Vaucouleurs, R. Buta.
Astron. J., Vol. 89, No. 8, p. 1264 (1984). See Abstr. 34.155.002.

155.006 **Distribution and properties of molecular clouds in the Galaxy.**
N. Z. Scoville, D. B. Sanders, D. P. Clemens.
Occas. Rep. R. Obs., Edinb., No. 13, p. 1 – 14 (1984). – See Abstr. 012.001.
The large scale distribution of the molecular clouds as a function of galactic radius and azimuth is discussed with particular emphasis on the molecular cloud ring at 5 to 8 kpc radius. A summary is provided of what is known about the properties of the emission regions – their size, mass and density. The issue of CO spiral structure is dealt with.

155.007 **The number and distribution of molecular clouds in the inner Galaxy.**
P. Thaddeus, T. M. Dame.
Occas. Rep. R. Obs., Edinb., No. 13, p. 15 – 26 (1984). – See Abstr. 012.001.
In attempting to determine the number and location of the largest molecular clouds in the inner Galaxy the authors' main exhibits are the two CO l, v diagrams they have obtained covering much of the first and second galactic quadrants, and the distribution of the largest clouds on the plane of the Galaxy derived from a first quadrant diagram. All observations are of the $1 \rightarrow 0$ rotational transition of the normal CO isotopic species at 115 GHz.

155.008 **A deep optical survey of a small region in Aquarius. I. Stellar photometry and galactic structure.**
K. P. Tritton, D. C. Morton.
Mon. Not. R. Astron. Soc., Vol. 209, No. 2, p. 429 – 447 (1984). With microfiche MN 209/1.
The authors have identified and classified as accurately as possible all objects with $B \leqslant 20.0$ in 0.31 deg$^2$ centred on the 1950.0 coordinates $\alpha = 22^h03^m04^s.5$, $\delta = -18°55'03''$ ($l = 36°.5$, $b = -51°.1$). They determined the $B$ and $V$ magnitudes of 601 normal stars and 2 white dwarfs to this magnitude limit and used these data to investigate galactic structure in this direction. Within 2 kpc of the sun a disc scale height of about 300 pc is found for $4 \leqslant M_v \leqslant 6$, consistent with determinations in other directions. However, at greater distances out to 5 kpc, where the counts are complete for only $4 \leqslant M_v \leqslant 5$, the density decreases much less steeply with distance, implying a rather flat spheroid component or a thick disc.

155.009 **Large, highly organized radio structures near the galactic centre.**
F. Yusef–Zadeh, M. Morris, D. Chance.
Nature, Vol. 310, No. 5978, p. 557 – 561 (1984).
A radio map of the galactic centre made at a wavelength of 20 cm with the VLA reveals that most of the radio emission arising within 50 pc of the galactic centre is not associated with ionizing radiation from recently formed stars. Rather, the large–scale geometry and the measured polarization of the radio emission strongly suggest that the nonthermally emitting gas is arranged along magnetic structures indicative of a substantial poloidal component to the magnetic field in the central region of the Galaxy.

155.010 **A radio lobe over the galactic centre.**
Y. Sofue, T. Handa.
Nature, Vol. 310, No. 5978, p. 568 – 569 (1984).
The authors report a prominent off–plane radio lobe structure above the galactic nucleus, which has been revealed during a 10–GHz continuum survey of the galactic plane.

155.011 **Features of the position of the sun in the Galaxy.**
L. S. Marochnik.
Astrofizika, Tom 19, Vyp. 3, p. 494 – 504 (1983). In Russian. English translation in Astrophysics, Vol. 19, No. 3.
The solar system position in the Galaxy is an exclusive one since the sun is close to the corotation circle which is the place where the angular velocity of the galactic differential rotation is equal to that of density waves displaying as the galactic spiral arms. The special conditions of the sun's position in the Galaxy explain the fundamental cosmogony time scales detected by radioactive decay of various nuclides.

155.012 **Far–infrared survey of the galactic disc in the southern hemisphere. I. The large–scale structure.**
E. Caux, G. Serra, R. Gispert, J. L. Puget, C. Ryter, N. Coron.
Astron. Astrophys., Vol. 137, No. 1, p. 1 – 5 (1984).
The authors present a FIR survey (114 – 196 μm) of the galactic disc made from a balloon borne instrument launched in Aus-

tralia. The emission of the galactic disc has been measured from $l = 250°$ to $l = 20°$. This emission appears as an intense ridge with a typical FIR brightness of $2 \times 10^{-5}$W m$^{-2}$ sterad$^{-1}$ with bright spots superimposed on an unresolved component. The authors have built a spiral structure model of the FIR source function spatial distribution which represents the large scale emission of the Galaxy. It is suggested that the FIR emission could be due to very large complexes of H II regions giant molecular clouds and lower density gas which are distributed along galactic spiral arms.

155.013 **The program of studying the main meridional section of the Galaxy: astrometric data.**
N. V. Kharchenko.
Tartu Astrofüüs. Obs. Teated, Nr. 72, p. 12 – 19 (1984).
Astrometric data to be used in the complex program of studying the main meridional section of the Galaxy are outlined and efforts to increase their accuracy are described.

155.014 **H I shells, supershells, shell–like objects, and "worms".**
C. Heiles.
Astrophys. J., Suppl. Ser., Vol. 55, No. 4, p. 585 – 595 (1984). With plates 10 – 34.
The author combines existing H I surveys and lists new shell–like objects that cross survey boundaries at $b = 10°$. In addition, spatial filtering reveals wormlike structures in the inner galaxy. Stellar winds and supernovae in star clusters are almost certainly not sufficiently energetic to form the largest supershells in our own Galaxy. The "worms" are probably parts of shells that are open at the top and should be good sources of hot gas for the galactic halo.

155.015 **On the statistical similarity of high–velocity clouds and stars of the spherical component of the Galaxy.**
R. B. Shatsova.
Astrofizika, Tom 19, Vyp. 4, p. 779 – 790 (1983). In Russian. English translation in Astrophysics, Vol. 19, No. 4.
The dependence of high–velocity clouds (HVC) radial velocities on galactic longitude is similar to the Strömberg asymmetry. The galactic rotation velocity and the velocity dispersion of HVC are the same as those of spherical component objects (subdwarfs, RR Lyrae–type variables, globular clusters). The HVC similar to the nearby subdwarfs do not reveal any concentration to the galactic centre. Stellar groups, including Kapteyn's group, projecting over (or in the vicinity) C II complexes, the "Magellanic Stream" and others with the same radial velocities as these HVC have are found. Most of the stars in these groups have distances of the order of 100 pc.

155.016 **Interstellar nitrogen isotope ratios.**
R. Güsten, H. Ungerechts.
MPA Rep., No. 90, p. 66 – 67 (1983). – See Abstr. 012.020.
The authors report first results of a systematic study of the galactic interstellar nitrogen isotope ratios, based upon measurements of the $(J, K) = (1, 1)$ and $(2, 2)$ lines of the $^{14}NH_3$ and $^{15}NH_3$ (ammonia) molecules. They find significant deviations of the $[^{14}N]/[^{15}N]$ abundance ratio from the solar system value ($\sim 270$), with the most extreme enrichment in the galactic center region, where $[^{14}N]/[^{15}N] \sim 1500$.

155.017 **Galactic oxygen gradient: airplane measurements of $O^{++}$.**
J. Schmid–Burgk.
MPA Rep., No. 90, p. 68 (1983). Abstract. – See Abstr. 012.020.

155.018 **Can the ultrahigh–energy cosmic rays stem from the galactic center?**
V. S. Berezinskij, A. A. Mikhajlov.
Sov. Astron. Lett., Vol. 10, No. 2, p. 112 – 114 (1984). English translation of 37.155.048.

155.019 **Faraday rotation and the galactic magnetic field – a review.**
P. A. H. Seymour.
Q. J. R. Astron. Soc., Vol. 25, No. 3, p. 293 – 300 (1984).
This paper starts by reviewing work on the interpretation of rotation measures from extra–galactic radio sources up to 1983, and then concentrates on the use of spherical harmonics in the analysis and interpretation of the data.

155.020 **Very deep survey of the Galaxy at 7.6 cm.**
A. B. Berlin, E. V. Bulaenko, V. V. Vitkovskij, V. K. Kononov, D. V. Korol'kov, Yu. N. Parijskij, S. A. Trushkin.
Spets. Astrofiz. Obs. Akad. Nauk SSSR. Prepr., No. 7, 21 pp. (1983). Abstr. in Ref. Zh., 51. Astron., 8.51.558 (1984).

155.021 **Journey to the centre of the Galaxy.**
H. Couper.
New Sci., Vol. 102, No. 1407, p. 32 – 39 (1984). Abstr. in Phys. Abstr., Vol. 87, No. 1242, Entry 94970 (1984).

155.022 **The galactic rotation outside the solar circle.**
R. Chini, J. E. Wink.
Astron. Astrophys., Vol. 139, No. 1, p. L5 – L8 (1984).
New photometric and spectroscopic observations of the exciting stars of galactic H II regions are presented. Combining the derived distances with the radial velocities of the regions gives evidence for a rising rotation curve beyond the solar circle.

155.023 **Shocked molecular hydrogen emission from the centre of the Galaxy.**
I. Gatley, T. J. Jones, A. R. Hyland, D. H. Beattie, T. J. Lee.
Mon. Not. R. Astron. Soc., Vol. 210, No. 3, p. 565 – 575 (1984).
Observations of molecular hydrogen emission lines near 2 μm show that the nucleus of the Galaxy is encircled by a ring of shocked gas; this ring has a radius of 2 parsecs, lies in the plane of the Galaxy, is symmetric about the centre of mass, and rotates in the sense of galactic rotation. Gas is being shocked at a rate of $> 10^{-2} M_{\odot} yr^{-1}$, to a temperature about 2000K, in a region of mean molecular density $5 \times 10^3 cm^{-3}$. The momentum needed to shock the gas cannot be provided radiatively. Mass loss from the nucleus can account naturally for the central density minimum and for the shocked gas; a mass loss rate of $3 \times 10^{-3} M_{\odot} yr^{-1}$ is required. Simple time–scale arguments suggest that observable molecular hydrogen emission from the galactic centre may be a long–lived phenomenon. A model involving a single central engine is suggested.

155.024 **Rotation curve of our Galaxy for $R > R_0$.**
U. Haud.
Astrophys. Space Sci., Vol. 104, No. 2, p. 337 – 345 (1984).
It is shown that the rise of rotation velocities in the outer regions of our Galaxy may be illusive, and is caused by an excessive simplification of the methods of processing the observational data. Taking into consideration the possibility of the radial motions of the gas, the author demonstrates that the existing observational data do not conflict with the hypothesis of the flat rotation curve of our Galaxy. The bending of the outer regions of the galactic disk has no significant effect on the observed radial velocities of the gaseous complexes. It is also shown that the wavy form of the rotation curve may be evoked by the spiral density waves in the Galaxy.

155.025 **Observations of the stellar radiation field at 975 Å.**
C. B. Opal, C. S. Weller.
Astrophys. J., Vol. 282, No. 2, p. 445 – 451 (1984). With plate 9.
Approximately two–thirds of the celestial sphere was scanned with a photometer sensitive over the 911 – 1050 Å band which was carried on the STP 72–1 satellite. Most of the observed emission originated from late O and early B stars in Gould's belt. Because of the 10° resolution of the instrument, only a few stars were free from source confusion. Simulated sky maps generated by taking fluxes predicted by stellar model atmospheres combined with star catalog integrations and folding them through the

instrument spectral and spatial response; good agreement was found if some allowance for reddening was made.

155.026 **Distribution of OB stars and absorbing matter in the region around P Cyg.**
A. T. Garibdzhanyan, K. G. Gasparyan, R. Kh. Oganesyan.
Astrofizika, Tom 20, Vyp. 2, p. 245 – 261 (1984). In Russian. English translation in Astrophysics, Vol. 20, No. 2.

The results of a spectrophotometric study of 277 OB stars in the region around P Cyg are presented. The absolute spectrophotometric blue (4000 – 4800 Å) gradients, the colour excesses and the distances of these stars are obtained.

155.027 **The H I halo in the inner Galaxy.**
F. J. Lockman.
Astrophys. J., Vol. 283, No. 1, p. 90 – 97 (1984).

There is 21 cm emission from corotating H I in the inner Galaxy to $|z| \geqslant 1000$ pc from the plane. Over most of the inner Galaxy more than 10% of the H I emission at the subcentral points comes from $|z| > 500$ pc. High–$z$ H I is not present $\lesssim 3$ kpc from the galactic center, but appears suddenly near $R = 3.5$ kpc, in qualitative agreement with Bregman's model for a wind from the bulge. The scale height and surface density of the halo are approximately constant over $4 \leqslant R \leqslant 8$ kpc. In this region $n_H(z)$ has a long exponential tail with a scale height $\sim 500$ pc and a density of 0.012 $cm^{-3}$ at $|z| = 750$ pc.

155.028 **Radio evidence for mass outflow in the Milky Way and external galaxies.**
R. D. Davies.
Mass loss from astronomical objects, p. 101 – 107 (1984). – See Abstr. 012.032.

This paper discusses observational evidence for mass outflow from the galactic center, for outflowing H I gas in the galactic halo, and for the ejection of matter from ordinary galaxies.

155.029 **A *B,V* sequence in Baade's window.**
V. M. Blanco, B. M. Blanco.
Publ. Astron. Soc. Pac., Vol. 96, No. 582, p. 603 – 608 (1984).

*B,V* measurements are presented of a 70–star sequence in Baade's window. Magnitudes in this extremely crowded region were measured with a photomultiplier, by reducing PDS scanned density–intensity calibrated photographic plates, and with a CCD detector. The degree of agreement obtained with these different methods suggests that systematic errors are negligible up to $V \sim 18^m.8$, $B \sim 19^m.5$.

155.030 **Far–infrared and submillimeter survey of the galactic plane from $l = 11°.5$ to $l = 17°.5$.**
M. F. Campbell, R. F. Silverberg, W. F. Hoffmann,
M. G. Hauser, D. W. Niles, M. T. Stier, H. A. Thronson Jr.,
T. Kelsall.
Astrophys. J., Vol. 283, No. 2, p. 566 – 572 (1984).

Medium resolution (11′) maps of the galactic plane are presented from $l = 11°.5$ to $l = 17°.5$ at wavelengths of 93 μm, 154 μm, and 190 μm. The maps are interpreted in terms of the temperature and spatial structure of diffuse far–infrared/submillimeter sources associated with evolved H II regions and a continuous ridge of galactic emission. The emission regions are found to be more extended at the longer wavelengths which implies that there must be a range of dust temperatures in the sources.

155.031 **A far–infrared survey of the galactic center.**
S. F. Odenwald, G. G. Fazio.
Astrophys. J., Vol. 283, No. 2, p. 601 – 614 (1984). With plate 17.

This paper presents the first large–scale, high–resolution, far–infrared survey of the galactic center region covering an 8.5 $deg^2$ field centered on Sgr A at a resolution of 1′ in the spectral range 40 – 250 μm. A total of 47 far–infrared sources were detected, of which 37 are probably indigenous to the galactic center region. For those 27 sources associated with H II regions luminosities range from $5 \times 10^5$ to $10^7 L_\odot$. They are located in a narrow band along the galactic plane and are centered on Sgr A. Each far–infrared source appears to be powered by small numbers of early–type O stars. The extended thermal emission of the galactic center region is consistent with a population of $3 - 5 \times 10^4$ B0 ZAMS stars embedded in an interstellar medium with a Lyman–continuum optical depth of 1 – 1.5. The ratio of the inferred O and B star populations suggests that there are fewer early O type stars relative to B type stars in the galactic center than are predicted by the Salpeter initial luminosity function.

155.032 **The distribution of CO in the Galaxy for longitudes 294° to 86°.**
B. J. Robinson, R. N. Manchester, J. B. Whiteoak,
D. B. Sanders, N. Z. Scoville, D. P. Clemens,
W. H. McCutcheon, P. M. Solomon.
Astrophys. J., Lett. Ed., Vol. 283, No. 2, p. L31 – L35 (1984). With plate L2.

Observations of 115 GHz ($J = 1 - 0$) emission from the CO molecule for the southern galactic plane from longitude 294° to the galactic center are compared with northern galactic plane observations to longitude 86°. Data from both hemispheres are used to derive a global average radial distribution of molecular gas out to $R_\odot$. The molecular cloud ring seen in earlier northern surveys remains the dominant feature in the galactic distribution, but differences in the width of the ring between north and south could be due to the effects of spiral arms.

155.033 **On the structure of the nonthermal radio radiation of the Galaxy at a frequency of 150 MHz.**
Ch. Seitnepesov, A. Kh. Khanberdiev.
Izv. Akad. Nauk TSSR. Ser. fiz.–tekh., khim. i geol. nauk., No. 6, p. 30 – 37 (1983). In Russian. Abstr. in Ref. Zh., 51. Astron., 9.51.714 (1984).

155.034 **The distribution of luminous stars at $(l,b) = (245°,0°)$ in Puppis.**
B. C. Reed, M. P. FitzGerald.
Mon. Not. R. Astron. Soc., Vol 211, No. 1, p. 235 – 241 (1984). = Contrib. Univ. Waterloo Obs., No. 107.

Spectroscopic and photometric parallaxes are presented for 108 O – B5 stars ($V \leqslant 12.5$) in the vicinity of $(l,b) = (245°,0°)$. The galactic plane as defined by these stars is found to be essentially flat to $R \geqslant 15$ kpc (galactocentric; $R_0 = 10$ kpc), in agreement with the distribution of neutral hydrogen. The stellar density distribution of O – B0 stars is found to consist of an inter–arm region for $r \leqslant 2.8$ kpc, followed by a rise to a peak density of $\sim 100$ stars $kpc^{-3}$ at $r = 3$ kpc ($R = 11.7$ kpc), declining exponentially thereafter with a scale length in $R$ of $1.6 \pm 0.1$ kpc, or greater. There is no evidence for the existence of Puppis OB1 or OB2.

155.035 **The stellar distribution in Puppis. I. Distribution of B5 to M5 stars to $V \cong 12.5$.**
B. C. Reed, M. P. FitzGerald.
Mon. Not. R. Astron. Soc., Vol 211, No. 1, p. 243 – 253 (1984). = Contrib. Univ. Waterloo Obs., No. 108. With microfiche MN 211/2.

BV photographic photometry and objective–prism spectral classifications for 3339 B5 – M5 stars in a 7.75 square degree field at $(l,b) = (245°,0°)$ have been determined. Interstellar absorption in this direction sets in at $r \sim 400$ pc, and rises steadily to $A_V \sim 1.6$ mag at $r \sim 8.8$ kpc. The spatial distributions of various spectral groups are consistent in number and structure with those in other Galactic plane fields. The A2 – A5 stars exhibit a density maximum at $r \sim 1.2$ kpc which may be associated with the Carina bridge. The G – K giants appear more populous in this direction than at other Galactic longitudes.

155.036 **Observations of broad helium and hydrogen lines in the very center of the Galaxy.**
T. R. Geballe, K. Krisciunas, T. J. Lee, I. Gatley, R. Wade,
W. D. Duncan, R. Garden, E. E. Becklin.
Astrophys. J., Vol. 284, No. 1, p. 118 – 125 (1984).

Measurements of the 2.06 μm $2\,^1P - 2\,^1S$ line of He I and the 4.05 μm Brackett–α line of H I at IRS 16 in the galactic center show that high velocity emission extends to $\pm 700$ km $s^{-1}$ in both lines. A map of Brα spectra demonstrates that this high velocity line emission is concentrated within a few arc seconds of IRS 16. When interpreted as outflow from IRS 16, the broad line emission implies $1 \times 10^{-4} \lesssim \dot{M} \lesssim 3 \times 10^{-2} M_{\odot} yr^{-1}$, consistent with that needed to produce the shocked $H_2$ seen 2 pc distant from IRS 16. IRS 16 or the compact radio source appears to supply the ionizing radiation responsible for the line emission in its flow. The line intensities suggest that He/H in IRS 16 is near normal.

155.037 **Small-scale variations in the galactic magnetic field: the rotation measure structure function and birefringence in interstellar scintillations.**
J. H. Simonetti, J. M. Cordes, S. R. Spangler.
Astrophys. J., Vol. 284, No. 1, p. 126 – 134 (1984).
The structure function of rotation measures of extragalactic sources and birefringence in interstellar scintillations are used to investigate variations in the interstellar magnetic field on length scales of ~0.01 – 100 pc and $\sim 10^{11}$cm, respectively. Model structure functions are derived for the case of a power-law power spectrum of irregularities in the quantity ($n_e$**B**), and an estimate for the structure function is computed for several regions of the sky using data on extragalactic sources. The effect of a random, Faraday-active medium on the diffraction of radio waves is derived, and an upper limit to the variations in $n_e$**B** on a length scale of $\sim 10^{11}$cm is obtained from available observations.

155.038 **Galactic rotation, correction to the constant of precession and coordinates of the solar apex according to proper motions of bright stars.**
E. V. Khrutskaya.
Glav. Astron. Obs. Akad. Nauk SSSR. Leningrad, 26 pp. (1984). In Russian. Abstr. in Ref. Zh., 51. Astron., 10.51.150 (1984).

155.039 **Sagittarius A and its environment.**
R. L. Brown, H. S. Liszt.
Annu. Rev. Astron. Astrophys., Vol. 22, p. 223 – 265 (1984). – See Abstr. 003.024.
Contents: (1) Introduction. (2) The environment of Sgr A: Large-scale structure of the ionized gas. X-ray emission from the Sgr A complex. The relationship between neutral and ionized gas in the Sgr A complex. Associations between neutral and ionized gas near Sgr A. (3) An overview of neutral gas patterns observed in the inner Galaxy: The geometry and kinematics of the inner few kiloparsecs. A framework for the 50 km $s^{-1}$ cloud. (4) The central 2 parsecs of the Galaxy: The stellar distribution. The $R \sim 2$ pc ring of dust and gas. (5) The region interior to the 2-pc ring: Ionized gas associated with the ring. Ionized gas in the bar. (6) IRS 16 and Sgr A*. (7) Evidence for an energetic object. (8) Summary.

155.040 **The center of the Galaxy.**
N. S. Kardashev, I. D. Novikov, B. E. Shtern.
Zemlya Vselennaya, No. 5, p. 19 – 26 (1984). In Russian.

155.041 **The initial mass function for massive stars in the Galaxy and the Magellanic Clouds.**
R. M. Humphreys, D. B. McElroy.
Astrophys. J., Vol. 284, No. 2, p. 565 – 577 (1984).
The authors have determined the "present-day" initial mass function for stars greater than 15 $M_{\odot}$ in our Galaxy and the Large and Small Magellanic Clouds. The data set includes O type stars, supergiants of all types, and the less luminous B stars for a total of over 5000 stars in our Galaxy and nearly 1300 stars in the LMC and almost 500 stars in the SMC. It is found that incompleteness may be a serious problem for determining the IMF for massive stars. The IMFs for the outer region of our Galaxy, the LMC, and SMC are all consistent with the Miller-Scalo IMF for the solar neighborhood. The authors find no observational evidence that the slope of the massive star IMF differs from the normal IMF for the solar neighborhood or that it varies with galactocentric distance.

155.042 **Naar het hart van de Melkweg.**
J. H. Oort.
Zenit, 11. Jaarg., No. 12, p. 512 – 521 (1984).

155.043 **Dissecting the hub of our Galaxy.**
D. H. Smith, L. J. Robinson.
Sky Telesc., Vol. 68, No. 6, p. 494 – 497 (1984).

155.044 **The distribution of near-infrared obscuration in the innermost region of the Galaxy.**
N. Hiromoto, T. Maihara, K. Mizutani, H. Takami, H. Shibai, H. Okuda.
Astron. Astrophys., Vol. 139, No. 2, p. 309 – 312 (1984).
A 2.4 µm map of the central 10° of the Galaxy has been obtained through a balloon observation with a 0°.4 resolution in order to reveal the non-uniform distribution of obscuration in the innermost region of the Galaxy. The map exhibits a prominent peak at the galactic centre as well as many areas with reduced 2.4 µm flux near the galactic centre. After applying a new correction for local interstellar extinction in the galactic disk, the authors have found three noticeable absorption features in the 2.4 µm surface brightness map of the bulge, which may indicate the presence of dust clouds within 300 pc of the centre.

155.045 **Einige neue Trends in der Erforschung des Milchstraßensystems.**
B. J. Bok.
Sterne, 60. Band, Heft 6, p. 331 – 346 (1984).

155.046 **Submillimeter wavelength survey of the galactic plane from $l = -5°$ to $l = +62°$: structure and energetics of the inner disk.**
M. G. Hauser, R. F. Silverberg, M. T. Stier, T. Kelsall, D. Y. Gezari, E. Dwek, D. Walser, J. C. Mather, L. H. Cheung.
Astrophys. J., Vol. 285, No. 1, p. 74 – 88 (1984).
The authors present results from a new large-scale survey of the first quadrant of the galactic plane at wavelengths of 150, 250, and 300 µm, with a 10′ × 10′ beam. The emission detected in the survey arises from compact sources, most of which are identified with known peaks of 5 GHz or CO emission, or both, and from an underlying diffuse background with a typical angular width of ~0°.9 (FWHM) which accounts for most of the emission. A total of 80 prominent discrete sources are identified and characterized, of which about half have not previously been reported at far-infrared wavelengths.

155.047 **The location of infrared sources in the galactic center from a deep 1 micron CCD image.**
J. P. Henry, D. L. DePoy, E. E. Becklin.
Astrophys. J., Lett. Ed., Vol. 285, No. 1, p. L27 – L30 (1984).
The authors have obtained a CCD image of the galactic center region which, for the first time, is deep enough that the bright discrete sources seen at longer wavelengths are detectable. It is found that the source IRS 16 Center is within 0″.3 ± 0″.5 of the nonthermal radio source usually identified with the galactic center. The position, color, and absence of emission lines indicate that IRS 16 Center is unique in the galactic center. The extinction to IRS 12, a late-type star, is 14.5 mag at 0.98 µm. This value extrapolates to a visual extinction of $A_V = 38$ mag.

155.048 **Kinematics of field K giants in the outer galactic halo.**
K. U. Ratnatunga.
Publ. Astron. Soc. Pac., Vol. 96, No. 584, p. 793 – 794 (1984). Abstract. – See Abstr. 010.281.

155.049 **Velocity dispersion and metallicity distribution of K giants in the galactic nuclear bulge.**
R. M. Rich, A. E. Whitford.
Publ. Astron. Soc. Pac., Vol. 96, No. 584, p. 794 (1984). Abstract. – See Abstr. 010.281.

155.050 **New color-magnitude diagrams for two galactic bulge fields.**
D. M. Terndrup, R. M. Rich, A. E. Whitford.
Publ. Astron. Soc. Pac., Vol. 96, No. 584, p. 796 (1984). Abstract. – See Abstr. 010.281.

155.051 **Where massive stars form: a correlative study of molecular and ionized gas emission in the Milky Way.**
W. H. Waller, D. B. Sanders.
Publ. Astron. Soc. Pac., Vol. 96, No. 584, p. 797 – 798 (1984). Abstract. – See Abstr. 010.281.

155.052 **K giants and the total amount of matter near the sun.**
J. N. Bahcall.
Bull. Am. Astron. Soc., Vol. 16, No. 3, Part I, p. 733 (1984). Abstract. – See Abstr. 010.065.

155.053 **The stellar distribution in Puppis.**
B. C. Reed.
J. R. Astron. Soc. Can., Vol. 78, No. 5, p. 208 (1984). Abstract. – See Abstr. 010.401.

155.054 **Distribution of CO in the Galaxy, $l$ = 294° to $l$ = 86°.**
B. J. Robinson, R. N. Manchester, J. B. Whiteoak, D. B. Sanders, N. Z. Scoville, D. P. Clemens, W. H. McCutcheon, P. M. Solomon.
J. R. Astron. Soc. Can., Vol. 78, No. 5, p. 211 (1984). Abstract. – See Abstr. 010.401.

155.055 **$^{13}$CO in the inner galactic plane.**
H. S. Liszt, W. B. Burton, D.-L. Xiang.
Astron. Astrophys., Vol. 140, No. 2, p. 303 – 313 (1984).

The authors have mapped $^{13}$CO emission at λ2.7 mm along the galactic plane from $l$ = 20°.5 to 40°.0. These observations are used to discuss three topics: (1) the ratio of emissivities $A\ ^{12}CO/A\ ^{13}CO$ varies by more than a factor 2 across the galactic disk. The variations mimic those seen for extinction across local clouds, indicating a lower mean hydrogen column density for clouds observed at larger galactocentric radii. (2) Analysis of terminal velocities measured on $^{13}$CO and H I profiles is used to constrain the cloud mean free path and cloud–cloud random velocity dispersion. (3) The authors model the effect of cold H I in molecular clouds on low–latitude inner–galaxy 21–cm H I profiles.

155.056 **A comparison of our Galaxy to NGC 891 and NGC 4565 and the structure of their spheroids.**
P. C. van der Kruit.
Astron. Astrophys., Vol. 140, No. 2, p. 470 – 475 (1984).

A discussion is given of the large–scale properties of the disks and spheroids of our Galaxy, NGC 891 and NGC 4565. It is concluded that the disks of the Galaxy and NGC 4565 are very similar, while that of NGC 891 may be less luminous but otherwise similar. The spheroid of NGC 4565 is brighter and structurally different from those in NGC 891 and in our Galaxy, which are very closely comparable. Surface photometry of the spheroid of NGC 891 can be fitted equally satisfactorily by an "$R^{1/4}$–spheroid" with changing axis ratio of the isophotes and by a fainter $R^{1/4}$–spheroid with uniform flattening plus an intermediately flattened component with exponential drop–offs in $R$ and $z$.

155.057 **A radio continuum survey of the galactic plane at 11 cm wavelength. I. The area 357°.4 ⩽ $l$ ⩽ 76°, –1°.5 ⩽ $b$ ⩽ 1°.5.**
W. Reich, E. Fürst, P. Steffen, K. Reif, C. G. T. Haslam.
Astron. Astrophys., Suppl. Ser., Vol. 58, No. 2, p. 197 – 248 (1984).

The paper shows the 11 cm radio continuum radiation of the galactic plane in the form of an atlas of contour maps. The angular resolution is about 4'.3 and the sensitivity is 50 mK $T_B$ (or 20 mJy/beam area). Additionally, the authors have compiled a catalogue of 1212 small diameter radio sources.

155.058 **Comparison between *UBV*– and *RGU*–photometrically determined density–functions for the (photometric) disc– and halo–populations in SA 82.**
R. P. Fenkart, F. Esin–Yilmaz.
Astron. Astrophys., Suppl. Ser., Vol. 58, No. 2, p. 357 – 364 (1984).

A further field of the Basle Halo Program (BHP), SA 82, already treated in *RGU* by Becker and Steppe (1977), has been investigated applying analogously the BHP methods in *UBV* in order to compare the performance of both systems in this respect. The overall space–density–functions of both systems are compared. They agree almost completely for both (disc and halo) populations.

155.059 **A survey of the 21 cm line of neutral hydrogen in the Cas–Per region 120° ⩽ $l$ ⩽ 142° and –5° ⩽ $b$ ⩽ +5° with the 100 m telescope at Effelsberg.**
E. Braunsfurth, K. Reif.
Astron. Astrophys., Suppl. Ser., Vol. 58, No. 3, p. 503 – 505 (1984). With microfiches 1/2 and 2/2.

The 100 m telescope at Effelsberg has been used to survey the 21 cm line radiation of neutral hydrogen in the Cas–Per region in the velocity range $-180\ km\ s^{-1} \leqslant v_{LSR} \leqslant +80\ km\ s^{-1}$. The sensitivity of the survey is about 0.3K.

155.060 **The Milky Way in the ultraviolet spectral region.**
C. Winkler, J. Pfleiderer, T. Schmidt-Kaler.
Astron. Astrophys., Suppl. Ser., Vol. 58, No. 3, p. 705 – 728 (1984).

A complete isophote map of the Galaxy in the ultraviolet spectral region (λ = 356 nm; $\Delta\lambda_{1/2}$ = 60 nm) is presented in graphical and in tabular form: 0° ⩽ $l$ < 360°, –40° ⩽ $b$ ⩽ +40°, resolution 1°. The brightest regions ($I$ > 250 S10 $U$ = 250 stars of 10th magnitude per square degree in $U$) are located at $l$ ≅ 290°, $b$ = 0° (Carina), whereas the minimum of surface brightness along the galactic equator appears at $l$ ≅ 150° ($I$ < 40 S10 $U$). The overall picture of the Galaxy is the base for an extensive study of the galactic stellar and interstellar environment.

155.061 **Very deep cross–cut of the Galaxy at 7.6 cm.**
A. B. Berlin, E. V. Bulaenko, V. V. Vitkovskij, V. K. Kononov, D. V. Korol'kov, Yu. N. Parijskij, S. A. Trushkin.
Soobshch. Spets. Astrofiz. Obs., Vyp. 43, p. 43 – 55 (1984).

The authors present the results of the "deep cross–cut" of the Galaxy at 7.6 cm wavelength. For the first time the background radio emission from the Galaxy at high latitudes and the emission in the anticentre region at centimeter wavelength were recorded. It is shown that the spectral index of the background radio emission of the Galaxy is constant in a wide range of galactic latitudes. The spectrum of the brightness temperature in the bright part of the Milky Way as well as in the intermediate component and in almost all complexes is nonthermal.

155.062 **The outer Galaxy study at 7.6 cm wavelength.**
T. B. Pyatunina.
Soobshch. Spets. Astrofiz. Obs., Vyp. 43, p. 63 – 65 (1984).

155.063 **The local system of early type stars – spatial extent and kinematics.**
T. N. G. Westin.
Stockholms Obs., Rep., No. 25, 122 pp. (1984).

The present paper presents the results from an investigation of the distribution and kinematics of early type stars, both the bright stars mainly in the vicinity of the sun, and fainter stars at greater distances at four galactic longitudes. These directions are particularly well suited for the determination of galactic rotation parameters and the expansion age from radial velocities for a system which started to expand from a small volume. Individual ages for the stars have been determined, which makes it possible to study the behavior of stars in different age groups. The author deals with the study of the distribution of nearby early type stars as a function of age, and the orientation and extent of the Gould belt. The kinematic behavior of the stars is examined, starting with a description of a general velocity field. Also included is the determination of the solar motion and galactic rotation parameters.

155.064 **The Galaxy as a self-regulated star-forming system: the case of the OB associations.**
J. Franco, S. N. Shore.
Astrophys. J., Vol. 285, No. 2, p. 813 – 817 (1984).
It is proposed that the energy and momentum injected from OB associations can act as the regulators of high-mass star formation throughout the disk of a spiral galaxy. The mechanism, interacting blast waves driven by supernovae and stellar winds, leads to a generalized Schmidt law for the star formation rate and could explain in a natural way the galactic gradients of metals and star formation rate. The formation of massive stars is determined by the average ambient density in the disk while low-mass star formation is regulated by the density in molecular clouds.

155.065 **On the age of the galactic disk.**
V. A. Marsakov, A. A. Suchkov.
Astron. Tsirk., No. 1300, p. 2 – 5 (1983). In Russian.
The age of the galactic disk can be determined by comparing the positions of field giants with respect to the ascending giant branch of the oldest open clusters (e.g., NGC 188) in the HR diagram. However, the position of giants depends not only on age, but on abundance as well. The authors have shown that the spread of the heavy element abundance among field giants of the disk leaves the possibility to no more than a few per cent of these giants to be older than the open cluster NGC 188.

155.066 **Chemical evolution of the Galaxy: theory and facts.**
V. A. Marsakov, A. A. Suchkov.
Astron. Tsirk., No. 1303, p. 3 – 6 (1983). In Russian.

155.067 **The luminosity gradient of young objects across the spiral arms of the Galaxy.**
P. G. Kovachev, A. D. Stoev.
Astron. Tsirk., No. 1312, p. 5 – 8 (1984). In Russian.

155.068 **The surface brightness of the Milky Way in the V system.**
Yu. M. Zavarzin.
Astron. Tsirk., No. 1313, p. 4 – 6 (1984). In Russian.

155.069 **Die Milchstraße im ultravioletten Spektralbereich.**
C. Winkler, T. Schmidt-Kaler, J. Pfleiderer.
Mitt. Astron. Ges., Nr. 62, p. 305 – 306 (1984). Abstract. – See Abstr. 010.341.

155.070 **Vergleich zwischen neuen X-ray Messungen und radioastronomischen Beobachtungen an prominenten galaktischen Strukturen.**
W. Hirth, U. Mebold, P. Müller.
Mitt. Astron. Ges., Nr. 62, p. 315 – 317 (1984). – See Abstr. 010.341.

155.071 **Eine Durchmusterung der galaktischen Ebene bei 11 cm Wellenlänge: Das Gebiet 357°.4 ≤ *l* ≤ 76°, –1°.5 ≤ *b* ≤ 1°.5.**
W. Reich, E. Fürst, P. Steffen, K. Reif, C. G. T. Haslam.
Mitt. Astron. Ges., Nr. 62, p. 331 – 332 (1984). – See Abstr. 010.341.

155.072 **Kinematical evidence for the local galactic shingle.**
P. Brosche, H. Schwan.
Mitt. Astron. Ges., Nr. 62, p. 332 (1984). Abstract. – See Abstr. 010.341.

155.073 **The correlation between the stars and the continuous radio emission intensity distributions in the north polar spur region.**
G. B. Anisimova.
Astron. Zh., Tom 61, Vyp. 6, p. 1226 – 1228 (1984). In Russian. English translation in Sov. Astron., Vol. 28, No. 6.
The correlation between the stars and the continuous 240 MHz radio emission intensity distributions in the Loop I main ridge region is calculated. The correlation factors $r = 0.2 - 0.98$ with high confidence level are obtained for different stellar groups from the catalogue SAO and supplementary data.

155.074 **Extremely large Faraday rotation at the radio arc in the galactic center region.**
M. Inoue, T. Takahashi, H. Tabara, T. Kato, M. Tsuboi.
Publ. Astron. Soc. Jpn., Vol. 36, No. 3, p. 633 – 638 (1984).
A multifrequency measurement system of linear polarization has been developed to measure Faraday rotation without the $n\pi$ ambiguity of the position angle of the polarization at 3 cm. Both the polarization and Faraday rotation of the galactic center region were observed by the "Faraday polarimeter" attached to the Nobeayma 45 m telescope. The authors found an extremely large amount of Faraday rotation of 1660 rad $m^{-2}$ near the center. The depolarization properties of the source indicate that the Faraday rotation occurs within the polarized source itself.

155.075 **Determinations of the molecular mass of the Galaxy from observations of CO.**
H. S. Liszt.
Comments Astrophys., Vol. 10, No. 4, p. 137 – 147 (1984).

155.076 **Probing the outer galactic halo.**
B. W. Carney.
Publ. Astron. Soc. Pac., Vol. 96, No. 585, p. 841 – 855 (1984). Invited paper presented at the Symposium on Stellar Populations at the 95th Annual Scientific Meeting of the Astronomical Society of the Pacific, Santa Cruz, Calif., July 1984.
The paper discusses recent and continuing studies of the outer halo of our galaxy. A brief inventory is conducted of the stellar systems lying at distances exceeding 25 kpc from the galactic center. The spatial distributions of such systems and the field stars are reviewed, as well as the galactic mass estimates that follow from considerations of their kinematics. The question of a gradient in the halo's metallicity is addressed, plus the scant information available on the chemical abundance pattern of outer halo stars. Finally, the probability of differences in ages of outer halo systems is discussed.

155.077 **Are halo carbon stars associated with dwarf spheroidal galaxies?**
S. van den Bergh, A. Lafontaine.
Publ. Astron. Soc. Pac., Vol. 96, No. 585, p. 869 – 870 (1984).
Star counts have been made in the vicinity of a carbon star $l = 69°$, $b = +55°$ that is located at a distance of $\sim 60$ kpc. These counts show that this object is not situated in a dwarf spheroidal galaxy brighter than $M_V = -5.7$.

155.078 **Simulation of the apparent anomalies of the galactic structure.**
T. Jaakkola, N. Holsti, E. Laurikainen, P. Teerikorpi.
Astrophys. Space Sci., Vol. 107, No. 1, p. 85 – 107 (1984).
In the maps of the galactic structure based on the kinematical method, several systematic heliocentric anomalies are found. All of these anomalies can be explained with a simple model involving a non-velocity redshift field within the Galaxy, with an enhancement within the spiral arms. This is demonstrated by numerical simulations of the structural anomalies. Reducing the redshift effect from the kinematic data, the Galaxy's structure and kinematics appear symmetric. The significance of the result for the redshift problem is discussed.

155.079 **Kinematics of nearby gas and stars.**
T. Goulet, W. L. H. Shuter.
NASA Conf. Publ., NASA CP-2345, p. 319 – 325 (1984). – See Abstr. 012.089 (IAU Colloq. No. 81).
The kinematic properties of gas and stars in the solar neighborhood are described in terms of the line-of-sight component of a three-dimensional first order Taylor series expansion of the local velocity field. The types of object analysed are (1) 21 cm absorbing clouds, (2) intercloud medium, (3) main sequence B stars closer than 200 pc, (4) B stars of luminosity class ranging from I to IV, (5) main sequence A stars, (6) K giants. Marked departures from circular motion are found in most cases, but the only systematic trend is a correlation between $\partial u/\partial x$ (u being the velocity component along the x-axis directed towards the Galactic center) and stellar spectral type, where the gas behaves like a medium 'younger' than the early type stars. The standard plane-

parallel model provides a good description for the intercloud medium, but is inadequate for the absorbing clouds.

155.080 **Dark matter near the Sun: simulated star counts and the Oort limit.**
D. Gilden, J. N. Bahcall.
NASA Conf. Publ., NASA CP-2345, p. 326 – 329 (1984). – See Abstr. 012.089 (IAU Colloq. No. 81).

An ensemble of orbits passing through the solar position have been generated for a specific mass model of the galaxy. These orbits are randomly sampled to form simulated density distributions of tracer stars perpendicular to the galactic disk. The simulated distributions are analyzed in order to determine the sampling errors in a self-consistent derivation of the total amount of matter near the Sun (the Oort limit).

155.081 **Limits on soft X-ray flux from distant emission regions.**
D. N. Burrows, D. McCammon, W. T. Sanders, W. L. Kraushaar.
Astrophys. J., Vol. 287, No. 1, p. 208 – 218 (1984).

The authors use the all-sky soft X-ray data of McCammon *et al.* and the new $N_H$ survey (Stark *et al.*) to place limits on the amount of the soft X-ray diffuse background that can originate beyond the neutral gas of the galactic disk. The X-ray data for two regions of the sky near the galactic poles are shown to be uncorrelated with the 21 cm column densities. Almost all of the observed X-ray flux must therefore originate on the near side of the most distant neutral gas.

155.082 **Ultraviolet radiation from white dwarfs and the carbon ionization in the galactic halo.**
N. Panagia, Y. Terzian.
Astrophys. J., Vol. 287, No. 1, p. 315 – 319 (1984).

The radiation provided by the whole white dwarf population in our Galaxy has been estimated. At frequencies near the Lyman edge ($h\nu \approx 13.6$ eV), white dwarfs are found to contribute a minor fraction ($\sim 2\%$) to the total galactic radiation, whereas they dominate the emission at higher UV frequencies ($h\nu \gtrsim 40$ eV). Moreover, most of the radiation from white dwarfs may leak out of the galactic plane and ionize the halo gas. In particular, the ionization equilibrium of C III – C IV can entirely be determined by white dwarf radiation.

155.083 **The electron density in the galactic center as derived from the S III 18.71/33.47 micron line ratio.**
T. Herter, J. R. Houck, M. Shure, G. E. Gull, P. Graf.
Astrophys. J., Lett. Ed., Vol. 287, No. 1, p. L15 – L18 (1984).

The authors report high signal-to-noise ratio measurements of the S III lines 18.71 and 33.47 μm from the galactic center, Sgr A West. The average electron density is calculated for the S III-emitting region and is found to be $1800 \pm 500$ cm$^{-3}$. This value is much less than the rms electron density of the clumps as derived from radio observations. The implications for various models are discussed.

155.084 **K giants and the total amount of matter near the Sun.**
J. N. Bahcall.
Astrophys. J., Vol. 287, No. 2, p. 926 – 944 (1984).

The Poisson and Boltzmann equations are solved numerically for realistic Galaxy models which include multiple disk components (between 14 and 28), a Population II spheroid, and an unseen massive halo. As a guide in constructing the theoretical models, a population of normal-metallicity (disk) K giants is isolated using existing data and is shown to be well described by an isothermal distribution function with a velocity dispersion perpendicular to the plane of 20 km s$^{-1}$. The total amount of matter in the vicinity of the Sun is determined by comparing the observed distributions of K giants with the predictions of detailed Galaxy models. The volume density of *unobserved* material near the Sun is about 0.1 $M_\odot$ pc$^{-3}$; the corresponding column density is about 30 $M_\odot$ pc$^{-2}$.

**Radial velocities of 232 stars belonging to two fields of 4° × 4°. Third list.**
See Abstr. 002.035.

**Gamma ray astronomy.**
See Abstr. 003.026.

**Magnetic fields in astrophysics.**
See Abstr. 003.045.

**Astronomical scrapbook: the discovery of the Milky Way's spiral arms.**
See Abstr. 004.053.

**On the validity of the local approximation for the *s*-process in the Os region, and implications for the $^{187}$Re–$^{187}$Os cosmochronology.**
See Abstr. 061.008.

**On a lower limit to the rest mass of ions in the halo of our Galaxy.**
See Abstr. 061.040.

**The $^{187}$Re – $^{187}$Os chronology and chemical evolution of the Galaxy.**
See Abstr. 061.043.

**Actinide *r*-process chronometers and β-delayed fission.**
See Abstr. 061.045.

**Numerical simulation of the plasma and gravitational dynamics of a Galactic magnetic monopole halo.**
See Abstr. 061.082.

**Magnetic field generation in celestial bodies.**
See Abstr. 062.003.

**Carbon deflagrating supernovae and the chemical history of the solar neighbourhood.**
See Abstr. 065.138.

**Possible scenarios for the [O/Fe] ratio in metal-poor stars.**
See Abstr. 065.139.

**The models of annihilation γ-line and continuous gamma-spectrum radiation in the center of the Galaxy.**
See Abstr. 067.085.

**The boundary of the Solar System.**
See Abstr. 091.005.

**Comets and the Galaxy.**
See Abstr. 102.059.

**A wave-like effect in the motions of nearby young stars.**
See Abstr. 111.019.

**Sur l'efficacité des mesures systématiques de mouvements propres.**
See Abstr. 111.023.

**Reinvestigation of O, B star kinematics and determination of corrections to precession.**
See Abstr. 111.025.

**Spatial distribution and luminosity function of OH/IR maser sources.**
See Abstr. 112.078.

***RGU*-photometry of a starfield in the direction to the Praesepe cluster.**
See Abstr. 113.004.

***RGU* three-colour photometry of a starfield in anticentre direction (A5).**
See Abstr. 113.005.

***RGU*-photometry of a starfield close to M67 and space density behaviour in the anticentre-northern galactic meridian.**
See Abstr. 113.012.

***RGU* photometry of a field in the Carina region near the cluster IC 2581.**
See Abstr. 113.019.

**Photometry of K giants in the nuclear bulge of the Galaxy.**
See Abstr. 113.027.

**Light element abundances in F stars and the chemical evolution of the galactic disk.**
See Abstr. 114.086.

**Studies of late–type dwarfs. VI. Identification of Population II main–sequence stars at $M_v = +14$.**
See Abstr. 114.121.

**Abundances of distant classical Cepheid variables.**
See Abstr. 122.085.

**A search for distant halo RR Lyrae stars.**
See Abstr. 122.097.

**Novae and galactic chemical evolution.**
See Abstr. 124.007.

**G357.7 + 0.3 and G359.1–0.5 – two shell–type supernova remnants in the galactic centre region.**
See Abstr. 125.002.

**Supernova rate and the number of supernova remnants in M31, M33 and the Galaxy.**
See Abstr. 125.059.

**Stars too small to burn.**
See Abstr. 126.019.

**Kinematics of anti–center subdwarfs.**
See Abstr. 126.062.

**A survey of the 21 cm line of neutral hydrogen in the galactic region $-3° \leqslant l \leqslant 21°$ and $-4° \leqslant b \leqslant +3°$.**
See Abstr. 131.009.

**Inflow of halo gas from the direction of the galactic North Pole.**
See Abstr. 131.010.

**Star formation and the galactic magnetic field.**
See Abstr. 131.011.

**Kinematics of local interstellar clouds.**
See Abstr. 131.016.

**High–latitude molecular clouds.**
See Abstr. 131.021.

**The size and mass distribution of Galactic molecular clouds.**
See Abstr. 131.043.

**A large–scale dark cloud system in the northern sky.**
See Abstr. 131.067.

**On the statistics of galactic H I clouds.**
See Abstr. 131.072.

**UV studies of the interstellar medium.**
See Abstr. 131.075.

**On the distribution of interstellar matter around the sun.**
See Abstr. 131.076.

**The interstellar absorption in the direction of some nearby A stars.**
See Abstr. 131.077.

**Use of high resolution IUE spectra of Mg II in cool stars to probe the local interstellar medium.**
See Abstr. 131.078.

**Interstellar lithium and the $^7Li/^6Li$ ratio in diffuse clouds.**
See Abstr. 131.086.

**Interstellar scattering in the inner parts of the Galaxy.**
See Abstr. 131.162.

**The Cygnus X region. XV. A 4.8 GHz continuum survey with the 100 m–telescope.**
See Abstr. 131.178.

**UV studies of the interstellar medium.**
See Abstr. 131.197.

**Parameter für Hochgeschwindigkeitswolken.**
See Abstr. 131.225.

**Radial distribution of giant molecular clouds. II. Comparison of the viscous cloud system with observations.**
See Abstr. 131.228.

**On the random motion of giant molecular clouds.**
See Abstr. 131.229.

**Coagulation of interstellar clouds in spiral gravitational potential and formation of giant molecular clouds.**
See Abstr. 131.230.

**Dust far infrared emission at high galactic latitudes.**
See Abstr. 131.236.

**Dust distribution nearby the solar system and its possible implications on cosmic radiation background anisotropy.**
See Abstr. 131.237.

**Optical emission line studies and the warm, ionized component of the local interstellar medium.**
See Abstr. 131.254.

**A wide latitude CO survey of molecular clouds in the northern Milky Way.**
See Abstr. 131.268.

**Large–scale bubble structure of the interstellar medium (ISM) and properties of the local spiral arm (LSA).**
See Abstr. 131.281.

**H II–Regionen, Radiosterne und galaktisches Radiokontinuum. Beobachtungen am Effelsberger 100–m–Teleskop.**
See Abstr. 132.005.

**Filling in the IRAS blanks.**
See Abstr. 133.003.

**A survey of faint, near–infrared sources towards the centre of the Galaxy.**
See Abstr. 133.010.

**Survey à 2.2 µm du plan galactique austral.**
See Abstr. 133.019.

**Galactic far infrared astronomy.**
See Abstr. 133.021.

**A 10–GHz radio continuum survey of the galactic plane region. I. A complex region at $l = 21° - 26°$.**
See Abstr. 141.008.

**Infrared observations of galactic bulge X–ray sources.**
See Abstr. 142.008.

**HEAO–1 diffuse soft X–ray sky maps.**
See Abstr. 142.087.

**SAS 3 survey of the soft X–ray background.**
See Abstr. 142.091.

**De gammahemel geeft enkele geheimen prijs (2).**
See Abstr. 143.008.

**Implications of the observation of ultra–high energy γ–rays. I. The production of diffuse X–rays.**
See Abstr. 143.032.

**Origin and propagation of cosmic rays.**
See Abstr. 144.013.

**Cosmic rays in the Galaxy.**
See Abstr. 144.016.

**Local conditions of cosmic ray propagation in the Galaxy.**
See Abstr. 144.029.

**Importance of transverse diffusion for cosmic–ray propagation.**
See Abstr. 144.032.

**Cosmic ray gradients in the Galaxy.**
See Abstr. 144.045.

**The galactic distribution of cosmic–ray particles.**
See Abstr. 144.072.

**Galactic cosmic rays.**
See Abstr. 144.074.

**Stability of star clusters as galactic satellites. III. Preliminary applications to the system of the globular clusters.**
See Abstr. 151.025.

**The distribution model of the velocity dispersion of stars within the Galaxy.**
See Abstr. 151.096.

**Phase–mixing of newly born stars in the galactic disk.**
See Abstr. 151.110.

***VBLUW* photometry of the association Sco OB1 (containing the open cluster NGC 6231). A discussion on the evolutionary status of the hypergiant $\zeta^1$ Sco (B1Ia$^+$).**
See Abstr. 152.007.

**Wolf–Rayet stars in open clusters and associations.**
See Abstr. 153.012.

**The rotation of the globular cluster system.**
See Abstr. 154.013.

**Kinematics of galactic globular clusters.**
See Abstr. 154.029.

**Interaction of the Large Magellanic Cloud with the halo population and the companions of the Galaxy.**
See Abstr. 156.018.

**Flat rotation curves and mass distribution for spiral galaxies with a disk model.**
See Abstr. 157.042.

**Velocity dispersions of disk stars in spiral galaxies.**
See Abstr. 157.120.

**Galactic chemical evolution and nucleocosmochronology: standard model with terminated infall.**
See Abstr. 157.177.

**Structural properties of giant H II regions in nearby galaxies.**
See Abstr. 157.234.

# 156 Magellanic Clouds

156.001 **Infrared photometry of normal and peculiar A–type supergiants, and the law of reddening in the Small Magellanic Cloud.**
M. W. Feast, P. A. Whitelock.
Observatory, Vol. 104, No. 1061, p. 193 – 196 (1984).
The A–type supergiants in the SMC with abnormally strong Balmer lines do not have markedly different infrared fluxes from normal SMC A–type supergiants. Infrared observations of some moderately reddened A supergiants show that the ratio of total to selective absorption in the SMC is 3.1, close to the galactic value.

156.002 **Compact emission–line nebulae in the Small Magellanic Cloud.**
D. H. Morgan.
Mon. Not. R. Astron. Soc., Vol. 209, No. 2, p. 241 – 244 (1984).
A systematic search of an objective prism plate of the SMC taken with the UK 1.2–m Schmidt telescope using the 2°11′ objective prism has revealed six more very–low–excitation compact nebulae bringing the total known for the SMC to eight. A new planetary nebula has also been found.

156.003 **A curious, high velocity, component within a small shell in the Large Magellanic Cloud.**
C. Goudis, J. Meaburn.
Astron. Astrophys., Vol. 137, No. 1, p. 152 – 153 (1984).
A faint velocity component at –120 km s$^{-1}$ with respect to the dominant one, in profiles of the Hα line, has been found within a small shell in the Large Magellanic Cloud. The close association of this high velocity feature with the central star implies that it is being blown off in some way from its vicinity.

156.004 **A neutral hydrogen line survey of the Large Magellanic Cloud.**
K. Rohlfs, J. Kreitschmann, B. C. Siegman, J. V. Feitzinger.
Astron. Astrophys., Vol. 137, No. 2, p. 343 – 357 (1984).
A new 21 cm line survey of the LMC with low noise ($\sigma_{TA} \cong 0.08$K) and high frequency resolution (channel width 0.8 km s$^{-1}$) has been made with the Parkes 64 m telescope. About 1/2 of the profiles show two clearly separated peaks, and often the shape is clearly non–gaussian. Both, the channel maps and the large–scale velocity field of the weighted mean radial velocity of the profiles are characteristic of that of a flat differentially rotating disk. Neutral hydrogen shows several large scale structures that are correlated in one or the other way with those of other cloud members. Finally, the rotation of the LMC is considered.

156.005 **Complex motions within the giant LMC shell N59A (DEM 241).**
C. Goudis, J. Meaburn.
Astron. Astrophys., Vol. 138, No. 1, p. 57 – 61 (1984).
New observations of the profiles of the Hα line over the face of N59A (DEM 241) have been combined with previous profiles, at lower angular resolution, of the [O II] lines. A "bubble" of 16 pc in diameter, which is expanding at ≅ 50 km s$^{-1}$, is apparently present within the 40 pc diameter of this LMC giant shell. Several theoretical models are considered for the formation of this, and other, LMC giant shells.

156.006 **The diffuse interstellar medium in the Magellanic Clouds.**
K. S. de Boer.
ESA Spec. Publ., ESA SP–218, p. 179 – 182 (1984). – See Abstr. 012.024.

156.007 **The spatial distribution of young objects in the Large Magellanic Cloud: a problem of pattern recognition.**
J. V. Feitzinger, E. Braunsfurth.
Astron. Astrophys., Vol. 139, No. 1, p. 104 – 114 (1984).

Methods used in pattern recognition and cluster analysis are applied to investigate the spiral distribution of OB associations and emission regions in the LMC. The LMC is an irregular system with some suspected spiral structure. Until now this classification has been done by visual inspection. The authors try to quantify this and to give objective criteria for orderness and clustering. At a high confidence level (>95%) associations and emission regions are not distributed at random. Several methods of cluster analytic partitions of the system reveal intermediate scale structures of 700 – 1500 pc being identical with the spiral arm filaments responsible for the flocculent appearance of spiral galaxies.

156.008 **Interstellar matter and star formation in the Large Magellanic Cloud.**
J. Isserstedt, W. Kohl.
Astron. Astrophys., Vol. 139, No. 1, p. 115 – 122 (1984). In German.

The distributions of luminous stars, atomic hydrogen and interstellar dust in the Large Magellanic Cloud (LMC) are compared. For this purpose a quantitative map of the distribution of interstellar reddenings in the LMC is constructed for the first time. The authors find a close correlation between the average areal densities of supergiants and H I only for the small number of stars in dense H I-regions. In general the distributions of young stars and neutral hydrogen are rather dissimilar. No correlation was found between the distributions of supergiants and interstellar dust. This is not compatible with the theory of stochastic self propagating star formation.

156.009 **Ultraviolet interstellar absorption toward stars in the Small Magellanic Cloud. II. Sk 159.**
E. L. Fitzpatrick.
Astrophys. J., Vol. 282, No. 2, p. 436 – 444 (1984).

Multiple high–dispersion IUE spectra of Sk 159 in the Small Magellanic Cloud (SMC) have been obtained and averaged in order to study interstellar gas associated with the SMC. Sk 159, spectral type B0.5 Iaw, is relatively cool and is located in a region of the SMC free of strong nebulosity. The weak low–ionization stage lines show SMC absorption at 150 km $s^{-1}$ (heliocentric) corresponding to SMC interstellar gas. The strong low–ionization stage lines show additional absorption near 100 km $s^{-1}$ and 215 km $s^{-1}$. The gas producing these features may be associated with recently detected large SMC neutral hydrogen masses. The origin of SMC absorption systems of C IV, Si IV and Al III is discussed in detail.

156.010 **The period–luminosity relation in the Magellanic Clouds.**
P. A. Wayman.
Ir. Astron. J., Vol. 16, No. 3, p. 188 – 194 (1984). – See Abstr. 012.036.

The Dunsink photometric data on Cepheid variable stars in the Magellanic Clouds are discussed in terms of period, luminosity and colour. The average differences between stars in three regions, one in the Small Cloud and two in the Large Cloud, are tentatively interpreted in terms of abundance and reddening differences. The difference of true distance modulus between the two Clouds is evaluated as $0\overset{m}{.}25 \pm 0\overset{m}{.}06$.

156.011 **The evolution of asymptotic giant branch stars in the Large Magellanic Cloud.**
N. Reid, J. Mould.
Astrophys. J., Vol. 284, No. 1, p. 98 – 107 (1984).

The authors present a bolometric luminosity function for stars on the asymptotic giant branch (AGB) in a 15 $deg^2$ area of the Large Magellanic Cloud. The luminosity function was derived by COSMOS scanning of multiple visual and infrared UK Schmidt plates and is photometrically complete for $V–I > 1.6$. The deficiency of luminous AGB stars which was apparent in earlier spectroscopic studies is seen in this study also. A likely explanation is that mass loss more severe than is predicted by standard scaling relations sets in on the upper AGB, terminating the evolution of these stars at quite modest luminosities.

156.012 **Comparative stellar field studies in the Magellanic Clouds.**
L. L. Stryker.
Publ. Astron. Soc. Pac., Vol. 96, No. 584, p. 795 (1984). Abstract. – See Abstr. 010.281.

156.013 **Current research in the Magellanic Clouds.**
J. A. Graham.
Ir. Astron. J., Vol. 16, No. 4, p. 243 – 250 (1984).

156.014 **The identification of HDE objects with Large Magellanic Cloud clusters and nebulae.**
M. Morel.
Bull. Inf. Cent. Données Stellaires, No. 27, p. 57 – 61 (1984).

A complete listing is presented here of all entries in the Henry Draper Extension (HDE) which are clearly identifiable with star clusters and/or emission nebulae in the Large Magellanic Cloud. The list also includes many individual HDE stars whose position near or within clusters requires clarification.

156.015 **A distance to the Large Magellanic Cloud by main–sequence fitting.**
R. A. Schommer, E. W. Olszewski, M. Aaronson.
Astrophys. J., Lett. Ed., Vol. 285, No. 2, p. L53 – L57 (1984).

Deep photometry of unevolved main sequences for two clusters in the Large Magellanic Cloud permits their distances to be directly measured by fits to (1) the semiempirical ZAMS of VandenBerg and Bridges, (2) Yale isochrone models, and (3) galactic open clusters (and specifically the Hyades). A conservative choice of the relevant reddening, helium abundance, and metallicity parameters leads to a LMC distance modulus of $(m–M)_0 = 18.2 \pm 0.2$. This result suggests the necessity of a zero–point revision in the galactic Cepheid scale. Implications for the value of the Hubble constant are briefly indicated.

156.016 **A distance to the Large Magellanic Cloud by main sequence fitting.**
R. A. Schommer, E. W. Olszewski, M. Aaronson.
Prepr. Steward Obs., No. 532, 19 pp. (1984). To appear in Astrophys. J., Lett. Ed.

156.017 **The stellar population of the Magellanic Clouds: a selective review.**
J. A. Frogel.
Publ. Astron. Soc. Pac., Vol. 96, No. 585, p. 856 – 868 (1984). Invited paper presented at the Symposium on Stellar Populations at the 95th Annual Scientific Meeting of the Astronomical Society of the Pacific, Santa Cruz, Calif., July 1984.

The article considers evidence for an old metal–poor population in the Magellanic Clouds corresponding to that of the Galactic halo, and evidence for subsequent star–forming episodes. Particular emphasis is placed on cluster research and on observations of bright giants in clusters and the field as a means to understand the past and present stellar content of the Clouds. The color–magnitude diagrams and luminosity functions for bright giants in clusters and in a selected field of the LMC are compared to see if the stellar content of the field can be reproduced by a superposition of clusters. The small, but growing body of data on the chemical enrichment history of the Clouds is considered. Finally, some remarks on the gaseous content of the Clouds, and the implication for present–day star formation, are made.

156.018 **Interaction of the Large Magellanic Cloud with the halo population and the companions of the Galaxy.**
M. J. Valtonen, K. A. Innanen, L. Tähtinen.
Astrophys. Space Sci., Vol. 107, No. 2, p. 209 – 222 (1984).

Numerical simulations have been performed in order to study the gravitational interaction between the outlying population of the Galaxy and the Large Magellanic Cloud (LMC). The following effects have been noted: (1) halo objects are partially removed from the orbital distance of the LMC such that a local minimum is created in the radial density profile; (2) a large number of halo objects go into escape orbits. This possibility makes the determination of the mass of the Galaxy by virial theorem arguments unreliable; (3) the orbit of the LMC decays faster than what one would expect on the basis of the classical dynamical friction calculations. The excess friction derives from three–body effects, which are not accounted for in the classical dynamical friction. Finally, the fates of globular clusters and dwarf galaxies near the Galaxy and the LMC are discussed.

**Synthetic H–R diagrams as an observational test of stellar evolution theory.**
See Abstr. 065.022.

**On two WR stars in the Large Magellanic Cloud: the high velocity of BR 52 and a newly discovered WN star.**
See Abstr. 111.026.

**The hybrid spectrum of the LMC hypergiant R 126.**
See Abstr. 112.087.

**10–micron observations of Magellanic Cloud supergiants.**
See Abstr. 113.015.

**Spectrophotometry of early type supergiants in the Large Magellanic Cloud.**
See Abstr. 114.014.

**CASPEC (*Cassegrain Echelle Spectrograph*) and IUE: a perfect match.**
See Abstr. 114.084.

**Variable blue supergiants in the Large Magellanic Cloud: R 84, R 85, and R 99.**
See Abstr. 114.090.

**Optical spectra of R136.**
See Abstr. 114.137.

**A superluminous object in the Large Cloud of Magellan.**
See Abstr. 115.002.

**Massive stars in the Galaxy and the Magellanic Clouds: clues to stellar evolution.**
See Abstr. 115.014.

**R136: multiple or monster?**
See Abstr. 115.020.

**Optical photometry of massive X–ray binaries: SMC X–1/Sk 160.**
See Abstr. 117.051.

**LMC X–3: a black hole in a neighbor galaxy.**
See Abstr. 117.053.

**Ultraviolet high–resolution spectroscopy of the X–ray binary Sk 160/SMC X–1.**
See Abstr. 117.096.

**SMC X–1 variability observed from *HEAO 1*.**
See Abstr. 117.110.

**R31: a massive O + W–R star binary in the Small Magellanic Cloud.**
See Abstr. 120.016.

**Spectra of seven long–period Cepheids in the Small Magellanic Cloud.**
See Abstr. 122.121.

**The period–luminosity relation for Mira variables and the distance of the Large Magellanic Cloud.**
See Abstr. 122.137.

**BVRI photoelectric photometry of Cepheids in the Small Magellanic Cloud.**
See Abstr. 122.196.

**The luminosity scale of Cepheid variable stars: a revision.**
See Abstr. 122.199.

**Die hellsten blauen Veränderlichen in der Großen Magellanschen Wolke.**
See Abstr. 122.213.

**Spectrophotometry of young supernova remnants.**
See Abstr. 125.005.

**A new look at radio supernova remnants.**
See Abstr. 125.017.

**An optical synchrotron nebula around the X–ray pulsar 0540–693 in the Large Magellanic Cloud.**
See Abstr. 125.082.

**Pulsar in Large Magellanic Cloud.**
See Abstr. 126.067.

**A new Crab–like pulsar in the Large Magellanic Cloud.**
See Abstr. 126.088.

**Infrared observations of recent star formation regions in the Large Magellanic Cloud: N 160 A and N 105 A.**
See Abstr. 131.157.

**Interstellar dust.**
See Abstr. 131.187.

**Interstellar Ca II lines in SMC stars.**
See Abstr. 131.190.

**Infrared extinction in the Small Magellanic Cloud.**
See Abstr. 131.193.

**UV studies of the interstellar medium.**
See Abstr. 131.197.

**Properties of R136a as derived from its optical light distribution.**
See Abstr. 132.022.

**High velocities within extensive regions of the 30 Doradus nebula.**
See Abstr. 132.035.

**The interstellar lines and the energetics of the inner 30 Doradus nebula from the ultraviolet spectrum of R 136a.**
See Abstr. 132.046.

**Detection of molecular hydrogen in the SMC H II region N 81.**
See Abstr. 132.048.

**Very–wide–field ultraviolet sky survey.**
See Abstr. 142.006.

**LMC X–4: the optical 30–day cycle and its implications.**
See Abstr. 142.049.

**Optical counterparts of the Large Magellanic Cloud X–ray point sources.**
See Abstr. 142.064.

**Discovery of a 50 millisecond pulsar in the Large Magellanic Cloud.**
See Abstr. 142.090.

**Optical observations of the bubble nebula N 157C (= 0536.6–6914) and the tight clusters in the association LH 90 in LMC.**
See Abstr. 152.006.

**Wolf–Rayet stars in open clusters and associations.**
See Abstr. 153.012.

**The clusters NGC 419 and NGC 416 in the Wing of the Small Magellanic Cloud.**
See Abstr. 153.021.

**A photographic color–magnitude diagram for the remote LMC cluster NGC 1777.**
See Abstr. 153.031.

**Does the ellipticity of clusters in the LMC correlate with age or luminosity?**
See Abstr. 154.033.

**Color–magnitude diagram photometry of the LMC red cluster NGC 1978.**
See Abstr. 154.035.

**Colour–magnitude diagrams of star clusters in the Magellanic Clouds from wide–field electronography – I. Hodge 11.**
See Abstr. 154.055.

***IUE* observations of the clusters of the Magellanic Clouds.**
See Abstr. 154.068.

**The extended giant branches of intermediate age globular clusters in the Magellanic Clouds. IV.**
See Abstr. 154.071.

**Main–sequence photometry of the SMC globular cluster Kron 3.**
See Abstr. 154.076.

**The Large Magellanic Cloud intermediate–age globular cluster NGC 1831.**
See Abstr. 154.079.

**The initial mass function for massive stars in the Galaxy and the Magellanic Clouds.**
See Abstr. 155.041.

**The underlying exponential disc in the Im galaxy NGC 4214.**
See Abstr. 157.153.

**Endpoints of stellar evolution: X–ray surveys of the Local Group.**
See Abstr. 160.146.

# 157 Normal Galaxies (Structure, Evolution, Pairs, etc.)

157.001 **Cigar–shaped ring galaxies.**
B. A. Vorontsov–Vel'yaminov, V. A. Dostal', V. G. Metlov.
Pis'ma Astron. Zh., Tom 10, No. 7, p. 494 – 500 (1984). In Russian. English translation in Sov. Astron. Lett., Vol. 10.

To solve the existing discussion on whether the central bodies of cigar–shaped ring galaxies, discovered by Vorontsov–Vel'yaminov in 1960, are oblate or prolate, photographic and spectroscopic observational data are collected and extended. Their analysis leads to the conclusion that there exist central bodies of both types.

157.002 **The nature of blue compact galaxies.**
P. M. Gondhalekar, D. H. Morgan, M. Dopita, A. P. Phillips.
Mon. Not. R. Astron. Soc., Vol. 209, No. 1, p. 59 – 67 (1984).

The near infrared (*J*, *H*, *K* and where possible *L*) photometry of two H II regions in external galaxies and of eight blue galaxies is presented. The ultraviolet, visible and near–infrared colours are used to demonstrate that some of the blue compact galaxies may be genuinely young objects.

157.003 **Recent star formation in interacting galaxies – I. Evidence from *JHKL* photometry.**
R. D. Joseph, W. P. S. Meikle, N. A. Robertson, G. S. Wright.
Mon. Not. R. Astron. Soc., Vol. 209, No. 1, p. 111 – 122 (1984).

The authors have carried out a survey using *JHKL* photometry to investigate recent star formation in interacting galaxies. The objective was to look for a *K–L* excess produced by "warm" dust heated by a putative burst of star formation. The authors find *K–L* excesses suggesting that interactions induce starbursts with an efficiency approaching 100 per cent. The appearance of these inferred starbursts in interacting systems of different morphological types is qualitatively consistent with dynamical studies of galaxy interactions.

157.004 **H I observations in the Virgo Cluster area. II. A complete, magnitude–limited sample of spiral galaxies.**
G. Helou, G. L. Hoffman, E. E. Salpeter.
Astrophys. J., Suppl. Ser., Vol. 55, No. 3, p. 433 – 453 (1984).

The authors report Arecibo H I observations of 74 spiral galaxies; they compute H I diameters and spin orientations, as well as velocity widths, systemic redshifts, and H I masses. Uniform H I data are now available for all spirals with $B_T^0 \leqslant 12.8$ in an area $\sim 250 \deg^2$ centered on the Virgo Cluster. Approximate redshifts are also given for five objects detected by chance.

157.005 **Galaxy evolution at late epochs.**
M. S. Longair.
Nature, Vol. 310, No. 5972, p. 13 – 14 (1984).

157.006 **Masers in the nuclei of galaxies.**
J. Moran.
Nature, Vol. 310, No. 5975, p. 270 – 271 (1984).

157.007 **Water–vapour maser emission from galactic nuclei.**
M. J. Claussen, G. M. Heiligman, K. Y. Lo.
Nature, Vol. 310, No. 5975, p. 298 – 300 (1984).

The authors report the progress of a survey of galactic nuclei for water–vapour maser emission at 22.235 GHz. They observed 29 late–type galaxies and detected maser emission from four nuclei: NGC 4258 (M106), NGC 1068 (M77), NGC 3034 (M82), and NGC 6946. The masers in NGC 4258 and NGC 1068 are extremely luminous. It is suggested that the extremely bright nuclear masers indicate an ongoing burst of star formation.

157.008 **M31: an extragalactic source of 1000 GeV gamma rays.**
J. C. Dowthwaite, A. B. Harrison, I. W. Kirkman, H. J. Macrae, K. J. Orford, K. E. Turver, M. Walmsley.
Astron. Astrophys., Vol. 136, No. 1, p. L14 – L15 (1984).

The Dugway array of γ–ray telescopes has been used to observe M31 at energies above 1000 GeV. An excess of γ–ray counts was seen from the direction of M31 with a conservative maximum likelihood significance of 1% to be due to a random fluctuation. The estimated flux of $(2.2 \pm 0.7) \times 10^{-10}$ photon $cm^{-2}s^{-1}$ corresponds to a luminosity of $4 \times 10^{40}$ erg $s^{-1}$ for energies above 1000 GeV. The measured spatial distribution is consistent with the angular dimension of the region of emission being about 1 degree.

157.009 **UBV photometry of some double galaxies.**
A. N. Tomov.
Astron. Tsirk., No. 1296, p. 1 – 2 (1983). In Russian.

157.010 **UBV photometry of triple galaxies.**
A. N. Tomov, N. A. Tomov.
Astron. Tsirk., No. 1296, p. 2 – 4 (1983). In Russian.

157.011 **On new data on the distribution of galaxies according to the main morphological types.**
V. A. Dostal'.
Astron. Tsirk., No. 1297, p. 5 – 6 (1983). In Russian.

157.012 **Optical properties of spiral galaxies with strong nuclear radio sources: active nuclei, starbursts, and composite objects.**
W. C. Keel.
Astrophys. J., Vol. 282, No. 1, p. 75 – 84 (1984). With plate 3.

New optical spectrophotometry, accurate positions, and narrow–band emission–line images have been obtained for the nuclei of spiral and irregular galaxies containing strong radio sources. The optical data support a general division of these objects into those powered by star formation and those dominated by nonthermal processes. In several cases, however, nuclei are classified differently from radio and optical properties, suggesting a composite nature for some nuclei. Correlations between optical emission–line and 6 cm luminosities are explored for each class of nucleus.

157.013 **The families of elliptical–like galaxies.**
A. Wirth, J. S. Gallagher III.
Astrophys. J., Vol. 282, No. 1, p. 85 – 94 (1984). With plate 4.

The authors hypothesize that M32–like classical dwarf ellipticals, and not the more diffuse NGC 205–like systems, represent the true low–luminosity extension of the classical giant elliptical galaxy family. A review of fundamental properties illustrates significant structural and stellar population characteristics which differentiate these two types of low–luminosity elliptical galaxies. While the well–known classical dwarf ellipticals are all companions to giant galaxies, a survey of nearby galaxy groups has produced a few relatively isolated classical elliptical dwarfs, indicating that their structure properties are not due solely to the tidal effects of near neighbors. A morphological selection technique is used to produce a magnitude–limited sample of compact ellipticals in the Fornax Cluster.

157.014 **A galactic disk is not a true exponential.**
P. E. Seiden, L. S. Schulman, B. G. Elmegreen.
Astrophys. J., Vol. 282, No. 1, p. 95 – 100 (1984).

Stochastic self–propagating star formation predicts that the distribution of atomic hydrogen in a galactic disk should be relatively independent of radius. The molecular hydrogen density should therefore vary as the total gas density minus a constant. The most likely distribution for the total gas density is approximately $1/r$, the same as the distribution of column density in a halo that produces a flat rotation curve. The authors propose that the approximately exponential surface brightness distribution observed in disk galaxies results from the intrinsic $r^{-1}$–constant distribution produced by propagating star formation. The exponential fit has no intrinsic physical meaning.

157.015 **The distance to NGC 2403 based on near–infrared observations of Cepheids.**
C. W. McAlary, B. F. Madore.
Astrophys. J., Vol. 282, No. 1, p. 101 – 105 (1984).

The authors have obtained near–infrared measurements of five Cepheids in the galaxy NGC 2403, which is thought to be a member of the M81 Group. Application of the $J$ band period–luminosity relation yields a distance modulus of $(m–M)_0 = 28.15 \pm 0.20$ (internal error) assuming a foreground reddening of $E(B–V) = 0.06$ mag. This value is approximately 0.5 mag larger than that derived by Tammann and Sandage from blue observations.

157.016 **Erratum: "X–ray observations of M33 with the high resolution imager on the Einstein Observatory" [Astrophys. J., Vol. 275, No. 2, p. 571 – 577 (1983)].**
T. H. Markert, A. D. Rallis.
Astrophys. J., Vol. 282, No. 1, p. 359 (1984). See Abstr. 34.157.190.

157.017 **A kinematic method for measuring the pattern speed of barred galaxies.**
S. Tremaine, M. D. Weinberg.
Astrophys. J., Lett. Ed., Vol. 282, No. 1, p. L5 – L7 (1984).

The purpose of this letter is to describe a novel procedure for measuring the pattern speed of a barred disk galaxy, using kinematical arguments. The main assumption required is that the luminosity density of the tracer (stars, neutral hydrogen, etc.) obey a continuity equation, i.e., that the total mass of tracer remain fixed and the apparent luminosity per unit mass be constant. In its simplest form, the method requires measuring the surface brightness and radial velocity along a strip parallel to the line of nodes. The method is applied to H I observations of the barred spiral galaxy NGC 5383.

157.018 **The relation between the colour of galaxies in pairs.**
V. V. Demin, A. V. Zasov, Eh. A. Dibaj, A. N. Tomov.
Astron. Zh., Tom 61, Vyp. 4, p. 625 – 633 (1984). In Russian. English translation in Sov. Astron., Vol. 28, No. 4.

Proceeding from the data of photoelectric photometry by Tomov the colours of galaxies in double systems are studied.

157.019 **Star formation bursts in the central regions of Sc–galaxies.**
O. K. Sil'chenko.
Astron. Zh., Tom 61, Vyp. 4, p. 634 – 638 (1984). In Russian. English translation in Sov. Astron., Vol. 28, No. 4.

Integral spectra of galaxies with star formation bursts of various duration and intensity are calculated by using the evolutionary synthesis. Two Sc–galaxies (NGC 1084 and 2903) are found to have star formation bursts in their central regions. Characteristics of these bursts are determined.

157.020 **Observations of the Andromeda nebula at a frequency of 103 MHz.**
V. S. Artyukh, M. A. Ogannisyan.
Astron. Zh., Tom 61, Vyp. 4, p. 639 – 644 (1984). In Russian. English translation in Sov. Astron., Vol. 28, No. 4.

Interplanetary scintillation observations of the region of M31 have been made at 103 MHz. 21 scintillating sources have been observed in the region $6° \times 4°$ around the nebula. One of them, placed 13′ to the north of the centre of M31, is a Cas A–type supernova remnant.

157.021 **Dwarf satellite galaxies and radiative decay of elementary particles composing massive galactic halos.**
A. L. Melott.
Astron. Zh., Tom 61, Vyp. 4, p. 817 – 819 (1984). In Russian. English translation in Sov. Astron., Vol. 28, No. 4.

It is shown that radiative decay of finite–rest–mass particles possibly composing galactic halos to ultraviolet photons may explain a correlation in the morphology of dwarf galaxies with their distance from large galaxies.

157.022 **Evolutionary modelling of the $M/L$ ratio for different types of galaxies.**
O. K. Sil'chenko, M. V. Tarasova.
Sov. Astron., Vol. 28, No. 1, p. 27 – 30 (1984). English translation of 37.157.001.

157.023 **M 83.**
F. Bertola.
Sterne Weltraum, 23. Jahrg., Nr. 8–9, p. 433 (1984).

157.024 **Photoelectric observations of interacting and compact galaxies.**
V. P. Arkhipova, M. V. Savel'eva.
Tr. Gos. Astron. Inst. Shternberg, Tom 54, p. 33 – 45 (1984). In Russian.
Results of photoelectric UBV observations of 70 interacting and 49 compact galaxies are presented.

157.025 **The luminosities of the M supergiants in M33.**
R. M. Humphreys, T. J. Jones, M. L. Sitko.
Astron. J., Vol. 89, No. 8, p. 1155 – 1159 (1984).
*JHK* photometry of ten spectroscopically confirmed M supergiants in M33 is used to derive their intrinsic colors, the interstellar extinction, and their resulting infrared, visual, and bolometric luminosities. The visual extinction affecting the M supergiants is highly variable ranging from 0.3 to 1.4 mag.The visual luminosity of the three brightest red supergiants is –8.1 mag, in agreement with their luminosities in other nearby spiral and irregular galaxies. The most luminous M supergiant in M33 has a bolometric luminosity of –9.2 mag, also in accord with the observed upper luminosity boundary for late–type supergiants.

157.026 **Leo A: color–magnitude diagram of its brightest stars.**
S. Demers, E. J. Kibblewhite, M. J. Irwin, P. S. Bunclark, M. T. Bridgeland.
Astron. J., Vol. 89, No. 8, p. 1160 – 1165 (1984). With plates 66 – 68.
The authors present the color–magnitude diagram for the bright stars of the dwarf irregular galaxy Leo A. Plates obtained with the CFH telescope were measured with the APM system at Cambridge. No secondary images are used for the faint–magnitude calibration. Leo A has a population of bright blue stars but no bright red supergiants at the same magnitude level are present. A distance of 2.3 Mpc ($\mu_0 = 26.8$) is determined from the apparent magnitude of its brightest blue stars. Leo A with $M_v = -14.4$ has the same brightness as Sextans A. Leo A is, however, bluer. Very recent star formation could explain the morphology of its color–magnitude diagram.

157.027 **IRAS observations of external galaxies.**
J. P. Emerson.
Occas. Rep. R. Obs., Edinb., No. 13, p. 159 – 175 (1984). – See Abstr. 012.001.
Some of the first IRAS results on normal galaxies are reviewed. The results include maps of M31 and a discussion of the IR properties of both an optically selected and an infrared selected sample of galaxies. Late type spirals are most readily detected, whereas normal ellipticals are not seen at all in the far infrared. The ratio of far infrared (80 micron) to visual (B filter) luminosity ranges from 0.04 to 50, the large range reflecting both a variety of current star formation activity and a variety of dust optical depths. The galaxies with extremely high L(80)/L(B) are only a small fraction of all galaxies. Interactions between galaxies may be stimulating star burst activity. Lists of IRAS circular objects identified with galaxies are given.

157.028 **Star formation in interacting galaxies.**
R. D. Joseph, W. P. S. Meikle, N. A. Robertson, G. S. Wright.
Occas. Rep. R. Obs., Edinb., No. 13, p. 177 – 181 (1984). – See Abstr. 012.001.
The authors have investigated the possibility that the nuclei of interacting galaxies might be the sites of recent star formation activity. A *JHKL* survey of 45 interacting galaxies reveals that, for 85% of the sample, one member of an interacting pair has a *K–L* colour index significantly redder than normal galaxy colours. These *K–L* excesses imply steeply–rising infrared spectra which are characteristic of thermal emission by dust and similar to the infrared spectra found for the nuclei of canonical star–burst galaxies. This inference is supported by 10 – 20 micron photometry obtained for a few of these galaxies. Thus, interactions apparently trigger bursts of star formation in the nuclei of galaxies with efficiency approaching 100%.

157.029 **Infrared mapping of galaxian centers: star forming regions with kiloparsec scale sizes.**
C. M. Telesco.
Occas. Rep. R. Obs., Edinb., No. 13, p. 183 – 198 (1984). – See Abstr. 012.001.
The author reviews 1 – 20 μm mapping of the centers of galaxies encompassing the regions that extend out to several kiloparsecs from the nuclei. The power emerging at $\lambda \geqslant 5$ μm is a direct measure of the number of young stars present. At shorter wavelengths, a variety of processes, including bremsstrahlung, are evident. These and other data constrain the properties of regions of extragalactic star formation.

157.030 **Molecules in external galaxies.**
N. Z. Scoville.
Occas. Rep. R. Obs., Edinb., No. 13, p. 199 – 210 (1984). – See Abstr. 012.001.

157.031 **Kinematical tests for the intrinsic shapes of galaxies.**
M. Capaccioli, G. Fasano, G. Lake.
Mon. Not. R. Astron. Soc., Vol. 209, No. 2, p. 317 – 334 (1984).
The authors study the theoretical correlation of velocity dispersion with observed eccentricity for elliptical galaxies, when the known dependence of velocity dispersion on luminosity is removed. Using a compilation of published data, they first determine accurately the relation between luminosity and velocity dispersion. The residuals are examined as a function of axis ratio in order to construct a test for the intrinsic shape of galaxies. The effects of projection are modelled and possible intrinsic variations are examined. Model catalogues, generated from the theory by a Monte Carlo technique, are compared to the observational data.

157.032 **Evidence for a spike of light in the centre of NGC 3379.**
J.-L. Nieto, J.-L. Vidal.
Mon. Not. R. Astron. Soc., Vol. 209, No. 2, p. 21P – 25P (1984).
The authors present evidence for a spike of light in the centre of the elliptical galaxy NGC 3379. This is based on the fact that the central surface brightness increases with increasing resolution, independent of any model assumed to fit the luminosity profile of the rest of the galaxy. The central luminosity profile does not fit an isothermal model. It is well described by a Young model of a star cluster having an isotropic velocity dispersion and containing a black hole of mass $1 - 2 \times 10^8$ $M_\odot$, although other interpretations of the data are possible.

157.033 **Inner structure of the lenticular galaxy NGC 3384.**
E. Davoust, G. Lelièvre, A. Maury, J.-L. Nieto.
Mon. Not. R. Astron. Soc., Vol. 209, No. 3, p. 503 – 509 (1984).
The inner region of the lenticular galaxy NGC 3384 presents a complex structure. The high resolution (FWHM $\sim$ 0.7 arcsec) obtained in the present photometric study allows the authors to argue that the faint elongated feature crossing the bulge perpendicularly to the major axis of the galaxy is probably not a bar, but a small polar ring of non uniform surface brightness, 800 pc in radius. The latter interpretation is dynamically more plausible and provides an explanation of the shape of the inner isophotes, in particular their off–centring.

157.034 **Radio emission from three face–on spiral galaxies.**
J. I. Harnett.
Mon. Not. R. Astron. Soc., Vol. 210, No. 1, p. 13 – 22 (1984).
Maps of NGC 1566, 2442 and 5236 (M83) at 843 MHz have been made with the Molonglo Observatory Synthesis Telescope. Emission extends over the nuclei and discs of the three galaxies. M83 has a significant thermal component in its emission at 10 GHz; this enabled the electron density and mass of H II to be estimated. The radial distributions of non–thermal emission from NGC 1566 and M83 show an exponential dependence similar to that of the blue light distributions. Scale lengths of non–thermal

emission are used to investigate the distributions of cosmic ray electrons and magnetic fields in NGC 1566 and M83.

157.035 **Spectroscopic survey of southern compact and bright-nucleus galaxies. VI.**
A. P. Fairall.
Mon. Not. R. Astron. Soc., Vol. 210, No. 1, p. 69 – 77 (1984).
Spectroscopy for 440 galaxies is presented. Most show absorption line spectra but 16 new narrow emission line galaxies and a Seyfert 2 galaxy are included.

157.036 **Galaxies in collision.**
M. G. Edmunds.
Nature, Vol. 311, No. 5981, p. 10 – 11 (1984).

157.037 **A four-colour photometry of spiral galaxies. A comparison with Markarian galaxies.**
A. T. Kalloglyan, F. Börngen.
Astrofizika, Tom 19, Vyp. 3, p. 403 – 410 (1983). In Russian. English translation in Astrophysics, Vol. 19, No. 3.
The results of *UBVR* surface photometry of seven normal spiral galaxies and a comparison of some photometric parameters of these galaxies with galaxies having ultraviolet excess of radiation are given.

157.038 **Investigation of morphological properties of 6 interacting galaxies. I.**
A. A. Korovyakovskaya.
Astrofizika, Tom 19, Vyp. 3, p. 419 – 428 (1983). In Russian. English translation in Astrophysics, Vol. 19, No. 3.
The morphology of six groups of interacting galaxies from Vorontsov-Vel'yaminov's atlas is investigated. It has been shown that only one among five groups of galaxies which could be considered as chains is, apparently, a real chain. Only two spiral galaxies are detected in the VV 551 group classified earlier as a blue nest consisting of 4 – 5 members.

157.039 **On the spectrum of the irregular galaxy NGC 2968.**
Eh. E. Khachikyan, N. K. Andreasyan.
Astrofizika, Tom 19, Vyp. 3, p. 575 – 577 (1983). In Russian. English translation in Astrophysics, Vol. 19, No. 3.
Results of a spectroscopic study of the galaxy NGC 2968 are presented. The dominance of type II stellar population in this galaxy is determined.

157.040 **On the irregularity of the MCG data caused by variability of observing conditions.**
N. G. Kogoshvili.
Astrofizika, Tom 19, Vyp. 3, p. 584 – 588 (1983). In Russian. English translation in Astrophysics, Vol. 19, No. 3.
The degree of irregularity in the apparent galaxy distribution arising in the Palomar Sky Atlas print-to-print transition was estimated. This irregularity is caused by the variability of observing conditions. Counts of galaxies within the separate Palomar Atlas prints were carried out on the basis of data taken from the Morphological Catalogue of Galaxies by Vorontsov-Vel'yaminov et al.

157.041 **The large-scale radio continuum structure of the edge-on spiral galaxy NGC 253.**
E. Hummel, P. Smith, J. M. van der Hulst.
Astron. Astrophys., Vol. 137, No. 1, p. 138 – 144 (1984).
VLA observations of NGC 253 are presented. The main results are: (1) The radio continuum emission consists of three main components, a central source ($\sim 50\%$), an inner disk component containing a bar ($\sim 30\%$) and an outer disk component ($\sim 20\%$). (2) The central source has a radio power $P_{4.9\mathrm{GHz}} \cong 10^{21}\mathrm{W\ Hz^{-1}}$ and a linear half power size of $\sim 130 \times 60$ pc. The spectral index is $-0.43 \pm 0.07$. (3) The inner disk emission has a half power diameter of $\sim 5$ kpc and has a ridge of radio emission at position angle $75° \pm 5°$ coinciding with the optical bar. The outer disk component has a total extent of $\sim 15$ kpc. There is some suggestion of the presence of a radio spiral arm. (4) The radio disk (inner + outer disk) has a half power $z$-thickness of $1.4 \pm 0.2$ kpc and the 1% level $z$-thickness is 4.5 kpc. No emission at a level $T_b > 0.1$K at 1.46 GHz was detected at $\geqslant 5$ kpc above the plane of the galaxy.

157.042 **Flat rotation curves and mass distribution for spiral galaxies with a disk model.**
M. Ubriaco, J. Tharrats.
Astrophys. Space Sci., Vol. 102, No. 1, p. 81 – 95 (1984).
A method to fit flat rotation curves is presented, wherein the galactic density for a disk model is expressed in terms of a Dirichlet polynomial. This procedure allows to obtain the total galactic mass and to predict the circular velocity at large galactocentric distances. Application of the method to the Galaxy, M31 and four Sc galaxies shows that a significant galactic mass is located beyond the optical radius although it is considerably smaller than the integral mass values obtained from current models with a massive corona included. Observed rotation curves and convergent total mass are obtained, thus the total mass for the Milky Way Galaxy is $5.69 \times 10^{11}\ M_{\odot}$.

157.043 **Rotation and masses of double galaxies.**
I. D. Karachentsev, V. A. Mineva.
Pis'ma Astron. Zh., Tom 10, No. 8, p. 563 – 573 (1984). In Russian. English translation in Sov. Astron. Lett., Vol. 10.
Results of the mass determination from rotation curves are presented for 37 galaxies which are members of isolated pairs. The majority of double galaxies have rotation curves of the flat type. Angular momenta of members in the pairs are randomly orientated with respect to each other and also to the orbital angular momentum.

157.044 **The dispersion velocity of the bulge of spiral galaxy NGC 4736.**
M. Rego, J. Gorgas.
An. Fis., Ser. B, Vol. 80, No. 1, p. 67 – 72 (1984). In Spanish. Abstr. in Phys. Abstr., Vol. 87, No. 1241, Entry 90244 (1984).

157.045 **Detection of molecular hydrogen in two merging galaxies.**
R. D. Joseph, G. S. Wright, R. Wade.
Nature, Vol. 311, No. 5982, p. 132 – 133 (1984).
NGC 6240 and Arp 220 (IC 4553) are two of the most luminous IR galaxies known. They are also both thought to be examples of galaxy-galaxy mergers. The authors have obtained IR spectra in the 2-μm region of NGC 6240 and Arp 220 and detected the $\upsilon = 1\text{–}0\ S(1)$ quadrupole rotation-vibration emission line of $H_2$ in both galaxies. These detections suggest that the merger of two galaxies results in the production of massive quantities of shocked molecular gas.

157.046 **IR observations of the peculiar galaxy Arp 220.**
J. P. Emerson, P. E. Clegg, G. Gee,
C. T. Cunningham, M. J. Griffin, L. M. J. Brown,
E. I. Robson, A. J. Longmore.
Nature, Vol. 311, No. 5983, p. 237 – 238 (1984).
Observations from IRAS have shown that the peculiar galaxy Arp 220 (= IC 4553) has extreme IR properties. Its IR luminosity is 100 times greater than normal spirals. The authors report here 20-, 350- and 760-μm observations which constrain the mass and size of the emitting region and compare Arp 220 with the archetypal starburst galaxy M82.

157.047 **Photometric investigation of the peculiar galaxy NGC 2685.**
V. A. Hagen-Thorn, I. I. Popov, V. A. Yakovleva.
Astrofizika, Tom 19, Vyp. 4, p. 599 – 612 (1983). In Russian. English translation in Astrophysics, Vol. 19, No. 4.
The results of a detailed *UBV* photometry of the highly peculiar galaxy NGC 2685 are given. Consideration of all the available data – photometric, spectroscopic and polarimetric – lead to the suggestion that NGC 2685 is a pair of colliding galaxies.

157.048 **Radial velocities and virial masses of some isolated triplets of galaxies.**
V. E. Karachentseva, I. D. Karachentsev.
Astrofizika, Tom 19, Vyp. 4, p. 613 – 617 (1983). In Russian. English translation in Astrophysics, Vol. 19, No. 4.
Spectral observations of isolated triple systems of galaxies have been continued. Radial velocities for 31 components of the triplets have been obtained. For 7 triplets the values of virial masses and mass–to–luminosity ratios were calculated.

157.049 **Superassociations in spiral galaxies with UV excess. I.**
A. R. Petrosyan, K. A. Saakyan, Eh. E. Khachikyan.
Astrofizika, Tom 19, Vyp. 4, p. 619 – 630 (1983). In Russian. English translation in Astrophysics, Vol. 19, No. 4.
The results of a morphological investigation of spiral galaxies with UV excess containing bright condensations–superassociations (SA) are given. A sample of 56 galaxies are listed containing 151 SA. For 26 galaxies out of 56 reproductions and descriptions are presented. The dimensions and brightnesses of SA and nuclei are determined and the distances of SA from the centre are measured.

157.050 **Morphology of galaxies with UV excess.**
A. A. Egiazaryan.
Astrofizika, Tom 19, Vyp. 4, p. 631 – 638 (1983). In Russian. English translation in Astrophysics, Vol. 19, No. 4.
A description of 43 galaxies with UV excess is presented. Morphological details and condensations are brought to light.

157.051 **Observations of galaxies of the Byurakan classification at a frequency of 102 MHz.**
V. S. Artyukh, M. A. Ogannisyan.
Astrofizika, Tom 19, Vyp. 4, p. 655 – 663 (1983). In Russian. English translation in Astrophysics, Vol. 19, No. 4.
Interplanetary scintillation observations of 131 galaxies of the Byurakan classification have been made at a frequency of 102 MHz. The galaxies which have brighter nuclei are more radioactive and they have more frequently scintillation components. The scintillation components have steep spectra.

157.052 **On statistics of morphological types of single galaxies and of the components of pairs.**
M. A. Arakelyan.
Astrofizika, Tom 19, Vyp. 4, p. 665 – 671 (1983). In Russian. English translation in Astrophysics, Vol. 19, No. 4.
The influence of observational selection on the distribution of morphological types in the catalogues of single galaxies and pairs of galaxies is investigated. It has been shown that the difference of morphological contents of two catalogues as well as the indication on non–random contents of pairs can be interpreted as results of selection.

157.053 **Distribution of the apparent magnitudes and luminosities of double galaxies.**
M. A. Arakelyan.
Astrofizika, Tom 19, Vyp. 4, p. 673 – 682 (1983). In Russian. English translation in Astrophysics, Vol. 19, No. 4.
The influence of observational selection on the distribution of the apparent magnitudes and the luminosities of objects in the catalogue of double galaxies is investigated. The method of determination of the luminosity function of the components of double galaxies utilizing an arbitrary sample of pairs of galaxies is proposed. The results are applied to data of Karachentsev's catalogue of isolated pairs. The existence of a correlation between the absolute magnitudes of the components of pairs is proved.

157.054 **On the origin of tightly wound spiral features in galactic nuclei.**
P. Pismis, E. Moreno.
Astrofizika, Tom 20, Vyp. 1, p. 7 – 20 (1984).
A mechanism for the origin and development of spiral arms proposed earlier by one of the authors (P. P.) is recalled to show that spiral structure in the nucleus of a galaxy can be produced by matter ejected from a rotating nucleus. The mechanism can account for the observations of outward and/or inward motions in the nuclear regions of spirals. In particular the model applied to the tight nuclear spiral of NGC 4736 shows satisfactory agreement with observations.

157.055 **Spectrophotometry and morphology of galaxies with UV excess. V.**
M. A. Kazaryan.
Astrofizika, Tom 20, Vyp. 1, p. 35 – 42 (1984). In Russian. English translation in Astrophysics, Vol. 20, No. 1.
The results of spectrophotometry and morphology of galaxies No. 198 and 199 from the list of the author (1979) are presented. The mass of the galaxy No. 199 and the mass of its gaseous component are obtained, which are $6.4 \times 10^9 M_\odot$ and $3.2 \times 10^4 M_\odot$, respectively. A dependence has been established between absolute photographic magnitude and the linear diameters as well as between surface brightness and diameters.

157.056 **UBV photometry of two physical systems of galaxies with UV excess.**
V. S. Tamazyan.
Astrofizika, Tom 20, Vyp. 1, p. 43 – 50 (1984). In Russian. English translation in Astrophysics, Vol. 20, No. 1.
Detailed photographic UBV photometry of the galaxies No. 27, 28, 29 and 346, 347, 348 from lists by Kazaryan is carried out. The distribution of surface brightness $B/\square''$ and colour indices $U{-}B$, $B{-}V$ along the major axes of the galaxies No. 28 and 348 as well as ($U{-}B$, $B{-}V$) diagrams of these galaxies are presented. The UV excess in the investigated galaxies probably has stellar origin and is concentrated in nuclear regions. Seven bright condensations in the galaxy No. 346 are noted.

157.057 **Galactic winds and fountains.**
F. D. Kahn.
Mass loss from astronomical objects, p. 108 (1984). Abstract. – See Abstr. 012.032.

157.058 **Investigation of galaxies of high surface brightness at 3.95 GHz.**
V. G. Malumyan.
Astrofizika, Tom 20, Vyp. 1, p. 182 – 186 (1984). In Russian. English translation in Astrophysics, Vol. 20, No. 1.
The results of observations of 31 galaxies of high optical surface brightness at 3.95 GHz are presented.

157.059 **Slit spectra of galaxies of the second Byurakan Survey. I.**
B. E. Markarian, V. A. Lipovetskij, D. A. Stepanyan.
Astrofizika, Tom 20, Vyp. 2, p. 213 – 222 (1984). In Russian. English translation in Astrophysics, Vol. 20, No. 2.
The results of spectral investigations of 63 peculiar (by their physical nature) galaxies, representing the majority of objects of the first list of the second Byurakan Survey (1983) are presented. The galaxies are distributed in a field $4° \times 4°$, centered at $\alpha = 11^h30^m$, $\delta = +59°00'$. Emission lines observed in the spectra of all of them served to determine the redshift of the galaxies and thus their luminosities. Seven of them have Seyfert properties.

157.060 **Comparison of M33 and NGC 7793: stochastic models of spiral galaxies modulated by density waves.**
G. Smith, B. G. Elmegreen, D. M. Elmegreen.
Mon. Not. R. Astron. Soc., Vol. 210, No. 2, p. 399 – 413 (1984).
Two late–type spiral galaxies with similar kinematic and photometric properties but different spiral arm structures, M33 and NGC 7793, are compared to model galaxies with stochastic self–propagating star formation. The spontaneous probability, $P_{sp}$, representing the rate of primary star formation, is modulated by a smooth, density wave–like spiral pattern in the models of M33. When propagating star formation is included, these models show no age gradients in the underlying spiral arms. Models which have no imposed spiral modulation to $P_{sp}$ resemble the observed structure of NGC 7793.

157.061 **Photographic photometry of galaxies using the INMP. III. The isophotal structure of the two early-type galaxies NGC 1600 and NGC 4261.**
R. Barbon, L. Benacchio, M. Capaccioli, R. Rampazzo.
Astron. Astrophys., Vol. 137, No. 2, p. 166 – 176 (1984).

A study of the photometric and geometric properties of the two early-type galaxies NGC 1600 and NGC 4261, using an interactive numerical mapping technique, is presented. The data for NGC 1600 confirm quantitatively its morphological classification as a giant E galaxy. The structural parameters indicate a triaxial configuration. The luminosity profile of NGC 4261 does not fit the $r^{1/4}$ law at faint levels where a light excess is present. This suggests that this galaxy is a misclassified lenticular. The contrasting results on the light distribution and structure of NGC 4261 reported in earlier studies are critically analyzed. The influence of the environment on both objects is also discussed.

157.062 **The stellar population of the nucleus of M33 from an analysis of its λλ1200 – 3000 Å spectrum.**
A. Ciani, S. D'Odorico, P. Benvenuti.
Astron. Astrophys., Vol. 137, No. 2, p. 223 – 234 (1984).

The λλ1200 – 3000 Å spectrum of the nucleus of M33 is used in conjunction with *U*, *B*, *V* photometry as input to a linear programming method to determine the stellar population of the system. No astrophysical constraints but modifications in the stellar library (which includes Population II objects) were used to explore different evolutionary histories of star formation in the nucleus. A multigeneration model, where a young component (age $\sim 10^7$years) and an old one (age $\sim 5 \times 10^9$ years) are superimposed, gives the best fit to the data and is favoured by the results at other wavelengths as well. The effect of reddening on the modelling and the interpretation of the nuclear X-ray source are briefly discussed.

157.063 **Past star formation rates in spiral galaxies.**
R. Mochkovitch, B. Rocca-Volmerange.
Astron. Astrophys., Vol. 137, No. 2, p. 298 – 306 (1984).

The authors discuss the determination of the history of the star formation rate in spiral galaxies using visible and far-UV photometric data. The results of theoretical evolutionary models indicate that the dereddened colors can be related to the ratio *R* of the present to the integrated past star formation rate (ISFR), provided that a metallicity and an initial mass function of stars have been assumed. An important problem in such a study is the correction of the observed fluxes for internal extinction. The authors consider a sample of spiral galaxies taken from those observed in UV light by Code and Welch (1982) with the OAO satellite. They demonstrate that the ISFR can be obtained in spite of the uncertainty in the internal reddening estimation.

157.064 **Shock-induced star formation in IC 2153. A colliding pair of galaxies.**
K. Olofsson, N. Bergvall, A. Ekman.
Astron. Astrophys., Vol. 137, No. 2, p. 327 – 334 (1984).

The authors present direct photography, photometry and spectroscopy of the interacting galaxy-pair IC 2153 = ESO 364-IG22. Data are based on observations with the ESO 3.6-m, 1.5-m and 1-m telescopes at La Silla, Chile. The mean radial velocity of the system is $2630 \pm 30$ km s$^{-1}$, which corresponds to a distance of 35 Mpc ($H_0$ = 75 km s$^{-1}$Mpc$^{-1}$). The absolute magnitude of the system is $M_B = -18.4$ mag. Spectral tracings of the nuclei of the main components show an early-type absorption spectrum and an emission spectrum respectively. The authors find support for shock-heating, possibly as an outcome of a direct collision between the gaseous components of the two galaxies. A synthesis of the underlying stellar population is made, which indicates that a burst of star formation started $15 \times 10^6$yr ago.

157.065 **Radio observations of H I and OH in the center of the galaxy M82.**
L. Weliachew, E. B. Fomalont, E. W. Greisen.
Astron. Astrophys., Vol. 137, No. 2, p. 335 – 342 (1984).

Using the VLA, the H I and OH absorption have been mapped across the central 1 kpc of the galaxy M82. The H I and OH are concentrated in a rotating ring of radius 250 pc. The radio and optical kinematic center coincide within 20 pc with the near infrared peak. The VLBI continuum point source is displaced by 140 pc from the kinematic center of this ring. The H I mass in the ring is $1.2 \times 10^7$ $M_\odot$. From arguments of kinematics and metal abundances, lower limits to the age of the H I ring are smaller than the time elapsed since the gravitational interaction of M82 with M81. It is therefore impossible to determine whether or not the ring is a consequence of that interaction. Several OH masers near the continuum point source were also mapped. For them the authors used higher spatial and velocity resolution to determine their characteristics.

157.066 **The radio source associated with the elliptical galaxy NGC 5322.**
L. Feretti, G. Giovannini, E. Hummel, C. G. Kotanyi.
Astron. Astrophys., Vol. 137, No. 2, p. 362 – 364 (1984).

The authors present high resolution radio observations at 1.4 and 5 GHz of the nearby elliptical galaxy NGC 5322, which is associated with a weak radio source ($P_{1.4\ \mathrm{GHz}}$ = $4 \times 10^{21}$ WHz$^{-1}$). The radio structure shows two symmetric jets emanating from a compact unresolved core and extending over the linear scale of ~1 kpc. These features are surrounded by a low brightness region, of total linear extent ~14 kpc, completely embedded within the optical galaxy.

157.067 **Shells at the periphery of elliptical galaxies.**
B. V. Komberg.
Priroda, No. 8, p. 103 – 104 (1984). In Russian.

157.068 **Tidal corrections to the orbital velocity of paired galaxies.**
G. S. Bisnovatyj-Kogan.
Sov. Astron. Lett., Vol. 10, No. 2, p. 75 – 76 (1984). English translation of 37.157.118.

157.069 **Rotation and mass of 21 isolated galaxies.**
I. D. Karachentsev, V. A. Mineva.
Sov. Astron. Lett., Vol. 10, No. 2, p. 105 – 108 (1984). English translation of 37.157.136.

157.070 **Detection of H I in the elliptical galaxy NGC 5173.**
G. R. Knapp, E. Raimond.
Astron. Astrophys., Vol. 138, No. 1, p. 77 – 84 (1984).

The pair of galaxies NGC 5169 (Sb) and NGC 5173 (E0) was observed in the 21 cm line of H I using the Westerbork Synthesis Radio Telescope. H I was clearly detected in NGC 5173, with $M$(H I) = $1.2 \times 10^9$ $M_\odot$ ($D$ = 33 Mpc). No H I bridge between the galaxies was seen. The available evidence suggests that the acquisition of gas by NGC 5173 is fairly recent.

157.071 **Rotation axes of gas and stars in elliptical galaxies.**
D. Bettoni.
Messenger, No. 37, p. 17 – 19 (1984).

157.072 **Wolf-Rayet stars in "lazy" galaxies.**
M. Joubert, D. Kunth.
Messenger, No. 37, p. 21 – 23 (1984).

157.073 **Colour gradients in elliptical galaxies – some results from CCD photometry.**
E. Sadler.
Messenger, No. 37, p. 36 – 39 (1984).

157.074 **Ultraviolet observations of normal galaxies.**
M. A. J. Snijders.
ESA Spec. Publ., ESA SP-218, p. 3 – 9 (1984). – See Abstr. 012.024.

Ultraviolet observations of normal galaxies are reviewed with respect to the knowledge of their stellar populations. UV data show that both elliptical galaxies and the bulges of spiral galaxies possess a hot star component of uncertain nature. The importance of a better understanding of the late stages of stellar evolution, both from an observational and a theoretical point, is stressed. In those spiral and elliptical galaxies undergoing a star-burst it is clear that young O- and B-type stars are responsible for the ultraviolet flux. Some recent results for these objects are briefly considered.

157.075 **Study of hot spot galaxies.**
K. Nandy, G. I. Thompson.
ESA Spec. Publ., ESA SP-218, p. 65 - 68 (1984). - See Abstr. 012.024.

A comparative study of UV flux distributions of several hot spot galaxies is presented. The UV spectra do not suggest the presence of O type stars in marked contrast to Markarian galaxies. Also, the N III λ1750 and C III λ1909 lines from high excitation H II regions or planetary nebulae are not present. The UV flux distributions have been used to determine the population of early type stars. They show evidence of reddening but the λ2200 dust feature is clearly absent.

157.076 **Far-UV excess and gas content of S0 galaxies.**
B. Rocca-Volmerange, C. Balkowski.
ESA Spec. Publ., ESA SP-218, p. 69 - 72 (1984). - See Abstr. 012.024.

Preliminary results about the analysis of IUE spectra of S0 galaxies in terms of far-UV excess are given. For the two S0 galaxies considered: NGC 5102 (gas-rich) and NGC 3115 (gas poor), a comparison with optical as well as UV stellar continua gives a uniform stellar population respectively A3 and G2. The UV excesses, normalized to the V band, represent a very low fraction (1 - 2%) of the total emitted energy.

157.077 **Evidence for dark halos around spiral galaxies.**
V. C. Rubin.
Large-scale structure of the universe, cosmology and fundamental physics, p. 204 - 205 (1984). - See Abstr. 012.025.

The polar ring galaxy, A0136-0801, consists of a rotating disk surrounded by a rotating ring which runs almost over the poles. Velocities in the ring serve as a probe of the vertical potential of the system. The agreement of the velocities in the ring with those in the disk indicates that (1) the mass of the system increases linearly with radius to the outer edge of the polar ring; and (2) the halo is more nearly spherical than flat.

157.078 **The $M_T/L$ ratio and hidden mass in dwarf galaxies.**
R. D. Davies.
Large-scale structure of the universe, cosmology and fundamental physics, p. 206 - 207 (1984). - See Abstr. 012.025.

Neutral hydrogen observations are used to measure the total mass of spiral and dwarf galaxies. It is found that the $M_T/L$ ratio does not change between spiral and dwarf galaxies. The dimensions and velocity field of dwarfs would require a neutrino mass of $\sim 150$ eV from phase space density arguments. The implications of this result are discussed.

157.079 **A simple model for the dynamical and chemical evolution of elliptical galaxies.**
B. J. T. Jones, R. F. G. Wyse.
Large-scale structure of the universe, cosmology and fundamental physics, p. 267 - 271 (1984). - See Abstr. 012.025.

The authors present a simple model for the formation of elliptical galaxies. The model is based on a clustering hierarchy of dark matter, the chemical enrichment of the gas being controlled by supernovae. The observed relationships between luminosity and velocity dispersion, between metallicity and velocity dispersion and between rotation and surface brightness are easily explained within such a scheme.

157.080 **The potentiality of the galaxy autocorrelation function.**
R. Fong.
Large-scale structure of the universe, cosmology and fundamental physics, p. 344 - 348 (1984). - See Abstr. 012.025.

This paper discusses the properties of the two-point autocorrelation function $\xi(r)$ for galaxies determined from the Durham-AAT redshift survey.

157.081 **The Andromeda Galaxy III. Tidal interaction with its companions.**
T. Sawa, N. Sato.
Astron. Her., Vol. 77, No. 9, p. 216 - 222 (1984). In Japanese.

157.082 **Recent star-formation in interacting galaxies.**
G. S. Wright.
Observatory, Vol. 104, No. 1062, p. 212 (1984). Abstract. - See Abstr. 010.682.

157.083 **Infrared observations of two blue early-type galaxies.**
I. S. Glass, A. F. M. Moorwood.
Observatory, Vol. 104, No. 1062, p. 231 - 233 (1984).

Infrared examination of the S0 galaxy NGC 5102 and the apparent dwarf elliptical NGC 1510 confirms their abnormally blue colours to be due to star-burst events.

157.084 **Emission line ratios for some southern compact and bright-nucleus galaxies.**
A. P. Fairall.
Mon. Notes Astron. Soc. S. Afr., Vol. 43, Nos. 5 - 8, p. 30 - 31 (1984).

157.085 **Rotation curve of the sodium D absorption line in the central region of M82.**
M. Saitō, M. Sasaki, N. Kaneko, M. Nishimura, K. Toyama.
Publ. Astron. Soc. Jpn., Vol. 36, No. 2, p. 305 - 311 (1984).

The authors show that the rotation curve of the Na D absorption line has a smaller gradient compared with rotation curves of the emission lines of Hα and [N II] λλ6548, 6584.

157.086 **Neutral hydrogen in M31. V. The formation rate of H II regions and a correction factor for the star formation rate.**
N. Nakai, Y. Sofue.
Publ. Astron. Soc. Jpn., Vol. 36, No. 2, p. 313 - 318 (1984).

The relation among the densities of H II regions, young stars, and H I gas in M31 is reexamined. The authors point out that the "star formation rate" derived from the densities of H II regions and H I gas does not represent the true rate.

157.087 **Star formation in a dwarf elliptical of the Virgo cluster.**
L. Vigroux, J. Souviron, J. P. Vader.
Astron. Astrophys., Vol. 139, No. 1, p. L9 - L12 (1984).

High resolution CCD photometry was performed at the Canada-France-Hawaii Telescope on 3 dwarf elliptical galaxies in the Virgo cluster. One of these exhibits a small excentric blue nucleus. The properties of this nucleus are very similar to those of the active star formation zone in NGC 205, a low luminosity elliptical companion of M31.

157.088 **Neutral hydrogen associated with the S0 galaxy NGC 1023.**
R. Sancisi, H. van Woerden, R. D. Davies, L. Hart.
Mon. Not. R. Astron. Soc., Vol. 210, No. 3, p. 497 - 514 (1984).

NGC 1023 has been observed with the Westerbork Synthesis Radio Telescope. The hydrogen, amounting to $1.5 \times 10^9 M_\odot$, has remarkable density distribution and kinematics, suggesting probably an intergalactic origin from an encounter of NGC 1023 with one or more late-type hydrogen-rich systems. One of the densest H I clumps is associated with the small optical companion visible near the eastern edge of NGC 1023. In addition to the Eastern companion, two other dwarf-like systems have been detected in H I in the neighbourhood of NGC 1023.

157.089 **Gas dynamics in barred spiral galaxies. II. NGC 7496 and 289.**
W. D. Pence, C. P. Blackman.
Mon. Not. R. Astron. Soc., Vol. 210, No. 3, p. 547 - 563 (1984).

This paper continues the study of the gas dynamics in barred spiral galaxies in which the authors have been looking for observational verification of recent theoretical models of the behaviour of gas in response to a rotating stellar bar. The presented results on two more barred spiral galaxies, NGC 7496 and 289, provide further support for the models.

157.090 **Direct determination of extragalactic distances from the 21 centimeter line width and the method of "sosies".**
G. Paturel.
Astrophys. J., Vol. 282, No. 2, p. 382 – 386 (1984).
The basic concept behind the Tully–Fisher relation is applied to "sosie" (or "look–alike") galaxies having the same morphological type, inclination, and 21 cm line width. The ratio of the distances of two "sosies" is then simply given by the ratio of their apparent diameters or deduced from the difference of their apparent magnitudes. This is independent of any assumption on the specific form of the T–F relation and without questionable corrections for turbulent velocities and inclination. An application is made to four galaxies usable as calibrators. These galaxies have a total of 13 "sosies". The corresponding values of the mean Hubble ratio are $H = 102 \pm 8$ km s$^{-1}$Mpc$^{-1}$ if the "short" scale values are adopted for the calibrating galaxies and $85 \pm 5$ if the "long" scale values are used.

157.091 **Angular correlations of galaxies to $B \approx 24$: another probe of cosmology and galaxy evolution.**
D. C. Koo, A. S. Szalay.
Astrophys. J., Vol. 282, No. 2, p. 390 – 397 (1984).
The authors investigate the two–point angular autocorrelation function, $\omega(\theta)$, of over 11,000 very faint galaxies using new 4 m data in two fields of 0.2 deg$^2$ and models that incorporate the best available estimates of the spectral evolution of galaxies. Within the framework of the models, it is concluded that the luminosity evolution of galaxies has been mild, that is, with galaxy formation having occurred in a universe of low density at redshifts beyond $z = 5$ and that clustering evolution has been negligible to the average distance of the faintest galaxies in the sample.

157.092 **Aperture synthesis observations of CO emission from the nucleus of IC 342.**
K. Y. Lo, G. L. Berge, M. J. Claussen, G. M. Heiligman, R. B. Leighton, C. R. Masson, A. T. Moffet, T. G. Phillips, A. I. Sargent, S. L. Scott, P. G. Wannier, D. P. Woody.
Astrophys. J., Lett. Ed., Vol. 282, No. 2, p. L59 – L63 (1984). With plates L5 – L6.
The authors present the first aperture synthesis maps of $\lambda$2.6 mm CO ($J = 1$–$0$) emission from an external galaxy, IC 342. The 7″ resolution maps of the nuclear region were made with the Owens Valley Millimeter–Wave Interferometer. They reveal that the CO source is distributed in a bar, 300 pc × $\gtrsim$1500 pc, with a velocity gradient across the width of the bar. The observations suggest that the molecular gas in the nucleus is moving in response to an oval gravitational potential.

157.093 **Superassociations in spiral galaxies with UV excess. II.**
A. R. Petrosyan, K. A. Saakyan, Eh. E. Khachikyan.
Astrofizika, Tom 20, Vyp. 1, p. 51 – 60 (1984). In Russian. English translation in Astrophysics, Vol. 20, No. 1.
Descriptions and data on 30 out of 56 galaxies containing superassociations are presented. Reproductions of 28 galaxies are given.

157.094 **Faint galaxy number–magnitude counts at high galactic latitude. I.**
P. Hall, C. D. Mackay.
Mon. Not. R. Astron. Soc., Vol. 210, No. 4, p. 979 – 992 (1984).
A new observing technique has been developed that avoids the limits normally set by detector non–uniformities and achieves shot–noise limited performance with a CCD detector. The technique is described together with its application in determining the galaxy number–magnitude counts in three fields at high galactic latitude, each in two passbands. Counts are presented to an integrated density of 150000 per degree$^2$. The faintest objects that the authors detect at the 4.5 sigma level have a magnitude of $KG = 25.9$, and the counts are substantially complete to $KG3 = 25.0$. The counts are found to have a slope of 0.4, and the average galaxy colour is constant for all magnitudes with an overall average of $(KG3–I) = 0.89$.

157.095 **On the correlation of CO line and radio continuum emission in nearby galaxies and the efficiency of OB star formation.**
F. P. Israel, M. Rowan–Robinson.
Astrophys. J., Vol. 283, No. 1, p. 81 – 89 (1984).
A comparison of published CO line strengths of nearby Sb and Sc galaxies with (21 cm) radio continuum disk surface brightness $\sigma_{21}$ suggests the quantitative correlation $T_A{}^*\Delta V/\sigma_{21} = 0.2 – 0.3$K km s$^{-1}$mJy$^{-1}$ in a 1′ beam. This correlation can be used to predict CO line strengths of hitherto undetected galaxies. The authors present a possible interpretation of this result in terms of similar OB star formation efficiencies in Sb and Sc galaxy disks.

157.096 **An empirical model of the M87 jet.**
I. S. Shklovskij.
Astron. Zh., Tom 61, Vyp. 5, p. 833 – 836 (1984). In Russian. English translation in Sov. Astron., Vol. 28, No. 5.
Recent observations of the M87 jet in radio, optical and X–ray regions have revealed an identity of the brightness distribution at all frequencies. This points to a decisive role of the regular magnetic field variations along the jet for its overall structure. The bright knots of the jet form in the places where the field is enhanced. The velocity of the plasma ejected from the nucleus of M87 should be 0.1 $c$. Thus, the M87 jet is one–sided.

157.097 **Rotation curves of normal galaxies. The catalogue.**
G. A. Kyazumov.
Astron. Zh., Tom 61, Vyp. 5, p. 846 – 853 (1984). In Russian. English translation in Sov. Astron., Vol. 28, No. 5.
The author presents a catalogue which includes a quantitative description of curves of rotation of 116 normal and irregular galaxies based on published data. Only those galaxies were chosen whose rotational velocities are known up to distances $R_1 \geqslant D_0/4$ from the centre ($D_0$ being the photometric diameter of the galaxy).

157.098 **Results from the Durham–AAT redshift survey.**
T. Shanks.
Ann. N.Y. Acad. Sci., Vol. 422, p. 382 (1984). Abstract. – See Abstr. 012.033.

157.099 **An unsuccessful search for very extended halos around NGC 1300 and M61.**
G. D. Bothun, B. Margon, B. Balick.
Publ. Astron. Soc. Pac., Vol. 96, No. 582, p. 583 – 587 (1984).
The authors have undertaken a sensitive search for H and K absorption arising of two spiral galaxies that are projected against background QSOs, using a blue–sensitive spectrograph attached to the Multiple Mirror Telescope. Despite good sensitivity (S/N $\sim$ 20) and high spectral resolution (1 Å), this search program produced a null result. In the best case, the authors can set an upper limit on the equivalent width of H and K absorption of 0.1 Å. A 21–cm search for a very extended halo around NGC 1300 was also negative, restricting H I emission column densities to $<10^{20}$cm$^{-2}$ at distances $>50$ kpc from the nucleus.

157.100 **H–alpha observations of spiral galaxies in Cancer, A1367, and Coma.**
R. C. Kennicutt Jr., G. D. Bothun, R. A. Schommer.
Astron. J., Vol. 89, No. 9, p. 1279 – 1287 (1984). With plates 70 – 71.
The authors have used large aperture H$\alpha$ photometry of 65 spiral galaxies in the Cancer, Coma, and Abell 1367 clusters to compare the ionized–gas contents and star–formation rates in cluster and field spirals. Overall, the authors do not observe any significant deficiency of H$\alpha$ emission in the cluster members. Emission strength correlates strongly with integrated galaxy colors, but only weakly with H I content. All three clusters contain several galaxies with unusually strong H$\alpha$ emission, including several H I–poor objects in Coma and A1367. Thus, spirals which appear "anemic" in their morphology or exhibit weak H$\alpha$ emission are not necessarily H I poor; conversely, H I poor spirals can show strong H$\alpha$ emission, indicating relatively high current star–formation rates. Gas depletion time scales for some

objects in the core of Coma are significantly shorter than the field, indicating rapid stellar and gaseous evolution.

157.101 **A possible relationship between metal abundance and luminosity for disk galaxies.**
G. D. Bothun, W. Romanishin, S. E. Strom, K. M. Strom.
Astron. J., Vol. 89, No. 9, p. 1300 – 1309 (1984).

Near–infrared colors have been measured for a sample of 31 late–type galaxies in the Pegasus I and Pisces clusters; system luminosities in the sample cover the range $-19 < M_H < -23.5$. The color index ($J$–$K$) correlates strongly with the absolute $H$ magnitude; lower–luminosity systems have bluer colors. These observations are consistent with the assumption that the mean metal abundance of the old disk population decreases systematically with luminosity. The systematic variation of ($B$–$H$) with absolute $H$ magnitude reported recently by Tully et al. derives in part from this proposed systematic change of metallicity with luminosity. However, one must still posit a relative increase in the number of newly formed stars and/or a systematic smaller age for lower–luminosity disks in order to fully explain the observed ($B$–$H$), $H$ relation.

157.102 **Redshifts for 228 southern galaxies.**
L. N. Da Costa, P. S. Pellegrini, M. A. Nunes, C. Willmer, D. W. Latham.
Astron. J., Vol. 89, No. 9, p. 1310 – 1318 (1984).

The authors present new redshifts for 228 galaxies located south of declination –30°. The observations were made with a photon–counting Reticon detector on the Observatório Nacional 60–in. telescope. The detector is identical to the one used at Mount Hopkins for the CfA Redshift Survey, and the redshifts were derived using the same data–analysis system. A preliminary comparison with published 21–cm redshifts indicates that the authors' velocities have a zero–point offset of about $-4$ km s$^{-1}$, with a typical uncertainty of 40 km s$^{-1}$.

157.103 **The extended hydrogen envelope around the dwarf galaxy DDO 154.**
N. Krumm, D. Burstein.
Astron. J., Vol. 89, No. 9, p. 1319 – 1326 (1984). With plate 73.

A 21–cm line map made with the Arecibo telescope reveals that the irregular galaxy DDO 154 is embedded in an H I disk at least seven times the optical (RC2) diameter. The H I mass–to–light ratio is 6.1 and the H I fraction by mass is 0.3 to 0.5 within four optical radii, making DDO 154 one of the most gas–rich galaxies known. The H I gas distribution is nonaxisymmetric, but it can be roughly approximated by an exponential disk with a scale length of 7 kpc (assuming a distance to the galaxy of 10 Mpc). The position angles of the H I and optical disks are similar, but the H I disk is significantly more elliptical than the much smaller optical disk. A linearly rising rotation curve gives an adequate fit to the H I line profiles out to a radius of 15 kpc.

157.104 **The Cepheid variables in NGC 300.**
J. A. Graham.
Astron. J., Vol. 89, No. 9, p. 1332 – 1342 (1984). With plates 74 – 77.

The southern spiral galaxy NGC 300 has been searched for variable stars. Coordinates and finding charts are given for 34 variables. Most are Cepheids, but also included are three red supergiants and two eclipsing variables, as well as a few objects whose light variation can not yet be classified. For the purpose of estimating their magnitudes, faint photometric sequences have been set up in $V$ and $B$ which extend to $21^m.7$ and $22^m.5$, respectively. Periods have been found for 18 Cepheids ranging from $9^d.67$ to $126^d.9$. Period–luminosity relations are shown with both mean $V$ and mean $B$ magnitudes plotted against log period. A comparison is made with the Cepheid variables of the Large Magellanic Cloud. An absorption–corrected distance modulus of $26^m.09 \pm 0^m.2$ is derived for NGC 300. This corresponds to a linear distance of $1.65 \pm 0.15$ Mpc.

157.105 **Evidence for a central mass concentration in M32.**
J. L. Tonry.
Astrophys. J., Lett. Ed., Vol. 283, No. 2, p. L27 – L30 (1984).

Observations of the velocity and velocity dispersion profiles of the galaxy M32 (NGC 221) reveal an unresolved jump in the rotation curve and a central bump in the velocity dispersion. The amplitude and sharpness of the jump in the rotation of M32 are explained by an increase in mass–to–luminosity ratio ($M/L$) in the center. This increase of $M/L$ in the center could be caused by a change in the stellar population, or, alternatively, the presence of central dark object of mass $5 \times 10^6 M_\odot$.

157.106 **X–ray observations of spiral galaxies. I. Integrated properties.**
G. Fabbiano, G. Trinchieri, A. Macdonald.
Astrophys. J., Vol. 284, No. 1, p. 65 – 74 (1984).

Fourteen nearby normal spiral galaxies were observed with the Einstein Observatory. Thirteen of them were detected with (0.5 – 3.0 keV) $L_x \sim 10^{38} - 10^{40}$ ergs s$^{-1}$. The authors report the global properties of the X–ray emission from these galaxies. They find that the X–ray luminosity is correlated with both the optical (blue) and the radio (1.4 GHz) luminosities. The linear relationship between the X–ray and the optical blue luminosities suggests a relationship between the X–ray luminosity and the stellar content of spiral galaxies responsible for the blue luminosity. The X–ray–radio relationship could have an impact on the current theories of the radio emission in spiral galaxies.

157.107 **A comparison of distance scales for early–type galaxies.**
G. de Vaucouleurs, D. W. Olson.
Astrophys. J., Suppl. Ser., Vol. 56, No. 1, p. 91 – 104 (1984).

The distance scales of elliptical and lenticular galaxies derived from several photometric and colorimetric distance indicators by Michard and by Sandage and Visvanathan have been intercompared and reduced to the scale previously derived by de Vaucouleurs and Olson from the velocity dispersion indicator via the revised Faber–Jackson relation. The standard errors of distance moduli derived from the different indicators range from 0.44 to 0.65 mag after reduction to the same scale. A general catalog of weighted mean distance moduli of 424 early–type galaxies based on one or more indicators is included.

157.108 **CCD surface photometry of field galaxies. I. Observations.**
S. M. Kent.
Astrophys. J., Suppl. Ser., Vol. 56, No. 1, p. 105 – 141 (1984).

Major and minor axis profiles for 105 intrinsically luminous galaxies have been obtained by a method in which elliptical contours whose position angle and ellipticity are allowed to vary with radius are fitted to the true isophotes of a galaxy. Additional photometric parameters derived include an isophotal radius and magnitude, an approximate total magnitude, the effective radius containing one–half the total light, and the mean surface brightness inside this radius.

157.109 **Surface photometry of edge–on galaxies. V. The vertical structure of the bar in NGC 4762.**
K.–I. Wakamatsu, M. Hamabe.
Astrophys. J., Suppl. Ser., Vol. 56, No. 2, p. 283 – 294 (1984). With plates 6 – 7.

From surface photometry, evidence is presented that the edge–on S0 galaxy NGC 4762 is a *barred* galaxy with a prominent lens and an outer ring. The vertical luminosity profiles of the lens decrease less steeply than isothermal sheet models and have a constant scale height $\bar{z} = 560 \pm 40$ pc. The profiles of the bar can be fitted to an isothermal model with a scale height $\bar{z} = 155 \pm 15$ pc.

157.110 ***JHK* colours of "ordinary" galaxies.**
I. S. Glass.
Mon. Not. R. Astron. Soc., Vol. 211, No. 2, p. 461 – 469 (1984).

Accurate $JHK$ photometry is presented for 19 elliptical, 10 lenticular, 29 "ordinary" spiral galaxies, and 1 Im type on the SAAO photometric system which has well–defined transformations to the Caltech/CTIO and AAO systems. The groups having the highest degree of colour uniformity are the ellipticals and lenticulars. "Ordinary", or inactive spirals, show slightly more scatter; late types are bluer than average in $J-H$ and emission-line galaxies are redder in $H-K$. In general, the infrared and $V-K$ colours determined here agree well with previous work. However, no evidence for significant change of infrared colour with radius has been seen, in contrast to the results of some other workers.

157.111 **Structure and evolution of irregular galaxies.**
J. S. Gallagher III, D. A. Hunter.
Annu. Rev. Astron. Astrophys., Vol. 22, p. 37 – 74 (1984). – See Abstr. 003.024.

Contents: (1) Introduction. (2) Basic properties: Light. Masses and kinematics. Abundances. Stellar populations. (3) Star formation properties: Current global star formation rates. Distributions of star-forming regions. Star-forming complexes and the interstellar medium. The initial mass function. Usable gas. (4) Star-forming histories: Constant star formation rates. Bursts of star formation. SSPSF: a possible model. (5) Evolutionary status.

157.112 **The influence of environment on the H I content of galaxies.**
M. P. Haynes, R. Giovanelli, G. L. Chincarini.
Annu. Rev. Astron. Astrophys., Vol. 22, p. 445 – 470 (1984). – See Abstr. 003.024.

Contents: (1) Introduction. (2) The H I content of normal galaxies: The definition of H I content. The distribution of H I gas. (3) Groups of galaxies: The group environment. The streams in nearby groups. The fate of tidal debris. (4) Clusters of galaxies: Comparison samples and definition of H I deficiency. Cluster samples. The Virgo cluster. Gas removal mechanisms. Discussion. (5) Concluding remarks.

157.113 **Spheroidal luminosities: a quantitative basis for the galaxian sequence.**
A. Meisels, J. P. Ostriker.
Astron. J., Vol. 89, No. 10, p. 1451 – 1460 (1984).

A quantitative one-dimensional scheme of galaxy classification is proposed based on the luminosity of the spheroidal component $L_{sph}$. After noting previous work showing the correlations between color, velocity dispersion, scale, metallicity, and degree of rotational support with $L_{sph}$, the authors show that the Hubble type correlates as well. The authors find that (in terms of blue magnitudes, with $H_0 = 100$ km/s/Mpc), if $M_{sph} < -20$ the system is almost always an elliptical, if $-20 < M_{sph} < -17$ comparable frequencies of elliptical, S0, and spirals are found, and if $-17 < M_{sph} < -16$ the systems are typically spiral or irregular. Due to incompleteness of the observational data, little conclusive can be said about systems with $M_{sph} > -16$.

157.114 **Far-infrared observations of galactic nuclei.**
L. J. Rickard, P. M. Harvey.
Astron. J., Vol. 89, No. 10, p. 1520 – 1530 (1984).

The authors report measurements of far-infrared continuum emission from the central regions of 30 late-type galaxies, 24 of which were previously unreported. They have obtained photometry covering wavelengths from 40 to 160 μ, yielding information on luminosities, temperatures, and dust masses for most. The authors find good correlations of the far-infrared luminosity with blue colors, reddening, and molecular line radiation, supporting the identification of the sources with the molecular cloud component in these galaxies, and strengthening the identification of the energy sources as the young massive stars formed from those clouds.

157.115 **Enhanced radio emission in binary spiral galaxies.**
D. R. Altschuler, C. A. Pantoja.
Astron. J., Vol. 89, No. 10, p. 1531 – 1535 (1984).

A comparison of the radio emission of a sample of isolated *pairs* of spiral galaxies with those of a sample of isolated spiral galaxies for which complete distance information is available shows that: (1) the total radio emission of the double galaxies is, on average, a factor about 2 more powerful than that of isolated spiral galaxies of comparable luminosity and (2) the radio emission enhancement in the double galaxies is correlated with a "tidal index", consistent with a tidal origin for this enhancement.

157.116 **Radio emission in isolated and cluster spiral galaxies.**
D. R. Altschuler, R. Giovanelli, M. P. Haynes.
Astron. J., Vol. 89, No. 11, p. 1695 – 1701 (1984).

The authors present observations of the radio-continuum emission at 2380 MHz of spiral galaxies found in the highest and lowest galaxian-density regions of space. Of comparable galaxy population, the two samples consist of spirals in the central regions of Abell clusters, and isolated ones, respectively. The results of this detection survey provide an initial indication of the occurrence of radio emission in these two samples.

157.117 **Metallicity estimates for H II regions in M81.**
J. R. Stauffer, G. D. Bothun.
Astron. J., Vol. 89, No. 11, p. 1702 – 1704 (1984).

Low-dispersion spectra for ten H II regions in the Sab galaxy M81 have been obtained. Oxygen abundances for those H II regions were estimated from the observed flux ratio of ([O II] + [O III])/Hβ. All the H II regions have fairly similar spectra, with a mean abundance near solar and a weak abundance gradient $d[\log(O/H)]/dr = -0.045$ dex kpc$^{-1}$. The derived abundances and abundance gradients are nearly identical to those observed in the Milky Way, considered by most to be of Hubble type Sb or later.

157.118 **Star formation histories of irregular galaxies.**
J. S. Gallagher III, D. A. Hunter, A. V. Tutukov.
Astrophys. J., Vol. 284, No. 2, p. 544 – 556 (1984).

The authors explore the star formation histories of a selection of irregular and spiral galaxies by using three parameters that sample the star formation rate (SFR) at different epochs: (1) the mass of a galaxy in the form of stars measures the SFR integrated over a galaxy's lifetime: (2) the blue luminosity is dominated primarily by stars formed over the past few billion years; and (3) Lyman continuum photon fluxes derived from Hα luminosities give the current SFR. It is found that for most irregulars these parameters are consistent with a nearly constant SFR and Salpeter initial mass function over the galaxies' lifetimes. The spirals and a few high-mass irregulars, on the other hand, appear to have had a constant SFR over the past few billion years but a high total mass indicates either an early phase of more intense star formation or the presence of dark matter.

157.119 **Ongoing star formation in NGC 3310: an infrared perspective.**
C. M. Telesco, I. Gatley.
Astrophys. J., Vol. 284, No. 2, p. 557 – 564 (1984).

Infrared observations of NGC 3310 demonstrate that the extremely blue *UBV* colors and intense radio, ultraviolet, and X-ray emission of this galaxy are attributable to an exceptionally luminous burst of star formation. The flux at 1 – 20 μm is separable into contributions from young and old stars, ionized gas, and warm dust. Simple models imply that the starburst is very young ($\sim 10^7$ yr) and that it will proceed at its present level of activity for a time which is only a small fraction of the age of the galaxy.

157.120 **Velocity dispersions of disk stars in spiral galaxies.**
J. N. Bahcall, S. Casertano.
Astrophys. J., Lett. Ed., Vol. 284, No. 2, p. L35 – L38 (1984).

For multicomponent spiral disks, the authors derive a relation for individual stellar components between the total surface density of mass and the observed luminosity scale height and vertical velocity dispersion. A criterion is given for the validity of the relation in the presence of spheroidal mass distributions and an inhomogeneous disk. Mass models for the Milky Way and NGC 891, 4565, and 5907 are used to illustrate what velocity dispersions observers might expect to measure in nearly face-on spiral galaxies.

157.121 **M82 – NGC 3034.**
F. Bertola.
Sterne Weltraum, 23. Jahrg., Nr. 11, p. 568 – 569 (1984).

157.122 **Haben Spiralgalaxien Halos hochenergetischer kosmischer Teilchen?**
K. Hummel, U. Klein.
Sterne Weltraum, 23. Jahrg., Nr. 11, p. 570 – 574 (1984).

157.123 **IRAS observes galaxies and quasars.**
L. I. Gurvits.
Priroda, No. 10, p. 58 – 61 (1984). In Russian.

157.124 **CO observations of M31 with 100 pc resolution.**
F. Boulanger, J. Bystedt, F. Casoli, F. Combes.
Astron. Astrophys., Vol. 140, No. 1, p. L5 – L8 (1984).

Results of CO observations with the Onsala 20 m telescope towards one spiral arm of the Andromeda galaxy are presented. Inside the spiral arm two components are observed: CO complexes giving narrow bright lines and a diffuse component showing up when adjacent spectra are averaged. This diffuse emission, very broad in velocity, can be interpreted as coming from a lot of small clouds. The narrow bright emission is just resolved by the beam and corresponds to giant molecular complexes of at least 100 pc in size and of total mass $\sim 10^6 M_\odot$. These complexes are located at the center of the spiral arm, are associated with maxima of H I emission and are gravitationally bound.

157.125 **Neutral hydrogen observations of the interacting galaxies NGC 4725 and NGC 4747.**
B. M. H. R. Wevers, P. N. Appleton, R. D. Davies, L. Hart.
Astron. Astrophys., Vol. 140, No. 1, p. 125 – 140 (1984).

New high sensitivity single dish and aperture synthesis H I observations are presented of the galaxies NGC 4725 and NGC 4747. The kinematics and distribution of H I in the two galaxies strongly suggest that they are tidally interacting. A 45 – 50 kpc long H I plume extends from the centre of the small peculiar galaxy NGC 4747 but is offset in position angle by 11° from an associated optical plume. To the south of the galaxy a weaker clumpy H I plume is observed which points back towards NGC 4725. This plume has a small velocity gradient along its length which is in the opposite sense to that of the main rotation of the galaxy. The giant spiral galaxy NGC 4725 is dominated in H I by a main outer arm. Evidence is presented which shows that the arm is probably warped out of the plane of the galaxy. A number of other inner spiral arms and an inner H I ring are discussed.

157.126 **Nuclear structure of NGC 3379.**
O. Bendinelli, G. Parmeggiani, F. Zavatti.
Astron. Astrophys., Vol. 140, No. 1, p. 174 – 176 (1984).

The deconvolution of the high–resolution photographic profile of NGC 3379 (Nieto, 1983) from its three Gaussian components *PSF* is in very good agreement with a previously restored photoelectric profile (Bendinelli et al., 1981). The authors argue that ground based observations do not contain any evidence of a point–like source in NGC 3379 such as those found at the centers of M32 and M87.

157.127 **NGC 5194/5195.**
F. Bertola.
Sterne Weltraum, 23. Jahrg., Nr. 12, p. 622 – 623 (1984).

157.128 **The bivariant luminosity function for isolated galaxy pairs.**
A. R. Petrosyan.
Sov. Astron., Vol. 28, No. 3, p. 258 – 263 (1984). English translation of 37.157.187.

157.129 **Chemical evolution of bulges and halos.**
J. R. Mould.
Publ. Astron. Soc. Pac., Vol. 96, No. 584, p. 773 – 778 (1984). Invited paper presented at the Symposium on Stellar Populations, at the 95th Annual Scientific Meeting of the Astronomical Society of the Pacific, University of California, Santa Cruz, July 1984.

The chemical properties of elliptical galaxies and the bulges of spiral galaxies are reviewed. The mass–metallicity relation for ellipticals appears to span the full range of these galaxies from dwarf spheroidal to gE. Explanations seem possible in terms of galaxy formation models with dissipation and mass loss, but other solutions such as variable yield are discussed. The properties of globular cluster systems and their "parent" galaxies are compared.

157.130 **CNO abundance variations in the ISM of galaxies.**
R. J. Dufour.
Publ. Astron. Soc. Pac., Vol. 96, No. 584, p. 787 (1984). Abstract. – See Abstr. 010.281.

157.131 **The stellar content of the nuclei of late–type spirals.**
J. A. Frogel.
Publ. Astron. Soc. Pac., Vol. 96, No. 584, p. 788 (1984). Abstract. – See Abstr. 010.281.

157.132 **Evolutionary status of luminous blue galaxies.**
J. S. Gallagher III, D. A. Hunter, H. Bushouse.
Publ. Astron. Soc. Pac., Vol. 96, No. 584, p. 788 (1984). Abstract. – See Abstr. 010.281.

157.133 **Wolf–Rayet stars in nearby galaxies: tracers of massive stars.**
P. Massey, T. Armandroff.
Publ. Astron. Soc. Pac., Vol. 96, No. 584, p. 791 (1984). Abstract. – See Abstr. 010.281.

157.134 **Color–magnitude diagrams of the dwarf spheroidal galaxies Andromeda I, II, and III.**
E. W. Olszewski, N. B. Suntzeff, R. Zinn, G. Da Costa.
Publ. Astron. Soc. Pac., Vol. 96, No. 584, p. 792 (1984). Abstract. – See Abstr. 010.281.

157.135 **Luminous late–type stars in NGC 205.**
H. Richer, D. R. Crabtree, C. Pritchet.
Publ. Astron. Soc. Pac., Vol. 96, No. 584, p. 794 (1984). Abstract. – See Abstr. 010.281.

157.136 **Metallicities of the galactic system dwarf spheroidal galaxies.**
N. B. Suntzeff, E. W. Olszewski, R. P. Kraft, E. Friel, M. Aaronson, K. Cook.
Publ. Astron. Soc. Pac., Vol. 96, No. 584, p. 795 (1984). Abstract. – See Abstr. 010.281.

157.137 **Gas dynamics of flow past galaxies: radiative cooling.**
T. J. Gaetz.
Bull. Am. Astron. Soc., Vol. 16, No. 3, Part I, p. 732 (1984). Abstract. – See Abstr. 010.065.

157.138 **The submillimetre and millimetre spectrum of NGC 5128.**
C. T. Cunningham, P. A. R. Ade, E. I. Robson, J. V. Radostitz.
Mon. Not. R. Astron. Soc., Vol. 211, No. 3, p. 543 – 547 (1984).

The authors' observations define the millimetre and submillimetre spectrum of the central 80 arcsec of NGC 5128 and show that in the submillimetre region the spectrum displays a radical departure from what, at longer wavelengths, appears to be non–thermal synchrotron emission. The observed submillimetre excess is probably the Rayleigh–Jeans tail of thermal emission from warm dust in the dark lane, which may be heated by an extended distribution of young stars and possibly partly by the nuclear source itself.

157.139 **The jets in NGC 1097.**
D. Carter, D. A. Allen, D. F. Malin.
Mon. Not. R. Astron. Soc., Vol. 211, No. 3, p. 707 – 712 (1984).

The authors have measured the surface brightness of the broad 'dog–leg jet' which appears to be associated with the bright barred spiral galaxy NGC 1097, in the *B*, *R*, *J* and *H* wavebands. The colours of the jet, together with published radio upper limits, appear to rule out both synchrotron and thermal Bremsstrahlung as the source of the emission. The authors suggest that the optical emission comes from stars near type G. The most promising explanation of this feature is that it is a tidal remnant; alternatively it might consist of stars formed from cooling plasma in a jet.

157.140 **Abundances in LINERs.**
L. Binette.
Messenger, No. 38, p. 13 – 16 (1984).

157.141 **A detailed study of peculiar galaxy pair Arp 205.**
L. Noreau, P. P. Kronberg.
J. R. Astron. Soc. Can., Vol. 78, No. 5, p. 204 (1984). Abstract. – See Abstr. 010.401.

157.142 **NGC 3109 – photometry of its bright stars.**
S. Demers, M. J. Irwin, W. E. Kunkel.
J. R. Astron. Soc. Can., Vol. 78, No. 5, p. 208 (1984). Abstract. – See Abstr. 010.401.

157.143 **Comparison of the ellipticity profiles of E and S0 galaxies.**
R. Michard.
Astron. Astrophys., Vol. 140, No. 2, p. L39 – L42 (1984).
The ellipticity profiles (run of the ellipticity ε of the isophotes against their semi-major axis a) for S0 and E galaxies have been intercompared using literature data and original observations. A striking similarity has been found between the profiles of E-objects flatter than E3 and those pertaining to S0 galaxies of similar apparent flattening. In both cases, and generally speaking, ε first increases with a, goes through a maximum, then decreases again. Relative to the effective isophote $a_e$, the maximum tends to occur farther out for S0 than for E objects.

157.144 **Radio continuum observations of the edge-on galaxies NGC 3628 and NGC 4244.**
R. Schlickeiser, W. Werner, R. Wielebinski.
Astron. Astrophys., Vol. 140, No. 2, p. 277 – 283 (1984).
The edge-on galaxies NGC 3628 and NGC 4244 were detected at 1.43, 2.68 and 4.75 GHz (λλ21, 11 and 6 cm) with the 100-m radio telescope, and sensitive maps of areas centred on the galaxies are presented. The observations are examined in light of the galactic wind theory. For NGC 3628 a spectral break is found indicating the presence of a galactic wind in this object. The detection of NGC 4244 is at a flux level which is in agreement with the previous negative results for this object.

157.145 **Ultraviolet luminosities and colors of spiral and irregular galaxies. The present-day star formation rate.**
J. Donas, J. M. Deharveng.
Astron. Astrophys., Vol. 140, No. 2, p. 325 – 333 (1984).
The OAO-2 observations by Code and Welch of 40 spiral and irregular galaxies in UV light are used to obtain the intrinsic 1910 Å flux and (1910-B) color of the stellar population of these galaxies. According to evolutionary models, the intrinsic UV fluxes are expressed in terms of recent star formation rates. These rates correlate with the total gas content but no significant dependence on the average gas surface density is found. The relatively small dispersion of the star formation rate from average trends is interpreted as a lack of significant variation in the initial mass function. The two-color (1910-B, *B-V*) diagram is presented and discussed in order to set limits on the importance of bursts in the star formation process.

157.146 ***UBVRI* photoelectric photometry of 191 southern galaxies.**
A. Lauberts.
Astron. Astrophys., Suppl. Ser., Vol. 58, No. 2, p. 249 – 260 (1984).
Multiaperture *UBVRI* photometry with the ESO 1-m reflector for 191 southern early type galaxies is presented. Out of this number, 114 galaxies had no previous photometry.

157.147 **Rotation and masses of some spiral galaxies: NGC 3726, NGC 3810, NGC 4559, NGC 6118.**
N. Meyssonnier.
Astron. Astrophys., Suppl. Ser., Vol. 58, No. 2, p. 351 – 356 (1984). In French.
The author has observed the rotational curves and derived the masses of four spiral galaxies. The emission line radial velocities were obtained principally from plates, covering the range from 6450 Å to 6800 Å, taken with the nebular spectrograph at the Cassegrain focus of the 193 cm telescope at the Observatoire de Haute-Provence. The dispersion was 32 Å $mm^{-1}$.

157.148 **A high resolution hydrogen line survey of Messier 31. II. The warped flaring hydrogen layer.**
E. Brinks, W. B. Burton.
Astron. Astrophys., Vol. 141, No. 1, p. 195 – 214 (1984).
Some of the global aspects of the H I distribution in M31 are discussed on the basis of the high-resolution survey of the 21-cm line made with the Westerbork Synthesis Radio Telescope by Brinks and Shane (1984). The morphology of the H I kinematics and structure in M31 can be modelled by a flat disk extending over the region at $R < 16$ kpc, and, at larger radii, a flaring warped gas distribution. The scale height of the gas in the warped part of M31 increases approximately linearly with radius to a maximum scale height of about 1.7 kpc. The maximum deviation of the warp from the plane of the extended inner disk is about 3.8 kpc. These and other parameters of the warp in M31 are similar to those in our Galaxy. The H I mass in the disk is about $2.4 \times 10^9 M_\odot$; the H I mass in the warp is about $1.5 \times 10^9 M_\odot$. As in our Galaxy, the flaring gas layer and the rather flat rotation curve in the regions of M31 require that the total mass there resides in a halo. The $2\pi$-symmetry and the general regularity of the warp in M31 argue against it being a consequence of interaction with the small elliptical companion M32.

157.149 **The velocity field of NGC 1365. I. The measurements and data reduction techniques.**
S. Jörsäter, C. J. Peterson, P. O. Lindblad, A. Boksenberg.
Astron. Astrophys., Suppl. Ser., Vol. 58, No. 3, p. 507 – 527 (1984).
The velocity field of the large barred spiral NGC 1365 has been mapped in considerable detail from optical emission lines and, in the case of the bar, also from stellar absorption lines. A part of the material has been obtained using a digital detector (IPCS) allowing estimates of velocity dispersions and line intensities to be made. In this paper the authors present the data and discuss the data reduction techniques employed.

157.150 **Radial velocities of galaxies: some considerations on the comparison between radio and optical data.**
G. Tanzella-Nitti.
Astron. Astrophys., Suppl. Ser., Vol. 58, No. 3, p. 549 – 557 (1984).
The H I-optical content of the Catalogue of Radial Velocities of Galaxies (Palumbo et al., 1983) and its up-dated extension have been investigated in order to study the cross-comparison of these two kinds of measurements over the whole range of measured velocities. The statistical behaviour of the radio-optical residuals have been surveyed for double galaxies, Seyfert galaxies, different morphological types and inclinations. The consistency among various sources of 21 cm data has been investigated and a self-adjusted average H I velocity available on each galaxy has been used to evaluate possible external errors of 25 optical references. The outcoming results have been compared with those obtained considering the H I standard proposed by Rood (1982).

157.151 **Study of a complete sample of galaxies. I. *UBV* aperture photometry.**
M. P. Véron-Cetty.
Astron. Astrophys., Suppl. Ser., Vol. 58, No. 3, p. 665 – 677 (1984).
The author has measured *UBV* photometry through a 15.54 arc second diaphragm of 249 galaxies (493 measurements) from a subsample of the Revised Shapley-Ames catalogue (Sandage and Tammann, 1981) and 111 other galaxies (166 measurements), mainly galaxies with starbursts or Seyfert nuclei.

157.152 **The distribution of neutral hydrogen in the Sombrero galaxy, NGC 4594.**
E. Bajaja, G. van der Burg, S. M. Faber, J. S. Gallagher, G. R. Knapp, W. W. Shane.
Astron. Astrophys., Vol. 141, No. 2, p. 309 – 317 (1984).

Fan beam observations with an E–W resolution of 11″.4 of NGC 4594 in the 21–cm line, made with the Westerbork Synthesis Radio Telescope, are reported. The results are presented in the form of a position–velocity contour map. A model is proposed which reproduces the gross properties of the H I distribution. The relative H I content is very much lower for this galaxy than for other Sa galaxies and is lower than for H I–rich ellipticals. The H I distribution suggests that only gas outside the optical bulge has been able to survive. Mechanisms causing the removal of gas from the spheroid of the galaxy are briefly discussed.

157.153 **The underlying exponential disc in the Im galaxy NGC 4214.**
R. Schild.
Astrophys. Lett., Vol. 24, No. 2, p. 85 – 89 (1984).

CCD images of the Magellanic Irregular Galaxy NGC 4214 have been combined to create a color photograph which shows that a disc of old population stars underlies the bright blue population which has previously been studied comprehensively. The underlying disc is offset from the distribution of bright blue stars and contains 10% of the total galaxy luminosity measured at B. The brightness profile of the old stellar population is exponential, like that in spiral galaxies. The transitional IAB(s)m galaxy NGC 4214 differs from the normal IB(s)m galaxies in a number of ways. No bar of old population stars appears to be present. The exponential disc is not displaced perpendicular to the center of the bar defined by young H II regions, and the large trailing spiral arm characteristic of the barred Magellanic irregulars is not present.

157.154 **Star formation in late type galaxies.**
F. Viallefond.
Spectral evolution of galaxies, p. 28 – 49 (1984). – See Abstr. 012.066.

Contents: Introduction. Large–scale rate of star formation in galaxies. Blue compact galaxies. H I – H II associations in M101. A dependence on the metallicity for the upper cut–off of the IMF? Extinction in H II regions.

157.155 **Chemical evolution of galaxies with H II regions.**
B. E. J. Pagel.
Spectral evolution of galaxies, p. 50 – 61 (1984). – See Abstr. 012.066.

The determination of abundances in H II regions of external galaxies is briefly described. The resulting radial abundance gradients in heavy elements typified by oxygen are related to local galactic parameters (disk surface density and gas fraction). It is suggested that they can be explained by a model in which inflow is neglected, but the yield is an increasing function of local surface density.

157.156 **Stellar synthesis models of galaxies.**
G. Bruzual A.
Spectral evolution of galaxies, p. 67 – 85 (1984). – See Abstr. 012.066.

The main purpose of building stellar synthesis models of galaxies is to derive the fractions of different stellar groups present in a stellar system at a given epoch. The paper summarizes the techniques used in building stellar synthesis models of galaxies, as well as some of the most relevant results derived from these models.

157.157 **Theoretical foundations of evolutionary population synthesis: a progress report.**
A. Renzini, A. Buzzoni.
Spectral evolution of galaxies, p. 86 – 94 (1984). – See Abstr. 012.066.

This paper summarizes some recent results in the attempts to describe the evolution of stellar populations in galaxies by models based on the theory of stellar evolution.

157.158 **An empirical study of star formation in galaxies.**
P. M. Gondhalekar.
Spectral evolution of galaxies, p. 95 – 116 (1984). – See Abstr. 012.066.

Comparison from ultraviolet to radio frequencies between stellar evolutionary models and galaxies suggest that global star formation in galaxies may be quasi–continuous although a single valued initial mass function may be unrealistic. The unknown reddening and extinction laws for spiral and irregular galaxies are major barriers to our understanding of star formation. The elliptical galaxies present more complex problems. Although the post–giant evolution has not been considered very vigorously in the models, it does seem likely that low level star formation is possible in elliptical galaxies.

157.159 **Infrared spectra of interacting galaxies.**
N. A. Robertson, R. D. Joseph, W. P. S. Meikle, G. S. Wright.
Spectral evolution of galaxies, p. 117 – 121 (1984). – See Abstr. 012.066.

Infrared observations of interacting galaxies provide strong evidence that tidal interactions are extremely efficient in inducing massive bursts of star formation in the nuclei of galaxies. It is therefore evident that interactions will play a major role in the spectral evolution of galaxies.

157.160 **Observational aspects of galaxy evolution.**
R. S. Ellis.
Spectral evolution of galaxies, p. 122 – 144 (1984). – See Abstr. 012.066.

This paper reviews recent progress in the two traditional tests for models of galaxy evolution. The first is to confront the models with spectroscopic data on nearby stellar populations (elliptical galaxies, globular clusters, etc). The second test is to study galaxies at high redshift making some (questionable) assumptions about the relationship these have with present day objects, e.g. members of distant clusters are normal elliptical galaxies.

157.161 **Colours as galaxy distance indicators.**
S. A. Butchins.
Spectral evolution of galaxies, p. 145 – 178 (1984). – See Abstr. 012.066.

This paper reports work on two–colour high–precision photographic photometry of galaxies by using plates obtained with the UK Schmidt telescope. The aim is to derive accurate redshifts and types for the observed distant galaxies.

157.162 **Unravelling the cosmic code.**
J. Silk.
Spectral evolution of galaxies, p. 205 – 210 (1984). – See Abstr. 012.066.

This note presents the author's view on the current state of theoretical and observational knowledge of galaxy evolution.

157.163 **UBV observations of double galaxies.**
A. N. Tomov.
Dokl. Bolg. Akad. Nauk, Vol. 37, No. 4, p. 415 – 418 (1984). Abstr. in Ref. Zh., 51. Astron., 11.51.728 (1984).

157.164 **Electrophotometry of isolated triple galaxies.**
A. N. Tomov, N. A. Tomov.
Dokl. Bolg. Akad. Nauk, Vol. 37, No. 4, p. 419 – 422 (1984). Abstr. in Ref. Zh., 51. Astron., 11.51.731 (1984).

157.165 **Warps, spiral structure and star bursts caused by perturbed flattened haloes in disc galaxies.**
S. D. Mathur.
Mon. Not. R. Astron. Soc., Vol. 211, No. 4, p. 901 – 910 (1984).

Rotation curve studies indicate that large amounts of dark matter may be present in disc galaxies. It is assumed that this dark matter is distributed in a flattened spheroidal form. Perturbations due to nearby galaxies, clusters, etc. would then cause changes in the orientation of the axis of symmetry of this dark halo. As a result, a torque is exerted by the halo on the orbiting elements in the disc. Different orbits precess at different rates, since the exerted torques and orbital angular frequencies vary. Consequently, a warp develops. It is shown that at the nodal points of the warp, the "crowding of orbits" leads to an increase in the pressure and density of the gas, which is significant even for small warps. This increase of pressure can lead to enhanced star formation. A mechanism is also proposed for the phenomenon of "star bursts" based on the density and pressure increases due to the warp.

157.166 **Stellar and gas kinematics in disk galaxies.**
B. C. Whitmore, V. C. Rubin, W. K. Ford Jr.
Space Telesc. Sci. Inst., Prepr. Ser., No. 19, 46 pp. (1984). To appear in Astrophys. J.

157.167 **Distribution and kinematics of CO in the spiral galaxy M51.**
G. Rydbeck, Å. Hjalmarson, O. E. H. Rydbeck.
Onsala Space Obs., Prepr., No. 84:48, 47 pp. (1984). To appear in Astron. Astrophys.

157.168 **The spiral arm structure of molecular gas in M51.**
G. Rydbeck, Å. Hjalmarson, L. E. B. Johansson, O. E. H. Rydbeck.
Onsala Space Obs., Prepr., No. 84:58, 4 pp. (1984).
The authors present the latest observational results from a continuing study of the distribution and dynamics of CO in the spiral galaxy M51, with the Onsala 20 m antenna.

157.169 **The stellar content of the A496 cD galaxy.**
E. A. Valentijn, A. F. M. Moorwood.
ESO Sci. Prepr., No. 334, 28 pp. (1984). Submitted to Astron. Astrophys.

157.170 **Star formation in early-type galaxies.**
P. Véron, M.-P. Véron-Cetty.
ESO Sci. Prepr., No. 340, 17 pp. (1984). Submitted to Astron. Astrophys.

157.171 **New carbon stars in spheroidal galaxies: I. Sculptor, Carina, Leo I and Leo II systems.**
M. Azzopardi, J. Lequeux, B. E. Westerlund.
ESO Sci. Prepr., No. 345, 19 pp. (1984). To appear in Astron. Astrophys.

157.172 **The flattening distribution of lenticular galaxies.**
R. E. De Souza, G. Vettolani, G. Chincarini.
ESO Sci. Prepr., No. 346, 24 pp. (1984). To appear in Astron. Astrophys.

157.173 **Photoionisation models for Liners: gas distribution and abundances.**
L. Binette.
ESO Sci. Prepr., No. 350, 44 pp. (1984). Submitted to Astron. Astrophys.

157.174 **Dust in elliptical galaxies - how often, how much?**
E. M. Sadler, O. E. Gerhard.
ESO Sci. Prepr., No. 354, 4 pp. (1984). To be published in the Proceedings of the specialized colloquium on New Aspects of Galaxy Photometry, Toulouse 1984.

157.175 **Massive neutrino halos and the rotation curves of galaxies.**
J.-L. Basdevant.
Inst. Astrophys. Paris, Pré-Publ., No. 63, 18+A5+7 pp. (1984). To appear in Astron. Astrophys.

157.176 **Far-UV excess and gas content of S0 galaxies.**
B. Rocca-Volmerange, C. Balkowski.
Inst. Astrophys. Paris, Pré-Publ., No. 68, 5 pp. (1984). To appear in Proceedings of the Fourth European IUE Conference, Rome 1984.

157.177 **Galactic chemical evolution and nucleocosmochronology: standard model with terminated infall.**
D. D. Clayton.
Astrophys. J., Vol. 285, No. 2, p. 411 - 425 (1984).
The author presents techniques for generating exact solutions to the differential equations for the chemical evolution of galactic gas $M_G(t)$, the star mass $S(t)$, the metallicity $Z(t)$, and the abundances of the radioactive chronometers $Z_\lambda(t)$ in the face of galactic infall $f(t)$. The author assumes instantaneous recycling and a linear dependence of star formation rate upon gas mass. Because some of these solutions are simple analytic functions, they can serve as standard models for the usual tests of galactic evolution. This paper emphasizes cases where infall $f(t)$ terminates at some time $t_0$ and is zero thereafter.

157.178 **Face-on galaxies.**
B. M. Lewis.
Astrophys. J., Vol. 285, No. 2, p. 453 - 457 (1984).
An H I survey of $\sim$200 face-on galaxies has tripled the number of galaxies known to have half-power widths $<36$ km s$^{-1}$. The H I profiles of these objects are consistent with integral line-of-sight velocity dispersions $\sigma \approx 8$ km s$^{-1}$, a value that is equally characteristic of all spiral morphological types. When combined with other data, the velocity dispersion of the H I cloud ensemble appears to be isotropic and independent of position, absolute magnitude, and morphological type.

157.179 **The age(s?) of the Sculptor dwarf galaxy.**
G. S. Da Costa.
Astrophys. J., Vol. 285, No. 2, p. 483 - 494 (1984). With plates 5 - 6.
A color-magnitude diagram reaching to the main-sequence turnoff in the Sculptor dwarf spheroidal galaxy is presented. The *C-M* diagram shows a small population of "blue stragglers" which if interpreted as normal main-sequence stars have ages as young as 5 billion yr. Comparison with theoretical isochrones indicates that the age of the bulk of the stellar population in the field studied may be 2 - 3 billion yr younger than that derived from similar isochrone fits to *C-M* diagrams of galactic globular clusters. This age difference could explain the predominantly red horizontal branch of the Sculptor system.

157.180 **Extended gaseous emission in normal elliptical galaxies.**
M.-H. Demoulin-Ulrich, H. R. Butcher, A. Boksenberg.
Astrophys. J., Vol. 285, No. 2, p. 527 - 546 (1984).
The authors report observations of the spatial distribution, the velocity field, and the line intensities of the extended gaseous emission in a small sample of nearby normal elliptical galaxies. Ionized gas regions extending 1 - 3 kpc in diameter are found in six out of 12 ellipticals selected for having gas in their nuclei. The velocity field of the gas is in every case consistent with the gas being in a disk in rotation around the nucleus, but turbulent motions are dominant in the center. In NGC 4278, the best studied case, it is found that the position angle of the rotation axis of the ionized gas differs from that of the neutral gas and of the stellar population. Possible origins of the ionized gas and the total energy requirements for its ionization are discussed.

157.181 **UBV observations of four isolated galaxies.**
V. G. Metlov.
Astron. Tsirk., No. 1303, p. 1 (1983). In Russian.

157.182 **Infrared photometry and the comparative stellar content of dwarf spheroidals in the galactic halo.**
M. Aaronson, J. Mould.
Prepr. Steward Obs., No. 544, 83 pp. (1984).
The authors report infrared JHK photometry for carbon stars in the Ursa Minor, Draco, Leo I, and Leo II dwarf spheroidal galaxies, and for a number of oxygen-rich giants in these systems as well. In combination with previously published data, the results enable a detailed comparison to be made of the red stellar content in all seven nearby spheroidals.

157.183 **Were ellipticals formed from spirals?**
S. Veeraraghavan, S. D. M. White.
Prepr. Steward Obs., No. 558, 17 pp. (1984).

The authors investigate the possibility that all elliptical galaxies could have formed from the merger of spiral galaxies similar to those currently observed. They use a simple semi-empirical model based on energy and mass conservation and on the observed scaling relations between velocities and luminosities. The results confirm previous arguments against the merger hypothesis based on the relative binding energies of spirals and ellipticals, but the authors find that the discrepancy is not large. Nevertheless it seems unlikely that most ellipticals were formed in this way.

157.184 **Structure of ring-like galaxies from a multi-colour photometry.**
V. L. Afanas'ev, I. P. Kostyuk.
Astron. Tsirk., No. 1316, p. 1 – 3 (1984). In Russian.

157.185 **Photo-red magnitudes of interacting galaxies.**
V. A. Dostal', G. M. Lyapunov, V. G. Metlov.
Astron. Tsirk., No. 1324, p. 6 – 8 (1984). In Russian.

157.186 **Non-physical pair of galaxies VV320.**
V. P. Arkhipova, R. I. Noskova.
Astron. Tsirk., No. 1329, p. 1 – 2 (1984). In Russian.

157.187 **OB stars in the M33 galaxy.**
P. Kynchev, N. Nikolov.
Astron. Tsirk., No. 1330, p. 1 – 2 (1984). In Russian.

157.188 **Rotation and surface brightness in elliptical galaxies.**
R. F. G. Wyse, B. J. T. Jones.
Astrophys. J., Vol. 286, No. 1, p. 88 – 96 (1984).

The authors present and discuss correlations between the mean surface brightness and level of rotational support, for the sample of elliptical galaxies presented by Davies et al. (1983). The importance of these distance-independent correlations lies in the possibility to isolate the effects of projection, and hence to define an intrinsic, edge-on correlation. Thus the position of an elliptical galaxy in this plane contains information about its true shape. The authors interpret the correlation as a manifestation of varying amounts of dissipation during the formation epoch of a galaxy, with higher surface brightness, rotating ellipticals having dissipated more binding energy.

157.189 **Studying the internal kinematics of galaxies using the calcium infrared triplet.**
A. Dressler.
Astrophys. J., Vol. 286, No. 1, p. 97 – 105 (1984).

The Ca II IR triplet lines 8498, 8542, and 8662 Å are strong in a wide variety of stellar types, making this an ideal spectral region for studying the internal kinematics of neighboring galaxies. With the calcium triplet it is possible to measure systemic velocity, velocity dispersion, and line strength even in galaxies whose light is dominated by young stars, hot gas, or nonthermal emission. Results are given for some sample galaxies, including radial profiles of $v$, $\sigma$, and $\gamma$ for M31, M32, and NGC 1068. Both M31 and M32 show a cusplike rise in velocity dispersion of 30% – 50% in the central few parsecs. These data suggest a rising $M/L$ in each nucleus perhaps due to a black hole of order $10^{6-7}$ solar masses.

157.190 **The velocity dispersion in the disk of the S0 galaxy NGC 1553.**
J. Kormendy.
Astrophys. J., Vol. 286, No. 1, p. 116 – 131 (1984). With plate 1.

Stellar rotation velocities and velocity dispersions have been measured along the major, minor, and mean axes of the normal SA(lens)0 galaxy NGC 1553. These are used to study the kinematics and dynamics of the disk as a function of radius. The lens is very hot. A significant fraction of the orbits in the inner part of the lens may be retrograde. The large excursions of stars dilute any radial gradients in disk properties, and phase-mix away any small-scale structure. The velocity dispersion appears to be large enough for stability against spiral and other structure in the absence of gas. If NGC 1553 is representative, S0 galaxies may lack spiral structure partly because they are too hot.

157.191 **The velocity dispersion in the disk of the SB0 galaxy NGC 936.**
J. Kormendy.
Astrophys. J., Vol. 286, No. 1, p. 132 – 143 (1984).

The effects of the velocity dispersion on the structure and stability of the disk of the SB0 galaxy NGC 936 are studied as a function of radius. The bar provides NGC 936 with an engine with which it can heat the disk. The results are therefore compared with those for three unbarred or weakly barred galaxies studied to date, i.e., the Galaxy (SABbc), NGC 488 (SAb), and NGC 1553 (SA0).

157.192 **Photometry of resolved galaxies. IV. Holmberg I and Holmberg II.**
J. G. Hoessel, G. E. Danielson.
Astrophys. J., Vol. 286, No. 1, p. 159 – 170 (1984). With plates 2 – 13.

The authors present colors and magnitudes for 279 resolved stars in the Holmberg I dwarf galaxy and 468 resolved stars in Holmberg II. Both systems are Magellanic type dwarf members of the M81-NGC 2403 Group, which lies at approximately 3 Mpc from the Local Group. The photometry was done in the *GRI* passbands using CCD detectors. Color-magnitude diagrams and luminosity functions are constructed; these are compared with results for several Local Group galaxies and with theoretical work.

157.193 **Nuclear H II regions in galaxies with emission lines.**
G. T. Petrov, I. M. Yankulova, V. K. Golev.
Dokl. Bolg. Akad. Nauk, Vol. 37, No. 4, p. 411 – 414 (1984). Abstr. in Ref. Zh., 51. Astron., 12.51.1029 (1984).

157.194 **Zur Struktur der irregulären Zwerggalaxien.**
M. Hoffmann.
Mitt. Astron. Ges., Nr. 62, p. 333 (1984). Abstract. – See Abstr. 010.341.

157.195 **Kugelsternhaufen in nahen irregulären Galaxien.**
E. H. Geyer, M. Hoffmann.
Mitt. Astron. Ges., Nr. 62, p. 334 – 335 (1984). – See Abstr. 010.341.

157.196 **Die räumliche Struktur von Zwerggalaxien anhand eines neu erstellten Objektkataloges.**
J. V. Feitzinger, T. Galinski.
Mitt. Astron. Ges., Nr. 62, p. 336 (1984). Abstract. – See Abstr. 010.341.

157.197 **Infrared observations of dwarf elliptical galaxies in the Virgo cluster.**
H. Zinnecker, R. D. Cannon, T. G. Hawarden, H. T. MacGillivray.
Mitt. Astron. Ges., Nr. 62, p. 337 (1984). Abstract. – See Abstr. 010.341.

157.198 **Kinematik von prolaten Staubstreifengalaxien.**
C. Möllenhoff, G. Marenbach.
Mitt. Astron. Ges., Nr. 62, p. 340 (1984). Abstract. – See Abstr. 010.341.

157.199 **The photometric peculiarities of E and S0 galaxies in double systems.**
V. V. Demin.
Astron. Zh., Tom 61, Vyp. 6, p. 1063 – 1073 (1984). In Russian. English translation in Sov. Astron., Vol. 28, No. 6.

The properties of EE and ES double systems have been studied from the data of photoelectric observations made by Tomov. The difference between the color – absolute magnitude diagrams for early-type galaxies in pairs and groups has been demonstrated. The evolution of galaxies in pairs differs from that of group and cluster members. This conclusion follows also from a comparison of integral photometric properties of E and S0 galaxies in EE and ES pairs.

157.200 **Near-infrared observations of NGC 253.**
K. Uyama, T. Matsumoto, J. A. Thomas.
Publ. Astron. Soc. Jpn., Vol. 36, No. 3, p. 477 – 489 (1984).
Isophotes of the surface brightness of NGC 253 were obtained at 0.7 – 0.9 μm, 1.25 μm, and 2.2 μm by means of the photographic and photoelectric methods. The observed color, [0.8 μm] – [2.2 μm], of the disk ranges from 2.0 to 2.8 mag, corresponding to the colors of M0 III and M4 III stars, respectively. The significant difference of colors between the near and far sides is indicative of the local distribution of dust and stars associated with the galactic arm. Dust-to-gas ratios are also obtained in the disk and bulge of NGC 253 and found to be a little smaller than that in the disk of the Galaxy.

157.201 **En tung halo omkring NGC 4378.**
J. Sørensen.
Astron. Tidsskr., Årg. 17, Nr. 4, p. 160 – 165 (1984).

157.202 **Three-dimensional mass distribution in discoid galaxies with both bulge and disk.**
Z. Tang.
Acta Astrophys. Sin., Vol. 4, No. 1, p. 43 – 48 (1984). In Chinese.
According to the distribution of surface density and the rotation curve in discoid galaxies with bulge, the steady hydrodynamic equations doubled with the Poission equation are numerically solved, from which the three-dimensional mass distribution and velocity distribution in discoid galaxies are obtained. The results show that the central spheroid of the discoid galaxy is a high density region. The stars and clusters in this region have the same dynamic characteristics as in the spheroidal galaxies. These results are consistent with observations.

157.203 **The spatial distribution of galaxies in the southern sky.**
A. P. Fairall, H. Winkler.
Clusters and groups of galaxies, p. 23 – 28 (1984). – See Abstr. 012.088.

157.204 **A redshift survey in the Lynx Gemini region.**
P. Focardi, B. Marano, G. Vettolani.
Clusters and groups of galaxies, p. 55 – 61 (1984). – See Abstr. 012.088.
The study of the redshift distribution of a complete sample of galaxies brighter than 14.5 mph has been performed over an area encompassing about 1800 square degrees in Lynx and Gemini. The main result is the discovery of a new filament of galaxies in Gemini, at a radial velocity of 4800 km/s, mainly composed of spirals. The possible connection of a cloud of galaxies surrounding the cluster A569, the new filament in Gemini and the Lynx-Ursa Major supercluster with the Perseus supercluster is briefly discussed.

157.205 **The geography of nearby space.**
P. Focardi, B. Marano, G. Vettolani.
Clusters and groups of galaxies, p. 69 – 70 (1984). – See Abstr. 012.088.
A series of maps showing the distribution of galaxies in different radial velocity ranges have been prepared.

157.206 **Galaxy redshifts from the U.K. Schmidt objective prisms (medium dispersion).**
J. B. Palmer.
Clusters and groups of galaxies, p. 71 – 72 (1984). – See Abstr. 012.088.
Two objective prisms with dispersions of 2480 Å/mm and 840 Å/mm (at Hγ) are available on the U.K. Schmidt telescope. Combinations give dispersions of 1240 Å/mm and 620 Å/mm at Hγ. This note discusses the accuracy of preliminary galaxy redshift surveys using the two intermediate dispersions.

157.207 **The spatial and peculiar velocity distribution of galaxies within 80 Mpc.**
L. Staveley-Smith, R. D. Davies.
Clusters and groups of galaxies, p. 75 – 76 (1984). – See Abstr. 012.088.

157.208 **Les régions externes des galaxies ne sont pas orientées au hasard.**
H. H. Fliche, J. M. Souriau.
Clusters and groups of galaxies, p. 77 – 78 (1984). – See Abstr. 012.088.

157.209 **Redshift estimation by colors.**
D. C. Koo, R. G. Kron, D. Nanni, D. Trevese, A. Vignato.
Clusters and groups of galaxies, p. 159 – 162 (1984). – See Abstr. 012.088.
Redshift estimation by colors could prove to be valuable for cosmological investigations, since colors for faint galaxies can be produced in large quantities by automatic techniques. In this contribution the authors explore this idea in practice by analysis of the colors of galaxies in the field of a rich, compact cluster at known redshift.

157.210 **Comparison of H I sizes of Virgo Cluster and field galaxies.**
R. H. Warmels, H. van Woerden.
Clusters and groups of galaxies, p. 251 – 259 (1984). – See Abstr. 012.088.
With the Westerbork Synthesis Radio Telescope the authors have measured the H I distributions of 35 galaxies in the region of the Virgo Cluster. For 19 Sab to Sd-type galaxies the authors present the ratio of the face-on corrected H I diameter $D_{HI}$ to the face-on corrected optical diameter $D_o$. Comparison with a sample of field galaxies suggests that the relative H I diameter of Virgo Cluster galaxies is smaller by about 20%. Within the present Virgo data set, the authors find no correlation between $D_{HI}/D_o$ and the projected distance to M87.

157.211 **The formation of H I rings around S0 galaxies.**
P. N. Appleton.
Clusters and groups of galaxies, p. 273 – 274 (1984). – See Abstr. 012.088.

157.212 **Investigations on the H I-content of field galaxies, Virgo cluster galaxies and UMa-I group galaxies.**
H.-D. Bohnenstengel.
Clusters and groups of galaxies, p. 275 – 276 (1984). – See Abstr. 012.088.

157.213 **Hot gaseous coronae around early-type galaxies.**
W. Forman, C. Jones, W. Tucker.
Clusters and groups of galaxies, p. 297 – 305 (1984). – See Abstr. 012.088.
The authors discuss the nature of the extended X-ray emission around three elliptical galaxies in the Virgo cluster. Based on energy spectra obtained with the Einstein Observatory, the authors argue that this emission is hot (~1 keV) gas. Interactions of these coronae with their environment of hot intracluster gas allows their detection at X-ray wavelengths.

157.214 **Diffuse Lyα emission around the cD galaxy in A1795.**
H. U. Nørgaard-Nielsen, H. E. Jørgensen, L. Hansen.
Clusters and groups of galaxies, p. 347 – 349 (1984). – See Abstr. 012.088.
The authors report briefly the discovery of diffuse Lyα emission coinciding with the optical filaments around the cD galaxy in the X-ray cluster A1795. The derived Lyα/Hβ ratio is close to the recombination value. From this ratio the authors find that the content of smoothly distributed H I is less than 100 solar masses. The non-detection of a stellar UV continuum implies that the formation rate of stars more massive than $3M_{\odot}$ is less than 1 per cent of the accretion rate. Combining UV, optical and radio data for the nonthermal nuclear source the authors find $\alpha \sim .8$.

157.215 **The distribution of faint galaxies detected on deep UKST plates.**
H. T. MacGillivray, R. J. Dodd.
Clusters and groups of galaxies, p. 589 – 594 (1984). – See Abstr. 012.088.

A programme is underway to investigate the large–scale projected distribution of faint galaxies from objective, machine–determined samples. Data are obtained from measurements made with COSMOS on deep photographs taken with the UK 1.2 m Schmidt Telescope. The ultimate intention is to obtain a complete Lick–type galaxy catalogue from machine data for a large area of the southern sky. In this paper, the authors describe the methods and data reduction techniques used, and present some preliminary results.

157.216 **An objective–prism galaxy redshift survey at the south galactic pole.**
Q. A. Parker, H. T. MacGillivray, R. J. Dodd.
Clusters and groups of galaxies, p. 595 – 600 (1984). – See Abstr. 012.088.

A project has been initiated to obtain objective–prism redshifts for large numbers of faint galaxies in several fields around the South Galactic Pole. Measurements are made with the COSMOS machine on deep photographs taken with the UK 1.2 m Schmidt Telescope. Preliminary results from one particular field of the southern sky are presented together with an assessment of the usefulness of the redshifts obtained for this type of analysis.

157.217 **Companions to S0 galaxies.**
A. Busch, B. F. Madore, E. R. Anderson.
Clusters and groups of galaxies, p. 601 – 602 (1984). – See Abstr. 012.088.

The authors present statistical results on the distribution of dwarf companions to S0 galaxies. They find that the slope of the galaxy covariance function $\gamma = 1.60 \pm 0.10$, consistent with the value $\gamma = 1.8$ obtained at larger separations.

157.218 **Strömgren photometry of galaxies.**
K. D. Rakos, N. Fiala.
Clusters and groups of galaxies, p. 603 – 604 (1984). – See Abstr. 012.088.

In preparation for a program of observations of different types of galaxies in distant clusters, the authors have calculated synthetic colors from the spectrophotometry of radio galaxies published by Yee and Oke (1978). The monochromatic magnitudes were calculated for the Strömgren v, b and y bandpasses shifted into the rest frame of each galaxy. The colors calculated were (b–y) and (v–b)–(b–y) = $m_1$.

157.219 **Dynamical environment and reliability of redshift measurements in statistical studies of binary galaxies.**
G. Tanzella–Nitti.
Clusters and groups of galaxies, p. 605 – 606 (1984). – See Abstr. 012.088.

157.220 **The Second Byurakan Sky Survey. Spectra of galaxies. II.**
B. E. Markarian, V. A. Lipovetskij, D. A. Stepanyan.
Astrofizika, Tom 21, Vyp. 1, p. 35 – 42 (1984). In Russian. English translation in Astrophysics, Vol. 21, No. 1.

The results of spectroscopic observations of 53 galaxies from the third list of the Second Byurakan Spectral Sky Survey centered at $\alpha = 08^h00^m$, $\delta = +59°00'$, are presented. The presence of emission lines in red and sometimes in green parts of the spectra is established for all of the investigated galaxies. The redshifts and luminosities of all galaxies have been determined.

157.221 **Investigation of morphological properties of some interacting galaxies. II.**
A. A. Korovyakovskaya.
Astrofizika, Tom 21, Vyp. 1, p. 43 – 48 (1984). In Russian. English translation in Astrophysics, Vol. 21, No. 1.

The morphology of the six groups of interacting galaxies earlier classified as chain galaxies is investigated from direct photographs by means of the digital filtration method. It is shown that only three among the six systems may be considered as chain galaxies.

157.222 **Isodensitometry of selected interacting galaxies. I.**
Yu. P. Korovyakovskij.
Astrofizika, Tom 21, Vyp. 1, p. 49 – 56 (1984). In Russian. English translation in Astrophysics, Vol. 21, No. 1.

The results of digital processing of large–scale photographs of interacting galaxies are presented. The processing results of the investigated objects are presented as two–dimensional maps, which contain quantitative information on the brightness distribution in the objects.

157.223 **Spectrophotometry of two emission regions in the irregular galaxy NGC 3448.**
N. K. Andreasyan.
Astrofizika, Tom 21, Vyp. 1, p. 73 – 80 (1984). In Russian. English translation in Astrophysics, Vol. 21, No. 1.

The isodensities of the galaxy NGC 3448 and results of spectrophotometry of two regions in it are presented. The physical parameters of ionized gas which appear to resemble those in H II regions and chemical abundances were estimated.

157.224 **Unusual properties of bar–within–ring galaxies and the projection effect model for barred spiral galaxies.**
R. L. Oldershaw.
Astrophys. Space Sci., Vol. 107, No. 2, p. 403 – 408 (1984).

Five anomalous properties of bar–within–ring galaxies, which comprise the majority of barred spiral galaxies, are reported. These properties appear to be in conflict with conventional ideas about galactic morphology, but are in keeping with the projection effect model for barred spiral galaxies.

157.225 **Isodensitometry of selected interacting galaxies. II.**
Yu. P. Korovyakovskij.
Astrofizika, Tom 21, Vyp. 2, p. 255 – 261 (1984). In Russian. English translation in Astrophysics, Vol. 21, No. 2.

The morphological investigation of large–scale photographs of selected interacting galaxies is continued. High–frequency photoemulsion noises have been decreased and the angular resolution of the photographs is improved with the help of the digital filtration method.

157.226 **Are cells visible in a deep sample of galaxies?**
B. I. Fesenko.
Astrofizika, Tom 21, Vyp. 2, p. 269 – 274 (1984). In Russian. English translation in Astrophysics, Vol. 21, No. 2.

The significance of apparent distribution analysis of galaxies for the investigation of the large–scale structure of the Universe is found and an analytical method for the calculation of the fluctuations of galaxy numbers in the case of cells is derived. The application of this method to a deep sample of galaxies in the case of cells was not confirmed. The observational data are easily explainable by small variations in the value of interstellar extinction.

157.227 **Double galaxy redshifts and dynamical analyses.**
N. A. Sharp.
Astrophys. J., Vol. 286, No. 2, p. 437 – 441 (1984).

Several samples of binary galaxy data are processed by the dynamical analysis method of White et al. Particular attention is paid to the distributions of velocity difference in order to assess the periodicity found by Tifft. While the high apparent significance of the effect is confirmed, the samples do not seem otherwise incompatible with conventional dynamics.

157.228 **High–resolution observations of the H I content of the barred spiral galaxies NGC 3992 and NGC 4731.**
S. T. Gottesman, R. Ball, J. H. Hunter Jr., J. M. Huntley.
Astrophys. J., Vol. 286, No. 2, p. 471 – 490 (1984). = Contrib. Dep. Astron., Univ. Florida, No. 67. With plates 14 – 17.

The atomic hydrogen properties of the barred spiral galaxies NGC 3992 and NGC 4731 have been observed at $<30''$ resolution with the VLA telescope. The kinematics and gas density have been mapped with a velocity resolution of $\sim 25$ km s$^{-1}$. The H I properties of NGC 3992 appear to change at the edge of the optical disk. This behavior may be the density and kinematic signatures of a truncated disk. Substantial noncircular phenomena are observed in the galaxy NGC 4731. These appear to be associated with the spiral structure of the galaxy. Four satellite systems were also found, three of which are associated with NGC 3992.

157.229 **The complex X-ray emission of NGC 253.**
G. Fabbiano, G. Trinchieri.
Astrophys. J., Vol. 286, No. 2, p. 491 – 497 (1984).
High-resolution X-ray observations of the starburst galaxy NGC 253 with the Einstein Observatory show that there are at least four components to its X-ray emission. Eight bright point-like sources were detected. Diffuse emission is visible in the inner disk of NGC 253. A more intense emission region is associated with the nuclear region and with an extension along the southern portion of the minor axis. This last component, in particular, suggests the presence of shock-heated gas associated with outflow from the starburst nucleus.

157.230 **Compact-core-dominated radio emission from bright E/S0 galaxies.**
J. M. Wrobel, D. S. Heeschen.
Astrophys. J., Vol. 287, No. 1, p. 41 – 65 (1984).
The continuum emission from 11 bright E/S0 galaxies that contain compact radio cores was surveyed with the VLA at frequencies of 1.5, 4.9, and 15 GHz. Information on the total intensity and, in some cases, the linearly polarized intensity was obtained on angular scales between $\sim 7'$ and $\sim 0''.1$. The compact radio cores investigated here each possess flat or inverted radio spectra at frequencies $\gtrsim 5$ GHz, and most exhibit time-variable flux densities and display structure on milliarcsec scales. The linearly polarized emission detected from four galaxies was found to be significantly displaced from the compact-core location. Extended emission on arcmin scales was found in four (possibly five) cases, and steep-spectrum components are associated with three galaxies.

157.231 **Stellar and gas kinematics in disk galaxies.**
B. C. Whitmore, V. C. Rubin, W. K. Ford Jr.
Astrophys. J., Vol. 287, No. 1, p. 66 – 79 (1984).
Extended absorption- and emission-line rotation curves and stellar velocity dispersion curves have been obtained for NGC 3898 (Sab), NGC 2841 (Sb), NGC 4062 (Sc), and NGC 2998 (Sc), along with a few observations of NGC 3200 (Sab) and M82 (I). In the disk, the rotation of the gas and stellar components are similar. In the bulge, stellar rotational velocities are only about half of the rotation of the disk. The authors have performed numerical experiments to determine values of the intrinsic rotation of the bulge in regions where significant fractions of the light come from both the bulge and the disk.

157.232 **Infrared observations of interacting/merging galaxies.**
C. J. Lonsdale, S. E. Persson, K. Matthews.
Astrophys. J., Vol. 287, No. 1, p. 95 – 107 (1984).
Observations from 1 to 10 μm are presented of a sample of 20 galaxy systems selected on the basis of morphological evidence of an ongoing tidal interaction, or merger, of two galaxies. The 11 systems detected at 10 μm have on average a significantly higher infrared luminosity than noninteracting galaxies; the enhanced infrared radiation is due to star formation bursts. Recent results from the *IRAS* on the far-infrared emission of a sample of galaxies are analyzed. It is found that the massive star formation rates in the interacting and merging galaxies, as a class, are $\sim 3$ times higher than those of noninteracting systems.

157.233 **On the structure of the nucleus of M31.**
J.-L. Nieto.
Astrophys. J., Vol. 287, No. 1, p. 108 – 111 (1984).
This paper discusses the nature of the nucleus of M31 and more particularly the possible evidence for a strong triaxiality-generated anisotropy on the basis of the insights that the projected geometry and rotation velocity can bring. First, it considers the nucleus of M31 as an elliptical stellar system for which the tensor virial theorem can be used. It then discusses on the basis of the projected geometry of the nucleus (and with the assumption of one axis perpendicular to the disk), the barlike structure of its three-dimensional shape, also suggesting a high degree of anisotropy.

157.234 **Structural properties of giant H II regions in nearby galaxies.**
R. C. Kennicutt Jr.
Astrophys. J., Vol. 287, No. 1, p. 116 – 130 (1984).
Balmer emission-line and radio continuum maps of nearby giant extragalactic and galactic H II regions have been used to define their basic physical properties. Even within the relatively narrow range of galaxy types studied here, the luminosities, masses, and densities of the largest H II regions range over several orders of magnitude. By combining maps of the extragalactic H II regions with radio continuum maps of the largest galactic objects, it has been possible to compare these objects on comparable spatial resolutions. While analogs to the largest known galactic objects have been found in nearby galaxies, the apparent absence of very bright 30 Doradus-type H II regions in the Galaxy is almost certainly a real effect.

157.235 **Spatially resolved high-velocity outflow from the nucleus of M51.**
G. Cecil, J. A. Rose.
Astrophys. J., Vol. 287, No. 1, p. 131 – 137 (1984).
Long-slit optical spectra were obtained at 3.5 Å resolution with a charge-coupled device (CCD) along four position angles through the nucleus of the spiral galaxy M51. Forbidden-line profiles reveal that a significant fraction of the ionized gas in the spatially resolved region within a 6″ radius of the center cannot be bound to the nucleus but is expanding outward with velocities up to 700 km $s^{-1}$.

157.236 **Luminous late-type stars in NGC 205.**
H. B. Richer, D. R. Crabtree, C. J. Pritchet.
Astrophys. J., Vol. 287, No. 1, p. 138 – 147 (1984).
Broad-band *VRI* and narrow-band CCD frames were obtained at the CFH telescope of the central region of NGC 205, the dwarf elliptical companion of the Andromeda galaxy. The narrow-band images isolate TiO and CN, and colors formed from the photometry of these images allow carbon and M stars to be unambiguously separated. Seven carbon stars were found together with 21 late-M stars with spectral types as cool as $\sim$M 5. Calibrating against the Magellanic Cloud carbon stars, a true distance modulus to NGC 205 of 24.5 is derived. The presence of two Cepheids is also noted.

157.237 **On the existence of a dark halo around the high-velocity compact object near NGC 1199.**
M. F. Walker, J. Andersen.
Astrophys. J., Vol. 287, No. 1, p. 148 – 152 (1984). = Lick Obs. Bull., No. 985. With plates 1 – 2.
Analysis of a blue light electrograph of the field of NGC 1199, taken with the 9 cm McMullan camera on the 1.5 m Danish reflector at La Silla, is presented. That analysis indicates that the dark halo, reported by Arp to surround the high-velocity compact object SW of NGC 1199, does not exist down to an intensity level of $\approx 1\%$ of the background light from NGC 1199 plus the night sky. In consequence, there is no reason to suppose that the compact object is not located beyond NGC 1199 at a distance given by its redshift.

157.238 **Limits on diffuse X-ray emission from M101.**
D. McCammon, W. T. Sanders.
Astrophys. J., Vol. 287, No. 1, p. 167 – 174 (1984).
Observed limits on diffuse X-ray emission from M101 require that the temperature of any coronal or matrix hot gas that is radiating an appreciable part ($\gtrsim 10\%$) of the average supernova power be less than $10^{5.7}$K. Furthermore, the fraction of the galactic plane occupied by hot bubbles similar to the one that apparently surrounds the Sun is at most 25% in the region between 10 kpc and 20 kpc from the galactic center.

157.239 **Do giant H II regions preferentially form in the outer disks of galaxies?**
R. C. Kennicutt Jr., P. W. Hodge.
Publ. Astron. Soc. Pac., Vol. 96, No. 586, p. 944 – 946 (1984).
The authors have used recently published data on the H II region distributions in 65 galaxies to test the commonly quoted observation that the largest star–forming regions in galaxies preferentially occur in the outer disk. Although a few nearby and well–studied galaxies do exhibit this effect, simple statistical tests indicate that in general there is no significant difference between the radial locations of the large star–forming regions and the general disk population. It suggests that radially varying parameters such as metal abundance and dust content may not be the key parameters influencing the number and size of star–forming regions in a galaxy.

157.240 **IRAS observations of NGC 1052.**
G. Neugebauer, B. T. Soifer, W. Rice, M. Rowan–Robinson.
Publ. Astron. Soc. Pac., Vol. 96, No. 586, p. 973 – 976 (1984).
The galaxy NGC 1052 has been observed with the IRAS satellite. The infrared emission at 100 μm is substantially larger than a smooth extrapolation of the radio data. Because of the large diaphragm size of IRAS, it is impossible to decide uniquely if the infrared radiation represents a self–absorbed nonthermal reradiation by heated dust.

157.241 **How to find galaxies at high redshift.**
L. W. Hartmann, J. P. Huchra, M. J. Geller.
Astrophys. J., Vol. 287, No. 2, p. 487 – 491 (1984).
The authors present ultraviolet and optical spectra of the blue star–forming galaxies Mrk 357, Mrk 496, and Mrk 26, which have sufficiently large redshifts to make observation of Lyα possible. Weak Lyα emission is observed in Mrk 357 and Mrk 496 with a flux approximately equal to Hβ. There is no detectable Lyα emission from Mrk 26. All three objects have strong ultraviolet continuum emission characteristic of hot stars. There is little evidence for dust in these galaxies. The results indicate that the best way to detect primeval galaxies is to look for the redshifted continuum emission from massive stars.

157.242 **Global redshift quantization.**
W. G. Tifft, W. J. Cocke.
Astrophys. J., Vol. 287, No. 2, p. 492 – 502 (1984).
The new (1981) Fisher–Tully survey of 21 cm redshifts of late–type galaxies shows sharp periodicities at exact submultiples (1/3 and 1/2) of 72.45 km $s^{-1}$. The periodicities allow a precise determination of the solar motion of $(\theta,\pi,z) = (231\pm2,-35\pm3,1\pm2)$ km $s^{-1}$. The 1/3 periodicity at 24.1 km $s^{-1}$ involves galaxies with narrow 21 cm profiles; and the 1/2 resonance at 36.2 km $s^{-1}$, those with wide profiles. The two data sets independently give the same solar motion velocity components.

157.243 **CO emission from *IRAS* galaxies.**
J. S. Young, J. Kenney, S. D. Lord, F. P. Schloerb.
Astrophys. J., Lett. Ed., Vol. 287, No. 2, p. L65 – L68 (1984). With plates L6 – L8.
The authors have searched for CO emission in 20 galaxies listed in the *IRAS* circulars using the 14 m telescope of the FCRAO (*Five College Radio Astronomy Observatory*) (HPBW = 45″). CO was detected in 10 of these galaxies, including Arp 220 and NGC 6240. The authors find a rough correlation between the CO and 100 μm luminosities, although a wide scatter is observed in this relation. The very luminous far–infrared galaxies Arp 220 and NGC 6240 contain $3\times10^{10} M_{\odot}$ of $H_2$.

157.244 **Neutral hydrogen in the Fourcade–Figueroa galaxy (A 1332–45).**
F. R. Colomb, N. Loiseau, J. C. Testori.
Astrophys. Lett., Vol. 24, No. 3, p. 139 – 147 (1984).
The H I 21–cm line has been observed around the Fourcade–Figueroa galaxy, a late spiral–type galaxy seen edge–on, with a systematic velocity of 823 km $s^{-1}$. Its distance is estimated as 3.2 Mpc, and the mass and other parameters of the galaxy are derived from the observed 21–cm profiles. The possibility of a physical association with the Centaurus Group is discussed.

**The second Byurakan spectral sky survey. II. The results of the area centred at $\alpha = 09^h50^m$, $\delta = +55°00'$.**
See Abstr. 002.007.

**Cambridge Society – TIFR workshop on galaxy interactions.**
See Abstr. 011.025.

**Workshop über wechselwirkende Galaxien.**
See Abstr. 011.030.

**Optimum signal to noise ratios for observations of faint galaxies.**
See Abstr. 036.002.

**On precision of measurement of galaxy images at Zô–Sè.**
See Abstr. 036.075.

**Galaxy photometry in the infrared.**
See Abstr. 036.155.

**Does wave–particle duality apply to galaxies?**
See Abstr. 061.099.

**Galactic neutrino halos.**
See Abstr. 061.116.

**Four *UBVRI* sequences near bright Virgo galaxies.**
See Abstr. 113.003.

**The infrared Ca II triplet: a luminosity indicator for stellar population synthesis.**
See Abstr. 114.056.

**Association of type II supernovae with giant H II regions: the case of M83 and M101.**
See Abstr. 125.044.

**The radio emission of galaxies and their supernova rate.**
See Abstr. 125.047.

**Supernova rate and the number of supernova remnants in M31, M33 and the Galaxy.**
See Abstr. 125.059.

**On the frequency of supernovae in Sb galaxies.**
See Abstr. 125.078.

**The rate of star formation in a galaxy derived from its integrated spectrum.**
See Abstr. 131.024.

**The origin of the infrared luminosity in violent star–formation regions.**
See Abstr. 131.130.

**Water maser in NGC 3079.**
See Abstr. 131.184.

**OH megamasers.**
See Abstr. 131.185.

**Radial distribution of giant molecular clouds. II. Comparison of the viscous cloud system with observations.**
See Abstr. 131.228.

**IUE UV spectra of extragalactic H II regions. I. The catalogue and the atlas.**
See Abstr. 132.006.

**One–millimeter continuum observations of galactic and extragalactic sources.**
See Abstr. 132.010.

**Ultraviolet study of the stellar population of the giant extragalactic H II region NGC 604.**
See Abstr. 132.013.

**On the composition of H II regions in southern galaxies – III. NGC 2997 and 7793.**
See Abstr. 132.034.

**IRAS circulars 14 and 15.**
See Abstr. 133.002.

**IRAS circular No. 14.**
See Abstr. 133.007.

**IRAS circular No. 15.**
See Abstr. 133.008.

**Optical counterparts for X–ray sources in M31.**
See Abstr. 142.038.

**How bar strength and pattern speed affect galactic spiral structure.**
See Abstr. 151.003.

**Orbital characteristics of polar rings of galaxies.**
See Abstr. 151.004.

**Spiral instabilities provoked by accretion and star formation.**
See Abstr. 151.007.

**Spiral structure and star formation. II. Stellar lifetimes and cloud kinematics.**
See Abstr. 151.008.

**Observable properties of non–axisymmetric galaxies. II. Photometric parameters.**
See Abstr. 151.017.

**Galactic detonation waves: numerical models illustrating the transition from deterministic to stochastic.**
See Abstr. 151.032.

**Violent relaxation and dissipationless collapse.**
See Abstr. 151.037.

**Bursts of star formation in computer simulations of dwarf galaxies.**
See Abstr. 151.038.

**Can the ballistic particle model explain the Hubble sequence?**
See Abstr. 151.047.

**Stellar and gas kinematics in disk galaxies.**
See Abstr. 151.059.

**Tidal encounters of elliptical galaxies.**
See Abstr. 151.060.

**The response of a spheroid to a disc field or were bulges ever ellipticals?**
See Abstr. 151.069.

**Making warps from bars and haloes.**
See Abstr. 151.070.

**A general method for constructing spherical galaxy models.**
See Abstr. 151.079.

**Angular momentum growth in protogalaxies.**
See Abstr. 151.080.

**Preferred orbit planes in triaxial galaxies. II. Tumbling about a nonprincipal axis.**
See Abstr. 151.081.

**Gravitational spurs and resonances: effects of small mass disturbers in spiral galaxy disks.**
See Abstr. 151.082.

**Integral characteristics of binary galaxies with tidal forces.**
See Abstr. 151.102.

**Dissipative formation of an elliptical galaxy.**
See Abstr. 151.103.

**Dissipative models for the sequence of elliptical galaxies.**
See Abstr. 151.104.

**Stochastic models for the evolution of open stellar systems.**
See Abstr. 151.105.

**On Schwarzschild's method for the construction of model galaxies.**
See Abstr. 151.106.

**Interacting binary galaxies. I. A numerical model and preliminary results.**
See Abstr. 151.107.

**Recognizing merger remnants among normal elliptical galaxies: NGC 5813.**
See Abstr. 151.109.

**The infrared color–luminosity relation for globular clusters in M31.**
See Abstr. 154.050.

**The metallicity of globular clusters associated with the giant elliptical galaxy NGC 4486 (M87).**
See Abstr. 154.065.

**Reddening of globular clusters in M31.**
See Abstr. 154.067.

**Infrared photometry of globular clusters in M31.**
See Abstr. 154.072.

**Old stellar populations. I. A spectroscopic comparison of galactic globular clusters, M31 globular clusters, and elliptical galaxies.**
See Abstr. 154.081.

**Effects of interstellar magnetic bubbles and of galactic tides on galactic magnetic fields.**
See Abstr. 155.001.

**Radio evidence for mass outflow in the Milky Way and external galaxies.**
See Abstr. 155.028.

**A comparison of our Galaxy to NGC 891 and NGC 4565 and the structure of their spheroids.**
See Abstr. 155.056.

**The Galaxy as a self–regulated star–forming system: the case of the OB associations.**
See Abstr. 155.064.

**Are halo carbon stars associated with dwarf spheroidal galaxies?**
See Abstr. 155.077.

**NGC 3256: an emerging elliptical galaxy.**
See Abstr. 158.006.

**Radio and optical investigation of UGC 6697 in Abell 1367.**
See Abstr. 158.055.

**The atypical galaxy NGC 4258 at high radio frequencies.**
See Abstr. 158.068.

**Cosmological evolution of radio sources found at 1.4 GHz.**
See Abstr. 158.106.

**Active inner regions of isolated galaxies.**
See Abstr. 158.118.

**Continuous transition from star burst galaxies to Seyfert galaxies?**
See Abstr. 158.121.

**A comparison of some photometric parameters of active and normal spiral galaxies.**
See Abstr. 158.126.

**Radio continuum observations of blue compact dwarf galaxies.**
See Abstr. 158.190.

**On the relative orientation of radio source axes and parent elliptical galaxies.**
See Abstr. 158.258.

**A search for changing structure in low–luminosity compact radio sources.**
See Abstr. 158.267.

**Low–redshift quasars as the active nuclei of cosmologically distant interacting galaxies: a spectroscopic investigation.**
See Abstr. 159.005.

**A search for QSOs in the fields of nearby galaxies. I. NGC 253, NGC 5236 and NGC 6744.**
See Abstr. 159.029.

**Two quasars seen near the spiral galaxy NGC 470.**
See Abstr. 159.032.

**Statistical evidence for possible association between QSOs and bright galaxies.**
See Abstr. 159.038.

**Strong Na I D absorption in the spectrum of the QSO PKS 1327–206 due to the galaxy MCG 03–34–085.**
See Abstr. 159.045.

**Two newly discovered quasars closely spaced across a galaxy.**
See Abstr. 159.048.

**Q 1114–2846: a quasar–galaxy binary.**
See Abstr. 159.066.

**A statistical analysis on the association of quasars and bright galaxies.**
See Abstr. 159.110.

**Studies of the Virgo cluster. III. A classification system and an illustrated atlas of Virgo cluster dwarf galaxies.**
See Abstr. 160.001.

**Nature of blue galaxies in the cluster Cl1447 + 2619.**
See Abstr. 160.002.

**H I synthesis observations of the high–redshift galaxies in Stephan's Quintet.**
See Abstr. 160.022.

**The effect of the perturbation of the local velocity field by Virgo on the calculation of differential luminosity functions.**
See Abstr. 160.028.

**Interaction of galaxy and quasar winds with the intergalactic medium.**
See Abstr. 160.031.

**The development of galaxy clusters.**
See Abstr. 160.032.

**The evolution of galaxies in clusters.**
See Abstr. 160.038.

**Spin statistics in binary galaxies: implications for formation and evolution.**
See Abstr. 160.042.

**The "excess" light in brightest cluster galaxies.**
See Abstr. 160.053.

**Extragalactic dust I: NGC 7070A.**
See Abstr. 160.066.

**H I deficiency and photometric evolution of spiral galaxies in the Virgo cluster.**
See Abstr. 160.067.

**The evolution of galaxies in clusters. V. A study of populations since $z \sim 0.5$.**
See Abstr. 160.068.

**The structure of first–ranked cluster galaxies and the radius–magnitude relation.**
See Abstr. 160.073.

**The orientations of spiral galaxies in the Local Supercluster.**
See Abstr. 160.083.

**Galaxy morphology in distant clusters.**
See Abstr. 160.092.

**The evolution of galaxies in clusters.**
See Abstr. 160.093.

**Radio continuum properties of spiral galaxies in the Coma/A1367 supercluster. Comparison between isolated and cluster galaxies.**
See Abstr. 160.109.

**Redshift differentials in a complete sample of galaxy groups.**
See Abstr. 160.142.

**CCD observations of galaxies in clusters at high redshift.**
See Abstr. 160.143.

**A complete sample of southern E and S0 galaxies: clustering properties.**
See Abstr. 160.144.

**The Shane–Wirtanen counts.**
See Abstr. 160.147.

**Firm bounds on the neutrino mass from the distribution of dark matter in galaxies.**
See Abstr. 161.009.

**The spatial correlation function of galaxies confronted with theoretical scenarios.**
See Abstr. 161.187.

**Adiabatic theories of galaxy formation and pancakes.**
See Abstr. 161.195.

**The statistics of gravitational lenses: the distributions of image angular separations and lens redshifts.**
See Abstr. 161.218.

**Constraints on the mass of muon neutrinos and their possible role in the galaxy formation.**
See Abstr. 161.266.

**Distant galaxies and the determination of the deceleration parameter.**
See Abstr. 161.310.

# 158 Active Galaxies (Seyfert Galaxies, BL Lacertae Objects Radio Galaxies)

158.001 **Experience in observing rapid variations of the emission spectrum of the nucleus of the peculiar galaxy NGC 1275.**
I. Pronik, N. Merkulova.
Astron. Tsirk., No. 1285, p. 1 - 3 (1983). In Russian.

158.002 **Redshifts of faint 3CR radio sources.**
M. A. C. Perryman, S. J. Lilly, M. S. Longair, A. J. B. Downes.
Mon. Not. R. Astron. Soc., Vol. 209, No. 1, p. 159 - 167 (1984).

Low-resolution optical spectra of nine faint objects associated with 3CR radio sources in the right ascension range 22 - 07 hr have resulted in the determination of seven new redshifts (4C 12.03, 3C 14, 22 and 41, 4C 14.11, 3C 441 and 457). One further redshift (3C 172) is provisional; for the remaining object (3C 65) no redshift has been determined.

158.003 **Companions of Seyfert galaxies – a statistical survey.**
O. Dahari.
Astron. J., Vol. 89, No. 7, p. 966 - 974 (1984). = Lick Obs. Bull., No. 978.

A search for close companion galaxies in a redshift-limited sample of Seyfert galaxies is reported. It is found that there is a definite excess of companions in that sample, compared with a control sample of field galaxies. The calculated percentage of Seyferts with *physical* companions is about 15%, while the upper limit on the percentage of the control-sample galaxies with physical companions is only 3.1%. Furthermore, about 7.5% of the Seyferts have companions that are very close and of comparable size, compared with only about 1% of the control sample.

158.004 **Sidedness, field configuration, and collimation of extragalactic radio jets.**
A. H. Bridle.
Astron. J., Vol. 89, No. 7, p. 979 - 986 (1984).

Data on 125 extragalactic radio jets are used to show that those in sources with total powers $\lesssim 10^{24.5}$ W/Hz at 1.4 GHz are generally two sided and dominated by perpendicular magnetic field components over most of their lengths. In contrast, those in sources with larger total and core powers are generally one sided and dominated by parallel magnetic field components over most of their lengths. The resolved jets in the weaker sources generally have faster lateral expansion rates than those in the more powerful sources.

158.005 **A three-dimensional model of the fluid dynamics of radio-trail sources.**
A. G. Williams, S. F. Gull.
Nature, Vol. 310, No. 5972, p. 33 - 36 (1984).

It is now generally accepted that radio-trail galaxies are produced by the motion of active galactic nuclei through the intergalactic medium in clusters of galaxies. Present evidence supports mass ejection from the active nucleus in a quasi-continuous jet. The authors adopt the jet model and study the formation of twin-tail sources by the motion of the active nucleus through intra-cluster gas, using three dimensional fluid-dynamical simulations of a supersonic jet in a cross-wind.

158.006 **NGC 3256: an emerging elliptical galaxy.**
J. R. Graham, G. S. Wright, W. P. S. Meikle, R. D. Joseph, M. F. Bode.
Nature, Vol. 310, No. 5974, p. 213 - 214 (1984).

NGC 3256 is a spectacular peculiar galaxy, with a highly chaotic nuclear region consisting of several bright knots. Two diffuse tidal tails are visible evidence of violent past history. The authors have discovered bright and exceptionally extended 10 μm emission that is evidence for an extremely luminous starburst extending over several kiloparsecs, which will leave the system severely gas-depleted. This depletion, in combination with the violent stellar relaxation which accompanies galaxy mergers, suggests that NGC 3256 will become an elliptical galaxy.

158.007 **Quick-look spectra of Markaryan galaxies.**
M. Dennefeld, F. Sèvre.
Astron. Astrophys., Suppl. Ser., Vol. 57, No. 2, p. 253 - 256 (1984).

Blue spectra have been obtained for 74 galaxies from the lists of Arakelyan or Markaryan. Velocities are presented for the 57 objects for which emission or absorption lines have been detected, together with an estimate of the excitation whenever possible. Comments are made on the possible Seyfert nature of a few of them.

158.008 **Distribution and kinematics of CO in the galaxy M82.**
H. Olofsson, G. Rydbeck.
Astron. Astrophys., Vol. 136, No. 1, p. 17 - 30 (1984).

The CO ($J$ = 1-0) emission towards the peculiar galaxy M82 has been mapped with 15" spacing. The mass of molecular gas within a radius of 500 pc is estimated to be $\sim 6 \times 10^7\ M_\odot$. This is $\sim$10% of the total mass, and about three times lower than the calculated mass of newborn stars. Thus, the recent conversion of gas into stars must have been extremely efficient. The current molecular cloud ensemble of M82 appears to be severely disrupted into a large number of small, high-temperature, low-mass clouds, which are not favourable to star formation. The kinematics of the molecular gas appears complicated. The data may be interpreted in terms of a warped galactic plane, which gradually becomes asymmetric with respect to the nucleus of the galaxy. The systemic velocity of M82 is estimated to be $v_{LSR} = 220 \pm 10$ km s$^{-1}$.

158.009 **Markarian Seyfert galaxies: optical and radio luminosity functions and other statistical studies.**
E. J. A. Meurs, A. S. Wilson.
Astron. Astrophys., Vol. 136, No. 2, p. 206 - 226 (1984).

This paper is devoted to a derivation of the (differential) optical, bivariate optical-radio and radio luminosity functions for Seyfert galaxies, as well as an investigation of the relationships between the radio emission and other properties of this class of galaxy. The basic sample comprises the Seyfert galaxies in the first nine lists of Markarian and almost all the radio data come from the Westerbork survey of Seyfert galaxies at 1415 MHz.

158.010 **An explanation of abrupt cutoffs in the optical-infrared spectra of non-thermal sources. A new pile-up mechanism for relativistic electron spectra.**
R. Schlickeiser.
Astron. Astrophys., Vol. 136, No. 2, p. 227 - 236 (1984).

An evolutionary model for relativistic electron momentum spectra is developed to explain abrupt cutoffs in the infrared-optical spectrum of non-thermal variable compact sources like BL Lacertae object AO 0235+164 and QSO 1413+135. The model is based on the interplay of Fermi acceleration at shock waves, second-order Fermi acceleration by moving magnetized fluid elements, radiation losses due to synchrotron and inverse Compton emission, and particle escape.

158.011 **Contribution of turbulent cascades to the luminosity of extragalactic radio-jets and Mach number constraints.**
G. Pelletier, L. Zaninetti.
Astron. Astrophys., Vol. 136, No. 2, p. 313 - 318 (1984).

The authors investigate the efficiency of turbulence produced by large hydrodynamic instabilities such as Kelvin-Helmholtz and internal shear to make radio-jets shine. Assuming that most of the power of a turbulent cascade is absorbed by radiating particles, the contribution of hydrodynamic instabilities to luminosity is deduced. Analysis of stochastic acceleration with synchrotron losses allow to predict a high turnover frequency in the millimeter band. The efficiency of the turbulent cascade is very sensitive to variation of the Mach number and this could explain the morphological differences between low luminosity sources with supersonic flow and high luminosity sources with hypersonic flow.

158.012 **The Hubble diagram and luminosity distribution of Arakelian galaxies.**
V. K. Golev, G. T. Petrov, Z. I. Tsvetanov.
Astron. Tsirk., No. 1292, p. 1 – 3 (1983). In Russian.

158.013 **Linearly polarized emission of variable radio sources. I. Background component.**
V. N. Kuril'chik, N. V. Kulakova.
Astron. Tsirk., No. 1295, p. 1 – 3 (1983). In Russian.

158.014 **Interpretation of the optical emission spectrum of radio galaxies.**
A. S. Zentsova.
Astron. Tsirk., No. 1295, p. 3 – 6 (1983). In Russian.

The optical emission spectrum of the broad–line radio galaxies arises as result of the photoionization of the gas by the radiation of central sources, while the spectrum of the narrow–line radio galaxies appears owing to heating of gas by shocks. It is shown that the typical size of shock waves is of the order of some kpc in the narrow radio galaxies. The velocity of the shock wave is determined.

158.015 **Linearly polarized emission of variable radio sources. II. Variable component.**
V. N. Kuril'chik, N. V. Kulakova.
Astron. Tsirk., No. 1298, p. 4 – 7 (1983). In Russian.

158.016 **Wavelength–dependent polarization in the optical flux of the BL Lacertae object B2 1308+326.**
M. L. Sitko, W. A. Stein, G. D. Schmidt.
Astrophys. J., Vol. 282, No. 1, p. 29 – 32 (1984).

Strong wavelength dependence of degree and position angle of polarization has been observed during an outburst of the BL Lacertae object B2 1308+326. Changes in the observed polarization characteristics with time suggest that they may be explained through models involving the changing orientation of a relativistic electron beam with respect to the line of sight.

158.017 **IRAS exposes a remarkable infrared galaxy.**
Phys. Today, Vol. 37, No. 8, p. 18 – 20 (1984).

158.018 ***U, B, V, R* photometry of 37 galaxies.**
É. A. Dibaj (*Eh. A. Dibaj*), V. T. Doroshenko, K. A. Postnov.
Sov. Astron., Vol. 28, No. 1, p. 1 – 6 (1984). English translation of 37.158.011.

158.019 **Optical variability parameters of active galactic nuclei.**
É. A. Dibaj (*Eh. A. Dibaj*), V. M. Lyutyj.
Sov. Astron., Vol. 28, No. 1, p. 7 – 11 (1984). English translation of 37.158.012.

158.020 **Thermal equilibrium of the emission–line clouds in quasar and Seyfert–nucleus envelopes.**
N. A. Katysheva.
Sov. Astron., Vol. 28, No. 1, p. 21 – 26 (1984). English translation of 37.158.013.

158.021 **RATAN–600 observations of 15 Bologna radio galaxies.**
V. G. Malumyan.
Sov. Astron. Lett., Vol. 10, No. 1, p. 35 – 36 (1984). English translation of 37.158.025.

158.022 **VLA observations of Markarian 8 at 6 cm wavelength.**
Q. Yin, D. S. Heeschen, J. Heidmann.
Sci. Sin., Ser. A, Vol. 27, No. 7, p. 750 – 756 (1984).

VLA observations show that the irregular galaxy Markarian 8 consists of three radio components coincident with optical clumps. The radio spectra of the components are flat and their emission is probably thermal. The radio components are imbedded in a diffused envelope of nonthermal emission. The radio and optical structures of Markarian 8 are similar. Under the assumption of thermal emission, electron densities and ionized hydrogen masses contained in the components are calculated. The results indicate that the clumps may indeed be supergiant H II regions.

158.023 **Spectral observations of Seyfert galaxies.**
V. T. Doroshenko, V. Yu. Terebizh.
Tr. Gos. Astron. Inst. Shternberg, Tom 55, p. 64 – 83 (1983). In Russian.

Results of spectral observations of 15 galaxies including 14 Seyfert galaxies with an image tube are presented. Observations were made during 1973 – 1981.

158.024 **Photoelectric comparison sequences in the fields of B2 1308+326 and 1418+54.**
H. R. Miller, J. W. Wilson, J. L. Africano, R. J. Quigley.
Astron. Astrophys., Suppl. Ser., Vol. 57, No. 3, p. 353 – 355 (1984). = Astron. Contrib. Georgia State Univ., No. 77.

Photoelectric *UBV* comparison sequences have been established in the fields of two violently variable BL Lacertae objects: B2 1308+326 and 1418+54 (=OQ 530). The sequences will provide accurate standards in order to calibrate the photographic photometry of each of these objects.

158.025 **Flickering of extragalactic radio sources.**
D. S. Heeschen.
Astron. J., Vol. 89, No. 8, p. 1111 – 1123 (1984).

Variability on time scales of 2 – 20 days, called "flickering" in the paper, has been studied at 9–cm wavelength. Measurements of 226 sources were made on about 25 consecutive days in each of three observing sessions. The observing list includes roughly equal numbers of steep spectrum (SC) and flat or complex spectrum (F) sources. The SC sources do not flicker while F sources do, with an average amplitude of about 1.5%. Observations of a smaller number of sources at 6–cm wavelength suggest that the amplitude of flickering is about the same at 6 and 9 cm. Flickering may be caused by scintillation in the interstellar or near–source medium, or it may be intrinsic to the source.

158.026 **Infrared polarimetry and photometry of BL Lac objects. II.**
C. D. Impey, P. W. J. L. Brand, R. D. Wolstencroft, P. M. Williams.
Mon. Not. R. Astron. Soc., Vol. 209, No. 2, p. 245 – 269 (1984).

This paper presents results from a continuing program to monitor the infrared polarization of BL Lac objects. Infrared (*JHK*) polarimetry and photometry have now been obtained for 19 BL Lac objects, with photometry alone for another six. The sample now represents a considerable fraction of the number of known BL Lac objects. It is used for a statistical analysis of the polarization properties of these objects.

158.027 **Continuum emission from the nucleus of NGC 1275.**
A. J. Longmore, R. M. Sharples, A. T. Tokunaga, R. J. Rudy, E. I. Robson, P. A. R. Ade, J. Radostitz.
Mon. Not. R. Astron. Soc., Vol. 209, No. 2, p. 373 – 385 (1984).

The authors present sub–millimetre and multi–aperture near–infrared observations of NGC 1275. The luminosity of the extended stellar component within the near–infrared apertures has been determined, and consequently also the 1.25 – 3.5 μm energy distribution of the unresolved nucleus. The 1 μm – 30 cm energy distribution is reviewed. It is argued that the most likely origin for the 100 μm – 30 cm radiation is synchrotron emission, with this non–thermal component also contributing significant flux at 10 μm.

158.028 **Surface photometry of powerful radio galaxies: their relation to Abell cluster cD galaxies.**
S. J. Lilly, I. S. McLean, M. S. Longair.
Mon. Not. R. Astron. Soc., Vol. 209, No. 2, p. 401 – 413 (1984).

Surface brightness profiles of 10 powerful 3C radio galaxies at moderate redshift ($0.06 < z < 0.3$) yield lower values of the structure parameter $\alpha$ than are typically found in first ranked Abell cluster galaxies. This difference has consequences for studies in observational cosmology that mix high-redshift radio galaxies and low-redshift Abell cluster galaxies. Comparisons with earlier qualitative work on lower-redshift radio galaxies suggests that this difference may reflect a change in environment and that it may be related to differences in the radio morphology.

158.029 **Erratum: "Detailed observations of NGC 4151 with *IUE*. III. Variability of the strong emission lines from 1978 February to 1980 May" [Mon. Not. R. Astron. Soc., Vol. 206, No. 1, p. 221 - 237 (1984)].**
M. H. Ulrich, A. Boksenberg, G. E. Bromage, J. Clavel, A. Elvius, M. V. Penston, G. C. Perola, M. Pettini, M. A. J. Snijders, E. G. Tanzi, M. Tarenghi.
Mon. Not. R. Astron. Soc., Vol. 209, No. 2, p. 479 (1984). See Abstr. 37.158.016.

158.030 **Sub-arcsec radio structure of the core-dominated Seyfert galaxy Markarian 348 (NGC 262).**
S. W. Unger, A. Pedlar, S. G. Neff, A. G. de Bruyn.
Mon. Not. R. Astron. Soc., Vol. 209, No. 2, p. 15P - 19P (1984).
The authors present 18-cm and 6-cm MERLIN observations of sub-arcsec structure in the centre of Mkn 348. The radio structure at high frequencies is dominated by an unresolved, highly variable, inverted-spectrum core. Although it is extremely unusual to find such a core in Seyfert galaxies, more normal steep-spectrum emission is also present extending about 50 pc to the north and south. Equipartition calculations indicate that the lobes must currently be in a phase of expansion.

158.031 **CCD observations of the bright radio sources 3C 292, 4C 14.27, 4C 74.16, 4C 13.66 and 4C 51.40.**
J. M. Riley, S. A. Eales, J. E. Baldwin.
Mon. Not. R. Astron. Soc., Vol. 209, No. 3, p. 641 - 643 (1984).
Optical identifications have been found for the bright radio sources 4C 74.16, 3C 292 and 4C 51.40, using a CCD camera on the Kitt Peak 0.9-m telescope. The faint galaxy associated with 4C 14.27 was confirmed. The field around 4C 13.66 was empty to a limiting magnitude of $R \sim 21.5$.

158.032 **A complete sample of extragalactic radio sources at 1 Jy at 408 MHz. III. Estimated redshifts and radio luminosities.**
J. R. Allington-Smith.
Mon. Not. R. Astron. Soc., Vol. 209, No. 3, p. 665 - 681 (1984).
Two previous papers presented measurements of the properties of a complete radio-selected sample of 59 B2 radio sources with flux densities between 1 and 2 Jy at 0.4 GHz. Here, the radio and optical identification data are discussed and interpreted. In order to estimate redshifts from the optical data in Paper II, the magnitude-redshift relations for radio galaxies and quasars in the bright-source sample of Laing, Riley & Longair were examined. The estimated redshifts are used in conjunction with existing data (mainly from bright-source samples) to examine the space-density evolution of the extragalactic radio-source population. Quasars appear to occupy the same range of radio luminosity at all redshifts, indicating that the shape of the quasar radio luminosity function may be independent of redshift.

158.033 **A deep radio/optical survey near the North Galactic Pole. III. A 4.85-GHz survey of the 5C12 area.**
C. R. Benn, J. V. Wall, G. Grueff, M. Vigotti.
Mon. Not. R. Astron. Soc., Vol. 209, No. 3, p. 683 - 695 (1984).
The 100-m Effelsberg telescope of MPIfR has been used to carry out a 4.85-GHz survey of the 5C12 area. The survey is complete to 18 mJy, and 93 sources are catalogued. Positions and flux densities are compared with other available data to check both random and systematic errors, and in particular to test for radial polar-diagram errors in the 5C12 survey. Source counts from the catalogue confirm the drop below a Euclidean law towards mJy levels, and provide no evidence of anisotropy on scales of degrees. The spectral-index distribution shows that by mJy levels the flat-spectrum source population, so predominant at 5-GHz Jy levels, has largely disappeared.

158.034 **Optical monitoring of radio sources.**
C. Lloyd.
Mon. Not. R. Astron. Soc., Vol. 209, No. 3, p. 697 - 718 (1984). With microfiche MN 290/2.
This paper presents new results from the Herstmonceux Optical Monitoring Programme for a selection of quasars, BL Lac objects, and different types of active galaxies. The light curves given cover the period ~1966 - 1980. Comparison with variations at other wavelengths indicates a similar type of behaviour in the optical and X-ray regions and a poor connection between optical and radio variations. An analysis of a large body of data from the literature shows that the Hubble diagrams of QSOs with flat and steep radio spectra are indistinguishable (in contrast to earlier work). It is also shown that neither the simple shot noise nor the periodic model provides an adequate description of optical variability.

158.035 **A multi-frequency radio study of Cygnus A.**
P. Alexander, M. T. Brown, P. F. Scott.
Mon. Not. R. Astron. Soc., Vol. 209, No. 4, p. 851 - 868 (1984).
New observations of Cygnus A made with the Cambridge 5-km radio telescope at 2.7, 5, 15.4 and 31.4 GHz are presented. The increased dynamic range over earlier 5-km maps at these frequencies has enabled the authors to map spectral variations over a large proportion of the source. The spectra are found to steepen away from the hotspots, with the spectral index reaching a maximum of about 2 in the most extended structure. The authors have interpreted these variations as due to synchrotron ageing and thus have estimated the speed of advance of the hotspots. Despite having polarization maps at three frequencies the authors are unable to determine unambiguously the distribution of rotation measure across the source, although the results are in broad agreement with earlier work.

158.036 **A catapult model for the narrow-line region in Seyferts and radio galaxies.**
M. D. Smith.
Mon. Not. R. Astron. Soc., Vol. 209, No. 4, p. 913 - 921 (1984).
The author studies the kinematics and stability of clouds falling radially into a supersonic wind. He finds a critical parameter, the ejection coefficient, which separates clouds which continue to gravitate inwards from those which are catapulted out by the ram pressure of the wind. This leads to a maximum size for ejected clouds. The clouds are partially broken up by fluid dynamic instabilities and the fragments expelled with enhanced velocities. The author applies this model to the narrow-line region of Seyferts and radio galaxies.

158.037 **Discovery of asymmetric optical structure in the radio-galaxies IC 708 and NGC 5352.**
R. A. Cordey.
Mon. Not. R. Astron. Soc., Vol. 210, No. 1, p. 183 - 187 (1984).
Red CCD images have been obtained of the radio galaxies IC 708 and NGC 5352 which reveal significant asymmetric structures within each. In IC 708 the structure coincides with a radio jet, but is too bright and too extended to be caused solely by synchrotron radiation and may be predominantly line radiation. Two large plumes are detected in NGC 5352, the longer being 62 kpc in projected length, but their nature and origin are unclear.

158.038 **Broad-line high-excitation gas in the elliptical galaxy NGC 5128.**
M. M. Phillips, K. Taylor, D. J. Axon, P. D. Atherton, R. N. Hook.
Nature, Vol. 310, No. 5978, p. 554 - 557 (1984).
A faint, but extensive component of broad-line ionized gas has been discovered in the peculiar giant elliptical galaxy NGC 5128. This component has a radically different spatial distribution from the well-studied rotating photoionized gas associated with the dust lane although the velocity fields of the two components are similar. The origin of the broad-line gas is considered and its possible relation to the active nucleus and the X-ray jet discussed.

158.039 **Infrared spectropolarimetry of Seyfert galaxy NGC 1068.**
D. K. Aitken, J. A. Bailey, G. Briggs, J. H. Hough, P. F. Roche.
Nature, Vol. 310, No. 5979, p. 660 – 662 (1984).
The authors present spectropolarimetric measurements of NGC 1068, between 8 and 13 μm, which show that the polarized flux at these wavelengths is intrinsic to the emission mechanism and not due to absorption by aligned grains.

158.040 **Black hole weighs in at one billion suns.**
D. H. Smith.
Sky Telesc., Vol. 68, No. 2, p. 127 – 128 (1984).

158.041 **Spectral investigation of the galaxy Markarian 201.**
A. N. Burenkov, A. R. Petrosyan, Eh. E. Khachikyan.
Astrofizika, Tom 19, Vyp. 3, p. 393 – 402 (1983). In Russian. English translation in Astrophysics, Vol. 19, No. 3.
The results of spectroscopic and spectrophotometric observations of Markarian 201 are presented. The spectrum of the nucleus shows both emission and absorption lines and in addition the first members of Balmer lines are in emission, and the highest in absorption. By means of intensity ratios of emission lines $T_e \sim 6700$K and $n_e \sim 440$ cm$^{-3}$ are evaluated. The abundances of O, N, He and S are estimated.

158.042 **A new galaxy of the Seyfert type.**
M. A. Kazaryan.
Astrofizika, Tom 19, Vyp. 3, p. 411 – 417 (1983). In Russian. English translation in Astrophysics, Vol. 19, No. 3.
The results of spectrophotometric studies of the galaxy No. 163 from the list of galaxies with *UV* excess (1979) are presented. The galaxy consists of two parts S and N. In the spectra S the broad and strong emission lines $H_\alpha$, [O III] λλ5007, 4959, $H_\beta$, $H_\gamma$ and $H_\delta$ are observed which are typical for the galaxies Sy 1. In the spectra N weak lines [N II] λ6584 and $H_\alpha$ in emission, $H_\beta$ and $H_\gamma$ in absorption are observed. It is concluded that the parts S and N of the galaxy strongly differ from each other in activity.

158.043 **Observations of galaxies with ultraviolet continuum with the radio telescope RATAN–600.**
V. A. Sanamyan, R. A. Kandalyan, G. A. Oganyan.
Astrofizika, Tom 19, Vyp. 3, p. 429 – 439 (1983). In Russian. English translation in Astrophysics, Vol. 19, No. 3.
The results of the observations of 101 galaxies with ultraviolet continuum at λλ3.9 – 31.0 cm are presented.

158.044 **On the configuration of radio galaxies.**
R. R. Andreasyan.
Astrofizika, Tom 19, Vyp. 3, p. 441 – 448 (1983). In Russian. English translation in Astrophysics, Vol. 19, No. 3.
It is assumed that galaxies possess dipole magnetic fields. The study of the motion of charged relativistic particles ejected from the nucleus of the parent galaxy in the dipole magnetic field offers a chance to find the main forms of radio galaxies. The obtained results are in good agreement with observations.

158.045 **Kinematic models of powerful radio galaxies and statistics of extended components.**
Yu. V. Baryshev.
Astrofizika, Tom 19, Vyp. 3, p. 461 – 469 (1983). In Russian. English translation in Astrophysics, Vol. 19, No. 3.
A statistical analysis of a sample of double radio galaxies from the paper of Longair and Riley (1979) is made. It is shown that the ratio of separations of the components from the nucleus do not depend on the projection of linear sizes of radio galaxies. This means that for the Ryle and Longair model (1967) there is no evolution of separation velocities.

158.046 **A correlation between ellipticity and core–strength in extended radio galaxies.**
T. P. Prabhu, R. K. Kochhar.
Astrophys. Space Sci., Vol. 102, No. 1, p. 29 – 39 (1984).
The authors show that in the case of extended radio sources a correlation exists between the fraction of the radio flux retained in the core component and the ellipticity of the underlying galaxy. The correlation is in the sense that stronger cores occur in flatter galaxies. It seems that there exists a class of intrinsically rounder, redder, massive ellipticals with larger velocity dispersions and metallicities that can form extended radio sources more efficiently. Thus the occurrence of a radio source appears to be related to the dynamical and chemical evolution of the galaxy.

158.047 **Far infrared emission from galaxies.**
K. V. K. Iyengar, R. P. Verma.
Astrophys. Space Sci., Vol. 103, No. 1, p. 61 – 70 (1984).
Until recently far infrared (FIR) observations of galaxies were limited to about a dozen bright and/or active galaxies. New photometric data has become available from Infrared Astronomical Satellite (IRAS) on 33 galaxies. The FIR spectra of these galaxies are similar. The far infrared flux in the wavelength interval 9 – 118 μm of the brighter galaxies is seen to be correlated with the integrated optical magnitude $B_T{}^0$. The 12 and 25 μm fluxes of these galaxies exhibit the same dependence on $B_T{}^0$ as the 10 and 21 μm fluxes from Seyferts and other emission–line galaxies. This suggests that the galaxies detected by IRAS are some type of active galaxies in accord with the high percentage of these galaxies predicted by Lock and Rowan–Robinson (1983).

158.048 **Search for a one–day variability of the broad components of emission line profiles of the NGC 4151 Seyfert galaxy nucleus.**
N. G. Bochkarev.
Pis'ma Astron. Zh., Tom 10, No. 8, p. 574 – 582 (1984). In Russian. English translation in Sov. Astron. Lett., Vol. 10.
A portion of the spectrum of NGC 4151 Seyfert galaxy nucleus was observed during 9 nights of the periods March 31 – April 6 and May 29 – June 3, 1983. A (1 – 2) – day variability of the broad components of the lines mentioned seems to exist. Increasing of the whole Hβ blue wing during the course of 2 months was observed.

158.049 **Fluxes in the lines Hβ, [O III] λ5007 and UBVR magnitudes of 10 Markarian galaxies with active nuclei.**
V. T. Doroshenko, V. A. Lipovetskij, V. Yu. Terebizh, A. I. Shapovalova.
Astrofizika, Tom 19, Vyp. 4, p. 823 – 826 (1983). In Russian. English translation in Astrophysics, Vol. 19, No. 4.
Fluxes in two lines and UBVR magnitudes of 10 Markarian galaxies mainly of Sy 1.5 – 2 types are given.

158.050 **Radial velocity field of Markarian 7.**
A. N. Burenkov, Eh. E. Khachikyan.
Astrofizika, Tom 19, Vyp. 4, p. 826 – 830 (1983). In Russian. English translation in Astrophysics, Vol. 19, No. 4.
The results of a detailed spectroscopy of Mark 7 are presented. Although the east and west parts of the galaxy have different kinematic properties, the area where both parts come into contact have the same radial velocity. It is concluded that Mark 7 is a single rather than a double galaxy.

158.051 **Polarimetric and photometric investigation of BL Lac. Results of observations.**
V. A. Hagen–Thorn, S. G. Marchenko, V. A. Yakovleva.
Astron. Zh., Tom 61, Vyp. 5, p. 925 – 934 (1984). In Russian. English translation in Sov. Astron., Vol. 28, No. 5.
The results of photoelectric observations of the optical brightness and polarization of BL Lac for 1969 – 1982 are given. These data are in good agreement with the results of other observers.

158.052 **Radio galaxies with broad and narrow emission lines.**
R. A. Kandalyan.
Astrofizika, Tom 20, Vyp. 1, p. 186 – 193 (1984). In Russian. English translation in Astrophysics, Vol. 20, No. 1.
The results of observations of radio galaxies with broad and narrow emission lines at 0.97 – 7.7 GHz are presented. Some properties of these galaxies are discussed. In particular, there is a difference between the average spectral indices for these two types of galaxies. This difference is due to the nucleus activity of these galaxies.

158.053 **Relativistic electron–photon showers in the nuclei of active galaxies and quasars.**
F. A. Agaronyan, V. V. Vardanyan, V. G. Kirillov–Ugryumov.
Astrofizika, Tom 20, Vyp. 2, p. 223 – 237 (1984). In Russian. English translation in Astrophysics, Vol. 20, No. 2.
A hypothesis is proposed that the observed gamma radiation from the nuclei of active galaxies and QSOs is due to the development of a high–energy electromagnetic cascade initiated by relativistic nonthermal particles in the sources of activity.

158.054 **Polarimetry and photometry of M87: is the jet fading?**
R. F. Warren–Smith, D. J. King, S. M. Scarrott.
Mon. Not. R. Astron. Soc., Vol. 210, No. 2, p. 415 – 424 (1984).
Optical linear polarization mapping and photometry of M87 is presented in $B$ and $R$ wavebands. The results indicate significant polarization in the galactic nucleus. Polarization in the jet in $B$ is consistent with other recent maps obtained for this area in blue light and no significant variation with wavelength between $B$ and $R$ wavebands is found. The spectral index of the jet radiation is measured to be $S = -1.65 \pm 0.2$ over the wavelength range observed. Comparison of the integrated $B$ magnitude of the jet with previous independent measurements over the period 1934 – 80 suggests that the jet is variable and has been fading more or less uniformly by about 0.8 mag per decade between 1964 and 1980. The time–scale of this variation is unexpectedly short for an object of this size and possible interpretations are discussed.

158.055 **Radio and optical investigation of UGC 6697 in Abell 1367.**
G. Gavazzi, M. Tarenghi, W. Jaffe, H. Butcher, A. Boksenberg.
Astron. Astrophys., Vol. 137, No. 2, p. 235 – 244 (1984).
The authors present radio maps of the irregular galaxy UGC 6697, in A1367, obtained at 0.6, 1.4, and 5 GHz with the WSRT and at 1.4 GHz with the VLA. These are combined with multicolor 4 m KPNO observations to establish that the marked asymmetry of the radio source (resembling a head–tail source) reflects an intrinsic asymmetry of the parent galaxy. Narrow band Hα observations reveal a ring–like structure in this object which indicates that the galaxy is not seen edge–on. The peculiar rotation curve of UGC 6697, as derived from IPCS spectra, can be interpreted as due to a rotating ring with strong non–circular motions. The data are discussed in the context of a possible interaction between UGC 6697 and the IGM or with its companion galaxy.

158.056 **High dynamic range observations of 3C123 at 1400 MHz. Is it a giant radio galaxy?**
R. J. Hanisch.
Astron. Astrophys., Vol. 137, No. 2, p. 250 – 254 (1984).
Observations have been made of the strong radio galaxy 3C123 at 21 cm wavelength with the 305–m telescope at Arecibo and the Westerbork Synthesis Radio Telescope in its high dynamic range configuration. The author has searched for evidence that 3C123 may indeed be associated with several of the nearby weak sources first observed by Reich et al. (1976). Such an association, if real, would imply a total source extent in excess of 10 Mpc and would make 3C123 the largest known radio source. The author's observations, however, do not support this interpretation. If 3C123 is indeed a giant radio galaxy, its properties are very unlike those of the other well–known giant radio sources.

158.057 **Optical spectra of southern Molonglo radio sources – II.**
H. S. Murdoch, R. W. Hunstead, G. L. White.
Proc. Astron. Soc. Aust., Vol. 5, No. 3, p. 341 – 354 (1984).
This is the second of a series of papers presenting detailed optical spectra for QSOs and strong–emission–line galaxies identified with radio sources from a selected area (–15° to –23° in declination) of the Molonglo Reference Catalogue. The aim is to produce a complete sample of QSOs and galaxies with strong emission lines to a limiting flux density of 1 Jy at 408 MHz and a limiting magnitude of 19.5 on the O plate of the Palomar Sky Survey.

158.058 **Multifrequency observations of the complex radio galaxy 1919+479 (4C 47.51).**
J. G. Robertson.
Astron. Astrophys., Vol. 138, No. 1, p. 41 – 48 (1984).
Observations of the radio galaxy 1919+479 using the Westerbork Synthesis Radio Telescope at 0.61, 1.4, and 5.0 GHz are presented. This source is located in a cluster of galaxies at redshift 0.10 and is a wide–angle–tail galaxy of very unusual morphology. Maps of both total and polarised intensity at all three frequencies are shown. Spectral index maps are derived from these multifrequency observations, and some inferences are drawn regarding the development of this unusual source.

158.059 **Optical structures of the M87 and 3C 273 jets.**
G. Lelièvre, J.–L. Nieto, D. Horville, L. Renard, B. Servan.
Astron. Astrophys., Vol. 138, No. 1, p. 49 – 56 (1984).
New high–resolution observations of the M87 and the 3C 273 jets with the wide–field electronographic camera attached to the CFH telescope reveal new features in the optical structures of these jets. In M87 there is evidence for a bent nuclear extension, very similar to that shown by VLBI maps, and oscillations in the inner part of the jet. The jet in 3C 273 shows a structure quite different at radio and optical wavelengths; in particular there is no strong luminosity gradient outwards in the optical such as in radio. The authors discuss in detail the possible origins for the optical morphologies of both jets.

158.060 **The broad absorption line region in NGC 4151.**
G. E. Bromage, A. Boksenberg, J. Clavel, M. H. Demoulin–Ulrich, A. Elvius, M. V. Penston, G. C. Perola, M. Pettini, M. A. J. Snijders.
ESA Spec. Publ., ESA SP–218, p. 77 – 79 (1984). – See Abstr. 012.024.
Extensive monitoring of NGC 4151 with IUE has yielded much new information on the absorption line region in the active galactic nucleus. The paper summarises the results and their implications for models of the absorption line region in the active nucleus – its structure, density, ionisation conditions, location and size.

158.061 **Component structure in the C IV line in NGC 4151.**
M. V. Penston, A. Boksenberg, M. Pettini, M. A. J. Snijders, G. E. Bromage, J. Clavel, A. Elvius, G. C. Perola, A. Altamore, M.–H. Demoulin–Ulrich, M. Tarenghi.
ESA Spec. Publ., ESA SP–218, p. 81 (1984). Abstract. – See Abstr. 012.024.

158.062 **The Seyfert I galaxy NGC 4593. III. Very rapid variations of its far UV spectrum and their implications.**
J. Clavel.
ESA Spec. Publ., ESA SP–218, p. 83 – 89 (1984). – See Abstr. 012.024.
New IUE observations of the Seyfert I galaxy NGC 4593 have revealed variations by a factor 2.3 in 69.7 hours of the intensity of its nuclear continuum at 1450 Å. This result sets an upper limit of $2.2 \pm 0.3$ light–days to the diameter of its far–UV continuum emitting region. In the same time–interval, the intensities and the profiles of the main UV emission lines did not vary. These data, together with results obtained previously, are entirely consistent with the hypothesis that a $10^7\ M_\odot$ black hole accretes at 4% of

its Eddington limit in the nucleus of NGC 4593. Only 3% of the accretion flow is in the form of broad line emitting gas.

158.063 **IUE observations of multiple nucleus galaxies.**
W. Kollatschny, K. J. Fricke.
ESA Spec. Publ., ESA SP–218, p. 91 – 95 (1984). – See Abstr. 012.024.

The authors describe the UV spectra of the 4 double nucleus Markarian galaxies Mkn 266, 463, 673, 1027. The first 3 of these objects have nonthermal nuclei and Seyfert or Liner characteristics in the line spectra of the component nuclei. Mkn 1027 has an absorption line spectrum typical for extragalactic H II regions.

158.064 **The 1978 – 1984 fading of the Seyfert I galaxy F–9.**
W. Wamsteker, R. Gilmozzi, P. Benvenuti,
A. Cassatella, B. Hassall, A. Talavera, A. W. Harris, H. Netzer,
D. Wills, B. J. Wills, P. Barr, C. Boisson, J. Clavel, J. Lub,
A. C. Danks, W. Kollatschny, K. J. Fricke, L. Colina,
J. W. Menzies, C. Gry.
ESA Spec. Publ., ESA SP–218, p. 97 – 99 (1984). – See Abstr. 012.024.

Results of simultaneous optical and UV spectra for the Seyfert I galaxy F–9 are presented. The UV variability is shown to be very large and to decrease with wavelength. It is shown that at one epoch a good fit to the continuum can be obtained with a single power law spectrum, Ba continuum and Fe II emission.

158.065 **MCG–2–58–22 = Mkn 926. A high luminosity Seyfert galaxy with two distinct broad line components.**
M. Jones, M. Ward.
ESA Spec. Publ., ESA SP–218, p. 101 – 105 (1984). – See Abstr. 012.024.

The authors use single epoch near simultaneous optical and ultraviolet spectra to derive a "snapshot" list of emission line fluxes. A composite spectrum containing many epochs of data is used to study the line profiles. Using data ranging from 21 cm radio continuum to 30 keV X–ray, the $\nu F_\nu$ energy distribution is discussed. Variations in the Lyman alpha emission line are reported.

158.066 **Multifrequency observations of three active galaxies.**
L. Bassani, M. J. Coe, R. Clement, A. J. Dean,
G. Di Cocco, B. Elsmore, M. Ferrari–Toniolo, F. Giovannelli,
J. Macdougall, P. Persi, F. Perotti, S. Sembay, L. Spinoglio,
P. Ubertini.
ESA Spec. Publ., ESA SP–218, p. 107 – 110 (1984). – See Abstr. 012.024.

As part of a multiwaveband observational program of active galactic nuclei, ultraviolet spectra have been obtained of Mk 501, MCG 8–11–11 and NGC 1275 in late 1983. The IUE data of the BL Lac object Mk 501 and the Seyfert galaxy MCG 8–11–11 are presented together with contemporaneous radio and near–infrared observations. For NGC 1275, a galaxy of uncertain classification, near–infrared data have also been obtained within two weeks of the IUE measurement.

158.067 **Multifrequency observations of active galactic nuclei.**
E. G. Tanzi, L. Chiappetti, J. Danziger, R. Falomo,
D. Maccagni, L. Maraschi, A. Treves, W. Wamsteker.
ESA Spec. Publ., ESA SP–218, p. 111 – 117 (1984). – See Abstr. 012.024.

A number of active galactic nuclei were observed simultaneously with EXOSAT, IUE and ground–based telescopes in order to determine the overall energy distribution over a wide frequency range. In this paper the authors describe the preliminary analysis of repeated observations of the BL Lac object PKS 2155–304 and of the Seyfert galaxies 3C 120 and MCG 8–11–11.

158.068 **The atypical galaxy NGC 4258 at high radio frequencies.**
M. Krause, R. Beck, U. Klein.
Astron. Astrophys., Vol. 138, No. 2, p. 385 – 395 (1984).

Radio continuum observations of NGC 4258 in total intensity and linear polarization obtained with the Effelsberg 100–m telescope at λ6.3 cm and λ2.8 cm are presented. The strongest linearly polarized intensity with degrees of polarization up to ≅50% at λ2.8 cm is found at the termination of the anomalous arms. The observations suggest that the magnetic field is aligned along both, the normal and anomalous arms. Equipartition considerations are used to calculate the magnetic field strengths. The observed Faraday rotation is compatible with a rotation measure of $5 \pm 4$ rad $m^{-2}$. This result supports the view that the anomalous radio arms lie within the disk of the galaxy.

158.069 **Emission–line variability in Seyfert galaxies.**
A. Robinson.
Observatory, Vol. 104, No. 1062, p. 215 – 216 (1984). Abstract. – See Abstr. 010.682.

158.070 **Jets in extragalactic radio sources.**
D. S. De Young.
Science, Vol. 225, No. 4663, p. 677 – 682 (1984).

Observations now require that there be a continuous supply of energy to the giant extragalactic radio sources. These observations also suggest that this energy input may be in the form of streams or jets of gas emanating from the centers of galaxies and quasi–stellar objects. Current data indicate that the large–scale jet structures are not moving with relativistic speeds, as previously proposed. Slow–moving jets, which possess turbulent interiors and are dominated by relatively cool gas, can account for the observed jet properties at optical and radio wavelengths. Extremely small–scale jets observed adjacent to the central energy source may or may not be in relativistic motion.

158.071 **UBV photometry of Seyfert galaxy nuclei.**
A. D. Mallama.
J. Am. Assoc. Variable Star Obs., Vol. 12, No. 2, p. 69 – 72 (1983).

158.072 **Galactic jets: two exotic cases.**
R. A. Schorn.
Sky Telesc., Vol. 68, No. 5, p. 410 – 411 (1984).

158.073 **Statistical properties of the radio cores in elliptical galaxies.**
L. Feretti, G. Giovannini, L. Gregorini, P. Parma,
G. Zamorani.
Astron. Astrophys., Vol. 139, No. 1, p. 55 – 63 (1984).

A statistical study of the properties of the radio cores in elliptical galaxies has been performed on a sample of 138 radio galaxies taken from the B2 and 3CR catalogues. About two thirds of these objects have a detected core at 5 GHz. A statistically significant correlation is found between the core and the total radio power. However, the large dispersion around the best fit regression line implies that the core radio power is not a good indicator of the total radio power. Moreover, the authors find a correlation between the core radio power and the optical absolute magnitude: optically brighter galaxies tend to have higher core radio powers. Within the limited redshift range of the sample, no redshift dependence of the two correlations is found.

158.074 **Variations of the linear sizes of extragalactic radio sources with radio luminosity and redshift.**
J. R. Allington–Smith.
Mon. Not. R. Astron. Soc., Vol. 210, No. 3, p. 611 – 631 (1984).

The variations with radio luminosity and redshift of the characteristic linear size of extragalactic radio sources are investigated in two ways: (1) by examining bright flux–density–limited samples in order to determine the dependence explicitly, and (2) by comparing angular size–flux density data from faint samples with numerical predictions, using individual sources in bright samples to define the local population. These predictions divide the total population into flat– and steep–spectrum components, thereby allowing populations found at different frequencies to be related. These investigations suggest that the apparent local distribution of linear sizes is determined by instrumental and physical selection effects which result in a weak inverse correlation of linear size with radio luminosity. For populations at earlier cosmological epochs, these effects cause an apparent decrease in linear size with redshift.

158.075 **Nuclear activity in two spiral galaxies with jets: NGC 1097 and 1598.**
M. M. Phillips, B. E. J. Pagel, M. G. Edmunds, A. Díaz.
Mon. Not. R. Astron. Soc., Vol. 210, No. 3, p. 701 – 710 (1984).

The barred spiral galaxies NGC 1097 and 1598 are unusual in being known to have optical jet–like features aligned with their nuclei; they also have conspicuous nuclear emission–line spectra which the authors have studied in some detail in an attempt to throw light on the underlying activity. The very nuclei are either classic 'Liners' (NGC 1097) or quite similar to 'Liners' (NGC 1598), and are surrounded by giant H II complexes with typical low–excitation H II region spectra. The presence of near–solar abundances of N and O in these H II regions, and thus presumably also in the nuclei themselves, supports power–law photoionization models for the excitation of 'Liners' in general, in preference to available shock models.

158.076 **Time variability of the 87 GHz linear polarization and flux density of active extragalactic objects.**
R. Barvainis, C. R. Predmore.
Astrophys. J., Vol. 282, No. 2, p. 402 – 411 (1984).

Measurements at 87 GHz of the linear polarization and total flux density of 16 active extragalactic sources are presented. The observations were made over a 7 month period with time resolution of days to weeks for some sources. Two rapid and closely spaced outbursts with time scales of about 1 week were seen in OJ 287. For 3C 345, variations in the polarization may indicate the propagation of radiating particles along a jet, in which the order and direction of the magnetic field changes with distance along the jet. No source was seen to undergo large–amplitude linear rotation of polarization position angle with time.

158.077 **A luminous 3 kiloparsec infrared disk in NGC 1068.**
C. M. Telesco, E. E. Becklin, C. G. Wynn–Williams, D. A. Harper.
Astrophys. J., Vol. 282, No. 2, p. 427 – 435 (1984).

The authors present a 10 μm map of the Seyfert galaxy NGC 1068 and airborne measurements of its angular extent in the far–infrared. They show that the infrared emission originates primarily from two physically distinct regions; approximately half of the total infrared luminosity of $3 \times 10^{11} L_{\odot}$ is associated with the Seyfert nucleus and half with a 3 kpc diameter disk surrounding it. It is argued that the disk component of infrared emission originates from an extended but heavily obscured burst of star formation which resembles those seen in some non–Seyfert galaxies.

158.078 **IS 3C 310 blowing bubbles?**
W. van Breugel, E. B. Fomalont.
Astrophys. J., Lett. Ed., Vol. 282, No. 2, p. L55 – L58 (1984). With plates L3 – L4.

Radio observations are presented of the nearby, wide double source 3C 310 which show intricate fine–scale radio structure (shells and filaments) embedded in diffuse, low–brightness lobes. These features have longitudinal magnetic fields and an exceptionally high percentage of linear polarization. The balloon–shaped morphology of some of these features suggests that they may be the edge–brightened boundaries of "bubbles" of hot gas which are blown up by weak jets and are surrounded by magnetic fields.

158.079 **The bridges of classical double radio sources.**
J. P. Leahy, A. G. Williams.
Mon. Not. R. Astron. Soc., Vol. 210, No. 4, p. 929 – 951 (1984).

New observations with the VLA of a complete sample of 39 3C sources with angular sizes larger than 45 arcsec are discussed. Nearly all of the classical double sources at redshifts less than 0.5 have bridges linking the central galaxy with the outer edge of the source, and the maps of these objects show that two–thirds of the bridges are significantly distorted near the central galaxy. Such distortions may be predicted from the jet model for double radio sources if the jet is much lighter than the confining medium, since the dynamics of the bridge region are then dominated by back flow of material that has passed through the hotspots. There is a correlation between axial ratio and luminosity which may be physically related to the compactness relation of Jenkins and McEllin.

158.080 **Infrared polarimetry and photometry of BL Lac objects. III.**
P. A. Holmes, P. W. J. L. Brand, C. D. Impey, P. M. Williams.
Mon. Not. R. Astron. Soc., Vol. 210, No. 4, p. 961 – 977 (1984).

The data presented here is part of a continuing monitoring programme of BL Lac objects with *J, H*,and *K* photometry and polarimetry. A total of 30 BL Lac objects have now been observed photometrically. The authors also have infrared polarimetry for 24 of these objects. Internight variations and wavelength dependence of polarization indicate that BL Lac objects, as a class, may be understood in terms of a relatively simple two–component model. Luminosity calculations suggest that relativistic beaming or non–spherical accretion is important in the most luminous objects. The $p(\lambda)$–$p$ relation allows to examine the magnetic–field structure of the emitting region.

158.081 **The spectra of extended radio sources with a nonuniform magnetic field and hard injection of particles.**
A. E. Kochanov.
Astrophys. Space Sci., Vol. 105, No. 1, p. 1 – 21 (1984).

On the basis of an analytical solution of the diffusion–type kinetic equation for electrons, electron distributions and radiation spectra have been found which result from a hard injection of particles in sources of the 'core halo' type, characterized by spatially nonuniform magnetic fields and diffusion parameters. Such radio sources are shown to possess nonlinear radiation spectra containing universal ($\alpha = 0.5$) and diffusion–controlled power–law sections shaped by synchrotron losses, spatial diffusion and radiation conditions of the electrons. The diffusion–controlled sections can be described by spectral indices $0.5 \leqslant \alpha < 1$, if the magnetic field decreases towards the source edge, and by $\alpha < 0.5$ where the magnetic field increases.

158.082 **Erratum: "An X–ray survey of a complete sample of 3CR radio galaxies" [Astrophys. J., Vol. 277, No. 1, p. 115 – 131 (1984)].**
G. Fabbiano, L. Miller, G. Trinchieri, M. Longair, M. Elvis.
Astrophys. J., Vol. 283, No. 1, p. 469 (1984). See Abstr. 37.158.090.

158.083 **The remarkable infrared galaxy Arp 220 = IC 4553.**
B. T. Soifer, G. Helou, C. J. Lonsdale, G. Neugebauer, P. Hacking, J. R. Houck, F. J. Low, W. Rice, M. Rowan–Robinson.
Astrophys. J., Lett. Ed., Vol. 283, No. 1, p. L1 – L4 (1984).

*IRAS* observations of the peculiar galaxy Arp 220 ≡ IC 4553 show that it is extremely luminous in the far–infrared, with a total luminosity of $\sim 2 \times 10^{12} L_{\odot}$. The infrared–to–blue luminosity ratio of this galaxy is $\sim 80$, which is the largest value of the ratio for galaxies in the UGC catalog. Other observations of Arp 220, combined with the luminosity in the infrared, allow either a Seyfert–like or starburst origin for this luminosity.

158.084 **Line variability in active nuclei and the structure of the broad–line region.**
M. H. Ulrich.
Ann. N.Y. Acad. Sci., Vol. 422, p. 291 – 302 (1984). – See Abstr. 012.033.

This paper reviews the observations of the time variations of the emission lines in Seyfert nuclei and in particular in NGC 4151, the galaxy that has been the most extensively studied in the ultraviolet. In this galaxy, the variations of the line intensities and of the line profiles allow insight into the structure of the broad–line region. The variations are consistent with a model where the optical thickness, the distance from the central continuum source, and the velocity of the clouds forming the broad–line region are correlated, the clouds with the highest velocities and smaller optical thickness being closer to the center.

158.085 **Models of active galactic nuclei.**
R. D. Blandford.
Ann. N.Y. Acad. Sci., Vol. 422, p. 303 – 318 (1984). – See Abstr. 012.033.

An attempt towards a unified model of active galactic nuclei, based on massive accreting black holes, is presented. It is argued that black hole models could account for both the diversity of extragalactic phenomena and the correlations between different wave bands in a particular source.

158.086 **The radio structure of the X–ray jet in Centaurus A.**
J. O. Burns, E. D. Feigelson, E. J. Schreier.
Ann. N.Y. Acad. Sci., Vol. 422, p. 334 (1984). Abstract. – See Abstr. 012.033.

158.087 **Broad–line clouds in active galactic nuclei: physical conditions and motions.**
C. M. Gaskell.
Ann. N.Y. Acad. Sci., Vol. 422, p. 350 (1984). Abstract. – See Abstr. 012.033.

158.088 **The gravitationally bound current loop model for quasars and galactic nuclei.**
H. D. Greyber.
Ann. N.Y. Acad. Sci., Vol. 422, p. 353 – 354 (1984). Abstract. – See Abstr. 012.033.

158.089 **X–ray spectral variability of BL Lac objects.**
D. Maccagni, T. Maccacaro, M. Tarenghi.
Ann. N.Y. Acad. Sci., Vol. 422, p. 367 (1984). Abstract. – See Abstr. 012.033.

158.090 **Black hole nuclear cluster model of quasars and active galactic nuclei.**
A. G. Pacholczyk, W. R. Stoeger.
Ann. N.Y. Acad. Sci., Vol. 422, p. 373 – 374 (1984). Abstract. – See Abstr. 012.033.

158.091 **X–ray constraints on relativistic beams in BL Lacertae objects.**
D. A. Schwartz, G. M. Madejski.
Ann. N.Y. Acad. Sci., Vol. 422, p. 381 (1984). Abstract. – See Abstr. 012.033.

158.092 **Improved collimation for radio sources.**
P. J. Wiita, M. J. Siah.
Ann. N.Y. Acad. Sci., Vol. 422, p. 393 (1984). Abstract. – See Abstr. 012.033.

158.093 **Relativistic explosion and activity in galactic nuclei.**
M. Yokosawa.
Astron. Her., Vol. 77, No. 10, p. 256 – 260 (1984). In Japanese.

158.094 **Very low luminosity active galaxies and the X–ray background.**
M. Elvis, A. Soltan, W. C. Keel.
Astrophys. J., Vol. 283, No. 2, p. 479 – 485 (1984).

Using the complete sample of bright spiral galaxy Hα luminosities of Keel and a mean $H\alpha/L_x$ ratio, the authors derive a bivariate X–ray luminosity function for active galaxies in the range $10^{39} - 10^{42.5}$ ergs $s^{-1}$. This X–ray luminosity function is then integrated to estimate the contribution of these very low luminosity active galaxies to the diffuse X–ray background.

158.095 **X–ray selected quasars and Seyfert galaxies: cosmological evolution, luminosity function, and contribution to the X–ray background.**
T. Maccacaro, I. M. Gioia, J. T. Stocke.
Astrophys. J., Vol. 283, No. 2, p. 486 – 494 (1984).

The cosmological evolution and the X–ray luminosity function of quasars and Seyfert galaxies (active galactic nuclei [AGNs]) are derived and discussed. The sample used consists of 56 objects extracted from the expanded Einstein Observatory Medium Sensitivity Survey, and it is exclusively defined by its X–ray properties. The X–ray luminosity function is derived. The simultaneous determination of the cosmological evolution and of the X–ray luminosity function of AGNs is then used to estimate the contribution to the extragalactic diffuse X–ray background.

158.096 **Radio jets in classical double radio sources with strong cores.**
J. O. Burns, J. P. Basart, D. S. De Young, D. C. Ghiglia.
Astrophys. J., Vol. 283, No. 2, p. 515 – 528 (1984).

Fifteen luminous radio galaxies and quasars with strong cores were observed with high dynamic range using the VLA at 6 cm. These observations were intended to determine if classical double (class II) sources have a significantly lower probability of jet emission, as had been suggested from lower contrast maps. About 50% of the sample contains jets, significantly higher than that seen for doubles with weaker cores. The properties of the jets are examined and it is discussed whether these properties are compatible with relativistic beaming models.

158.097 **MHD Kelvin–Helmholtz instability in extended radio jets.**
R. Fiedler, T. W. Jones.
Astrophys. J., Vol. 283, No. 2, p. 532 – 539 (1984).

The linearized MHD dispersion relation is exactly solved numerically for a cylindrical vortex sheet under a wide range of conditions appropriate for extragalactic radio jets. In the zero–field, hydrodynamic, limit empirical expressions are derived which allow one to estimate simply the growth rates of instabilities. The expressions for maximally disruptive growth rates can be used to evaluate the propagation of jets before disruption, to compare the physical properties of different radio sources, and to study instability induced eddies for application to particle reacceleration schemes.

158.098 **Double Faraday rotation toward 3C 27.**
S. J. Goldstein Jr., J. A. Reed.
Astrophys. J., Vol. 283, No. 2, p. 540 – 545 (1984).

From observations of the integrated flux of 3C 27 with the NRAO 140 foot (43 m) telescope at 40 frequencies between 1250 and 1445 MHz, the authors deduce rotation measures of $165 \pm 15$ and $-104 \pm 4$ rad $m^{-2}$. Since the source (assumed to be a radio galaxy) has components 45″ apart, it is concluded that the net magnetic field reverses between these directions. One explanation is that a large magnetic field surrounding the central galaxy of the distant source covers one component but not the other. Another explanation is that our Galaxy contains a dipole field with a scale of order 1 pc. One component of the distant source is seen inside the current loop associated with the dipole field, while the other is seen outside the loop.

158.099 **On the space density of Arakelian galaxies.**
V. K. Golev, Z. I. Tsvetanov, G. T. Petrov.
Dokl. Bolg. Akad. Nauk, Vol. 37, No. 1, p. 15 – 18 (1984). Abstr. in Ref. Zh., 51. Astron., 9.51.752 (1984).

158.100 **On the nature of variable spectra of radio radiation of some extragalactic objects.**
Yu. A. Kovalev.
Inst. kosm. issled. Akad. Nauk SSSR. Prepr., No. 879, 22 pp. (1984). In Russian. Abstr. in Ref. Zh., 51. Astron., 9.51.757 (1984).

158.101 **Photographic *B*–band monitoring of NGC 4151.**
T. R. Gill, C. Lloyd, M. V. Penston, M. A. J. Snijders.
Mon. Not. R. Astron. Soc., Vol. 211, No. 1, p. 31 – 37 (1984). With microfiche MN 211/1.

Results of a photographic monitoring programme of the Seyfert galaxy NGC 4151 for the period 1968 February to 1983 March are presented. An optical light curve which covers the time during which this Seyfert galaxy was observed in the ultraviolet, 1978 – 83, is constructed.

158.102 **The tail of 3C129: observations at 2.7 GHz.**
A. Downes.
Mon. Not. R. Astron. Soc., Vol 211, No. 1, p. 215 – 220 (1984).
Observations of the tailed radio galaxy 3C129 at 2.7 GHz with an angular resolution of 4.3 arcmin are presented. These observations allow the outer regions of the tail to be investigated. The tail is shorter than at low frequencies because of spectral steepening, and shows polarization as strong as 50 per cent of the total power. The magnetic field lies along the tail.

158.103 **An evolutionary link between Seyfert I and II galaxies?**
M. V. Penston, E. Pérez.
Mon. Not. R. Astron. Soc., Vol. 211, No. 1, p. 33P – 39P (1984).
First spectra from the newly sited INT show NGC 4151 and 3C 390.3 to have taken on a classification very close to Seyfert II. The authors propose that Seyfert II galaxies are Seyfert Is in which the continuum source is temporarily off.

158.104 **He I λ10830 emission in Seyfert galaxies and QSOs.**
P. D. LeVan, R. C. Puetter, H. E. Smith, R. J. Rudy.
Astrophys. J., Vol. 284, No. 1, p. 23 – 32 (1984).
He I λ10830 line fluxes are reported for eight type 1 Seyfert galaxies, three QSOs, and two type 2 Seyfert galaxies, and combined with published He I λ5876 line fluxes to form the ratio λ10830/λ5876. The observed values of λ10830/λ5876 are compared to values calculated for and measured in gaseous nebulae. These "dustless gaseous nebula" values are very similar to values observed in the type 2 Seyfert galaxies but are considerably larger than, and incompatible for any amount of reddening with, values measured in the type 1 Seyfert galaxies and QSOs.

158.105 **Effects of dust on the infrared emission, selected line ratios, and polarization of Seyfert 1 galaxies, broad–line radio galaxies, and quasars.**
R. J. Rudy.
Astrophys. J., Vol. 284, No. 1, p. 33 – 43 (1984).
This paper investigates the effects of dust on the infrared emission, the Balmer decrement, the [O III]/Hβ (broad) value, the Hβ luminosity, and the optical polarization for active extragalactic objects with broad emission lines. Selection effects pertaining to dust content which are inherent to optical surveys are also discussed. The apparent lack of observational evidence for dust in quasars in past studies is a consequence of both selection effects and the gradual weakening of the narrow–line region with increased luminosity. Dust does exist in detectable amounts in quasars with strong narrow–line regions.

158.106 **Cosmological evolution of radio sources found at 1.4 GHz.**
J. J. Condon.
Astrophys. J., Vol. 284, No. 1, p. 44 – 53 (1984).
Radio source counts, the strong–source redshift distribution, and the local luminosity functions of elliptical and spiral galaxies are used to constrain the evolution of radio sources selected at 1.4 GHz. These observables are expressed in terms of "visibility functions" to show quantitatively (1) the insensitivity of statistical tests on faint–source samples to possible cutoffs at high redshifts, (2) the precision with which source evolution must be specified by physical theory before the density parameter Ω can be determined from source counts, and (3) differential evolution is not required by the data – all sources can evolve in the same way, regardless of luminosity.

158.107 **Hard X–ray variability of three active galactic nuclei.**
P. Ubertini, A. Bazzano, C. La Padula, V. F. Polcaro, R. K. Manchanda.
Astrophys. J., Vol. 284, No. 1, p. 54 – 59 (1984).
The authors report spectral observation of three active galactic nuclei in the hard X–ray range 20 – 100 keV. The data include the two Seyfert galaxies MCG 8–11–11 and NGC 4151 and BL Lac object Mrk 421. The results have been obtained with a large–area balloon–borne telescope consisting of a 1 $m^2$ array of multiwire Spectroscopic Proportional Counters. The very flat hard X–ray spectra are fitted by power laws. A comparison of the present spectral measurements with earlier data shows the long–term variability of these AGNs, in terms of integral intensity, above 20 keV.

158.108 **The sub–milliarcsecond radio structure of AO 0235+164.**
D. L. Jones, L. B. Bååth, M. M. Davis, S. C. Unwin.
Astrophys. J., Vol. 284, No. 1, p. 60 – 64 (1984).
Very high resolution (<0.2 milli–arcsec) VLBI hybrid maps of the BL Lacertae object AO 0235+164 have been produced at a wavelength of 1.3 cm. These maps show the object's radio structure to be dominated by a strong, nearly unresolved core. The small angular size of this core, combined with previous X–ray observations, implies that bulk relativistic motion is occurring in the core. In addition, a weak component is detected 0.7 milli–arcsec from the core in position angle +7°.

158.109 **The $\Psi_{IPS}$–LAS (*largest angular size*) relation for extragalactic radio sources.**
D. G. Banhatti.
Mon. Not. R. Astron. Soc., Vol. 211, No. 2, p. 369 – 373 (1984).
Metre–wavelength interplanetary scintillation (IPS) observations give the overall angular sizes ψ of scintillating compact structures in radio sources. From 326.5–MHz IPS data for a sample of faint (Ooty) radio sources, log ψ versus log LAS is seen, on average, to have a slope 0.2. A similar trend is seen from 81.5–MHz IPS data for a sample of strong, powerful (3CR) double sources, although the slope is 0.4 and the mean ψ about four times larger. If ψ represents the hotspot size, the log ψ–log LAS correlation has a bearing on the collimation of extended jets in powerful double radio sources. A similar relation derived from direct observations of hotspots in such sources chosen to minimize projection effects has been found to have a slope 0.7. The author shows that projection effects are partly responsible for the small value 0.4 of the slope of the log ψ–log LAS correlation. Blending effects in the IPS observations probably account for the remaining flattening.

158.110 **NGC 4151: Sy2 in a deep photometric minimum.**
V. M. Lyutyj, V. L. Oknyanskij, K. K. Chuvaev.
Pis'ma Astron. Zh., Tom 10, No. 11, p. 803 – 807 (1984). In Russian. English translation in Sov. Astron. Lett., Vol. 10.
Coordinated *UBV* and spectral observations of the nucleus of the Seyfert galaxy NGC 4151 revealed that the brightness decrease in April – May 1984 was accompanied by a ~2.5–fold decrease of the $W_\lambda(H\beta)/W_\lambda(N_2)$ ratio during ~3 months. This minimum is the deepest one since 1967. The color indices $B-V = +0^m92$, $U-B = -0^m05$ and the value $W_\lambda(H\beta)/W_\lambda(N_2) \approx 0.3$ in this minimum correspond to a Sy2 galaxy.

158.111 **Distant radio galaxies of moderate luminosity in the vicinity of 3C 273.**
V. V. Vitkovskij, W. Reich, N. S. Soboleva, A. V. Temirova.
Pis'ma Astron. Zh., Tom 10, No. 11, p. 808 – 814 (1984). In Russian. English translation in Sov. Astron. Lett., Vol. 10.
The coordinates and spectra of sources detected at 7.6–cm wavelength in the region of the sky $\alpha_{1950.0} = 12^h19^m \pm 5^m$ and $\delta_{1950.0} = 2°22' \pm 20'$ are presented. Four objects with flux density more than 30 mJy are most likely radio galaxies of moderate luminosity.

158.112 ***UBV* photometry of a galaxy of Seyfert type.**
M. A. Kazaryan, V. S. Tamazyan.
Pis'ma Astron. Zh., Tom 10, No. 11, p. 815 – 819 (1984). In Russian. English translation in Sov. Astron. Lett., Vol. 10.
Detailed photographic *UBV* photometry of the double galaxy with *UV* excess No. 163 from the list by Kazaryan (1979) was made. The integral *B* magnitudes and color indices *U–B*, *B–V* for the whole galaxy as well as for different components and their nuclei are presented. The location of the components on the two–color diagram is considered.

158.113 **Extragalactic radio jets.**
A. H. Bridle, R. A. Perley.
Annu. Rev. Astron. Astrophys., Vol. 22, p. 319 – 358 (1984). – See Abstr. 003.024.
Contents: (1) Introduction: Why "jets"? What makes a "narrow feature" a "jet"? (2) The incidence of extragalactic radio jets: A list of known jets. Occurrence rates of jets. Jets in weak sources? (3) Trends with luminosity: Sidedness. Magnetic configuration. Size, curvature, and misalignments. (4) Structural details: Collimation, freedom, and confinement. Radio spectra. Intensity evolution. (5) Radio jets at other wavelengths: Optical and infrared wavelengths. X-ray wavelengths. (6) Jet lag–missing physics: Velocity estimates. The velocity dilemma. Jet/hot-spot symmetries and the sidedness dilemma. Stability. Unified models. A broader unified model. (7) Summary and some key experiments.

158.114 **Optical and UV emission of active galactic nuclei.**
M. H. Ulrich.
MPE Rep., No. 184, p. 1 – 10 (1984). – See Abstr. 012.039.
The study of the optical and ultraviolet properties of active galactic nuclei – quasars, BL Lac objects, Seyfert and radio galaxies – has been very active in the last ten years. This paper summarizes the most recent progress and outlines some of the outstanding unsolved problems.

158.115 **BL Lacertae objects: multifrequency observations and model applications.**
D. M. Worrall.
MPE Rep., No. 184, p. 11 – 13 (1984). – See Abstr. 012.039.
Coordinated multifrequency observations at radio, infrared, optical and ultraviolet wavelengths have been made for 8 BL Lacertae-type and related objects. In order to obtain an adequate fit to spectral and variability data, a moderate relativistic beaming factor, $\delta \lesssim 30$, is required.

158.116 **Radio Seyferts & the forbidden line region.**
A. Pedlar, J. E. Dyson.
MPE Rep., No. 184, p. 14 – 17 (1984). – See Abstr. 012.039.
The authors describe a model in which the expansion of the radio emitting components in Seyfert 2 nuclei compress and accelerate the ambient medium by means of shock waves. When the expansion velocity of the component has decreased to $<400$ km/s the compressed gas cools and radiates producing the forbidden line region. This readily accounts for the intimate relationship between the radio continuum emission and the properties of the forbidden line region.

158.117 **The relationship between radio structures and stellar dynamics of nearby elliptical galaxies.**
M. Birkinshaw, R. L. Davies.
MPE Rep., No. 184, p. 18 – 20 (1984). – See Abstr. 012.039.
Preliminary results of a project to compare the radio and rotation axes of nearby elliptical galaxies are presented. The results so far suggest that the radio axes generally lie close to the rotation axes, and that the galaxies hosting radio sources are not dynamically unusual.

158.118 **Active inner regions of isolated galaxies.**
N. Brosch.
MPE Rep., No. 184, p. 21 – 24 (1984). – See Abstr. 012.039.
A sample of very isolated galaxies has been observed in a number of spectral regions, from 5 GHz to $2.5 \times 10^{15}$Hz. Nuclear UV excesses were found to be a general phenomenon. This is interpreted as indicating that in isolated galaxies the inner regions are more active in forming stars than in non-isolated galaxies.

158.119 **A new estimate of the local luminosity function of Seyfert 1 nuclei.**
F. Z. Cheng, L. Danese, G. De Zotti, A. Franceschini.
MPE Rep., No. 184, p. 25 – 27 (1984). – See Abstr. 012.039.
A new optical luminosity function for Seyfert 1 nuclei has been derived from a sample of 88 galaxies.

158.120 **Optical studies of X-ray selected AGN in the direction of the LMC.**
M. W. Pakull.
MPE Rep., No. 184, p. 28 – 31 (1984). – See Abstr. 012.039.
Optical follow-up studies of 19 X-ray point sources detected with the Einstein HRI detector have revealed that the majority (17) are not associated with our neighbouring galaxy being either galactic foreground stars or background interlopers. It is found that 80% of this sample are coronal-type stellar X-ray emitters. Only 4 objects are identified with active galactic nuclei showing broad Balmer- and narrow forbidden emission. Their X-ray and optical luminosity classify them as Seyfert 1 galaxies at low redshift ($z < 0.2$). Two AGN at $z = 0.063$ and 0.075 are optically resolved on deep Schmidt plates.

158.121 **Continuous transition from star burst galaxies to Seyfert galaxies?**
G. F. O. Schnur, T. Schmidt-Kaler, J. V. Feitzinger.
MPE Rep., No. 184, p. 32 – 34 (1984). – See Abstr. 012.039.
The authors present high-resolution ($2 \times 10^4$) spectra of the nuclei of 5 galaxies, including the Seyfert galaxy NGC 2110 and the star burst galaxy NGC 2993, and discuss the possible relation between the Seyfert phenomenon and star formation activity in galactic nuclei.

158.122 **The ultraviolet spectra of the narrow line active galaxies NGC 3081 and NGC 4507.**
F. Durret, J. Bergeron.
MPE Rep., No. 184, p. 35 – 39 (1984). – See Abstr. 012.039.
Ultraviolet and optical spectroscopy of the narow line active galaxies NGC 3081 and NGC 4507 has been obtained. The power law index for the continuum between the UV and the X-ray range is derived for the galactic nuclei and emission line intensities are presented.

158.123 **Preliminary results on the ultraviolet observations of a complete X-ray selected sample of AGNs.**
C. Boisson.
MPE Rep., No. 184, p. 40 – 42 (1984). – See Abstr. 012.039.
UV spectra from a sample of X-ray selected Seyfert galaxies and BL Lacertae objects are summarily described.

158.124 **Spatially resolved optical spectroscopy of the radio jet of the X-ray galaxy NGC 2110.**
G. F. O. Schnur, J. V. Feitzinger, T. Schmidt-Kaler.
MPE Rep., No. 184, p. 55 – 57 (1984). – See Abstr. 012.039.

158.125 **Extragalactic cosmic rays from active galaxies.**
M. M. Shapiro, R. Silberberg.
MPE Rep., No. 184, p. 58 – 60 (1984). – See Abstr. 012.039.
This paper discusses the feasibility of detecting neutrinos from active galactic nuclei with the DUMAND (deep underwater muon and neutrino detection) array.

158.126 **A comparison of some photometric parameters of active and normal spiral galaxies.**
A. T. Kalloghlian (*A. T. Kalloglyan*).
MPE Rep., No. 184, p. 61 – 62 (1984). Abstract. – See Abstr. 012.039.

158.127 **X-ray emission from active galactic nuclei.**
R. Mushotzky.
MPE Rep., No. 184, p. 73 – 87 (1984). – See Abstr. 012.039.
It is often held that the X-ray emission from active galactic nuclei (AGN) arises from a region close to the central energy source. Thus X-ray observations may provide our best constraints on the central engine. In particular, the shape of the X-ray continuum gives information about the mechanism for photon generation, X-ray time variability data can constrain the "size" and "mass" of the continuum source, and X-ray "occultation" data gives constraints on the relative sizes of the continuum source and the intervening absorbing material (often assumed to be the broad line clouds).

158.128 **EXOSAT observations of active galactic nuclei.**
G. Branduardi-Raymont, S. J. Bell-Burnell, B. Kellett, H. Fink, D. Molteni, I. McHardy.
MPE Rep., No. 184, p. 88 – 91 (1984). – See Abstr. 012.039.
The authors report EXOSAT observations of the type 1 Seyfert galaxies NGC 3783 and NGC 5548.

158.129 **Soft X-ray emission from BL Lac objects.**
D. Maccagni, L. Chiappetti, B. Garilli, P. Giommi, D. Molteni, M. Morini, A. Rampini.
MPE Rep., No. 184, p. 92 – 95 (1984). – See Abstr. 012.039.
EXOSAT X-ray imaging observations of the BL Lacertae objects PKS 0548–322, AP Librae and BL Lacertae are reported.

158.130 **X-ray emission from PKS 2155–304.**
H. Kunieda.
MPE Rep., No. 184, p. 99 – 101 (1984). – See Abstr. 012.039.
Observations of the BL Lacertae object PKS 2155–304 with the Japanese X-ray satellite TENMA are summarized.

158.131 **X-ray emission from Centaurus A.**
J. Terrell.
MPE Rep., No. 184, p. 102 – 104 (1984). – See Abstr. 012.039.
The X-ray strength of Cen A (NGC 5128) was monitored for 10 years by Vela 5B, beginning in 1979. The 1973 – 1975 period of exceptional activity and strength was especially well covered, and was characterized by numerous rapid changes in intensity, in agreement with other data, indicating that the nucleus of Cen A is small but massive.

158.132 **X-ray emission from Centaurus A observed by TENMA.**
F. Nagase, T. Hirano.
MPE Rep., No. 184, p. 105 – 107 (1984). – See Abstr. 012.039.
The radio galaxy Centaurus A was observed from 30 March to 4 April 1984 by TENMA. The average X-ray flux was $1.3 \times 10^{-10} \mathrm{erg\,s^{-1}cm^{-2}}$ in the 2 – 6 keV range, suggesting the low intensity state during the observation. The observed pulse height spectrum indicates an excess in the soft X-ray emission below $\sim$3 keV.

158.133 **The X-ray spectra of NGC 4151.**
M. Matsuoka, T. Ikegami.
MPE Rep., No. 184, p. 108 – 110 (1984). – See Abstr. 012.039.
TENMA observed NGC 4151 in January and March 1984. During these two intervals the X-ray intensity was as weak as a few μJy, showing variability on a time scale of a few days. Spectra from NGC 4151 reveal a strong iron emission line with an energy of $\sim$6.4 keV; the equivalent width for both periods was 300 – 350 eV.

158.134 **UV and X-ray observations of AGN: two comparisons with other wavelengths.**
M. A. Malkan.
MPE Rep., No. 184, p. 121 – 128 (1984). – See Abstr. 012.039.
Ultraviolet and X-ray continuum observations have proven especially useful in the study of active galactic nuclei when they are combined with measurements at lower frequencies. This paper discusses correlations between blue and UV colors and between infrared and X-ray spectral slopes for AGN.

158.135 **On the surface density of X-ray selected BL Lacertae objects.**
I. M. Gioia, T. Maccacaro, D. Maccagni, J. Stocke.
MPE Rep., No. 184, p. 135 – 137 (1984). – See Abstr. 012.039.
This paper derives an estimate of the X-ray number flux relation of BL Lac objects and discusses the implications of the results with respect to their luminosity function, cosmological evolution and X-ray to optical flux ratio.

158.136 **Comments on the physics of active galactic nuclei.**
M. J. Rees.
MPE Rep., No. 184, p. 138 – 151 (1984). – See Abstr. 012.039.
Some physical processes relevant to the central parsec of active galactic nuclei (AGNs) are discussed.

158.137 **On the broad band energy spectra of quasars and BL Lac objects.**
L. Maraschi.
MPE Rep., No. 184, p. 164 – 176 (1984). – See Abstr. 012.039.
This paper discusses emission mechanisms operating in the various wavelength ranges for quasars and BL Lacertae objects and describes attempts to relate these mechanisms to a unified model of active galactic nuclei.

158.138 **A universal spectrum for AGNs and QSOs.**
D. Kazanas.
MPE Rep., No. 184, p. 177 – 180 (1984). – See Abstr. 012.039.
The effects of the feedback of $e^+$–$e^-$ pair reinjection in a plasma due to photon photon absorption of its own radiation is examined. Under the assumption of continuous electron injection with a power law spectrum $E^{-\Gamma}$ and Compton losses only, it is shown that for $\Gamma < 2$ the steady state electron distribution function has a unique form independent of the primary injection spectrum. This electron distribution function can by then reproduce the general characteristics of AGN, QSO spectra from radio to X-rays.

158.139 **An inverse Compton model of the X-ray emission in Seyfert galaxies.**
G. C. Perola, L. Piro.
MPE Rep., No. 184, p. 181 – 184 (1984). – See Abstr. 012.039.
Inverse Compton (IC) scattering of low energy photons by relativistic electrons is one of the radiation mechanism which may be responsable for the X-ray emission in Seyfert galaxies. This contribution summarizes the radiative properties of IC models as they apply to the case of NGC 4151, which takes into account all the observational constraints available from the extensive data on this object.

158.140 **Theoretical studies of emission line variability in type 1 Seyfert galaxies.**
A. Robinson.
MPE Rep., No. 184, p. 224 – 228 (1984). – See Abstr. 012.039.
The emission line variability observed in type 1 Seyfert galaxies is investigated by means of a time dependent photoionization model of the broad line emitting gas.

158.141 **On the stability of broad line clouds in active galactic nuclei.**
M. Elitzur, G. Ferland.
MPE Rep., No. 184, p. 229 – 231 (1984). – See Abstr. 012.039.
The radiation pressure caused by the trapped line photons in clouds producing the broad line emission around active galactic nuclei is analyzed and found to be important for the stability of the clouds. It may provide a natural explanation for the great uniformity in values of density and ionization parameter found in quasars and Seyfert galaxies.

158.142 **Active galactic nuclei and their cosmological evolution.**
G. Setti.
MPE Rep., No. 184, p. 243 – 253 (1984). – See Abstr. 012.039.
This paper reviews evidence for the cosmological evolution of quasars and of Seyfert type 1 nuclei as indicated mainly from optical observations.

158.143 **VLBI and X-ray observations of complete samples of radioselected active galactic nuclei.**
P. Biermann, J. A. Zensus.
MPE Rep., No. 184, p. 275 – 280 (1984). – See Abstr. 012.039.
With the purpose to improve the physical understanding of active galactic nuclei the authors discuss their campaign to obtain VLBI and X-ray observations of complete samples, selected by flat radio spectrum. They concentrate here on the sample from the 5 GHz 1-Jy catalogue in the declination range $+10°$ to $+70°$. The inverse Compton predictions for the expected X-ray flux densities are derived; excess over observed flux density is usually interpreted as evidence for bulk relativistic motion.

158.144 **Dust in the nuclei of the Seyfert galaxies Markarian 231 and NGC 4151.**
B. Jones, D. M. Worrall, J. M. Rodriguez–Espinosa, W. A. Stein, F. C. Gillett.
Publ. Astron. Soc. Pac., Vol. 96, No. 583, p. 692 – 698 (1984).
The authors report 8 – 13 μm grating–spectrometer observations of Mrk 231 and NGC 4151. The Mrk 231 data can be fitted to various thermal dust emission models or a single power law, with dust extinction. In all the model fits, except for that of graphite and silicon carbide grain emission, a component of silicate absorption of optical depth $\tau_{9.7\mu m} \geqslant 0.7$ is required. Confirming published work, the authors rule out the absorption being at the redshift of the low–redshift absorption–line system. The high values of $\tau_{9.7}$ do not give ratios to the galaxy's visual extinction which are comparable to those of galactic H II regions. The authors also report weak evidence for a 10 μm absorption feature in NGC 4151.

158.145 **Optical monitoring of OJ 287 during the 1982 – 83 observing season.**
G. J. Corso, B. Purcell, M. Giroux, J. Schultz.
Publ. Astron. Soc. Pac., Vol. 96, No. 583, p. 705 – 706 (1984).
Iris photometry of photographic plates of OJ 287 taken in blue light with the 1–meter reflector of Northwestern's Lindheimer Astronomical Research Center between mid–November 1982 and early May 1983 are reported.

158.146 **VLA observations of radio galaxies with extended lobe emission: 3C 79 and 3C 430.**
S. R. Spangler, S. T. Myers, J. J. Pogge.
Astron. J., Vol. 89, No. 10, p. 1478 – 1486 (1984).
Dual–frequency (1.446 and 4.885 GHz) total intensity and polarimetric observations have been made with the VLA of the radio galaxies 3C 79 and 3C 430. These sources are characterized by extended lobe and/or radio bridge emission. 3C 430 possesses an asymmetric structure which may indicate large–scale inhomogeneities in a confining medium. This source also displays Faraday rotation fluctuations which are apparently unrelated to depolarization, suggesting an origin in an external medium. The polarization observations suggest that the magnetic fields in both sources possess large–amplitude, random components. In the case of 3C 79, the authors observe a correlation between depolarization and position–angle rotation. It is unclear whether this is due to internal Faraday rotation, indicating the presence of a large–scale magnetic field, or a Faraday–rotating screen external to the radio–emitting material.

158.147 **Kinematics and morphological type of the Seyfert galaxy NGC 4507.**
K. J. Adney, A. S. Wilson, J. A. Baldwin.
Astron. J., Vol. 89, No. 10, p. 1514 – 1519 (1984). With plate 82.
Long–slit, medium–high dispersion optical spectroscopy and direct imaging of the class 2 Seyfert galaxy NGC 4507 are reported. The results are consistent with the morphological classification SBab(rs)I given in the Revised Shapley–Ames Catalog. Evidence has been found that the nuclear [O III] line emitting region is structurally and kinematically distinct from the galaxy disk. The distribution of line profiles over this nuclear region is presented and its kinematics discussed.

158.148 **Radio observations of the jet in Fornax A.**
B. J. Geldzahler, E. B. Fomalont.
Astron. J., Vol. 89, No. 11, p. 1650 – 1657 (1984).
The radiation near the center of Fornax A (NGC 1316) has been mapped at the VLA at 1.5, 4.9, and 15.0 GHz with arcsecond and subarcsecond resolution. The radio structure consists of a steep–spectrum core and dual–opposing jets which extend about 5 kpc before abruptly terminating. The inner 300 pc of the jets contain discrete knots, the outer parts of the jets are continuous. The jets show significant departures from linearity which appear to be produced by interactions with the galactic medium. The jets point in the general vicinity of the extended radio lobes with which there is no visible connection. The lack of hot spots in the lobes may be caused by short–term variation in the direction of the energy flow to the lobes.

158.149 **H I absorption in the direction of galactic–plane variables: evidence for new, rapidly variable extragalactic radio sources.**
A. R. Taylor, E. R. Seaquist.
Astron. J., Vol. 89, No. 11, p. 1692 – 1694 (1984).
The authors have carried out H I absorption observations of three radio sources in the galactic plane that vary on time scales of a few days. The absorption spectra of all three sources contain components corresponding to the full range of H I emission velocities. Although the authors cannot strictly rule out a galactic origin, their results strongly suggest that these objects are rapidly variable extragalactic radio sources. One source, GT2100 + 468, is among the most highly variable of known extragalactic radio sources.

158.150 **Cen A – galactic cannibal.**
J. Milogradov–Turin.
Vasiona, Année 32, No. 3, p. 45 – 48 (1984). In Serbo–Croatian.
A review of the observational results and the recent model based on galactic cannibalism are presented.

158.151 **Multifrequency observations of the BL Lacertae objects OQ 530 and ON 325.**
D. M. Worrall, J. J. Puschell, F. C. Bruhweiler, M. L. Sitko, W. A. Stein, M. F. Aller, H. D. Aller, P. E. Hodge, R. J. Rudy, H. R. Miller, W. Z. Wiśniewski, F. A. Córdova, K. O. Mason.
Astrophys. J., Vol. 284, No. 2, p. 512 – 518 (1984).
The authors present spectral measurements of the two BL Lac objects OQ 530 and ON 325, from radio to ultraviolet frequencies. Both sources exhibit spectral curvature between infrared and ultraviolet frequencies. The sources differ from one another in optical variability behavior. The intensity and the spectral shape of OQ 530 vary, but in an uncorrelated manner. The spectral shape of ON 325, however, appears to remain constant during intensity variability. A nonthermal jet emission model is applied to the data, incorporating both the spectral and time variability information.

158.152 **A search at the millijansky level for milli–arcsecond cores in a complete sample of radio galaxies.**
A. E. Wehrle, R. A. Preston, D. L. Meier, M. V. Gorenstein, I. I. Shapiro, A. E. E. Rogers, A. Rius.
Astrophys. J., Vol. 284, No. 2, p. 519 – 522 (1984).
A complete sample of 26 extended radio galaxies was observed at 2.29 GHz with the Mark III VLBI system. The fringe spacing was ~3 milli–arcsec, and the detection limit was ~2 millijanskys. Half of the galaxies were found to possess milli–arcsec radio cores. In all but three sources, the nuclear flux density was less than 0.04 of the total flux density. Galaxies with high optical luminosity ($M_v \leqslant -21.2$) were more likely than less luminous galaxies to contain a detectable milli–arcsec radio core (69% vs. 20%).

158.153 **Reddening in the broad–line radio galaxy 3C 234.**
N. P. Carleton, S. P. Willner, R. J. Rudy, A. T. Tokunaga.
Astrophys. J., Vol. 284, No. 2, p. 523 – 530 (1984).
A measurement of the Paschen–α flux from the broad–line radio galaxy 3C 234 shows a line strong enough to demand reddening ($A_V \approx 2.6$) as a crucial part of the explanation of the observed hydrogen line ratios. Since the narrow–line region appears strongly photoionized and only slightly reddened, the obscuring material is neither in our line of sight to this region nor between this region and the central source. A high–resolution profile of [O III] λ5007 shows attenuation of the red wing, consistent with obscuration of outflowing material that is moving directly away from us. A continuum corrected for reddening can provide all of the ionizing photons and sufficient luminosity to explain the infrared emission.

158.154 **Radio continuum activity in the elliptical galaxy NGC 1052.**
J. M. Wrobel.
Astrophys. J., Vol. 284, No. 2, p. 531 – 543 (1984).

Aperture synthesis maps of the continuum emission from the nucleus of the elliptical galaxy NGC 1052 (0238–084) are presented at two frequencies near 1.5 GHz. The radio emission is dominated by a compact, flat-spectrum core of extent $\lesssim 1''$ (93 pc). A 30" (2.8 kpc) double source straddles the compact core, in contrast with the usual compact core plus jet morphology. The 30" source shows a collimated, bent, and asymmetric structure. Embedded within each lobe of the double source is an active region which exhibits a relatively flat spectral index (–0.4), perhaps indicating the occurrence of in situ particle acceleration. The predicted inverse Compton X–ray flux from both active regions is four orders of magnitude below the observed level, making it unlikely that this emission mechanism is the cause of the galaxy's elongated X–ray image. The active regions are in pressure balance with the postshock gas associated with the "clouds" of gas which form the extended optical line–emitting region.

158.155 **On the surface density of X–ray selected BL Lacertae objects.**
T. Maccacaro, I. M. Gioia, D. Maccagni, J. T. Stocke.
Astrophys. J., Lett. Ed., Vol. 284, No. 2, p. L23 – L27 (1984).

Only a handful of BL Lac objects have been found as a result of systematic optical identification of serendipitous Einstein X–ray sources. By combining the data from two flux–limited complete X–ray surveys the authors have evaluated the surface density of X–ray emitting BL Lac objects as a function of their X–ray flux. They find that a single power law is not an acceptable representation of the BL Lac objects' X–ray log $N$–log $S$. The number–flux relationship is consistent with the Euclidean slope at "high" flux levels but shows a drastic flattening below fluxes of the order of $10^{-12}$ergs cm$^{-2}$s$^{-1}$.

158.156 **Luminosity–dependent effects on the hot spots in extended double radio sources.**
Gopal–Krishna, L. Saripalli.
Astron. Astrophys., Vol. 139, No. 2, p. L19 – L21 (1984).

Following a simple approach based on the beam model for the formation of extended extragalactic double radio sources and assuming that the usual equipartition condition prevails inside their hot spots which are roughly spherical in shape, the authors attempt to explain the observed trend that the hot spots found in sources of lower luminosities are generally more diffuse. It is further argued, that non–relativistic beam velocities ($\leqslant 0.1c$) can only give rise to hot spots of luminosities up to $\sim 10^{44}$erg s$^{-1}$. For hot spots powered by relativistic beams, much higher luminosities are attainable but a qualitative change in morphology should occur above a luminosity of around $10^{46}$erg s$^{-1}$.

158.157 **A new estimate of the mass of M87.**
J.–L. Nieto, G. Monnet.
Astron. Astrophys., Vol. 139, No. 2, p. 464 – 466 (1984).

The luminosity profile of the compact elliptical galaxy NGC 4486B, companion of M87, follows with a reasonable accuracy a King profile having a tidal radius of 27". An estimate of its mass can be derived from that of M32. The use of the revised classical von Hoerner's formula in which M87 has a distributed mass $M(r) \propto r^{+1.3}$, leads to a mass $M > 5.2 \times 10^{12} M_\odot$ in the central 7:3 of M87 and to a corresponding $M/L_v > 58$ for an adopted distance of 15 Mpc. A conservative assumption of a point mass would have given $M > 3.3 \times 10^{12} M_\odot$ and $M/L_v > 38$. These results are in excellent agreement with estimates of the M87 mass from recent X–ray observations.

158.158 **The recent two magnitude decline of OJ287 in blue light.**
G. J. Corso, J. Schultz, T. Pfaff, B. Purcell.
Astron. Astrophys., Vol. 140, No. 1, p. L23 (1984).

Iris photometry is reported for blue photographic observations of OJ287 made between early March and late May 1984 with the 1–meter reflector of Northwestern's Lindheimer Astronomical Research Center.

158.159 **Seyfert nuclei as short–lived stochastic accretion events.**
R. H. Sanders.
Astron. Astrophys., Vol. 140, No. 1, p. 52 – 54 (1984).

The minimum statistical lifetime of Seyfert activity in a particular galaxy is $3 - 7 \times 10^8$yr. The recent discovery of radio continuum jets in the central regions of Seyferts may constrain the lifetime of individual Seyfert episodes to be shorter than $10^6$yr implying that nuclei evolve through at least 100 recurring Seyfert episodes. If the energy source of Seyfert activity is accretion onto a central black hole, then random jet orientations suggest that each event is accompanied by its own independent accretion disk in no preferred plane (stochastic fueling). The two different types of Seyferts may result from very different time scales for the fueling events.

158.160 **Multiple hotspots in radio galaxies.**
E. Valtaoja.
Astron. Astrophys., Vol. 140, No. 1, p. 148 – 154 (1984).

The properties of multiple hotspots in a sample of radio sources are investigated. The morphologies and energetics of the more diffuse "secondary" hotspots do not support the hypothesis, common to both the beam and the slingshot models of radio sources, that they are simply remnants of previous activity now observed in the compact "primary" hotspots. If the secondary hotspots are instead formed by directed outflow of plasma from the primary hotspots, their properties can be better understood. Moderately collimated ejection along the axis of a supermassive object embedded in the primary hotspot is suggested to be the underlying mechanism.

158.161 **Spectrophotometric scans of galaxies and Strömgren *uvby* photometry.**
M. J. Stift.
Astron. Astrophys., Vol. 140, No. 1, p. 155 – 158 (1984).

The author investigates recent claims that there exist certain intrinsic relations between Strömgren indices and redshift values of elliptical galaxies and of N–galaxies. Estimating synthetic colours in the rest system of the respective galaxies from spectrophotometric surveys, the author cannot confirm any of the alleged trends in the ($b$–$y$) and the $m_1$ indices of these galaxies. At present no discrimination appears possible between N–galaxies, "active ellipticals" and "standard candles" with the help of Strömgren indices only. Finally, problems arising from the calibration of absolute spectrophotometry at near–infrared wavelengths and from chance effects such as light variability and reddening are pointed out.

158.162 **The ionization mechanism for gas in the nuclei of NGC 1052 and NGC 4278.**
J. A. Rose, M. J. Tripicco.
Astrophys. J., Vol. 285, No. 1, p. 55 – 63 (1984).

Photographic image–tube spectra at a resolution of 2.5 Å in the blue have been obtained for the nuclei of the active elliptical galaxies NGC 1052 and NGC 4278 and for several emission–free galaxy nuclei. The purpose of this study was to determine the ionization mechanism for low–ionization active galactic nuclei. A search for the diagnostic emission line [O III]λ4363 in NGC 1052 yields a lower limit of 47 to the ratio [O III] (λ5007 + λ4959)/λ4363, which implies a temperature for the [O III]–emitting gas less than 17,600K. This temperature is incompatible with shock ionization, and it is concluded that the nuclear gas in NGC 1052 is photoionized by a nonstellar UV continuum.

158.163 **The variable component of the NGC 4151 *IUE* spectrum: evidence for an expanding spherical emission shell.**
R. Stoner, R. Ptak, S. Gregory.
Astrophys. J., Vol. 285, No. 1, p. 69 – 73 (1984).

Low–resolution SWP spectra of the nucleus of the Seyfert galaxy NGC 4151 obtained with the *IUE* in 1980 November and 1982 December are compared. In the difference spectrum, which is interpreted to be the spectrum of a time–variable component, C IV λ1550 is the only strong emission feature. The profile of this line closely resembles that of a spherical shell surrounding a less dense gas uniformly expanding from a central opaque object. The authors infer that the shell has a radius of about $10^{15}$cm, a mass of 0.25 $M_\odot$, an expansion velocity of 13,000 km $s^{-1}$, and an electron density of $10^{11}cm^{-3}$.

158.164 **Submillimeter continuum observations of M82.**
D. T. Jaffe, E. E. Becklin, R. H. Hildebrand.
Astrophys. J., Lett. Ed., Vol. 285, No. 1, p. L31 – L33 (1984).
The authors have mapped the submillimeter continuum emission from the core of the irregular galaxy M82. The 400 μm flux density toward the emission peak is 30 Jy. The source size is ⩽40″ (630 pc). The position is consistent with the 2.2 μm nuclear position, with that of the compact nonthermal radio source near the nucleus and with the center of the 10 μm continuum and $J = 1 \rightarrow 0$ CO distributions. Assuming a normal gas–to–dust ratio, the submillimeter emission implies a mass of $8 \times 10^7 M_\odot$ for the interstellar medium in the inner 650 pc diameter region.

158.165 **The jet and filaments in Cygnus A.**
R. A. Perley, J. W. Dreher, J. J. Cowan.
Astrophys. J., Lett. Ed., Vol. 285, No. 1, p. L35 – L38 (1984). With plates L1 – L2.
Multiconfiguration VLA observations of Cygnus A at 1.4 and 5 GHz have been used to produce detailed maps of exceptional dynamic range. A radio jet extends from the core into the northwest lobe. The broad radio lobes reveal an unexpected wealth of filamentary structure. Two new hot spots are identified. The pressure of the cluster gas in which Cyg A is embedded appears to be insufficient to confine either the jet or the filaments. The jet is probably relativistic.

158.166 **Rapid variability in optical polarization of the quasar–like object OJ287.**
A. K. Kulshrestha, U. C. Joshi, M. R. Deshpande.
Nature, Vol. 311, No. 5988, p. 733 – 734 (1984).
The authors report observations of rapid and strong variations in the degree of linear polarization of the optical flux of the BL Lacertae object OJ287.

158.167 **Parameters of active galactic nuclei and the approach to critical luminosity.**
É. A. Dibaj (*Eh. A. Dibaj*).
Sov. Astron., Vol. 28, No. 2, p. 123 – 127 (1984). English translation of 37.158.101.

158.168 **An empirical model for active galactic nuclei. I. The catalog.**
É. A. Dibai (*Eh. A. Dibaj*).
Sov. Astron., Vol. 28, No. 3, p. 245 – 253 (1984). English translation of 37.158.201.

158.169 **An analysis of the narrow–line profiles in high–ionization Seyfert galaxies.**
M. M. De Robertis, D. E. Osterbrock.
Publ. Astron. Soc. Pac., Vol. 96, No. 584, p. 787 (1984). Abstract. – See Abstr. 010.281.

158.170 **Spectra of emission–line galaxies in Wasilewski's north galactic pole field.**
D. E. Osterbrock.
Publ. Astron. Soc. Pac., Vol. 96, No. 584, p. 792 (1984). Abstract. – See Abstr. 010.281.

158.171 **Imaging and spectroscopy of very distant radio galaxies.**
H. Spinrad, S. Djorgovski.
Publ. Astron. Soc. Pac., Vol. 96, No. 584, p. 795 (1984). Abstract. – See Abstr. 010.281.

158.172 **Distorted structure in a $z$ = 1.9 radio source.**
C. J. Lonsdale, P. D. Barthel.
Bull. Am. Astron. Soc., Vol. 16, No. 3, Part I, p. 727 (1984). Abstract. – See Abstr. 010.065.

158.173 **Time variations in ultraviolet emission lines of the Seyfert galaxies F9 and MKN1095.**
R. Ptak, S. McCracken, R. Stoner.
Bull. Am. Astron. Soc., Vol. 16, No. 3, Part I, p. 731 (1984). Abstract. – See Abstr. 010.065.

158.174 **A polarization flare in OJ 287.**
P. A. Holmes, P. W. J. L. Brand, C. D. Impey, P. M. Williams, P. Smith, R. Elston, T. Balonek, M. Zeilik, J. Burns, P. Heckert, R. Barvainis, J. Kenny, G. Schmidt, J. Puschell.
Mon. Not. R. Astron. Soc., Vol. 211, No. 3, p. 497 – 506 (1984).
Observations of the BL Lac object OJ 287 were made simultaneously at infrared *JHK* wavelengths and in optical and near infrared wavelengths at *UBVRI* in 1983 January. These observations, of total flux reaching a peak and then decaying, and of complex polarization changes, are explained in terms of a model of a jet emanating from an accretion disc.

158.175 **A VLBI map at 1661 MHz of the brightest radio source in M82.**
P. N. Wilkinson, A. G. de Bruyn.
Mon. Not. R. Astron. Soc., Vol. 211, No. 3, p. 593 – 605 (1984).
A map of 41.9 + 58, the brightest radio source in M82, made with data from the European VLBI network at 1.66 GHz, shows an elongated structure of length (FWHM) ~20 milliarcsec (~0.30 pc) in pa 47°. Transatlantic VLBI data at the same frequency reveal that there is a more compact core (~4 milliarcsec) embedded within the source, with the extended structure rather symmetrically disposed on either side of it. 5–GHz data yield roughly the same estimate of overall size and position angle as at 1.66 GHz, showing that the effects of interstellar scattering on the more extended component are negligible. If the event that created the source dates back to at least 1965, it is possible to set upper limits of about 2000 km $s^{-1}$ and 10000 km $s^{-1}$ to the expansion velocity of the core and extended components respectively.

158.176 **Variations in the optical continuum of OJ 287.**
M. Moles, J. Garcia–Pelayo, J. Masegosa.
Mon. Not. R. Astron. Soc., Vol. 211, No. 3, p. 621 – 627 (1984).
Optical differential photometry of OJ 287 shows rapid variability of this object, amounting to 0.05 mag in an hour in the *B* band. There are no signs of periodicity from the corresponding power spectrum. The continua calculated from *UBVRI* photometry are all very well fitted by power laws, indicating a very small dust content. A correlation is found between flux and spectral index; the spectrum steepens as the luminosity decreases.

158.177 **CCD astrometry and photometry of IAU reference frame benchmark radio sources.**
A. N. Argue.
Mon. Not. R. Astron. Soc., Vol. 211, No. 3, p. 713 – 718 (1984).
CCD astrometry and photometry have been carried out on 11 radio sources taken from the IAU lists of extragalactic objects selected as suitable for radio and optical astrometry, and which had previously been either unidentified or of doubtful status. Nine have now been positively identified. A discussion is given on the accuracy of measurement by CCD, based on King's treatment.

158.178 **Blue compact galaxies: infants of the Universe?**
N. Bergvall, K. Olofsson.
Messenger, No. 38, p. 2 – 6 (1984).

158.179 **Studies of extragalactic sources with the VLA.**
A. H. Bridle.
J. R. Astron. Soc. Can., Vol. 78, No. 5, p. 202 (1984). Abstract. – See Abstr. 010.401.

158.180 **A first look at the engine room in the nucleus of an active galaxy.**
P. P. Kronberg, R. A. Sramek, P. Biermann.
J. R. Astron. Soc. Can., Vol. 78, No. 5, p. 203 (1984). Abstract. – See Abstr. 010.401.

158.181 **The radio spectrum of 3C 84 at the epoch 1981.8.**
V. M. Bogod, M. N. Naugol'naya, Yu. N. Pariiskii (*Yu. N. Parijskij*), N. S. Soboleva, N. A. Yudaeva.
Bull. Spec. Astrophys. Obs. – North Caucasus, Vol. 17, p. 54 – 57 (1984). English translation of 34.158.197.

158.182 **The kinematics of the hot gas in the nuclear region of NGC 1365.**
S. Jörsäter, P. O. Lindblad, A. Boksenberg.
Astron. Astrophys., Vol. 140, No. 2, p. 288 – 294 (1984).
A number of new, high quality digital spectra confirm the peculiar velocity field previously derived from the [O III] lines in the nuclear region of NGC 1365. The [Ne III] line is found to behave similarly, and the hot medium producing these lines appears to form a separate dynamic entity. A model is presented which assumes perpendicular outflow from the plane of the galaxy. This model fits the data well, but the physical reasons for such a flow are not clear.

158.183 **Composite models for the narrow line regions of active galactic nuclei. II. The UV lines.**
S. M. V. Aldrovandi, M. Contini.
Astron. Astrophys., Vol. 140, No. 2, p. 368 – 372 (1984).
UV line calculations are presented for the narrow line regions of active galactic nuclei, assuming the infalling of matter towards the nucleus. A composite model, which accounts for ionizing radiation as well as the shock wave effect, is adopted. UV line ratios are calculated for different values of the radiation flux, the spectral index, the shock velocity and the cut–off energy. The results are given with the aim of explaining future observations and of supplying, in particular, the narrow line core contribution to the lines which permits a more reliable fitting of the calculated broad line ratios to observations.

158.184 **Inclination effects and reddening of Seyfert 1 nuclei.**
L. Danese, G. De Zotti.
Astron. Astrophys., Vol. 140, No. 2, p. 393 – 394 (1984).
The authors show that the reddening law for Seyfert 1 nuclei, recently proposed, quantitatively accounts for the deficiency of nearly edge–on optically selected Seyferts, and for the correlation with inclination of the ratio $L_{H\beta}/L_x$.

158.185 **The radio emission of Markarian galaxies at λ11 cm.**
H. M. Tovmassian (*G. M. Tovmasyan*), W. A. Sherwood, V. E. Sherwood, G. V. Schultz, C. J. Salter, H. E. Matthews.
Astron. Astrophys., Suppl. Ser., Vol. 58, No. 2, p. 317 – 325 (1984).
Four hundred sixty–four objects from Markarian lists 7 to 11 were observed at λ11 cm with a 3σ detection limit of 30 mJy. Fourteen Markarian galaxies were detected. Eighteen objects from earlier studies were also reobserved to determine spectral indices and to search for possible variability. The measurements show that at least four and possibly as many as eight objects are variable.

158.186 **Strong $H_2O$ maser emission from the peculiar galaxy NGC 3079.**
C. Henkel, R. Güsten, D. Downes, C. Thum, T. L. Wilson, P. Biermann.
Astron. Astrophys., Vol. 141, No. 1, p. L1 – L3 (1984).
$H_2O$ maser emission with an (isotropic) total luminosity of $\sim 500\ L_\odot$ has been detected in the peculiar galaxy NGC 3079 within 5″ of the nucleus; this is the most powerful maser yet detected. The peak flux density is 11 Jy. The $H_2O$ features cover a velocity range of at least 140 km $s^{-1}$. A possible detection of a $\sim 200\ L_\odot$ $H_2O$ maser in the distant interacting galaxy NGC 6240 is also reported.

158.187 **Possible relationship between the radio and optical structures associated with the active galaxy Centaurus A.**
Gopal–Krishna, L. Saripalli.
Astron. Astrophys., Vol. 141, No. 1, p. 61 – 66 (1984).
The authors present a detailed comparison of the recent deep optical and radio photographs of the nearby radio galaxy Centaurus A (NGC 5128). This has provided evidence for dynamical interaction between the relativistic plasma and optically detected shell–type structures. Based on this, a new interpretation is suggested for the formation of multiple radio peaks observed in the two giant radio lobes of this active galaxy and also for their inversion symmetric morphology. Also proposed is a method for estimating magnetic field strength inside the radio lobes, based on dynamical considerations of the optical structures interacting with the radio components.

158.188 **Jets and emission–line regions.**
C. Norman, G. Miley.
Astron. Astrophys., Vol. 141, No. 1, p. 85 – 90 (1984).
The authors propose that the low excitation broad line regions in active galaxies are closely associated with the jets responsible for compact radio sources. Various consequences of this hypothesis are considered for emission line structure, including broad and narrow line regions, radio morphology and active galaxy type.

158.189 **The correlation between distortion of radio structure and linear size in low luminosity radio galaxies.**
H. R. de Ruiter, P. Parma.
Astron. Astrophys., Vol. 141, No. 1, p. 189 – 194 (1984).
Two samples of low luminosity B2 radio galaxies have been observed with the VLA at 20 cm. The angular resolution (about 3.5 arcsec) allows the study of distortions of the radio structure well inside the galaxy. The distortion angle at 5 kpc from the nucleus correlates with both the linear size and radio power. Small radio sources are highly distorted at 5 kpc from the nucleus; larger sources are usually very straight inside the parent galaxy. A possible explanation for the correlation involving interstellar gas/dust clouds is briefly discussed.

158.190 **Radio continuum observations of blue compact dwarf galaxies.**
U. Klein, R. Wielebinski, T. X. Thuan.
Astron. Astrophys., Vol. 141, No. 1, p. 241 – 247 (1984).
A sample of 52 blue compact dwarf and low surface brightness galaxies has been observed at λ6.3 cm. Blue compact dwarf galaxies exhibit about 10 times higher radio–to–optical luminosity than normal spirals. There is evidence that the relation between radio and optical luminosity of blue compact dwarf galaxies is different from that found for normal spirals. The average radio continuum spectra of blue compact dwarf galaxies appear to be significantly flatter suggesting a low fraction of synchrotron radiation. Since these blue galaxies show clear evidence for violent star formation taking place at present, nonthermal emission is expected to be present due to supernova events. It is suggested that weak magnetic fields in blue compact dwarf galaxies are the reason for this weak synchrotron emission.

158.191 **A 4850 MHz survey of a part of the "GB region" and the combined counts of sources.**
J. Maslowski, I. I. K. Pauliny–Toth, A. Witzel, H. Kühr.
Astron. Astrophys., Vol. 141, No. 2, p. 376 – 382 (1984).
The results of a survey of a part of the "GB region", made at a frequency of 4850 MHz are presented. This region had been surveyed earlier at a frequency of 1400 MHz. The present survey includes 99 sources above a limiting flux density of 0.030 Jy in a solid angle of $9.18 \times 10^{-3}$ sterad. Two–point spectral indices for sources selected at 1400 and 4850 MHz are derived and discussed. The combined source counts of the present and other MPIfR–NRAO surveys near 5 GHz are presented, in tabular and graphical form. Power–law representations of the combined source counts are derived.

158.192 **Four double galaxies with ultraviolet–excess components.**
A. R. Petrosyan, K. A. Saakyan, É. E. Khachikyan (*Eh. E. Khachikyan*), A. N. Burenkov.
Sov. Astron. Lett., Vol. 10, No. 3, p. 169 – 171 (1984). English translation of 37.158.260.

158.193 **Strong hydrogen emission from Arp 220 and NGC 6240.**
IAU Circ., No. 3968 (1984).

158.194 **Changes in the stellar populations of radio galaxies with epoch.**
S. J. Lilly.
Spectral evolution of galaxies, p. 179 – 189 (1984). – See Abstr. 012.066.
The radio galaxies associated with 3CR radio sources at $z \sim 1$ are the most distant objects known in which the optical and infrared light is dominated by starlight. By studying the colours and magnitudes of these galaxies, one can follow changes in the stellar populations over look-back times of more than half the present age of the Universe.

158.195 **Eleven new emission-line galaxies.**
J. Liu, Y. Huang, X. Feng.
Kexue Tongbao (Beijing), Vol. 29, No. 7, p. 910 – 911 (1984).
The authors have found eleven new emission-line galaxies on their objective prism survey plates. Basic parameters and spectral characteristics of these galaxies from the observations are briefly presented.

158.196 **On the physical state in the narrow-line region of the classical Seyfert galaxy NGC 7469.**
V. K. Golev, I. M. Yankulova, G. T. Petrov.
Dokl. Bolg. Akad. Nauk, Vol. 37, No. 5, p. 549 – 551 (1984). Abstr. in Ref. Zh., 51. Astron., 11.51.705 – 706.

158.197 **Multifrequency light curves of low-frequency variable radio sources.**
D. R. Altschuler, J. J. Broderick, J. J. Condon, B. Dennison, K. J. Mitchell, S. L. O'Dell, H. E. Payne.
Astron. J., Vol. 89, No. 12, p. 1784 – 1798 (1984).
The authors present detailed multifrequency light curves, spanning the range 0.3 – 1.4 GHz, for the low-frequency variable sources AO 0235+16, NRAO 140, PKS 1117+14, DA 406, CTA 102, and 3C 454.3. They argue that the variations observed in AO 0235+16 are due to changes in relativistically moving synchrotron components, but that the low-frequency variations observed in NRAO 140, PKS 1117+14, DA 406, CTA 102, and 3C 454.3 are of a quite different character, which suggests a distinct low-frequency variability mechanism.

158.198 **High-resolution observations of radio-source hot spots at 329 MHz.**
R. Linfield, R. S. Simon.
Astron. J., Vol. 89, No. 12, p. 1799 – 1801 (1984).
The hot spots of several luminous double radio sources have been observed at 329 MHz with very long baseline interferometry at fringe spacings ranging from 0.05" to 0.70". The authors detected two out of two high-redshift ($z \sim 0.5$) sources, but only one of six low-redshift ($z \sim 0.05$) sources. For the low-redshift source (3C 234) which was detected, either the hot spot is larger at 329 MHz than at 15 GHz, or else the brightness contrast between it and the surrounding lobe is lower at the lower frequency.

158.199 **Jet deflection and hairpin bends.**
M. D. Smith.
Mon. Not. R. Astron. Soc., Vol. 211, No. 4, p. 767 – 773 (1984).
The deflection of a hydrodynamic jet via an oblique internal shock is studied. The pressure increase across the shock and the confinement of the shocked jet are maintained by the ramming of the bend through the ambient medium. A steady flow pattern is found in the reference frame co-moving with the bend: the external ram pressure gradually bends the deflected jet by up to 180°. The shock polar method is used to quantify this model and the author finds that changes in the initial angle of deflection and fluid dynamic instabilities will destroy this flow pattern for heavy jets. The model is applied to the U-shaped channels in the radio galaxies 3C 401 and 4C 32.69 and the head-tail source IC 708.

158.200 **VLA observations of the wide-angle tailed radio source 1313+073.**
A. R. Patnaik, D. G. Banhatti, C. R. Subrahmanya.
Mon. Not. R. Astron. Soc., Vol. 211, No. 4, p. 775 – 781 (1984).
The authors present VLA observations at 20 and 6 cm of the wide-angle tailed source 1313+073. It has an asymmetric structure with a sharp bend in the eastern tail and a gradual bend in the western. Both tails become diffuse after the bends. The authors explore several models to explain this structure and conclude that the tails bend because of the motion of the parent cD galaxy through the intracluster medium, which is possible if the cluster containing the source is dynamically young. Examination of the optical fields of 26 similar sources shows that in 20 of them the parent galaxies have a fainter companion (by ~2 mag), or are D, cD or db. This indicates that cannibalism may be important in the formation and powering of such sources.

158.201 **Young supernovae in the starburst galaxy M82.**
S. W. Unger, A. Pedlar, D. J. Axon, P. N. Wilkinson, P. N. Appleton.
Mon. Not. R. Astron. Soc., Vol. 211, No. 4, p. 783 – 792 (1984).
MERLIN observations at frequencies of 408, 1666 and 4995 MHz reveal 18 compact (<0.25 arcsec) radio sources within the central few hundred parsecs of M82. M82 is well known as a "starburst" galaxy, and it is probable that these objects are either radio supernovae or young supernova remnants. The distribution of radio sources appears asymmetric with respect to the centre of the galaxy, and the authors suggest that this is due to the current region of star formation lying within a molecular ring.

158.202 **Stellar populations in distant radio galaxies.**
S. J. Lilly, M. S. Longair.
Mon. Not. R. Astron. Soc., Vol. 211, No. 4, p. 833 – 855 (1984).
A homogeneous data set of infrared observations of 83 3CR galaxies with redshifts $0 < z < 1.6$, selected from a statistically complete sample of 90 radio sources, is used to study the colours and magnitudes of these galaxies as a function of their redshifts. It is shown that the infrared colours do not deviate from the predicted relations with redshift for a standard giant elliptical galaxy spectrum, indicating that the Giant Branch is basically unchanged at redshifts of unity. The optical to infrared colours, however, show substantial deviations at high redshift. No galaxies have been found that are significantly redder than a passively evolving galaxy, and there is a significant scatter of colours bluewards from this model. The excess of ultraviolet light responsible for these colours is not concentrated at the nucleus, and is interpreted as resulting from bursts of star formation throughout the galaxy that may be associated with the radio activity.

158.203 **Energetic particles in radio sources.**
P. A. G. Scheuer.
Adv. Space Res., Vol. 4, No. 2 – 3, p. 337 – 343 (1984). – See Abstr. 012.075.
Acceleration mechanisms for high-energy particles operating in supernova remnants and in the radio jets of extragalactic sources are briefly reviewed.

158.204 **Particle acceleration in active galactic nuclei.**
A. Ferrari.
Adv. Space Res., Vol. 4, No. 2 – 3, p. 345 – 355 (1984). – See Abstr. 012.075.
Multifrequency observations of active galactic nuclei (AGN) from radio to the X-ray band have now provided evidence that they are site of production of high temperature plasmas (kT > 100 MeV) with strong tails of suprathermal particles. A considerable part of high-energy cosmic rays may therefore originate in AGN. This review discusses the physical processes which have been suggested to operate in AGN, with special attention for the aspects connected with the generation of specific and/or bulk relativistic energies.

158.205 **The environment of active nuclei.**
G. Miley.
Space Telesc. Sci. Inst., Prepr. Ser., No. 24, 6 pp. (1984). To be published in Proceedings of the ESA Workshop on "QUASAT", June 1984, Vienna, Austria.

158.206 **The near ultraviolet spectrum of the high-redshift BL Lacertae object 0215+015.**
J. C. Blades, R. W. Hunstead, H. S. Murdoch, M. Pettini.
Space Telesc. Sci. Inst., Prepr. Ser., No. 27, 43 pp. (1984). To appear in Astrophys. J.

158.207 **Nano-quasars next door?**
C. Jenkins.
Gemini, No. 12, p. 2 - 3 (1984).

158.208 **The variable extragalactic object 3C 446.**
C. Barbieri, S. Cristiani, S. Omizzolo, G. Romano.
ESO Sci. Prepr., No. 339, 18 pp. (1984). Submitted to Astron. Astrophys.

158.209 **The ultraviolet absorption spectrum of NGC 4151.**
P. Véron, M.-P. Véron-Cetty, M. Tarenghi.
ESO Sci. Prepr., No. 341, 17 pp. (1984). Submitted to Astron. Astrophys.

158.210 **Wolf-Rayet stars in "lazy" galaxies: a statistical approach.**
D. Kunth, M. Joubert.
Inst. Astrophys. Paris, Pré-Publ., No. 71, 29 pp. (1984). To appear in Astron. Astrophys.

158.211 **The nucleus of NGC 7469: physical conditions and structure.**
B. A. M. Westin.
Stockholms Obs., Rep., No. 24, 24 pp. (1984).
Optical data together with data from the literature, including IUE observations, have been used to describe the physical conditions in the nucleus of the Seyfert 1 galaxy NGC 7469. The ultraviolet excess emission, over a power-law component, cannot be explained by Balmer continuum emission alone, suggesting also emission from a blackbody with T = 20000K, respresenting the accretion disk, probably the source of variable X-ray emission, around a black hole of mass $0.7 \times 10^9 M_\odot$ in the very center. A two-component model for the broad-line region could explain the differences in line width between the broad Balmer components. Outflow of gas from the center of NGC 7469 seems to be the most probable explanation to the observed emission-line profiles.

158.212 **Radio structures of Seyfert galaxies. VI. VLA observations of a nearby sample.**
J. S. Ulvestad, A. S. Wilson.
Astrophys. J., Vol. 285, No. 2, p. 439 - 452 (1984).
This paper is concerned with the radio properties of a distance-limited sample of 25 Seyfert and relatively narrow-line X-ray galaxies (NLXGs) having recession velocities with respect to the centroid of the Local Group of less than 3100 km $s^{-1}$. All but one of these galaxies have been observed with the VLA and all 24 of the observed galaxies have been detected. Analysis confirms the previous conclusions that Seyfert 2 galaxies have radio sources larger and more powerful than Seyfert 1 galaxies. The five NLXGs seem to contain radio sources similar to those in Seyfert 2 galaxies. This implies that type 2 Seyferts and probably NLXGs cannot be merely type 1 Seyferts with central regions heavily obscured by dust.

158.213 **NGC 7213: a key to the nature of Liners?**
A. V. Filippenko, J. P. Halpern.
Astrophys. J., Vol. 285, No. 2, p. 458 - 474 (1984).
New optical spectra show that the nucleus of the S0 galaxy NGC 7213 resembles those of type 1 Seyferts: emission lines such as [Ne V] λ3426, He II λ4686, and broad Hα are visible, and a nonstellar component accounts for a substantial portion of the continuum at blue and ultraviolet wavelengths. Many of the emission-line intensity ratios suggest heating due to shocks, since they are similar to those in "Liners". After subtracting the stellar continuum careful measurements of various emission lines are presented and a number of important physical parameters are derived. The Seyfert characteristics of NGC 7213 are reviewed, and its shock features are reconciled with photoionization models. The arguments may apply to classical Liners as well. Finally, theoretical models are explored that can help explain the relationships discovered among the emission lines.

158.214 **The nonstellar continuum of the Seyfert galaxy NGC 7213.**
J. P. Halpern, A. V. Filippenko.
Astrophys. J., Vol. 285, No. 2, p. 475 - 482 (1984).
Optical spectrophotometry of the nearby Seyfert 1 galaxy NGC 7213 reveals that nonstellar radiation accounts for at least 50% of the observed continuum at λ3300. A formal decomposition of the continuum into a power law of index 1.1 and the spectrum of a giant elliptical galaxy suggests that the true nonstellar flux may be substantially larger if reddening is allowed. X-ray spectra obtained with the IPC and MPC on the Einstein Observatory are also presented. Published observations at radio, IR, and UV wavelengths are combined with the new data to produce the overall continuum, which is similar to that of other type 1 Seyfert galaxies.

158.215 **Near-infrared photometry of faint radio galaxies in selected areas.**
T. X. Thuan, R. A. Windhorst, J. J. Puschell, R. B. Isaacman, F. N. Owen.
Astrophys. J., Vol. 285, No. 2, p. 515 - 526 (1984).
The authors present near-infrared *JHK* photometry for 40 radio galaxies and 8 quasars from the complete sample of mJy radio sources in the Leiden-Berkeley Deep Survey. The detection rate is 48 out of 52 attempted objects down to $K \sim 17.5$ mag. By combining these measurements with optical photometry and spectral evolution models it is found that the optical-infrared colors of faint radio galaxies are independent of their radio flux and power and are due to their stellar populations.

158.216 **Coordinated multifrequency observations of the BL Lacertae Objects Mrk 180 and Mrk 501.**
S. L. Mufson, D. J. Hutter, K. R. Hackney, R. L. Hackney, C. M. Urry, R. F. Mushotzky, Y. Kondo, W. Z. Wiśniewski, H. D. Aller, M. F. Aller, P. E. Hodge.
Astrophys. J., Vol. 285, No. 2, p. 571 - 579 (1984).
Coordinated observations of the nearby BL Lac objects MrK 180 and MrK 501 have been obtained at X-ray, ultraviolet, optical and radio wavelengths. The authors present for each of these sources a set of observations which gives a time-frozen picture of its continuous spectrum from the X-ray to the radio. The spectra have been fitted with two different models: spherically symmetric, synchrotron self-Compton models and relativistic jet models. It is found that both models can account for the spectral observations.

158.217 **Position-dependent infrared spectroscopy of M82: 100,000 Orion nebulae?**
B. Jones, J. M. Rodríguez-Espinosa.
Astrophys. J., Vol. 285, No. 2, p. 580 - 586 (1984).
The authors report 8-13 μm spectra obtained at several positions across the extended emission region of M82. They do not find major differences among the spectra. There is a strong similarity between these spectra and the infrared spectrum of the Orion ionization front. The data are interpreted using a model for M82 in which the Orion nebula is taken as typical of a region of active star formation. A star burst model for M82 is constructed using 100,000 regions similar to Orion.

158.218 **Spectroscopy of extremely distant radio galaxies.**
H. Spinrad, S. Djorgovski.
Astrophys. J., Lett. Ed., Vol. 285, No. 2, p. L49 - L52 (1984). With plates L3 - L4.
The authors report the firm identification of two faint radio galaxies, 3C 13 and 3C 256, and a spectroscopic survey of 13 faint

3CR galaxies. The emission–line redshifts now extend up to $z = 1.82$ (3C 256), and the authors present a list and a preliminary spectrum summary of the character of 11 new radio galaxies with $z > 0.77$. These are the most distant (non–QSO) galaxies known, and the cosmological and galaxy evolution tests can now be extended into previously unprobed territory.

158.219 **The brightness of NGC 4151 between 1906 and 1911.**
V. L. Oknyanskij.
Astron. Tsirk., No. 1300, p. 1 – 2 (1983). In Russian.

158.220 **Linearly polarized emission of variable radio sources. III. Comparison with observations.**
N. V. Kulakova, V. N. Kuril'chik.
Astron. Tsirk., No. 1302, p. 1 – 3 (1983). In Russian.

158.221 **Linearly polarized emission of variable radio sources. IV. Comparison with observations.**
N. V. Kulakova, V. N. Kuril'chik.
Astron. Tsirk., No. 1307, p. 1 – 4 (1984). In Russian.

158.222 **Linearly polarized emission of variable radio sources. V. The sign of Faraday rotation.**
V. N. Kuril'chik, N. V. Kulakova.
Astron. Tsirk., No. 1309, p. 1 – 4 (1984). In Russian.

158.223 **Determination of optical positions of fourteen extragalactic radio sources.**
R. Ya. Inasaridze.
Astron. Tsirk., No. 1309, p. 6 – 7 (1984). In Russian.

158.224 **Similar structures in the outbursts of OJ 287 in 1972 and 1983.**
A. Sillanpää, P. Teerikorpi, S. Haarala, T. Korhonen, Yu. S. Efimov, N. M. Shakhovskoy (*N. M. Shakhovskoj*).
Rep. Ser., Dep. Phys. Sci., Univ. Turku, No. FTL–R69, 11 pp. (1984). = Turku Univ. Obs., Informo No. 85. Submitted to Astron. Astrophys.

158.225 **Relativistic electrons and magnetic fields in extragalactic radio sources.**
E. Valtaoja.
Rep. Ser., Dep. Phys. Sci., Univ. Turku, No. D12, 5+142 pp. (1984). = Turku Univ. Obs., Informo No. 88.

Different aspects of the synchrotron–emitting relativistic plasma in extragalactic radio sources are studied. Theoretical models for the structure and evolution of the nonthermal components are constructed, based on the observed radio properties of active galaxies. The existence of luminosity–dependent excess circular polarization in compact radio sources is demonstrated, and used to predict the magnetic field structure. The observed spatial variation of the spectral index in extended sources is explained with a rapid diffusion model for the motion of electrons.

158.226 **Optical outburst in OJ 287.**
S. Haarala, T. Korhonen, A. Sillanpää, P. Teerikorpi.
Rep. Ser., Dep. Phys. Sci., Univ. Turku, No. FTL–R68, p. 7 – 11 (1984). – See Abstr. 012.079.

Optical light curve of BL Lacertae object OJ 287 during its outburst is presented. There seems to be an interesting sequence of secondary maxima in the time scales of a few days in the declining tail of the outburst.

158.227 **Polarization properties of compact radio sources.**
E. Valtaoja.
Rep. Ser., Dep. Phys. Sci., Univ. Turku, No. FTL–R68, p. 37 (1984). – See Abstr. 012.079.

158.228 **The luminous host galaxy and anomalous Brackett–gamma line of Markarian 231.**
R. M. Cutri, G. H. Rieke, M. J. Lebofsky.
Prepr. Steward Obs., No. 530, 22 pp. (1984). To appear in Astrophys. J.

158.229 **Circular polarization of compact, extragalactic radio sources – synchrotron emission and circular repolarization.**
A. Y. S. Cheng, A. G. Pacholczyk, K. H. Cook.
Prepr. Steward Obs., No. 537, 14 pp. (1984). Submitted to Astrophys. J., Lett. Ed.

158.230 **Echelle spectroscopy of the Seyfert 1 galaxy Markarian 231.**
R. J. Rudy, C. B. Foltz, J. T. Stocke.
Prepr. Steward Obs., No. 540, 22 pp. (1984).

This paper presents high resolution (~10 km/sec) spectroscopy of the Na D absorptions associated with two of the three absorption line systems present in the spectrum of the Seyfert 1 galaxy Markarian 231. These features are blueshifted by several thousand km/sec relative to the emission lines and show the shape, displacement, and extent in velocity space characteristic of broad–absorption–line QSOs (BAL–QSOs). This indicates that the mechanism which accelerates the gas in those high luminosity objects may be operating in Mrk 231 as well.

158.231 **$10^{12}$ $L_\odot$ starbursts and shocked molecular hydrogen in the colliding galaxies Arp 220 (= IC 4553) and NGC 6240.**
G. H. Rieke, R. M. Cutri, J. H. Black, W. F. Kailey, C. W. McAlary, M. J. Lebofsky, R. Elston.
Prepr. Steward Obs., No. 541, 36 pp. (1984).

The authors present infrared spectroscopy and photometry and optical spectroscopy of the exceedingly luminous interacting galaxies Arp 220 (= IC 4553) and NGC 6240. These galaxies are the sites of exceedingly powerful bursts of star formation, as shown by (1) the luminosity and extent of the stellar component seen near 2 μm; (2) the large population of red supergiants required to account for the depth of the stellar CO absorption; (3) the absorption and emission features in the spectrum longward of 3 μm; and (4) the extended luminosity source indicated by the spatial extent of the galaxies at 10 μm. Scaled–up models similar to those developed for less luminous starburst galaxies are consistent with this hypothesis.

158.232 **UV and X–ray observations of AGN: two comparisons with other wavelengths.**
M. A. Malkan.
Prepr. Steward Obs., No. 542, 8 pp. (1984). To be published in the proceedings of: X–ray and UV emission from active galactic nuclei, July 9 – 13, 1984, Garching, West Germany.

158.233 **The narrow line region of active galaxies. I. Nuclear [O III] profiles.**
M. Whittle.
Prepr. Steward Obs., No. 548, 48 pp. (1984). Submitted to Mon. Not. R. Astron. Soc.

158.234 **The narrow line region of active galaxies. II. Relations between [O III] profile shape and other properties.**
M. Whittle.
Prepr. Steward Obs., No. 549, 37 pp. (1984). Submitted to Mon. Not. R. Astron. Soc.

158.235 **Active galaxies viewed from the infrared.**
G. H. Rieke.
Prepr. Steward Obs., No. 553, 13 pp. (1984). Review prepared for the Santa Cruz Workshop on Active Galaxies and QSOs, July, 1984.

158.236 **Mass–to–luminosity ratio for Seyfert galaxies.**
A. V. Zasov.
Astron. Tsirk., No. 1338, p. 1 – 3 (1984). In Russian.

158.237 **Method of estimate of the physical parameters of double radio sources with hot spots.**
Yu. V. Baryshev.
Astron. Tsirk., No. 1334, p. 1 – 3 (1984). In Russian.

A method of estimation of jet velocity, jet density, and mass outflow for relativistic and nonrelativistic cases is given. These physical parameters are evaluated for known linear size and synchrotron luminosity of hot spots. The relation between hot spot radius and luminosity is predicted in the form $L_s \sim R_{hs}^{2/3}$.

158.238 **Cosmic rays and the emission–line regions of active galactic nuclei.**
G. J. Ferland, R. F. Mushotzky.
Astrophys. J., Vol. 286, No. 1, p. 42 – 52 (1984).
The authors discuss the effects that the synchrotron emitting relativistic electrons, which are known to exist throughout many distance scales in active nuclei, could have on the emission–line regions that characterize these objects. Detailed models of both the inner, dense, broad–line region and the outer, lower density, narrow–line region are presented, together with the first models of the optically emitting gas often found within extended radio lobes. The implications of the results for correlations between radio and optical properties of nuclei are discussed. The effects of the synchrotron emitting electrons on filaments in the Crab Nebula are discussed in an appendix.

158.239 **A model for the surface brightness of a turbulent low Mach number jet. I. Theoretical development and application to 3C 31.**
G. V. Bicknell.
Astrophys. J., Vol. 286, No. 1, p. 68 – 87 (1984).
It is proposed that radio jets "turn on" as a result of shocks induced by their becoming underexpanded a few seconds of arc from the core. A sequence of shocks and/or turbulent entrainment causes some jets to decelerate to a Mach number near unity and to spread laterally at a rate comparable to that of laboratory jets. A model is developed to infer the velocity and surface brightness of the jet in this region. The analysis makes extensive use of Favre's (1969) statistical mass–averaging technique for compressible turbulent flow and the existing experimental and theoretical literature on laboratory jets. This model is used to describe the surface brightness of the 3C 31 jet and to study the physical state of the interstellar medium in the optical counterpart NGC 383.

158.240 **X–ray emission from M82.**
M. G. Watson, V. Stanger, R. E. Griffiths.
Astrophys. J., Vol. 286, No. 1, p. 144 – 158 (1984).
The authors present an X–ray study of M82 with the Einstein Observatory imaging instruments. Observations with the HRI show a luminous, extended X–ray source centered on M82. The most intense feature in the X–ray image has dimensions $\sim 30'' - 60''$ and corresponds quite closely to the optically brightest region near the nucleus of the galaxy. Several discrete unresolved sources are detected in this central region. Outside the central region a diffuse X–ray halo is detected which extends for several arc minutes along the minor axis of the galaxy both north and south of the nucleus. These results are interpreted in the context of the current "starburst" models for the nuclear activity in M82.

158.241 **An analysis of the narrow–line profiles in high ionization Seyfert galaxies.**
M. M. De Robertis, D. E. Osterbrock.
Astrophys. J., Vol. 286, No. 1, p. 171 – 185 (1984). = Lick Obs. Bull., No. 983.
Narrow optical emission–line profiles representing a wide range in ionization potential for 12 high–ionization Seyfert galaxies are analyzed using deconvolution techniques. The data reduction process is described in detail. A good correlation is found between line width and ionization potential in many of the objects. For some galaxies a good correlation is also found between line width and critical density. For a given object, the ratios of line widths at various fractions of full intensity are remarkably similar over the entire range of ionization potential and critical density. The possible implication of these and other correlations, and the absence of still other correlations, are discussed. Descriptions of the details of the individual spectra are presented.

158.242 **Kernaktivität in wechselwirkenden Galaxien.**
W. Kollatschny, K. J. Fricke.
Mitt. Astron. Ges., Nr. 62, p. 151 (1984). – See Abstr. 010.341.

158.243 **Gaskinematik in der aktiven Galaxie NGC 4151.**
H. Schulz.
Mitt. Astron. Ges., Nr. 62, p. 339 – 340 (1984). – See Abstr. 010.341.

158.244 **The variability of emission–line spectra and the spatial structure of broad–line regions of Seyfert galaxies.**
I. I. Shevchenko.
Pis'ma Astron. Zh., Tom 10, No. 12, p. 896 – 902 (1984). In Russian. English translation in Sov. Astron. Lett., Vol. 10.
Conditions on which the maximum of the flare of an active galactic nucleus in emission lines lags behind the maximum of its flare in the continuum are deduced under the assumption of isotropic distribution of line–emitting matter around the central source of ionizing radiation. In particular, these conditions constrain the allowed radial (in relation to the central source) distributions of line–emitting matter; thus they may be used to investigate the spatial structure of broad–line regions of active galactic nuclei. Applications to NGC 4151 and other Seyfert galaxies are discussed.

158.245 **Magnetic field configurations of radio jets in the double source of a radio galaxy.**
W. R. Hu.
Sci. Sin., Ser. A, Vol. 27, No. 11, p. 1190 – 1201 (1984).
In the present paper, a two dimensional magnetohydrodynamical theory is applied to the analysis of the magnetic field configurations in radio jets. Two–dimensional results not only are consistent theoretically, but also explain the fine structures of observations. One of the theoretical models is discussed in detail, and is in good agreement as compared with the observed radio jets of NGC 6251.

158.246 **The nature of the IRAS "warm" galaxies.**
D. Carter.
Astron. Express, Vol. 1, Issue 2, p. 61 – 74 (1984).
The IRAS satellite has detected as strong infra–red sources a number of otherwise unspectacular galaxies. Here the author presents optical and near infra–red spectra of a sample of 13 such galaxies from the IRAS preliminary lists, flux limited at 1.2 Jy at 60 μm. All of the galaxies show strong emission lines, 10 are Seyfert galaxies and three have the spectra of giant H II regions. For the Seyferts there is a strong correlation between far infra–red luminosity and the luminosity in the forbidden lines, supporting the hypothesis that the far infra–red emission mechanism is thermal re–radiation by dust in the narrow line region.

158.247 **Shock wave from a galactic nucleus into a halo and intergalactic space.**
Y. Sofue.
Publ. Astron. Soc. Jpn., Vol. 36, No. 3, p. 539 – 550 (1984).
The propagation of a shock wave associated with an explosion at the nucleus of a disk galaxy through the halo and intergalactic gas is studied. An initially spherical shock front expands into the disk and halo, being elongated in the direction perpendicular to the disk plane. In the halo the front attains an Ω–shaped shell structure, which is observed as a large radio lobe in edge–on galaxies.

158.248 **Aktiviteten i Seyfertgalaxer. Kommer den från svarta hål?**
A. Elvius.
Astron. Tidsskr., Årg. 17, Nr. 4, p. 166 – 174 (1984).

158.249 **The radio luminosity functions of cD galaxies; the fuelling of radio sources.**
W. Bijleveld, E. A. Valentijn.
Clusters and groups of galaxies, p. 133 – 138 (1984). – See Abstr. 012.088.

158.250 **The extended radio source near Coma A.**
H. Andernach, L. Feretti, G. Giovannini.
Clusters and groups of galaxies, p. 215 – 220 (1984). – See Abstr. 012.088.

Maps at 2.7 and 4.75 GHz are presented for the elongated radio source SE of Coma A (3C 277.3). No radio bridge to Coma A could be found nor can any other optical object be easily identified to account for the emission. Spectral index variations across the strongly polarized source suggest a particle diffusion along the major axis from the centre to the outer ends of the source, parallel to the magnetic field direction. The authors argue against a galactic foreground feature, but an association with the Coma cluster or its galaxies is also uncertain.

158.251 **Observation of a large halo around M87.**
C. Ceccarelli, G. Dall'Oglio, P. de Bernardis, N. Mandolesi, S. Masi, B. Melchiorri, F. Melchiorri, G. Moreno, G. Morigi, L. Pietranera.
Clusters and groups of galaxies, p. 287 – 288 (1984). – See Abstr. 012.088.

158.252 **High resolution radio and X–ray observations of 3C 28 in A115.**
L. Feretti, I. M. Gioia, G. Giovannini, L. Gregorini, L. Padrielli.
Clusters and groups of galaxies, p. 289 – 290 (1984). – See Abstr. 012.088.

158.253 **The angular dimensions of extragalactic radio sources.**
J. C. Carvalho.
Astrophys. Space Sci., Vol. 107, No. 1, p. 199 – 205 (1984).

The angular dimensions of compact components of extragalactic radio sources are calculated theoretically on three hypotheses: (1) magnetic field and relativistic particles energy equipartition; (2) equilibrium between synchrotron and Compton losses; (3) radiative lifetime of the same order as the source's light–travel time. Using angular measurements of 73 radio components, divided into two groups (multi–component and single–component sources), a correcting formula is derived to reduce the theoretical values to the observed values.

158.254 **Spectral observations of Markarian galaxies. IV.**
Eh. K. Denisyuk, V. A. Lipovetskij.
Astrofizika, Tom 20, Vyp. 3, p. 525 – 537 (1984). In Russian. English translation in Astrophysics, Vol. 20, No. 3.

The spectral observations of 109 Markarian galaxies are presented. The radial velocities, absolute magnitudes, linear sizes and visual estimates of relative intensities of emission lines are obtained. Five objects are classified as Seyfert galaxies, they are Markarian 928, 957, 984, 1058 and 1066. The masses and mass–to–light ratios are evaluated for four pairs and groups of galaxies. The authors found no emission lines in the spectra of 25 Markarian galaxies. Their list is given.

158.255 **Evolutionary stages and extended radio structures of active galactic nuclei.**
I. I. Shevchenko.
Astrofizika, Tom 20, Vyp. 3, p. 539 – 546 (1984). In Russian. English translation in Astrophysics, Vol. 20, No. 3.

The distribution of linear sizes of extended radio structures of broad–line radio galaxies strongly differs from that of most radio galaxies. The shape of the distribution suggests that broad–line radio galaxies with extended radio structures exhibit a stage of galactic nuclear activity which does not occur in the beginning, but nearer the end of the existence of structures. In the class of quasars with extended radio structures, a subclass of objects which have on the average low optical luminosity, high amplitude of optical variability and very broad emission lines in optical spectra seem to relate to the same stage of nuclear activity.

158.256 **Spectrophotometry of the Seyfert galaxy NGC 4151. I. Search for and identification of faint lines.**
V. A. Lipovetskij, A. I. Shapovalova.
Astrofizika, Tom 21, Vyp. 1, p. 5 – 19 (1984). In Russian. English translation in Astrophysics, Vol. 21, No. 1.

Spectrophotometric results for the nucleus of the Seyfert galaxy NGC 4151 are presented. The broadened spectra are obtained in the range λλ4100 – 7300 Å. The method of the search and identification of faint emission lines is described. 77 lines are identified in the spectra and good accordance is shown with the results of other authors. The authors have discovered many narrow permitted lines of iron Fe II 37, 38, 40, 74, 210. The presence of the line [Fe X] λ6374 is shown distinctly and the assumption about its intensity variability with time scale of 3 – 5 years is suggested.

158.257 **Superassociations in spiral galaxies with UV excess. III.**
A. R. Petrosyan, K. A. Saakyan, Eh. E. Khachikyan.
Astrofizika, Tom 21, Vyp. 1, p. 57 – 71 (1984). In Russian. English translation in Astrophysics, Vol. 21, No. 1.

On the basis of data on UV spiral galaxies containing superassociations (SA) a statistical investigation is carried out. The following results are obtained: (1) The frequency of occurrence of spiral galaxies with SA among UV galaxies is twice as high as in normal spirals. (2) SA are found in dwarf and giant as well as in supergiant galaxies, but in most cases in barred galaxies.

158.258 **On the relative orientation of radio source axes and parent elliptical galaxies.**
R. R. Andreasyan.
Astrofizika, Tom 21, Vyp. 1, p. 93 – 102 (1984). In Russian. English translation in Astrophysics, Vol. 21, No. 1.

A relation between the major axes of radio galaxies and parent elliptical galaxies is discussed. At the value of the ratio between the sizes of radio axes and radio components beyond 2.5 a significant correlation between the orientations of radio axes and optical minor axes is found. Otherwise, at the value of this ratio below 2.5 a good correlation between the orientations of radio axes and optical major axes exists.

158.259 **Mass accretion by the nuclei of disk galaxies.**
A. Saha, T. Bhattacharyya, B. Basu.
Astrophys. Space Sci., Vol. 107, No. 2, p. 255 – 270 (1984).

In this work a model has been proposed to explain how the nucleus of a galaxy can accumulate mass and becomes supermassive – ultimately giving way to gravitational instability leading to an explosion in the nucleus. The process may be repeated many times during the life–span of a galaxy.

158.260 **Investigation of the circumnuclear region of the Seyfert galaxy NGC 1275.**
L. P. Metik, I. I. Pronik.
Astrofizika, Tom 21, Vyp. 2, p. 233 – 244 (1984). In Russian. English translation in Astrophysics, Vol. 21, No. 2.

New data on the structure of the circumnuclear region and nucleus of the Seyfert galaxy NGC 1275 have been obtained.

158.261 **The preferred orientation of Seyfert galaxies as the result of the selection effect.**
V. P. Reshetnikov.
Astrofizika, Tom 21, Vyp. 2, p. 263 – 268 (1984). In Russian. English translation in Astrophysics, Vol. 21, No. 2.

The observed preferred orientation of Seyfert galaxies "face on" is shown to be presumably the result of the selection effect, that is the choice of these galaxies from Markarian objects for which this effect is naturally explained by self–absorption in the ultraviolet.

158.262 **Interferenzfilter–Beobachtung der Irr II/Sc Galaxie M82.**
U. Görze.
Diplomarbeit, Landessternwarte Heidelberg–Königstuhl, Heidelberg, F.R.Germany, 3 + 89 pp. (1984).

158.263 **The ionization mechanism of the "Liner" galaxies Pictor A and PKS 1718–65.**
R. F. Carswell, J. A. Baldwin, B. Atwood, M. M. Phillips.
Astrophys. J., Vol. 286, No. 2, p. 464 – 470 (1984).

High–resolution optical spectra are presented of two galaxies with Liner (low–ionization nuclear emission–line region) nuclei,

Pictor A and PKS 1718–65. Pictor A is shown to have a strong featureless optical continuum and He II λ4686 emission, making it the best case to date of a power–law photoionized Liner. The spectrum of PKS 1718–65 shows neither of these features. If the Liner in this galaxy is photoionized rather than shock–excited, the ionizing spectrum must be more like a blackbody than a power–law spectrum.

158.264 **Emission–line gas associated with the radio lobes of the high–luminosity radio source 3C 171.**
T. M. Heckman, W. J. M. van Breugel, G. K. Miley.
Astrophys. J., Vol. 286, No. 2, p. 509 – 516 (1984). With plates 18 – 19.

The authors report the discovery of spatially extended optical emission–line gas associated with the radio lobes of the high–redshift ($z = 0.2384$) radio galaxy 3C 171. This is the first known case of such an association in a high–luminosity radio source. The emission–line gas occurs adjacent to (along the boundaries of) the bright regions of radio emission; the emission–line region is relatively small ($\sim 50$ kpc); the gas velocities range up to almost $10^3$km s$^{-1}$; the emission–line spectrum of the gas associated with the radio lobes bears a strong spectroscopic resemblance to gas in the narrow–line region of active nuclei.

158.265 **Multifrequency spectral behavior of the BL Lacertae objects OI 90.4 and 3C 66A.**
D. M. Worrall, J. J. Puschell, J. M. Rodriguez–Espinosa, F. C. Bruhweiler, H. R. Miller, M. F. Aller, H. D. Aller.
Astrophys. J., Vol. 286, No. 2, p. 711 – 717 (1984).

The authors present single–epoch multifrequency data for the BL Lacertae objects OI 90.4 and 3C 66A, spanning radio through ultraviolet wavelengths. Other measurements relevant to intensity and spectral–shape variability are also presented. Both BL Lac objects exhibit spectral curvature in the infrared to ultraviolet region. Variability in the ultraviolet spectral index of OI 90.4 is found. A synchrotron self–Compton jet model is applied to each BL Lacertae object.

158.266 **The extragalactic nature of H0323+022.**
B. Margon, G. H. Jacoby.
Astrophys. J., Lett. Ed., Vol. 286, No. 2, p. L31 – L33 (1984). With plate L3.

The bright X–ray source H0323+022 has been identified by Doxsey and colleagues with a 16th mag ultraviolet–excess object with a featureless optical spectrum and radio characteristics of a BL Lac object. However, on two occasions those authors found the object to display an absorption spectrum of a G star at zero velocity. This letter presents imaging and spectroscopic observations showing that their G spectra were due to misidentification of the object. H0323+022 is extragalactic and is probably quite similar to several other X–ray selected BL Lac objects.

158.267 **A search for changing structure in low–luminosity compact radio sources.**
D. L. Jones.
Astrophys. J., Vol. 287, No. 1, p. 33 – 40 (1984).

Low–luminosity compact radio sources in the nuclei of several nearby galaxies were observed during two 6 cm VLBI experiments. The experiments were separated by 1.75 yr and were designed to detect changes in the structure of the sources. With the possible exception of NGC 4278, no evidence for any changes in structure was seen. This could be due to an absence of relativistic motion in these sources, but it is also consistent with the standard relativistic beaming models if the lower luminosity sources are oriented at larger angles to our line of sight than the high–luminosity sources.

158.268 **Photon–photon absorption and the uniqueness of the spectra of active galactic nuclei.**
D. Kazanas.
Astrophys. J., Vol. 287, No. 1, p. 112 – 115 (1984).

The effects of the feedback of $e^+ - e^-$ pair reinjection in a plasma due to photon–photon absorption of its own radiation are examined. Under the assumption of continuous electron injection with a power–law spectrum $E^{-\Gamma}$ and Compton losses only, it is shown that for $\Gamma < 2$ the steady state electron distribution function has a unique form independent of the primary injection spectrum. This electron distribution function can, by synchrotron emission, reproduce the general characteristics of the observed radio to optical active galactic nuclei spectra.

158.269 **Molecular clouds in M82.**
J. S. Young, N. Z. Scoville.
Astrophys. J., Vol. 287, No. 1, p. 153 – 166 (1984).

CO observations of the peculiar galaxy M82 have been made at 79 positions using the 14 m FCRAO telescope (HPBW = 50″) in order to obtain a fully sampled map of the central 2 kpc as well as the radial distribution of molecular gas out to 3 kpc along the major axis. The velocity field of the molecular gas is derived and it is suggested that the systemic velocity of M82 is close to a value of 175 km s$^{-1}$. The CO intensity peaks in the center of the galaxy and falls off smoothly by two orders of magnitude out to radii of 2 – 3 kpc. If the relation between CO emission and the mass of molecules is the same in M82 as in our Galaxy, the total $H_2$ mass for the inner 6 kpc of M82 is $\sim 2 \times 10^9 M_\odot$ as compared to $2 \times 10^8 M_\odot$ in H I for the same region. These observations are consistent with the previously proposed models that M82 was once a normal Sc/Irr galaxy which is now accreting interstellar material.

158.270 **The NGC 5128 globular cluster system: star counts in *U*, *V*, and *R*.**
G. L. H. Harris, J. E. Hesser, H. C. Harris, P. J. Curry.
Astrophys. J., Vol. 287, No. 1, p. 175 – 184 (1984).

Star counts in $U$, $V$, and $R$ within $\sim 23$ arcmin of the nearby peculiar elliptical galaxy, NGC 5128, show 800 – 900 counted objects in excess of background in the region studied. The cluster system inferred from these data seems to follow an $r^{1/4}$ law distribution for $r \gtrsim 4'$. A cluster luminosity function over a range of 5 mag has been constructed. The analysis of the counts suggests a total cluster population of $1200 \lesssim N_t \lesssim 1900$, a specific frequency of $3.0 \lesssim S \lesssim 3.2$ and a distance of $\sim 3 - 4$ Mpc.

158.271 **The NGC 5128 globular cluster system: a complete analysis of images using COSMOS.**
H. C. Harris, G. L. H. Harris, J. E. Hesser, H. T. MacGillivray.
Astrophys. J., Vol. 287, No. 1, p. 185 – 193 (1984). = Contrib. Univ. Waterloo Obs., No. 30. With plate 3.

A large region around the peculiar elliptical galaxy NGC 5128 (Centaurus A) has been analyzed using prime focus plates taken in four colors to search for globular clusters. The plates have been scanned with COSMOS and processed to detect and analyze all images (7656 images in the region studied). Nonstellar images are identified, and a visual classification is used to obatin a sample of likely cluster candidates. Most bright clusters have probably been found, so the bright tail of the luminosity function is now fairly well known. In most respects, the cluster system of NGC 5128 appears to be normal for a large elliptical galaxy.

158.272 **Far–infrared photometry of compact extragalactic sources: OJ 287 and BL Lacertae.**
P. M. Harvey, M. Joy, D. F. Lester, B. A. Wilking.
Astrophys. J., Lett. Ed., Vol. 287, No. 1, p. L9 – L10 (1984).

The authors have detected the 50 and 100 μm emission of OJ 287 and obtained upper limits for BL Lac. These first measurements of two BL Lac objects in the far–infrared show them to be similar to the few quasars previously observed in the far–infrared. In particular there is no evidence for significant dust emission, and the $\lambda \approx 100$ μm flux density fits on a smooth line joining the near–infrared and millimeter continuum fluxes.

158.273 **The electron density in M82 from the S III mid–infrared line ratio.**
J. R. Houck, M. A. Shure, G. E. Gull, T. Herter.
Astrophys. J., Lett. Ed., Vol. 287, No. 1, p. L11 – L13 (1984).

Strong detections have been made of both the 18.7 μm and 33.4 μm lines of S III in the irregular galaxy M82. This represents the first detection of the 33.4 μm line in an extragalactic object. These measurements sample the ionized gas component which dominates the observed infrared forbidden line and infrared hydrogen recombination line emission. From the ratio of the line fluxes an electron density of 120 (+280,–120) $cm^{-3}$ is deduced.

158.274 **Identifications and spectra of extragalactic radio sources.**
M. S. Longair, S. J. Lilly.
J. Astrophys. Astron., Vol. 5, No. 4, p. 349 – 368 (1984).
The intention of this brief review is to concentrate on the identification and redshifts of extragalactic radio sources but it is worthwhile beginning with a historical survey of some of the more remarkable products of the process of identification of radio sources in both its galactic and extragalactic contexts. The authors then look in more detail at some of their recent work on the identification of extragalactic radio sources at optical and infrared wavelengths and at the problems of measuring their redshifts and spectra.

158.275 **Extragalactic sources with asymmetric radio structure. I. Observations of 17 sources.**
D. J. Saikia, P. Shastri, R. P. Sinha, V. K. Kapahi, G. Swarup.
J. Astrophys. Astron., Vol. 5, No. 4, p. 429 – 473 (1984).
The authors present total–intensity and linear–polarization observations with the Very Large Array (VLA) at λ6 and 2 cm of 17 sources, almost all of which were suspected to have extended emission only on one side of the nucleus. Five of them are still one–sided, three appear unresolved, while seven have radio lobes on both sides of the nucleus. The outer components in the double–lobed sources, however, have significantly different surface brightness or are very asymmetrically located with respect to the nucleus.

158.276 **Radio observations of the BL Lac object 1400+162.**
D. J. Saikia, G. Swarup, R. P. Sinha.
J. Astrophys. Astron., Vol. 5, No. 4, p. 475 – 482 (1984).
The authors present total–intensity and linear–polarization observations made with the VLA at 5 GHz of 1400+162, a BL Lac object in a group of galaxies. It has a misaligned triple structure with a prominent radio jet towards the east. There is evidence of a weak counter–jet towards the western component which also has the more prominent warm–spot. The authors discuss possible explanations for some of the observed features of this source.

158.277 **Cosmological evolution of radio sources.**
J. J. Condon.
Astrophys. J., Vol. 287, No. 2, p. 461 – 474 (1984).
Source counts at 0.408, 0.61, 1.4, 2.7, and 5 GHz; redshift distributions of strong sources selected at 1.4, 2.7, and 5 GHz; spectral–index distributions of sources in samples complete above different flux–density limits at 1.4, 2.7, and 5 GHz; and the 1.4 GHz local luminosity function were used to constrain the epoch–dependent spectral luminosity function of extragalactic radio sources. A simple evolutionary model for the spectral luminosity function was found which is consistent with these data and the magnitude distribution of galaxies optically identified with faint sources. This model evolves all sources in the same way.

158.278 **On the origin of power–law X–ray spectra of active galactic nuclei.**
I. Schlosman, J. Shaham, G. Shaviv.
Astrophys. J., Vol. 287, No. 2, p. 534 – 537 (1984).
A simple analytical model for a power–law X–ray continuum production in active galactic nuclei is presented. Dissipation of turbulent energy flux above the accretion disk forms an optically thin transition layer with an inverted temperature gradient. The authors show that the thermal emitted radiation has a power–law spectrum in the range 0.1 – 100 (300) keV with a photon energy spectral index $\gamma \approx 0.4 - 1$. The value $\gamma = 0.75$ appears to be a "natural" power–law index for Seyfert galaxies and QSOs.

158.279 **The luminous host galaxy and anomalous Brackett–gamma line of Markarian 231.**
R. M. Cutri, G. H. Rieke, M. J. Lebofsky.
Astrophys. J., Vol. 287, No. 2, p. 566 – 570 (1984).
New infrared observations have been made of the Seyfert galaxy Markarian 231. Multiaperture photometry reveals that the underlying galaxy is extremely luminous, with $M(K) = -24.7$. This implies that the Seyfert nucleus likely resides in a giant elliptical galaxy, or possibly in a galaxy which is undergoing a massive starburst. The strengths of the Paschen–alpha and Brackett–gamma lines of hydrogen have been measured using Fourier transform spectroscopy. The authors examine the possible origin of the exceptional activity and the unusual spectroscopic properties in the nucleus of Mrk 231 in the context of the nature of the host galaxy.

158.280 **Fine structure of the radio source 4C 21.53 in the decameter waveband.**
V. P. Bovkun, I. N. Zhuk.
Astron. Zh., Tom 61, Vyp. 6, p. 1108 – 1112 (1984). In Russian. English translation in Sov. Astron., Vol. 28, No. 6.
The results of interplanetary scintillation observations of the radio source 4C 21.53 at the frequencies 20 and 25 MHz are presented.

**The Case low–dispersion northern sky survey. II.**
See Abstr. 002.001.

**The second Byurakan spectral sky survey. I. The results of the area centred at $\alpha = 11^h30^m$, $\delta = 59°00'$.**
See Abstr. 002.006.

**Galaxies: activity and environment. Summary of the RAS specialist discussion, held 1984 April 13 at the University of Leicester.**
See Abstr. 010.683.

**Assembling the cloudy jigsaw.**
See Abstr. 011.020.

**Workshop über wechselwirkende Galaxien.**
See Abstr. 011.030.

**High resolution observations of galactic nuclei.**
See Abstr. 035.062.

**On the detection of positrons via the optical lines of positronium.**
See Abstr. 061.002.

**The hotspots from a time–varying jet.**
See Abstr. 062.013.

**On the viability of acceleration mechanisms for high–velocity clouds and quasar emission–line clouds.**
See Abstr. 062.014.

**Helical twisting on an adiabatically expanding jet. I. Propagation and temporal growth.**
See Abstr. 062.171.

**Physical processes in relativistic plasmas.**
See Abstr. 063.045.

**Power law X– and γ–ray emission from relativistic thermal plasmas.**
See Abstr. 063.046.

**Pair production at moderate optical depths.**
See Abstr. 063.047.

**Nonadiabatic self–consistent spherical accretion as a model for quasars and active galactic nuclei.**
See Abstr. 067.003.

**Clusters of black holes as models for active galactic nuclei.**
See Abstr. 067.034.

**Thick discs around black holes as energy sources for quasars and other astrophysical objects.**
See Abstr. 067.047.

**Black hole models for active galactic nuclei.**
See Abstr. 067.079.

**Shock acceleration in active galactic nuclei and quasars.**
See Abstr. 067.080.

**Collimated winds from active galactic nuclei.**
See Abstr. 067.082.

**An alternative to relativistic beaming in AGN.**
See Abstr. 067.083.

**Models of non–Keplerian accretion disks based on dynamo magnetic field: application to AGN.**
See Abstr. 067.084.

**Centrifugally driven MHD–winds and the non–thermal emission regions in AGNs.**
See Abstr. 067.086.

**Macroscopic currents and wind perturbations in the frame of the wind and shock model.**
See Abstr. 067.087.

**Duelling winds: a model for broad–line regions.**
See Abstr. 067.088.

**X–ray spectra of accretion disks.**
See Abstr. 067.089.

**Extragalactic gravitational collapse.**
See Abstr. 067.141.

**High resolution observations of extended radio sources near $l = 94°$.**
See Abstr. 125.020.

**H I absorption observations of two radio sources near the supernova remnant G 127.1 + 0.5.**
See Abstr. 125.085.

**Dynamical evolution of fragmenting gas clouds.**
See Abstr. 131.093.

**The origin of the infrared luminosity in violent star–formation regions.**
See Abstr. 131.130.

**Some comments on models of photoionized nebulae.**
See Abstr. 132.021.

**Nature of the infrared sources in IRAS Circulars 2 – 4.**
See Abstr. 133.012.

**A deep 6–centimeter radio source survey.**
See Abstr. 141.002.

**A deep Westerbork survey of areas with multicolor Mayall 4 m plates. I. The 1412 MHz catalogue, source counts and angular size statistics.**
See Abstr. 141.004.

**A deep Westerbork survey of areas with multicolor Mayall 4 m plates. II. Optical identifications.**
See Abstr. 141.005.

**A confusion limited radio survey from the Molonglo Cross.**
See Abstr. 141.007.

**Variability of SS433 and extragalactic compact radio sources.**
See Abstr. 141.011.

**A VLBI survey of peaked spectrum radio sources.**
See Abstr. 141.013.

**The nature of BG2107 + 49 – galactic or extragalactic?**
See Abstr. 141.015.

**Infrared study and classification of optically faint steep spectrum radio sources.**
See Abstr. 141.021.

**Very high resolution observations of compact sources in the direction of supernova remnants: G74.87 + 1.22.**
See Abstr. 141.024.

**The Medium Sensitivity Survey: a new sample of X–ray sources with optical identifications and the revised extragalactic log *N*–log *S*.**
See Abstr. 142.026.

**The extragalactic nature of H0323 + 022.**
See Abstr. 142.032.

**Characteristics of X–ray spectra from galactic and extra–galactic X–ray sources.**
See Abstr. 142.033.

**EXO 1102.8 + 2539, an X–ray variable AGN.**
See Abstr. 142.034.

**The HEAO–1 diffuse component spectrum from 15 to above 500 keV.**
See Abstr. 142.035.

**The origin of the X–ray background.**
See Abstr. 142.036.

**Active galaxies and the diffuse gamma–ray background.**
See Abstr. 143.016.

**Astrophysical boosters.**
See Abstr. 144.025.

**Astrophysical boosters.**
See Abstr. 144.027.

**On the formation of the superhigh–energy spectrum of cosmic rays in the nuclei of active galaxies.**
See Abstr. 144.078.

**Constructing distribution functions for spherical galaxies.**
See Abstr. 151.029.

**New stellar associations in M33 galaxy.**
See Abstr. 152.008.

**X–ray sources in globular clusters.**
See Abstr. 154.028.

**The nature of blue compact galaxies.**
See Abstr. 157.002.

**Masers in the nuclei of galaxies.**
See Abstr. 157.006.

**Water–vapour maser emission from galactic nuclei.**
See Abstr. 157.007.

**Optical properties of spiral galaxies with strong nuclear radio sources: active nuclei, starbursts, and composite objects.**
See Abstr. 157.012.

**Spectroscopic survey of southern compact and bright–nucleus galaxies. VI.**
See Abstr. 157.035.

**A four–colour photometry of spiral galaxies. A comparison with Markarian galaxies.**
See Abstr. 157.037.

**The large–scale radio continuum structure of the edge–on spiral galaxy NGC 253.**
See Abstr. 157.041.

**Superassociations in spiral galaxies with UV excess. I.**
See Abstr. 157.049.

**On the origin of tightly wound spiral features in galactic nuclei.**
See Abstr. 157.054.

**The radio source associated with the elliptical galaxy NGC 5322.**
See Abstr. 157.066.

**Study of hot spot galaxies.**
See Abstr. 157.075.

**Superassociations in spiral galaxies with UV excess. II.**
See Abstr. 157.093.

**Radial velocities of galaxies: some considerations on the comparison between radio and optical data.**
See Abstr. 157.150.

**The complex X–ray emission of NGC 253.**
See Abstr. 157.229.

**Compact–core–dominated radio emission from bright E/S0 galaxies.**
See Abstr. 157.230.

**A search for QSOs in the fields of nearby galaxies. I. NGC 253, NGC 5236 and NGC 6744.**
See Abstr. 159.029.

**Statistical evidence for possible association between QSOs and bright galaxies.**
See Abstr. 159.038.

**A third quasar close to NGC 3842.**
See Abstr. 159.042.

**Photoionization models and diagnosis of physical properties of the broad–line emission gas in quasars and Seyfert nuclei.**
See Abstr. 159.050.

**The properties of X–ray emission of quasars.**
See Abstr. 159.051.

**CCD imaging and neutral hydrogen emission in IZw1 and other low–redshift QSOs/AGNs.**
See Abstr. 159.061.

**Spectroscopy of 26 QSO candidates from the Jodrell Bank 966–MHz survey.**
See Abstr. 159.065.

**One to eight month X–ray light curves for 4 active extragalactic objects.**
See Abstr. 159.074.

**Complete quasar search in the NGC 1097 field.**
See Abstr. 159.088.

**Radiation pressure and the stability of broad–line region clouds.**
See Abstr. 159.090.

**Radiation pressure and the stability of broad–line region clouds.**
See Abstr. 159.094.

**Identifications of blue stellar objects from the Case Low–Dispersion Northern Sky Survey.**
See Abstr. 159.095.

**The radio–to–UV continuum in 3C 273.**
See Abstr. 159.105.

**A small–size (less than 1 min of arc) decametric radio source in the Perseus cluster.**
See Abstr. 160.008.

**High resolution radio and X–ray observations of A 115.**
See Abstr. 160.023.

**1.4 GHz continuum sources in the Hercules Cluster.**
See Abstr. 160.041.

**The intracluster gas around Cygnus–A.**
See Abstr. 160.061.

**Nonthermal radio sources in clusters of galaxies.**
See Abstr. 160.107.

**Westerbork observations of 37 poor clusters of galaxies at 1415 MHz.**
See Abstr. 160.108.

**Radio continuum properties of spiral galaxies in the Coma/A1367 supercluster. Comparison between isolated and cluster galaxies.**
See Abstr. 160.109.

**The radio luminosity function of first ranked galaxies in Abell clusters.**
See Abstr. 160.114.

**The clustering of galaxies about powerful radio sources.**
See Abstr. 160.135.

**CCD observations of galaxies in clusters at high redshift.**
See Abstr. 160.143.

**Is there really a large scale geometric anisotropy in extragalactic radio sources?**
See Abstr. 161.289.

# 159 Quasi-stellar Objects

159.001 **On the redshift relation of the density of absorption systems in the spectra of QSOs.**
E. A. Nazarov.
Astron. Tsirk., No. 1284, p. 1 – 3 (1983). In Russian.

159.002 **A metre-wave outburst from the QSO 1055+018.**
O. B. Slee.
Mon. Not. R. Astron. Soc., Vol. 209, No. 1, p. 215 – 226 (1984).
Observations of the QSO 1055+018 with the Culgoora circular array show that the source flux density doubled at 160 MHz and tripled at 80 MHz during the first seven months of 1983, while complementary observations between 0.843 and 15 GHz revealed no significant changes. The time variability at metre wavelengths may naturally be explained by the brightening of a coherent source. Slow interstellar scintillations provide another possible explanation.

159.003 **Wide-angle radio tail QSOs as members of clusters of galaxies. II. Direct optical observations and spectroscopy of QSO fields.**
P. Hintzen.
Astrophys. J., Suppl. Ser., Vol. 55, No. 3, p. 533 – 550 (1984). With plates 4 – 9.
Deep direct observations ($R \approx 23$) have been obtained for the fields of 19 radio QSOs ($0.3 < z < 1.5$), including both wide-angle radio tail QSOs (RT QSOs) and undistorted comparison QSOs. Counts of excess galaxies in these fields confirm the "cosmological" interpretation of QSO redshifts at the 99% confidence level and also support with 97% confidence the hypothesis that RT QSOs lie in richer concentrations of galaxies than do undistorted QSOs. Spectroscopic redshifts for galaxies in the field of 3C 275.1 provide preliminary confirmation that this RT QSO lies in a first-ranked galaxy at the center of a rich cluster of galaxies.

159.004 **VLA observations of Jodrell Bank radio quasars.**
F. N. Owen, J. J. Puschell.
Astron. J., Vol. 89, No. 7, p. 932 – 957 (1984).
VLA observations are presented of 26 radio quasars from the Jodrell Bank 966-MHz survey. The observations were made at 6 cm in the A array and thus have about 0.5-arcsec resolution. Jets are found in at least seven of the sources indicating that powerful jets are a very common feature in radio quasars. Central components are also detected in all the quasars at levels at least an order of magnitude higher than is found in similar samples of radio galaxies.

159.005 **Low-redshift quasars as the active nuclei of cosmologically distant interacting galaxies: a spectroscopic investigation.**
T. M. Heckman, G. D. Bothun, B. Balick, E. P. Smith.
Astron. J., Vol. 89, No. 7, p. 958 – 965 (1984).
The authors present CCD spectra obtained with the 4-m telescope at Kitt Peak of 21 apparent companion galaxies to 15 low-redshift quasars. With one exception, each galaxy is located at an apparent projected distance of $\leqslant 50$ kpc from its quasar. Redshifts have been determined for 19 of these galaxies. In 18 of the 19 total cases (95%), the galaxy and quasar radial velocities agree to $\leqslant 10^3 \mathrm{km\,s^{-1}}$ in the quasar rest frame. Combining the results with similar data in the literature yields the following: 86% of the apparent close companions ($\leqslant 50$ kpc projected quasar/galaxy separation) share the quasar redshift, compared to only $\sim$40% of the more distant (in projection) apparent companions. The authors believe the present data represent the most convincing evidence yet for a cosmological origin of quasar redshifts. The results also support the hypothesis that galaxy interactions are an important means of triggering quasar activity.

159.006 **What triggers a quasar?**
C. M. Gaskell.
Nature, Vol. 310, No. 5973, p. 102 (1984).
This note comments on a recent paper by R. C. Kennicutt and W. C. Keel presenting observational support to the hypothesis that interactions between companion galaxies are an important way of stimulating quasar-like activity. (See Abstr. 037.159.148).

159.007 **Multiple imaging of quasars by galaxies and clusters.**
R. Narayan, R. Blandford, R. Nityananda.
Nature, Vol. 310, No. 5973, p. 112 – 115 (1984).
Multiply-imaged quasars have generally been explained as a combination of gravitational lensing by galaxies and magnification by clusters. The observational difficulties with this view suggest that some such quasars may be more naturally interpreted as images split by the cores of very rich clusters. The observational distinctions between the two models are discussed.

159.008 **Quasar polarization properties.**
R. Carswell.
Nature, Vol. 310, No. 5974, p. 186 (1984).
This note discusses the main results of a recent systematic investigation of quasar polarization properties published by R. L. Moore and H. Stockman (see Abstract 37.159.135) and by H. Stockman, R. L. Moore and J. R. P. Angel (see Abstract 37.159.136).

159.009 **The clustering of quasars.**
P. A. Shaver.
Astron. Astrophys., Vol. 136, No. 1, p. L9 – L10 (1984).
A new method of studying the clustering of quasars is applied to a large sample of quasars with known redshifts. Tentative first results reveal possible clustering, which may be comparable to that of galaxies at the current epoch.

159.010 **The [O III]-emission associated with the QSO MR2251–178.**
L. Hansen, H. U. Nørgaard-Nielsen, H. E. Jørgensen.
Astron. Astrophys., Vol. 136, No. 1, p. L11 – L13 (1984).
The authors have observed MR2251–178 through narrow-band filters isolating the redshifted [O III] λ4959/5007 lines. The [O III]-image shows that the underlying galaxy is filled up by hot gas. Furthermore it shows three shell-like filamentary structures surrounding the QSO-galaxy reaching distances up to 110 kpc. The authors are not able to confirm the presence of a vast H II region extending 200 kpc to the south-east announced by Bergeron et al. (1983). From the off-band continuum images they estimate that the QSO-galaxy is fainter than the central cD of the surrounding cluster by $0\overset{m}{.}8$, but brighter than the remaining galaxies. No [O III]-emission was found in the cD galaxy or in the two elliptical galaxies close to the cD.

159.011 **The high energy X-ray spectrum of 3C 273.**
M. Bezler, E. Kendziorra, R. Staubert, G. Hasinger, W. Pietsch, C. Reppin, J. Trümper, W. Voges.
Astron. Astrophys., Vol. 136, No. 2, p. 351 – 354 (1984).
The quasar 3C 273 was observed in the hard X-ray range (18 – 240 keV) with a balloon-borne detector on 1981 September 28. The source was found to be quite strong and to emit a very hard spectrum. The photon number spectrum is well described by a single power law of index $\alpha = -1.2$, with no apparent break or cut-off. The integrated hard X-ray luminosity (20 – 200 keV) is $1.2 \times 10^{47} \mathrm{erg\,s^{-1}}$. Compared to other observations in 1978/79 this corresponds to an increase in luminosity (18 – 240 keV) by a factor of three, representing the first observation of time variability of 3C 273 in this energy range. The authors briefly discuss this behaviour with respect to Synchrotron-self-Compton and thermal Comptonization models.

159.012 **Nine new quasi–stellar objects with broad absorption lines.**
C. Hazard, D. C. Morton, R. Terlevich, R. McMahon.
Astrophys. J., Vol. 282, No. 1, p. 33 – 52 (1984). With plates 1 – 2.

The authors present slit spectra of nine new broad absorption line (BAL) QSOs selected from objective prism plates. These QSOs have a wide variety of properties. The emission redshifts range from 2.20 to 3.07 with Lyα apparently absent in six of them. The absorption troughs have widths up to 25,000 km $s^{-1}$ and blue edges displaced by as much as 36,000 km $s^{-1}$ from the emission lines. Some narrow absorption lines are also present. A new hypothesis for the origin of the BAL phenomenon is presented.

159.013 **Discovery of a new gravitational lens.**
S. Djorgovski, H. Spinrad.
Astrophys. J., Lett. Ed., Vol. 282, No. 1, p. L1 – L4 (1984).

The faint QSO pair 1635+267 A+B, discovered in a slitless spectroscopy survey by Weedman, are found to be at practically the same redshift, $z = 1.961 \pm 0.003$. Their redshift difference, determined with a cross–correlation technique, is found to be $(3.2 \pm 8.5) \times 10^{-4}$. This low value and the similarity of the spectra strongly argue that this is another case of a gravitational lens. Deep CCD images do not reveal the presence of a lensing object down to a red magnitude of 23.5, if it is more than an arc second away from either QSO image.

159.014 **Neutrino objects and quasar redshifts.**
L. Fang, S. Xiang.
Kexue Tongbao (Beijing), Vol. 28, No. 5, p. 644 – 648 (1983). Abstr. in Phys. Abstr., Vol. 87, No. 1237, Entry 73097 (1984).

159.015 **The anisotropic Compton effect in quasars and pulsars.**
V. M. Charugin.
Sov. Astron., Vol. 28, No. 1, p. 67 – 70 (1984). English translation of 37.159.005.

159.016 **Spectra of the quasar OQ 172.**
V. B. Nebelitskij, A. F. Fomenko, S. A. Levshakov, D. A. Varshalovich.
Sov. Astron. Lett., Vol. 10, No. 1, p. 32 – 35 (1984). English translation of 37.159.014.

159.017 **On the separation ratio of the outer radio components in quasars.**
D. J. Saikia.
Mon. Not. R. Astron. Soc., Vol. 209, No. 3, p. 525 – 531 (1984).

If the observed asymmetry in the radio jets is due to relativistic beaming, then the approaching outer component can be identified and the ratio $r_\theta$ of the core separations of the components can be calculated. The expected value of this ratio, *F*, should always be $\geqslant 1$ for an intrinsically symmetric, collinear source. The author finds, however, from a sample of quasars observed with high resolution that $r_\theta$ is $\lesssim F$ for most of them and $<1$ for about half of them. The author investigates possible explanations for such an effect and suggests that intrinsic non–collinearity and the existence of extremely dissipative beams probably play an important role.

159.018 **Gravitational lens model of the double QSO 0957+561 A,B incorporating *VLBI* features.**
D. Narasimha, K. Subramanian, S. M. Chitre.
Mon. Not. R. Astron. Soc., Vol. 210, No. 1, p. 79 – 88 (1984).

The twin quasar QSO 0957+561 A,B along with bright radio arches and *VLBI* structures is modelled using a gravitational lens consisting of the giant elliptical galaxy and the background cluster. The effective time–delay between components A and B comes out to be around one yr, which is significantly different from the value obtained by Young et al. for similar set of lens parameters. It is demonstrated that the observed intensity variation in component B between 1979 and mid–1981, reported by Keel (1982) can result from minilensing by low mass stars.

159.019 **An attempt of identifying QSO spectral absorption lines with small redshifts.**
I. E. Val'tts, B. V. Komberg.
Astrofizika, Tom 19, Vyp. 3, p. 449 – 460 (1983). In Russian. English translation in Astrophysics, Vol. 19, No. 3.

High–resolution spectral data available in literature for 38 quasi–stellar objects (QSO) have been used in a statistical analysis of the reliability of narrow absorption line identification with small redshifts ($z$ from –0.005 to +0.05). Emphasis was given to QSOs located within the region $\pm 10°$ to the supergalactic equator, at short angular distances from near galaxies, or coincident as to the location on the celestial sphere, with high–latitude high–speed hydrogen clouds. In most cases identifications with Ca II (H and K, 3969.65 and 3934.83) and Na I (3303.98 and 3303.37) are shown to be random. The only exceptions are the objects 3C 232, PKS 0457+024 and may be PHL 957.

159.020 **Deconvolution of quasar images from point–spread function: application to 3C 273, PKS 2135+147, and PKS 0812+020.**
O. Bendinelli, G. Parmeggiani, F. Zavatti, S. Lorenzutta.
Astrophys. Space Sci., Vol. 103, No. 1, p. 165 – 187 (1984).

The photometric structure of the resolved quasars 3C 273, PKS 2135+147, and PKS 0812+020 has been restored by an effective deconvolution procedure. Three well–defined components have been found in quasars: (1) the central point–like source, containing about 75% of the total light, (2) an intermediate region of about 10 kpc radius, whose luminosity seems to depend on the quasar's absolute luminosity, and (3) an external region with dimensions and behaviour of the brightness profile typical of giant elliptical and cD galaxies. An extensive discussion is devoted to the numerical aspects of the deconvolution.

159.021 **The quasar PKS 0237–233: chance coincidences and the alleged CO redshift systems.**
Y. P. Varshni.
Astrophys. Space Sci., Vol. 103, No. 2, p. 203 – 206 (1984).

It is shown that the number of redshifted CO absorption–line systems which are found in the spectrum of the quasar PKS 0237–233, using the rules of Varshalovich and Levshakov (1981), differ insignificantly from that that would be expected from chance coincidences. Consequently, the CO systems proposed by Varshalovich and Levshakov have no physical reality.

159.022 **Some recent developments in problems of quasar like objects.**
S. Tsuruta.
Proceedings of the Third Marcel Grossmann Meeting on General Relativity, Part A, p. 609 – 623 (1983). – See Abstr. 012.002.

After a review of the pertinent properties of 'quasar like objects' this paper presents models based on supermassive black holes and discusses the radiation mechanisms operating in these objects.

159.023 **Some recent development on the study of QSOs absorption lines.**
J. Chen, Z. Zou, Z. Cui, Y. Bian, X. Tang, S. Sun.
Proceedings of the Third Marcel Grossmann Meeting on General Relativity, Part A, p. 855 – 868 (1983). – See Abstr. 012.002.

This review paper discusses the methods of identifying redshift systems in quasar absorption spectra and examines current physical models attempting to explain these absorption systems.

159.024 **The periodicity in the distribution of quasar redshifts – an analysis for quasars around selected directions.**
Z. Deng.
Proceedings of the Third Marcel Grossmann Meeting on General Relativity, Part B, p. 1309 – 1313 (1983). – See Abstr. 012.002.

159.025 **The exponential law for superluminal expansion of the quasar 3C 273.**
S. Cao, Y. Liu.
Proceedings of the Third Marcel Grossmann Meeting on General Relativity, Part B, p. 1319 – 1324 (1983). – See Abstr. 012.002.

159.026 **Possible existence of two populations of quasi-stellar objects (QSO).**
B. V. Komberg.
Astrofizika, Tom 20, Vyp. 1, p. 73 – 83 (1984). In Russian. English translation in Astrophysics, Vol. 20, No. 1.

Observational data on QSO belonging to groups and clusters poor in galaxies and theoretical concepts of a possible merging of their members into a giant central cD galaxy are used to discuss the question about the existence of two populations of QSOs with considerably different properties. Weaker QSO I are associated with the nuclei of primary galaxies belonging to distant clusters while brighter QSO II with nuclei of secondary (merged) cD galaxies that form in the centers of groups and poor clusters. Possible properties of QSO I and II are discussed, some observational data are reported which may be interpreted as favoring the idea that there exist two QSO populations.

159.027 **Spectrophotometry of the QSO NRAO 530.**
V. Junkkarinen.
Publ. Astron. Soc. Pac., Vol. 96, No. 581, p. 539 – 542 (1984).

Spectrophotometry of the QSO NRAO 530 with the KPNO Cryogenic Camera is presented. A redshift of 0.902 is determined from Hβ, Hγ, and [O II] λ3727 emission lines. The Hβ equivalent width is smaller than in normal QSOs which may be characteristic of the optically violent variable QSOs.

159.028 **Binary quasi-stellar object Q 0957+561 A, B – a close pair of galaxies.**
B. V. Komberg.
Astrofizika, Tom 20, Vyp. 2, p. 351 – 355 (1984). In Russian. English translation in Astrophysics, Vol. 20, No. 2.

It has been shown that the assumption on the system Q 0957+561 A, B being a true binary rather than a gravitational-lens-split image of a single quasi-stellar object does not contradict the available observational data. A number of observational tests are mentioned which may prove that the system is in fact a binary one.

159.029 **A search for QSOs in the fields of nearby galaxies. I. NGC 253, NGC 5236 and NGC 6744.**
A. S. Pocock, J. C. Blades, M. V. Penston, M. Pettini.
Mon. Not. R. Astron. Soc., Vol. 210, No. 2, p. 373 – 380 (1984).

The authors give positions and magnitudes of QSO and background-galaxy candidates identified on UKST prism plates centred on the galaxies NGC 253, NGC 5236 and NGC 6744. Follow-up low-resolution spectroscopy with the 1.9-m telescope of the South African Astronomical Observatory confirmed that four of the candidates are QSOs (finding charts are given), and a further seven are galaxies sufficiently bright in the region of the Ca II λλ3933, 3968 lines to permit future high-resolution spectroscopy.

159.030 **The distribution of the true ratio between component size and overall size of double quasar sources.**
A. D. Fokker.
Astron. Astrophys., Vol. 137, No. 2, p. 360 – 361 (1984).

From the distribution of the ratio $\Phi$ = component angular diameter/apparent overall angular size, as determined for quasars that have a double structure, the true distribution of the ratio $\varphi = d/D$ (where $d$ is the component diameter and $D$ the overall dimension of the system) can be derived under the assumption of isotropy of the directions of source axis. The $\Phi$-distribution as derived by Nottale (1982) from published data on the structure of quasar doubles is used. It turns out that the quantity $\varphi$ varies by a factor of the order of 3. The mean value of $1/\varphi$ is found to be 7.1.

159.031 **Absorption lines in the spectra of the QSO PKS 1448-232.**
J.-s. Chen, D. C. Morton, B. A. Peterson, A. E. Wright, D. L. Jauncey.
Proc. Astron. Soc. Aust., Vol. 5, No. 3, p. 355 – 359 (1984).

Spectra of the QSO PKS 1448-232 at a resolution of 33 Å mm$^{-1}$ were obtained with the RGO spectrograph and IPCS at the f/8 focus of the AAT. An emission line redshift $z_{em} = 2.2176$ was derived from the strong symmetric Lα profile. Two absorption redshift systems with $z_{ab} = 1.72349$ and $z_{ab} = 1.58462$ could be identified with a high confidence level.

159.032 **Two quasars seen near the spiral galaxy NGC 470.**
H. Arp, J. Surdej, J.-P. Swings.
Astron. Astrophys., Vol. 138, No. 1, p. 179 – 182 (1984).

The discovery of two quasars with 16″ separation from each other on the sky and within 95″ of the nucleus of the spiral galaxy NGC 470 is reported. The quasars were discovered in a search for ultraviolet-excess objects in four overlapping fields of about 25 square degrees each. Preliminary spectroscopy of the quasars yields $z = 1.875$ ($m_v = 19.9$ mag) and $z = 1.533$ ($m_v = 18.2$ mag).

159.033 **Study of the overall spectrum of the QSO 3C273.**
T. J.-L. Courvoisier.
Messenger, No. 37, p. 8 – 10 (1984).

159.034 **UV observations of the variable quasar PG 1351+64.**
A. Treves, J. Drew, R. Falomo, L. Maraschi, E. G. Tanzi, R. Wilson.
ESA Spec. Publ., ESA SP-218, p. 119 – 124 (1984). – See Abstr. 012.024.

PG 1351+64 was repeatedly observed in 1979 – 83 with IUE. The spectra exhibit strong emission lines superposed to a continuum with a spectral index of $\alpha \cong -0.3$. Variations of a factor 2 in line and continuum intensity were observed with a minimum time scale of a month. The nature of the continuum emission is shortly discussed and an upper limit to the mass within the line emitting region is deduced from the observed line variability.

159.035 **Quasars as probes of the intergalactic medium.**
R. F. Green, J. Bechtold.
ESA Spec. Publ., ESA SP-218, p. 125 – 128 (1984). – See Abstr. 012.024.

The authors have obtained spectral energy distributions for 10 high-redshift quasars in the observed wavelength range 1250 to 10000 Å. A pronounced steepening is detected shortward of about 1200 Å rest, which correlates well with quasar redshift, but poorly with luminosity. The authors can account for the observed slope change by Lyman continuum absorption from the strongest 5% of the Lyman alpha forest lines, identified with the low column density metal containing systems. Reddening may also contribute to the steepening, but a much lower neutral gas to dust ratio than that found in the Galaxy is required. Optically thick Lyman limit discontinuities are seen with $z_{abs} \ll z_{em}$.

159.036 **Determination of difference in light travel time for QSO 0957+561A, B.**
R. Florentin-Nielsen.
Astron. Astrophys., Vol. 138, No. 2, p. L19 – L20 (1984).

The time delay, Δt, between light variations in the two images of QSO 0957+561 has been established observationally: $\Delta t = 1.55 \pm 0.1$ yrs. This places the QSO at its cosmological distance. Most probable value of Hubble parameter is $H_0 \cong 77$ km s$^{-1}$Mpc$^{-1}$.

159.037 **Deconvolution of photographic images of quasars from point spread functions.**
O. Bendinelli, S. Lorenzutta, G. Parmeggiani, F. Zavatti.
Astron. Astrophys., Vol. 138, No. 2, p. 337 – 342 (1984).

The authors present a reliable method, based on the use of a priori information and on Tikhonov's regularization of unstable inverse problems, for the deconvolution of quasar and distant galactic nuclei from point spread function. The application to four quasars: 3C273, PKS2135+147, 3C206, and PKS0812+020, chosen as representative members of the low red-shift quasar sample of Wyckoff et al., 1981, shows that the deconvolved photometric structure of the region surrounding the central point-like source is very similar, if suitably corrected, to that of a set of well studied gE and cD galaxies.

159.038 **Statistical evidence for possible association between QSOs and bright galaxies.**
Y. Chu, X. Zhu, G. Burbidge, A. Hewitt.
Astron. Astrophys., Vol. 138, No. 2, p. 408 – 414 (1984).

A statistical study has been made of QSO–bright galaxy correlations using the QSO catalogue of Hewitt and Burbidge (1980) and the Second Reference Catalogue of Bright Galaxies (de Vaucouleurs et al., 1976). Both cross–correlation function techniques and the nearest neighbor method have been used. The results from all of these analyses provide strong statistical evidence for the association of QSOs at all redshifts with bright galaxies which have redshifts $\leqslant 0.05$.

159.039 **The search for high–redshift QSOs.**
B. J. Boyle.
Observatory, Vol. 104, No. 1062, p. 216 (1984). Abstract. – See Abstr. 010.682.

159.040 **Some evidence of the axial symmetry of emitting regions in quasars with broad absorption lines.**
V. P. Grinin.
Pis'ma Astron. Zh., Tom 10, No. 9, p. 643 – 653 (1984). In Russian. English translation in Sov. Astron. Lett., Vol. 10.

Formation of resonance lines in QSOs with broad absorptions is considered. It is shown that the axisymmetric decelerated outflow model with the number density of the scattering atoms being maximal in the equatorial plane and minimal in the directions of the poles explains both the PHL 5200–type profiles (with a sharp transition from emission to absorption and $W_e/W_a \lesssim 1$) and the Q1246–057–type profiles (with separated emission and absorption components and $W_e/W_a \gtrsim 1$). Some other arguments in favor of the disc–like geometry of the outflows in QSOs are given.

159.041 **A negative search for CO absorption lines ($J$ = 0–1) toward high–redshift quasars.**
F. Takahara, Y. Sofue, M. Inoue, N. Nakai, H. Tabara, T. Kato.
Publ. Astron. Soc. Jpn., Vol. 36, No. 2, p. 387 – 390 (1984).

A search has been made for absorption lines of CO molecules in the direction of six high–redshift quasars with strong continuum emission in the millimeter–wave range. The 45–m telescope at the Nobeyama Radio Observatory (NRO) combined with wide–band spectrometers was used to investigate redshifted CO ($J$ = 0–1) absorption lines in the 40–GHz band with a 0.5 – 3.75–GHz band width. No significant features are found in the obtained spectra.

159.042 **A third quasar close to NGC 3842.**
H. Arp, G. Gavazzi.
Astron. Astrophys., Vol. 139, No. 1, p. 240 – 242 (1984).

Radio and X–ray observations have identified quasar candidates in the area of Abell 1367, a cluster of galaxies in the Coma supercluster. NGC 3842 is one of the largest galaxies in A 1367 and was previously reported as having two quasars within 73″ of the center of the galaxy. The present paper reports the confirmation of a third quasar at a distance of 73″ from NGC 3842. The new quasar has a redshift of $z = 2.20$.

159.043 **A deconvolved structure of the nebulosity associated with 3C 206.**
O. Bendinelli, G. Parmeggiani, F. Zavatti.
Astrophys. Space Sci., Vol. 104, No. 2, p. 199 – 204 (1984).

The published photographic profile of 3C 206 (reported in the low redshift sample of quasars by Wyckoff et al., 1981) has been deconvolved from the point–spread–function by means of an effective restoration procedure. The deconvolved photometric structure of the quasar consists of a central point–like source, containing 68% of the integrated luminosity, an intermediate region of about 10 kpc radius ($H_0 = 60$ km s$^{-1}$ Mpc$^{-1}$, $q_0 = 0$) and an external region with nearly–linear slope and brightness level of the profile similar to those of the corresponding regions in giant elliptical and cD galaxies. The result confirms the previous findings in 3C 273, PKS 2135+147, and PKS 0812+020 obtained in the same way.

159.044 **Relativistic star clusters with large central redshifts.**
P. K. Das.
Mon. Not. R. Astron. Soc., Vol. 210, No. 4, p. 753 – 761 (1984).

Previous calculations of the velocity distribution function for a star cluster scenario of a class of core–envelope type static, stable, massive equilibrium configurations, capable of giving high internal gravitational redshifts, have been extended to show that positive distribution functions can be obtained for these objects by suitably altering the envelope equation of state, without affecting the overall structure significantly. Limits on the central redshift obtainable from such configurations, subject to conditions of pulsational stability and positivity of the distribution function, have been obtained. Finally, the relevance of the new composite configurations as models for high redshift quasi–stellar objects is discussed.

159.045 **Strong Na I D absorption in the spectrum of the QSO PKS 1327–206 due to the galaxy MCG 03–34–085.**
D. Kunth, J. Bergeron.
Mon. Not. R. Astron. Soc., Vol. 210, No. 4, p. 873 – 882 (1984).

The authors have discovered a strong Na I D absorption line in the spectrum of the QSO PKS 1327–206 ($z_e = 1.1696$) at a redshift $z_a = 0.0180$, very similar to that of the galaxy MCG 03–34–085 which lies 38 arcsec away in the plane of the sky. The lower limit on the column density of the Na I D absorption is $7.4 \times 10^{12}$cm$^{-2}$. Such an unusually large column density may imply an extinction of $E(B-V) = 0.18(+0.04, -0.03)$. Only an upper limit of 0.7 Å is given for the Ca II K line leading to a low Ca II/Na I ratio. From this ratio and the inclination of 62° of the galaxy one can argue that the absorption lines are produced in the outer disc at a distance of 21.0 $h_0^{-1}$kpc from the nucleus, although a contribution from the outer halo cannot be ruled out. It is shown that the galaxy itself is gaseous at least in its nucleus since its spectrum presents emission lines typical of a Liner.

159.046 **A similarity solution for a QSO envelope with radiation driving.**
L. T. Hildyard.
Mon. Not. R. Astron. Soc., Vol. 210, No. 4, p. 953 – 959 (1984).

For some time now it has been recognized that the wind from a QSO central source is appreciably affected by radiative driving. This paper outlines an unsteady solution, in terms of a similarity variable, for a model in which the radiative acceleration is either directly proportional to the local fluid density, or is inversely proportional to the square of the radial distance from the (massive) source. The flow is assumed in this model to be spherically symmetric and externally bounded by a shock wave, expanding into a medium at rest. In view of the restrictive assumptions, no attempt has been made to relate the results to any practical case, the emphasis always being to attempt to predict how a given flow is modified by the additional effect.

159.047 **A complete sample of quasars at $B$ = 19.80.**
H. L. Marshall, Y. Avni, A. Braccesi, J. P. Huchra, H. Tananbaum, G. Zamorani, V. Zitelli.
Astrophys. J., Vol. 283, No. 1, p. 50 – 58 (1984).

The authors present spectroscopy for quasar candidates in a complete UV excess survey with $B < 19.80$. Including the 10 previously known quasars, there are 35 quasars in the 1.72 deg$^2$ BF region. Fifteen quasars are detected in X–rays by deep exposures with the Einstein Observatory. The X–ray data for the 35 quasars are used to show that quasars with $B < 19.80$ and $z < 2.2$ contribute $\sim$24% of the observed X–ray background at 2 keV. By combining the present sample with a sample of 22 quasars having $B < 18.25$ (in the 37.2 deg$^2$ AB region), the authors find that fits of evolution and luminosity functions are substantially similar to results reported in an earlier paper.

159.048 **Two newly discovered quasars closely spaced across a galaxy.**
H. Arp.
Astrophys. J., Vol. 283, No. 1, p. 59 – 63 (1984).

Two stellar appearing objects, closely situated on either side of a large galaxy in the cluster A1367, were recently identified, as

X-ray sources by Bechtold and colleagues. Observations reported in the present paper establish that both objects are quasars of nearly identical apparent magnitude and redshifts of $z = 0.33$ and $z = 0.95$. At distances of $r = 73''$ and $r = 59''$ from the galaxy NGC 3842 ($z = 0.026$), the chance of accidentally encountering two quasars of this apparent magnitude is less than $7 \times 10^{-6}$.

159.049 **Images and spectra of the host galaxy of the QSO Markarian 1014.**
J. W. MacKenty, A. Stockton.
Astrophys. J., Vol. 283, No. 1, p. 64 – 69 (1984). With plates 1 – 2.

The authors present images and spectra of the extended luminous material associated with the low-redshift QSO Markarian 1014. Stellar absorption lines indicative of an admixture of early- and late-type stars are present at the same redshift as that of the QSO nucleus. The distribution of the luminous material is highly asymmmetric, appears spiral-like, and on a deep CCD image, its total extent is $\sim 90$ kpc ($H_0 = 75$, $q_0 = 1/2$). Markarian 1014 is one of the most striking examples yet found of a galaxy with a fairly luminous QSO nucleus.

159.050 **Photoionization models and diagnosis of physical properties of the broad-line emission gas in quasars and Seyfert nuclei.**
J. Kwan.
Astrophys. J., Vol. 283, No. 1, p. 70 – 80 (1984).

The earlier photoionization calculations of Kwan and Krolik are extended and improved. The present models all pertain to slabs of finite column densities. The O I and Fe II multilevel systems are modeled better, and the dependence of different line ratios on the physical conditions is illustrated graphically. The quantitative dependence of the line ratios on the physical parameters provides tests of the photoionization model, a framework for modeling an individual quasar spectrum, and, in conjunction with kinematic models, allows specific calculations of both the relative intensities and profiles for the different lines.

159.051 **The properties of X-ray emission of quasars.**
H.-s. Yang, R.-l. Liu, H.-q. Zhang.
Chin. Phys., Vol. 4, No. 2, p. 331 – 340 (1984). Abstr. in Phys. Abstr., Vol. 87, No. 1245, Entry 110770 (1984).

159.052 **Mass loss from quasars – observational evidence from their absorption-line spectra.**
M. G. Smith.
Mass loss from astronomical objects, p. 109 – 116 (1984). – See Abstr. 012.032.

A brief account is given of the absorption-line spectra of QSOs. Evidence for mass loss at velocities of up to 0.1c (perhaps 0.18c) is found in the form of broad-absorption-line systems. Further observational generalisations about these systems, needed in order to construct a useful mass-loss model, are fraught with difficulties. Some sharp-lined absorption systems may also be associated with mass loss from the associated quasar.

159.053 **Mass loss from quasars – more observational evidence, energetics and model constraints.**
R. F. Carswell.
Mass loss from astronomical objects, p. 117 – 124 (1984). – See Abstr. 012.032.

The observational evidence for mass flows in the material which gives rise to the emission lines seen in quasar spectra is discussed briefly. Observed velocity differences between high-ionization and low-ionization material argue strongly that some mass flow is occurring but the form of this is very uncertain.

159.054 **QSO winds and the broad emission line regions.**
J. E. Dyson.
Mass loss from astronomical objects, p. 125 – 127 (1984). Abstract. – See Abstr. 012.032.

159.055 **The emission-line regions of quasars.**
J. Kwan.
Ann. N.Y. Acad. Sci., Vol. 422, p. 284 – 290 (1984). – See Abstr. 012.033.

This paper summarizes the present understanding of the physical conditions of the emission-line gas in quasars and addresses the question of the cloud kinematics.

159.056 **Quasars as supermassive binaries.**
C. M. Gaskell.
Ann. N.Y. Acad. Sci., Vol. 422, p. 349 (1984). Abstract. – See Abstr. 012.033.

159.057 **An accretion disk origin for the ultraviolet excess of luminous quasars: the confrontation of theory with observation.**
M. A. Malkan.
Ann. N.Y. Acad. Sci., Vol. 422, p. 368 (1984). Abstract. – See Abstr. 012.033.

159.058 **A complete sample of quasars at $B = 19.80$.**
H. L. Marshall, Y. Avni, A. Braccesi, J. P. Huchra, H. Tananbaum, G. Zamorani, V. Zitelli.
Ann. N.Y. Acad. Sci., Vol. 422, p. 370 (1984). Abstract. – See Abstr. 012.033.

159.059 **United Kingdom Schmidt Telescope objective prism surveys for QSOs.**
A. Savage.
Ann. N.Y. Acad. Sci., Vol. 422, p. 380 (1984). Abstract. – See Abstr. 012.033.

159.060 **Alternating ejection in radio-loud quasars?**
L. M. Ensman, J. S. Ulvestad.
Astron. J., Vol. 89, No. 9, p. 1275 – 1278 (1984).

A sample of radio quasars mapped at the Very Large Array has been used to test the hypothesis that ejection of material in extragalactic radio sources occurs on only one side at a time. This hypothesis, most recently put forward by Rudnick (1982), predicts a deficit of double sources having their two outer radio lobes at equal distances from the radio core. Such a deficit is not confirmed in the authors' sample of 47 quasars containing triple radio sources. The statistics imply that alternation of ejection direction is not the dominant effect in determining relative distances from the core to the radio lobes in the majority of quasars; factors such as nonuniformities in the intergalactic medium may be more important.

159.061 **CCD imaging and neutral hydrogen emission in IZw1 and other low-redshift QSOs/AGNs.**
G. D. Bothun, T. M. Heckman, R. A. Schommer, B. Balick.
Astron. J., Vol. 89, No. 9, p. 1293 – 1299 (1984). With plate 72.

The authors present the results of a survey for H I emission from a sample of 14 low-redshift ($z < 0.07$) QSOs and active galactic nuclei. In addition, they have obtained high-quality CCD images for most of the sample objects. Six of the 14 objects in the sample have been detected in H I, with a bias toward detecting objects that exhibited nebulosity on the Palomar Sky Survey. The range in detected H I mass is similar to that seen in previous studies of Seyfert galaxies. In general, neither active galactic nuclei or low-redshift QSOs are immersed in host galaxies that are overabundant in H I with respect to normal spirals. Furthermore, the optical luminosities of the host galaxies are not large in comparison with normal spirals. The most significant result is the strong H I detection of the QSO/Seyfert I galaxy IZw1 at a velocity that is within 50 km $s^{-1}$ of the optical redshift.

159.062 **Gravitational lens models for the triple radio source MG 2016+112.**
D. Narasimha, K. Subramanian, S. M. Chitre.
Astrophys. J., Vol. 283, No. 2, p. 512 – 514 (1984).

The authors present two sets of models for the newly reported gravitational lens system MG 2016+112, one involving a galaxy in a rich cluster and the other made up of two galaxies at different redshifts. They suggest possible tests of the model, specifically in relation to the location and characteristics of the third image and the time delay between images.

159.063 **Are forbidden lines present in the optical spectrum of the QSO 3C 273?**
B. M. Peterson, K. A. Meyers, E. R. Capriotti.
Astrophys. J., Vol. 283, No. 2, p. 529 – 531 (1984).
The emission lines in the vicinity of Hβ in the spectrum of 3C 273 are reexamined. It is found that while most of the emission redward of Hβ is attributable to strong Fe II lines, [O III] λλ4959, 5007 also appears to be present. The [O III] lines are broad compared to those in other active galaxies, but they are narrower than the permitted lines and presumably arise in a distinct region. The large forbidden line widths account for the unusual appearance of the 3C 273 spectrum.

159.064 **Relative orientation of core polarization at λ6 cm and the overall radio axes of quasars.**
D. J. Saikia, P. Shastri.
Mon. Not. R. Astron. Soc., Vol. 211, No. 1, p. 47 – 56 (1984).
For a sample of 127 quasars observed with high angular resolution at the VLA, the authors have investigated the relative orientation of the core polarization vector at λ6 cm and the overall radio axis. While the core polarization **E**–vector tends to be perpendicular to the radio axis in the two–sided quasars with lobes of radio emission on opposite sides of the nucleus, no correlation is found for the one–sided sources which appear to have only a single extended lobe. The authors have also investigated the dependence of the angle, $\phi$, between the polarization vector and the radio axis, on the fraction of emission from the core, $f_c$. They find that, for sources with weak cores, $\phi$ tends to be close to 90° while, for those with the strongest cores, $\phi$ appears to be more or less uniformly distributed between 0° and 90°.

159.065 **Spectroscopy of 26 QSO candidates from the Jodrell Bank 966–MHz survey.**
D. Walsh, J. M. Beckers, R. F. Carswell, R. J. Weymann.
Mon. Not. R. Astron. Soc., Vol. 211, No. 1, p. 105 – 109 (1984).
Twenty–six QSO candidates from the Jodrell Bank 966–MHz survey have been observed spectroscopically at the Multiple Mirror Telescope. Nineteen are emission–line QSOs and redshifts are given for these. Four have smooth, featureless spectra and are classified as BL Lacertae–type objects. The remaining three objects are galactic stars.

159.066 **Q 1114–2846: a quasar–galaxy binary.**
G. Gilmore.
Mon. Not. R. Astron. Soc., Vol. 211, No. 1, p. 25P – 27P (1984).
A quasar–galaxy pair with identical redshifts and an angular separation of six arcsec is reported.

159.067 **Infrared spectrophotometry of the most luminous quasar S5 0014+81.**
H. Kühr, C. W. McAlary, R. J. Rudy, P. A. Strittmatter, G. H. Rieke.
Astrophys. J., Lett. Ed., Vol. 284, No. 1, p. L5 – L8 (1984).
Spectrophotometric and broad–band infrared observations are reported for the most luminous quasar: S5 0014+81. Hβ and the [O III] λλ4959 + 5007 emission lines are detected at a redshift of 3.41, in agreement with the optically determined redshift. S5 0014+81 is the first object at such high redshift in which Hβ and [O III] have been measured separately, showing rest frame equivalent widths of 37 and 31 Å, respectively.

159.068 **A search for quasars in the Virgo cluster region.**
X.–T. He, R. D. Cannon, J. A. Peacock, M. G. Smith, J. B. Oke.
Mon. Not. R. Astron. Soc., Vol. 211, No. 2, p. 443 – 459 (1984).
Using objective–prism plates taken with the 44–arcmin objective prism mounted on the UK Schmidt telescope, the authors have found 53 emission–line quasar candidates and 29 ultraviolet–excess objects (possible low–redshift quasars) in a 5 × 5 degree² field centred on $12^h27^m$, +13°30′ (1950) in the Virgo cluster of galaxies. Eighteen of these 82 candidates were observed using the double spectrograph on the Palomar 5–metre telescope; 13 of the observed objects proved to be quasars. The broad–absorption–line QSO Q1232+134 is the first example of the class to show broad low–ionization absorption lines – such as Mg II 2798 Å – in addition to the usual high–excitation lines such as N V λ1240 Å.

159.069 **Detection of lines by calibrated algorithm in a close pair of quasars.**
A. S. Trew, P. W. J. L. Brand.
Mon. Not. R. Astron. Soc., Vol. 211, No. 2, p. 485 – 495 (1984).
Comparison of the current quasar line detection techniques shows that the results do not always agree and as a consequence a debate has arisen as the true redshift distribution of the narrow quasar absorption lines. The authors have developed a method for finding such features and have subjected it to rigorous testing on model spectra. The results are presented showing that this method is very efficient and has no significant biases. The algorithm has also been applied to a close pair of quasars in a grouping discussed by Arp & Hazard which have been observed at high resolution. The authors find only marginal evidence for absorption common to both objects and that the redshift distributions of the absorption systems are otherwise normal.

159.070 **1159+123: a bright high–redshift QSO selected from a UK Schmidt IIIaF plate.**
C. Hazard, R. Terlevich, R. McMahon, D. Turnshek, C. Foltz, J. Stocke, R. Weymann.
Mon. Not. R. Astron. Soc., Vol. 211, No. 2, p. 45P – 50P (1984).
Using IIIaF, IIaD and IIIaJ objective prism plates from the UK Schmidt telescope the authors have discovered a strong emission–line QSO having $z_{em} \cong 3.5$. The only known QSOs with higher redshifts were discovered by radio or colour selection techniques. The discovery indicates that, up to redshifts of 3.5, QSOs are not conspicuously less frequent than in the range $3.0 < z_{em} < 3.3$. Spectroscopic studies at Steward Observatory show a strong absorption system displaced to the red of the emission lines, indicating an inflow velocity of 500 km s$^{-1}$ with respect to the emission lines.

159.071 **Low excitation absorptions in the spectrum of the QSO PKS 1327–206 due to the galaxy MCG 03–34–085.**
D. Kunth, J. Bergeron.
MPE Rep., No. 184, p. 43 – 46 (1984). – See Abstr. 012.039.
The authors have discovered a strong Na I D absorption line in the spectrum of the QSO PKS 1327–206 ($z_e$ = 1.1696) at a redshift $z_a$ = 0.0180, very similar to that of the galaxy MCG 03–34–085 which lies 38 arcsec away in the plane of the sky. It is argued that the absorbing material actually belongs to that galaxy.

159.072 **Morphology and optical polarization of the 3C 273 jet.**
K. Meisenheimer, H.–J. Röser.
MPE Rep., No. 184, p. 51 – 54 (1984). – See Abstr. 012.039.

159.073 **X–ray surveys of quasars with the Einstein Observatory.**
T. Maccacaro.
MPE Rep., No. 184, p. 63 – 72 (1984). – See Abstr. 012.039.
A review of quasar observations with the Einstein Observatory is given. In particular, luminosity correlations in quasars and the relations between X–ray selected samples and optically selected quasar samples are discussed.

159.074 **One to eight month X–ray light curves for 4 active extragalactic objects.**
W. A. Snyder, K. S. Wood.
MPE Rep., No. 184, p. 114 – 117 (1984). – See Abstr. 012.039.
The authors present the light curves of 4 extragalactic objects using the 0.5 – 20 keV LASS detectors on HEAO 1. Two of the objects are quasi–stellar sources, one is a BL Lacertae object, while the fourth, although not unambiguously identified, is certainly extragalactic. In each case the light curves have a temporal resolution of a few days, and extend at least 30 days in length. Variability has definitely been detected in 3 of the 4 objects.

159.075 **The high energy X-ray spectrum of 3C 273.**
M. Bezler, R. Staubert, E. Kendziorra, G. Hasinger, W. Pietsch, C. Reppin, J. Trümper, W. Voges.
MPE Rep., No. 184, p. 118 – 120 (1984). – See Abstr. 012.039.
This note reports on X-ray observations of the quasar 3C 273 in the energy range 20 – 200 keV.

159.076 **An objective prism quasar survey with the Schmidt telescope on Calar Alto.**
U. Haug.
MPE Rep., No. 184, p. 268 – 270 (1984). – See Abstr. 012.039.

159.077 **Multifrequency observations of the quasi-stellar object Ton 1542.**
D. M. Worrall, J. J. Puschell, F. C. Bruhweiler, H. R. Miller, M. F. Aller, H. D. Aller.
Publ. Astron. Soc. Pac., Vol. 96, No. 583, p. 699 – 704 (1984).
The authors present single-epoch multifrequency data for the radio-quiet low-redshift QSO Ton 1542, spanning radio through ultraviolet wavelengths. They measure a $3\sigma$ upper limit of 9.2 mJy for the 20.3 GHz flux density. The first ultraviolet spectrum of this object is presented and discussed. The authors find a particularly strong excess of flux density at $\sim 3000$ Å, which is likely to be dominated by Balmer continuum emission. They also find a $L\alpha/H\beta$ flux ratio of $\sim 17$, which is higher than that usually observed in QSOs.

159.078 **Radio and X-ray emission in radio-selected quasars.**
E. D. Feigelson, T. Isobe, A. Kembhavi.
Astron. J., Vol. 89, No. 10, p. 1464 – 1477 (1984).
In order to investigate the physical relationship between radio and X-ray emission in quasars, the authors have observed with the Very Large Array 31 radio-loud quasars and a comparable number of radio-quiet quasars. The sample was selected to include quasars with known X-ray luminosities but poorly-known radio properties. Maps with polarization data of 15 extended radio-loud quasars at 20 cm and 6 cm are presented. They show jet magnetic fields to be aligned along the source axis while lobe fields have misaligned orientations. The authors have sought statistical associations between X-ray luminosity and extended radio luminosity, spectrum, morphology, and size in an extended sample of 44 radio-loud quasars. Only one statistically significant correlation, between X-ray and 178-MHz power, is found.

159.079 **Radio and optical morphology of low-redshift quasars.**
A. C. Gower, J. B. Hutchings.
Astron. J., Vol. 89, No. 11, p. 1658 – 1687 (1984).
Snapshot VLA radio maps are presented of a representative sample of low-redshift quasars, and radio properties summarized for 40 objects. Resolutions are typically 1.4″ at 20 cm and 0.4″ at 6 cm wavelength. Half the sources have a bright core and two lobes, while 20% have a single lobe. With data of dynamic range $\sim 100$, 55% have connecting luminous bridges between components, and more than half of these are curved. 25% of the sources are significantly nonlinear, and over 40% have cores with resolved structure. 20% of the sources are unresolved. Optical imaging data, principally from Hutchings, Crampton, and Campbell (1983), exist for 24 of the quasars. The authors discuss possible correlations between optical and radio morphology.

159.080 **The radio-to-visual spectral-flux distribution of very red QSOs.**
W. A. Stein, M. L. Sitko.
Astron. J., Vol. 89, No. 11, p. 1688 – 1691 (1984).
Very red QSOs exhibit flattening of the spectral-flux distributions at infrared wavelengths. The objects in this study are associated with radio sources with spectra that are generally described by power laws, and such spectra are characteristic of extended lobes of radio emission. One source that has been studied with high angular resolution at $\lambda = 6$ cm, 3CR68.1, exhibits only the emission from the lobes, and a low upper limit of 2 mJy has been obtained on the 6-cm flux of the visually identified QSO. This object appears to be a relatively radio-weak (or radio-quiet?) QSO. High angular resolution radio-frequency studies of other objects in the sample are encouraged.

159.081 **An X-ray survey of variable radio bright quasars.**
M. J. Henriksen, F. E. Marshall, R. F. Mushotzky.
Astrophys. J., Vol. 284, No. 2, p. 491 – 496 (1984).
A sample consisting primarily of radio bright quasars has been observed in X-rays with the Einstein Observatory for times ranging from 1500 to 5000 seconds. Detected sources have luminosities ranging from 0.2 to $41.0 \times 10^{45}$ ergs s$^{-1}$ in the 0.5 to 4.5 keV band. Three of the 14 objects which have been reobserved show flux increases greater than a factor of 2 on a time scale greater than six months. The optical and X-ray luminosities are correlated, which suggests a common origin. From the comparison of the calculated Compton luminosity and the observed levels, it is found that the Compton component is suppressed in at least some of the strong radio quasars, perhaps due to near alignment of a jet.

159.082 **Spectrophotometry of intermediate-redshift quasars.**
A. Uomoto.
Astrophys. J., Vol. 284, No. 2, p. 497 – 511 (1984).
Measurements of the spectra of six intermediate-redshift quasars were made with a self-scanned Digicon detector. Emission lines were used to measure the C:N:O ratio with the result that this ratio is solar except for a slight nitrogen enhancement in two of the quasars. Intensity measurements of weak emission lines, including some due to iron, were also made. Gaskell's discovery of a redshift discrepancy between the high- and low-ionization lines is confirmed, but an unusual narrow-lined object, PKS 1148-001, appears not to show it.

159.083 **Optical variability of 14 quasi-stellar objects.**
C. Barbieri, G. Romano.
Acta Astron., Vol. 34, No. 1, p. 117 – 123 (1984).
The optical variability of 14 QSOs has been examined on plates taken with the Asiago Observatory telescopes. Five of the QSOs show large variations while 9 do not appear to vary.

159.084 **The variability of blue objects in the field of M3.**
G. Romano.
Acta Astron., Vol. 34, No. 2, p. 315 – 316 (1984).
The author has examined 48 blue objects discovered by Richter et al. in the field of M3. Six of these show variations of light.

159.085 **3C 273: a gravitationally lensed quasar?**
S. M. Chitre, D. Narasimha, J. V. Narlikar, K. Subramanian.
Astron. Astrophys., Vol. 139, No. 2, p. 289 – 295 (1984).
It is suggested that the abnormal brightness of the quasar 3C 273, coupled with the observations of superluminal motions in its radio core B as well as the misalignment of its VLBI jet with the optical jet, can be understood by postulating a faint spheroidal lensing galaxy located about halfway along the line of sight to the quasar. The lens model also helps to understand why the quasar is found slightly off-centre with respect to its nebulosity. The probability for such a lens system to arise for 3C 273 by chance is shown to be no less than that computed on the basis of the relativistic beaming hypothesis. Further tests of the lens model are discussed.

159.086 **The exceptionally large flux variability of the quasar 1055+018 at metre wavelengths.**
Gopal-Krishna, A. K. Singal, S. Krishnamohan.
Astron. Astrophys., Vol. 140, No. 1, p. L19 – L22 (1984).
By combining the published VLBI and lunar occultation data the authors have delineated the quiescent radio structure of the quasar 1055+018 in terms of 4 components of which two are rather compact ($< 0\overset{''}{.}1$). If an apparent variation of either of these two components is responsible for the recently reported outburst of flux density of the quasar, this would imply a rise in flux density of that component in the course of $\lesssim 4$ years by factors of $\gtrsim 20$ at 80 MHz and $\gtrsim 3$ at 160 MHz. The probability of such a variation occurring due to slow interstellar scintillations is shown to be exceedingly small ($< 10^{-5}$). An attempt to explain the flux variation in terms of a newly created relativistically approaching weak component of incoherent synchrotron emission

seems to require for the bulk motion an extremely large Doppler factor $\delta \gtrsim 10^2$.

159.087 **Quasars et lentilles gravitationnelles.**
J. Surdej.
Ciel Terre, Vol. 100, No. 6, p. 153 – 162 (1984).
The author reviews the most important characteristics of quasars such as redshift, peculiar spectrum, high luminosity, variability, etc. and shows that the evidence for a cosmological origin of quasars has been recently strengthened by the discovery of gravitational lenses.

159.088 **Complete quasar search in the NGC 1097 field.**
H. Arp, R. D. Wolstencroft, X. T. He.
Astrophys. J., Vol. 285, No. 1, p. 44 – 54 (1984). With plates 1–4.
An area $2.85 \times 2.85$ deg$^2$ around the jet galaxy NGC 1097 has been completely searched for quasars. Two objective prism plates taken with the UK Schmidt telescope yielded 43 candidates reaching as faint as $V_{5000} = 20.5$ mag. Thirty–three of these candidates have been examined with aperture spectrophotometry. The candidate list appears to be about 94% true quasars. The quasar density in the field increases strongly toward NGC 1097. It reaches its greatest enhancement in a small region centered between the two strongest jets, which emerge on the northern side of NGC 1097.

159.089 **Constraints on continuous beam models of the jet in 3C 273.**
D. Roberts.
Astrophys. J., Vol. 285, No. 1, p. 64 – 68 (1984).
The one–sidedness, "superluminal" motion, and detailed spectra observed for 3C 273 are used to put constraints on continuous beam models for the source of the extended emission in 3C 273 A. The one–sidedness of the extended emission implies that the age of the radio–frequency cloud is probably ~3–5 Myr. The "superluminal" motion observed with VLBI constrains the bulk Lorentz factor of the jet to be $\gtrsim 10$. Finally, the observed X–ray emission from the jet provides a method for deciding the question of beam composition.

159.090 **Radiation pressure and the stability of broad–line region clouds.**
G. J. Ferland, M. Elitzur.
Astrophys. J., Lett. Ed., Vol. 285, No. 1, p. L11 – L14 (1984).
The authors discuss two constraints on the existence and stability of quasar emission–line gas. The case of a constant pressure photoionized cloud in pressure equilibrium with a surrounding hot gas is considered. It is shown that the radiation pressure of the internally generated line emission makes a significant contribution to the total pressure in standard broad–line region models, and that models with either a higher ionization parameter or lower gas density are dominated by radiation pressure and hence unstable. Further, clouds with ionization parameters lower than the standard value cannot be supported by any hot gas which is heated only by the observed radiation field.

159.091 **The identification of the quasar PKS 0335–122 with redshift 3.45.**
Y. Chu, H. Butcher.
Kexue Tongbao (Beijing), Vol. 29, No. 4, p. 498 – 500 (1984).

159.092 **Quasar absorption spectra and the structure of the universe.**
A. G. Doroshkevich.
Sov. Astron., Vol. 28, No. 2, p. 127 – 132 (1984). English translation of 37.159.071.

159.093 **Observations of Parkes quasars 102 MHz.**
V. S. Artyukh.
Sov. Astron., Vol. 28, No. 2, p. 140 – 143 (1984). English translation of 37.159.072.

159.094 **Radiation pressure and the stability of broad–line region clouds.**
G. J. Ferland, M. Elitzur.
Publ. Astron. Soc. Pac., Vol. 96, No. 584, p. 787 (1984). Abstract. – See Abstr. 010.281.

159.095 **Identifications of blue stellar objects from the Case Low–Dispersion Northern Sky Survey.**
N. V. Zotov.
Bull. Am. Astron. Soc., Vol. 16, No. 3, Part I, p. 726 (1984). Abstract. – See Abstr. 010.065.

159.096 **Effects of optical variability on X–ray optical correlations.**
J. R. Webb, A. G. Smith, R. J. Leacock.
Bull. Am. Astron. Soc., Vol. 16, No. 3, Part I, p. 727 (1984). Abstract. – See Abstr. 010.065.

159.097 **The nature of the underlying galaxy in the QSO 3C 273.**
T. A. Boroson, J. B. Oke.
Bull. Am. Astron. Soc., Vol. 16, No. 3, Part I, p. 731 (1984). Abstract. – See Abstr. 010.065.

159.098 **The X–ray to optical luminosity ratio of optically and X–ray selected quasars.**
G. A. Kriss, C. R. Canizares.
Bull. Am. Astron. Soc., Vol. 16, No. 3, Part I, p. 731 (1984). Abstract. – See Abstr. 010.065.

159.099 **3 mm VLBI observations of 3C273B with transcontinental baselines.**
C. R. Predmore, D. C. Backer, R. L. Dickman, C. R. Masson, A. T. Moffet, J. M. Moran, T. J. Pearson, R. Plambeck, A. E. E. Rogers, D. Woody, M. C. H. Wright.
Bull. Am. Astron. Soc., Vol. 16, No. 3, Part I, p. 732 (1984). Abstract. – See Abstr. 010.065.

159.100 **High resolution spectroscopy in the Lyman alpha forest; possible evidence for absorption line clustering.**
J. K. Webb, R. F. Carswell, M. J. Irwin.
Bull. Am. Astron. Soc., Vol. 16, No. 3, Part I, p. 733 (1984). Abstract. – See Abstr. 010.065.

159.101 **Companion galaxies match quasar redshifts: the debate goes on.**
B. M. Schwarzschild.
Phys. Today, Vol. 37, No. 12, p. 17 – 19 (1984).

159.102 **A complete optical survey of candidate quasars down to B = 22.0 with the ESO 3.6 m telescope.**
B. Marano, G. Zamorani, V. Zitelli.
Messenger, No. 38, p. 6 – 8 (1984).

159.103 **The quasar 0805+046 as a helium–rich shell star.**
Y. P. Varshni.
J. R. Astron. Soc. Can., Vol. 78, No. 5, p. 211 – 212 (1984). Abstract. – See Abstr. 010.401.

159.104 **Image separation statistics for multiply imaged quasars.**
C. C. Dyer.
J. R. Astron. Soc. Can., Vol. 78, No. 5, p. 214 (1984). Abstract. – See Abstr. 010.401.

159.105 **The radio–to–UV continuum in 3C 273.**
M. Camenzind, T. J.-L. Courvoisier.
Astron. Astrophys., Vol. 140, No. 2, p. 341 – 344 (1984).
The authors decompose the observed radio–to–infrared spectrum of 3C 273 into a self–absorbed synchrotron spectrum S of a partially opaque shell source and radio emission from 5 moving knots $C_1$–$C_5$, the first 4 being already detected by VLBI observations. The VLBI core–component D is just a superposition of S and the undetected knot $C_5$. The observed UV–, X–ray and $\gamma$–ray luminosities are used to constrain the physical parameters of the shocked region which is proposed to be the source of all nonthermal radiation in 3C 273.

159.106 **Compact radio cores in extended quasars.**
P. D. Barthel, G. K. Miley, R. T. Schilizzi, E. Preuss.
Astron. Astrophys., Vol. 140, No. 2, p. 399 – 404 (1984).

A complete sample of quasars with extended radio structure and bright radio cores has been surveyed for ultra compact core components. Using a transatlantic VLBI system at 5 GHz the authors found a high detection rate for compact radio core components, implying linear sizes $\lesssim 10$ pc. In some quasar cores multiple component structure was detected, with separations (projected on the overall source axis) of about 10 pc.

159.107 **The distribution of absorption line redshifts of quasars and its origin.**
Y.–q. Chu, L.–z. Fang, Y.–z. Liu.
Astrophys. Lett., Vol. 24, No. 2, p. 95 – 102 (1984).

Available data on quasar absorption line redshifts have been studied by means of statistical analysis. All of the statistical results, including the relationship between relative velocity and emission line redshift; the peaks in the distribution of relative velocity; and the power–spectrum of the absorption–system redshift distribution, favor the cosmological hypothesis for the origin of the absorption lines of quasars. Especially, it is pointed out that the observed peaks in the distribution of relative velocity are not uniquely evidence for the intrinsic hypothesis of the origin of absorption lines in quasar spectra, because they might also be rather well explained by the model of density wave perturbations in the early universe (Fang, Liu, 1981).

159.108 **On the statistics of quasar absorption lines.**
E. J. Zuiderwijk.
Astron. J., Vol. 89, No. 12, p. 1808 – 1810 (1984).

The distribution of absorption lines in 13 quasar spectra is analyzed and shown to be fully consistent with the hypothesis of randomly, but not uniformly, distributed absorption features. The analysis by Libby et al. (1984), in which it is claimed that the number of wavelength coincidences among absorption lines in different quasar spectra (as measured in the rest frame of the quasars) is much larger than expected, implying absorbers in the quasars themselves, is totally invalid. Instead, the number of these coincidences is fully commensurate with the expected one on the assumption of randomness.

159.109 **Optical spectroscopy of two broad absorption line QSOs and implications for spherical–symmetric absorbing wind models.**
J. E. Drew, A. Boksenberg.
Mon. Not. R. Astron. Soc., Vol. 211, No. 4, p. 813 – 831 (1984).

Low resolution spectroscopic observations of two broad absorption line QSOs, spanning 3200 Å to 1 µm, are described and briefly discussed. Consistency is sought with a model for broad absorption line QSOs, in which the absorbing medium is a spherically–symmetric wind. It is found that the spherically-symmetric wind model is only viable for QSOs emitting weakly at X–ray energies and that the mass–loss rate needs to be of the order of 40 $M_{\odot} yr^{-1}$.

159.110 **A statistical analysis on the association of quasars and bright galaxies.**
Y.–q. Chu, X.–f. Zhu.
Chin. Astron. Astrophys., Vol. 8, No. 3, p. 238 – 242 (1984). English translation of Acta Astrophys. Sin., Vol. 4, No. 1, p. 49 – 55 (1984).

The authors applied the nearest neighbour test in a statistical analysis on the association of quasars and bright galaxies. They found that the radio quasars, especially the radio quasars with large redshifts are clearly associated with bright galaxies.

159.111 **Photographic observations of 3C 273.**
Y.–h. Chu, C.–j. Shen, C.–d. Wang, H.–w. Wu, S.–c. Wang.
Chin. Astron. Astrophys., Vol. 8, No. 3, p. 243 – 244 (1984). English translation of Acta Astrophys. Sin., Vol. 4, No. 1, p. 56 – 64 (1984).

159.112 **The spectrum of the QSO 1442+101 (OQ172) at intermediate dispersion.**
B. A. Peterson, J.–S. Chen, D. C. Morton, A. E. Wright, D. L. Jauncey.
Anglo–Aust. Obs., Prepr., No. 196, 33 pp. (1984). Submitted to Mon. Not. R. Astron. Soc.

159.113 **The close QSO pair Q1548+114A,B.**
P. A. Shaver, J. G. Robertson.
ESO Sci. Prepr., No. 344, 16 pp. (1984). To appear in Mon. Not. R. Astron. Soc.

159.114 **High redshift Mg II and Fe II absorptions in QSO spectra from known C IV systems.**
P. Boissé, J. Bergeron.
Inst. Astrophys. Paris, Pré–Publ., No. 80, 35 pp. (1984). To appear in Astron. Astrophys.

159.115 **Nineteen newly discovered quasars in the declination –35° zone I.**
H. Arp.
Astrophys. J., Vol. 285, No. 2, p. 547 – 554 (1984). With plates 7 – 9.

Redshifts are reported for 19 new quasars in the declination –35° zone between R.A. = $23^h20^m$ and $1^h40^m$. These quasars were identified by Cecilia Aguirre on Curtis Schmidt objective prism plates taken by Malcolm Smith. Flux–calibrated spectra are presented for these 19 new quasars. Equivalent widths for all major lines in the spectra are derived. The spectra of the two unusual quasars QSO 0218–367 and QSO 0134–376 are discussed in some detail.

159.116 **Properties of the quasars in the Sculptor region II.**
H. Arp.
Astrophys. J., Vol. 285, No. 2, p. 555 – 566 (1984).

Evidence for a concentration of quasars in the Sculptor Group region is reviewed. It is shown that quasars in this association have different properties from quasars in adjoining regions. In particular they contain a higher percentage of quasars with strong lines and high redshifts. They show a well–defined Baldwin relation (apparent magnitude vs. equivalent width). In contrast, the apparent magnitude–equivalent width plot for quasars on either side of the Sculptor Group region shows relatively poor correlation. These differences are interpreted as evidence against a cosmological or redshift distance of the quasars in the Sculptor group.

159.117 **[O III] emission surrounding the quasar MR 2251–178.**
S. di Serego Alighieri, M. A. C. Perryman, F. Macchetto.
Astrophys. J., Vol. 285, No. 2, p. 567 – 570 (1984). With plates 10 – 11.

Narrow–band images in [O III] λλ4959, 5007 and Hβ of the field of the low–redshift quasar MR 2251–178 have revealed, in addition to nebulosity closely associated with the quasar, diffuse [O III]–emitting regions separated by up to 100 kpc from the nucleus but kinematically associated with it. These regions of enhanced [O III] emission are attributed to density perturbations of the gaseous envelope, or disk, which rotates about the quasar.

159.118 **The synchro–compton reactor in the quasar 3C 345.**
A. P. Mosin, V. M. Charugin.
Astron. Tsirk., No. 1319, p. 1 – 3 (1984). In Russian.

159.119 **Optische Morphologie, Farbe und Polarisation des Jets von 3C 273.**
K. Meisenheimer, H.–J. Röser.
Mitt. Astron. Ges., Nr. 62, p. 341 (1984). Abstract. – See Abstr. 010.341.

159.120 **Neuere Beobachtungsergebnisse von S5 0014+81.**
H. Kühr.
Mitt. Astron. Ges., Nr. 62, p. 342 (1984). Abstract. – See Abstr. 010.341.

159.121 **Study of quasar candidates on objective prism plates.**
D. Groote, U. Haug, H. Kühr.
Mitt. Astron. Ges., Nr. 62, p. 342 (1984). Abstract. – See Abstr. 010.341.

159.122 **Interpretation des Quasarspektrums von GD 1339.**
G. Schönknecht, I. Bues, G. Rupprecht.
Mitt. Astron. Ges., Nr. 62, p. 343 – 344 (1984). – See Abstr. 010.341.

159.123 **Are extended nebulae around quasars normal galaxies?**
B. V. Komberg.
Astron. Zh., Tom 61, Vyp. 6, p. 1048 – 1053 (1984). In Russian. English translation in Sov. Astron., Vol. 28, No. 6.
An attempt is made to reveal the nature of extended nebulae detected in recent years around all nearby quasi-stellar objects (QSO). Basic data are briefly described that may help us to estimate, for all of these nebulae, the dependences between their luminosities, velocity dispersions and dimensions. Similar dependences for normal galaxies are then discussed and used as a basis for comparison of the nebulae enveloping QSOs with normal galaxies.

159.124 **Photometry of the active nuclei of galaxies – photographic photometry of the quasar 3C 273.**
Y.-h. Chu, C.-j. Shen, C.-d. Wang, H.-w. Wu, S.-c. Wang.
Acta Astrophys. Sin., Vol. 4, No. 1, p. 56 – 64 (1984). In Chinese. English translation of the part concerning the quasar 3C 273 see Abstr. 159.111.
The photographic photometry of some stars of the open cluster NGC 2287 as well as the optical variations of the quasar 3C 273 observed in 1980 are presented.

159.125 **The redshift distribution and the selection effects.**
Y.-y. Zhou, Z.-g. Deng, Z.-l. Zhou.
Acta Astron. Sin., Vol. 25, No. 3, p. 241 – 249 (1984). In Chinese.
The main purpose of the paper is to investigate the influence of the emission line identification on the redshift distribution of QSOs. Using the sample from the Hewitt and Burbidge QSO catalogue, choosing 7, 11, 13 spectrum lines as key lines in the identification of QSO redshifts, respectively, and assuming the observational response function in the 3400 – 7000 Å range to be independent upon the wavelength, the authors calculate the influence of the selection effect caused by line identification on the redshift distribution.

159.126 **Progress report on a new survey to study the clustering properties of quasars at z ~ 2.**
P. S. Osmer.
Clusters and groups of galaxies, p. 579 – 581 (1984). – See Abstr. 012.088.

159.127 **Properties of groups and clusters of galaxies associated with quasars.**
H. K. C. Yee, R. F. Green.
Clusters and groups of galaxies, p. 583 – 587 (1984). – See Abstr. 012.088.
Images of fields around 108 quasars are used to study the association of galaxies with quasars. The authors find that the excess objects in the quasar fields are consistent with being galaxies at the cosmological distances implied by the redshifts of the quasars, and that quasars tend to prefer regions having higher than normal galaxy density.

159.128 **Properties of the optical variability of the quasar 3C 345.**
M. K. Babadzhanyants, E. T. Belokon'.
Astrofizika, Tom 21, Vyp. 2, p. 217 – 231 (1984). In Russian. English translation in Astrophysics, Vol. 21, No. 2.
The light curve (B) of 3C 345 is given containing all available data for 1965 – 1983. A systematic increase of the mean luminosity ~$0^m.05$/year in B during the last 15 years was found. A connection is assumed between the "slow" optical flashes of 3C 345 and the arising of the compact "superluminal" radio components in their *VLBI*-structure. With these assumptions the authors predict the appearance of a new radio component in 3C 345 connected with the optical flash in 1982. The "fast" optical flashes appear to show a 327-day periodicity in 1972 – 1975 that coincide with the period ($P = 321^d$) obtained by Kinman et al. (1968) for 1965 – 1967. This suggests that the periodicity of the "fast" flashes may occur at limited time intervals.

159.129 **Redshifts of southern radio sources. VI.**
D. L. Jauncey, M. J. Batty, A. E. Wright,
B. A. Peterson, A. Savage.
Astrophys. J., Vol. 286, No. 2, p. 498 – 502 (1984).
Redshifts are presented for 30 radio-selected QSOs drawn from the Parkes 2.7 GHz survey, and also for three optically selected QSOs. For two radio-selected QSOs, PKS 1935-692 and PKS 2000-330, the authors measure redshifts greater than 3.0 and confirm the redshift of 3.614 for the optically selected QSO DHM 0054-284.

159.130 **VLBI observations of the polarized radio emission from the quasar 3C 454.3.**
W. D. Cotton, B. J. Geldzahler, J. M. Marcaide, I. I. Shapiro, M. Sanromá, A. Rius.
Astrophys. J., Vol. 286, No. 2, p. 503 – 508 (1984).
Polarization-sensitive VLBI observations of 3C 454.3 at a radio wavelength of 13 cm with a synthesized beam of dimension $0''.006 \times 0''.002$ disclosed a small core with a jet emerging to the southeast and extending about $0''.01$ from the core. Near the core, the electric field vector is well ordered and nearly parallel to the jet axis, but it rotates with distance such that, far from the core, the electric field vector is nearly perpendicular to the jet axis.

159.131 **Image separation statistics for multiply imaged quasars.**
C. C. Dyer.
Astrophys. J., Vol. 287, No. 1, p. 26 – 32 (1984).
The gravitational lens effect for multiply imaged quasars is considered for various mass distributions in the lensing galaxies. It is shown that the presently observed statistics of maximum separations are not in strong disagreement with predictions based on these realistic lens models after some account is taken of the observational limitations of present searches. It is also shown how these image separations may be used to place certain constraints on various properties of the lensing objects and their distribution in space.

159.132 **A medium-bright quasar sample: new quasar surface densities in the magnitude range 16.4 < $B$ < 17.65.**
K. J. Mitchell, A. Warnock III, P. D. Usher.
Astrophys. J., Lett. Ed., Vol. 287, No. 1, p. L3 – L7 (1984).
A new medium-bright quasar sample (MBQS) is constructed from spectroscopic observations of 140 bright objects selected for varying degrees of blue and ultraviolet excess in five Palomar 1.2 m Schmidt fields. The MBQS contains 32 quasars with $B < 17.65$ mag. The new integral surface densities in the range $16.45 < B < 17.65$ mag are $\gtrsim 40\%$ higher than expected. The MBQS and its redshift distribution increase the area of the quasar Hubble diagram covered by complete samples of quasars.

159.133 **Faster-than-light motion in quasars.**
J. V. Narlikar, S. M. Chitre.
J. Astrophys. Astron., Vol. 5, No. 4, p. 495 – 506 (1984).
Observations of some quasars have shown that the angular separation between pairs of radio-emitting regions in their cores is increasing year after year. If the quasars are indeed as far away as implied by Hubble's law, then these angular motions translate into linear speeds several times the speed of light. Several theoretical scenarios have been proposed to show that the observed motions are illusory. The leading contender in this field – the relativistic beam model – and an alternative offered by the concept of a gravitational screen are described and compared in the light of recent observational data.

159.134 **The milli–arcsecond images of Q0957+561.**
M. V. Gorenstein, I. I. Shapiro, A. E. E. Rogers, N. L. Cohen, B. E. Corey, R. W. Porcas, E. E. Falco, R. J. Bonometti, R. A. Preston, A. Rius, A. R. Whitney.
Astrophys. J., Vol. 287, No. 2, p. 538 – 548 (1984).

The authors detected similar compact radio structures in the A and B images of the quasar Q0957+561 from observations in 1980 February that utilized a three–element VLBI array operating at 13 cm (2292 MHz). The array included antennas at Goldstone, California, Effelsberg, Federal Republic of Germany, and Madrid, Spain. Data from the long baselines show that the A and B images contain core components with flux densities of $22 \pm 1$ and $18 \pm 1$ mJy, respectively. Data from the short Effelsberg–Madrid baseline suggest that additional correlated flux density, distributed over an angular scale of $\sim 20$ mas, accompanies both cores. The observed brightness distributions of these cores are related by a linear magnification matrix, consistent with conservation of surface brightness, as expected for images produced by a transparent gravitational lens.

159.135 **The broad absorption lines in the spectrum of the QSO PKS 1157+014: a possible link between broad absorption line QSOs, metal enrichment, and the formation of galaxies.**
F. H. Briggs, D. A. Turnshek, A. M. Wolfe.
Astrophys. J., Vol. 287, No. 2, p. 549 – 554 (1984).

The broad absorption trough to the blue of the Lyα emission redshift in the spectrum of the QSO PKS 1157+014 is largely accounted for by a high column density, low velocity dispersion absorption system ($z = 1.94362$) that is detected in the H I λ21 cm transition. The apparent deficiency of Lyα emission at $z_{em} = 1.978$ is due to attenuation by the damping wings of the $z = 1.94362$ system rather than an intrinsic absence. The morphology of the optical spectrum of PKS 1157+014 places it midway between the radio–weak BAL class and a group of radio–bright QSOs with "associated" absorption complexes.

159.136 **The underlying galaxies of quasars. II. Imaging of a radio–loud sample.**
M. A. Malkan.
Astrophys. J., Vol. 287, No. 2, p. 555 – 565 (1984).

The author presents new deep *v*, *g*, and *r* SIT frames of 23 low–redshift quasars, mostly selected by their strong radio emission. Seventeen quasars with redshifts up to 0.4 are resolved. Assuming that the extended emission is starlight, it is found that, when the underlying galaxy is associated with a radio–quiet quasar, it has the color, size, surface brightness, and scale length of a normal spiral galaxy. In contrast, the host galaxies of radio–loud quasars are larger and appear to be bright elliptical galaxies.

**The Case low–dispersion northern sky survey. II.**
See Abstr. 002.001.

**The second Byurakan spectral sky survey. I. The results of the area centred at $\alpha = 11^h30^m$, $\delta = 59°00'$.**
See Abstr. 002.006.

**The second Byurakan spectral sky survey. II. The results of the area centred at $\alpha = 09^h50^m$, $\delta = +55°00'$.**
See Abstr. 002.007.

**A search for faint blue stars in high galactic latitudes. III. Twelve PSS fields at declinations –6° and –12° near the South Galactic Pole.**
See Abstr. 002.036.

**Faint blue objects at high galactic latitude. IV. Palomar Schmidt fields centered on Selected Areas 55 and 94.**
See Abstr. 002.047.

**The IIIa–F emulsion and the quasar redshift cut–off.**
See Abstr. 034.144.

**Uses of very large telescopes for galaxy research.**
See Abstr. 036.097.

**On the viability of acceleration mechanisms for high–velocity clouds and quasar emission–line clouds.**
See Abstr. 062.014.

**Cerenkov line emission: basic theory.**
See Abstr. 063.058.

**Gravitational lensing by stars in a galaxy halo: theory of combined weak and strong scattering.**
See Abstr. 066.186.

**Gravitational lensing by cosmic strings.**
See Abstr. 066.215.

**Nonadiabatic self–consistent spherical accretion as a model for quasars and active galactic nuclei.**
See Abstr. 067.003.

**Clusters of black holes as models for active galactic nuclei.**
See Abstr. 067.034.

**Thick discs around black holes as energy sources for quasars and other astrophysical objects.**
See Abstr. 067.047.

**Black hole models for active galactic nuclei.**
See Abstr. 067.079.

**Shock acceleration in active galactic nuclei and quasars.**
See Abstr. 067.080.

**Collimated winds from active galactic nuclei.**
See Abstr. 067.082.

**An alternative to relativistic beaming in AGN.**
See Abstr. 067.083.

**Models of non–Keplerian accretion disks based on dynamo magnetic field: application to AGN.**
See Abstr. 067.084.

**Centrifugally driven MHD–winds and the non–thermal emission regions in AGNs.**
See Abstr. 067.086.

**Macroscopic currents and wind perturbations in the frame of the wind and shock model.**
See Abstr. 067.087.

**Duelling winds: a model for broad–line regions.**
See Abstr. 067.088.

**X–ray spectra of accretion disks.**
See Abstr. 067.089.

**Extragalactic gravitational collapse.**
See Abstr. 067.141.

**Sco X–1: a mini quasar?**
See Abstr. 117.007.

**The properties of molecular clouds with high redshifts ($z = 2 - 3$).**
See Abstr. 131.238.

**A deep Westerbork survey of areas with multicolor Mayall 4 m plates. II. Optical identifications.**
See Abstr. 141.005.

**Observations of weak radio sources with flat or inverted spectra at 10 GHz.**
See Abstr. 141.009.

**A VLBI survey of peaked spectrum radio sources.**
See Abstr. 141.013.

**Infrared study and classification of optically faint steep spectrum radio sources.**
See Abstr. 141.021.

**The Medium Sensitivity Survey: a new sample of X–ray sources with optical identifications and the revised extragalactic log *N*–log *S*.**
See Abstr. 142.026.

**The origin of the X–ray background.**
See Abstr. 142.036.

**Redshifts of faint 3CR radio sources.**
See Abstr. 158.002.

**Sidedness, field configuration, and collimation of extragalactic radio jets.**
See Abstr. 158.004.

**An explanation of abrupt cutoffs in the optical–infrared spectra of non–thermal sources. A new pile–up mechanism for relativistic electron spectra.**
See Abstr. 158.010.

**Thermal equilibrium of the emission–line clouds in quasar and Seyfert–nucleus envelopes.**
See Abstr. 158.020.

**Flickering of extragalactic radio sources.**
See Abstr. 158.025.

**A complete sample of extragalactic radio sources at 1 Jy at 408 MHz. III. Estimated redshifts and radio luminosities.**
See Abstr. 158.032.

**Optical monitoring of radio sources.**
See Abstr. 158.034.

**Relativistic electron–photon showers in the nuclei of active galaxies and quasars.**
See Abstr. 158.053.

**Optical spectra of southern Molonglo radio sources – II.**
See Abstr. 158.057.

**Optical structures of the M87 and 3C 273 jets.**
See Abstr. 158.059.

**Time variability of the 87 GHz linear polarization and flux density of active extragalactic objects.**
See Abstr. 158.076.

**Models of active galactic nuclei.**
See Abstr. 158.085.

**The gravitationally bound current loop model for quasars and galactic nuclei.**
See Abstr. 158.088.

**Black hole nuclear cluster model of quasars and active galactic nuclei.**
See Abstr. 158.090.

**X–ray selected quasars and Seyfert galaxies: cosmological evolution, luminosity function, and contribution to the X–ray background.**
See Abstr. 158.095.

**Radio jets in classical double radio sources with strong cores.**
See Abstr. 158.096.

**He I λ10830 emission in Seyfert galaxies and QSOs.**
See Abstr. 158.104.

**Effects of dust on the infrared emission, selected line ratios, and polarization of Seyfert 1 galaxies, broad–line radio galaxies, and quasars.**
See Abstr. 158.105.

**Extragalactic radio jets.**
See Abstr. 158.113.

**Optical and UV emission of active galactic nuclei.**
See Abstr. 158.114.

**Preliminary results on the ultraviolet observations of a complete X–ray selected sample of AGNs.**
See Abstr. 158.123.

**X–ray emission from active galactic nuclei.**
See Abstr. 158.127.

**UV and X–ray observations of AGN: two comparisons with other wavelengths.**
See Abstr. 158.134.

**On the broad band energy spectra of quasars and BL Lac objects.**
See Abstr. 158.137.

**Theoretical studies of emission line variability in type 1 Seyfert galaxies.**
See Abstr. 158.140.

**On the stability of broad line clouds in active galactic nuclei.**
See Abstr. 158.141.

**Active galactic nuclei and their cosmological evolution.**
See Abstr. 158.142.

**VLBI and X–ray observations of complete samples of radioselected active galactic nuclei.**
See Abstr. 158.143.

**Luminosity–dependent effects on the hot spots in extended double radio sources.**
See Abstr. 158.156.

**An empirical model for active galactic nuclei. I. The catalog.**
See Abstr. 158.168.

**Distorted structure in a $z = 1.9$ radio source.**
See Abstr. 158.172.

**Near–infrared photometry of faint radio galaxies in selected areas.**
See Abstr. 158.215.

**Active galaxies viewed from the infrared.**
See Abstr. 158.235.

**Cosmological evolution of radio sources.**
See Abstr. 158.277.

**On the origin of power–law X–ray spectra of active galactic nuclei.**
See Abstr. 158.278.

**Superclusters at large redshifts. Can protosuperclusters and the birth of "pancakes" be observed?**
See Abstr. 160.024.

**Interaction of galaxy and quasar winds with the intergalactic medium.**
See Abstr. 160.031.

**Distribution functions of intergalactic clouds.**
See Abstr. 160.040.

**On a cosmology explaining Sandage's high–z–"cutoff".**
See Abstr. 161.133.

**Manifestations of a cosmological density of compact objects in quasar light.**
See Abstr. 161.202.

**The statistics of gravitational lenses: the distributions of image angular separations and lens redshifts.**
See Abstr. 161.218.

**A possibility to use C I infrared absorption lines for determination of the background radiation temperature at earlier cosmological epochs.**
See Abstr. 161.221.

**Clustering in a two–component universe.**
See Abstr. 161.239.

# 160 Groups of Galaxies, Clusters of Galaxies, Superclusters, Intergalactic Matter

160.001 **Studies of the Virgo cluster. III. A classification system and an illustrated atlas of Virgo cluster dwarf galaxies.**
A. Sandage, B. Binggeli.
Astron. J., Vol. 89, No. 7, p. 919 – 931 (1984). With plates 41 – 63.

Photographs enlarged to a common scale are given for 138 dwarf galaxies in the region of the Virgo cluster. Most are cluster members, as judged either from their uniquely low surface brightness and/or morphology, or occasionally from velocity data. This Atlas and its descriptions are meant to form a preliminary classification system that organizes all known types of low–luminosity galaxies into a coherent scheme. It is also meant as a preliminary candidate list of late–type, low–SB dwarfs that can be usefully observed with Space Telescope to resolve the brightest stars in a next step to establish the extragalactic distance scale.

160.002 **Nature of blue galaxies in the cluster Cl1447 + 2619.**
H. R. Butcher, A. Oemler Jr.
Nature, Vol. 310, No. 5972, p. 31 – 33 (1984).

Spectroscopy is presented that confirms a previous photometric result that the distant ($z = 0.38$) cluster Cl1447+2619 contains the largest excess of blue galaxies yet encountered. The galaxy surface density profile of the cluster shows it to be open and irregular, making it the first such cluster to be studied at high redshift. The spectra indicate that the blue colours are predominantly the result of recent star formation.

160.003 **The Horologium region. I. The data.**
G. Chincarini, M. Tarenghi, H. Sol, P. Crane, I. Manousoyannaki, J. Materne.
Astron. Astrophys., Suppl. Ser., Vol. 57, No. 1, p. 1 – 21 (1984).

The authors present observations, spectroscopic and photometric, of a subsample of the Shapley catalogue in Horologium. The clumpy distribution of objects is dominated by a structure at $<V> = 13436$ km/s.

160.004 **The north extension of the Perseus supercluster.**
P. Focardi, B. Marano, G. Vettolani.
Astron. Astrophys., Vol. 136, No. 1, p. 178 – 180 (1984).

The authors discuss the possible extension of the Perseus supercluster in the northern galactic hemisphere. They show the existence of a large cloud of galaxies around the cluster Abell 569 at the same radial velocity as the Perseus supercluster and the presence of an excess of galaxies behind the galactic plane in the region where the Perseus supercluster crosses the Milky Way.

160.005 **Velocity distribution of Virgo Cluster galaxies.**
C. Ftaclas, M. N. Fanelli, M. F. Struble.
Astrophys. J., Vol. 282, No. 1, p. 19 – 28 (1984).

The authors have analyzed several effects of the general population of the Local Supercluster on observed properties of the Virgo Cluster. They identify two subgroups in the field of the Virgo Cluster that are most likely beyond the Virgo distance. Combined with an intrusion of the W cloud into the field of the cluster, this clumped component accounts for $\sim 9\%$ of the galaxies in the complete sample. Evidence is also presented of a near Hubble expansion of galaxies in the outer 2° of the cluster suggesting that the flow within the supercluster may be generally retarded, but not very disturbed otherwise. Several different approaches suggest a cluster mean velocity in the range of 960 – 1000 km $s^{-1}$.

160.006 **Cooling flows in clusters of galaxies.**
A. C. Fabian, P. E. J. Nulsen, C. R. Canizares.
Nature, Vol. 310, No. 5980, p. 733 – 740 (1984).

Cooling gas in the centre of many clusters of galaxies is compressed and pushed inward by the thermal pressure of the hot, uncooled outer gas. Gravity acts as a focusing agent until the gas has cooled to galactic temperatures. The cooling process is thermally unstable, giving rise to optical filamentation. Once the gas cools below $10^4$K it presumably forms stars which add to the mass of a central galaxy. Accretion rates of up to $400 M_\odot$ $yr^{-1}$ are inferred from X–ray measurements so that the mass of such central galaxies may be due to cooling flows. Their optical and UV appearance emphasize that the gas predominantly turns into low–mass stars with an initial mass function unlike that of the solar neighbourhood.

160.007 **Intergalactic dust.**
K. Rudnicki.
Sky Telesc., Vol. 68, No. 2, p. 123 – 126 (1984).

160.008 **A small–size (less than 1 min of arc) decametric radio source in the Perseus cluster.**
A. V. Megn (*A. V. Men'*), S. Ya. Braude, S. L. Rashkovski, I. S. Falkovich, N. K. Sharykin, V. A. Shepelev, A. D. Khristenko.
Astrophys. Space Sci., Vol. 102, No. 1, p. 155 – 160 (1984).

Results of observations of decametric radio emission from the Perseus cluster with the aid of URAN–1 interferometer are presented. The 42.2 km baseline of the interferometer is oriented along the parallel. The object is shown to contain a radio source with dimensions which do not exceed 60″ and whose flux is not less than 95 Jy over the frequency range from 16.7 to 25 MHz. Several source models are considered and a comparison with higher–frequency measurements is carried out. It seems possible that the source is a component of 3C84A which was reportedly observed earlier by the interferometry technique at 408 and 1415 MHz.

160.009 **Infrared emission of dust grains in the clusters of galaxies.**
N. V. Voshchinnikov, V. K. Khersonskij.
Astrophys. Space Sci., Vol. 103, No. 2, p. 301 – 319 (1984).

The major mechanisms of the expected infrared emission from the galaxy clusters are considered. These mechanisms are the dust emission (if the dust is present in the intergalactic medium of the clusters) as well as the Sunyaev–Zel'dovich effect. The dust grains' temperature has been determined. The expected infrared emission intensities and fluxes from Coma and Perseus clusters are calculated for the wavelength region of 50 – 300 μ. It is shown that in the wavelength region under consideration, the contribution of the Sunyaev–Zel'dovich effect to the total emission is very small and can be neglected. The total fluxes from the clusters constitute $10^5$ – $10^6$ Jy and can be registered with the aid of modern infrared telescopes.

160.010 **Superclusters: giants of the Cosmos.**
L. A. Marshall.
Astronomy, Vol. 12, No. 4, p. 6 – 15 (1984). Abstr. in Phys. Abstr., Vol. 87, No. 1238, Entry 77943 (1984).

160.011 **Cosmic gamma rays from galaxy clusters.**
B. P. Houston, A. W. Wolfendale, E. C. M. Young.
J. Phys. G, Vol. 10, No. 6, p. L147 – L150 (1984). Abstr. in Phys. Abstr., Vol. 87, No. 1239, Entry 82554 (1984).

160.012 **A list of nearby groups of galaxies.**
J. Vennik.
Tartu Astrofüüs. Obs. Teated, Nr. 73, 67 pp. (1984).

A new list of groups of galaxies located in the northern hemisphere has been compiled on the basis of the catalogue of radial velocities of galaxies. The list includes 116 small groups of galaxies having mostly one center of concentration. Some groups consist of two subcondensations – A and B components. A short description is added for every group. The list of the groups of galaxies is compared with similar lists by other investigators.

160.013 **On a group of peripherical type.**
L. Rosino, R. K. Shakhbazyan.
Astrofizika, Tom 19, Vyp. 4, p. 834 – 835 (1983). In Russian. English translation in Astrophysics, Vol. 19, No. 4.

The description of a group of spiral galaxies having a peripherical structure is given. The images of the brightest galaxies are greatly stretched and directed toward the empty central region of the system.

160.014 **On thermal energy sources for intergalactic gas in clusters of galaxies.**
V. G. Gorbatskij.
Astrofizika, Tom 20, Vyp. 1, p. 61 – 71 (1984). In Russian. English translation in Astrophysics, Vol. 20, No. 1.

Possible mechanisms of heating of intergalactic gas in clusters are discussed. It is shown that energy released in supernova outbursts in the form of cosmic rays in early stages of evolution of galaxies is great enough to be an important source of observed high temperature of the intracluster gas. The heating rate of gas in the intracluster space due to the action of cosmic rays is estimated.

160.015 **Redshifts in Klemola 27.**
O.–G. Richter.
Astron. Astrophys., Suppl. Ser., Vol. 58, No. 1, p. 131 – 143 (1984).

Observational data for galaxies in the galaxy cluster Klemola 27 gathered over the last few years are presented. These are mostly new redshifts and further, accurate coordinates and morphological types for a large number of "prominent" galaxies in the ESO/SRC–Survey fields 444 and 445. A few preliminary H I radial velocities are also reported. The data available for this cluster are briefly discussed.

160.016 **Submillimeter spectrophotometry for measuring the peculiar radial velocities of galaxy clusters.**
G. B. Sholomitskij.
Sov. Astron. Lett., Vol. 10, No. 2, p. 108 – 112 (1984). English translation of 37.160.081.

160.017 **The Coma/A 1367 filament of galaxies.**
P. Fontanelli.
Astron. Astrophys., Vol. 138, No. 1, p. 85 – 92 (1984).

In the Coma/A 1367 supercluster the surface distribution of galaxies from Zwicky and UGC catalogs outlines a filamentary structure in which the Coma cluster itself is embedded. In its densest part the filament extends for approximately 3.5 h in right ascension and shows an average density contrast of about 2.7 with respect to the nearby regions. A sample of new 21–cm redshifts is used together with the redshifts data available in the literature to study the structure of the filament: the majority of the galaxies on it falls into the velocity range of Coma indicating that the filament is a coherent structure in the three–dimensional space and represents an important element of the supercluster.

160.018 **Suspected rotation of an X–ray cluster of galaxies.**
J. Materne.
Messenger, No. 37, p. 19 – 21 (1984).

160.019 **Clusters, superclusters and their distribution.**
J. H. Oort.
Large–scale structure of the universe, cosmology and fundamental physics, p. 209 – 224 (1984). – See Abstr. 012.025.

Galaxies are distributed in a very irregular manner; they are clumped on scales ranging from doubles to clusters and superclusters. The clusters, which are the largest bound systems, are half–way on the path to dynamical evolution. Superclusters are entirely unevolved along their major diameters, which range up to several hundred Mpc. Rough statistics are given on diameters and shapes. The superclusters may be interconnected, and form a sort of network. Current theories on their origin are briefly reviewed.

160.020 **Tidal torques and the orientations of galaxies in clusters.**
P. S. Wesson.
Astron. Astrophys., Vol. 138, No. 2, p. 253 – 257 (1984).

Aspherical galaxies in the outer regions of a cluster are acted upon by a tidal torque due predominantly to the cluster core. The stability of the orientations of galaxies under the action of this torque is investigated. It is found that a typical spiral galaxy, by virtue of its spin, has a mean orientation that is stable over times greater than the age of the Universe. An aspherical galaxy with little or no spin has a mean orientation that is unstable, but in this case the timescale of the motion is at least comparable to the age of the Universe. These results tend to confirm the widespread belief that the orientations of galaxies in the outer parts of clusters are primeval. Alignments of galaxies in clusters can therefore be interpreted as evidence of an asymmetric or pancake mode of cluster formation.

160.021 **Intergalactic matter model for the missing mass in the Coma cluster of galaxies.**
Y. Fujishima, M. Tosa.
Publ. Astron. Soc. Jpn., Vol. 36, No. 2, p. 333 – 345 (1984).

The authors propose a model of the Coma cluster of galaxies which is composed of the observed visible galaxies and hypothetical invisible intergalactic objects, the latter accounting for the "missing mass". Numerical simulations are performed of the motion of the member galaxies in the gravitational field of the invisible objects. The dynamical friction due to the invisible objects on the galaxy causes a segregation of galaxies by mass. It is shown by the simulations that this model can well explain the observed mass segregation of the member galaxies.

160.022 **H I synthesis observations of the high–redshift galaxies in Stephan's Quintet.**
G. S. Shostak, W. T. Sullivan III, R. J. Allen.
Astron. Astrophys., Vol. 139, No. 1, p. 15 – 24 (1984).

The authors have used the Westerbork Synthesis Telescope to map the high–velocity neutral hydrogen in the region of Stephan's Quintet with a resolution of $\sim 35''$. Three distinct velocity systems are seen, at 5700, 6000, and 6600 km $s^{-1}$, and what is remarkable is that the bulk of the H I at all velocities lies outside the optical bounds of the Quintet galaxies. The authors interpret this result in terms of a collision among the high-velocity objects, including NGC 7320c, in which tidal and collisional stripping have succeeded in removing the majority of the disk gas from the interacting galaxies. Detection of H I in several galaxies near the Quintet provides evidence that the high-redshift objects are at distances commensurate with their redshifts. In this view, the apparent redshift discrepancy of Stephan's Quintet arises from the chance juxtaposition of two groupings of galaxies, at 800 and 6000 km $s^{-1}$.

160.023 **High resolution radio and X-ray observations of A 115.**
L. Feretti, I. M. Gioia, G. Giovannini, L. Gregorini, L. Padrielli.
Astron. Astrophys., Vol. 139, No. 1, p. 50 – 54 (1984).
High resolution radio and X-ray observations are presented for the cluster of galaxies Abell 115, which is known in the literature for being characterized by a double X-ray structure and a multiple optical structure. The brightest galaxy of the cluster is associated with the radio source 3C28, which shows a very unusual radio morphology in the high resolution map presented here. A tail-type radio source is also found at $\approx 4'$ from 3C28, in NE direction. A resolved X-ray structure is detected around 3C28 and is interpreted to originate from a hot gas which is likely to be accreting onto the galaxy. The complex radio structure of 3C28 is understood by invoking a combination of external ram pressure plus buoyancy effects of light radio plasmoids.

160.024 **Superclusters at large redshifts. Can protosuperclusters and the birth of "pancakes" be observed?**
J. H. Oort.
Astron. Astrophys., Vol. 139, No. 1, p. 211 – 214 (1984).
The article starts from the working hypothesis that throughout the universe the luminous matter is largely concentrated in supercluster structures. It then follows that the Lyman α absorption lines in quasars are connected with this structure, and that they can give information on superclustering at large redshifts. Some direct evidence for the existence of (super)clustering at large redshifts is discussed. The supposition of a relation between superclusters and Ly α lines is supported by the observation.

160.025 **Limits on the infrared and visual luminosity of the intergalactic H I cloud in Leo.**
M. F. Skrutskie, M. A. Shure, S. Beckwith.
Astrophys. J., Lett. Ed., Vol. 282, No. 2, p. L65 – L68 (1984).
Low surface brightness photometry of the intergalactic neutral hydrogen cloud in Leo sets upper limits on the intensity of this object of 28.0 mag $arcsec^{-2}$ in the $V$ photometric band and 22.8 mag $arcsec^{-2}$ in the $K$ band. The corresponding ratio of the H I mass to blue luminosity in the H I emission peaks of the cloud must be greater than 2.9 (solar units) depending on the star formation history of the object.

160.026 **On the morphology of two systems of galaxies.**
R. K. Shakhbazyan, A. I. Shapovalova.
Astrofizika, Tom 20, Vyp. 1, p. 179 – 182 (1984). In Russian. English translation in Astrophysics, Vol. 20, No. 1.
Using plates taken with the 6-m telescope of the Special Astrophysical Observatory a morphological investigation of two groups is carried out.

160.027 **The virial masses of clusters of galaxies and the effects of contamination.**
J. A. Fernley, S. P. Bhavsar.
Mon. Not. R. Astron. Soc., Vol. 210, No. 4, p. 883 – 890 (1984).
The authors have used model clusters, selected from $n$-body simulations of clustering, to determine the accuracy of using virial methods for mass determination. All the clusters which have spuriously high virial masses, as a result of contamination, can be identified by using a $\chi^2$ test which compares their line-of-sight velocities with a normal distribution. For the remaining clusters, with good $\chi^2$ fits, the ratio of virial mass to the actual mass has a mean value of $0.97 \pm 0.36$ ($1\sigma$). The authors have calculated the mass-to-light ratios for three clusters (Coma, A2151 and Hercules Supercluster) in an analogous manner. They conclude Hercules Supercluster is seriously contaminated. The $M/L$ ratios of the other two clusters could be accounted for by 'massive halo' galaxies and intracluster gas.

160.028 **The effect of the perturbation of the local velocity field by Virgo on the calculation of differential luminosity functions.**
R. C. Kraan-Korteweg, A. Sandage, G. A. Tammann.
Astrophys. J., Vol. 283, No. 1, p. 24 – 32 (1984).
The authors have repeated the determination of $\varphi(M)$ for galaxies in the *Revised Shapley-Ames Catalog* using Virgo-centric perturbation fields of different strength to counter criticism that past calculations of the luminosity functions for galaxies of various Hubble types and van den Bergh luminosity classes are significantly in error as a result of neglect of this local velocity perturbation. Differential luminosity functions are given for 18 different Hubble types and luminosity classes for perturbation models with infall velocities of our Galaxy to Virgo of $v_{V_c} = 0$, 220, and 440 km $s^{-1}$. The conclusion is that any realistic perturbation of the local velocity field does not introduce large errors in the calculation of $\varphi(M)$.

160.029 **Seven poor clusters of galaxies.**
T. C. Beers, M. J. Geller, J. P. Huchra, D. W. Latham, R. J. Davis.
Astrophys. J., Vol. 283, No. 1, p. 33 – 49 (1984).
The authors have measured 83 new redshifts for galaxies in the region of seven of the poor clusters of galaxies identified by Morgan, Kayser, and White and Albert, White, and Morgan. For three systems (MKW 1s, AWM 1, and AWM 7) complete redshift samples were obtained for galaxies brighter than $m_{B(0)} = 15.7$ within 1° of the D or cD galaxy. The authors estimate masses for the clusters by applying both the virial theorem and the projected mass method. For the two clusters with the highest X-ray luminosities, the line-of-sight velocity dispersions are $\sim 700$ km $s^{-1}$, and mass-to-light ratios $M/L_{B(0)} \gtrsim 400\ M_\odot/L_\odot$. For the five other clusters the velocity dispersions are $\lesssim 370$ km $s^{-1}$, and four of the five have mass-to-light ratios $\lesssim 250\ M_\odot/L_\odot$. The D or cD galaxy in each poor cluster is at the kinematic center of the system.

160.030 **Rich clusters of galaxies as the result of large-scale motions inside superclusters.**
S. F. Shandarin, A. A. Klypin.
Astron. Zh., Tom 61, Vyp. 5, p. 837 – 845 (1984). In Russian. English translation in Sov. Astron., Vol. 28, No. 5.
Formation of rich clusters of galaxies is studied in the framework of the adiabatic scenario. A three-dimensional numerical model gives evidence for the following three-step process. First, flattened pancakes formed. After this, inside the pancakes filaments formed. Finally, inside the filaments rich clusters formed.

160.031 **Interaction of galaxy and quasar winds with the intergalactic medium.**
J. Silk.
Mass loss from astronomical objects, p. 128 – 146 (1984). – See Abstr. 012.032.
This paper gives estimates for the mean power and rate of mass loss likely to be expended in the form of steady or non-steady winds from galaxies and quasars. The ensuing rates of heating and enrichment of the intergalactic medium are derived, and implications for galaxy formation and pregalactic enrichment described. The author reviews a recent theory of galaxy formation based on explosive energy release from quasars and radio galaxies, and finally compares this approach to galaxy formation with more conventional theories.

160.032 **The development of galaxy clusters.**
E. E. Salpeter.
Ann. N.Y. Acad. Sci., Vol. 422, p. 95 – 105 (1984). – See Abstr. 012.033.

Recent work on the dynamical evolution of galaxy clusters is reviewed. The nature of the "missing matter" is considered and the question whether it is attached to individual galaxies in halos or is more loosely distributed in clusters is discussed. The problem whether galaxies were formed earlier or later than the first contraction of their respective clusters is also examined. The influence of the cluster environment on the mass to light ratio of single galaxies is studied and the evidence for deviations in the Hubble flow due to the Virgo and Coma clusters is reviewed.

160.033 **Optical studies of southern clusters of galaxies. III. Photoelectric photometry of galaxies in 15 clusters.**
J. Melnick, H. Quintana.
Astron. J., Vol. 89, No. 9, p. 1288 – 1292 (1984).

Multiaperture *g* and *r* photoelectric photometry is presented for 48 galaxies in the clusters: A85, A119, SC0316–344, SC0342–538, Sersic 40/6, SC0626–547, A754, A1146, Klemola 21, SC1329–314, SC1326–311, A2052, SC2008–569, SC2059–247, Klemola 44. The data are mainly intended for the calibration of photographic plates and the existence of a linear transformation to the *UBV* system make them useful to workers using plates taken in the *B* or *V* bands. A compilation of published photoelectric photometry on these clusters is given and a detailed comparison between these and the authors' observations is presented.

160.034 **On the spatial correlations of Abell clusters.**
N. Kaiser.
Astrophys. J., Lett. Ed., Vol. 284, No. 1, p. L9 – L12 (1984).

If rich clusters formed where the primordial density enhancement, when averaged over an appropriate volume, was unusually large, then they give a biased measure of the large-scale density correlation function: $\xi_{\text{clusters}}(r) \approx A\xi_{\text{density}}(r)$. The factor $A$ is determined by the probability distribution of the density fluctuations on a rich cluster mass scale, and if this distribution was Gaussian the correlation function is amplified. The amplification for rich $R \geqslant 1$ clusters is estimated to be $A \approx 10$, and the predicted trend of $A$ with richness agrees qualitatively with that observed.

160.035 **Distant clusters of galaxies. I. Uniform photometry of 14 rich clusters.**
W. J. Couch, E. B. Newell.
Astrophys. J., Suppl. Ser., Vol. 56, No. 1, p. 143 – 192 (1984). With plates 1 – 2.

The authors present *JF* photographic photometry of 14 rich clusters of galaxies in the redshift range $0.18 \leqslant z \leqslant 0.39$. The photometry, which includes the inner 60% of each cluster's population together with a large number of field galaxies measured in several areas close to each cluster, is presented in the form of color–magnitude arrays, color–magnitude tables, and red luminosity functions. The authors also present redshift measurements for 11 of the clusters, nine of which did not previously have redshift information.

160.036 **Monte-Carlo simulations of galaxy systems. VI. The angular correlation function for faint galaxy samples.**
H. T. MacGillivray, R. J. Dodd.
Astrophys. Space Sci., Vol. 105, No. 2, p. 331 – 337 (1984).

The angular correlation function for the sample of faint galaxies described in MacGillivray and Dodd (1980), obtained from objective measures made with the COSMOS machine on a deep UKST photograph, is examined and compared with that for the model fields described in Paper V of this series (MacGillivray and Dodd, 1982). The results support the presence of clustering of galaxies in these machine-determined samples on scales up to $7\,\text{hr}^{-1}$Mpc and are consistent with both the hierarchical and adiabatic schemes for galaxy clustering.

160.037 **Statistical masses of Abell clusters.**
J. A. de Freitas Pacheco.
Astrophys. Space Sci., Vol. 105, No. 2, p. 393 – 400 (1984).

An analysis of a homogeneous sample of 108 Abell clusters has led to an average peculiar velocity for the center of mass motion of these clusters of $610 \pm 750$ km s$^{-1}$. From this result, an upper limit for the average mass of the Abell clusters of $(1.6 \pm 2.4) \times 10^{15} M_\odot$ was obtained under the assumption that the peculiar motion is due to the excess of neighbours with respect to a uniform background. A lower limit of $(2.4 \pm 2.9) \times 10^{14} h^{-1} \Omega^{0.4} M_\odot$ was derived if one assumes that the peculiar velocity results from the mutual acceleration with the nearest neighbour.

160.038 **The evolution of galaxies in clusters.**
A. Dressler.
Annu. Rev. Astron. Astrophys., Vol. 22, p. 185 – 222 (1984). – See Abstr. 003.024.

Contents: (1) Introduction. (2) The laboratory. (3) Mergers and tidal stripping: the evolution of cD galaxies: Criteria for identifying cD galaxies. Simulations of cluster evolution and cD models. Optical observations relevant to merging, accretion and tidal stripping. Summary. (4) The evolution of different morphological types: A basic description of galaxies. Three classes of models. (5) Looking forward to looking back: Galaxy color as a function of look-back time. Verifying and interpreting the Butcher–Oemler effect by spectroscopy.

160.039 **Morphological classification (revised RS) of Abell clusters in $D = 5$ and an analysis of observed correlations.**
M. F. Struble, H. J. Rood.
Astron. J., Vol. 89, No. 10, p. 1487 – 1513 (1984).

A catalogue is presented of Rood–Sastry (RS) types for the 1087 Abell clusters in distance class $D = 5$. The structural information was determined from visual inspection of the KPNO glass copy of the National Geographic Society Palomar Observatory Sky Survey (POSS). The catalogue also includes the Hubble morphological type of the three brightest cluster members and the separations between them. The separation between the components of each binary galaxy (counted as one galaxy) among the three brightest is given; the observed number of other binaries is also given. Abell counts $N_A$, Bautz–Morgan (BM) types, redshifts $z$, and line-of-sight velocity dispersions gathered from the literature are also tabulated. The authors describe several empirical results derived from their data. These results are discussed primarily in relation to selection effects. The identification of connections between empirical results and theoretical predictions are left primarily to the reader. The authors mention three current modeling efforts directed towards understanding overall cluster dynamics; they differ from one another depending on how the missing mass is distributed.

160.040 **Distribution functions of intergalactic clouds.**
R. D. Sherman.
Astrophys. J., Vol. 284, No. 2, p. 457 – 460 (1984).

The rest equivalent width spectrum data derived from high-redshift QSO Ly$\alpha$ absorption line systems is a sensitive indicator of intergalactic cloud statistical properties. Assuming that the lines are due to clouds (randomly distributed, primordially constituted, photoionized, and uniform), model distribution functions for size (radius) and density are tested against those data for two types of line broadening mechanisms: thermal and cosmological. Optimal compatibility of model and data is found if the clouds are *thermally* broadened, and if radius and density distributions are power laws.

160.041 **1.4 GHz continuum sources in the Hercules Cluster.**
J. M. Dickey, E. E. Salpeter.
Astrophys. J., Vol. 284, No. 2, p. 461 – 470 (1984).

The authors have made a continuum survey near 1.4 GHz with the VLA D configuration of three fields, centered on the Hercules Cluster of galaxies (A2151). Positions and fluxes for 65 sources are given. There are 21 – 25 sources associated with galaxies (mostly bright galaxies). The three fields contain some optically bright galaxies and one strongly radio–emitting galaxy each. Two of the three strongly radio–emitting galaxies are head–tail sources, which indicates an intergalactic medium. The authors find five sources with 1.4 GHz flux >20 mJy, which are not associated with any galaxy. It is conjectured that these five galaxyless sources are not very distant background objects, but are produced by some plasma effect in the subcore medium of the Hercules Cluster.

160.042 **Spin statistics in binary galaxies: implications for formation and evolution.**
G. Helou.
Astrophys. J., Vol. 284, No. 2, p. 471 – 478 (1984).

New data are presented defining unambiguously the spin vectors of spiral galaxies in binaries and small groups. A strong anticorrelation is found, whereby the spins of companion galaxies avoid being parallel and favor being antiparallel. The anticorrelation is stronger for pairs with low indicative mass–to–light ratio. Clues to the origin of spin in galaxies are also direct clues to the mechanism of galaxy formation. The evidence so far is clearly against a simple picture where primeval turbulence is the source of spin. The data are consistent with the hypothesis that spins were acquired via tidal torquing.

160.043 **X–ray emission from Stephan's Quintet and other compact groups.**
N. A. Bahcall, D. E. Harris, H. J. Rood.
Astrophys. J., Lett. Ed., Vol. 284, No. 2, p. L29 – L33 (1984). With plate L1.

A search for X–ray emission from five compact groups of galaxies with the Einstein Observatory revealed detections from three groups. Soft, extended X–ray emission was observed in Stephan's Quintet, which is most likely caused by hot intracluster gas. This provides evidence for dynamical interaction among the group galaxies. X–ray emission from the group Arp 330 may also originate in hot intracluster gas. X–ray emission from Arp 318 may originate in its member galaxies.

160.044 **The galaxy group Klemola 25.**
A. C. Danks, J. Materne.
Astron. Astrophys., Vol. 139, No. 2, p. 455 – 458 (1984).

High resolution photographs of the compact group of galaxies Klemola 25 show that this group forms a ring. This ring resembles very much model simulations of rings undergoing bead instabilities. The mass of the system is equivalent to the mass of a giant elliptical galaxy.

160.045 **The prevalence of cooling flows in clusters of galaxies.**
G. C. Stewart, A. C. Fabian, C. Jones, W. Forman.
Astrophys. J., Vol. 285, No. 1, p. 1 – 6 (1984).

The central density and temperature of the intracluster gas in 36 clusters of galaxies have been determined by deprojection of the X–ray surface brightness distribution. The cooling time of the gas in half of these clusters is less than a Hubble time ($H_0$ = 50 km s$^{-1}$Mpc$^{-1}$). For these clusters mass inflow rates between 2 and 400 $M_\odot$ year$^{-1}$ are derived. These cooling flows may provide a significant fraction of the mass of the central cD galaxies.

160.046 **Globular cluster swapping in clusters of galaxies.**
J. C. Muzzio, R. E. Martínez, M. Rabolli.
Astrophys. J., Vol. 285, No. 1, p. 7 – 15 (1984).

The authors use numerical models that simulate the dynamical evolution of the globular cluster systems of galaxies which, in turn, belong to clusters of galaxies. The results show that, for conditions similar to those of the Virgo cluster and over a time interval comparable to the age of the universe, one–third of the globulars are lost by their parent galaxies; half of that number are clusters captured by other galaxies, while the rest end up as intergalactic clusters. The globulars are lost and captured by galaxies of any mass, but the net result is that small galaxies lose clusters and large galaxies gain them, the latter being able to capture amounts of clusters similar to the quantities they originally had.

160.047 **On the characteristics of clusters of galaxies with unseen mass.**
H. Smith Jr.
Astrophys. J., Vol. 285, No. 1, p. 16 – 24 (1984). = Dep. Astron., Univ. Florida, Contrib. No. 68.

*N*–body calculations have been used to study the effects of different background distributions of unseen mass on the characteristics (surface–density and radial–velocity profiles and virial mass) of simulated clusters of galaxies. The computed systems had an inverse–square mass spectrum, some were initially contracting, and some were calculated with ratios of background to total mass of 1/2 and 3/4 in addition to the previously used 9/10. The results are generally consistent with those already reported.

160.048 **Relations between spatial correlations of rich clusters of galaxies.**
H. D. Politzer, M. B. Wise.
Astrophys. J., Lett. Ed., Vol. 285, No. 1, p. L1 – L3 (1984).

The authors express the *N*–point rich cluster correlation function in terms of the two–point rich cluster correlation function in a model where rich clusters formed wherever suitably averaged primordial energy density fluctuations are unusually large. The validity of the results is *not* restricted to regions where the connected (reduced) *N*–point correlation functions are small compared with unity.

160.049 **The intergalactic H I cloud in Leo: a Spitzer–Baade collision product?**
H. J. Rood, B. A. Williams.
Astrophys. J., Lett. Ed., Vol. 285, No. 1, p. L5 – L9 (1984).

Results of a controlled experiment suggest that the recently discovered intergalactic H I cloud in Leo is the product of a Spitzer–Baade (gas–sweeping) collision between the large spiral galaxy NGC 3368 and the spiral predecessor of NGC 3384. These galaxies are members of the group Materne GWa (= de Vaucouleurs G11) and lie at equal angular distances on either side of the cloud center. Traces of two optical rings in NGC 3368 and an arc or spiral arm in NGC 3384 have been detected. These are precisely the features found in computer–simulated collisions by Theys and Spiegel and by Toomre.

160.050 **H I synthesis of the Virgo cluster.**
J. H. van Gorkom, C. G. Kotanyi, C. Balkowski.
Bull. Am. Astron. Soc., Vol. 16, No. 3, Part I, p. 720 (1984).
Abstract. – See Abstr. 010.064.

160.051 **High resolution imaging of distant galaxy clusters.**
L. Thompson.
Bull. Am. Astron. Soc., Vol. 16, No. 3, Part I, p. 728 (1984).
Abstract. – See Abstr. 010.065.

160.052 **A survey of galaxies in the field of Abell 194.**
R. J. Nemiroff, C. Ftaclas, M. F. Struble.
Bull. Am. Astron. Soc., Vol. 16, No. 3, Part I, p. 730 (1984).
Abstract. – See Abstr. 010.065.

160.053 **The "excess" light in brightest cluster galaxies.**
C. Ftaclas, M. F. Struble.
Bull. Am. Astron. Soc., Vol. 16, No. 3, Part I, p. 730 (1984).
Abstract. – See Abstr. 010.065.

160.054 **CCD camera observations of nearby rich clusters – II. Techniques and results for A85.**
H. P. Murphy.
Mon. Not. R. Astron. Soc., Vol. 211, No. 3, p. 637 – 658 (1984).

A CCD camera survey has been conducted of the X–ray emitting regions of a sample of clusters of galaxies. This sample includes at least one cluster from each of the classes in the new Forman–Jones classification scheme (Forman & Jones 1982). In this paper, the techniques used to obtain, reduce and analyse the data are presented, along with results for the cluster Abell 85. For the central X–ray emitting region of this cluster, a core radius of between 1.2 and 1.6 Mpc is estimated from the galaxy radial number density distribution. The properties of the galaxies in the core of this cluster are significantly different from those of the galaxies in the outer regions of the area covered in the survey, in terms of luminosity function and colour distribution.

160.055 **From the fringe of the Virgo supercluster.**
A. P. Fairall.
Mon. Notes Astron. Soc. S. Afr., Vol. 43, Nos. 9 – 10, p. 55 – 67 (1984).

160.056 **A photographic study of the cluster of galaxies A2052.**
J. Couture, E. Hardy.
J. R. Astron. Soc. Can., Vol. 78, No. 5, p. 211 (1984). Abstract. – See Abstr. 010.401.

160.057 **Gas in cosmic voids.**
N. Brosch, P. M. Gondhalekar.
Astron. Astrophys., Vol. 140, No. 2, p. L43 – L46 (1984).
Absorption lines at the redshifts of cosmic voids in Perseus–Pisces and Bootes have been detected in the ultraviolet spectra of background quasars. The detection of metal lines suggests chemical enrichment of gas in the voids in support of explosion dominated model of formation of galaxies.

160.058 **Galaxy clusters: dynamical times and brightest member magnitudes.**
G. Giuricin, F. Mardirossian, M. Mezzetti.
Astron. Astrophys., Vol. 141, No. 2, p. 419 – 421 (1984).
In contrast with earlier findings, the authors' analysis of the data of $\sim 30$ galaxy clusters reveals no significant correlations between cluster dynamical time scales and the magnitude differences between the first– and second– (or third–) ranked cluster galaxies. This result can be regarded as an indication that the luminosity of the brightest cluster members, which are often thought to undergo considerable ongoing evolution, cannot be appreciably affected by evolutionary processes related to cluster dynamical time scales. Furthermore, unlike previous assertions, the authors find no correlation between the cluster virial radii and velocity dispersions.

160.059 **Intracluster gravitational separation of deuterium and helium in rich galaxy clusters.**
M. R. Gil'fanov, R. A. Syunyaev.
Sov. Astron. Lett., Vol. 10, No. 3, p. 137 – 140 (1984). English translation of 37.160.100.

160.060 **On detecting intergalactic dispersion.**
P. J. Wiita, J. J. Mitteldorf.
Observatory, Vol. 104, No. 1063, p. 270 – 272 (1984).

160.061 **The intracluster gas around Cygnus–A.**
K. A. Arnaud, A. C. Fabian, S. A. Eales, C. Jones, W. Forman.
Mon. Not. R. Astron. Soc., Vol. 211, No. 4, p. 981 – 989 (1984).
Einstein Observatory X–ray observations show that the radio source Cygnus–A lies at the centre of about $10^{14} M_{\odot}$ of hot gas. In one direction this gas extends to over 1 Mpc from the radio source, a size comparable with previously observed large X–ray clusters. Gas out to a radius of 125 kpc has a cooling time of less than a Hubble time leading to a cooling flow of about 90 $M_{\odot}\mathrm{yr}^{-1}$. The hot gas has an $r^{-1}$ density profile giving an age of less than $2 \times 10^7$yr for a constant power input from the beam. The hotspot advance speed is found to be greater than 0.011 $c$, using recently published estimates of the pressure in the radio hotspots. Present X–ray observations are not able to detect the shocked gas confining the radio plasma.

160.062 **An X–ray survey of clusters of glaxies. IV. A survey of southern clusters and a compilation of upper limits for both Abell and southern clusters.**
M. P. Kowalski, M. P. Ulmer, R. G. Cruddace, K. S. Wood.
Astrophys. J., Suppl. Ser., Vol. 56, No. 4, p. 403 – 506 (1984).
Using HEAO 1 A–1, the authors have obtained X–ray error boxes and intensities or explicit upper limits for all clusters of galaxies in the Abell catalog and in the catalog of southern clusters and groups of Duus and Newell, a total of more than 3600 clusters. Calculated X–ray luminosity functions are valid to $10^{42}$ergs s$^{-1}$ and show that clusters and groups contribute 9.3% (+1.9, –1.5, 95% confidence) to the diffuse X–ray background. The authors have obtained upper limits to cluster volume emissivity evolution and have examined correlations between X–ray luminosity and other cluster properties.

160.063 **X–ray emission from Stephan's Quintet and other compact groups.**
N. A. Bahcall, D. E. Harris, H. J. Rood.
Space Telesc. Sci. Inst., Prepr. Ser., No. 21, 15 pp. (1984). To appear in Astrophys. J.

160.064 **The galaxy group Klemola 25.**
A. C. Danks, J. Materne.
ESO Sci. Prepr., No. 335, 17 pp. (1984). Submitted to Astron. Astrophys.

160.065 **The distribution of voids.**
G. Vettolani, R. E. de Souza, B. Marano, G. Chincarini.
ESO Sci. Prepr., No. 352, 35 pp. (1984). To appear in Astron. Astrophys.

160.066 **Extragalactic dust I: NGC 7070A.**
N. Brosch, J. M. Greenberg, P. J. Grosbøl.
ESO Sci. Prepr., No. 353, 28 pp. (1984). Submitted to Astron. Astrophys.

160.067 **H I deficiency and photometric evolution of spiral galaxies in the Virgo cluster.**
B. Guiderdoni, B. Rocca–Volmerange.
Inst. Astrophys. Paris, Pré–Publ., No. 79, 9 pp. (1984). To be published in Proceedings of the ESO Workshop on "The Virgo cluster of galaxies", ESO Garching, 4 – 7 September 1984.

160.068 **The evolution of galaxies in clusters. V. A study of populations since $z \sim 0.5$.**
H. Butcher, A. Oemler Jr.
Astrophys. J., Vol. 285, No. 2, p. 426 – 438 (1984).
The authors analyze photometry of 33 clusters of galaxies, with redshifts between 0.003 (the Virgo Cluster) and 0.54 (Cl 0016+16) to search for evolution of the colors of cluster populations. They determine the blue galaxy population in both compact, concentrated clusters and in open, irregular clusters. Evidence for an increase of the blue galaxy population with $z$ for both types of clusters is presented. Most of the data are consistent with the hypothesis that, in some fraction ($\sim 25\%$) of galaxies, the rate of star formation has decayed rapidly with epoch, *independent of the galaxy's environment*. One possible cause is simply a general depletion of gas in at least some regions of the universe.

160.069 **Origin of redshift differentials in galaxy groups.**
G. G. Byrd, M. J. Valtonen.
Rep. Ser., Dep. Phys. Sci., Univ. Turku, No. FTL–R63, 19 pp. (1984). = Turku Univ. Obs., Informo No. 82.
Sulentic has found a statistically significant excess number of higher redshifts companions relative to the brightest galaxy in spiral dominated groups in Huchra and Geller's catalog of galaxy groups. He proposed that the excess is caused by non–Doppler redshifts in the spiral groups. Instead, the authors propose that the excess results from most of the group population being composed of unbound expanding members and a tendency to mistakenly pick a galaxy of brighter apparent magnitude from the nearer portion of the expanding group population rather than the true primary which would generally be in the middle of the population. Using the known galaxy number versus luminosity distribution, and the assumption that most spiral group members are unbound, the authors calculate the expected redshift excess for the spiral dominated groups.

160.070 **Simulations of clusters of galaxies with massive cores.**
S. Saarinen, M. J. Valtonen.
Rep. Ser., Dep. Phys. Sci., Univ. Turku, No. FTL–R64, 33 pp. (1984). = Turku Univ. Obs., Informo No. 83.

Clusters of galaxies with strong concentration of mass in their cores can be very far from dynamical equilibrium during their early evolution. Then a straightforward application of the virial theorem can lead to a totally wrong idea of the mass distribution of the cluster, hiding away the very central mass concentration. Numerical simulations have been carried out of clusters of galaxies which separate out of the Hubble flow and collapse on their massive cores.

160.071 **The infrared color–magnitude and color–gas content relations for cluster spirals.**
G. D. Bothun, J. Mould, R. A. Schommer, M. Aaronson.
Prepr. Steward Obs., No. 545, 48 pp. (1984).

Using multiaperture optical and infrared magnitudes of 164 spirals in nearby galaxy clusters, the authors explore the utility of the spiral color–magnitude relation as an accurate distance indicator. In combination with H I line widths, the authors find there to be a well defined relationship between the observed $(B–H)_{-0.5}$ color of spirals and their observed 20% line width. This relation demands there to be a wavelength dependent slope in the Tully–Fisher relation. However, there are important variations in the zero point of the color–line width relation for individual clusters which lead directly to zero–point differences in their corresponding color–magnitude relation. Thus, while individual clusters are shown to yield color–magnitude relations which have a scatter of 0.4 – 0.5 mag, the authors show that the spiral color–magnitude relation may be impossible to calibrate due to the lack of a well–defined zero point.

160.072 **The brightest clusters of galaxies and the cosmological constant.**
V. E. Yakimov.
Astron. Tsirk., No. 1330, p. 2 – 4 (1984). In Russian.

160.073 **The structure of first–ranked cluster galaxies and the radius–magnitude relation.**
P. M. Lugger.
Astrophys. J., Vol. 286, No. 1, p. 106 – 115 (1984).

To investigate theoretical predictions for the dynamical evolution of first–ranked galaxies, a quantitative study of their properties, as a function of cluster morphology, has been carried out using photographic plates obtained with the Palomar Schmidt telescope. Surface brightness profiles to radii of several hundred kpc first–ranked cluster galaxies have been analyzed. The dispersion in the metric magnitudes of first–ranked galaxies is quite small ($\sim$0.4 mag). For the cD (supergiant elliptical) galaxy sample, the mean metric magnitude is $\sim$0.5 mag brighter than for the non–cD galaxies. The relation between de Vaucouleurs effective radius and magnitude determined in the present study for first–ranked galaxies is consistent with the relations found for fainter galaxies by Strom and Strom as well as Wirth.

160.074 **X–ray measurements of the nonspherical mass distribution in the cluster of galaxies A2256.**
D. Fabricant, G. Rybicki, P. Gorenstein.
Astrophys. J., Vol. 286, No. 1, p. 186 – 195 (1984).

The authors describe a technique for the analysis of X–ray surface brightness maps of rich clusters of galaxies with a smooth X–ray morphology and appropriate bilateral symmetry to yield an accurate estimate of the underlying cluster mass distribution. This method assumes that the X–ray emitting gas is in hydrostatic equilibrium in the gravitational potential of the cluster, and that the cluster possesses an axis of symmetry that lies in the plane of the sky. The errors associated with the latter assumption are discussed. The authors apply this technique to the cluster of galaxies A2256 and find that the mass distribution departs significantly from spherical symmetry, with axial ratios of approximately 1.6. This is to be compared with the axial ratio of 2 – 3.3 found for the galaxy distribution. They find that the surfaces of constant total density are more complex than similar oblate or prolate spheroids with a fixed center. The radially integrated mass derived from the X–ray data is consistent with that estimated from the galaxy surface density and velocity dispersion.

160.075 **Two–dimensional numerical modelling of formation of superclusters from collisionless particles: two–point correlation function.**
A. Melott.
Astron. Zh., Tom 61, Vyp. 6, p. 1079 – 1086 (1984). In Russian. English translation in Sov. Astron., Vol. 28, No. 6.

The two–point correlation function and the velocity field are investigated in the two–dimensional numerical modelling of the adiabatic theory of formation of galaxies under the presence of collisionless particles. The correlation function during nonlinear clustering remains small by amplitude at radii comparable to the scale of the cluster. It is revealed that the behaviour of the correlation function in the small–amplitude region leads to a sharp break or to a change of the slope of the power asymptotics. It is shown that the velocities associated with large voids are much smaller than the diameter of the void divided by the Hubble time–scale.

160.076 **Stephan's quintet, superclusters of galaxies and giant voids.**
P. Focardi, B. Marano, G. Vettolani.
Astron. Express, Vol. 1, Issue 2, p. 75 – 79 (1984).

The high redshift members and the low redshift members of the Stephan Quintet appear to belong to different superclusters of galaxies separated by a large "void" in radial velocity. The authors suggest that any non–classical interpretation of the cosmological redshift must deal with the evidence of associations of galaxies in large scale structures.

160.077 **Luminosities of first–ranked galaxies.**
C. L. Morbey.
Publ. Astron. Soc. Pac., Vol. 96, No. 585, p. 874 – 879 (1984).

A number of first–ranked galaxies in poor and rich clusters have absolute magnitudes which correlate well with the richness classes of their parent clusters. The correlation is consistent with the statistical theory for the luminosities of first–ranked galaxies in clusters advocated by Geller and Peebles. This does not rule out special process theories but suggests that, if special processes do occur, their effects may be somewhat nonuniform from cluster to cluster. It is shown that the commonly observed small dispersion in the absolute magnitudes of giant elliptical galaxies is a result of (1) the size of the sampling aperture used and (2) the intrinsic gradients of their observed surface brightness profiles. Parameters associated with the dominant galaxies do not correlate well with Bautz–Morgan class.

160.078 **Superclusters.**
J. H. Oort.
Clusters and groups of galaxies, p. 1 – 16 (1984). – See Abstr. 012.088.

160.079 **Three–dimensional structure in the Indus supercluster.**
S. M. Beard, J. A. Cooke, D. Emerson, B. D. Kelly.
Clusters and groups of galaxies, p. 17 – 22 (1984). – See Abstr. 012.088.

Redshifts of large numbers of galaxies can be obtained to an accuracy of $\pm0.01$ in Z by measurement of their spectra on UK Schmidt Telescope objective–prism plates using the COSMOS plate–scanning machine. 2294 redshifts have been measured in a $4.7 \times 5.2$ deg$^2$ area of the sky which includes part of the Indus supercluster. The redshifts enable the three–dimensional distribution of the galaxies within the supercluster to be investigated, and a subset of this data is presented.

160.080 **Recent studies of the Local Supercluster.**
G. de Vaucouleurs.
Clusters and groups of galaxies, p. 29 – 42 (1984). – See Abstr. 012.088.

After a brief historical review of the discovery of the Local Supercluster, this paper reports on recent research on the space distribution of nearby galaxies, on the space motions of the Sun, Galaxy and Local Group relative to these nearby galaxies, and on a first empirical mapping of the velocity field near the plane of the Local Supercluster.

160.081 **A catalog of candidate superclusters and voids.**
J. O. Burns, D. J. Batuski.
Clusters and groups of galaxies, p. 43 – 48 (1984). – See Abstr. 012.088.

160.082 **Statistical analysis of the shapes of rich superclusters.**
R. Ciardullo, R. Harms, H. Ford.
Clusters and groups of galaxies, p. 49 – 54 (1984). – See Abstr. 012.088.

The authors outline a technique for estimating the large scale shape of rich superclusters using complete samples of objects extending out to moderately large redshifts. The authors describe two tests, a principal axes test and a velocity gradient test, which are especially useful for this purpose. They apply these tests to samples of clusters from the Abell catalog and search for evidence of disk–like or filamentary structures.

160.083 **The orientations of spiral galaxies in the Local Supercluster.**
E. R. Anderson, C. Norman, G. Illingworth.
Clusters and groups of galaxies, p. 63 – 64 (1984). – See Abstr. 012.088.

Studies of the frequency distribution of supergalactic positon angles for spiral galaxies of the Local Supercluster show no preferred orientation. However, for galaxies in the magnitude range $14.0 \leqslant m_B \leqslant 15.5$ positioned over the whole sky, there is a preferred supergalactic position angle of 94° with a dispersion of 40°.

160.084 **On the orientation of galaxies in the Local Supercluster.**
P. Flin, W. Godłowski.
Clusters and groups of galaxies, p. 65 – 66 (1984). – See Abstr. 012.088.

160.085 **Superclusters: a new insight into the problem of groups with discrepant redshifts.**
P. Focardi, B. Marano, G. Vettolani.
Clusters and groups of galaxies, p. 67 – 68 (1984). – See Abstr. 012.088.

160.086 **On the large–scale distribution of clusters of galaxies.**
K.–H. Schmidt.
Clusters and groups of galaxies, p. 73 – 74 (1984). – See Abstr. 012.088.

The redshifts of more than 600 clusters of galaxies have been compiled from the literature in order to study the large–scale distribution of these objects.

160.087 **The core of the Virgo cluster.**
J. P. Huchra, R. J. Davis, D. W. Latham.
Clusters and groups of galaxies, p. 79 – 87 (1984). – See Abstr. 012.088.

160.088 **The structure and dynamics of the Coma cluster.**
J. V. Peach.
Clusters and groups of galaxies, p. 89 – 92 (1984). – See Abstr. 012.088.

160.089 **Fibre optic spectra and the dynamics of Sersic 40/6.**
D. Carter, P. F. Teague, P. M. Gray.
Clusters and groups of galaxies, p. 93 – 98 (1984). – See Abstr. 012.088.

160.090 **Clusters of galaxies with very large redshifts.**
J. B. Oke.
Clusters and groups of galaxies, p. 99 – 107 (1984). – See Abstr. 012.088.

160.091 **CCD observations of distant clusters of galaxies.**
R. S. Ellis.
Clusters and groups of galaxies, p. 109 – 111 (1984). – See Abstr. 012.088.

160.092 **Galaxy morphology in distant clusters.**
L. A. Thompson.
Clusters and groups of galaxies, p. 113 – 116 (1984). – See Abstr. 012.088.

Direct information on the evolution of galaxies and of rich clusters may be available in the images of clusters at high redshift. A program is underway at Mauna Kea Observatory to obtain subarcsecond images of distant clusters. CCD images of clusters out to $z = 0.4$ have been obtained under good seeing conditions through a filter that mimics standard photographic bandpasses used in studies of nearby clusters (IIIa–J + GG 385).

160.093 **The evolution of galaxies in clusters.**
A. Dressler.
Clusters and groups of galaxies, p. 117 – 132 (1984). – See Abstr. 012.088.

Clusters of galaxies are laboratories for the study of galaxy evolution. Predictions of the effects of tidal stripping, merging, and accretion are compared to optical observations, particularly in connection with the formation of cD galaxies in rich clusters. The development of the different morphological types is discussed in terms of models that depend entirely on later evolution and those that rely on initial conditions.

160.094 **Globular clusters and galaxy formation in clusters of galaxies.**
S. van den Bergh.
Clusters and groups of galaxies, p. 139 – 143 (1984). – See Abstr. 012.088.

Taken at face value globular cluster counts indicate that: (1) most elliptical galaxies in rich clusters were not formed by merging spirals and (2) central galaxies in rich clusters were already very special at the time when they were forming globular clusters. These results are difficult to reconcile with models in which clusters of galaxies are formed entirely by gravitational clustering of pre–existing galaxies.

160.095 **Infall into the Virgo cluster and some cosmological constraints.**
R. B. Tully, E. J. Shaya.
Clusters and groups of galaxies, p. 145 – 146 (1984). – See Abstr. 012.088.

160.096 **Matter distributions of various components in the Coma cluster.**
D. Gerbal, G. Mathez, A. Mazure, J. L. Monin.
Clusters and groups of galaxies, p. 147 – 151 (1984). – See Abstr. 012.088.

160.097 **The two–component cluster in Centaurus.**
J. R. Lucey, M. J. Currie, R. J. Dickens, J. A. Dawe.
Clusters and groups of galaxies, p. 153 – 158 (1984). – See Abstr. 012.088.

160.098 **The ellipticity of galaxy clusters.**
P. Flin.
Clusters and groups of galaxies, p. 163 – 168 (1984). – See Abstr. 012.088.

The paper deals with changes of ellipticity with radial distance from the cluster centre. The Dressler sample of galaxy clusters is analysed. It is shown that the ellipticity changes are statistically significant. In many clusters they are not caused by subclustering. For this reason ellipticity changes ought to be regarded as an intrinsic independent cluster feature.

160.099 **Structure of the Hydra–I cluster of galaxies.**
F. W. Baier, H. Oleak.
Clusters and groups of galaxies, p. 169 – 170 (1984). – See Abstr. 012.088.

160.100 **Redshifts for Zwicky's near clusters.**
G. C. Baiesi–Pillastrini, G. G. C. Palumbo, G. Vettolani.
Clusters and groups of galaxies, p. 171 – 172 (1984). – See Abstr. 012.088.

160.101 **Tidal stripping of galaxies in clusters revisited.**
M. J. Currie.
Clusters and groups of galaxies, p. 173 – 174 (1984). – See Abstr. 012.088.

The Stroms (1978, 1979) claimed that they had found evidence for stripping of galaxies in cluster cores because galaxian sizes varied with environment, however this work has not been corroborated.

160.102 **Cluster elliptical and lenticular colour–magnitude relationships.**
M. J. Currie.
Clusters and groups of galaxies, p. 175 – 176 (1984). – See Abstr. 012.088.

160.103 **On the perceptibility of distant clusters of galaxies.**
H.-E. Frölich, K.-H. Schmidt, R. Schmidt.
Clusters and groups of galaxies, p. 177 – 178 (1984). – See Abstr. 012.088.

160.104 **Luminosity segregation and anti–segregation in clusters of galaxies.**
A. Mazure, H. Capelato, G. des Forêts.
Clusters and groups of galaxies, p. 179 – 180 (1984). – See Abstr. 012.088.

160.105 **An objective prism survey of emission line galaxies in nearby Abell clusters.**
C. Moss, J. F. Goeke, M. Whittle.
Clusters and groups of galaxies, p. 181 (1984). Abstract. – See Abstr. 012.088.

160.106 **Studies of galaxy clusters using APM.**
S. Phillipps.
Clusters and groups of galaxies, p. 183 – 184 (1984). – See Abstr. 012.088.

The Automatic Plate Measuring System (APM) at the Institute of Astronomy, Cambridge, has been used to obtain surface photometry of a large sample of galaxies. Several studies involving galaxies in the Fornax cluster are currently being undertaken.

160.107 **Nonthermal radio sources in clusters of galaxies.**
R. Fanti.
Clusters and groups of galaxies, p. 185 – 201 (1984). – See Abstr. 012.088.

The paper deals mostly with properties of nonthermal radio sources in rich Abell clusters, with particular attention to radio morphology and to the radio luminosity functions.

160.108 **Westerbork observations of 37 poor clusters of galaxies at 1415 MHz.**
R. J. Hanisch.
Clusters and groups of galaxies, p. 203 – 208 (1984). – See Abstr. 012.088.

Observations of a sample of 37 poor clusters of galaxies north of declination +35° have been made at 1415 MHz with the Westerbork Synthesis Radio Telescope. 204 radio sources with 1415 MHz flux densities greater than 10 mJy were detected. However, only 14 of these sources could be identified with galaxies that are poor cluster members. The morphologies of the identified poor cluster radio sources are studied.

160.109 **Radio continuum properties of spiral galaxies in the Coma/A1367 supercluster. Comparison between isolated and cluster galaxies.**
G. Gavazzi, W. Jaffe, E. Valentijn.
Clusters and groups of galaxies, p. 209 – 214 (1984). – See Abstr. 012.088.

160.110 **H I content of groups and clusters of galaxies.**
W. K. Huchtmeier.
Clusters and groups of galaxies, p. 221 – 242 (1984). – See Abstr. 012.088.

160.111 **Cluster evolution and sweeping: a comparative study of nearby clusters.**
R. Giovanelli.
Clusters and groups of galaxies, p. 243 – 249 (1984). – See Abstr. 012.088.

160.112 **H I deficiency in the Virgo Cluster: VLA observations.**
J. H. Van Gorkom, C. Balkowski, C. Kotanyi.
Clusters and groups of galaxies, p. 261 – 265 (1984). – See Abstr. 012.088.

H I line observations of 10 bright spiral galaxies within 6° of the centre of the Virgo Cluster were made at a resolution of about 45″ using the VLA. H I disks within 2° of the cluster centre are H I deficient and much smaller than the optical disks, whereas disks outside 3° are normal. NGC 4254 and NGC 4654 situated at about 2° show asymmetric H I and optical structures suggesting the action of ram–pressure stripping. It is concluded that stripping is nearly complete and instantaneous and may account for the Hubble type segregation of spiral galaxies within rich clusters.

160.113 **Microwave decrement measurements in galaxy clusters.**
R. D. Davies, A. N. Lasenby.
Clusters and groups of galaxies, p. 267 – 272 (1984). – See Abstr. 012.088.

Photons from the 3K cosmic microwave background originating at $z \sim 1500$ are inverse Compton scattered on passing through the hot ($10^7$ to $10^8$K) gas in galaxy clusters. Photons from the Rayleigh–Jeans ($\lambda \gg 1$ mm) part of the spectrum are scattered to $\lambda \ll 1$ mm, thereby producing a reduction in radiation intensity at $\lambda > 1$ mm. This microwave decrement of "cooling" in clusters has been actively pursued observationally in the 10 years since its prediction by Sunyaev and Zeldovich (1970, 1972). The authors describe the significance of such observations, the present observational status and the prospects for the future.

160.114 **The radio luminosity function of first ranked galaxies in Abell clusters.**
L. Feretti, G. Giovannini, L. Gregorini.
Clusters and groups of galaxies, p. 291 – 292 (1984). – See Abstr. 012.088.

160.115 **Extended radio sources and the diffuse medium in the Coma–A1367 supercluster.**
W. J. Jaffe, G. Gavazzi, E. A. Valentijn.
Clusters and groups of galaxies, p. 293 – 294 (1984). – See Abstr. 012.088.

Studies of extended, thermally confined radio sources in rich clusters provided estimates of the pressure and temperature in the cluster medium that confirmed the values obtained from X–ray studies. The authors have now found some extended sources in the Coma–A1367 supercluster that may yield similar information on the supercluster medium.

160.116 **Preliminary results of the 20–cm VLA survey of Abell clusters.**
F. N. Owen, J. O. Burns, R. A. White.
Clusters and groups of galaxies, p. 295 – 296 (1984). – See Abstr. 012.088.

160.117 **Implications of HEAO A–1 and Einstein satellite observations of clusters of galaxies.**
M. P. Ulmer, M. P. Kowalski, R. G. Cruddace, K. S. Wood, G. Fritz.
Clusters and groups of galaxies, p. 307 – 312 (1984). – See Abstr. 012.088.

The authors present a review of X–ray observations they have made over the past several years. This work is a survey of all rich clusters with the HEAO A–1 experiment, plus some follow–up work with the Einstein Observatory.

160.118 **Dark matter in poor clusters of galaxies.**
G. A. Kriss.
Clusters and groups of galaxies, p. 313 – 318 (1984). – See Abstr. 012.088.

X–ray observations of 16 poor clusters containing central dominant galaxies have been performed with the Imaging Proportional Counter of the Einstein Observatory. Twelve clusters were detected, and in each case the X–ray emission is centered on the dominant galaxy. For the six brightest clusters the author finds extended X–ray emission that is smooth, centrally peaked, and reasonably symmetric. The X–ray surface brightness implies density profiles that are roughly inversely proportional to radius. The author finds total binding masses of $\sim 10^{13} M_{\odot}$ within 0.5 Mpc and mean mass to light ratios of 100 $M/L_V$. Comparison of the deduced distribution of binding mass with the light distribution of the central galaxies of four clusters shows that $M/L_V$ rises to over 200 in the galaxy halos.

160.119 **Einstein images of clusters of galaxies: galaxy haloes, the intracluster medium, and the intercluster gas.**
C. Jones, W. Forman.
Clusters and groups of galaxies, p. 319 – 340 (1984). – See Abstr. 012.088.

Analyses of the X–ray images obtained with the Einstein Observatory have provided a new and different view of clusters of galaxies. The primary topics discussed in this review are the role of the central dominant galaxy, the origin of S0 galaxies, cluster classification and the dynamically young state of most clusters, the large scale height of the gas compared to the galaxy distribution, and the effects of a hot intercluster gas on the intracluster gas.

160.120 **Cluster X–ray luminosity and first ranked galaxy size.**
A. Porter.
Clusters and groups of galaxies, p. 351 – 352 (1984). – See Abstr. 012.088.

The author examines the relation between the metric X–ray luminosity of a cluster of galaxies and the optical core radius of its brightest member. Implications for theories of galaxy growth are briefly discussed.

160.121 **Groups of galaxies.**
M. J. Geller.
Clusters and groups of galaxies, p. 353 – 366 (1984). – See Abstr. 012.088.

The author concentrates on the analysis of statistically complete group catalogs rather than on detailed studies of individual systems. She focuses on four aspects of these analyses: (1) the definition and construction of group catalogs, (2) the extraction of physical results (particularly mass–to–light ratios) from the study of groups of galaxies, (3) the statistical problems involved in the study of groups, and (4) the overlap or "continuum" of properties of "groups" and "clusters" of galaxies.

160.122 **The structure of compact groups of galaxies.**
P. Hickson, Z. Ninkov, J. P. Huchra, G. A. Mamon.
Clusters and groups of galaxies, p. 367 – 373 (1984). – See Abstr. 012.088.

A survey was undertaken of the Palomar Observatory Sky Survey prints for groups satisfying three selection criteria of population, isolation, and mean surface brightness. The resulting catalog of 100 groups forms the study sample.

160.123 **A dynamical study of two groups in the Coma/A1367 supercluster.**
B. A. Williams.
Clusters and groups of galaxies, p. 375 – 380 (1984). – See Abstr. 012.088.

Located $\sim 12°$ southwest of A1367, the IC 698 group and the NGC 3825 group are two systems of galaxies within the Coma/A1367 supercluster. All of the spiral galaxies in both groups were observed and detected at 21 cm with the 305–m Arecibo radio telescope. The dynamical stability of each group was analyzed by using the accurate radial velocities and the direct estimates of the masses obtained from these 21–cm line observations.

160.124 **A neutral hydrogen study of Seyfert's very compact group.**
J. R. Dickel, H. J. Rood, B. A. Williams.
Clusters and groups of galaxies, p. 389 – 394 (1984). – See Abstr. 012.088.

A cloud of neutral hydrogen with an extent greater than that of any individual galaxy is present in the direction of Seyfert's very compact group of galaxies. At each position within this cloud, the velocity width is about 160 km $s^{-1}$ and the radial velocity lies in the range 4530 km $s^{-1}$ to more than 4600 km $s^{-1}$. The largest radial velocity of any galaxy in the group is 4503 km $s^{-1}$.

160.125 **Hot gas in groups of galaxies.**
P. Biermann, P. P. Kronberg.
Clusters and groups of galaxies, p. 395 – 397 (1984). – See Abstr. 012.088.

The properties of sparse groups of galaxies dominated by early Hubble type galaxies are a natural extension of those of larger clusters. The evolutionary model by Tinsley and Larson for the formation of elliptical galaxies is strongly supported by the detection of $\sim 10^{11} M_{\odot}$ of hot gas around the elliptical galaxy NGC 5846 which sits in only a sparse group. The virial mass required to contain the gas is $\sim 10^{13} M_{\odot}$ which is about the same mass as required to hold the entire group.

160.126 **On merging in galaxy groups.**
M. Mezzetti, G. Giuricin, F. Mardirossian.
Clusters and groups of galaxies, p. 399 – 400 (1984). – See Abstr. 012.088.

160.127 **The blue Tully–Fisher relation for the Local Group of galaxies and the Virgo cluster of galaxies.**
O.–G. Richter.
Clusters and groups of galaxies, p. 401 – 403 (1984). – See Abstr. 012.088.

160.128 **Hierarchical clustering.**
P. J. E. Peebles.
Clusters and groups of galaxies, p. 405 – 414 (1984). – See Abstr. 012.088.

Studies of how galaxies and clusters of galaxies have formed have reached an interesting stage where good arguments can be adduced in favor of quite different scenarios. The author explains why he thinks the evidence for the "top–down" picture, where galaxies are fragments of pre–existing protoclusters, seems a little questionable, and then presents his assessment of the evidence on the "bottom–up" picture where galaxies form as bound objects before clusters.

160.129 **Percolation studies of galaxy clustering.**
S. P. Bhavsar, J. D. Barrow.
Clusters and groups of galaxies, p. 415 – 421 (1984). – See Abstr. 012.088.

A percolation cluster analysis is performed on equivalent regions of the CFA redshift survey of galaxies and the 4000 body simulations of gravitational clustering made by Aarseth, Gott and Turner (1979). The observed and simulated percolation properties are compared.

160.130 **The spatial distribution of dark matter in the Coma cluster.**
Y. Rephaeli.
Clusters and groups of galaxies, p. 423 – 428 (1984). – See Abstr. 012.088.

Dark matter in clusters can be either more or less centrally concentrated than galaxies, depending on whether it was once bound to galaxies in halos. If the distributions of galaxies and dark matter are represented by similar profiles, but with different core radii, then the radial variation of the velocity dispersion in the Coma cluster and the degree of efficiency of gravitational drag on galaxies are somewhat more consistent with the dark matter being more centrally concentrated than galaxies.

160.131 **The dynamical evolution of clusters and groups of galaxies.**
A. Cavaliere, P. Santangelo, G. Tarquini, N. Vittorio.
Clusters and groups of galaxies, p. 499 – 504 (1984). – See Abstr. 012.088.

Computer simulations of the evolution of clusters and of groups (including degrees of freedom internal to galaxies) test how the dissipationless clustering scenario meets the challenge posed by the observed variety of morphologies and structures.

160.132 **Mass estimation of clusters with unseen mass.**
H. Smith Jr.
Clusters and groups of galaxies, p. 523 – 524 (1984). – See Abstr. 012.088.

160.133 **Estimation of the velocity law of infalling galaxies in a spherically symmetric cluster from observational data.**
N. Voglis.
Clusters and groups of galaxies, p. 527 – 528 (1984). – See Abstr. 012.088.

160.134 **Information on galaxy clustering from gravitational lenses.**
E. L. Turner.
Clusters and groups of galaxies, p. 553 – 558 (1984). – See Abstr. 012.088.

Observations of gravitational lens systems can provide information on the distribution and perhaps the nature of dark matter in clusters of galaxies. Lenses may prove to be one of the most powerful tools for the study of a variety of cosmological problems.

160.135 **The clustering of galaxies about powerful radio sources.**
R. M. Prestage.
Clusters and groups of galaxies, p. 559 – 564 (1984). – See Abstr. 012.088.

The author has investigated the clustering environment of a sample of ~300 powerful radio sources. He finds that on average flat spectrum sources have clustering properties similar to those of galaxies in general. Steep spectrum sources are on average in regions of rather enhanced galaxy density; more so for FR I than FR II sources. The mean galaxy density around classical double sources is consistent with that of FR II sources as a whole.

160.136 **Correlation analyses of galaxy clustering to B ~ 24.**
P. R. F. Stevenson, T. Shanks, R. Fong.
Clusters and groups of galaxies, p. 565 – 571 (1984). – See Abstr. 012.088.

Estimates of the angular two–point galaxy correlation function, w(θ), are presented as obtained from COSMOS machine measurements of 1.2 m UK Schmidt telescope (UKST) and 4 m Anglo–Australian telescope (AAT) plates.

160.137 **Clustering of early–type galaxies in the southern hemisphere.**
E. M. Sadler.
Clusters and groups of galaxies, p. 573 – 577 (1984). – See Abstr. 012.088.

The results of a large observational study of early–type galaxies are used to investigate the large–scale clustering of galaxies in a region of sky only poorly covered by previous surveys. Redshifts and magnitudes have been measured for all elliptical and S0 galaxies south of declination –32° to a limiting magnitude of 14.0, and the data can be used to study both angular (2D) and spatial (3D) clustering. The two–point correlation function for southern galaxies agrees closely with that already derived for the north (δ > 0°).

160.138 **Two–colour photometry of clusters of galaxies. II. A1213.**
A. G. Egikyan, A. T. Kalloglyan, D. Nanni, D. Trevese, A. Vignato.
Astrofizika, Tom 21, Vyp. 1, p. 21 – 34 (1984).

A two–colour photometry of more than 420 galaxies in the cluster A1213 has been carried out. The luminosity function, radial and colour distributions in the cluster have been investigated. The existence of a core of about 0.5 Mpc radius consisting of mainly bright and red galaxies is suggested. About half of the total luminosity of the cluster is included within the core radius.

160.139 **The interaction of high–energy particles with intergalactic gas.**
N. Ya. Sotnikova.
Astrofizika, Tom 21, Vyp. 1, p. 103 – 109 (1984). In Russian. English translation in Astrophysics, Vol. 21, No. 1.

It is shown that the energy of cosmic rays emitted from galaxies into the intracluster medium may be the substantial source of its heating. The formation of non–Maxwellian distribution with cut–off "tail" is emphasized to be possible. The interaction of electrons with cosmic rays cannot provide the existence of this "tail". The X–ray spectrum peculiarities of gas in the A2142 cluster in the high–energy region may be attributed to deviations from Maxwellian distribution.

160.140 **Emission–line galaxies in the cluster A 634.**
D. A. Stepanyan.
Astrofizika, Tom 21, Vyp. 2, p. 245 – 254 (1984). In Russian. English translation in Astrophysics, Vol. 21, No. 2.

The results of investigations of the cluster of galaxies A 634 are presented. 35 emission–line galaxies are ascribed to the members of the cluster. The average radial velocity of the cluster is 8010 km/s, velocity dispersion 309 km/s. The lower limit of relative number of emission–line galaxies brighter than $15^m.7$ in the cluster A 634 is 0.29.

160.141 **The A2197 and A2199 galaxy clusters.**
S. A. Gregory, L. A. Thompson.
Astrophys. J., Vol. 286, No. 2, p. 422 – 436 (1984).

New redshifts are presented for 74 galaxies in and near the rich clusters A2197 and A2199. When added to results in the literature, the authors have a total sample of 136 galaxies with known redshifts in the region. They determine the systematic redshift and line–of–sight velocity dispersions for both clusters. By applying the virial theorem they find mass–to–light ($M/L$) ratios for each cluster core in the range 200 – 400. The authors model the binary cluster with a two–body infall analysis and thereby determine an independent estimate for the $M/L$ ratio of the entire system. Surprisingly, the double cluster method indicates that $M/L < 100$.

160.142 **Redshift differentials in a complete sample of galaxy groups.**
J. W. Sulentic.
Astrophys. J., Vol. 286, No. 2, p. 442 – 449 (1984).

A test is made of the hypothesis that companion galaxies in spiral–dominated groups contain an excess positive redshift. The sample of galaxy groups was rigorously selected by Huchra and Geller (1982). A sample of 196 companion galaxies in 60 groups shows a statistically significant positive redshift relative to the brightest spiral group members. A sample of 62 companions in 21 E/S0–dominated groups shows no such excess. The positive excess redshift reported here is not expected within the framework of conventional dynamical theory. Conventional explanations for the excess of higher redshift companions are examined.

160.143 **CCD observations of galaxies in clusters at high redshift.**
R. Schild.
Astrophys. J., Vol. 286, No. 2, p. 450 – 463 (1984).

Observations of magnitudes of galaxies obtained with a CCD camera and filters approximating the photographic *J*, *F*, and *I* are transformed to standard color indices ($g$–$r$) and ($R$–$I$) to permit studies of galaxies in clusters around radio sources with known redshift. The author presents magnitudes and colors of galaxies associated with high–redshift radio galaxies, and discusses the existence of foreground clusters on the basis of the color–color diagram. Although many galaxies are identified as probably at the distance of 3C 352 ($z$ = 0.81) and 3C 265 ($z$ = 0.81), none have the color of a redshifted giant elliptical galaxy. The relative importance of luminosity evolution due to evolution of stellar populations as compared to galaxy mergers is discussed.

160.144 **A complete sample of southern E and S0 galaxies: clustering properties.**
E. M. Sadler, N. A. Sharp.
Astrophys. J., Vol. 287, No. 1, p. 80 – 88 (1984).
The authors analyze the angular and spatial clustering of a sample of southern elliptical and S0 galaxies, complete to a limiting apparent magnitude of 13.8 and all having measured radial velocities. The sample has an angular covariance function $w = 5.6\theta^{-1.1}$, in agreement with previous work. The spatial density field is derived, and thence the three-dimensional covariance function.

160.145 **The radial X-ray brightness profile of the Coma cluster.**
G. A. Chanan, F. Abramopoulos.
Astrophys. J., Vol. 287, No. 1, p. 89 – 94 (1984).
The authors present radial brightness profiles of the Coma cluster of galaxies in the 0.15 – 4.0 keV X-ray band, along both major and minor axes, from data obtained with the *Einstein Observatory* imaging proportional counter. These are compared to various theoretical brightness distributions predicted by Strimpel and Binney in 1979. The observed profiles are closest to the isothermal oblate model (although there are some systematic differences), and may therefore lend support to "pancake" theories of cluster origin.

160.146 **Endpoints of stellar evolution: X-ray surveys of the Local Group.**
D. J. Helfand.
Publ. Astron. Soc. Pac., Vol. 96, No. 586, p. 913 – 931 (1984). = Columbia Astrophys. Lab. Contrib., No. 275.
The author reviews the subject of Local Group X-ray astronomy – from the first discovery of a source in the Large Magellanic Cloud 15 years ago, to the most recent analysis of the Einstein Observatory data in which over 200 X-ray sources were detected in Local Group members. Following a detailed presentation of the latest results, the author discusses two examples of the constraints these data impose on stellar and galactic evolution: a study of the distribution of properties for the 30 detected supernova remnants in the Large Magellanic Cloud, and a comparison of the populations of classical X-ray binaries in M31, M33, the Magellanic Clouds, and the Galaxy.

160.147 **The Shane-Wirtanen counts.**
M. J. Geller, V. de Lapparent, M. J. Kurtz.
Astrophys. J., Lett. Ed., Vol. 287, No. 2, p. L55 – L58 (1984). With plates L3 – L5.
The authors show that the 2°5 break in the galaxy correlation function derived from the Shane-Wirtanen counts is indistinguishable from an artifact introduced by residual systematic variations in the effective magnitude limit from plate to plate. In order to avoid introduction of a break, the systematic errors from plate to plate must be $\lesssim 0.05$ mag.

160.148 **A statistical comparison of voids in the galaxy distribution and *n*-body simulations.**
B. S. Ryden, E. L. Turner.
Astrophys. J., Lett. Ed., Vol. 287, No. 2, p. L59 – L63 (1984).
A quantitative statistical comparison is made between the voids in the three-dimensional galaxy distribution and several *n*-body simulations. The statistic used is the distribution of the distances to the nearest neighboring galaxy from each point in a regular grid filling space. It is found, using a Kolmogorov-Smirnov test, that the voids in the galaxy distribution are significantly larger than voids in the *n*-body simulations. The largest voids in the volume-limited galaxy samples have radii equal to about 2 – 3 times the inverse cube root of the mean galaxy density.

**Faint object detection on photographic plates.**
See Abstr. 036.008.

**A bootstrap resampling analysis of galaxy clustering.**
See Abstr. 036.018.

**The nature of galaxies and clusters of galaxies at very large distances.**
See Abstr. 036.098.

**EXOSAT observations.**
See Abstr. 051.076.

**Does the missing mass problem signal the breakdown of Newtonian gravity?**
See Abstr. 066.266.

**Radio lines of heavy elements from the hot gas in supernova remnants and clusters of galaxies.**
See Abstr. 125.001.

**X-ray gas scintillation spectrometer experiment.**
See Abstr. 142.005.

**The Medium Sensitivity Survey: a new sample of X-ray sources with optical identifications and the revised extragalactic log *N*-log *S*.**
See Abstr. 142.026.

**N-body problem of clusters of galaxies.**
See Abstr. 151.011.

**Computer modelling of the gas dynamics in small groups of interacting galaxies.**
See Abstr. 151.098.

**H I observations in the Virgo Cluster area. II. A complete, magnitude-limited sample of spiral galaxies.**
See Abstr. 157.004.

**Galaxy evolution at late epochs.**
See Abstr. 157.005.

**The potentiality of the galaxy autocorrelation function.**
See Abstr. 157.080.

**Star formation in a dwarf elliptical of the Virgo cluster.**
See Abstr. 157.087.

**Angular correlations of galaxies to $B \approx 24$: another probe of cosmology and galaxy evolution.**
See Abstr. 157.091.

**Results from the Durham-AAT redshift survey.**
See Abstr. 157.098.

**H-alpha observations of spiral galaxies in Cancer, A1367, and Coma.**
See Abstr. 157.100.

**A possible relationship between metal abundance and luminosity for disk galaxies.**
See Abstr. 157.101.

**The influence of environment on the H I content of galaxies.**
See Abstr. 157.112.

**Radio emission in isolated and cluster spiral galaxies.**
See Abstr. 157.116.

**Gas dynamics of flow past galaxies: radiative cooling.**
See Abstr. 157.137.

**Infrared observations of dwarf elliptical galaxies in the Virgo cluster.**
See Abstr. 157.197.

**The spatial distribution of galaxies in the southern sky.**
See Abstr. 157.203.

**A redshift survey in the Lynx Gemini region.**
See Abstr. 157.204.

**The geography of nearby space.**
See Abstr. 157.205.

**Comparison of H I sizes of Virgo Cluster and field galaxies.**
See Abstr. 157.210.

**Investigations on the H I–content of field galaxies, Virgo cluster galaxies and UMa–I group galaxies.**
See Abstr. 157.212.

**Hot gaseous coronae around early–type galaxies.**
See Abstr. 157.213.

**Diffuse Lyα emission around the cD galaxy in A1795.**
See Abstr. 157.214.

**Double galaxy redshifts and dynamical analyses.**
See Abstr. 157.227.

**A three–dimensional model of the fluid dynamics of radio–trail sources.**
See Abstr. 158.005.

**Surface photometry of powerful radio galaxies: their relation to Abell cluster cD galaxies.**
See Abstr. 158.028.

**Radio and optical investigation of UGC 6697 in Abell 1367.**
See Abstr. 158.055.

**Multifrequency observations of the complex radio galaxy 1919+479 (4C 47.51).**
See Abstr. 158.058.

**VLA observations of the wide–angle tailed radio source 1313+073.**
See Abstr. 158.200.

**Shock wave from a galactic nucleus into a halo and intergalactic space.**
See Abstr. 158.247.

**The extended radio source near Coma A.**
See Abstr. 158.250.

**High resolution radio and X–ray observations of 3C 28 in A115.**
See Abstr. 158.252.

**Wide–angle radio tail QSOs as members of clusters of galaxies. II. Direct optical observations and spectroscopy of QSO fields.**
See Abstr. 159.003.

**Multiple imaging of quasars by galaxies and clusters.**
See Abstr. 159.007.

**The [O III]–emission associated with the QSO MR2251–178.**
See Abstr. 159.010.

**Quasars as probes of the intergalactic medium.**
See Abstr. 159.035.

**A third quasar close to NGC 3842.**
See Abstr. 159.042.

**A search for quasars in the Virgo cluster region.**
See Abstr. 159.068.

**High resolution spectroscopy in the Lyman alpha forest; possible evidence for absorption line clustering.**
See Abstr. 159.100.

**Properties of the quasars in the Sculptor region II.**
See Abstr. 159.116.

**Properties of groups and clusters of galaxies associated with quasars.**
See Abstr. 159.127.

**Non–linear gravitational clustering.**
See Abstr. 161.001.

**The origin of galaxies and clusters of galaxies in the neutrino universe.**
See Abstr. 161.024.

**Cosmological observations of galaxies: the observational map.**
See Abstr. 161.029.

**Numerical experiments on galaxy clustering in open universes.**
See Abstr. 161.126.

**Topology of the supercluster–void structure.**
See Abstr. 161.167.

**Reheating the intergalactic medium under extremal conditions.**
See Abstr. 161.180.

**The spatial correlation function of galaxies confronted with theoretical scenarios.**
See Abstr. 161.187.

**Adiabatic theories of galaxy formation and pancakes.**
See Abstr. 161.195.

**Pancakes and the formation of galaxies. Neutrinos and other 'inos.**
See Abstr. 161.196.

**The texture of the universe.**
See Abstr. 161.197.

**Nature and significance of the mass distribution in the universe.**
See Abstr. 161.199.

**The statistics of gravitational lenses: the distributions of image angular separations and lens redshifts.**
See Abstr. 161.218.

**Superpancakes and the cluster correlation function.**
See Abstr. 161.236.

**Statistical implications of a neutrino universe.**
See Abstr. 161.285.

**Distant galaxies and the determination of the deceleration parameter.**
See Abstr. 161.310.

**Gravitational clustering for a multicomponent system with a distribution of temperatures.**
See Abstr. 161.321.

**The voids and the cosmic microwave photons.**
See Abstr. 161.342.

**The dependence on distance and redshift of the velocity vectors of the Sun, the Galaxy, and the Local Group with respect to different extragalactic frames of reference.**
See Abstr. 161.363.

# 161 Universe, Cosmology, Background Radiation

161.001 **Non–linear gravitational clustering.**
R. Juszkiewicz, D. H. Sonoda, J. D. Barrow.
Mon. Not. R. Astron. Soc., Vol. 209, No. 1, p. 139 – 144 (1984).
The authors show that non–linear gravitational effects can produce a significant shortening of the characteristic scale of galaxy clustering during the period from last scattering to the present. These results can be compared with recent numerical simulations of galaxy clustering.

161.002 **Isotropic singularities and isotropization in a class of Bianchi type–$VI_h$ cosmologies.**
J. Wainwright, P. J. Anderson.
Gen. Relativ. Gravitation, Vol. 16, No. 7, p. 609 – 624 (1984).
The evolution of a class of exact spatially homogeneous cosmological models of Bianchi type $VI_h$ is discussed. It is known that solutions of type $VI_h$ cannot approach isotropy asymptotically at large times. Nevertheless, the initial anisotropy can decay, leading to a stage of finite duration in which the model is close to isotropy. This quasi–isotropic stage can commence at the initial singularity, in which case the singularity is of the type known as "isotropic" or "Friedmann–like".

161.003 **Power law singularities in orthogonal spatially homogeneous cosmologies.**
J. Wainwright.
Gen. Relativ. Gravitation, Vol. 16, No. 7, p. 657 – 674 (1984).
The author investigates the structure of the so–called power asymptote singularities in orthogonal spatially homogenous solutions of the Einstein field equations with perfect fluid source.

161.004 **Cosmic strings and galaxies.**
C. Hogan.
Nature, Vol. 310, No. 5976, p. 365 – 366 (1984).

161.005 **Microwave anisotropy due to cosmic strings.**
N. Kaiser, A. Stebbins.
Nature, Vol. 310, No. 5976, p. 391 – 393 (1984).
There has been much interest in the cosmological consequences of line–like topological defects ('cosmic strings') which may be produced in a phase transition in the early Universe. Such strings would produce density fluctuations on a broad range of scales and may have been responsible for the formation of galaxies. The authors show here that cosmic strings would generate anisotropy of the microwave background temperature with a very distinctive feature; the temperature would have steplike discontinuities on curves on the sky. Present anisotropy measurements limit the line density parameter, $G\mu$, to be $<10^{-5}$.

161.006 **Die ferne Zukunft des Universums. Kann Leben im Universum unbegrenzt existieren?**
J. N. Islam.
Umschau, 84. Jahrg., Nr. 14/15, p. 441 – 443 (1984).

161.007 **Large–scale structure in the Universe.**
M. Umemura, S. Ikeuchi.
Astron. Her., Vol. 77, No. 7, p. 171 – 176 (1984). In Japanese.

161.008 **The redshift structure of the big bang in inhomogeneous cosmological models. I. Spherical dust solutions.**
C. Hellaby, K. Lake.
Astrophys. J., Vol. 282, No. 1, p. 1 – 10 (1984).
The redshift from the big bang in the standard model is always infinite, but in inhomogeneous cosmological models infinite blueshifts are also possible. To avoid such divergent energy fluxes, the authors require that all realistic cosmological models must not display infinite blueshifts. They apply this requirement to the Tolman model (spherically symmetric dust), using the geometrical optics approximation, and assuming that the geodesic tangent vectors may be expanded in power series. It is concluded that the bang time must be simultaneous. The stronger requirement, that only infinite redshifts from the big bang may occur, does not lead to a stronger condition on the metric. Consequences of simultaneity are that no decaying mode fluctuations are possible, and that the only acceptable model which is homogeneous at late times is the Robertson–Walker model.

161.009 **Firm bounds on the neutrino mass from the distribution of dark matter in galaxies.**
J. Madsen, R. I. Epstein.
Astrophys. J., Vol. 282, No. 1, p. 11 – 18 (1984).
The authors explore the possibility that the dark or missing matter in galaxies is due to massive neutrinos. Two approaches are described which give lower mass limits for a single neutrino type in terms of the observable properties of galaxies. In the first, the distribution of dark matter is compared to that of the most compact mass distribution which is possible for relic neutrinos. In the second, the deduced pressure for the dark matter in galaxies is compared with the equation of state for relic neutrinos. From the well studied galaxies M87 and M31 it is found that the neutrino mass must exceed 8 eV. The preliminary reports of dark matter in dwarf galaxies, if confirmed, would imply that the neutrino mass exceeds 125 eV.

161.010 **Exact solutions for inhomogeneous generalisations of some vacuum Bianchi models.**
J. L. Hanquin, J. Demaret.
Classical Quantum Gravity, Vol. 1, No. 3, p. 291 – 300 (1984). Abstr. in Phys. Abstr., Vol. 87, No. 1237, Entry 68059 (1984).

161.011 **Riemannian approach and cosmological singularity.**
E. Calzetta, M. Castagnino.
Phys. Rev. D, Vol. 29, No. 8, p. 1609 – 1617 (1984). Abstr. in Phys. Abstr., Vol. 87, No. 1237, Entry 68097 (1984).

161.012 **Gravitational theory in atomic scale units in Dirac cosmology.**
W. Davidson.
Proc. R. Soc. Edinb., Sect. A, Vol. 96, Part 1 – 2, p. 143 – 162 (1984). Abstr. in Phys. Abstr., Vol. 87, No. 1237, Entry 73106 (1984).

161.013 **Quark–hadron and chiral transitions and their relation to the early Universe.**
D. N. Schramm, K. A. Olive.
Nucl. Phys. A, Vol. A418, p. 289 – 300 (1984). Abstr. in Phys. Abstr., Vol. 87, No. 1237, Entry 73107 (1984). – See Abstr. 012.007.

161.014 **The smallness of the cosmological constant and the principle of naturalness.**
J.-J. Huang, M.-J. Wang.
Nuovo Cimento B, Vol. 80B, Ser. 11, No. 1, p. 13 – 20 (1984). Abstr. in Phys. Abstr., Vol. 87, No. 1237, Entry 73108 (1984).

161.015 **What made the world?**
E. P. Tryon.
New Sci., Vol. 101, No. 1400, p. 14 – 16 (1984). Abstr. in Phys. Abstr., Vol. 87, No. 1237, Entry 73109 (1984).

161.016 **Gravitational creation of particles by a vacuum bubble.**
W. A. Hiscock.
Phys. Lett. B, Vol. 139B, No. 1 – 2, p. 31 – 34 (1984). Abstr. in Phys. Abstr., Vol. 87, No. 1237, Entry 73111 (1984).

161.017 **Strongly–coupled nature of the primordial SUSY SU(5) plasma.**
H. Goldberg.
Phys. Lett. B, Vol. 139B, No. 1 – 2, p. 45 – 50 (1984). Abstr. in Phys. Abstr., Vol. 87, No. 1237, Entry 73112 (1984).

161.018 **Stretching cosmic strings.**
N. Turok, P. Bhattacharjee.
Phys. Rev. D, Vol. 29, No. 8, p. 1557 – 1562 (1984). Abstr. in Phys. Abstr., Vol. 87, No. 1237, Entry 73113 (1984).

161.019 **Causal cosmological perturbations and implications for the Sachs–Wolfe effect.**
J. Traschen.
Phys. Rev. D, Vol. 29, No. 8, p. 1563 – 1574 (1984). Abstr. in Phys. Abstr., Vol. 87, No. 1237, Entry 73114 (1984).

161.020 **Equation of motion for bubble boundaries.**
K. Lake.
Phys. Rev. D, Vol. 29, No. 8, p. 1861 – 1862 (1984). Abstr. in Phys. Abstr., Vol. 87, No. 1237, Entry 73115 (1984).

161.021 **Spontaneously broken global symmetries and cosmology.**
Q. Shafi, A. Vilenkin.
Phys. Rev. D, Vol. 29, No. 8, p. 1870 – 1871 (1984). Abstr. in Phys. Abstr., Vol. 87, No. 1237, Entry 73116 (1984).

161.022 **Foundations and working pictures in microphysical cosmology.**
F. Wilczek.
Phys. Rep., Vol. 104, No. 2 – 4, p. 143 – 157 (1984). Abstr. in Phys. Abstr., Vol. 87, No. 1237, Entry 73117 (1984). – See Abstr. 012.009.

161.023 **Modification of the Reissner–Nordström metric in an Einstein–de Sitter cosmology.**
L. Nunez, H. Rago, L. Aulestia.
Rev. Mex. Fis., Vol. 30, No. 1, p. 83 – 89 (1983). In Spanish. Abstr. in Phys. Abstr., Vol. 87, No. 1237, Entry 73118 (1984).

161.024 **The origin of galaxies and clusters of galaxies in the neutrino universe.**
A. G. Doroshkevich.
Astron. Zh., Tom 61, Vyp. 4, p. 645 – 653 (1984). In Russian. English translation in Sov. Astron., Vol. 28, No. 4.
The analysis of 27 clusters of galaxies shows that in addition to the hidden mass (neutrinos, approximately 90% of mass) about 5 – 7% of mass of the cluster is maintained by the galaxies and 3 – 5% by the hot gas with very high entropy. The comparison of the density of the clusters and parameters of the neutrino–gaseous pancakes permits to estimate the redshift corresponding to the formation of first pancakes. A model of the later pancake evolution is proposed.

161.025 **The role of massive neutrinos in the evolution of gravitational cosmological perturbations.**
I. K. Rozgacheva.
Astron. Zh., Tom 61, Vyp. 4, p. 654 – 664 (1984). In Russian. English translation in Sov. Astron., Vol. 28, No. 4.
Gravitational perturbations of a standard model universe filled by weakly interacting neutrinos and radiation–dominated plasma are considered. Accounting for density perturbations of the neutrino gas alters the law of evolution of plasma adiabatic perturbations.

161.026 **Numerical simulation of protogalactic eddies.**
A. Yu. Ushakov, A. D. Chernin.
Sov. Astron., Vol. 28, No. 1, p. 11 – 14 (1984). English translation of 37.161.010.

161.027 **Kinetics of $H_2$ formation in the primordial gas.**
Yu. I. Izotov, I. G. Kolesnik.
Sov. Astron., Vol. 28, No. 1, p. 15 – 21 (1984). English translation of 37.161.011.

161.028 **Mystery of large numbers and harmony of the Universe.**
T. Matsuda.
Astron. Her., Vol. 77, No. 8, p. 203 – 206 (1984). In Japanese.

161.029 **Cosmological observations of galaxies: the observational map.**
G. F. R. Ellis, J. J. Perry, A. W. Sievers.
Astron. J., Vol. 89, No. 8, p. 1124 – 1154 (1984).
The theoretical foundations of established methods using astronomical observations of distant astronomical sources to determine the structure of the universe are reviewed. The authors give a general description of the relation between source and image properties, and apply it specifically to optical observations of galaxies. The concept of the observational map is developed and illustrated for well–defined families of galaxies for which the authors map source intrinsic magnitude and size onto the image apparent magnitude and angle. The properties of the resulting map, which is highly nonlinear, depend critically on the limiting surface brightness and point–spread characteristics of the particular observation. Using these maps as a basis, the authors discuss the nature and effects of several important selection effects.

161.030 **Is the Universe near the state of maximal expansion?**
A. K. Raychaudhuri, G. Mukherjee.
Mon. Not. R. Astron. Soc., Vol. 209, No. 2, p. 353 – 358 (1984).
A quadratic relation between the redshift and distance as has apparently emerged from the analysis of Nicoll & Segal, may be consistent with relativistic cosmology if the present deceleration parameter has a high value and the Hubble constant a low value. Consequences of these values are discussed.

161.031 **The size of clusters in a neutrino–dominated universe.**
S. D. M. White, M. Davis, C. S. Frenk.
Mon. Not. R. Astron. Soc., Vol. 209, No. 2, p. 27P – 31P (1984).
Quite soon after the first collapse of structure almost half the matter in a neutrino–dominated universe is expected to reside in clusters. The masses and binding energies of these neutrino clusters are too large for them to be identified with observed galaxy clusters. Even if such objects were able to suppress all galaxy formation, their X–ray emission would, however, make them highly visible if more than 2.5 per cent of their mass was in ordinary matter. A neutrino–dominated universe appears to conflict with observation irrespective of the details of the processes which govern galaxy formation.

161.032 **Hunting for the missing mass.**
J. Maddox.
Nature, Vol. 310, No. 5979, p. 627 (1984).

161.033 **Can a neutrino–dominated Universe be rejected?**
P. Hut, S. D. M. White.
Nature, Vol. 310, No. 5979, p. 637 – 640 (1984).
The conservative hypothesis that the Universe is dominated by nonbaryonic matter made up of particles that have already been observed in the laboratory is not tenable within a standard big–bang cosmology if all neutrino species are stable. It can be salvaged only if one neutrino species has a mass in the range 0.1 – 250 MeV and decays on a short mass–dependent time scale, while a second species has a mass in the range 10 – 50 eV and is stable.

161.034 **Stability of nonrelativistic configurations in the Jordan and Brans–Dicke theory.**
S. M. Kozyrev.
Astrofizika, Tom 19, Vyp. 3, p. 592 – 594 (1983). In Russian. English translation in Astrophysics, Vol. 19, No. 3.
The stability limits for nonrelativistic polytropes in the theory of Jordan and Brans–Dicke are obtained.

161.035 **Some general results in self–creation cosmologies.**
T. Singh, T. Singh.
Astrophys. Space Sci., Vol. 102, No. 1, p. 67 – 70 (1984).
Raychaudhuri–type equations are written for a cosmological model filled with perfect fluid and obeying the equations of a self–creation theory recently proposed by G. A. Barber. In addition some general results on spatially homogeneous cosmological models are obtained. The Hawking–Penrose energy condition suggests that the singularity can be avoided in this theory.

161.036 **The neutrino in the Friedmann universe: non-zero rest-mass effects.**
G. S. Bisnovatyi-Kogan (*G. S. Bisnovatyj-Kogan*), Z. F. Seidov.
Astrophys. Space Sci., Vol. 102, No. 1, p. 131 – 154 (1984).

The authors have considered the problem of the propagation of non-zero rest-mass neutrinos in the Friedmann dust universes of three types: open, flat, and closed, as well as in the radiation-dominated epoch of the hot universe. The total Lagrangian path of the particle has been calculated, and this is shown to be finite for all three universes – contrary to the total path of the photon, which is infinite in the open and flat universes. The particle horizon is found as a function of the relativistic parameter at the emission moment and at the moment of observation. The extreme relativistic and nonrelativistic particle motion and its difference from the photon motion have been investigated.

161.037 **Some results in the general scalar tensor theory.**
T. Singh, T. Singh.
Astrophys. Space Sci., Vol. 102, No. 2, p. 223 – 228 (1984).

Raychaudhuri-type equations are written for a cosmological model filled with a perfect fluid and obeying the equations of general scalar-tensor theory proposed by Nordtvedt. It is found that a non-rotating cosmological model may, in general, have a minimum volume. The Hawking-Penrose energy condition suggests that the singularity can be avoided in this theory.

161.038 **Exact Bianchi type-I cosmological model with matter in Brans-Dicke theory.**
S. Ram, D. K. Singh.
Astrophys. Space Sci., Vol. 102, No. 2, p. 229 – 234 (1984).

In this paper an exact solution of the Brans-Dicke field equations in the presence of stiff matter is obtained for the Bianchi type-I cosmological space-time. The new solution represents an anisotropic homogeneous cosmological model which tends to anisotropic expansion. The behaviour of the solution near the singularities and late stages of the expansion is discussed.

161.039 **Exact Bianchi type $VI_0$ cosmological solutions with matter in Brans-Dicke theory.**
S. Ram, D. K. Singh.
Astrophys. Space Sci., Vol. 103, No. 1, p. 21 – 26 (1984).

The authors present an exact solution of the Brans-Dicke equations for cosmological models of Bianchi type $VI_0$ with stiff matter. The solution represents anisotropic universe which has its analogy in Einstein's theory. The corresponding result for a plane symmetry Bianchi type I model is obtained as a special case.

161.040 **The evolution of nonlinear hydrodynamical density fluctuations of the photon-plasma gas during the recombination era of the Universe.**
E. Nowotny.
Astrophys. Space Sci., Vol. 103, No. 2, p. 249 – 270 (1984).

On the basis of the hydrodynamical equations of a two-component gas (photons and hydrogen with coupling via Thomson scattering) in the recombination era of the Universe (standard model), the evolution of the density perturbations up to second order is calculated. This calculation indicates that the linear perturbation analysis is very restricted, especially at wavelengths near the lower limit of the Jeans length.

161.041 **Size of ambiplasma domains in the Universe.**
G. L. Kalra, B. Singh.
Astrophys. Space Sci., Vol. 103, No. 2, p. 321 – 329 (1984).

Assuming that the initial state of the Universe can be simulated by a thin large-scale homogeneous ambiplasma described by fluid type equations for charged particles (both matter and antimatter particles), stability calculations have been carried out of the various normal modes which might be supported by such a system. The present simplified analysis does not lead to the formation of matter-antimatter symmetric domains smaller than the size of the proto-metagalaxy (Alfvén, 1981) from the initial state of the Universe.

161.042 **Large-scale inhomogeneities in the Universe – some recent developments.**
L. Fang.
Proceedings of the Third Marcel Grossmann Meeting on General Relativity, Part A, p. 383 – 404 (1983). – See Abstr. 012.002.

This paper discusses the large-scale inhomogeneities in the distribution of galaxies and quasars on scales up to 100 Mpc. The question whether there are periodic components in the three-dimensional network of superclusters is studied. Observational constraints on the distribution of dark matter in the Universe are considered.

161.043 **Cluster and void in the expanding universe.**
H. Sato.
Proceedings of the Third Marcel Grossmann Meeting on General Relativity, Part A, p. 405 – 414 (1983). – See Abstr. 012.002.

The origin of cluster-void structure in the universe is discussed in terms of the neutrino dominated model. The cellular structure in the distribution of galaxies might be explained by the enlargement of the negatively perturbed region rather than the collapse of "pancakes". The resolution of the horizon problem by the topologically closed universe is also suggested.

161.044 **Finite temperature quantum processes in the early universe.**
B. L. Hu.
Proceedings of the Third Marcel Grossmann Meeting on General Relativity, Part A, p. 415 – 430 (1983). – See Abstr. 012.002.

Finite temperature quantum field theory is presented here as an approximate theory for the study of quantum statistical processes in the early universe. The author starts with a discussion of the conditions for the establishment of thermal equilibrium in dynamic spacetimes and presents the adiabatic approximation for treating quantum systems in quasi-equilibrium. The method is then used to compute the finite temperature quantum energy densities for scalar fields arbitrarily coupled to the Robertson-Walker universe. The results are applied to the analysis of the problem of constructing cosmological solutions with back reactions from finite temperature quantum sources, and to the problem of symmetry restoration at finite temperature.

161.045 **Some recent developments in the massive neutrino cosmology.**
T. Lu, L. Fang.
Proceedings of the Third Marcel Grossmann Meeting on General Relativity, Part A, p. 461 – 480 (1983). – See Abstr. 012.002.

In this paper the recent developments and achievements in the field of massive neutrino cosmology are reviewed. Especially, the clustering processes of massive neutrinos and their role in the formation mechanism of galaxies and clusters of galaxies are discussed in some detail.

161.046 **Black hole and wormhole creation by phase transition and monopole problem.**
K. Sato, M. Izawa.
Proceedings of the Third Marcel Grossmann Meeting on General Relativity, Part A, p. 481 – 494 (1983). – See Abstr. 012.002.

If GUT-phase transition is of strongly first order, the false vacuum regions surrounded by bubbles become black holes and wormholes eventually. By the evaporation of these holes, baryons and the entropy of the universe are produced and 't Hooft-Polyakov type monopoles remain after the evaporation. The authors calculate the ratios of monopole/entropy and baryon/entropy in this scenario and obtain the result that these ratios can be consistent with observations, but very fine-tuning of the nucleation rate of the true vacuum is necessary.

161.047 **Self-consistent generalized Hawking thermal states for FRW universes.**
G. Horwitz.
Proceedings of the Third Marcel Grossmann Meeting on General Relativity, Part A, p. 495 – 509 (1983). – See Abstr. 012.002.

A novel microcanonical entropy definition in cosmology is exploited to determine thermodynamic equilibrium configurations for a Friedmann universe ($\Lambda \neq 0$) filled with conformal, massless, scalar bosons. The Gibbons–Hawking de Sitter entropy is a second solution, which is, however, found to be unstable. The entropy is calculated in a fully self–consistent treatment of quantum bosons and the quasi–classical cosmic scale function of the RW metric. The dynamics of the cosmological solution is determined by analytic continuation to real time.

161.048 **Cosmology, inflation, and supersymmetry.**
A. Albrecht, S. Dimopoulos, W. Fischler, E. W. Kolb, S. Raby, P. J. Steinhardt.
Proceedings of the Third Marcel Grossmann Meeting on General Relativity, Part A, p. 511 – 525 (1983). – See Abstr. 012.002.
Cosmological consequences of supersymmetric grand unified models based on the Witten–O'Raifeartaigh potential are discussed. The authors study the development of the phase transition in the spontaneous breaking of supersymmetry. They find that in realistic models where light fields feel supersymmetry breaking only through coupling to massive fields, e.g., the Geometric Hierarchy model, the transition does not produce sufficient entropy. Thus, the standard cosmological flatness, monopole, and horizon problems remain. In addition, it is found that the transition is never completed, in the sense that the universe remains dominated by coherent Higgs field energy.

161.049 **Early universe and GUT.**
M. Yoshimura.
Proceedings of the Third Marcel Grossmann Meeting on General Relativity, Part A, p. 527 – 544 (1983). – See Abstr. 012.002.
Recent developments on applications of grand unified theories (GUT) to the early universe are reviewed. The classical puzzles of the standard cosmology (baryon asymmetry, horizon–homogeneity, flatness, monopole overproduction) may be solved by GUT cosmology.

161.050 **On the structure of semidegenerate massive neutrino halos.**
R. Ruffini, L. Stella.
Proceedings of the Third Marcel Grossmann Meeting on General Relativity, Part A, p. 545 – 559 (1983). – See Abstr. 012.002.
This study investigates the properties of self–gravitating massive neutrino halos in structures ranging from galaxies to clusters of galaxies and attempts to obtain testable predictions. In deriving the basic features of the model, the authors use concepts and approximations typical of the theory of dynamical models for star clusters.

161.051 **On Cosmic Censorship: do compact Cauchy horizons imply symmetry?**
J. Isenberg, V. Moncrief.
Proceedings of the Third Marcel Grossmann Meeting on General Relativity, Part A, p. 561 – 573 (1983). – See Abstr. 012.002.
The authors discuss some new results which provide strong support for one of the statements of Cosmic Censorship. The statement, which Penrose has labeled the "Strong Cosmic Censorship Principle", asserts that a physically reasonable spacetime should be globally hyperbolic.

161.052 **Cosmological consequences of the Langmuir condensation and Zakharov collapse.**
A. V. Byalko.
Proceedings of the Third Marcel Grossmann Meeting on General Relativity, Part A, p. 823 – 826 (1983). – See Abstr. 012.002.
The aim of this paper is to outline the principal possibility of small density perturbations in the early universe resulting from phenomena of the condensation of plasma oscillations and their collapse.

161.053 **Bifurcation in the early cosmos.**
M. Novello, L. M. C. S. Rodriguez.
Proceedings of the Third Marcel Grossmann Meeting on General Relativity, Part A, p. 883 – 887 (1983). – See Abstr. 012.002.
It is shown that, due to the non–linear character of Einstein's equations of motion for the metric of space–time, bifurcation does occur in the early universe.

161.054 **Is our universe an open system?**
D. N. Page.
Proceedings of the Third Marcel Grossmann Meeting on General Relativity, Part B, p. 1153 – 1155 (1983). – See Abstr. 012.002.

161.055 **Density fluctuation models and low–order harmonics in the background radiation anisotropy.**
R. Fabbri, I. Guidi, V. Natale.
Proceedings of the Third Marcel Grossmann Meeting on General Relativity, Part B, p. 1301 – 1304 (1983). – See Abstr. 012.002.
Recent experiments on the large–scale angular distribution of the cosmic background radiation suggest that anisotropies of cosmological origin may have been detected. The authors consider linear perturbations of the cosmic matter density with size larger than the particle horizon at the last scattering of the radiation and discuss their implications for the observed microwave background.

161.056 **The cosmic density wave and its observable vestige.**
Y. Liu.
Proceedings of the Third Marcel Grossmann Meeting on General Relativity, Part B, p. 1305 – 1308 (1983). – See Abstr. 012.002.
The existence of the periodic components in the distribution of emission–line redshifts of quasars means that perhaps there is a certain periodic structure in the large–scale distribution of quasars in space. As a possible explanation the author suggests a tentative theory of the cosmic density wave.

161.057 **Systematic differences in the determination of $q_0$ and its implication.**
F. Hu, L. Fang.
Proceedings of the Third Marcel Grossmann Meeting on General Relativity, Part B, p. 1315 – 1318 (1983). – See Abstr. 012.002.
The Hubble constant $H_0$ and the deceleration parameter $q_0$ are two most important parameters in geometrical cosmology. The former has now been narrowed down to the range from 50 to 100 km $s^{-1}Mpc^{-1}$, yet the various results about the deceleration parameter $q_0$ do not show any tendency to converge.

161.058 **The cosmic density wave and the Lemaitre model.**
L. Yang, S. Cao, C. Xu, P. Yang.
Proceedings of the Third Marcel Grossmann Meeting on General Relativity, Part B, p. 1337 – 1340 (1983). – See Abstr. 012.002.
This paper presents the theory of the cosmic density wave in the expanding universe of the Lemaitre model.

161.059 **The first–order phase transition and the origin of clusters of galaxies.**
H. Kodama, M. Sasaki, K. Sato.
Proceedings of the Third Marcel Grossmann Meeting on General Relativity, Part B, p. 1341 – 1342 (1983). – See Abstr. 012.002.

161.060 **Gravitational effects on vacuum decay at finite temperature.**
R. Su, P. Bi.
Proceedings of the Third Marcel Grossmann Meeting on General Relativity, Part B, p. 1343 – 1347 (1983). – See Abstr. 012.002.

161.061 **Cosmological consequences of conformal noninvariance.**
L. H. Ford.
Proceedings of the Third Marcel Grossmann Meeting on General Relativity, Part B, p. 1349 – 1351 (1983). – See Abstr. 012.002.
When quantum fields in the early universe are coupled to the cosmological background in such a way as to break conformal invariance, the effects of the gravitational coupling can be especially large. Several physical effects which can be traced to the loss of conformal symmetry are discussed.

161.062 **On the role of fermion condensates for phase transitions in the early Universe.**
C.–G. Källman, C. Montonen.
Proceedings of the Third Marcel Grossmann Meeting on General Relativity, Part B, p. 1353 – 1357 (1983). – See Abstr. 012.002.
The Grand Unified phase transition in the early Universe is discussed. It is argued that the dynamics of fermion condensates

is complex and that only a qualitative picture of their role is possible at present.

161.063 **Particle creation and the cosmological constant.**
L. Parker.
Proceedings of the Third Marcel Grossmann Meeting on General Relativity, Part B, p. 1359 – 1362 (1983). – See Abstr. 012.002.

161.064 **The metric of space–time with two bubbles' collision.**
Z. Wu.
Proceedings of the Third Marcel Grossmann Meeting on General Relativity, Part B, p. 1363 – 1367 (1983). – See Abstr. 012.002.
It is believed that a first order phase transition occurred in the very early universe. Coleman described this process as the formation and expansion of bubbles. This paper investigates the physical consequences of collisions of bubbles.

161.065 **A singularity avoided oscillating bouncing cosmological model.**
D. Xu, L. Liu.
Proceedings of the Third Marcel Grossmann Meeting on General Relativity, Part B, p. 1369 – 1375 (1983). – See Abstr. 012.002.
The one–loop contribution of the quantum effect of matter field to the effective action is evaluated and a semiclassical Einstein equation of the Friedmann universe is derived. With reasonable boundary condition one obtains a time symmetric, oscillating, bouncing solution with no singularity.

161.066 **Quantum effects in the Gowdy $T^3$ cosmological model.**
B. K. Berger.
Proceedings of the Third Marcel Grossmann Meeting on General Relativity, Part B, p. 1377 – 1380 (1983). – See Abstr. 012.002.

161.067 **Horizon in Dirac cosmology and neutrino mass.**
R. Su, S. Zhang.
Proceedings of the Third Marcel Grossmann Meeting on General Relativity, Part B, p. 1381 – 1385 (1983). – See Abstr. 012.002.

161.068 **The cosmic neutrino clustering and the isotropy of background radiation.**
T. Lu, P. Zhu.
Proceedings of the Third Marcel Grossmann Meeting on General Relativity, Part B, p. 1387 – 1390 (1983). – See Abstr. 012.002.

161.069 **The expansion of void inhomogeneous cosmological model.**
S. Zhu.
Proceedings of the Third Marcel Grossmann Meeting on General Relativity, Part B, p. 1391 – 1398 (1983). – See Abstr. 012.002.
The low–density region in a Tolman inhomogeneous cosmic model is called Tolman void. In this paper the expansions of a mass invariable void and a mass variable void are discussed. It is concluded that the expansion parameter of a Tolman void is smaller than the Hubble parameter of the cosmological background.

161.070 **Some cosmological aspects of quantum gravity.**
L. Parker.
Quantum theory of gravity, p. 89 – 102 (1984). – See Abstr. 003.005.
This paper discusses briefly the theory of particle production by the expansion of the universe. A description of the use of Bogolubov transformations to analyze particle production by an expanding universe is given and the creation of gravitons on a Robertson–Walker background metric is studied. A brief discussion of the reaction back of particle creation on the expansion of the universe is included.

161.071 **Vacuum energy in a squashed Einstein universe.**
J. S. Dowker.
Quantum theory of gravity, p. 103 – 124 (1984). – See Abstr. 003.005.

161.072 **Primordial cosmological inflation versus local supersymmetry breaking in SUSY GUTs coupled to $N = 1$ supergravity.**
B. Gato, J. Leon, M. Quiros, M. Ramon–Medrano.
Z. Phys., C, Vol. 22, No. 4, p. 345 – 349 (1984). Abstr. in Phys. Abstr., Vol. 87, No. 1238, Entry 73324 (1984).

161.073 **More fluctuations in the inflationary universe.**
I. G. Moss.
Nucl. Phys. B, Part. Phys., Vol. B238, No. 2, p. 436 – 452 (1984). Abstr. in Phys. Abstr., Vol. 87, No. 1238, Entry 73689 (1984).

161.074 **Particle physics and cosmology: the inner space/outer space connection.**
M. S. Turner.
AIP Conf. Proc., No. 112, p. 51 – 89 (1984). Abstr. in Phys. Abstr., Vol. 87, No. 1238, Entry 77986 (1984). – See Abstr. 012.013.

161.075 **Large scale structure in the Universe and its connection to particle physics.**
M. Davis.
AIP Conf. Proc., No. 112, p. 90 – 92 (1984). Abstr. in Phys. Abstr., Vol. 87, No. 1238, Entry 77987 (1984). – See Abstr. 012.013.

161.076 **Supersymmetric relics from the big bang.**
J. Ellis, J. S. Hagelin, D. V. Nanopoulos, K. Olive, M. Srednicki.
Nucl. Phys. B, Part. Phys., Vol. B238, No. 2, p. 453 – 476 (1984). Abstr. in Phys. Abstr., Vol. 87, No. 1238, Entry 77988 (1984).

161.077 **Some general results in a new scalar–tensor theory.**
T. Singh, T. Singh.
Indian J. Phys., Part B, Vol. 58B, No. 1, p. 41 – 47 (1984). Abstr. in Phys. Abstr., Vol. 87, No. 1238, Entry 77989 (1984).

161.078 **Quantum creation of the inflationary Universe.**
A. D. Linde.
Lett. Nuovo Cimento, Vol. 39, Ser. 2, No. 17, p. 401 – 405 (1984). Abstr. in Phys. Abstr., Vol. 87, No. 1238, Entry 77990 (1984).

161.079 **Can entropy and "order" increase together?**
P. T. Landsberg.
Phys. Lett. A, Vol. 102A, No. 4, p. 171 – 173 (1984). Abstr. in Phys. Abstr., Vol. 87, No. 1238, Entry 77991 (1984).

161.080 **A note on compact Cauchy horizons.**
A. Borde.
Phys. Lett. A, Vol. 102A, No. 5 – 6, p. 224 – 226 (1984). Abstr. in Phys. Abstr., Vol. 87, No. 1238, Entry 77992 (1984).

161.081 **Quantum mechanics of the Robertson–Walker geometry.**
T. Christodoulakis, J. Zanelli.
Phys. Lett. A, Vol. 102A, No. 5 – 6, p. 227 – 230 (1984). Abstr. in Phys. Abstr., Vol. 87, No. 1238, Entry 77993 (1984).

161.082 **The dynamics of dimensional reduction.**
I. G. Moss.
Phys. Lett. B, Vol. 140B, No. 1 – 2, p. 29 – 32 (1984). Abstr. in Phys. Abstr., Vol. 87, No. 1238, Entry 77994 (1984).

161.083 **Supersymmetry and the entropy crisis.**
G. D. Coughlan, R. Holman, P. Ramond, G. G. Ross.
Phys. Lett. B, Vol. 140B, No. 1 – 2, p. 44 – 48 (1984). Abstr. in Phys. Abstr., Vol. 87, No. 1238, Entry 77995 (1984).

161.084 **Exact Brans–Dicke Bianchi type–I solutions with a cosmological constant.**
D. Lorenz–Petzold.
Phys. Rev. D, Vol. 29, No. 10, p. 2399 – 2401 (1984). Abstr. in Phys. Abstr., Vol. 87, No. 1238, Entry 77996 (1984).

161.085 **Closed, homogeneous cosmology with massive magnetic field.**
J. L. C. Costa, I. Wolk, A. F. da F. Teixeira.
Phys. Rev. D, Vol. 29, No. 10, p. 2402 – 2403 (1984). Abstr. in Phys. Abstr., Vol. 87, No. 1238, Entry 77997 (1984).

161.086 **Neutrino cosmology.**
J. Bernstein.
Report CERN 84–06, CERN, Geneva, Switzerland, 5+73 pp. (1984). Abstr. in Phys. Abstr., Vol. 87, No. 1238, Entry 77999 (1984).

161.087 **The quantum state of the universe.**
S. W. Hawking.
Nucl. Phys. B, Part. Phys., Vol. B239, No. 1, p. 257 – 276 (1984). Abstr. in Phys. Abstr., Vol. 87, No. 1239, Entry 78202 (1984).

161.088 **Inflation in SU(5) GUT models coupled to gravity.**
A. Albrecht, L. G. Jensen, P. J. Steinhardt.
Nucl. Phys. B, Part. Phys., Vol. B239, No. 1, p. 290 – 300 (1984). Abstr. in Phys. Abstr., Vol. 87, No. 1239, Entry 78212 (1984).

161.089 **Supersymmetric cosmology and running masses and coupling constants.**
D. Bailin, A. Love.
Nucl. Phys. B, Part. Phys., Vol. B239, No. 1, p. 277 – 289 (1984). Abstr. in Phys. Abstr., Vol. 87, No. 1239, Entry 78502 (1984).

161.090 **Thermal production of superheavy magnetic monopoles in the new inflationary–Universe scenario.**
W. Collins, M. S. Turner.
Phys. Rev. D, Vol. 29, No. 10, p. 2158 – 2161 (1984). Abstr. in Phys. Abstr., Vol. 87, No. 1239, Entry 78512 (1984).

161.091 **The general field equations in Friedmann spacetime.**
D. Xu.
Acta Sci. Nat. Univ. Pekin., No. 4, p. 19 – 23 (1983). In Chinese. Abstr. in Phys. Abstr., Vol. 87, No. 1239, Entry 82569 (1984).

161.092 **The vanishing likelihood of a space–time singularity in quantum conformal cosmology.**
J. V. Narlikar.
Found. Phys., Vol. 14, No. 5, p. 443 – 456 (1984). Abstr. in Phys. Abstr., Vol. 87, No. 1239, Entry 82570 (1984).

161.093 **New evidence for the principle of self–similarity.**
R. L. Oldershaw.
Int. J. Gen. Syst., Vol. 9, No. 1, p. 37 – 42 (1982). Abstr. in Phys. Abstr., Vol. 87, No. 1239, Entry 82574 (1984).

161.094 **Friedmann cosmology and hadron properties.**
A. Italiano, E. Recami.
Lett. Nuovo Cimento, Vol. 40, Ser. 2, No. 5, p. 140 – 144 (1984). Abstr. in Phys. Abstr., Vol. 87, No. 1239, Entry 82575 (1984).

161.095 **On three equivalent definitions for the spin 1 massive particles in a Robertson–Walker universe.**
M. Castagnino, L. Chimento, C. Laciana.
Prog. Theor. Phys., Vol. 71, No. 2, p. 309 – 319 (1984). Abstr. in Phys. Abstr., Vol. 87, No. 1239, Entry 82576 (1984).

161.096 **Prescription for successful new inflation.**
P. J. Steinhardt, M. S. Turner.
Phys. Rev. D, Vol. 29, No. 10, p. 2162 – 2171 (1984). Abstr. in Phys. Abstr., Vol. 87, No. 1239, Entry 82579 (1984).

161.097 **Cosmological perturbations in inflationary–universe models.**
R. Brandenberger, R. Kakn.
Phys. Rev. D, Vol. 29, No. 10, p. 2172 – 2190 (1984). Abstr. in Phys. Abstr., Vol. 87, No. 1239, Entry 82580 (1984).

161.098 **On Planckian seconds of the universe.**
H.–J. Treder.
Ann. Phys. (Leipzig), Vol. 40, No. 6, p. 375 – 377 (1983). In German. Abstr. in Phys. Abstr., Vol. 87, No. 1240, Entry 86093 (1984).

161.099 **The Dirac equation in Friedmann spacetime.**
D. Xu.
Acta Sci. Nat. Univ. Pekin., No. 5, p. 17 – 24 (1983). In Chinese. Abstr. in Phys. Abstr., Vol. 87, No. 1240, Entry 86095 (1984).

161.100 **The thermodynamical history of the Universe.**
G. Marx.
Fortschr. Phys., Vol. 32, No. 4, p. 185 – 206 (1984). Abstr. in Phys. Abstr., Vol. 87, No. 1240, Entry 86096 (1984).

161.101 **Exact Brans–Dicke–Bianchi type–V solutions.**
D. Lorenz–Petzold.
Math. Proc. Camb. Philos. Soc., Vol. 96, Part 1, p. 183 – 189 (1984). Abstr. in Phys. Abstr., Vol. 87, No. 1240, Entry 86098 (1984).

161.102 **May cosmological heavy photons form a Bose condensate?**
D. Fargion.
Phys. Lett. B, Vol. 141B, No. 1 – 2, p. 53 – 56 (1984). Abstr. in Phys. Abstr., Vol. 87, No. 1240, Entry 86099 (1984).

161.103 **Nonlinear photons in the universe. II. The anisotropic case.**
M. Novello, J. M. Salim, E. Ruckert.
Lett. Nuovo Cimento, Vol. 40, Ser. 2, No. 8, p. 232 – 234 (1984). Abstr. in Phys. Abstr., Vol. 87, No. 1241, Entry 86279 (1984).

161.104 **Does the Newton's world model revive?**
A. Meszaros.
Report KFKI–1984–34. Hungarian Acad. Sci., Budapest, Hungary, 5 pp. (1984). Abstr. in Phys. Abstr., Vol. 87, No. 1241, Entry 86292 (1984).

161.105 **Quantization of Robertson–Walker geometry coupled to fermionic matter.**
T. Christodoulakis, J. Zanelli.
Phys. Rev. D, Vol. 29, No. 12, p. 2738 – 2745 (1984). Abstr. in Phys. Abstr., Vol. 87, No. 1241, Entry 86302 (1984).

161.106 **Measurements of the cosmic background radiation temperature at 3.3 and 9.1 mm.**
G. De Amici, C. Witebsky, G. F. Smoot, S. D. Friedman.
Phys. Rev. D, Vol. 29, No. 12, p. 2673 – 2679 (1984). Abstr. in Phys. Abstr., Vol. 87, No. 1241, Entry 90258 (1984).

161.107 **Measurement of the cosmic background radiation temperature at 3.0 cm.**
S. D. Friedman, G. F. Smoot, G. De Amici, C. Witebsky.
Phys. Rev. D, Vol. 29, No. 12, p. 2677 – 2679 (1984). Abstr. in Phys. Abstr., Vol. 87, No. 1241, Entry 90259 (1984).

161.108 **Measurement of the cosmic background radiation temperature at 6.3 cm.**
N. Mandolesi, P. Calzolari, S. Cortiglioni, G. Morigi.
Phys. Rev. D, Vol. 29, No. 12, p. 2680 – 2682 (1984). Abstr. in Phys. Abstr., Vol. 87, No. 1241, Entry 90260 (1984).

161.109 **Measurement of the cosmic background radiation at 12 cm.**
G. Sironi, P. Inzani, A. Ferrari.
Phys. Rev. D, Vol. 29, No. 12, p. 2686 – 2689 (1984). Abstr. in Phys. Abstr., Vol. 87, No. 1241, Entry 90261 (1984).

161.110 **Cosmological baryon generation at low temperatures.**
M. Claudson, L. J. Hall, I. Hinchliffe.
Nucl. Phys. B, Part. Phys., Vol. B241, No. 1, p. 309 – 332 (1984). Abstr. in Phys. Abstr., Vol. 87, No. 1241, Entry 90264 (1984).

161.111 **Helium formation in a Bianchi type V cosmological model with tilt.**
D. R. Matravers, D. L. Vogel, M. S. Madsen.
Classical Quantum Gravity, Vol. 1, No. 4, p. 407 – 415 (1984). Abstr. in Phys. Abstr., Vol. 87, No. 1241, Entry 90265 (1984).

161.112 **A fractal set of perpetually bouncing universe?**
D. N. Page.
Classical Quantum Gravity, Vol. 1, No. 4, p. 417 – 427 (1984). Abstr. in Phys. Abstr., Vol. 87, No. 1241, Entry 90266 (1984).

161.113 **An exact nonghost solution for a plane–symmetric cosmology containing a classical spinor field.**
R. A. Matzner, M. P. Ryan Jr.
J. Math. Phys., Vol. 25, No. 7, p. 2236 – 2239 (1984). Abstr. in Phys. Abstr., Vol. 87, No. 1241, Entry 90267 (1984).

161.114 **Cosmology, galactic astronomy and elementary particle physics.**
D. W. Sciama.
Proc. R. Soc. London, Ser. A, Vol. 394, No. 1806, p. 1 – 17 (1984). Abstr. in Phys. Abstr., Vol. 87, No. 1241, Entry 90268 (1984).

161.115 **Entropy from extra dimensions.**
S. M. Barr, L. S. Brown.
Phys. Rev. D, Vol. 29, No. 12, p. 2779 – 2782 (1984). Abstr. in Phys. Abstr., Vol. 87, No. 1241, Entry 90269 (1984).

161.116 **Coleman–Weinberg symmetry breaking in an anisotropic spacetime.**
T. Futamase.
Phys. Rev. D, Vol. 29, No. 12, p. 2783 – 2790 (1984). Abstr. in Phys. Abstr., Vol. 87, No. 1241, Entry 90270 (1984).

161.117 **Models which allow a "new inflationary" universe history.**
S. Gupta, H. R. Quinn.
Phys. Rev. D, Vol. 29, No. 12, p. 2791 – 2797 (1984). Abstr. in Phys. Abstr., Vol. 87, No. 1241, Entry 90271 (1984).

161.118 **Imbedding a Schwarzschild mass into cosmology.**
N. Van den Bergh, P. Wils.
Phys. Rev. D, Vol. 29, No. 12, p. 3002 – 3003 (1984). Abstr. in Phys. Abstr., Vol. 87, No. 1241, Entry 90272 (1984).

161.119 **Non–equilibrium thermodynamics of cosmological event horizons.**
R. E. Zimmermann.
Z. Angew. Math. Mech., Vol. 64, No. 5, p. T402 – T404 (1984). Abstr. in Phys. Abstr., Vol. 87, No. 1241, Entry 90273 (1984).

161.120 **Plane symmetric cosmologies with a Synge gas and radiation.**
Abdussattar.
Gen. Relativ. Gravitation, Vol. 16, No. 8, p. 705 – 710 (1984).
A family of plane symmetric models is obtained for a mixture of electrons, protons, and photons expanding in thermal equilibrium, characterized by two independent parameters: the initial photon/baryon number ratio in the universe and the anisotropy parameter.

161.121 **Gravitational interactions of cosmic strings.**
C. J. Hogan, M. J. Rees.
Nature, Vol. 311, No. 5982, p. 109 – 114 (1984).
Breaking a Yang–Mills symmetry at very high energy could lead to macroscopic structures in the vacuum called 'strings', which may be the cause of galaxy formation and clustering. Dissipative gravitational interactions of string loops can produce unusual astrophysical side effects. Such scenarios generally predict a stochastic background of gravitational radiation which should be observable using the millisecond pulsar 1937+21. Gravitational radiation recoil accelerates loops, but dynamical friction on ordinary matter tends to slow them down so they may be able to accrete matter. Internal oscillations of loops produce a time–varying shear perturbation which may lead to an observable heating of stellar or gaseous systems in their immediate neighbourhood.

161.122 **Stellar He production in a cold Universe?**
W. W. Ober, H. J. Falk.
MPA Rep., No. 90, p. 62 – 65 (1983). – See Abstr. 012.020.
A hot Big–Bang model of the Universe is quite successful in explaining the observed helium abundance together with the 2.7K microwave background radiation. An alternative scenario is the mass–energy conversion by nucleosynthesis in stars formed in a cold (no photons or particle pairs) Universe out of density fluctuations. The authors have investigated the evolution of pure hydrogen stars in the mass range of 10 to 500 solar masses including mass loss by stellar winds. The enrichment of the pregalactic medium with helium and heavier elements has been calculated using a nuclear reaction network to follow in detail the chemical evolution.

161.123 **An Einstein–Strauss–de Sitter model of the Universe.**
Z. Stuchlík.
Bull. Astron. Inst. Czech., Vol. 35, No. 4, p. 205 – 215 (1984).
The possibility of constructing an Einstein–Strauss–de Sitter model of the Universe (i.e. a model of Schwarzschild–de Sitter mass condensations immersed in a Friedmann universe with the cosmological constant $\Lambda$ present) is proved. Geodesic motion is considered in both the Schwarzschild–de Sitter and Friedmann parts of the model, and the matching of geodesics across the matching hypersurface of these parts is given. Some consequences of this geodesics–matching, which should be relevant in the process of formation and evolution of mass condensations due to gravitational instability of massive neutrinos in a neutrino–dominated universe, are discussed.

161.124 **Variation of the gravitational constant, *G*, and the age of the Sun in the Brans–Dicke cosmology.**
A. D. Pinotsis.
Bull. Astron. Inst. Czech., Vol. 35, No. 4, p. 220 – 223 (1984).
The present solar position in the H–R diagram is compared with the isochrones of three and five billion years and with the models of $M = 1\ M_{\odot}$, which lie on these last curves, obtained under the assumption of the Brans–Dicke cosmological theory with $\omega = 2$. Since the Sun cannot be as young (from geological evidence) as this comparison shows, this cosmological theory is ruled out, at least for such small values of $\omega$, and indicates that $G$ decreases at a rate of less than $\dot{G}/G = -2\times10^{-11}$/yr, if the mass of the model remains constant.

161.125 **On the large–scale distribution of faint blue galaxies.**
I. D. Karachentsev.
Sov. Astron. Lett., Vol. 10, No. 2, p. 72 – 74 (1984). English translation of 37.161.118.

161.126 **Numerical experiments on galaxy clustering in open universes.**
R. H. Miller.
Astron. Astrophys., Vol. 138, No. 1, p. 121 – 130 (1984).
Open universe models have been explored in numerical experiments undertaken with the goal of understanding how the dynamics is affected by changing $\Omega_F$. Configurations clump into a "strings connecting neighboring clumps, with large empty regions" form that looks very much like the "filaments and voids" structure of galaxy clustering reported observationally on the scale of superclusters. The configurations that develop in each of these experiments are strikingly similar in spite of quite different values of $\Omega_F$ (ranging from $\Omega_F = 1$ to 0.03). Major galaxy clusters are dynamically young objects; very few galaxies have crossed a cluster more than twice at presently observed clumping strengths.

161.127 **Intérêt cosmologique de l'étude du lithium dans l'univers.**
F. Spite.
Astronomie, Vol. 98, p. 371 – 380 (1984).

161.128 **An unconventional view of the missing mass problem.**
W. Tucker, K. Tucker.
Mercury, Vol. 13, No. 4, p. 118, 120 (1984).

161.129 **Comment on the Brans–Dicke–Kasner solution.**
D. Lorenz-Petzold.
Aust. J. Phys., Vol. 37, No. 3, p. 317 – 319 (1984).

The author points out that the Brans–Dicke–Kasner solution recently given by Johri and Goswami (1983) is nothing but the general Bianchi type–I vacuum solution first given by Ruban and Finkelstein (1972).

161.130 **Urknall und Evolution des Kosmos. I. Einstein–Friedmann–Kosmos und das Neutrino–Problem.**
H. J. Blome, W. Priester.
Naturwissenschaften, 71. Jahrg., Heft 9, p. 456 – 467 (1984).

Recent progress in particle physics has an important impact on cosmology. The large–scale structure of the universe and its expansion rate suggest an Einstein–Friedmann world with an age of $(18 \pm 4) \times 10^9$ years. It was assumed that neutrinos do not contribute appreciably to the matter density. If, however, the average mass of the three species of neutrinos is above 2 eV, they would dominate the dynamics of the universe. From the primordial nucleosynthesis of helium and deuterium and its observed values follows an upper limit for the number of neutrino species (or quark–lepton families).

161.131 **Introductory survey (First ESO–CERN Symposium on "Large–scale structure of the universe, cosmology and fundamental physics, CERN, Geneva, 21 – 25 November 1983).**
D. W. Sciama.
Large–scale structure of the universe, cosmology and fundamental physics, p. 3 – 15 (1984). – See Abstr. 012.025.

161.132 **The dynamical parameters of the universe: $H_0$, $q_0$, $\Omega_0$, $\Lambda$, and $K$.**
A. Sandage, G. A. Tammann.
Large–scale structure of the universe, cosmology and fundamental physics, p. 127 – 149 (1984). – See Abstr. 012.025.

The simplest Friedmann model ($\Lambda = 0$, $\Omega_0 \cong 0$) fits all available data if $H_0 = 50$ km s$^{-1}$ Mpc$^{-1}$) and the age of the universe is $\sim 18 \times 10^9$years, but the model is not established by the data. If $H_0 \gtrsim 60$ keeping the age fixed, then $\Lambda > 0$ by amounts shown in the diagrams. A review is given of the methods to determine $H_0$, $\Omega_0$, and the space–time curvature, with the result that the preferred values are $H_0 = 50$, $\Omega_0 = 0.12$, $\Lambda = 0$, and $K < 0$.

161.133 **On a cosmology explaining Sandage's high–$z$–"cutoff".**
R. V. Popić.
Large–scale structure of the universe, cosmology and fundamental physics, p. 150 – 151 (1984). – See Abstr. 012.025.

A generalisation of Friedmann cosmology is introduced. It is capable to explain Sandage's high–$z$–rareness ("cutoff") of QSO, by the behaviour of the cosmological scale factor.

161.134 **Background radiation in the universe.**
D. T. Wilkinson.
Large–scale structure of the universe, cosmology and fundamental physics, p. 153 – 169 (1984). – See Abstr. 012.025.

Diffuse background radiation in several spectral regions may be carrying detailed information about conditions and processes in the early universe. Some conclusions can already be drawn from observations of the 2.7K microwave background. Diffuse X–ray background radiation has been measured but its cosmological implications are unclear. The reverse is true for visible–infrared background radiation. It hasn't yet been seen, but most theories of galaxy formation predict background radiation in this part of the spectrum. The models and observations of the 2.7K and visible–infrared radiations are described in this paper.

161.135 **Microwave sky patterns and the density of the universe.**
J. D. Barrow, R. Juszkiewicz, D. H. Sonoda.
Large–scale structure of the universe, cosmology and fundamental physics, p. 170 – 176 (1984). – See Abstr. 012.025.

The authors predict the possible microwave background temperature distributions in anisotropic and inhomogeneous universes close to the Friedmann model. Homogeneous, anisotropic universes are found to possess either quadrupole or hot–spot temperature variations. The detailed appearance of the hot–spot is predicted. The discovery of such a feature would demonstrate the universe to be open and allow the total density to be measured directly.

161.136 **Intrinsic fluctuations in the cosmic microwave background on a large angular scale.**
R. D. Davies.
Large–scale structure of the universe, cosmology and fundamental physics, p. 177 – 179 (1984). – See Abstr. 012.025.

A determination of the motion of the Sun and Local Group relative to the backdrop of galaxies extending to 70 Mpc is used to correct the observed dipole anisotropy in the cosmic microwave background. The resultant intrinsic dipole component of $\Delta T/T = 3 \pm 2 \times 10^{-4}$ is consistent with no detected anisotropy.

161.137 **Massive neutrinos, deviations from Hubble flow and radiation background fluctuations.**
E. E. Salpeter.
Large–scale structure of the universe, cosmology and fundamental physics, p. 180 – 185 (1984). – See Abstr. 012.025.

The spatial distribution of mass to light ratio is likely to be quite different in two extreme cosmological models: (A) dark matter in the form of massive neutrinos, giving a cosmological density parameter $\Omega$ close to unity and (B) only baryon dark matter in the form of "population III" stars, giving $\Omega \lesssim 0.05$. Models (A) and (B) make different predictions for two observable quantities.

161.138 **Galaxy formation and cosmology.**
S. M. Faber.
Large–scale structure of the universe, cosmology and fundamental physics, p. 187 – 203 (1984). – See Abstr. 012.025.

Roughly 90% of the universe appears to consist of dark matter (DM), the nature of which has not yet been identified. An attractive hypothesis is that DM consists of some type of non–interacting elementary particle. Particle candidates for DM are conveniently classified as "hot", "warm", and "cold" depending on their mass and the epoch at which they become non–relativistic. The observed properties of galaxies are compared with each of these possibilities.

161.139 **Elementary particles and the large–scale structure of the universe.**
J. Silk.
Large–scale structure of the universe, cosmology and fundamental physics, p. 225 – 256 (1984). – See Abstr. 012.025.

Inflationary scenarios for the early universe predict the value of the cosmological density parameter and the primordial fluctuation spectrum. Popular options for dark matter include massive neutrinos and cold supersymmetric relics such as massive photinos. Constraints on these predictions are reviewed in the harsh light of astronomical reality. Observations of the isotropy of the cosmic background radiation and of the large–scale galaxy distribution provide unique and important constraints on elementary particle physics.

161.140 **Massive neutrinos and galaxy formation.**
C. S. Frenk, S. D. M. White, M. Davis.
Large–scale structure of the universe, cosmology and fundamental physics, p. 257 – 266 (1984). – See Abstr. 012.025.

The hypothesis is that the mass density of the universe today is dominated by massive neutrinos results in a definite prediction for the spectrum of density fluctuations at recombination. Since at later epochs these fluctuations give rise to non–linear large–scale structures, a comparison of the observed clustering scale of galaxies with the predicted scale for the neutrino distribution is an important test of this hypothesis. If galaxies are to form at the relatively high redshifts implied by observation, then the predicted amplitude of large–scale density fluctuations considerably exceeds that observed in the galaxy distribution for any acceptable set of cosmological parameters. This discrepancy appears to rule out the conventional neutrino–dominated picture.

161.141 **Early nucleosynthesis.**
J. Audouze.
Large–scale structure of the universe, cosmology and fundamental physics, p. 293 – 318 (1984). – See Abstr. 012.025.

Primordial nucleosynthesis responsible for the formation of the lightest elements provides a unique opportunity to probe the very early phases of the Universe. This paper first reviews standard Big Bang nucleosynthesis. It then discusses a hot and dense Big Bang where D is not synthesized together with $^4$He but is subsequently formed by partial photodisintegration of $^4$He induced by energetic photons coming from the decay of unstable massive neutrinos. Finally, scenarios are considered where He is produced in a population of massive pregalactic stars.

161.142 **Helium production by pregalactic stars.**
B. J. Carr, J. R. Bond, W. D. Arnett.
Large–scale structure of the universe, cosmology and fundamental physics, p. 324 – 330 (1984). – See Abstr. 012.025.

The dark matter in galactic halos could plausibly be contained in the black hole remnants of stars with mass in the range $10^2 - 10^5\ M_\odot$. These stars would naturally generate a helium abundance of around 25%, providing they ejected their envelopes during hydrogen–shell burning. Cosmological nucleosynthesis is therefore not the only possible explanation of the "primordial" helium abundance.

161.143 **The quark–hadron transition in cosmology.**
S. A. Bonometto.
Large–scale structure of the universe, cosmology and fundamental physics, p. 331 – 333 (1984). – See Abstr. 012.025.

The way in which lattice QCD results can be used to study the cosmological quark–hadron transition is discussed. The relevance of the pressure is outlined. The possible existence of a temperature plateau of a few $10^{-5}$ sec is shown and its impact on cosmological limits for photino mass is briefly analyzed.

161.144 **Observational evidence for the evolution of the universe.**
L. Woltjer.
Large–scale structure of the universe, cosmology and fundamental physics, p. 335 – 343 (1984). – See Abstr. 012.025.

Galaxies appear to evolve with redshift, although not necessarily in accordance with simple models. Powerful radio sources and quasars evolve strongly, being very numerous at redshifts of 2 – 3, with possibly a cutoff just under $z = 4$. In addition, they provide evidence for evolution in the intergalactic medium with cosmic epoch.

161.145 **Unified field theories and the early universe.**
D. V. Nanopoulos.
Large–scale structure of the universe, cosmology and fundamental physics, p. 349 – 391 (1984). – See Abstr. 012.025.

Recent developments on the interface between grand unification, supersymmetry and cosmology are discussed. Emphasis is given to the baryon asymmetry of the Universe, the problem of magnetic monopoles and to the cosmic inflation scenario. Primordial inflation is analyzed in considerable detail.

161.146 **The quantum mechanics of the universe.**
S. W. Hawking.
Large–scale structure of the universe, cosmology and fundamental physics, p. 415 – 422 (1984). – See Abstr. 012.025.

Classical general relativity predicts that the universe had a singular origin. The author shows that the singularity can be removed by quantum mechanics, just as in the case of the classical model of the atom. It is proposed that the quantum state of the universe is defined by a path integral over compact positive definite metrics. In a simple model this boundary condition leads to a wave function which can be regarded as a superposition of wave functions peaked around classical oscillating solutions with a long inflationary period.

161.147 **Concluding comments: a cosmologist's view. (First ESO–CERN Symposium on "Large–scale structure of the universe, cosmology and fundamental physics", CERN, Geneva, 21 – 25 November 1983).**
M. J. Rees.
Large–scale structure of the universe, cosmology and fundamental physics, p. 423 – 434 (1984). – See Abstr. 012.025.

161.148 **A brilliant past in front of us.**
J. Ellis.
Large–scale structure of the universe, cosmology and fundamental physics, p. 435 – 450 (1984). – See Abstr. 012.025.

This paper presents the text of a concluding lecture given at the First ESO–CERN Symposium on "Large–scale structure of the universe, cosmology and fundamental physics", CERN, Geneva, 21 – 25 November 1983.

161.149 **Evolution of adiabatic fluctuations in massive–neutrino–dominated Universe models.**
S. A. Bonometto, F. Lucchin, F. Occhionero, N. Vittorio.
Astron. Astrophys., Vol. 138, No. 2, p. 477 – 484 (1984).

The authors investigate the influence of "hot", dark, non–baryonic matter, upon the growth of linear adiabatic fluctuations in the early Universe by considering the "canonical" case of three equally massive neutrinos of $\sim 30$ eV. The background model is assumed to have a density parameter $\Omega = 1$, with baryonic matter having a density parameter $\Omega_m$ much less than unity; the results reported here refer to $\Omega_m = 0.03$. The authors follow the evolution of fluctuations in the range of masses from $10^{14}$ to $10^{19}\ M_\odot$, from well before their entry into the horizon to the epoch when they become non–linear.

161.150 **On the physical nature of the hidden mass of the universe.**
A. G. Doroshkevich, M. Yu. Khlopov.
Yad. Fiz., Tom 39, No. 4, p. 869 – 872 (1984). In Russian. Abstr. in Ref. Zh., 51. Astron., 8.51.621 (1984).

161.151 **Attenuation of gravitational waves in the early universe.**
Yu. G. Ignat'ev, V. Yu. Shulikovskij.
Kazan. univ. Kazan', 10 pp. (1984). In Russian. Abstr. in Ref. Zh., 51. Astron., 8.51.627 (1984).

161.152 **"Large numbers" problem in the Brans–Dicke theory with variable cosmological term.**
B. L. Al'tshuler.
Dokl. Akad. Nauk SSSR. Ser. Mat. Fiz., Tom 274, No. 5, p. 1076 – 1079 (1984). In Russian. Abstr. in Ref. Zh., 51. Astron., 8.51.628 (1984).

161.153 **The effect of scalar particle creation in isotropic models of the universe with torsion.**
A. V. Minkevich, Nguen Hong Tyong.
Dokl. Akad. Nauk BSSR, Tom 28, No. 3, p. 211 – 214 (1984). In Russian. Abstr. in Ref. Zh., 51. Astron., 8.51.633 (1984).

161.154 **On the possibility of an oscillating stage in the evolution of the early universe with non–ideal equation of state.**
Yu. A. Beletskij.
Inst. teor. fiz. Akad. Nauk USSR. Prepr., No. 7R, 16 pp. (1984). In Russian. Abstr. in Ref. Zh., 51. Astron., 8.51.635 (1984).

161.155 **On a regular model of an isotropic universe.**
V. S. Kiriya.
Tbilis. univ. Tbilisi, 55 pp. (1984). In Russian. Abstr. in Ref. Zh., 51. Astron., 8.51.645 (1984).

161.156 **Urknall und Evolution des Kosmos. II. Inflationär modifizierter Urknall und Eschatologie des Kosmos.**
H. J. Blome, W. Priester.
Naturwissenschaften, 71. Jahrg., Heft 10, p. 515 – 527 (1984).

For the very early times ("Urblitz $< 10^{-6}$ s) cosmology requires knowledge about the fundamental particles and their interactions at very high energies. The Grand Unified Theories (GUTs) provide insight into the open problems of the cosmological standard model: The absence of antimatter, the number ratio of photons and baryons and the problem of the large–scale homogeneity and isotropy. The importance of an extremely high energy density of the "vacuum" is discussed for the inflationary expansion in the very early universe and the related origin of matter. Finally, ideas about the future evolution of the cosmos are considered.

161.157 **Vacuum energy in a Friedmann–Lemaître cosmos.**
H.–J. Blome, W. Priester.
Naturwissenschaften, 71. Jahrg., Heft 10, p. 528 – 531 (1984).

161.158 **Dark matter in the universe.**
J. O. Burns.
Sky Telesc., Vol. 68, No. 5, p. 396 – 399 (1984).

161.159 **Bianchi type–I models with conformally invariant scalar field with trace–free energy–momentum tensor.**
S. Ram.
Nuovo Cimento B, Vol. 81B, Ser. 11, No. 1, p. 1 – 6 (1984). Abstr. in Phys. Abstr., Vol. 87, No. 1242, Entry 90447 (1984).

161.160 **Inhomogeneous cosmologies with cosmological constant.**
J. D. Barrow, J. Stein–Schabes.
Phys. Lett. A, Vol. 103A, No. 6 – 7, p. 315 – 317 (1984). Abstr. in Phys. Abstr., Vol. 87, No. 1242, Entry 90450 (1984).

161.161 **Some higher dimensional vacuum solutions of Einstein equations with a cosmological constant.**
H. Ishihara, K. Tomita, H. Nariai.
Prog. Theor. Phys., Vol. 71, No. 4, p. 859 – 861 (1984). Abstr. in Phys. Abstr., Vol. 87, No. 1242, Entry 90451 (1984).

161.162 **Black–hole formation in the new inflationary universe.**
Z. C. Wu.
Phys. Rev. D, Vol. 30, No. 2, p. 286 – 294 (1984). Abstr. in Phys. Abstr., Vol. 87, No. 1242, Entry 90452 (1984).

161.163 **Primordial Coleman–Weinberg inflation.**
P. F. Gonzalez–Diaz.
Phys. Lett. B, Vol. 141B, No. 5 – 6, p. 314 – 316 (1984). Abstr. in Phys. Abstr., Vol. 87, No. 1242, Entry 90459 (1984).

161.164 **Fluctuations in models with primordial inflation.**
R. Kahn, R. Brandenberger.
Phys. Lett. B, Vol. 141B, No. 5 – 6, p. 317 – 322 (1984). Abstr. in Phys. Abstr., Vol. 87, No. 1242, Entry 90460 (1984).

161.165 **Effective action and cosmological evolution of scale factors in higher–dimensional curved spacetime.**
M. Yoshimura.
Phys. Rev. D, Vol. 30, No. 2, p. 344 – 356 (1984). Abstr. in Phys. Abstr., Vol. 87, No. 1242, Entry 90465 (1984).

161.166 **Cosmic separation of phases.**
E. Witten.
Phys. Rev. D, Vol. 30, No. 2, p. 272 – 285 (1984). Abstr. in Phys. Abstr., Vol. 87, No. 1242, Entry 90803 (1984).

161.167 **Topology of the supercluster–void structure.**
T. Matsuda, E. Shima.
Prog. Theor. Phys., Vol. 71, No. 4, p. 855 – 858 (1984). Abstr. in Phys. Abstr., Vol. 87, No. 1242, Entry 94977 (1984).

161.168 **Problems of a perpetually oscillating universe.**
M. A. Markov.
Ann. Phys. (N.Y.), Vol. 155, No. 2, p. 333 – 357 (1984). Abstr. in Phys. Abstr., Vol. 87, No. 1242, Entry 94998 (1984).

161.169 **Bifurcation in the early cosmos.**
M. Novello, L. M. C. S. Rodriguez.
Lett. Nuovo Cimento, Vol. 40, Ser. 2, No. 10, p. 317 – 320 (1984). Abstr. in Phys. Abstr., Vol. 87, No. 1242, Entry 95003 (1984).

161.170 **Cosmology and pilot wave interpretation of quantum mechanics.**
F. J. Tipler.
Phys. Lett. A, Vol. 103A, No. 4, p. 188 – 192 (1984). Abstr. in Phys. Abstr., Vol. 87, No. 1242, Entry 95006 (1984).

161.171 **The cosmological peeling–off property of gravity.**
M. Carmeli, A. Feinstein.
Phys. Lett. A, Vol. 103A, No. 6 – 7, p. 318 – 320 (1984). Abstr. in Phys. Abstr., Vol. 87, No. 1242, Entry 95007 (1984).

161.172 **Massive neutrinos and cosmology.**
A. A. Natale.
Phys. Lett. B, Vol. 141B, No. 5 – 6, p. 323 – 328 (1984). Abstr. in Phys. Abstr., Vol. 87, No. 1242, Entry 95008 (1984).

161.173 **On the relation of the cosmological constraints on neutrino flavors to the width of the $Z^0$.**
D. N. Schramm, G. Steigman.
Phys. Lett. B, Vol. 141B, No. 5 – 6, p. 337 – 341 (1984). Abstr. in Phys. Abstr., Vol. 87, No. 1242, Entry 95009 (1984).

161.174 **Evolution of density perturbations through cosmological phase transitions.**
J. A. Frieman, M. S. Turner.
Phys. Rev. D, Vol. 30, No. 2, p. 265 – 271 (1984). Abstr. in Phys. Abstr., Vol. 87, No. 1242, Entry 95010 (1984).

161.175 **Cooling pancakes.**
J. R. Bond, J. Centrella, A. S. Szalay, J. R. Wilson.
Mon. Not. R. Astron. Soc., Vol. 210, No. 3, p. 515 – 545 (1984).
In theories of galaxy formation with a damping cut–off in the density fluctuation spectrum, the first non–linear structures to form are Zeldovich pancakes in which dissipation separates gas from any collisionless dark matter then present. One–dimensional numerical simulations of the collapse, shock heating, and subsequent thermal evolution of pancakes are described. Neutrinos (or any other cool collisionless particles) are followed by direct $N$–body methods and the gas by Eulerian hydrodynamics with conduction as well as cooling included.

161.176 **The formation of shock waves in the recombination era of the Universe.**
E. Nowotny.
Astrophys. Space Sci., Vol. 104, No. 1, p. 73 – 82 (1984).
The formation and the evolution of shock waves in a two–component fluid during the recombination era of the Universe is studied. The fluid consists of photons and ionized hydrogen gas with interaction due to Thomson scattering. The author gives the analytical solution which answers the questions as to when the discontinuities appear in the physical quantities themselves. Of the two possible waves (a radiation– and a matter–dominated wave) only the matter–dominated wave shows the shock phenomena. However, the shock is induced by the radiation due to Thomson scattering. Therefore, the time for the shock formation depends on the initial amplitude of the wave (as in usual shock phenomena) and on the collision frequency of the photons.

161.177 **Bianchi type–$VI_0$ conformally invariant scalar fields.**
S. Ram.
Astrophys. Space Sci., Vol. 104, No. 1, p. 151 – 155 (1984).
An exact solution of Einstein's equations corresponding to the conformally invariant scalar field with tracefree energy–momentum tensor as source is obtained in Bianchi type $VI_0$ class of metrics. The solution represents a spatially homogeneous but anisotropic universe which admits anisotropic expansions. Some properties of the cosmological model are discussed.

161.178 **Physical processes in the universe at the epoch of the quark–hadron transition.**
S. A. Bonometto, M. Sakellariadou.
Astrophys. J., Vol. 282, No. 2, p. 370 – 373 (1984).
Using QCD results on the lattice the authors determine the time dependence of the scale factor and of the temperature at the end of the quark era. It is found that processes leading to the

quark–hadron transition take a fairly long time ($\sim 2\times10^{-5}$ s), during which the temperature keeps almost constant, while a great many quarks and gluons turn into photons and leptons. The duration of a possible phase transition is likely to be only a minor fraction of this period ($\sim 10^{-7}$ s).

161.179 **Minimal anisotropy of the microwave background radiation in the gravitational instability picture.**
N. Kaiser.
Astrophys. J., Vol. 282, No. 2, p. 374 – 381 (1984).
This paper examines the residual radiation anisotropy arising if the universe were optically thick to electron scattering at some time after the standard epoch of recombination. This would be the state of affairs in any scenario in which the first luminous objects form at redshifts $\gtrsim 20$, and photoionize the unclustered material. The results of this analysis are presented as a linear relation between the angular power spectrum of the sky brightness fluctuations and the spatial power spectrum of the density fluctuations at the epoch of last scattering. It is relatively straightforward to relate these quantities to a range of observed statistics.

161.180 **Reheating the intergalactic medium under extremal conditions.**
R. D. Sherman.
Astrophys. J., Vol. 282, No. 2, p. 387 – 389 (1984).
Previously published studies on the recombination epoch of a hypothetical closure density IGM are extended to the reheating phase. Extremal heating rates are sought and compared to normal ones. The minimal rate is $\sim 10^{-31}$ergs s$^{-1}$cm$^{-3}$, and it must commence heating the IGM at a redshift no greater than $\sim 28$. Normal galactic heating rates are $\sim 10^{-30}$ergs s$^{-1}$cm$^{-3}$ and must be initiated at a redshift no later than $\sim 23$. At heating rates $\gtrsim 10^{-27}$ergs s$^{-1}$cm$^{-3}$ over redshift intervals $\lesssim 10^{-5}$ ionization equilibrium is not maintained; such impulsive heating lacks credible sources and yields line emission radiation far in excess of observations.

161.181 **Large–scale anisotropy of the microwave background and the amplitude of energy density fluctuations in the early universe.**
L. F. Abbott, M. B. Wise.
Astrophys. J., Lett. Ed., Vol. 282, No. 2, p. L47 – L50 (1984).
The spectrum of energy density fluctuations predicted by the inflationary cosmology is used to compute multipole moments of the microwave background temperature and the probability distributions for these moments. Comparison with the observed dipole moment gives a 90% confidence level upper bound of $3.4\times10^{-6}$ on the amplitude of these fluctuations when they cross the horizon.

161.182 **Cosmic strings as gravitational lenses.**
A. Vilenkin.
Astrophys. J., Lett. Ed., Vol. 282, No. 2, p. L51 – L53 (1984).
Superheavy strings produced at a phase transition in the early universe can be detected through their gravitational lens effect. Long strings can give rise to lines of double galaxies with relatively large angular separations between the images. It is argued that some of the known double quasars may be due to strings.

161.183 **What was, before "nothing" was?**
I. S. Shklovskij.
Zemlya Vselennaya, No. 4, p. 34 – 39 (1984). In Russian.
Discussion on the problem of the earliest stage of evolution of the Universe.

161.184 **Characteristic masses in dynamics of a gravitating gas.**
D. I. Barausov, A. D. Chernin.
Astrofizika, Tom 20, Vyp. 1, p. 208 – 211 (1984). In Russian. English translation in Astrophysics, Vol. 20, No. 1.
Dynamics of massive cooling gravitating gaseous clouds is characterized by values of their sizes and masses which are expressed by combinations of universal physical constants and close in magnitude to the parameters of the large–scale structures in the Universe.

161.185 **A Bianchi type–III cosmological model in Brans–Dicke theory for vacuum field.**
R. Tiwari, D. K. Singh.
Astrophys. Space Sci., Vol. 105, No. 1, p. 67 – 71 (1984).
An exact solution of Brans–Dicke field equations for the metric tensor of a spatially homogeneous Bianchi type–III configuration has been obtained for vacuum field. It is shown that in the limiting case the solution reduces to that of Einstein field equations in vacuum.

161.186 **Gravitational collapse of cosmic gas clouds and formation of star clusters.**
B. Basu, T. Bhattacharyya.
Astrophys. Space Sci., Vol. 105, No. 1, p. 85 – 97 (1984).
In this paper the gravitational collapse of cosmic gas clouds and formation of star clusters has been considered. Hoyle's view of successive fragmentation has been taken as the basic mechanism in the present work. It has been shown that the fragmentation process is greatly inhibited by the presence of a strong magnetic field. A model has been constructed showing how a protostar grows in mass by accretion from the surrounding medium.

161.187 **The spatial correlation function of galaxies confronted with theoretical scenarios.**
A. Dekel, S. J. Aarseth.
Astrophys. J., Vol. 283, No. 1, p. 1 – 23 (1984).
The authors study the two–point spatial correlation function of galaxies, $\xi(r)$, in a series of numerical simulations of cosmological clustering scenarios in comparison with $\xi(r)$ obtained from redshift samples. The evolution of $\xi(r)$ is evaluated in new large–scale 10000–body simulations of the pancake scenarios. The observed excess of correlation in the range 2 – 10 Mpc $h^{-1}$, which does not agree with a pure hierarchical clustering scenario, is reproduced in a scenario where $\lambda = 30\pm5$ Mpc $h^{-1}$ superclusters collapse first to flat (or elongated) pancakes. The shape of $\xi(r)$ is reproduced provided that the first pancaking occurred at $z \leqslant 0.5$ if $\Omega = 1$ and $z < 2$ if $\Omega_0 = 0.1$. This result indicates that galaxy formation preceded supercluster formation. Constraints on the spectrum of density perturbations after decoupling are obtained and the origin of such a spectrum with respect to the particles that dominate the mass of the universe is considered.

161.188 **The detectability of population III "Jupiters".**
H. Karimabadi, L. Blitz.
Astrophys. J., Vol. 283, No. 1, p. 169 – 173 (1984).
Under the assumption that the closure density of the universe is provided by low–mass stars below the hydrogen burning limit (i.e., "Jupiters"), the contribution of these objects to the electromagnetic background is calculated. The resultant spectrum is found to deviate significantly from a blackbody spectrum, thus providing a distinctive radiation signature. The peak of the spectrum is broad and depends only weakly on mass, and is found to lie between $\lambda$ = 10 μm to 100 μm for the mass ranges considered. For certain cases the peak brightness is found to approach $10^4$Jy sr$^{-1}$.

161.189 **Cosmologists in the dark.**
J. Silk.
Nature, Vol. 311, No. 5986, p. 508 – 509 (1984).

161.190 **Formation of galaxies and large–scale structure with cold dark matter.**
G. R. Blumenthal, S. M. Faber, J. R. Primack, M. J. Rees.
Nature, Vol. 311, No. 5986, p. 517 – 525 (1984). With a correction in Vol. 313, No. 5997, p. 72 (1985).
The dark matter that appears to be gravitationally dominant on all scales larger than galactic cores may consist of axions, stable photinos, or other collisionless particles whose velocity dispersion in the early Universe is so small that fluctuations of galactic size or larger are not damped by free streaming. An attractive feature of this cold dark matter hypothesis is its considerable predictive power: the post–recombination fluctuation spectrum is calculable, and it in turn governs the formation of

galaxies and clusters. Good agreement with the data is obtained for a Zeldovich spectrum of primordial fluctuations.

161.191 **Structural stability properties of Friedmann cosmology.**
M. Szydlowski, M. Heller, Z. Golda.
Gen. Relativ. Gravitation, Vol. 16, No. 9, p. 877 – 890 (1984).
A dynamical system with Robertson–Walker symmetries and the equation of the state $p = \gamma\varepsilon$, $0 \leqslant \gamma \leqslant 1$, considered both as a conservative and nonconservative system, is studied with respect to its structural stability properties.

161.192 **The new inflationary universe.**
A. H. Guth.
Ann. N.Y. Acad. Sci., Vol. 422, p. 1 – 14 (1984). – See Abstr. 012.033.
This review considers the development of the basic ideas of grand unified theories (GUT), which led to the concept of the new inflationary universe as an alternative to the standard big–bang scenario.

161.193 **Microcosmology. New particles and cosmology.**
H. R. Pagels.
Ann. N.Y. Acad. Sci., Vol. 422, p. 15 – 32 (1984). – See Abstr. 012.033.
This paper reviews the role of cosmology and astrophysics in providing information on the properties of new hypothetical particles that emerge as field quanta from unified relativistic quantum field theories. Emphasis is on supersymmetric field theories and the gravitino and photino, but the axion and majoron are also discussed. Other topics include the role of heavy quanta, galaxy formation, and origin problems.

161.194 **Massive magnetic monopoles in cosmology and astrophysics.**
E. W. Kolb.
Ann. N.Y. Acad. Sci., Vol. 422, p. 33 – 44 (1984). – See Abstr. 012.033.
This paper discusses the production of heavy magnetic monopoles in the early universe, considers astrophysical limits on their current density in the Galaxy, and examines some consequences of monopole–induced proton decay.

161.195 **Adiabatic theories of galaxy formation and pancakes.**
J. R. Bond, A. S. Szalay.
Ann. N.Y. Acad. Sci., Vol. 422, p. 82 – 89 (1984). – See Abstr. 012.033.
The adiabatic theory of galaxy formation in universes dominated by massive neutrinos or other more weakly interacting particles such as gravitinos or photinos is reviewed. Collisionless damping leads to a density fluctuation spectrum with a cutoff, the nonlinear evolution of which naturally results in the formation of pancakes, strings, voids, galaxies, and galactic halos.

161.196 **Pancakes and the formation of galaxies. Neutrinos and other 'inos.**
P. R. Shapiro, C. Struck–Marcell.
Ann. N.Y. Acad. Sci., Vol. 422, p. 90 (1984). Abstract. – See Abstr. 012.033.

161.197 **The texture of the universe.**
R. P. Kirshner, A. Oemler Jr., P. L. Schechter, S. A. Shectman.
Ann. N.Y. Acad. Sci., Vol. 422, p. 91 – 94 (1984). – See Abstr. 012.033.
This paper discusses the large–scale space distribution of galaxies in the universe as revealed by recent extensive redshift surveys. In particular, the discovery of large galaxy voids is considered.

161.198 **Big bang nucleosynthesis. Gateway to the very early universe.**
M. S. Turner.
Ann. N.Y. Acad. Sci., Vol. 422, p. 106 – 117 (1984). – See Abstr. 012.033.
This paper reviews the status of the concordance of the standard model of primordial nucleosynthesis with current observational data. At present the standard model is consistent with all the observational data provided that the baryon–to–photon ratio $\eta \cong (3-7) \times 10^{-10}$ and the number of light neutrino species (or the equivalent number of other light species) $N_v \leqslant 4$. Should the primordial mass fraction of $^4$He be found to be $\leqslant 0.22$, the standard model is in serious trouble.

161.199 **Nature and significance of the mass distribution in the universe.**
P. J. E. Peebles.
Ann. N.Y. Acad. Sci., Vol. 422, p. 118 – 125 (1984). – See Abstr. 012.033.
Measures of the mass distribution on scales greater than a few kpc have considerably improved in the last few years. The author reviews the status of some of the measures and the interpretations the results suggest to him.

161.200 **Trace anomaly in the "massive" Gowdy $T^3$ cosmology.**
B. K. Berger.
Ann. N.Y. Acad. Sci., Vol. 422, p. 331 (1984). Abstract. – See Abstr. 012.033.

161.201 **Cosmology in General Relativity with higher derivatives.**
K. Brecher.
Ann. N.Y. Acad. Sci., Vol. 422, p. 332 (1984). Abstract. – See Abstr. 012.033.

161.202 **Manifestations of a cosmological density of compact objects in quasar light.**
C. R. Canizares.
Ann. N.Y. Acad. Sci., Vol. 422, p. 335 – 336 (1984). Abstract. – See Abstr. 012.033.

161.203 **FRW models as exact viscous magnetohydrodynamical cosmologies.**
A. A. Coley, B. O. J. Tupper.
Ann. N.Y. Acad. Sci., V 422, p. 338 (1984). Abstract. – See Abstr. 012.033.

161.204 **A single expression for the general solution of Einstein's vacuum and electrovac field equations with cosmological constant for Petrov type D admitting a nonsingular aligned Maxwell field.**
R. Debever, N. Kamram, R. G. McLenaghan.
Ann. N.Y. Acad. Sci., Vol. 422, p. 339 (1984). Abstract. – See Abstr. 012.033.

161.205 **Curvature coordinates in cosmology.**
R. Gautreau.
Ann. N.Y. Acad. Sci., Vol. 422, p. 351 (1984). Abstract. – See Abstr. 012.033.

161.206 **The evolution of voids in the expanding universe.**
M. A. Hausman, D. W. Olson, B. D. Roth.
Ann. N.Y. Acad. Sci., Vol. 422, p. 357 (1984). Abstract. – See Abstr. 012.033.

161.207 **Three–millimeter anisotropy measurement.**
P. M. Lubin, G. L. Epstein, G. F. Smoot.
Ann. N.Y. Acad. Sci., Vol. 422, p. 366 (1984). Abstract. – See Abstr. 012.033.

161.208 **The unobservability of higher–order moments in the cosmic background radiation.**
D. W. Marks.
Ann. N.Y. Acad. Sci., Vol. 422, p. 369 (1984). Abstract. – See Abstr. 012.033.

161.209 **Nonlinear density fluctuations in a two-component fluid during the recombination era of the universe.**
E. Nowotny.
Ann. N.Y. Acad. Sci., Vol. 422, p. 372 (1984). Abstract. – See Abstr. 012.033.

161.210 **A gravitation theory with *G* determined by retarded cosmic potential.**
K. Philberth.
Ann. N.Y. Acad. Sci., Vol. 422, p. 375 – 376 (1984). Abstract. – See Abstr. 012.033.

161.211 **Spectrum of the cosmic background radiation.**
G. Sironi, N. Mandolesi, G. De Amici, S. Friedman, G. Smoot, C. Witebsky, R. B. Partridge, L. Danese.
Ann. N.Y. Acad. Sci., Vol. 422, p. 383 (1984). Abstract. – See Abstr. 012.033.

161.212 **A closed pulsating universe with inner dimensionality.**
F. R. Tangherlini.
Ann. N.Y. Acad. Sci., Vol. 422, p. 385 – 389 (1984). – See Abstr. 012.033.

This paper describes the model of a pulsating universe where the gravitational collapse rebounds at some minimum value of the expansion parameter. Thus an ultimate singularity is avoided.

161.213 **Die Explosion des Universums. Materie, die aus dem Nichts kommt.**
H. J. Fahr.
Umschau, 84. Jahrg., Nr. 22, p. 678 – 683 (1984).

161.214 **Small-scale isotropy of the cosmic microwave background at 19.5 GHz.**
J. M. Uson, D. T. Wilkinson.
Astrophys. J., Vol. 283, No. 2, p. 471 – 478 (1984).

The 140-foot (43 m) telescope of the NRAO, with the *K*-band maser front-end, has a 50K system temperature with a bandwidth of 370 MHz centered at 19.5 GHz. The authors have used this instrument for measurements of small-scale anisotropy in the cosmic microwave background. They place an upper limit of $\Delta T_{rms}/T < 4.5 \times 10^{-5}$ at the 95% confidence level on fluctuations at a scale of 4'.5. The sensitivity attained comes within a factor of 2 of the theoretical value for 112 hours of integration.

161.215 **Calibration theory of gravitation when localizing a group of motions of the Einstein Universe.**
V. N. Tunyak.
Izv. vuzov. Fiz., Tomsk, 13 pp. (1984). In Russian. Abstr. in Ref. Zh., 51. Astron., 9.51.824 (1984).

161.216 **Estimate of the causality region of the Gödel cosmological model.**
V. I. Denisov.
Ukr. geom. Sb., Khar'kov, No. 27, p. 26 – 31 (1984). In Russian. Abstr. in Ref. Zh., 51. Astron., 9.51.827 (1984).

161.217 **Helium formation in cosmologies with anisotropic curvature.**
J. D. Barrow.
Mon. Not. R. Astron. Soc., Vol 211, No. 1, p. 221 – 227 (1984).

Primordial nucleosynthesis is calculated in three classes of exact cosmological model possessing anisotropic three-curvature. The calculations lead to primordial helium-4 abundances within the observed range, $Y = 0.21 - 0.26$, in many cases. This is unlike the behaviour found in the anisotropic cosmological models used in previous nucleosynthesis studies. This difference arises because the curvature anisotropy gives rise to a more complicated expansion anisotropy than in the Kasner-like models studied previously. One of the exact models studied possesses non-zero vorticity as well as shear and curvature anisotropy. These examples demonstrate that the range of acceptable cosmological models for the early Universe may be considerably wider than is usually believed.

161.218 **The statistics of gravitational lenses: the distributions of image angular separations and lens redshifts.**
E. L. Turner, J. P. Ostriker, J. R. Gott III.
Astrophys. J., Vol. 284, No. 1, p. 1 – 22 (1984).

The authors compute analytically the probability $d\tau$ that the light from a distant quasar at redshift $z_Q$ will be lensed by objects in redshift range $dz_L$ and find that for nearby objects $\tau$ goes as $z_Q^2$ for point lenses and $z_Q^3$ for galaxy lenses (modeled as singular isothermal spheres). At fixed $z_Q$ the most probable lens redshift is typically small, $z_L \approx \mathrm{Min}(z_Q/2, 0.8)$, so the lensing objects (if they are galaxies) should be detectable. A variety of numerical simulations are made to determine the expected properties of double optical or radio quasars. These simulations take into account realistic selection biases which affect the expected properties of observed lens systems. The results are insensitive to the assumed cosmology but highly dependent on the central velocity dispersions of bright elliptical galaxies and to the maximum surface densities in great clusters. The inferred luminosity of an imaged quasar is usually much greater than its true luminosity, with the average flux amplification a factor of 4 in an unbiased sample.

161.219 **Microwave background anisotropy and hydrodynamic formation of large-scale structure.**
C. J. Hogan.
Astrophys. J., Lett. Ed., Vol. 284, No. 1, p. L1 – L4 (1984).

A test is proposed of the hypothesis that the large-scale distribution of galaxies is generated by large-scale pressure gradients in protogalactic gas, due to pregalactic explosions or other inhomogeneous energy sources at redshifts $z \lesssim 10$. Pressure inhomogeneities in hot gas lead to anisotropy in the microwave background via the Sunyaev-Zel'dovich Compton heating effect. The amplitude of the predicted anisotropy is estimated using simplified models and is found to conflict with current upper limits. Several possible complications are considered and are found to increase the predicted anisotropy.

161.220 **Formation of structure in a universe with unstable neutrinos.**
A. G. Doroshkevich, M. Yu. Khlopov.
Mon. Not. R. Astron. Soc., Vol. 211, No. 2, p. 277 – 282 (1984).

Possible influence of the instability of massive neutrinos relative to decays $\nu_H \rightarrow \nu_L + \mu(f)$ with $\mu(f)$ majoron (familon) on the processes of the formation of the structure of the universe is discussed. It is shown that it is possible to decrease the density of neutrino "pancakes" by 10 – 20 times as compared to the standard model. It makes possible: (1) the stopping of the process of decay of the structure and thus it makes compatible the formation of first objects at $z \cong 4.5$ with the existence of evolved structure at $z \cong 0$, (2) a compatibility of the predicted and observed value of the density of the hidden mass, and (3) a better compatibility of the predicted and the observed correlation function. It is possible in some models to decrease by 5 – 10 times the predicted value of the amplitude of large-scale thermal background fluctuations.

161.221 **A possibility to use C I infrared absorption lines for determination of the background radiation temperature at earlier cosmological epochs.**
I. E. Waltz (*I. E. Val'tts*), V. K. Khersonskij.
Astrophys. Space Sci., Vol. 105, No. 2, p. 215 – 226 (1984).

A method for the determination of the background radiation temperature in earlier cosmological epochs at redshifts $z > 2$ is discussed in detail. The method is based on the fact that in the clouds situated in cosmological distances at redshift $z$ the background radiation temperature must be $(1+z)$ times more than in the modern epoch ($z = 0$). This shall affect the level populations of the atoms, ions and molecules and, consequently, the parameters of the absorption lines observed in quasar absorption spectra. It is proposed to use the transition $^3P_0$–$^3P_1$ of C I ($\lambda = 610\ \mu$) for the measurement of the background radiation temperature. At $z > 2$ this absorption line shifts to the millimeter radio region and can be observed by radioastronomical methods.

161.222 **The microwave background produced by particle decay in the steady state – a possible model?**
P. R. Marchant.
Astrophys. Space Sci., Vol. 105, No. 2, p. 227 – 235 (1984).

The conjecture that the cosmic microwave background can be produced by particle decay in steady-state cosmology is developed. A fit to the measurements would require that the particles (fermions), have a mass $2 \times 10^{-3}$eV/$c^2$, and decay into much lighter particles, one of which is a photon, on a time-scale $\sim 100\ \tau_H$. These particles would contribute $\sim 1\%$ to the density parameter of the Universe.

161.223 **Nonapplicability of the viscous approximation method in homogeneous cosmological models.**
M. Ostrowski.
Astrophys. Space Sci., Vol. 105, No. 2, p. 253 – 257 (1984).

By considering two conceptually important instances it is shown that one cannot use the viscous approximation method to estimate the effects of nonequilibrium processes in homogeneous cosmological models. Despite the widespread acceptance of this method, it is argued that not even qualitative estimates can be obtained by its use.

161.224 **Formation of the structure of the universe in models with nonstable neutrinos.**
A. G. Doroshkevich, M. Yu. Khlopov.
Inst. prikl. mat. Akad. Nauk SSSR. Prepr., No. 47, 14 pp. (1984). In Russian. Abstr. in Ref. Zh., 51. Astron., 10.51.760 (1984).

161.225 **The Syunyaev – Zel'dovich effect experimentally confirmed.**
Yu. Eh. Lyubarskij.
Priroda, No. 11, p. 76 – 77 (1984). In Russian.

161.226 **Alternatives to the big bang.**
G. F. R. Ellis.
Annu. Rev. Astron. Astrophys., Vol. 22, p. 157 – 184 (1984). – See Abstr. 003.024.

Contents: (1) Introduction. (2) Basic results: The nature of redshift. Basic equations. The standard models. The cosmological constant. Implications. (3) Alternative views of redshift: Alternative theories of light. Local redshift theories. (4) Alternative physics: Alternative matter. Alternative gravity. (5) Alternative geometries: Anisotropies. Inhomogeneities. (6) Conclusion.

161.227 **Axion cosmology.**
L. F. Abbott.
Relativity, cosmology, topological mass and supergravity, p. 100 – 107 (1983). – See Abstr. 012.040.

The role of axions in big-bang and inflationary cosmology is discussed. A cosmological bound on the axion mass is derived and the possibility that axions comprise the dark matter of the universe is explored.

161.228 **On the causal Green function in a Robertson–Walker universe with linear expansion.**
E. Calzetta, M. Castagnino.
Relativity, cosmology, topological mass and supergravity, p. 242 – 247 (1983). – See Abstr. 012.040.

It is demonstrated that three reasonable physical criteria yield to a unique causal Green function for a conformally coupled neutral scalar field in a linearly expanding Robertson–Walker universe without spatial curvature.

161.229 **A class of Szekeres space–times with cosmological constant.**
G. M. Covarrubias.
Relativity, cosmology, topological mass and supergravity, p. 251 – 254 (1983). – See Abstr. 012.040.

The author presents a generalization of a class of Szekeres space–times. The new solutions satisfy Einstein's equations with a cosmological constant and have the same geometrical properties as the corresponding class found by Szekeres.

161.230 **Gravitational instability in the presence of a magnetic field in the expanding universe.**
S. Hacyan.
Relativity, cosmology, topological mass and supergravity, p. 255 – 257 (1983). – See Abstr. 012.040.

Closed-form solutions for the evolution of linear density and magnetic field perturbations in a model universe with a uniform magnetic field are presented.

161.231 **An exact solution for a quantum Taub model.**
S. E. Martínez, M. P. Ryan Jr.
Relativity, cosmology, topological mass and supergravity, p. 262 – 265 (1983). – See Abstr. 012.040.

There exists a number of exact solutions to quantized cosmological models in a formulation due to Misner. The authors present an exact solution to the vacuum Taub model in the Misner formalism. Such a solution is of interest because the Taub model is of Bianchi type IX.

161.232 **Quantum fluctuations in the early universe.**
E. Nowotny.
Relativity, cosmology, topological mass and supergravity, p. 272 – 277 (1983). – See Abstr. 012.040.

161.233 **Hamiltonian formulation of diagonal Bianchi type IX electromagnetic cosmology.**
S. M. Waller.
Relativity, cosmology, topological mass and supergravity, p. 285 – 290 (1983). – See Abstr. 012.040.

The dynamical effects of spatially homogeneous electromagnetic fields on anisotropic Bianchi type IX cosmological models are studied. The potentials affecting the evolution of the anisotropy for a diagonal metric are obtained by means of the Arnowitt, Deser, and Misner Hamiltonian formalism. A Hamiltonian for the gravitational variables in which the electromagnetic fields appear only as constants of motion is constructed.

161.234 **Neutral hydrogen in cosmic voids.**
N. Krumm, N. Brosch.
Astron. J., Vol. 89, No. 10, p. 1461 – 1463 (1984).

The authors have searched unsuccessfully for H I clouds in two voids in the cosmic distribution of galaxies. The search employed the Green Bank 300-ft telescope, and was sensitive to clouds of mass greater than about $10^{10}h^{-2}M_{\odot}$. Roughly 7% of the Perseus–Pisces void and 19% of the Hercules void were searched. The authors conclude that the voids do not contain a cosmologically significant population of H I protogalaxies, and they summarize evidence consistent with the hypothesis that the voids are indeed regions of low matter density in the Universe.

161.235 **Tests of cosmological models constrained by inflation.**
P. J. E. Peebles.
Astrophys. J., Vol. 284, No. 2, p. 439 – 444 (1984).

The inflationary scenario requires that the universe have negligible curvature along constant-density surfaces. In the Friedmann–Lemaitre cosmology that leaves us with two free parameters, Hubble's constant $H_0$ and the density parameter $\Omega_0$ (or, equivalently, the cosmological constant $\Lambda$). The author discusses here tests of this set of models from local and high-redshift observations. The data agree reasonably well with $\Omega_0 \sim 0.2$.

161.236 **Superpancakes and the cluster correlation function.**
A. Dekel.
Astrophys. J., Vol. 284, No. 2, p. 445 – 456 (1984).

It is proposed that the apparent difference between the clustering lengths of galaxies and of rich clusters reflects the fact that galaxies populate $\lambda_g \sim 20 -$ 30 Mpc $h^{-1}$ *pancakes* (flat or elongated) while clusters are distributed in $\lambda_c \sim 100 -$ 150 Mpc $h^{-1}$ *superpancakes*. The correlation functions near unity are assumed to have grown because of the large-scale pancaking process rather than by hierarchical clustering due to local gravity. Such a structure of flattened, first order and second order superclusters arises if the linear spectrum of perturbations had two coherence lengths associated with amplitude jumps on the corresponding scales. The rms amplitude ratio $\delta(\lambda_g)/\delta(\lambda_c)$ should be 3:1.

161.237 **Limits on arcsecond–scale fluctuations in the cosmic microwave background.**
J. E. Knoke, R. B. Partridge, M. I. Ratner, I. I. Shapiro.
Astrophys. J., Vol. 284, No. 2, p. 479 – 490 (1984).
The authors used the NRAO Very Large Array in its C configuration at a wavelength of 6 cm to set upper limits on the rms fluctuation of sky brightness on angular scales of 6″ – 18″ from sources too weak to be detected individually. At the highest resolution, they establish a limit of 8 μJy per beam area on the rms sky fluctuation. If this fluctuation level is the result of a Poisson distribution of unresolved sources, each of flux density $S_0$ Jy, then the number density of such sources per steradian must be less than 0.08 $S_0{}^{-2}\mathrm{sr}^{-1}$. These limits also establish limits on the rms temperature fluctuation for simple models of fluctuations in the cosmic microwave background.

161.238 **Photon mass and cosmic microwave background anisotropy.**
P. De Bernardis, S. Masi, F. Melchiorri, A. Moleti.
Astrophys. J., Lett. Ed., Vol. 284, No. 2, p. L21 – L22 (1984).
Following the hypothesis of Georgi et al. on the existence of massive photons, the authors have investigated their effect on the spectral behavior of the cosmic background dipole anisotropy. They have found a distortion increasing with wavelength to such an extent that the direction of the dipole anisotropy can be reversed. The best fit of the available experimental data gives a value for the photon mass of $(2.9 \pm 0.1) \times 10^{-51}$g with $\chi^2/\mathrm{DOF} = 0.53$, while the best fit with a pure blackbody has $\chi^2/\mathrm{DOF} = 1.32$.

161.239 **Clustering in a two–component universe.**
L. Z. Fang, S. X. Li, S. P. Xiang.
Astron. Astrophys., Vol. 140, No. 1, p. 77 – 81 (1984).
The density distributions of visible objects and invisible matter in the universe are rather different from each other. Visible objects show marked clustering on the scales of galaxies, clusters, superclusters, while the dark matter is distributed fairly uniformly at least on scales smaller than that of superclusters. A possible origin of this difference has been studied. The model used implies that the dark matter contains not only massive neutrinos as a dominant component, but also some type of more massive, noninteracting particles as a non–dominant component. The scenario of structure formation in a universe with two such components of dark matter is discussed. A testable prediction from this scenario is that no strong clustering on large scale exists in the distribution of quasars. Several statistical analyses of quasar distributions tend to be in favour of this prediction.

161.240 **Grand unified strings and galaxy formation.**
N. Turok.
Nucl. Phys. B, Part. Phys., Vol. B242, No. 2, p. 520 – 541 (1984). Abstr. in Phys. Abstr., Vol. 87, No. 1243, Entry 95264 (1984).

161.241 **A closed Friedmann universe with a scalar field as source.**
S. Gottlöber.
Ann. Phys. (Leipzig), Vol. 41, No. 1, p. 45 – 53 (1984). Abstr. in Phys. Abstr., Vol. 87, No. 1243, Entry 101543 (1984).

161.242 **Why is the Universe so big?**
H.–J. Treder.
Ann. Phys. (Leipzig), Vol. 41, No. 2, p. 81 – 87 (1984). Abstr. in Phys. Abstr., Vol. 87, No. 1243, Entry 101544 (1984).

161.243 **Mass scales and the cosmological coincidences.**
P. T. Landsberg.
Ann. Phys. (Leipzig), Vol. 41, No. 2, p. 88 – 92 (1984). Abstr. in Phys. Abstr., Vol. 87, No. 1243, Entry 101545 (1984).

161.244 **Quantum effects in the Gowdy $T^3$ cosmology.**
B. K. Berger.
Ann. Phys. (N.Y.), Vol. 156, No. 1, p. 155 – 193 (1984). Abstr. in Phys. Abstr., Vol. 87, No. 1243, Entry 101546 (1984).

161.245 **Formation of galaxies in G–variable cosmologies. II. The Hoyle–Narlikar and the Brans–Dicke gravitational theories.**
J. Krempec–Krygier.
Acta Phys. Pol., Ser. B, Vol. B15, No. 6, p. 455 – 464 (1984). Abstr. in Phys. Abstr., Vol. 87, No. 1243, Entry 101547 (1984).

161.246 **Early evolution of the Universe and the primordial nucleosynthesis.**
N. C. Rana.
Curr. Sci., Vol. 52, No. 24, p. 1161 – 1168 (1983). Abstr. in Phys. Abstr., Vol. 87, No. 1243, Entry 101548 (1984).

161.247 **Axially symmetric cosmological models for the perfect fluid.**
N. F. Dandach (*N. F. Dandash*), N. V. Mitskievic (*N. V. Mitskevich*).
J. Phys. A, Vol. 17, No. 11, p. 2335 – 2343 (1984). Abstr. in Phys. Abstr., Vol. 87, No. 1243, Entry 101550 (1984).

161.248 **A nonequilibrium universe of stars and radiation.**
D. G. Long.
Lett. Nuovo Cimento, Vol. 40, Ser. 2 No. 11, p. 325 – 328 (1984). Abstr. in Phys. Abstr., Vol. 87, No. 1243, Entry 101551 (1984).

161.249 **Nature of dark matter and pancake mass.**
S. A. Bonometto, R. Valdarnini.
Phys. Lett. A, Vol. 103A, No. 8, p. 369 – 373 (1984). Abstr. in Phys. Abstr., Vol. 87, No. 1243, Entry 101553 (1984).

161.250 **Dynamical evolution of a Higgs field in the inflationary stage of the Universe.**
K. Sato, H. Kodama.
Phys. Lett. B, Vol. 142B, No. 1 – 2, p. 18 – 22 (1984). Abstr. in Phys. Abstr., Vol. 87, No. 1243, Entry 101554 (1984).

161.251 **Quantum creation of universes.**
A. Vilenkin.
Phys. Rev. D, Vol. 30, No. 2, p. 509 – 511 (1984). Abstr. in Phys. Abstr., Vol. 87, No. 1243, Entry 101555 (1984).

161.252 **Improved limits on small–scale anisotropy in cosmic microwave background.**
J. M. Uson, D. T. Wilkinson.
Nature, Vol. 312, No. 5993, p. 427 – 429 (1984).
The authors' search for small–scale anisotropy in the background radiation has now resulted in a lowering of the upper limit on root–mean–square fluctuations observed at an angular scale of ~4 arc min to $\Delta T_{\mathrm{r.m.s.}}/T < 2.1 \times 10^{-5}$ at the 95% confidence level. The actual limits deduced from the experiment depend on the model assumed for the unseen fluctuations. Several possibilities are discussed as well as the implications this new measurement has for various cosmological models.

161.253 **Inflation in the universe and time asymmetry.**
P. C. W. Davies.
Nature, Vol. 312, No. 5994, p. 524 – 527 (1984).
The author previously argued that the origin of the time asymmetry is to be sought in the processes that characterized the very early cosmological epochs, and that the recent inflationary scenario provides a crucial element in our understanding of this origin. Here, he develops this thesis further, in the light of criticisms by Page. It is argued in particular that any of the "initial states" that appear to be permitted by quantum gravity, if succeeded by a period of inflation, would have wound up the universe to the state from which, as indicated by increasing entropy, it is now winding down.

161.254 **Strings and the origins of galaxies.**
N. Turok, D. N. Schramm.
Nature, Vol. 312, No. 5995, p. 598 – 599 (1984).
This note discusses some new aspects of the recent hypothesis that superheavy strings, formed at a phase transition in the very early universe, may be responsible for the origin of galaxies.

161.255 **Exact solutions of plane symmetric cosmological models.**
R. Hojman, J. Santamarina.
J. Math. Phys., Vol. 25, No. 6, p. 1973 – 1974 (1984). Abstr. in Phys. Abstr., Vol. 87, No. 1244, Entry 101719 (1984).

161.256 **Kaluza–Klein cosmologies at late times.**
D. Bailin, A. Love, C. E. Vayonakis.
Phys. Lett. B, Vol. 142B, No. 5 – 6, p. 344 – 348 (1984). Abstr. in Phys. Abstr., Vol. 87, No. 1244, Entry 101730 (1984).

161.257 **Rapid phase transitions in primordial inflation with flat supergravity potentials.**
K. Enqvist, D. V. Nanopoulos.
Phys. Lett. B, Vol. 142B, No. 5 – 6, p. 349 – 354 (1984). Abstr. in Phys. Abstr., Vol. 87, No. 1244, Entry 101731 (1984).

161.258 **The isotropy of the Universe.**
S. W. Hawking, J. C. Luttrell.
Phys. Lett. B, Vol. 143B, No. 1 – 3, p. 83 – 86 (1984). Abstr. in Phys. Abstr., Vol. 87, No. 1244, Entry 102039 (1984).

161.259 **The inflationary Universe.**
A. D. Linde.
Rep. Prog. Phys., Vol. 47, No. 8, p. 925 – 986 (1984). Abstr. in Phys. Abstr., Vol. 87, No. 1244, Entry 106088 (1984).

161.260 **Inflation from quantum gravity.**
T. Padmanabhan.
Phys. Lett. A, Vol. 104A, No. 4, p. 196 – 199 (1984). Abstr. in Phys. Abstr., Vol. 87, No. 1245, Entry 110792 (1984).

161.261 **On the problems of a very early Universe.**
M. A. Markov, V. F. Mukhanov.
Phys. Lett. A, Vol. 104A, No. 4, p. 200 – 203 (1984). Abstr. in Phys. Abstr., Vol. 87, No. 1245, Entry 110793 (1984).

161.262 **Recombination of the primeval plasma and light–ions.**
P. Salati, J. C. Wallet.
Phys. Lett. B, Vol. 144B, No. 1 – 2, p. 61 – 68 (1984). Abstr. in Phys. Abstr., Vol. 87, No. 1245, Entry 110794 (1984).

161.263 **Universe and entropy.**
H. Sato.
Solid State Phys., Vol. 19, No. 6, p. 366 – 372 (1984). In Japanese. Abstr. in Phys. Abstr., Vol. 87, No. 1245, Entry 110795 (1984).

161.264 **The cosmological constant as a canonical variable.**
M. Henneaux, C. Teitelboim.
Phys. Lett. B, Vol. 143B, No. 4 – 6, p. 415 – 420 (1984). Abstr. in Phys. Abstr., Vol. 87, No. 1246, Entry 110973 (1984).

161.265 **Cosmological gravitino regeneration and decay.**
J. Ellis, J. E. Kim, D. V. Nanopoulos.
Phys. Lett. B, Vol. 145B, No. 3 – 4, p. 181 – 186 (1984). Abstr. in Phys. Abstr., Vol. 87, No. 1246, Entry 110975 (1984).

161.266 **Constraints on the mass of muon neutrinos and their possible role in the galaxy formation.**
M. Fukugita, T. Yanagida.
Phys. Lett. B, Vol. 144B, No. 5 – 6, p. 386 – 390 (1984). Abstr. in Phys. Abstr., Vol. 87, No. 1246, Entry 115826 (1984).

161.267 **Phase transitions and fluctuations in inflationary Universe models based on (nearly) Coleman–Weinberg GUT models.**
P. J. Steinhardt.
AIP Conf. Proc., No. 116, p. 147 – 165 (1984). Abstr. in Phys. Abstr., Vol. 87, No. 1246, Entry 115838 (1984). – See Abstr. 012.053.

161.268 **A new inhomogeneous cosmological model in general relativity.**
D. Kramer.
Classical Quantum Gravity, Vol. 1, No. 5, p. 611 – 618 (1984). Abstr. in Phys. Abstr., Vol. 87, No. 1246, Entry 115839 (1984).

161.269 **A class one cosmological model of electrified viscous fluid.**
K. P. Singh, O. N. Singh.
Proc. Natl. Acad. Sci. India, Sect. A, Vol. 53, Part 2, p. 135 – 141 (1983). Abstr. in Phys. Abstr., Vol. 87, No. 1246, Entry 115841 (1984).

161.270 **Matter–antimatter domains: a possible solution to the CP domain wall problem in the early Universe.**
A. K. Mohanty, F. W. Stecker.
Phys. Lett. B, Vol. 143B, No. 4 – 6, p. 351 – 356 (1984). Abstr. in Phys. Abstr., Vol. 87, No. 1246, Entry 115842 (1984).

161.271 **Formation of monopole–proton bound states in the hot Universe.**
L. Bracci, G. Fiorentini, G. Mezzorani, P. Quarati.
Phys. Lett. B, Vol. 143B, No. 4 – 6, p. 357 – 362 (1984). Abstr. in Phys. Abstr., Vol. 87, No. 1246, Entry 115843 (1984).

161.272 **Galaxy formation in a monopole dominated universe.**
J. Madsen.
Phys. Lett. B, Vol. 143B, No. 4 – 6, p. 363 – 367 (1984). Abstr. in Phys. Abstr., Vol. 87, No. 1246, Entry 115844 (1984).

161.273 **The inflationary universe in the Coleman–Weinberg theory.**
J. Okada.
Phys. Lett. B, Vol. 144B, No. 3 – 4, p. 183 – 186 (1984). Abstr. in Phys. Abstr., Vol. 87, No. 1246, Entry 115845 (1984).

161.274 **Kaluza–Klein cosmologies and inflation.**
R. B. Abbott, S. M. Barr, S. D. Ellis.
Phys. Rev. D, Vol. 30, No. 4, p. 720 – 727 (1984). Abstr. in Phys. Abstr., Vol. 87, No. 1246, Entry 115846 (1984).

161.275 **Search for higher–dimensional cosmologies.**
M. Gleiser, S. Rajpoot, J. G. Taylor.
Phys. Rev. D, Vol. 30, No. 4, p. 756 – 759 (1984). Abstr. in Phys. Abstr., Vol. 87, No. 1246, Entry 115847 (1984).

161.276 **Perfect fluid models of Bianchi type–I in modified Brans–Dicke cosmology.**
T. Singh, T. Singh.
J. Math. Phys., Vol. 25, No. 9, p. 2800 – 2802 (1984). Abstr. in Phys. Abstr., Vol. 88, No. 1247, Entry 157 (1985).

161.277 **Cosmology in *D*–dimensional gravity coupled to a rank–three–field $F_{MNP}$.**
Y. S. Myung, Y. D. Kim.
Phys. Lett. B, Vol. 145B, No. 1 – 2, p. 45 – 48 (1984). Abstr. in Phys. Abstr., Vol. 88, No. 1247, Entry 159 (1985).

161.278 **Higgs scalar densities and cosmology.**
A. Love, J. Stow.
Nucl. Phys. B, Part. Phys., Vol. B243, No. 3, p. 537 – 549 (1984). Abstr. in Phys. Abstr., Vol. 88, No. 1247, Entry 4897 (1985).

161.279 **One step on from the moment of creation.**
J. Gribbin.
New Sci., Vol. 103, No. 1416, p. 30 – 31 (1984). Abstr. in Phys. Abstr., Vol. 88, No. 1247, Entry 4899 (1985).

161.280 **Expansion of voids in a matter–dominated universe.**
Y. Suto, K. Sato, H. Sato.
Prog. Theor. Phys., Vol. 71, No. 5, p. 938 – 945 (1984). Abstr. in Phys. Abstr., Vol. 88, No. 1247, Entry 4900 (1985).

161.281 **Generation and evolution of the density fluctuation in the evolving Universe. I. The effect of weak transient phenomena.**
H. Kodama.
Prog. Theor. Phys., Vol. 71, No. 5, p. 946 – 959 (1984). Abstr. in Phys. Abstr., Vol. 88, No. 1247, Entry 4901 (1985).

161.282 **The homogeneous, isotropic Universe in new general relativity.**
M. Fukui, T. Shirafuji.
Prog. Theor. Phys., Vol. 71, No. 5, p. 1063 – 1073 (1984). Abstr. in Phys. Abstr., Vol. 88, No. 1247, Entry 4902 (1985).

161.283 **Contributions to evolutive mechanics and to its cosmological extension.**
L. Herscu.
Rev. Roum. Phys., Vol. 29, No. 6, p. 503 – 507 (1984). Abstr. in Phys. Abstr., Vol. 88, No. 1247, Entry 4905 (1985).

161.284 **Cosmic density wave and voids in the universe.**
Y. Liu.
Kexue Tongbao (Beijing), Vol. 29, No. 4, p. 490 – 492 (1984).
The author draws the following conclusions: 1. The voids in the universe should appear near or between the superclusters, and they correspond to the vestige of the cosmic density wave pattern which may have happened in the early universe. 2. The sizes of the observed voids and superclusters should be of the same order of magnitude. 3. For the big–bang Friedmann model of the universe the upper limit scale of the observable voids and superclusters is about 170 – 250 Mpc, corresponding to the range of the redshift $0.5 > z > 0$.

161.285 **Statistical implications of a neutrino universe.**
A. G. Doroshkevich.
Sov. Astron., Vol. 28, No. 3, p. 253 – 257 (1984). English translation of 37.161.156.

161.286 **Quasilinear relaxation of the distribution function for collisionless, relativistic gravitating particles in an expanding universe.**
B. V. Vajner, P. D. Nasel'skij.
Sov. Astron., Vol. 28, No. 3, p. 264 – 269 (1984). English translation of 37.161.157.

161.287 **Molecular hydrogen as a probe of the early universe.**
Yu. A. Shchekinov, M. B. Éntél' (*M. B. Ehntehl'*).
Sov. Astron., Vol. 28, No. 3, p. 270 – 274 (1984). English translation of 37.161.158.

161.288 **Another cosmological redshift theory.**
D. C. Wilson.
Publ. Astron. Soc. Pac., Vol. 96, No. 584, p. 798 (1984). Abstract. – See Abstr. 010.281.

161.289 **Is there really a large scale geometric anisotropy in extragalactic radio sources?**
M. F. Bietenholz, P. P. Kronberg.
J. R. Astron. Soc. Can., Vol. 78, No. 5, p. 203 – 204 (1984). Abstract. – See Abstr. 010.401.

161.290 **The red–shift structure of the big bang in inhomogeneous cosmological models.**
C. Hellaby, K. Lake.
J. R. Astron. Soc. Can., Vol. 78, No. 5, p. 213 (1984). Abstract. – See Abstr. 010.401.

161.291 **Relativistic cosmology: its nature, aims and problems.**
G. F. R. Ellis.
General relativity and gravitation, p. 215 – 288 (1984). – See Abstr. 012.060.

161.292 **Voids in the expanding universe.**
H. Sato.
General relativity and gravitation, p. 289 – 312 (1984). – See Abstr. 012.060.
Enlargement of a spherical void in the expanding universe is discussed. In a nonlinear regime of density perturbation, the low–density region or a void behaves like an explosive source and the dense thin shell is formed by this explosion of the void. Israel's formalism for an infinitely thin shell is applied to derive a relativistic relation, which is analysed for various models of the universe.

161.293 **Cosmological parameters and cosmological models. Report of workshop B3 (*10th International Conference on General Relativity and Gravitation, Padua, July 3 – 8, 1983*).**
I. D. Novikov.
General relativity and gravitation, p. 337 – 341 (1984). – See Abstr. 012.060.

161.294 **On the stochastic properties of relativistic cosmological models near the singularity.**
I. M. Khalatnikov, E. M. Lifshitz (*E. M. Lifshits*), K. M. Khanin, L. N. Shchur, Ya. G. Sinai.
General relativity and gravitation, p. 343 – 349 (1984). – See Abstr. 012.060.

161.295 **The early universe. Report of workshop B2 (*10th International Conference on General Relativity and Gravitation, Padua, July 3 – 8, 1983*).**
J. B. Hartle.
General relativity and gravitation, p. 351 – 354 (1984). – See Abstr. 012.060.

161.296 **The quantum mechanics of the universe.**
S. W. Hawking.
General relativity and gravitation, p. 491 – 498 (1984). – See Abstr. 012.060.
Classical general relativity predicts that the universe had a singular origin. The author shows that the singularity can be removed by quantum mechanics, just as in the case of the classical model of the atom. He proposes that the quantum state of the universe is defined by a path integral over compact positive definite metrics. It is shown that in a simple model this boundary condition leads to a wave function which can be regarded as a superposition of wave functions peaked around classical oscillating solutions with a long inflationary period.

161.297 **Small–scale anisotropies of the microwave background in model universes dominated by dark matter.**
S. A. Bonometto, F. Lucchin, R. Valdarnini.
Astron. Astrophys., Vol. 140, No. 2, p. L27 – L30 (1984).
The authors give the results of detailed computations of small–scale fluctuations in the MW background for flat models of the universe, dominated either by massive 'neutrinos' (m = 32 eV) or by low–mass 'photino' or 'gravitinos' (m = 880 eV). They consider different values for the spectral index of primeval fluctuations and normalize the results to the mean mass variance $\delta M/M$. The authors find that a 'neutrino' dominated model can be in conflict with MW observational limits, if $\delta M/M$, on the smallest scales, is 1 at a redshift $z_{nl} > 3.4$. The 'photino' or 'gravitino' dominated cases are in conflict with observations only if $z_{nl} > 10$ (for spectral index $n = -1$) or for even higher $z_{nl}$ if $n > -1$. In this case the dependence on n is much stronger than in the neutrino case.

161.298 **Homogeneous cosmological models and non–linear electrodynamics.**
S. Hacyan.
Astron. Astrophys., Vol. 140, No. 2, p. 363 – 367 (1984).
A Bianchi Type I cosmological model with a homogeneous electromagnetic field is studied using the Born–Infeld non–linear electrodynamics. With this approach, it is possible to describe the behaviour of the Universe near the initial singularity, even when the magnetic induction becomes infinite. It is found that the electromagnetic field does not influence in any appreciable way

the expansion of the Universe. Implications for other Bianchi models are briefly discussed.

161.299 **The virialized regime of non-linear clustering in an expanding universe.**
F. R. Bouchet, R. Pellat.
Astron. Astrophys., Vol. 141, No. 1, p. 77 – 84 (1984).
Starting with the Vlasov set of BBGKY hierarchy equations, the authors give an explicit connection between the work of Mandelbrot and that of Davis and Peebles concerning the observed mass-clustering in our universe. They obtain a simple reduced set of exact equations including the two-point density, the velocity and the temperature tensor correlation functions. Solutions based on a virialization hypothesis at short distances are derived.

161.300 **Gravitational amplification of light at relativistic distance.**
L. Nottale, F. Hammer.
Astron. Astrophys., Vol. 141, No. 1, p. 144 – 150 (1984).
The optical scalar equations (O.S.E.) are used to derive solutions for the gravitational amplification of light by the contribution of matter in the beam from a transparent lens lying at relativistic distance. The authors first obtain the general solution of the null shear O.S.E. in Friedmann universes in terms of a new distance, and are able to describe the distance dependence of amplification without any hypothesis on the lens. Then detailed analytic solutions are found for a thin lens.

161.301 **The Hubble parameter: an upper limit from QSO 0957+561 A, B.**
U. Borgeest, S. Refsdal.
Astron. Astrophys., Vol. 141, No. 2, p. 318 – 322 (1984).
It is shown that the cluster of galaxies which contributes to the light deflection by the double quasar can only reduce the time delay $\Delta t$ between light variations of the images. Assuming a King type deflecting galaxy the authors give an upper limit for the Hubble parameter of $H_0 \leqslant 200\ \mathrm{km\ s^{-1} Mpc^{-1}}/\Delta t(\mathrm{yr})$ or a lower limit for the Hubble age of $H_0^{-1} \geqslant 5 \times 10^9 \Delta t$.

161.302 **Malmquist bias in a relation of the form $M = a \cdot p + b$.**
P. Teerikorpi.
Astron. Astrophys., Vol. 141, No. 2, p. 407 – 410 (1984).
Malmquist bias is discussed in the case where absolute magnitudes are derived from a relation $M = a \cdot p + b$ where $p$ is a distance-independent measurable parameter. It is shown that the amount of bias at constant observed $p$ depends on (1) the intrinsic scatter of the relation, (2) the observational random errors in $p$, and (3) the luminosity function of the objects. A generalized Malmquist formula for the case where the luminosity function is gaussian, is given. The importance of the observational error in $p$ is emphasized. The distance-dependence of the bias as averaged over $p$ is discussed.

161.303 **Quantum creation of a universe with nontrivial topology.**
Ya. B. Zel'dovich, A. A. Starobinskij.
Sov. Astron. Lett., Vol. 10, No. 3, p. 135 – 137 (1984). English translation of 37.161.146.

161.304 **Cosmic-ray antiprotons as a probe of a photino-dominated universe.**
J. Silk, M. Srednicki.
Phys. Rev. Lett., Vol. 53, No. 6, p. 624 – 627 (1984).
Observational tests of the hypothesis that the universe is flat and dominated by dark matter in the form of massive photinos include the production of significant fluxes of cosmic rays and gamma rays in the galactic halo. Specification of the cosmological photino density and the masses of scalar quarks and leptons determines the present annihilation rate. The predicted number of low-energy cosmic-ray antiprotons is comparable to the observed flux.

161.305 **Inflationary cosmology and the mass hierarchy in locally supersymmetric theories.**
B. A. Ovrut, P. J. Steinhardt.
Phys. Rev. Lett., Vol. 53, No. 7, p. 732 – 735 (1984).
The authors present an $N = 1$ locally supersymmetric theory in which all constraints for a successful new-inflationary-universe scenario are satisfied and the gravitino mass is on the order of the electroweak scale. Sufficient baryon asymmmerty is generated and the overproduction of entropy due to gravitinos and/or coherent field oscillations (which is typical of supersymmetric models) is avoided.

161.306 **String-dominated universe.**
A. Vilenkin.
Phys. Rev. Lett., Vol. 53, No. 10, p. 1016 – 1018 (1984).
The evolution of a universe with $\Omega = 1$ which becomes dominated by relatively light, nonintercommuting strings at redshift $z_s$ is similar to that of a matter-dominated universe with $\Omega \sim (2+z_s)^{-1} < 1$. The existence of such strings can resolve the conflict between the $\Omega = 1$ prediction of inflationary cosmology and observations.

161.307 **Implications of a class of grand-unified theories for large-scale structure in the universe.**
Q. Shafi, F. W. Stecker.
Phys. Rev. Lett., Vol. 53, No. 13, p. 1292 – 1295 (1984).
The authors consider a class of grand-unified theories in which cosmologically significant axion and neutrino energy densities arise naturally. To obtain large-scale structure they consider (1) an inflationary scenario, (2) inflation followed by string production, and (3) a noninflationary scenario with density fluctuations caused solely by strings. The authors show that inflation may be compatible with the recent observational indications that $\Omega < 1$ on the scale of superclusters, particularly if strings are present.

161.308 **Cosmic strings and galaxy formation.**
J. Silk, A. Vilenkin.
Phys. Rev. Lett., Vol. 53, No. 17, p. 1700 – 1703 (1984).
Accretion onto closed loops of vacuum string results in the formation of compact objects of subgalactic mass which are comparable in number to the frequency of active galactic nuclei and quasars. Such structure develops in baryon- and neutrino-dominated, as well as in axion-dominated, cosmological models. Planar wakes also develop behind relativistically moving open strings and generate fluctuations of order unity in the galaxy density on scales comparable to those of galaxy superclusters.

161.309 **Smoothing out spatially closed cosmologies.**
M. Carfora, A. Marzuoli.
Phys. Rev. Lett., Vol. 53, No. 25, p. 2445 – 2448 (1984).
The authors discuss a smoothing-out procedure that deforms a family of locally inhomogeneous and anisotropic spatially closed cosmological models into closed Friedmann-Robertson-Walker universes. This gives a precise content to the averaging hypothesis tacitly assumed in cosmology by providing explicitly the correction terms to the physical sources induced on smoothing out the space-time geometry. The consequences of such terms on the dynamics of the Universe are discussed.

161.310 **Distant galaxies and the determination of the deceleration parameter.**
R. G. Kron.
Spectral evolution of galaxies, p. 190 – 204 (1984). – See Abstr. 012.066.
This paper reviews recent progress in the attempt to determine the cosmological deceleration parameter $q_0$ from studies of distant galaxies in clusters. It is concluded that at present no reliable measurement of $q_0$ is available, but that a useful upper limit may be established.

161.311 **A radiation- and matter-filled universe in Brans–Dicke cosmology.**
P. Chauvet.
Astrophys. Space Sci., Vol. 106, No. 1, p. 207 – 209 (1984).
In attempting to obtain the solution to a more realistic isotropic flat space model of the Universe filled with noninteracting "radiation and nonrelativistic matter (dust)" in the Brans–Dicke cosmological theory (BDT), several general properties of this theory, are obtained. In contrast to general relativity theory (GRT), where the solution to a mixture of noninteracting "radiation and dust" is the sum of the "radiation" plus "dust" solutions taken separately, in BDT this is not so. Yet, the numerical value of the integration constant ν can make the BDT solution almost indistinguishable from GRT, in the radiation dominated epoch.

161.312 **Some special non-flat BDT–FRW solutions with a cosmological constant.**
D. Lorenz-Petzold.
Astrophys. Space Sci., Vol. 106, No. 1, p. 211 – 212 (1984).
The author considers the non-flat $k^2 = 1$ Friedmann–Robertson–Walker models with a non-vanishing cosmological constant in the Brans–Dicke theory of gravitation. Some special solutions are given in case of $\omega = 0$, which may be considered as exact solutions of the projective theory of Jordan and Thiry.

161.313 **Exact Brans–Dicke–Bianchi type-VII$_h$ solutions.**
D. Lorenz-Petzold.
Astrophys. Space Sci., Vol. 106, No. 2, p. 409 – 413 (1984).

161.314 **Some remarks on certain special BDT–FRW (*Brans–Dicke theory – Friedmann–Robertson–Walker models*) solutions.**
D. Lorenz-Petzold.
Astrophys. Space Sci., Vol. 106, No. 2, p. 415 – 417 (1984).

161.315 **Dynamical friction due to cosmological background. II. Newtonian cosmology.**
P. Hadrava, V. Karas.
Bull. Astron. Inst. Czech., Vol. 35, No. 6, p. 343 – 346 (1984).
The perturbation of the Newtonian cosmological model by a moving gravitating body is calculated. The narrow, infinitely massive tail of background dust, condensed in the trace of the body, is shown to cause dynamical friction, which is negligible in the course of cosmological evolution, if the present velocity of the body is sufficiently high.

161.316 **Big bang cosmology, relic neutrinos, and absorption of neutrino cosmic rays.**
T. Weiler.
Astrophys. J., Vol. 285, No. 2, p. 495 – 500 (1984).
The proposal to search for relic neutrinos through their absorption effect on the neutrino cosmic-ray flux is carefully examined. The author shows that detection of the neutrino flux in the absorption region requires an improbable power output for a *single* source. A summation over a distribution of sources is then performed, using the quasar distribution as a model. It is concluded that within the context of big bang cosmology, detection of relic neutrinos through their absorption effects depends crucially on whether or not there exist sources at large $z$, emitting a sufficient flux of ultraenergy neutrinos.

161.317 **Fine-scale anisotropy of the cosmic microwave background in a universe dominated by cold dark matter.**
N. Vittorio, J. Silk.
Astrophys. J., Lett. Ed., Vol. 285, No. 2, p. L39 – L43 (1984).
The fine-scale anisotropy of the cosmic microwave background radiation has been studied in cosmological models with a scale-invariant primordial adiabatic density fluctuation spectrum that are dominated by cold, weakly interacting particles such as axions or photinos. Normalization of the present fluctuation spectrum to the observed galaxy distribution results in excessive temperature anisotropy when compared to a recent upper limit on $4\overset{'}{.}5$ unless the density parameter $\Omega_0$ exceeds 0.4 ($50\ \mathrm{km\ s^{-1} Mpc^{-1}}/H_0$).

161.318 **Cosmic background radiation anisotropies in universes dominated by nonbaryonic dark matter.**
J. R. Bond, G. Efstathiou.
Astrophys. J., Lett. Ed., Vol. 285, No. 2, p. L45 – L48 (1984).
The authors present detailed calculations of the temperature fluctuations in the cosmic background radiation for universes dominated by massive collisionless relics of the big bang. They assume an initially adiabatic constant curvature perturbation spectrum. In models with cold dark matter, the simplest hypothesis – that galaxies follow the mass distribution – leads to small-scale anisotropies which exceed current observational limits if $\Omega < 0.2\, h^{-4/3}$. Since low values of $\Omega$ are indicated by dynamical studies of galaxy clustering, cold particle models in which light traces mass are probably incorrect.

161.319 **The Malmquist bias and the value of $H_0$ from the Tully–Fisher relation.**
L. Bottinelli, L. Gouguenheim, G. Paturel, P. Teerikorpi.
Rep. Ser., Dep. Phys. Sci., Univ. Turku, No. FTL-R71, 45 pp. (1984). = Turku Univ. Obs., Informo No. 87. To appear in Astron. Astrophys.

161.320 **Local structure, isotropy, and stability in a static, steady-state universe.**
T. Jaakkola.
Rep. Ser., Dep. Phys. Sci., Univ. Turku, No. FTL-R68, p. 13 (1984). – See Abstr. 012.079.

161.321 **Gravitational clustering for a multicomponent system with a distribution of temperatures.**
H. E. Kandrup.
Astrophys. J., Vol. 286, No. 1, p. 15 – 26 (1984).
This paper examines the onset of gravitational clustering, in the context of a homogeneous and isotropic Newtonian cosmology, for a complicated system of point mass "particles", e.g., galaxies, characterized by a distribution of masses and temperatures. The principal conclusion is that, to the extent that nonlinear and collective effects may be neglected, the spatial clustering of any two species of particles may be described as if the two constituents were characterized by a single effective temperature, the value of which depends only upon the temperatures of the two constituents, and the ratio of their masses.

161.322 **Attenuation of the spectrum of relic radiation by disturbances at an early stage of expansion of the universe with a magnetic field.**
A. S. Il'chenko.
Vopr. teor. i yader fiz., Saratov, No. 9, p. 101 – 110 (1983). In Russian. Abstr. in Ref. Zh., 51. Astron., 12.51.1114 (1984).

161.323 **Quantum generation of the inflationary universe.**
A. D. Linde.
Zh. Ehksp. Teor. Fiz., Tom 87, No. 2, p. 369 – 374 (1984). In Russian. Abstr. in Ref. Zh., 51. Astron., 12.51.1124 (1984).

161.324 **Cosmological transitions with change of the sign of metrics.**
A. D. Sakharov.
Zh. Ehksp. Teor. Fiz., Tom 87, No. 2, p. 375 – 383 (1984). In Russian. Abstr. in Ref. Zh., 51. Astron., 12.51.1138 (1984).

161.325 **Cosmological model of a rotating universe with magnetic field.**
V. G. Agakov.
Modern theoretical and experimental problems of the theory of relativity and gravitation (1984). In Russian. Abstr. in Ref. Zh., 51. Astron., 12.51.1145 (1984). – See Abstr. 012.080.

161.326 **The homogeneity of the universe and anisotropy of the background radiation.**
V. F. Mukhanov, G. V. Chibisov.
Pis'ma Astron. Zh., Tom 10, No. 12, p. 890 – 895 (1984). In Russian. English translation in Sov. Astron. Lett., Vol. 10.
The background radiation fluctuations due to scalar type cosmological perturbations are calculated using relativistic potential

gauge. The observational upper limit of the quadrupole anisotropy of the background radiation restricts the characteristic scale of perturbations which are larger than the transparency region. The sizes of perturbations with amplitude of the order of unity must be at least forty–five times as large as the observational horizon.

161.327 **Multipole harmonics of the relic background and the unresolved–sources background.**
V. A. Korolev, M. V. Sazhin.
Astron. Zh., Tom 61, Vyp. 6, p. 1054 – 1062 (1984). In Russian. English translation in Sov. Astron., Vol. 28, No. 6.

The fluctuations of the unresolved–sources background are considered. A method of calculation of the correlation function of temperature fluctuations produced by radio sources is developed. Multipole components of the flux and of the effective temperature at the wavelength 8 mm are calculated. This wavelength is the one for the experiment RELIC. The dispersion of the temperature is 6 μK. The quadrupole component is approximately ~1 μK.

161.328 **Friedmann's cosmology.**
O. Godart, M. Heller.
Astron. Q., Vol. 5, No. 17, p. 5 – 12 (1984).

161.329 **Logarithmic minus infinity.**
M. Heller, O. Godart.
Astron. Q., Vol. 5, No. 17, p. 13 – 25 (1984).

161.330 **Determination of cosmic time intervals.**
K. Rudnicki.
Astron. Q., Vol. 5, No. 17, p. 27 – 38 (1984).

161.331 **Some outstanding problems in cosmology.**
A. K. Raychaudhuri.
Gravitation and relativistic astrophysics, p. 113 – 119 (1984). – See Abstr. 012.083.

161.332 **Axial and skew–axial homogeneous cosmologies.**
M. A. Melvin.
Gravitation and relativistic astrophysics, p. 121 – 124 (1984). – See Abstr. 012.083.

161.333 **Rotating systems in cosmological background.**
P. C. Vaidya.
Gravitation and relativistic astrophysics, p. 125 – 131 (1984). – See Abstr. 012.083.

One outstanding problem in gravitational collapse is "the pre-collapse configuration with rotation". As a first step, this paper presents a mathematical treatment of a rotating system in a cosmological background.

161.334 **Quantum fluctuations near the classical space–time singularity.**
J. V. Narlikar.
Gravitation and relativistic astrophysics, p. 135 – 143 (1984). – See Abstr. 012.083.

161.335 **Stationary states in quantum cosmology.**
T. Padmanabhan.
Gravitation and relativistic astrophysics, p. 145 – 151 (1984). – See Abstr. 012.083.

161.336 **Cosmology, elementary particles and the regularity of the universe.**
J. D. Barrow.
Relativistic astrophysics and cosmology, p. 137 – 199 (1984). – See Abstr. 012.084.

An introductory survey is given of the standard hot big bang model with special reference to new developments in elementary particle physics and the structure of the microwave background radiation in anisotropic and inhomogeneous cosmological models.

161.337 **Einstein's equations and some cosmological solutions.**
S. T. C. Siklos.
Relativistic astrophysics and cosmology, p. 201 – 248 (1984). – See Abstr. 012.084.

This paper gives a very general discussion of Einstein's equations, followed by a more detailed examination of a few exact solutions. It discusses in particular spatially homogeneous solutions and the singularities in these solutions with emphasis on the cosmological singularity. Finally, some diverse topics including supergravity, inflation and gravitational instantons are reviewed.

161.338 **Observational cosmology and large scale structure.**
J. Silk.
Relativistic astrophysics and cosmology, p. 249 – 293 (1984). – See Abstr. 012.084.

Observational cosmology is reviewed, with emphasis on determination of two parameters that are of especial interest to studies of the very early universe. These are the density parameter Ω and the amplitude of the primordial density fluctuations at horizon crossing. The role of weakly interacting particles such as massive neutrinos, photinos, or axions that may dominate the present mass density is evaluated. Implications for the large–scale galaxy distribution are examined.

161.339 **An analysis on the clustering time of the relic neutrinos.**
X.–h. Xiao, S.–l. Cao, G.–l. Zhu.
Acta Astron. Sin., Vol. 25, No. 3, p. 250 – 254 (1984). In Chinese.

161.340 **New methods of investigation for extragalactic backgrounds in the infrared.**
C. Ceccarelli, G. Dall'Oglio, P. de Bernardis, S. Masi, B. Melchiorri, F. Melchiorri, G. Moreno, L. Pietranera.
Clusters and groups of galaxies, p. 277 – 278 (1984). – See Abstr. 012.088.

161.341 **Thermal radiation noise and the 3K background.**
C. Ceccarelli, G. Dall'Oglio, P. de Bernardis, S. Masi, B. Melchiorri, F. Melchiorri, G. Moreno, L. Pietranera.
Clusters and groups of galaxies, p. 283 – 284 (1984). – See Abstr. 012.088.

161.342 **The voids and the cosmic microwave photons.**
C. Ceccarelli, G. Dall'Oglio, P. de Bernardis, S. Masi, B. Melchiorri, F. Melchiorri, G. Moreno, L. Pietranera.
Clusters and groups of galaxies, p. 285 – 286 (1984). – See Abstr. 012.088.

161.343 **Nonlinear evolution of large–scale density structure.**
A. R. Rivolo, A. Yahil.
Clusters and groups of galaxies, p. 429 – 434 (1984). – See Abstr. 012.088.

161.344 **Growth of perturbations between horizon crossing and matter dominance: implications for galaxy formation.**
J. R. Primack, G. R. Blumenthal.
Clusters and groups of galaxies, p. 435 – 440 (1984). – See Abstr. 012.088.

161.345 **Radiation perturbations in a three component universe.**
N. Vittorio, F. Occhionero, F. Lucchin, S. Bonometto.
Clusters and groups of galaxies, p. 441 – 446 (1984). – See Abstr. 012.088.

The authors study numerically the evolution of initial adiabatic perturbations in a universe with critical density containing three components: massive neutrinos or the like, baryonic matter and radiation. The evaluation at decoupling of the residual perturbation in the radiation field shows that the common expectations on the neutrino influence on the monopole and the dipole components are likely to be modified. Thus, the lack of detection of distortions in the cosmic microwave background may remain an unresolved problem.

161.346 **Pancakes and galaxy formation.**
P. R. Shapiro.
Clusters and groups of galaxies, p. 447 – 478 (1984). – See Abstr. 012.088.
The pancake scenario for the formation of large–scale structure in the Universe by the growth of a spectrum of primordial, adiabatic density perturbations is reviewed.

161.347 **Curvature fluctuations as progenitors of large scale holes.**
N. Vittorio, P. Santangelo, F. Occhionero.
Clusters and groups of galaxies, p. 479 – 484 (1984). – See Abstr. 012.088.
The authors extend previous work to study the formation and evolution of deep holes, under the assumption that they arise from curvature or energy perturbations in the Hubble flow. The algorithm, which makes use of the spherically symmetric and pressureless Tolman–Bondi solution, can embed a perturbation in any cosmological background. After recalling previous results on the central depth of the hole and its radial dimension, the authors give specific examples of density and peculiar velocity profiles, which may have a bearing on whether galaxy formation is a dissipative or dissipationless process.

161.348 **Clusters and their precursors.**
M. J. Rees.
Clusters and groups of galaxies, p. 485 – 497 (1984). – See Abstr. 012.088.
Contents: Three possible forms for the post–recombination spectrum of fluctuations. The first bound baryonic systems: population III. Does baryonic matter contribute all the "unseen" mass? Feedback effects on uncondensed gas. Formation of "secondary" large–scale fluctuations by pregalactic objects. Is primordial nucleosynthesis of light elements compatible with $\Omega_b \cong 1$?

161.349 **The non–linear evolution of the clustering length in the "pancake" scenario.**
K. Górski, R. Juszkiewicz.
Clusters and groups of galaxies, p. 505 – 507 (1984). – See Abstr. 012.088.

161.350 **The inhomogeneity of invisible matter in an expanding universe.**
L. Z. Fang.
Clusters and groups of galaxies, p. 509 – 513 (1984). – See Abstr. 012.088.

161.351 **Cosmic string–loops and galaxy formation – angular momentum and the Hubble sequence.**
P. Corbyn.
Clusters and groups of galaxies, p. 517 – 519 (1984). – See Abstr. 012.088.
The cosmic string–loop model is developed whereby supermassive string–loops form nucleation centres for galaxies. Estimation of the interaction of string–loops with matter gives acceptable decay times and radiative power for (proto) galaxy cores. The modes of formation and angular momentum properties of string–loops suggest an explanation for the elliptical/spiral discontinuity and a modified Hubble tuning–fork sequence.

161.352 **The influence of environment on galaxy formation.**
E. J. Shaya, R. B. Tully.
Clusters and groups of galaxies, p. 521 – 522 (1984). – See Abstr. 012.088.

161.353 **Adiabatic perturbations in a photino dominated universe.**
R. Valdarnini.
Clusters and groups of galaxies, p. 525 – 526 (1984). – See Abstr. 012.088.

161.354 **The Hubble diagram.**
G. A. Tammann.
Clusters and groups of galaxies, p. 529 – 552 (1984). – See Abstr. 012.088.
Contents: The importance of the Hubble diagram. Early development. Deceptive standard candles. Hubble diagram, peculiar motions, and Hubble constant.

161.355 **Magneto–viscous fluid cosmological model of plane symmetry in general relativity.**
R. Bali.
Astrophys. Space Sci., Vol. 107, No. 1, p. 155 – 165 (1984).
The object of this paper is to investigate the behavior of magnetic field in a viscous fluid cosmological model. Various physical and geometrical properties of the model have also been discussed.

161.356 **Relation between physical and metric field in general relativity and cosmology.**
S. Selak.
Astrophys. Space Sci., Vol. 107, No. 2, p. 357 – 371 (1984).
By assuming that the whole matter of the Universe possesses a dilatational degree of freedom, the author attempted to show the equivalence between the curved metric of space–time and the flat metric of dilated space–time. He supposed that the metrical field and the physical space and time change their roles. The basic result suggests that gravitation and dilatation are interrelated phenomena. In addition the author discusses the possibility of the Universe which is of a hybrid type: it possesses at the same time properties of the evolutionary and stationary Universe. Finally, the lenghtening of day in time is discussed.

161.357 **Validity of Mach's principle in a hybrid Universe model.**
S. Selak.
Astrophys. Space Sci., Vol. 107, No. 2, p. 409 – 411 (1984).
The author discusses the validity of Mach's principle using its non–relativistic interpretation according to Sciama (1961). The procedure was performed in the framework of a hybrid Universe model proposed by Selak (1979, 1980, 1984). This model possesses at the same time properties of the evolutionary and the stationary Universe.

161.358 **Cosmological constant and rotation of the Universe.**
R. M. Muradyan.
Astrofizika, Tom 21, Vyp. 2, p. 396 – 398 (1984). In Russian. English translation in Astrophysics, Vol. 21, No. 2.

161.359 **Behavior of higher modes of gravitational waves and gauge–invariant density perturbations in Bianchi IX cosmological models.**
R. J. Slagter.
Astrophys. J., Vol. 286, No. 2, p. 379 – 386 (1984).
High–frequency gravitational waves considered as "ripples" in a given background cosmological model are studied in a time–dependent perturbation expansion. The effects of back–reaction of gravitational waves on the background metric are investigated and serve as a preliminary description of quantum effects in anisotropic universes. The asymptotic features of the gravitational perturbations are derived. Finally, a gauge–invariant density perturbation is constructed in the anisotropic Bianchi IX model.

161.360 **Limits on a stochastic gravitational wave background from observations of terrestrial and solar oscillations.**
S. P. Boughn, J. R. Kuhn.
Astrophys. J., Vol. 286, No. 2, p. 387 – 391 (1984).
Realistic models of the Earth and Sun are used to calculate their respective responses to a homogeneous, isotropic background of gravitational radiation. Solar velocity data constrain the energy density of such a background at a frequency of $4 \times 10^{-4}$Hz to be less than $10^2$ times the closure density of the universe, as does Earth seismic data at frequencies of $2 \times 10^{-3}$Hz and $2 \times 10^{-2}$Hz. Errors in previous analyses of the Earth are pointed out.

161.361 **Some observable effects in relativistic cosmology with non–perfect fluids.**
L. J. Goicoechea, J. L. Sanz.
Astrophys. J., Vol. 286, No. 2, p. 392 – 396 (1984).
The authors consider non–perfect fluid distributions as sources of Friedmann–Robertson–Walker (FRW) geometries. They study the propagation of cosmic background radiation (CBR) with the FRW geometries, obtaining as observable effects a dipole and a quadrupole component in the temperature of the CBR due to Doppler shift. In the case of a flat geometry, the authors obtain the conditions satisfied by a homogeneous velocity field in order to have viscous and heat conduction terms, and they give different values and upper limits on the shear–viscosity coefficient, the relative shear, and the relative acceleration.

161.362 **Supermassive monopole stars.**
J. N. Fry, G. M. Fuller.
Astrophys. J., Vol. 286, No. 2, p. 397 – 402 (1984).
The authors reexamine the evolution of a superheavy monopole–dominated universe; in particular, they examine the evolution of rotation–stabilized monopole–burning "stars", balancing effects of radiation and annihilation. In contrast to earlier results, the authors find that this can lead to complete burning of all poles present in such an object within finite time. If monopoles can be brought within acceptable limits by dynamical processes, an inflationary epoch of the early universe need not be associated with grand unification.

161.363 **The dependence on distance and redshift of the velocity vectors of the Sun, the Galaxy, and the Local Group with respect to different extragalactic frames of reference.**
G. de Vaucouleurs, W. L. Peters.
Astrophys. J., Vol. 287, No. 1, p. 1 – 16 (1984).
A new analysis of the solar motion and Hubble ratio derived from a sample of 600 "best observed" galaxies of all types having distance moduli estimated by one to four independent methods confirms a phenomenon first reported in 1968: the solar apex $S$ moves steadily from $S'$ ($V' = 336\ \mathrm{km\,s^{-1}}$ toward supergalactic coordinates $L' = 338°$, $B' = +27°$) to $S^*$($V^* = 366\ \mathrm{km\,s^{-1}}$ toward $L^* = 112°$, $B^* = -17°$) when the mean redshift of the reference frame increases from small values (Local Group, $z = 0$) to large values (background radiation). The mean values of the Hubble ratio derived from galaxies in six different distance intervals ranging from $\sim 6$ to $\sim 36$ Mpc are shown to be constant and independent of distance. The overall mean value $\langle H^* \rangle = 101 \pm 2$ (internal m.e.) or $\pm 10$, including zero point errors, should be a close approximation of the Hubble constant.

161.364 **Is there really evidence for universal rotation?**
M. F. Bietenholz, P. P. Kronberg.
Astrophys. J., Lett. Ed., Vol. 287, No. 1, p. L1 – L2 (1984).
The authors investigate claims in the recent literature for the existence of a large–scale (out to $z \geqslant 1$) anisotropy in the properties of extragalactic radio sources in the universe and conclude, from an independent analysis of a new, enhanced sample of radio sources, that there is no evidence for a large–scale anisotropy of this type, or for universal rotation.

161.365 **Is the universe flat?**
M. J. Rees.
J. Astrophys. Astron., Vol. 5, No. 4, p. 331 – 348 (1984).
The evidence for unseen mass (which is briefly reviewed) suggests that the cosmological density parameter $\Omega$ is at least 0.1 – 0.2. An Einstein–de Sitter "flat" universe with $\Omega = 1$ – which is appealing for theoretical reasons – can only be reconciled with the data if the galaxies are more "clumped" than the overall mass distribution, and are poor tracers of the unseen mass even on scales of several Mpc. Possible forms for the unseen mass are discussed; and feedback processes are outlined whereby galaxy formation can be suppressed in underdense regions.

161.366 **Some rotating, time–dependent Bianchi type VIII cosmologies with heat flow.**
J. M. Bradley, E. Sviestins.
Gen. Relativ. Gravitation, Vol. 16, No. 12, p. 1119 – 1133 (1984).
A number of explicit, rotating and expanding cosmological solutions of the Einstein field equations are presented. They have homogeneous hypersurfaces of Bianchi type VIII, and they are locally rotationally symmetric. The thermodynamic properties of the models are discussed.

161.367 **Solitonic gravitational waves in Bianchi II cosmologies. 2. One solitonic perturbations.**
V. Belinskii (*V. A. Belinskij*), M. Francaviglia.
Gen. Relativ. Gravitation, Vol. 16, No. 12, p. 1189 – 1209 (1984).
This paper is a continuation of a previous one, in which the general framework for constructing solitonic perturbations on Bianchi II background was described. Here the authors calculate the exact solitonic perturbation corresponding to one (real) pole and investigate in detail some of its main features.

161.368 **Comment on the Brans–Dicke–Bianchi type–I solution.**
D. Lorenz–Petzold.
Gen. Relativ. Gravitation, Vol. 16, No. 12, p. 1211 – 1213 (1984).
The Brans–Dicke–Bianchi type–I vacuum solution given in a recent paper by Shri Ram is shown to be identical with the general solution first given by Ruban and Finkelstein.

**Die Physik der Zeitrichtung.**
See Abstr. 003.016.

**Kosmologie heute. Eine Einführung in ihre philosophischen und naturwissenschaftlichen Problemkreise.**
See Abstr. 003.020.

**O universo inflacionário.**
See Abstr. 003.022.

**Urknall und Evolution des Kosmos – Fortschritte in der Kosmologie.**
See Abstr. 003.044.

**The amazing universe.**
See Abstr. 003.088.

**The catastrophic universe. An essay in the philosophy of cosmology.**
See Abstr. 003.131.

**Atoms of silence: an exploration of cosmic evolution.**
See Abstr. 003.137.

**Elementary particles and the structure of the universe.**
See Abstr. 003.145.

**The influence of radio astronomy on cosmology.**
See Abstr. 004.030.

**Symposium in Pamporovo, Bulgaria, 1983, July 18 – 22.**
See Abstr. 011.008.

**Cosmology of the early universe.**
See Abstr. 012.091.

**Entropy.**
See Abstr. 014.012.

**Line strengths and positions of the submillimeter magnetic dipole transitions of $O_2$.**
See Abstr. 022.079.

**Space Telescope and cosmology.**
See Abstr. 035.021.

**Renormalization–group analysis of grand unified theories in curved spacetime.**
See Abstr. 061.003.

**Intermediate vector bosons and neutrino cosmology.**
See Abstr. 061.004.

**Is it safe to disturb the vacuum?**
See Abstr. 061.005.

**Subquark pregeometry and the possible origin of our Universe.**
See Abstr. 061.009.

**Unstable photino mass bound from cosmology.**
See Abstr. 061.036.

**Time development of Higgs field at finite temperature.**
See Abstr. 061.037.

**Deuterium and helium–3 production from massive neutrino decay.**
See Abstr. 061.061.

**Unified field theories.**
See Abstr. 061.063.

**Naturally vanishing cosmological constant and dynamical generation of $m_W$, $m_{3/2}/m_p$ hierarchy.**
See Abstr. 061.064.

**Magnetic monopoles.**
See Abstr. 061.066.

**Cosmological and experimental constraints on the tau neutrino.**
See Abstr. 061.069.

**Deuterium, helium, and the Big–Bang nucleosynthesis.**
See Abstr. 061.070.

**Detection of massive cosmic neutrinos by second–order in $G$ coherent scattering.**
See Abstr. 061.074.

**Supernovae, neutrino rest mass, and the middle–energy neutrino background in the universe.**
See Abstr. 061.075.

**Superunification from eleven dimensions.**
See Abstr. 061.085.

**Conflicts between B–violating phenomena in supergravity models.**
See Abstr. 061.105.

**A simple linear description of gauge interactions.**
See Abstr. 061.106.

**Effective Lagrangian for $\lambda\phi^4$ theory in curved spacetime with varying background fields: quasilocal approximation.**
See Abstr. 061.107.

**Radiative breaking of cosmologically acceptable grand unified theories.**
See Abstr. 061.108.

**Coloured scalars and cosmology.**
See Abstr. 061.109.

**Cosmological equations in Kaluza–Klein $d = 11$ supergravity theory.**
See Abstr. 061.110.

**More dimensions – less entropy.**
See Abstr. 061.111.

**Cosmological solutions of $N$–1 supergravity in 11 dimensions.**
See Abstr. 061.113.

**On compressional modes in radiative plasmas.**
See Abstr. 062.045.

**A current generation by Compton scattering in a relativistic plasma with velocity shear and temperature gradient.**
See Abstr. 062.100.

**Kerr metric in the de Sitter background.**
See Abstr. 066.005.

**Gravitational gauge fields and the cosmological constant.**
See Abstr. 066.008.

**Chaotic inflation in supergravity.**
See Abstr. 066.009.

**Comment on the LRS–BDT–Bianchi type–V vacuum solution.**
See Abstr. 066.020.

**The general non–LRS (*locally rotationally–symmetric*)–BDT (*Brans–Dicke theory*)–Bianchi type–V radiation solution.**
See Abstr. 066.021.

**A class of Szekeres space–times with cosmological constant.**
See Abstr. 066.022.

**Gauge freedom in perfect fluid spatially homogeneous spacetimes.**
See Abstr. 066.044.

**Aspects of interacting quantum field theory in curved space–time: renormalization and symmetry breaking.**
See Abstr. 066.127.

**A class of spherically symmetric perfect fluid distributions in general relativity.**
See Abstr. 066.141.

**Exact Brans–Dicke–Bianchi type–II solutions.**
See Abstr. 066.143.

**Electromagnetic fields invariant up to a duality rotation under a group of isometries.**
See Abstr. 066.146.

**Segal's mechanism for the redshift and its physical bases.**
See Abstr. 066.148.

**Einstein–Weyl field equations in a Bianchi type–IX space–time.**
See Abstr. 066.155.

**Cosmic censorship and conformal transformations.**
See Abstr. 066.173.

**Cosmological solution with an energy flux.**
See Abstr. 066.188.

**The amplification caused by gravitational bending of light.**
See Abstr. 066.196.

**Tetrads, broken symmetries, and the gravitational constant.**
See Abstr. 066.198.

**Fourth–order gravity as general relativity plus matter.**
See Abstr. 066.207.

**On the potential of induced gravity.**
See Abstr. 066.208.

**On the identification of gravitation with the massless spin 2 field.**
See Abstr. 066.209.

**Gravitational lensing by cosmic strings.**
See Abstr. 066.215.

**Exact non-flat Friedmann–Robertson–Walker solutions in Barker's theory of gravitation.**
See Abstr. 066.244.

**Milgrom's revision of Newton's laws: dynamical and cosmological consequences.**
See Abstr. 066.265.

**Does the missing mass problem signal the breakdown of Newtonian gravity?**
See Abstr. 066.266.

**Censorship, strong curvature, and asymptotic causal pathology.**
See Abstr. 066.289.

**Persistent curvature and cosmic censorship.**
See Abstr. 066.290.

**Massive black holes generated by cosmic strings.**
See Abstr. 067.103.

**Mining the Universe.**
See Abstr. 067.108.

**An upper limit to the beryllium abundance in the population II star HD 140283 from a high resolution IUE spectrum.**
See Abstr. 114.035.

**An upper limit to the abundance of $^9$Be in the Population II star HD 76932 from a high resolution spectrum with IUE.**
See Abstr. 114.070.

**Abundance of lithium in another sample of halo dwarfs, and in the spectroscopic binary BD –0°4234.**
See Abstr. 114.091.

**The luminosity scale of Cepheid variable stars: a revision.**
See Abstr. 122.199.

**Dust far infrared emission at high galactic latitudes.**
See Abstr. 131.236.

**Dust distribution nearby the solar system and its possible implications on cosmic radiation background anisotropy.**
See Abstr. 131.237.

**The properties of molecular clouds with high redshifts ($z = 2 - 3$).**
See Abstr. 131.238.

**Search for the radiative decay of a cosmic relic particle with I.U.E.**
See Abstr. 142.020.

**The anisotropy of the diffuse cosmic X–ray background.**
See Abstr. 142.057.

**Anti–gravity and galaxy rotation curves.**
See Abstr. 151.006.

**Galaxy clustering in two dimensions: some analytic results.**
See Abstr. 151.030.

**Flat rotation curves according to the fragmentation/shear flow model of galaxy formation.**
See Abstr. 151.033.

**Dissipative formation of an elliptical galaxy.**
See Abstr. 151.103.

**Dissipative models for the sequence of elliptical galaxies.**
See Abstr. 151.104.

**A distance to the Large Magellanic Cloud by main–sequence fitting.**
See Abstr. 156.015.

**Galaxy evolution at late epochs.**
See Abstr. 157.005.

**A simple model for the dynamical and chemical evolution of elliptical galaxies.**
See Abstr. 157.079.

**The potentiality of the galaxy autocorrelation function.**
See Abstr. 157.080.

**Direct determination of extragalactic distances from the 21 centimeter line width and the method of "sosies".**
See Abstr. 157.090.

**Angular correlations of galaxies to $B \approx 24$: another probe of cosmology and galaxy evolution.**
See Abstr. 157.091.

**Results from the Durham–AAT redshift survey.**
See Abstr. 157.098.

**Unravelling the cosmic code.**
See Abstr. 157.162.

**The spatial and peculiar velocity distribution of galaxies within 80 Mpc.**
See Abstr. 157.207.

**Global redshift quantization.**
See Abstr. 157.242.

**Surface photometry of powerful radio galaxies: their relation to Abell cluster cD galaxies.**
See Abstr. 158.028.

**Variations of the linear sizes of extragalactic radio sources with radio luminosity and redshift.**
See Abstr. 158.074.

**Cosmological evolution of radio sources found at 1.4 GHz.**
See Abstr. 158.106.

**Active galactic nuclei and their cosmological evolution.**
See Abstr. 158.142.

**Cosmological evolution of radio sources.**
See Abstr. 158.277.

**The clustering of quasars.**
See Abstr. 159.009.

**Neutrino objects and quasar redshifts.**
See Abstr. 159.014.

**The periodicity in the distribution of quasar redshifts – an analysis for quasars around selected directions.**
See Abstr. 159.024.

**Quasar absorption spectra and the structure of the universe.**
See Abstr. 159.092.

**Companion galaxies match quasar redshifts: the debate goes on.**
See Abstr. 159.101.

**Progress report on a new survey to study the clustering properties of quasars at $z \sim 2$.**
See Abstr. 159.126.

**Nature of blue galaxies in the cluster Cl1447 + 2619.**
See Abstr. 160.002.

**Superclusters: giants of the Cosmos.**
See Abstr. 160.010.

**Clusters, superclusters and their distribution.**
See Abstr. 160.019.

**Superclusters at large redshifts. Can protosuperclusters and the birth of "pancakes" be observed?**
See Abstr. 160.024.

**Rich clusters of galaxies as the result of large-scale motions inside superclusters.**
See Abstr. 160.030.

**The development of galaxy clusters.**
See Abstr. 160.032.

**On the spatial correlations of Abell clusters.**
See Abstr. 160.034.

**Distribution functions of intergalactic clouds.**
See Abstr. 160.040.

**Relations between spatial correlations of rich clusters of galaxies.**
See Abstr. 160.048.

**The evolution of galaxies in clusters. V. A study of populations since $z \sim 0.5$.**
See Abstr. 160.068.

**The brightest clusters of galaxies and the cosmological constant.**
See Abstr. 160.072.

**Stephan's quintet, superclusters of galaxies and giant voids.**
See Abstr. 160.076.

**A catalog of candidate superclusters and voids.**
See Abstr. 160.081.

**On the large-scale distribution of clusters of galaxies.**
See Abstr. 160.086.

**Infall into the Virgo cluster and some cosmological constraints.**
See Abstr. 160.095.

**Hierarchical clustering.**
See Abstr. 160.128.

**Percolation studies of galaxy clustering.**
See Abstr. 160.129.

**The Shane-Wirtanen counts.**
See Abstr. 160.147.

# Author Index

The authors are listed in alphabetical order according to the initial letter following the first names. Author names which are not transliterated according to our scheme are indicated by †.

**Zavody**

# Subject Index

The Subject Index provides an alphabetical subject key to the papers abstracted in *Astronomy and Astrophysics Abstracts*. In order to explain the structure of this Index some general remarks on the construction principles of our key words will follow.

Whenever possible a key word was formed in such a way as to combine two different descriptors characterizing an abstracted article. Such a combination represents the contents of a paper more precisely than both components regarded as single entries. As an example we take the two terms Minor Planets and Orbits and the combination

Minor Planets
  Orbits

Efforts were made to choose combinations which can be inverted in order to increase the usefulness of the Index. That means the two components represent terms of equal status which both serve as primary entries. In the given example we find the two entries

Minor Planets
  Orbits

and

Orbits
  Minor Planets

But there are also two-term key words which cannot be reversed. The conditions for non-inversibility are: the secondary descriptor is either a very specific term (e.g. LTE Analyses) or a very general one (e.g. Models). Thus we distinguish two kinds of descriptors, the primary ones, serving as direct entries and the secondary ones as auxiliary entries to the Index. Our intention is to choose comprehensive as well as significant formulations.

Some further aids to the use of this Subject Index are given by the following concordances:

| | |
|---|---|
| Asteroids | see Minor Planets |
| Earth: Magnetic Field | see Geomagnetic Field |
| Galactic... | see also Galaxies, Galaxy |
| Lunar... | see also Moon |
| Planetary... | see also Planets |
| Solar... | see also Sun |
| Stellar... | see also Stars |

Efforts were made to standardize the key words. This idea led to the development of our vocabulary of astronomical and astrophysical terms. From time to time we enlarged our vocabulary and at present version 07.84 consists of 2273 items.

The list of key words was examined in order to standardize the key words and to confine some of the synonymous entries. For example, according to our vocabulary the key words Catalogues, Positions, Star Catalogues, and Star Positions were admissible terms, likewise the combinations

Catalogues
  Star Positions

Star Catalogues
  Positions

Star Catalogues
  Star Positions

The user is – in any case – kindly requested to look for synonymous entries, because further references to the topic looked for might exist elsewhere in the Index under another current astronomical term.

Some key words concerning comets, novae, and solar eclipses are sorted by a special routine. Comets are listed according to their Roman numeral designation or preliminary designation, novae according to their stellar constellation and solar eclipses according to the date of the event.

Starting with Volume 30 of *Astronomy and Astrophysics Abstracts*, this Index is given in a four-column arrangement. The overall size further has been optimized by suppression of a repetition of the first key word in a two-term combination. Thus, the primary term is printed only once (in semi-boldface) for all the following secondary key words which belong to this header.

The following List of Key Words might serve as a further guidance to the use of this Subject Index. This list, a subset of our vocabulary 07.84, consists of 1176 entries and comprises all one-term key words and all primary terms of the two-term ones actually used in this Subject Index.

# List of Key Words

A Dwarfs
A Stars
A Supergiants
Absolute Magnitudes
Absorption
Absorption Lines
Absorption Spectra
Accretion
Accretion Disk
Accretion Disks
Achondrites
Acoustic Waves
Active Galactic Nuclei
Active Galaxies
Adaptive Optics
Ae Stars
Aerosols
Air Showers
Airglow
Albedo
Alfven Waves
Algol
Algol Systems
AM Herculis
AM Herculis Stars
Am Stars
Amor Objects
Andromeda Nebula
Angular Diameters
Angular Momentum
Antennas
Antimatter
Antineutrinos
Ap Stars
Aperture Synthesis
Apollo Objects
Apparent Magnitudes
Apsidal Motion
Arakelian Galaxies
Archaeoastronomy
Artificial Satellites
Asteroidal Belt
Astrodynamics
Astrographs
Astrolabes
Astrometric Binaries
Astrometric Plates
Astrometric Reflectors
Astrometry
Astronomical Constants
Astronomical Geodesy
Astronomical Instruments
Astronomical Optics
Atlases
Atmospheres
Atmospheric Optics
Atomic Clocks
Atomic Processes
Aurorae
Auroral Arcs
Auxiliary Instrumentation
Azimuth Determination

B Dwarfs
B Stars
B Subdwarfs
B Supergiants
Background Radiation
Balloons
Balmer Lines
Barium Stars
Barred Spirals
Baryons
Be Stars
Beta Cephei Stars
Beta CMa Stars
Beta Lyrae Stars
Binaries
Binary Pulsars
Bipolar Nebulae
BL Lacertae
BL Lacertae Objects
Black Dwarfs
Black Holes
Black-Body Radiation
Blue Galaxies
Blue Objects
Blue Stars
Blue Stragglers
Bolometers
Bolometric Corrections
Bp Stars
Bright Galaxies
Bright Stars
Brightness Temperature
Brightness Temperatures
Broad-Band Photometry
BV Photometry
BY Draconis Stars

C-M Diagrams
C-M Relation
Callisto
Cameras
Carbon Burning
Carbon Stars
Carina Nebula
Cassegrain Spectrographs
Cassegrain Telescopes
Cataclysmic Binaries
Cataclysmic Variables
Catalogues
CCD Cameras
CCD Detectors
CCD Observations
CCD Photometry
Celestial Bodies
Celestial Mechanics
Central Stars
Cepheids
Cerenkov Radiation
Chandler Wobble
Chemical Elements
Chondrites
Chondrules
Circumpolar Stars
Circumstellar Clouds
Circumstellar Disks
Circumstellar Dust
Circumstellar Grains
Circumstellar Matter
Circumstellar Shell
Circumstellar Shells
Clocks
Close Binaries
Clusters of Galaxies
CN Stars
CNO Cycle
Coalsack
Collapse
Collapsed Stars
Collapsing Clouds
Collapsing Stars
Collimators
Color Excesses
Color Indices
Colors
Cometary Atmospheres
Cometary Comae
Cometary Dust
Cometary Ionospheres
Cometary Nebulae
Cometary Nuclei
Cometary Plasma
Cometary Tails
Comets
Compact Galaxies
Compact Objects
Compton Effect
Compton Scattering
Contact Binaries
Continental Drift
Convection
Convective Shells
Convective Zones
Cool Giants
Cool Stars
Coronographs
Cosmic Dust
Cosmic Rays
Cosmochemistry
Cosmochronology
Cosmogony
Cosmological Constant
Cosmology
Coude Spectrographs
Crab Nebula
Curves-of-Growth
Cyclotron Radiation
Cygnus Loop

D Galaxies
Dark Clouds
Dark Matter
Dark Nebulae
Data Centers
Data Processing
DDO Photometry
Degenerate Dwarfs
Degenerate Stars
Delta Scuti Stars
Dense Clouds
Dense Matter
Densities
Density
Density Waves
Detector Arrays
Detectors
Diameter
Diameters
Differential Rotation
Diffuse Nebulae
Disk Galaxies
Disk Stars
Distance
Distance Indicators
Distance Scale
Distances
Domes
Dust
Dust Clouds
Dust Grains
Dust Shells
Dwarf Cepheids
Dwarf Galaxies
Dwarf Novae
Dwarfs
Dynamical Parallaxes
Dynamo Theory

Early Universe
Early-Stage Stars
Early-Type Galaxies
Early-Type Stars
Early-Type Supergiants
Earth
Earth Atmosphere
Earth Interior
Earth Ionosphere
Earth Magnetosphere
Earth Mantle
Earth Mesosphere
Earth Plasmasphere
Earth Radiation Belts
Earth Rotation
Earth Stratosphere
Earth Thermosphere
Earth Troposphere
Earth-Moon System
Echelle Spectra
Echelle Spectrographs
Echelle Spectrometers
Eclipses
Eclipsing Binaries
Ecliptic
Effective Temperatures
Electron Densities
Electron Temperatures
Electronographic Cameras
Elementary Particles
Elliptical Galaxies
Emission Lines
Emission Nebulae
Emission Spectra
Emission-Line Galaxies
Emission-Line Objects
Emission-Line Stars
Enceladus
Ephemerides
Ephemeris Time
Equidensitometry
Equilibrium Figures
Eruptive Binaries
Eruptive Stars
Eta Carinae
Europa
Europa Surface
Extinction
Extragalactic Objects
Extraterrestrial Intelligence
Extraterrestrial Life
Extreme UV
Extreme UV Astronomy
Extreme UV Spectra

F Dwarfs
F Stars
F Supergiants
Fabry-Perot Interferometers
Fabry-Perot Interferometry
Fabry-Perot Spectrometers
Faint Galaxies
Faint Stars
Filamentary Nebulae
Filters
Fireballs
FK4
FK5
FK5 System
Flare Stars
Fluid Dynamics
Fluid Spheres
Flux Densities
Forbidden Lines
Four-Body Problem
Four-Color Photometry
Fourier Spectrometers
Fourier Spectroscopy
Fraunhofer Lines
Frequency Standards
FU Orionis Stars
Fundamental Astrometry
Fundamental Constants
Fundamental Systems

G Dwarfs
G Stars
Galactic Anticenter
Galactic Bars
Galactic Center
Galactic Centers
Galactic Coronae
Galactic Disk
Galactic Disks
Galactic Halo
Galactic Halos
Galactic Magnetic Field
Galactic Nebulae
Galactic Nuclei
Galactic Nucleus

# ASTRONOMY AND ASTROPHYSICS ABSTRACTS

A Publication of the Astronomisches Rechen-Institut Heidelberg

Editors: S. Böhme, W. Fricke, H. Hefele, I. Heinrich, W. Hofmann, D. Krahn, V. R. Matas, L. D. Schmadel, G. Zech

| | | | |
|---|---|---|---|
| **Volume 1** | Literature 1969, Part 1 | X+ 435 pp. | (1969) |
| **Volume 2** | Literature 1969, Part 2 | X+ 516 pp. | (1970) |
| **Volume 3** | Literature 1970, Part 1 | X+ 490 pp. | (1970) |
| **Volume 4** | Literature 1970, Part 2 | X+ 562 pp. | (1971) |
| **Volume 5** | Literature 1971, Part 1 | X+ 505 pp. | (1971) |
| **Volume 6** | Literature 1971, Part 2 | X+ 560 pp. | (1972) |
| **Volume 7** | Literature 1972, Part 1 | X+ 526 pp. | (1972) |
| **Volume 8** | Literature 1972, Part 2 | X+ 594 pp. | (1973) |
| **Volume 9** | Literature 1973, Part 1 | X+ 610 pp. | (1973) |
| **Volume 10** | Literature 1973, Part 2 | X+ 661 pp. | (1974) |
| **Volume 11** | Literature 1974, Part 1 | X+ 579 pp. | (1974) |
| **Volume 12** | Literature 1974, Part 2 | X+ 699 pp. | (1975) |
| **Volume 13** | Literature 1975, Part 1 | X+ 632 pp. | (1975) |
| **Volume 14** | Literature 1975, Part 2 | X+ 747 pp. | (1976) |
| **Volume 15/16** | Author and Subject Indexes to Volumes 1–10<br>Literature 1969–1973 | VII+ 655 pp. | (1976) |
| **Volume 17** | Literature 1976, Part 1 | XII+ 645 pp. | (1976) |
| **Volume 18** | Literature 1976, Part 2 | X+ 859 pp. | (1977) |
| **Volume 19** | Literature 1977, Part 1 | X+ 732 pp. | (1977) |
| **Volume 20** | Literature 1977, Part 2 | X+ 786 pp. | (1978) |
| **Volume 21** | Literature 1978, Part 1 | X+ 834 pp. | (1978) |
| **Volume 22** | Literature 1978, Part 2 | X+ 849 pp. | (1979) |
| **Volume 23/24** | Author and Subject Indexes to Volumes 11–14 and 17–22<br>Literature 1974–1978 | VIII+1127 pp. | (1979) |
| **Volume 25** | Literature 1979, Part 1 | X+ 872 pp. | (1979) |
| **Volume 26** | Literature 1979, Part 2 | X+ 794 pp. | (1980) |
| **Volume 27** | Literature 1980, Part 1 | X+ 939 pp. | (1980) |
| **Volume 28** | Literature 1980, Part 2 | X+ 841 pp. | (1981) |
| **Volume 29** | Literature 1981, Part 1 | X+ 853 pp. | (1981) |
| **Volume 30** | Literature 1981, Part 2 | X+ 792 pp. | (1982) |
| **Volume 31** | Literature 1982, Part 1 | X+ 776 pp. | (1982) |
| **Volume 32** | Literature 1982, Part 2 | X+ 848 pp. | (1983) |
| **Volume 33** | Literature 1983, Part 1 | X+ 815 pp. | (1983) |
| **Volume 34** | Literature 1983, Part 2 | X+ 921 pp. | (1984) |
| **Volume 35/36** | Author and Subject Indexes to Volumes 25–34<br>Literature 1979–1983 | XIII+ 891 pp. | (1984) |
| **Volume 37** | Literature 1984, Part 1 | X+ 937 pp. | (1984) |
| **Volume 38** | Literature 1984, Part 2 | X+ 919 pp. | (1985) |

Published for Astronomisches Rechen-Institut by
Springer-Verlag Berlin Heidelberg New York Tokyo

The manufacturer's authorised representative in the EU is Springer
Nature Customer Service Centre GmbH, Europaplatz 3, 69115 Heidelberg,
Germany. If you have any concerns regarding our products, please
contact ProductSafety@springernature.com

Printed and bound by CPI Group (UK) Ltd, Croydon, CR0 4YY
16/07/2026
02168263-0007